# LEARNSMART®

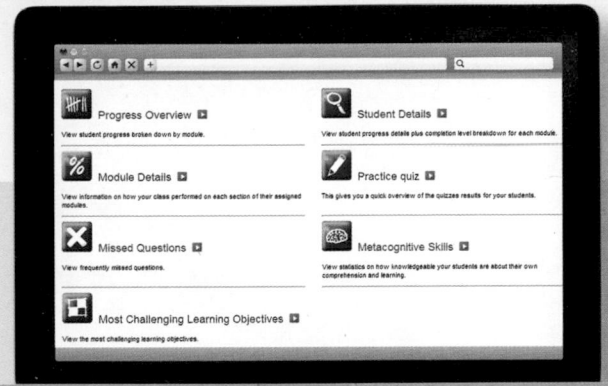

The market leading **adaptive study tool** proven to strengthen memory recall, increase class retention and boost grades.

> Moves students beyond memorizing

> Allows instructors to align content with their goals

> Allows instructors to spend more time teaching higher level concepts

# SMARTBOOK™

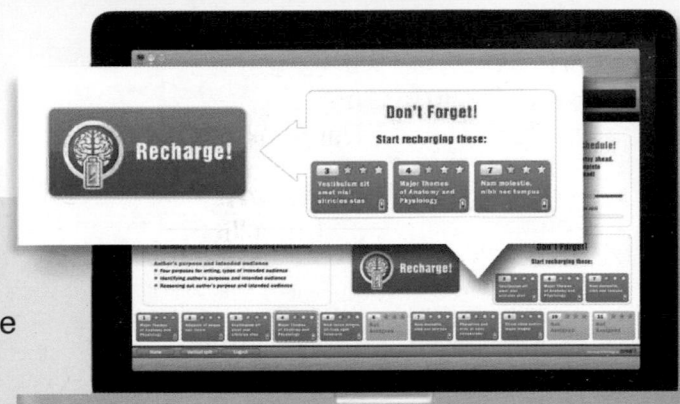

The first—and only—adaptive reading experience designed to transform the way students read.

> Engages students with a personalized reading experience

> Ensures students retain knowledge

# LEARNSMART PREP™

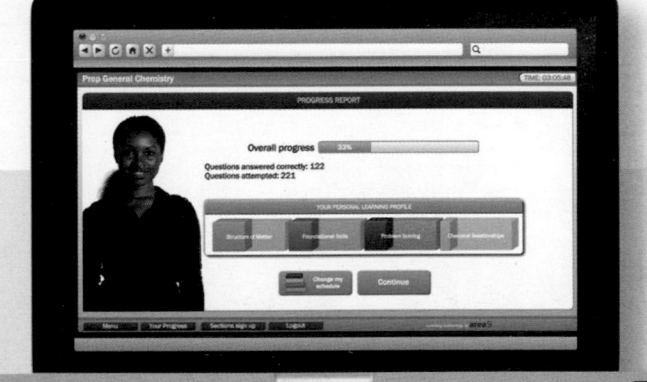

An adaptive course preparation tool that quickly and efficiently helps students prepare for college level work.

> Levels out student knowledge

> Keeps students on track

# WWW.LEARNSMARTADVANTAGE.COM

# ANATOMY &

# PHYSIOLOGY

## The Unity of Form and Function

### Seventh Edition

## KENNETH S. SALADIN

*Georgia College & State University*

**Digital Authors**

## STEPHEN J. SULLIVAN

*Bucks County Community College*

## CHRISTINA A. GAN

*Highline Community College*

Mc
Graw
Hill
Education

ANATOMY & PHYSIOLOGY: THE UNITY OF FORM AND FUNCTION, SEVENTH EDITION
International Edition 2015

10  09  08  07  06  05  04  03  02  01
20  16  15  14
CTP  SLP

**When ordering this title, use ISBN 978-981-4646-43-7 or MHID 981-4646-43-1.**

www.mhhe.com

# BRIEF CONTENTS

# CONTENTS

**KENNETH S. SALADIN** has taught since 1977 at Georgia College & State University in Milledgeville, Georgia. He earned a B.S. in zoology at Michigan State University and a Ph.D. in parasitology at Florida State University, with interests especially in the sensory ecology of freshwater invertebrates. In addition to human anatomy and physiology, his teaching experience includes histology, parasitology, animal behavior, sociobiology, introductory biology, general zoology, biological etymology, and study abroad in the Galápagos Islands. Ken has been recognized as "most significant undergraduate mentor" nine times over the years by outstanding students inducted into Phi Kappa Phi. He received the university's Excellence in Research and Publication Award for the first edition of this book, and was named Distinguished Professor in 2001.

Ken is a member of the Human Anatomy and Physiology Society, the Society for Integrative and Comparative Biology, the American Association of Anatomists, American Physiological Society, and the American Association for the Advancement of Science. He served as a developmental reviewer and wrote supplements for several other McGraw-Hill anatomy and physiology textbooks for a number of years before becoming a textbook writer.

Ken's outside interests include the Big Brothers/Big Sisters program for single-parent children, the Galápagos Conservancy, and student scholarships. Ken is married to Diane Saladin, a registered nurse. They have two adult children.

**STEPHEN J. SULLIVAN,** digital coauthor for Connect, has been teaching anatomy and physiology at Bucks County Community College in Pennsylvania since 2002. Steve started consulting with McGraw-Hill on the development of digital tools in 2009. His goal for Connect is to create digital assessments that directly reflect the content and style of Ken Saladin's text, provide student access, and foster student success. Steve is a member of the Human Anatomy and Physiology Society and the American Association of Anatomists, and is a 2013 recipient of the Lindback Award for Distinguished Teaching.

**CHRISTINA A. GAN,** digital coauthor for Connect, has been teaching anatomy and physiology, microbiology, and general biology at Highline Community College in Des Moines, WA since 1994. Before that she taught at Rogue Community College in Medford, OR for six years. She earned her MA in biology from Humboldt State University, researching the genetic variation of mitochondrial DNA in various salmonid species, and is a member of the Human Anatomy and Physiology Society. When she is not in the classroom or developing digital media, she is climbing, mountaineering, skiing, kayaking, sailing, cycling, and mountain biking throughout the Pacific Northwest.

# THE EVOLUTION OF A
# STORYTELLER

Ken in 1964

Ken Saladin's first step into authoring was a 318-page paper on the ecology of hydras written for his 10th-grade biology class. With his "first book," featuring 53 original India ink drawings and photomicrographs, a true storyteller was born.

*When I first became a textbook writer, I found myself bringing the same enjoyment of writing and illustrating to this book that I first discovered when I was 15.*

—Ken Saladin

Ken's "first book," *Hydra Ecology*, 1965

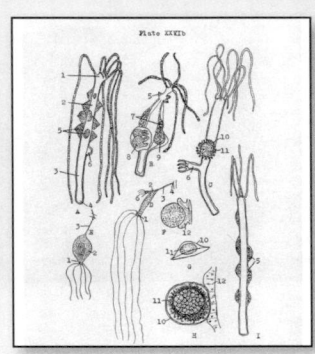

One of Ken's drawings from *Hydra Ecology*

Ken began working on his first book for McGraw-Hill in 1993, and in 1997 the first edition of *The Unity of Form and Function* was published. In 2014 the story continues with the seventh edition of Ken's best-selling A&P textbook.

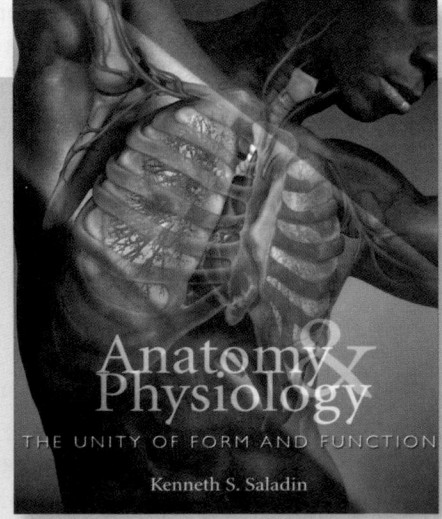

The first edition (1997)

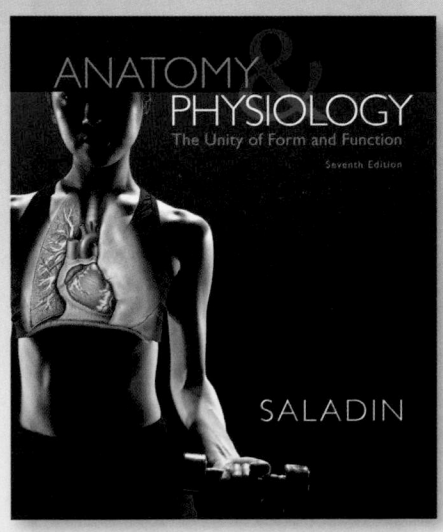

The story continues (2014)

# PREFACE

***Anatomy & Physiology: The Unity of Form and Function*** tells a story made of many layers including the core science, clinical applications, the history of medicine, and the evolution of the human body. Saladin combines this humanistic perspective on anatomy and physiology with vibrant photos and art to convey the beauty and excitement of the subject to beginning students.

To help students manage the tremendous amount of information in this introductory course, the narrative is broken into short segments, each framed by expected learning outcomes and self-testing review questions. This presentation strategy works as a whole to create a more efficient and effective way for students to learn A&P.

## WRITING STYLE AND LEVEL

Saladin's text is written using plain language for A&P students who may be taking this course early in their curricula. Careful attention has been given to word selection and paragraph structure to maintain the appropriate level for students, while avoiding the appearance of dumbed down content. This writing level makes the text accessible for *all* audiences (international readers, English-as-a-second-language students, and nontraditional students).

Students say the enlightening analogies, clinical applications, historical notes, biographical vignettes, and evolutionary insights make the book not merely informative, but a pleasure to read. Even instructors say they often learn something new and interesting from Saladin's innovative perspectives.

> **Analogies** explain tough scientific content in a way students can understand.

The cytoskeleton is composed of *microfilaments, intermediate filaments,* and *microtubules.* If you think of intermediate filaments as being like the stiff rods of uncoooked spaghetti, you could, by comparison, think of microfilaments as being like fine angel-hair pasta and microtubles as being like tubular penne pasta.

> **Medical History**
>
> Saladin "puts the human in human A&P" with his occasional vignettes on the people behind the science. Students say these stories make learning A&P more fun and stimulating.

 **DEEPER INSIGHT 2.1**

**MEDICAL HISTORY**

### *Radiation and Madame Curie*

In 1896, French scientist Henri Becquerel (1852–1908) discovered that uranium darkened photographic plates through several thick layers of paper. Marie Curie (1867–1934) and Pierre Curie (1859–1906), her husband, discovered that polonium and radium did likewise. Marie Curie coined the term *radioactivity* for the emission of energy by these elements. Becquerel and the Curies shared a Nobel Prize in 1903 for this discovery.

Marie Curie (fig. 2.3) was not only the first woman in the world to receive a Nobel Prize but also the first woman in France even to receive a Ph.D. She received a second Nobel Prize in 1911 for further work in radiation. Curie crusaded to train women for careers in science, and in World War I, she and her daughter, Irène Joliot-Curie (1897–1956), trained physicians in the use of X-ray machines. Curie pioneered radiation therapy for breast and uterine cancer.

In the wake of such discoveries, radium was regarded as a wonder drug. Unaware of its danger, people drank radium tonics and flocked to health spas to bathe in radium-enriched waters. Marie herself suffered extensive damage to her hands from handling radioactive minerals and died of radiation poisoning at age 67. The following year, Irène and her husband, Frédéric Joliot (1900–1958), were awarded a Nobel Prize for work in artificial radioactivity and synthetic radioisotopes. Apparently also a martyr to her science, Irène died of leukemia, possibly induced by radiation exposure.

**FIGURE 2.3** **Marie Curie (1867–1934).** This portrait was made in 1911, when Curie received her second Nobel Prize.

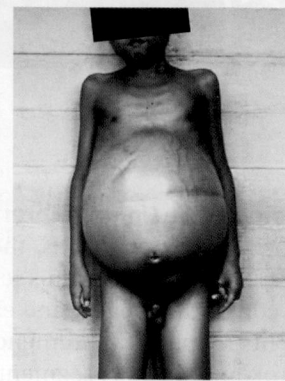

**Clinical Applications** make the abstract science more relevant.

*More than a few distinguished scientists and clinicians say they found their inspiration in reading of the lives of their predecessors. Maybe these stories will inspire some of our own students to go on to do great things.*

—Ken Saladin

## DEEPER INSIGHT 8.4
### EVOLUTIONARY MEDICINE

#### Skeletal Adaptations for Bipedalism

Some mammals can stand, hop, or walk briefly on their hind legs, but humans are the only mammals that are habitually bipedal. Footprints preserved in a layer of volcanic ash in Tanzania indicate that hominids walked upright as early as 3.6 million years ago. This bipedal locomotion is possible only because of several adaptations of the human feet, legs, spine, and skull (fig. 8.43). These features are so distinctive that paleoanthropologists (those who study human fossil remains) can tell with considerable certainty whether a fossil species was able to walk upright.

As important as the hand has been to human evolution, the foot may be an even more significant adaptation. Unlike other mammals, humans support their entire body weight on two feet. While apes are flat-footed, humans have strong, springy foot arches that absorb shock as the body jostles up and down during walking and running. The tarsal bones are tightly articulated with one another, and the calcaneus is strongly developed. The hallux (great toe) is not opposable

as it is in most Old World monkeys and apes, but it is highly developed so that it provides the "toe-off" that pushes the body forward in the last phase of the stride (fig. 8.43a). For this reason, loss of the hallux has a more crippling effect than the loss of any other toe.

While the femurs of apes are nearly vertical, in humans they angle medially from the hip to the knee (fig. 8.43b). This places our knees closer together, beneath the body's center of gravity. We lock our knees when standing, allowing us to maintain an erect posture with little muscular effort. Apes cannot do this, and they cannot stand on two legs for very long without tiring—much as you would if you tried to maintain an erect posture with your knees slightly bent.

In apes and other quadrupedal (four-legged) mammals, the abdominal viscera are supported by the muscular abdominal wall. In humans, the viscera bear down on the floor of the pelvic cavity, and a bowl-shaped pelvis

**Evolutionary Medicine** provides novel and intriguing ways of looking at:

- menopause
- the sweet tooth
- bipedalism
- the origin of mitochondria
- skin color
- body hair
- lactose intolerance
- the kidney and life on dry land
- the palate
- theories of aging and death

**(a) Foot**

**(b) Knee**

**(c) Gluteal muscles**

Chimpanzee

Chimpanzee

Human

Chimpanzee

Human

Human

**FIGURE 8.43 Skeletal Adaptations for Bipedalism.** Human adaptations for bipedalism are best understood by comparison to our close living relative, the chimpanzee, which is not adapted for a comfortable or sustained erect stance. See the text for the relevance of each comparison.

# INTERACTIVE MATERIAL

Review activities are integrated into each chapter. Self-teaching prompts and simple experiments are liberally seeded throughout the narrative. Learning aids such as pronunciation guides and insights into the origins and root meanings of medical terms are sprinkled liberally throughout the text to help students retain meaning and spelling of terms.

> **Self-teaching prompts** make reading more active.

> **Pro-NUN-see-AY-shun guides** help beginning students master A&P.

> **Word origins** are footnoted.

## The Ulna

At the proximal end of the **ulna** (fig. 8.33) is a deep, C-shaped **trochlear notch** that wraps around the trochlea of the humerus. The posterior side of this notch is formed by a prominent **olecranon**—the bony point where you rest your elbow on a table. The anterior side is formed by a less prominent **coronoid process.** Laterally, the head of the ulna has a less conspicuous **radial notch,** which accommodates the edge of the head of the radius. At the distal end of the ulna is a medial **styloid process.** The bony lumps you can palpate on each side of your wrist are the styloid processes of the radius and ulna.

The radius and ulna are attached along their shafts by a ligament called the **interosseous** (IN-tur-OSS-ee-us) **membrane (IM),** which is attached to an angular ridge called the **interosseous margin** on each bone. Most fibers of the IM are oriented obliquely, slanting upward from the ulna to the radius. If you lean forward on a table supporting your weight on your hands, about 80% of the force is borne by the radius. This tenses the IM, which pulls the ulna upward and

The primary feature of the shaft is a posterior ridge called the **linea aspera**[68] (LIN-ee-uh ASS-peh-ruh) at its midpoint. At its upper end, the linea aspera forks into a medial **spiral (pectineal) line** and a lateral **gluteal tuberosity.** The gluteal tuberosity is a rough ridge (sometimes a depression) that

> **Familiarity with word origins** helps students retain meaning and spelling.

[66]*fovea* = pit; *capitis* = of the head
[67]*trochanter* = to run
[68]*linea* = line; *asper* = rough

# WHAT'S NEW IN THE SEVENTH EDITION?

## New Content

Several sections include new content, especially

- Gradients and flow as another unifying concept of physiology (chapter 1)
- Pseudopods as another class of cell surface features (chapter 3)
- New table of art and descriptions of the hierarchy of muscle structure (chapter 11)
- New Deeper Insight on portal hypertension and ascites (chapter 20)

- Updated treatment of chronic obstructive pulmonary disease (COPD) (chapter 22)
- New section on magnesium homeostasis (chapter 24)
- History's youngest mother, pregnant at age 4 (chapter 28)
- Deeper Insight on endometriosis (chapter 28)

## New Science

Other new content in Saladin's *Anatomy & Physiology* updates the book to stay abreast of new developments in science. Some of the more significant examples are listed here.

- The high frequency of everyday DNA damage (chapter 4)
- The proteome as a new frontier in cytology (chapter 4)
- Epigenetic inheritance and disease (chapter 4)

> Despite all this intricate packaging, the DNA of the average mammalian cell is damaged an astonishing 10,000 to 100,000 times per day! The consequences would be catastrophic were it not for DNA repair enzymes that detect and undo most of the damage.

- Umbrella cells and protective role of transitional epithelium (chapter 5)
- Reinterpretation of pelvic floor muscle compartments based on new medical imaging techniques (chapter 10)
- Updates in muscle physiology on the length–tension relationship, the size principle in recruitment, tetanus, fatigue, muscle fiber types, and smooth muscle caveolae (chapter 11)
- Autoimmune pathogenesis of pernicious anemia (chapter 18)
- New biomechanical interpretations of papillary muscles and trabeculae carneae (chapter 19)
- New insights on functions of the spleen (chapter 21)
- Update on the status of reproductive effects of endocrine disrupting chemicals (chapter 27)
- New understanding of the 209-day timetable of ovarian follicle and oocyte development (chapter 28)
- Update on free radical–DNA damage theory of aging (chapter 29)

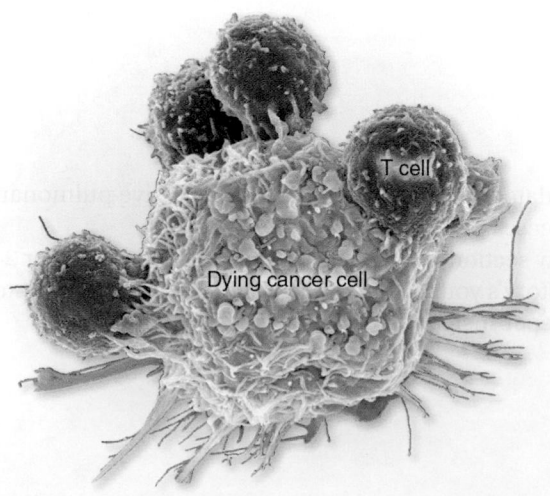

## New Photographs

- Figure A.8: intestinal mesentery
- Figure 17.11: histology of the adrenal cortex
- Figure 21.7: macrophages attacking bacteria
- Figure 21.24: T cells attacking cancer cell
- Figure 28.9d: SEM of acini and myoepithelial cells of the mammary gland

## New Art

- Figure 3.13: pseudopods as cell surface features
- Figure 3.31: the proteasome as an organelle
- Figure 4.23: cancer metastasis
- Figure 5.32: comparison of merocrine, apocrine, and holocrine secretion
- Figure 14.39: full-page summary flowchart of cranial nerve pathways
- Figure 17.18: hormone actions on their target cells
- Figure 17.23: antagonistic effects of insulin and glucagon on a hepatocyte
- Figure 18.22: enhanced presentation of blood-clotting cascade
- Figure 25.31: full-page summary figure of macronutrient digestion and absorption
- Figure 26.1: simplified flowchart of gut–brain peptides and appetite regulation

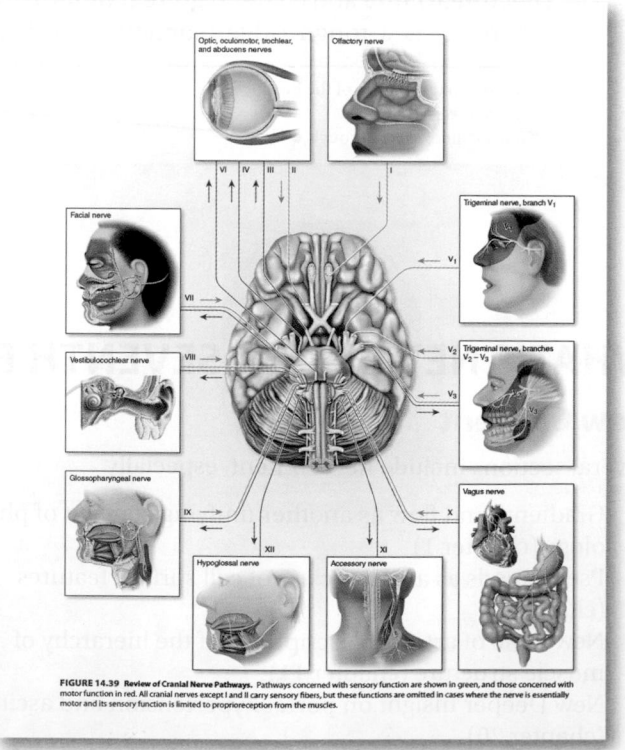

FIGURE 14.39 **Review of Cranial Nerve Pathways.** Pathways concerned with sensory function are shown in green, and those concerned with motor function in red. All cranial nerves except I and II carry sensory fibers, but these functions are omitted in cases where the nerve is essentially motor and its sensory function is limited to proprioception from the muscles.

## New Pedagogy

- Appendix E: a table of the complete genetic code

# MAKING
# *Anatomy & Physiology*
## INTRIGUING AND INSPIRING

## INNOVATIVE CHAPTER SEQUENCING

Some chapters and topics are presented in a sequence that is more instructive than the conventional order.

### Early Presentation of Heredity

Fundamental principles of heredity are presented in the last few pages of chapter 4 rather than at the back of the book to better integrate molecular and Mendelian genetics. This organization also prepares students to learn about such genetic traits and conditions as cystic fibrosis, color blindness, blood types, hemophilia, cancer genes, and sickle-cell disease by first teaching them about dominant and recessive alleles, genotype and phenotype, and sex linkage.

### Urinary System Presented Close to Circulatory and Respiratory Systems

Most textbooks place this system near the end of the book because of its anatomical and developmental relationships with the reproductive system. However, its physiological ties to the circulatory and respiratory systems are much more important. Except for a necessary digression on lymphatics and immunity, the circulatory system is followed almost immediately with the respiratory and urinary systems, which regulate blood composition and whose functional mechanisms rely on recently covered principles of blood flow and capillary exchange.

### Muscle Anatomy and Physiology Follow Skeleton and Joints

The functional morphology of the skeleton, joints, and muscles is treated in three consecutive chapters, 8 through 10, so when students learn muscle origins and insertions, these come only two chapters after the names of the relevant bone features. When they learn muscle actions, it is in the first chapter after learning the terms for the joint movements. This order brings another advantage: the physiology of muscle and nerve cells is treated in two consecutive chapters (11 and 12), which are thus closely integrated in their treatment of synapses, neurotransmitters, and membrane electrophysiology.

## BRIEF CONTENTS

iii

# LEARNING TOOLS

## Engaging Chapter Layouts

- Chapters are structured around the way students learn.
- Frequent subheadings and expected learning outcomes help students plan their study time and review strategies.

**Chapter Outlines** provide quick overviews of the content.

**Deeper Insights** highlight areas of interest and career relevance for students.

# BONE TISSUE

Spongy bone of the human femur

Anatomy & Physiology REVEALED
aprevealed.com

Module 5: Skeletal System

---

**BRUSHING UP**

- Bone histology was introduced on page 156 and is detailed in this chapter.
- The hyaline cartilage histology introduced on page 156 is important for understanding bone development and certain features of the mature skeleton.
- The introduction on stem cells on page 169 will be helpful for understanding some types of bone cells.

In art and history, nothing has so often symbolized death as a skull or skeleton.[1] Bones and teeth are the most durable remains of a once-living body and the most vivid reminder of the impermanence of life.

The dry bones presented for laboratory study may wrongly suggest that the skeleton is a nonliving scaffold for the body, like the steel girders of a building. Seeing it in such a sanitized form makes it easy to forget that the living skeleton is made of dynamic tissues, full of cells—that it continually remodels itself and interacts physiologically with all of the other organ systems of the body. The skeleton is permeated with nerves and blood vessels, which attests to its sensitivity and metabolic activity.

**Osteology,**[2] the study of bone, is the subject of these next three chapters. In this chapter, we study bone as a tissue—its composition, its functions, how it develops and grows, how its metabolism is regulated, and some of its disorders. This will provide a basis for understanding the skeleton, joints, and muscles in the chapters that follow.

**7.1 Tissues and Organs of the Skeletal System**

**Expected Learning Outcomes**

When you have completed this section, you should be able to

a. name the tissues and organs that compose the skeletal system;

b. state several functions of the skeletal system;

c. distinguish between bone as a tissue and as an organ;

d. describe four types of bones classified by shape; and

e. describe the general features of a long bone and a flat bone.

The **skeletal system** is composed of bones, cartilages, and ligaments joined tightly to form a strong, flexible framework for the body. Cartilage, the forerunner of most bones in embryonic and childhood development, covers many joint surfaces in the mature skeleton. Ligaments hold bones together at the joints and are discussed in chapter 9. Tendons are structurally similar to ligaments but attach muscle to bone; they are discussed with the muscular system in chapter 10. Here, we focus on the bones.

[1] skelet = dried up
[2] osteo = bone; logy = study of

### Functions of the Skeleton

The skeleton plays at least six roles:

1. **Support.** Bones of the limbs and vertebral column support the body; the mandible and maxilla support the teeth; and some viscera are supported by nearby bones.

2. **Protection.** Bones enclose and protect the brain, spinal cord, heart, lungs, pelvic viscera, and bone marrow.

3. **Movement.** Limb movements, breathing, and other movements are produced by the action of muscles on the bones.

4. **Electrolyte balance.** The skeleton stores calcium and phosphate ions and releases them into the tissue fluid and blood according to the body's physiological needs.

5. **Acid–base balance.** Bone tissue buffers the blood against excessive pH changes by absorbing or releasing alkaline phosphate and carbonate salts.

6. **Blood formation.** Red bone marrow is the major producer of blood cells, including cells of the immune system.

### Bones and Osseous Tissue

Bone, or **osseous tissue,** is a connective tissue in which the matrix is hardened by the deposition of calcium phosphate and other minerals. The hardening process is called **mineralization** or **calcification.** (Bone is not the hardest substance in the body; that distinction goes to tooth enamel.) Osseous tissue is only one of the tissues that make up a bone. Also present are blood, bone marrow, cartilage, adipose tissue, nervous tissue, and fibrous connective tissue. The word *bone* can denote an organ composed of all these tissues, or it can denote just the osseous tissue.

### General Features of Bones

Bones have a wide variety of shapes correlated with their varied protective and locomotor functions. Most of the cranial bones are in the form of thin curved plates called **flat bones,** such as the paired parietal bones that form the dome of the top of the head. The sternum (breastbone), scapula (shoulder blade), ribs, and hip bones are also flat bones. The most important bones in body movement are the **long bones** of the limbs—the humerus, radius, and ulna of the arm and forearm; the femur, tibia, and fibula of the thigh and leg; and the metacarpals, metatarsals, and phalanges of the hands and feet. Like crowbars, long bones serve as rigid levers that are acted upon by skeletal muscles to produce the major body movements. The wrists and ankles have a total of 30 **short bones** (carpal and tarsal bones), which are approximately equal in length and width and which produce relatively limited gliding movements. The patella is also a short bone. Many bones, however, do not fit any of these categories and are collectively

[3] os, ossa, oste = bone

---

Each chapter begins with **Brushing Up** to emphasize the interrelatedness of concepts, and serves as an aid for instructors when teaching chapters out of order.

## Tiered Assessments Based on Key Learning Outcomes

- Chapters are divided into easily manageable chunks, which help students budget study time effectively.
- Section-ending questions allow students to check their understanding before moving on.

Each numbered section begins with **Expected Learning Outcomes** to help focus the reader's attention on the larger concepts and make the course outcome-driven. This also assists instructors in structuring their courses around expected learning outcomes.

**Questions** in figure legends and **Apply What You Know** items prompt students to think more deeply about the implications and applications of what they have learned.

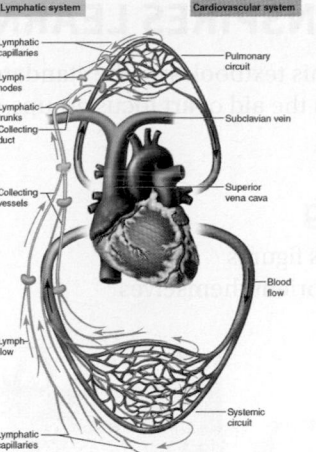

left subclavian vein. Collectively, this duct therefore drains all of the body below the diaphragm, and the left upper limb and left side of the head, neck, and thorax.

### Flow of Lymph

Lymph flows under forces similar to those that govern venous return, except that the lymphatic system has no pump like the heart, and lymph flows at even lower pressure and speed than venous blood. The primary mechanism of flow is rhythmic contractions of the lymphatic vessels themselves, which contract when the fluid stretches them. The valves of lymphatic vessels, like those of veins, prevent the fluid from flowing backward. Lymph flow is also produced by skeletal muscles squeezing the lymphatic vessels, like the skeletal muscle pump that moves venous blood. Since lymphatic vessels are often wrapped with an artery in a common connective tissue sheath, arterial pulsation may also rhythmically squeeze the lymphatic vessels and contribute to lymph flow. A thoracic (respiratory) pump promotes the flow of lymph from the abdominal to the thoracic cavity as one inhales, just as it does in venous return. Finally, at the point where the collecting ducts empty into the subclavian veins, the rapidly flowing bloodstream draws the lymph into it. Considering these mechanisms of lymph flow, it should be apparent that physical exercise significantly increases the rate of lymphatic return.

**▶▶▶ APPLY WHAT YOU KNOW**

*Why does it make more functional sense for the collecting ducts to connect to the subclavian veins than it would for them to connect to the subclavian arteries?*

### Lymphatic Cells AP|R

Another component of the lymphatic system is lymphatic tissue, which ranges from loosely scattered cells in the mucous membranes of the respiratory, digestive, urinary, and reproductive tracts to compact cell populations encapsulated in lymphatic organs. These tissues are composed of a variety of lymphocytes and other cells with various roles in defense and immunity:

1. **Natural killer (NK) cells** are large lymphocytes that attack and destroy bacteria, transplanted tissues, and *host cells* (cells of one's own body) that have either become infected with viruses or turned cancerous.

2. **T lymphocytes (T cells)** are lymphocytes that mature in the thymus and later depend on thymic hormones; the *T* stands for *thymus-dependent*. There are several subclasses of T cells that will be introduced later.

3. **B lymphocytes (B cells)** are lymphocytes that differentiate into *plasma cells*—connective tissue cells that secrete antibodies. They are named for an organ in chickens (the *bursa of Fabricius*[1]) in which they were first discovered.

[1]Hieronymus Fabricius (Girolamo Fabrizzi) (1537–1619), Italian anatomist

**FIGURE 21.5  Fluid Exchange Between the Circulatory and Lymphatic Systems.** Blood capillaries lose fluid to the tissue spaces. The lymphatic system picks up excess tissue fluid and returns it to the bloodstream.

❓ *Identify two benefits in having lymphatic capillaries pick up tissue fluid that is not reclaimed by the blood capillaries.*

drainage from the right arm and right side of the thorax and head and empties into the right subclavian vein.

2. The **thoracic duct,** on the left, is larger and longer. It begins just below the diaphragm anterior to the vertebral column at the level of the second lumbar vertebra. Here, the two lumbar trunks and the intestinal trunk join and form a prominent sac called the **cisterna chyli** (sis-TUR-nuh KY-lye), named for the large amount of *chyle* (fatty intestinal lymph) that it collects after a meal. The thoracic duct then passes through the diaphragm with the aorta and ascends the mediastinum, adjacent to the vertebral column. As it passes through the thorax, it receives additional lymph from the left bronchomediastinal, left subclavian, and left jugular trunks, then empties into the

---

The end-of-chapter **Study Guide** offers several methods for assessment that are useful to both students and instructors.

**Assess Your Learning Outcomes** provides students a study outline for review, and addresses the needs of instructors whose colleges require outcome-oriented syllabi and assessment of student achievement of the expected learning outcomes.

**End-of-chapter questions** build on all levels of Bloom's taxonomy in sections to
1. assess learning outcomes
2. test simple recall and analytical thought
3. build medical vocabulary
4. apply the basic knowledge to new clinical problems and other situations

**True or False** questions further address Bloom's taxonomy by asking the student to explain *why* the false statements are untrue.

**Testing Your Comprehension** questions address Bloom's Taxonomy in going beyond recall to application of ideas.

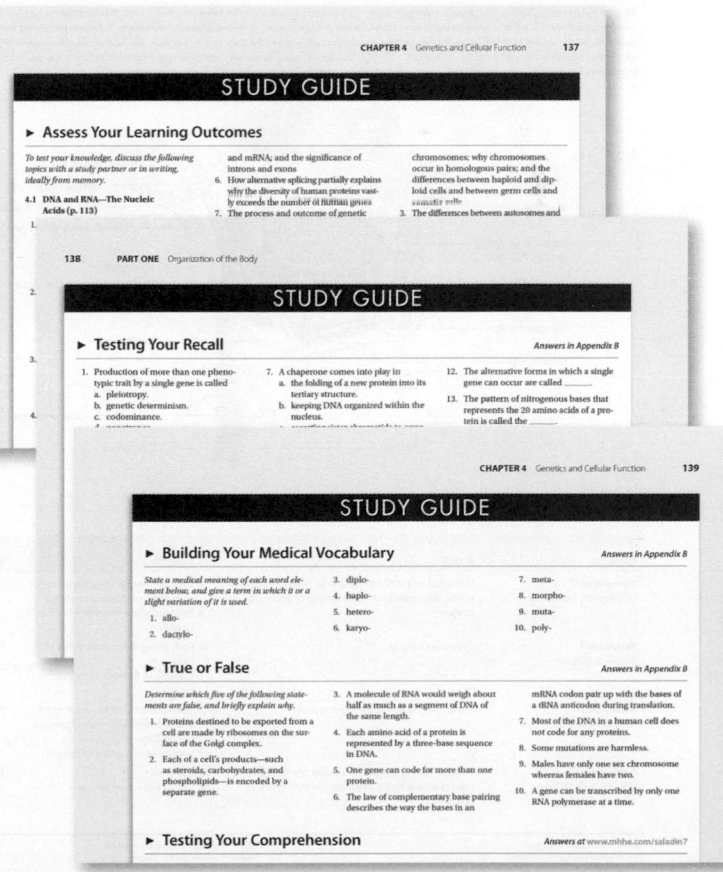

# ARTWORK THAT INSPIRES LEARNING

The incredible art program in this textbook sets the standard in A&P. The stunning portfolio of art and photos was created with the aid of art focus groups, and with feedback from hundreds of accuracy reviews.

## Conducive to Learning

- Easy-to-understand process figures
- Tools for students to easily orient themselves

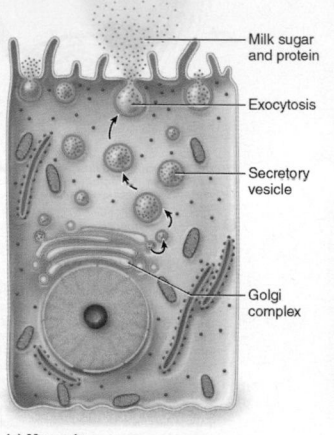
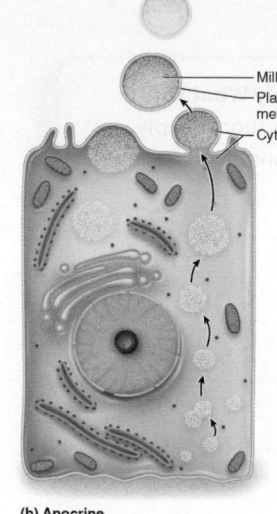
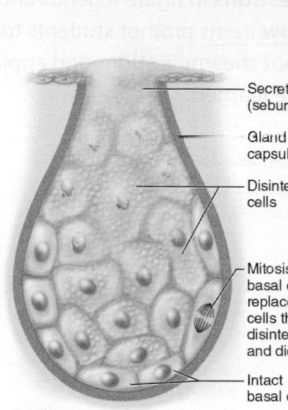

Milk sugar and protein
Exocytosis
Secretory vesicle
Golgi complex

Milk fat
Plasma membrane
Cytoplasm

Secretion (sebum)
Gland capsule
Disintegrating cells
Mitosis in basal cells replaces cells that disintegrate and die
Intact basal cells

(a) Merocrine     (b) Apocrine     (c) Holocrine

> **Vivid Illustrations**
>
> Rich textures and shading and bold, bright colors bring structures to life.

---

| TABLE 10.2 | Muscles of Chewing and Swallowing |
|---|---|

The following muscles contribute to facial expression and speech but are primarily concerned with the manipulation of food, including tongue movements, chewing, and swallowing.

**Extrinsic Muscles of the Tongue.** The tongue is a very agile organ. It pushes food between the molars for chewing (mastication) and later forces the food into the pharynx for swallowing (deglutition); it is also, of course, of crucial importance to speech. Both intrinsic and extrinsic muscles are responsible for its complex movements. The intrinsic muscles consist of a variable number of vertical fascicles that extend from the superior to the inferior sides of the tongue, transverse fascicles that extend from right to left, and longitudinal fascicles that extend from root to tip (see figs. 10.1c and 25.5b, p. 953). The extrinsic muscles listed here connect the tongue to other structures in the head (fig. 10.9). Three of these are innervated by the hypoglossal nerve (CN XII), whereas the fourth is innervated by both the vagus (CN X) and accessory (CN XI) nerves.

Palatoglossus
Styloglossus
Inferior longitudinal muscle of tongue
Genioglossus
Mylohyoid (cut)
Geniohyoid
Larynx
Trachea

Styloid process
Mastoid process
Posterior belly of digastric (cut)
Superior pharyngeal constrictor (cut)
Stylohyoid
Middle pharyngeal constrictor
Hyoglossus
Hyoid bone
Inferior pharyngeal constrictor
Esophagus

**FIGURE 10.9 Muscles of the Tongue and Pharynx.** [AP][R]

| Name | Action | O: Origin / I: Insertion | Innervation |
|---|---|---|---|
| **Genioglossus**[29] (JEE-nee-oh-GLOSS-us) | Unilateral action draws tongue to one side; bilateral action depresses midline of tongue or protrudes tongue | O: Superior mental spine on posterior surface of mental protuberance / I: Inferior surface of tongue from root to apex | Hypoglossal nerve |
| **Hyoglossus**[30] (HI-oh-GLOSS-us) | Depresses tongue | O: Body and greater horn of hyoid bone / I: Lateral and inferior surfaces of tongue | Hypoglossal nerve |
| **Styloglossus**[31] (STY-lo-GLOSS-us) | Draws tongue upward and posteriorly | O: Styloid process of temporal bone and ligament from styloid process to mandible / I: Superolateral surface of tongue | Hypoglossal nerve |
| **Palatoglossus**[32] (PAL-a-toe-GLOSS-us) | Elevates root of tongue and closes oral cavity off from pharynx; forms palatoglossal arch at rear of oral cavity | O: Soft palate / I: Lateral surface of tongue | Accessory and vagus nerves |

[29]*genio* = chin; *gloss* = tongue
[30]*hyo* = hyoid bone; *gloss* = tongue
[31]*stylo* = styloid process; *gloss* = tongue
[32]*palato* = palate; *gloss* = tongue

> **Muscle Tables**
>
> Muscle tables are organized into columnar format and enhanced with shading for easier reading and learning.

> *The visual appeal of nature is immensely important in motivating one to study it. We certainly see this at work in human anatomy—in the countless students who describe themselves as visual learners", in the many laypeople who find anatomy atlases so intriguing, and in the enormous popularity of Body Worlds and similar exhibitions of human anatomy.*
>
> —Ken Saladin

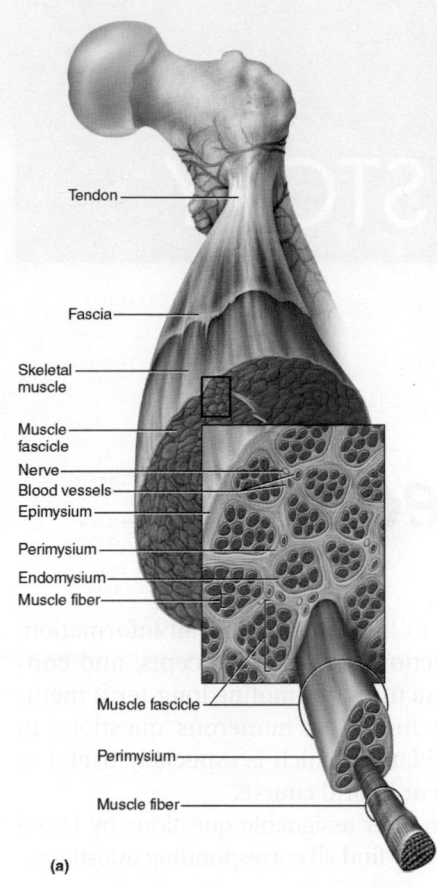

(a)

Tendon
Fascia
Skeletal muscle
Muscle fascicle
Nerve
Blood vessels
Epimysium
Perimysium
Endomysium
Muscle fiber
Muscle fascicle
Perimysium
Muscle fiber

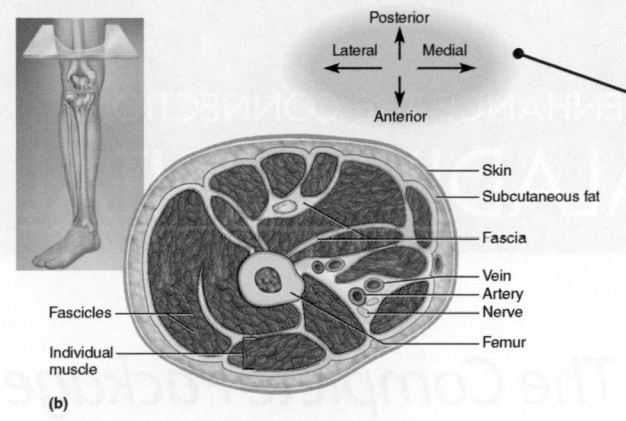

(b)

Posterior
Lateral — Medial
Anterior

Skin
Subcutaneous fat
Fascia
Vein
Artery
Nerve
Femur
Fascicles
Individual muscle

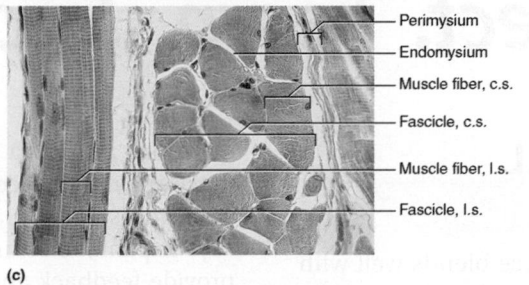

(c)

Perimysium
Endomysium
Muscle fiber, c.s.
Fascicle, c.s.
Muscle fiber, l.s.
Fascicle, l.s.

## Orientation Tools

Saladin art integrates tools to help students quickly orient themselves within a figure and make connections between ideas.

## Process Figures

Saladin breaks complicated physiological processes into numbered steps for a manageable introduction to difficult concepts.

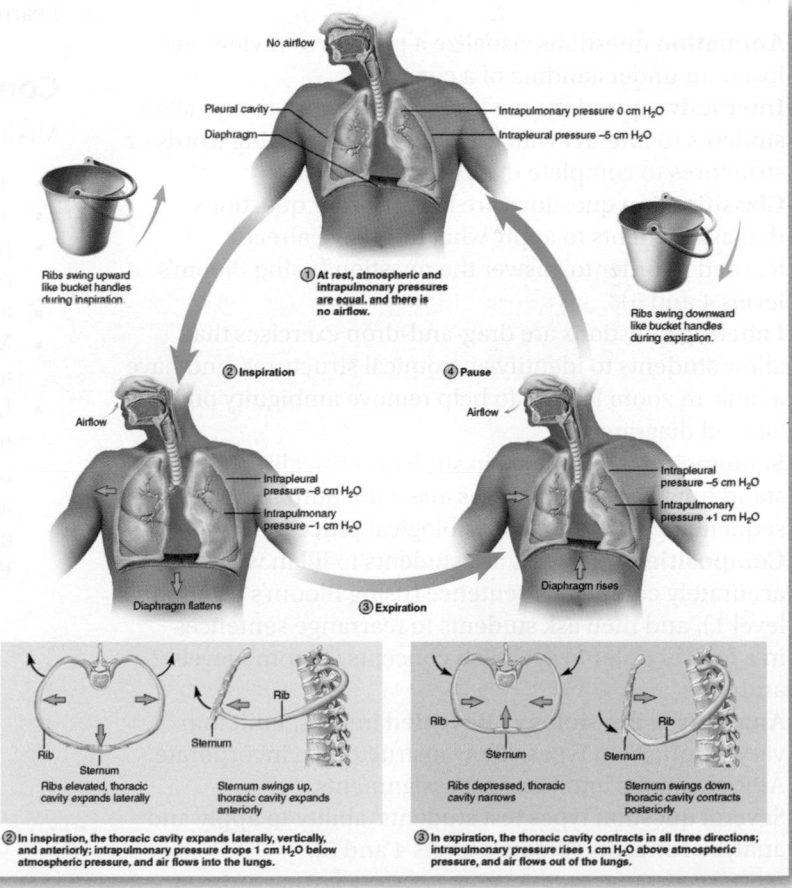

No airflow

Pleural cavity
Diaphragm

Intrapulmonary pressure 0 cm $H_2O$
Intrapleural pressure −5 cm $H_2O$

Ribs swing upward like bucket handles during inspiration.

① At rest, atmospheric and intrapulmonary pressures are equal, and there is no airflow.

Ribs swing downward like bucket handles during expiration.

② Inspiration

④ Pause

Airflow

Airflow

Intrapleural pressure −8 cm $H_2O$
Intrapulmonary pressure −1 cm $H_2O$

Intrapleural pressure −5 cm $H_2O$
Intrapulmonary pressure +1 cm $H_2O$

Diaphragm flattens

③ Expiration

Diaphragm rises

Rib
Sternum
Ribs elevated, thoracic cavity expands laterally

Rib
Sternum
Sternum swings up, thoracic cavity expands anteriorly

Rib
Sternum
Ribs depressed, thoracic cavity narrows

Rib
Sternum
Sternum swings down, thoracic cavity contracts posteriorly

② In inspiration, the thoracic cavity expands laterally, vertically, and anteriorly; intrapulmonary pressure drops 1 cm $H_2O$ below atmospheric pressure, and air flows into the lungs.

③ In expiration, the thoracic cavity contracts in all three directions; intrapulmonary pressure rises 1 cm $H_2O$ above atmospheric pressure, and air flows out of the lungs.

## The Complete Package
# Connect. Learn. Succeed.

## CONNECT WITH DIGITAL PRODUCTS

To ensure that the student study experience blends well with the learning outcomes in this text, digital authors Steve Sullivan and Chris Gan prepared Connect® questions in multiple formats.

- **Animation** questions visualize a process to review and foster an understanding of a concept.
- **Interactive** questions are kinesthetic in nature and allow students to interact with the question by moving words or structures to complete the exercise.
- **Classification** questions are higher-order questions that ask students to apply what they have already learned in order to answer the question (using Bloom's levels 4 and 5).
- **Labeling** questions are drag-and-drop exercises that allow students to identify anatomical structures, and have a built-in zoom feature to help remove ambiguity on labeled diagrams.
- **Sequencing** questions help students critically understand how disease processes affect the human body by sequencing events in a physiological process.
- **Composition** questions ask students to fill in words to accurately complete a sentence (using Bloom's taxonomy level 1), and then ask students to rearrange sentences in a logical order to establish concepts (Bloom's levels 2 and 3).
- **Anatomy & Physiology | Revealed** images, built into varying question types, allow instructors to incorporate APR questions into Connect assignments.
- Several question types test students' ability to apply and analyze concepts (Bloom's levels 4 and 5).

Other question types, such as multiple choice, true/false, fill-in-the-blank, and "Before You Go On" essay, are included to ensure consistent and adequate depth and breadth of coverage for each Learning Outcome in the text. Many Connect questions include a hint to help students recall information, make meaningful connections between concepts, and connect words and visual structures, promoting long-term memory. Explanations are included for numerous questions to provide feedback and guidance, which is especially useful to students taking an online or hybrid course.

Instructors are able to filter assignable questions by HAPS Learning Objectives to quickly find all corresponding questions.

### ConnectPlus® Anatomy & Physiology

McGraw-Hill ConnectPlus interactive learning platform provides

- auto-graded assignments
- adaptive diagnostic tools
- powerful reporting against learning outcomes and level of difficulty
- an easy-to-use interface
- McGraw-Hill Tegrity® Campus, which digitally records and distributes your lectures with a click of a button
- Learning Objectives from the textbook that are tied to interactive questions in Connect, to ensure that all parts of the chapters have adequate coverage within Connect assignments. ConnectPlus includes the full textbook as an integrated, dynamic eBook, which you can also assign. Everything you need—in one place!

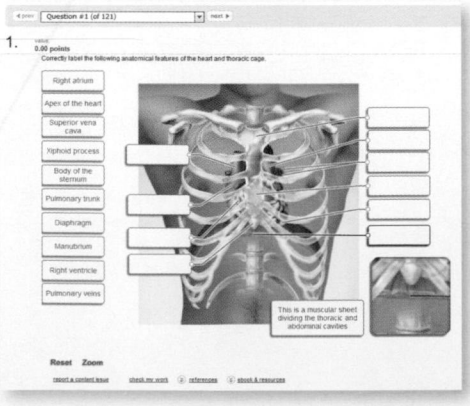

## LearnSmart™ Advantage
### The Evolution of Learning

LearnSmart Advantage is a new series of adaptive learning products fueled by LearnSmart—the most widely used and adaptive learning resource proven to strengthen memory recall, increase retention, and boost grades. Each product in the series helps students study smarter and retain more knowledge. www.learnsmartadvantage.com

### McGraw-Hill Smartbook™

This is the first—and only—adaptive reading experience designed to transform the way students read.

- Engages students with a personalized reading experience
- Ensures that students retain knowledge

### McGraw-Hill LearnSmart®

Unlike static flashcards or rote memorization, LearnSmart ensures that your students have mastered course concepts before taking the exam, thereby saving you time and increasing student success.

- The only truly adaptive learning system
- Intelligently identifies course content students have not yet mastered
- Maps out personalized study plans for student success

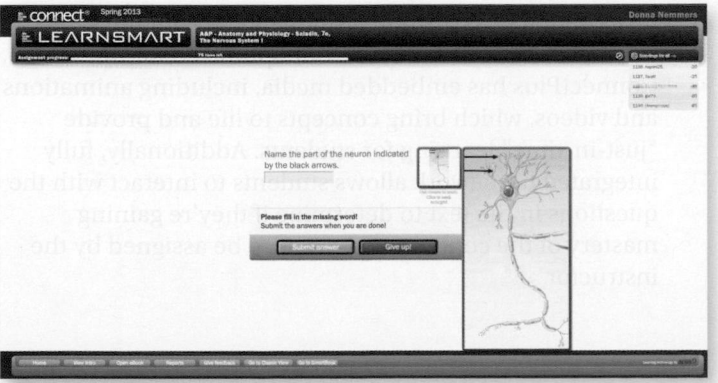

## McGraw-Hill LearnSmart Labs™

 THE Virtual Lab Experience. Based on the same world-class super-adaptive technology as LearnSmart, McGraw-Hill's LearnSmart Labs are must-see, outcomes-based lab simulations. LearnSmart Labs assess a student's knowledge and adaptively correct deficiencies, allowing the student to learn faster and retain more knowledge with greater success.

First, a student's knowledge is adaptively leveled on core learning outcomes: Questioning reveals knowledge deficiencies that are corrected by the delivery of content that is conditional on a student's response. Then, a simulated lab experience requires the student to think and act like a scientist: recording, interpreting, and analyzing data using simulated equipment found in labs and clinics. The student is allowed to make mistakes—a powerful part of the learning experience! A virtual coach provides subtle hints when needed, asks questions about the student's choices, and allows the student to reflect upon and correct those mistakes. Whether your need is to overcome the logistical challenges of a traditional lab, provide better lab prep, improve student performance, or make your online experience one that rivals the real world, LearnSmart Labs accomplish it al!.

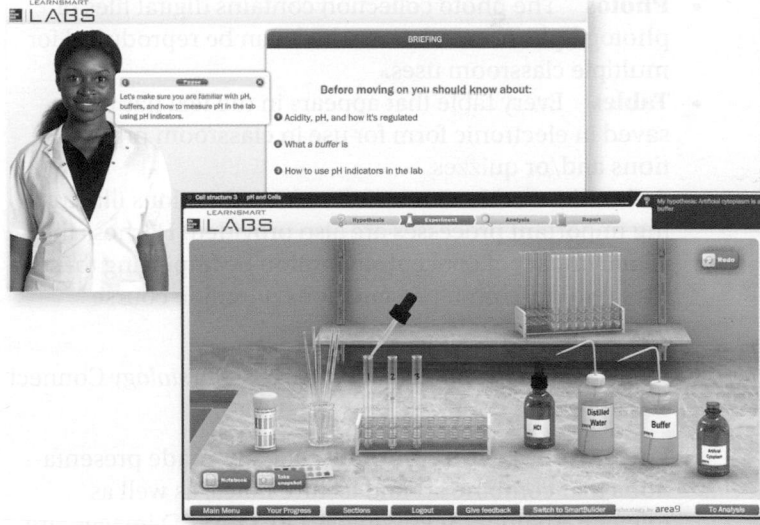

### LearnSmart Prep™

This adaptive course preparation tool quickly and efficiently helps students prepare for college-level work.

- Levels out student knowledge
- Keeps students on track

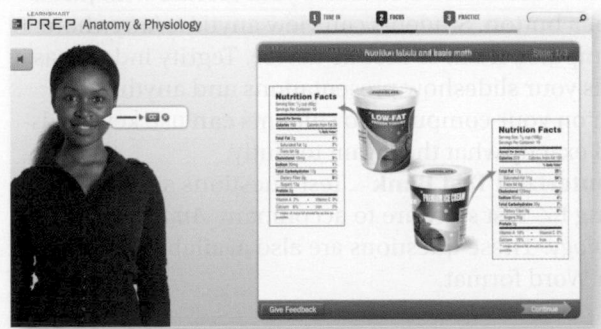

## Textbook Website: www.mhhe.com/saladin7e

Accessed from the *Anatomy & Physiology* Connect website or through the URL shown above, Presentation Center is an online digital library containing photos, artwork, animations, and other media types that can be used to create customized lectures, visually enhanced tests and quizzes, compelling course websites, or attractive printed support materials. All assets are copyrighted by McGraw-Hill but can be used by instructors for classroom purposes. The visual resources in this collection include

- **Art**   Full-color digital files (both labeled and unlabeled versions) of all illustrations in the book can be readily incorporated into lecture presentations, exams, or custom-made classroom materials. In addition, all files are pre-inserted into PowerPoint slides for ease of lecture preparation.
- **Photos**   The photo collection contains digital files of photographs from the text, which can be reproduced for multiple classroom uses.
- **Tables**   Every table that appears in the text has been saved in electronic form for use in classroom presentations and/or quizzes.
- **Animations**   Numerous full-color animations illustrating important processes are also provided. Harness the visual impact of concepts in motion by importing these files into classroom presentations or online course materials.

Also accessed through the *Anatomy & Physiology* Connect website are

- **PowerPoint Lecture Outlines**   Ready-made presentations that combine art and lecture notes, as well as relevant Anatomy & Physiology | REVEALED images, are provided for each chapter of the text.
- **PowerPoint Slides**   For instructors who prefer to create their lectures from scratch, all illustrations, photos, tables, and animations are pre-inserted by chapter into blank PowerPoint slides.
- **Digital Lecture Capture: Tegrity®**   McGraw-Hill Tegrity Campus records and distributes your lecture with just a click of a button. Students can view anytime/anywhere via computer, iPod, or mobile device. Tegrity indexes as it records your slideshow presentations and anything shown on your computer, so students can use keywords to find exactly what they want to study.
- **Computerized Test Bank**   Test questions are served up utilizing EZ Test software to accompany *Anatomy & Physiology*. These questions are also available to instructors in Word format.

*Instructors:* To access Connect, request registration information from your McGraw-Hill sales representative.

## Content Delivery Flexibility

*Anatomy & Physiology* is available in many formats in addition to the traditional textbook to give instructors and students more choices. Choices include

- **Customizable Textbooks: Create™**   Introducing McGraw-Hill Create, a self-service website that allows you to create custom course materials—print and eBooks—by drawing upon McGraw-Hill's comprehensive, cross-disciplinary content. Add your own content quickly and easily. Tap into other rights-secured, third-party sources as well. Then, arrange the content in a way that makes the most sense for your course. Even personalize your book with your course name and information. Choose the best format for your course: color print, black and white print, or eBook. The eBook is now viewable on an iPad! When you are finished customizing, you will receive a free PDF review copy in just minutes. Visit McGraw-Hill Create—**www.mcgrawhillcreate.com**—today and begin building your perfect book.
- **ConnectPlus eBook**   McGraw-Hill's ConnectPlus eBook takes digital texts beyond a simple PDF. With the same content as the printed book, but optimized for the screen, ConnectPlus has embedded media, including animations and videos, which bring concepts to life and provide "just-in-time" learning for students. Additionally, fully integrated homework allows students to interact with the questions in the text to determine if they're gaining mastery of the content, and can also be assigned by the instructor.

# An Interactive Cadaver Dissection Experience

## Anatomy & Physiology | REVEALED
www.aprevealed.com

APR 3.0 is an interactive cadaver dissection tool to enhance lectures and labs that students can use anytime, anywhere. Instructors may customize APR 3.0 to their course by selecting the specific structures they require in their course, and APR 3.0 does the rest. Once the structure list is generated, APR highlights these selected structures for students. APR contains all the material covered in an A&P course, including these three new modules:

- Tissues
- Cells and Chemistry
- Body Orientation

An APR application also makes APR accessible on Apple® and Android™ tablets.

*Anatomy & Physiology's* text, artwork, and photos are integrated with APR 3.0, so that wherever students see the APR 3.0 logo **AP|R** in their eBook, they can simply click the logo and they will be taken specifically to the dissection photos, animations, histology slides, and radiological images in APR that support and enrich their understanding of the text.

## APR Is Now Available in Two New Versions

**Anatomy & Physiology Revealed | Cat®** and **Anatomy & Physiology Revealed | Fetal Pig®** are online interactive cat dissection and fetal pig dissection experiences that use cat photos or fetal pig photos, combined with a layering technique that allows you to peel away layers to reveal structures beneath the surface. Both **Anatomy & Physiology Revealed | Cat** and **Anatomy & Physiology Revealed | Fetal Pig** offer animations, histologic and radiologic imaging, audio pronunciations, and comprehensive quizzing.

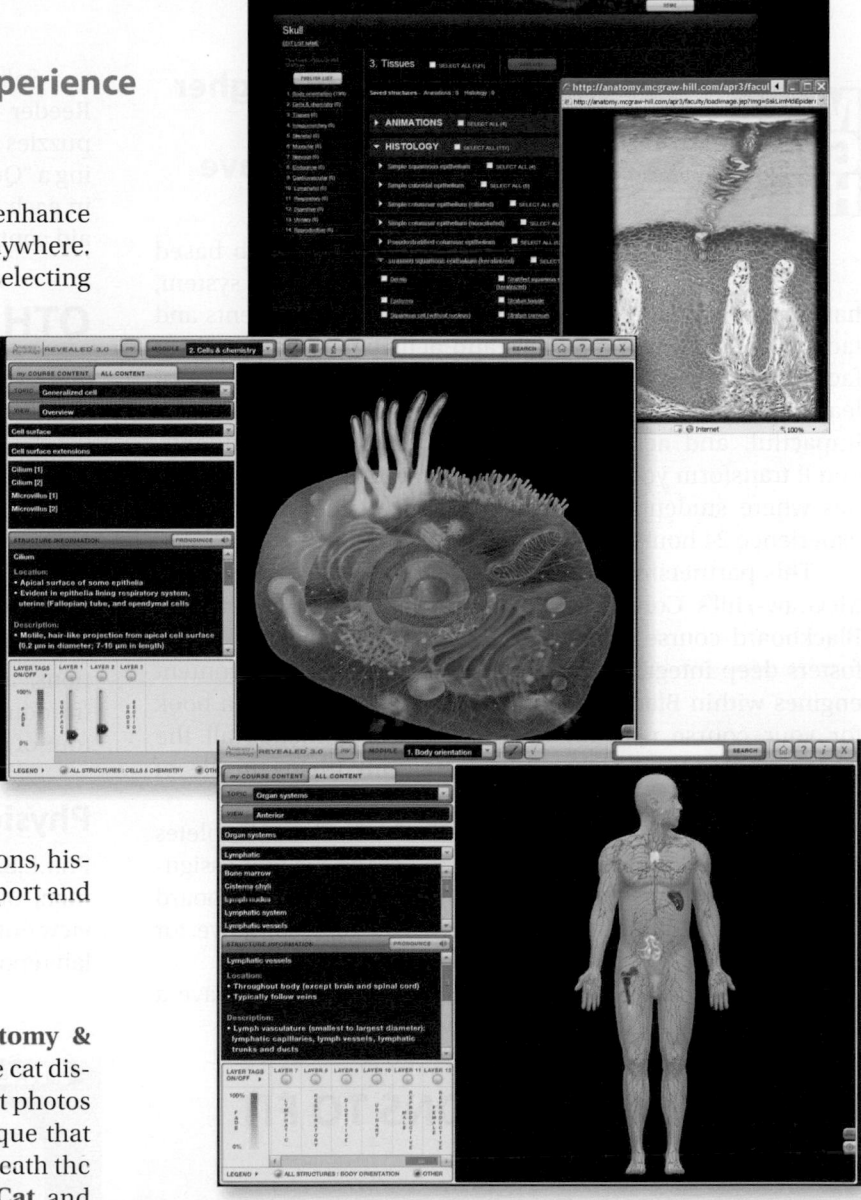

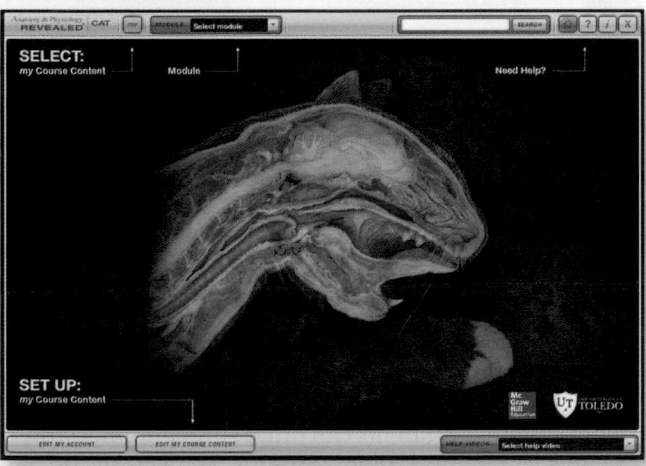

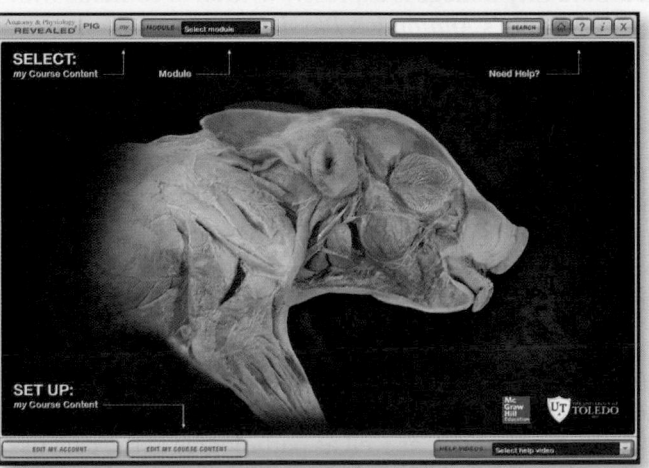

## McGraw-Hill Higher Education and Blackboard® Have Teamed Up.

## Do More

Blackboard, the Web-based course-management system, has partnered with McGraw-Hill to better allow students and faculty to use online materials and activities to complement face-to-face teaching. Blackboard features exciting social learning and teaching tools that foster more logical, visually impactful, and active learning opportunities for students. You'll transform your closed-door classrooms into communities where students remain connected to their educational experience 24 hours a day.

This partnership allows you and your students access to McGraw-Hill's Connect and Create right from within your Blackboard course—all with one single sign-on. Blackboard fosters deep integration of McGraw-Hill content and content engines within Blackboard. Whether you're choosing a book for your course or building Connect assignments, all the tools you need are right where you want them—inside of Blackboard.

Gradebooks are now seamless. When a student completes an integrated Connect assignment, the grade for that assignment automatically (and instantly) feeds your Blackboard grade center. Ask your local McGraw-Hill representative for details.

Even if your school is not using Blackboard, we have a solution for you. Learn more at **www.domorenow.com.**

## LAB MANUAL OPTIONS TO FIT YOUR COURSE

The *Anatomy & Physiology Laboratory Manual* by Eric Wise of Santa Barbara City College is expressly written to coincide with chapters of Saladin's *Anatomy & Physiology.*

The *Laboratory Manual for Human Anatomy & Physiology* by Terry Martin of Kishwaukee College is written to coincide with Saladin or any A&P textbook.

- Three versions available including the main version, cat, and fetal pig
- Includes Ph.I.L.S. 4.0
- Outcomes and assessments format
- Clear, concise writing style

The *Clinical Applications Manual to Accompany Anatomy and Physiology* is correlated to the Saladin text. The first six chapters introduce general aspects of patient examination, diagnosis, and treatment, as well as pathologies at the cell, tissue, and genetic levels. All later chapters explain disorders and procedures specific to each body system, elaborating on some clinical conditions described in the text, and introducing additional conditions that are especially common or interesting.

*Solve Saladin Crossword Puzzle Supplement* by Greg Reeder of Broward Community College contains crossword puzzles tailored to each section of Saladin's chapters, including a "Quiz" puzzle to summarize the most important concepts in each chapter. This excellent anatomy and physiology study aid contains all puzzle answers in the back of the book.

## OTHER RESOURCES AVAILABLE

### Student Study Guide

This comprehensive study guide written by experienced instructor Jacque Homan in collaboration with Ken Saladin contains vocabulary-building and content-testing exercises, labeling exercises, and practice exams.

### Physiology Tutorials

MediaPhys offers detailed explanations, high-quality illustrations, and animations to provide students with a thorough introduction to the world of physiology—giving them a virtual tour of physiological processes.

### Physiology Interactive Lab Simulations

Ph.I.L.S. offers 42 lab simulations that may be used to supplement or substitute for wet labs. Users may adjust variables, view outcomes, make predictions, draw conclusions, and print lab reports.

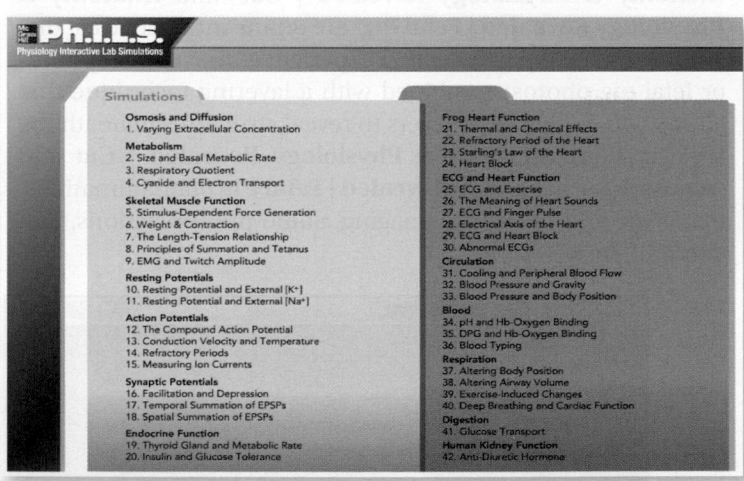

McGraw-Hill offers various tools and technology products to support the textbook. Instructors can obtain teaching aids to accompany this textbook by visiting our online catalog at **www.mhhe.com,** by calling McGraw-Hill Customer Service at 1-800-338-3987, or by contacting their local McGraw-Hill sales representative. Students can order supplemental study materials by contacting their campus bookstore or online at **www.shopmcgraw-hill.com.**

# ACKNOWLEDGMENTS

Our grateful thanks are extended to the reviewers of this text, who provided instructive comments to help refine and update the content within these pages.

Michele D. Alexandre
*Durham Technical Community College*

Tami Asplin
*North Dakota State University*

Seher Atamturktur
*Bronx Community College*

Tim Ballard
*University of North Carolina-Wilmington*

Patricia Bernard
*Trocaire College*

Christopher Brandon Jr.
*Georgia Gwinnett College, School of Science & Technology*

Winnifred Bryant
*University of Wisconsin-Eau Claire*

Wayne Busch
*Riverland Community College*

Juliann Chmielecki
*Bunker Hill Community College*

Roger Choate
*Oklahoma City Community College*

Thomas F. Chubb
*Villanova University*

Tracy Curtis
*Broome Community College*

Patricia DeLeon
*BMCC/CUNY*

Mathew Edick
*Grand Valley State University*

Amy Fenech Sandy
*Columbus Technical College*

Julie Fischer
*Wallace Community College*

Teresa Fischer
*Indian River State College*

Ted Fleming
*Bradley University*

Clifford Fontenot
*Southeastern Louisiana University*

Seth Gardner
*Bowling Green State University-Firelands Campus*

Carrie Gaynor
*Macomb Community College*

Mike Gehner
*Xavier University*

Christine Griffiths
*Jackson State Community College*

Matthew Halter
*Cabrillo College*

Steven B. Hammer
*Indian River State College*

Judy Y. Harris
*Kennebec Valley Community College*

DeLoris Hesse
*University of Georgia*

William Hopkins
*Whatcom Community College*

Melinda Hutton
*McNeese State University*

Kathy Jo Ann Jackson
*McNeese State University*

Judy Jiang
*Triton College*

Astagik Kalashnikova
*Broward College*

Steve Kash
*Oklahoma City Community College*

Raymond Larsen
*Bowling Green State University*

Deborah Lawson
*Maryville University-St. Louis*

Carlos Liachovitzky
*Bronx Community College of The City University of New York*

Sarah Liechty
*Ivy Tech Community College*

David Barry Markillie
*Cape Fear Community College*

Tiffany McFalls
*Elizabethtown Community & Technical College*

Abraham Miller
*University of Tampa*

Robert Moldenhauer
*St. Clair County Community College*

April Murphy
*Columbus Technical College*

Keith (Dan) Murray
*South Texas College*

Ellen Ott-Reeves
*Blinn College-Bryan*

Kjerstin Owens
*Trocaire College*

James Palmer
*Southeastern Louisiana University*

Maria Pereira
*Hunter College, CUNY*

Jocelyn Ramos
*Ivy Tech Community College-Franklin*

Cynthia P. Robison
*Wallace Community College-Dothan*

Susan Rohde
*Triton College*

Donald Shaw
*University of Tennessee-Martin*

Gidi Shemer
*University of North Carolina-Chapel Hill*

Hollis Smith
*Kennebec Valley Community College*

Michelle Smith
*Columbus Technical College*

Kimberly Sonanstine
*Wallace Community College*

Christopher Sorenson
*St. Cloud Technical and Community College*

Jeff Spencer
*The University of Akron*

William Stewart
*Middle Tennessee State University*

Justin St. Juliana
*Ivy Tech Community College-Wabash Valley*

Diana Sturges
*Georgia Southern University*

Pete Sutliff
*Wayne County Community College District*

Mohamad Termos
*South Texas College*

Diane Tice
*Morrisville State College*

Todd Tolar
*Wallace Community College*

Janice Webster
*Ivy Tech Community College*

Andrzej Wieraszko
*The College of Staten Island/CUNY*

Daniel B. Williams
*Winston Salem State University*

Sonya J. Williams
*Oklahoma City Community College*

Vanessa Williams
*University of Georgia*

Madelyn Wilson
*Three Rivers College*

Kerby Winters
*Treasure Valley Community College*

Jen Wortham
*University of Tampa*

Catherine Young
*Bunker Hill Community College*

Jeffrey Zuiderveen
*Columbus State University*

When I was a young boy, I became interested in what I then called "nature study" for two reasons. One was the sheer beauty of nature. I reveled in children's books with abundant, colorful drawings and photographs of animals, plants, minerals, and gems. It was this esthetic appreciation of nature that made me want to learn more about it and made me happily surprised to discover I could make a career of it. At a slightly later age, another thing that drew me still deeper into biology was to discover writers who had a way with words—who could captivate my imagination and curiosity with their elegant prose. Once I was old enough to hold part-time jobs, I began buying zoology and anatomy books that mesmerized me with their gracefulness of writing and fascinating art and photography. I wanted to write and draw like that myself, and I began teaching myself by learning from "the masters." I spent many late nights in my room peering into my microscope and jars of pond water, typing page after page of manuscript, and trying pen and ink as a medium. My "first book" was a 318-page paper on some little pond animals called hydras, with 53 India ink illustrations that I wrote for my tenth-grade biology class when I was 16.

Fast-forward about 30 years, to when I became a textbook writer, and I found myself bringing that same enjoyment of writing and illustrating to the first edition of this book you are now holding. Why? Not only for its intrinsic creative satisfaction, but because I'm guessing that you're like I was—you can appreciate a book that does more than simply give you the information you need. You appreciate, I trust, a writer who makes it enjoyable for you through his scientific, storytelling prose and his concept of the way things should be illustrated to spark interest and facilitate understanding.

I know from my own students, however, that you need more than captivating illustrations and enjoyable reading. Let's face it—A&P is a complex subject and it may seem a formidable task to acquire even a basic knowledge of the human body. It was difficult even for me to learn (and the learning never ends). So in addition to simply writing this book, I've given a lot of thought to its pedagogy—the art of teaching. I've designed my chapters to make them easier for you to study and to give you abundant opportunity to check whether you've understood what you read—to test yourself (as I advise my own students) before the instructor tests you.

Each chapter is broken down into short, digestible bits with a set of Expected Learning Outcomes at the beginning of each section, and self-testing questions (Before You Go On) just a few pages later. Even if you have just 30 minutes to read during a lunch break or a bus ride, you can easily read or review one of these brief sections. There are also numerous self-testing questions in a Study Guide at the end of each chapter, in some of the figure legends, and the occasional Apply What You Know questions dispersed throughout each chapter. The questions cover a broad range of cognitive skills, from simple recall of a term to your ability to evaluate, analyze, and apply what you've learned to new clinical situations or other problems.

I hope you enjoy your study of this book, but I know there are always ways to make it even better. Indeed, what quality you may find in this edition owes a great deal to feedback I've received from students all over the world. If you find any typos or other errors, if you have any suggestions for improvement, if I can clarify a concept for you, or even if you just want to comment on something you really like about the book, I hope you'll feel free to write to me. I correspond quite a lot with students and would enjoy hearing from you.

Ken Saladin
Georgia College & State University
Milledgeville, GA 31061 (USA)
ken.saladin@gcsu.edu

*Dedicated to the memory of*
Libbie Henrietta Hyman
*who patiently indulged my adolescent zoological questions*
*and, unknowingly, taught me how to write*

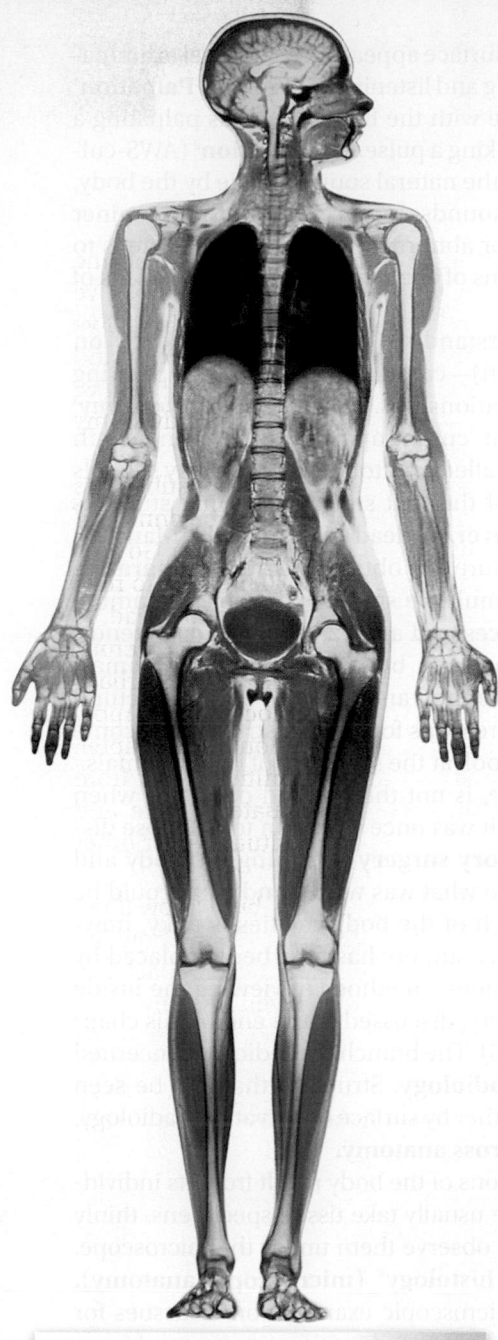

A colorized MRI scan of the human body

# CHAPTER 1

# MAJOR THEMES OF ANATOMY AND PHYSIOLOGY

aprevealed.com

**Volume 1 Skeletal and Muscular Systems**

No branch of science hits as close to home as the science of our own bodies. We're grateful for the dependability of our hearts; we're awed by the capabilities of muscles and joints displayed by Olympic athletes; and we ponder with philosophers the ancient mysteries of mind and emotion. We want to know how our body works, and when it malfunctions, we want to know what is happening and what we can do about it. Even the most ancient writings of civilization include medical documents that attest to humanity's timeless drive to know itself. You are embarking on a subject that is as old as civilization, yet one that grows by thousands of scientific publications every week.

This book is an introduction to human structure and function, the biology of the human body. It is meant primarily to give you a foundation for advanced study in health care, exercise physiology, pathology, and other fields related to health and fitness. Beyond that purpose, however, it can also provide you with a deeply satisfying sense of self-understanding.

As rewarding and engrossing as this subject is, the human body is highly complex, and understanding it requires us to comprehend a great deal of detail. The details will be more manageable if we relate them to a few broad, unifying concepts. The aim of this chapter, therefore, is to introduce such concepts and put the rest of the book into perspective. We consider the historical development of anatomy and physiology, the thought processes that led to the knowledge in this book, the meaning of human life, some central concepts of physiology, and how to better understand medical terminology.

## 1.1   The Scope of Anatomy and Physiology

### Expected Learning Outcomes

When you have completed this section, you should be able to

a. define *anatomy* and *physiology* and relate them to each other;

b. describe several ways of studying human anatomy; and

c. define a few subdisciplines of human physiology.

**Anatomy** is the study of structure, and **physiology** is the study of function. These approaches are complementary and never entirely separable. Together, they form the bedrock of the health sciences. When we study a structure, we want to know, What does it do? Physiology thus lends meaning to anatomy; and, conversely, anatomy is what makes physiology possible. This *unity of form and function* is an important point to bear in mind as you study the body. Many examples of it will be apparent throughout the book—some of them pointed out for you, and others you will notice for yourself.

### Anatomy—The Study of Form

There are several ways to examine the structure of the human body. The simplest is **inspection**—simply looking at the body's appearance, as in performing a physical examination or making a clinical diagnosis from surface appearance. Physical examinations also involve touching and listening to the body. **Palpation**[1] means feeling a structure with the hands, such as palpating a swollen lymph node or taking a pulse. **Auscultation**[2] (AWS-cul-TAY-shun) is listening to the natural sounds made by the body, such as heart and lung sounds. In **percussion,** the examiner taps on the body, feels for abnormal resistance, and listens to the emitted sound for signs of abnormalities such as pockets of fluid or air.

But a deeper understanding of the body depends on **dissection** (dis-SEC-shun)—carefully cutting and separating tissues to reveal their relationships. The very words *anatomy*[3] and *dissection*[4] both mean "cutting apart"; until the nineteenth century, dissection was called "anatomizing." In many schools of health science, one of the first steps in training students is dissection of the **cadaver,**[5] a dead human body. Many insights into human structure are obtained from **comparative anatomy**—the study of multiple species in order to examine similarities and differences and analyze evolutionary trends. Anatomy students often begin by dissecting other animals with which we share a common ancestry and many structural similarities. Many of the reasons for human structure become apparent only when we look at the structure of other animals.

Dissection, of course, is not the method of choice when studying a living person! It was once common to diagnose disorders through **exploratory surgery**—opening the body and taking a look inside to see what was wrong and what could be done about it. Any breach of the body cavities is risky, however, and most exploratory surgery has now been replaced by **medical imaging** techniques—methods of viewing the inside of the body without surgery, discussed at the end of this chapter (see Deeper Insight 1.5). The branch of medicine concerned with imaging is called **radiology.** Structure that can be seen with the naked eye—whether by surface observation, radiology, or dissection—is called **gross anatomy.**

Ultimately, the functions of the body result from its individual cells. To see those, we usually take tissue specimens, thinly slice and stain them, and observe them under the microscope. This approach is called **histology**[6] (**microscopic anatomy**). **Histopathology** is the microscopic examination of tissues for signs of disease. **Cytology**[7] is the study of the structure and function of individual cells. **Ultrastructure** refers to fine detail, down to the molecular level, revealed by the electron microscope.

### Physiology—The Study of Function

Physiology[8] uses the methods of experimental science discussed later. It has many subdisciplines such as *neurophysiology* (physiology of the nervous system), *endocrinology* (physiology

---

[1]*palp* = touch, feel; *ation* = process
[2]*auscult* = listen; *ation* = process
[3]*ana* = apart; *tom* = cut
[4]*dis* = apart; *sect* = cut
[5]from *cadere* = to fall down or die
[6]*histo* = tissue; *logy* = study of
[7]*cyto* = cell; *logy* = study of
[8]*physio* = nature; *logy* = study of

of hormones), and *pathophysiology* (mechanisms of disease). Partly because of limitations on experimentation with humans, much of what we know about bodily function has been gained through **comparative physiology,** the study of how different species have solved problems of life such as water balance, respiration, and reproduction. Comparative physiology is also the basis for the development of new drugs and medical procedures. For example, a cardiac surgeon may learn animal surgery before practicing on humans, and a vaccine cannot be used on human subjects until it has been demonstrated through animal research that it confers significant benefits without unacceptable risks.

**BEFORE YOU GO ON**

Answer the following questions to test your understanding of the preceding section:

1. What is the difference between anatomy and physiology? How do these two sciences support each other?

2. Name the method that would be used for each of the following: listening to a patient for a heart murmur; studying the microscopic structure of the liver; microscopically examining liver tissue for signs of hepatitis; learning the blood vessels of a cadaver; and performing a breast self-examination.

## 1.2 The Origins of Biomedical Science

### Expected Learning Outcomes

When you have completed this section, you should be able to

a. give examples of how modern biomedical science emerged from an era of superstition and authoritarianism; and

b. describe the contributions of some key people who helped to bring about this transformation.

Any science is more enjoyable if we consider not just the current state of knowledge, but how it compares to past understandings of the subject and how our knowledge was gained. Of all sciences, medicine has one of the most fascinating histories. Medical science has progressed far more in the last 50 years than in the 2,500 years before that, but the field did not spring up overnight. It is built upon centuries of thought and controversy, triumph and defeat. We cannot fully appreciate its present state without understanding its past—people who had the curiosity to try new things, the vision to look at human form and function in new ways, and the courage to question authority.

### The Greek and Roman Legacy

As early as 3,000 years ago, physicians in Mesopotamia and Egypt treated patients with herbal drugs, salts, physical therapy, and faith healing. The "father of medicine," however, is usually considered to be the Greek physician **Hippocrates** (c. 460–c. 375 BCE). He and his followers established a code of ethics for physicians, the Hippocratic Oath, that is still recited in modern form by graduating physicians at some medical schools. Hippocrates urged physicians to stop attributing disease to the activities of gods and demons and to seek their natural causes, which could afford the only rational basis for therapy.

**Aristotle** (384–322 BCE) was one of the first philosophers to write about anatomy and physiology. He believed that diseases and other natural events could have either supernatural causes, which he called *theologi,* or natural ones, which he called *physici* or *physiologi.* We derive such terms as *physician* and *physiology* from the latter. Until the nineteenth century, physicians were called "doctors of physic." In his anatomy book, *On the Parts of Animals,* Aristotle tried to identify unifying themes in nature. Among other points, he argued that complex structures are built from a smaller variety of simple components—a perspective that we will find useful later in this chapter.

### ►►►APPLY WHAT YOU KNOW

*When you have completed this chapter, discuss the relevance of Aristotle's philosophy to our current thinking about human structure.*

**Claudius Galen** (c. 130–c. 200), physician to the Roman gladiators, wrote the most influential medical textbook of the ancient era—a book worshipped to excess by medical professors for centuries to follow. Cadaver dissection was banned in Galen's time because of some horrid excesses that preceded him, including public dissection of living slaves and prisoners. Aside from what he could learn by treating gladiators' wounds, Galen was therefore limited to dissecting pigs, monkeys, and other animals. Because he was not permitted to dissect cadavers, he had to guess at much of human anatomy and made some incorrect deductions from animal dissections. He described the human liver, for example, as having five fingerlike lobes, somewhat like a baseball glove, because that is what he had seen in baboons. But Galen saw science as a method of discovery, not as a body of fact to be taken on faith. He warned that even his own books could be wrong and advised his followers to trust their own observations more than any book. Unfortunately, his advice was not heeded. For nearly 1,500 years, medical professors dogmatically taught what they read in Aristotle and Galen, seldom daring to question the authority of these "ancient masters."

### The Birth of Modern Medicine

In the Middle Ages, the state of medical science varied greatly from one religious culture to another. Science was severely repressed in the Christian culture of Europe until about the sixteenth century, although some of the most famous medical schools of Europe were founded during this era. Their professors, however, taught medicine primarily as a dogmatic commentary on Galen and Aristotle, not as a field of original research. Medieval medical illustrations were crude representations of the body intended more to decorate a page than to depict the body realistically (fig. 1.1a). Some were astrological charts that showed which sign of the zodiac was thought to influence each organ of the body. From such pseudoscience came the word *influenza,* Italian for "influence."

Free inquiry was less inhibited in Jewish and Muslim culture during this time. Jewish physicians were the most esteemed practitioners of their art—and none more famous than *Moses ben Maimon* (1135–1204), known in Christendom as **Maimonides.** Born in Spain, he fled to Egypt at age 24 to escape antisemitic persecution. There he served the rest of his life as physician to the court of the sultan, Saladin. A highly admired rabbi, Maimonides wrote voluminously on Jewish law and theology, but also wrote 10 influential medical books and numerous treatises on specific diseases.

Among Muslims, probably the most highly regarded medical scholar was *Ibn Sina* (980–1037), known in the West as **Avicenna** or "the Galen of Islam." He studied Galen and Aristotle, combined their findings with original discoveries, and questioned authority when the evidence demanded it. Medicine in the Mideast soon became superior to European medicine. Avicenna's textbook, *The Canon of Medicine,* was the leading authority in European medical schools for over 500 years.

Chinese medicine had little influence on Western thought and practice until relatively recently; the medical arts evolved in China quite independently of European medicine. Later chapters of this book describe some of the insights of ancient China and India.

Modern Western medicine began around the sixteenth century in the innovative minds of such people as the anatomist Andreas Vesalius and the physiologist William Harvey.

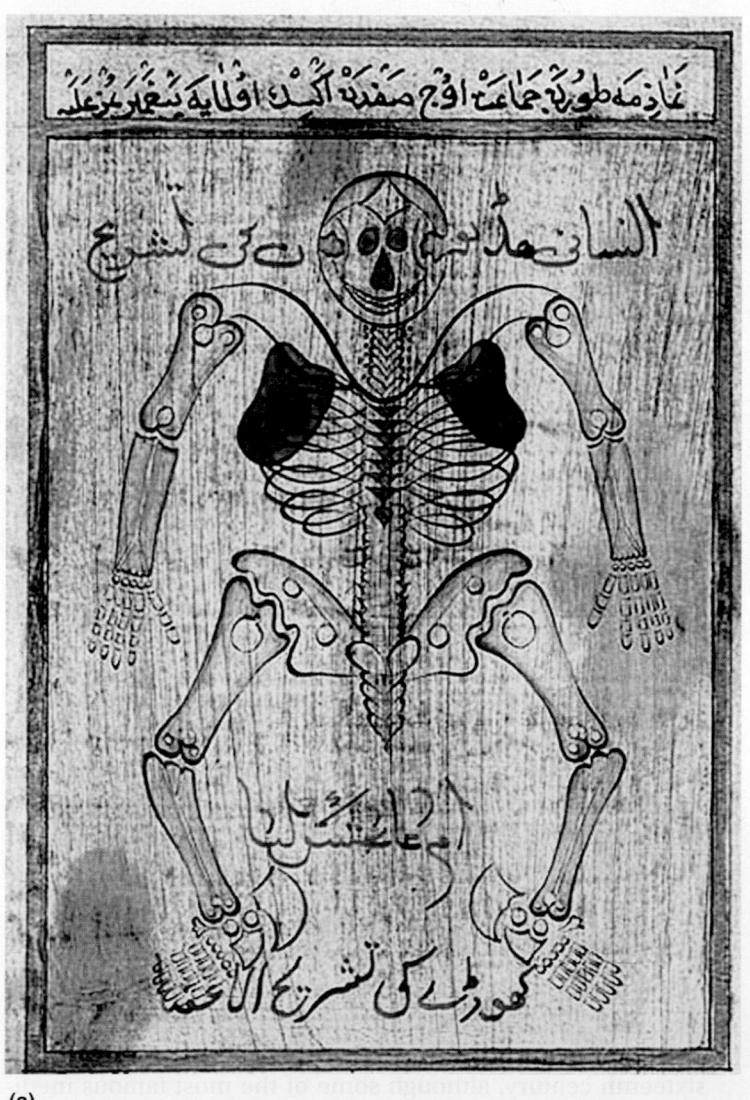

(a)

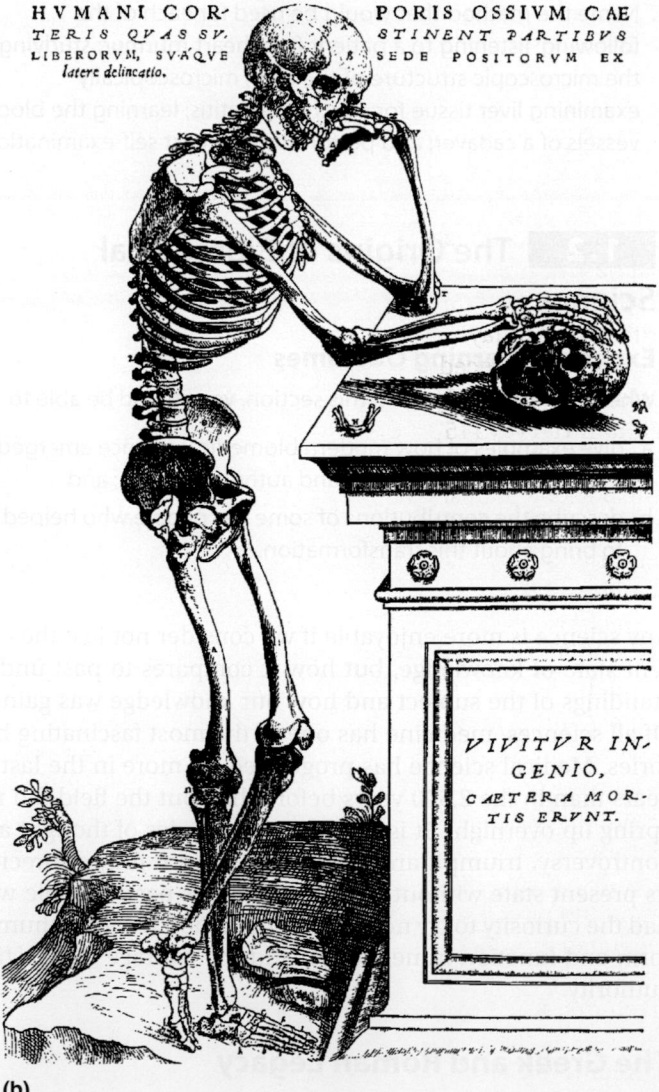

HVMANI COR-
*TERIS QVAS SV.*
*LIBERORVM, SVÆQVE*
*latere delincatio.*

PORIS OSSIVM CAE
*STINENT PARTIBVS*
*SEDE POSITORVM EX*

*VIVITVR IN-
GENIO,
CÆTERA MOR-
TIS ERVNT.*

(b)

**FIGURE 1.1 Evolution of Medical Art.** Two illustrations of the skeletal system made about 500 years apart. (a) From an eleventh-century work attributed to Persian physician Avicenna. (b) From *De Humani Corporis Fabrica* by Andreas Vesalius, 1543.

**Andreas Vesalius** (1514–64) taught anatomy in Italy. In his time, the Catholic Church relaxed its prohibition against cadaver dissection, in part to allow autopsies in cases of suspicious death. Furthermore, the Italian Renaissance created an environment more friendly to innovative scholarship. Dissection gradually found its way into the training of medical students throughout Europe. It was an unpleasant business, however, and most professors considered it beneath their dignity. In those days before refrigeration or embalming, the odor from the decaying cadaver was unbearable. Dissections were a race against decay. Bleary medical students had to fight the urge to vomit, lest they incur the wrath of an overbearing professor. Professors typically sat in an elevated chair, the cathedra, reading dryly in Latin from Galen or Aristotle while a lower-ranking *barber–surgeon* removed putrefying organs from the cadaver and held them up for the students to see. Barbering and surgery were considered to be "kindred arts of the knife"; today's barber poles date from this era, their red and white stripes symbolizing blood and bandages.

Vesalius broke with tradition by coming down from the cathedra and doing the dissections himself. He was quick to point out that much of the anatomy in Galen's books was wrong, and he was the first to publish accurate illustrations for teaching anatomy (fig. 1.1b). When others began to plagiarize them, Vesalius published the first atlas of anatomy, *De Humani Corporis Fabrica (On the Structure of the Human Body),* in 1543. This book began a rich tradition of medical illustration that has been handed down to us through such milestones as *Gray's Anatomy* (1856) and the vividly illustrated atlases and textbooks of today.

Anatomy preceded physiology and was a necessary foundation for it. What Vesalius was to anatomy, the Englishman **William Harvey** (1578–1657) was to physiology. Harvey is remembered especially for his studies of blood circulation and a little book he published in 1628, known by its abbreviated title *De Motu Cordis (On the Motion of the Heart).* He and **Michael Servetus** (1511–53) were the first Western scientists to realize that blood must circulate continuously around the body, from the heart to the other organs and back to the heart again. This flew in the face of Galen's belief that the liver converted food to blood, the heart pumped blood through the veins to all other organs, and those organs consumed it. Harvey's colleagues, wedded to the ideas of Galen, ridiculed him for his theory, though we now know he was correct (see p. 744). Despite persecution and setbacks, Harvey lived to a ripe old age, served as physician to the kings of England, and later did important work in embryology. Most importantly, Harvey's contributions represent the birth of experimental physiology—the method that generated most of the information in this book.

Modern medicine also owes an enormous debt to two inventors from this era, Robert Hooke and Antony van Leeuwenhoek, who extended the vision of biologists to the cellular level.

**Robert Hooke** (1635–1703), an Englishman, designed scientific instruments of various kinds, including the compound microscope. This is a tube with a lens at each end—an *objective lens* near the specimen, which produces an initial magnified image, and an *ocular lens (eyepiece)* near the observer's eye, which magnifies the first image still further. Although crude compound microscopes had existed since 1595, Hooke improved the optics and invented several of the helpful features found in microscopes today—a stage to hold the specimen, an illuminator, and coarse and fine focus controls. His microscopes magnified only about 30 times, but with them, he was the first to see and name cells. In 1663, he observed thin shavings of cork and observed that they "consisted of a great many little boxes," which he called *cellulae* (little cells) after the cubicles of a monastery (fig. 1.2). He later observed living cells "filled with juices." Hooke became particularly interested in microscopic examination of such material as insects, plant tissues, and animal parts. He published the first comprehensive book of microscopy, *Micrographia,* in 1665.

**Antony van Leeuwenhoek** (an-TOE-nee vahn LAY-wenhook) (1632–1723), a Dutch textile merchant, invented a *simple* (single-lens) *microscope,* originally for the purpose of examining the weave of fabrics. His microscope was a beadlike lens mounted in a metal plate equipped with a movable

**FIGURE 1.2 Hooke's Compound Microscope.** (a) The compound microscope had a lens at each end of a tubular body. (b) Hooke's drawing of cork cells, showing the thick cell walls characteristic of plants.

(a)                    (b)

specimen clip. Even though his microscopes were simpler than Hooke's, they achieved much greater useful magnification (up to 200×) owing to Leeuwenhoek's superior lens-making technique. Out of curiosity, he examined a drop of lake water and was astonished to find a variety of microorganisms—"little animalcules," he called them, "very prettily a-swimming." He went on to observe practically everything he could get his hands on, including blood cells, blood capillaries, sperm, muscular tissue, and bacteria from tooth scrapings. Leeuwenhoek began submitting his observations to the Royal Society of London in 1673. He was praised at first, and his observations were eagerly read by scientists, but enthusiasm for the microscope did not last. By the end of the seventeenth century, it was treated as a mere toy for the upper classes, as amusing and meaningless as a kaleidoscope. Leeuwenhoek and Hooke had even become the brunt of satire. But probably no one in history had looked at nature in such a revolutionary way. By taking biology to the cellular level, the two men had laid an entirely new foundation for the modern medicine to follow centuries later.

The Hooke and Leeuwenhoek microscopes produced poor images with blurry edges *(spherical aberration)* and rainbowlike distortions *(chromatic aberration)*. These problems had to be solved before the microscope could be widely used as a biological tool. In the nineteenth century, German inventors greatly improved the compound microscope, adding the condenser and developing superior optics. With improved microscopes, biologists began eagerly examining a wider variety of specimens. By 1839, botanist **Matthias Schleiden** (1804–81) and zoologist **Theodor Schwann** (1810–82) concluded that all organisms were composed of cells. Although it took another century for this idea to be generally accepted, it became the first tenet of the **cell theory,** added to by later biologists and summarized in chapter 3. The cell theory was perhaps the most important breakthrough in biomedical history; all functions of the body are now interpreted as the effects of cellular activity.

Although the philosophical foundation for modern medicine was largely established by the time of Leeuwenhoek, Hooke, and Harvey, clinical practice was still in a dismal state. Few doctors attended medical school or received any formal education in basic science or human anatomy. Physicians tended to be ignorant, ineffective, and pompous. Their practice was heavily based on expelling imaginary toxins from the body by bleeding their patients or inducing vomiting, sweating, or diarrhea. They performed operations with filthy hands and instruments, spreading lethal infections from one patient to another and refusing, in their vanity, to believe that they themselves were the carriers of disease. Countless women died of infections acquired during childbirth from their obstetricians. Fractured limbs often became gangrenous and had to be amputated, and there was no anesthesia to lessen the pain. Disease was still widely attributed to demons and witches, and many people felt they would be interfering with God's will if they tried to treat it.

## Living in a Revolution

This short history brings us only to the threshold of modern biomedical science; it stops short of such momentous discoveries as the germ theory of disease, the mechanisms of heredity, and the structure of DNA. In the twentieth century, basic biology and biochemistry yielded a much deeper understanding of how the body works. Advances in medical imaging enhanced our diagnostic ability and life-support strategies. We witnessed monumental developments in chemotherapy, immunization, anesthesia, surgery, organ transplants, and human genetics. By the close of the twentieth century, we had discovered the chemical "base sequence" of every human gene and begun attempting gene therapy to treat children born with diseases recently considered incurable. As future historians look back on the turn of this century, they may exult about the Genetic Revolution in which you are now living.

Several discoveries of the nineteenth and twentieth centuries, and the men and women behind them, are covered in short historical sketches in later chapters. Yet, the stories told in this chapter are different in a significant way. The people discussed here were pioneers in establishing the scientific way of thinking. They helped to replace superstition with an appreciation of natural law. They bridged the chasm between mystery and medication. Without this intellectual revolution, those who followed could not have conceived of the right questions to ask, much less a method for answering them.

**BEFORE YOU GO ON**

Answer the following questions to test your understanding of the preceding section:

3. In what way did the followers of Galen disregard his advice? How does Galen's advice apply to you and this book?

4. Describe two ways in which Vesalius improved medical education and set standards that remain relevant today.

5. How is our concept of human form and function today affected by inventors from the seventeenth to the nineteenth century?

## 1.3   Scientific Method

### Expected Learning Outcomes

When you have completed this section, you should be able to

a. describe the inductive and hypothetico–deductive methods of obtaining scientific knowledge;

b. describe some aspects of experimental design that help to ensure objective and reliable results; and

c. explain what is meant by *hypothesis, fact, law,* and *theory* in science.

Prior to the seventeenth century, science was done in a haphazard way by a small number of isolated individuals.

The philosophers **Francis Bacon** (1561–1626) in England and **René Descartes** (1596–1650) in France envisioned science as a far greater, systematic enterprise with enormous possibilities for human health and welfare. They detested those who endlessly debated ancient philosophy without creating anything new. Bacon argued against biased thinking and for more objectivity in science. He outlined a systematic way of seeking similarities, differences, and trends in nature and drawing useful generalizations from observable facts. You will see echoes of Bacon's philosophy in the discussion of scientific method that follows.

Though the followers of Bacon and Descartes argued bitterly with one another, both men wanted science to become a public, cooperative enterprise, supported by governments and conducted by an international community of scholars rather than a few isolated amateurs. Inspired by their vision, the French and English governments established academies of science that still flourish today. Bacon and Descartes are credited with putting science on the path to modernity, not by discovering anything new in nature or inventing any techniques—for neither man was a scientist—but by inventing new habits of scientific thought.

When we say "scientific," we mean that such thinking is based on assumptions and methods that yield reliable, objective, testable information about nature. The assumptions of science are ideas that have proven fruitful in the past—for example, the idea that natural phenomena have natural causes and nature is therefore predictable and understandable. The methods of science are highly variable. **Scientific method** refers less to observational procedures than to certain habits of disciplined creativity, careful observation, logical thinking, and honest analysis of one's observations and conclusions. It is especially important in health science to understand these habits. This field is littered with more fads and frauds than any other. We are called upon constantly to judge which claims are trustworthy and which are bogus. To make such judgments depends on an appreciation of how scientists think, how they set standards for truth, and why their claims are more reliable than others.

## The Inductive Method

The **inductive method,** first prescribed by Bacon, is a process of making numerous observations until one feels confident in drawing generalizations and predictions from them. What we know of anatomy is a product of the inductive method. We describe the normal structure of the body based on observations of many bodies.

This raises the issue of what is considered proof in science. We can never prove a claim beyond all possible refutation. We can, however, consider a statement as proven *beyond reasonable doubt* if it was arrived at by reliable methods of observation, tested and confirmed repeatedly, and not falsified by any credible observation. In science, all truth is tentative; there is no room for dogma. We must always be prepared to abandon yesterday's truth if tomorrow's facts disprove it.

## The Hypothetico–Deductive Method

Most physiological knowledge was obtained by the **hypothetico–deductive method.** An investigator begins by asking a question and formulating a **hypothesis**—an educated speculation or possible answer to the question. A good hypothesis must be (1) consistent with what is already known and (2) capable of being tested and possibly falsified by evidence. **Falsifiability** means that if we claim something is scientifically true, we must be able to specify what evidence it would take to prove it wrong. If nothing could possibly prove it wrong, then it is not scientific.

**►►►APPLY WHAT YOU KNOW**

*The ancients thought that gods or invisible demons caused epilepsy. Today, epileptic seizures are attributed to bursts of abnormal electrical activity in nerve cells of the brain. Explain why one of these claims is falsifiable (and thus scientific), whereas the other claim is not.*

The purpose of a hypothesis is to suggest a method for answering a question. From the hypothesis, a researcher makes a deduction, typically in the form of an "if–then" prediction: *If* my hypothesis on epilepsy is correct and I record the brain waves of patients during seizures, *then* I should observe abnormal bursts of activity. A properly conducted experiment yields observations that either support a hypothesis or require the scientist to modify or abandon it, formulate a better hypothesis, and test that one. Hypothesis testing operates in cycles of conjecture and disproof until one is found that is supported by the evidence.

## Experimental Design

Doing an experiment properly involves several important considerations. What shall I measure and how can I measure it? What effects should I watch for and which ones should I ignore? How can I be sure my results are due to the variables that I manipulate and not due to something else? When working on human subjects, how can I prevent the subject's expectations or state of mind from influencing the results? How can I eliminate my own biases and be sure that even the most skeptical critics will have as much confidence in my conclusions as I do? Several elements of experimental design address these issues:

- **Sample size.** The number of subjects (animals or people) used in a study is the sample size. An adequate sample size controls for chance events and individual variations in response and thus enables us to place more confidence in the outcome. For example, would you rather trust your health to a drug that was tested on 5 people or one tested on 5,000? Why?

- **Controls.** Biomedical experiments require comparison between treated and untreated individuals so that we can judge whether the treatment has any effect. A **control group** consists of subjects that are as much like the **treatment group** as possible except with respect to the variable being tested. For example, there is evidence that garlic lowers blood cholesterol levels. In one study, volunteers with high cholesterol were each given 800 mg of garlic powder daily for 4 months and exhibited an average 12% reduction in cholesterol. Was this a significant reduction, and was it due to the garlic? It is impossible to say without comparison to a control group of similar people who received no treatment. In this study, the control group averaged only a 3% reduction in cholesterol, so garlic *seems* to have made a difference.

- **Psychosomatic effects.** Psychosomatic effects (effects of the subject's state of mind on his or her physiology) can have an undesirable effect on experimental results if we do not control for them. In drug research, it is therefore customary to give the control group a **placebo** (pla-SEE-bo)—a substance with no significant physiological effect on the body. If we were testing a drug, for example, we could give the treatment group the drug and the control group identical-looking sugar tablets. Neither group must know which tablets it is receiving. If the two groups showed significantly different effects, we could feel confident that it did not result from a knowledge of what they were taking.

- **Experimenter bias.** In the competitive, high-stakes world of medical research, experimenters may want certain results so much that their biases, even subconscious ones, can affect their interpretation of the data. One way to control for this is the **double-blind method.** In this procedure, neither the subject to whom a treatment is given nor the person giving it and recording the results knows whether that subject is receiving the experimental treatment or placebo. A researcher might prepare identical-looking tablets, some with the drug and some with placebo; label them with code numbers; and distribute them to participating physicians. The physicians themselves do not know whether they are administering drug or placebo, so they cannot give the subjects even accidental hints of which substance they are taking. When the data are collected, the researcher can correlate them with the composition of the tablets and determine whether the drug had more effect than the placebo.

- **Statistical testing.** If you tossed a coin 100 times, you would expect it to come up about 50 heads and 50 tails. If it actually came up 48:52, you would probably attribute this to random error rather than bias in the coin. But what if it came up 40:60? At what point would you begin to suspect bias? This type of problem is faced routinely in research—how great a difference must there be between control and experimental groups before we feel confident that it was due to the treatment and not merely random variation? What if a treatment group exhibited a 12% reduction in cholesterol level and the placebo group a 10% reduction? Would this be enough to conclude that the treatment was effective? Scientists are well grounded in **statistical tests** that can be applied to the data—the chi-square test, the *t* test, and analysis of variance, for example. A typical outcome of a statistical test might be expressed, "We can be 99.5% sure that the difference between group A and group B was due to the experimental treatment and not to random variation." Science is grounded not in statements of absolute truth, but in statements of probability.

## Peer Review

When a scientist applies for funds to support a research project or submits results for publication, the application or manuscript is submitted to **peer review**—a critical evaluation by other experts in that field. Even after a report is published, if the results are important or unconventional, other scientists may attempt to reproduce them to see if the author was correct. At every stage from planning to postpublication, scientists are therefore subject to intense scrutiny by their colleagues. Peer review is one mechanism for ensuring honesty, objectivity, and quality in science.

## Facts, Laws, and Theories

The most important product of scientific research is understanding how nature works—whether it be the nature of a pond to an ecologist or the nature of a liver cell to a physiologist. We express our understanding as *facts, laws,* and *theories* of nature. It is important to appreciate the differences among these.

A scientific **fact** is information that can be independently verified by any trained person—for example, the fact that an iron deficiency leads to anemia. A **law of nature** is a generalization about the predictable ways in which matter and energy behave. It is the result of inductive reasoning based on repeated, confirmed observations. Some laws are expressed as concise verbal statements, such as the *law of complementary base pairing:* In the double helix of DNA, a chemical base called adenine always pairs with one called thymine, and a base called guanine always pairs with cytosine (see p. 114). Other laws are expressed as mathematical formulae, such as *Boyle's law,* used in respiratory physiology: Under specified conditions, the volume of a gas $(V)$ is inversely proportional to its pressure $(P)$—that is, $V \propto 1/P$.

A **theory** is an explanatory statement or set of statements derived from facts, laws, and confirmed hypotheses. Some theories have names, such as the *cell theory,* the *fluid-mosaic theory* of cell membranes, and the *sliding filament theory* of muscle contraction. Most, however, remain unnamed. The purpose of a theory is not only to concisely summarize what we already know but, moreover, to suggest directions for further study and to help predict what the findings should be if the theory is correct.

*Law* and *theory* mean something different in science than they do to most people. In common usage, a law is a rule created and enforced by people; we must obey it or risk a penalty. A law of nature, however, is a description; laws do not *govern* the universe—they *describe* it. Laypeople tend to use the word *theory* for what a scientist would call a hypothesis—for example, "I have a theory why my car won't start." The difference in meaning causes significant confusion when it leads people to think that a scientific theory (such as the theory of evolution) is merely a guess or conjecture, instead of recognizing it as a summary of conclusions drawn from a large body of observed facts. The concepts of gravity and electrons are theories, too, but this does not mean they are merely speculations.

### ▶▶▶ APPLY WHAT YOU KNOW

*Was the cell theory proposed by Schleiden and Schwann more a product of the hypothetico–deductive method or of the inductive method? Explain your answer.*

### BEFORE YOU GO ON

Answer the following questions to test your understanding of the preceding section:

6. Describe the general process involved in the inductive method.

7. Describe some sources of potential bias in biomedical research. What are some ways of minimizing such bias?

8. Is there more information in an individual scientific fact or in a theory? Explain.

## 1.4 Human Origins and Adaptations

### Expected Learning Outcomes

When you have completed this section, you should be able to

a. explain why evolution is relevant to understanding human form and function;

b. define *evolution* and *natural selection;*

c. describe some human characteristics that can be attributed to the tree-dwelling habits of earlier primates; and

d. describe some human characteristics that evolved later in connection with upright walking.

If any two theories have the broadest implications for understanding the human body, they are probably the *cell theory* and the *theory of natural selection.* As an explanation of how species originate and change through time, natural selection was the brainchild of **Charles Darwin** (1809–82)—certainly the most influential biologist who ever lived. His book, *On the Origin of Species by Means of Natural Selection* (1859), has been called "the book that shook the world." In presenting the first well-supported theory of how evolution works, *On the Origin of Species* not only caused the restructuring of all of biology but also profoundly changed the prevailing view of our origin, nature, and place in the universe. In *The Descent of Man* (1871), Darwin directly addressed the issue of human evolution and emphasized features of anatomy and behavior that reveal our relationship to other animals. No understanding of human form and function is complete without an understanding of our evolutionary history. Here we will touch just briefly on how natural selection helps explain some of the distinctive characteristics seen in *Homo sapiens* today.

## Evolution, Selection, and Adaptation

**Evolution** simply means change in the genetic composition of a population of organisms. Examples include the evolution of bacterial resistance to antibiotics, the appearance of new strains of the AIDS virus, and the emergence of new species of organisms.

**Natural selection** is the principal theory of how evolution works. It states essentially this: Some individuals within a species have hereditary advantages over their competitors—for example, better camouflage, disease resistance, or ability to attract mates—that enable them to produce more offspring. They pass these advantages on to their offspring, and such characteristics therefore become more and more common in successive generations. This brings about the genetic change in a population that constitutes evolution.

Natural forces that promote the reproductive success of some individuals more than others are called **selection pressures.** They include such things as climate, predators, disease, competition, and the availability of food. **Adaptations** are features of anatomy, physiology, and behavior that have evolved in response to these selection pressures and enable the organism to cope with the challenges of its environment.

Darwin could scarcely have predicted the overwhelming mass of genetic, molecular, fossil, and other evidence of human evolution that would accumulate in the twentieth century and further substantiate his theory. A technique called DNA hybridization, for example, suggests a difference of only 1.6% in DNA structure between humans and chimpanzees. Chimpanzees and gorillas differ by 2.3%. DNA structure thus suggests that a chimpanzee's closest living relative is not the gorilla or any other ape—it is us, *Homo sapiens.*

Several aspects of our anatomy make little sense without an awareness that the human body has a history (see Deeper Insight 1.1). Our evolutionary relationship to other species is also important in choosing animals for biomedical research. If there were no issues of cost, availability, or ethics, we might test drugs on our close living relatives, the chimpanzees, before

 **DEEPER INSIGHT 1.1**

### EVOLUTIONARY MEDICINE

#### *Vestiges of Human Evolution*

One of the classic lines of evidence for evolution, debated even before Darwin was born, is *vestigial organs.* These structures are the remnants of organs that apparently were better developed and more functional in the ancestors of a species. They now serve little or no purpose or, in some cases, have been converted to new functions.

Our bodies, for example, are covered with millions of hairs, each equipped with a useless little muscle called a *piloerector.* In other mammals, these muscles fluff the hair and conserve heat. In humans, they merely produce goose bumps. Above each ear, we have three *auricularis muscles.* In other mammals, they move the ears to receive sounds better or to repel flies and other pests, but most people cannot contract them at all. As Darwin said, it makes no sense that humans would have such structures were it not for the fact that we came from ancestors in which they were functional.

approving them for human use. Their genetics, anatomy, and physiology are most similar to ours, and their reactions to drugs therefore afford the best prediction of how the human body would react. On the other hand, if we had no kinship with any other species, the selection of a test species would be arbitrary; we might as well use frogs or snails. In reality, we compromise. Rats and mice are used extensively for research because they are fellow mammals with a physiology similar to ours, but they present fewer of the aforementioned issues than chimpanzees or other mammals do. An animal species or strain selected for research on a particular problem is called a **model**—for example, a mouse model for leukemia.

## Our Basic Primate Adaptations

We belong to an order of mammals called the Primates, which also includes the monkeys and apes. Some of our anatomical and physiological features can be traced to the earliest primates, which descended from certain squirrel-sized, insect-eating, African mammals that took up life in the trees 55 to 60 million years ago. This **arboreal**[9] (treetop) habitat probably afforded greater safety from predators, less competition, and a rich food supply of leaves, fruit, insects, and lizards. But the forest canopy is a challenging world, with dim and dappled sunlight, swaying branches, and prey darting about in the dense foliage. Any new feature that enabled arboreal animals to move about more easily in the treetops would have been strongly favored by natural selection. Thus, the shoulder became more mobile and enabled

primates to reach out in any direction (even overhead, which few other mammals can do). The thumbs became fully **opposable**—they could cross the palm to touch the fingertips—and enabled primates to hold small objects and manipulate them more precisely than other mammals can. Opposable thumbs made the hands **prehensile**[10]—able to grasp branches by encircling them with the thumb and fingers (fig. 1.3a). The thumb is so important that it receives highest priority in the repair of hand injuries. If the thumb can be saved, the hand can be reasonably functional; if it is lost, hand functions are severely diminished.

The eyes of primates moved to a more forward-facing position (fig. 1.3b), which allowed for **stereoscopic**[11] vision (depth perception). This adaptation provided better hand–eye coordination in catching and manipulating prey, with the added advantage of making it easier to judge distances accurately in leaping from tree to tree. Color vision, rare among mammals, is also a primate hallmark. Primates eat mainly fruit and leaves. The ability to distinguish subtle shades of orange and red enables them to distinguish ripe, sugary fruits from unripe ones. Distinguishing subtle shades of green helps them to differentiate between tender young leaves and tough, more toxic older foliage.

Various fruits ripen at different times and in widely separated places in the tropical forest. This requires a good memory of what will be available, when, and how to get there. Larger brains may have evolved in response to the challenge of efficient food finding and, in turn, laid the foundation for more sophisticated social organization.

---

[9]*arbor* = tree; *eal* = pertaining to

[10]*prehens* = to seize
[11]*stereo* = solid; *scop* = vision

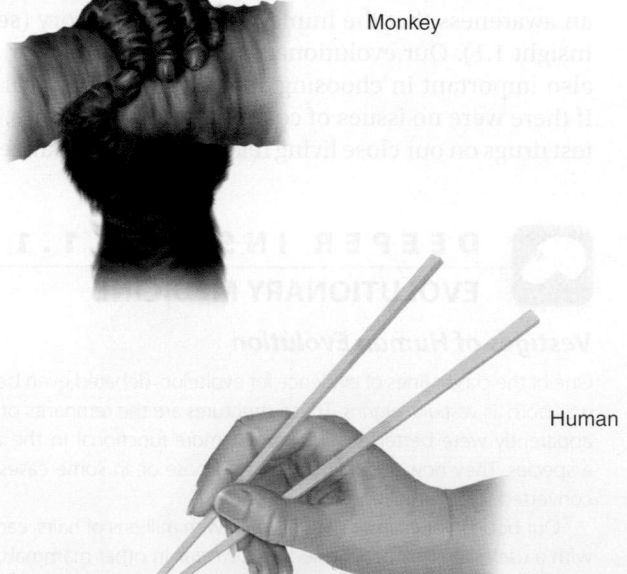

Monkey

Human

(a)

(b)

**FIGURE 1.3 Human Adaptations Shared With Other Primates.** (a) The opposable thumb makes the hand prehensile, able to encircle and grasp objects. (b) Chimpanzees exhibit the prehensile hands and forward-facing eyes typical of primates. Such traits endow primates with stereoscopic vision (depth perception) and good hand–eye coordination, two factors of enormous importance in human evolution.

None of this is meant to imply that humans evolved from monkeys or apes—a common misconception about evolution that no biologist believes. Monkeys, apes, and humans do, however, share common ancestors. Our relationship is not like parent and child, but more like cousins who have the same grandparents. Observations of monkeys and apes provide insight into how primates adapt to the arboreal habitat and therefore how certain human adaptations probably originated.

## Walking Upright

About 4 to 5 million years ago, parts of Africa became hotter and drier, and much of the forest was replaced by savanna (grassland). Some primates adapted to living on the savanna, but this was a dangerous place with more predators and less protection. Just as squirrels and monkeys stand briefly on their hind legs to look around for danger, so would these early ground dwellers. Being able to stand up not only helps an animal stay alert, but also frees the forelimbs for purposes other than walking. Chimpanzees sometimes walk upright to carry food, infants, or weapons (sticks and rocks), and it is reasonable to suppose that our early ancestors did so too.

These advantages are so great that they favored skeletal modifications that made **bipedalism**[12]—standing and walking on two legs—easier. Fossil evidence indicates that bipedalism was firmly established more than 4 million years ago. The anatomy of the human pelvis, femur, knee, great toe, foot arches, spinal column, skull, arms, and many muscles became adapted for bipedal locomotion, as did many aspects of human family life and society. As the skeleton and muscles became adapted for bipedalism, brain volume increased dramatically, from 400 mL around 4 million years ago to an average of 1,350 mL today. It must have become increasingly difficult for a fully developed, large-brained infant to pass through the mother's pelvic outlet at birth. This may explain why humans are born in a relatively immature, helpless state compared with other mammals, before their nervous systems have matured and the bones of the skull have fused. The helplessness of human young and their extended dependence on parental care may help to explain why humans have such exceptionally strong family ties.

Most of the oldest bipedal primates are classified in the genus *Australopithecus* (aus-TRAL-oh-PITH-eh-cus). About 2.5 million years ago, hominids appeared with taller stature, greater brain volumes, simple stone tools, and probably articulate speech. These are the earliest members of the genus *Homo*. By at least 1.8 million years ago, *Homo erectus* migrated from Africa to parts of Asia. Anatomically modern *Homo sapiens*, our own species, originated in Africa about 200,000 years ago and is the sole surviving hominid species.

This brief account barely begins to explain how human anatomy, physiology, and behavior have been shaped by ancient selection pressures. Later chapters further demonstrate that the evolutionary perspective provides a meaningful understanding of why humans are the way we are. Evolution is the basis

for comparative anatomy and physiology, which have been so fruitful for the understanding of human biology. If we were not related to any other species, those sciences would be pointless.

The emerging science of **evolutionary medicine** analyzes how human disease and dysfunctions can be traced to differences between the artificial environment in which we now live, and the prehistoric environment to which *Homo sapiens* was biologically adapted. For example, we can relate sleep and mood disorders to artificial lighting and night-shift work, and the rise of asthma to our modern obsession with sanitation. Other examples in this book will relate evolution to obesity, diabetes, low-back pain, skin cancer, and other health issues.

### BEFORE YOU GO ON

Answer the following questions to test your understanding of the preceding section:

9. Define *adaptation* and *selection pressure*. Why are these concepts important in understanding human anatomy and physiology?

10. Select any two human characteristics and explain how they may have originated in primate adaptations to an arboreal habitat.

11. Select two other human characteristics and explain how they may have resulted from adaptation to a grassland habitat.

---

**1.5** **Human Structure**

### Expected Learning Outcomes

When you have completed this section, you should be able to

a. list the levels of human structure from the most complex to the simplest;

b. discuss the value of both reductionistic and holistic viewpoints to understanding human form and function; and

c. discuss the clinical significance of anatomical variation among humans.

Earlier in this chapter, we observed that human anatomy is studied by a variety of techniques—dissection, palpation, and so forth. In addition, anatomy is studied at several levels of detail, from the whole body down to the molecular level.

## The Hierarchy of Complexity

Consider for the moment an analogy to human structure: The English language, like the human body, is very complex, yet an infinite variety of ideas can be conveyed with a limited number of words. All words in English are, in turn, composed of various combinations of just 26 letters. Between an essay and an alphabet are successively simpler levels of organization: paragraphs, sentences, words, and syllables. We can say that

---

[12]*bi* = two; *ped* = foot

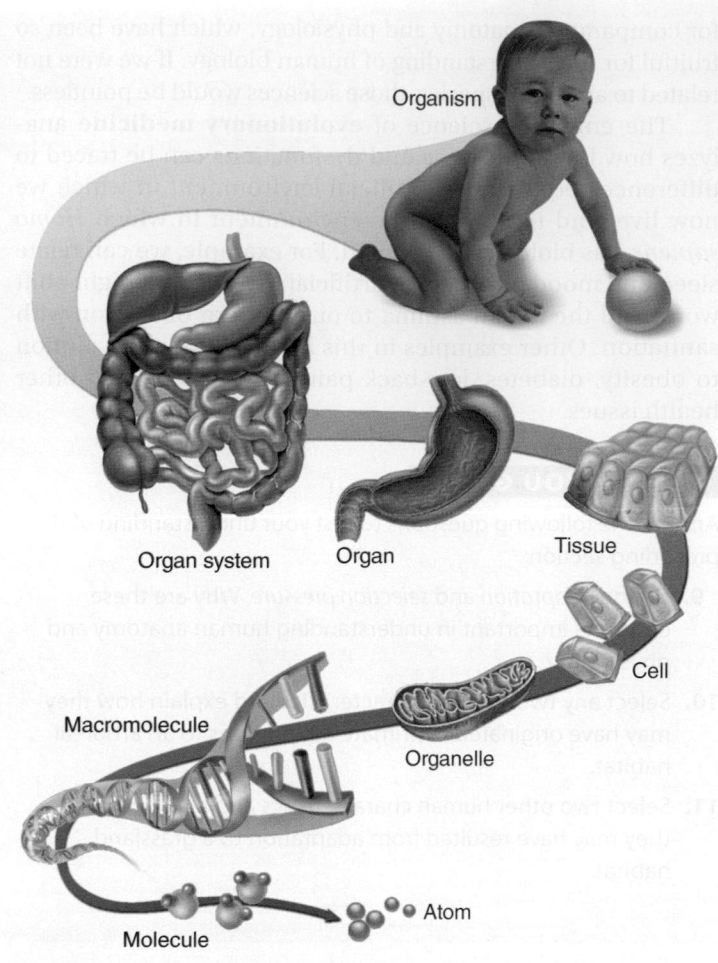

**FIGURE 1.4 The Body's Structural Hierarchy.**

language exhibits a hierarchy of complexity, with letters, syllables, words, and so forth being successive levels of the hierarchy. Humans have an analogous hierarchy of complexity, as follows (fig. 1.4):

The organism is composed of organ systems,

     organ systems are composed of organs,

         organs are composed of tissues,

            tissues are composed of cells,

                cells are composed partly of organelles,

                   organelles are composed of molecules, and

                       molecules are composed of atoms.

     The **organism** is a single, complete individual.

     An **organ system** is a group of organs with a unique collective function, such as circulation, respiration, or digestion. The human body has 11 organ systems, illustrated in atlas A immediately following this chapter: the integumentary, skeletal, muscular, nervous, endocrine, circulatory, lymphatic, respiratory, urinary, digestive, and reproductive systems. Usually, the organs of one system are physically interconnected, such as the kidneys, ureters, urinary bladder, and urethra, which compose the urinary system. Beginning with chapter 6, this book is organized around the organ systems.

An **organ** is a structure composed of two or more tissue types that work together to carry out a particular function. Organs have definite anatomical boundaries and are visibly distinguishable from adjacent structures. Most organs and higher levels of structure are within the domain of gross anatomy. However, there are organs within organs—the large organs visible to the naked eye often contain smaller organs visible only with the microscope. The skin, for example, is the body's largest organ. Included within it are thousands of smaller organs: Each hair, nail, gland, nerve, and blood vessel of the skin is an organ in itself. A single organ can belong to two organ systems. For example, the pancreas belongs to both the endocrine and digestive systems.

A **tissue** is a mass of similar cells and cell products that forms a discrete region of an organ and performs a specific function. The body is composed of only four primary classes of tissue: epithelial, connective, nervous, and muscular tissue. Histology, the study of tissues, is the subject of chapter 5.

**Cells** are the smallest units of an organism that carry out all the basic functions of life; nothing simpler than a cell is considered alive. A cell is enclosed in a *plasma membrane* composed of lipids and proteins. Most cells have one nucleus, an organelle that contains its DNA. *Cytology,* the study of cells and organelles, is the subject of chapters 3 and 4.

**Organelles**[13] are microscopic structures in a cell that carry out its individual functions. Examples include mitochondria, centrioles, and lysosomes.

Organelles and other cellular components are composed of **molecules.** The largest molecules, such as proteins, fats, and DNA, are called *macromolecules.* A molecule is a particle composed of at least two **atoms,** the smallest particles with unique chemical identities.

The theory that a large, complex system such as the human body can be understood by studying its simpler components is called **reductionism.** First espoused by Aristotle, this has proven to be a highly productive approach; indeed, it is essential to scientific thinking. Yet the reductionistic view is not the only way of understanding human life. Just as it would be very difficult to predict the workings of an automobile transmission merely by looking at a pile of its disassembled gears and levers, one could never predict the human personality from a complete knowledge of the circuitry of the brain or the genetic sequence of DNA. **Holism**[14] is the complementary theory that there are "emergent properties" of the whole organism that cannot be predicted from the properties of its separate parts—human beings are more than the sum of their parts. To be most effective, a health-care provider treats not merely a disease or an organ system, but a whole person. A patient's perceptions, emotional responses to life, and confidence in the nurse, therapist, or physician profoundly affect the outcome of treatment. In fact, these psychological factors often play a greater role in a patient's recovery than the physical treatments administered.

[13]*elle* = little
[14]*holo* = whole, entire

## Anatomical Variation

A quick look around any classroom is enough to show that no two humans are exactly alike; on close inspection, even identical twins exhibit differences. Yet anatomy atlases and textbooks can easily give the impression that everyone's internal anatomy is the same. This simply is not true. Books such as this one can teach you only the most common structure—the anatomy seen in about 70% or more of people. Someone who thinks that all human bodies are the same internally would make a very confused medical student or an incompetent surgeon.

Some people lack certain organs. For example, most of us have a *palmaris longus* muscle in the forearm and a *plantaris* muscle in the leg, but these are absent from others. Most of us have five lumbar vertebrae (bones of the lower spine), but some people have six and some have four. Most of us have one spleen and two kidneys, but some have two spleens or only one kidney. Most kidneys are supplied by a single *renal artery* and are drained by one *ureter,* but some have two renal arteries or ureters. Figure 1.5 shows some common variations in human anatomy, and Deeper Insight 1.2 describes a particularly dramatic and clinically important variation.

### ▶▶▶APPLY WHAT YOU KNOW

*People who are allergic to aspirin or penicillin often wear MedicAlert bracelets or necklaces that note this fact in case they need emergency medical treatment and are unable to communicate. Why would it be important for a person with situs inversus (see Deeper Insight 1.2) to have this noted on a MedicAlert bracelet?*

### DEEPER INSIGHT 1.2
#### CLINICAL APPLICATION

**Situs Inversus and Other Unusual Anatomy**

In most people, the spleen, pancreas, sigmoid colon, and most of the heart are on the left, while the appendix, gallbladder, and most of the liver are on the right. The normal arrangement of these and other internal organs is called *situs* (SITE-us) *solitus.* About 1 in 8,000 people, however, is born with an abnormality called *situs inversus*—the organs of the thoracic and abdominal cavities are reversed between right and left. A selective right–left reversal of the heart is called *dextrocardia.* In *situs perversus,* a single organ occupies an atypical position—for example, a kidney located low in the pelvic cavity instead of high in the abdominal cavity.

Conditions such as dextrocardia in the absence of complete situs inversus can cause serious medical problems. Complete situs inversus, however, usually causes no functional problems because all of the viscera, though reversed, maintain their normal relationships to one another. Situs inversus is often discovered in the fetus by sonography, but many people remain unaware of their condition for decades until it is discovered by medical imaging, on physical examination, or in surgery. You can easily imagine the importance of such conditions in diagnosing appendicitis, performing gallbladder surgery, interpreting an X-ray, auscultating the heart valves, or recording an electrocardiogram.

**BEFORE YOU GO ON**

Answer the following questions to test your understanding of the preceding section:

12. In the hierarchy of human structure, what is the level between organ system and tissue? Between cell and molecule?
13. How are tissues relevant to the definition of an organ?
14. Why is reductionism a necessary but not sufficient point of view for fully understanding a patient's illness?
15. Why should medical students observe multiple cadavers and not be satisfied to dissect only one?

## 1.6 Human Function

### Expected Learning Outcomes

When you have completed this section, you should be able to

a. state the characteristics that distinguish living organisms from nonliving objects;
b. explain the importance of physiological variation among persons;
c. define *homeostasis* and explain why this concept is central to physiology;
d. define *negative feedback,* give an example of it, and explain its importance to homeostasis;
e. define *positive feedback* and give examples of its beneficial and harmful effects; and
f. define *gradient,* describe the variety of gradients in human physiology, and identify some forms of matter and energy that flow down gradients.

## Characteristics of Life

Why do we consider a growing child to be alive, but not a growing crystal? Is abortion the taking of a human life? If so, what about a contraceptive foam that kills only sperm? As a patient is dying, at what point does it become ethical to disconnect life-support equipment and remove organs for donation? If these organs are alive, as they must be to serve someone else, then why isn't the donor considered alive? Such questions have no easy answers, but they demand a concept of what life is—a concept that may differ with one's biological, medical, legal, or religious perspective.

From a biological viewpoint, life is not a single property. It is a collection of properties that help to distinguish living from nonliving things:

- **Organization.** Living things exhibit a far higher level of organization than the nonliving world around them. They expend a great deal of energy to maintain order, and a breakdown in this order is accompanied by disease and often death.
- **Cellular composition.** Living matter is always compartmentalized into one or more cells.

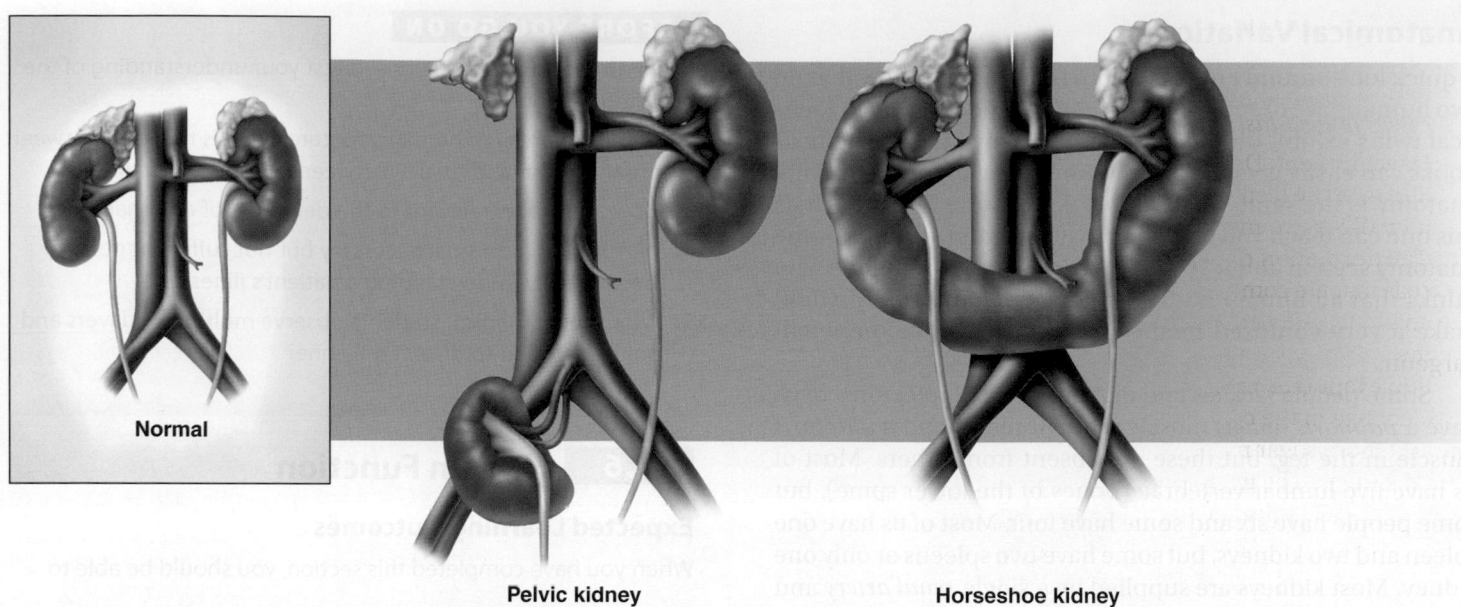

Normal

Pelvic kidney

Horseshoe kidney

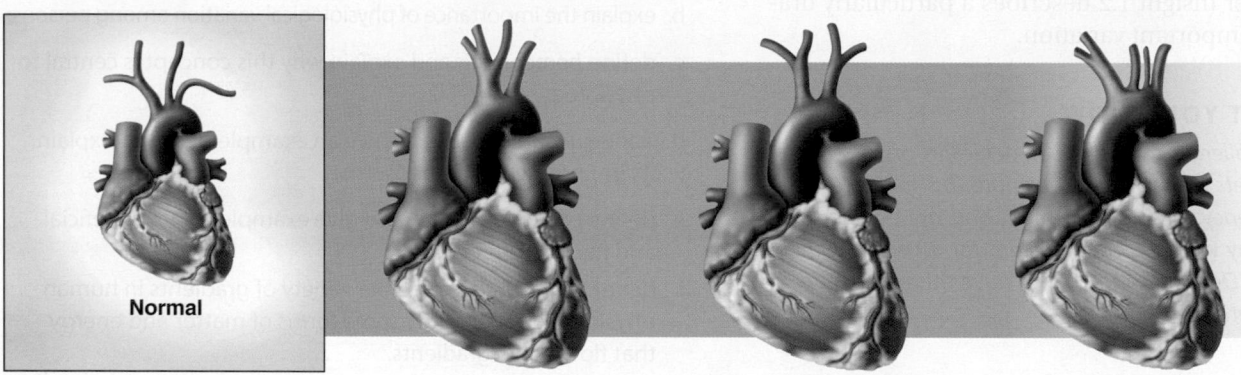

Normal

Variations in branches of the aorta

**FIGURE 1.5  Variation in Anatomy of the Kidneys and Major Arteries Near the Heart.**

- **Metabolism.** Living things take in molecules from the environment and chemically change them into molecules that form their own structures, control their physiology, or provide them with energy. **Metabolism**[15] is the sum of all this internal chemical change. It consists of two classes of reactions: *anabolism,*[16] in which relatively complex molecules are synthesized from simpler ones (for example, protein synthesis), and *catabolism,*[17] in which relatively complex molecules are broken down into simpler ones (for example, protein digestion). Metabolism inevitably produces chemical wastes, some of which are toxic if they accumulate. Metabolism therefore requires **excretion,** the separation of wastes from the tissues and their elimination from the body. There is a constant turnover of molecules in the body; few of the molecules now in your body have been there for

more than a year. It is food for thought that although you sense a continuity of personality and experience from your childhood to the present, nearly all of your body has been replaced within the past year.

- **Responsiveness** and **movement.** The ability of organisms to sense and react to **stimuli** (changes in their environment) is called *responsiveness, irritability,* or *excitability.* It occurs at all levels from the single cell to the entire body, and it characterizes all living things from bacteria to you. Responsiveness is especially obvious in animals because of nerve and muscle cells that exhibit high sensitivity to environmental stimuli, rapid transmission of information, and quick reactions. Most living organisms are capable of self-propelled movement from place to place, and all organisms and cells are at least capable of moving substances internally, such as moving food along the digestive tract or moving molecules and organelles from place to place within a cell.

[15]*metabol* = change; *ism* = process
[16]*ana* = up
[17]*cata* = down

- **Homeostasis.** Although the environment around an organism changes, the organism maintains relatively stable internal conditions. This ability to maintain internal stability, called *homeostasis,* is explored in more depth shortly.

- **Development.** Development is any change in form or function over the lifetime of the organism. In most organisms, it involves two major processes: (1) **differentiation,** the transformation of cells with no specialized function into cells that are committed to a particular task; and (2) **growth,** an increase in size. Some nonliving things grow, but not in the way your body does. If you let a saturated sugar solution evaporate, crystals will grow from it, but not through a change in the composition of the sugar. They merely add more sugar molecules from the solution to the crystal surface. The growth of the body, by contrast, occurs through chemical change (metabolism); for the most part, your body is not composed of the molecules you ate but of molecules made by chemically altering your food.

- **Reproduction.** All living organisms can produce copies of themselves, thus passing their genes on to new, younger containers—their offspring.

- **Evolution.** All living species exhibit genetic change from generation to generation and therefore evolve. This occurs because *mutations* (changes in DNA structure) are inevitable and because environmental selection pressures favor the transmission of some genes more than others. Unlike the other characteristics of life, evolution is a characteristic seen only in the population as a whole. No single individual evolves over the course of its life.

Clinical and legal criteria of life differ from these biological criteria. A person who has shown no brain waves for 24 hours, and has no reflexes, respiration, or heartbeat other than what is provided by artificial life support, can be declared legally dead. At such time, however, most of the body is still biologically alive and its organs may be useful for transplant.

## Physiological Variation

Earlier we considered the clinical importance of variations in human anatomy, but physiology is even more variable. Physiological variables differ with sex, age, weight, diet, degree of physical activity, and environment, among other things. Failure to consider such variation leads to medical mistakes such as overmedication of the elderly or medicating women on the basis of research that was done on young men. If a textbook states a typical human heart rate, blood pressure, red blood cell count, or body temperature, it is generally assumed, unless otherwise stated, that such values refer to a healthy 22-year-old weighing 58 kg (128 lb) for a female and 70 kg (154 lb) for a male, and a lifestyle of light physical activity and moderate caloric intake (2,000 and 2,800 kcal/day, respectively).

## Homeostasis and Negative Feedback

The human body has a remarkable capacity for self-restoration. Hippocrates commented that it usually returns to a state of equilibrium by itself, and people recover from most illnesses

even without the help of a physician. This tendency results from **homeostasis**[18] (HO-me-oh-STAY-sis), the body's ability to detect change, activate mechanisms that oppose it, and thereby maintain relatively stable internal conditions.

French physiologist **Claude Bernard** (1813–78) observed that the internal conditions of the body remain quite constant even when external conditions vary greatly. For example, whether it is freezing cold or swelteringly hot outdoors, the internal temperature of the body stays within a range of about 36° to 37°C (97°–99°F). American physiologist **Walter Cannon** (1871–1945) coined the term *homeostasis* for this tendency to maintain internal stability. Homeostasis has been one of the most enlightening theories in physiology. We now see physiology as largely a group of mechanisms for maintaining homeostasis, and the loss of homeostatic control as the cause of illness and death. Pathophysiology is essentially the study of unstable conditions that result when our homeostatic controls go awry.

Do not, however, overestimate the degree of internal stability. Internal conditions are not absolutely constant but fluctuate within a limited range, such as the range of body temperatures noted earlier. The internal state of the body is best described as a **dynamic equilibrium** (balanced change), in which there is a certain **set point** or average value for a given variable (such as 37°C for body temperature) and conditions fluctuate slightly around this point.

The fundamental mechanism that keeps a variable close to its set point is **negative feedback**—a process in which the body senses a change and activates mechanisms that negate or reverse it. By maintaining stability, negative feedback is the key mechanism for maintaining health.

These principles can be understood by comparison to a home heating system (fig. 1.6a). Suppose it is a cold winter day and you have set your thermostat for 20°C (68°F)—the set point. If the room becomes too cold, a temperature-sensitive switch in the thermostat turns on the furnace. The temperature rises until it is slightly above the set point, and then the switch breaks the circuit and turns off the furnace. This is a negative feedback process that reverses the falling temperature and restores it to the set point. When the furnace turns off, the temperature slowly drops again until the switch is reactivated—thus, the furnace cycles on and off all day. The room temperature does not stay at exactly 20°C but *fluctuates* a few degrees either way—the system maintains a state of dynamic equilibrium in which the temperature averages 20°C and deviates only slightly from the set point. Because feedback mechanisms alter the original changes that triggered them (temperature, for example), they are often called **feedback loops.**

Body temperature is similarly regulated by a "thermostat"—a group of nerve cells in the base of the brain that monitor the temperature of the blood. If you become overheated, the thermostat triggers heat-losing mechanisms (fig. 1.6b). One of these is **vasodilation** (VAY-zo-dy-LAY-shun), the widening of blood vessels. When blood vessels of the skin dilate, warm blood flows closer to the body surface and loses heat to the surrounding

[18]*homeo* = the same; *stas* = to place, stand, stay

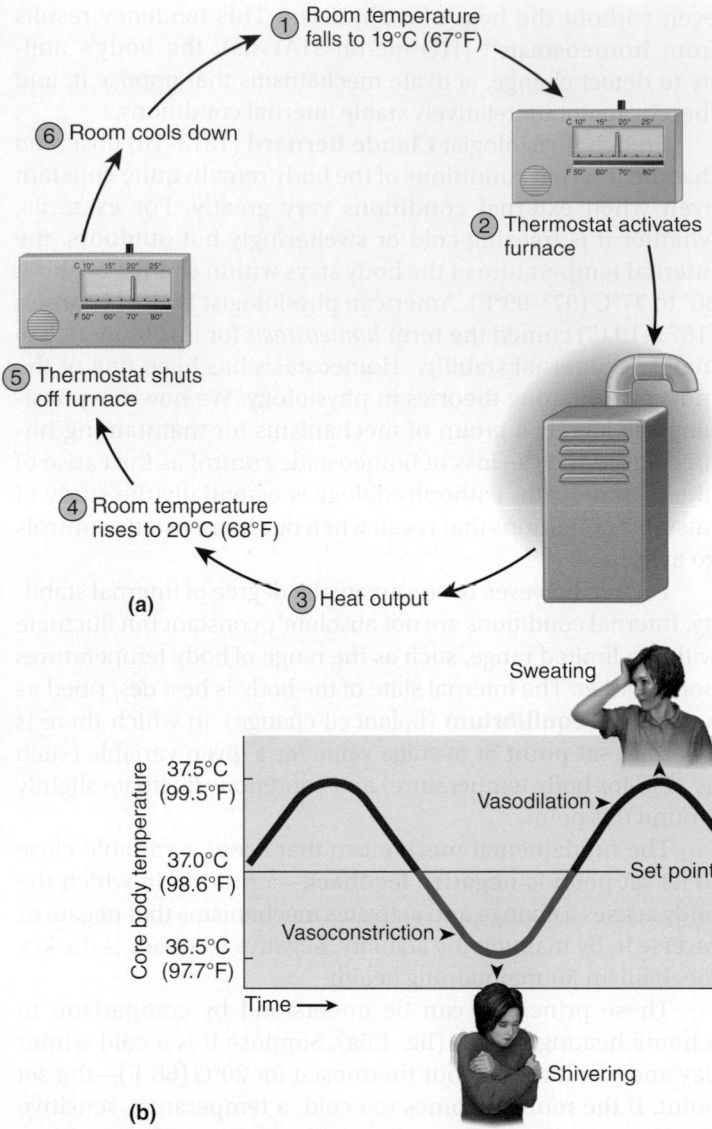

(a)

Sweating

37.5°C
(99.5°F)

Vasodilation ▶

37.0°C
(98.6°F)

Set point

Vasoconstriction ▶

36.5°C
(97.7°F)

Time ⟶

Core body temperature

Shivering

(b)

**FIGURE 1.6 Negative Feedback in Thermoregulation.** (a) The negative feedback loop that maintains room temperature. (b) Negative feedback usually keeps the human body temperature within about 0.5°C of a 37°C set point. Cutaneous vasoconstriction and shivering set in when the body temperature falls too low, and soon raise it. Cutaneous vasodilation and sweating set in when body temperature rises too high, and soon lower it.

❓ *How does vasodilation reduce the body temperature?*

air. If this is not enough to return your temperature to normal, sweating occurs; the evaporation of water from the skin has a powerful cooling effect (see Deeper Insight 1.3). Conversely, if it is cold outside and your body temperature drops much below 37°C, these nerve cells activate heat-conserving mechanisms. The first to be activated is **vasoconstriction,** a narrowing of the blood vessels in the skin, which serves to retain warm blood deeper in your body and reduce heat loss. If this is not enough, the brain activates shivering—muscle tremors that generate heat.

 **DEEPER INSIGHT 1.3**

## MEDICAL HISTORY

### Men in the Oven

English physician Charles Blagden (1748–1820) staged a rather theatrical demonstration of homeostasis long before Cannon coined the word. In 1775, Blagden spent 45 minutes in a chamber heated to 127°C (260°F)—along with a dog, a beefsteak, and some research associates. Being dead and unable to maintain homeostasis, the beefsteak was cooked. But being alive and capable of evaporative cooling, the dog panted, the men sweated, and all of them survived. History does not record whether the men ate the steak in celebration or shared it with the dog.

Let's consider one more example—a case of homeostatic control of blood pressure. When you first rise from bed in the morning, gravity causes some of your blood to drain away from your head and upper torso, resulting in falling blood pressure in this region—a local imbalance in your homeostasis (fig. 1.7). This is detected by sensory nerve endings called *baroreceptors* in large arteries near the heart. They transmit nerve signals to the brainstem, where we have a *cardiac center* that regulates the heart rate. The cardiac center responds by transmitting

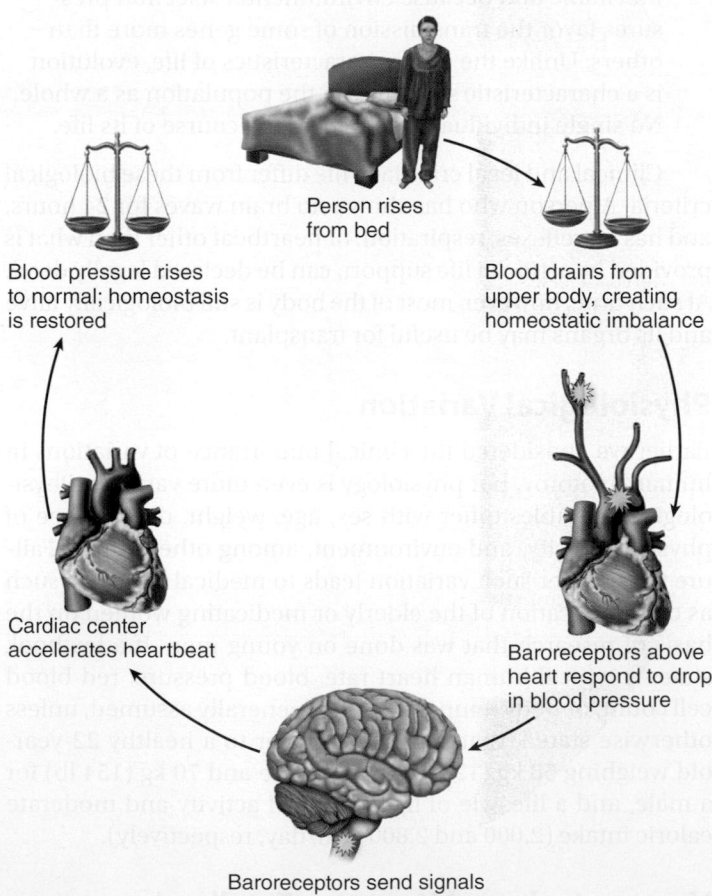

Person rises
from bed

Blood pressure rises
to normal; homeostasis
is restored

Blood drains from
upper body, creating
homeostatic imbalance

Cardiac center
accelerates heartbeat

Baroreceptors above
heart respond to drop
in blood pressure

Baroreceptors send signals
to cardiac center of brainstem

**FIGURE 1.7 Homeostatic Compensation for a Postural Change in Blood Pressure.**

nerve signals to the heart, which speed it up. The faster heart rate quickly raises the blood pressure and restores normal homeostasis. In elderly people, this feedback loop is sometimes insufficiently responsive, and they may feel dizzy as they rise from a reclining position and their cerebral blood pressure falls. This sometimes causes fainting.

This reflexive correction of blood pressure *(baroreflex)* illustrates three common, although not universal, components of a feedback loop: a receptor, an integrating center, and an effector. The **receptor** is a structure that senses a change in the body, such as the stretch receptors that monitor blood pressure. The **integrating (control) center,** such as the cardiac center of the brain, is a mechanism that processes this information, relates it to other available information (for example, comparing what the blood pressure is with what it should be), and "makes a decision" about what the appropriate response should be. The **effector** is the cell or organ that carries out the final corrective action. In the foregoing example, it is the heart. The response, such as the restoration of normal blood pressure, is then sensed by the receptor, and the feedback loop is complete.

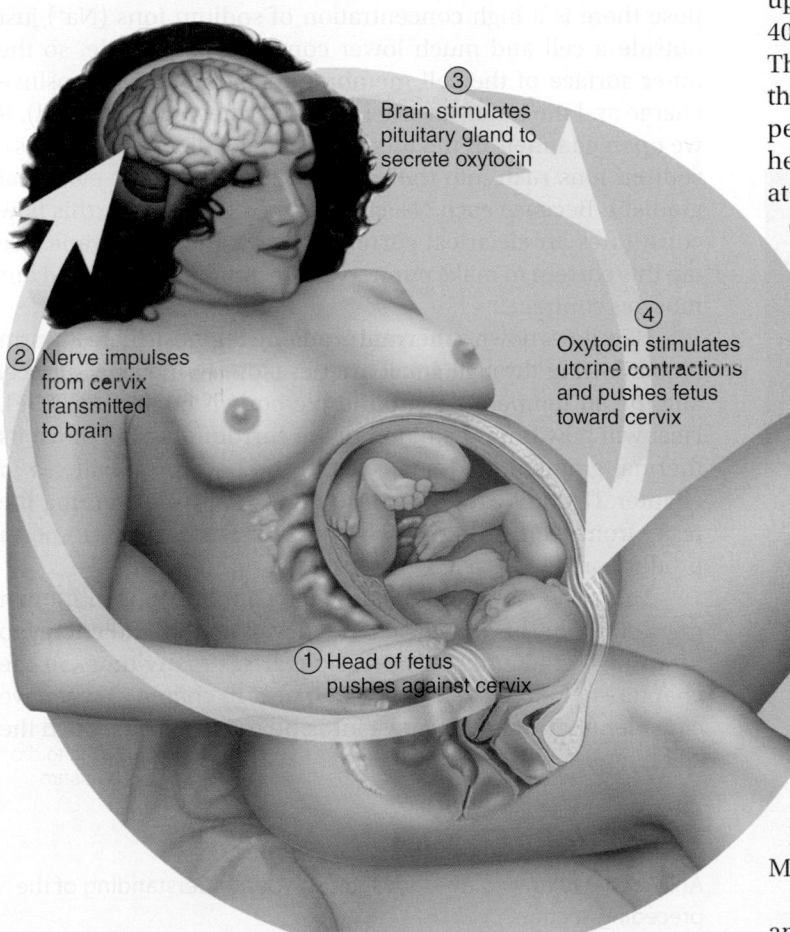

**FIGURE 1.8 Positive Feedback in Childbirth.**

❓ *Could childbirth as a whole be considered a negative feedback event? Discuss.*

③ Brain stimulates pituitary gland to secrete oxytocin

② Nerve impulses from cervix transmitted to brain

④ Oxytocin stimulates uterine contractions and pushes fetus toward cervix

① Head of fetus pushes against cervix

## Positive Feedback and Rapid Change

**Positive feedback** is a self-amplifying cycle in which a physiological change leads to even greater change in the same direction, rather than producing the corrective effects of negative feedback. Positive feedback is often a normal way of producing rapid change. When a woman is giving birth, for example, the head of the fetus pushes against her cervix (the neck of the uterus) and stimulates its nerve endings (fig. 1.8). Nerve signals travel to the brain, which, in turn, stimulates the pituitary gland to secrete the hormone oxytocin. Oxytocin travels in the blood and stimulates the uterus to contract. This pushes the fetus downward, stimulating the cervix still more and causing the positive feedback loop to be repeated. Labor contractions therefore become more and more intense until the fetus is expelled. Other cases of beneficial positive feedback are seen later in the book in, for example, blood clotting, protein digestion, and the generation of nerve signals.

Frequently, however, positive feedback is a harmful or even life-threatening process. This is because its self-amplifying nature can quickly change the internal state of the body to something far from its homeostatic set point. Consider a high fever, for example. A fever triggered by infection is beneficial up to a point, but if the body temperature rises much above 40°C (104°F), it may create a dangerous positive feedback loop. This high temperature raises the metabolic rate, which makes the body produce heat faster than it can get rid of it. Thus, temperature rises still further, increasing the metabolic rate and heat production still more. This "vicious circle" becomes fatal at approximately 45°C (113°F). Thus, positive feedback loops often create dangerously out-of-control situations that require emergency medical treatment.

## Gradients and Flow

Another fundamental concept that will arise repeatedly in this book is that matter and energy tend to *flow down gradients*. This simple principle underlines processes as diverse as blood circulation, respiratory airflow, urine formation, nutrient absorption, body water distribution, temperature regulation, and the action of nerves and muscles.

A physiological **gradient** is a difference in chemical concentration, electrical charge, physical pressure, temperature, or other variable between one point and another. If matter or energy moves from the point where this variable has a higher value to the point with a lower value, we say it flows **down the gradient**— for example, from a warmer to a cooler point, or a place of high chemical concentration to one of lower concentration. Movement in the opposite direction is **up the gradient.**

Outside of biology, *gradient* can mean a hill or slope, and this affords us a useful analogy to biological processes (fig. 1.9a). A wagon released at the top of a hill will roll down it ("flow") spontaneously, without need for anyone to exert energy to move it. Similarly, matter and energy in the body spontaneously flow down gradients, without the expenditure

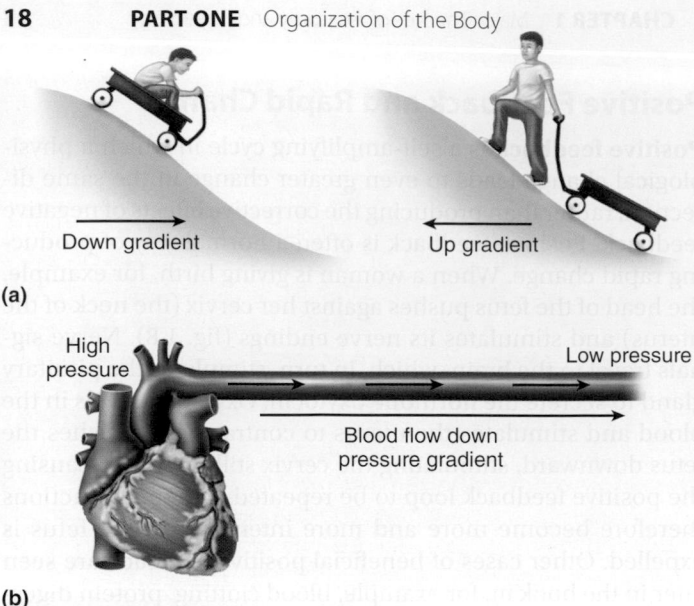

(a)

(b)

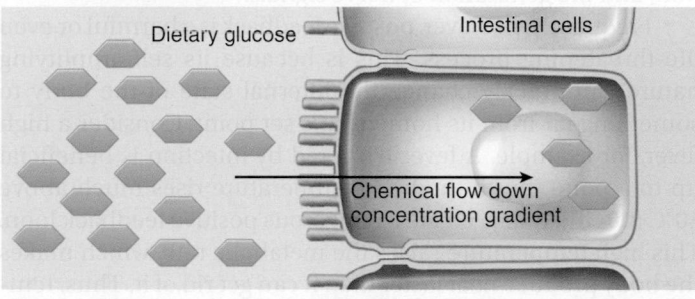

(c)

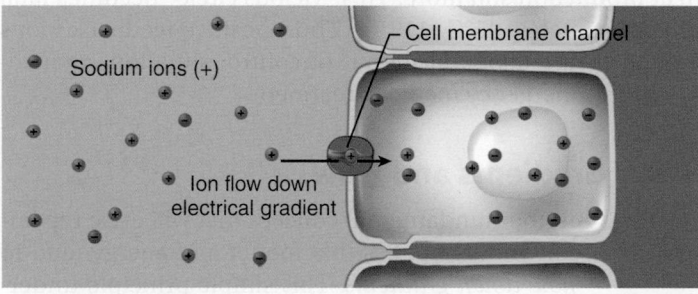

(d)

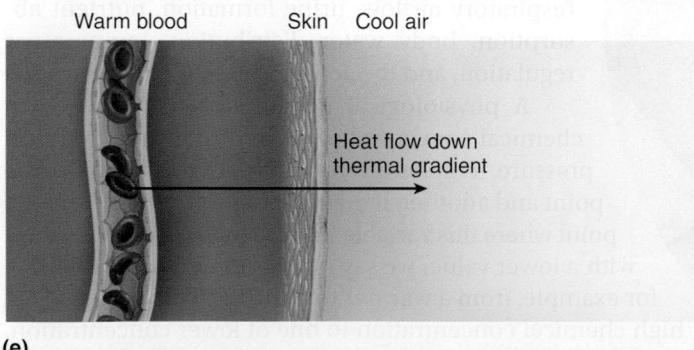

(e)

**FIGURE 1.9  Flow Down Gradients.** (a) A wagon rolling downhill (down a gradient) is a useful analogy to spontaneous, gradient-driven physiological processes. (b) Blood flowing down a pressure gradient. (c) Dietary sugars flowing down a concentration gradient into an intestinal cell. (d) Sodium ions flowing down an electrical gradient into a cell. (e) Heat flowing down a thermal gradient to leave the body through the skin.

of metabolic energy. Movement up a gradient does require an energy expenditure, just as we would have to push or pull a wagon to move it uphill.

Consider some examples and analogies. If you open a water tap with a garden hose on it, you create a **pressure gradient;** water flows down the hose from the high-pressure point at the tap to the low-pressure point at the open end. Each heartbeat is like that, creating a gradient from high blood pressure near the heart to low pressure farther away; blood flows down this gradient away from the heart (fig. 1.9b). When we inhale, air flows down a pressure gradient from the surrounding atmosphere to pulmonary air passages where the pressure is lower. A pressure gradient also drives the process in which the kidneys filter water and waste products from the blood.

Chemicals flow down **concentration gradients.** When we digest starch, a high concentration of free sugars accumulates in the small intestine. The cells lining the intestine contain only a low concentration of sugars, so sugars flow from the intestinal space into these cells, thus becoming absorbed into the body's tissues (fig. 1.9c). Water flows through cell membranes and epithelia by *osmosis,* from the side where it is more concentrated to the side where it is less so.

Charged particles flow down **electrical gradients.** Suppose there is a high concentration of sodium ions ($Na^+$) just outside a cell and much lower concentration inside, so the outer surface of the cell membrane has a relatively positive charge and the inner surface is relatively negative (fig. 1.9d). If we open channels in the membrane that will let sodium pass, sodium ions rush into the cell, flowing down their electrical gradient. Because each $Na^+$ carries a positive charge, this flow constitutes an electrical current through the membrane. We tap this current to make our nerves fire, our heart beat, and our muscles contract.

Heat flows down a **thermal gradient.** Suppose there is warm blood flowing through small arteries close to the skin surface, and the air temperature around the body is cooler (fig. 1.9e). Heat will flow from the blood to the surrounding air, down its thermal gradient, and be lost from the body. You will see in chapter 27 that heat flow is also important in preventing the testes from overheating, which would otherwise prevent sperm production.

Thus, you can see there are many applications in human physiology for this universal tendency of matter and energy to flow down gradients. This principle arises many times in the chapters to follow. We will revisit it next in chapter 3 when we consider how materials move into and out of cells through the cell membrane.

**BEFORE YOU GO ON**

Answer the following questions to test your understanding of the preceding section:

16. List four biological criteria of life and one clinical criterion. Explain how a person could be clinically dead but biologically alive.

17. What is meant by dynamic equilibrium? Why would it be wrong to say homeostasis prevents internal change?

18. Explain why stabilizing mechanisms are called *negative* feedback.

19. Explain why positive feedback is more likely than negative feedback to disturb homeostasis.

20. Active tissues generate carbon dioxide, which diffuses out of the tissue into the bloodstream, to be carried away. Is this diffusion into the blood a case of flow up a gradient, or down? Explain.

## 1.7 The Language of Medicine

### Expected Learning Outcomes

When you have completed this section, you should be able to

a. explain why modern anatomical terminology is so heavily based on Greek and Latin;

b. recognize eponyms when you see them;

c. describe the efforts to achieve an internationally uniform anatomical terminology;

d. break medical terms down into their basic word elements;

e. state some reasons why the literal meaning of a word may not lend insight into its definition;

f. relate singular noun forms to their plural and adjectival forms; and

g. discuss why precise spelling is important in anatomy and physiology.

One of the greatest challenges faced by students of anatomy and physiology is the vocabulary. In this book, you will encounter such Latin terms as *corpus callosum* (a brain structure), *ligamentum arteriosum* (a small fibrous band near the heart), and *extensor carpi radialis longus* (a forearm muscle). You may wonder why structures aren't named in "just plain English," and how you will ever remember such formidable names. This section will give you some answers to these questions and some useful tips on mastering anatomical terminology.

### The History of Anatomical Terminology

The major features of human gross anatomy have standard international names prescribed by a book titled the *Terminologia Anatomica (TA)*. The *TA* was codified in 1998 by an international committee of anatomists and approved by professional associations of anatomists in more than 50 countries.

About 90% of today's medical terms are formed from just 1,200 Greek and Latin roots. Why those two languages? Scientific investigation began in ancient Greece and soon spread to Rome. The Greeks and Romans coined many of the words still used in human anatomy today: *duodenum, uterus, prostate, cerebellum, diaphragm, sacrum, amnion,* and others. In the Renaissance, the fast pace of discovery required a profusion of new terms to describe things. Anatomists in different countries began giving different names to the same structures. Adding to the confusion, they often named new structures and diseases in honor of

their esteemed teachers and predecessors, giving us such non-descriptive terms as *fallopian tube* and *duct of Santorini.* Terms coined from the names of people, called **eponyms,**[19] afford little clue as to what a structure or condition is.

In hopes of resolving this growing confusion, anatomists began meeting as early as 1895 to devise a uniform international terminology. After several false starts, they agreed on a list of terms that rejected all eponyms and gave each structure a unique Latin name to be used worldwide. Even if you were to look at an anatomy atlas in Japanese or Arabic, the illustrations may be labeled with the same Latin terms as in an English-language atlas. That list served for many decades until recently replaced by the *TA*, which prescribes both Latin names and accepted English equivalents. The terminology in this book conforms to the *TA* except where undue confusion would result from abandoning widely used, yet unofficial, terms.

### Analyzing Medical Terms

The task of learning medical terminology seems overwhelming at first, but it is a simple skill to become more comfortable with the technical language of medicine. People who find scientific terms confusing and difficult to pronounce, spell, and remember often feel more confident once they realize the logic of how terms are composed. A term such as *hyponatremia* is less forbidding once we recognize that it is composed of three common word elements: *hypo-* (below normal), *natr-* (sodium), and *-emia* (blood condition). Thus, hyponatremia is a deficiency of sodium in the blood. Those word elements appear over and over in many other medical terms: *hypothermia, natriuretic, anemia,* and so on. Once you learn the meanings of *hypo-, natri-,* and *-emia,* you already have the tools to at least partially understand hundreds of other biomedical terms. Inside the back cover, you will find a lexicon of the 400 word elements most commonly footnoted in this book.

Scientific terms are typically composed of one or more of the following elements:

- At least one *root (stem)* that bears the core meaning of the word. In *cardiology,* for example, the root is *cardi-* (heart). Many words have two or more roots. In *cardiomyopathy,* for example, the roots are *cardi-* (heart), *my-* (muscle), and *path-* (disease).

- *Combining vowels* that are often inserted to join roots and make the word easier to pronounce. In *cardiomyopathy,* each *o* is a combining vowel. Although *o* is the most common combining vowel, all vowels of the alphabet are used in this way, such as *a* in *ligament, e* in *vitreous, i* in *fusiform, u* in *ovulation,* and *y* in *tachycardia.* Some words, such as *intervertebral,* have no combining vowels. A combination of a root and combining vowel is called a *combining form;* for example, *chrom-* (color) + *o* (a combining vowel) make the combining form *chromo-,* as in *chromosome.*

- A *prefix* may be present to modify the core meaning of the word. For example, *gastric* (pertaining to the stomach or to the belly of a muscle) takes on a variety of new meanings when prefixes are added to it: *epigastric* (above the stomach), *hypogastric* (below the stomach), *endogastric* (within the stomach), and *digastric* (a muscle with two bellies).

- A *suffix* may be added to the end of a word to modify its core meaning. For example, *microscope, microscopy, microscopic,* and *microscopist* have different meanings because of their suffixes alone. Often two or more suffixes, or a root and suffix, occur together so often that they are treated jointly as a *compound suffix;* for example, *log* (study) + *y* (process) form the compound suffix *-logy* (the study of).

To summarize these basic principles, consider the word *gastroenterology,* a branch of medicine dealing with the stomach and small intestine. It breaks down into *gastro/entero/logy:*

gastro     =   a combining form meaning "stomach"

entero    =   a combining form meaning "small intestine"

logy      =   a compound suffix meaning "the study of"

"Dissecting" words in this way and paying attention to the word-origin footnotes throughout this book will help you become more comfortable with the language of anatomy. Knowing how a word breaks down and knowing the meaning of its elements make it far easier to pronounce a word, spell it, and remember its definition. There are a few unfortunate exceptions, however. The path from original meaning to current usage has often become obscured by history (see Deeper Insight 1.4). The foregoing approach also is no help with eponyms or **acronyms**—words composed of the first letter, or first few letters, of a series of words. For example, a common medical imaging method is the PET scan, an acronym for *positron emission tomography.* Note that PET is a pronounceable word, hence a true acronym. Acronyms are not to be confused with simple abbreviations such as DNA and MRI, in which each letter must be pronounced separately.

## Plural, Adjectival, and Possessive Forms

A point of confusion for many beginning students is how to recognize the plural forms of medical terms. Few people would fail to recognize that *ovaries* is the plural of *ovary,* but the connection is harder to make in other cases: For example, the plural of *cortex* is *cortices* (COR-ti-sees), the plural of *corpus* is *corpora,* and the plural of *epididymis* is *epididymides* (EP-ih-DID-ih-MID-eze). Table 1.1 will help you make the connection between common singular and plural noun terminals.

In some cases, what appears to the beginner to be two completely different words may be only the noun and adjectival forms of the same word. For example, *brachium* denotes the arm, and *brachii* (as in the muscle name *biceps brachii*) means "of the arm." *Carpus* denotes the wrist, and *carpi,* a word used in several muscle names, means "of the wrist." Adjectives can also take different forms for the singular and plural and for different degrees of comparison. The *digits* are the fingers and toes. The word *digiti* in a muscle name means "of a single finger (or toe)," whereas *digitorum* is the plural, meaning "of multiple fingers (or toes)." Thus, the *extensor digiti minimi* muscle

## DEEPER INSIGHT 1.4
## MEDICAL HISTORY

### Obscure Word Origins

The literal translation of a word doesn't always provide great insight into its modern meaning. The history of language is full of twists and turns that are fascinating in their own right and say much about the history of human culture, but they can create confusion for students.

For example, the *amnion* is a transparent sac that forms around the developing fetus. The word is derived from *amnos,* from the Greek for "lamb." From this origin, *amnos* came to mean a bowl for catching the blood of sacrificial lambs, and from there the word found its way into biomedical usage for the membrane that emerges (quite bloody) as part of the afterbirth. The *acetabulum,* the socket of the hip joint, literally means "vinegar cup." Apparently the hip socket reminded an anatomist of the little cups used to serve vinegar as a condiment on dining tables in ancient Rome. The word *testicles* can be translated "little pots" or "little witnesses." The history of medical language has several amusing conjectures as to why this word was chosen to name the male gonads.

| TABLE 1.1 | Singular and Plural Forms of Some Noun Terminals | |
|---|---|---|
| **Singular Ending** | **Plural Ending** | **Examples** |
| -a | -ae | axilla, axillae |
| -ax | -aces | thorax, thoraces |
| -en | -ina | lumen, lumina |
| -ex | -ices | cortex, cortices |
| -is | -es | diagnosis, diagnoses |
| -is | -ides | epididymis, epididymides |
| -ix | -ices | appendix, appendices |
| -ma | -mata | carcinoma, carcinomata |
| -on | -a | ganglion, ganglia |
| -um | -a | septum, septa |
| -us | -era | viscus, viscera |
| -us | -i | villus, villi |
| -us | -ora | corpus, corpora |
| -x | -ges | phalanx, phalanges |
| -y | -ies | ovary, ovaries |
| -yx | -yces | calyx, calyces |

extends only the little finger, whereas the *extensor digitorum* muscle extends all fingers except the thumb.

The English words *large, larger,* and *largest* are examples of the positive, comparative, and superlative degrees of comparison. In Latin, these are *magnus, major* (from *maior*), and *maximus.* We find these in the muscle names *adductor magnus* (a *large* muscle of the thigh), the *pectoralis major* (the *larger* of two pectoralis muscles of the chest), and *gluteus maximus* (the *largest* of the three gluteal muscles of the buttock).

Some noun variations indicate the possessive, such as the *rectus abdominis,* a straight *(rectus)* muscle of the abdomen *(abdominis,* "of the abdomen"), and the *erector spinae,* a muscle that straightens *(erector)* the spinal column *(spinae,* "of the spine").

Anatomical terminology also frequently follows the Greek and Latin practice of placing the adjective after the noun. Thus, we have such names as the *stratum lucidum* for a clear *(lucidum)* layer *(stratum)* of the epidermis, the *foramen magnum* for a large *(magnum)* hole *(foramen)* in the skull, and the aforementioned *pectoralis major* muscle of the chest.

This is not to say that you must be conversant in Latin or Greek grammar to proceed with your study of anatomy. These few examples, however, may alert you to some patterns to watch for in the terminology you study and, ideally, will make your encounters with anatomical terminology less confusing.

## Pronunciation

Pronunciation is another stumbling block for many beginning anatomy and physiology students. This book gives simple pro-NUN-see-AY-shun guides for many terms when they are first introduced. You can also hear pronunciations of most of the anatomical terms within the *Anatomy & Physiology | Revealed* product, which can be accessed from this book's website.

## The Importance of Precision

A final word of advice for your study of anatomy and physiology: Be precise in your use of terms. It may seem trivial if you misspell *trapezius* as *trapezium,* but in doing so, you would be changing the name of a back muscle to the name of a wrist bone. Similarly, changing *occipitalis* to *occipital* or *zygomaticus* to *zygomatic* changes other muscle names to bone names. Changing *malleus* to *malleolus* changes the name of a middle-ear bone to the name of a bony protuberance of the ankle. And there is only a one-letter difference between *ileum* (the final portion of the small intestine) and *ilium* (part of the hip bone), and between *gustation* (the sense of taste) and *gestation* (pregnancy).

The health professions demand the utmost attention to detail and precision—people's lives may one day be in your hands. The habit of carefulness must extend to your use of language as well. Many patients have died simply because of tragic written and oral miscommunication in the hospital.

### BEFORE YOU GO ON

Answer the following questions to test your understanding of the preceding section:

21. Explain why modern anatomical terminology is so heavily based on Greek and Latin.

22. Distinguish between an eponym and an acronym, and explain why both of these present difficulties for interpreting anatomical terms.

23. Break each of the following words down into its roots, prefixes, and suffixes, and state their meanings, following the example of *gastroenterology* analyzed earlier: *pericardium, appendectomy, subcutaneous, phonocardiogram, otorhinolaryngology.* Consult the list of word elements inside the back cover for help.

24. Write the singular form of each of the following words: *pleurae, gyri, ganglia, fissures.* Write the plural form of each of the following: *villus, tibia, encephalitis, cervix, stoma.*

## 1.8 Review of Major Themes

To close this chapter, let's distill a few major points from it. These themes can provide you with a sense of perspective that will make the rest of the book more meaningful and not just a collection of disconnected facts. These are some key unifying principles behind all study of human anatomy and physiology:

- **Unity of form and function.** *Form and function complement each other; physiology cannot be divorced from anatomy.* This unity holds true even down to the molecular level. Our very molecules, such as DNA and proteins, are structured in ways that enable them to carry out their functions. Slight changes in molecular structure can destroy their activity and threaten life.

- **Cell theory.** *All structure and function result from the activity of cells.* Every physiological concept in this book ultimately must be understood from the standpoint of how cells function. Even anatomy is a result of cellular function. If cells are damaged or destroyed, we see the results in disease symptoms of the whole person.

- **Evolution.** *The human body is a product of evolution.* Like every other living species, we have been molded by millions of years of natural selection to function in a changing environment. Many aspects of human anatomy and physiology reflect our ancestors' adaptations to their environment. Human form and function cannot be fully understood except in light of our evolutionary history.

- **Hierarchy of complexity.** *Human structure can be viewed as a series of levels of complexity.* Each level is composed of a smaller number of simpler subunits than the level above it. These subunits are arranged in different ways to form diverse structures of higher complexity. Understanding the simpler components is the key to understanding higher levels of structure.

- **Homeostasis.** *The purpose of most normal physiology is to maintain stable conditions within the body.* Human physiology is essentially a group of homeostatic mechanisms that produce stable internal conditions favorable to cellular function. Any serious departure from these conditions can be harmful or fatal to cells and thus to the whole body.

- **Gradients and flow.** Matter and energy tend to flow down gradients such as differences in chemical concentration, pressure, temperature, and electrical charge. This accounts for much of their movement in human physiology.

▶▶▶ **APPLY WHAT YOU KNOW**

*Architect Louis Henri Sullivan coined the phrase, "Form ever follows function." What do you think he meant by this? Discuss how this idea could be applied to the human body and cite a specific example of human anatomy to support it.*

# DEEPER INSIGHT 1.5

## CLINICAL APPLICATION

### Medical Imaging

The development of techniques for looking into the body without having to do exploratory surgery has greatly accelerated progress in medicine. A few of these techniques are described here.

### Radiography

*X-rays,* a form of high-energy radiation, were discovered by William Roentgen in 1885. X-rays can penetrate soft tissues of the body and darken photographic film on the other side. They are absorbed, however, by dense tissues such as bone, teeth, tumors, and tuberculosis nodules, which leave the film lighter in these areas (fig. 1.10a). The process of examining the body with X-rays is called

*radiography.* The term *X-ray* also applies to an image *(radiograph)* made by this method. Radiography is commonly used in dentistry, mammography, diagnosis of fractures, and examination of the chest. Hollow organs can be visualized by filling them with a *radiopaque* substance that absorbs X-rays. Barium sulfate is given orally for examination of the esophagus, stomach, and small intestine or by enema for examination of the large intestine. Other substances are given by injection for *angiography,* the examination of blood vessels (fig. 1.10b). Some disadvantages of radiography are that images of overlapping organs can be confusing, slight differences in tissue density are not easily detected, and X-rays can cause mutations leading to cancer and birth defects. Radiography

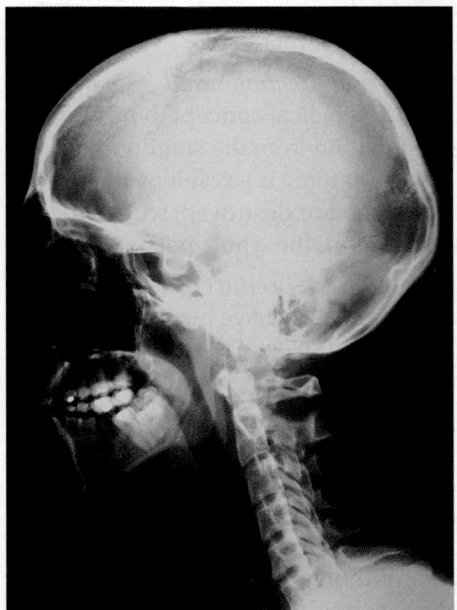

**(a) X-ray (radiograph)**

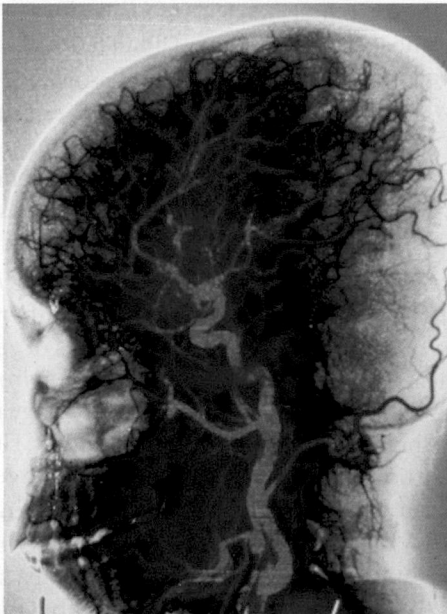

**(b) Cerebral angiogram**

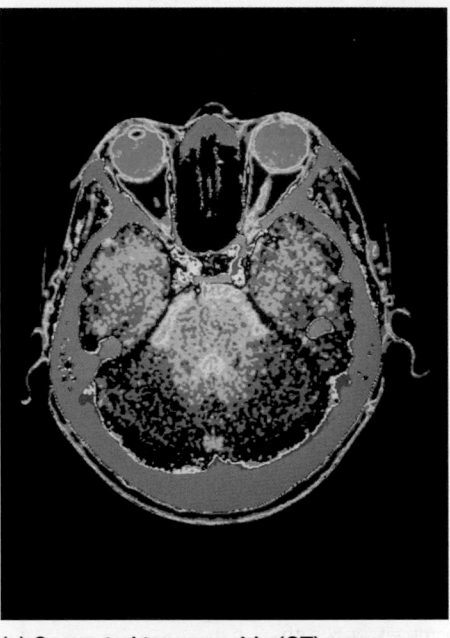

**(c) Computed tomographic (CT) scan**

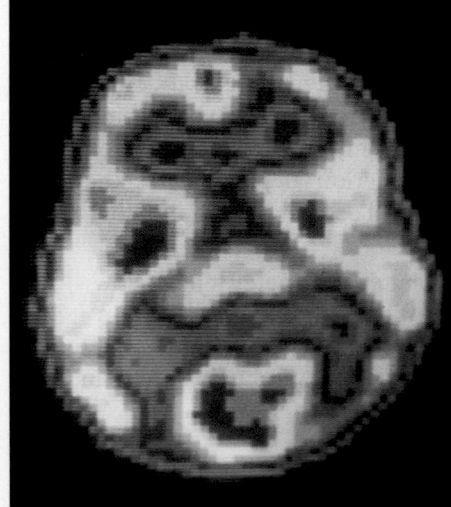

**(d) Positron emission tomographic (PET) scan**

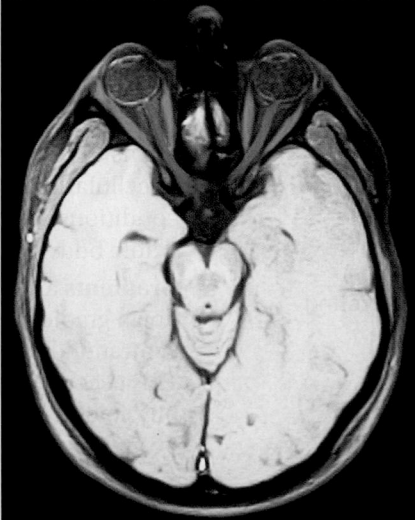

**(e) Magnetic resonance image (MRI)**

**FIGURE 1.10  Radiologic Images of the Head.**  (a) X-ray (radiograph) showing the bones and teeth. (b) An angiogram of the cerebral blood vessels. The arteries are enhanced with false color. (c) A CT scan. The eyes and skin are shown in blue, bone in pink, and the brain in green. (d) A PET scan of the brain of an unmedicated schizophrenic patient. Red areas indicate regions of high metabolic rate. In this patient, the visual center of the brain at the rear of the head (bottom of photo) was especially active during the scan. (e) An MRI scan at the level of the eyes. The optic nerves appear in red and the muscles that move the eyes appear in green.

therefore cannot be used indiscriminately. Nevertheless, radiography still accounts for over half of all clinical imaging. Until the 1960s, it was the only method widely available.

### Computed Tomography (CT)

The *CT (computerized tomographic[20]) scan* is a more sophisticated application of X-rays developed in 1972. The patient is moved through a ring-shaped machine that emits low-intensity X-rays on one side and receives them with a detector on the opposite side. A computer analyzes signals from the detector and produces an image of a "slice" of the body about as thin as a coin (fig. 1.10c). The computer can "stack" a series of these images to construct a three-dimensional image of the body. CT scanning has the advantage of imaging thin sections of the body, so there is little overlap of organs and the image is much sharper than a conventional X-ray. It requires extensive knowledge of cross-sectional anatomy to interpret the images. CT scanning is useful for identifying tumors, aneurysms, cerebral hemorrhages, kidney stones, and other abnormalities. It has virtually eliminated exploratory surgery.

### Positron Emission Tomography (PET)

The *PET scan,* developed in the 1970s, is used to assess the metabolic state of a tissue and distinguish which tissues are most active at a given moment (fig. 1.10d). The procedure begins with an injection of radioactively labeled glucose, which emits positrons (electron-like particles with a positive charge). When a positron and electron meet, they annihilate each other and give off a pair of gamma rays that can be detected by sensors and analyzed by computer. The computer displays a color image that shows which tissues were using the most glucose at the moment. In cardiology, PET scans can show the extent of damaged heart tissue. Since it consumes little or no glucose, the damaged tissue appears dark. PET scans are also widely used to diagnose cancer and evaluate tumor status. The PET scan is an example of *nuclear medicine*—the use of radioactive isotopes to treat disease or to form diagnostic images of the body.

### Magnetic Resonance Imaging (MRI)

*MRI* was developed in the 1970s as a technique superior to CT scanning for visualizing some soft tissues. The patient lies in a cylindrical chamber surrounded by a large electromagnet that creates a very strong magnetic field. Hydrogen atoms in the tissues align themselves with this field. The technologist then turns on a field of radio waves, causing the hydrogen atoms to absorb additional energy and align in a different direction. When the radio waves are turned off, the hydrogen atoms realign themselves to the magnetic field, giving off their excess energy at different rates that depend on the type of tissue. A computer analyzes the emitted energy to produce an image of the body. MRI can "see" clearly through the skull and spinal column to produce images of the nervous tissue (fig. 1.10e). Moreover, it is better than CT for distinguishing between soft tissues such as the white and gray matter of the nervous system. MRI also eliminates exposure to harmful X-rays.

One disadvantage of MRI is that it requires a patient to lie still in the enclosed space for up to 45 minutes to scan one region of the body and may entail 90 minutes to scan multiple regions such as the abdominal and pelvic cavities. Some patients find they cannot do this. Also because of the long exposures involved, MRI is not as good as CT for scanning the digestive tract; stomach and intestinal motility produce blurred images over such long exposures.

*Functional MRI (fMRI)* is a variation that visualizes moment-to-moment changes in tissue function. fMRI scans of the brain, for example, show shifting patterns of activity as the brain applies itself to a specific sensory, mental, or motor task. fMRI has lately replaced the PET scan as the most important method for visualizing brain function. The use of fMRI in brain imaging is further discussed in Deeper Insight 14.4 (p. 553).

### Sonography

*Sonography[21]* is the second oldest and second most widely used method of imaging. It is an outgrowth of sonar technology developed in World War II. A handheld device held firmly to the skin produces high-frequency ultrasound waves and receives the signals that echo back from internal organs. Although sonography was first used medically in the 1950s, images of significant clinical value had to wait until computer technology had developed enough to analyze differences in the way tissues reflect ultrasound. Sonography is not very useful for examining bones or lungs, but it is the method of choice in obstetrics, where the image *(sonogram)* can be used to locate the placenta and evaluate fetal age, position, and development. *Echocardiography* is the use of sonography to visualize motion of the heart wall and valves and blood flow through the heart chambers. Sonography avoids the harmful effects of X-rays, and the equipment is inexpensive and portable. Its primary disadvantage is that it does not produce a very sharp image (fig. 1.11).

---

[21]*sono* = sound; *graphy* = recording process

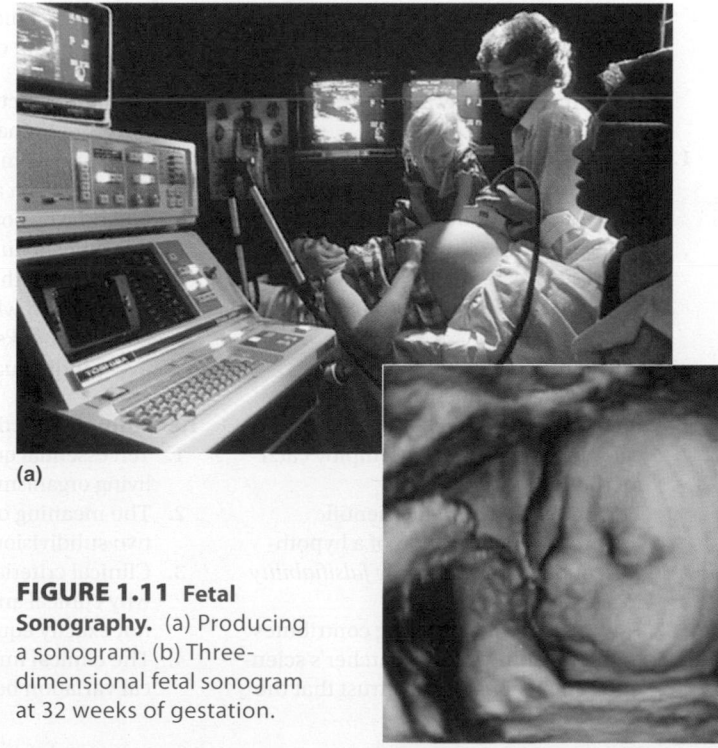

**(a)**

**FIGURE 1.11 Fetal Sonography.** (a) Producing a sonogram. (b) Three-dimensional fetal sonogram at 32 weeks of gestation.

**(b)**

---

[20]*tomo* = section, cut, slice; *graphic* = pertaining to a recording

# STUDY GUIDE

## ► Assess Your Learning Outcomes

*To test your knowledge, discuss the following topics with a study partner or in writing, ideally from memory.*

### 1.1 The Scope of Anatomy and Physiology (p. 2)

1. The meaning of *anatomy* and *physiology* and what it means to say these two sciences are complementary
2. Methods of study in anatomy and clinical examination
3. Branches of anatomy that study the body at different levels of detail
4. How comparative physiology advances the understanding human function

### 1.2 The Origins of Biomedical Science (p. 3)

1. Greek and Roman scholars who first gave medicine a scientific basis
2. Ways in which the work of Maimonides, Avicenna, Vesalius, and Harvey were groundbreaking in the context of their time and culture
3. Why medical science today owes such a great debt to Hooke, Leeuwenhoek, and other inventors
4. How Schleiden and Schwann revolutionized and unified the understanding of biological structure, ultimately including human anatomy and physiology

### 1.3 Scientific Method (p. 6)

1. How the philosophers Bacon and Descartes revolutionized society's view of science, even though neither of them was a scientist
2. The essential qualities of the scientific method
3. The nature of the inductive and hypothetico–deductive methods, how they differ, and which areas of biomedical science most heavily employ each method
4. The qualities of a valid scientific hypothesis, the function of a hypothesis, and what is meant by *falsifiability* in science
5. How each of the following contributes to the reliability of a researcher's scientific conclusions and the trust that the public may place in science: sample size, control groups, the double-blind method, statistical testing, and peer review
6. The distinctions between scientific facts, laws, and theories; the purpose of a theory; and how the scientific meanings of *law* and *theory* differ from the common lay meanings

### 1.4 Human Origins and Adaptations (p. 9)

1. The meaning of *evolution, natural selection, selection pressure,* and *adaptation,* with examples of each
2. The historical origin of the theory of natural selection and how this theory is relevant to a complete understanding of human anatomy and physiology
3. How the kinship among all species is relevant to the choice of model animals for biomedical research
4. Ecological conditions thought to have selected for such key characteristics of *Homo sapiens* as opposable thumbs, shoulder mobility, prehensile hands, stereoscopic vision, color vision, and bipedal locomotion
5. A description of evolutionary medicine

### 1.5 Human Structure (p. 11)

1. Levels of human structural complexity from organism to atom
2. Reductionism and holism; how they differ and why both ideas are relevant to the study of human anatomy and physiology and to the clinical care of patients
3. Examples of why the anatomy presented in textbooks is not necessarily true of every individual

### 1.6 Human Function (p. 13)

1. Ten essential qualities that distinguish living organisms from nonliving things
2. The meaning of *metabolism* and of its two subdivisions
3. Clinical criteria for life and death, and why clinical and biological death are not exactly equivalent
4. The clinical importance of physiological variation between people, and the assumptions that underlie typical values given in textbooks
5. The meaning of *homeostasis;* its importance for survival; and the historical origin of this concept
6. How negative feedback contributes to homeostasis; the meaning of *negative feedback loop;* how a receptor, integrating center, and effector are involved in many negative feedback loops; and at least one example of such a loop
7. How positive feedback differs from negative feedback; examples of beneficial and harmful cases of positive feedback
8. The concept of matter and energy flowing down gradients and how this applies to various areas of human physiology

### 1.7 The Language of Medicine (p. 19)

1. The origin and purpose of the *Terminologia Anatomica (TA)* and its relevance for anatomy textbooks and students
2. How to break biomedical terms into familiar roots, prefixes, and suffixes, and why the habit of doing so aids in learning
3. Acronyms and eponyms, and why they cannot be understood by trying to analyze their roots
4. How to recognize when two or more words are singular and plural versions of one another; when one word is the possessive form of another; and when medical terms built on the same root represent different degrees of comparison (such as terms denoting *large, larger,* and *largest*)
5. Why precision in spelling and usage of medical terms can be a matter of life or death in a hospital or clinic, and how seemingly trivial spelling errors can radically alter meaning

### 1.8 Review of Major Themes (p. 21)

1. A description of five core themes of this book: cell theory, homeostasis, evolution, hierarchy of structure, and complementarity of form and function

# STUDY GUIDE

## ▶ Testing Your Recall

*Answers in Appendix B*

1. Structure that can be observed with the naked eye is called
   a. gross anatomy.
   b. ultrastructure.
   c. microscopic anatomy.
   d. histology.
   e. cytology.

2. The word root *homeo-* means
   a. tissue.
   b. metabolism.
   c. change.
   d. human.
   e. same.

3. The simplest structures considered to be alive are
   a. organisms.
   b. organs.
   c. tissues.
   d. cells.
   e. organelles.

4. Which of the following people revolutionized the teaching of gross anatomy?
   a. Vesalius
   b. Aristotle
   c. Hippocrates
   d. Leeuwenhoek
   e. Cannon

5. Which of the following embodies the greatest amount of scientific information?
   a. a fact
   b. a law of nature
   c. a theory
   d. a deduction
   e. a hypothesis

6. An informed, uncertain, but testable conjecture is
   a. a natural law.
   b. a scientific theory.
   c. a hypothesis.
   d. a deduction.
   e. a scientific fact.

7. A self-amplifying chain of physiological events is called
   a. positive feedback.
   b. negative feedback.
   c. dynamic constancy.
   d. homeostasis.
   e. metabolism.

8. Which of the following is *not* a human organ system?
   a. integumentary
   b. muscular
   c. epithelial
   d. nervous
   e. endocrine

9. _____ means studying anatomy by touch.
   a. Gross anatomy
   b. Auscultation
   c. Osculation
   d. Palpation
   e. Percussion

10. The prefix *hetero-* means
    a. same.
    b. different.
    c. both.
    d. solid.
    e. below.

11. Cutting and separating tissues to reveal structural relationships is called _____.

12. A difference in chemical concentration between one point and another is called a concentration _____.

13. By the process of _____, a scientist predicts what the result of a certain experiment will be if his or her hypothesis is correct.

14. Physiological effects of a person's mental state are called _____ effects.

15. The tendency of the body to maintain stable internal conditions is called _____.

16. Blood pH averages 7.4 but fluctuates from 7.35 to 7.45. A pH of 7.4 can therefore be considered the _____ for this variable.

17. Self-corrective mechanisms in physiology are called _____ loops.

18. A/an _____ is the simplest body structure to be composed of two or more types of tissue.

19. Depth perception, or the ability to form three-dimensional images, is also called _____ vision.

20. Our hands are said to be _____ because they can encircle an object such as a branch or tool. The presence of an _____ thumb is important to this ability.

# STUDY GUIDE

## ▶ Building Your Medical Vocabulary

*Answers in Appendix B*

*State a medical meaning of each word element below, and give a term in which it or a slight variation of it is used.*

1. auscult-
2. dis-
3. homeo-
4. metabolo-
5. palp-
6. physio-
7. -sect
8. -stasis
9. stereo-
10. tomo-

## ▶ True or False

*Answers in Appendix B*

*Determine which five of the following statements are false, and briefly explain why.*

1. The technique for listening to the sounds of the heart valves is auscultation.

2. Techniques of radiology have made exploratory surgery much rarer than it once was.

3. Abnormal skin color or dryness could be one piece of diagnostic information gained by auscultation.

4. There are more organelles than cells in the body.

5. Matter does not generally move up a gradient in the body unless the body expends metabolic energy to move it.

6. Leeuwenhoek was a biologist who invented the simple microscope in order to examine organisms in lake water.

7. A scientific theory is just a speculation until someone finds the evidence to prove it.

8. In a typical clinical research study, volunteer patients are in the treatment group and the physicians and scientists who run the study constitute the control group.

9. The great mobility of the primate shoulder joint is an adaptation to the arboreal habitat.

10. Negative feedback usually has a negative (harmful) effect on the body.

## ▶ Testing Your Comprehension

*Answers at www.mhhe.com/saladin7*

1. Ellen is pregnant and tells Janet, one of her coworkers, that she is scheduled to get a fetal sonogram. Janet expresses alarm and warns Ellen about the danger of exposing a fetus to X-rays. Discuss why you think Janet's concern is warranted or unwarranted.

2. Which of the characteristics of living things are possessed by an automobile? What bearing does this have on our definition of life?

3. About 1 out of every 120 live-born infants has a structural defect in the heart such as a hole between two heart chambers. Such infants often suffer pulmonary congestion and heart failure, and about one-third of them die as a result. Which of the major themes in this chapter does this illustrate? Explain your answer.

4. How might human anatomy be different today if the forerunners of humans had never inhabited the forest canopy?

5. Suppose you have been doing heavy yard work on a hot day and sweating profusely. You become very thirsty, so you drink a tall glass of lemonade. Explain how your thirst relates to the concept of homeostasis. Which type of feedback—positive or negative—does this illustrate?

## ▶ Improve Your Grade at www.mhhe.com/saladin7

*Download animations to study when it fits your schedule. Practice quizzes, labeling activities, games, and flashcards offer fun ways to master the chapter concepts. Or, download image PowerPoint files for each chapter to create a study guide or for taking notes during lecture.*

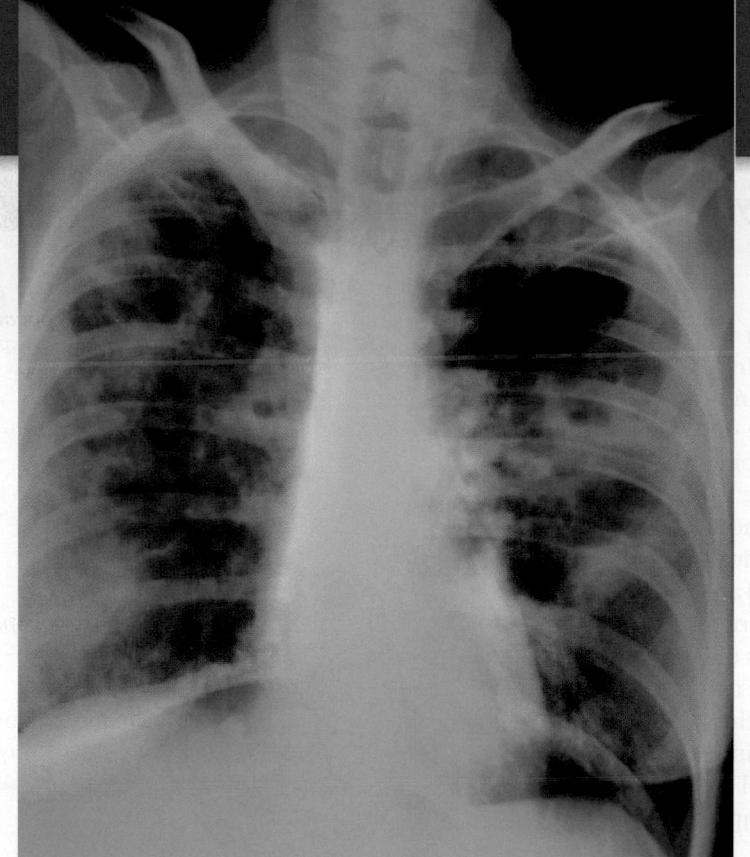

Colorized chest X-ray showing lung damage from tuberculosis.

ATLAS A

# GENERAL ORIENTATION TO HUMAN ANATOMY

Anatomy & Physiology | **REVEALED®**
aprevealed.com

**Module 1: Body Orientation**

## A.1   General Anatomical Terminology

### Anatomical Position

In describing the human body, anatomists assume that it is in **anatomical position** (fig. A.1)—that of a person standing upright with the feet flat on the floor, arms at the sides, and the palms and face directed forward. Without such a frame of reference, to say that a structure such as the sternum, thyroid gland, or aorta is "above the heart" would be vague, since it would depend on whether the subject was standing, lying face down *(prone)*, or lying face up *(supine)*. From the perspective of anatomical position, however, we can describe the thyroid as *superior* to the heart, the sternum as *anterior* to it, and the aorta as *posterior* to it. These descriptions remain valid regardless of the subject's position. Even if the body is lying down, such as a cadaver on the medical student's dissection table, to say the sternum is anterior to the heart invites the viewer to imagine the body is standing in anatomical position and not to call it "above the heart" simply because that is the way the body happens to be lying.

Unless stated otherwise, assume that all anatomical descriptions refer to anatomical position. Bear in mind that if a subject is facing you, the subject's left will be on your right

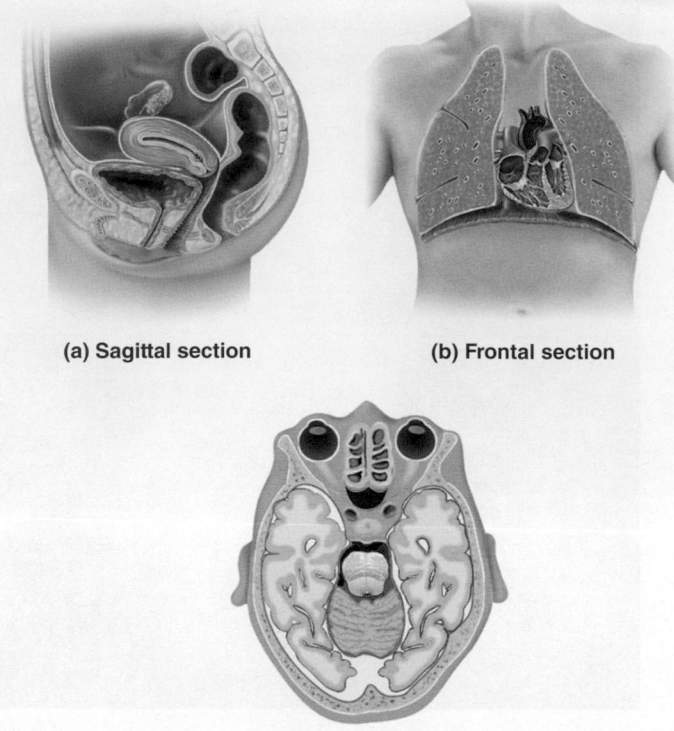

**(a) Sagittal section**          **(b) Frontal section**

**(c) Transverse section**

**FIGURE A.2  Body Sections Cut Along the Three Primary Anatomical Planes.**  (a) Sagittal section of the pelvic region. (b) Frontal section of the thoracic region. (c) Transverse section of the head at the level of the eyes.

and vice versa. In most anatomical illustrations, for example, the left atrium of the heart appears toward the right side of the page, and although the appendix is located in the right lower quadrant of the abdomen, it appears on the left side of most illustrations.

### Anatomical Planes

Many views of the body are based on real or imaginary "slices" called sections or planes. *Section* implies an actual cut or slice to reveal internal anatomy, whereas *plane* implies an imaginary flat surface passing through the body. The three major anatomical planes are *sagittal, frontal,* and *transverse* (fig. A.1).

A **sagittal**[1] (SADJ-ih-tul) **plane** passes vertically through the body or an organ and divides it into right and left portions. The sagittal plane that divides the body or organ into equal halves is also called the **median (midsagittal) plane** (fig. A.2a). The head and pelvic organs are commonly illustrated on the median plane. Other sagittal planes parallel to this (off center) divide the body into unequal right and left portions. Such planes are sometimes called *parasagittal*[2] planes.

A **frontal (coronal) plane** also extends vertically, but it is perpendicular to the sagittal plane and divides the body into anterior (front) and posterior (back) portions. A frontal section

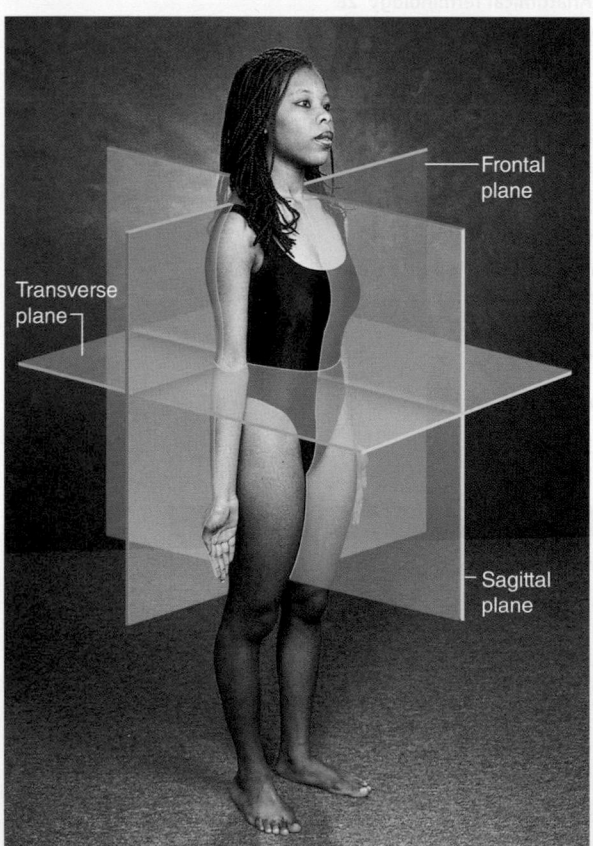

Frontal plane

Transverse plane

Sagittal plane

**FIGURE A.1  Anatomical Position and the Three Primary Anatomical Planes.**

---

[1] *sagitta* = arrow
[2] *para* = next to

| TABLE A.1 | Directional Terms in Human Anatomy  AP|R | |
|---|---|---|
| **Term** | **Meaning** | **Examples of Usage** |
| Ventral | Toward the front* or belly | The aorta is ventral to the vertebral column. |
| Dorsal | Toward the back or spine | The vertebral column is dorsal to the aorta. |
| Anterior | Toward the ventral side* | The sternum is anterior to the heart. |
| Posterior | Toward the dorsal side* | The esophagus is posterior to the trachea. |
| Cephalic | Toward the head or superior end | The cephalic end of the embryonic neural tube develops into the brain. |
| Rostral | Toward the forehead or nose | The forebrain is rostral to the brainstem. |
| Caudal | Toward the tail or inferior end | The spinal cord is caudal to the brain. |
| Superior | Above | The heart is superior to the diaphragm. |
| Inferior | Below | The liver is inferior to the diaphragm. |
| Medial | Toward the median plane | The heart is medial to the lungs. |
| Lateral | Away from the median plane | The eyes are lateral to the nose. |
| Proximal | Closer to the point of attachment or origin | The elbow is proximal to the wrist. |
| Distal | Farther from the point of attachment or origin | The fingernails are at the distal ends of the fingers. |
| Ipsilateral | On the same side of the body | The liver is ipsilateral to the appendix. |
| Contralateral | On opposite sides of the body | The spleen is contralateral to the liver. |
| Superficial | Closer to the body surface | The skin is superficial to the muscles. |
| Deep | Farther from the body surface | The bones are deep to the muscles. |

*In humans only; definition differs for other animals.

of the head, for example, would divide it into one portion bearing the face and another bearing the back of the head. Contents of the thoracic and abdominal cavities are most commonly shown in the frontal section (fig. A.2b).

A **transverse (horizontal) plane** passes across the body or an organ perpendicular to its long axis; it divides the body or organ into superior (upper) and inferior (lower) portions (fig. A.2c). CT scans are typically transverse sections (see fig. 1.10c, p. 22).

## Directional Terms

Words that describe the location of one structure relative to another are generally called the **directional terms** of anatomy. Table A.1 summarizes those most frequently used. Most of these terms exist in pairs with opposite meanings: *anterior* versus *posterior, rostral* versus *caudal, superior* versus *inferior, medial* versus *lateral, proximal* versus *distal, ipsilateral* versus *contralateral,* and *superficial* versus *deep.* Intermediate directions are often indicated by combinations of these terms. For example, one's cheeks may be described as *inferolateral* to the eyes (below and to the side).

The terms *proximal* and *distal* are used especially in the anatomy of the limbs, with *proximal* used to denote something relatively close to the limb's point of attachment (the shoulder or hip) and *distal* to denote something farther away. These terms do have some applications to anatomy of the trunk, however—for example, in referring to certain aspects of the intestines and microscopic anatomy of the kidneys. But when describing the trunk and referring to a structure that lies above or below another, *superior* and *inferior* are the preferred terms. These terms are not usually used for the limbs. Although it may be technically correct, one would not generally say that the elbow is superior to the wrist; rather, one would describe the elbow as proximal to the wrist.

Because of the bipedal, upright stance of humans, some directional terms have different meanings for humans than they do for other animals. *Anterior,* for example, denotes the region of the body that leads the way in normal locomotion. For a four-legged animal such as a cat, this is the head end of the body; for a human, however, it is the front of the chest and abdomen. Thus, *anterior* has the same meaning as *ventral* for a human but not for a cat. *Posterior* denotes the region of the body that comes last in normal locomotion—the tail end of a cat but the dorsal side (back) of a human. In the anatomy of most other animals, *ventral* denotes the surface of the body closest to the ground and *dorsal* denotes the surface farthest away from the ground. These two words are too entrenched in human anatomy to completely ignore them, but we will minimize their use in this book to avoid confusion. You must keep such differences in mind, however, when dissecting other animals for comparison to human anatomy.

One vestige of the term *dorsal* is **dorsum,** used to denote the upper surface of the foot and the back of the hand. If you consider how a cat stands, the corresponding surfaces of its paws are uppermost, facing the same direction as the dorsal side of its trunk. Although these surfaces of the human hand and foot face entirely different directions in anatomical position, the term *dorsum* is still used.

## A.2   Major Body Regions

Knowledge of the external anatomy and landmarks of the body is important in performing a physical examination and many other clinical procedures. For purposes of study, the body is divided into two major regions called the *axial* and *appendicular* regions. Smaller areas within the major regions are described in the following paragraphs and illustrated in figure A.3.

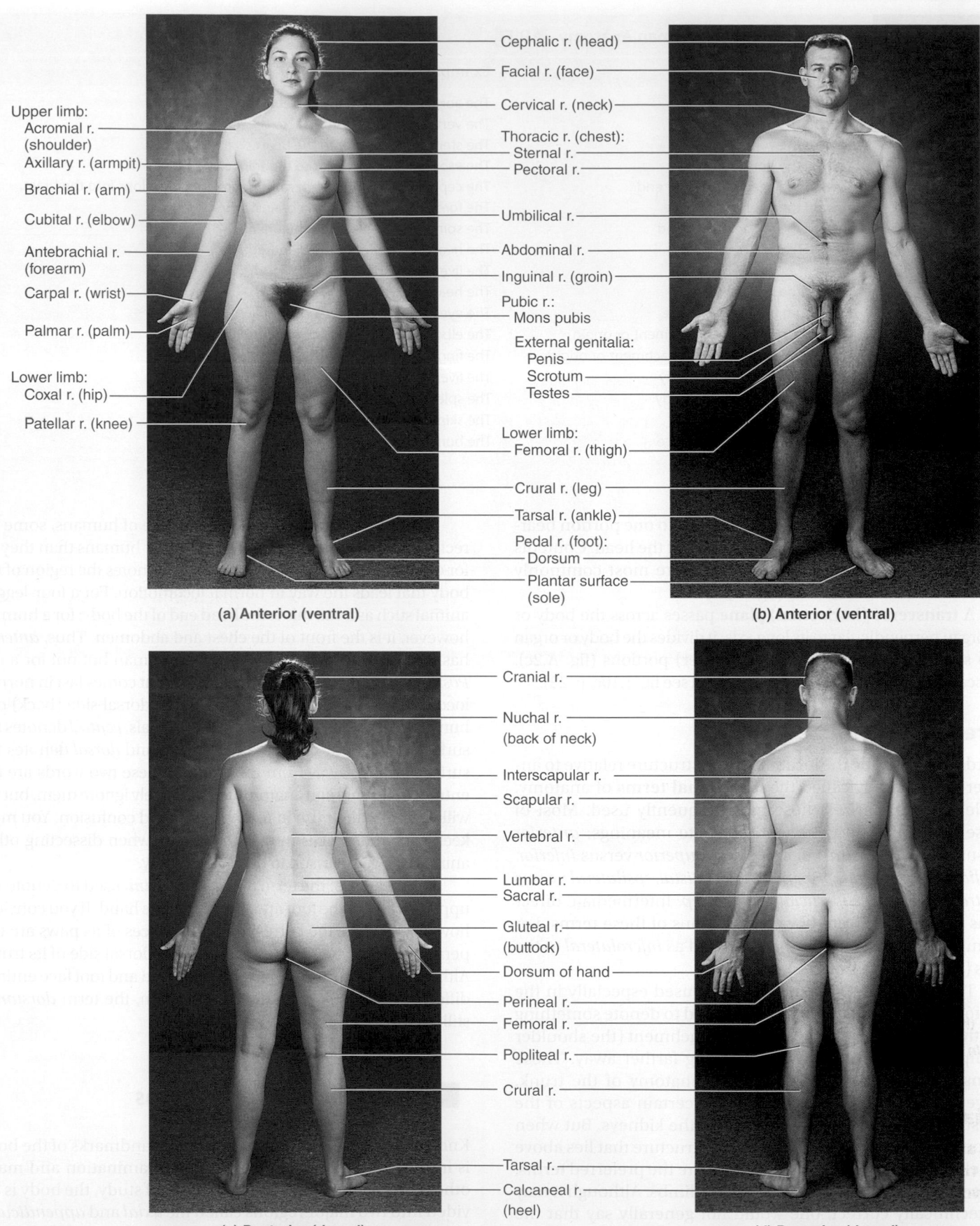

Upper limb:
Acromial r. (shoulder)
Axillary r. (armpit)
Brachial r. (arm)
Cubital r. (elbow)
Antebrachial r. (forearm)
Carpal r. (wrist)
Palmar r. (palm)

Lower limb:
Coxal r. (hip)
Patellar r. (knee)

Cephalic r. (head)
Facial r. (face)
Cervical r. (neck)
Thoracic r. (chest):
Sternal r.
Pectoral r.
Umbilical r.
Abdominal r.
Inguinal r. (groin)
Pubic r.:
Mons pubis
External genitalia:
Penis
Scrotum
Testes
Lower limb:
Femoral r. (thigh)
Crural r. (leg)
Tarsal r. (ankle)
Pedal r. (foot):
Dorsum
Plantar surface (sole)

**(a) Anterior (ventral)**

**(b) Anterior (ventral)**

Cranial r.
Nuchal r. (back of neck)
Interscapular r.
Scapular r.
Vertebral r.
Lumbar r.
Sacral r.
Gluteal r. (buttock)
Dorsum of hand
Perineal r.
Femoral r.
Popliteal r.
Crural r.
Tarsal r.
Calcaneal r. (heel)

**(c) Posterior (dorsal)**

**(d) Posterior (dorsal)**

**FIGURE A.3 The Adult Female and Male Bodies (r. = region).**

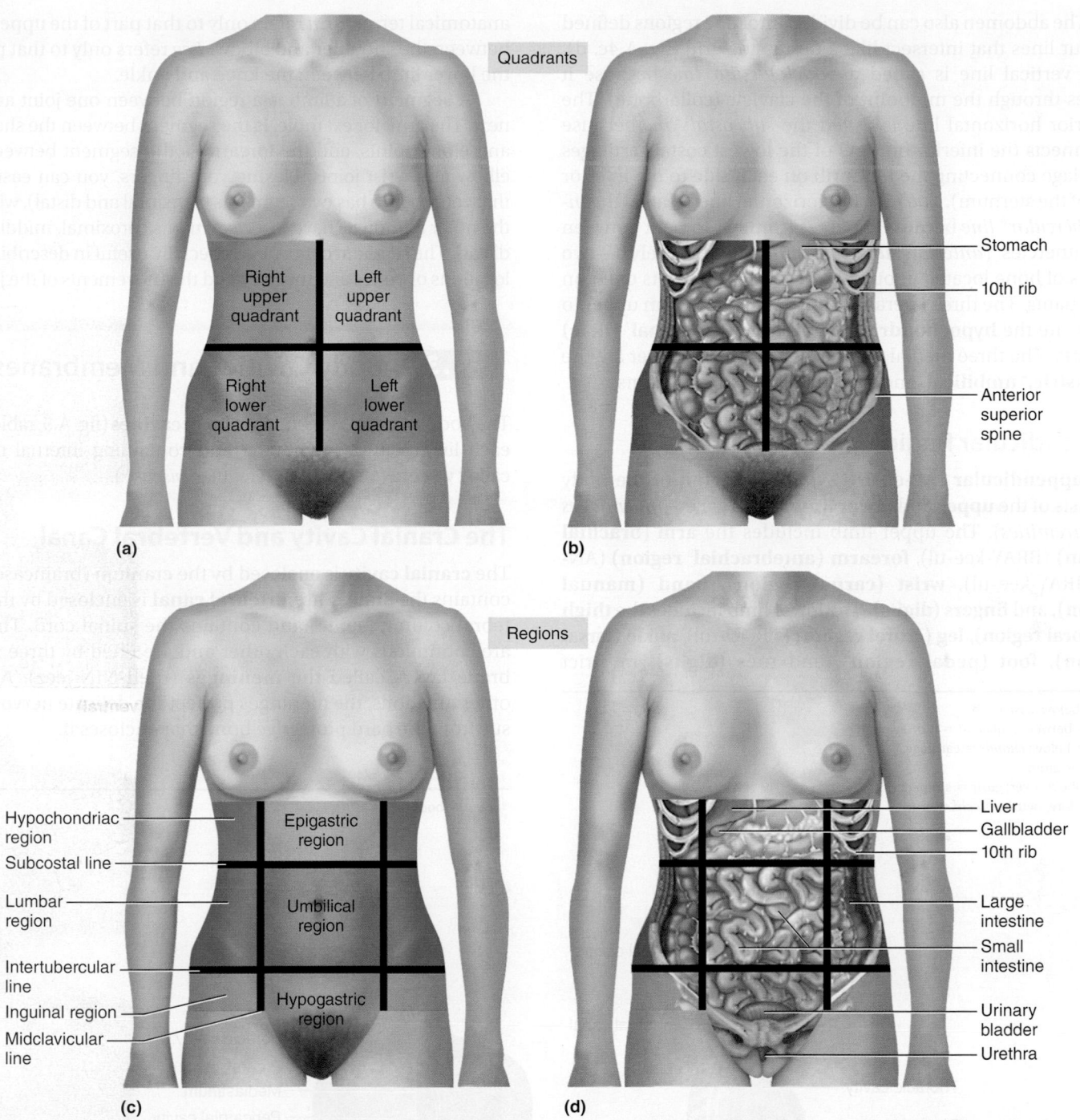

**Quadrants**

Right upper quadrant

Left upper quadrant

Right lower quadrant

Left lower quadrant

(a)

Stomach

10th rib

Anterior superior spine

(b)

**Regions**

Hypochondriac region

Subcostal line

Lumbar region

Intertubercular line

Inguinal region

Midclavicular line

Epigastric region

Umbilical region

Hypogastric region

(c)

Liver

Gallbladder

10th rib

Large intestine

Small intestine

Urinary bladder

Urethra

(d)

**FIGURE A.4  The Four Quadrants and Nine Regions of the Abdomen.**  (a) External division into four quadrants. (b) Internal anatomy correlated with the four quadrants. (c) External division into nine regions. (d) Internal anatomy correlated with the nine regions.  AP|R

❓ *In what quadrant would the pain of appendicitis usually be felt?*

## Axial Region

The **axial region** consists of the **head, neck (cervical[3] region),** and **trunk.** The trunk is further divided into the **thoracic region** above the diaphragm and the **abdominal region** below it.

---

[3]*cervic* = neck

One way of referring to the locations of abdominal structures is to divide the region into quadrants. Two perpendicular lines intersecting at the umbilicus (navel) divide the abdomen into a **right upper quadrant (RUQ), right lower quadrant (RLQ), left upper quadrant (LUQ),** and **left lower quadrant (LLQ)** (fig. A.4a, b). The quadrant scheme is often used to describe the site of an abdominal pain or abnormality.

The abdomen also can be divided into nine regions defined by four lines that intersect like a tic-tac-toe grid (fig. A.4c, d). Each vertical line is called a *midclavicular line* because it passes through the midpoint of the clavicle (collarbone). The superior horizontal line is called the *subcostal*[4] *line* because it connects the inferior borders of the lowest costal cartilages (cartilage connecting the tenth rib on each side to the inferior end of the sternum). The inferior horizontal line is called the *intertubercular*[5] *line* because it passes from left to right between the tubercles *(anterior superior spines)* of the pelvis—two points of bone located about where the front pockets open on most pants. The three lateral regions of this grid, from upper to lower, are the **hypochondriac,**[6] **lumbar,** and **inguinal**[7] **(iliac) regions.** The three medial regions from upper to lower are the **epigastric,**[8] **umbilical,** and **hypogastric (pubic) regions.**

## Appendicular Region

The **appendicular** (AP-en-DIC-you-lur) **region** of the body consists of the **upper** and **lower limbs** (also called *appendages* or *extremities*). The upper limb includes the **arm (brachial region)** (BRAY-kee-ul), **forearm (antebrachial**[9] **region)** (AN-teh-BRAY-kee-ul), **wrist (carpal region), hand (manual region),** and **fingers (digits).** The lower limb includes the **thigh (femoral region), leg (crural region)** (CROO-rul), **ankle (tarsal region), foot (pedal region),** and **toes (digits).** In strict

---

[4]*sub* = below; *cost* = rib
[5]*inter* = between; *tubercul* = little swelling
[6]*hypo* = below; *chondr* = cartilage
[7]*inguin* = groin
[8]*epi* = above, over; *gastr* = stomach
[9]*ante* = fore, before; *brachi* = arm

anatomical terms, *arm* refers only to that part of the upper limb between the shoulder and elbow. *Leg* refers only to that part of the lower limb between the knee and ankle.

A **segment** of a limb is a region between one joint and the next. The arm, for example, is the segment between the shoulder and elbow joints, and the forearm is the segment between the elbow and wrist joints. Flexing your fingers, you can easily see that your thumb has two segments (proximal and distal), whereas the other four digits have three segments (proximal, middle, and distal). The segment concept is especially useful in describing the locations of bones and muscles and the movements of the joints.

### A.3  Body Cavities and Membranes

The body wall encloses multiple **body cavities** (fig. A.5, table A.2), each lined with a membrane and containing internal organs called **viscera** (VISS-er-uh) (singular, *viscus*[10]).

## The Cranial Cavity and Vertebral Canal

The **cranial cavity** is enclosed by the cranium (braincase) and contains the brain. The **vertebral canal** is enclosed by the vertebral column (spine) and contains the spinal cord. The two are continuous with each other and are lined by three membrane layers called the **meninges** (meh-NIN-jeez). Among other functions, the meninges protect the delicate nervous tissue from the hard protective bone that encloses it.

---

[10]*viscus* = body organ

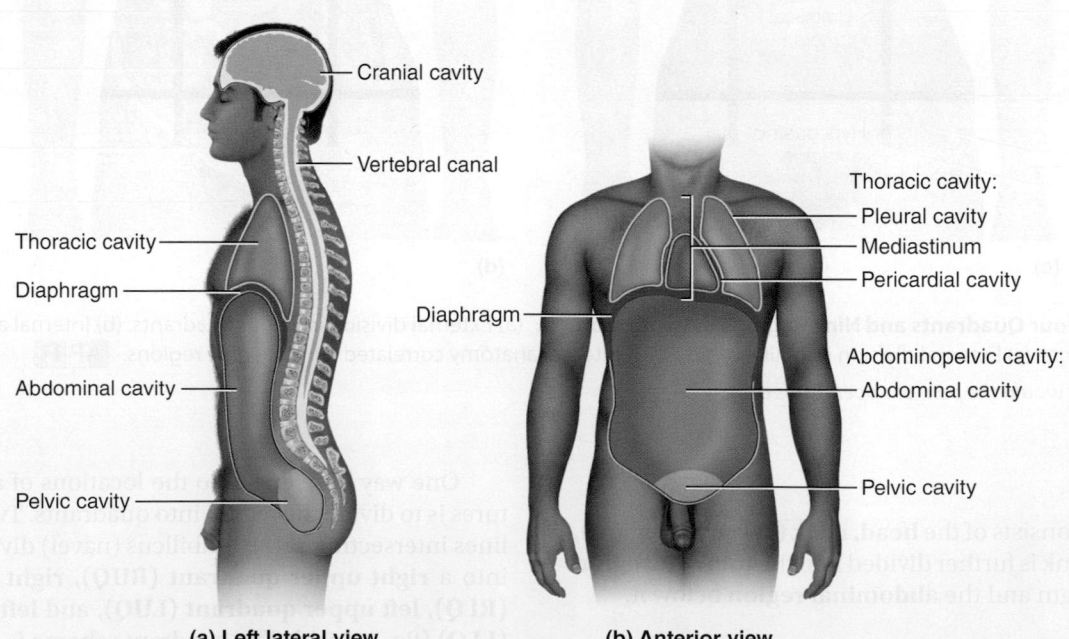

**(a) Left lateral view**                    **(b) Anterior view**

**FIGURE A.5  The Major Body Cavities.**

| TABLE A.2 | Directional Terms in Human Anatomy AP|R | |
|---|---|---|
| **Name of Cavity** | **Associated Viscera** | **Membranous Lining** |
| Cranial cavity | Brain | Meninges |
| Vertebral canal | Spinal cord | Meninges |
| Thoracic cavity | | |
|    Pleural cavities (2) | Lungs | Pleurae |
|    Pericardial cavity | Heart | Pericardium |
| Abdominopelvic cavity | | |
|    Abdominal cavity | Digestive organs, spleen, kidneys | Peritoneum |
|    Pelvic cavity | Bladder, rectum, reproductive organs | Peritoneum |

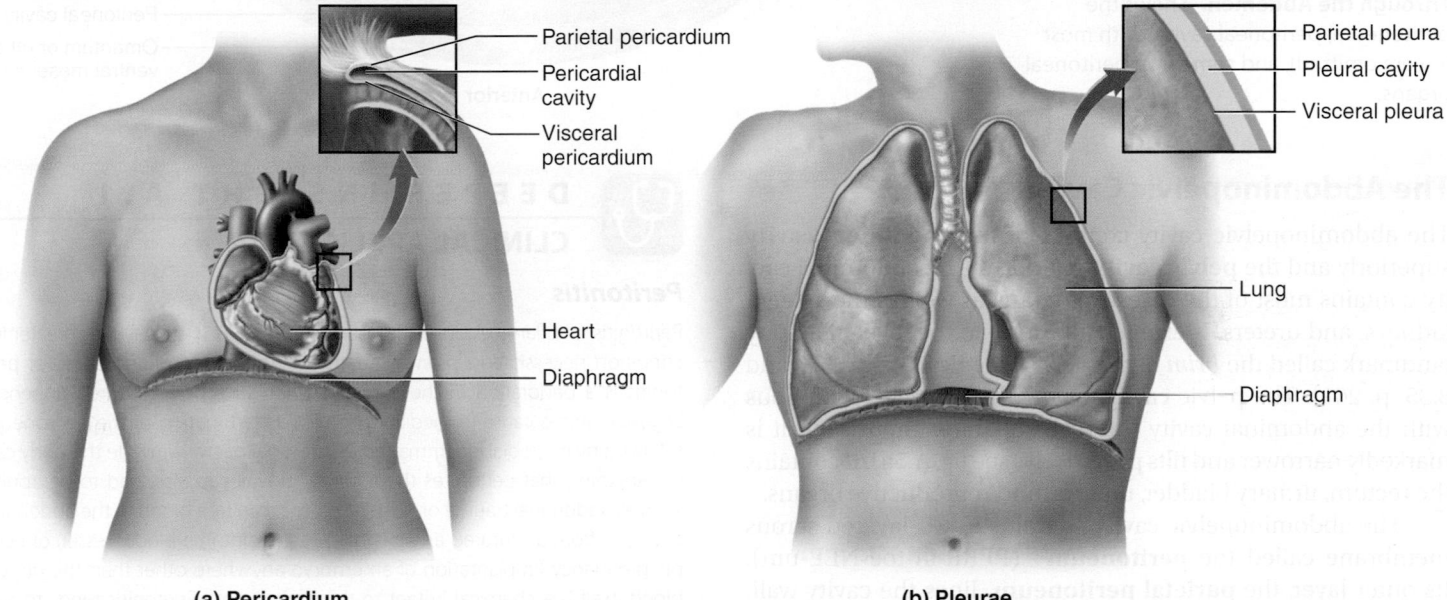

(a) Pericardium
(b) Pleurae

**FIGURE A.6** Parietal and Visceral Layers of Double-Walled Membranes.

## The Thoracic Cavity

During embryonic development, a space called the **coelom** (SEE-loam) forms within the trunk. It subsequently becomes partitioned by a muscular sheet, the **diaphragm**, into a superior **thoracic cavity** and an inferior **abdominopelvic cavity.** Both cavities are lined with thin **serous membranes**, which secrete a lubricating film of moisture similar to blood serum (hence their name).

The thoracic cavity is divided by a thick wall called the **mediastinum**[11] (ME-dee-ah-STY-num) (fig. A.5). This is the region between the lungs, extending from the base of the neck to the diaphragm. It is occupied by the heart, the major blood vessels connected to it, the esophagus, the trachea and bronchi, and a gland called the thymus.

The heart is enfolded in a two-layered membrane called the **pericardium.**[12] The inner layer of the pericardium forms the surface of the heart itself and is called the **visceral** (VISS-er-ul) **pericardium.** The outer layer is called the **parietal** (pa-RYE-eh-tul) **pericardium (pericardial sac).** It is separated from the visceral pericardium by a space called the **pericardial cavity** (fig. A.6a), which is lubricated by **pericardial fluid.**

The right and left sides of the thoracic cavity contain the lungs. Each lung is enfolded by a serous membrane called the **pleura**[13] (PLOOR-uh) (fig. A.6b). Like the pericardium, the pleura has visceral (inner) and parietal (outer) layers. The **visceral pleura** forms the external surface of the lung, and the **parietal**[14] **pleura** lines the inside of the rib cage. The narrow space between them is called the **pleural cavity** (see fig. B.11, page 386). It is lubricated by slippery **pleural fluid.**

Note that in both the pericardium and the pleura, the visceral layer of the membrane *covers* an organ surface and the parietal layer *lines* the inside of a body cavity. We will see this pattern elsewhere, including the abdominopelvic cavity.

---

[11]*mediastinum* = in the middle
[12]*peri* = around; *cardi* = heart

[13]*pleur* = rib, side
[14]*pariet* = wall

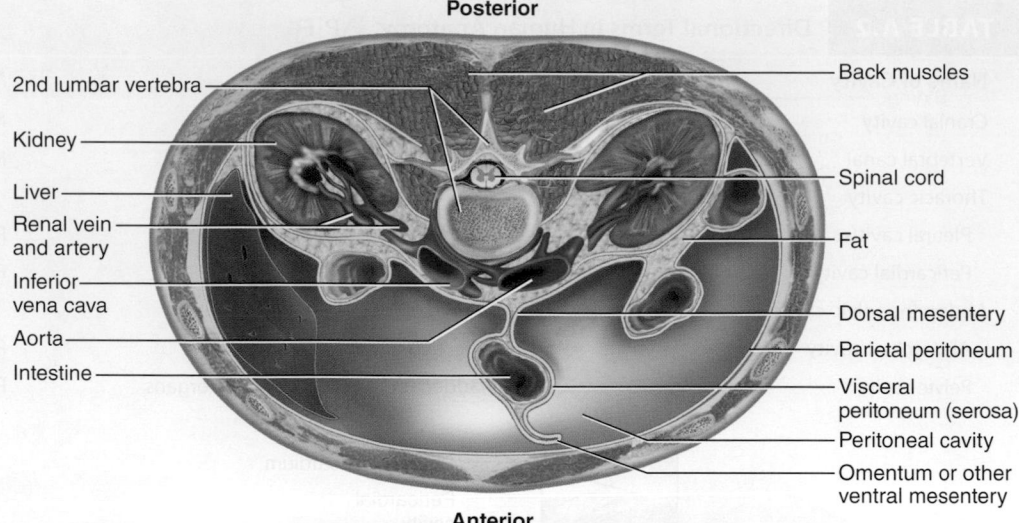

**Posterior**

2nd lumbar vertebra
Kidney
Liver
Renal vein and artery
Inferior vena cava
Aorta
Intestine

Back muscles
Spinal cord
Fat
Dorsal mesentery
Parietal peritoneum
Visceral peritoneum (serosa)
Peritoneal cavity
Omentum or other ventral mesentery

**Anterior**

**FIGURE A.7 Transverse Section Through the Abdomen.** Shows the peritoneum, peritoneal cavity (with most viscera omitted), and some retroperitoneal organs.

## The Abdominopelvic Cavity

The abdominopelvic cavity consists of the **abdominal cavity** superiorly and the **pelvic cavity** inferiorly. The abdominal cavity contains most of the digestive organs as well as the spleen, kidneys, and ureters. It extends inferiorly to the level of a bony landmark called the *brim* of the pelvis (see figs. B.7, p. 383, and 8.35, p. 261). The pelvic cavity, below the brim, is continuous with the abdominal cavity (no wall separates them), but it is markedly narrower and tilts posteriorly (see fig. A.5a). It contains the rectum, urinary bladder, urethra, and reproductive organs.

The abdominopelvic cavity contains a two-layered serous membrane called the **peritoneum**[15] (PERR-ih-toe-NEE-um). Its outer layer, the **parietal peritoneum,** lines the cavity wall. Along the posterior midline, it turns inward and becomes another layer, the **visceral peritoneum,** suspending certain abdominal viscera from the body wall, covering their outer surfaces, and holding them in place. The **peritoneal cavity** is the space between the parietal and visceral layers. It is lubricated by **peritoneal fluid.**

Some organs of the abdominal cavity lie against the posterior body wall and are covered by peritoneum only on the side facing the peritoneal cavity. They are said to have a **retroperitoneal**[16] position (fig. A.7). These include the kidneys, ureters, adrenal glands, most of the pancreas, and abdominal portions of two major blood vessels—the aorta and inferior vena cava (see fig. B.6, p. 382). Organs that are encircled by peritoneum and connected to the posterior body wall by peritoneal sheets are described as **intraperitoneal.**[17]

The visceral peritoneum is also called a **mesentery**[18] (MEZ-en-tare-ee) at points where it forms a translucent, membranous curtain suspending and anchoring the viscera (fig. A.8), and a **serosa** (seer-OH-sa) at points where it enfolds and covers the outer surfaces of organs such as the stomach and small intestine. The intestines are suspended

## DEEPER INSIGHT A.1

### CLINICAL APPLICATION

#### Peritonitis

Peritonitis is inflammation of the peritoneum. It is a critical, life-threatening condition necessitating prompt treatment. The most serious cause of peritonitis is a perforation in the digestive tract, such as a ruptured appendix. Digestive juices cause immediate chemical inflammation of the peritoneum, followed by microbial inflammation as intestinal bacteria invade the body cavity. Anything that perforates the abdominal wall can also lead to peritonitis, such as abdominal trauma or surgery. So, too, can free blood in the abdominal cavity, as from a ruptured aneurysm (a weak point in a blood vessel) or ectopic pregnancy (implantation of an embryo anywhere other than the uterus); blood itself is a chemical irritant to the peritoneum. Peritonitis tends to shift fluid from the circulation into the abdominal cavity. Death can follow within a few days from severe electrolyte imbalance, respiratory distress, kidney failure, and widespread blood clotting called *disseminated intravascular coagulation*.

from the posterior (dorsal) abdominal wall by the **posterior mesentery**. The posterior mesentery of the large intestine is called the **mesocolon.** In some places, after wrapping around the intestines or other viscera, the mesentery continues toward the anterior body wall as the **anterior mesentery.** The most significant example of this is a fatty membrane called the **greater omentum,**[19] which hangs like an apron from the inferolateral margin of the stomach and overlies the intestines (figs. A.8a and B.4, p. 380). The greater omentum is unattached at its inferior border and can be lifted to reveal the intestines. A smaller **lesser omentum** extends from the superomedial margin of the stomach to the liver.

## Potential Spaces

Some of the spaces between body membranes are considered to be **potential spaces,** so named because under normal conditions, the membranes are pressed firmly together and there

---

[15]*peri* = around; *tone* = stretched

[16]*retro* = behind

[17]*intra* = within

[18]*mes* = in the middle; *enter* = intestine

[19]*omentum* = covering

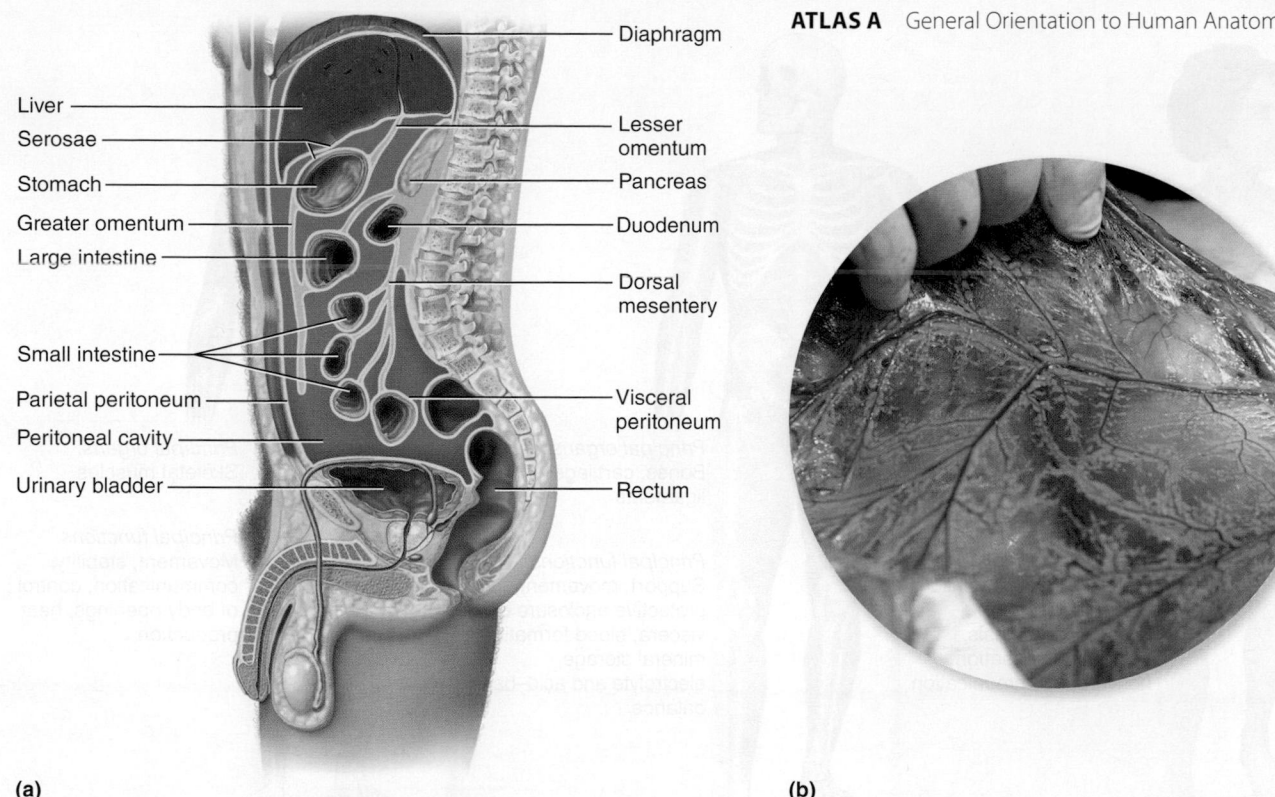

Liver
Serosae
Stomach
Greater omentum
Large intestine
Small intestine
Parietal peritoneum
Peritoneal cavity
Urinary bladder

Diaphragm
Lesser omentum
Pancreas
Duodenum
Dorsal mesentery
Visceral peritoneum
Rectum

(a)

(b)

**FIGURE A.8  Serous Membranes of the Abdominal Cavity.**  (a) Sagittal section, left lateral view. (b) Photo of the mesentery of the small intestine. Mesenteries contain blood vessels, lymphatic vessels, and nerves supplying the viscera.  AP|R

❓ *Is the urinary bladder in the peritoneal cavity?*

is no actual space between them. The membranes are not physically attached, however, and under unusual conditions, they may separate and create a space filled with fluid or other matter. Thus there is normally no actual space, but only a potential for membranes to separate and create one.

The pleural cavity is one example. Normally the parietal and visceral pleurae are pressed together without a gap between them, but under pathological conditions, air or serous fluid can accumulate between the membranes and open up a space. The internal cavity *(lumen)* of the uterus is another. In a nonpregnant uterus, the mucous membranes of opposite walls are pressed together so that there is no open space in the organ. In pregnancy, of course, a growing fetus occupies this space and pushes the mucous membranes apart.

### A.4   Organ Systems

The human body has 11 **organ systems** (fig. A.9) and an immune system, which is better described as a population of cells that inhabit multiple organs rather than as an organ system. These systems are classified in the following list by their principal functions, but this is an unavoidably flawed classification. Some organs belong to two or more systems—for example, the male urethra is part of both the urinary and reproductive systems; the pharynx is part of the respiratory and digestive systems; and the mammary glands can be considered part of the integumentary and female reproductive systems. The organ systems are as follows:

**Systems of protection, support, and movement**

Integumentary system

Skeletal system

Muscular system

**Systems of internal communication and integration**

Nervous system

Endocrine system

**Systems of fluid transport**

Circulatory system

Lymphatic system

**Systems of intake and output**

Respiratory system

Urinary system

Digestive system

**Systems of reproduction**

Male reproductive system

Female reproductive system

Some medical terms combine the names of two systems—for example, the *musculoskeletal system, cardiopulmonary system,* and *urogenital (genitourinary) system.* These terms serve to call attention to the close anatomical or physiological relationships between two systems, but these are not literally individual organ systems.

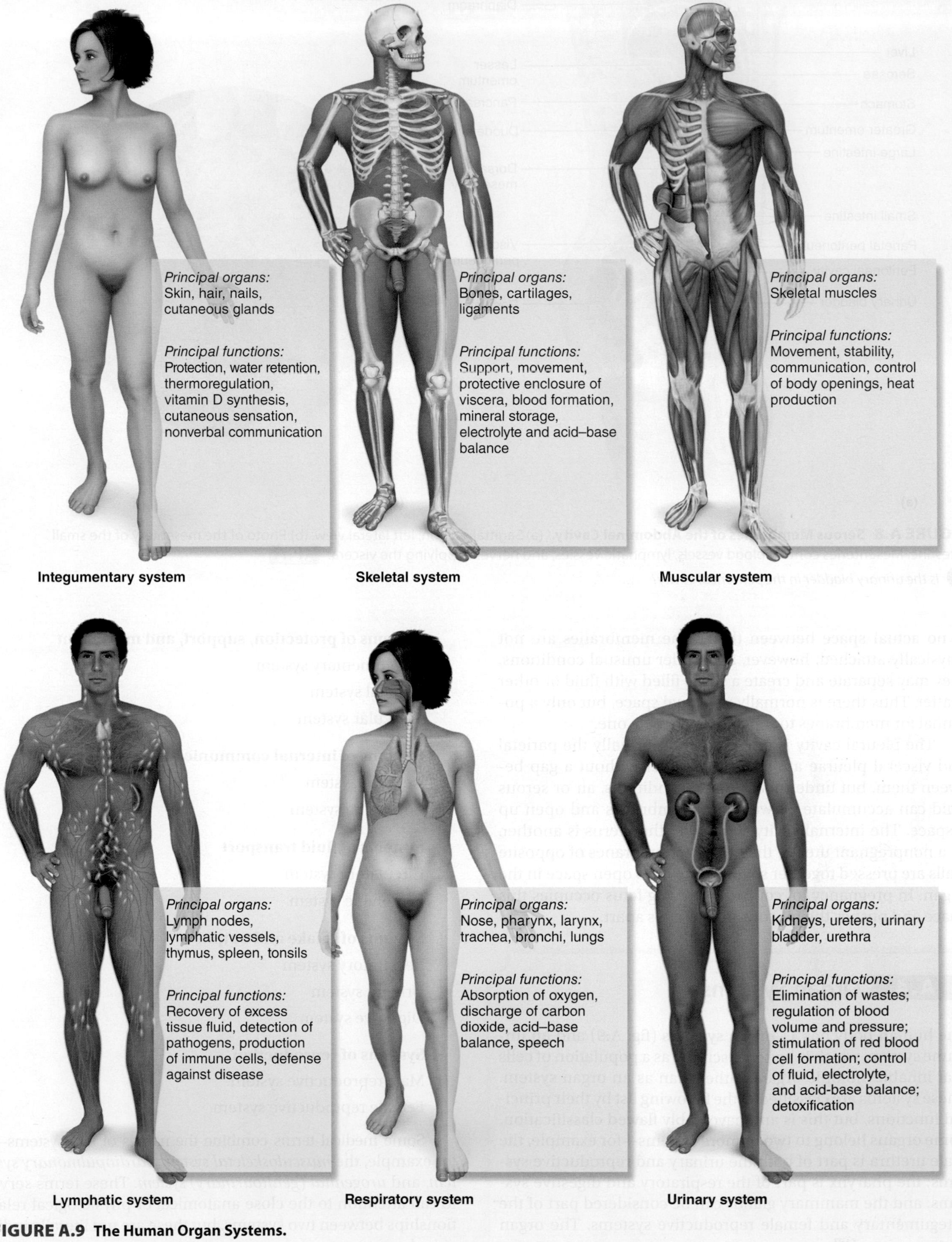

Principal organs:
Skin, hair, nails,
cutaneous glands

Principal functions:
Protection, water retention,
thermoregulation,
vitamin D synthesis,
cutaneous sensation,
nonverbal communication

**Integumentary system**

Principal organs:
Bones, cartilages,
ligaments

Principal functions:
Support, movement,
protective enclosure of
viscera, blood formation,
mineral storage,
electrolyte and acid–base
balance

**Skeletal system**

Principal organs:
Skeletal muscles

Principal functions:
Movement, stability,
communication, control
of body openings, heat
production

**Muscular system**

Principal organs:
Lymph nodes,
lymphatic vessels,
thymus, spleen, tonsils

Principal functions:
Recovery of excess
tissue fluid, detection of
pathogens, production
of immune cells, defense
against disease

**Lymphatic system**

Principal organs:
Nose, pharynx, larynx,
trachea, bronchi, lungs

Principal functions:
Absorption of oxygen,
discharge of carbon
dioxide, acid–base
balance, speech

**Respiratory system**

Principal organs:
Kidneys, ureters, urinary
bladder, urethra

Principal functions:
Elimination of wastes;
regulation of blood
volume and pressure;
stimulation of red blood
cell formation; control
of fluid, electrolyte,
and acid-base balance;
detoxification

**Urinary system**

**FIGURE A.9 The Human Organ Systems.**

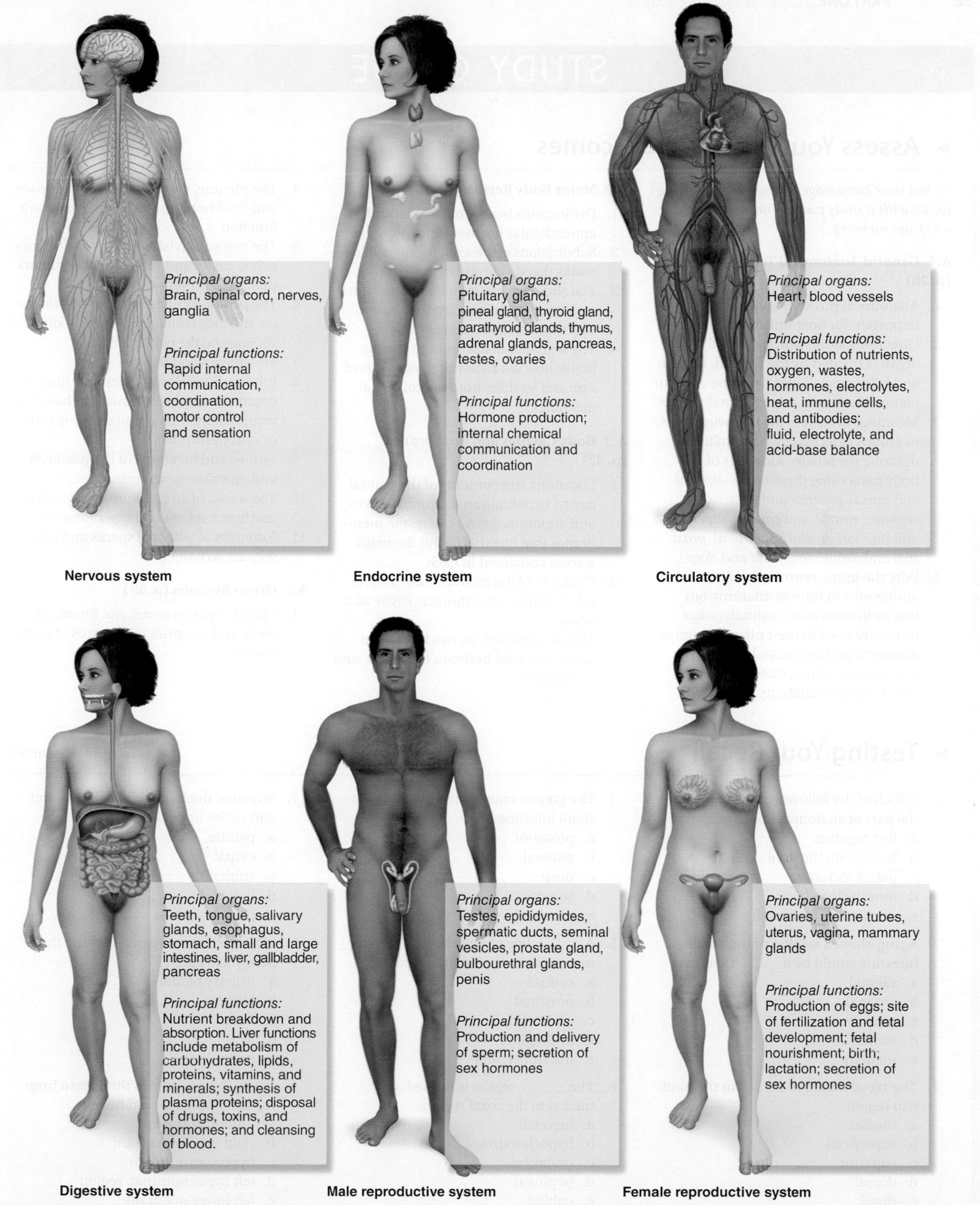

*Principal organs:*
Brain, spinal cord, nerves, ganglia

*Principal functions:*
Rapid internal communication, coordination, motor control and sensation

**Nervous system**

*Principal organs:*
Pituitary gland, pineal gland, thyroid gland, parathyroid glands, thymus, adrenal glands, pancreas, testes, ovaries

*Principal functions:*
Hormone production; internal chemical communication and coordination

**Endocrine system**

*Principal organs:*
Heart, blood vessels

*Principal functions:*
Distribution of nutrients, oxygen, wastes, hormones, electrolytes, heat, immune cells, and antibodies; fluid, electrolyte, and acid-base balance

**Circulatory system**

*Principal organs:*
Teeth, tongue, salivary glands, esophagus, stomach, small and large intestines, liver, gallbladder, pancreas

*Principal functions:*
Nutrient breakdown and absorption. Liver functions include metabolism of carbohydrates, lipids, proteins, vitamins, and minerals; synthesis of plasma proteins; disposal of drugs, toxins, and hormones; and cleansing of blood.

**Digestive system**

*Principal organs:*
Testes, epididymides, spermatic ducts, seminal vesicles, prostate gland, bulbourethral glands, penis

*Principal functions:*
Production and delivery of sperm; secretion of sex hormones

**Male reproductive system**

*Principal organs:*
Ovaries, uterine tubes, uterus, vagina, mammary glands

*Principal functions:*
Production of eggs; site of fertilization and fetal development; fetal nourishment; birth; lactation; secretion of sex hormones

**Female reproductive system**

**FIGURE A.9  The Human Organ Systems** *(continued).*

# STUDY GUIDE

## ▶ Assess Your Learning Outcomes

*To test your knowledge, discuss the following topics with a study partner or in writing, ideally from memory.*

### A.1 General Anatomical Terminology (p. 28)

1. Anatomical position and why it is important for anatomical description
2. Directions along which the body or an organ is divided by the sagittal, frontal, and transverse planes; how the median plane differs from other sagittal planes
3. Meanings of each of the following pairs or groups of terms, and the ability to describe the relative locations of two body parts using these terms: *ventral* and *dorsal; anterior* and *posterior; cephalic, rostral,* and *caudal; superior* and *inferior; medial* and *lateral; proximal* and *distal; superficial* and *deep*
4. Why the terms *ventral* and *dorsal* are ambiguous in human anatomy but less so in most other animals; what terms are used in their place in human anatomy; and reasons why they are occasionally appropriate or unavoidable in human anatomy

### A.2 Major Body Regions (p. 29)

1. Distinctions between the axial and appendicular regions of the body
2. Subdivisions of the axial region and landmarks that divide and define them
3. The abdomen's four quadrants and nine regions; their defining landmarks; and why this scheme is clinically useful
4. The segments of the upper and lower limbs; how the anatomical meanings of *arm* and *leg* differ from the colloquial meanings

### A.3 Body Cavities and Membranes (p. 32)

1. Locations and contents of the cranial cavity, vertebral canal, thoracic cavity, and abdominopelvic cavity; the membranes that line them; and the main viscera contained in each
2. Contents of the mediastinum and its relationship to the thoracic cavity as a whole
3. The pericardium, its two layers, the space and fluid between the layers, and its function

4. The pleurae, their two layers, the space and fluid between the layers, and their function
5. The two subdivisions of the abdominopelvic cavity and the skeletal landmark that divides them
6. The peritoneum; its functions; its two layers and their relationship to the abdominal viscera; and the peritoneal fluid
7. Mesenteries and serosae
8. Intraperitoneal versus retroperitoneal organs, examples of both, and how one would identify an organ as being intra- or retroperitoneal
9. Names and locations of the posterior and anterior mesenteries
10. The serosa of an abdominopelvic organ and how it relates to the peritoneum
11. Examples of potential spaces and why they are so named

### A.4 Organ Systems (p. 35)

1. The 11 organ systems, the functions of each, and the principal organs of each system

## ▶ Testing Your Recall

*Answers in Appendix B*

1. Which of the following is *not* an essential part of anatomical position?
   a. feet together
   b. feet flat on the floor
   c. palms forward
   d. mouth closed
   e. arms down to the sides

2. A ring-shaped section of the small intestine would be a _____ section.
   a. sagittal
   b. coronal
   c. transverse
   d. frontal
   e. median

3. The tarsal region is _____ to the popliteal region.
   a. medial
   b. superficial
   c. superior
   d. dorsal
   e. distal

4. The greater omentum is _____ to the small intestine.
   a. posterior
   b. parietal
   c. deep
   d. superficial
   e. proximal

5. A _____ line passes through the sternum, umbilicus, and mons pubis.
   a. central
   b. proximal
   c. midclavicular
   d. midsagittal
   e. intertubercular

6. The _____ region is immediately medial to the coxal region.
   a. inguinal
   b. hypochondriac
   c. umbilical
   d. popliteal
   e. cubital

7. Which of the following regions is not part of the upper limb?
   a. plantar
   b. carpal
   c. cubital
   d. brachial
   e. palmar

8. Which of these organs is within the peritoneal cavity?
   a. urinary bladder
   b. kidneys
   c. heart
   d. liver
   e. brain

9. In which area do you think pain from the gallbladder would be felt?
   a. umbilical region
   b. right upper quadrant
   c. hypogastric region
   d. left hypochondriac region
   e. left lower quadrant

# STUDY GUIDE

10. Which organ system regulates blood volume, controls acid–base balance, and stimulates red blood cell production?
    a. digestive system
    b. lymphatic system
    c. nervous system
    d. urinary system
    e. circulatory system

11. The translucent membranes that suspend the intestines and hold them in place are called _____.

12. The superficial layer of the pleura is called the _____ pleura.

13. The right and left pleural cavities are separated by a thick wall called the _____.

14. The back of the neck is the _____ region.

15. The manus is more commonly known as the _____ and the pes is more commonly known as the _____.

16. The cranial cavity is lined by membranes called the _____.

17. Organs that lie within the abdominal cavity but not within the peritoneal cavity are said to have a _____ position.

18. The sternal region is _____ to the pectoral region.

19. The pelvic cavity can be described as _____ to the abdominal cavity in position.

20. The anterior pit of the elbow is the _____ region, and the corresponding (but posterior) pit of the knee is the _____ region.

## ▶ Building Your Medical Vocabulary

*Answers in Appendix B*

*State a medical meaning of each word element below, and give a term in which it or a slight variation of it is used.*

1. ante-
2. cervico-
3. epi-
4. hypo-
5. inguino-
6. intra-
7. parieto-
8. peri-
9. retro-
10. sagitto-

## ▶ True or False

*Answers in Appendix B*

*Determine which five of the following statements are false, and briefly explain why.*

1. A single sagittal section of the body can pass through one lung but not through both.

2. It would be possible to see both eyes in one frontal section of the head.

3. The knee is both superior and proximal to the tarsal region.

4. The diaphragm is posterior to the lungs.

5. The esophagus is inferior to the stomach.

6. The liver is in the lumbar region.

7. The heart is in the mediastinum.

8. Both kidneys could be shown in a single coronal section of the body.

9. The peritoneum lines the inside of the stomach and intestines.

10. The sigmoid colon is in the lower right quadrant of the abdomen.

## ▶ Testing Your Comprehension

*Answers at www.mhhe.com/saladin7*

1. Identify which anatomical plane—sagittal, frontal, or transverse—is the only one that could *not* show (a) both the brain and tongue, (b) both eyes, (c) both the hypogastric and gluteal regions, (d) both kidneys, (e) both the sternum and vertebral column, and (f) both the heart and uterus.

2. Laypeople often misunderstand anatomical terminology. What do you think people really mean when they say they have "planter's warts"?

3. Name one structure or anatomical feature that could be found in each of the following locations relative to the ribs: medial, lateral, superior, inferior, deep, superficial, posterior, and anterior. Try not to use the same example twice.

4. Based on the illustrations in this atlas, identify an internal organ that is (a) in the upper left quadrant and retroperitoneal, (b) in the lower right quadrant of the peritoneal cavity, (c) in the hypogastric region, (d) in the right hypochondriac region, and (e) in the pectoral region.

5. Why do you think people with imaginary illnesses came to be called hypochondriacs?

# STUDY GUIDE

## ▶ Improve Your Grade at www.mhhe.com/saladin7

*Download animations to study when it fits your schedule. Practice quizzes, labeling activities, games, and flashcards offer fun ways to master the chapter concepts. Or, download image PowerPoint files for each chapter to create a study guide or for taking notes during lecture.*

Cholesterol crystals seen through a polarizing microscope

# THE CHEMISTRY OF LIFE

Anatomy & Physiology | REVEALED®
aprevealed.com

**Module 2: Cells and Chemistry**

## BRUSHING UP...

Beginning here, each chapter builds on information presented in earlier ones. If you do not clearly remember these concepts, you may find that brushing up on them before you proceed will enable you to get more out of the new chapter.

- Chapter 1 discusses metabolism as one of the fundamental characteristics of life, and mentions that it is divided into catabolism and anabolism (p. 14). Here we will delve more deeply into the chemical meaning of these two divisions. Metabolism, in turn, provides the foundation for perhaps the most important of all concepts in chapter 1, homeostasis (p. 15).

**W**hy is too much sodium or cholesterol harmful? Why does an iron deficiency cause anemia and an iodine deficiency cause a goiter? Why does a pH imbalance make some drugs less effective? Why do some pregnant women suffer convulsions after several days of vomiting? How can radiation cause cancer as well as cure it?

None of these questions can be answered, nor would the rest of this book be intelligible, without understanding the chemistry of life. A little knowledge of chemistry can help you choose a healthy diet, use medications more wisely, avoid worthless health fads and frauds, and explain treatments and procedures to your patients or clients. Thus, we begin our study of the human body with basic chemistry, the simplest level of the body's structural organization.

We will progress from general chemistry to **biochemistry,** study of the molecules that compose living organisms—especially molecules unique to living things, such as carbohydrates, fats, proteins, and nucleic acids. Most people have at least heard of these and it is common knowledge that we need proteins, fats, carbohydrates, vitamins, and minerals in our diet, and that we should avoid consuming too much saturated fat and cholesterol. But most people have only a vague concept of what these molecules are, much less how they function in the body. Such knowledge is very helpful in matters of personal fitness and patient education and is essential to the comprehension of the rest of this book.

## 2.1  Atoms, Ions, and Molecules

### Expected Learning Outcomes

When you have completed this section, you should be able to

a. name the chemical elements of the body from their chemical symbols;

b. distinguish between chemical elements and compounds;

c. state the functions of minerals in the body;

d. explain the basis for radioactivity and the types and hazards of ionizing radiation;

e. distinguish between ions, electrolytes, and free radicals; and

f. define the types of chemical bonds.

### The Chemical Elements

A chemical **element** is the simplest form of matter to have unique chemical properties. Water, for example, has unique properties, but it can be broken down into two elements, hydrogen and oxygen, that have unique chemical properties of their own. If we carry this process any further, however, we find that hydrogen and oxygen are made of protons, neutrons, and electrons—and none of these are unique. A proton of gold is identical to a proton of oxygen. Hydrogen and oxygen are the simplest chemically unique components of water and are thus elements.

Each element is identified by an *atomic number,* the number of protons in its nucleus. The atomic number of carbon is 6 and that of oxygen is 8, for example. The periodic table of the elements (see appendix A) arranges the elements in order by their atomic numbers. The elements are represented by one- or two-letter symbols, usually based on their English names: C for carbon, Mg for magnesium, Cl for chlorine, and so forth. A few symbols are based on Latin names, such as K for potassium *(kalium),* Na for sodium *(natrium),* and Fe for iron *(ferrum).*

There are 91 naturally occurring elements on earth, 24 of which play normal physiological roles in humans. Table 2.1 groups these 24 according to their abundance in the body. Six of them account for 98.5% of the body's weight: oxygen, carbon, hydrogen, nitrogen, calcium, and phosphorus. The next 0.8% consists of another 6 elements: sulfur, potassium, sodium, chlorine, magnesium, and iron. The remaining 12 elements account for 0.7% of body weight, and no one of them accounts for more than 0.02%; thus they are known as **trace elements.** Despite their minute quantities, trace elements play vital roles in physiology. Other elements without

| **TABLE 2.1** | **Elements of the Human Body** | |
|---|---|---|
| **Name** | **Symbol** | **Percentage of Body Weight** |
| **Major Elements (Total 98.5%)** | | |
| Oxygen | O | 65.0 |
| Carbon | C | 18.0 |
| Hydrogen | H | 10.0 |
| Nitrogen | N | 3.0 |
| Calcium | Ca | 1.5 |
| Phosphorus | P | 1.0 |
| **Lesser Elements (Total 0.8%)** | | |
| Sulfur | S | 0.25 |
| Potassium | K | 0.20 |
| Sodium | Na | 0.15 |
| Chlorine | Cl | 0.15 |
| Magnesium | Mg | 0.05 |
| Iron | Fe | 0.006 |
| **Trace Elements (Total 0.7%)** | | | |
| Chromium | Cr | Molybdenum | Mo |
| Cobalt | Co | Selenium | Se |
| Copper | Cu | Silicon | Si |
| Fluorine | F | Tin | Sn |
| Iodine | I | Vanadium | V |
| Manganese | Mn | Zinc | Zn |

natural physiological roles can contaminate the body and severely disrupt its functions, as in heavy-metal poisoning with lead or mercury.

Several of these elements are classified as **minerals**—inorganic elements that are extracted from the soil by plants and passed up the food chain to humans and other organisms. Minerals constitute about 4% of the human body by weight. Nearly three-quarters of this is Ca and P; the rest is mainly Cl, Mg, K, Na, and S. Minerals contribute significantly to body structure. The bones and teeth consist partly of crystals of calcium, phosphate, magnesium, fluoride, and sulfate ions. Many proteins include sulfur, and phosphorus is a major component of nucleic acids, ATP, and cell membranes. Minerals also enable enzymes and other organic molecules to function. Iodine is a component of thyroid hormone; iron is a component of hemoglobin; and some enzymes function only when manganese, zinc, copper, or other minerals are bound to them. The electrolytes needed for nerve and muscle function are mineral salts. The biological roles of minerals are discussed in more detail in chapters 24 and 26.

## Atomic Structure

In the fifth century BCE, the Greek philosopher Democritus reasoned that we can cut matter such as a gold nugget into smaller and smaller pieces, but there must ultimately be particles so small that nothing could cut them. He called these imaginary particles atoms[1] ("indivisible"). Atoms were only a philosophical concept until 1803, when English chemist John Dalton began to develop an atomic theory based on experimental evidence. In 1913, Danish physicist Niels Bohr proposed a model of atomic structure similar to planets orbiting the sun (figs. 2.1 and 2.2). Although this *planetary model* is too simple to account for many of the properties of atoms, it remains useful for elementary purposes.

---

[1] *a* = not; *tom* = cut

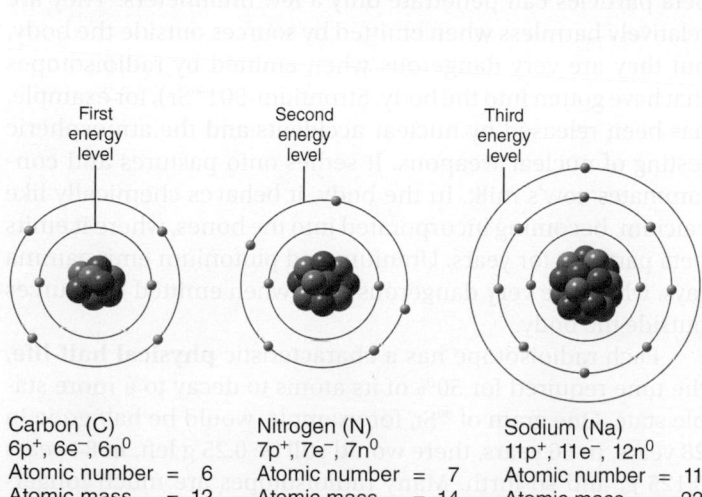

First energy level | Second energy level | Third energy level

Carbon (C)
$6p^+$, $6e^-$, $6n^0$
Atomic number = 6
Atomic mass = 12

Nitrogen (N)
$7p^+$, $7e^-$, $7n^0$
Atomic number = 7
Atomic mass = 14

Sodium (Na)
$11p^+$, $11e^-$, $12n^0$
Atomic number = 11
Atomic mass = 23

**FIGURE 2.1 Bohr Planetary Models of Three Representative Elements.** Note the filling of electron shells as atomic number increases ($p^+$ = protons; $e^-$ = electrons; $n^0$ = neutrons).

❓ *Will sodium have a greater tendency to give up an electron or to take one away from another atom? (See hint later in this chapter.)*

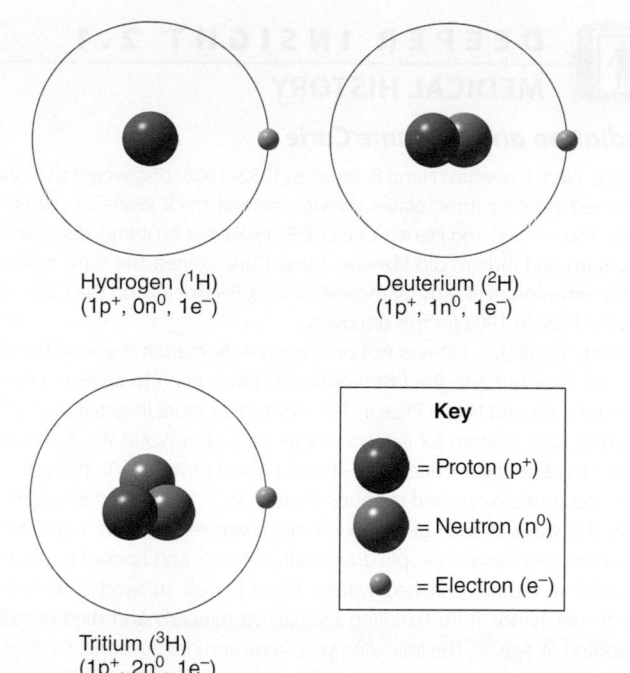

Hydrogen ($^1$H)
($1p^+$, $0n^0$, $1e^-$)

Deuterium ($^2$H)
($1p^+$, $1n^0$, $1e^-$)

Tritium ($^3$H)
($1p^+$, $2n^0$, $1e^-$)

**Key**

● = Proton ($p^+$)

● = Neutron ($n^0$)

• = Electron ($e^-$)

**FIGURE 2.2 Isotopes of Hydrogen.** The three isotopes differ only in the number of neutrons present.

At the center of an atom is the *nucleus,* composed of protons and neutrons. **Protons ($p^+$)** have a single positive charge and **neutrons ($n^0$)** have no charge. Each proton or neutron weighs approximately 1 *atomic mass unit (amu)*. The *atomic mass* of an element is approximately equal to its total number of protons and neutrons.

Around the nucleus are one or more concentric clouds of **electrons ($e^-$)**, tiny particles with a single negative charge and very low mass. It takes 1,836 electrons to equal the mass of one proton, so for most purposes we can disregard their mass. A person who weighs 64 kg (140 lb) contains less than 24 g (1 oz) of electrons. This hardly means that we can ignore electrons, however. They determine the chemical properties of an atom, thereby governing what molecules can exist and what chemical reactions can occur. The number of electrons equals the number of protons, so their charges cancel each other and an atom is electrically neutral.

Electrons swarm about the nucleus in concentric regions called *electron shells (energy levels)*. The more energy an electron has, the farther away from the nucleus its orbit lies. Each shell holds a limited number of electrons (fig. 2.1). The elements known to date have up to seven electron shells, but those ordinarily involved in human physiology do not exceed four. Electrons of the outermost shell, called **valence electrons,** determine the chemical bonding properties of an atom.

## Isotopes and Radioactivity

Dalton believed that every atom of an element was identical. We now know, however, that all elements have varieties called **isotopes,**[2] which differ from one another only in number of

---

[2] *iso* = same; *top* = place (same position in the periodic table)

# DEEPER INSIGHT 2.1

## MEDICAL HISTORY

### Radiation and Madame Curie

In 1896, French scientist Henri Becquerel (1852–1908) discovered that uranium darkened photographic plates through several thick layers of paper. Marie Curie (1867–1934) and Pierre Curie (1859–1906), her husband, discovered that polonium and radium did likewise. Marie Curie coined the term *radioactivity* for the emission of energy by these elements. Becquerel and the Curies shared a Nobel Prize in 1903 for this discovery.

Marie Curie (fig. 2.3) was not only the first woman in the world to receive a Nobel Prize but also the first woman in France even to receive a Ph.D. She received a second Nobel Prize in 1911 for further work in radiation. Curie crusaded to train women for careers in science, and in World War I, she and her daughter, Irène Joliot-Curie (1897–1956), trained physicians in the use of X-ray machines. Curie pioneered radiation therapy for breast and uterine cancer.

In the wake of such discoveries, radium was regarded as a wonder drug. Unaware of its danger, people drank radium tonics and flocked to health spas to bathe in radium-enriched waters. Marie herself suffered extensive damage to her hands from handling radioactive minerals and died of radiation poisoning at age 67. The following year, Irène and her husband, Frédéric Joliot (1900–1958), were awarded a Nobel Prize for work in artificial radioactivity and synthetic radioisotopes. Apparently also a martyr to her science, Irène died of leukemia, possibly induced by radiation exposure.

**FIGURE 2.3  Marie Curie (1867–1934).** This portrait was made in 1911, when Curie received her second Nobel Prize.

neutrons and therefore in atomic mass. Hydrogen atoms, for example, have only one proton. In the most common isotope, symbolized $^1H$, that is all there is to the nucleus. Hydrogen has two other isotopes, however: *deuterium* ($^2H$) with one proton and one neutron, and *tritium* ($^3H$) with one proton and two neutrons (fig. 2.2). Over 99% of carbon atoms have an atomic mass of 12 ($6p^+$, $6n^0$) and are called carbon-12 ($^{12}C$), but a small percentage of carbon atoms are $^{13}C$, with seven neutrons, and $^{14}C$, with eight. All isotopes of a given element behave the same chemically. Deuterium ($^2H$), for example, reacts with oxygen the same way $^1H$ does to produce water.

The *atomic weight (relative atomic mass)* of an element accounts for the fact that an element is a mixture of isotopes. If all carbon were $^{12}C$, the atomic weight of carbon would be the same as its atomic mass, 12.000. But since a sample of carbon also contains small amounts of the heavier isotopes $^{13}C$ and $^{14}C$, the atomic weight is slightly higher, 12.011.

Although different isotopes of an element exhibit identical chemical behavior, they differ in physical behavior. Many of them are unstable and *decay* (break down) to more stable isotopes by giving off radiation. Unstable isotopes are therefore called **radioisotopes,** and the process of decay is called **radioactivity** (see Deeper Insight 2.1). Every element has at least one radioisotope. Oxygen, for example, has three stable isotopes and five radioisotopes. All of us contain radioisotopes such as $^{14}C$ and $^{40}K$—that is, we are all mildly radioactive!

Many forms of radiation, such as light and radio waves, have low energy and are harmless. High-energy radiation, however, ejects electrons from atoms, converting atoms to ions; thus, it is called **ionizing radiation.** It destroys molecules

and produces dangerous free radicals and ions in human tissues. In high doses, ionizing radiation is quickly fatal. In lower doses, it can be *mutagenic* (causing mutations in DNA) and *carcinogenic* (triggering cancer as a result of mutation).

Examples of ionizing radiation include ultraviolet rays, X-rays, and three kinds of radiation produced by nuclear decay: *alpha (α) particles, beta (β) particles,* and *gamma (γ) rays.* An alpha particle is composed of two protons and two neutrons (equivalent to a helium nucleus), and a beta particle is a free electron. Alpha particles are too large to penetrate the skin, and beta particles can penetrate only a few millimeters. They are relatively harmless when emitted by sources outside the body, but they are very dangerous when emitted by radioisotopes that have gotten into the body. Strontium-90 ($^{90}Sr$), for example, has been released by nuclear accidents and the atmospheric testing of nuclear weapons. It settles onto pastures and contaminates cow's milk. In the body, it behaves chemically like calcium, becoming incorporated into the bones, where it emits beta particles for years. Uranium and plutonium emit gamma rays, which are very dangerous even when emitted by sources outside the body.

Each radioisotope has a characteristic **physical half-life,** the time required for 50% of its atoms to decay to a more stable state. One gram of $^{90}Sr$, for example, would be half gone in 28 years. In 56 years, there would still be 0.25 g left, in 84 years 0.125 g, and so forth. Many radioisotopes are much longer-lived. The half-life of $^{40}K$, for example, is 1.3 billion years. Nuclear power plants produce hundreds of radioisotopes that will be intensely radioactive for at least 10,000 years—longer than the life of any disposal container yet conceived.

The **biological half-life** of a radioisotope is the time required for half of it to disappear from the body. Some of it is lost by radioactive decay and even more of it by excretion from the body. Cesium-137, for example, has a physical half-life of 30 years but a biological half-life of only 17 days. Chemically, it behaves like potassium; it is quite mobile and rapidly excreted by the kidneys.

There are several ways to measure the intensity of ionizing radiation, the amount absorbed by the body, and its biological effects. To understand the units of measurement requires a grounding in physics beyond the scope of this book, but the standard international (SI) unit of radiation dosage is the *sievert*[3] *(Sv),* which takes into account the type and intensity of radiation and its biological effect. Doses of 5 Sv or more are usually fatal. The average person worldwide receives about 2.4 millisieverts (mSv) per year in *background radiation* from natural sources and another 0.6 mSv from artificial sources. The most significant natural source is *radon,* a gas produced by the decay of uranium in the earth; it can accumulate in buildings to unhealthy levels. Artificial sources of radiation exposure include medical X-rays, radiation therapy, and consumer products such as color televisions, smoke detectors, and luminous watch dials. Such voluntary exposure must be considered from the standpoint of its risk-to-benefit ratio. The benefits of a smoke detector or mammogram far outweigh the risk from the low levels of radiation involved. Radiation therapists and radiologists face a greater risk than their patients, however, and astronauts and airline flight crews receive more than average exposure. U.S. federal standards set a limit of 50 mSv/year as acceptable occupational exposure to ionizing radiation. A typical mammogram exposes one to 0.4 to 0.6 mSv and a full-body CT scan, 10 to 30 mSv.

## Ions, Electrolytes, and Free Radicals

**Ions** are charged particles with unequal numbers of protons and electrons. An ion can consist of a single atom with a positive or negative charge, or it can be as large as a protein with many charges on it.

Ions form because elements with one to three valence electrons tend to give them up, and those with four to seven electrons tend to gain more. If an atom of the first kind is exposed to an atom of the second, electrons may transfer from one to the other and turn both of them into ions. This process is called *ionization.* The particle that gains electrons acquires a negative charge and is called an **anion** (AN-eye-on). The one that loses electrons acquires a positive charge (because it then has a surplus of protons) and is called a **cation** (CAT-eye-on).

Consider, for example, what happens when sodium and chlorine meet (fig. 2.4). Sodium has three electron shells with a total of 11 electrons: 2 in the first shell, 8 in the second, and 1 in the third. If it gives up the electron in the third shell, its second shell becomes the valence shell and has a stable configuration of 8 electrons. Chlorine has 17 electrons: 2 in the first shell, 8 in

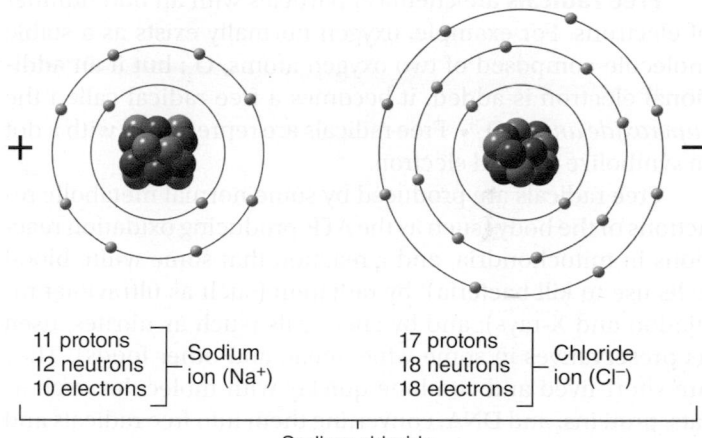

① Transfer of an electron from a sodium atom to a chlorine atom

| 11 protons | | 17 protons | |
| 12 neutrons | Sodium | 18 neutrons | Chlorine |
| 11 electrons | atom (Na) | 17 electrons | atom (Cl) |

| 11 protons | | 17 protons | |
| 12 neutrons | Sodium | 18 neutrons | Chloride |
| 10 electrons | ion (Na⁺) | 18 electrons | ion (Cl⁻) |

Sodium chloride

② The charged sodium ion (Na⁺) and chloride ion (Cl⁻) that result

**FIGURE 2.4  Ionization.**

the second, and 7 in the third. If it can gain one more electron, it can fill the third shell with 8 electrons and become stable. Sodium and chlorine seem "made for each other"—one needs to lose an electron and the other needs to gain one. This is just what they do. When they interact, an electron transfers from sodium to chlorine. Now, sodium has 11 protons in its nucleus but only 10 electrons. This imbalance gives it a positive charge, so we symbolize the sodium ion Na⁺. Chlorine has been changed to the chloride ion with a surplus negative charge, symbolized Cl⁻.

Some elements exist in two or more ionized forms. Iron, for example, has ferrous ($Fe^{2+}$) and ferric ($Fe^{3+}$) ions. Note that some ions have a single positive or negative charge, whereas others have charges of $\pm2$ or $\pm3$ because they gain or lose more than one electron. The charge on an ion is called its *valence.* Ions are not always single atoms that have become charged; some are groups of atoms—phosphate ($PO_4^{3-}$) and bicarbonate ($HCO_3^-$) ions, for example.

Ions with opposite charges are attracted to each other and tend to follow each other through the body. Thus, when Na⁺ is excreted in the urine, Cl⁻ tends to follow it. The attraction of

---

[3]Rolf Maximillian Sievert (1896–1966), Swedish radiologist

cations and anions to each other is important in maintaining the excitability of muscle and nerve cells, as we shall see in chapters 11 and 12.

**Electrolytes** are substances that ionize in water (acids, bases, or salts) and form solutions capable of conducting electricity (table 2.2). We can detect electrical activity of the muscles, heart, and brain with electrodes on the skin because electrolytes in the body fluids conduct electrical currents from these organs to the skin surface. Electrolytes are important for their chemical reactivity (as when calcium phosphate becomes incorporated into bone), osmotic effects (influence on water content and distribution in the body), and electrical effects (which are essential to nerve and muscle function). Electrolyte balance is one of the most important considerations in patient care. Electrolyte imbalances have effects ranging from muscle cramps and brittle bones to coma and cardiac arrest.

**Free radicals** are chemical particles with an odd number of electrons. For example, oxygen normally exists as a stable molecule composed of two oxygen atoms, $O_2$; but if an additional electron is added, it becomes a free radical called the *superoxide anion,* $O_2^-\bullet$. Free radicals are represented with a dot to symbolize the odd electron.

Free radicals are produced by some normal metabolic reactions of the body (such as the ATP-producing oxidation reactions in mitochondria, and a reaction that some white blood cells use to kill bacteria); by radiation (such as ultraviolet radiation and X-rays); and by chemicals (such as nitrites, used as preservatives in some wine, meat, and other foods). They are short-lived and combine quickly with molecules such as fats, proteins, and DNA, converting them into free radicals and triggering chain reactions that destroy still more molecules. Among the damages caused by free radicals are some forms of cancer and myocardial infarction, the death of heart tissue. One theory of aging is that it results in part from lifelong cellular damage by free radicals.

Because free radicals are so common and destructive, we have multiple mechanisms for neutralizing them. An **antioxidant** is a chemical that neutralizes free radicals. The body produces an enzyme called *superoxide dismutase (SOD),* for example, that converts superoxide into oxygen and hydrogen peroxide. Selenium, vitamin E (α-tocopherol), vitamin C (ascorbic acid), and carotenoids (such as β-carotene) are some

antioxidants obtained from the diet. Dietary deficiencies of antioxidants have been associated with increased risk of heart attacks, sterility, muscular dystrophy, and other disorders.

## Molecules and Chemical Bonds

**Molecules** are chemical particles composed of two or more atoms united by a chemical bond. The atoms may be identical, as in nitrogen ($N_2$), or different, as in glucose ($C_6H_{12}O_6$). Molecules composed of two or more elements are called **compounds.** Oxygen ($O_2$) and carbon dioxide ($CO_2$) are both molecules, because they consist of at least two atoms; but only $CO_2$ is a compound, because it has atoms of two different elements.

Molecules can be represented by *molecular formulae* that identify their constituent elements and show how many atoms of each are present. Molecules with identical molecular formulae but different arrangements of their atoms are called **isomers**[4] of each other. For example, both ethanol (grain alcohol) and ethyl ether have the molecular formula $C_2H_6O$, but they are certainly not interchangeable! To show the difference between them, we use *structural formulae* that show the location of each atom (fig. 2.5).

The **molecular weight (MW)** of a compound is the sum of the atomic weights of its atoms. Rounding to whole numbers, we can calculate the approximate MW of glucose ($C_6H_{12}O_6$), for example, as

| 6 | C atoms | × | 12 amu each | = | 72 amu |
|---|---------|---|-------------|---|--------|
| 12 | H atoms | × | 1 amu each | = | 12 amu |
| 6 | O atoms | × | 16 amu each | = | 96 amu |
| | Molecular weight (MW) | | | = | 180 amu |

Molecular weight is needed to compute some measures of concentration discussed later.

---

[4]*iso* = same; *mer* = part

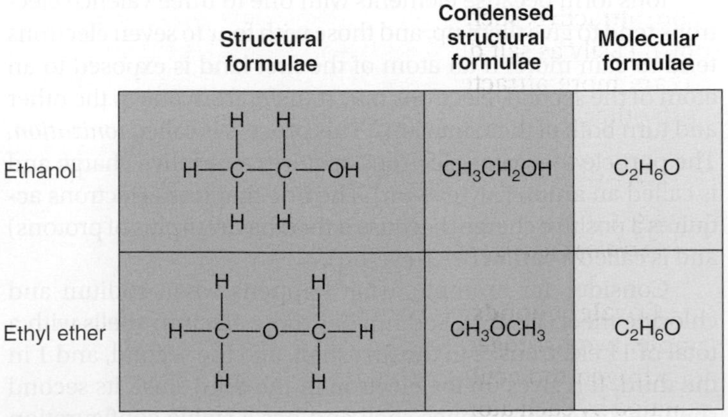

| **TABLE 2.2** | **Major Electrolytes and the Ions Released by their Dissociation** | | |
|---|---|---|---|
| **Electrolyte** | | **Cation** | **Anion** |
| Calcium chloride (CaCl$_2$) | → | Ca$^{2+}$ | 2 Cl$^-$ |
| Disodium phosphate (Na$_2$HPO$_4$) | → | 2 Na$^+$ | HPO$_4^{2-}$ |
| Magnesium chloride (MgCl$_2$) | → | Mg$^{2+}$ | 2 Cl$^-$ |
| Potassium chloride (KCl) | → | K$^+$ | Cl$^-$ |
| Sodium bicarbonate (NaHCO$_3$) | → | Na$^+$ | HCO$_3^-$ |
| Sodium chloride (NaCl) | → | Na$^+$ | Cl$^-$ |

| | Structural formulae | Condensed structural formulae | Molecular formulae |
|---|---|---|---|
| Ethanol | | $CH_3CH_2OH$ | $C_2H_6O$ |
| Ethyl ether | | $CH_3OCH_3$ | $C_2H_6O$ |

**FIGURE 2.5  Structural Isomers, Ethanol and Ethyl Ether.**  The molecular formulae are identical, but the structures and chemical properties are different.

| TABLE 2.3 | Types of Chemical Bonds |
|---|---|
| **Bond Type** | **Definition and Remarks** |
| Ionic bond | Relatively weak attraction between an anion and a cation. Easily disrupted in water, as when salt dissolves. |
| Covalent bond | Sharing of one or more pairs of electrons between nuclei. |
| Single covalent | Sharing of one electron pair. |
| Double covalent | Sharing of two electron pairs. Often occurs between carbon atoms, between carbon and oxygen, and between carbon and nitrogen. |
| Nonpolar covalent | Covalent bond in which electrons are equally attracted to both nuclei. May be single or double. Strongest type of chemical bond. |
| Polar covalent | Covalent bond in which electrons are more attracted to one nucleus than to the other, resulting in slightly positive and negative regions in one molecule. May be single or double. |
| Hydrogen bond | Weak attraction between polarized molecules or between polarized regions of the same molecule. Important in the three-dimensional folding and coiling of large molecules. Easily disrupted by temperature and pH changes. |
| Van der Waals force | Weak, brief attraction due to random disturbances in the electron clouds of adjacent atoms. Weakest of all bonds. |

A molecule is held together, and molecules are attracted to one another, by forces called **chemical bonds.** The bonds of greatest physiological interest are *ionic bonds, covalent bonds, hydrogen bonds,* and *van der Waals forces* (table 2.3).

An **ionic bond** is the attraction of a cation to an anion. Sodium ($Na^+$) and chloride ($Cl^-$) ions, for example, are attracted to each other and form the compound sodium chloride (NaCl), common table salt. Ionic compounds can be composed of more than two ions. Calcium has two valence electrons. It can become stable by donating one electron to one chlorine atom and the other electron to another chlorine, thus producing a calcium ion ($Ca^{2+}$) and two chloride ions. The result is calcium chloride, $CaCl_2$. Ionic bonds are weak and easily dissociate (break up) in the presence of something more attractive, such as water. The ionic bonds of NaCl break down easily as salt dissolves in water, because both $Na^+$ and $Cl^-$ are more attracted to water molecules than they are to each other.

#### ▶▶▶APPLY WHAT YOU KNOW

*Do you think ionic bonds are common in the human body? Explain your answer.*

**Covalent bonds** form by the sharing of electrons. For example, two hydrogen atoms share valence electrons to form a hydrogen molecule, $H_2$ (fig. 2.6a). The two electrons, one donated by each atom, swarm around both nuclei in a dumbbell-shaped cloud. A *single covalent bond* is the sharing of a single pair of electrons. It is symbolized by a single line between atomic symbols, for example H—H. A *double covalent*

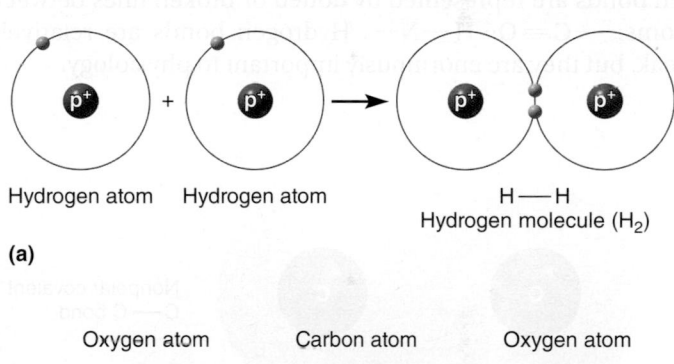

Hydrogen atom    Hydrogen atom          H—H
Hydrogen molecule ($H_2$)

**(a)**

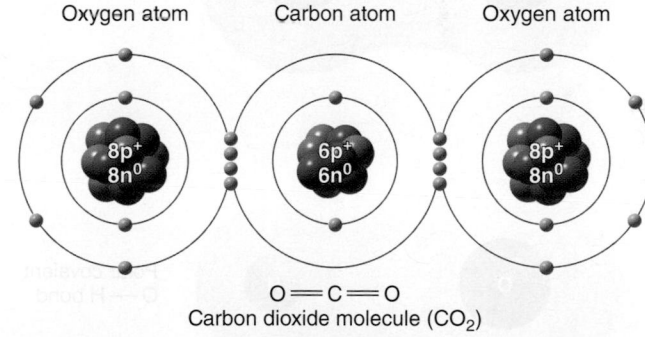

Oxygen atom        Carbon atom        Oxygen atom

O=C=O
Carbon dioxide molecule ($CO_2$)

**(b)**

**FIGURE 2.6 Covalent Bonding.** (a) Two hydrogen atoms share a single pair of electrons to form a hydrogen molecule. (b) A carbon dioxide molecule, in which a carbon atom shares two pairs of electrons with each oxygen atom, forming double covalent bonds.

*bond* is the sharing of two pairs of electrons. In carbon dioxide, for example, a central carbon atom shares two electron pairs with each oxygen atom. Such bonds are symbolized by two lines—for example, O=C=O (fig. 2.6b).

When shared electrons spend approximately equal time around each nucleus, they form a *nonpolar covalent bond* (fig. 2.7a), the strongest of all chemical bonds. Carbon atoms bond to each other with nonpolar covalent bonds. If shared electrons spend significantly more time orbiting one nucleus than they do the other, they lend their negative charge to the region where they spend the most time, and they form a *polar covalent bond* (fig. 2.7b). When hydrogen bonds with oxygen, for example, the electrons are more attracted to the oxygen nucleus and orbit it more than they do the hydrogen. This makes the oxygen region of the molecule slightly negative and the hydrogen regions slightly positive. The Greek delta (δ) is used to symbolize a charge less than that of one electron or proton. A slightly negative region of a molecule is represented δ− and a slightly positive region is represented δ+.

A **hydrogen bond** is a weak attraction between a slightly positive hydrogen atom in one molecule and a slightly negative oxygen or nitrogen atom in another. Water molecules, for example, are weakly attracted to each other by hydrogen bonds (fig. 2.8). Hydrogen bonds also form between different regions of the same molecule, especially in very large molecules such as proteins and DNA. They cause such molecules to fold or coil into precise three-dimensional shapes. Hydrogen bonds are represented by dotted or broken lines between atoms: —C═O···H—N—. Hydrogen bonds are relatively weak, but they are enormously important to physiology.

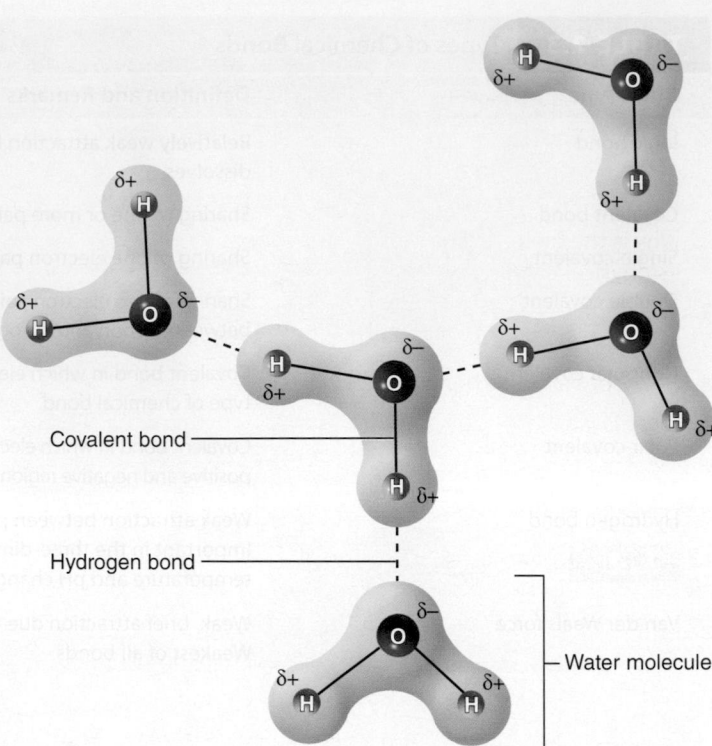

**FIGURE 2.8  Hydrogen Bonding of Water.** The polar covalent bonds of water molecules enable each oxygen to form a hydrogen bond with a hydrogen of a neighboring molecule. Thus, the water molecules are weakly attracted to each other.

**?** *Why would this behavior raise the boiling point of water above that of a nonpolar liquid?*

**Van der Waals[5] forces** are weak, brief attractions between neutral atoms. When electrons orbit an atom's nucleus, they don't maintain a uniform distribution but show random fluctuations in density. If the electrons briefly crowd toward one side of an atom, they render that side slightly negative and the other side slightly positive for a moment. If another atom is close enough to this one, the second atom responds with disturbances in its own electron cloud. Oppositely charged regions of the two atoms then attract each other for a very brief moment.

A single van der Waals force is only about 1% as strong as a covalent bond, but when two surfaces or large molecules meet, the van der Waals forces between large numbers of atoms can create a very strong attraction. This is how plastic wrap clings to food and dishes; flies and spiders walk across a ceiling; and even a 100 g lizard, the Tokay gecko, can run up a windowpane. Van der Waals forces also have a significant effect on the boiling points of liquids. In human structure, they are especially important in protein folding, the binding of proteins to each other and to other molecules such as hormones, and the association of lipid molecules with each other. Some of these molecular behaviors are described later in this chapter.

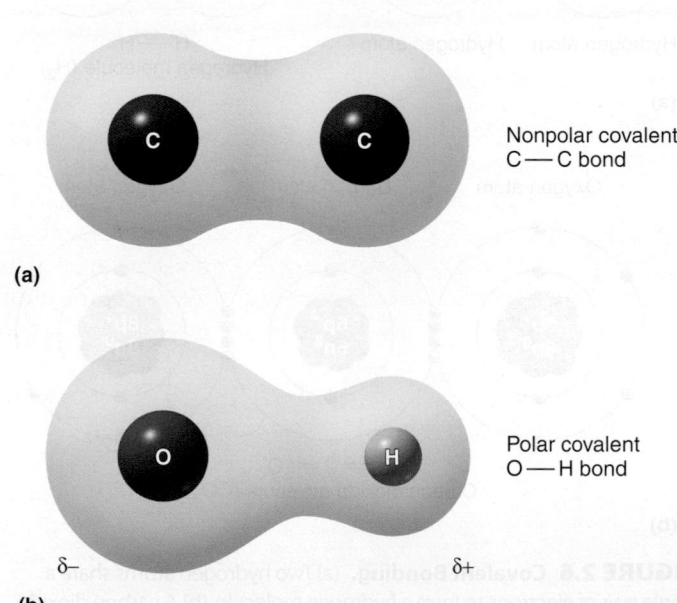

Nonpolar covalent
C — C bond

**(a)**

Polar covalent
O — H bond

**(b)**

**FIGURE 2.7  Nonpolar and Polar Covalent Bonds.** (a) A nonpolar covalent bond between two carbon atoms, formed by electrons that spend an equal amount of time around each nucleus, as represented by the symmetric blue cloud. (b) A polar covalent bond, in which electrons orbit one nucleus significantly more than the other, as represented by the asymmetric cloud. This results in a slight negative charge (δ−) in the region where the electrons spend most of their time, and a slight positive charge (δ+) at the other pole.

Answer the following questions to test your understanding of the preceding section:

1. Consider iron (Fe), hydrogen gas ($H_2$), and ammonia ($NH_3$). Which of them is or are atoms? Which of them is or are molecules? Which of them is or are compounds? Explain each answer.

2. Why is the biological half-life of a radioisotope shorter than its physical half-life?

3. Where do free radicals come from? What harm do they do? How is the body protected from free radicals?

4. How does an ionic bond differ from a covalent bond?

5. What is a hydrogen bond? Why do hydrogen bonds depend on the existence of polar covalent bonds?

## 2.2 Water and Mixtures

### Expected Learning Outcomes

When you have completed this section, you should be able to

a. define *mixture* and distinguish between mixtures and compounds;

b. describe the biologically important properties of water;

c. show how three kinds of mixtures differ from each other;

d. discuss some ways in which the concentration of a solution can be expressed, and explain why different expressions of concentration are used for different purposes; and

e. define *acid* and *base* and interpret the pH scale.

Our body fluids are complex mixtures of chemicals. A **mixture** consists of substances that are physically blended but not chemically combined. Each substance retains its own chemical properties. To contrast a mixture with a compound, consider sodium chloride again. Sodium is a lightweight metal that bursts into flame if exposed to water, and chlorine is a yellow-green poisonous gas that was used for chemical warfare in World War I. When these elements chemically react, they form common table salt. Clearly, the compound has properties much different from the properties of its elements. But if you were to put a little salt on your watermelon, the watermelon would taste salty and sweet because the sugar of the melon and the salt you added would merely form a mixture in which each compound retained its individual properties.

### Water

Most mixtures in our bodies consist of chemicals dissolved or suspended in water. Water constitutes 50% to 75% of your body weight, depending on age, sex, fat content, and other factors. Its structure, simple as it is, has profound biological effects. Two aspects of its structure are particularly important:

(1) Its atoms are joined by polar covalent bonds, and (2) the molecule is V-shaped, with a 105° bond angle (fig. 2.9a). This makes the molecule as a whole polar, because there is a slight negative charge ($\delta-$) on the oxygen at the apex of the V and a slight positive charge ($\delta+$) on each hydrogen. Like little magnets, water molecules are attracted to one another by hydrogen bonds (see fig. 2.8). This gives water a set of properties that account for its ability to support life: *solvency, cohesion, adhesion, chemical reactivity,* and *thermal stability.*

*Solvency* is the ability to dissolve other chemicals. Water is sometimes called the *universal solvent* because it dissolves a broader range of substances than any other liquid. Substances that dissolve in water, such as sugar, are said to be **hydrophilic**[6] (HY-dro-FILL-ic); the relatively few substances that do not, such as fats, are **hydrophobic**[7] (HY-dro-FOE-bic). Virtually all metabolic reactions depend on the solvency of water. Biological molecules must be dissolved in water to move freely, come together, and react. The solvency of water also makes it the body's primary means of transporting substances from place to place.

To be soluble in water, a molecule must be polarized or charged so that its charges can interact with those of water. When NaCl is dropped into water, for example, the ionic bonds between $Na^+$ and $Cl^-$ are overpowered by the attraction of each ion to water molecules. Water molecules form a cluster, or *hydration sphere,* around each sodium ion with the $O^{\delta-}$ pole of each water molecule facing the sodium ion. They also form a hydration sphere around each chloride ion, with the $H^{\delta+}$ poles facing it. This isolates the sodium ions from the chloride ions and keeps them dissolved (fig. 2.9b).

*Adhesion* is the tendency of one substance to cling to another, whereas *cohesion* is the tendency of molecules of the same substance to cling to each other. Water adheres to the body's tissues and forms a lubricating film on membranes such as the pleura and pericardium. This reduces friction as the lungs and heart contract and expand and rub against these membranes. Water also is a very cohesive liquid because of its hydrogen bonds. This is why, when you spill water on the floor, it forms a puddle and evaporates slowly. By contrast, if you spill a nonpolar substance such as liquid nitrogen, it dances about and evaporates in seconds, like a drop of water in a hot dry skillet. This is because nitrogen molecules have no attraction for each other, so the little bit of heat provided by the floor is enough to disperse them into the air. The cohesion of water is especially evident at its surface, where it forms an elastic layer called the *surface film* held together by a force called *surface tension.* This force causes water to hang in drops from a leaky faucet and travel in rivulets down a window.

The *chemical reactivity* of water is its ability to participate in chemical reactions. Not only does water ionize many other chemicals such as acids and salts, but water itself ionizes into

---

[6]*hydro* = water; *philic* = loving, attracted to
[7]*phobic* = fearing, avoiding

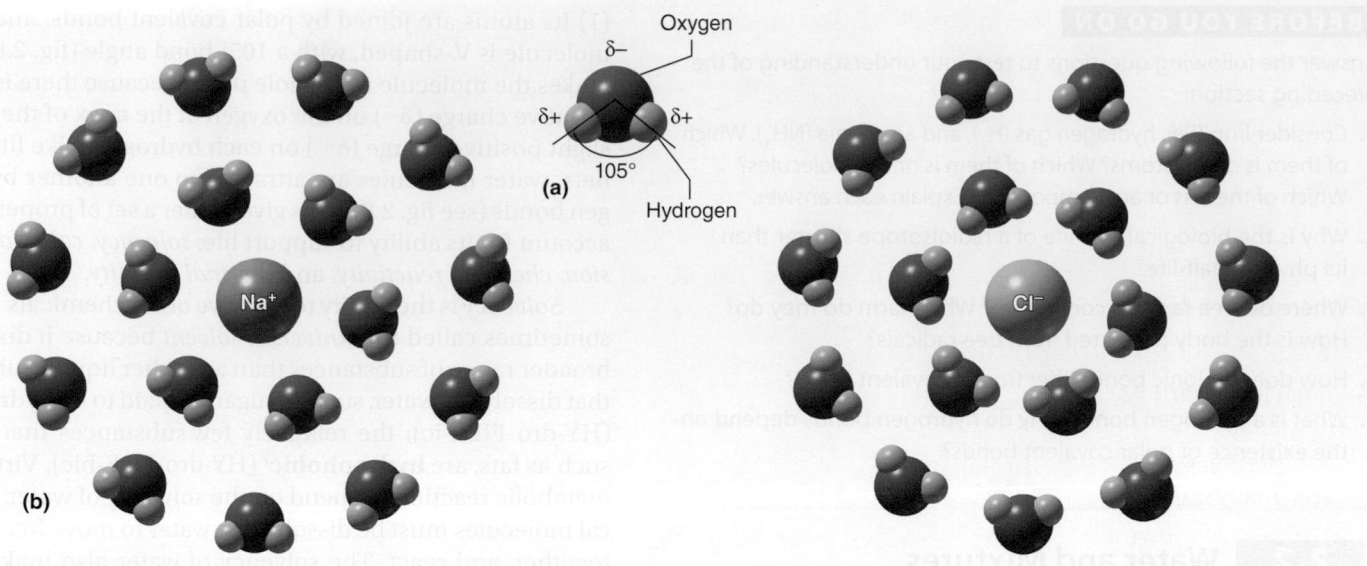

**FIGURE 2.9 Water and Hydration Spheres.** (a) A water molecule showing its bond angle and polarity. (b) Water molecules aggregate around a sodium ion with their negatively charged oxygen poles facing the $Na^+$ and aggregate around a chloride ion with their positively charged hydrogen poles facing the $Cl^-$.

$H^+$ and $OH^-$. These ions can be incorporated into other molecules, or released from them, in the course of chemical reactions such as *hydrolysis* and *dehydration synthesis,* described later in this chapter.

The *thermal stability* of water helps to stabilize the internal temperature of the body. It results from the high *heat capacity* of water—the amount of heat required to raise the temperature of 1 g of a substance by 1°C. The base unit of heat is the **calorie**[8] **(cal)**—1 cal is the amount of heat that raises the temperature of 1 g of water 1°C. The same amount of heat would raise the temperature of a nonpolar substance such as nitrogen about four times as much. The difference stems from the presence or absence of hydrogen bonding. To increase in temperature, the molecules of a substance must move around more actively. The hydrogen bonds of water molecules inhibit their movement, so water can absorb a given amount of heat without changing temperature (molecular motion) as much.

The high heat capacity of water also makes it a very effective coolant. When it changes from a liquid to a vapor, water carries a large amount of heat with it. One milliliter of perspiration evaporating from the skin removes about 500 cal of heat from the body. This effect is very apparent when you are sweaty and stand in front of a fan.

▶▶▶**APPLY WHAT YOU KNOW**

*Why are heat and temperature not the same thing?*

## Solutions, Colloids, and Suspensions

Mixtures of other substances in water can be classified as *solutions, colloids,* and *suspensions.*

A **solution** consists of particles of matter called the **solute** mixed with a more abundant substance (usually water) called the **solvent.** The solute can be a gas, solid, or liquid—as in a solution of oxygen, sodium chloride, or alcohol in water, respectively. Solutions are defined by the following properties:

- The solute particles are under 1 nanometer (nm) in size. The solute and solvent therefore cannot be visually distinguished from each other, even with a microscope.
- Such small particles do not scatter light noticeably, so solutions are usually transparent (fig. 2.10a).
- The solute particles will pass through most selectively permeable membranes, such as dialysis tubing and cell membranes.
- The solute does not separate from the solvent when the solution is allowed to stand.

The most common **colloids**[9] in the body are mixtures of protein and water, such as the albumin in blood plasma. Many colloids can change from liquid to gel states—gelatin desserts, agar culture media, and the fluids within and between our cells, for example. Colloids are defined by the following physical properties:

- The colloidal particles range from 1 to 100 nm in size.

---

[8]*calor* = heat

[9]*collo* = glue; *oid* = like, resembling

- Particles this large scatter light, so colloids are usually cloudy (fig. 2.10b).
- The particles are too large to pass through most selectively permeable membranes.
- The particles are still small enough, however, to remain permanently mixed with the solvent when the mixture stands.

The blood cells in our blood plasma exemplify a **suspension.** Suspensions are defined by the following properties:

- The suspended particles exceed 100 nm in size.
- Such large particles render suspensions cloudy or opaque.
- The particles are too large to penetrate selectively permeable membranes.
- The particles are too heavy to remain permanently suspended, so suspensions separate on standing. Blood cells, for example, form a suspension in the blood plasma and settle to the bottom of a tube when blood is allowed to stand without mixing (fig. 2.10c, d).

An **emulsion** is a suspension of one liquid in another, such as oil-and-vinegar salad dressing. The fat in breast milk is an emulsion, as are medications such as Kaopectate and milk of magnesia.

A single mixture can fit into more than one of these categories. Blood is a perfect example—it is a solution of sodium chloride, a colloid of protein, and a suspension of cells. Milk is a solution of calcium, a colloid of protein, and an emulsion of fat. Table 2.4 summarizes the types of mixtures and provides additional examples.

## Measures of Concentration

Solutions are often described in terms of their concentration—how much solute is present in a given volume of solution. Concentration is expressed in different ways for different purposes, some of which are explained here. The table of symbols and measures in appendix C may be helpful as you study this section.

### Weight per Volume

A simple way to express concentration is the weight of solute in a given volume of solution. For example, intravenous (I.V.) saline typically contains 8.5 grams of NaCl per liter of solution (8.5 g/L). For many biological purposes, however, we deal with smaller quantities such as milligrams per deciliter (mg/dL; 1 dL = 100 mL). For example, a typical serum cholesterol concentration may be 200 mg/dL, also expressed as 200 mg/100 mL or 200 milligram-percent (mg-%).

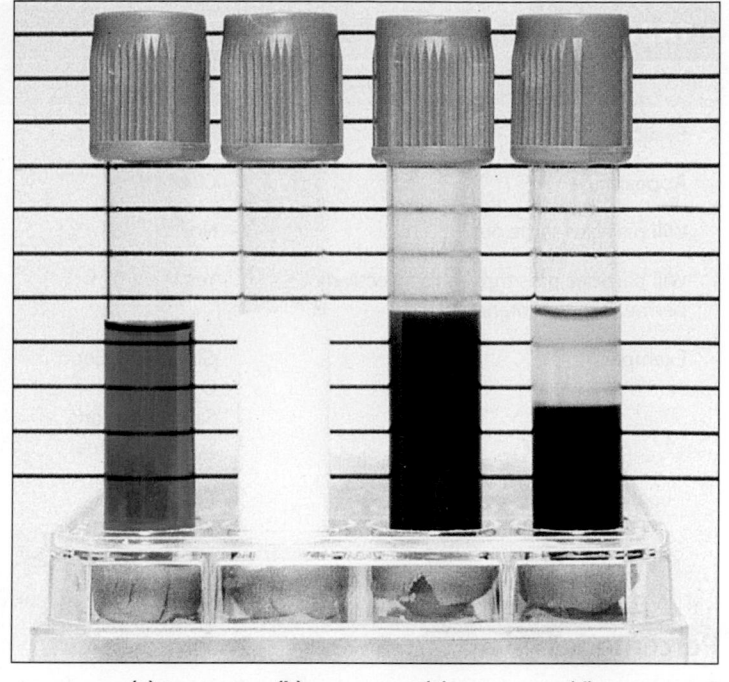

(a)          (b)          (c)          (d)

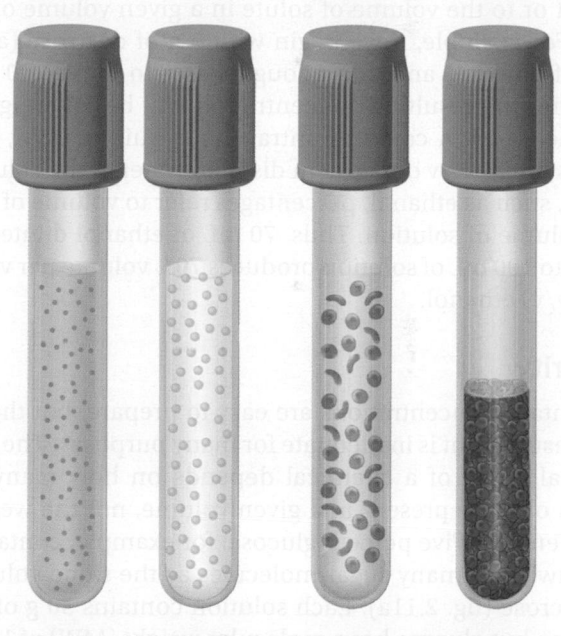

Solution          Colloid          Suspension

**FIGURE 2.10 A Solution, a Colloid, and a Suspension.**
Top row: Photographs of a representative solution, colloid, and suspension. Bottom row: Symbolic representation of the particle sizes in each mixture. (a) In a copper sulfate solution, the solute particles are so small they remain permanently mixed and the solution is transparent. (b) In milk, the protein molecules are small enough to remain permanently mixed, but large enough to scatter light, so the mixture is opaque. (c) In blood, the red blood cells scatter light and make the mixture opaque. (d) Red blood cells are too large to remain evenly mixed, so they settle to the bottom as in this blood specimen that stood overnight.

| TABLE 2.4 | Types of Mixtures | | |
|---|---|---|---|
| | **Solution** | **Colloid** | **Suspension** |
| Particle Size | <1 nm | 1–100 nm | >100 nm |
| Appearance | Clear | Often cloudy | Cloudy-opaque |
| Will particles settle out? | No | No | Yes |
| Will particles pass through a selectively permeable membrane? | Yes | No | No |
| Examples | Glucose in blood<br>$O_2$ in water<br>Saline solutions<br>Sugar in coffee | Proteins in blood<br>Intracellular fluid<br>Milk protein<br>Gelatin | Blood cells<br>Cornstarch in water<br>Fats in blood<br>Kaopectate |

## Percentages

Percentage concentrations are also simple to compute, but it is necessary to specify whether the percentage refers to the weight or to the volume of solute in a given volume of solution. For example, if we begin with 5 g of dextrose (an isomer of glucose) and add enough water to make 100 mL of solution, the resulting concentration will be 5% weight per volume (w/v). A common intravenous fluid is D5W, which stands for 5% w/v dextrose in distilled water. If the solute is a liquid, such as ethanol, percentages refer to volume of solute per volume of solution. Thus, 70 mL of ethanol diluted with water to 100 mL of solution produces 70% volume per volume (70% v/v) ethanol.

## Molarity

Percentage concentrations are easy to prepare, but that unit of measurement is inadequate for many purposes. The physiological effect of a chemical depends on how many molecules of it are present in a given volume, not the weight of the chemical. Five percent glucose, for example, contains almost twice as many sugar molecules as the same volume of 5% sucrose (fig. 2.11a). Each solution contains 50 g of sugar per liter, but glucose has a molecular weight (MW) of 180 and sucrose has a MW of 342. Since each molecule of glucose is lighter, 50 g of glucose contains more molecules than 50 g of sucrose.

To produce solutions with a known number of molecules per volume, we must factor in the molecular weight. If we know the MW and weigh out that many grams of the substance, we have a quantity known as 1 *mole*. **Molarity (M)** is the number of moles of solute per liter of solution. A *one-molar* (1.0 M) solution of glucose contains 180 g/L, and 1.0 M solution of sucrose contains 342 g/L. Both have the same number of solute molecules in a given volume (fig. 2.11b). Body fluids and laboratory solutions usually are less concentrated than

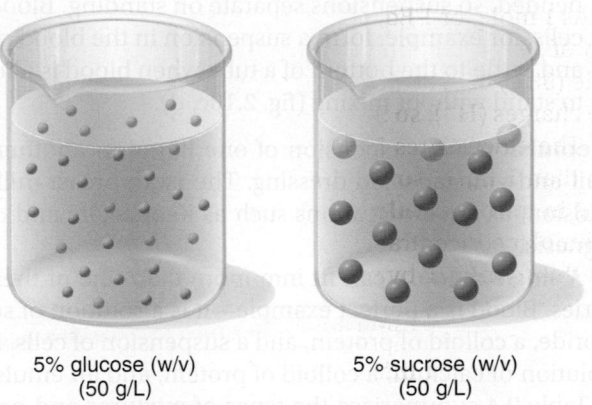

5% glucose (w/v)　　5% sucrose (w/v)
(50 g/L)　　　　　　(50 g/L)

**(a) Solutions of equal percentage concentration**

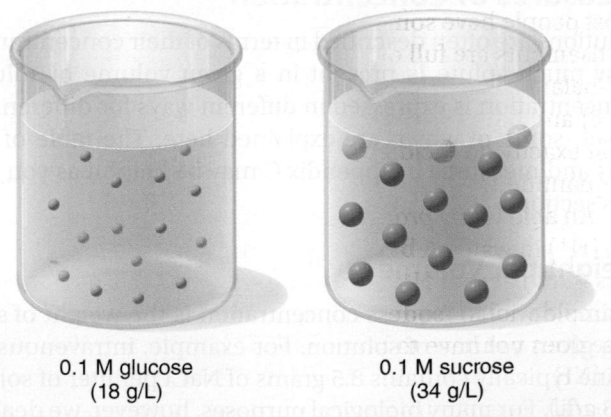

0.1 M glucose　　0.1 M sucrose
(18 g/L)　　　　　(34 g/L)

**(b) Solutions of equal molar concentration**

**FIGURE 2.11 Comparison of Percentage and Molar Concentrations.** (a) Solutions with the same percentage concentrations can differ greatly in the number of molecules per volume because of differences in molecular weights of the solutes. Fifty grams of sucrose has about half as many molecules as 50 g of glucose, for example. (b) Solutions with the same molarity have the same number of molecules per volume because molarity takes differences in molecular weight into account.

1 M, so biologists and clinicians more often work with *milli-molar (mM)* and *micromolar (μM)* concentrations—$10^{-3}$ and $10^{-6}$ M, respectively.

## Electrolyte Concentrations

Electrolytes are important for their chemical, physical (osmotic), and electrical effects on the body. Their electrical effects, which determine such things as nerve, heart, and muscle actions, depend not only on their concentration but also on their electrical charge. A calcium ion ($Ca^{2+}$) has twice the electrical effect of a sodium ion ($Na^+$), for example, because it carries twice the charge. When we measure electrolyte concentrations, we must therefore take the charges into account.

One *equivalent (Eq)* of an electrolyte is the amount that would electrically neutralize 1 mole of hydrogen ions ($H^+$) or hydroxide ions ($OH^-$). For example, 1 mole (58.4 g) of NaCl yields 1 mole, or 1 Eq, of $Na^+$ in solution. Thus, an NaCl solution of 58.4 g/L contains 1 Eq of $Na^+$ per liter (1 Eq/L). One mole (98 g) of sulfuric acid ($H_2SO_4$) contains 2 moles of positive charges ($H^+$), so 98 g/L of $H_2SO_4$ would be 2 Eq/L.

The electrolytes in our body fluids have concentrations less than 1 Eq/L, so we more often express their concentrations in **milliequivalents per liter (mEq/L).** If you know the millimolar concentration of an electrolyte, you can easily convert this to mEq/L by multiplying it by the valence of the ion:

$$1 \text{ mM } Na^+ = 1 \text{ mEq/L}$$
$$1 \text{ mM } Ca^{2+} = 2 \text{ mEq/L}$$
$$1 \text{ mM } Fe^{3+} = 3 \text{ mEq/L}$$

## Acids, Bases, and pH

Most people have some sense of what acids and bases are. Advertisements are full of references to excess stomach acid and pH-balanced shampoo. We know that drain cleaner (a strong base) and battery acid can cause serious chemical burns. But what exactly do "acidic" and "basic" mean, and how can they be quantified?

An **acid** is any *proton donor,* a molecule that releases a proton ($H^+$) in water. A **base** is a proton acceptor. Since hydroxide ions ($OH^-$) accept $H^+$, many bases are substances that release hydroxide ions—sodium hydroxide (NaOH), for example. A base does not have to be a hydroxide donor, however. Ammonia ($NH_3$) is also a base. It does not release hydroxide ions, but it readily accepts hydrogen ions to become the ammonium ion ($NH_4^+$).

Acidity is expressed in terms of **pH,** a measure derived from the molarity of $H^+$. Molarity is represented by square brackets, so the molarity of $H^+$ is symbolized $[H^+]$. pH is the negative logarithm of hydrogen ion molarity—that is, pH = $-\log [H^+]$. In pure water, 1 in 10 million molecules ionizes into

hydrogen and hydroxide ions: $H_2O \leftrightharpoons H^+ + OH^-$. (The symbol $\leftrightharpoons$ denotes a reversible chemical reaction.) Pure water has a neutral pH because it contains equal amounts of $H^+$ and $OH^-$. Since 1 in 10 million molecules ionize, the molarity of $H^+$ and the pH of water are

$$[H^+] = 0.0000001 \text{ molar} = 10^{-7} \text{ M}$$
$$\log [H^+] = -7$$
$$pH = -\log [H^+] = 7$$

The pH scale (fig. 2.12) was invented in 1909 by Danish biochemist and brewer Sören Sörensen to measure the acidity of beer. The scale extends from 0.0 to 14.0. A solution with a pH of 7.0 is **neutral;** solutions with pH below 7 are **acidic;** and solutions with pH above 7 are **basic (alkaline).** The lower the pH value, the more hydrogen ions a solution has and the more acidic it is. Since the pH scale is logarithmic, a change of one whole number on the scale represents a 10-fold change in $H^+$ concentration. In other words, a solution with pH 4 is 10 times as acidic as one with pH 5 and 100 times as acidic as one with pH 6.

Slight disturbances of pH can seriously disrupt physiological functions and alter drug actions (see Deeper Insight 2.2), so it is important that the body carefully control its pH. Blood, for example, normally has a pH ranging from 7.35 to 7.45. Deviations from this range cause tremors, fainting, paralysis, or even death. Chemical solutions that resist changes in pH are called **buffers.** Buffers and pH regulation are considered in detail in chapter 24.

### ▶▶▶ APPLY WHAT YOU KNOW

*A pH of 7.20 is slightly alkaline, yet a blood pH of 7.20 is called acidosis. Why do you think it is called this?*

## DEEPER INSIGHT 2.2
### CLINICAL APPLICATION

### *pH and Drug Action*

The pH of our body fluids has a direct bearing on how we react to drugs. Depending on pH, drugs such as aspirin, phenobarbital, and penicillin can exist in charged (ionized) or uncharged forms. Whether a drug is charged or not can determine whether it will pass through cell membranes. When aspirin is in the acidic environment of the stomach, for example, it is uncharged and passes easily through the stomach lining into the bloodstream. Here it encounters a basic pH, whereupon it ionizes. In this state, it is unable to pass back through the membrane, so it accumulates in the blood. This effect, called ion trapping or pH partitioning, can be controlled to help clear poisons from the body. The pH of the urine, for example, can be manipulated so that poisons become trapped there and thus are rapidly excreted from the body.

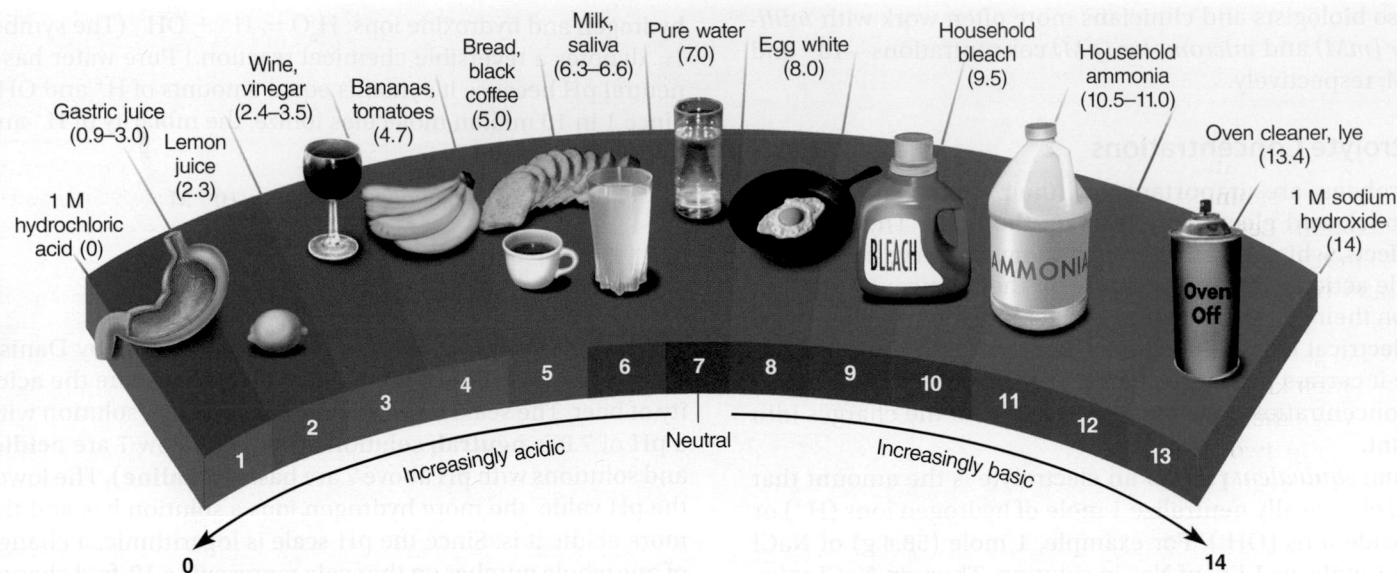

**FIGURE 2.12   The pH Scale.** The pH is shown within the colored bar. $H^+$ molarity increases 10-fold for every step down the scale.

Answer the following questions to test your understanding of the preceding section:

6. What is the difference between a mixture and a compound?

7. What are hydrophilic and hydrophobic substances? Give an example of each.

8. Why would the cohesion and thermal stability of water be less if water did not have polar covalent bonds?

9. How do solutions, colloids, and suspensions differ from each other? Give an example of each in the human body.

10. What is one advantage of percentage over molarity as a measure of solute concentration? What is one advantage of molarity over percentage?

11. If solution A had an $H^+$ concentration of $10^{-8}$ M, what would be its pH? If solution B had 1,000 times this $H^+$ concentration, what would be its pH? Would solution A be acidic or basic? What about solution B?

## 2.3   Energy and Chemical Reactions

### Expected Learning Outcomes

When you have completed this section, you should be able to

a. define *energy* and *work*, and describe some types of energy;

b. understand how chemical reactions are symbolized by chemical equations;

c. list and define the fundamental types of chemical reactions;

d. identify the factors that govern the speed and direction of a reaction;

e. define *metabolism* and its two subdivisions; and

f. define *oxidation* and *reduction* and relate these to changes in the energy content of a molecule.

## Energy and Work

**Energy** is the capacity to do work. To do **work** means to move something, whether it is a muscle or a molecule. Some examples of physiological work are breaking chemical bonds, building molecules, pumping blood, and contracting skeletal muscles. All of the body's activities are forms of work.

Energy is broadly classified as *potential* or *kinetic energy*. *Potential energy* is energy contained in an object because of its position or internal state but that is not doing work at the time. *Kinetic energy* is energy of motion, energy that is doing work. It is observed in musculoskeletal movements, the flow of ions into a cell, and vibration of the eardrum, for example. The water behind a dam has potential energy because of its position. Let the water flow through, and it exhibits kinetic energy that can be tapped for generating electricity. Like water behind a dam, ions concentrated on one side of a cell membrane have potential energy that can be released by opening gates in the membrane. As the ions flow through the gates, their kinetic energy can be tapped to create a nerve signal or make the heart beat.

Within the two broad categories of potential and kinetic energy, several forms of energy are relevant to human physiology. *Chemical energy* is potential energy stored in the bonds of molecules. Chemical reactions release this energy and make it available for physiological work. *Heat* is the kinetic energy of molecular motion. The temperature of a substance is a measure of rate of this motion, and adding heat to a substance increases this rate. *Electromagnetic energy* is the kinetic energy of moving "packets" of radiation called *photons*. The most familiar form of electromagnetic energy is light. *Electrical energy* has both potential and kinetic forms. It is potential energy when charged particles have accumulated at a point such as a battery terminal or on one side of a cell membrane; it becomes kinetic energy when these particles begin to move and create an electrical current—for example, when electrons move through your household wiring or sodium ions move through a cell membrane.

**Free energy** is the potential energy available in a system to do useful work. In human physiology, the most relevant free energy is the energy stored in the chemical bonds of organic molecules.

## Classes of Chemical Reactions

A **chemical reaction** is a process in which a covalent or ionic bond is formed or broken. The course of a chemical reaction is symbolized by a **chemical equation** that typically shows the *reactants* on the left, the *products* on the right, and an arrow pointing from the reactants to the products. For example, consider this common occurrence: If you open a bottle of wine and let it stand for several days, it turns sour. Wine "turns to vinegar" because oxygen gets into the bottle and reacts with ethanol to produce acetic acid and water. Acetic acid gives the tart flavor to vinegar and spoiled wine. The equation for this reaction is

$$CH_3CH_2OH \ + \ O_2 \ \longrightarrow \ CH_3COOH \ + \ H_2O$$

Ethanol $\qquad$ Oxygen $\qquad$ Acetic acid $\qquad$ Water

Ethanol and oxygen are the reactants, and acetic acid and water are the products of this reaction. Not all reactions are shown with the arrow pointing from left to right. In complex biochemical equations, reaction chains are often written vertically or even in circles.

Chemical reactions can be classified as *decomposition, synthesis,* or *exchange reactions*. In **decomposition reactions,** a large molecule breaks down into two or more smaller ones (fig. 2.13a); symbolically, AB $\longrightarrow$ A + B. When you eat a potato, for example, digestive enzymes decompose its starch into thousands of glucose molecules, and most cells further decompose glucose to water and carbon dioxide. Starch, a very large molecule, ultimately yields about 36,000 molecules of $H_2O$ and $CO_2$.

**Synthesis reactions** are just the opposite—two or more small molecules combine to form a larger one; symbolically, A + B $\longrightarrow$ AB (fig. 2.13b). When the body synthesizes proteins, for example, it combines several hundred amino acids into one protein molecule.

In **exchange reactions,** two molecules exchange atoms or groups of atoms; AB + CD $\longrightarrow$ AC + BD (fig. 2.13c). For example, when stomach acid (HCl) enters the small intestine, the pancreas secretes sodium bicarbonate ($NaHCO_3$) to neutralize it. The reaction between the two is $NaHCO_3$ + HCl $\longrightarrow$ NaCl + $H_2CO_3$. We could say the sodium atom has exchanged its bicarbonate group ($-HCO_3$) for a chlorine atom.

**Reversible reactions** can go in either direction under different circumstances. For example, carbon dioxide combines with water to produce carbonic acid, which in turn decomposes into bicarbonate ions and hydrogen ions:

$$CO_2 \ + \ H_2O \ \rightleftharpoons \ H_2CO_3 \ \rightleftharpoons \ HCO_3^- \ + \ H^+$$

Carbon $\quad$ Water $\quad$ Carbonic $\quad$ Bicarbonate $\quad$ Hydrogen
dioxide $\qquad\qquad$ acid $\qquad$ ion $\qquad$ ion

This reaction appears in this book more often than any other, especially as we discuss respiratory, urinary, and digestive physiology.

The direction in which a reversible reaction goes is determined by the relative abundance of substances on each side of the equation. If there is a surplus of $CO_2$, this reaction proceeds to the right and produces bicarbonate and hydrogen ions. If bicarbonate and hydrogen ions are present in excess, the reaction proceeds to the left and generates $CO_2$ and $H_2O$. Reversible reactions follow the **law of mass action:** They proceed from the reactants in greater quantity to the substances with the lesser quantity. This law will help to explain processes discussed in later chapters, such as why hemoglobin binds oxygen in the lungs yet releases it to muscle tissue.

In the absence of upsetting influences, reversible reactions exist in a state of **equilibrium,** in which the ratio of

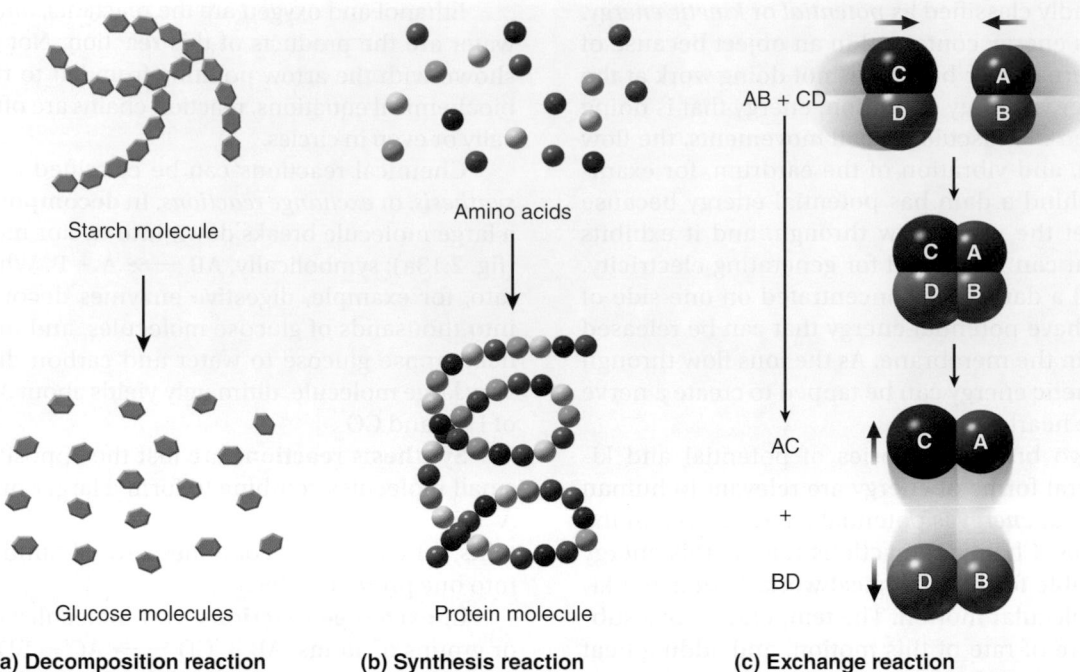

Starch molecule

Amino acids

Glucose molecules

Protein molecule

(a) **Decomposition reaction**          (b) **Synthesis reaction**          (c) **Exchange reaction**

**FIGURE 2.13 Decomposition, Synthesis, and Exchange Reactions.** (a) In a decomposition reaction, large molecules are broken down into simpler ones. (b) In a synthesis reaction, smaller molecules are joined to form larger ones. (c) In an exchange reaction, two molecules exchange atoms.

**?** *To which of these categories does the digestion of food belong?*

products to reactants is stable. The carbonic acid reaction, for example, normally maintains a 20:1 ratio of bicarbonate ions to carbonic acid molecules. This equilibrium can be upset, however, by a surplus of hydrogen ions, which drive the reaction to the left, or adding carbon dioxide and driving it to the right.

## Reaction Rates

Chemical reactions are based on molecular motion and collisions. All molecules are in constant motion, and reactions occur when mutually reactive molecules collide with sufficient force and the right orientation. The rate of a reaction depends on the nature of the reactants and on the frequency and force of these collisions. Some factors that affect reaction rates are

- **Concentration.** Reaction rates increase when the reactants are more concentrated. This is because the molecules are more crowded and collide more frequently.

- **Temperature.** Reaction rate increases as the temperature rises. This is because heat causes molecules to move more rapidly and collide with greater force and frequency.

- **Catalysts** (CAT-uh-lists). These are substances that temporarily bind to reactants, hold them in a favorable position to react with each other, and may change the shapes of reactants in ways that make them more likely to react. By reducing the element of chance in molecular collisions, a catalyst speeds up a reaction. It then releases the products and is available to repeat the process with more reactants. The catalyst itself is not consumed or changed by the reaction. The most important biological catalysts are enzymes, discussed later in this chapter.

## Metabolism, Oxidation, and Reduction

All the chemical reactions in the body are collectively called **metabolism.** Metabolism has two divisions: *catabolism* and *anabolism.* **Catabolism**[10] (ca-TAB-oh-lizm) consists of energy-releasing decomposition reactions. Such reactions break covalent bonds, produce smaller molecules from larger ones, and release energy that can be used for other physiological work. Energy-releasing reactions are called

---

[10]*cato* = down, to break down

*exergonic*[11] reactions. If you hold a beaker of water in your hand and pour sulfuric acid into it, for example, the beaker will get so hot you may have to put it down. If you break down energy-storage molecules to run a race, you too will get hot. In both cases, the heat signifies that exergonic reactions are occurring.

**Anabolism**[12] (ah-NAB-oh-lizm) consists of energy-storing synthesis reactions, such as the production of protein or fat. Reactions that require an energy input, such as these, are called *endergonic*[13] reactions. Anabolism is driven by the energy that catabolism releases, so endergonic and exergonic processes, anabolism and catabolism, are inseparably linked.

**Oxidation** is any chemical reaction in which a molecule gives up electrons and releases energy. A molecule is *oxidized* by this process, and whatever molecule takes the electrons from it is an *oxidizing agent* (electron acceptor). The term *oxidation* stems from the fact that oxygen is often involved as the electron acceptor. Thus, we can sometimes recognize an oxidation reaction from the fact that oxygen has been added to a molecule. The rusting of iron, for example, is a slow oxidation process in which oxygen is added to iron to form iron oxide ($Fe_2O_3$). Many oxidation reactions, however, do not involve oxygen at all. For example, when yeast ferments glucose to alcohol, no oxygen is required; indeed, the alcohol *contains less oxygen* than the sugar originally did, but it is *more oxidized* than the sugar:

$$C_6H_{12}O_6 \longrightarrow 2\,CH_3CH_2OH + 2\,CO_2$$

Glucose　　　　　　Ethanol　　　　　Carbon dioxide

---

[11]*ex, exo* = out; *erg* = work
[12]*ana* = up, to build up
[13]*end, endo* = in; *erg* = work

**Reduction** is a chemical reaction in which a molecule gains electrons and energy. When a molecule accepts electrons, it is said to be *reduced;* a molecule that donates electrons to another is therefore called a *reducing agent* (electron donor). The oxidation of one molecule is always accompanied by the reduction of another, so these electron transfers are known as *oxidation–reduction (redox) reactions.*

It is not necessary that *only* electrons be transferred in a redox reaction. Often, the electrons are transferred in the form of hydrogen atoms. The fact that a proton (the hydrogen nucleus) is also transferred is immaterial to whether we consider a reaction oxidation or reduction.

Table 2.5 summarizes these energy-transfer reactions. We can symbolize oxidation and reduction as follows, letting A and B symbolize arbitrary molecules and letting e⁻ represent one or more electrons:

$$Ae^- + B \longrightarrow A + Be^-$$

High-energy　　Low-energy　　Low-energy　　High-energy
reduced　　　　oxidized　　　　oxidized　　　　reduced
state　　　　　　state　　　　　　state　　　　　　state

$Ae^-$ is a reducing agent because it reduces B, and B is an oxidizing agent because it oxidizes $Ae^-$.

**BEFORE YOU GO ON**

Answer the following questions to test your understanding of the preceding section:

12. Define *energy.* Distinguish potential energy from kinetic energy.

13. Define *metabolism, catabolism,* and *anabolism.*

14. What does *oxidation* mean? What does *reduction* mean? Which of them is endergonic and which is exergonic?

15. When sodium chloride forms, which element—sodium or chlorine—is oxidized? Which one is reduced?

| TABLE 2.5 | Energy-Transfer Reactions in the Human Body |
|---|---|
| Exergonic Reactions | Reactions in which there is a net release of energy. The products have less total free energy than the reactants did. |
| Oxidation | An exergonic reaction in which electrons are removed from a reactant. Electrons may be removed one or two at a time and may be removed in the form of hydrogen atoms (H or $H_2$). The product is then said to be oxidized. |
| Decomposition | A reaction such as digestion and cell respiration, in which larger molecules are broken down into smaller ones. |
| Catabolism | The sum of all decomposition reactions in the body. |
| Endergonic Reactions | Reactions in which there is a net input of energy. The products have more total free energy than the reactants did. |
| Reduction | An endergonic reaction in which electrons are donated to a reactant. The product is then said to be reduced. |
| Synthesis | A reaction such as protein and glycogen synthesis, in which two or more smaller molecules are combined into a larger one. |
| Anabolism | The sum of all synthesis reactions in the body. |

## 2.4    Organic Compounds

### Expected Learning Outcomes

When you have completed this section, you should be able to

a.  explain why carbon is especially well suited to serve as the structural foundation of many biological molecules;

b.  identify some common functional groups of organic molecules from their formulae;

c.  discuss the relevance of polymers to biology and explain how they are formed and broken by dehydration synthesis and hydrolysis;

d.  discuss the types and functions of carbohydrates;

e.  discuss the types and functions of lipids;

f.  discuss protein structure and function;

g.  explain how enzymes function;

h.  describe the structure, production, and function of ATP;

i.  identify other nucleotide types and their functions; and

j.  identify the principal types of nucleic acids.

### Carbon Compounds and Functional Groups

*Organic chemistry* is the study of compounds of carbon. By 1900, biochemists had classified the organic molecules of life into four primary categories: *carbohydrates, lipids, proteins,* and *nucleic acids.* We examine the first three in this chapter but describe the details of nucleic acids, which are concerned with genetics, in chapter 4.

Carbon is an especially versatile atom that serves as the basis of a wide variety of structures. It has four valence electrons, so it bonds with other atoms that can provide it with four more to complete its valence shell. Carbon atoms readily bond with each other and can form long chains, branched molecules, and rings—an enormous variety of **carbon backbones** for organic molecules. Carbon also commonly forms covalent bonds with hydrogen, oxygen, nitrogen, and sulfur.

Carbon backbones carry a variety of **functional groups**— small clusters of atoms that determine many of the properties of an organic molecule. For example, organic acids bear a **carboxyl** (car-BOC-sil) **group,** and ATP is named for its three **phosphate groups.** Other common functional groups include **hydroxyl, methyl,** and **amino groups** (fig. 2.14).

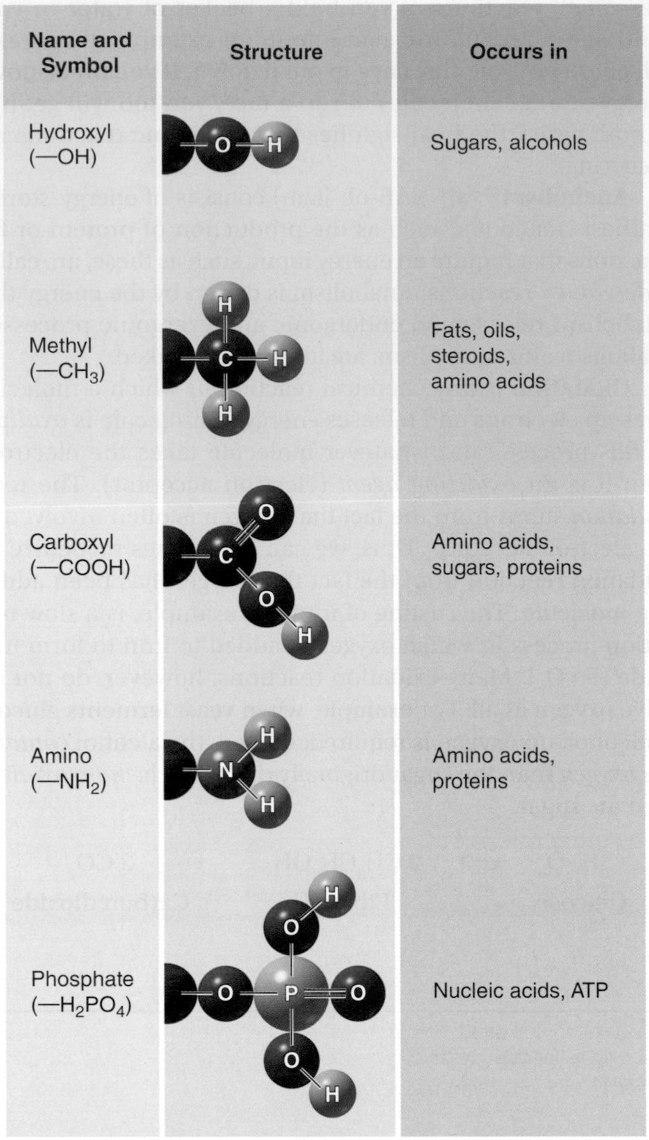

### Monomers and Polymers

Since carbon can form long chains, some organic molecules are gigantic *macromolecules* with molecular weights that range from the thousands (as in starch and proteins) to the millions (as in DNA). Most macromolecules are **polymers**[14]— molecules made of a repetitive series of identical or similar

---

[14]*poly* = many; *mer* = part

| Name and Symbol | Structure | Occurs in |
|---|---|---|
| Hydroxyl (—OH) | | Sugars, alcohols |
| Methyl (—CH₃) | | Fats, oils, steroids, amino acids |
| Carboxyl (—COOH) | | Amino acids, sugars, proteins |
| Amino (—NH₂) | | Amino acids, proteins |
| Phosphate (—H₂PO₄) | | Nucleic acids, ATP |

**FIGURE 2.14    Functional Groups of Organic Molecules.**

subunits called **monomers** (MON-oh-murs). Starch, for example, is a polymer of about 3,000 glucose monomers. In starch, the monomers are identical, whereas in other polymers they have a basic structural similarity but differ in detail. DNA, for example, is made of 4 different kinds of monomers (nucleotides), and proteins are made of 20 kinds (amino acids).

The joining of monomers to form a polymer is called *polymerization.* Living cells achieve this by means of a reaction called **dehydration synthesis (condensation)** (fig. 2.15a). A hydroxyl (–OH) group is removed from one monomer and a hydrogen (–H) from another, producing water as a by-product. The two monomers become joined by a covalent bond, forming a *dimer.* This is repeated for each monomer added to the chain, potentially leading to a chain long enough to be considered a polymer.

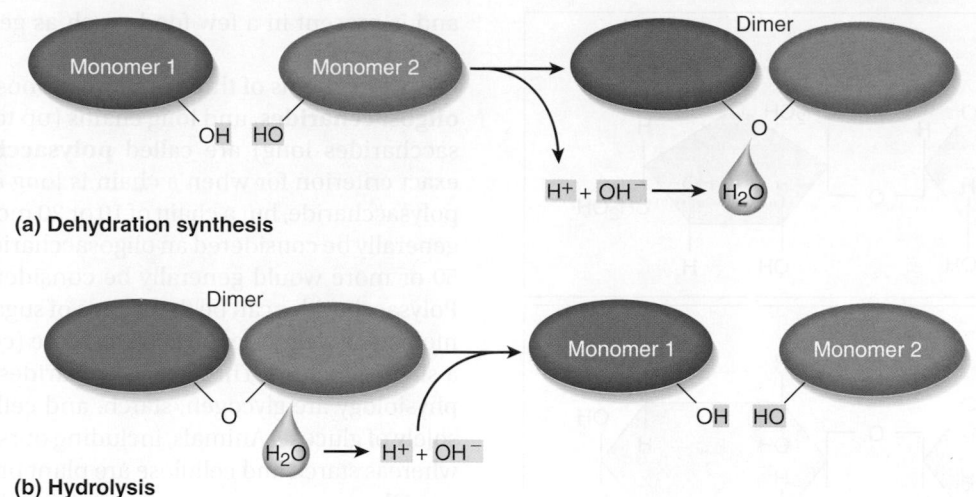

**(a) Dehydration synthesis**

**(b) Hydrolysis**

**FIGURE 2.15 Synthesis and Hydrolysis Reactions.** (a) In dehydration synthesis, a hydrogen atom is removed from one monomer and a hydroxyl group is removed from another. These combine to form water as a by-product. The monomers become joined by a covalent bond to form a dimer. (b) In hydrolysis, a covalent bond between two monomers is broken. Water donates a hydrogen atom to one monomer and a hydroxyl group to the other.

The opposite of dehydration synthesis is **hydrolysis**[15] (fig. 2.15b). In hydrolysis, a water molecule ionizes into $OH^-$ and $H^+$. A covalent bond linking one monomer to another is broken, the $OH^-$ is added to one monomer, and the $H^+$ is added to the other one. All digestion consists of hydrolysis reactions.

## Carbohydrates

A **carbohydrate**[16] is a hydrophilic organic molecule with the general formula $(CH_2O)_n$, where $n$ represents the number of carbon atoms. In glucose, for example, $n = 6$ and the formula is $C_6H_{12}O_6$. As the generic formula shows, carbohydrates have a 2:1 ratio of hydrogen to oxygen.

### ▶▶▶APPLY WHAT YOU KNOW

*Why is carbohydrate an appropriate name for this class of compounds? Relate this name to the general formula of carbohydrates.*

The names of individual carbohydrates are often built on the word root *sacchar-* or the suffix *-ose,* both of which mean "sugar" or "sweet." The most familiar carbohydrates are the sugars and starches.

The simplest carbohydrates are monomers called **monosaccharides**[17] (MON-oh-SAC-uh-rides), or simple sugars. The three of primary importance are **glucose, fructose,** and **galactose,** all with the molecular formula $C_6H_{12}O_6$; they are isomers of each other (fig. 2.16). We obtain these sugars mainly by the digestion of more complex carbohydrates. Glucose is the "blood sugar" that provides energy to most of our cells. Two other monosaccharides, ribose and deoxyribose, are important components of DNA and RNA.

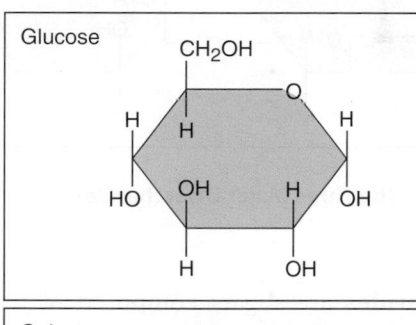

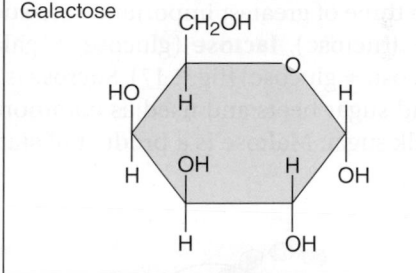

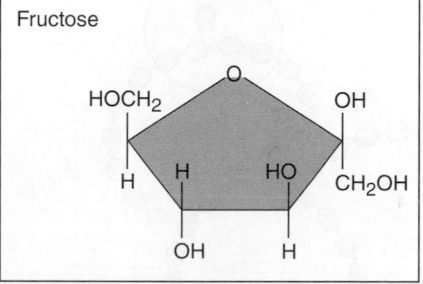

**FIGURE 2.16 The Three Major Monosaccharides.** All three have the molecular formula $C_6H_{12}O_6$. Each angle in the rings represents a carbon atom except the one where oxygen is shown. This is a conventional way of representing carbon in the structural formulae of organic compounds.

---

[15]*hydro* = water; *lysis* = splitting apart
[16]*carbo* = carbon; *hydr* = water
[17]*mono* = one; *sacchar* = sugar

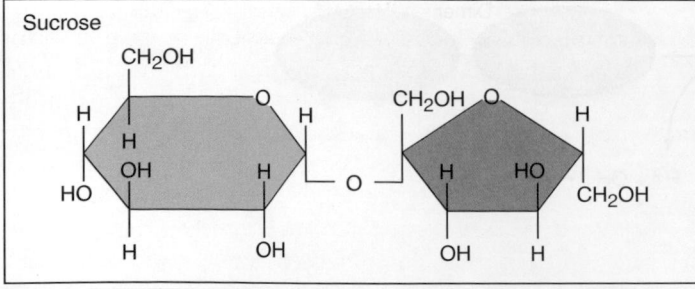

Sucrose

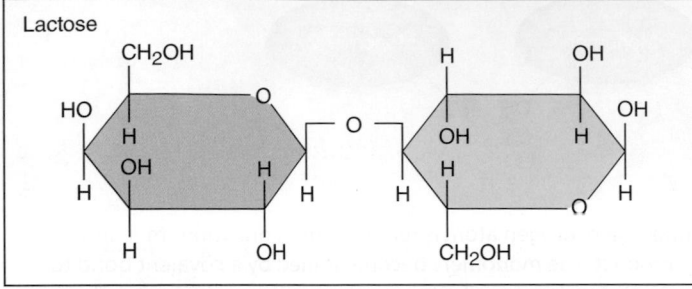

Lactose

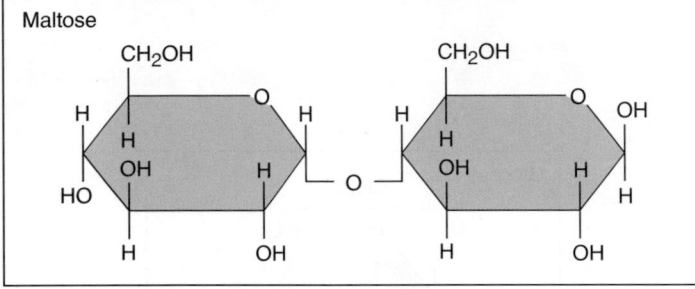

Maltose

**FIGURE 2.17   The Three Major Disaccharides.**

**Disaccharides** are sugars composed of two monosaccharides. The three of greatest importance are **sucrose** (made of glucose + fructose), **lactose** (glucose + galactose), and **maltose** (glucose + glucose) (fig. 2.17). Sucrose is produced by sugarcane and sugar beets and used as common table sugar. Lactose is milk sugar. Maltose is a product of starch digestion

and is present in a few foods such as germinating wheat and malt beverages.

Short chains of three or more monosaccharides are called **oligosaccharides,** and long chains (up to thousands of monosaccharides long) are called **polysaccharides.** There is no exact criterion for when a chain is long enough to be called a polysaccharide, but a chain of 10 or 20 monosaccharides would generally be considered an oligosaccharide, whereas a chain of 50 or more would generally be considered a polysaccharide. Polysaccharides can be thousands of sugars long and may have molecular weights of 500,000 or more (compared with 180 for a single glucose). Three polysaccharides of interest to human physiology are glycogen, starch, and cellulose—all composed solely of glucose. Animals, including ourselves, make glycogen, whereas starch and cellulose are plant products.

**Glycogen**[18] is an energy-storage polysaccharide made by cells of the liver, muscles, brain, uterus, and vagina. It is a long branched glucose polymer (fig. 2.18). The liver produces glycogen after a meal, when the blood glucose level is high, and then breaks it down between meals to maintain blood glucose levels when there is no food intake. Muscle stores glycogen for its own energy needs, and the uterus uses it in pregnancy to nourish the embryo.

**Starch** is the corresponding energy-storage polysaccharide of plants. They store it when sunlight and nutrients are available and draw from it when photosynthesis is not possible (for example, at night and in winter, when a plant has shed its leaves). Starch is the only significant digestible polysaccharide in the human diet.

**Cellulose** is a structural polysaccharide that gives strength to the cell walls of plants. It is the principal component of wood, cotton, and paper. It consists of a few thousand glucose monomers joined together, with every other monomer "upside down" relative to the next. (The $CH_2OH$ groups all face in the same direction in glycogen and starch, but alternate between facing up and down in cellulose.) Cellulose is the most

---

[18]*glyco* = sugar; *gen* = producing

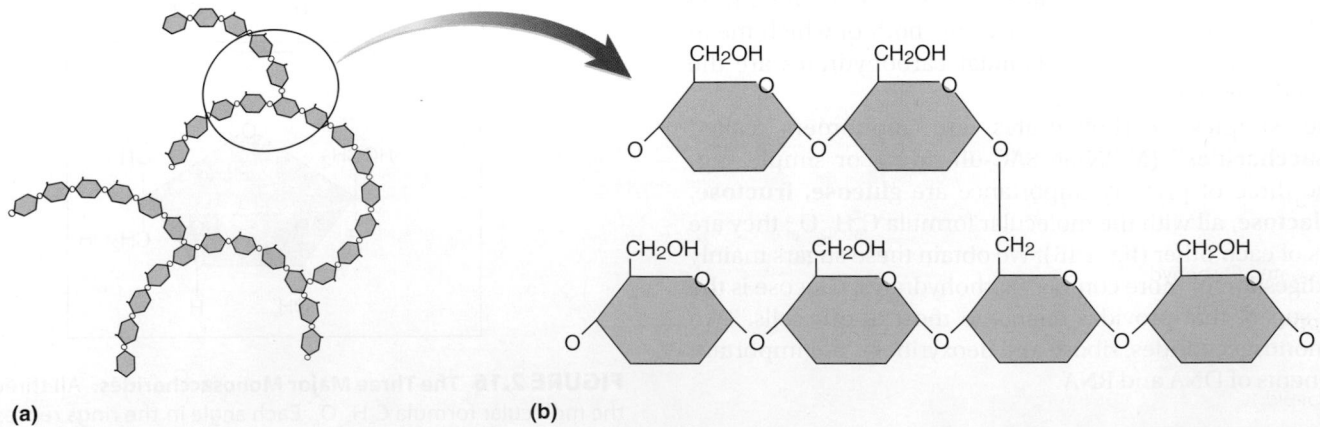

(a)                    (b)

**FIGURE 2.18   Glycogen.** This is the only polysaccharide found in human tissues. (a) Part of a glycogen molecule showing the chain of glucose monomers and branching pattern. (b) Detail of the molecule at a branchpoint.

abundant organic compound on earth and it is a common component of the diets of humans and other animals—yet we have no enzymes to digest it and thus derive no energy or nutrition from it. Nevertheless, it is important as dietary fiber ("bulk" or "roughage"). It swells with water in the digestive tract and helps move other materials through the intestine.

Carbohydrates are, above all, a source of energy that can be quickly mobilized. All digested carbohydrate is ultimately converted to glucose, and glucose is oxidized to make ATP, a high-energy compound discussed later. But carbohydrates have other functions as well (table 2.6). They are often **conjugated**[19] with (covalently bound to) proteins and lipids. Many of the lipid and protein molecules at the external surface of the cell membrane have chains of up to 12 sugars attached to them, thus forming **glycolipids** and **glycoproteins,** respectively. Among other functions, glycoproteins are a major component of mucus, which traps particles in the respiratory system, resists infection, and protects the digestive tract from its own acid and enzymes.

[19]*con* = together; *jug* = join

**Proteoglycans** are macromolecules in which the carbohydrate component is dominant and a peptide or protein forms a smaller component. Proteoglycans form gels that help hold cells and tissues together, form a gelatinous filler in the umbilical cord and eye, lubricate the joints of the skeletal system, and account for the tough rubbery texture of cartilage. Their structure and functions are further considered in chapter 5.

When discussing conjugated macromolecules, it is convenient to refer to each chemically different component as a **moiety**[20] (MOY-eh-tee). Proteoglycans have a protein moiety and a carbohydrate moiety, for example.

## Lipids

A **lipid** is a hydrophobic organic molecule, usually composed only of carbon, hydrogen, and oxygen, with a high ratio of hydrogen to oxygen. A fat called *tristearin* (tri-STEE-uh-rin), for example, has the molecular formula $C_{57}H_{110}O_6$—more than 18 hydrogens for every oxygen. Lipids are less oxidized than carbohydrates, and thus have more calories per gram. Beyond these criteria, it is difficult to generalize about lipids; they are much more variable in structure than the other macromolecules we are considering. We consider here the five primary types of lipids in humans—*fatty acids, triglycerides, phospholipids, eicosanoids,* and *steroids* (table 2.7).

A **fatty acid** is a chain of usually 4 to 24 carbon atoms with a carboxyl group at one end and a methyl group at the other. Fatty acids and the fats made from them are classified as *saturated* or *unsaturated*. A **saturated fatty acid** such as palmitic acid has as much hydrogen as it can carry. No more could be

[20]*moiet* = half

| TABLE 2.6 | Carbohydrate Functions |
|---|---|
| **Type** | **Function** |
| Monosaccharides | |
| Glucose | Blood sugar—energy source for most cells |
| Galactose | Converted to glucose and metabolized |
| Fructose | Fruit sugar—converted to glucose and metabolized |
| Disaccharides | |
| Sucrose | Cane sugar—digested to glucose and fructose |
| Lactose | Milk sugar—digested to glucose and galactose; important in infant nutrition |
| Maltose | Malt sugar—product of starch digestion, further digested to glucose |
| Polysaccharides | |
| Cellulose | Structural polysaccharide of plants; dietary fiber |
| Starch | Energy storage in plant cells |
| Glycogen | Energy storage in animal cells (liver, muscle, brain, uterus, vagina) |
| Conjugated Carbohydrates | |
| Glycoprotein | Component of the cell surface coat and mucus, among other roles |
| Glycolipid | Component of the cell surface coat |
| Proteoglycan | Cell adhesion; lubrication; supportive filler of some tissues and organs |

| TABLE 2.7 | Lipid Functions |
|---|---|
| **Type** | **Function** |
| Bile acids | Steroids that aid in fat digestion and nutrient absorption |
| Cholesterol | Component of cell membranes; precursor of other steroids |
| Eicosanoids | Chemical messengers between cells |
| Fat-soluble vitamins (A, D, E, and K) | Involved in a variety of functions including blood clotting, wound healing, vision, and calcium absorption |
| Fatty acids | Precursor of triglycerides; source of energy |
| Phospholipids | Major component of cell membranes; aid in fat digestion |
| Steroid hormones | Chemical messengers between cells |
| Triglycerides | Energy storage; thermal insulation; filling space; binding organs together; cushioning organs |

added without exceeding four covalent bonds per carbon atom; thus it is "saturated" with hydrogen. In **unsaturated fatty acids** such as linoleic acid, however, some carbon atoms are joined by double covalent bonds (fig. 2.19). Each of these could potentially share one pair of electrons with another hydrogen atom instead of the adjacent carbon, so hydrogen could be added to this molecule. **Polyunsaturated fatty acids** are those with many C=C bonds. Most fatty acids can be synthesized by the human body, but a few, called **essential fatty acids,** must be obtained from the diet because we cannot synthesize them (see chapter 26).

A **triglyceride** (try-GLISS-ur-ide) is a molecule consisting of a three-carbon alcohol called **glycerol** linked to three fatty acids; triglycerides are more correctly, although less widely, also known as *triacylglycerols.* Each bond between a fatty acid and glycerol is formed by dehydration synthesis (fig. 2.19).

Once joined to glycerol, a fatty acid can no longer donate a proton to solution and is therefore no longer an acid. For this reason, triglycerides are also called **neutral fats.** Triglycerides are broken down by hydrolysis reactions, which split each of these bonds apart by the addition of water.

Triglycerides that are liquid at room temperature are also called *oils,* but the difference between a fat and oil is fairly arbitrary. Coconut oil, for example, is solid at room temperature. Animal fats are usually made of saturated fatty acids, so they are called *saturated fats.* They are solid at room or body temperature. Most plant triglycerides are *polyunsaturated fats,* which generally remain liquid at room temperature. Examples include peanut, olive, corn, and linseed oils. Saturated fats contribute more to cardiovascular disease than unsaturated fats, and for this reason it is healthier to cook with vegetable oils than with lard, bacon fat, or butter (see Deeper Insight 2.3).

## DEEPER INSIGHT 2.3

## CLINICAL APPLICATION

### Trans *Fats and Cardiovascular Health*

There has been a great deal of public interest lately in *trans* fats and cardiovascular health, with some U.S. states even regulating or banning the use of *trans* fats in restaurants. What, then, are *trans* fats and why have they gotten such a bad reputation?

A *trans* fat is a triglyceride containing one or more **trans**-fatty acids. In such fatty acids, there is at least one unsaturated C=C double bond. On each side of the C=C bond, the single covalent C—C bonds angle in opposite directions (*trans* means "across from") like a pair of bicycle pedals (see arrows in figure 2.20a). This is in contrast to *cis*-fatty acids, in which the two C—C bonds adjacent to the C=C bond both angle in the same direction (*cis* means "on the same side"), as indicated by the arrows in figure 2.20b. As you can see, the *cis* configuration creates a kink in the chain, whereas the *trans* configuration results in a relatively straight chain.

In the early 1900s, the food industry developed a method for converting *cis*-fatty acids to *trans*-fatty acids. Fats with straight-chain *trans*-fatty acids pack more densely together and consequently are solids at room temperature. This has several advantages for cooking. Solids such as vegetable shortening are easier to use in making such baked goods as pie crusts and biscuits; they have a longer shelf life; and vegetable products are more acceptable than animal fat to vegetarians. *Trans* fats are now used abundantly in snack foods, baked goods, fast foods such as french fries, and many other foods. *Trans* fats constitute about 30% of the fat in shortening, but only 4% of the animal fat in butter.

The disadvantage of trans fats is that they resist enzymatic breakdown in the human body, remain in circulation longer, and have more tendency to deposit in the arteries than saturated and cis-unsaturated fats do. Therefore, they raise the risk of coronary heart disease (CHD). From 1980 to 1994, medical scientists tracked a cohort of 80,082 female nurses (the Nurses' Health Study II) and, among other things, correlated their incidence of CHD with their self-reported diets. They concluded that for every 2% increase in calories from trans fats as compared to carbohydrate calories, the women had a 93% higher risk of CHD.

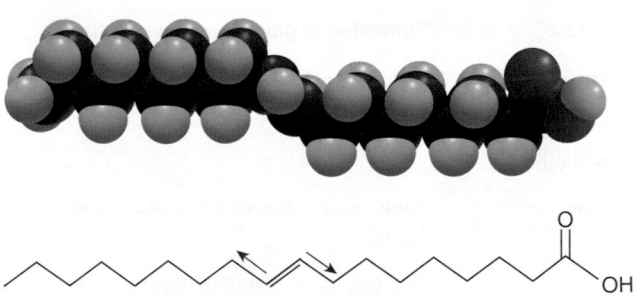

**(a) A *trans*-fatty acid (elaidic acid)**

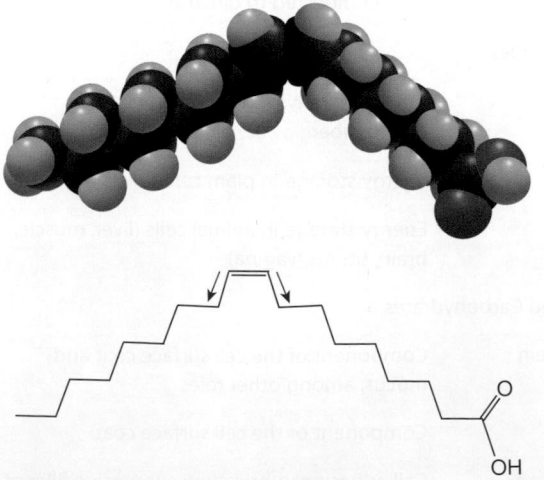

**FIGURE 2.20** *Trans- and Cis-*Fatty Acids. Each example is unsaturated at the C=C double bond in red. (a) In a *trans*-fatty acid, the bonds on opposite sides of the C=C bond angle in opposite directions (arrows). (b) In a *cis*-fatty acid, the bonds on opposite sides of the C=C bond angle in the same direction. The straight chains of *trans*-fatty acids allow fat molecules to pack more tightly together, thus to remain solid (greasy) at room temperature. Oleic acid melts at 13.5°C (56.3°F), whereas elaidic acid melts at 46.5°C (115.7°F).

**(b) A *cis*-fatty acid (oleic acid)**

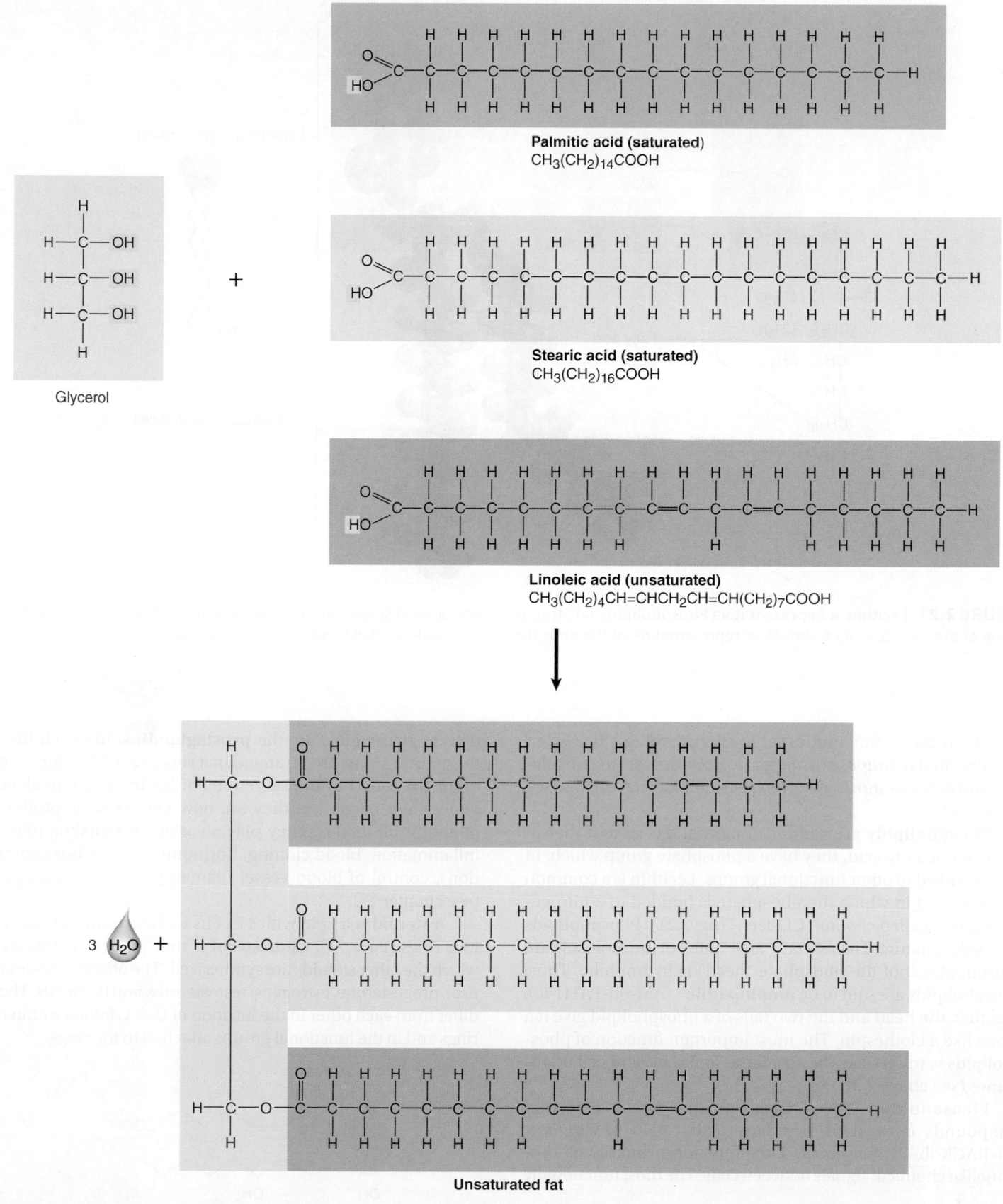

**Palmitic acid (saturated)**
CH$_3$(CH$_2$)$_{14}$COOH

**Stearic acid (saturated)**
CH$_3$(CH$_2$)$_{16}$COOH

**Linoleic acid (unsaturated)**
CH$_3$(CH$_2$)$_4$CH=CHCH$_2$CH=CH(CH$_2$)$_7$COOH

Glycerol

**Unsaturated fat**

**FIGURE 2.19 Triglyceride (Fat) Synthesis.** Note the difference between saturated and unsaturated fatty acids and the production of 3 H$_2$O as a by-product of this dehydration synthesis reaction.

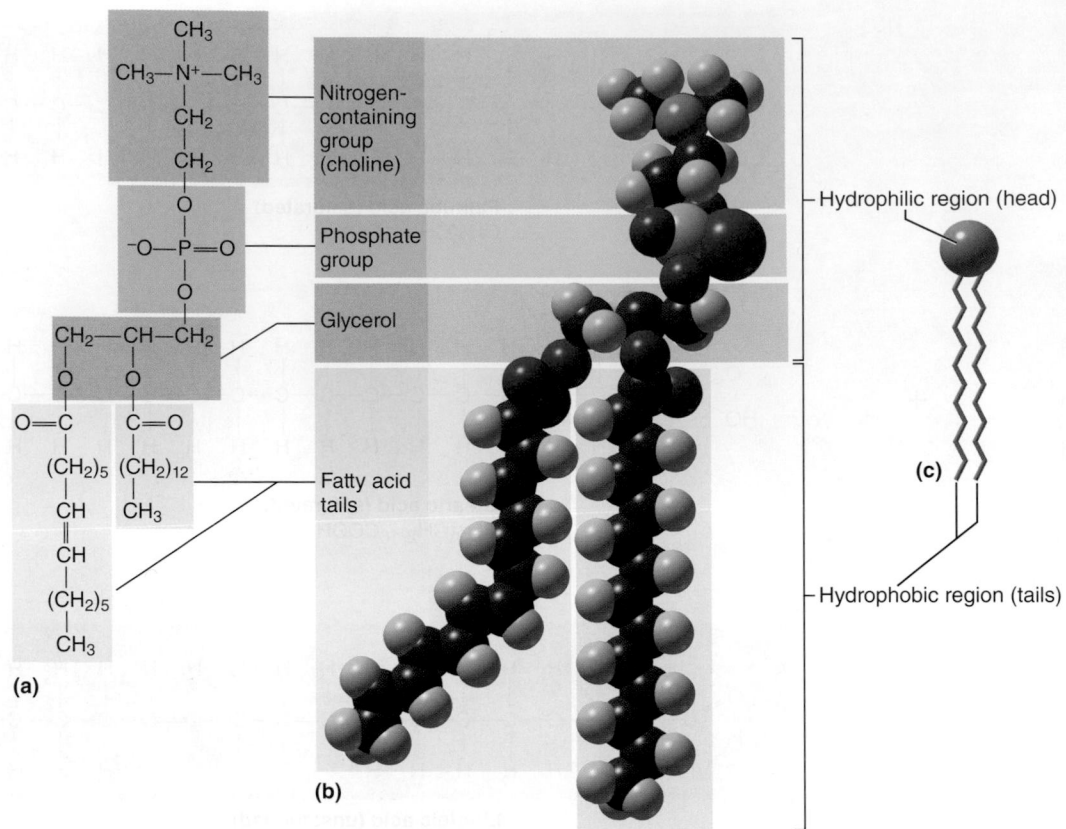

**FIGURE 2.21  Lecithin, a Representative Phospholipid.**  (a) Structural formula. (b) A space-filling model that gives some idea of the actual shape of the molecule. (c) A simplified representation of the phospholipid molecule used in diagrams of cell membranes.

The primary function of fat is energy storage, but when concentrated in *adipose tissue,* it also provides thermal insulation and acts as a shock-absorbing cushion for vital organs (see chapter 5).

**Phospholipids** are similar to neutral fats except that in place of one fatty acid, they have a phosphate group which, in turn, is linked to other functional groups. Lecithin is a common phospholipid in which the phosphate is bonded to a nitrogenous group called *choline* (CO-leen) (fig. 2.21). Phospholipids have a dual nature. The two fatty acid "tails" of the molecule are hydrophobic, but the phosphate "head" is hydrophilic. Thus, phospholipids are said to be **amphipathic**[21] (AM-fih-PATH-ic). Together, the head and the two tails of a phospholipid give it a shape like a clothespin. The most important function of phospholipids is to serve as the structural foundation of cell membranes (see chapter 3).

**Eicosanoids**[22] (eye-CO-sah-noyds) are 20-carbon compounds derived from a fatty acid called *arachidonic* (ah-RACK-ih-DON-ic) *acid.* They function primarily as hormonelike chemical signals between cells. The most functionally diverse eicosanoids are the **prostaglandins,** in which five of the carbon atoms are arranged in a ring (fig. 2.22). These were originally found in the secretions of bovine prostate glands, hence their name, but they are now known to be produced in almost all tissues. They play a variety of signaling roles in inflammation, blood clotting, hormone action, labor contractions, control of blood vessel diameter, and other processes (see chapter 17).

A **steroid** is a lipid with 17 of its carbon atoms arranged in four rings (fig. 2.23). **Cholesterol** is the "parent" steroid from which the other steroids are synthesized. The others include cortisol, progesterone, estrogens, testosterone, and bile acids. These differ from each other in the location of C=C bonds within the rings and in the functional groups attached to the rings.

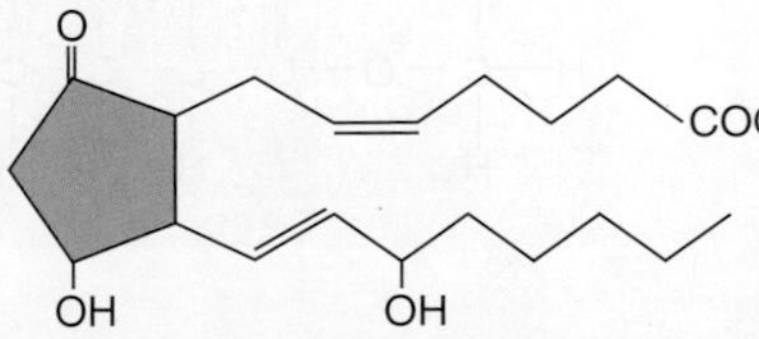

**FIGURE 2.22  Prostaglandin.**  This is a modified fatty acid with five of its carbon atoms arranged in a ring.

---

[21]*amphi* = both; *pathic* = feeling
[22]*eicosa* = 20

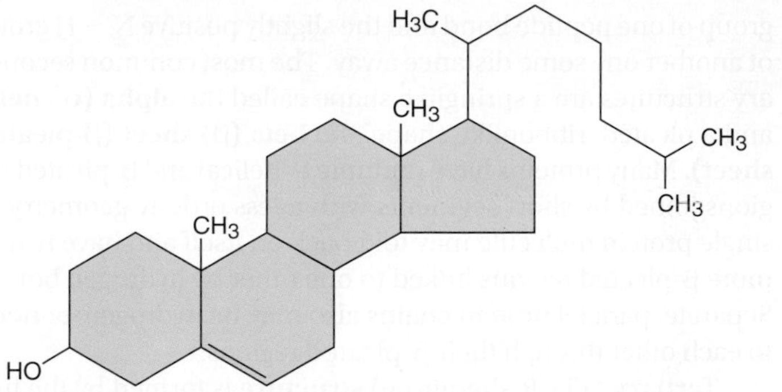

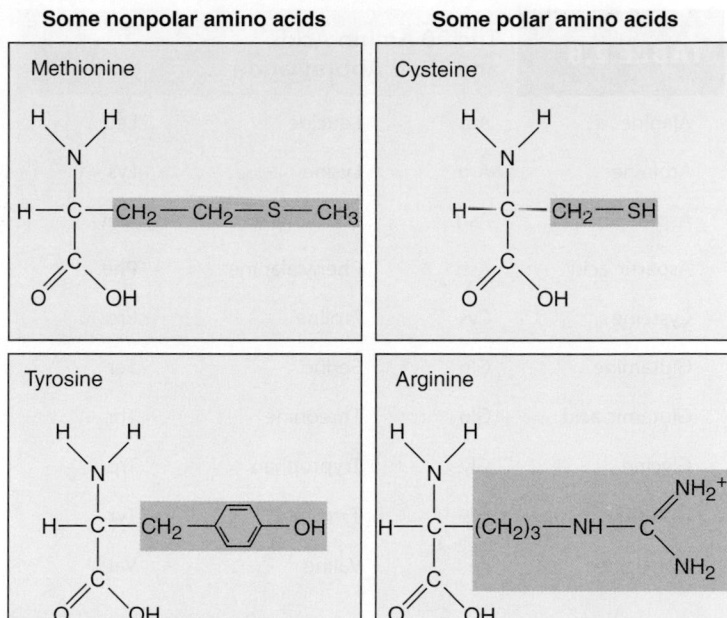

FIGURE 2.23 **Cholesterol.** All steroids have this basic four-ringed structure, with variations in the functional groups and locations of double bonds within the rings.

 **DEEPER INSIGHT 2.4**

## CLINICAL APPLICATION

### *"Good" and "Bad" Cholesterol*

There is only one kind of cholesterol, and it does far more good than harm. When the popular press refers to "good" and "bad" cholesterol, it is actually referring to droplets in the blood called *lipoproteins,* which are a complex of cholesterol, fat, phospholipids, and protein. So-called bad cholesterol refers to low-density lipoprotein (LDL), which has a high ratio of lipid to protein and contributes to cardiovascular disease. So-called good cholesterol refers to high-density lipoprotein (HDL), which has a lower ratio of lipid to protein and may help to prevent cardiovascular disease.

Even when food products are advertised as cholesterol-free, they may be high in saturated fat, which stimulates the body to produce more cholesterol. Palmitic acid seems to be the greatest culprit in stimulating elevated cholesterol levels, while linoleic acid has a cholesterol-lowering effect. Both are shown in figure 2.19. Cardiovascular disease is further discussed at the end of chapter 19, and LDLs and HDLs are more fully explained in chapter 26.

We obtain dietary cholesterol only from foods of animal origin; plants make only trace amounts of no dietary importance. The average adult contains over 200 g (half a pound) of cholesterol. Cholesterol has a bad reputation as a factor in cardiovascular disease (see Deeper Insight 2.4), and it is true that hereditary and dietary factors can elevate blood cholesterol to dangerously high levels. Nevertheless, cholesterol is a natural product of the body and is necessary for human health. In addition to being the precursor of other steroids, it is an important component of cell membranes and is required for proper nervous system function. Only about 15% of our cholesterol comes from the diet; the other 85% is internally synthesized, primarily by the liver.

The principal lipids and their functions are summarized in table 2.7 (p. 61).

## Proteins

The word *protein* is derived from the Greek word *proteios,* meaning "of first importance." Proteins are the most versatile molecules in the body, and many discussions in this book will draw on your understanding of protein structure and behavior.

FIGURE 2.24 **Amino Acids and Peptides.** (a) Four representative amino acids. Note that they differ only in the R group, shaded in pink. (b) The joining of two amino acids by a peptide bond, forming a dipeptide. Side groups $R_1$ and $R_2$ could be the groups indicated in pink in part (a), among other possibilities.

## Amino Acids and Peptides

A **protein** is a polymer of **amino acids.** An amino acid has a central carbon atom with an amino ($-NH_2$) and a carboxyl ($-COOH$) group bound to it (fig. 2.24a). The 20 amino acids used to make proteins are identical except for a third functional group called the *radical* (R group) attached to the central carbon. In the simplest amino acid, glycine, R is merely a hydrogen atom, whereas in the largest amino acids it includes rings of carbon. Some radicals are hydrophilic and some are hydrophobic.

| TABLE 2.8 | The 20 Amino Acids and Their Abbreviation | | |
|---|---|---|---|
| Alanine | Ala | Leucine | Leu |
| Arginine | Arg | Lysine | Lys |
| Asparagine | Asn | Methionine | Met |
| Aspartic acid | Asp | Phenylalanine | Phe |
| Cysteine | Cys | Proline | Pro |
| Glutamine | Gln | Serine | Ser |
| Glutamic acid | Glu | Threonine | Thr |
| Glycine | Gly | Tryptophan | Trp |
| Histidine | His | Tyrosine | Tyr |
| Isoleucine | Ile | Valine | Val |

Being composed of many amino acids, proteins as a whole are therefore often amphiphilic. The 20 amino acids involved in proteins are listed in table 2.8 along with their abbreviations.

A **peptide** is any molecule composed of two or more amino acids joined by **peptide bonds.** A peptide bond, formed by dehydration synthesis, joins the amino group of one amino acid to the carboxyl group of the next (fig. 2.24b). Peptides are named for the number of amino acids they have—for example, dipeptides have two and tripeptides have three. Chains of fewer than 10 or 15 amino acids are called **oligopeptides,**[23] and chains larger than that are called **polypeptides.** An example of an oligopeptide is the childbirth-inducing hormone oxytocin, composed of 9 amino acids. A representative polypeptide is adrenocorticotropic hormone (ACTH), which is 39 amino acids long. A protein is a polypeptide of 50 amino acids or more. A typical amino acid has a molecular weight of about 80 amu, and the molecular weights of the smallest proteins are around 4,000 to 8,000 amu. The average protein weighs in at about 30,000 amu, and some of them have molecular weights in the hundreds of thousands.

## Protein Structure

Proteins have complex coiled and folded structures that are critically important to the roles they play. Even slight changes in their **conformation** (three-dimensional shape) can destroy protein function. Protein molecules have three to four levels of complexity, from primary through quaternary structure (fig. 2.25).

**Primary structure** is the protein's sequence of amino acids, which is encoded in the genes (see chapter 4).

**Secondary structure** is a coiled or folded shape held together by hydrogen bonds between the slightly negative C=O

group of one peptide bond and the slightly positive N—H group of another one some distance away. The most common secondary structures are a springlike shape called the **alpha (α) helix** and a pleated, ribbonlike shape, the **beta (β) sheet (β-pleated sheet).** Many proteins have multiple α-helical and β-pleated regions joined by short segments with a less orderly geometry. A single protein molecule may fold back on itself and have two or more β-pleated regions linked to one other by hydrogen bonds. Separate, parallel protein chains also may be hydrogen-bonded to each other through their β-pleated regions.

**Tertiary**[24] (TUR-she-air-ee) **structure** is formed by the further bending and folding of proteins into various globular and fibrous shapes. It results from hydrophobic radicals associating with each other and avoiding water, while the hydrophilic radicals are attracted to the surrounding water. Van der Waals forces play a significant role in stabilizing tertiary structure. *Globular proteins*, somewhat resembling a wadded ball of yarn, have a compact tertiary structure well suited for proteins embedded in cell membranes and proteins that must move around freely in the body fluids, such as enzymes and antibodies. *Fibrous proteins* such as myosin, keratin, and collagen are slender filaments better suited for such roles as muscle contraction and providing strength to skin, hair, and tendons.

The amino acid cysteine (Cys), whose radical is —CH$_2$—SH (see fig. 2.24a), often stabilizes a protein's tertiary structure by forming covalent **disulfide bridges.** When two cysteines align with each other, each can release a hydrogen atom, leaving the sulfur atoms to form a disulfide (—S—S—) bridge. Disulfide bridges hold separate polypeptide chains together in such molecules as antibodies and insulin (fig. 2.26).

**Quaternary**[25] (QUA-tur-nare-ee) **structure** is the association of two or more polypeptide chains by noncovalent forces such as ionic bonds and hydrophilic–hydrophobic interactions. It occurs in only some proteins. Hemoglobin, for example, consists of four polypeptides: two identical alpha chains and two identical, slightly longer beta chains (see fig. 2.25).

One of the most important properties of proteins is their ability to change conformation, especially tertiary structure. This can be triggered by such influences as voltage changes on a cell membrane during the action of nerve cells, the binding of a hormone to a protein, or the dissociation of a molecule from a protein. Subtle, reversible changes in conformation are important to processes such as enzyme function, muscle contraction, and the opening and closing of pores in cell membranes. **Denaturation** is a more drastic conformational change in response to conditions such as extreme heat or pH. It is seen, for example, when you cook an egg and the egg white (albumen) turns from clear and runny to opaque and stiff. Denaturation makes a protein unable to perform its normal function. It is sometimes reversible, but more often it permanently destroys protein function.

---

[23]*oligo* = a few

[24]*tert* = third
[25]*quater* = fourth
[26]*prosthe* = appendage, addition

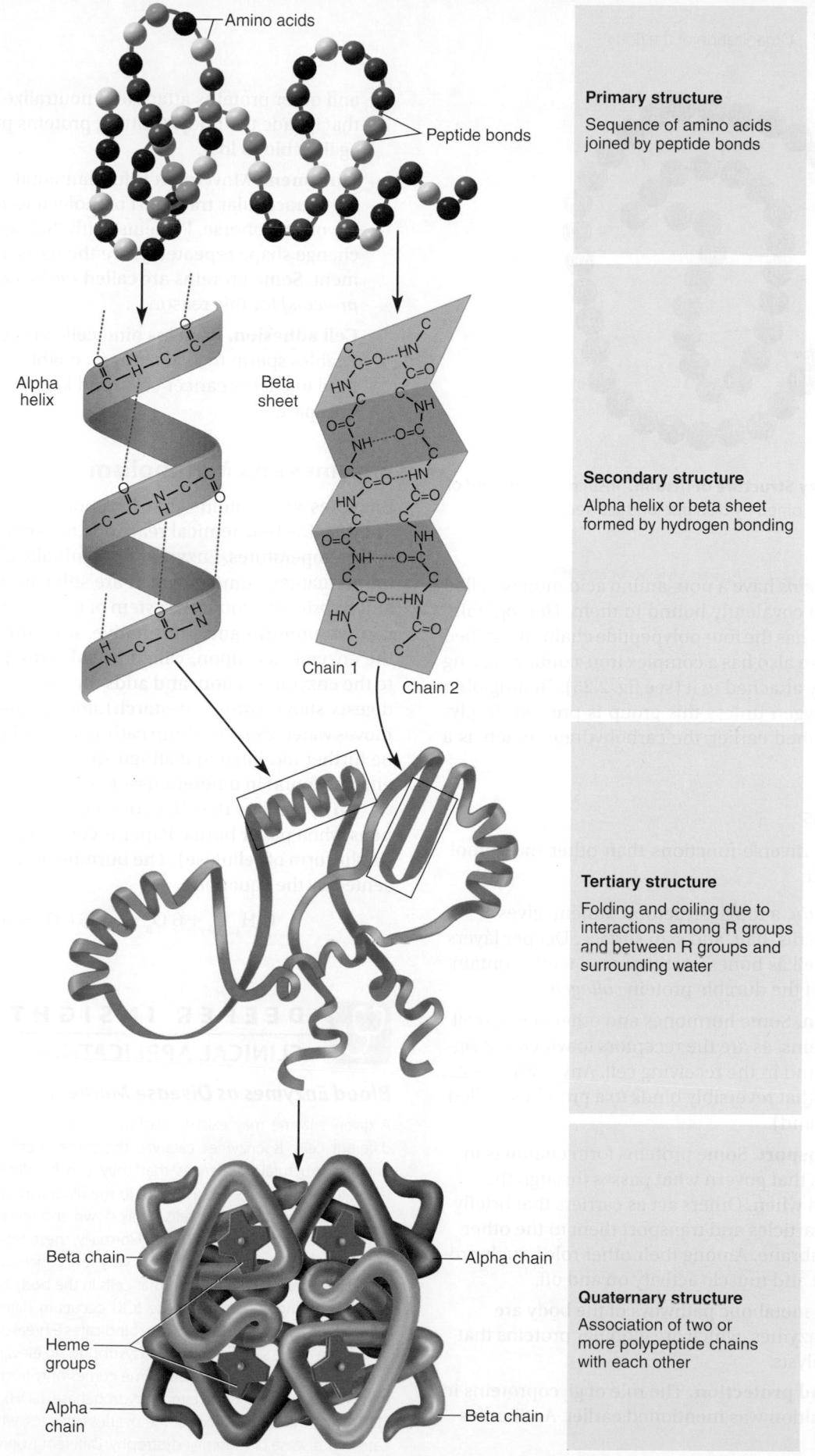

**Primary structure**

Sequence of amino acids joined by peptide bonds

**Secondary structure**

Alpha helix or beta sheet formed by hydrogen bonding

**Tertiary structure**

Folding and coiling due to interactions among R groups and between R groups and surrounding water

**Quaternary structure**

Association of two or more polypeptide chains with each other

Amino acids

Peptide bonds

Alpha helix

Beta sheet

Chain 1    Chain 2

Beta chain — — Alpha chain

Heme groups

Alpha chain — — Beta chain

**FIGURE 2.25 Four Levels of Protein Structure.** The molecule shown for quaternary structure is hemoglobin, which is composed of four polypeptide chains. The heme groups are iron-containing nonprotein moieties.

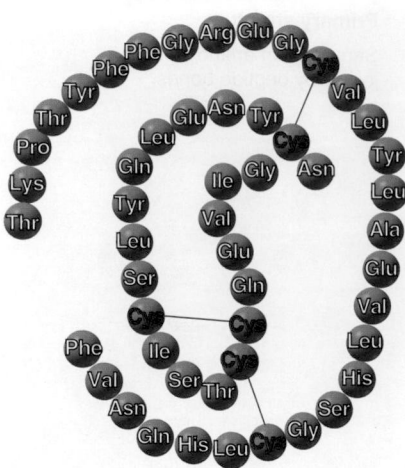

**FIGURE 2.26 Primary Structure of Insulin.** Insulin is composed of two polypeptide chains joined by disulfide bridges (red).

*Conjugated proteins* have a non–amino acid moiety called a **prosthetic**[26] **group** covalently bound to them. Hemoglobin, for example, not only has the four polypeptide chains described earlier, but each chain also has a complex iron-containing ring called a *heme* moiety attached to it (see fig. 2.25). Hemoglobin cannot transport oxygen unless this group is present. In glycoproteins, as described earlier, the carbohydrate moiety is a prosthetic group.

## Protein Functions

Proteins have more diverse functions than other macromolecules. These include

- **Structure.** *Keratin,* a tough structural protein, gives strength to the nails, hair, and skin surface. Deeper layers of the skin, as well as bones, cartilage, and teeth, contain an abundance of the durable protein *collagen.*

- **Communication.** Some hormones and other cell-to-cell signals are proteins, as are the receptors to which the signal molecules bind in the receiving cell. Any hormone or other molecule that reversibly binds to a protein is called a **ligand**[27] (LIG-and).

- **Membrane transport.** Some proteins form channels in cell membranes that govern what passes through the membranes and when. Others act as carriers that briefly bind to solute particles and transport them to the other side of the membrane. Among their other roles, such proteins turn nerve and muscle activity on and off.

- **Catalysis.** Most metabolic pathways of the body are controlled by enzymes, which are globular proteins that function as catalysts.

- **Recognition and protection.** The role of glycoproteins in immune recognition was mentioned earlier. Antibodies

and other proteins attack and neutralize organisms that invade the body. Clotting proteins protect the body against blood loss.

- **Movement.** Movement is fundamental to all life, from the intracellular transport of molecules to the galloping of a racehorse. Proteins, with their special ability to change shape repeatedly, are the basis for all such movement. Some proteins are called *molecular motors (motor proteins)* for this reason.

- **Cell adhesion.** Proteins bind cells to each other, which enables sperm to fertilize eggs, enables immune cells to bind to enemy cancer cells, and keeps tissues from falling apart.

## Enzymes and Metabolism

**Enzymes** are proteins that function as biological catalysts. They enable biochemical reactions to occur rapidly at normal body temperatures. Enzymes were initially given somewhat arbitrary names, some of which are still with us, such as *pepsin* and *trypsin.* The modern system of naming enzymes, however, is more uniform and informative. It identifies the substance the enzyme acts upon, called its **substrate;** sometimes refers to the enzyme's action; and adds the suffix *-ase.* Thus, *amylase* digests starch (*amyl-* = starch) and *carbonic anhydrase* removes water (*anhydr-*) from carbonic acid. Enzyme names may be further modified to distinguish different forms of the same enzyme found in different tissues (see Deeper Insight 2.5).

To appreciate the effect of an enzyme, think of what happens when paper burns. Paper is composed mainly of glucose (in the form of cellulose). The burning of glucose can be represented by the equation

$$C_6H_{12}O_6 + 6\ O_2 \longrightarrow 6\ CO_2 + 6\ H_2O$$

 **DEEPER INSIGHT 2.5**

### CLINICAL APPLICATION

#### Blood Enzymes as Disease Markers

A given enzyme may exist in slightly different forms, called *isoenzymes,* in different cells. Isoenzymes catalyze the same chemical reactions but have enough structural differences that they can be distinguished by standard laboratory techniques. This is useful in the diagnosis of disease. When organs are diseased, some of their cells break down and release specific isoenzymes that can be detected in the blood. Normally, these isoenzymes would not be present in the blood or would have very low concentrations. An elevation in blood levels can help pinpoint what cells in the body have been damaged.

For example, creatine kinase (CK) occurs in different forms in different cells. An elevated serum level of CK-1 indicates a breakdown of skeletal muscle and is one of the signs of muscular dystrophy. An elevated CK-2 level indicates heart disease, because this isoenzyme comes only from cardiac muscle. There are five isoenzymes of lactate dehydrogenase (LDH). High serum levels of LDH-1 may indicate a tumor of the ovaries or testes, whereas LDH-5 may indicate liver disease or muscular dystrophy. Different isoenzymes of phosphatase in the blood may indicate bone or prostate disease.

[27]*lig* = to bind

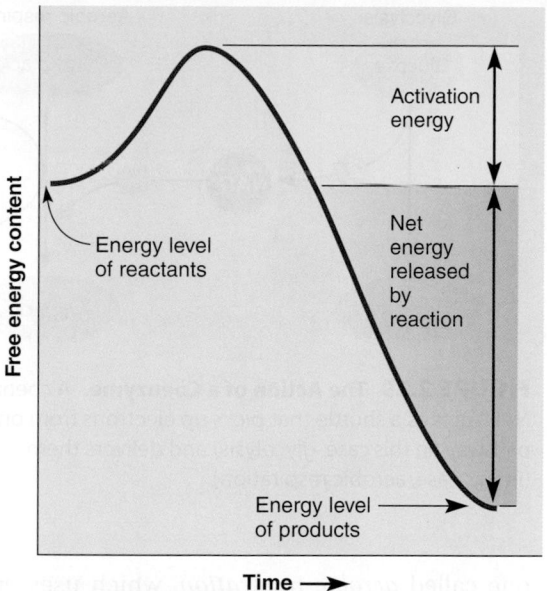
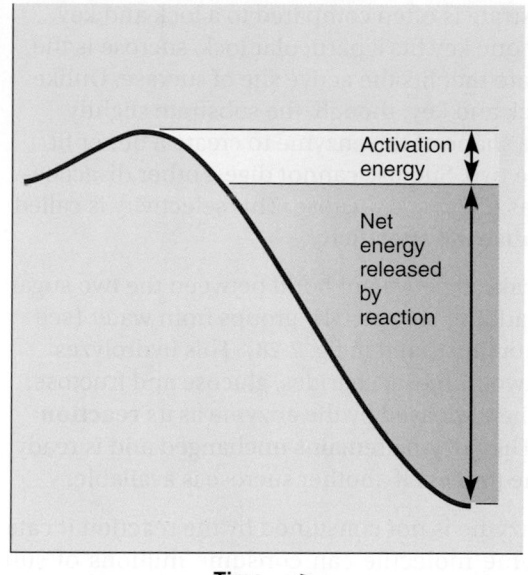

**FIGURE 2.27  Effect of an Enzyme on Activation Energy.**  (a) Without catalysts, some chemical reactions proceed slowly because of the high activation energy needed to get molecules to react. (b) A catalyst facilitates molecular interaction, thus lowering the activation energy and making the reaction proceed  more rapidly.

❓ *Does an enzyme release more energy from its substrate than an uncatalyzed reaction would release?*

Paper does not spontaneously burst into flame because few of its molecules have enough kinetic energy to react. Lighting the paper with a match, however, raises the kinetic energy enough to initiate combustion (rapid oxidation). The energy needed to get the reaction started, supplied by the match, is called the **activation energy** (fig. 2.27a).

In the body, we carry out the same reaction and oxidize glucose to water and carbon dioxide to extract its energy. We could not tolerate the heat of combustion in our bodies, however, so we must oxidize glucose in a more controlled way at a biologically feasible and safe temperature. Enzymes make this happen by lowering the activation energy—that is, by reducing the barrier to glucose oxidation (fig. 2.27b)—and by releasing the energy in small steps rather than a single burst of heat.

## Enzyme Structure and Action

Figure 2.28 illustrates the action of an enzyme, using the example of *sucrase,* an enzyme that breaks sucrose down to glucose and fructose. The process occurs in three principal steps:

① A substrate molecule approaches a pocket on the enzyme surface called the **active site.** Amino acid side groups in this region of the enzyme are arranged so as to bind functional groups on the substrate molecule. Many enzymes have two active sites, enabling them to bind two different substrates and bring them together in a way that makes them react more readily with each other.

② The substrate binds to the enzyme, forming an **enzyme-substrate complex.** The fit between a particular enzyme

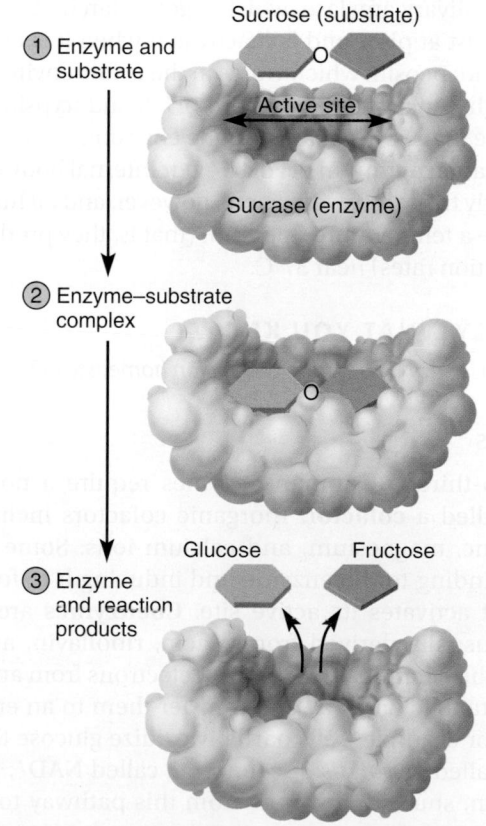

**FIGURE 2.28  The Three Steps of an Enzymatic Reaction.**

and its substrate is often compared to a lock and key. Just as only one key fits a particular lock, sucrose is the only substrate that fits the active site of sucrase. Unlike a simple lock and key, though, the substrate slightly changes the shape of the enzyme to create a better fit between the two. Sucrase cannot digest other disaccharides such as maltose or lactose. This selectivity is called **enzyme–substrate specificity.**

③ Sucrase breaks the covalent bond between the two sugars of sucrose, adding $H^+$ and $OH^-$ groups from water (see fig. 2.15b; not illustrated in fig. 2.28). This hydrolyzes sucrose to two monosaccharides, glucose and fructose, which are then released by the enzyme as its **reaction products.** The enzyme remains unchanged and is ready to repeat the process if another sucrose is available.

Since an enzyme is not consumed by the reaction it catalyzes, one enzyme molecule can consume millions of substrate molecules, and at astonishing speed. A single molecule of carbonic anhydrase, for example, breaks carbonic acid ($H_2CO_3$) down to $H_2O$ and $CO_2$ at a rate of 36 million molecules per minute.

Factors that change the shape of an enzyme—notably temperature and pH—tend to alter or destroy the ability of the enzyme to bind its substrate. They disrupt the hydrogen bonds and other weak forces that hold the enzyme in its proper conformation, essentially changing the shape of the "lock" (active site) so that the "key" (substrate) no longer fits. Enzymes vary in optimum pH according to where in the body they normally function. Thus salivary amylase, which digests starch in the mouth, functions best at pH 7 and is inactivated when it is exposed to stomach acid; pepsin, which works in the acidic environment of the stomach, functions best around pH 2; and trypsin, a digestive enzyme that works in the alkaline environment of the small intestine, has an optimum pH of 9.5. Our internal body temperature is nearly the same everywhere, however, and all human enzymes have a temperature optimum (that is, they produce their fastest reaction rates) near 37°C.

#### ▶▶▶ APPLY WHAT YOU KNOW

*Why does enzyme function depend on homeostasis?*

## Cofactors

About two-thirds of human enzymes require a nonprotein partner called a **cofactor.** Inorganic cofactors include iron, copper, zinc, magnesium, and calcium ions. Some of these work by binding to the enzyme and inducing it to fold into a shape that activates its active site. **Coenzymes** are organic cofactors usually derived from niacin, riboflavin, and other water-soluble vitamins. They accept electrons from an enzyme in one metabolic pathway and transfer them to an enzyme in another. For example, cells partially oxidize glucose through a pathway called *glycolysis.* A coenzyme called $NAD^+$,[28] derived from niacin, shuttles electrons from this pathway to another

---

[28]nicotinamide adenine dinucleotide

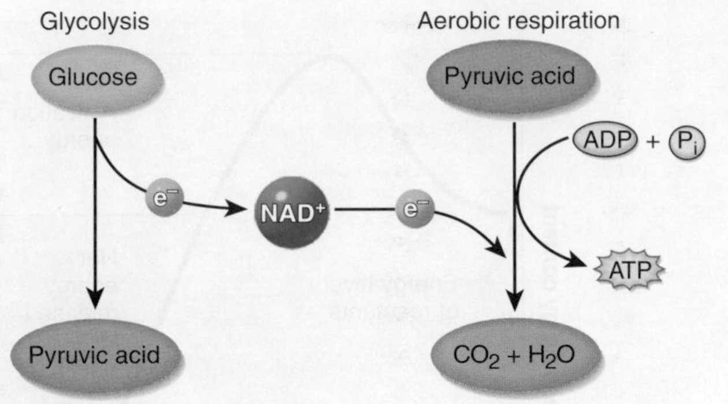

**FIGURE 2.29  The Action of a Coenzyme.**  A coenzyme such as $NAD^+$ acts as a shuttle that picks up electrons from one metabolic pathway (in this case, glycolysis) and delivers them to another (in this case, aerobic respiration).

one called *aerobic respiration,* which uses energy from the electrons to make ATP (fig. 2.29). If $NAD^+$ is unavailable, the glycolysis pathway shuts down.

## Metabolic Pathways

A **metabolic pathway** is a chain of reactions with each step usually catalyzed by a different enzyme. A simple metabolic pathway can be symbolized

$$A \xrightarrow{\alpha} B \xrightarrow{\beta} C \xrightarrow{\gamma} D$$

where A is the initial *reactant,* B and C are *intermediates,* and D is the *end product.* The Greek letters above the reaction arrows represent enzymes that catalyze each step of the reaction. A is the substrate for enzyme $\alpha$, B is the substrate for enzyme $\beta$, and C for enzyme $\gamma$. Such a pathway can be turned on or off by altering the conformation of any of these enzymes, thereby activating or deactivating them. This can be done by such means as the binding or dissociation of a cofactor, or by an end product of the pathway binding to an enzyme at an earlier step (product D binding to enzyme $\alpha$ and shutting off the reaction chain at that step, for example). In these and other ways, cells are able to turn on metabolic pathways when their end products are needed and shut them down when the end products are not needed.

## ATP, Other Nucleotides, and Nucleic Acids

**Nucleotides** are organic compounds with three principal components: a single or double carbon–nitrogen ring called a *nitrogenous base,* a monosaccharide, and one or more phosphate groups. One of the best-known nucleotides is ATP (fig. 2.30a), in which the nitrogenous base is a double ring called *adenine,* the sugar is *ribose,* and there are three phosphate groups.

### Adenosine Triphosphate

**Adenosine triphosphate (ATP)** is the body's most important energy-transfer molecule. It briefly stores energy gained from exergonic reactions such as glucose oxidation and releases it

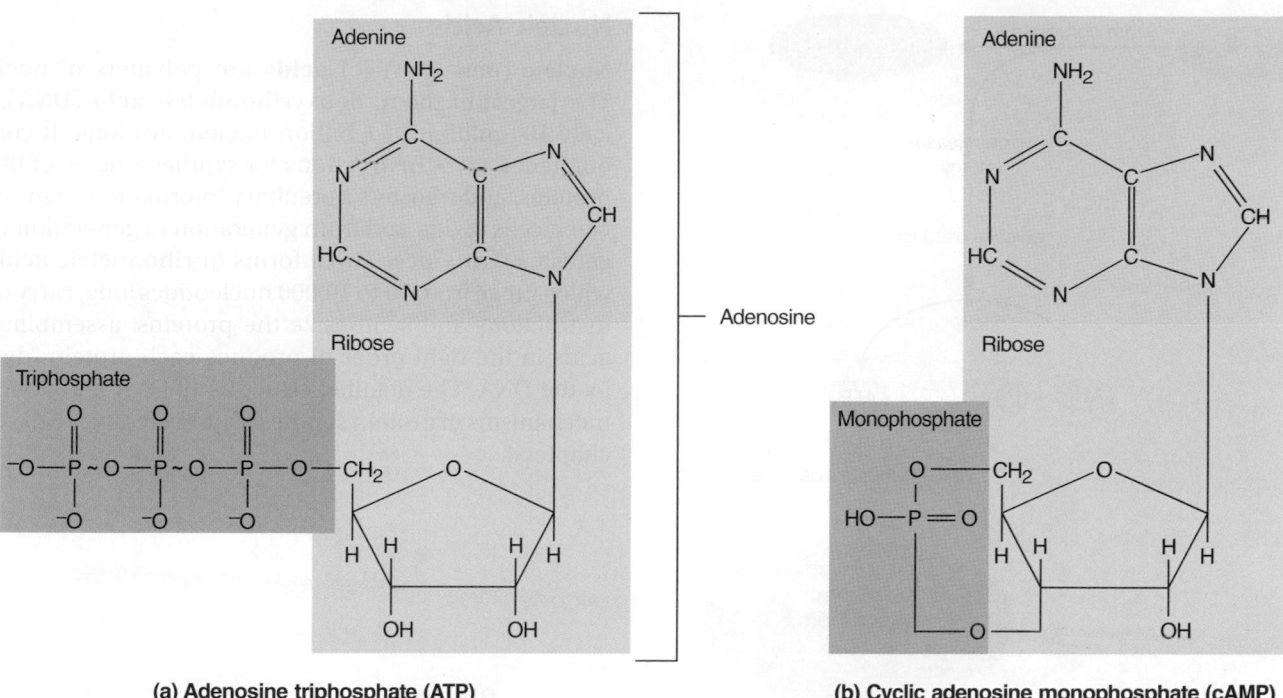

(a) Adenosine triphosphate (ATP)

(b) Cyclic adenosine monophosphate (cAMP)

**FIGURE 2.30 Adenosine Triphosphate (ATP) and Cyclic Adenosine Monophosphate (cAMP).** The last two P~O bonds in ATP, indicated by wavy lines, are high-energy bonds.

within seconds for physiological work such as polymerization reactions, muscle contraction, and pumping ions through cell membranes. The second and third phosphate groups of ATP are attached to the rest of the molecule by high-energy covalent bonds traditionally indicated by a wavy line (~) in the structural formula. Since phosphate groups are negatively charged, they repel each other. It requires a high-energy bond to overcome that repellent force and hold them together—especially to add the third phosphate group to a chain that already has two negatively charged phosphates. Most energy transfers to and from ATP involve adding or removing that third phosphate.

Enzymes called **adenosine triphosphatases (ATPases)** are specialized to hydrolyze the third phosphate bond, producing **adenosine diphosphate (ADP)** and an inorganic phosphate group ($P_i$). This reaction releases 7.3 kilocalories (kcal) of energy for every mole (505 g) of ATP. Most of this energy escapes as heat, but we live on the portion of it that does useful work. We can summarize this as follows:

$$ATP + H_2O \xrightarrow{\text{ATPase}} ADP + P_i + Energy \begin{array}{l} \nearrow \text{Heat} \\ \searrow \text{Work} \end{array}$$

The free phosphate groups released by ATP hydrolysis are often added to enzymes or other molecules to activate them. This addition of $P_i$, called **phosphorylation,** is carried out by enzymes called **kinases.** The phosphorylation of an enzyme is sometimes the "switch" that turns a metabolic pathway on or off.

ATP is a short-lived molecule, usually consumed within 60 seconds of its formation. The entire amount in the body would support life for less than 1 minute if it were not continually replenished. At a moderate rate of physical activity, a full day's supply of ATP would weigh twice as much as you do. Even if you never got out of bed, you would need about 45 kg (99 lb) of ATP to stay alive for a day. The reason cyanide is so lethal is that it halts ATP synthesis.

ATP synthesis is explained in detail in chapter 26, but you will find it necessary to become familiar with the general idea of it before you reach that chapter—especially in understanding muscle physiology (chapter 11). Much of the energy for ATP synthesis comes from glucose oxidation (fig. 2.31). The first stage in glucose oxidation (fig. 2.32) is the reaction pathway known as **glycolysis** (gly-COLL-ih-sis). This literally means "sugar splitting," and indeed its major effect is to split the six-carbon glucose molecule into two three-carbon molecules of *pyruvic acid.* A little ATP is produced in this stage (a net yield of 2 ATP per glucose), but most of the chemical energy of the glucose is still in the pyruvic acid.

What happens to pyruvic acid depends on how much oxygen is available relative to ATP demand. When the demand for ATP outpaces the oxygen supply, excess pyruvic acid is converted to lactic acid by a pathway called **anaerobic**[29] (AN-err-OH-bic) **fermentation.** This pathway has two noteworthy

---

[29]*an* = without; *aer* = air, oxygen; *obic* = pertaining to life

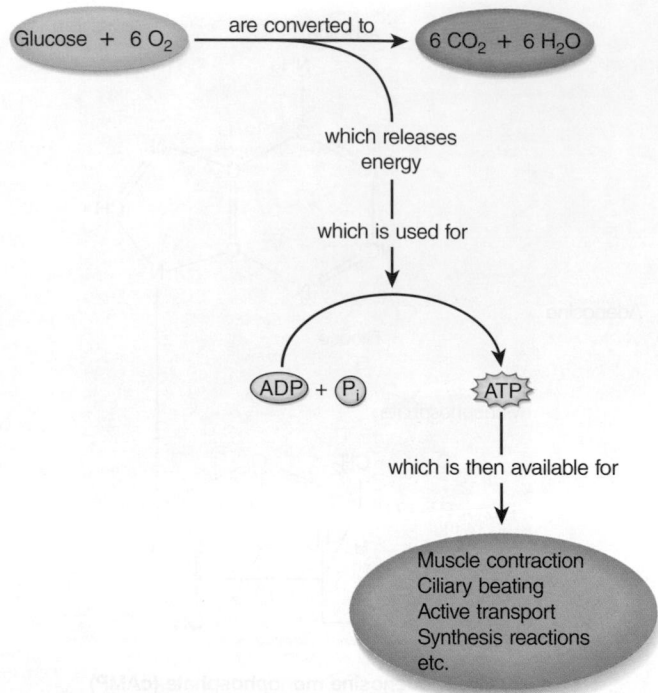

**FIGURE 2.31  The Source and Uses of ATP.**

disadvantages: First, it does not extract any more energy from pyruvic acid; second, the lactic acid it produces is toxic, so most cells can use anaerobic fermentation only as a temporary measure. The only advantage to this pathway is that it enables glycolysis to continue (for reasons explained in chapter 26) and thus enables a cell to continue producing a small amount of ATP.

If enough oxygen is available, a more efficient pathway called **aerobic respiration** occurs. This breaks pyruvic acid down to carbon dioxide and water and generates up to 36 more molecules of ATP for each of the original glucose molecules. The reactions of aerobic respiration are carried out in the cell's *mitochondria* (described in chapter 3), so mitochondria are regarded as a cell's principal "ATP factories."

## Other Nucleotides

**Guanosine** (GWAH-no-seen) **triphosphate (GTP)** is another nucleotide involved in energy transfers. In some reactions, it donates phosphate groups to other molecules. For example, it can donate its third phosphate group to ADP to regenerate ATP.

**Cyclic adenosine monophosphate (cAMP)** (see fig. 2.30b) is a nucleotide formed by the removal of both the second and third phosphate groups from ATP. In many cases, when a hormone or other chemical signal ("first messenger") binds to a cell surface, it triggers an internal reaction that converts ATP to cAMP. The cAMP then acts as a "second messenger" to activate metabolic effects within the cell.

## Nucleic Acids

**Nucleic** (new-CLAY-ic) **acids** are polymers of nucleotides. The largest of them, **deoxyribonucleic acid (DNA),** is typically 100 million to 1 billion nucleotides long. It constitutes our genes, gives instructions for synthesizing all of the body's proteins, and transfers hereditary information from cell to cell when cells divide and from generation to generation when organisms reproduce. Three forms of **ribonucleic acid (RNA),** which range from 70 to 10,000 nucleotides long, carry out those instructions and synthesize the proteins, assembling amino acids in the right order to produce each protein "described" by the DNA. The detailed structure of DNA and RNA and the mechanisms of protein synthesis and heredity are described in chapter 4.

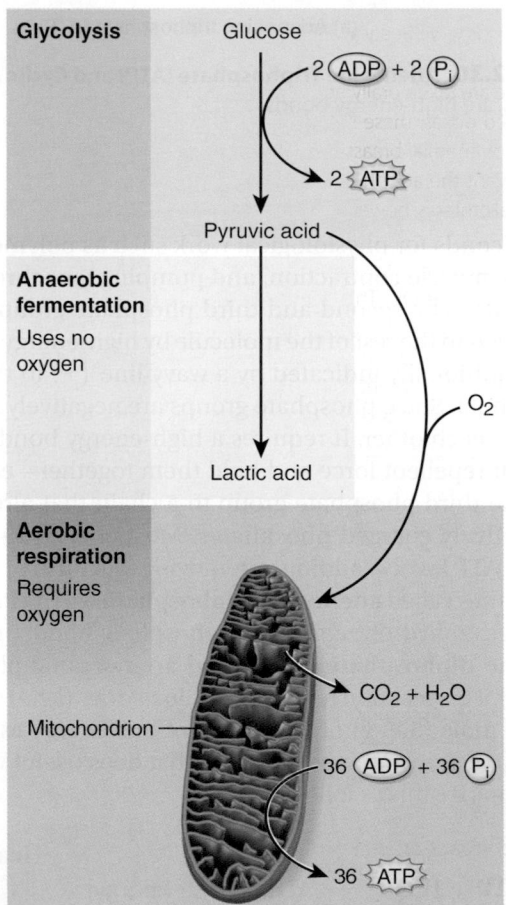

**FIGURE 2.32  ATP Production.** Glycolysis produces pyruvic acid and a net gain of two ATPs. Anaerobic fermentation converts pyruvic acid to lactic acid and permits glycolysis to continue producing ATP in the absence of oxygen. Aerobic respiration produces a much greater ATP yield but requires oxygen.

# DEEPER INSIGHT 2.6

## CLINICAL APPLICATION

### Anabolic–Androgenic Steroids

It is routine news to hear of sports celebrities suspended or stripped of their honors for the use of anabolic steroids. Magazines of physical culture carry many tragic reports of the deaths of amateur athletes, or violent crimes committed by them, attributed to steroid abuse.

Anabolic steroids, as they are known on the street, are more properly called anabolic–androgenic[30] steroids. They are hormones derived from testosterone that stimulate muscle growth (the anabolic effect) and masculinize the body (the androgenic effect). Perhaps the earliest notion to put them to use arose in Nazi Germany, where testosterone was given to SS troops in an effort to make them more aggressive—but with no proven success. In the 1950s, however, when Soviet weight-lifting teams were routinely defeating American teams, it came to light that the Soviets were using testosterone as a performance enhancer. American team physician John Ziegler began experimenting with this back in the United States. He disliked the androgenic side effects and approached the Ciba Pharmaceutical Company to develop a testosterone analog (a molecule of slightly altered structure) that would enhance the anabolic effect and weaken the androgenic effect. Ciba soon developed Dianabol. It produced spectacular effects in weight lifters and by the 1960s, several testosterone analogs were freely and legally available, designed to enhance anabolic potency, reduce androgenic effects, and prolong the half-life of the drug in the body. Some are taken orally and others by intramuscular (I.M.) injection.

In limited doses, these steroids have legitimate medical uses such as the treatment of anemia, breast cancer, osteoporosis, and some muscle diseases and to prevent the atrophy of muscles in immobilized patients. Amateur and professional athletes, however, sometimes use them in doses 10 to 1,000 times stronger than therapeutic doses. Such doses have devastating effects on the body. They raise cholesterol levels, which promotes fatty degeneration of the arteries (atherosclerosis). This can lead to coronary artery disease, heart and kidney failure, and stroke. Deteriorating circulation also sometimes results in gangrene, and many have suffered amputation of the lower extremities as a result. As the liver attempts to dispose of the high concentration of steroids, liver cancer and other liver diseases may ensue. In addition, steroids suppress the immune system, so the user is more subject to infection and cancer. They cause a premature end to bone elongation, so people who use anabolic steroids in adolescence may never attain normal adult height.

Paradoxically, anabolic–androgenic steroids can have masculinizing effects on women and feminizing effects on men. In women, who are especially sensitive to the androgenic effect, the steroids commonly produce growth of facial hair, enlargement of the clitoris, atrophy of the breasts and uterus, and irregularities of ovulation and menstruation. An enzyme called aromatase converts androgens to estrogens, and in men these often induce breast enlargement (gynecomastia), atrophy of the testes, impotence (inability to achieve or maintain an erection), low sperm count, and infertility.

Especially in men, steroid abuse can be linked to severe emotional disorders. Individuals vary in susceptibility, but the androgenic effects include heightened aggressiveness and unpredictable mood swings, so some abusers vacillate between depression and violence ("roid rage"), including physical abuse of family members and crimes as serious as homicide.

As the heavy recreational use of anabolic–androgenic steroids became widespread, so did the tragic side effects of such heavy use—an outcome that Dr. Ziegler deeply regretted as the low point of his career. The U.S. Congress classified anabolic–androgenic steroids as a controlled substance in 1991. Their use in sports has been condemned by the American Medical Association and American College of Sports Medicine and banned by the International Olympic Committee, National Football League, Major League Baseball Players' Association, National Basketball Association, and National Collegiate Athletic Association.

Yet in spite of such warnings and bans, many continue to use steroids and related performance-enhancing drugs, which remain available through unscrupulous coaches, physicians, Internet sources, and foreign mail-order suppliers under a cloud of confusing trade names (Durabolin, Anadrol, Oxandrin, Dianabol, Winstrol, Primobolan, and others). By some estimates, as many as 80% of competitive weight lifters, 30% of college and professional athletes, and 20% of male high school athletes now use anabolic–androgenic steroids. The National Institutes of Health find increasing use among high school students and increasing denial that the steroids present a significant health hazard.

---

[30] andro = male; genic = producing

---

**BEFORE YOU GO ON**

Answer the following questions to test your understanding of the preceding section:

16. Which reaction—dehydration synthesis or hydrolysis—converts a polymer to its monomers? Which one converts monomers to a polymer? Explain your answer.

17. What is the chemical name of blood sugar? What carbohydrate is polymerized to form starch and glycogen?

18. What is the main chemical similarity between carbohydrates and lipids? What are the main differences between them?

19. Explain the statement, All proteins are polypeptides but not all polypeptides are proteins.

20. Which is more likely to be changed by heating a protein, its primary structure or its tertiary structure? Explain.

21. Use the lock-and-key analogy to explain why excessively acidic body fluids (acidosis) could destroy enzyme function.

22. How does ATP change structure in the process of releasing energy?

23. What advantage and disadvantage does anaerobic fermentation have compared with aerobic respiration?

24. How is DNA related to nucleotides?

# STUDY GUIDE

## ▶ Assess Your Learning Outcomes

*To test your knowledge, discuss the following topics with a study partner or in writing, ideally from memory.*

### 2.1 Atoms, Ions, and Molecules (p. 42)

1. The definition of *chemical element;* the six most abundant elements in the human body; and trace elements important in human physiology
2. The structure of an atom and the special functional relevance of its valence electrons
3. How isotopes of the same element differ from each other, and how radioisotopes differ from other isotopes
4. The clinical relevance of ionizing radiation; its three forms; the difference between the physical and biological half-life of a radioisotope; and the clinical relevance of that difference
5. The difference between an ion and an atom; how ions form; and the two types of ions and examples of each
6. How an electrolyte differs from an atom and from an ion; the most common ions that constitute electrolytes; and the functions and medical relevance of electrolytes
7. The definition of *free radical;* the medical relevance of free radicals; and how the body is partially protected against them
8. The definitions of *molecule* and *compound*
9. How isomers resemble and differ from each other
10. How a molecule's molecular weight is determined
11. The nature and distinguishing characteristics of ionic bonds, covalent bonds, hydrogen bonds, and van der Waals forces; how polar and nonpolar covalent bonds differ from each other; and how polar covalent bonds can give rise to hydrogen bonds

### 2.2 Water and Mixtures (p. 49)

1. How the biologically important properties of water arise from the polarity and bond angle of its molecules
2. The difference between a mixture and a compound
3. The differences between solutions, colloids, and suspensions, and examples of body fluids in each category

4. How various measures of chemical concentration differ from each other—weight per volume, percentage, molarity, and milliequivalents per liter—and why each unit of measure may be the most appropriate for different purposes
5. How pH is mathematically defined; the pH scale; and the meanings of *acid* and *base*
6. The action and physiological function of buffers

### 2.3 Energy and Chemical Reactions (p. 54)

1. The definition of *energy,* and the two basic forms of energy
2. The differences between decomposition, synthesis, and exchange reactions
3. What determines the direction of a reversible chemical reaction; the nature of a chemical equilibrium
4. Factors that determine the rate of a chemical reaction
5. The definition of *metabolism* and its two subdivisions
6. The difference between oxidation and reduction

### 2.4 Organic Compounds (p. 58)

1. The criterion for considering a compound to be organic
2. The difference between the carbon backbone and the functional group(s) of an organic molecule; the physiological relevance of functional groups
3. The structures of hydroxyl, methyl, carboxyl, amino, and phosphate functional groups
4. The difference between monomers and polymers; how dehydration synthesis and hydrolysis reactions convert one to the other; and the role of water in both types of reactions
5. The defining characteristics of carbohydrates and their principal roles in the body
6. The names of and basic structural differences between the three monosaccharides, three disaccharides, and three polysaccharides that are most abundant in the diet and most relevant in human physiology
7. The defining characteristics of lipids and their principal roles in the body
8. The major categories of lipids in human physiology, and the roles of each
9. How fatty acids and triglycerides are related through dehydration synthesis and hydrolysis reactions

10. The defining characteristics of amino acids and how the 20 amino acids involved in protein structure differ from each other
11. How amino acids are polymerized, and the structure of peptide bonds
12. Differences between a dipeptide, oligopeptide, polypeptide, and protein
13. The nature of the primary through quaternary levels of protein structure, and how an alpha helix differs from a beta sheet
14. Why protein function depends so strongly on the shape (conformation) of the molecule; how and why functionality is affected by denaturation; and the most common causes of protein denaturation
15. What defines a conjugated protein; the general term for the nonprotein component of such a molecule; and examples of conjugated proteins
16. The functions of proteins in human anatomy and physiology
17. How enzymes differ from other proteins, and the general role played by all enzymes
18. The general term for substances acted upon by enzymes; the relevance of active sites to enzyme action; why the active sites limit the range of substances on which an enzyme can act; and the name for this principle of selective enzyme action
19. The nature of cofactors and coenzymes, with examples
20. The term for a chain of linked enzymatic reactions; for its input and output; and for the intermediate steps between the input and output molecules
21. The basic structural components of adenosine triphosphate (ATP) and their organization in the molecule; the function of ATP and why life instantly ceases without it; and where in its molecular structure ATP carries the energy that is transferred to other chemicals
22. Differences between the aerobic and anaerobic mechanisms of producing ATP
23. How guanosine triphosphate (GTP) and cyclic adenosine monophosphate (cAMP) are related to ATP, and their functions
24. The basic chemical nature of the nucleic acids, DNA and RNA; their fundamental function; and how they are structurally related to ATP

# STUDY GUIDE

## ▶ Testing Your Recall

*Answers in Appendix B*

1. A substance that _____ is considered to be a chemical compound.
   a. contains at least two different elements
   b. contains at least two atoms
   c. has a chemical bond
   d. has a stable valence shell
   e. has covalent bonds

2. An ionic bond is formed when
   a. two anions meet.
   b. two cations meet.
   c. an anion meets a cation.
   d. electrons are unequally shared between nuclei.
   e. electrons transfer completely from one atom to another.

3. The ionization of a sodium atom to produce $Na^+$ is an example of
   a. oxidation.
   b. reduction.
   c. catabolism.
   d. anabolism.
   e. decomposition.

4. The weakest and most temporary of the following chemical bonds are
   a. polar covalent bonds.
   b. nonpolar covalent bonds.
   c. hydrogen bonds.
   d. ionic bonds.
   e. double covalent bonds.

5. A substance capable of dissolving freely in water is
   a. hydrophilic.
   b. hydrophobic.
   c. hydrolyzed.
   d. hydrated.
   e. amphipathic.

6. A carboxyl group is symbolized
   a. –OH.
   b. $-NH_2$.
   c. $-CH_3$.
   d. $-CH_2OH$.
   e. –COOH.

7. The only polysaccharide synthesized in the human body is
   a. cellulose.
   b. glycogen.
   c. cholesterol.
   d. starch.
   e. prostaglandin.

8. The arrangement of a polypeptide into a fibrous or globular shape is called its
   a. primary structure.
   b. secondary structure.
   c. tertiary structure.
   d. quaternary structure.
   e. conjugated structure.

9. Which of the following functions is more characteristic of carbohydrates than of proteins?
   a. contraction
   b. energy storage
   c. catalyzing reactions
   d. immune defense
   e. intercellular communication

10. The feature that most distinguishes a lipid from a carbohydrate is that a lipid has
    a. more phosphate.
    b. more sulfur.
    c. a lower ratio of carbon to oxygen.
    d. a lower ratio of oxygen to hydrogen.
    e. a greater molecular weight.

11. When an atom gives up an electron and acquires a positive charge, it is called a/an _____.

12. Dietary antioxidants are important because they neutralize _____.

13. Any substance that increases the rate of a reaction without being consumed by it is a/an _____. In the human body, _____ serve this function.

14. All the synthesis reactions in the body form a division of metabolism called _____.

15. A chemical reaction that joins two organic molecules into a larger one and produces water as a by-product is called _____.

16. The suffix _____ denotes a sugar, while the suffix _____ denotes an enzyme.

17. The amphipathic lipids of cell membranes are called _____.

18. A chemical named _____ is derived from ATP and widely employed as a "second messenger" in cellular signaling.

19. When oxygen is too limited to meet a cell's ATP demand, a cell can employ a metabolic pathway called _____ to produce ATP.

20. A substance acted upon and changed by an enzyme is called the enzyme's _____.

# STUDY GUIDE

## ▶ Building Your Medical Vocabulary

*Answers in Appendix B*

*State a medical meaning of each word element below, and give a term in which it or a slight variation of it is used.*

1. a-
2. aero-
3. amphi-
4. caloro-
5. collo-
6. hydro-
7. -mer
8. mono-
9. oligo-
10. -philic

## ▶ True or False

*Answers in Appendix B*

*Determine which five of the following statements are false, and briefly explain why.*

1. The monomers of a polysaccharide are called amino acids.

2. An emulsion is a mixture of two liquids that separate from each other on standing.

3. Two molecules with the same atoms arranged in a different order are called isotopes.

4. If a pair of shared electrons is more attracted to one nucleus than to the other, they form a polar covalent bond.

5. Amino acids are joined by a unique type of bond called a peptide bond.

6. A saturated fat is defined as a fat to which no more carbon can be added.

7. Organic compounds get their unique chemical characteristics more from their functional groups than from their carbon backbones.

8. The higher the temperature is, the faster an enzyme works.

9. Two percent sucrose and 2% sodium bicarbonate have the same number of molecules per liter of solution.

10. A solution of pH 8 has one-tenth the hydrogen ion concentration of a solution with pH 7.

## ▶ Testing Your Comprehension

*Answers at www.mhhe.com/saladin7*

1. Suppose a pregnant woman with severe morning sickness has been vomiting steadily for several days. How will her loss of stomach acid affect the pH of her body fluids? Explain.

2. Suppose a person with a severe anxiety attack hyperventilates and exhales $CO_2$ faster than his body produces it. Consider the carbonic acid reaction on page 55 and explain what effect this hyperventilation will have on his blood pH. (*Hint:* Remember the law of mass action.)

3. In one form of nuclear decay, a neutron breaks down into a proton and electron and emits a gamma ray. Is this an endergonic or exergonic reaction, or neither? Is it an anabolic or catabolic reaction, or neither? Explain both answers.

4. How would the body's metabolic rate be affected if there were no such thing as enzymes? Explain.

5. Some metabolic conditions such as diabetes mellitus cause disturbances in the acid–base balance of the body, which gives the body fluids an abnormally low pH. Explain how this could affect enzyme–substrate reactions and metabolic pathways in the body.

## ▶ Improve Your Grade at www.mhhe.com/saladin7

*Download animations to study when it fits your schedule. Practice quizzes, labeling activities, games, and flashcards offer fun ways to master the chapter concepts. Or, download image PowerPoint files for each chapter to create a study guide or for taking notes during lecture.*

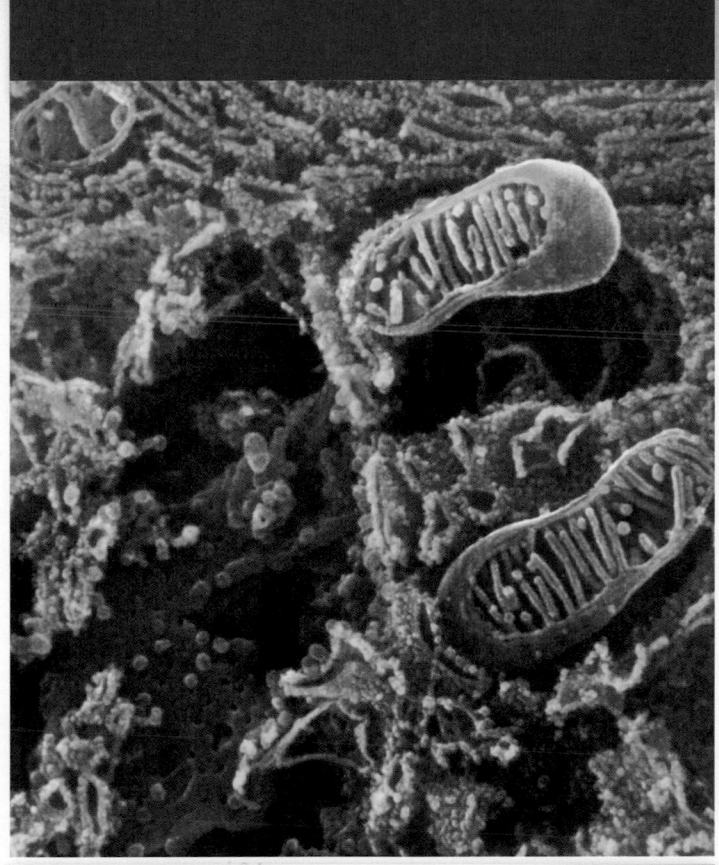

Mitochondria (blue) and rough endoplasmic reticulum (orange) in a pancreatic cell (SEM)

# CELLULAR FORM AND FUNCTION

aprevealed.com

**Module 2: Cells and Chemistry**

All organisms, from the simplest to the most complex, are composed of cells—whether the single cell of a bacterium or the trillions of cells that constitute the human body. These cells are responsible for all structural and functional properties of a living organism. A knowledge of cells is therefore indispensable to any true understanding of the workings of the human body, the mechanisms of disease, and the rationale of therapy. Thus, this chapter and the next one introduce the basic cell biology of the human body, and subsequent chapters expand upon this information as we examine the specialized cellular structure and function of specific organs.

## 3.1    Concepts of Cellular Structure

### Expected Learning Outcomes

When you have completed this section, you should be able to

a. discuss the development and modern tenets of the cell theory;

b. describe cell shapes from their descriptive terms;

c. state the size range of human cells and discuss factors that limit their size;

d. discuss the way that developments in microscopy have changed our view of cell structure; and

e. outline the major components of a cell.

## Development of the Cell Theory

**Cytology,**[1] the scientific study of cells, was born in 1663 when Robert Hooke observed the empty cell walls of cork and coined the word *cellulae* ("little cells") to describe them. Soon he studied thin slices of fresh wood and saw living cells "filled with juices"—a fluid later named *protoplasm.*[2] Two centuries later, Theodor Schwann studied a wide range of animal tissues and concluded that all animals are made of cells.

Schwann and other biologists originally believed that cells came from nonliving body fluid that somehow congealed and acquired a membrane and nucleus. This idea of *spontaneous generation*—that living things arise from nonliving matter—was

rooted in the scientific thought of the times. For centuries, it seemed to be simple common sense that decaying meat turned into maggots, stored grain into rodents, and mud into frogs. Schwann and his contemporaries merely extended this idea to cells. The idea of spontaneous generation wasn't discredited until some classic experiments by French microbiologist Louis Pasteur in 1859.

By the end of the nineteenth century, it was established beyond all reasonable doubt that cells arise only from other cells and that every living organism is composed of cells and cell products. The cell came to be regarded, and still is, as the simplest structural and functional unit of life. There are no smaller subdivisions of a cell or organism that, in themselves, are alive. An enzyme molecule, for example, is not alive, although the life of a cell depends on the activity of numerous enzymes.

The development of biochemistry from the late nineteenth to the twentieth century made it further apparent that all physiological processes of the body are based on cellular activity and that the cells of all species exhibit remarkable biochemical unity. These various generalizations now constitute the modern **cell theory.**

## Cell Shapes and Sizes

We will shortly examine the structure of a generic cell, but the generalizations we draw should not blind you to the diversity of cellular form and function in humans. There are about 200 kinds of cells in the human body, with a variety of shapes, sizes, and functions.

Descriptions of organ and tissue structure often refer to the shapes of cells by the following terms (fig. 3.1):

- **Squamous**[3] (SQUAY-mus)—a thin, flat, scaly shape, often with a bulge where the nucleus is, much like the shape of a fried egg "sunny side up." Squamous cells line the esophagus and form the surface layer (epidermis) of the skin.

- **Cuboidal**[4] (cue-BOY-dul)—squarish-looking in frontal tissue sections and about equal in height and width; liver cells are a good example.

- **Columnar**—distinctly taller than wide, such as the inner lining cells of the stomach and intestines.

- **Polygonal**[5]—having irregularly angular shapes with four, five, or more sides.

- **Stellate**[6]—having multiple pointed processes projecting from the body of a cell, giving it a somewhat starlike shape. The cell bodies of many nerve cells are stellate.

- **Spheroidal** to **ovoid**—round to oval, as in egg cells and white blood cells.

- **Discoid**—disc-shaped, as in red blood cells.

---

[1]*cyto* = cell; *logy* = study of
[2]*proto* = first; *plasm* = formed

[3]*squam* = scale; *ous* = characterized by
[4]*cub* = cube; *oidal* = like, resembling
[5]*poly* = many; *gon* = angles
[6]*stell* = star; *ate* = characterized by

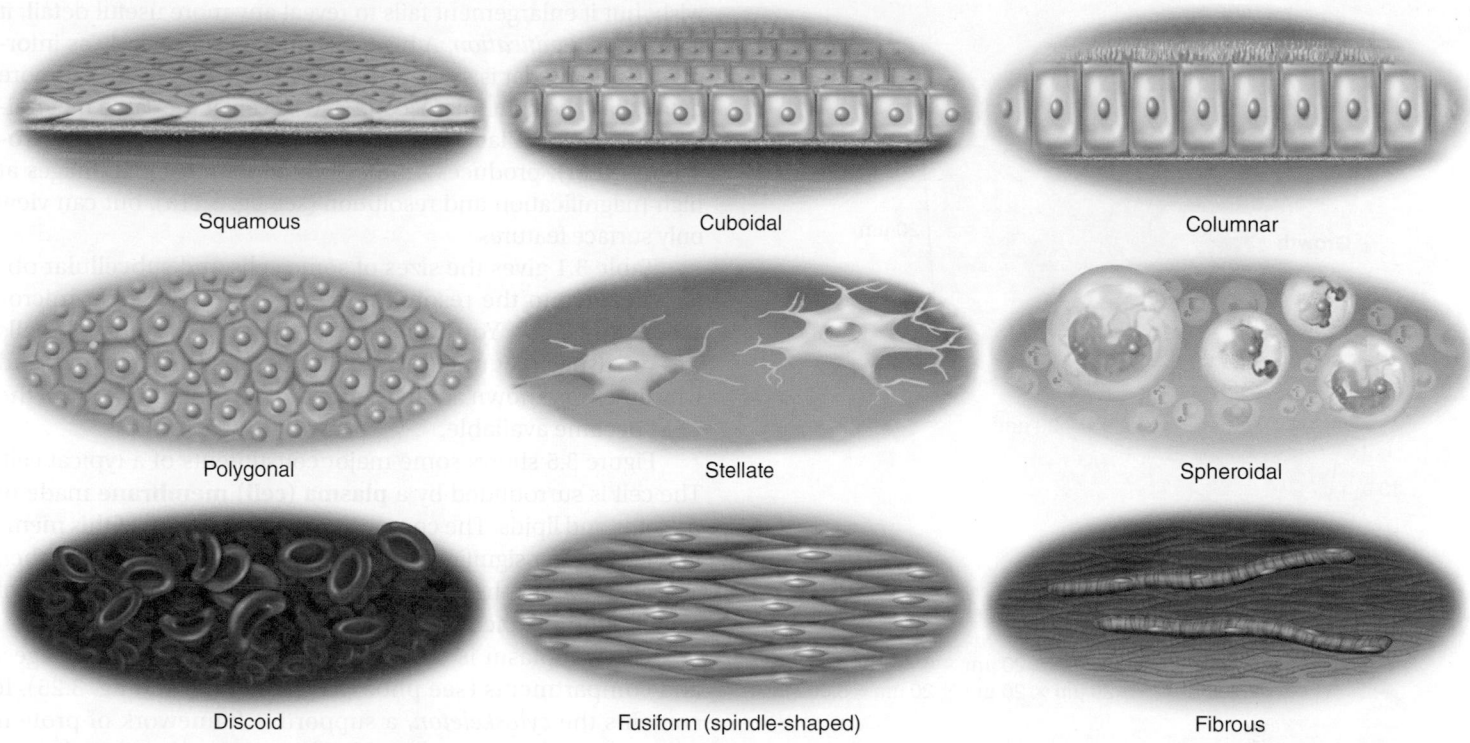

**FIGURE 3.1 Common Cell Shapes.** AP|R

- **Fusiform**[7] (FEW-zih-form)—spindle-shaped; elongated, with a thick middle and tapered ends, as in smooth muscle cells.

- **Fibrous**—long, slender, and threadlike, as in skeletal muscle cells and the axons (nerve fibers) of nerve cells.

Some of these shapes refer to the way a cell looks in typical tissue sections, not to the complete three-dimensional shape of the cell. A cell that looks squamous, cuboidal, or columnar in a tissue section, for example, usually looks polygonal if viewed from its upper surface.

The most useful unit of measurement for designating cell sizes is the **micrometer (µm),** formerly called the micron—one-millionth ($10^{-6}$) of a meter, one-thousandth ($10^{-3}$) of a millimeter. (See appendix C for units of measurement.) The smallest objects most people can see with the naked eye are about 100 µm, which is about one-quarter the size of the period at the end of this sentence. A few human cells fall within this range, such as the egg cell and some fat cells, but most human cells are about 10 to 15 µm wide. The longest human cells are nerve cells (sometimes over a meter long) and muscle cells (up to 30 cm long), but both are usually too slender to be seen with the naked eye.

There is a limit to how large a cell can be, partly due to the relationship between its volume and surface area. The surface area of a cell is proportional to the square of its diameter, while volume is proportional to the cube of its diameter. Thus, for a given increase in diameter, volume increases much more than surface area. Picture a cuboidal cell 10 µm on each side (fig 3.2). It would have a surface area of 600 µm² (10 µm × 10 µm × 6 sides) and a volume of 1,000 µm³ (10 × 10 × 10 µm). Now, suppose it grew by another 10 µm on each side. Its new surface area would be 20 µm × 20 µm × 6 = 2,400 µm², and its volume would be 20 × 20 × 20 µm = 8,000 µm³. The 20 µm cell has eight times as much protoplasm needing nourishment and waste removal, but only four times as much membrane surface through which wastes and nutrients can be exchanged. A cell that is too big cannot support itself. In addition, an overly large cell is at risk of rupturing like an overfilled water balloon.

#### ▶▶▶ APPLY WHAT YOU KNOW

*Can you conceive of some other reasons for an organ to consist of many small cells rather than fewer larger ones?*

---

[7]*fusi* = spindle; *form* = shape

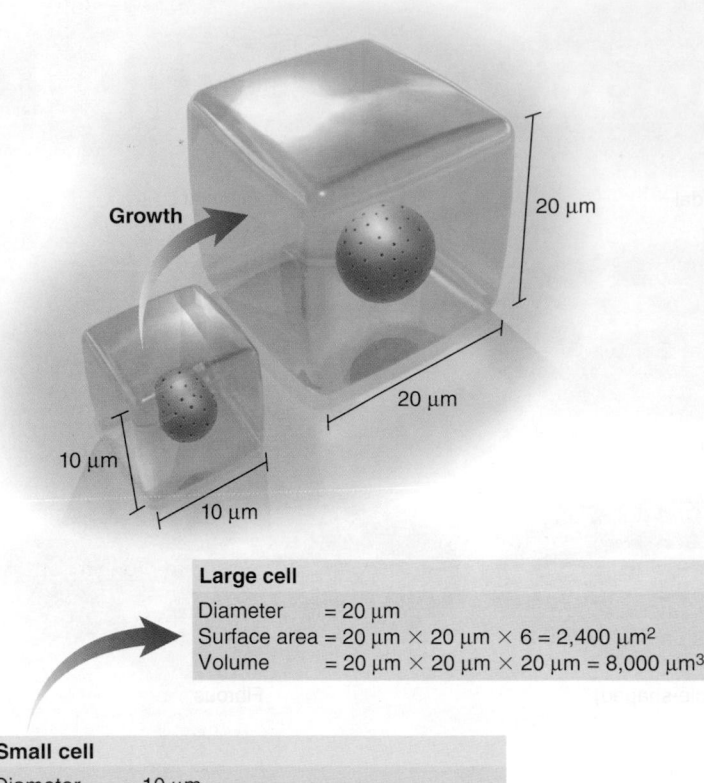

**Growth**

20 μm

20 μm

10 μm

10 μm

**Large cell**

Diameter = 20 μm
Surface area = 20 μm × 20 μm × 6 = 2,400 μm²
Volume = 20 μm × 20 μm × 20 μm = 8,000 μm³

**Small cell**

Diameter = 10 μm
Surface area = 10 μm × 10 μm × 6 = 600 μm²
Volume = 10 μm × 10 μm × 10 μm = 1,000 μm³

**Effect of cell growth:**

Diameter ($D$) increased by a factor of 2
Surface area increased by a factor of 4 (= $D^2$)
Volume increased by a factor of 8 (= $D^3$)

**FIGURE 3.2  The Relationship Between Cell Surface Area and Volume.** As a cell doubles in diameter, its volume increases eightfold, but its surface area increases only fourfold. A cell that is too large may have too little plasma membrane to serve the metabolic needs of its increased volume of cytoplasm.

## Basic Components of a Cell

In Schwann's time, little was known about cells except that they were enclosed in a membrane and contained a nucleus. The fluid between the nucleus and surface membrane, called **cytoplasm,**[8] was thought to be little more than a gelatinous mixture of chemicals and vaguely defined particles. The **transmission electron microscope (TEM),** invented in the mid-twentieth century, radically changed this concept. Using a beam of electrons in place of light, the TEM enabled biologists to see a cell's *ultrastructure* (fig. 3.3), a fine degree of detail extending even to the molecular level. The most important thing about a good microscope is not magnification but **resolution**—the ability to reveal detail. Any image can be photographed and enlarged as much as we

[8]*cyto* = cell; *plasm* = formed, molded

wish, but if enlargement fails to reveal any more useful detail, it is *empty magnification.* A big fuzzy image is not nearly as informative as one that is small and sharp. The TEM reveals far more detail than the light microscope (LM), even at the same magnification (fig. 3.4). A later invention, the **scanning electron microscope (SEM),** produces dramatic three-dimensional images at high magnification and resolution (see fig. 3.11a), but can view only surface features.

Table 3.1 gives the sizes of some cells and subcellular objects relative to the resolution of the naked eye, light microscope, and TEM. You can see why the very existence of cells was unsuspected until the light microscope was invented, and why little was known about their internal components until the TEM became available.

Figure 3.5 shows some major constituents of a typical cell. The cell is surrounded by a **plasma (cell) membrane** made of proteins and lipids. The composition and functions of this membrane can differ significantly from one region of a cell to another, especially among the basal, lateral, and apical (upper) surfaces of cells like the one pictured.

The cytoplasm is crowded with fibers, tubules, passages, and compartments (see photograph on p. 77 and fig. 3.25). It contains the *cytoskeleton,* a supportive framework of protein filaments and tubules; an abundance of *organelles,* diverse structures that perform various metabolic tasks for the cell; and *inclusions,* which are foreign matter or stored cell products.

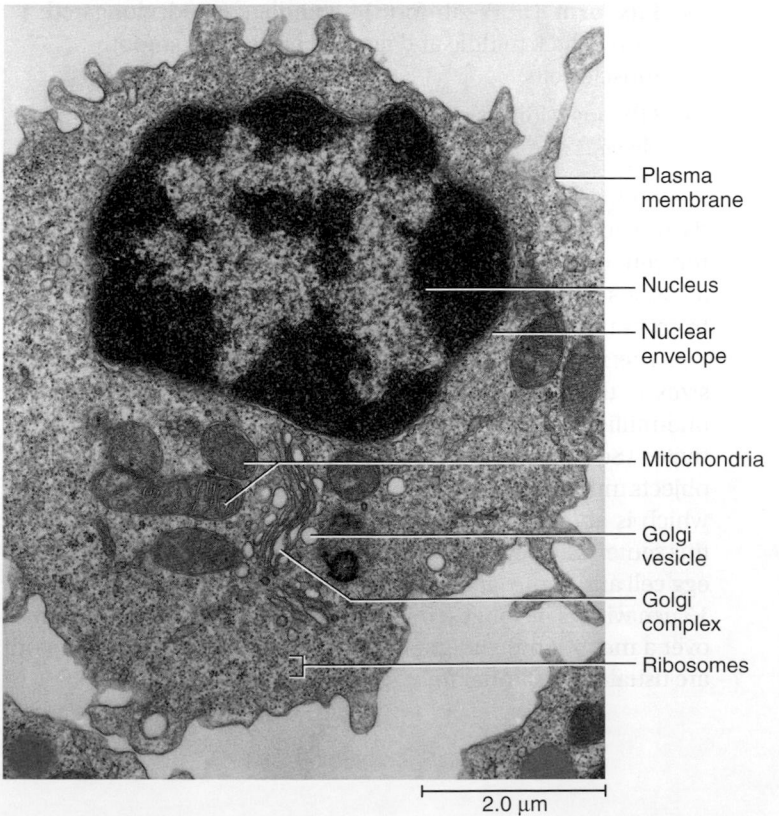

Plasma membrane

Nucleus

Nuclear envelope

Mitochondria

Golgi vesicle

Golgi complex

Ribosomes

2.0 μm

**FIGURE 3.3  Ultrastructure of a White Blood Cell (TEM).** AP|R

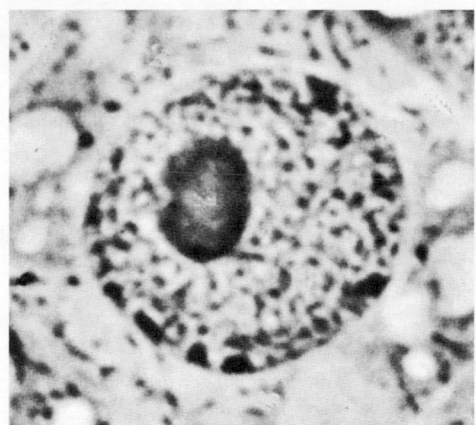

**(a) Light microscope (LM)**

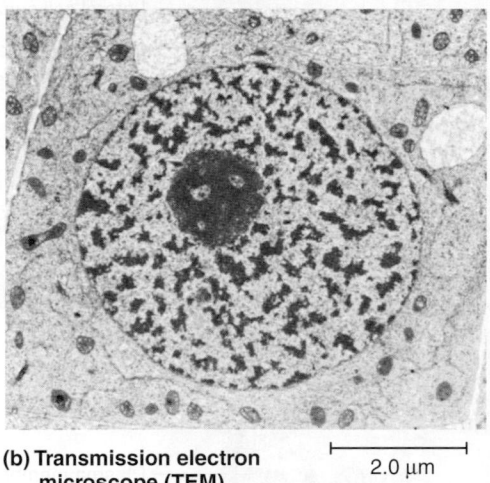

2.0 μm

**(b) Transmission electron microscope (TEM)**

**FIGURE 3.4  Magnification Versus Resolution.** These cells were photographed at the same magnification (about ×750) through (a) a light microscope and (b) a transmission electron microscope.

| TABLE 3.1 | Sizes of Biological Structures in Relation to the Resolution of the Eye, Light Microscope, and Transmission Electron Microscope |
|---|---|
| **Object** | **Size** |
| **Visible to the Naked Eye (Resolution 70–100 μm)** | |
| Human egg, diameter | 100 μm |
| **Visible with the Light Microscope (Resolution 200 nm)** | |
| Most human cells, diameter | 10–15 μm |
| Cilia, length | 7–10 μm |
| Mitochondria, width × length | 0.2 × 4 μm |
| Bacteria (*Escherichia coli*), length | 1–3 μm |
| Microvilli, length | 1–2 μm |
| Lysosomes, diameter | 0.5 μm = 500 nm |
| **Visible with the Transmission Electron Microscope (Resolution 0.5 nm)** | |
| Nuclear pores, diameter | 30–100 nm |
| Centriole, diameter × length | 20 × 50 nm |
| Poliovirus, diameter | 30 nm |
| Ribosomes, diameter | 15 nm |
| Globular proteins, diameter | 5–10 nm |
| Plasma membrane, thickness | 7.5 nm |
| DNA molecule, diameter | 2.0 nm |
| Plasma membrane channels, diameter | 0.8 nm |

A cell may have 10 billion protein molecules, including potent enzymes with the potential to destroy the cell if they are not contained and isolated from other cellular components. You can imagine the enormous problem of keeping track of all this material, directing molecules to the correct destinations, and maintaining order against the incessant trend toward disorder. Cells maintain order partly by compartmentalizing their contents in the organelles.

The cytoskeleton, organelles, and inclusions are embedded in a clear gel called the **cytosol** or **intracellular fluid (ICF).** All body fluids not contained in the cells are collectively called the **extracellular fluid (ECF).** The ECF located amid the cells is also called **tissue (interstitial) fluid.** Some other extracellular fluids include blood plasma, lymph, and cerebrospinal fluid.

**BEFORE YOU GO ON**

Answer the following questions to test your understanding of the preceding section:

1. What are the basic principles of the cell theory?

2. What does it mean to say a cell is squamous, stellate, columnar, or fusiform?

3. Why can cells not grow to unlimited size?

4. What is the difference between cytoplasm and cytosol?

5. Define *intracellular fluid (ICF)* and *extracellular fluid (ECF)*.

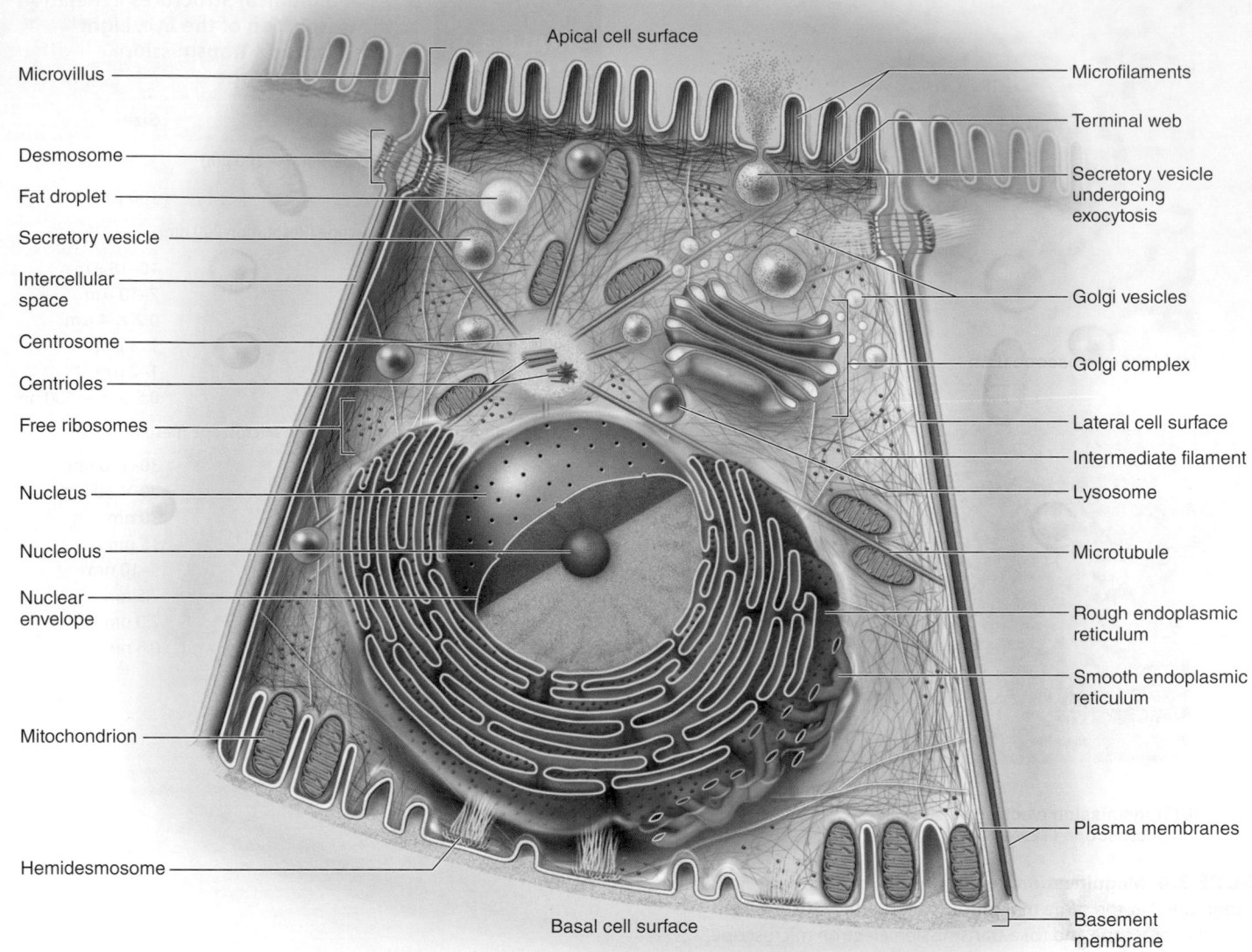

Microvillus

Desmosome

Fat droplet

Secretory vesicle

Intercellular space

Centrosome

Centrioles

Free ribosomes

Nucleus

Nucleolus

Nuclear envelope

Mitochondrion

Hemidesmosome

Apical cell surface

Microfilaments

Terminal web

Secretory vesicle undergoing exocytosis

Golgi vesicles

Golgi complex

Lateral cell surface

Intermediate filament

Lysosome

Microtubule

Rough endoplasmic reticulum

Smooth endoplasmic reticulum

Plasma membranes

Basal cell surface

Basement membrane

**FIGURE 3.5   Structure of a Representative Cell.**

## 3.2   The Cell Surface

### Expected Learning Outcomes

When you have completed this section, you should be able to

a. describe the structure of the plasma membrane;

b. explain the functions of the lipid, protein, and carbohydrate components of the plasma membrane;

c. describe a second-messenger system and discuss its importance in human physiology;

d. explain the composition and functions of the glycocalyx that coats cell surfaces; and

e. describe the structure and functions of microvilli, cilia, flagella, and pseudopods.

Many of the most physiologically important processes occur at the surface of a cell—such events as immune responses, the binding of egg and sperm, cell-to-cell signaling by hormones, and the detection of tastes and smells, for example. A substantial part of this chapter is therefore concerned with the cell surface. Before we venture into the interior of the cell, we will examine the structure of the plasma membrane, surface features such as cilia and microvilli, and methods of transport through the membrane.

### The Plasma Membrane

The electron microscope reveals that the cell and many of the organelles within it are bordered by a **plasma membrane,** which appears as a pair of dark parallel lines with a total thickness of about 7.5 nm (fig. 3.6a). It defines the boundaries of the cell, governs its interactions with other cells, and controls the passage of materials into and out of the cell. The side that faces the cytoplasm is the *intracellular face* of the membrane, and the side that faces outward is the *extracellular face*.

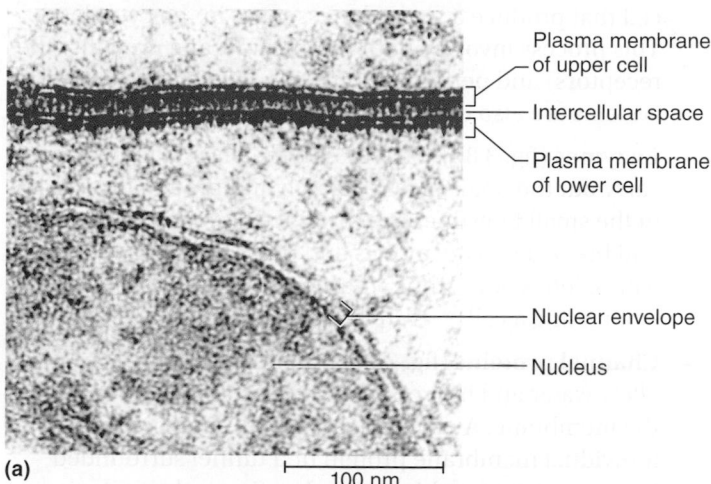

Plasma membrane of upper cell

Intercellular space

Plasma membrane of lower cell

Nuclear envelope

Nucleus

**(a)**

100 nm

## Membrane Lipids

Figure 3.6b shows our current concept of the molecular structure of the plasma membrane—an oily film of lipids with proteins embedded in it. Typically about 98% of the membrane molecules are lipids, and about 75% of those are phospholipids. These amphipathic molecules arrange themselves into a bilayer, with their hydrophilic phosphate-containing heads facing the water on each side and their hydrophobic tails directed toward the center, avoiding the water. The phospholipids drift laterally from place to place, spin on their axes, and flex their tails. These movements keep the membrane fluid.

### ▶▶▶ APPLY WHAT YOU KNOW

*What would happen if the plasma membrane were made primarily of a hydrophilic substance such as carbohydrate? Which of the major themes at the end of chapter 1 does this point best exemplify?*

Cholesterol molecules, found near the membrane surfaces amid the phospholipids, constitute about 20% of the membrane lipids. By interacting with the phospholipids and

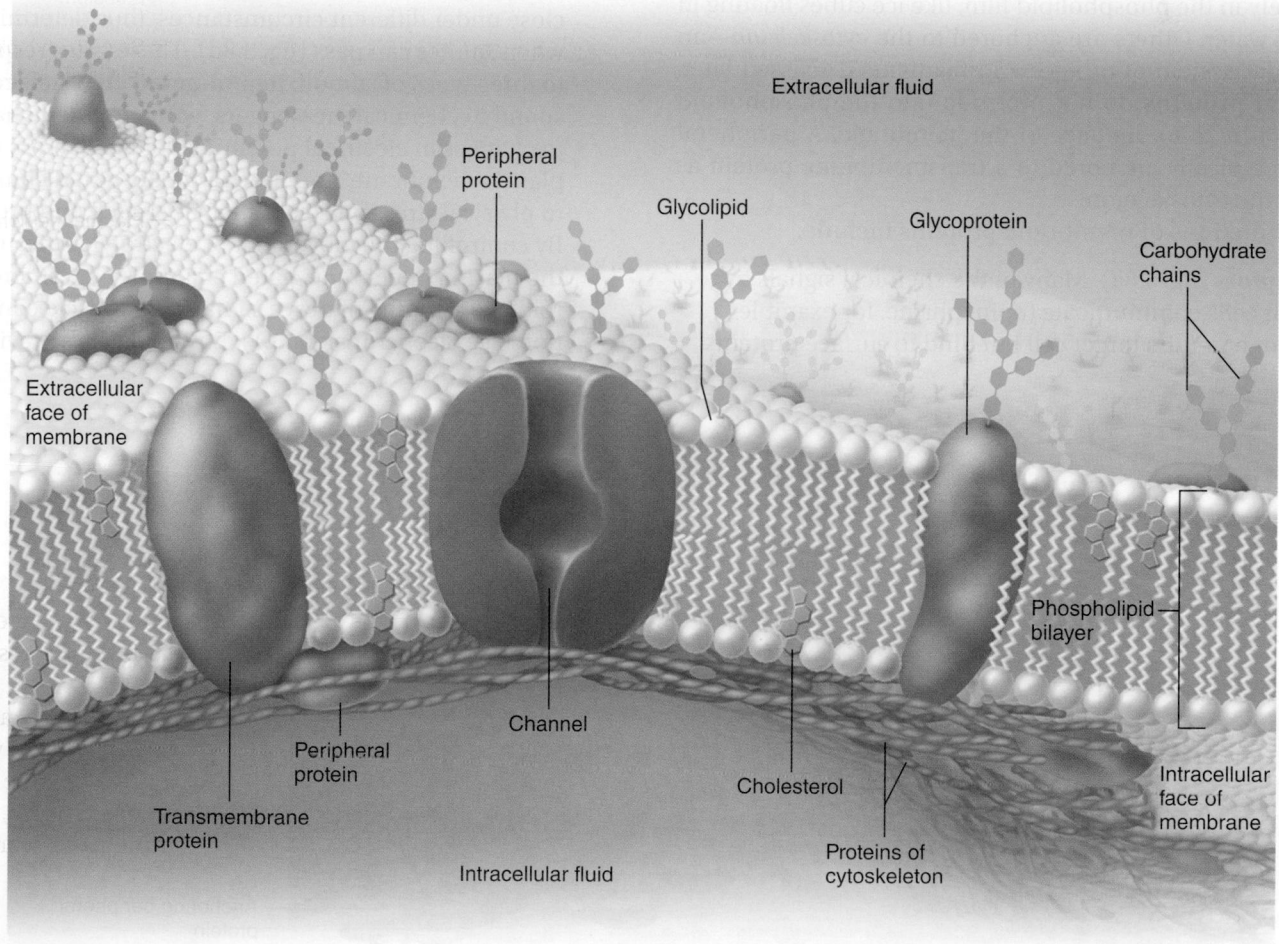

Extracellular fluid

Peripheral protein

Glycolipid

Glycoprotein

Carbohydrate chains

Extracellular face of membrane

Phospholipid bilayer

Channel

Peripheral protein

Cholesterol

Intracellular face of membrane

Transmembrane protein

Proteins of cytoskeleton

Intracellular fluid

**(b)**

**FIGURE 3.6 The Plasma Membrane.** (a) Plasma membranes of two adjacent cells (TEM). (b) Molecular structure of the plasma membrane.

holding them still, cholesterol can stiffen the membrane (make it less fluid) in spots. Higher concentrations of cholesterol, however, can increase membrane fluidity by preventing phospholipids from becoming packed closely together.

The remaining 5% of the membrane lipids are glycolipids—phospholipids with short carbohydrate chains on the extracellular face of the membrane. They help to form the *glycocalyx,* a carbohydrate coating on the cell surface with multiple functions described shortly.

## Membrane Proteins

Although proteins are only about 2% of the molecules of the plasma membrane, they are larger than lipids and constitute about 50% of the membrane weight. There are two broad classes of membrane proteins: integral and peripheral. **Integral proteins** penetrate into the phospholipid bilayer or all the way through it. Those that pass completely through are also called **transmembrane proteins.** These have hydrophilic regions in contact with the water on both sides of the membrane, and hydrophobic regions that pass back and forth through the lipid of the membrane (fig. 3.7). Most transmembrane proteins are glycoproteins, bound to oligosaccharides on the extracellular side of the membrane. Many integral proteins drift about freely in the phospholipid film, like ice cubes floating in a bowl of water. Others are anchored to the *cytoskeleton*—an intracellular system of tubules and filaments discussed later. **Peripheral proteins** do not protrude into the phospholipid layer but adhere to one face of the membrane. A peripheral protein is typically anchored to a transmembrane protein as well as to the cytoskeleton.

The functions of membrane proteins include:

- **Receptors** (fig. 3.8a). Many of the chemical signals by which cells communicate (epinephrine, for example) cannot enter the target cell but bind to surface proteins called receptors. Receptors are usually specific for one particular messenger, much like an enzyme that is specific for one substrate. Plasma membranes also have receptor proteins that bind chemicals and transport them into the cell, as discussed later in this chapter.

- **Second-messenger systems.** When a messenger binds to a surface receptor, it may trigger changes within the cell that produce a second messenger in the cytoplasm. This process involves both transmembrane proteins (the receptors) and peripheral proteins. Second-messenger systems are also discussed later in more detail.

- **Enzymes** (fig. 3.8b). Enzymes in the plasma membrane carry out the final stages of starch and protein digestion in the small intestine, help produce second messengers, and break down hormones and other signaling molecules whose job is done, thus stopping them from excessively stimulating a cell.

- **Channel proteins** (fig. 3.8c). Channels are passages that allow water and hydrophilic solutes to move through the membrane. A channel can be a tunnel through an individual membrane protein or a tunnel surrounded by a complex of multiple proteins. Some channels are always open, whereas others are **gates** that open and close under different circumstances, thus determining when solutes can pass (fig. 3.8d). These gates respond to three types of stimuli: **ligand-gated channels** respond to chemical messengers, **voltage-gated channels** to changes in electrical potential (voltage) across the plasma membrane, and **mechanically gated channels** to physical stress on a cell, such as stretch and pressure. By controlling the movement of electrolytes through the plasma membrane, gated channels play an important role in the timing of nerve signals and muscle contraction (see Deeper Insight 3.1). Physiologists are avidly

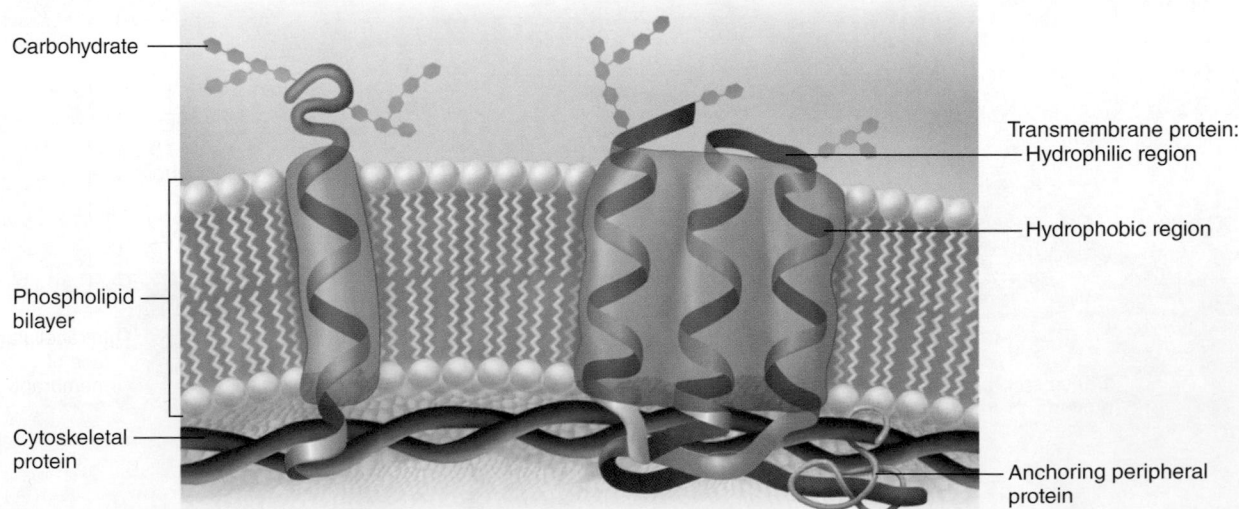

Carbohydrate

Transmembrane protein:
Hydrophilic region

Hydrophobic region

Phospholipid bilayer

Cytoskeletal protein

Anchoring peripheral protein

**FIGURE 3.7 Transmembrane Proteins.** A transmembrane protein has hydrophobic regions embedded in the phospholipid bilayer and hydrophilic regions projecting into the intracellular and extracellular fluids. The protein may cross the membrane once (left) or multiple times (right). The intracellular regions are often anchored to the cytoskeleton by peripheral proteins.

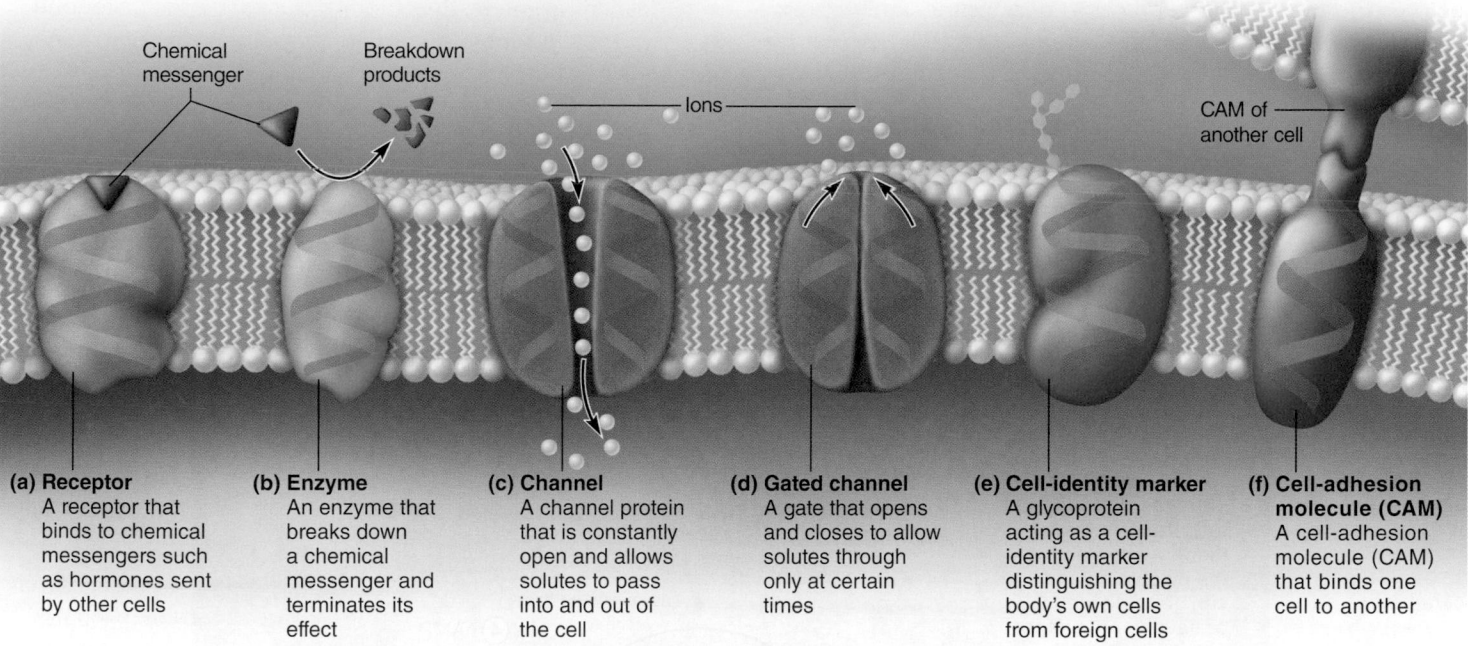

(a) **Receptor**
A receptor that binds to chemical messengers such as hormones sent by other cells

(b) **Enzyme**
An enzyme that breaks down a chemical messenger and terminates its effect

(c) **Channel**
A channel protein that is constantly open and allows solutes to pass into and out of the cell

(d) **Gated channel**
A gate that opens and closes to allow solutes through only at certain times

(e) **Cell-identity marker**
A glycoprotein acting as a cell-identity marker distinguishing the body's own cells from foreign cells

(f) **Cell-adhesion molecule (CAM)**
A cell-adhesion molecule (CAM) that binds one cell to another

**FIGURE 3.8 Some Functions of Membrane Proteins.**

investigating a family of diseases called *channelopathies* that result from defects in channel proteins.

- **Carriers** (see fig. 3.18). Carriers are transmembrane proteins that bind to glucose, electrolytes, and other solutes and transfer them to the other side of the membrane. Some carriers, called **pumps,** consume ATP in the process.

- **Cell-identity markers** (fig. 3.8e). Glycoproteins contribute to the glycocalyx, which acts like an "identification tag" that enables our bodies to tell which cells belong to it and which are foreign invaders.

 **DEEPER INSIGHT 3.1**

**CLINICAL APPLICATION**

### *Calcium Channel Blockers*

*Calcium channel blockers* are a class of drugs that show the therapeutic relevance of understanding gated membrane channels. The walls of the arteries contain smooth muscle that contracts or relaxes to change their diameter. These changes modify the blood flow and strongly influence blood pressure. Blood pressure rises when the arteries constrict and falls when they relax and dilate. Excessive, widespread vasoconstriction can cause hypertension (high blood pressure), and vasoconstriction in the coronary blood vessels of the heart can cause pain (angina) due to inadequate blood flow to the cardiac muscle. In order to contract, a smooth muscle cell must open calcium channels in its plasma membrane and allow calcium to enter from the extracellular fluid. Calcium channel blockers prevent these channels from opening, thereby relaxing the arteries, relieving angina, and lowering blood pressure.

- **Cell-adhesion molecules** (fig. 3.8f). Cells adhere to one another and to extracellular material through membrane proteins called cell-adhesion molecules (CAMs). With few exceptions (such as blood cells and metastasizing cancer cells), cells do not grow or survive normally unless they are mechanically linked to the extracellular material. Special events such as sperm–egg binding and the binding of an immune cell to a cancer cell also require CAMs.

## Second Messengers

**Second messengers** are of such importance that they require a closer look. You will find this information essential for your later understanding of hormone and neurotransmitter action. Let's consider how the hormone epinephrine stimulates a cell. Epinephrine, the "first messenger," cannot pass through the plasma membrane, so it binds to a surface receptor. The receptor is linked on the intracellular side to a peripheral **G protein** (fig. 3.9). G proteins are named for the ATP-like chemical, guanosine triphosphate (GTP), from which they get their energy. When activated by the receptor, a G protein relays the signal to another membrane protein, **adenylate cyclase** (ah-DEN-ih-late SY-clase). Adenylate cyclase removes two phosphate groups from ATP and converts it to **cyclic AMP (cAMP),** the second messenger (see fig. 2.30b, p. 71). Cyclic AMP then activates cytoplasmic enzymes called **kinases** (KY-nace-es), which add phosphate groups to other cellular enzymes. This activates some enzymes and deactivates others, but either way, it triggers a great variety of physiological changes within the cell. Up to 60% of drugs work by altering the activity of G proteins.

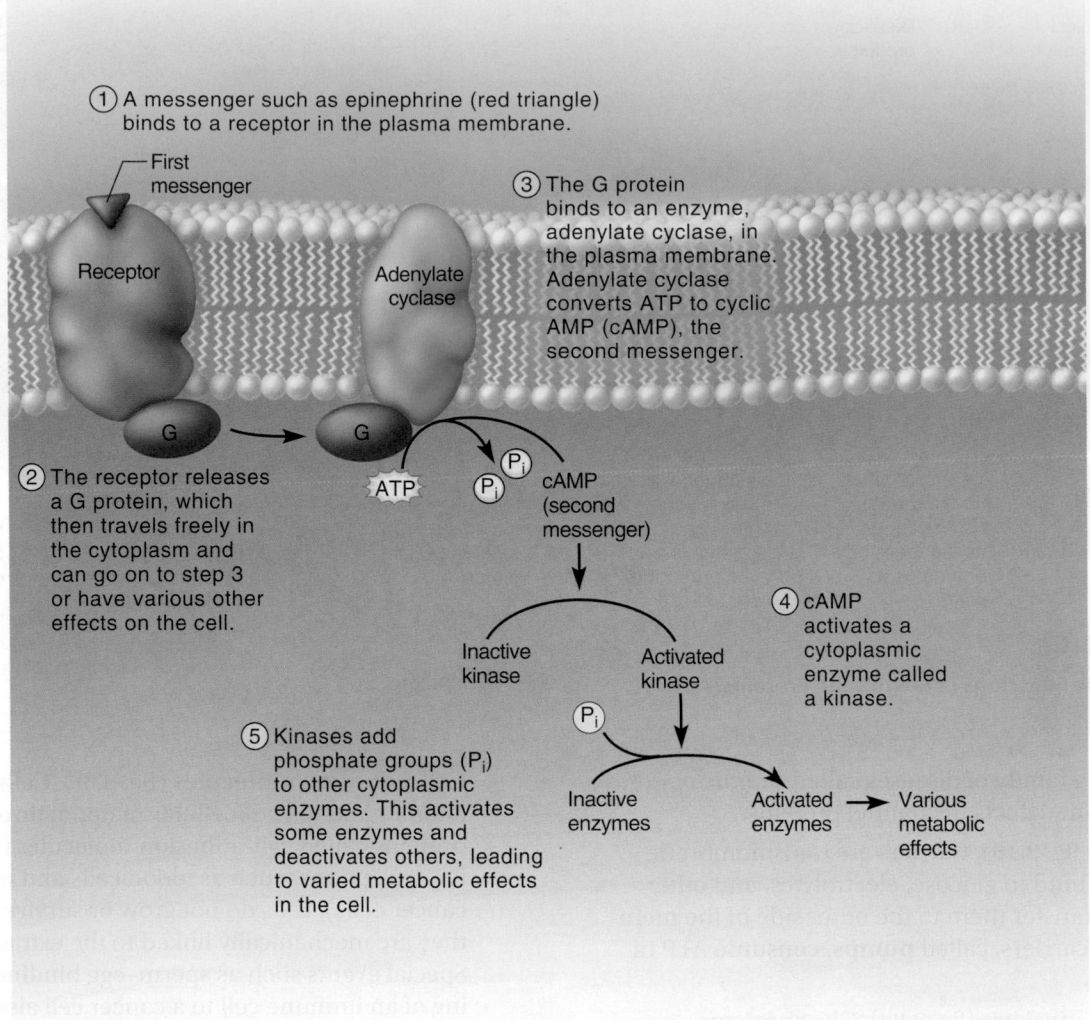

**FIGURE 3.9  A Second-Messenger System.**

❓ *Is adenylate cyclase a transmembrane protein or a peripheral protein? What about the G protein?*

## The Glycocalyx

External to the plasma membrane, all animal cells have a fuzzy coat called the **glycocalyx**[9] (GLY-co-CAY-licks) (fig. 3.10), composed of the carbohydrate moieties of membrane glycolipids and glycoproteins. It is chemically unique in everyone but identical twins, and acts like an identification tag that enables the body to distinguish its own healthy cells from transplanted tissues, invading organisms, and diseased cells. Human blood types and transfusion compatibility are determined by glycolipids. Functions of the glycocalyx are summarized in table 3.2.

| TABLE 3.2 | Functions of the Glycocalyx |
|---|---|
| Protection | Cushions the plasma membrane and protects it from physical and chemical injury |
| Immunity to infection | Enables the immune system to recognize and selectively attack foreign organisms |
| Defense against cancer | Changes in the glycocalyx of cancerous cells enable the immune system to recognize and destroy them |
| Transplant compatibility | Forms the basis for compatibility of blood transfusions, tissue grafts, and organ transplants |
| Cell adhesion | Binds cells together so tissues do not fall apart |
| Fertilization | Enables sperm to recognize and bind to eggs |
| Embryonic development | Guides embryonic cells to their destinations in the body |

[9]*glyco* = sugar; *calyx* = cup, vessel

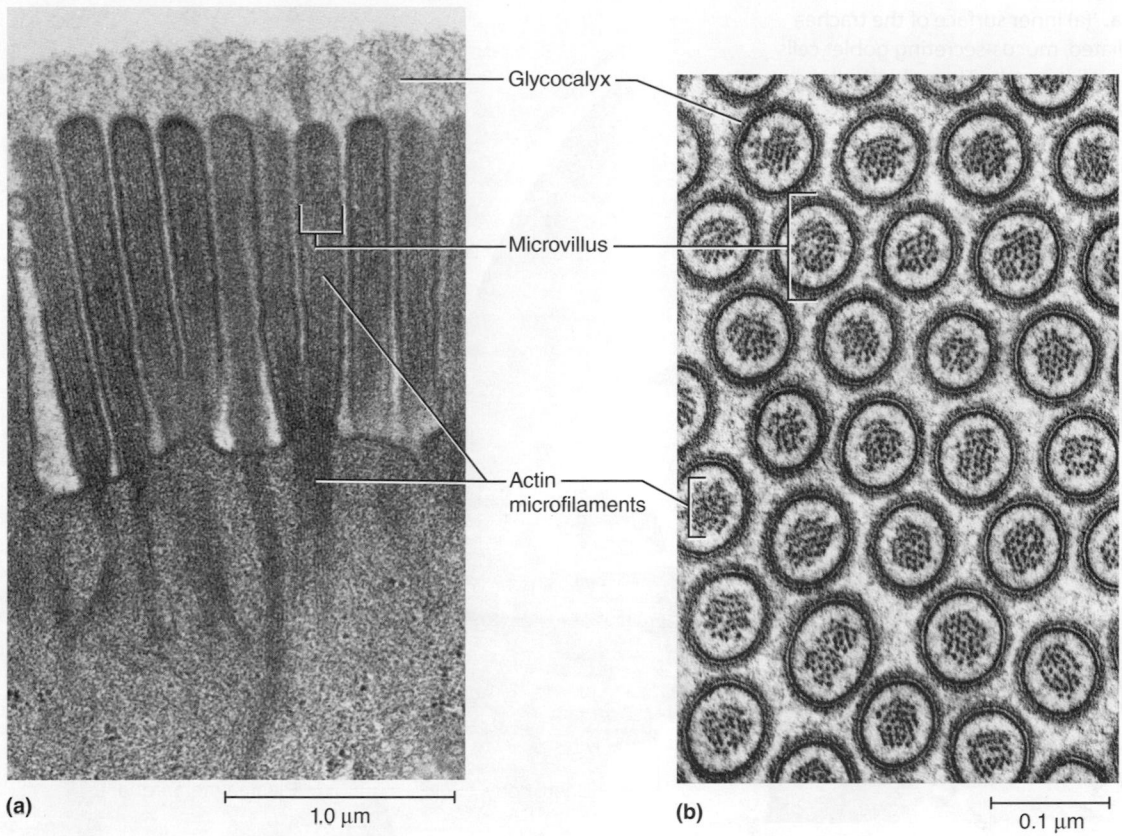

**FIGURE 3.10 Microvilli and the Glycocalyx (TEM).** The microvilli are anchored by microfilaments of actin, which occupy the core of each microvillus and project into the cytoplasm. (a) Longitudinal section, perpendicular to the cell surface. (b) Cross section.

## Microvilli, Cilia, Flagella, and Pseudopods

Many cells have surface extensions called *microvilli, cilia, flagella,* and *pseudopods.* These aid in absorption, movement, and sensory processes.

### Microvilli

**Microvilli**[10] (MY-cro-VIL-eye; singular, *microvillus*) are extensions of the plasma membrane that serve primarily to increase a cell's surface area (figs. 3.10 and 3.11a, c). They are best developed in cells specialized for absorption, such as the epithelial cells of the intestines and kidney tubules. They give such cells 15 to 40 times as much absorptive surface area as they would have if their apical surfaces were flat.

Individual microvilli cannot be distinguished very well with the light microscope because they are only 1 to 2 μm long. On some cells, they are very dense and appear as a fringe called the **brush border** at the apical cell surface. With the scanning electron microscope, they resemble a deep-pile carpet. With the transmission electron microscope, microvilli typically look like finger-shaped projections of the cell surface. They show little internal structure, but some have a bundle of stiff filaments of a protein called *actin.* Actin filaments attach to the inside of

the plasma membrane at the tip of the microvillus, and at its base they extend a little way into the cell and anchor the microvillus to a protein mesh called the *terminal web.* When tugged by another protein in the cytoplasm, actin can shorten a microvillus to milk its absorbed contents downward into the cell.

In contrast to the long, shaggy microvilli of absorptive cells, those on many other cells are little more than tiny bumps on the surface. On cells of the taste buds and inner ear, they are well developed but serve sensory rather than absorptive functions.

### Cilia

Cilia (SIL-ee-uh; singular, *cilium*[11]) (fig. 3.11) are hairlike processes about 7 to 10 μm long. Nearly every human cell has a single, nonmotile *primary cilium* a few micrometers long. Its function in some cases is still a mystery, but apparently many of them are sensory, serving as the cell's "antenna" for monitoring nearby conditions. In the inner ear, they play a role in the sense of balance; in the retina of the eye, they are highly elaborate and form the light-absorbing part of the receptor cells; and in the kidney, they are thought to monitor the flow of fluid as it is processed into urine. In some cases, they open calcium gates in the plasma membrane, activating an informative signal in

---

[10]*micro* = small; *villi* = hairs

[11]*cilium* = eyelash

**FIGURE 3.11  Cilia.** (a) Inner surface of the trachea (SEM). Several nonciliated, mucus-secreting goblet cells are visible among the ciliated cells. The goblet cells have short microvilli on their surface. (b) Three-dimensional structure of a cilium. (c) Cross section of a few cilia and microvilli (TEM). (d) Cross-sectional structure of a cilium. Note the relative sizes of cilia and microvilli in parts (a) and (c).

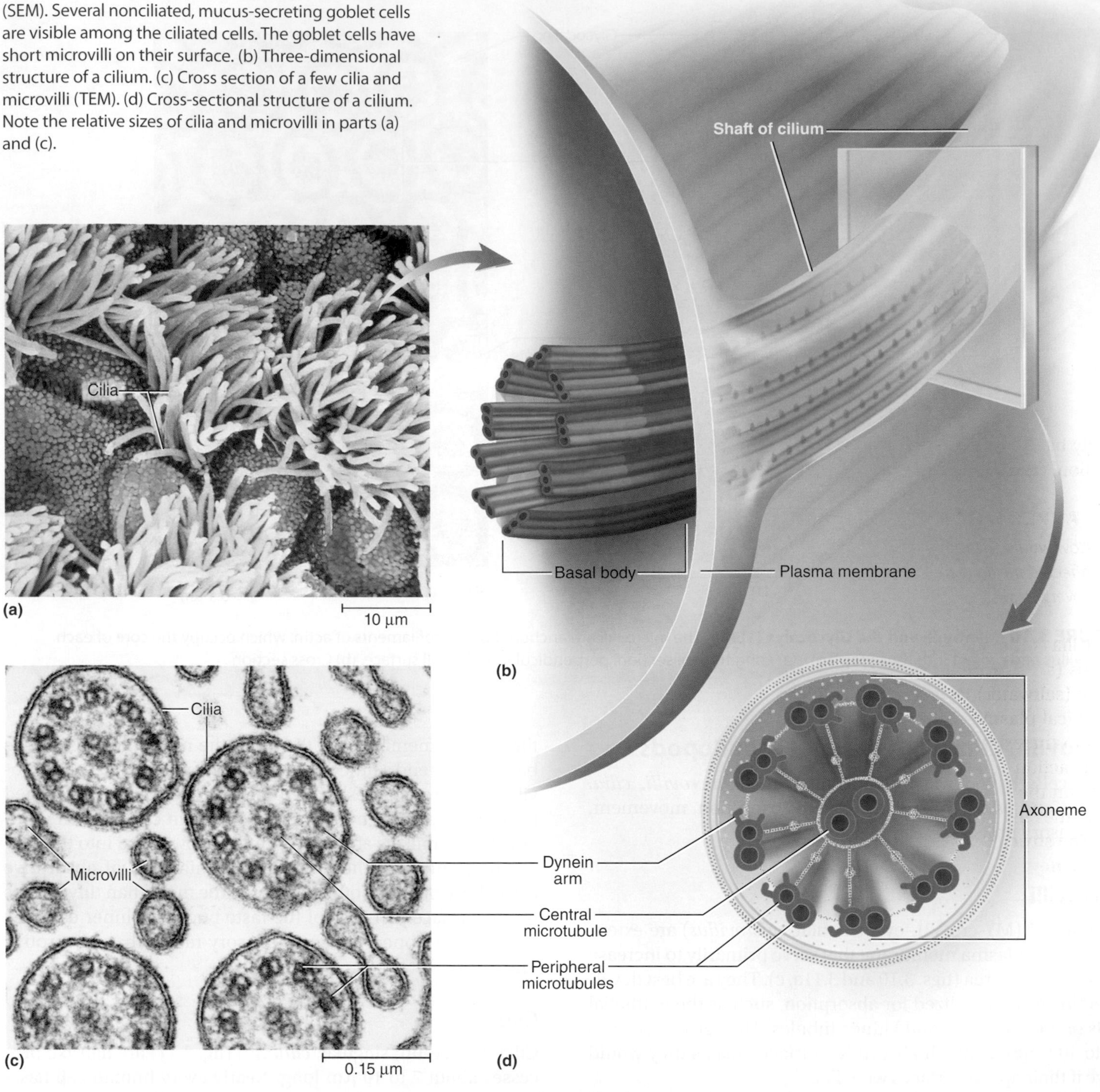

(a)

Cilia

10 μm

Shaft of cilium

Basal body      Plasma membrane

(b)

Cilia

Microvilli

0.15 μm

(c)                  (d)

Dynein arm

Central microtubule

Peripheral microtubules

Axoneme

the cell. Sensory cells in the nose have multiple nonmotile cilia that bind odor molecules. Defects in the development, structure, or function of cilia—especially these nonmotile primary cilia—are responsible for several hereditary diseases called *ciliopathies.* (See Testing Your Comprehension question 5 on p. 111.)

Motile cilia are less widespread, occurring in the respiratory tract, uterine (fallopian) tubes, internal cavities *(ventricles)* of the brain, and short ducts *(efferent ductules)* associated with

the testes. There may be 50 to 200 cilia on the surface of one cell. They beat in waves that sweep across the surface of an epithelium, always in the same direction (fig. 3.12), propelling such materials as mucus, an egg cell, or cerebrospinal fluid. Each cilium bends stiffly forward and produces a *power stroke* that pushes along the mucus or other matter. Shortly after a cilium begins its power stroke, the one just ahead of it begins, and the next and the next—collectively producing a wavelike motion. After a cilium completes its power stroke, it is pulled

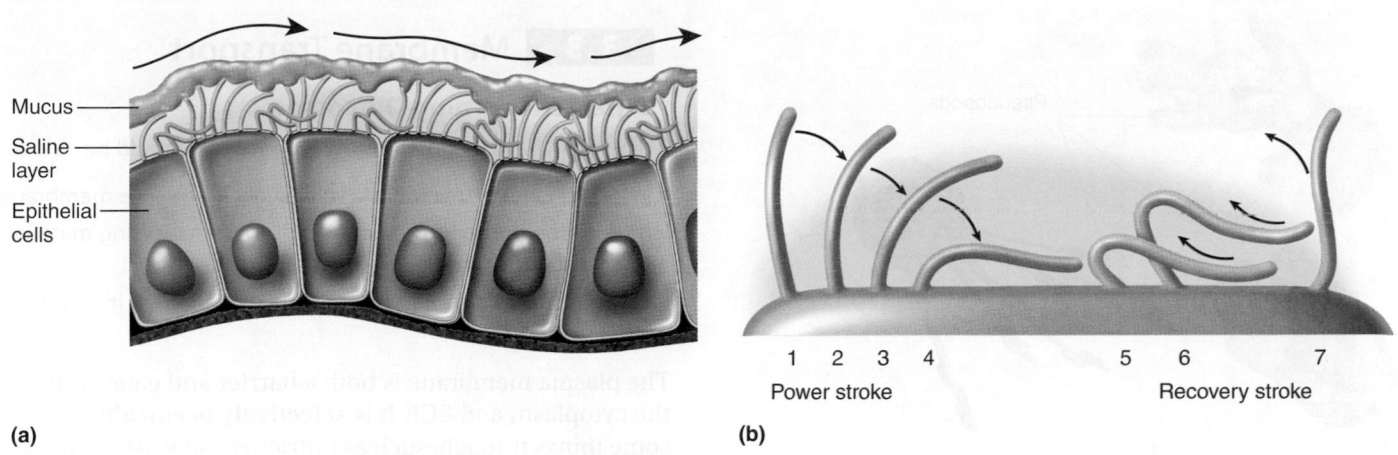

**(a)**    **(b)**

**FIGURE 3.12 Ciliary Action.** (a) Cilia of an epithelium moving mucus along a surface layer of saline. (b) Power and recovery strokes of an individual cilium.

limply back by a *recovery stroke* that restores it to the upright position, ready to flex again.

**►►►APPLY WHAT YOU KNOW**

*How would the movement of mucus in the respiratory tract be affected if cilia were equally stiff on both their power and recovery strokes?*

Cilia could not beat freely if they were embedded in sticky mucus (see Deeper Insight 3.2). Instead, they beat within a saline (saltwater) layer at the cell surface. *Chloride pumps* in the apical plasma membrane produce this layer by pumping Cl⁻ into the extracellular fluid. Sodium ions follow by electrical attraction and water follows by osmosis. Mucus essentially floats on the surface of this layer and is pushed along by the tips of the cilia.

The structural basis for ciliary movement is a core called the **axoneme**[12] (ACK-so-neem), which consists of an array of

---

[12]*axo* = axis; *neme* = thread

 **DEEPER INSIGHT 3.2**

**CLINICAL APPLICATION**

**Cystic Fibrosis**

The significance of chloride pumps becomes especially evident in *cystic fibrosis (CF)*, a hereditary disease affecting primarily white children of European descent. CF is usually caused by a defect in which cells make chloride pumps but fail to install them in the plasma membrane. Consequently, there is an inadequate saline layer on the cell surface and the mucus is dehydrated and overly sticky. This thick mucus plugs the ducts of the pancreas and prevents it from secreting digestive enzymes into the small intestine, so digestion and nutrition are compromised. In the respiratory tract, the mucus clogs the cilia and prevents them from beating freely. The respiratory tract becomes congested with thick mucus, often leading to chronic infection and pulmonary collapse. The mean life expectancy of people with CF is about 30 years.

thin protein cylinders called *microtubules.* There are two central microtubules surrounded by a ring of nine microtubule pairs—an arrangement called the *9 + 2 structure*, reminiscent of a Ferris wheel in cross section (fig. 3.11d). The central microtubules stop at the cell surface, but the peripheral microtubules continue a short distance into the cell as part of a **basal body** that anchors the cilium. In each pair of peripheral microtubules, one tubule has two little **dynein**[13] (DINE-een) **arms.** Dynein, a motor protein, uses energy from ATP to "crawl" up the adjacent pair of microtubules. When microtubules on the front of the cilium crawl up the microtubules behind them, the cilium bends toward the front. The primary cilia, which cannot move, lack the two central microtubules but still have the nine peripheral pairs; they are said to have a *9 + 0 structure.*

### Flagella

There is only one functional **flagellum**[14] (fla-JEL-um) in humans—the whiplike tail of a sperm. It is much longer than a cilium and has an identical axoneme, but between the axoneme and plasma membrane it also has a complex sheath of coarse fibers that stiffen the tail and give it more propulsive power. A flagellum does not beat with power and recovery strokes like those of a cilium, but in a more undulating, snakelike or corkscrew fashion. It is described in further detail on page 1047.

### Pseudopods

**Pseudopods**[15] (SOO-do-pods) are cytoplasm-filled extensions of the cell varying in shape from fine, filamentous to blunt fingerlike processes (fig. 3.13). Unlike the other three kinds of surface extensions, they change continually. Some form anew as the cell surface bubbles outward and cytoplasm flows into a

---

[13]*dyn* = power, energy; *in* = protein
[14]*flagellum* = whip
[15]*pseudo* = false; *pod* = foot

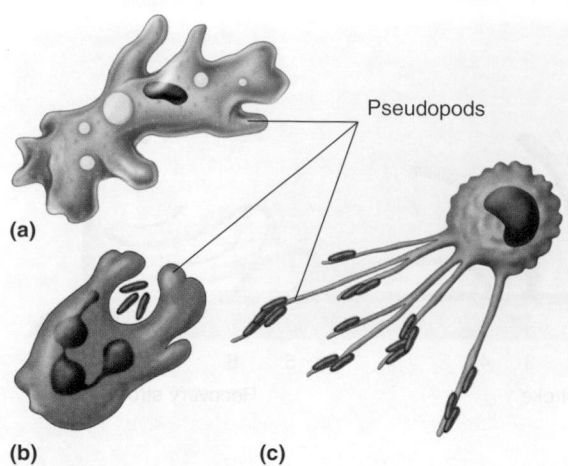

Pseudopods

(a)

(b)          (c)

**FIGURE 3.13  Pseudopods.** (a) *Amoeba,* a freshwater organism that crawls and captures food by means of pseudopods. (b) A neutrophil (white blood cell) that similarly uses pseudopods for locomotion and capturing bacteria. (c) A macrophage extending filamentous pseudopods to snare and "reel in" bacteria.

lengthening pseudopod, while others are retracted into the cell by disassembling protein filaments that supported them like a scaffold.

The freshwater organism *Amoeba* furnishes a familiar example of pseudopods, which it uses for locomotion and food capture. White blood cells called *neutrophils* crawl about like amebae by means of fingerlike pseudopods, and when they encounter a bacterium or other foreign particle, they reach out with their pseudopods to surround and engulf it. *Macrophages*—tissue cells derived from certain white blood cells—reach out with thin filamentous pseudopods to snare bacteria and cell debris and "reel them in" to be digested by the cell. Like little janitors, macrophages thereby keep our tissues cleaned up. Blood platelets reach out with thin pseudopods to adhere to each other and to the walls of damaged blood vessels, forming plugs that temporarily stop bleeding (see fig. 18.21, p. 698).

**BEFORE YOU GO ON**

Answer the following questions to test your understanding of the preceding section:

6. How does the structure of a plasma membrane depend on the amphipathic nature of phospholipids?

7. Distinguish between integral and peripheral proteins.

8. Explain the differences between a receptor, pump, and cell-adhesion molecule.

9. How does a gate differ from other channel proteins? What three factors open and close membrane gates?

10. What roles do cAMP, adenylate cyclase, and kinases play in cellular function?

11. Identify several reasons why the glycocalyx is important to human survival.

12. How do microvilli and cilia differ in structure and function?

**3.3**   **Membrane Transport**

**Expected Learning Outcomes**

When you have completed this section, you should be able to

a. explain what is meant by a selectively permeable membrane;

b. describe the various mechanisms for transporting material through the plasma membrane; and

c. define *osmolarity* and *tonicity* and explain their importance.

The plasma membrane is both a barrier and gateway between the cytoplasm and ECF. It is **selectively permeable**—it allows some things through, such as nutrients and wastes, but usually prevents other things, such as proteins and phosphates, from entering or leaving the cell.

The methods of moving substances through the membrane can be classified in two overlapping ways: as passive or active mechanisms and as *carrier-mediated* or not. *Passive* mechanisms require no energy (ATP) expenditure by the cell. In most cases, the random molecular motion of the particles themselves provides the necessary energy. Passive mechanisms include filtration, diffusion, and osmosis. *Active* mechanisms, however, consume ATP. These include active transport and vesicular transport. *Carrier-mediated* mechanisms use a membrane protein to transport substances from one side of the membrane to the other.

**Filtration**

**Filtration** is a process in which a physical pressure forces fluid through a selectively permeable membrane. A coffee filter provides an everyday example. The weight of the water drives water and dissolved matter through the filter, while the filter holds back larger particles (the coffee grounds). In physiology, the most important case of filtration is seen in the blood capillaries, where blood pressure forces fluid through gaps in the capillary wall (fig. 3.14). This is how water, salts, nutrients, and other solutes are transferred from the bloodstream to the tissue fluid and how the kidneys filter wastes from the blood. Capillaries hold back larger particles such as blood cells and proteins. In most cases, water and solutes filter through narrow gaps between the capillary cells. In some capillaries, however, the cells have large *filtration pores* through them, like the holes in a slice of Swiss cheese, allowing for more rapid filtration of large solutes such as protein hormones.

**Simple Diffusion**

**Simple diffusion** is the net movement of particles from a place of high concentration to a place of lower concentration as a result of their constant, spontaneous motion. In other words, substances diffuse *down their concentration gradients* (see Gradients and Flow, p. 17). Molecules move at astonishing speeds. At body temperature, the average water molecule moves about

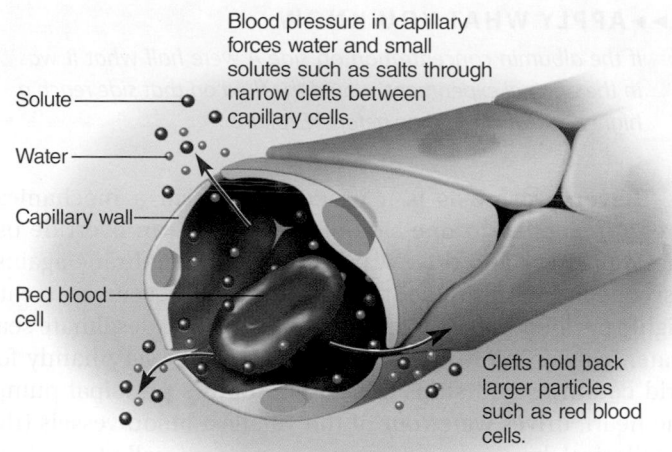

Solute

Water

Capillary wall

Red blood cell

Blood pressure in capillary forces water and small solutes such as salts through narrow clefts between capillary cells.

Clefts hold back larger particles such as red blood cells.

**FIGURE 3.14 Filtration Through the Wall of a Blood Capillary.** Water and small solutes pass through gaps between cells, while blood cells and other large particles are held back.

2,500 km/h (1,500 mi./h)! However, a molecule can travel only a very short distance before colliding with another and careening off in a new direction, like colliding billiard balls. The rate of diffusion, therefore, is much slower than the rate of molecular motion.

Diffusion occurs readily in air or water and doesn't necessarily need a membrane. However, if there is a membrane in the path of the diffusing molecules, and if it is permeable to that substance, the molecules will pass from one side of the membrane to the other. This is how oxygen passes from the air we inhale into the bloodstream. Dialysis treatment for kidney patients is based on diffusion of solutes through artificial *dialysis membranes.*

Diffusion rates are very important to cell survival because they determine how quickly a cell can acquire nutrients or rid itself of wastes. Some factors that affect the rate of diffusion through a membrane are as follows:

- **Temperature.** Diffusion is driven by the kinetic energy of the particles, and temperature is a measure of that kinetic energy. The warmer a substance is, the more rapidly its particles diffuse. This is why sugar diffuses more quickly through hot tea than through iced tea.

- **Molecular weight.** Heavy molecules such as proteins move more sluggishly and diffuse more slowly than light particles such as electrolytes and gases. Small molecules also pass through membrane pores more easily than large ones.

- **"Steepness" of the concentration gradient.** The steepness of a gradient refers to the concentration difference between two points. Particles diffuse more rapidly if there is a greater concentration difference.

- **Membrane surface area.** As noted earlier, the apical surface of cells specialized for absorption (for example, in the small intestine) is often extensively folded into

microvilli. This makes more membrane available for particles to diffuse through.

- **Membrane permeability.** Diffusion through a membrane depends on how permeable it is to the particles. For example, potassium ions diffuse more rapidly than sodium ions through a plasma membrane. Nonpolar, hydrophobic, lipid-soluble substances such as oxygen, nitric oxide, alcohol, and steroids diffuse through the phospholipid regions of a plasma membrane. Water and small charged, hydrophilic solutes such as electrolytes do not mix with lipids but diffuse primarily through channel proteins in the membrane. Cells can adjust their permeability to such a substance by adding channel proteins to the membrane, by taking them away, or by opening and closing membrane gates.

## Osmosis

**Osmosis**[16] is the net flow of water from one side of a selectively permeable membrane to the other. It is crucial to the body's water distribution (fluid balance). Imbalances in osmosis underlie such problems as diarrhea, constipation, and edema (tissue swelling); osmosis also is a vital consideration in intravenous (I.V.) fluid therapy.

Osmosis occurs through nonliving membranes, such as cellophane and dialysis membranes, and through the plasma membranes of cells. The usual direction of net movement is from the side with the higher concentration of water molecules (less dissolved matter) to the side with the lower water concentration (more dissolved matter)—that is, down the water concentration gradient. The reason for this is that when water molecules encounter a solute particle, they tend to associate with it to form a *hydration sphere* (see fig. 2.9, p. 50). Even though this is a loose, reversible attraction, it does make those water molecules less available to diffuse back across the membrane to the side from which they came. In essence, solute particles on one side of the membrane draw water away from the other side. Thus, water accumulates on the side with the most solute. All of this assumes that the solute molecules cannot pass through the membrane, but stay on one side. The rate and direction of osmosis depend on the relative concentration of these nonpermeating solutes on the two sides of the membrane.

Significant amounts of water diffuse even through the hydrophobic, phospholipid regions of a plasma membrane, but it diffuses more easily through the channels of transmembrane proteins called **aquaporins,** specialized for the passage of water. Cells can increase the rate of osmosis by installing more aquaporins in the membrane or decrease the rate by removing them. Certain cells of the kidney, for example, regulate the rate of urinary water loss by adding or removing aquaporins.

A cell can exchange a tremendous amount of water by osmosis. In red blood cells, for example, the amount of water passing through the plasma membrane every second is 100 times the volume of the cell.

[16]*osm* = push, thrust; *osis* = condition, process

Figure 3.15 is a conceptual model of osmosis. Imagine a chamber divided by a selectively permeable membrane. Side A contains large particles that cannot pass through the membrane pores—a *nonpermeating* solute such as albumin (egg white protein). Side B contains distilled water. Water passes down its concentration gradient from B to A (fig. 3.15a) and associates with the solute molecules on side A, hindering water movement back to side B.

Under such conditions, the water level in side B would fall and the level in side A would rise. It might seem as if this would continue indefinitely until side B dried up. This would not happen, however, because as water accumulated in side A, it would become heavier and exert more force, called **hydrostatic pressure,** on that side of the membrane. This would cause some filtration of water from A back to B. At some point, the rate of filtration would equal the rate of "forward" osmosis, water would pass through the membrane equally in both directions, and net osmosis would slow down and stop. At this point, an equilibrium (balance between opposing forces) would exist. The hydrostatic pressure required on side A to halt osmosis is called **osmotic pressure.** The more nonpermeating solute there is in A, the greater the osmotic pressure.

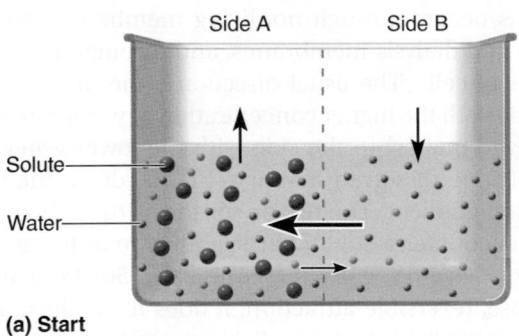

**(a) Start**

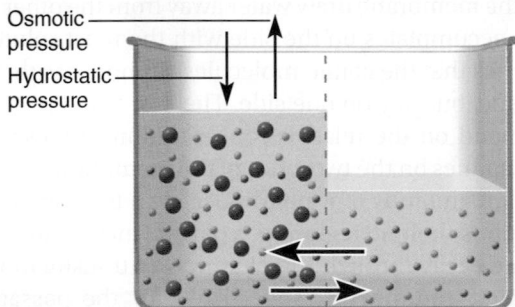

**(b) 30 minutes later**

**FIGURE 3.15 Osmosis.** The dashed line represents a selectively permeable membrane dividing the chamber in half. The large particles on side A represent any solute, such as albumin, too large to pass through the membrane. The small particles are water molecules. (a) Water diffuses from side B, where it is relatively concentrated, to side A, where it is less concentrated. Fluid level rises in side A and falls in side B. (b) Net diffusion stops when the weight (hydrostatic pressure) of the fluid in side A balances the osmotic pressure. At this point, water passes at equal rates from A to B by filtration and from B to A by osmosis. The two processes are then in equilibrium.

▶▶▶**APPLY WHAT YOU KNOW**

*If the albumin concentration on side A were half what it was in the original experiment, would the fluid on that side reach a higher or lower level than before? Explain.*

**Reverse osmosis** is a process in which a mechanical pressure applied to one side of the system can override osmotic pressure and drive water through a membrane against its concentration gradient. This principle is used to create highly purified water for laboratory use and to desalinate seawater, converting it to drinkable freshwater—very handy for arid countries and ships at sea. The body's principal pump, the heart, drives water out of the smallest blood vessels (the capillaries) by reverse osmosis—a process called *capillary filtration.* The equilibrium between osmosis and filtration will be an important consideration when we study fluid exchange by the capillaries in chapter 20. Blood plasma also contains albumin. In the preceding discussion, side A is analogous to the bloodstream and side B to the tissue fluid surrounding the capillaries. Water leaves the capillaries by filtration, but this is approximately balanced by water reentering the capillaries by osmosis.

## Osmolarity and Tonicity

The osmotic concentration of body fluids has such a great effect on cellular function that it is important to understand the units in which it is measured. One **osmole** is 1 mole of dissolved particles. If a solute does not ionize in water, then 1 mole of the solute yields 1 osmole (osm) of dissolved particles. A solution of 1-molar (1 M) glucose, for example, is also 1 osm/L. If a solute does ionize, it yields two or more dissolved particles in solution. A 1 M solution of NaCl, for example, contains 1 mole of sodium ions and 1 mole of chloride ions per liter. Both ions affect osmosis and must be separately counted in a measure of osmotic concentration. Thus, 1 M NaCl = 2 osm/L.

**Osmolality** is the number of osmoles of solute *per kilogram of water,* and **osmolarity** is the number of osmoles *per liter of solution.* Most clinical calculations are based on osmolarity, since it is easier to measure the volume of a solution than the weight of water it contains. At the concentrations of human body fluids, there is less than 1% difference between osmolality and osmolarity, and the two terms are nearly interchangeable. All body fluids and many clinical solutions are mixtures of many chemicals. The osmolarity of such a solution is the total osmotic concentration of all of its dissolved particles.

A concentration of 1 osm/L is much higher than we find in most body fluids, so physiological concentrations are usually expressed in **milliosmoles per liter** (mOsm/L) (1 mOsm/L = $10^{-3}$ osm/L). Blood plasma, tissue fluid, and intracellular fluid measure about 300 mOsm/L.

**Tonicity** is the ability of a solution to affect the fluid volume and pressure in a cell. If a solute cannot pass through a plasma membrane but remains more concentrated on one side of the membrane than on the other, it causes osmosis.

**(a) Hypotonic**   **(b) Isotonic**   **(c) Hypertonic**

**FIGURE 3.16 Effects of Tonicity on Red Blood Cells (RBCs).** (a) RBC swelling in a hypotonic medium such as distilled water. (b) Normal RBC size and shape in an isotonic medium such as 0.9% NaCl. (c) RBC shriveling in a hypertonic medium such as 2% NaCl.

A **hypotonic**[17] solution has a lower concentration of nonpermeating solutes than the intracellular fluid (ICF). Cells in a hypotonic solution absorb water, swell, and may burst *(lyse)* (fig. 3.16a). Distilled water is the extreme example; a sufficient quantity given to a person intravenously would lyse the blood cells, with dire consequences. A **hypertonic**[18] solution is one with a higher concentration of nonpermeating solutes than the ICF. It causes cells to lose water and shrivel *(crenate)* (fig. 3.16c). Such cells may die of torn membranes and cytoplasmic loss. In **isotonic**[19] solutions, the total concentration of nonpermeating solutes is the same as in the ICF—hence, isotonic solutions cause no change in cell volume or shape (fig. 3.16b).

It is essential for cells to be in a state of osmotic equilibrium with the fluid around them, and this requires that the ECF have the same concentration of nonpermeating solutes as the ICF. Intravenous fluids given to patients are usually isotonic solutions, but hypertonic or hypotonic fluids are given for special purposes. A 0.9% solution of NaCl, called *normal saline,* is isotonic to human blood cells.

It is important to note that osmolarity and tonicity are not the same. Urea, for example, is a small organic molecule that easily penetrates plasma membranes. If cells are placed in 300 mOsm/L urea, urea diffuses into them (down its concentration gradient), water follows by osmosis, and the cells swell and burst. Thus, 300 mOsm/L urea is not isotonic to the cells. Sodium chloride, by contrast, penetrates plasma membranes poorly. In 300 mOsm/L NaCl, there is little change in cell volume; this solution is isotonic to cells.

[17]*hypo* = less; *ton* = tension
[18]*hyper* = more; *ton* = tension
[19]*iso* = equal; *ton* = tension

## Carrier-Mediated Transport

The processes of membrane transport described up to this point do not necessarily require a cell membrane; they can occur as well through artificial membranes. Now, however, we come to processes for which a cell membrane is necessary, because they employ transport proteins, or carriers. Thus, the next three processes are classified as **carrier-mediated transport.** In these cases, a solute binds to a carrier in the plasma membrane, which then changes shape and releases the solute to the other side. Carriers can move substances into or out of a cell, and into or out of organelles within the cell. The process is very rapid; for example, one carrier can transport 1,000 glucose molecules per second across the membrane.

Carriers act like enzymes in some ways: The solute is a ligand that binds to a specific receptor site on the carrier, like a substrate binding to the active site of an enzyme. The carrier exhibits **specificity** for its ligand, just as an enzyme does for its substrate. A glucose carrier, for example, cannot transport fructose. Carriers also exhibit **saturation;** as the solute concentration rises, its rate of transport increases, but only up to a point. When every carrier is occupied, adding more solute cannot make the process go any faster. The carriers are saturated—no more are available to handle the increased demand, and transport levels off at a rate called the **transport maximum (T$_m$)** (fig. 3.17). You could think of carriers as analogous to buses. If all the buses on a given line are full ("saturated"), they cannot carry any more passengers, regardless of how many people are waiting at the bus stop. As we'll see later in the book, the T$_m$ explains why glucose appears in the urine of people with diabetes mellitus.

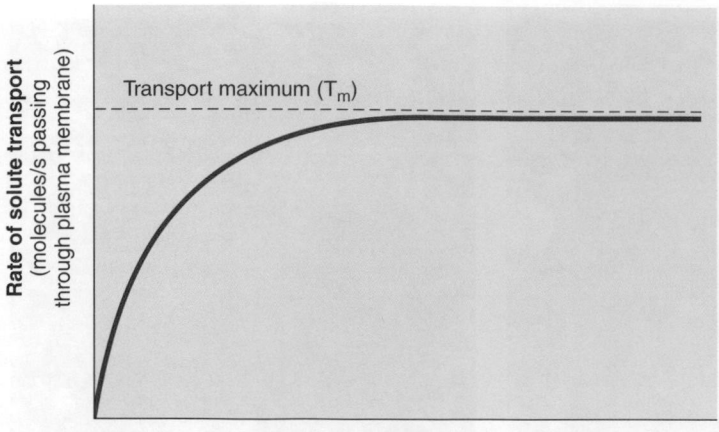

**FIGURE 3.17  Carrier Saturation and Transport Maximum.**  Up to a point, increasing the solute concentration increases the rate of transport through a membrane. At the transport maximum ($T_m$), however, all carrier proteins are busy and cannot transport the solute any faster, even if more solute is added.

An important difference between a carrier and an enzyme is that carriers do not chemically change their ligands; they simply pick them up on one side of the membrane and release them, unchanged, on the other.

There are three kinds of carriers: uniports, symports, and antiports. A **uniport**[20] carries only one type of solute. For example, most cells pump out calcium by means of a uniport, maintaining a low intracellular concentration so calcium salts don't crystallize in the cytoplasm. Some carriers move two or more solutes through a membrane simultaneously in the same

direction; this process is called **cotransport**[21] and the carrier protein that performs it is called a **symport.**[22] For example, absorptive cells of the small intestine and kidneys have a symport that takes up sodium and glucose simultaneously. Other carriers move two or more solutes in opposite directions; this process is called **countertransport** and the carrier protein is called an **antiport.**[23] For example, nearly all cells have an antiport called the *sodium–potassium pump* that continually removes Na⁺ from the cell and brings in K⁺.

There are three mechanisms of carrier-mediated transport: facilitated diffusion, primary active transport, and secondary active transport. **Facilitated**[24] **diffusion** (fig. 3.18) is the carrier-mediated transport of a solute through a membrane *down its concentration gradient*. It does not require any expenditure of metabolic energy (ATP) by the cell. It transports solutes such as glucose that cannot pass through the membrane unaided. The solute attaches to a binding site on the carrier, then the carrier changes conformation and releases the solute on the other side of the membrane.

**Primary active transport** is a process in which a carrier moves a substance through a cell membrane *up its concentration gradient* using energy provided by ATP. Just as rolling a ball up a ramp would require you to push it (an energy input), this mechanism requires energy to move material up its concentration gradient. ATP supplies this energy by transferring a phosphate group to the transport protein. The calcium pump mentioned previously uses this mechanism. Even though Ca²⁺ is already more concentrated in the ECF than within the cell, this carrier pumps still more of it out. Active transport also

---

[20]*uni* = one; *port* = carry

[21]*co* = together; *trans* = across; *port* = carry
[22]*sym* = together; *port* = carry
[23]*anti* = opposite; *port* = carry
[24]*facil* = easy

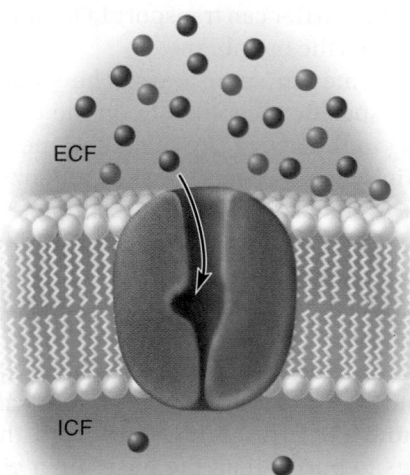

① A solute particle enters the channel of a membrane protein (carrier).

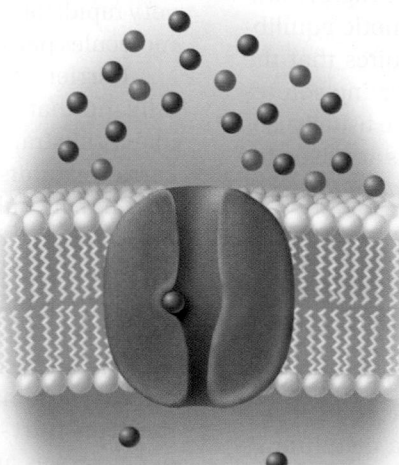

② The solute binds to a receptor site on the carrier and the carrier changes conformation.

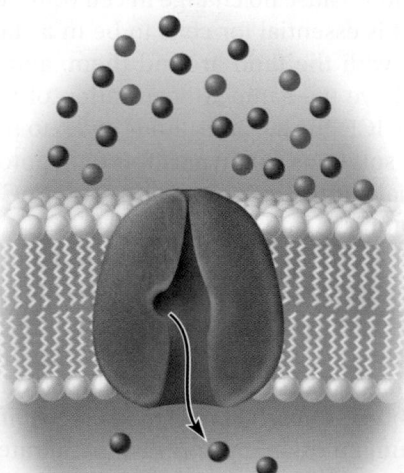

③ The carrier releases the solute on the other side of the membrane.

**FIGURE 3.18  Facilitated Diffusion.**  Note that the solute moves down its concentration gradient.

enables cells to absorb amino acids that are already more concentrated in the cytoplasm than in the ECF.

**Secondary active transport** also requires an energy input, but depends only indirectly on ATP. For example, certain kidney tubules have proteins called *sodium–glucose transporters (SGLTs)* that simultaneously bind sodium ions ($Na^+$) and glucose molecules and transport them into the tubule cells, saving glucose from being lost in the urine (fig. 3.19). An SGLT itself does not use ATP. However, it depends on the fact that the cell actively maintains a low internal $Na^+$ concentration, so $Na^+$ will diffuse down its gradient into the cell. Glucose "hitches a ride" with the ingoing $Na^+$. But what keeps the intracellular $Na^+$ concentration low is that the basal membrane of the cell has an ATP-driven **sodium–potassium ($Na^+$–$K^+$) pump** that

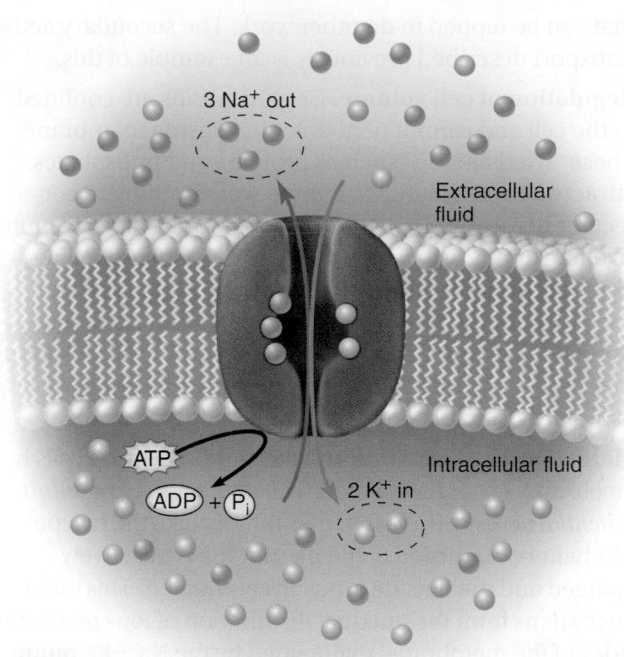

**FIGURE 3.20 The Sodium–Potassium Pump ($Na^+$–$K^+$ ATPase).** In each cycle of action, this membrane carrier removes three sodium ions ($Na^+$) from the cell, brings two potassium ions ($K^+$) into the cell, and hydrolyzes one molecule of ATP.

❓ *Why would the $Na^+$–$K^+$ pump, but not osmosis, cease to function after a cell dies?*

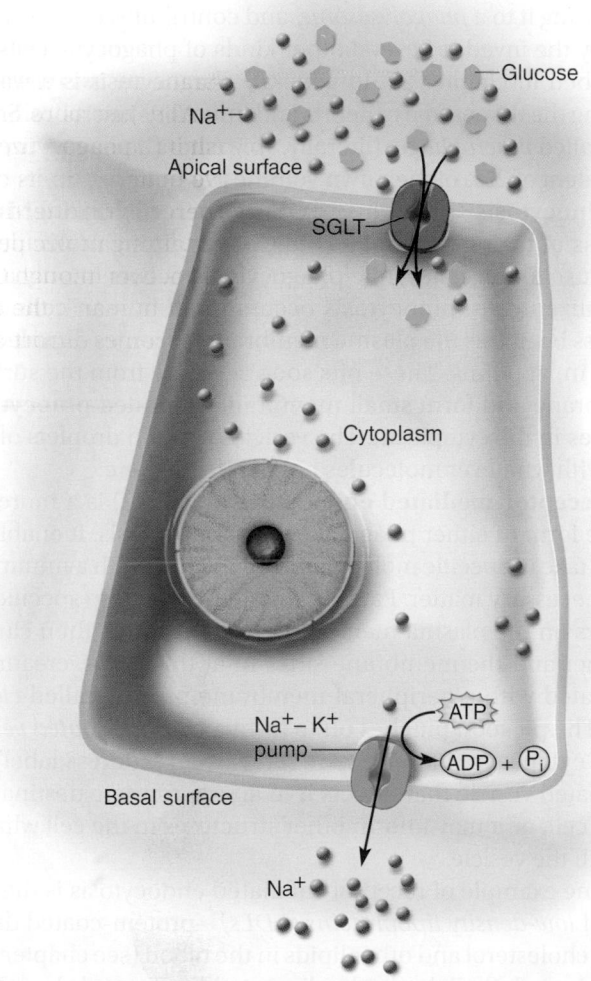

**FIGURE 3.19 Secondary Active Transport.** At the basal surface of the cell, an $Na^+$–$K^+$ pump removes sodium ions ($Na^+$) from the cytoplasm, maintaining a low sodium concentration within the cell. At the apical surface, $Na^+$ enters the cell by facilitated diffusion, following its concentration gradient. It can gain entry to the cell only by binding to a carrier, the sodium–glucose transporter (SGLT), which simultaneously binds and transports glucose. The SGLT does not consume ATP, but does depend on the ATP-consuming pump at the base of the cell.

constantly removes $Na^+$ from the cell. If not for this, the $Na^+$ and glucose inflow via the SGLT would soon cease. Therefore, the SGLT does not use ATP directly, but depends on ATP to drive the $Na^+$–$K^+$ pump; it is therefore a secondary active transport protein. (*Secondary active transport* is an unfortunate name for this, as the SGLT is actually carrying out facilitated diffusion, but its dependence on a primary active transport pump has led to this name.)

The $Na^+$–$K^+$ pump itself (fig. 3.20) is a good example of primary active transport. It is also known as *$Na^+$–$K^+$ ATPase* because it is an enzyme that hydrolyzes ATP. The $Na^+$–$K^+$ pump binds three $Na^+$ simultaneously on the cytoplasmic side of the membrane, releases these to the ECF, binds two $K^+$ simultaneously from the ECF, and releases these into the cell. Each cycle of the pump consumes one ATP and exchanges three $Na^+$ for two $K^+$. This keeps the $K^+$ concentration higher and the $Na^+$ concentration lower within the cell than they are in the ECF. These ions continually leak through the membrane, and the $Na^+$–$K^+$ pump compensates like bailing out a leaky boat.

Lest you question the importance of the $Na^+$–$K^+$ pump, about half of the calories you use each day go to this purpose alone. Beyond compensating for a leaky plasma membrane, the $Na^+$–$K^+$ pump has at least four functions:

1. **Secondary active transport.** It maintains a steep $Na^+$ concentration gradient across the membrane. Like water behind a dam, this gradient is a source of potential energy

that can be tapped to do other work. The secondary active transport described previously is an example of this.

2. **Regulation of cell volume.** Certain anions are confined to the cell and cannot penetrate the plasma membrane. These "fixed anions," such as proteins and phosphates, attract and retain cations. If there were nothing to correct for it, the retention of these ions would cause osmotic swelling and possibly lysis of the cell. Cellular swelling, however, elevates activity of the $Na^+-K^+$ pumps. Since each cycle of the pump removes one ion more than it brings in, the pumps are part of a negative feedback loop that reduces intracellular ion concentration, controls osmolarity, and prevents cellular swelling.

3. **Maintenance of a membrane potential.** All living cells have an electrical charge difference called the *resting membrane potential* across the plasma membrane. Like the two poles of a battery, the inside of the membrane is negatively charged and the outside is positively charged. This difference stems from the unequal distribution of ions on the two sides of the membrane, maintained by the $Na^+-K^+$ pump. The membrane potential is essential to the excitability of nerve and muscle cells.

4. **Heat production.** When the weather turns chilly, we turn up not only the furnace in our home but also the "furnace" in our body. Thyroid hormone stimulates cells to produce more $Na^+-K^+$ pumps. As these pumps consume ATP, they release heat, compensating for the body heat we lose to the cold air around us.

#### ▶▶▶ APPLY WHAT YOU KNOW

*An important characteristic of proteins is their ability to change conformation in response to the binding or dissociation of a ligand (chapter 2). Explain how this characteristic is essential to carrier-mediated transport.*

In summary, carrier-mediated transport is any process in which solute particles move through a membrane by means of a transport protein. The protein is a uniport if it transports only one solute, a symport if it carries two types of solutes at once in the same direction, and an antiport if it carries two or more solutes in opposite directions. If the carrier does not depend on ATP at all and it moves solutes down their concentration gradient, the process is called facilitated diffusion. If the carrier itself consumes ATP and moves solutes up their concentration gradient, the process is called primary active transport. If the carrier does not directly use ATP, but depends on a concentration gradient produced by ATP-consuming $Na^+-K^+$ pumps elsewhere in the plasma membrane, the process is called secondary active transport.

## Vesicular Transport

So far, we have considered processes that move one or a few ions or molecules at a time through the plasma membrane. **Vesicular transport** processes, by contrast, move large particles, droplets of fluid, or numerous molecules at once through the membrane, contained in bubblelike **vesicles** of membrane. Vesicular processes that bring matter into a cell are called **endocytosis**[25] (EN-doe-sy-TOE-sis) and those that release material from a cell are called **exocytosis**[26] (EC-so-sy-TOE-sis). These processes employ motor proteins whose movements are energized by ATP.

There are three forms of endocytosis: phagocytosis, pinocytosis, and receptor-mediated endocytosis. **Phagocytosis**[27] (FAG-oh-sy-TOE-sis), or "cell eating," is the process of engulfing particles such as bacteria, dust, and cellular debris—particles large enough to be seen with a microscope. For example, neutrophils (a class of white blood cells) protect the body from infection by phagocytizing and killing bacteria. A neutrophil spends most of its life crawling about in the connective tissues by means of its pseudopods. When a neutrophil encounters a bacterium, it surrounds it with pseudopods and traps it in a vesicle called a **phagosome**[28]—a vesicle in the cytoplasm surrounded by a unit membrane (fig. 3.21). A lysosome merges with the phagosome, converting it to a *phagolysosome,* and contributes enzymes that destroy the invader. Several other kinds of phagocytic cells are described in chapter 21. In general, phagocytosis is a way of keeping the tissues free of debris and infectious microbes. Some cells called *macrophages* (literally, "big eaters") phagocytize the equivalent of 25% of their own volume per hour.

**Pinocytosis**[29] (PIN-oh-sy-TOE-sis), or "cell drinking," is the process of taking in droplets of ECF containing molecules of some use to the cell. While phagocytosis occurs in only a few specialized cells, pinocytosis occurs in all human cells. The process begins as the plasma membrane becomes dimpled, or caved in, at points. These pits soon separate from the surface membrane and form small membrane-bounded **pinocytotic vesicles** in the cytoplasm. The vesicles contain droplets of the ECF with whatever molecules happen to be there.

**Receptor-mediated endocytosis** (fig. 3.22) is a more selective form of either phagocytosis or pinocytosis. It enables a cell to take in specific molecules from the ECF with a minimum of unnecessary matter. Particles in the ECF bind to specific receptors on the plasma membrane. The receptors then cluster together and the membrane sinks in at this point, creating a pit coated with a peripheral membrane protein called *clathrin*.[30] The pit soon pinches off to form a *clathrin-coated vesicle* in the cytoplasm. Clathrin may serve as an "address label" on the coated vesicle that directs it to an appropriate destination in the cell, or it may inform other structures in the cell what to do with the vesicle.

One example of receptor-mediated endocytosis is the uptake of *low-density lipoproteins (LDLs)*—protein-coated droplets of cholesterol and other lipids in the blood (see chapter 26). The thin endothelial cells that line our blood vessels have LDL receptors on their surfaces and absorb LDLs in clathrin-coated vesicles. Inside the cell, the LDL is freed from the vesicle and

---

[25]*endo* = into; *cyt* = cell; *osis* = process
[26]*exo* = out of; *cyt* = cell; *osis* = process
[27]*phago* = eating; *cyt* = cell; *osis* = process
[28]*phago* = eaten; *some* = body
[29]*pino* = drinking; *cyt* = cell; *osis* = process
[30]*clathr* = lattice; *in* = protein

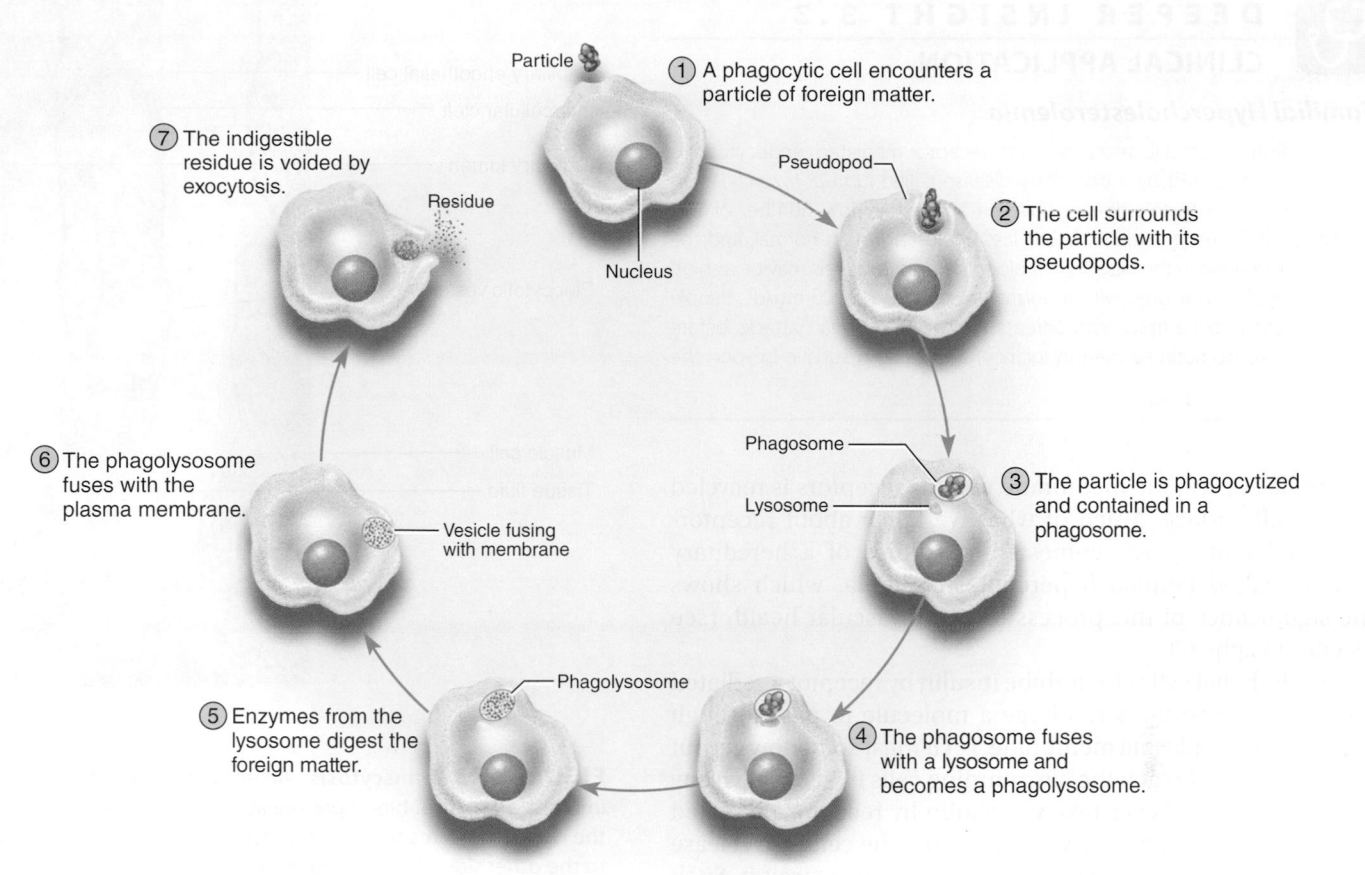

**FIGURE 3.21**  **Phagocytosis, Intracellular Digestion, and Exocytosis.**

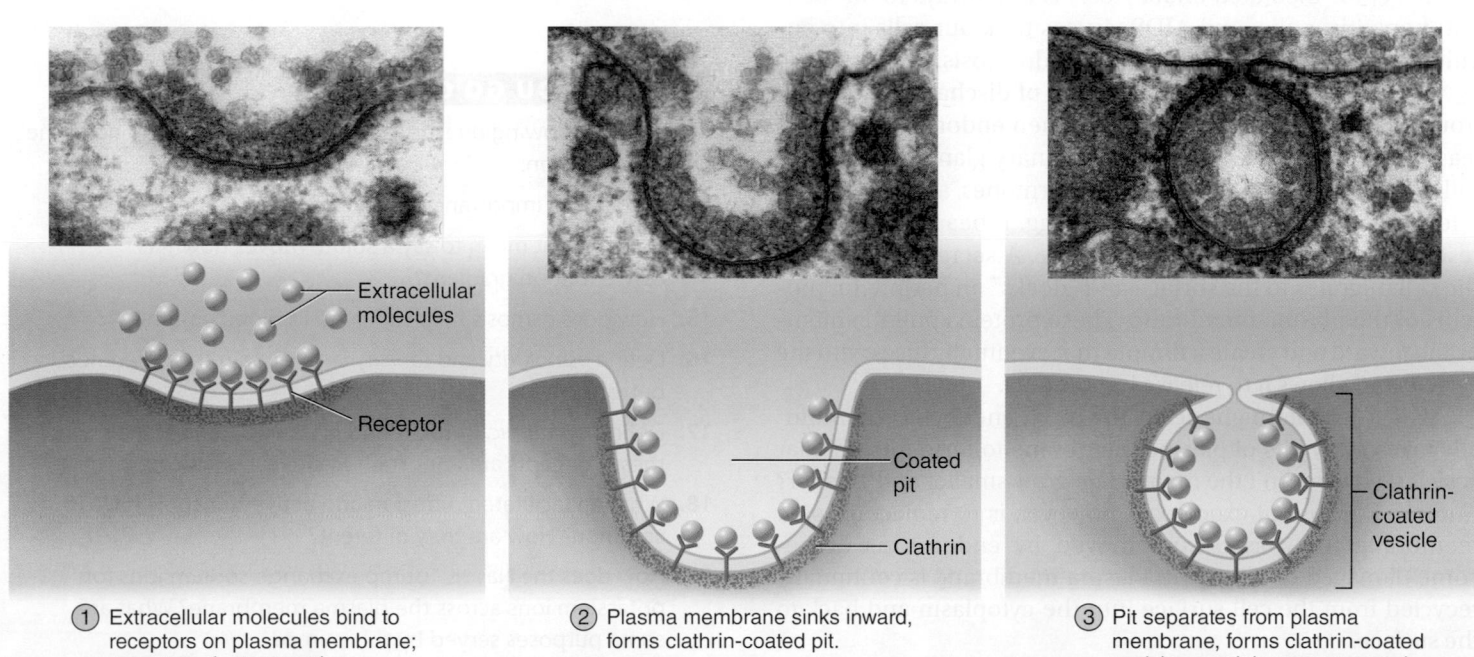

**FIGURE 3.22**  **Receptor-Mediated Endocytosis.**

# DEEPER INSIGHT 3.3

## CLINICAL APPLICATION

### Familial Hypercholesterolemia

The significance of LDL receptors and receptor-mediated endocytosis is dramatically illustrated by a hereditary disease called *familial hypercholesterolemia*.[31] People with this disease have an abnormally low number of LDL receptors. Their cells therefore absorb less cholesterol than normal, and the cholesterol remains in the blood. Their blood cholesterol levels may be as high as 1,200 mg/dL, compared with a normal level of about 200 mg/dL. People who inherit the gene from both parents typically have heart attacks before the age of 20 (sometimes even in infancy) and seldom survive beyond the age of 30.

metabolized, and the membrane with its receptors is recycled to the cell surface. Much of what we know about receptor-mediated endocytosis comes from studies of a hereditary disease called familial hypercholesterolemia, which shows the significance of this process to cardiovascular health (see Deeper Insight 3.3).

Endothelial cells also imbibe insulin by receptor-mediated endocytosis. Insulin is too large a molecule to pass through channels in the plasma membrane, yet it must somehow get out of the blood and reach the surrounding cells if it is to have any effect. Endothelial cells take up insulin by receptor-mediated endocytosis, transport the vesicles across the cell, and release the insulin on the other side, where tissue cells await it. Such transport of material across a cell (capture on one side and release on the other side) is called **transcytosis**[32] (fig. 3.23). This process is especially active in muscle capillaries and transfers a significant amount of blood albumin into the tissue fluid.

Receptor-mediated endocytosis is not always to our benefit; hepatitis, polio, and AIDS viruses trick our cells into engulfing them by receptor-mediated endocytosis.

**Exocytosis** (fig. 3.24) is a process of discharging material from a cell. It occurs, for example, when endothelial cells release insulin to the tissue fluid, mammary gland cells secrete milk sugar, other gland cells release hormones, and sperm cells release enzymes for penetrating an egg. It bears a superficial resemblance to endocytosis in reverse. A secretory vesicle in the cell migrates to the surface and "docks" on peripheral proteins of the plasma membrane. These proteins pull the membrane inward and create a dimple that eventually fuses with the vesicle and allows it to release its contents.

The question might occur to you, If endocytosis continually takes away bits of plasma membrane to form intracellular vesicles, why doesn't the membrane grow smaller and smaller? Another purpose of exocytosis, however, is to replace plasma membrane that has been removed by endocytosis or become damaged or worn out. Plasma membrane is continually recycled from the cell surface into the cytoplasm and back to the surface.

Table 3.3 summarizes these mechanisms of transport.

[31]*familial* = running in the family; *hyper* = above normal; *emia* = blood condition
[32]*trans* = across; *cyt* = cell; *osis* = process

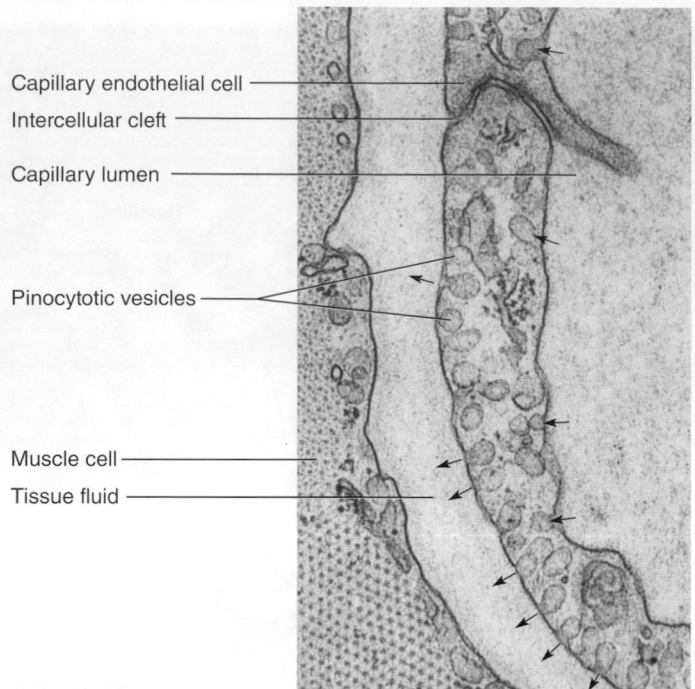

Capillary endothelial cell
Intercellular cleft
Capillary lumen
Pinocytotic vesicles
Muscle cell
Tissue fluid

0.25 µm

**FIGURE 3.23  Transcytosis.** An endothelial cell of a capillary imbibes droplets of blood plasma at sites indicated by arrows along the right. This forms pinocytotic vesicles, which the cell transports to the other side. Here, it releases the contents by exocytosis at sites indicated by arrows along the left side of the cell. This process is especially active in muscle capillaries and transfers a significant amount of blood albumin into the tissue fluid.

 *Why isn't transcytosis listed as a separate means of membrane transport, in addition to pinocytosis and the others?*

### BEFORE YOU GO ON

Answer the following questions to test your understanding of the preceding section:

13. What is the importance of filtration to human physiology?

14. What does it mean to say a solute moves down its concentration gradient?

15. How does osmosis help to maintain blood volume?

16. Define *osmolarity* and *tonicity,* and explain the difference between them.

17. Define *hypotonic, isotonic,* and *hypertonic,* and explain why these concepts are important in clinical practice.

18. What do facilitated diffusion and active transport have in common? How are they different?

19. How does the $Na^+$–$K^+$ pump exchange sodium ions for potassium ions across the plasma membrane? What are some purposes served by this pump?

20. How does phagocytosis differ from pinocytosis?

21. Describe the process of exocytosis. What are some of its purposes?

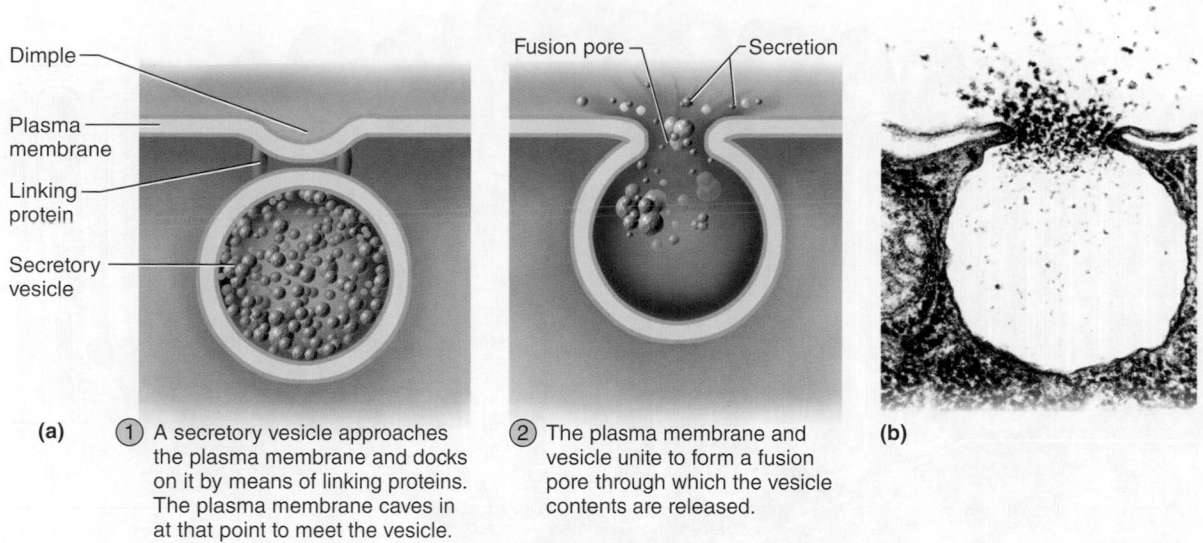

Dimple

Plasma membrane

Linking protein

Secretory vesicle

Fusion pore — Secretion

**(a)**

① A secretory vesicle approaches the plasma membrane and docks on it by means of linking proteins. The plasma membrane caves in at that point to meet the vesicle.

② The plasma membrane and vesicle unite to form a fusion pore through which the vesicle contents are released.

**(b)**

**FIGURE 3.24 Exocytosis.** (a) Stages of exocytosis. (b) Electron micrograph of exocytosis.

| **TABLE 3.3** | **Methods of Membrane Transport** |
|---|---|
| **Transport Without Carriers** | Movement of material without the aid of carrier proteins |
| Filtration | Movement of water and solutes through a selectively permeable membrane as a result of hydrostatic pressure |
| Simple diffusion | Diffusion of particles through water or air or through a living or artificial membrane, down their concentration gradient, without the aid of membrane carriers |
| Osmosis | Net flow of water through a selectively permeable membrane, driven by either a difference in solute concentration or a mechanical force |
| **Carrier-Mediated Transport** | Movement of material through a cell membrane with the aid of carrier proteins |
| Facilitated diffusion | Transport of particles through a selectively permeable membrane, down their concentration gradient, by a carrier that does not directly consume ATP |
| Active transport | Transport of particles through a selectively permeable membrane, up their concentration gradient, with the aid of a carrier that consumes ATP |
| Primary active transport | Direct transport of solute particles by an ATP-using membrane pump |
| Secondary active transport | Transport of solute particles by a carrier that does not in itself use ATP but depends on concentration gradients produced by primary active transport |
| Cotransport | Transport of two or more solutes simultaneously in the same direction through a membrane by either facilitated diffusion or active transport |
| Countertransport | Transport of two or more different solutes in opposite directions through a membrane by either facilitated diffusion or active transport |
| Uniport | A carrier that transports only one solute, using either facilitated diffusion or active transport |
| Symport | A carrier that performs cotransport |
| Antiport | A carrier that performs countertransport |
| **Vesicular (Bulk) Transport** | Movement of fluid and particles through a plasma membrane by way of membrane vesicles; consumes ATP |
| Endocytosis | Vesicular transport of particles into a cell |
| Phagocytosis | Process of engulfing large particles by means of pseudopods; "cell eating" |
| Pinocytosis | Process of imbibing extracellular fluid in which the plasma membrane sinks in and pinches off small vesicles containing droplets of fluid |
| Receptor-mediated endocytosis | Phagocytosis or pinocytosis in which specific solute particles bind to receptors on the plasma membrane, and are then taken into the cell in clathrin-coated vesicles with a minimal amount of extraneous matter |
| Exocytosis | Process of eliminating material from a cell by means of a vesicle approaching the cell surface, fusing with the plasma membrane, and expelling its contents; used to release cell secretions, replace worn-out plasma membrane, and replace membrane that has been internalized by endocytosis |

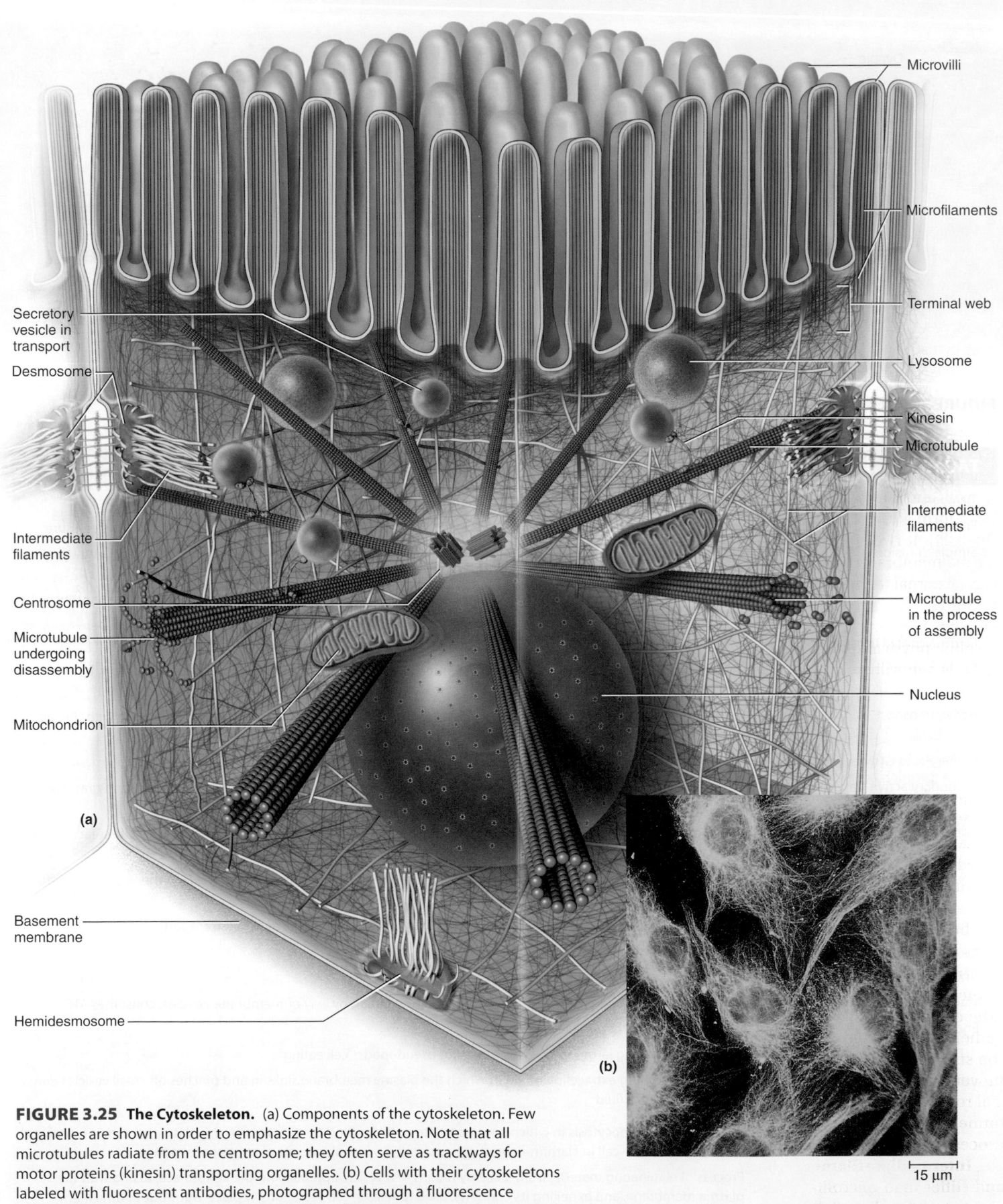

Microvilli

Microfilaments

Terminal web

Lysosome

Kinesin

Microtubule

Intermediate filaments

Microtubule in the process of assembly

Nucleus

Secretory vesicle in transport

Desmosome

Intermediate filaments

Centrosome

Microtubule undergoing disassembly

Mitochondrion

Basement membrane

Hemidesmosome

(a)

(b)

15 μm

**FIGURE 3.25  The Cytoskeleton.** (a) Components of the cytoskeleton. Few organelles are shown in order to emphasize the cytoskeleton. Note that all microtubules radiate from the centrosome; they often serve as trackways for motor proteins (kinesin) transporting organelles. (b) Cells with their cytoskeletons labeled with fluorescent antibodies, photographed through a fluorescence microscope. The density of a typical cytoskeleton far exceeds even that shown in part (a).

## 3.4  The Cell Interior

### Expected Learning Outcomes

When you have completed this section, you should be able to

a. list the main organelles of a cell, describe their structure, and explain their functions;

b. describe the cytoskeleton and its functions; and

c. give some examples of cell inclusions and explain how inclusions differ from organelles.

We now probe more deeply into the cell to study its internal structures. These are classified into three groups—*cytoskeleton, organelles,* and *inclusions*—all embedded in the clear, gelatinous cytosol.

## The Cytoskeleton

The **cytoskeleton** is a network of protein filaments and cylinders that structurally support a cell, determine its shape, organize its contents, direct the movement of materials within the cell, and contribute to movements of the cell as a whole. It can form a very dense supportive scaffold in the cytoplasm (fig. 3.25). It is connected to transmembrane proteins of the plasma membrane, and they in turn are connected to protein fibers external to the cell, creating a strong structural continuity from extracellular material to the cytoplasm. Cytoskeletal elements may even connect to chromosomes in the nucleus, enabling physical tension on a cell to move nuclear contents and mechanically stimulate genetic function.

The cytoskeleton is composed of *microfilaments, intermediate filaments,* and *microtubules.* If you think of intermediate filaments as being like the stiff rods of uncoooked spaghetti, you could, by comparison, think of microfilaments as being like fine angel-hair pasta and microtubles as being like tubular penne pasta.

**Microfilaments (thin filaments)** are about 6 nm thick and are made of the protein *actin.* They form a fibrous mat called the **terminal web (membrane skeleton)** on the cytoplasmic side of the plasma membrane. The phospholipids of the plasma membrane spread out over the terminal web like butter on a slice of bread. The web, like the bread, provides physical support, whereas the lipids, like butter, provide a permeability barrier. It is thought that without the support of the terminal web, the phospholipids would break up into little droplets and the plasma membrane would not hold together. As described earlier, actin microfilaments also form the supportive cores of the microvilli and play a role in cell movement. Through its role in cell motility, actin plays a crucial role in embryonic development, muscle contraction, immune function, wound healing, cancer metastasis, and other processes that involve cell migration.

**Intermediate filaments** (8–10 nm in diameter) are thicker and stiffer than microfilaments. They give the cell its shape, resist stress, and participate in junctions that attach cells to

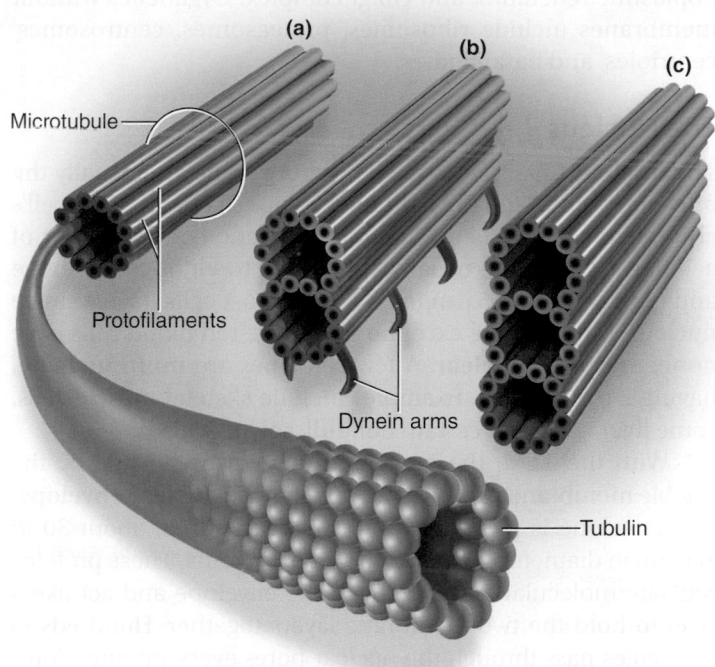

**FIGURE 3.26  Microtubules.** (a) A microtubule is composed of 13 protofilaments. Each protofilament is a helical chain of globular proteins called tubulin. (b) One of the nine microtubule pairs that form the axonemes of cilia and flagella. (c) One of the nine microtubule triplets that form a centriole.

their neighbors. In epidermal cells, they are made of the tough protein *keratin* and occupy most of the cytoplasm. They are responsible for the strength of hair and fingernails.

A **microtubule** (25 nm in diameter) is a cylinder made of 13 parallel strands called *protofilaments.* Each protofilament is a long chain of globular proteins called *tubulin* (fig. 3.26). Microtubules radiate from an area of the cell called the *centrosome* and hold organelles in place, form bundles that maintain cell shape and rigidity, and act somewhat like railroad tracks. Motor proteins "walk" along these tracks carrying organelles and macromolecules to specific destinations in the cell. Microtubules form the axonemes of cilia and flagella and are responsible for their beating movements, and form the mitotic spindle that guides chromosome movement during cell division. Microtubules are not permanent structures. They come and go moment by moment as tubulin molecules assemble into a tubule and then suddenly break apart again to be used somewhere else in the cell. The microtubules in cilia, flagella, basal bodies, and centrioles, however, are more stable.

## Organelles

**Organelles** are internal structures of a cell that carry out specialized metabolic tasks. Some are surrounded by membranes and are therefore referred to as *membranous organelles.* These are the nucleus, mitochondria, lysosomes, peroxisomes, endoplasmic reticulum, and Golgi complex. Organelles without membranes include ribosomes, proteasomes, centrosomes, centrioles, and basal bodies.

## The Nucleus

The **nucleus** (fig. 3.27) is the largest organelle and usually the only one visible with the light microscope. It contains the cell's chromosomes and is therefore the genetic control center of cellular activity. It is usually spheroidal to elliptical in shape and typically about 5 μm in diameter. Most cells have a single nucleus, but there are exceptions. Mature red blood cells have none; they are **anuclear.** A few cell types are **multinucleate,** having 2 to 50 nuclei. Examples include skeletal muscle cells, some liver cells, and certain bone-dissolving cells.

With the TEM, the nucleus can be distinguished by the double membrane surrounding it, called the **nuclear envelope.** The envelope is perforated with **nuclear pores,** about 30 to 100 nm in diameter, formed by a ring of proteins. These proteins regulate molecular traffic through the envelope and act like a rivet to hold the two membrane layers together. Hundreds of molecules pass through the nuclear pores every minute. Coming into the nucleus are raw materials for DNA and RNA synthesis, enzymes that are made in the cytoplasm but function in the nucleus, and hormones that activate certain genes. Going the other way, RNA is made in the nucleus but leaves to perform its job in the cytoplasm.

Immediately inside the nuclear envelope is a narrow but densely fibrous zone called the **nuclear lamina,** composed of a web of intermediate filaments. It supports the nuclear envelope and pores, provides points of attachment and

organization for the chromosomes inside the nucleus, and plays a role in regulating the cell life cycle. Abnormalities of its structure or function are associated with certain genetic diseases and premature cell death.

The material in the nucleus is called **nucleoplasm.** This includes **chromatin**[33] (CRO-muh-tin)—fine threadlike matter composed of DNA and protein—and one or more dark-staining masses called **nucleoli** (singular, *nucleolus*), where ribosomes are produced. The genetic function of the nucleus is described in chapter 4.

## Endoplasmic Reticulum

**Endoplasmic reticulum (ER)** literally means "little network within the cytoplasm." It is a system of interconnected channels called **cisternae**[34] (sis-TUR-nee) enclosed by a unit membrane (fig. 3.28). In areas called **rough endoplasmic reticulum,** the cisternae are parallel, flattened sacs covered with granules called *ribosomes.* The rough ER is continuous with the outer membrane of the nuclear envelope, and cisternae adjacent to each other are connected by bridges. In areas called **smooth endoplasmic reticulum,** the cisternae are more tubular, branch more extensively, and lack ribosomes. The cisternae of the smooth ER are continuous with those of the rough ER, so the two are different parts of the same network.

The ER synthesizes steroids and other lipids, detoxifies alcohol and other drugs, and manufactures all membranes of the cell. Rough ER produces the phospholipids and proteins of the plasma membrane and synthesizes proteins that are either

[33]*chromat* = color
[34]*cistern* = reservoir

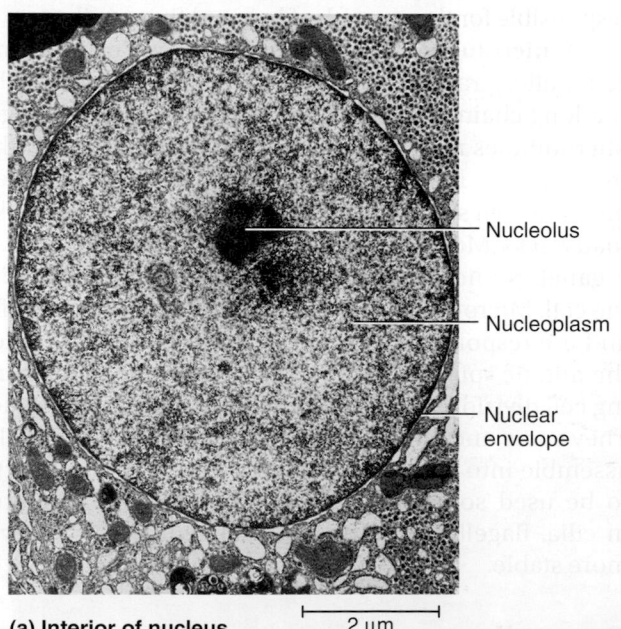

(a) Interior of nucleus    2 μm

— Nucleolus

— Nucleoplasm

— Nuclear envelope

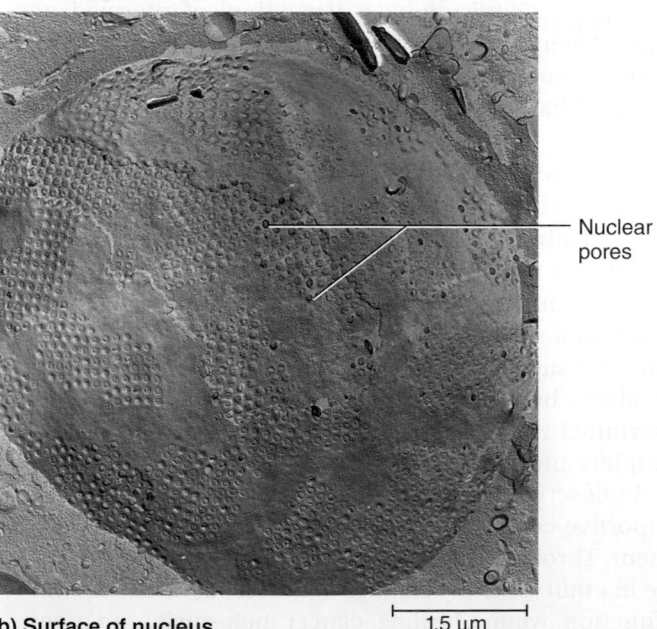

(b) Surface of nucleus    1.5 μm

— Nuclear pores

**FIGURE 3.27  The Nucleus.**  These photomicrographs were made by different TEM methods to show the internal structure of the nucleus and surface of the nuclear envelope.

**?** *Why do these nuclear pores have to be larger in diameter than the channels in the cell's plasma membrane? (See table 3.1.)*

packaged in other organelles such as lysosomes or secreted from the cell. Rough ER is most abundant in cells that synthesize large amounts of protein, such as antibody-producing cells and cells of the digestive glands. This role is discussed further in chapter 4.

Most cells have only a scanty smooth ER, but it is relatively abundant in cells that engage extensively in detoxification, such as liver and kidney cells. Long-term abuse of alcohol, barbiturates, and other drugs leads to tolerance partly because the smooth ER proliferates and detoxifies the drugs more quickly. Smooth ER is also abundant in cells of the testes and ovaries that synthesize steroid hormones. Skeletal and cardiac muscle contain extensive networks of smooth ER that store calcium and release it to trigger muscle contraction.

## Ribosomes

**Ribosomes** are small granules of protein and RNA found in the nucleoli, in the cytosol, and on the outer surfaces of the rough ER and nuclear envelope. They "read" coded genetic messages (messenger RNA) and assemble amino acids into proteins specified

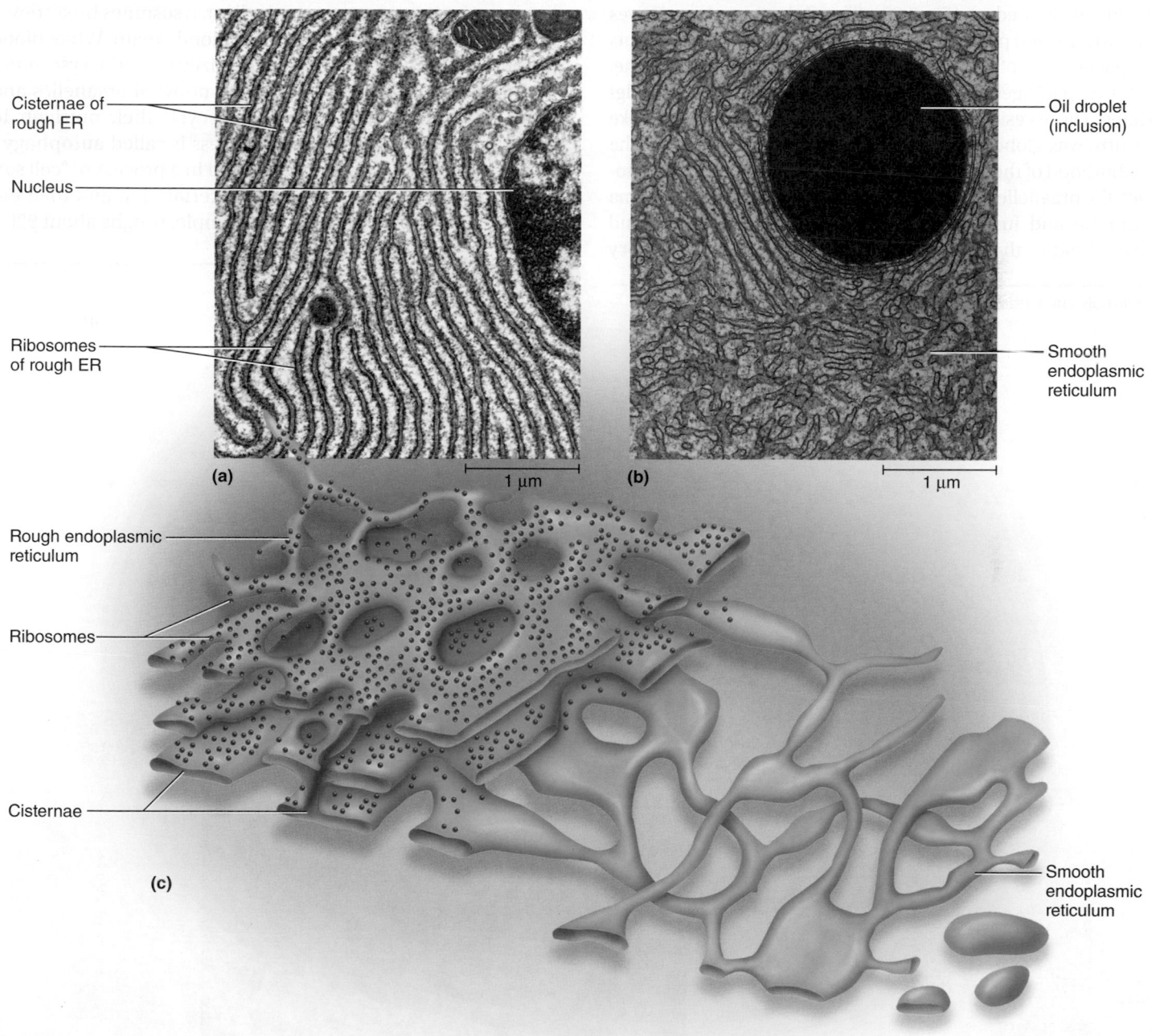

**FIGURE 3.28 Endoplasmic Reticulum (ER).** (a) Rough ER. (b) Smooth ER and an inclusion (oil droplet). (c) Structure of the endoplasmic reticulum, with rough and smooth regions.

by the code (see chapter 4). The unattached ribosomes scattered throughout the cytoplasm make enzymes and other proteins for use within the cell. Ribosomes attach to the rough ER when they make proteins destined to be packaged in lysosomes or, as in cases such as digestive enzymes, to be secreted from the cell.

## Golgi Complex

The **Golgi**[35] (GOAL-jee) **complex** is a small system of cisternae that synthesize carbohydrates and put the finishing touches on protein and glycoprotein synthesis. The complex resembles a stack of pita bread. It consists of only a few cisternae, slightly separated from each other; each cisterna is a flattened, often curved sac with swollen edges (fig. 3.29). The Golgi complex receives newly synthesized proteins from the rough ER. It sorts them, cuts and splices some of them, adds carbohydrate moieties to some, and finally packages the proteins in membrane-bounded **Golgi vesicles.** These vesicles bud off the swollen rim of a cisterna like the warm wax globules in a lava lamp, and are abundant in the neighborhood of the Golgi complex. Some vesicles become *lysosomes,* the organelle discussed next; some migrate to the plasma membrane and fuse with it, contributing fresh protein and phospholipid to the membrane; and some become **secretory**

vesicles that store a cell product, such as breast milk or digestive enzymes, for later release. The role of the Golgi complex in protein synthesis and secretion is detailed in chapter 4.

## Lysosomes

A **lysosome**[36] (LY-so-some) (fig. 3.30a) is a package of enzymes bounded by a membrane. Although often round or oval, lysosomes are extremely variable in shape. When viewed with the TEM, they often exhibit dark gray contents devoid of structure, but sometimes show crystals or parallel layers of protein. At least 50 lysosomal enzymes have been identified. They hydrolyze proteins, nucleic acids, complex carbohydrates, phospholipids, and other substrates. In the liver, lysosomes break down glycogen to release glucose into the bloodstream. White blood cells use lysosomes to digest phagocytized bacteria. Lysosomes also digest and dispose of surplus or nonvital organelles and other cell components in order to recycle their nutrients to more important cell needs; this process is called **autophagy**[37] (aw-TOFF-uh-jee). Lysosomes also aid in a process of "cell suicide." Some cells are meant to do a certain job and then destroy themselves. The uterus, for example, weighs about 900 g

---

[35]Camillo Golgi (1843–1926), Italian histologist

[36]*lyso* = loosen, dissolve; *some* = body
[37]*auto* = self; *phagy* = eating

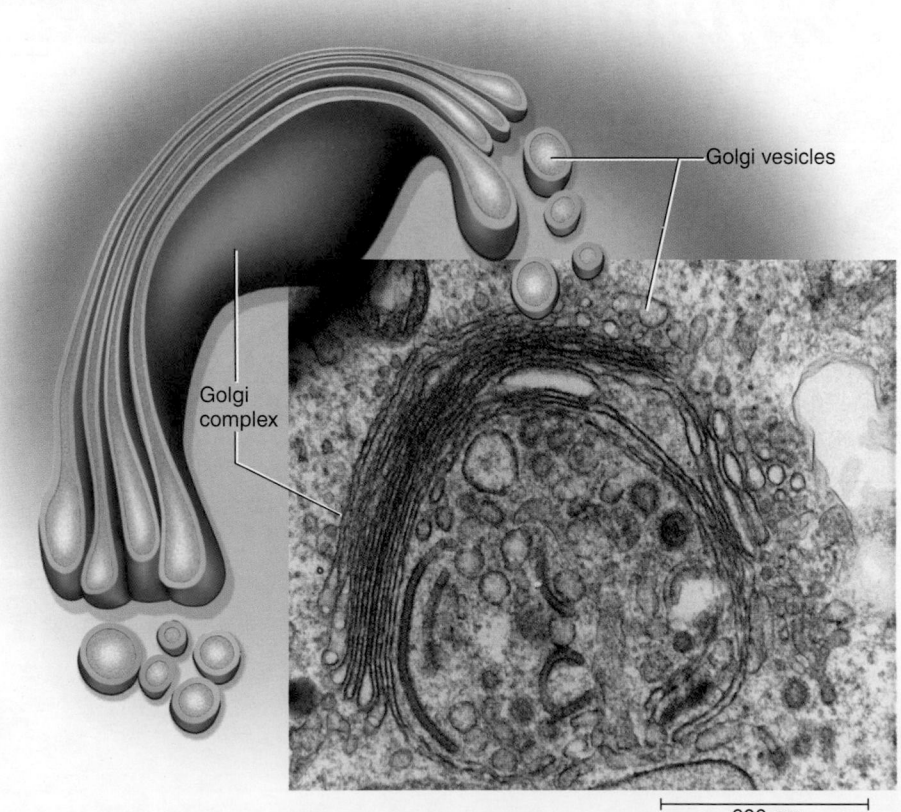

Golgi vesicles

Golgi complex

600 nm

**FIGURE 3.29  The Golgi Complex.**

at full-term pregnancy and shrinks to 60 g within 5 or 6 weeks after birth. This shrinkage is due to **autolysis,**[38] the digestion of surplus cells by their own lysosomal enzymes.

## Peroxisomes

**Peroxisomes** (fig. 3.30b) resemble lysosomes but contain different enzymes and are produced by the endoplasmic reticulum rather than the Golgi complex. Their general function is to use molecular oxygen ($O_2$) to oxidize organic molecules. These reactions produce hydrogen peroxide ($H_2O_2$)—hence, the name of the organelle. $H_2O_2$ is then used to oxidize other molecules, and the excess is broken down to water and oxygen by an enzyme called *catalase.*

Peroxisomes occur in nearly all cells but are especially abundant in liver and kidney cells. They neutralize free radicals and detoxify alcohol, other drugs, and a variety of blood-borne toxins. Peroxisomes also decompose fatty acids into two-carbon fragments that the mitochondria use as an energy source for ATP synthesis.

## Proteasomes

Cells must tightly control the concentration of proteins in their cytoplasm. Therefore, they must not only synthesize new proteins, but also dispose of those that are no longer needed. Cells

also need to rid themselves of damaged and nonfunctional proteins and foreign proteins introduced by such events as viral infections. Protein synthesis, we have seen, is the domain of the ribosomes; protein disposal is the function of another structurally simple organelle called a **proteasome.**

Proteasomes are hollow, cylindrical complexes of proteins located in both the cytoplasm and nucleus (fig. 3.31). A cell tags undesirable proteins for destruction and transports them to a proteasome. As the undesirable protein passes through the core of this organelle, the proteasome's enzymes unfold it and break it down into short peptides and free amino acids. These can be used to synthesize new proteins or presented to the immune system for further degradation. Proteasomes degrade more than 80% of a cell's proteins.

## Mitochondria

**Mitochondria**[39] (MY-toe-CON-dree-uh) (singular, *mitochondrion*) are organelles specialized for synthesizing ATP. They have a variety of shapes—spheroidal, rod-shaped, kidney-shaped, or threadlike (fig. 3.32)—but they are quite mobile and squirm and change shape continually. Like the nucleus, a mitochondrion is surrounded by a double membrane. The inner membrane usually has folds called **cristae**[40] (CRIS-tee), which

---

[38]*auto* = self; *lysis* = dissolving

[39]*mito* = thread; *chondr* = grain
[40]*crista* = crest

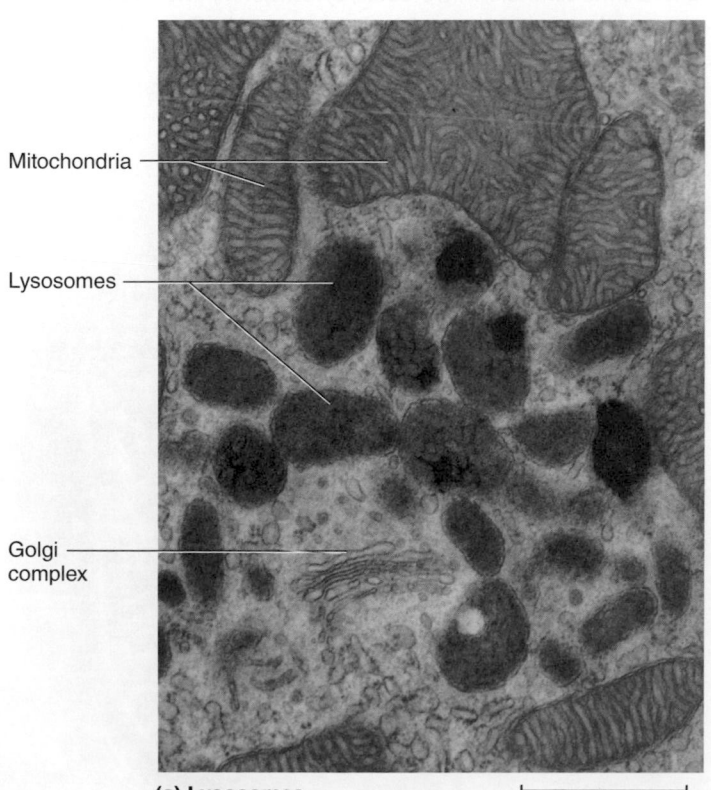

**(a) Lysosomes**

1 μm

**FIGURE 3.30  Lysosomes and Peroxisomes.**

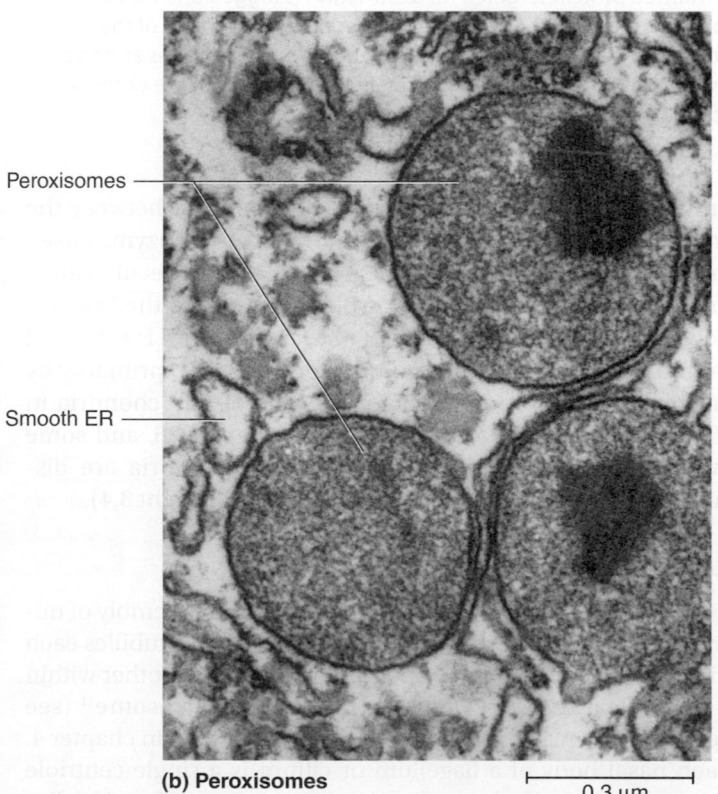

**(b) Peroxisomes**

0.3 μm

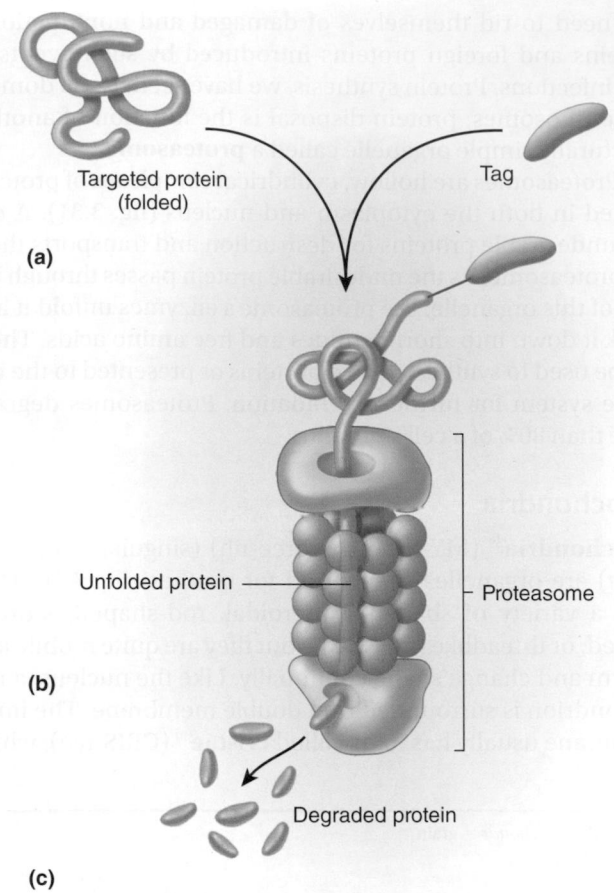

Targeted protein
(folded)

Tag

**(a)**

Unfolded protein

Proteasome

**(b)**

Degraded protein

**(c)**

**FIGURE 3.31 Protein Degradation by a Proteasome.** (a) The unwanted protein targeted for destruction is tagged and transported to a proteasome. (b) As the protein passes down the core of the proteasome, it is unfolded and cleaved into small peptides and free amino acids. (c) The degraded protein is released from the other end of the proteasome.

project like shelves across the organelle. The space between the cristae, called the **matrix,** contains ribosomes; enzymes used in ATP synthesis; and a small, circular DNA molecule called *mitochondrial DNA (mtDNA).* Mitochondria are the "powerhouses" of the cell. Energy is not *made* here, but it is extracted from organic compounds and transferred to ATP, primarily by enzymes located on the cristae. The role of mitochondria in ATP synthesis is explained in detail in chapter 26, and some evolutionary and clinical aspects of mitochondria are discussed at the end of this chapter (see Deeper Insight 3.4).

## Centrioles

A **centriole** (SEN-tree-ole) is a short cylindrical assembly of microtubules, arranged in nine groups of three microtubules each (fig. 3.33). Two centrioles lie perpendicular to each other within a small clear area of cytoplasm called the **centrosome**[41] (see fig. 3.5). They play a role in cell division, described in chapter 4. Each basal body of a flagellum or cilium is a single centriole oriented perpendicular to the plasma membrane. Basal bodies

---

[41]*centro* = central; *some* = body

originate in a *centriolar organizing center* and migrate to the plasma membrane. Two microtubules of each triplet then elongate to form the nine pairs of peripheral microtubules of the axoneme. A cilium can grow to its full length in less than an hour.

## Inclusions

**Inclusions** are of two kinds: accumulated cell products such as glycogen granules, pigments, and fat droplets (see fig. 3.28b); and foreign bodies such as viruses, bacteria, and dust particles and other debris phagocytized by a cell. Inclusions are never enclosed in a membrane, and unlike the organelles and cytoskeleton, they are not essential to cell survival.

The major features of a cell are summarized in table 3.4.

---

**BEFORE YOU GO ON**

Answer the following questions to test your understanding of the preceding section:

22. Distinguish between organelles and inclusions. State two examples of each.

23. Briefly state how each of the following cell components can be recognized in electron micrographs: the nucleus, a mitochondrion, a lysosome, and a centriole. What is the primary function of each?

24. What three organelles are involved in protein synthesis?

25. In what ways do rough and smooth endoplasmic reticulum differ?

26. Define *centriole, microtubule, cytoskeleton,* and *axoneme.* How are these structures related to one another?

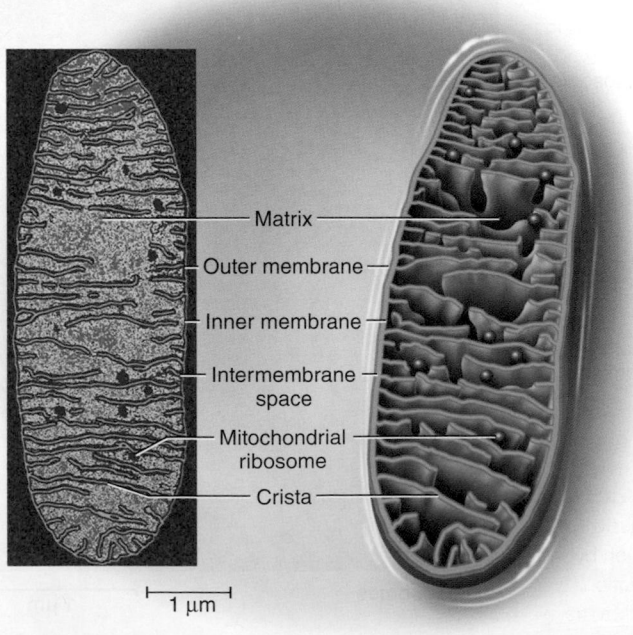

Matrix

Outer membrane

Inner membrane

Intermembrane space

Mitochondrial ribosome

Crista

1 µm

**FIGURE 3.32  A Mitochondrion.**

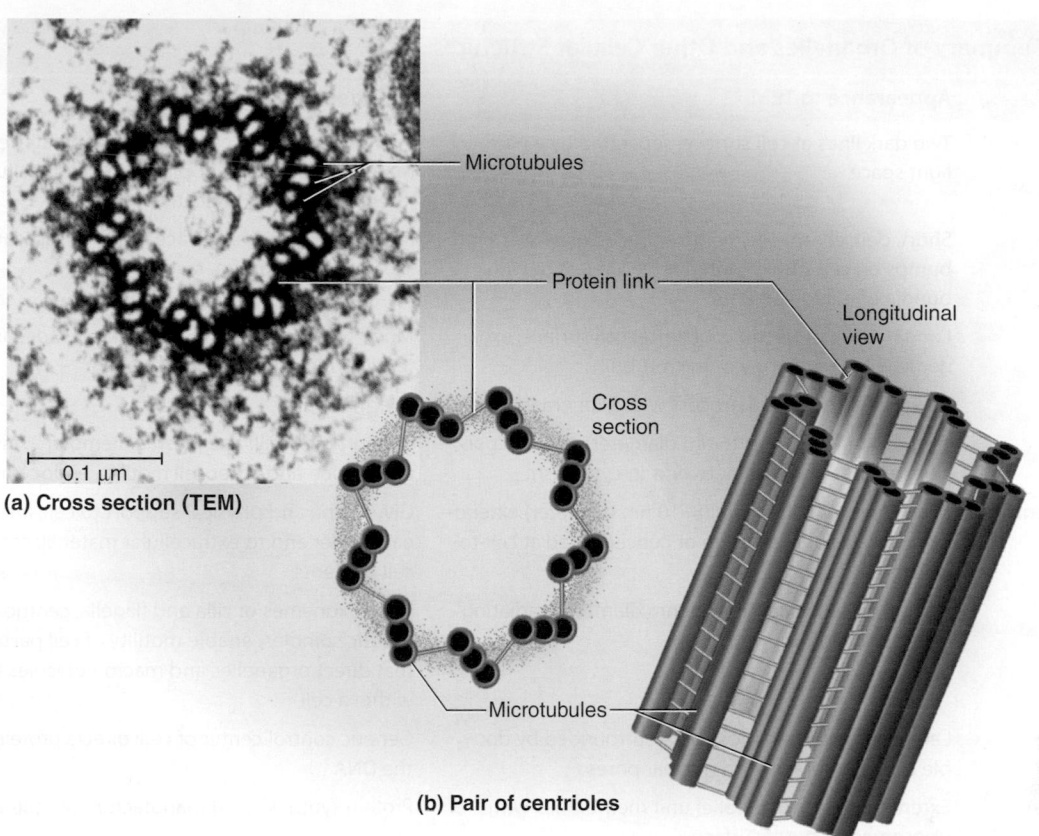

Microtubules

Protein link

Longitudinal view

Cross section

0.1 μm

**(a) Cross section (TEM)**

Microtubules

**(b) Pair of centrioles**

**FIGURE 3.33 Centrioles.** (a) Electron micrograph of a centriole as seen in cross section. (b) A pair of perpendicular centrioles.

 *How does a centriole resemble the axoneme of a cilium? How does it differ?*

 **DEEPER INSIGHT 3.4**

## EVOLUTIONARY MEDICINE

### Mitochondria—Evolution and Clinical Significance

It is virtually certain that mitochondria evolved from bacteria that invaded another primitive cell, survived in its cytoplasm, and became permanent residents. Certain modern bacteria called *rickettsii* live in the cytoplasm of other cells, showing that this mode of life is feasible. The two unit membranes around the mitochondrion suggest that the original bacterium provided the inner membrane, and the host cell's phagosome provided the outer membrane when the bacterium was phagocytized.

Several comparisons show the apparent relationship of mitochondria to bacteria. Their ribosomes are more like bacterial ribosomes than those of eukaryotic (nucleated) cells. Mitochondrial DNA (mtDNA) is a small, circular molecule that resembles the circular DNA of bacteria, not the linear DNA of the cell nucleus. It replicates independently of nuclear DNA. Mitochondrial DNA codes for some of the enzymes employed in ATP synthesis. It consists of 16,569 *base pairs* (explained in chapter 4), comprising 37 genes, compared with over 3 billion base pairs and about 20,000 to 25,000 genes in nuclear DNA.

When a sperm fertilizes an egg, any mitochondria introduced by the sperm are usually destroyed and only those provided by the egg are passed on to the developing embryo. Therefore, mitochondrial DNA is inherited almost exclusively through the mother. While nuclear DNA is reshuffled in every generation by sexual reproduction, mtDNA remains unchanged except by random mutation. Biologists and anthropologists have used mtDNA as a "molecular clock" to trace evolutionary lineages in humans and other species. Mitochondrial DNA has also been used as evidence in criminal law and to identify the remains of soldiers killed in combat. It was used in 2001 to identify the remains of the

famed bandit Jesse James, who was killed in 1882. Anthropologists have gained evidence from mtDNA that of all the women who lived in Africa 200,000 years ago, only one has any descendants still living today. This "mitochondrial Eve" is ancestor to us all.

Mitochondrial DNA is very exposed to damage from free radicals normally generated in mitochondria by aerobic respiration. Yet unlike nuclear DNA, mtDNA has no effective mechanism for repairing damage. Therefore, it mutates about 10 times as rapidly as nuclear DNA. Some of these mutations are responsible for various rare hereditary diseases and death in early childhood. Tissues and organs with the highest energy demands are the most vulnerable to mitochondrial dysfunctions—nervous tissue, the heart, the kidneys, and skeletal muscles, for example.

*Mitochondrial myopathy* is a degenerative muscle disease in which the muscle displays "ragged red fibers," cells with abnormal mitochondria that stain red with a particular histological stain. Another mtDNA disease is *MELAS (mitochondrial encephalomyopathy, lactic acidosis and strokelike episodes)*, a syndrome involving seizures, paralysis, dementia, muscle deterioration, and a toxic accumulation of lactic acid in the blood. *Leber hereditary optic neuropathy (LHON)* is a form of blindness that usually appears in young adulthood as a result of damage to the optic nerve. *Kearns–Sayre syndrome (KSS)* involves paralysis of the eye muscles, degeneration of the retina, heart disease, hearing loss, diabetes, and kidney failure. Damage to mtDNA has also been implicated as a possible factor in Alzheimer disease, Huntington disease, and other degenerative diseases of old age.

| TABLE 3.4 | Summary of Organelles and Other Cellular Structures | |
|---|---|---|
| **Structure** | **Appearance to TEM** | **Function** |
| Plasma membrane (figs. 3.3 and 3.6) | Two dark lines at cell surface, separated by a narrow light space | Prevents escape of cell contents; regulates exchange of materials between cytoplasm and extracellular fluid; involved in intercellular communication |
| Microvilli (fig. 3.10) | Short, densely spaced, hairlike processes or scattered bumps on cell surface; interior featureless or with bundle of microfilaments | Increase absorptive surface area; widespread sensory roles (hearing, equilibrium, taste) |
| Cilia (fig. 3.11) | Long hairlike projections of apical cell surface; axoneme with 9 + 2 array of microtubules | Move substances along cell surface; widespread sensory roles (equilibrium, smell, vision) |
| Flagellum | Long, single, whiplike process with axoneme | Sperm motility |
| Microfilaments (figs. 3.10 and 3.25) | Thin protein filaments (6 nm diameter), often in parallel bundles or dense networks in cytoplasm | Support microvilli and plasma membrane; involved in muscle contraction and other cell motility, endocytosis, and cell division |
| Intermediate filaments (fig. 3.25) | Thicker protein filaments (8–10 nm diameter) extending throughout cytoplasm or concentrated at cell-to-cell junctions | Give shape and physical support to cell; anchor cells to each other and to extracellular material; compartmentalize cell contents |
| Microtubules (figs. 3.25 and 3.26) | Hollow protein cylinders (25 nm diameter) radiating from centrosome | Form axonemes of cilia and flagella, centrioles, basal bodies, and mitotic spindles; enable motility of cell parts; form trackways that direct organelles and macromolecules to their destinations within a cell |
| Nucleus (figs. 3.3 and 3.27) | Largest organelle in most cells, surrounded by double unit membrane with nuclear pores | Genetic control center of cell; directs protein synthesis; shelters the DNA |
| Rough ER (fig. 3.28a) | Extensive sheets of parallel unit membranes with ribosomes on outer surface | Protein synthesis and manufacture of cellular membranes |
| Smooth ER (fig. 3.28b) | Branching network of tubules with smooth surface (no ribosomes); usually broken into numerous small segments in TEM photos | Lipid synthesis, detoxification, calcium storage |
| Ribosomes (fig. 3.28a) | Small dark granules free in cytosol or on surface of rough ER and nuclear envelope | Interpret the genetic code and synthesize polypeptides |
| Golgi complex (fig. 3.29) | Several closely spaced, parallel cisternae with thick edges, usually near nucleus, often with many Golgi vesicles nearby | Receives and modifies newly synthesized polypeptides, synthesizes carbohydrates, adds carbohydrates to glycoproteins; packages cell products into Golgi vesicles |
| Golgi vesicles (fig. 3.29) | Round to irregular sacs near Golgi complex, usually with light, featureless contents | Become secretory vesicles and carry cell products to apical surface for exocytosis, or become lysosomes |
| Lysosomes (fig. 3.30a) | Round to oval sacs with single unit membrane, often a dark featureless interior but sometimes with protein layers or crystals | Contain enzymes for intracellular digestion, autophagy, programmed cell death, and glucose mobilization |
| Peroxisomes (fig. 3.30b) | Similar to lysosomes; often lighter in color | Contain enzymes for detoxification of free radicals, alcohol, and other drugs; oxidize fatty acids |
| Proteasomes (fig. 3.31) | Small cytoplasmic granules composed of a cylindrical array of proteins | Degrade proteins that are undesirable or no longer needed by a cell |
| Mitochondria (fig. 3.32) | Round, rod-shaped, bean-shaped, or threadlike structures with double unit membrane and shelflike infoldings called cristae | ATP synthesis |
| Centrioles (fig. 3.33) | Short cylindrical bodies, each composed of a circle of nine triplets of microtubules | Form mitotic spindle during cell division; unpaired centrioles form basal bodies of cilia and flagella |
| Centrosome (fig. 3.5) | Clear area near nucleus containing a pair of centrioles | Organizing center for formation of microtubules of cytoskeleton and mitotic spindle |
| Basal body (fig. 3.11b) | Unpaired centriole at the base of a cilium or flagellum | Point of origin, growth, and anchorage of a cilium or flagellum; produces axoneme |
| Inclusions (fig. 3.28b) | Highly variable—fat droplets, glycogen granules, protein crystals, dust, bacteria, viruses; never enclosed in unit membranes | Storage products or other products of cellular metabolism, or foreign matter retained in cytoplasm |

# STUDY GUIDE

## ► Assess Your Learning Outcomes

*To test your knowledge, discuss the following topics with a study partner or in writing, ideally from memory.*

### 3.1 Concepts of Cellular Structure (p. 78)

1. The scope of cytology
2. Basic tenets of the cell theory
3. The nine common cell shapes
4. The size range of most human cells; some extremes outside this range; and some factors that limit cells from growing indefinitely large
5. The two kinds of electron microscopes; why they have enhanced the modern understanding of cells; and the distinction between magnification and resolution in microscopy
6. Basic structural components of a cell
7. The distinction between intracellular fluid (ICF) and extracellular fluid (ECF)

### 3.2 The Cell Surface (p. 82)

1. The molecules of the plasma membrane and how they are organized
2. The distinctive roles of phospholipids, glycolipids, cholesterol, integral proteins, peripheral proteins, and glycoproteins in membrane structure
3. Seven roles played by membrane proteins
4. Distinctions between membrane gates and other channels, and between ligand-gated, voltage-gated, and mechanically gated channels
5. The function of second-messenger systems associated with the plasma membrane; the specific roles of membrane receptors, G proteins, adenylate cyclase, cyclic adenosine monophosphate, and kinases in the cAMP second-messenger system
6. Composition and functions of the glycocalyx
7. Structure and functions of microvilli, and where they are found
8. Structure and functions of cilia, and where they are found
9. Structure and function of the only human flagellum
10. Structure and function of pseudopods

### 3.3 Membrane Transport (p. 90)

1. What it means to say that a plasma membrane is selectively permeable, and why this property is important for human survival
2. Filtration, where it occurs in the body, and why it depends on hydrostatic pressure
3. Simple diffusion, factors that determine its speed, and examples of its physiological and clinical relevance
4. Osmosis, examples of its physiological and clinical relevance, factors that determine its speed and direction, and the role of aquaporins
5. Reverse osmosis, where it occurs in the body, and the purpose it serves
6. In relation to osmosis, the meaning of *osmotic pressure, osmole, osmolarity, tonicity,* and *milliosmoles per liter (mOsm/L)*
7. Distinctions between hypotonic, hypertonic, and isotonic solutions; their effects on cells; and how this relates to intravenous fluid therapy
8. How carrier-mediated transport differs from other types of transport, and the relevance of specificity to this process
9. How carrier-mediated transport is limited by carrier saturation and the transport maximum ($T_m$)
10. Distinctions between a uniport, symport, and antiport; the meanings of *cotransport* and *countertransport;* and examples of where each is relevant in human physiology
11. Similarities and differences between facilitated diffusion and active transport
12. The distinction between primary and secondary active transport
13. The mechanism and roles of the sodium–potassium ($Na^+$–$K^+$) pump
14. How vesicular transport differs from other modes of membrane transport; the difference between endocytosis and exocytosis; different forms of endocytosis; and examples of the physiological relevance of each kind of vesicular transport

15. Of the preceding mechanisms of transport, which ones require a membrane; which ones require a plasma membrane and which ones can also occur through artificial membranes; which ones require ATP and which ones cease if ATP is unavailable, as upon death

### 3.4 The Cell Interior (p. 101)

1. Distinctions between cytoplasm, cytosol, cytoskeleton, organelles, and inclusions; and the respective, general roles of each in the internal organization of a cell
2. Overall functions of the cytoskeleton and the differences between microfilaments, intermediate filaments, and microtubules
3. Which organelles are considered membranous and why, and which of these are enclosed in single or double membranes
4. Structure and function of the nucleus, especially the nuclear envelope and nuclear lamina
5. General structure of the endoplasmic reticulum (ER); the two types of ER and the structural and functional differences between them
6. The composition, location, and function of ribosomes
7. The structure and functions of the Golgi complex; the origin and destiny of Golgi vesicles
8. The structures and functions of lysosomes and peroxisomes, and the similarities and differences between them
9. Structure and function of proteasomes
10. Structure and function of mitochondria
11. Structures and functions of centrioles, the centrosome, and basal bodies; and how these relate to each other
12. How inclusions differ from organelles; the origins and types of inclusions

# STUDY GUIDE

## ▶ Testing Your Recall

*Answers in Appendix B*

1. The clear, structureless gel in a cell is its
   a. nucleoplasm.
   b. protoplasm.
   c. cytoplasm.
   d. neoplasm.
   e. cytosol.

2. The $Na^+$–$K^+$ pump is
   a. a peripheral protein.
   b. a transmembrane protein.
   c. a G protein.
   d. a glycolipid.
   e. a phospholipid.

3. Which of the following processes could occur *only* in the plasma membrane of a living cell?
   a. facilitated diffusion
   b. simple diffusion
   c. filtration
   d. active transport
   e. osmosis

4. Cells specialized for absorption of matter from the ECF are likely to show an abundance of
   a. lysosomes.
   b. microvilli.
   c. mitochondria.
   d. secretory vesicles.
   e. ribosomes.

5. Aquaporins are transmembrane proteins that promote
   a. pinocytosis.
   b. carrier-mediated transport.
   c. active transport.
   d. facilitated diffusion.
   e. osmosis.

6. Membrane carriers resemble enzymes except for the fact that carriers
   a. are not proteins.
   b. do not have binding sites.
   c. are not selective for particular ligands.
   d. change conformation when they bind a ligand.
   e. do not chemically change their ligands.

7. The cotransport of glucose derives energy from
   a. a $Na^+$ concentration gradient.
   b. the glucose being transported.
   c. a $Ca^{2+}$ gradient.
   d. the membrane voltage.
   e. body heat.

8. The function of cAMP in a cell is
   a. to activate a G protein.
   b. to remove phosphate groups from ATP.
   c. to activate kinases.
   d. to bind to the first messenger.
   e. to add phosphate groups to enzymes.

9. Most cellular membranes are made by
   a. the nucleus.
   b. the cytoskeleton.
   c. enzymes in the peroxisomes.
   d. the endoplasmic reticulum.
   e. replication of existing membranes.

10. Matter can leave a cell by any of the following means *except*
    a. active transport.
    b. pinocytosis.
    c. an antiport.
    d. simple diffusion.
    e. exocytosis.

11. Most human cells are 10 to 15 _____ in diameter.

12. When a hormone cannot enter a cell, it activates the formation of a/an _____ inside the cell.

13. _____ gates in the plasma membrane open or close in response to changes in the electrical charge difference across the membrane.

14. The force exerted on a membrane by water is called _____.

15. A concentrated solution that causes a cell to shrink is _____ to the cell.

16. Fusion of a secretory vesicle with the plasma membrane and release of the vesicle's contents is a process called _____.

17. _____ and _____ are two granular organelles (enzyme complexes) that, respectively, synthesize and degrade proteins.

18. Liver cells can detoxify alcohol with two organelles, the _____ and _____.

19. An ion gate in the plasma membrane that opens or closes when a chemical binds to it is called a/an _____.

20. The space enclosed by the membranes of the Golgi complex and endoplasmic reticulum is called the _____.

# STUDY GUIDE

## ▶ Building Your Medical Vocabulary

*Answers in Appendix B*

*State a medical meaning of each word element below, and give a term in which it or a slight variation of it is used.*

1. anti-
2. chromato-
3. co-
4. cyto-
5. endo-
6. facil-
7. fusi-
8. -logy
9. -osis
10. phago-

## ▶ True or False

*Answers in Appendix B*

*Determine which five of the following statements are false, and briefly explain why.*

1. If a cell were poisoned so it could not make ATP, osmosis through its membrane would cease.

2. Material can move either into or out of a cell by means of active transport.

3. A cell's second messengers serve mainly to transport solutes through the membrane.

4. The Golgi complex makes lysosomes but not peroxisomes.

5. Some membrane channels are peripheral proteins.

6. The plasma membrane consists primarily of protein molecules.

7. The brush border of a cell is composed of cilia.

8. Human cells swell or shrink in any solution other than an isotonic one.

9. Osmosis is not limited by the transport maximum ($T_m$).

10. It is very unlikely for a cell to have more centrosomes than ribosomes.

## ▶ Testing Your Comprehension

*Answers at www.mhhe.com/saladin7*

1. If someone bought a saltwater fish in a pet shop and put it in a freshwater aquarium at home, what would happen to the fish's cells? What would happen if someone put a freshwater fish in a saltwater aquarium? Explain.

2. A farmer's hand and forearm are badly crushed in a hay baler. When examined at the hospital, his blood potassium level is found to be abnormal. Would you expect it to be higher or lower than normal? Explain.

3. Many children worldwide suffer from a severe deficiency of dietary protein. As a result, they have very low levels of blood albumin. How do you think this affects the water content and volume of their blood? Explain.

4. It is often said that mitochondria make energy for a cell. Why is this statement false?

5. Kartagener syndrome is a hereditary disease in which dynein arms are lacking from the axonemes of cilia and flagella. Predict the effect of Kartagener syndrome on a man's ability to father a child. Predict its effect on his respiratory health. Explain both answers.

## ▶ Improve Your Grade at www.mhhe.com/saladin7

*Download animations to study when it fits your schedule. Practice quizzes, labeling activities, games, and flashcards offer fun ways to master the chapter concepts. Or, download image PowerPoint files for each chapter to create a study guide or for taking notes during lecture.*

# GENETICS AND CELLUAR FUNCTION

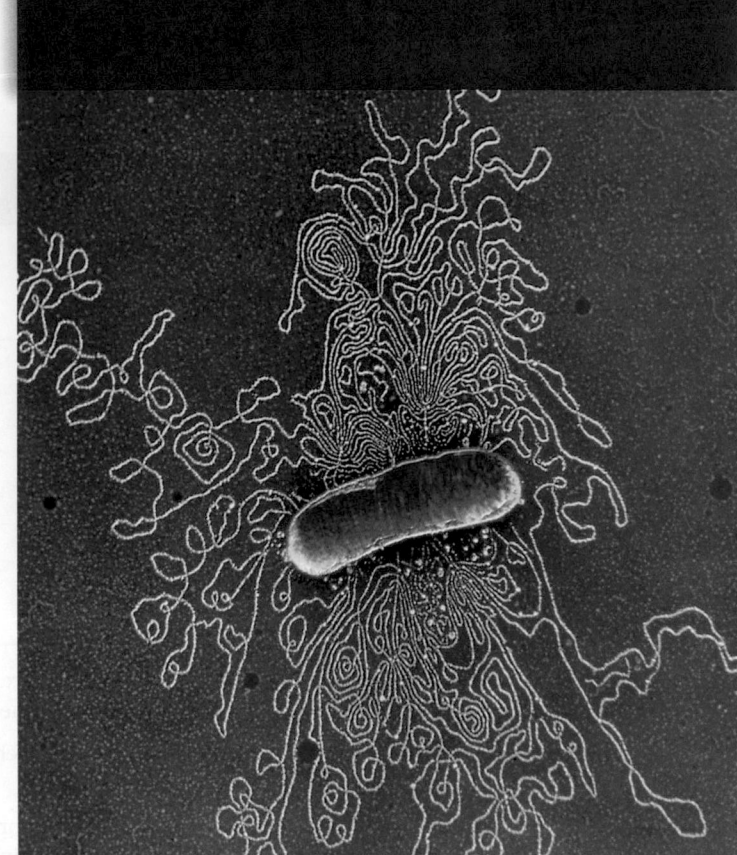

A single DNA molecule spilling from a ruptured bacterial cell (TEM)

Anatomy & Physiology | REVEALED®
aprevealed.com

**Module 2: Cells and Chemistry**

**S**everal chapters in this book discuss hereditary traits such as blood types and hair color and genetic disorders such as color blindness, cystic fibrosis, diabetes mellitus, and hemophilia. To understand such conditions, it is necessary to have some familiarity with DNA and genes. This chapter is intended to provide such preparation.

Heredity has been a matter of human interest dating even to biblical writings, but a scientific understanding of how traits pass from parent to offspring began with the Austrian monk Gregor Mendel (1822–84) and his famous experiments on garden peas. By the late nineteenth century, biologists had observed chromosomes and their behavior during cell division. In the early twentieth century, they rediscovered Mendel's work and realized the correlation between chromosome behavior and his laws of heredity. From that simple but insightful beginning, genetics has grown into a highly diverse science with several subdisciplines and is arguably the most dynamic of all natural sciences in these early years of the twenty-first century.

*Mendelian genetics* deals with parent–offspring and larger family relationships to discern and predict patterns of inheritance within a family line. *Cytogenetics* uses the techniques of cytology and microscopy to study chromosomes and their relationship to hereditary traits. *Molecular genetics* uses the techniques of biochemistry to study the structure and function of DNA. *Genomic medicine* comprehensively studies the entire DNA endowment of an individual (the *genome*), how it influences health and disease, and how this knowledge can be used to prevent, treat, or cure diseases. We will examine all four of these perspectives in this chapter.

## 4.1 DNA and RNA—The Nucleic Acids

### Expected Learning Outcomes
When you have completed this section, you should be able to

a. describe the structure of DNA and relate this to its function;

b. explain how DNA and proteins are organized to form the chromosomes; and

c. describe the types of RNA, their structural and functional differences, and how they compare with DNA.

With improvements in the microscope, biologists of the late nineteenth century saw that cell division is immediately preceded by nuclear division, and during nuclear division, the chromosomes split neatly in two and distribute their halves to the two daughter cells. They came to suspect that the nucleus was the center of heredity and cellular control, and they began probing it for the biochemical secrets of heredity. Swiss biochemist Johann Friedrich Miescher (1844–95) studied the nuclei of white blood cells extracted from pus in used hospital bandages, and later the nuclei of salmon sperm, since both cell types offered large nuclei with minimal amounts of contaminating cytoplasm. In 1869, he discovered an acidic, phosphorus-rich substance he named *nuclein*. He correctly believed this to be the cell's hereditary matter, although he was never able to convince other scientists of this. We now call this substance **deoxyribonucleic acid (DNA)** and know it to be the repository of our genes.

By 1900, biochemists knew the basic components of DNA—sugar, phosphate groups, and organic rings called nitrogenous bases—but they didn't have the technology to determine how these were put together. That understanding didn't come until 1953, in one of the twentieth century's most dramatic and important stories of scientific discovery (see Deeper Insight 4.1). The following description of DNA is largely the outcome of that work.

## DNA Structure and Function

DNA is a long threadlike molecule with a uniform diameter of 2 nm, although its length varies greatly from the smallest to the largest chromosomes. Most human cells have 46 molecules of DNA totaling 2 m in length. This makes the average DNA molecule about 43 mm (almost 2 in.) long. To put this in perspective, imagine that an average DNA molecule was scaled up to the diameter of a utility pole (about 20 cm, or 8 in.). At this diameter, a pole proportionate to DNA would rise about 4,400 km (2,700 mi.) into space—far higher than the orbits of space shuttles (320–390 km) and the Hubble Space Telescope (600 km).

At the molecular level, DNA and other nucleic acids are polymers of **nucleotides** (NEW-clee-oh-tides). A nucleotide consists of a sugar, a phosphate group, and a single- or double-ringed **nitrogenous** (ny-TRODJ-eh-nus) **base** (fig. 4.1a). Two of the bases in DNA—**cytosine (C)** and **thymine (T)**—have a single carbon–nitrogen ring and are classified as *pyrimidines* (py-RIM-ih-deens). The other two bases—**adenine (A)** and **guanine (G)**—have double rings and are classified as *purines* (fig. 4.1b). The bases of RNA are somewhat different, as explained later, but still fall into the purine and pyrimidine classes.

The structure of DNA, commonly described as a *double helix*, resembles a spiral staircase (fig. 4.2). Each sidepiece is a backbone composed of phosphate groups alternating with the sugar *deoxyribose*. The steplike connections between the backbones are pairs of nitrogenous bases. The bases face the inside of the helix and hold the two backbones together with hydrogen bonds. Across from a purine on one backbone, there is a pyrimidine on the other. The pairing of each small, single-ringed pyrimidine with a large, double-ringed purine gives the DNA molecule its uniform 2 nm width.

A given purine cannot arbitrarily bind to just any pyrimidine. Adenine and thymine form two hydrogen bonds with each other, and guanine and cytosine form three, as shown in figure 4.2b. Therefore, where there is an A on one backbone, there is normally a T across from it, and every C is normally paired with a G. A–T and C–G are called the **base pairs.** The fact that one strand governs the base sequence of the other is called the **law of complementary base pairing.** It enables us to predict the base sequence of one strand if we know the sequence of the complementary strand.

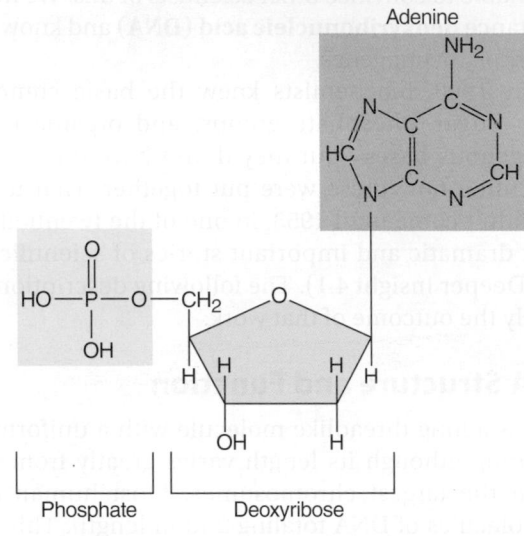

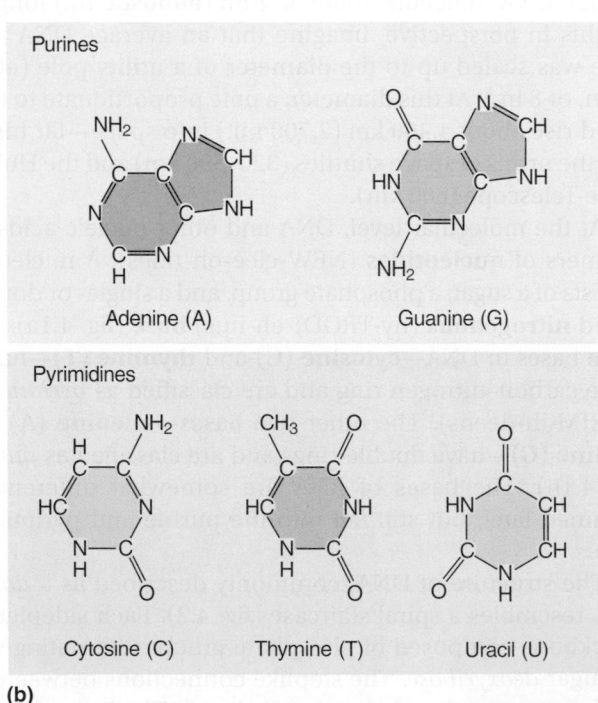

**FIGURE 4.1  Nucleotides and Nitrogenous Bases.** (a) The structure of a nucleotide, one of the monomers of DNA and RNA. In RNA, the sugar is ribose. (b) The five nitrogenous bases found in DNA and RNA.

❓ *How would the uniform 2 nm diameter of DNA be affected if two purines or two pyrimidines could pair with each other?*

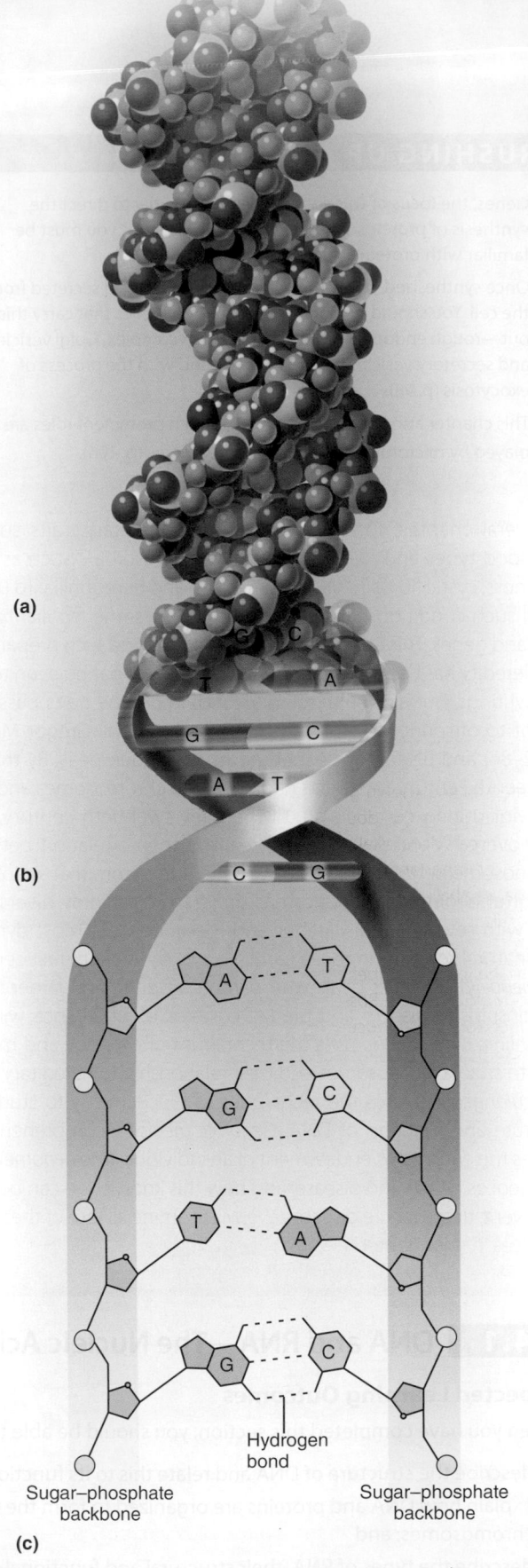

**FIGURE 4.2  DNA Structure.** (a) A molecular space-filling model of DNA giving some impression of its molecular geometry. (b) The "spiral staircase" structure. The two sugar–phosphate backbones twine around each other while complementary bases (colored bars) face each other on the inside of the double helix. (c) A small segment of DNA showing the composition of the backbone and complementary pairing of the nitrogenous bases.

## DEEPER INSIGHT 4.1

### MEDICAL HISTORY

### Discovery of the Double Helix

Credit for determining the double-helical structure of DNA has gone mainly to James Watson and Francis Crick (fig. 4.3). The events surrounding their discovery form one of the most dramatic stories of modern science—the subject of many books and at least one movie. When Watson and Crick came to share a laboratory at Cambridge University in 1951, both had barely begun their careers. Watson, age 23, had just completed his Ph.D. in the United States, and Crick, 11 years older, was a doctoral candidate in England. Yet the two were about to become the most famous molecular biologists of the twentieth century, and the discovery that won them such acclaim came without a single laboratory experiment of their own.

Others were fervently at work on DNA, including Rosalind Franklin and Maurice Wilkins at King's College in London. Using a technique called X-ray diffraction, Franklin had determined that DNA had a repetitious helical structure with sugar and phosphate on the outside of the helix. Without her permission, Wilkins showed one of Franklin's best X-ray photographs to Watson. Watson said, "The instant I saw the picture my mouth fell open and my pulse began to race." It provided a flash of insight that allowed the Watson and Crick team to beat Franklin to the goal. Combining what they already knew with the molecular geometry revealed by Franklin's photo, they were quickly able to piece together a scale model from cardboard and sheet metal that fully accounted for the known geometry of DNA. They rushed a paper into print in 1953 describing the double helix, barely mentioning the importance of Franklin's 2 years of painstaking X-ray diffraction work in unlocking the mystery of life's most important molecule. Franklin published her findings in a separate paper back to back with theirs.

For this discovery and the ensuing decade of research on DNA that it opened up, Watson, Crick, and Wilkins shared the Nobel Prize for Physiology or Medicine in 1962. Nobel Prizes are awarded only to the living, and in the final irony of her career, Rosalind Franklin had died in 1958, at the age of 37, of a cancer possibly induced by the X-rays that were her window on DNA architecture.

(a)

(b)

(c)

**FIGURE 4.3 Discoverers of the Double Helix.** (a) Rosalind Franklin (1920–58), whose painstaking X-ray diffraction photographs revealed important information about the molecular geometry of DNA. (b) One of Franklin's X-ray photographs. (c) James Watson (1928–) (left) and Francis Crick (1916–2004) with their model of the double helix.

▶▶▶**APPLY WHAT YOU KNOW**

*What would be the base sequence of the DNA strand across from ATTGACTCG? If a DNA molecule was known to be 20% adenine, predict its percentage of cytosine and explain your answer.*

The essential function of DNA is to carry instructions, called *genes,* for the synthesis of proteins. At this point in the chapter, we will provisionally regard a gene as a segment of DNA that codes for a protein. Later, we shall have to confront the fact that this is an inadequate definition, and we'll examine the meaning of the word more deeply.

Humans are estimated to have about 20,000 genes. These constitute only about 2% of the DNA. The other 98% is noncoding DNA, most of which apparently plays various roles in chromosome structure and regulation of gene activity.

## Chromatin and Chromosomes

DNA does not exist as a naked double helix in the nucleus of a cell, but is complexed with proteins to form a fine filamentous material called **chromatin.** In most cells, the chromatin occurs as 46 long filaments called **chromosomes.** There is a stupendous amount of DNA in one nucleus—about 2 m (6 ft) of it in the first half of a cell's life cycle and twice as much when a cell has replicated its DNA in preparation for cell division. It is a prodigious feat to pack this much DNA into a nucleus only about 5 μm in diameter—and in such an orderly fashion that it does not become tangled, broken, and damaged beyond use. Here, we will examine how this is achieved.

In nondividing cells, the chromatin is so slender that it usually cannot be seen with the light microscope. With a high-resolution electron microscope, however, it has a granular appearance, like beads on a string (fig. 4.4a). Each "bead" is a disc-shaped cluster of eight proteins called **histones.** A DNA molecule winds around the cluster for a little over one and a half turns, like a ribbon around a spool, and then continues on its way until it reaches the next histone cluster and winds around that one (fig. 4.4b). The average chromatin thread repeats this pattern almost 800,000 times, and thus appears divided into segments called **nucleosomes.** Each nucleosome consists of a *core particle* (the spool of histones

with the DNA ribbon around them) and a short segment of *linker DNA* leading to the next core particle. Winding the DNA around the histones makes the chromatin thread more than five times as thick (11 nm) and one-third shorter than the DNA alone.

But even at this degree of compaction, a single chromosome would cross the entire nucleus hundreds of times. There are higher orders of structure that make the chromosome still more compact. First, the nucleosomes are arranged in a zigzag pattern, folding the chromatin like an accordion. This produces a strand 30 nm wide, but still 100 times as long as the nuclear diameter. Then, the 30 nm strand is thrown into complex, irregular loops and coils that make the chromosome 300 nm thick and 1,000 times shorter than the DNA molecule. Finally, each chromosome is packed into its own sphcroidal region of the nucleus, called a *chromosome territory.* A chromosome territory is permeated with channels that allow regulatory chemicals to have access to the genes.

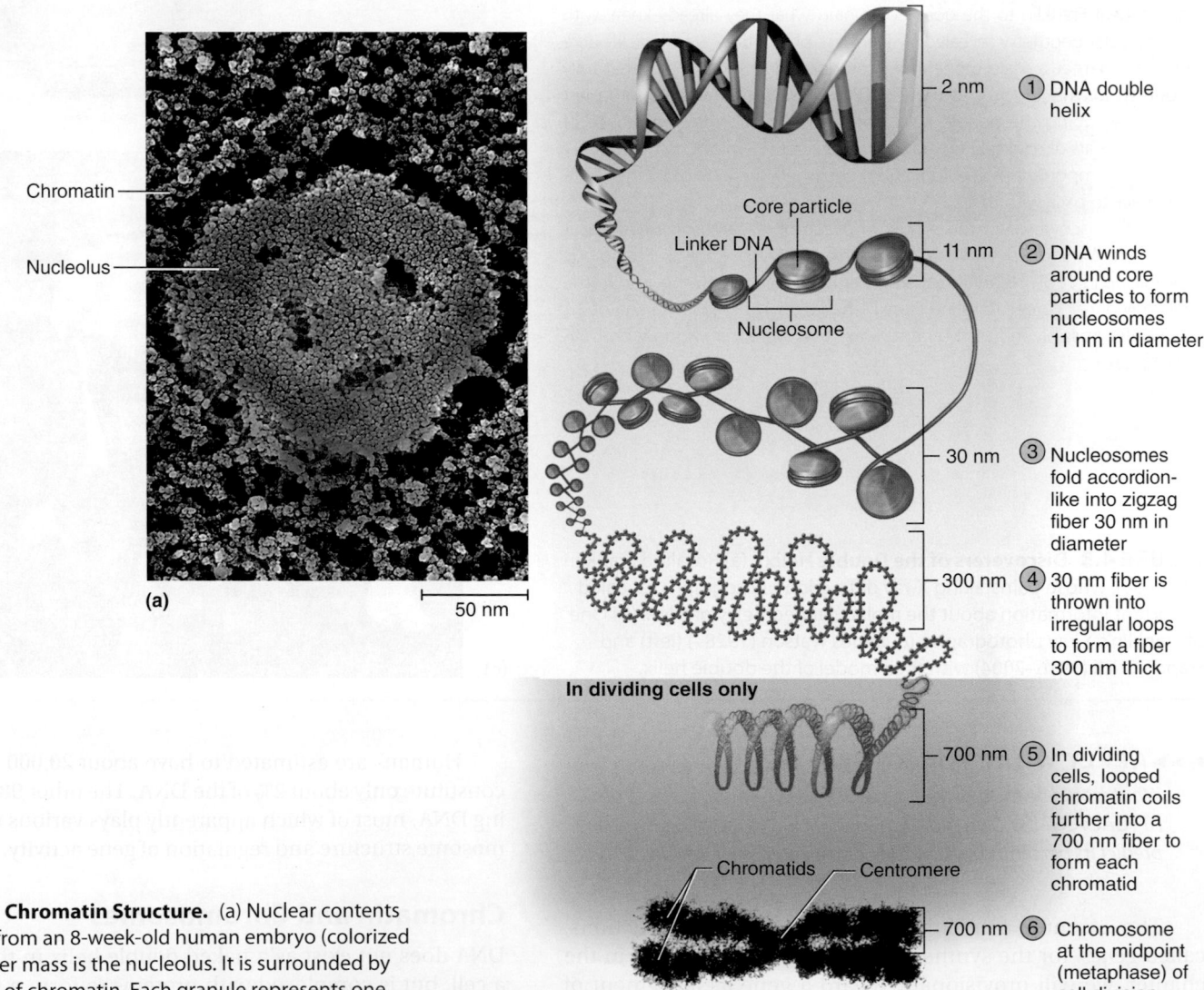

**FIGURE 4.4 Chromatin Structure.** (a) Nuclear contents of a germ cell from an 8-week-old human embryo (colorized SEM). The center mass is the nucleolus. It is surrounded by granular fibers of chromatin. Each granule represents one nucleosome. (b) The coiling of chromatin and its relationship to the nucleosomes.

This is the state of the DNA in a nondividing cell. It is not a static structure, but changes from moment to moment according to the genetic activity of the cell as individual genes are turned on and off. Whole chromosomes migrate to new territories as a cell develops—for example, moving from the edge to the core of a nucleus as its genes are activated for a certain developmental task. This allows genes on different chromosomes to partner with each other in bringing about developmental changes in the cell.

When a cell is preparing to divide, it makes an exact copy of all its nuclear DNA by a process described later, increasing its allotment to about 4 m of DNA. Each chromosome then consists of two parallel filaments called **sister chromatids.** In the early stage of cell division *(prophase),* these chromatids coil some more until each one becomes another 10 times shorter and about 700 nm wide. Thus, at its most compact, each thread of chromatin is 10,000 times shorter but 350 times thicker than the DNA double helix. Only now are the chromosomes thick enough to be seen with a light microscope. This compaction not only allows the 4 m of DNA to fit in the nucleus, but also enables the two sister chromatids to be pulled apart and carried to separate daughter cells.

Despite all this intricate packaging, the DNA of the average mammalian cell is damaged an astonishing 10,000 to 100,000 times per day! The consequences would be catastrophic were it not for DNA repair enzymes that detect and undo most of the damage.

Figure 4.5 shows the structure of a chromosome in early cell division, when it is compacted to its maximum extent. It consists of two genetically identical, rodlike sister chromatids joined together at a pinched spot called the **centromere.** On each side of the centromere, there is a protein plaque called a **kinetochore**[1] (kih-NEE-to-core), which has a role in cell division explained later.

---

[1]*kineto* = motion; *chore* = place

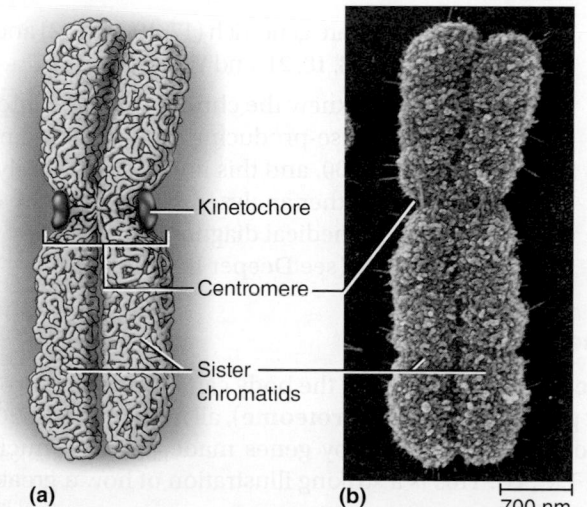

(a)  (b)  700 nm

**FIGURE 4.5 Chromosome Structure at Metaphase.** (a) Drawing of a metaphase chromosome. (b) Scanning electron micrograph.

| Feature | DNA | RNA |
|---|---|---|
| Sugar | Deoxyribose | Ribose |
| Types of nitrogenous bases | A, T, C, G | A, U, C, G |
| Number of nitrogenous bases | Averages $10^8$ base pairs | 70–10,000 bases, mostly unpaired |
| Number of nucleotide chains | Two (double helix) | One |
| Site of action | Functions in nucleus; cannot leave | Leaves nucleus; functions mainly in cytoplasm |
| Function | Codes for synthesis of RNA and protein | Carries out the instructions in DNA; assembles proteins |

**TABLE 4.1 Comparison of DNA and RNA**

## RNA Structure and Function

Before we leave this section, we mustn't forget those smaller cousins of DNA, the ribonucleic acids (RNAs). There are several forms of RNA in a cell, but we will focus on the three that are directly involved in producing proteins: *messenger RNA (mRNA), ribosomal RNA (rRNA),* and *transfer RNA (tRNA).* DNA cannot produce proteins without their help. Other RNA types play various regulatory roles that we will not consider.

What do mRNA, rRNA, and tRNA have in common, and how do they differ from DNA? The most significant difference is that RNA is much smaller, ranging from about 70 to 90 bases in tRNA to slightly over 10,000 bases in the largest mRNA. DNA, by contrast, averages more than 100 million base pairs long (table 4.1). Also, whereas DNA is a double helix, RNA consists of only one nucleotide chain, not held together by complementary base pairs except in certain short regions where the molecule folds back on itself. The sugar in RNA is ribose instead of deoxyribose. RNA contains three of the same nitrogenous bases as DNA—adenine, cytosine, and guanine—but it has no thymine; a base called **uracil (U)** takes its place (see fig. 4.1b). Transfer RNA has more than 50 different nitrogenous bases, but for our purposes, we don't need to consider any except A, U, C, and G.

The essential function of the three principal RNAs is to interpret the code in DNA and use those instructions to synthesize proteins. RNA is a disposable molecule that works mainly in the cytoplasm, while DNA is irreplaceable and remains safely behind in the nucleus, "giving orders" from there. This process is described in the next section of the chapter.

### BEFORE YOU GO ON

Answer the following questions to test your understanding of the preceding section:

1. What are the three components of a nucleotide? Which component varies from one nucleotide to another in DNA?

2. What governs the pattern of base pairing in DNA?

3. What is the difference between DNA and chromatin?

4. Summarize the structural and functional differences between DNA and RNA.

## 4.2 Genes and Their Action

### Expected Learning Outcomes
When you have completed this section, you should be able to

a. give a working definition of the *gene* and explain why new discoveries in genetics have changed our concept of what a gene is;

b. explain what the human genome is and what relationship it has to the health sciences;

c. define *genetic code* and describe how DNA codes for protein structure;

d. describe the process of assembling amino acids to form a protein;

e. explain what happens to a protein after its amino acid sequence has been synthesized;

f. describe some ways that a gene can be turned on or off; and

g. explain how DNA indirectly regulates the synthesis of nonprotein molecules.

## What Is a Gene?

As much as biologists talk about genes, the term is devilishly difficult to define. The classical concept of the gene, rooted in Mendel's studies of heredity in peas, was that it is an abstract "unit of heredity" by which a trait passes from parent to offspring. Following discovery of the double helix, molecular biologists worked out the *genetic code* contained in the four bases of DNA and considered a gene to be a segment of the DNA that carries the code for a particular protein. Now, however, we know that the human body has millions of different proteins but only 20,000 genes; obviously there isn't a separate gene for every protein. In addition, we know now that several human genes produce only RNA molecules that never go on to direct the synthesis of a protein; RNA is their final product. Other recent discoveries have complicated our concept of the gene still more—genes overlapping each other, so some segments of DNA belong to two different genes; short genes embedded within longer ones; multiple related proteins encoded by a single gene; and other unexpected arrangements. As molecular biologists have learned more and more about DNA, the definition of the gene has become more and more frayed around the edges.

For the purposes of this introductory book, however, we can settle for an approximate meaning. We will define *gene* as an information-containing segment of DNA that codes for the production of a molecule of RNA, which in most cases goes on to play a role in the synthesis of one or more proteins. The amino acid sequence of a protein is determined by a nucleotide sequence in the DNA.

## The Genome

The 46 human chromosomes come in two sets of 23 each, one set from each parent. All the DNA, both coding and noncoding, in one 23-chromosome set is called the **genome.** The human genome consists of about 3.1 billion nucleotide pairs. Owing to a massive multinational undertaking called the **Human Genome Project (HGP),** carried out from 1990 to 2003, biologists now know the base sequence (A, T, C, G) of more than 99% of the genome.

Sequencing the human genome has been hailed as a technological accomplishment as momentous as splitting the atom and landing explorers on the moon. It has opened a new field of biology called **genomics,** the comprehensive study of the whole genome and how its genes and noncoding DNA interact to affect the structure and function of the whole organism. Among the revolutionary findings to come from the Human Genome Project are these:

- *Homo sapiens* has far fewer genes than the 100,000 previously believed.

- These genes generate millions of different proteins, so gone is the old idea of one gene for each protein. In its place is the realization that a single gene can code for many different proteins through alternative splicing of mRNA (see p. 120) and other means.

- Genes average about 3,000 bases long, but range up to 2.4 million bases.

- All humans, worldwide, are at least 99.99% genetically identical, but even the 0.01% variation means that we can differ from one another in more than 3 million base pairs. Various combinations of these **single-nucleotide polymorphisms**[2] account for all human genetic variation.

- Some chromosomes are gene-rich (17, 19, and 22) and some are gene-poor (4, 8, 13, 18, 21, and Y; see fig. 4.17).

- Before the HGP, we knew the chromosomal locations of fewer than 100 disease-producing mutations; we now know more than 1,400, and this number will surely rise as the genome is further analyzed. This opens the door to a new branch of medical diagnosis and therapy called **genomic medicine** (see Deeper Insight 4.2).

## The Genetic Code

It seems remarkable that the body can make millions of different proteins (called the **proteome**), all from the same 20 amino acids and all encoded by genes made of just 4 nucleotides (A, T, C, G). This is a striking illustration of how a great variety

---

[2]*poly* = multiple; *morph* = form

of complex structures can be made from a small variety of simpler components. The **genetic code** is a system that enables these 4 nucleotides to code for the amino acid sequences of all proteins.

It is not unusual for simple codes to represent complex information. Computers store and transmit complex information, including pictures and sounds, in a binary code with only the symbols 1 and 0. Thus, it should not be surprising that a mere 20 amino acids can be represented by a code of 4 nucleotides; all that is required is to combine these symbols in varied ways. It requires more than 2 nucleotides to code for each amino acid, because A, U, C, and G can combine in only 16 different pairs (AA, AU, AC, AG, UA, UU, and so on). The minimum code to symbolize 20 amino acids is 3 nucleotides per amino acid, and indeed, this is the case in DNA. A sequence of 3 DNA nucleotides that stands for 1 amino acid is called a **base triplet.** When messenger RNA is produced, it carries a coded message based on these DNA triplets. A 3-base sequence in mRNA is called a **codon.** The genetic code is expressed in terms of codons.

For the purpose of this book, it isn't necessary to tabulate the entire genetic code, but sufficient to see a few illuminating examples of how the DNA triplets relate to the mRNA codons and how they relate to the amino acids of a protein (table 4.2). You can see from the table that sometimes two or more codons represent the same amino acid. The reason for this is easy to explain mathematically. Four symbols *(N)* taken three at a time *(x)* can be combined in $N^x$ different ways; that is, there are $4^3 = 64$ possible codons available to represent the 20 amino acids.

Only 61 of these, however, code for amino acids. The other 3—UAG, UGA, and UAA—are called **stop codons;** they signal "end of message," like the period at the end of a sentence. A stop codon enables the cell's protein-synthesizing machinery to sense that it has reached the end of the instruction for a

| TABLE 4.2 | Examples of the Genetic Code* | | |
|---|---|---|---|
| Base Triplet of DNA | Codon of mRNA | Name of Amino Acid | Abbreviation for Amino Acid |
| CCT | GGA | Glycine | Gly |
| CCA | GGU | Glycine | Gly |
| CCC | GGG | Glycine | Gly |
| CGT | GCU | Alanine | Ala |
| CGT | GCA | Alanine | Ala |
| TGG | ACC | Threonine | Thr |
| TGC | ACG | Threonine | Thr |
| GTA | CAU | Histidine | His |
| TAC | AUG | Methionine | Met |

*For the complete genetic code, see appendix B.

particular protein. The codon AUG plays two roles: It serves as a code for methionine and as a **start codon.** This dual function is explained shortly.

## Protein Synthesis

We can now move on to an understanding of how DNA and RNA collaborate to produce proteins. Before studying the details, however, it will be helpful to consider the big picture. In brief, the genetic code in DNA specifies which proteins a cell can make. All the body's cells except the sex cells and some immune cells contain identical genes. However, different genes are activated in different cells; for example, genes for digestive enzymes are active in stomach cells but not in muscle cells.

---

# DEEPER INSIGHT 4.2

## CLINICAL APPLICATION

### Genomic Medicine

*Genomic medicine* is the application of our knowledge of the genome to the prediction, diagnosis, and treatment of disease. It is relevant to disorders as diverse as cancer, Alzheimer disease, schizophrenia, obesity, and even a person's susceptibility to nonhereditary diseases such as AIDS and tuberculosis.

Genomic technology has advanced to the point that for less than $1,000, one can already have one's entire genome scanned for markers of disease risk. Why would anyone want to? Because knowing one's genome could dramatically change clinical care. It may allow clinicians to forecast a person's risk of disease and to predict its course; mutations in a single gene can affect the severity of such diseases as hemophilia, muscular dystrophy, cancer, and cystic fibrosis. Genomics should also allow for earlier detection of diseases and for earlier, more effective clinical intervention. Drugs that are safe for most people can have serious side effects in others, owing to genetic variations in drug metabolism. Genomics has begun providing a basis for choosing the safest or most effective drugs and for adjusting dosages for different patients on the basis of their genetic makeup.

Knowing the sites of disease-producing mutations expands the potential for *gene-substitution therapy.* This is a procedure in which cells are removed from a patient with a genetic disorder, supplied with a normal gene in place of the defective one, and reintroduced to the body. The hope is that these genetically modified cells will proliferate and provide the patient with a gene product that he or she was lacking—perhaps insulin for a patient with diabetes or a blood-clotting factor for a patient with hemophilia. Attempts at gene therapy have been marred by some tragic setbacks, however, and still face great technical difficulties that have, as yet, seldom been surmounted.

Genomics is introducing new problems in medical ethics and law. Should your genome be a private matter between you and your physician? Or should an insurance company be entitled to know your genome before issuing health or life insurance to you so it can know your risk of contracting a catastrophic illness, adjust the cost of your coverage, or even deny coverage? Should a prospective employer have the right to know your genome before offering employment? These are areas in which biology, politics, and law converge to shape public policy.

Any given cell uses only one-third to two-thirds of its genes; the rest remain dormant in that cell, but may be functional in other types of cells.

When a gene is activated, a **messenger RNA (mRNA)** is made—a mirror image of the gene, more or less. Most mRNA migrates from the nucleus to the cytoplasm, where it serves as a code for assembling amino acids in the right order to make a particular protein. In summary, you can think of the process of protein synthesis as DNA ⟶ mRNA ⟶ protein, with each arrow reading as "codes for the production of." The step from DNA to mRNA is called *transcription,* and the step from mRNA to protein is called *translation.* Transcription occurs in the nucleus, where the DNA is. Most translation occurs in the cytoplasm, but 10% to 15% of proteins are synthesized in the nucleus, with both steps occurring there.

## Transcription

DNA is too large to leave the nucleus and participate directly in cytoplasmic protein synthesis. It is necessary, therefore, to make a small mRNA copy that can migrate through a nuclear pore into the cytoplasm. Just as we might transcribe (copy) a document, **transcription** in genetics means the process of copying genetic instructions from DNA to RNA. An enzyme called **RNA polymerase** (po-LIM-ur-ase) binds to the DNA and assembles the RNA. Certain distinctive base sequences (often TATATA or TATAAA) inform the polymerase where to begin.

RNA polymerase opens up the DNA helix about 17 base pairs at a time. It reads the bases from one strand of the DNA and makes a corresponding RNA. Where it finds a C on the DNA, it adds a G to the RNA; where it finds an A, it adds a U; and so forth. The enzyme then rewinds the DNA helix behind it. Another RNA polymerase may follow closely behind the first one; thus, a gene may be transcribed by several polymerase molecules at once, and numerous copies of the same RNA are made. At the end of the gene is a base sequence that serves as a terminator, which signals the polymerase to stop.

The RNA produced by transcription is an "immature" form called *pre-mRNA.* This molecule contains "sense" portions called *exons* that will be translated into a protein, and "nonsense" portions called *introns* that must be removed before translation. Enzymes remove and degrade the introns and splice the exons together into a functional mRNA molecule, which then leaves the nucleus. It may help you in remembering these if you think of *in*trons being removed while still *in* the nucleus and the *ex*ons being *ex*ported from the nucleus to undergo translation in the cytoplasm. Another way of remembering these is that *ex*ons are *ex*pressed in protein synthesis, whereas *in*trons are *in*tervening segments of DNA between the exons.

Through a mechanism called *alternative splicing,* one gene can code for more than one protein. Suppose a gene produced a pre-mRNA containing six exons separated by noncoding introns. As shown in figure 4.6, these exons can be spliced together in various combinations to yield codes for two or more proteins. This is a partial explanation of how the body can produce millions of different proteins with no more than 20,000 genes.

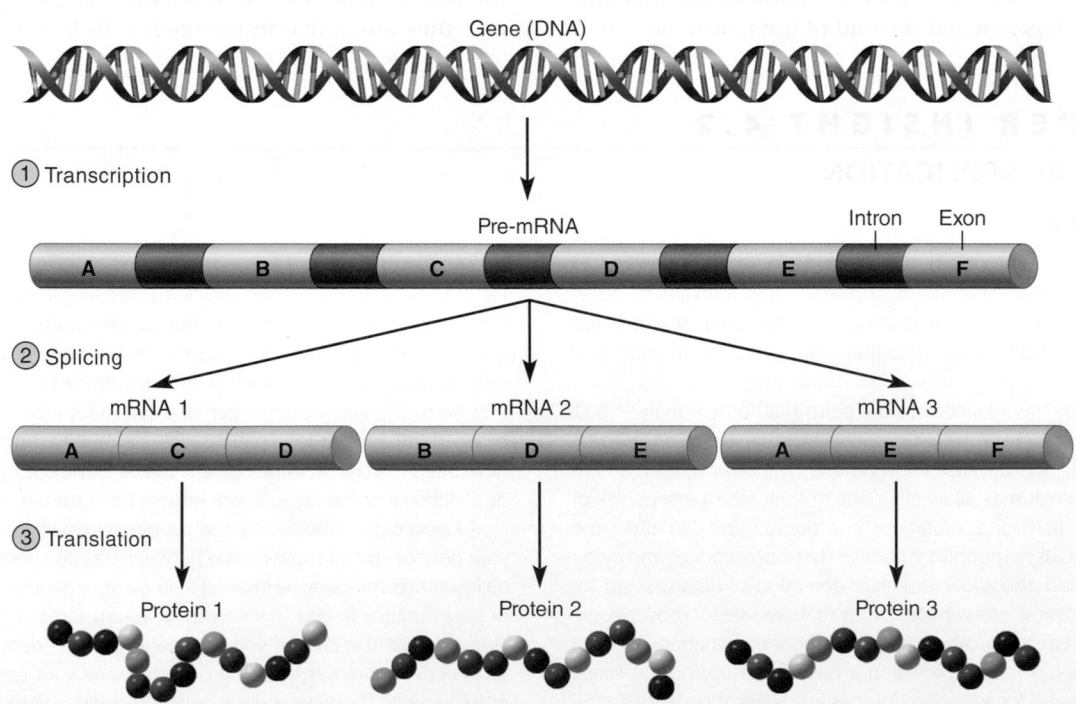

**FIGURE 4.6 Alternative Splicing of mRNA.** By splicing together different combinations of exons from a single pre-mRNA, a cell can generate multiple proteins from a single gene.

## Translation

Just as we might translate a work from Spanish into English, genetic **translation** converts the language of nucleotides into the language of amino acids. This job is carried out by the following participants.

1. **Messenger RNA (mRNA),** which carries the genetic code from the nucleus to the cytoplasm. During its synthesis in the nucleus, mRNA acquires a protein cap that acts like a passport, permitting it to pass through a nuclear pore into the cytosol. The cap also acts as a recognition site that tells a ribosome where to begin translation.

2. **Transfer RNA (tRNA),** a relatively small RNA whose job is to bind a free amino acid in the cytosol and deliver it to the ribosome to be added to a growing protein chain. tRNA is a single-stranded molecule that turns back and coils on itself to form an angular L shape (fig. 4.7). One loop of the molecule includes an **anticodon,** a series of three nucleotides complementary to a specific codon of mRNA. For the codon AUG, for example, the anticodon is UAC. The other end of the tRNA has a binding site for a specific amino acid corresponding to that codon. The first tRNA to bind to a ribosome at the start of translation is called the *initiator tRNA.* It always has the anticodon UAC and always carries the amino acid methionine.

3. **Ribosomes,** the little "reading machines" found in the cytosol and on the outside of the rough ER and nuclear envelope. Inactive ribosomes occur in the cytosol in two pieces—a **small subunit** and a **large subunit.** Each is composed of several enzymes and ribosomal RNA (rRNA) molecules. The two subunits come together only when translating mRNA. A ribosome has three pockets that serve as binding sites for tRNA. In the course of translation, a tRNA usually binds first to the *A site* on one side of the ribosome, then shifts to the *P site* in the middle, and finally to the *E site* on the other side. To remember the order of these sites, it may help you to think of *A* for the site that *accepts* a new amino acid; *P* for the site that carries the growing *protein;* and *E* for *exit.* (*A* and *P* actually stand for *aminoacyl* and *peptidyl* sites.)

Translation occurs in three steps called *initiation, elongation,* and *termination.* These are shown in figure 4.8, panels 1 through 3; panel 4 illustrates a further aspect of the production of proteins destined to be packaged into lysosomes or secretory vesicles.

(1) **Initiation.** mRNA passes through a nuclear pore into the cytosol and forms a loop. A small ribosomal subunit binds to a *leader sequence* of bases on the mRNA near the cap, then slides along the mRNA until it recognizes the start codon AUG. An initiator tRNA with the anticodon UAC pairs with the start codon and settles into the P (middle) site of the ribosome with its cargo of methionine (Met). The large subunit of the ribosome then joins the complex. The assembled ribosome now embraces the mRNA in a groove between the subunits and begins sliding along it, reading its bases.

(2) **Elongation.** The next tRNA arrives, carrying another amino acid; it binds to the A site of the ribosome and its anticodon pairs with the second codon of the mRNA— GCU, for example. A tRNA with the anticodon CGA would bind here, and according to the genetic code, it would carry alanine (Ala). An enzyme in the ribosome transfers the Met of the initiator tRNA to the Ala delivered by the second tRNA and creates a peptide bond between them, giving us a dipeptide, Met–Ala. The ribosome then slides down to read the next codon. This shifts the initiator tRNA (now carrying no amino acid) into the E site, where it leaves the ribosome. The second tRNA (now carrying Met–Ala) shifts into the P site. The now-vacant A site binds a third tRNA. Suppose the next codon is ACG. A tRNA with anticodon UGC would bind here, and would carry threonine (Thr). (This is the state shown in the figure.) The ribosome transfers the Met–Ala to the Thr, creates another peptide bond, and we now have a tripeptide, Met–Ala–Thr. By repetition of this process, a larger and larger protein is produced. As the protein elongates, it folds into its three-dimensional shape.

Each time a tRNA leaves the E site, it goes off to pick up another amino acid from a pool of free amino acids in the cytosol. One ATP molecule is used in binding the

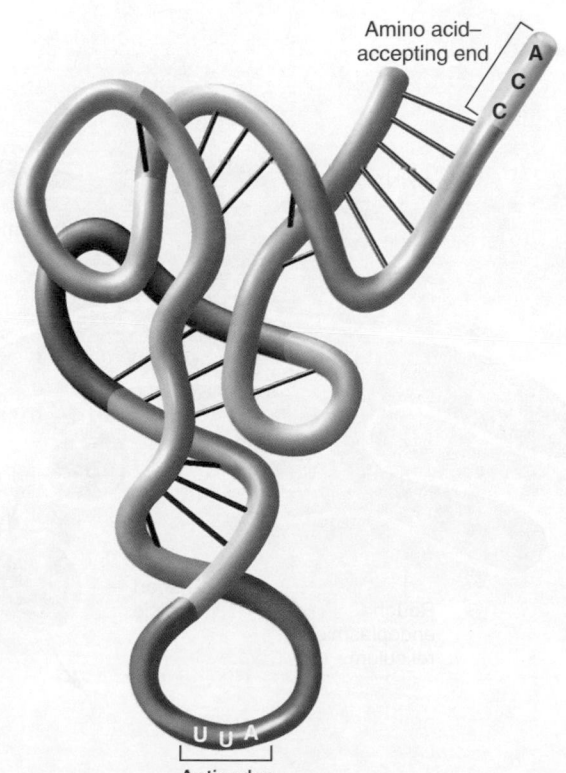

Amino acid–accepting end

A
C
C

U U A
Anticodon

**FIGURE 4.7 Transfer RNA (tRNA).** tRNA has an amino acid–accepting end that binds to one specific amino acid, and an anticodon at the other end that binds to a complementary codon of mRNA.

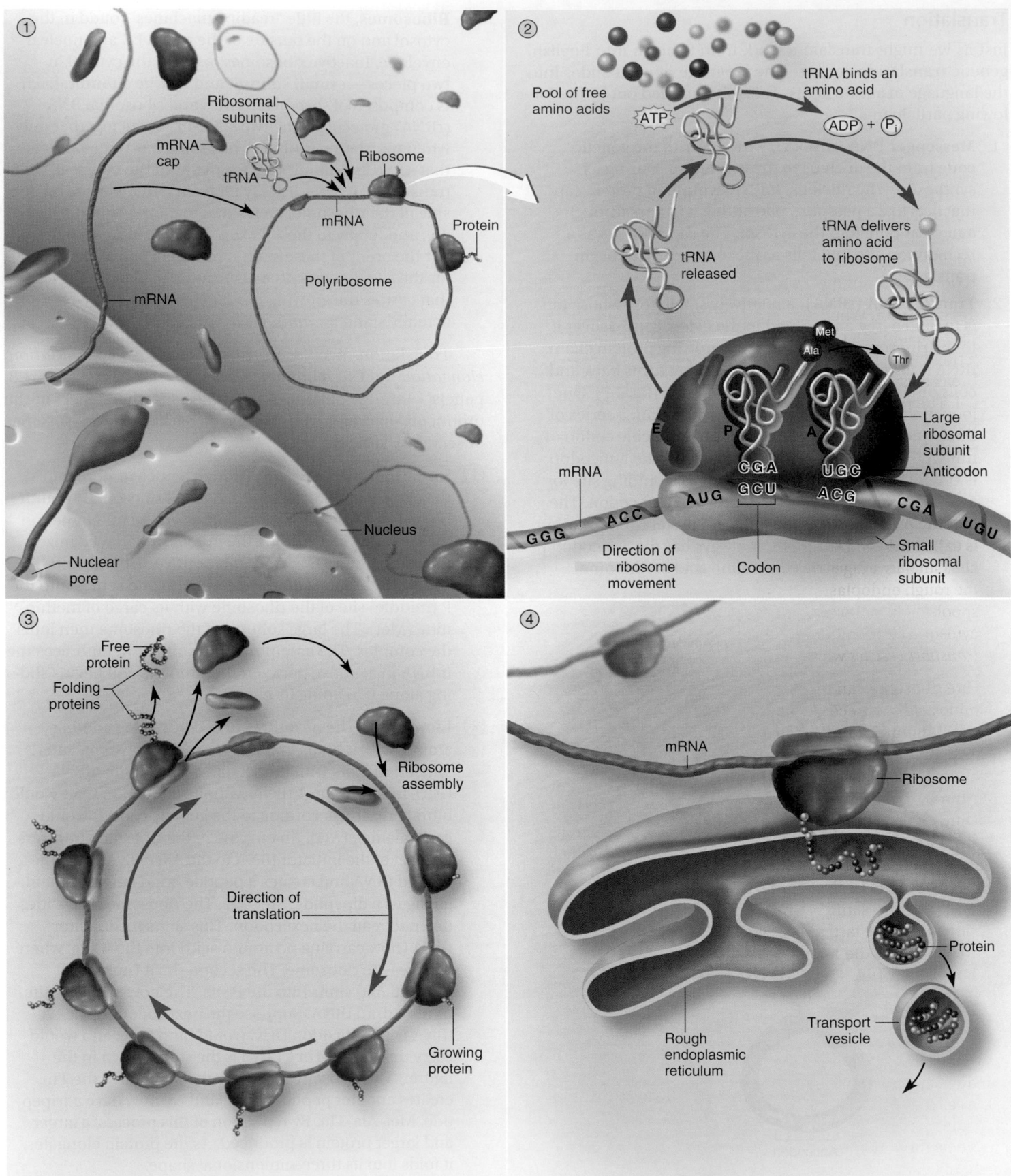

**FIGURE 4.8 Translation of mRNA.** See the text for explanation of the processes occurring at each step.

❓ *Why would translation not work if ribosomes could bind only one tRNA at a time?*

amino acid to the tRNA; therefore protein synthesis consumes one ATP for every amino acid added to the chain.

All new proteins, as we can see, begin with the amino acid methionine, carried by the initiator tRNA. This is often cleaved off in later processing, however, so not every finished protein starts with methionine.

Codon–anticodon pairing is less precise than just depicted; it tolerates some mismatches, especially at the third base of the codon. Therefore, UGC isn't necessarily the only anticodon that can pair with ACG. Due to this imprecision or "wobble" in the system, as few as 48 different tRNAs are needed to pair up with the 61 codons that represent the amino acids.

③ **Termination.** When the ribosome reaches a stop codon, its A site binds a protein called a *release factor* instead of a tRNA. The release factor causes the finished protein to break away from the ribosome and go off into the cytosol. The ribosome then dissociates into its two subunits, but since these are now so close to the mRNA's leader sequence, they often reassemble on the same mRNA and repeat the process, making another copy of the same protein.

④ **Making proteins for packaging or export.** If a protein is to be packaged into a lysosome or secreted from the cell (such as a digestive enzyme), the ribosome docks on the rough endoplasmic reticulum and the new protein spools off into the cisterna of the ER instead of into the cytosol. The ER modifies this protein and packages it into *transport vesicles* whose destiny we will examine later.

One ribosome can work very rapidly, adding about two to six amino acids per second. Most proteins take from 20 seconds to several minutes to make. But a ribosome does not work at the task alone. After one ribosome moves away from the leader sequence, another one often binds there and begins the process, following along behind the first—and then another and another, so that a single mRNA is commonly translated by 10 or 20 ribosomes at once. This cluster of ribosomes, all translating the same mRNA, is called a **polyribosome.** The farther along the mRNA each ribosome is, the longer is the protein spooling from it (fig. 4.9). Not only is each mRNA translated by all these ribosomes at once, but a cell may have 300,000 identical mRNA molecules, each being simultaneously translated by some 20 ribosomes. With so many "factory workers" doing the same task, a cell may produce over 100,000 protein molecules per second—a remarkably productive factory! As much as 25% of the dry weight of liver cells, which are highly active in protein synthesis, consists of ribosomes.

Figure 4.10 summarizes transcription and translation and shows how a nucleo-

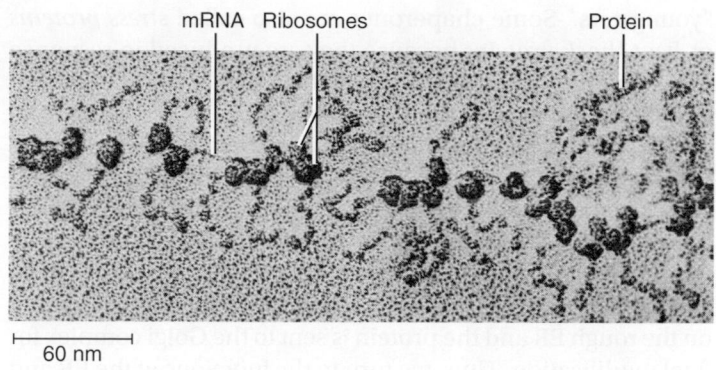

mRNA  Ribosomes  Protein

⊢ 60 nm ⊣

**FIGURE 4.9  Several Ribosomes Attached to a Single mRNA Molecule, Forming a Polyribosome.**

tide sequence translates to a hypothetical peptide of 6 amino acids. An average protein is about 400 amino acids long; it would have to be represented, at a minimum, by a sequence of 1,203 nucleotides (3 for each amino acid, plus a stop codon).

## Protein Processing and Secretion

Protein synthesis is not finished when its amino acid sequence (primary structure) has been assembled. To be functional, it must coil or fold into precise secondary and tertiary structures; in some cases, it associates with other protein chains (quaternary structure) or binds with a nonprotein such as a vitamin or carbohydrate. As a new protein is assembled by a ribosome, it is often bound by an older protein called a **chaperone.** The chaperone guides the new protein in folding into the proper shape and helps to prevent improper associations between different proteins. As in the colloquial sense of the word, a chaperone is an older protein that escorts and regulates the behavior of the

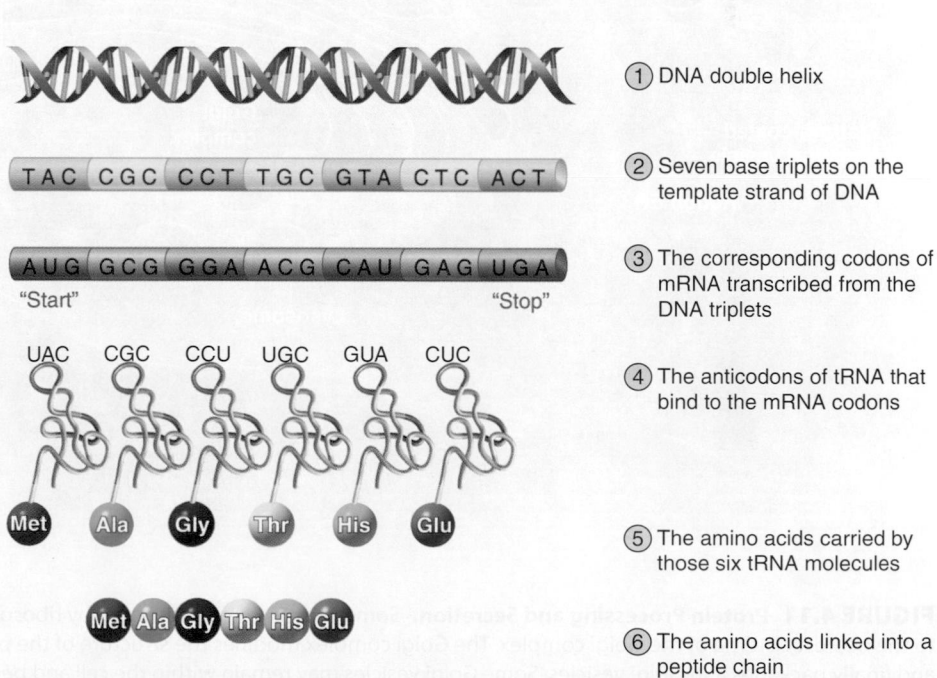

① DNA double helix

② Seven base triplets on the template strand of DNA

③ The corresponding codons of mRNA transcribed from the DNA triplets

④ The anticodons of tRNA that bind to the mRNA codons

⑤ The amino acids carried by those six tRNA molecules

⑥ The amino acids linked into a peptide chain

**FIGURE 4.10  Relationship of a DNA Base Sequence to Peptide Structure.**

"youngsters." Some chaperones are also called *stress proteins* or *heat shock proteins* because they are produced in response to heat or other stress on a cell and help damaged proteins fold back into their correct functional shapes.

If a protein is going to be used in the cytosol (for example, the enzymes of glycolysis), it is likely to be made by free ribosomes in the cytosol. However, if it is going to be packaged into a lysosome or secreted from the cell (for example, insulin), the entire polyribosome migrates to the rough ER and docks on its surface. Assembly of the amino acid chain is then completed on the rough ER and the protein is sent to the Golgi complex for final modification. Thus, we turn to the functions of the ER and Golgi in the processing and secretion of a protein. Compare the following description to figure 4.11.

(1) As a protein is assembled on the ER surface, it threads itself through a pore in the ER membrane and into the cisterna. Enzymes in the cisterna modify the new protein in a variety of ways—removing some amino acid segments, folding the protein and stabilizing it with disulfide bridges, adding carbohydrates, and so forth. Such changes are called **posttranslational modification.** Insulin, for example, is first synthesized as a protein 86 amino acids long. In posttranslational modification, the chain folds back on itself, three disulfide bridges are formed, and 35 amino acids are removed from the middle of the protein. The final insulin molecule is therefore made of two chains of 21 and 30 amino acids held together by disulfide bridges (see fig. 2.26, p. 68).

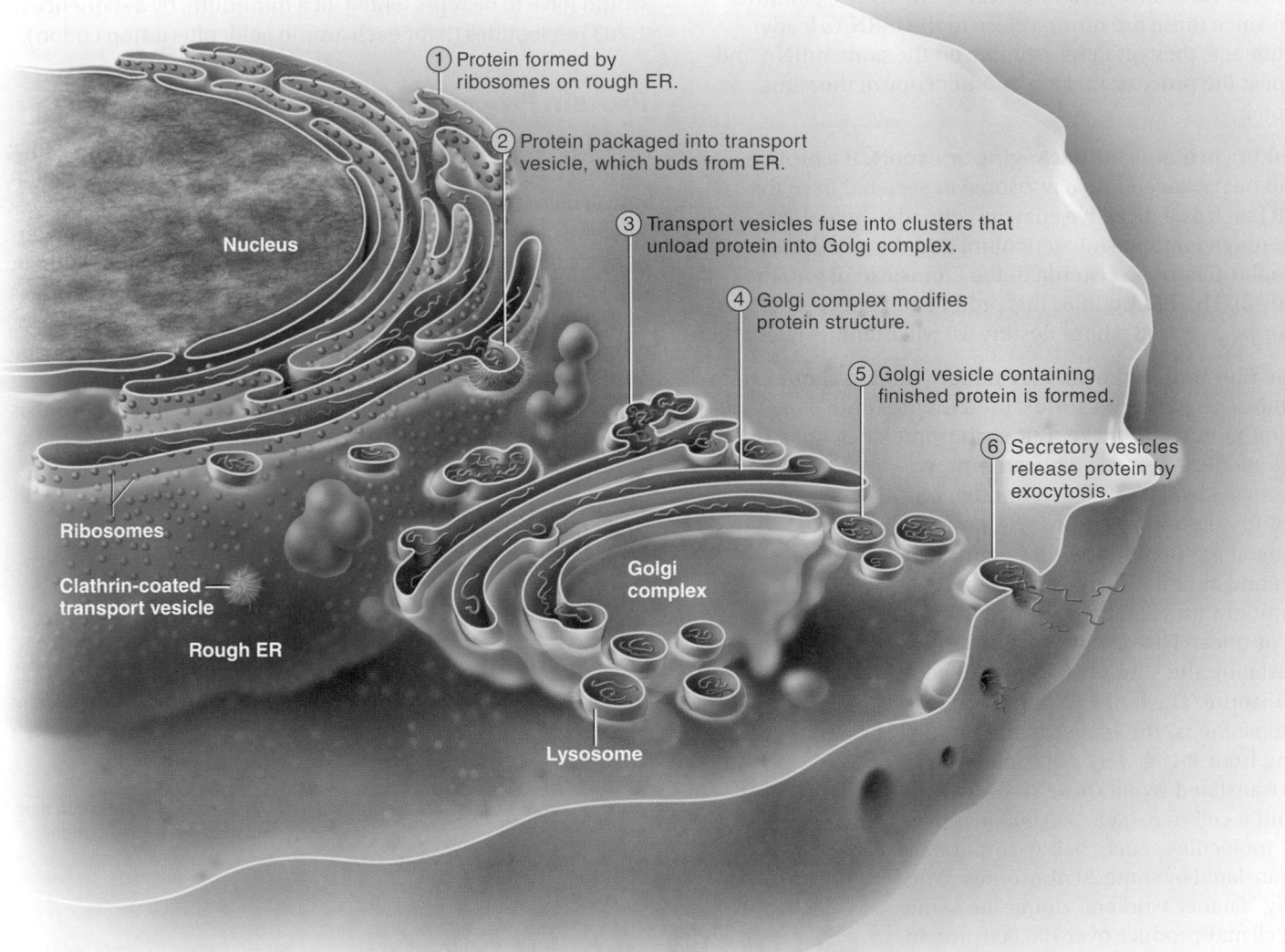

**FIGURE 4.11 Protein Processing and Secretion.** Some proteins are synthesized by ribosomes on the rough ER and carried in transport vesicles to the nearest cisterna of the Golgi complex. The Golgi complex modifies the structure of the protein, transferring it from one cisterna to the next, and finally packages it in Golgi vesicles. Some Golgi vesicles may remain within the cell and become lysosomes, while others migrate to the plasma membrane and release the cell product by exocytosis.

②  When the rough ER is finished with a protein, it pinches off bubblelike **transport vesicles** coated with a protein called *clathrin.* Clathrin apparently helps to select the proteins to be transported in the vesicles, and as a basketlike cage, it helps to mold the forming vesicles. Soon after the vesicles detach from the ER, they fuse into irregularly shaped clusters that carry their cargo to the nearest cisterna of the Golgi complex.

③  Upon contact with the Golgi complex, the cluster fuses with it and unloads its protein cargo into the Golgi cisterna.

④  The Golgi complex further modifies the protein, often by adding carbohydrate chains. This is where the *glycoproteins* mentioned in chapter 2 are assembled. There is still some disagreement about how the Golgi cisternae behave. Some say the maturing protein is passed by transport vesicles from one cisterna to the next, as shown in the figure. Others say the whole cisterna migrates from one side of the complex to the other and then breaks up into vesicles.

⑤  The final Golgi cisterna, farthest from the ER, either buds off new coated **Golgi vesicles** containing the finished protein, or may simply break up into vesicles, to be replaced by a younger cisterna behind it.

⑥  Some of the Golgi vesicles become lysosomes, while others become **secretory vesicles** that migrate to the plasma membrane and fuse with it, releasing the cell product by exocytosis. This is how a cell of a salivary gland, for example, secretes mucus and digestive enzymes.

Table 4.3 summarizes the diverse destinations and functions of newly synthesized proteins.

## Gene Regulation

Genes don't simply produce their products at a steady, incessant pace, like a 24-hour nonstop manufacturing plant. They are turned on and off from day to day, even hour to hour, as their products are needed or not, and many genes are permanently turned off in any given cell. The genes for hemoglobin and digestive enzymes, for example, are present but inactive in liver cells.

There are several ways to turn genes on or off. We cannot consider all of them here, but an example can convey the general principle. Consider a woman who has just given birth to her first baby. In the ensuing days, the hormone *prolactin* stimulates cells of her mammary glands to synthesize the various components of breast milk, including the protein *casein*—something her body has never synthesized before. How is the gene for casein turned on at this point in her life? Figure 4.12 shows the steps leading from prolactin stimulation to casein secretion.

①  Prolactin binds to its receptor, a pair of proteins in the plasma membrane of the mammary cell.

| Destination or Function | Proteins (Examples) |
|---|---|
| Deposited as a structural protein within cells | Actin of cytoskeleton; keratin of epidermis |
| Used in the cytosol as a metabolic enzyme | ATPase; kinases |
| Returned to the nucleus for use in nuclear metabolism | Histones of chromatin; RNA polymerase |
| Packaged in lysosomes for autophagy, intracellular digestion, and other functions | Numerous lysosomal enzymes |
| Delivered to other organelles for intracellular use | Catalase of peroxisomes; mitochondrial enzymes |
| Delivered to plasma membrane to serve transport and other functions | Hormone receptors; sodium–potassium pumps |
| Secreted by exocytosis for extracellular functions | Digestive enzymes; casein of breast milk |

**TABLE 4.3  Some Destinations and Functions of Newly Synthesized Proteins**

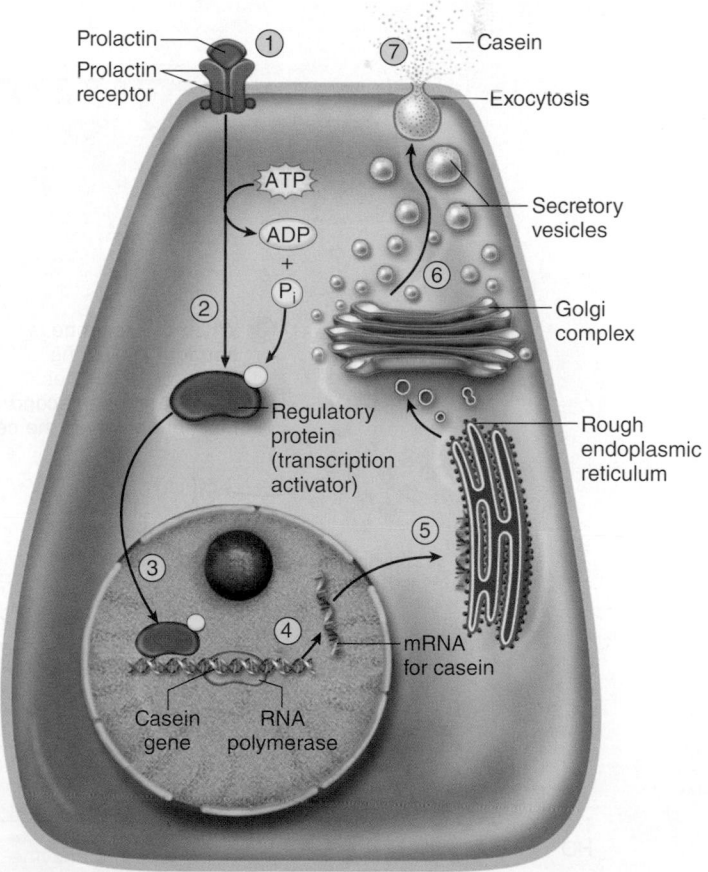

**FIGURE 4.12  A Mechanism of Gene Activation.** The hormone prolactin triggers intracellular reactions that activate a regulatory protein. This protein binds to the DNA near the casein gene, enabling RNA polymerase to produce mRNA for casein. The mRNA is then translated to synthesize the casein itself. See the text for further explanation of the numbered steps.

(2) The receptor triggers the activation of a **regulatory protein (transcription activator)** in the cytoplasm.

(3) The regulatory protein moves into the nucleus and binds to the DNA near the casein gene.

(4) This binding enables RNA polymerase to bind to the gene and transcribe it, producing the mRNA for casein.

(5) The casein mRNA moves into the cytoplasm and is translated by ribosomes on the rough endoplasmic reticulum.

(6) The Golgi complex packages casein into secretory vesicles.

(7) The secretory vesicles release the casein by exocytosis, and it becomes part of the milk.

At step 4, there are multiple ways that regulatory proteins can activate gene transcription. Some of them attract and position RNA polymerase so it can begin transcribing the gene. Others modify the coiling of DNA in a nucleosome in a way that makes specific genes more accessible to RNA polymerase. To turn off a gene, a regulatory protein can coil the chromatin in a different way that makes the gene less accessible, thus preventing transcription. There are several additional ways, beyond the scope of this book, for inducing or halting the production of a gene product, but the casein example shows how a certain gene may lie dormant in a person until, only a few times in one's life (and only if one bears children), it is activated by a stimulus such as a hormonal signal.

## Synthesizing Compounds Other Than Proteins

Cells, of course, make more than proteins—they also synthesize glycogen, fat, steroids, phospholipids, pigments, and many other compounds. There are no genes for these cell products, yet their synthesis is under indirect genetic control. How? They are produced by enzymatic reactions, and enzymes are proteins encoded by genes.

Consider the production of testosterone, for example (fig. 4.13). This is a steroid; there is no gene for testosterone. But to make it, a cell of the testis takes in cholesterol and enzymatically converts it to testosterone. This can occur only if the genes for the enzymes are active. Yet a further implication of this is that genes may greatly affect such complex outcomes as

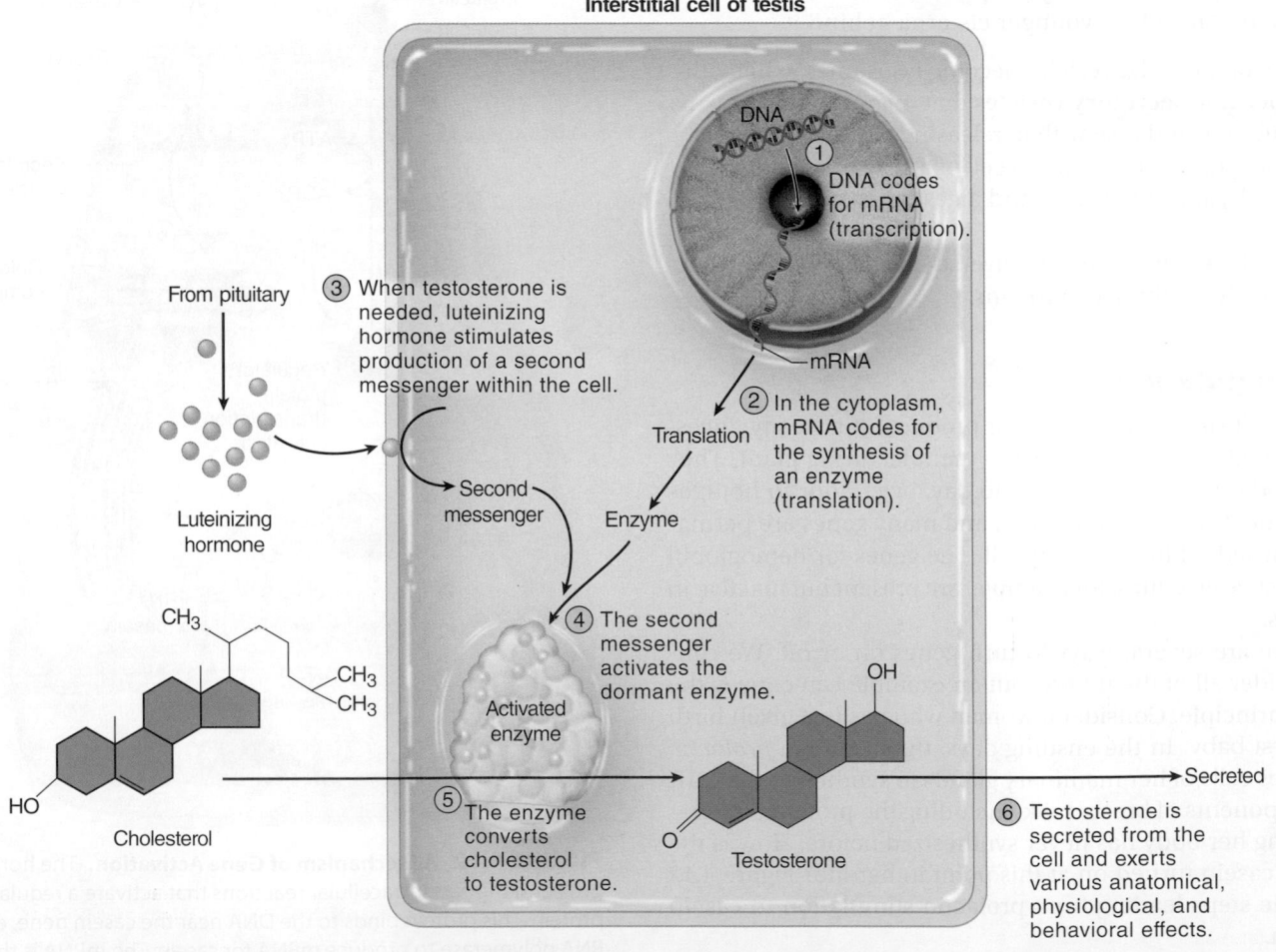

**FIGURE 4.13  Indirect Control of Testosterone Synthesis by DNA.** There is no gene for testosterone, but DNA regulates its production by coding for enzymes that convert cholesterol to testosterone.

behavior, since testosterone strongly influences such behaviors as aggression and sex drive. In short, DNA codes only for RNA and protein synthesis, yet it indirectly controls the synthesis of a much wider range of substances concerned with all aspects of anatomy, physiology, and behavior.

**BEFORE YOU GO ON**

Answer the following questions to test your understanding of the preceding section:

5. Define *gene, genetic code, codon,* and *anticodon.*

6. Describe the roles of RNA polymerase, ribosomes, and tRNA in producing a protein.

7. What is the difference between genetic transcription and translation?

8. Summarize the processing of a protein from the time a ribosome finishes its work to the time a protein is secreted from the cell. What roles do the endoplasmic reticulum and Golgi complex play in this?

9. Describe a way that a gene can be turned on or off at different times in a person's life.

10. Considering that genes can code only for RNA or proteins, how can the synthesis of nonprotein substances such as carbohydrates or steroids be under genetic control?

## 4.3 DNA Replication and the Cell Cycle

### Expected Learning Outcomes

When you have completed this section, you should be able to

a. describe how DNA is replicated;

b. discuss the consequences of replication errors;

c. describe the life history of a cell, including the events of mitosis; and

d. explain how the timing of cell division is regulated.

Before a cell divides, it must duplicate its DNA so it can give complete and identical copies of all of its genes to each daughter cell. Since DNA controls all cellular function, this replication process must be very exact. We now examine how it is accomplished and consider the consequences of mistakes.

### DNA Replication

The law of complementary base pairing shows that we can predict the base sequence of one DNA strand if we know the sequence of the other. More importantly, it enables a cell to reproduce one strand based on information in the other. The fundamental steps of the replication process are as follows (fig. 4.14):

1. The double helix unwinds from the histones.

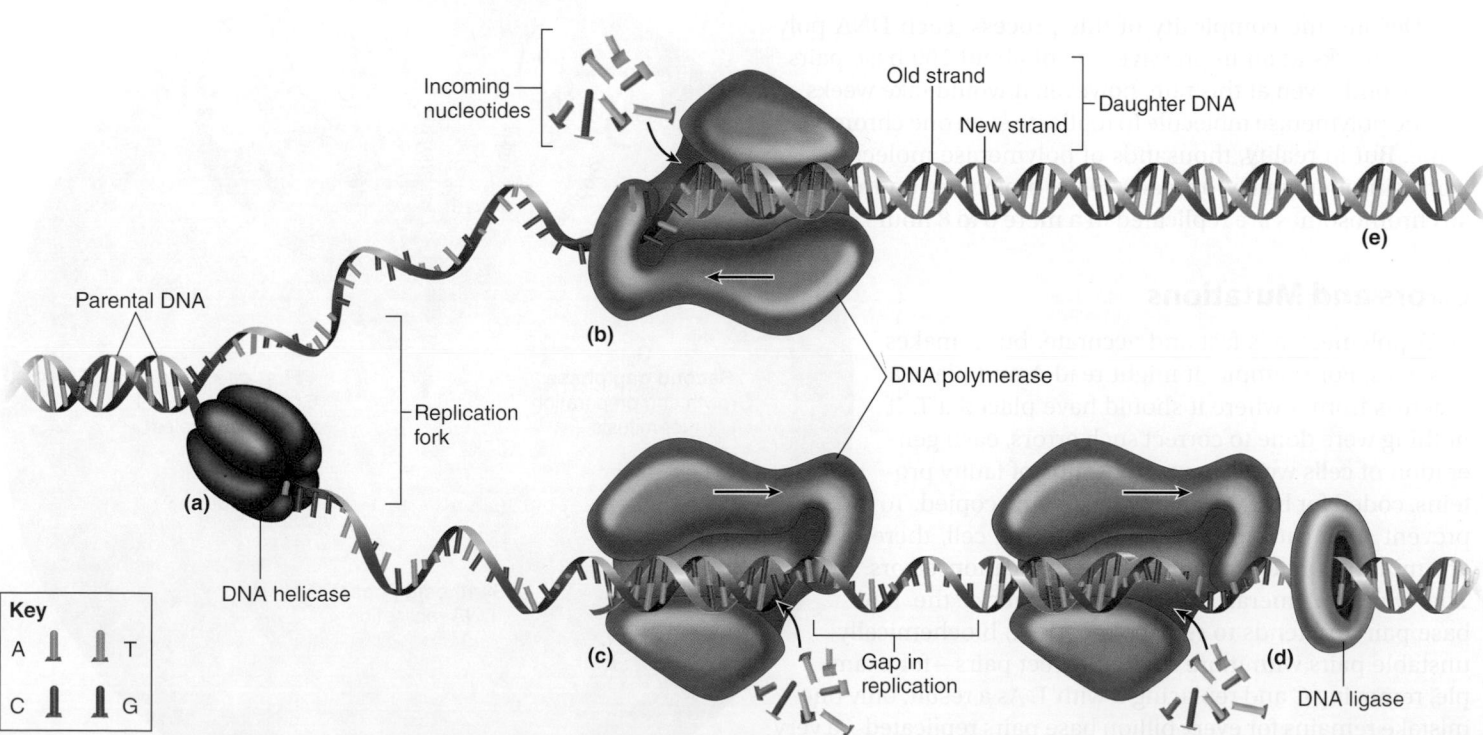

**FIGURE 4.14 Semiconservative DNA Replication.** (a) At the replication fork, DNA helicase unwinds the double helix and exposes the bases. DNA polymerases begin assembling new bases across from the existing ones. (b) On one strand, DNA polymerase moves toward the replication fork and makes one long, continuous new DNA strand. (c) On the other strand, DNA polymerases begin at the fork and move away from it, replicating the DNA in short segments with gaps between them. (d) DNA ligase closes the gaps to join the segments into a continuous double helix. (e) The ultimate result is two DNA double helices, each composed of one strand of the original DNA and one newly synthesized strand.

2. Like a zipper, an enzyme called **DNA helicase** opens up one short segment of the helix at a time, exposing its nitrogenous bases. The point where the DNA is opened up, like the two halves of a zipper separating, is called the **replication fork.**

3. Molecules of the enzyme **DNA polymerase** move along each strand; read the exposed bases; and like a matchmaker, arrange "marriages" with complementary free nucleotides. If the polymerase finds the sequence TCG, for example, it assembles AGC across from it. The two separated strands of DNA are copied by separate polymerase molecules, proceeding in opposite directions. On one strand (top of figure 4.14), the DNA polymerase moves toward the replication fork and makes a long, continuous, new strand of DNA to complement the old one. On the other strand (bottom of the figure), DNA polymerase moves away from the replication fork and copies only a short segment of DNA at a time. The segments are then joined together by another enzyme called **DNA ligase.** Ultimately, from the old *parental DNA* molecule, two new *daughter DNA* molecules are made. Each daughter DNA consists of one new helix synthesized from free nucleotides and one old helix conserved from the parental DNA. The process is therefore called **semiconservative replication.**

4. While DNA is synthesized in the nucleus, new histones are synthesized in the cytoplasm. Millions of histones are transported into the nucleus within a few minutes after DNA replication, and each new DNA helix wraps around them to make new nucleosomes.

Despite the complexity of this process, each DNA polymerase works at an impressive rate of about 100 base pairs per second. Even at this rate, however, it would take weeks for one polymerase molecule to replicate even one chromosome. But in reality, thousands of polymerase molecules work simultaneously on each DNA molecule, and all 46 chromosomes are replicated in a mere 6 to 8 hours.

## Errors and Mutations

DNA polymerase is fast and accurate, but it makes mistakes. For example, it might read A and place a C across from it where it should have placed a T. If nothing were done to correct such errors, each generation of cells would have thousands of faulty proteins, coded for by DNA that had been miscopied. To prevent such catastrophic damage to the cell, there are multiple modes of correcting replication errors. The DNA polymerase itself double-checks the new base pair and tends to replace incorrect, biochemically unstable pairs with more stable, correct pairs—for example, removing C and replacing it with T. As a result, only one mistake remains for every billion base pairs replicated—a very high degree of replication accuracy, if not completely flawless.

Changes in DNA structure, called **mutations,**[3] can result from replication errors or from environmental factors such as radiation, chemicals, and viruses. Uncorrected mutations can be passed on to the descendants of that cell, but many of them have no adverse effect. One reason is that a new base sequence sometimes codes for the same thing as the old one. For example, TGG and TGC both code for threonine (see table 4.2), so a mutation from G to C in the third place would not necessarily change protein structure. Another reason is that a change in protein structure is not always critical to its function. For example, the beta chain of hemoglobin is 146 amino acids long in both humans and horses, but 25 of these amino acids differ between the two species. Nevertheless, the hemoglobin is fully functional in both species. Furthermore, since 98% of the DNA does not code for any proteins, the great majority of mutations do not affect protein structure at all. Other mutations, however, may kill a cell, turn it cancerous, or cause genetic defects in future generations. When a mutation changes the sixth amino acid of β-hemoglobin from glutamic acid to valine, for example, the result is a crippling disorder called sickle-cell disease. Clearly some amino acid substitutions are more critical than others, and this affects the severity of a mutation.

## The Cell Cycle

Most cells periodically divide into two daughter cells, so a cell has a life cycle extending from one division to the next. This **cell cycle** is divided into four main phases: $G_1$, $S$, $G_2$, and $M$ (fig. 4.15).

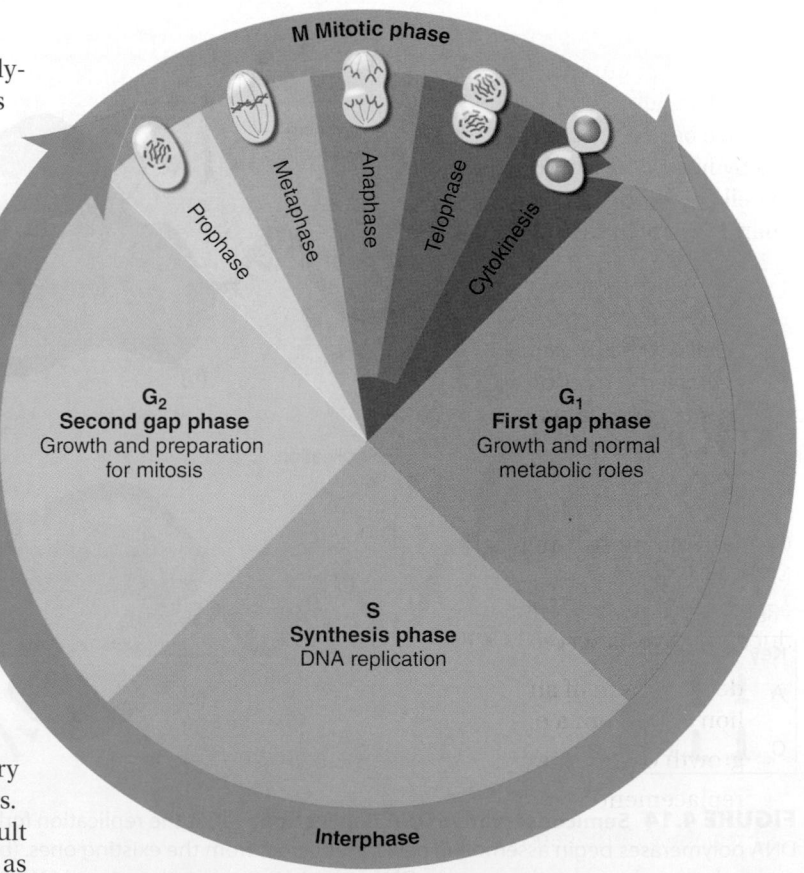

**FIGURE 4.15  The Cell Cycle.**

---

[3]*muta* = change

$G_1$ is the **first gap phase,** an interval between cell division and DNA replication. During this time, a cell synthesizes proteins, grows, and carries out its preordained tasks for the body. Almost all of the discussion in this book relates to what cells do in the $G_1$ phase. Cells in $G_1$ also accumulate the materials needed to replicate their DNA in the next phase. In cultured cells called fibroblasts, which divide every 18 to 24 hours, $G_1$ lasts 8 to 10 hours.

**S** is the **synthesis phase,** in which a cell makes a duplicate copy of its centrioles and all of its nuclear DNA. The two identical sets of DNA molecules are then available to be divided up between daughter cells at the next cell division. This phase takes 6 to 8 hours in cultured fibroblasts.

$G_2$, the **second gap phase,** is a relatively brief interval (4–6 hours) between DNA replication and cell division. In $G_2$, a cell exhibits further growth, makes more organelles, finishes replicating its centrioles, and synthesizes enzymes that control cell division. It also checks the fidelity of DNA replication and usually repairs any errors that are detected.

**M** is the **mitotic phase,** in which a cell replicates its nucleus and then pinches in two to form two new daughter cells. In cultured fibroblasts, the M phase takes 1 to 2 hours. The details of this phase are considered in the next section. Phases $G_1$, S, and $G_2$ are collectively called **interphase**—the time between M phases.

The length of the cell cycle varies greatly from one cell type to another. Stomach and skin cells divide rapidly, whereas bone and cartilage cells divide slowly. Some cells leave the cell cycle for a "rest" and cease to divide for days, years, or the rest of one's life—mature neurons, skeletal muscle cells, and adipocytes, for example. Such cells are said to be in the $G_0$ **(G-zero) phase.** The balance between cells that are actively cycling and those standing by in $G_0$ is an important factor in determining the number of cells in the body. An inability to stop cycling and enter $G_0$ is characteristic of cancer cells (see Deeper Insight 4.3).

#### ▶▶▶ APPLY WHAT YOU KNOW

*What is the maximum number of DNA molecules ever contained in a cell over the course of its life cycle? (Assume the cell has only one nucleus, and disregard mitochondrial DNA.)*

## Mitosis

Cells divide by two mechanisms called mitosis and meiosis. Meiosis, however, is restricted to one purpose, the production of eggs and sperm, and is therefore treated in chapter 27 on reproduction. **Mitosis** serves all the other functions of cell division:

- development of an individual, composed of some 50 trillion cells, from a one-celled fertilized egg;

- growth of all tissues and organs after birth;

- replacement of cells that die; and

- repair of damaged tissues.

Four phases of mitosis are recognizable: *prophase, metaphase, anaphase,* and *telophase* (fig. 4.16).

① **Prophase.**[4] At the outset of mitosis, the chromosomes shorten and thicken, eventually coiling into compact rods that are easier to distribute to daughter cells than the long, delicate chromatin of interphase. There are 46 chromosomes, each with two chromatids and one molecule of DNA per chromatid. The nuclear envelope disintegrates during prophase and releases the chromosomes into the cytosol. The centrioles begin to sprout elongated microtubules called **spindle fibers,** which push the centrioles apart as they grow. Eventually, a pair of centrioles comes to lie at each pole of the cell. Some spindle fibers grow toward the chromosomes and become attached to the kinetochore on each side of the centromere (see fig. 4.5). The spindle fibers then tug the chromosomes back and forth until they line up along the midline of the cell.

② **Metaphase.**[5] The chromosomes are aligned on the cell equator, oscillating slightly and awaiting a signal that stimulates each of them to split in two at the centromere. The spindle fibers now form a lemon-shaped array called the **mitotic spindle.** Long microtubules reach out from each centriole to the chromosomes, and shorter microtubules form a starlike *aster,*[6] which anchors the assembly to the inside of the plasma membrane at each end of the cell.

③ **Anaphase.**[7] This phase begins with activation of an enzyme that cleaves the two sister chromatids from each other at the centromere. Each chromatid is now regarded as a separate, single-stranded *daughter chromosome.* One daughter chromosome migrates to each pole of the cell, with its centromere leading the way and the arms trailing behind. Migration is achieved by means of motor proteins in the kinetochore crawling along the spindle fiber as the fiber itself is "chewed up" and disassembled at the chromosomal end. Since sister chromatids are genetically identical, and since each daughter cell receives one chromatid from each chromosome, the daughter cells of mitosis are genetically identical.

④ **Telophase.**[8] The chromatids cluster on each side of the cell. The rough ER produces a new nuclear envelope around each cluster, and the chromatids begin to uncoil and return to the thinly dispersed chromatin form. The mitotic spindle breaks up and vanishes. Each new nucleus forms nucleoli, indicating it has already begun making RNA and is preparing for protein synthesis.

Telophase is the end of nuclear division but overlaps with **cytokinesis**[9] (SY-toe-kih-NEE-sis), division of the cytoplasm

---

[4]*pro* = first
[5]*meta* = next in a series
[6]*aster* = star
[7]*ana* = apart
[8]*telo* = end, final
[9]*cyto* = cell; *kinesis* = action, motion

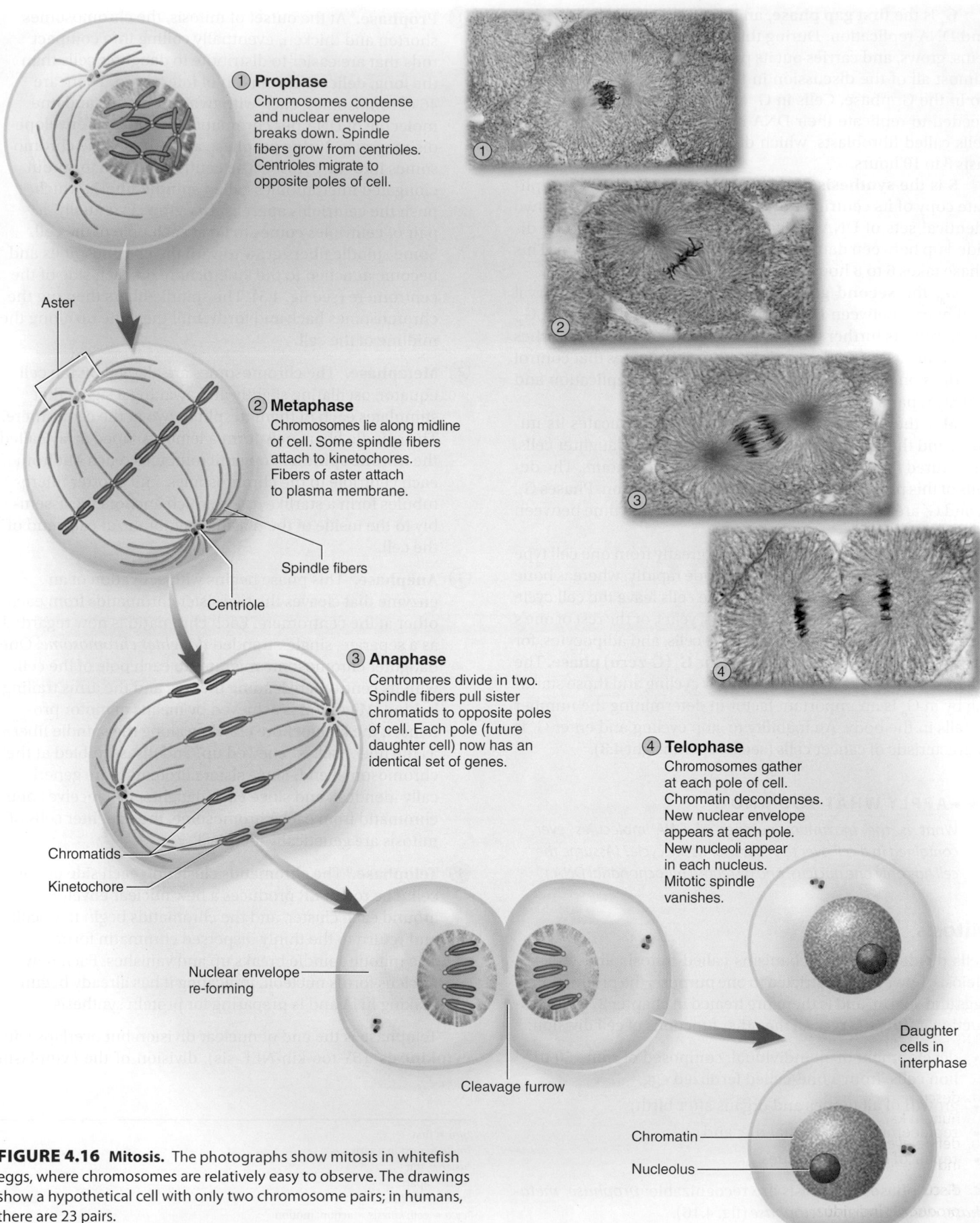

**① Prophase**
Chromosomes condense
and nuclear envelope
breaks down. Spindle
fibers grow from centrioles.
Centrioles migrate to
opposite poles of cell.

Aster

**② Metaphase**
Chromosomes lie along midline
of cell. Some spindle fibers
attach to kinetochores.
Fibers of aster attach
to plasma membrane.

Spindle fibers

Centriole

**③ Anaphase**
Centromeres divide in two.
Spindle fibers pull sister
chromatids to opposite poles
of cell. Each pole (future
daughter cell) now has an
identical set of genes.

Chromatids

Kinetochore

**④ Telophase**
Chromosomes gather
at each pole of cell.
Chromatin decondenses.
New nuclear envelope
appears at each pole.
New nucleoli appear
in each nucleus.
Mitotic spindle
vanishes.

Nuclear envelope
re-forming

Cleavage furrow

Daughter
cells in
interphase

Chromatin

Nucleolus

**FIGURE 4.16 Mitosis.** The photographs show mitosis in whitefish
eggs, where chromosomes are relatively easy to observe. The drawings
show a hypothetical cell with only two chromosome pairs; in humans,
there are 23 pairs.

into two cells. Early traces of cytokinesis are visible even at anaphase. It is achieved by the motor protein myosin pulling on microfilaments of actin in the terminal web of the cytoskeleton. This creates a crease called the *cleavage furrow* around the equator of the cell, and the cell eventually pinches in two. Interphase has now begun for these new cells.

## The Timing of Cell Division

One of the most important questions in biology is what signals cells when to divide and when to stop. The activation and inhibition of cell division are subjects of intense research for obvious reasons such as management of cancer and tissue repair. Cells divide when (1) they grow large enough to have enough cytoplasm to distribute to their two daughter cells; (2) they have replicated their DNA, so they can give each daughter cell a duplicate set of genes; (3) they receive an adequate supply of nutrients; (4) they are stimulated by **growth factors,** chemical signals secreted by blood platelets, kidney cells, and other sources; or (5) neighboring cells die, opening up space in a tissue to be occupied by new cells. Growth factors and their receptors are a central issue in understanding the uncontrolled growth of cancer (see Deeper Insight 4.3, p. 136). Cells stop dividing when they snugly contact neighboring cells or when nutrients or growth factors are withdrawn. The cessation of cell division in response to contact with other cells is called **contact inhibition.**

### BEFORE YOU GO ON

Answer the following questions to test your understanding of the preceding section:

11. Describe the genetic roles of DNA helicase and DNA polymerase. Contrast the function of DNA polymerase with that of RNA polymerase.

12. Explain why DNA replication is called *semiconservative.*

13. Define *mutation.* Explain why some mutations are harmless and others can be lethal.

14. List the stages of the cell cycle and summarize what occurs in each one.

15. List the stages of mitosis and the main processes that occur in each one.

### 4.4 Chromosomes and Heredity

#### Expected Learning Outcomes

When you have completed this section, you should be able to

a. describe the paired arrangement of chromosomes in the human karyotype;

b. define *allele* and discuss how alleles affect the traits of an individual; and

c. discuss the interaction of heredity and environment in producing individual traits.

**Heredity** is the transmission of genetic characteristics from parent to offspring. In the following discussion, we will examine a few basic principles of normal heredity, thus establishing a basis for understanding hereditary traits in later chapters. Hereditary defects are described in chapter 29 along with nonhereditary birth defects.

## The Karyotype

As we have seen, the agents of heredity are the genes, and the genes are carried on the chromosomes. Nearly all human cells—with the exception of eggs, sperm, and a few others—have 46 chromosomes. When we lay the chromosomes out in order by size and other physical features, we get a chart called a **karyotype**[10] (fig. 4.17). These chromosomes occur in 23 pairs; the two members of each pair are called **homologous**[11] (ho-MOLL-uh-gus) **chromosomes.** One member of each pair is inherited from the individual's mother and one from the father. Except for the X and Y chromosomes, two homologous chromosomes look alike and carry the same genes, although they may have different varieties of those genes. Chromosomes X and Y are called the **sex chromosomes** because they determine the individual's sex; all the others are called **autosomes.** A female normally has a homologous pair of X chromosomes, whereas a male has one X and a much smaller Y.

---

[10]*karyo* = nucleus
[11]*homo* = same; *log* = relation

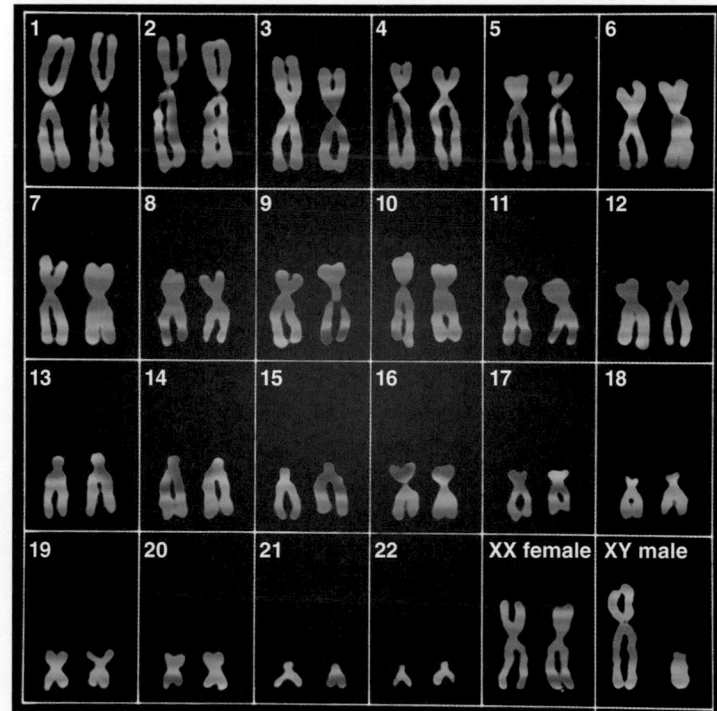

**FIGURE 4.17 Karyotype of a Normal Human Male.** This is a false-color micrograph of chromosomes stained to accentuate their banding patterns. The two chromosomes of each homologous pair exhibit similar size, shape, and banding. Each chromosome at this stage has two chromatids.

▶▶▶**APPLY WHAT YOU KNOW**

*Why would a cell in metaphase be more useful than a cell in interphase for producing a karyotype?*

Any cell with 23 pairs of chromosomes is said to be **diploid.**[12] Sperm and egg cells, however, are **haploid,**[13] meaning they contain only 23 unpaired chromosomes. Sperm and eggs, and cells on their way to becoming sperm and eggs, are called **germ cells.** All other cells of the body are called **somatic cells.** In meiosis (see chapter 27), the two homologous chromosomes of each pair become *segregated* into separate daughter cells leading to the haploid germ cells. At fertilization, one set of *paternal* (sperm) chromosomes unites with a set of *maternal* (egg) chromosomes, restoring the diploid number to the fertilized egg and the somatic cells that arise from it.

## Genes and Alleles

The location of a particular gene on a chromosome is called its **locus.** Homologous chromosomes have the same gene at the same locus, although they may carry different forms of that gene, called **alleles**[14] (ah-LEELS), which produce alternative forms of a particular trait. Frequently, one allele is **dominant** and the other one **recessive.** If at least one chromosome carries the dominant allele, the corresponding trait is usually detectable in the individual. A dominant allele masks the effect of any recessive allele that may be present. Typically, but not always, dominant alleles code for a normal, functional protein and recessive alleles for a nonfunctional variant of the protein.

A feature of the chin presents an example of dominant and recessive genetic effects. Some people have a *cleft chin,* with a deep dimple in the middle, whereas most people do not (fig. 4.18a). The allele for cleft chin is dominant; we will represent it with a capital *C.* Those who inherit this from one or both parents will normally have a cleft chin. The allele for an uncleft chin is recessive, here represented with a lowercase *c.* (It is customary to represent a dominant allele with a capital letter and a recessive allele with its lowercase equivalent.) One must inherit the recessive allele from both parents to have an uncleft chin (fig. 4.18b).

Individuals with two identical alleles, such as *CC* or *cc,* are said to be **homozygous**[15] (HO-mo-ZY-gus) for that trait. If the homologous chromosomes have different alleles for that gene *(Cc),* the individual is **heterozygous**[16] (HET-er-oh-ZY-gus). The paired alleles that an individual possesses for a particular trait constitute the **genotype** (JEE-no-type). An observable trait such as cleft or uncleft chin is called the **phenotype**[17] (FEE-no-type).

---

[12]*diplo* = double
[13]*haplo* = half
[14]*allo* = different
[15]*homo* = same; *zygo* = union, joined
[16]*hetero* = different
[17]*pheno* = showing, evident

Cleft chin
*CC, Cc*

Uncleft chin
*cc*

**(a)**

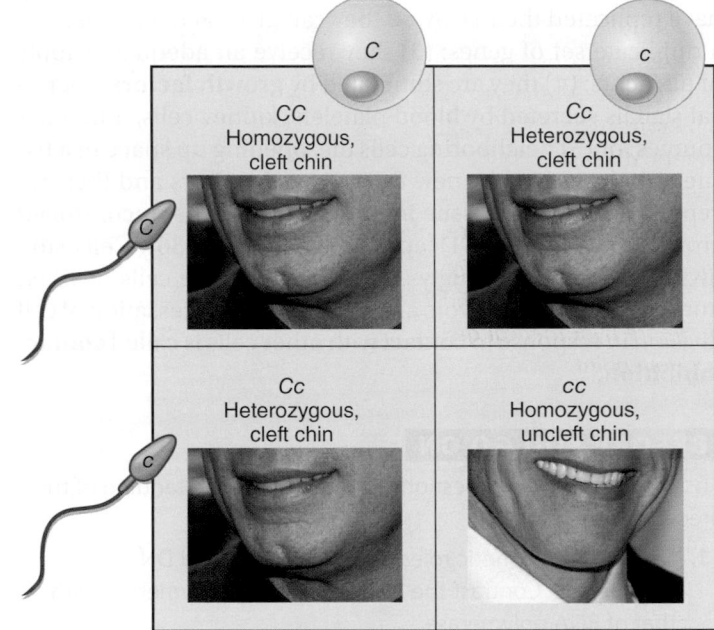

**(b)**

**FIGURE 4.18 Genetics of Cleft Chin.** (a) A cleft chin occurs if even one allele of the pair is dominant (*C*). The more common uncleft chin occurs only when both alleles are recessive (*c*). (b) A Punnett square shows why such a trait can "skip a generation." Both parents in this case have heterozygous genotypes (*Cc*) and cleft chins. An egg from the mother can carry either allele *C* or *c* (top), as can a sperm from the father (left). This gives their offspring a one-in-four chance of having an uncleft chin (*cc*).

We say that an allele is **expressed** if it shows in the phenotype of an individual. Chin allele *c* is expressed only when it is present in a homozygous state *(cc);* allele *C* is expressed whether it is homozygous *(CC)* or heterozygous *(Cc).* The only way most recessive alleles can be expressed is for an individual to inherit them from both parents.

Recessive traits can "skip" one or more generations. A diagram called a *Punnett square* (fig. 4.18b) shows how two heterozygous parents with cleft chins can produce a child with an uncleft chin. Across the top are the two genetically possible types of eggs the mother could produce, and on the left side are the possible types of sperm from the father. The four cells of the square show the genotypes and phenotypes that would result from each possible combination of sperm and egg. You can see that three of the possible combinations would produce a child with cleft chin (genotypes *CC* and *Cc*), but one combination *(cc)* would produce a child with an uncleft chin. Therefore, the uncleft chin trait skipped the parental generation in this case but could be expressed in their child.

This phenomenon becomes more significant when parents are heterozygous **carriers** of hereditary diseases such as cystic fibrosis—individuals who carry a recessive allele and may pass it on, but do not phenotypically express it in themselves. For some hereditary diseases, tests are available to detect carriers and allow couples to weigh their risk of having children with genetic disorders. *Genetic counselors* perform genetic testing or refer clients for tests, advise couples on the probability of transmitting genetic diseases, and assist people in coping with genetic disease.

▶▶▶**APPLY WHAT YOU KNOW**

*Would it be possible for a woman with an uncleft chin to have children with cleft chins? Use a Punnett square and one or more hypothetical genotypes for the father to demonstrate your point.*

## Multiple Alleles, Codominance, and Incomplete Dominance

Some genes exist in more than two allelic forms—that is, there are **multiple alleles** within the collective genetic makeup, or **gene pool,** of the population as a whole. For example, there are over 100 alleles responsible for cystic fibrosis, and there are 3 alleles for ABO blood types. Two of the ABO blood type alleles are dominant and symbolized with a capital $I$ (for *immunoglobulin*) and a superscript: $I^A$ and $I^B$. There is one recessive allele, symbolized with a lowercase $i$. Which two alleles one inherits determine the blood type, as follows:

| Genotype | Phenotype |
|----------|-----------|
| $I^AI^A$ | Type A |
| $I^Ai$ | Type A |
| $I^BI^B$ | Type B |
| $I^Bi$ | Type B |
| $I^AI^B$ | Type AB |
| $ii$ | Type O |

▶▶▶**APPLY WHAT YOU KNOW**

*Why can't one person have all three of the ABO alleles?*

Some alleles are equally dominant, or **codominant.** When both of them are present, both are phenotypically expressed. For example, a person who inherits allele $I^A$ from one parent and $I^B$ from the other has blood type AB. These alleles code for enzymes that produce the surface glycolipids of red blood cells. Type AB means that both A and B glycolipids are present, and type O means that neither of them is present.

Other alleles exhibit **incomplete dominance.** When two different alleles are present, the phenotype is intermediate between the traits that each allele would produce alone. Familial hypercholesterolemia, the disease discussed in Deeper Insight 3.3 (p. 98), is a good example. Individuals with two abnormal alleles die of heart attacks in childhood, those with only one abnormal allele typically die as young adults, and those with two normal alleles have normal life expectancies. Thus, the heterozygous individuals suffer an effect between the two extremes.

## Polygenic Inheritance and Pleiotropy

**Polygenic (multiple-gene) inheritance** (fig. 4.19) is a phenomenon in which genes at two or more loci, or even on different chromosomes, contribute to a single phenotypic trait. Human eye and skin colors are normal polygenic traits, for example. They result from the combined expression of all the genes for each trait. Several diseases are also thought to stem from polygenic inheritance, including some forms of alcoholism, mental illness, cancer, and heart disease.

**Pleiotropy**[18] (ply-OT-roe-pee) is a phenomenon in which one gene produces multiple phenotypic effects. For example, about 1 in 200,000 people has a genetic disorder called *alkaptonuria,* caused by a mutation on chromosome 3. The

---

[18]*pleio* = more; *trop* = changes

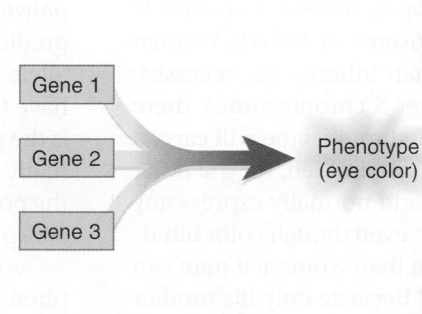

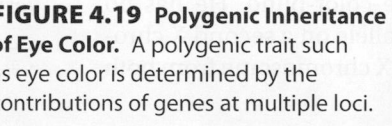

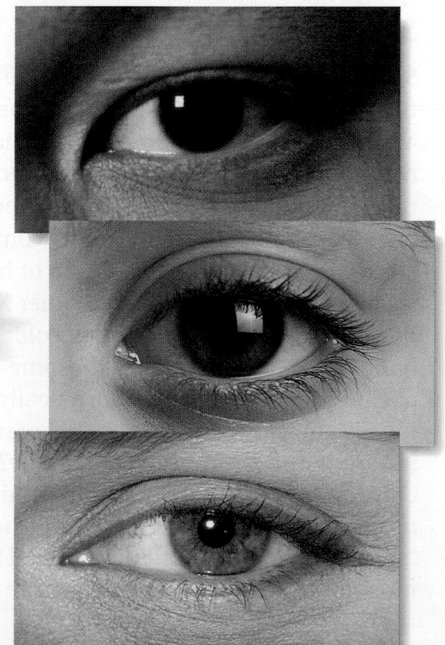

**FIGURE 4.19 Polygenic Inheritance of Eye Color.** A polygenic trait such as eye color is determined by the contributions of genes at multiple loci.

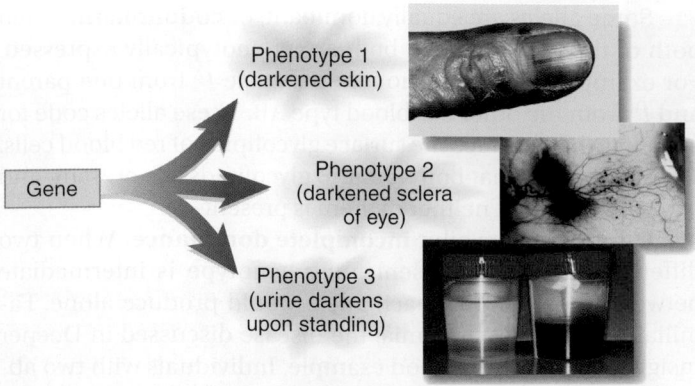

**FIGURE 4.20  Pleiotropy in Alkaptonuria.** Multiple phenotypic effects can result from a single gene mutation. In alkaptonuria, a single mutation leads to darkening of the skin, dark patches in the sclera ("white") of the eye, and darkening of the urine as it stands and oxidizes.

mutation blocks the normal breakdown of an amino acid named tyrosine, resulting in the accumulation of an intermediate breakdown product, homogentisic acid, in the body fluids and connective tissues. When homogentisic acid oxidizes, it binds to collagen and turns the tissues gray to bluish black; and for unknown reasons, it causes degeneration of cartilages and other connective tissues. Among the multiple phenotypic effects of this disorder (fig. 4.20) are darkening of the skin; darkening and degeneration of the cartilages in places such as the ears, knees, and intervertebral discs; darkening of the urine when it stands long enough for the homogentisic acid to oxidize; discoloration of the teeth and the whites of the eyes; arthritis of the shoulders, hips, and knees; and damage to the heart valves, prostate gland, and other internal organs. Another well-known example of pleiotropy is sickle-cell disease, detailed in chapter 18.

## Sex Linkage

**Sex-linked traits** are carried on the X or Y chromosome and therefore tend to be inherited by one sex more than the other. Men are more likely than women to have red–green color blindness or hemophilia, for example, because the allele for each is recessive and located on the X chromosome *(X-linked)*. Women have two X chromosomes. If a woman inherits the recessive color-blindness allele *(c)* on one of her X chromosomes, there is still a good chance that her other X chromosome will carry a dominant allele *(C)* for normal color vision. Men, on the other hand, have only one X chromosome and normally express any allele found there (fig. 4.21). Ironically, even though color blindness is far more common among men than women, a man can inherit it only from his mother. Why? Because only his mother contributes an X chromosome to him. If he inherits *c* on his mother's X chromosome, he will be color-blind. He has no "second chance" to inherit a normal allele on a second X chromosome. A woman, however, gets an X chromosome from both

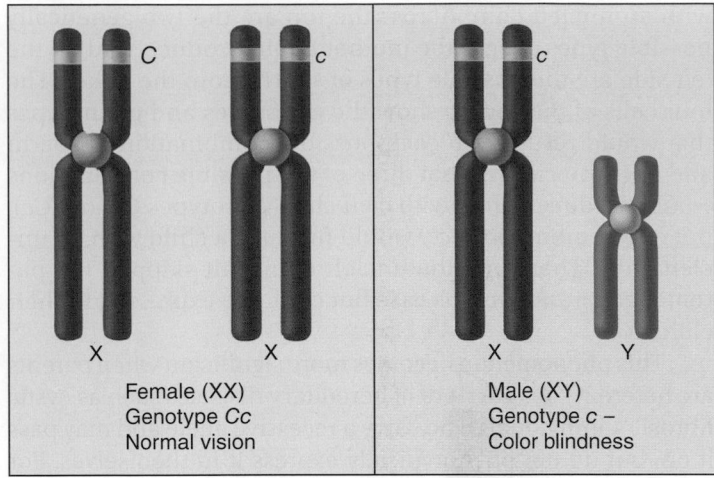

**FIGURE 4.21  Sex-Linked Inheritance of Red–Green Color Blindness.** Left: A female who inherits a recessive allele (*c*) for color blindness from one parent may not exhibit the trait, because she is likely to inherit the dominant allele (*C*) for normal vision from her other parent. Right: A male who inherits c from his mother will exhibit red–green color blindness, because the Y chromosome inherited from his father does not have a corresponding gene locus and therefore has no ability to mask the effect of *c*.

parents. Even if one parent transmits the recessive allele to her, the chances are high that she will inherit a normal allele from her other parent. She would have to have the extraordinarily bad luck to inherit it from both parents in order for her to have a trait such as red–green color blindness.

The X chromosome is thought to carry about 260 genes, most of which have nothing to do with determining an individual's sex. There are so few functional genes on the Y chromosome—concerned mainly with development of the testes—that virtually all sex-linked traits are associated with the X chromosome.

## Penetrance and Environmental Effects

A given genotype does not inevitably produce the expected phenotype. For example, a dominant allele causes *polydactyly,*[19] the presence of extra fingers or toes. We might predict that since it is dominant, anyone who inherited the allele would exhibit this trait. Most do, but others known to have the allele have the normal number of digits. **Penetrance** is the percentage of a population with a given genotype that actually exhibits the predicted phenotype. If 80% of people with the polydactyly allele actually exhibit extra digits, the allele has 80% penetrance.

Another reason the connection between genotype and phenotype is not inevitable is that environmental factors play an important role in the expression of all genes. At the very least, all gene expression depends on nutrition. Brown

[19]*epi* = many; *dactyl* = fingers, toes

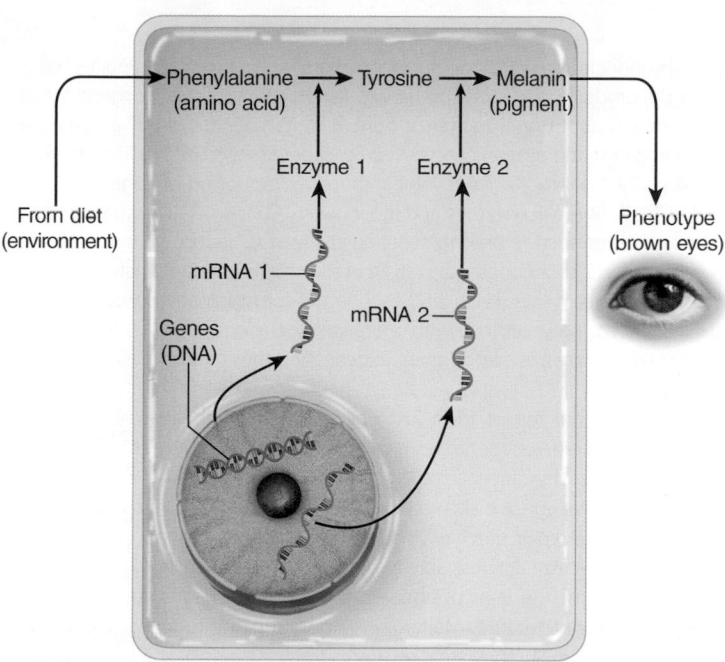

**FIGURE 4.22 The Roles of Environment and Heredity in Producing a Phenotype.** Brown eye color requires phenylalanine from the diet (environment) and two genetically coded (hereditary) enzymes to convert phenylalanine to melanin, the eye pigment.

eyes, for example, require not only genes for the enzymes that synthesize the pigment melanin, but also the dietary raw material, phenylalanine, from which the melanin is made (fig. 4.22).

No gene can produce a phenotypic effect without nutritional and other environmental input, and no nutrients can produce a body or specific phenotype without genetic instructions that tell cells what to do with them. Just as you need both a recipe and ingredients to make a cake, it takes both heredity and environment to make a phenotype.

## Dominant and Recessive Alleles at the Population Level

It is a common misconception that dominant alleles must be more common in the gene pool than recessive alleles. The truth is that dominance and recessiveness have little to do with how common an allele is. For example, type O is the most common ABO blood type in North America, but it is caused by the recessive allele *i*. Blood type AB, caused by the two dominant ABO alleles, is the rarest. Polydactyly, caused by a dominant allele, also is rare in the population. We also saw earlier that people

with cleft chins are in a minority, even though the allele for that is dominant.

## Epigenetics

We have long thought of heredity as involving only the genes passed from parent to offspring. There is a growing awareness, however, of nongenetic changes that can occur in one generation and be inherited by the next. These are the subject of a new, astonishing, and fast-growing field called **epigenetics**[20]—heritable changes beyond the level of the genome. There are a dozen or so ways of altering gene expression other than mutation. One of the most common is **DNA methylation**—the addition of methyl ($-CH_3$) groups to DNA, especially its cytosine bases. This often silences a gene, "shuts it off." This is a normal and necessary part of human development, because not all genes can be active in every cell; as embryonic cells become specialized for one task, the genes for other tasks need to be inactivated by methylation or other means.

Inappropriate DNA methylation, however, is becoming increasingly implicated in diseases ranging from cancer (see Deeper Insight 4.3) to the hereditary disorder *Prader–Willi syndrome*[21], in which either mutations or gene silencing leads to diverse effects including poor muscle tone, short stature, overeating, and childhood behavioral problems. There is speculation that epigenetic changes can persist over multiple generations. If so, some aspects of your phenotype and health could stem even from what your grandparents ate, whether they smoked, and other environmental influences in their generation. Things you do or experience today might affect even your children or grandchildren through epigenetic changes. We must now view heredity as entailing more than just the genes passed from one generation to the next.

### BEFORE YOU GO ON

Answer the following questions to test your understanding of the preceding section:

16. Why must the carrier of a genetic disease be heterozygous?

17. State at least three reasons why a person's phenotype can't always be determined from the genotype.

18. A man can inherit color blindness only from his mother, whereas a woman must inherit it from both her father and mother to show the trait. Explain this apparent paradox.

---

[20]*epi* = above, beyond
[21]Andrea Prader (1919–2001), Swiss pediatrician; Heinrich Willi (1900–1971), Swiss pediatrician

### Cancer

Anyone awaiting the results of a tumor biopsy hopes for the good news: benign! This means the tumor is slow-growing and contained in a fibrous capsule, so it will not metastasize and in most cases it is relatively easy to treat. The dreaded news is that it's malignant, meaning that it tends to grow rapidly and to *metastasize*—to give off cells that seed the growth of multiple tumors elsewhere, such as colon cancer metastasizing to the lungs and brain (fig. 4.23).

*Oncology* is a medical specialty that deals with both benign and malignant tumors, but only malignancies are called *cancer*. The word literally means "crab." Hippocrates was the first to use the word this way, upon seeing a breast tumor with a tangle of blood vessels that reminded him of a crab's outstretched legs. Energy-hungry tumors often stimulate such ingrowth of blood vessels—a phenomenon called *tumor angiogenesis*.

Cancers are named for the tissue of origin: *carcinomas* originate in epithelial tissue; *lymphomas* in the lymph nodes; *melanomas* in pigment cells (melanocytes) of the epidermis; *leukemias* in blood-forming tissues such as bone marrow; and *sarcomas* in bone, other connective tissue, or muscle. About 90% of malignancies are carcinomas, probably because epithelial cells have a high rate of mitosis, making them especially subject to mutation, and because epithelia are more exposed than other tissues to *carcinogens* (environmental cancer-causing agents).

Only 5% to 10% of cancers are hereditary, but cancer is always a genetic disease. This is not as contradictory as it might seem. Most cases are due to mutations arising anew in the affected individual, not to genes received from a parent. Mutations can arise simply through errors in DNA replication, or from exposure to carcinogens—radiation such as ultraviolet rays and X-rays; chemicals such as cigarette tar; and viruses such as human papillomavirus (HPV), hepatitis C, and type 2 herpes simplex.

Oncologists are especially interested in two families of cancer genes called oncogenes and tumor-suppressor genes. An *oncogene* is analogous to a stuck accelerator on a car—it causes cell division to accelerate out of control, some-times by inducing the excessive secretion of growth factors that stimulate mitosis, or the production of excessive growth-factor receptors. An oncogene called *ras* underlies about one-quarter of human cancers, and *erbB2* is a common factor in breast and ovarian cancer. A *tumor-suppressor (TS) gene* is like a broken brake pedal. Healthy TS genes inhibit cancer by opposing oncogene action, coding for DNA-repair enzymes, and other means. Consequently, mutations that destroy their protective "braking" function can lead to cancer. Mutation of a TS gene called *p53*, for example, is involved in about 50% of cases of leukemia and colon, lung, breast, liver, brain, and esophageal tumors. Many human cancers are associated, however, not with mutation but aberrant DNA methylation, which can, for example, silence one's TS genes and thus turn off their protective function.

Cancer seldom results from just one mutation. It usually requires 5 to 10 mutations at different gene loci. It takes time for so many mutations to accumulate, which is why cancer is more common in the elderly than in the young. In addition, as we age, we accrue more lifetime exposure to carcinogens, our DNA- and tissue-repair mechanisms become less efficient, and our immune system grows weaker and less able to detect and destroy malignant cells.

Cancer claims the lives of about one in five Americans, and is almost always fatal if not treated. Malignant tumors replace functional tissue in vital organs (fig. 4.24); they steal nutrients from the rest of the body, some-times causing a severe wasting away called *cachexia* (ka-KEX-ee-ah); they weaken one's immunity, opening the door to *opportunistic infections* that a healthier person could ward off; and they often invade blood vessels, lung tissue, or brain tissue, with such consequences as hemorrhage, pulmonary collapse, seizures, or coma. Mortality usually results not from the original (primary) tumor, but from metastasis.

Cancer is usually treated with surgery, chemotherapy, or radiation therapy. A lively area of cancer research today is the development of drugs to starve tumors by blocking tumor angiogenesis.

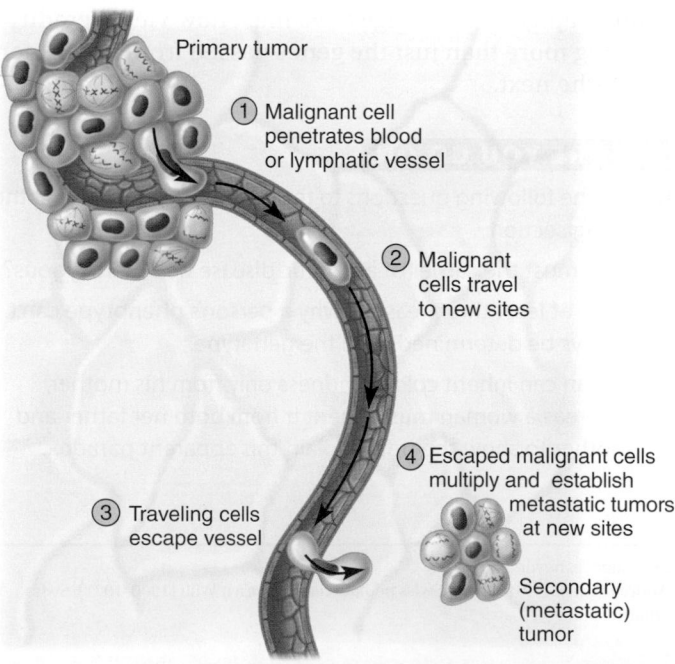

**FIGURE 4.23 Metastasis.** The process by which malignant cells escape from a primary tumor, travel in the blood or lymph, and seed the growth of new (metastatic) tumors in other localities. Colon cancer, for example, can metastasize to the liver or brain by this method.

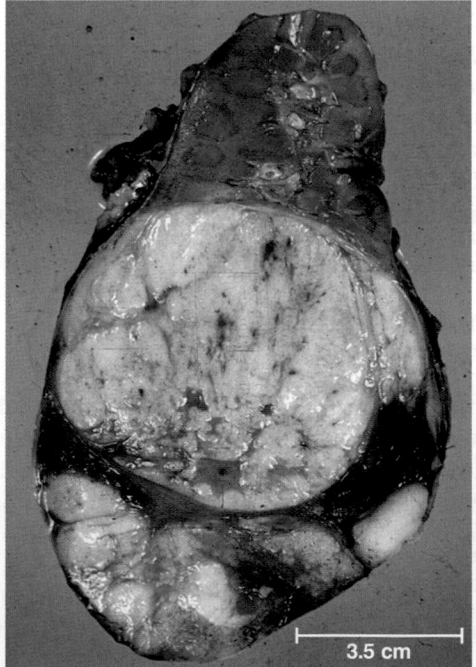

**FIGURE 4.24 Wilms Tumor.** This is a malignant tumor of the kidney occurring especially in children. (From the University of Alabama at Birmingham Department of Pathology PEIR Digital Library © http://peir.net)

# STUDY GUIDE

## ► Assess Your Learning Outcomes

*To test your knowledge, discuss the following topics with a study partner or in writing, ideally from memory.*

### 4.1   DNA and RNA—The Nucleic Acids (p. 113)

1. The general name of the monomers that compose DNA and RNA; the three universal structural components of each monomer; the purines and pyrimidines involved in DNA and RNA structure; and how a purine differs from a pyrimidine
2. Structure of the double helix of DNA; what holds the two strands of the helix together; why DNA normally has only A–T and C–G base pairs (not A–C or G–T); and the law of complementary base pairing
3. The approximate number of genes in the human genome; what percentage of the DNA consists of genes; and to what extent, and in what ways, the remaining DNA may serve a purpose
4. How DNA and protein are combined to form chromatin; how the chromatin is coiled and supercoiled to fit into the compact cell nucleus and minimize damage; and the role of histones and nucleosomes in the organization of chromatin
5. The structure of a metaphase chromosome and how many chromosomes are in a typical cell
6. How RNA differs from DNA in structure and function, and the three types of RNA involved in protein synthesis

### 4.2   Genes and Their Action (p. 118)

1. The definition of *gene;* how genes relate to the amino acid sequence of proteins; why it cannot be said that every gene codes for protein structure; and why not every protein is encoded by its own unique gene
2. Definitions of *genome, genomics,* and *genomic medicine*
3. The organization of nucleotides into DNA triplets; how these relate to but differ from the codons of mRNA; the number of codons; and stop codons and their function
4. How the genetic code relates mRNA codons to protein structure
5. The process and outcome of genetic transcription; the enzyme that carries it out; the difference between pre-mRNA

and mRNA; and the significance of introns and exons
6. How alternative splicing partially explains why the diversity of human proteins vastly exceeds the number of human genes
7. The process and outcome of genetic translation; the roles of mRNA, tRNA, and ribosomes in translation
8. Posttranslational modification of proteins, where it occurs, and how new proteins destined for extracellular use are packaged and released from a cell
9. How gene expression can be turned on or off according to the functions of different kinds of cells or physiological needs that change over time, such as a temporary need for breast milk
10. How the synthesis of nonproteins such as carbohydrates and steroids is regulated by the genes, even though genes code only for RNA or protein

### 4.3   DNA Replication and the Cell Cycle (p. 127)

1. Why every generation of cells must synthesize new DNA even though the chromosome number remains constant from generation to generation
2. Semiconservative replication, the enzymes that carry it out, and why it results in two DNA molecules that each contain one old and one new nucleotide polymer
3. What a mutation is and how a cell detects and corrects most mutations that arise during DNA replication; the varied consequences of uncorrected mutations
4. The four stages of the cell cycle, what occurs in each stage, and which stages occur during interphase
5. The four stages of mitosis; its bodily functions; what occurs in each stage; and why the amount of DNA in a cell is halved by mitosis, yet the number of chromosomes remains the same
6. Cytokinesis and how it overlaps but differs from telophase
7. How mitosis can be either stimulated or inhibited according to the need for tissue maintenance or growth

### 4.4   Chromosomes and Heredity (p. 131)

1. The definition of *heredity*
2. Organization of the karyotype; the number of homologous pairs of human

chromosomes; why chromosomes occur in homologous pairs; and the differences between haploid and diploid cells and between germ cells and somatic cells
3. The differences between autosomes and sex chromosomes, and which sex chromosomes occur in males and females
4. Dominant and recessive alleles, the difference between homozygous and heterozygous individuals, and why a person can be homozygous for some alleles and heterozygous for others
5. The relationship of genotype to phenotype, the difference between the two, and why phenotype is not determined solely by genotype
6. Why a recessive trait can skip a generation, with examples; what is meant by a *carrier*
7. The differences between the genotype, genome, and gene pool; why a genotype can never be composed of more than two alleles of the same gene, but the gene pool can contain three or more alleles of one gene
8. How codominance and incomplete dominance differ from simple dominance of an allele
9. What polygenic inheritance and pleiotropy imply about the relationship between a certain phenotype and its associated genes; examples of both
10. Why some traits occur more in males than in females, the name of this phenomenon, and examples
11. Why some individuals may not exhibit the phenotype that one would predict from their genotype; the name for this incomplete expression of certain genotypes
12. Why it cannot be said that any human trait is exclusively the result of the genes or of the environment; an example that demonstrates this
13. Why it cannot be said that dominant alleles are the most common ones in a population and recessive alleles are less common; examples that support this point
14. How hereditary and health are affected not only by genetic changes (mutations) but also by epigenetic ones

# STUDY GUIDE

▶ **Testing Your Recall**                                              *Answers in Appendix B*

1. Production of more than one pheno-typic trait by a single gene is called
   a. pleiotropy.
   b. genetic determinism.
   c. codominance.
   d. penetrance.
   e. genetic recombination.

2. When a ribosome reads a codon on mRNA, it must bind to the _____ of a corresponding tRNA.
   a. start codon
   b. stop codon
   c. intron
   d. exon
   e. anticodon

3. The normal functions of a liver cell—synthesizing proteins, detoxifying wastes, storing glycogen, and so forth—are done during its
   a. anaphase.
   b. telophase.
   c. $G_1$ phase.
   d. $G_2$ phase.
   e. synthesis phase.

4. Two genetically identical strands of a metaphase chromosome, joined at the centromere, are its
   a. kinetochores.
   b. centrioles.
   c. sister chromatids.
   d. homologous chromatids.
   e. nucleosomes.

5. Which of the following is *not* found in DNA?
   a. thymine
   b. phosphate
   c. cytosine
   d. deoxyribose
   e. uracil

6. Genetic transcription is performed by
   a. ribosomes.
   b. RNA polymerase.
   c. DNA polymerase.
   d. helicase.
   e. chaperones.

7. A chaperone comes into play in
   a. the folding of a new protein into its tertiary structure.
   b. keeping DNA organized within the nucleus.
   c. escorting sister chromatids to oppo-site daughter cells during mitosis.
   d. repairing DNA that has been damaged by mutagens.
   e. preventing malignant cells from metastasizing.

8. An allele that is not phenotypically expressed in the presence of an alternative allele of the same gene is said to be
   a. codominant.
   b. lacking penetrance.
   c. heterozygous.
   d. recessive.
   e. subordinate.

9. Semiconservative replication occurs during
   a. transcription.
   b. translation.
   c. posttranslational modification.
   d. the S phase of the cell cycle.
   e. mitosis.

10. Mutagens sometimes cause no harm to cells for all of the following reasons *except*
    a. some mutagens are natural, harm-less products of the cell itself.
    b. most of the human DNA does not code for any proteins.
    c. the body's DNA repair mechanisms detect and correct genetic damage.
    d. change in a codon does not always change the amino acid encoded by it.
    e. some mutations change protein structure in ways that are not critical to normal function.

11. The cytoplasmic division at the end of mitosis is called _____.

12. The alternative forms in which a single gene can occur are called _____.

13. The pattern of nitrogenous bases that represents the 20 amino acids of a pro-tein is called the _____.

14. Several ribosomes attached to one mRNA, which they are all transcribing, form a cluster called a/an _____.

15. The enzyme that produces pre-mRNA from the instructions in DNA is _____.

16. All the DNA in a haploid set of chromo-somes is called a person's _____.

17. At prophase, a cell has _____ chromo-somes, _____ chromatids, and _____ molecules of DNA.

18. The cytoplasmic granule of RNA and protein that reads the message in mRNA is a/an _____.

19. Cells are stimulated to divide by chemi-cal signals called _____.

20. All chromosomes except the sex chro-mosomes are called _____.

# STUDY GUIDE

## ▶ Building Your Medical Vocabulary

*Answers in Appendix B*

*State a medical meaning of each word element below, and give a term in which it or a slight variation of it is used.*

1. allo-
2. dactylo-
3. diplo-
4. haplo-
5. hetero-
6. karyo-
7. meta-
8. morpho-
9. muta-
10. poly-

## ▶ True or False

*Answers in Appendix B*

*Determine which five of the following statements are false, and briefly explain why.*

1. Proteins destined to be exported from a cell are made by ribosomes on the surface of the Golgi complex.

2. Each of a cell's products—such as steroids, carbohydrates, and phospholipids—is encoded by a separate gene.

3. A molecule of RNA would weigh about half as much as a segment of DNA of the same length.

4. Each amino acid of a protein is represented by a three-base sequence in DNA.

5. One gene can code for more than one protein.

6. The law of complementary base pairing describes the way the bases in an mRNA codon pair up with the bases of a tRNA anticodon during translation.

7. Most of the DNA in a human cell does not code for any proteins.

8. Some mutations are harmless.

9. Males have only one sex chromosome whereas females have two.

10. A gene can be transcribed by only one RNA polymerase at a time.

## ▶ Testing Your Comprehension

*Answers at* www.mhhe.com/saladin7

1. Why would the supercoiled, condensed form of chromosomes seen in metaphase not be suitable for the $G_1$ phase of the cell cycle? Why would the finely dispersed chromatin of the $G_1$ phase not be suitable for mitosis?

2. Suppose a woman was heterozygous for blood type A and her husband had blood type AB. With the aid of a Punnett square, explain what blood type(s) their children could possibly have.

3. Given the information in this chapter, present an argument that evolution is not merely possible but inevitable. (*Hint:* Review the definition of evolution in chapter 1.)

4. What would be the minimum length (approximate number of bases) of an mRNA that coded for a protein 300 amino acids long?

5. Until recently, textbooks taught the concept of "one gene, one protein"—that a gene is a segment of DNA that codes for just one protein, and every protein is represented by a separate gene. Discuss the evidence that shows that this can no longer be regarded as true.

## ▶ Improve Your Grade at www.mhhe.com/saladin7

*Download animations to study when it fits your schedule. Practice quizzes, labeling activities, games, and flashcards offer fun ways to master the chapter concepts. Or, download image PowerPoint files for each chapter to create a study guide or for taking notes during lecture.*

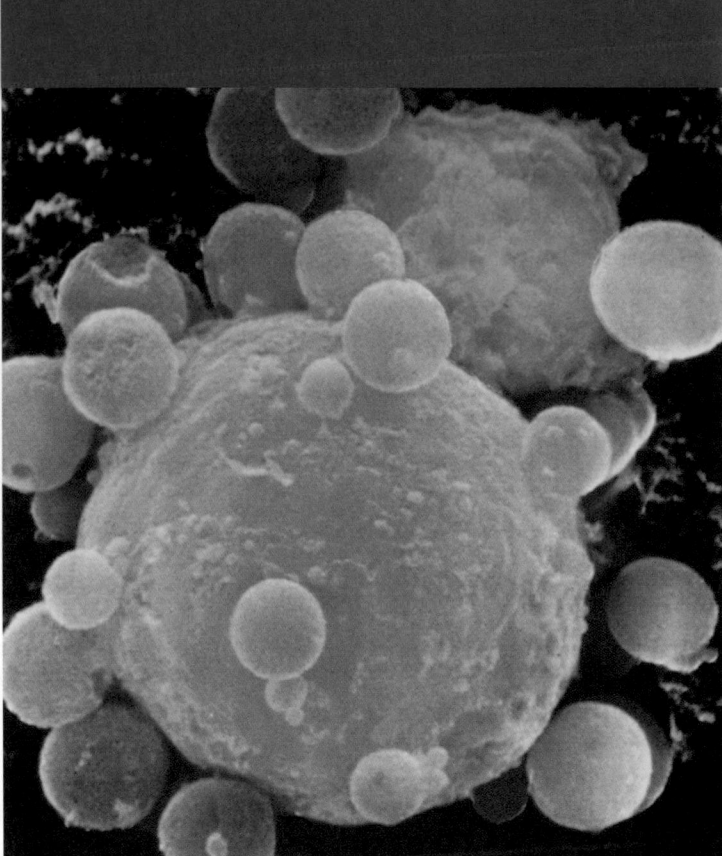

A cancer cell (pink) undergoing apoptosis (cell suicide) under attack by an immune cell (orange) (SEM)

Anatomy & Physiology REVEALED®
aprevealed.com

**Module 3: Tissues**

## BRUSHING UP

- Membranes that line the body cavities were introduced on pages 32 to 34. This chapter returns to these membranes to describe their histological structure.
- Connective tissues are characterized by a large amount of ground substance, two of the chief components of which are the glycoproteins and proteoglycans introduced on page 61.
- The terminology of cell shapes (p. 78) is used in the naming of epithelial tissue in this chapter.
- To understand the gland types introduced in this chapter, you should be familiar with secretory vesicles and exocytosis (p. 98).

W ith its 50 trillion cells and thousands of organs, the human body may seem to be a structure of forbidding complexity. Fortunately for our health, longevity, and self-understanding, biologists of the past were not discouraged by this complexity, but discovered patterns that made it more understandable. One such pattern is the fact that these trillions of cells belong to only 200 different types or so, and they are organized into tissues that fall into just four broad categories: *epithelial, connective, nervous,* and *muscular tissue.*

Organs derive their function not from their cells alone but from how the cells are organized into tissues. Cells are specialized for certain tasks: muscle contraction, defense, enzyme secretion, and so forth. No one cell type can carry out all of the body's vital functions. Cells therefore work together at certain tasks and form tissues that carry out a particular function, such as nerve signaling or nutrient digestion. An organ is a structure with discrete boundaries that is composed of two or more tissue types.

The study of tissues and how they are arranged into organs is called **histology,**[1] or **microscopic anatomy**—the subject of this chapter. Here we study the four tissue classes; the variations within each class; how to recognize tissue types microscopically and relate their microscopic anatomy to their function; how tissues are arranged

to form an organ; how tissues change as they grow, shrink, or change from one tissue type to another over the life of the individual; and modes of tissue degeneration and death. Histology bridges the gap between the cytology of the preceding chapters and the organ system approach of the chapters that follow.

### 5.1  The Study of Tissues

#### Expected Learning Outcomes

When you have completed this section, you should be able to

a. name the four primary classes into which all adult tissues are classified;

b. name the three embryonic germ layers and some adult tissues derived from each; and

c. visualize the three-dimensional shape of a structure from a two-dimensional tissue section.

### The Primary Tissue Classes

A **tissue** is a group of similar cells and cell products that arise from the same region of the embryo and work together to perform a specific structural or physiological role in an organ. The four *primary tissues*—epithelial, connective, nervous, and muscular—are summarized in table 5.1. They differ in the types and functions of their cells, the characteristics of the **matrix (extracellular material)** that surrounds the cells, and the relative amount of space occupied by the cells and matrix. In muscle and epithelium, the cells are so close together that the matrix is scarcely visible, but in connective tissues, the matrix usually occupies much more space than the cells do.

The matrix is composed of fibrous proteins and, usually, a clear gel variously known as **ground substance, tissue fluid,**

[1] *histo* = tissue; *logy* = study of

| TABLE 5.1 | The Four Primary Tissue Classes | |
|---|---|---|
| **Type** | **Definition** | **Representative Locations** |
| Epithelial | Tissue composed of layers of closely spaced cells that cover organ surfaces, form glands, and serve for protection, secretion, and absorption | Epidermis<br>Inner lining of digestive tract<br>Liver and other glands |
| Connective | Tissue with usually more matrix than cell volume, often specialized to support, bind, and protect organs | Tendons and ligaments<br>Cartilage and bone<br>Blood |
| Nervous | Tissue containing excitable cells specialized for rapid transmission of coded information to other cells | Brain<br>Spinal cord<br>Nerves |
| Muscular | Tissue composed of elongated, excitable muscle cells specialized for contraction | Skeletal muscles<br>Heart (cardiac muscle)<br>Walls of viscera (smooth muscle) |

**extracellular fluid (ECF),** or **interstitial**[2] **fluid.** In cartilage and bone, it can be rubbery or stony in consistency. The ground substance contains water, gases, minerals, nutrients, wastes, hormones, and other chemicals. This is the medium from which all cells obtain their oxygen, nutrients, and other needs, and into which cells release metabolic wastes, hormones, and other products.

In summary, a tissue is composed of cells and matrix, and the matrix is composed of fibers and ground substance.

## Embryonic Tissues

Human development begins with a single cell, the fertilized egg, which soon divides to produce scores of identical, smaller cells. The first tissues appear when these cells start to organize themselves into layers—first two, and soon three strata called the **primary germ layers,** which give rise to all of the body's mature tissues. The three layers are called *ectoderm, mesoderm,* and *endoderm.* The **ectoderm**[3] is an outer layer that gives rise to the epidermis and nervous system. The innermost layer, the **endoderm,**[4] gives rise to the mucous membranes of the digestive and respiratory tracts and to the digestive glands, among other things. Between these two is the **mesoderm,**[5] a layer of more loosely organized cells. Mesoderm eventually turns to a

---

[2]*inter* = between; *stit* = to stand
[3]*ecto* = outer; *derm* = skin
[4]*endo* = inner; *derm* = skin
[5]*meso* = middle; *derm* = skin

gelatinous tissue called **mesenchyme,** composed of fine, wispy collagen (protein) fibers and branching *mesenchymal cells* embedded in a gelatinous ground substance. Mesenchyme gives rise to muscle, bone, and blood among other tissues. The development of the three primary tissues in the embryo is detailed in chapter 29 (p. 1102). Most organs are composed of tissues derived from two or more primary germ layers. The rest of this chapter concerns the "mature" tissues that exist from infancy through adulthood.

## Interpreting Tissue Sections

In your study of histology, you may be presented with various tissue preparations mounted on microscope slides. Most such preparations are thin slices called **histological sections,** and are artificially colored to bring out detail. The best anatomical insight depends on an ability to deduce the three-dimensional structure of an organ from these two-dimensional sections (fig. 5.1). This ability, in turn, depends on an awareness of how tissues are prepared for study.

Histologists use a variety of techniques for preserving, sectioning (slicing), and staining tissues to show their structural details as clearly as possible. Tissue specimens are preserved in a **fixative**—a chemical such as formalin that prevents decay. After fixation, most tissues are cut into sections typically only one or two cells thick. Sectioning is necessary to allow the light of a microscope to pass through and so the image is not confused by too many layers of overlapping cells. The sections are

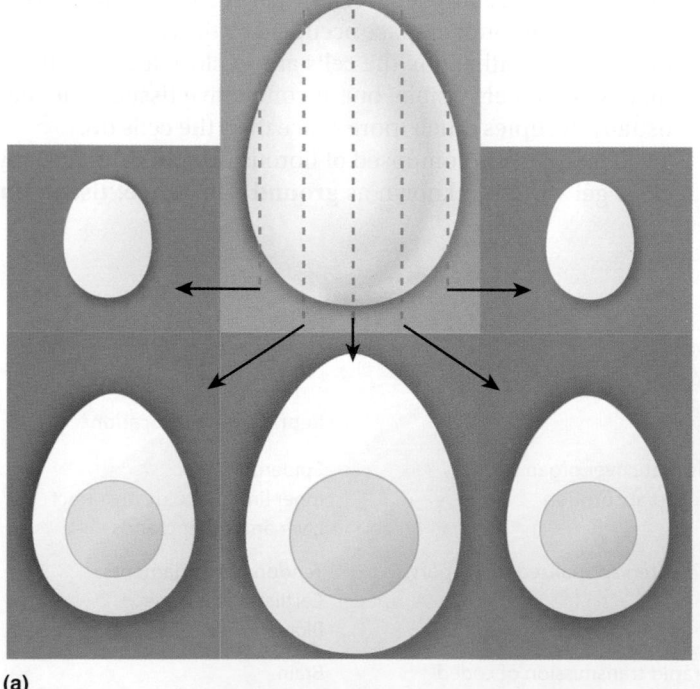
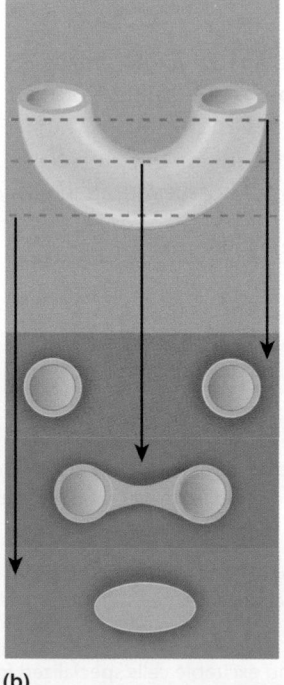
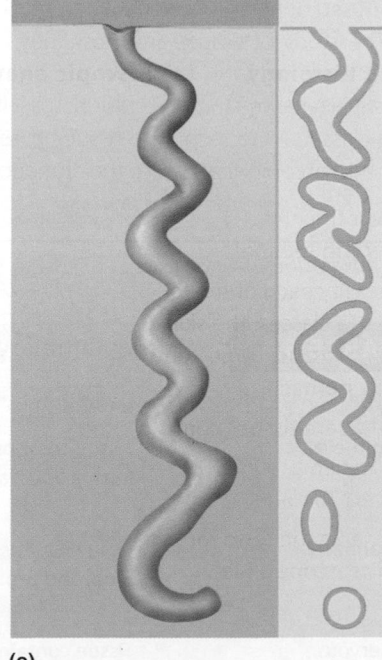

(a)                                      (b)                    (c)

**FIGURE 5.1 Three-Dimensional Interpretation of Two-Dimensional Images.** (a) A boiled egg. Note that grazing sections (top left and right) would miss the yolk, just as a tissue section may miss a nucleus or other structure. (b) Elbow macaroni, which resembles many curved ducts and tubules. A section far from the bend would give the impression of two separate tubules; a section near the bend would show two interconnected lumina (cavities); and a section still farther down could miss the lumen completely. (c) A coiled gland in three dimensions and as it would look in a vertical tissue section of a tissue such as the lining of the uterus.

then mounted on slides and artificially colored with histological **stains** to enhance detail. If they were not stained, most tissue sections would appear pale gray. With stains that bind to different components of a tissue, however, you may see pink cytoplasm; violet nuclei; and blue, green, or golden-brown protein fibers, depending on the stain used.

Sectioning a tissue reduces a three-dimensional structure to a series of two-dimensional slices. You must keep this in mind and try to translate the microscopic image into a mental image of the whole structure. Like the boiled egg and elbow macaroni in figure 5.1, an object may look quite different when it is cut at various levels, or *planes of section.* A coiled tube, such as a gland of the uterus (fig. 5.1c), is often broken up into multiple portions since it meanders in and out of the plane of section. An experienced viewer, however, recognizes that the separated pieces are parts of a single tube winding its way to the organ surface. Note that a grazing slice through a boiled egg might miss the yolk, just as a tissue section might miss the nucleus of a cell even though it was present.

Many anatomical structures are longer on one axis than another—the humerus and esophagus, for example. A tissue cut on its long axis is called a **longitudinal section (l.s.),** and one cut perpendicular to this is a **cross section (c.s. or x.s.)** or **transverse section (t.s.).** A section cut on a slant between a longitudinal and cross section is an **oblique section.** Figure 5.2 shows how certain organs look when sectioned on each of these planes.

Not all histological preparations are sections. Liquid tissues such as blood and soft tissues such as spinal cord may be prepared as **smears,** in which the tissue is rubbed or spread across the slide rather than sliced. Some membranes and cobwebby tissues like the *areolar tissue* in figure 5.14 are sometimes mounted as **spreads,** in which the tissue is laid out on the slide, like placing a small square of tissue paper or a tuft of lint on a sheet of glass.

### BEFORE YOU GO ON

Answer the following questions to test your understanding of the preceding section:

1. Classify each of the following into one of the four primary tissue classes: the skin surface, fat, the spinal cord, most heart tissue, bone, tendons, blood, and the inner lining of the stomach.

2. What are tissues composed of in addition to cells?

3. What embryonic germ layer gives rise to nervous tissue? To the liver? To muscle?

4. What is the term for a thin, stained slice of tissue mounted on a microscope slide?

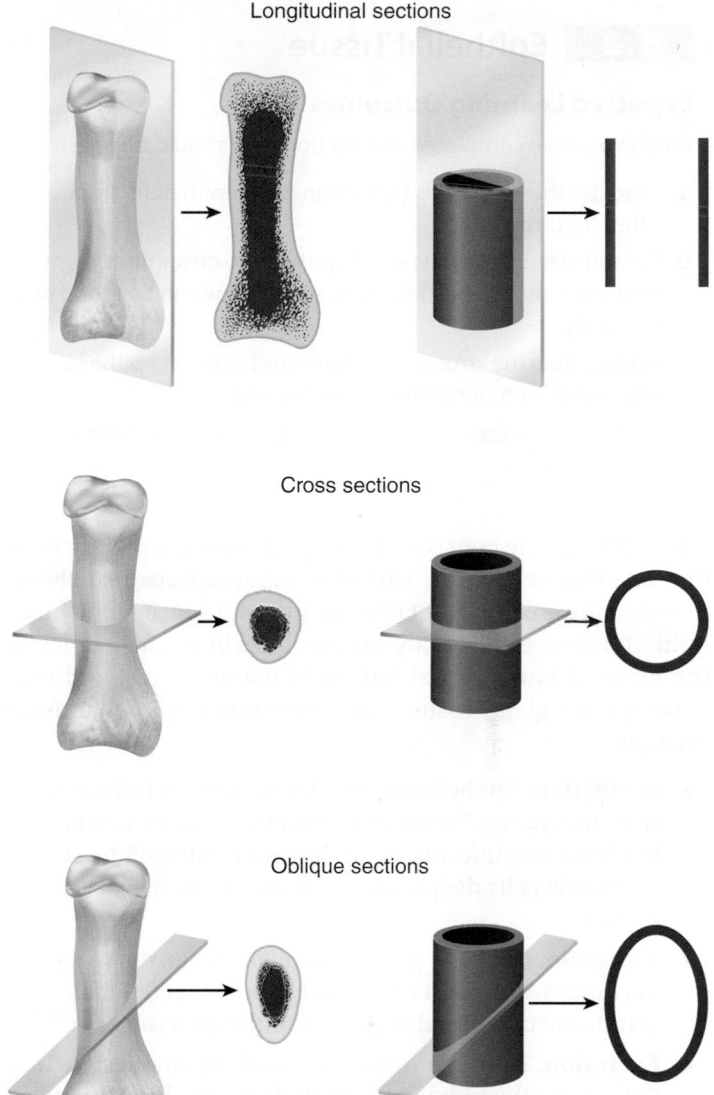

Longitudinal sections

Cross sections

Oblique sections

**FIGURE 5.2 Longitudinal, Cross, and Oblique Sections.** Note the effect of the plane of section on the two-dimensional appearance of elongated structures such as bones and blood vessels.

*Would you classify the egg sections in the previous figure as longitudinal, cross, or oblique sections? How would the egg look if sectioned in the other two planes?*

## 5.2 Epithelial Tissue

### Expected Learning Outcomes

When you have completed this section, you should be able to

a. describe the properties that distinguish epithelium from other tissue classes;

b. list and classify eight types of epithelium, distinguish them from each other, and state where each type can be found in the body;

c. explain how the structural differences between epithelia relate to their functional differences; and

d. visually recognize each epithelial type from specimens or photographs.

**Epithelial**[6] **tissue** consists of a sheet of closely adhering cells, one or more cells thick, with the upper surface usually exposed to the environment or to an internal space in the body. Epithelium covers the body surface, lines body cavities, forms the external and internal linings of many organs, and constitutes most gland tissue. The functions of epithelial tissue include

- **Protection.** Epithelia protect deeper tissues from invasion and injury. The epidermis of the skin, for example, is a barrier to infection, and the inner lining of the stomach protects its deeper tissues from stomach acid and enzymes.

- **Secretion.** Epithelia produce mucus, sweat, enzymes, hormones, and most of the body's other secretions; glands are composed largely of epithelial tissue.

- **Excretion.** Epithelia void wastes from the tissues, such as $CO_2$ across the pulmonary epithelium and bile from the epithelium of the liver.

- **Absorption.** Epithelia absorb chemicals from the adjacent medium; nutrients, for example, are absorbed through the epithelium of the small intestine.

- **Filtration.** All substances leaving the blood are selectively filtered through the epithelium that lines the blood vessels; all urinary waste is filtered through epithelia of the kidneys.

- **Sensation.** Epithelia are provided with nerve endings that sense stimulation ranging from a touch on the skin to irritation of the stomach.

The cells and extracellular material of an epithelium can be loosely compared to the bricks and mortar of a wall. The extracellular material ("mortar") is so thin, however, that it is barely visible with a light microscope, and the cells appear pressed very close together. Epithelia are *avascular*[7]—there is no room between the cells for blood vessels. Epithelia,

however, usually lie on a vessel-rich layer of connective tissue, which furnishes them with nutrients and waste removal. Epithelial cells closest to the connective tissue typically exhibit a high rate of mitosis. This allows epithelia to repair themselves quickly—an ability of special importance in protective epithelia that are highly vulnerable to such injuries as skin abrasions and erosion by stomach acid.

Between an epithelium and the underlying connective tissue is a layer called the **basement membrane.** It contains collagen, glycoproteins, and other protein–carbohydrate complexes, and blends into other proteins of the connective tissue. The basement membrane serves to anchor an epithelium to the connective tissue; it controls the exchange of materials between the epithelium and the underlying tissues; and it binds growth factors from below that regulate epithelial development. The surface of an epithelial cell that faces the basement membrane is its *basal surface,* and the one that faces away from it toward the internal cavity (lumen) of an organ is the *apical surface.*

Epithelia are classified into two broad categories—*simple* and *stratified*—with four types in each category. In a simple epithelium, every cell touches the basement membrane, whereas in a stratified epithelium, some cells rest on top of other cells and do not contact the basement membrane (fig. 5.3).

Table 5.2 (containing figs. 5.4 to 5.7) summarizes the structural and functional differences between the four simple epithelia, and table 5.3 (figs. 5.8 to 5.11) compares the major types of stratified epithelia. In these and subsequent tables, each photograph is accompanied by a line drawing with labels. The drawings clarify cell boundaries and other relevant features that may otherwise be difficult to see or identify in photographs or through the microscope. Each figure indicates the approximate magnification at which the original photograph was made. Each is enlarged much more than this when printed in the book, but selecting the closest magnification on a microscope should enable you to see a comparable level of detail (resolution).

### Simple Epithelia

Generally, a **simple epithelium** has only one layer of cells, although this is a somewhat debatable point in the *pseudostratified columnar* type. Three types of simple epithelia are named for the shapes of their cells: **simple squamous**[8] (thin scaly cells), **simple cuboidal** (squarish or round cells), and **simple columnar** (tall narrow cells). In the fourth type, **pseudostratified columnar,** not all cells reach the surface; the shorter cells are covered by the taller ones. This epithelium looks stratified in most tissue sections, but careful examination, especially with the electron microscope, shows that every cell reaches the basement membrane—like trees in a forest, where some grow taller than others but all are anchored in the soil below.

---

[6]*epi* = upon; *theli* = nipple, female
[7]*a* = without; *vas* = vessels

[8]*squam* = scale

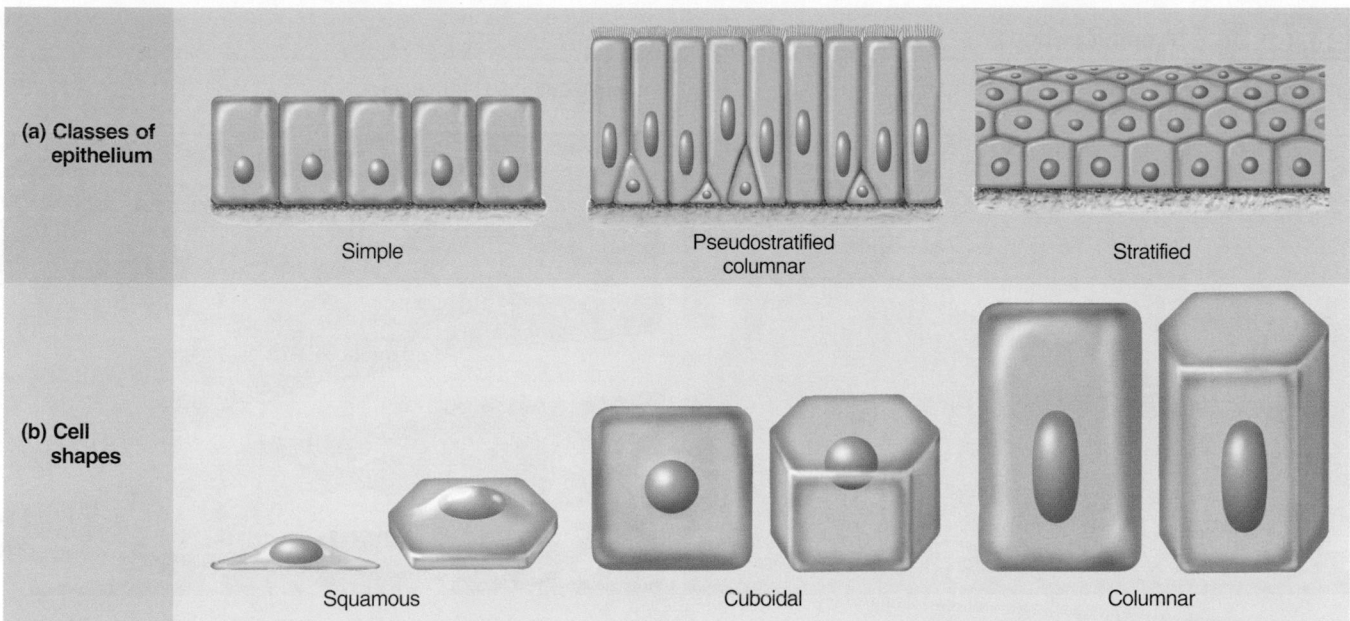

**FIGURE 5.3 Cell Shapes and Epithelial Types.** Pseudostratified columnar epithelium is a special type of simple epithelium that gives a false appearance of multiple cell layers.

Simple columnar and pseudostratified columnar epithelia often have wineglass-shaped **goblet cells** that produce protective mucous coatings over the mucous membranes. These cells have an expanded apical end filled with secretory vesicles; their product becomes mucus when it is secreted and absorbs water. The basal part of the cell is a narrow stem, like that of a wineglass, that reaches to the basement membrane.

## Stratified Epithelia

**Stratified epithelia** range from 2 to 20 or more layers of cells, with some cells resting directly on others and only the deepest layer attached to the basement membrane. Three of the stratified epithelia are named for the shapes of their surface cells: **stratified squamous, stratified cuboidal,** and **stratified columnar epithelia.** The deeper cells, however, may be of a different shape than the surface cells. The fourth type, **transitional epithelium,** was named when it was thought to represent a transitional stage between stratified squamous and stratified columnar epithelium. This is now known to be untrue, but the name has persisted.

Stratified columnar epithelium is rare and of relatively minor importance—seen only in places where two other epithelial types meet, as in limited regions of the pharynx, larynx, anal canal, and male urethra. We will not consider this type any further.

The most widespread epithelium in the body is stratified squamous epithelium, which deserves further discussion. Its deepest layer of cells are cuboidal to columnar, and include mitotically active stem cells. Their daughter cells push toward the surface and become flatter (more scaly) as they migrate

farther upward, until they finally die and flake off. Their loss is called **exfoliation (desquamation)** (see fig. 5.12); the study of exfoliated cells is called *exfoliate cytology.* You can easily study exfoliated cells by scraping your gums with a toothpick, smearing this material on a slide, and staining it with iodine. A similar procedure is used in the *Pap smear,* an examination of exfoliated cells from the cervix for signs of uterine cancer (see fig. 28.5, p. 1064).

Stratified squamous epithelia are of two kinds—keratinized and nonkeratinized. A **keratinized (cornified)** epithelium, found in the epidermis, is covered with a layer of dead compressed cells. These cells are packed with the durable protein keratin and coated with a water-repellent glycolipid. The skin surface is therefore relatively dry; it retards water loss from the body; and it resists penetration by disease organisms. (Keratin is also the protein of which animal horns are made, hence its name.[9]) The tongue, esophagus, vagina, and a few other internal membranes are covered with the **nonkeratinized** type, which lacks the surface layer of dead cells. This type provides a surface that is, again, abrasion-resistant, but also moist and slippery. These characteristics are well suited to resist stress produced by chewing and swallowing food and by sexual intercourse and childbirth.

Transitional epithelium is another particularly interesting type of stratified epithelium. Why is it limited to the urinary tract? The answer relates to the fact that urine is usually acidic and hypertonic to the intracellular fluid. It would tend

*(text continued on p. 150)*

[9]*kerat* = horn

| TABLE 5.2 | Simple Epithelia |
|---|---|

| **Simple Squamous Epithelium** | **Simple Cuboidal Epithelium** |
|---|---|

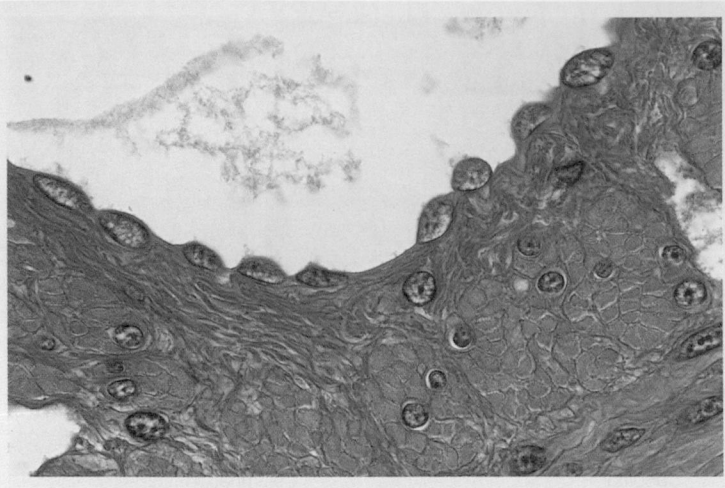

(a)

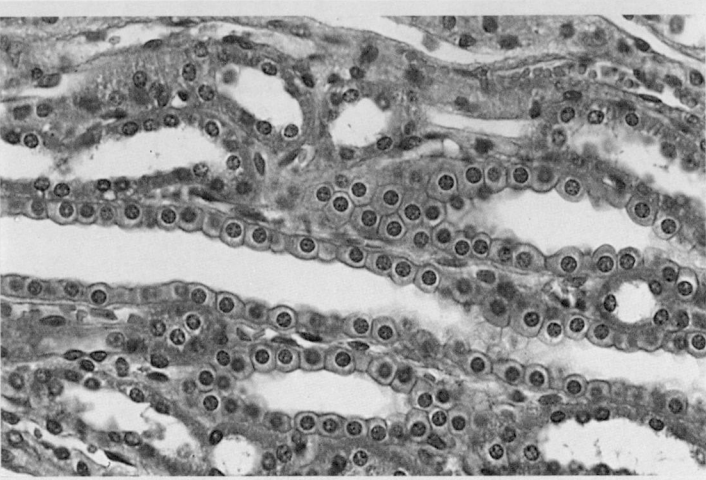

(a)

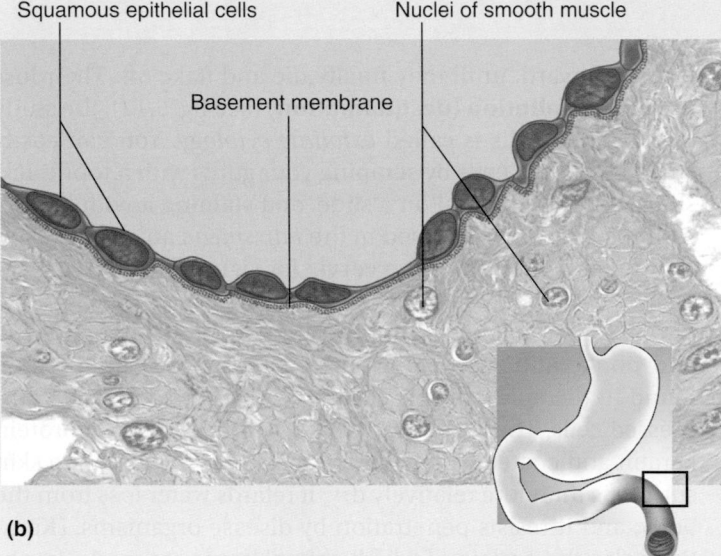

Squamous epithelial cells    Nuclei of smooth muscle

Basement membrane

(b)

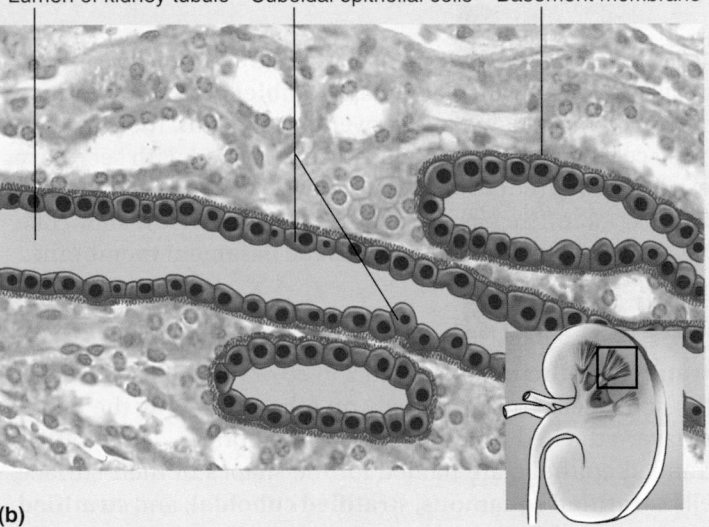

Lumen of kidney tubule    Cuboidal epithelial cells    Basement membrane

(b)

**FIGURE 5.4  Simple Squamous Epithelium on the External Surface (Serosa) of the Small Intestine (×400).** AP|R

Microscopic appearance: Single layer of thin cells, shaped like fried eggs with bulge where nucleus is located; nucleus flattened in the plane of the cell, like an egg yolk; cytoplasm may be so thin it is hard to see in tissue sections; in surface view, cells have angular contours and nuclei appear round

Representative locations: Air sacs (alveoli) of lungs; glomerular capsules of kidneys; some kidney tubules; inner lining (endothelium) of heart and blood vessels; serous membranes of stomach, intestines, and some other viscera; surface mesothelium of pleura, pericardium, peritoneum, and mesenteries

Functions: Allows rapid diffusion or transport of substances through membrane; secretes lubricating serous fluid

**FIGURE 5.5  Simple Cuboidal Epithelium in the Kidney Tubules (×400).** AP|R

Microscopic appearance: Single layer of square or round cells; in glands, cells often pyramidal and arranged like segments of an orange around a central space; spherical, centrally placed nuclei; with a brush border of microvilli in some kidney tubules; ciliated in bronchioles of lung

Representative locations: Liver, thyroid, mammary, salivary, and other glands; most kidney tubules; bronchioles

Functions: Absorption and secretion; production of protective mucous coat; movement of respiratory mucus

| **TABLE 5.2** | Simple Epithelia (continued) |
|---|---|

| **Simple Columnar Epithelium** | **Pseudostratified Columnar Epithelium** |
|---|---|

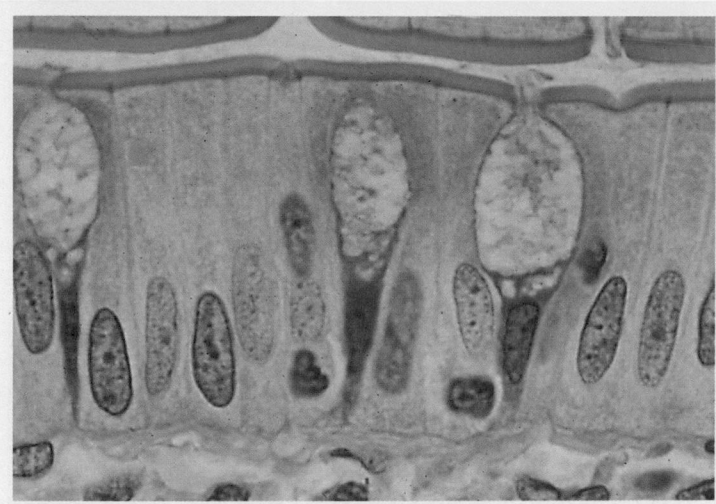

(a)

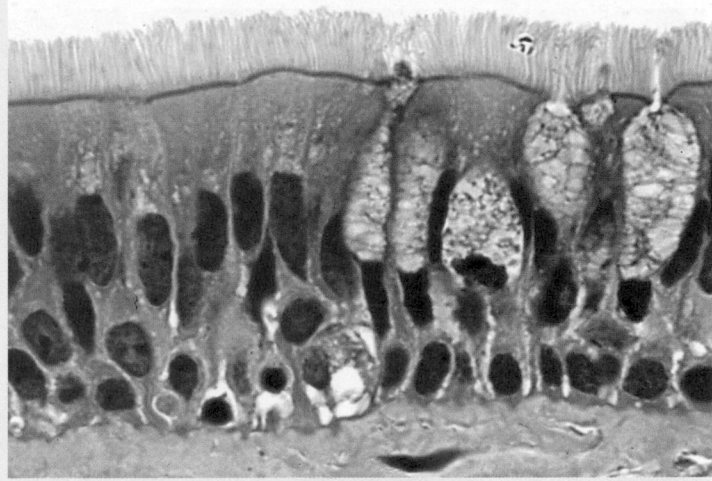

(a)

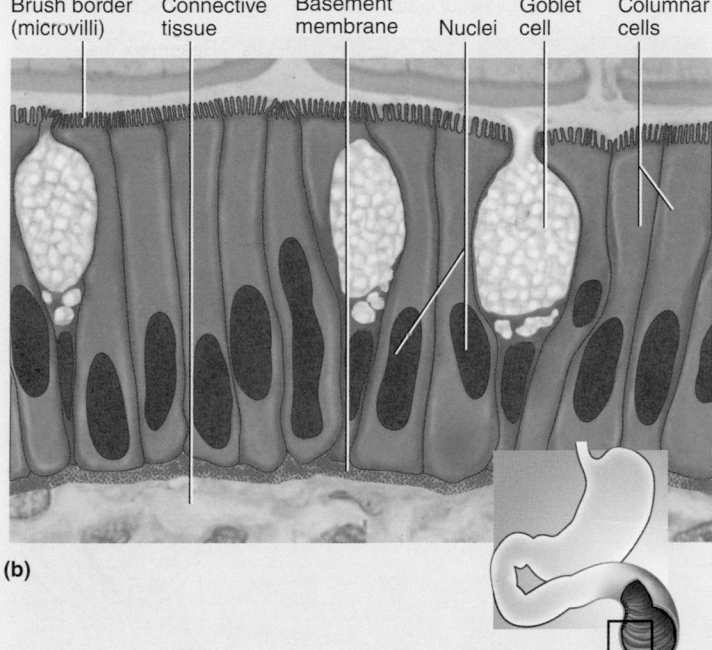

Brush border (microvilli)  Connective tissue  Basement membrane  Nuclei  Goblet cell  Columnar cells

(b)

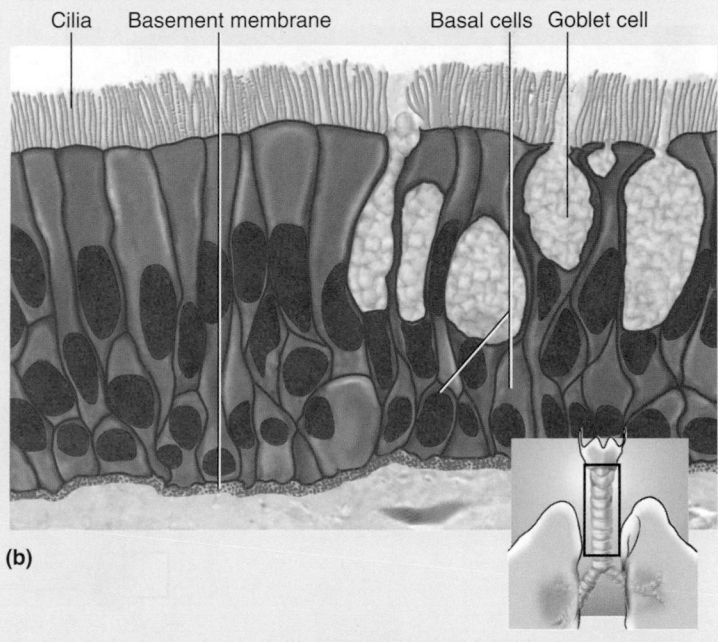

Cilia  Basement membrane  Basal cells  Goblet cell

(b)

**FIGURE 5.6 Simple Columnar Epithelium on the Internal Surface (Mucosa) of the Small Intestine (×400).** AP|R

Microscopic appearance: Single layer of tall, narrow cells; oval or sausage-shaped nuclei, vertically oriented, usually in basal half of cell; apical portion of cell often shows secretory vesicles visible with TEM; often shows a brush border of microvilli; ciliated in some organs; may possess goblet cells

Representative locations: Inner lining of stomach, intestines, gallbladder, uterus, and uterine tubes; some kidney tubules

Functions: Absorption; secretion of mucus and other products; movement of egg and embryo in uterine tube

**FIGURE 5.7 Ciliated Pseudostratified Columnar Epithelium in the Mucosa of the Trachea (×400).** AP|R

Microscopic appearance: Looks multilayered; some cells do not reach free surface, but all cells reach basement membrane; nuclei at several levels in deeper half of epithelium; often with goblet cells; often ciliated

Representative locations: Respiratory tract from nasal cavity to bronchi; portions of male urethra

Functions: Secretes and propels mucus

| TABLE 5.3 | Stratified Epithelia |
|---|---|

| **Stratified Squamous Epithelium—Keratinized** | **Stratified Squamous Epithelium—Nonkeratinized** |
|---|---|

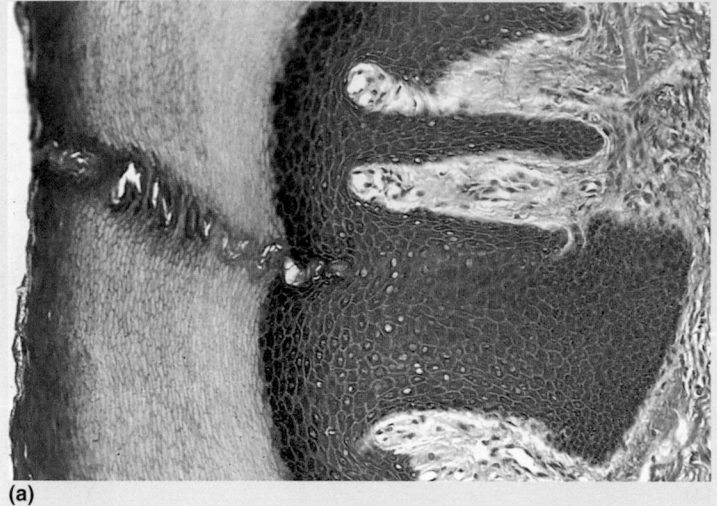

(a)

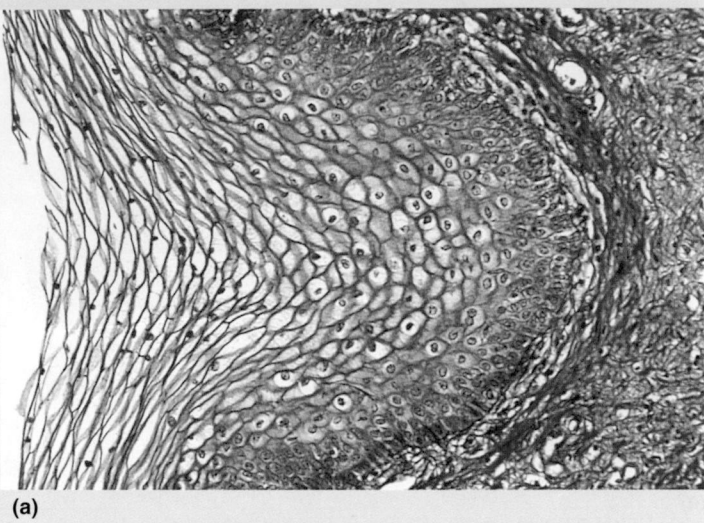

(a)

Dead squamous cells   Living epithelial cells   Dense irregular connective tissue

Living epithelial cells   Connective tissue

Areolar tissue

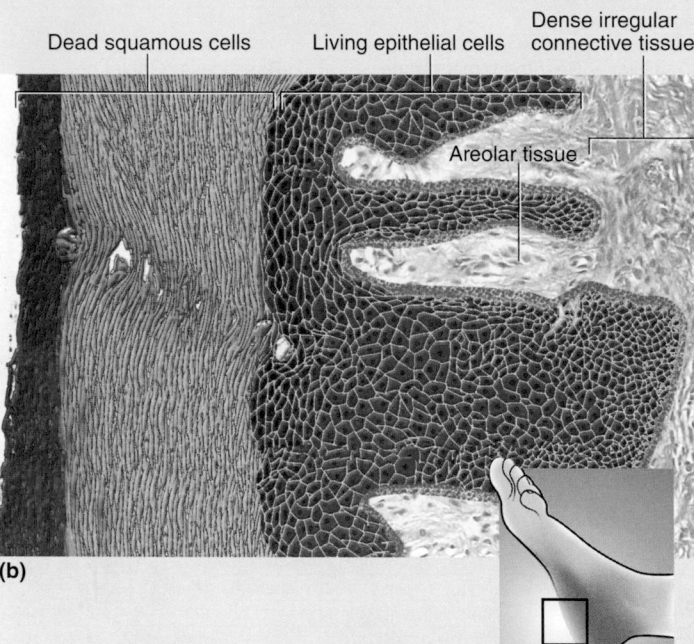

(b)

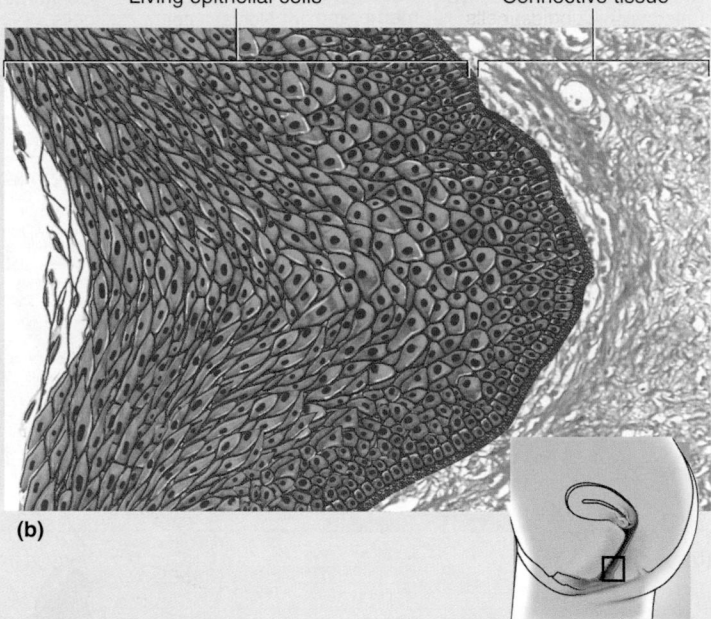

(b)

**FIGURE 5.8  Keratinized Stratified Squamous Epithelium on the Sole of the Foot (×400).**  AP|R

Microscopic appearance:  Multiple cell layers with cells becoming increasingly flat and scaly toward surface; surface covered with a layer of compact dead cells without nuclei; basal cells may be cuboidal to columnar

Representative locations:  Epidermis; palms and soles are especially heavily keratinized

Functions:  Resists abrasion and penetration by pathogenic organisms; retards water loss through skin

**FIGURE 5.9  Nonkeratinized Stratified Squamous Epithelium in the Mucosa of the Vagina (×400).**  AP|R

Microscopic appearance:  Same as keratinized epithelium but without the surface layer of dead cells

Representative locations:  Tongue, oral mucosa, esophagus, anal canal, vagina

Functions:  Resists abrasion and penetration by pathogenic organisms

**TABLE 5.3** Stratified Epithelia (continued)

| Stratified Cuboidal Epithelium | Transitional Epithelium |
|---|---|

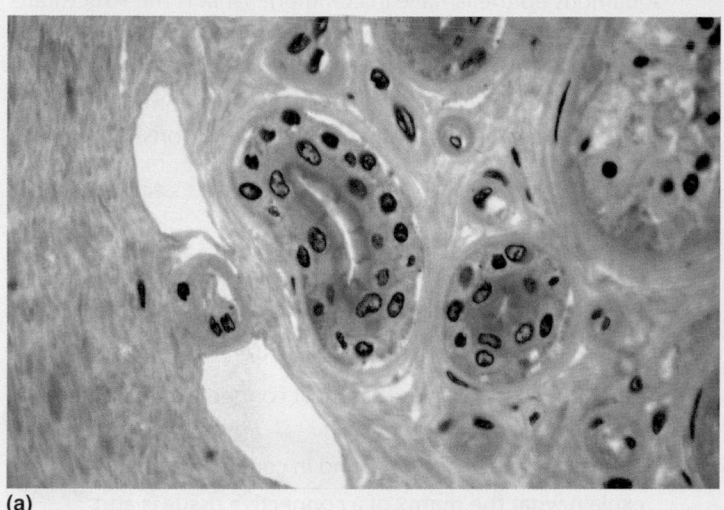

(a)

Cuboidal cells　Epithelium　Connective tissue

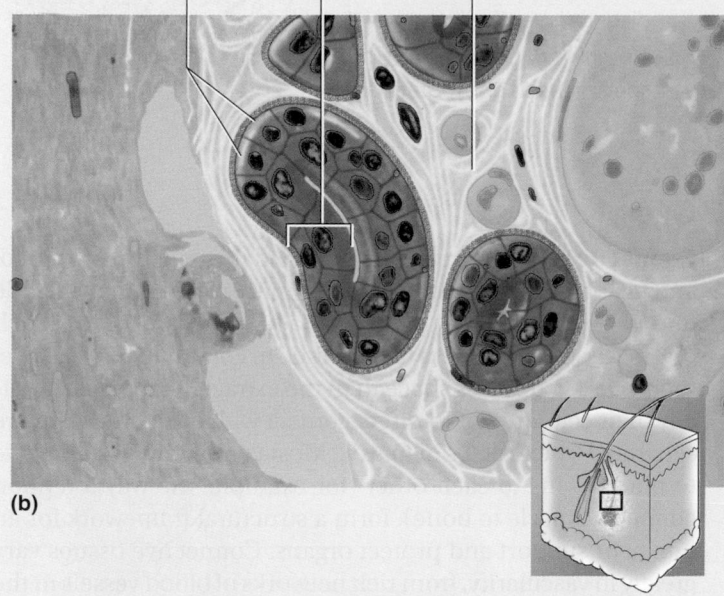

(b)

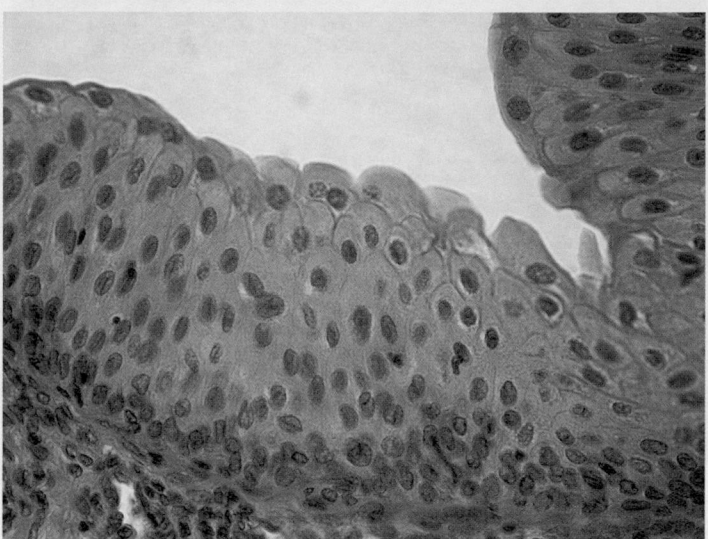

(a)

Basement membrane　Connective tissue　Binucleate epithelial cell

(b)

**FIGURE 5.10** Stratified Cuboidal Epithelium in the Duct of a Sweat Gland (×400). AP|R

Microscopic appearance: Two or more layers of cells; surface cells square or round

Representative locations: Sweat gland ducts; egg-producing vesicles (follicles) of ovaries; sperm-producing ducts (seminiferous tubules) of testis

Functions: Contributes to sweat secretion; secretes ovarian hormones; produces sperm

**FIGURE 5.11** Transitional Epithelium in the Kidney (×400). AP|R

Microscopic appearance: Somewhat resembles stratified squamous epithelium, but surface cells are rounded, not flattened, and often bulge at surface; typically five or six cells thick when relaxed and two or three cells thick when stretched; cells may be flatter and thinner when epithelium is stretched (as in a distended bladder); some cells have two nuclei

Representative locations: Urinary tract—part of kidney, ureter, bladder, part of urethra; umbilical cord

Functions: Stretches to allow filling of urinary tract; protects underlying tissues from osmotic damage by urine

**FIGURE 5.12 Exfoliation of Squamous Cells from the Mucosa of the Vagina.** © Dr. Kessel & Dr. Kardon/Tissues and Organs/Visuals Unlimited, Inc.

❓ *Aside from the gums and vagina, name another epithelium in the body that would look like this to the scanning electron microscope.*

to draw water out of the cells by osmosis and kill them if there were nothing to protect the cells. The domed surface cells of transitional epithelium, however, have a unique protective property. They are called **umbrella cells.** On the upper surface of an umbrella cell, the outer phospholipid layer of the plasma membrane is thicker than usual and has dense patches called *lipid rafts* with embedded proteins called *uroplakins.* Uroplakins are impermeable to urine and protect the epithelium, including the cytoplasm of the umbrella cell itself. Lipid rafts are connected to each other by hinges of ordinary plasma membrane. When the bladder is empty and relaxed, these plaques fold at the hinges and drop into the cell interior for storage (like folding a laptop computer and putting it away), and the cell bulges upward as seen in figure 5.11. As the bladder fills with urine, the hinges open (like opening the computer), the plaques spread out over the surface to protect the cell, and the umbrella cells become thinner and flatter. Not surprisingly, this type of epithelium is best developed in the bladder, where it is subject to prolonged contact with stored urine.

**BEFORE YOU GO ON**

Answer the following questions to test your understanding of the preceding section:

5. Distinguish between simple and stratified epithelia, and explain why pseudostratified columnar epithelium belongs in the former category despite its superficial appearance.

6. Explain how to distinguish a stratified squamous epithelium from a transitional epithelium.

7. What function do keratinized and nonkeratinized stratified squamous epithelia have in common? What is the structural difference between these two? How is this structural difference related to a functional difference between them?

8. How do the epithelia of the esophagus and stomach differ? How does this relate to their respective functions?

## 5.3   Connective Tissue

### Expected Learning Outcomes

When you have completed this section, you should be able to

a. describe the properties that most connective tissues have in common;

b. discuss the types of cells found in connective tissue;

c. explain what the matrix of a connective tissue is and describe its components;

d. name and classify 10 types of connective tissue, describe their cellular components and matrix, and explain what distinguishes them from each other; and

e. visually recognize each connective tissue type from specimens or photographs.

### Overview

**Connective tissues** are the most abundant, widely distributed, and histologically variable of the primary tissues. They include fibrous tissue, adipose tissue, cartilage, bone, and blood. Such diverse tissues may seem to have little in common, but as a rule, their cells occupy less space than the extracellular matrix. Usually their cells are not in direct contact with each other, but are separated by expanses of matrix. Most connective tissues serve to bind organs to each other (for example, the way a tendon connects muscle to bone), form a structural framework for an organ, or support and protect organs. Connective tissues vary greatly in vascularity, from rich networks of blood vessels in the loose connective tissues to few or no blood vessels in cartilage.

The functions of connective tissue include

- **Binding of organs.** Tendons bind muscle to bone, ligaments bind one bone to another, fat holds the kidneys and eyes in place, and fibrous tissue binds the skin to underlying muscle.

- **Support.** Bones support the body; cartilage supports the ears, nose, larynx, and trachea; fibrous tissues form the framework of organs such as the spleen.

- **Physical protection.** The cranium, ribs, and sternum protect delicate organs such as the brain, lungs, and heart; fatty cushions protect the kidneys and eyes.

- **Immune protection.** Connective tissue cells attack foreign invaders, and connective tissue fiber forms a "battlefield" under the skin and mucous membranes where immune cells can be quickly mobilized against disease agents.

- **Movement.** Bones provide the lever system for body movement, cartilages are involved in movement of the vocal cords, and cartilages on bone surfaces ease joint movements.

- **Storage.** Fat is the body's major energy reserve; bone is a reservoir of calcium and phosphorus that can be drawn upon when needed.

- **Heat production.** Metabolism of brown fat generates heat in infants and children.

- **Transport.** Blood transports gases, nutrients, wastes, hormones, and blood cells.

The mesenchyme described earlier in this chapter is a form of embryonic connective tissue. The mature connective tissues fall into four broad categories: fibrous connective tissue, adipose tissue, supportive connective tissues (cartilage and bone), and fluid connective tissue (blood).

## Fibrous Connective Tissue

Fibrous connective tissue is the most diverse type. It is also called *fibroconnective tissue.* Nearly all connective tissues contain fibers, but the tissues considered here are classified together because the fibers are so conspicuous. Fibers are, of course, just one component of the tissue, which also includes cells and ground substance. Before examining specific types of fibrous connective tissue, let's examine these components.

### Components of Fibrous Connective Tissue

*Cells*  The cells of fibrous connective tissue include the following types:

- **Fibroblasts.**[10] These are large, fusiform or stellate cells that often show slender, wispy branches. They produce the fibers and ground substance that form the matrix of the tissue.

- **Macrophages.**[11] These are large phagocytic cells that wander through the connective tissues, where they engulf and destroy bacteria, other foreign particles, and dead or dying cells of our own body. They also activate the immune system when they sense foreign matter called *antigens.* They arise from certain white blood cells called *monocytes* or from the same stem cells that produce monocytes.

- **Leukocytes,**[12] or **white blood cells (WBCs).** WBCs travel briefly in the bloodstream, then crawl out through the walls of small blood vessels and spend most of their time in the connective tissues. The two most common types are *neutrophils,* which wander about attacking bacteria, and *lymphocytes,* which react against bacteria, toxins, and other foreign agents. Lymphocytes often form dense patches in the mucous membranes.

- **Plasma cells.** Certain lymphocytes turn into plasma cells when they detect foreign agents. The plasma cells then synthesize disease-fighting proteins called *antibodies.* Plasma cells are rarely seen except in the wall of the intestines and in inflamed tissue.

- **Mast cells.** These cells, found especially alongside blood vessels, secrete a chemical called *heparin* that inhibits blood clotting, and one called *histamine* that increases blood flow by dilating blood vessels.

- **Adipocytes** (AD-ih-po-sites), or **fat cells.** These appear in small clusters in some fibrous connective tissues. When they dominate an area, the tissue is called *adipose tissue.*

*Fibers*  Three types of protein fibers are found in fibrous connective tissues:

- **Collagenous** (col-LADJ-eh-nus) **fibers.** These fibers, made of collagen, are tough and flexible and resist stretching. Collagen is the body's most abundant protein, constituting about 25% of the total. It is the base of such animal products as gelatin, leather, and glue.[13] In fresh tissue, collagenous fibers have a glistening white appearance, as seen in tendons and some cuts of meat (fig. 5.13); thus, they are often called *white fibers.* In tissue sections, collagen forms coarse, wavy bundles, often dyed pink, blue, or green by the most common histological stains. Tendons, ligaments, and the dermis of the skin are made mainly of collagen. Less visibly, collagen pervades the matrix of cartilage and bone.

- **Reticular**[14] **fibers.** These are thin collagen fibers coated with glycoprotein. They form a spongelike framework for such organs as the spleen and lymph nodes.

- **Elastic fibers.** These are thinner than collagenous fibers, and they branch and rejoin each other along their course. They are made of a protein called **elastin,** whose coiled structure allows it to stretch and recoil like a rubber band. Elastic fibers account for the ability of the skin, lungs, and arteries to spring back after they are stretched. (Elasticity is not the ability to stretch, but the tendency to recoil when tension is released.)

*Ground Substance*  Amid the cells and fibers in some tissue sections, there appears to be a lot of empty space. In life, this space is occupied by the featureless **ground substance.**

---

[10]*fibro* = fiber; *blast* = producing
[11]*macro* = big; *phage* = eater
[12]*leuko* = white; *cyte* = cell

[13]*colla* = glue; *gen* = producing
[14]*ret* = network; *icul* = little

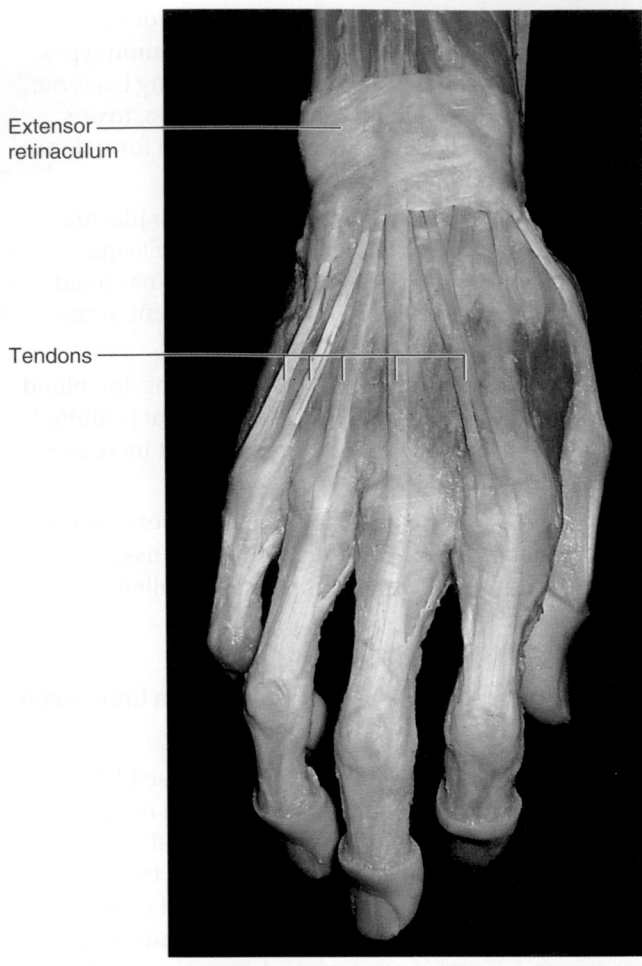

Extensor retinaculum

Tendons

**FIGURE 5.13  Tendons and Ligament.** The tendons of the hand and the ligamentous band (extensor retinaculum) of the wrist are composed of collagen, which has a white glistening appearance.

Ground substance usually has a gelatinous to rubbery consistency resulting from three classes of large molecules: glycosaminoglycans, proteoglycans, and adhesive glycoproteins. It absorbs compressive forces and, like the styrofoam packing in a shipping carton, protects the more delicate cells from mechanical injury.

A **glycosaminoglycan (GAG)** (gly-COSE-ah-MEE-no-GLY-can) is a long polysaccharide composed of unusual disaccharides called *amino sugars* and *uronic acid.* GAGs are negatively charged and thus tend to attract sodium and potassium ions, which in turn cause the GAGs to absorb and retain water. Thus, GAGs play an important role in regulating the water and electrolyte balance of tissues. The most common GAG is **chondroitin** (con-DRO-ih-tin) **sulfate.** It is abundant in blood vessels and bones and gives cartilage its relative stiffness. Other GAGs that you will read of in this book are *heparin* (an anticoagulant) and *hyaluronic* (HY-uh-loo-RON-ic) *acid.* The latter is a gigantic molecule up to 20 μm long, as large as most cells. It is a viscous, slippery substance that forms a lubricant in the joints and constitutes much of the jellylike *vitreous body* of the eyeball.

A **proteoglycan** is another gigantic molecule. It is shaped somewhat like a bottle brush, with a central core of protein and bristlelike outgrowths composed of GAGs. The entire proteoglycan may be attached to hyaluronic acid, thus forming an enormous molecular complex. Proteoglycans form thick colloids similar to those of gravy, gelatin, and glue. This gel slows the spread of pathogenic organisms through the tissues. Some proteoglycans are embedded in the plasma membranes of cells, attached to the cytoskeleton on the inside and to other extracellular molecules in the matrix. They create a strong structural bond between cells and extracellular macromolecules and help to hold tissues together.

**Adhesive glycoproteins** are protein–carbohydrate complexes that bind plasma membrane proteins to extracellular collagen and proteoglycans. They bind the components of a tissue together and mark paths that guide migrating embryonic cells to their destinations in a tissue.

## Types of Fibrous Connective Tissue

Fibrous connective tissue is divided into two broad categories according to the relative abundance of fiber: *loose* and *dense connective tissue.* In **loose connective tissue,** much of the space is occupied by ground substance, which dissolves out of the tissue during histological fixation and leaves empty space in prepared tissue sections. The loose connective tissues we will discuss are *areolar* and *reticular tissue* (table 5.4). In **dense connective tissue,** fiber occupies more space than the cells and ground substance, and appears closely packed in tissue sections. We will discuss two types: *dense regular* and *dense irregular connective tissue* (table 5.5).

**Areolar**[15] (AIR-ee-OH-lur) **tissue** exhibits loosely organized fibers, abundant blood vessels, and a lot of seemingly empty space. It possesses all six of the aforementioned cell types. Its fibers run in random directions and are mostly collagenous, but elastic and reticular fibers are also present. Areolar tissue is highly variable in appearance. In many serous membranes, it looks like figure 5.14, but in the skin and mucous membranes, it is more compact (see fig. 5.8) and sometimes difficult to distinguish from dense irregular connective tissue. Some advice on how to tell them apart is given after the discussion of dense irregular connective tissue.

Areolar tissue is found in tissue sections from almost every part of the body. It surrounds blood vessels and nerves and penetrates with them even into the small spaces of muscle, tendon, and other tissues. Nearly every epithelium rests on a layer of areolar tissue, whose blood vessels provide the epithelium with nutrition, waste removal, and a ready supply of infection-fighting leukocytes in times of need. Because of the abundance of open, fluid-filled space, leukocytes can move about freely in areolar tissue and can easily find and destroy pathogens.

**Reticular tissue** (fig. 5.15) is a mesh of reticular fibers and fibroblasts. It forms the framework (stroma) of such organs as

---

[15]*areola* = little space

| **TABLE 5.4** | **Loose Connective Tissues** |
| --- | --- |

| **Areolar Tissue** | **Reticular Tissue** |
| --- | --- |

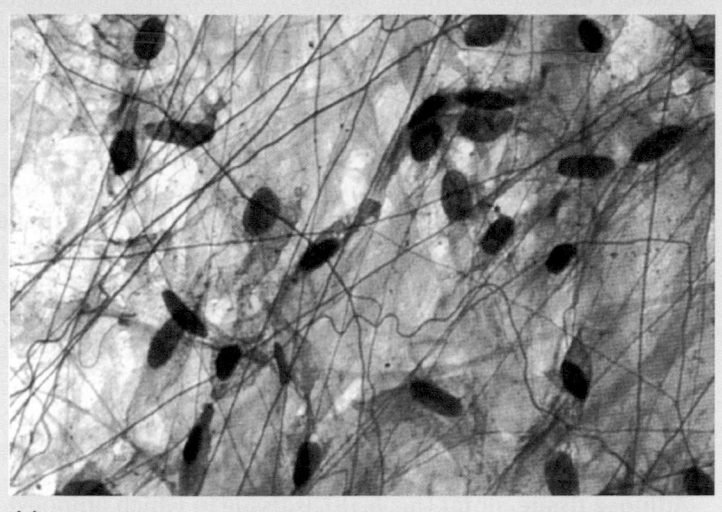

(a)

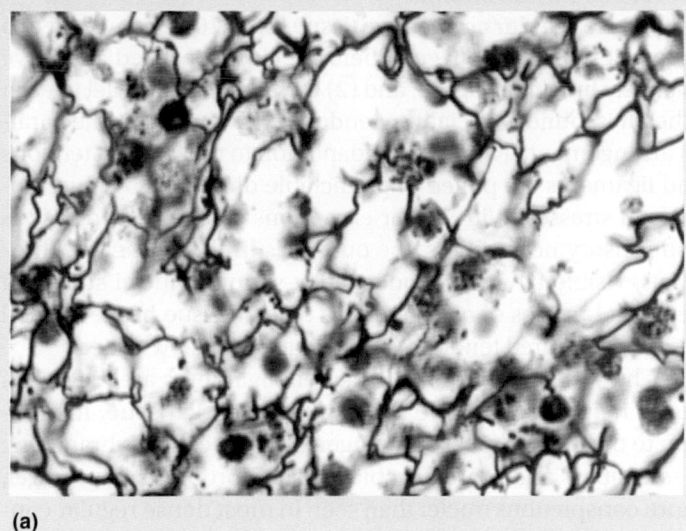

(a)

Ground substance    Elastic fibers    Collagenous fibers    Fibroblasts

Leukocytes    Reticular fibers

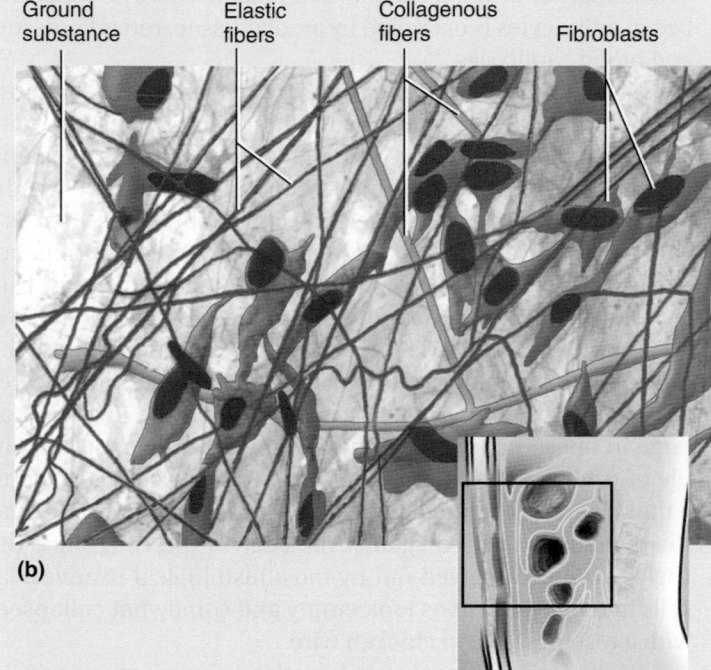

(b)

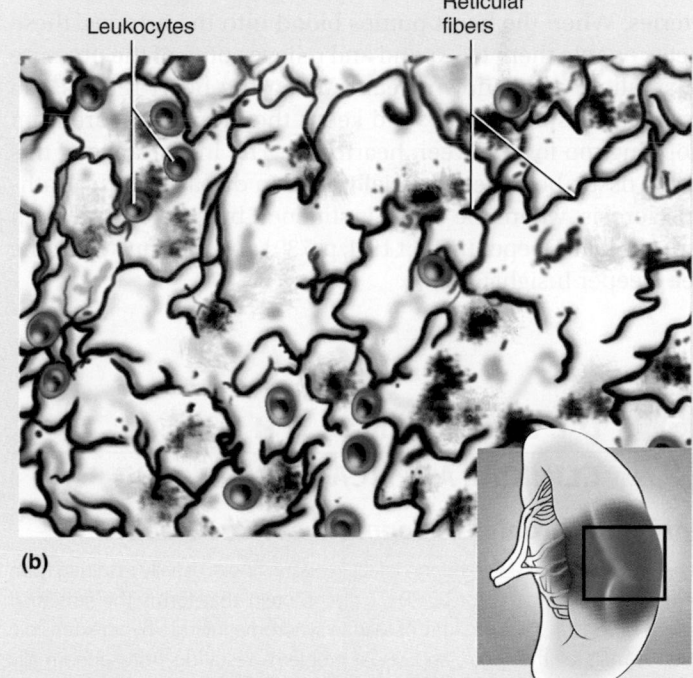

(b)

**FIGURE 5.14 Areolar Tissue in a Spread of the Mesentery (×400).** AP|R

**FIGURE 5.15 Reticular Tissue of the Spleen (×400).** AP|R

**Microscopic appearance:** Loose arrangement of collagenous and elastic fibers; scattered cells of various types; abundant ground substance; numerous blood vessels

**Representative locations:** Underlying nearly all epithelia; surrounding blood vessels, nerves, esophagus, and trachea; fascia between muscles; mesenteries; visceral layers of pericardium and pleura

**Functions:** Loosely binds epithelia to deeper tissues; allows passage of nerves and blood vessels through other tissues; provides an arena for immune defense; blood vessels provide nutrients and waste removal for overlying epithelia

**Microscopic appearance:** Loose network of reticular fibers and cells, infiltrated with numerous leukocytes, especially lymphocytes

**Representative locations:** Lymph nodes, spleen, thymus, bone marrow

**Function:** Forms supportive stroma (framework) for lymphatic organs

the lymph nodes, spleen, thymus, and bone marrow. The space amid the fibers is filled with blood cells. If you imagine a kitchen sponge soaked with blood, the sponge fibers are analogous to the reticular tissue stroma.

**Dense regular connective tissue** (fig. 5.16) is named for two properties: (1) the collagen fibers are closely packed and leave relatively little open space, and (2) the fibers are parallel to each other. It is found especially in tendons and ligaments. The parallel arrangement of fibers is an adaptation to the fact that tendons and ligaments are pulled in predictable directions by musculoskeletal stresses. With minor exceptions such as blood vessels and sensory nerve fibers, the only cells in this tissue are fibroblasts, visible by their slender, violet-staining nuclei squeezed between bundles of collagen. This type of tissue has few blood vessels, so injured tendons and ligaments are slow to heal.

The vocal cords and some spinal ligaments are made of a dense regular connective tissue called **elastic tissue.** In addition to the densely packed collagen fibers, it exhibits branching elastic fibers and more fibroblasts. The fibroblasts have larger, more conspicuous nuclei than seen in most dense regular connective tissue.

Elastic tissue also forms wavy sheets in large and medium arteries. When the heart pumps blood into the arteries, these sheets enable them to expand and relieve some of the pressure on smaller vessels downstream. When the heart relaxes, the arterial wall springs back and keeps the blood pressure from dropping too low between heartbeats. The importance of this elastic tissue becomes especially clear in diseases such as atherosclerosis, where the tissue is stiffened by lipid and calcium deposits (see Deeper Insight 19.4, p. 739) and Marfan syndrome (see Deeper Insight 5.1).

## DEEPER INSIGHT 5.1
### CLINICAL APPLICATION

#### Marfan Syndrome—A Connective Tissue Disease

*Marfan*[16] syndrome is a hereditary defect in elastin fibers, usually resulting from a mutation in the gene for *fibrillin,* a glycoprotein that forms the structural scaffold for elastin. Clinical signs of Marfan syndrome include hyperextensible joints, hernias of the groin, and visual problems resulting from abnormally elongated eyes and deformed lenses. People with Marfan syndrome typically show unusually tall stature, long limbs, spidery fingers, abnormal spinal curvature, and a protruding "pigeon breast." More serious problems are weakened heart valves and arterial walls. The aorta, where blood pressure is highest, is sometimes enormously dilated close to the heart and may rupture. Marfan syndrome is present in about 1 out of 20,000 live births, and most victims die by their mid-30s. Abraham Lincoln's tall, gangly physique and spindly fingers led some authorities to suspect that he had Marfan syndrome, but the evidence is inconclusive. Some star athletes have died at a young age of Marfan syndrome, including Olympic volleyball champion Flo Hyman (1954–86), who died at the age of 31 of a ruptured aorta during a game in Japan.

[16]Antoine Bernard-Jean Marfan (1858–1942), French physician

**Dense irregular connective tissue** (fig. 5.17) also has thick bundles of collagen and relatively little room for cells and ground substance, but the collagen bundles run in seemingly random directions. This arrangement enables the tissue to resist unpredictable stresses. This tissue constitutes most of the dermis, where it binds the skin to the underlying muscle and connective tissue. It forms a protective capsule around organs such as the kidneys, testes, and spleen and a tough fibrous sheath around the bones, nerves, and most cartilages.

It is sometimes difficult to judge whether a tissue is areolar or dense irregular. In the dermis, for example, these tissues occur side by side, and the transition from one to the other is not at all obvious (see fig. 5.8). A relatively large amount of clear space suggests areolar tissue, and thicker bundles of collagen and relatively little clear space suggest dense irregular tissue.

## Adipose Tissue

**Adipose tissue,** or **fat** (fig. 5.18), is tissue in which adipocytes are the dominant cell type (table 5.6). Adipocytes may also occur singly or in small clusters in areolar tissue. The space between adipocytes is occupied by areolar tissue, reticular tissue, and blood capillaries.

Fat is the body's primary energy reservoir. The quantity of stored triglyceride and the number of adipocytes are quite stable in a person, but this doesn't mean stored fat is stagnant. New triglycerides are constantly synthesized and stored as others are hydrolyzed and released into circulation. Thus, there is a constant turnover of stored triglyceride, with an equilibrium between synthesis and hydrolysis, energy storage and energy use.

There are two kinds of fat in humans—white (or yellow) fat and brown fat. **White fat** is the more abundant and is the only significant adipose tissue of the adult body. Its adipocytes are usually 70 to 120 $\mu$m in diameter, but may be five times as large in obese people. They have a single large, central globule of triglyceride. Their cytoplasm is otherwise restricted to a thin layer immediately beneath the plasma membrane, and the nucleus is pushed against the edge of the cell. Since the triglyceride is dissolved out by most histological fixatives, fat cells in most specimens look empty and somewhat collapsed, with a resemblance to chicken wire.

White fat provides thermal insulation, anchors and cushions such organs as the eyeballs and kidneys, and contributes to body contours such as the female breasts and hips. On average, women have more fat relative to body weight than men do. It helps to meet the caloric needs of pregnancy and nursing an infant, and having too little fat can reduce female fertility.

**Brown fat** is found mainly in fetuses, infants, and children; it accounts for up to 6% of an infant's weight, and is concentrated especially in fat pads in the shoulders, upper back, and around the kidneys. It stores lipid in the form of multiple globules rather than one large one. It gets its color from an unusual abundance of blood vessels and certain enzymes in its mitochondria. Brown fat is a heat-generating tissue. It has numerous mitochondria, but their oxidative pathway is not linked to

| **TABLE 5.5** | **Dense Connective Tissues** |
| --- | --- |

| **Dense Regular Connective Tissue** | **Dense Irregular Connective Tissue** |
| --- | --- |

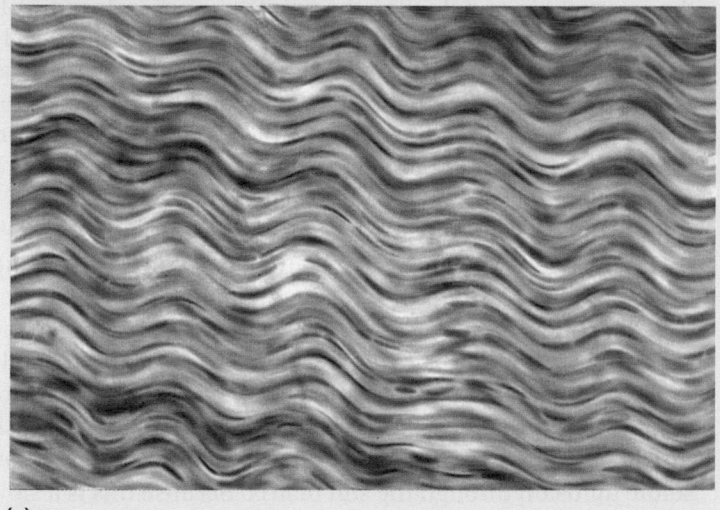

(a)

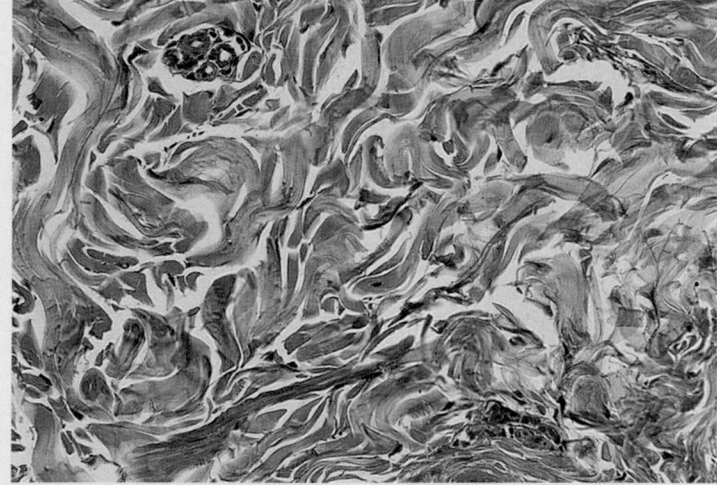

(a)

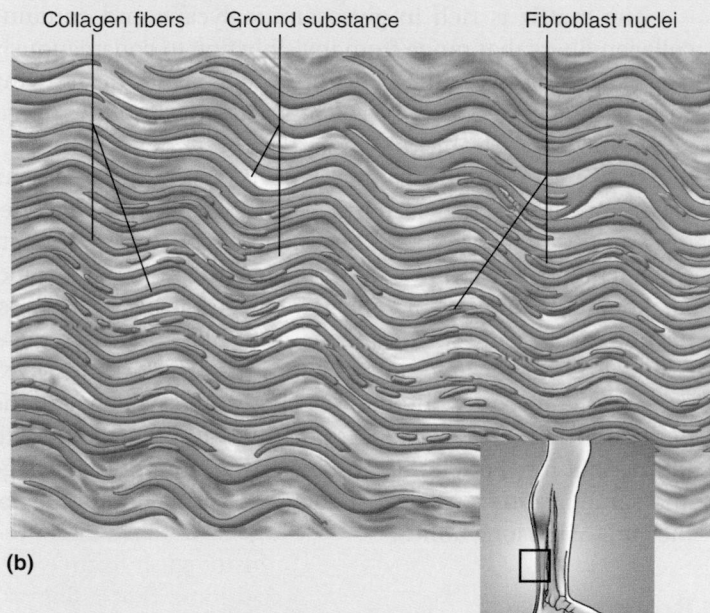

Collagen fibers    Ground substance    Fibroblast nuclei

(b)

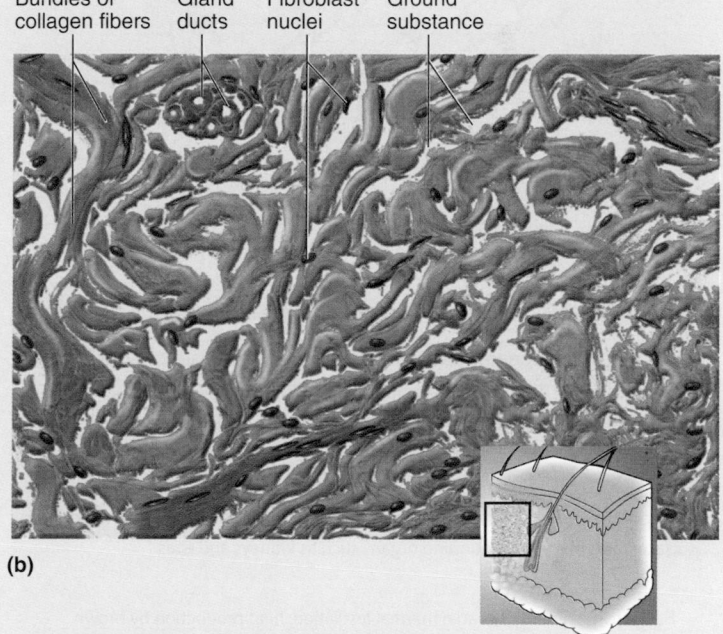

Bundles of collagen fibers    Gland ducts    Fibroblast nuclei    Ground substance

(b)

**FIGURE 5.16** Dense Regular Connective Tissue of a Tendon (×400). AP|R

Microscopic appearance: Densely packed, parallel, often wavy collagen fibers; slender fibroblast nuclei compressed between collagen bundles; scanty open space (ground substance); scarcity of blood vessels

Representative locations: Tendons and ligaments

Functions: Ligaments tightly bind bones together and resist stress; tendons attach muscle to bone and transfer muscular tension to bones

**FIGURE 5.17** Dense Irregular Connective Tissue in the Dermis of the Skin (×400). AP|R

Microscopic appearance: Densely packed collagen fibers running in random directions; scanty open space (ground substance); few visible cells; scarcity of blood vessels

Representative locations: Deeper portion of dermis of skin; capsules around viscera such as liver, kidney, spleen; fibrous sheaths around cartilages and bones

Functions: Withstands stresses applied in unpredictable directions; imparts durability to tissues

| TABLE 5.6 | Adipose Tissue |
|---|---|

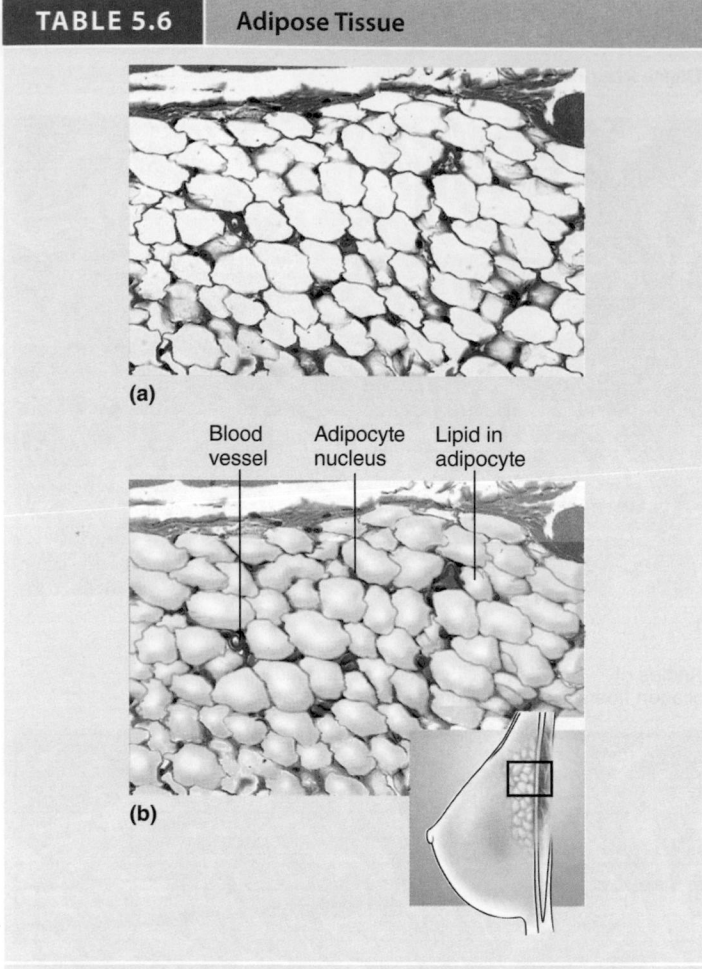

**FIGURE 5.18** Adipose Tissue of the Breast (x100). AP|R

Microscopic appearance: Dominated by adipocytes—large, empty-looking cells with thin margins; tissue sections often very pale because of scarcity of stained cytoplasm; adipocytes shrunken; nucleus pressed against plasma membrane; blood vessels present

Representative locations: Subcutaneous fat beneath skin; breast; heart surface; mesenteries; surrounding organs such as kidneys and eyes

Functions: Energy storage; thermal insulation; heat production by brown fat; protective cushion for some organs; filling space, shaping body

ATP synthesis. Therefore, when these cells oxidize fats, they release all of the energy as heat. Hibernating animals accumulate brown fat in preparation for winter.

▶▶▶ **APPLY WHAT YOU KNOW**

*Why would infants and children have more need for brown fat than adults do? (Hint: Smaller bodies have a higher ratio of surface area to volume than larger bodies do.)*

## Cartilage

**Cartilage** is a relatively stiff connective tissue with a flexible rubbery matrix; you can feel its texture by folding and releasing the external ear or palpating the tip of your nose or your "Adam's apple" (the *thyroid cartilage* of the larynx). It is also easily seen in many grocery items—it is the milky-colored gristle at the ends of pork ribs and on chicken leg and breast bones, for example. Among other functions, cartilages shape and support the nose and ears and partially enclose the larynx (voice box), trachea (windpipe), and thoracic cavity.

Cartilage is produced by cells called **chondroblasts**[17] (CON-dro-blasts), which secrete the matrix and surround themselves with it until they become trapped in little cavities called **lacunae**[18] (la-CUE-nee). Once enclosed in lacunae, the cells are called **chondrocytes** (CON-dro-sites). Cartilage only rarely exhibits blood vessels, and even when it does, they are just passing through without giving off capillaries to nourish the tissue. Therefore, nutrition and waste removal depend on solute diffusion through the stiff matrix. Because this is a slow process, chondrocytes have low rates of metabolism and cell division, and injured cartilage heals slowly.

The matrix is rich in glycosaminoglycans and contains collagen fibers that range from invisibly fine to conspicuously coarse. Differences in the fibers provide a basis for classifying cartilage into three types: *hyaline cartilage, elastic cartilage,* and *fibrocartilage* (table 5.7, figs. 5.19 to 5.21).

**Hyaline**[19] (HY-uh-lin) **cartilage** is named for its clear, glassy appearance, which stems from the usually invisible fineness of its collagen fibers. **Elastic cartilage** is named for its conspicuous elastic fibers, and **fibrocartilage** for its coarse, readily visible bundles of collagen. Elastic cartilage and most hyaline cartilage are surrounded by a sheath of dense irregular connective tissue called the **perichondrium**[20] (PAIR-ih-CON-dree-um). A reserve population of chondroblasts between the perichondrium and cartilage contributes to cartilage growth throughout life. Perichondrium is lacking from fibrocartilage and some hyaline cartilage, such as the cartilaginous caps at the ends of the long bones.

## Bone

**Bone,** or **osseous tissue** (table 5.8, fig. 5.22) is a hard, calcified connective tissue that composes the skeleton. The term *bone* has two meanings in anatomy—an entire organ such as the femur and mandible, or just the osseous tissue. Bones are composed of not only **osseous tissue,** but also cartilage, bone marrow, dense irregular connective tissue, and other tissue types.

---

[17]*chondro* = cartilage, gristle; *blast* = forming
[18]*lacuna* = lake, cavity
[19]*hyal* = glass
[20]*peri* = around; *chondri* = cartilage

| **TABLE 5.7** | **Cartilage** |
|---|---|

| **Hyaline Cartilage** | **Elastic Cartilage** | **Fibrocartilage** |
|---|---|---|

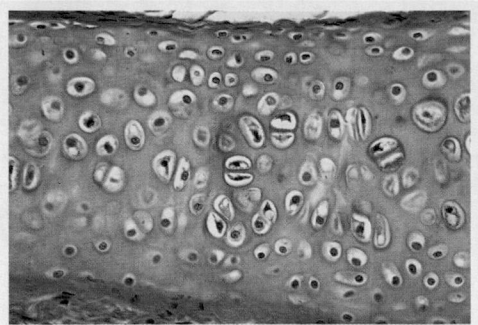

(a)

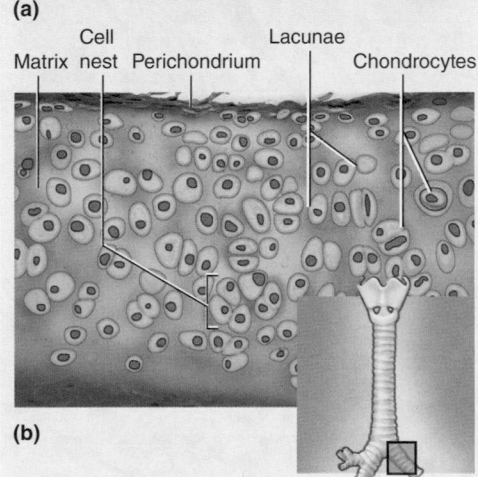

Matrix — Cell nest — Perichondrium — Lacunae — Chondrocytes

(b)

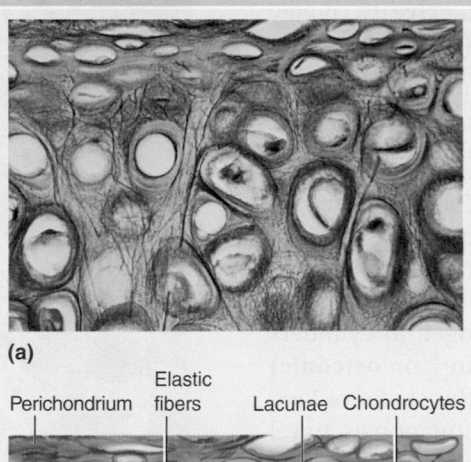

(a)

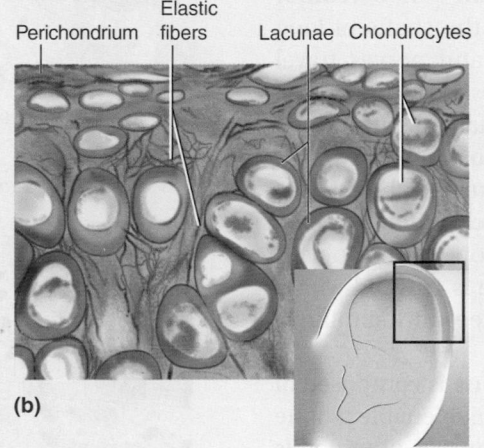

Perichondrium — Elastic fibers — Lacunae — Chondrocytes

(b)

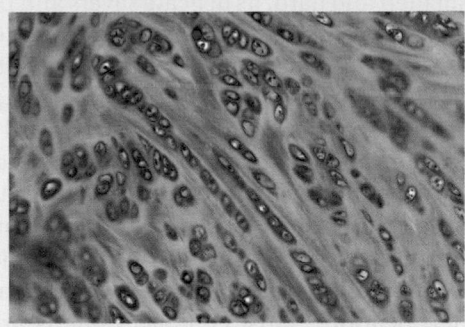

(a)

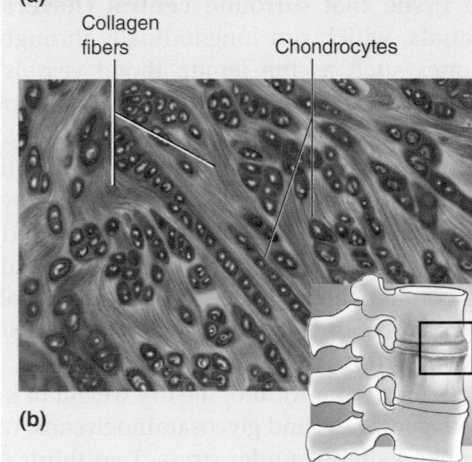

Collagen fibers — Chondrocytes

(b)

**FIGURE 5.19 Hyaline Cartilage of a Bronchus (×400).** AP|R

Microscopic appearance: Clear, glassy matrix, often stained light blue or pink in tissue sections; fine, dispersed collagen fibers, not usually visible; chondrocytes enclosed in lacunae, often in small clusters of three or four cells *(cell nests);* usually covered by perichondrium

Representative locations: A thin *articular cartilage,* lacking perichondrium, over the ends of bones at movable joints; supportive rings and plates around trachea and bronchi; a boxlike enclosure around the larynx; much of the fetal skeleton; and a *costal cartilage* attaches the end of a rib to the breastbone

Functions: Eases joint movements; holds airway open during respiration; moves vocal cords during speech; a precursor of bone in the fetal skeleton and the growth zones of long bones of children

**FIGURE 5.20 Elastic Cartilage of the External Ear (×1,000).** AP|R

Microscopic appearance: Elastic fibers form weblike mesh amid lacunae; always covered by perichondrium

Representative locations: External ear; epiglottis

Functions: Provides flexible, elastic support

**FIGURE 5.21 Fibrocartilage of an Intervertebral Disc (×400).** AP|R

Microscopic appearance: Parallel collagen fibers similar to those of tendon; rows of chondrocytes in lacunae between collagen fibers; never has a perichondrium

Representative locations: Pubic symphysis (anterior joint between two halves of pelvic girdle); intervertebral discs, which separate bones of vertebral column; menisci, or pads of shock-absorbing cartilage, in knee joint; at points where tendons insert on bones near articular hyaline cartilage

Functions: Resists compression and absorbs shock in some joints; often a transitional tissue between dense connective tissue and hyaline cartilage (for example, at some tendon–bone junctions)

There are two forms of osseous tissue: (1) **Spongy bone** fills the heads of the long bones and forms the middle layer of flat bones such as the sternum and cranial bones. Although it is calcified and hard, its delicate slivers and plates give it a spongy appearance (fig. 7.4a, p. 208). (2) **Compact (dense) bone** is a denser calcified tissue with no spaces visible to the naked eye. It forms the external surfaces of all bones, so spongy bone, when present, is always covered by a shell of compact bone.

Further differences between compact and spongy bone are described in chapter 7. Here, we examine only compact bone. Most specimens you study will probably be chips of dead, dried bone ground to microscopic thinness. In such preparations, the cells are absent but spaces reveal their former locations. Most compact bone is arranged in cylinders of tissue that surround **central (haversian**[21] **or osteonic) canals,** which run longitudinally through the shafts of long bones such as the femur. Blood vessels and nerves travel through these canals. The bone matrix is deposited in **concentric lamellae**—onionlike layers around each canal. A central canal and its surrounding lamellae are called an **osteon.** Tiny lacunae between the lamellae are occupied by mature bone cells, or **osteocytes.**[22] Delicate channels called **canaliculi** radiate from each lacuna to its neighbors and allow the osteocytes to contact each other. The bone as a whole is covered with a tough fibrous **periosteum** (PAIR-ee-OSS-tee-um) similar to the perichondrium of cartilage.

About one-third of the dry weight of bone is composed of collagen fibers and glycosaminoglycans, which enable a bone to bend slightly under stress. Two-thirds consists of minerals (mainly calcium and phosphate salts) that enable bones to withstand compression by the weight of the body.

## Blood

**Blood** (table 5.9, fig. 5.23) is a fluid connective tissue that travels through tubular blood vessels. Its primary function is to transport cells and dissolved matter from place to place. It may seem odd that a tissue as fluid as blood and another as rock hard as bone are both considered connective tissues, but they have more in common than first meets the eye. Like other connective tissues, blood is composed of more ground substance than cells. Its ground substance is the **blood plasma** and its cellular components are collectively called the **formed elements.** Unlike other connective tissues, blood does not exhibit fibers except when it clots. Another factor placing blood in the connective tissue category is that it is produced by the connective tissues of the bone marrow and lymphatic organs.

The formed elements are of three kinds—erythrocytes, leukocytes, and platelets. **Erythrocytes**[23] (eh-RITH-ro-sites), or **red blood cells (RBCs),** are the most abundant. In stained blood films, they look like pink discs with thin, pale centers and no nuclei. Erythrocytes transport oxygen and carbon dioxide. **Leukocytes,** or **white blood cells (WBCs),** serve various roles

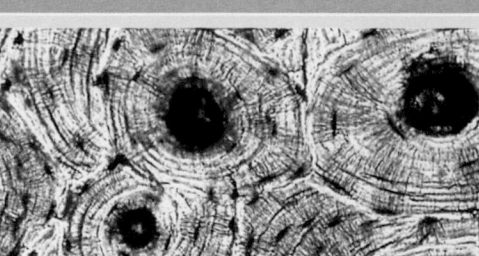

| TABLE 5.8 | Bone |
| --- | --- |

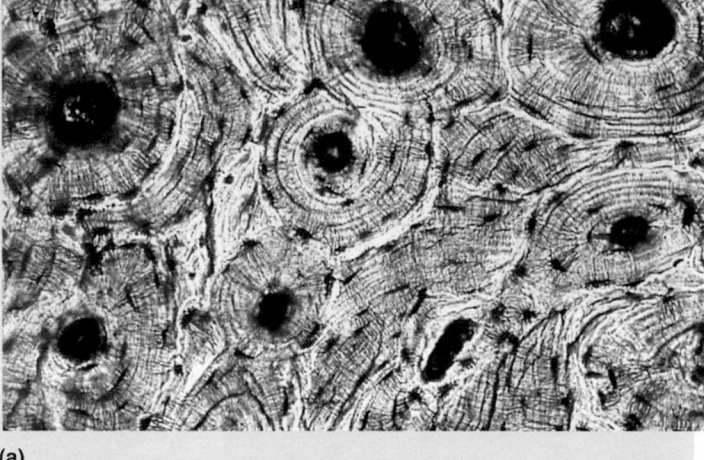

(a)

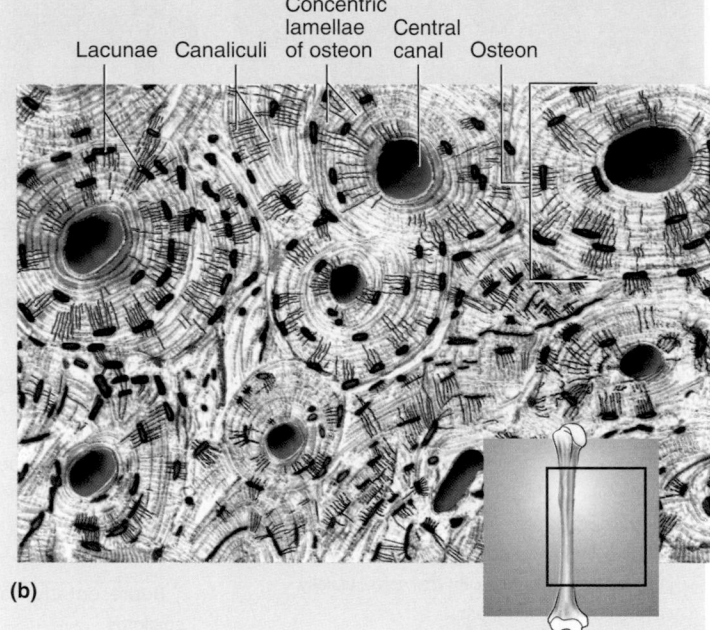

Lacunae   Canaliculi   Concentric lamellae of osteon   Central canal   Osteon

(b)

**FIGURE 5.22**  **Compact Bone (×100).**  **AP|R**

Microscopic appearance (compact bone):  Calcified matrix arranged in concentric lamellae around central canals; osteocytes in lacunae between adjacent lamellae; lacunae interconnected by delicate canaliculi

Representative location:  Skeleton

Functions:  Physical support of body; leverage for muscle action; protective enclosure of viscera; reservoir of calcium and phosphorus

---

[21]Clopton Havers (1650–1702), English anatomist
[22]*osteo* = bone; *cyte* = cell
[23]*erythro* = red; *cyte* = cell

| TABLE 5.9 | Blood |
|---|---|

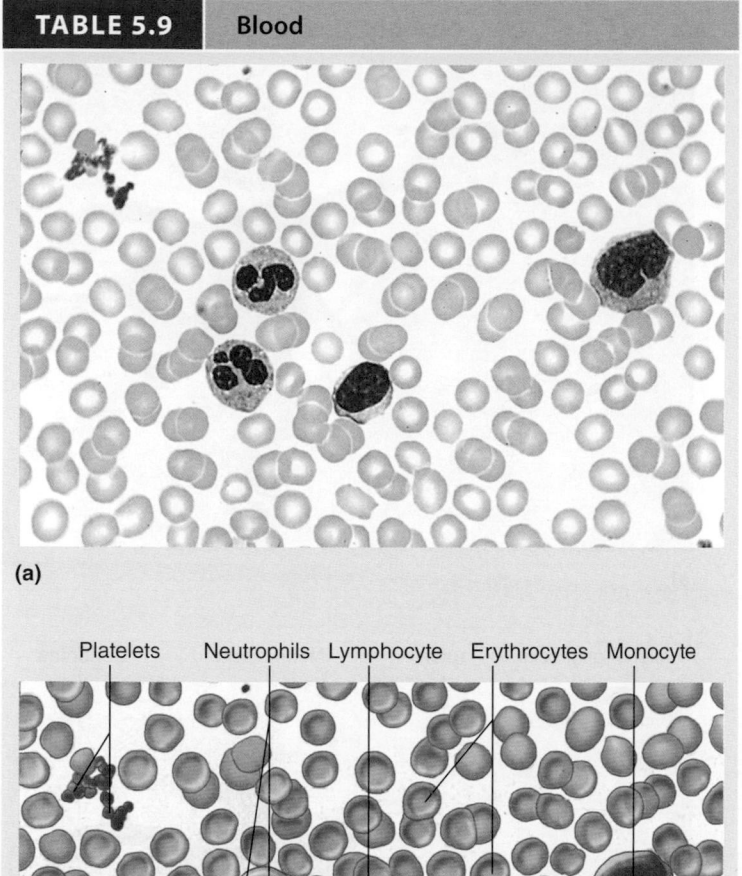

(a)

Platelets   Neutrophils   Lymphocyte   Erythrocytes   Monocyte

(b)

**FIGURE 5.23 Blood Smear (x1,000).** AP|R

Microscopic appearance: Erythrocytes appear as pale pink discs with light centers and no nuclei; leukocytes are slightly larger, are much fewer, and have variously shaped nuclei, usually stained violet; platelets are cell fragments with no nuclei, about one-quarter the diameter of erythrocytes

Representative locations: Contained in heart and blood vessels

Functions: Transports gases, nutrients, wastes, chemical signals, and heat throughout body; provides defensive leukocytes; contains clotting agents to minimize bleeding; platelets secrete growth factors that promote tissue maintenance and repair

in defense against infection and other diseases. They travel from one organ to another in the bloodstream and lymph but spend most of their lives in the connective tissues. Leukocytes are somewhat larger than erythrocytes and have conspicuous nuclei that usually appear violet in stained preparations. There are five kinds, distinguished partly by variations in nuclear shape: *neutrophils, eosinophils, basophils, lymphocytes,* and *monocytes.* Their individual characteristics are considered in detail in chapter 18. **Platelets** are small cell fragments scattered amid the blood cells. They are involved in clotting and other mechanisms for minimizing blood loss, and in secreting growth factors that promote blood vessel growth and maintenance.

### BEFORE YOU GO ON

Answer the following questions to test your understanding of the preceding section:

9. What features do most or all connective tissues have in common to set this class apart from nervous, muscular, and epithelial tissue?

10. List the cell and fiber types found in fibrous connective tissues and state their functional differences.

11. What substances account for the gelatinous consistency of connective tissue ground substance?

12. What is areolar tissue? How can it be distinguished from any other kind of connective tissue?

13. Discuss the difference between dense regular and dense irregular connective tissue as an example of the relationship between form and function.

14. Describe some similarities, differences, and functional relationships between hyaline cartilage and bone.

15. What are the three basic kinds of formed elements in blood, and what are their respective functions?

## 5.4  Nervous and Muscular Tissues—Excitable Tissues

### Expected Learning Outcomes

When you have completed this section, you should be able to

a. explain what distinguishes excitable tissues from other tissues;

b. name the cell types that compose nervous tissue;

c. identify the major parts of a nerve cell;

d. visually recognize nervous tissue from specimens or photographs;

e. name the three kinds of muscular tissue and describe the differences between them; and

f. visually identify any type of muscular tissue from specimens or photographs.

Excitability is a characteristic of all living cells, but it is developed to its highest degree in nervous and muscular tissues, which are therefore described as **excitable tissues.** The basis for their excitation is an electrical charge difference (voltage) called the **membrane potential,** which occurs across the plasma membranes of all cells. Nervous and muscular tissues respond quickly to outside stimuli by means of changes in membrane potential. In nerve cells, these changes result in the rapid transmission of signals to other cells. In muscle cells, they result in contraction, or shortening of the cell.

## Nervous Tissue

**Nervous tissue** (table 5.10, fig. 5.24) is specialized for communication by means of electrical and chemical signals. It consists of **neurons** (NOOR-ons), or nerve cells, and a much greater number of **neuroglia** (noo-ROG-lee-uh), or **glial** (GLEE-ul) **cells,** which protect and assist the neurons. Neurons detect stimuli, respond quickly, and transmit coded information rapidly to other cells. Each neuron has a prominent **neurosoma,** or cell body, that houses the nucleus and most other organelles. This is the cell's center of genetic control and protein synthesis. Neurosomas are usually round, ovoid, or stellate in shape. Extending from the neurosoma, there are usually multiple short, branched processes called **dendrites,**[24] which receive signals from other cells and conduct messages to the neurosoma; and a single, much longer **axon,** or **nerve fiber,** which sends outgoing signals to other cells. Some axons are more than a meter long and extend from the brainstem to the foot.

Glial cells constitute most of the volume of the nervous tissue. They are usually much smaller than neurons. There are six types of glial cells, described in chapter 12, which provide a variety of supportive, protective, and "housekeeping" functions for the nervous system. Although they communicate with neurons and each other, they do not transmit long-distance signals.

Nervous tissue is found in the brain, spinal cord, nerves, and ganglia, which are knotlike swellings in nerves. Local variations in the structure of nervous tissue are described in chapters 12 to 16.

## Muscular Tissue

**Muscular tissue** is specialized to contract when stimulated, and thus to exert a physical force on other tissues, organs, or fluids—for example, a skeletal muscle pulls on a bone, the heart contracts and expels blood, and the bladder contracts and expels urine. Not only do movements of the body and its limbs depend on muscle, but so do such processes as digestion, waste elimination, breathing, speech, and blood circulation. The muscles are also an important source of body heat.

There are three types of muscular tissue—*skeletal, cardiac,* and *smooth*—which differ in appearance, physiology, and

---

[24]*dendr* = tree; *ite* = little

| TABLE 5.10 | Nervous Tissue |
| --- | --- |

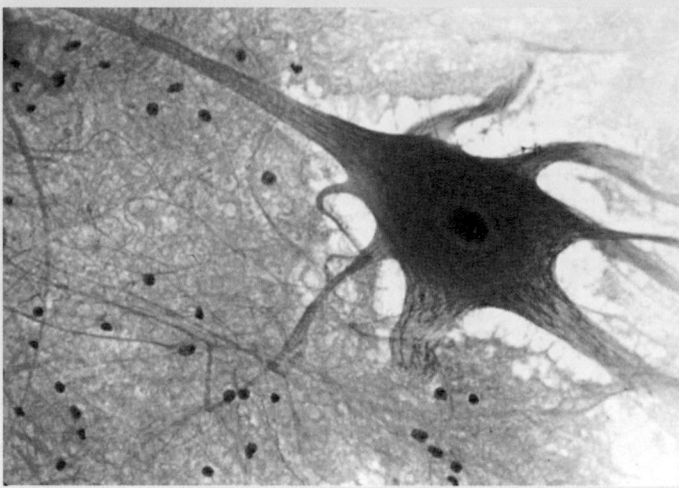

(a)

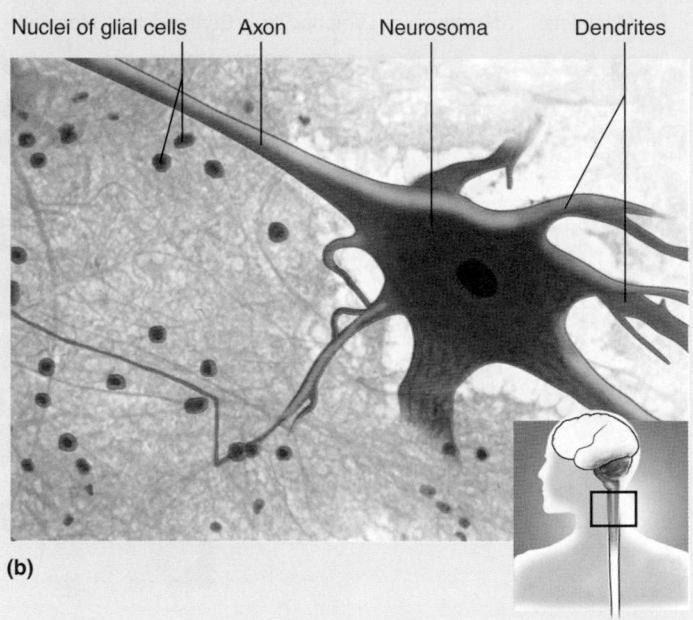

Nuclei of glial cells   Axon   Neurosoma   Dendrites

(b)

**FIGURE 5.24  A Neuron and Glial Cells of a Spinal Cord Smear (×400).** AP|R

Microscopic appearance:  Most sections show a few large neurons, usually with rounded or stellate cell bodies (neurosomas) and fibrous processes (axon and dendrites) extending from the neurosomas; neurons are surrounded by a greater number of much smaller glial cells, which lack dendrites and axons.

Representative locations:  Brain, spinal cord, nerves, ganglia

Function:  Internal communication

function (table 5.11, figs. 5.25 to 5.27). **Skeletal muscle** consists of long threadlike cells called **muscle fibers.** Most skeletal muscle is attached to bones, but there are exceptions in the tongue, upper esophagus, some facial muscles, and some **sphincter**[25] (SFINK-tur) muscles (muscular rings or cuffs that open and close body passages). Each cell contains multiple nuclei adjacent to the plasma membrane. Skeletal muscle is described as striated and voluntary. The first term refers to alternating light and dark bands, or **striations** (stry-AY-shuns), created by the overlapping pattern of cytoplasmic protein filaments that cause muscle contraction. The second term, *voluntary,* refers to the fact that we usually have conscious control over skeletal muscle.

**Cardiac muscle** is limited to the heart. It too is striated, but it differs from skeletal muscle in its other features. Its cells are much shorter, so they are commonly called **myocytes**[26] or **cardiocytes** rather than fibers. The myocytes are branched or notched at the ends. They contain only one nucleus, which is located near the center and often surrounded by a light-staining region of glycogen. Cardiac myocytes are joined end to end by junctions called **intercalated**[27] (in-TUR-kuh-LAY-ted) **discs.** Electrical connections at these junctions enable a wave of excitation to travel rapidly from cell to cell, and mechanical connections keep the myocytes from pulling apart when the heart contracts. The electrical junctions allow a wave of electrical excitation to travel rapidly from cell to cell so that all the myocytes of a heart chamber are stimulated and contract almost simultaneously. Intercalated discs appear as dark transverse lines separating each myocyte from the next. They may be only faintly visible, however, unless the tissue has been specially stained for them. Cardiac muscle is considered *involuntary* because it is not usually under conscious control; it contracts even if all nerve connections to it are severed.

**Smooth muscle** lacks striations and is involuntary. Smooth muscle cells, also called myocytes, are fusiform and relatively short. They have only one, centrally placed nucleus. Small amounts of smooth muscle are found in the iris of the eye and in the skin, but most of it, called **visceral muscle,** forms layers in the walls of the digestive, respiratory, and urinary tracts; blood vessels; the uterus; and other viscera. In locations such as the esophagus and small intestine, smooth muscle forms adjacent layers, with the cells of one layer encircling the organ and the cells of the other layer running longitudinally. When the circular smooth muscle contracts, it may propel contents such as food through the organ. When the longitudinal layer contracts, it makes the organ shorter and thicker. By regulating the diameter of blood vessels, smooth muscle is very important in controlling blood pressure and flow. Both smooth and skeletal muscle form sphincters that control the emptying of the bladder and rectum.

▶▶▶ **APPLY WHAT YOU KNOW**

*How does the meaning of the word* fiber *differ in the following uses:* muscle fiber, nerve fiber, *and* connective tissue fiber?

---

[25]*sphinc* = squeeze, bind tightly
[26]*myo* = muscle; *cyte* = cell
[27]*inter* = between; *calated* = inserted

**BEFORE YOU GO ON**

Answer the following questions to test your understanding of the preceding section:

16. What do nervous and muscular tissue have in common? What is the primary function of each?

17. What kinds of cells compose nervous tissue, and how can they be distinguished from each other?

18. Name the three kinds of muscular tissue, describe how to distinguish them from each other in microscopic appearance, and state a location and function for each.

## 5.5 Cell Junctions, Glands, and Membranes

### Expected Learning Outcomes

When you have completed this section, you should be able to

a. describe the junctions that hold cells and tissues together;

b. describe or define different types of glands;

c. describe the typical anatomy of a gland;

d. name and compare different modes of glandular secretion;

e. describe the way tissues are organized to form the body's membranes; and

f. name and describe the major types of membranes in the body.

### Cell Junctions

Most cells, with the exception of blood cells, macrophages, and metastatic cancer cells, must be anchored to each other and to the matrix if they are to grow and divide normally. The connections between one cell and another are called **cell junctions.** They enable the cells to resist stress, communicate with each other, and control the movement of substances through the gaps between cells. Without them, cardiac muscle cells would pull apart when they contracted, and every swallow of food would scrape away the lining of your esophagus. The main types of cell junctions are shown in figure 5.28.

### Tight Junctions

A **tight junction** completely encircles an epithelial cell near its apical surface and joins it tightly to the neighboring cells, somewhat like the plastic harness on a six-pack of soda cans. At a tight junction, the plasma membranes of two adjacent cells come very close together and are linked by transmembrane cell-adhesion proteins. These proteins seal off the intercellular space and make it difficult or impossible for substances to pass between cells.

In the stomach and intestines, tight junctions prevent digestive juices from seeping between epithelial cells and digesting the underlying connective tissue. They also help to prevent

| TABLE 5.11 | Muscular Tissue | |
|---|---|---|
| **Skeletal Muscle** | **Cardiac Muscle** | **Smooth Muscle** |

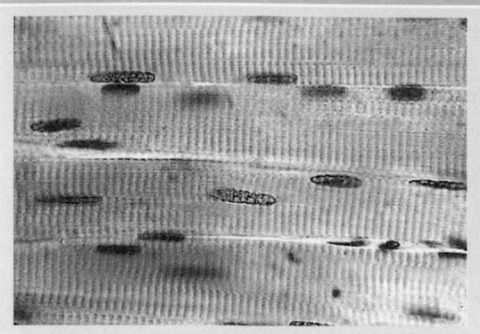

(a)

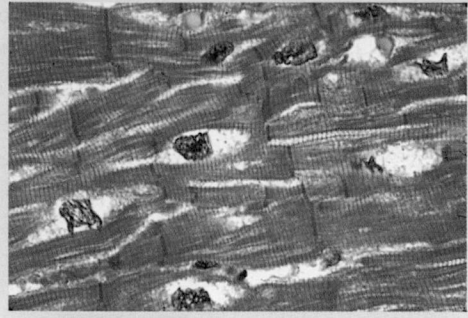

(a)

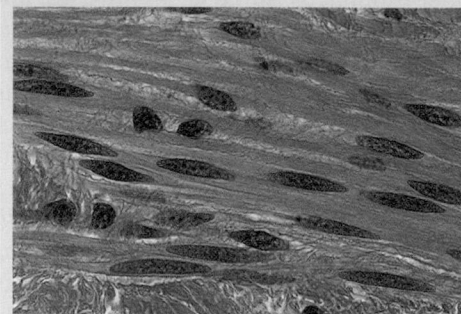

(a)

Nuclei    Striations  Muscle fiber

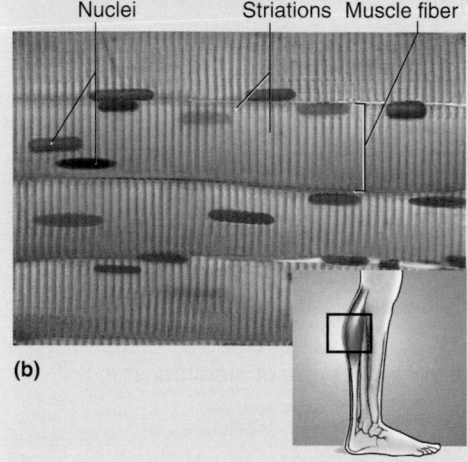

(b)

Intercalated discs   Striations   Glycogen

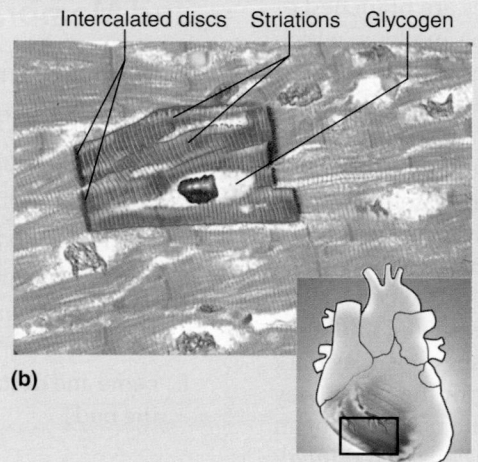

(b)

Nuclei          Muscle cells

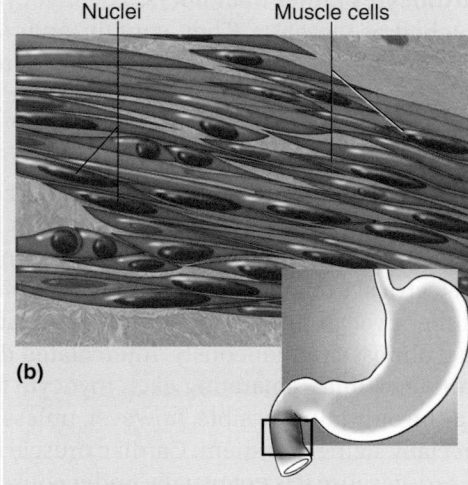

(b)

**FIGURE 5.25  Skeletal Muscle (×400).**
AP|R

Microscopic appearance:  Long, threadlike, unbranched cells (fibers), relatively parallel in longitudinal tissue sections; striations; multiple nuclei per cell, near plasma membrane

Representative locations:  Skeletal muscles, mostly attached to bones but also in the tongue, esophagus, and encircling the lips, eyelids, urethra, and anus

Functions:  Body movements, facial expression, posture, breathing, speech, swallowing, control of urination and defecation, and assistance in childbirth; under voluntary control

**FIGURE 5.26  Cardiac Muscle (×400).**
AP|R

Microscopic appearance:  Short cells (myocytes) with notched or slightly branched ends; less parallel appearance in tissue sections; striations; intercalated discs; one nucleus per cell, centrally located and often surrounded by a light zone

Representative location:  Heart

Functions:  Pumping of blood; under involuntary control

**FIGURE 5.27  Smooth Muscle of the Intestinal Wall (x1,000).**  AP|R

Microscopic appearance:  Short fusiform cells overlapping each other; nonstriated; one nucleus per cell, centrally located

Representative locations:  Usually found as sheets of tissue in walls of blood vessels and viscera such as the digestive tract; also in iris and associated with hair follicles; involuntary sphincters of urethra and anus

Functions:  Swallowing; contractions of stomach and intestines; expulsion of feces and urine; labor contractions; control of blood pressure and flow; control of respiratory airflow; control of pupillary diameter; erection of hairs; under involuntary control

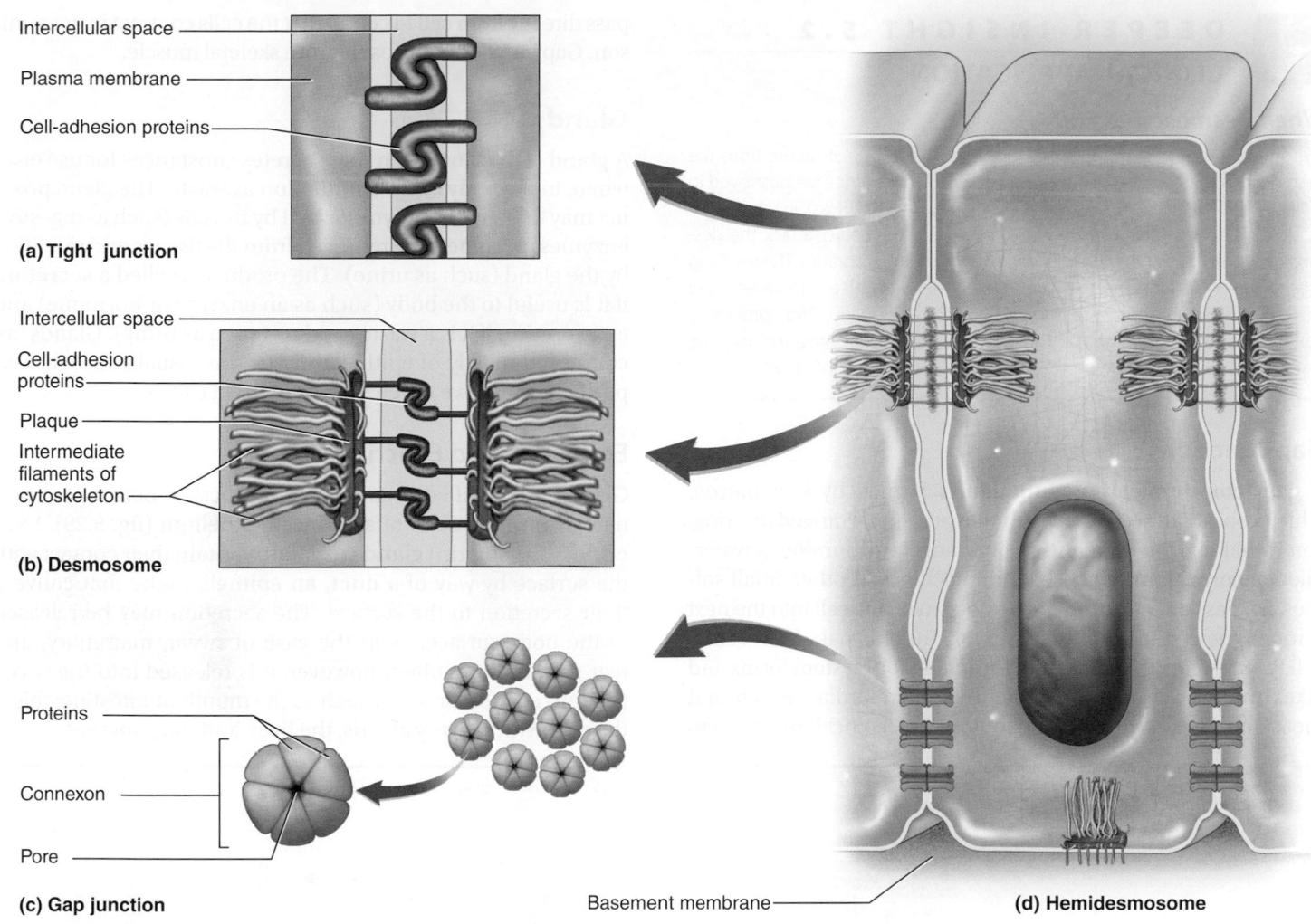

Intercellular space
Plasma membrane
Cell-adhesion proteins

**(a) Tight junction**

Intercellular space
Cell-adhesion proteins
Plaque
Intermediate filaments of cytoskeleton

**(b) Desmosome**

Proteins
Connexon
Pore

**(c) Gap junction**

Basement membrane

**(d) Hemidesmosome**

**FIGURE 5.28 Types of Cell Junctions.**

❓ *Which of these junctions allows material to pass from one cell directly into the next?*

bacteria from invading the tissues, and they ensure that most nutrients pass *through* the epithelial cells and not *between* them. In addition, some membrane proteins function in the apical domain of the cell, and others in the lateral or basal domains; tight junctions limit how far drifting proteins can travel and keep them segregated in the appropriate domains of the membrane where they are needed to perform their tasks.

## Desmosomes

A **desmosome**[28] (DEZ-mo-some) is a patch that holds cells together somewhat like the snap on a pair of jeans. They are not continuous and cannot prevent substances from passing around them and going between the cells, but serve to keep cells from pulling apart and enable a tissue to resist mechanical stress. Desmosomes are common in the epidermis, the epithelium of the uterine cervix, other epithelia, and cardiac muscle. Hooklike

J-shaped proteins arise from the cytoskeleton, approach the cell surface from within, and penetrate into a thick protein plaque on the inner face of the plasma membrane; then the short arm of the J turns back into the cell—thus anchoring the cytoskeleton to the membrane plaque. Proteins of the plaque are linked to transmembrane proteins that, in turn, are linked to transmembrane proteins of the next cell, forming a zone of strong cell adhesion. Each cell mirrors the other and contributes half of the desmosome. Such connections create a strong structural network that binds cells together throughout the tissue. The basal cells of an epithelium are similarly linked to the underlying basement membrane by half desmosomes called **hemidesmosomes,** so an epithelium cannot easily peel away from the underlying tissue.

### ▶▶▶ APPLY WHAT YOU KNOW

*Why would desmosomes not be suitable as the sole type of cell junction between epithelial cells of the stomach?*

---

[28]*desmo* = band, bond, ligament; *som* = body

## DEEPER INSIGHT 5.2

### CLINICAL APPLICATION

#### When Desmosomes Fail

We often get our best insights into the importance of a structure from the dysfunctions that occur when it breaks down. Desmosomes are destroyed in a disease called *pemphigus vulgaris*[29] (PEM-fih-gus vul-GAIR-iss), in which misguided antibodies (defensive proteins) called *autoantibodies* attack the desmosome proteins, especially in the skin and mucous membranes. The resulting breakdown of desmosomes between the epithelial cells leads to widespread blistering of the skin and oral mucosa, loss of tissue fluid, and sometimes death. The condition can be controlled with drugs that suppress the immune system, but such drugs compromise the body's ability to fight off infections.

## Gap Junctions

A **gap (communicating) junction** is formed by a *connexon,* which consists of six transmembrane proteins arranged in a ring, somewhat like the segments of an orange, surrounding a water-filled channel. Ions, glucose, amino acids, and other small solutes can pass directly from the cytoplasm of one cell into the next through the channel. In the embryo, nutrients pass from cell to cell through gap junctions until the circulatory system forms and takes over the role of nutrient distribution. In cardiac muscle and most smooth muscle, gap junctions allow electrical excitation to pass directly from cell to cell so that the cells contract in near unison. Gap junctions are absent from skeletal muscle.

## Glands

A **gland** is a cell or organ that secretes substances for use elsewhere in the body or for elimination as waste. The gland product may be something synthesized by its cells (such as digestive enzymes) or something removed from the tissues and modified by the gland (such as urine). The product is called a **secretion** if it is useful to the body (such as an enzyme or hormone) and an **excretion** if it is a waste product (such as urine). Glands are composed mostly of epithelial tissue, but usually have a supportive connective tissue framework and capsule.

## Endocrine and Exocrine Glands

Glands are classified as endocrine or exocrine. Both types originate as invaginations of a surface epithelium (fig. 5.29). **Exocrine**[30] (EC-so-crin) **glands** usually maintain their contact with the surface by way of a **duct,** an epithelial tube that conveys their secretion to the surface. The secretion may be released to the body surface, as in the case of sweat, mammary, and tear glands. More often, however, it is released into the cavity (lumen) of another organ such as the mouth or intestine; this is the case with salivary glands, the liver, and the pancreas.

[29]*pemphigus* = blistering; *vulgaris* = common

[30]*exo* = out; *crin* = to separate, secrete

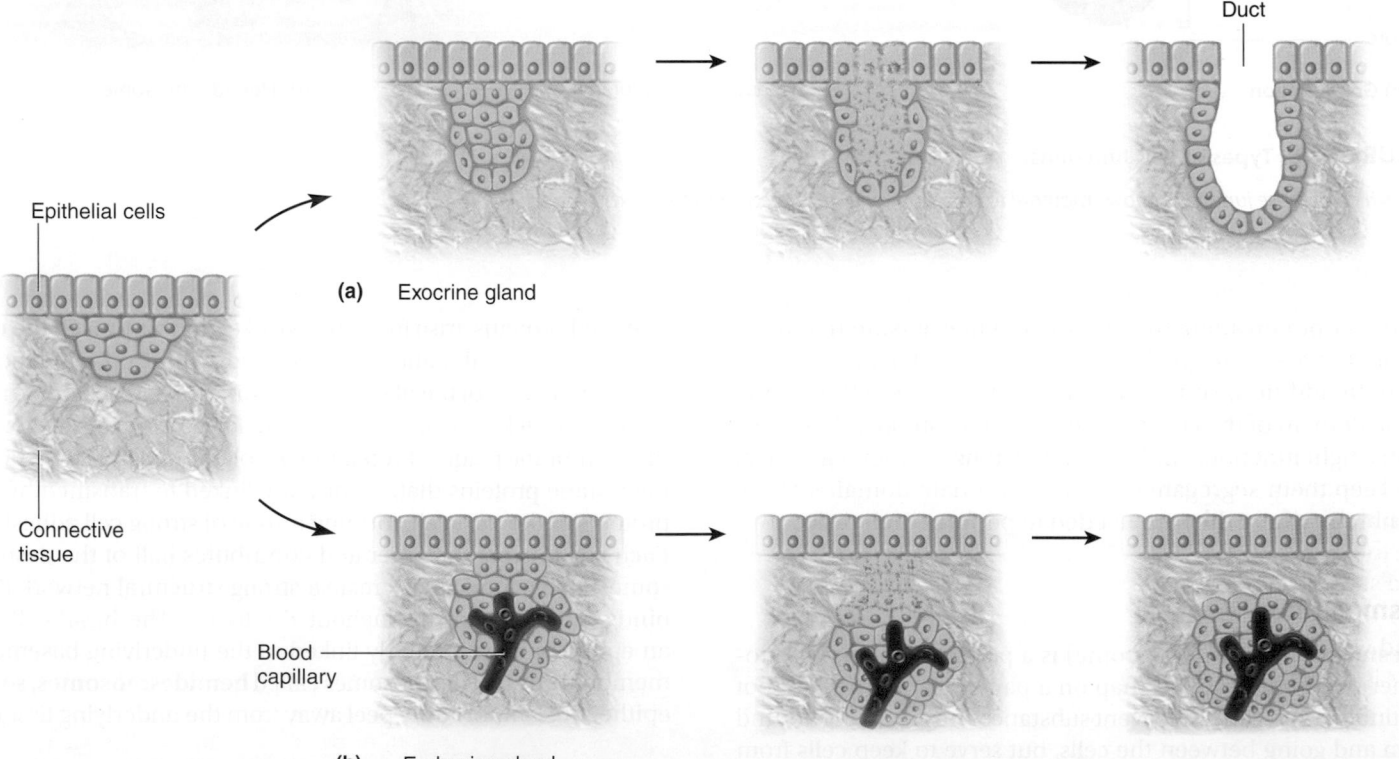

**FIGURE 5.29 Development of Exocrine and Endocrine Glands.** (a) An exocrine gland begins with epithelial cells proliferating into the connective tissue below. A form of cell death called *apoptosis* (see p. 172) hollows out the core and creates a duct to the surface. The gland remains connected to the surface for life by way of this duct and releases its secretions onto the epithelial surface. (b) An endocrine gland begins similarly, but the cells connecting it to the surface degenerate while the secretory tissue becomes infiltrated with blood capillaries. The secretory cells will secrete their products (hormones) into the blood.

**Endocrine**[31] **glands** lose contact with the surface and have no ducts. They do, however, have a high density of blood capillaries and secrete their products directly into the blood (fig. 5.30a). The secretions of endocrine glands, called *hormones,* function as chemical messengers to stimulate cells elsewhere in the body. Endocrine glands include the pituitary, thyroid, and adrenal glands.

The exocrine–endocrine distinction is not always clear. The liver is an exocrine gland that secretes one of its products, bile, through a system of ducts, but secretes hormones, albumin, and other products directly into the blood. Several glands, such as the pancreas and kidney, have both exocrine and endocrine components. Nearly all of the viscera have at least some cells that secrete hormones, even though most of these organs are not usually thought of as glands (for example, the brain and heart).

**Unicellular glands** are secretory cells found in an epithelium that is predominantly nonsecretory. They can be endocrine or exocrine. For example, the respiratory tract, which is lined mainly by ciliated cells, also has a liberal scattering of exocrine goblet cells (see figs. 5.6 and 5.7). The stomach and small intestine have scattered endocrine cells, which secrete hormones that regulate digestion.

Endocrine glands are the subject of chapter 17 and are not further considered here.

## Exocrine Gland Structure

Multicellular exocrine glands such as the pancreas and salivary glands are usually enclosed in a fibrous **capsule** (fig. 5.30b). The capsule often gives off extensions called **septa** (singular, *septum*), or **trabeculae** (trah-BEC-you-lee), that divide the interior of the gland into compartments called **lobes,** which are visible to the naked eye. Finer connective tissue septa may further subdivide each lobe into microscopic **lobules.** Blood vessels, nerves, and the gland's own ducts generally travel through these septa. The connective tissue framework of the gland, called its **stroma,** supports and organizes the glandular tissue. The cells that perform the tasks of synthesis and secretion are collectively called the **parenchyma** (pa-REN-kih-muh). This is typically simple cuboidal or simple columnar epithelium.

Exocrine glands are classified as **simple** if they have a single unbranched duct and **compound** if they have a branched duct. If the duct and secretory portion are of uniform diameter, the gland is called **tubular.** If the secretory cells form a dilated sac, the gland is called **acinar** and the sac is an **acinus**[32] (ASS-ih-nus), or **alveolus**[33] (AL-vee-OH-lus) (fig. 5.30c). A gland with secretory cells in both the tubular and acinar portions is called a **tubuloacinar gland** (fig. 5.31).

---

[31]*endo* = in, into; *crin* = to separate, secrete
[32]*acinus* = berry
[33]*alveol* = cavity, pit

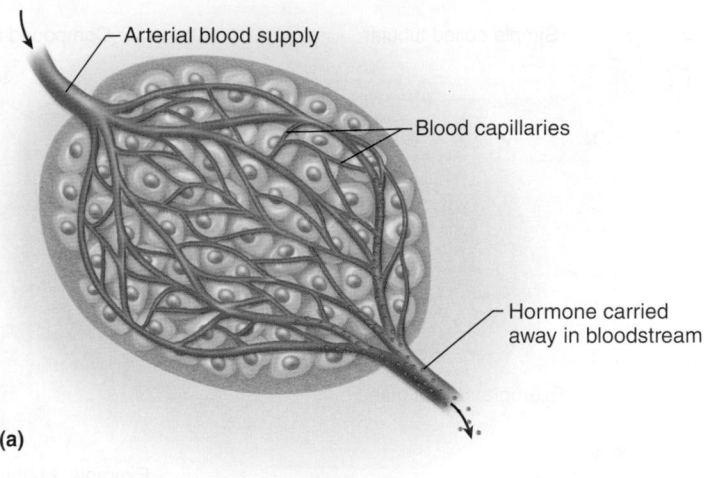

(a)

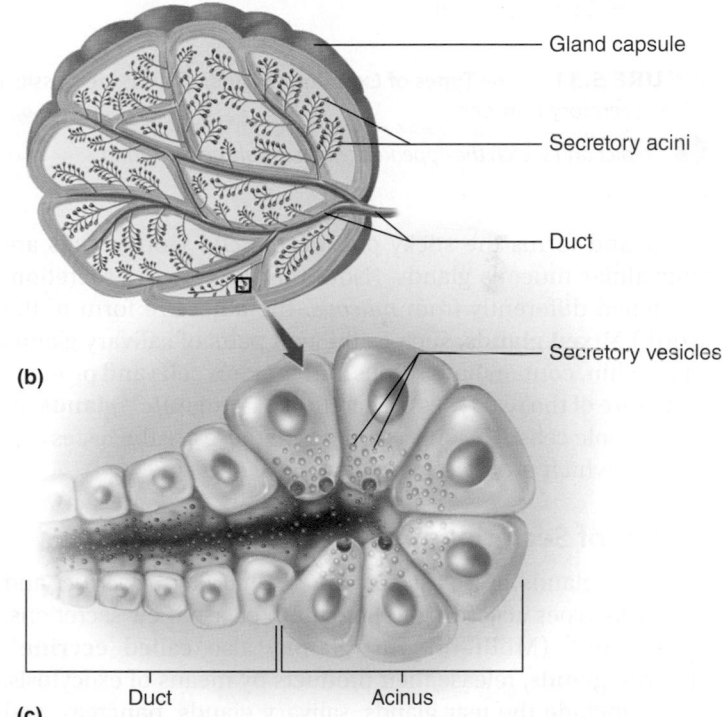

(b)

(c) Duct      Acinus

**FIGURE 5.30 General Structure of Endocrine and Exocrine Glands.** (a) Endocrine glands have no ducts but have a high density of blood capillaries and secrete their products (hormones) directly into the bloodstream. (b) Exocrine glands usually have a system of ducts, which often follow connective tissue septa, until their finest divisions end in saccular acini of secretory cells. (c) Detail of an acinus and the beginning of a duct.

 *What membrane transport process are the cells of this acinus carrying out? (Review pp. 90–98.)*

## Types of Secretions

Glands are classified not only by their structure but also by the nature of their secretions. **Serous** (SEER-us) **glands** produce relatively thin, watery fluids such as perspiration, milk, tears, and digestive juices. **Mucous glands,** found in the tongue and roof of the mouth among other places, secrete a glycoprotein called *mucin* (MEW-sin). After it is secreted, mucin absorbs

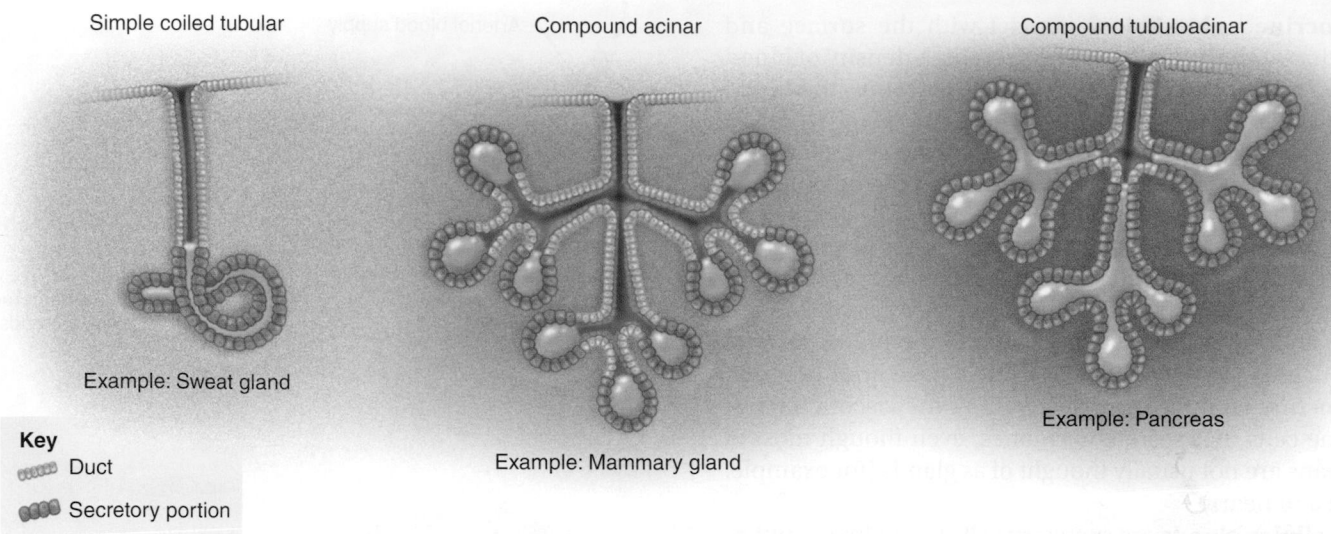

Simple coiled tubular

Example: Sweat gland

**Key**
Duct
Secretory portion

Compound acinar

Example: Mammary gland

Compound tubuloacinar

Example: Pancreas

**FIGURE 5.31   Some Types of Exocrine Glands.**  Glands are classified according to the branching of their ducts and the appearance and extent of the secretory portions.

**?** *Predict and sketch the appearance of a simple acinar gland.*

water and forms the sticky product *mucus.* Goblet cells are unicellular mucous glands. (Note that *mucus,* the secretion, is spelled differently from *mucous,* the adjective form of the word.) **Mixed glands,** such as the two pairs of salivary glands in the chin, contain both serous and mucous cells and produce a mixture of the two types of secretions. **Cytogenic**[34] **glands** release whole cells. The only examples of these are the testes and ovaries, which produce sperm and egg cells.

## Modes of Secretion

Exocrine glands are classified into *merocrine, apocrine,* and *holocrine* types according to how they release their secretions. **Merocrine**[35] (MERR-oh-crin) glands, also called **eccrine**[36] (EC-rin) **glands,** release their products by means of exocytosis. These include the tear glands, salivary glands, pancreas, and most others. Mammary glands secrete milk sugar (lactose) and proteins (casein, lactalbumin) by this method (fig. 5.32a), but secrete the milk fat by another method called **apocrine**[37] secretion (fig. 5.32b). Lipids coalesce from the cytosol into a droplet that buds from the cell surface, covered by a layer of plasma membrane and a very thin film of cytoplasm. Sweat glands of the axillary (armpit) region were once thought to use the apocrine method as well. Closer study showed this to be untrue; they are merocrine, but they are nevertheless different from other merocrine glands in function and histological appearance (see chapter 6) and are still referred to as apocrine sweat glands.

In **holocrine**[38] **glands,** cells accumulate a product and then the entire cell disintegrates, *becoming* the secretion instead of *releasing* one (fig. 5.32c). Holocrine secretions tend

to be relatively thick and oily, composed of cell fragments and the substances the cells had synthesized before disintegrating. Only a few glands use this method, such as the oil-producing glands of the scalp and other areas of skin, and certain glands of the eyelid.

## Membranes

Atlas A describes the major body cavities and the membranes that line them and cover their viscera (p. 32). We now consider some histological aspects of these membranes. Membranes may be composed of epithelial tissue only; connective tissue only; or epithelial, connective, and muscular tissue.

The largest membrane of the body is the **cutaneous membrane**—or more simply, the skin (detailed in chapter 6). It consists of a stratified squamous epithelium (epidermis) resting on a layer of connective tissue (dermis). Unlike the other membranes to be considered, it is relatively dry. It resists dehydration of the body and provides an inhospitable environment for the growth of infectious organisms.

The two principal kinds of internal membranes are mucous and serous membranes. A **mucous membrane (mucosa)** (fig. 5.33a) lines passages that open to the exterior environment: the digestive, respiratory, urinary, and reproductive tracts. A mucous membrane consists of two to three layers: (1) an epithelium; (2) an areolar connective tissue layer called the **lamina propria**[39] (LAM-ih-nuh PRO-pree-uh); and often (3) a layer of smooth muscle called the **muscularis mucosae** (MUSK-you-LAIR-iss mew-CO-see). Mucous membranes have absorptive, secretory, and protective functions. They are often covered with mucus secreted by goblet cells, multicellular mucous glands, or both. The mucus traps bacteria and foreign particles, which

---

[34]*cyto* = cell; *genic* = producing
[35]*mero* = part; *crin* = to separate, secrete
[36]*ec* = *ex* = out; *crin* = to separate, secrete
[37]*apo* = from, off, away; *crin* = to separate, secrete
[38]*holo* = whole, entire; *crin* = to separate, secrete

[39]*lamina* = layer; *propria* = of one's own

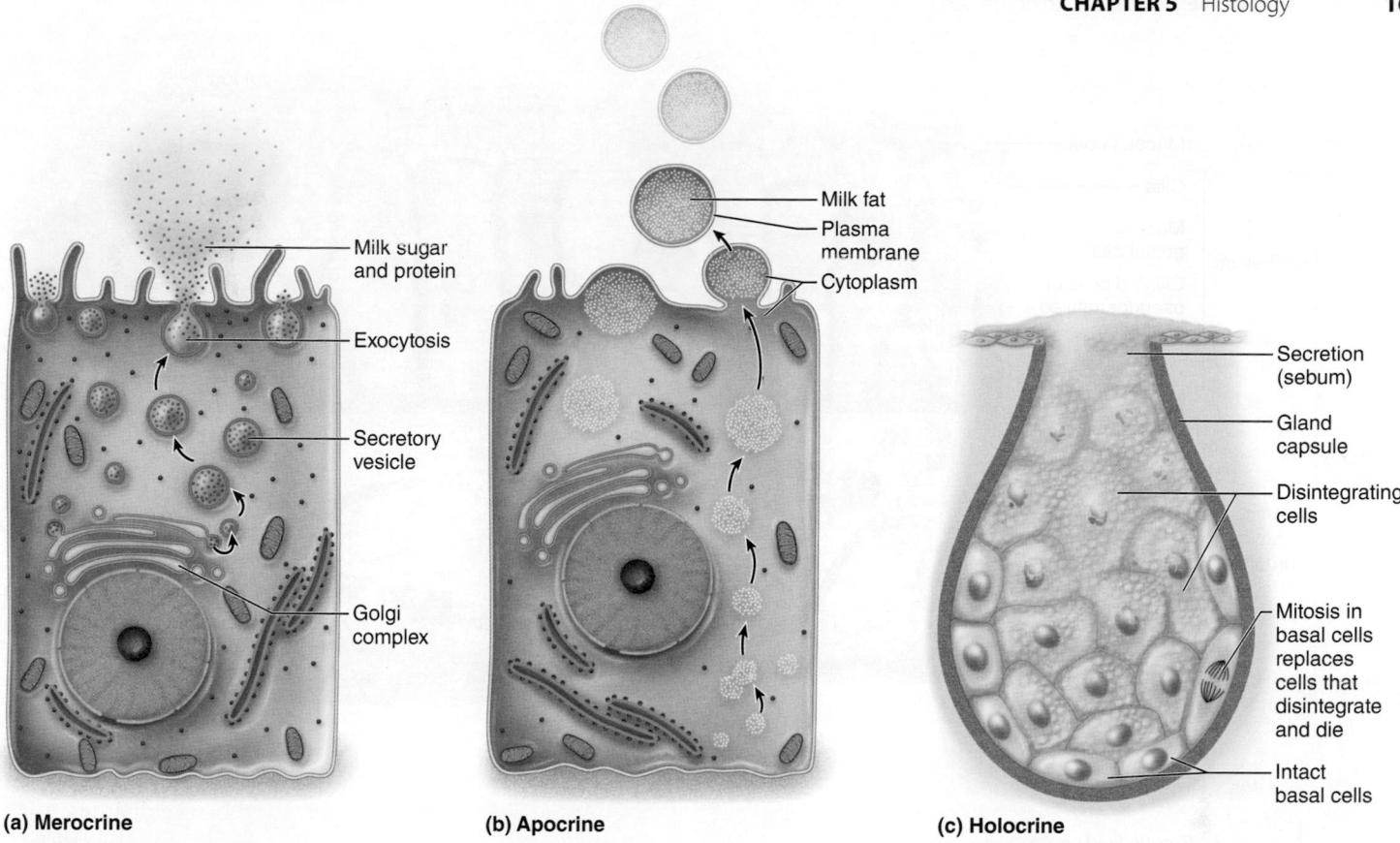

**(a) Merocrine**

Milk sugar and protein
Exocytosis
Secretory vesicle
Golgi complex

**(b) Apocrine**

Milk fat
Plasma membrane
Cytoplasm

**(c) Holocrine**

Secretion (sebum)
Gland capsule
Disintegrating cells
Mitosis in basal cells replaces cells that disintegrate and die
Intact basal cells

**FIGURE 5.32 The Three Modes of Exocrine Secretion.** (a) Merocrine secretion in a cell of the mammary gland, secreting milk sugar (lactose) and proteins (casein, lactalbumin) by exocytosis. (b) Apocrine secretion of fat by a cell of the mammary gland. Fat droplets coalesce in the cytosol, then bud from the cell surface with a thin coat of cytoplasm and plasma membrane. (c) Holocrine secretion by a sebaceous (oil) gland of the scalp. In this method, entire gland cells break down and become the secretion (sebum).

❓ *Which of these three glands would require the highest rate of mitosis? Why?*

keeps them from invading the tissues and aids in their removal from the body. The epithelium of a mucous membrane may also include absorptive, ciliated, and other types of cells.

A **serous membrane (serosa)** is composed of a simple squamous epithelium resting on a thin layer of areolar connective tissue (fig. 5.33b). Serous membranes produce watery **serous fluid,** which arises from the blood and derives its name from the fact that it is similar to blood serum in composition. Serous membranes line the insides of some body cavities and form a smooth outer surface on some of the viscera, such as the digestive tract. The pleurae, pericardium, and peritoneum described in atlas A are serous membranes. Their epithelial component is called **mesothelium.**

The circulatory system is lined with a simple squamous epithelium called **endothelium,** derived from mesoderm. The endothelium rests on a thin layer of areolar tissue, which often rests in turn on an elastic sheet. Collectively, these tissues make up a membrane called the *tunica interna* of the blood vessels and *endocardium* of the heart.

The foregoing membranes are composed of two to three tissue types. By contrast, some membranes composed only of connective tissue include the *dura mater* around the brain, *synovial membranes* that enclose joints of the skeletal system,

and the *periosteum* that covers each bone. Some membranes composed only of epithelium include the anterior surfaces of the lens and cornea of the eye. All of these are described in later chapters.

**BEFORE YOU GO ON**

Answer the following questions to test your understanding of the preceding section:

19. Compare the structure of tight junctions and gap junctions. Relate their structural differences to their functional differences.

20. Distinguish between a simple gland and a compound gland, and give an example of each. Distinguish between a tubular gland and an acinar gland, and give an example of each.

21. Contrast the merocrine, apocrine, and holocrine methods of secretion, and name a gland product produced by each method.

22. Describe the differences between a mucous and a serous membrane.

23. Name the layers of a mucous membrane, and state which of the four primary tissue classes composes each layer.

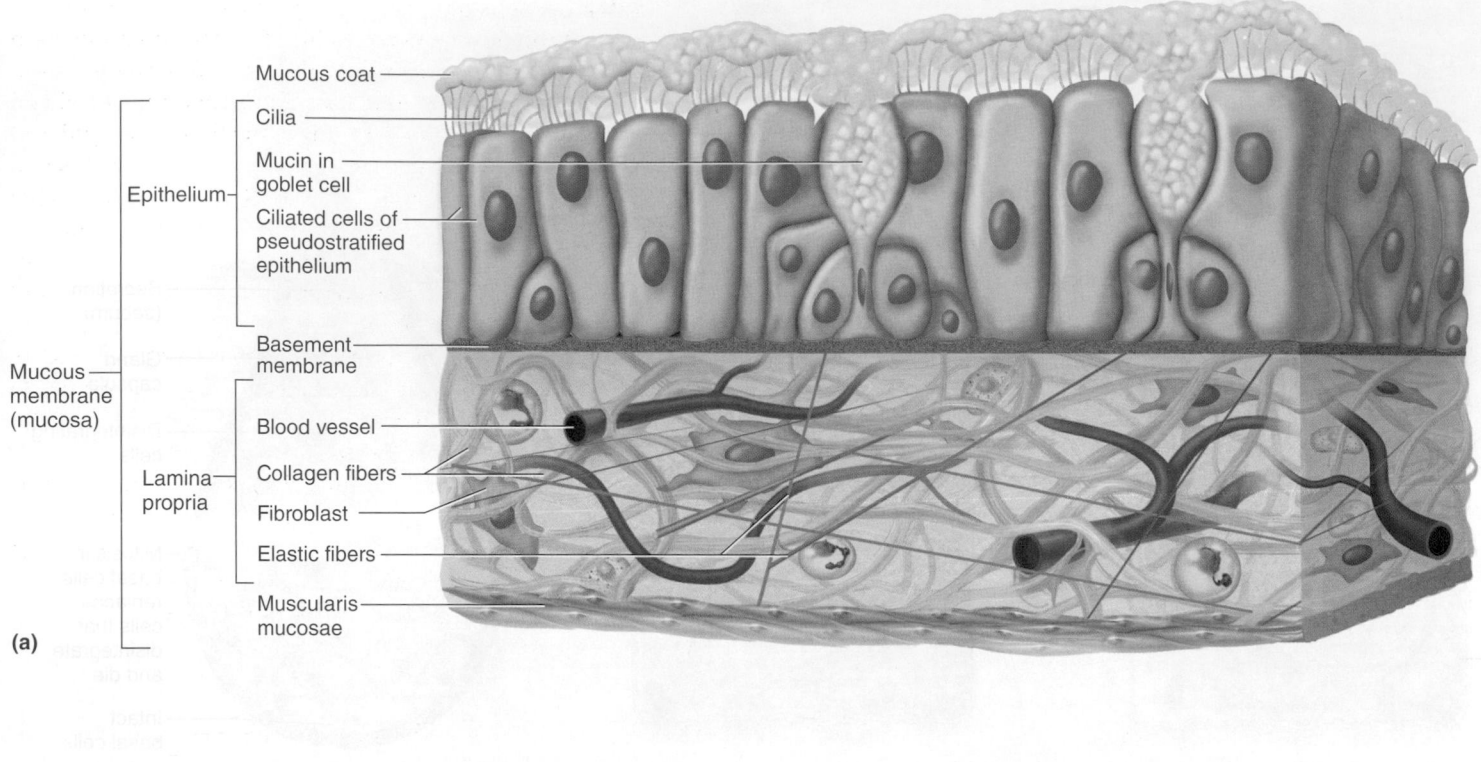

**FIGURE 5.33   Histology of Mucous and Serous Membranes.**   (a) A mucous membrane such as the inner lining of the trachea. (b) A serous membrane such as the external surface of the small intestine.

**5.6**   Tissue Growth, Development, Repair, and Degeneration

## 5.6   Tissue Growth, Development, Repair, and Degeneration

### Expected Learning Outcomes

When you have completed this section, you should be able to

a. name and describe the modes of tissue growth;

b. define *adult* and *embryonic stem cells* and their varied degrees of developmental plasticity;

c. name and describe the ways that a tissue can change from one type to another;

d. name and describe the modes and causes of tissue shrinkage and death; and

e. name and describe the ways the body repairs damaged tissues.

## Tissue Growth

Tissues grow because their cells increase in number or size. Most embryonic and childhood growth occurs by **hyperplasia**[40] (HY-pur-PLAY-zhuh)—tissue growth through cell multiplication. Skeletal muscles and adipose tissue grow, however, through **hypertrophy**[41] (hy-PUR-truh-fee)—the enlargement of preexisting cells. Even a very muscular or fat adult has essentially the same number of muscle fibers or adipocytes as he or she had in childhood, but the cells may be substantially larger. **Neoplasia**[42] (NEE-oh-PLAY-zhuh) is the development of a tumor (*neoplasm*)—whether benign or malignant—composed of abnormal, nonfunctional tissue.

---

[40]*hyper* = excessive; *plas* = growth
[41]*hyper* = excessive; *trophy* = nourishment
[42]*neo* = new; *plas* = form, growth

## Tissue Development

You have studied the form and function of more than two dozen discrete types of human tissue in this chapter. You should not leave this subject, however, with the impression that once these tissue types are established, they never change. Tissues are, in fact, capable of changing from one type to another within certain limits. Most obviously, unspecialized tissues of the embryo develop into more diverse and specialized types of mature tissue—mesenchyme to muscle, for example. This development of a more specialized form and function is called **differentiation.**

Epithelia sometimes exhibit **metaplasia,**[43] a change from one type of mature tissue to another. For example, the vagina of a young girl is lined with a simple cuboidal epithelium. At puberty, it changes to a stratified squamous epithelium, better adapted to the future demands of intercourse and childbirth. The nasal cavity is lined with ciliated pseudostratified columnar epithelium. However, if we block one nostril and breathe through the other one for several days, the epithelium in the unblocked passage changes to stratified squamous. In smokers, the pseudostratified columnar epithelium of the bronchi may transform into a stratified squamous epithelium.

#### ►►►APPLY WHAT YOU KNOW

*What functions of a ciliated pseudostratified columnar epithelium could not be served by a stratified squamous epithelium? In light of this, what might be some consequences of bronchial metaplasia in heavy smokers?*

## Stem Cells

The growth and differentiation of tissues depend upon a supply of reserve **stem cells.** These are undifferentiated cells that are not yet performing any specialized function, but that have the potential to differentiate into one or more types of mature functional cells, such as liver, brain, cartilage, or skin cells. Such cells have various degrees of **developmental plasticity,** or diversity of mature cell types to which they can give rise.

There are two types of stem cells: *embryonic* and *adult.* **Embryonic stem cells** compose the early human embryo. In the early stages of development, these are called **totipotent** stem cells, because they have the potential to develop into any type of fully differentiated human cell—not only cells of the later embryonic, fetal, or adult body, but also cells of the temporary structures of pregnancy, such as the placenta and amniotic sac. Totipotency is unlimited developmental plasticity. About 4 days after fertilization, the developing embryo enters the *blastocyst* stage. The blastocyst is a hollow ball with an *outer cell mass* that helps form the placenta and other accessory organs of pregnancy, and an *inner cell mass* that becomes the embryo itself (see fig. 29.4, p. 1100). Cells of the inner cell mass are called **pluripotent** stem cells; they can still develop into any cell type of the embryo, but not into the accessory organs of pregnancy. Thus their developmental plasticity is already somewhat limited.

**Adult stem cells** occur in small numbers in mature organs and tissues throughout a person's life. Typically an adult stem cell divides mitotically; one of its daughter cells remains a stem cell and the other one differentiates into a mature specialized cell. The latter cell may replace another that has grown old and died, contribute to the development of growing organs (as in a child), or help to repair damaged tissue. Some adult stem cells are **multipotent**—able to develop into two or more different cell lines, but not just any type of body cell. Certain multipotent bone marrow stem cells, for example, can give rise to red blood cells, five kinds of white blood cells, and platelet-producing cells. **Unipotent** stem cells have the most limited plasticity, as they can produce only one mature cell type. Examples include the cells that give rise to sperm, eggs, and keratinocytes (the majority cell type of the epidermis).

Both embryonic and adult stem cells have enormous potential for therapy, but stem-cell research has been embroiled in great political controversy in the past several years. Deeper Insight 5.4 (p. 172) addresses the clinical potential of stem cells and the ethical and political issues surrounding stem-cell research.

## Tissue Repair

Damaged tissues can be repaired in two ways: *regeneration* or *fibrosis*. **Regeneration** is the replacement of dead or damaged cells by the same type of cells as before; it restores normal function to the organ. Most skin injuries (cuts, scrapes, and minor burns) heal by regeneration. The liver also regenerates remarkably well. **Fibrosis** is the replacement of damaged tissue with scar tissue, composed mainly of collagen produced by fibroblasts. Scar tissue helps to hold an organ together, but it does not restore normal function. Examples include the healing of severe cuts and burns, the healing of muscle injuries, and scarring of the lungs in tuberculosis.

Figure 5.34 illustrates the following stages in the healing of a cut in the skin, where both regeneration and fibrosis are involved:

1. Severed blood vessels bleed into the cut. Mast cells and cells damaged by the cut release histamine, which dilates blood vessels, increases blood flow to the area, and makes blood capillaries more permeable. Blood plasma seeps into the wound, carrying antibodies and clotting proteins.

2. A blood clot forms in the tissue, loosely knitting the edges of the cut together and inhibiting the spread of pathogens from the site of injury into healthy tissues. The surface of the blood clot dries and hardens in the air, forming a scab that temporarily seals the wound and blocks infection. Beneath the scab, macrophages begin to phagocytize and digest tissue debris.

3. New blood capillaries sprout from nearby vessels and grow into the wound. The deeper portions of the clot become infiltrated by capillaries and fibroblasts and transform into a soft mass called **granulation tissue.** Macrophages remove the blood clot while fibroblasts deposit new collagen to replace it. This *fibroblastic (reconstructive) phase* of repair begins 3 to 4 days after the injury and lasts up to 2 weeks.

---

[43]*meta* = change; *plas* = form, growth

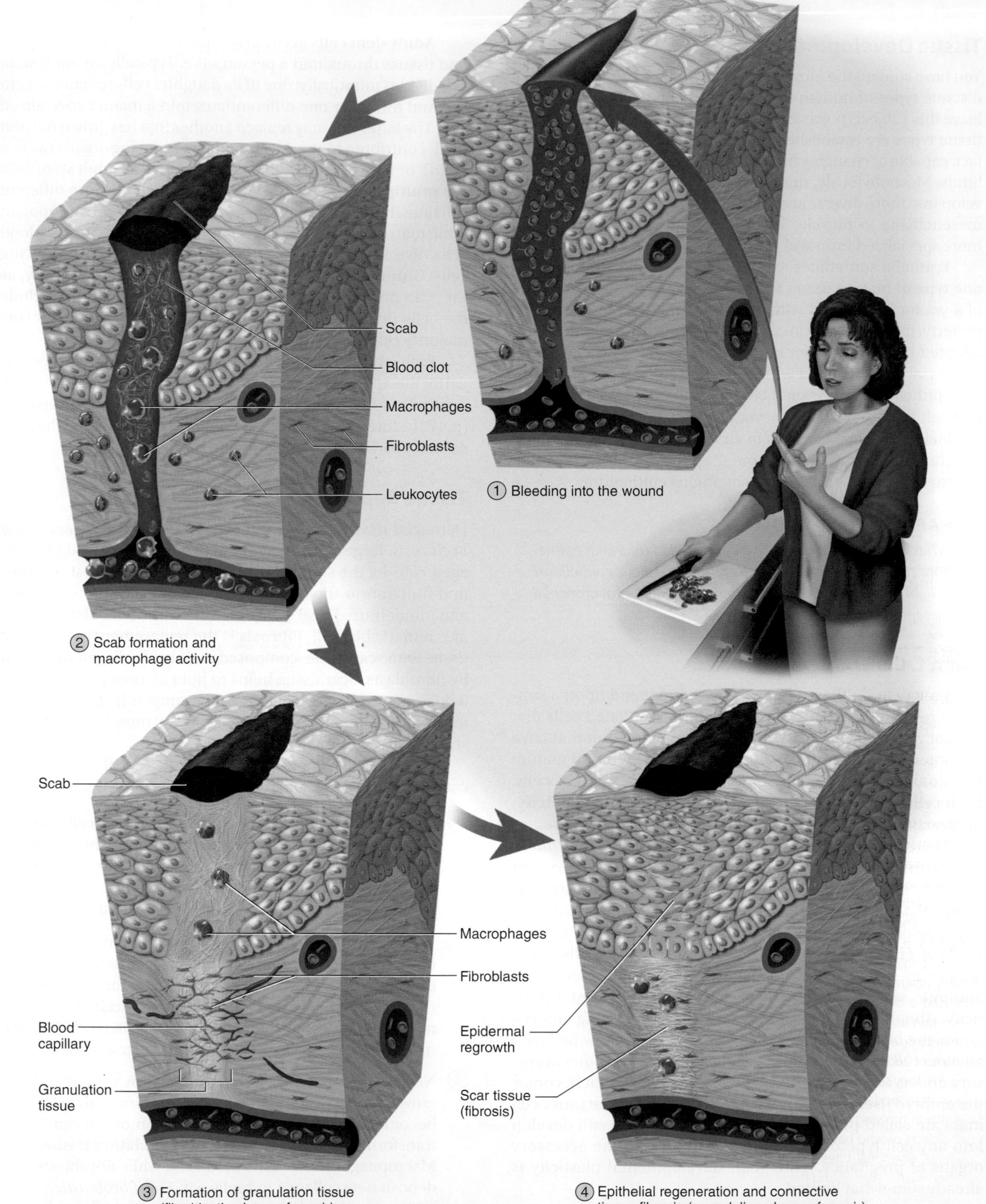

Scab

Blood clot

Macrophages

Fibroblasts

Leukocytes

① Bleeding into the wound

② Scab formation and
macrophage activity

Scab

Macrophages

Fibroblasts

Blood
capillary

Epidermal
regrowth

Granulation
tissue

Scar tissue
(fibrosis)

③ Formation of granulation tissue
(fibroblastic phase of repair)

④ Epithelial regeneration and connective
tissue fibrosis (remodeling phase of repair)

**FIGURE 5.34 Stages in the Healing of a Skin Wound.**

(4) Surface epithelial cells around the wound multiply and migrate into the wounded area, beneath the scab. The scab loosens and eventually falls off, and the epithelium grows thicker. Thus, the epithelium *regenerates* while the underlying connective tissue undergoes *fibrosis,* or scarring. Capillaries withdraw from the area as fibrosis progresses. The scar tissue may or may not show through the epithelium, depending on the severity of the wound. The wound may exhibit a depressed area at first, but this is often filled in by continued fibrosis and remodeling from below, until the scar becomes unnoticeable. This *remodeling (maturation) phase* of tissue repair begins several weeks after injury and may last as long as 2 years.

## Tissue Degeneration and Death

**Atrophy**[44] (AT-ruh-fee) is the shrinkage of a tissue through a loss in cell size or number. It results from both normal aging *(senile atrophy)* and lack of use of an organ *(disuse atrophy).* Muscles that are not exercised exhibit disuse atrophy as their cells become smaller. This was a serious problem for the first astronauts who participated in prolonged microgravity space flights. Upon return to normal gravity, they were sometimes too

weak from muscular atrophy to walk. Space stations and shuttles now include exercise equipment to maintain the crew's muscular condition. Disuse atrophy also occurs when a limb is immobilized in a cast or by paralysis.

**Necrosis**[45] (neh-CRO-sis) is premature, pathological tissue death due to trauma, toxins, infection, and so forth. **Infarction** is the sudden death of tissue, such as cardiac muscle *(myocardial infarction)* or brain tissue *(cerebral infarction),* that occurs when its blood supply is cut off. **Gangrene** is any tissue necrosis resulting from an insufficient blood supply, usually involving infection. *Dry gangrene* often occurs in diabetics, especially in the feet. A lack of sensation due to diabetic nerve damage can make a person oblivious to injury and infection, and poor blood circulation due to diabetic arterial damage results in slow healing and rapid spread of infection. This often necessitates the amputation of toes, feet, or legs. A **decubitus ulcer (bed sore** if **pressure sore)** is a form of dry gangrene that occurs when immobilized persons, such as those confined to a hospital bed or wheelchair, are unable to move, and continual pressure on the skin cuts off blood flow to an area. Pressure sores occur most often in places where a bone comes close to the body surface, such as the hips, sacral region, and ankles. Here, the thin layer of skin and connective

---

[44]*a* = without; *trophy* = nourishment

[45]*necr* = death; *osis* = process

---

## DEEPER INSIGHT 5.3
### CLINICAL APPLICATION

### Tissue Engineering

Tissue repair is not only a natural process but also a lively area of research in biotechnology. **Tissue engineering** is the artificial production of tissues and organs in the laboratory for implantation in the human body. The process commonly begins with building a scaffold (supportive framework) of collagen or biodegradable polyester, sometimes in the shape of a desired organ such as a blood vessel or an ear. The scaffold is seeded with human cells and put in a "bioreactor" to grow. The bioreactor supplies nutrients, oxygen, and growth factors. It may be an artificial chamber, or the body of a human patient or laboratory animal. When a lab-grown tissue reaches a certain point, it is implanted into the desired location in the patient.

Tissue-engineered skin grafts have long been on the market (see Deeper Insight 6.5, p. 197), and artificial tissues have since grown into a billion-dollar industry. Engineers are working on heart components such as valves, coronary arteries, patches of cardiac tissue, and whole heart chambers, and some have produced a beating rodent heart from a cell-seeded scaffold. Others have grown human liver, bone, ureter, tendon, intestinal, and breast tissue in the laboratory. Scientists at the University of Massachusetts and the Massachusetts Institute of Technology grew a "human" outer ear on the back of a mouse (fig. 5.35). They seeded a polymer scaffold with human cartilage cells and grew it in an immunodeficient mouse unable to reject the human tissue. They see potential in growing ears and noses for cosmetic treatment of children with birth defects or who have suffered disfiguring injuries from playground fights, accidents, or animal bites. One of the most daunting problems in growing artificial organs is producing the microvascular blood supply needed to sustain an organ such as a liver. Nevertheless, tissue engineers have constructed new urinary bladders in several patients and a new bronchus in one, seeding a nonliving protein scaffold with cells taken from elsewhere in the patients' bodies.

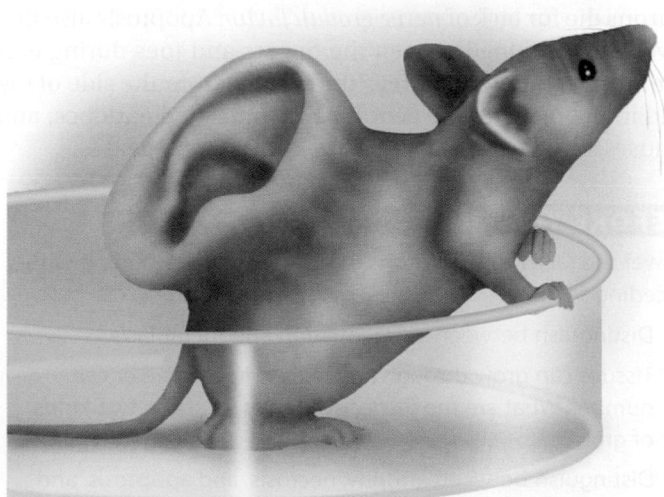

**FIGURE 5.35 Tissue Engineering.** Scientists have grown external ears from human tissue on the backs of immunodeficient mice. The ear can be removed without killing the mouse. In the future, such artificial organs might be used to improve the facial appearance of patients with missing parts.

tissue is especially prone to compression between the bone and a bed or wheelchair.

*Wet gangrene* typically occurs in internal organs and involves neutrophil invasion, liquefaction of the tissue, pus, and a foul odor. It can result from appendicitis or an obstructed colon, for example. *Gas gangrene* is necrosis of a wound resulting from infection with certain bacteria of the genus *Clostridium,* usually introduced when a wound is contaminated with soil. The disorder is named for bubbles of gas (mainly hydrogen) that accumulate in the tissues. This is a deadly condition that requires immediate intervention, often including amputation.

Cells dying by necrosis usually swell, exhibit *blebbing* (bubbling) of their plasma membranes, and then rupture. The cell contents released into the tissues trigger an inflammatory response in which macrophages phagocytize the cellular debris.

**Apoptosis**[46] (AP-op-TOE-sis), or **programmed cell death,** is the normal death of cells that have completed their function and best serve the body by dying and getting out of the way. Cells undergoing apoptosis shrink and are quickly phagocytized by macrophages and other cells. The cell contents never escape, so there is no inflammatory response. Although billions of cells die every hour by apoptosis, they are engulfed so quickly that they are almost never seen except within macrophages.

Apparently, nearly every cell has a built-in "suicide program" that enables the body to dispose of it when necessary. In some cases, an extracellular suicide signal binds to a receptor protein in the plasma membrane called *Fas.* Fas then activates intracellular enzymes that destroy the cell, including an *endonuclease* that chops up its DNA and a *protease* that destroys cellular proteins. In other cases, cells seem to undergo apoptosis automatically if they stop receiving growth factors from other cells. For example, in embryonic development we produce about twice as many neurons as we need. Those that make connections with target cells survive, while the excess neurons die for lack of *nerve growth factor.* Apoptosis also dissolves the webbing between the fingers and toes during embryonic development; it frees the earlobe from the side of the head in people with the genotype for detached earlobes; and it causes shrinkage of the breasts after lactation ceases.

---

### BEFORE YOU GO ON

Answer the following questions to test your understanding of the preceding section:

24. Distinguish between differentiation and metaplasia.

25. Tissues can grow through an increase in cell size or cell number. What are the respective terms for these two kinds of growth?

26. Distinguish between atrophy, necrosis, and apoptosis, and describe a circumstance under which each of these forms of tissue loss may occur.

27. Distinguish between regeneration and fibrosis. Which process restores normal cellular function? What good is the other process if it does not restore function?

---

[46]*apo* = away; *ptosis* = falling

## DEEPER INSIGHT 5.4

### CLINICAL APPLICATION

#### The Stem-Cell Controversy

Stem-cell research has been one of the most politically controversial areas of biological science at the dawn of this century. At least 18 countries have recently debated or enacted laws to regulate it. Politicians, scientists, bioethicists, philosophers, and theologians have joined in the debate; legions of lay citizens have voiced their opinions in newspaper editorial pages; and stem cells have been a contentious issue in U.S. presidential politics. In 2001, President George W. Bush signed into law an act that prohibited federal funding for research on any embryonic stem (ES) cell lines created after that date, on the grounds that he regarded the harvesting of the cells to be the taking of a human life. However, California voters flouted the federal ban in 2005 by approving a bill to provide state funds—$300 million per year for 10 years—for embryonic stem-cell research. President Barack Obama reversed the federal ban promptly upon taking office in 2009.

Not surprisingly, biologists see stem cells as a possible treatment for diseases that result from the loss of functional cells. Skin and bone marrow stem cells have been used in therapy for many years. Scientists hope that with a little coaxing, stem cells might replace cardiac muscle damaged by heart attack; restore function to an injured spinal cord; cure parkinsonism by replacing lost brain cells; or cure diabetes mellitus by replacing lost insulin-secreting cells. *Adult stem cells (AS cells)* have narrower developmental potential and may be unable to make all the cell types needed to treat a broad range of degenerative diseases. In addition, they are present in very small numbers and are difficult to harvest and culture in the quantities needed for therapy.

ES cells, however, may hold greater potential. New laboratory methods have made them easier to culture than AS cells and have greatly accelerated stem-cell research in recent years. In animal studies, ES cells have already proven effective for treating degenerative disease. In rats, for example, neurons have been produced from ES cells, implanted into the animal, and shown to reverse the signs of Parkinson disease.

The road to therapy with ES cells remains full of technical, ethical, and legal speed bumps. Will ES cells be rejected by the recipient's immune system? Can the ES cells or the growth media in which they are cultured introduce viruses or other pathogens into the recipient? How can the ES cells be made to lodge and grow in the right place in the patient's body? Could they grow into tumors instead of healthy tissue? Can ES cell therapy ever be economical enough to be affordable to any but the very rich? Scientists can scarcely begin to tackle these problems, however, unless and until a bioethical question is resolved: Can we balance the benefits of stem-cell therapy against the destruction of early human embryos from which the ES cells are harvested?

Where do these embryos come from? Most are donated by couples using *in vitro fertilization (IVF)* to conceive a child. IVF entails collecting numerous eggs from the prospective mother, fertilizing them in glassware with the father's sperm, letting them develop into embryos (technically, preembryos) of about 8 to 16 cells, and then transplanting *some* of these into the mother's uterus (see Deeper Insight 29.4, p. 1124). To overcome the low odds of success, excess embryos are always produced and some are always left over. The excess embryos are often destroyed, but many couples choose instead to donate them for research that may ultimately benefit other patients. It would seem sensible to use the embryos for beneficial purposes rather than to simply destroy and discard them. Opponents of stem-cell research argue, however, that potential medical benefits cannot justify the destruction of a human embryo. Understandably, such issues have aroused an intense debate that is likely to restrain stem-cell research for some time to come.

Even stem-cell researchers admit that IVF clinics alone could not meet the demand for ES cells as research and therapy move ahead. In hopes of reducing the need for embryonic cells, another line of intensive research has sought and achieved ways of biochemically inducing adult stem cells to revert to an embryonic level of developmental plasticity. These manipulated cells, called *induced pluripotent stem cells (iPS cells),* have begun to show great promise.

# STUDY GUIDE

## ▶ Assess Your Learning Outcomes

*To test your knowledge, discuss the following topics with a study partner or in writing, ideally from memory.*

### 5.1 The Study of Tissues (p. 141)

1. Two names for the branch of biology concerned with tissue structure
2. The four primary tissue classes that constitute the body
3. The roles of cells, matrix, fibers, and ground substance in tissue composition, and how these terms relate to each other
4. Primary germ layers of the embryo and their relevance to the histology of mature tissues
5. How and why tissues are prepared as stained histological sections; the three common planes of section; and some ways that tissues are prepared other than sectioning

### 5.2 Epithelial Tissue (p. 144)

1. Characteristics that distinguish epithelium from the other three primary tissue classes
2. Functions of the basement membrane and its relationship to an epithelium
3. Defining characteristics of a simple epithelium
4. Four types of simple epithelium and the appearance, functions, and representative locations of each
5. Defining characteristics of a stratified epithelium
6. Four types of stratified epithelium and the appearance, functions, and representative locations of each
7. Distinctions between keratinized and nonkeratinized stratified squamous epithelium, including differences in histology, locations, and functions
8. Epithelial exfoliation and its clinical relevance

### 5.3 Connective Tissue (p. 150)

1. Characteristics that distinguish connective tissue from the other three primary tissue classes
2. Functions of connective tissues
3. Cell types found in fibrous connective tissue, and the functions of each

4. Fiber types found in fibrous connective tissue, their composition, and the functions of each
5. The composition and variations in the ground substance of fibrous connective tissue
6. The appearance, functions, and representative locations of areolar, reticular, dense irregular, and dense regular connective tissue
7. The appearance, functions, and representative locations of adipose tissue, including the differences between white fat and brown fat
8. Defining characteristics of cartilage as a class; the three types of cartilage and how they differ in histology, function, and location; the relationship of the perichondrium to cartilage; and where perichondrium is absent
9. Defining characteristics of osseous tissue as a class; the distinction between spongy and compact bone; and the relationship of the periosteum to bone
10. The histological appearance of cross sections of compact bone
11. Why blood is classified as a connective tissue; the term for its matrix; and the three major categories of formed elements in blood

### 5.4 Nervous and Muscular Tissues—Excitable Tissues (p. 159)

1. Why nervous and muscular tissue are called *excitable tissues* even though excitability is a universal property of all living cells
2. The two basic types of cells in nervous tissue and their functional differences
3. The general structure of neurons
4. Defining characteristics of muscular tissue as a class
5. Three types of muscle and how they differ in histology, function, and location

### 5.5 Cell Junctions, Glands, and Membranes (p. 161)

1. The general function of cell junctions
2. Differences in the structure and function of tight junctions, desmosomes, hemidesmosomes, and gap junctions

3. The definition of a *gland* and the two basic functions of glands
4. The developmental, structural, and functional distinctions between exocrine and endocrine glands; examples of each; and why some glands cannot be strictly classified into one category or the other
5. Examples of unicellular glands in both the exocrine and endocrine categories
6. General histology of a typical exocrine gland
7. The scheme for classifying exocrine glands according to the anatomy of their duct systems and their distribution of secretory cells
8. Differences between serous, mucous, mixed, and cytogenic glands, and examples of each
9. Comparison of the mode of secretion of merocrine, apocrine, and holocrine glands
10. The variety of serous, mucous, and other membranes in the body, and names of some specialized membranes of the skin, blood vessels, and joints

### 5.6 Tissue Growth, Development, Repair, and Degeneration (p. 168)

1. Differences between hyperplasia, hypertrophy, and neoplasia as normal and pathological modes of tissue growth
2. Differences between differentiation and metaplasia as modes of transformation from one tissue type to another
3. What stem cells are and how they relate to developmental plasticity; differences between embryonic and adult stem cells; and differences between totipotent, pluripotent, multipotent, and unipotent stem cells
4. Differences between regeneration and fibrosis as modes of tissue repair
5. Steps in the healing of a wound such as a cut in the skin
6. The general meaning of tissue *atrophy* and two forms or causes of atrophy
7. Differences between necrosis and apoptosis as modes of cell death and tissue shrinkage; some normal functions of apoptosis
8. Varieties of necrosis including infarction, dry gangrene, and gas gangrene

# STUDY GUIDE

## ► Testing Your Recall

*Answers in Appendix B*

1. Transitional epithelium is found in
   a. the urinary system.
   b. the respiratory system.
   c. the digestive system.
   d. the reproductive system.
   e. all of the above.

2. The external surface of the stomach is covered by
   a. a mucosa.
   b. a serosa.
   c. the parietal peritoneum.
   d. a lamina propria.
   e. a basement membrane.

3. Which of these is a primary germ layer?
   a. epidermis
   b. mucosa
   c. ectoderm
   d. endothelium
   e. epithelium

4. A seminiferous tubule of the testis is lined with _____ epithelium.
   a. simple cuboidal
   b. pseudostratified columnar ciliated
   c. stratified squamous
   d. transitional
   e. stratified cuboidal

5. _____ prevent fluids from seeping between epithelial cells.
   a. Glycosaminoglycans
   b. Hemidesmosomes
   c. Tight junctions
   d. Communicating junctions
   e. Basement membranes

6. A fixative serves to
   a. stop tissue decay.
   b. improve contrast.
   c. repair a damaged tissue.
   d. bind epithelial cells together.
   e. bind cardiac myocytes together.

7. The collagen of areolar tissue is produced by
   a. macrophages.
   b. fibroblasts.
   c. mast cells.
   d. leukocytes.
   e. chondrocytes.

8. Tendons are composed of _____ connective tissue.
   a. skeletal
   b. areolar
   c. dense irregular
   d. yellow elastic
   e. dense regular

9. The shape of the external ear is due to
   a. skeletal muscle.
   b. elastic cartilage.
   c. fibrocartilage.
   d. articular cartilage.
   e. hyaline cartilage.

10. The most abundant formed element(s) of blood is/are
    a. plasma.
    b. erythrocytes.
    c. platelets.
    d. leukocytes.
    e. proteins.

11. Any form of pathological tissue death is called _____.

12. The simple squamous epithelium that lines the peritoneal cavity is called _____.

13. Osteocytes and chondrocytes occupy little cavities called _____.

14. Muscle cells and axons are often called _____ because of their shape.

15. Tendons and ligaments are made mainly of the protein _____.

16. The only type of muscle that lacks gap junctions is _____.

17. An epithelium rests on a layer called the _____ between its deepest cells and the underlying connective tissue.

18. Fibers and ground substance make up the _____ of a connective tissue.

19. A/An _____ adult stem cell can differentiate into two or more mature cell types.

20. Any epithelium in which every cell touches the basement membrane is called a/an _____ epithelium.

## ► Building Your Medical Vocabulary

*Answers in Appendix B*

*State a medical meaning of each word element below, and give a term in which it or a slight variation of it is used.*

1. apo-

2. chondro-

3. ecto-

4. -gen

5. histo-

6. holo-

7. hyalo-

8. necro-

9. plas-

10. squamo-

# STUDY GUIDE

## ► True or False

*Answers in Appendix B*

*Determine which five of the following statements are false, and briefly explain why.*

1. The esophagus is protected from abrasion by a keratinized stratified squamous epithelium.

2. All cells of a pseudostratified columnar epithelium contact the basement membrane.

3. Not all skeletal muscle is attached to bones.

4. The stroma of a gland does not secrete anything.

5. In all connective tissues, the matrix occupies more space than the cells do.

6. Adipocytes are limited to adipose tissue.

7. Tight junctions function primarily to prevent cells from pulling apart.

8. Metaplasia is a normal, healthy tissue transformation but neoplasia is not.

9. Nerve and muscle cells are not the body's only electrically excitable cells.

10. Cartilage is always covered by a fibrous perichondrium.

## ► Testing Your Comprehension

*Answers at* www.mhhe.com/saladin7

1. A woman in labor is often told to push. In doing so, is she consciously contracting her uterus to expel the baby? Justify your answer based on the muscular composition of the uterus.

2. A major tenet of the cell theory is that all bodily structure and function are based on cells. The structural properties of bone, cartilage, and tendons, however, are due more to their extracellular material than to their cells. Is this an exception to the cell theory? Why or why not?

3. When cartilage is compressed, water is squeezed out of it, and when pressure is taken off, water flows back into the matrix. This being the case, why do you think cartilage at weight-bearing joints such as the knees can degenerate from lack of exercise?

4. The epithelium of the respiratory tract is mostly of the pseudostratified columnar ciliated type, but in the alveoli—the tiny air sacs where oxygen and carbon dioxide are exchanged between the blood and inhaled air—the epithelium is simple squamous. Explain the functional significance of this histological difference. That is, why don't the alveoli have the same kind of epithelium as the rest of the respiratory tract?

5. Which do you think would heal faster, cartilage or bone? Stratified squamous or simple columnar epithelium? Why?

## ► Improve Your Grade at www.mhhe.com/saladin7

*Download animations to study when it fits your schedule. Practice quizzes, labeling activities, games, and flashcards offer fun ways to master the chapter concepts. Or, download image PowerPoint files for each chapter to create a study guide or for taking notes during lecture.*

CHAPTER

# 6

# THE INTEGUMENTARY SYSTEM

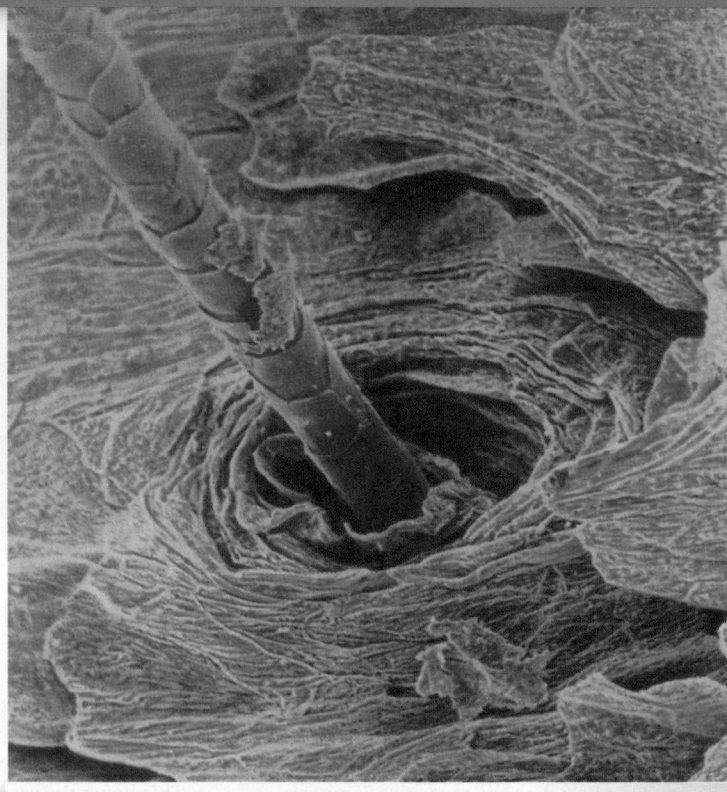

**A human hair emerging from its follicle (SEM)**

Anatomy & Physiology | **REVEALED**®
aprevealed.com

**Module 4: Integumentary System**

The skin is also known as the **integument,**[1] whereas the **integumentary system** consists of the skin and its *accessory organs*—the hair, nails, and cutaneous glands. We pay more attention to this organ system than to any other. It is, after all, the most visible one, and its appearance strongly affects our social interactions. Few people venture out of the house without first looking in a mirror to see if their skin and hair are presentable. Social considerations aside, the integumentary system is important to one's self-image, and a positive self-image is important to the attitudes that promote overall good health. Care of the integumentary system is thus a particularly important part of the total plan of patient care.

Inspection of the skin, hair, and nails is a significant part of a physical examination. The integumentary system provides clues not only to its own health, but also to deeper disorders such as liver cancer, anemia, and heart failure. The skin also is the most vulnerable of our organs, exposed to radiation, trauma, infection, and injurious chemicals. Consequently, it needs and receives more medical attention than any other organ system. The scientific study and medical treatment of the integumentary system are called **dermatology.**[2]

## 6.1 The Skin and Subcutaneous Tissue

### Expected Learning Outcomes

When you have completed this section, you should be able to

a. list the functions of the skin and relate them to its structure;

b. describe the histological structure of the epidermis, dermis, and subcutaneous tissue;

c. describe the normal and pathological colors that the skin can have, and explain their causes; and

d. describe the common markings of the skin.

The **skin** is the body's largest and heaviest organ. In adults, it covers an area of 1.5 to 2.0 m² and accounts for about 15% of the body weight. It consists of two layers: a stratified

---

[1] *integument* = covering
[2] *dermat* = skin; *logy* = study of

squamous epithelium called the *epidermis* and a deeper connective tissue layer called the *dermis* (fig. 6.1). Below the dermis is another connective tissue layer, the *hypodermis,* which is not part of the skin but is customarily studied in conjunction with it.

Most of the skin is 1 to 2 mm thick, but it ranges from less than 0.5 mm on the eyelids to 6 mm between the shoulder blades. The difference is due mainly to variation in thickness of the dermis, although skin is classified as thick or thin based on the relative thickness of the epidermis alone. **Thick skin** covers the palms, soles, and corresponding surfaces of the fingers and toes. Its epidermis alone is about 0.5 mm thick, due to a very thick surface layer of dead cells called the *stratum corneum* (see fig. 6.3). Thick skin has sweat glands but no hair follicles or sebaceous (oil) glands. The rest of the body is covered with **thin skin,** which has an epidermis about 0.1 mm thick, with a thin stratum corneum (see fig. 6.6). It possesses hair follicles, sebaceous glands, and sweat glands.

## Functions of the Skin

The skin is much more than a container for the body. It has a variety of important functions that go well beyond appearance, as we shall see here.

1. **Resistance to trauma and infection.** The skin suffers the most physical injuries to the body, but it resists and recovers from trauma better than other organs do. The epidermal cells are packed with the tough protein **keratin** and linked by strong desmosomes that give this epithelium its durability. Few infectious organisms can penetrate the intact skin. Bacteria and fungi colonize the surface, but their numbers are kept in check by its relative dryness, its slight acidity (pH 4–6), and certain defensive antimicrobial peptides called *dermcidin* and *defensins.* The protective acidic film is called the *acid mantle.*

2. **Other barrier functions.** The skin is important as a barrier to water. It prevents the body from absorbing excess water when you are swimming or bathing, but even more importantly, it prevents the body from losing excess water. The epidermis is also a barrier to ultraviolet (UV) rays, blocking much of this cancer-causing radiation from reaching deeper tissue layers; and it is a barrier to many potentially harmful chemicals. It is, however, permeable to several drugs and poisons (see Deeper Insight 6.1).

3. **Vitamin D synthesis.** The skin carries out the first step in the synthesis of vitamin D, which is needed for bone development and maintenance. The liver and kidneys complete the process. This is further detailed in figure 7.14 (p. 218).

4. **Sensation.** The skin is our most extensive sense organ. It is equipped with a variety of nerve endings that react to heat, cold, touch, texture, pressure, vibration, and tissue injury (see chapter 16). These sensory receptors are especially abundant on the face, palms, fingers, soles, nipples, and genitals. There are relatively few on the back and in skin overlying joints such as the knees and elbows.

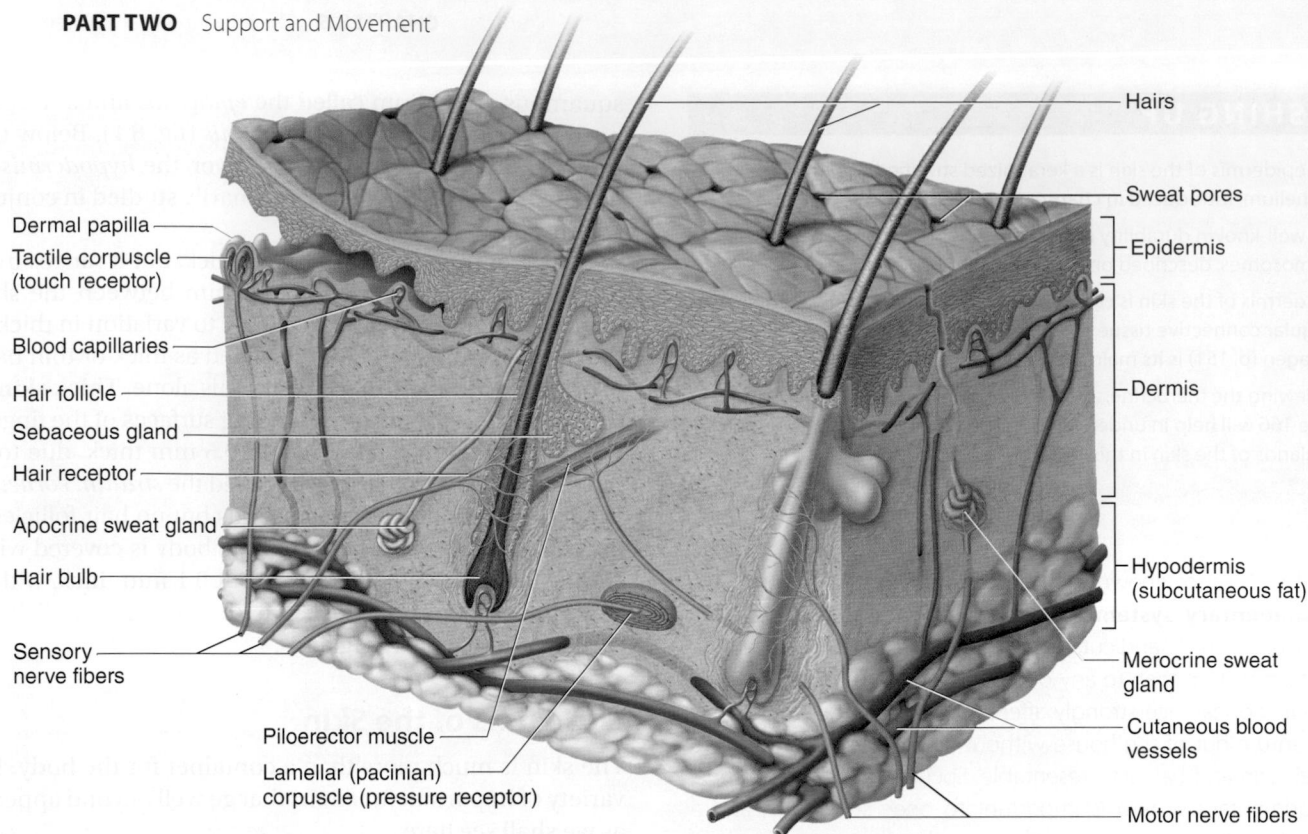

Dermal papilla

Tactile corpuscle (touch receptor)

Blood capillaries

Hair follicle

Sebaceous gland

Hair receptor

Apocrine sweat gland

Hair bulb

Sensory nerve fibers

Piloerector muscle

Lamellar (pacinian) corpuscle (pressure receptor)

Hairs

Sweat pores

Epidermis

Dermis

Hypodermis (subcutaneous fat)

Merocrine sweat gland

Cutaneous blood vessels

Motor nerve fibers

**FIGURE 6.1  Structure of the Skin and Subcutaneous Tissue.**  The epidermis is peeled up at the upper left corner to show the dermal–epidermal boundary.

5. **Thermoregulation.** Cutaneous nerve endings called **thermoreceptors** monitor the body surface temperature. In response to chilling, the body retains heat by constricting blood vessels of the dermis *(cutaneous vasoconstriction)*, keeping warm blood deeper in the body. In response to overheating, it loses excess heat by dilating those vessels *(cutaneous vasodilation)*, allowing more blood to flow close to the surface and lose heat through the skin. If this is insufficient to restore normal temperature, sweat glands secrete perspiration. The evaporation of sweat can have a powerful cooling effect. Thus, the skin plays roles in both warming and cooling the body.

6. **Nonverbal communication.** The skin is an important means of nonverbal communication. Humans, like most other primates, have much more expressive faces than other mammals (fig. 6.2). Complex skeletal muscles insert on dermal collagen fibers and pull on the skin to create subtle and varied facial expressions. The general appearance of the skin, hair, and nails is also important to social acceptance and to a person's self-image and emotional state—whether the ravages of adolescent acne, the presence of a birthmark or scar, or just a "bad hair day."

 **DEEPER INSIGHT 6.1**

## CLINICAL APPLICATION

### *Transdermal Absorption*

The ability of the skin to absorb chemicals makes it possible to administer several medicines as ointments or lotions, or by means of adhesive patches that release the medicine steadily through a membrane. For example, inflammation can be treated with a hydrocortisone ointment, nitroglycerine patches are used to relieve heart pain, nicotine patches are used to help overcome tobacco addiction, and other medicated patches are used to control high blood pressure and motion sickness.

Unfortunately, the skin can also be a route for absorption of poisons. These include toxins from poison ivy and other plants; metals such as mercury, arsenic, and lead; and solvents such as acetone (nail polish remover), paint thinner, and pesticides. Some of these can cause brain damage, liver failure, or kidney failure, which is good reason for using protective gloves when handling such substances.

## The Epidermis

The **epidermis**[3] is a keratinized stratified squamous epithelium, as described in chapter 5. That is, its surface consists of dead cells packed with the tough protein keratin. Like other epithelia, the epidermis lacks blood vessels and depends on the diffusion of nutrients from the underlying connective tissue. It has sparse nerve endings for touch and pain, but most sensations of the skin are due to nerve endings in the dermis.

---

[3]*epi* = above, upon; *derm* = skin

**(a)**

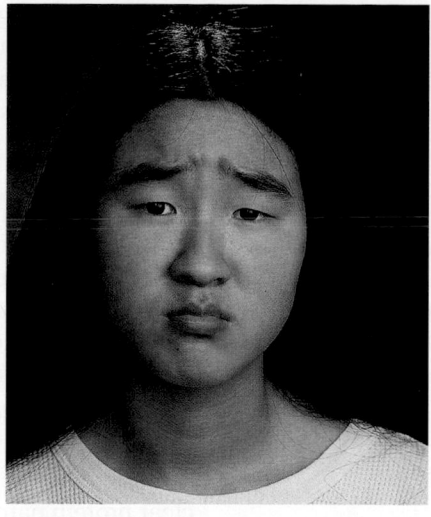

**(b)**

**FIGURE 6.2 Importance of the Skin in Nonverbal Expression.** Primates differ from other mammals in having very expressive faces due to facial muscles that insert on the collagen fibers of the dermis and move the skin.

## Cells of the Epidermis

The epidermis is composed of five types of cells (fig. 6.3):

1. **Stem cells** are undifferentiated cells that divide and give rise to the keratinocytes (described next). They are found only in the deepest layer of the epidermis, called the *stratum basale* (described later).

2. **Keratinocytes** (keh-RAT-ih-no-sites) are the great majority of epidermal cells. They are named for their role in synthesizing keratin. In ordinary histological specimens, nearly all visible epidermal cells are keratinocytes.

3. **Melanocytes** also occur only in the stratum basale, amid the stem cells and deepest keratinocytes. They synthesize the brown to black pigment *melanin*. They have branching processes that spread among the keratinocytes and continually shed melanin-containing fragments from their tips. The keratinocytes phagocytize these fragments and accumulate melanin granules on the "sunny side" of the nucleus. The ultraviolet (UV) component of sunlight has the potential to mutate DNA, causing skin cancer and other consequences. But like a parasol, this aggregation of melanin granules shields the DNA from UV radiation.

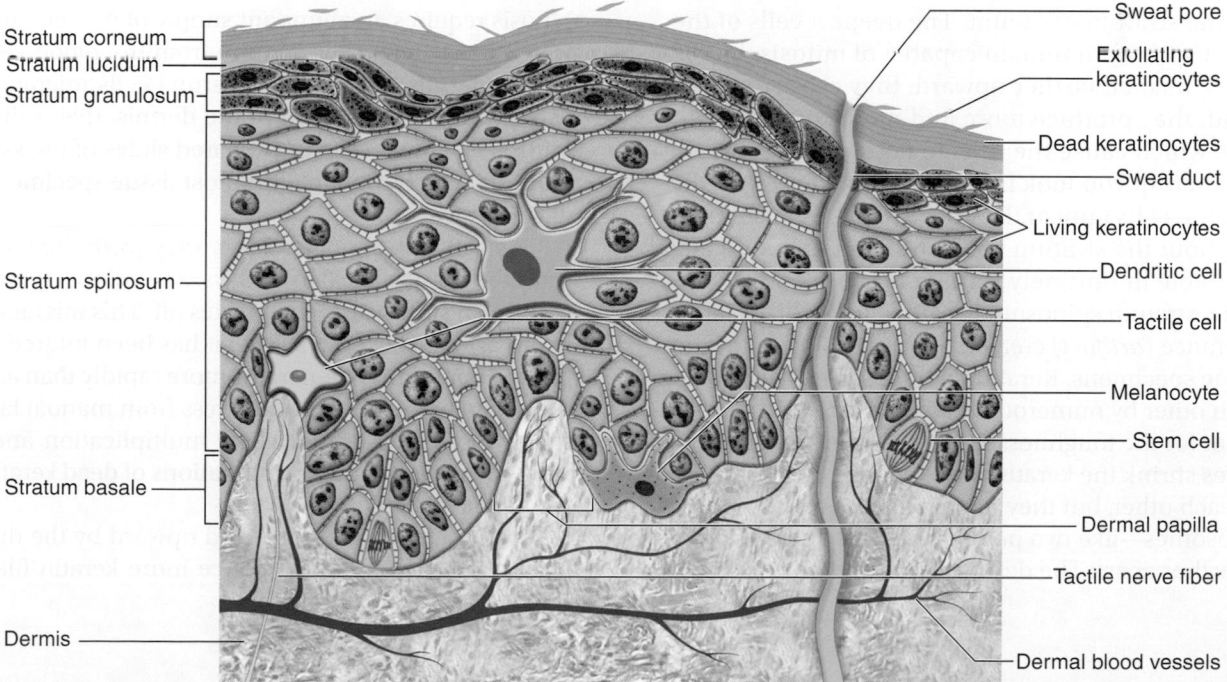

Stratum corneum
Stratum lucidum
Stratum granulosum
Stratum spinosum
Stratum basale
Dermis

Sweat pore
Exfoliating keratinocytes
Dead keratinocytes
Sweat duct
Living keratinocytes
Dendritic cell
Tactile cell
Melanocyte
Stem cell
Dermal papilla
Tactile nerve fiber
Dermal blood vessels

**FIGURE 6.3 Strata and Cell Types of the Epidermis.** AP|R

4. **Tactile cells,** relatively few in number, are receptors for touch. They, too, are found in the basal layer of the epidermis and are associated with an underlying dermal nerve fiber. The tactile cell and its nerve fiber are collectively called a *tactile disc.*

5. **Dendritic**[4] **cells** are found in two layers of the epidermis called the *stratum spinosum* and *stratum granulosum* (described in the next section). They are immune cells that originate in the bone marrow but migrate to the epidermis and epithelia of the oral cavity, esophagus, and vagina. The epidermis has as many as 800 dendritic cells per square millimeter. They stand guard against toxins, microbes, and other pathogens that penetrate into the skin. When they detect such invaders, they alert the immune system so the body can defend itself.

## Layers of the Epidermis

Cells of the epidermis are arranged in four to five zones, or strata (five in thick skin) (fig. 6.3). The following description progresses from deep to superficial, and from the youngest to the oldest keratinocytes.

1. The **stratum basale** (bah-SAY-lee) consists mainly of a single layer of cuboidal to low columnar stem cells and keratinocytes resting on the basement membrane. Scattered among these are the melanocytes, tactile cells, and stem cells. As the stem cells divide, they give rise to keratinocytes that migrate toward the skin surface and replace lost epidermal cells. The life history of these cells is described in the next section.

2. The **stratum spinosum** (spy-NO-sum) consists of several layers of keratinocytes. In most skin, this is the thickest stratum, but in thick skin it is usually exceeded by the stratum corneum. The deepest cells of the stratum spinosum remain capable of mitosis, but as they are pushed farther upward, they cease dividing. Instead, they produce more and more keratin filaments, which cause the cells to flatten. Therefore, the higher up you look in the stratum spinosum, the flatter the cells appear. Dendritic cells are also found throughout the stratum spinosum, but are not usually identifiable in routinely stained tissue sections.

   The stratum spinosum is named for an artificial appearance *(artifact)* created by the histological fixation of tissue specimens. Keratinocytes are firmly attached to each other by numerous desmosomes, which partly account for the toughness of the epidermis. Histological fixatives shrink the keratinocytes so they pull away from each other, but they remain attached by the desmosomes—like two people holding hands while they step farther apart. The desmosomes thus create bridges

from cell to cell, giving each cell a spiny appearance from which we derive the word *spinosum.*

Epidermal keratinocytes are also bound to each other by tight junctions, which make an essential contribution to water retention by the skin. This is further discussed in the next section.

3. The **stratum granulosum** consists of three to five layers of flat keratinocytes—more in thick skin than in thin skin. The keratinocytes of this layer contain coarse, dark-staining *keratohyalin granules* that give the layer its name. The functional significance of these granules will be explained shortly.

4. The **stratum lucidum**[5] (LOO-sih-dum) is a thin zone superficial to the stratum granulosum, seen only in thick skin. Here, the keratinocytes are densely packed with a clear protein named *eleidin* (ee-LEE-ih-din). The cells have no nuclei or other organelles. This zone has a pale, featureless appearance with indistinct cell boundaries.

5. The **stratum corneum** consists of up to 30 layers of dead, scaly, keratinized cells that form a durable surface layer. It is especially resistant to abrasion, penetration, and water loss.

## The Life History of a Keratinocyte

Dead cells constantly flake off the skin surface. They float around as tiny white specks in the air, settling on household surfaces and forming much of the house dust that accumulates there (see Deeper Insight 6.2). Because we constantly lose these epidermal cells, they must be continually replaced.

Keratinocytes are produced deep in the epidermis by the mitosis of stem cells in the stratum basale. Some of the deepest keratinocytes in the stratum spinosum also continue dividing. Mitosis requires an abundant supply of oxygen and nutrients, which these deep cells acquire from the blood vessels in the nearby dermis. Once the epidermal cells migrate more than two or three cells away from the dermis, their mitosis ceases. Mitosis is seldom seen in prepared slides of the skin, because it occurs mainly at night and most tissue specimens are taken during the day.

As new keratinocytes form, they push the older ones toward the surface. In 30 to 40 days, a keratinocyte makes its way to the skin surface and then flakes off. This migration is slower in old age and faster in skin that has been injured or stressed. Injured epidermis regenerates more rapidly than any other tissue in the body. Mechanical stress from manual labor or tight shoes accelerates keratinocyte multiplication and results in *calluses* or *corns,* thick accumulations of dead keratinocytes on the hands or feet.

As keratinocytes are shoved upward by the dividing cells below, they flatten and produce more keratin filaments and

---

[4]*dendr* = tree, branch

[5]*lucid* = light, clear

## DEEPER INSIGHT 6.2

### CLINICAL APPLICATION

#### Dead Skin and Dust Mites

In the beams of late afternoon sun that shine aslant through a window, you may see tiny white specks floating through the air. Most of these are flakes of dander; the dust on top of your bookshelves is largely a film of dead human skin. Composed of protein, this dust in turn supports molds and other microscopic organisms that feed on the skin cells and each other. One of these organisms is the house dust mite, *Dermatophagoides*[6] (der-MAT-oh-fah-GOY-deez) (fig. 6.4). (What wonders may be found in humble places!)

*Dermatophagoides* thrives abundantly in pillows, mattresses, and upholstery—warm, humid places that are liberally sprinkled with edible flakes of keratin. No home is without these mites, and it is impossible to entirely exterminate them. What was once regarded as "house dust allergy" has been identified as an allergy to the inhaled feces of these mites.

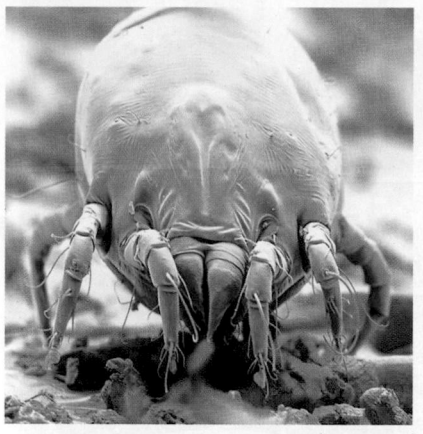

**FIGURE 6.4** The House Dust Mite, *Dermatophagoides*.

0.1 mm

---

lipid-filled **membrane-coating vesicles.** In the stratum granulosum, four important developments occur: (1) Keratohyalin granules release a protein called *filaggrin* that binds the cytoskeletal keratin filaments together into coarse, tough bundles. (2) The cells produce a tough layer of *envelope proteins* just beneath the plasma membrane, resulting in a nearly indestructible protein sac around the keratin bundles. (3) Membrane-coating vesicles release a lipid mixture that spreads out over the cell surface and waterproofs it. (4) Finally, as these barriers cut the keratinocytes off from the supply of nutrients from below, their organelles degenerate and the cells die, leaving just the tough waterproof sac enclosing coarse bundles of keratin. These processes, along with the tight junctions between keratinocytes, result in an **epidermal water barrier** that is crucial to the retention of body water.

The stratum corneum consists of compact layers of dead keratinocytes and keratinocyte fragments. Dead keratinocytes soon **exfoliate** (flake off) from the epidermal surface as specks called **dander.** *Dandruff* is composed of clumps of dander stuck together by sebum (oil).

A curious effect of the epidermal water barrier is the way our skin wrinkles when we linger in the bath or a lake. The keratin of the stratum corneum absorbs water and swells, but the deeper layers of the skin do not. The thickening of the stratum corneum forces it to wrinkle. This is especially conspicuous on the fingers and toes ("prune fingers") because they have such a thick stratum corneum and they lack the sebaceous glands that produce water-resistant oil elsewhere on the body. There may be more to the story than this, however, because the wrinkles do not form when the nerves to the fingers are severed, indicating some role for the nervous system. It has recently been hypothesized that

this buckling of the skin may serve a function similar to treads on a car tire, to channel water away and improve our grip when we press our fingertips to wet surfaces.

## The Dermis

Beneath the epidermis is a connective tissue layer, the **dermis.** It ranges from 0.2 mm thick in the eyelids to about 4 mm thick in the palms and soles. It is composed mainly of collagen, but also contains elastic and reticular fibers, fibroblasts, and the other cells typical of fibrous connective tissue (described in chapter 5). It is well supplied with blood vessels, cutaneous glands, and nerve endings. The hair follicles and nail roots are embedded in the dermis. In the face, skeletal muscles attach to dermal collagen fibers and produce such expressions as a smile, a wrinkle of the forehead, or the lifting of an eyebrow.

The boundary between the epidermis and dermis is histologically conspicuous and usually wavy. The upward waves are fingerlike extensions of the dermis called **dermal papillae**[7] (see fig. 6.3), and the downward waves are extensions of the epidermis called **epidermal ridges.** The dermal and epidermal boundaries thus interlock like corrugated cardboard, an arrangement that resists slippage of the epidermis across the dermis. If you look closely at your hand and wrist, you will see delicate furrows that divide the skin into tiny rectangular to rhomboidal areas. The dermal papillae produce the raised areas between the furrows. On the fingertips, this wavy boundary forms the *friction ridges* that produce fingerprints. In highly sensitive areas such as the lips and genitals, exceptionally tall dermal papillae allow nerve fibers to reach close to the surface.

---

[6]*dermato* = skin; *phag* = eat

[7]*pap* = nipple; *illa* = little

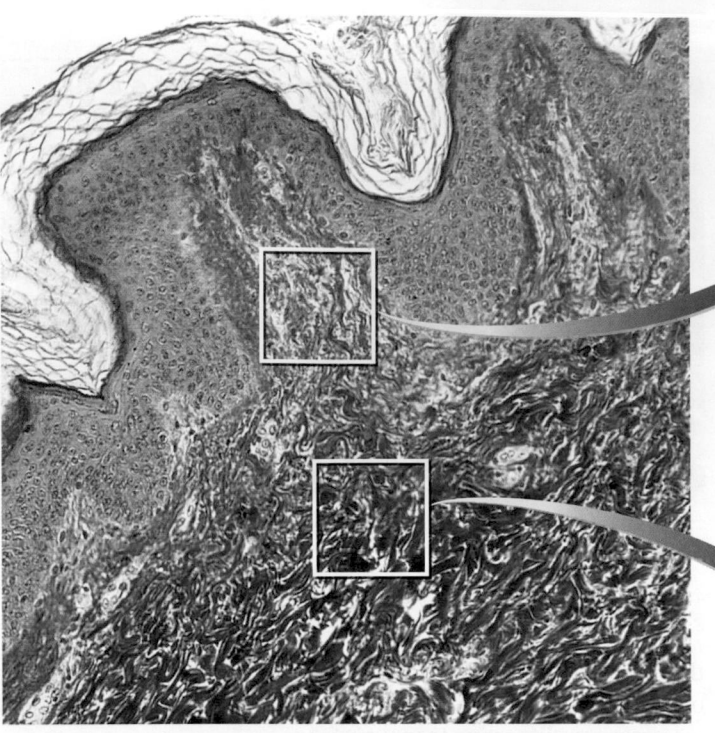

(a)

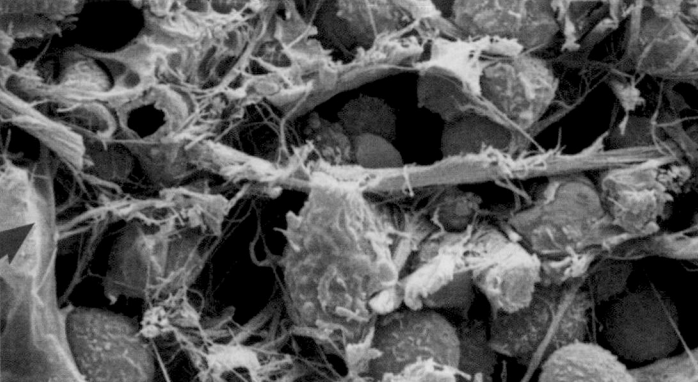

**(b) Papillary layer of dermis**

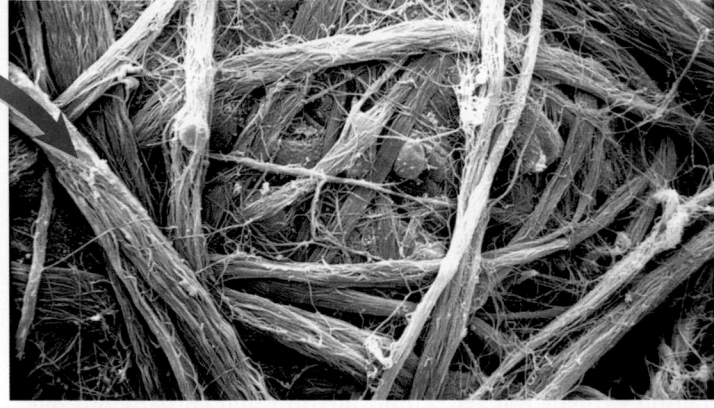

**(c) Reticular layer of dermis**

**FIGURE 6.5  The Dermis.** (a) Light micrograph of axillary skin, with the collagen stained blue. (b) The papillary layer, made of loose (areolar) tissue, forms the dermal papillae. (c) The reticular layer, made of dense irregular connective tissue, forms the deeper four-fifths of the dermis.  © Dr. Kessel & Dr. Kardon/Tissues and Organs/Visuals Unlimited, Inc.  **AP|R**

▶▶▶ **APPLY WHAT YOU KNOW**

*Dermal papillae are relatively high and numerous in palmar and plantar skin but low and few in number in the skin of the face and abdomen. What do you think is the functional significance of this difference?*

There are two zones of dermis called the papillary and reticular layers (fig. 6.5). The **papillary** (PAP-ih-lerr-ee) **layer** is a thin zone of areolar tissue in and near the dermal papillae. This loosely organized tissue allows for mobility of leukocytes and other defenses against organisms introduced through breaks in the epidermis. This layer is especially rich in small blood vessels.

The **reticular**[8] **layer** of the dermis is deeper and much thicker. It consists of dense irregular connective tissue. The boundary between the papillary and reticular layers is often vague. In the reticular layer, the collagen forms thicker bundles with less room for ground substance, and there are often small clusters of adipocytes. Stretching of the skin in obesity and pregnancy can tear the collagen fibers and produce *striae* (STRY-ee), or stretch marks. These occur especially in areas most stretched by weight gain: the thighs, buttocks, abdomen, and breasts.

## The Hypodermis

Beneath the skin is a layer called the **hypodermis**[9] or **subcutaneous tissue** (see fig. 6.1). The boundary between the dermis and hypodermis is indistinct, but the hypodermis generally has more areolar and adipose tissue. It pads the body and binds the skin to the underlying tissues. Drugs are introduced into the hypodermis by injection because the subcutaneous tissue is highly vascular and absorbs them quickly.

**Subcutaneous fat** is hypodermis composed predominantly of adipose tissue. It serves as an energy reservoir and thermal insulation. It is not uniformly distributed; for example, it is virtually absent from the scalp but relatively abundant in the breasts, abdomen, hips, and thighs. The subcutaneous fat averages about 8% thicker in women than in men, and varies with age. Infants and elderly people have less subcutaneous fat than other people and are therefore more sensitive to cold.

Table 6.1 summarizes the layers of the skin and hypodermis.

---

[8]*reti* = network; *cul* = little

[9]*hypo* = below; *derm* = skin

| TABLE 6.1 | Stratification of the Skin and Hypodermis |
|-----------|--------------------------------------------|
| **Layer** | **Description** |
| *Epidermis* | Keratinized stratified squamous epithelium |
| Stratum corneum | Dead, keratinized cells of the skin surface |
| Stratum lucidum | Clear, featureless, narrow zone seen only in thick skin |
| Stratum granulosum | Two to five layers of cells with dark-staining keratohyalin granules; scanty in thin skin |
| Stratum spinosum | Many layers of keratinocytes, typically shrunken in fixed tissues but attached to each other by desmosomes, which give them a spiny look; progressively flattened the farther they are from the dermis. Dendritic cells are abundant here but are not distinguishable in routinely stained preparations. |
| Stratum basale | Single layer of cuboidal to columnar cells resting on basement membrane; site of most mitosis; consists of stem cells, keratinocytes, melanocytes, and tactile cells, but these are difficult to distinguish with routine stains. Melanin is conspicuous in keratinocytes of this layer in black to brown skin. |
| *Dermis* | Fibrous connective tissue, richly endowed with blood vessels and nerve endings. Sweat glands and hair follicles originate here and in hypodermis. |
| Papillary layer | Superficial one-fifth of dermis; composed of areolar tissue; often extends upward as dermal papillae |
| Reticular layer | Deeper four-fifths of dermis; dense irregular connective tissue |
| *Hypodermis* | Areolar or adipose tissue between skin and muscle |

## Skin Color

The most significant factor in skin color is **melanin.** This is produced by the melanocytes but accumulates in the keratinocytes of the stratum basale and stratum spinosum (fig. 6.6). There are two forms of melanin: a brownish black **eumelanin**[10] and a reddish yellow sulfur-containing pigment, **pheomelanin.**[11] People of different skin colors have essentially the same number of melanocytes, but in dark skin, the melanocytes produce greater quantities of melanin, the melanin granules in the keratinocytes are more spread out than tightly clumped, and the melanin breaks down more slowly. Thus, melanized cells may be seen throughout the epidermis, from stratum basale to stratum corneum. In light skin, the melanin is clumped near the keratinocyte nucleus, so it imparts less color to the cells. It also breaks down more rapidly, so little of it is seen beyond the stratum basale, if even there.

Skin color also varies with exposure to the UV rays of sunlight, which stimulate melanin synthesis and darken the skin. A suntan fades as melanin is degraded in older keratinocytes and as keratinocytes migrate to the surface and exfoliate. The amount of melanin also varies from place to place on the body. It is relatively concentrated in freckles and moles, on the dorsal surfaces of the hands and feet as compared with the palms and soles, on the nipple and surrounding area (areola) of the breast, around the anus, on the scrotum and penis, and on the lateral surfaces of the female genital folds (labia majora). The contrast between heavily melanized and lightly melanized regions of the skin is more pronounced in some people than others, but it exists to some extent in nearly everyone. Variation in ancestral exposure to UV is the primary reason for the geographic and ethnic variation in skin color today (see Deeper Insight 6.3).

Other factors in skin color are hemoglobin and carotene. **Hemoglobin,** the red pigment of blood, imparts reddish to pinkish hues as blood vessels show through the skin. Its color is lightened by the white of the dermal collagen. The skin is redder in places such as the lips, where blood capillaries come closer to the surface and the hemoglobin shows through more vividly. **Carotene**[12] is a yellow pigment acquired from egg yolks and yellow and orange vegetables. Depending on the diet, carotene or related compounds can become concentrated to various degrees in the stratum corneum and subcutaneous fat, imparting a yellow color. This is often most conspicuous in skin of the heel and in calluses of the feet, because this is where the stratum corneum is thickest.

The skin may also exhibit abnormal colors of diagnostic value:

- **Cyanosis**[13] is blueness of the skin resulting from a deficiency of oxygen in the circulating blood. Oxygen deficiency turns the hemoglobin a reddish violet color, which is lightened to blue-violet as it shows through the white dermal collagen. Oxygen deficiency can result from conditions that prevent the blood from picking up a normal load of oxygen in the lungs, such as airway obstructions in drowning and choking, lung diseases such as emphysema,

---

[10]*eu* = true; *melan* = black; *in* = substance
[11]*pheo* = dusky; *melan* = black; *in* = substance

[12]*carot* = carrot
[13]*cyan* = blue; *osis* = condition

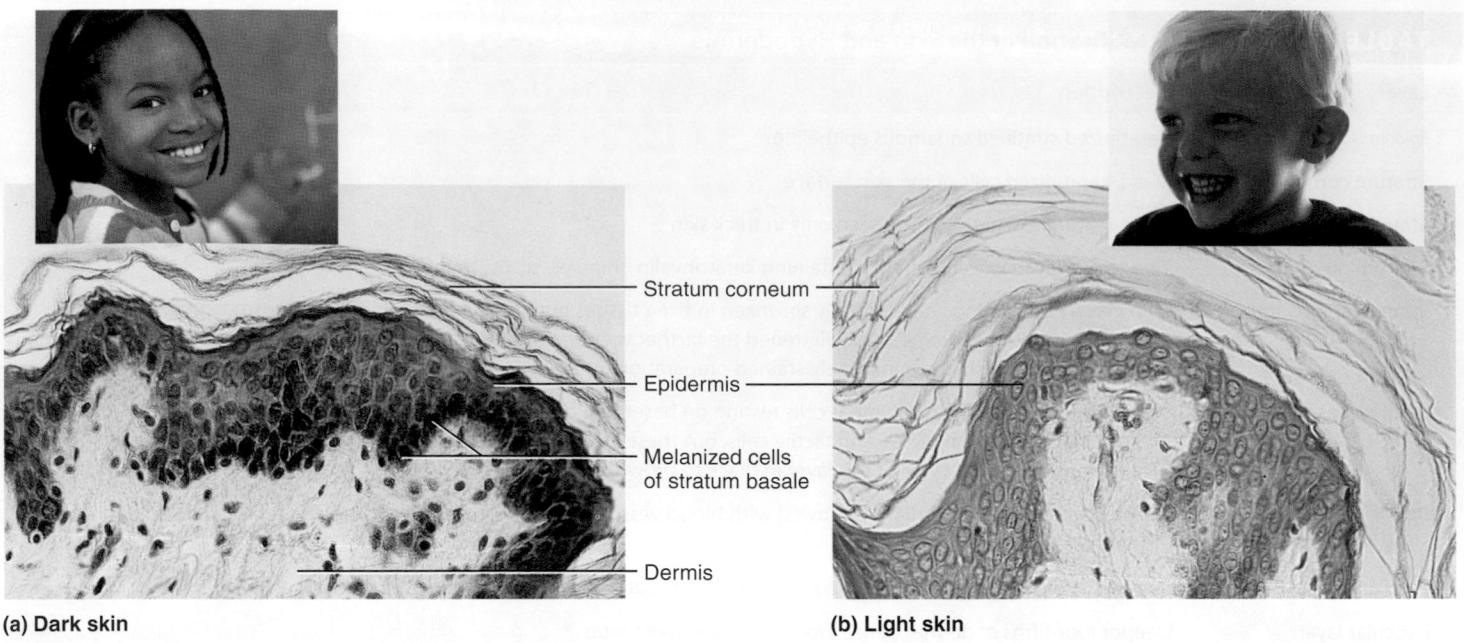

**(a) Dark skin**

Stratum corneum

Epidermis

Melanized cells
of stratum basale

Dermis

**(b) Light skin**

**FIGURE 6.6  Variations in Skin Pigmentation.** (a) The stratum basale shows heavy deposits of melanin in dark skin. (b) Light skin shows little to no visible melanin.

 *Which of the five types of epidermal cells are the melanized cells in part (a)?*

 **DEEPER INSIGHT 6.3**

**EVOLUTIONARY MEDICINE**

### The Evolution of Skin Color

One of the most conspicuous signs of human variation is skin color, which can range from the color of espresso or milk chocolate to café au lait or light peach. Such variation results from a combination of evolutionary selection pressures, especially differences in exposure to ultraviolet (UV) radiation.

UV can have two adverse effects: It causes skin cancer and it breaks down folic acid, a B vitamin needed for normal cell division, fertility, and fetal development. It also has a desirable effect: It stimulates keratinocytes to synthesize vitamin D, which is needed for the absorption of dietary calcium and thus for healthy bone development. Too much UV and one is at risk of infertility and fetal deformities such as spina bifida; too little and one is at risk of bone deformities such as rickets. Consequently, populations native to the tropics and people descended from them tend to have well-melanized skin to screen out excessive UV. Populations native to far northern and southern latitudes, where the sunlight is weak, tend to have light skin to allow for adequate UV penetration. Ancestral skin color is thus partly a compromise between vitamin D and

folic acid requirements. Worldwide, women have skin averaging about 4% lighter than men do, perhaps because of their greater need for vitamin D and calcium to support pregnancy and lactation.

But for multiple reasons, there are exceptions to this trend. UV exposure is determined by more than latitude. It increases at higher elevations and in dry air, because the thinner, drier atmosphere filters out less UV. This helps to explain the dark skin of people in such localities as the Andes and the high plateaus of Tibet and Ethiopia. UV levels account for up to 77% of the variation in human skin color. Some other exceptions may be the result of human migrations from one latitude to another occurring too recently for their skin color to have adapted to the new level of UV exposure. Variation may also result from cultural differences in clothing and shelter, intermarriage among people of different geographic ancestries, and darwinian sexual selection—a preference in mate choice for partners of light or dark complexion.

and respiratory arrest. Cyanosis also occurs in situations such as cold weather and cardiac arrest, when blood flows so slowly through the skin that the tissues consume most of the blood's oxygen faster than freshly oxygenated blood arrives.

- **Erythema**[14] (ERR-ih-THEE-muh) is abnormal redness of the skin. It occurs in such situations as exercise, hot

weather, sunburn, anger, and embarrassment. Erythema is caused by increased blood flow in dilated cutaneous blood vessels or by dermal pooling of red blood cells that have escaped from abnormally permeable capillaries, as in sunburn.

- **Pallor** is a pale or ashen color that occurs when there is so little blood flow through the skin that the white of the dermal collagen shows through. It can result from emotional stress, low blood pressure, circulatory shock, cold temperatures, or severe anemia.

---

[14]*eryth* = red; *em* = blood

- **Albinism**[15] is a genetic lack of melanin that usually results in milky white hair and skin, and blue-gray eyes. Melanin is synthesized from the amino acid tyrosine by the enzyme tyrosinase. People with albinism have inherited a recessive, nonfunctional tyrosinase allele from both parents.

- **Jaundice**[16] is yellowing of the skin and whites of the eyes resulting from high levels of bilirubin in the blood. Bilirubin is a hemoglobin breakdown product. When erythrocytes get old, they disintegrate and release their hemoglobin. The liver and spleen convert hemoglobin to bilirubin, which the liver excretes in the bile. Bilirubin can accumulate enough to discolor the skin, however, in such situations as a rapid rate of erythrocyte destruction; when diseases such as cancer, hepatitis, and cirrhosis compromise liver function; and in premature infants, where the liver is not well enough developed to dispose of bilirubin efficiently.

- A **hematoma**,[17] or bruise, is a mass of clotted blood showing through the skin. It is usually due to accidental trauma (blows to the skin), but it may indicate hemophilia, other metabolic or nutritional disorders, or physical abuse.

### ▶▶▶ APPLY WHAT YOU KNOW

*An infant brought to a clinic shows abnormally yellow skin. What sign could you look for to help decide whether this was due to jaundice or to a large amount of carotene from strained vegetables in the diet?*

## Skin Markings

The skin is marked by many lines, creases, ridges, and patches of accentuated pigmentation. **Friction ridges** are the markings on the fingertips that leave distinctive oily fingerprints on surfaces we touch. They are characteristic of most primates, though their function has long been obscure. They enhance one's sensitivity to texture by vibrating and stimulating sense organs called *lamellar corpuscles* deeper in the skin when the fingertips stroke an uneven surface. They are also thought to improve one's grasp and aid in the manipulation of small and rough-surfaced objects. Friction ridges form during fetal development and remain essentially unchanged for life. Everyone has a unique pattern of friction ridges; not even identical twins produce identical fingerprints.

**Flexion lines (flexion creases)** are the lines on the flexor surfaces of the digits, palms, wrists, elbows, and other places (see fig. B.19, p. 391). They mark sites where the skin folds during flexion of the joints. The skin is tightly bound to deeper connective tissues along these lines.

Freckles and moles are tan to black aggregations of melanocytes. **Freckles** are flat melanized patches that vary with heredity and exposure to the sun. A **mole (nevus)** is an elevated patch of melanized skin, often with hair. Moles are harmless and sometimes even regarded as "beauty marks," but they should be watched for changes in color, diameter, or contour that may suggest malignancy (skin cancer) (see p. 193).

Birthmarks, or **hemangiomas**,[18] are patches of skin discolored by benign tumors of the blood capillaries. *Capillary hemangiomas* (strawberry birthmarks) usually develop about a month after birth. They become bright red to deep purple and develop small capillary-dense elevations that give them a strawberry-like appearance. About 90% of capillary hemangiomas disappear by the age of 5 or 6 years. *Cavernous hemangiomas* are flatter and duller in color. They are present at birth, enlarge up to 1 year of age, and then regress. About 90% disappear by the age of 9 years. A *port-wine stain* is flat and pinkish to dark purple in color. It can be quite large and remains for life.

### BEFORE YOU GO ON

Answer the following questions to test your understanding of the preceding section:

1. What is the major histological difference between thick and thin skin? Where on the body is each type of skin found?
2. How does the skin help to adjust body temperature?
3. List the five cell types of the epidermis. Describe their locations and functions.
4. List the five layers of epidermis from deep to superficial. What are the distinctive features of each layer? Which layer is often absent?
5. What are the two layers of the dermis? What type of tissue composes each layer?
6. Name the pigments responsible for normal skin colors and explain how certain conditions can produce discolorations of the skin.

## 6.2 Hair and Nails

### Expected Learning Outcomes

When you have completed this section, you should be able to

a. distinguish between three types of hair;
b. describe the histology of a hair and its follicle;
c. discuss some theories of the purposes served by various kinds of hair; and
d. describe the structure and function of nails.

The hair, nails, and cutaneous glands are the **accessory organs (appendages)** of the skin. Hair and nails are composed mostly of dead, keratinized cells. The stratum corneum of the skin is

---

[15]*alb* = white; *ism* = state, condition
[16]*jaun* = yellow
[17]*hemat* = blood; *oma* = mass

[18]*hem* = blood; *angi* = vessels; *oma* = mass, tumor

made of pliable **soft keratin,** but the hair and nails are composed mostly of **hard keratin.** This is more compact than soft keratin and is toughened by numerous cross-linkages between the keratin molecules.

## Hair

A hair is also known as a **pilus** (PY-lus); in the plural, *pili* (PY-lye). It is a slender filament of keratinized cells that grows from an oblique tube in the skin called a **hair follicle** (fig. 6.7).

### Distribution and Types

Hair occurs almost everywhere on the body except the palms and soles; palmar, plantar, and lateral surfaces and distal segments of the fingers and toes; and the lips, nipples, and parts of the genitals. The limbs and trunk have about 55 to 70 hairs per square centimeter, and the face has about 10 times as many. There are about 30,000 hairs in a man's beard and about 100,000 hairs in the average person's scalp. The density of hair does not differ much from one person to another or even between the sexes; indeed, it is virtually the same in humans,

chimpanzees, and gorillas. Differences in apparent hairiness are due mainly to differences in texture and pigmentation.

Not all hair is alike, even on one person. Over the course of our lives, we grow three kinds of hair: lanugo, vellus, and terminal hair. **Lanugo**[19] is fine, downy, unpigmented hair that appears on the fetus in the last 3 months of development. By the time of birth, most of it is replaced by **vellus,**[20] similarly fine, pale hair. Vellus constitutes about two-thirds of the hair of women, one-tenth of the hair of men, and all the hair of children except for the eyebrows, eyelashes, and hair of the scalp. **Terminal hair** is longer, coarser, and usually more heavily pigmented. It forms the eyebrows and eyelashes and covers the scalp; after puberty, it forms the axillary and pubic hair, the male facial hair, and some of the hair on the trunk and limbs.

### Structure of the Hair and Follicle

A hair is divisible into three zones along its length: (1) the **bulb,** a swelling at the base where the hair originates in the dermis or hypodermis; (2) the **root,** which is the remainder of the hair within the follicle; and (3) the **shaft,** which is the portion above the skin surface. The only living cells of a hair are in and near the bulb. The bulb grows around a bud of vascular connective

---

[19]*lan* = down, wool
[20]*vellus* = fleece

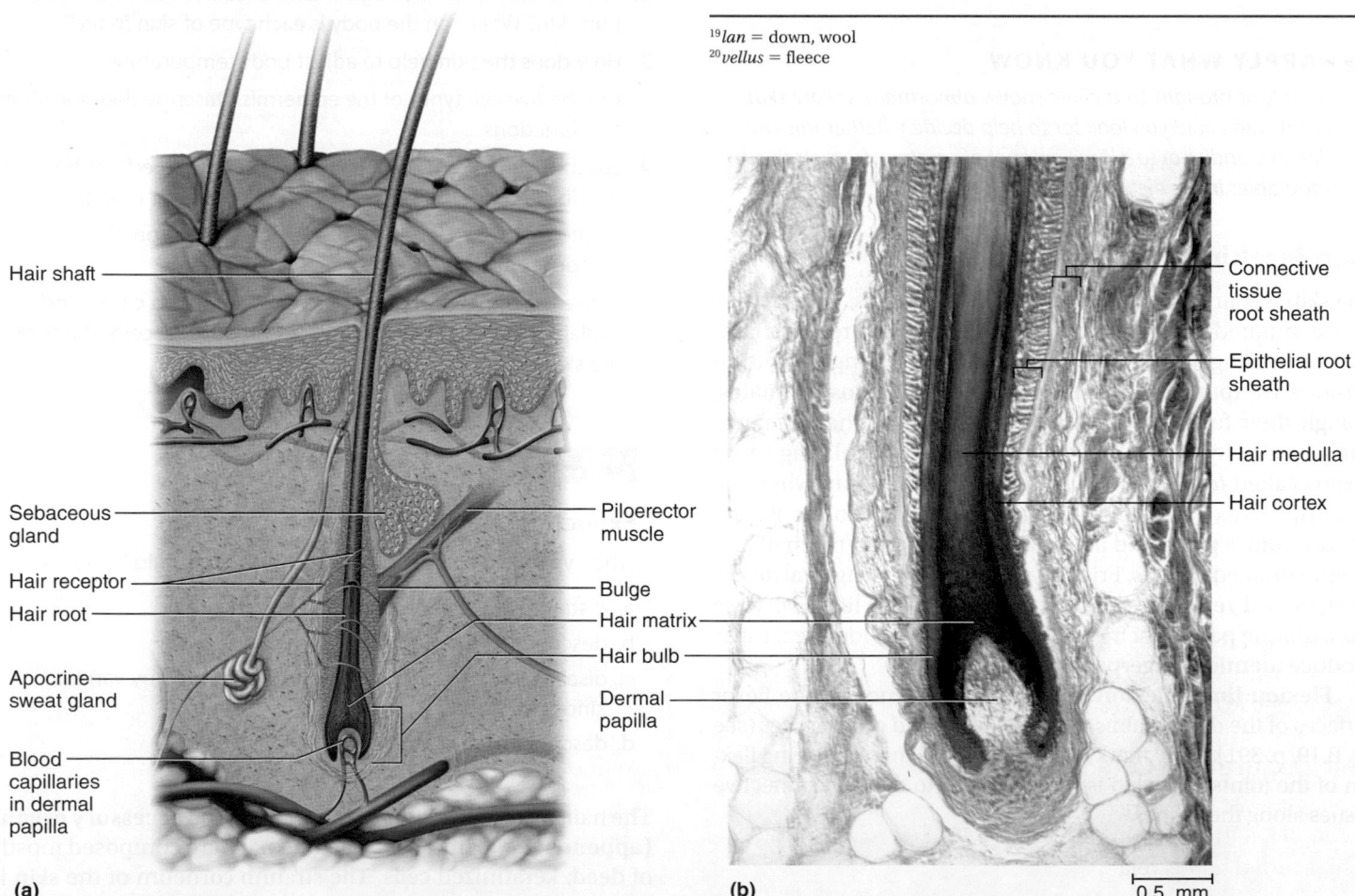

Hair shaft

Sebaceous gland

Hair receptor

Hair root

Apocrine sweat gland

Blood capillaries in dermal papilla

Piloerector muscle

Bulge

Hair matrix

Hair bulb

Dermal papilla

(a)

Connective tissue root sheath

Epithelial root sheath

Hair medulla

Hair cortex

(b)                    0.5 mm

**FIGURE 6.7  Structure of a Hair and Its Follicle.**  (a) Anatomy of the follicle and associated structures. (b) Light micrograph of the base of a hair follicle.  **AP|R**

tissue called the **dermal papilla,** which provides the hair with its sole source of nutrition. Immediately above the papilla is a region of mitotically active cells, the **hair matrix,** which is the hair's growth center. All cells higher up are dead.

In cross section, a hair reveals up to three layers. From the inside out, these are the medulla, cortex, and cuticle. The **medulla** is a core of loosely arranged cells and air spaces. It is most prominent in thick hairs such as those of the eyebrows, but narrower in hairs of medium thickness and absent from the thinnest hairs of the scalp and elsewhere. The **cortex** constitutes most of the bulk of a hair. It consists of several layers of elongated keratinized cells that appear cuboidal to flattened in cross sections. The **cuticle** is composed of multiple layers of very thin, scaly cells that overlap each other like roof shingles with their free edges directed upward (see photo on p. 176). Cells lining the follicle are like shingles facing in the opposite direction. They interlock with the scales of the hair cuticle and resist pulling on the hair. When a hair is pulled out, this layer of follicle cells comes with it.

The follicle is a diagonal tube that dips deeply into the dermis and sometimes extends as far as the hypodermis. It has two principal layers: an **epithelial root sheath** and a **connective tissue root sheath.** The epithelial root sheath, which is an extension of the epidermis, is composed of stratified squamous epithelium and lies immediately adjacent to the hair root. Toward the deep end of the follicle, it widens to form a **bulge,** a source of stem cells for follicle growth. The connective tissue root sheath, which is derived from the dermis and composed of collagenous connective tissue, surrounds the epithelial sheath and is somewhat denser than the adjacent dermal tissue.

Associated with the follicle are nerve and muscle fibers. Nerve fibers called **hair receptors** entwine each follicle and respond to hair movements. You can feel their effect by carefully moving a single hair with a pin or by lightly running your finger over the hairs of your forearm without touching the skin. Each hair has a **piloerector muscle**—also known as a *pilomotor muscle* or *arrector pili*[21]—a bundle of smooth muscle cells extending from dermal collagen fibers to the connective tissue root sheath of the follicle (see figs. 6.1 and 6.7a). In response to cold, fear, touch, or other stimuli, the sympathetic nervous system stimulates the piloerector to contract, making the hair stand on end and wrinkling the skin in such areas as the scrotum and areola. In other mammals, piloerection traps an insulating layer of warm air next to the skin or makes the animal appear larger and less vulnerable to a potential enemy. In humans, it pulls the follicles into a vertical position and causes "goose bumps," but serves no useful purpose.

## Hair Texture and Color

The texture of hair is related to differences in cross-sectional shape (fig. 6.8)—straight hair is round, wavy hair is oval, and tightly curled hair is relatively flat. Hair color is due to pigment granules in the cells of the cortex. Brown and black hair are rich in eumelanin. Red hair has a slight amount of eumelanin but a high concentration of pheomelanin. Blond hair has an intermediate amount of pheomelanin but very little eumelanin. Gray and white hair result from a scarcity or absence of melanins in the cortex and the presence of air in the medulla.

## Hair Growth and Loss

A given hair goes through a **hair cycle** consisting of three developmental stages: anagen, catagen, and telogen (fig. 6.9). At any given time, about 90% of the scalp follicles are in the **anagen**[22] stage. In this stage, stem cells from the bulge in the follicle multiply and travel downward, pushing the dermal papilla deeper into the skin and forming the epithelial root sheath. Root sheath cells directly above the dermal papilla form the hair matrix. Here, sheath cells transform into hair cells, which synthesize keratin and then die as they are pushed upward away from the papilla. The new hair grows up the follicle, often alongside an old *club hair* left from the previous cycle.

In the **catagen**[23] stage, mitosis in the hair matrix ceases and sheath cells below the bulge die. The follicle shrinks and the dermal papilla draws up toward the bulge. The base of the hair keratinizes into a hard club and the hair, now known as a **club hair,** loses its anchorage. Club hairs are easily pulled out by brushing the hair, and the hard club can be felt at the hair's end. When the papilla reaches the bulge, the hair goes into a resting period called the **telogen**[24] stage. Eventually, anagen begins anew and the cycle repeats itself. A club hair may fall out during catagen or telogen, or as it is pushed out by the new hair in the next anagen phase. We lose about 50 to 100 scalp hairs daily.

In a young adult, scalp follicles typically spend 6 to 8 years in anagen, 2 to 3 weeks in catagen, and 1 to 3 months in telogen. Scalp hairs grow at a rate of about 1 mm per 3 days (10–18 cm/yr) in the anagen phase.

Hair grows fastest from adolescence until the 40s. After that, an increasing percentage of follicles are in the catagen and telogen phases rather than the growing anagen phase. Follicles also shrink and begin producing wispy vellus hairs instead of thicker terminal hairs. Thinning of the hair, or baldness, is called **alopecia**[25] (AL-oh-PEE-she-uh). It occurs to some degree in both sexes and may be worsened by disease, poor nutrition, fever, emotional stress, radiation, or chemotherapy. In the great majority of cases, however, it is simply a matter of aging.

**Pattern baldness** is the condition in which hair is lost from select regions of the scalp rather than thinning uniformly across the entire scalp. It results from a combination of genetic and hormonal influences. The relevant gene has two alleles: one for uniform hair growth and a baldness allele for patchy

---

[21]*arrect* = erect; *pili* = of hair

[22]*ana* = up; *gen* = build, produce
[23]*cata* = down
[24]*telo* = end
[25]*alopecia* = fox mange

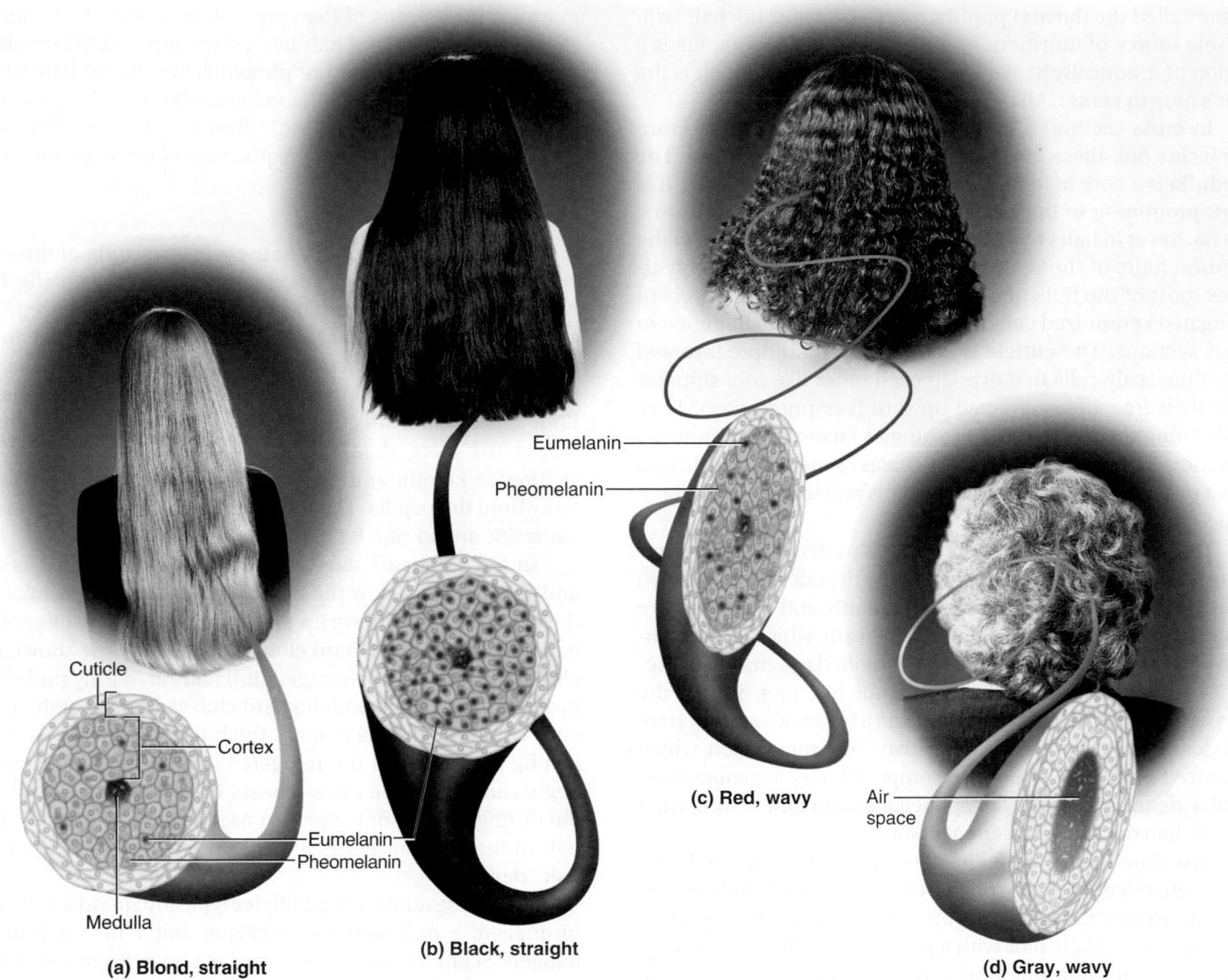

**FIGURE 6.8 The Basis of Hair Color and Texture.** Straight hair (a and b) is round in cross section, whereas curly hair (c and d) is flatter. Blond hair (a) has scanty eumelanin and a moderate amount of pheomelanin. Eumelanin predominates in black and brown hair (b). Red hair (c) derives its color predominantly from pheomelanin. Gray and white hair (d) lack pigment and have air in the medulla.

*Which of the hair layers illustrated here corresponds to the scales seen on the hair in the photo on page 176?*

hair growth. The baldness allele is dominant in males and is expressed only in the presence of the high level of testosterone characteristic of men. In men who are either heterozygous or homozygous for the baldness allele, testosterone causes the terminal hair of the scalp to be replaced by thinner vellus, beginning on top of the head and later the sides. In women, the baldness allele is recessive. Homozygous dominant and heterozygous women show normal hair distribution; only homozygous recessive women are at risk of pattern baldness. Even then, they exhibit the trait only if their testosterone levels are abnormally high for a woman (for example, because of a tumor of the adrenal gland, a woman's principal source

of testosterone). Such characteristics in which an allele is dominant in one sex and recessive in the other are called *sex-influenced traits.*

Excessive or undesirable hairiness in areas that are not usually hairy, especially in women and children, is called **hirsutism.**[26] It tends to run in families and usually results from either masculinizing ovarian tumors or hypersecretion of testosterone by the adrenal cortex. It is often associated with menopause.

---

[26]*hirsut* = shaggy

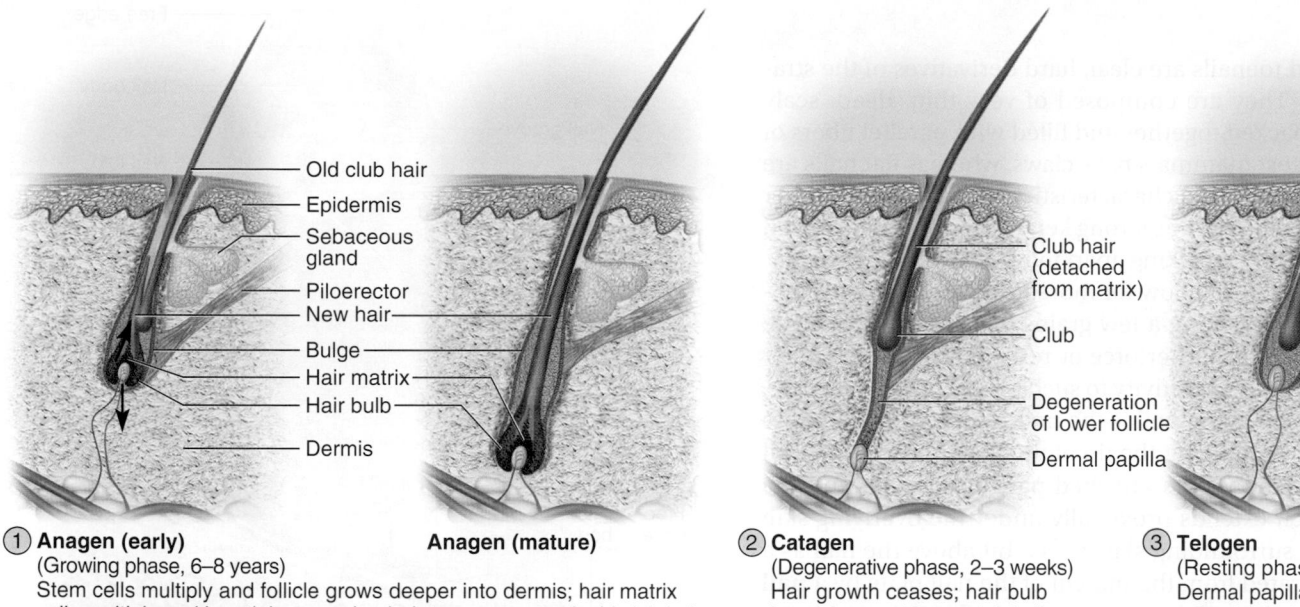

① **Anagen (early)**
(Growing phase, 6–8 years)
Stem cells multiply and follicle grows deeper into dermis; hair matrix cells multiply and keratinize, causing hair to grow upward; old club hair may persist temporarily alongside newly growing hair.

**Anagen (mature)**

② **Catagen**
(Degenerative phase, 2–3 weeks)
Hair growth ceases; hair bulb keratinizes and forms club hair; lower follicle degenerates.

③ **Telogen**
(Resting phase, 1–3 months)
Dermal papilla has ascended to level of bulge; club hair falls out, usually in telogen or next anagen.

**FIGURE 6.9** **The Hair Cycle.**

Contrary to popular misconceptions, hair and nails do not continue to grow after a person dies, cutting hair does not make it grow faster or thicker, and emotional stress cannot make the hair turn white overnight.

## Functions of Hair

Compared with other mammals, the relative hairlessness of humans is so unusual that it raises the question, Why do we have any hair at all? What purpose does it serve? There are different answers for the different types of hair; furthermore, some of the answers would make little sense if we limited our frame of reference to industrialized societies, where barbers and hairdressers are engaged to alter the natural state of the hair. It is more useful to take a comparative approach to this question and consider the purposes hair serves in other species of mammals.

Most hair of the human trunk and limbs is probably best interpreted as vestigial, with little present function. Body hair undoubtedly served to keep our ancestors warm, but in modern humans it is too scanty. Stimulation of the hair receptors, however, alerts people to parasites crawling on the skin, such as lice and fleas.

The scalp is normally the only place where the hair is thick enough to retain heat. Heat loss from a bald scalp can be substantial and quite uncomfortable. The brain receives a rich supply of warm blood, and most of the scalp lacks an insulating fat layer. Heat is easily conducted through the bones of the skull and lost to the surrounding air. In addition, without hair there is nothing to break the wind and stop it from carrying away heat. Hair also protects the scalp from sunburn, since the scalp is otherwise most directly exposed to the sun's rays. These may

be the reasons humans have retained thick hair on their heads while losing most of it from the rest of the body.

Tufts and patches of hair, sometimes with contrasting colors, are important among mammals for advertising species, age, sex, and individual identity. For the less groomed members of the human species, scalp hair may play a similar role. The indefinitely growing hair of a man's scalp and beard, for example, could provide a striking contrast to a face that is otherwise almost hairless. It creates a badge of recognition instantly visible at a distance.

The beard and pubic and axillary hair signify sexual maturity and aid in the transmission of sexual scents. We will further reflect on this in a later discussion of apocrine sweat glands, whose distribution and function add significant evidence to support this theory.

Stout protective **guard hairs,** or **vibrissae** (vy-BRISS-ee), guard the nostrils and ear canals and prevent foreign particles from entering easily. The eyelashes and blink reflex shield the eye from windblown debris. In windy or rainy conditions, we can squint so that the eyelashes protect the eyes without completely obstructing our vision.

The eyebrows are often presumed to keep sweat or debris out of the eyes, but this seems negligible. More likely, they function mainly to enhance facial expression. Movements of the eyebrows are an important means of nonverbal communication in humans of all cultures, and we even have special *frontalis* muscles for this purpose. Eyebrow expressiveness is not unique to humans; even monkeys and apes use quick eyebrow flashes to greet each other, assert their dominance, and break up quarrels.

## Nails

Fingernails and toenails are clear, hard derivatives of the stratum corneum. They are composed of very thin, dead, scaly cells, densely packed together and filled with parallel fibers of hard keratin. Most mammals have claws, whereas flat nails are one of the distinguishing characteristics of humans and other primates. Flat nails serve as strong keratinized "tools" that can be used for grooming, picking apart food, and other manipulations. In addition, they allow for more fleshy and sensitive fingertips. Imagine touching a few grains of salt on a table. The nail, by providing a counterforce or resistance from the other side, enhances one's sensitivity to such tiny objects.

The hard part of the nail is the **nail plate,** which includes the **free edge** overhanging the tip of the finger or toe; the **nail body,** which is the visible attached part of the nail; and the **nail root,** which extends proximally under the overlying skin (fig. 6.10). The surrounding skin rises a bit above the nail as a **nail fold,** separated from the margin of the nail plate by a **nail groove.** The groove and the space under the free edge accumulate dirt and bacteria and require special attention when scrubbing for duty in an operating room or nursery.

The skin underlying the nail plate is the **nail bed;** its epidermis is called the **hyponychium**[27] (HIPE-o-NICK-ee-um). At the proximal end of the nail, the stratum basale thickens into a growth zone called the **nail matrix.** Mitosis in the matrix accounts for the growth of the nail—about 1 mm per week in the fingernails and slightly slower in the toenails. The thickness of the matrix obscures the underlying dermal blood vessels and is the reason why an opaque white crescent, the **lunule**[28] (LOON-yule), often appears at the proximal end of a nail. A narrow zone of dead skin, the **cuticle** or **eponychium**[29] (EP-o-NICK-ee-um), commonly overhangs this end of the nail.

The appearance of the fingertips and nails can be valuable in medical diagnosis. The fingertips become swollen or *clubbed* in response to long-term hypoxemia—a deficiency of oxygen in the blood stemming from conditions such as congenital heart defects and emphysema. Dietary deficiencies may be reflected in the appearance of the nails. An iron deficiency, for example, may cause them to become flat or concave (spoonlike) rather than convex. Contrary to popular belief, adding gelatin to the diet has no effect on the growth or hardness of the nails.

**BEFORE YOU GO ON**

Answer the following questions to test your understanding of the preceding section:

7.  What is the difference between vellus and terminal hair?

8.  State the functions of the hair papilla, hair receptors, and piloerector.

---

[27]*hypo* = below; *onych* = nail
[28]*lun* = moon; *ule* = little
[29]*ep* = above; *onych* = nail

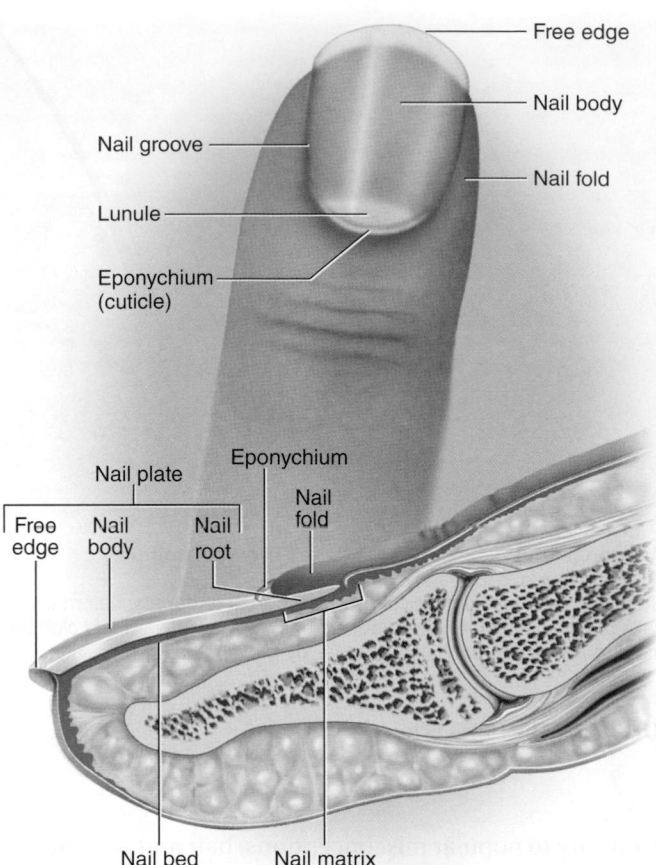

**FIGURE 6.10  Anatomy of a Fingernail.  AP|R**

9.  Describe what happens in the anagen, catagen, and telogen phases of the hair cycle.

10.  State some reasonable theories for the different functions of hair of the eyebrows, eyelashes, scalp, nostrils, and axilla.

11.  Define or describe the nail plate, nail fold, eponychium, hyponychium, and nail matrix.

## 6.3    Cutaneous Glands

### Expected Learning Outcomes

When you have completed this section, you should be able to

a.  name two types of sweat glands, and describe the structure and function of each;

b.  describe the location, structure, and function of sebaceous and ceruminous glands; and

c.  discuss the distinction between breasts and mammary glands, and explain their respective functions.

The skin has five kinds of glands: *merocrine sweat glands, apocrine sweat glands, sebaceous glands, ceruminous glands,* and *mammary glands.*

## Sweat Glands

Sweat glands, or **sudoriferous**[30] (soo-dor-IF-er-us) **glands,** are of two kinds: apocrine and merocrine (compare figure 5.32, p. 167). **Apocrine sweat glands** (fig. 6.11a) occur in the groin, anal region, axilla, and areola, and in mature males, in the beard area. They are absent from the axillary region of Koreans and sparse in the Japanese. Their ducts lead into nearby hair follicles rather than directly to the skin surface. Apocrine glands, unfortunately, are misnamed because they do not use the apocrine mode of secretion; they use exocytosis, the same as merocrine sweat glands do (see p. 166). The secretory part of an apocrine gland, however, has a much larger lumen than that of a merocrine gland, so these glands continue to be called apocrine glands to distinguish them functionally and histologically

[30] *sudor* = sweat; *fer* = carry, bear

from the merocrine type. Apocrine sweat is thicker and more milky than merocrine sweat because it has more fatty acids in it.

Apocrine sweat glands are scent glands that respond especially to stress and sexual stimulation. They are not active until puberty, and in women, they enlarge and shrink in phase with the menstrual cycle. These facts, as well as experimental evidence, suggest that their function is to secrete *sex pheromones*—chemicals that exert subtle effects on the sexual behavior and physiology of other people (see Deeper Insight 16.1, p. 591). They apparently correspond to the scent glands that develop in other mammals as they approach sexual maturity. Fresh apocrine sweat does not have a disagreeable odor; indeed, it is considered attractive or arousing in some cultures, where it is as much a part of courtship as artificial perfume is to others. Clothing, however, traps stale sweat long enough for bacteria to degrade the secretion and release free fatty acids with a rancid odor. Disagreeable body odor is called

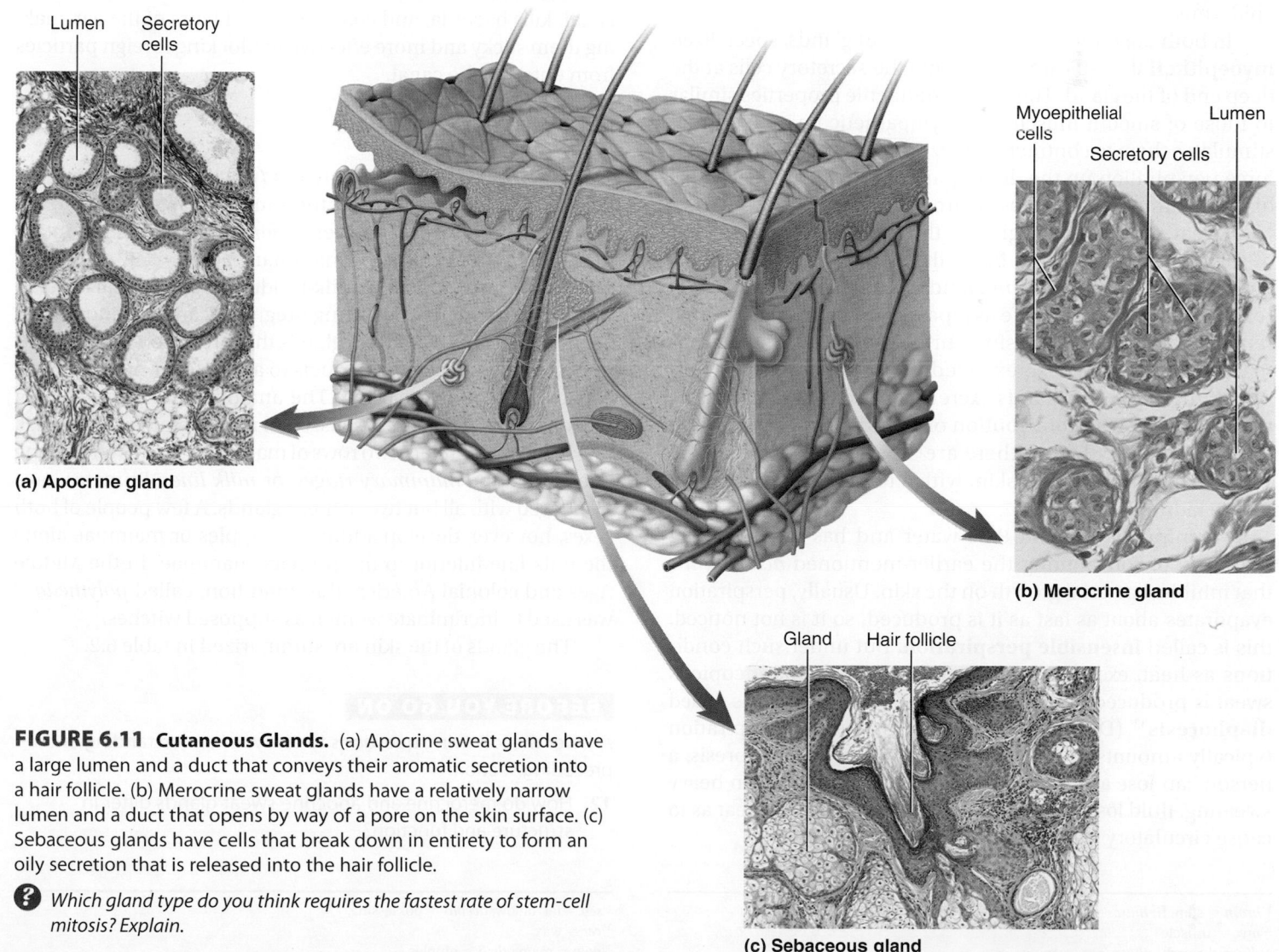

**FIGURE 6.11 Cutaneous Glands.** (a) Apocrine sweat glands have a large lumen and a duct that conveys their aromatic secretion into a hair follicle. (b) Merocrine sweat glands have a relatively narrow lumen and a duct that opens by way of a pore on the skin surface. (c) Sebaceous glands have cells that break down in entirety to form an oily secretion that is released into the hair follicle.

❓ *Which gland type do you think requires the fastest rate of stem-cell mitosis? Explain.*

(a) Apocrine gland

(b) Merocrine gland

(c) Sebaceous gland

*bromhidrosis.*[31] It occasionally indicates a metabolic disorder, but more often reflects inadequate hygiene.

Many mammals have apocrine scent glands associated with specialized tufts of hair. In humans, apocrine glands are found mainly in the regions covered by the pubic hair, axillary hair, and beard. This supports the interpretation that they are pheromone glands. The hair serves to retain the aromatic secretion and regulate its rate of evaporation from the skin. Thus, it seems no mere coincidence that women's faces lack both apocrine scent glands and a beard.

**Merocrine (eccrine) sweat glands** (fig. 6.11b) are widely distributed over the entire body, but are especially abundant on the palms, soles, and forehead. Their primary function is to cool the body. Overheating activates sweating through the sympathetic nervous system. Then, because of the high heat capacity of water (see p. 50), the evaporation of sweat carries away a substantial amount of body heat, cooling one down. Each is a simple tubular gland with a twisted coil in the dermis or hypodermis, and an undulating or coiled duct leading to a sweat pore on the skin surface. The duct is lined by a stratified cuboidal epithelium in the dermis and by keratinocytes in the epidermis.

In both apocrine and merocrine sweat glands, specialized **myoepithelial**[32] **cells** are found amid the secretory cells at the deep end of the gland. They have contractile properties similar to those of smooth muscle. The sympathetic nervous system stimulates them to contract, squeeze the base of the gland, and force perspiration up the duct—particularly under conditions of overheating, nervousness, or arousal.

Sweat production begins in the deep secretory portion of the gland. Protein-free fluid filters from the blood capillaries into the lumen of the gland. Most sodium chloride is reabsorbed from the filtrate as it passes up the duct, but some remains, along with potassium, urea, lactic acid, and ammonia. Some drugs are also excreted in the perspiration. Thus, the merocrine sweat glands excrete some of the same wastes as the kidneys. The contribution of an individual sweat gland to this is minuscule, but there are 3 or 4 million merocrine sweat glands in the adult skin, with a total mass about equal to one kidney.

Perspiration is about 99% water and has a pH ranging from 4 to 6, contributing the earlier-mentioned *acid mantle* that inhibits bacterial growth on the skin. Usually, perspiration evaporates about as fast as it is produced, so it is not noticed; this is called **insensible perspiration.** But under such conditions as heat, exercise, and circulatory shock, more copious sweat is produced and noticeably wets the skin; this is called **diaphoresis**[33] (DY-uh-foe-REE-sis). Insensible perspiration typically amounts to about 500 mL/day, but in diaphoresis, a person can lose as much as a liter of sweat per hour. In heavy sweating, fluid loss from the bloodstream can be so great as to cause circulatory shock.

## Sebaceous Glands

**Sebaceous**[34] (see-BAY-shus) **glands** produce an oily secretion called **sebum** (SEE-bum). They are flask-shaped, with short ducts that usually open into a hair follicle (fig. 6.11c), although some of them open directly onto the skin surface. These are holocrine glands with little visible lumen. Their secretion consists of broken-down cells that are replaced by mitosis at the base of the gland. Sebum keeps the skin and hair from becoming dry, brittle, and cracked. The sheen of well-brushed hair is due to sebum distributed by the hairbrush. Ironically, we go to great lengths to wash sebum from the skin, only to replace it with various skin creams and hand lotions made of little more than lanolin, which is sheep sebum.

## Ceruminous Glands

**Ceruminous** (seh-ROO-mih-nus) **glands** are found only in the external ear canal, where their secretion combines with sebum and dead epidermal cells to form earwax, or **cerumen.**[35] They are simple, coiled, tubular glands with ducts leading to the skin surface. Cerumen keeps the eardrum pliable, waterproofs the canal, kills bacteria, and coats the guard hairs of the ear, making them sticky and more effective in blocking foreign particles from entering the canal.

## Mammary Glands

The **mammary glands** and breasts *(mammae)* are often mistakenly regarded as one and the same. Breasts, however, are present in both sexes, and even in females they rarely contain more than small traces of mammary gland. The mammary glands, by contrast, are the milk-producing glands that develop within the female breast during pregnancy and lactation. They are modified apocrine sweat glands that produce a richer secretion and channel it through ducts to a nipple for more efficient conveyance to the offspring. The anatomy and physiology of the mammary gland are discussed in more detail in chapter 28.

In most mammals, two rows of mammary glands form along lines called the *mammary ridges,* or *milk lines.* Primates have dispensed with all but two of these glands. A few people of both sexes, however, develop additional nipples or mammae along the milk line inferior to the primary mammae. In the Middle Ages and colonial America, this condition, called *polythelia,*[36] was used to incriminate women as supposed witches.

The glands of the skin are summarized in table 6.2.

**BEFORE YOU GO ON**

Answer the following questions to test your understanding of the preceding section:

12. How do merocrine and apocrine sweat glands differ in structure and function?

---

[31]*brom* = stench; *hidros* = sweat
[32]*myo* = muscle
[33]*dia* = through; *phoresis* = carrying

[34]*seb* = fat, tallow; *aceous* = possessing
[35]*cer* = wax
[36]*poly* = many; *theli* = nipples

| TABLE 6.2 | Cutaneous Glands |
|-----------|------------------|
| **Gland Type** | **Definition** |
| Sudoriferous glands | Sweat glands |
| Merocrine glands | Sweat glands that function in evaporative cooling; widely distributed over the body surface; open by ducts onto the skin surface |
| Apocrine glands | Sweat glands that function as scent glands; found in the regions covered by the pubic, axillary, and male facial hair; open by ducts into hair follicles |
| Sebaceous glands | Oil glands associated with hair follicles |
| Ceruminous glands | Glands of the ear canal that contribute to the cerumen (earwax) |
| Mammary glands | Milk-producing glands located in the breasts |

13. What other type of gland is associated with hair follicles? How does its mode of secretion differ from that of sweat glands?

14. What is the difference between the breast and mammary gland?

## 6.4   Skin Disorders

### Expected Learning Outcomes

When you have completed this section, you should be able to

a. describe the three most common forms of skin cancer; and

b. describe the three classes of burns and the priorities in burn treatment.

Because it is the most exposed of all our organs, skin is not only the most vulnerable to injury and disease, but is also the one place where we are most likely to notice anything out of the ordinary. Skin diseases become increasingly common in old age, and most people over age 70 have complaints about their integumentary system. Aging of the skin is discussed more fully on page 1118. The healing of cuts and other injuries to the skin occurs by the process described at the end of chapter 5. We focus here on two particularly common and serious disorders: skin cancer and burns. Other skin diseases are briefly summarized in table 6.3.

### Skin Cancer

Skin cancer befalls about one out of five people in the United States at some time in their lives. Most cases are caused by UV radiation from the sun, which damages DNA and disables protective tumor suppressor genes in the epidermal cells. Consequently, most tumors occur on the head, neck, and hands, where exposure to the sun is greatest. It is most common in fair-skinned people and the elderly, who have had the longest lifetime UV exposure (see Deeper Insight 6.4). The ill-advised popularity of suntanning, however, has caused an alarming increase in skin cancer among younger people. Skin cancer is one of the most common cancers, but it is also one of the easiest to treat and has one of the highest survival rates when it is detected and treated early.

#### ▶▶▶ APPLY WHAT YOU KNOW

*Skin cancer is relatively rare in people with dark skin. Other than possible differences in behavior, such as less intentional suntanning, why do you think this is so?*

There are three types of skin cancer named for the epidermal cells in which they originate: *basal cell carcinoma, squamous cell carcinoma,* and *melanoma.* The three types are also distinguished from each other by the appearance of their **lesions**[37] (zones of tissue injury).

**Basal cell carcinoma**[38] is the most common type. It is the least deadly because it seldom metastasizes, but if neglected, it can severely disfigure the face. It arises from cells of the stratum basale and eventually invades the dermis. On the surface, the lesion first appears as a small, shiny bump. As the bump enlarges, it often develops a central depression and a beaded "pearly" edge (fig. 6.12a).

**Squamous cell carcinoma** arises from keratinocytes of the stratum spinosum. Lesions usually appear on the scalp, ears, lower lip, or back of the hand. They have a raised, reddened, scaly appearance, later forming a concave ulcer with raised edges (fig. 6.12b). The chance of recovery is good with early detection and surgical removal, but if it goes unnoticed or is neglected, this cancer tends to metastasize to the lymph nodes and can be lethal.

**Melanoma** is a skin cancer that arises from the melanocytes. It accounts for no more than 5% of skin cancers, but it is extremely aggressive and drug-resistant. It can be treated surgically if caught early, but if it metastasizes—which it does quickly—it is unresponsive to chemotherapy and is usually fatal. The average person with metastatic melanoma lives only 6 months from diagnosis, and only 5% to 14% of patients survive with it for 5 years. The greatest risk factor for melanoma is a family history of the disease. It has a relatively high incidence in men, in redheads, and in people who experienced severe sunburns in childhood.

About two-thirds of cases of melanoma in men result from an oncogene called *BRAF.* In women, *BRAF* does not appear to trigger melanoma, but it has been linked to some breast and ovarian cancers. *BRAF* mutations are commonly found in moles. *BRAF* was recently discovered in the course of the new Cancer Genome Project, a multinational effort to identify cancer genes.

---

[37]*lesio* = injure
[38]*carcin* = cancer; *oma* = tumor

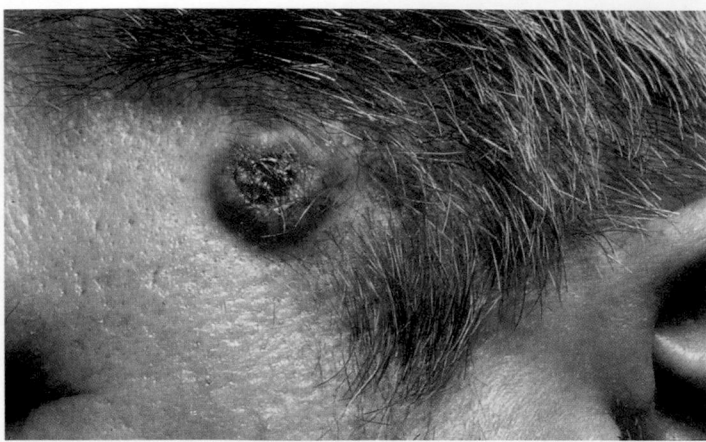

**(a) Basal cell carcinoma**

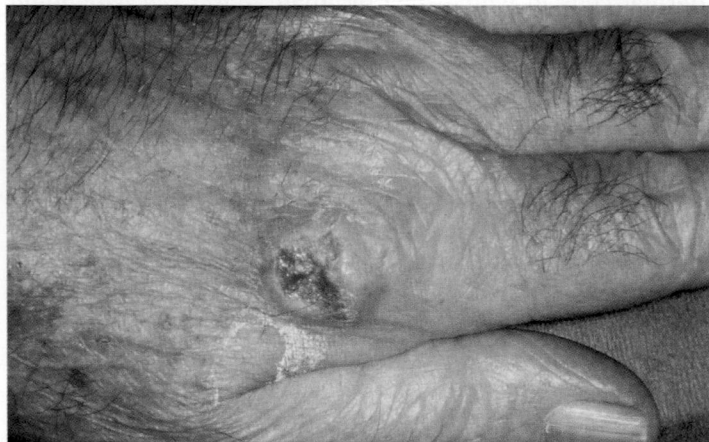

**(b) Squamous cell carcinoma**

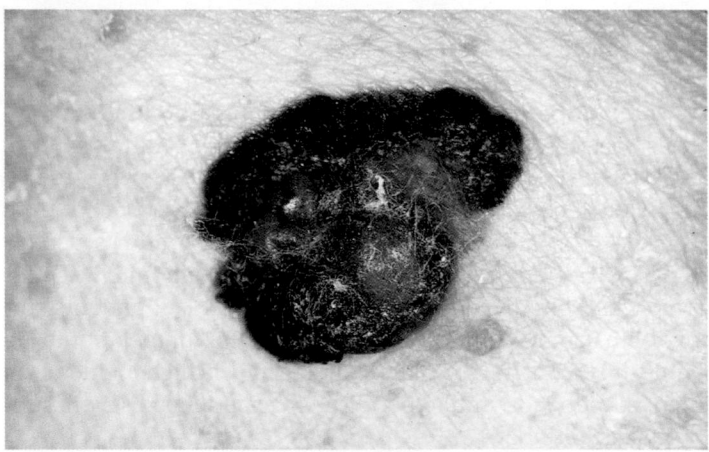

**(c) Melanoma**

**FIGURE 6.12 Skin Cancer.**

❓ *Which of the ABCD rules can you identify in part (c)?*

It is important to distinguish a mole from a melanoma. A mole usually has a uniform color and even contour, and it is no larger in diameter than the end of a pencil eraser (about 6 mm). If it becomes malignant, however, it forms a large, flat, spreading lesion with a scalloped border (fig. 6.12c). The American

Cancer Society suggests an "ABCD rule" for recognizing melanoma: *A* for asymmetry (one side of the lesion looks different from the other); *B* for border irregularity (the contour is not uniform but wavy or scalloped); *C* for color (often a mixture of brown, black, tan, and sometimes red and blue); and *D* for diameter (greater than 6 mm).

Skin cancer is treated by surgical excision, radiation therapy, or destruction of the lesion by heat (electrodesiccation) or cold (cryosurgery).

## Burns

**Burns** are the leading cause of accidental death. They are usually caused by fires, kitchen spills, or excessively hot bath water, but they also can be caused by sunlight, ionizing radiation, strong acids and bases, or electrical shock. Burn deaths result primarily from fluid loss, infection, and the toxic effects of **eschar** (ESS-car)—the burned, dead tissue.

Burns are classified according to the depth of tissue involvement (fig. 6.13). **First-degree burns** involve only the epidermis and are marked by redness, slight edema, and pain. They heal in a few days and seldom leave scars. Most sunburns are first-degree burns.

**Second-degree burns** involve the epidermis and part of the dermis but leave at least some of the dermis intact. First- and second-degree burns are therefore also known as **partial-thickness burns.** A second-degree burn may be red, tan, or white and is blistered and very painful. It may take from 2 weeks to several months to heal and may leave scars. The epidermis regenerates by division of epithelial cells in the hair follicles and sweat glands and around the edges of the lesion. Severe sunburns and many scalds are second-degree burns.

**Third-degree burns** are also called **full-thickness burns** because the epidermis, all of the dermis, and often some deeper tissues (muscle and bone) are destroyed. (Some authorities call burns that extend to the bone *fourth-degree burns.*) Since no dermis remains, the skin can regenerate only from the edges of the wound. Third-degree burns often require skin grafts (see Deeper Insight 6.5). If a third-degree burn is left to itself to heal, contracture (abnormal connective tissue fibrosis) and severe disfigurement may result.

▶▶▶ **APPLY WHAT YOU KNOW**

*A third-degree burn may be surrounded by painful areas of first- and second-degree burns, but the region of the third-degree burn is painless. Explain the reason for this lack of pain.*

The two most urgent considerations in treating a burn patient are fluid replacement and infection control. A patient can lose several liters of water, electrolytes, and protein each day from the burned area. As fluid is lost from the tissues, more is transferred from the bloodstream to replace it, and the volume of circulating blood declines. A patient may lose up to 75% of the blood plasma within a few hours, potentially leading to circulatory shock and cardiac arrest—the principal cause of death in burn patients. Intravenous fluid must be given to make up

Partial-thickness burns                                   Full-thickness burns

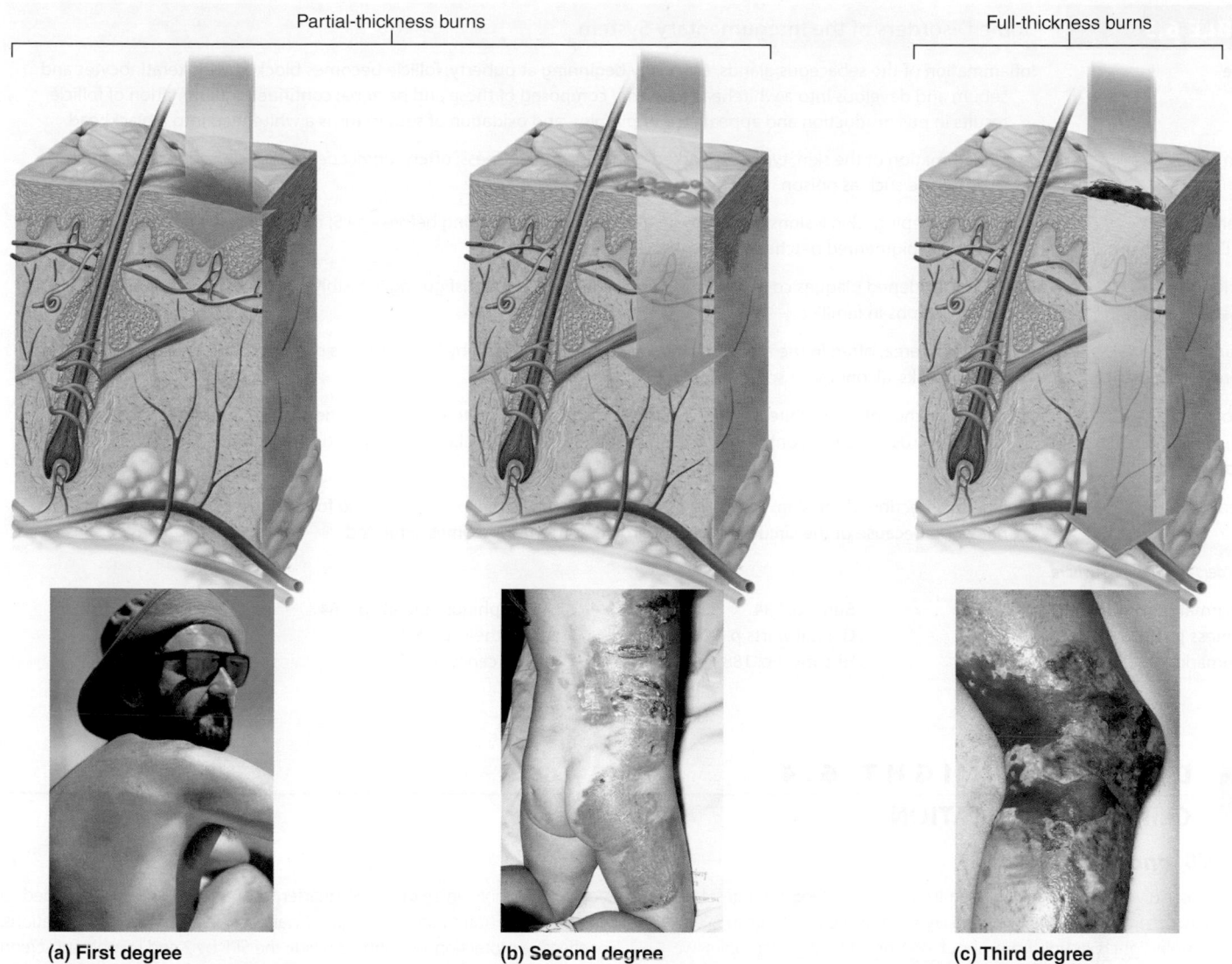

**(a) First degree**          **(b) Second degree**          **(c) Third degree**

**FIGURE 6.13  Burns.** (a) First-degree burn, involving only the epidermis. (b) Second-degree burn, involving the epidermis and part of the dermis. (c) Third-degree burn, extending through the entire dermis and often involving even deeper tissue.

for this loss. A severely burned patient may also require thousands of extra calories daily to compensate for protein loss and the demands of tissue repair. Supplemental nutrients are given intravenously or via gastric tube.

Infection is controlled by keeping the patient in an aseptic (germ-free) environment and administering antibiotics. The eschar is sterile for the first 24 hours, but then it quickly becomes infected and may have toxic effects on the digestive, respiratory, and other systems. Its removal, called **debridement**[39] (deh-BREED-ment), is essential to infection control.

---

**BEFORE YOU GO ON**

Answer the following questions to test your understanding of the preceding section:

15. What types of cells are involved in each type of skin cancer?

16. Which type of skin cancer is most dangerous? What are its early warning signs?

17. What are the differences between first-, second-, and third-degree burns?

18. What are the two most urgent priorities in treating a burn victim? How are these needs addressed?

---

[39]*de* = un; *bride* = bridle

| TABLE 6.3 | Some Disorders of the Integumentary System |
|-----------|---------------------------------------------|
| Acne | Inflammation of the sebaceous glands, especially beginning at puberty; follicle becomes blocked with keratinocytes and sebum and develops into a whitehead *(comedo)* composed of these and bacteria; continued inflammation of follicle results in pus production and appearance of pimples, and oxidation of sebum turns a whitehead into a blackhead. |
| Dermatitis | Any inflammation of the skin, typically marked by itching and redness; often *contact dermatitis,* caused by exposure to toxic foliage such as poison ivy. |
| Eczema (ECK-zeh-mah) | Itchy, red, "weeping" skin lesions caused by an allergy, usually beginning before age 5; may progress to thickened, leathery, darkly pigmented patches of skin. |
| Psoriasis (so-RY-ah-sis) | Recurring, reddened plaques covered with silvery scale; sometimes disfiguring; possibly caused by an autoimmune response; runs in families. |
| Rosacea (ro-ZAY-she-ah) | A red rashlike area, often in the area of the nose and cheeks, marked by fine networks of dilated blood vessels; worsened by hot drinks, alcohol, and spicy food. |
| Seborrheic (seb-oh-REE-ik) dermatitis | Recurring patches of scaly white or yellowish inflammation often on the head, face, chest, and back; called *cradle cap* (a yellow, crusty scalp lesion) in infants. Cause unknown, but correlated with genetic and climatic factors. |
| Tinea | Any fungal infection of the skin; common in moist areas such as the axilla, groin, and foot *(athlete's foot).* Misnamed *ringworm* because of the circular, wormlike growth pattern sometimes exhibited. |

*Disorders described elsewhere*

| | | |
|---|---|---|
| Abnormal skin coloration p. 183 | Burns p. 194 | Pemphigus vulgaris p. 164 |
| Baldness p. 187 | Genital warts p. 1054 | Polythelia p. 192 |
| Birthmarks p. 185 | Hirsutism p. 188 | Skin cancer p. 193 |

## DEEPER INSIGHT 6.4

## CLINICAL APPLICATION

### *UVA, UVB, and Sunscreens*

Photobiologists divide ultraviolet radiation into two wavelength ranges: UVA with wavelengths from 320 to 400 nm and UVB with wavelengths from 290 to 320 nm. (Visible light extends from about 400 nm, the deepest violet we can see, to about 700 nm, the deepest red.) The shorter the wavelength, the higher the energy of electromagnetic radiation, so UVB has a higher energy than UVA. UVA and UVB are sometimes called "tanning rays" and "burning rays," respectively. Tanning salons often advertise that they only use "safe" UVA rays, but public health authorities are more skeptical. UVA can burn as well as tan, it is responsible for most of the undesirable *photoaging* effects on the skin (see p. 1118), and it inhibits the immune system; both UVA and UVB are now thought to initiate skin cancer. As dermatologists like to say, there is no such thing as a healthy tan.

Many health-conscious people buy sunscreen according to its sun protection factor (SPF), but may be misled by misunderstandings of the concept. SPF is a laboratory measure of protection from UVB radiation. It is advisable to use a minimum SPF of 15 for meaningful protection. It might seem that an SPF of 30 would give twice as much protection as SPF 15, but the relationship between protection and SPF is not linear. An SPF 15 sunscreen protects the skin from 93% of UVB, but SPF 30 protects only slightly more—97%. Although manufacturers compete to produce high-SPF products, higher numbers tend to be misleading, giving people a false sense of much greater protection. The U.S. government considers any SPF above 50 to be misleading, whereas the government of Australia sets the bounds at SPF 30.

Further, many people use sunscreen ineffectively. Effectiveness varies with skin type, the amount of sweating, amount applied, and frequency of reapplication. No sunscreen is waterproof; it must be reapplied after swimming.

People commonly apply only one-quarter to one-half as much as needed to provide the SPF rating on the label, and wait too long between reapplications. For effective protection, one should divide the SPF by 2 and reapply sunscreen that many minutes after the onset of sun exposure—that is, every 15 minutes for an SPF of 30. Some products claim "broad spectrum UVA/UVB protection," but there is no good evidence that they protect adequately against UVA.

It was once assumed that sunscreens furnish protection against skin cancer, but more careful studies have cast doubt on this and made the issue increasingly puzzling and controversial. Sunburn and skin cancer are caused by different mechanisms, and although sunscreen can protect against burning, it provides little protection against cancer. Ironically, as the sale of sunscreen has risen in recent decades, so has the incidence of skin cancer—perhaps because people falsely assume that when they use sunscreen, they can safely stay in the sun longer. Sunscreen does provide some protection against squamous cell carcinoma, but apparently not against basal cell carcinoma or melanoma. Indeed, people who use sunscreen have a higher incidence of basal cell carcinoma than people who do not. Some of the chemicals used in sunscreens damage DNA and generate harmful free radicals when exposed to UV—chemicals such as zinc oxide and titanium dioxide—but much is still unknown about how sunscreen reacts on and with the skin. Epidemiological data also are yet inconclusive; skin cancer is most prevalent in older people, and not enough data are available on older people who have used sunscreen their entire lives.

Given the uncertainty, it is best not to assume that merely because sunscreens protect you from sunburn, they also protect against skin cancer. Above all, one should not assume that using sunscreen makes it safe (with respect to cancer) to spend longer hours on the beach or tanning by the pool.

# DEEPER INSIGHT 6.5
## CLINICAL APPLICATION

### Skin Grafts and Artificial Skin

Third-degree burns leave no dermal tissue to regenerate what was lost, therefore they generally require skin grafts. The ideal graft is an *autograft*[40]—tissue taken from another location on the same person's body—because it is not rejected by the immune system. An autograft is performed by taking epidermis and part of the dermis from an undamaged area such as the thigh or buttock and grafting it to a burned area. This method is called a *split-skin* graft because part of the dermis is left behind to proliferate and replace the epidermis that was removed—the same way a second-degree burn heals.

An autograft may not be possible, however, if the burns are too extensive. The best treatment option in this case is an *isograft*,[41] which uses skin from an identical twin. Because the donor and recipient are genetically identical, the recipient's immune system is unlikely to reject the graft. Since identical twins are rare, however, the best one can hope for in most cases is donor skin from another close relative.

An *allograft*,[42] or *homograft*,[43] is a graft from any other person. Skin banks provide skin from deceased persons for this purpose. The immune system attempts to reject allografts, but they suffice as temporary coverings for the burned area. They can be replaced by autografts when the patient is well enough for healthy skin to be removed from an undamaged area of the body.

Pig skin is sometimes used on burn patients but presents the same problem of immune rejection. A graft of tissue from a different species is called a *heterograft*,[44] or *xenograft*.[45] This is a special case of a *heterotransplant*, which also includes transplantation of organs such as baboon hearts or livers into humans. Heterografts and heterotransplants are short-term methods of maintaining a patient until a better, long-term solution is possible. The immune reaction can be suppressed by drugs called *immunosuppressants*. This procedure is risky, however, because it lowers a person's resistance to infection, which is already compromised in a burn patient.

Some alternatives to skin grafts are also being used. Burns are sometimes temporarily covered with amnion (the membrane that surrounds a developing fetus) obtained from afterbirths. In addition, tiny keratinocyte patches cultured with growth stimulants have produced sheets of epidermal tissue as large as the entire body surface. These can replace large areas of burned tissue. Dermal fibroblasts also have been successfully cultured and used for autografts. A drawback to these approaches is that the culture process requires 3 or 4 weeks, which is too long a wait for some patients with severe burns.

Various kinds of artificial skin have also been developed as a temporary burn covering. One concept is a sheet with an upper layer of silicone and a lower layer of collagen and chondroitin sulfate. It stimulates growth of connective tissue and blood vessels from the patient's underlying tissue. The artificial skin can be removed after about 3 weeks and replaced with a thin layer of cultured or grafted epidermis. The manufacture of one such product begins by culturing fibroblasts on a collagen gel to produce a dermis, then culturing keratinocytes on this dermal substrate to produce an epidermis. Such products are used to treat burn patients as well as leg and foot ulcers that result from diabetes mellitus. This is one aspect of the larger field of *tissue engineering*, which biotechnology companies hope will lead, within a few decades, even to engineering replacement livers and other organs (see Deeper Insight 5.3, p. 171).

---

[40] *auto* = self
[41] *iso* = same
[42] *allo* = different, other
[43] *homo* = same
[44] *hetero* = different
[45] *xeno* = strange, alien

# CONNECTIVE ISSUES

## Effects of the INTEGUMENTARY SYSTEM
## On Other Organ Systems

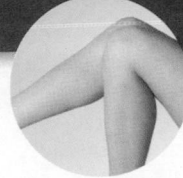

### ALL SYSTEMS
Skin covers the body and provides a barrier to pathogens and to excessive water loss; epidermal keratinocytes initiate synthesis of calcitriol, with effects on multiple other organ systems as noted.

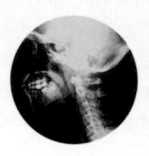

### SKELETAL SYSTEM
Bone growth and maintenance depend on calcium, which is absorbed from the diet under the influence of calcitriol.

### MUSCULAR SYSTEM
Muscle contraction depends on calcium, which is absorbed from the diet under the influence of calcitriol.

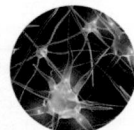

### NERVOUS SYSTEM
The transmission of nerve signals across synapses depends on calcium, which is absorbed from the diet under the influence of calcitriol.

### ENDOCRINE SYSTEM
Hormone secretion depends on calcium as a trigger for exocytosis and, therefore, on calcitriol; the role of epidermal keratinocytes in synthesizing calcitriol is itself an endocrine function.

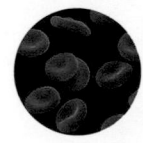

### CIRCULATORY SYSTEM
The skin is a major blood reservoir; cutaneous vasoconstriction diverts blood to other organs; skin supports blood volume by retarding fluid loss; dermal vasoconstriction and vasodilation help to regulate blood temperature.

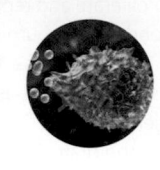

### LYMPHATIC/ IMMUNE SYSTEM
Dendritic cells of the skin alert the immune system when pathogens breach the epidermal barrier.

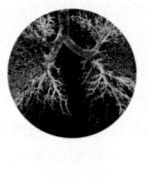

### RESPIRATORY SYSTEM
Nasal guard hairs block some airborne debris from being inhaled; calcium is required for the secretion of respiratory mucus, which therefore depends on calcitriol.

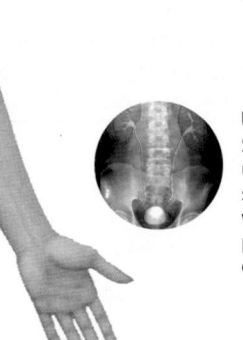

### URINARY SYSTEM
Skin complements the urinary system by excreting salts and some nitrogenous waste in the sweat; calcitriol promotes reabsorption of calcium by the kidneys.

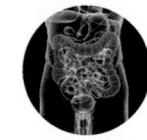

### DIGESTIVE SYSTEM
By their role in calcitriol synthesis, keratinocytes influence intestinal absorption of calcium; calcium is needed for the secretion of all digestive enzymes and mucus.

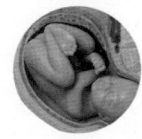

### REPRODUCTIVE SYSTEM
Cutaneous nerve endings are important in sexual stimulation; mammary glands produce milk; apocrine sweat glands secrete pheromones that affect sexual behavior and physiology; skin stretches to accommodate abdominal growth in pregnancy.

# STUDY GUIDE

## ▶ Assess Your Learning Outcomes

*To test your knowledge, discuss the following topics with a study partner or in writing, ideally from memory.*

### 6.1 The Skin and Subcutaneous Tissue (p. 177)

1. The name of the branch of medicine that deals with the integumentary system
2. The difference between the terms *integumentary system* and *integument;* organs that belong to the integumentary system
3. The two principal layers of the skin, and the name of the connective tissue layer that lies between the skin and the deeper muscle or other tissue
4. Normal thickness of the skin, and differences in histology and location between thick and thin skin
5. Functions of the skin
6. Five histological layers of thick skin and which of them is lacking from thin skin
7. Five kinds of epidermal cells, their respective functions, and the epidermal layers in which they occur
8. The life history of a keratinocyte from the time it is "born" by mitosis at the base of the epidermis to the time it exfoliates from the surface
9. Fiber and cell types of the dermis, other dermal structures, and the typical thickness of the dermis

10. Dermal papillae, epidermal ridges, their function, and their relationship to the surface appearance of skin
11. The two layers of the dermis, and how they differ histologically and functionally
12. Composition and functions of the hypodermis, and an alternative name for it when it is composed predominantly of adipose tissue
13. Factors that account for the variety of normal skin colors; abnormal skin colors and their causes
14. Friction ridges, flexion lines, freckles, moles, and hemangiomas

### 6.2 Hair and Nails (p. 185)

1. How the keratin of hair and nails differs from keratin of the epidermis
2. Three kinds of hair, including fetal and adult types
3. The three regions of a hair from base to tip, and the three layers of a hair from core to surface
4. Location of a hair's growth zone and of its source of nourishment
5. The two layers of a hair follicle; their composition; and the specialized nerve endings and smooth muscle associated with a follicle
6. The basis for differences between straight, wavy, and curly hair and for differences in hair color

7. Events of the anagen, catagen, and telogen stages of a hair's life; the typical life span and growth rate of scalp hairs
8. Alopecia, pattern baldness, and hirsutism
9. Functions and body distribution of the various kinds of hair
10. The anatomy of fingernails and toenails; location of their growth zone; and a typical rate of nail growth

### 6.3 Cutaneous Glands (p. 190)

1. Apocrine and merocrine sweat gland distribution, development, structure, and function
2. The same characteristics of sebaceous glands; the name of their product; and how their mode of secretion differs from that of the sweat glands
3. Ceruminous gland distribution, development, structure, and function
4. Mammary gland structure and development and how mammary glands relate to a type of sudoriferous glands

### 6.4 Skin Disorders (p. 193)

1. Three forms of skin cancer and differences in their appearance, the cells in which they originate, their frequency of occurrence, and their severity
2. Three degrees of burns and how they are treated

## ▶ Testing Your Recall

*Answers in Appendix B*

1. Cells of the _____ are keratinized and dead.
   a. papillary layer
   b. stratum spinosum
   c. stratum basale
   d. stratum corneum
   e. stratum granulosum

2. The epidermal water barrier is formed at the point where epidermal cells
   a. pass from stratum basale to stratum spinosum.
   b. enter the telogen phase.
   c. pass from stratum spinosum to stratum granulosum.
   d. form the epidermal ridges.
   e. exfoliate.

3. Which of the following skin conditions or appearances would most likely result from liver failure?
   a. pallor
   b. erythema
   c. seborrheic dermatitis
   d. jaundice
   e. melanization

4. All of the following interfere with microbial invasion of the body *except*
   a. the acid mantle.
   b. melanization.
   c. dendritic cells.
   d. keratinization.
   e. sebum.

5. The hair on a 6-year-old's arms is
   a. vellus.
   b. lanugo.
   c. alopecia.
   d. terminal hair.
   e. rosacea.

6. Which of the following terms is *least* related to the rest?
   a. lunule
   b. nail plate
   c. hyponychium
   d. free edge
   e. cortex

# STUDY GUIDE

7. Which of the following is a scent gland?
   a. eccrine gland
   b. sebaceous gland
   c. apocrine gland
   d. ceruminous gland
   e. merocrine gland

8. _____ are skin cells with a sensory role.
   a. Tactile cells
   b. Dendritic cells
   c. Stem cells
   d. Melanocytes
   e. Keratinocytes

9. Which of the following glands produce the acid mantle?
   a. merocrine sweat glands
   b. apocrine sweat glands
   c. mammary glands
   d. ceruminous glands
   e. sebaceous glands

10. Which of the following skin cells alert the immune system to pathogens?
    a. fibroblasts
    b. melanocytes
    c. keratinocytes
    d. dendritic cells
    e. tactile cells

11. _____ is sweating without noticeable wetness of the skin.

12. A muscle that causes a hair to stand on end is called a/an _____.

13. The process of removing burned skin from a patient is called _____.

14. Blueness of the skin due to low oxygen concentration in the blood is called _____.

15. Projections of the dermis toward the epidermis are called _____.

16. Cerumen is more commonly known as _____.

17. The holocrine glands that secrete into a hair follicle are called _____.

18. Hairs grow only during the _____ phase of the hair cycle.

19. A hair is nourished by blood vessels in a connective tissue projection called the _____.

20. A _____ burn destroys the entire dermis.

## ▶ Building Your Medical Vocabulary

*Answers in Appendix B*

*State a medical meaning of each word element below, and give a term in which it or a slight variation of it is used.*

1. -in
2. albo-
3. dermato-
4. dia-
5. homo-
6. lesio-
7. melano-
8. -oma
9. onycho-
10. pilo-

## ▶ True or False

*Answers in Appendix B*

*Determine which five of the following statements are false, and briefly explain why.*

1. Dander consists of dead keratinocytes.

2. The term *integument* means only the skin, but *integumentary system* includes the skin, hair, nails, and cutaneous glands.

3. The dermis is composed mainly of keratin.

4. Vitamin D synthesis begins in certain cutaneous glands.

5. Cells of the stratum granulosum cannot undergo mitosis.

6. Dermal papillae are better developed in skin subjected to a lot of mechanical stress than in skin subjected to less stress.

7. The three layers of the skin are the epidermis, dermis, and hypodermis.

8. People of African descent have a much higher density of epidermal melanocytes than do people of northern European descent.

9. Pallor indicates a genetic lack of melanin.

10. Apocrine sweat glands develop at the same time in life as the pubic and axillary hair.

# STUDY GUIDE

## ▶ Testing Your Comprehension

*Answers at* www.mhhe.com/saladin7

1. Many organs of the body contain numerous smaller organs, perhaps even thousands. Describe an example of this in the integumentary system.

2. Certain aspects of human form and function are easier to understand when viewed from the perspective of comparative anatomy and evolution. Discuss examples of this in the integumentary system.

3. Explain how the complementarity of form and function is reflected in the fact that the dermis has two histological layers and not just one.

4. Cold weather does not normally interfere with oxygen uptake by the blood, but it can cause cyanosis anyway. Why?

5. Why is it important for the epidermis to be effective, but not *too* effective, in screening out UV radiation?

## ▶ Improve Your Grade at www.mhhe.com/saladin7

*Download animations to study when it fits your schedule. Practice quizzes, labeling activities, games, and flashcards offer fun ways to master the chapter concepts. Or, download image PowerPoint files for each chapter to create a study guide or for taking notes during lecture.*

# 7

# BONE TISSUE

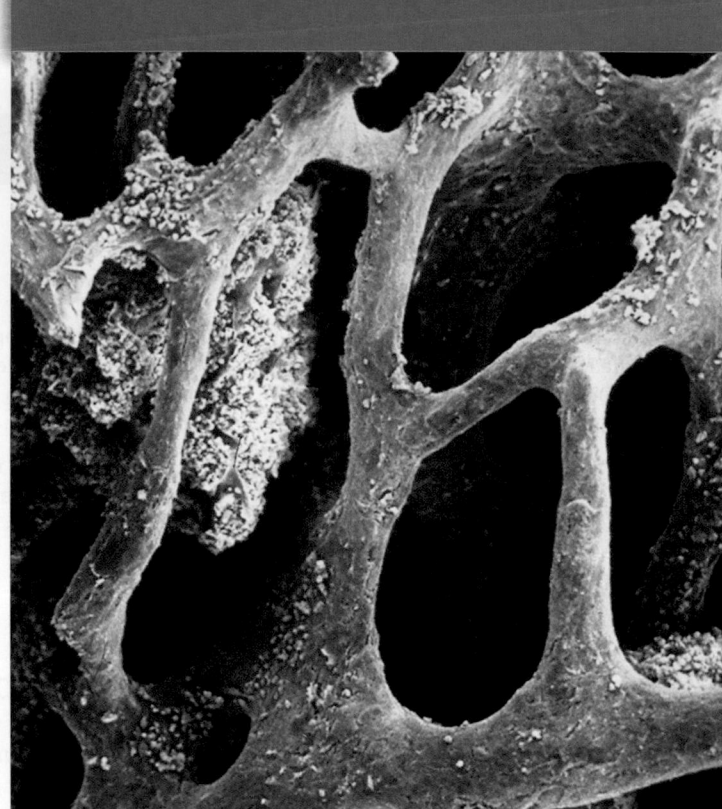

**Spongy bone of the human femur**

Anatomy & Physiology | **REVEALED**®
aprevealed.com

**Module 5: Skeletal System**

In art and history, nothing has so often symbolized death as a skull or skeleton.[1] Bones and teeth are the most durable remains of a once-living body and the most vivid reminder of the impermanence of life.

The dry bones presented for laboratory study may wrongly suggest that the skeleton is a nonliving scaffold for the body, like the steel girders of a building. Seeing it in such a sanitized form makes it easy to forget that the living skeleton is made of dynamic tissues, full of cells—that it continually remodels itself and interacts physiologically with all of the other organ systems of the body. The skeleton is permeated with nerves and blood vessels, which attests to its sensitivity and metabolic activity.

**Osteology,**[2] the study of bone, is the subject of these next three chapters. In this chapter, we study bone as a tissue—its composition, its functions, how it develops and grows, how its metabolism is regulated, and some of its disorders. This will provide a basis for understanding the skeleton, joints, and muscles in the chapters that follow.

## 7.1 Tissues and Organs of the Skeletal System

### Expected Learning Outcomes

When you have completed this section, you should be able to

a. name the tissues and organs that compose the skeletal system;
b. state several functions of the skeletal system;
c. distinguish between bone as a tissue and as an organ;
d. describe four types of bones classified by shape; and
e. describe the general features of a long bone and a flat bone.

The **skeletal system** is composed of bones, cartilages, and ligaments joined tightly to form a strong, flexible framework for the body. Cartilage, the forerunner of most bones in embryonic and childhood development, covers many joint surfaces in the mature skeleton. Ligaments hold bones together at the joints and are discussed in chapter 9. Tendons are structurally similar to ligaments but attach muscle to bone; they are discussed with the muscular system in chapter 10. Here, we focus on the bones.

## Functions of the Skeleton

The skeleton plays at least six roles:

1. **Support.** Bones of the limbs and vertebral column support the body; the mandible and maxilla support the teeth; and some viscera are supported by nearby bones.
2. **Protection.** Bones enclose and protect the brain, spinal cord, heart, lungs, pelvic viscera, and bone marrow.
3. **Movement.** Limb movements, breathing, and other movements are produced by the action of muscles on the bones.
4. **Electrolyte balance.** The skeleton stores calcium and phosphate ions and releases them into the tissue fluid and blood according to the body's physiological needs.
5. **Acid–base balance.** Bone tissue buffers the blood against excessive pH changes by absorbing or releasing alkaline phosphate and carbonate salts.
6. **Blood formation.** Red bone marrow is the major producer of blood cells, including cells of the immune system.

## Bones and Osseous Tissue

Bone, or **osseous**[3] **tissue,** is a connective tissue in which the matrix is hardened by the deposition of calcium phosphate and other minerals. The hardening process is called **mineralization** or **calcification.** (Bone is not the hardest substance in the body; that distinction goes to tooth enamel.) Osseous tissue is only one of the tissues that make up a bone. Also present are blood, bone marrow, cartilage, adipose tissue, nervous tissue, and fibrous connective tissue. The word *bone* can denote an organ composed of all these tissues, or it can denote just the osseous tissue.

## General Features of Bones

Bones have a wide variety of shapes correlated with their varied protective and locomotor functions. Most of the cranial bones are in the form of thin curved plates called **flat bones,** such as the paired parietal bones that form the dome of the top of the head. The sternum (breastbone), scapula (shoulder blade), ribs, and hip bones are also flat bones. The most important bones in body movement are the **long bones** of the limbs—the humerus, radius, and ulna of the arm and forearm; the femur, tibia, and fibula of the thigh and leg; and the metacarpals, metatarsals, and phalanges of the hands and feet. Like crowbars, long bones serve as rigid levers that are acted upon by skeletal muscles to produce the major body movements. The wrists and ankles have a total of 30 **short bones** (carpal and tarsal bones), which are approximately equal in length and width and which produce relatively limited gliding movements. The patella is also a short bone. Many bones, however, do not fit any of these categories and are collectively

---

[1] *skelet* = dried up
[2] *osteo* = bone; *logy* = study of

[3] *os, osse, oste* = bone

considered **irregular bones**—the vertebrae and three tiny middle-ear bones, for example.

Figure 7.1 shows the general anatomy of a long bone. Much of it is composed of an outer shell of dense white osseous tissue called **compact (dense) bone.** The shell encloses a space called the **medullary** (MED-you-lerr-ee) **cavity,** or **marrow cavity,** which contains bone marrow. At the ends of the bone, the central space is occupied by a more loosely organized form of osseous tissue called **spongy (cancellous) bone.** A narrow zone of spongy bone also occurs just inside the compact bone of the shaft and in the middle of most flat, irregular, and short bones. The skeleton is about three-quarters compact bone and one-quarter spongy bone by weight. Spongy bone is always enclosed by more durable compact bone.

The principal features of a long bone are its shaft, called the **diaphysis**[4] (dy-AF-ih-sis), and an expanded head at each end called the **epiphysis**[5] (eh-PIF-ih-sis). The diaphysis provides leverage, and the epiphysis is enlarged to strengthen the joint and provide added surface area for the attachment of tendons and ligaments. The joint surface where one bone meets another is covered with a layer of hyaline cartilage called the **articular cartilage.** Together with a lubricating fluid secreted between the bones, this cartilage enables a joint to move far more easily than it would if one bone rubbed directly against the other. Blood vessels penetrate into the bone through minute holes called **nutrient foramina** (for-AM-ih-nuh); we will trace where they go when we consider the histology of bone.

Externally, a bone is covered with a sheath called the **periosteum.**[6] This has a tough, outer *fibrous layer* of collagen and an inner *osteogenic layer* of bone-forming cells described later in the chapter. Some collagen fibers of the outer layer are continuous with the tendons that bind muscle to bone, and some penetrate into the bone matrix as **perforating fibers.** The periosteum thus provides strong attachment and continuity from muscle to tendon to bone. The osteogenic layer is important to the growth of bone and healing of fractures. There is no periosteum over the articular cartilage.

A thin layer of reticular connective tissue called the **endosteum**[7] lines the internal marrow cavity, covers all the honeycombed surfaces of spongy bone, and lines the canal system, described later, in compact bone.

In children and adolescents, an **epiphyseal** (EP-ih-FIZZ-ee-ul) **plate** of hyaline cartilage separates the marrow spaces of the epiphysis and diaphysis (fig. 7.1a). On X-rays, it appears as a transparent line at the end of a long bone (see fig. 7.11). The epiphyseal plate is a zone where the bones grow in length. In adults, the epiphyseal plate is depleted and the bones can grow no longer, but an *epiphyseal line* marks where the plate used to be.

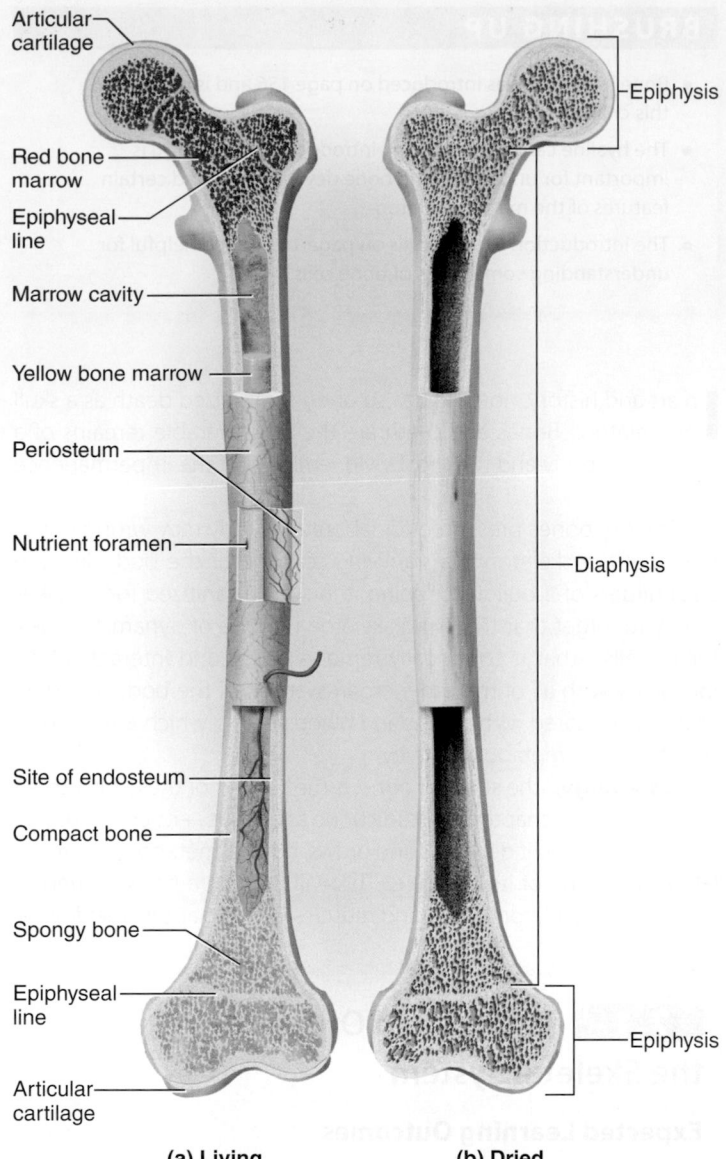

**(a) Living**          **(b) Dried**

**FIGURE 7.1  Anatomy of a Long Bone.** (a) The femur, with its soft tissues including bone marrow, articular cartilage, blood vessels, and periosteum. (b) A dried femur in longitudinal section.

 *What is the functional significance of a long bone being wider at the epiphyses than at the diaphysis?*

Figure 7.2 shows a flat bone of the cranium. It has a sandwichlike construction with two layers of compact bone enclosing a middle layer of spongy bone. The spongy layer in the cranium is called **diploe**[8] (DIP-lo-ee). A moderate blow to the skull can fracture the outer layer of compact bone, but the diploe may absorb the impact and leave the inner layer of compact bone unharmed. Both surfaces of a flat bone are covered with periosteum, and the marrow spaces amid the spongy bone are lined with endosteum.

---

[4]*dia* = across; *physis* = growth; originally named for a ridge on the shaft of the tibia
[5]*epi* = upon, above; *physis* = growth
[6]*peri* = around; *oste* = bone
[7]*endo* = within; *oste* = bone

[8]*diplo* = double

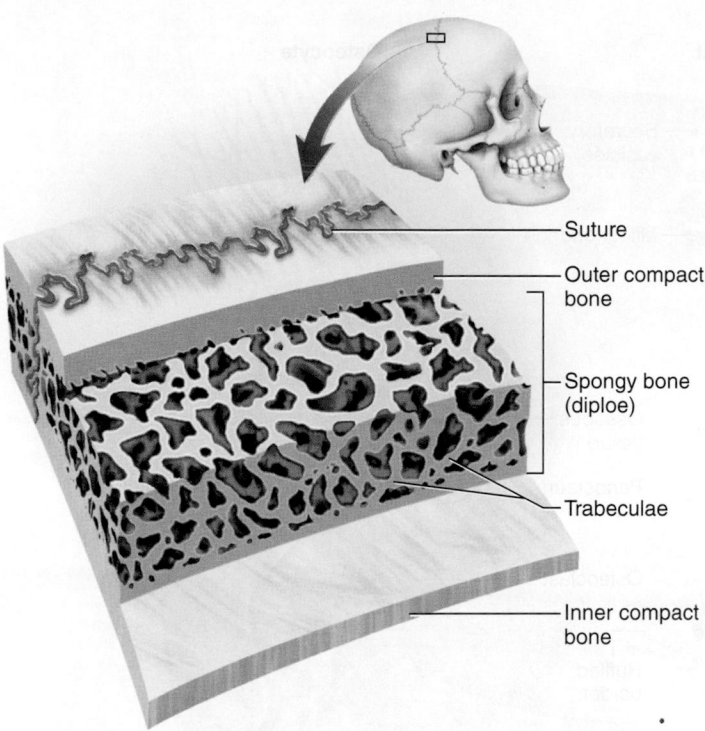

Suture

Outer compact bone

Spongy bone (diploe)

Trabeculae

Inner compact bone

**FIGURE 7.2 Anatomy of a Flat Bone.**

Answer the following questions to test your understanding of the preceding section:

1. Name at least five tissues found in a bone.

2. List three or more functions of the skeletal system other than supporting the body and protecting some of the internal organs.

3. Name the four bone shapes and give an example of each.

4. Explain the difference between compact and spongy bone, and describe their spatial relationship to each other.

5. State the anatomical terms for the shaft, head, growth zone, and fibrous covering of a long bone.

## 7.2 Histology of Osseous Tissue

### Expected Learning Outcomes

When you have completed this section, you should be able to

a. list and describe the cells, fibers, and ground substance of bone tissue;

b. state the importance of each constituent of bone tissue;

c. compare the histology of the two types of bone tissue; and

d. distinguish between the two types of bone marrow.

## Bone Cells

Like any other connective tissue, bone consists of cells, fibers, and ground substance. There are four principal types of bone cells (fig. 7.3):

1. **Osteogenic**[9] **cells** are stem cells that develop from embryonic mesenchymal cells (see p. 142) and then give rise to most other bone cell types. They are found in the endosteum and inner layer of the periosteum. They multiply continually, and some go on to become the *osteoblasts* described next.

2. **Osteoblasts**[10] are bone-forming cells. They are roughly cuboidal or angular, and line up in a single layer on the bone surface under the endosteum and periosteum and resemble a cuboidal epithelium (see fig. 7.8). Osteoblasts are nonmitotic, so the only way new ones can be generated is by mitosis and differentiation of osteogenic cells. They synthesize the soft organic matter of the bone matrix, which then hardens by mineral deposition. Stress and fractures stimulate osteogenic cells to multiply more rapidly and quickly generate increased numbers of osteoblasts, which reinforce or rebuild the bone. Osteoblasts also have an endocrine function: They secrete a hormone, *osteocalcin,* previously thought to be only a structural protein of bone. Osteocalcin stimulates insulin secretion by the pancreas, increases insulin sensitivity in adipocytes, and limits the growth of adipose tissue.

3. **Osteocytes** are former osteoblasts that have become trapped in the matrix they deposited. They reside in tiny cavities called **lacunae,**[11] which are interconnected by slender channels called **canaliculi**[12] (CAN-uh-LIC-you-lye). Each osteocyte has delicate cytoplasmic processes that reach into the canaliculi to contact the processes from neighboring osteocytes. Some of them also contact osteoblasts on the bone surface. Neighboring osteocytes are connected by gap junctions where their processes meet, so they can pass nutrients and chemical signals to one another and pass their metabolic wastes to the nearest blood vessel for disposal.

   Osteocytes have multiple functions. Some resorb bone matrix and others deposit it, so they contribute to the homeostatic maintenance of both bone density and blood concentrations of calcium and phosphate ions. Perhaps even more importantly, they are strain sensors. When a load is applied to a bone, it produces a flow in the extracellular fluid of the lacunae and canaliculi. This apparently stimulates the sensory primary cilia on the osteocytes and induces the cells to secrete signals that regulate bone remodeling—adjustments in bone shape and density to adapt to stress.

---

[9]*osteo* = bone; *genic* = producing
[10]*osteo* = bone; *blast* = form, produce
[11]*lac* = lake, hollow; *una* = little
[12]*canal* = canal, channel; *icul* = little

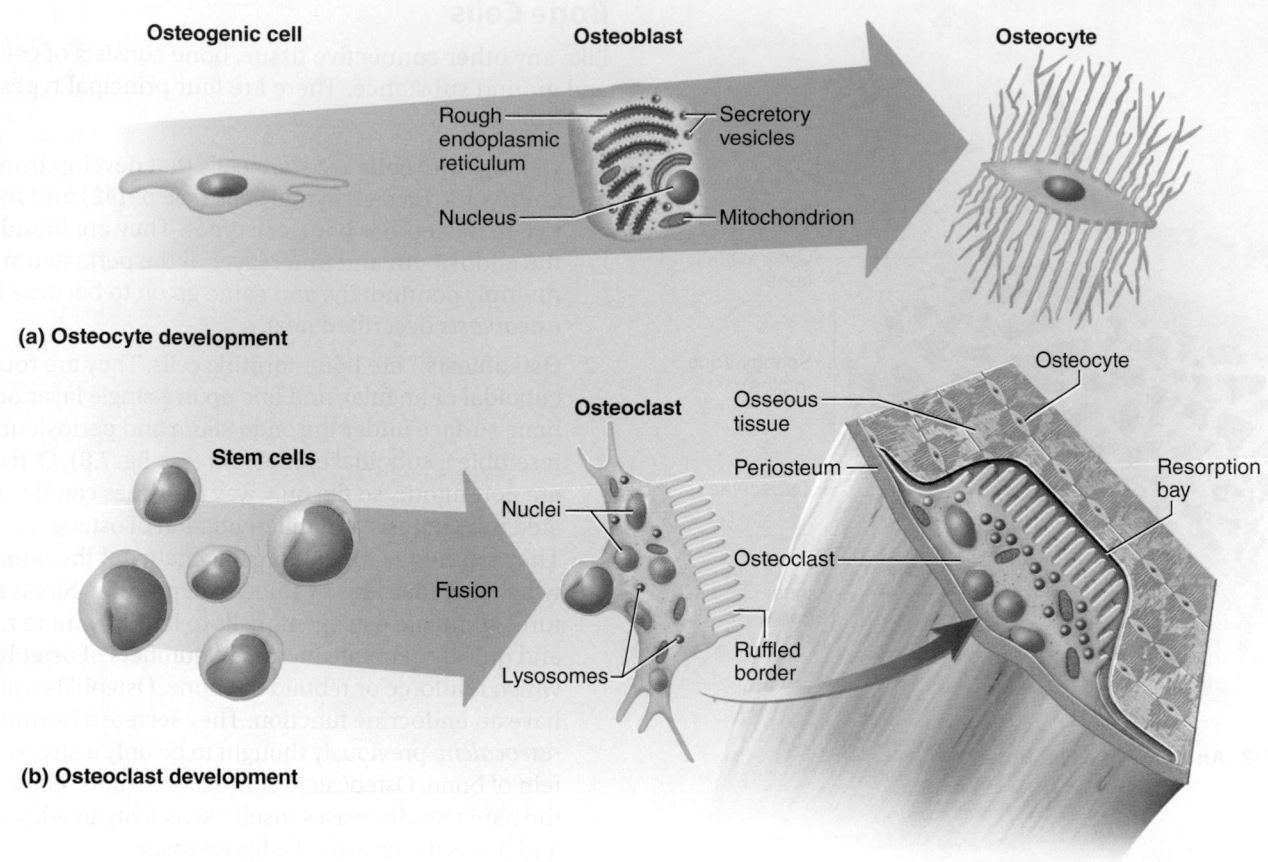

**Osteogenic cell**

**Osteoblast**

**Osteocyte**

Rough endoplasmic reticulum

Secretory vesicles

Nucleus

Mitochondrion

**(a) Osteocyte development**

**Stem cells**

**Osteoclast**

Nuclei

Fusion

Lysosomes

Ruffled border

Osteocyte

Osseous tissue

Periosteum

Osteoclast

Resorption bay

**(b) Osteoclast development**

**FIGURE 7.3** **Bone Cells and Their Development.** (a) Osteogenic cells give rise to osteoblasts, which deposit matrix around themselves and transform into osteocytes. (b) Bone marrow stem cells fuse to form osteoclasts.

4. **Osteoclasts**[13] are bone-dissolving cells on the bone surfaces. They develop from the same bone marrow stem cells that give rise to the blood cells. Thus, osteogenic cells, osteoblasts, and osteocytes all belong to one cell lineage, but osteoclasts have an independent origin (fig. 7.3). Each osteoclast is formed by the fusion of several stem cells, so osteoclasts are unusually large (up to 150 μm in diameter, visible to the naked eye). They typically have 3 or 4 nuclei, but sometimes up to 50, each contributed by one stem cell. The side of the osteoclast facing the bone surface has a *ruffled border* with many deep infoldings of the plasma membrane. This increases the cell surface area and thus enhances the efficiency of bone resorption. Osteoclasts often reside in pits called *resorption bays (Howship*[14] *lacunae)* that they have etched into the bone surface. Bone remodeling results from the combined action of these bone-dissolving osteoclasts and bone-depositing osteoblasts.

[13]*osteo* = bone; *clast* = destroy, break down
[14]J. Howship (1781–1841), English surgeon

## The Matrix

The matrix of osseous tissue is, by dry weight, about one-third organic and two-thirds inorganic matter. The organic matter, synthesized by the osteoblasts, includes collagen and various protein–carbohydrate complexes such as glycosaminoglycans, proteoglycans, and glycoproteins. The inorganic matter is about 85% **hydroxyapatite,** a crystallized calcium phosphate salt $[Ca_{10}(PO_4)_6(OH)_2]$; 10% calcium carbonate $(CaCO_3)$; and lesser amounts of magnesium, sodium, potassium, fluoride, sulfate, carbonate, and hydroxide ions. Several foreign elements behave chemically like bone minerals and become incorporated into osseous tissue as contaminants, sometimes with deadly results (see Deeper Insight 7.1).

### ▶▶▶ APPLY WHAT YOU KNOW

*What two organelles do you think are especially prominent in osteoblasts? (Hint: Consider the major substances that osteoblasts synthesize.)*

Bone is in a class of materials that engineers call a **composite**, a combination of two basic structural materials—in this case a ceramic and a polymer. A composite can combine the optimal mechanical properties of each component. Consider a fiberglass fishing rod, for example, made of a ceramic (glass fibers) embedded in a polymer (resin). The resin alone would be too brittle and the fibers alone too flexible and limp to serve the purpose of a fishing rod, but together they produce a material of great strength and flexibility.

In bone, the polymer is collagen and the ceramic is hydroxyapatite and other minerals. The ceramic component enables a bone to support the weight of the body without sagging. When the bones are deficient in calcium salts, they are soft and bend easily. One way to demonstrate this is to soak a clean dried bone, such as a chicken bone, in vinegar for a few days. As the mild acid of the vinegar dissolves the minerals away, the bone becomes flexible and rubbery. Such mineral deficiency and flexibility are the central problems in the childhood disease, *rickets,* in which the soft bones of the lower limbs bend under the body's weight and become permanently deformed.

The protein component gives bone a degree of flexibility. Without protein, a bone is excessively brittle, as in *osteogenesis imperfecta,* or *brittle bone disease* (see table 7.2). Without collagen, a jogger's bones would shatter under the impact of running. But normally, when a bone bends slightly toward one side, the tensile strength of the collagen fibers on the opposite side holds the bone together and prevents it from snapping like a stick of chalk. Collagen molecules have *sacrificial bonds* that break under stress, protecting a bone from fracture by dissipating some of the shock. The bonds re-form when the collagen is relieved of stress.

Unlike fiberglass, bone varies from place to place in its ratio of minerals to collagen. Osseous tissue is thus adapted to different amounts of tension and compression exerted on different parts of the skeleton.

## Compact Bone

The histological study of compact bone usually uses slices that have been dried, cut with a saw, and ground to translucent thinness. This procedure destroys the cells but reveals fine details of the matrix (fig. 7.4). Such sections show onionlike **concentric lamellae**—layers of matrix concentrically arranged around a **central (haversian[15]) canal** and connected with each other by canaliculi. A central canal and its lamellae constitute an **osteon (haversian system)**—the basic structural unit of compact bone. In longitudinal views and three-dimensional reconstructions, we can see that an osteon is actually a cylinder of tissue surrounding a central canal. Along their length, central canals are joined by transverse or diagonal passages called **perforating (Volkmann[16]) canals.** The central and perforating canals are lined with endosteum.

Collagen fibers "corkscrew" down the matrix of a given lamella in a helical arrangement like the threads of a screw. The helices coil in one direction in one lamella and in the opposite direction in the next lamella (fig. 7.4b). This enhances the strength of bone on the same principle as plywood, made of thin layers of wood with the grain running in different directions from one layer to the next. In areas where the bone must resist tension (bending), the helix is loosely coiled like the threads on a wood screw and the fibers are more stretched out on the longitudinal axis of the bone. In weight-bearing areas, where the bone must resist compression, the helix is more tightly coiled like the closely spaced threads on a bolt, and the fibers are more nearly transverse.

The skeleton receives about half a liter of blood per minute. Blood vessels, along with nerves, enter the bone tissue through nutrient foramina on the surface. These foramina open into the perforating canals that cross the matrix and feed into the central canals. The innermost osteocytes around each central canal receive nutrients from these blood vessels and pass them along through their gap junctions to neighboring osteocytes. They also receive wastes from their neighbors and convey them to the central canal for removal by the bloodstream. Thus, the cytoplasmic processes of the osteocytes maintain a two-way flow of nutrients and wastes between the central canal and the outermost cells of the osteon.

Not all of the matrix is organized into osteons. The inner and outer boundaries of dense bone are arranged in *circumferential lamellae* that run parallel to the bone surface. Between

## DEEPER INSIGHT 7.1
## MEDICAL HISTORY

### Bone Contamination

When Marie and Pierre Curie and Henri Becquerel received their 1903 Nobel Prize for the discovery of radioactivity (see Deeper Insight 2.1, p. 44), radiation captured the public imagination. Not for several decades did anyone realize its dangers. For example, factories employed women to paint luminous numbers on watch and clock dials with radium paint. The women moistened their paint brushes with their tongues to keep them finely pointed and ingested radium in the process. The radium accumulated in their bones and caused many of them to develop a form of bone cancer called osteosarcoma.

Even more horrific, in the wisdom of hindsight, was a deadly health fad in which people drank "tonics" made of radium-enriched water. One famous enthusiast was the millionaire playboy and championship golfer Eben Byers (1880–1932), who drank several bottles of radium tonic each day and praised its virtues as a wonder drug and aphrodisiac. Like the factory women, Byers contracted osteosarcoma. By the time of his death, holes had formed in his skull and doctors had removed his entire upper jaw and most of his mandible in an effort to halt the spreading cancer. Byers' bones and teeth were so radioactive they could expose photographic film in the dark. Brain damage left him unable to speak, but he remained mentally alert to the bitter end. His tragic decline and death shocked the world and helped put an end to the radium tonic fad.

---

[15]Clopton Havers (1650–1702), English anatomist
[16]Alfred Volkmann (1800–1877), German physiologist

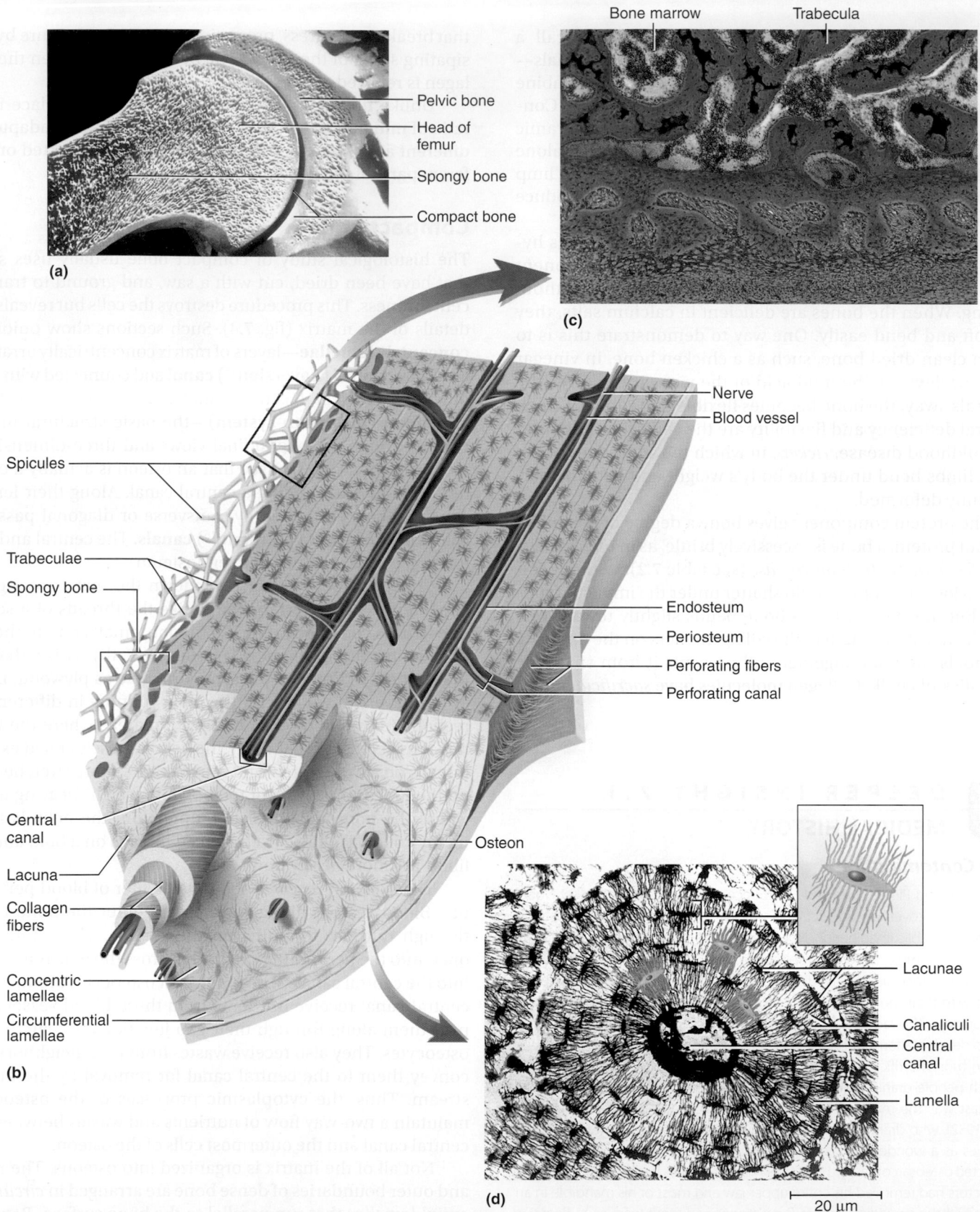

Bone marrow   Trabecula

Pelvic bone
Head of femur
Spongy bone
Compact bone

(a)

(c)

Nerve
Blood vessel

Spicules

Trabeculae
Spongy bone

Endosteum
Periosteum
Perforating fibers
Perforating canal

Central canal
Lacuna
Collagen fibers
Concentric lamellae
Circumferential lamellae

Osteon

(b)

Lacunae
Canaliculi
Central canal
Lamella

(d)

20 μm

**FIGURE 7.4  The Histology of Osseous Tissue.**  (a) Compact and spongy bone in a frontal section of the hip joint. (b) The three-dimensional structure of compact bone. Lamellae of one osteon are telescoped to show their alternating arrangement of collagen fibers. (c) Microscopic appearance of decalcified spongy bone. (d) Microscopic appearance of a cross section of an osteon of dried compact bone.  **AP|R**

❓ *Which type of bone, spongy or compact, has more surface area exposed to osteoclast action?*

osteons, we can find irregular regions called *inter-stitial lamellae,* the remains of old osteons that broke down as the bone grew and remodeled itself.

## Spongy Bone

Spongy bone (fig. 7.4a) consists of a lattice of delicate slivers of bone called **spicules**[17] (rods or spines, as in the photo on p. 202) and **trabeculae**[18] (thin plates). Although calcified and hard, it is named for its spongelike appearance. It is covered with endosteum and permeated by spaces filled with bone marrow. The matrix is arranged in lamellae like those of compact bone, but there are few osteons. Central canals are not needed here because no osteocyte is very far from the marrow. Spongy bone is well designed to impart strength to a bone while adding a minimum of weight. Its trabeculae are not randomly arranged as they might seem at a glance, but develop along the bone's lines of stress (fig. 7.5).

## Bone Marrow

**Bone marrow** is a general term for soft tissue that occupies the marrow cavity of a long bone, the spaces amid the trabeculae of spongy bone, and the larger central canals. There are two kinds of marrow—red and yellow. We can best appreciate their differences by considering how marrow changes over a person's lifetime.

In a child, the marrow cavity of nearly every bone is filled with **red bone marrow (myeloid tissue).** This is often described as *hemopoietic*[19] (HE-mo-poy-ET-ic) *tissue*—tissue that produces blood cells—but it is actually composed of multiple tissues in a delicate but intricate arrangement, and is properly considered an organ unto itself. Its structure is further described in chapter 21.

In adults, most of the red marrow turns to fatty **yellow bone marrow,** like the fat at the center of a ham bone. Red marrow is then limited to the skull, vertebrae, ribs, sternum, part of the pelvic (hip) girdle, and the proximal heads of the humerus and femur (fig. 7.6). Yellow bone marrow no longer produces blood, although in the event of severe or chronic anemia, it can transform back into red marrow and resume its hemopoietic function.

### BEFORE YOU GO ON

Answer the following questions to test your understanding of the preceding section:

6. Suppose you had unlabeled electron micrographs of the four kinds of bone cells and their neighboring tissues. Name the four cells and explain how you could visually distinguish each one from the other three.

7. Name three organic components of the bone matrix.

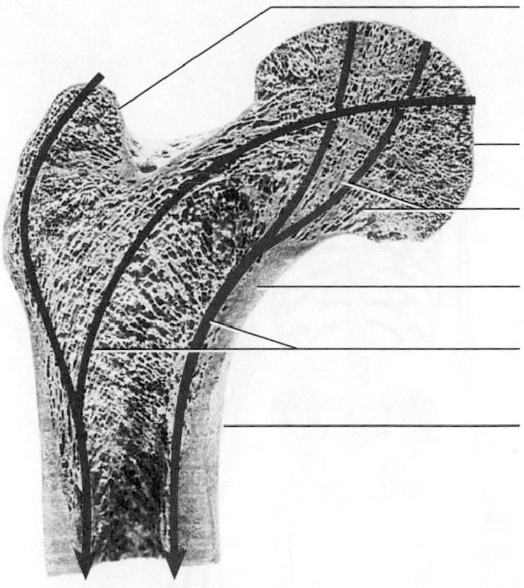

**FIGURE 7.5  Spongy Bone Structure in Relation to Mechanical Stress.**  In this frontal section of the femur (thighbone), the trabeculae of spongy bone can be seen oriented along lines of mechanical stress applied by the weight of the body or the pull of a muscle.

8. What are the mineral crystals of bone called, and what are they made of?

9. Sketch a cross section of an osteon and label its major parts.

10. What are the two kinds of bone marrow? What does hemopoietic tissue mean? Which type of bone marrow fits this description?

## 7.3  Bone Development

### Expected Learning Outcomes

When you have completed this section, you should be able to

a. describe two mechanisms of bone formation; and

b. explain how mature bone continues to grow and remodel itself.

The formation of bone is called **ossification** (OSS-ih-fih-CAY-shun) or **osteogenesis.** There are two methods of ossification— *intramembranous* and *endochondral.* Both begin with a soft embryonic tissue called *mesenchyme* (MEZ-en-kime).

## Intramembranous Ossification

**Intramembranous**[20] (IN-tra-MEM-bra-nus) **ossification** produces the flat bones of the skull and most of the clavicle (collarbone). Follow its stages in figure 7.7 as you read the correspondingly numbered descriptions here.

---

[17]*spic* = dart, point; *ule* = little
[18]*trabe* = plate; *cul* = little
[19]*hemo* = blood; *poietic* = forming

[20]*intra* = within; *membran* = membrane

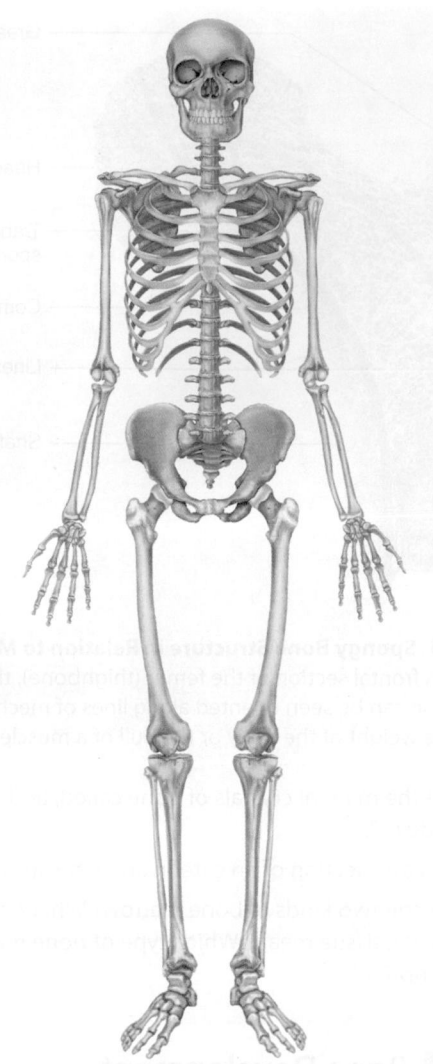

**FIGURE 7.6  Distribution of Red and Yellow Bone Marrow.**
In an adult, red bone marrow occupies the marrow cavities of the axial skeleton and proximal heads of the humerus and femur. Yellow bone marrow occurs in the long bones of the limbs.

❓ *What would be the most accessible places to draw red bone marrow from an adult?*

(1) Mesenchyme first condenses into a soft sheet of tissue permeated with blood vessels—the *membrane* to which *intramembranous* refers. Mesenchymal cells line up along the blood vessels, become osteoblasts, and secrete a soft collagenous *osteoid*[21] *tissue (prebone)* (fig. 7.8) in the direction away from the vessel. Osteoid tissue resembles bone but is not yet calcified.

(2) Calcium phosphate and other minerals crystallize on the collagen fibers of the osteoid tissue and harden the matrix. Continued osteoid deposition and mineralization squeeze the blood vessels and future bone marrow into narrower

and narrower spaces. As osteoblasts become trapped in their own hardening matrix, they become osteocytes.

(3) While the foregoing processes are going on, more of the mesenchyme adjacent to the developing bone condenses and forms a fibrous periosteum on each surface. The spongy bone becomes a honeycomb of slender calcified trabeculae.

(4) At the surfaces, osteoblasts beneath the periosteum deposit layers of bone, fill in the spaces between trabeculae, and create a zone of compact bone on each side, as well as thickening the bone overall. This process gives rise to the sandwichlike structure typical of a flat cranial bone—a layer of spongy bone between two layers of compact bone.

Intramembranous ossification also plays an important role in the lifelong thickening, strengthening, and remodeling of the long bones discussed next. Throughout the skeleton, it is the method of depositing new tissue on the bone surface even past the age where the bones can no longer grow in length.

## Endochondral Ossification

**Endochondral**[22] (EN-doe-CON-drul) **ossification** is a process in which a bone develops from a preexisting model composed of hyaline cartilage. It begins around the sixth week of fetal development and continues into a person's 20s. Most bones of the body develop in this way, including the vertebrae, ribs, sternum, scapula, pelvic girdle, and bones of the limbs.

Figure 7.9 shows the following steps in endochondral ossification. This figure uses a metacarpal bone from the palmar region of the hand as an example because of its relative simplicity, having only one *epiphyseal plate* (growth center). Many other bones develop in more complex ways, having an epiphyseal plate at both ends or multiple plates at each end, but the basic process is the same.

(1) Mesenchyme develops into a body of hyaline cartilage, covered with a fibrous perichondrium, in the location of a future bone. For a time, the perichondrium produces chondrocytes and the cartilage model grows in thickness.

(2) In a **primary ossification center** near the middle of this cartilage, chondrocytes begin to inflate and die, while the thin walls between them calcify. The perichondrium stops producing chondrocytes and begins producing osteoblasts. These deposit a thin collar of bone around the middle of the cartilage model, reinforcing it like a napkin ring. The former perichondrium is now considered to be a periosteum. As chondrocytes in the middle of the model die, their lacunae merge into a single cavity.

(3) Osteoclasts arrive in the blood and digest calcified tissue in the shaft, hollowing it out and creating the **primary marrow cavity.** Osteoblasts also arrive and deposit layers of bone lining the cavity, thickening the shaft. As the

---

[21]*oste* = bone; *oid* = like, resembling

[22]*endo* = within; *chondr* = cartilage

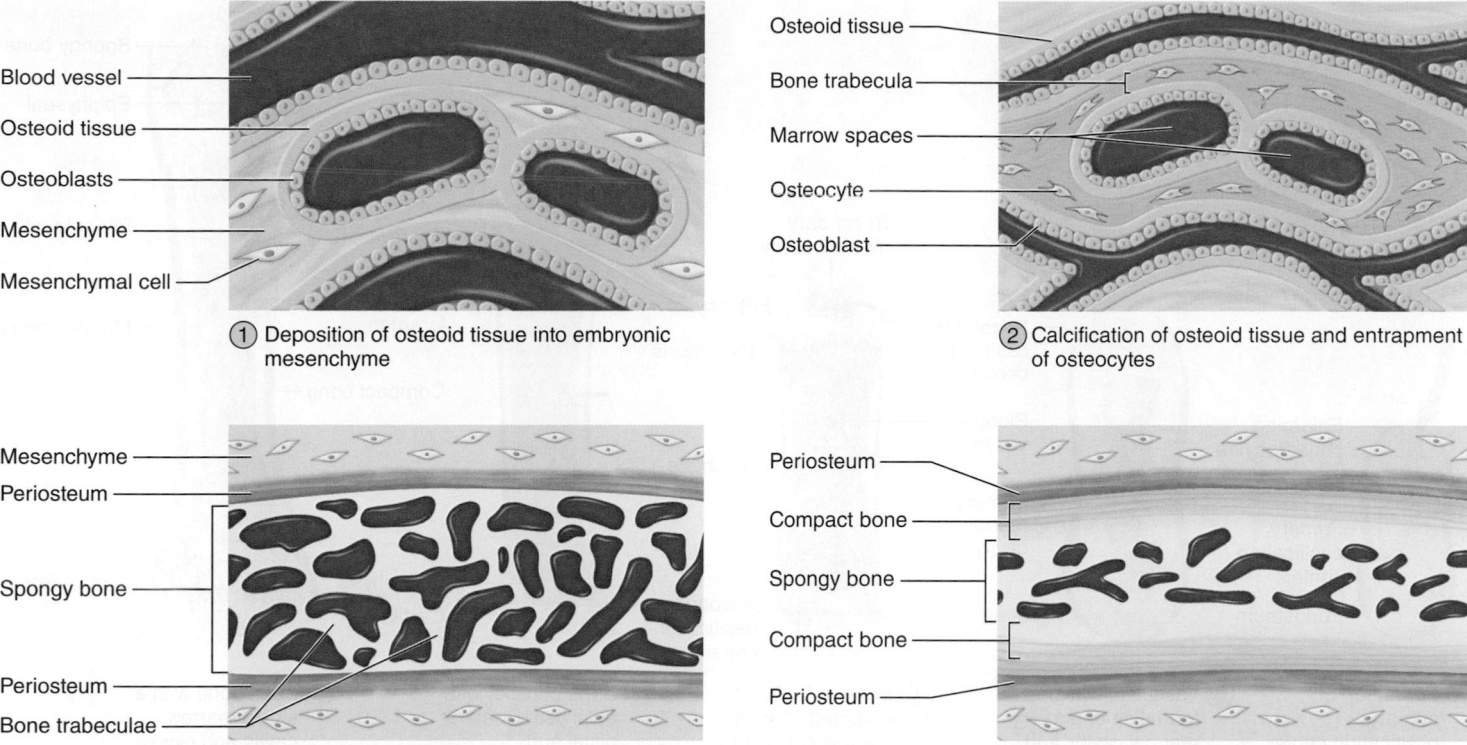

Blood vessel

Osteoid tissue

Osteoblasts

Mesenchyme

Mesenchymal cell

① Deposition of osteoid tissue into embryonic mesenchyme

Osteoid tissue

Bone trabecula

Marrow spaces

Osteocyte

Osteoblast

② Calcification of osteoid tissue and entrapment of osteocytes

Mesenchyme

Periosteum

Spongy bone

Periosteum

Bone trabeculae

③ Honeycomb of spongy bone with developing periosteum

Periosteum

Compact bone

Spongy bone

Compact bone

Periosteum

④ Filling of space to form compact bone at surfaces, leaving spongy bone in middle

**FIGURE 7.7  Intramembranous Ossification.** The figures are drawn to different scales, with the highest magnification and detail at the beginning and backing off for a broader overview at the end of the process.

❓ *With the aid of chapter 8, name at least two specific bones other than the clavicle that would form by this process.*

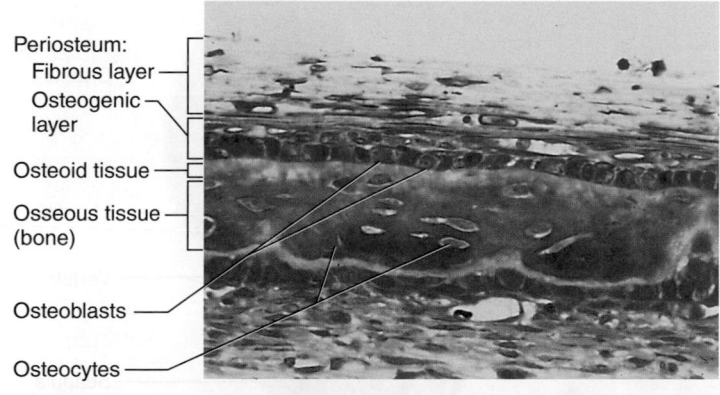

Periosteum:
  Fibrous layer
  Osteogenic layer
Osteoid tissue
Osseous tissue (bone)

Osteoblasts

Osteocytes

**FIGURE 7.8  Intramembranous Ossification of a Cranial Bone of the Human Fetus.** Note the layers of osteoid tissue, osteoblasts, and fibrous periosteum on both sides of the bone.

bony collar under the periosteum thickens and elongates, a wave of cartilage death progresses toward the ends of the bone. Osteoclasts in the marrow cavity follow this wave, dissolving calcified cartilage remnants and enlarging the marrow cavity of the diaphysis. The region of transition from cartilage to bone at each end of the primary marrow cavity is called a **metaphysis.**

Soon, chondrocyte enlargement and death occur in the epiphysis of the model as well, creating a **secondary ossification center.** In the metacarpal bones, as illustrated in figure 7.9, this occurs in only one epiphysis. In longer bones of the arms, forearms, legs, and thighs, it occurs at both ends.

④ The secondary ossification center hollows out by the same process as the diaphysis, generating a **secondary marrow cavity** in the epiphysis. This cavity expands outward from the center, in all directions. At the time of birth, the bone typically looks like step 4 in figure 7.9. In bones with two secondary ossification centers, one center lags behind the other, so at birth there is a secondary marrow cavity at one end while chondrocyte growth has just begun at the other. The joints of the limbs are still cartilaginous at birth, much as they are in the 12-week fetus in figure 7.10.

⑤ During infancy and childhood, the epiphyses fill with spongy bone. Cartilage is then limited to the articular cartilage covering each joint surface, and to an **epiphyseal** (EP-ih-FIZ-ee-ul) **plate,** a thin wall of cartilage separating the primary and secondary marrow cavities at one or both ends of the bone. The plate persists through childhood and adolescence and serves as a growth zone for bone elongation.

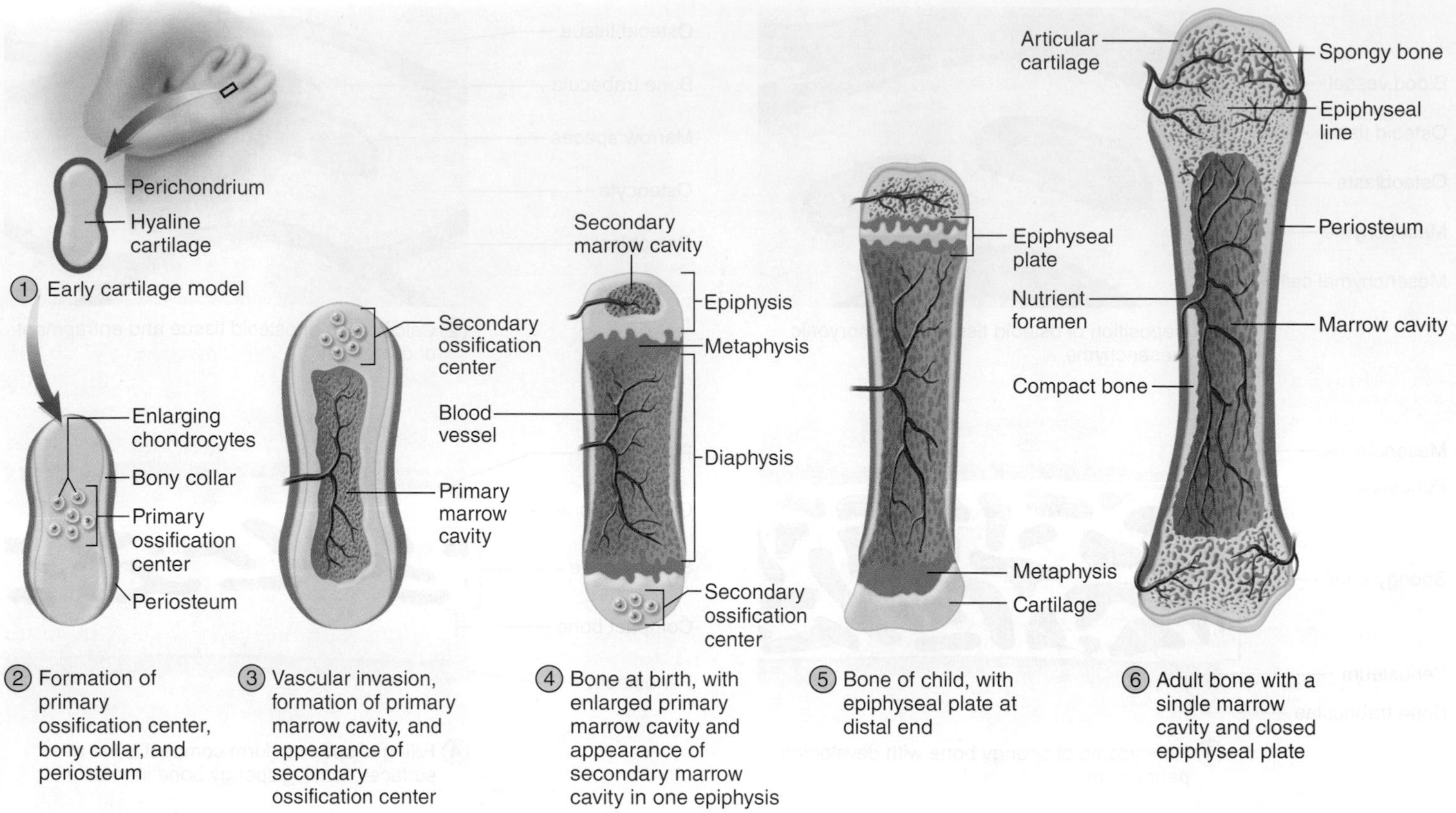

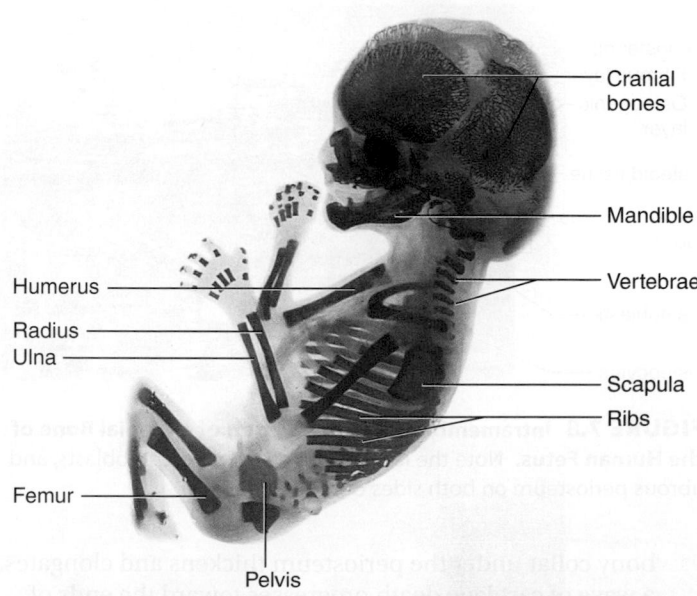

**FIGURE 7.9  Stages of Endochondral Ossification.**  A metacarpal bone of the hand.

❓ *With the aid of chapter 8, name at least two specific bones that would have two epiphyseal plates (proximal and distal) at stage 5.*

⑥ By the late teens to early twenties, all remaining cartilage in the epiphyseal plate is generally consumed and the gap between the epiphysis and diaphysis closes. The primary and secondary marrow cavities then unite into a single cavity.

## Bone Growth and Remodeling

Ossification does not end at birth, but continues throughout life with the growth and remodeling of bones. Bones grow in two directions: length and width.

## Bone Elongation

To understand growth in length, we must return to the epiphyseal plates mentioned earlier (see fig. 7.9, step 5). From infancy through adolescence, an epiphyseal plate is present at one or both ends of a long bone, at the junction between the diaphysis and epiphysis. On X-rays, it appears as a translucent line across the end of a bone, since it is not yet ossified (fig. 7.11; compare the X-ray of an adult hand in fig. 8.34, p. 260). The epiphyseal plate is a region of transition from cartilage to bone, and functions as a growth zone where the bones elongate. Growth here is responsible for a person's increase in height.

**FIGURE 7.10  The Fetal Skeleton at 12 Weeks.**  The red-stained regions are calcified at this age, whereas the elbow, wrist, knee, and ankle joints appear translucent because they are still cartilaginous.

❓ *Why are the joints of an infant weaker than those of an older child?*

The epiphyseal plate consists of typical hyaline cartilage in the middle, with a transitional zone on each side where cartilage is being replaced by bone. The transitional zone, facing the marrow cavity, is called the **metaphysis** (meh-TAF-ih-sis). In figure 7.9, step 4, the cartilage is the blue region and each metaphysis is violet. Figure 7.12 shows the histological structure of the metaphysis and the following steps in the conversion of cartilage to bone.

1. **Zone of reserve cartilage.** This region, farthest from the marrow cavity, consists of typical hyaline cartilage with resting chondrocytes, not yet showing any sign of transformation into bone.

2. **Zone of cell proliferation.** A little closer to the marrow cavity, chondrocytes multiply and arrange themselves into longitudinal columns of flattened lacunae.

3. **Zone of cell hypertrophy.** Next, the chondrocytes cease to multiply and begin to hypertrophy (enlarge), much like they do in the primary ossification center of the fetus. The walls of matrix between lacunae become very thin.

4. **Zone of calcification.** Minerals are deposited in the matrix between the columns of lacunae and calcify the cartilage. These are not the permanent mineral deposits of bone, but only a temporary support for the cartilage that would otherwise soon be weakened by the breakdown of the enlarged lacunae.

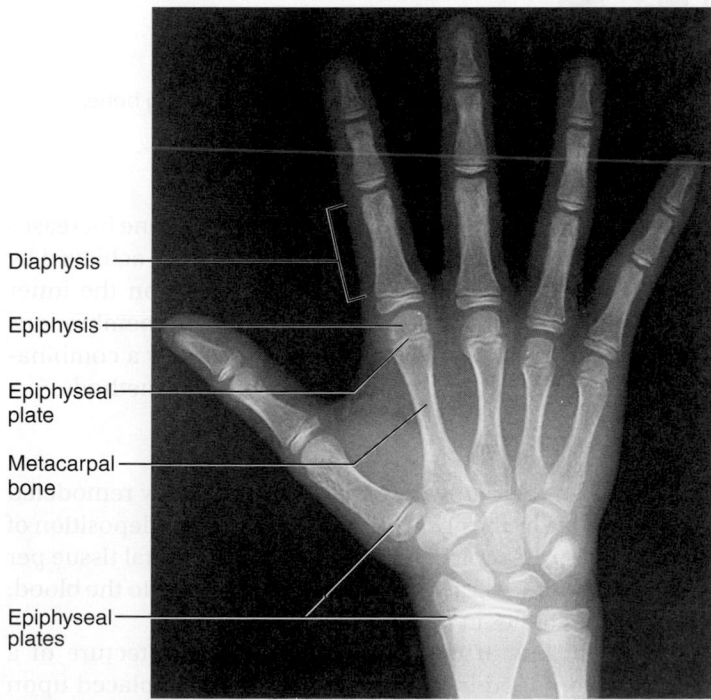

**FIGURE 7.11  X-Ray of a Child's Hand.** The cartilaginous epiphyseal plates are evident at the ends of the long bones. By adulthood, these will disappear and the epiphyses will fuse with the diaphyses. Long bones of the hand and fingers develop only one epiphyseal plate.

5. **Zone of bone deposition.** Within each column, the walls between the lacunae break down and the chondrocytes die. This converts each column into a longitudinal channel (white spaces in the figure), which is immediately invaded by blood vessels and marrow from the marrow cavity. Osteoblasts line up along the walls of these channels and begin depositing concentric lamellae of matrix, while osteoclasts dissolve the temporarily calcified cartilage.

The process of bone deposition in zone 5 creates a region of spongy bone at the end of the marrow cavity facing the metaphysis. This spongy bone remains for life, although with extensive lifelong remodeling. But around the perimeter of the marrow cavity, continuing ossification converts this spongy bone to compact bone. Osteoblasts lining the aforementioned channels deposit layer after layer of bone matrix, so the channel grows narrower and narrower. These layers become the concentric lamellae of an osteon. Finally only a slender channel persists, the central canal of a new osteon. Osteoblasts trapped in the matrix become osteocytes.

#### ▶▶▶ APPLY WHAT YOU KNOW

*In a given osteon, which lamellae are the oldest—those immediately adjacent to the central canal or those around the perimeter of the osteon? Explain your answer.*

How does a child or adolescent grow in height? Chondrocyte multiplication in zone 2 and hypertrophy in zone 3 continually push the zone of reserve cartilage (1) toward the ends of the bone, so the bone elongates. In the lower limbs, this process causes a person to grow in height, while bones of the upper limbs grow proportionately.

Thus, bone elongation is really a result of cartilage growth. Cartilage growth from within, by the multiplication of chondrocytes and deposition of new matrix in the interior, is called **interstitial growth.**[23] The most common form of dwarfism results from a failure of cartilage growth in the long bones (see Deeper Insight 7.2).

In the late teens to early twenties, all the cartilage of the epiphyseal plate is depleted. The primary and secondary marrow cavities now unite into one cavity. The junctional region where they meet is filled with spongy bone, and the site of the original epiphyseal plate is marked with a line of slightly denser spongy bone called the **epiphyseal line** (see figs. 7.1; 7.5; and 7.9, step 6). Often a delicate ridge on the bone surface marks the location of this line. When the epiphyseal plate is depleted, we say that the epiphyses have "closed" because no gap between the epiphysis and diaphysis is visible on an X-ray. Once the epiphyses have all closed in the lower limbs, a person can grow no taller. The epiphyseal plates close at different ages in different bones and in different regions of the same bone. The

Diaphysis

Epiphysis

Epiphyseal plate

Metacarpal bone

Epiphyseal plates

---

[23]*inter* = between; *stit* = to place, stand

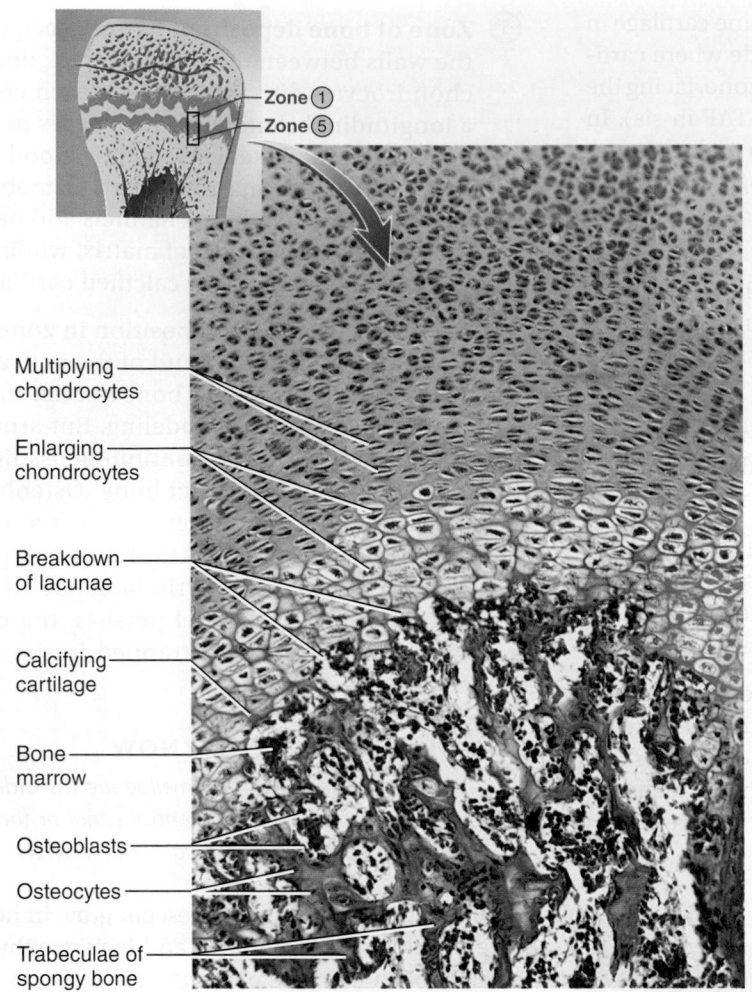

Zone ①
Zone ⑤

Multiplying chondrocytes

Enlarging chondrocytes

Breakdown of lacunae

Calcifying cartilage

Bone marrow

Osteoblasts

Osteocytes

Trabeculae of spongy bone

① **Zone of reserve cartilage**
Typical histology of resting hyaline cartilage

② **Zone of cell proliferation**
Chondrocytes multiplying and lining up in rows of small flattened lacunae

③ **Zone of cell hypertrophy**
Cessation of mitosis; enlargement of chondrocytes and thinning of lacuna walls

④ **Zone of calcification**
Temporary calcification of cartilage matrix between columns of lacunae

⑤ **Zone of bone deposition**
Breakdown of lacuna walls, leaving open channels; death of chondrocytes; bone deposition by osteoblasts, forming trabeculae of spongy bone

**FIGURE 7.12 Zones of the Metaphysis.** This micrograph shows the transition from cartilage to bone in the growth zone of a long bone.

*Which two zones in this figure account for a child's growth in height?*

state of closure in various bones of a subadult skeleton is often used in forensic science to estimate the **bone age** at death.

## Bone Widening and Thickening

Bones also continually grow in diameter and thickness. This involves a process called **appositional growth,**[24] the deposition of new tissue at the surface. Cartilages can enlarge by both interstitial and appositional growth. In bone, however, osteocytes embedded in calcified matrix have little room to spare for the deposition of more matrix internally. Bone is therefore limited to appositional growth.

Appositional growth occurs by intramembranous ossification at the bone surface. Osteoblasts in the inner layer of periosteum deposit osteoid tissue on the bone surface, calcify it, and become trapped in it as osteocytes—much like the process in figure 7.8. They lay down matrix in layers parallel to the surface, not in cylindrical osteons like those deeper in the bone. This process produces the surface layers of bone called

*circumferential lamellae,* described earlier. As a bone increases in diameter, its marrow cavity also widens. This is achieved by osteoclasts of the endosteum dissolving tissue on the inner bone surface. Thus, flat bones develop by intramembranous ossificaton alone, whereas long bones develop by a combination of the intramembranous and endochondral methods.

## Bone Remodeling

In addition to their growth, bones are continually remodeled throughout life by the absorption of old bone and deposition of new. This process replaces about 10% of the skeletal tissue per year. It repairs microfractures, releases minerals into the blood, and reshapes bones in response to use and disuse.

**Wolff's**[25] **law of bone** states that the architecture of a bone is determined by the mechanical stresses placed upon it, and the bone thereby adapts to withstand those stresses. Wolff's law is a fine example of the complementarity of form and function, showing that the form of a bone is shaped by

---

[24]*ap = ad* = to, near; *posit* = to place

[25]Julius Wolff (1836–1902), German anatomist and surgeon

# DEEPER INSIGHT 7.2

## CLINICAL APPLICATION

### Achondroplastic Dwarfism

Achondroplastic[26] (a-con-dro-PLAS-tic) *dwarfism* is a condition in which the long bones of the limbs stop growing in childhood, while the growth of other bones is unaffected. As a result, a person has a short stature but a normal-size head and trunk (fig. 7.13). As its name implies, achondroplastic dwarfism results from a failure of cartilage growth—specifically, failure of the chondrocytes in zones 2 and 3 of the metaphysis to multiply and enlarge. This is different from *pituitary dwarfism,* in which a deficiency of growth hormone stunts the growth of all of the bones, and a person has short stature but normal proportions throughout the skeletal system.

Achondroplastic dwarfism results from a spontaneous mutation that can arise any time DNA is replicated. Two people of normal height with no family history of dwarfism can therefore have a child with achondroplastic dwarfism. The mutant allele is dominant, so the offspring of a heterozygous achondroplastic dwarf have at least a 50% chance of exhibiting dwarfism, depending on the genotype of the other parent. Persons homozygous for the trait (those who inherit it from both parents) are usually stillborn or die soon after birth.

**FIGURE 7.13 Achondroplastic Dwarfism.** The student on the right, pictured with her roommate of normal height, is an achondroplastic dwarf with a height of about 122 cm (48 in.). Her parents were of normal height. Note the normal proportion of head to trunk but shortening of the limbs.

[26]*a* = without; *chondr* = cartilage; *plast* = growth

its functional experience. It is admirably demonstrated by figure 7.5, in which we see that the trabeculae of spongy bone have developed along the lines of stress placed on the femur. Wolff observed that these stress lines were very similar to the ones that engineers knew of in mechanical cranes. The effect of stress on bone development is quite evident in tennis players, in whom the bones of the racket arm are more robust than those of the other arm. Long bones of the limbs are thickest at about midshaft, because this is where they are subjected to the greatest stress.

Bone remodeling comes about through the collaborative action of osteoblasts and osteoclasts. If a bone is little used, osteoclasts remove matrix and get rid of unnecessary mass. If a bone is heavily used, or a stress is consistently applied to a particular region of a bone, osteoblasts deposit new osseous tissue and thicken the bone. Consequently, the comparatively smooth bones of an infant or toddler develop a variety of surface bumps, ridges, and spines (described in chapter 8) as the child begins to walk. The greater trochanter of the femur, for example (see fig. 7.5; also see fig. 8.38, p. 265), is a massive outgrowth of bone stimulated by the pull of tendons from several powerful hip muscles employed in walking.

On average, bones have a greater density and mass in athletes and people engaged in heavy manual labor than they do in sedentary people. Anthropologists who study ancient skeletal remains use evidence of this sort to help distinguish between members of different social classes, such as distinguishing aristocrats from laborers. Even in studying modern skeletal remains, as in investigating a suspicious death, Wolff's law comes into play as the bones give evidence of a person's sex, race, height, weight, work or exercise habits, nutritional status, and medical history.

The orderly remodeling of bone depends on a precise balance between deposition and resorption, between osteoblasts and osteoclasts. If one process outpaces the other, or both processes occur too rapidly, various bone deformities, developmental abnormalities, and other disorders occur, such as *osteitis deformans* (Paget disease), *osteogenesis imperfecta* (brittle bone disease), and *osteoporosis* (see table 7.2 and Deeper Insight 7.4).

### BEFORE YOU GO ON

Answer the following questions to test your understanding of the preceding section:

11. Describe the stages of intramembranous ossification. Name a bone that is formed in this way.
12. Describe how a cartilage model is transformed into a long bone in endochondral ossification.
13. Describe the five zones of a metaphysis and the major distinctions between them.
14. How does Wolff's law explain some of the structural differences between the bones of a young child and the bones of a young adult?

## 7.4 Physiology of Osseous Tissue

### Expected Learning Outcomes

When you have completed this section, you should be able to

a. describe the processes by which minerals are added to and removed from bone tissue;

b. discuss the role of the bones in regulating blood calcium and phosphate levels; and

c. name several hormones that regulate bone physiology, and describe their effects.

Even after a bone is fully formed, it remains a metabolically active organ with many roles to play. Not only is it involved in its own maintenance, growth, and remodeling; it also exerts a profound influence on the rest of the body by exchanging minerals with the tissue fluid. Disturbances of calcium homeostasis in the skeleton can disrupt the functioning of other organ systems, especially the nervous and muscular systems. For reasons explained later, such disturbances can even cause death by suffocation.

### Mineral Deposition and Resorption

**Mineral deposition (mineralization)** is a crystallization process in which calcium, phosphate, and other ions are taken from the blood plasma and deposited in bone tissue, mainly as needlelike crystals of hydroxyapatite. Deposition begins in fetal ossification and continues throughout life.

Osteoblasts begin the process by laying down collagen fibers in a helical pattern along the length of the osteon. These fibers then become encrusted with minerals that harden the matrix. Hydroxyapatite crystals form only when the product of calcium and phosphate concentration in the tissue fluids, represented $[Ca^{2+}] \cdot [PO_4^{3-}]$, reaches a critical value called the **solubility product.** Most tissues have inhibitors to prevent this so they do not become calcified. Osteoblasts, however, apparently neutralize these inhibitors and thus allow the salts to precipitate in the bone matrix. The first few crystals to form act as "seed crystals" that attract more calcium and phosphate from solution. The more hydroxyapatite that forms, the more it attracts additional minerals from the tissue fluid, until the matrix is thoroughly calcified.

Osseous tissue sometimes forms in the lungs, brain, eyes, muscles, tendons, arteries, and other organs. Such abnormal calcification of tissues is called **ectopic**[27] **ossification.** One example of this is arteriosclerosis, or "hardening of the arteries," which results from calcification of the arterial walls. A calcified mass in an otherwise soft organ such as the lungs is called a **calculus.**[28]

▶▶▶ **APPLY WHAT YOU KNOW**

*What positive feedback process can you recognize in bone deposition?*

**Mineral resorption** is the process of dissolving bone. It releases minerals into the blood and makes them available for other uses. Resorption is carried out by osteoclasts. They have surface receptors for calcium and respond to falling levels of calcium in the tissue fluid. Hydrogen pumps in the ruffled border of the osteoclast secrete hydrogen ions into the tissue fluid, and chloride ions follow by electrical attraction. The space between the osteoclast and the bone thus becomes filled with concentrated hydrochloric acid with a pH of about 4. The acid dissolves the bone minerals. The osteoclast also secretes an enzyme called **acid phosphatase** that digests the collagen of the bone matrix. This enzyme is named for its ability to function in a highly acidic environment.

When orthodontic appliances (braces) are used to reposition teeth, a tooth moves because osteoclasts dissolve bone ahead of the tooth (where the appliance creates greater pressure of the tooth against the bone) and osteoblasts deposit bone in the low-pressure zone behind it.

### Calcium Homeostasis

Calcium and phosphate are used for much more than bone structure. Phosphate groups are a component of DNA, RNA, ATP, phospholipids, and many other compounds. Phosphate ions also help to correct acid–base imbalances in the body fluids (see Deeper Insight 7.3). Calcium plays roles in communication among neurons, and in muscle contraction, blood clotting, and exocytosis. It is also a second or third messenger in many cell-signaling processes and a cofactor for some enzymes. The skeleton is a reservoir for these minerals. Minerals are deposited in the skeleton when the supply is ample and withdrawn when they are needed for other purposes.

The adult body contains about 1,100 g of calcium, with 99% of it in the bones. Bone has two calcium reserves: (1) a stable pool of calcium, which is incorporated into hydroxyapatite and is not easily exchanged with the blood; and (2) exchangeable calcium, which is 1% or less of the total but is easily released to the tissue fluid. The adult skeleton exchanges about 18% of its calcium with the blood each year.

The calcium concentration in the blood plasma is normally 9.2 to 10.4 mg/dL. This is a rather narrow margin of safety, as we shall soon see. About 45% of it is in the ionized form ($Ca^{2+}$), which can diffuse through capillary walls and affect neighboring cells. The rest of it is bound to plasma proteins and other solutes. It is not physiologically active, but it serves as a reserve from which free $Ca^{2+}$ can be obtained as needed.

Even slight changes in blood calcium concentration can have serious consequences. A calcium deficiency is called **hypocalcemia**[29] (HY-po-cal-SEE-me-uh). It causes excessive excitability of the nervous system and leads to muscle tremors, spasms, or **tetany**—the inability of the muscle to relax. Tetany begins to occur as the plasma $Ca^{2+}$ concentration falls to 6 mg/dL. One sign of hypocalcemia is strong spasmodic flexion of the wrist and thumb and extension of the other fingers,

[27]*ec* = out of; *top* = place
[28]*calc* = stone; *ulus* = little

[29]*hypo* = below normal; *calc* = calcium; *emia* = blood condition

called the *Trousseau*[30] *sign*—often induced by the inflation of a blood pressure cuff putting pressure on the brachial nerve. At 4 mg/dL, muscles of the larynx contract tightly, a condition called *laryngospasm,* which can shut off airflow and cause suffocation.

The reason for this hypocalcemic excitability is this: Calcium ions normally bind to and mask negatively charged groups on glycoproteins of the cell surface, contributing to the difference between the relatively positive charge on the outer face of the membrane and the negative charge on the inner face. In hypocalcemia, fewer calcium ions are available to mask the external negative charges, so there is less charge difference between the two sides of the membrane. Voltage-gated sodium channels in the plasma membrane are sensitive to this charge difference, and when the difference is diminished, they open more easily and stay open longer. This allows sodium ions to enter the cell too freely. As you will see in chapters 11 and 12, an inflow of sodium is the normal process that excites nerve and muscle cells. In hypocalcemia, this excitation is excessive and results in the aforementioned tetany.

A blood calcium excess is called **hypercalcemia.**[31] In this condition, excessive amounts of calcium bind to the cell surface, increasing the charge difference across the membrane and making sodium channels less responsive. In addition, calcium ions bind to membrane proteins that serve as sodium channels and inhibit them from opening. Both actions render nerve and muscle cells less excitable than normal. At 12 mg/dL and higher, hypercalcemia causes depression of the nervous system, emotional disturbances, muscle weakness, sluggish reflexes, and sometimes cardiac arrest.

You can see how critical blood calcium level is, but what causes it to deviate from the norm, and how does the body correct such imbalances? Hypercalcemia is rare, but hypocalcemia can result from a wide variety of causes including vitamin D deficiency, diarrhea, thyroid tumors, or underactive parathyroid glands. Pregnancy and lactation put women at

---

### DEEPER INSIGHT 7.3
#### CLINICAL APPLICATION

#### Osseous Tissue and pH Balance

The urinary, respiratory, and skeletal systems cooperate to maintain the body's acid–base balance (pH). pH balance is threatened by such conditions as kidney diseases that impair hydrogen ion excretion in the urine. The accumulation of $H^+$ in the blood lowers its pH, potentially causing a state of acidosis (pH < 7.35). Urinary and respiratory responses to acidosis are discussed in chapter 24. The role of the skeleton in responding to acidosis is to dissolve bone and release calcium carbonate into circulation. Carbonate ions neutralize some of the acid in the blood. The withdrawal of calcium carbonate from the skeleton, however, can lead to osteomalacia, a softening of the bones. Bone strength can be preserved by treating acidosis with intravenous bicarbonate.

---

risk of hypocalcemia because of the calcium demanded by ossification of the fetal skeleton and synthesis of milk. The leading cause of hypocalcemic tetany is accidental removal of the parathyroid glands during thyroid surgery or damage to their blood supply by head and neck surgery. Without hormone replacement therapy, the lack of parathyroid glands can lead to fatal tetany within 4 days.

Calcium homeostasis depends on a balance between dietary intake, urinary and fecal losses, and exchanges with the osseous tissue. It is regulated by three hormones: *calcitriol, calcitonin,* and *parathyroid hormone.*

## Calcitriol

**Calcitriol** (CAL-sih-TRY-ol) is a form of vitamin D produced by the sequential action of the skin, liver, and kidneys (fig. 7.14):

1. Epidermal keratinocytes use ultraviolet radiation from sunlight to convert a steroid, 7-dehydrocholesterol, to previtamin $D_3$. Over another 3 days, the warmth of sunlight on the skin further converts this to vitamin $D_3$, and a transport protein carries this to the bloodstream.

2. The liver adds a hydroxyl group to the molecule, converting it to *calcidiol.*

3. The kidney then adds another hydroxyl group, converting calcidiol to calcitriol, the most active form of vitamin D.

Calcitriol behaves as a hormone—a blood-borne chemical messenger from one organ to another. It is called a vitamin only because it is added to the diet, mainly in fortified milk, as a safeguard for people who do not get enough sunlight to initiate its synthesis in the skin.

The principal function of calcitriol is to raise the blood calcium concentration. It does this in three ways (fig. 7.15), especially the first of these:

1. It increases calcium absorption by the small intestine, using mechanisms detailed in chapter 25.

2. It increases calcium resorption from the skeleton. Calcitriol binds to osteoblasts, which release another chemical messenger called RANKL. This is the ligand (L) for a receptor named RANK[32] on the surfaces of osteoclast-producing stem cells. This messenger stimulates the stem cells to differentiate into osteoclasts. The new osteoclasts then liberate calcium and phosphate ions from bone.

3. It weakly promotes the reabsorption of calcium ions by the kidneys, so less calcium is lost in the urine.

Although calcitriol promotes bone resorption, it is also necessary for bone deposition. Without it, calcium and phosphate levels in the blood are too low for normal deposition. The result is a softness of the bones called **rickets** in children and **osteomalacia**[33] in adults.

---

[30]Armand Trousseau (1801–67), French physician
[31]*hyper* = above normal; *calc* = calcium; *emia* = blood condition

[32]RANK = receptor activator of nuclear factor kappa B
[33]*osteo* = bone; *malacia* = softening

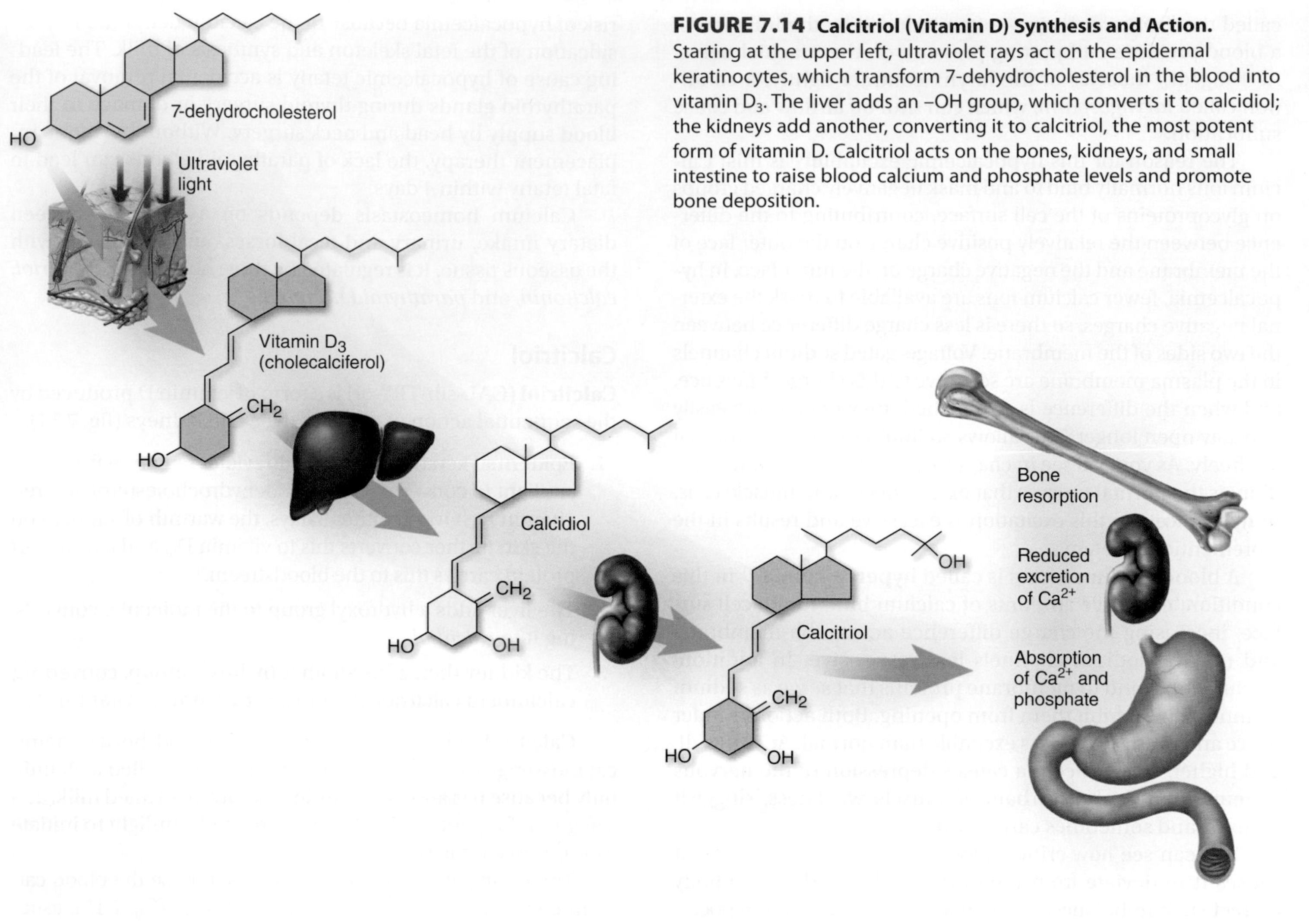

**FIGURE 7.14 Calcitriol (Vitamin D) Synthesis and Action.**
Starting at the upper left, ultraviolet rays act on the epidermal keratinocytes, which transform 7-dehydrocholesterol in the blood into vitamin $D_3$. The liver adds an –OH group, which converts it to calcidiol; the kidneys add another, converting it to calcitriol, the most potent form of vitamin D. Calcitriol acts on the bones, kidneys, and small intestine to raise blood calcium and phosphate levels and promote bone deposition.

## Calcitonin

**Calcitonin** is secreted by *C cells (clear cells)* of the thyroid gland (see fig. 17.9, p. 642). It is secreted when the blood calcium concentration rises too high, and it lowers the concentration by two principal mechanisms (figs. 7.15 and 7.16a):

1. **Osteoclast inhibition.** Within 15 minutes after it is secreted, calcitonin reduces osteoclast activity by as much as 70%, so osteoclasts liberate less calcium from the skeleton.

2. **Osteoblast stimulation.** Within an hour, calcitonin increases the number and activity of osteoblasts, which deposit calcium into the skeleton.

Calcitonin plays an important role in children but has only a weak effect in most adults. The osteoclasts of children are highly active in skeletal remodeling and release 5 g or more of calcium into the blood each day. By inhibiting this activity, calcitonin can significantly lower the blood calcium level in children. In adults, however, the osteoclasts release only about 0.8 g of calcium per day. Calcitonin cannot change adult blood calcium very much by suppressing this lesser contribution.

Calcitonin deficiency is not known to cause any adult disease. Calcitonin may, however, inhibit bone loss in pregnant and lactating women.

## Parathyroid Hormone

**Parathyroid hormone (PTH)** is secreted by the parathyroid glands, which adhere to the posterior surface of the thyroid gland. These glands release PTH when blood calcium is low. A mere 1% drop in the blood calcium level doubles the secretion of PTH. PTH raises the blood calcium level by four mechanisms (figs. 7.15 and 7.16b):

1. PTH binds to receptors on the osteoblasts, stimulating them to secrete RANKL, which in turn raises the osteoclast population and promotes bone resorption.

2. PTH promotes calcium reabsorption by the kidneys, so less calcium is lost in the urine.

3. PTH promotes the final step of calcitriol synthesis in the kidneys, thus enhancing the calcium-raising effect of calcitriol.

4. PTH inhibits collagen synthesis by osteoblasts, thus inhibiting bone deposition.

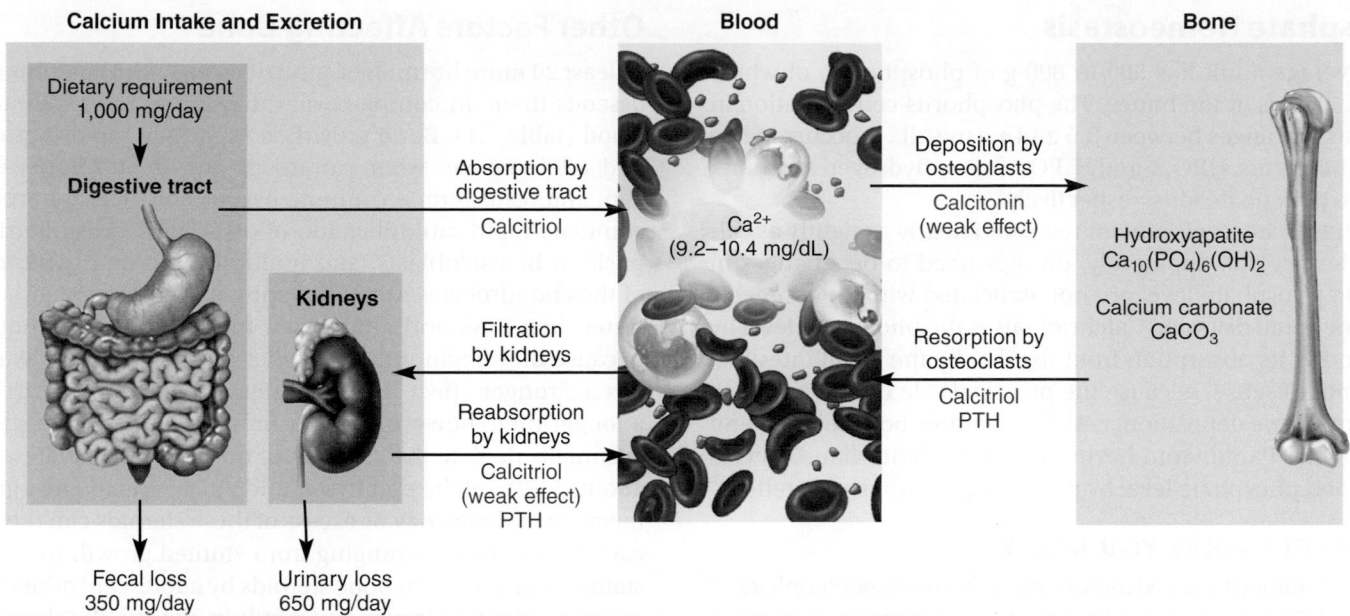

**FIGURE 7.15 Hormonal Control of Calcium Balance.** The central panel represents the blood reservoir of calcium and shows its normal (safe) range. Calcitriol and PTH regulate calcium exchanges between the blood and the small intestine and kidneys (left). Calcitonin, calcitriol, and PTH regulate calcium exchanges between blood and bone (right).

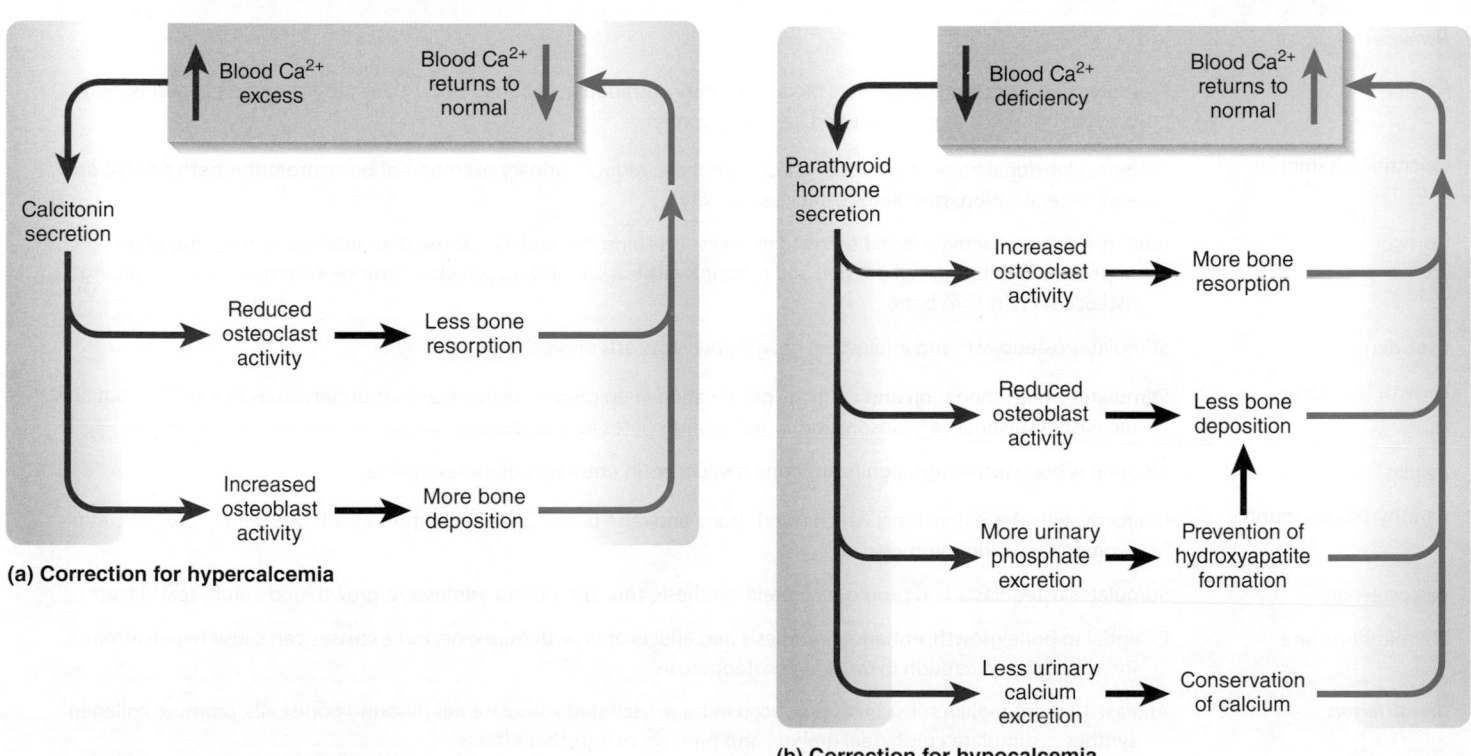

**FIGURE 7.16 Negative Feedback Loops in Calcium Homeostasis.** (a) The correction of hypercalcemia by calcitonin. (b) The correction of hypocalcemia by parathyroid hormone.

## Phosphate Homeostasis

The average adult has 500 to 800 g of phosphorus, of which 85% to 90% is in the bones. The phosphorus concentration in the plasma ranges between 3.5 and 4.0 mg/dL. It occurs in two principal forms, $HPO_4^{2-}$ and $H_2PO_4^-$ (monohydrogen and dihydrogen phosphate ions, respectively).

Phosphate levels are not regulated nearly as tightly as calcium levels. Nor, apparently, do they need to be; changes in plasma phosphate level are not associated with any immediate functional disorder. Calcitriol raises the phosphate level by promoting its absorption from the diet by the small intestine. This makes sense, because the principal role of calcitriol is to promote bone deposition, and that requires both calcium and phosphate. Parathyroid hormone, on the other hand, lowers the blood phosphate level by promoting its urinary excretion.

### ▶▶▶ APPLY WHAT YOU KNOW

*While raising the blood calcium level, PTH lowers the phosphate level. Explain why this is important for achieving the purpose of PTH.*

## Other Factors Affecting Bone

At least 20 more hormones, growth factors, and vitamins affect osseous tissue in complex ways that are still not well understood (table 7.1). Bone growth is especially rapid in puberty and adolescence, when surges of growth hormone, estrogen, and testosterone promote ossification. These hormones stimulate rapid multiplication of osteogenic cells, matrix deposition by osteoblasts, and multiplication and hypertrophy of the chondrocytes in the metaphyses. Adolescent girls grow faster than boys and attain their full height earlier, not only because they begin puberty earlier but also because estrogen has a stronger effect than testosterone. Since males grow for a longer time, however, they usually grow taller. Sex steroids eventually deplete the cartilage of the epiphyseal plates, bring about closure of the epiphyses, and put an end to one's growth in height. A deficiency or excess of these steroids can therefore cause abnormalities ranging from stunted growth to very tall stature. The use of anabolic steroids by adolescent athletes can cause premature closure and result in abnormally short adult stature (see p. 73). The excessive consumption of cola (more

| TABLE 7.1 | Agents Affecting Calcium and Bone Metabolism |
|---|---|
| **Name** | **Effect** |
| *Hormones* | |
| Calcitonin | Promotes mineralization and lowers blood $Ca^{2+}$ concentration in children, but usually has little effect in adults; may prevent bone loss in pregnant and lactating women |
| Calcitriol (vitamin D) | Promotes intestinal absorption of $Ca^{2+}$ and phosphate; reduces urinary excretion of both; promotes both resorption and mineralization; stimulates osteoclast activity |
| Cortisol | Inhibits osteoclast activity, but if secreted in excess (Cushing disease), can cause osteoporosis by reducing bone deposition (inhibiting cell division and protein synthesis), inhibiting growth hormone secretion, and stimulating osteoclasts to resorb bone |
| Estrogen | Stimulates osteoblasts and adolescent growth; prevents osteoporosis |
| Growth hormone | Stimulates bone elongation and cartilage proliferation at epiphyseal plate; increases urinary excretion of $Ca^{2+}$ but also increases intestinal $Ca^{2+}$ absorption, which compensates for the loss |
| Insulin | Stimulates bone formation; significant bone loss occurs in untreated diabetes mellitus |
| Parathyroid hormone | Indirectly activates osteoclasts, which resorb bone and raise blood $Ca^{2+}$ concentration; inhibits urinary $Ca^{2+}$ excretion; promotes calcitriol synthesis |
| Testosterone | Stimulates osteoblasts and promotes protein synthesis, thus promoting adolescent growth and epiphyseal closure |
| Thyroid hormone | Essential to bone growth; enhances synthesis and effects of growth hormone, but excesses can cause hypercalcemia, increased $Ca^{2+}$ excretion in urine, and osteoporosis |
| *Growth factors* | At least 12 hormonelike substances produced in bone itself that stimulate neighboring bone cells, promote collagen synthesis, stimulate epiphyseal growth, and produce many other effects |
| *Vitamins* | |
| Vitamin A | Promotes glycosaminoglycan (chondroitin sulfate) synthesis |
| Vitamin C (ascorbic acid) | Required for collagen synthesis, bone growth, and fracture repair |
| Vitamin D | Normally functions as a hormone (see calcitriol) |

than three 12-ounce servings per day) is associated with loss of bone density in women, but not in men. The effect is thought to be due to the phosphoric acid in cola, which binds intestinal calcium and interferes with its absorption. Other soft drinks do not contain phosphoric acid and show no effect on bone density.

**BEFORE YOU GO ON**

Answer the following questions to test your understanding of the preceding section:

15. Describe the role of collagen and seed crystals in bone mineralization.

16. Why is it important to regulate blood calcium concentration within such a narrow range?

17. What effect does calcitonin have on blood calcium concentration, and how does it produce this effect? Answer the same questions for parathyroid hormone.

18. How is vitamin D synthesized, and what effect does it have on blood calcium concentration?

## 7.5 Bone Disorders

### Expected Learning Outcomes

When you have completed this section, you should be able to

a. name and describe several bone diseases;

b. name and describe the types of fractures;

c. explain how a fracture is repaired; and

d. discuss some clinical treatments for fractures and other skeletal disorders.

Most people probably give little thought to their skeletal systems unless they break a bone. This section describes bone fractures, their healing, and their treatment, followed by a summary of other bone diseases. Bone disorders are among the concerns of the branch of medicine known as **orthopedics.**[34] As the name implies, this field originated as the treatment of skeletal deformities in children. It is now much more extensive and deals with the prevention and correction of injuries and disorders of the bones, joints, and muscles. It includes the design of artificial joints and limbs and the treatment of athletic injuries.

### Fractures and Their Repair

There are multiple ways of classifying bone fractures. A **stress fracture** is a break caused by abnormal trauma to a bone, such as fractures incurred in falls, athletics, and military combat. A **pathological fracture** is a break in a bone weakened

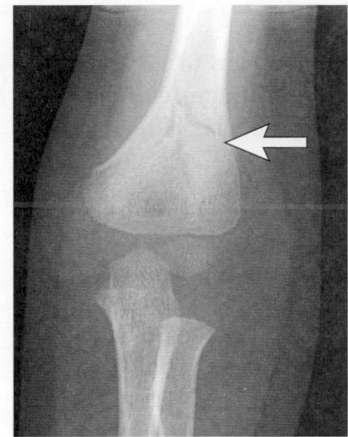
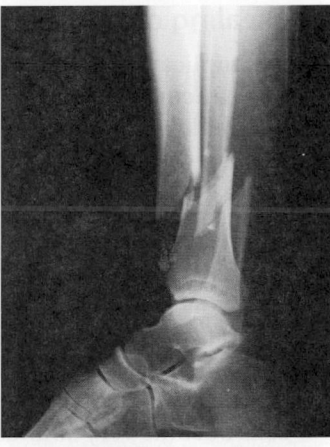

**(a) Nondisplaced**      **(b) Displaced**

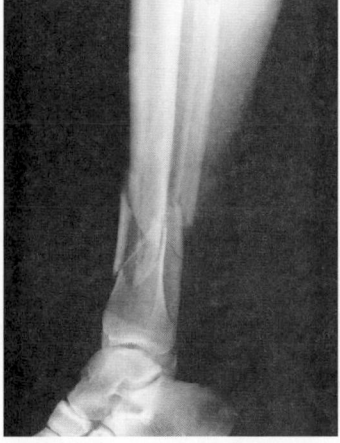
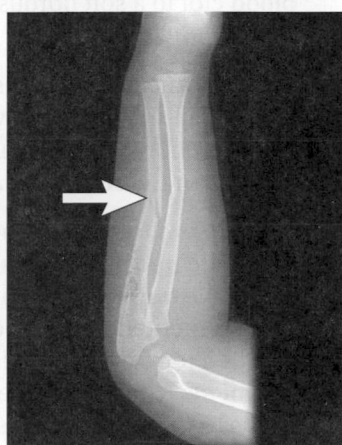

**(c) Comminuted**      **(d) Greenstick**

**FIGURE 7.17** **X-Rays of Representative Fracture Types.**
(a) Nondisplaced fracture of the distal humerus in a 3-year-old.
(b) Displaced fracture of the tibia and fibula. (c) Comminuted fracture of the tibia and fibula. (d) Greenstick fracture of the ulna.

by some other disease, such as bone cancer or osteoporosis, usually caused by a stress that would not normally fracture a bone. Fractures are also classified according to the direction of the fracture line, whether the skin is broken, and whether a bone is merely cracked or broken into separate pieces. For example, a *nondisplaced* fracture is one in which the bone pieces remain in proper anatomical alignment, whereas a *displaced* fracture is one in which at least one piece is shifted out of alignment with the other (fig. 7.17a, b). A *comminuted* fracture is one in which a bone is broken into three or more pieces (fig. 7.17c). A *greenstick* fracture is one in which the bone is incompletely broken on one side but merely bent on the opposite side (fig. 7.17d), the way a green twig breaks only partially and not into separate pieces. Many other types of fractures are routinely taught in clinical and first aid courses.

---

[34]*ortho* = straight; *ped* = child, foot

## The Healing of Fractures

An uncomplicated fracture heals in about 8 to 12 weeks, but complex fractures take longer and all fractures heal more slowly in older people. The healing process occurs in the following stages (fig. 7.18):

1. **Formation of hematoma and granulation tissue.**
A bone fracture severs blood vessels of the bone and periosteum, causing bleeding and the formation of a blood clot *(fracture hematoma)*. Blood capillaries soon grow into the clot, while fibroblasts, macrophages, osteoclasts, and osteogenic cells invade the tissue from both the periosteal and medullary sides of the fracture. Osteogenic cells become very abundant within 48 hours of the injury. All of this capillary and cellular invasion converts the blood clot to a soft fibrous mass called **granulation tissue.**

2. **Formation of a soft callus.**[35] Fibroblasts deposit collagen in the granulation tissue, while some osteogenic cells become chondroblasts and produce patches of fibrocartilage called the **soft callus.**

3. **Conversion to hard callus.** Other osteogenic cells differentiate into osteoblasts, which produce a bony collar called the **hard callus** around the fracture. The hard callus is cemented to dead bone around the injury site and acts as a temporary splint to join the broken ends or bone fragments together. It takes about 4 to 6 weeks for a hard callus to form. During this period, it is important that

---

[35]*call* = hard, tough

a broken bone be immobilized by traction or a cast to prevent reinjury.

4. **Remodeling.** The hard callus persists for 3 to 4 months. Meanwhile, osteoclasts dissolve small fragments of broken bone, and osteoblasts deposit spongy bone to bridge the gap between the broken ends. This spongy bone gradually fills in to become compact bone, in a manner similar to intramembranous ossification. Usually the fracture leaves a slight thickening of the bone visible by X-ray, but in some cases, healing is so complete that no trace of the fracture can be found.

## The Treatment of Fractures

Most fractures are set by **closed reduction,** a procedure in which the bone fragments are manipulated into their normal positions without surgery. **Open reduction** involves the surgical exposure of the bone and the use of plates, screws, or pins to realign the fragments (fig. 7.19). To stabilize the bone during healing, fractures are often set in casts. Traction is used to treat fractures of the femur in children. It aids in the alignment of the bone fragments by overriding the force of the strong thigh muscles. Traction is rarely used for elderly patients, however, because the risks from long-term confinement to bed outweigh the benefits. Hip fractures are usually pinned, and early ambulation (walking) is encouraged because it promotes blood circulation and healing. Fractures that take longer than 2 months to heal may be treated with electrical stimulation, which accelerates repair by suppressing the effects of parathyroid hormone.

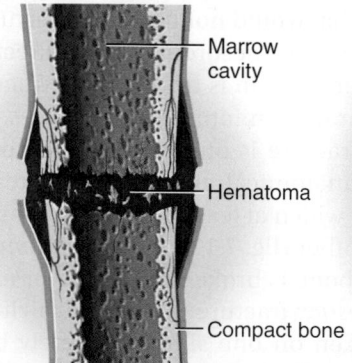

① **Hematoma formation**
The hematoma is converted to granulation tissue by invasion of cells and blood capillaries.

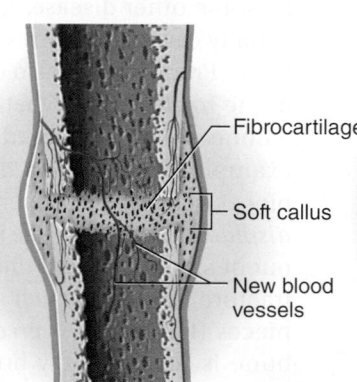

② **Soft callus formation**
Deposition of collagen and fibrocartilage converts granulation tissue to a soft callus.

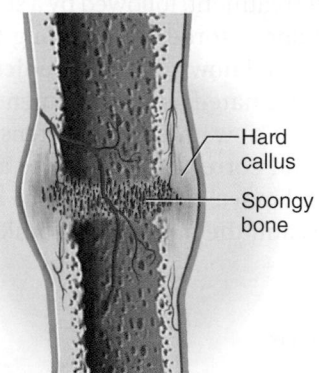

③ **Hard callus formation**
Osteoblasts deposit a temporary bony collar around the fracture to unite the broken pieces while ossification occurs.

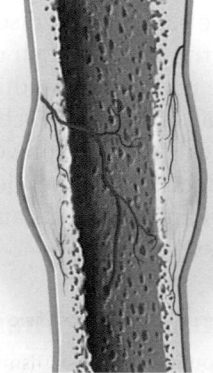

④ **Bone remodeling**
Small bone fragments are removed by osteoclasts, while osteoblasts deposit spongy bone and then convert it to compact bone.

**FIGURE 7.18   The Healing of a Bone Fracture.**

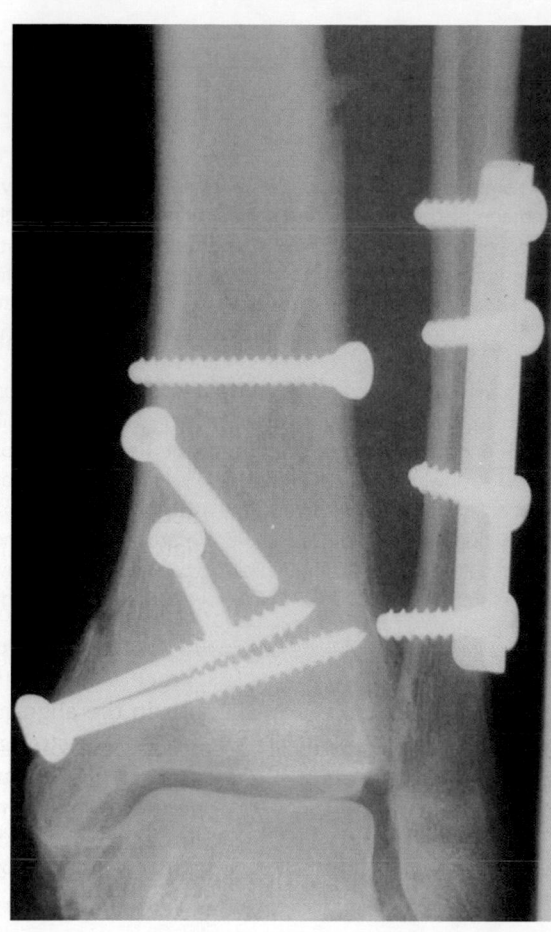

## Other Bone Disorders

Several additional bone disorders are summarized in table 7.2. The most common bone disease, **osteoporosis,**[36] receives special consideration in Deeper Insight 7.4. The effects of aging on the skeletal system are described on page 1119.

### BEFORE YOU GO ON

Answer the following questions to test your understanding of the preceding section:

**19.** Name and describe four types of bone fractures.

**20.** Why would osteomyelitis be more likely to occur in an open fracture than in a closed fracture?

**21.** What is a callus? How does it contribute to fracture repair?

**FIGURE 7.19 Open Reduction of an Ankle Fracture.** This fracture has been set by surgically exposing the bones and realigning the fragments with plates and screws.

| TABLE 7.2 | Bone Diseases |
|-----------|---------------|
| Osteitis deformans (Paget[37] disease) | Excessive proliferation of osteoclasts and resorption of excess bone, with osteoblasts attempting to compensate by depositing extra bone. This results in rapid, disorderly bone remodeling and weak, deformed bones. Osteitis deformans usually passes unnoticed, but in some cases it causes pain, disfiguration, and fractures. It is most common in males over the age of 50. |
| Osteomyelitis[38] | Inflammation of osseous tissue and bone marrow as a result of bacterial infection. This disease was often fatal before the discovery of antibiotics and is still very difficult to treat. |
| Osteogenesis imperfecta (brittle bone disease) | A defect in collagen deposition that renders bones exceptionally brittle, resulting in fractures present at birth or occurring with extraordinary frequency during childhood; also causing tooth deformity, and hearing loss due to deformity of middle-ear bones. |
| Osteosarcoma[39] (osteogenic sarcoma) | The most common and deadly form of bone cancer. It occurs most often in the tibia, femur, and humerus of males between the ages of 10 and 25. In 10% of cases, it metastasizes to the lungs or other organs; if untreated, death occurs within 1 year. |

*Disorders described elsewhere*

| | | | |
|---|---|---|---|
| Achondroplastic dwarfism p. 215 | Fractured clavicle p. 256 | Joint diseases p. 303 | Osteomalacia p. 217 |
| Cleft palate p. 243 | Fractured skull p. 242 | Kyphosis p. 248 | Osteoporosis p. 223 |
| Ectopic ossification p. 216 | Fractures in general p. 221 | Lordosis p. 248 | Rickets p. 217 |
| Fallen arches p. 268 | Herniated disc p. 248 | Mastoiditis p. 239 | Scoliosis p. 248 |

[36] *osteo* = bone; *por* = porous; *osis* = condition
[37] Sir James Paget (1814–99), English surgeon
[38] *osteo* = bone; *myel* = marrow; *itis* = inflammation
[39] *osteo* = bone; *sarc* = flesh; *oma* = tumor

# DEEPER INSIGHT 7.4

## CLINICAL APPLICATION

### Osteoporosis

The most common bone disease is osteoporosis (literally, "porous bones"), a severe loss of bone density. It affects especially spongy bone, since this is more metabolically active than dense bone and has more surface area exposed to bone-dissolving osteoclasts (fig. 7.20a). As a result, the bones become brittle and highly subject to pathological fractures from stresses as slight as sitting down too quickly. Fractures occur especially in the hip, wrist, and vertebral column. As the vertebrae lose bone, they become compressed and the spine is often deformed into a condition called *kyphosis* ("widow's hump") (fig. 7.20b, c). Hip fractures are especially serious. Each year, about 275,000 elderly Americans fracture their hips, and about 1 in 5 of these die within a year from complications of the resulting loss of mobility, such as pneumonia and thrombosis.

Osteoporosis occurs in both sexes and at all ages from adolescence to old age. Postmenopausal white women, however, are at the greatest risk. In contrast to men and black women, they have less bone density; white women begin losing it earlier (as young as age 35); and they lose it more rapidly. By age 70, the average white woman loses 30% of her bone tissue. Until menopause, estrogen maintains bone density by inhibiting osteoclasts. At menopause, the ovaries cease to secrete estrogen and osteoclast activity begins to outpace bone deposition by osteoblasts. Black women also lose bone density after menopause, but having denser bones to begin with, they usually do not lose enough to suffer osteoporosis. About 20% of osteoporosis patients are men. For most, the testes and adrenal glands secrete enough estrogen even in old age to maintain adequate bone density. Other than age, race, and sex, some other risk factors for osteoporosis include smoking, diabetes mellitus, poor diet, and inadequate weight-bearing exercise. Osteoporosis is surprisingly common among young female runners, dancers, and gymnasts. Their percentage of body fat is often so low that they stop ovulating and ovarian estrogen secretion is low. In early long-term space flights, astronauts developed osteoporosis because in a microgravity environment, their bones were subjected to too little of the stress that normally would stimulate bone deposition. This and the prevention of muscle atrophy are reasons that exercise equipment is now standard on space shuttles and stations.

Osteoporosis is now diagnosed with *dual-energy X-ray absorptiometry (DEXA)*, which uses low-dose X-rays to measure bone density. DEXA allows for early diagnosis and more effective drug treatment. Treatments for osteoporosis are aimed at slowing the rate of bone resorption. Estrogen-replacement therapy is out of favor because it increases the risk of breast cancer, stroke, and coronary artery disease. The *bis-phosphonates* (Fosamax, Actonel), among the current preferred treatments, act by inhibiting osteoclasts. They have been shown to increase bone mass by 5% to 10% over 3 years, and to reduce the incidence of fractures by 50%. Parathyroid hormone and derivatives such as *teriparatide* (Forteo) also are highly effective, but present a risk of bone cancer if used too long. The quest for safer drugs continues.

But as is so often true, an ounce of prevention is worth a pound of cure. Drug treatments for osteoporosis cost a patient thousands of dollars annually; exercise and a good bone-building diet are far less costly. The optimal means of preventing osteoporosis is with good diet and exercise habits between the ages of 25 and 40, when bone density is on the rise. The greater bone density a person has going into middle age, the less he or she will be affected by osteoporosis later.

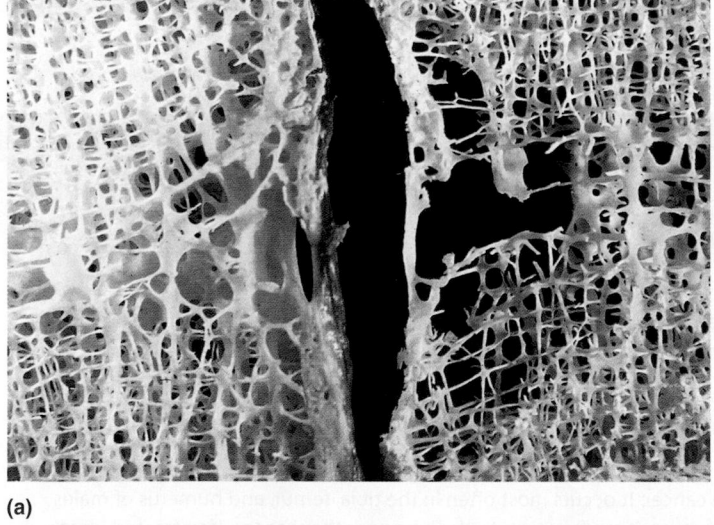

(a)

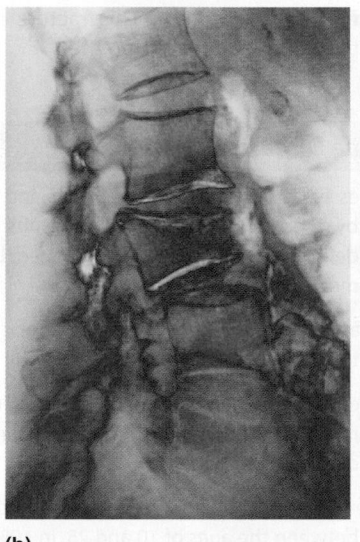

(b)

(c)

**FIGURE 7.20 Spinal Osteoporosis.** (a) Spongy bone in the body of a vertebra in good health (left) and with osteoporosis (right). (b) Colorized X-ray of lumbar vertebrae severely damaged by osteoporosis. (c) Abnormal thoracic spinal curvature (kyphosis) due to compression of thoracic vertebrae with osteoporosis.

# CONNECTIVE ISSUES

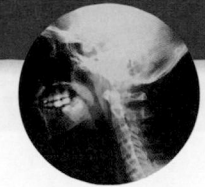

## Effects of the SKELETAL SYSTEM
## On Other Organ Systems

**INTEGUMENTARY SYSTEM**

Bones lying close to the body surface support and shape the skin.

**MUSCULAR SYSTEM**
Bones are the attachment sites for most skeletal muscles and provide leverage for muscle action; calcium homeostasis, important for muscle contraction, is achieved partly through a balance between bone deposition and resorption.

**NERVOUS SYSTEM**
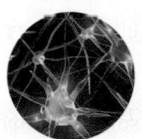
The cranium and vertebral column protect the brain and spinal cord; osseous tissue provides the calcium homeostasis needed for nerve function.

**ENDOCRINE SYSTEM**

Bones protect endocrine glands in the head, thorax, and pelvis; bones secrete the hormone osteocalcin, which promotes insulin action.

**CIRCULATORY SYSTEM**
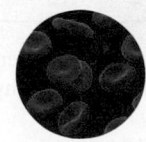
Bone marrow forms blood cells and platelets; osseous tissue provides the calcium homeostasis needed for cardiac function and blood clotting.

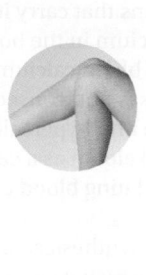

**LYMPHATIC/ IMMUNE SYSTEM**
White blood cells produced in the bone marrow carry out the body's immune functions.

**RESPIRATORY SYSTEM**
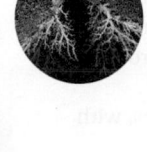
Ventilation of the lungs is achieved by musculoskeletal actions of the thoracic cage; the thoracic cage protects the delicate lungs from trauma; bones support and shape the nasal cavity.

**URINARY SYSTEM**
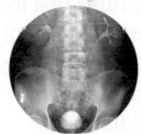
The thoracic cage partially protects the kidneys, and the pelvic girdle protects the lower urinary tract.

**DIGESTIVE SYSTEM**
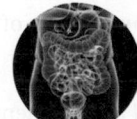
Osseous tissue interacts with the digestive system in maintaining calcium homeostasis; the thoracic cage and pelvic girdle protect portions of the digestive tract; musculoskeletal movements are necessary for chewing.

**REPRODUCTIVE SYSTEM**
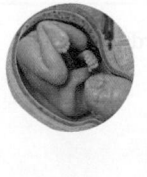
The pelvic girdle protects the internal reproductive organs; childbirth is adapted to the anatomy of the female pelvic girdle; ligaments anchor the penis and clitoris to the pelvic girdle.

# STUDY GUIDE

## ► Assess Your Learning Outcomes

*To test your knowledge, discuss the following topics with a study partner or in writing, ideally from memory.*

### 7.1 Tissues and Organs of the Skeletal System (p. 203)

1. The branch of medicine and biology that deals with the skeleton and bone tissue
2. Organs and tissues that constitute the skeletal system
3. Functions of the skeletal system
4. Which primary tissue category includes bone, and how bone differs from other tissues in that category
5. The categories of bone shapes, with examples of each
6. The relationship of compact bone, spongy bone, and the marrow cavity in the anatomy of a long bone
7. Other anatomical features of a long bone including the diaphysis, epiphysis, epiphyseal plate, articular cartilage, periosteum, and endosteum
8. Structure of a typical flat bone

### 7.2 Histology of Osseous Tissue (p. 205)

1. The four cell types in bone tissue; their functions, origins, and locations in the tissue
2. Organic and inorganic components of the bone matrix; their respective contributions to bone strength; and the significance of the helical arrangement of collagen fibers in bone

3. Osteon structure and the relationship of osteonic bone to interstitial and circumferential lamellae
4. The route by which nerves and blood vessels penetrate throughout a bone
5. Comparisons of the histology of spongy bone with that of compact bone; where spongy bone is found; and why bones are not composed solely of compact bone
6. Location and functions of the bone marrow; the composition and childhood versus adult distribution of the two types of marrow

### 7.3 Bone Development (p. 209)

1. Stages of intramembranous ossification; some bones that form in this way; and how far this process has progressed by birth
2. The same points concerning endochondral ossification
3. Histology, cell transformations, and tissue zones of the metaphysis; which zones and processes account for a child's or adolescent's growth in height
4. How stresses on bones remodel them throughout life; the difference between interstitial and appositional growth

### 7.4 Physiology of Osseous Tissue (p. 216)

1. The purpose and process of mineralization of osseous tissue, and the identity of the cells that carry it out
2. The purpose and process of bone resorption, and the identity of the cells

and cell secretions that carry it out
3. Functions of calcium in the body; the normal range of blood calcium concentration; and causes and consequences of hypocalcemia and hypercalcemia
4. The role of the skeleton as a calcium reservoir in regulating blood calcium levels
5. How calcitriol is synthesized and the mechanisms by which it supports or raises blood calcium level
6. The source of calcitonin and how it corrects hypercalcemia
7. The source of parathyroid hormone and multiple mechanisms by which it corrects hypocalcemia
8. Two forms of phosphate ions in the blood; the bodily functions of phosphate; and how calcitriol and parathyroid hormone affect blood phosphate levels
9. Effects of dietary vitamins A, C, and D on bone metabolism
10. Effects of cortisol, estrogen, testosterone, growth hormone, insulin, and thyroid hormone on bone metabolism

### 7.5 Bone Disorders (p. 221)

1. The difference between a stress fracture and a pathological fracture; stages in the healing of a fractured bone; and approaches to the clinical treatment of fractures
2. Causes of osteoporosis; its risk factors, pathological effects, diagnosis, treatment, and prevention

## ► Testing Your Recall

*Answers in Appendix B*

1. Which cells have a ruffled border and secrete hydrochloric acid?
   a. C cells
   b. osteocytes
   c. osteogenic cells
   d. osteoblasts
   e. osteoclasts

2. The marrow cavity of an adult bone may contain
   a. myeloid tissue.
   b. hyaline cartilage.
   c. periosteum.
   d. osteocytes.
   e. articular cartilages.

3. The spurt of growth in puberty results from cell proliferation and hypertrophy in
   a. the epiphysis.
   b. the epiphyseal line.
   c. compact bone.
   d. the epiphyseal plate.
   e. spongy bone.

# STUDY GUIDE

4. Osteoclasts are most closely related, by common descent, to
   a. osteoprogenitor cells.
   b. osteogenic cells.
   c. blood cells.
   d. fibroblasts.
   e. osteoblasts.

5. The walls between cartilage lacunae break down in the zone of
   a. cell proliferation.
   b. calcification.
   c. reserve cartilage.
   d. bone deposition.
   e. cell hypertrophy.

6. Which of these is *not* an effect of PTH?
   a. rise in blood phosphate level
   b. reduction of calcium excretion
   c. increased intestinal calcium absorption
   d. increased number of osteoclasts
   e. increased calcitriol synthesis

7. A child jumps to the ground from the top of a playground "jungle gym." His leg bones do not shatter mainly because they contain
   a. an abundance of glycosaminoglycans.
   b. young, resilient osteocytes.

   c. an abundance of calcium phosphate.
   d. collagen fibers.
   e. hydroxyapatite crystals.

8. One long bone meets another at its
   a. diaphysis.
   b. epiphyseal plate.
   c. periosteum.
   d. metaphysis.
   e. epiphysis.

9. Calcitriol is made from
   a. calcitonin.
   b. 7-dehydrocholesterol.
   c. hydroxyapatite.
   d. estrogen.
   e. PTH.

10. One sign of osteoporosis is
    a. osteosarcoma.
    b. osteomalacia.
    c. osteomyelitis.
    d. a wrist fracture.
    e. hypocalcemia.

11. Calcium phosphate crystallizes in bone as a mineral called _____.

12. Osteocytes contact each other through channels called _____ in the bone matrix.

13. A bone increases in diameter only by _____ growth, the addition of new surface lamellae.

14. Seed crystals of hydroxyapatite form only when the levels of calcium and phosphate in the tissue fluid exceed the _____.

15. A calcium deficiency called _____ can cause death by suffocation.

16. _____ are cells that secrete collagen and stimulate calcium phosphate deposition.

17. The most active form of vitamin D, produced mainly by the kidneys, is _____.

18. The most common bone disease is _____.

19. The transitional region between epiphyseal cartilage and the primary marrow cavity of a young bone is called the _____.

20. A pregnant, poorly nourished woman may suffer a softening of the bones called _____.

## ▶ Building Your Medical Vocabulary                    *Answers in Appendix B*

*State a medical meaning of each word element below, and give a term in which it or a slight variation of it is used.*

1. calc-

2. -clast

3. -malacia

4. myelo-

5. ortho-

6. osse-

7. osteo-

8. -physis

9. spic-

10. topo-

## ▶ True or False                    *Answers in Appendix B*

*Determine which five of the following statements are false, and briefly explain why.*

1. Spongy bone is always covered by compact bone.

2. Most bones develop from hyaline cartilage.

3. Fractures are the most common bone disorder.

4. The growth zone of the long bones of adolescents is the articular cartilage.

5. Osteoclasts develop from osteoblasts.

6. Osteocytes develop from osteoblasts.

7. The protein of the bone matrix is called hydroxyapatite.

8. Blood vessels travel through the central canals of compact bone.

9. Vitamin D promotes bone deposition, not resorption.

10. Parathyroid hormone promotes bone resorption and raises blood calcium concentration.

# STUDY GUIDE

## ▶ Testing Your Comprehension

*Answers at* www.mhhe.com/saladin7

1. Most osteocytes of an osteon are far removed from blood vessels, but still receive blood-borne nutrients. Explain how this is possible.

2. A 50-year-old business executive decides he has not been getting enough exercise for the last several years. He takes up hiking and finds that he really loves it. Within 2 years, he is spending many of his weekends hiking with a heavy backpack and camping in the mountains. Explain what changes in his anatomy could be predicted from Wolff's law of bone.

3. How does the regulation of blood calcium concentration exemplify negative feedback and homeostasis?

4. Describe how the arrangement of trabeculae in spongy bone demonstrates the unity of form and function.

5. Identify two bone diseases you would expect to see if the epidermis were a completely effective barrier to UV radiation and a person took no dietary supplements to compensate for this. Explain your answer.

## ▶ Improve Your Grade at www.mhhe.com/saladin7

*Download animations and movies to study when it fits your schedule. Practice quizzes, labeling activities, games, and flashcards offer fun ways to master the chapter concepts. Or, download image PowerPoint files for each chapter to create a study guide or for taking notes during lecture.*

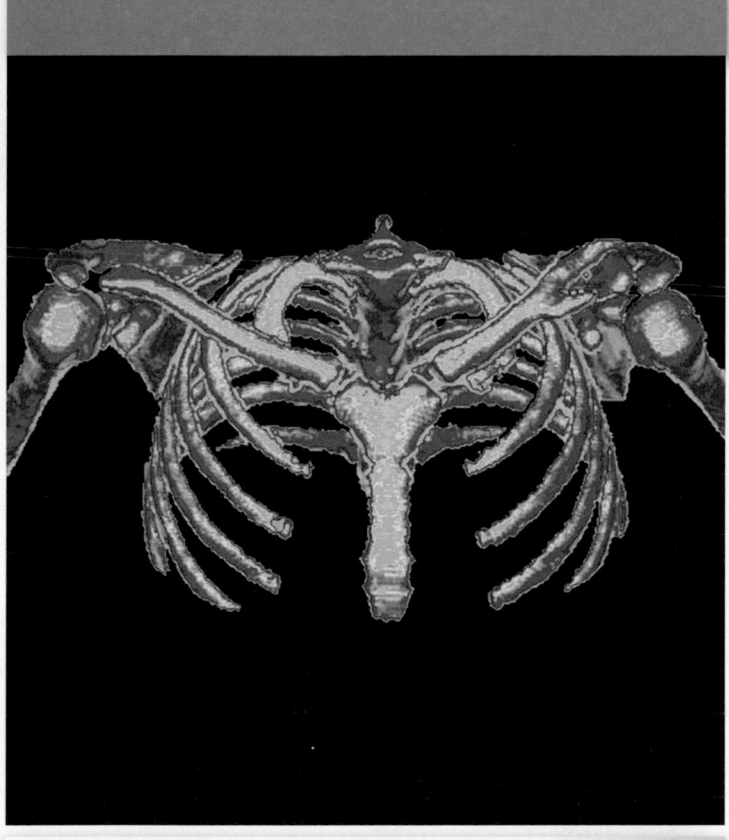

Colorized CT scan of the thoracic cage and pectoral girdle

**Module 5: Skeletal System**

**BRUSHING UP**

- Anatomical descriptions of the skeletal system depend heavily on the directional terminology introduced in table A.1 (p. 29).
- Understanding skeletal anatomy also depends on familiarity with the terminology of the body regions and cavities described in atlas A (pp. 29–34).

Knowledge of skeletal anatomy will be useful as you study later chapters. It provides a point of reference for studying the gross anatomy of other organ systems because many organs are named for their relationships to nearby bones. The subclavian artery and vein, for example, lie adjacent to the clavicle; the temporalis muscle is attached to the temporal bone; the ulnar nerve and radial artery travel beside the ulna and radius of the forearm, respectively; and the frontal, parietal, temporal, and occipital lobes of the brain are named for adjacent bones of the cranium. Understanding how the muscles produce body movements also depends on knowledge of skeletal anatomy. Additionally, the positions, shapes, and processes of bones can serve as landmarks for clinicians in determining where to give an injection or record a pulse, what to look for in an X-ray, and how to perform physical therapy and other clinical procedures.

## 8.1 Overview of the Skeleton

### Expected Learning Outcomes

When you have completed this section, you should be able to

a. define the two subdivisions of the skeleton;
b. state the approximate number of bones in the adult body;
c. explain why this number varies with age and from one person to another; and
d. define several terms that denote surface features of bones.

The skeleton (fig. 8.1) is divided into two regions: the axial skeleton and appendicular skeleton. The **axial skeleton,** which forms the central supporting axis of the body, includes the skull, auditory ossicles, hyoid bone, vertebral column, and thoracic cage (ribs and sternum). The **appendicular skeleton** includes the bones of the upper limb and pectoral girdle and the bones of the lower limb and pelvic girdle.

### Bones of the Skeletal System

It is often stated that there are 206 bones in the skeleton, but this is only a typical adult count, not an invariable number. At birth there are about 270, and even more bones form during childhood. With age, however, the number decreases as separate bones gradually fuse. For example, each side of a child's pelvic girdle has three bones—the *ilium, ischium,* and *pubis*—but in

adults, these are fused into a single *hip bone* on each side. The fusion of several bones, completed by late adolescence to the mid-20s, brings about the average adult number of 206. These bones are listed in table 8.1.

This number varies even among adults. One reason is the development of **sesamoid**[1] **bones**—bones that form within some tendons in response to strain. The patella (kneecap) is the largest of these; most of the others are small, rounded bones in such locations as the hands and feet (see fig. 8.34c). Another reason for adult variation is that some people have extra bones in the skull called **sutural** (SOO-chur-ul), or **wormian,**[2] **bones** (see fig. 8.6).

---

[1]*sesam* = sesame seed; *oid* = resembling
[2]Ole Worm (1588–1654), Danish physician

| **TABLE 8.1** | | **Bones of the Adult Skeletal System** |
|---|---|---|
| **Axial Skeleton** | | |
| *Skull (22 bones)* | | *Auditory ossicles (6 bones)* |
| Cranial bones | | Malleus (2) |
|   Frontal bone (1) | | Incus (2) |
|   Parietal bone (2) | | Stapes (2) |
|   Occipital bone (1) | | *Hyoid bone (1 bone)* |
|   Temporal bone (2) | | |
|   Sphenoid bone (1) | | *Vertebral column (26 bones)* |
|   Ethmoid bone (1) | | Cervical vertebrae (7) |
| Facial bones | | Thoracic vertebrae (12) |
|   Maxilla (2) | | Lumbar vertebrae (5) |
|   Palatine bone (2) | | Sacrum (1) |
|   Zygomatic bone (2) | | Coccyx (1) |
|   Lacrimal bone (2) | | |
|   Nasal bone (2) | | *Thoracic cage (25 bones plus* |
|   Vomer (1) | | *thoracic vertebrae)* |
|   Inferior nasal concha (2) | | Ribs (24) |
|   Mandible (1) | | Sternum (1) |
| **Appendicular Skeleton** | | |
| *Pectoral girdle (4 bones)* | | *Hip bones (2)* |
| Scapula (2) | | |
| Clavicle (2) | | *Lower limb (60 bones)* |
| | | Femur (2) |
| *Upper limb (60 bones)* | | Patella (2) |
| Humerus (2) | | Tibia (2) |
| Radius (2) | | Fibula (2) |
| Ulna (2) | | Tarsals (14) |
| Carpals (16) | | Metatarsals (10) |
| Metacarpals (10) | | Phalanges (28) |
| Phalanges (28) | | |
| **Grand Total: 206 Bones** | | |

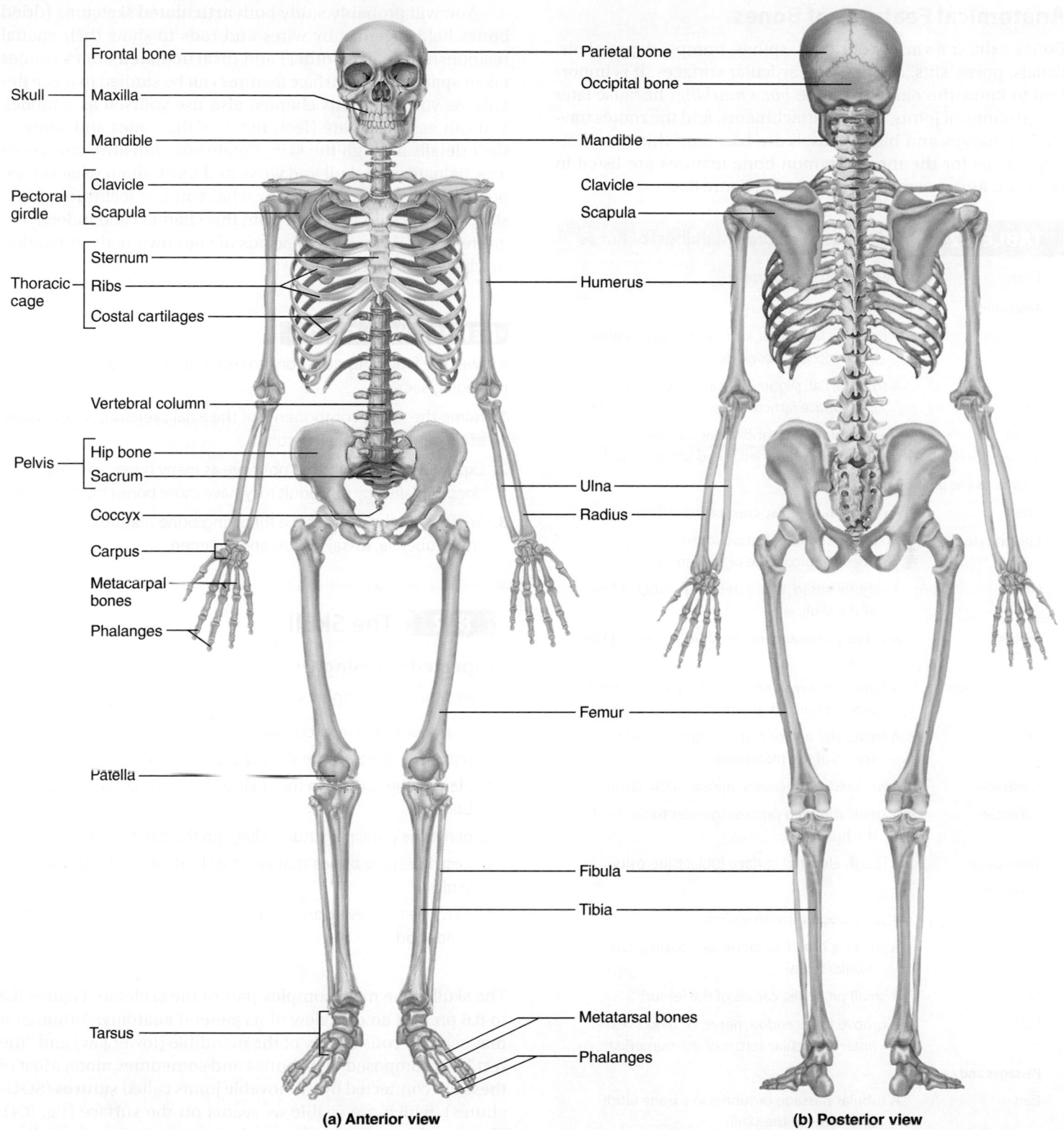

Frontal bone
Maxilla
Mandible
Skull

Clavicle
Scapula
Pectoral girdle

Sternum
Ribs
Costal cartilages
Thoracic cage

Vertebral column

Hip bone
Sacrum
Pelvis

Coccyx

Carpus

Metacarpal bones

Phalanges

Patella

Tarsus

Parietal bone
Occipital bone

Mandible

Clavicle
Scapula

Humerus

Ulna
Radius

Femur

Fibula

Tibia

Metatarsal bones
Phalanges

(a) Anterior view

(b) Posterior view

**FIGURE 8.1 The Adult Skeleton.** The appendicular skeleton is colored green, and the rest is axial skeleton.  **AP|R**

## Anatomical Features of Bones

Bones exhibit a variety of ridges, spines, bumps, depressions, canals, pores, slits, cavities, and articular surfaces. It is important to know the names of these *bone markings* because later descriptions of joints, muscle attachments, and the routes traveled by nerves and blood vessels are based on this terminology. Terms for the most common bone features are listed in table 8.2, and several are illustrated in figure 8.2.

| TABLE 8.2 | Anatomical Features (Markings) of Bones |
|---|---|
| **Term** | **Description and Example** |
| **Articulations** | |
| Condyle | A rounded knob that articulates with another bone (occipital condyles of the skull) |
| Facet | A smooth, flat, slightly concave or convex articular surface (articular facets of the vertebrae) |
| Head | The prominent expanded end of a bone, sometimes rounded (head of the femur) |
| **Extensions and projections** | |
| Crest | A narrow ridge (iliac crest of the pelvis) |
| Epicondyle | An expanded region superior to a condyle (medial epicondyle of the femur) |
| Line | A slightly raised, elongated ridge (nuchal lines of the skull) |
| Process | Any bony prominence (mastoid process of the skull) |
| Protuberance | A bony outgrowth or protruding part (mental protuberance of the chin) |
| Spine | A sharp, slender, or narrow process (mental spines of the mandible) |
| Trochanter | Two massive processes unique to the femur |
| Tubercle | A small, rounded process (greater tubercle of the humerus) |
| Tuberosity | A rough elevated surface (tibial tuberosity) |
| **Depressions** | |
| Alveolus | A pit or socket (tooth socket) |
| Fossa | A shallow, broad, or elongated basin (mandibular fossa) |
| Fovea | A small pit (fovea capitis of the femur) |
| Sulcus | A groove for a tendon, nerve, or blood vessel (intertubercular sulcus of the humerus) |
| **Passages and cavities** | |
| Canal | A tubular passage or tunnel in a bone (auditory canal of the skull) |
| Fissure | A slit through a bone (orbital fissures behind the eye) |
| Foramen | A hole through a bone, usually round (foramen magnum of the skull) |
| Meatus | A canal (external acoustic meatus of the ear) |
| Sinus | An air-filled space in a bone (frontal sinus of the forehead) |

You will probably study both **articulated** skeletons (dried bones held together by wires and rods to show their spatial relationships to each other) and **disarticulated** bones (bones taken apart so their surface features can be studied in more detail). As you study this chapter, also use yourself as a model. You can easily palpate (feel) many of the bones and some of their details through the skin. Rotate your forearm, cross your legs, palpate your skull and wrist, and think about what is happening beneath the surface or what you can feel through the skin. You will gain the most from this chapter (and indeed, the entire book) if you are conscious of your own body in relation to what you are studying.

### BEFORE YOU GO ON

Answer the following questions to test your understanding of the preceding section:

1. Name the major components of the axial skeleton. Name those of the appendicular skeleton.

2. Explain why an adult does not have as many bones as a child does. Explain why one adult may have more bones than another.

3. Briefly describe each of the following bone features: a condyle, crest, tubercle, fossa, sulcus, and foramen.

### 8.2 The Skull

#### Expected Learning Outcomes

When you have completed this section, you should be able to

a. distinguish between cranial and facial bones;

b. name the bones of the skull and their anatomical features;

c. identify the cavities in the skull and in some of its individual bones;

d. name the principal sutures that join the bones of the skull;

e. describe some bones that are closely associated with the skull;

f. describe the development of the skull from infancy through childhood.

The skull is the most complex part of the skeleton. Figures 8.3 to 8.6 present an overview of its general anatomy. Although it may seem to consist only of the mandible (lower jaw) and "the rest," it is composed of 22 bones and sometimes more. Most of these are connected by immovable joints called **sutures** (SOO-chures), which are visible as seams on the surface (fig. 8.4). These are important landmarks in the descriptions that follow.

The skull contains several prominent cavities (fig. 8.7). The largest, with an adult volume of about 1,350 mL, is the **cranial cavity,** which encloses the brain. Other cavities include the **orbits** (eye sockets), **nasal cavity, oral (buccal) cavity,**

*(text continued on p. 236)*

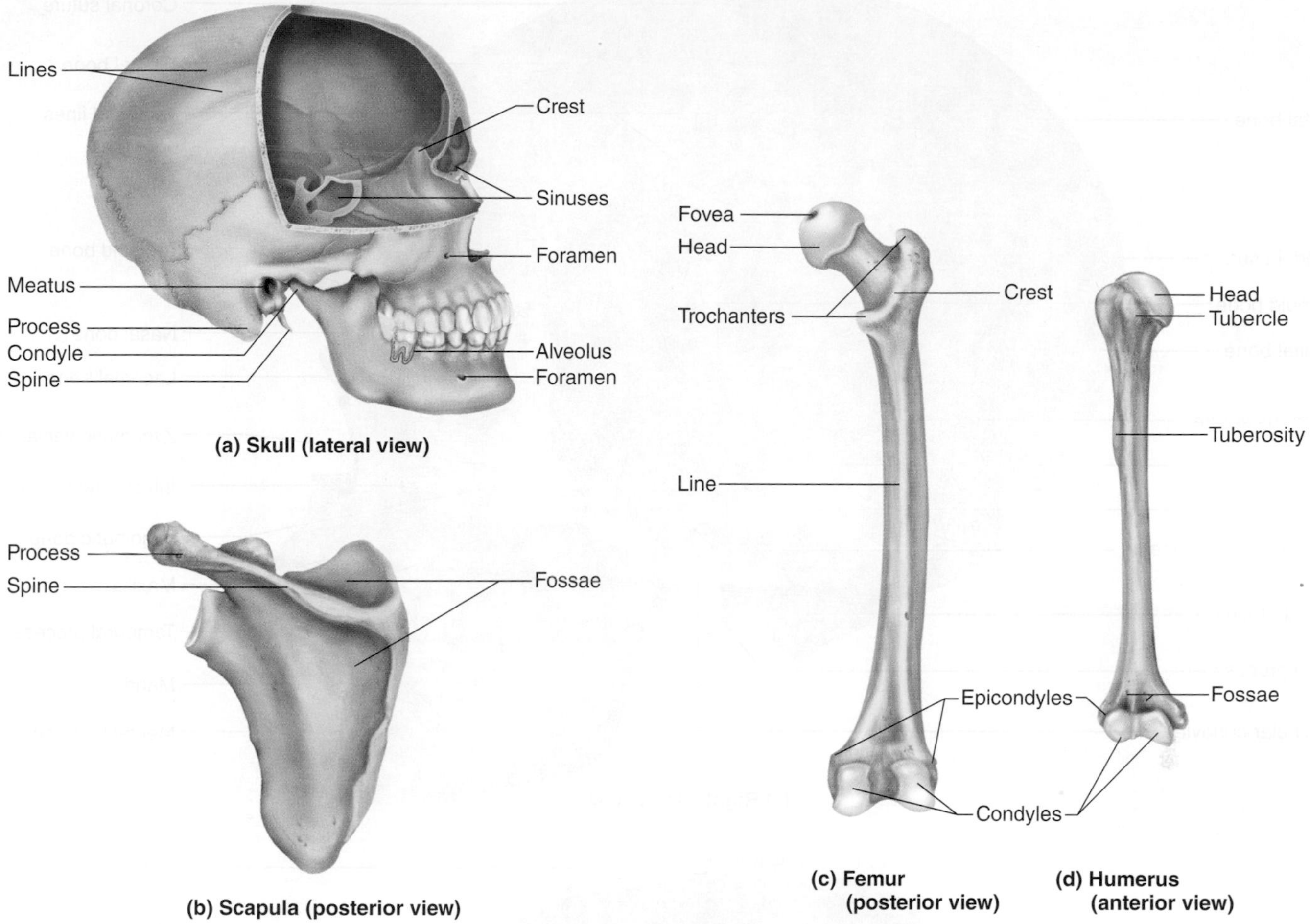

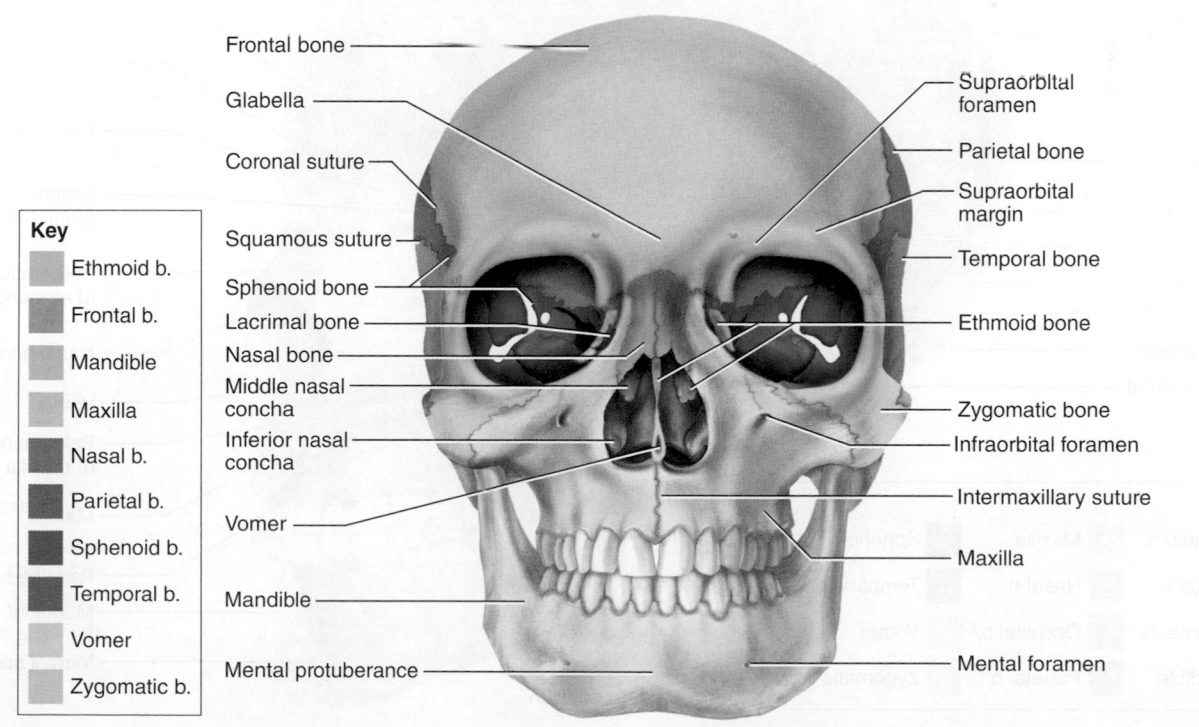

**(a) Skull (lateral view)**

**(b) Scapula (posterior view)**

**(c) Femur (posterior view)**

**(d) Humerus (anterior view)**

**FIGURE 8.2 Anatomical Features of Bones.** Most of these features also occur on many other bones of the body.

**FIGURE 8.3 The Skull (Anterior View).** AP|R

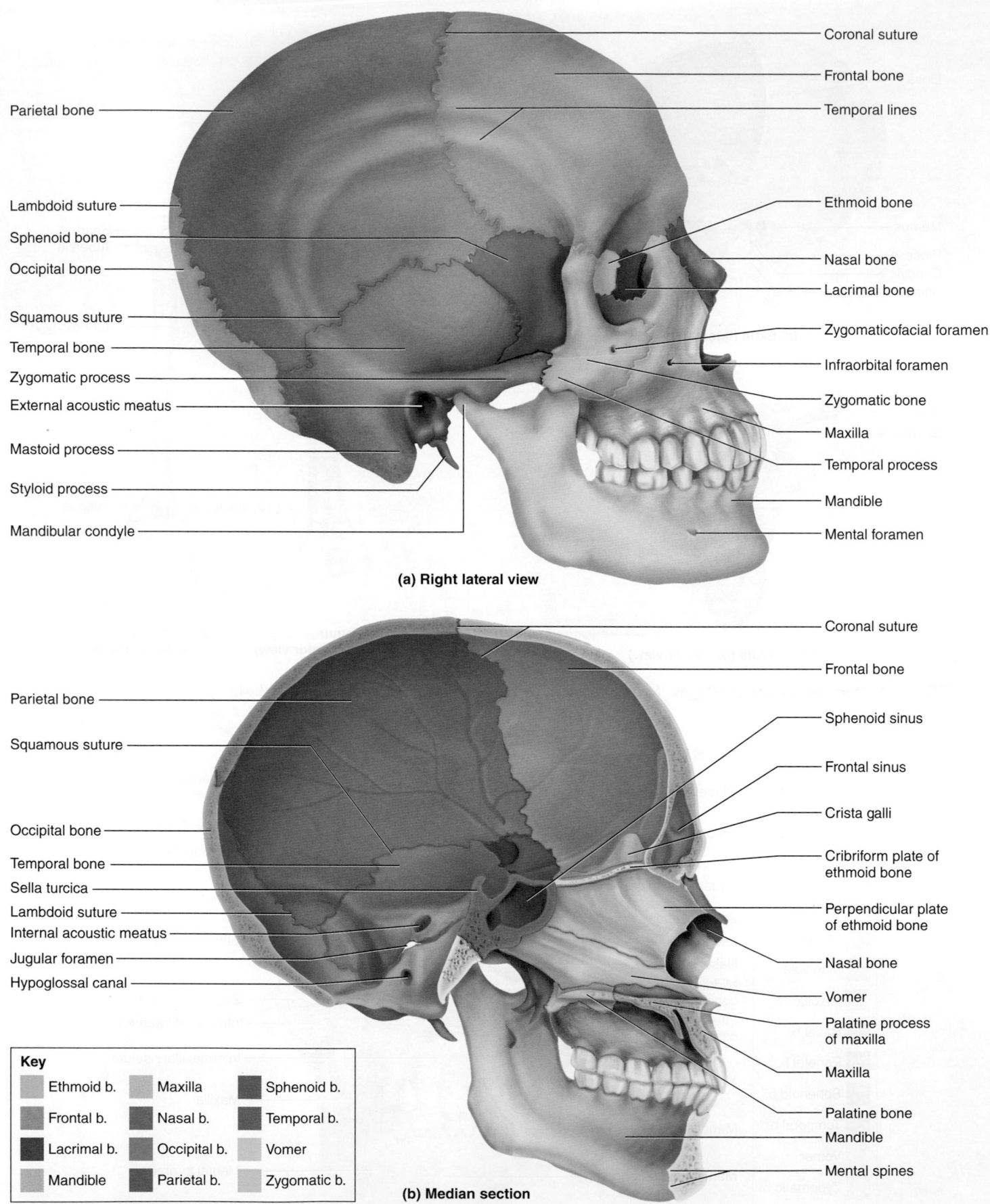

Coronal suture

Frontal bone

Temporal lines

Parietal bone

Ethmoid bone

Lambdoid suture

Sphenoid bone

Occipital bone

Nasal bone

Lacrimal bone

Squamous suture

Temporal bone

Zygomatic process

External acoustic meatus

Zygomaticofacial foramen

Infraorbital foramen

Zygomatic bone

Maxilla

Mastoid process

Temporal process

Styloid process

Mandible

Mandibular condyle

Mental foramen

**(a) Right lateral view**

Coronal suture

Frontal bone

Parietal bone

Squamous suture

Sphenoid sinus

Frontal sinus

Crista galli

Cribriform plate of ethmoid bone

Occipital bone

Temporal bone

Sella turcica

Lambdoid suture

Internal acoustic meatus

Jugular foramen

Hypoglossal canal

Perpendicular plate of ethmoid bone

Nasal bone

Vomer

Palatine process of maxilla

Maxilla

Palatine bone

Mandible

Mental spines

**Key**

| Ethmoid b. | Maxilla | Sphenoid b. |
| Frontal b. | Nasal b. | Temporal b. |
| Lacrimal b. | Occipital b. | Vomer |
| Mandible | Parietal b. | Zygomatic b. |

**(b) Median section**

**FIGURE 8.4 The Skull (Lateral Surface and Median Section).** AP|R

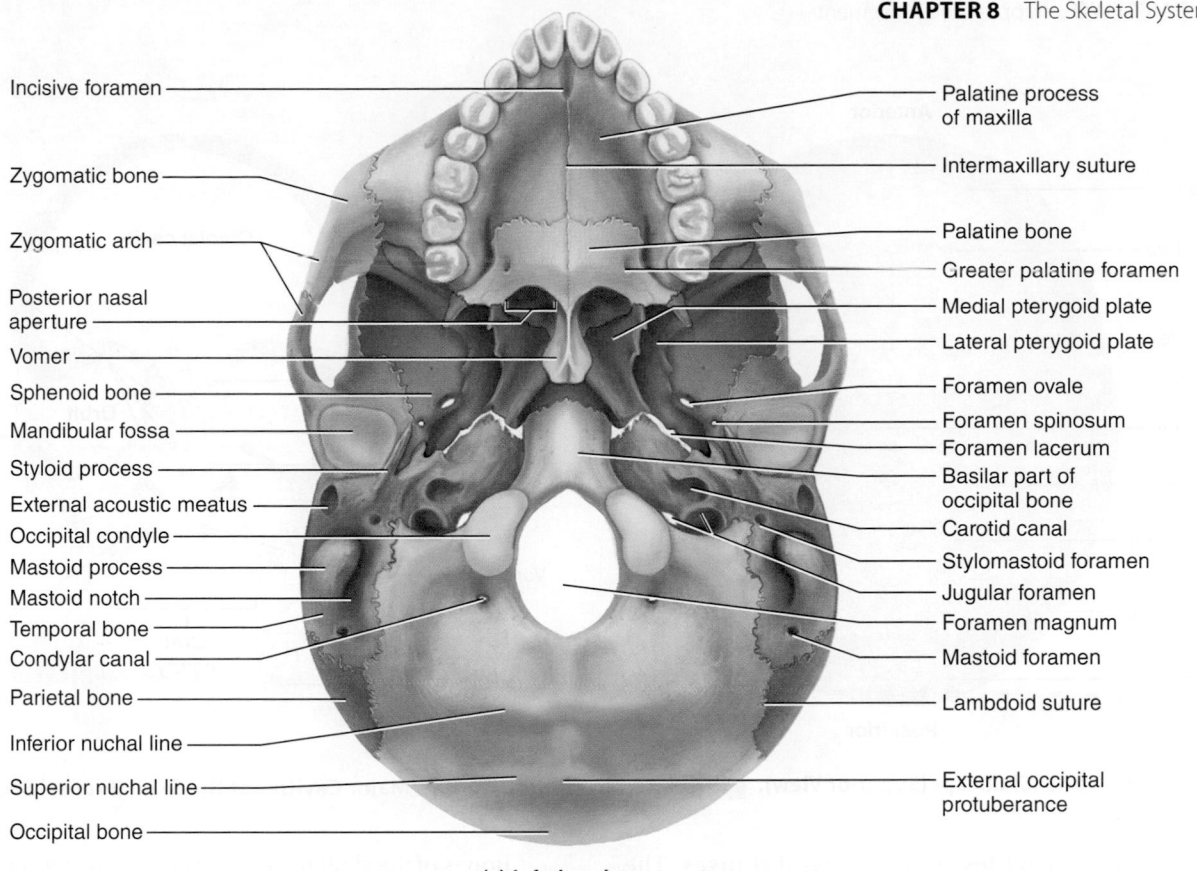

Incisive foramen

Zygomatic bone

Zygomatic arch

Posterior nasal aperture

Vomer

Sphenoid bone

Mandibular fossa

Styloid process

External acoustic meatus

Occipital condyle

Mastoid process

Mastoid notch

Temporal bone

Condylar canal

Parietal bone

Inferior nuchal line

Superior nuchal line

Occipital bone

Palatine process of maxilla

Intermaxillary suture

Palatine bone

Greater palatine foramen

Medial pterygoid plate

Lateral pterygoid plate

Foramen ovale

Foramen spinosum

Foramen lacerum

Basilar part of occipital bone

Carotid canal

Stylomastoid foramen

Jugular foramen

Foramen magnum

Mastoid foramen

Lambdoid suture

External occipital protuberance

**(a) Inferior view**

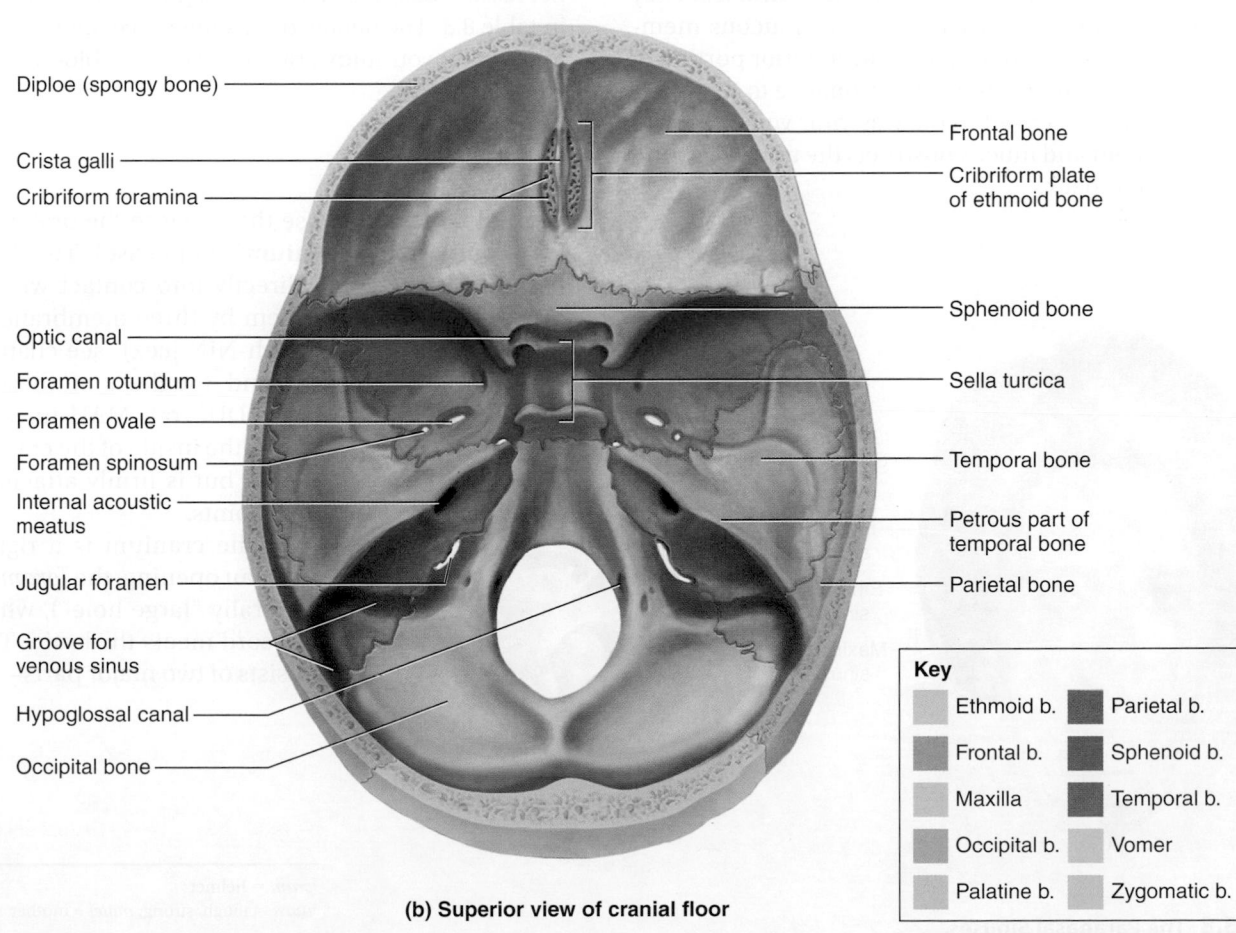

Diploe (spongy bone)

Crista galli

Cribriform foramina

Optic canal

Foramen rotundum

Foramen ovale

Foramen spinosum

Internal acoustic meatus

Jugular foramen

Groove for venous sinus

Hypoglossal canal

Occipital bone

Frontal bone

Cribriform plate of ethmoid bone

Sphenoid bone

Sella turcica

Temporal bone

Petrous part of temporal bone

Parietal bone

**Key**

Ethmoid b.

Frontal b.

Maxilla

Occipital b.

Palatine b.

Parietal b.

Sphenoid b.

Temporal b.

Vomer

Zygomatic b.

**(b) Superior view of cranial floor**

**FIGURE 8.5  The Base of the Skull.**  AP|R

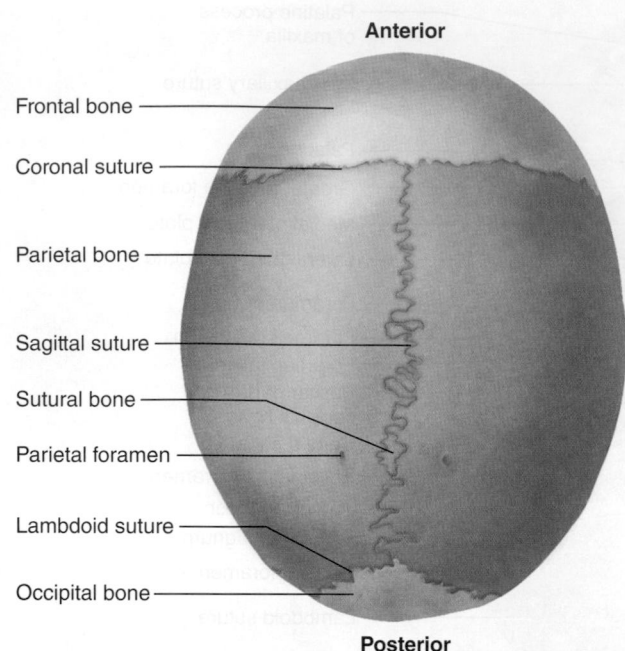

**FIGURE 8.6** The Calvaria (Skullcap) (Superior View). AP|R

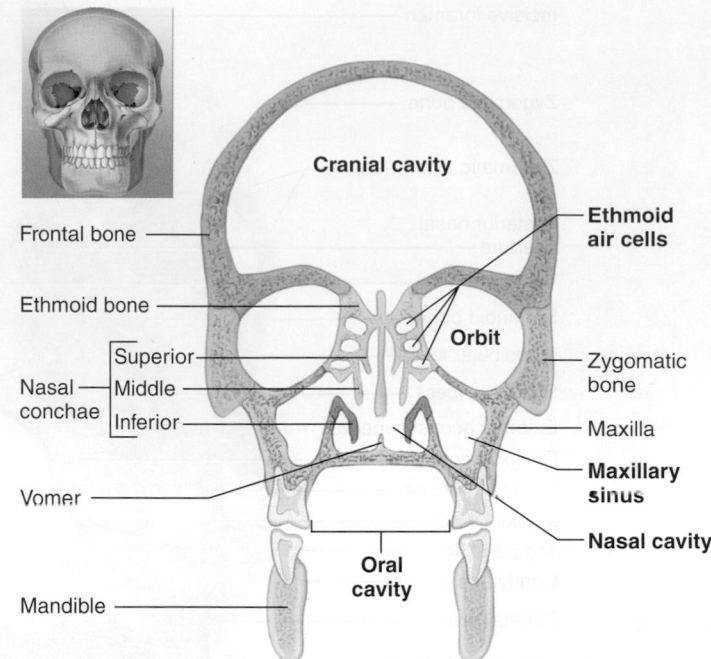

**FIGURE 8.7** Major Cavities of the Skull (Frontal Section).

middle- and **inner-ear cavities,** and **paranasal sinuses.** The sinuses are named for the bones in which they occur (fig. 8.8)—the **frontal, sphenoid, ethmoid,** and **maxillary sinuses.** They are connected with the nasal cavity, lined by mucous membranes, and filled with air. They lighten the anterior portion of the skull and act as chambers that add resonance to the voice. The latter effect can be sensed in the way your voice changes when you have a cold and mucus obstructs the travel of sound into the sinuses and back.

Bones of the skull have especially conspicuous **foramina**—singular, *foramen* (fo-RAY-men)—holes that allow passage for nerves and blood vessels. The major foramina are summarized in table 8.3. The details of this reference table will mean more to you after you study cranial nerves and blood vessels in later chapters.

## Cranial Bones

**Cranial bones** are those that enclose the brain; collectively, they compose the **cranium**[3] (braincase). The delicate brain tissue does not come directly into contact with the bones, but is separated from them by three membranes called the *meninges* (meh-NIN-jeez) (see chapter 14). The thickest and toughest of these, the *dura mater*[4] (DUE-rah MAH-tur), lies loosely against the inside of the cranium in most places but is firmly attached to it at a few points.

The cranium is a rigid structure with an opening, the *foramen magnum* (literally "large hole"), where the spinal cord meets the brain. The cranium consists of two major parts—the calvaria

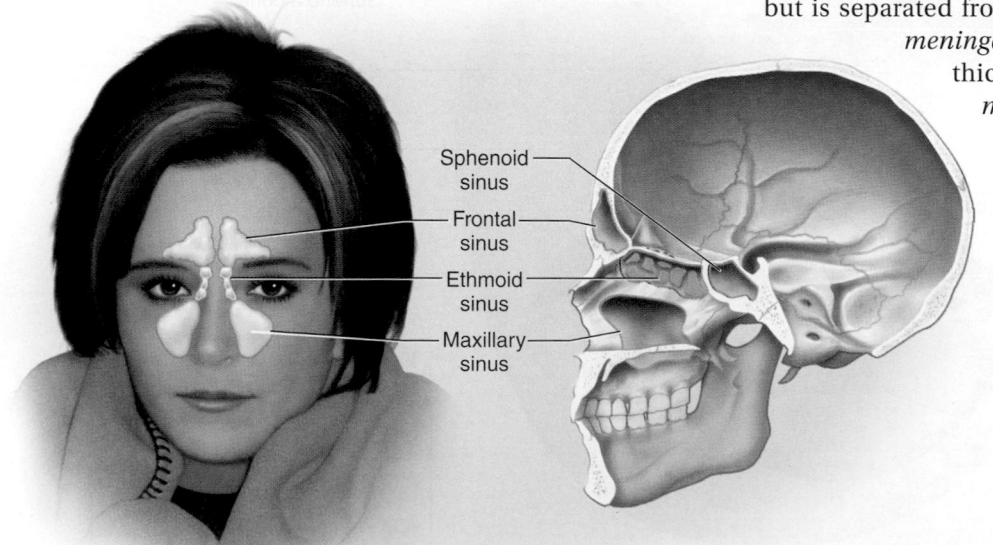

**FIGURE 8.8** The Paranasal Sinuses.

---

[3]*crani* = helmet
[4]*dura* = tough, strong; *mater* = mother

| **TABLE 8.3** | Foramina of the Skull and the Nerves and Blood Vessels Transmitted Through Them |
|---|---|
| **Bones and Their Foramina*** | **Structures Transmitted** |
| **Frontal bone** | |
| Supraorbital foramen or notch | Supraorbital nerve, artery, and vein; ophthalmic nerve |
| **Parietal bone** | |
| Parietal foramen | Emissary vein of superior sagittal sinus |
| **Temporal bone** | |
| Carotid canal | Internal carotid artery |
| External acoustic meatus | Sound waves to eardrum |
| Internal acoustic meatus | Vestibulocochlear nerve; internal auditory vessels |
| Stylomastoid foramen | Facial nerve |
| Mastoid foramen | Meningeal artery; vein from sigmoid sinus |
| **Temporal–occipital region** | |
| Jugular foramen | Internal jugular vein; glossopharyngeal, vagus, and accessory nerves |
| **Temporal–occipital–sphenoid region** | |
| Foramen lacerum | Closed by cartilage; not completely penetrated by any major nerves or vessels |
| **Occipital bone** | |
| Foramen magnum | Spinal cord; accessory nerve; vertebral arteries |
| Hypoglossal canal | Hypoglossal nerve to muscles of tongue |
| Condylar canal | Vein from transverse sinus |
| **Sphenoid bone** | |
| Foramen ovale | Mandibular division of trigeminal nerve; accessory meningeal artery |
| Foramen rotundum | Maxillary division of trigeminal nerve |
| Foramen spinosum | Middle meningeal artery; spinosal nerve; part of trigeminal nerve |
| Optic canal | Optic nerve; ophthalmic artery |
| Superior orbital fissure | Oculomotor, trochlear, and abducens nerves; ophthalmic division of trigeminal nerve; ophthalmic veins |
| **Ethmoid bone** | |
| Cribriform foramina | Olfactory nerves |
| **Maxilla** | |
| Infraorbital foramen | Infraorbital nerve and vessels |
| Incisive foramen | Nasopalatine nerves |
| **Maxilla–sphenoid region** | |
| Inferior orbital fissure | Infraorbital nerve; zygomatic nerve; infraorbital vessels |
| **Lacrimal bone** | |
| Lacrimal foramen | Tear duct leading to nasal cavity |
| **Palatine bone** | |
| Greater palatine foramen | Palatine nerves |
| **Zygomatic bone** | |
| Zygomaticofacial foramen | Zygomaticofacial nerve |
| Zygomaticotemporal foramen | Zygomaticotemporal nerve |
| **Mandible** | |
| Mental foramen | Mental nerve and vessels |
| Mandibular foramen | Inferior alveolar nerves and vessels to the lower teeth |

*When two or more bones are listed together (for example, temporal–occipital), it indicates that the foramen passes between them.

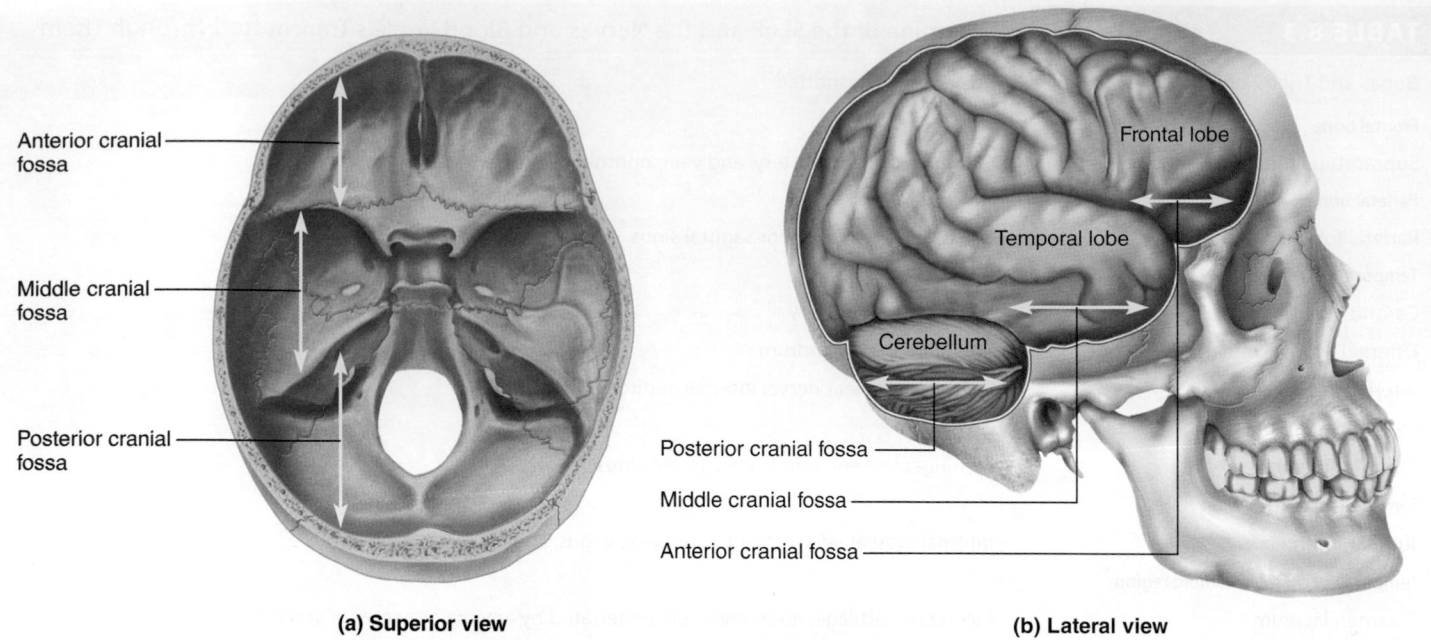

Anterior cranial fossa

Middle cranial fossa

Posterior cranial fossa

Frontal lobe

Temporal lobe

Cerebellum

Posterior cranial fossa

Middle cranial fossa

Anterior cranial fossa

**(a) Superior view**

**(b) Lateral view**

**FIGURE 8.9 Cranial Fossae.** The three fossae conform to the contours of the base of the brain.

and the base. The **calvaria**[5] (skullcap) is not a single bone but simply the dome of the top of the skull; it is composed of parts of multiple bones that form the roof and walls (see fig. 8.6). In study skulls, the calvaria is often sawed so that part of it can be lifted off for examination of the interior. This reveals the **base** (floor) of the cranial cavity (see fig. 8.5b), which exhibits three paired depressions called cranial fossae. These correspond to the contour of the inferior surface of the brain (fig. 8.9). The relatively shallow **anterior cranial fossa** is crescent-shaped and accommodates the frontal lobes of the brain. The **middle cranial fossa,** which drops abruptly deeper, is shaped like a pair of outstretched bird's wings and accommodates the temporal lobes. The **posterior cranial fossa** is deepest and houses a large posterior division of the brain called the cerebellum.

There are eight cranial bones:

| | |
|---|---|
| 1 frontal bone | 1 occipital bone |
| 2 parietal bones | 1 sphenoid bone |
| 2 temporal bones | 1 ethmoid bone |

## The Frontal Bone

The **frontal bone** extends from the forehead back to a prominent *coronal suture,* which crosses the crown of the head from right to left and joins the frontal bone to the parietal bones (see figs. 8.3 and 8.4). The frontal bone forms the anterior wall and about one-third of the roof of the cranial cavity, and it turns inward to form nearly all of the anterior cranial fossa and the roof of the orbit. Deep to the eyebrows it has a ridge called the **supraorbital margin.** Each margin is perforated by a single **supraorbital foramen** (see figs. 8.3 and 8.14), which provides passage for a nerve, artery, and veins. In some people, the edge of this foramen breaks through the margin of the orbit and forms a *supraorbital notch.* A person may have a foramen on one supraorbital margin and a notch on the other. The smooth area of the frontal bone just above the root of the nose is called the **glabella.**[6] The frontal bone also contains the frontal sinus. You may not see this sinus on all study skulls. On some, the calvaria is cut too high to show it, and some people simply do not have one. Along the cut edge of the calvaria, you can also see the diploe (DIP-lo-ee)—the layer of spongy bone in the middle of the cranial bones (see fig. 8.5b).

## The Parietal Bones

The right and left **parietal** (pa-RY-eh-tul) **bones** form most of the cranial roof and part of its walls (see figs. 8.4 and 8.6). Each is bordered by four sutures that join it to the neighboring bones: (1) a **sagittal suture** between the parietal bones; (2) the **coronal**[7] **suture** at the anterior margin; (3) the **lambdoid**[8] (LAM-doyd) **suture** at the posterior margin; and (4) the **squamous suture** laterally. Small sutural (wormian) bones are often seen along the sagittal and lambdoid sutures, like little

[5]*calvar* = bald, skull

[6]*glab* = smooth
[7]*corona* = crown
[8]Shaped like the Greek letter lambda (λ)

islands of bone with the suture lines passing around them. Internally, the parietal and frontal bones have markings that look a bit like aerial photographs of river tributaries (see fig. 8.4b). These represent places where the bone has been molded around blood vessels of the meninges.

Externally, the parietal bones have few features. A **parietal foramen** sometimes occurs near the corner of the lambdoid and sagittal sutures (see fig. 8.6). A pair of slight thickenings, the superior and inferior **temporal lines,** form an arc across the parietal and frontal bones (see fig. 8.4a). They mark the attachment of the large, fan-shaped *temporalis muscle,* a chewing muscle that inserts on the mandible.

## The Temporal Bones

If you palpate your skull just above and anterior to the ear—that is, the temporal region—you can feel the **temporal bone,** which forms the lower wall and part of the floor of the cranial cavity (fig. 8.10). The temporal bone derives its name from the fact that people often develop their first gray hairs on the temples with the passage of time.[9] The relatively complex shape of the temporal bone is best understood by dividing it into four parts:

1. The **squamous**[10] **part** (which you just palpated) is relatively flat and vertical. It is encircled by the squamous suture. It bears two prominent features: (a) the **zygomatic process,** which extends anteriorly to form part of the zygomatic arch (cheekbone), and (b) the **mandibular fossa,** a depression where the mandible articulates with the cranium.

2. The **tympanic**[11] **part** is a small ring of bone that borders the opening of the **external acoustic meatus** (me-AY-tus), or ear canal. It has a pointed spine on its inferior surface, the **styloid process,** named for its resemblance to the stylus used by ancient Greeks and Romans to write on wax tablets. The styloid process provides attachment for muscles of the tongue, pharynx, and hyoid bone.

3. The **mastoid**[12] **part** lies posterior to the tympanic part. It bears a heavy **mastoid process,** which you can palpate as a prominent lump behind the earlobe. It is filled with small air sinuses that communicate with the middle-ear cavity. These sinuses are subject to infection and inflammation *(mastoiditis),* which can erode the bone and spread to the brain. A groove called the **mastoid notch** lies medial to the mastoid process (see fig. 8.5a). It is the

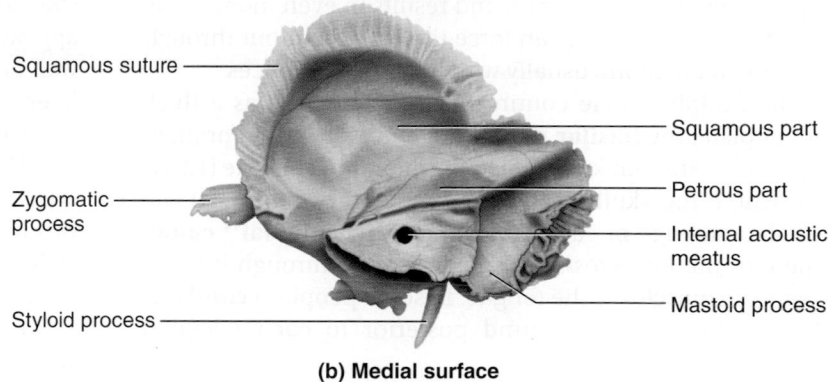

**(a) Lateral surface**

Squamous suture
Squamous part
Mastoid part
Zygomatic process
Mandibular fossa
Mastoid notch
Mastoid process
External acoustic meatus
Styloid process
Tympanic part

Squamous suture
Squamous part
Petrous part
Zygomatic process
Internal acoustic meatus
Styloid process
Mastoid process

**(b) Medial surface**

**FIGURE 8.10 The Right Temporal Bone.** The lateral surface faces the scalp and external ear; the medial surface faces the brain. AP|R

❓ *List five bones that articulate with the temporal bone.*

origin of the *digastric muscle,* which opens the mouth. The notch is perforated by the **stylomastoid foramen** at its anterior end and the **mastoid foramen** at its posterior end.

4. The **petrous**[13] **part** can be seen in the cranial floor, where it resembles a little mountain range separating the middle cranial fossa from the posterior fossa (fig. 8.10b). It houses the middle- and inner-ear cavities. The **internal acoustic meatus,** an opening on its posteromedial surface, allows passage of the *vestibulocochlear* (vess-TIB-you-lo-COC-lee-ur) *nerve,* which carries signals for hearing and balance from the inner ear to the brain. On the inferior surface of the petrous part are two prominent foramina named for the major blood vessels that pass through them (see fig. 8.5a): (a) The **carotid canal** is a passage for the internal carotid artery, a major blood supply to the brain. This artery is so close to the inner ear that one can some-

---

[9]*tempor* = time
[10]*squam* = flat; *ous* = characterized by
[11]*tympan* = drum (eardrum); *ic* = pertaining to
[12]*mast* = breast; *oid* = resembling

[13]*petr* = stone, rock; *ous* = like

times hear the pulsing of its blood when the ear is resting on a pillow or the heart is beating hard. (b) The **jugular foramen** is a large, irregular opening just medial to the styloid process, between the temporal and occipital bones. Blood from the brain drains through this foramen into the internal jugular vein of the neck. Three cranial nerves also pass through this foramen (see table 8.3).

## The Occipital Bone

The **occipital** (oc-SIP-ih-tul) **bone** forms the rear of the skull *(occiput)* and much of its base (see fig. 8.5). Its most conspicuous feature, the **foramen magnum,** admits the spinal cord to the cranial cavity and provides a point of attachment for the dura mater. An important consideration in head injuries is swelling of the brain. Since the cranium cannot expand, swelling puts pressure on the brain and results in even more tissue damage. Severe swelling can force the brainstem out through the foramen magnum, usually with fatal consequences.

The occipital bone continues anterior to this as a thick median plate, the **basilar part.** On either side of the foramen magnum is a smooth knob called the **occipital condyle** (CON-dile), where the skull rests on the vertebral column. At the anterolateral edge of each condyle is a **hypoglossal**[14] **canal,** named for the *hypoglossal nerve* that passes through it to innervate the muscles of the tongue. In some people, a **condylar** (CON-dih-lur) **canal** is found posterior to each occipital condyle.

Internally, the occipital bone displays impressions left by large venous sinuses that drain blood from the brain (see fig. 8.5b). One of these grooves travels along the midsagittal line. Just before reaching the foramen magnum, it branches into right and left grooves that wrap around the occipital bone like outstretched arms before terminating at the jugular foramina. The venous sinuses that occupy these grooves are described in chapter 20.

Other features of the occipital bone can be palpated on the back of your head. One is a prominent medial bump called the **external occipital protuberance**—the attachment for the **nuchal**[15] (NEW-kul) **ligament,** which binds the skull to the vertebral column. A ridge, the **superior nuchal line,** can be traced horizontally from this protuberance toward the mastoid process (see fig. 8.5a). It defines the superior limit of the neck and provides attachment to the skull for several neck and back muscles. It forms the boundary where, in palpating the upper neck, you feel the transition from muscle to bone. By pulling down on the occipital bone, some of these muscles help to keep the head erect. The deeper **inferior nuchal line** provides attachment for some of the deep neck muscles. This inconspicuous ridge cannot be palpated on the living body but is visible on an isolated skull.

## The Sphenoid Bone

The **sphenoid**[16] (SFEE-noyd) **bone** has a complex shape with a thick median **body** and outstretched **greater** and **lesser wings,** which give the bone as a whole a ragged mothlike shape. Most of it is best seen from the superior perspective (fig. 8.11a). In this view, the lesser wings form the posterior margin of the anterior cranial fossa and end at a sharp bony crest, where the sphenoid drops abruptly to the greater wings. The greater wings form about half of the middle cranial fossa (the temporal bone forming the rest) and are perforated by several foramina to be discussed shortly.

The greater wing also forms part of the lateral surface of the cranium just anterior to the temporal bone (see fig. 8.4a). The lesser wing forms the posterior wall of the orbit and contains the **optic canal,** which permits passage of the optic nerve and ophthalmic artery (see fig. 8.14). Superiorly, a pair of bony spines of the lesser wing called the **anterior clinoid processes** appears to guard the optic foramina. A gash in the posterior wall of the orbit, the **superior orbital fissure,** angles upward lateral to the optic canal. It serves as a passage for three nerves that supply the muscles of eye movement.

The body of the sphenoid bone contains a pair of sphenoid sinuses and has a saddlelike surface feature named the **sella turcica**[17] (SEL-la TUR-sih-ca). The sella consists of a deep pit called the *hypophyseal fossa,* which houses the pituitary gland (hypophysis); a raised anterior margin called the *tuberculum sellae* (too-BUR-cu-lum SEL-lee); and a posterior margin called the *dorsum sellae.* In life, the dura mater stretches over the sella turcica and attaches to the anterior clinoid processes. A stalk penetrates the dura to connect the pituitary gland to the base of the brain.

Lateral to the sella turcica, the sphenoid is perforated by several foramina (see fig. 8.5a). The **foramen rotundum** and **foramen ovale** (oh-VAY-lee) are passages for two branches of the trigeminal nerve. The **foramen spinosum,** about the diameter of a pencil lead, provides passage for an artery of the meninges. An irregular gash called the **foramen lacerum**[18] (LASS-eh-rum) occurs at the junction of the sphenoid, temporal, and occipital bones. It is filled with cartilage in life and transmits no major vessels or nerves.

In an inferior view of the skull, the sphenoid can be seen just anterior to the basilar part of the occipital bone. The internal openings of the nasal cavity seen here are called the **posterior nasal apertures,** or **choanae**[19] (co-AH-nee). Lateral to each aperture, the sphenoid bone exhibits a pair of parallel plates—the **medial** and **lateral pterygoid**[20] (TERR-ih-goyd) **plates** (see fig. 8.5a). Each plate has a narrower inferior extension called the **pterygoid process.** These plates and processes provide attachment for some of the jaw muscles. The sphenoid sinus occurs within the body of the sphenoid bone.

---

[14]*hypo* = below; *gloss* = tongue
[15]*nucha* = back of the neck

[16]*sphen* = wedge; *oid* = resembling
[17]*sella* = saddle; *turcica* = Turkish
[18]*lacerum* = torn, lacerated
[19]*choana* = funnel
[20]*pteryg* = wing; *oid* = resembling

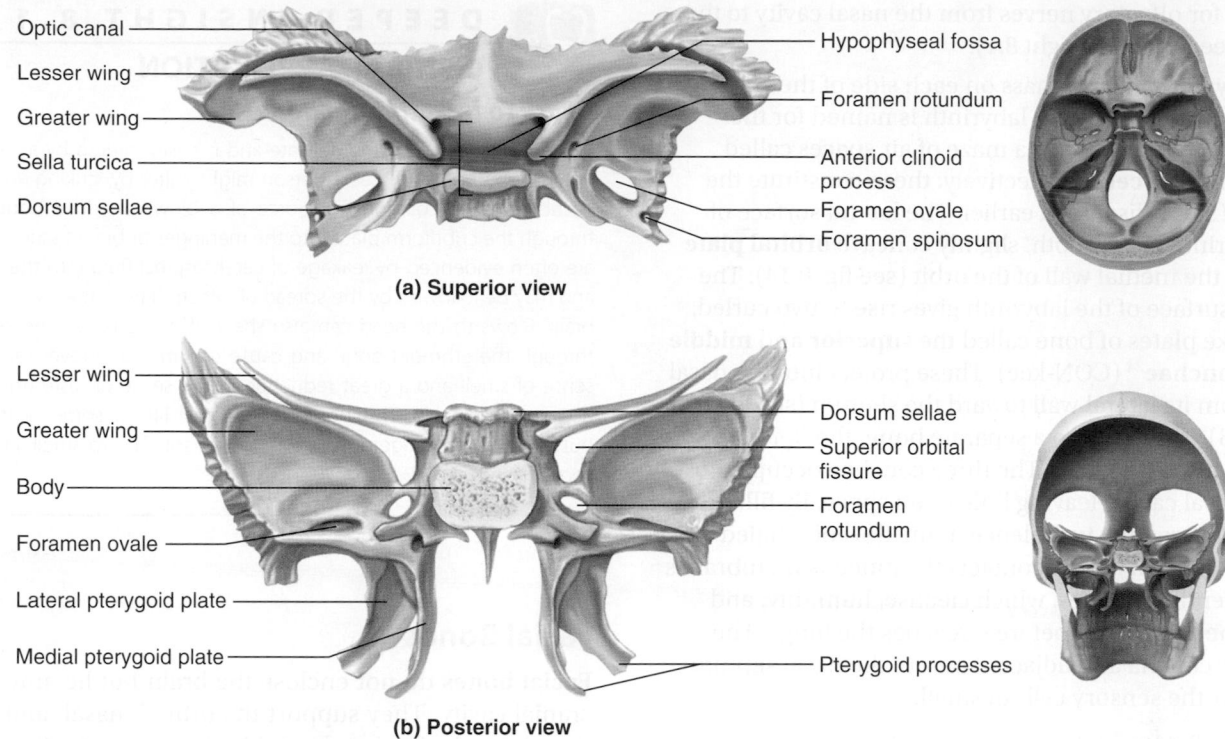

Optic canal

Lesser wing

Greater wing

Sella turcica

Dorsum sellae

Hypophyseal fossa

Foramen rotundum

Anterior clinoid process

Foramen ovale

Foramen spinosum

**(a) Superior view**

Lesser wing

Greater wing

Body

Foramen ovale

Lateral pterygoid plate

Medial pterygoid plate

Dorsum sellae

Superior orbital fissure

Foramen rotundum

Pterygoid processes

**(b) Posterior view**

**FIGURE 8.11 The Sphenoid Bone.** AP|R

## The Ethmoid Bone

The **ethmoid**[21] (ETH-moyd) **bone** is an anterior cranial bone located between the eyes (figs. 8.7 and 8.12). It contributes to the medial wall of the orbit, the roof and walls of the nasal cavity, and the nasal septum. It is a very porous and delicate bone, with three major portions:

1. The vertical **perpendicular plate,** a thin median plate of bone that forms the superior two-thirds of the nasal septum (see fig. 8.4b). (The lower part is formed by the *vomer,* discussed later.) The septum divides the nasal cavity into right and left air spaces called the **nasal fossae** (FOSS-ee). The septum is often curved, or deviated, toward one nasal fossa or the other.

2. A horizontal **cribriform**[22] (CRIB-rih-form) **plate,** which forms the roof of the nasal cavity. This plate has a median blade called the **crista galli**[23] (GAL-eye), an attachment point for the dura mater. On each side of the crista is an elongated depressed area perforated with numerous holes, the **cribriform (olfactory) foramina.** A pair of *olfactory bulbs* of the brain, concerned with the sense of smell, rests in these depressions, and the foramina allow

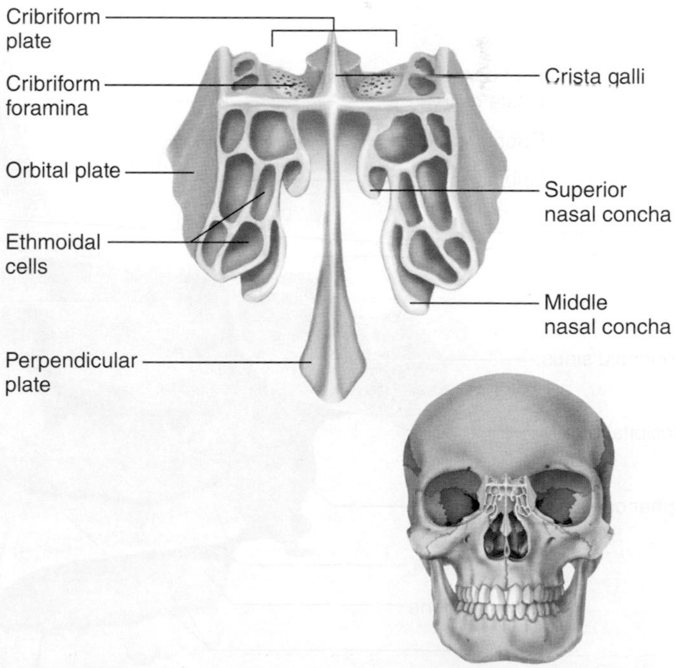

Cribriform plate

Cribriform foramina

Orbital plate

Ethmoidal cells

Perpendicular plate

Crista galli

Superior nasal concha

Middle nasal concha

**FIGURE 8.12 The Ethmoid Bone (Anterior View).**

❓ *List five bones that articulate with the ethmoid bone.* AP|R

---

[21]*ethmo* = sieve, strainer; *oid* = resembling
[22]*cribri* = sieve; *form* = in the shape of
[23]*crista* = crest; *galli* = of a rooster

passage for olfactory nerves from the nasal cavity to the bulbs (see Deeper Insight 8.1).

3. The **labyrinth,** a large mass on each side of the perpendicular plate. The labyrinth is named for the fact that internally, it has a maze of air spaces called the **ethmoidal cells.** Collectively, these constitute the *ethmoid sinus* discussed earlier. The lateral surface of the labyrinth is a smooth, slightly concave **orbital plate** seen on the medial wall of the orbit (see fig. 8.14). The medial surface of the labyrinth gives rise to two curled, scroll-like plates of bone called the **superior** and **middle nasal conchae**[24] (CON-kee). These project into the nasal fossa from its lateral wall toward the septum (see figs. 8.7 and 8.13). There is also a separate bone, the *inferior nasal concha,* discussed later. The three conchae occupy most of the nasal cavity, leaving little open space. By filling space and creating turbulence in the flow of inhaled air, they ensure that the air contacts the mucous membranes that cover these bones, which cleanse, humidify, and warm the inhaled air before it reaches the lungs. The superior concha and adjacent part of the nasal septum also bear the sensory cells of smell.

Usually, all that can be seen of the ethmoid is the perpendicular plate, by looking into the nasal cavity (see fig. 8.3); the orbital plate, by looking at the medial wall of the orbit (fig. 8.14); and the crista galli and cribriform plate, seen from within the cranial cavity (see fig. 8.5b).

---

[24]*concha* = conch (large marine snail)

## DEEPER INSIGHT 8.1

### CLINICAL APPLICATION

#### *Injury to the Ethmoid Bone*

The ethmoid bone is very delicate and is easily injured by a sharp upward blow to the nose, such as a person might suffer by striking an automobile dashboard in a collision. The force of a blow can drive bone fragments through the cribriform plate into the meninges or brain tissue. Such injuries are often evidenced by leakage of cerebrospinal fluid into the nasal cavity, and may be followed by the spread of infection from the nasal cavity to the brain. Blows to the head can also shear off the olfactory nerves that pass through the ethmoid bone and cause *anosmia,* an irreversible loss of the sense of smell and a great reduction in the sense of taste (most of which depends on smell). This not only deprives life of some of its pleasures, but can also be dangerous, as when a person fails to smell smoke, gas, or spoiled food.

---

## Facial Bones

**Facial bones** do not enclose the brain but lie anterior to the cranial cavity. They support the orbital, nasal, and oral cavities, shape the face, and provide attachment for the muscles of facial expression and mastication. There are 14 facial bones:

| | |
|---|---|
| 2 maxillae | 2 nasal bones |
| 2 palatine bones | 2 inferior nasal conchae |
| 2 zygomatic bones | 1 vomer |
| 2 lacrimal bones | 1 mandible |

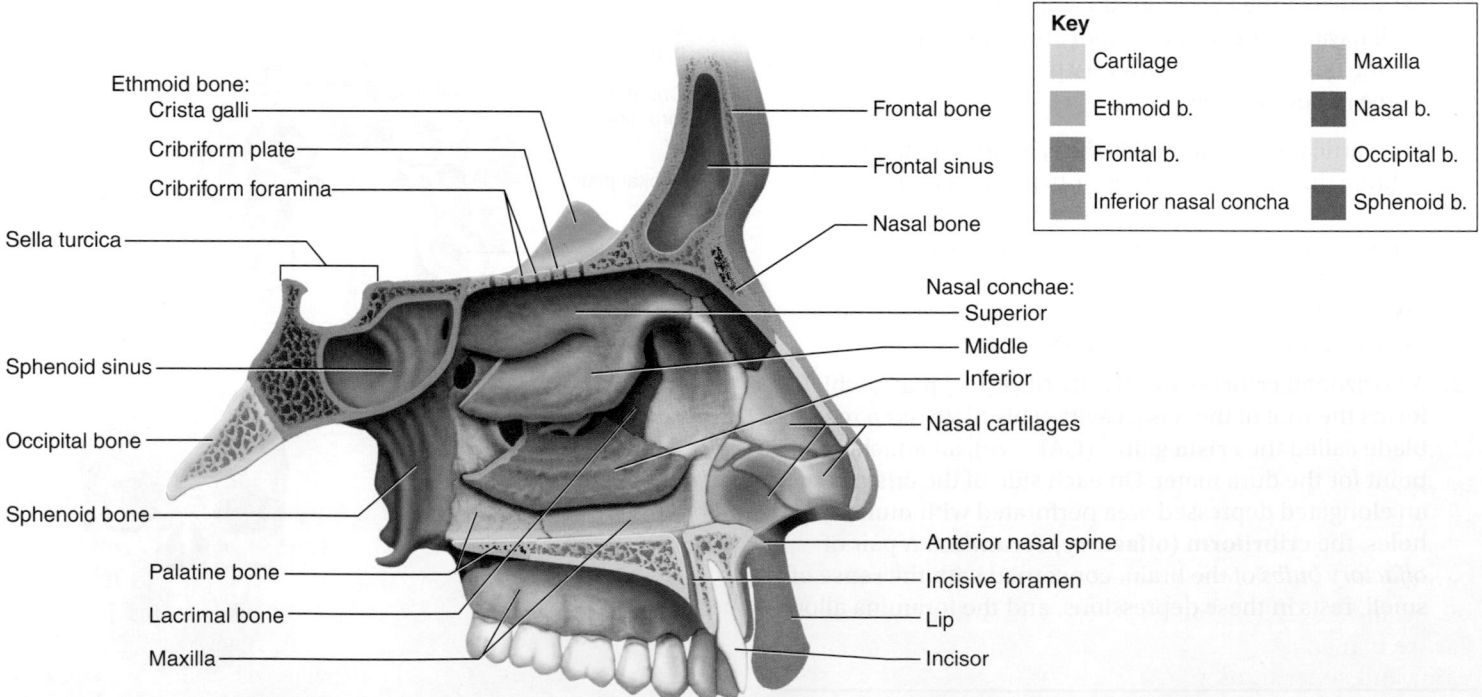

**Key**

| | | | |
|---|---|---|---|
| ▢ | Cartilage | ▢ | Maxilla |
| ▢ | Ethmoid b. | ▢ | Nasal b. |
| ▢ | Frontal b. | ▢ | Occipital b. |
| ▢ | Inferior nasal concha | ▢ | Sphenoid b. |

Ethmoid bone:
- Crista galli
- Cribriform plate
- Cribriform foramina

Sella turcica

Sphenoid sinus

Occipital bone

Sphenoid bone

Palatine bone

Lacrimal bone

Maxilla

Frontal bone

Frontal sinus

Nasal bone

Nasal conchae:
- Superior
- Middle
- Inferior

Nasal cartilages

Anterior nasal spine

Incisive foramen

Lip

Incisor

**FIGURE 8.13 The Left Nasal Cavity, Sagittal Section with Nasal Septum Removed.**

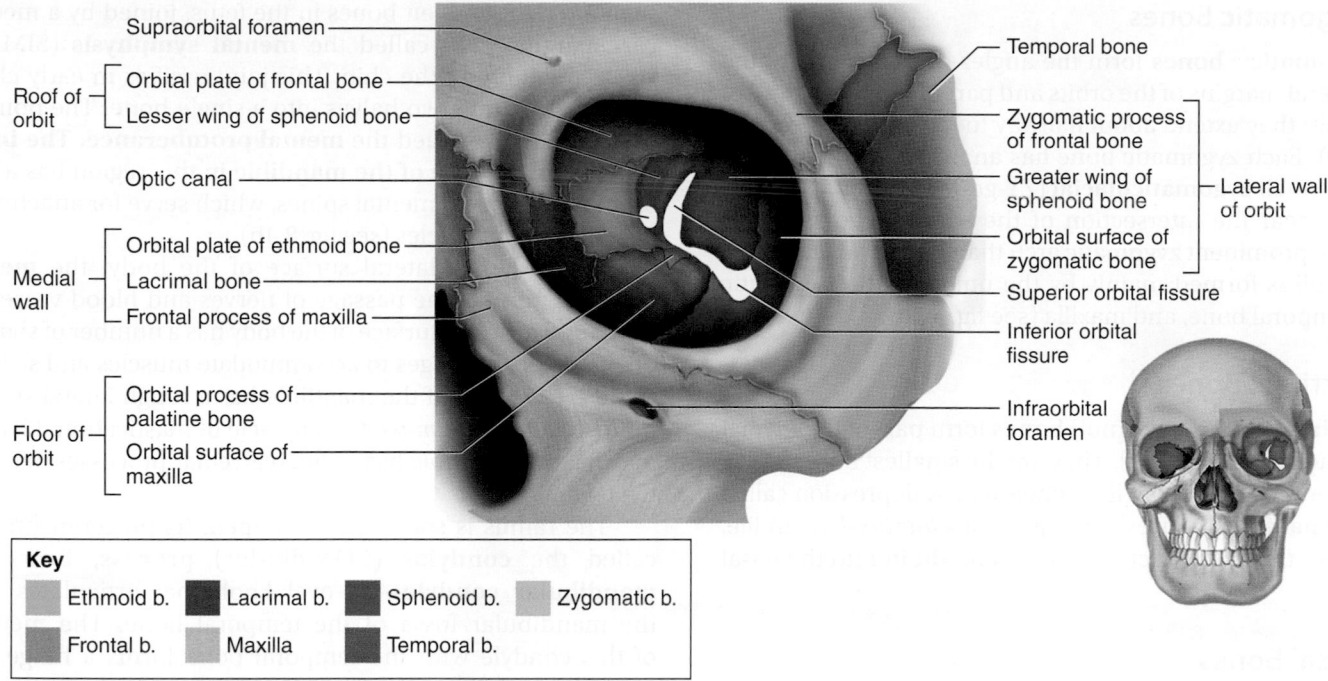

Key
- Ethmoid b.
- Lacrimal b.
- Sphenoid b.
- Zygomatic b.
- Frontal b.
- Maxilla
- Temporal b.

**FIGURE 8.14 The Left Orbit (Anterior View). AP|R**

## The Maxillae

The **maxillae** (mac-SILL-ee) are the largest facial bones. They form the upper jaw and meet each other at a median *intermaxillary suture* (see figs. 8.3, 8.4a, and 8.5a). Small points of maxillary bone called **alveolar processes** grow into the spaces between the bases of the teeth. The root of each tooth is inserted into a deep socket, or **alveolus.** If a tooth is lost or extracted so that chewing no longer puts stress on the maxilla, the alveolar processes are resorbed and the alveolus fills in with new bone, leaving a smooth area on the maxilla.

Although they are preserved with the skull, the teeth are not bones. They are discussed in detail in chapter 25.

### ▶▶▶ APPLY WHAT YOU KNOW

*Suppose you were studying a skull with some teeth missing. How could you tell whether the teeth had been lost after the person's death or years before it?*

Each maxilla extends from the teeth to the inferomedial wall of the orbit. Just below the orbit, it exhibits an **infraorbital foramen,** which provides passage for a blood vessel to the face and a nerve that receives sensations from the nasal region and cheek. This nerve emerges through the foramen rotundum into the cranial cavity. The maxilla forms part of the floor of the orbit, where it exhibits a gash called the **inferior orbital fissure** that angles downward and medially (fig. 8.14). The inferior and superior orbital fissures form a sideways V whose apex lies near the optic canal. The inferior orbital fissure is a passage for blood vessels and sensory nerves from the face.

The **palate** forms the roof of the mouth and floor of the nasal cavity. Its function is to separate the nasal cavity from the oral cavity, enabling us (and other mammals) to continue breathing while chewing. The high metabolic rate of humans requires rapid digestion of food, which in turn is aided by prolonged and thorough mastication into small, easily digested particles. This would be difficult if such prolonged mastication required an interruption of airflow.

The palate consists of a bony **hard palate** anteriorly and a fleshy **soft palate** posteriorly. Most of the hard palate is formed by horizontal extensions of the maxilla called **palatine** (PAL-uh-tine) **processes** (see fig. 8.5a). Just behind the incisors (front teeth) is a pair of **incisive foramina.** The palatine processes normally meet at the intermaxillary suture at about 12 weeks of gestation. Failure to join results in a *cleft palate,* often accompanied by a *cleft lip* lateral to the midline. A cleft palate and lip can be surgically corrected with good cosmetic results, but a cleft palate makes it difficult for an infant to generate the suction needed for nursing.

## The Palatine Bones

The **palatine bones** are located in the posterior nasal cavity (fig. 8.13). Each has an L shape formed by a *horizontal plate* and a *perpendicular plate.* The horizontal plates form the posterior one-third of the bony palate. Each is marked by a large **greater palatine foramen,** a nerve passage to the palate. The perpendicular plate is a thin, delicate, irregularly shaped plate that forms part of the wall between the nasal cavity and the orbit (see figs. 8.5a and 8.13).

## The Zygomatic Bones

The **zygomatic**[25] **bones** form the angles of the cheeks at the inferolateral margins of the orbits and part of the lateral wall of each orbit; they extend about halfway to the ear (see figs. 8.4a and 8.5a). Each zygomatic bone has an inverted T shape and usually a small **zygomaticofacial** (ZY-go-MAT-ih-co-FAY-shul) **foramen** near the intersection of the stem and crossbar of the T. The prominent zygomatic arch that flares from each side of the skull is formed mainly by the union of the zygomatic bone, temporal bone, and maxilla (see fig. 8.4a).

## The Lacrimal Bones

The **lacrimal**[26] (LACK-rih-mul) **bones** form part of the medial wall of each orbit (fig. 8.14). They are the smallest bones of the skull, about the size of the little fingernail. A depression called the **lacrimal fossa** houses a membranous *lacrimal sac* in life. Tears from the eye collect in this sac and drain into the nasal cavity.

## The Nasal Bones

Two small rectangular **nasal bones** form the bridge of the nose (see fig. 8.3) and support cartilages that shape its lower portion. If you palpate the bridge, you can easily feel where the nasal bones end and the cartilages begin. The nasal bones are often fractured by blows to the nose.

## The Inferior Nasal Conchae

There are three conchae in the nasal cavity. The superior and middle conchae, as discussed earlier, are parts of the ethmoid bone. The **inferior nasal concha**—the largest of the three—is a separate bone (see fig. 8.13).

## The Vomer

The **vomer** forms the inferior half of the nasal septum (see figs. 8.3 and 8.4b). Its name literally means "plowshare," which refers to its resemblance to the blade of a plow. The superior half of the nasal septum is formed by the perpendicular plate of the ethmoid bone, as mentioned earlier. The vomer and perpendicular plate support a wall of *septal cartilage* that forms most of the anterior part of the septum.

## The Mandible

The **mandible** (fig. 8.15) is the strongest bone of the skull and the only one that can move significantly. It supports the lower teeth and provides attachment for muscles of mastication and facial expression. The horizontal portion, bearing the teeth, is called the **body;** the vertical to oblique posterior portion is the **ramus** (RAY-mus) (plural, *rami*); and these two portions meet at a corner called the **angle.** The mandible develops as

---

separate right and left bones in the fetus, joined by a median cartilaginous joint called the **mental symphysis** (SIM-fih-sis) at the point of the chin. This joint ossifies in early childhood, uniting the two halves into a single bone. The point of the chin itself is called the **mental protuberance. The inner (posterior) surface of the mandible** in this region has a pair of small points, the mental spines, which serve for attachment of certain chin muscles (see fig. 8.4b).

On the anterolateral surface of the body, the **mental foramen** permits the passage of nerves and blood vessels of the chin. The inner surface of the body has a number of shallow depressions and ridges to accommodate muscles and salivary glands. The angle of the mandible has a rough lateral surface for insertion of the *masseter,* a muscle of mastication. Like the maxilla, the mandible has pointed alveolar processes between the teeth.

The ramus is somewhat Y-shaped. Its posterior branch, called the **condylar** (CON-dih-lur) **process,** bears the **mandibular condyle**—an oval knob that articulates with the mandibular fossa of the temporal bone. The meeting of this condyle with the temporal bone forms a hinge, the **temporomandibular joint (TMJ).** The anterior branch of the ramus is a blade called the **coronoid process.** It is the point of insertion for the temporalis muscle, which pulls the mandible upward when you bite. The U-shaped arch between the two processes is the **mandibular notch.** Just below the notch, on the medial surface of the ramus, is the **mandibular foramen.** The nerve and blood vessels that supply the lower teeth enter this foramen and then travel through the bone of the mandibular body, giving off branches to each tooth along the way. Dentists commonly inject lidocaine near the mandibular foramen to deaden sensation from the lower teeth.

## Bones Associated with the Skull

Seven bones are closely associated with the skull but not considered part of it. These are the three auditory ossicles in each middle-ear cavity and the hyoid bone beneath the chin. The

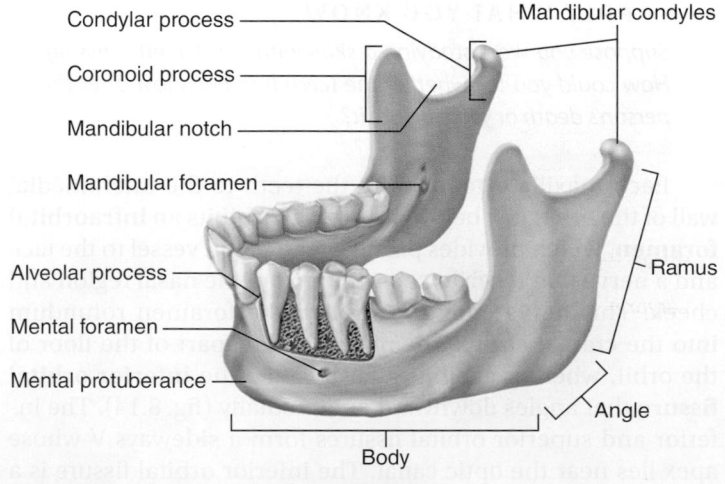

**FIGURE 8.15  The Mandible.**  AP|R

**auditory ossicles**[27]—named the **malleus** (hammer), **incus** (anvil), and **stapes** (STAY-peez) (stirrup)—are discussed in connection with hearing in chapter 16. The **hyoid**[28] **bone** is a slender U-shaped bone between the chin and larynx (fig. 8.16). It is one of the few bones that does not articulate with any other. It is suspended from the styloid processes of the skull, somewhat like a hammock, by the small *stylohyoid muscles* and *stylohyoid ligaments.* The median **body** of the hyoid is flanked on either side by hornlike projections called the **greater** and **lesser horns (cornua).** The larynx (voice box) is suspended from the hyoid bone by a broad ligament (see fig. 22.4a, p. 853), and the hyoid serves for attachment of several muscles that control the mandible, tongue, and larynx. Forensic pathologists look for a fractured hyoid as evidence of strangulation.

## The Skull in Infancy and Childhood

The head of an infant could not fit through the mother's pelvic outlet at birth were it not for the fact that the bones of its skull are not yet fused. The shifting of the skull bones during birth may cause the infant to appear deformed, but the head soon assumes a more normal shape. Spaces between the unfused cranial bones are called **fontanels,**[29] after the fact that pulsation of the infant's blood can be felt there. The bones are joined at these points only by fibrous membranes, in which intramembranous ossification is completed later. Four of these sites are especially prominent and regular in location: the **anterior, posterior, sphenoid (anterolateral),** and **mastoid (posterolateral) fontanels** (fig. 8.17). Most fontanels ossify by the time the infant is a year old, but the largest one—the anterior fontanel—can still be palpated 18 to 24 months after birth.

---

[27]*os* = bone; *icle* = little
[28]*hy* = the letter U; *oid* = resembling
[29]*fontan* = fountain; *el* = little

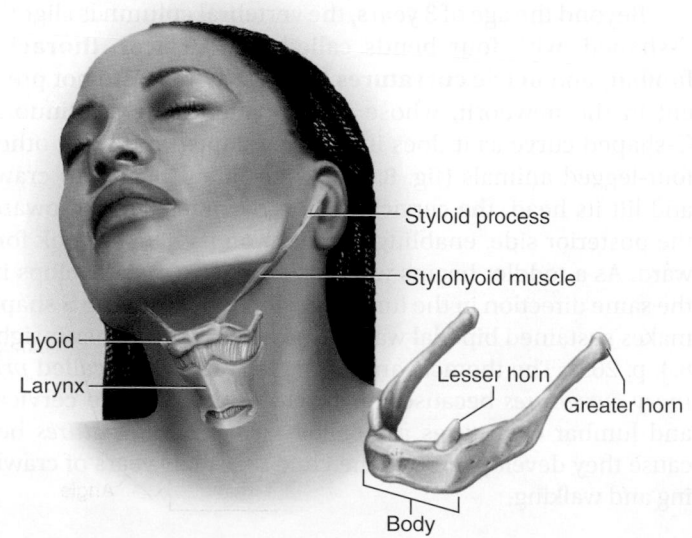

**FIGURE 8.16 The Hyoid Bone.**

# DEEPER INSIGHT 8.2
## CLINICAL APPLICATION
### Cranial Assessment of the Newborn

Obstetric nurses routinely assess the fontanels of newborns by palpation. In a difficult delivery, one cranial bone may override another along a suture line, which calls for close monitoring of the infant. Abnormally wide sutures may indicate hydrocephalus, the accumulation of excessive amounts of cerebrospinal fluid, which causes the cranium to swell. Bulging fontanels suggest abnormally high intracranial pressure, while depressed fontanels indicate dehydration.

---

The frontal bone and mandible are separate right and left bones at birth, but fuse medially in early childhood. The frontal bones usually fuse by age 5 or 6, but in some children a *metopic*[30] *suture* persists between them. Traces of this suture are evident in some adult skulls.

The face of a newborn is flat and the cranium relatively large. To accommodate the growing brain, the skull grows more rapidly than the rest of the skeleton during childhood. It reaches about half its adult size by 9 months of age, three-quarters by age 2, and nearly final size by 8 or 9 years. The heads of babies and children are therefore much larger in proportion to the trunk than the heads of adults—an attribute thoroughly exploited by cartoonists and advertisers who draw big-headed characters to give them a more endearing or immature appearance. In humans and other animals, the large rounded heads of the young are thought to promote survival by stimulating parental caregiving instincts.

### BEFORE YOU GO ON

Answer the following questions to test your understanding of the preceding section:

4. Name the paranasal sinuses and state their locations. Name any four other cavities in the skull.

5. Explain the difference between a cranial bone and a facial bone. Give four examples of each.

6. Draw an oval representing a superior view of the calvaria. Draw lines representing the coronal, lambdoid, and sagittal sutures. Label the four bones separated by these sutures.

7. State which bone has each of these features: a squamous part, hypoglossal foramen, greater horn, greater wing, condylar process, and cribriform plate.

8. Determine which of the following structures cannot normally be palpated on a living person: the mastoid process, crista galli, superior orbital fissure, palatine processes, zygomatic bone, mental protuberance, and stapes. You may find it useful to palpate some of these on your own skull as you try to answer.

---

[30]*met* = beyond; *op* = the eyes

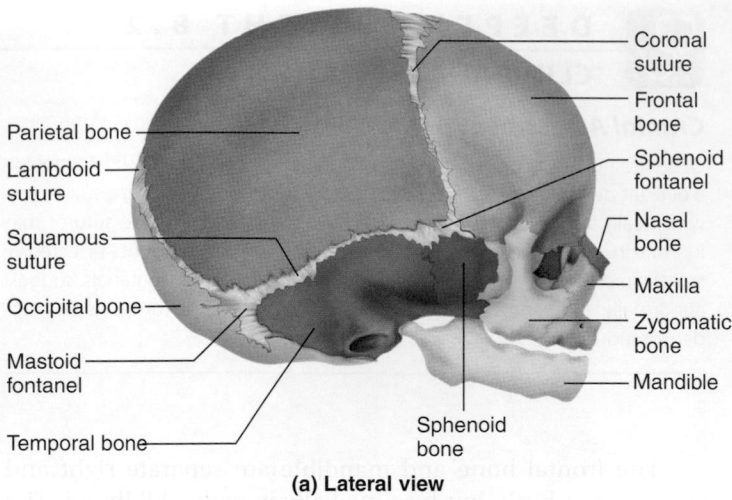

Coronal suture
Frontal bone
Parietal bone
Lambdoid suture
Sphenoid fontanel
Squamous suture
Nasal bone
Occipital bone
Maxilla
Mastoid fontanel
Zygomatic bone
Mandible
Temporal bone
Sphenoid bone

**(a) Lateral view**

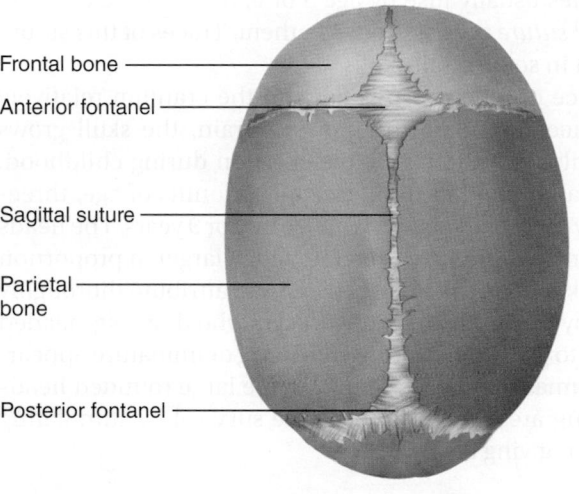

Frontal bone
Anterior fontanel
Sagittal suture
Parietal bone
Posterior fontanel

**(b) Superior view**

**FIGURE 8.17    The Fetal Skull Near the Time of Birth.**

---

### 8.3    The Vertebral Column and Thoracic Cage

#### Expected Learning Outcomes

When you have completed this section, you should be able to

a. describe the general features of the vertebral column and those of a typical vertebra;

b. describe the structure of the intervertebral discs and their relationship to the vertebrae;

c. describe the special features of vertebrae in different regions of the vertebral column, and discuss the functional significance of the regional differences; and

d. describe the anatomy of the sternum and ribs and how the ribs articulate with the thoracic vertebrae.

---

## General Features of the Vertebral Column

The **vertebral** (VUR-teh-brul) **column (spine)** physically supports the skull and trunk, allows for their movement, protects the spinal cord, and absorbs stresses produced by walking, running, and lifting. It also provides attachment for the limbs, thoracic cage, and postural muscles. Although commonly called the backbone, it consists not of a single bone but a flexible chain of 33 **vertebrae** with **intervertebral discs** of fibrocartilage between most of them. The adult vertebral column averages about 71 cm (28 in.) long, with the intervertebral discs accounting for about one-quarter of the length.

Most people are about 1% shorter when they go to bed at night than when they first rise in the morning. This is because during the day, the weight of the body compresses the intervertebral discs and squeezes water out of them. When one is sleeping, with the weight off the spine, the discs reabsorb water and swell.

As shown in figure 8.18, the vertebrae are divided into five groups, usually numbering 7 *cervical* (SUR-vih-cul) *vertebrae* in the neck, 12 *thoracic vertebrae* in the chest, 5 *lumbar vertebrae* in the lower back, 5 *sacral vertebrae* at the base of the spine, and 4 tiny *coccygeal* (coc-SIDJ-ee-ul) *vertebrae*. To help remember the numbers of cervical, thoracic, and lumbar vertebrae—7, 12, and 5—you might think of a typical workday: Go to work at 7, have lunch at 12, and go home at 5. All mammals have 7 cervical vertebrae, even in the famously long necks of giraffes.

Variations in this arrangement occur in about 1 person in 20. For example, the last lumbar vertebra is sometimes incorporated into the sacrum, producing four lumbar and six sacral vertebrae. In other cases, the first sacral vertebra fails to fuse with the second, producing six lumbar and four sacral vertebrae. The coccyx usually has four but sometimes five vertebrae. The cervical and thoracic vertebrae are more constant in number.

Beyond the age of 3 years, the vertebral column is slightly S-shaped, with four bends called the **cervical, thoracic, lumbar,** and **pelvic curvatures** (fig. 8.19). These are not present in the newborn, whose spine exhibits one continuous C-shaped curve as it does in monkeys, apes, and most other four-legged animals (fig. 8.20). As an infant begins to crawl and lift its head, the cervical region becomes curved toward the posterior side, enabling an infant on its belly to look forward. As a toddler begins walking, another curve develops in the same direction in the lumbar region. The resulting S shape makes sustained bipedal walking possible (see Deeper Insight 8.4, p. 269). The thoracic and pelvic curvatures are called *primary curvatures* because they exist from birth. The cervical and lumbar curvatures are called *secondary curvatures* because they develop later, in the child's first few years of crawling and walking.

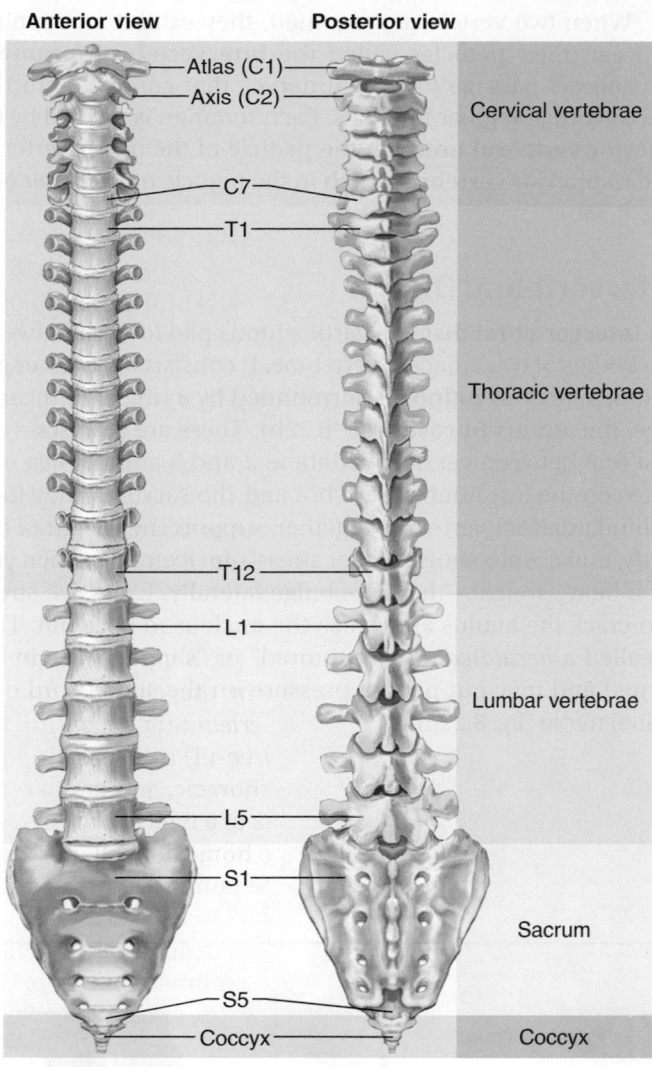

**FIGURE 8.18 The Vertebral Column.** AP|R

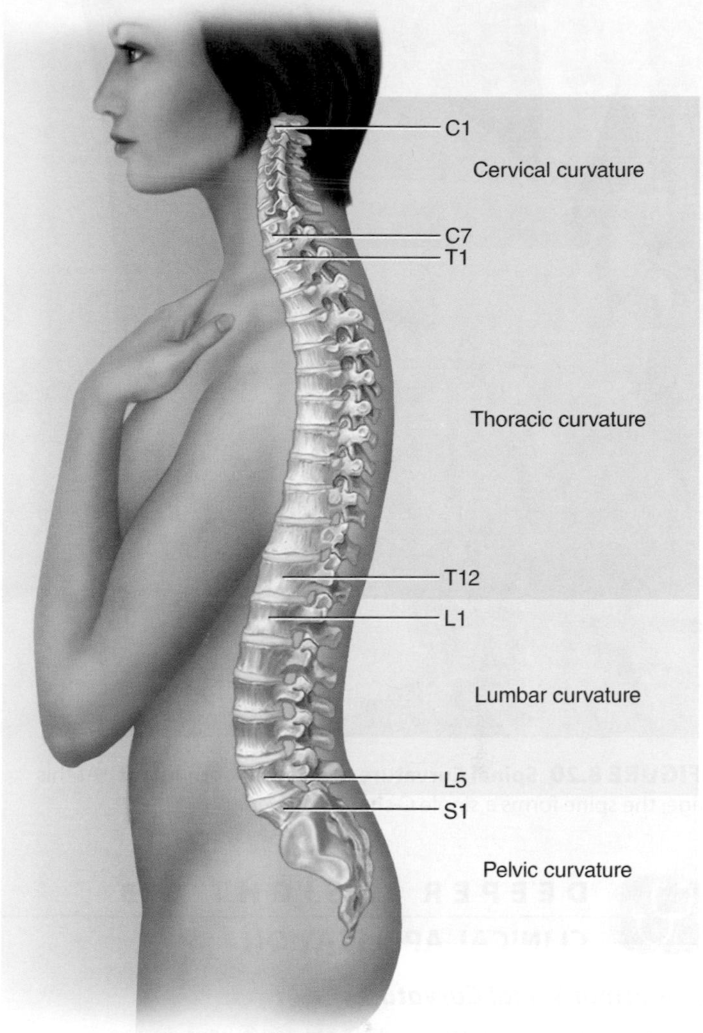

**FIGURE 8.19 Curvatures of the Adult Vertebral Column.**

## General Structure of a Vertebra

A representative vertebra and intervertebral disc are shown in figure 8.22. The most obvious feature of a vertebra is the **body (centrum)**—a mass of spongy bone and red bone marrow covered with a thin shell of compact bone. This is the weight-bearing portion of the vertebra. Its rough superior and inferior surfaces provide firm attachment to the intervertebral discs.

### ►►►APPLY WHAT YOU KNOW

*The vertebral bodies and intervertebral discs get progressively larger as we look lower and lower on the vertebral column. What is the functional significance of this trend?*

Posterior to the body of each vertebra is a triangular space called the **vertebral foramen.** The vertebral foramina collectively form the **vertebral canal,** a passage for the spinal cord. Each foramen is bordered by a bony **vertebral arch** composed of two parts on each side: a pillarlike **pedicle**[31] and platelike **lamina.**[32] Extending from the apex of the arch, a projection called the **spinous process** is directed posteriorly and downward. You can see and feel the spinous processes on a living person as a row of bumps along the spine. A **transverse process** extends laterally from the point where the pedicle and lamina meet. The spinous and transverse processes provide points of attachment for ligaments, ribs, and spinal muscles.

A pair of **superior articular processes** projects upward from one vertebra and meets a similar pair of inferior articular processes that projects downward from the vertebra above (fig. 8.23a). Each process has a flat articular surface (facet) facing that of the adjacent vertebra. These processes restrict twisting of the vertebral column, which could otherwise severely damage the spinal cord.

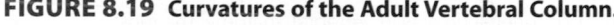

[31]*ped* = foot; *icle* = little
[32]*lamina* = plate

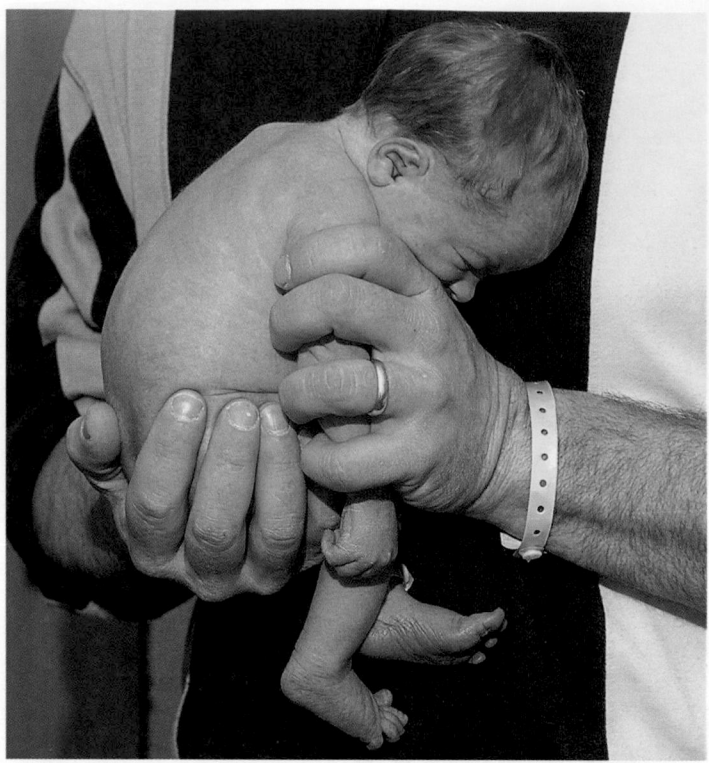

**FIGURE 8.20  Spinal Curvature of the Newborn Infant.** At this age, the spine forms a single C-shaped curve.

When two vertebrae are joined, they exhibit an opening between their pedicles called the **intervertebral foramen.** This allows passage for spinal nerves that connect with the spinal cord at regular intervals. Each foramen is formed by an **inferior vertebral notch** in the pedicle of the upper vertebra and a **superior vertebral notch** in the pedicle of the lower one (fig. 8.23b).

## Intervertebral Discs

An **intervertebral disc** is a cartilaginous pad located between the bodies of two adjacent vertebrae. It consists of an inner gelatinous **nucleus pulposus** surrounded by a ring of fibrocartilage, the **anulus fibrosus** (fig. 8.22b). There are 23 discs—the first one between cervical vertebrae 2 and 3 and the last one between the last lumbar vertebra and the sacrum. They help to bind adjacent vertebrae together, support the weight of the body, and absorb shock. Under stress—for example, when you lift a heavy weight—the discs bulge laterally. Excessive stress can crack the anulus and cause the nucleus to ooze out. This is called a *herniated disc* ("ruptured" or "slipped" disc in lay terms) and may put painful pressure on the spinal cord or a spinal nerve (fig. 8.22c).

 **DEEPER INSIGHT 8.3**

### CLINICAL APPLICATION

#### Abnormal Spinal Curvatures

Abnormal spinal curvatures (fig. 8.21) can result from disease; weakness or paralysis of the trunk muscles; poor posture; pregnancy; or congenital defects in vertebral anatomy. The most common deformity is an abnormal lateral curvature called *scoliosis*. It occurs most often in the thoracic region, particularly among adolescent girls. It sometimes results from a developmental abnormality in which the body and arch fail to develop on one side of a vertebra. If the person's skeletal growth is not yet complete, scoliosis can be corrected with a back brace.

An exaggerated thoracic curvature is called *kyphosis* (hunchback, in lay language). It is usually a result of osteoporosis, but it also occurs in people with osteomalacia or spinal tuberculosis and in adolescent boys who engage heavily in such spine-loading sports as wrestling and weight lifting. An exaggerated lumbar curvature is called *lordosis* (swayback, in lay language). It may have the same causes as kyphosis, or it may result from added abdominal weight in pregnancy or obesity.

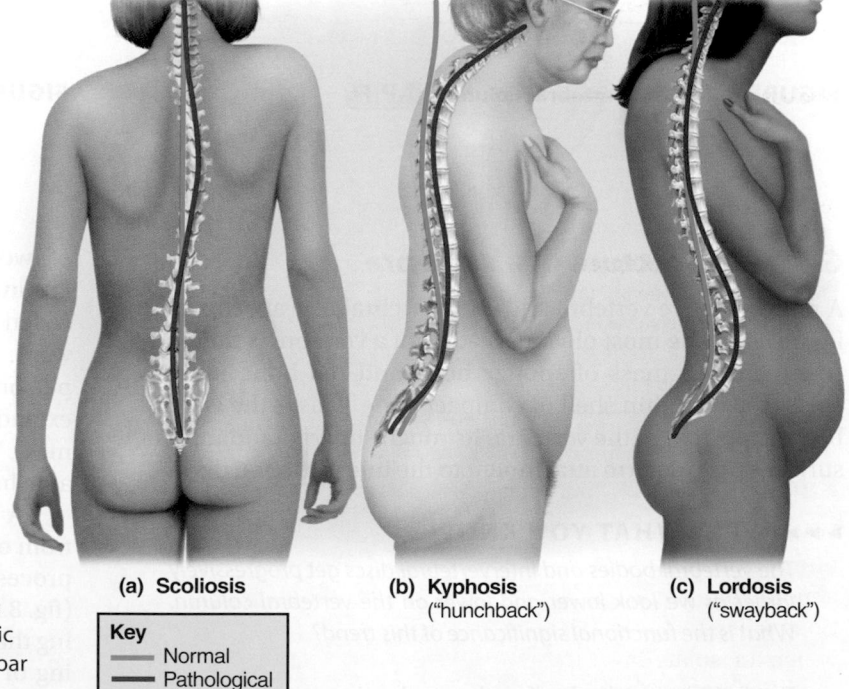

(a) Scoliosis    (b) Kyphosis ("hunchback")    (c) Lordosis ("swayback")

**Key**
— Normal
— Pathological

**FIGURE 8.21  Abnormal Spinal Curvatures.** (a) Scoliosis, an abnormal lateral deviation. (b) Kyphosis, an exaggerated thoracic curvature common in old age. (c) Lordosis, an exaggerated lumbar curvature common in pregnancy and obesity.

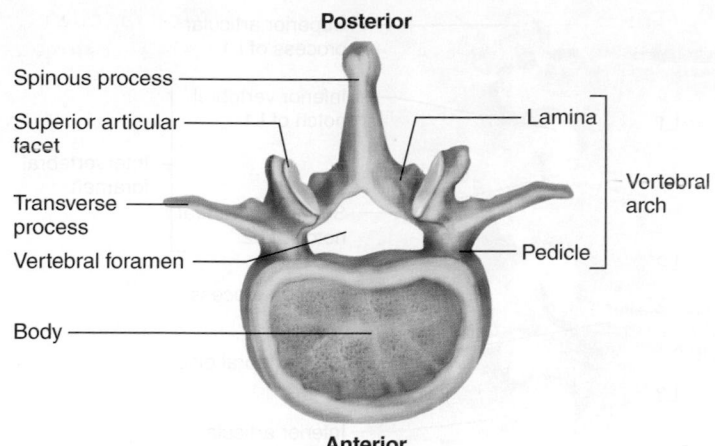

**Posterior**

Spinous process

Superior articular facet

Transverse process

Vertebral foramen

Body

Lamina

Vertebral arch

Pedicle

**Anterior**

**(a) 2nd lumbar vertebra (L2)**

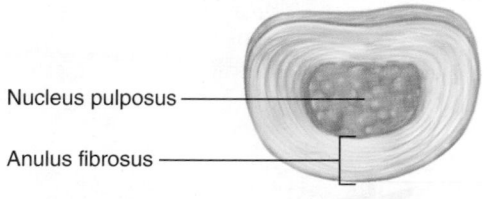

Nucleus pulposus

Anulus fibrosus

**(b) Intervertebral disc**

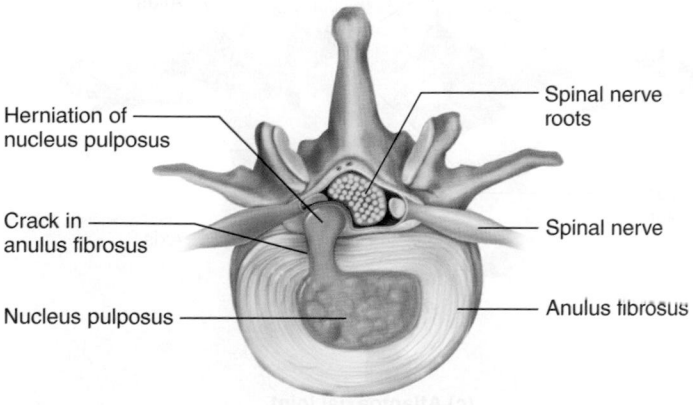

Herniation of nucleus pulposus

Crack in anulus fibrosus

Nucleus pulposus

Spinal nerve roots

Spinal nerve

Anulus fibrosus

**(c) Herniated disc**

**FIGURE 8.22 A Representative Vertebra and Intervertebral Disc (Superior Views).** (a) A typical vertebra. (b) An intervertebral disc, oriented the same way as the vertebral body in part (a) for comparison. (c) A herniated disc, showing compression of the spinal nerve roots by the nucleus pulposus oozing from the disc.

## Regional Characteristics of Vertebrae

We are now prepared to consider how vertebrae differ from one region of the vertebral column to another and from the generalized anatomy just described. Knowing these variations will enable you to identify the region of the spine from which an isolated vertebra was taken. More importantly, these modifications in form reflect functional differences among the vertebrae.

## The Cervical Vertebrae

The cervical vertebrae (C1–C7) are relatively small. Their function is to support the head and allow for its movements. The first two (C1 and C2) have unique structures for this purpose (fig. 8.24). Vertebra C1 is called the **atlas** because it supports the head in a manner reminiscent of the Titan of Greek mythology who was condemned by Zeus to carry the heavens on his shoulders. It scarcely resembles the typical vertebra; it has no body, and is little more than a delicate ring surrounding a large vertebral foramen. On each side is a **lateral mass** with a deeply concave **superior articular facet** that articulates with the occipital condyle of the skull. A nodding motion of the skull, as in gesturing "yes," causes the occipital condyles to rock back and forth on these facets. The **inferior articular facets,** which are comparatively flat or only slightly concave, articulate with C2. The lateral masses are connected by an **anterior arch** and a **posterior arch,** which bear slight protuberances called the **anterior** and **posterior tubercle,** respectively.

Vertebra C2, the **axis,** allows rotation of the head as in gesturing "no." Its most distinctive feature is a prominent anterior knob called the **dens** (pronounced "denz"), or **odontoid**[33] **process,** on its anterosuperior side. No other vertebra has a dens. It begins to form as an independent ossification center during the first year of life and fuses with the axis by the age of 3 to 6 years. It projects into the vertebral foramen of the atlas, where it is nestled in a facet and held in place by a **transverse ligament** (fig. 8.24c). A heavy blow to the top of the head can cause a fatal injury in which the dens is driven through the foramen magnum into the brainstem.

The articulation between the atlas and the cranium is called the **atlanto–occipital joint;** the one between the atlas and axis is called the **atlantoaxial joint.**

The axis is the first vertebra that exhibits a spinous process. In vertebrae C2 through C6, the process is forked, or *bifid,*[34] at its tip (fig. 8.25a). This fork provides attachment for the *nuchal ligament* of the back of the neck. All seven cervical vertebrae have a prominent round **transverse foramen** in each transverse process. These foramina provide passage and protection for the *vertebral arteries,* which supply blood to the brain, and *vertebral veins,* which drain blood from various neck structures (but not from the brain). Transverse foramina occur in no other vertebrae and thus provide an easy means of recognizing a cervical vertebra.

### ▶▶▶ APPLY WHAT YOU KNOW

*How would head movements be affected if vertebrae C1 and C2 had the same structure as C3? What is the functional advantage of the lack of a spinous process in C1?*

---

[33]*dens = odont* = tooth; *oid* = resembling
[34]*bifid* = cleft into two parts

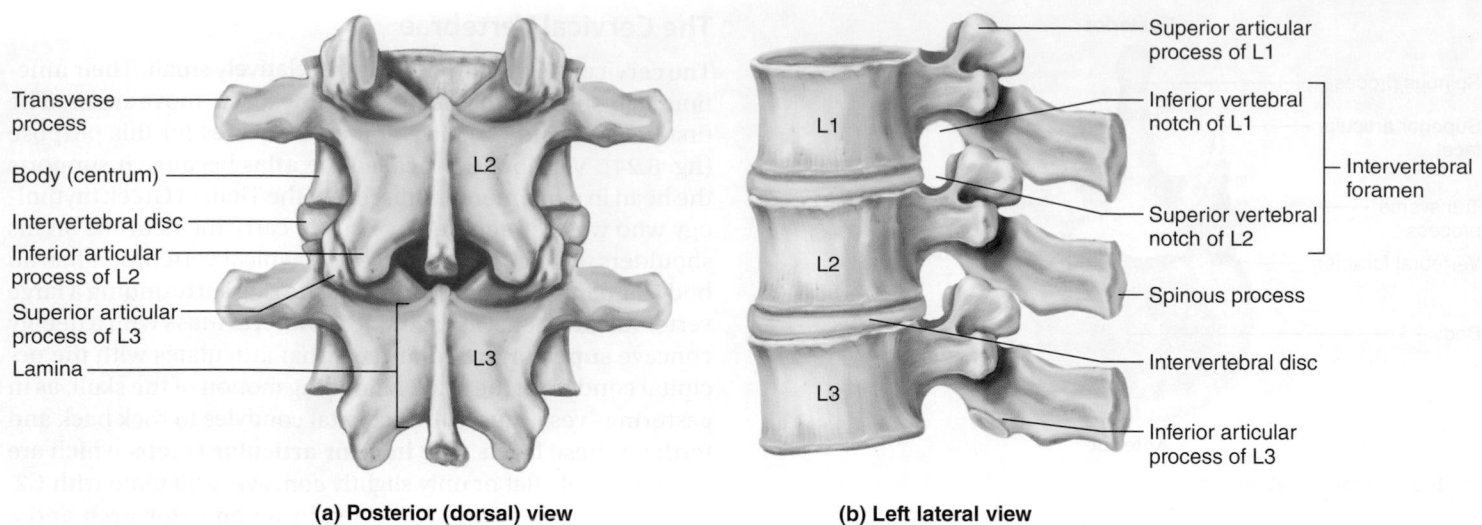

**FIGURE 8.23  Articulated Vertebrae.**

(a) Posterior (dorsal) view

(b) Left lateral view

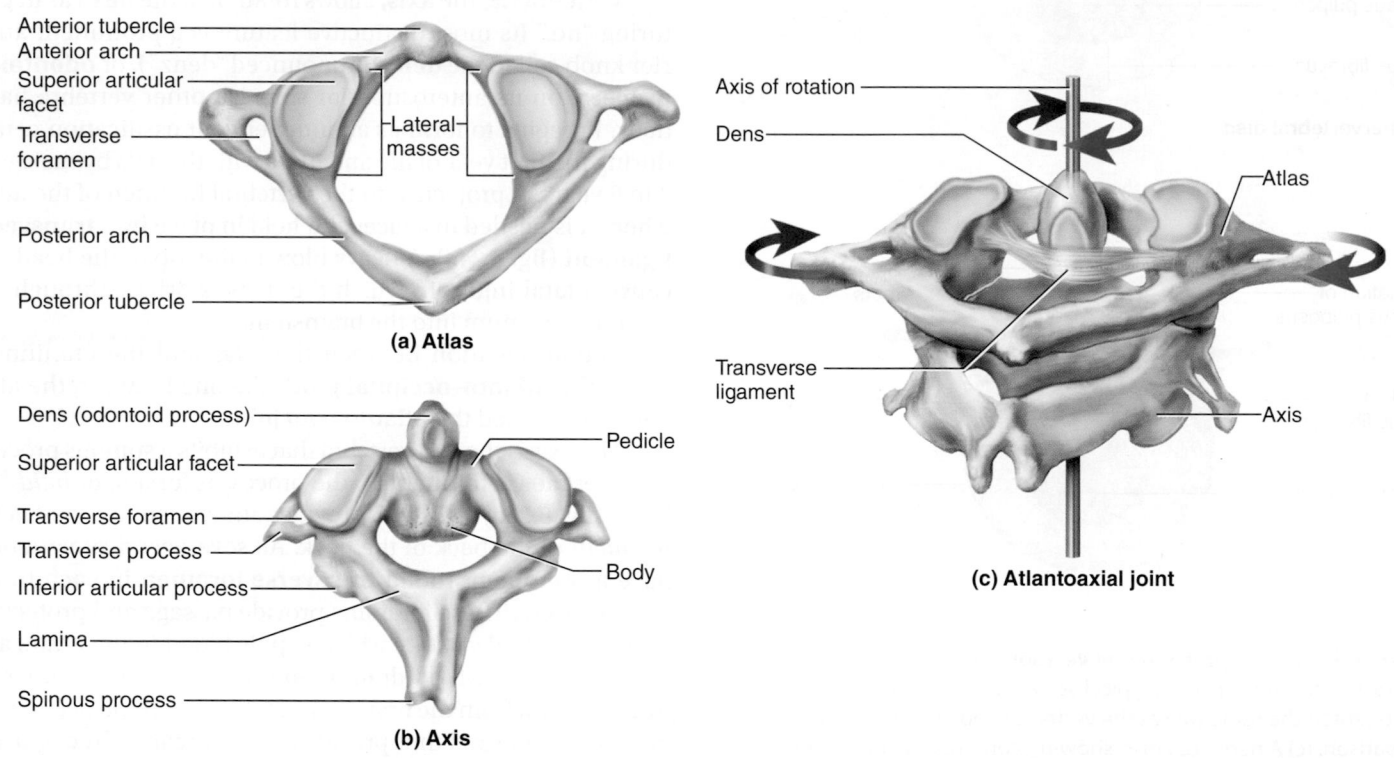

(a) Atlas

(b) Axis

(c) Atlantoaxial joint

**FIGURE 8.24  The Atlas and Axis, Cervical Vertebrae C1 and C2.**  (a) The atlas, superior view. (b) The axis, posterosuperior view. (c) Articulation of the atlas and axis and rotation of the atlas. This movement turns the head from side to side, as in gesturing "no." Note the transverse ligament holding the dens of the axis in place.  AP|R

❓ *What serious consequence could result from a rupture of the transverse ligament?*

Cervical vertebrae C3 through C6 are similar to the typical vertebra described earlier, with the addition of the transverse foramina and bifid spinous processes. Vertebra C7 is a little different—its spinous process is not bifid, but it is especially long and forms a prominent bump on the lower back of the neck. C7 is sometimes called the *vertebra prominens* because of this especially conspicuous spinous process. This feature is a convenient landmark for counting vertebrae. One can easily identify the largest bump on the neck as C7, then count up or down from there to identify others.

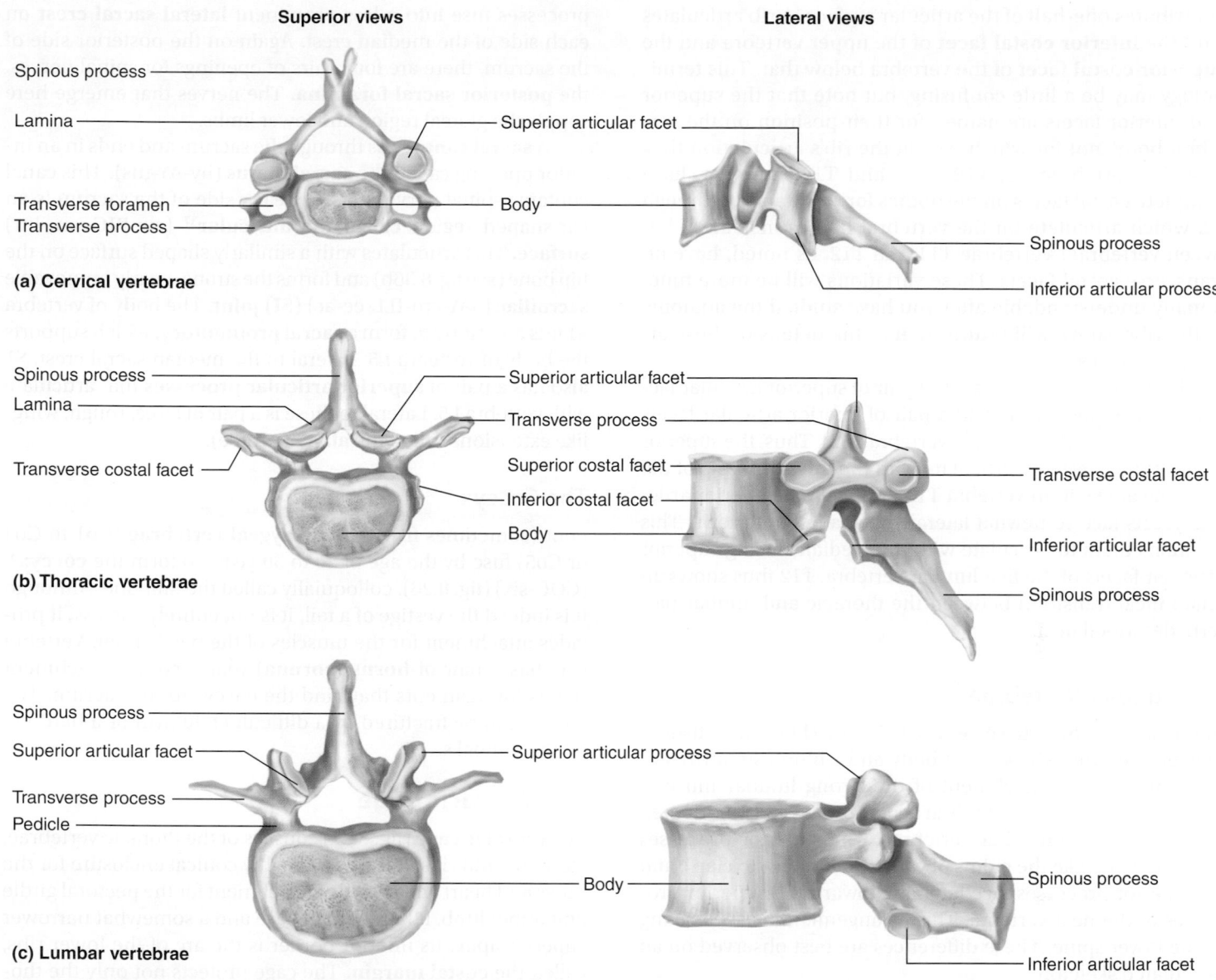

**Superior views**

**Lateral views**

**(a) Cervical vertebrae**

- Spinous process
- Lamina
- Transverse foramen
- Transverse process
- Superior articular facet
- Body
- Spinous process
- Inferior articular process

**(b) Thoracic vertebrae**

- Spinous process
- Lamina
- Transverse costal facet
- Superior articular facet
- Transverse process
- Superior costal facet
- Inferior costal facet
- Body
- Transverse costal facet
- Inferior articular facet
- Spinous process

**(c) Lumbar vertebrae**

- Spinous process
- Superior articular facet
- Transverse process
- Pedicle
- Superior articular process
- Body
- Spinous process
- Inferior articular facet

**FIGURE 8.25 Typical Cervical, Thoracic, and Lumbar Vertebrae.** The left-hand figures are superior views and the right-hand figures are left lateral views. **AP|R**

## The Thoracic Vertebrae

There are 12 **thoracic vertebrae** (T1–T12), corresponding to the 12 pairs of ribs attached to them; no other vertebrae have ribs. The function of the thoracics is to support the thoracic cage enclosing the heart and lungs. They lack the transverse foramina and bifid processes that distinguish the cervicals, but possess the following distinctive features of their own (fig. 8.25b):

- The spinous processes are relatively pointed and angle sharply downward.
- The body is somewhat heart-shaped, more massive than in the cervical vertebrae but less than in the lumbar vertebrae.

- The body has small, smooth, slightly concave spots called *costal facets* for attachment of the ribs.
- Vertebrae T1 through T10 have a shallow, cuplike **transverse costal**[35] **facet** at the end of each transverse process. These provide a second point of articulation for ribs 1 to 10. There are no transverse costal facets on T11 and T12 because ribs 11 and 12 attach only to the bodies of the vertebrae.

Thoracic vertebrae vary among themselves mainly in the mode of articulation with the ribs (see fig. 8.29). In most cases, a rib inserts between two vertebrae, so each vertebra

[35]*costa* = rib; *al* = pertaining to

contributes one-half of the articular surface. A rib articulates with the **inferior costal facet** of the upper vertebra and the **superior costal facet** of the vertebra below that. This terminology may be a little confusing, but note that the superior and inferior facets are named for their position on the vertebral body, not for which part of the rib's articulation they provide. Vertebrae T1, T10, T11, and T12, however, have complete costal facets on the bodies for ribs 1 and 10 through 12, which articulate on the vertebral bodies instead of between vertebrae. Vertebrae T11 and T12, as noted, have no transverse costal facets. These variations will be more functionally understandable after you have studied the anatomy of the ribs, so we will return then to the details of these articular surfaces.

Each thoracic vertebra has a pair of superior articular facets that face posteriorly and a pair of inferior articular facets that face anteriorly (except in vertebra T12). Thus, the superior facets of one vertebra articulate with the inferior facets of the next one above it. In vertebra T12, however, the inferior articular facets face somewhat laterally instead of anteriorly. This positions them to articulate with the medially facing superior articular facets of the first lumbar vertebra. T12 thus shows an anatomical transition between the thoracic and lumbar pattern, described next.

## The Lumbar Vertebrae

There are five **lumbar vertebrae** (L1–L5). Their most distinctive features are a thick, stout body and a blunt, squarish spinous process for attachment of the strong lumbar muscles (fig. 8.25c). In addition, their articular processes are oriented differently than on other vertebrae. The superior processes face medially (like the palms of your hands about to clap), and the inferior processes face laterally, toward the superior processes of the next vertebra. This arrangement resists twisting of the lower spine. These differences are best observed on an articulated skeleton.

## The Sacrum

The **sacrum** (SACK-rum, SAY-krum) is a bony plate that forms the posterior wall of the pelvic girdle (fig. 8.26). It was named *sacrum* for its prominence as the largest and most durable bone of the vertebral column.[36] In children, there are five separate **sacral vertebrae** (S1–S5). They begin to fuse around age 16 and are fully fused by age 26.

The anterior surface of the sacrum is relatively smooth and concave and has four transverse lines that indicate where the five vertebrae have fused. This surface exhibits four pairs of large **anterior sacral (pelvic) foramina,** which allow for passage of nerves and arteries to the pelvic organs. The posterior surface is very rough. The spinous processes of the vertebrae fuse into a ridge called the **median sacral crest.** The transverse

processes fuse into a less prominent **lateral sacral crest** on each side of the median crest. Again on the posterior side of the sacrum, there are four pairs of openings for spinal nerves, the **posterior sacral foramina.** The nerves that emerge here supply the gluteal region and lower limbs.

A **sacral canal** runs through the sacrum and ends in an inferior opening called the **sacral hiatus** (hy-AY-tus). This canal contains spinal nerve roots. On each side of the sacrum is an ear-shaped region called the **auricular**[37] (aw-RIC-you-lur) **surface.** This articulates with a similarly shaped surface on the hip bone (see fig. 8.36b) and forms the strong, nearly immovable **sacroiliac** (SAY-cro-ILL-ee-ac) **(SI) joint.** The body of vertebra S1 juts anteriorly to form a **sacral promontory,** which supports the body of vertebra L5. Lateral to the median sacral crest, S1 also has a pair of **superior articular processes** that articulate with vertebra L5. Lateral to these is a pair of large, rough, winglike extensions called the **alae**[38] (AIL-ee).

## The Coccyx

Four (sometimes five) tiny **coccygeal vertebrae** (Co1 to Co4 or Co5) fuse by the age of 20 to 30 years to form the **coccyx**[39] (COC-six) (fig. 8.26), colloquially called the tailbone. Although it is indeed the vestige of a tail, it is not entirely useless; it provides attachment for the muscles of the pelvic floor. Vertebra Co1 has a pair of **horns (cornua)** that serve as attachment points for ligaments that bind the coccyx to the sacrum. The coccyx can be fractured by a difficult childbirth or a hard fall on the buttocks.

## The Thoracic Cage

The **thoracic cage** (fig. 8.27) consists of the thoracic vertebrae, sternum, and ribs. It forms a roughly conical enclosure for the lungs and heart and provides attachment for the pectoral girdle and upper limb. It has a broad base and a somewhat narrower superior apex. Its inferior border is the arc of the lower ribs, called the **costal margin.** The cage protects not only the thoracic organs but also the spleen, most of the liver, and to some extent the kidneys. Most important is its role in breathing; it is rhythmically expanded by the respiratory muscles to create a vacuum that draws air into the lungs, and then compressed to expel air.

## The Sternum

The **sternum** (breastbone) is a bony plate anterior to the heart. It is subdivided into three regions: the manubrium, body, and xiphoid process. The **manubrium**[40] (ma-NOO-bree-um) is the broad superior portion, shaped like the knot of a necktie. It lies at the level of vertebrae T3 to T4. It has a median **suprasternal**

---

[36]*sacr* = great, prominent

[37]*auri* = ear; *cul* = little; *ar* = pertaining to
[38]*alae* = wings
[39]*coccyx* = cuckoo (named for resemblance to a cuckoo's beak)
[40]*manubrium* = handle

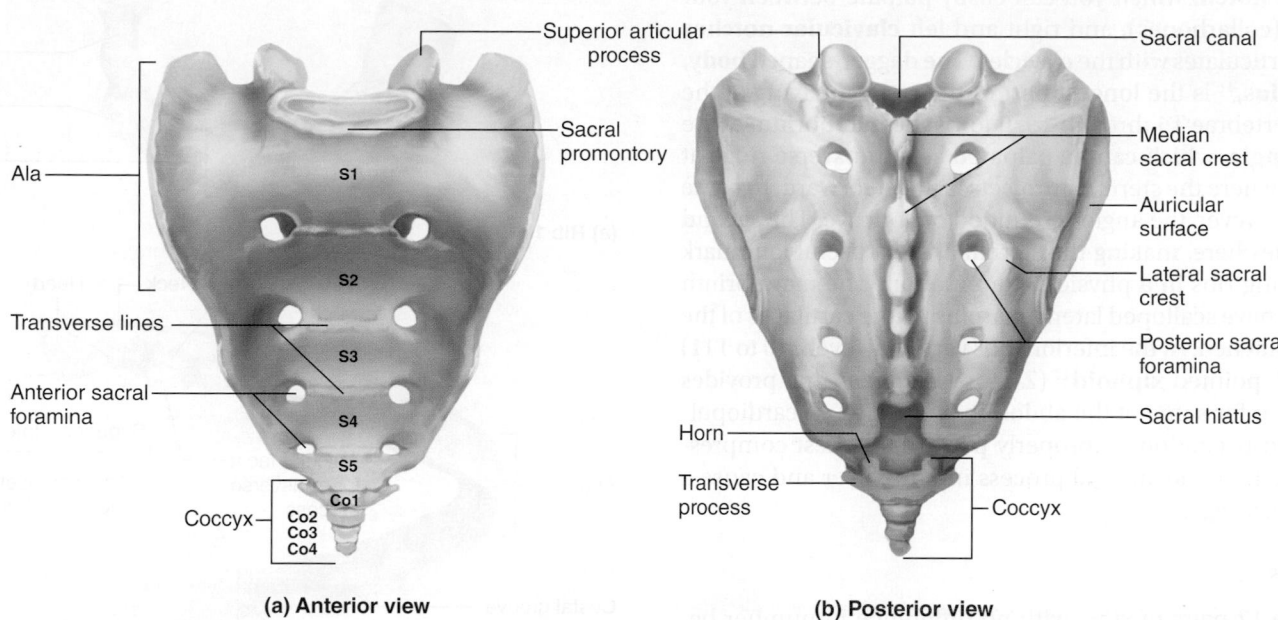

**(a) Anterior view**

**(b) Posterior view**

**FIGURE 8.26 The Sacrum and Coccyx.** (a) The anterior surface, which faces the viscera of the pelvic cavity. (b) The posterior surface. The processes of this surface can be palpated in the sacral region. AP|R

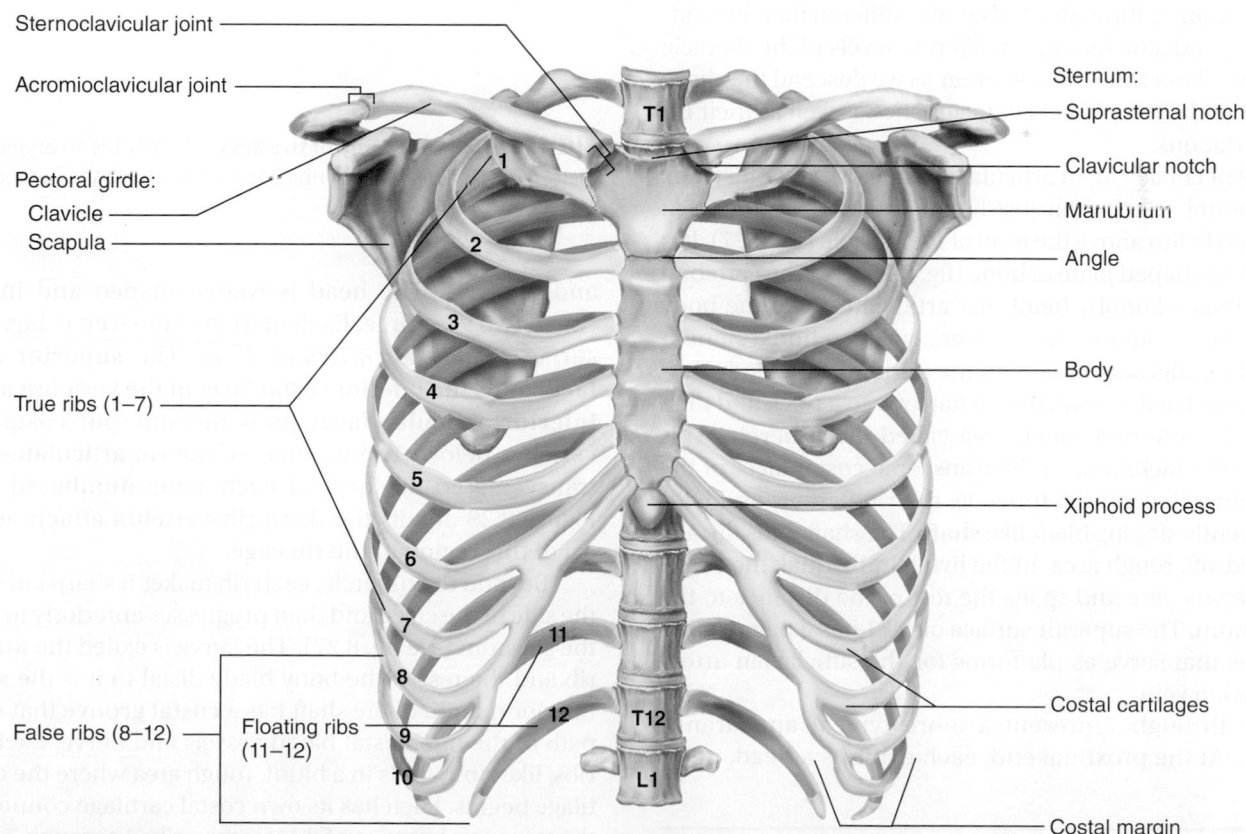

**FIGURE 8.27 The Thoracic Cage and Pectoral Girdle (Anterior View).** AP|R

**(jugular) notch,** which you can easily palpate between your clavicles (collarbones), and right and left **clavicular notches** where it articulates with the clavicles. The dagger-shaped **body,** or **gladiolus,**[41] is the longest part of the sternum, lying at the level of vertebrae T5 through T9. It joins the manubrium at the **sternal angle,** which can be palpated as a transverse ridge at the point where the sternum projects farthest forward. In some people, however, the angle is rounded or concave. The second rib attaches here, making the sternal angle a useful landmark for counting ribs in a physical examination. The manubrium and body have scalloped lateral margins where cartilages of the ribs are attached. At the inferior end (vertebral level T10 to T11) is a small, pointed **xiphoid**[42] (ZIF-oyd) **process** that provides attachment for some of the abdominal muscles. In cardiopulmonary resuscitation, improperly performed chest compressions can drive the xiphoid process into the liver and cause a fatal hemorrhage.

## The Ribs

There are 12 pairs of **ribs,** with no difference in number between the sexes. Each is attached at its posterior (proximal) end to the vertebral column, and most of them are also attached at the anterior (distal) end to the sternum. The anterior attachment is by way of a long strip of hyaline cartilage called the **costal cartilage.**

As a rule, the ribs increase in length from 1 through 7 and become progressively smaller again through rib 12. They are increasingly oblique (slanted) in orientation from 1 through 9, then less so from 10 through 12. They also differ in their individual structure and attachments at different levels of the thoracic cage, so we will examine them in order as we descend the chest, taking note of their universal characteristics as well as their individual variations.

Rib 1 is peculiar. On an articulated skeleton, you must look for its vertebral attachment just below the base of the neck; much of this rib lies above the level of the clavicle (fig. 8.27). It is a short, flat, C-shaped plate of bone (fig. 8.28a). At the vertebral end, it exhibits a knobby **head** that articulates with the body of vertebra T1. On an isolated vertebra, you can find a smooth costal facet for this attachment on the middle of the body. Immediately distal to the head, the rib narrows to a **neck** and then widens again to form a rough area called the **tubercle.** This is its point of attachment to the transverse costal facet of the same vertebra. Beyond the tubercle, the rib flattens and widens into a gently sloping bladelike **shaft.** The shaft ends distally in a squared-off, rough area. In the living individual, the costal cartilage begins here and spans the rest of the distance to the upper sternum. The superior surface of rib 1 has a pair of shallow grooves that serve as platforms for the subclavian artery and subclavian vein.

Ribs 2 through 7 present a more typical appearance (fig. 8.28b). At the proximal end, each exhibits a head, neck,

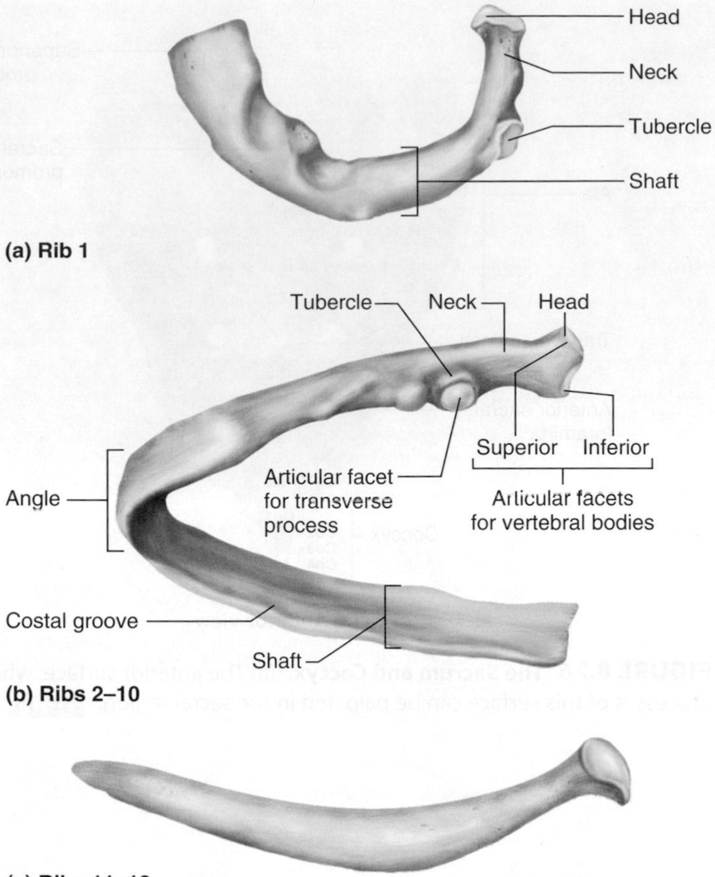

**(a) Rib 1**

**(b) Ribs 2–10**

**(c) Ribs 11–12**

**FIGURE 8.28  Anatomy of the Ribs.**  (a) Rib 1 is an atypical flat plate. (b) Typical features of ribs 2 to 10. (c) Appearance of the floating ribs, 11 and 12.

and tubercle. The head is wedge-shaped and inserts between two vertebrae. Each margin of the wedge has a smooth surface called an *articular facet.* The **superior articular facet** joins the inferior costal facet of the vertebra above; the **inferior articular facet** joins the superior costal facet of vertebra below. The tubercle of the rib articulates with the transverse costal facet of each same-numbered vertebra. Figure 8.29 details the three rib–vertebra attachments typical of this region of the rib cage.

Beyond the tubercle, each rib makes a sharp curve around the side of the chest and then progresses anteriorly to approach the sternum (see fig. 8.27). The curve is called the **angle** of the rib and the rest of the bony blade distal to it is the **shaft.** The inferior margin of the shaft has a **costal groove** that marks the path of the intercostal blood vessels and nerve. Each of these ribs, like rib 1, ends in a blunt, rough area where the costal cartilage begins. Each has its own costal cartilage connecting it to the sternum; because of this feature, ribs 1 through 7 are called **true ribs.**

---

[41]*gladiolus* = sword
[42]*xipho* = sword; *oid* = resembling

Ribs 8 through 12 are called **false ribs** because they lack independent cartilaginous connections to the sternum. In 8 through 10, the costal cartilages sweep upward and end on the costal cartilage of rib 7 (see fig. 8.27). Rib 10 also differs from 2 through 9 in that it attaches to the body of a single vertebra (T10) rather than between vertebrae. Thus, vertebra T10 has a complete costal facet on its body for rib 10.

Ribs 11 and 12 are again unusual (fig. 8.28c). Posteriorly, they articulate with the bodies of vertebrae T11 and T12, but they do not have tubercles and do not attach to the transverse processes of the vertebrae. Those two vertebrae therefore have no transverse costal facets. At the distal end, these two relatively small, delicate ribs taper to a point and are capped by a small cartilaginous tip, but there is no cartilaginous connection to the sternum or to any of the higher costal cartilages. The ribs are merely embedded in lumbar muscle at this end. Consequently, 11 and 12 are also called **floating ribs.** Among the Japanese and some other people, rib 10 is also usually floating. Table 8.4 summarizes these variations in rib anatomy and their vertebral and sternal attachments.

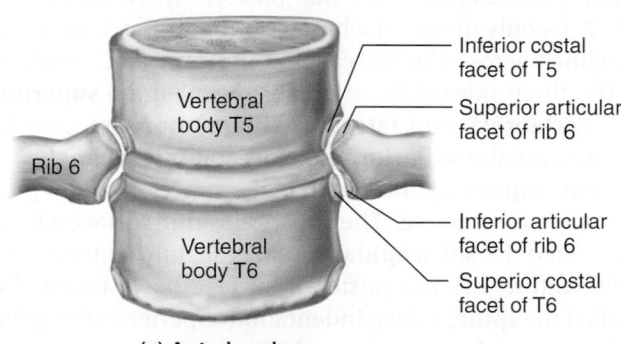

**(a) Anterior view**

Inferior costal facet of T5
Superior articular facet of rib 6
Vertebral body T5
Rib 6
Inferior articular facet of rib 6
Vertebral body T6
Superior costal facet of T6

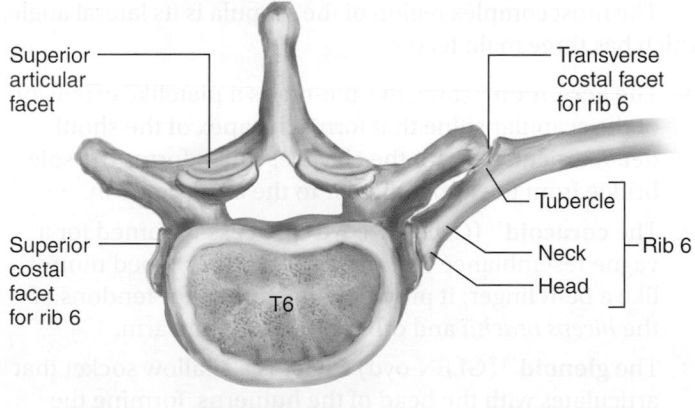

**(b) Superior view**

Superior articular facet
Transverse costal facet for rib 6
Tubercle
Superior costal facet for rib 6
Neck
Rib 6
Head
T6

**FIGURE 8.29 Articulation of Rib 6 with Vertebrae T5 and T6.**
(a) Anterior view. Note the relationship of the articular facets of the rib with the costal facets of the two vertebrae. (b) Superior view. Note that the rib articulates with a vertebra at two points: the costal facet on the vertebral body and the transverse costal facet on the transverse process.

┌─────────────────────────────────────┐
**BEFORE YOU GO ON**
└─────────────────────────────────────┘

Answer the following questions to test your understanding of the preceding section:

9. Discuss the contribution of the intervertebral discs to the length and flexibility of the spine.

10. Construct a three-column table headed C4, T4, and L4. In each column, list all anatomical features that would distinguish that vertebra from the other two.

11. Name the three parts of the sternum. How many ribs attach (directly or indirectly) to each part?

12. Describe how rib 5 articulates with the spine. How do ribs 1 and 12 differ from this and from each other in their modes of articulation?

13. Distinguish between true, false, and floating ribs. Which ribs fall into each category?

14. Name the three divisions of the sternum and list the sternal features that can be palpated on a living person.

## **8.4** The Pectoral Girdle and Upper Limb

### Expected Learning Outcome

When you have completed this section, you should be able to

a. identify and describe the features of the clavicle, scapula, humerus, radius, ulna, and bones of the wrist and hand.

### The Pectoral Girdle

The **pectoral girdle** (shoulder girdle) supports the arm and links it to the axial skeleton. It consists of two bones on each side of the body: the *clavicle* (collarbone) and *scapula* (shoulder blade). The medial end of the clavicle articulates with the sternum at the **sternoclavicular joint,** and its lateral end articulates with the scapula at the **acromioclavicular joint** (see fig. 8.27). The scapula also articulates with the humerus at the **glenohumeral joint.** These are loose attachments that result in a shoulder far more flexible than that of most other mammals, but they also make the shoulder joint easy to dislocate.

#### ▶▶▶ APPLY WHAT YOU KNOW

*How is the unusual flexibility of the human shoulder joint related to the habitat of our primate ancestors?*

### The Clavicle

The **clavicle**[43] (fig. 8.30) is slightly S-shaped, somewhat flattened from the upper to lower surface, and easily seen and palpated on the upper thorax (see fig. B.1b, p. 377). The superior surface is relatively smooth and rounded, whereas the inferior

---
[43]*clav* = hammer, club, key; *icle* = little

| TABLE 8.4 | | Articulations of the Ribs | | | |
|---|---|---|---|---|---|
| Rib | Type | Costal Cartilage | Articulating Vertebral Bodies | Articulating with a Transverse Costal Facet? | Rib Tubercle |
| 1 | True | Individual | T1 | Yes | Present |
| 2 | True | Individual | T1 and T2 | Yes | Present |
| 3 | True | Individual | T2 and T3 | Yes | Present |
| 4 | True | Individual | T3 and T4 | Yes | Present |
| 5 | True | Individual | T4 and T5 | Yes | Present |
| 6 | True | Individual | T5 and T6 | Yes | Present |
| 7 | True | Individual | T6 and T7 | Yes | Present |
| 8 | False | Shared with rib 7 | T7 and T8 | Yes | Present |
| 9 | False | Shared with rib 7 | T8 and T9 | Yes | Present |
| 10 | False | Shared with rib 7 | T10 | Yes | Present |
| 11 | False, floating | None | T11 | No | Absent |
| 12 | False, floating | None | T12 | No | Absent |

surface is flatter and marked by grooves and ridges for muscle attachment. The medial **sternal end** has a rounded, hammer-like head, and the lateral **acromial end** is markedly flattened. Near the acromial end is a rough tuberosity called the **conoid tubercle**—a ligament attachment that faces toward the rear and slightly downward.

The clavicle braces the shoulder, keeping the upper limb away from the midline of the body. It also transfers force from the arm to the axial region of the body, as when doing push-ups. It is thickened in people who do heavy manual labor, and because most people are right-handed, the right clavicle is usually stronger and shorter than the left. Without the clavicles, the pectoralis major muscles would pull the shoulders forward and medially—which indeed happens when a clavicle is fractured. The clavicle is the most commonly fractured bone in the body because it is so close to the surface and because people often reach out with their arms to break a fall.

## The Scapula

The **scapula**[44] (fig. 8.31), named for its resemblance to a spade or shovel, is a triangular plate that posteriorly overlies ribs 2 through 7. Its only direct attachment to the thorax is by muscles; it glides across the rib cage as the arm and shoulder move. The three sides of the triangle are called the **superior, medial (vertebral),** and **lateral (axillary) borders,** and its three angles are the **superior, inferior,** and **lateral angles.** A conspicuous **suprascapular notch** in the superior border provides passage for a nerve. The broad anterior surface of the scapula, called the **subscapular fossa,** is slightly concave and relatively featureless. The posterior surface has a transverse ridge called the **spine,** a deep indentation superior to the spine called the **supraspinous fossa,** and a broad surface inferior to it called the **infraspinous fossa.**[45]

The most complex region of the scapula is its lateral angle, which has three main features:

1. The **acromion**[46] (ah-CRO-me-on) is a platelike extension of the scapular spine that forms the apex of the shoulder. It articulates with the clavicle, which forms the sole bridge from the appendicular to the axial skeleton.

2. The **coracoid**[47] (COR-uh-coyd) **process** is named for a vague resemblance to a crow's beak, but shaped more like a bent finger; it provides attachment for tendons of the *biceps brachii* and other muscles of the arm.

3. The **glenoid**[48] (GLEN-oyd) **cavity** is a shallow socket that articulates with the head of the humerus, forming the glenohumeral joint.

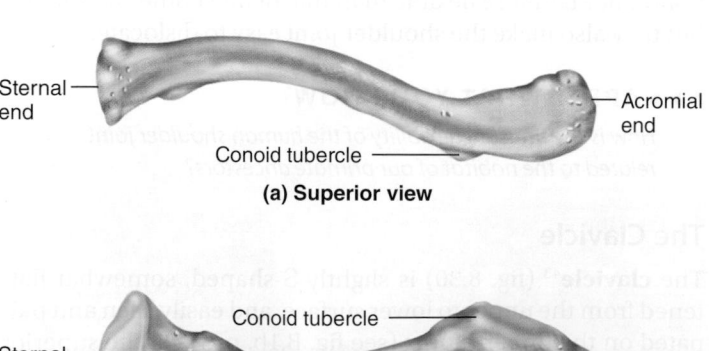

Sternal end

Acromial end

Conoid tubercle

**(a) Superior view**

Conoid tubercle

Sternal end

Acromial end

**(b) Inferior view**

**FIGURE 8.30 The Right Clavicle.** AP|R

---

[44]*scap* = spade, shovel; *ula* = little
[45]*supra* = above; *infra* = below
[46]*acr* = extremity, point; *omi* = shoulder
[47]*corac* = crow; *oid* = resembling
[48]*glen* = pit, socket; *oid* = resembling

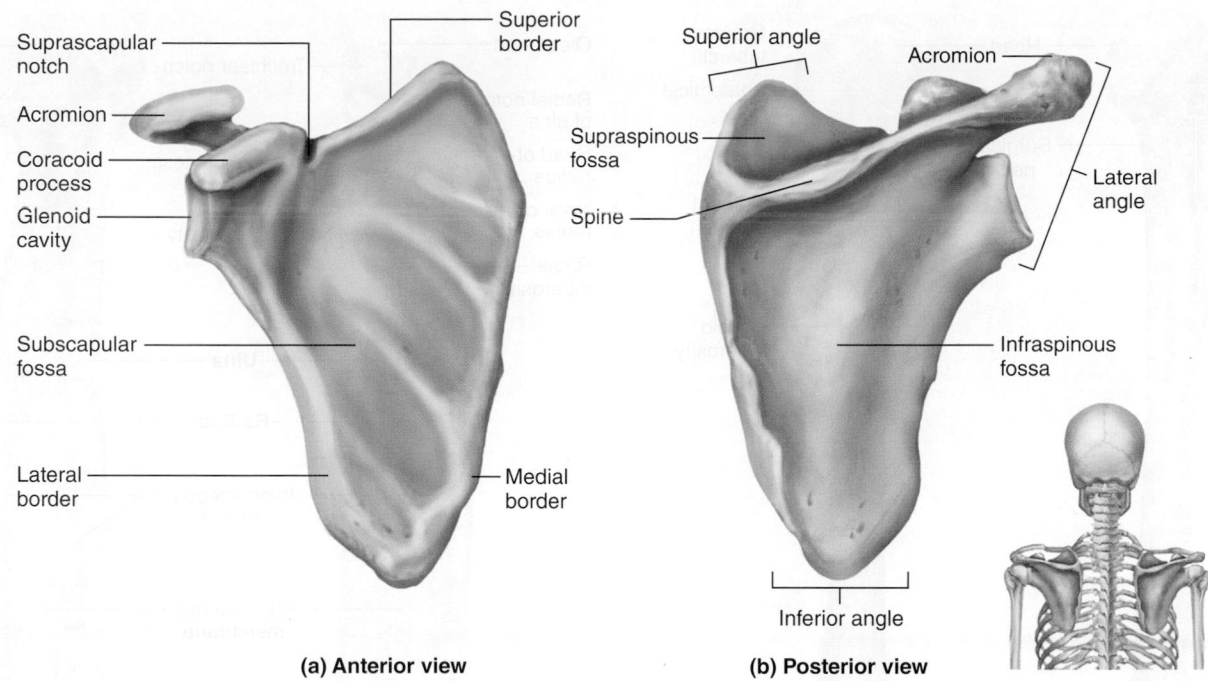

**FIGURE 8.31 The Right Scapula.** AP|R

## The Upper Limb

The upper limb is divided into four segments containing a total of 30 bones per limb:

1. The **brachium**[49] (BRAY-kee-um), or arm proper, extends from shoulder to elbow. It contains only one bone, the *humerus.*

2. The **antebrachium,**[50] or forearm, extends from elbow to wrist and contains two bones: the *radius* and *ulna.* In anatomical position, these bones are parallel and the radius is lateral to the ulna.

3. The **carpus,**[51] or wrist, contains eight small *carpal bones* arranged in two rows.

4. The **manus,**[52] or hand, contains 19 bones in two groups: 5 *metacarpals* in the palm and 14 *phalanges* in the fingers.

## The Humerus

The **humerus** has a hemispherical **head** that articulates with the glenoid cavity of the scapula (fig. 8.32). The smooth surface of the head (covered with articular cartilage in life) is bordered by a groove called the **anatomical neck.** Other prominent features of the proximal end are muscle attachments called the **greater** and **lesser tubercles** and an **intertubercular sulcus** between them that accommodates a tendon of the biceps

muscle. The **surgical neck,** a common fracture site, is a narrowing of the bone just distal to the tubercles, at the transition from the head to the shaft. The shaft has a rough area called the **deltoid tuberosity** on its lateral surface. This is an insertion for the *deltoid muscle* of the shoulder.

The distal end of the humerus has two smooth condyles. The lateral one, called the **capitulum**[53] (ca-PIT-you-lum), is shaped like a wide tire and articulates with the radius. The medial one, called the **trochlea**[54] (TROCK-lee-uh), is pulleylike and articulates with the ulna. Immediately proximal to these condyles, the humerus flares out to form two bony processes, the **lateral** and **medial epicondyles.** The medial epicondyle protects the *ulnar nerve,* which passes close to the surface across the back of the elbow. This epicondyle is popularly known as the "funny bone" because striking the elbow on the edge of a table stimulates the ulnar nerve and produces a sharp tingling sensation. Immediately proximal to the epicondyles, the margins of the humerus are called the **lateral** and **medial supracondylar ridges.** These are attachments for certain forearm muscles.

The distal end of the humerus also shows three deep pits: two anterior and one posterior. The posterior pit, called the **olecranon** (oh-LEC-ruh-non) **fossa,** accommodates a process of the ulna called the *olecranon* when the elbow is extended. On the anterior surface, a medial pit called the **coronoid fossa** accommodates the *coronoid process* of the ulna when the forearm is flexed. The lateral pit is the **radial fossa,** named for the nearby head of the radius.

---

[49]*brachi* = arm

[50]*ante* = before

[51]*carp* = wrist

[52]*man* = hand

[53]*capit* = head; *ulum* = little

[54]*troch* = wheel, pulley

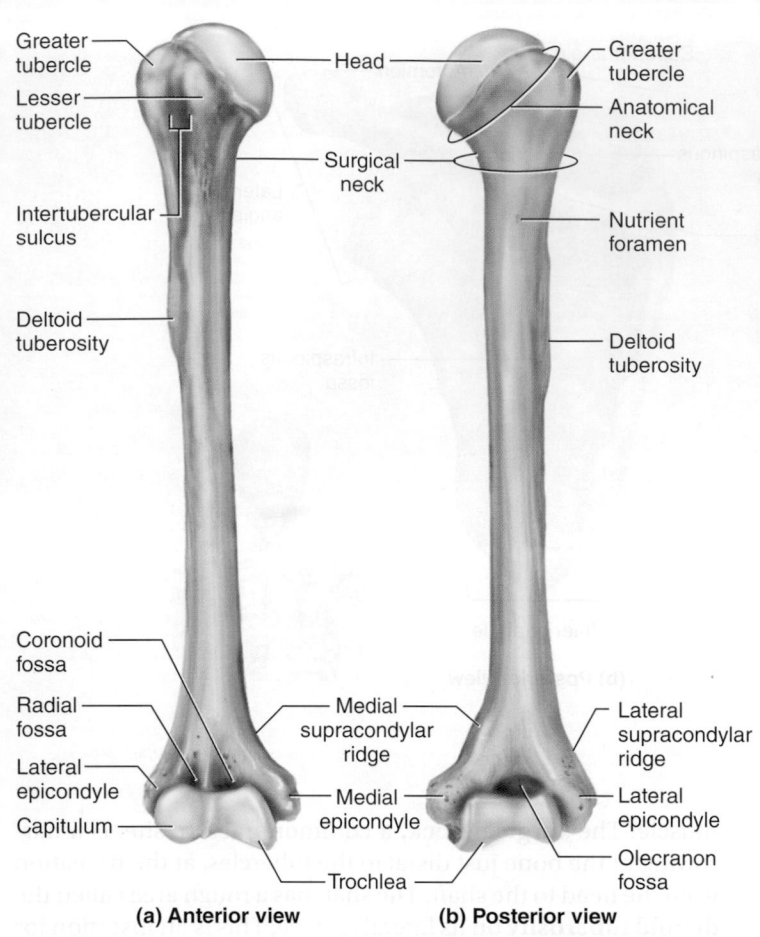

Greater tubercle
Lesser tubercle
Intertubercular sulcus
Deltoid tuberosity
Coronoid fossa
Radial fossa
Lateral epicondyle
Capitulum
Head
Surgical neck
Medial supracondylar ridge
Medial epicondyle
Trochlea
Greater tubercle
Anatomical neck
Nutrient foramen
Deltoid tuberosity
Lateral supracondylar ridge
Lateral epicondyle
Olecranon fossa

(a) Anterior view    (b) Posterior view

**FIGURE 8.32** The Right Humerus.

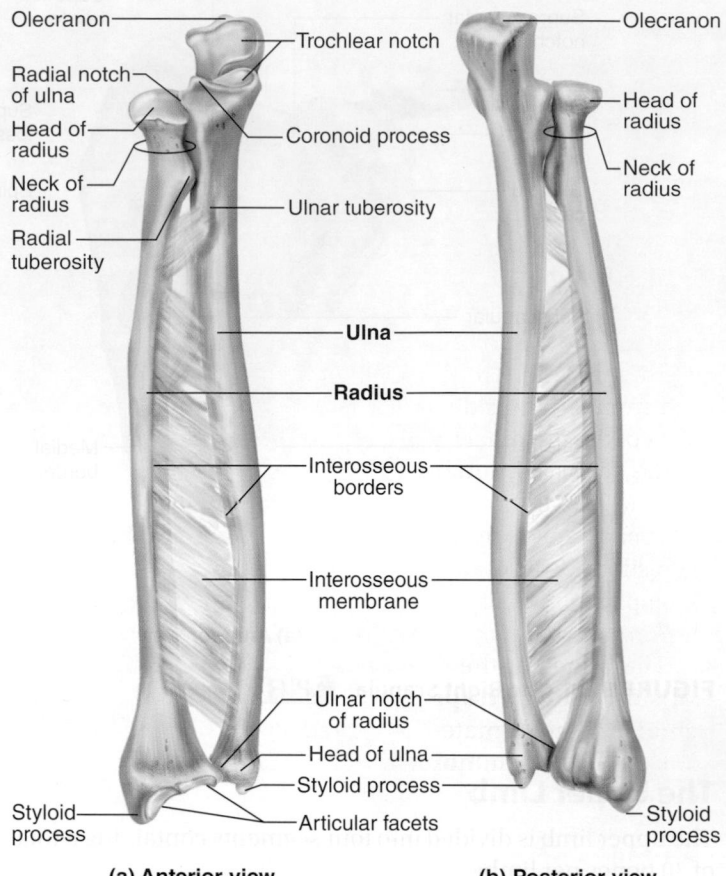

Olecranon
Radial notch of ulna
Head of radius
Neck of radius
Radial tuberosity
Trochlear notch
Coronoid process
Ulnar tuberosity
**Ulna**
**Radius**
Interosseous borders
Interosseous membrane
Ulnar notch of radius
Head of ulna
Styloid process
Articular facets
Styloid process
Olecranon
Head of radius
Neck of radius
Styloid process

(a) Anterior view    (b) Posterior view

**FIGURE 8.33** The Right Radius and Ulna. AP|R

## The Radius

The **radius** has a distinctive discoidal **head** at its proximal end (fig. 8.33). When the forearm is rotated so the palm turns forward and back, the circular superior surface of this disc spins on the capitulum of the humerus, and the edge of the disc spins on the radial notch of the ulna. Immediately distal to the head, the radius has a narrower **neck** and then widens to a rough prominence, the **radial tuberosity,** on its medial surface. The distal tendon of the biceps muscle terminates on this tuberosity.

The distal end of the radius has the following features, from lateral to medial:

1. a bony point, the **styloid process,** which can be palpated proximal to the thumb;

2. two shallow depressions (articular facets) that articulate with the scaphoid and lunate bones of the wrist; and

3. the **ulnar notch,** which articulates with the end of the ulna.

## The Ulna

At the proximal end of the **ulna** (fig. 8.33) is a deep, C-shaped **trochlear notch** that wraps around the trochlea of the humerus. The posterior side of this notch is formed by a prominent **olecranon**—the bony point where you rest your elbow on a table. The anterior side is formed by a less prominent **coronoid process.** Laterally, the head of the ulna has a less conspicuous **radial notch,** which accommodates the edge of the head of the radius. At the distal end of the ulna is a medial **styloid process.** The bony lumps you can palpate on each side of your wrist are the styloid processes of the radius and ulna.

The radius and ulna are attached along their shafts by a ligament called the **interosseous** (IN-tur-OSS-ee-us) **membrane (IM),** which is attached to an angular ridge called the **interosseous margin** on each bone. Most fibers of the IM are oriented obliquely, slanting upward from the ulna to the radius. If you lean forward on a table supporting your weight on your hands, about 80% of the force is borne by the radius. This tenses the IM, which pulls the ulna upward and transfers some of this force through the ulna to the humerus. The IM thereby enables two elbow joints (humeroradial and humeroulnar) to share the load and reduces the wear and tear that one joint would otherwise have to bear alone. The IM also serves as an attachment for several forearm muscles.

## The Carpal Bones

The **carpal bones** are arranged in two rows of four bones each (fig. 8.34). Although they are colloquially called wrist bones, the narrow point where one might wear a wristwatch is at the distal end of the radius and ulna. The carpal bones are in the base of the hand. These short bones allow movements of the wrist from side to side and anterior to posterior. The carpal bones of the proximal row, starting at the lateral (thumb) side, are the **scaphoid, lunate, triquetrum** (tri-QUEE-trum), and **pisiform** (PY-sih-form)—Latin for "boat-," "moon-," "triangle-," and "pea-shaped," respectively. Unlike the other carpal bones, the pisiform is a sesamoid bone; it is not present at birth but develops around the age of 9 to 12 years within the tendon of the *flexor carpi ulnaris muscle.*

The bones of the distal row, again starting on the lateral side, are the **trapezium,**[55] **trapezoid, capitate,**[56] and **hamate.**[57] The hamate can be recognized by a prominent hook called the **hamulus** on the palmar side (fig. 8.34b). The hamulus is an attachment for the *flexor retinaculum,* a fibrous sheet in the wrist that covers the carpal tunnel (see fig. 10.30, p. 352).

## The Metacarpal Bones

Bones of the palm are called **metacarpals.**[58] Metacarpal I is located proximal to the base of the thumb and metacarpal V proximal to the base of the little finger. On a skeleton, the metacarpals look like extensions of the fingers, making the fingers seem much longer than they really are. The proximal end of a metacarpal bone is called the **base,** the shaft is called the **body,** and the distal end is called the **head.** The heads of the metacarpals form knuckles when you clench your fist.

## The Phalanges

The bones of the fingers are called **phalanges** (fah-LAN-jeez), in the singular, *phalanx* (FAY-lanks). There are two phalanges in the **pollex** (thumb) and three in each of the other digits. Phalanges are identified by roman numerals preceded by *proximal, middle,* and *distal.* For example, proximal phalanx I

is in the basal segment of the thumb (the first segment beyond the web between the thumb and palm); the left proximal phalanx IV is where people usually wear wedding rings; and distal phalanx V forms the tip of the little finger. The three parts of a phalanx are the same as in a metacarpal: base, body, and head. The anterior (palmar) surface of a phalanx is slightly concave from end to end and flattened from side to side; the posterior surface is rounder and slightly convex.

### BEFORE YOU GO ON

Answer the following questions to test your understanding of the preceding section:

15. Describe how to distinguish the medial and lateral ends of the clavicle from each other, and how to distinguish its superior and inferior surfaces.
16. Name the three fossae of the scapula and describe the location of each.
17. What three bones meet at the elbow? Identify the fossae, articular surfaces, and processes of this joint and state to which bone each of these features belongs.
18. Name the four carpal bones of the proximal row from lateral to medial, then the four bones of the distal row in the same order.
19. Name the four bones from the tip of the little finger to the base of the hand on that side.

## 8.5 The Pelvic Girdle and Lower Limb

### Expected Learning Outcomes
When you have completed this section, you should be able to

a. identify and describe the features of the pelvic girdle, femur, patella, tibia, fibula, and bones of the foot; and
b. compare the anatomy of the male and female pelvic girdles and explain the functional significance of the differences.

## The Pelvic Girdle

The terms *pelvis* and *pelvic girdle* are used in contradictory ways by various anatomical authorities. Here we will follow the practice of *Gray's Anatomy* and the *Terminologia Anatomica* in considering the **pelvic girdle** to consist of a complete ring composed of three bones (fig. 8.35)—two **hip (coxal) bones** and the sacrum (which of course is also part of the vertebral column). The hip bones are also frequently called the *ossa*

---

[55]*trapez* = table, grinding surface
[56]*capit* = head; *ate* = possessing
[57]*ham* = hook; *ate* = possessing
[58]*meta* = beyond; *carp* = wrist

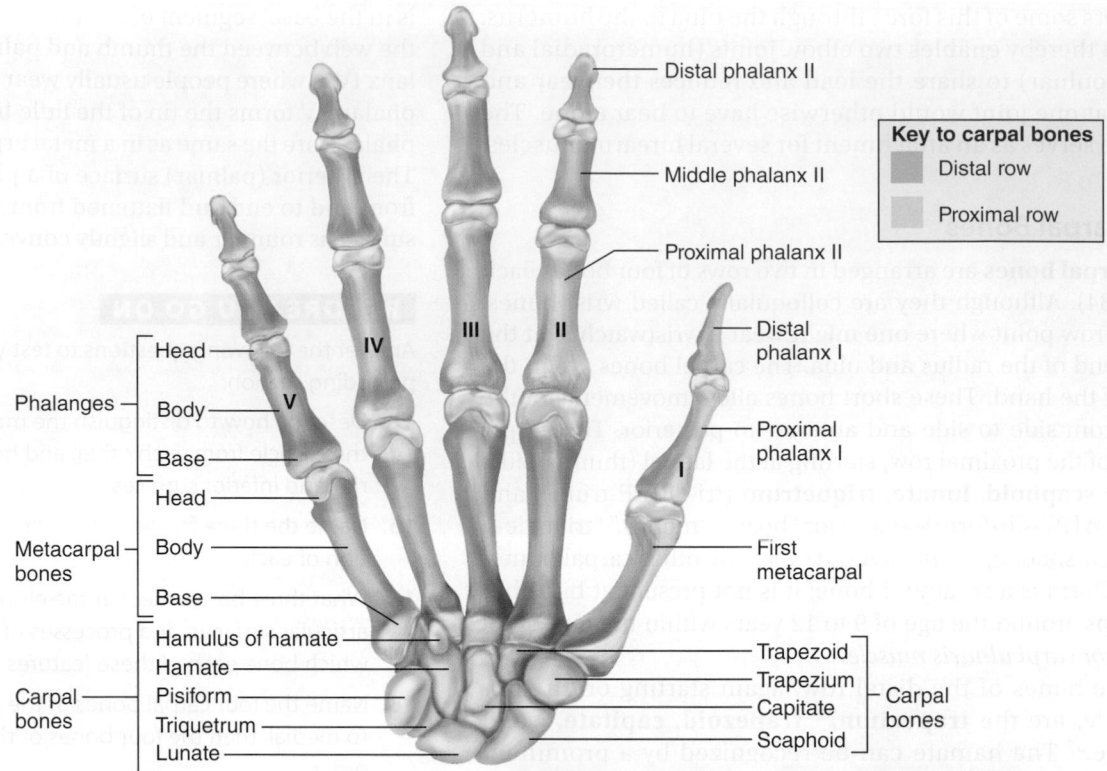

**(a) Anterior view**

Distal phalanx II
Middle phalanx II
Proximal phalanx II
Distal phalanx I
Proximal phalanx I
Phalanges
Head
Body
Base
Metacarpal bones
Head
Body
Base
Hamulus of hamate
Hamate
Pisiform
Triquetrum
Lunate
Carpal bones
First metacarpal
Trapezoid
Trapezium
Capitate
Scaphoid
Carpal bones

**Key to carpal bones**
Distal row
Proximal row

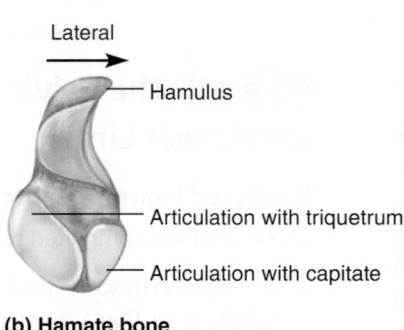

**(b) Hamate bone**

Lateral
Hamulus
Articulation with triquetrum
Articulation with capitate

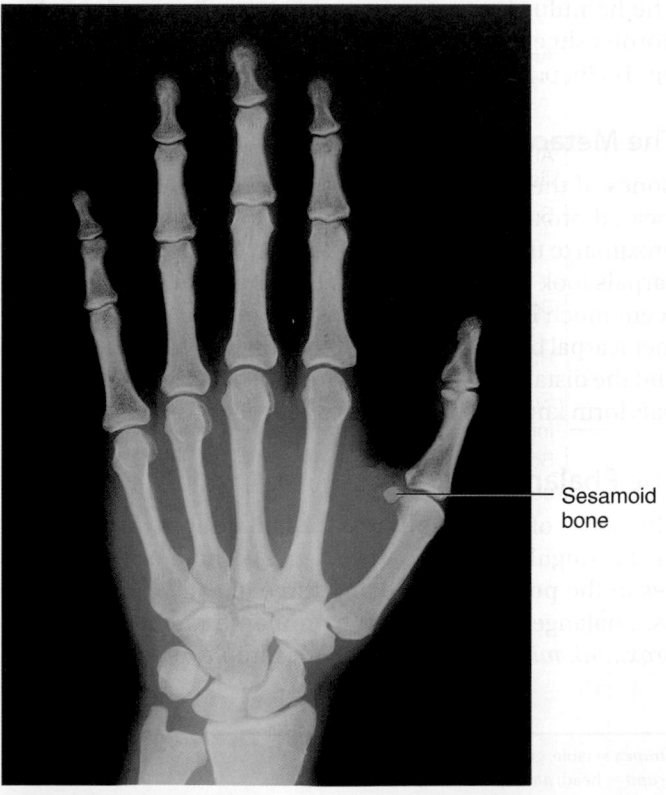

Sesamoid bone

**(c) X-ray of adult hand**

**FIGURE 8.34 The Right Wrist and Hand (Anterior, or Palmar, View).** (a) Carpal bones are color-coded to distinguish the proximal (yellow) and distal (green) rows. Some people remember the names of the carpal bones with the mnemonic, "Sally left the party to take Charlie home." The first letters of these words correspond to the first letters of the carpal bones, from lateral to medial, proximal row first. (b) The right hamate bone, viewed from the proximal side of the wrist to show its distinctive hook. This unique bone is a useful landmark for locating the others when studying the skeleton. (c) Color-enhanced X-ray of an adult hand. Identify the unlabeled bones in the X-ray by comparing it with the drawing in part (a). A small sesamoid bone is visible at the base of the thumb. The pisiform bone in part (a) is also a sesamoid bone. **AP|R**

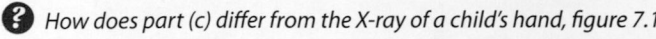 *How does part (c) differ from the X-ray of a child's hand, figure 7.11?*

*coxae*[59] (OS-sa COC-see) or *innominate*[60] (ih-NOM-ih-nate) *bones;* the latter is arguably the most self-contradictory term in human anatomy (the "bones with no name"). The **pelvis**[61] is a bowl-shaped structure composed of these bones as well as their ligaments and the muscles that line the pelvic cavity and form its floor. The pelvic girdle supports the trunk on the lower limbs and encloses and protects the viscera of the pelvic cavity—mainly the lower colon, urinary bladder, and internal reproductive organs.

Each hip bone is joined to the vertebral column at one point, the sacroiliac joint, where its **auricular surface** matches the auricular surface of the sacrum. The two hip bones articulate with each other on the anterior side of the pelvis, where they are joined by a pad of fibrocartilage called the **interpubic**

**disc.** The interpubic disc and the adjacent region of each pubic bone constitute the **pubic symphysis**,[62] which can be palpated as a hard prominence immediately above the genitalia.

The pelvis has a bowl-like shape with the broad **greater (false) pelvis** between the flare of the hips, and the narrower **lesser (true) pelvis** below. The two are separated by a round margin called the **pelvic brim.** The opening circumscribed by the brim is called the **pelvic inlet**—an entry into the lesser pelvis through which an infant's head passes during birth. The lower margin of the lesser pelvis is called the **pelvic outlet.**

The hip bones have three distinctive features that will serve as landmarks for further description. These are the **iliac**[63] **crest** (superior crest of the hip); **acetabulum**[64] (ASS-eh-TAB-you-lum) (the hip socket—named for its resemblance to

---

[59]*os* = bone; *coxae* = of the hip
[60]*in* = without; *nomin* = name; *ate* = having
[61]*pelv* = basin, bowl

[62]*sym* = together; *physis* = growth
[63]*ili* = flank, loin; *ac* = pertaining to
[64]*acetabulum* = vinegar cup

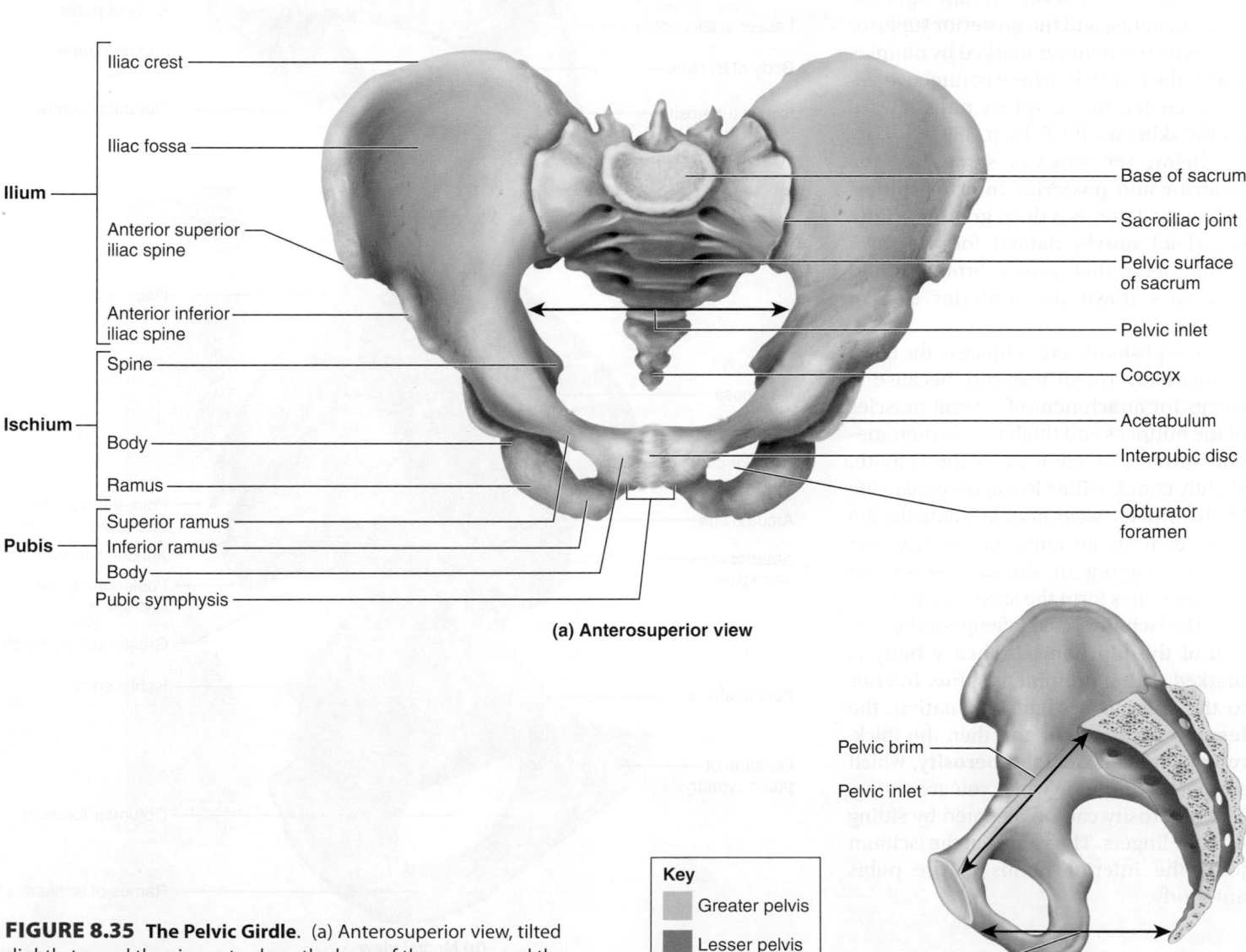

(a) Anterosuperior view

**Key**
Greater pelvis
Lesser pelvis

Pelvic brim
Pelvic inlet
Pelvic outlet

(b) Median section

**FIGURE 8.35 The Pelvic Girdle**. (a) Anterosuperior view, tilted slightly toward the viewer to show the base of the sacrum and the pelvic inlet. (b) Median section, to show the greater and lesser pelvis and the pelvic inlet and outlet. AP|R

vinegar cups used on ancient Roman dining tables); and **obturator**[65] **foramen** (a large round-to-triangular hole below the acetabulum, closed by a ligament called the *obturator membrane* in life).

The adult hip bone forms by the fusion of three childhood bones called the *ilium* (ILL-ee-um), *ischium* (ISS-kee-um), and *pubis* (PEW-biss), identified by color in figure 8.36. The largest of these is the **ilium,** which extends from the iliac crest to the center of the acetabulum. The iliac crest extends from an anterior point or angle called the **anterior superior spine** to a sharp posterior angle called the **posterior superior spine.** In a lean person, the anterior superior spines form visible anterior protrusions at a point where the front pockets usually open on a pair of pants, and the posterior superior spines are sometimes marked by dimples above the buttocks where connective tissue attached to the spines pulls inward on the skin (see fig. B.15, p. 389).

Below the superior spines are the **anterior** and **posterior inferior spines.** Below the latter is a deep **greater sciatic** (sy-AT-ic) **notch,** named for the thick sciatic nerve that passes through it and continues down the posterior side of the thigh.

The posterolateral surface of the ilium is relatively rough-textured because it serves for attachment of several muscles of the buttocks and thighs. The anteromedial surface, by contrast, is the smooth, slightly concave **iliac fossa,** covered in life by the broad *iliacus muscle.* Medially, the ilium exhibits an auricular surface that matches the one on the sacrum, so that the two bones form the sacroiliac joint.

The **ischium** is the inferoposterior portion of the hip bone. Its heavy **body** is marked with a prominent **spine.** Inferior to the spine is a slight indentation, the **lesser sciatic notch,** and then the thick, rough-surfaced **ischial tuberosity,** which supports your body when you are sitting. The tuberosity can be palpated by sitting on your fingers. The **ramus** of the ischium joins the inferior ramus of the pubis anteriorly.

[65]*obtur* = to close, stop up; *ator* = that which

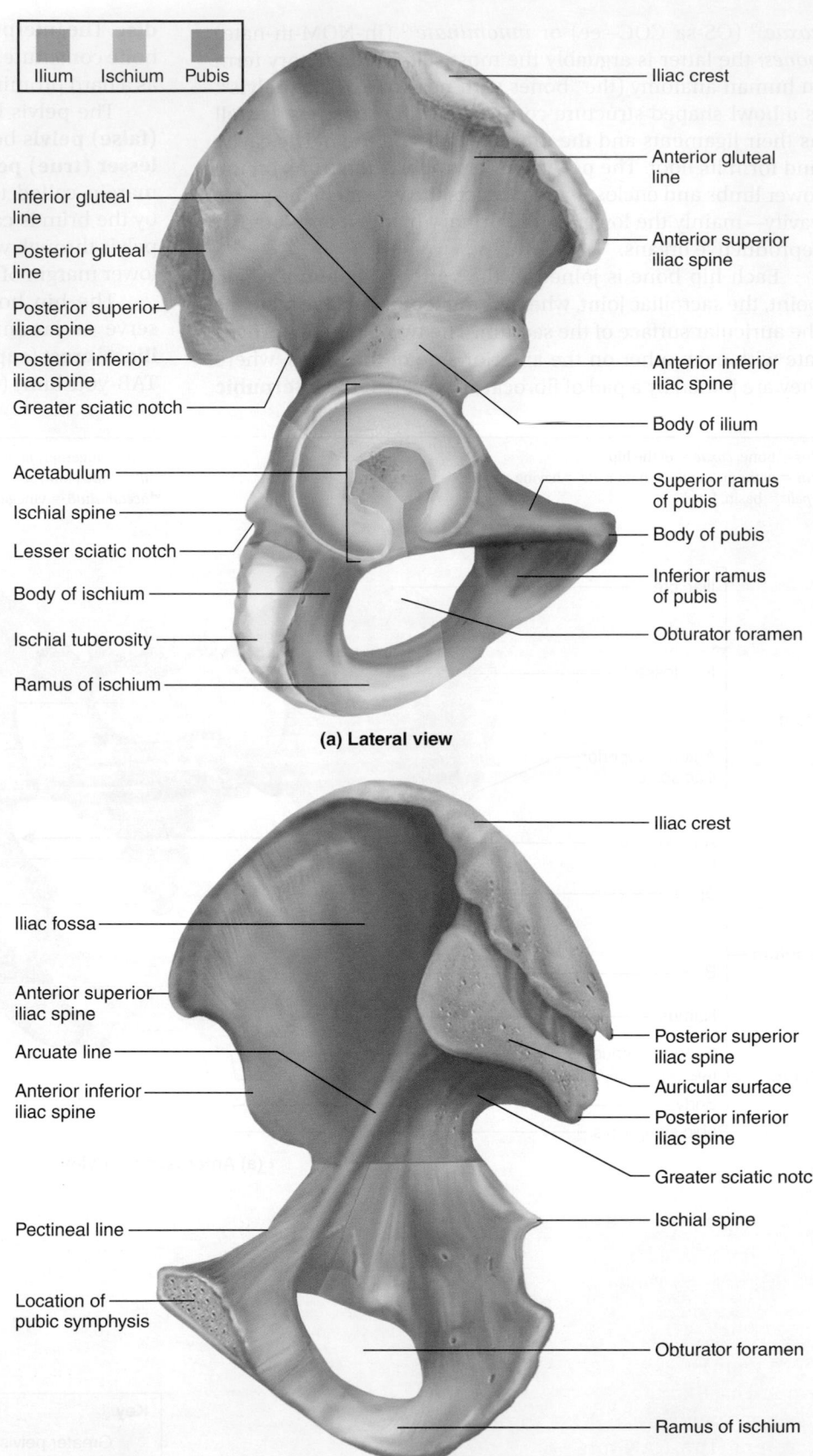

Ilium    Ischium    Pubis

**(a) Lateral view**

Iliac crest
Anterior gluteal line
Anterior superior iliac spine
Anterior inferior iliac spine
Body of ilium
Superior ramus of pubis
Body of pubis
Inferior ramus of pubis
Obturator foramen

Inferior gluteal line
Posterior gluteal line
Posterior superior iliac spine
Posterior inferior iliac spine
Greater sciatic notch
Acetabulum
Ischial spine
Lesser sciatic notch
Body of ischium
Ischial tuberosity
Ramus of ischium

**(b) Medial view**

Iliac crest
Posterior superior iliac spine
Auricular surface
Posterior inferior iliac spine
Greater sciatic notch
Ischial spine
Obturator foramen
Ramus of ischium

Iliac fossa
Anterior superior iliac spine
Arcuate line
Anterior inferior iliac spine
Pectineal line
Location of pubic symphysis

**FIGURE 8.36 The Right Hip Bone.** The three childhood bones that fuse to form the adult hip bone are identified by color according to the key at the top. **AP|R**

The **pubis (pubic bone)** is the most anterior portion of the hip bone. In anatomical position, it is nearly horizontal and serves as a platform for the urinary bladder. It has a **superior** and **inferior ramus** and a triangular **body.** The body of one pubis meets the body of the other at the pubic symphysis. The pubis and ischium encircle the obturator foramen. The pubis is often fractured when the pelvis is subjected to violent antero-posterior compression, as in seat-belt injuries.

The pelvis is the most *sexually dimorphic* part of the skeleton—that is, the one whose anatomy most differs between the sexes. In identifying the sex of skeletal remains, forensic scientists focus especially on the pelvis. The average male pelvis is more robust (heavier and thicker) than the female's owing to the forces exerted on the bones by stronger muscles. The female pelvis is adapted to the needs of pregnancy and childbirth. It is wider and shallower and has a larger pelvic inlet and outlet for passage of the infant's head. Table 8.5 and figure 8.37 summarize the most useful features of the pelvis in sex identification.

## The Lower Limb

The number and arrangement of bones in the lower limb are similar to those of the upper limb. In the lower limb, however, they are adapted for weight bearing and locomotion and are therefore shaped and articulated differently. The femur and tibia are essentially pillars for supporting the weight of the body. The lower limb is divided into four regions containing a total of 30 bones per limb:

1. The **femoral region,** or thigh, extends from hip to knee and contains the *femur.* The *patella* (kneecap) is a sesamoid bone at the junction of the femoral and crural regions.

2. The **crural** (CROO-rul) **region,** or leg proper, extends from knee to ankle and contains two bones, the medial *tibia* and lateral *fibula.*

3. The **tarsal region (tarsus),** or ankle, is the union of the crural region with the foot. The tarsal bones are treated as part of the foot.

4. The **pedal region (pes),** or foot, is composed of 7 *tarsal bones,* 5 *metatarsals,* and 14 *phalanges* in the toes.

## The Femur

The **femur** (FEE-mur) is the longest and strongest bone of the body, measuring about one-quarter of one's height (fig. 8.38). It has a hemispherical head that articulates with the acetabulum of the pelvis, forming a quintessential *ball-and-socket joint.* A ligament extends from the acetabulum to a pit, the **fovea capitis**[66] (FOE-vee-uh CAP-ih-tiss), in the head of the femur. Distal to the head is a constricted **neck** and then two massive, rough processes called the **greater** and **lesser trochanters**[67] (tro-CAN-turs), which are insertions for the powerful muscles of the hip. The trochanters are connected on the posterior side by a thick oblique ridge of bone, the **intertrochanteric crest,** and on the anterior side by a more delicate **intertrochanteric line.**

The primary feature of the shaft is a posterior ridge called the **linea aspera**[68] (LIN-ee-uh ASS-peh-ruh) at its midpoint. At its upper end, the linea aspera forks into a medial **spiral (pectineal) line** and a lateral **gluteal tuberosity.** The gluteal tuberosity is a rough ridge (sometimes a depression) that serves for attachment of the powerful *gluteus maximus* muscle of the buttock. At its lower end, the linea aspera forks into **medial** and **lateral supracondylar lines,** which continue down to the respective epicondyles.

The **medial** and **lateral epicondyles** are the widest points of the femur at the knee. These and the supracondylar lines are attachments for certain thigh and leg muscles and knee ligaments. At the distal end of the femur are two smooth round surfaces of the knee joint, the **medial** and **lateral condyles,** separated by a groove called the **intercondylar** (IN-tur-CON-dih-lur) **fossa.** During knee flexion and extension, the condyles rock on the superior surface of the tibia. On the anterior side of the femur, a smooth medial depression called the **patellar surface** articulates with the patella. On the posterior side is a flat or slightly depressed area called the **popliteal surface.**

## The Patella

The **patella,**[69] or kneecap (fig. 8.38), is a roughly triangular sesamoid bone embedded in the tendon of the knee. It is cartilaginous at birth and ossifies at 3 to 6 years of age. It has a broad superior **base,** a pointed inferior **apex,** and a pair of shallow **articular facets** on its posterior surface where it articulates with the femur. The lateral facet is usually larger than the medial. The *quadriceps femoris tendon* extends from the anterior *quadriceps femoris muscle* of the thigh to the patella, and it continues as the *patellar ligament* from the patella to the tibia. This is a change in terminology more than a change in structure or function, as a tendon connects muscle to bone and a ligament connects bone to bone. Because of the way the quadriceps tendon loops over the patella, the patella modifies the direction of pull by the quadriceps muscle and improves its efficiency in extending the knee.

## The Tibia

The leg has two bones: a thick strong tibia (TIB-ee-uh) on the medial side and a slender fibula (FIB-you-luh) on the lateral side (fig. 8.39). The **tibia** is the only weight-bearing bone of the crural region. Its broad superior head has two fairly flat articular surfaces, the **medial** and **lateral condyles,** separated by a ridge called the **intercondylar eminence.** The condyles of the tibia articulate with those of the femur. The rough anterior surface of the tibia, the **tibial tuberosity,** can be palpated just below the patella. This is an attachment for the powerful thigh muscles that extend (straighten) the knee. Distal to this, the shaft has a sharply angular **anterior border,** which can be palpated in the shin. At the ankle, just above the rim of a standard dress shoe, you can palpate

---

[66]*fovea* = pit; *capitis* = of the head
[67]*trochanter* = to run
[68]*linea* = line; *asper* = rough
[69]*pat* = pan; *ella* = little

| TABLE 8.5 | Comparison of the Male and Female Pelvic Girdles | |
|---|---|---|
| **Feature** | **Male** | **Female** |
| General appearance | More massive; rougher; heavier processes | Less massive; smoother; more delicate processes |
| Tilt | Upper end of pelvis relatively vertical | Upper end of pelvis tilted forward |
| Depth of greater pelvis | Deeper; ilium projects farther above sacroiliac joint | Shallower; ilium does not project as far above sacroiliac joint |
| Width of greater pelvis | Hips less flared; anterior superior spines closer together | Hips more flared; anterior superior spines farther apart |
| Pelvic inlet | Heart-shaped | Round or oval |
| Pelvic outlet | Smaller | Larger |
| Subpubic angle | Narrower, usually 90° or less | Wider, usually 100° or more |
| Pubic symphysis | Taller | Shorter |
| Body of pubis | More triangular | More rectangular |
| Greater sciatic notch | Narrower | Wider |
| Obturator foramen | Rounder | More oval to triangular |
| Acetabulum | Larger, faces more laterally | Smaller, faces slightly anteriorly |
| Sacrum | Narrower and deeper | Wider and shallower |
| Coccyx | Less movable; more vertical | More movable; tilted posteriorly |

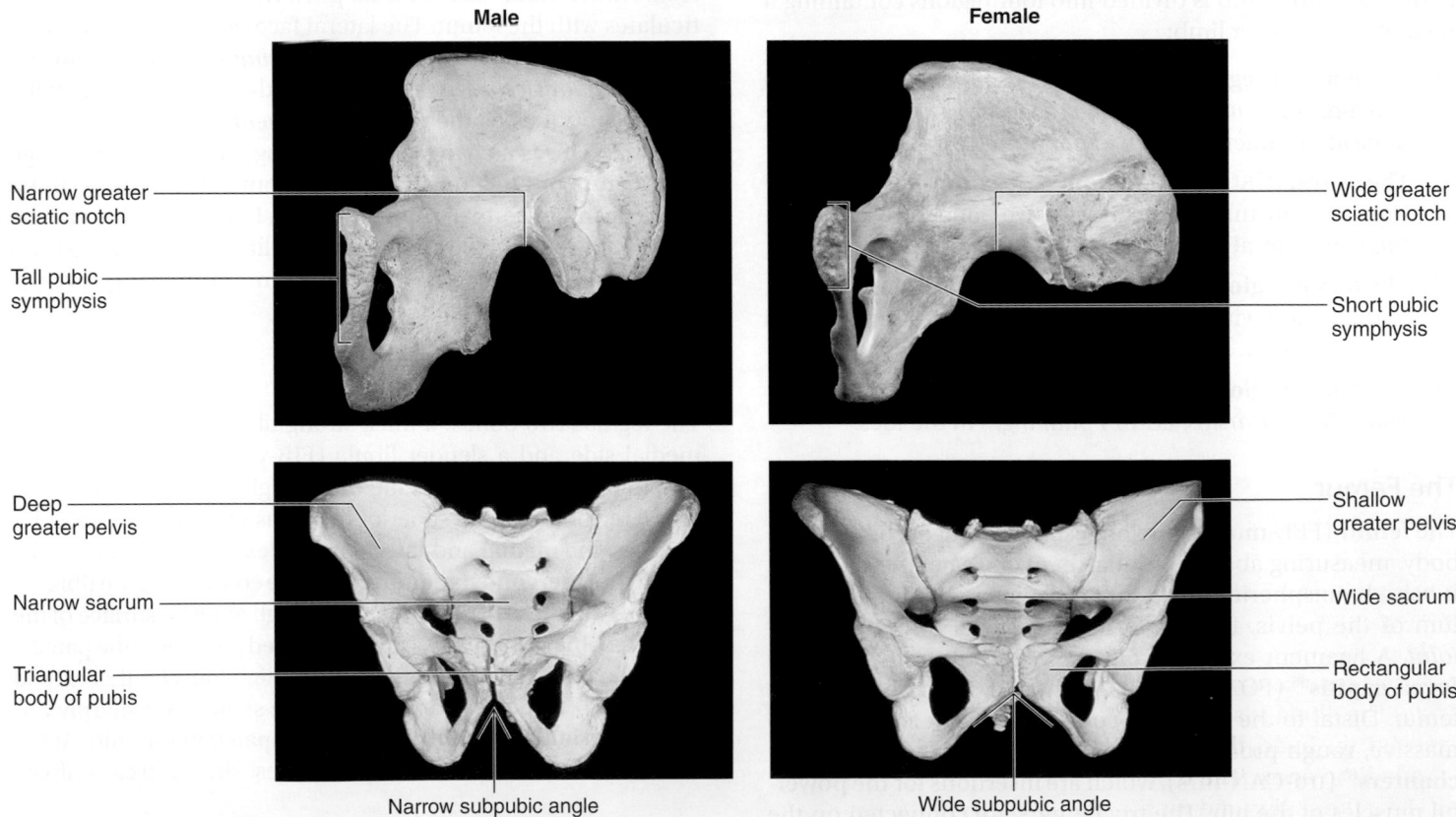

**FIGURE 8.37  Comparison of the Male and Female Pelvic Girdles.** Compare with table 8.5.

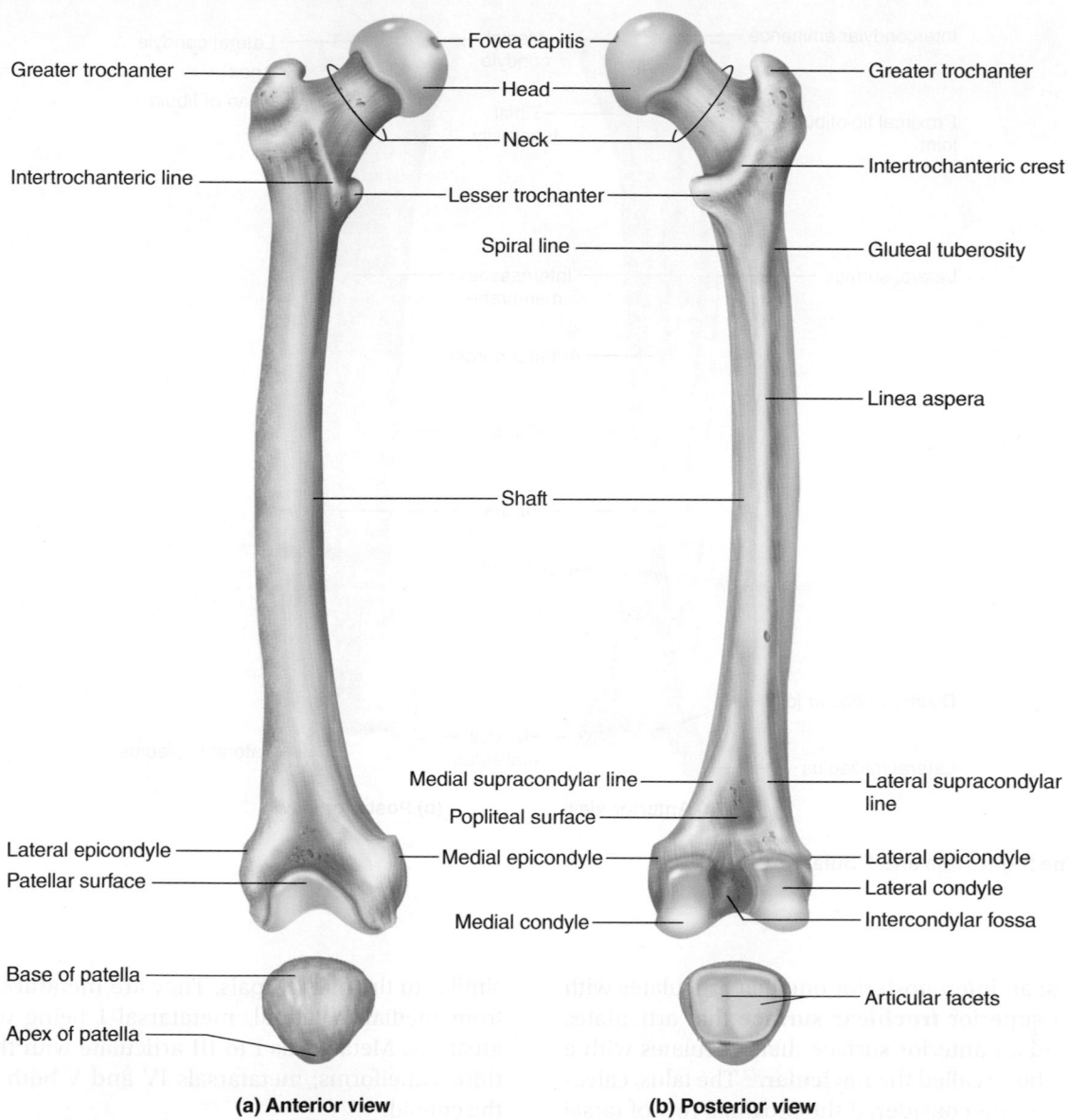

Greater trochanter — Fovea capitis — — Greater trochanter
— Head —
Neck — — Intertrochanteric crest
Intertrochanteric line — — Lesser trochanter —
Spiral line — — Gluteal tuberosity
— Linea aspera
Shaft — — Shaft
Medial supracondylar line — — Lateral supracondylar line
Popliteal surface —
Lateral epicondyle — — Medial epicondyle — — Lateral epicondyle
Patellar surface — — Lateral condyle
Medial condyle — — Intercondylar fossa
Base of patella — — Articular facets
Apex of patella —

**(a) Anterior view**          **(b) Posterior view**

**FIGURE 8.38 The Right Femur and Patella.** AP|R

a prominent bony knob on each side. These are the **medial** and **lateral malleoli**[70] (MAL-ee-OH-lie). The medial malleolus is part of the tibia, and the lateral malleolus is the part of the fibula.

## The Fibula

The **fibula** (fig. 8.39) is a slender lateral strut that helps to stabilize the ankle. It does not bear any of the body's weight; indeed, orthopedic surgeons sometimes remove part of the fibula and use it to replace damaged or missing bone elsewhere in the body. The fibula is somewhat thicker and broader at its proximal end, the **head,** than at the distal end. The point of the head is called the **apex.** The distal expansion is the lateral malleolus. Like the radius and ulna, the tibia and fibula are joined by an interosseous membrane along their shafts, and by shorter

ligaments at the superior and inferior ends where the fibular head and apex contact the tibia.

## The Ankle and Foot

The **tarsal bones** of the ankle are arranged in proximal and distal groups somewhat like the carpal bones of the wrist (fig. 8.40). Because of the load-bearing role of the ankle, however, their shapes and arrangement are conspicuously different from those of the carpal bones, and they are thoroughly integrated into the structure of the foot. The largest tarsal bone is the **calcaneus**[71] (cal-CAY-nee-us), which forms the heel. Its posterior end is the point of attachment for the **calcaneal (Achilles) tendon** from the calf muscles. The second-largest tarsal bone, and the most superior, is the **talus.** It has three

---

[70]*malle* = hammer; *olus* = little

[71]*calc* = stone, chalk

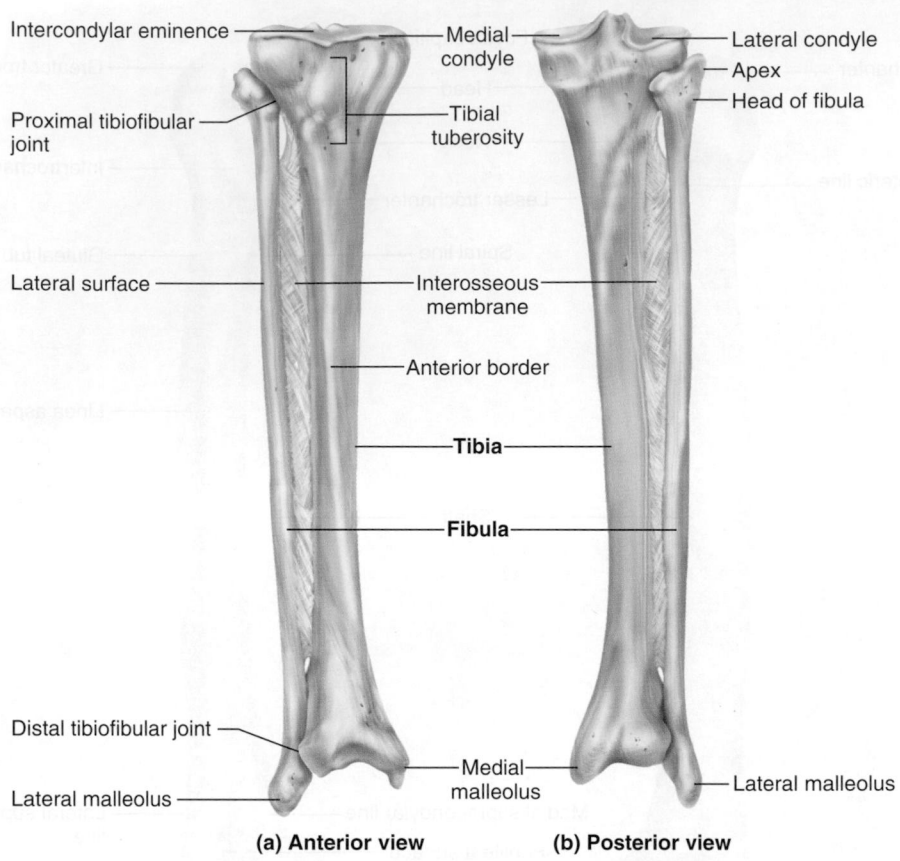

Intercondylar eminence

Medial condyle

Lateral condyle

Apex

Head of fibula

Proximal tibiofibular joint

Tibial tuberosity

Lateral surface

Interosseous membrane

Anterior border

**Tibia**

**Fibula**

Distal tibiofibular joint

Lateral malleolus

Medial malleolus

Lateral malleolus

**(a) Anterior view**     **(b) Posterior view**

**FIGURE 8.39 The Right Tibia and Fibula.** AP|R

articular surfaces: an inferoposterior one that articulates with the calcaneus, a superior **trochlear surface** that articulates with the tibia, and an anterior surface that articulates with a short, wide tarsal bone called the **navicular.**[72] The talus, calcaneus, and navicular are considered the proximal row of tarsal bones.

The distal group forms a row of four bones. Proceeding from the medial to lateral, these are the **medial, intermediate,** and **lateral cuneiforms**[73] (cue-NEE-ih-forms) and the **cuboid.**[74] The cuboid is the largest.

#### ▶▶▶ APPLY WHAT YOU KNOW

*The upper and lower limbs each contain 30 bones, yet we have 8 carpal bones in the upper limb and only 7 tarsal bones in the lower limb. What makes up the difference in the lower limb?*

The remaining bones of the foot are similar in arrangement and name to those of the hand. The proximal **metatarsals**[75] are

similar to the metacarpals. They are **metatarsals I** through **V** from medial to lateral, metatarsal I being proximal to the great toe. Metatarsals I to III articulate with the first through third cuneiforms; metatarsals IV and V both articulate with the cuboid.

Bones of the toes, like those of the fingers, are called phalanges. The great toe is the **hallux** and contains only two bones, the proximal and distal phalanx I. The other toes each contain a proximal, middle, and distal phalanx, and are numbered II through V from medial to lateral. Thus middle phalanx V, for example, would be the middle bone of the smallest toe. The metatarsal and phalangeal bones each have a base, body, and head, like the bones of the hand. All of them, especially the phalanges, are slightly concave on the inferior (plantar) side.

Note that roman numeral I represents the *medial* group of bones in the foot but the *lateral* group in the hand. In both cases, however, it refers to the largest digit of the limb. The reason for the difference between the hand and foot lies in a rotation of the limbs that occurs in the seventh week of embryonic development. Early in the seventh week, the limbs extend anteriorly from the body, the foot is a paddlelike *foot plate,* and the hand is also more or less paddlelike with the finger buds showing early separation (fig. 8.41a). The future thumb and great toe

---

[72]*navi* = boat; *cul* = little; *ar* = like
[73]*cunei* = wedge; *form* = in the shape of
[74]*cub* = cube; *oid* = resembling
[75]*meta* = beyond; *tars* = ankle

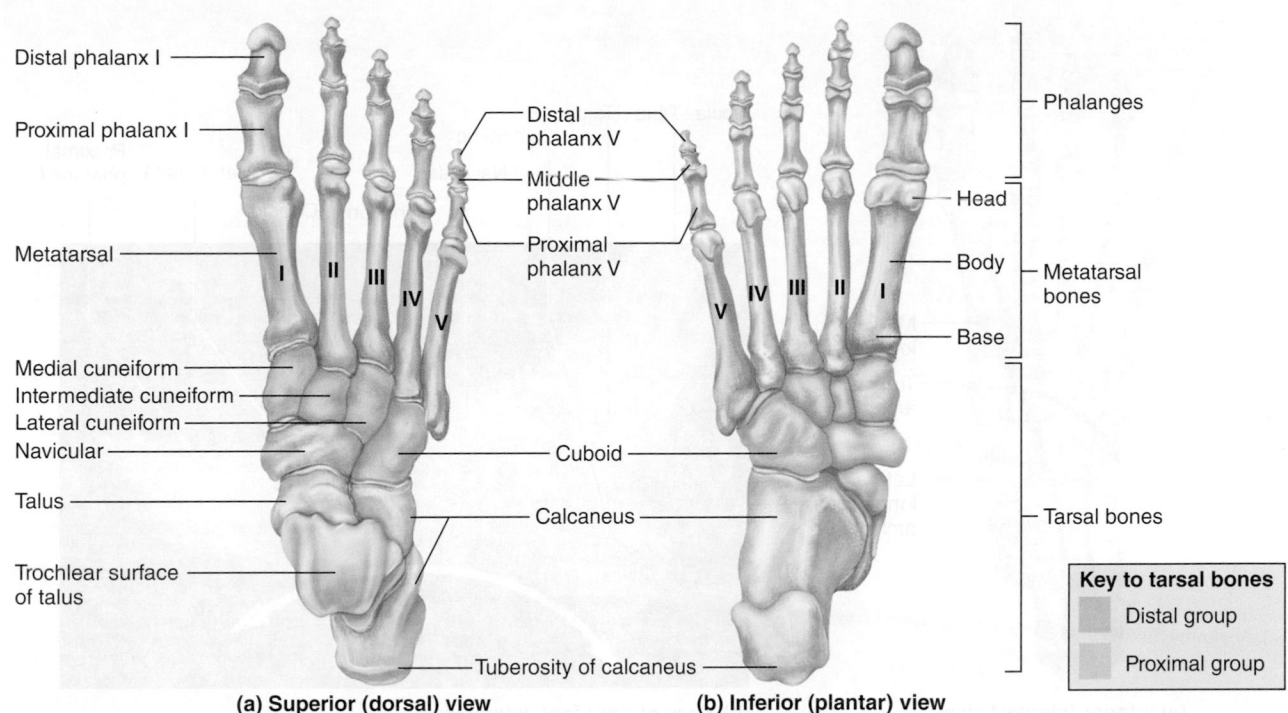

**FIGURE 8.40** **The Right Foot.** AP|R

❓ *Contrast the tarsal bones with the carpal bones. Which ones are similar in location? Which ones are different?*

are both directed superiorly, and the future palms and soles face each other medially. But then each limb rotates about 90° in opposite directions. The upper limb rotates laterally. To visualize this, hold your hands straight out in front of you with the palms facing each other as if you were about to clap. Then rotate your forearms so the thumbs face away from each other (laterally) and the palms face upward. The lower limbs rotate in the opposite direction, medially, so that the soles face downward and the great toes become medial. So even though the thumb and great toe (digit I of the hand and foot) start out facing in the same direction, these opposite rotations result in their being on opposite sides of the hand and foot (fig. 8.41b). This rotation also explains why the elbow flexes posteriorly and the knee flexes anteriorly, and why (as you will see in chapter 10) the muscles that flex the elbow are on the anterior side of the arm, whereas those that flex the knee are on the posterior side of the thigh.

The foot normally does not rest flat on the ground, but has three springy arches that distribute the body's weight between the heel and the heads of the metatarsal bones and

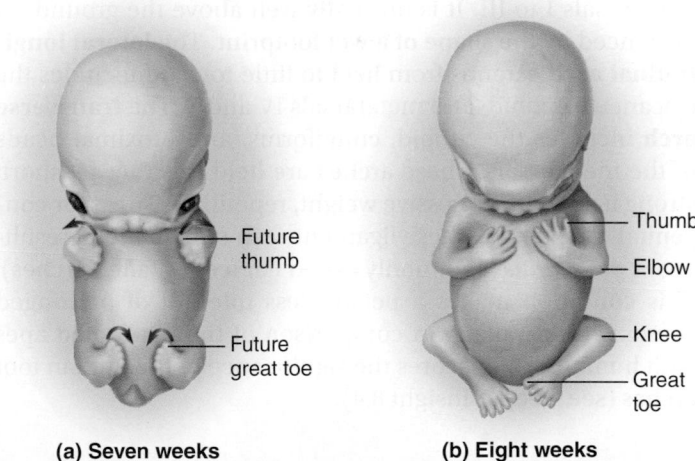

**FIGURE 8.41** **Embryonic Limb Rotation.** In the seventh week of development, the forelimbs and hindlimbs of the embryo rotate about 90° in opposite directions. This explains why the largest digits (digit I) are on opposite sides of the hand and foot, and why the elbow and knee flex in opposite directions.

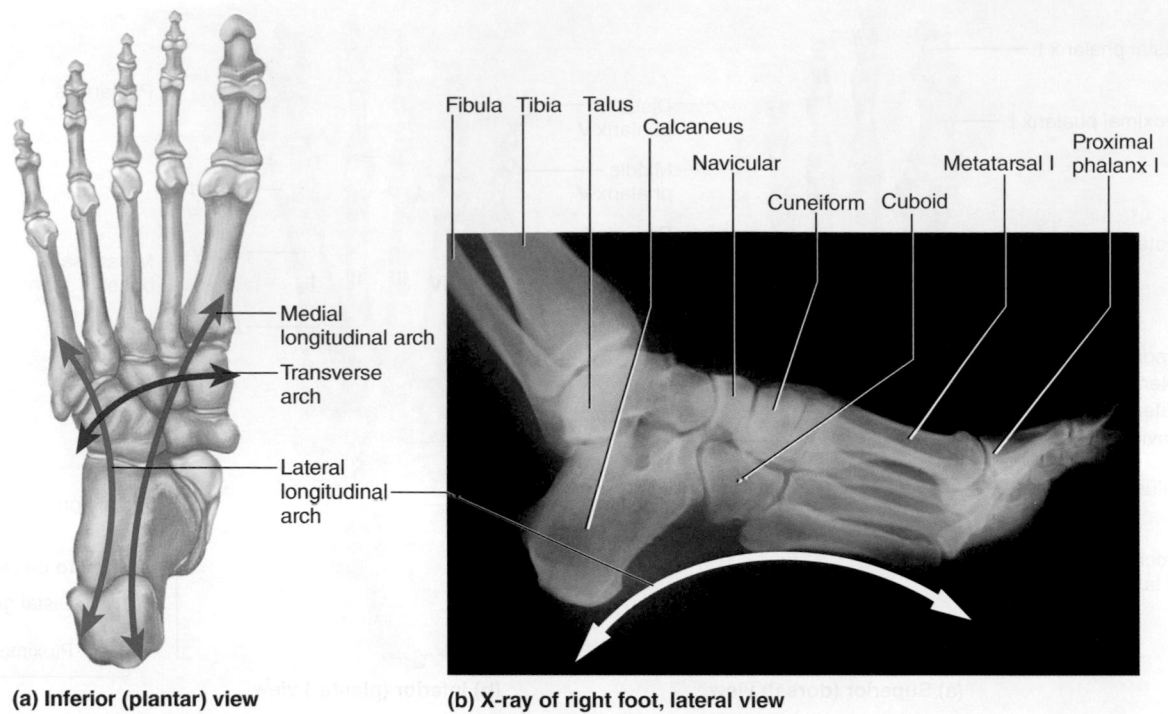

(a) Inferior (plantar) view

Labels: Medial longitudinal arch, Transverse arch, Lateral longitudinal arch

(b) X-ray of right foot, lateral view

Labels: Fibula, Tibia, Talus, Calcaneus, Navicular, Cuneiform, Cuboid, Metatarsal I, Proximal phalanx I

**FIGURE 8.42  Arches of the Foot.**

absorb the stress of walking (fig. 8.42). The **medial longitudinal arch,** which essentially extends from heel to hallux, is formed from the calcaneus, talus, navicular, cuneiforms, and metatarsals I to III. It is normally well above the ground, as evidenced by the shape of a wet footprint. The **lateral longitudinal arch** extends from heel to little toe and includes the calcaneus, cuboid, and metatarsals IV and V. The **transverse arch** includes the cuboid, cuneiforms, and proximal heads of the metatarsals. These arches are held together by short, strong ligaments. Excessive weight, repetitive stress, or congenital weakness of these ligaments can stretch them, resulting in *pes planus* (commonly called flat feet or fallen arches). This condition makes a person less tolerant of prolonged standing and walking. A comparison of the flat-footed apes with humans underscores the significance of the human foot arches (see Deeper Insight 8.4).

**BEFORE YOU GO ON**

Answer the following questions to test your understanding of the preceding section:

20. Name the bones of the adult pelvic girdle. What three bones of a child fuse to form the hip bone of an adult?

21. Name any four structures of the pelvis that you can palpate and describe where to palpate them.

22. Describe several ways in which the male and female pelvic girdles differ.

23. What parts of the femur are involved in the hip joint? What parts are involved in the knee joint?

24. Name the prominent knobs on each side of your ankle. What bones contribute to these structures?

25. Name all the bones that articulate with the talus and describe the location of each.

## DEEPER INSIGHT 8.4

## EVOLUTIONARY MEDICINE

### Skeletal Adaptations for Bipedalism

Some mammals can stand, hop, or walk briefly on their hind legs, but humans are the only mammals that are habitually bipedal. Footprints preserved in a layer of volcanic ash in Tanzania indicate that hominids walked upright as early as 3.6 million years ago. This bipedal locomotion is possible only because of several adaptations of the human feet, legs, spine, and skull (fig. 8.43). These features are so distinctive that paleoanthropologists (those who study human fossil remains) can tell with considerable certainty whether a fossil species was able to walk upright.

As important as the hand has been to human evolution, the foot may be an even more significant adaptation. Unlike other mammals, humans support their entire body weight on two feet. While apes are flat-footed, humans have strong, springy foot arches that absorb shock as the body jostles up and down during walking and running. The tarsal bones are tightly articulated with one another, and the calcaneus is strongly developed. The hallux (great toe) is not opposable

as it is in most Old World monkeys and apes, but it is highly developed so that it provides the "toe-off" that pushes the body forward in the last phase of the stride (fig. 8.43a). For this reason, loss of the hallux has a more crippling effect than the loss of any other toe.

While the femurs of apes are nearly vertical, in humans they angle medially from the hip to the knee (fig. 8.43b). This places our knees closer together, beneath the body's center of gravity. We lock our knees when standing, allowing us to maintain an erect posture with little muscular effort. Apes cannot do this, and they cannot stand on two legs for very long without tiring—much as you would if you tried to maintain an erect posture with your knees slightly bent.

In apes and other quadrupedal (four-legged) mammals, the abdominal viscera are supported by the muscular abdominal wall. In humans, the viscera bear down on the floor of the pelvic cavity, and a bowl-shaped pelvis

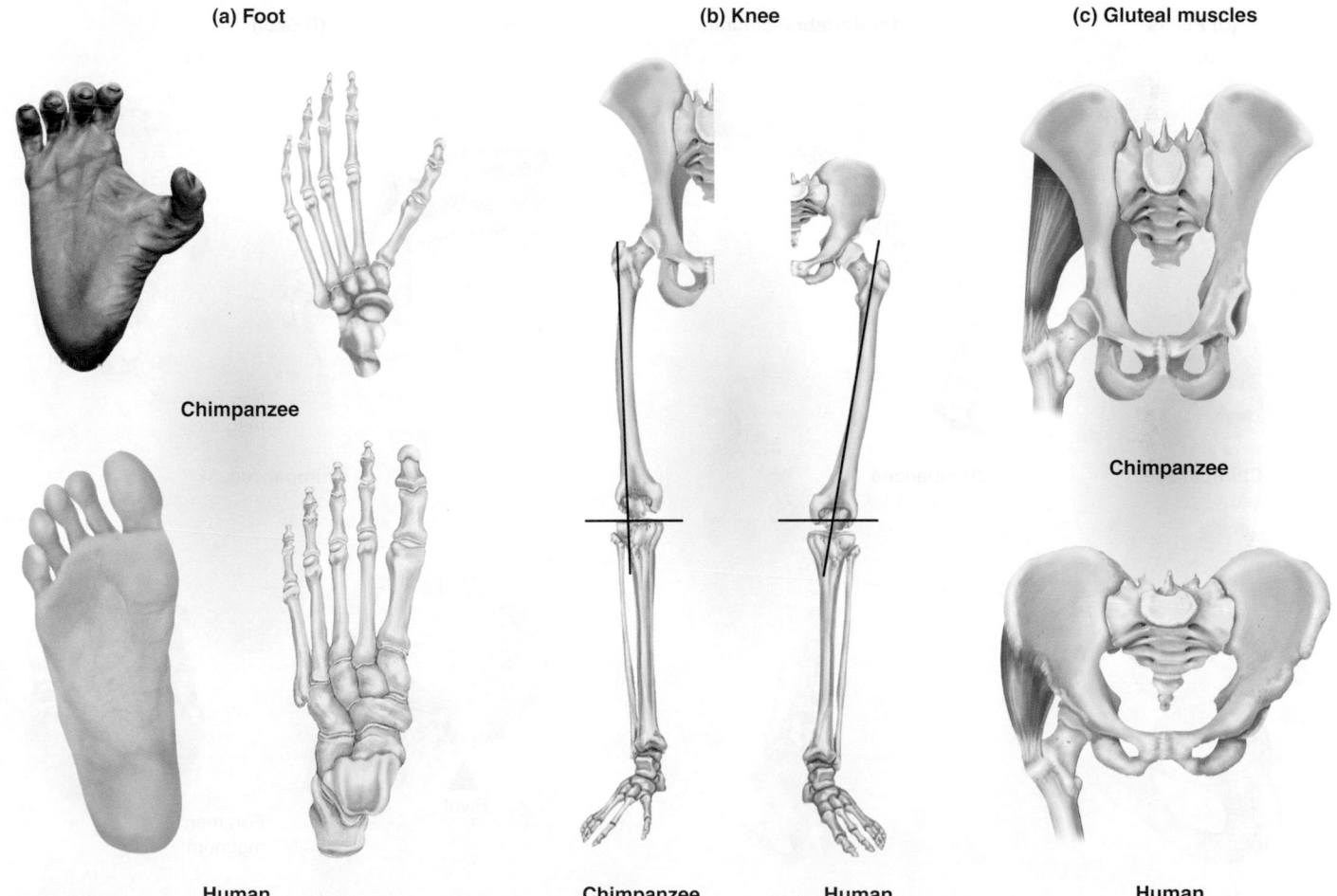

**(a) Foot**  **(b) Knee**  **(c) Gluteal muscles**

Chimpanzee

Chimpanzee

Human  Chimpanzee  Human  Human

**FIGURE 8.43 Skeletal Adaptations for Bipedalism.** Human adaptations for bipedalism are best understood by comparison to our close living relative, the chimpanzee, which is not adapted for a comfortable or sustained erect stance. See the text for the relevance of each comparison.

*(figure continued on p. 270)*

is necessary to support their weight. This has resulted in a narrower pelvic outlet—a condition that creates pain and difficulty in giving birth to such large-brained infants. The pain of childbirth seems unique to humans and, one might say, is a price we must pay for having both a large brain and a bipedal stance.

The largest muscle of the buttock, the *gluteus maximus,* serves in apes primarily as an abductor of the thigh—that is, it moves the leg laterally. In humans, however, the ilium has expanded posteriorly, so the gluteus maximus originates behind the hip joint. This changes the function of the muscle—instead of abducting the thigh, it pulls the thigh back in the second half of a stride (pulling back on your right thigh, for example, when your left foot is off the ground and swinging forward). Two other buttock muscles, the *gluteus medius* and *gluteus minimus,* extend laterally in humans from the surface of the ilium to the greater trochanter of the femur (fig. 8.43c). In walking, when one foot is lifted from the ground, these muscles shift the body weight over the other foot so we do not fall over. The actions of all three gluteal muscles, and the corresponding evolutionary remodeling of the pelvis, account for the smooth, efficient stride of a human as compared with the awkward, shuffling gait of a chimpanzee or gorilla when it is walking upright. The posterior growth of the ilium (fig. 8.43d) is the reason the greater sciatic notch is so deeply concave.

The lumbar curvature of the human spine allows for efficient bipedalism by shifting the body's center of gravity to the rear, above and slightly behind the hip joint (fig. 8.43e). Because of their C-shaped spines, chimpanzees cannot stand as easily. Their center of gravity is anterior to the hip joint when they stand; they must exert a continual muscular effort to keep from falling forward, and fatigue sets in relatively quickly. Humans, by contrast, require little muscular effort to keep their balance. Our australopithecine ancestors probably could travel all day with relatively little fatigue.

The human head is balanced on the vertebral column with the gaze directed forward. The cervical curvature of the spine and remodeling of the skull have made this possible. The foramen magnum has moved to a more inferior and anterior location, and the face is much flatter than an ape's face (fig. 8.43f), so there is less weight anterior to the occipital condyles. Being balanced on the spine, the head does not require strong muscular attachments to hold it erect.

The forelimbs of apes are longer than the hindlimbs; indeed, some species such as the orangutan and gibbons hold their long forelimbs over their heads when they walk on their hind legs. By contrast, our forelimbs are shorter than our hindlimbs and far less muscular than the forelimbs of apes. No longer needed for locomotion, our forelimbs have become better adapted for carrying objects, holding things closer to the eyes, and manipulating them more precisely.

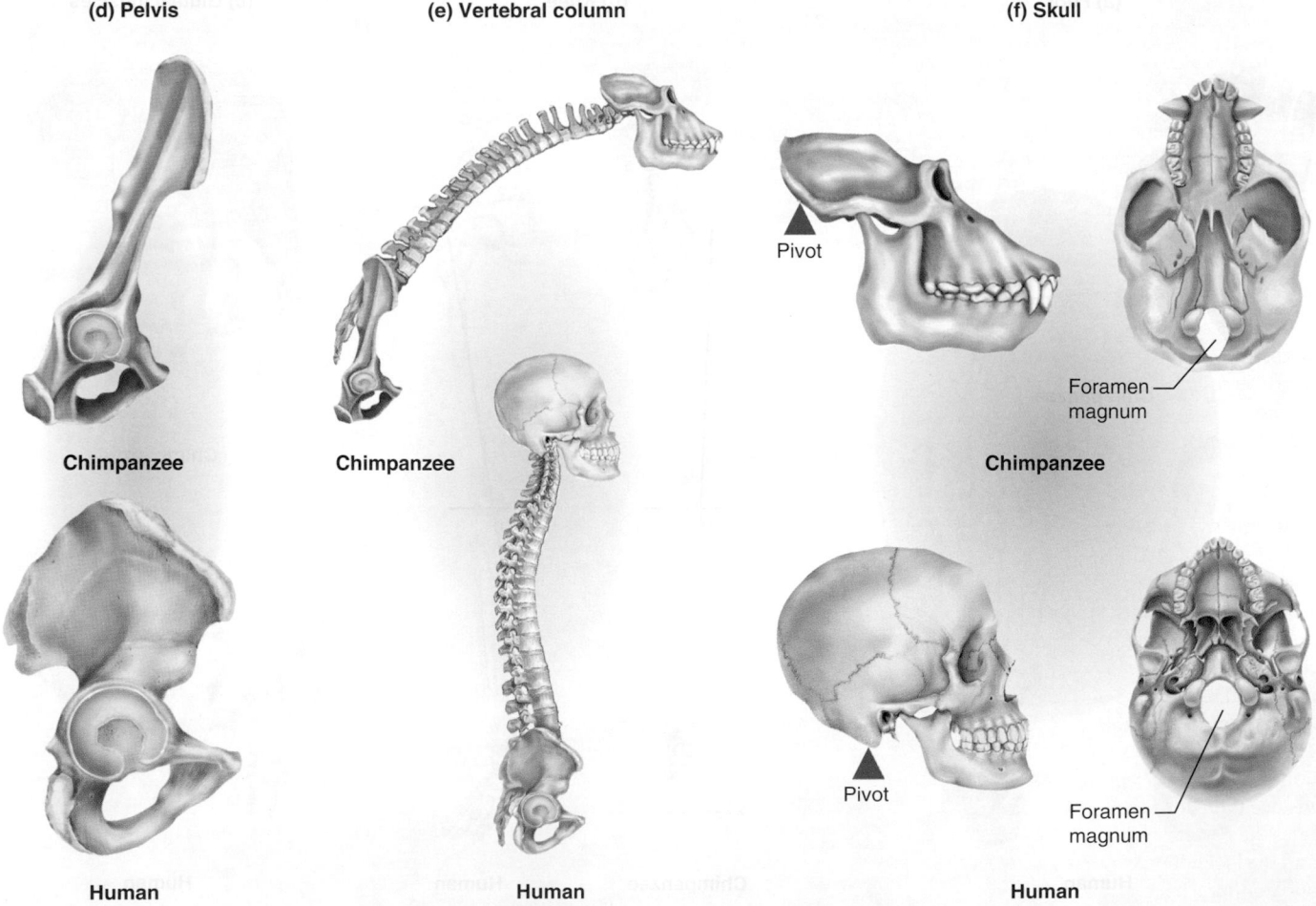

**(d) Pelvis**   **(e) Vertebral column**   **(f) Skull**

Chimpanzee   Chimpanzee   Pivot   Foramen magnum   Chimpanzee

Human   Human   Pivot   Foramen magnum   Human

**FIGURE 8.43** **Skeletal Adaptations for Bipedalism (continued).**

# STUDY GUIDE

## ▶ Assess Your Learning Outcomes

*Answers in Appendix B*

*To test your knowledge, discuss the following topics with a study partner or in writing, ideally from memory.*

### 8.1 Overview of the Skeleton (p. 230)

1. The difference between the axial and appendicular skeletons, and the bones in each category
2. The typical number of named bones in an adult; why this number differs in newborns and children; and why the number varies among adults
3. Sutural and sesamoid bones, and examples of the latter
4. Names of the various outgrowths, depressions, articular surfaces, cavities, and passages in bones

### 8.2 The Skull (p. 232)

1. The usual number of bones in the adult skull, and the collective name of the seams or joints that bind most of them together
2. Names and locations of the cavities that enclose the brain, nose, ears, and eyes, and of the paranasal sinuses
3. The collective function of the skull foramina; the location and function of the largest one, the foramen magnum
4. Major features of the cranium; the difference between its base and calvaria; the three cranial fossae and how they relate to brain anatomy
5. Names of the six kinds of cranial bones; which ones are solitary and which are bilaterally paired; and what distinguishes a cranial bone from a facial bone
6. The location and extent of the frontal bone; the suture that binds it to the parietal bones; and the locations of its supraorbital margin and foramen, glabella, frontal sinus, and diploe
7. The location and extent of the parietal bones; the suture formed where they meet the occipital bone and the one that separates the parietal bones from each other; and the locations of the parietal foramen and temporal lines
8. The location and extent of the temporal bone; the suture that borders it; and the names and boundaries of its four major parts
9. The location and extent of the occipital bone, its basilar part, and the names and locations of its foramina, canals, and surface protrusions

10. The location and extent of the sphenoid bone; its wings, body, pterygoid plates, clinoid processes, and foramina; its relationships with the pituitary gland and nasal apertures
11. The location and extent of the ethmoid bone; its part in defining the nasal fossae; and the locations of its plates, foramina, air cells, and nasal conchae
12. Names of the eight kinds of facial bones; which ones are solitary and which are bilaterally paired
13. The location and extent of the maxilla; its foramina, alveoli, and alveolar and palatine processes; and the suture that joins the right and left maxillae
14. The location and extent of the palatine bones; their foramina; and their part in partially defining the walls of the nasal cavity and orbit
15. Structure of the palate, including the hard and soft regions and the contributions of the palatine processes and palatine bones
16. The location and extent of the zygomatic bones, and the temporal process and main foramen of each
17. The three parts of the zygomatic arch
18. The locations and structures of the tiny lacrimal and nasal bones
19. The inferior nasal concha and why it is distinguished from the superior and middle conchae
20. The location and extent of the vomer; contributions of the vomer and ethmoid bone to the nasal septum
21. Structure of the mandible, including the body, ramus, and angle; its two main processes and the notch between them; its foramina, symphysis, protuberance, and spines
22. The bones, and their specific features, that form the temporomandibular joint
23. The locations and names of the auditory ossicles; location and features of the hyoid bone; and functions of these bones
24. Names and locations of the fontanels of the neonatal skull; why they exist; and how a child's skull changes between birth and 9 years of age

### 8.3 The Vertebral Column and Thoracic Cage (p. 246)

1. The number of vertebrae and intervertebral discs in the vertebral column (spine)

2. Four curvatures of the adult spine; which ones are present at birth; and when and how the others develop
3. Features of a typical vertebra
4. The five classes of vertebrae and the number of vertebrae in each class; the system of numbering them; and why the number of vertebrae in a child differs from the number at 30 years of age and beyond
5. Features that identify an isolated vertebra as cervical, thoracic, or lumbar
6. How the anatomy of the first two vertebrae (C1–C2) relates to movements of the head
7. How the anatomy of the thoracic vertebrae (T1–T12) relates to the articulations of the ribs
8. The structure and function of intervertebral discs; which vertebrae have discs between them and which ones do not
9. Anatomical features of the sacrum, including its foramina, crests, canal and hiatus, auricular surface and sacroiliac joint, promontory, and alae
10. Features of the coccyx
11. Components and general shape of the thoracic cage
12. Three main regions of the sternum; its notches and sternal angle
13. The number of ribs; which ones are true, false, and floating ribs
14. All features seen in most of the ribs
15. Which ribs differ from that typical anatomy, and how
16. How the ribs articulate with the vertebrae, including variations from the top to bottom of the rib cage

### 8.4 The Pectoral Girdle and Upper Limb (p. 255)

1. Names and locations of the 4 bones of the pectoral girdle and 30 bones of each upper limb
2. Names of the joints at which the humerus articulates with the scapula, the scapula with the clavicle, and the clavicle with the axial skeleton
3. Features of the clavicle, including the sternal and acromial ends and conoid tubercle; function of the clavicles
4. Features of the scapula, including its borders and angles, fossae, suprascapular notch, acromion, coracoid process, and glenoid cavity
5. Names of the four segments of the upper limb and the bones contained in each

# STUDY GUIDE

6. Features of the humerus, including the head, necks, tubercles, intertubercular sulcus, deltoid tuberosity, capitulum, trochlea, epicondyles, supracondylar ridges, and three fossae

7. Features of the radius, including the head, neck, radial tuberosity, styloid process, and ulnar notch

8. Features of the ulna, including the trochlear and radial notches, coronoid and styloid processes, and olecranon; and the relationship of the radius and ulna to the interosseous membrane

9. Names of the carpal bones, in order, from lateral to medial in the proximal row and lateral to medial in the distal row; the unusual structure of the hamate bone

10. The system of naming and numbering the 5 metacarpal bones of the palmar region and 5 sets of phalanges in the digits; why there are 5 digits but only 14 phalanges; the base, body, and head of all 19 metacarpals and phalanges; and the special anatomical name of the thumb

### 8.5 The Pelvic Girdle and Lower Limb (p. 259)

1. Names and locations of the 3 bones of the pelvic girdle and 30 bones of each lower limb

2. Names of the joints at which the lower limb articulates with the pelvic girdle and the pelvic girdle articulates with the axial skeleton

3. The distinction between the pelvic girdle and pelvis

4. Three childhood bones that fuse to form each adult hip (coxal) bone, and the boundaries of each on the hip bone

5. Features of the hip (coxal) bones and pelvic girdle including the auricular surfaces; interpubic disc and pubic symphysis; greater and lesser pelves; pelvic brim, inlet, and outlet; iliac crest; acetabulum; obturator foramen; four spines; two sciatic notches; iliac fossa; and parts of the ischium and pubis

6. Differences between the male and female pelvic girdles and the essential reason for these differences

7. Names of the four segments of the lower limb and the bones contained in each

8. Features of the femur, including the head, neck, fovea capitis, trochanters, intertro-

chanteric crest, gluteal tuberosity, condyles, intercondylar fossa, epicondyles, lines, and patellar and popliteal surfaces

9. Features of the patella, including the base, apex, and articular facets

10. Features of the tibia, including the lateral and medial condyles and intercondylar eminence; tibial tuberosity; anterior crest; and medial malleolus

11. Features of the fibula, including the head, apex, and lateral malleolus

12. Names of the tarsal bones from posterior to anterior, and from lateral to medial in the distal row; why they are more fully integrated into the foot than the carpal bones are into the hand

13. The system of naming and numbering the 5 metatarsal bones of the foot and 5 sets of phalanges in the digits; why there are 5 digits but only 14 phalanges; the base, body, and head of all 19 metatarsals and phalanges; and the special anatomical name of the great toe

14. Why the elbows and knees flex in opposite directions, and why the largest digit is lateral in the hand but medial in the foot

15. Names and locations of the three foot arches

## ▶ Testing Your Recall

*Answers in Appendix B*

1. Which of these is *not* a paranasal sinus?
   a. frontal      d. ethmoid
   b. temporal     e. maxillary
   c. sphenoid

2. Which of these is a facial bone?
   a. frontal      d. temporal
   b. ethmoid      e. lacrimal
   c. occipital

3. Which of these *cannot* be palpated on a living person?
   a. the crista galli
   b. the mastoid process
   c. the zygomatic arch
   d. the superior nuchal line
   e. the hyoid bone

4. All of the following are groups of vertebrae *except* for _____, which is a spinal curvature.
   a. thoracic      d. pelvic
   b. cervical      e. sacral
   c. lumbar

5. Thoracic vertebrae do *not* have
   a. transverse foramina.
   b. costal facets.

   c. spinous processes.
   d. transverse processes.
   e. pedicles.

6. The tubercle of a rib articulates with
   a. the sternal notch.
   b. the margin of the gladiolus.
   c. the costal facets of two vertebrae.
   d. the body of a vertebra.
   e. the transverse process of a vertebra.

7. The disc-shaped head of the radius articulates with the _____ of the humerus.
   a. radial tuberosity    d. olecranon
   b. trochlea             e. glenoid cavity
   c. capitulum

8. All of the following are carpal bones, *except* the _____, which is a tarsal bone.
   a. trapezium      d. triquetrum
   b. cuboid         e. pisiform
   c. trapezoid

9. The bone that supports your body weight when you are sitting down is
   a. the acetabulum.    d. the coccyx.
   b. the pubis.         e. the ischium.
   c. the ilium.

10. Which of these is the bone of the heel?
    a. cuboid       d. trochlear
    b. calcaneus    e. talus
    c. navicular

11. Gaps between the cranial bones of an infant are called _____.

12. The external auditory canal is a passage in the _____ bone.

13. Bones of the skull are joined along lines called _____.

14. The _____ bone has greater and lesser wings and protects the pituitary gland.

15. A herniated disc occurs when a ring called the _____ cracks.

16. The transverse ligament of the atlas holds the _____ of the axis in place.

17. The sacroiliac joint is formed where the _____ surface of the sacrum articulates with that of the ilium.

18. The _____ processes of the radius and ulna form bony protuberances on each side of the wrist.

# STUDY GUIDE

19. The thumb is also known as the _____ and the great toe is also known as the _____.

20. The _____ arch of the foot extends from the heel to the great toe.

## ▶ Building Your Medical Vocabulary

Answers in Appendix B

*State a medical meaning of each word element below, and give a term in which it or a slight variation of it is used.*

1. costo-

2. cranio-

3. dura

4. glosso-

5. -icle

6. masto-

7. pedo-

8. pterygo-

9. supra-

10. tarso-

## ▶ True or False

Answers in Appendix B

*Determine which five of the following statements are false, and briefly explain why.*

1. Not everyone has a frontal sinus.

2. The hands have more phalanges than the feet.

3. As an adaptation to pregnancy, the female's pelvis is deeper than the male's.

4. There are more carpal bones than tarsal bones.

5. On a living person, it would be possible to palpate the muscles in the infraspinous fossa but not those of the subscapular fossa.

6. If you rest your chin on your hands and your elbows on a table, the olecranon of the ulna rests on the table.

7. The lumbar vertebrae do not articulate with any ribs and therefore do not have transverse processes.

8. The most frequently broken bone is the humerus.

9. In strict anatomical terminology, the words *arm* and *leg* both refer to regions with only one bone.

10. The pisiform bone and patella are both sesamoid bones.

## ▶ Testing Your Comprehension

Answers at www.mhhe.com/saladin7

1. A child was involved in an automobile collision. She was not wearing a safety restraint, and her chin struck the dashboard hard. When the physician looked into her ear, he could see into her throat. What do you infer from this about the nature of her injury?

2. By palpating the hind leg of a cat or dog or by examining a laboratory skeleton, you can see that cats and dogs stand on the heads of their metatarsal bones; the calcaneus does not touch the ground. How is this similar to the stance of a woman wearing high-heeled shoes? How is it different?

3. Between any two of the unfused vertebrae (cervical through lumbar), there is an intervertebral disc—except between C1 and C2. Give some reasons for the unique absence of a disc at that location.

4. In adolescents, trauma sometimes separates the head of the femur from the neck. Why do you think this is more common in adolescents than in adults?

5. Andy, a 55-year-old, 85 kg (187 lb) roofer, is shingling the steeply pitched roof of a new house when he loses his footing and slides down the roof and over the edge, feet first.

He braces himself for the fall, and when he hits the ground he cries out and doubles up in excruciating pain. Emergency medical technicians called to the scene tell him he has broken his hips. Describe, more specifically, where his fractures most likely occurred. On the way to the hospital, Andy says, "You know it's funny, when I was a kid, I used to jump off roofs that high, and I never got hurt." Why do you think Andy was more at risk of a fracture as an adult than he was as a boy?

## ▶ Improve Your Grade at www.mhhe.com/saladin7

*Download animations and movies to study when it fits your schedule. Practice quizzes, labeling activities, games, and flashcards offer fun ways to master the chapter concepts. Or, download image PowerPoint files for each chapter to create a study guide or for taking notes during lecture.*

# 9

# JOINTS

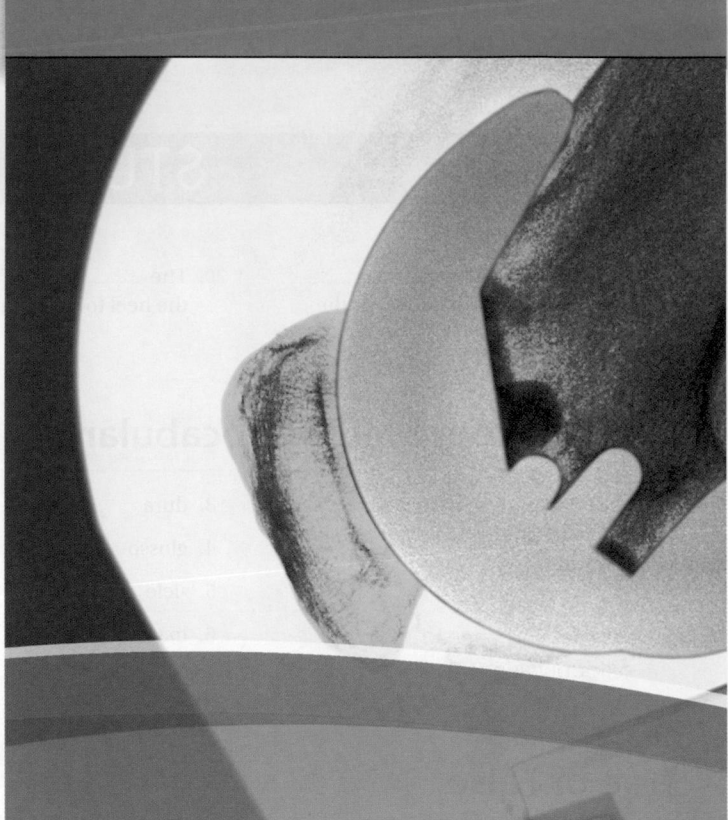

**X-ray of an artifical knee joint (prosthesis)**

**Anatomy & Physiology | REVEALED®**
aprevealed.com

**Module 5: Skeletal System**

Joints, or articulations, link the bones of the skeletal system into a functional whole—a system that supports the body, permits effective movement, and protects the softer organs. Joints such as the shoulder, elbow, and knee are remarkable specimens of biological design—self-lubricating, almost frictionless, and able to bear heavy loads and withstand compression while executing smooth and precise movements (fig. 9.1). Yet it is equally important that other joints be less movable or even immobile. Such joints are better able to support the body and protect delicate organs. The vertebral column, for example, is only moderately mobile, for it must allow for flexibility of the torso and yet protect the delicate spinal cord and support much of the body's weight. Bones of the cranium must protect the brain and sense organs, but need not allow for movement (except during birth); thus, they are locked together by immobile joints, the sutures studied in chapter 8.

In everyday life, we take the greatest notice of the most freely mobile joints of the limbs, and it is here that people feel most severely compromised by such disabling diseases as arthritis. Much of the work of physical therapists focuses on limb mobility. In this chapter, we will survey all types of joints, from the utterly immobile to the most mobile, but with an emphasis on the latter. This survey of joint anatomy and movements will provide a foundation for the study of muscle actions in chapter 10.

**FIGURE 9.1  Joint Flexibility.** This gymnast demonstrates the flexibility, precision, and weight-bearing capacity of the body's joints.

## 9.1   Joints and Their Classification

### Expected Learning Outcomes

When you have completed this section, you should be able to

a. explain what joints are, how they are named, and what functions they serve;

b. name and describe the four major classes of joints;

c. describe the three types of fibrous joints and give an example of each;

d. distinguish between the three types of sutures;

e. describe the two types of cartilaginous joints and give an example of each; and

f. name some joints that become synostoses as they age.

Any point where two bones meet is called a **joint (articulation),** whether or not the bones are mobile at that interface. The science of joint structure, function, and dysfunction is called **arthrology.**[1] The study of musculoskeletal movement is **kinesiology**[2] (kih-NEE-see-OL-oh-jee). This is a branch of **biomechanics,** which deals with a broad variety of movements and mechanical processes in the body, including the physics of blood circulation, respiration, and hearing.

The name of a joint is typically derived from the names of the bones involved. For example, the *atlanto–occipital joint* is where the atlas meets the occipital condyles; the *gleno-humeral joint* is where the glenoid cavity of the scapula meets the humerus; and the *radioulnar joint* is where the radius meets the ulna.

---

[1]*arthro* = joint; *logy* = study of
[2]*kinesio* = movement; *logy* = study of

Joints can be classified according to the manner in which the adjacent bones are bound to each other, with corresponding differences in how freely the bones can move. Authorities differ in their classification schemes, but one common view places the joints in four major categories: *bony, fibrous, cartilaginous,* and *synovial joints.* This section will describe the first three of these and the subclasses of each. The remainder of the chapter will then be concerned primarily with synovial joints.

## Bony Joints

A **bony joint,** or **synostosis**[3] (SIN-oss-TOE-sis), is an immobile joint formed when the gap between two bones ossifies and they become, in effect, a single bone. Bony joints can form by ossification of either fibrous or cartilaginous joints. An infant is born with right and left frontal and mandibular bones, for example, but these soon fuse seamlessly into a single frontal bone and mandible. In old age, some cranial sutures become obliterated by ossification and the adjacent cranial bones, such as the parietal bones, fuse. The epiphyses and diaphyses of the long bones are joined by cartilaginous joints in childhood and adolescence, and these become synostoses in early adulthood. The attachment of the first rib to the sternum also becomes a synostosis with age.

## Fibrous Joints

A **fibrous joint** is also called a **synarthrosis**[4] (SIN-ar-THRO-sis) or **synarthrodial joint.** It is a point at which adjacent bones are bound by collagen fibers that emerge from one bone, cross the space between them, and penetrate into the other (fig. 9.2). There are three kinds of fibrous joints: *sutures, gomphoses,* and

---

[3]*syn* = together; *ost* = bone; *osis* = condition

[4]*syn* = together; *arthr* = joined; *osis* = condition

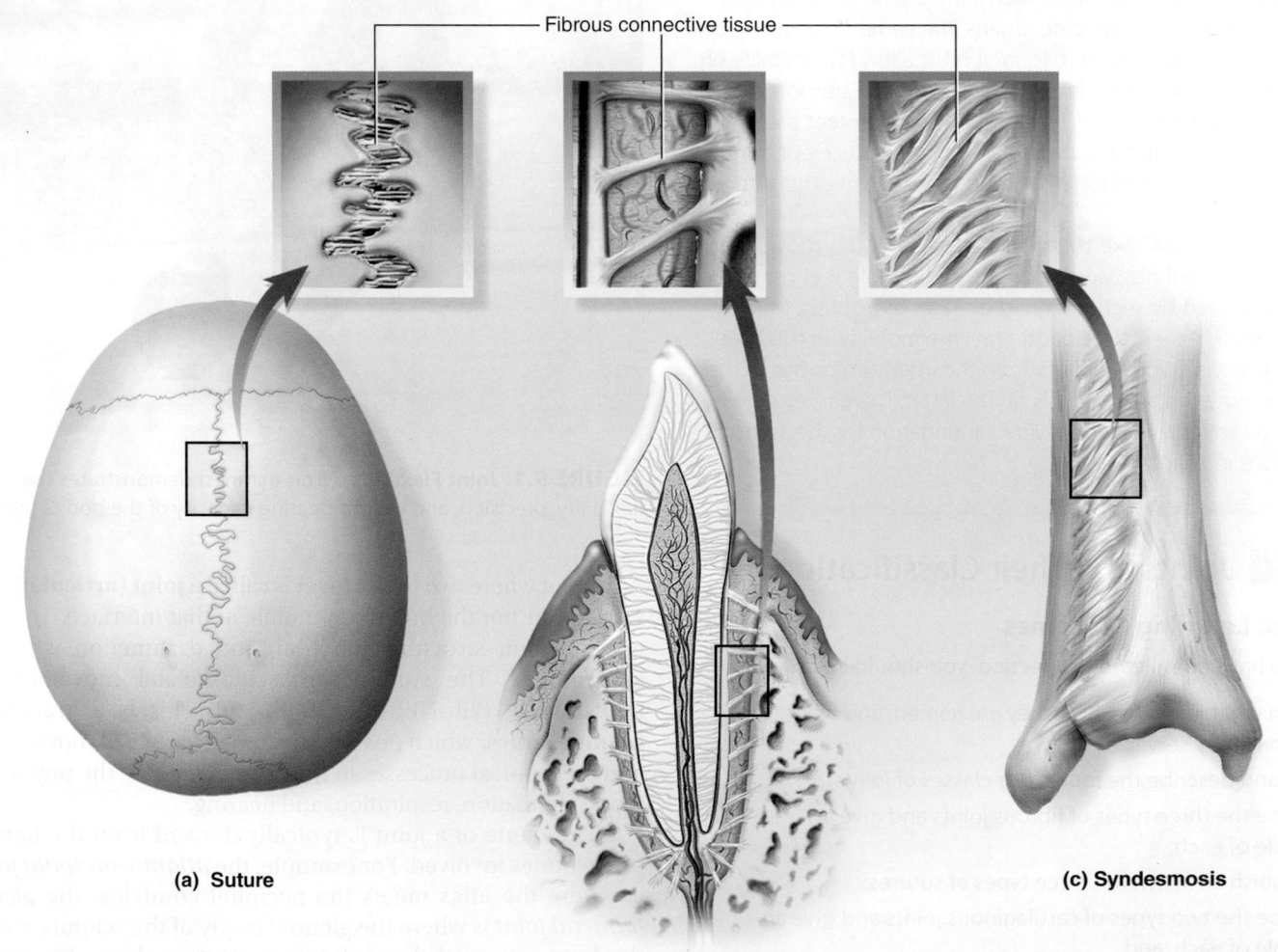

Fibrous connective tissue

**(a) Suture**

**(b) Gomphosis**

**(c) Syndesmosis**

**FIGURE 9.2 Fibrous Joints.** (a) A suture between the parietal bones. (b) A gomphosis between a tooth and the jaw. (c) A syndesmosis between the tibia and fibula.

❓ *Which of these is not a joint between two bones? Why?*

*syndesmoses.* In sutures and gomphoses, the fibers are very short and allow for little or no movement. In syndesmoses, the fibers are longer and the attached bones are more mobile.

## Sutures

**Sutures** are immobile or only slightly mobile fibrous joints that closely bind the bones of the skull to each other; they occur nowhere else. In chapter 8 we did not take much notice of the differences between one suture and another, but some differences may have caught your attention as you studied the diagrams in that chapter or examined laboratory specimens. Sutures can be classified as *serrate, lap,* and *plane sutures.* Readers with some knowledge of woodworking may recognize that the structures and functional properties of these sutures have something in common with basic types of carpentry joints (fig. 9.3).

**Serrate sutures** appear as wavy lines along which the adjoining bones firmly interlock with each other by their serrated margins, like pieces of a jigsaw puzzle. Serrate sutures are analogous to a dovetail wood joint. Examples include the coronal, sagittal, and lambdoid sutures that border the parietal bones.

**Lap (squamous) sutures** occur where two bones have overlapping beveled edges, like a miter joint in carpentry. On the surface, a lap suture appears as a relatively smooth (nonserrated) line. An example is the squamous suture between the temporal and parietal bones. The beveled edge of the temporal bone can be seen in figure 8.10b (p. 239).

**Plane (butt) sutures** occur where two bones have straight nonoverlapping edges. The two bones merely border on each other, like two boards glued together in a butt joint. This type of joint is represented by the intermaxillary suture in the roof of the mouth (see fig. 8.5a, p. 235).

## Gomphoses

Even though the teeth are not bones, the attachment of a tooth to its socket is classified as a joint called a **gomphosis** (gom-FOE-sis). The term refers to its similarity to a nail hammered into wood.[5] The tooth is held firmly in place by a fibrous **periodontal ligament,** which consists of collagen fibers that extend from the bone matrix of the jaw into the dental tissue (see fig. 9.2b). The periodontal ligament allows the tooth to move or give a little under the stress of chewing. Along with associated nerve endings, this slight tooth movement allows us to sense how hard we are biting or to sense a particle of food stuck between the teeth.

---

[5]*gomph* = nail, bolt; *osis* = condition

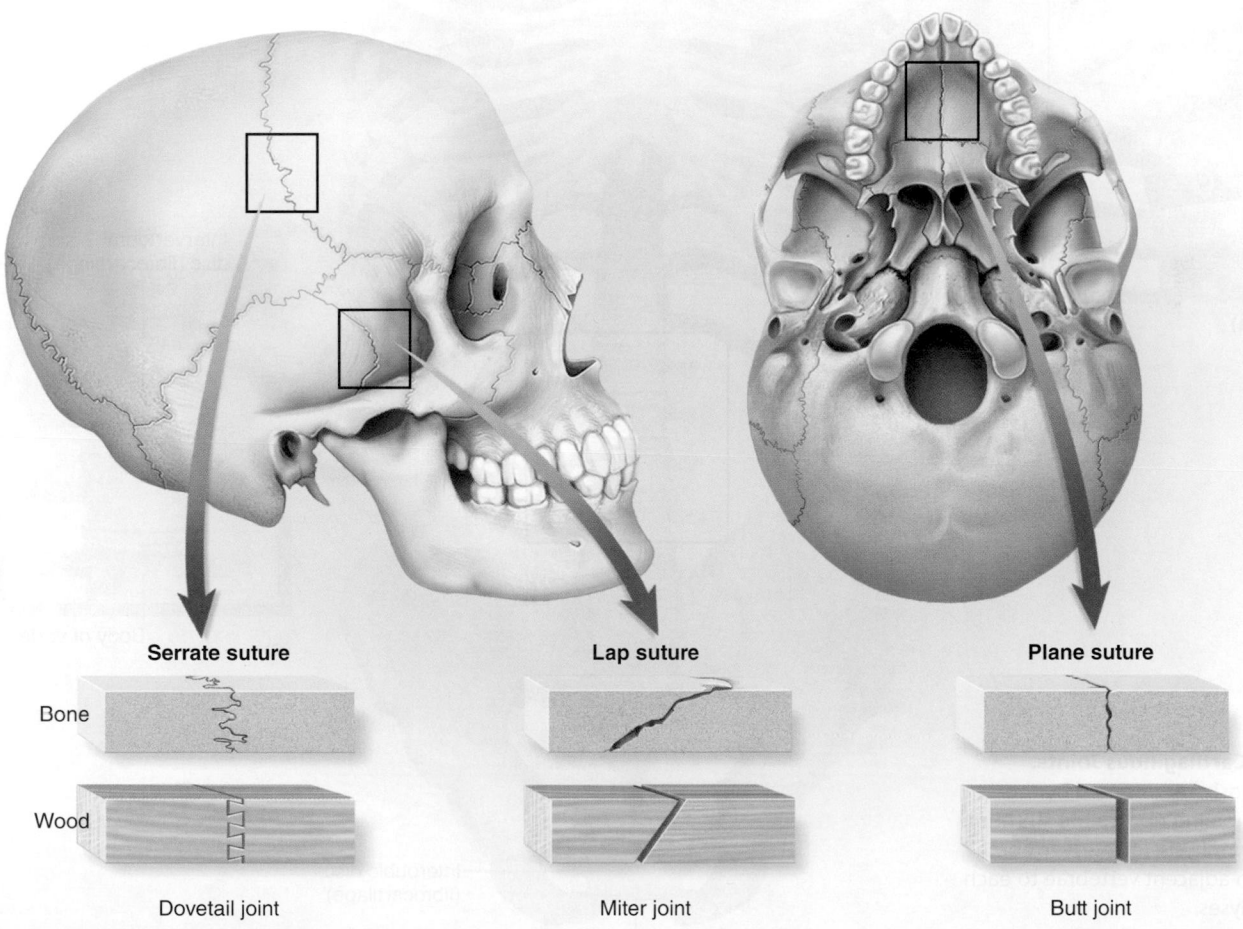

**FIGURE 9.3  Sutures.** Serrate, lap, and plane sutures compared to some common wood joints.

## Syndesmoses

A **syndesmosis**[6] (SIN-dez-MO-sis) is a fibrous joint at which two bones are bound by relatively long collagenous fibers. The separation between the bones and length of the fibers give these joints more mobility than a suture or gomphosis has. An especially mobile syndesmosis exists between the shafts of the radius and ulna, which are joined by a broad fibrous *interosseous membrane*. This permits such movements as pronation and supination of the forearm. A less mobile syndesmosis is the one that binds the distal ends of the tibia and fibula together, side by side (see fig. 9.2c).

---

[6]*syn* = together; *desm* = band; *osis* = condition

## Cartilaginous Joints

A **cartilaginous joint** is also called an **amphiarthrosis**[7] (AM-fee-ar-THRO-sis) or **amphiarthrodial joint.** In these joints, two bones are linked by cartilage (fig. 9.4). The two types of cartilaginous joints are *synchondroses* and *symphyses.*

### Synchondroses

A **synchondrosis**[8] (SIN-con-DRO-sis) is a joint in which the bones are bound by hyaline cartilage. An example is the temporary joint between the epiphysis and diaphysis of a long bone in a child, formed by the cartilage of the epiphyseal plate.

---

[7]*amphi* = on all sides; *arthr* = joined; *osis* = condition
[8]*syn* = together; *chondr* = cartilage; *osis* = condition

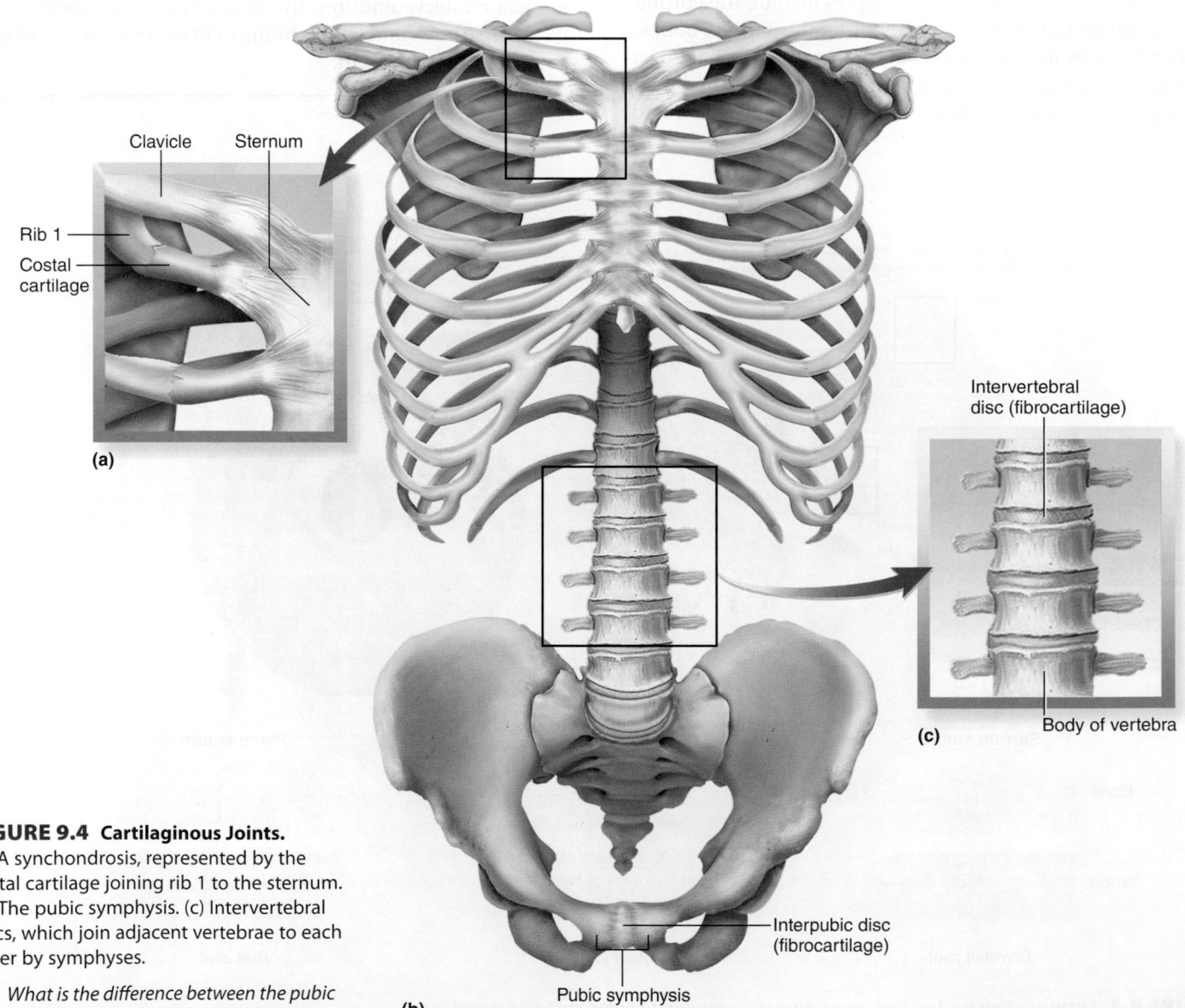

**FIGURE 9.4  Cartilaginous Joints.**
(a) A synchondrosis, represented by the costal cartilage joining rib 1 to the sternum. (b) The pubic symphysis. (c) Intervertebral discs, which join adjacent vertebrae to each other by symphyses.

❓ *What is the difference between the pubic symphysis and the interpubic disc?*

Another is the attachment of the first rib to the sternum by a hyaline costal cartilage (fig. 9.4a). (The other costal cartilages are joined to the sternum by synovial joints.)

## Symphyses

In a **symphysis**[9] (SIM-fih-sis), two bones are joined by fibrocartilage (fig. 9.4b, c). One example is the pubic symphysis, in which the right and left pubic bones are joined anteriorly by the cartilaginous interpubic disc. Another is the joint between the bodies of two vertebrae, united by an intervertebral disc. The surface of each vertebral body is covered with hyaline cartilage. Between the vertebrae, this cartilage becomes infiltrated with collagen bundles to form fibrocartilage. Each intervertebral disc permits only slight movement between adjacent vertebrae, but the collective effect of all 23 discs gives the spine considerable flexibility.

▶▶▶**APPLY WHAT YOU KNOW**

*The intervertebral joints are symphyses only in the cervical through the lumbar region. How would you classify the intervertebral joints of the sacrum and coccyx in a middle-aged adult?*

**BEFORE YOU GO ON**

Answer the following questions to test your understanding of the preceding section:

1. What is the difference between arthrology and kinesiology?
2. Distinguish between a synostosis, synarthrosis, and amphiarthrosis.
3. Define suture, gomphosis, and syndesmosis, and explain what these three joints have in common.
4. Name the three types of sutures and describe how they differ.
5. Name two synchondroses and two symphyses.
6. Give some examples of joints that become synostoses with age.

## 9.2  Synovial Joints

### Expected Learning Outcomes

When you have completed this section, you should be able to

a. identify the anatomical components of a typical synovial joint;
b. classify any given joint action as a first-, second-, or third-class lever;
c. explain how mechanical advantage relates to the power and speed of joint movement;
d. discuss the factors that determine a joint's range of motion;
e. describe the primary axes of rotation that a bone can have and relate this to a joint's degrees of freedom;

f. name and describe six classes of synovial joints; and
g. use the correct standard terminology for various joint movements.

The most familiar type of joint is the **synovial** (sih-NO-vee-ul) **joint,** also called a **diarthrosis**[10] (DY-ar-THRO-sis) or **diarthrodial joint.** Ask most people to point out any joint in the body, and they are likely to point to a synovial joint such as an elbow, knee, or knuckle. Many synovial joints, like these examples, are freely mobile. Others, such as the joints between the wrist and ankle bones and between the articular processes of the vertebrae, have more limited mobility.

Synovial joints are the most structurally complex type of joint and are the type most likely to develop uncomfortable and crippling dysfunctions. They are the most important joints for such professionals as physical and occupational therapists, athletic coaches, nurses, and fitness trainers to understand well. Their mobility makes the synovial joints especially important to the quality of life. Reflect, for example, on the performance extremes of a young athlete, the decline in flexibility that comes with age, and the crippling effect of rheumatoid arthritis. The rest of this chapter is concerned with synovial joints.

### General Anatomy

In synovial joints, the facing surfaces of the two bones are covered with **articular cartilage,** a layer of hyaline cartilage usually 2 or 3 mm thick. These surfaces are separated by a narrow space, the **joint (articular) cavity,** containing a slippery lubricant called **synovial fluid** (fig. 9.5). This fluid, for which the joint is named, is rich in albumin and hyaluronic acid, which give it a viscous, slippery texture similar to raw egg white.[11] It nourishes the articular cartilages, removes their wastes, and makes movements at synovial joints almost friction-free. A connective tissue **joint (articular) capsule** encloses the cavity and retains the fluid. It has an outer **fibrous capsule** continuous with the periosteum of the adjoining bones, and an inner, cellular **synovial membrane.** The synovial membrane is composed mainly of fibroblast-like cells that secrete the fluid, and is populated by macrophages that remove debris from the joint cavity.

In a few synovial joints, fibrocartilage grows inward from the joint capsule and forms a pad between the articulating bones. In the jaw (temporomandibular) joint, at both ends of the clavicle (sternoclavicular and acromioclavicular joints), and between the ulna and carpal bones, the pad crosses the entire joint capsule and is called an **articular disc** (see fig. 9.23c). In the knee, two cartilages extend inward from the left and right but do not entirely cross the joint (see fig. 9.29d). Each is called a **meniscus**[12] because of its crescent-moon shape. These

---

[9]*sym* = together; *physis* = growth

[10]*dia* = separate, apart; *arthr* = joint; *osis* = condition
[11]*ovi* = egg
[12]*men* = moon, crescent; *isc* = little

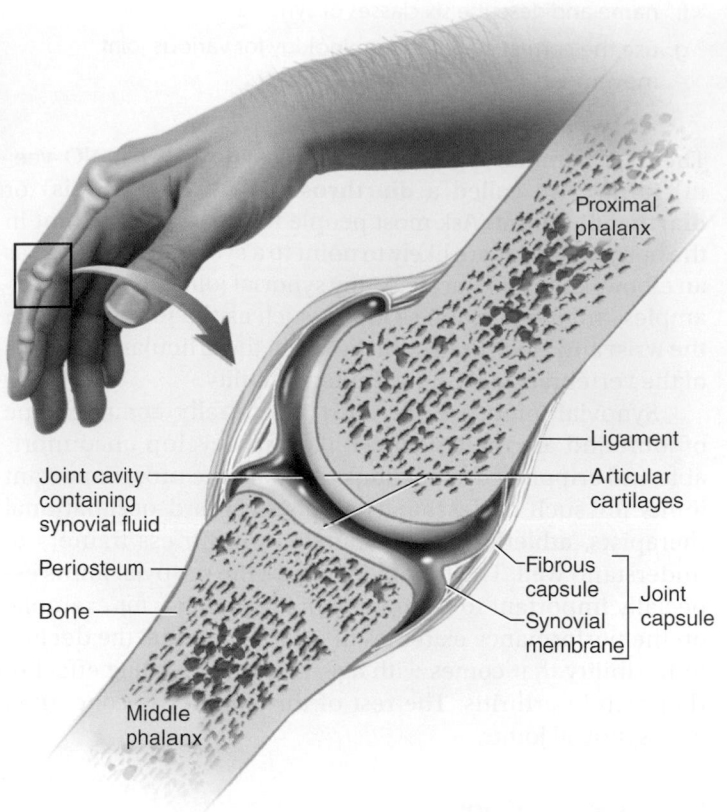

**FIGURE 9.5 Structure of a Simple Synovial Joint.**

❓ *Why is a meniscus unnecessary in an interphalangeal joint?*

cartilages absorb shock and pressure, guide the bones across each other, improve the fit between the bones, and stabilize the joint, reducing the chance of dislocation.

Accessory structures associated with a synovial joint include tendons, ligaments, and bursae. A **tendon** is a strip or sheet of tough collagenous connective tissue that attaches a muscle to a bone. Tendons are often the most important structures in stabilizing a joint. A **ligament** is a similar tissue that attaches one bone to another. Several ligaments are named and illustrated in our later discussion of individual joints, and tendons are more fully considered in chapter 10 along with the gross anatomy of muscles.

A **bursa**[13] is a fibrous sac filled with synovial fluid, located between adjacent muscles, where a tendon passes over a bone, or between bone and skin (see fig. 9.24). Bursae cushion muscles, help tendons slide more easily over the joints, and sometimes enhance the mechanical effect of a muscle by modifying the direction in which its tendon pulls. **Tendon (synovial) sheaths** are elongated cylindrical bursae wrapped around a tendon, seen especially in the hand and foot (fig. 9.6). They enable tendons to move back and forth more freely in such tight spaces as the wrist and ankle.

[13]*bursa* = purse

## DEEPER INSIGHT 9.1
### CLINICAL APPLICATION

#### Exercise and Articular Cartilage

When synovial fluid is warmed by exercise, it becomes thinner (less viscous) and more easily absorbed by the articular cartilage. The cartilage then swells and provides a more effective cushion against compression. For this reason, a warm-up period before vigorous exercise helps protect the articular cartilage from undue wear and tear.

Because cartilage is nonvascular, repetitive compression during exercise is important to its nutrition and waste removal. Each time a cartilage is compressed, fluid and metabolic wastes are squeezed out of it. When weight is taken off the joint, the cartilage absorbs synovial fluid like a sponge, and the fluid carries oxygen and nutrients to the chondrocytes. Without exercise, articular cartilages deteriorate more rapidly from inadequate nutrition, oxygenation, and waste removal.

Weight-bearing exercise builds bone mass and strengthens the muscles that stabilize many of the joints, thus reducing the risk of joint dislocations. Excessive joint stress, however, can hasten the progression of osteoarthritis by damaging the articular cartilage (see Deeper Insight 9.5, p. 303). Swimming is a good way of exercising the joints with minimal damage.

## Joints and Lever Systems

Many bones, especially the long bones, act as levers to enhance the speed or power of limb movements. A lever is any elongated, rigid object that rotates around a fixed point called the fulcrum (fig. 9.7). Rotation occurs when an effort applied to one point on the lever overcomes a resistance (load) at some other point. The portion of a lever from the fulcrum to the point of effort is called the **effort arm,** and the part from the fulcrum to the point of resistance is called the **resistance arm.** In skeletal anatomy, the fulcrum is a joint; the effort is applied by a muscle; and the resistance can be an object against which the body is working (as in weight lifting), the weight of the limb itself, or the tension in an opposing muscle.

## Mechanical Advantage

The function of a lever is to produce a gain in the speed, distance, or force of a motion—either to exert more force against a resisting object than the force applied to the lever (for example, in moving a heavy object with a crowbar), or to move the resisting object farther or faster than the effort arm is moved (as in rowing a boat, where the blade of the oar moves much farther and faster than the handle). A single lever cannot confer both advantages. There is a trade-off between force on one hand and speed or distance on the other—as one increases, the other decreases.

The **mechanical advantage (MA)** of a lever is the ratio of its output force to its input force. If $L_E$ is the length of the effort arm and $L_R$ is the length of the resistance arm, $MA = L_E/L_R$. If MA is greater than 1.0, the lever produces more force, but less

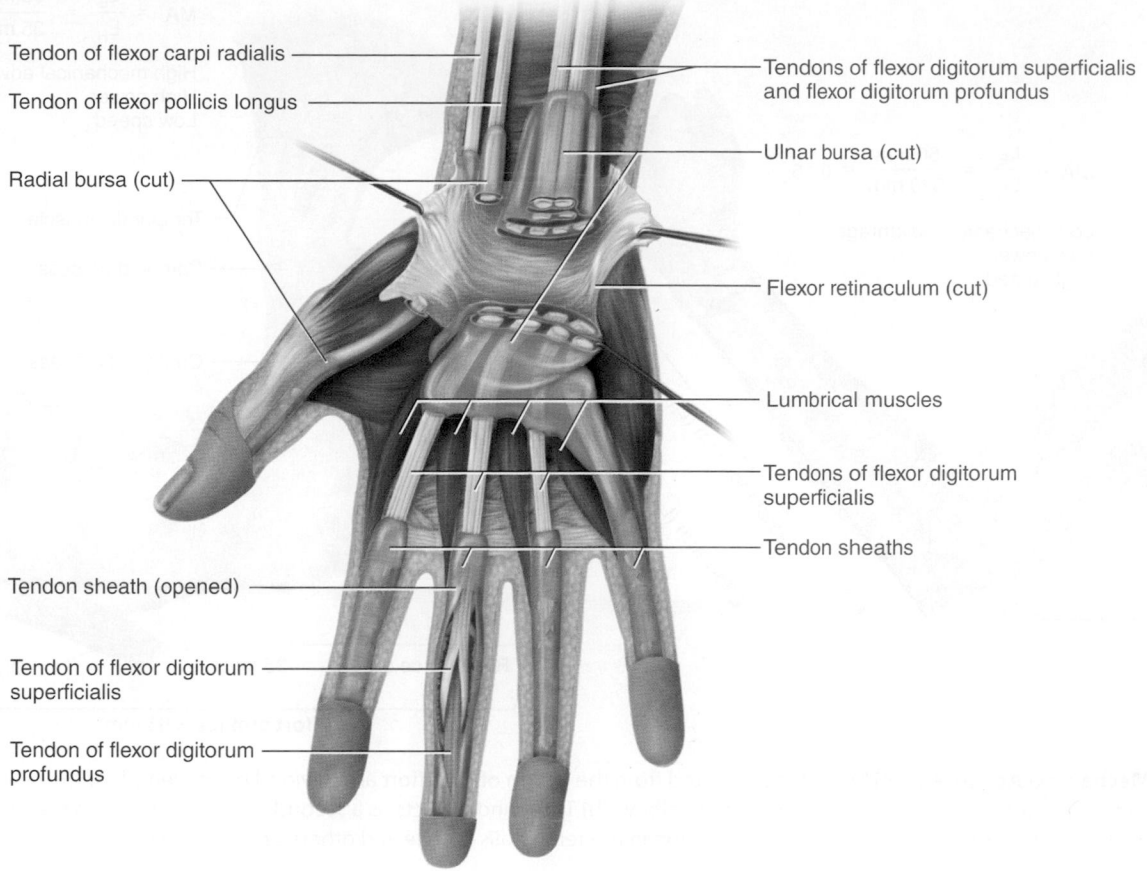

**FIGURE 9.6 Tendon Sheaths and Other Bursae in the Hand and Wrist.**

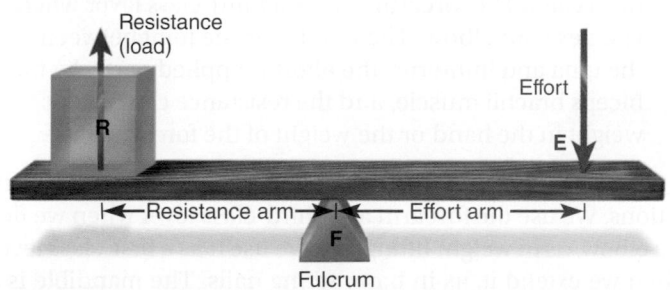

**FIGURE 9.7 The Basic Components of a Lever.** This example is a first-class lever.

❓ *What would be the mechanical advantage of the lever shown here? Where would you put the fulcrum to increase the mechanical advantage without changing the lever class?*

speed or distance, than the force exerted on it. If MA is less than 1.0, the lever produces more speed or distance, but less force, than the input.

Consider the forearm, for example (fig. 9.8a). The resistance arm of the ulna is longer than its effort arm, so we know

from the preceding formula that MA is less than 1.0. The figure shows some representative values for $L_E$ and $L_R$ that yield MA = 0.15. The biceps muscle puts more power into the lever than we get out of it, but the hand moves farther and faster than the point where the biceps tendon inserts on the radius. Most musculoskeletal levers operate with an MA much less than 1, but figure 9.8b shows a case with MA greater than 1.

In chapter 10, you will often find that two or more muscles act on the same joint, seemingly producing the same effect. This may seem redundant, but it makes sense if the tendinous insertions of the muscles are at different points on a bone and produce different mechanical advantages. A sprinter taking off from the starting line, for example, uses "low-gear" (high-MA) muscles that do not generate maximum speed, but have the power to overcome the inertia of the body. The runner then "shifts into high gear" by using muscles with different insertions that have a lower mechanical advantage but produce more speed. This is analogous to the way an automobile transmission uses one gear to get a car moving and other gears to cruise at higher speeds.

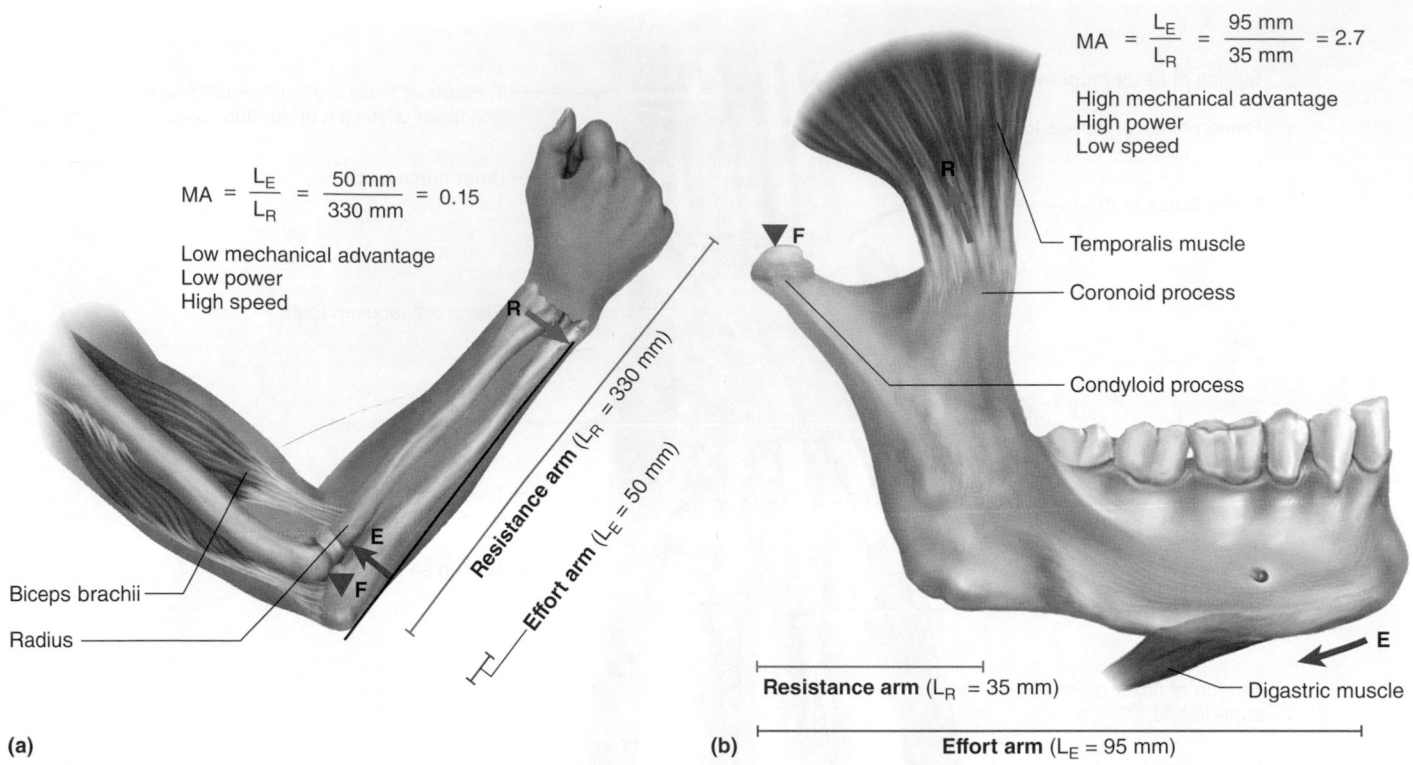

**FIGURE 9.8 Mechanical Advantage (MA).** MA is calculated from the length of the effort arm divided by the length of the resistance arm. (a) The forearm acts as a third-class lever during flexion of the elbow. (b) The mandible acts as a second-class lever when the jaw is forcibly opened. The digastric muscle and others provide the effort, while tension in the temporalis muscle and others provide resistance.

## Types of Levers

There are three classes of levers that differ with respect to which component is in the middle—the fulcrum (F), effort (E), and resistance (R) (fig. 9.9).

1. A **first-class lever** is one with the fulcrum in the middle (EFR), such as a seesaw. An anatomical example is the atlanto–occipital joint of the neck, where the muscles of the back of the neck pull down on the occipital bone of the skull and oppose the tendency of the head to tip forward. Loss of muscle tone here can be embarrassing if you nod off in class. Rocking of the foot on the tibia as the toes are raised and lowered also exemplifies a first-class lever. (It is often misinterpreted as a second-class lever because of a superficial resemblance between standing on tiptoes and the wheelbarrow example that follows.)

2. A **second-class lever** has the resistance in the middle (FRE). Lifting the handles of a wheelbarrow, for example, causes it to pivot on the axle of the wheel at the opposite end and lift a load in the middle. If you sit in a chair and raise one thigh, like bouncing a small child on your knee, the femur pivots on the hip joint (the fulcrum), the quadriceps femoris muscle of the anterior thigh elevates the tibia like the wheelbarrow handles, and the resistance is the weight of the child or of the thigh itself.

3. In a **third-class lever,** the effort is applied between the fulcrum and resistance (REF). For example, in paddling a canoe, the relatively stationary grip at the upper end of the paddle is the fulcrum, the effort is applied to the middle of the shaft, and the resistance is produced by the water against the blade. Most musculoskeletal levers are third class. The forearm acts as a third-class lever when you flex your elbow. The fulcrum is the joint between the ulna and humerus, the effort is applied partly by the biceps brachii muscle, and the resistance can be any weight in the hand or the weight of the forearm itself.

The classification of a lever changes as it makes different actions. We use the forearm as a third-class lever when we flex the elbow, as in weight lifting, but we use it as a first-class lever when we extend it, as in hammering nails. The mandible is a second-class lever when we open the mouth and a third-class lever when we close it to bite off a piece of food.

## Range of Motion

One aspect of joint performance and physical assessment of a patient is a joint's flexibility, or **range of motion (ROM)**—the degrees through which a joint can move. The knee, for example, can flex through an arc of 130° to 140°, the metacarpophalangeal joint of the index finger about 90°, and the ankle about 74°. ROM obviously affects a person's functional independence and quality of life. It is also an important consideration in training for athletics or dance, in clinical diagnosis, and in monitoring the progress of rehabilitation. The ROM of a joint is normally determined by the following factors:

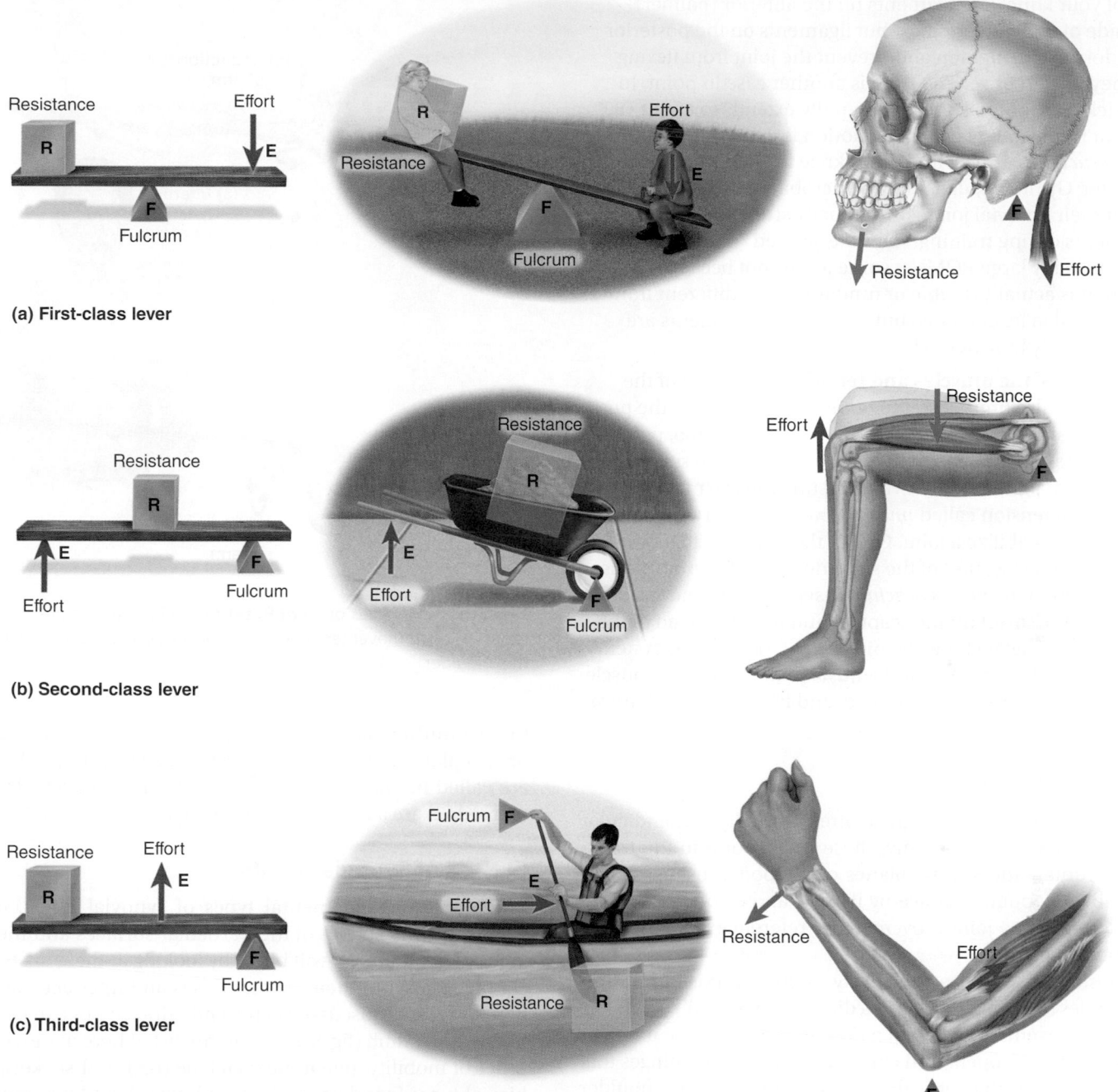

**(a) First-class lever**

**(b) Second-class lever**

**(c) Third-class lever**

**FIGURE 9.9 The Three Classes of Levers.** Left: The lever classes defined by the relative positions of the resistance (load), fulcrum, and effort. Center: Mechanical examples. Right: Anatomical examples. (a) Muscles of the back of the neck pull down on the occipital bone to oppose the tendency of the head to drop forward. The fulcrum is the occipital condyles. (b) The quadriceps muscle of the anterior thigh elevates the knee. The fulcrum is the hip joint. (c) In flexing the elbow, the biceps brachii muscle exerts an effort on the radius. Resistance is provided by the weight of the forearm or anything held in the hand. The fulcrum is the elbow joint.

- **Structure of the articular surfaces of the bones.** In many cases, joint movement is limited by the shapes of the bone surfaces. For example, you cannot straighten your elbow beyond 180° or so because, as it straightens, the olecranon of the ulna swings into the olecranon fossa of the humerus and the fossa prevents it from moving any farther.

- **Strength and tautness of ligaments and joint capsules.** Some bone surfaces impose little if any limitation on joint movement. The articulations of the phalanges are an example; as one can see by examining a dry skeleton, an interphalangeal joint can bend through a broad arc. In the living body, however, these bones are joined by ligaments that limit their movement. As you flex one of your knuckles, ligaments on the anterior (palmar) side of the joint go slack, but ligaments on the posterior (dorsal) side tighten and prevent the joint from flexing beyond 90° or so. The knee is another case in point. In kicking a football, the knee rapidly extends to about 180°, but it can go no farther. Its motion is limited in part by a *cruciate ligament* and other knee ligaments described later. Gymnasts, dancers, and acrobats increase the ROM of their synovial joints by gradually stretching their ligaments during training. "Double-jointed" people have unusually large ROMs at some joints, not because the joint is actually double or fundamentally different from normal in its anatomy, but because the ligaments are unusually long or slack.

- **Action of the muscles and tendons.** Extension of the knee is also limited by the *hamstring muscles* on the posterior side of the thigh. In many other joints, too, pairs of muscles oppose each other and moderate the speed and range of joint motion. Even a resting muscle maintains a state of tension called *muscle tone,* which serves in many cases to stabilize a joint. One of the major factors preventing dislocation of the shoulder joint, for example, is tension in the *biceps brachii* muscle, whose tendons cross the joint, insert on the scapula, and hold the head of the humerus against the glenoid cavity. The nervous system continually monitors and adjusts joint angles and muscle tone to maintain joint stability and limit unwanted movements.

## Axes of Rotation

In solid geometry, we recognize three mutually perpendicular axes, *x, y,* and *z.* In anatomy, these correspond to the transverse, frontal, and sagittal planes of the body. Just as we can describe any point in space by its *x, y,* and *z* coordinates, we can describe any joint movement by reference to these three anatomical planes.

A moving bone has a relatively stationary **axis of rotation** that passes through the bone in a direction perpendicular to the plane of movement. Think of a door for comparison; it moves horizontally as it opens and closes and it rotates on hinges that are oriented on the vertical axis. Now consider the shoulder joint (fig. 9.10), where the convex head of the humerus inserts into the concave glenoid cavity of the scapula. If you raise your arm to one side of your body, the head of the humerus rotates on an axis that passes from anterior to posterior; the arm rises in the frontal plane whereas its axis of rotation is in the sagittal plane. If you lift your arm to point at something straight in front of you, it moves through the sagittal plane whereas its axis of rotation is on the frontal plane, passing through the shoulder from lateral to medial. And if you swing your arm in a horizontal arc, for example to fold it across your chest, the humeral head rotates in the transverse plane and its axis of rotation passes vertically through the joint.

Because the arm can move in all three anatomical planes, the shoulder joint is said to have three **degrees of freedom,** or

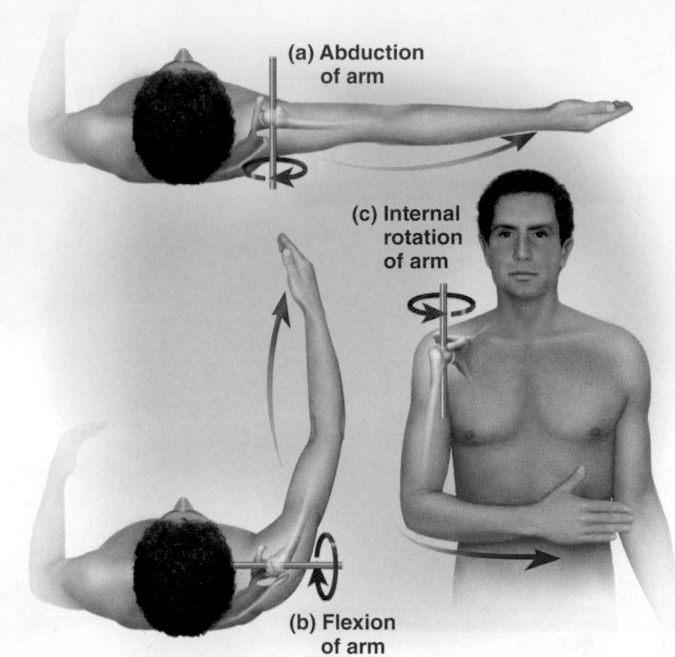

**FIGURE 9.10  Axes of Joint Rotation.**  All three axes are represented in movements of the multiaxial ball-and-socket joint of the shoulder.

to be a **multiaxial** joint. Other joints move through only one or two planes; they have one or two degrees of freedom and are called **monaxial** and **biaxial** joints, respectively. Degrees of freedom are a factor used in classifying the synovial joints.

## Classes of Synovial Joints

There are six fundamental types of synovial joints, distinguished by the shapes of their articular surfaces and their degrees of freedom. We will begin by looking at these six types in simple terms, but then see that this is an imperfect classification for reasons discussed at the end. All six types can be found in the upper limb (fig. 9.11). They are listed here in descending order of mobility: one multiaxial type (ball-and-socket), three biaxial types (condylar, saddle, and plane), and two monaxial types (hinge and pivot).

1. **Ball-and-socket joints.** These are the shoulder and hip joints—the only multiaxial joints in the body. In both cases, one bone (the humerus or femur) has a smooth hemispherical head that fits into a cuplike socket on the other (the glenoid cavity of the scapula or the acetabulum of the hip bone).

2. **Condylar (ellipsoid) joints.** These joints exhibit an oval convex surface on one bone that fits into a complementary-shaped depression on the other. The radiocarpal joint of the wrist and metacarpophalangeal (MET-uh-CAR-po-fah-LAN-jee-ul) joints at the bases of the fingers are examples. They are biaxial joints, capable

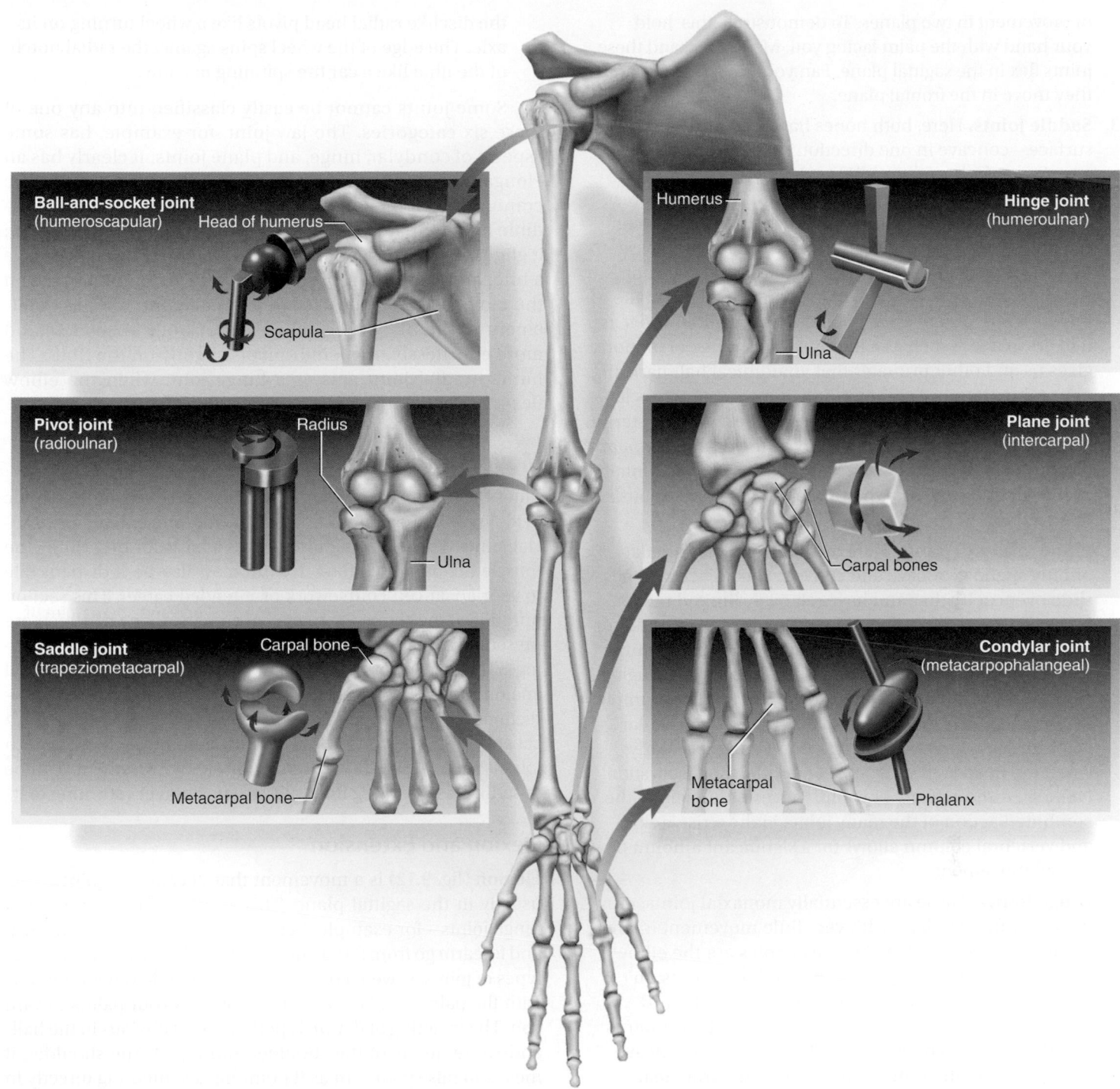

**FIGURE 9.11  The Six Types of Synovial Joints.**  All six have representatives in the forelimb. Mechanical models show the types of motion possible at each joint.

of movement in two planes. To demonstrate this, hold your hand with the palm facing you. Make a fist, and these joints flex in the sagittal plane. Fan your fingers apart, and they move in the frontal plane.

3. **Saddle joints.** Here, both bones have a saddle-shaped surface—concave in one direction (like the front-to-rear curvature of a horse's saddle) and convex in the other (like the left-to-right curvature of a saddle). The clearest example of this is the trapeziometacarpal joint between the trapezium of the wrist and metacarpal I at the base of the thumb. Saddle joints are biaxial. The thumb, for example, moves in a frontal plane when you spread the fingers apart, and in a sagittal plane when you move it as if to grasp a tool such as a hammer. This range of motion gives us and other primates that anatomical hallmark, the opposable thumb. Another saddle joint is the sternoclavicular joint, where the clavicle articulates with the sternum. The clavicle moves vertically in the frontal plane at this joint when you lift a suitcase, and moves horizontally in the transverse plane when you reach forward to push open a door.

4. **Plane (gliding) joints.** Here the bone surfaces are flat or only slightly concave and convex. The adjacent bones slide over each other and have relatively limited movement. Plane joints are found between the carpal bones of the wrist, the tarsal bones of the ankle, and the articular processes of the vertebrae. Their movements, although slight, are complex. They are usually biaxial. For example, when the head is tilted forward and back, the articular facets of the vertebrae slide anteriorly and posteriorly; when the head is tilted from side to side, the facets slide laterally. Although any one joint moves only slightly, the combined action of the many joints in the wrist, ankle, and vertebral column allows for a significant amount of overall movement.

5. **Hinge joints.** These are essentially monaxial joints, moving freely in one plane with very little movement in any other, like a door hinge. Some examples are the elbow, knee, and interphalangeal (finger and toe) joints. In these cases, one bone has a convex (but not hemispherical) surface, such as the trochlea of the humerus and the condyles of the femur. This fits into a concave depression on the other bone, such as the trochlear notch of the ulna and the condyles of the tibia.

6. **Pivot joints.** These are monaxial joints in which a bone spins on its longitudinal axis. There are two principal examples: the atlantoaxial joint between the first two vertebrae, and the radioulnar joint at the elbow. At the atlantoaxial joint, the dens of the axis projects into the vertebral foramen of the atlas and is held against the anterior arch of the atlas by the transverse ligament (see fig. 8.24, p. 250). As the head rotates left and right, the skull and atlas pivot around the dens. At the radioulnar joint, the anular ligament of the ulna wraps around the neck of the radius. During pronation and supination of the forearm, the disclike radial head pivots like a wheel turning on its axle. The edge of the wheel spins against the radial notch of the ulna like a car tire spinning in snow.

Some joints cannot be easily classified into any one of these six categories. The jaw joint, for example, has some aspects of condylar, hinge, and plane joints. It clearly has an elongated condyle where it meets the temporal bone of the cranium, but it moves in a hingelike fashion when the mandible moves up and down in speaking, biting, and chewing; it glides slightly forward when the jaw juts (protracts) to take a bite; and it glides from side to side to grind food between the molars. The knee is a classic hinge joint, but has an element of the pivot type; when we lock our knees to stand more effortlessly, the femur pivots slightly on the tibia. The humeroradial joint acts as a hinge joint when the elbow flexes and a pivot joint when the forearm pronates.

## Movements of Synovial Joints

Kinesiology, physical therapy, and other medical and scientific fields have a specific vocabulary for the movements of synovial joints. The following terms form a basis for describing the muscle actions in chapter 10 and may also be indispensable to your advanced coursework or intended career. This section introduces the terms for joint movements, many of which are presented in pairs or groups with opposite or contrasting meanings. This section relies on familiarity with the three cardinal anatomical planes and the directional terms in atlas A. All directional terms used here refer to a person in standard anatomical position. When one is standing in anatomical position, each joint is said to be in its **zero position.** Joint movements can be described as deviating from the zero position or returning to it.

### Flexion and Extension

**Flexion** (fig. 9.12) is a movement that decreases a joint angle, usually in the sagittal plane. This is particularly common at hinge joints—for example, bending the elbow so that the arm and forearm go from a 180° angle to 90° or less. It occurs in other types of joints as well. For example, if you hold out your hands with the palms up, flexion of the wrist tips your palms toward you. The meaning of *flexion* is perhaps least obvious in the ball-and-socket joints of the shoulder and hip. At the shoulder, it means to raise your arm as if pointing at something directly in front of you or to continue in that arc and point toward the sky. At the hip, it means to raise the thigh, for example to place your foot on the next higher step when ascending a flight of stairs.

**Extension** (fig. 9.12) is a movement that straightens a joint and generally returns a body part to the zero position—for example, straightening the elbow, wrist, or knee, or returning the arm or thigh back to zero position. In stair climbing, both the hip and knee extend when lifting the body to the next higher step.

Further extension of a joint beyond the zero position is called **hyperextension.**[14] For example, if you hold your hand in

---

[14]*hyper* = excessive, beyond normal

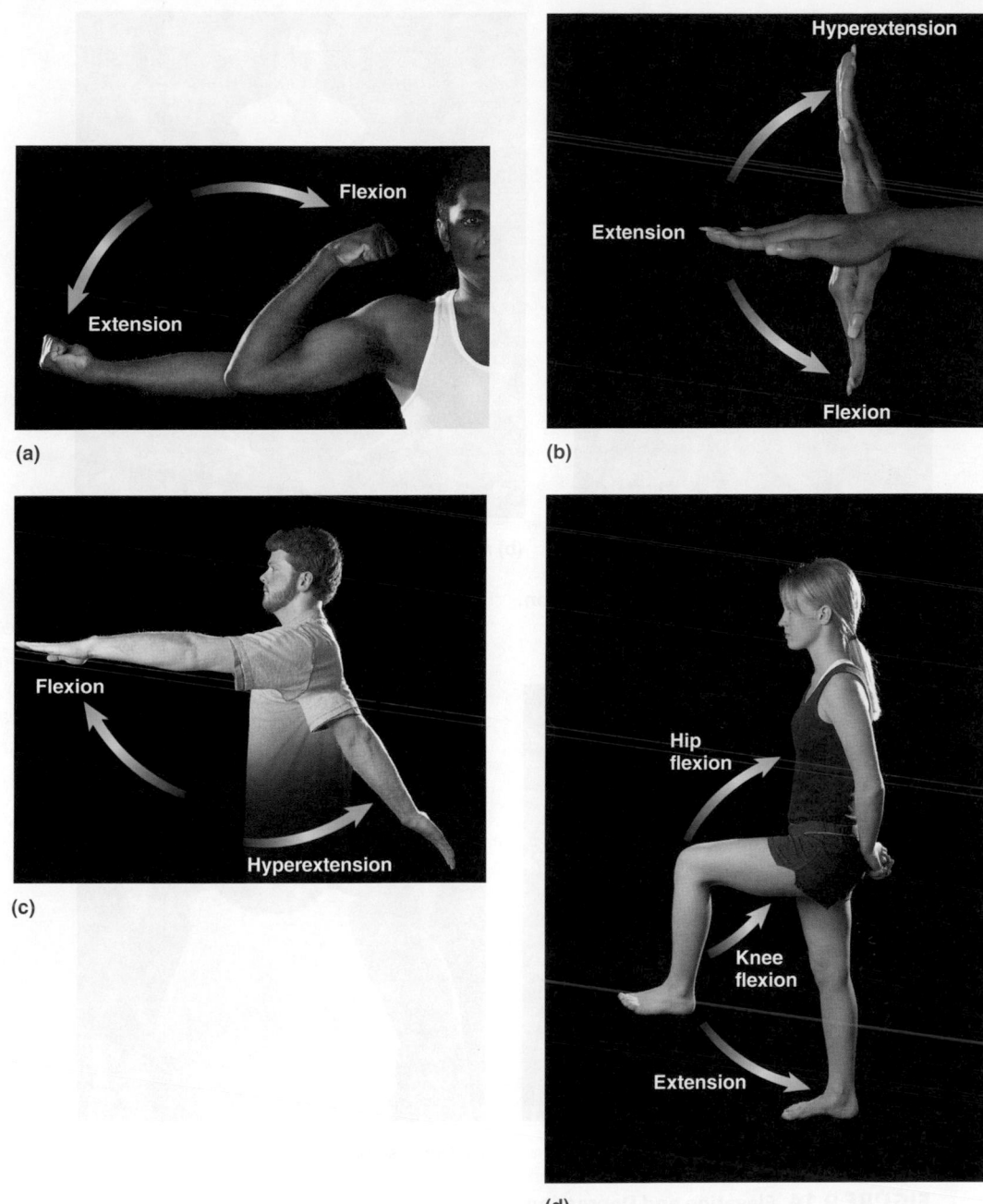

**FIGURE 9.12 Flexion and Extension.** (a) Flexion and extension of the elbow. (b) Flexion, extension, and hyperextension of the wrist. (c) Flexion and hyperextension of the shoulder. (d) Flexion and extension of the hip and knee.

front of you with the palm down, then raise the back of your hand as if you were admiring a new ring, you hyperextend the wrist. Hyperextension of the upper or lower limb means to move the limb to a position behind the frontal plane of the trunk, as if reaching around with your arm to scratch your back. Each backswing of the lower limb when you walk hyperextends the hip.

Flexion and extension occur at nearly all diarthroses, but hyperextension is limited to only a few. At most diarthroses, ligaments or bone structure prevents hyperextension.

## Abduction and Adduction

**Abduction**[15] (ab-DUC-shun) (fig. 9.13a) is the movement of a body part in the frontal plane away from the midline of the body—for example, moving the feet apart to stand spread-legged, or raising an arm to one side of the body. **Adduction**[16] (fig. 9.13b) is movement in the frontal plane back toward the

---

[15]*ab* = away; *duc* = to lead or carry
[16]*ad* = toward; *duc* = to lead or carry

 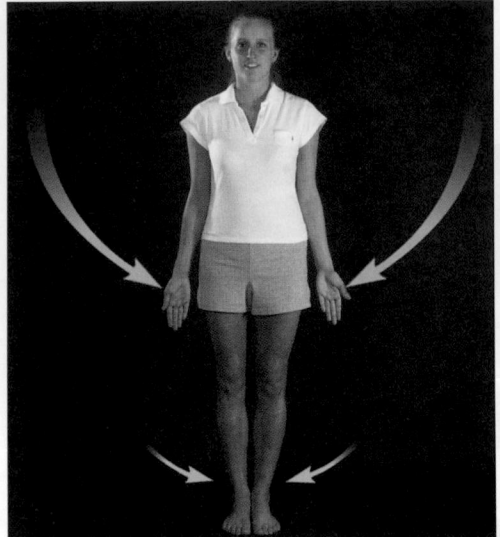

(a) Abduction                              (b) Adduction

**FIGURE 9.13**  Abduction and Adduction.

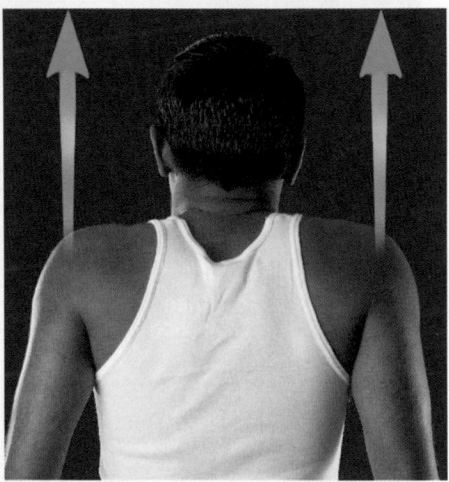

(a) Elevation                              (b) Depression

**FIGURE 9.14**  Elevation and Depression.

midline. Some joints can be **hyperadducted,** as when you stand with your ankles crossed, cross your fingers, or hyperadduct the shoulder to stand with your elbows straight and your hands clasped below your waist. You **hyperabduct** the arm if you raise it high enough to cross slightly over the front or back of your head.

## Elevation and Depression

**Elevation** (fig. 9.14a) is a movement that raises a body part vertically in the frontal plane. **Depression** (fig. 9.14b) lowers a body part in the same plane. For example, to lift a suitcase

from the floor, you elevate your scapula; in setting it down again, you depress the scapula. These are also important jaw movements in biting.

## Protraction and Retraction

**Protraction**[17] (fig. 9.15a) is the anterior movement of a body part in the transverse (horizontal) plane, and **retraction**[18] (fig. 9.15b) is posterior movement. Your shoulder protracts, for

[17]*pro* = forward; *trac* = to pull or draw
[18]*re* = back; *trac* = to pull or draw

(a) Protraction

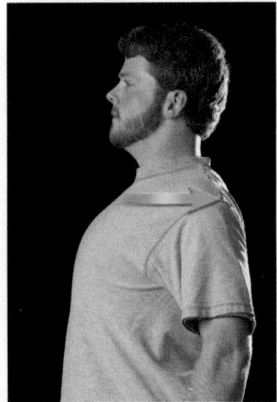

(b) Retraction

**FIGURE 9.15** Protraction and Retraction.

example, when you reach in front of you to push a door open. It retracts when you return it to the resting (zero) position or pull the shoulders back to stand at military attention. Such exercises as rowing a boat, bench presses, and push-ups involve repeated protraction and retraction of the shoulders.

## Circumduction

In **circumduction**[19] (fig. 9.16), one end of an appendage remains fairly stationary while the other end makes a circular motion. If an artist standing at an easel reaches forward and draws a circle on a canvas, she circumducts the upper limb; the shoulder remains stationary while the hand moves in a circle. A baseball player winding up for the pitch circumducts the upper limb in a more extreme "windmill" fashion. One can also circumduct an individual finger, the hand, the thigh, the foot, the trunk, and the head.

▶▶▶ **APPLY WHAT YOU KNOW**

*Choose any example of circumduction and explain why this motion is actually a sequence of flexion, abduction, extension, and adduction.*

---

[19]*circum* = around; *duc* = to carry, lead

**FIGURE 9.16** Circumduction.

## Rotation

In one sense, the term *rotation* applies to any bone turning around a fixed axis, as described earlier. But in the terminology of specific joint movements, **rotation** (fig. 9.17) is a movement in which a bone spins on its longitudinal axis. For example, if you stand with bent elbow and move your forearm to place your palm against your abdomen, your humerus spins in a motion called **medial (internal) rotation** (fig. 9.17a). If you make the opposite motion, so the forearm points away from the body, your humerus exhibits **lateral (external) rotation** (fig. 9.17b). A tennis player's forehand and backhand strokes

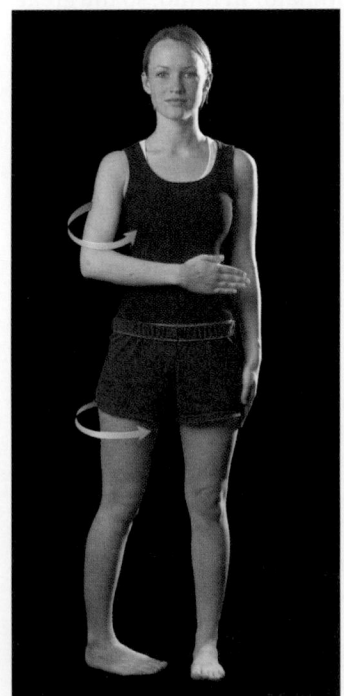

(a) Medial (internal) rotation

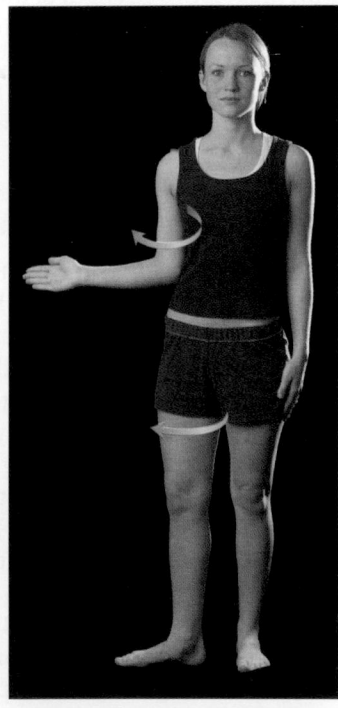

(b) Lateral (external) rotation

**FIGURE 9.17** Medial (Internal) and Lateral (External) Rotation. Shows rotation of both the humerus and femur.

entail vigorous rotation of the humerus. If you are standing and turn your right foot so your toes are pointing away from your left foot, and then turn it so your toes are pointing toward your left foot, your femur undergoes lateral and medial rotation, respectively.

## Supination and Pronation

Supination and pronation are known primarily as forearm movements, but see also the later discussion of foot movements. **Supination**[20] (SOO-pih-NAY-shun) (fig. 9.18a) of the forearm is a movement that turns the palm to face anteriorly or upward; in anatomical position, the forearm is supinated and the radius is parallel to the ulna. **Pronation**[21] (fig. 9.18b) is the opposite movement, causing the palm to face posteriorly or downward, and the radius to cross the ulna like an X. During these movements, the concave end of the disc-shaped head of the radius spins on the capitulum of the humerus, and the edge of the disc spins in the radial notch of the ulna. The ulna remains relatively stationary.

As an aid to remembering these terms, think of it this way: You are *prone* to stand in the most comfortable position, which is with the forearm *pronated*. But if you were holding a cup of *soup* in your palm, you would need to *supinate* the forearm to keep from spilling it.

Chapter 10 describes the muscles that perform these actions. Of these, the *supinator* is the most powerful. Supination is the type of movement you would usually make with your right hand to turn a doorknob clockwise or to drive a screw into a piece of wood. The threads of screws and bolts are designed with the relative strength of the supinator in mind, so the greatest power can be applied when driving them with a screwdriver in the right hand.

We will now consider a few body regions that combine the foregoing motions, or that have unique movements and terminology.

## Special Movements of the Head and Trunk

*Flexion* of the vertebral column produces forward-bending movements, as in tilting the head forward or bending at the waist in a toe-touching exercise (fig. 9.19a). *Extension* of the vertebral column straightens the trunk or the neck, as in standing up or returning the head to a forward-looking (zero) position. *Hyperextension* is employed in looking up toward the sky or bending over backward (fig. 9.19b).

**Lateral flexion** is tilting the head or trunk to the right or left of the midline (fig. 9.19c). Twisting at the waist or turning of the head is called **right rotation** or **left rotation** when the chest or the face turns to the right or left of the forward-facing zero position (fig. 9.19d, e). Powerful right and left rotation at the waist is important in baseball pitching, golf, discus throwing, and other sports.

[20]*supin* = to lay back
[21]*pron* = to bend forward

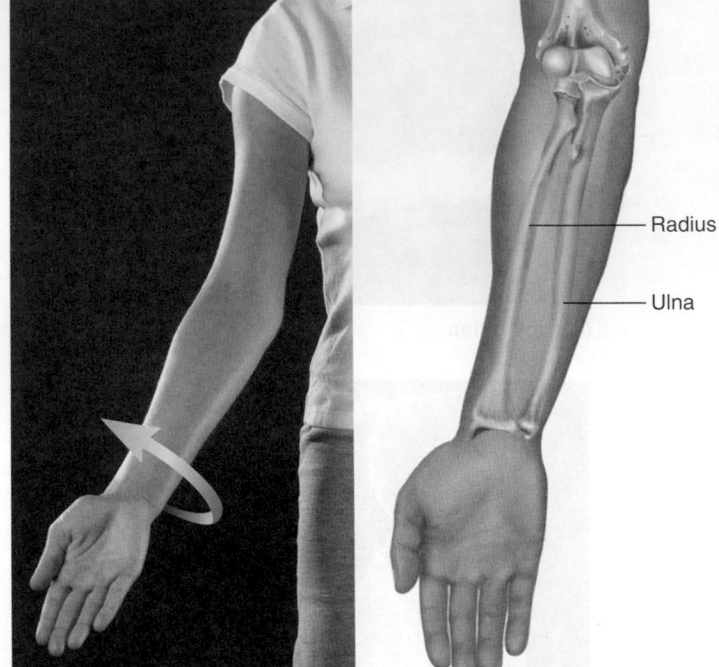

Radius

Ulna

**(a) Supination**

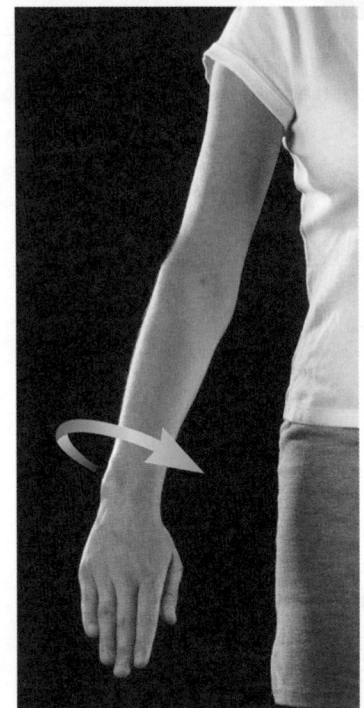
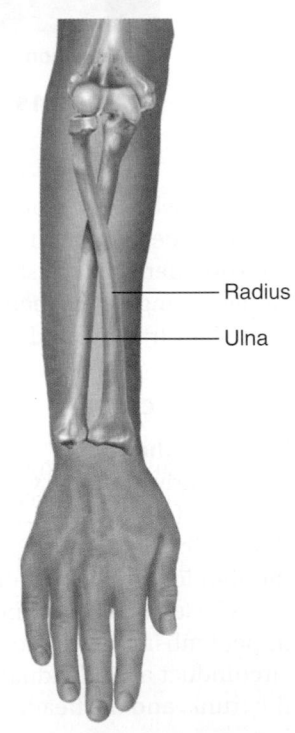

Radius

Ulna

**(b) Pronation**

**FIGURE 9.18  Supination and Pronation.**  Note the way these forearm rotations affect the relationship of the radius and ulna. Relative positions of muscles, nerves, and blood vessels are similarly affected.

## Special Movements of the Mandible

Movements of the mandible are concerned especially with biting and chewing. Imagine taking a bite of raw carrot. Most people have some degree of overbite; at rest, the upper incisors

(a) Flexion

(b) Hyperextension

(c) Lateral flexion

(d) Right rotation

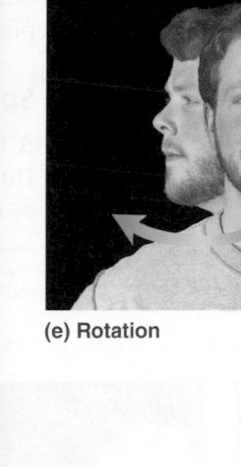

(e) Rotation

**FIGURE 9.19  Movements of the Head and Trunk.**

❓ *In rotation of the head (e), what bone spins on its axis?*

(front teeth) overhang the lower ones. For effective biting, however, the chisel-like edges of the incisors must meet. In preparation to bite, we therefore *protract* the mandible to bring the lower incisors forward. After the bite is taken, we *retract* it (fig. 9.20a, b). To actually take the bite, we must *depress* the mandible to open the mouth, then *elevate* it so the incisors cut off the piece of food.

Next, to chew the food, we do not simply raise and lower the mandible as if hammering away at the food between the teeth; rather, we exercise a grinding action that shreds the food between the broad, bumpy surfaces of the premolars and molars. This entails a side-to-side movement of the mandible called **lateral excursion** (movement to the left or right of the zero position) and **medial excursion** (movement back to the median, zero position) (fig. 9.20c, d).

## Special Movements of the Hand and Digits

The hand moves anteriorly and posteriorly by flexion and extension of the wrist. It can also move in the frontal plane. **Ulnar** **flexion** tilts the hand toward the little finger, and **radial flexion** tilts it toward the thumb (fig. 9.21a, b). We often use such motions when waving hello to someone with a side-to-side wave of the hand, or when washing windows, polishing furniture, or keyboarding.

Movements of the digits are more varied, especially those of the thumb. *Flexion* of the fingers is curling them; *extension* is straightening them. Most people cannot hyperextend their fingers. Spreading the fingers apart is abduction (fig. 9.21c), and bringing them together again so they touch along their surfaces is *adduction* (as in fig. 9.21a, b).

The thumb is different, however, because in embryonic development it rotates nearly 90° from the rest of the hand. If you hold your hand in a completely relaxed position (but not resting on a table), you will probably see that the plane that contains your thumb and index finger is about 90° to the plane that contains the index through little fingers. Much of the terminology of thumb movement therefore differs from that of the other four fingers. *Flexion* of the thumb is bending

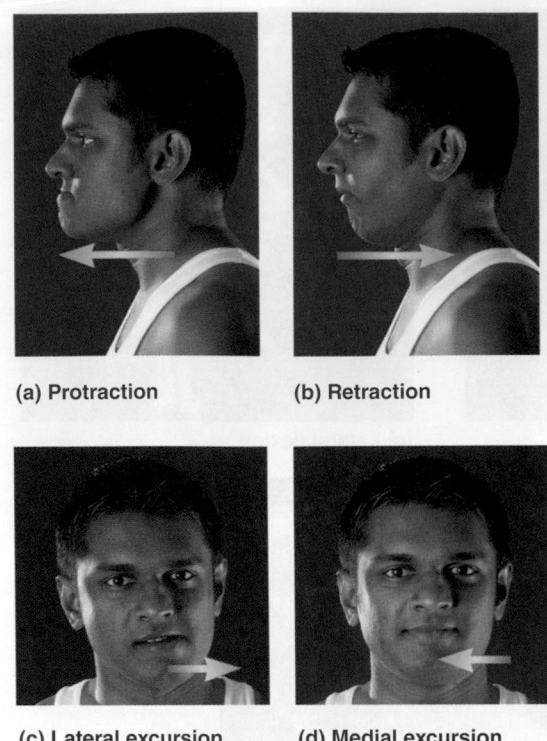

**(a) Protraction**          **(b) Retraction**

**(c) Lateral excursion**    **(d) Medial excursion**

**FIGURE 9.20   Movements of the Mandible.**

the joints so the tip of the thumb is directed toward the palm, and *extension* is straightening it. If you now place the palm of your hand on a tabletop with all five digits parallel and touching, the thumb is extended. Keeping your hand there, if you move your thumb away from the index finger so they form a 90° angle (but both are on the plane of the table), the thumb movement is called **radial abduction** (as in fig. 9.21c). Another movement, **palmar abduction,** moves the thumb away from the plane of the hand so it points anteriorly, as you would do if you were about to wrap your hand around a tool handle (fig. 9.21d). From either position—radial or palmar abduction—*adduction* of the thumb means to bring it back to zero position, touching the base of the index finger.

Two terms are unique to the thumb: **Opposition**[22] means to move the thumb to approach or touch the tip of any of the other four fingers (fig. 9.21e). **Reposition**[23] is the return to zero position.

## Special Movements of the Foot

A few additional movement terms are unique to the foot. **Dorsiflexion** is a movement in which the toes are elevated, as one might do in applying toenail polish (fig. 9.22a). In each

---

[22]*op* = against; *posit* = to place
[23]*re* = back; *posit* = to place

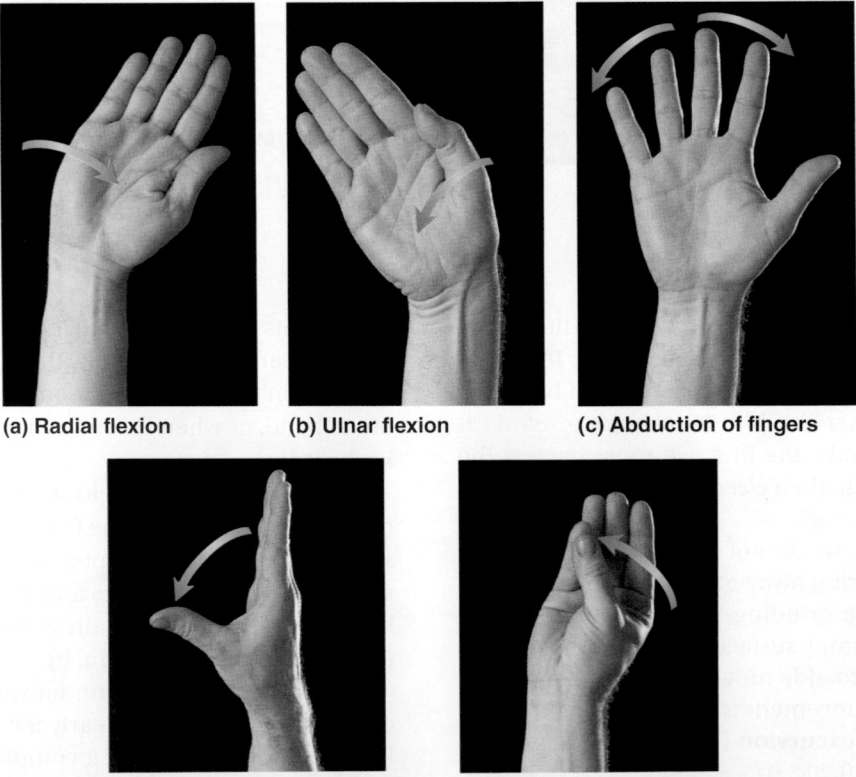

**(a) Radial flexion**       **(b) Ulnar flexion**       **(c) Abduction of fingers**

**(d) Palmar abduction of thumb**       **(e) Opposition of thumb**

**FIGURE 9.21   Movements of the Hand and Digits.**   (a) Radial flexion of the wrist. (b) Ulnar flexion of the wrist. (c) Abduction of the fingers. The thumb position in this figure is called *radial abduction*. Parts (a) and (b) show adduction of the fingers. (d) Palmar abduction of the thumb. (e) Opposition of the thumb; reposition is shown in parts (a) and (b).

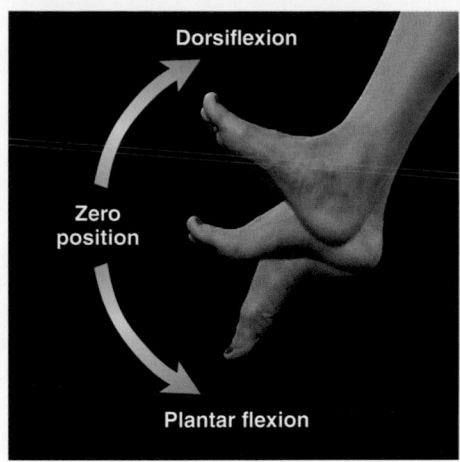

(a) Flexion of ankle

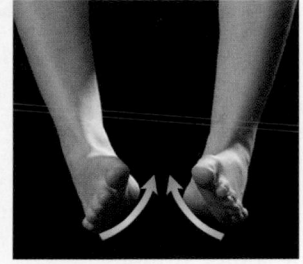

(b) Inversion

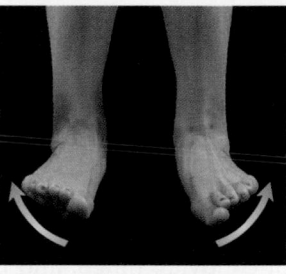

(c) Eversion

**FIGURE 9.22** Movements of the Foot.

step you take, the foot dorsiflexes as it comes forward. This prevents you from scraping your toes on the ground and results in the characteristic *heel strike* of human locomotion when the foot touches down in front of you. **Plantar flexion** is movement of the foot so the toes point downward, as in pressing the gas pedal of a car or standing on tiptoes. This motion also produces the *toe-off* in each step you take, as the heel of the foot behind you lifts off the ground. Plantar flexion can be a very powerful motion, epitomized by high jumpers and the jump shots of basketball players.

**Inversion**[24] is a foot movement that tips the soles medially, somewhat facing each other, and **eversion**[25] is a movement that tips the soles laterally, away from each other (fig. 9.22b, c). These movements are important in walking on uneven surfaces such as a rocky trail. They are common in fast sports such as tennis and football, and sometimes cause ankle sprains. These terms also refer to congenital deformities of the feet, which are often corrected by orthopedic shoes or braces.

*Pronation* and *supination,* referring mainly to forearm movements, also apply to the feet but refer here to a more complex combination of movements. Pronation of the foot is a combination of dorsiflexion, eversion, and abduction—that is, the toes are elevated and turned away from the other foot and the sole is tilted away from the other foot. Supination of the foot is a combination of plantar flexion, inversion, and adduction—the toes are lowered and turned toward the other foot and the sole is tilted toward it. These may seem a little difficult to visualize and perform, but they are common motions in walking, running, ballet, and crossing uneven surfaces.

You can perhaps understand why these terms apply to the feet if you place the palms of your hands on a table and pretend they are your soles. Tilt your hands so the inner edge (thumb side) of each is raised from the table. This is like raising the medial edge of your foot from the ground, and as you

can see, it involves a slight supination of your forearms. Resting your hands palms down on a table, your forearms are already pronated; but if you raise the outer edges of your hands (the little finger side), like pronating the feet, you will see that it involves a continuation of the pronation movement of the forearm.

**BEFORE YOU GO ON**

Answer the following questions to test your understanding of the preceding section:

7. Describe the roles of articular cartilage and synovial fluid in joint mobility.

8. Give an anatomical example of each class of levers and explain why each example belongs in that class.

9. Give an example of each of the six classes of synovial joints and state how many axes of rotation each example has.

10. Suppose you reach overhead and screw a light bulb into a ceiling fixture. Name each joint that would be involved and the joint actions that would occur.

11. Where are the effort, fulcrum, and resistance in the act of dorsiflexion? What class of lever does the foot act as during dorsiflexion? Would you expect it to have a mechanical advantage greater or less than 1.0? Why?

## 9.3   Anatomy of Selected Diarthroses

### Expected Learning Outcomes

When you have completed this section, you should be able to

a. identify the major anatomical features of the jaw, shoulder, elbow, hip, knee, and ankle joints; and

b. explain how the anatomical differences between these joints are related to differences in function.

We now examine the gross anatomy of certain diarthroses. It is beyond the scope of this book to discuss all of them, but the ones selected here most often require medical attention and

---

[24]*in* = inward; *version* = turning
[25]*e* = outward; *version* = turning

many of them have a strong bearing on athletic performance and everyday mobility.

## The Jaw Joint

The **temporomandibular (jaw) joint (TMJ)** is the articulation of the condyle of the mandible with the mandibular fossa of the temporal bone (fig. 9.23). You can feel its action by pressing your fingertips against the jaw immediately anterior to the ear while opening and closing your mouth. This joint combines elements of condylar, hinge, and plane joints. It functions in a hingelike fashion when the mandible is elevated and depressed, it glides slightly forward whenever the mouth is opened or the jaw is protracted to take a bite, and it glides from side to side to grind food between the molars. To observe the importance of the forward glide, try to open your mouth while pushing the jaw posteriorly with the heel of your hand; it is difficult to open the mouth more than 1 or 2 cm when there is resistance to protraction of the mandible.

The synovial cavity of the TMJ is divided into superior and inferior chambers by an articular disc, which permits lateral and medial excursion of the mandible. Two ligaments support the joint. The **lateral ligament** prevents posterior displacement of the mandible. If the jaw receives a hard blow,

**DEEPER INSIGHT 9.2**

## CLINICAL APPLICATION

### TMJ Syndrome

Temporomandibular joint (TMJ) syndrome has received medical recognition only recently, although it may affect as many as 75 million Americans. It can cause moderate intermittent facial pain, clicking sounds in the jaw, limitation of jaw movement, and in some people, more serious symptoms—severe headaches, vertigo (dizziness), tinnitus (ringing in the ears), and pain radiating from the jaw down the neck, shoulders, and back. It seems to be caused by a combination of psychological tension and malocclusion (misalignment of the teeth). Treatment may involve psychological management, physical therapy, analgesic and anti-inflammatory drugs, and sometimes corrective dental appliances to align the teeth properly.

this ligament normally prevents the condylar process from being driven upward and fracturing the base of the skull. The **sphenomandibular ligament** on the medial side of the joint extends from the sphenoid bone to the ramus of the mandible. A *stylomandibular ligament* extends from the styloid process to the angle of the mandible but is not part of the TMJ proper.

A deep yawn or other strenuous depression of the mandible can dislocate the TMJ by making the condyle pop out of the

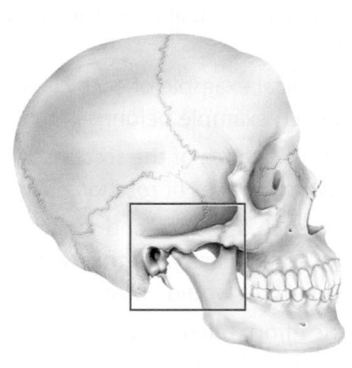

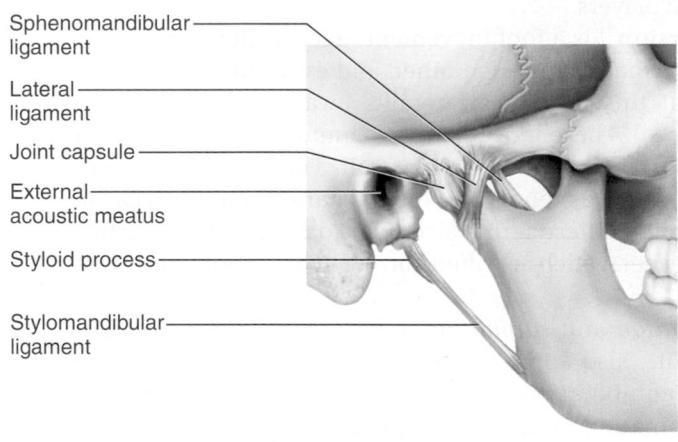

Sphenomandibular ligament
Lateral ligament
Joint capsule
External acoustic meatus
Styloid process
Stylomandibular ligament

**(a) Lateral view**

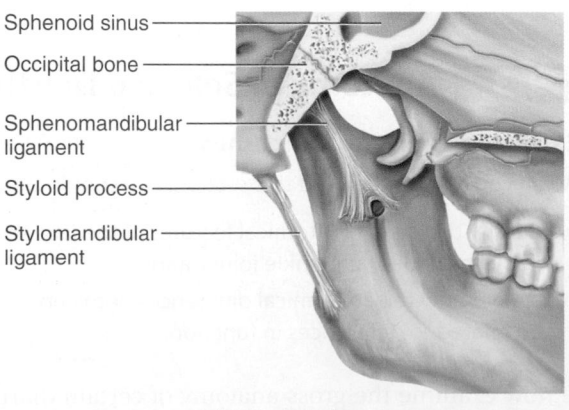

Sphenoid sinus
Occipital bone
Sphenomandibular ligament
Styloid process
Stylomandibular ligament

**(b) Medial view**

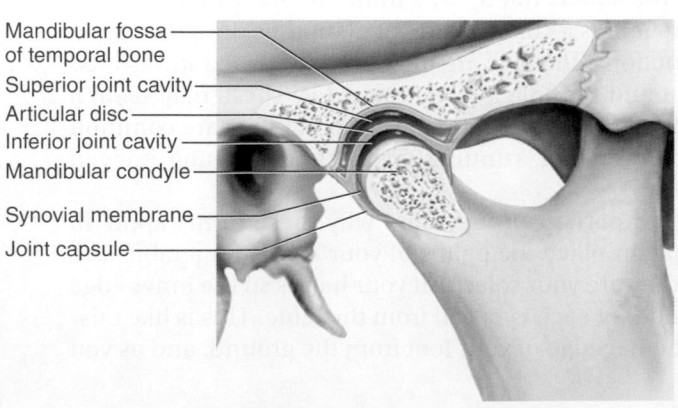

Mandibular fossa of temporal bone
Superior joint cavity
Articular disc
Inferior joint cavity
Mandibular condyle
Synovial membrane
Joint capsule

**(c) Sagittal section**

**FIGURE 9.23  The Temporomandibular Joint (TMJ).  AP|R**

fossa and slip forward. The joint is relocated by pressing down on the molars while pushing the jaw posteriorly.

## The Shoulder Joint

The **glenohumeral (humeroscapular) joint,** or shoulder joint, is where the hemispherical head of the humerus articulates with the glenoid cavity of the scapula (fig. 9.24). Together, the shoulder and elbow joints serve to position the hand for the performance of a task; without a hand, shoulder and elbow movements are almost useless. The relatively loose shoulder joint capsule and shallow glenoid cavity sacrifice joint stability for freedom of movement (see Deeper Insight 9.3). The

cavity, however, has a ring of fibrocartilage called the **glenoid labrum**[26] around its margin, making it somewhat deeper than it looks on a dried skeleton.

The shoulder is stabilized mainly by the biceps brachii muscle on the anterior side of the arm. One of its tendons arises from the *long head* of the muscle (see chapter 10), passes through the intertubercular groove of the humerus, and inserts on the superior margin of the glenoid cavity. It acts as a taut strap that presses the humeral head against the glenoid cavity. Four additional muscles help to stabilize this joint: the *supraspinatus, infraspinatus, teres minor,* and

[26]*labrum* = lip

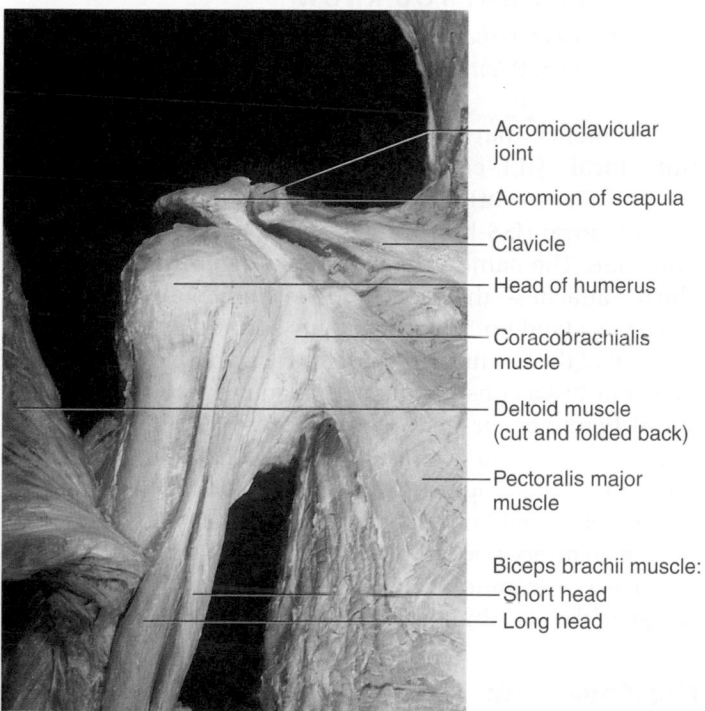

**(a) Anterior dissection**

- Acromioclavicular joint
- Acromion of scapula
- Clavicle
- Head of humerus
- Coracobrachialis muscle
- Deltoid muscle (cut and folded back)
- Pectoralis major muscle
- Biceps brachii muscle:
  - Short head
  - Long head

**(b) Anterior view**

- Acromion
- Subacromial bursa
- Supraspinatus tendon
- Coracohumeral ligament
- Subdeltoid bursa
- Subscapularis tendon
- Transverse humeral ligament
- Tendon sheath
- Biceps brachii tendon (long head)
- Humerus
- Acromioclavicular ligament
- Clavicle
- Coraco-clavicular ligament
- Coraco-acromial ligament
- Coracoid process
- Subcoracoid bursa
- Subscapular bursa
- Glenohumeral ligaments

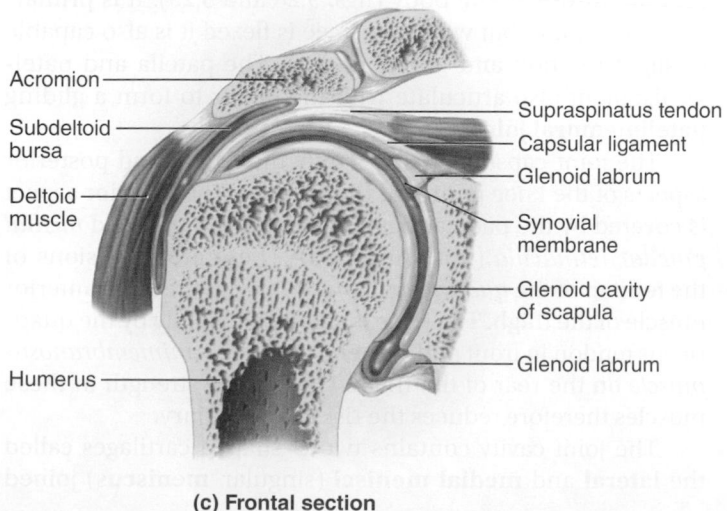

**(c) Frontal section**

- Acromion
- Subdeltoid bursa
- Deltoid muscle
- Humerus
- Supraspinatus tendon
- Capsular ligament
- Glenoid labrum
- Synovial membrane
- Glenoid cavity of scapula
- Glenoid labrum

**(d) Lateral view, humerus removed**

- Acromion
- Supraspinatus tendon
- Subdeltoid bursa
- Infraspinatus tendon
- Glenoid cavity (articular cartilage)
- Teres minor tendon
- Synovial membrane (cut)
- Coracoid process
- Coracohumeral ligament
- Superior glenohumeral ligament
- Biceps brachii tendon (long head)
- Subscapular bursa
- Subscapularis tendon
- Middle glenohumeral ligament
- Inferior glenohumeral ligament

**FIGURE 9.24 The Glenohumeral (Shoulder) Joint.**

### Shoulder Dislocation

The anatomy and mobility of the shoulder joint make it especially susceptible to dislocation. Over 95% of cases are classified as *anterior dislocation* (displacement of the humeral head in the anterior direction). Such dislocations are usually caused when the arm is abducted and receives a blow from above—for example, by heavy objects falling from a shelf. A complex of nerves and blood vessels traverses the axillary region and shoulder dislocation can easily damage the axillary nerve or artery (see figs. 13.16, p. 492, and 20.35, p. 787). Left untreated, this can lead to muscle atrophy, weakness, or paralysis. Greek physician Hippocrates taught students to treat shoulder dislocation by placing a heel in the patient's axilla and pulling on the arm, but this can cause even worse nerve damage and is never done anymore by professionals. Because the shoulder is so easily dislocated, one also should never attempt to move an unconscious or immobilized person by pulling on his or her arm.

subscapularis. Their tendons form the **rotator cuff,** which is fused to the joint capsule on all sides except the inferior (see fig. 10.25, p. 345). Chapter 10 further describes the rotator cuff and its injuries (p. 370).

Five principal ligaments also support this joint. Three of them, called the **glenohumeral ligaments,** are relatively weak and sometimes absent. The other two are the **coracohumeral ligament,** which extends from the coracoid process of the scapula to the greater tubercle of the humerus, and the **transverse humeral ligament,** which extends from the greater to the lesser tubercle of the humerus and forms a tunnel housing the tendon from the long head of the biceps.

Four bursae occur at the shoulder. Their names describe their locations: the **subdeltoid, subacromial, subcoracoid,** and **subscapular bursae.** The *deltoid* is the large muscle that caps the shoulder, and the other bursae are named for parts of the scapula described in chapter 8.

## The Elbow Joint

The elbow is a hinge joint composed of two articulations: the **humeroulnar joint** where the trochlea of the humerus joins the trochlear notch of the ulna, and the **humeroradial joint** where the capitulum of the humerus meets the head of the radius (fig. 9.25). Both are enclosed in a single joint capsule. On the posterior side of the elbow, there is a prominent **olecranon bursa** to ease the movement of tendons over the joint. Side-to-side motions of the elbow joint are restricted by a pair of ligaments: the **radial (lateral) collateral ligament** and **ulnar (medial) collateral ligament.**

Another joint occurs in the elbow region, the **proximal radioulnar joint,** but it is not involved in the hinge. At this joint, the edge of the disclike head of the radius fits into the radial notch of the ulna. It is held in place by the **anular ligament,** which encircles the radial head and is attached at each end to the ulna. The radial head rotates like a wheel against the ulna as the forearm is pronated or supinated.

## The Hip Joint

The **coxal (hip) joint** is the point where the head of the femur inserts into the acetabulum of the hip bone (fig. 9.26). Because the coxal joints bear much of the body's weight, they have deep sockets and are much more stable than the shoulder joint. The depth of the socket is somewhat greater than you see on dried bones because of a horseshoe-shaped ring of fibrocartilage, the **acetabular labrum,** attached to its rim. Dislocations of the hip are rare, but some infants suffer congenital dislocations because the acetabulum is not deep enough to hold the head of the femur in place. If detected early, this condition can be treated with a harness, worn for 2 to 4 months, that holds the head of the femur in the proper position until the joint is stronger (fig. 9.27).

### ▶▶▶ APPLY WHAT YOU KNOW

*Where else in the body is there a structure similar to the acetabular labrum? What do those two locations have in common?*

Ligaments that support the coxal joint include the **iliofemoral** (ILL-ee-oh-FEM-oh-rul) and **pubofemoral** (PYU-bo-FEM-or-ul) **ligaments** on the anterior side and the **ischiofemoral** (ISS-kee-oh-FEM-or-ul) **ligament** on the posterior side. The name of each ligament refers to the bones to which it attaches—the femur and the ilium, pubis, or ischium. When you stand up, these ligaments become twisted and pull the head of the femur tightly into the acetabulum. The head of the femur has a conspicuous pit called the **fovea capitis.** The **round ligament,** or **ligamentum teres**[27] (TERR-eez), arises here and attaches to the lower margin of the acetabulum. This is a relatively slack ligament, so it is doubtful that it plays a significant role in holding the femur in its socket. It does, however, contain an artery that supplies blood to the head of the femur. A **transverse acetabular ligament** bridges a gap in the inferior margin of the acetabular labrum.

## The Knee Joint

The **tibiofemoral (knee) joint** is the largest and most complex diarthrosis of the body (figs. 9.28 and 9.29). It is primarily a hinge joint, but when the knee is flexed it is also capable of slight rotation and lateral gliding. The patella and patellar ligament also articulate with the femur to form a gliding **patellofemoral joint.**

The joint capsule encloses only the lateral and posterior aspects of the knee joint, not the anterior. The anterior aspect is covered by the patellar ligament and the *lateral* and *medial patellar retinacula* (not illustrated). These are extensions of the tendon of the *quadriceps femoris muscle,* the large anterior muscle of the thigh. The knee is stabilized mainly by the quadriceps tendon in front and the tendon of the *semimembranosus muscle* on the rear of the thigh. Developing strength in these muscles therefore reduces the risk of knee injury.

The joint cavity contains two C-shaped cartilages called the **lateral** and **medial menisci** (singular, **meniscus**) joined

---

[27]*teres* = round

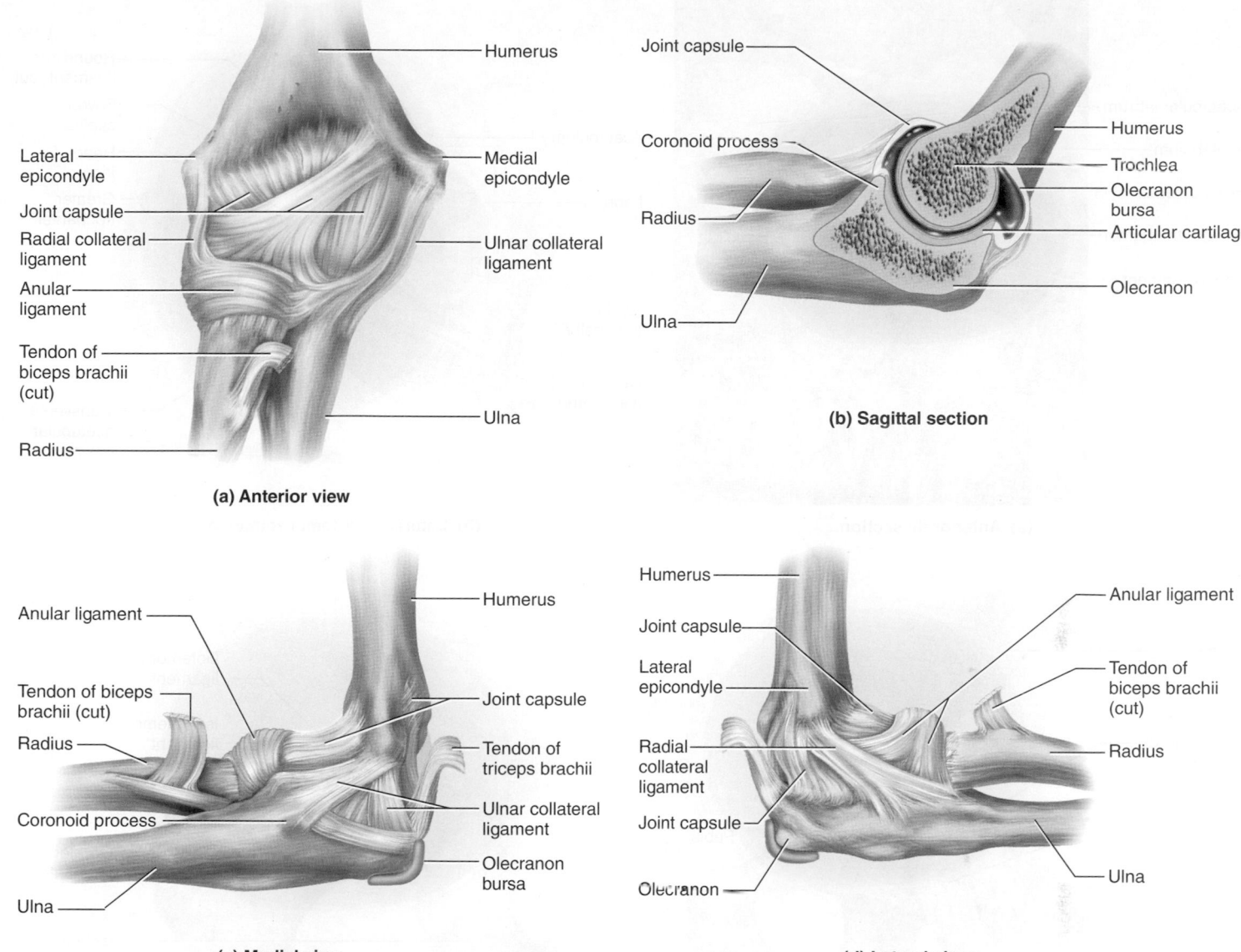

**FIGURE 9.25 The Elbow Joint.** This region includes two joints that form the elbow hinge—the humeroulnar and humeroradial—and one joint, the radioulnar, not involved in the hinge. **AP|R**

by a **transverse ligament.** The menisci absorb the shock of the body weight jostling up and down on the knee and prevent the femur from rocking from side to side on the tibia.

The posterior **popliteal** (pop-LIT-ee-ul) **region** of the knee is supported by a complex array of *extracapsular ligaments* external to the joint capsule and two *intracapsular ligaments* within it. The extracapsular ligaments include two collateral ligaments that prevent the knee from rotating when the joint is extended—the **fibular (lateral) collateral ligament** and the **tibial (medial) collateral ligament**—and other ligaments not illustrated.

The two intracapsular ligaments lie deep within the joint. The synovial membrane folds around them, however, so that they are excluded from the fluid-filled synovial cavity. These ligaments cross each other in the form of an X; hence, they are

called the **anterior cruciate**[28] (CROO-she-ate) **ligament (ACL)** and **posterior cruciate ligament (PCL).** These are named according to whether they attach to the anterior or posterior side of the tibia, not for their attachments to the femur. When the knee is extended, the ACL is pulled tight and prevents hyperextension. The PCL prevents the femur from sliding off the front of the tibia and prevents the tibia from being displaced backward. The ACL is one of the most common sites of knee injury (see Deeper Insight 9.4).

**▶▶▶ APPLY WHAT YOU KNOW**

*What structure in the elbow joint serves the same function as the ACL of the knee?*

[28]*cruci* = cross; *ate* = characterized by

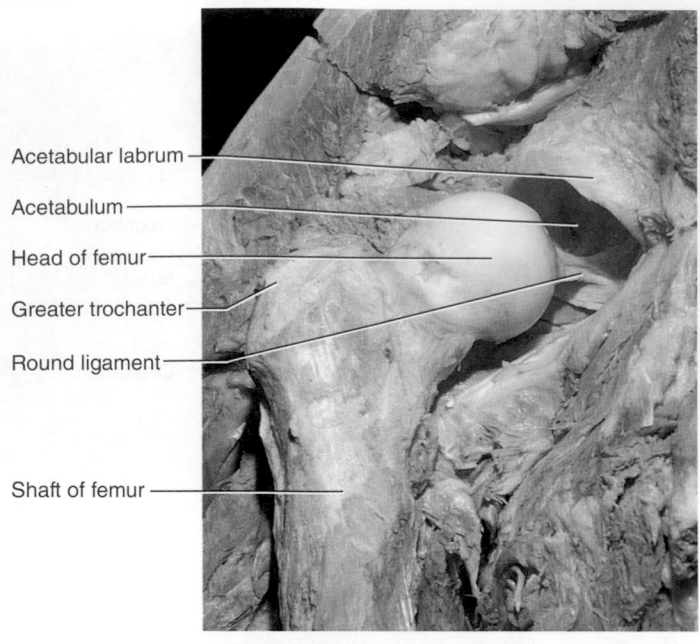

Acetabular labrum
Acetabulum
Head of femur
Greater trochanter
Round ligament
Shaft of femur

**(a) Anterior dissection**

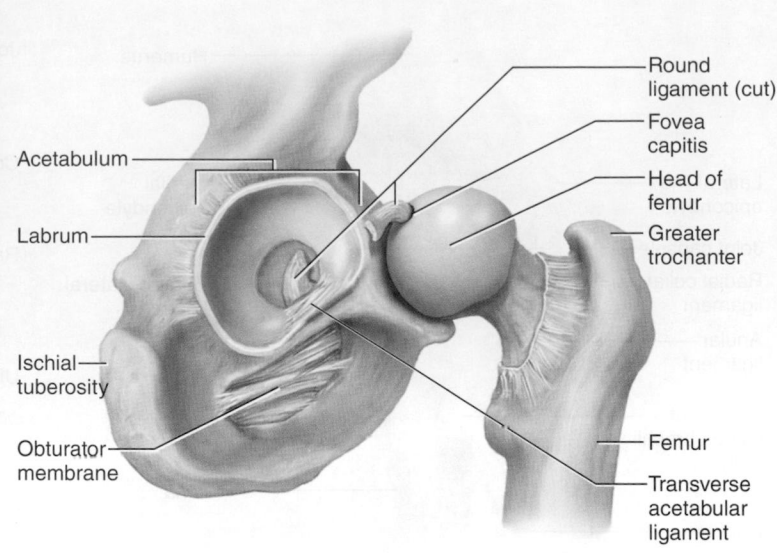

Acetabulum
Labrum
Ischial tuberosity
Obturator membrane
Round ligament (cut)
Fovea capitis
Head of femur
Greater trochanter
Femur
Transverse acetabular ligament

**(b) Lateral view, femur retracted**

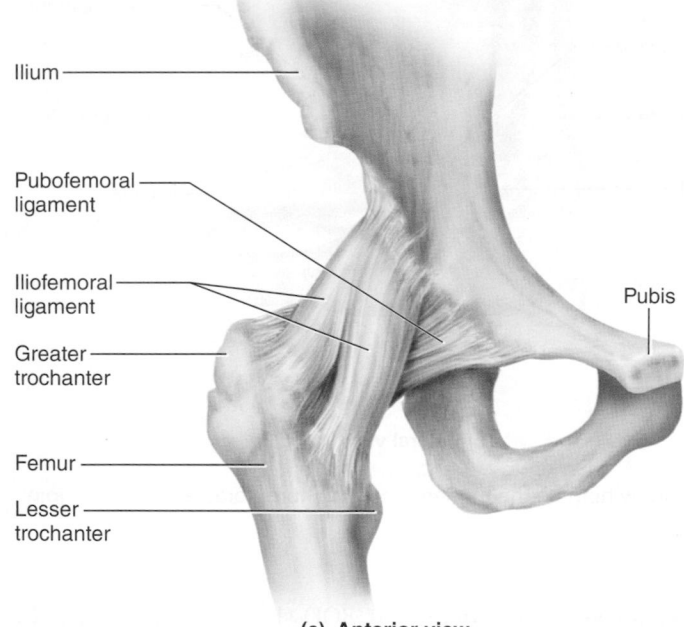

Ilium
Pubofemoral ligament
Iliofemoral ligament
Greater trochanter
Femur
Lesser trochanter
Pubis

**(c) Anterior view**

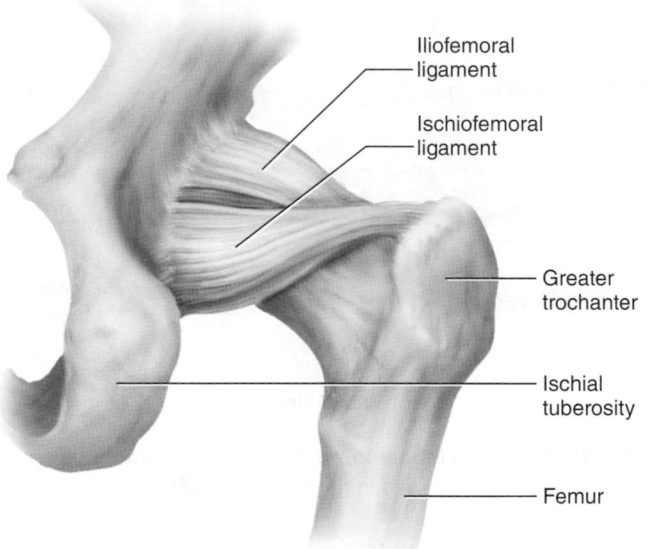

Iliofemoral ligament
Ischiofemoral ligament
Greater trochanter
Ischial tuberosity
Femur

**(d) Posterior view**

**FIGURE 9.26 The Coxal (Hip) Joint.**

An important aspect of human bipedalism is the ability to lock the knees and stand erect without tiring the extensor muscles of the leg. When the knee is extended to the fullest degree allowed by the ACL, the femur rotates medially on the tibia. This action locks the knee, and in this state, all the major knee ligaments are twisted and taut. To unlock the knee, the *popliteus* muscle rotates the femur laterally and untwists the ligaments.

The knee joint has at least 13 bursae. Four of these are anterior: the **superficial infrapatellar, suprapatellar,** **prepatellar,** and **deep infrapatellar.** Located in the popliteal region are the *popliteal bursa* and *semimembranosus bursa* (not illustrated). At least seven more bursae are found on the lateral and medial sides of the knee joint. From figure 9.29c, your knowledge of the relevant word elements *(infra-, supra-, pre-),* and the terms *superficial* and *deep,* you should be able to work out the reasoning behind most of these names and develop a system for remembering the locations of these bursae.

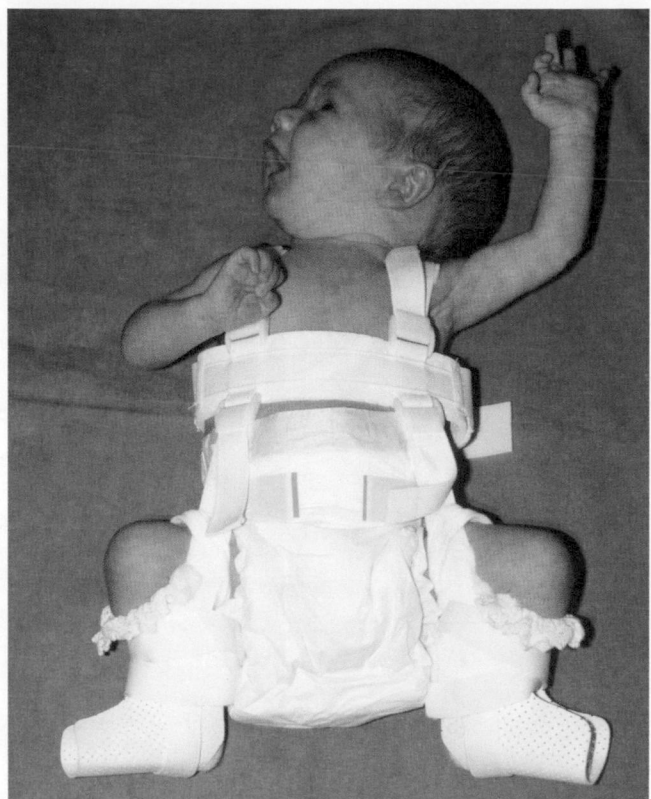

**FIGURE 9.27  Treatment of Congenital Hip Dislocation.** This infant is wearing a harness that holds the head of the femur in the acetabulum.

## The Ankle Joint

The **talocrural**[29] **(ankle) joint** includes two articulations—a medial joint between the tibia and talus and a lateral joint between the fibula and talus, both enclosed in one joint capsule (fig. 9.31). The malleoli of the tibia and fibula overhang the talus on each side like a cap and prevent most side-to-side motion. The ankle therefore has a more restricted range of motion than the wrist.

The ligaments of the ankle include (1) **anterior** and **posterior tibiofibular ligaments,** which bind the tibia to the fibula; (2) a multipart **medial (deltoid**[30]**) ligament,** which binds the tibia to the foot on the medial side; and (3) a multipart **lateral (collateral) ligament,** which binds the fibula to the foot on the lateral side. The **calcaneal (Achilles) tendon** extends from the calf muscles to the calcaneus. It plantarflexes the foot and limits dorsiflexion. Plantar flexion is limited by extensor tendons on the anterior side of the ankle and by the anterior part of the joint capsule.

Sprains (torn ligaments and tendons) are common at the ankle, especially when the foot is suddenly inverted or everted to excess. They are painful and usually accompanied by immediate swelling. They are best treated by immobilizing the joint and reducing swelling with an ice pack, but in extreme

[29]*talo* = ankle; *crural* = pertaining to the leg
[30]*delt* = triangular, Greek letter delta (Δ); *oid* = resembling

Lateral ← | → Medial

Femur:
 Shaft
 Patellar surface
 Medial condyle
 Lateral condyle

Joint capsule

Joint cavity:
 Anterior cruciate ligament
 Medial meniscus
 Lateral meniscus

Tibia:
 Lateral condyle
 Medial condyle
 Tuberosity

Patellar ligament

Patella (posterior surface)

Articular facets

Quadriceps tendon (reflected)

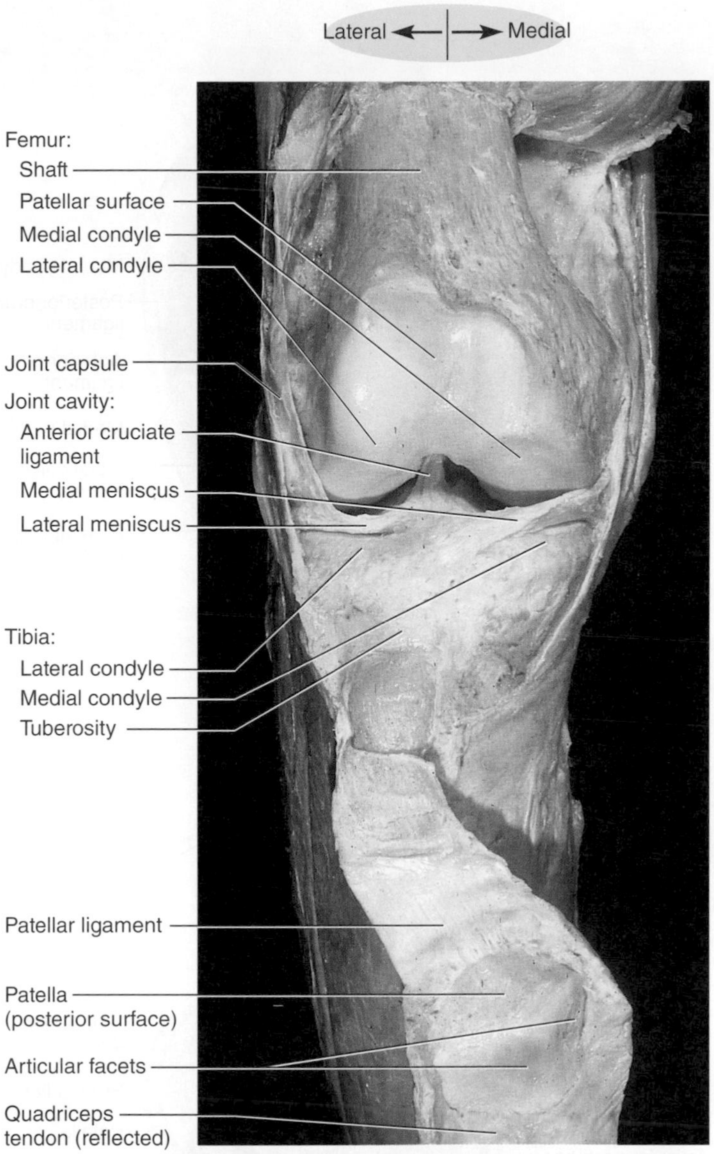

**FIGURE 9.28  The Right Knee, Anterior Dissection.** The quadriceps tendon has been cut and folded (reflected) downward to expose the joint cavity and the posterior surface of the patella.

cases may require a cast or surgery. Sprains and other joint disorders are briefly described in table 9.1.

### BEFORE YOU GO ON

Answer the following questions to test your understanding of the preceding section:

**12.** What keeps the mandibular condyle from slipping out of its fossa in a posterior direction?

**13.** Explain how the biceps tendon braces the shoulder joint.

**14.** Identify the three joints found at the elbow and name the movements in which each joint is involved.

**15.** What keeps the femur from slipping backward off the tibia?

**16.** What keeps the tibia from slipping sideways off the talus?

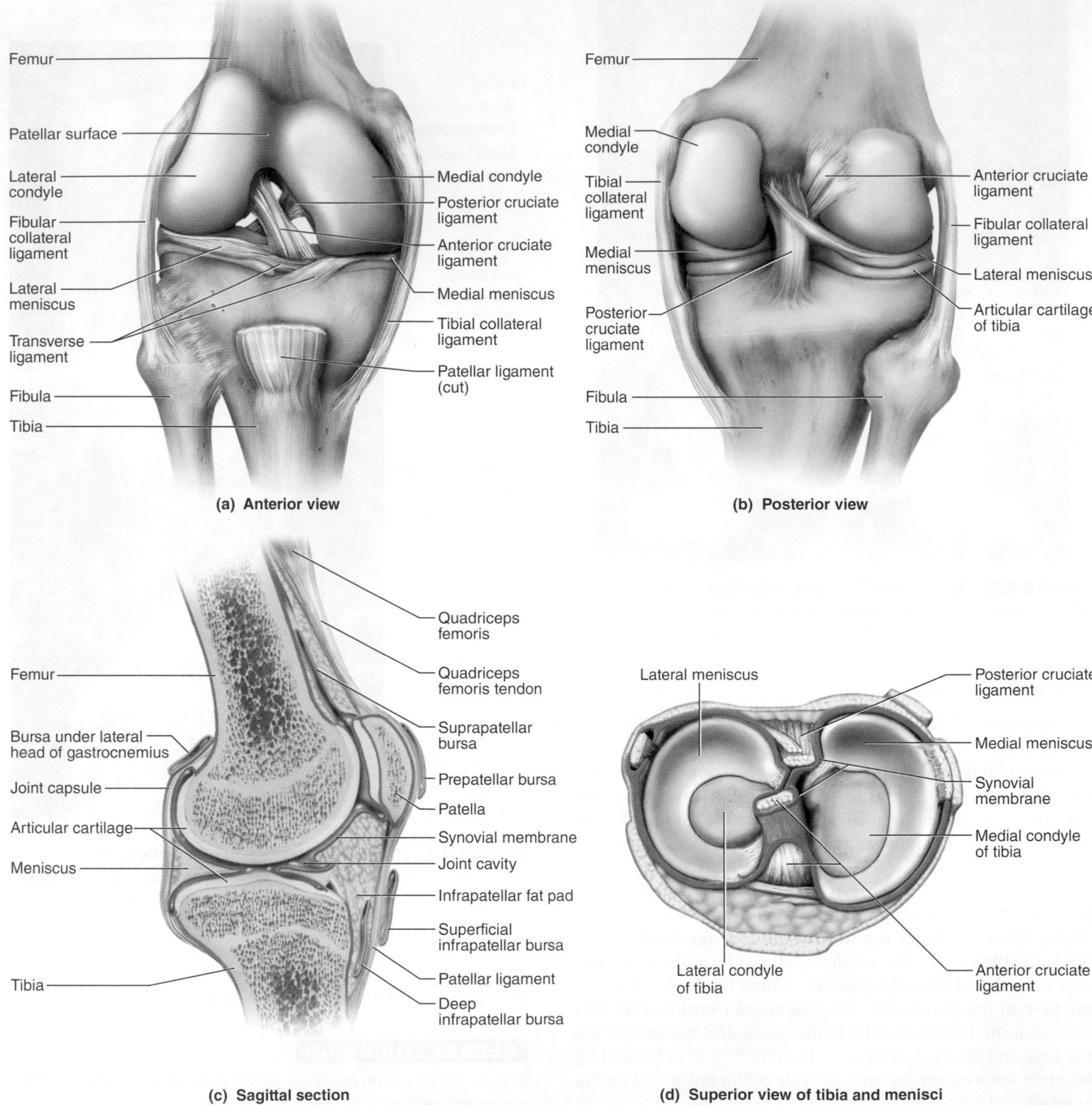

(a) Anterior view

Femur

Patellar surface

Lateral condyle

Fibular collateral ligament

Lateral meniscus

Transverse ligament

Fibula

Tibia

Medial condyle

Posterior cruciate ligament

Anterior cruciate ligament

Medial meniscus

Tibial collateral ligament

Patellar ligament (cut)

(b) Posterior view

Femur

Medial condyle

Tibial collateral ligament

Medial meniscus

Posterior cruciate ligament

Fibula

Tibia

Anterior cruciate ligament

Fibular collateral ligament

Lateral meniscus

Articular cartilage of tibia

(c) Sagittal section

Femur

Bursa under lateral head of gastrocnemius

Joint capsule

Articular cartilage

Meniscus

Tibia

Quadriceps femoris

Quadriceps femoris tendon

Suprapatellar bursa

Prepatellar bursa

Patella

Synovial membrane

Joint cavity

Infrapatellar fat pad

Superficial infrapatellar bursa

Patellar ligament

Deep infrapatellar bursa

(d) Superior view of tibia and menisci

Lateral meniscus

Lateral condyle of tibia

Posterior cruciate ligament

Medial meniscus

Synovial membrane

Medial condyle of tibia

Anterior cruciate ligament

**FIGURE 9.29  The Right Tibiofemoral (Knee) Joint.  AP|R**

# DEEPER INSIGHT 9.4

## CLINICAL APPLICATION

### Knee Injuries and Arthroscopic Surgery

Although the knee can bear a lot of weight, it is highly vulnerable to rotational and horizontal stress, especially when the knee is flexed (as in skiing or running) and receives a blow from behind or from the side. The most common injuries are to a meniscus or the anterior cruciate ligament (ACL) (fig. 9.30). Knee injuries heal slowly because ligaments and tendons have a scanty blood supply and cartilage usually has no blood vessels at all.

The diagnosis and surgical treatment of knee injuries have been greatly improved by *arthroscopy,* a procedure in which the interior of a joint is viewed with a pencil-thin instrument, the *arthroscope,* inserted through a small incision. The arthroscope has a light, a lens, and fiber optics that allow a viewer to see into the cavity and take photographs or video recordings. A surgeon can also withdraw samples of synovial fluid by arthroscopy or inject saline into the joint cavity to expand it and provide a clearer view. If surgery is required, additional small incisions can be made for the surgical instruments and the procedures can be observed through the arthroscope or on a monitor. Arthroscopic surgery produces much less tissue damage than conventional surgery and enables patients to recover more quickly.

Orthopedic surgeons now often replace a damaged ACL with a graft from the patellar ligament or a hamstring tendon. The surgeon "harvests" a strip from the middle of the patient's ligament (or tendon), drills a hole into the femur and tibia within the joint cavity, threads the ligament through the holes, and fastens it with biodegradable screws. The grafted ligament is more taut and "competent" than the damaged ACL. It becomes ingrown with blood vessels and serves as a substrate for the deposition of more collagen, which further strengthens it in time. Following arthroscopic ACL reconstruction, a patient typically must use crutches for 7 to 10 days and undergo physical therapy for 6 to 10 weeks, followed by self-directed exercise therapy. Healing is completed in about 9 months.

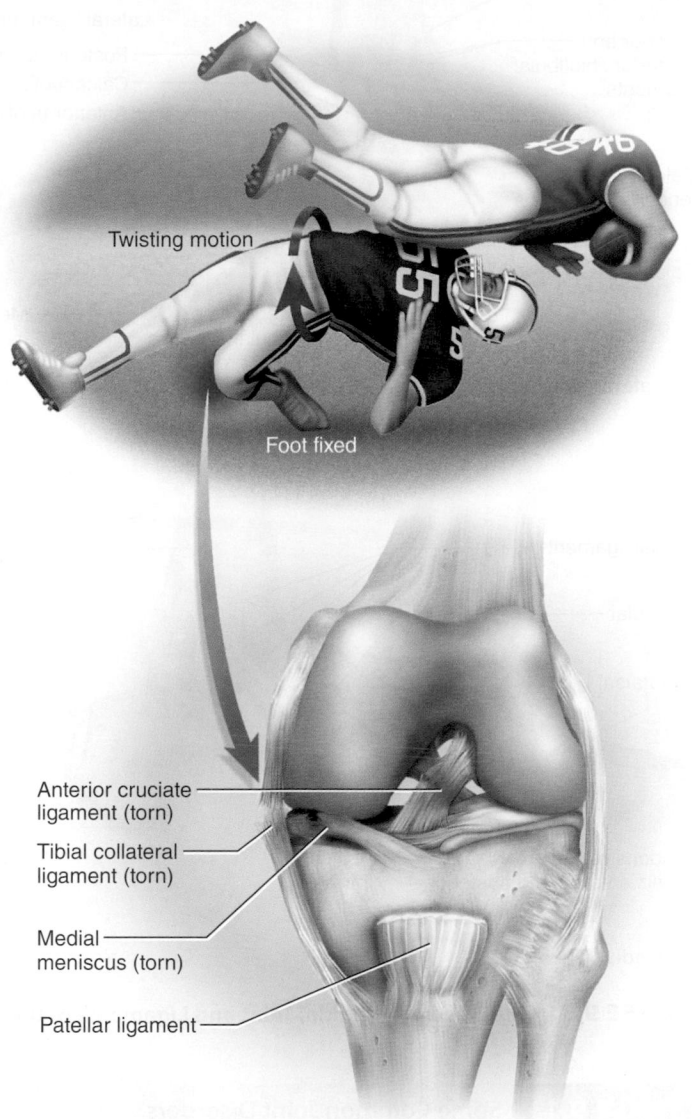

Twisting motion

Foot fixed

Anterior cruciate ligament (torn)

Tibial collateral ligament (torn)

Medial meniscus (torn)

Patellar ligament

**FIGURE 9.30  Knee Injuries.**

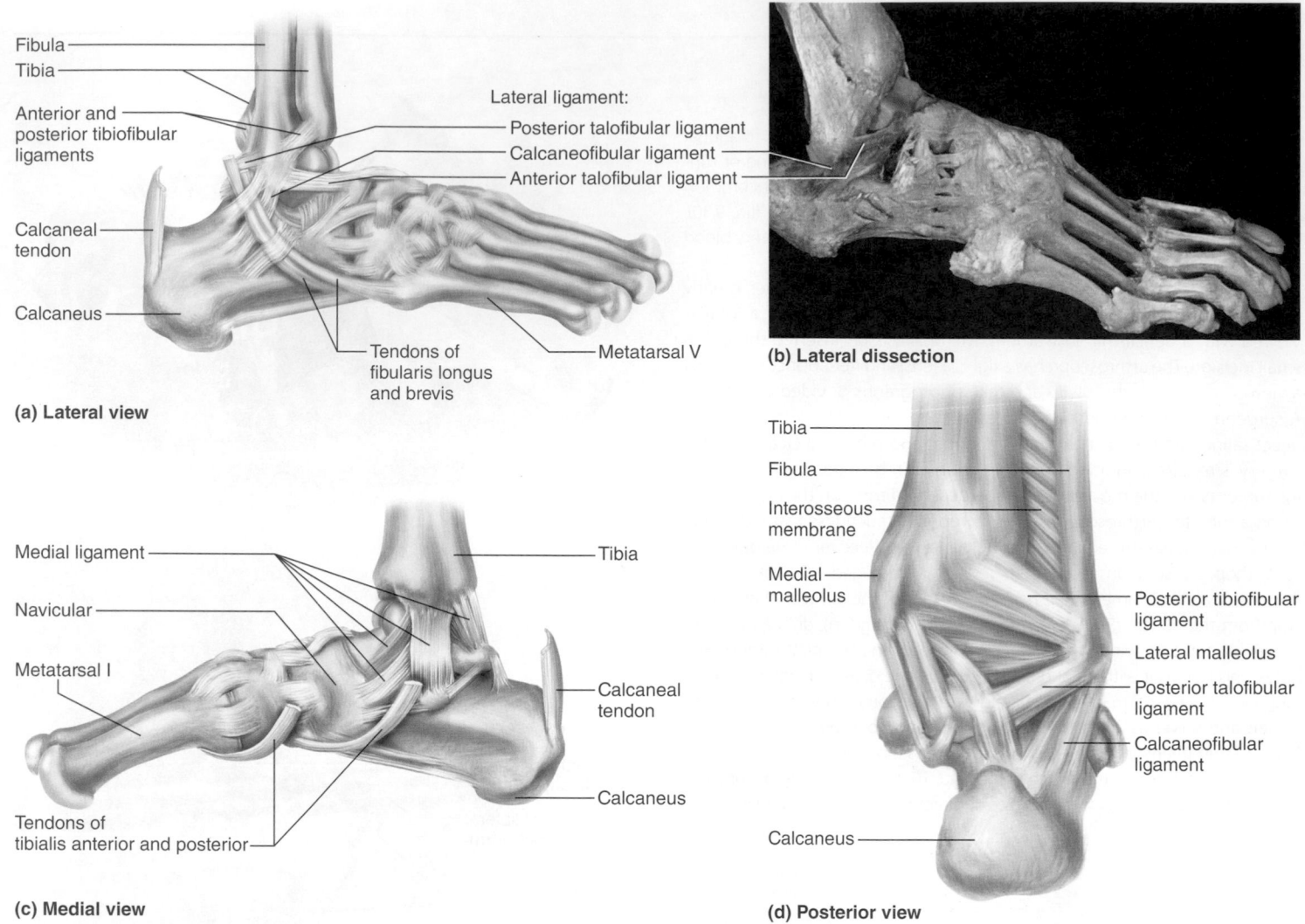

**FIGURE 9.31** The Talocrural (Ankle) Joint and Ligaments of the Right Foot.   AP|R

| TABLE 9.1 | Some Common Joint Disorders |
|---|---|
| Arthritis | Broad term embracing more than 100 types of joint rheumatism. |
| Bursitis | Inflammation of a bursa, usually due to overexertion of a joint. |
| Dislocation | Displacement of a bone from its normal position at a joint, usually accompanied by a sprain of the adjoining connective tissues. Most common at the fingers, thumb, shoulder, and knee. |
| Gout | A hereditary disease, most common in men, in which uric acid crystals accumulate in the joints and irritate the articular cartilage and synovial membrane. Causes gouty arthritis, with swelling, pain, tissue degeneration, and sometimes fusion of the joint. Most commonly affects the great toe. |
| Rheumatism | Broad term for any pain in the supportive and locomotory organs of the body, including bones, ligaments, tendons, and muscles. |
| Sprain | Torn ligament or tendon, sometimes with damage to a meniscus or other cartilage. |
| Strain | Painful overstretching of a tendon or muscle without serious tissue damage. Often results from inadequate warm-up before exercise. |
| Synovitis | Inflammation of a joint capsule, often as a complication of a sprain. |
| Tendinitis | A form of bursitis in which a tendon sheath is inflamed. |

*Disorders described elsewhere*

Hip dislocation p. 296          Osteoarthritis p. 303          Rotator cuff injury p. 370

Knee injuries p. 301          Rheumatoid arthritis p. 303          Shoulder dislocation p. 296

## DEEPER INSIGHT 9.5
## CLINICAL APPLICATION

### Arthritis and Artificial Joints

**Arthritis**[31] is a broad term for pain and inflammation of a joint and embraces more than a hundred different diseases of largely obscure or unknown causes. In all of its forms, it is the most common crippling disease in the United States; nearly everyone past middle age develops arthritis to some degree. Physicians who treat arthritis and other joint disorders are called *rheumatologists*.

The most common form of arthritis is *osteoarthritis (OA)*, also called "wear-and-tear arthritis" because it is apparently a normal consequence of years of wear on the joints. As joints age, the articular cartilages soften and degenerate. As the cartilage becomes roughened by wear, joint movement may be accompanied by crunching or crackling sounds called *crepitus*. OA affects especially the fingers, intervertebral joints, hips, and knees. As the articular cartilage wears away, exposed bone tissue often develops spurs that grow into the joint cavity, restrict movement, and cause pain. OA rarely occurs before age 40, but it affects about 85% of people older than 70, especially those who are overweight. It usually does not cripple, but in severe cases it can immobilize the hip.

*Rheumatoid arthritis (RA)*, which is far more severe than osteoarthritis, results from an autoimmune attack against the joint tissues. It begins when the body produces antibodies to fight an infection. Failing to recognize the body's own tissues, a misguided antibody known as *rheumatoid factor* also attacks the synovial membranes. Inflammatory cells accumulate in the synovial fluid and produce enzymes that degrade the articular cartilage. The synovial membrane thickens and adheres to the articular cartilage, fluid accumulates in the joint capsule, and the capsule is invaded by fibrous connective tissue. As articular cartilage degenerates, the joint begins to ossify, and sometimes the bones become solidly fused and immobilized, a condition called *ankylosis*[32] (fig. 9.32). The disease tends to develop symmetrically—if the right wrist or hip develops RA, so does the left.

Rheumatoid arthritis is named for the fact that symptoms tend to flare up and subside (go into remission) periodically.[33] It affects women far more often than men, and because RA typically begins as early as age 30 to 40, it can cause decades of pain and disability. There is no cure, but joint damage can be slowed with hydrocortisone or other steroids. Because long-term use of steroids weakens the bone, however, aspirin is the treatment of first choice to control the inflammation. Physical therapy is also used to preserve the joint's range of motion and the patient's functional ability.

*Arthroplasty*,[34] a treatment of last resort, is the replacement of a diseased joint with an artificial device called a *prosthesis*.[35] Joint prostheses were first developed to treat injuries in World War II and the Korean War. Total hip replacement (THR), first performed in 1963 by English orthopedic surgeon Sir John Charnley, is now the most common orthopedic procedure for the elderly. The first knee replacements were performed in the 1970s. Joint prostheses are now available for finger, shoulder, and elbow joints, as well as the hip and knee. Arthroplasty is performed on over 250,000 patients per year in the United States, primarily to relieve pain and restore function in elderly people with OA or RA.

Arthroplasty presents ongoing challenges for biomedical engineering. An effective prosthesis must be strong, nontoxic, and corrosion-resistant. In addition, it must bond firmly to the patient's bones and enable a normal

---

[31]*arthr* = joint; *itis* = inflammation
[32]*ankyl* = bent, crooked; *osis* = condition
[33]*rheumat* = tending to change
[34]*arthro* = joint; *plasty* = surgical repair
[35]*prosthe* = something added

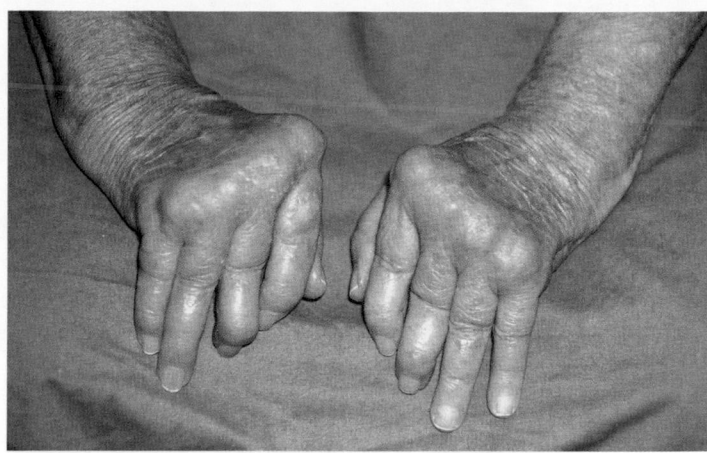

**(a)**

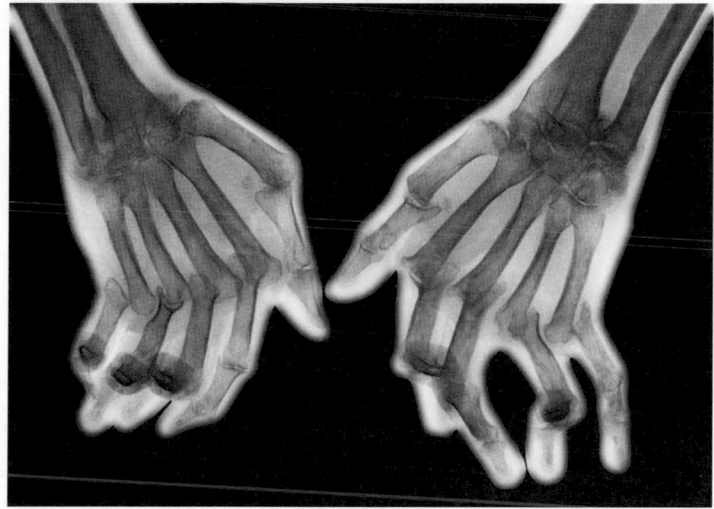

**(b)**

**FIGURE 9.32 Rheumatoid Arthritis (RA).** (a) A severe case with ankylosis of the joints. (b) Colorized X-ray of severe RA of the hands.

range of motion with a minimum of friction. The heads of long bones are usually replaced with prostheses made of a metal alloy such as cobalt–chrome, titanium alloy, or stainless steel. Joint sockets are made of polyethylene (fig. 9.33). Prostheses are bonded to the patient's bone with screws or bone cement.

Improvements in technology have resulted in long-lasting prostheses. Over 75% of artificial knees last 20 years, nearly 85% last 15 years, and over 90% last 10 years. The most common form of failure is detachment of the prosthesis from the bone. This problem has been reduced by using *porous-coated prostheses*, which become infiltrated by the patient's own bone and

*(continued)*

## DEEPER INSIGHT 9.5 (continued)

### CLINICAL APPLICATION

create a firmer bond. A prosthesis is not as strong as a natural joint, however, and is not an option for many young, active patients.

Arthroplasty has been greatly improved by *computer-assisted design and manufacture (CAD/CAM)*. A computer scans X-rays from the patient and presents several design possibilities for review. Once a design is selected, the computer generates a program to operate the machinery that produces the prosthesis. CAD/CAM has reduced the waiting period for a prosthesis from 12 weeks to about 2 weeks and has lowered the cost dramatically.

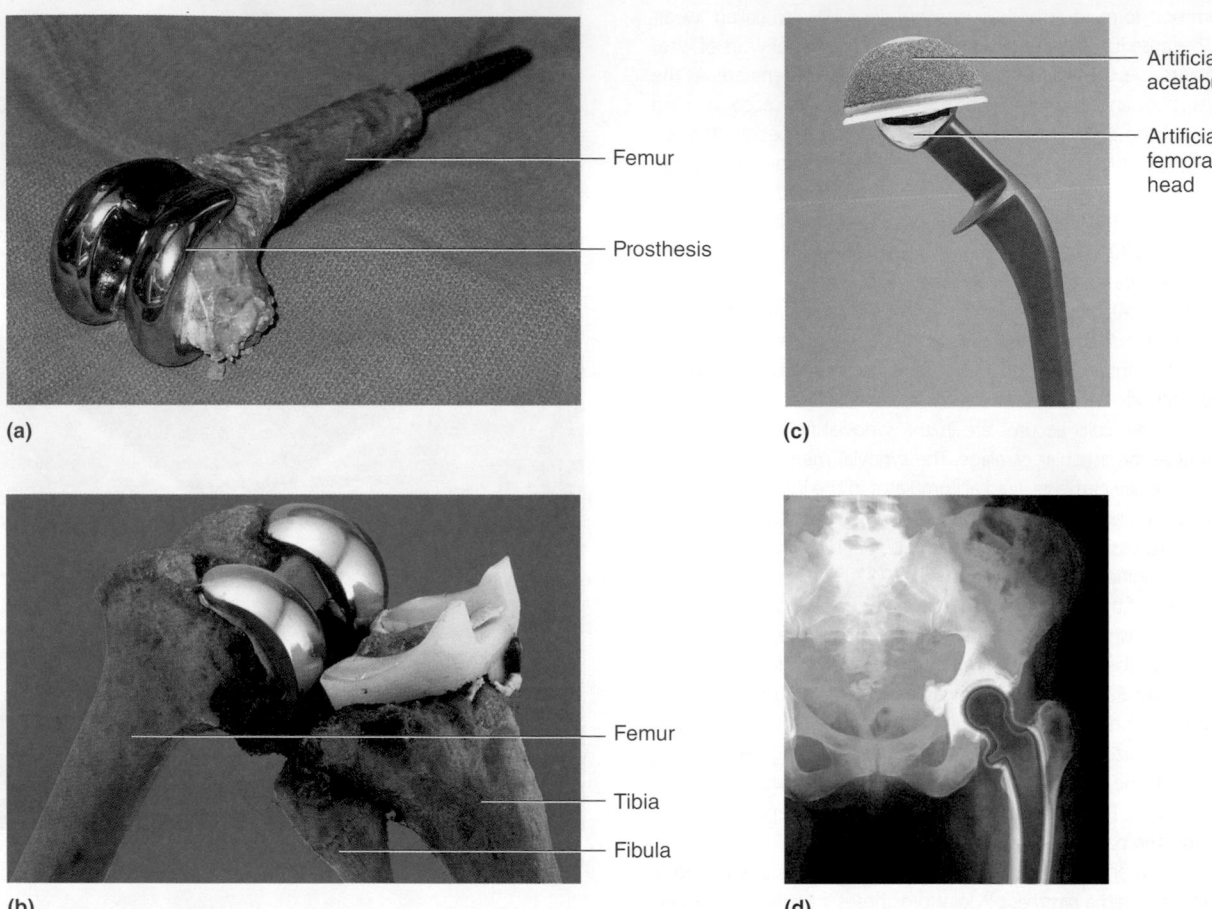

**FIGURE 9.33  Joint Prostheses.** (a) Artificial femoral condyles affixed to the distal end of the femur. (b) An artificial knee joint bonded to a natural femur and tibia. (c) A porous-coated hip prosthesis. The caplike portion replaces the acetabulum of the hip bone, and the ball and shaft below it are bonded to the proximal end of the femur. (d) X-ray of a patient with a total hip replacement.

# STUDY GUIDE

## ▶ Assess Your Learning Outcomes

*To test your knowledge, discuss the following topics with a study partner or in writing, ideally from memory.*

### 9.1 Joints and Their Classification (p. 275)

1. The fundamental definition of *joint (articulation)* and why it cannot be defined as a point at which one bone moves relative to an adjacent bone
2. Relationships and differences between the sciences of arthrology, kinesiology, and biomechanics
3. The typical system for naming most joints after the bones they involve; examples of this
4. Basic criteria for classifying joints
5. Characteristics and examples of bony joints (synostoses)
6. Characteristics of fibrous joints (synarthroses) and each of their subclasses, with examples
7. Characteristics of cartilaginous joints (amphiarthroses) and each of their subclasses, with examples

### 9.2 Synovial Joints (p. 279)

1. The definition and anatomical features of a synovial joint (diarthrosis), examples of this type, and why this type is of greatest interest for kinesiology
2. General anatomy of tendons, ligaments, bursae, and tendon sheaths, and their contributions to joint function
3. Three essential components of a lever
4. The meaning of mechanical advantage (MA); how the MA of a lever can be determined from measurements of

its effort and resistance arms; and the respective advantages of levers in which the MA is greater than or less than 1.0
5. Comparison of first-, second-, and third-class levers, and anatomical examples of each
6. Variables that determine a joint's range of motion (ROM), and the clinical relevance of ROM
7. Axes of rotation and degrees of freedom in joint movement, and how this relates to the classification of joints as monaxial, biaxial, or multiaxial
8. Six kinds of synovial joints; how each is classified as monaxial, biaxial, or multiaxial; imperfections in this classification; and examples of each type in the body
9. The concept of *zero position* and how it relates to the description of joint function
10. Examples of each of the following limb movements, including an ability to describe or demonstrate them: flexion, extension, hyperextension, abduction, adduction, hyperabduction, hyperadduction, circumduction, medial rotation, and lateral rotation
11. The same for supination, pronation, ulnar flexion, and radial flexion of the forearm and hand, and opposition, reposition, abduction, and adduction of the thumb
12. The same for flexion, extension, hyperextension, and lateral flexion of the spine, and right and left rotation of the trunk

13. The same for elevation, depression, protraction, retraction, and lateral and medial excursion of the mandible
14. The same for dorsiflexion, plantar flexion, inversion, eversion, pronation, and supination of the foot

### 9.3 Anatomy of Selected Diarthroses (p. 293)

1. Features of the jaw (temporomandibular) joint including the mandibular condyle, mandibular fossa, synovial cavity, articular disc, and principal ligaments
2. Features of the shoulder (glenohumeral) joint including the humeral head, glenoid cavity and labrum, five major ligaments and four bursae, and tendons of the biceps brachii and four rotator cuff muscles
3. Features of the elbow; the three joints that occur here; the olecranon bursa and four major ligaments
4. Features of the hip (coxal) joint including the femoral head, fovea capitis, acetabulum and labrum, and five principal ligaments
5. Features of the knee (tibiofemoral and patellofemoral joints), including the menisci, cruciate and other ligaments, and four major bursae around the patella
6. Features of the ankle (talocrural) joint, including the malleoli, calcaneal tendon, and major ligaments

## ▶ Testing Your Recall

*Answers in Appendix B*

1. Internal and external rotation of the humerus are made possible by a _____ joint.
   a. pivot
   b. condylar
   c. ball-and-socket
   d. saddle
   e. hinge

2. Which of the following is the least movable?
   a. a diarthrosis
   b. a synostosis
   c. a symphysis
   d. a synovial joint
   e. a condylar joint

3. Which of the following movements are unique to the foot?
   a. dorsiflexion and inversion
   b. elevation and depression
   c. circumduction and rotation
   d. abduction and adduction
   e. opposition and reposition

# STUDY GUIDE

4. Which of the following joints cannot be circumducted?
   a. carpometacarpal
   b. metacarpophalangeal
   c. glenohumeral
   d. coxal
   e. interphalangeal

5. Which of the following terms denotes a general condition that includes the other four?
   a. gout
   b. arthritis
   c. rheumatism
   d. osteoarthritis
   e. rheumatoid arthritis

6. In the adult, the ischium and pubis are united by
   a. a synchondrosis.
   b. a diarthrosis.
   c. a synostosis.
   d. an amphiarthrosis.
   e. a symphysis.

7. In a second-class lever, the effort
   a. is applied to the end opposite the fulcrum.
   b. is applied to the fulcrum itself.
   c. is applied between the fulcrum and resistance.

   d. always produces an MA less than 1.0.
   e. is applied on one side of the fulcrum to move a resistance on the other side.

8. Which of the following joints has anterior and posterior cruciate ligaments?
   a. the shoulder
   b. the elbow
   c. the hip
   d. the knee
   e. the ankle

9. To bend backward at the waist involves _____ of the vertebral column.
   a. rotation
   b. hyperextension
   c. dorsiflexion
   d. abduction
   e. flexion

10. The rotator cuff includes the tendons of all of the following muscles *except*
    a. the subscapularis.
    b. the supraspinatus.
    c. the infraspinatus.
    d. the biceps brachii.
    e. the teres minor.

11. The lubricant of a diarthrosis is called _____.

12. A fluid-filled sac that eases the movement of a tendon over a bone is called a/an _____.

13. A _____ joint allows one bone to swivel on another.

14. _____ is the science of movement.

15. The joint between a tooth and the mandible is called a/an _____.

16. In a _____ suture, the articulating bones have interlocking wavy margins, somewhat like a dovetail joint in carpentry.

17. In kicking a football, what type of action does the knee joint exhibit?

18. The angle through which a joint can move is called its _____.

19. The menisci of the knee are functionally similar to the _____ of the temporomandibular joint.

20. At the ankle, both the tibia and fibula articulate with what tarsal bone?

## ▶ Building Your Medical Vocabulary

*Answers in Appendix B*

*State a medical meaning of each word element below, and give a term in which it or a slight variation of it is used.*

1. ab-
2. arthro-

3. -ate
4. cruci-
5. cruro-
6. -duc

7. kinesio-
8. men-
9. supin-
10. -trac

## ▶ True or False

*Answers in Appendix B*

*Determine which five of the following statements are false, and briefly explain why.*

1. More people get rheumatoid arthritis than osteoarthritis.

2. A doctor who treats arthritis is called a kinesiologist.

3. Synovial joints are also known as synarthroses.

4. There is no meniscus in the elbow joint.

5. Reaching behind you to take something out of your hip pocket involves hyperextension of the shoulder.

6. The anterior cruciate ligament normally prevents hyperextension of the knee.

7. The femur is held tightly in the acetabulum mainly by the round ligament.

8. The knuckles are diarthroses.

9. Synovial fluid is secreted by the bursae.

10. Unlike most ligaments, the periodontal ligaments do not attach one bone to another.

# STUDY GUIDE

## ▶ Testing Your Comprehension

*Answers at* www.mhhe.com/saladin7

1. All second-class levers produce a mechanical advantage greater than 1.0 and all third-class levers produce a mechanical advantage less than 1.0. Explain why.

2. For each of the following joint movements, state what bone the axis of rotation passes through and which of the three anatomical planes contains the axis of rotation. You might find it helpful to produce some of these actions on an articulated laboratory skeleton so you can more easily visualize the axis of rotation. (a) Plantar flexion; (b) flexion of the hip; (c) adduction of the thigh; (d) flexion of the knee; (e) flexion of the first interphalangeal joint of the index finger. (Do not bend the fingers of a wired laboratory skeletal hand, because they can break off.)

3. In order of occurrence, list the joint actions (flexion, pronation, etc.) and the joints where they would occur as you (a) sit down at a table, (b) reach out and pick up an apple, (c) take a bite, and (d) chew it. Assume that you start in anatomical position.

4. The deltoid muscle inserts on the deltoid tuberosity of the humerus and abducts the arm. Imagine a person holding a weight in the hand and abducting the arm. On a laboratory skeleton, identify the fulcrum; measure the effort arm and resistance arm; determine the mechanical advantage of this movement; and determine which of the three lever types the upper limb acts as when performing this movement.

5. List the six types of synovial joints, and for each one, if possible, identify a joint in the upper limb and a joint in the lower limb that falls into each category. Which of these six joints has/have no examples in the lower limb?

## ▶ Improve Your Grade at www.mhhe.com/saladin7

*Download animations to study when it fits your schedule. Practice quizzes, labeling activities, games, and flashcards offer fun ways to master the chapter concepts. Or, download image PowerPoint files for each chapter to create a study guide or for taking notes during lecture.*

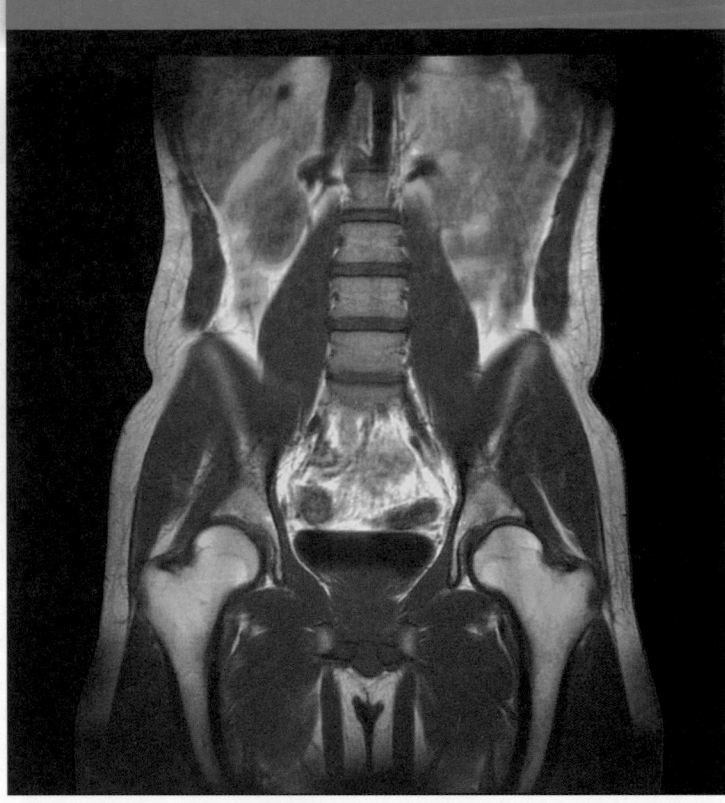

Colorized MRI scan showing muscles of the lumbar, pelvic, and upper femoral regions

Anatomy &
Physiology | **REVEALED**®
aprevealed.com

**Module 6: Muscular System**

## BRUSHING UP

- Understanding of the skeletal muscles depends on a thorough knowledge of skeletal anatomy (chapter 8), including not just the bones but also their surface features, many of which are muscle attachments.
- The movements produced by muscles, called their actions, are described in terms of the joint movements in chapter 9 (pp. 286–293).

Muscles constitute nearly half of the body's weight and occupy a place of central interest in several fields of health care and fitness. Physical and occupational therapists must be well acquainted with the muscular system to plan and carry out rehabilitation programs. Athletes and trainers, dancers and acrobats, and amateur fitness enthusiasts follow programs of resistance training to strengthen individual muscle groups through movement regimens based on knowledge of muscle, bone, and joint anatomy. Nurses employ their knowledge of the muscular system to give intramuscular injections correctly and to safely and effectively move patients who are physically incapacitated. Gerontological nurses are keenly aware of how deeply a person's muscular condition affects the quality of life in old age. The muscular system is highly important to biomedical disciplines even beyond the scope of the movement sciences. For example, it is the primary source of body heat in the moving individual, and loss of muscle mass can be a contributing factor in diabetes mellitus.

The muscular system is closely related to what we have covered in the preceding chapters. After this, we will examine the mechanisms of muscle contraction at the cellular and molecular levels in chapter 11, and chapter 12 will shed light on the relationship of the muscles to the nerves that control them.

## 10.1 The Structural and Functional Organization of Muscles

### Expected Learning Outcomes

When you have completed this section, you should be able to

a. describe the varied functions of muscles;

b. describe the connective tissue components of a muscle and their relationship to the bundling of muscle fibers;

c. describe the various shapes of skeletal muscles and relate this to their functions;

d. explain what is meant by the origin, insertion, belly, action, and innervation of a muscle;

e. describe the ways that muscles work in groups to aid, oppose, or moderate each other's actions;

f. distinguish between intrinsic and extrinsic muscles;

g. describe in general terms the nerve supply to the muscles and where these nerves originate; and

h. explain how the Latin names of muscles can aid in visualizing and remembering them.

As we saw in chapter 5, there are three kinds of muscular tissue in the human body—skeletal, cardiac, and smooth. All types, however, are specialized for one fundamental purpose: to convert the chemical energy of ATP into the mechanical energy of motion. Muscle cells exert a useful force on other tissues and organs, either to produce desirable movements or to prevent undesirable ones.

Although we examine all three muscle types in this chapter, most of our attention will be on the **muscular system,** composed of the skeletal muscles only. There are about 600 muscles in the human muscular system, but we survey fewer than one-third of them in this chapter, and most introductory courses cover even fewer. The study of this system is called **myology.**[1] The word *muscle*[2] means "little mouse," apparently referring to the appearance of muscles rippling under the skin.

### The Functions of Muscles

The functions of muscles are as follows:

- **Movement.** Muscles enable us to move from place to place and to move individual body parts; they move body contents in the course of breathing, blood circulation, feeding and digestion, defecation, urination, and childbirth; and they serve various roles in communication—speech, writing, facial expressions, and other body language.

- **Stability.** Muscles maintain posture by preventing unwanted movements. Some are called *antigravity muscles* because, at least part of the time, they resist the pull of gravity and prevent us from falling or slumping over. Many muscles also stabilize the joints by maintaining tension on tendons and bones.

- **Control of body openings and passages.** Muscles encircling the mouth serve not only for speech but also for food intake and retention of food while chewing. In the eyelid and pupil, they regulate the admission of light to the eye. Internal muscular rings control the movement of food, bile, blood, and other materials within the body. Muscles encircling the urethra and anus control the elimination of waste. (Some of these muscles are called *sphincters,* but not all; this is clarified later.)

- **Heat production.** The skeletal muscles produce as much as 85% of one's body heat, which is vital to the functioning of enzymes and therefore to all metabolism.

---

[1]*myo* = muscle; *logy* = study of
[2]*mus* = mouse; *cle* = little

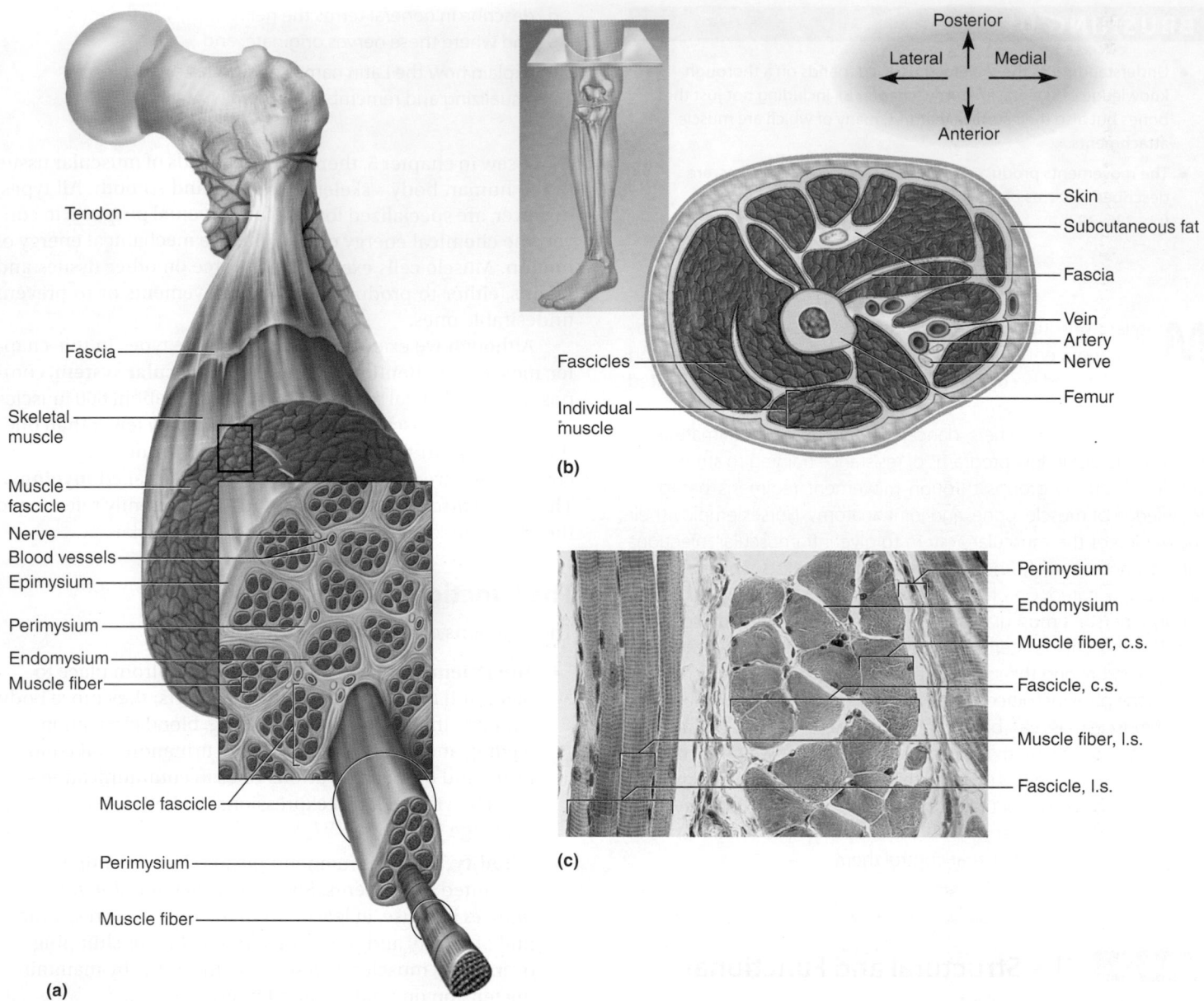

**FIGURE 10.1 Connective Tissues of a Muscle.** (a) The muscle–bone attachment. (b) A cross section of the thigh showing the relationship of neighboring muscles to fasciae and bone. (c) Muscle fascicles in the tongue. Vertical fascicles passing between the superior and inferior surfaces of the tongue are seen alternating with cross-sectioned horizontal fascicles that pass from the tip to the rear of the tongue. A fibrous perimysium can be seen between the fascicles, and endomysium can be seen between the muscle fibers within each fascicle (c.s. = cross section; l.s. = longitudinal section).

- **Glycemic control.** This means the regulation of blood glucose concentration within its normal range. The skeletal muscles absorb, store, and use a large share of one's glucose and play a highly significant role in stabilizing its blood concentration. In old age, in obesity, and when muscles become deconditioned and weakened, people suffer an increased risk of type 2 diabetes mellitus because of the decline in this glucose-buffering function.

## Connective Tissues and Fascicles

A skeletal muscle is more than muscular tissue. It also contains connective tissue, nerves, and blood vessels. The connective tissue components, from the smallest to largest and from deep to superficial, are as follows (fig. 10.1):

- **Endomysium**[3] (EN-doe-MIZ-ee-um). This is a thin sleeve of loose connective tissue that surrounds each muscle fiber. It creates room for blood capillaries and nerve fibers

---

[3]*endo* = within; *mys* = muscle

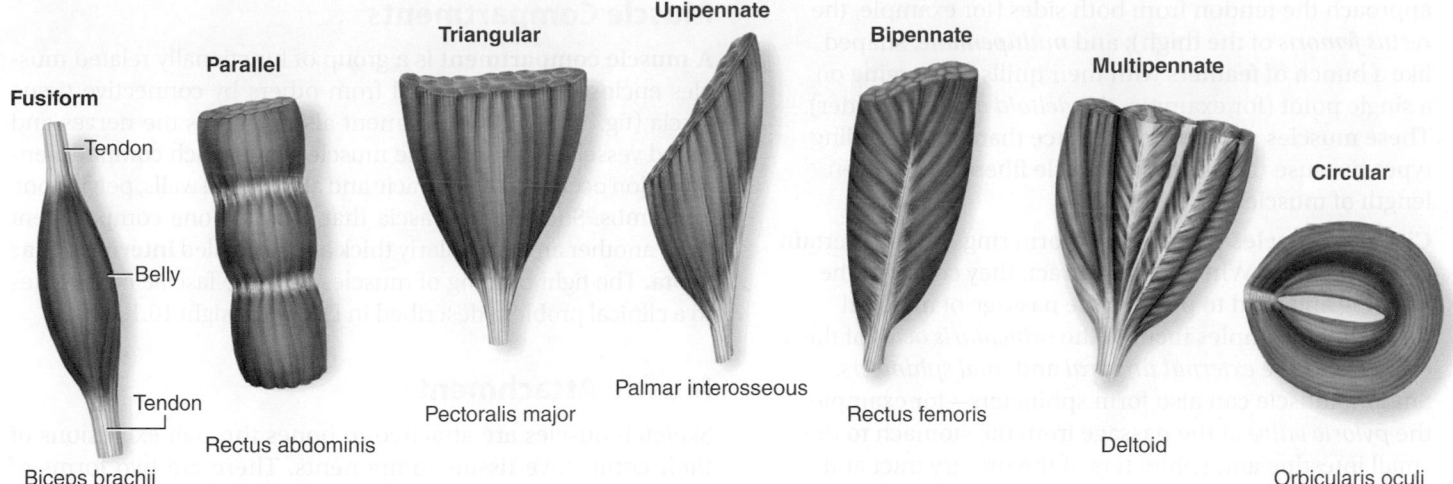

**FIGURE 10.2 Classification of Muscles According to Fascicle Orientation.** The fascicles are the "grain" visible in each illustration.

to reach every muscle fiber, ensuring that no muscle cell is without stimulation and nourishment. The endomysium also provides the extracellular chemical environment for the muscle fiber and its associated nerve ending. Excitation of a muscle fiber is based on the exchange of calcium, sodium, and potassium ions between the endomysial tissue fluid and the nerve and muscle fibers.

- **Perimysium.**[4] This is a thicker connective tissue sheath that wraps muscle fibers together in bundles called **fascicles**[5] (FASS-ih-culs). Fascicles are visible to the naked eye as parallel strands—the "grain" in a cut of meat; tender roast beef is easily pulled apart along its fascicles. The perimysium carries the larger nerves and blood vessels as well as stretch receptors called muscle spindles (see p. 497).

- **Epimysium.**[6] This is a fibrous sheath that surrounds the entire muscle. On its outer surface, the epimysium grades into the fascia, and its inner surface issues projections between the fascicles to form the perimysium.

- **Fascia**[7] (FASH-ee-uh). This is a sheet of connective tissue that separates neighboring muscles or muscle groups from each other and from the subcutaneous tissue. Muscles are grouped in *compartments* separated from each other by fascia.

## Fascicles and Muscle Shapes

The strength of a muscle and the direction of its pull are determined partly by the orientation of its fascicles. Muscles can be classified according to fascicle orientation as follows (fig. 10.2):

- **Fusiform**[8] **muscles** are thick in the middle and tapered at each end. The *biceps brachii* of the arm and *gastrocnemius* of the calf are examples of this type. Muscle strength is proportional to the diameter of a muscle at its thickest point, and fusiform muscles are relatively strong.

- **Parallel muscles** have a fairly uniform width and parallel fascicles. Some of these are elongated straps, such as the *rectus abdominis* of the abdomen, *sartorius* of the thigh, and *zygomaticus major* of the face. Others are more squarish and are called *quadrilateral* (four-sided) muscles, such as the *masseter* of the jaw. Parallel muscles can span long distances, such as from hip to knee, and they shorten more than other muscle types. However, having fewer muscle fibers than a fusiform muscle of the same mass, they produce less force.

- **Triangular (convergent) muscles** are fan-shaped—broad at one end and narrower at the other. Examples include the *pectoralis major* in the chest and the *temporalis* on the side of the head. Despite their small localized insertions on a bone, these muscles are relatively strong because they contain a large number of fibers in the wider part of the muscle.

- **Pennate**[9] **muscles** are feather-shaped. Their fascicles insert obliquely on a tendon that runs the length of the muscle, like the shaft of a feather. There are three types of pennate muscles: *unipennate,* in which all fascicles approach the tendon from one side (for example, the *palmar interosseous muscles* of the hand and *semimembranosus* of the thigh); *bipennate,* in which fascicles

---

[4]*peri* = around; *mys* = muscle
[5]*fasc* = bundle; *icle* = little
[6]*epi* = upon, above; *mys* = muscle
[7]*fascia* = band

[8]*fusi* = spindle; *form* = shape
[9]*penna* = feather; *ate* = characterized by

approach the tendon from both sides (for example, the *rectus femoris* of the thigh); and *multipennate,* shaped like a bunch of feathers with their quills converging on a single point (for example, the *deltoid* of the shoulder). These muscles generate more force than the preceding types because they fit more muscle fibers into a given length of muscle.

- **Circular muscles (sphincters)** form rings around certain body openings. When they contract, they constrict the opening and tend to prevent the passage of material through it. Examples include the *orbicularis oculi* of the eyelids and the *external urethral* and *anal sphincters.* Smooth muscle can also form sphincters—for example, the *pyloric valve* at the passage from the stomach to the small intestine and sphincters of the urinary tract and anal canal.

## Muscle Compartments

A **muscle compartment** is a group of functionally related muscles enclosed and separated from others by connective tissue fascia (fig. 10.3). A compartment also contains the nerves and blood vessels that supply the muscle group. Such compartmentalization occurs in the thoracic and abdominal walls, pelvic floor, and limbs. Some of the fascia that separate one compartment from another are particularly thick and are called **intermuscular septa.** The tight binding of muscles by these fasciae contributes to a clinical problem described in Deeper Insight 10.1.

## Muscle Attachments

Skeletal muscles are attached to bones through extensions of their connective tissue components. There are two forms of attachment—*indirect* and *direct.*

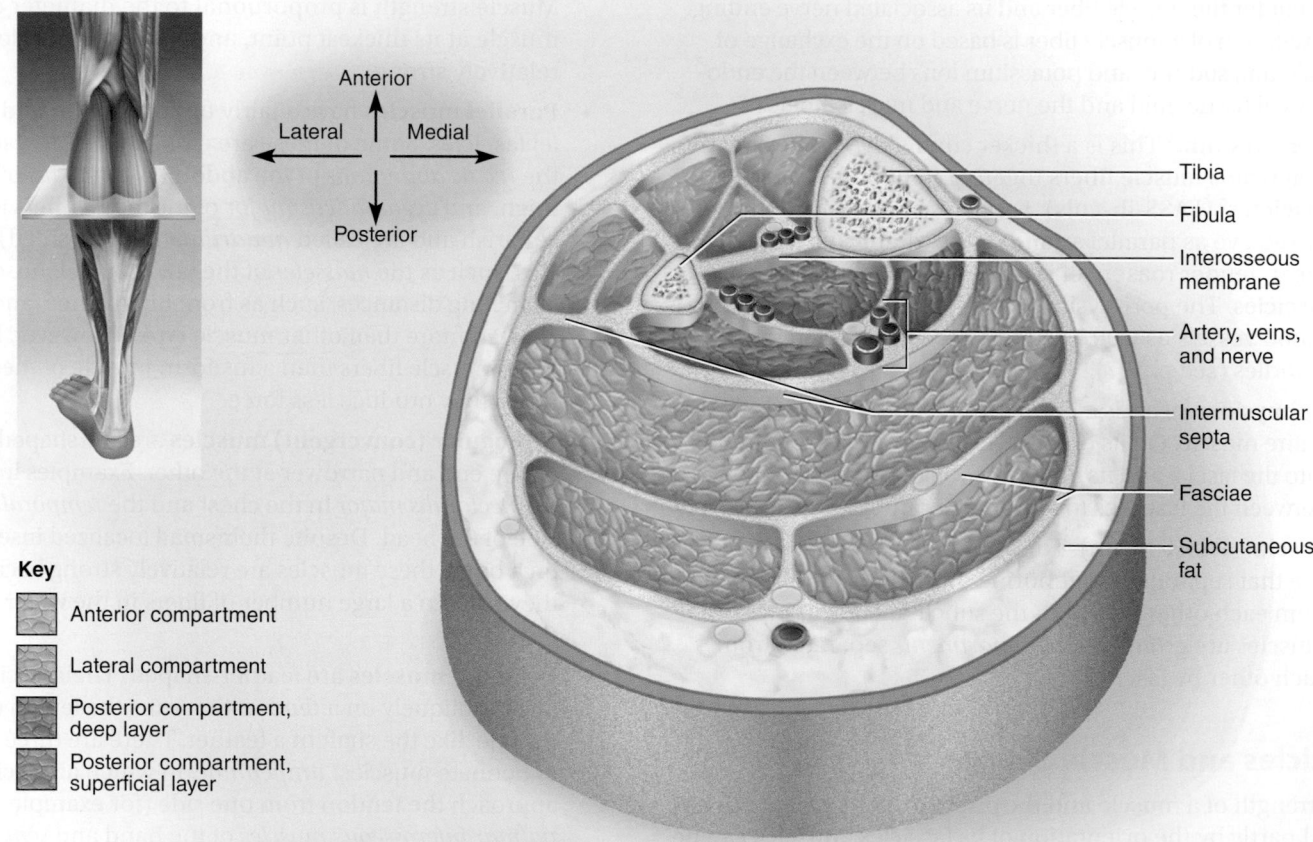

**Key**

- Anterior compartment
- Lateral compartment
- Posterior compartment, deep layer
- Posterior compartment, superficial layer

Anterior

Lateral — Medial

Posterior

- Tibia
- Fibula
- Interosseous membrane
- Artery, veins, and nerve
- Intermuscular septa
- Fasciae
- Subcutaneous fat

**FIGURE 10.3 Muscle Compartments.** A cross section of the left leg slightly above midcalf, oriented the same way as the reader's.

## DEEPER INSIGHT 10.1
### CLINICAL APPLICATION

#### Compartment Syndrome

Muscle compartments are very snugly contained in their fasciae. If a blood vessel in a compartment is damaged by overuse or contusion (a bruising injury), blood and tissue fluid accumulate in the compartment. The inelastic fascia prevents the compartment from expanding to relieve the pressure. Mounting pressure on the muscles, nerves, and blood vessels triggers a sequence of degenerative events called *compartment syndrome*. Blood flow to the compartment is obstructed by pressure on its arteries. If *ischemia* (poor blood flow) persists for more than 2 to 4 hours, nerves begin to die, and after 6 hours, so does muscle tissue. Nerves can regenerate after the pressure is relieved, but muscle necrosis is irreversible. The breakdown of muscle releases myoglobin into the blood. *Myoglobinuria,* the presence of myoglobin in the urine, gives the urine a dark color and is one of the key signs of compartment syndrome and some other degenerative muscle disorders. Compartment syndrome is treated by immobilizing and resting the limb and, if necessary, making an incision *(fasciotomy)* to relieve the pressure.

In an **indirect attachment,** the muscle ends conspicuously short of its bony destination, and the gap is bridged by a fibrous band or sheet called a **tendon.** See, for example, the two ends of the biceps brachii in figure 10.4 and the photographs of tendons in figures 10.31b and 10.36. You can easily palpate tendons and feel their texture just above the heel (your *calcaneal* or *Achilles tendon*) and on the anterior side of the wrist (tendons of the *palmaris longus* and *flexor carpi radialis* muscles). Collagen fibers of the muscle (the endo-, peri-, and epimysium) continue into the tendon and from there into the periosteum and matrix of the bone, creating very strong structural continuity from muscle to bone.

In some cases, the tendon is a broad sheet called an **aponeurosis**[10] (AP-oh-new-RO-sis). This term originally referred to the tendon located beneath the scalp, but now it also refers to similar tendons associated with certain abdominal, lumbar, hand, and foot muscles. For example, the palmaris longus tendon passes through the wrist and then expands into a fanlike *palmar aponeurosis* beneath the skin of the palm (see fig. 10.28a).

In some places, groups of tendons from separate muscles pass under a band of connective tissue called a **retinaculum.**[11] One of these covers each surface of the wrist like a bracelet, for example. The tendons of several forearm muscles pass under them on their way to the hand (see fig. 5.13, p. 152, and fig. 10.28a).

In a **direct (fleshy) attachment,** there is so little separation between muscle and bone that to the naked eye, the red muscular tissue seems to emerge directly from the bone—for

[10] *apo* = upon, above; *neuro* = nerve
[11] *retinac* = retainer, bracelet; *cul* = little

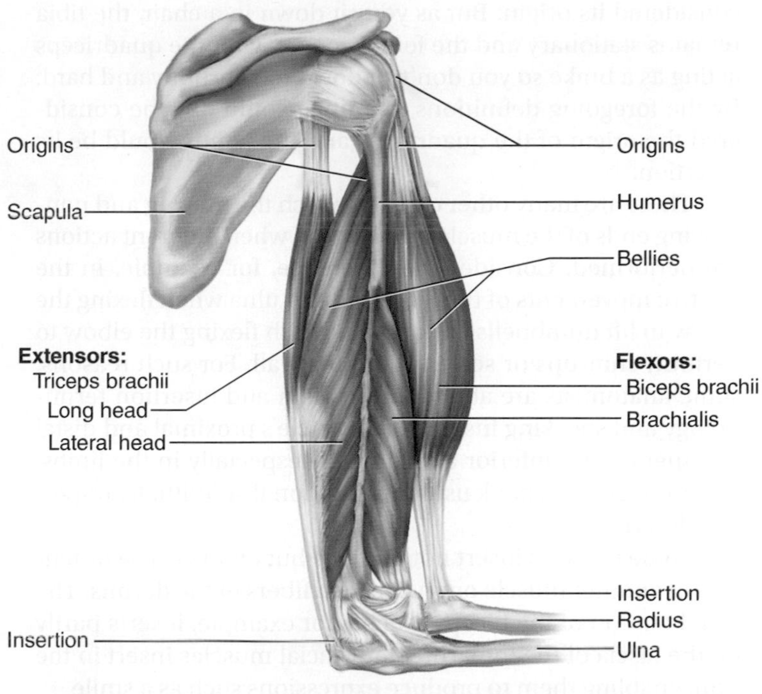

**FIGURE 10.4  Synergistic and Antagonistic Muscle Pairs.**
The biceps brachii and brachialis muscles are synergists in elbow flexion. The triceps brachii is an antagonist of those two muscles and is the prime mover in elbow extension.

*Which of these muscles have direct attachments to the bones, and which have indirect attachments?*

example, along the margins of the *brachialis* and lateral head of the *triceps brachii* in figure 10.4. At a microscopic level, however, the muscle fibers stop slightly short of the bone, and the gap between muscle and bone is spanned by collagen fibers.

### Muscle Origins and Insertions

Most skeletal muscles are attached to a different bone at each end, so either the muscle or its tendon spans at least one joint. When the muscle contracts, it moves one bone relative to the other. The bony site of attachment at the relatively stationary end is called its **origin.** The attachment site at its more mobile end is called the **insertion.** For the biceps brachii, for example, the origin is on the scapula and the insertion is on the radius (fig. 10.4). The middle, usually thicker region is called the **belly.**

The terminology of origins and insertions, however, is imperfect and sometimes misleading. One end of a muscle might function as its stationary origin during one action, but as its moving insertion during a different action. For example, consider the *quadriceps femoris* muscle on the anterior side of the thigh. It is a powerful extensor of the knee, connected at its proximal end mainly to the femur and at its distal end to the tibia, just below the knee. If you kick a soccer ball, the tibia moves more than the femur, so the tibia would be considered the insertion of the quadriceps and the femur would be considered its origin. But as you sit down in a chair, the tibia remains stationary and the femur moves, with the quadriceps acting as a brake so you don't sit down too abruptly and hard. By the foregoing definitions, the tibia would now be considered the origin of the quadriceps and the femur would be its insertion.

There are many other cases in which the moving and non-moving ends of the muscle are reversed when different actions are performed. Consider the difference, for example, in the relative movements of the humerus and ulna when flexing the elbow to lift dumbbells as compared with flexing the elbow to perform chin-ups or scale a climbing wall. For such reasons, some anatomists are abandoning origin and insertion terminology and speaking instead of a muscle's proximal and distal or superior and inferior attachments, especially in the limbs. Nevertheless, this book uses the traditional, admittedly imperfect descriptions.

Some muscles insert not on bone but on the fascia or tendon of another muscle or on collagen fibers of the dermis. The distal tendon of the biceps brachii, for example, inserts partly on the fascia of the forearm. Many facial muscles insert in the skin, enabling them to produce expressions such as a smile.

## Functional Groups of Muscles

The effect produced by a muscle, whether it is to produce or prevent a movement, is called its **action.** Skeletal muscles seldom act independently; instead, they function in groups whose combined actions produce the coordinated control of a joint. Muscles can be classified into four categories according to their actions, but it must be stressed that a particular muscle can act in a certain way during one joint action and in a different way during other actions of the same joint. Furthermore, the action of a given muscle depends on what other muscles are doing. For example, the *gastrocnemius* of the posterior calf usually flexes the knee, but if the *quadriceps* of the anterior thigh prevents knee flexion, the gastrocnemius flexes the ankle, causing plantar flexion. The four categories of muscle action are illustrated in figure 10.4:

1. The **prime mover (agonist)** is the muscle that produces most of the force during a particular joint action. In flexing the elbow, for example, the prime mover is the *brachialis.*

2. A **synergist**[12] (SIN-ur-jist) is a muscle that aids the prime mover. Two or more synergists acting on a joint can produce more power than a single larger muscle. The biceps brachii, for example, overlies the brachialis and works with it as a synergist to flex the elbow. The actions of a prime mover and its synergist are not necessarily identical and redundant. If the prime mover worked alone at a

joint, it might cause rotation or other undesirable movements of a bone. A synergist may stabilize a joint and restrict these movements, or modify the direction of a movement so that the action of the prime mover is more coordinated and specific.

3. An **antagonist**[13] is a muscle that opposes the prime mover. In some cases, it relaxes to give the prime mover almost complete control over an action. More often, however, the antagonist maintains some tension on a joint and thus limits the speed or range of the prime mover, preventing excessive movement, joint injury, or inappropriate actions. If you extend your arm to reach out and pick up a cup of tea, for example, your *triceps brachii* serves as the prime mover of elbow extension, and your brachialis acts as an antagonist to slow the extension and stop it at the appropriate point. If you extend your arm rapidly to throw a dart, however, the brachialis must be quite relaxed. The brachialis and triceps represent an **antagonistic pair** of muscles that act on opposite sides of a joint. We need antagonistic pairs at a joint because a muscle can only pull, not push—for example, a single muscle cannot flex *and* extend the elbow. Which member of the pair acts as the prime mover depends on the motion under consideration. In flexion of the elbow, the brachialis is the prime mover and the triceps is the antagonist; when the elbow is extended, their roles are reversed.

4. A **fixator** is a muscle that prevents a bone from moving. To *fix* a bone means to hold it steady, allowing another muscle attached to it to pull on something else. For example, consider again the flexion of the elbow by the biceps brachii. The biceps originates on the scapula, crosses both the shoulder and elbow joints, and inserts on the radius. The scapula is loosely attached to the axial skeleton, so when the biceps contracts, it seems that it would pull the scapula laterally. However, there are fixator muscles (the *rhomboids*) that attach the scapula to the vertebral column. They contract at the same time as the biceps, holding the scapula firmly in place and ensuring that the force generated by the biceps moves the radius rather than the scapula.

## Intrinsic and Extrinsic Muscles

In places such as the tongue, larynx, back, hand, and foot, anatomists distinguish between intrinsic and extrinsic muscles. An **intrinsic muscle** is entirely contained within a particular region, having both its origin and insertion there. An **extrinsic muscle** acts upon a designated region but has its origin elsewhere. For example, some movements of the fingers are

---

[12]*syn* = together; *erg* = work

[13]*ant* = against; *agonist* = actor, competitor

produced by extrinsic muscles in the forearm, whose long tendons reach to the phalanges; other finger movements are produced by the intrinsic muscles of the hand, located between the metacarpal bones.

## Muscle Innervation

The **innervation** of a muscle refers to the identity of the nerve that stimulates it. Knowing the innervation to each muscle enables clinicians to diagnose nerve, spinal cord, and brainstem injuries from their effects on muscle function, and to set realistic goals for rehabilitation. The innervations described in this chapter will be more meaningful after you have studied the peripheral nervous system (chapters 13 and 14), but a brief orientation will be helpful here. The muscles are innervated by two groups of nerves:

- **Spinal nerves** arise from the spinal cord, emerge through the intervertebral foramina, and innervate muscles below the neck. Spinal nerves are identified by letters and numbers that refer to the adjacent vertebrae—for example, T6 for the sixth thoracic nerve and S2 for the second sacral nerve. Immediately after emerging from an intervertebral foramen, each spinal nerve branches into a *posterior* and *anterior ramus*. You will note references to nerve numbers and rami in many of the muscle tables. The term *plexus* in some of the tables refers to weblike networks of spinal nerves adjacent to the vertebral column. All of the spinal nerves named here are illustrated, and most are also discussed, in chapter 13 (see tables 13.3–13.6).

- **Cranial nerves** arise from the base of the brain, emerge through the skull foramina, and innervate muscles of the head and neck. Cranial nerves are identified by roman numerals (CN I to CN XII) and by names given in chapter 14 (see table 14.1), although not all 12 of them innervate skeletal muscles.

## Blood Supply

The muscular system as a whole receives about 1.24 L of blood per minute at rest—which is about one-quarter of the blood pumped by the heart. During heavy exercise, total cardiac output rises and the muscular system's share is more than three-quarters, or 11.6 L/min. Working muscles have a great demand for glucose, fatty acids, and oxygen. Blood capillaries branch extensively through the endomysium to reach every muscle fiber, sometimes so intimately associated with the muscle fibers that the fibers have surface indentations to accommodate them. The capillaries of skeletal muscle undulate or coil when the muscle is contracted, allowing them enough slack to stretch

out straight, without breaking, when the muscle lengthens. Chapter 20 describes some special physiological properties of muscle circulation and names the major arteries that supply the skeletal muscle groups.

## How Muscles Are Named

Figure 10.5 shows an overview of the major superficial muscles. Learning the names of these and other muscles may seem a forbidding task at first, especially when some of them have such long Latin names as *depressor labii inferioris* and *flexor digiti minimi brevis*. Such names, however, typically describe some distinctive aspects of the structure, location, or action of a muscle, and become very helpful once we grow familiar with a few common Latin words. For example, the depressor labii inferioris is a muscle that lowers (depresses) the bottom (inferior) lip (labium), and the flexor digiti minimi brevis is a short (brevis) muscle that flexes the smallest (minimi) finger (digit). Muscle names are interpreted in footnotes throughout the chapter. Familiarity with these terms and attention to the footnotes will help you translate muscle names and remember the location, appearance, and action of the muscles. You can listen to pronunciations of these muscle names online by visiting *Anatomy & Physiology|Revealed* (www.aprevealed.com).

## A Learning Strategy

In the remainder of this chapter, we consider about 160 muscles; most courses cover considerably fewer according to the choices of individual instructors. The following suggestions may help you develop a rational strategy for learning the muscular system:

- Examine models, cadavers, dissected animals, or a photographic atlas as you read about these muscles. Visual images are often easier to remember than words, and direct observation of a muscle may stick in your memory better than descriptive text or two-dimensional drawings.

- When studying a particular muscle, palpate it on yourself if possible. Contract the muscle to feel it bulge and sense its action. This makes muscle locations and actions less abstract. Atlas B, following this chapter, shows where you can see and palpate several of these muscles on the living body.

- Locate the origins and insertions of muscles on an articulated skeleton. Some study skeletons are painted and labeled to show these. This helps you visualize the locations of muscles and understand how they produce particular joint actions.

- Study the derivation of each muscle name; the name usually describes the muscle's location, appearance, origin, insertion, or action.

Superficial | Deep
← | →

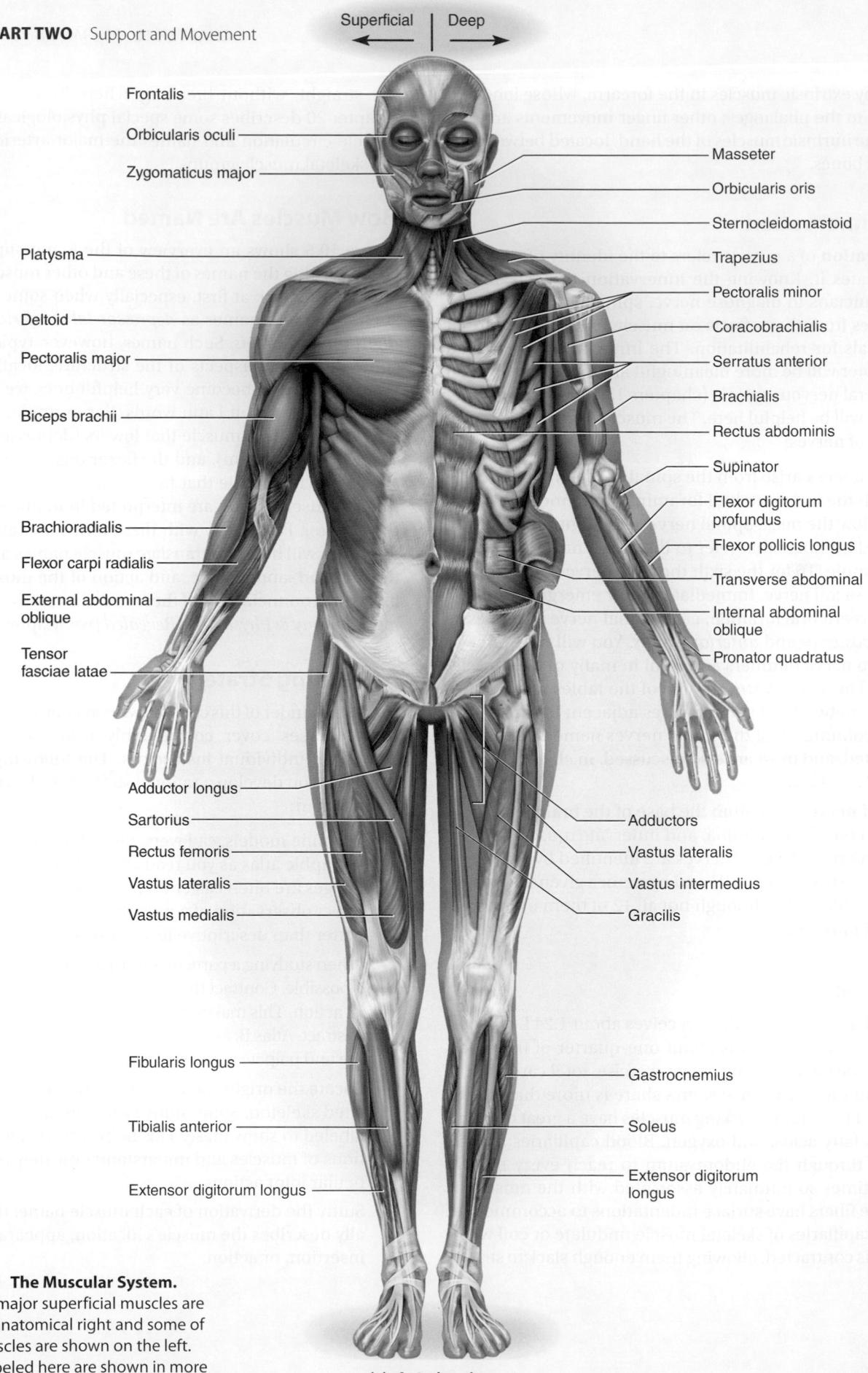

Frontalis

Orbicularis oculi

Zygomaticus major

Masseter

Orbicularis oris

Sternocleidomastoid

Platysma

Trapezius

Pectoralis minor

Coracobrachialis

Deltoid

Serratus anterior

Pectoralis major

Brachialis

Biceps brachii

Rectus abdominis

Supinator

Flexor digitorum profundus

Brachioradialis

Flexor pollicis longus

Flexor carpi radialis

Transverse abdominal

External abdominal oblique

Internal abdominal oblique

Tensor fasciae latae

Pronator quadratus

Adductor longus

Sartorius

Adductors

Rectus femoris

Vastus lateralis

Vastus lateralis

Vastus intermedius

Vastus medialis

Gracilis

Fibularis longus

Gastrocnemius

Tibialis anterior

Soleus

Extensor digitorum longus

Extensor digitorum longus

**FIGURE 10.5  The Muscular System.**
In each figure, major superficial muscles are shown on the anatomical right and some of the deeper muscles are shown on the left. Muscles not labeled here are shown in more detail in later figures.

**(a) Anterior view**

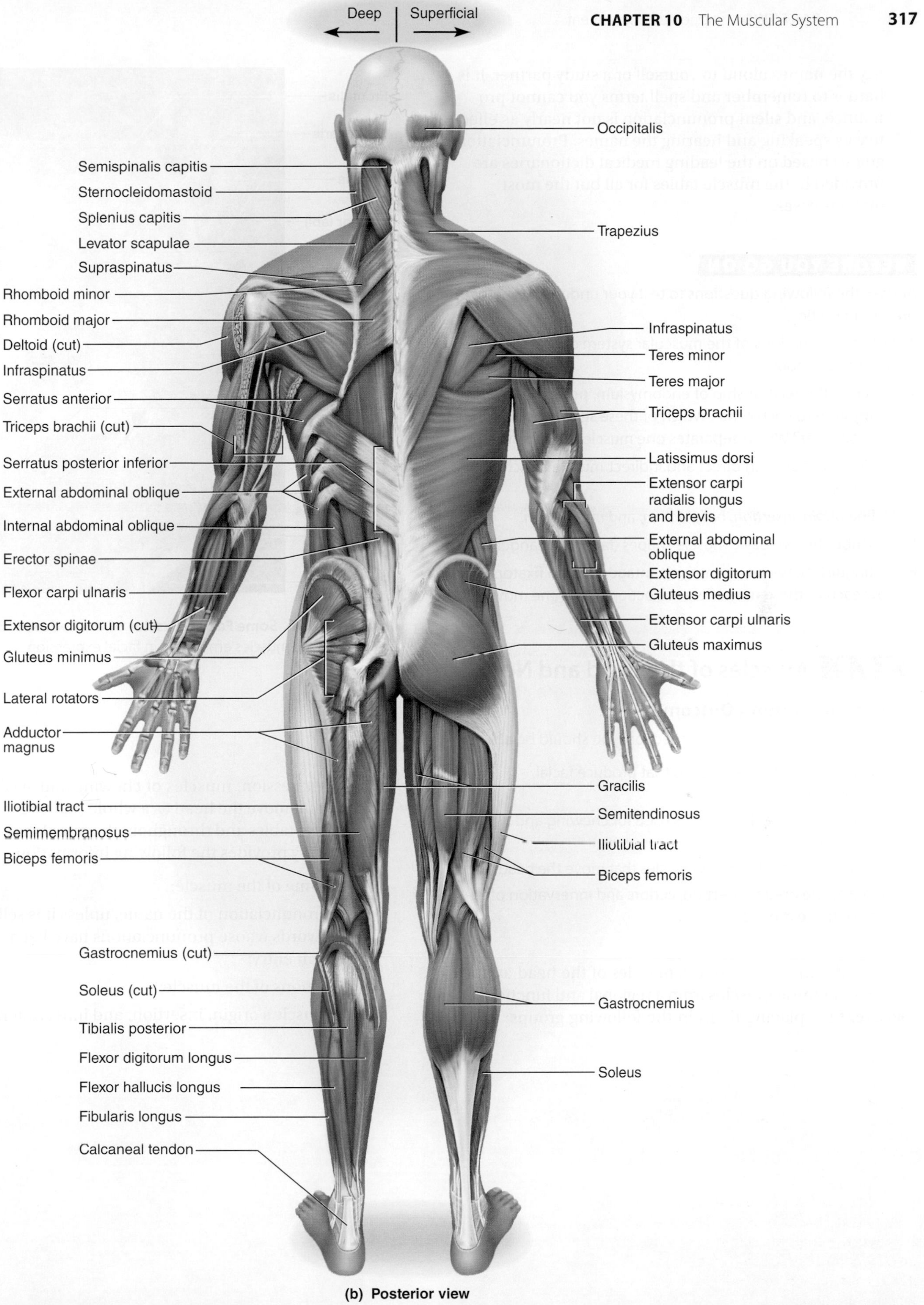

Semispinalis capitis

Sternocleidomastoid

Splenius capitis

Levator scapulae

Supraspinatus

Rhomboid minor

Rhomboid major

Deltoid (cut)

Infraspinatus

Serratus anterior

Triceps brachii (cut)

Serratus posterior inferior

External abdominal oblique

Internal abdominal oblique

Erector spinae

Flexor carpi ulnaris

Extensor digitorum (cut)

Gluteus minimus

Lateral rotators

Adductor magnus

Iliotibial tract

Semimembranosus

Biceps femoris

Gastrocnemius (cut)

Soleus (cut)

Tibialis posterior

Flexor digitorum longus

Flexor hallucis longus

Fibularis longus

Calcaneal tendon

Occipitalis

Trapezius

Infraspinatus

Teres minor

Teres major

Triceps brachii

Latissimus dorsi

Extensor carpi radialis longus and brevis

External abdominal oblique

Extensor digitorum

Gluteus medius

Extensor carpi ulnaris

Gluteus maximus

Gracilis

Semitendinosus

Iliotibial tract

Biceps femoris

Gastrocnemius

Soleus

**(b) Posterior view**

**FIGURE 10.5 The Muscular System (continued).**

- Say the names aloud to yourself or a study partner. It is harder to remember and spell terms you cannot pronounce, and silent pronunciation is not nearly as effective as speaking and hearing the names. Pronunciation guides based on the leading medical dictionaries are provided in the muscle tables for all but the most obvious cases.

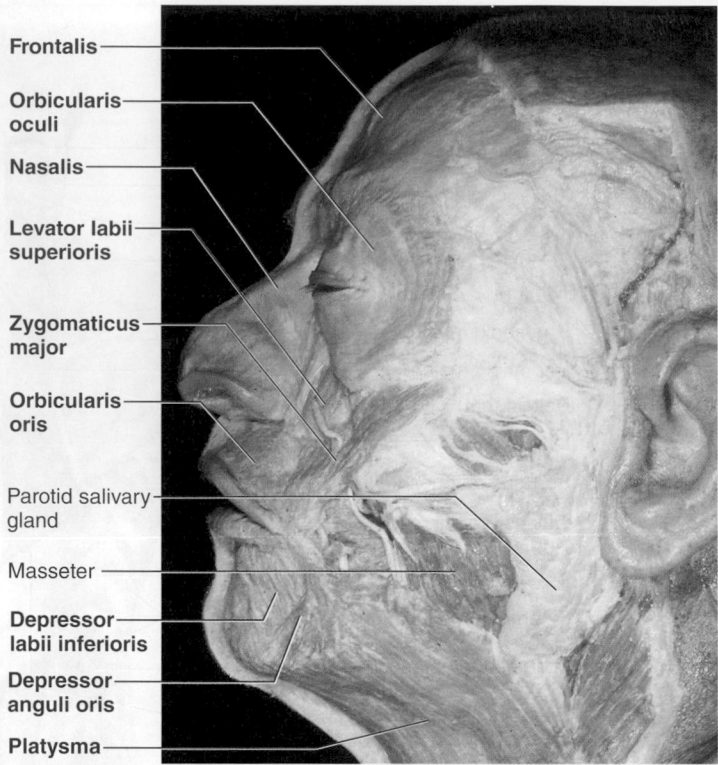

**FIGURE 10.6 Some Facial Muscles of the Cadaver.** Boldface labels indicate muscles employed in facial expression.

Labels in figure: Frontalis, Orbicularis oculi, Nasalis, Levator labii superioris, Zygomaticus major, Orbicularis oris, Parotid salivary gland, Masseter, Depressor labii inferioris, Depressor anguli oris, Platysma

---

**BEFORE YOU GO ON**

Answer the following questions to test your understanding of the preceding section:

1. List some functions of the muscular system other than movement of the body.

2. Describe the relationship of endomysium, perimysium, and epimysium to each other. Which of these separates one fascicle from another? Which separates one muscle from another?

3. Distinguish between direct and indirect muscle attachments to bones.

4. Define *origin, insertion, belly, action,* and *innervation.*

5. Describe the five basic muscle shapes (fascicle arrangements).

6. Distinguish between a synergist, antagonist, and fixator. Explain how each of these may affect the action of a prime mover.

## 10.2 Muscles of the Head and Neck

### Expected Learning Outcomes

When you have completed this section, you should be able to

a. name and locate the muscles that produce facial expressions;

b. name and locate the muscles used for chewing and swallowing;

c. name and locate the neck muscles that move the head; and

d. identify the origin, insertion, action, and innervation of any of these muscles.

Figure 10.6 shows some of the muscles of the head and neck. We will treat these muscles from a regional and functional perspective, thus placing them in the following groups: muscles of facial expression, muscles of chewing and swallowing, and muscles that move the head as a whole (tables 10.1–10.3).

In these tables and throughout the rest of the chapter, each muscle entry provides the following information:

- the name of the muscle;

- the pronunciation of the name, unless it is self-evident or uses words whose pronunciations have been provided in a recent entry;

- the actions of the muscle;

- the muscle's origin, insertion, and innervation.

| **TABLE 10.1** | **Muscles of Facial Expression** |
| --- | --- |

Humans have much more expressive faces than other mammals because of a complex array of muscles that insert in the dermis and subcutaneous tissues (figs. 10.6 and 10.7). These muscles tense the skin and produce such expressions as a pleasant smile, a threatening scowl, a puzzled frown, or a flirtatious wink. They add subtle shades of meaning to our spoken words. Facial muscles also contribute directly to speech, chewing, and other oral functions. All but one of these muscles are innervated by the facial nerve (cranial nerve VII). This nerve is especially vulnerable to injury from lacerations and skull fractures, which can paralyze the muscles and cause parts of the face to sag. The only muscle in this table not innervated by the facial nerve is the levator palpebrae superioris, innervated by the oculomotor nerve (CN III).

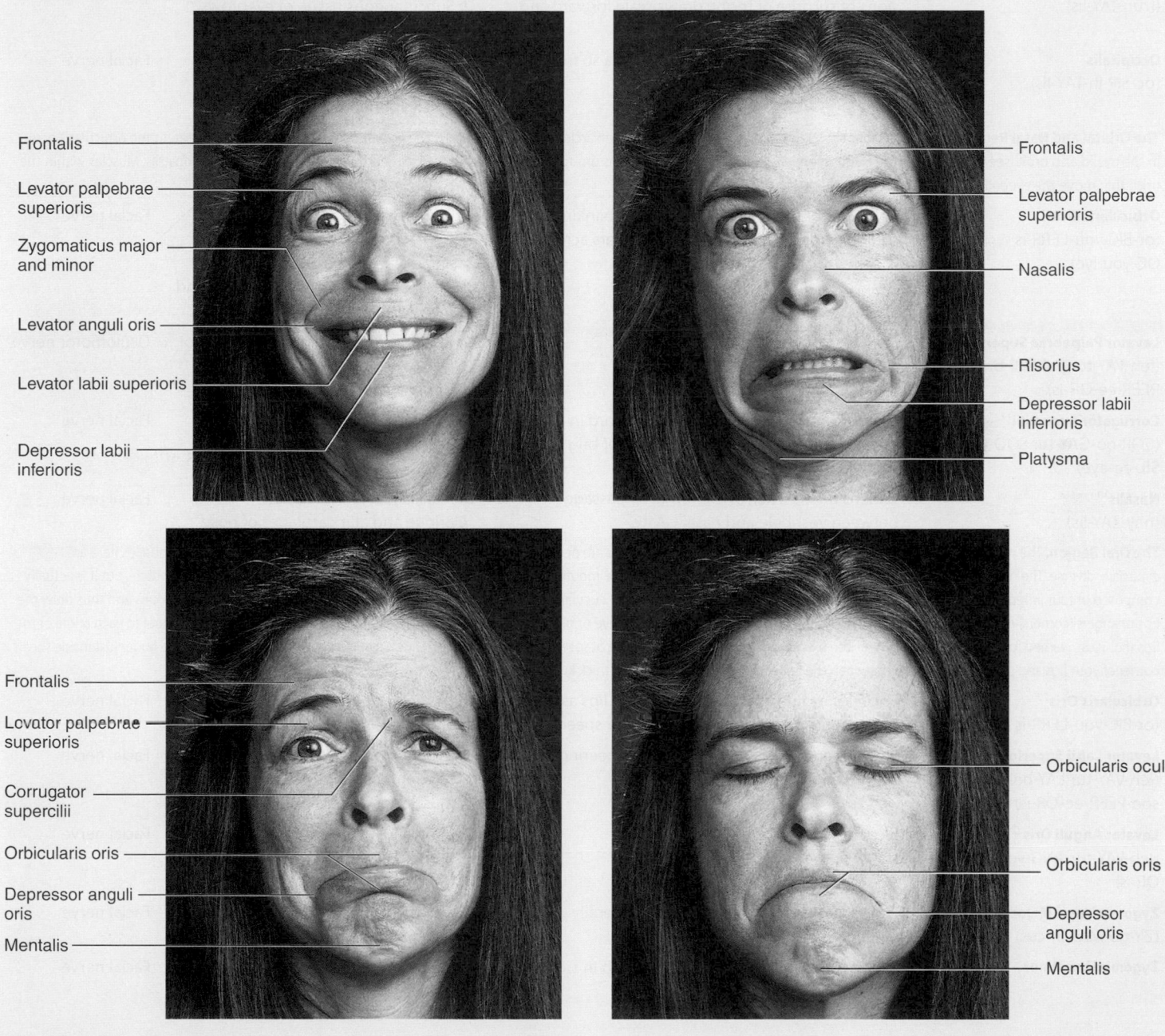

**FIGURE 10.7 Expressions Produced by Several of the Facial Muscles.** The ordinary actions of these muscles are usually more subtle than these demonstrations.

❓ *Name an antagonist of each of these muscles: the depressor anguli oris, orbicularis oculi, and levator labii superioris.*

| TABLE 10.1 | Muscles of Facial Expression (continued) | | |
|---|---|---|---|
| **Name** | **Action** | **O: Origin** **I: Insertion** | **Innervation** |
| **The Scalp.** The *occipitofrontalis* overlies the dome of the cranium. It is divided into the *frontalis* of the forehead and *occipitalis* at the rear of the head, named for the frontal and occipital bones underlying them. They are connected to each other by a broad aponeurosis, the **galea aponeurotica**[14] (GAY-lee-uh APO-oh-new-ROT-ih-cuh). | | | |
| **Frontalis** (frun-TAY-lis) | Elevates eyebrows in glancing upward and expressions of surprise or fright; draws scalp forward and wrinkles skin of forehead | *O:* Galea aponeurotica *I:* Subcutaneous tissue of eyebrows | Facial nerve |
| **Occipitalis** (oc-SIP-ih-TAY-lis) | Retracts scalp; fixes galea aponeurotica so frontalis can act on eyebrows | *O:* Superior nuchal line and temporal bone *I:* Galea aponeurotica | Facial nerve |
| **The Orbital and Nasal Regions.** The *orbicularis oculi* is a sphincter of the eyelid that encircles and closes the eye. The *levator palpebrae superioris* lies deep to the orbicularis oculi, in the eyelid and orbit (see fig. 16.23a, p. 607), and opens the eye. Other muscles in this group move the eyelids and skin of the forehead and dilate the nostrils. Muscles within the orbit that move the eyeball itself are discussed in chapter 16. | | | |
| **Orbicularis Oculi**[15] (or-BIC-you-LERR-is OC-you-lye) | Sphincter of the eyelids; closes eye in blinking, squinting, and sleep; aids in flow of tears across eye | *O:* Lacrimal bone; adjacent regions of frontal bone and maxilla; medial angle of eyelids *I:* Upper and lower eyelids; skin around margin of orbit | Facial nerve |
| **Levator Palpebrae Superioris**[16] (leh-VAY-tur pal-PEE-bree soo-PEER-ee-OR-is) | Elevates upper eyelid; opens eye | *O:* Lesser wing of sphenoid in posterior wall of orbit *I:* Upper eyelid | Oculomotor nerve |
| **Corrugator Supercilii**[17] (COR-oo-GAY-tur SOO-per-SIL-ee-eye) | Draws eyebrows medially and downward in frowning and concentration; reduces glare of bright sunlight | *O:* Medial end of supraorbital margin *I:* Skin of eyebrow | Facial nerve |
| **Nasalis**[18] (nay-ZAY-lis) | Widens nostrils; narrows internal air passage between vestibule and nasal cavity | *O:* Maxilla just lateral to nose *I:* Bridge and alar cartilages of nose | Facial nerve |
| **The Oral Region.** The mouth is the most expressive part of the face, and lip movements are necessary for intelligible speech; thus, it is not surprising that the muscles here are especially diverse. The *orbicularis oris* is a complex of muscles in the lips that encircles the mouth; until recently it was misinterpreted as a sphincter, or circular muscle, but it is actually composed of four independent quadrants that interlace and give only an appearance of circularity. Other muscles in this region approach the lips from all directions and thus draw the lips or angles (corners) of the mouth upward, laterally, and downward. Some of these have origins or insertions in a complex cord called the **modiolus**[19] just lateral to each angle of the lips (fig. 10.8). Named for the hub of a cartwheel, the modiolus is a point of convergence of several muscles of the lower face. You can palpate it by inserting one finger just inside the corner of your lips and pinching the corner between the finger and thumb, feeling for a thick knot of tissue. | | | |
| **Orbicularis Oris**[20] (or-BIC-you-LERR-is OR-is) | Encircles mouth, closes lips, protrudes lips as in kissing; uniquely developed in humans for speech | *O:* Modiolus of mouth *I:* Submucosa and dermis of lips | Facial nerve |
| **Levator Labii Superioris**[21] (leh-VAY-tur LAY-bee-eye soo-PEER-ee-OR-is) | Elevates and everts upper lip in sad, sneering, or serious expressions | *O:* Zygomatic bone and maxilla near inferior margin of orbit *I:* Muscles of upper lip | Facial nerve |
| **Levator Anguli Oris**[22] (leh-VAY-tur ANG-you-lye OR-is) | Elevates angle of mouth as in smiling | *O:* Maxilla just below infraorbital foramen *I:* Muscles at angle of mouth | Facial nerve |
| **Zygomaticus**[23] **Major** (ZY-go-MAT-ih-cus) | Draws angle of mouth upward and laterally in laughing | *O:* Zygomatic bone *I:* Superolateral angle of mouth | Facial nerve |
| **Zygomaticus Minor** | Elevates upper lip, exposes upper teeth in smiling or sneering | *O:* Zygomatic bone *I:* Muscles of upper lip | Facial nerve |

---

[14]*galea* = helmet; *apo* = above; *neuro* = nerves, the brain
[15]*orb* = circle; *ocul* = eye
[16]*levator* = that which raises; *palpebr* = eyelid; *superior* = upper
[17]*corrug* = wrinkle; *supercilii* = of the eyebrow
[18]*nas* = of the nose

[19]*modiolus* = hub
[20]*orb* = circle; *oris* = of the mouth
[21]*levat* = to raise; *labi* = lip; *superior* = upper
[22]*angul* = angle, corner; *oris* = of the mouth
[23]*zygo* = join, unite (refers to zygomatic bone)

**TABLE 10.1** Muscles of Facial Expression (continued)

**FIGURE 10.8 Muscles of Facial Expression.**
Boldface labels indicate muscles employed in facial
expression. AP|R

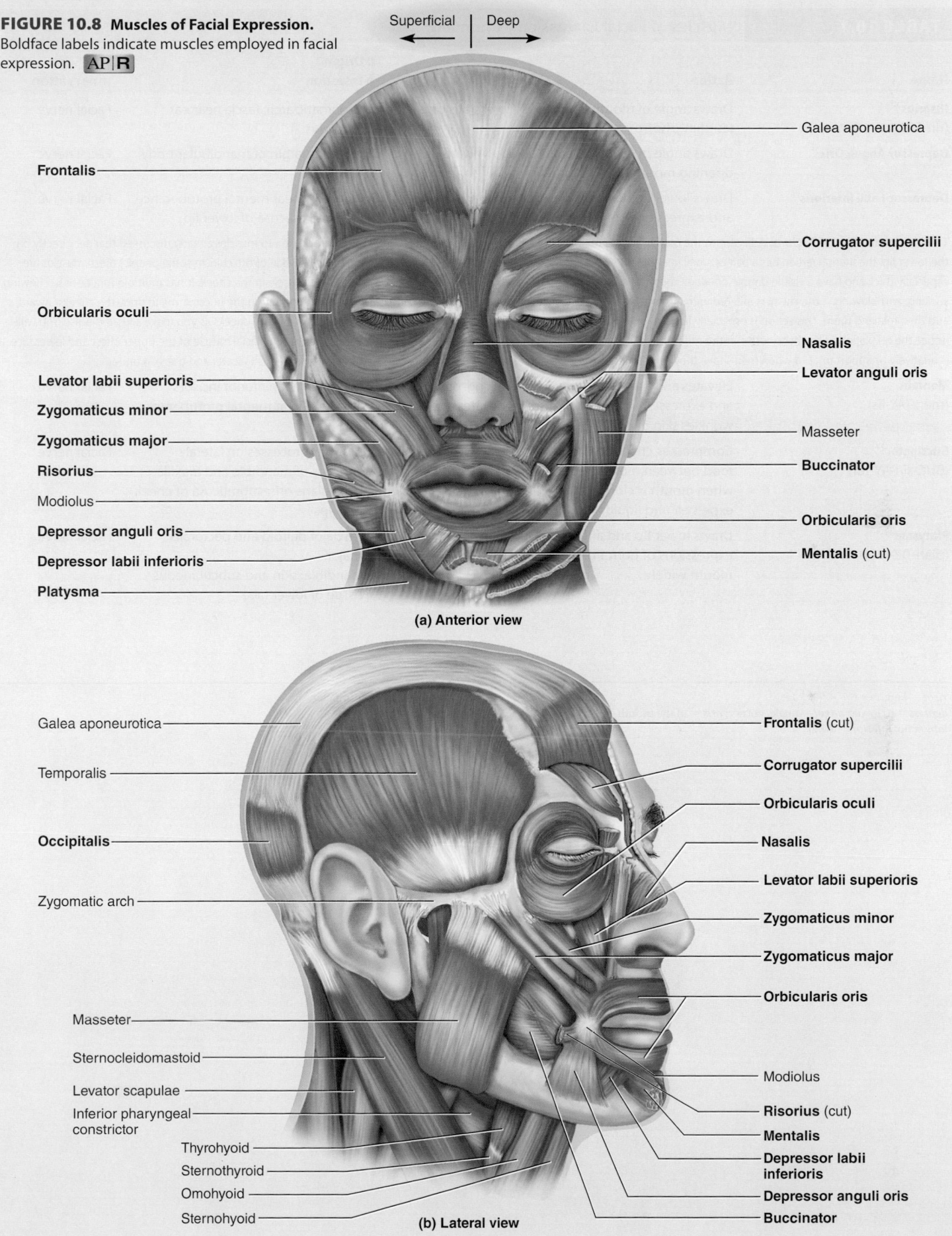

Superficial ← | → Deep

**(a) Anterior view**

Galea aponeurotica
**Corrugator supercilii**
**Nasalis**
**Levator anguli oris**
Masseter
**Buccinator**
**Orbicularis oris**
**Mentalis** (cut)

**Frontalis**
**Orbicularis oculi**
**Levator labii superioris**
**Zygomaticus minor**
**Zygomaticus major**
**Risorius**
Modiolus
**Depressor anguli oris**
**Depressor labii inferioris**
**Platysma**

**(b) Lateral view**

Galea aponeurotica
Temporalis
**Occipitalis**
Zygomatic arch
Masseter
Sternocleidomastoid
Levator scapulae
Inferior pharyngeal
constrictor
Thyrohyoid
Sternothyroid
Omohyoid
Sternohyoid

**Frontalis** (cut)
**Corrugator supercilii**
**Orbicularis oculi**
**Nasalis**
**Levator labii superioris**
**Zygomaticus minor**
**Zygomaticus major**
**Orbicularis oris**
Modiolus
**Risorius** (cut)
**Mentalis**
**Depressor labii
inferioris**
**Depressor anguli oris**
**Buccinator**

| TABLE 10.1 | Muscles of Facial Expression (continued) | | |
|---|---|---|---|
| **Name** | **Action** | **O: Origin** <br> **I: Insertion** | **Innervation** |
| **Risorius**[24] <br> (rih-SOR-ee-us) | Draws angle of mouth laterally in expressions of laughing, horror, or disdain | **O:** Zygomatic arch; fascia near ear <br> **I:** Modiolus | Facial nerve |
| **Depressor Anguli Oris**[25] | Draws angle of mouth laterally and downward in opening mouth or sad expressions | **O:** Inferior margin of mandibular body <br> **I:** Modiolus | Facial nerve |
| **Depressor Labii Inferioris**[26] | Draws lower lip downward and laterally in chewing and expressions of melancholy or doubt | **O:** Mandible near mental protuberance <br> **I:** Skin and mucosa of lower lip | Facial nerve |
| **The Mental and Buccal Regions.** Adjacent to the oral orifice are the mental region (chin) and buccal region (cheek). In addition to muscles already discussed that act directly on the lower lip, the mental region has a pair of small *mentalis muscles* extending from the upper margin of the mandible to the skin of the chin. In some people, these muscles are especially thick and have a visible dimple between them called the *mental cleft* (see fig. 4.18, p. 132). The *buccinator* is the muscle in the cheek. It has multiple functions in chewing, sucking, and blowing. If the cheek is inflated with air, compression of the buccinator blows it out. Sucking is achieved by contracting the buccinators to draw the cheeks inward, and then relaxing them. This action is especially important to nursing infants. To feel this action, hold your fingertips lightly on your cheeks as you make a kissing noise. You will notice the relaxation of the buccinators at the moment air is sharply drawn in through the pursed lips. The *platysma* is a thin superficial muscle of the upper chest and lower face. It is relatively unimportant, but when men shave they tend to tense the platysma to make the concavity between the jaw and neck shallower and the skin tauter. | | | |
| **Mentalis** <br> (men-TAY-lis) | Elevates and protrudes lower lip in drinking, pouting, and expressions of doubt or disdain; elevates and wrinkles skin of chin | **O:** Mandible near inferior incisors <br> **I:** Skin of chin at mental protuberance | Facial nerve |
| **Buccinator**[27] <br> (BUC-sin-AY-tur) | Compresses cheek against teeth and gums; directs food between molars; retracts cheek from teeth when mouth is closing to prevent biting cheek; expels air and liquid | **O:** Alveolar processes on lateral surfaces of mandible and maxilla <br> **I:** Orbicularis oris; submucosa of cheek and lips | Facial nerve |
| **Platysma**[28] <br> (plah-TIZ-muh) | Draws lower lip and angle of mouth downward in expressions of horror or surprise; may aid in opening mouth widely | **O:** Fascia of deltoid and pectoralis major <br> **I:** Mandible; skin and subcutaneous tissue of lower face | Facial nerve |

---

[24] *risor* = laughter
[25] *depress* = to lower; *angul* = angle, corner; *oris* = of the mouth
[26] *labi* = lip; *inferior* = lower

[27] *buccinator* = trumpeter
[28] *platy* = flat

| TABLE 10.2 | Muscles of Chewing and Swallowing |
|---|---|

The following muscles contribute to facial expression and speech but are primarily concerned with the manipulation of food, including tongue movements, chewing, and swallowing.

**Extrinsic Muscles of the Tongue.** The tongue is a very agile organ. It pushes food between the molars for chewing (mastication) and later forces the food into the pharynx for swallowing (deglutition); it is also, of course, of crucial importance to speech. Both intrinsic and extrinsic muscles are responsible for its complex movements. The intrinsic muscles consist of a variable number of vertical fascicles that extend from the superior to the inferior sides of the tongue, transverse fascicles that extend from right to left, and longitudinal fascicles that extend from root to tip (see figs. 10.1c and 25.5b, p. 953). The extrinsic muscles listed here connect the tongue to other structures in the head (fig. 10.9). Three of these are innervated by the hypoglossal nerve (CN XII), whereas the fourth is innervated by both the vagus (CN X) and accessory (CN XI) nerves.

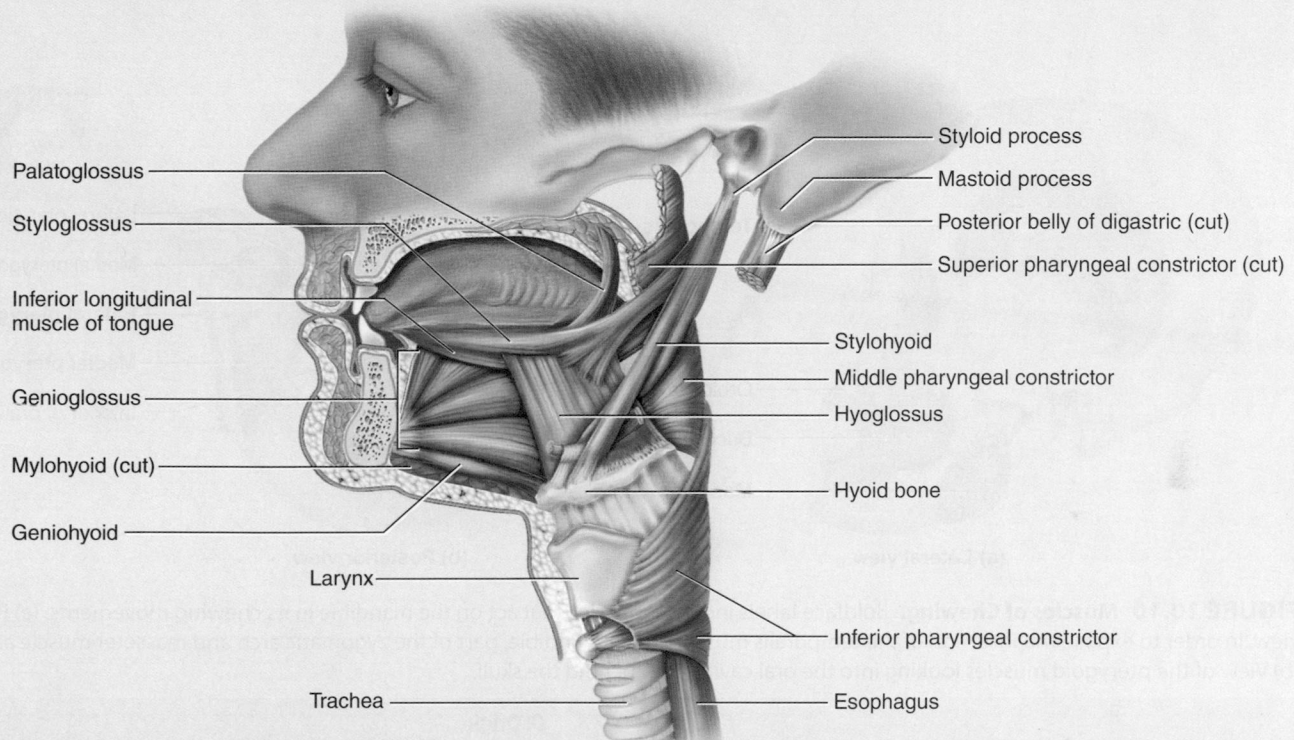

**FIGURE 10.9  Muscles of the Tongue and Pharynx.** AP|R

| Name | Action | *O:* Origin<br>*I:* Insertion | Innervation |
|---|---|---|---|
| **Genioglossus**[29]<br>(JEE-nee-oh-GLOSS-us) | Unilateral action draws tongue to one side; bilateral action depresses midline of tongue or protrudes tongue | *O:* Superior mental spine on posterior surface of mental protuberance<br>*I:* Inferior surface of tongue from root to apex | Hypoglossal nerve |
| **Hyoglossus**[30]<br>(HI-oh-GLOSS-us) | Depresses tongue | *O:* Body and greater horn of hyoid bone<br>*I:* Lateral and inferior surfaces of tongue | Hypoglossal nerve |
| **Styloglossus**[31]<br>(STY-lo-GLOSS-us) | Draws tongue upward and posteriorly | *O:* Styloid process of temporal bone and ligament from styloid process to mandible<br>*I:* Superolateral surface of tongue | Hypoglossal nerve |
| **Palatoglossus**[32]<br>(PAL-a-toe-GLOSS-us) | Elevates root of tongue and closes oral cavity off from pharynx; forms palatoglossal arch at rear of oral cavity | *O:* Soft palate<br>*I:* Lateral surface of tongue | Accessory and vagus nerves |

[29]*genio* = chin; *gloss* = tongue
[30]*hyo* = hyoid bone; *gloss* = tongue
[31]*stylo* = styloid process; *gloss* = tongue
[32]*palato* = palate; *gloss* = tongue

| TABLE 10.2 | Muscles of Chewing and Swallowing (continued) |
| --- | --- |

**Muscles of Chewing.** Four pairs of muscles produce the biting and chewing movements of the mandible: the *temporalis, masseter,* and two pairs of *pterygoid muscles* (fig. 10.10). Their actions include *depression* to open the mouth for receiving food; *elevation* for biting off a piece of food or crushing it between the teeth; *protraction* so that the incisors meet in cutting off a piece of food; *retraction* to draw the lower incisors behind the upper incisors and make the rear teeth meet; and *lateral* and *medial excursion,* the side-to-side movements that grind food between the rear teeth. The last four of these movements are shown in figure 9.20 (p. 292). All of these muscles are innervated by the mandibular nerve, which is a branch of the trigeminal (CN V).

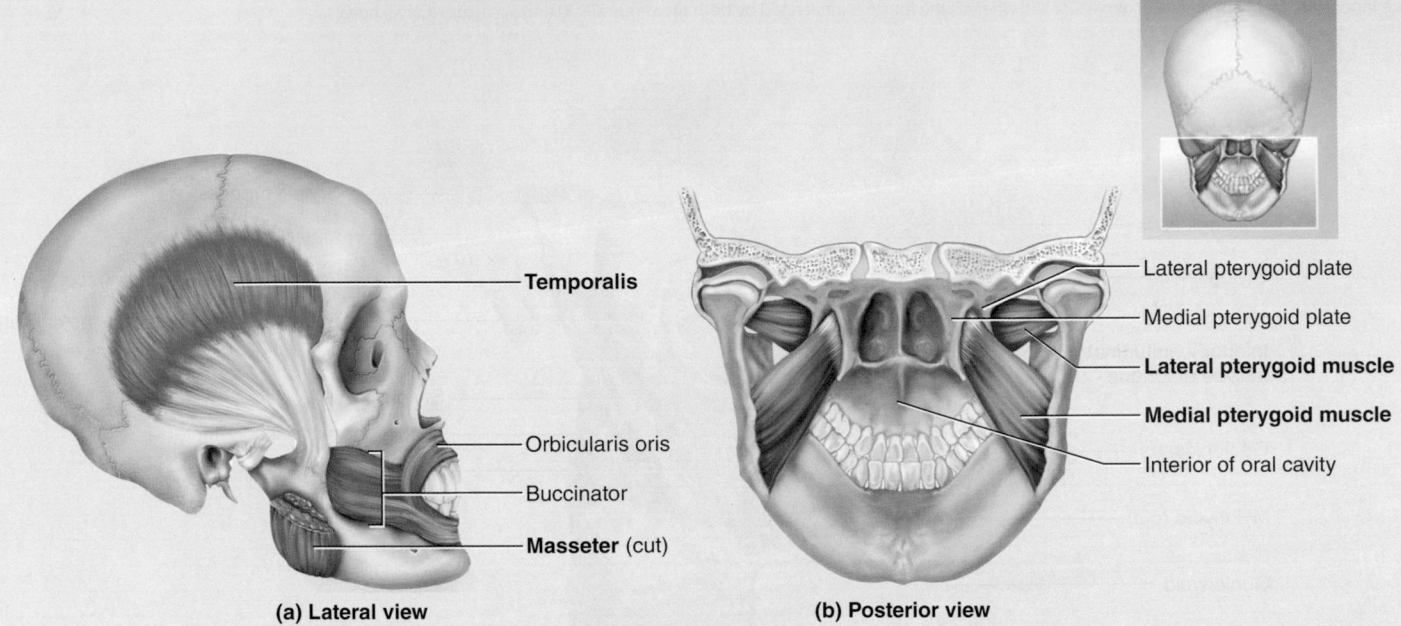

(a) Lateral view          (b) Posterior view

**FIGURE 10.10 Muscles of Chewing.** Boldface labels indicate muscles that act on the mandible in its chewing movements. (a) Right lateral view. In order to expose the insertion of the temporalis muscle on the mandible, part of the zygomatic arch and masseter muscle are removed. (b) View of the pterygoid muscles looking into the oral cavity from behind the skull.

| Name | Action | O: Origin<br>I: Insertion | Innervation |
| --- | --- | --- | --- |
| **Temporalis**[33]<br>(TEM-po-RAY-liss) | Elevation, retraction, and lateral and medial excursion of the mandible | **O:** Temporal lines and temporal fossa of cranium<br>**I:** Coronoid process and anterior border of mandibular ramus | Mandibular nerve |
| **Masseter**[34]<br>(ma-SEE-tur) | Elevation of the mandible, with smaller roles in protraction, retraction, and lateral and medial excursion | **O:** Zygomatic arch<br>**I:** Lateral surface of mandibular ramus and angle | Mandibular nerve |
| **Medial Pterygoid**[35]<br>(TERR-ih-goyd) | Elevation, protraction, and lateral and medial excursion of the mandible | **O:** Medial surface of lateral pterygoid plate; palatine bone; lateral surface of maxilla near molar teeth<br>**I:** Medial surface of mandibular ramus and angle | Mandibular nerve |
| **Lateral Pterygoid** | Depression (in wide opening of the mouth), protraction, and lateral and medial excursion of the mandible | **O:** Lateral surfaces of lateral pterygoid plate; greater wing of sphenoid<br>**I:** Neck of mandible (just below condyle); articular disc and capsule of temporomandibular joint | Mandibular nerve |

**Hyoid Muscles—Suprahyoid Group.** Several aspects of chewing, swallowing, and vocalizing are aided by eight pairs of *hyoid muscles* associated with the hyoid bone (fig. 10.11). The *suprahyoid group* is composed of the four pairs superior to the hyoid—the *digastric, geniohyoid, mylohyoid,* and *stylohyoid.* The digastric is an unusual muscle, named for its two bellies. Its *posterior belly* arises from the mastoid notch of the cranium and slopes downward and forward. The *anterior belly* arises from a trench called the *digastric fossa* on the inner surface of the mandibular body. It slopes downward and backward. The two bellies meet at a constriction, the *intermediate tendon.* This tendon passes through a connective tissue loop, the *fascial sling,* attached to the hyoid bone. Thus, when the two bellies of the digastric contract, they pull upward on the hyoid; but if the hyoid is fixed from below, the digastric aids in wide opening of the mouth. The lateral pterygoids are more important in wide mouth opening, with the digastrics coming into play only in extreme opening, as in yawning or taking a large bite of an apple. Cranial nerves V (trigeminal), VII (facial), and XII (hypoglossal) innervate these muscles; the trigeminal nerve gives rise to the mylohyoid nerve of the digastric and mylohyoid muscles.

---

[33]*temporalis* = of the temporal region of the head
[34]*masset* = chew
[35]*pteryg* = wing; *oid* = resembling (refers to pterygoid plate of sphenoid)

| TABLE 10.2 | Muscles of Chewing and Swallowing (continued) | | |
|---|---|---|---|
| **Name** | **Action** | **O: Origin**<br>**I: Insertion** | **Innervation** |
| **Digastric**[36] | Depresses mandible when hyoid is fixed; opens mouth widely, as when ingesting food or yawning; elevates hyoid when mandible is fixed | **O:** Mastoid notch of temporal bone; digastric fossa of mandible<br>**I:** Hyoid bone via fascial sling | Posterior belly: facial nerve<br>Anterior belly: mylohyoid nerve |
| **Geniohyoid**[37]<br>(JEE-nee-oh-HY-oyd) | Depresses mandible when hyoid is fixed; elevates and protracts hyoid when mandible is fixed | **O:** Inferior mental spine of mandible<br>**I:** Hyoid bone | Spinal nerve C1 via hypoglossal nerve |
| **Mylohyoid**[38] | Spans mandible from side to side and forms floor of mouth; elevates floor of mouth in initial stage of swallowing | **O:** Mylohyoid line near inferior margin<br>**I:** Hyoid bone | Mylohyoid nerve of mandible |
| **Stylohyoid** | Elevates and retracts hyoid, elongating floor of mouth; roles in speech, chewing, and swallowing are not yet clearly understood | **O:** Styloid process of temporal bone<br>**I:** Hyoid bone | Facial nerve |

**Hyoid Muscles—Infrahyoid Group.** The infrahyoid muscles are inferior to the hyoid bone. By fixing the hyoid from below, they enable the suprahyoid muscles to open the mouth. The *omohyoid* is unusual in that it arises from the shoulder, passes under the sternocleidomastoid, and then ascends to the hyoid bone. Like the digastric, it has two bellies. The *thyrohyoid,* named for the hyoid bone and the large shield-shaped *thyroid cartilage* of the larynx, helps prevent choking. It elevates the larynx during swallowing so that its superior opening is sealed by a flap of tissue, the *epiglottis.* You can feel this effect by placing your finger on the "Adam's apple" (the anterior prominence of the thyroid cartilage) and feeling it bob up as you swallow. The *sternothyroid* muscle then pulls the larynx down again so you can resume breathing; it is the only infrahyoid muscle with no connection to the hyoid bone. The sternohyoid lowers the hyoid bone after it has been elevated.

The infrahyoid muscles that act on the larynx are regarded as its *extrinsic muscles.* The *intrinsic muscles,* considered in chapter 22, are concerned with control of the vocal cords and laryngeal opening. The *ansa cervicalis,*[39] which innervates three of these muscles, is a loop of nerve on the side of the neck formed by certain fibers from cervical nerves 1 to 3. Cranial nerves IX (glossopharyngeal), X (vagus), and XII (hypoglossal) also innervate these muscles.

| | | | |
|---|---|---|---|
| **Omohyoid**[40] | Depresses hyoid after it has been elevated | **O:** Superior border of scapula<br>**I:** Hyoid bone | Ansa cervicalis |
| **Sternohyoid**[41] | Depresses hyoid after it has been elevated | **O:** Manubrium of sternum; medial end of clavicle<br>**I:** Hyoid bone | Ansa cervicalis |
| **Thyrohyoid**[42] | Depresses hyoid; with hyoid fixed, elevates larynx as in singing high notes | **O:** Thyroid cartilage of larynx<br>**I:** Hyoid bone | Spinal nerve C1 via hypoglossal nerve |
| **Sternothyroid** | Depresses larynx after it has been elevated in swallowing and vocalization; aids in singing low notes | **O:** Manubrium of sternum; costal cartilage<br>**I:** Thyroid cartilage of larynx | Ansa cervicalis |

**Muscles of the Pharynx.** The three pairs of *pharyngeal constrictors* encircle the pharynx on its posterior and lateral sides, forming a muscular funnel (see fig. 10.9).

| | | | |
|---|---|---|---|
| **Superior, Middle, and Inferior Pharyngeal Constrictors** | During swallowing, contract in order from *superior* to *middle* to *inferior constrictor* to drive food into esophagus | **O:** Medial pterygoid plate; mandible; hyoid; stylohyoid ligament; cartilages of larynx<br>**I:** Posteromedial seam of pharynx; basilar part of occipital bone | Glossopharyngeal and vagus nerves |

---

[36] *di* = two; *gastr* = bellies
[37] *genio* = chin
[38] *mylo* = mill, molar tooth
[39] *ansa* = handle; *cervic* = neck; *alis* = of, belonging to

[40] *omo* = shoulder
[41] *sterno* = chest, sternum
[42] *thyro* = shield, thyroid cartilage

| **TABLE 10.2** | Muscles of Chewing and Swallowing (continued) |
|---|---|

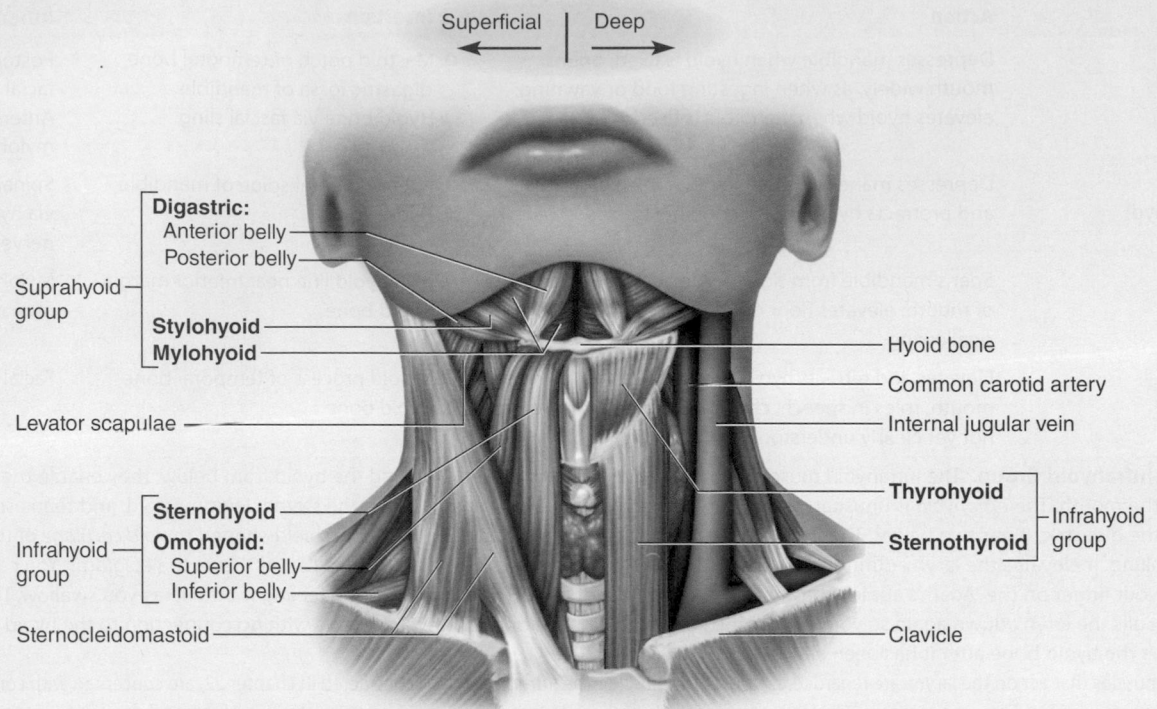

(a) Anterior view

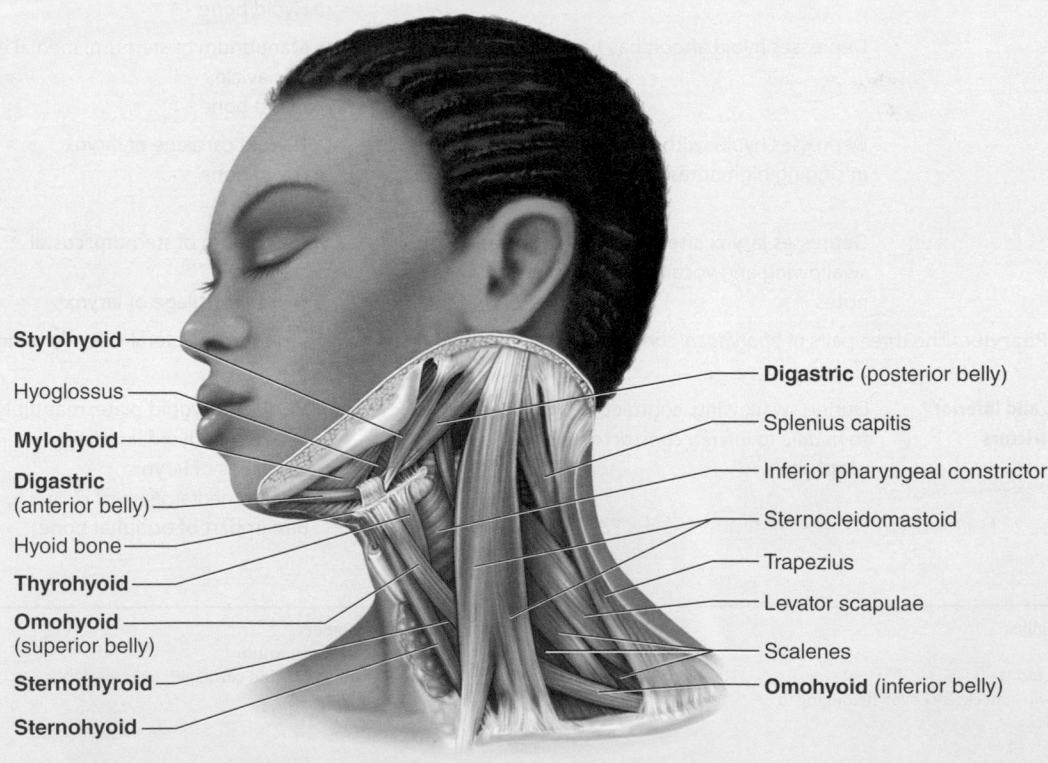

(b) Lateral view

**FIGURE 10.11 Muscles of the Neck.** Boldface labels indicate muscles of the suprahyoid and infrahyoid groups. Another muscle of the suprahyoid group, the geniohyoid, lies deep to the mylohyoid and can be seen in figure 10.8. **AP|R**

| TABLE 10.3 | Muscles Acting on the Head |
|---|---|

Muscles that move the head originate on the vertebral column, thoracic cage, and pectoral girdle and insert on the cranial bones. Their actions include flexion (tipping the head forward), extension (holding the head erect), hyperextension (as in looking upward), lateral flexion (tilting the head to one side), and rotation (turning the head to look left or right). Flexion, extension, and hyperextension involve simultaneous action of the right and left muscles of a pair; the other actions require the muscle on one side to contract more strongly than its mate. Many head actions result from a combination of these movements—for example, looking up over the shoulder involves a combination of rotation and hyperextension.

Depending on the relations of the muscle origins and insertions, a muscle may cause a **contralateral** movement of the head (toward the opposite side, as when contraction of a muscle on the left turns the face toward the right) or an **ipsilateral** movement (toward the same side as the muscle, as when contraction of a muscle on the left tilts the head to the left).

**Flexors of the Neck.** The prime mover of neck flexion is the *sternocleidomastoid,* a thick muscular cord that extends from the upper chest (sternum and clavicle) to the mastoid process behind the ear (fig. 10.11). This is most easily seen when the head is rotated to one side and slightly extended. To visualize the action of a single sternocleidomastoid, place the index finger of your left hand on your left mastoid process and the index finger of your right hand on your suprasternal notch. Now contract your left sternocleidomastoid to bring your two fingertips as close together as possible. You will see that this tilts your face upward and to the right.

The three *scalenes,*[43] located on the side of the neck, are named for being arranged somewhat like a staircase. Their actions are similar so they are considered collectively.

| Name | Action | O: Origin<br>I: Insertion | Innervation |
|---|---|---|---|
| **Sternocleidomastoid**[44]<br>(STIR-no-CLY-do-MAST-oyd) | Unilateral action tilts head slightly upward and toward the opposite side, as in looking over one's contralateral shoulder. The most common action is probably rotating the head to the left or right. Bilateral action draws the head straight forward and down, as when eating or reading. Aids in deep breathing when head is fixed. | **O:** Manubrium of sternum; medial one-third of clavicle<br>**I:** Mastoid process; lateral half of superior nuchal line | Accessory nerve; spinal nerves C2–C3 |
| **Anterior, Middle, and Posterior Scalenes**<br>(SCAY-leens) | Unilateral contraction causes ipsilateral flexion or contralateral rotation (tilts head toward same shoulder, or rotates face away), depending on action of other muscles. Bilateral contraction flexes neck. If spine is fixed, scalenes elevate ribs 1–2 and aid in breathing. | **O:** Transverse processes of all cervical vertebrae (C1–C7)<br>**I:** Ribs 1–2 | Anterior rami of spinal nerves C3–C8 |

**Extensors of the Neck.** The extensors are located mainly in the nuchal region (back of the neck; fig. 10.12) and therefore tend to hold the head erect or draw it back. The *trapezius* is the most superficial of these. It extends from the nuchal region over the shoulders and halfway down the back. It is named for the fact that the right and left trapezii together form a diamond or trapezoidal shape (see fig. 10.5b). The *splenius* is a deeper, elongated muscle with *splenius capitis* and *splenius cervicis* regions in the head and neck, respectively. It is nicknamed the "bandage muscle" because of the way it wraps around still deeper neck muscles. One of those deeper muscles is the semispinalis, another elongated muscle with head, neck, and thoracic regions. Only the *semispinalis capitis* and *cervicis* are tabulated here; the semispinalis thoracis does not act on the neck, but is included in table 10.6.

| Name | Action | O: Origin<br>I: Insertion | Innervation |
|---|---|---|---|
| **Trapezius**[45]<br>(tra-PEE-zee-us) | Extends and laterally flexes neck. See also roles in scapular movement in table 10.8. | **O:** External occipital protuberance; medial one-third of superior nuchal line; nuchal ligament; spinous processes of vertebrae C7–T3 or T4<br>**I:** Acromion and spine of scapula; lateral one-third of clavicle | Accessory nerve; anterior rami of spinal nerves C3–C4 |
| **Splenius Capitis**[46] and **Splenius Cervicis**[47]<br>(SPLEE-nee-us CAP-ih-tiss and SIR-vih-sis) | Acting unilaterally, produce ipsilateral flexion and slight rotation of head; extend head when acting bilaterally. | **O:** Inferior half of nuchal ligament; spinous processes of vertebrae C7–T6<br>**I:** Mastoid process and occipital bone just inferior to superior nuchal line; cervical vertebrae C1–C2 or C3 | Posterior rami of middle cervical nerves |
| **Semispinalis Capitis and Semispinalis Cervicis**<br>(SEM-ee-spy-NAY-lis) | Extend and contralaterally rotate head | **O:** Articular processes of vertebrae C4–C7; transverse processes of T1–T6<br>**I:** Occipital bone between nuchal lines; spinous processes of vertebrae C2–C5 | Posterior rami of cervical and thoracic nerves |

[43] *scal* = staircase
[44] *sterno* = chest, sternum; *cleido* = hammer, clavicle; *masto* = breastlike, mastoid process

[45] *trapez* = table, trapezoid
[46] *splenius* = bandage; *capitis* = of the head
[47] *cervicis* = of the neck

| TABLE 10.3 | Muscles Acting on the Head (continued) |
|---|---|

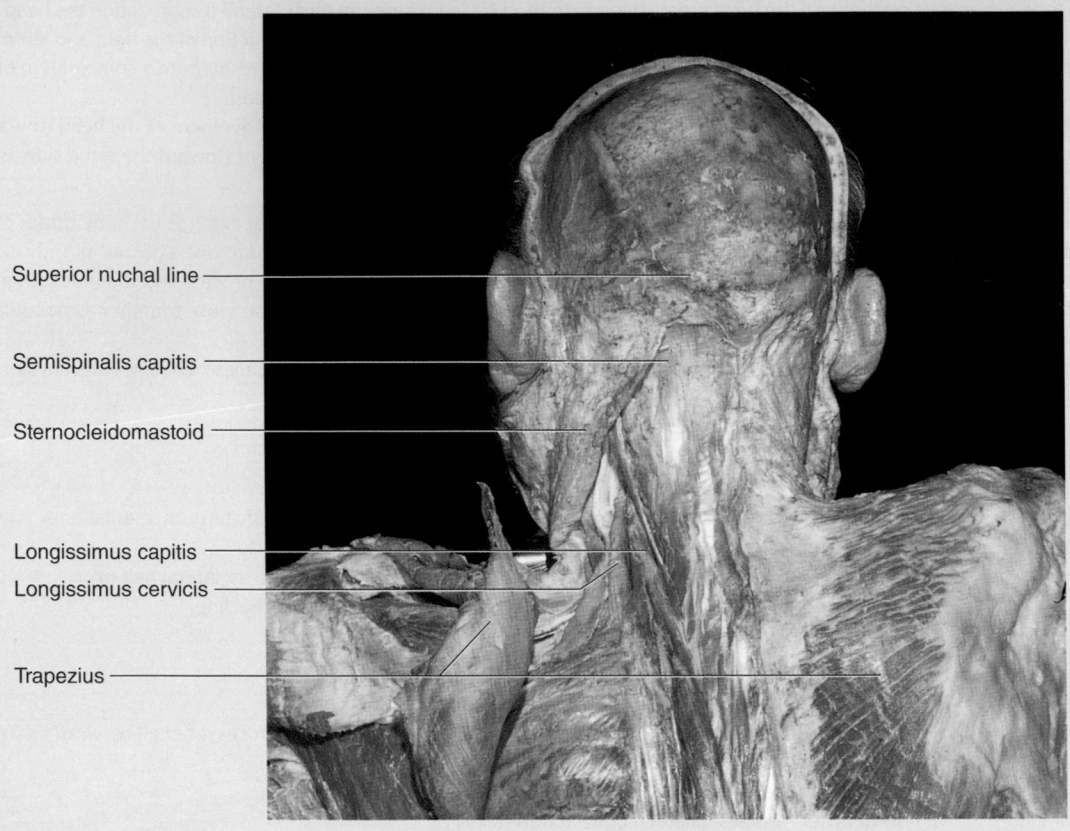

Superior nuchal line

Semispinalis capitis

Sternocleidomastoid

Longissimus capitis

Longissimus cervicis

Trapezius

**FIGURE 10.12**  **Muscles of the Shoulder and Nuchal Regions.**

### ►►►APPLY WHAT YOU KNOW

*Of the muscles you have studied so far, name three that you would consider intrinsic muscles of the head and three that you would classify as extrinsic. Explain your reason for each.*

### BEFORE YOU GO ON

Answer the following questions to test your understanding of the preceding section:

7. Name two muscles that elevate the upper lip and two that depress the lower lip.

8. Name the four paired muscles of mastication and state where they insert on the mandible.

9. Distinguish between the functions of the suprahyoid and infrahyoid muscles.

10. List the muscles of neck extension and flexion.

### 10.3 | Muscles of the Trunk

#### Expected Learning Outcomes

When you have completed this section, you should be able to

a. name and locate the muscles of respiration and explain how they affect airflow and abdominal pressure;

b. name and locate the muscles of the abdominal wall, back, and pelvic floor; and

c. identify the origin, insertion, action, and innervation of any of these muscles.

In this section, we will examine muscles of the trunk of the body in three functional groups concerned with respiration, support of the abdominal wall and pelvic floor, and movement of the vertebral column (tables 10.4–10.7). In the illustrations, you will note some major muscles that are not discussed in the associated tables—for example, the pectoralis major and serratus anterior. Although they are *located* in the trunk, they *act upon* the limbs and limb girdles, and are further discussed in tables 10.8 and 10.9.

| TABLE 10.4 | Muscles of Respiration |
| --- | --- |

We breathe primarily by means of muscles that enclose the thoracic cavity—the diaphragm, external intercostal, internal intercostal, and innermost intercostal muscles (fig. 10.13).

The *diaphragm* is a muscular dome between the thoracic and abdominal cavities, bulging upward against the base of the lungs. It has openings for passage of the esophagus, major blood and lymphatic vessels, and nerves between the two cavities. Its fibers converge from the margins toward a fibrous central tendon. When the diaphragm contracts, it flattens slightly and enlarges the thoracic cavity, causing air intake *(inspiration);* when it relaxes, it rises and shrinks the thoracic cavity, expelling air *(expiration).*

Three layers of muscle lie between the ribs: the external, internal, and innermost intercostal muscles. The 11 pairs of *external intercostal muscles* constitute the most superficial layer. They extend from the rib tubercle posteriorly almost to the beginning of the costal cartilage anteriorly. Each one slopes downward and anteriorly from one rib to the next inferior one. The 11 pairs of *internal intercostal muscles* lie deep to the external intercostals and extend from the margin of the sternum to the angles of the ribs. They are thickest in the region between the costal cartilages and grow thinner in the region where they overlap the internal intercostals. Their fibers slope downward and posteriorly from each rib to the one below, at nearly right angles to the external intercostals. Each is divided into an *intercartilaginous part* between the costal cartilages and an *interosseous part* between the bony part of the ribs. The two parts differ in their respiratory roles. The *innermost intercostal muscles* vary in number, as they are sometimes absent from the upper thoracic cage. Their fibers run in the same direction as the internal intercostals, and they are presumed to serve the same function. The internal and innermost intercostals are separated by a fascia that allows passage for intercostal nerves and blood vessels (see fig. 13.13b, p. 489).

The primary function of the intercostal muscles is to stiffen the thoracic cage during respiration so that it does not cave inward when the diaphragm descends. However, they also contribute to enlargement and contraction of the thoracic cage and thus add to the air volume that ventilates the lungs.

Many other muscles of the chest and abdomen contribute significantly to breathing: the sternocleidomastoid and scalenes of the neck; pectoralis major and serratus anterior of the chest; latissimus dorsi of the lower back; internal and external abdominal obliques and transverse abdominal muscle; and even some of the anal muscles. The respiratory actions of all these muscles are described in chapter 22.

| Name | Action | O: Origin<br>I: Insertion | Innervation |
| --- | --- | --- | --- |
| **Diaphragm**[48]<br>(DY-ah-fram) | Prime mover of inspiration (responsible for about two-thirds of air intake); contracts in preparation for sneezing, coughing, crying, laughing, and weight lifting; contraction compresses abdominal viscera and aids in childbirth and expulsion of urine and feces. | *O:* Xiphoid process of sternum; ribs and costal cartilages 7–12; lumbar vertebrae<br>*I:* Central tendon of diaphragm | Phrenic nerves |
| **External Intercostals**[49]<br>(IN-tur-COSS-tulz) | When scalenes fix rib 1, external intercostals elevate and protract ribs 2–12, expanding the thoracic cavity and creating a partial vacuum causing inflow of air; exercise a braking action during expiration so that expiration is not overly abrupt. | *O:* Inferior margins of ribs 1–11<br>*I:* Superior margin of next lower rib | Intercostal nerves |
| **Internal Intercostals** | In inspiration, the intercartilaginous part aids in elevating the ribs and expanding the thoracic cavity; in expiration, the interosseous part depresses and retracts the ribs, compressing the thoracic cavity and expelling air; the latter occurs only in forceful expiration, not in relaxed breathing. | *O:* Superior margins and costal cartilages of ribs 2–12; margin of sternum<br>*I:* Inferior margin of next higher rib | Intercostal nerves |

---

[48]*dia* = across; *phragm* = partition
[49]*inter* = between; *costa* = rib

| TABLE 10.4 | Muscles of Respiration (continued) | | |
|---|---|---|---|
| **Name** | **Action** | ***O:* Origin** <br> ***I:* Insertion** | **Innervation** |
| **Innermost Intercostals** | Presumed to have the same action as the internal intercostals | ***O:*** Superomedial surface of ribs 2–12; may be absent from upper ribs <br> ***I:*** Medial edge of costal groove of next higher rib | Intercostal nerves |

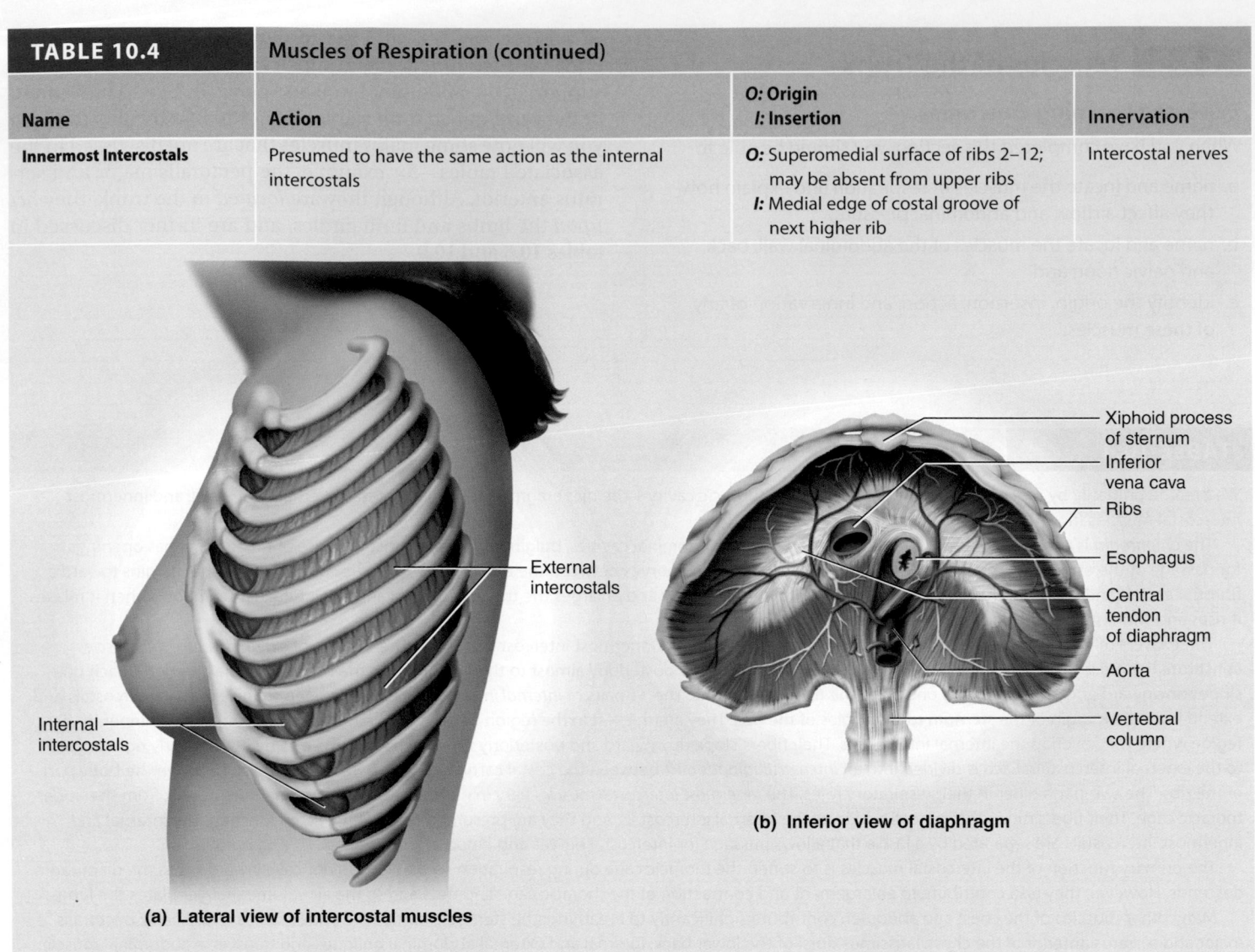

(a) Lateral view of intercostal muscles

(b) Inferior view of diaphragm

**FIGURE 10.13  Muscles of Respiration.** AP|R

▶▶▶ **APPLY WHAT YOU KNOW**

*A young thoracic surgery resident performing an operation for esophageal cancer accidentally severs the patient's left phrenic nerve. Predict the effect of this accident on the patient's respiration.*

▶▶▶ **APPLY WHAT YOU KNOW**

*What muscles are eaten as "spare ribs"? What is the tough fibrous membrane between the meat and the bone?*

| TABLE 10.5 | Muscles of the Anterior Abdominal Wall |
|---|---|

Unlike the thoracic cavity, the abdominal cavity has little skeletal support. It is enclosed, however, in layers of broad flat muscles whose fibers run in different directions, strengthening the abdominal wall on the same principle as the alternating layers of plywood. Three layers of muscle enclose the lumbar region and extend about halfway across the anterior abdomen (fig. 10.14). The most superficial layer is the *external abdominal oblique*. Its fibers pass downward and anteriorly. The next deeper layer is the *internal abdominal oblique,* whose fibers pass upward and anteriorly, roughly perpendicular to those of the external oblique. The deepest layer is the *transverse abdominal (transversus abdominis),* with horizontal fibers. Anteriorly, a pair of vertical *rectus abdominis* muscles extend from sternum to pubis. These are divided into segments by three transverse tendinous intersections, giving them an appearance that body builders nickname the "six pack."

The tendons of the oblique and transverse muscles are *aponeuroses*—broad fibrous sheets that continue medially and inferiorly (figs. 10.15 and 10.16). At the rectus abdominis, they diverge and pass around its anterior and posterior sides, enclosing the muscle in a vertical sleeve called the **rectus sheath.** They meet again at a median line called the **linea alba** between the rectus muscles. Another line, the **linea semilunaris,** marks the lateral boundary where the rectus sheath meets the aponeurosis. The aponeurosis of the external oblique also forms a cordlike **inguinal ligament** at its inferior margin. This extends obliquely from the anterior superior spine of the ilium to the pubis. The linea alba, linea semilunaris, and inguinal ligament are externally visible on a person with good muscle definition (see fig. B.8, p. 384).

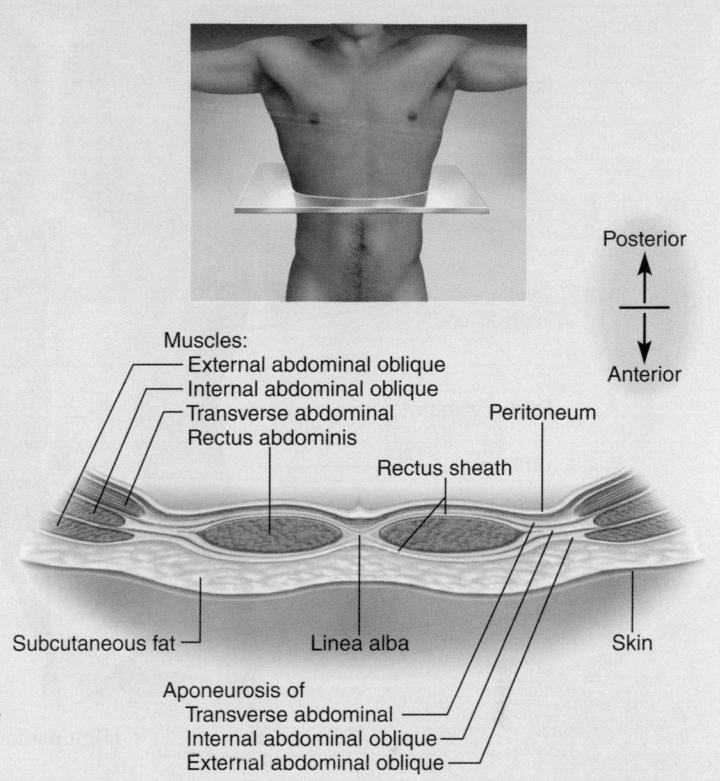

**FIGURE 10.14  Cross Section of the Anterior Abdominal Wall.**

| Name | Action | O: Origin<br>I: Insertion | Innervation |
|---|---|---|---|
| **External Abdominal Oblique** | Supports abdominal viscera against pull of gravity; stabilizes vertebral column during heavy lifting; maintains posture; compresses abdominal organs, thus aiding in deep breathing, loud vocalizations such as singing and public speaking, and in expulsion of abdominopelvic contents during childbirth, urination, defecation, and vomiting; unilateral contraction causes contralateral rotation of the spine, as in twisting at the waist. | O: Ribs 5–12<br>I: Anterior half of iliac crest; symphysis and superior margin of pubis | Anterior rami of spinal nerves T7–T12 |
| **Internal Abdominal Oblique** | Same as external oblique except that unilateral contraction causes ipsilateral rotation of waist | O: Inguinal ligament; iliac crest; thoracolumbar fascia<br>I: Ribs 10–12; costal cartilages 7–10; pubis | Anterior rami of spinal nerves T7–L1 |
| **Transverse Abdominal** | Compresses abdominal contents, with same effects as external oblique, but does not contribute to movements of vertebral column | O: Inguinal ligament; iliac crest; thoracolumbar fascia; costal cartilages 7–12<br>I: Linea alba; pubis; aponeurosis of internal oblique | Anterior rami of spinal nerves T7–L1 |
| **Rectus[50] Abdominis**<br>(REC-tus ab-DOM-ih-nis) | Flexes waist, as in bending forward or doing sit-ups; stabilizes pelvic region during walking; and compresses abdominal viscera | O: Pubic symphysis and superior margin of pubis<br>I: Xiphoid process; costal cartilages 5–7 | Anterior rami of spinal nerves T6–T12 |

[50]*rectus* = straight

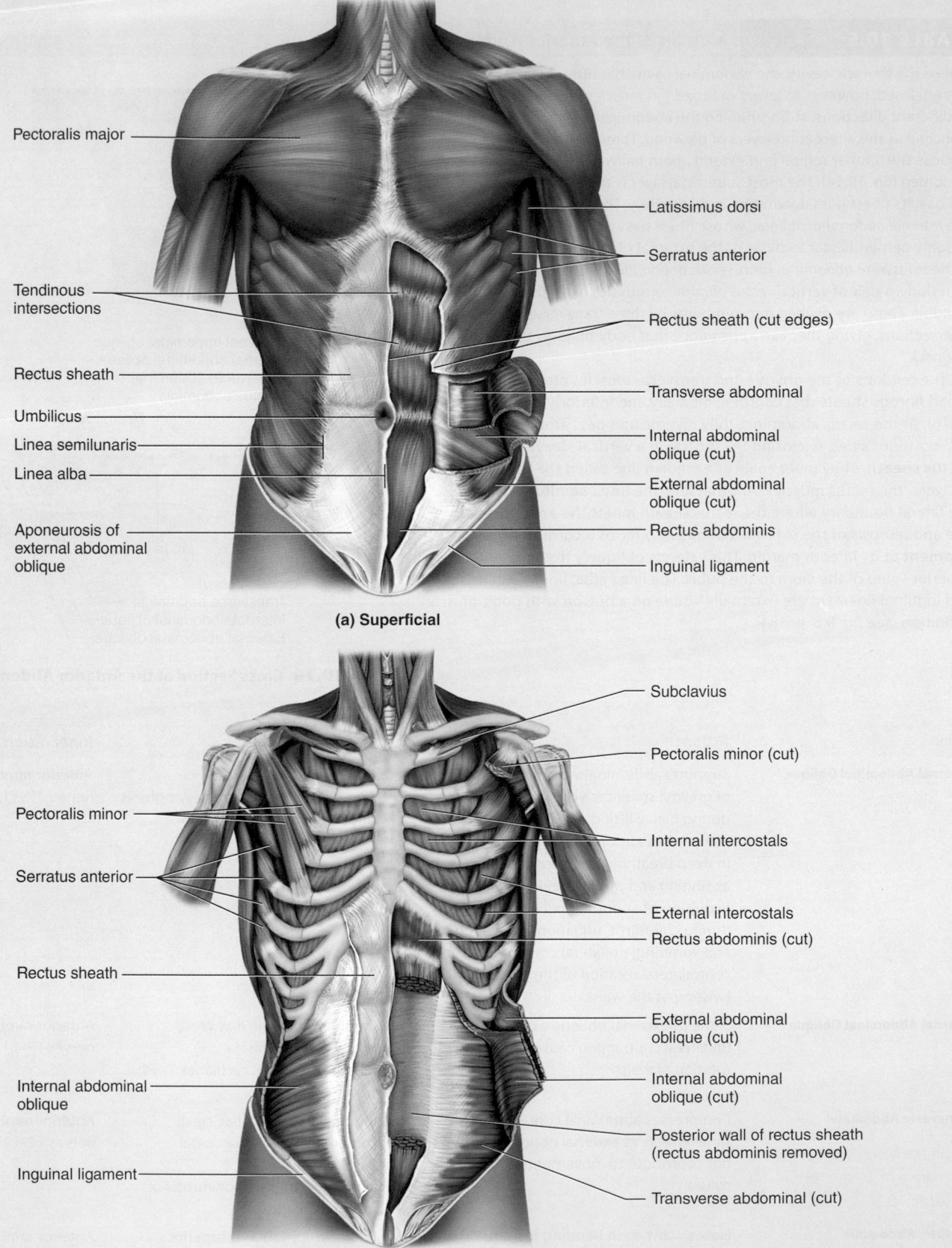

(a) Superficial

(b) Deep

**FIGURE 10.15** **Thoracic and Abdominal Muscles.** (a) Superficial muscles. The left rectus sheath is cut away to expose the rectus abdominis muscle. (b) Deep muscles. On the anatomical right, the external oblique has been removed to expose the internal oblique and the pectoralis major has been removed to expose the pectoralis minor. On the anatomical left, the internal oblique has been cut to expose the transverse abdominal, and the middle of the rectus abdominis has been cut out to expose the posterior rectus sheath. **AP|R**

*Name at least three muscles that lie deep to the pectoralis major.*

| TABLE 10.5 | Muscles of the Anterior Abdominal Wall (*continued*) |
| --- | --- |

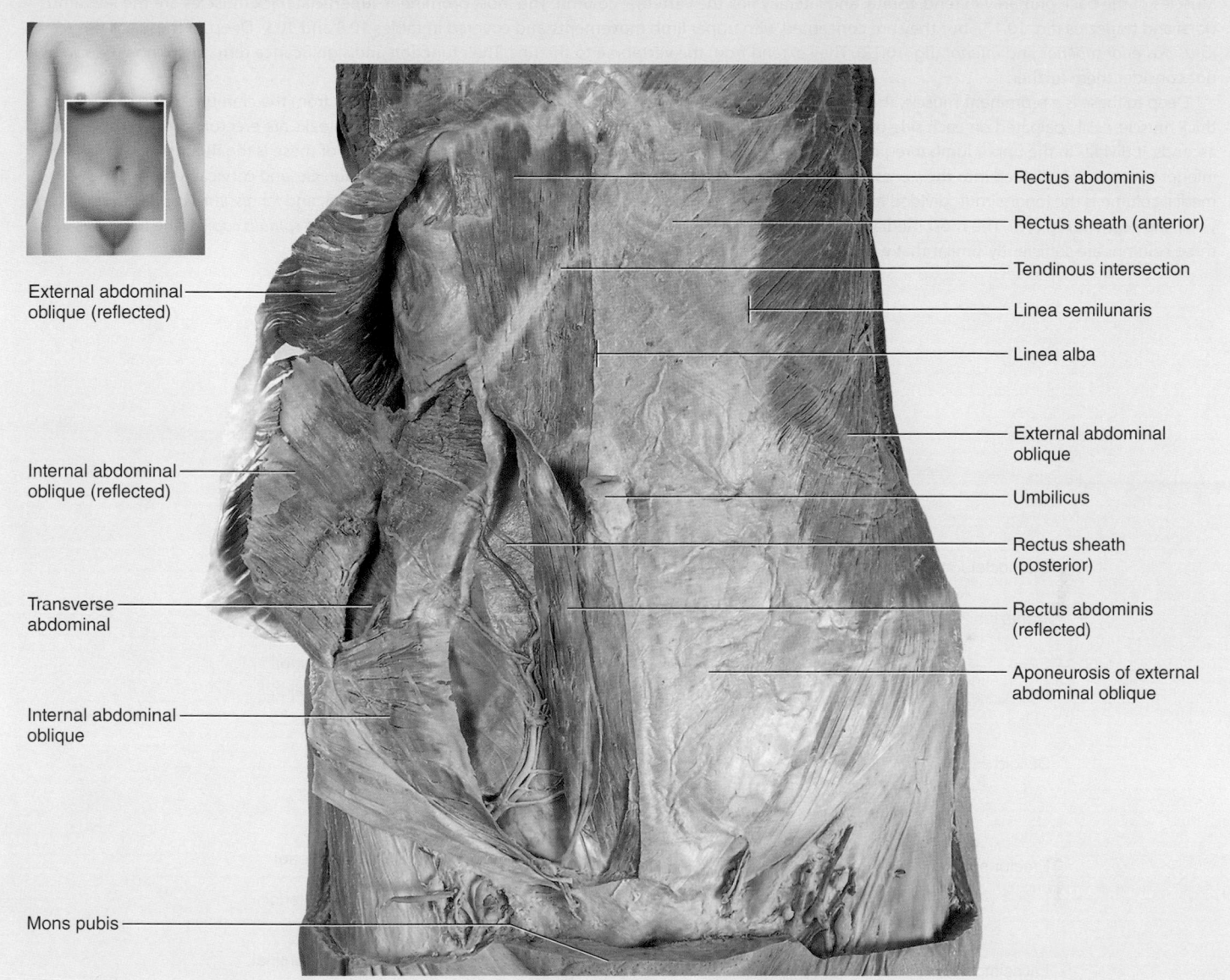

External abdominal oblique (reflected)

Internal abdominal oblique (reflected)

Transverse abdominal

Internal abdominal oblique

Mons pubis

Rectus abdominis

Rectus sheath (anterior)

Tendinous intersection

Linea semilunaris

Linea alba

External abdominal oblique

Umbilicus

Rectus sheath (posterior)

Rectus abdominis (reflected)

Aponeurosis of external abdominal oblique

**FIGURE 10.16 Thoracic and Abdominal Muscles of the Cadaver.** The rectus sheath has been removed on the anatomical right to expose the right rectus abdominis muscle. Inset shows area of dissection.

▶▶▶**APPLY WHAT YOU KNOW**

*Alice works out at a fitness center three times a week doing weight lifting and abdominal crunches. Martha prefers to sit on the sofa eating potato chips and watching TV. Both become pregnant. Other things being equal, give one reason related to table 10.5 why Alice may have an easier time with her childbirth than Martha will.*

| **TABLE 10.6** | **Muscles of the Back** |
|---|---|

Muscles of the back primarily extend, rotate, and laterally flex the vertebral column. The most prominent superficial back muscles are the latissimus dorsi and trapezius (fig. 10.17), but they are concerned with upper limb movements and covered in tables 10.8 and 10.9. Deep to these are the *serratus posterior superior* and *inferior* (fig. 10.18). They extend from the vertebrae to the ribs. Their function and significance remain unknown, so we will not consider them further.

Deep to these is a prominent muscle, the erector spinae, which runs vertically for the entire length of the back from the cranium to the sacrum. It is a thick muscle, easily palpated on each side of the vertebral column in the lumbar region. (Pork chops and T-bone steaks are erector spinae muscles.) As it ascends, it divides in the upper lumbar region into three parallel columns (figs. 10.18 and 10.19). The most lateral of these is the **iliocostalis,** which from inferior to superior is divided into the *iliocostalis lumborum, iliocostalis thoracis,* and *iliocostalis cervicis* (lumbar, thoracic, and cervical regions). The next medial column is the longissimus, divided from inferior to superior into the *longissimus thoracis, longissimus cervicis,* and *longissimus capitis* (thoracic, cervical, and cephalic regions). The most medial column is the **spinalis,** divided into *spinalis thoracis, spinalis cervicis,* and *spinalis capitis.* The functions of all three columns are sufficiently similar that we will treat them collectively as the erector spinae.

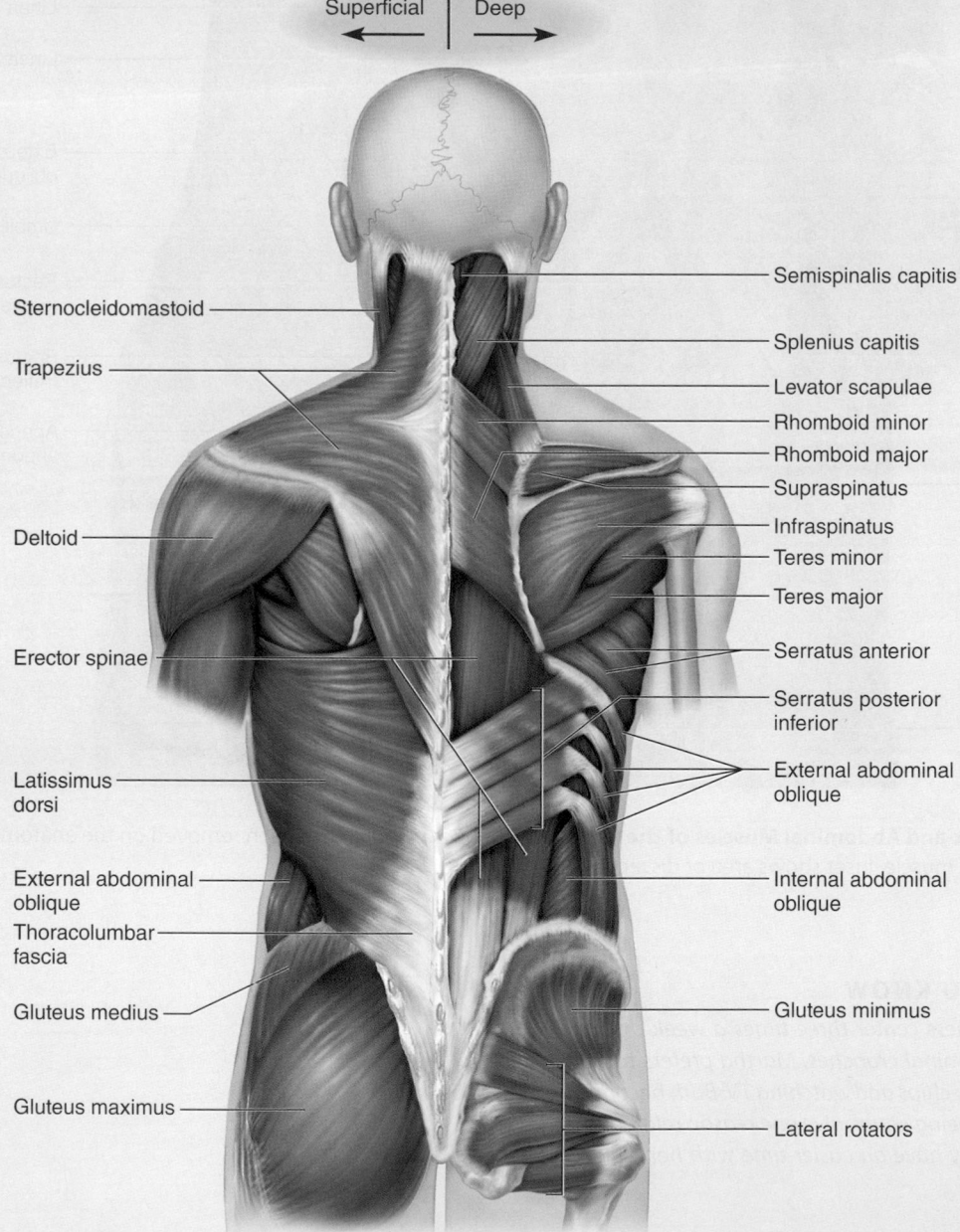

**FIGURE 10.17  Neck, Back, and Gluteal Muscles.**  The most superficial muscles are shown on the left and the next deeper layer on the right.  AP|R

| TABLE 10.6 | Muscles of the Back (continued) |
|---|---|

The major deep muscles are the *semispinalis thoracis* in the thoracic region and *quadratus lumborum* in the lumbar region. The erector spinae and quadratus lumborum are enclosed in a fibrous sheath called the thoracolumbar fascia, which is the origin of some of the abdominal and lumbar muscles. The *multifidus* is a collective name for a series of tiny muscles that connect adjacent vertebrae to each other from the cervical to lumbar region.

| Name | Action | O: Origin<br>I: Insertion | Innervation |
|---|---|---|---|
| **Erector Spinae**<br>(eh-REC-tur SPY-nee) | Aids in sitting and standing erect; straightens back after one bends at waist, and is employed in arching the back; unilateral contraction flexes waist laterally; the longissimus capitis also produces ipsilateral rotation of the head. | **O:** Nuchal ligament; ribs 3–12; thoracic and lumbar vertebrae; median and lateral sacral crests; thoracolumbar fascia<br>**I:** Mastoid process; cervical and thoracic vertebrae; all ribs | Posterior rami of cervical to lumbar spinal nerves |
| **Semispinalis Thoracis**<br>(SEM-ee-spy-NAY-liss tho-RA-sis) | Extension and contralateral rotation of vertebral column | **O:** Vertebrae T6–T10<br>**I:** Vertebrae C6–T4 | Posterior rami of cervical and thoracic spinal nerves |

**FIGURE 10.18 Muscles Acting on the Vertebral Column.** These are deeper than the muscles in figure 10.17, and those on the right are deeper than those on the left. AP|R

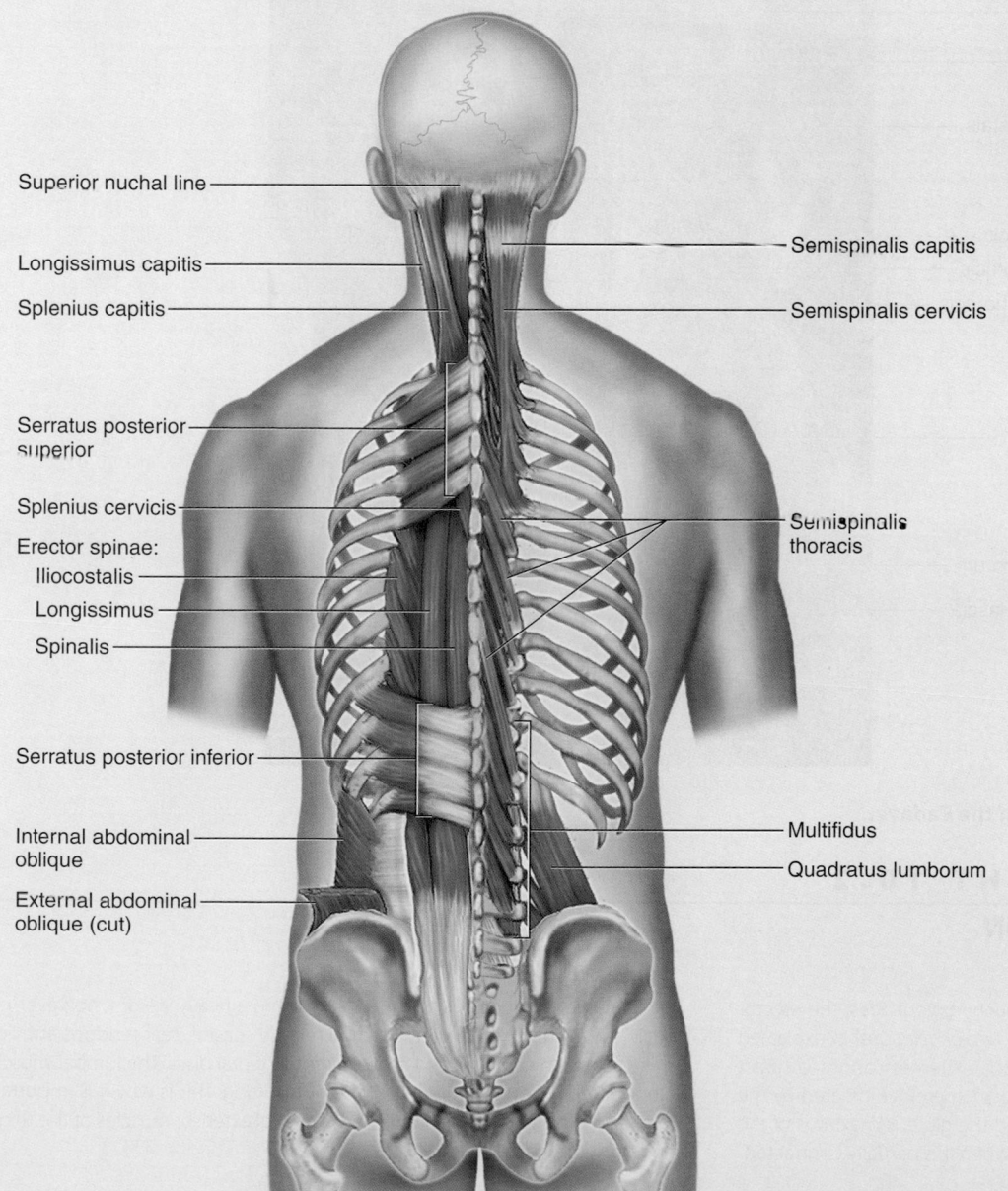

Superior nuchal line

Longissimus capitis

Splenius capitis

Serratus posterior superior

Splenius cervicis

Erector spinae:
  Iliocostalis
  Longissimus
  Spinalis

Serratus posterior inferior

Internal abdominal oblique

External abdominal oblique (cut)

Semispinalis capitis

Semispinalis cervicis

Semispinalis thoracis

Multifidus

Quadratus lumborum

| TABLE 10.6 | Muscles of the Back (continued) |

| Name | Action | O: Origin<br>I: Insertion | Innervation |
|---|---|---|---|
| **Quadratus Lumborum**[51]<br>(quad-RAY-tus lum-BORE-um) | Aids respiration by fixing rib 12 and stabilizing inferior attachments of diaphragm. Unilateral contraction causes ipsilateral flexion of lumbar vertebral column; bilateral contraction extends lumbar vertebral column. | O: Iliac crest; iliolumbar ligament<br>I: Rib 12; vertebrae L1–L4 | Anterior rami of spinal nerves T12–L4 |
| **Multifidus**[52]<br>(mul-TIFF-ih-dus) | Stabilization of adjacent vertebrae, maintenance of posture, control of vertebral movement when erector spinae acts on vertebral column | O: Vertebrae C4–L5; posterior superior iliac spine; sacrum; aponeurosis of erector spinae<br>I: Laminae and spinous processes of vertebrae superior to origins | Posterior rami of cervical to lumbar spinal nerves |

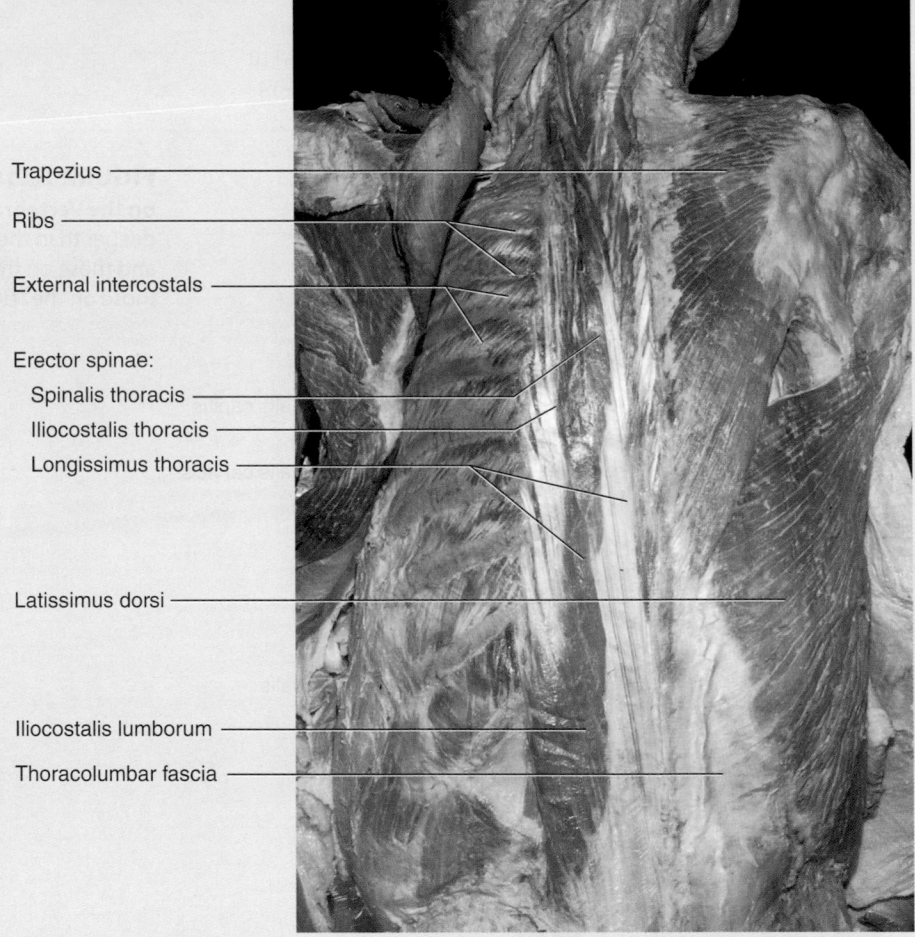

Trapezius

Ribs

External intercostals

Erector spinae:
  Spinalis thoracis
  Iliocostalis thoracis
  Longissimus thoracis

Latissimus dorsi

Iliocostalis lumborum

Thoracolumbar fascia

**FIGURE 10.19 Deep Back Muscles of the Cadaver.**

 **DEEPER INSIGHT 10.2**

## CLINICAL APPLICATION

### Heavy Lifting and Back Injuries

When you are fully bent over forward, as in touching your toes, the erector spinae is fully stretched. Because of the *length–tension relationship* explained in chapter 11, muscles that are stretched to such extremes cannot contract very effectively. Standing up from such a position is therefore initiated by the hamstring muscles on the back of the thigh and the gluteus maximus of the buttocks. The erector spinae joins in the action when it is partially contracted.

Standing too suddenly or improperly lifting a heavy weight, however, can strain the erector spinae, cause painful muscle spasms, tear tendons and ligaments of the lower back, and rupture intervertebral discs. The lumbar muscles are adapted for maintaining posture, not for lifting. This is why it is important, in heavy lifting, to crouch and use the powerful extensor muscles of the thighs and buttocks to lift the load.

---

[51]*quadrat* = four-sided; *lumborum* = of the lumbar region

[52]*multi* = many; *fid* = branched, sectioned

| TABLE 10.7 | Muscles of the Pelvic Floor |
|---|---|

The floor of the pelvic cavity is formed mainly by an extensive muscle called the *levator ani.* Inferior to this is the **perineum** (PERR-ih-NEE-um), a diamond-shaped area between the thighs bordered by four bony landmarks: the pubic symphysis anteriorly, the coccyx posteriorly, and the ischial tuberosities laterally. The pelvic floor and perineum are penetrated by the anal canal, urethra, and vagina. The anterior half of the perineum is the **urogenital triangle** and the posterior half is the **anal triangle** (fig. 10.20b). These are especially important landmarks in obstetrics. The urogenital triangle is divided into two muscle compartments separated by a strong fibrous **perineal membrane.** The muscle compartment between this membrane and the skin is called **superficial perineal space,** and the compartment between the perineal membrane and levator ani is the **deep perineal space.** We will examine these structures beginning inferiorly, just beneath the skin, and progressing superiorly to the pelvic floor.

**Superficial Perineal Space.** The superficial perineal space (fig. 10.20a) contains three pairs of muscles: the ischiocavernosus, bulbospongiosus, and superficial transverse perineal. In females, this space also contains the clitoris; various glands and erectile tissues of the genitalia (see fig. 28.8, p. 1066); and adipose tissue, which extends into and fattens the mons pubis and labia majora. In males, it contains the root of the penis. The *ischiocavernosus* muscles converge like a V from the ischial tuberosities toward the penis or clitoris. In males, the *bulbospongiosus* (*bulbocavernosus*) muscles form a sheath around the root of the penis, and in females they enclose the vagina like a pair of parentheses. *Cavernosus* in these names refers to the spongy, cavernous structure of tissues in the penis and clitoris. The *superficial transverse perineal* muscles extend from the ischial tuberosities to a strong median fibromuscular anchorage, the **perineal body,** and a median seam called the **perineal raphe** (RAY-fee) that extends anteriorly from the perineal body. The muscles may help to anchor the perineal body, but they are weakly developed and not always present, therefore not tabulated below. The other two muscle pairs of this layer serve primarily sexual functions.

| Name | Action | O: Origin<br>I: Insertion | Innervation |
|---|---|---|---|
| **Ischiocavernosus**[53]<br>(ISS-kee-oh-CAV-er-NO-sus) | Maintains erection of the penis or clitoris by compressing deep structures of the organ and forcing blood forward into its body. | **O:** Ramus and tuberosity of ischium<br>**I:** Ensheaths internal structures of penis and clitoris | Pudendal nerve |
| **Bulbospongiosus**[54]<br>(BUL-bo-SPUN-jee-OH-sus) | Expels residual urine from urethra after bladder has emptied. Aids in erection of penis or clitoris. In male, spasmodic contractions expel semen during ejaculation. In female, contractions constrict vaginal orifice and expel secretions of greater vestibular glands. | **O:** Perineal body and median raphe<br>**I:** Male: Ensheaths root of penis.<br>Female: Pubic symphysis | Pudendal nerve |

**Deep Perineal Space.** The deep perineal space (fig. 10.20b) contains a pair of *deep transverse perineal* muscles and, in females only, the *compressor urethrae* muscles. The deep transverse perineal muscles anchor the perineal body on the median plane; the perineal body, in turn, anchors other pelvic muscles. The female *external urethral sphincter,* long thought to be part of the deep perineal space, is now regarded as part of the urethra itself and not part of the pelvic floor musculature.

| **Deep Transverse Perineal** | Anchors perineal body, which supports other pelvic muscles; supports vaginal and urethral canals | **O:** Ischiopubic rami<br>**I:** Perineal body | Pudendal nerve |
| **Compressor Urethrae**<br>(yu-REE-three) | Aids in urine retention; found in females only | **O:** Ischiopubic rami<br>**I:** Right and left compressor urethrae meet each other inferior to external urethral sphincter | Pudendal nerve; spinal nerves S2–S4; pelvic splanchnic nerve |

**Anal Triangle.** The anal triangle contains the *external anal sphincter* and *anococcygeal ligament* (fig. 10.20b). The external anal sphincter is a tubular muscle surrounding the lower anal canal. The anococcygeal ligament is the median insertion of the levator ani muscles, and the ligament, in turn, inserts on the coccyx. It is therefore a major anchorage for the structures that compose the pelvic floor.

| **External Anal Sphincter** | Retains feces in rectum until voluntarily voided | **O:** Coccyx, perineal body<br>**I:** Encircles anal canal and orifice | Pudendal nerve; spinal nerves S2–S4; pelvic splanchnic nerve |

**Pelvic Diaphragm.** The pelvic diaphragm (fig. 10.20c) is deep to the foregoing structures and is composed mainly of the right and left *levator ani muscles.* (The piriformis, also illustrated, is primarily a lower limb muscle.) The levator ani spans most of the pelvic outlet and forms the floor of the lesser (true) pelvis. It is divided into three portions that are sometimes regarded as separate muscles—the *ischiococcygeus* (or *coccygeus*), *iliococcygeus,* and *pubococcygeus.* The left and right levator ani muscles converge on the *anococcygeal ligament,* through which they are indirectly anchored to the coccyx.

| **Levator Ani**[55]<br>(leh-VAY-tur AY-nye) | Compresses anal canal and reinforces external anal and urethral sphincters; supports uterus and other pelvic viscera; aids in falling away of the feces; vertical movements affect pressure differences between abdominal and thoracic cavities and thus aid in deep breathing. | **O:** Inner surface of lesser pelvis from pubis through margin of obturator internus to spine of ischium<br>**I:** Coccyx via anococcygeal body; walls of urethra, vagina, and anal canal | Pudendal nerve; spinal nerves S2–S3 |

[53]*ischio* = ischium of hip bone; *cavernosus* = corpus cavernosum of the penis or clitoris

[54]*bulbo* = bulb of the penis; *spongiosus* = corpus spongiosum of the penis
[55]*levator* = that which elevates; *ani* = of the anus

| TABLE 10.7 | Muscles of the Pelvic Floor (continued) |
|---|---|

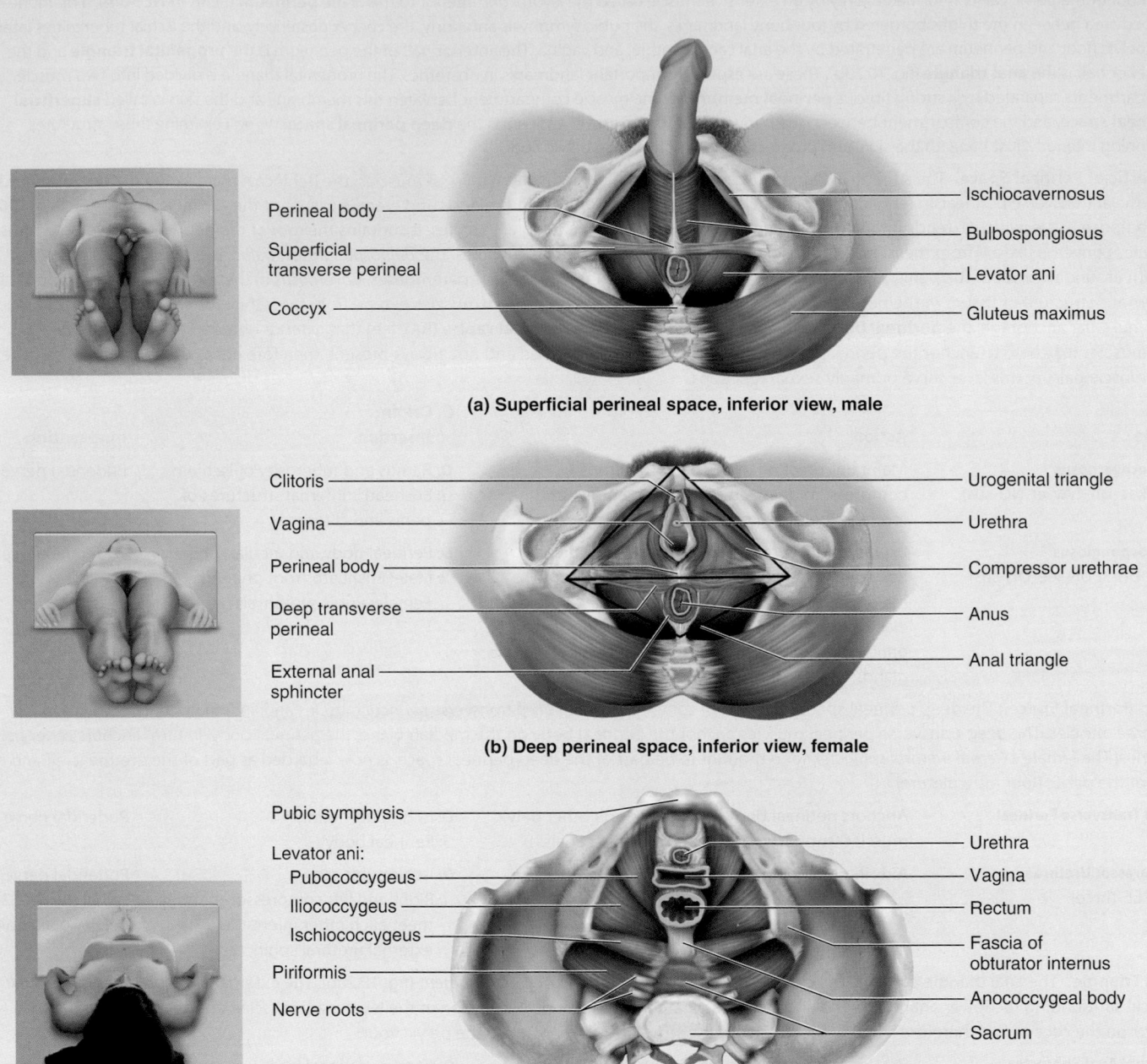

(a) **Superficial perineal space, inferior view, male**

Perineal body

Superficial transverse perineal

Coccyx

Ischiocavernosus

Bulbospongiosus

Levator ani

Gluteus maximus

(b) **Deep perineal space, inferior view, female**

Clitoris

Vagina

Perineal body

Deep transverse perineal

External anal sphincter

Urogenital triangle

Urethra

Compressor urethrae

Anus

Anal triangle

(c) **Pelvic diaphragm, superior view, female**

Pubic symphysis

Levator ani:
  Pubococcygeus
  Iliococcygeus
  Ischiococcygeus

Piriformis

Nerve roots

Urethra

Vagina

Rectum

Fascia of obturator internus

Anococcygeal body

Sacrum

**FIGURE 10.20 Muscles of the Pelvic Floor.** (a) The superficial perineal space, male, viewed from below (the inferior aspect). (b) The deep perineal space, female, viewed from the same perspective. Other than the vaginal canal, the sexes are nearly identical at this level, including the urogenital and anal triangles. The root of the penis does not extend to this level. (c) The pelvic diaphragm, female, viewed from above (from within the pelvic cavity).

# DEEPER INSIGHT 10.3

## CLINICAL APPLICATION

### Hernias

A hernia is any condition in which the viscera protrude through a weak point in the muscular wall of the abdominopelvic cavity. The most common type to require treatment is an *inguinal hernia* (fig. 10.21). In the male fetus, each testis descends from the pelvic cavity into the scrotum by way of a passage called the *inguinal canal* through the muscles of the groin. This canal remains a weak point in the pelvic floor, especially in infants and children. When pressure rises in the abdominal cavity, it can force part of the intestine or bladder into this canal or even into the scrotum. This also sometimes occurs in men who hold their breath while lifting heavy weights. When the diaphragm and abdominal muscles contract, pressure in the abdominal cavity can soar to 1,500 pounds per square inch—more than 100 times the normal pressure and quite sufficient to produce an inguinal hernia, or "rupture." Inguinal hernias rarely occur in women.

Two other sites of hernia are the diaphragm and navel. A *hiatal hernia* is a condition in which part of the stomach protrudes through the diaphragm into the thoracic cavity. This is most common in overweight people over 40. It may cause heartburn due to the regurgitation of stomach acid into the esophagus, but most cases go undetected. In an *umbilical hernia,* abdominal viscera protrude through the navel.

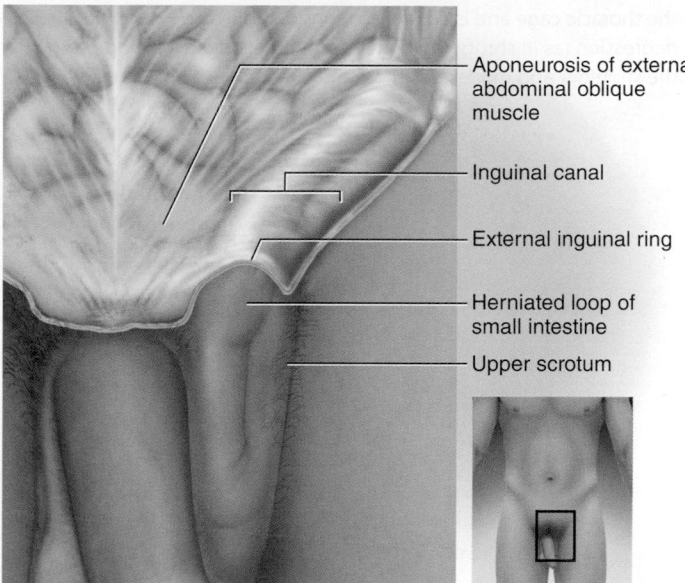

Aponeurosis of external abdominal oblique muscle

Inguinal canal

External inguinal ring

Herniated loop of small intestine

Upper scrotum

**FIGURE 10.21 Inguinal Hernia.** A loop of small intestine has protruded through the inguinal canal into a space beneath the skin.

---

### BEFORE YOU GO ON

Answer the following questions to test your understanding of the preceding section:

11. Which muscles are used more often, the external intercostals or internal intercostals? Explain.

12. Explain how pulmonary ventilation affects abdominal pressure and vice versa.

13. Name a major superficial muscle and two major deep muscles of the back.

14. Define perineum, urogenital triangle, and anal triangle.

15. Name one muscle in the superficial perineal space, one in the urogenital diaphragm, and one in the pelvic diaphragm. State the function of each.

## 10.4 Muscles Acting on the Shoulder and Upper Limb

### Expected Learning Outcomes

When you have completed this section, you should be able to

a. name and locate the muscles that act on the pectoral girdle, shoulder, elbow, wrist, and hand;

b. relate the actions of these muscles to the joint movements described in chapter 9; and

c. describe the origin, insertion, and innervation of each muscle.

The upper and lower limbs have numerous muscles that serve primarily for movement of the body and manipulation of objects. These muscles are organized into distinct compartments separated from each other by the interosseous membranes of the forearm and leg (see pp. 258 and 265) and by intermuscular septa. In the ensuing tables, you will find muscles of the upper limb divided into anterior and posterior compartments, and those of the lower limb divided into anterior, posterior, medial, and lateral compartments. In most limb regions, the muscle groups are further subdivided by thinner fasciae into superficial and deep layers.

The upper limb is used for a broad range of both powerful and subtle actions, ranging from climbing, grasping, and throwing to writing, playing musical instruments, and manipulating small objects. It therefore has an especially complex array of muscles, but the muscles fall into logical groups that make their functional relationships and names easier to understand. Tables 10.8 through 10.12 group these into muscles that act on the scapula, those that act on the humerus and shoulder joint, those that act on the forearm and elbow joint, extrinsic (forearm) muscles that act on the wrist and hand, and intrinsic (hand) muscles that act on the fingers.

| **TABLE 10.8** | Muscles Acting on the Shoulder |
| --- | --- |

Muscles that act on the pectoral girdle originate on the axial skeleton and insert on the clavicle and scapula. The scapula is only loosely attached to the thoracic cage and is capable of considerable movement (fig. 10.22)—rotation (as in raising and lowering the apex of the shoulder), elevation and depression (as in shrugging and lowering the shoulders), and protraction and retraction (pulling the shoulders forward and back). The clavicle braces the shoulder and moderates these movements.

**Lateral rotation**
Trapezius (superior part)
Serratus anterior

**Elevation**
Levator scapulae
Trapezius (superior part)
Rhomboid major
Rhomboid minor

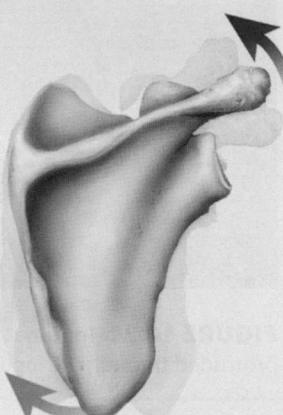

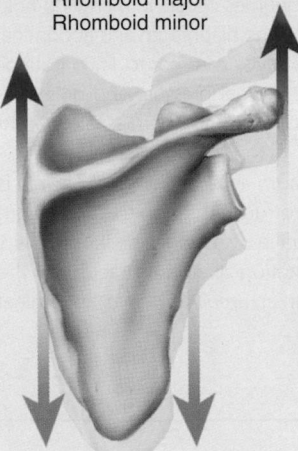

**Medial rotation**
Levator scapulae
Rhomboid major
Rhomboid minor

**Depression**
Trapezius (inferior part)
Serratus anterior

**Retraction**
Rhomboid major
Rhomboid minor
Trapezius

**Protraction**
Pectoralis minor
Serratus anterior

**FIGURE 10.22  Actions of Some Thoracic Muscles on the Scapula.**  Note that an individual muscle can contribute to multiple actions, depending on which fibers contract and what synergists act with it.

| TABLE 10.8 | Muscles Acting on the Shoulder (continued) |
|---|---|

**Anterior Group.** Muscles of the pectoral girdle fall into anterior and posterior groups (see figs. 10.23 and 10.24). The major muscles of the anterior group are the *pectoralis minor* and *serratus anterior* (see fig. 10.15b). The pectoralis minor arises by three heads from ribs 3 to 5 and converges on the coracoid process of the scapula. The serratus anterior arises from separate heads on all or nearly all of the ribs, wraps laterally around the chest, passes across the back between the rib cage and scapula, and inserts on the medial (vertebral) border of the scapula. Thus, when it contracts, the scapula glides laterally and slightly forward around the ribs. The serratus anterior is nicknamed the "boxer's muscle" because of its role in powerful thrusting movements of the arm such as a boxer's jab.

| Name | Action | O: Origin<br>I: Insertion | Innervation |
|---|---|---|---|
| **Pectoralis Minor**<br>(PECK-toe-RAY-liss) | With serratus anterior, draws scapula laterally and forward around chest wall; with other muscles, rotates scapula and depresses apex of shoulder, as in reaching down to pick up a suitcase | **O:** Ribs 3–5 and overlying fascia<br>**I:** Coracoid process of scapula | Medial and lateral pectoral nerves |
| **Serratus**[56] **Anterior**<br>(serr-AY-tus) | With pectoralis minor, draws scapula laterally and forward around chest wall; protracts scapula, and is the prime mover in all forward-reaching and pushing actions; aids in rotating scapula to elevate apex of shoulder; fixes scapula during abduction of arm | **O:** All or nearly all ribs<br>**I:** Medial border of scapula | Long thoracic nerve |

**Posterior Group.** The posterior muscles that act on the scapula include the large, superficial trapezius, already discussed (table 10.3), and three deep muscles: the *levator scapulae, rhomboid minor,* and *rhomboid major*. The action of the trapezius depends on whether its superior, middle, or inferior fibers contract and whether it acts alone or with other muscles. The levator scapulae and superior fibers of the trapezius rotate the scapula in opposite directions if either of them acts alone. If both act together, their opposite rotational effects balance each other and they elevate the scapula and shoulder, as when you lift a suitcase from the floor. Depression of the scapula occurs mainly by gravitational pull, but the trapezius and serratus anterior can depress it more rapidly and forcefully, as in swimming, hammering, and rowing.

| Name | Action | O: Origin<br>I: Insertion | Innervation |
|---|---|---|---|
| **Trapezius**<br>(tra-PEE-zee-us) | Stabilizes scapula and shoulder during arm movements; elevates and depresses apex of shoulder; acts with other muscles to rotate and retract scapula (see also roles in head and neck movements in table 10.3) | **O:** External occipital protuberance; medial one-third of superior nuchal line; nuchal ligament; spinous processes of vertebrae C7–T3 or T4<br>**I:** Acromion and spine of scapula; lateral one-third of clavicle | Accessory nerve; anterior rami of C3–C4 |
| **Levator Scapulae**<br>(leh-VAY-tur SCAP-you-lee) | Elevates scapula if cervical vertebrae are fixed; flexes neck laterally if scapula is fixed; retracts scapula and braces shoulder; rotates scapula and depresses apex of shoulder | **O:** Transverse processes of vertebrae C1–C4<br>**I:** Superior angle to medial border of scapula | Spinal nerves C3–C4, and C5 via posterior scapular nerve |
| **Rhomboid Minor**<br>(ROM-boyd) | Retracts scapula and braces shoulder; fixes scapula during arm movements | **O:** Spinous processes of vertebrae C7–T1; nuchal ligament<br>**I:** Medial border of scapula | Posterior scapular nerve |
| **Rhomboid Major** | Same as rhomboid minor | **O:** Spinous processes of vertebrae T2–T5<br>**I:** Medial border of scapula | Posterior scapular nerve |

[56]*serrate* = scalloped, zigzag

| **TABLE 10.9** | Muscles Acting on the Arm |
|---|---|

**Axial Muscles.** Nine muscles cross the shoulder joint and insert on the humerus. Two are considered **axial muscles** because they originate primarily on the axial skeleton—the *pectoralis major* and *latissimus dorsi* (figs. 10.15, 10.23, and 10.24). The pectoralis major is the thick, fleshy muscle of the mammary region, and the latissimus dorsi is a broad muscle of the back that extends from the waist to the axilla. These muscles bear the primary responsibility for attaching the arm to the trunk and are the prime movers of the shoulder joint.

| Name | Action | O: Origin<br>I: Insertion | Innervation |
|---|---|---|---|
| **Pectoralis Major**<br>(PECK-toe-RAY-liss) | Flexes, adducts, and medially rotates humerus, as in climbing or hugging; aids in deep inspiration | **O:** Medial half of clavicle; lateral margin of sternum; costal cartilages 1–7; aponeurosis of external oblique<br>**I:** Lateral lip of intertubercular sulcus of humerus | Medial and lateral pectoral nerves |
| **Latissimus Dorsi**[57]<br>(la-TISS-ih-mus DOR-sye) | Adducts and medially rotates humerus; extends the shoulder joint as in pulling on the oars of a rowboat; produces backward swing of arm in such actions as walking and bowling; with hands grasping overhead objects, pulls body forward and upward, as in climbing; aids in deep inspiration, sudden expiration such as sneezing and coughing, and prolonged forceful expiration as in singing or blowing a sustained note on a wind instrument | **O:** Vertebrae T7–L5; lower three or four ribs; iliac crest; thoracolumbar fascia<br>**I:** Floor of intertubercular sulcus of humerus | Thoracodorsal nerve |

**Scapular Muscles.** The other seven muscles of the shoulder are considered **scapular muscles** because they originate on the scapula. Four of them form the rotator cuff and are treated in the next section. The most conspicuous scapular muscle is the *deltoid,* the thick triangular muscle that caps the shoulder. This is a commonly used site of drug injections. Its anterior, lateral, and posterior fibers act like three different muscles.

| Name | Action | O: Origin / I: Insertion | Innervation |
|---|---|---|---|
| **Deltoid** | Anterior fibers flex and medially rotate arm; lateral fibers abduct arm; posterior fibers extend and laterally rotate arm; involved in arm swinging during such actions as walking or bowling, and in adjustment of hand height for various manual tasks | **O:** Acromion and spine of scapula; clavicle<br>**I:** Deltoid tuberosity of humerus | Axillary nerve |
| **Teres Major**<br>(TERR-eez) | Extends and medially rotates humerus; contributes to arm swinging | **O:** Inferior angle of scapula<br>**I:** Medial lip of intertubercular sulcus of humerus | Lower subscapular nerve |
| **Coracobrachialis**<br>(COR-uh-co-BRAY-kee-AY-lis) | Flexes and medially rotates arm; resists deviation of arm from frontal plane during abduction | **O:** Coracoid process<br>**I:** Medial aspect of humeral shaft | Musculocutaneous nerve |

[57]*latissimus* = broadest; *dorsi* = of the back

►►►**APPLY WHAT YOU KNOW**

*Perform an action as if lifting a cup to your mouth to take a sip of tea. Describe the contribution of your deltoid to this action, using the terminology of joint movement in chapter 9.*

**TABLE 10.9** Muscles Acting on the Arm (continued)

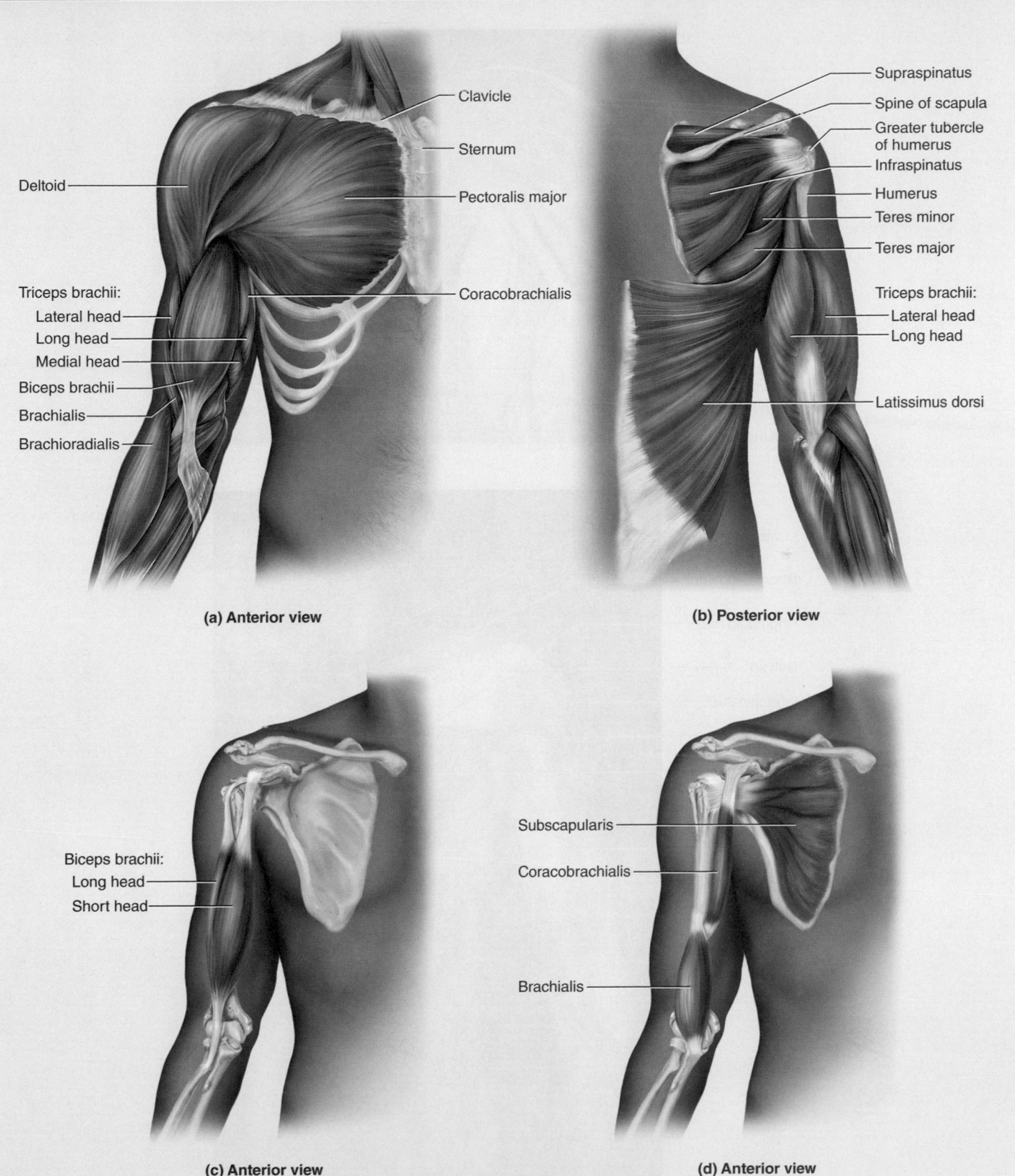

(a) Anterior view

(b) Posterior view

(c) Anterior view

(d) Anterior view

**FIGURE 10.23 Pectoral and Brachial Muscles.** (a) Superficial muscles, anterior view. (b) Superficial muscles, posterior view. (c) The biceps brachii, the superficial flexor of the elbow. (d) The brachialis, the deep flexor of the elbow, and the coracobrachialis and subscapularis, which act on the humerus. **AP|R**

**TABLE 10.9**    Muscles Acting on the Arm (continued)

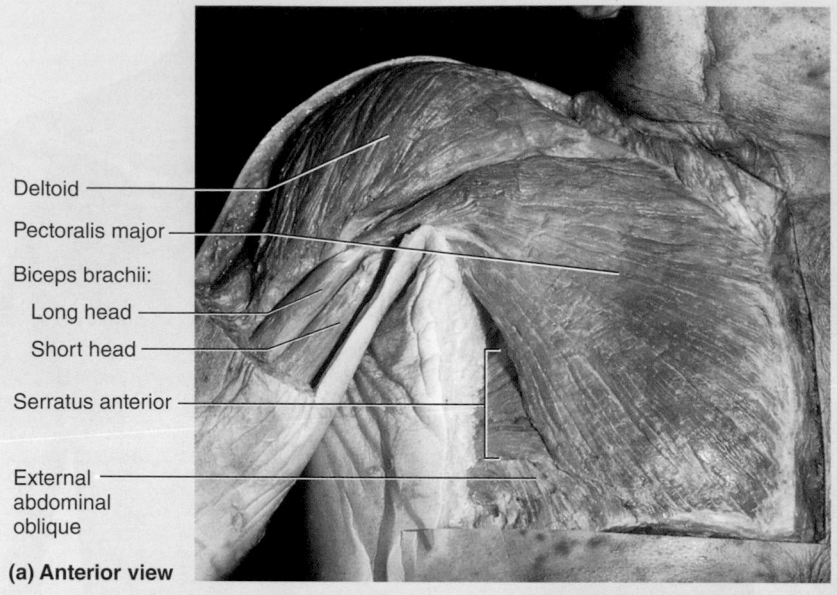

Deltoid

Pectoralis major

Biceps brachii:

Long head

Short head

Serratus anterior

External
abdominal
oblique

**(a) Anterior view**

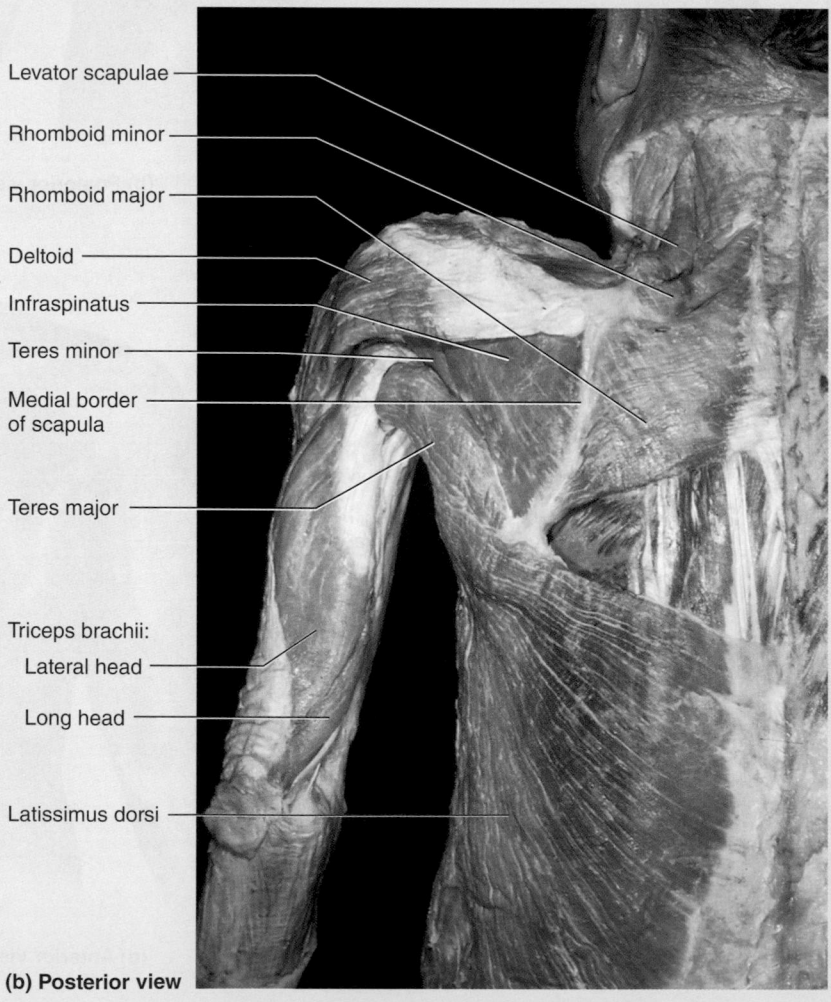

Levator scapulae

Rhomboid minor

Rhomboid major

Deltoid

Infraspinatus

Teres minor

Medial border
of scapula

Teres major

Triceps brachii:

Lateral head

Long head

Latissimus dorsi

**(b) Posterior view**

**FIGURE 10.24 Pectoral, Brachial, and Upper Back Muscles of the Cadaver.**

| TABLE 10.9 | Muscles Acting on the Arm (continued) |
|---|---|

**The Rotator Cuff.** Tendons of the remaining four scapular muscles form the **rotator cuff** (fig. 10.25). These muscles are nicknamed the "SITS muscles" for the first letters of their names—*supraspinatus, infraspinatus, teres minor,* and *subscapularis.* The first three muscles lie on the posterior side of the scapula (see fig. 10.23b). The supraspinatus and infraspinatus occupy the supraspinous and infraspinous fossae, above and below the scapular spine. The teres minor lies inferior to the infraspinatus. The subscapularis occupies the subscapular fossa on the anterior surface of the scapula, between the scapula and ribs (see fig. 10.23d). The tendons of these muscles merge with the joint capsule of the shoulder as they cross it en route to the humerus. They insert on the proximal end of the humerus, forming a partial sleeve around it. The rotator cuff reinforces the joint capsule and holds the head of the humerus in the glenoid cavity. The rotator cuff, especially the supraspinatus tendon, is easily damaged by strenuous circumduction or hard blows to the shoulder (see Deeper Insight 10.5).

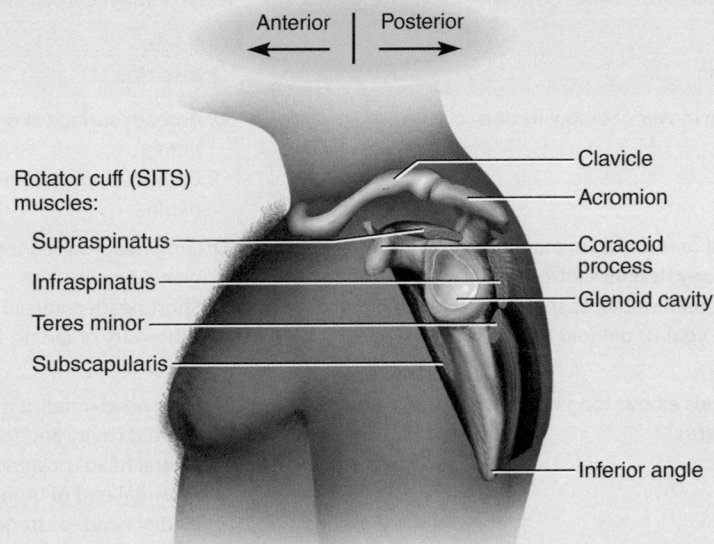

**FIGURE 10.25  Rotator Cuff Muscles in Relation to the Scapula (Lateral View).**  For posterior and anterior views of these muscles, see figure 10.23b, d.

| Name | Action | O: Origin<br>I: Insertion | Innervation |
|---|---|---|---|
| **Supraspinatus**[58]<br>(SOO-pra-spy-NAY-tus) | Aids deltoid in abduction of arm; resists downward slippage of humeral head when arm is relaxed or when carrying weight | O: Supraspinous fossa of scapula<br>I: Greater tubercle of humerus | Suprascapular nerve |
| **Infraspinatus**[59]<br>(IN-fra-spy-NAY-tus) | Modulates action of deltoid, preventing humeral head from sliding upward; rotates humerus laterally | O: Infraspinous fossa of scapula<br>I: Greater tubercle of humerus | Suprascapular nerve |
| **Teres Minor**<br>(TERR-eez) | Modulates action of deltoid, preventing humeral head from sliding upward as arm is abducted; rotates humerus laterally | O: Lateral border and adjacent posterior surface of scapula<br>I: Greater tubercle of humerus; posterior surface of joint capsule | Axillary nerve |
| **Subscapularis**[60]<br>(SUB-SCAP-you-LERR-iss) | Modulates action of deltoid, preventing humeral head from sliding upward as arm is abducted; rotates humerus medially | O: Subscapular fossa of scapula<br>I: Lesser tubercle of humerus; anterior surface of joint capsule | Upper and lower subscapular nerves |

[58] *supra* = above; *spin* = spine of scapula
[59] *infra* = below, under; *spin* = spine of scapula
[60] *sub* = below, under

| TABLE 10.10 | Muscles Acting on the Forearm |
|---|---|

The elbow and forearm are capable of four motions—flexion, extension, pronation, and supination—carried out by muscles in both the brachium and antebrachium (arm and forearm).

**Muscles with Bellies in the Arm (Brachium).** The principal elbow flexors are in the anterior compartment of the arm—the brachialis and biceps brachii (see fig. 10.23c–d). The *biceps brachii* appears as a large anterior bulge on the arm and commands considerable interest among body builders, but the *brachialis* underlying it generates about 50% more power and is thus the prime mover of elbow flexion. The biceps is not only a flexor but also a powerful forearm supinator. It is named for its two heads: a *short head* whose tendon arises from the coracoid process of the scapula, and a *long head* whose tendon originates on the superior margin of the glenoid cavity, loops over the shoulder, and braces the humerus against the glenoid cavity (see p. 295). The two heads converge close to the elbow on a single distal tendon that inserts on the radius and on the fascia of the medial side of the upper forearm. Note that *biceps* is the singular term; there is no such word as *bicep*. To refer to the biceps muscles of both arms, the plural is *bicipites* (by-SIP-ih-teez).

The triceps brachii is a three-headed muscle on the posterior side of the humerus, and is the prime mover of elbow extension (see fig. 10.23b).

| Name | Action | O: Origin<br>I: Insertion | Innervation |
|---|---|---|---|
| **Brachialis**<br>(BRAY-kee-AY-lis) | Prime mover of elbow flexion | **O:** Anterior surface of distal half of humerus<br>**I:** Coronoid process and tuberosity of ulna | Musculocutaneous nerve; radial nerve |
| **Biceps Brachii**<br>(BY-seps BRAY-kee-eye) | Rapid or forceful supination of forearm; synergist in elbow flexion; slight shoulder flexion; tendon of long head stabilizes shoulder by holding humeral head against glenoid cavity | **O:** Long head–superior margin of glenoid cavity<br>Short head–coracoid process<br>**I:** Tuberosity of radius; fascia of forearm | Musculocutaneous nerve |
| **Triceps Brachii**<br>(TRI-seps BRAY-kee-eye) | Extends elbow; long head extends and adducts humerus | **O:** Long head–inferior margin of glenoid cavity and joint capsule<br>Lateral head–posterior surface of proximal end of humerus<br>Medial head–posterior surface of entire humeral shaft<br>**I:** Olecranon; fascia of forearm | Radial nerve |

**Muscles with Bellies in the Forearm (Antebrachium).** Most forearm muscles act on the wrist and hand, but two of them are synergists in elbow flexion and extension and three of them function in pronation and supination. The *brachioradialis* is the large fleshy mass of the lateral (radial) side of the forearm just distal to the elbow (see figs. 10.23a and 10.28a). Its origin is on the distal end of the humerus and its insertion on the distal end of the radius. With the insertion so far from the fulcrum of the elbow, it does not generate as much force as the brachialis and biceps; it is effective mainly when those muscles have already partially flexed the elbow. The *anconeus* is a weak synergist of elbow extension on the posterior side of the elbow (see fig. 10.29). Pronation is achieved by the *pronator teres* near the elbow and *pronator quadratus* (the prime mover) near the wrist. Supination is usually achieved by the *supinator* of the upper forearm, with the biceps brachii aiding when additional speed or power is required (fig. 10.26).

| Name | Action | O: Origin / I: Insertion | Innervation |
|---|---|---|---|
| **Brachioradialis**<br>(BRAY-kee-oh-RAY-dee-AY-lis) | Flexes elbow | **O:** Lateral supracondylar ridge of humerus<br>**I:** Lateral surface of radius near styloid process | Radial nerve |
| **Anconeus**[61]<br>(an-CO-nee-us) | Extends elbow; may help to control ulnar movement during pronation | **O:** Lateral epicondyle of humerus<br>**I:** Olecranon and posterior surface of ulna | Radial nerve |
| **Pronator Quadratus**<br>(PRO-nay-tur quad-RAY-tus) | Prime mover of forearm pronation; also resists separation of radius and ulna when force is applied to forearm through wrist, as in doing push-ups | **O:** Anterior surface of distal ulna<br>**I:** Anterior surface of distal radius | Median nerve |
| **Pronator Teres**<br>(PRO-nay-tur TERR-eez) | Assists pronator quadratus in pronation, but only in rapid or forceful action; weakly flexes elbow | **O:** Humeral shaft near medial epicondyle; coronoid process of ulna<br>**I:** Lateral surface of radial shaft | Median nerve |
| **Supinator**<br>(SOO-pih-NAY-tur ) | Supinates forearm | **O:** Lateral epicondyle of humerus; supinator crest and fossa of ulna just distal to radial notch; anular and radial collateral ligaments of elbow<br>**I:** Proximal one-third of radius | Posterior interosseous nerve |

[61]*anconeus* = elbow

| **TABLE 10.10** | **Muscles Acting on the Forearm (continued)** |
| --- | --- |

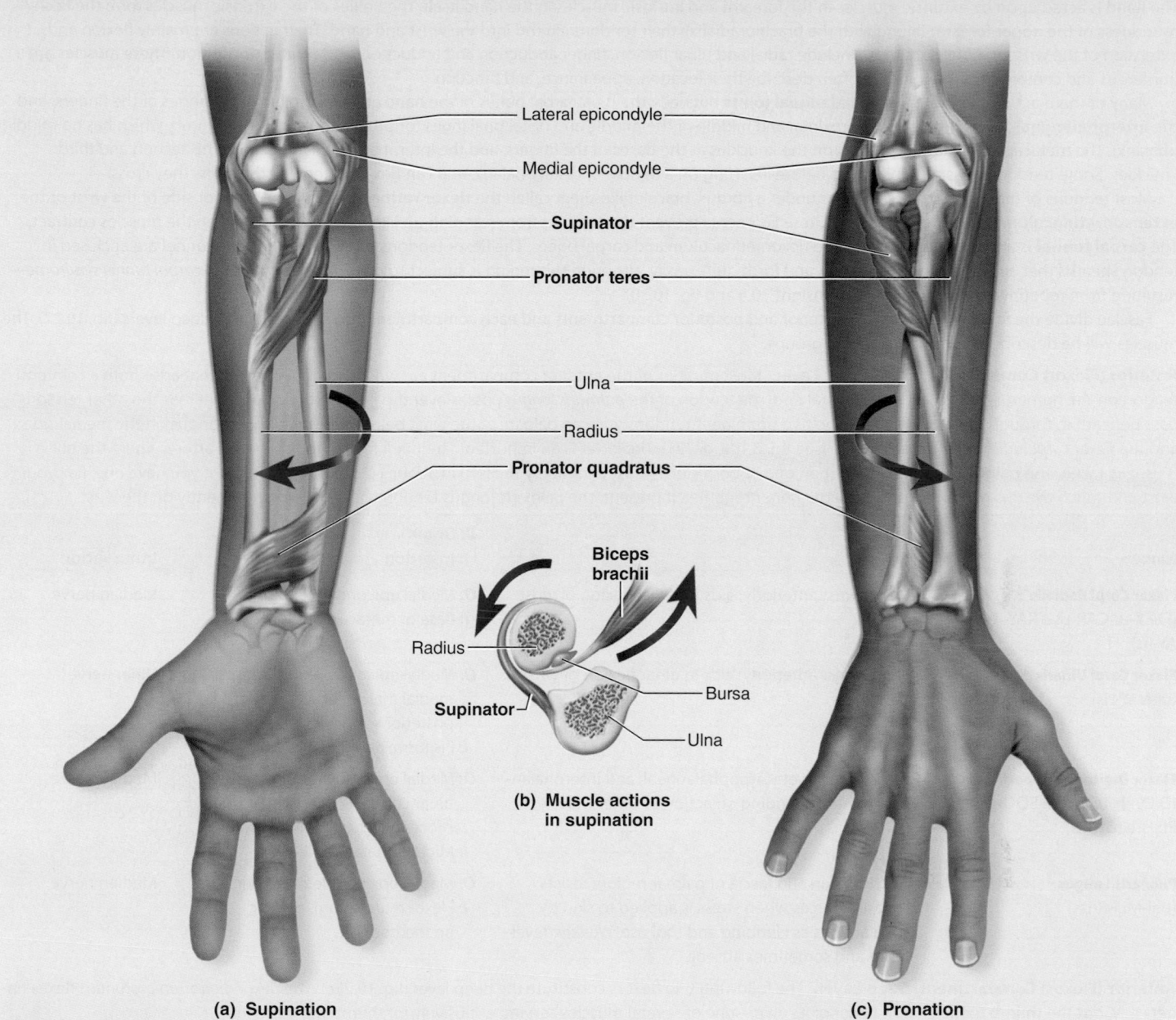

Lateral epicondyle

Medial epicondyle

**Supinator**

**Pronator teres**

Ulna

Radius

**Pronator quadratus**

**Biceps brachii**

Radius

**Supinator**

Bursa

Ulna

**(b) Muscle actions in supination**

**(a) Supination**

**(c) Pronation**

**FIGURE 10.26 Actions of the Rotator Muscles on the Forearm.** (a) Supination. (b) Cross section just distal to the elbow, showing the synergistic action of the biceps brachii and supinator. (c) Pronation.

❓ *What do the names of the pronator teres and pronator quadratus muscles indicate about their shapes?*

| TABLE 10.11 | Muscles Acting on the Wrist and Hand |
|---|---|

The hand is acted upon by extrinsic muscles in the forearm and intrinsic muscles in the hand itself. The bellies of the extrinsic muscles form the fleshy roundness of the upper forearm (along with the brachioradialis); their tendons extend into the wrist and hand. Their actions are mainly flexion and extension of the wrist and digits, but also include radial and ulnar flexion, finger abduction and adduction, and thumb opposition. These muscles are numerous and complex, but their names often describe their location, appearance, and function.

Many of them act on the **metacarpophalangeal joints** between the metacarpal bones of the hand and the proximal phalanges of the fingers, and the **interphalangeal joints** between the proximal and middle or the middle and distal phalanges (or proximal-distal in the thumb, which has no middle phalanx). The metacarpophalangeal joints form the knuckles at the bases of the fingers, and the interphalangeal joints form the second and third knuckles. Some tendons cross multiple joints before inserting on a middle or distal phalanx, and can flex or extend all the joints they cross.

Most tendons of the extrinsic muscles pass under a fibrous, braceletlike sheet called the **flexor retinaculum** on the anterior side of the wrist or the **extensor retinaculum** on the posterior side. These ligaments prevent the tendons from standing up like taut bowstrings when the muscles contract. The **carpal tunnel** is a tight space between the flexor retinaculum and carpal bones. The flexor tendons passing through the tunnel are enclosed in tendon sheaths that enable them to slide back and forth quite easily, although this region is subject to painful inflammation—*carpal tunnel syndrome*—resulting from repetitive motion (see Deeper Insight 10.4 and fig. 10.30).

Fasciae divide the forearm muscles into anterior and posterior compartments and each compartment into superficial and deep layers (fig. 10.27). The muscles will be described below in these four groups.

**Anterior (Flexor) Compartment, Superficial Layer.** Most muscles of the anterior compartment are wrist and finger flexors that arise from a common tendon on the humerus (fig. 10.28). At the distal end, the tendon of the *palmaris longus* passes over the flexor retinaculum, whereas the other tendons pass beneath it, through the carpal tunnel. The two prominent tendons you can palpate at the wrist belong to the palmaris longus on the medial side and the *flexor carpi radialis* on the lateral side (see fig. B.19a, p. 391). The latter is an important landmark for finding the radial artery, where the pulse is usually taken. The palmaris longus is absent on one or both sides (most commonly the left) in about 14% of people. To see if you have one, flex your wrist and touch the tips of your thumb and little finger together. If present, the palmaris longus tendon will stand up prominently on the wrist.

| Name | Action | O: Origin<br>I: Insertion | Innervation |
|---|---|---|---|
| **Flexor Carpi Radialis**[62]<br>(FLEX-ur CAR-pye RAY-dee-AY-lis) | Flexes wrist anteriorly; aids in radial flexion of wrist | **O:** Medial epicondyle of humerus<br>**I:** Base of metacarpals II–III | Median nerve |
| **Flexor Carpi Ulnaris**[63]<br>(ul-NAY-ris) | Flexes wrist anteriorly; aids in ulnar flexion of wrist | **O:** Medial epicondyle of humerus; medial margin of olecranon; posterior surface of ulna<br>**I:** Pisiform; hamate; metacarpal V | Ulnar nerve |
| **Flexor Digitorum Superficialis**[64]<br>(DIDJ-ih-TOE-rum SOO-per-FISH-ee-AY-lis) | Flexes wrist, metacarpophalangeal, and interphalangeal joints depending on action of other muscles | **O:** Medial epicondyle of humerus; ulnar collateral ligament; coronoid process; superior half of radius<br>**I:** Middle phalanges II–V | Median nerve |
| **Palmaris Longus**<br>(pal-MERR-iss) | Anchors skin and fascia of palmar region; resists shearing forces when stress is applied to skin by such actions as climbing and tool use. Weakly developed and sometimes absent. | **O:** Medial epicondyle of humerus<br>**I:** Flexor retinaculum, palmar aponeurosis | Median nerve |

**Anterior (Flexor) Compartment, Deep Layer.** The following two flexors constitute the deep layer (fig. 10.28c). The *flexor digitorum profundus* flexes fingers II–V, but the thumb (pollex) has a flexor of its own—one of several muscles serving exclusively for thumb movements.

| | | | |
|---|---|---|---|
| **Flexor Digitorum Profundus** | Flexes wrist, metacarpophalangeal, and interphalangeal joints; sole flexor of the distal interphalangeal joints | **O:** Proximal three-quarters of ulna; coronoid process; interosseous membrane<br>**I:** Distal phalanges II–V | Median nerve; ulnar nerve |
| **Flexor Pollicis Longus**<br>(PAHL-ih-sis) | Flexes phalanges of thumb | **O:** Radius; interosseous membrane<br>**I:** Distal phalanx I | Median nerve |

---

[62]*carpi* = of the wrist; *radialis* = of the radius
[63]*ulnaris* = of the ulna
[64]*digitorum* = of the digits; *superficialis* = shallow, near the surface

**TABLE 10.11**       **Muscles Acting on the Wrist and Hand (continued)**

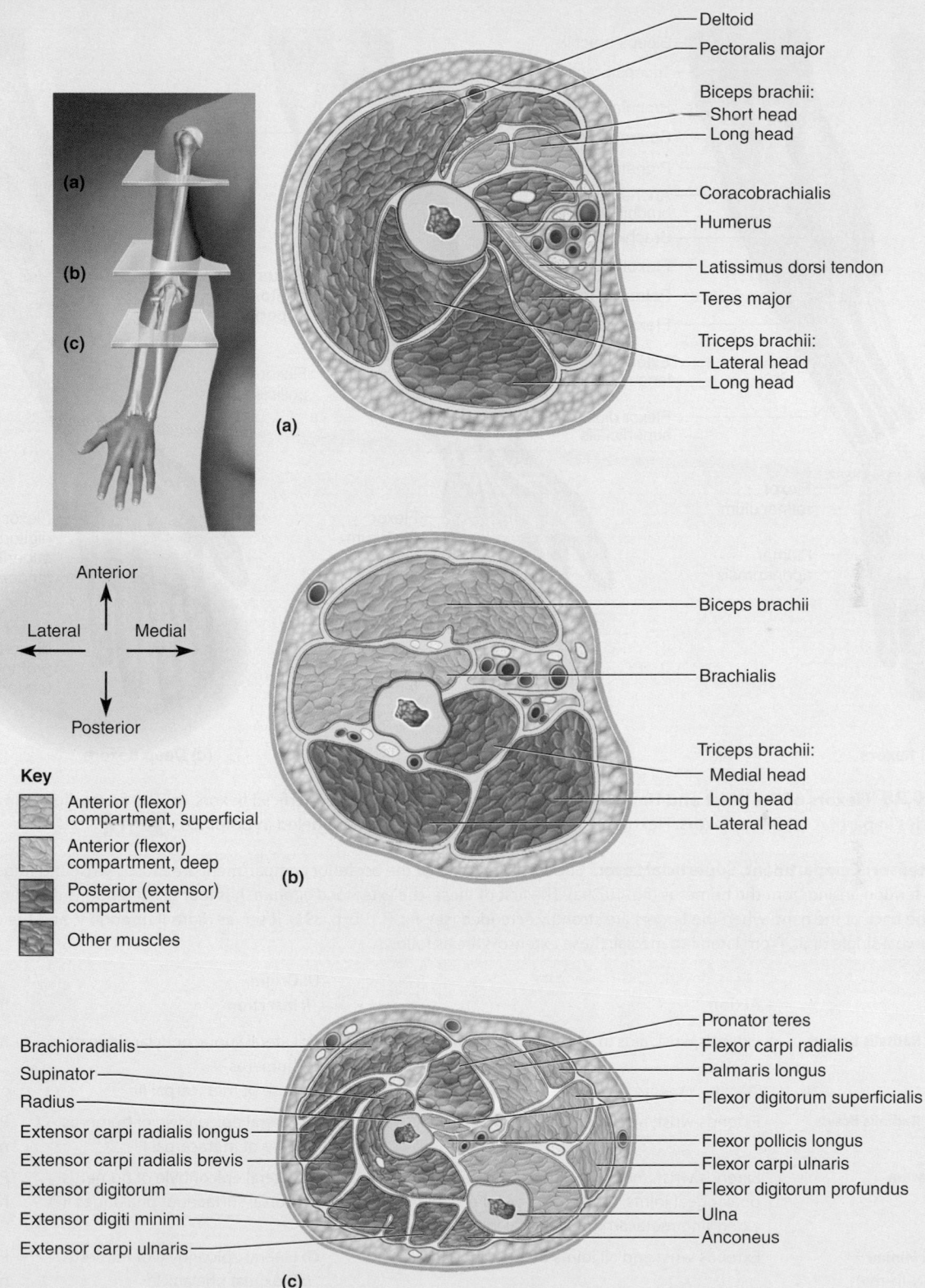

**FIGURE 10.27 Serial Cross Sections Through the Upper Limb.** Each section is taken at the correspondingly lettered level in the figure at the left and is pictured with the posterior muscle compartment facing the bottom of the page, as if viewing a person's right limb extended toward you with the palm up.

❓ *Why are the extensor pollicis longus and extensor indicis not seen in part (c)?*

| TABLE 10.11 | Muscles Acting on the Wrist and Hand (continued) |
|---|---|

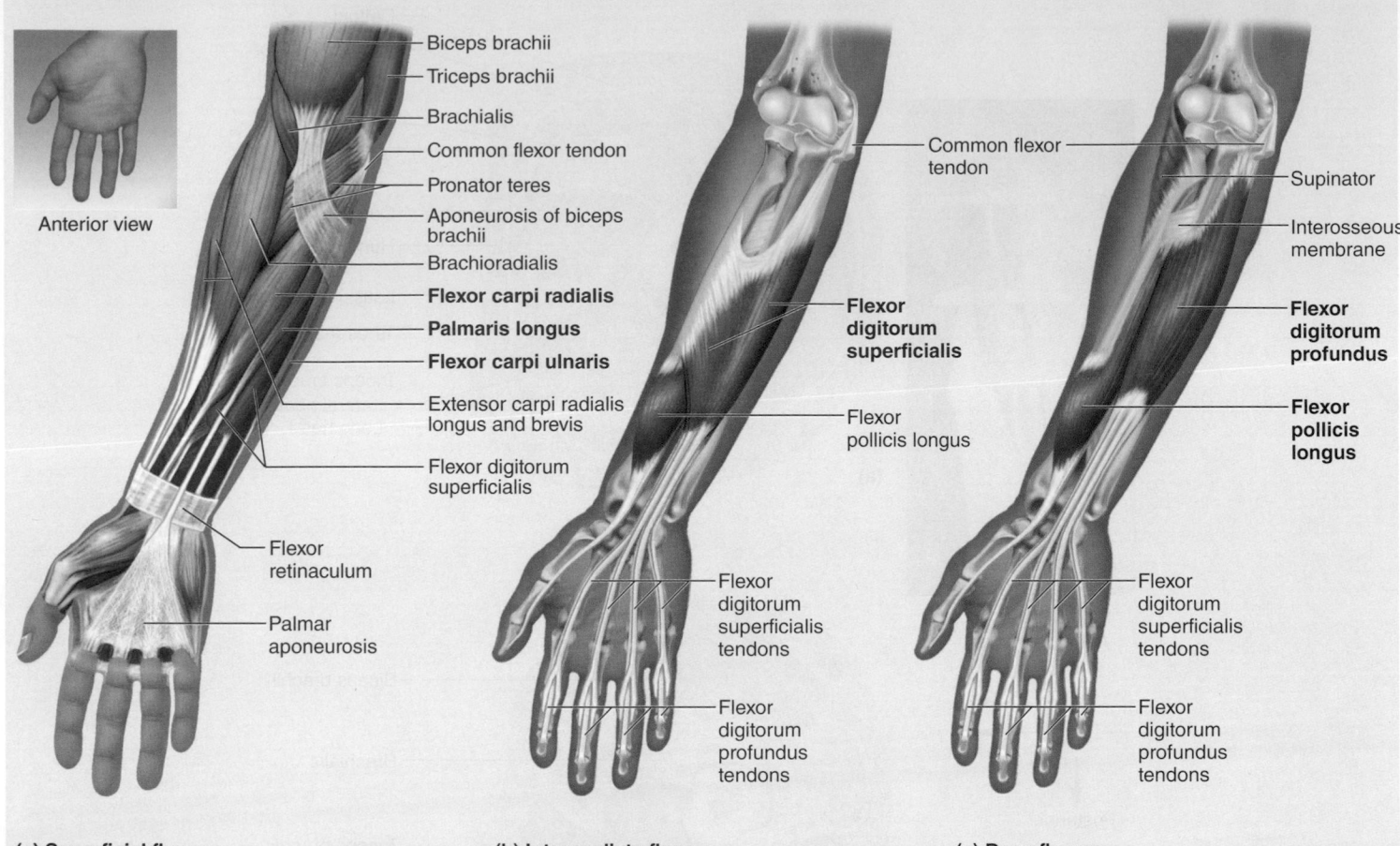

(a) Superficial flexors

(b) Intermediate flexor

(c) Deep flexors

**FIGURE 10.28 Flexors of the Wrist and Hand.** Anterior views of the forearm. (a) Superficial flexors. (b) The flexor digitorum superficialis, deep to the muscles in part (a). (c) Deep flexors. Flexor muscles of each compartment are labeled in boldface. **AP|R**

**Posterior (Extensor) Compartment, Superficial Layer.** Superficial muscles of the posterior compartment are mostly wrist and finger extensors, and share a single tendon arising from the humerus (fig. 10.29a). The first of these, the *extensor digitorum,* has four distal tendons that can easily be seen and palpated on the back of the hand when the fingers are strongly extended (see fig. B.19b, p. 391). It serves digits II through V, and the other muscles in this group each serve a single digit. From lateral to medial, these extensors are as follows.

| Name | Action | O: Origin<br>I: Insertion | Innervation |
|---|---|---|---|
| **Extensor Carpi Radialis Longus** | Extends wrist; aids in radial flexion of wrist | **O:** Lateral supracondylar ridge of humerus<br>**I:** Base of metacarpal II | Radial nerve |
| **Extensor Carpi Radialis Brevis**<br>(BREV-iss) | Extends wrist; aids in radial flexion of wrist | **O:** Lateral epicondyle of humerus<br>**I:** Base of metacarpal III | Posterior interosseous nerve |
| **Extensor Digitorum** | Extends wrist, metacarpophalangeal, and inter-phalangeal joints; tends to spread digits apart when extending metacarpophalangeal joints | **O:** Lateral epicondyle of humerus<br>**I:** Dorsal surfaces of phalanges II–V | Posterior interosseous nerve |
| **Extensor Digiti Minimi**[65]<br>(DIDJ-ih-ty MIN-ih-my) | Extends wrist and all joints of little finger | **O:** Lateral epicondyle of humerus<br>**I:** Proximal phalanx V | Posterior interosseous nerve |
| **Extensor Carpi Ulnaris** | Extends and fixes wrist when fist is clenched or hand grips an object; aids in ulnar flexion of wrist | **O:** Lateral epicondyle of humerus; posterior<br>**I:** Base of metacarpal V surface of ulnar shaft | Posterior interosseous nerve |

[65]*digit* = finger; *minim* = smallest

| **TABLE 10.11** | **Muscles Acting on the Wrist and Hand (continued)** |
|---|---|

**Posterior (Extensor) Compartment, Deep Layer.** The deep muscles that follow serve only the thumb and index finger (fig. 10.29b). By strongly abducting and extending the thumb into a hitchhiker's position, you may see a deep dorsolateral pit at the base of the thumb, with a taut tendon on each side of it (see fig. B.19b, p. 391). This depression is called the *anatomical snuffbox* because it was once fashionable to place a pinch of snuff here and inhale it. It is bordered laterally by the tendons of the *abductor pollicis longus* and *extensor pollicis brevis,* and medially by the tendon of the *extensor pollicis longus.* The extensor muscles from lateral to medial are as follows.

| Name | Action | O: Origin<br>I: Insertion | Innervation |
|---|---|---|---|
| **Abductor Pollicis Longus** | Abducts thumb in frontal (palmar) plane (radial abduction); extends thumb at carpometacarpal joint | **O:** Posterior surfaces of radius and ulna; interosseous membrane<br>**I:** Trapezium; base of metacarpal I | Posterior interosseous nerve |
| **Extensor Pollicis Brevis** | Extends metacarpal I and proximal phalanx of thumb | **O:** Shaft of radius; interosseous membrane<br>**I:** Proximal phalanx I | Posterior interosseous nerve |
| **Extensor Pollicis Longus** | Extends distal phalanx I; aids in extending proximal phalanx I and metacarpal I; adducts and laterally rotates thumb | **O:** Posterior surface of ulna; interosseous membrane<br>**I:** Distal phalanx I | Posterior interosseous nerve |
| **Extensor Indicis** (IN-dih-sis) | Extends wrist and index finger | **O:** Posterior surface of ulna; interosseous membrane<br>**I:** Middle and distal phalanges of index finger | Posterior interosseous nerve |

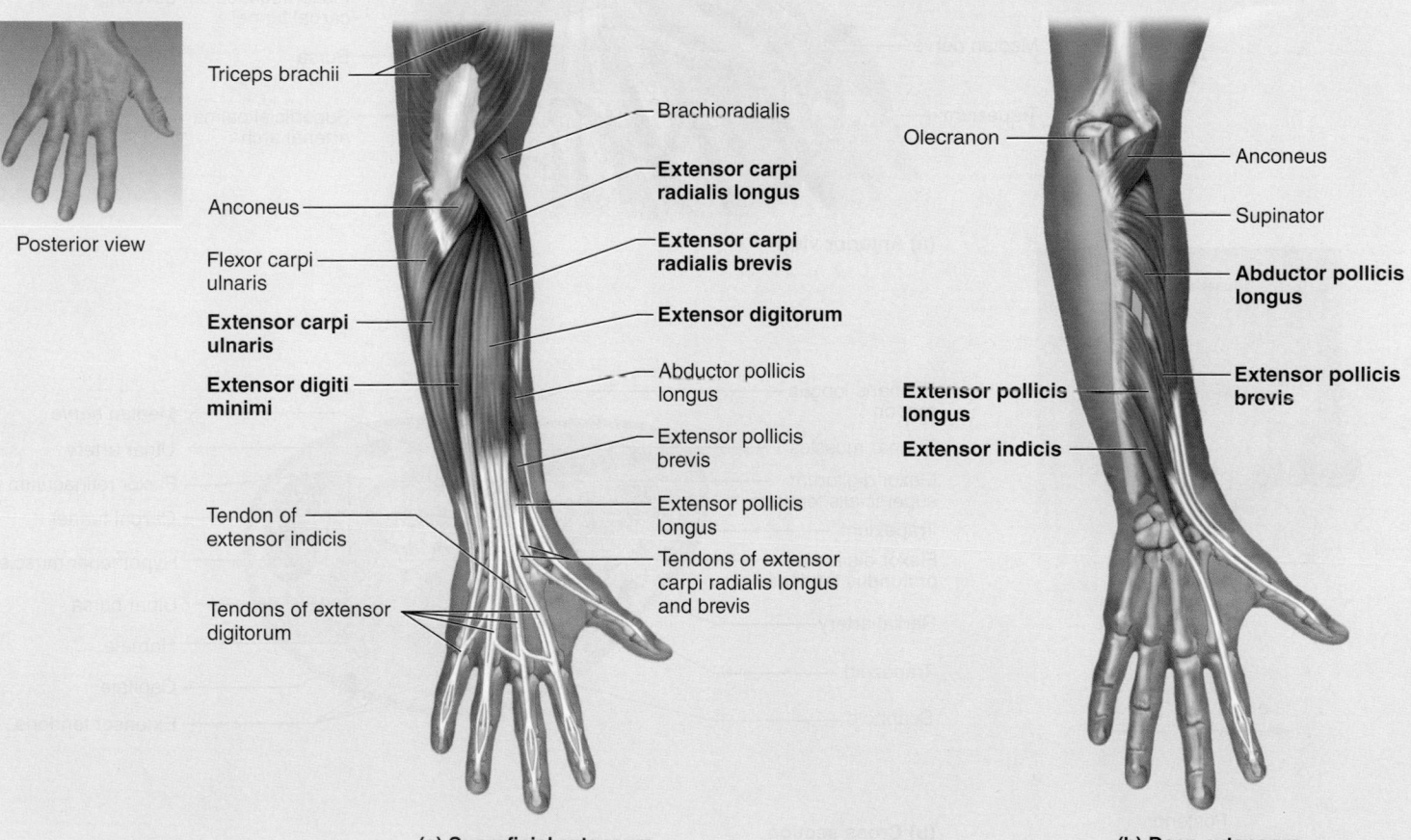

(a) Superficial extensors

(b) Deep extensors

**FIGURE 10.29 Extensors of the Wrist and Hand.** Posterior views of the forearm. Extensor muscles of each compartment are labeled in boldface. AP|R

# DEEPER INSIGHT 10.4

## CLINICAL APPLICATION

### Carpal Tunnel Syndrome

Prolonged, repetitive motions of the wrist and fingers can cause tissues in the carpal tunnel to become inflamed, swollen, or fibrotic. Since the carpal tunnel cannot expand, swelling puts pressure on the median nerve of the wrist, which passes through the carpal tunnel with the flexor tendons (fig. 10.30). This pressure causes tingling and muscular weakness in the palm and medial side of the hand and pain that may radiate to the arm and

shoulder. This condition, called *carpal tunnel syndrome,* is common among keyboard operators, pianists, meat cutters, and others who spend long hours making repetitive wrist motions. Carpal tunnel syndrome is treated with aspirin and other anti-inflammatory drugs, immobilization of the wrist, and sometimes surgical removal of part or all of the flexor retinaculum to relieve pressure on the nerve.

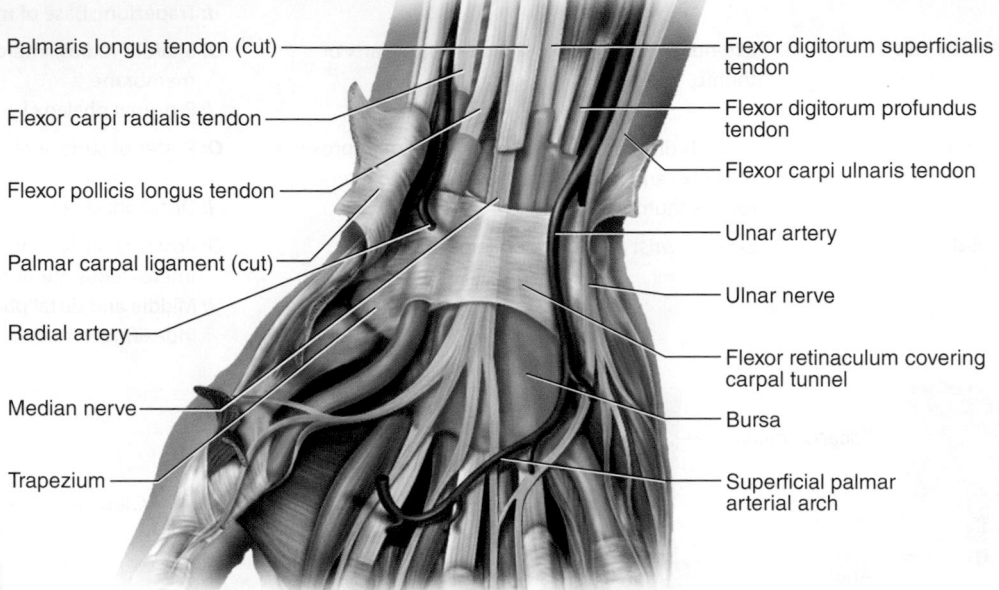

Palmaris longus tendon (cut)

Flexor carpi radialis tendon

Flexor pollicis longus tendon

Palmar carpal ligament (cut)

Radial artery

Median nerve

Trapezium

Flexor digitorum superficialis tendon

Flexor digitorum profundus tendon

Flexor carpi ulnaris tendon

Ulnar artery

Ulnar nerve

Flexor retinaculum covering carpal tunnel

Bursa

Superficial palmar arterial arch

**(a) Anterior view**

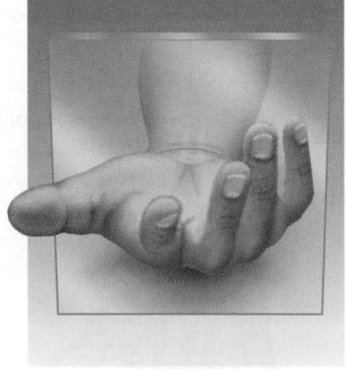

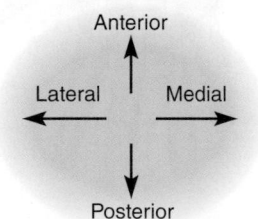

Anterior

Lateral — Medial

Posterior

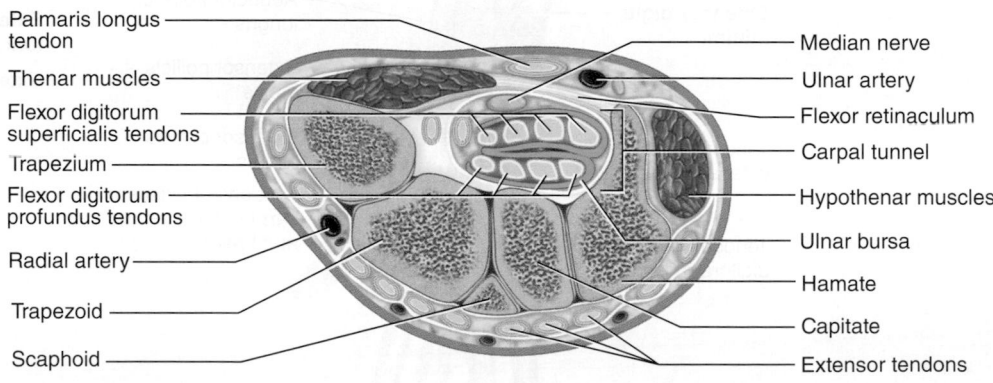

Palmaris longus tendon

Thenar muscles

Flexor digitorum superficialis tendons

Trapezium

Flexor digitorum profundus tendons

Radial artery

Trapezoid

Scaphoid

Median nerve

Ulnar artery

Flexor retinaculum

Carpal tunnel

Hypothenar muscles

Ulnar bursa

Hamate

Capitate

Extensor tendons

**(b) Cross section**

**FIGURE 10.30 The Carpal Tunnel.** (a) Dissection of the wrist (anterior aspect) showing the tendons, nerve, and bursae that pass under the flexor retinaculum. **AP|R** (b) Cross section of the wrist, viewed as if from the distal end of a person's right forearm extended toward you with the palm up. Note how the flexor tendons and median nerve are confined in the tight space between the carpal bones and flexor retinaculum. That tight packing and repetitive sliding of the flexor tendons through the tunnel contribute to carpal tunnel syndrome.

| TABLE 10.12 | Intrinsic Muscles of the Hand |
|---|---|

The intrinsic muscles of the hand assist the flexors and extensors in the forearm and make finger movements more precise. They are divided into three groups: the *thenar group* at the base of the thumb, the *hypothenar group* at the base of the little finger, and the *midpalmar group* between these (fig. 10.31).

**Thenar Group.** The thenar group of muscles forms the thick fleshy mass *(thenar eminence)* at the base of the thumb, and the *adductor pollicis* forms the web between the thumb and palm. All are concerned with thumb movements. The adductor pollicis has an *oblique head* that extends from the capitate bone of the wrist to the ulnar side of the base of the thumb, and a *transverse head* that extends from metacarpal III to the same insertion as the oblique head.

| Name | Action | O: Origin<br>I: Insertion | Innervation |
|---|---|---|---|
| **Adductor Pollicis** | Draws thumb toward palm as in gripping a tool | O: Capitate; bases of metacarpals II–III; anterior ligaments of wrist; tendon sheath of flexor carpi radialis<br>I: Medial surface of proximal phalanx I | Ulnar nerve |
| **Abductor Pollicis Brevis** | Abducts thumb in sagittal plane | O: Mainly flexor retinaculum; also scaphoid, trapezium, and abductor pollicis longus tendon<br>I: Lateral surface of proximal phalanx I | Median nerve |
| **Flexor Pollicis Brevis** | Flexes metacarpophalangeal joint of thumb | O: Trapezium; trapezoid; capitate; anterior ligaments of wrist; flexor retinaculum<br>I: Proximal phalanx I | Median nerve; ulnar nerve |
| **Opponens Pollicis**<br>(op-PO-nenz) | Flexes metacarpal I to oppose thumb to fingertips | O: Trapezium; flexor retinaculum<br>I: Metacarpal I | Median nerve |

**Hypothenar Group.** The hypothenar group forms the fleshy mass *(hypothenar eminence)* at the base of the little finger. All of these muscles are concerned with movement of that digit.

| Name | Action | O: Origin<br>I: Insertion | Innervation |
|---|---|---|---|
| **Abductor Digiti Minimi** | Abducts little finger, as in spreading fingers apart | O: Pisiform; tendon of flexor carpi ulnaris<br>I: Medial surface of proximal phalanx V | Ulnar nerve |
| **Flexor Digiti Minimi Brevis** | Flexes little finger at metacarpophalangeal joint | O: Hamulus of hamate bone; flexor retinaculum<br>I: Medial surface of proximal phalanx V | Ulnar nerve |
| **Opponens Digiti Minimi** | Flexes metacarpal V at carpometacarpal joint when little finger is moved into opposition with tip of thumb; deepens palm of hand | O: Hamulus of hamate bone; flexor retinaculum<br>I: Medial surface of metacarpal V | Ulnar nerve |

**Midpalmar Group.** The midpalmar group occupies the hollow of the palm. It has 11 small muscles divided into three groups.

| Name | Action | O: Origin<br>I: Insertion | Innervation |
|---|---|---|---|
| **Four Dorsal Interosseous**[66] **Muscles**<br>(IN-tur-OSS-ee-us) | Abduct fingers; strongly flex metacarpophalangeal joints but extend interphalangeal joints, depending on action of other muscles; important in grip strength | O: Each with two heads arising from facing surfaces of adjacent metacarpals<br>I: Proximal phalanges II–IV | Ulnar nerve |
| **Three Palmar Interosseous Muscles** | Adduct fingers; other actions same as for dorsal interosseous muscles | O: Metacarpals I, II, IV, V<br>I: Proximal phalanges II, IV, V | Ulnar nerve |
| **Four Lumbrical Muscles**[67]<br>(LUM-brih-cul) | Extend interphalangeal joints; contribute to ability to pinch objects between fleshy pulp of thumb and finger, instead of these digits meeting by the edges of their nails | O: Tendons of flexor digitorum profundus<br>I: Proximal phalanges II–V | Median nerve; ulnar nerve |

[66]*inter* = between; *osse* = bones
[67]*lumbrical* = resembling an earthworm

**TABLE 10.12**   Intrinsic Muscles of the Hand (continued)

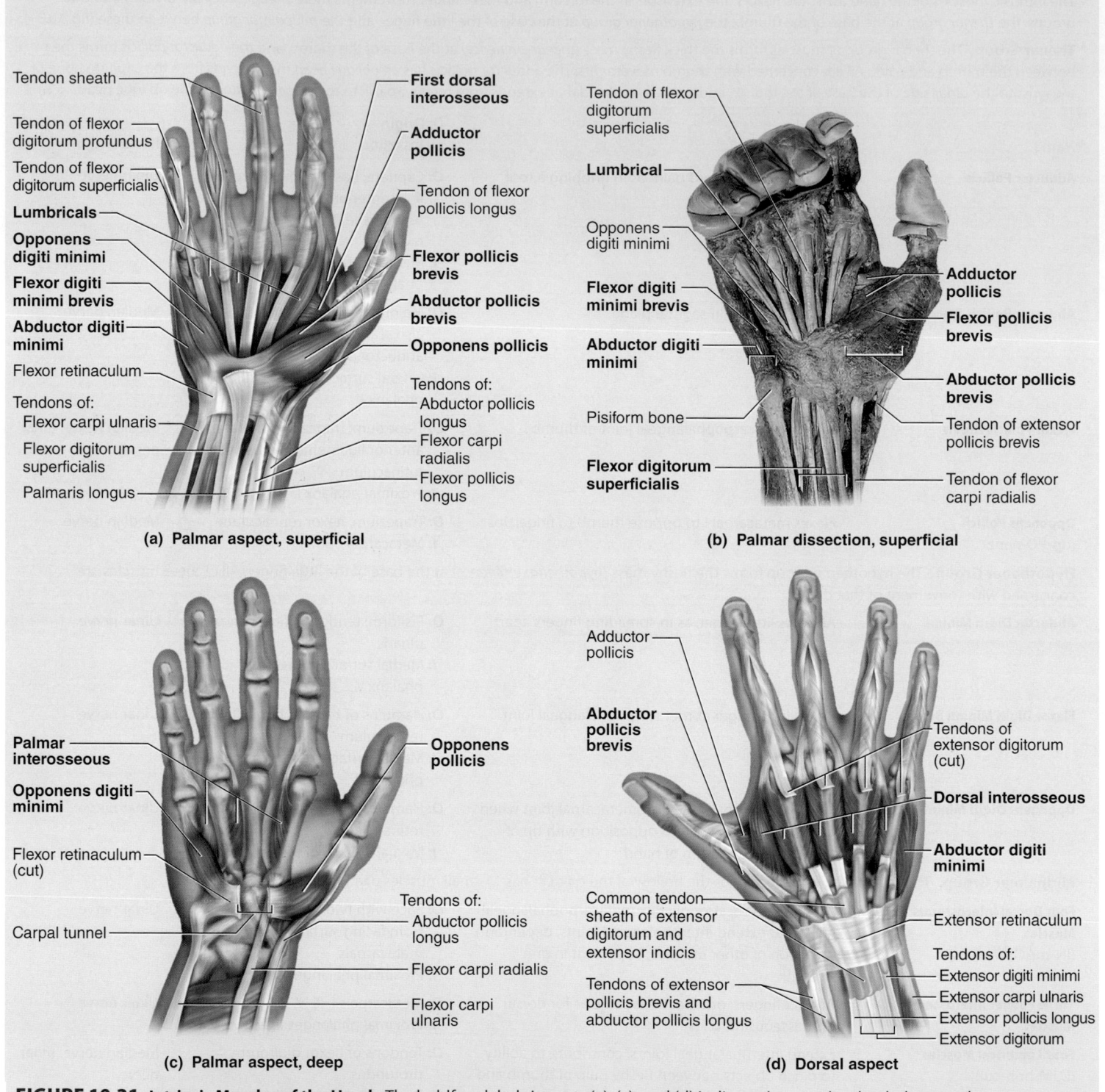

Tendon sheath

Tendon of flexor digitorum profundus

Tendon of flexor digitorum superficialis

**Lumbricals**

**Opponens digiti minimi**

**Flexor digiti minimi brevis**

**Abductor digiti minimi**

Flexor retinaculum

Tendons of:
Flexor carpi ulnaris
Flexor digitorum superficialis
Palmaris longus

**First dorsal interosseous**

**Adductor pollicis**

Tendon of flexor pollicis longus

**Flexor pollicis brevis**

**Abductor pollicis brevis**

**Opponens pollicis**

Tendons of:
Abductor pollicis longus
Flexor carpi radialis
Flexor pollicis longus

**(a) Palmar aspect, superficial**

Tendon of flexor digitorum superficialis

**Lumbrical**

Opponens digiti minimi

**Flexor digiti minimi brevis**

**Abductor digiti minimi**

Pisiform bone

**Flexor digitorum superficialis**

**Adductor pollicis**

**Flexor pollicis brevis**

**Abductor pollicis brevis**

Tendon of extensor pollicis brevis

Tendon of flexor carpi radialis

**(b) Palmar dissection, superficial**

**Palmar interosseous**

**Opponens digiti minimi**

Flexor retinaculum (cut)

Carpal tunnel

**Opponens pollicis**

Tendons of:
Abductor pollicis longus
Flexor carpi radialis
Flexor carpi ulnaris

**(c) Palmar aspect, deep**

Adductor pollicis

**Abductor pollicis brevis**

Common tendon sheath of extensor digitorum and extensor indicis

Tendons of extensor pollicis brevis and abductor pollicis longus

Tendons of extensor digitorum (cut)

**Dorsal interosseous**

**Abductor digiti minimi**

Extensor retinaculum

Tendons of:
Extensor digiti minimi
Extensor carpi ulnaris
Extensor pollicis longus
Extensor digitorum

**(d) Dorsal aspect**

**FIGURE 10.31  Intrinsic Muscles of the Hand.** The boldface labels in parts (a), (c), and (d) indicate the muscles that belong to the respective layer. AP|R

Answer the following questions to test your understanding of the preceding section:

16. Name a muscle that inserts on the scapula and plays a significant role in each of the following actions:

    a. pushing a stalled car,

    b. paddling a canoe,

    c. squaring the shoulders in military attention,

    d. lifting the shoulder to carry a heavy box on it, and

    e. lowering the shoulder to lift a suitcase.

17. Describe three contrasting actions of the deltoid muscle.

18. Name the four rotator cuff muscles and describe the scapular surfaces against which they lie.

19. Name the prime movers of elbow flexion and extension.

20. Identify three functions of the biceps brachii.

21. Name three extrinsic muscles and two intrinsic muscles that flex the phalanges.

## 10.5 Muscles Acting on the Hip and Lower Limb

### Expected Learning Outcomes

When you have completed this section, you should be able to

a. name and locate the muscles that act on the hip, knee, ankle, and toe joints;

b. relate the actions of these muscles to the joint movements described in chapter 9; and

c. describe the origin, insertion, and innervation of each muscle.

The largest muscles are found in the lower limb. Unlike those of the upper limb, they are adapted less for precision than for the strength needed to stand, maintain balance, walk, and run. Several of them cross and act upon two or more joints, such as the hip and knee. To avoid confusion in this discussion, remember that in the anatomical sense the word *leg* refers only to that part of the limb between the knee and ankle. The term *foot* includes the tarsal region (ankle), metatarsal region, and toes. Tables 10.13 through 10.16 group the muscles of the lower limb into those that act on the femur and hip joint, those that act on the leg and knee joint, extrinsic (leg) muscles that act on the foot and ankle joint, and intrinsic (foot) muscles that act on the arches and toes.

| TABLE 10.13 | Muscles Acting on the Hip and Femur |
|---|---|

**Anterior Muscles of the Hip.** Most muscles that act on the femur originate on the hip bone. The two principal anterior muscles are the *iliacus,* which fills most of the broad iliac fossa of the pelvis, and the *psoas major,* a thick rounded muscle that arises mainly from the lumbar vertebrae (fig. 10.32). Collectively, they are called the **iliopsoas** and share a common tendon to the femur.

| Name | Action | O: Origin<br>I: Insertion | Innervation |
|---|---|---|---|
| **Iliacus**[68]<br>(ih-LY-uh-cus) | Flexes thigh at hip when trunk is fixed; flexes trunk at hip when thigh is fixed, as in bending forward in a chair or sitting up in bed; balances trunk during sitting | **O:** Iliac crest and fossa; superolateral region of sacrum; anterior sacroiliac and iliolumbar ligaments<br>**I:** Lesser trochanter and nearby shaft of femur | Femoral nerve |
| **Psoas**[69] **major**<br>(SO-ass) | Same as iliacus | **O:** Bodies and intervertebral discs of vertebrae T12–L5; transverse processes of lumbar vertebrae<br>**I:** Lesser trochanter and nearby shaft of femur | Anterior rami of lumbar spinal nerves |

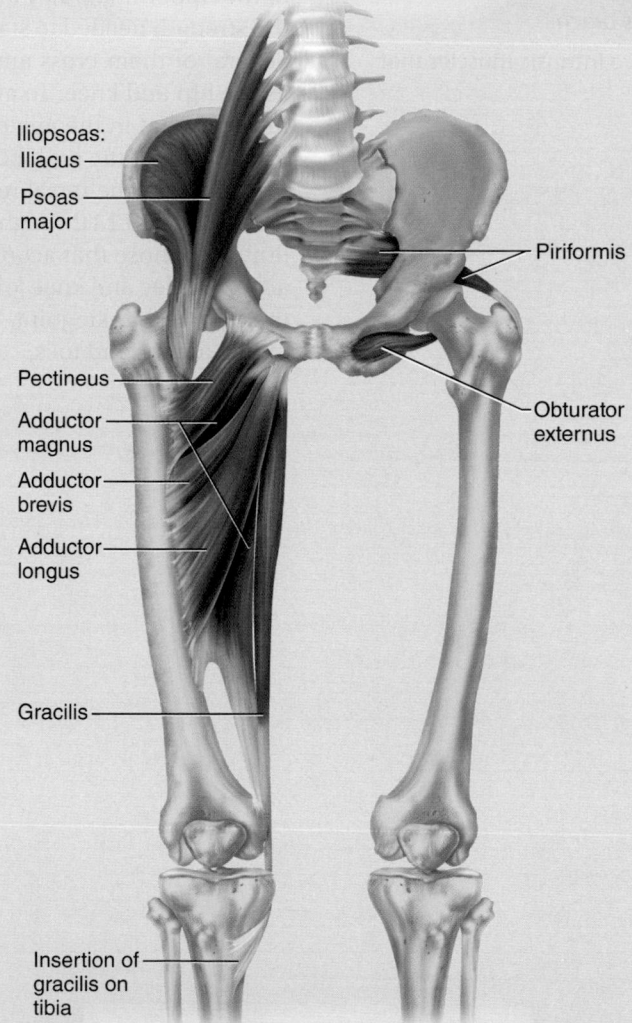

Iliopsoas:
Iliacus
Psoas major
Piriformis
Pectineus
Adductor magnus
Adductor brevis
Adductor longus
Obturator externus
Gracilis
Insertion of gracilis on tibia

**FIGURE 10.32  Muscles That Act on the Hip and Femur (Anterior View).**

---

[68]*ili* = loin, flank                    [69]*psoa* = loin

| **TABLE 10.13** | **Muscles Acting on the Hip and Femur (continued)** |
| --- | --- |

**Lateral and Posterior Muscles of the Hip.** On the lateral and posterior sides of the hip are the *tensor fasciae latae* and three gluteal muscles. The **fascia lata** is a fibrous sheath that encircles the thigh like a subcutaneous stocking and tightly binds its muscles. On the lateral surface, it combines with the tendons of the gluteus maximus and tensor fasciae latae to form the **iliotibial tract,** which extends from the iliac crest to the lateral condyle of the tibia (see fig. 10.34, table 10.14). The tensor fasciae latae tautens the iliotibial tract and braces the knee, especially when the opposite foot is lifted.

The gluteal muscles are the *gluteus maximus, gluteus medius,* and *gluteus minimus* (fig. 10.33). The gluteus maximus is the largest of these and forms most of the lean mass of the buttock. It is an extensor of the hip joint that produces the backswing of the leg in walking and provides most of the lift when you climb stairs. It generates its maximum force when the thigh is flexed at a 45° angle to the trunk. This is the advantage in starting a foot race from a crouched position. The gluteus medius is deep and lateral to the gluteus maximus. Its name refers to its size, not its position. The gluteus minimus is the smallest and deepest of the three.

| Name | Action | **O: Origin**<br>**I: Insertion** | Innervation |
| --- | --- | --- | --- |
| **Tensor Fasciae Latae**[70]<br>(TEN-sur FASH-ee-ee LAY-tee) | Extends knee, laterally rotates tibia, aids in abduction and medial rotation of femur; during standing, steadies pelvis on femoral head and steadies femoral condyles on tibia | **O:** Iliac crest; anterior superior spine; deep surface of fascia lata<br>**I:** Lateral condyle of tibia via iliotibial tract | Superior gluteal nerve |
| **Gluteus Maximus**[71] | Extends thigh at hip as in stair climbing (rising to next step) or running and walking (backswing of limb); abducts thigh; elevates trunk after stooping; prevents trunk from pitching forward during walking and running; helps stabilize femur on tibia | **O:** Posterior gluteal line of ilium, on posterior surface from iliac crest to posterior superior spine; coccyx; posterior surface of lower sacrum; aponeurosis of erector spinae<br>**I:** Gluteal tuberosity of femur; lateral condyle of tibia via iliotibial tract | Inferior gluteal nerve |
| **Gluteus Medius and Gluteus Minimus** | Abduct and medially rotate thigh; during walking, shift weight of trunk toward limb with foot on the ground as other foot is lifted | **O:** Most of lateral surface of ilium between crest and acetabulum<br>**I:** Greater trochanter of femur | Superior gluteal nerve |

**Lateral Rotators.** Inferior to the gluteus minimus and deep to the other two gluteal muscles are six muscles called the **lateral rotators,** named for their action on the femur (fig. 10.33). Their action is most clearly visualized when you cross your legs to rest an ankle on your knee, causing your femur to rotate and the knee to point laterally. Thus, they oppose medial rotation by the gluteus medius and minimus. Most of them also abduct or adduct the femur. The abductors are important in walking because when one lifts a foot from the ground, they shift the body weight to the other leg and prevent falling.

| | | | |
| --- | --- | --- | --- |
| **Gemellus**[72] **Superior**<br>(jeh-MEL-us) | Laterally rotates extended thigh; abducts flexed thigh; sometimes absent | **O:** Ischial spine<br>**I:** Greater trochanter of femur | Nerve to obturator internus |
| **Gemellus Inferior** | Same actions as gemellus superior | **O:** Ischial tuberosity<br>**I:** Greater trochanter of femur | Nerve to quadratus femoris |
| **Obturator**[73] **Externus**<br>(OB-too-RAY-tur) | Not well understood; thought to laterally rotate thigh in climbing | **O:** External surface of obturator membrane; rami of pubis and ischium<br>**I:** Femur between head and greater trochanter | Obturator nerve |
| **Obturator Internus** | Not well understood; thought to laterally rotate extended thigh and abduct flexed thigh | **O:** Ramus of ischium; inferior ramus of pubis; anteromedial surface of lesser pelvis<br>**I:** Greater trochanter of femur | Nerve to obturator internus |
| **Piriformis**[74]<br>(PIR-ih-FOR-mis) | Laterally rotates extended thigh; abducts flexed thigh | **O:** Anterior surface of sacrum; gluteal surface of ilium; capsule of sacroiliac joint<br>**I:** Greater trochanter of femur | Spinal nerves L5–S2 |
| **Quadratus Femoris**[75]<br>(quad-RAY-tus FEM-oh-ris) | Laterally rotates thigh | **O:** Ischial tuberosity<br>**I:** Intertrochanteric crest of femur | Nerve to quadratus femoris |

**Medial (Adductor) Compartment of the Thigh.** Fasciae divide the thigh into three compartments: the *anterior (extensor) compartment, posterior (flexor) compartment,* and *medial (adductor) compartment.* Muscles of the anterior and posterior compartments function mainly as extensors and flexors of the knee, respectively, and are treated in table 10.14. The five muscles of the medial compartment act primarily as adductors of the thigh (see fig. 10.32), but some of them cross both the hip and knee joints and have additional actions as follows.

| | | | |
| --- | --- | --- | --- |
| **Adductor Brevis** | Adducts thigh | **O:** Body and inferior ramus of pubis<br>**I:** Linea aspera and spiral line of femur | Obturator nerve |
| **Adductor Longus** | Adducts and medially rotates thigh; flexes thigh at hip | **O:** Body and inferior ramus of pubis<br>**I:** Linea aspera of femur | Obturator nerve |

---

[70]*fasc* = band; *lat* = broad
[71]*glut* = buttock; *maxim* = largest
[72]*gemellus* = twin

[73]*obtur* = to close, stop up
[74]*piri* = pear; *form* = shaped
[75]*quadrat* = four-sided; *femoris* = of the thigh or femur

| TABLE 10.13 | Muscles Acting on the Hip and Femur (continued) | | |
|---|---|---|---|
| Name | Action | O: Origin<br>I: Insertion | Innervation |
| **Adductor Magnus** | Adducts and medially rotates thigh; extends thigh at hip | **O:** Inferior ramus of pubis; ramus and tuberosity of ischium<br>**I:** Linea aspera, gluteal tuberosity, and medial supracondylar line of femur | Obturator nerve; tibial nerve |
| **Gracilis**[76]<br>(GRASS-ih-lis) | Flexes and medially rotates tibia at knee | **O:** Body and inferior ramus of pubis; ramus of ischium<br>**I:** Medial surface of tibia just below condyle | Obturator nerve |
| **Pectineus**[77]<br>(pec-TIN-ee-us) | Flexes and adducts thigh | **O:** Superior ramus of pubis<br>**I:** Spiral line of femur | Femoral nerve |

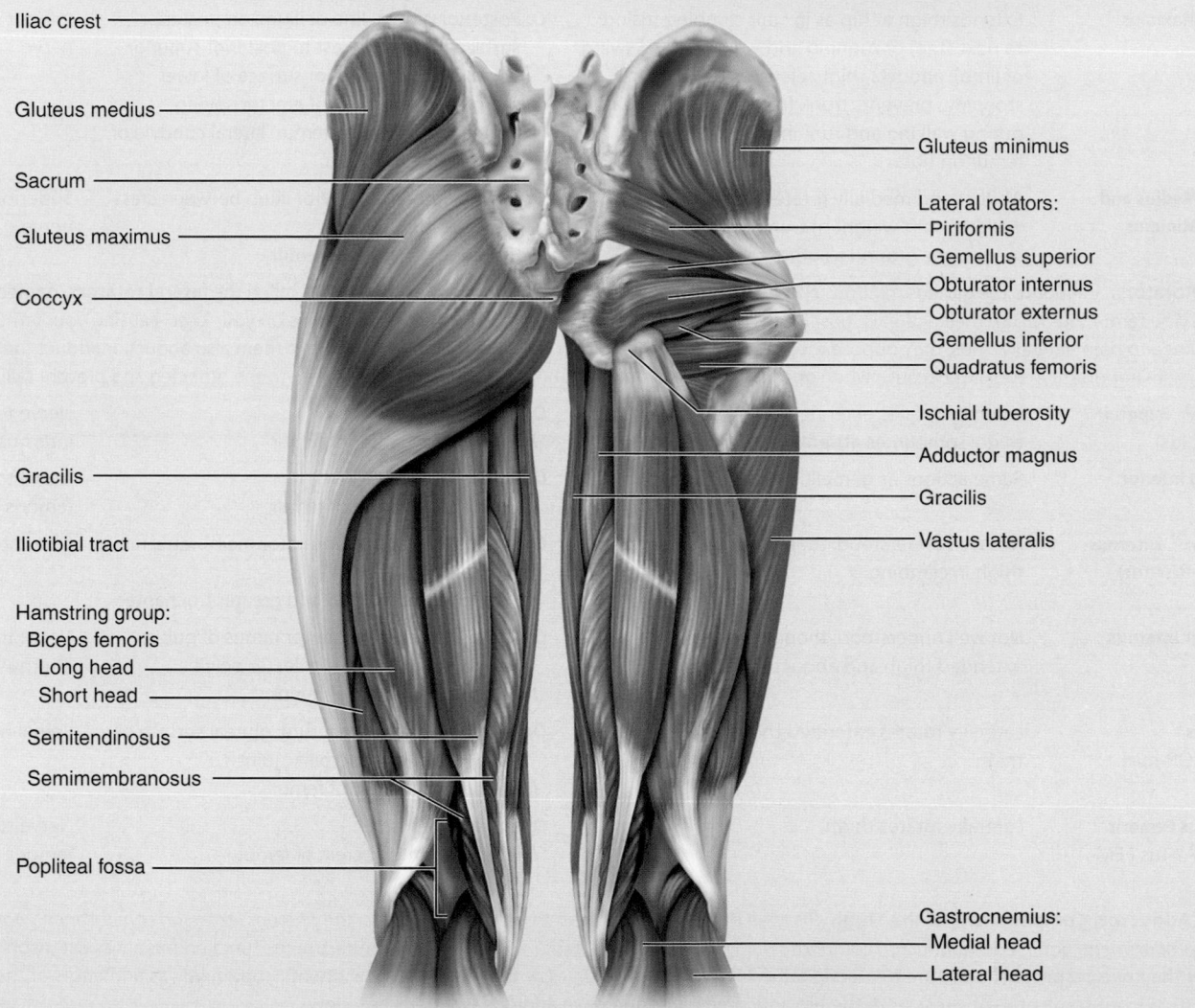

Iliac crest

Gluteus medius

Sacrum

Gluteus maximus

Coccyx

Gracilis

Iliotibial tract

Hamstring group:
  Biceps femoris
    Long head
    Short head
  Semitendinosus
  Semimembranosus

Popliteal fossa

Gluteus minimus

Lateral rotators:
  Piriformis
  Gemellus superior
  Obturator internus
  Obturator externus
  Gemellus inferior
  Quadratus femoris

Ischial tuberosity

Adductor magnus

Gracilis

Vastus lateralis

Gastrocnemius:
  Medial head
  Lateral head

**FIGURE 10.33  Posterior Gluteal and Thigh Muscles.**  The left side shows the superficial muscles. On the right, the gluteal medius and maximus are removed to show the deeper gluteus minimus, lateral rotator group, and hamstring origins.  AP|R

❓ *Describe two everyday movements of the body that employ the power of the gluteus maximus.*

[76]*gracil* = slender
[77]*pectin* = comb

| **TABLE 10.14** | Muscles Acting on the Knee and Leg |
|---|---|

The following muscles form most of the mass of the thigh and produce their most obvious actions on the knee joint. Some of them, however, cross both the hip and knee joints and produce actions at both, moving the femur, tibia, and fibula.

**Anterior (Extensor) Compartment of the Thigh.** The anterior compartment of the thigh contains the large *quadriceps femoris* muscle, the prime mover of knee extension and the most powerful muscle of the body (figs. 10.34 and 10.35). As the name implies, it has four heads: the *rectus femoris, vastus lateralis, vastus medialis,* and *vastus intermedius.* All four converge on a single **quadriceps (patellar) tendon,** which extends to the patella, then continues as the **patellar ligament** and inserts on the tibial tuberosity. (Remember that a tendon usually extends from muscle to bone, and a ligament from bone to bone.) The patellar ligament is struck with a rubber reflex hammer to test the knee-jerk reflex. The quadriceps extends the knee when you stand up, take a step, or kick a ball. One head, the rectus femoris, contributes to running by acting with the iliopsoas to flex the hip in each airborne phase of the leg's cycle of motion. The rectus femoris also flexes the hip in such actions as high kicks, stair climbing, or simply drawing the leg forward during a stride.

Crossing the quadriceps from the lateral side of the hip to the medial side of the knee is the narrow, straplike *sartorius,* the longest muscle of the body. It flexes the hip and knee joints and laterally rotates the thigh, as in crossing the legs. It is colloquially called the "tailor's muscle" after the cross-legged posture of a tailor supporting his work on the raised knee.

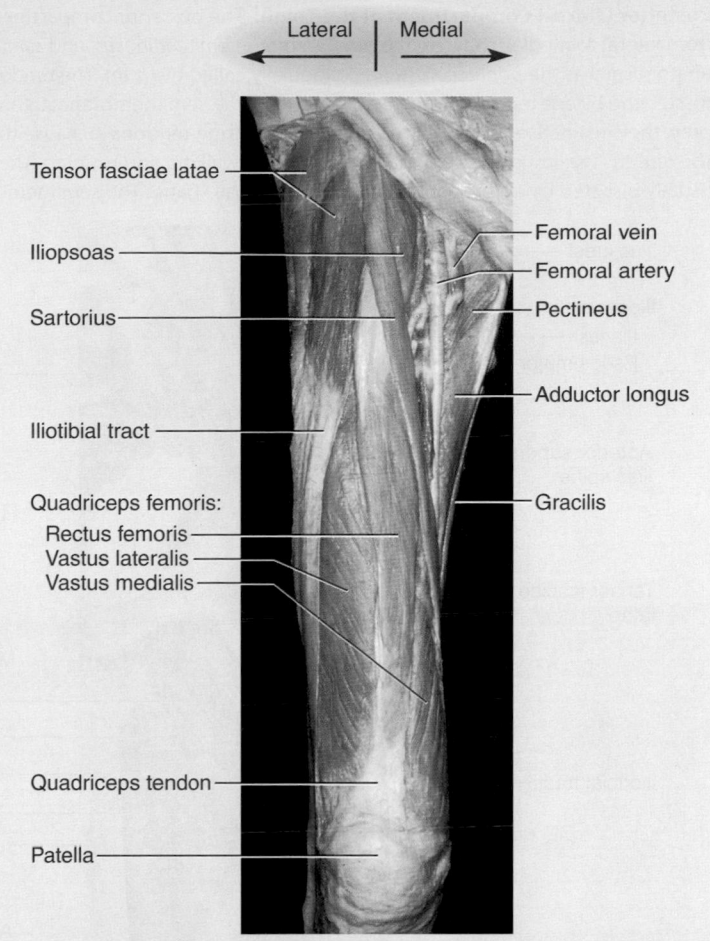

**FIGURE 10.34 Superficial Anterior Thigh Muscles of the Cadaver.** Right thigh.

| Name | Action | O: Origin<br>I: Insertion | Innervation |
|---|---|---|---|
| **Quadriceps Femoris**<br>(QUAD-rih-seps FEM-oh-ris) | Extends the knee, in addition to the actions of individual heads noted below | **O:** Varies; see individual heads below<br>**I:** Patella; tibial tuberosity; lateral and medial condyles of tibia | Femoral nerve |
| **Rectus Femoris** | Extends knee; flexes thigh at hip; flexes trunk on hip if thigh is fixed | **O:** Ilium at anterior inferior spine and superior margin of acetabulum; capsule of hip joint<br>**I:** See quadriceps femoris above | Femoral nerve |
| **Vastus**[78] **Lateralis** | Extends knee; retains patella in groove on femur during knee movements | **O:** Femur at greater trochanter and intertrochanteric line, gluteal tuberosity, and linea aspera<br>**I:** See quadriceps femoris above | Femoral nerve |
| **Vastus Medialis** | Same as vastus lateralis | **O:** Femur at intertrochanteric line, spiral line, linea aspera, and medial supracondylar line<br>**I:** See quadriceps femoris above | Femoral nerve |
| **Vastus Intermedius** | Extends knee | **O:** Anterior and lateral surfaces of femoral shaft<br>**I:** See quadriceps femoris above | Femoral nerve |
| **Sartorius**[79] | Aids in knee and hip flexion, as in sitting or climbing; abducts and laterally rotates thigh | **O:** On and near anterior superior spine of ilium<br>**I:** Medial surface of proximal end of tibia | Femoral nerve |

[78] *vastus* = large, extensive

[79] *sartor* = tailor

| **TABLE 10.14** | **Muscles Acting on the Knee and Leg (continued)** |

**Posterior (Flexor) Compartment of the Thigh.** The posterior compartment contains three muscles colloquially known as the **hamstring muscles;** from lateral to medial, they are the *biceps femoris, semitendinosus,* and *semimembranosus* (see fig. 10.33). The pit at the back of the knee, known anatomically as the *popliteal fossa,* is colloquially called the *ham.* The tendons of these muscles can be felt as prominent cords on both sides of the fossa—the biceps tendon on the lateral side and the semimembranosus and semitendinosus tendons on the medial side. When wolves attack large prey, they instinctively attempt to sever the hamstring tendons, because this renders the prey helpless. The hamstrings flex the knee, and aided by the gluteus maximus, they extend the hip during walking and running. The semitendinosus is named for its unusually long tendon. This muscle also is usually bisected by a transverse or oblique tendinous band. The semimembranosus is named for the flat shape of its superior attachment.

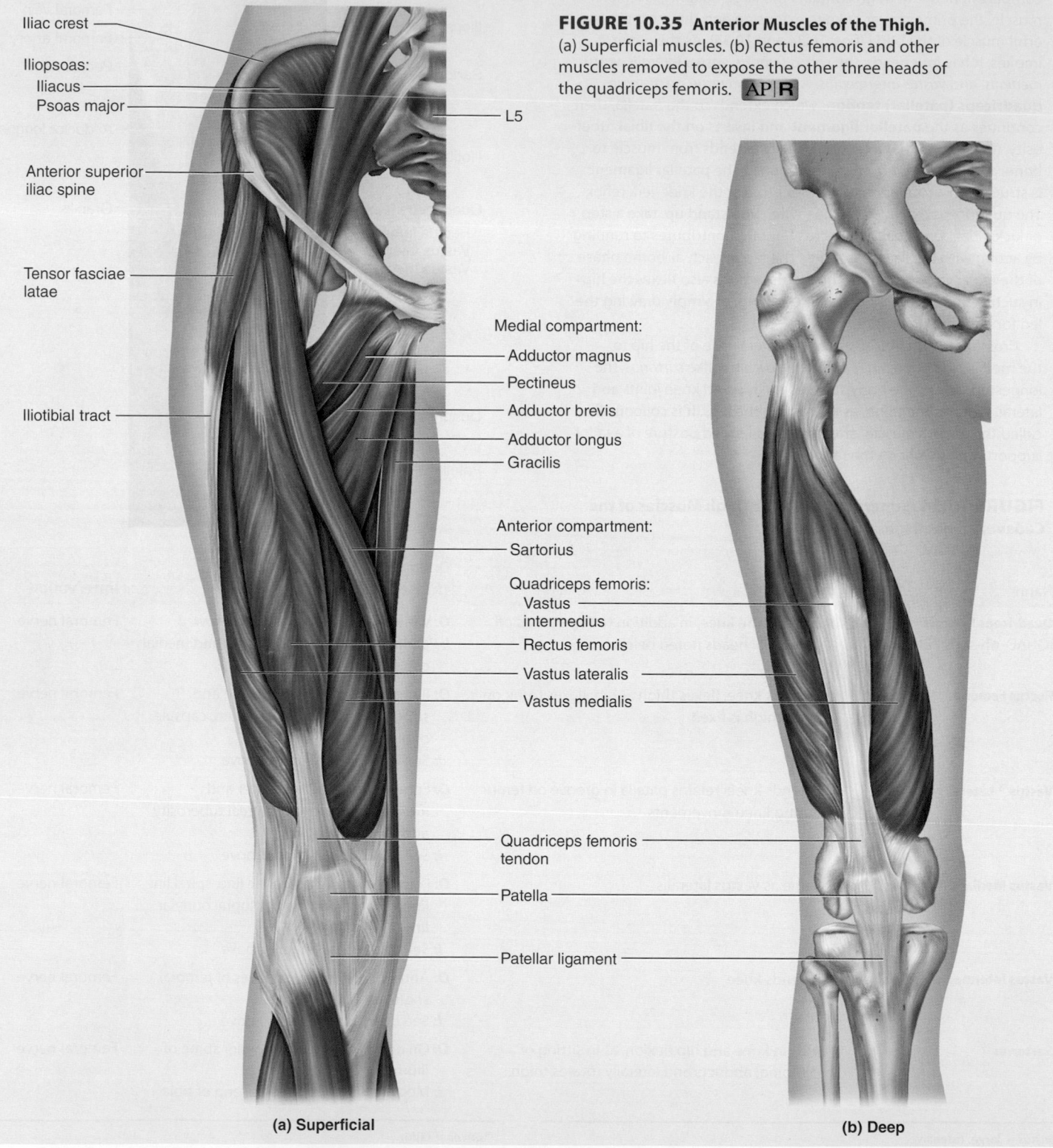

**FIGURE 10.35 Anterior Muscles of the Thigh.**
(a) Superficial muscles. (b) Rectus femoris and other muscles removed to expose the other three heads of the quadriceps femoris. AP|R

Iliac crest

Iliopsoas:
 Iliacus
 Psoas major

L5

Anterior superior iliac spine

Tensor fasciae latae

Iliotibial tract

Medial compartment:
 Adductor magnus
 Pectineus
 Adductor brevis
 Adductor longus
 Gracilis

Anterior compartment:
 Sartorius

Quadriceps femoris:
 Vastus intermedius
 Rectus femoris
 Vastus lateralis
 Vastus medialis

Quadriceps femoris tendon

Patella

Patellar ligament

(a) Superficial

(b) Deep

| TABLE 10.14 | Muscles Acting on the Knee and Leg (continued) | | |
|---|---|---|---|
| **Name** | **Action** | **O: Origin**<br>**I: Insertion** | **Innervation** |
| **Biceps Femoris** | Flexes knee; extends hip; elevates trunk from stooping posture; laterally rotates tibia on femur when knee is flexed; laterally rotates femur when hip is extended; counteracts forward bending at hips | **O:** Long head–ischial tuberosity<br>    Short head–linea aspera and lateral supracondylar line of femur<br>**I:** Head of fibula | Tibial nerve; common fibular nerve |
| **Semitendinosus**[80]<br>(SEM-ee-TEN-din-OH-sus) | Flexes knee; medially rotates tibia on femur when knee is flexed; medially rotates femur when hip is extended; counteracts forward bending at hips | **O:** Ischial tuberosity<br>**I:** Medial surface of upper tibia | Tibial nerve |
| **Semimembranosus**[81]<br>(SEM-ee-MEM-bran-OH-sus) | Same as semitendinosus | **O:** Ischial tuberosity<br>**I:** Medial condyle and nearby margin of tibia; intercondylar line and lateral condyle of femur; ligament of popliteal region | Tibial nerve |

**Posterior Compartment of the Leg.** Most muscles in the posterior compartment of the leg act on the ankle and foot and are reviewed in table 10.15, but the *popliteus* acts on the knee (see fig. 10.39a, d).

| | | | |
|---|---|---|---|
| **Popliteus**[82]<br>(pop-LIT-ee-us) | Rotates tibia medially on femur if femur is fixed (as in sitting down), or rotates femur laterally on tibia if tibia is fixed (as in standing up); unlocks knee to allow flexion; may prevent forward dislocation of femur during crouching | **O:** Lateral condyle of femur; lateral meniscus and joint capsule<br>**I:** Posterior surface of upper tibia | Tibial nerve |

---

[80] *semi* = half; *tendinosus* = tendinous
[81] *semi* = half; *membranosus* = membranous

[82] *poplit* = ham (pit) of the knee

| **TABLE 10.15** | **Muscles Acting on the Foot** |
|---|---|

The fleshy mass of the leg is formed by a group of crural muscles, which act on the foot (fig. 10.36). These muscles are tightly bound by fasciae that compress them and aid in the return of blood from the legs. The fasciae separate the crural muscles into anterior, lateral, and posterior compartments (see fig. 10.40b).

**Anterior (Extensor) Compartment of the Leg.**  Muscles of the anterior compartment dorsiflex the ankle and prevent the toes from scuffing the ground during walking. From lateral to medial, these muscles are the *fibularis tertius, extensor digitorum longus* (extensor of toes II–V), *extensor hallucis longus* (extensor of the great toe), and *tibialis anterior*. Their tendons are held tightly against the ankle and kept from bowing by two extensor retinacula similar to the one at the wrist (fig. 10.37a).

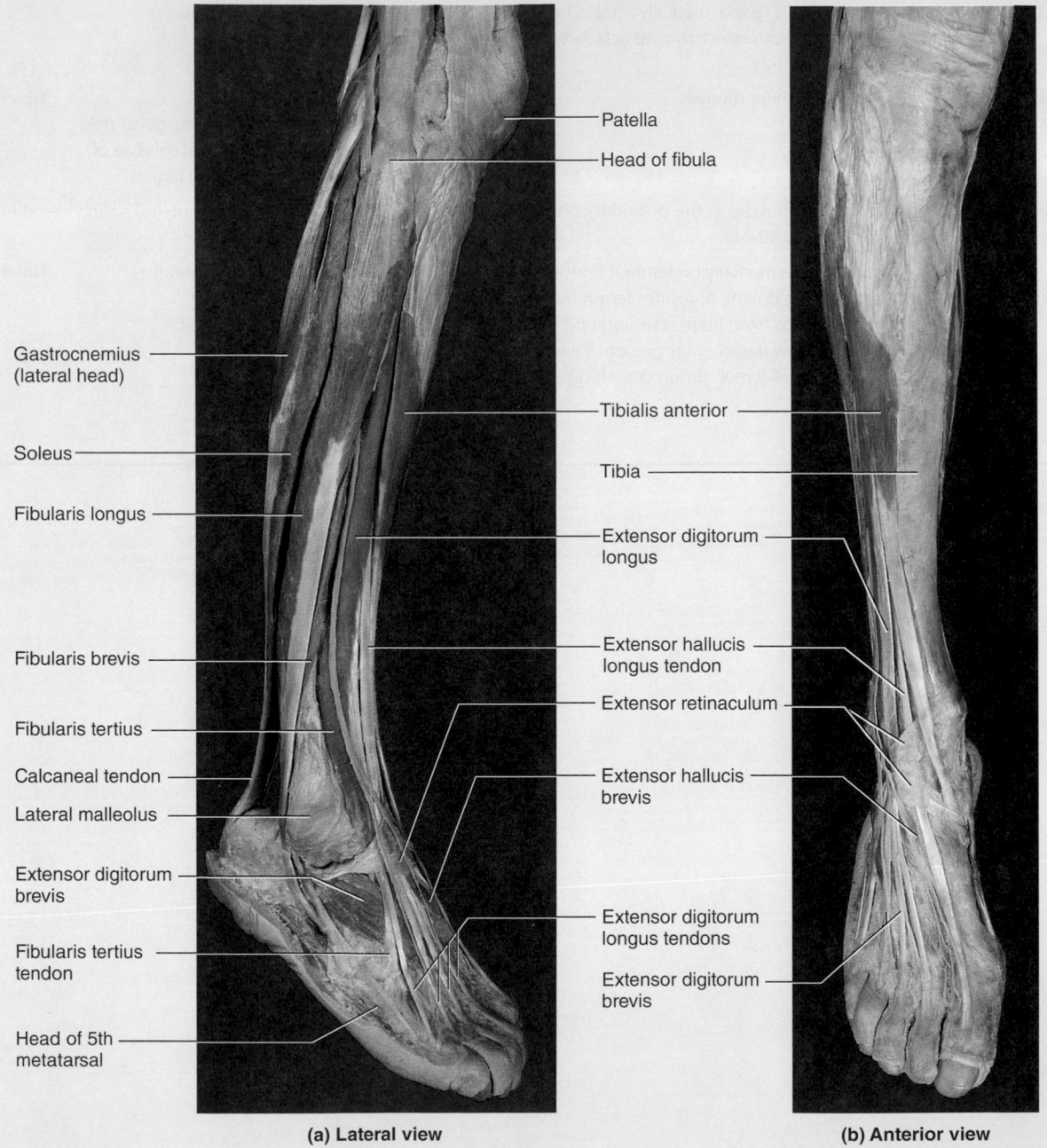

(a) Lateral view          (b) Anterior view

**FIGURE 10.36  Superficial Crural Muscles.**  Right leg of the cadaver.

| TABLE 10.15 | Muscles Acting on the Foot (continued) |
|---|---|

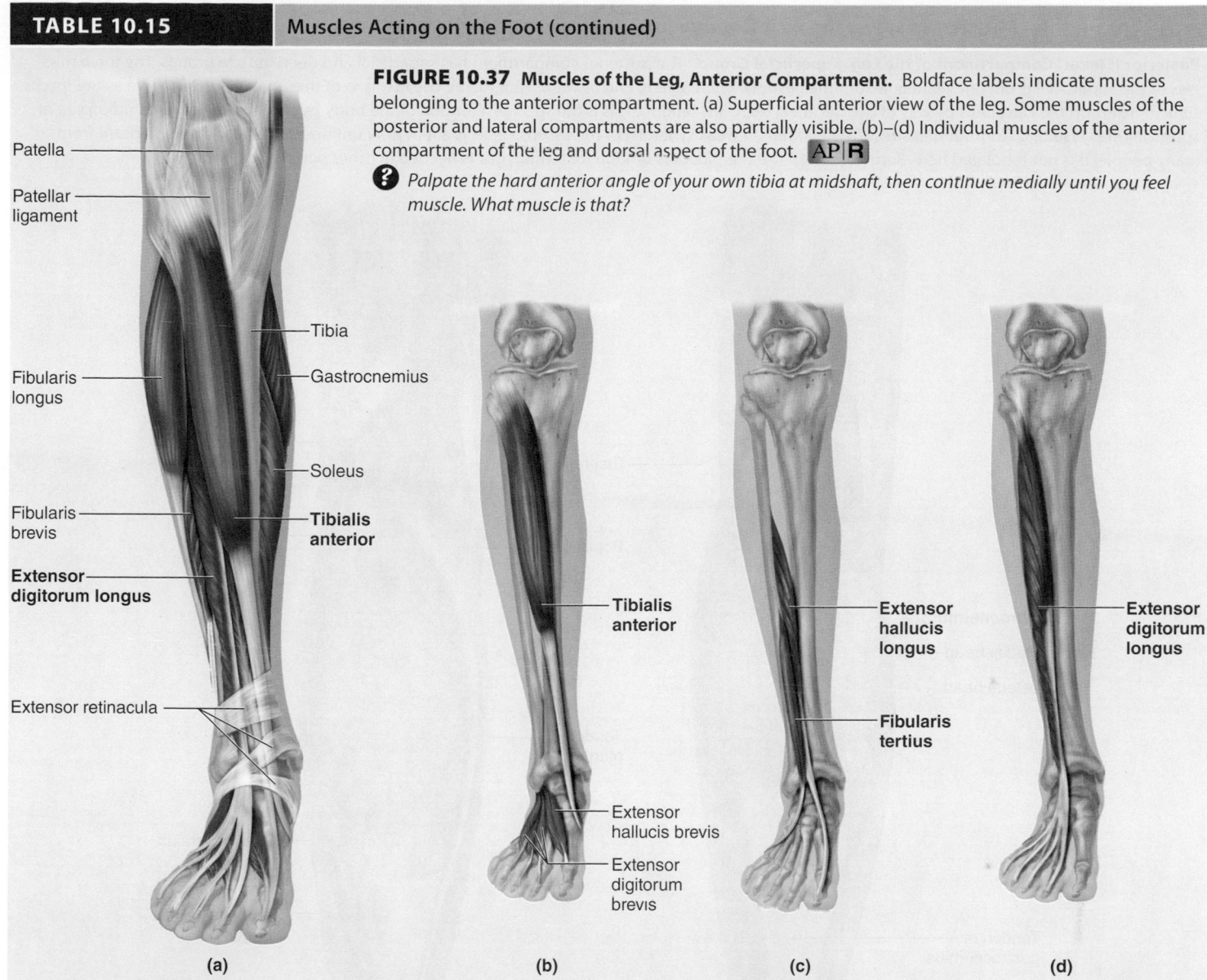

**FIGURE 10.37 Muscles of the Leg, Anterior Compartment.** Boldface labels indicate muscles belonging to the anterior compartment. (a) Superficial anterior view of the leg. Some muscles of the posterior and lateral compartments are also partially visible. (b)–(d) Individual muscles of the anterior compartment of the leg and dorsal aspect of the foot. AP|R

❓ *Palpate the hard anterior angle of your own tibia at midshaft, then continue medially until you feel muscle. What muscle is that?*

Labels in figure (a): Patella; Patellar ligament; Fibularis longus; Fibularis brevis; **Extensor digitorum longus**; Extensor retinacula; Tibia; Gastrocnemius; Soleus; **Tibialis anterior**

Labels in figure (b): **Tibialis anterior**; Extensor hallucis brevis; Extensor digitorum brevis

Labels in figure (c): **Extensor hallucis longus**; **Fibularis tertius**

Labels in figure (d): **Extensor digitorum longus**

(a)  (b)  (c)  (d)

| Name | Action | O: Origin<br>I: Insertion | Innervation |
|---|---|---|---|
| **Fibularis (Peroneus**[83]**) Tertius**[84]<br>(FIB-you-LERR-iss TUR-she-us) | Dorsiflexes and everts foot during walking; helps toes clear the ground during forward swing of leg | *O:* Medial surface of lower one-third of fibula; interosseous membrane<br>*I:* Metatarsal V | Deep fibular (peroneal) nerve |
| **Extensor Digitorum Longus**<br>(DIDJ-ih-TOE-rum) | Extends toes; dorsiflexes foot; tautens plantar aponeurosis | *O:* Lateral condyle of tibia; shaft of fibula; interosseous membrane<br>*I:* Middle and distal phalanges II–V | Deep fibular (peroneal) nerve |
| **Extensor Hallucis Longus**<br>(ha-LOO-sis) | Extends great toe; dorsiflexes foot | *O:* Anterior surface of middle of fibula, interosseous membrane<br>*I:* Distal phalanx I | Deep fibular (peroneal) nerve |
| **Tibialis**[85] **Anterior**<br>(TIB-ee-AY-lis) | Dorsiflexes and inverts foot; resists backward tipping of body (as when standing on a moving boat deck); helps support medial longitudinal arch of foot | *O:* Lateral condyle and lateral margin of proximal half of tibia; interosseous membrane<br>*I:* Medial cuneiform, metatarsal I | Deep fibular (peroneal) nerve |

[83] *perone* = pinlike (fibula)
[84] *fibularis* = of the fibula; *tert* = third

[85] *tibialis* = of the tibia

| **TABLE 10.15** | **Muscles Acting on the Foot (continued)** |
| --- | --- |

**Posterior (Flexor) Compartment of the Leg, Superficial Group.** The posterior compartment has superficial and deep muscle groups. The three muscles of the superficial group are plantar flexors: the *gastrocnemius, soleus,* and *plantaris* (fig. 10.38). The first two of these, collectively known as the *triceps surae,*[86] insert on the calcaneus by way of the calcaneal (Achilles) tendon. This is the strongest tendon of the body but is nevertheless a common site of sports injuries resulting from sudden stress. The plantaris, a weak synergist of the triceps surae, is a relatively unimportant muscle and is absent from many people; it is not tabulated here. Surgeons often use the plantaris tendon for tendon grafts needed in other parts of the body.

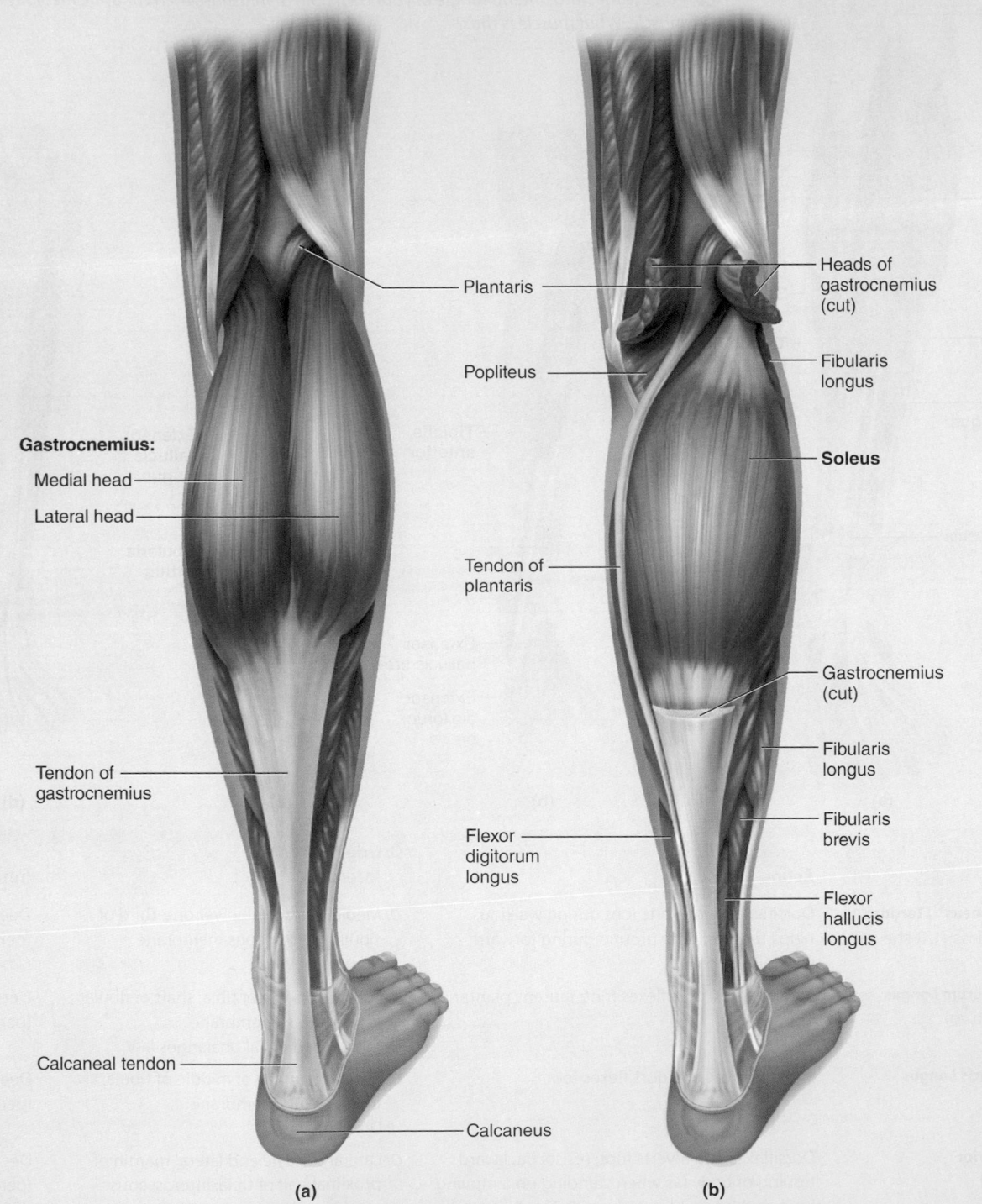

(a)                                              (b)

**FIGURE 10.38 Superficial Muscles of the Leg, Posterior Compartment.** (a) The gastrocnemius. (b) The soleus, deep to the gastrocnemius and sharing the calcaneal tendon with it. **AP|R**

[86]*sura* = calf of leg

| TABLE 10.15 | Muscles Acting on the Foot (continued) | | |
|---|---|---|---|
| **Name** | **Action** | **O: Origin**<br>**I: Insertion** | **Innervation** |
| **Gastrocnemius**[87]<br>(GAS-trock-NEE-me-us) | Plantar flexes foot, flexes knee; active in walking, running, and jumping | **O:** Condyles, popliteal surface, and lateral supracondylar line of femur; capsule of knee joint<br>**I:** Calcaneus | Tibial nerve |
| **Soleus**[88]<br>(SO-lee-us) | Plantar flexes foot; steadies leg on ankle during standing | **O:** Posterior surface of head and proximal one-fourth of fibula; middle one-third of tibia; interosseous membrane<br>**I:** Calcaneus | Tibial nerve |

**Posterior (Flexor) Compartment of the Leg, Deep Group.** There are four muscles in the deep group (fig. 10.39). The *flexor digitorum longus, flexor hallucis longus,* and *tibialis posterior* are plantar flexors. The fourth muscle, the *popliteus,* is described in table 10.14 because it acts on the knee rather than on the foot.

| | | | |
|---|---|---|---|
| **Flexor Digitorum Longus** | Flexes phalanges of digits II–V as foot is raised from ground; stabilizes metatarsal heads and keeps distal pads of toes in contact with ground in toe-off and tiptoe movements | **O:** Posterior surface of tibial shaft<br>**I:** Distal phalanges II–V | Tibial nerve |
| **Flexor Hallucis Longus** | Same actions as flexor digitorum longus, but for great toe (digit I) | **O:** Distal two-thirds of fibula and interosseous membrane<br>**I:** Distal phalanx I | Tibial nerve |
| **Tibialis Posterior** | Inverts foot; may assist in strong plantar flexion or control pronation of foot during walking | **O:** Posterior surface of proximal half of tibia, fibula, and interosseous membrane<br>**I:** Navicular, medial cuneiform, metatarsals II–IV | Tibial nerve |

**Lateral (Fibular) Compartment of the Leg.** The lateral compartment includes the *fibularis brevis* and *fibularis longus* (see figs. 10.36a, 10.37a, 10.40b). They plantar flex and evert the foot. Plantar flexion is important not only in standing on tiptoes but in providing lift and forward thrust each time you take a step.

| | | | |
|---|---|---|---|
| **Fibularis (Peroneus) Brevis** | Maintains concavity of sole during toe-off and tiptoeing; may evert foot and limit inversion and help steady leg on foot | **O:** Lateral surface of distal two-thirds of fibula<br>**I:** Base of metatarsal V | Superficial fibular (peroneal) nerve |
| **Fibularis (Peroneus) Longus** | Maintains concavity of sole during toe-off and tiptoeing; everts and plantar flexes foot | **O:** Head and lateral surface of proximal two-thirds of fibula<br>**I:** Medial cuneiform, metatarsal I | Superficial fibular (peroneal) nerve |

[87]*gastro* = belly; *cnem* = leg

[88]Named for its resemblance to a flatfish (sole)

▶▶▶**APPLY WHAT YOU KNOW**

*Suppose you tilt your foot upward to trim or paint your toenails. Identify as many muscles as you can that may contribute to this action.*

**TABLE 10.15** Muscles Acting on the Foot (continued)

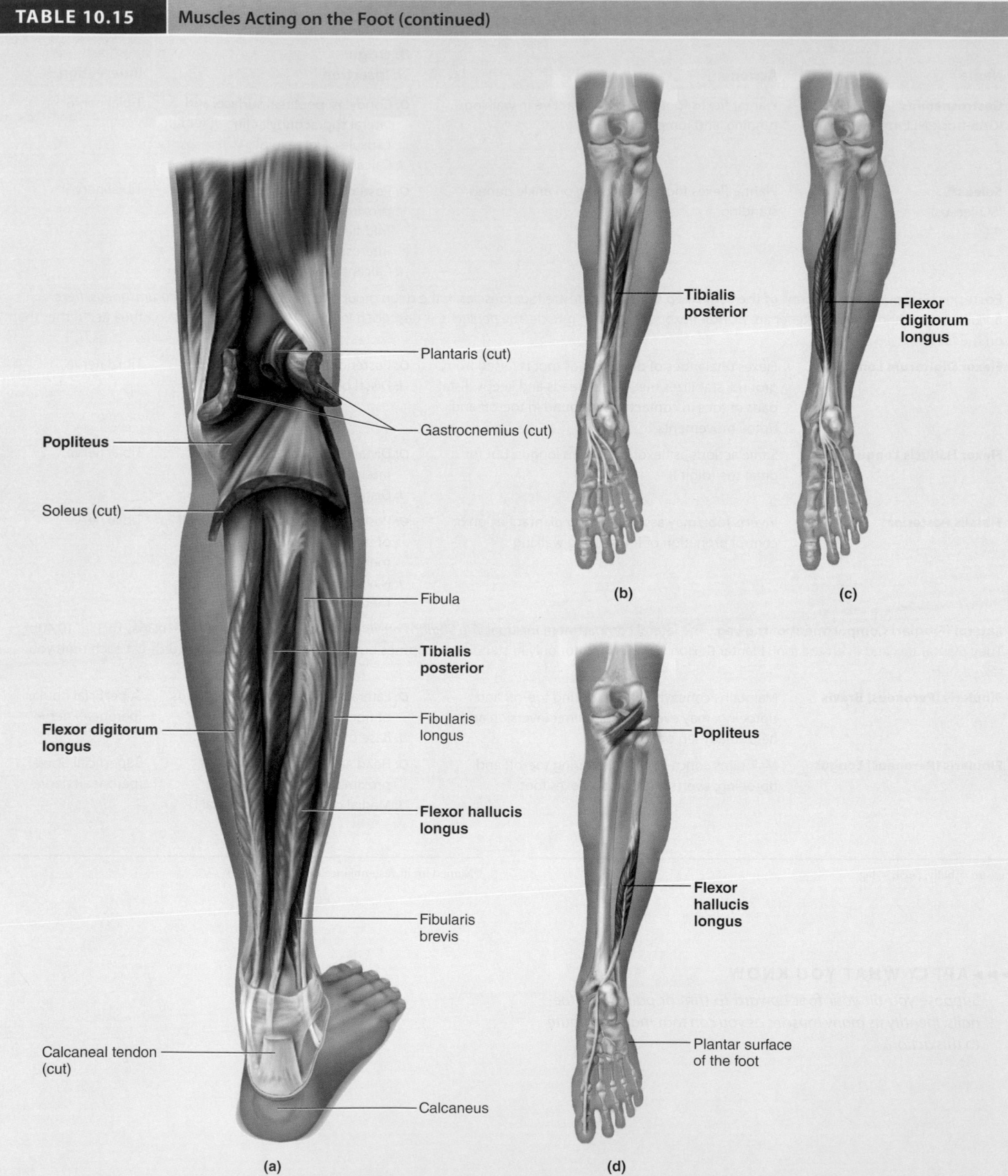

Plantaris (cut)

Gastrocnemius (cut)

**Popliteus**

Soleus (cut)

Fibula

**Tibialis posterior**

Fibularis longus

**Flexor digitorum longus**

**Flexor hallucis longus**

Fibularis brevis

Calcaneal tendon (cut)

Calcaneus

(a)

**Tibialis posterior**

(b)

**Flexor digitorum longus**

(c)

**Popliteus**

**Flexor hallucis longus**

Plantar surface of the foot

(d)

**FIGURE 10.39 Deep Muscles of the Leg, Posterior and Lateral Compartments.** (a) Muscles deep to the soleus. (b)–(d) Exposure of some individual deep muscles with the foot plantar flexed (sole facing viewer). **AP|R**

| TABLE 10.15 | Muscles Acting on the Foot (continued) |
| --- | --- |

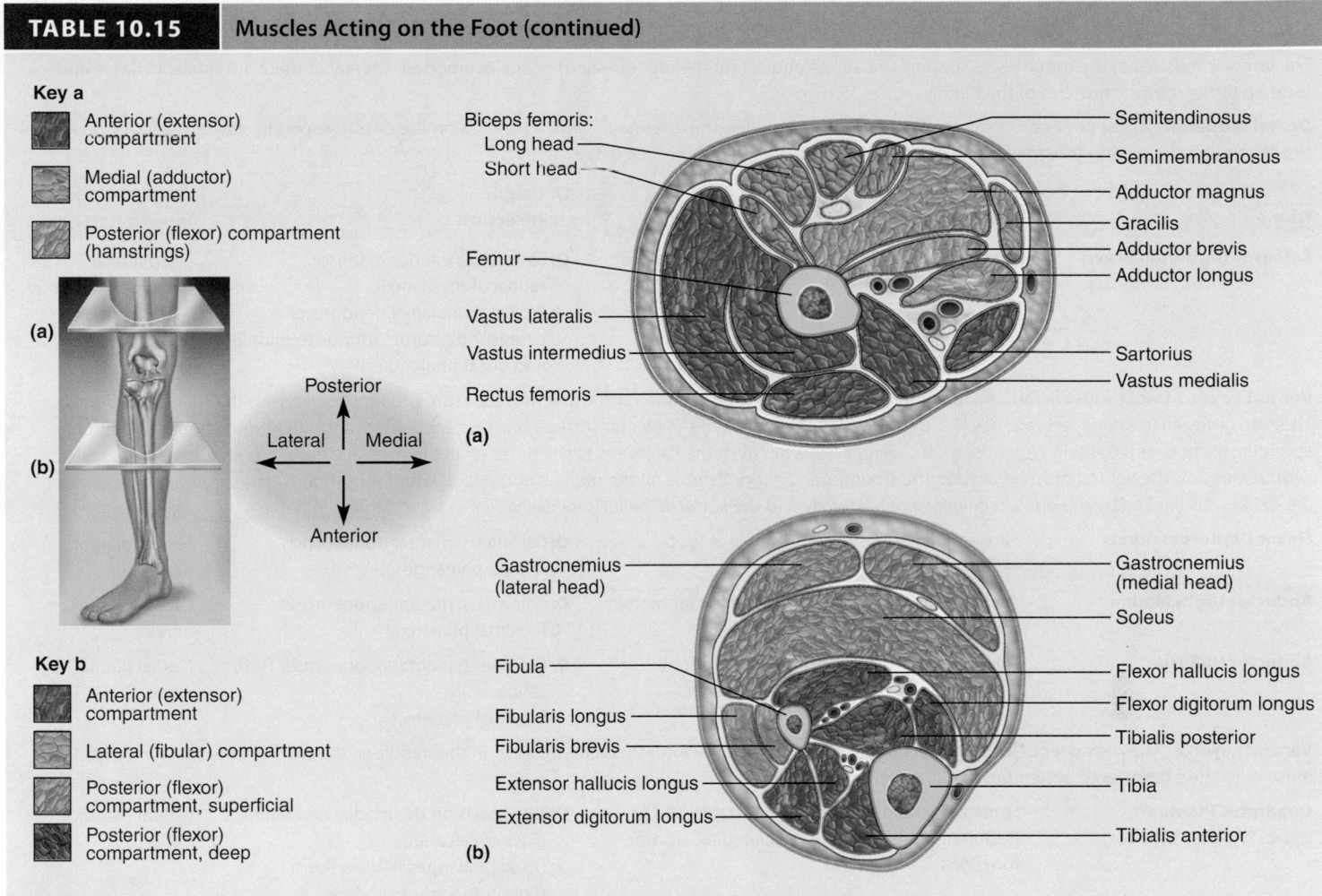

**Key a**

- Anterior (extensor) compartment
- Medial (adductor) compartment
- Posterior (flexor) compartment (hamstrings)

(a)

(b)

Posterior

Lateral | Medial

Anterior

**Key b**

- Anterior (extensor) compartment
- Lateral (fibular) compartment
- Posterior (flexor) compartment, superficial
- Posterior (flexor) compartment, deep

Biceps femoris:
Long head
Short head

Femur

Vastus lateralis
Vastus intermedius
Rectus femoris

(a)

Semitendinosus
Semimembranosus
Adductor magnus
Gracilis
Adductor brevis
Adductor longus

Sartorius
Vastus medialis

Gastrocnemius (lateral head)

Fibula
Fibularis longus
Fibularis brevis
Extensor hallucis longus
Extensor digitorum longus

(b)

Gastrocnemius (medial head)
Soleus

Flexor hallucis longus
Flexor digitorum longus
Tibialis posterior

Tibia

Tibialis anterior

**FIGURE 10.40 Serial Cross Sections Through the Lower Limb.** Each section is taken at the correspondingly lettered level in the figure at the left.

### ▶▶▶ APPLY WHAT YOU KNOW

*Suppose you are playing Frisbee and your Frisbee lands on the roof of your house. You have to climb a ladder to get it. Identify as many muscles as you can that would aid you in ascending from one rung of the ladder to the next.*

| TABLE 10.16 | Intrinsic Muscles of the Foot |
|---|---|

The intrinsic muscles of the foot help to support the arches and act on the toes in ways that aid locomotion. Several of them are similar in name and location to the intrinsic muscles of the hand.

**Dorsal (Superior) Aspect of Foot.** Only one of the intrinsic muscles, the *extensor digitorum brevis,* is on the dorsal (superior) side of the foot. The medial slip of this muscle, serving the great toe, is sometimes called the *extensor hallucis brevis.*

| Name | Action | O: Origin<br>I: Insertion | Innervation |
|---|---|---|---|
| **Extensor Digitorum Brevis** | Extends proximal phalanx I and all phalanges of digits II–IV | O: Calcaneus; inferior extensor retinaculum of ankle<br>I: Proximal phalanx I, tendons of extensor digitorum longus to middle and distal phalanges II–IV | Deep fibular (peroneal) nerve |

**Ventral Layer 1 (most superficial).** All remaining intrinsic muscles are on the ventral (inferior) aspect of the foot or between the metatarsal bones. They are grouped in four layers (fig. 10.41). Dissecting into the foot from the plantar surface, one first encounters a tough fibrous sheet, the plantar aponeurosis, between the skin and muscles. It diverges like a fan from the calcaneus to the bases of all the toes, and serves as an origin for several ventral muscles. The ventral muscles include the stout *flexor digitorum brevis* on the midline of the foot, with four tendons that supply all digits except the hallux. It is flanked by the *abductor digiti minimi* laterally and the *abductor hallucis* medially.

| | | | |
|---|---|---|---|
| **Flexor Digitorum Brevis** | Flexes digits II–IV; supports arches of foot | O: Calcaneus; plantar aponeurosis<br>I: Middle phalanges II–V | Medial plantar nerve |
| **Abductor Digiti Minimi**[89] | Abducts and flexes little toe; supports arches of foot | O: Calcaneus; plantar aponeurosis<br>I: Proximal phalanx V | Lateral plantar nerve |
| **Abductor Hallucis** | Abducts great toe; supports arches of foot | O: Calcaneus; plantar aponeurosis; flexor retinaculum<br>I: Proximal phalanx I | Medial plantar nerve |

**Ventral Layer 2.** The next deeper layer consists of the thick *quadratus plantae (flexor accessorius)* in the middle of the foot and the four *lumbrical* muscles located between the metatarsals.

| | | | |
|---|---|---|---|
| **Quadratus Plantae**[90]<br>(quad-RAY-tus PLAN-tee) | Same as flexor digitorum longus (table 10.15); flexion of digits II–V and associated locomotor functions | O: Two heads on the medial and lateral sides of calcaneus<br>I: Distal phalanges II–V via flexor digitorum longus tendons | Lateral plantar nerve |
| **Four Lumbrical Muscles**<br>(LUM-brih-cul) | Flex toes II–V | O: Tendon of flexor digitorum longus<br>I: Proximal phalanges II–V | Lateral and medial plantar nerves |

**Ventral Layer 3.** The muscles of this layer serve only the great and little toes. They are the *flexor digiti minimi brevis, flexor hallucis brevis,* and *adductor hallucis.* The adductor hallucis has an *oblique head* that extends diagonally from the midplantar region to the base of the great toe, and a *transverse head* that passes across the bases of digits II–IV and meets the long head at the base of the great toe.

| | | | |
|---|---|---|---|
| **Flexor Digiti Minimi Brevis** | Flexes little toe | O: Metatarsal V, sheath of fibularis longus<br>I: Proximal phalanx V | Lateral plantar nerve |
| **Flexor Hallucis Brevis** | Flexes great toe | O: Cuboid; lateral cuneiform; tibialis posterior tendon<br>I: Proximal phalanx I | Medial plantar nerve |
| **Adductor Hallucis** | Adducts great toe | O: Metatarsals II–IV; fibularis longus tendon; ligaments at bases of digits III–V<br>I: Proximal phalanx I | Lateral plantar nerve |

**Ventral Layer 4 (deepest).** This layer consists only of the small interosseous muscles located between the metatarsal bones—four dorsal and three plantar. Each dorsal interosseous muscle is bipennate and originates on two adjacent metatarsals. The plantar interosseous muscles are unipennate and originate on only one metatarsal each.

| | | | |
|---|---|---|---|
| **Four Dorsal Interosseous Muscles** | Abduct toes II–IV | O: Each with two heads arising from facing surfaces of two adjacent metatarsals<br>I: Proximal phalanges II–IV | Lateral plantar nerve |
| **Three Plantar Interosseous Muscles** | Adduct toes III–V | O: Medial aspect of metatarsals III–V<br>I: Proximal phalanges III–V | Lateral plantar nerve |

[89]*digit* = toe; *minim* = smallest

[90]*quadrat* = four-sided; *plantae* = of the plantar region

**TABLE 10.16**    Intrinsic Muscles of the Foot (continued)

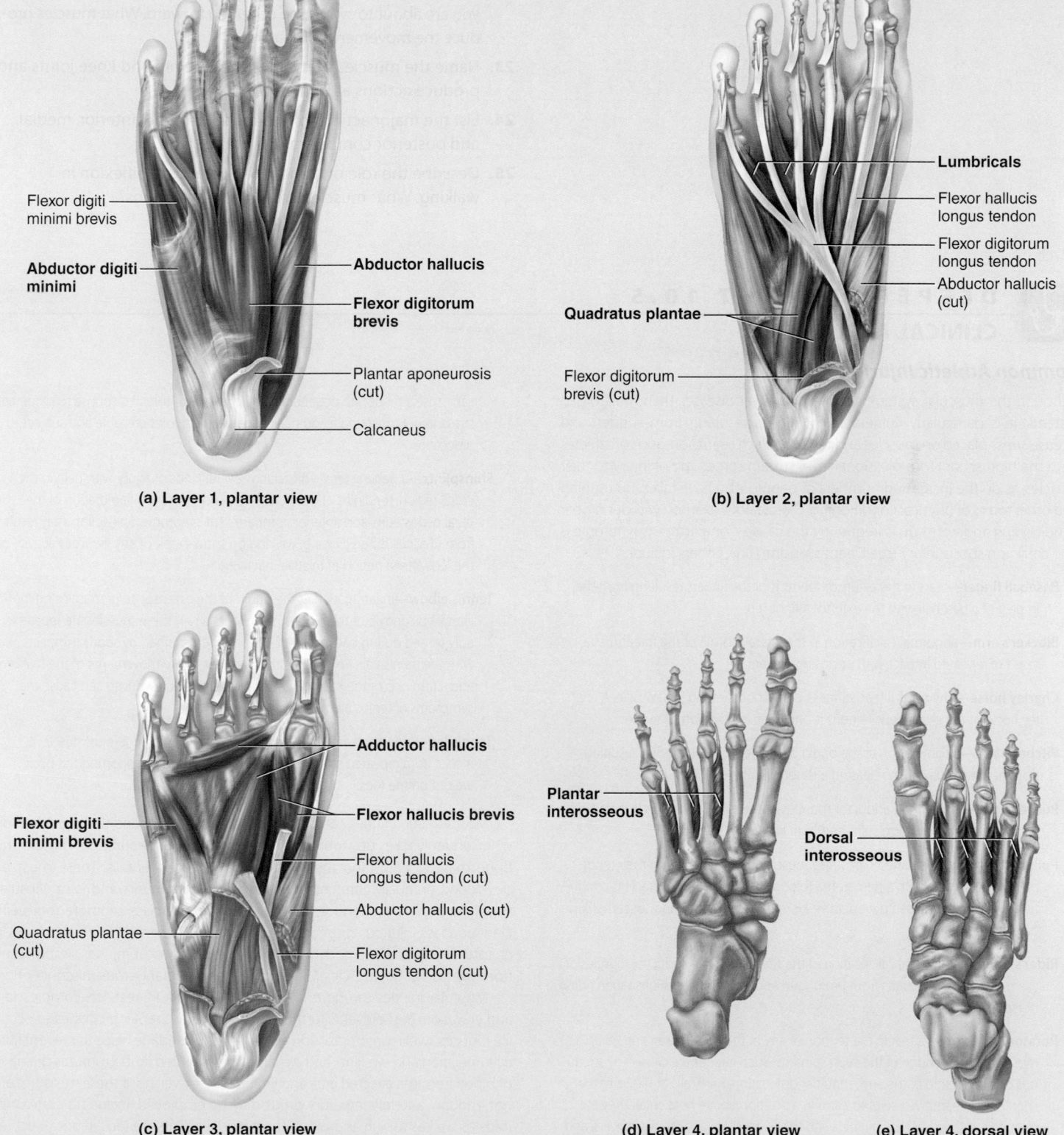

(a) Layer 1, plantar view

Flexor digiti minimi brevis
Abductor digiti minimi
Abductor hallucis
Flexor digitorum brevis
Plantar aponeurosis (cut)
Calcaneus

(b) Layer 2, plantar view

Lumbricals
Flexor hallucis longus tendon
Flexor digitorum longus tendon
Abductor hallucis (cut)
Quadratus plantae
Flexor digitorum brevis (cut)

(c) Layer 3, plantar view

Adductor hallucis
Flexor hallucis brevis
Flexor digiti minimi brevis
Flexor hallucis longus tendon (cut)
Abductor hallucis (cut)
Quadratus plantae (cut)
Flexor digitorum longus tendon (cut)

(d) Layer 4, plantar view

Plantar interosseous

(e) Layer 4, dorsal view

Dorsal interosseous

**FIGURE 10.41  Intrinsic Muscles of the Foot.**  (a)–(d) First through fourth layers, respectively, in ventral (plantar) views. (e) Fourth layer, dorsal view. The muscles belonging to each layer are shown in color and with boldface labels.  **AP|R**

▶▶▶**APPLY WHAT YOU KNOW**

*Not everyone has the same muscles. From the information provided in this chapter, identify at least three muscles that are lacking in some people.*

**BEFORE YOU GO ON**

Answer the following questions to test your understanding of the preceding section:

**22.** In the middle of a stride, you have one foot on the ground and you are about to swing the other leg forward. What muscles produce the movements of that leg?

**23.** Name the muscles that cross both the hip and knee joints and produce actions at both.

**24.** List the major actions of the muscles of the anterior, medial, and posterior compartments of the thigh.

**25.** Describe the role of plantar flexion and dorsiflexion in walking. What muscles produce these actions?

# DEEPER INSIGHT 10.5
## CLINICAL APPLICATION

### Common Athletic Injuries

Although the muscular system is subject to fewer diseases than most organ systems, it is particularly vulnerable to injuries resulting from sudden and intense stress placed on muscles and tendons. Each year, thousands of athletes from the high school to professional level sustain some type of injury to their muscles, as do the increasing numbers of people who have taken up running and other forms of physical conditioning. Overzealous exertion without proper conditioning and warm-up is frequently the cause. Compartment syndrome is one common sports injury (see Deeper Insight 10.1). Others include:

**Baseball finger**—tears in the extensor tendons of the fingers resulting from the impact of a baseball with the extended fingertip.

**Blocker's arm**—abnormal calcification in the lateral margin of the forearm as a result of repeated impact with opposing players.

**Charley horse**—any painful tear, stiffness, and blood clotting in a muscle. A charley horse of the quadriceps femoris is often caused by football tackles.

**Pitcher's arm**—inflammation at the origin of the flexor carpi muscles resulting from hard wrist flexion in releasing a baseball.

**Pulled groin**—strain in the adductor muscles of the thigh; common in gymnasts and dancers who perform splits and high kicks.

**Pulled hamstrings**—strained hamstring muscles or a partial tear in their tendinous origins, often with a hematoma (blood clot) in the fascia lata. This condition is frequently caused by repetitive kicking (as in football and soccer) or long, hard running.

**Rider's bones**—abnormal calcification in the tendons of the adductor muscles of the medial thigh. It results from prolonged abduction of the thighs when riding horses.

**Rotator cuff injury**—a tear in the tendon of any of the SITS (rotator cuff) muscles, most often the tendon of the supraspinatus. Such injuries are caused by strenuous circumduction of the arm, shoulder dislocation, hard falls or blows to the shoulder, or repetitive use of the arm in a position above horizontal. They are common among baseball pitchers and third basemen, bowlers, swimmers, and weight lifters, and in racquet sports. Recurrent inflammation of a SITS tendon

can cause a tendon to degenerate and then to rupture in response to moderate stress. Injury causes pain and makes the shoulder joint unstable and subject to dislocation.

**Shinsplints**—a general term embracing several kinds of injury with pain in the crural region: tendinitis of the tibialis posterior muscle, inflammation of the tibial periosteum, and anterior compartment syndrome. Shinsplints may result from unaccustomed jogging, walking on snowshoes, or any vigorous activity of the legs after a period of relative inactivity.

**Tennis elbow**—inflammation at the origin of the extensor carpi muscles on the lateral epicondyle of the humerus. It occurs when these muscles are repeatedly tensed during backhand strokes and then strained by sudden impact with the tennis ball. Any activity that requires rotary movements of the forearm and a firm grip of the hand (for example, using a screwdriver) can cause the symptoms of tennis elbow.

**Tennis leg**—a partial tear in the lateral origin of the gastrocnemius muscle. It results from repeated strains put on the muscle while supporting the body weight on the toes.

Most athletic injuries can be prevented by proper conditioning. A person who suddenly takes up vigorous exercise may not have sufficient muscle and bone mass to withstand the stresses such exercise entails. These must be developed gradually. Stretching exercises keep ligaments and joint capsules supple and therefore reduce injuries. Warm-up exercises promote more efficient and less injurious musculoskeletal function in several ways, discussed in chapter 11. Most of all, moderation is important, as most injuries simply result from overuse of the muscles. "No pain, no gain" is a dangerous misconception.

Muscular injuries can be treated initially with "RICE": rest, ice, compression, and elevation. Rest prevents further injury and allows repair processes to occur; ice reduces swelling; compression with an elastic bandage helps to prevent fluid accumulation and swelling; and elevation of an injured limb promotes drainage of blood from the affected area and limits further swelling. If these measures are not enough, anti-inflammatory drugs may be employed, including corticosteroids as well as aspirin and other nonsteroidal agents. Serious injuries, such as compartment syndrome, require emergency attention by a physician.

# STUDY GUIDE

## ▶ Assess Your Learning Outcomes

*Answers in Appendix B*

*To test your knowledge, discuss the following topics with a study partner or in writing, ideally from memory.*

### 10.1 The Structural and Functional Organization of Muscles (p. 309)

1. Which muscles are included in the muscular system and which ones are not; the name of the science that specializes in the muscular system
2. Functions of the muscular system
3. The relationship of muscle structure to the endomysium, perimysium, and epimysium; what constitutes a fascicle of skeletal muscle and how it relates to these connective tissues; and the relationship of a fascia to a muscle
4. Classification of muscles according to the orientation of their fascicles
5. Muscle compartments, interosseous membranes, and intermuscular septa
6. The difference between direct and indirect muscle attachments
7. The origin, belly, and insertion of a muscle; the imperfection in origin–insertion terminology
8. The action of a muscle; how it relates to the classification of muscles as prime movers, synergists, antagonists, or fixators; why these terms are not fixed for a given muscle but differ from one joint movement to another, and examples to illustrate this point
9. Intrinsic versus extrinsic muscles, with examples
10. The innervation of muscles
11. Features to which the Latin names of muscles commonly refer, with examples

### 10.2 Muscles of the Head and Neck (p. 318)

*Know the location, action, origin, insertion, and innervation of the named muscles in each of the following groups, and be able to recognize them on laboratory specimens or models to the extent required in your course.*

1. The frontalis and occipitalis muscles of the scalp, eyebrows, and forehead (table 10.1)
2. The orbicularis oculi, levator palpebrae superioris, and corrugator supercilii muscles, which move the eyelid and other tissues around the eye (table 10.1)
3. The nasalis muscle, which flares and compresses the nostrils (table 10.1)
4. The orbicularis oris, levator labii superioris, levator anguli oris, zygomaticus major and minor, risorius, depressor anguli oris, depressor labii inferioris, and mentalis muscles, which act on the lips (table 10.1)
5. The buccinator muscles of the cheeks (table 10.1)
6. The platysma, which acts upon the mandible and the skin of the neck (table 10.1)
7. The intrinsic muscles of the tongue in general, and specific extrinsic muscles: the genioglossus, hyoglossus, styloglossus, and palatoglossus muscles (table 10.2)
8. The temporalis, masseter, medial pterygoid, and lateral pterygoid muscles of biting and chewing (table 10.2)
9. The suprahyoid group: the digastric, geniohyoid, mylohyoid, and stylohyoid muscles (table 10.2)
10. The infrahyoid group: the omohyoid, sternohyoid, thyrohyoid, and sternothyroid muscles (table 10.2)
11. The superior, middle, and inferior pharyngeal constrictor muscles of the throat (table 10.2)
12. The sternocleidomastoid and three scalene muscles, which flex the neck, and the trapezius, splenius capitis, and semispinalis capitis muscles, which extend it (table 10.3)

### 10.3 Muscles of the Trunk (p. 329)

*For the following muscles, know the same information as for section 10.2.*

1. The diaphragm and the external intercostal, internal intercostal, and innermost intercostal muscles of respiration (table 10.4)
2. The external abdominal oblique, internal abdominal oblique, transverse abdominal, and rectus abdominis muscles of the anterior and lateral abdominal wall (table 10.5)
3. The superficial erector spinae muscle (and its subdivisions) and the deep semispinalis thoracis, quadratus lumborum, and multifidus muscles of the back (table 10.6)
4. The perineum, its two triangles, and their skeletal landmarks (table 10.7)
5. The ischiocavernosus and bulbospongiosus muscles of the superficial perineal space of the pelvic floor (table 10.7)
6. The deep transverse perineal muscle, and in females, the compressor urethrae, of the deep perineal space of the pelvic floor, and the external anal sphincter of the anal triangle (table 10.7)
7. The levator ani muscle of the pelvic diaphragm, the deepest compartment of the pelvic floor (table 10.7)

### 10.4 Muscles Acting on the Shoulder and Upper Limb (p. 339)

*For the following muscles, know the same information as for section 10.2.*

1. The pectoralis minor, serratus anterior, trapezius, levator scapulae, rhomboid major, and rhomboid minor muscles of scapular movement (table 10.8)
2. Muscles that act on the humerus, including the pectoralis major, latissimus dorsi, deltoid, teres major, coracobrachialis, and four rotator cuff (SITS) muscles—the supraspinatus, infraspinatus, teres minor, and subscapularis (table 10.9)
3. The brachialis, biceps brachii, triceps brachii, brachioradialis, anconeus, pronator quadratus, pronator teres, and supinator muscles of forearm movement (table 10.10)
4. The relationship of the flexor retinaculum, extensor retinaculum, and carpal tunnel to the tendons of the forearm muscles
5. The palmaris longus, flexor carpi radialis, flexor carpi ulnaris, and flexor digitorum superficialis muscles of the superficial anterior compartment of the forearm, and the flexor digitorum profundus and flexor pollicis longus muscles of the deep anterior compartment (table 10.11)
6. The extensor carpi radialis longus, extensor carpi radialis brevis, extensor digitorum, extensor digiti minimi, and extensor carpi ulnaris muscles of the superficial posterior compartment (table 10.11)
7. The abductor pollicis longus, extensor pollicis brevis, extensor pollicis longus, and extensor indicis muscles of the deep posterior compartment (table 10.11)
8. The thenar group of intrinsic hand muscles: adductor pollicis, abductor pollicis brevis, flexor pollicis brevis, and opponens pollicis (table 10.12)
9. The hypothenar group of intrinsic hand muscles: abductor digiti minimi, flexor

# STUDY GUIDE

digiti minimi brevis, and opponens digiti minimi (table 10.12)

10. The midpalmar group of intrinsic hand muscles: four dorsal interosseous muscles, three palmar interosseous muscles, and four lumbrical muscles (table 10.12)

**10.5 Muscles Acting on the Hip and Lower Limb (p. 355)**

*For the following muscles, know the same information as for section 10.2.*

1. The iliopsoas muscle of the hip, and its two subdivisions, the iliacus and psoas major (table 10.13)

2. The tensor fasciae latae, gluteus maximus, gluteus medius, and gluteus minimus muscles of the hip and buttock, and the relationship of the first two to the fascia lata and iliotibial tract (table 10.13)

3. The lateral rotators: gemellus superior, gemellus inferior, obturator externus, obturator internus, piriformis, and quadratus femoris muscles (table 10.13)

4. The compartments of the thigh muscles: anterior (extensor), medial (adductor), and posterior (flexor) compartments

5. Muscles of the medial compartment of the thigh: adductor brevis, adductor longus, adductor magnus, gracilis, and pectineus (table 10.13)

6. Muscles of the anterior compartment of the thigh: sartorius and quadriceps femoris, and the four heads of the quadriceps (table 10.14)

7. The hamstring muscles of the posterior compartment of the thigh: biceps femoris, semitendinosus, and semimembranosus (table 10.14)

8. The compartments of the leg muscles: anterior, posterior, and lateral (table 10.15)

9. Muscles of the anterior compartment of the leg: fibularis tertius, extensor digitorum longus, extensor hallucis longus, and tibialis anterior muscles of the anterior compartment (table 10.15)

10. Muscles of the superficial posterior compartment of the leg: popliteus and triceps

surae (gastrocnemius and soleus), and the relationship of the triceps surae to the calcaneal tendon and calcaneus (table 10.15)

11. Muscles of the deep posterior compartment of the leg: flexor digitorum longus, flexor hallucis longus, and tibialis posterior muscles of the deep posterior compartment

12. Muscles of the lateral compartment of the leg: fibularis brevis and fibularis longus (table 10.15)

13. The extensor digitorum brevis of the dorsal aspect of the foot (table 10.16)

14. The four muscle compartments (layers) of the ventral aspect of the foot, and the muscles in each: the flexor digitorum brevis, abductor digiti minimi, and abductor hallucis (layer 1); the quadratus plantae and four lumbrical muscles (layer 2); the flexor digiti minimi brevis, flexor hallucis brevis, and adductor hallucis (layer 3); and the four dorsal interosseous muscles and three plantar interosseous muscles (layer 4) (table 10.16)

► **Testing Your Recall**

*Answers in Appendix B*

1. Which of the following muscles is the prime mover in spitting out a mouthful of liquid?
   a. platysma
   b. buccinator
   c. risorius
   d. masseter
   e. palatoglossus

2. Each muscle fiber has a sleeve of areolar connective tissue around it called
   a. the fascia.
   b. the endomysium.
   c. the perimysium.
   d. the epimysium.
   e. the intermuscular septum.

3. Which of these is *not* a suprahyoid muscle?
   a. genioglossus
   b. geniohyoid
   c. stylohyoid
   d. mylohyoid
   e. digastric

4. Which of these muscles is an extensor of the neck?
   a. external oblique
   b. sternocleidomastoid
   c. splenius capitis
   d. iliocostalis
   e. latissimus dorsi

5. Which of these muscles of the pelvic floor is the deepest?
   a. superficial transverse perineal
   b. bulbospongiosus
   c. ischiocavernosus
   d. deep transverse perineal
   e. levator ani

6. Which of these actions is *not* performed by the trapezius?
   a. extension of the neck
   b. depression of the scapula
   c. elevation of the scapula
   d. rotation of the scapula
   e. adduction of the humerus

7. Both the hands and feet are acted upon by a muscle or muscles called
   a. the extensor digitorum.
   b. the abductor digiti minimi.
   c. the flexor digitorum profundus.
   d. the abductor hallucis.
   e. the flexor digitorum longus.

8. Which of the following muscles does *not* extend the hip joint?
   a. quadriceps femoris
   b. gluteus maximus
   c. biceps femoris
   d. semitendinosus
   e. semimembranosus

9. Both the gastrocnemius and _____ muscles insert on the heel by way of the calcaneal tendon.
   a. semimembranosus
   b. tibialis posterior
   c. tibialis anterior
   d. soleus
   e. plantaris

# STUDY GUIDE

10. Which of the following muscles raises the upper lip?
    a. levator palpebrae superioris
    b. orbicularis oris
    c. zygomaticus minor
    d. masseter
    e. mentalis

11. The _____ of a muscle is the point where it attaches to a relatively stationary bone.

12. A bundle of muscle fibers surrounded by perimysium is called a/an _____.

13. The _____ is the muscle that generates the most force in a given joint movement.

14. The three large muscles on the posterior side of the thigh are commonly known as the _____ muscles.

15. Connective tissue bands called _____ prevent flexor tendons of the forearm and leg from rising like bowstrings.

16. The anterior half of the perineum is a region called the _____.

17. The abdominal aponeuroses converge on a median fibrous band on the abdomen called the _____.

18. A muscle that works with another to produce the same or similar movement is called a/an _____.

19. A muscle somewhat like a feather, with fibers obliquely approaching its tendon from both sides, is called a/an _____ muscle.

20. A circular muscle that closes a body opening is called a/an _____.

## ▶ Building Your Medical Vocabulary

*Answers in Appendix B*

*State a medical meaning of each word element below, and give a term in which it or a slight variation of it is used.*

1. capito-
2. ergo-
3. fasc-
4. labio-
5. lumbo-
6. mus-
7. mys-
8. omo-
9. penn-
10. tert-

## ▶ True or False

*Answers in Appendix B*

*Determine which five of the following statements are false, and briefly explain why.*

1. Cutting the phrenic nerves would paralyze the prime mover of respiration.

2. The orbicularis oculi is a sphincter.

3. The origin of the sternocleidomastoid muscle is the mastoid process of the skull.

4. To push someone away from you, you would use the serratus anterior more than the trapezius.

5. Both the extensor digitorum and extensor digiti minimi extend the little finger.

6. Curling the toes employs the quadratus plantae.

7. The scalenes are superficial to the trapezius.

8. Exhaling requires contraction of the internal intercostal muscles.

9. Hamstring injuries often result from rapid flexion of the knee.

10. The tibialis anterior and tibialis posterior are synergists.

# STUDY GUIDE

## ► Testing Your Comprehension

*Answers at* www.mhhe.com/saladin7

1. Radical mastectomy, once a common treatment for breast cancer, involved removal of the pectoralis major along with the breast. What functional impairments would result from this? What synergists could a physical therapist train a patient to use to recover some lost function?

2. Removal of cancerous lymph nodes from the neck sometimes requires removal of the sternocleidomastoid on that side. How would this affect a patient's range of head movement?

3. Poorly conditioned, middle-aged people may suffer a rupture of the calcaneal tendon when the foot is suddenly dorsiflexed. Explain each of the following signs of a ruptured calcaneal tendon: (a) a prominent lump typically appears in the calf; (b) the foot can be dorsiflexed farther than usual; and (c) the patient cannot plantar flex the foot very effectively.

4. Women who habitually wear high heels may suffer painful "high heel syndrome" when they go barefoot or wear flat shoes. What muscle(s) and tendon(s) are involved? Explain.

5. A student moving out of a dormitory crouches, in correct fashion, to lift a heavy box of books. What prime movers are involved as he straightens his legs to lift the box?

## ► Improve Your Grade at www.mhhe.com/saladin7

*Download animations to study when it fits your schedule. Practice quizzes, labeling activities, games, and flashcards offer fun ways to master the chapter concepts. Or, download image PowerPoint files for each chapter to create a study guide or for taking notes during lecture.*

How many muscles can you identify from their surface appearance?

# REGIONAL AND SURFACE ANATOMY

## ATLAS OUTLINE

Anatomy & Physiology | REVEALED®

aprevealed.com

**Module 6: Muscular System**

## B.1 Regional Anatomy

On the whole, this book takes a **systems approach** to anatomy, examining the structure and function of each organ system, one at a time, regardless of which body regions it may traverse. Physicians and surgeons, however, think and act in terms of regional anatomy. If a patient presents with pain in the lower right quadrant (see fig. A.4a, p. 31), the source may be the appendix, an ovary, or an inguinal muscle, among other possibilities. The question is to think not of an entire organ system (the esophagus is probably irrelevant to that quadrant), but of what organs are present in that region and what possibilities must be considered as the cause of the pain. This atlas presents several views of the body region by region so that you can see some of the spatial relationships that exist among the organ systems considered in their separate chapters.

## B.2 The Importance of Surface Anatomy

In the study of human anatomy, it is easy to become so preoccupied with internal structure that we forget the importance of what we can see and feel externally. Yet external anatomy and appearance are major concerns in giving a physical examination and in many aspects of patient care. A knowledge of the body's surface landmarks is essential to one's competence in physical therapy, cardiopulmonary resuscitation, surgery, making X-rays and electrocardiograms, giving injections, drawing blood, listening to heart and respiratory sounds, measuring the pulse and blood pressure, and finding pressure points to stop arterial bleeding, among other procedures. A misguided attempt to perform some of these procedures while disregarding or misunderstanding external anatomy can be very harmful and even fatal to a patient.

Having just studied skeletal and muscular anatomy in the preceding chapters, this is an opportune time for you to study the body surface. Much of what we see there reflects the underlying structure of the superficial bones and muscles. A broad photographic overview of surface anatomy is given in atlas A (see fig. A.3, p. 30), where it is necessary for providing a vocabulary for reference in subsequent chapters. This atlas shows this surface anatomy in closer detail so you can relate it to the musculoskeletal anatomy of chapters 8 through 10.

## B.3 Learning Strategy

To make the most profitable use of this atlas, refer back to earlier chapters as you study these illustrations. Relate drawings of the clavicles in chapter 8 to the photograph in figure B.1, for example. Study the shape of the scapula in chapter 8 and see how much of it you can trace on the photographs in figure B.9. See if you can relate the tendons visible on the hand (see fig. B.19) to the muscles of the forearm illustrated in chapter 10, and the external markings of the pelvic girdle (see fig. B.15) to bone structure in chapter 8.

For learning surface anatomy, there is a resource available to you that is far more valuable than any laboratory model or textbook illustration—your own body. For the best understanding of human structure, compare the art and photographs in this book with your body or with structures visible on a study partner. In addition to bones and muscles, you can palpate a number of superficial arteries, veins, tendons, ligaments, and cartilages, among other structures. By palpating regions such as the shoulder, elbow, or ankle, you can develop a mental image of the subsurface structures better than the image you can obtain by looking at two-dimensional textbook images. And the more you can study with other people, the more you will appreciate the variations in human structure and be able to apply your knowledge to your future patients or clients, who will not look quite like any textbook diagram or photograph you have ever seen. Through comparisons of art, photography, and the living body, you will get a much deeper understanding of the body than if you were to study this atlas in isolation from the earlier chapters.

At the end of this atlas, you can test your knowledge of externally visible muscle anatomy. The two photographs in figure B.25 have 30 numbered muscles and a list of 26 names, some of which are shown more than once in the photographs and some of which are not shown at all. Identify the muscles to your best ability without looking back at the previous illustrations, and then check your answers in appendix B at the back of the book.

Throughout these illustrations, the following abbreviations apply: a. = artery; m. = muscle; n. = nerve; v. = vein. Double letters such as mm. or vv. represent the plurals.

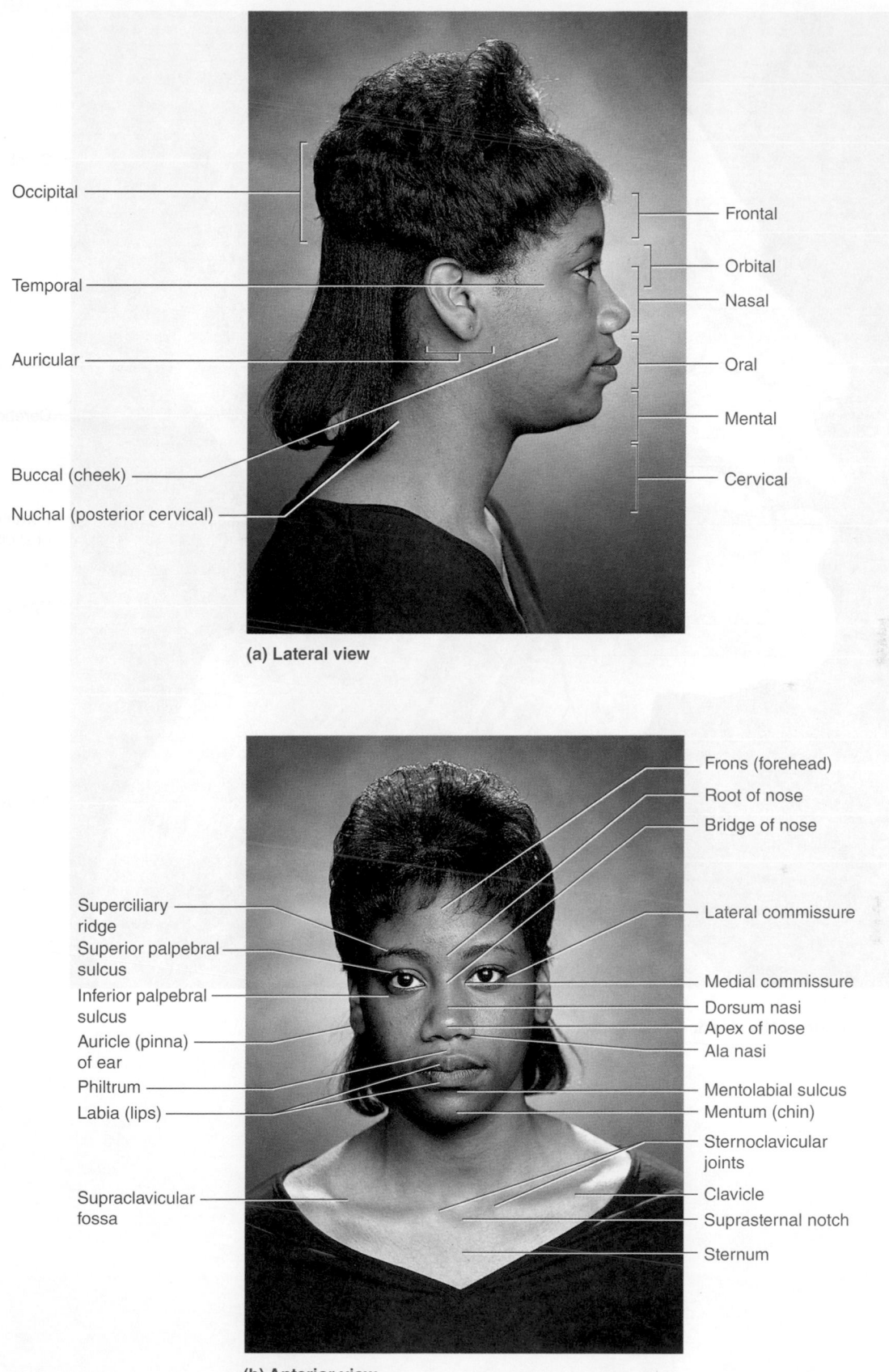

Occipital

Frontal

Temporal

Orbital

Nasal

Auricular

Oral

Mental

Buccal (cheek)

Cervical

Nuchal (posterior cervical)

**(a) Lateral view**

Frons (forehead)

Root of nose

Bridge of nose

Superciliary ridge

Lateral commissure

Superior palpebral sulcus

Medial commissure

Inferior palpebral sulcus

Dorsum nasi

Auricle (pinna) of ear

Apex of nose

Ala nasi

Philtrum

Mentolabial sulcus

Labia (lips)

Mentum (chin)

Sternoclavicular joints

Supraclavicular fossa

Clavicle

Suprasternal notch

Sternum

**(b) Anterior view**

**FIGURE B.1  The Head and Neck.**  (a) Anatomical regions of the head. (b) Features of the facial region and upper thorax.

❓ *What muscle underlies the region of the philtrum? What muscle forms the slope of the shoulder?*

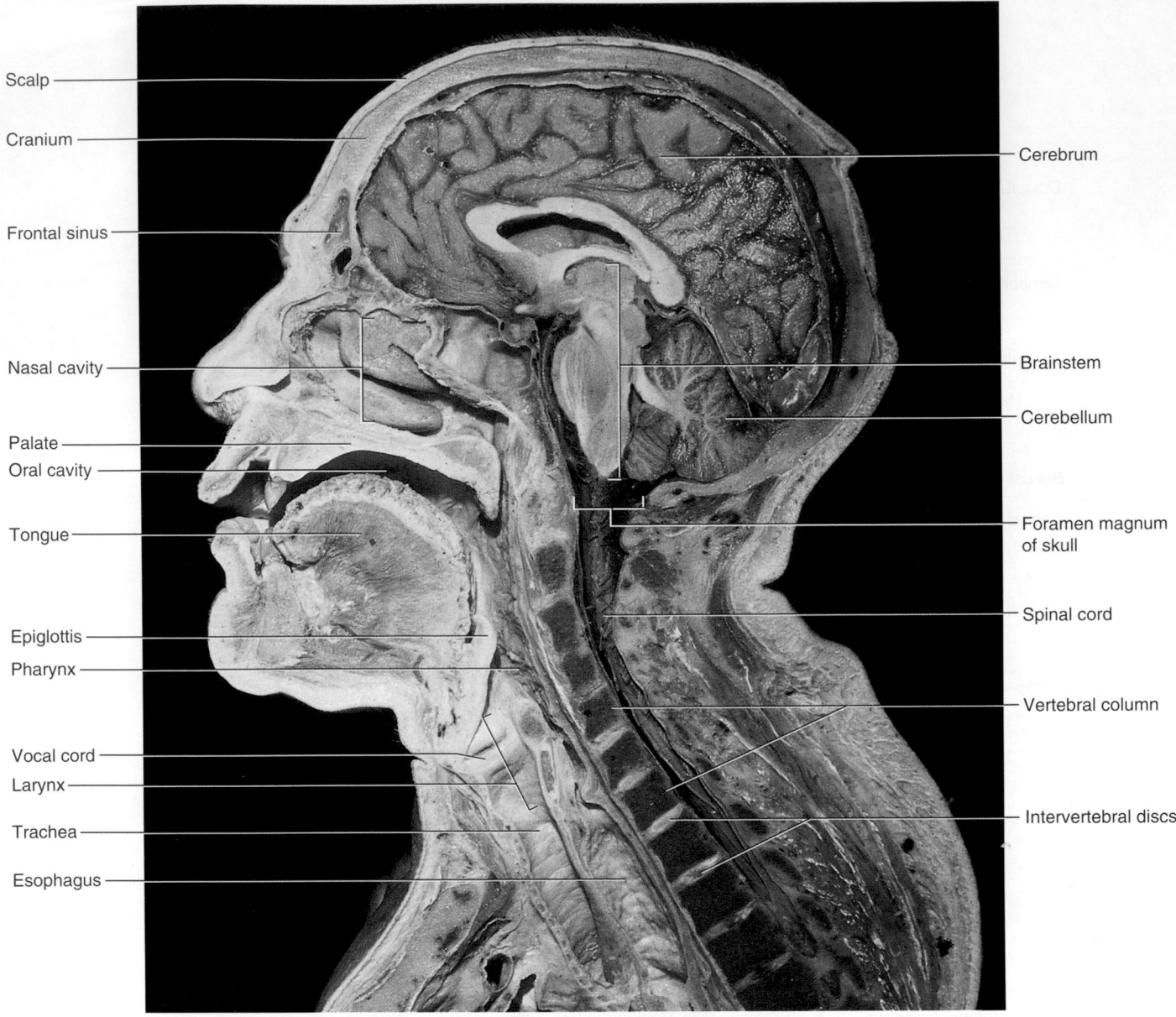

Scalp

Cranium

Frontal sinus

Nasal cavity

Palate

Oral cavity

Tongue

Epiglottis

Pharynx

Vocal cord

Larynx

Trachea

Esophagus

Cerebrum

Brainstem

Cerebellum

Foramen magnum of skull

Spinal cord

Vertebral column

Intervertebral discs

**FIGURE B.2   Median Section of the Head.**   Shows contents of the cranial, nasal, and oral cavities.

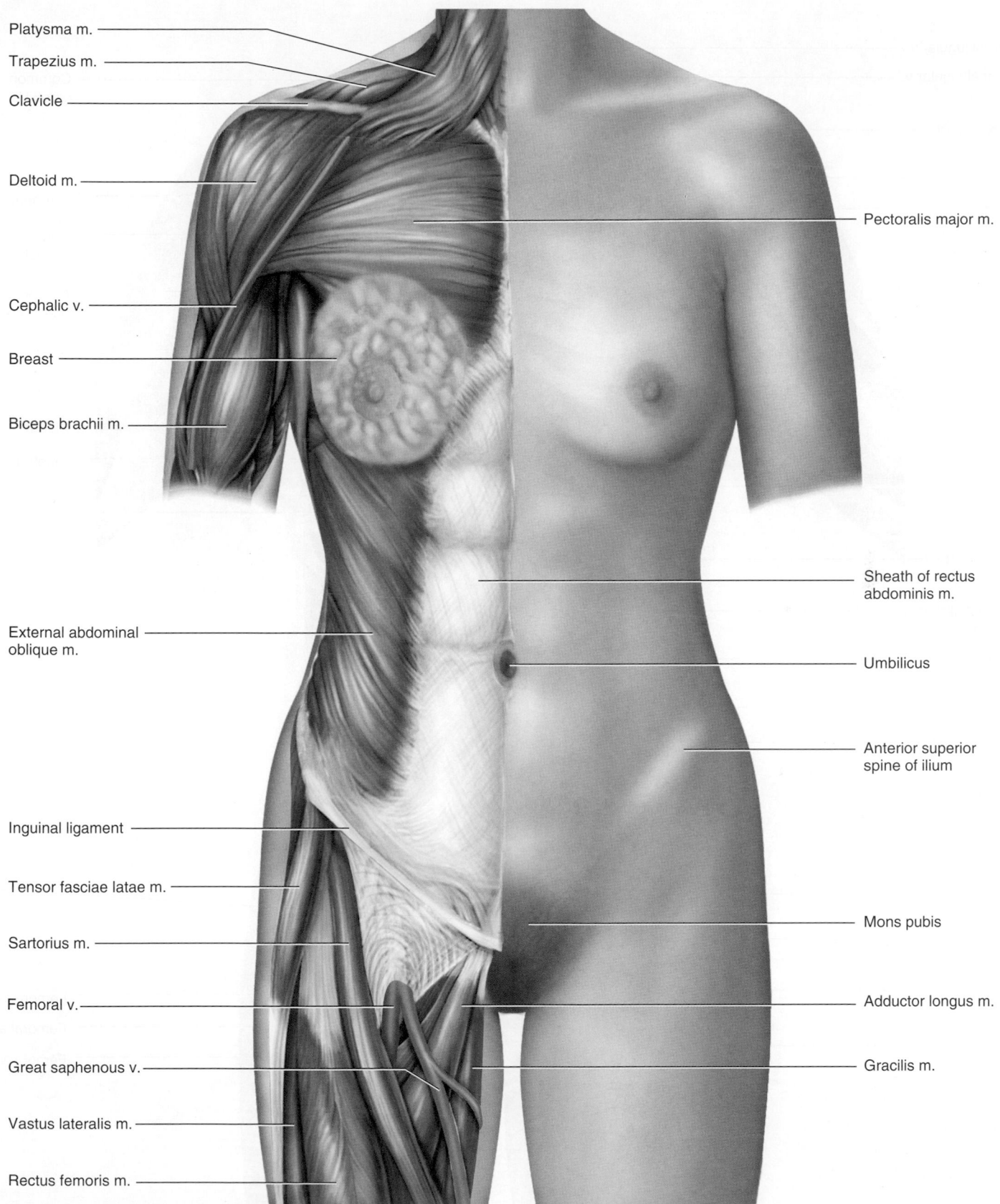

Platysma m.

Trapezius m.

Clavicle

Deltoid m.

Cephalic v.

Breast

Biceps brachii m.

External abdominal oblique m.

Inguinal ligament

Tensor fasciae latae m.

Sartorius m.

Femoral v.

Great saphenous v.

Vastus lateralis m.

Rectus femoris m.

Pectoralis major m.

Sheath of rectus abdominis m.

Umbilicus

Anterior superior spine of ilium

Mons pubis

Adductor longus m.

Gracilis m.

**FIGURE B.3 Superficial Anatomy of the Trunk (Female).** Surface anatomy is shown on the anatomical left, and structures immediately deep to the skin on the right.

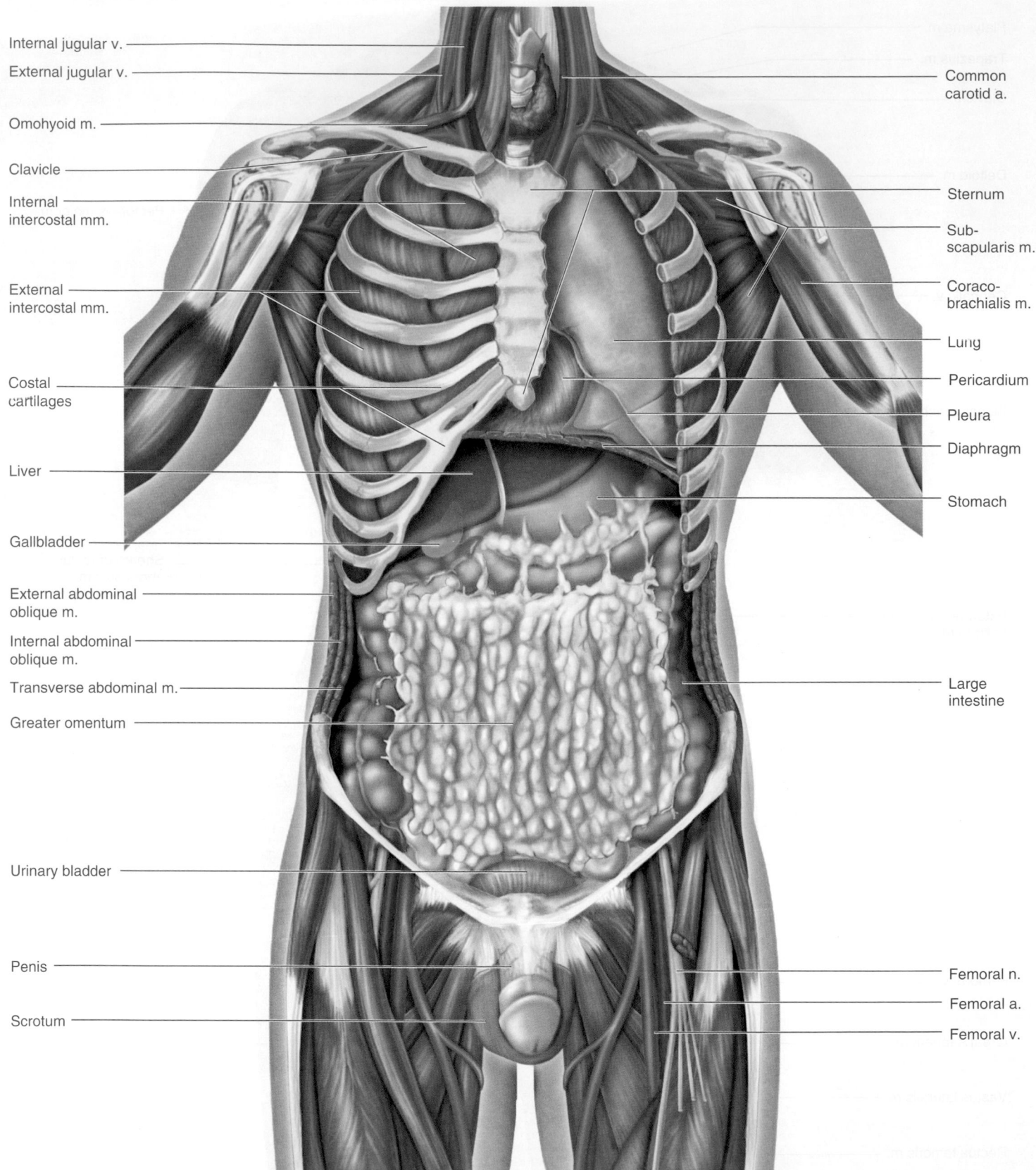

Internal jugular v.

External jugular v.

Omohyoid m.

Clavicle

Internal intercostal mm.

External intercostal mm.

Costal cartilages

Liver

Gallbladder

External abdominal oblique m.

Internal abdominal oblique m.

Transverse abdominal m.

Greater omentum

Urinary bladder

Penis

Scrotum

Common carotid a.

Sternum

Sub-scapularis m.

Coraco-brachialis m.

Lung

Pericardium

Pleura

Diaphragm

Stomach

Large intestine

Femoral n.

Femoral a.

Femoral v.

**FIGURE B.4** **Anatomy at the Level of the Rib Cage and Greater Omentum (Male).** The anterior body wall is removed, and the ribs, intercostal muscles, and pleura are removed from the anatomical left.

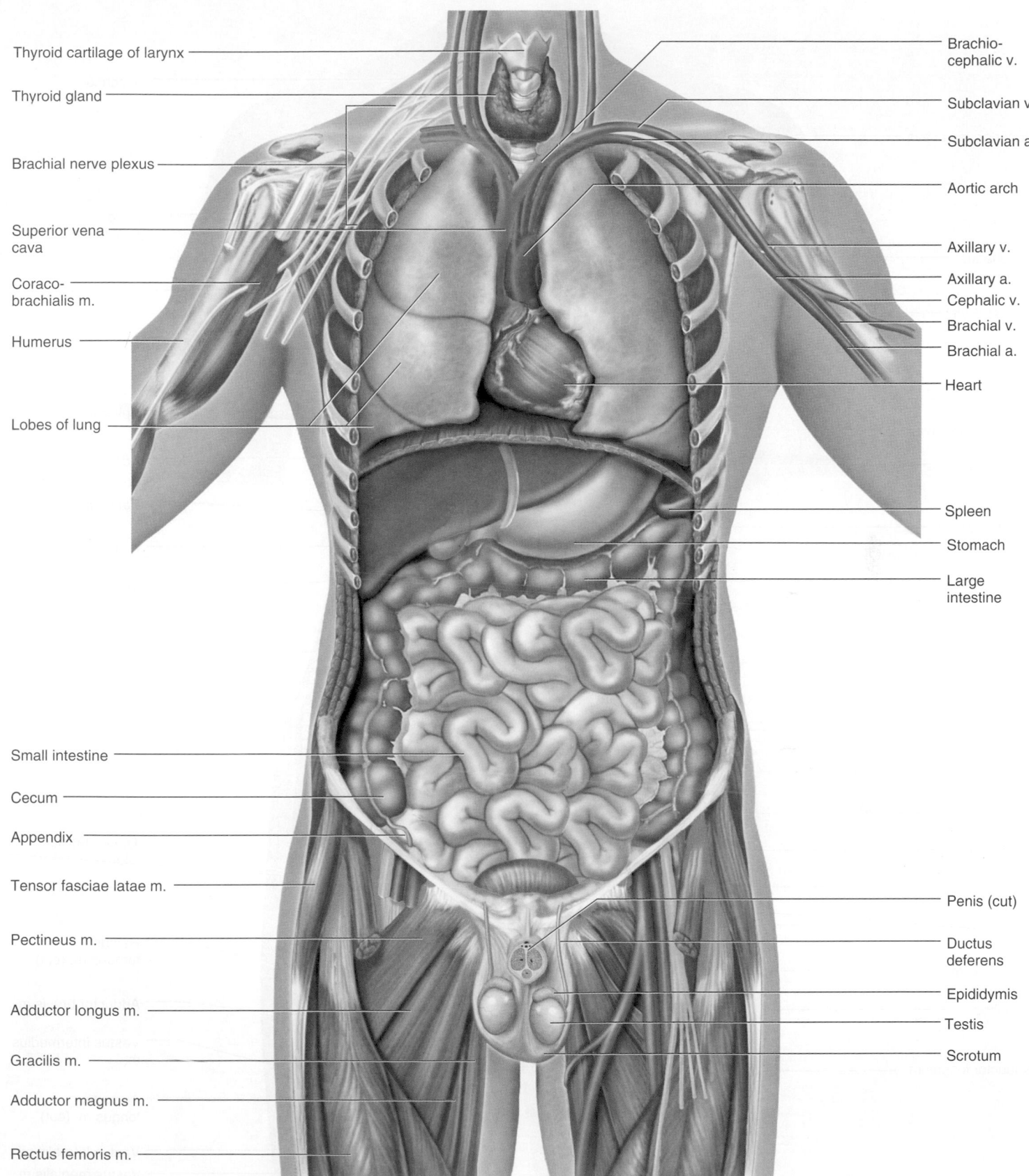

Thyroid cartilage of larynx

Thyroid gland

Brachial nerve plexus

Superior vena cava

Coraco-brachialis m.

Humerus

Lobes of lung

Small intestine

Cecum

Appendix

Tensor fasciae latae m.

Pectineus m.

Adductor longus m.

Gracilis m.

Adductor magnus m.

Rectus femoris m.

Brachio-cephalic v.

Subclavian v.

Subclavian a.

Aortic arch

Axillary v.

Axillary a.
Cephalic v.
Brachial v.
Brachial a.

Heart

Spleen

Stomach

Large intestine

Penis (cut)

Ductus deferens

Epididymis

Testis

Scrotum

**FIGURE B.5** **Anatomy at the Level of the Lungs and Intestines (Male).** The sternum, ribs, and greater omentum are removed.

❓ *Name several viscera that are protected by the rib cage.*

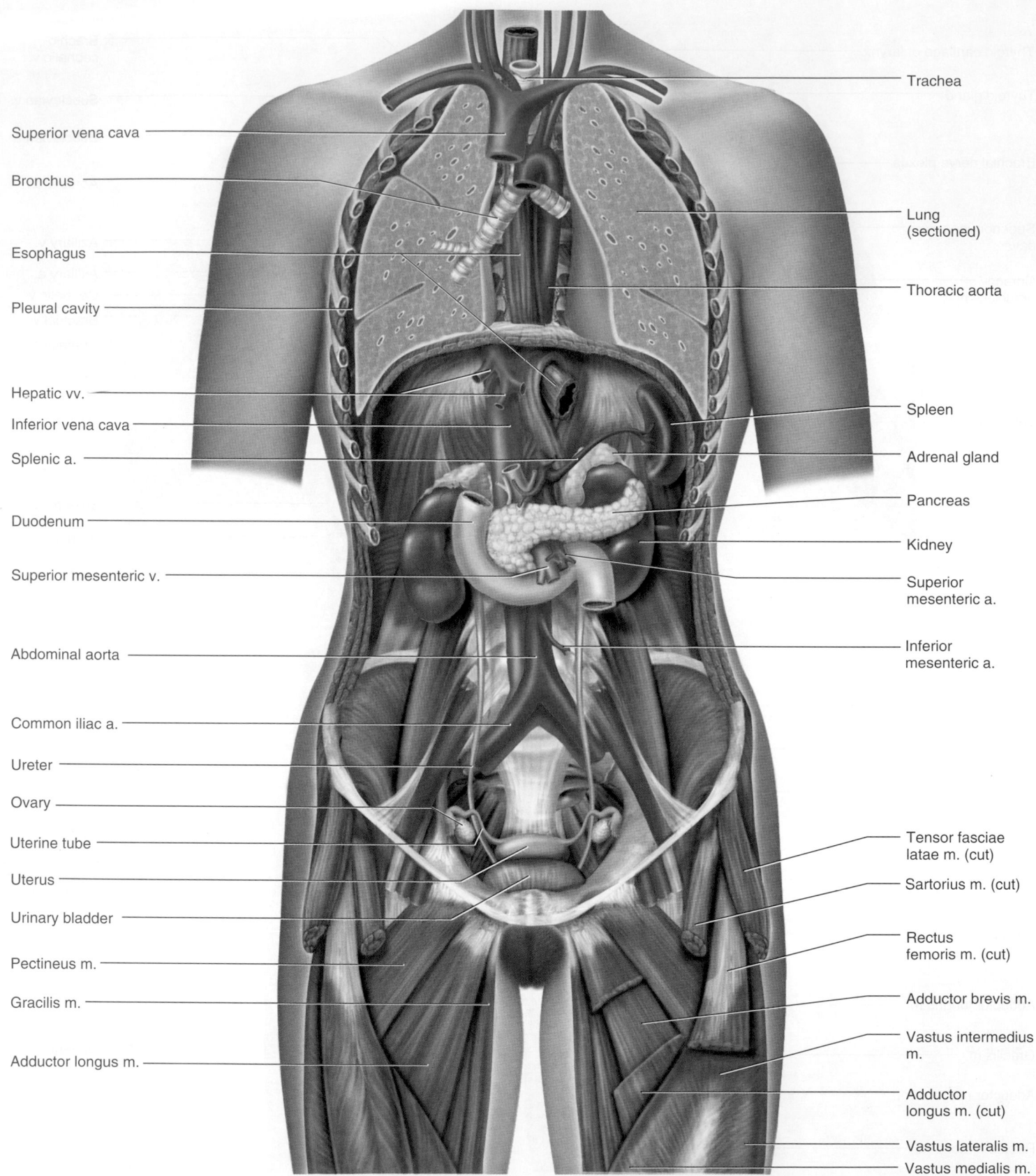

Superior vena cava

Bronchus

Esophagus

Pleural cavity

Hepatic vv.

Inferior vena cava

Splenic a.

Duodenum

Superior mesenteric v.

Abdominal aorta

Common iliac a.

Ureter

Ovary

Uterine tube

Uterus

Urinary bladder

Pectineus m.

Gracilis m.

Adductor longus m.

Trachea

Lung
(sectioned)

Thoracic aorta

Spleen

Adrenal gland

Pancreas

Kidney

Superior
mesenteric a.

Inferior
mesenteric a.

Tensor fasciae
latae m. (cut)

Sartorius m. (cut)

Rectus
femoris m. (cut)

Adductor brevis m.

Vastus intermedius
m.

Adductor
longus m. (cut)

Vastus lateralis m.

Vastus medialis m.

**FIGURE B.6** **Anatomy at the Level of the Retroperitoneal Viscera (Female).** The heart is removed, the lungs are frontally sectioned, and the viscera of the peritoneal cavity and the peritoneum itself are removed.

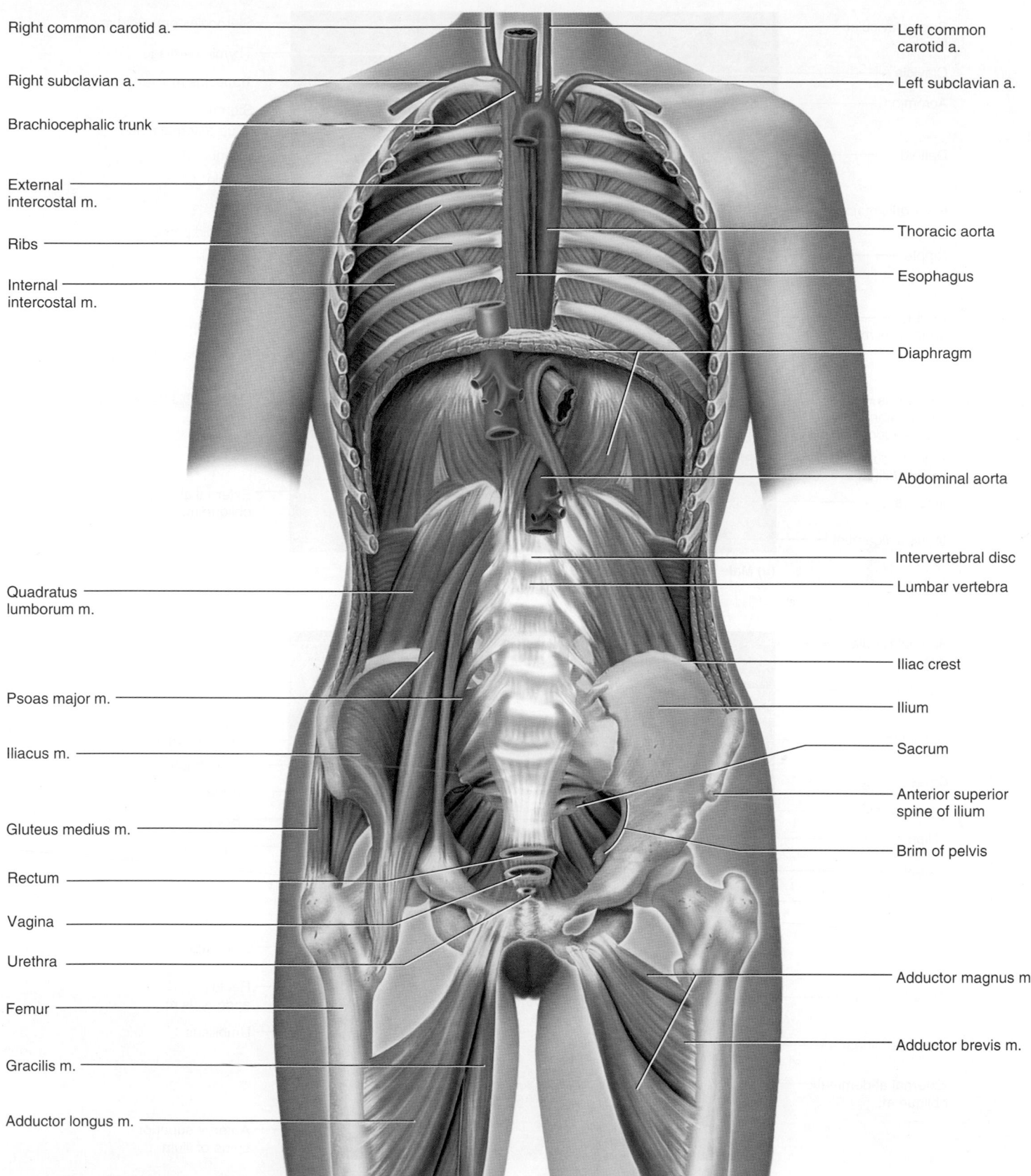

Right common carotid a.

Right subclavian a.

Brachiocephalic trunk

External intercostal m.

Ribs

Internal intercostal m.

Quadratus lumborum m.

Psoas major m.

Iliacus m.

Gluteus medius m.

Rectum

Vagina

Urethra

Femur

Gracilis m.

Adductor longus m.

Left common carotid a.

Left subclavian a.

Thoracic aorta

Esophagus

Diaphragm

Abdominal aorta

Intervertebral disc

Lumbar vertebra

Iliac crest

Ilium

Sacrum

Anterior superior spine of ilium

Brim of pelvis

Adductor magnus m.

Adductor brevis m.

**FIGURE B.7 Anatomy at the Level of the Posterior Body Wall (Female).** The lungs and retroperitoneal viscera are removed.

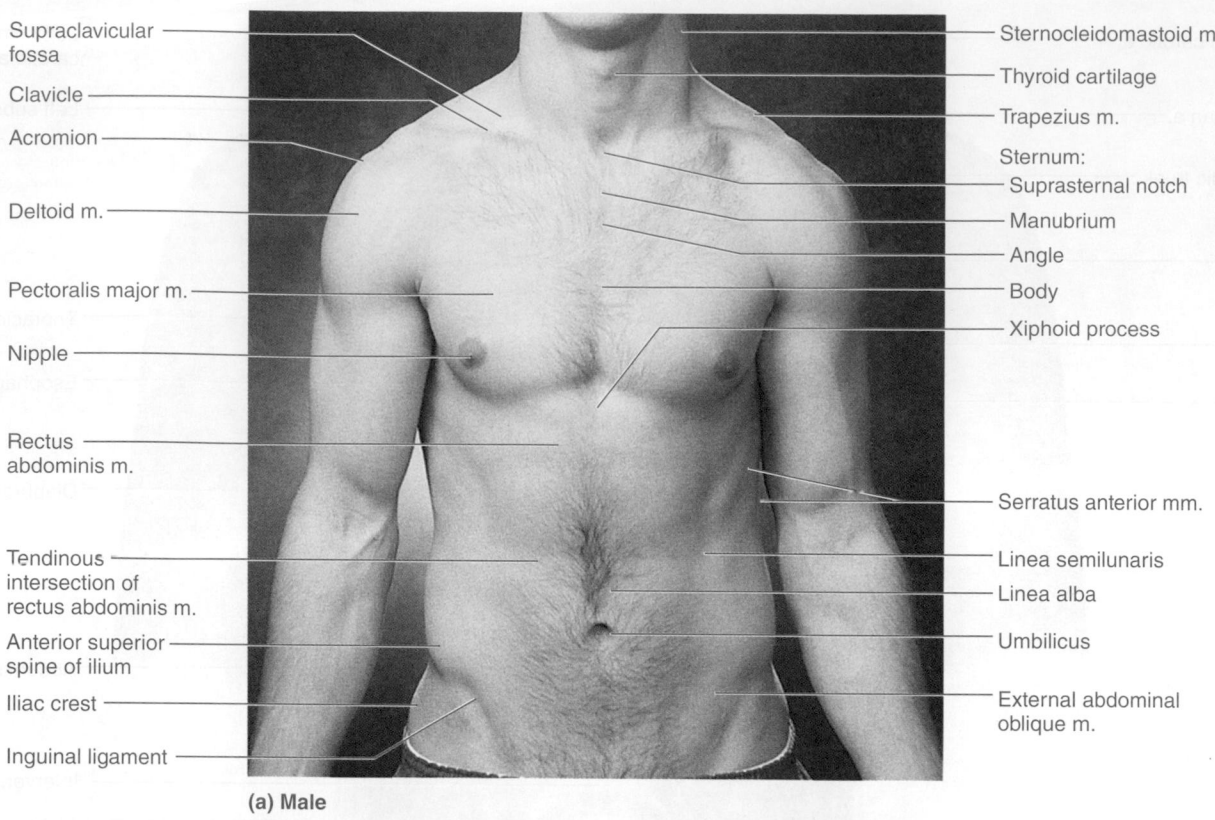

Supraclavicular fossa

Clavicle

Acromion

Deltoid m.

Pectoralis major m.

Nipple

Rectus abdominis m.

Tendinous intersection of rectus abdominis m.

Anterior superior spine of ilium

Iliac crest

Inguinal ligament

Sternocleidomastoid m.

Thyroid cartilage

Trapezius m.

Sternum:
Suprasternal notch

Manubrium

Angle

Body

Xiphoid process

Serratus anterior mm.

Linea semilunaris

Linea alba

Umbilicus

External abdominal oblique m.

**(a) Male**

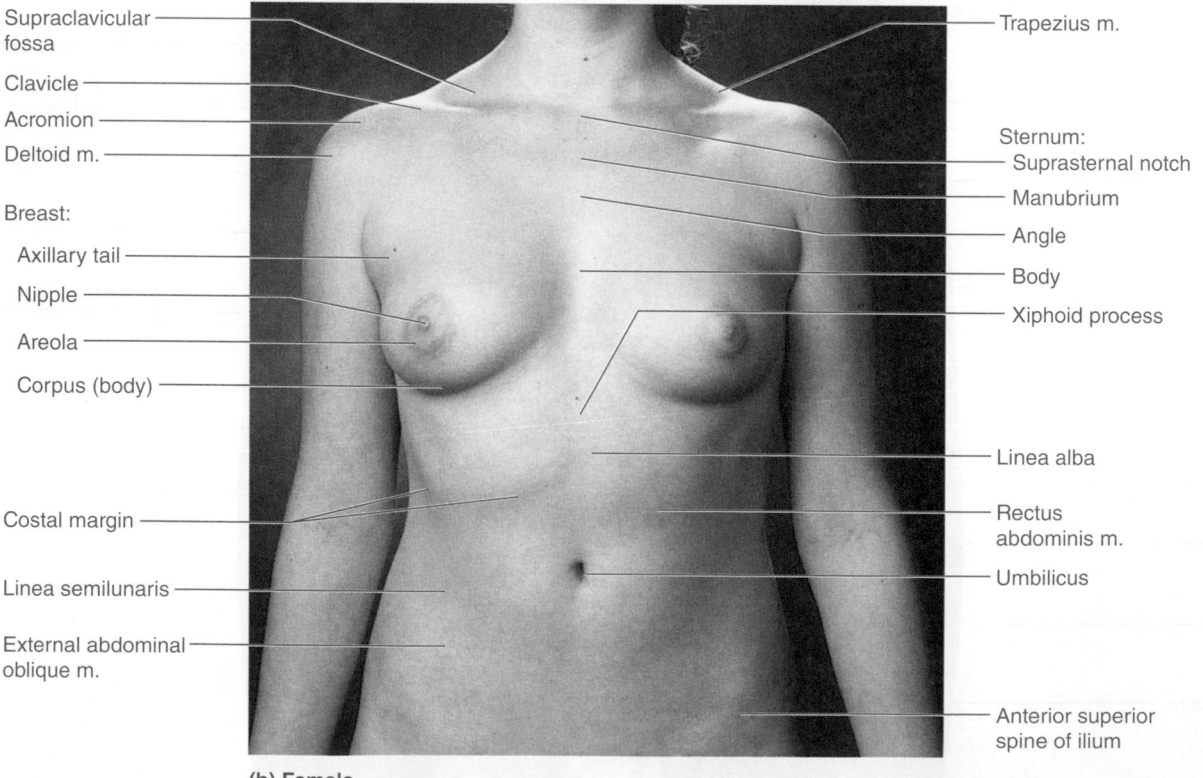

Supraclavicular fossa

Clavicle

Acromion

Deltoid m.

Breast:

Axillary tail

Nipple

Areola

Corpus (body)

Costal margin

Linea semilunaris

External abdominal oblique m.

Trapezius m.

Sternum:
Suprasternal notch

Manubrium

Angle

Body

Xiphoid process

Linea alba

Rectus abdominis m.

Umbilicus

Anterior superior spine of ilium

**(b) Female**

**FIGURE B.8  The Thorax and Abdomen (Anterior View).** All of the features labeled are common to both sexes, though some are labeled only on the photograph that shows them best.

❓ *The V-shaped tendons on each side of the suprasternal notch in part (a) belong to what muscles?*

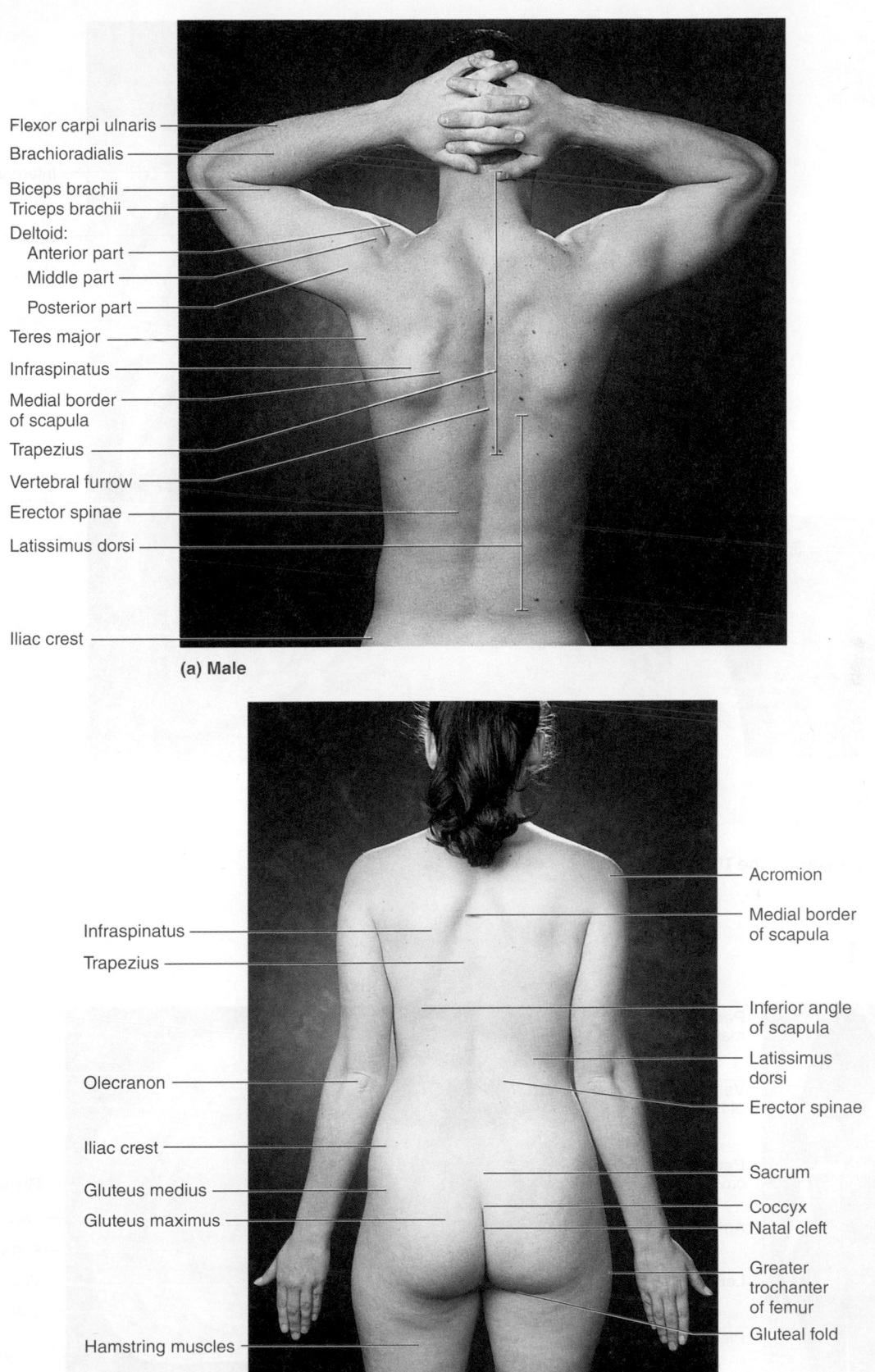

**(a) Male**

Flexor carpi ulnaris
Brachioradialis
Biceps brachii
Triceps brachii
Deltoid:
 Anterior part
 Middle part
 Posterior part
Teres major
Infraspinatus
Medial border
of scapula
Trapezius
Vertebral furrow
Erector spinae
Latissimus dorsi
Iliac crest

**(b) Female**

Infraspinatus
Trapezius
Olecranon
Iliac crest
Gluteus medius
Gluteus maximus
Hamstring muscles

Acromion
Medial border
of scapula
Inferior angle
of scapula
Latissimus
dorsi
Erector spinae
Sacrum
Coccyx
Natal cleft
Greater
trochanter
of femur
Gluteal fold

**FIGURE B.9  The Back and Gluteal Region.** All of the features labeled are common to both sexes, though some are labeled only on the photograph that shows them best.

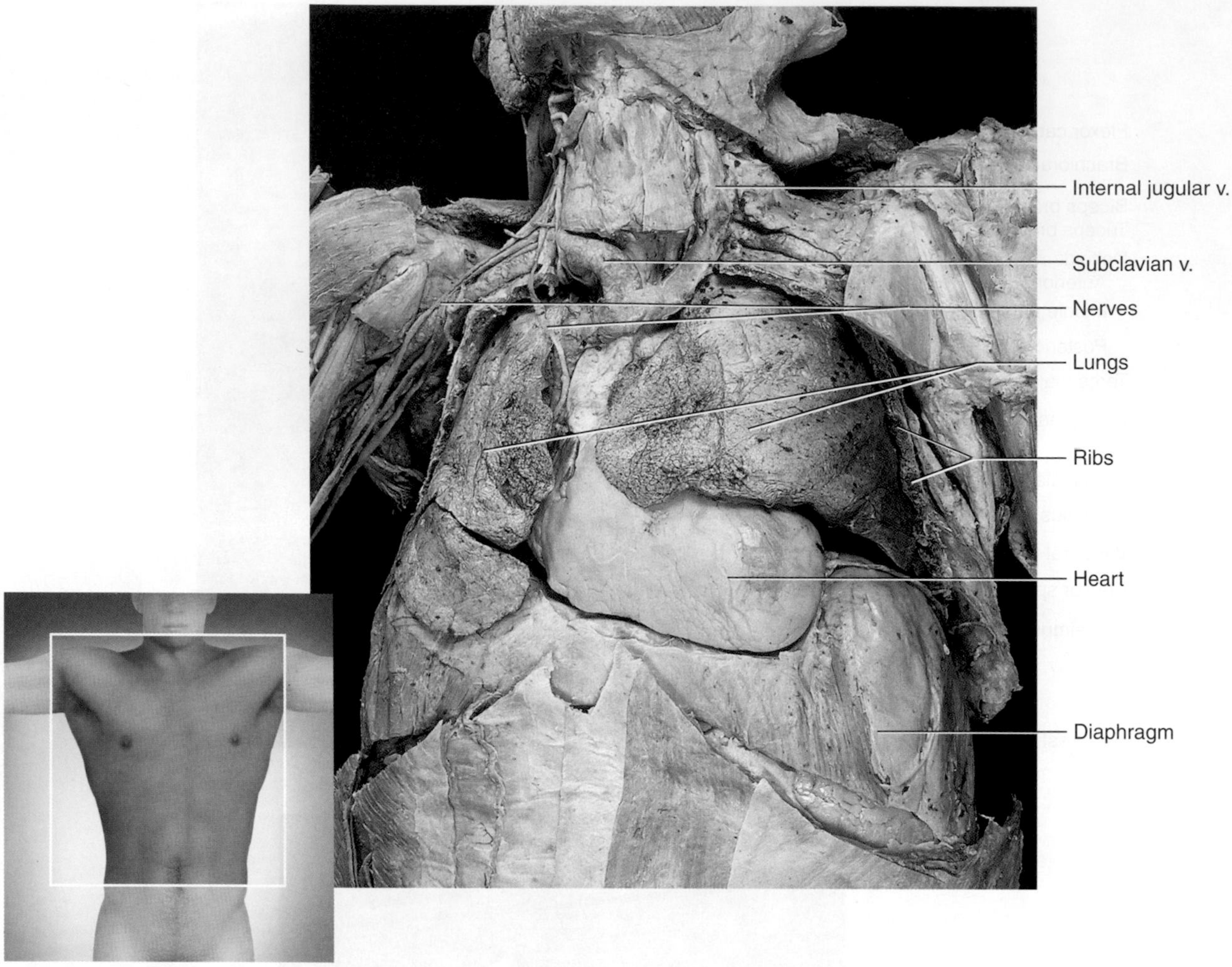

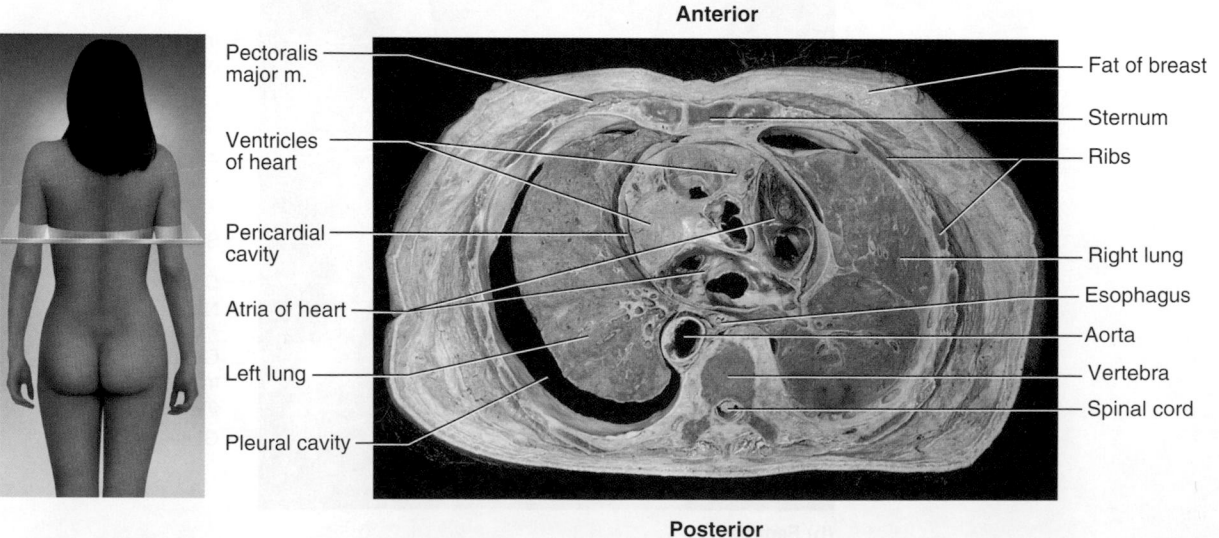

**FIGURE B.10** **Frontal View of the Thoracic Cavity.**

Internal jugular v.
Subclavian v.
Nerves
Lungs
Ribs
Heart
Diaphragm

**FIGURE B.11** **Transverse Section of the Thorax.** Section taken at the level shown by the inset and oriented the same as the reader's body.

Anterior
Pectoralis major m.
Ventricles of heart
Pericardial cavity
Atria of heart
Left lung
Pleural cavity
Fat of breast
Sternum
Ribs
Right lung
Esophagus
Aorta
Vertebra
Spinal cord
Posterior

**?** *In this section, which term best describes the position of the aorta relative to the heart: posterior, lateral, inferior, or proximal?*

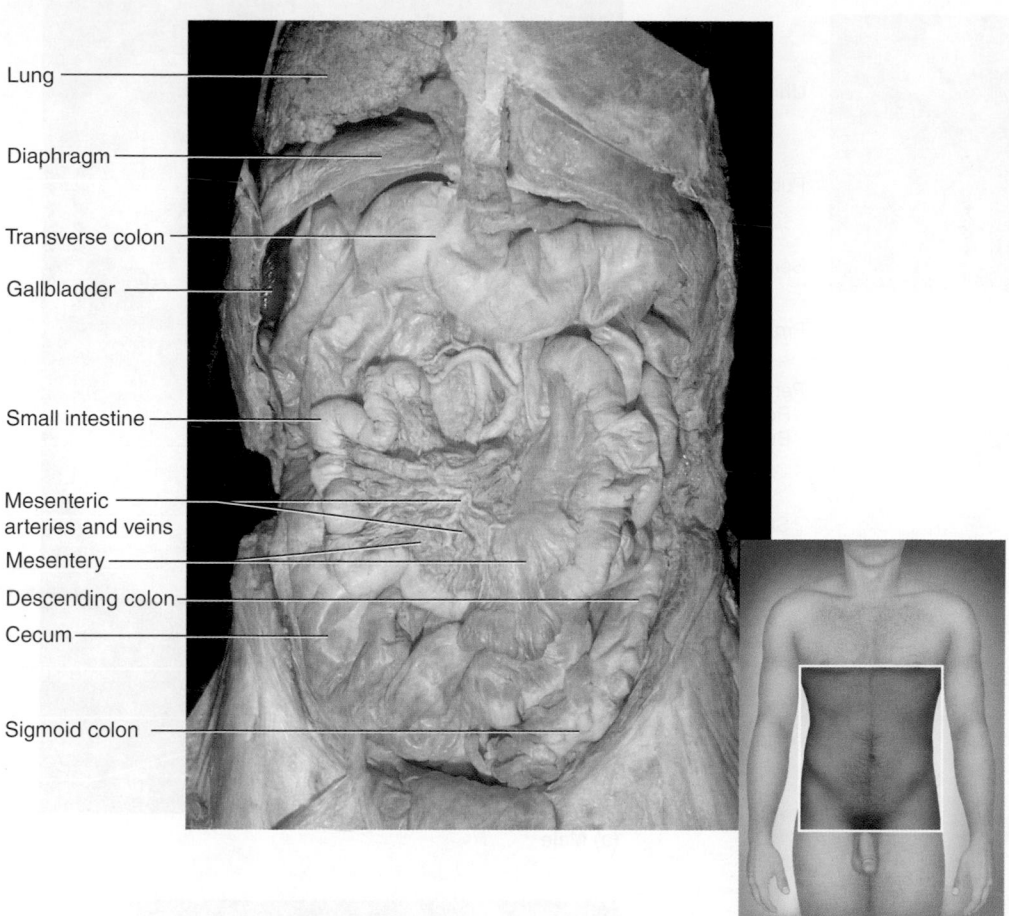

Lung

Diaphragm

Transverse colon

Gallbladder

Small intestine

Mesenteric
arteries and veins

Mesentery

Descending colon

Cecum

Sigmoid colon

**FIGURE B.12**  **Frontal View of the Abdominal Cavity.**

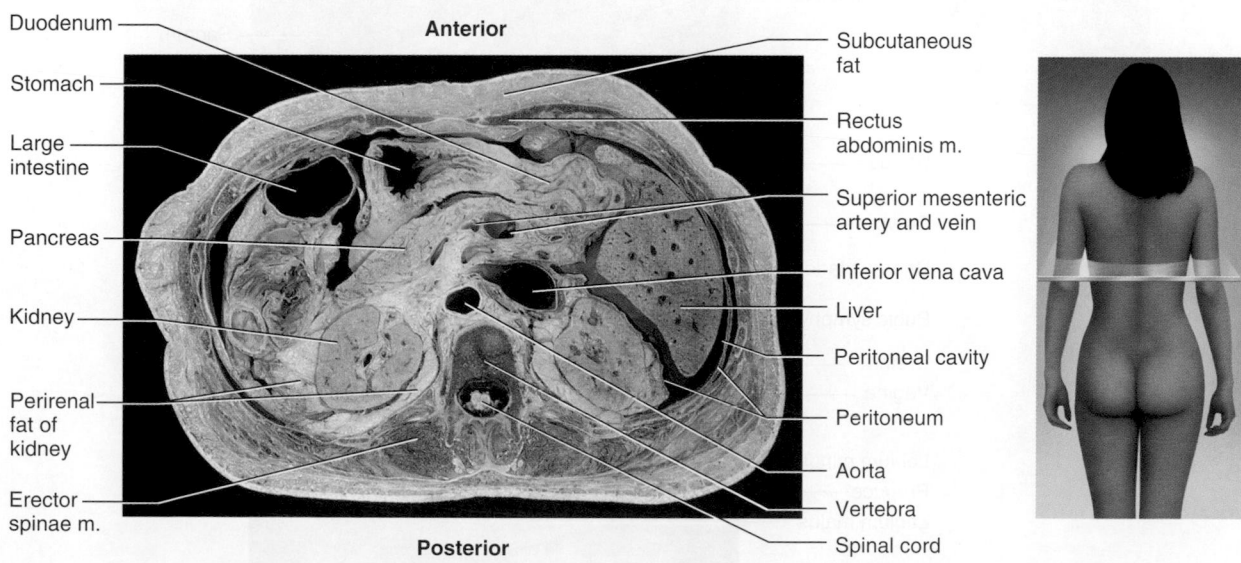

Duodenum

Stomach

Large
intestine

Pancreas

Kidney

Perirenal
fat of
kidney

Erector
spinae m.

**Anterior**

Subcutaneous
fat

Rectus
abdominis m.

Superior mesenteric
artery and vein

Inferior vena cava

Liver

Peritoneal cavity

Peritoneum

Aorta

Vertebra

Spinal cord

**Posterior**

**FIGURE B.13**  **Transverse Section of the Abdomen.**  Section taken at the level shown by the inset and oriented the same as the reader's body.

❓ *What tissue in this photograph is immediately superficial to the rectus abdominis muscle?*

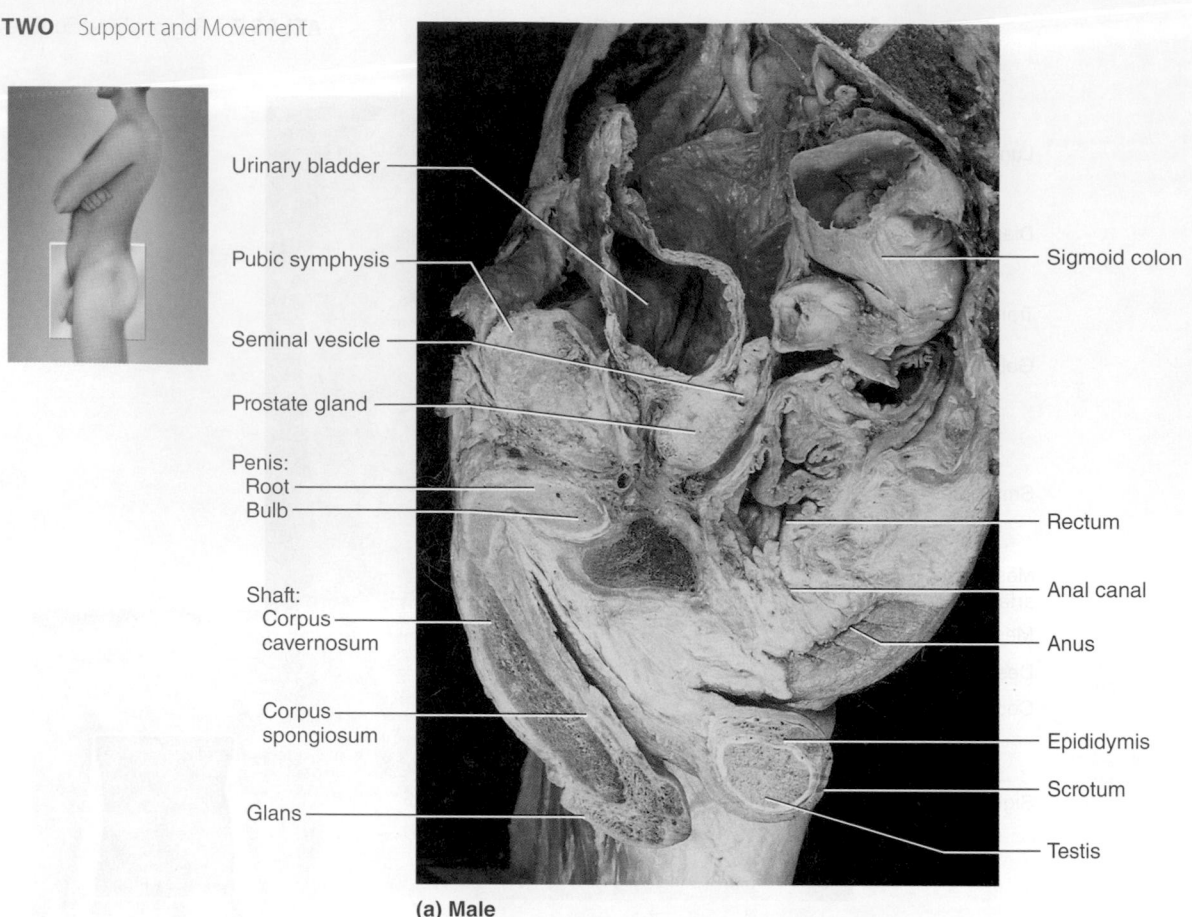

Urinary bladder

Pubic symphysis

Seminal vesicle

Prostate gland

Penis:
  Root
  Bulb

Shaft:
  Corpus
  cavernosum

Corpus
spongiosum

Glans

Sigmoid colon

Rectum

Anal canal

Anus

Epididymis

Scrotum

Testis

**(a) Male**

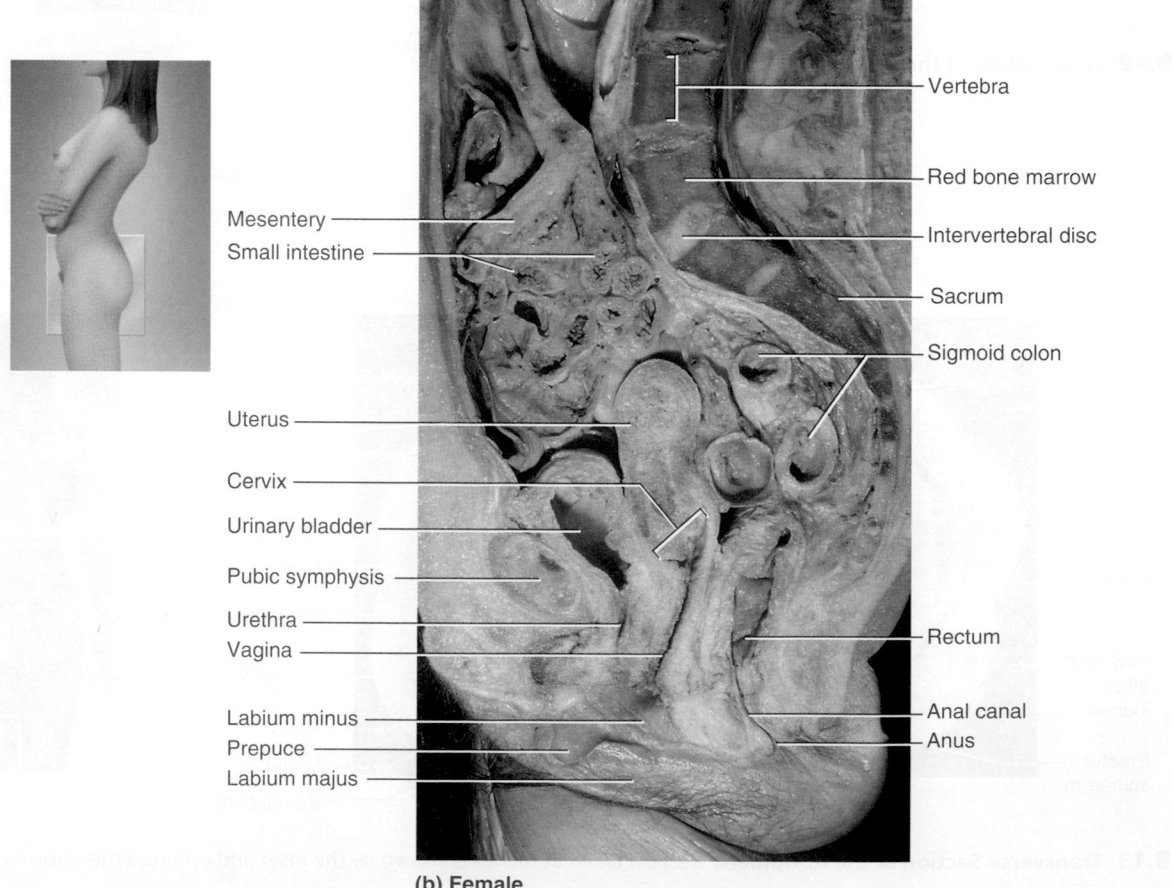

Vertebra

Red bone marrow

Intervertebral disc

Sacrum

Sigmoid colon

Mesentery

Small intestine

Uterus

Cervix

Urinary bladder

Pubic symphysis

Urethra

Vagina

Labium minus

Prepuce

Labium majus

Rectum

Anal canal

Anus

**(b) Female**

**FIGURE B.14   Median Sections of the Pelvic Cavity (Left Lateral Views).**

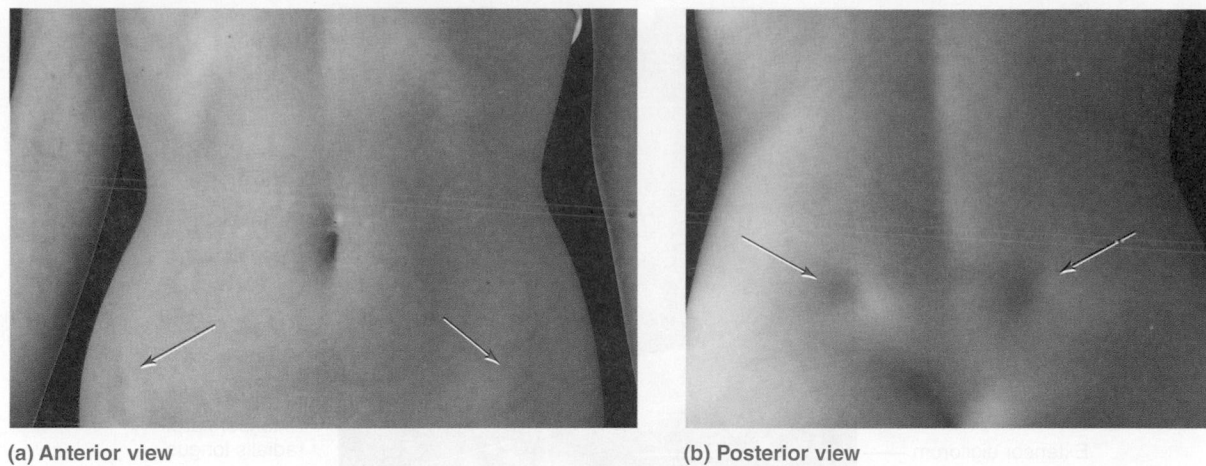

**(a) Anterior view**          **(b) Posterior view**

**FIGURE B.15  Pelvic Landmarks.**  (a) The anterior superior spines of the ilium are marked by anterolateral protuberances (arrows) at about the location where the front pockets usually open on a pair of pants. (b) The posterior superior spines are marked in some people by dimples in the sacral region (arrows).

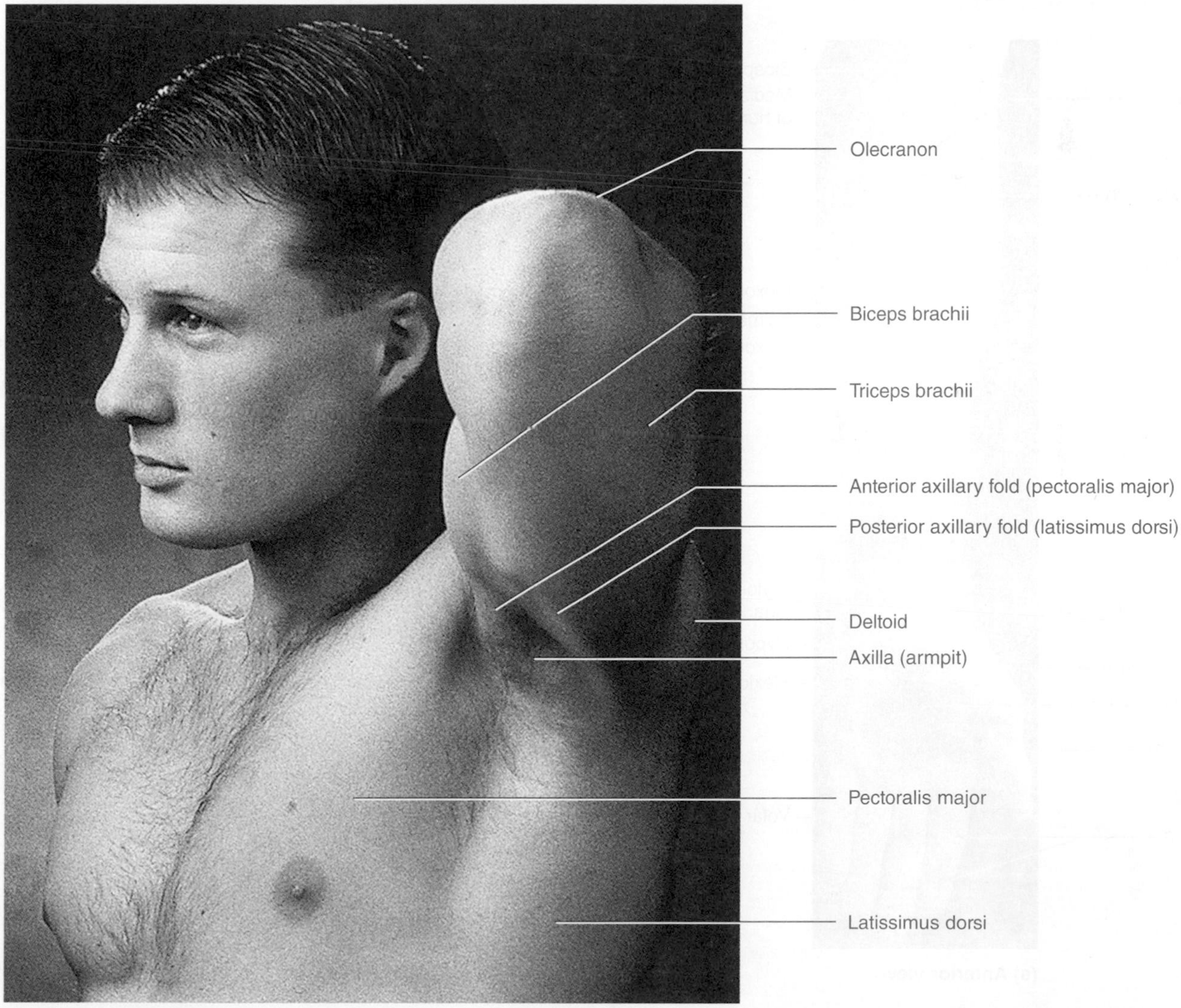

Olecranon

Biceps brachii

Triceps brachii

Anterior axillary fold (pectoralis major)

Posterior axillary fold (latissimus dorsi)

Deltoid

Axilla (armpit)

Pectoralis major

Latissimus dorsi

**FIGURE B.16  The Axillary Region.**

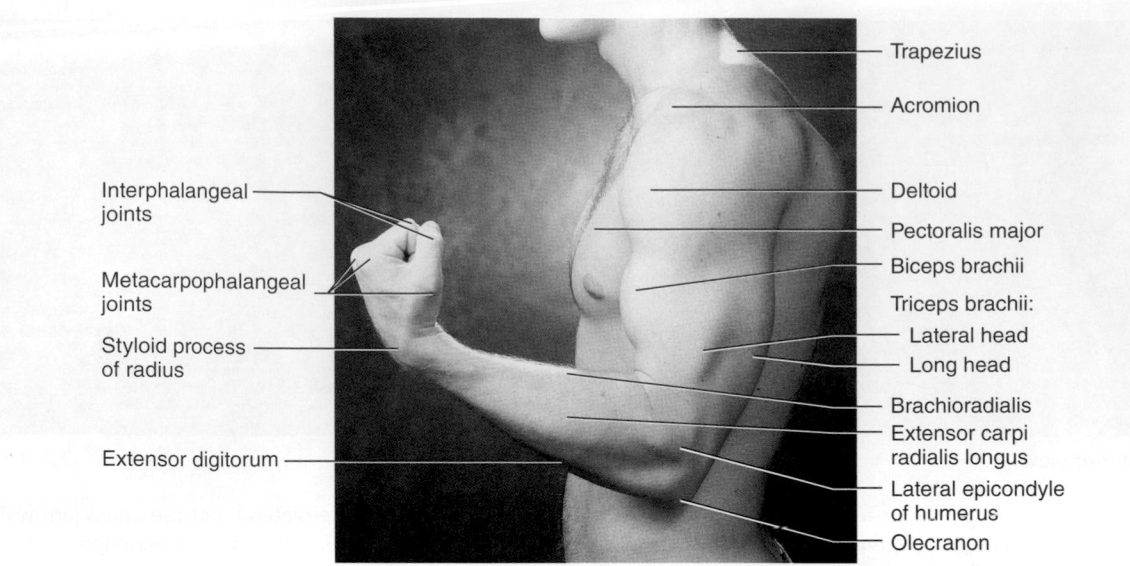

**FIGURE B.17** **The Upper Limb (Left Lateral View).**

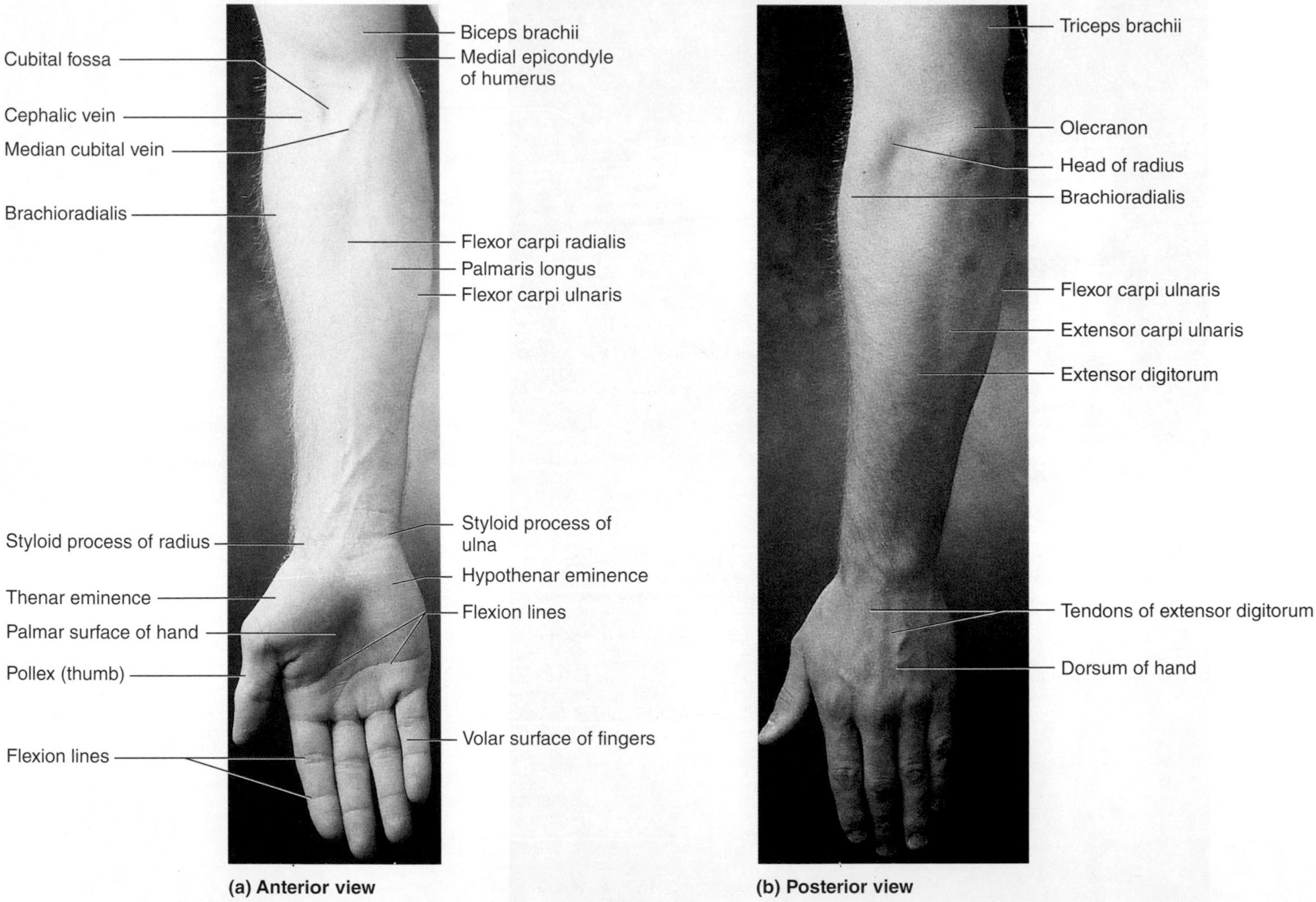

Trapezius

Acromion

Deltoid

Pectoralis major

Biceps brachii

Triceps brachii:
 Lateral head
 Long head

Brachioradialis

Extensor carpi radialis longus

Lateral epicondyle of humerus

Olecranon

Interphalangeal joints

Metacarpophalangeal joints

Styloid process of radius

Extensor digitorum

---

Biceps brachii

Medial epicondyle of humerus

Cubital fossa

Cephalic vein

Median cubital vein

Brachioradialis

Flexor carpi radialis

Palmaris longus

Flexor carpi ulnaris

Styloid process of radius

Thenar eminence

Palmar surface of hand

Pollex (thumb)

Flexion lines

Styloid process of ulna

Hypothenar eminence

Flexion lines

Volar surface of fingers

**(a) Anterior view**

Triceps brachii

Olecranon

Head of radius

Brachioradialis

Flexor carpi ulnaris

Extensor carpi ulnaris

Extensor digitorum

Tendons of extensor digitorum

Dorsum of hand

**(b) Posterior view**

**FIGURE B.18** **The Antebrachium (Forearm).**

❓ *Only two tendons of the extensor digitorum are labeled, but how many tendons does this muscle have in all?*

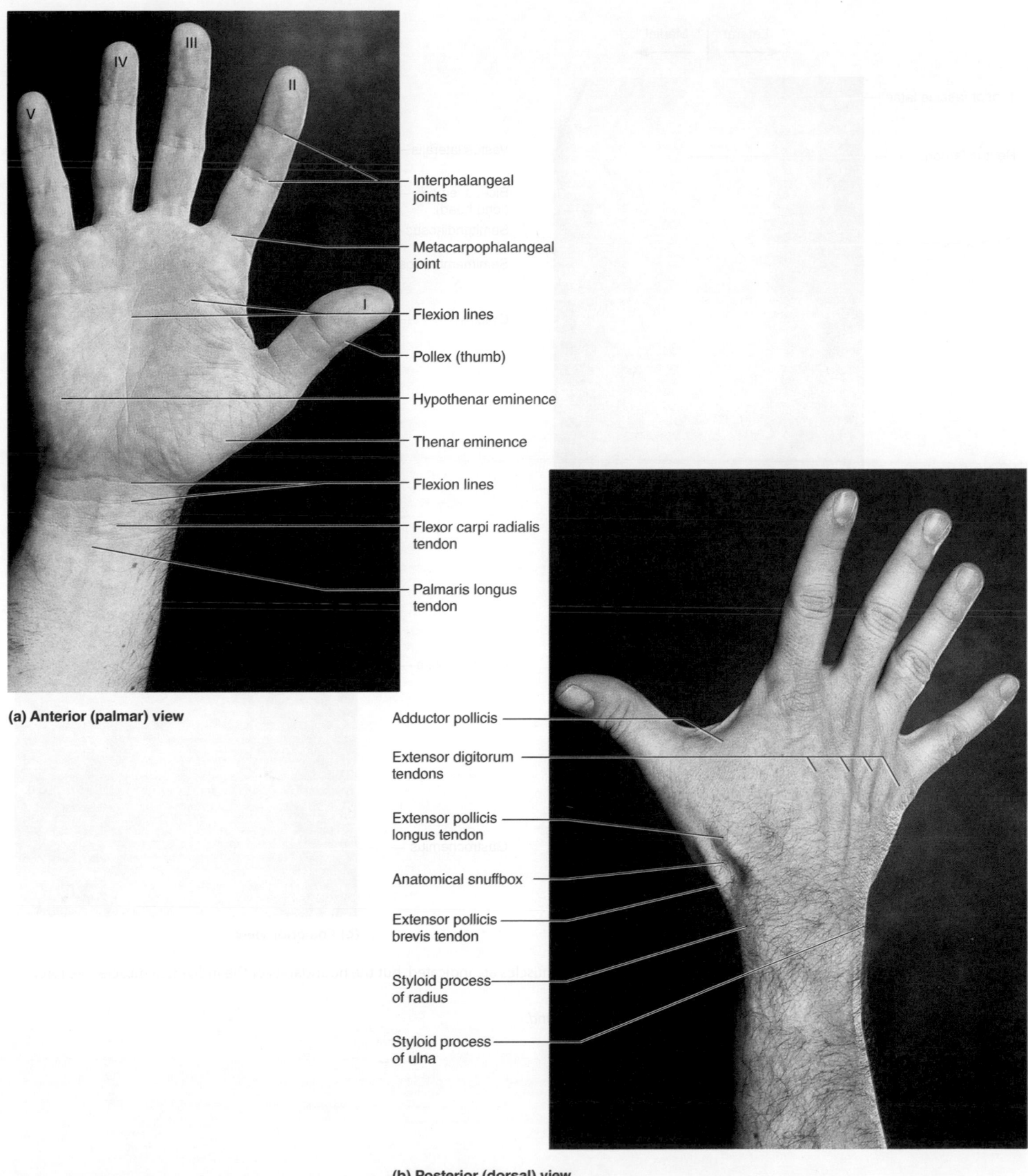

(a) Anterior (palmar) view

Interphalangeal joints

Metacarpophalangeal joint

Flexion lines

Pollex (thumb)

Hypothenar eminence

Thenar eminence

Flexion lines

Flexor carpi radialis tendon

Palmaris longus tendon

Adductor pollicis

Extensor digitorum tendons

Extensor pollicis longus tendon

Anatomical snuffbox

Extensor pollicis brevis tendon

Styloid process of radius

Styloid process of ulna

(b) Posterior (dorsal) view

**FIGURE B.19  The Wrist and Hand.**

❓ *Mark the spot on one or both photographs where a saddle joint can be found.*

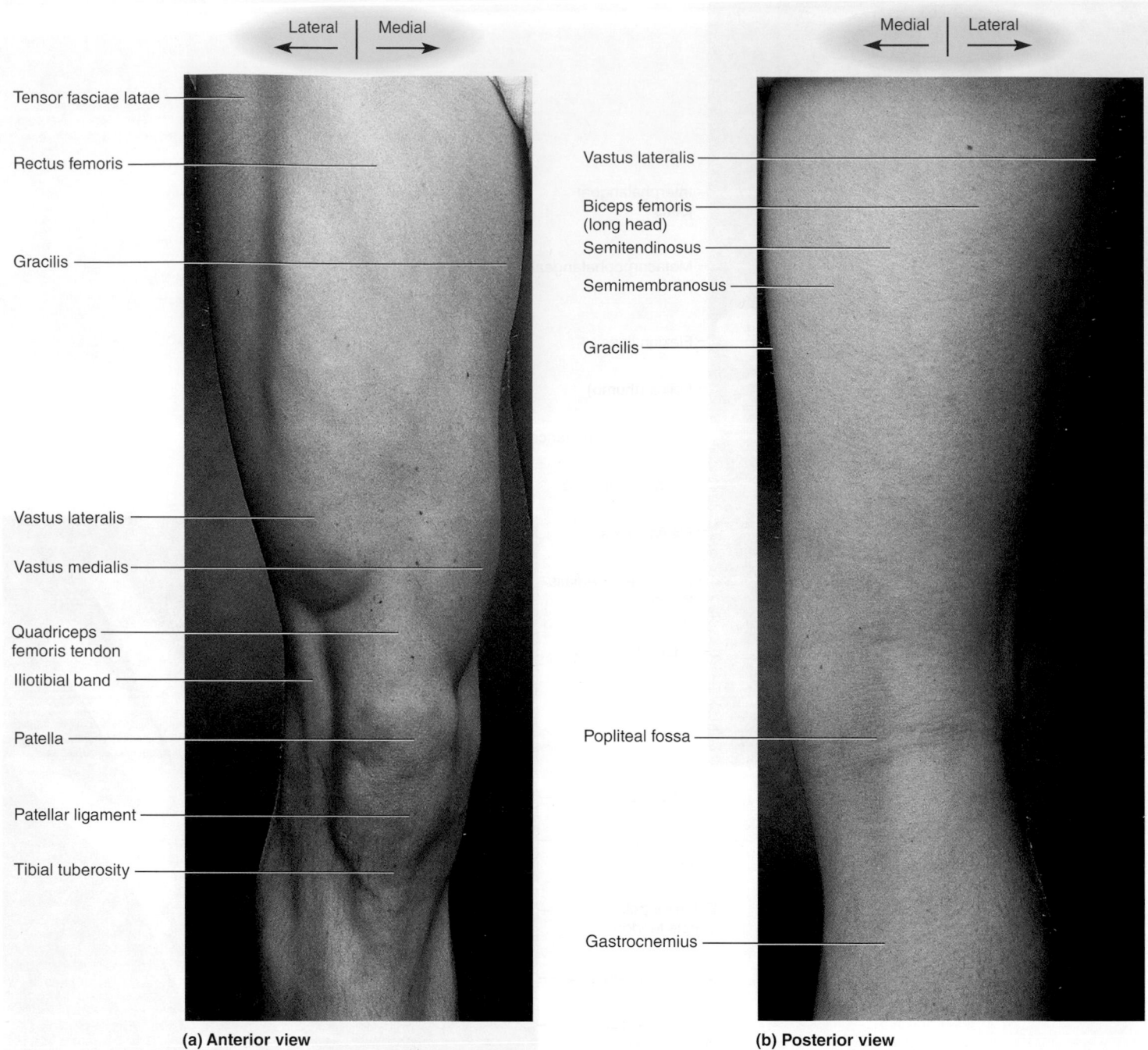

Lateral ← | → Medial

Tensor fasciae latae

Rectus femoris

Gracilis

Vastus lateralis

Vastus medialis

Quadriceps femoris tendon

Iliotibial band

Patella

Patellar ligament

Tibial tuberosity

**(a) Anterior view**

Medial ← | → Lateral

Vastus lateralis

Biceps femoris (long head)

Semitendinosus

Semimembranosus

Gracilis

Popliteal fossa

Gastrocnemius

**(b) Posterior view**

**FIGURE B.20    The Thigh and Knee.** Locations of posterior thigh muscles are indicated, but the boundaries of the individual muscles are rarely visible on a living person.

❓ *Mark the spot on part (a) where the vastus intermedius would be found.*

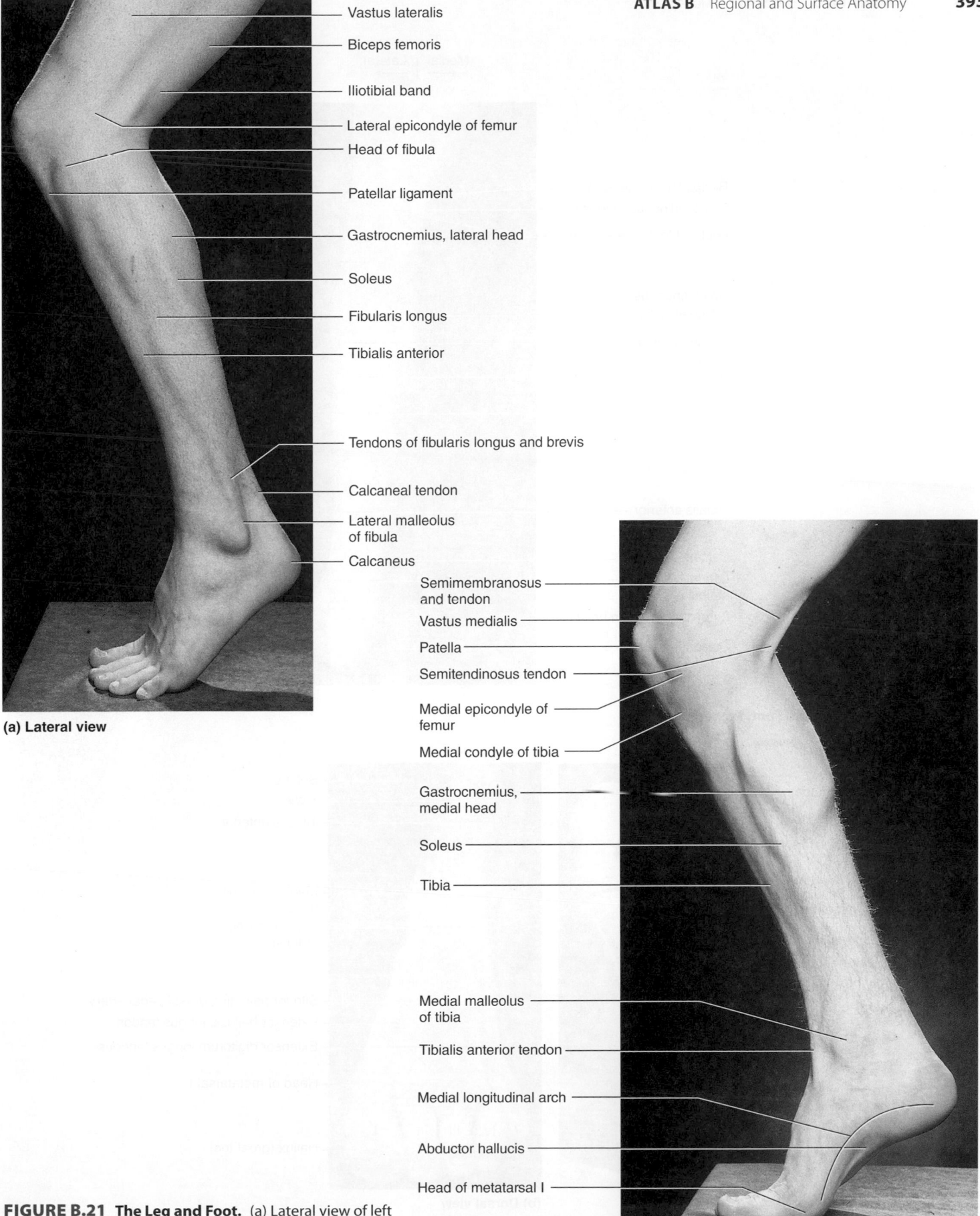

Vastus lateralis

Biceps femoris

Iliotibial band

Lateral epicondyle of femur

Head of fibula

Patellar ligament

Gastrocnemius, lateral head

Soleus

Fibularis longus

Tibialis anterior

Tendons of fibularis longus and brevis

Calcaneal tendon

Lateral malleolus of fibula

Calcaneus

**(a) Lateral view**

Semimembranosus and tendon

Vastus medialis

Patella

Semitendinosus tendon

Medial epicondyle of femur

Medial condyle of tibia

Gastrocnemius, medial head

Soleus

Tibia

Medial malleolus of tibia

Tibialis anterior tendon

Medial longitudinal arch

Abductor hallucis

Head of metatarsal I

**(b) Medial view**

**FIGURE B.21 The Leg and Foot.** (a) Lateral view of left limb. (b) Medial view of right limb.

**?** *The lateral malleolus is part of what bone?*

Medial | Lateral

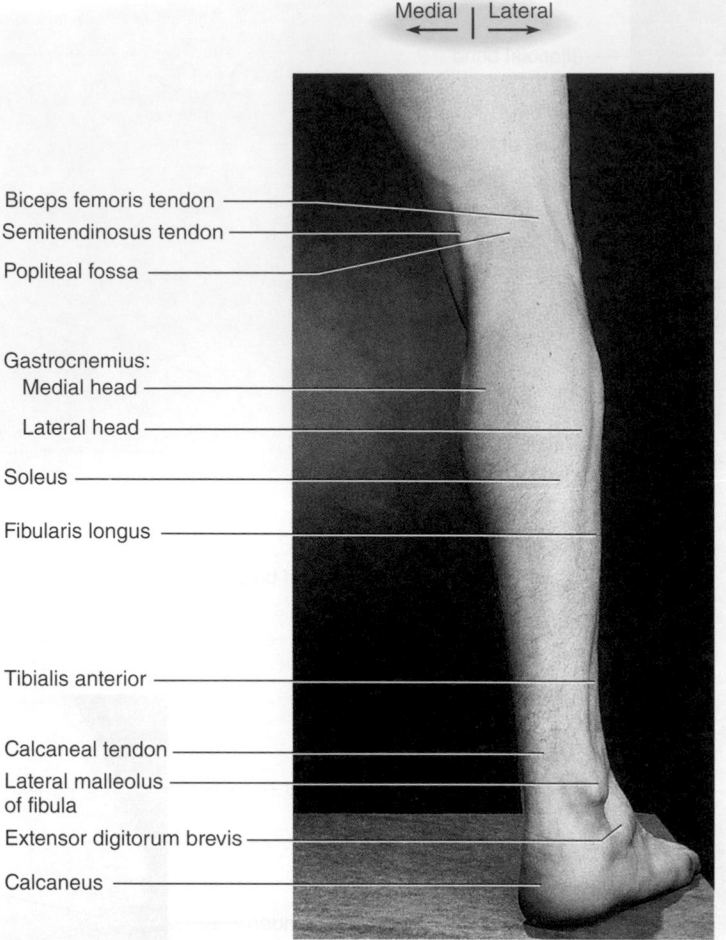

Biceps femoris tendon
Semitendinosus tendon
Popliteal fossa

Gastrocnemius:
Medial head
Lateral head
Soleus
Fibularis longus

Tibialis anterior

Calcaneal tendon
Lateral malleolus
of fibula
Extensor digitorum brevis
Calcaneus

**FIGURE B.22  The Leg and Foot, Posterior View.**

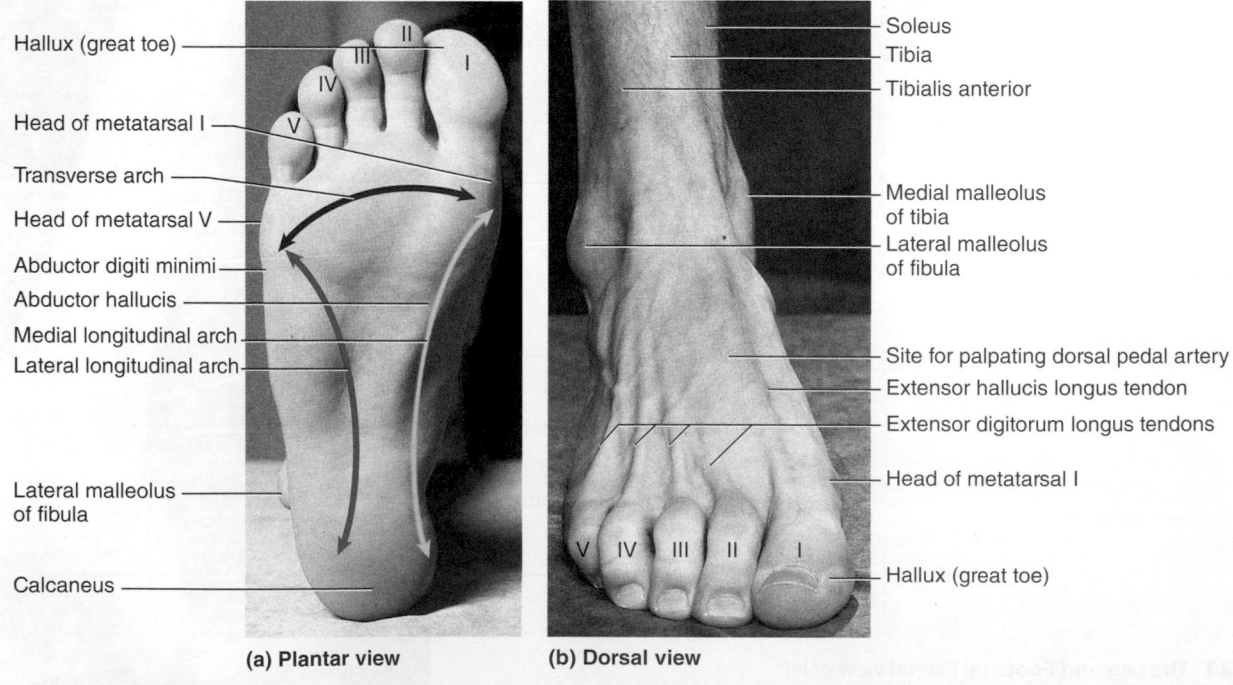

Hallux (great toe)

Head of metatarsal I
Transverse arch
Head of metatarsal V
Abductor digiti minimi
Abductor hallucis
Medial longitudinal arch
Lateral longitudinal arch

Lateral malleolus
of fibula

Calcaneus

Soleus
Tibia
Tibialis anterior

Medial malleolus
of tibia
Lateral malleolus
of fibula

Site for palpating dorsal pedal artery
Extensor hallucis longus tendon
Extensor digitorum longus tendons
Head of metatarsal I
Hallux (great toe)

**(a) Plantar view**        **(b) Dorsal view**

**FIGURE B.23  The Foot (Plantar and Dorsal Views).**  Compare the arches in part (a) to the skeletal anatomy in figure 8.42 (p. 268).

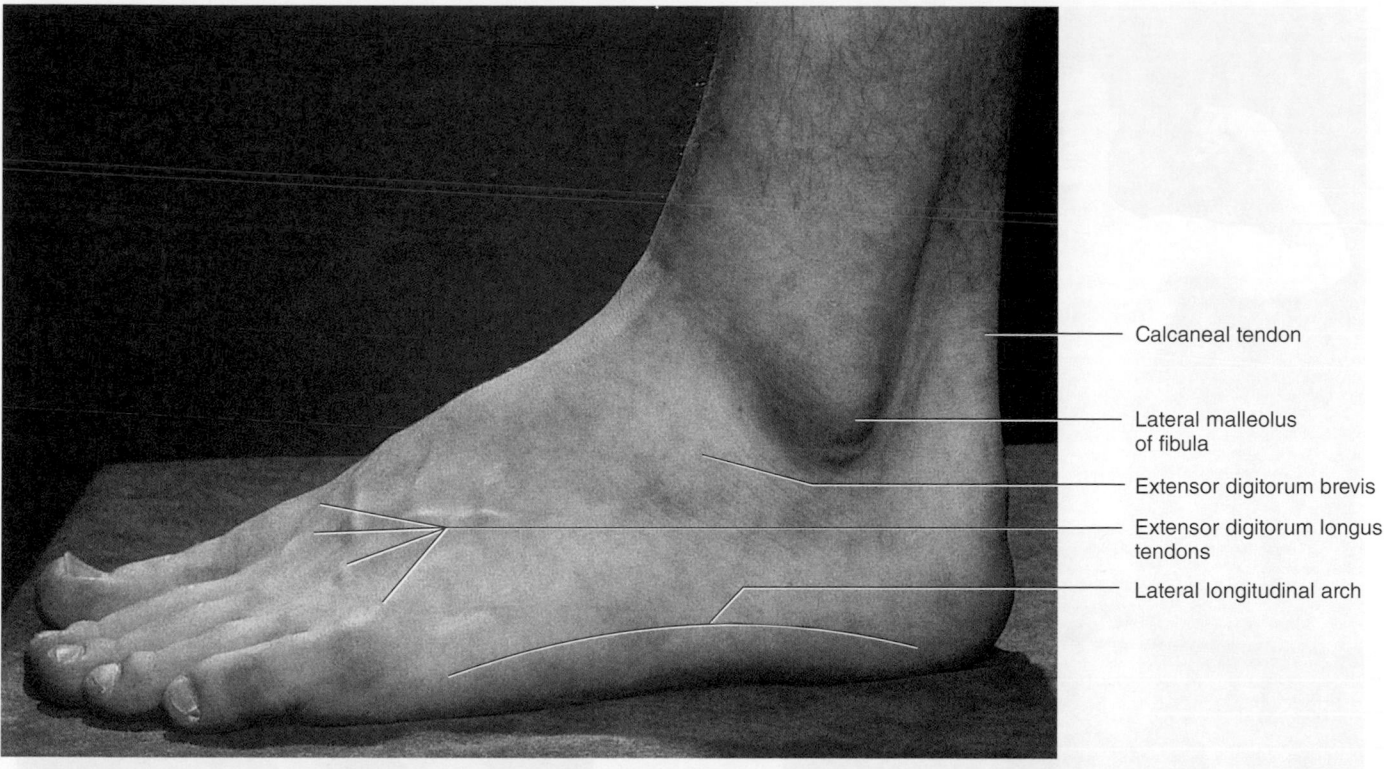

Calcaneal tendon

Lateral malleolus of fibula

Extensor digitorum brevis

Extensor digitorum longus tendons

Lateral longitudinal arch

**(a) Lateral view**

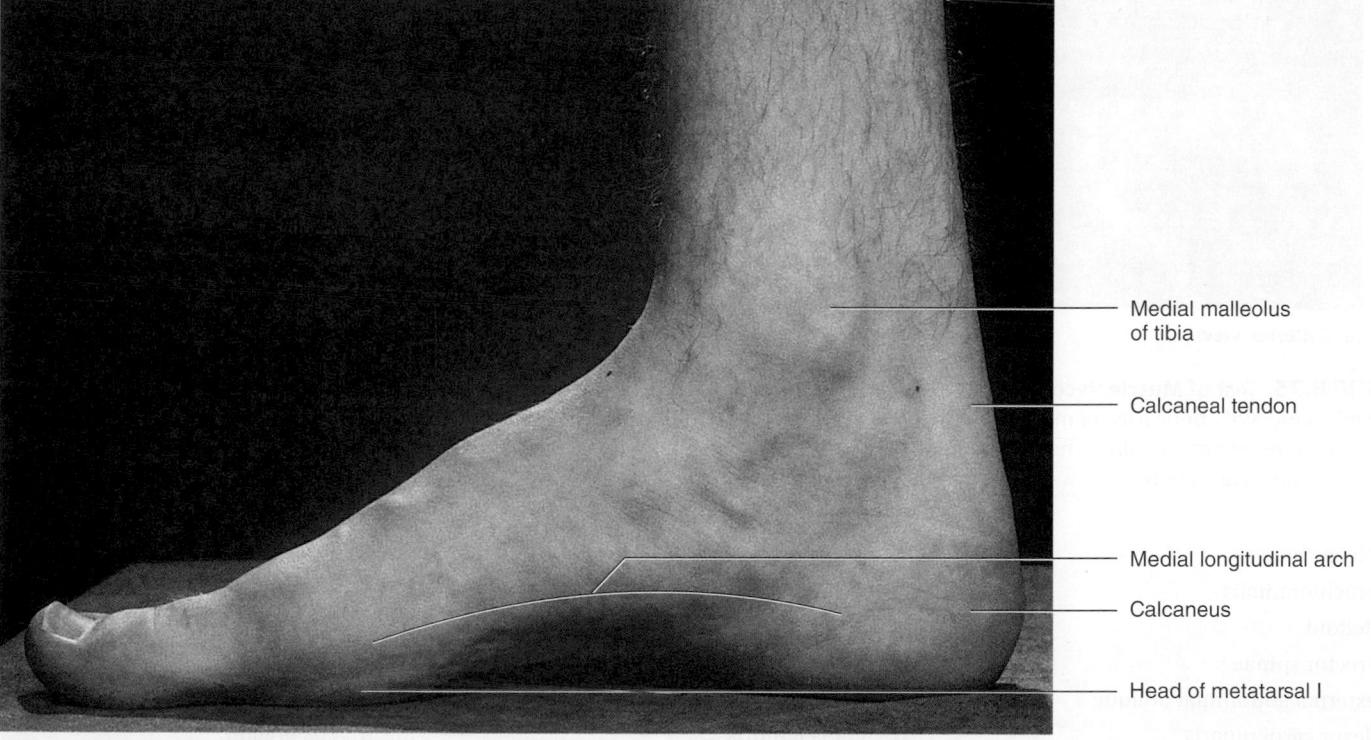

Medial malleolus of tibia

Calcaneal tendon

Medial longitudinal arch

Calcaneus

Head of metatarsal I

**(b) Medial view**

**FIGURE B.24** **The Foot (Lateral and Medial Views).**

*Indicate the position of middle phalanx I on each photograph.*

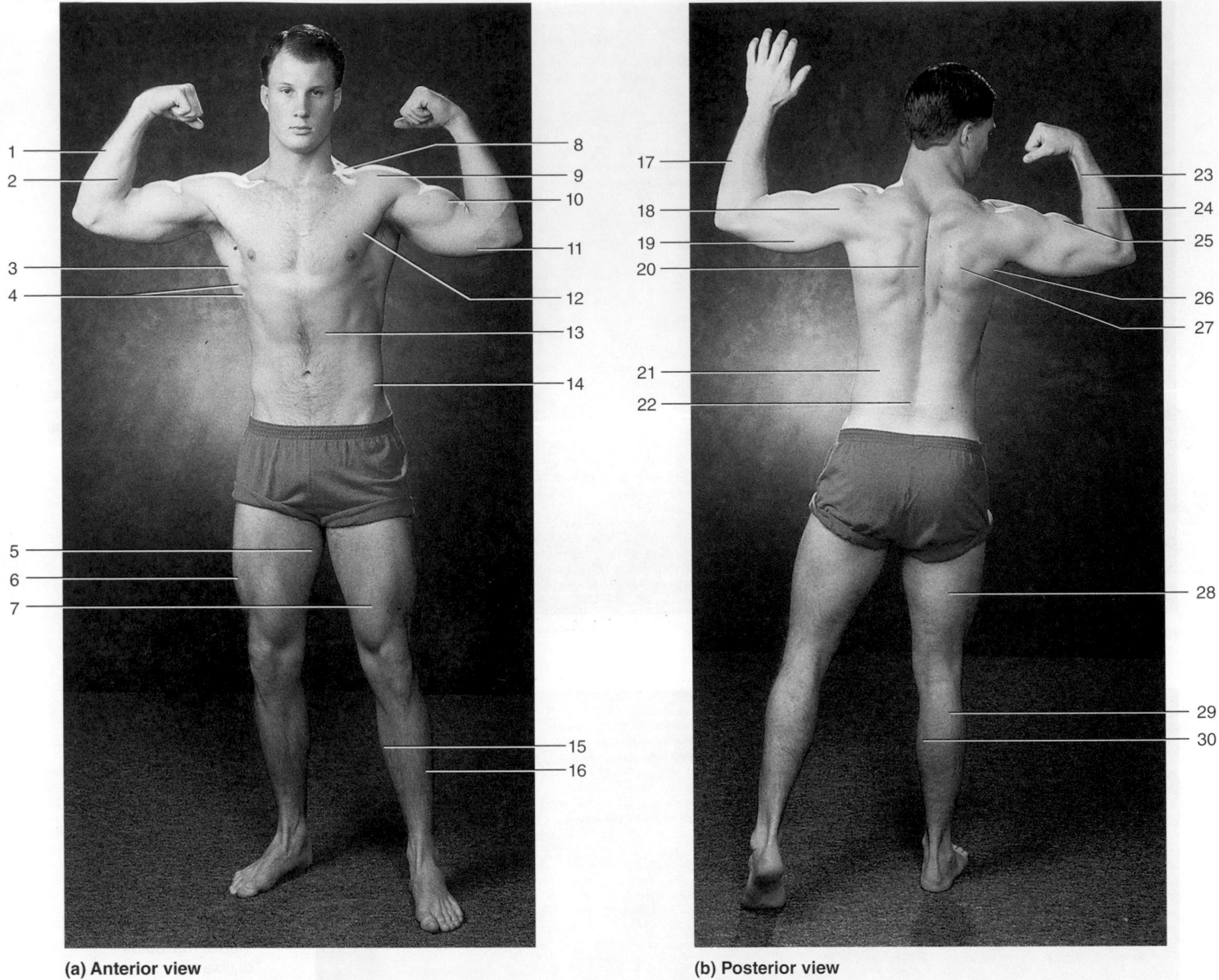

**(a) Anterior view**

**(b) Posterior view**

**FIGURE B.25  Test of Muscle Recognition.**  To test your knowledge of muscle anatomy, match the 30 labeled muscles on these photographs to the following alphabetical list of muscles. Answer as many as possible without referring back to the previous illustrations. Some of these names will be used more than once since the same muscle may be shown from different perspectives, and some of these names will not be used at all. The answers are in appendix B.

a. biceps brachii
b. brachioradialis
c. deltoid
d. erector spinae
e. external abdominal oblique
f. flexor carpi ulnaris
g. gastrocnemius
h. gracilis
i. hamstrings

j. infraspinatus
k. latissimus dorsi
l. pectineus
m. pectoralis major
n. rectus abdominis
o. rectus femoris
p. serratus anterior
q. soleus
r. splenius capitis

s. sternocleidomastoid
t. subscapularis
u. teres major
v. tibialis anterior
w. trapezius
x. triceps brachii
y. vastus lateralis
z. vastus medialis

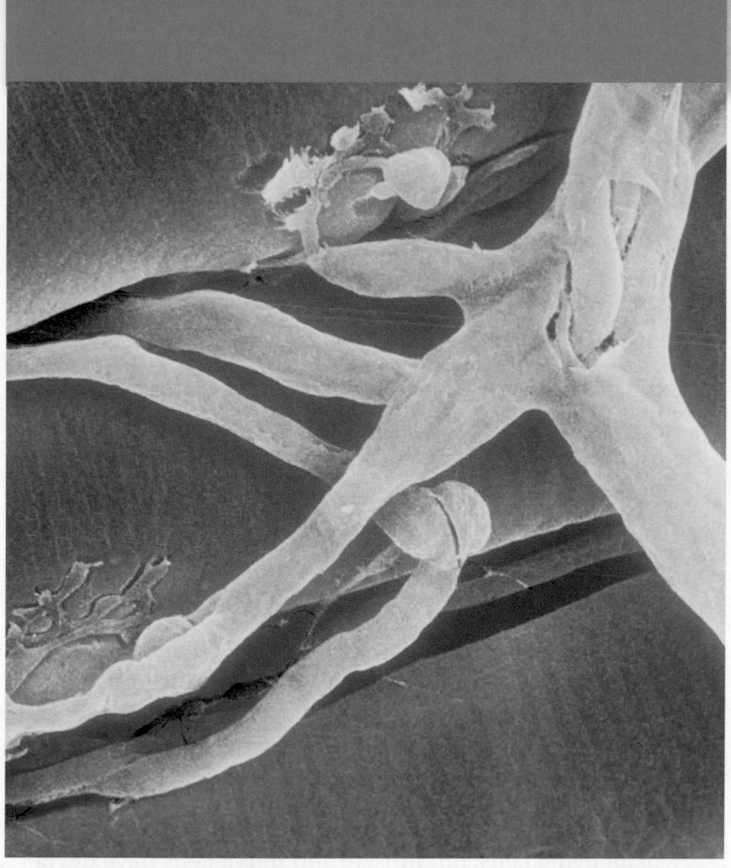

Neuromuscular junctions (SEM)

Anatomy & Physiology REVEALED®
aprevealed.com

**Module 6: Muscular System**

## BRUSHING UP

- Histological differences between the three muscle types were introduced on pages 160–161. Those differences will be more deeply explored in this chapter.
- You should be familiar with the organization of skeletal muscle and its associated connective tissues, introduced on page 310.
- This chapter deals with the relationship of neurons to muscle fibers. You should know the basic neuron structure introduced on page 160.
- The stimulation of a muscle fiber by a neuron is based on principles of plasma membrane proteins as receptors and as ligand- and voltage-gated ion channels (p. 84).
- The concept of flow down gradients (p. 17) is helpful for understanding muscle excitation and the activation of contraction.
- The differences between aerobic respiration and anaerobic fermentation (pp. 70–71) are central to understanding exercise physiology and the energy metabolism of muscle.
- Understanding cardiac and smooth muscle requires familiarity with desmosomes and gap junctions (pp. 163–164).

**M**ovement is a fundamental characteristic of all living organisms, from bacteria to humans. Even plants and other seemingly immobile organisms move cellular components from place to place. Across the entire spectrum of life, the molecular mechanisms of movement are very similar, involving motor proteins such as myosin and dynein. But in animals, movement has developed to the highest degree, with the evolution of **muscle cells** specialized for this function. A muscle cell is essentially a device for converting the chemical energy of ATP into the mechanical energy of movement.

The three types of muscular tissue—skeletal, cardiac, and smooth—were described and compared in chapter 5. Cardiac and smooth muscle are further described in this chapter, and cardiac muscle is discussed most extensively in chapter 19. Most of the present chapter, however, concerns skeletal muscle, the type that holds the body erect against the pull of gravity and produces its outwardly visible movements.

This chapter treats the structure, contraction, and metabolism of skeletal muscle at the molecular, cellular, and tissue levels of organization. Understanding muscle at these levels provides an indispensable basis for understanding such aspects of motor performance as warm-up, quickness, strength, endurance, and fatigue. Such factors have obvious relevance to athletic performance, and they become very important when a lack of physical conditioning, old age, or injury interferes with a person's ability to carry out everyday tasks or meet the extra demands for speed or strength that we all occasionally encounter.

## 11.1 Types and Characteristics of Muscular Tissue

### Expected Learning Outcomes
When you have completed this section, you should be able to

a. describe the physiological properties that all muscle types have in common;

b. list the defining characteristics of skeletal muscle; and

c. discuss the possible elastic functions of the connective tissue components of a muscle.

### Universal Characteristics of Muscle

The functions of the muscular system were detailed in the preceding chapter: movement, stability, communication, control of body openings and passages, heat production, and glycemic control (p. 309). To carry out those functions, all muscle cells have the following characteristics.

- **Excitability (responsiveness).** Excitability is a property of all living cells, but muscle and nerve cells have developed this property to the highest degree. When stimulated by chemical signals, stretch, and other stimuli, muscle cells respond with electrical changes across the plasma membrane.

- **Conductivity.** Stimulation of a muscle cell produces more than a local effect. The local electrical change triggers a wave of excitation that travels rapidly along the cell and initiates processes leading to contraction.

- **Contractility.** Muscle cells are unique in their ability to shorten substantially when stimulated. This enables them to pull on bones and other organs to create movement.

- **Extensibility.** In order to contract, a muscle cell must also be extensible—able to stretch again between contractions. Most cells rupture if they are stretched even a little, but skeletal muscle cells can stretch to as much as three times their contracted length.

- **Elasticity.** When a muscle cell is stretched and then released, it recoils to a shorter length. If it were not for this elastic recoil, resting muscles would be too slack.

### Skeletal Muscle

**Skeletal muscle** may be defined as voluntary striated muscle that is usually attached to one or more bones. A skeletal muscle exhibits alternating light and dark transverse bands, or **striations** (fig. 11.1), that reflect an overlapping arrangement of their

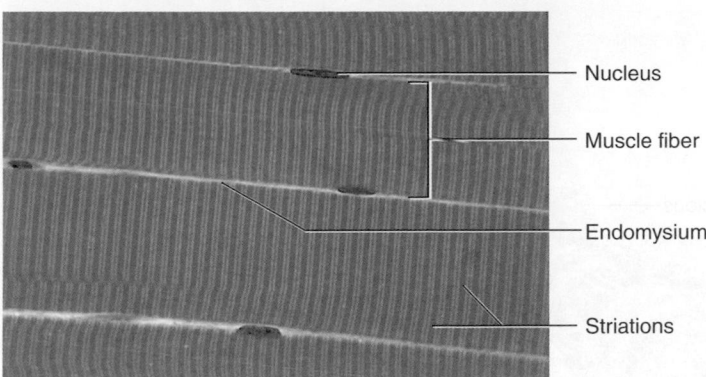

**FIGURE 11.1** **Skeletal Muscle Fibers.** AP|R

❓ *What tissue characteristics evident in this photo distinguish this from cardiac and smooth muscle?*

internal contractile proteins. Skeletal muscle is called **voluntary** because it is usually subject to conscious control. The other types of muscle are **involuntary** (not usually under conscious control), and they are never attached to bones.

A typical skeletal muscle cell is about 100 μm in diameter and 3 cm (30,000 μm) long; some are as thick as 500 μm and as long as 30 cm. Because of their extraordinary length, skeletal muscle cells are usually called **muscle fibers** or **myofibers.**

Recall from chapter 10 that a skeletal muscle is composed not only of muscular tissue, but also of fibrous connective tissue: the *endomysium* that surrounds each muscle fiber, the *perimysium* that bundles muscle fibers together into fascicles, and the *epimysium* that encloses the entire muscle (see fig. 10.1, p. 310). These connective tissues are continuous with the collagen fibers of tendons and those, in turn, with the collagen of the bone matrix. Thus, when a muscle fiber contracts, it pulls on these collagen fibers and often moves a bone.

Collagen is neither excitable nor contractile, but it is somewhat extensible and elastic. When a muscle lengthens, for example during extension of a joint, its collagenous components resist excessive stretching and protect the muscle from injury. When a muscle relaxes, elastic recoil of the collagen may help to return the muscle to its resting length and keep it from becoming too flaccid. Some authorities contend that recoil of the tendons and other collagenous tissues contributes significantly to the power output and efficiency of a muscle. When you are running, for example, recoil of the calcaneal tendon may help to lift the heel and produce some of the thrust as your toes push off from the ground. (Such recoil contributes significantly to the long, efficient leaps of a kangaroo.) Others feel that the elasticity of these components is negligible in humans and that the recoil is produced entirely by certain intracellular proteins of the muscle fibers themselves.

Answer the following questions to test your understanding of the preceding section:

1. Define *responsiveness, conductivity, contractility, extensibility,* and *elasticity.* State why each of these properties is necessary for muscle function.

2. How is skeletal muscle different from the other types of muscle?

3. Name and define the three layers of collagenous connective tissue in a skeletal muscle.

## 11.2 Microscopic Anatomy of Skeletal Muscle

### Expected Learning Outcomes

When you have completed this section, you should be able to

a. describe the structural components of a muscle fiber;

b. relate the striations of a muscle fiber to the overlapping arrangement of its protein filaments; and

c. name the major proteins of a muscle fiber and state the function of each.

### The Muscle Fiber

In order to understand muscle function, you must know how the organelles and macromolecules of a muscle fiber are arranged. Perhaps more than any other cell, a muscle fiber exemplifies the adage, Form follows function. It has a complex, tightly organized internal structure in which even the spatial arrangement of protein molecules is closely tied to its contractile function.

The plasma membrane of a muscle fiber is called the **sarcolemma,**[1] and its cytoplasm is called the **sarcoplasm.** The sarcoplasm is occupied mainly by long protein cords called **myofibrils** about 1 μm in diameter (fig. 11.2). It also contains an abundance of **glycogen,** a starchlike carbohydrate that provides energy for the cell during heightened levels of exercise, and the red oxygen-binding pigment **myoglobin,** which provides some of the oxygen needed for muscular activity.

Muscle fibers have multiple flattened or sausage-shaped nuclei pressed against the inside of the sarcolemma. This unusual multinuclear condition results from the embryonic development of a muscle fiber—several stem cells called **myoblasts**[2] fuse to produce each fiber, with each myoblast contributing one nucleus. Some myoblasts remain as unspecialized **satellite cells** between the muscle fiber and endomysium. These play an important role in the regeneration of damaged skeletal muscle.

Most other organelles of the cell, such as mitochondria, are packed into the spaces between the myofibrils. The smooth endoplasmic reticulum, here called the **sarcoplasmic reticulum (SR),**

[1] *sarco* = flesh, muscle; *lemma* = husk
[2] *myo* = muscle; *blast* = precursor

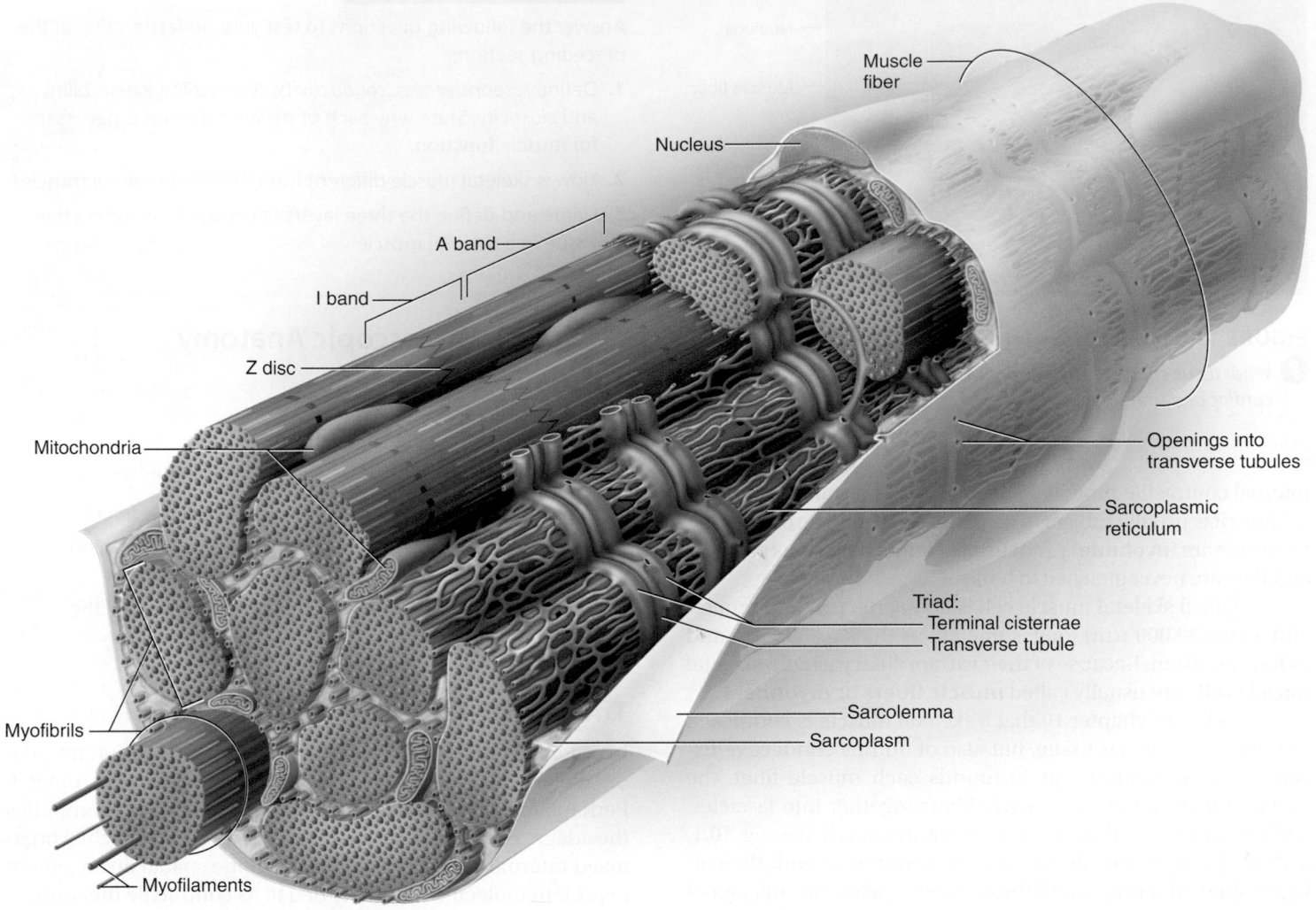

**FIGURE 11.2 Structure of a Skeletal Muscle Fiber.** This is a single cell containing 11 myofibrils (9 shown at the left end and 2 cut off at midfiber). A few myofilaments are shown projecting from the myofibril at the left. Most muscle fibers have from several dozen to a thousand or more myofibrils. Myofibril fine structure is shown in figure 11.3.

**?** *Why is it important for the transverse tubule to be so closely associated with the terminal cisternae?*

forms a network around each myofibril. It periodically exhibits dilated end-sacs called **terminal cisternae,** which cross the muscle fiber from one side to the other. The sarcolemma has tubular infoldings called **transverse (T) tubules,** which penetrate through the cell and emerge on the other side. Each T tubule is closely associated with two terminal cisternae running alongside it, one on each side. The T tubule and the two terminal cisternae associated with it constitute a **triad.** The SR is a reservoir of calcium ions; it has gated channels in its membrane that open at the right times to release a flood of calcium into the cytosol, which activates the muscle contraction process. The T tubule signals the SR when to release these calcium bursts.

## Myofilaments

Let's return to the myofibrils just mentioned—the long protein cords that fill most of the muscle cell—and look at their structure at a finer, molecular level. It is here that the key to

muscle contraction lies. Each myofibril is a bundle of parallel protein microfilaments called **myofilaments** (see the left end of fig. 11.2). There are three kinds of myofilaments:

1. **Thick filaments** (fig. 11.3a, b, d) are about 15 nm in diameter. Each is made of several hundred molecules of a protein called **myosin.** A myosin molecule is shaped like a golf club, with two chains intertwined to form a shaft-like *tail* and a double globular *head* projecting from it at an angle. A thick filament may be likened to a bundle of golf clubs, with their heads directed outward in a helical array around the bundle. The heads on one half of the thick filament angle to the left, and the heads on the other half angle to the right; in the middle is a *bare zone* with no heads.

2. **Thin filaments** (fig. 11.3c, d), 7 nm in diameter, are composed primarily of two intertwined strands of a protein called **fibrous (F) actin.** Each F actin is like a bead

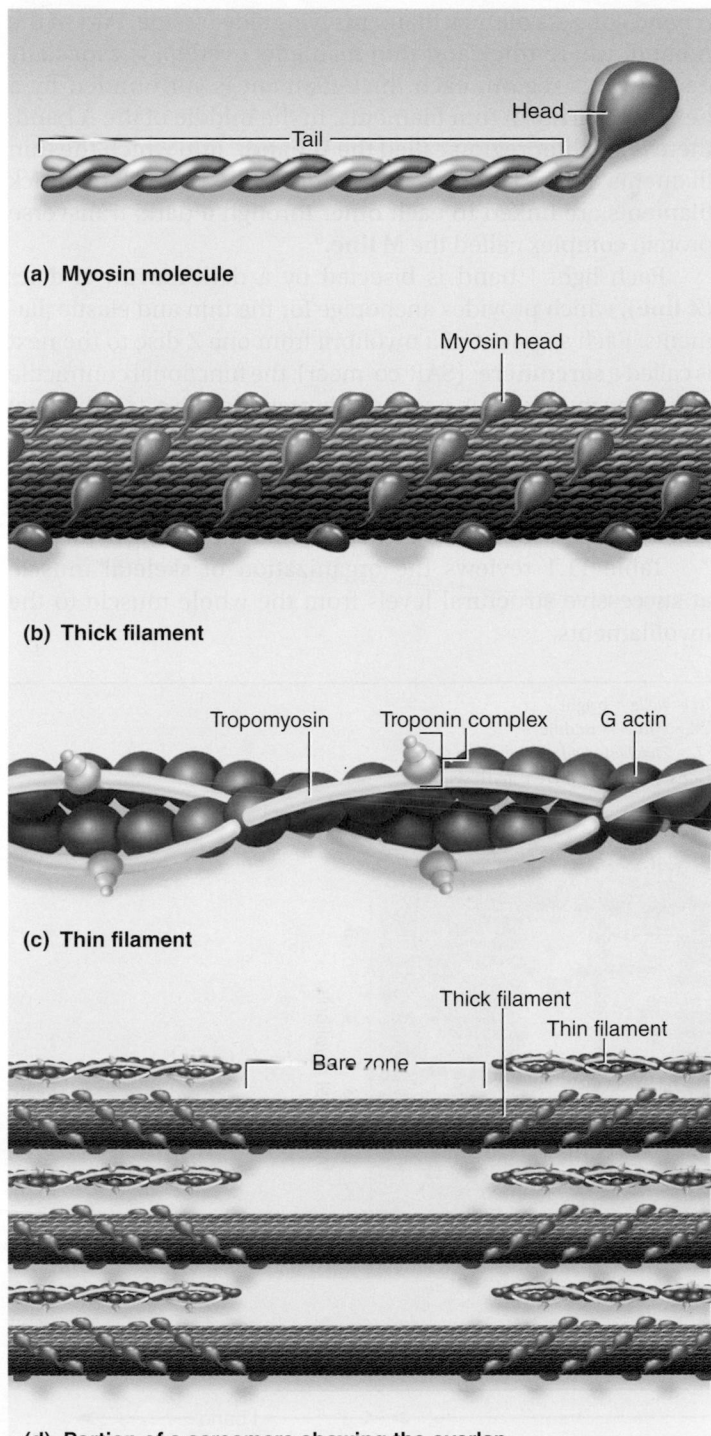

**(a) Myosin molecule**

**(b) Thick filament**

**(c) Thin filament**

**(d) Portion of a sarcomere showing the overlap of thick and thin filaments**

**FIGURE 11.3 Molecular Structure of Thick and Thin Filaments.** (a) A single myosin molecule consists of two intertwined polypeptides forming a twisted filamentous tail and a double globular head. (b) A thick filament consists of 200 to 500 myosin molecules bundled together with the heads projecting outward in a helical array. (c) A thin filament consists of two intertwined chains of G actin molecules, smaller filamentous tropomyosin molecules, and a calcium-binding protein called troponin associated with the tropomyosin. (d) A region of overlap between the thick and thin myofilaments.

necklace—a string of subunits called **globular (G) actin.** Each G actin has an **active site** that can bind to the head of a myosin molecule. A thin filament also has 40 to 60 molecules of yet another protein, **tropomyosin.** When a muscle fiber is relaxed, each tropomyosin blocks the active sites of six or seven G actins and prevents myosin from binding to them. Each tropomyosin molecule, in turn, has a smaller calcium-binding protein called **troponin** bound to it.

3. **Elastic filaments** (see fig. 11.5b), 1 nm in diameter, are made of a huge springy protein called **titin.**[3] They run through the core of each thick filament and anchor it to structures called the *Z disc* at one end and *M line* at the other. Titin stabilizes the thick filament, centers it between the thin filaments, prevents overstretching, and recoils like a spring after a muscle is stretched.

Myosin and actin are called **contractile proteins** because they do the work of shortening the muscle fiber. Tropomyosin and troponin are called **regulatory proteins** because they act like a switch to determine when the fiber can contract and when it cannot. Several clues as to how they do this may be apparent from what has already been said—calcium ions are released into the sarcoplasm to activate contraction; calcium binds to troponin; troponin is also bound to tropomyosin; and tropomyosin blocks the active sites of actin, so that myosin cannot bind to it when the muscle is not stimulated. Perhaps you are already forming some idea of the contraction mechanism to be explained shortly.

At least seven other accessory proteins occur in the thick and thin filaments or are associated with them. Among other functions, they anchor the myofilaments, regulate their length, and keep them aligned with each other for optimal contractile effectiveness. The most clinically important of these is **dystrophin,** an enormous protein located between the sarcolemma and the outermost myofilaments. It links actin filaments to a peripheral protein on the inner face of the sarcolemma. Through a series of links (fig. 11.4), this leads ultimately to the fibrous endomysium surrounding the muscle fiber. Therefore, when the thin filaments move, they pull on the dystrophin, and this ultimately pulls on the extracellular connective tissues leading to the tendon. Genetic defects in dystrophin are responsible for the disabling disease, *muscular dystrophy* (see Deeper Insight 11.4).

## Striations

Myosin and actin are not unique to muscle; they occur in all cells, where they function in cellular motility, mitosis, and transport of intracellular materials. In skeletal and cardiac muscle they are especially abundant, however, and are organized in a precise array that accounts for the striations of these muscle types (fig. 11.5).

Striated muscle has dark **A bands** alternating with lighter **I bands.** (*A* stands for *anisotropic* and *I* for *isotropic,* which refer to the way these bands affect polarized light. To help remember which band is which, think "dArk" and "lIght.") Each

[3]*tit* = giant; *in* = protein.

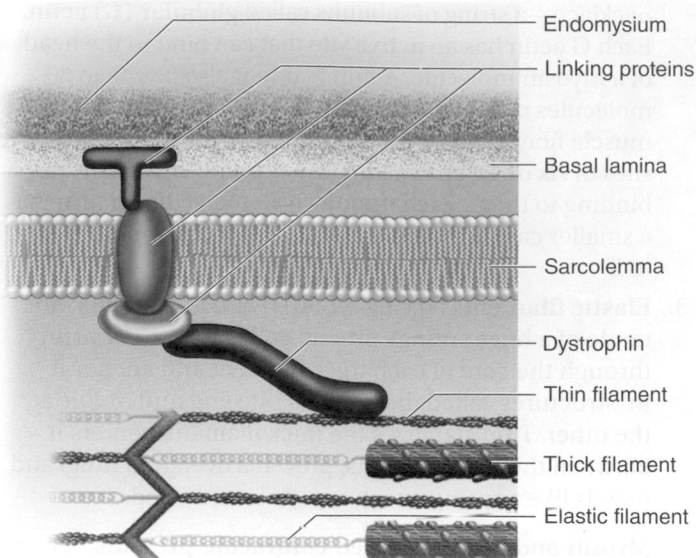

Endomysium

Linking proteins

Basal lamina

Sarcolemma

Dystrophin

Thin filament

Thick filament

Elastic filament

**FIGURE 11.4 Dystrophin.** One end of this large protein is linked to the actin of a thin myofilament near the surface of the muscle fiber. The other is linked to a peripheral protein on the inside of the sarcolemma. Through a complex of transmembrane and extracellular proteins, dystrophin transfers the force of myofilament movement to the basal lamina, endomysium, and other extracellular components of the muscle.

A band consists of thick filaments lying side by side. Part of the A band, where thick and thin filaments overlap, is especially dark. In this region, each thick filament is surrounded by a hexagonal array of thin filaments. In the middle of the A band, there is a lighter region called the **H band,**[4] into which the thin filaments do not reach. In the middle of the H band, the thick filaments are linked to each other through a dark, transverse protein complex called the **M line.**[5]

Each light I band is bisected by a dark narrow **Z disc**[6] **(Z line),** which provides anchorage for the thin and elastic filaments. Each segment of a myofibril from one Z disc to the next is called a **sarcomere**[7] (SAR-co-meer), the functional contractile unit of the muscle fiber. A muscle shortens because its individual sarcomeres shorten and pull the Z discs closer to each other, and dystrophin and the linking proteins pull on the extracellular proteins of the muscle. As the Z discs are pulled closer together, they pull on the sarcolemma to achieve overall shortening of the cell.

Table 11.1 reviews the organization of skeletal muscle at successive structural levels from the whole muscle to the myofilaments.

---

[4]H = *helle* = bright
[5]M = *Mittel* = middle
[6]Z = *Zwichenscheibe* = between disc
[7]*sarco* = muscle; *mere* = part, segment

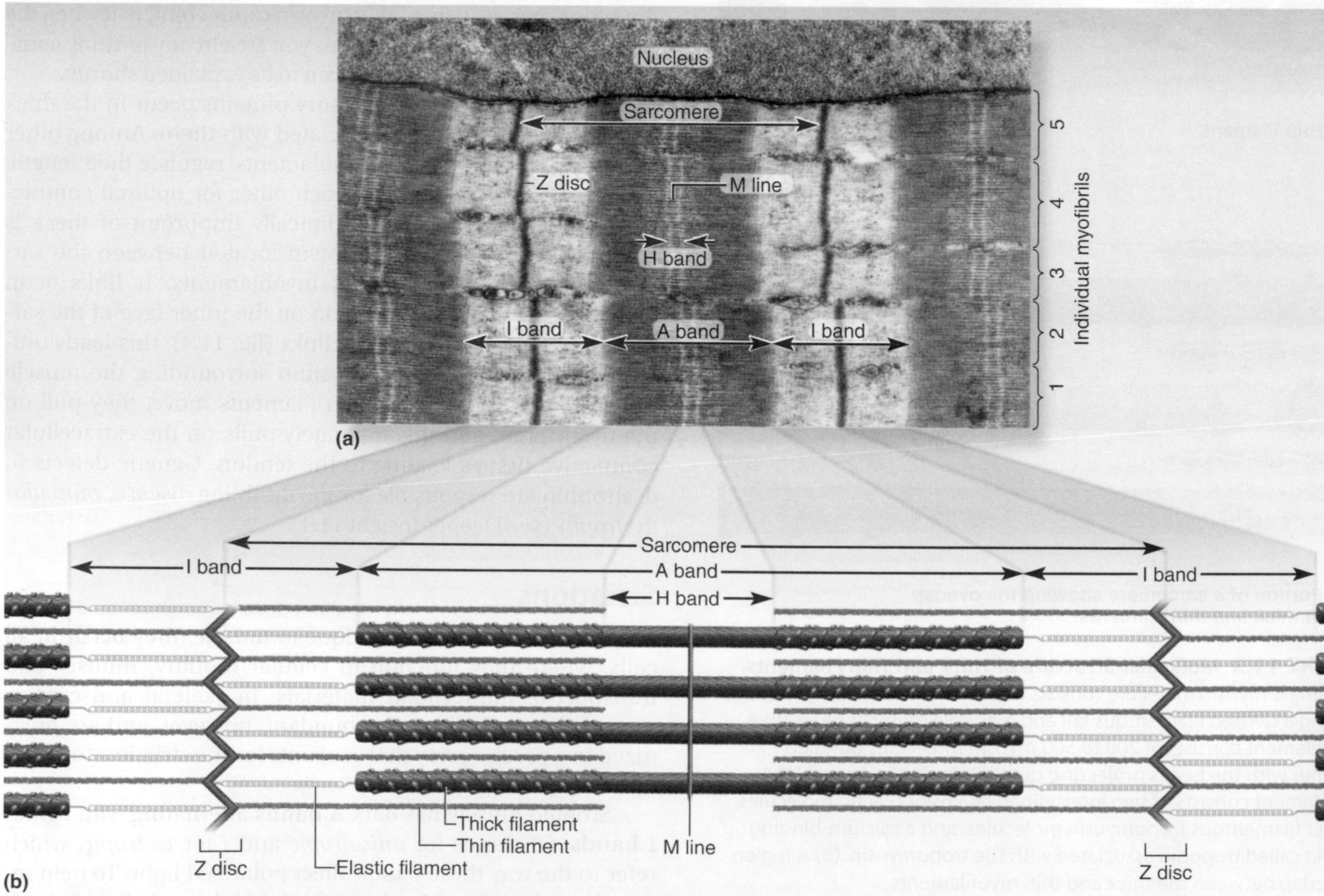

Nucleus

Sarcomere

Z disc

M line

H band

I band    A band    I band

Individual myofibrils

**(a)**

Sarcomere

I band    A band    I band

H band

Thick filament

Thin filament

M line

Z disc    Elastic filament    Z disc

**(b)**

**FIGURE 11.5 Muscle Striations and Their Molecular Basis.** (a) Five myofibrils of a single muscle fiber, showing the striations in the relaxed state (TEM). (b) The overlapping pattern of thick and thin myofilaments that accounts for the striations seen in part (a). **AP|R**

| TABLE 11.1 | The Structural Hierarchy of a Skeletal Muscle |
|---|---|

| Structural Level | Description |
|---|---|
| **Muscle** 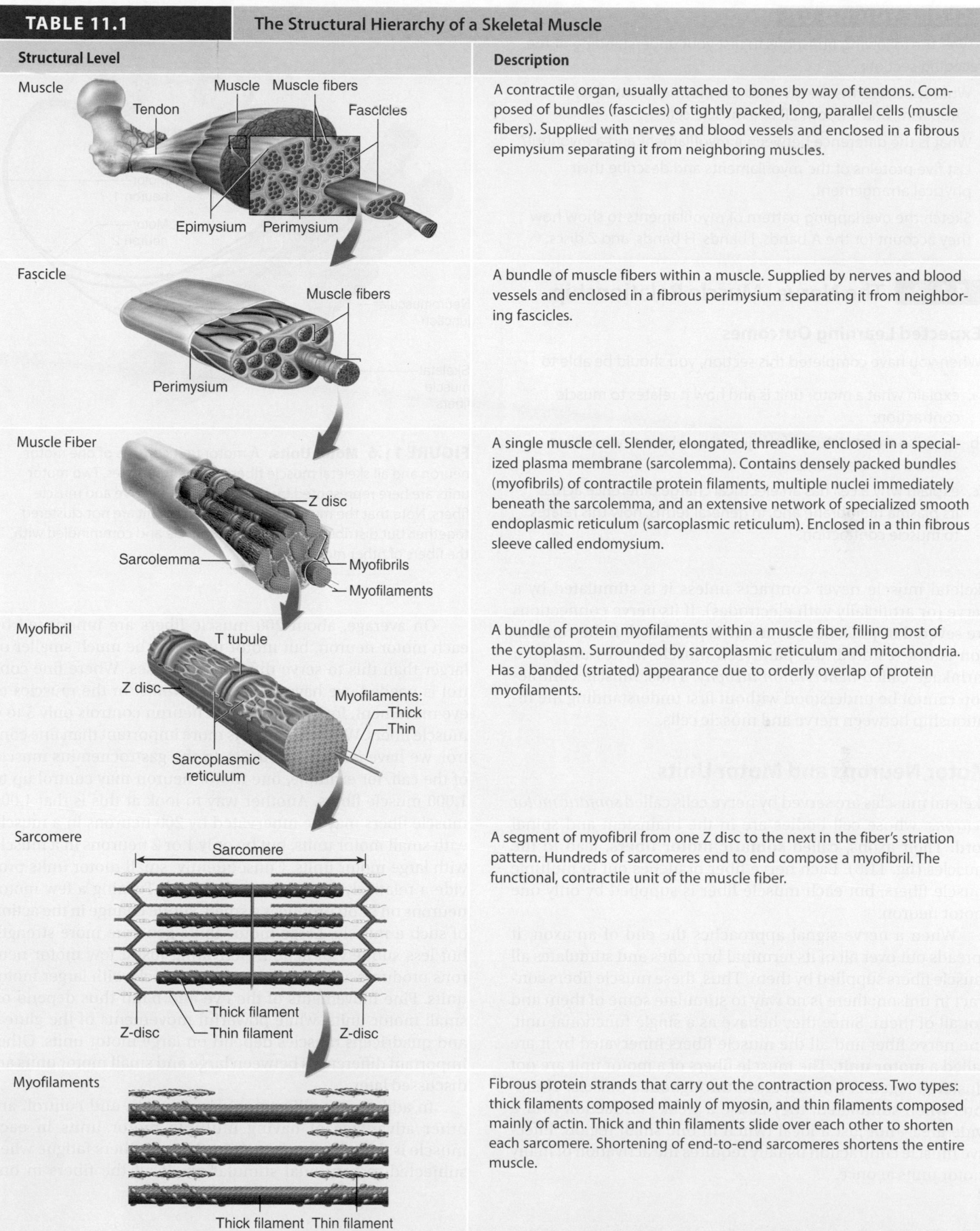 | A contractile organ, usually attached to bones by way of tendons. Composed of bundles (fascicles) of tightly packed, long, parallel cells (muscle fibers). Supplied with nerves and blood vessels and enclosed in a fibrous epimysium separating it from neighboring muscles. |
| **Fascicle** | A bundle of muscle fibers within a muscle. Supplied by nerves and blood vessels and enclosed in a fibrous perimysium separating it from neighboring fascicles. |
| **Muscle Fiber** | A single muscle cell. Slender, elongated, threadlike, enclosed in a specialized plasma membrane (sarcolemma). Contains densely packed bundles (myofibrils) of contractile protein filaments, multiple nuclei immediately beneath the sarcolemma, and an extensive network of specialized smooth endoplasmic reticulum (sarcoplasmic reticulum). Enclosed in a thin fibrous sleeve called endomysium. |
| **Myofibril** | A bundle of protein myofilaments within a muscle fiber, filling most of the cytoplasm. Surrounded by sarcoplasmic reticulum and mitochondria. Has a banded (striated) appearance due to orderly overlap of protein myofilaments. |
| **Sarcomere** | A segment of myofibril from one Z disc to the next in the fiber's striation pattern. Hundreds of sarcomeres end to end compose a myofibril. The functional, contractile unit of the muscle fiber. |
| **Myofilaments** | Fibrous protein strands that carry out the contraction process. Two types: thick filaments composed mainly of myosin, and thin filaments composed mainly of actin. Thick and thin filaments slide over each other to shorten each sarcomere. Shortening of end-to-end sarcomeres shortens the entire muscle. |

Labels in illustration:

**Muscle:** Tendon, Muscle, Muscle fibers, Fascicles, Epimysium, Perimysium

**Fascicle:** Muscle fibers, Perimysium

**Muscle Fiber:** Z disc, Sarcolemma, Myofibrils, Myofilaments

**Myofibril:** T tubule, Z disc, Sarcoplasmic reticulum, Myofilaments: Thick, Thin

**Sarcomere:** Sarcomere, Thick filament, Thin filament, Z-disc, Z-disc

**Myofilaments:** Thick filament, Thin filament

Answer the following questions to test your understanding of the preceding section:

4. What special terms are given to the plasma membrane, cytoplasm, and smooth ER of a muscle cell?

5. What is the difference between a myofilament and a myofibril?

6. List five proteins of the myofilaments and describe their physical arrangement.

7. Sketch the overlapping pattern of myofilaments to show how they account for the A bands, I bands, H bands, and Z discs.

## 11.3 The Nerve–Muscle Relationship

### Expected Learning Outcomes

When you have completed this section, you should be able to

a. explain what a motor unit is and how it relates to muscle contraction;

b. describe the structure of the junction where a nerve fiber meets a muscle fiber; and

c. explain why a cell has an electrical charge difference across its plasma membrane and, in general terms, how this relates to muscle contraction.

Skeletal muscle never contracts unless it is stimulated by a nerve (or artificially with electrodes). If its nerve connections are severed or poisoned, a muscle is paralyzed. If the connection is not restored, the paralyzed muscle wastes away in a shrinkage called *denervation atrophy*. Thus, muscle contraction cannot be understood without first understanding the relationship between nerve and muscle cells.

### Motor Neurons and Motor Units

Skeletal muscles are served by nerve cells called *somatic motor neurons,* whose cell bodies are in the brainstem and spinal cord. Their axons, called **somatic motor fibers,** lead to the muscles (fig. 11.6). Each nerve fiber branches out to multiple muscle fibers, but each muscle fiber is supplied by only one motor neuron.

When a nerve signal approaches the end of an axon, it spreads out over all of its terminal branches and stimulates all muscle fibers supplied by them. Thus, these muscle fibers contract in unison; there is no way to stimulate some of them and not all of them. Since they behave as a single functional unit, one nerve fiber and all the muscle fibers innervated by it are called a **motor unit.** The muscle fibers of a motor unit are not clustered together but dispersed throughout a muscle. Therefore, when stimulated, they cause a weak contraction over a wide area—not just a local twitch in one small region. Effective muscle contraction usually requires the activation of many motor units at once.

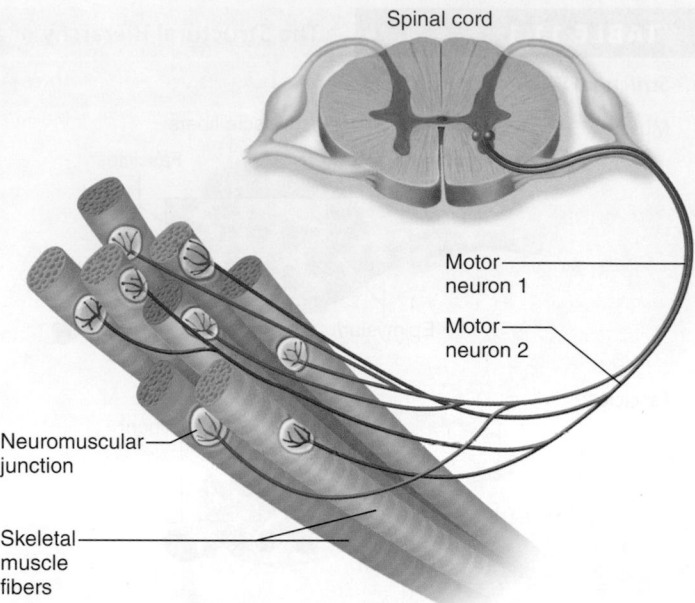

**FIGURE 11.6 Motor Units.** A motor unit consists of one motor neuron and all skeletal muscle fibers that it innervates. Two motor units are here represented by the red and blue nerve and muscle fibers. Note that the muscle fibers of a motor unit are not clustered together, but distributed through the muscle and commingled with the fibers of other motor units.

On average, about 200 muscle fibers are innervated by each motor neuron, but motor units can be much smaller or larger than this to serve different purposes. Where fine control is needed, we have *small motor units.* In the muscles of eye movement, for example, each neuron controls only 3 to 6 muscle fibers. Where strength is more important than fine control, we have *large motor units.* In the gastrocnemius muscle of the calf, for example, one motor neuron may control up to 1,000 muscle fibers. Another way to look at this is that 1,000 muscle fibers may be innervated by 200 neurons in a muscle with small motor units, but by only 1 or 2 neurons in a muscle with large motor units. Consequently, small motor units provide a relatively fine degree of control. Turning a few motor neurons on or off produces a small, subtle change in the action of such a muscle. Large motor units generate more strength but less subtlety of action. Activating just a few motor neurons produces a greater change in a muscle with larger motor units. Fine movements of the eye and hand thus depend on small motor units, while powerful movements of the gluteal and quadriceps muscles depend on large motor units. Other important differences between large and small motor units are discussed later.

In addition to adjustments in strength and control, another advantage of having multiple motor units in each muscle is that they work in shifts. Muscle fibers fatigue when subjected to continual stimulation. If all the fibers in one of your postural muscles fatigued at once, for example, you might collapse. To prevent this, other motor units take over while the fatigued ones recover, and the muscle as a whole can sustain long-term contraction. The role of motor units in muscular strength is further discussed later in the chapter.

## The Neuromuscular Junction

The point where a nerve fiber meets any target cell is called a **synapse** (SIN-aps). When the target cell is a muscle fiber, the synapse is also called a **neuromuscular junction (NMJ)** or **motor end plate** (fig. 11.7). Each terminal branch of the nerve fiber within the NMJ forms a separate synapse with the muscle fiber. The sarcolemma of the NMJ is irregularly indented, a little like a handprint pressed into soft clay. If you

imagine the nerve fiber to be like your forearm, branching out at the end like your fingers, the individual synapses would be like the dents where your fingertips press into the clay. Thus, one nerve fiber stimulates the muscle fiber at several points within the NMJ.

At each synapse, the nerve fiber ends in a bulbous swelling called a **synaptic knob.** The knob doesn't directly touch the muscle fiber but is separated from it by a narrow space called the **synaptic cleft,** about 60 to 100 nm wide (scarcely wider than the thickness of one plasma membrane). A third cell, called a **Schwann cell,** envelops the entire junction and isolates it from the surrounding tissue fluid.

The synaptic knob contains spheroidal organelles called **synaptic vesicles,** which are filled with a chemical, **acetylcholine (ACh)** (ASS-eh-tul-CO-leen)—one of many

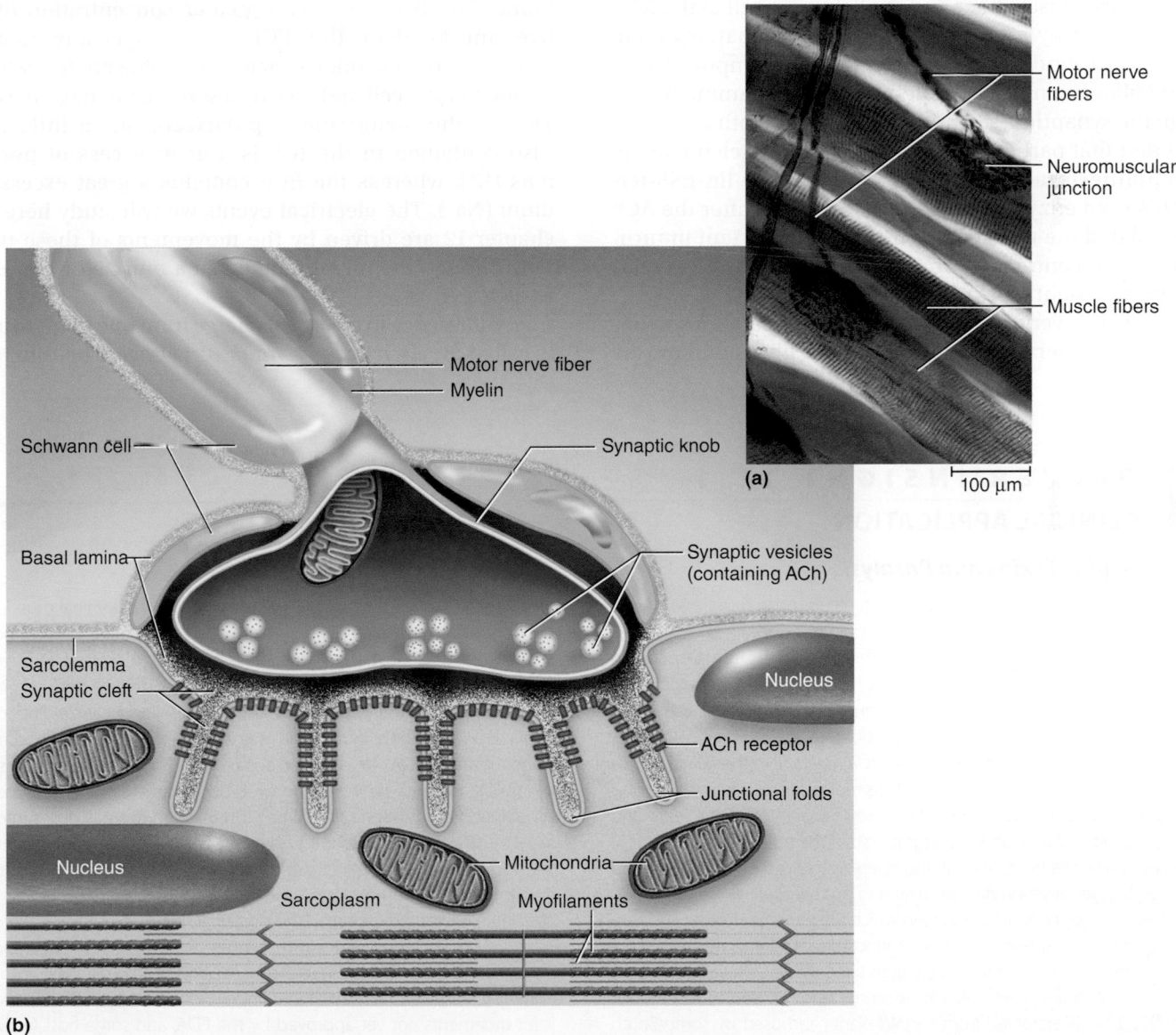

(a)

| 100 μm |

(b)

**FIGURE 11.7 Innervation of Skeletal Muscle.** (a) Neuromuscular junctions, with muscle fibers slightly teased apart (LM; compare the SEM on p. 397). (b) Structure of a single neuromuscular junction. **AP|R**

*neurotransmitters* to be introduced in chapter 12. The electrical signal (nerve impulse) traveling down a nerve fiber cannot cross the synaptic cleft like a spark jumping between two electrodes—rather, it causes the synaptic vesicles to undergo exocytosis, releasing ACh into the cleft. ACh thus functions as a chemical messenger from the nerve cell to the muscle cell.

To respond to ACh, the muscle fiber has about 50 million **ACh receptors**—proteins incorporated into its plasma membrane across from the synaptic knobs. To maximize the number of ACh receptors and thus its sensitivity to the neurotransmitter, the sarcolemma in this area has numerous infoldings, about 1 μm deep, called **junctional folds,** which increase the surface area of ACh-sensitive membrane. The muscle nuclei beneath the folds are specifically dedicated to the synthesis of ACh receptors and other proteins of the local sarcolemma. A deficiency of ACh receptors leads to muscle paralysis in the disease *myasthenia gravis* (see Deeper Insight 11.4, p. 429).

The entire muscle fiber and the Schwann cell of the NMJ are surrounded by a **basal lamina,** which separates them from the surrounding connective tissue. Composed partially of collagen and glycoproteins, the basal lamina passes through the synaptic cleft and virtually fills it. Both the sarcolemma and that part of the basal lamina in the cleft contain **acetylcholinesterase (AChE)** (ASS-eh-till-CO-lin-ESS-ter-ase). This is an enzyme that breaks down ACh after the ACh has stimulated the muscle cell; thus, it is important in turning off muscle contraction and allowing the muscle to relax (see Deeper Insight 11.1).

You must be very familiar with the foregoing terms to understand how a nerve stimulates a muscle fiber and how the fiber contracts. They are summarized in table 11.2 for your later reference.

## Electrically Excitable Cells

Muscle and nerve cells are regarded as *electrically excitable cells* because their plasma membranes exhibit voltage changes in response to stimulation. The study of the electrical activity of cells, called **electrophysiology,** is a key to understanding nervous activity, muscle contraction, the heartbeat, and other physiological phenomena. The details of electrophysiology are presented in chapter 12, but a few fundamental principles must be introduced here so you can understand muscle excitation.

The electrical activity of cells hinges on differences in the concentration of ions in the intracellular fluid (ICF) and extracellular fluid (ECF) adjacent to the plasma membrane. The ICF contains a greater concentration of negative anions than the ECF does—especially negatively charged proteins, nucleic acids, and phosphates, which are trapped in the cell and give its interior a net negative charge. That is, the membrane is **polarized,** like a little battery. Also contained in the ICF is a great excess of potassium ions ($K^+$), whereas the ECF contains a great excess of sodium ($Na^+$). The electrical events we will study here and in chapter 12 are driven by the movements of these two cations through the membrane when a muscle or nerve cell is excited.

A difference in electrical charge from one point to another is called an *electrical potential,* or *voltage.* The difference is

---

## DEEPER INSIGHT 11.1

### CLINICAL APPLICATION

#### Neuromuscular Toxins and Paralysis

Toxins that interfere with synaptic function can paralyze the muscles. Organophosphate pesticides such as malathion, for example, contain *cholinesterase inhibitors* that bind to AChE and prevent it from degrading ACh. Depending on the dose, this can prolong the action of ACh and produce *spastic paralysis,* a state in which the muscles contract and cannot relax; clinically, this is called a *cholinergic crisis.* Another example of spastic paralysis is *tetanus (lockjaw),* caused by the toxin of a soil bacterium, *Clostridium tetani.* In the spinal cord, a neurotransmitter called glycine normally stops motor neurons from producing unwanted muscle contractions. The tetanus toxin blocks glycine release and thus causes overstimulation and spastic paralysis of the muscles. (At the cost of possible confusion, the word *tetanus* also refers to a completely different and normal muscle phenomenon discussed later in this chapter.)

*Flaccid paralysis,* by contrast, is a state in which the muscles are limp and cannot contract. It poses a threat of death by suffocation if it affects the respiratory muscles. Among the causes of flaccid paralysis are poisons such as curare (cue-RAH-ree), which competes with ACh for receptor sites but does not stimulate the muscle. Curare is extracted from certain plants and used by some South American natives to poison blowgun darts. It has been used to treat muscle

spasms in some neurological disorders and to relax abdominal muscles for surgery, but other muscle relaxants have now replaced curare for most purposes.

Another cause of flaccid paralysis is *botulism,* a type of food poisoning caused by a neuromuscular toxin secreted by the bacterium *Clostridium botulinum.* Botulinum toxin blocks ACh release and causes flaccid muscle paralysis. Purified botulinum toxin was approved by the U.S. Food and Drug Administration (FDA) in 2002 for cosmetically treating "frown lines" caused by muscle tautness between the eyebrows. Marketed as Botox Cosmetic (a prescription drug despite the name), it is injected in small doses into specific facial muscles. The wrinkles gradually disappear as muscle paralysis sets in over the next few hours. The effect lasts about 4 months until the muscles retighten and the wrinkles return. Botox treatment has become the fastest growing cosmetic medical procedure in the United States, with many people going for treatment every few months in their quest for a youthful appearance. It has begun to have some undesirable consequences, however, as it is sometimes administered by unqualified practitioners. Even some qualified physicians use it for treatments not yet approved by the FDA, and some host festive "Botox parties" for treatment of patients in assembly-line fashion.

| TABLE 11.2 | Components of the Neuromuscular Junction |
|---|---|
| **Term** | **Definition** |
| Neuromuscular junction | A functional connection between the distal end of a nerve fiber and the middle of a muscle fiber |
| Synaptic knob | The dilated tip of a nerve fiber; contains synaptic vesicles |
| Synaptic cleft | A gap of about 60 to 100 nm between the synaptic knob and sarcolemma |
| Synaptic vesicle | A secretory vesicle in the synaptic knob; contains acetylcholine |
| Junctional folds | Invaginations of the sarcolemma where ACh receptors are especially concentrated |
| Acetylcholine (ACh) | The neurotransmitter released by a somatic motor fiber that stimulates a skeletal muscle fiber (also used elsewhere in the nervous system) |
| ACh receptor | A transmembrane protein in the sarcolemma of the neuromuscular junction that binds to ACh |
| Acetylcholinesterase (AChE) | An enzyme in the sarcolemma and basal lamina of the muscle fiber in the synaptic region; responsible for degrading ACh and stopping the stimulation of the muscle fiber |

typically 12 volts (V) for a car battery and 1.5 V for a flashlight battery, for example. On the sarcolemma of a muscle cell, the voltage is much smaller, about –90 millivolts (mV), but critically important to life. (The negative sign refers to the relatively negative charge on the intracellular side of the membrane.) This voltage is called the **resting membrane potential (RMP).** It is maintained by the sodium–potassium pump, as noted in chapter 3.

When a nerve or muscle cell is stimulated, dramatic things happen electrically, as we shall soon see in the excitation of muscle. Ion channels in the plasma membrane open and Na$^+$ instantly flows into the cell, driven both by its concentration difference across the membrane and by its attraction to the negative charge of the cell interior—that is, it flows down an **electrochemical gradient.** These Na$^+$ cations override the negative charge just inside the membrane, so the inside of the membrane briefly becomes positive. This is called **depolarization** of the membrane. Immediately, Na$^+$ channels close and K$^+$ channels open. K$^+$ rushes out of the cell, partly because it is repelled by the positive sodium charge and partly because it is more concentrated in the ICF than in the ECF—that is, it flows down its own electrochemical gradient in the direction opposite from the sodium movement. The loss of K$^+$ ions from the cell turns the inside of the membrane negative again **(repolarization).** This quick up-and-down voltage shift, from the negative RMP to a positive value and then back to the RMP again, is called an **action potential.** The RMP is a stable voltage seen in a "waiting" cell, whereas the action potential is a quickly fluctuating voltage seen in an active, stimulated cell. Chapter 12 explains the RMP and the mechanism of action potentials more fully.

Action potentials have a way of perpetuating themselves—an action potential at one point on a plasma membrane causes another one to happen immediately in front of it, which triggers another one a little farther along, and so forth. A wave of action potentials spreading along a nerve fiber like this is called a *nerve impulse* or *nerve signal.* Such signals also travel along the sarcolemma of a muscle fiber. We will see how this leads to muscle contraction.

**BEFORE YOU GO ON**

Answer the following questions to test your understanding of the preceding section:

8. What differences would you expect to see between a motor unit where muscular strength is more important than fine control and another motor unit where fine control is more important?

9. Distinguish between acetylcholine, an acetylcholine receptor, and acetylcholinesterase. State where each is found and describe the function it serves.

10. What accounts for the resting membrane potential seen in unstimulated nerve and muscle cells?

11. What is the difference between a resting membrane potential and an action potential?

## 11.4 Behavior of Skeletal Muscle Fibers

### Expected Learning Outcomes

When you have completed this section, you should be able to

a. explain how a nerve fiber stimulates a skeletal muscle fiber;

b. explain how stimulation of a muscle fiber activates its contractile mechanism;

c. explain the mechanism of muscle contraction;

d. explain how a muscle fiber relaxes; and

e. explain why the force of a muscle contraction depends on sarcomere length prior to stimulation.

The process of muscle contraction and relaxation has four major phases: (1) excitation, (2) excitation–contraction coupling, (3) contraction, and (4) relaxation. Each phase occurs in several smaller steps, which we now examine in detail. The steps are numbered in the following descriptions to correspond to those in figures 11.8 to 11.11.

# Excitation

**Excitation** is the process in which action potentials in the nerve fiber lead to action potentials in the muscle fiber. The steps in excitation are shown in figure 11.8.

①  A nerve signal arrives at the synaptic knob and opens voltage-gated calcium channels. Calcium ions enter the synaptic knob.

②  Calcium stimulates the synaptic vesicles to release acetylcholine (ACh) into the synaptic cleft. One action potential causes exocytosis of about 60 vesicles, and each vesicle releases about 10,000 molecules of ACh.

③  ACh diffuses across the synaptic cleft and binds to receptors on the sarcolemma.

④  These receptors are ligand-gated ion channels. Two ACh molecules must bind to each receptor to open the channel. When it opens, $Na^+$ flows quickly into the cell and $K^+$ flows out. As a result, the sarcolemma reverses polarity—its voltage quickly jumps from the RMP of $-90$ mV to a peak of $+75$ mV as $Na^+$ enters, and then falls back to a level close to the RMP as $K^+$ exits. This rapid fluctuation in voltage at the motor end plate is called the **end-plate potential (EPP).**

⑤  Areas of sarcolemma next to the end plate have voltage-gated ion channels that open in response to the EPP. Some of these are specific for $Na^+$ and admit it to the cell, while others are specific for $K^+$ and allow it to leave. These ion movements create an *action potential*. The muscle fiber is now excited.

## Excitation–Contraction Coupling

**Excitation–contraction coupling** refers to events that link action potentials on the sarcolemma to activation of the myofilaments, thereby preparing them to contract. The steps in the coupling process are shown in figure 11.9.

⑥  A wave of action potentials spreads from the motor end plate in all directions, like ripples on a pond. When this wave of excitation reaches the T tubules, it continues down them into the cell interior.

⑦  Action potentials open voltage-gated ion channels in the T tubules. These are linked to calcium channels in the terminal cisternae of the sarcoplasmic reticulum (SR). Thus, channels in the SR open as well and calcium diffuses out of the SR, down its concentration gradient into the cytosol.

⑧  Calcium binds to the troponin of the thin filaments.

⑨  The troponin–tropomyosin complex changes shape and exposes the active sites on the actin filaments. This makes them available for binding to myosin heads.

# Contraction

**Contraction** is the step in which the muscle fiber develops tension and may shorten. (Muscles often "contract," or develop tension, without shortening, as we see later.) The mechanism of contraction is called the **sliding filament theory.** It holds that the myofilaments do not become any shorter during contraction; rather, the thin filaments slide over the thick ones and pull the Z discs behind them, causing each sarcomere as a whole to shorten. The individual steps in this mechanism are shown in figure 11.10.

⑩  The myosin head must have an ATP molecule bound to it to initiate contraction. **Myosin ATPase,** an enzyme in the head, hydrolyzes this ATP into ADP and phosphate ($P_i$). The energy released by this process activates the head, which "cocks" into an extended, high-energy position. The head temporarily keeps the ADP and $P_i$ bound to it.

⑪  The cocked myosin binds to an exposed active site on the thin filament, forming a **cross-bridge** between the myosin and actin.

⑫  Myosin releases the ADP and $P_i$ and flexes into a bent, low-energy position, tugging the thin filament along with it. This is called the **power stroke.** The head remains bound to actin until it binds a new ATP.

⑬  The binding of ATP to myosin destabilizes the myosin–actin bond, breaking the cross-bridge. Myosin is now prepared to repeat the whole process—it will hydrolyze the new ATP, recock (the **recovery stroke**), attach to a new active site farther down the thin filament, and produce another power stroke.

It may seem as if releasing the thin filament at step 13 would simply allow it to slide back to its previous position, so nothing would have been accomplished. Think of the sliding filament mechanism, however, as being similar to the way you would pull in a boat anchor hand over hand. When the myosin head cocks, it is like your hand reaching out to grasp the anchor rope. When it flexes back into the low-energy position, it is like your elbow flexing to pull on the rope and draw the anchor up a little bit. When you let go of the rope with one hand, you hold onto it with the other, alternating hands until the anchor is pulled in. Similarly, when one myosin head releases actin in preparation for the recovery stroke, there are many other heads on the same thick filament holding onto the thin filament so it doesn't slide back. At any given moment during contraction, about half of the heads are bound to the thin filament and the other half are extending forward to grasp it farther down. That is, the myosin heads do not all stroke at once but contract sequentially.

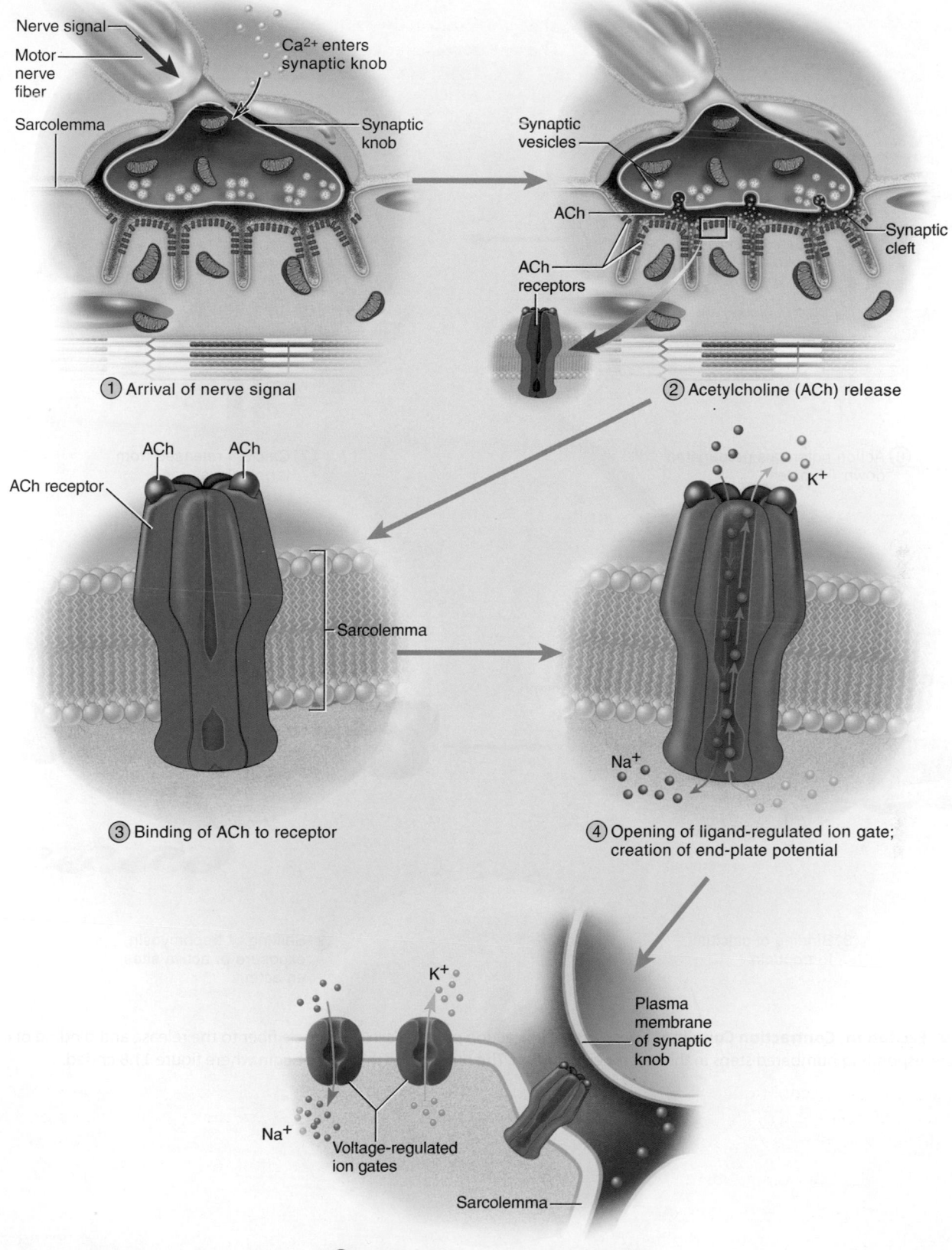

① Arrival of nerve signal

Nerve signal
Motor nerve fiber
Sarcolemma
Ca²⁺ enters synaptic knob
Synaptic knob

② Acetylcholine (ACh) release

Synaptic vesicles
ACh
ACh receptors
Synaptic cleft

③ Binding of ACh to receptor

ACh
ACh
ACh receptor
Sarcolemma

④ Opening of ligand-regulated ion gate; creation of end-plate potential

K⁺
Na⁺

⑤ Opening of voltage-regulated ion gates; creation of action potentials

K⁺
Na⁺
Voltage-regulated ion gates
Plasma membrane of synaptic knob
Sarcolemma

**FIGURE 11.8 Excitation of a Muscle Fiber.** These events link action potentials in a nerve fiber to the generation of action potentials in the muscle fiber. See the corresponding numbered steps in the text for explanation.

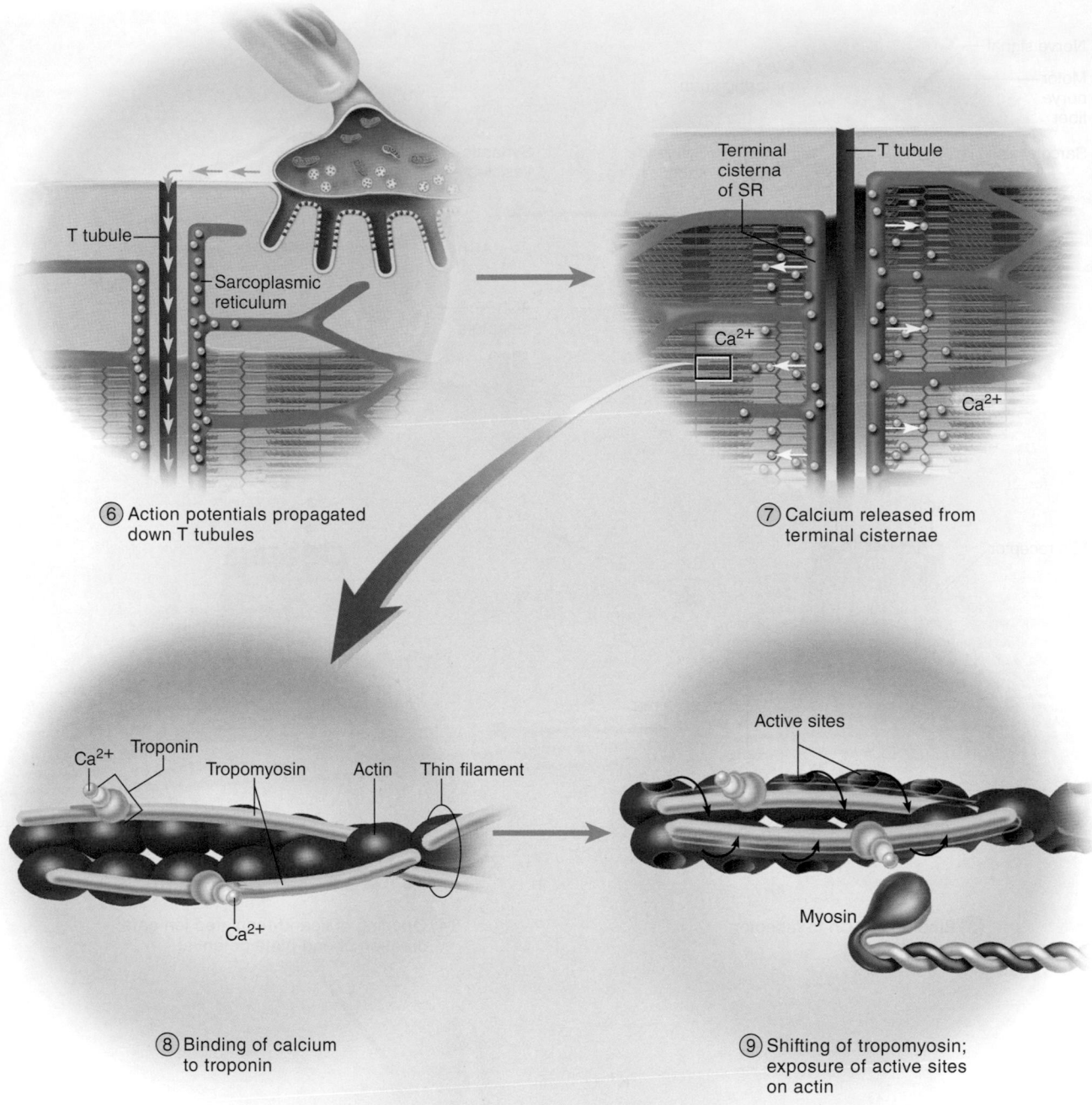

⑥ Action potentials propagated down T tubules

⑦ Calcium released from terminal cisternae

⑧ Binding of calcium to troponin

⑨ Shifting of tropomyosin; exposure of active sites on actin

**FIGURE 11.9 Excitation–Contraction Coupling.** These events link action potentials in the muscle fiber to the release and binding of calcium ions. See the corresponding numbered steps in the text for explanation. The numbers in this figure begin where figure 11.8 ended.

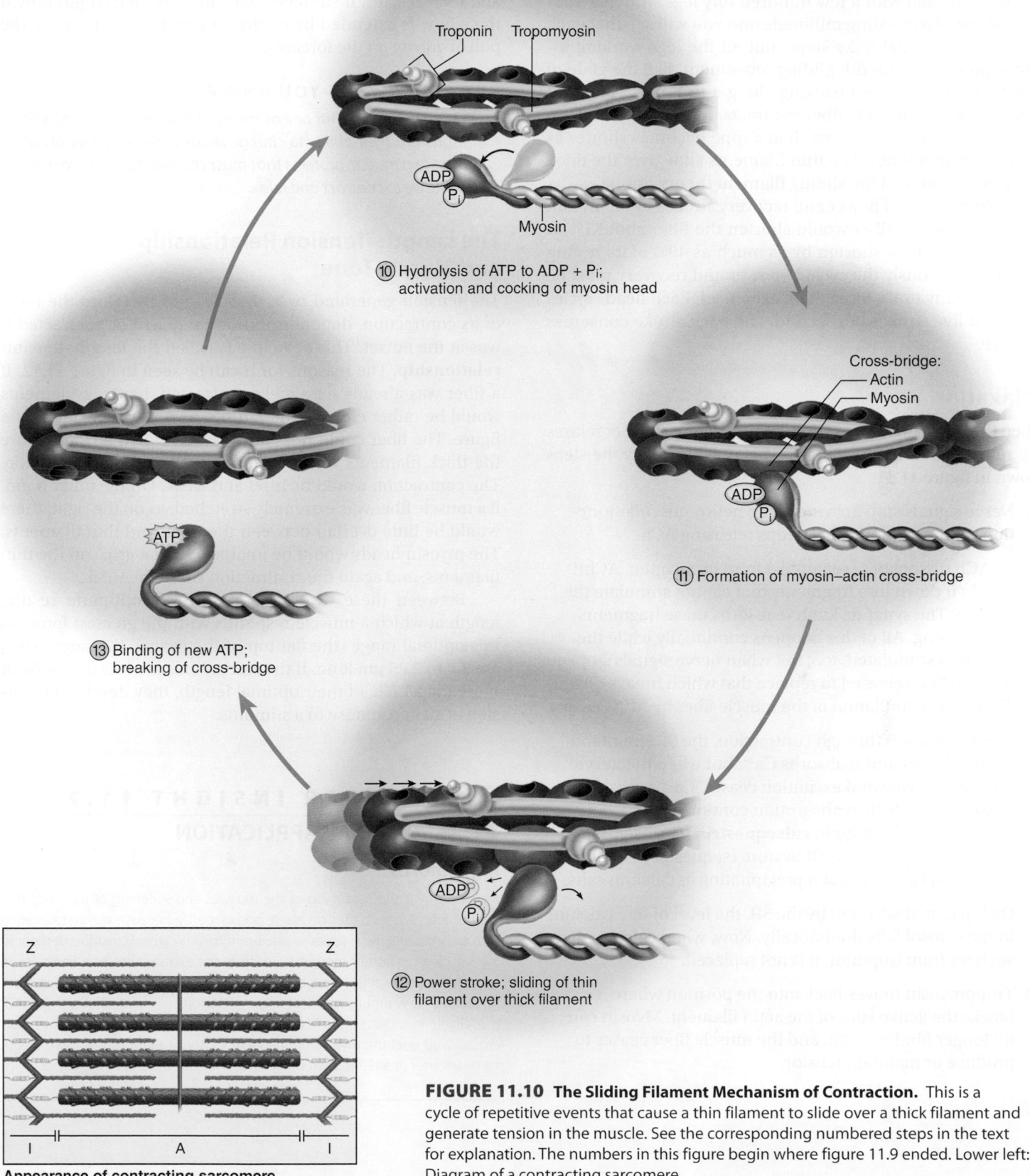

Troponin  Tropomyosin

ADP  Pi

Myosin

⑩ Hydrolysis of ATP to ADP + Pi;
activation and cocking of myosin head

Cross-bridge:
— Actin
— Myosin

ADP
Pi

⑪ Formation of myosin–actin cross-bridge

ATP

⑬ Binding of new ATP;
breaking of cross-bridge

ADP
Pi

⑫ Power stroke; sliding of thin
filament over thick filament

Z          Z

I   ⊢—⊣        A        ⊢—⊣  I

**Appearance of contracting sarcomere**

**FIGURE 11.10  The Sliding Filament Mechanism of Contraction.** This is a
cycle of repetitive events that cause a thin filament to slide over a thick filament and
generate tension in the muscle. See the corresponding numbered steps in the text
for explanation. The numbers in this figure begin where figure 11.9 ended. Lower left:
Diagram of a contracting sarcomere.

Each head acts in a jerky manner, but hundreds of them working together produce a smooth, steady pull on the thin filament. This is similar to the locomotion of a millipede—a wormlike animal with a few hundred tiny legs. Watch a YouTube video of a crawling millipede and you will see that each leg takes individual jerky steps, but all the legs working together produce a smooth gliding movement, like the glide of a thick muscle filament walking along a thin one. Note that even though the muscle fiber contracts, the *myofilaments do not become shorter* any more than a rope becomes shorter as you pull in an anchor. The thin filaments slide over the thick ones, as the name of the sliding filament theory implies.

A single cycle of power and recovery strokes by all myosin heads in a muscle fiber would shorten the fiber about 1%. A fiber, however, may shorten by as much as 40% of its resting length, so obviously the cycle of power and recovery must be repeated many times by each myosin head. Each head carries out about five strokes per second, and each stroke consumes one ATP.

## Relaxation

When the nerve fiber stops stimulating it, a muscle fiber relaxes and returns to its resting length. This is achieved by the steps shown in figure 11.11.

⑭ Nerve signals stop arriving at the neuromuscular junction, so the synaptic knob stops releasing ACh.

⑮ As ACh dissociates (separates) from its receptor, AChE breaks it down into fragments that cannot stimulate the muscle. The synaptic knob reabsorbs these fragments for recycling. All of this happens continually while the muscle is stimulated, too, but when nerve signals stop, no more ACh is released to replace that which breaks down. Therefore, stimulation of the muscle fiber by ACh ceases.

⑯ From excitation through contraction, the SR simultaneously releases and reabsorbs $Ca^{2+}$; but when the nerve fiber stops firing and excitation ceases, $Ca^{2+}$ release also ceases and only its reabsorption continues. In the SR, $Ca^{2+}$ binds to the protein **calsequestrin** (CAL-seh-QUES-trin), which allows the SR to store (sequester) a large amount of $Ca^{2+}$ without it precipitating as calcium salts.

⑰ Owing to reabsorption by the SR, the level of free calcium in the cytosol falls dramatically. Now, when calcium dissociates from troponin, it is not replaced.

⑱ Tropomyosin moves back into the position where it blocks the active sites of the actin filament. Myosin can no longer bind to actin, and the muscle fiber ceases to produce or maintain tension.

Relaxation alone does not return muscle to its resting length. That must be achieved by some force pulling the muscle and stretching it. For example, if the biceps flexes the elbow and then relaxes, it stretches back to its resting length only if the elbow is extended by contraction of the triceps or by the pull of gravity on the forearm.

▶▶▶ **APPLY WHAT YOU KNOW**

*Chapter 2 noted that one of the most important properties of proteins is their ability to change shape repeatedly. Identify at least two muscle proteins that must change shape in order for a muscle to contract and relax.*

## The Length–Tension Relationship and Muscle Tone

The tension generated by a muscle, and therefore the force of its contraction, depends on how stretched or contracted it was at the outset. This principle is called the **length–tension relationship.** The reasons for it can be seen in figure 11.12. If a fiber was already extremely contracted, its thick filaments would be rather close to the Z discs, as on the left side of the figure. The fiber could not contract very much farther before the thick filaments would butt against the Z discs and stop. The contraction would be brief and weak. On the other hand, if a muscle fiber was extremely stretched, as on the right, there would be little overlap between the thick and thin filaments. The myosin heads would be unable to "get a grip" on the thin filaments, and again the contraction would be weak.

Between these extremes, there is an optimum resting length at which a muscle responds with the greatest force. In this optimal range (the flat top of the curve), the sarcomeres are 2.0 to 2.25 μm long. If the sarcomeres are less than 60% or more than 175% of their optimal length, they develop no tension at all in response to a stimulus.

# DEEPER INSIGHT 11.2
## CLINICAL APPLICATION

### Rigor Mortis

*Rigor mortis*[8] is the hardening of the muscles and stiffening of the body that begins 3 to 4 hours after death. It occurs partly because the deteriorating sarcoplasmic reticulum releases calcium into the cytosol, and the deteriorating sarcolemma admits more calcium from the extracellular fluid. The calcium activates myosin–actin cross-bridging. Once bound to actin, myosin cannot release it without first binding an ATP molecule, and of course no ATP is available in a dead body. Thus, the thick and thin filaments remain rigidly cross-linked until the myofilaments begin to decay. Rigor mortis peaks about 12 hours after death and then diminishes over the next 48 to 60 hours.

---

[8] *rigor* = rigidity; *mortis* = of death

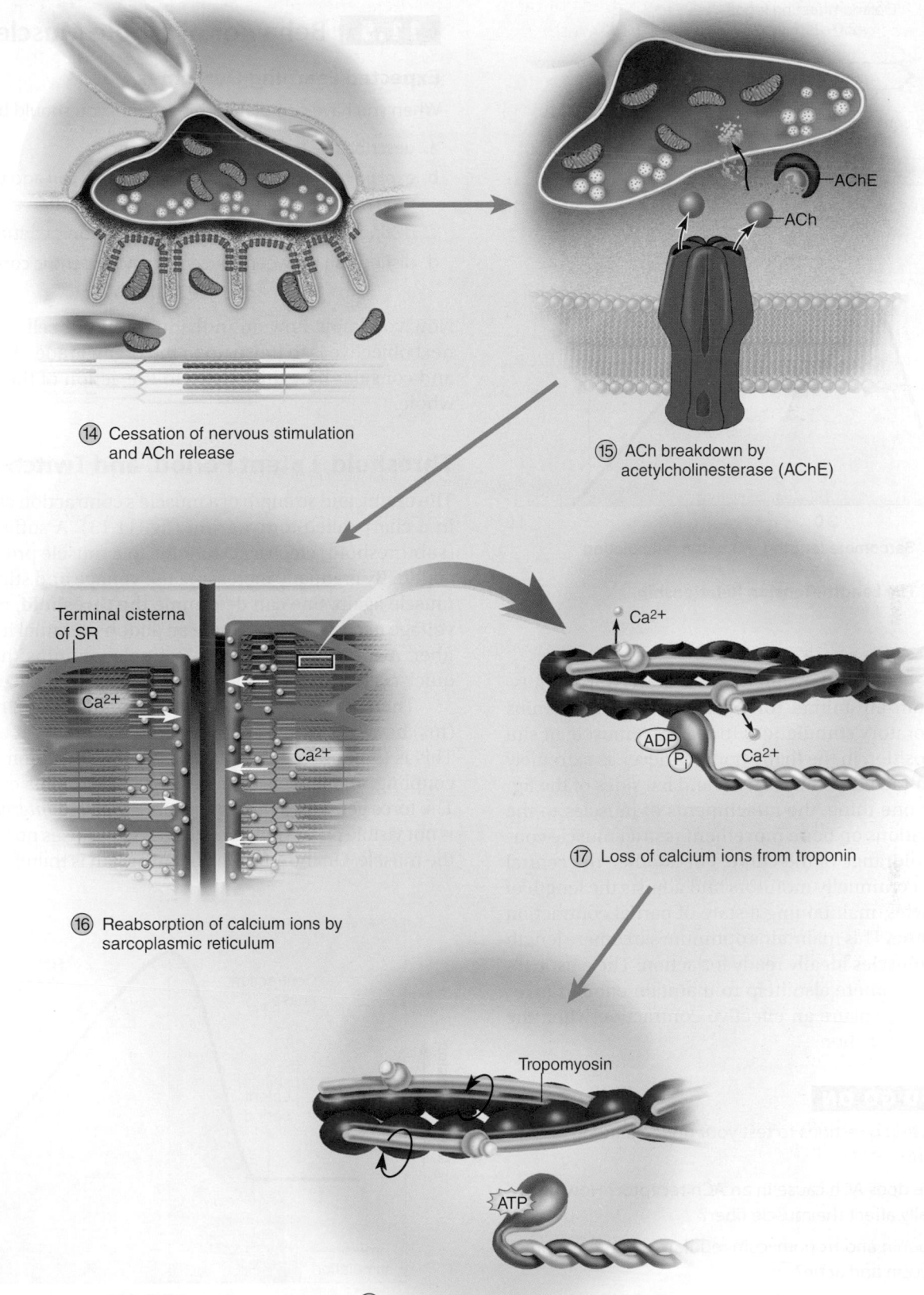

⑭ Cessation of nervous stimulation and ACh release

⑮ ACh breakdown by acetylcholinesterase (AChE)

AChE

ACh

Terminal cisterna of SR

Ca²⁺

Ca²⁺

⑯ Reabsorption of calcium ions by sarcoplasmic reticulum

Ca²⁺

ADP

Pᵢ

Ca²⁺

⑰ Loss of calcium ions from troponin

Tropomyosin

ATP

⑱ Return of tropomyosin to position blocking active sites of actin

**FIGURE 11.11 Relaxation of a Muscle Fiber.** These events lead from the cessation of a nerve signal to the release of thin filaments by myosin. See the corresponding numbered steps in the text for explanation. The numbers in this figure begin where figure 11.10 ended.

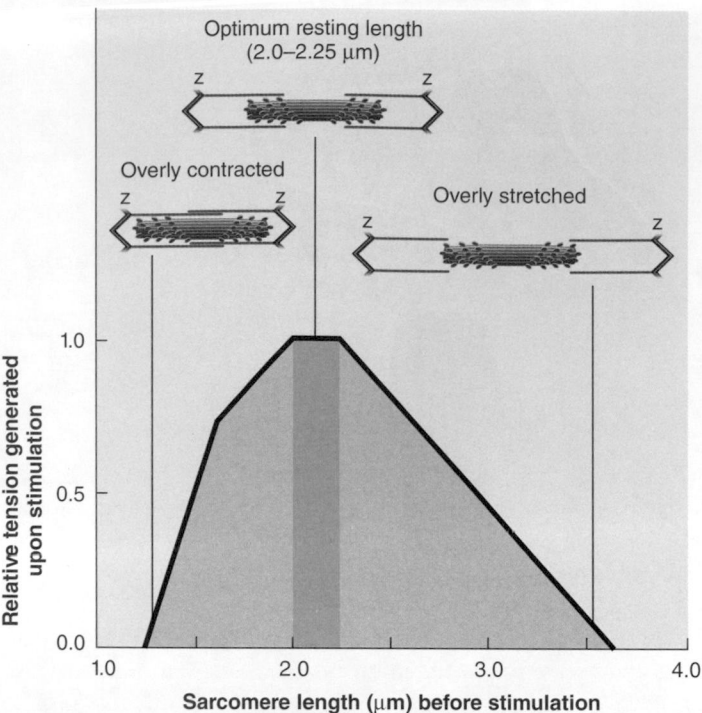

**FIGURE 11.12** **The Length–Tension Relationship.**

The complete length-tension curve is derived from muscles isolated from an animal (often the frog gastrocnemius muscle) for laboratory stimulation. In reality, a muscle in situ (in its natural position in the living body) is never as extremely stretched or contracted as the far right and left sides of the figure depict. For one thing, the attachments of muscles to the bones and limitations on bone movement restrict muscle contraction to the midrange of the curve. For another, the central nervous system continually monitors and adjusts the length of the resting muscles, maintaining a state of partial contraction called **muscle tone.** This maintains optimum sarcomere length and makes the muscles ideally ready for action. The elastic filaments of the sarcomere also help to maintain enough myofilament overlap to ensure an effective contraction when the muscle is called into action.

### BEFORE YOU GO ON

Answer the following questions to test your understanding of the preceding section:

12. What change does ACh cause in an ACh receptor? How does this electrically affect the muscle fiber?

13. How do troponin and tropomyosin regulate the interaction between myosin and actin?

14. Describe the roles played by ATP in the power and recovery strokes of myosin.

15. What steps are necessary for a contracted muscle to return to its resting length?

## 11.5    Behavior of Whole Muscles

### Expected Learning Outcomes

When you have completed this section, you should be able to

a. describe the stages of a muscle twitch;

b. explain how successive muscle twitches can add up to produce stronger muscle contractions;

c. distinguish between isometric and isotonic contraction; and

d. distinguish between concentric and eccentric contraction.

Now you know how an individual muscle cell shortens. Our next objective is to move up to the organ grade of construction and consider how this relates to the action of the muscle as a whole.

## Threshold, Latent Period, and Twitch

The timing and strength of a muscle's contraction can be shown in a chart called a **myogram** (fig. 11.13). A sufficiently weak (subthreshold) electrical stimulus to a muscle produces no reaction. By gradually increasing the voltage and stimulating the muscle again, one can determine the **threshold,** or minimum voltage necessary to generate an action potential in the muscle fiber. At threshold or higher, a single stimulus thus causes a quick cycle of contraction and relaxation called a **twitch.**

There is a delay, or **latent period,** of about 2 milliseconds (ms) between the onset of the stimulus and onset of the twitch. This is the time required for excitation, excitation–contraction coupling, and tensing of the elastic components of the muscle. The force generated during this time is called *internal tension.* It is not visible on the myogram because it causes no shortening of the muscle. On the left side, the myogram is therefore flat.

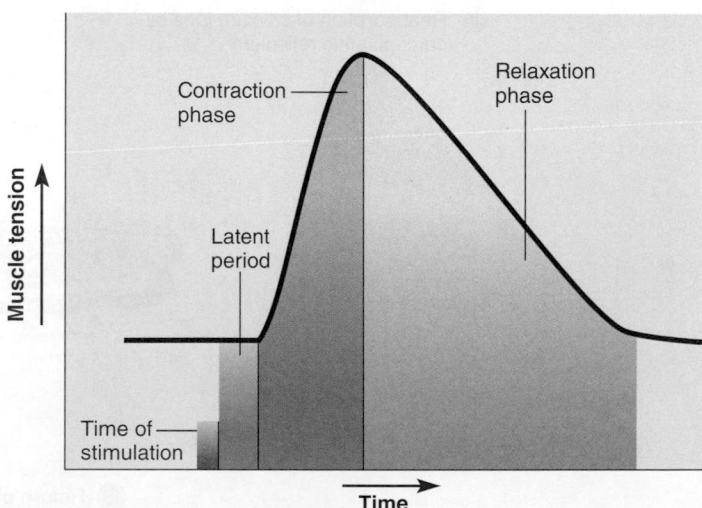

**FIGURE 11.13** **A Muscle Twitch.**

❓ *What role does ATP play during the relaxation phase?*

Once the elastic components are taut, the muscle begins to produce *external tension* and move a resisting object, or load, such as a bone or body limb. This is called the **contraction phase** of the twitch. By analogy, imagine lifting a weight from a table with a rubber band. At first, internal tension would only stretch the rubber band. Then as the rubber band became taut, external tension would lift the weight.

The contraction phase is short-lived, because the SR quickly reabsorbs $Ca^{2+}$ before the muscle develops maximal force. As the $Ca^{2+}$ level in the cytoplasm falls, myosin releases the thin filaments and muscle tension declines. This is seen in the myogram as the **relaxation phase.** As shown by the asymmetry of the myogram, the muscle is quicker to contract than it is to relax. The entire twitch lasts from about 7 to 100 ms, so a muscle could theoretically complete about 10 to 140 twitches per second (if only the math mattered).

## Contraction Strength of Twitches

We have seen that a subthreshold stimulus induces no muscle contraction at all, but at threshold intensity, a twitch is produced. Increasing the stimulus voltage still more, however, produces twitches no stronger than those at threshold. Superficially, the muscle fiber seems to be giving its maximum response once the stimulus intensity is at threshold or higher. However, even for a constant voltage, twitches vary in strength. This is so for a variety of reasons, some of which are causally linked to each other:

- Twitch strength depends on how stretched the muscle was just before it was stimulated, as we have just seen in the length–tension relationship.

- Twitches become weaker as a muscle fatigues, as discussed later in this chapter.

- Twitches vary with the temperature of the muscle; a warmed-up muscle contracts more strongly because enzymes such as the myosin heads work more quickly.

- Twitch strength varies with the muscle's state of hydration, which affects the spacing between thick and thin filaments and therefore the ability to form myosin–actin cross-bridges.

- Twitch strength varies with stimulus frequency; stimuli arriving close together produce stronger twitches than stimuli arriving at longer time intervals, as we will see shortly.

It should not be surprising that twitches vary in strength. Indeed, an individual twitch is not strong enough to do any useful work. Muscles must contract with variable strength for different tasks, such as lifting a glass of champagne compared with lifting barbells at the gym.

Let us examine more closely the contrasting effects of stimulus *intensity* versus stimulus *frequency* on contraction strength. Suppose we apply a stimulating electrode to a motor nerve that supplies a muscle, such as a laboratory preparation of a frog sciatic nerve connected to its gastrocnemius muscle. Subthreshold stimulus voltages produce no response (fig. 11.14). At threshold, we see a weak twitch (at *3* in the bottom row of the figure), and if we continue to raise the voltage, we see stronger twitches. The reason for this is that higher voltages excite more and more nerve fibers in the motor nerve (middle row of the figure), and thus stimulate more and more motor units to contract. The process of bringing more motor units into play is called **recruitment,** or **multiple motor unit (MMU) summation.** This is seen not just in artificial stimulation, but is part of the way the nervous system behaves naturally to produce varying muscle contractions. The neuromuscular system behaves according to the **size principle**—smaller, less powerful motor units with smaller, slower nerve fibers are activated first. This is sufficient for delicate tasks and refined movements, but if more power is needed, then larger motor units with larger, faster nerve fibers are subsequently activated.

But even when stimulus intensity (voltage) remains constant, twitch strength can vary with stimulus frequency. High-frequency stimulation produces stronger twitches than low-frequency stimulation. In figure 11.15a, we see that when a muscle is stimulated at low frequency, it produces an identical twitch for each stimulus and fully recovers between twitches.

At higher stimulus frequencies, say 20 to 40 stimuli/s, each new stimulus arrives before the previous twitch is over. Each new twitch "rides piggyback" on the previous one and generates higher tension (fig. 11.15b). This phenomenon goes by two names: **temporal[9] summation,** because it results

---

[9] *tempor* = time

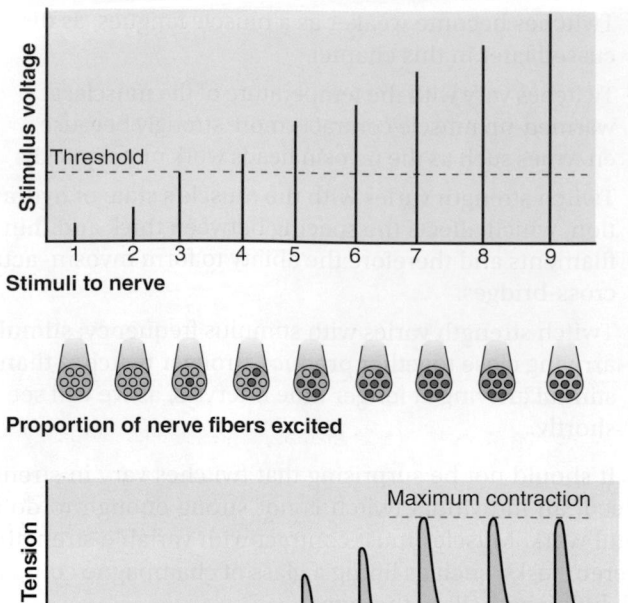

**Stimulus voltage**

Threshold

1  2  3  4  5  6  7  8  9
**Stimuli to nerve**

**Proportion of nerve fibers excited**

**Tension**

Maximum contraction

1  2  3  4  5  6  7  8  9
**Responses of muscle**

**FIGURE 11.14  The Relationship Between Stimulus Intensity (Voltage) and Muscle Tension.** Top row: Nine stimuli of increasing strength. The first two are subthreshold stimuli. Middle row: Cross section of a motor nerve with seven nerve fibers. The colored nerve fibers are the excited ones; note that none are excited by the subthreshold stimuli above. Bottom row: Graph of muscle tension. Subthreshold stimuli (1–2) produce no muscle contraction. When stimuli reach or exceed threshold (3–7), they excite more and more nerve fibers and motor units; thus, they produce stronger and stronger contractions. This is multiple motor unit summation (recruitment). Once all of the nerve fibers are stimulated (7–9), further increases in stimulus strength produce no further increase in muscle tension.

from two stimuli arriving close together, or **wave summation,** because it results from one wave of contraction added to another. Wave upon wave, each twitch reaches a higher level of tension than the one before, and the muscle relaxes only partially between stimuli. This effect produces a state of sustained fluttering contraction called **incomplete tetanus.**

In the laboratory, an isolated muscle can be stimulated at such high frequency that the twitches fuse into a single, nonfluctuating contraction called **complete (fused) tetanus.** This does not happen in the body, however, because motor neurons do not fire that fast. Indeed, there is an inhibitory mechanism in the spinal cord that prevents them from doing so. Complete tetanus is very injurious to muscle and associated soft tissues, so spinal inhibition protects the muscles by preventing complete tetanus. Muscle tetanus should not be confused with the disease of the same name caused by the tetanus toxin (see Deeper Insight 11.1).

Despite the fluttering contraction seen in incomplete tetanus, we know that a muscle taken as a whole can contract very smoothly. This is possible because motor units function asynchronously; when one motor unit relaxes, another contracts and takes over so that the muscle does not lose tension.

## Isometric and Isotonic Contraction

In muscle physiology, "contraction" doesn't always mean the shortening of a muscle—it may mean only that the muscle produces internal tension while an external resistance causes it to stay the same length or even become longer. Thus, physiologists speak of different kinds of muscle contraction as *isometric* versus *isotonic* and *concentric* versus *eccentric.*

Suppose you lift a heavy dumbbell. When you first contract the muscles of your arm, you can feel the tension building in them even though the dumbbell isn't yet moving. At this point, your muscles are contracting at a cellular level, but their tension is being absorbed by the elastic components and is resisted by the weight of the load; the muscle as a whole is not producing any external movement. This phase is called **isometric**[10] **contraction**—contraction without a change in length (fig. 11.16a). Isometric contraction is not merely a prelude to movement. The isometric contraction of antagonistic muscles at a single joint is important in maintaining joint stability at rest, and the isometric contraction of postural muscles is what keeps us from sinking in a heap to the floor. **Isotonic**[11] **contraction**—contraction with a change in length but no change in tension—begins when internal tension builds to the point that it overcomes the resistance. The muscle now shortens, moves the load, and maintains essentially the same tension from then on (fig. 11.16b). Isometric and isotonic contraction are both phases of normal muscular action (fig. 11.17).

There are two forms of isotonic contraction: concentric and eccentric. In **concentric contraction,** a muscle shortens as it maintains tension—for example, when the biceps contracts and flexes the elbow. In **eccentric contraction,** a muscle lengthens as it maintains tension. If you set that dumbbell down again (fig. 11.16c), your biceps lengthens as you extend your elbow, but it maintains tension to act as a brake and keep you from simply dropping the weight. A weight lifter uses concentric contraction when lifting a dumbbell and eccentric contraction when lowering it. When weight lifters suffer muscle injuries, it is usually during the eccentric phase, because the sarcomeres and connective tissues of the muscle are pulling in one direction while the weight is pulling the muscle in the opposite direction.

In summary, during isometric contraction, a muscle develops tension without changing length, and in isotonic contraction, it changes length while maintaining constant tension. In concentric contraction, a muscle maintains tension as it shortens, and in eccentric contraction, it maintains tension while it is lengthening.

---

[10]*iso* = same, uniform; *metr* = length
[11]*iso* = same, uniform; *ton* = tension

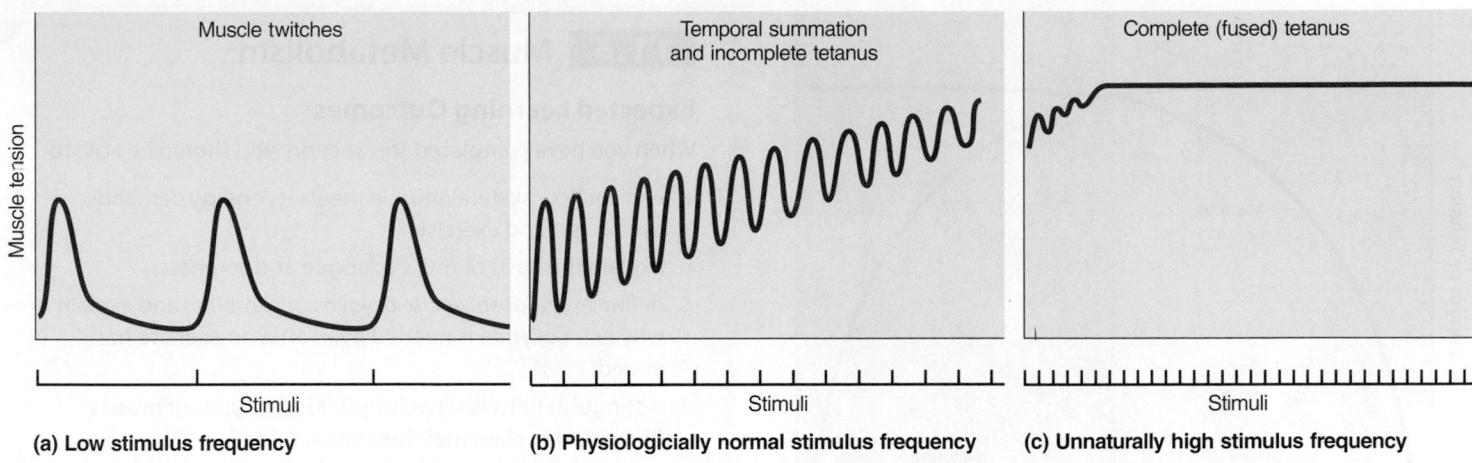

**FIGURE 11.15 The Relationship Between Stimulus Frequency and Muscle Tension.** (a) Twitch. At an unnaturally low stimulus frequency, as in laboratory preparations, the muscle relaxes completely between stimuli and shows twitches of uniform strength. (b) At a stimulus frequency within normal physiological range, the muscle does not have time to relax completely between twitches and the force of each twitch builds on the previous one, creating the state of incomplete tetanus. In this state, a muscle can attain three to four times as much tension, or force, as a single twitch produces. (c) At an unnaturally high stimulus frequency, attained only in laboratory preparations, the muscle cannot relax at all between twitches, and twitches fuse into a state of complete tetanus.

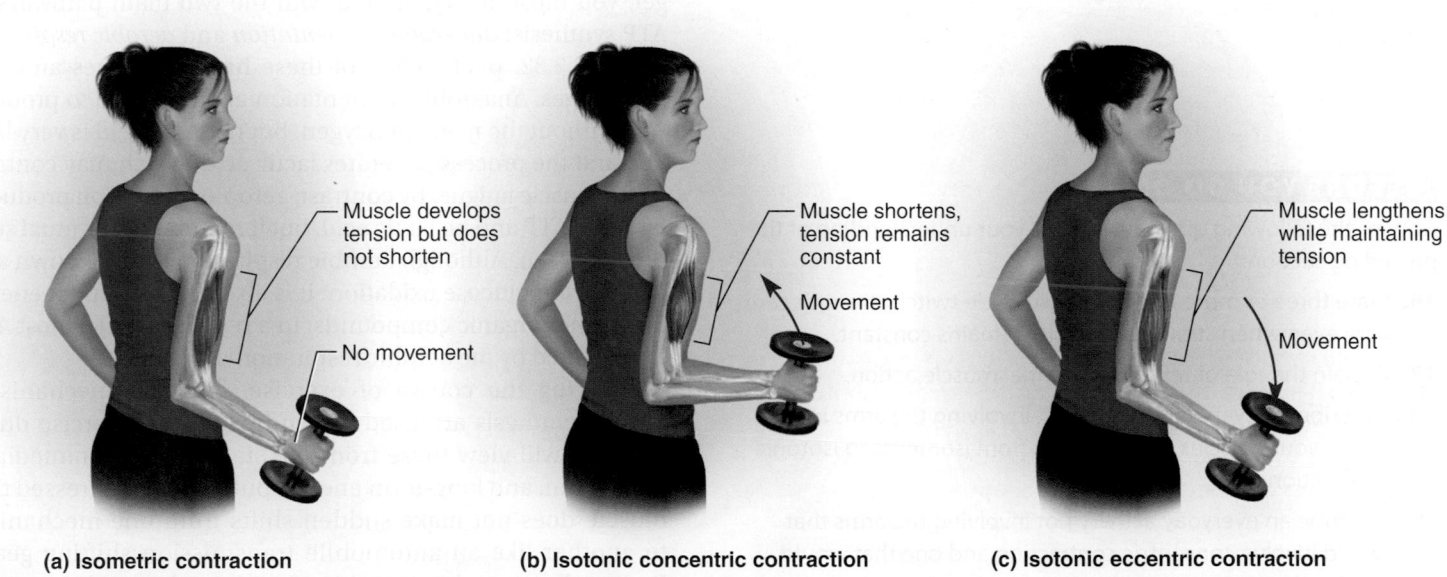

**FIGURE 11.16 Isometric and Isotonic Contraction.** (a) Isometric contraction, in which a muscle develops tension but does not shorten. This occurs at the beginning of any muscle contraction but is prolonged in actions such as lifting heavy weights. (b) Isotonic concentric contraction, in which the muscle shortens while maintaining a constant degree of tension. In this phase, the muscle moves a load. (c) Isotonic eccentric contraction, in which the muscle maintains tension while it lengthens, allowing a muscle to relax without going suddenly limp.

❓ *Name a muscle that undergoes eccentric contraction as you sit down in a chair.*

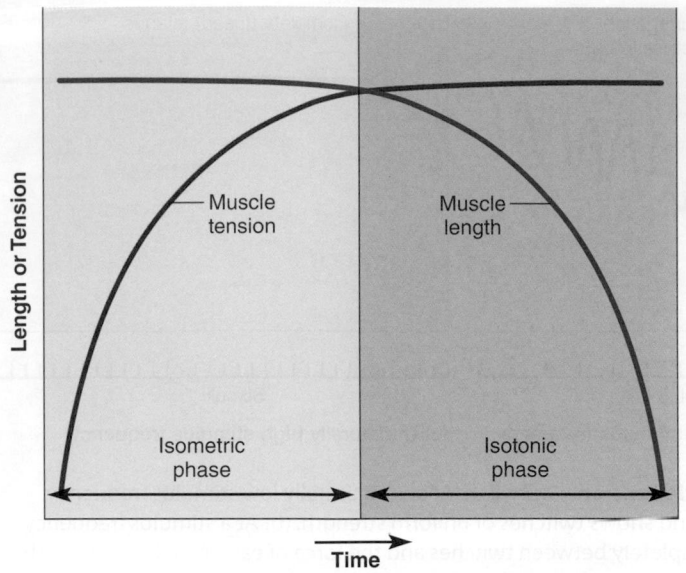

**FIGURE 11.17  Isometric and Isotonic Phases of Contraction.**
At the beginning of a contraction (isometric phase), muscle tension
rises but the length remains constant (the muscle does not shorten).
When tension overcomes the resistance of the load, the tension levels
off and the muscle begins to shorten and move the load (isotonic
phase).

 *How would you extend this graph in order to show eccentric
contraction?*

**BEFORE YOU GO ON**

Answer the following questions to test your understanding of the
preceding section:

16. State three or more reasons why muscle twitch strength can
    vary even when stimulus intensity remains constant.

17. Explain the role of tetanus in normal muscle action.

18. Describe an everyday activity not involving the arms in
    which your muscles would switch from isometric to isotonic
    contraction.

19. Describe an everyday activity not involving the arms that
    would involve concentric contraction and one that would
    involve eccentric contraction.

---

## 11.6    Muscle Metabolism

### Expected Learning Outcomes

When you have completed this section, you should be able to

a. explain how skeletal muscle meets its energy demands
   during rest and exercise;

b. explain the basis of muscle fatigue and soreness;

c. define *excess postexercise oxygen consumption* and explain
   why extra oxygen is needed even after an exercise has
   ended;

d. distinguish between two physiological types of muscle
   fibers, and explain their functional roles;

e. discuss the factors that affect muscular strength; and

f. discuss the effects of resistance and endurance exercises on
   muscles.

### ATP Sources

All muscle contraction depends on ATP; no other energy source
can serve in its place. The supply of ATP depends, in turn, on
the availability of oxygen and organic fuels such as glucose and
fatty acids. To understand how muscle manages its ATP budget, you must be acquainted with the two main pathways of
ATP synthesis: *anaerobic fermentation* and *aerobic respiration*
(see fig. 2.32, p. 71). Each of these has advantages and disadvantages. Anaerobic fermentation enables a cell to produce
ATP without the need for oxygen, but the ATP yield is very limited and the process generates lactic acid, which may contribute to muscle fatigue. By contrast, aerobic respiration produces
far more ATP and no lactic acid, but it requires a continual supply of oxygen. Although aerobic respiration is best known as a
pathway for glucose oxidation, it is also used to extract energy
from other organic compounds. In a resting muscle, most ATP
is generated by the aerobic respiration of fatty acids.

During the course of exercise, different mechanisms
of ATP synthesis are used depending on the exercise duration. We will view these from the standpoint of immediate,
short-term, and long-term energy, but it must be stressed that
muscle does not make sudden shifts from one mechanism
to another like an automobile transmission shifting gears.
Rather, these mechanisms blend and overlap as the exercise
continues (fig. 11.18).

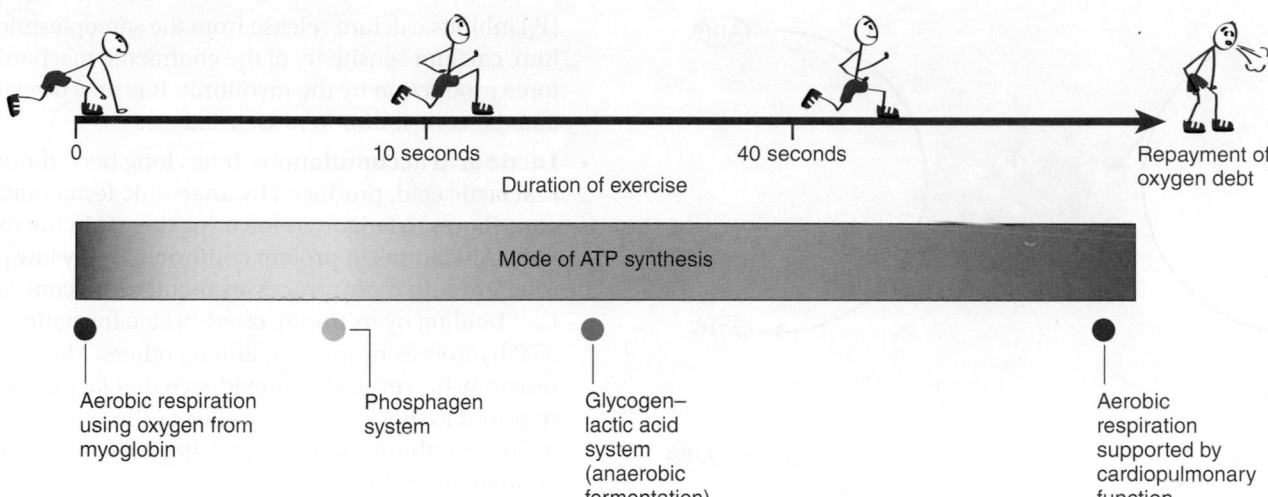

**FIGURE 11.18 Modes of ATP Synthesis During Exercise.**

## Immediate Energy

In a short, intense exercise such as a 100 m dash, the myoglobin in a muscle fiber supplies oxygen for a limited amount of aerobic respiration at the outset, but this oxygen supply is quickly depleted. Until the respiratory and cardiovascular systems catch up with the heightened oxygen demand, the muscle meets most of its ATP needs by borrowing phosphate groups ($P_i$) from other molecules and transferring them to ADP. Two enzyme systems control these phosphate transfers (fig. 11.19):

1. **Myokinase** (MY-oh-KY-nase) transfers $P_i$ from one ADP to another, converting the latter to ATP.

2. **Creatine kinase** (CREE-uh-tin KY-nase) obtains $P_i$ from a phosphate-storage molecule, **creatine phosphate (CP),** and donates it to ADP to make ATP. This is a fast-acting system that helps to maintain the ATP level while other ATP-generating mechanisms are being activated.

ATP and CP, collectively called the **phosphagen system,** provide nearly all the energy used for short bursts of intense activity. Muscle contains about 5 millimoles of ATP and 15 millimoles of CP per kilogram of tissue. Perhaps surprisingly, at the outset of an intense exercise, the amount of ATP in the muscle fibers changes very little, but the amount of CP drops rapidly. The total supply of ATP + CP is enough to power about 1 minute of brisk walking or 6 seconds of sprinting or fast swimming. The phosphagen system is especially important in activities requiring brief but maximal effort, such as football, baseball, and weight lifting.

## Short-Term Energy

As the phosphagen system is exhausted, the muscles transition to anaerobic fermentation to "buy time" until cardiopulmonary function can catch up with the muscles' oxygen demand. During this period, the muscles obtain glucose from the blood and their own stored glycogen. You may recall from chapter 2 that when oxygen is lacking, the pathway of glycolysis can generate a net gain of 2 ATP for every glucose molecule consumed, as it converts glucose to lactic acid. The pathway from glycogen to lactic acid, called the **glycogen–lactic acid system,** produces enough ATP for 30 to 40 seconds of maximum activity. Playing basketball or running completely around a baseball diamond, for example, depends heavily on this energy-transfer system.

## Long-Term Energy

After 40 seconds or so, the respiratory and cardiovascular systems "catch up" and deliver oxygen to the muscles fast enough for aerobic respiration to meet most of the ATP demand. Aerobic respiration produces much more ATP than glycolysis does—typically another 30 ATP per glucose. Thus it is a very efficient means of meeting the ATP demands of prolonged exercise. One's rate of oxygen consumption rises for 3 to 4 minutes and then levels off at a *steady state* in which aerobic ATP production keeps pace with the demand. In exercises lasting more than 10 minutes, over 90% of the ATP is produced aerobically. For up to 30 minutes, the energy for this comes about equally from glucose and fatty acids; then as glucose and glycogen are depleted, fatty acids become the more significant fuel.

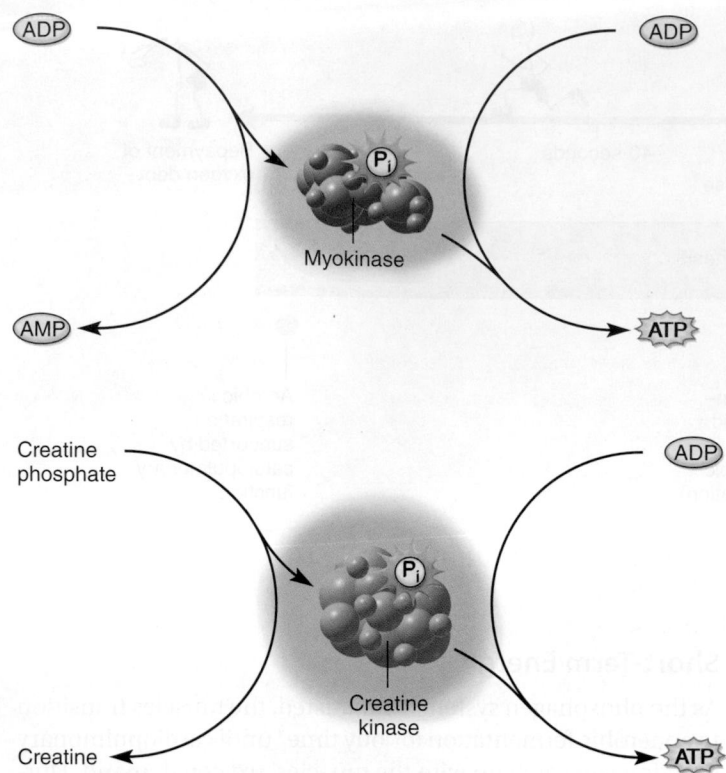

**FIGURE 11.19  The Phosphagen System.** Two enzymes, myokinase and creatine kinase, generate ATP in the absence of oxygen. Myokinase borrows phosphate groups from ADP, and creatine kinase borrows them from creatine phosphate, to convert an ADP to ATP.

## Fatigue and Endurance

Muscle **fatigue** is the progressive weakness and loss of contractility that results from prolonged use of the muscles. For example, if you hold this book at arm's length for a minute, you will feel your muscles growing weaker and eventually you will be unable to hold it up. Repeatedly squeezing a rubber ball, pushing a video game button, or trying to take lecture notes from a fast-talking professor produces fatigue in the hand muscles. In high-intensity, short-duration exercise, fatigue is thought to result from the following factors:

- **Potassium accumulation.**  Each action potential releases $K^+$ from the sarcoplasm to the extracellular fluid. This lowers the membrane potential (hyperpolarizes the cell) and makes the muscle fiber less excitable. This is especially significant in the T tubules, where the low volume of ECF enables the $K^+$ concentration to rise to a high level and interfere with the release of calcium from the sarcoplasmic reticulum.

- **ADP/$P_i$ accumulation.**  The hydrolysis of ATP generates an ever-growing pool of ADP + $P_i$. ADP slows the cross-bridge cycling mechanism of contraction. Free phosphate ($P_i$) inhibits calcium release from the sarcoplasmic reticulum, calcium sensitivity of the contractile mechanism, and force production by the myofibrils. It is now thought to be a major contributor to muscle fatigue.

- **Lactic acid accumulation.**  It has long been thought that lactic acid, produced by anaerobic fermentation, contributes to fatigue by lowering the pH in the muscle fiber. Alterations in protein conformation by low pH can interfere with many processes required for contraction—$Ca^{2+}$ binding by troponin, cross-bridge formation, and ATP hydrolysis by myosin, among others. More recent research, however, shows evidence that lactic acid is removed to the liver about as fast as the muscles produce it. Some authorities now argue that muscle fatigue is independent of lactic acid.

In low-intensity, long-duration exercise, fatigue may result partially from the preceding causes, but it results predominantly from the following:

- **Fuel depletion.**  Declining levels of muscle glycogen and blood glucose leave less fuel for ATP synthesis. Long-distance runners and cyclists call this "hitting the wall," and often endeavor to delay fatigue by means of high-carbohydrate diets before the race, loading the muscles with extra glycogen.

- **Electrolyte loss.**  The loss of electrolytes through sweating can alter the ion balance of the extracellular fluid enough to reduce muscle excitability.

- **Central fatigue.**  Exercising muscle generates ammonia, which is absorbed by the brain and inhibits motor neurons of the cerebrum. For this and other reasons not yet well understood, the central nervous system produces less signal output to the skeletal muscles. This is where psychological factors come into play, such as the will to complete a marathon.

ATP depletion per se is no longer thought to cause fatigue. The ATP level in fatigued muscle is almost as great as in rested muscle.

The ability to maintain high-intensity exercise for more than 4 to 5 minutes is determined in large part by one's **maximum oxygen uptake ($VO_{2max}$)**—the point at which the rate of oxygen consumption reaches a plateau and does not increase further with an added workload. $VO_{2max}$ is proportional to body size; it peaks around age 20; it is usually greater in males than in females; and it can be twice as great in a trained endurance athlete as in an untrained person. A typical sedentary adult uses a maximum of about 35 milliliters of oxygen per minute per kilogram of body weight. Such a person weighing 73 kg (160 lb) and exercising at maximum intensity might therefore "burn" oxygen at about 2.6 L/min., which sets a limit to his rate of ATP production. Elite endurance athletes can have a $VO_{2max}$ of about 70 mL/min./kg (5.2 L/min. for the same body weight).

## Excess Postexercise Oxygen Consumption

You have probably noticed that you breathe heavily not only during strenuous exercise but also for several minutes afterward (fig. 11.18). This is to meet a metabolic demand called **excess postexercise oxygen consumption (EPOC),** also known by an older popularized term, **oxygen debt.** EPOC is the difference between the elevated rate of oxygen consumption at the end of an exercise and the rate at rest. It occurs in part because exercise depletes your stores of ATP and creatine phosphate (CP). Oxygen is required to synthesize ATP aerobically, and some of that ATP is used to regenerate CP. A small amount of oxygen serves to reoxygenate the muscle myglobin, and the liver consumes oxygen in disposing of the lactic acid generated by exercise. In addition, exercise raises the body temperature and overall metabolic rate, which in itself consumes more oxygen.

ATP and CP are replenished in the early minutes of heaviest postexercise breathing, and oxygen consumption remains elevated for as much as an hour more as lactic acid is oxidized. EPOC can be as much as six times one's basal oxygen consumption, indicating that anaerobic mechanisms of ATP production during exercise allow six times as much physical exertion as would have been possible without those mechanisms.

## Physiological Classes of Muscle Fibers

Not all muscle fibers are alike or adapted for the same tasks. For example, the weight-bearing and postural muscles of the back and lower limbs react slowly to stimulation and take up to 100 ms to reach peak tension. By contrast, muscles that control eye and hand movements react quickly to stimulation and reach peak tension in as little as 7.5 ms. The predominant fibers that compose these muscles are therefore called *slow-twitch* and *fast-twitch* fibers, respectively. We can find reasons for their differences in response time by looking at their cellular structure and biochemistry (table 11.3).

**Slow-twitch fibers** are also called **slow oxidative (SO)** or **type I** fibers. They are well adapted for endurance and fatigue resistance, so they are particularly important in muscles that support the body and maintain posture, such as the erector spinae and quadratus lumborum of the back. Their fatigue resistance stems from their *oxidative* mode of ATP production—that is, aerobic respiration. Oxidative metabolism, of course, requires a liberal supply of oxygen and the means to use it efficiently. Therefore, these fibers are surrounded by a dense network of blood capillaries, they are rich in mitochondria, and they have a high concentration of myoglobin, the red pigment that facilitates diffusion of oxygen from the blood into the muscle fiber. Slow-twitch fibers are also relatively thin, which minimizes the distance that oxygen must diffuse to even the deepest mitochondria. The slowness of these muscles is due to a sarcoplasmic reticulum that is relatively slow to release and reabsorb calcium, and a form of myosin ATPase that is relatively slow in its ATP hydrolysis and cross-bridge cycling. The high

| TABLE 11.3 | Classification of Skeletal Muscle Fibers | |
|---|---|---|
| | **Fiber Type** | |
| **Properties** | **Slow-Twitch (Slow Oxidative)** | **Fast-Twitch (Fast Glycolytic)** |
| Twitch duration | As long as 100 ms | As short as 7.5 ms |
| Motor unit size | Smaller | Larger |
| Motor neurons | Smaller, more excitable | Larger, less excitable |
| Motor unit strength | Weaker | Stronger |
| Relative diameter | Smaller | Larger |
| ATP synthesis | Aerobic | Anaerobic |
| Fatigue resistance | Good | Poor |
| ATP hydrolysis | Slow | Fast |
| Glycolysis | Moderate | Fast |
| Myoglobin content | Abundant | Low |
| Glycogen content | Low | Abundant |
| Mitochondria | Abundant and large | Fewer and smaller |
| Capillaries | Abundant | Fewer |
| Color | Red | White, pale |
| *Representative muscles in which fiber type is predominant* | Soleus Erector spinae Quadratus lumborum | Gastrocnemius Biceps brachii Muscles of eye movement |

myoblogin concentration of these fibers gives them a relatively bright red color, so they are also called **red fibers.**

**Fast-twitch fibers** are also called **fast glycolytic (FG)** or **type IIb** fibers. They are well adapted for quick responses, so they are particularly important in the aforesaid eye and hand muscles, and in large muscles such as the gastrocnemius and biceps brachii, which we employ in such actions as jumping and elbow flexion. Their quickness stems from an especially extensive sarcoplasmic reticulum with fast release and reabsorption of calcium, and a form of myosin with very quick ATP hydrolysis and cross-bridge cycling. For energy, they depend primarily on glycolysis and anaerobic fermentation, which produces ATP more quickly (yet less efficiently) than aerobic respiration. To support this, they contain a high concentration of glycogen. They also have a high concentration of creatine phosphate, which, you will recall, aids in rapidly regenerating ATP. They have fewer mitochondria than slow-twitch fibers. Fast-twitch fibers are thicker than slow-twitch, because thick fibers are stronger and they have no need for especially rapid oxygen delivery to the deepest cytoplasm. Also, without need of such rapid oxygen uptake, they have less myoglobin. For this reason, fast-twitch fibers are relatively pale and are called **white fibers.**

But the price paid for these fast-twitch mechanisms and anaerobic ATP production is that these fibers fatigue more easily, as you may know from writer's cramp or doing rapid biceps curls.

There is also a fiber type called IIa, which is fast-twitch but oxidative and fatigue-resistant. However, type IIa fibers are uncommon in humans and known mainly from other species of mammals.

Every skeletal muscle is composed of a mixture of fast- and slow-twitch muscle fibers, but usually one fiber type or the other predominates according to the functions of that muscle—quick motility and reflexes, or sustained weight-bearing tension. In a histological section of a single muscle, we can recognize both fiber types from their relative sizes and with the aid of stains that distinguish their oxidative or glycolytic enzymes and other chemical constituents (fig. 11.20).

Muscles composed mainly of SO fibers are called *red muscles* and those composed mainly of FG fibers are called *white muscles* because of the color difference stemming from their difference in myoglobin content. Anyone who eats chicken or turkey may be unwittingly familiar with this distinction. The thighs are "red meat" composed of SO fibers adapted to long periods of standing, and the breast is "white meat" composed of FG fibers adapted for short bursts of power when the birds take flight. Duck breast, however, is "dark meat" (red muscle) adapted for long-distance flight.

In humans, small motor units are composed of relatively small SO muscle fibers and supplied by relatively small but easily excited motor neurons. These motor units are not as strong as large ones but produce more precise movements. Large motor units are composed of larger FG fibers; supplied by larger, less excitable neurons; and produce more power but less fine control. In the multiple motor unit summation discussed earlier, the nervous system recruits small SO motor units first, then larger FG motor units only if more strength is needed for a particular task.

People with different types and levels of physical activity differ in the proportion of one fiber type to another even in the same muscle, such as the *quadriceps femoris* of the anterior thigh (table 11.4). It is thought that people are born with a genetic predisposition for a certain ratio of fiber types. Those who go into competitive sports discover the sports at which they can excel and gravitate toward those for which heredity has

**FIGURE 11.20** **Muscle Stained to Distinguish Slow Oxidative (SO) from Fast Glycolytic (FG) Fibers.** Cross section. Note the differences in staining and fiber diameter.

best equipped them. One person might be a "born sprinter" and another a "born marathoner."

We saw in chapter 10 that sometimes two or more muscles act across the same joint and seem to have the same function. We have already seen some reasons why such muscles are not as redundant as they seem. Another reason is that they may differ in the proportion of SO to FG fibers. For example, the gastrocnemius and soleus muscles of the calf both insert on the calcaneus, so they exert the same pull on the heel. The gastrocnemius, however, is a white, predominantly FG muscle adapted for quick, powerful movements such as jumping, whereas the soleus is a red, SO muscle that does most of the work in standing and in endurance exercises such as jogging and skiing.

## Muscular Strength and Conditioning

We have far more muscular strength than we normally use. The gluteus maximus can generate 1,200 kg of tension, and all the muscles collectively can produce a total tension of 22,000 kg (nearly 25 tons). Indeed, the muscles can generate more tension than the bones and tendons can withstand—a fact that accounts for many injuries to the patellar and calcaneal tendons. Muscular strength depends on a variety of anatomical and physiological factors:

- **Muscle size.** The strength of a muscle depends primarily on its size; thicker muscles can form more myosin–actin cross-bridges and therefore generate more tension. A muscle can exert a tension of about 3 to 4 kg/cm$^2$ of cross-sectional area. This is why weight lifting increases both size and strength of a muscle.

- **Fascicle arrangement.** Pennate muscles such as the quadriceps femoris are stronger than parallel muscles such as the sartorius, which in turn are stronger than circular muscles such as the orbicularis oculi.

| TABLE 11.4 | Proportion of Slow Oxidative (SO) and Fast Glycolytic (FG) Fibers in the Quadriceps Femoris Muscle of Male Athletes | |
|---|---|---|
| **Sample Population** | **SO** | **FG** |
| Marathon runners | 82% | 18% |
| Swimmers | 74 | 26 |
| Average males | 45 | 55 |
| Sprinters and jumpers | 37 | 63 |

- **Size of active motor units.** Large motor units produce stronger contractions than small ones.
- **Multiple motor unit summation.** When a stronger muscle contraction is desired, the nervous system activates more and larger motor units. Getting "psyched up" for athletic competition is partly a matter of motor unit summation.
- **Temporal summation.** Nerve impulses usually arrive at a muscle in a series of closely spaced action potentials. Because of the *temporal summation* described earlier, the greater the frequency of stimulation, the more strongly a muscle contracts.
- **The length–tension relationship.** As noted earlier, a muscle resting at optimum length is prepared to contract more forcefully than a muscle that is excessively contracted or stretched.
- **Fatigue.** Rested muscles contract more strongly than fatigued ones.

**Resistance exercise,** such as weight lifting, is the contraction of muscles against a load that resists movement. A few minutes of resistance exercise at a time, a few times each week, is enough to stimulate muscle growth. Growth results primarily from cellular enlargement, not cellular division. The muscle fibers synthesize more myofilaments and the myofibrils grow thicker. Myofibrils split longitudinally when they reach a certain size, so a well-conditioned muscle has more myofibrils than a poorly conditioned one. Muscle fibers themselves are incapable of mitosis, but there is some evidence that as they enlarge, they too may split longitudinally. A small part of muscle growth may therefore result from an increase in the number of fibers, but most results from the enlargement of fibers that have existed since childhood.

▶▶▶**APPLY WHAT YOU KNOW**

*Is muscle growth mainly the result of hypertrophy or hyperplasia?*

**Endurance (aerobic) exercise,** such as jogging and swimming, improves the fatigue resistance of the muscles by enhancing the delivery and use of oxygen. Slow-twitch fibers, especially, produce more mitochondria and glycogen and acquire a greater density of blood capillaries as a result of conditioning. Endurance exercise also improves skeletal strength; increases the red blood cell count and the oxygen transport capacity of the blood; and enhances the function of the cardiovascular, respiratory, and nervous systems.

Endurance training does not significantly increase muscular strength, and resistance training does not improve endurance. Optimal performance and musculoskeletal health require **cross-training,** which incorporates elements of both types. If muscles are not kept sufficiently active, they become *deconditioned*—weaker and more easily fatigued.

**BEFORE YOU GO ON**

Answer the following questions to test your understanding of the preceding section:

20. From which two molecules can ADP borrow a phosphate group to become ATP? What is the enzyme that catalyzes each transfer?
21. In a long period of intense exercise, why does muscle generate ATP anaerobically at first and then switch to aerobic respiration?
22. List four causes of muscle fatigue.
23. List three causes of oxygen debt.
24. What properties of fast glycolytic and slow oxidative fibers adapt them for different physiological purposes?

## 11.7 Cardiac and Smooth Muscle

### Expected Learning Outcomes

When you have completed this section, you should be able to

a. describe the structural and physiological differences between cardiac muscle and skeletal muscle;
b. explain why these differences are important to cardiac function;
c. describe the structural and physiological differences between smooth muscle and skeletal muscle; and
d. relate the unique properties of smooth muscle to its locations and functions.

Cardiac and smooth muscle have special structural and physiological properties in common with each other, but different from those of skeletal muscle (table 11.5). These are related to their distinctive functions.

Any of the three types of muscle cells can be called **myocytes.** This term is preferable to *muscle fiber* for smooth and cardiac muscle because these two types of cells do not have the long fibrous shape of skeletal muscle cells. They are relatively short, and in further contrast to skeletal muscle fibers, they have only one nucleus. Cardiac myocytes are also called **cardiocytes.**

Cardiac and smooth muscle are *involuntary* muscle tissues, not usually subject to our conscious control. They receive no innervation from somatic motor neurons, but cardiac muscle and some smooth muscle receive nerves from the sympathetic and parasympathetic divisions of the autonomic nervous system (see chapter 15).

### Cardiac Muscle AP|R

**Cardiac muscle** is limited to the heart, where its function is to pump blood. Knowing that, we can predict the properties that it must have: (1) It must contract with a regular rhythm; (2) it

| TABLE 11.5 | Comparison of Skeletal, Cardiac, and Smooth Muscle | | |
|---|---|---|---|
| **Feature** | **Skeletal Muscle** | **Cardiac Muscle** | **Smooth Muscle** |
| Location | Associated with skeletal system | Heart | Walls of viscera and blood vessels, iris of eye, piloerector of hair follicles |
| Cell shape | Long threadlike fibers | Short, slightly branched cells | Short fusiform cells |
| Cell length | 100 μm–30 cm | 50–100 μm | 30–200 μm |
| Cell width | 10–500 μm | 10–20 μm | 5–10 μm |
| Striations | Present | Present | Absent |
| Nuclei | Multiple nuclei, adjacent to sarcolemma | Usually one nucleus, near middle of cell | One nucleus, near middle of cell |
| Connective tissues | Endomysium, perimysium, epimysium | Endomysium only | Endomysium only |
| Sarcoplasmic reticulum | Abundant | Present | Scanty |
| T tubules | Present, narrow | Present, wide | Absent |
| Gap junctions | Absent | Present in intercalated discs | Present in single-unit smooth muscle |
| Autorhythmicity | Absent | Present | Present in single-unit smooth muscle |
| Thin filament attachment | Z discs | Z discs | Dense bodies |
| Regulatory proteins | Tropomyosin, troponin | Tropomyosin, troponin | Calmodulin, myosin light-chain kinase |
| $Ca^{2+}$ source | Sarcoplasmic reticulum | Sarcoplasmic reticulum and extracellular fluid | Mainly extracellular fluid |
| $Ca^{2+}$ receptor | Troponin of thin filament | Troponin of thin filament | Calmodulin of thick filament |
| Innervation and control | Somatic motor fibers (voluntary) | Autonomic fibers (involuntary) | Autonomic fibers (involuntary) |
| Nervous stimulation required? | Yes | No | No |
| Effect of nervous stimulation | Excitatory only | Excitatory or inhibitory | Excitatory or inhibitory |
| Mode of tissue repair | Limited regeneration, mostly fibrosis | Limited regeneration, mostly fibrosis | Relatively good capacity for regeneration |

must function in sleep and wakefulness, without fail or need of conscious attention; (3) it must be highly resistant to fatigue; (4) the cardiocytes of a given heart chamber must contract in unison so that the chamber can effectively expel blood; and (5) each contraction must last long enough to expel blood from the chamber. These functional properties are the key to understanding how cardiac muscle differs structurally and physiologically from skeletal muscle (table 11.5).

Cardiac muscle is striated like skeletal muscle, but cardiocytes are shorter and thicker. Each is joined end to end with several others through linkages called **intercalated** (in-TUR-kuh-LAY-ted) **discs.** These appear as thick dark lines in stained tissue sections. An intercalated disc has electrical *gap junctions* that allow each cardiocyte to directly stimulate its neighbors, and mechanical junctions that keep the cardiocytes from pulling apart when the heart contracts. The sarcoplasmic reticulum is less developed than in skeletal

muscle, but the T tubules are larger and admit $Ca^{2+}$ from the extracellular fluid. Damaged cardiac muscle is repaired by fibrosis. Cardiac muscle has no satellite cells, and even though mitosis has recently been detected in cardiocytes following heart attacks, it does not produce a significant amount of regenerated functional muscle.

Unlike skeletal muscle, cardiac muscle can contract without the need of nervous stimulation. The heart has a built-in **pacemaker** that rhythmically sets off a wave of electrical excitation. This wave travels through the muscle and triggers the contraction of the heart chambers. The heart is said to be **autorhythmic**[12] because of this ability to contract rhythmically and independently. Stimulation by the autonomic nervous system, however, can increase or decrease

---

[12]*auto* = self

the heart rate and contraction strength. Cardiac muscle does not exhibit quick twitches like skeletal muscle. Rather, it maintains tension for about 200 to 250 ms, giving the heart time to expel blood.

Cardiac muscle uses aerobic respiration almost exclusively. It is very rich in myoglobin and glycogen, and it has especially large mitochondria that fill about 25% of the cell, compared with smaller mitochondria occupying about 2% of a skeletal muscle fiber. Cardiac muscle is very adaptable with respect to the fuel used, but very vulnerable to interruptions in oxygen supply. Because it makes little use of anaerobic fermentation, cardiac muscle is highly resistant to fatigue.

## Smooth Muscle  AP|R

**Smooth muscle** is named for the fact that it has no striations, for a reason to be described shortly. Its myocytes are relatively small, allowing for fine control of such tissues and organs as a single hair, the iris of the eye, and the tiniest arteries; yet in the pregnant uterus, the myocytes become quite large and produce the powerful contractions of childbirth.

Smooth muscle is not always innervated, but when it is, the nerve supply is autonomic, like that of the heart. Autonomic nerve fibers do not form precisely localized neuromuscular junctions with the myocytes. Rather, a nerve fiber has as many as 20,000 periodic swellings called **varicosities** along its length (fig. 11.21). Each varicosity contains synaptic vesicles from which it releases neurotransmitters—usually norepinephrine from the sympathetic fibers and acetylcholine

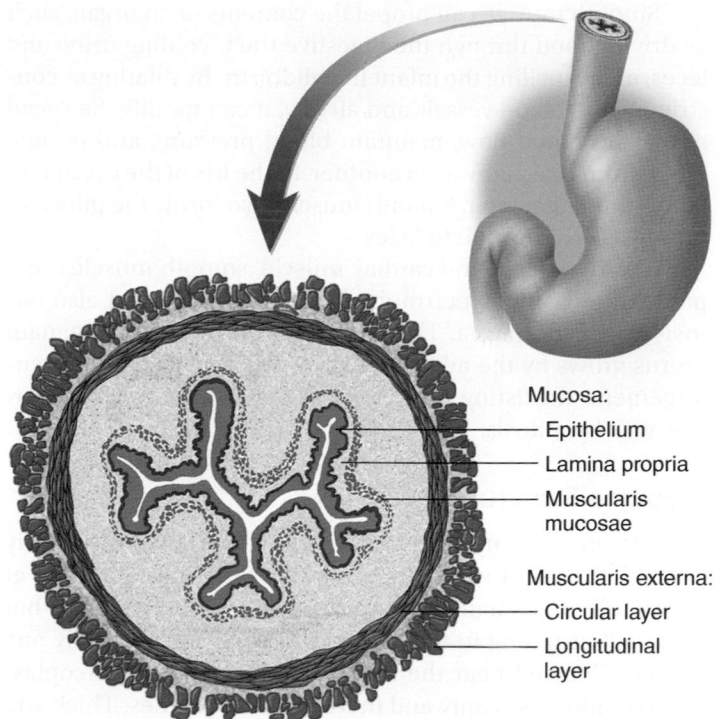

**FIGURE 11.22  Layers of Visceral Muscle in a Cross Section of the Esophagus.**  Many hollow organs have adjacent circular and longitudinal layers of smooth muscle.

Mucosa:
- Epithelium
- Lamina propria
- Muscularis mucosae

Muscularis externa:
- Circular layer
- Longitudinal layer

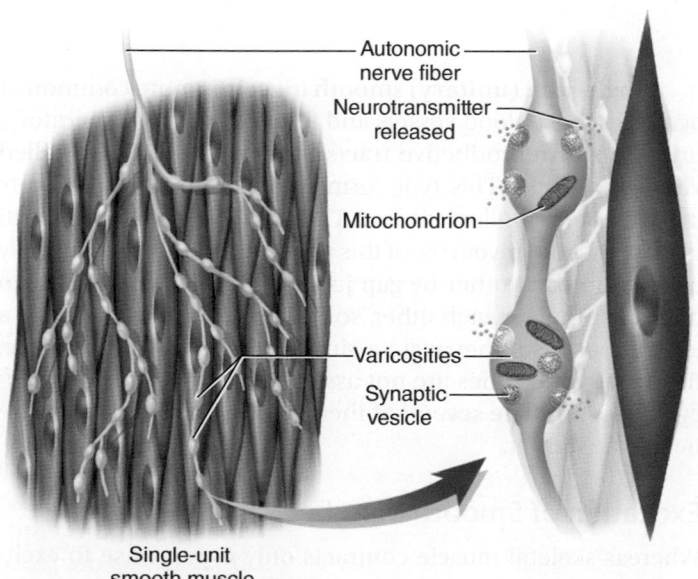

Autonomic nerve fiber
Neurotransmitter released
Mitochondrion
Varicosities
Synaptic vesicle
Single-unit smooth muscle

**FIGURE 11.21  Varicosities of an Autonomic Nerve Fiber in Single-Unit Smooth Muscle.**

from the parasympathetic fibers. The myocyte has no motor end plate, but instead has receptors for these neurotransmitters distributed over its surface. The varicosities simply release a flood of neurotransmitter into the tissue, and each myocyte may respond to more than one nerve fiber.

Whether innervated or not, smooth muscle responds to a wide variety of stimuli and often without any electrical excitation of the sarcolemma. It is much slower than skeletal and cardiac muscle to contract and relax, but it can remain contracted for a long time without fatigue and with minimal energy expenditure.

Smooth muscle does not form organs in itself, but usually forms layers in the walls of larger organs such as the stomach, intestines, uterus, and urinary bladder. In such cases, the muscle layer is quite variable in complexity. It can consist of as little as one cell in small arteries. The esophagus and intestines have a thick outer layer of longitudinal smooth muscle adjacent to a deeper, thick inner layer of circular muscle (fig. 11.22). When the longitudinal layer contracts, it shortens and dilates the organ; when the circular layer contracts, it constricts and lengthens the organ. In the stomach, urinary bladder, and uterus, smooth muscle forms three or more layers with bundles of myocytes running in multiple directions.

Smooth muscle can propel the contents of an organ, such as driving food through the digestive tract, voiding urine and feces, and expelling the infant in childbirth. By dilating or constricting the blood vessels and airway, it can modify the speed of air and blood flow, maintain blood pressure, and reroute blood from one pathway to another. In the iris of the eye, it regulates pupil diameter. Smooth muscle also forms the piloerector muscles of the hair follicles.

Unlike skeletal and cardiac muscle, smooth muscle is capable of not only hypertrophy (cellular growth) but also mitosis and hyperplasia. Thus, an organ such as the pregnant uterus grows by the addition of new myocytes as well as enlargement of existing ones. Injured smooth muscle regenerates well by mitosis.

## Myocyte Structure

Smooth muscle myocytes have a fusiform shape, typically about 5 to 10 μm wide at the middle, tapering to a point at each end, and usually ranging from 30 to 200 μm long—but up to 500 μm long in the pregnant uterus. There is only one nucleus, located near the middle of the cell. The sarcoplasmic reticulum is scanty and there are no T tubules. Thick and thin filaments are present, but there are no striations, sarcomeres, or myofibrils because the myofilaments are not bundled and aligned with each other the way they are in striated muscle. Z discs are absent. In their place are protein plaques called **dense bodies,** some associated with the inner face of the plasma membrane and others dispersed throughout the sarcoplasm (see fig. 11.24). The membrane-associated dense bodies of one cell are often directly across from those of another, with linkages between them so that contractile force can be transmitted from cell to cell. Associated with the dense bodies is an extensive cytoskeletal network of intermediate filaments. Actin filaments attach to the intermediate filaments as well as directly to the dense bodies, so their movement (powered by myosin) is transferred to the sarcolemma and shortens the cell.

## Types of Smooth Muscle

Smooth muscle tissue shows a range of types between two extremes called *multiunit* and *single-unit* types (fig. 11.23). **Multiunit smooth muscle** occurs in some of the largest arteries and pulmonary air passages, piloerector muscles, and eye muscles that control the iris and lens. Its innervation, although autonomic, is to some degree similar to that of skeletal muscle—the terminal branches of a nerve fiber synapse with individual myocytes and form a motor unit. The varicosities are specifically associated with a particular myocyte, and each myocyte responds independently of all the others—hence the name *multiunit.* Multiunit smooth muscle does not, however, generate action potentials. It contracts in response to variable (graded) electrical changes in the sarcolemma or even in the absence of electrical excitation.

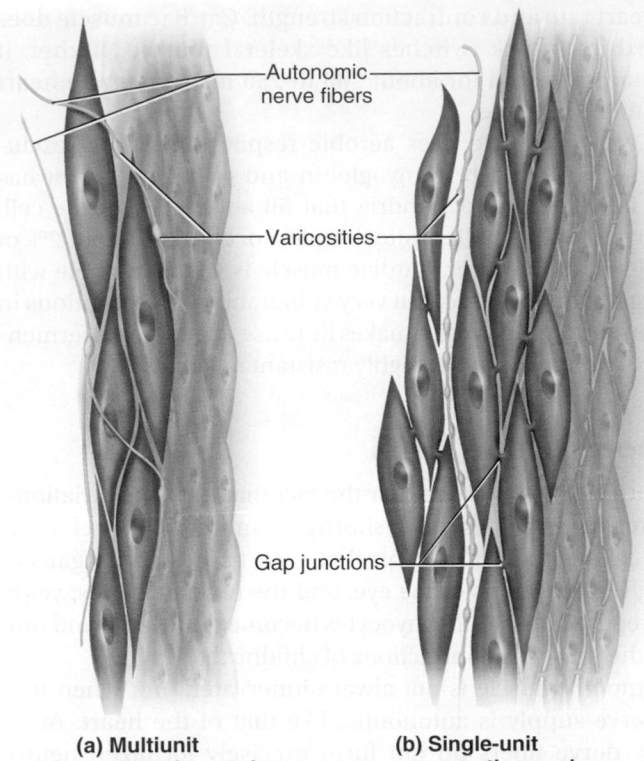

**(a) Multiunit smooth muscle**   **(b) Single-unit smooth muscle**

**FIGURE 11.23 Multiunit and Single-Unit Smooth Muscle.** (a) Multiunit smooth muscle, in which each muscle cell receives its own nerve supply and contracts independently. (b) Single-unit smooth muscle, in which a nerve fiber passes through the tissue without synapsing with any specific muscle cell, and muscle cells are coupled by electrical gap junctions.

**Single-unit (unitary) smooth muscle** is more common. It occurs in most blood vessels and in the digestive, respiratory, urinary, and reproductive tracts—therefore, it is also called **visceral muscle.** This type forms the aforementioned layers in many of the hollow viscera. The name *single-unit* refers to the fact that the myocytes of this type of muscle are electrically coupled to each other by gap junctions. This allows them to directly stimulate each other, so numerous cells contract as a unit, almost as if they were a single cell. In this muscle type, the nerve varicosities are not associated with a specific myocyte, but stimulate several of them at once when they release neurotransmitter.

## Excitation of Smooth Muscle

Whereas skeletal muscle contracts only in response to excitatory stimulation by a somatic motor fiber, smooth muscle can be stimulated in a multitude of ways. Some stimuli excite the myocyte and others inhibit it. Some produce action potentials in the sarcolemma, particularly in single-unit smooth muscle, whereas others stimulate the myocyte by nonelectrical means. Some smooth muscle has no nerve supply at all. Modes of smooth muscle stimulation and some examples include

- **Autonomic nerve fibers and neurotransmitters.** For example, parasympathetic nerves secrete acetylcholine and stimulate gastrointestinal motility; sympathetic nerves secrete norepinephrine and dilate the bronchioles of the lungs.

- **Chemicals.** Smooth muscle reacts to hormones, carbon dioxide, oxygen, nitric oxide, low pH, and other chemical stimuli. The hormone oxytocin, for example, stimulates the labor contractions of the uterus, and histamine relaxes the smooth muscle of arteries.

- **Temperature.** Cold induces contraction of smooth muscle resulting in erection of the hairs and tautening of the skin in such regions as the areola and scrotum, whereas warmth relaxes smooth muscle in arteries of the skin.

- **Stretch.** The stomach and urinary bladder contract when stretched by food or urine.

- **Autorhythmicity.** Some single-unit smooth muscle is autorhythmic, especially in the stomach and intestines. Some myocytes act as pacemaker cells, spontaneously depolarizing at regular time intervals and setting off waves of contraction throughout an entire layer of muscle. The rhythm is much slower than in cardiac muscle.

Regardless of how a myocyte is stimulated, however, the immediate trigger for contraction is the same as in skeletal and cardiac muscle—calcium ions. In some cases, the $Ca^{2+}$ comes from the sarcoplasmic reticulum (SR), as it does in skeletal muscle. With a relatively sparse SR, however, smooth muscle usually gets most of its $Ca^{2+}$ from the extracellular fluid by way of gated calcium channels in the sarcolemma. Compensating for the paucity of SR, the sarcolemma has numerous little pockets called **caveolae** (CAV-ee-OH-lee) where its calcium channels are concentrated. Calcium is 10,000 times as concentrated in the ECF as in the cytosol, so if these channels are opened, it diffuses quickly into the cell. Because smooth muscle cells are relatively small, the incoming $Ca^{2+}$ can quickly reach all of the myofilaments.

What opens these gated channels? Some are mechanically gated and open in response to physical disortion such as stretch. This is the case in organs that periodically fill and empty, such as the stomach and urinary bladder. Some are voltage-gated and open in response to electrical depolarization of the sarcolemma. Still others are ligand-gated and open in response to chemicals originating outside the cell or from the cytosol. Acetylcholine and norepinephrine, for example, bind to surface receptors and activate the formation of second messengers within the smooth muscle cell. This leads to a series of intracellular events that open the surface calcium channel from within.

►►►**APPLY WHAT YOU KNOW**

*How is smooth muscle contraction affected by the drugs called calcium channel blockers? (See p. 85.)*

## Contraction and Relaxation

Calcium ions are the immediate trigger for contraction, but unlike skeletal and cardiac muscle, smooth muscle has no troponin to bind it. Calcium binds instead to a similar protein called **calmodulin**[13] (cal-MOD-you-lin), associated with myosin. Calmodulin then activates an enzyme called **myosin light-chain kinase,** which adds a phosphate group to a small regulatory protein on the myosin head. This activates the myosin ATPase, enabling it to bind to actin and hydrolyze ATP. The myosin then produces repetitive power and recovery strokes like those of skeletal muscle.

As thick filaments pull on the thin ones, the thin filaments pull on the dense bodies and membrane plaques. Through the dense bodies and cytoskeleton, force is transferred to the plasma membrane and the entire cell shortens. When a smooth muscle cell contracts, it puckers and twists somewhat like wringing out a wet towel (fig. 11.24).

[13]acronym for *cal*cium *modul*ating prote*in*

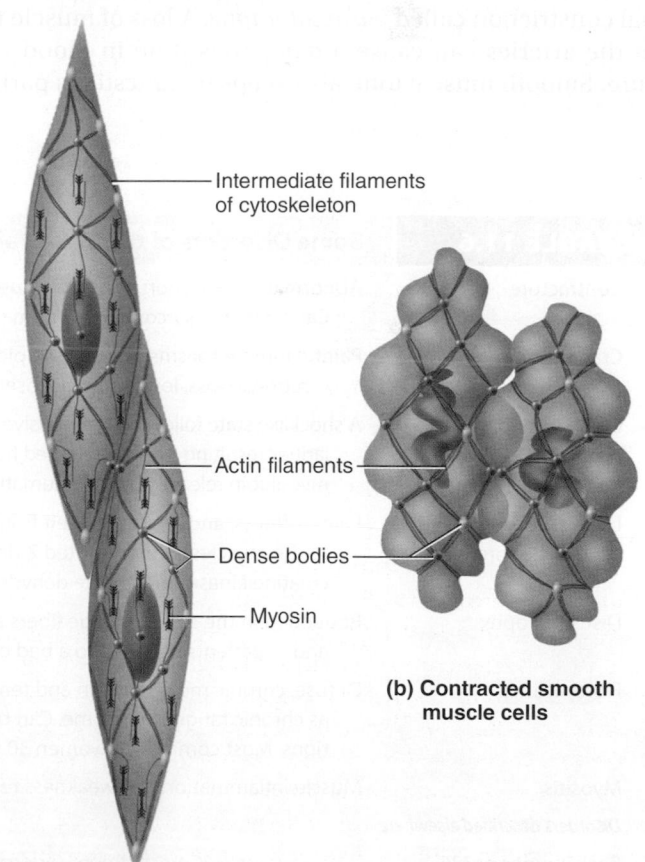

Intermediate filaments of cytoskeleton

Actin filaments

Dense bodies

Myosin

**(b) Contracted smooth muscle cells**

**(a) Relaxed smooth muscle cells**

**FIGURE 11.24 Smooth Muscle Contraction.** (a) Relaxed cells. Actin myofilaments are anchored to plaques on the plasma membrane and to dense bodies in the sarcoplasm, rather than to Z discs. (b) Contracted cells.

In skeletal muscle, there is typically a 2 ms latent period between stimulation and the onset of contraction. In smooth muscle, by contrast, the latent period is 50 to 100 ms long. Tension peaks about 500 ms (0.5 second) after the stimulus and then declines over a period of 1 to 2 seconds. The effect of all this is that compared with skeletal muscle, smooth muscle is very slow to contract and relax. It is slow to contract because its myosin ATPase is a slow enzyme. It is slow to relax because the pumps that remove $Ca^{2+}$ from the cytosol are also slow. As the $Ca^{2+}$ level falls, myosin is dephosphorylated and is no longer able to hydrolyze ATP and execute power strokes. However, it does not necessarily detach from actin immediately. It has a **latch-bridge mechanism** that enables it to remain attached to actin for a prolonged time without consuming more ATP.

Smooth muscle often exhibits tetanus and is very resistant to fatigue. It makes most of its ATP aerobically, but its ATP requirement is small and it has relatively few mitochondria. Skeletal muscle requires 10 to 300 times as much ATP as smooth muscle to maintain the same amount of tension. The fatigue resistance and latch-bridge mechanism of smooth muscle are important in enabling it to maintain a state of continual **smooth muscle tone** (tonic contraction). Tonic contraction keeps the arteries in a state of partial constriction called *vasomotor tone*. A loss of muscle tone in the arteries can cause a dangerous drop in blood pressure. Smooth muscle tone also keeps the intestines partially contracted. The intestines are much longer in a cadaver than they are in a living person because of the loss of muscle tone at death.

## Response to Stretch

Stretch alone sometimes causes smooth muscle to contract by opening mechanically gated calcium channels in the sarcolemma. Distension of the esophagus with food or the colon with feces, for example, evokes a wave of contraction called **peristalsis** (PERR-ih-STAL-sis) that propels the contents along the organ.

Smooth muscle exhibits a reaction called the **stress–relaxation** (or **receptive-relaxation**) **response.** When stretched, it briefly contracts and resists, but then relaxes. The significance of this response is apparent in the urinary bladder. If the stretched bladder contracted and did not soon relax, it would expel urine almost as soon as it began to fill, thus failing to store the urine until an opportune time.

Remember that skeletal muscle cannot contract very forcefully if it is overstretched. Smooth muscle, however, is less limited by the length–tension relationship. It contracts forcefully even when greatly stretched, so that hollow organs such as the stomach and bladder can fill and then expel their contents efficiently. Skeletal muscle must be within 30% of optimum length in order to contract strongly when stimulated. Smooth muscle, by contrast, can be anywhere from half to twice its

| TABLE 11.6 | Some Disorders of the Muscular System |
|---|---|
| Contracture | Abnormal muscle shortening not caused by nervous stimulation. Can result from failure of the calcium pump to remove $Ca^{2+}$ from the sarcoplasm or from contraction of scar tissue, as in burn patients. |
| Cramps | Painful muscle spasms caused by rapid firing of motor neurons; triggered by heavy exercise, extreme cold, dehydration, electrolyte loss, low blood glucose, or lack of blood flow. |
| Crush syndrome | A shocklike state following the massive crushing of muscles; associated with high and potentially fatal fever, cardiac irregularities resulting from $K^+$ released from the muscle, and kidney failure resulting from blockage of the renal tubules with myoglobin released by the traumatized muscle. Myoglobinuria (myoglobin in the urine) is a common sign. |
| Delayed-onset muscle soreness | Pain, stiffness, and tenderness felt from several hours to a day after strenuous exercise. Associated with microtrauma to the muscles; with disrupted Z discs, myofibrils, and plasma membranes; and with elevated levels of myoglobin, creatine kinase, and lactate dehydrogenase in the blood. |
| Disuse atrophy | Reduction in the size of muscle fibers as a result of nerve damage or muscular inactivity, for example in limbs in a cast and in patients confined to a bed or wheelchair. Muscle strength can be lost at a rate of 3% per day of bed rest. |
| Fibromyalgia | Diffuse, chronic muscular pain and tenderness, often associated with sleep disturbances and fatigue; often misdiagnosed as chronic fatigue syndrome. Can be caused by various infectious diseases, physical or emotional trauma, or medications. Most common in women 30 to 50 years old. |
| Myositis | Muscle inflammation and weakness resulting from infection or autoimmune disease. |

*Disorders described elsewhere*

| | | |
|---|---|---|
| Back injuries p. 336 | Compartment syndrome p. 313 | Pitcher's arm p. 370 |
| Baseball finger p. 370 | Hernia p. 339 | Pulled hamstrings p. 370 |
| Carpal tunnel syndrome p. 352 | Muscular dystrophy p. 429 | Rotator cuff injury p. 370 |
| | Myasthenia gravis p. 429 | Tennis elbow p. 370 |
| Charley horse p. 370 | Paralysis pp. 406, 503 | Tennis leg p. 370 |

resting length and still contract powerfully. There are three reasons for this: (1) There are no Z discs, so thick filaments cannot butt against them and stop the contraction; (2) since the thick and thin filaments are not arranged in orderly sarcomeres, stretching of the muscle does not cause a situation in which there is too little overlap for cross-bridges to form; and (3) the thick filaments of smooth muscle have myosin heads along their entire length (there is no bare zone), so cross-bridges can form anywhere, not just at the ends. Smooth muscle also exhibits **plasticity**—the ability to adjust its tension to the degree of stretch. Thus, a hollow organ such as the bladder can be greatly stretched yet not become flabby when it is empty.

The muscular system suffers fewer diseases than any other organ system, but several of its more common dysfunctions are listed in table 11.6. The effects of aging on the muscular system are described on page 1119.

### BEFORE YOU GO ON

Answer the following questions to test your understanding of the preceding section:

25. Explain why intercalated discs are important to cardiac muscle function.
26. Explain why it is important for cardiac muscle to have longer-lasting contractions than skeletal muscle.
27. How do single-unit and multiunit smooth muscle differ in innervation and contractile behavior?
28. How does smooth muscle differ from skeletal muscle with respect to its source of calcium and its calcium receptor?
29. Explain why the stress–relaxation response is an important factor in smooth muscle function.

## DEEPER INSIGHT 11.4

### CLINICAL APPLICATION

#### *Muscular Dystrophy and Myasthenia Gravis*

*Muscular dystrophy*[14] is a collective term for several hereditary diseases in which the muscles degenerate, weaken, and are gradually replaced by fat and fibrous scar tissue. The most common form of the disease is *Duchenne*[15] *muscular dystrophy (DMD)*, a sex-linked recessive trait affecting about 1 out of every 3,500 live-born boys.

DMD is not evident at birth, but begins to exhibit its effects as a child shows difficulty keeping up with other children, falls frequently, and finds it hard to stand again. It is typically diagnosed between the ages of 2 and 10 years. It affects the muscles of the hips first; then the legs; and then progresses to the abdominal, spinal, and respiratory muscles as well as cardiac muscle. The muscles shorten as they atrophy, causing postural abnormalities such as scoliosis. Persons with DMD are usually wheelchair-dependent by the age of 10 or 12, and seldom live past the age of 20. For obscure reasons, they also frequently suffer a progressive decline in mental ability. Death usually results from respiratory insufficiency, pulmonary infection, or heart failure. DMD is incurable, but is treated with exercise to slow the atrophy of the muscles and with braces to reinforce the weakened hips and maintain posture.

The underlying cause of DMD is a mutation in the gene for the muscle protein dystrophin (see p. 401 )—a large gene highly vulnerable to mutation. Without dystrophin, there is no coupling between the thin myofilaments and the sarcolemma. The sarcomeres move independently of the sarcolemma, creating tears in the membrane. The torn membrane admits excess $Ca^{2+}$ into the cell, which activates intracellular proteases (protein-digesting enzymes). These enzymes degrade the contractile proteins of the muscle, leading to weakness and cellular necrosis. Dying muscle fibers are replaced with scar tissue, which blocks blood circulation in the muscle and thereby contributes to still further necrosis. Muscle degeneration accelerates in a fatal spiral of positive feedback.

Genetic screening can identify heterozygous carriers of DMD, allowing for counseling of prospective parents on the risk of having a child with the disease. However, about one out of three cases arises by a new spontaneous mutation and therefore cannot be predicted by genetic testing.

A less severe form of muscular dystrophy is *facioscapulohumeral (Landouzy–Déjérine*[16]*) MD*, an autosomal dominant trait that begins in adolescence and affects both sexes equally. It involves the facial and shoulder muscles more than the pelvic muscles and cripples some individuals while it barely affects others. A third form, *limb-girdle dystrophy*, is a combination of several diseases of intermediate severity that affect the shoulder, arm, and pelvic muscles.

*Myasthenia gravis*[17] (MY-ass-THEE-nee-uh GRAV-is) *(MG)* usually occurs in women between the ages of 20 and 40. It is an autoimmune disease in which antibodies attack the neuromuscular junctions and bind ACh receptors together in clusters. The muscle fiber then removes the clusters from the sarcolemma by endocytosis. As a result, the muscle fibers become less and less sensitive to ACh. The effects often appear first in the facial muscles and commonly include drooping eyelids (*ptosis*, fig. 11.25) and double vision

*continued*

---

[14]*dys* = bad, abnormal; *trophy* = growth
[15] Guillaume B. A. Duchenne (1806–75), French physician

[16] Louis T. J. Landouzy (1845–1917) and Joseph J. Déjérine (1849–1917), French neurologists
[17]*my* = muscle; *asthen* = weakness; *grav* = severe

(due to *strabismus,* inability to fixate on the same point with both eyes). The initial symptoms are often followed by difficulty in swallowing, weakness of the limbs, and poor physical endurance. Some people with MG die quickly as a result of respiratory failure, but others have normal life spans. One method of assessing the progress of the disease is to use *bungarotoxin,* a protein from cobra venom that binds to ACh receptors. The amount that binds is proportional to the number of receptors that are still functional. The muscle of an MG patient sometimes binds less than one-third as much bungarotoxin as normal muscle does.

Myasthenia gravis is often treated with cholinesterase inhibitors. These drugs retard the breakdown of ACh in the neuromuscular junction and enable it to stimulate the muscle longer. Immunosuppressive agents such as prednisone and azathioprine (Imuran) may be used to suppress the production of the antibodies that destroy ACh receptors. Since certain immune cells are stimulated by hormones from the thymus, removal of the thymus *(thymectomy)* helps to dampen the overactive immune response that causes myasthenia gravis. Also, a technique called *plasmapheresis* may be used to remove harmful antibodies from the blood plasma.

**FIGURE 11.25  Test of Myasthenia Gravis.**  The subject is told to gaze upward (top panel). Within 60 seconds (middle panel) to 90 seconds (bottom), there is obvious sagging of the left eyelid (ptosis) due to inability to sustain stimulation of the orbicularis oculi muscle.

## Effects of the MUSCULAR SYSTEM
## On Other Organ Systems

**INTEGUMENTARY SYSTEM**
Facial expressions result from the action of muscles on the skin.

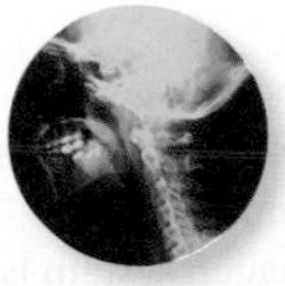

**SKELETAL SYSTEM**
Muscles move and stabilize joints and produce stresses that affect ossification, bone remodeling, and the shapes of bones.

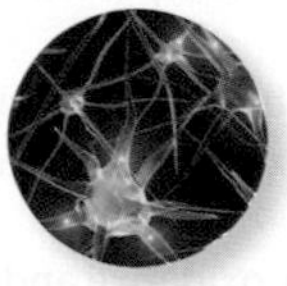

**NERVOUS SYSTEM**
Muscles give expression to thoughts, emotions, and motor commands that arise in the central nervous system.

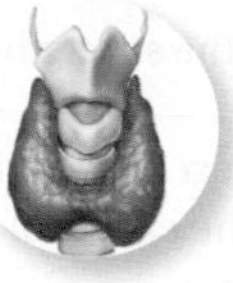

**ENDOCRINE SYSTEM**
Skeletal muscles provide protective cover for some endocrine glands; muscle mass affects insulin sensitivity.

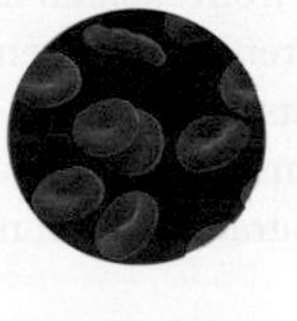

**CIRCULATORY SYSTEM**
Muscle contractions affect blood flow in many veins; exercise stimulates growth of new blood vessels.

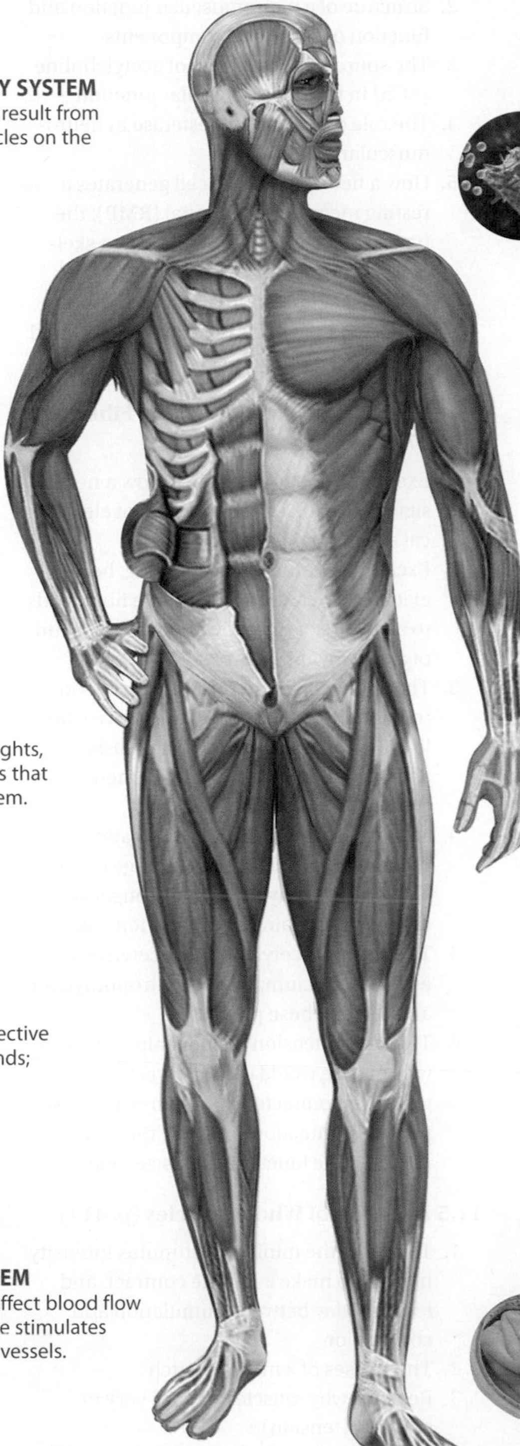

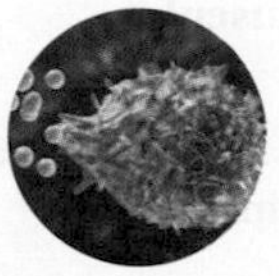

**LYMPHATIC/IMMUNE SYSTEM**
Muscle contractions aid flow of lymph; exercise elevates levels of immune cells and antibodies in circulation; excessive exercise can inhibit immunity.

**RESPIRATORY SYSTEM**
Muscle contractions ventilate the lungs; muscles of the larynx and pharynx regulate airflow; $CO_2$ generated by muscular activity stimulates respiration; abdominal muscles produce pressure bursts of coughing and sneezing and aid in deep breathing.

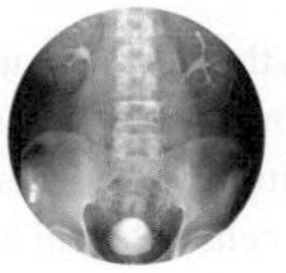

**URINARY SYSTEM**
A skeletal muscle sphincter retains urine in bladder until convenient for release; abdominal and pelvic muscles aid in compressing and emptying bladder; muscles of pelvic floor support bladder; bulbospongiosus muscle helps clear urine from male urethra.

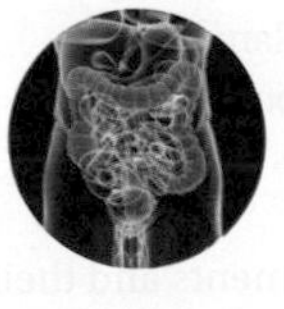

**DIGESTIVE SYSTEM**
Muscles enable chewing and swallowing; muscles control voluntary aspect of defecation; abdominal muscles produce vomiting; abdominal and lumbar muscles protect digestive organs.

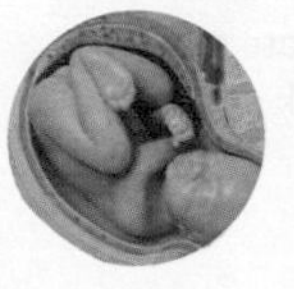

**REPRODUCTIVE SYSTEM**
Muscles are involved in sexual responses including erection and ejaculation; abdominal and pelvic muscles aid in childbirth.

# STUDY GUIDE

## ► Assess Your Learning Outcomes

*To test your knowledge, discuss the following topics with a study partner or in writing, ideally from memory.*

### 11.1 Types and Characteristics of Muscular Tissue (p. 398)

1. Five physiological properties of all muscular tissue and their relevance to muscle function
2. Distinguishing characteristics of skeletal muscle
3. Dimensions of a typical skeletal muscle fiber and of the longest fibers
4. Connective tissues associated with a muscle fiber and their relationship to muscle–bone attachments

### 11.2 Microscopic Anatomy of Skeletal Muscle (p. 399)

1. The sarcolemma and sarcoplasm, and the roles of glycogen and myoglobin in the sarcoplasm
2. The role of myoblasts in the development of a muscle fiber, and how they give rise to the multinucleate condition of the muscle fiber and to the satellite cells external to the fiber
3. Structure and function of the sarcoplasmic reticulum and transverse tubules
4. Types of myofilaments that constitute a myofibril
5. Composition and molecular organization of a thick myofilament, and the structure of a myosin molecule
6. Composition of a thin myofilament; the organization of its actin, tropomyosin, and troponin; and the active sites of its actin monomers
7. Composition of elastic filaments and their relationship to the thick filaments and Z discs
8. The position and function of dystrophin in the muscle fiber
9. Names of the striations of skeletal and cardiac muscle and how they relate to the overlapping arrangement of thick and thin myofilaments
10. The definition of *sarcomere*

### 11.3 The Nerve–Muscle Relationship (p. 404)

1. Motor units; the meaning of large and small motor units; and the respective advantages of the two types

2. Structure of a neuromuscular junction and function of each of its components
3. The source, role, and fate of acetylcholine (ACh) in the neuromuscular junction
4. The role of acetylcholinesterase in neuromuscular function
5. How a nerve or muscle cell generates a resting membrane potential (RMP); the typical voltage of this potential in a skeletal muscle fiber
6. How an action potential differs from the RMP, and the effects of an action potential on a nerve or muscle cell

### 11.4 Behavior of Skeletal Muscle Fibers (p. 407)

1. Excitation of a muscle fiber; how a nerve signal leads to a traveling wave of electrical excitation in a muscle fiber
2. Excitation–contraction coupling; how electrical excitation of a muscle fiber leads to exposure of the active sites on the actin of a thin myofilament
3. The sliding filament mechanism of contraction; how exposure of the active sites leads to repetitive binding of myosin to actin and sliding of the thin filaments over the thick filaments
4. Muscle relaxation; how the cessation of the nerve signal leads to blockage of the active sites so myosin can no longer bind to them and maintain muscle tension
5. The roles of acetylcholine, acetylcholinesterase, calcium, troponin, tropomyosin, and ATP in these processes
6. The length–tension relationship in muscle; why muscle would contract weakly if it were overcontracted or overstretched just prior to stimulation; and how this principle relates to the function of muscle tone

### 11.5 Behavior of Whole Muscles (p. 414)

1. Terms for the minimum stimulus intensity needed to make a muscle contract, and for the delay between stimulation and contraction
2. The phases of a muscle twitch
3. Reasons why muscle twitches vary in strength (tension)
4. How recruitment and tetanus are produced and how they affect muscle tension
5. Differences between isometric and isotonic contraction, and between the con-

centric and eccentric forms of isotonic contraction tension

### 11.6 Muscle Metabolism (p. 418)

1. Why a muscle cannot contract without ATP
2. Differences between aerobic respiration and anaerobic fermentation with respect to muscle function
3. The use of myoglobin and aerobic respiration to generate ATP at the outset of exercise
4. Two ways in which the phosphagen system generates ATP for continued exercise
5. How anaerobic fermentation generates ATP after the phosphagen system is depleted
6. Why a muscle is able to switch back to aerobic respiration to generate ATP after 40 seconds or so of exercise
7. Causes of muscle fatigue
8. $VO_2$max, why it partially determines one's ability to maintain high-intensity exercise, and why it differs from one person to another
9. Why exercise is followed by a prolonged state of elevated oxygen consumption, and the name of that state
10. Differences between slow oxidative and fast glycolytic muscle fibers; the respective advantages of each; how they relate to the power and recruitment of motor units; and examples of muscles in which each type predominates
11. Factors that determine the strength of a muscle
12. Examples of resistance exercise and endurance exercise, and the effects of each on muscle performance

### 11.7 Cardiac and Smooth Muscle (p. 423)

1. Reasons why cardiac muscle must differ physiologically from skeletal muscle
2. Structural differences between cardiocytes and skeletal muscle fibers
3. The autorhythmicity of the heart and its ability to contract without nervous stimulation
4. The unusual fatigue resistance of cardiac muscle; structural and biochemical properties that account for it
5. Functional differences between smooth muscle and the two forms of striated muscle

# STUDY GUIDE

6. How the innervation of smooth muscle differs from that of skeletal muscle
7. Variations in the complexity and anatomical organization of smooth muscle
8. Various functions of smooth muscle
9. Two modes of growth of smooth muscle tissue
10. The structure of smooth muscle myocytes and what takes the place of the absent Z discs and T tubules

11. Differences between multiunit and single-unit smooth muscle, and in the nerve–muscle relationship in each
12. Various modes of stimulation of smooth muscle
13. How excitation–contraction coupling in smooth muscle differs from that in skeletal muscle; the roles of calmodulin and myosin light-chain kinase in smooth muscle contraction

14. The nature and effect of the latch-bridge mechanism in smooth muscle
15. The role of smooth muscle in peristalsis
16. Benefits of the stress–relaxation response of smooth muscle, and of its absence of a length–tension relationship

## ▶ Testing Your Recall

*Answers in Appendix B*

1. To make a muscle contract more strongly, the nervous system can activate more motor units. This process is called
   a. recruitment.
   b. summation.
   c. incomplete tetanus.
   d. twitch.
   e. concentric contraction.

2. The functional unit of a muscle fiber is the _____, a segment from one Z disc to the next.
   a. myofibril
   b. I band
   c. sarcomere
   d. neuromuscular junction
   e. striation

3. Before a muscle fiber can contract, ATP must bind to
   a. a Z disc.
   b. the myosin head.
   c. tropomyosin.
   d. troponin.
   e. actin.

4. Before a muscle fiber can contract, $Ca^{2+}$ must bind to
   a. calsequestrin.
   b. the myosin head.
   c. tropomyosin.
   d. troponin.
   e. actin.

5. Which of the following muscle proteins is *not* intracellular?
   a. actin
   b. myosin
   c. collagen
   d. troponin
   e. dystrophin

6. Smooth muscle cells have _____, whereas skeletal muscle fibers do not.
   a. sarcoplasmic reticulum
   b. tropomyosin
   c. calmodulin
   d. Z discs
   e. myosin ATPase

7. ACh receptors are found mainly in
   a. synaptic vesicles.
   b. terminal cisternae.
   c. thick filaments.
   d. thin filaments.
   e. junctional folds.

8. Single-unit smooth muscle cells can stimulate each other because they have
   a. a latch-bridge.
   b. diffuse junctions.
   c. gap junctions.
   d. tight junctions.
   e. cross-bridges.

9. A person with a high $VO_2$max
   a. needs less oxygen than someone with a low $VO_2$max.
   b. has stronger muscles than someone with a low $VO_2$max.
   c. is less likely to show muscle tetanus than someone with a low $VO_2$max.
   d. has fewer muscle mitochondria than someone with a low $VO_2$max.
   e. experiences less muscle fatigue during exercise than someone with a low $VO_2$max.

10. Slow oxidative fibers have all of the following *except*
    a. an abundance of myoglobin.
    b. an abundance of glycogen.
    c. high fatigue resistance.
    d. a red color.
    e. a high capacity to synthesize ATP aerobically.

11. The minimum stimulus intensity that will make a muscle contract is called _____.

12. Red muscles consist mainly of slow _____, or type I, muscle fibers.

13. Parts of the sarcoplasmic reticulum called _____ lie on each side of a T tubule.

14. Thick myofilaments consist mainly of the protein _____.

15. The neurotransmitter that stimulates skeletal muscle is _____.

16. Muscle contains an oxygen-binding pigment called _____.

17. The _____ of skeletal muscle play the same role as dense bodies in smooth muscle.

18. In autonomic nerve fibers that stimulate single-unit smooth muscle, the neurotransmitter is contained in swellings called _____.

19. A state of continual partial muscle contraction is called _____.

20. _____ an increase in muscle tension without a change in length.

# STUDY GUIDE

## ▶ Building Your Medical Vocabulary

*Answers in Appendix B*

*State a medical meaning of each word element below, and give a term in which it or a slight variation of it is used.*

1. astheno-

2. auto-

3. dys-

4. iso-

5. metri-

6. myo-

7. sarco-

8. temporo-

9. tono-

10. -trophy

## ▶ True or False

*Answers in Appendix B*

*Determine which five of the following statements are false, and briefly explain why.*

1. Each motor neuron supplies just one muscle fiber.

2. To initiate muscle contraction, calcium ions must bind to the myosin heads.

3. Slow oxidative fibers are relatively resistant to fatigue.

4. Thin filaments are found in both the A bands and I bands of striated muscle.

5. Thin filaments do not change length when a muscle contracts.

6. Smooth muscle lacks striations because it does not have thick and thin myofilaments.

7. A muscle must contract to the point of complete tetanus if it is to move a load.

8. If no ATP were available to a muscle fiber, the excitation stage of muscle action could not occur.

9. For the first 30 seconds of an intense exercise, muscle gets most of its energy from lactic acid.

10. The heart and some smooth muscle are autorhythmic, but skeletal muscle is not.

## ▶ Testing Your Comprehension

*Answers at* www.mhhe.com/saladin7

1. Without ATP, relaxed muscle cannot contract and a contracted muscle cannot relax. Explain why.

2. Smooth muscle controls the curvature of the lens of the eye and the diameter of the pupil, but it would serve poorly for controlling eye movements as in tracking a flying bird or reading a page of print. Explain why.

3. Why would skeletal muscle be unsuitable for the wall of the urinary bladder? Explain how this illustrates the complementarity of form and function at a cellular and molecular level.

4. As skeletal muscle contracts, one or more bands of the sarcomere become narrower and disappear, and one or more of them remain the same width. Which bands will change—A, H, or I—and why?

5. Three physiology students are arguing about ATP and muscle function. Tom says ATP is needed for muscle contraction, not relaxation. Wendy says no, it is needed for relaxation, not contraction. Andrew says he thinks it is needed for both contraction and relaxation. Which of them is right, and why?

## ▶ Improve Your Grade at www.mhhe.com/saladin7

*Download animations to study when it fits your schedule. Practice quizzes, labeling activities, games, and flashcards offer fun ways to master the chapter concepts. Or, download image PowerPoint files for each chapter to create a study guide or for taking notes during lecture.*

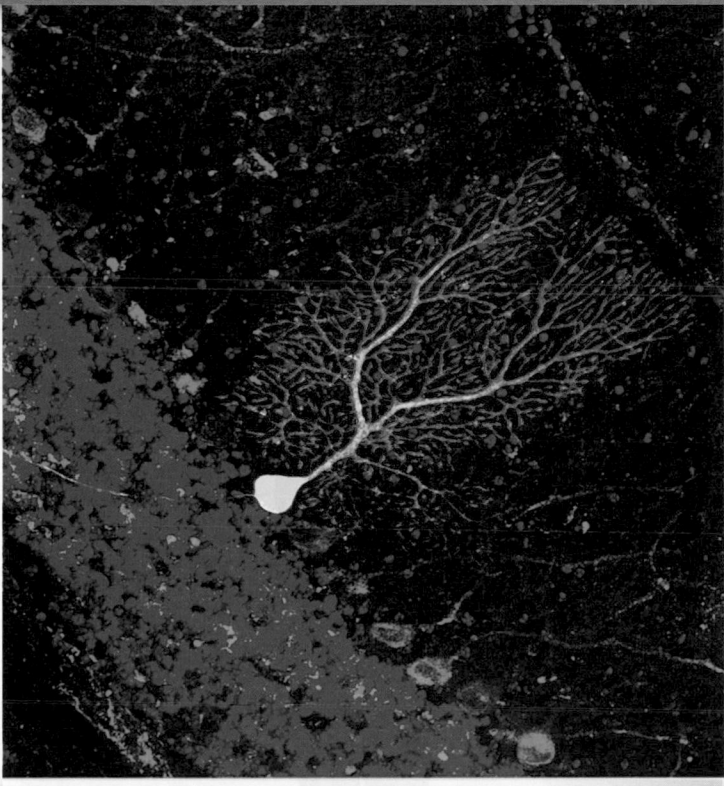

A Purkinje cell, a neuron from the cerebellum of the brain

CHAPTER

# 12

# NERVOUS TISSUE

Anatomy & Physiology | REVEALED®
aprevealed.com

**Module 7:  Nervous System**

## BRUSHING UP

- Although neuron structure is detailed here, you may find the brief introduction to neurons and neuroglia on page 160 helpful.

- Much of this chapter hinges on an understanding of the principles of electrophysiology introduced on page 406.

- Electrophysiology, in turn, depends on an understanding of the principles of diffusion (p. 90) and ligand- and voltage-gated ion channels (p. 84).

- The generation and conduction of nerve signals depend on the principles of gradients and flow introduced on page 17.

- You must be familiar with the sodium–potassium pump (p. 95), which maintains the membrane potential and excitability of neurons.

- Communication from one neuron to another depends in some cases on second-messenger systems such as cAMP (p. 85).

The nervous system is one of great complexity and mystery, and will absorb our attention for the next five chapters. It is the foundation of all our conscious experience, personality, and behavior. It profoundly intrigues biologists, physicians, psychologists, and even philosophers. Its scientific study, called **neurobiology,** is regarded by many as the ultimate challenge facing the behavioral and life sciences. We will begin at the simplest organizational level—the nerve cells *(neurons)* and cells called *neuroglia* that support their function in various ways. We will then progress to the organ level to examine the spinal cord (chapter 13), brain (chapter 14), autonomic nervous system (chapter 15), and sense organs (chapter 16).

### 12.1  Overview of the Nervous System

#### Expected Learning Outcomes

When you have completed this section, you should be able to

a.  describe the overall function of the nervous system; and

b.  describe its major anatomical and functional subdivisions.

If the body is to maintain homeostasis and function effectively, its trillions of cells must work together in a coordinated fashion. If each cell behaved without regard to what others were doing, the result would be physiological chaos and death. We have two organ systems dedicated to maintaining internal coordination—the **endocrine system,** which communicates by means of chemical messengers (hormones) secreted into the blood, and the **nervous system** (fig. 12.1), which employs electrical and chemical means to send messages very quickly from cell to cell.

The nervous system carries out its coordinating task in three basic steps: (1) Through sense organs and simple nerve endings, it receives information about changes in the body and

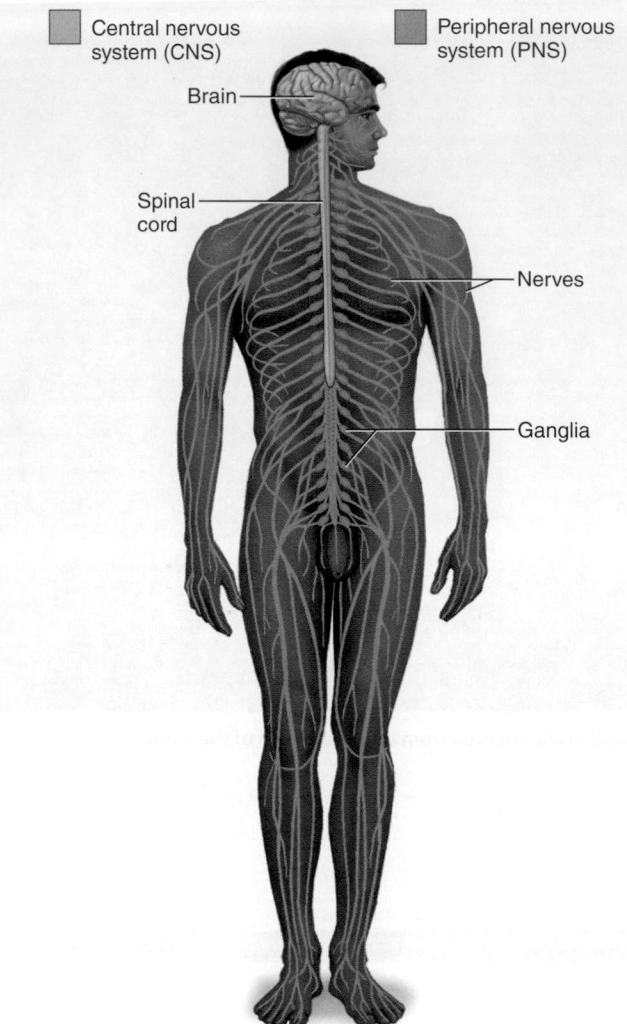

**FIGURE 12.1  The Nervous System.**

the external environment and transmits messages to the *central nervous system (CNS)*. (2) The CNS processes this information and determines what response, if any, is appropriate to the circumstances. (3) The CNS issues commands primarily to muscle and gland cells to carry out such responses.

The nervous system has two major anatomical subdivisions (fig. 12.2):

- The **central nervous system (CNS)** consists of the brain and spinal cord, which are enclosed and protected by the cranium and vertebral column.

- The **peripheral nervous system (PNS)** consists of all the rest; it is composed of nerves and ganglia. A **nerve** is a bundle of nerve fibers (axons) wrapped in fibrous connective tissue. Nerves emerge from the CNS through

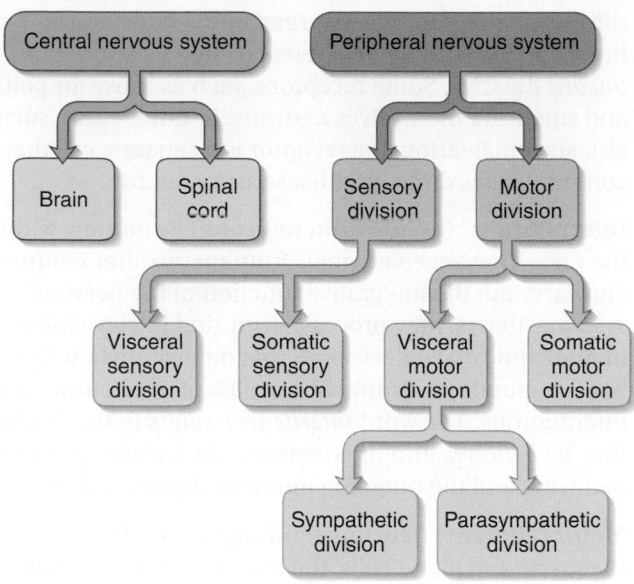

**FIGURE 12.2 Subdivisions of the Nervous System.**

foramina of the skull and vertebral column and carry signals to and from other organs of the body. A **ganglion**[1] (plural, *ganglia*) is a knotlike swelling in a nerve where the cell bodies of neurons are concentrated.

The peripheral nervous system is functionally divided into *sensory* and *motor* divisions, and each of these is further divided into *somatic* and *visceral* subdivisions.

- The **sensory (afferent**[2]**) division** carries signals from various **receptors** (sense organs and simple sensory nerve endings) to the CNS. This pathway informs the CNS of stimuli within and around the body.
  - The **somatic**[3] **sensory division** carries signals from receptors in the skin, muscles, bones, and joints.
  - The **visceral sensory division** carries signals mainly from the viscera of the thoracic and abdominal cavities, such as the heart, lungs, stomach, and urinary bladder.
- The **motor (efferent**[4]**) division** carries signals from the CNS to gland and muscle cells that carry out the body's responses. Cells and organs that respond to these signals are called **effectors.**
  - The **somatic motor division** carries signals to the skeletal muscles. This produces voluntary muscle contractions as well as involuntary *somatic reflexes.*

- The **visceral motor division (autonomic**[5] **nervous system, ANS)** carries signals to glands, cardiac muscle, and smooth muscle. We usually have no voluntary control over these effectors, and the ANS operates at an unconscious level. The responses of the ANS and its effectors are *visceral reflexes.* The ANS has two further divisions:
  - The **sympathetic division** tends to arouse the body for action—for example, by accelerating the heartbeat and increasing respiratory airflow—but it inhibits digestion.
  - The **parasympathetic division** tends to have a calming effect—slowing the heartbeat, for example—but it stimulates digestion.

The foregoing terms may give the impression that we have several nervous systems—central, peripheral, sensory, motor, somatic, and visceral. These are just terms of convenience, however. There is only one nervous system, and these subsystems are interconnected parts of the whole.

**BEFORE YOU GO ON**

Answer the following questions to test your understanding of the preceding section:

1. What is a receptor? Give two examples of effectors.
2. Distinguish between the central and peripheral nervous systems, and between visceral and somatic divisions of the sensory and motor systems.
3. What is another name for the visceral motor nervous system? What are its two subdivisions? What are their functions?

### 12.2 Properties of Neurons

**Expected Learning Outcomes**

When you have completed this section, you should be able to

a. describe three functional properties found in all neurons;
b. define the three most basic functional categories of neurons;
c. identify the parts of a neuron; and
d. explain how neurons transport materials between the cell body and tips of the axon.

## Universal Properties

The communicative role of the nervous system is carried out by **nerve cells,** or **neurons.** These cells have three fundamental physiological properties that enable them to communicate with other cells:

---

[1] *gangli* = knot
[2] *af* = *ad* = toward; *fer* = to carry
[3] *somat* = body; *ic* = pertaining to
[4] *ef* = *ex* = out, away; *fer* = to carry

[5] *auto* = self; *nom* = law, governance

1. **Excitability.** All cells are excitable—that is, they respond to environmental changes **(stimuli).** Neurons exhibit this property to the highest degree.

2. **Conductivity.** Neurons respond to stimuli by producing electrical signals that are quickly conducted to other cells at distant locations.

3. **Secretion.** When the signal reaches the end of a nerve fiber, the neuron secretes a *neurotransmitter* that crosses the gap and stimulates the next cell.

▶▶▶**APPLY WHAT YOU KNOW**

*What basic physiological properties do a nerve cell and a muscle cell have in common? Name a physiological property of each that the other one lacks.*

## Functional Classes

There are three general classes of neurons (fig. 12.3) corresponding to the three major aspects of nervous system function listed earlier:

① **Sensory (afferent) neurons** are specialized to detect stimuli such as light, heat, pressure, and chemicals, and transmit information about them to the CNS. Such neurons begin in almost every organ of the body and end in the CNS; the word *afferent* refers to signal conduction *toward* the CNS. Some receptors, such as those for pain and smell, are themselves neurons. In other cases, such as taste and hearing, the receptor is a separate cell that communicates directly with a sensory neuron.

② **Interneurons**[6] **(association neurons)** lie entirely within the CNS. They receive signals from many other neurons and carry out the integrative function of the nervous system—that is, they process, store, and retrieve information and "make decisions" that determine how the body responds to stimuli. About 90% of our neurons are interneurons. The word *interneuron* refers to the fact that they lie *between,* and interconnect, the incoming sensory pathways and the outgoing motor pathways of the CNS.

③ **Motor (efferent) neurons** send signals predominantly to muscle and gland cells, the effectors. They are called *motor* neurons because most of them lead to muscle cells, and *efferent* neurons to signify signal conduction *away from* the CNS.

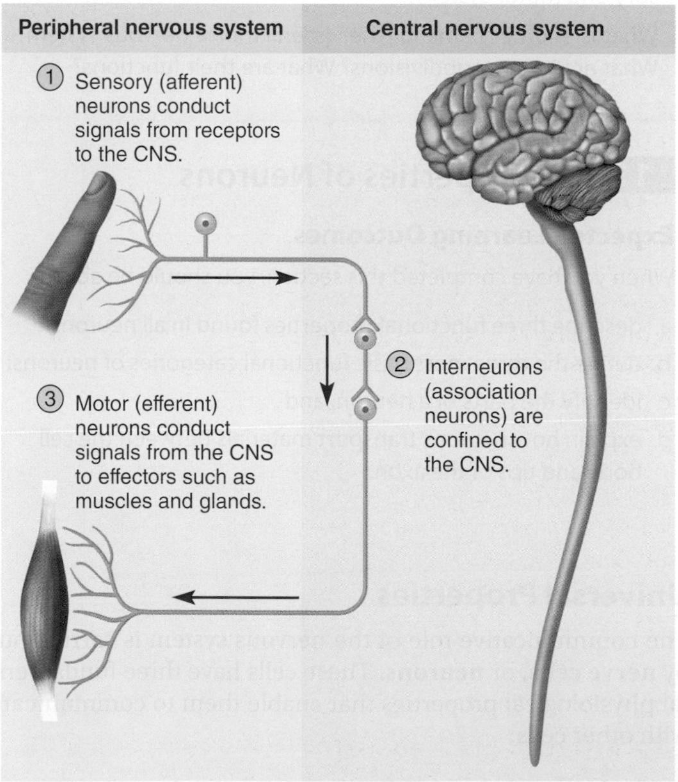

## Structure of a Neuron

There are several varieties of neurons, but a good starting point for discussion is a motor neuron of the spinal cord (fig. 12.4). The control center of the neuron is the **soma,**[7] also called the **neurosoma** or **cell body.** It has a centrally located nucleus with a large nucleolus. The cytoplasm contains mitochondria, lysosomes, a Golgi complex, numerous inclusions, and an extensive rough endoplasmic reticulum and cytoskeleton. The cytoskeleton consists of a dense mesh of microtubules and **neurofibrils** (bundles of actin filaments), which compartmentalize the rough ER into dark-staining regions called **Nissl**[8] **bodies** (fig. 12.4e). Nissl bodies are unique to neurons and a helpful clue to identifying them in tissue sections with mixed cell types. Mature neurons have no centrioles and apparently undergo no further mitosis after adolescence; however, they are unusually long-lived cells, capable of functioning for over a hundred years. Even into old age, however, there are unspecialized stem cells in the CNS that can divide and develop into new neurons.

The major inclusions in a neuron are glycogen granules, lipid droplets, melanin, and a golden brown pigment called *lipofuscin*[9] (LIP-oh-FEW-sin), produced when lysosomes degrade worn-out organelles and other products. Lipofuscin accumulates with age and pushes the nucleus to one side of the cell. Lipofuscin granules are also called "wear-and-tear granules" because they are most abundant in old neurons.

| Peripheral nervous system | Central nervous system |

① Sensory (afferent) neurons conduct signals from receptors to the CNS.

② Interneurons (association neurons) are confined to the CNS.

③ Motor (efferent) neurons conduct signals from the CNS to effectors such as muscles and glands.

**FIGURE 12.3  Functional Classes of Neurons.** All neurons can be classified as sensory, motor, or interneurons depending on their location and the direction of signal conduction.

---

[6]*inter* = between
[7]*soma* = body
[8]Franz Nissl (1860–1919), German neuropathologist
[9]*lipo* = fat, lipid; *fusc* = dusky, brown

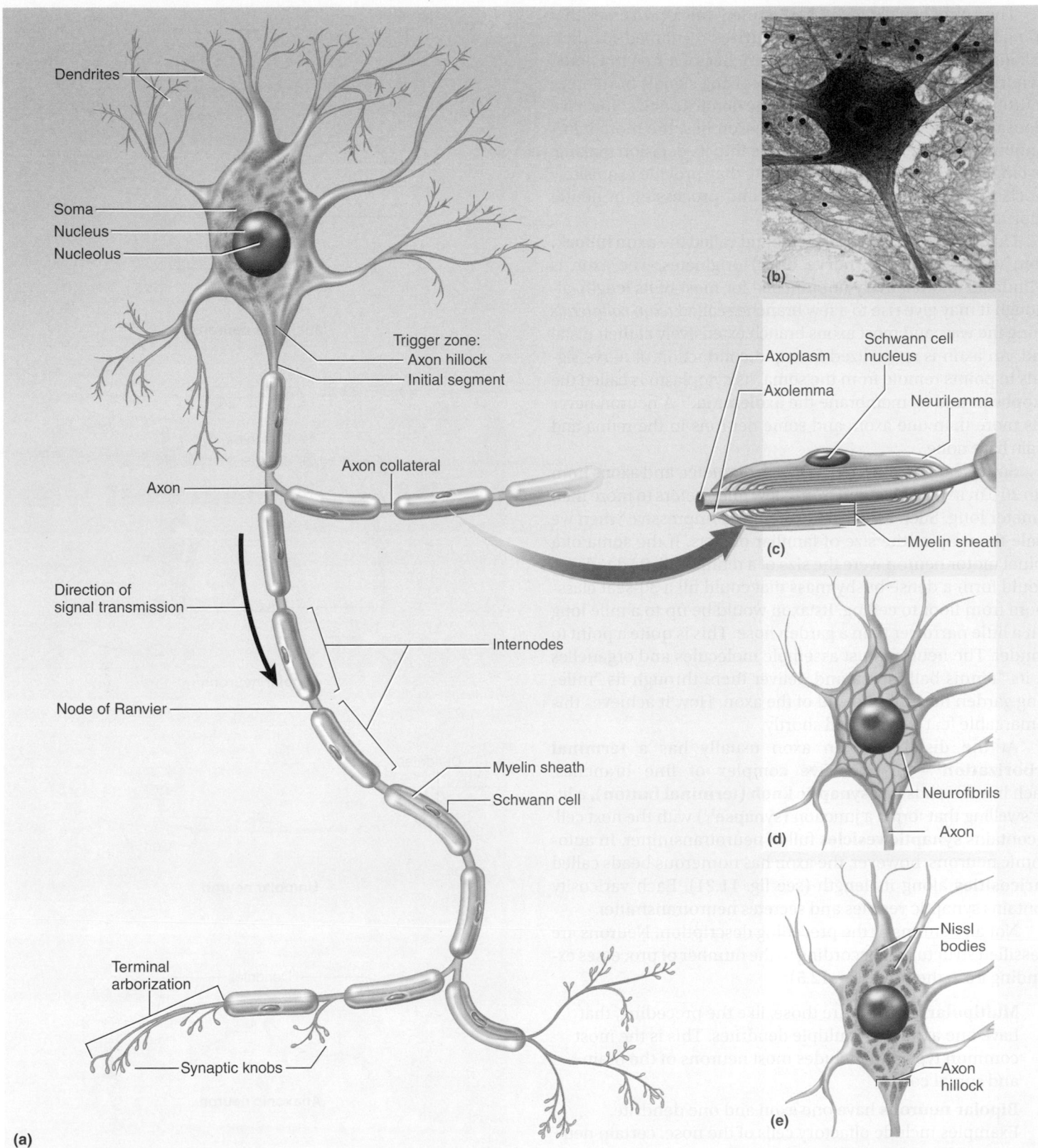

**(a)**

Dendrites

Soma
Nucleus
Nucleolus

Trigger zone:
Axon hillock
Initial segment

Axon

Axon collateral

Direction of
signal transmission

Internodes

Node of Ranvier

Myelin sheath
Schwann cell

Terminal
arborization

Synaptic knobs

**(b)**

**(c)**

Axoplasm
Axolemma

Schwann cell
nucleus

Neurilemma

Myelin sheath

**(d)**

Neurofibrils

Axon

**(e)**

Nissl
bodies

Axon
hillock

**FIGURE 12.4 A Representative Neuron.** The Schwann cells and myelin sheath are explained later in this chapter. (a) Structure of a multipolar neuron such as a spinal motor neuron. (b) Photograph of this neuron type. (c) Detail of the myelin sheath. (d) Neurofibrils of the soma. (e) Nissl bodies, stained masses of rough ER separated by the bundles of neurofibrils shown in part (d).

The soma usually gives rise to a few thick processes that branch into a vast number of **dendrites**[10]—named for their striking resemblance to the bare branches of a tree in winter. Dendrites are the primary site for receiving signals from other neurons. Some neurons have only one dendrite and some have thousands. The more dendrites a neuron has, the more information it can receive and incorporate into its decision making. As tangled as the dendrites may seem, they provide exquisitely precise pathways for the reception and processing of neural information.

On one side of the soma is a mound called the **axon hillock,** from which the **axon (nerve fiber)** originates. The axon is cylindrical and relatively unbranched for most of its length, although it may give rise to a few branches called *axon collaterals* along the way, and most axons branch extensively at their distal end. An axon is specialized for rapid conduction of nerve signals to points remote from the soma. Its cytoplasm is called the **axoplasm** and its membrane the **axolemma.**[11] A neuron never has more than one axon, and some neurons in the retina and brain have none.

Somas range from 5 to 135 $\mu$m in diameter, and axons from 1 to 20 $\mu$m in diameter and from a few millimeters to more than a meter long. Such dimensions are more impressive when we scale them up to the size of familiar objects. If the soma of a spinal motor neuron were the size of a tennis ball, its dendrites would form a dense bushy mass that could fill a 30-seat classroom from floor to ceiling. Its axon would be up to a mile long but a little narrower than a garden hose. This is quite a point to ponder. The neuron must assemble molecules and organelles in its "tennis ball" soma and deliver them through its "mile-long garden hose" to the end of the axon. How it achieves this remarkable feat is explained shortly.

At the distal end, an axon usually has a **terminal arborization**[12]—an extensive complex of fine branches. Each branch ends in a **synaptic knob (terminal button),** a little swelling that forms a junction (**synapse**[13]) with the next cell. It contains **synaptic vesicles** full of neurotransmitter. In autonomic neurons, however, the axon has numerous beads called **varicosities** along its length (see fig. 11.21). Each varicosity contains synaptic vesicles and secretes neurotransmitter.

Not all neurons fit the preceding description. Neurons are classified structurally according to the number of processes extending from the soma (fig. 12.5):

- **Multipolar neurons** are those, like the preceding, that have one axon and multiple dendrites. This is the most common type and includes most neurons of the brain and spinal cord.

- **Bipolar neurons** have one axon and one dendrite. Examples include olfactory cells of the nose, certain neurons of the retina, and sensory neurons of the ear.

[10]*dendr* = tree, branch; *ite* = little
[11]*axo* = axis, axon; *lemma* = husk, peel, sheath
[12]*arbor* = tree
[13]*syn* = together; *aps* = to touch, join

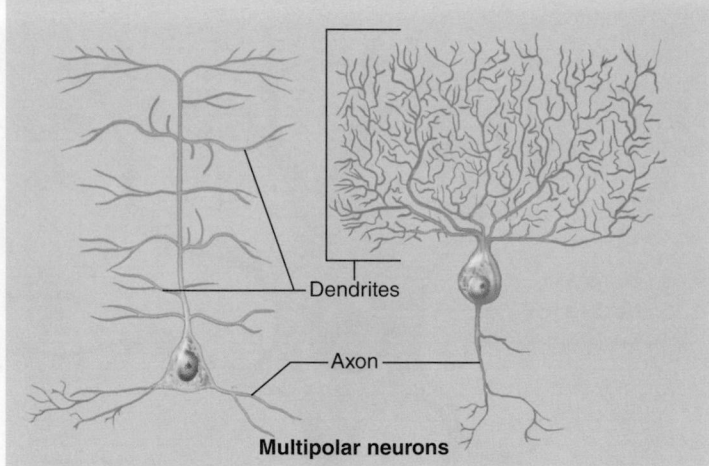

**Multipolar neurons**

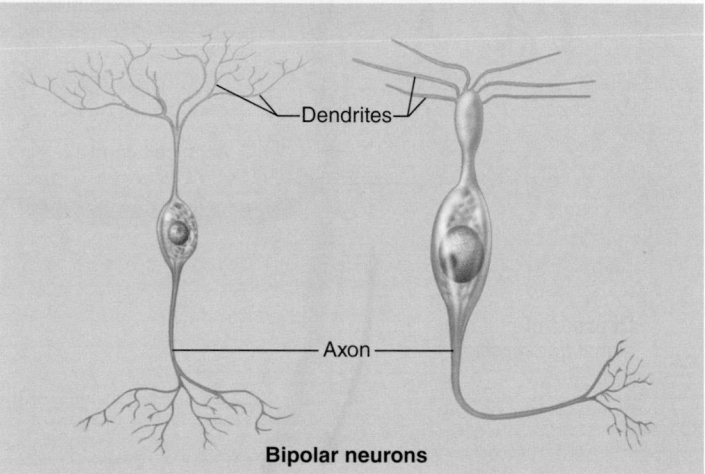

**Bipolar neurons**

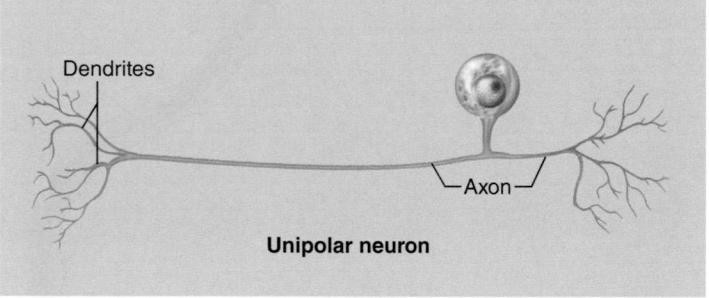

**Unipolar neuron**

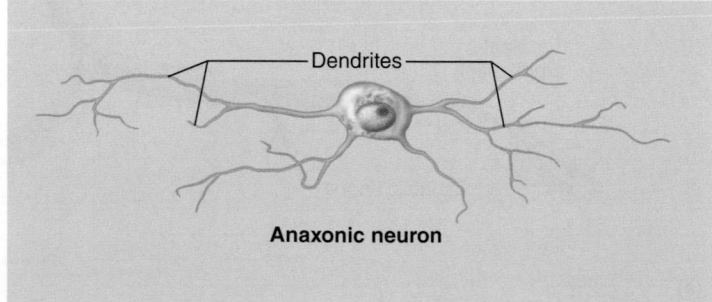

**Anaxonic neuron**

**FIGURE 12.5 Variation in Neuron Structure.** Top row, left to right: Two multipolar neurons of the brain—a pyramidal cell and a Purkinje cell. Second row, left to right: Two bipolar neurons—a bipolar cell of the retina and an olfactory neuron. Third row: A unipolar neuron of the type involved in the senses of touch and pain. Bottom row: An anaxonic neuron (amacrine cell) of the retina.

- **Unipolar neurons** have only a single process leading away from the soma. They are represented by the neurons that carry sensory signals to the spinal cord. They are also called *pseudounipolar* because they start out as bipolar neurons in the embryo, but their two processes fuse into one as the neuron matures. A short distance away from the soma, the process branches like a T into a *peripheral fiber* and a *central fiber.* The peripheral fiber begins with a sensory ending often far away from the soma—in the skin, for example. Its signals travel toward the soma, but bypass it and continue along the central fiber for a short remaining distance to the spinal cord. The dendrites are considered to be only the short receptive endings. The rest of the process, both peripheral and central, is the axon, defined by the presence of myelin and the ability to generate action potentials—two concepts explained later.

- **Anaxonic neurons** have multiple dendrites but no axon. They communicate locally through their dendrites and do not produce action potentials. Some anaxonic neurons are found in the brain, retina, and adrenal medulla. In the retina, they help in visual processes such as the perception of contrast.

## Axonal Transport

All of the proteins needed by a neuron must be made in the soma, where the protein-synthesizing organelles such as the nucleus, ribosomes, and rough endoplasmic reticulum are located. Yet many of these proteins are needed in the axon, for example to repair and maintain the axolemma, to serve as ion channels in the membrane, or to act in the synaptic knob as enzymes and signaling molecules. Other substances are transported from the axon terminals back to the soma for disposal or recycling. The two-way passage of proteins, organelles, and other materials along an axon is called **axonal transport.** Movement away from the soma down the axon is called **anterograde**[14] **transport** and movement up the axon toward the soma is called **retrograde**[15] **transport.**

Materials travel along axonal microtubules that act like monorail tracks to guide them to their destination. But what is the "engine" that drives them along the tracks? Anterograde transport employs a motor protein called *kinesin*[16] and retrograde transport uses one called *dynein*[17] (the same protein we encountered earlier in cilia and flagella; see chapter 3). These proteins carry materials "on their backs" while they reach out, like the myosin heads of muscle (see chapter 11), to bind repeatedly to the microtubules and walk along them.

There are two types of axonal transport: fast and slow.

1. **Fast axonal transport** occurs at a rate of 20 to 400 mm/day and may be either anterograde or retrograde:

   - *Fast anterograde transport* moves mitochondria; synaptic vesicles; other organelles; components of the axolemma; calcium ions; enzymes such as acetylcholinesterase; and small molecules such as glucose, amino acids, and nucleotides toward the distal end of the axon.

   - *Fast retrograde transport* returns used synaptic vesicles and other materials to the soma and informs the soma of conditions at the axon terminals. Some pathogens exploit this process to invade the nervous system. They enter the distal tips of an axon and travel to the soma by retrograde transport. Examples include tetanus toxin and the herpes simplex, rabies, and polio viruses. In such infections, the delay between infection and the onset of symptoms corresponds to the time needed for the pathogens to reach the somas.

2. **Slow axonal transport** is an anterograde process that works in a stop-and-go fashion. If we compare fast axonal transport to an express train traveling nonstop to its destination, slow axonal transport is like a local train that stops at every station. When moving, it goes just as fast as the express train, but the frequent stops result in an overall progress of only 0.5 to 10 mm/day. It moves enzymes and cytoskeletal components down the axon, renews worn-out axoplasmic components in mature neurons, and supplies new axoplasm for developing or regenerating neurons. Damaged nerves regenerate at a speed governed by slow axonal transport.

▶▶▶**APPLY WHAT YOU KNOW**

*The axon of a neuron has a dense cytoskeleton. Considering the functions of the cytoskeleton discussed in chapter 3, give two reasons why this is so important to neuron structure and function.*

**BEFORE YOU GO ON**

Answer the following questions to test your understanding of the preceding section:

4. Sketch a multipolar neuron and label its soma, dendrites, axon, terminal arborization, synaptic knobs, myelin sheath, and nodes of Ranvier.

5. Explain the difference between a sensory neuron, motor neuron, and interneuron.

6. What is the functional difference between a dendrite and an axon?

7. How do proteins and other chemicals synthesized in the soma get to the synaptic knobs? By what process can a virus that invades a peripheral nerve fiber get to the soma of that neuron?

---

[14]*antero* = forward; *grad* = to walk, to step
[15]*retro* = back; *grad* = to walk, to step
[16]*kines* = motion; *in* = protein
[17]*dyne* = force; *in* = protein

## 12.3   Supportive Cells (Neuroglia)

### Expected Learning Outcomes

When you have completed this section, you should be able to

a. name the six types of cells that aid neurons and state their respective functions;

b. describe the myelin sheath that is found around certain nerve fibers and explain its importance;

c. describe the relationship of unmyelinated nerve fibers to their supportive cells; and

d. explain how damaged nerve fibers regenerate.

There are about a trillion ($10^{12}$) neurons in the nervous system—10 times as many neurons in your body as there are stars in our galaxy! Because they branch so extensively, they make up about 50% of the volume of the nervous tissue. Yet they are outnumbered at least 10 to 1 by cells called **neuroglia** (noo-ROG-lee-uh), or **glial** (GLEE-ul) **cells.** Glial cells protect the neurons and help them function. The word *glia,* which means "glue," implies one of their roles—to bind neurons together and provide a supportive framework for the nervous tissue. In the fetus, they form a scaffold that guides young migrating neurons to their destinations. Wherever a mature neuron is not in synaptic contact with another cell, it is covered with glial cells. This prevents neurons from contacting each other except at points specialized for signal transmission, thus giving precision to their conduction pathways.

### Types of Neuroglia

There are six kinds of neuroglia, each with a unique function (table 12.1). Four types occur only in the central nervous system (fig. 12.6):

1. **Oligodendrocytes**[18] (OL-ih-go-DEN-dro-sites) somewhat resemble an octopus; they have a bulbous body with as many as 15 arms. Each arm reaches out to a nerve fiber and spirals around it like electrical tape wrapped repeatedly around a wire. This wrapping, called the *myelin sheath,* insulates the nerve fiber from the extracellular fluid. For reasons explained later, it speeds up signal conduction in the nerve fiber.

2. **Ependymal**[19] (ep-EN-dih-mul) **cells** resemble a cuboidal epithelium lining the internal cavities of the brain and spinal cord. Unlike true epithelial cells, however, they have no basement membrane and they exhibit rootlike processes that penetrate into the underlying tissue. Ependymal cells produce *cerebrospinal fluid (CSF),* a liquid that bathes the CNS and fills its internal cavities. They have patches of cilia on their apical surfaces that help to circulate the CSF. Ependymal cells and CSF are considered in more detail in chapter 14.

| TABLE 12.1 | Types of Glial Cells |
|---|---|
| **Types** | **Functions** |
| **Neuroglia of CNS** | |
| Oligodendrocytes | Form myelin in brain and spinal cord |
| Ependymal cells | Line cavities of brain and spinal cord; secrete and circulate cerebrospinal fluid |
| Microglia | Phagocytize and destroy microorganisms, foreign matter, and dead nervous tissue |
| Astrocytes | Cover brain surface and nonsynaptic regions of neurons; form supportive framework in CNS; induce formation of blood–brain barrier; nourish neurons; produce growth factors that stimulate neurons; communicate electrically with neurons and may influence synaptic signaling; remove $K^+$ and some neurotransmitters from ECF of brain and spinal cord; help to regulate composition of ECF; form scar tissue to replace damaged nervous tissue |
| **Neuroglia of PNS** | |
| Schwann cells | Form neurilemma around all PNS nerve fibers and myelin around most of them; aid in regeneration of damaged nerve fibers |
| Satellite cells | Surround somas of neurons in the ganglia; provide electrical insulation and regulate chemical environment of neurons |

3. **Microglia** are small macrophages that develop from white blood cells called monocytes. They wander through the CNS, putting out fingerlike extensions to constantly probe the tissue for cellular debris or other problems. They are thought to perform a complete checkup on the brain tissue several times a day, phagocytizing dead tissue, microorganisms, and other foreign matter. They become concentrated in areas damaged by infection, trauma, or stroke. Pathologists look for clusters of microglia in brain tissue as a clue to sites of injury. Microglia also aid in synaptic remodeling, changing the connections between neurons.

4. **Astrocytes**[20] are the most abundant glial cells in the CNS and constitute over 90% of the tissue in some areas of the brain. They cover the entire brain surface and most nonsynaptic regions of the neurons in the gray matter of the CNS. They are named for their many-branched, somewhat starlike shape. They have the most diverse functions of any glia:

   • They form a supportive framework for the nervous tissue.

---

[18]*oligo* = few; *dendro* = branches; *cyte* = cell
[19] *ependyma* = upper garment

[20]*astro* = star; *cyte* = cell

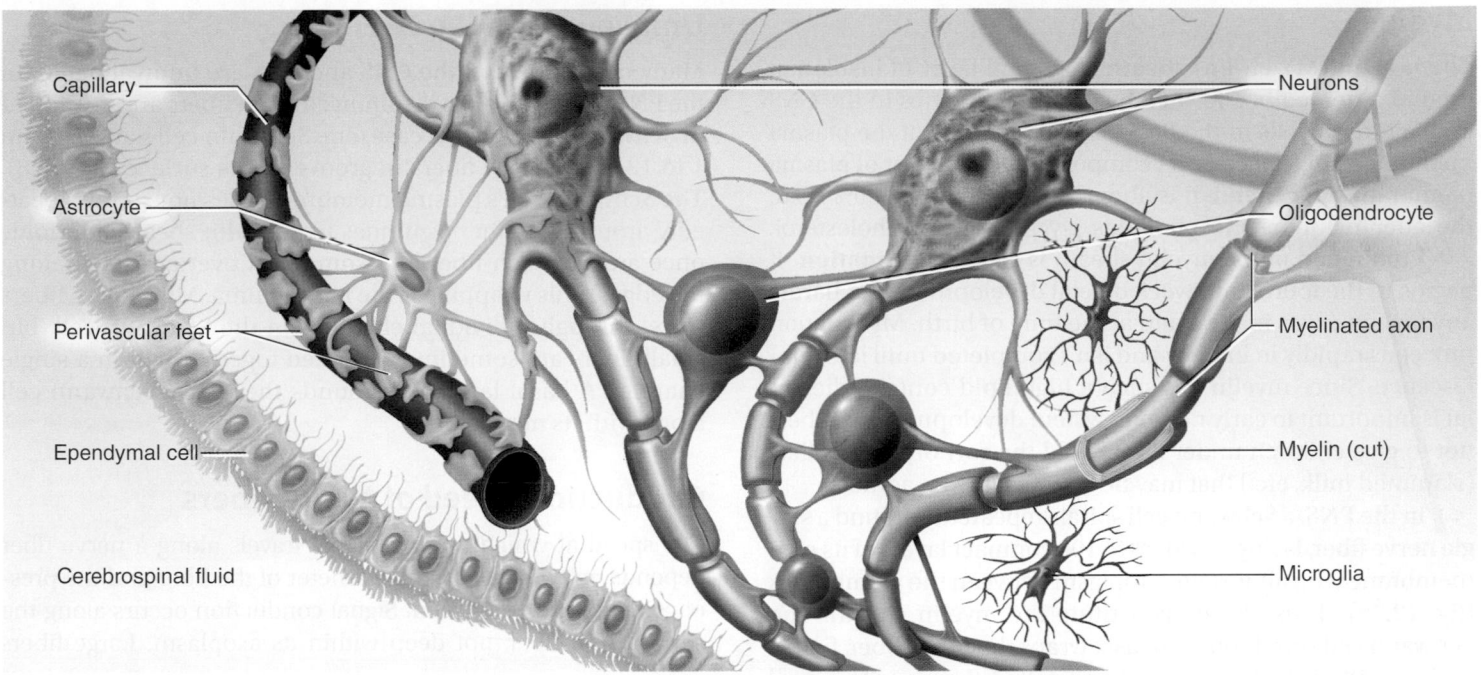

Capillary

Astrocyte

Perivascular feet

Ependymal cell

Cerebrospinal fluid

Neurons

Oligodendrocyte

Myelinated axon

Myelin (cut)

Microglia

**FIGURE 12.6  Neuroglia of the Central Nervous System.**

- They have extensions called *perivascular feet,* which contact the blood capillaries and stimulate them to form a tight, protective seal called the *blood–brain barrier* (see chapter 14).

- They monitor neuron activity and regulate blood flow in the brain tissue to meet changing needs for oxygen and nutrients.

- They convert blood glucose to lactate and supply this to the neurons for nourishment.

- They secrete *nerve growth factors* that regulate nerve development (see Deeper Insight 12.3).

- They communicate electrically with neurons and influence synaptic signaling between them.

- They regulate the composition of the tissue fluid. When neurons transmit signals, they release neurotransmitters and potassium ions. Astrocytes absorb these and prevent them from reaching excessive levels in the tissue fluid.

- When neurons are damaged, astrocytes form hardened scar tissue and fill space formerly occupied by the neurons. This process is called *astrocytosis* or *sclerosis.*

The other two types of glial cells occur only in the peripheral nervous system:

 **DEEPER INSIGHT 12.1**

**CLINICAL APPLICATION**

### Glial Cells and Brain Tumors

A tumor consists of a mass of rapidly dividing cells. Mature neurons, however, have little or no capacity for mitosis and seldom form tumors. Some brain tumors arise from the meninges (protective membranes of the CNS) or arise by metastasis from tumors elsewhere, such as malignant melanoma and colon cancer. Most adult brain tumors, however, are composed of glial cells, which are mitotically active throughout life. Such tumors are called *gliomas.*[21] Gliomas usually grow rapidly and are highly malignant. Because of the blood–brain barrier, brain tumors usually do not yield to chemotherapy and must be treated with radiation or surgery.

5. **Schwann**[22] (pronounced "shwon") **cells,** or **neurilemmocytes,** envelop nerve fibers of the PNS. In most cases, a Schwann cell winds repeatedly around a nerve fiber and produces a myelin sheath similar to the one produced by oligodendrocytes in the CNS. There are some important differences in myelin production between the CNS and PNS, which we consider shortly. Schwann cells also assist in the regeneration of damaged fibers, as described later.

6. **Satellite cells** surround the somas in ganglia of the PNS. They provide insulation around the soma and regulate the chemical environment of the neurons.

[21]*glia* = glial cells; *oma* = tumor
[22]Theodor Schwann (1810–82), German histologist

## Myelin

The **myelin** (MY-eh-lin) **sheath** is a spiral layer of insulation around a nerve fiber, formed by oligodendrocytes in the CNS and Schwann cells in the PNS. Since it consists of the plasma membranes of glial cells, its composition is like that of plasma membranes in general. It is about 20% protein and 80% lipid, the latter including phospholipids, glycolipids, and cholesterol.

Production of the myelin sheath is called **myelination.** It begins in the fourteenth week of fetal development, yet hardly any myelin exists in the brain at the time of birth. Myelination proceeds rapidly in infancy and isn't completed until late adolescence. Since myelin has such a high lipid content, dietary fat is important to early nervous system development. It is best not to give children under 2 years old the sort of low-fat diets (skimmed milk, etc.) that may be beneficial to an adult.

In the PNS, a Schwann cell spirals repeatedly around a single nerve fiber, laying down up to 100 compact layers of its own membrane with almost no cytoplasm between the membranes (fig. 12.7a). These layers constitute the myelin sheath. The Schwann cell spirals outward as it wraps the nerve fiber, finally ending with a thick outermost coil called the **neurilemma**[23] (noor-ih-LEM-ah). Here, the bulging body of the Schwann cell contains its nucleus and most of its cytoplasm. External to the neurilemma is a basal lamina and then a thin sleeve of fibrous connective tissue called the *endoneurium.* To visualize this myelination process, imagine that you wrap an almost-empty tube of toothpaste tightly around a pencil. The pencil represents the axon, and the spiral layers of toothpaste tube represent the myelin. The toothpaste would be forced to one end of the tube, which would form a bulge on the external surface of the wrapping, like the body of the Schwann cell.

In the CNS, each oligodendrocyte reaches out to myelinate several nerve fibers in its immediate vicinity (fig. 12.7b). Since it is anchored to multiple nerve fibers, it cannot migrate around any one of them like a Schwann cell does. It must push newer layers of myelin under the older ones, so myelination spirals inward toward the nerve fiber. Nerve fibers of the CNS have no neurilemma or endoneurium.

In both the PNS and CNS, a nerve fiber is much longer than the reach of a single glial cell, so it requires many Schwann cells or oligodendrocytes to cover one nerve fiber. Consequently, the myelin sheath is segmented. The gaps between the segments are called **nodes of Ranvier**[24] (RON-vee-AY), and the myelin-covered segments from one gap to the next are called **internodes** (see fig. 12.4a). The internodes are about 0.2 to 1.0 mm long. The short section of nerve fiber between the axon hillock and the first glial cell is called the **initial segment.** Since the axon hillock and initial segment play an important role in initiating a nerve signal, they are collectively called the **trigger zone.**

## Unmyelinated Nerve Fibers

Many nerve fibers in the CNS and PNS are unmyelinated. In the PNS, however, even the unmyelinated fibers are enveloped in Schwann cells. In this case, one Schwann cell harbors from 1 to 12 small nerve fibers in grooves in its surface (fig. 12.8). The Schwann cell's plasma membrane does not spiral repeatedly around the fiber as it does in a myelin sheath, but folds once around each fiber and somewhat overlaps itself along the edges. This wrapping is the neurilemma. Most nerve fibers travel through individual channels in the Schwann cell, but small fibers are sometimes bundled together within a single channel. A basal lamina surrounds the entire Schwann cell along with its nerve fibers.

## Conduction Speed of Nerve Fibers

The speed at which a nerve signal travels along a nerve fiber depends on two factors: the diameter of the fiber and the presence or absence of myelin. Signal conduction occurs along the surface of a fiber, not deep within its axoplasm. Large fibers

### DEEPER INSIGHT 12.2
### CLINICAL APPLICATION

#### Diseases of the Myelin Sheath

Multiple sclerosis and Tay–Sachs disease are degenerative disorders of the myelin sheath. In *multiple sclerosis*[25] *(MS),* the oligodendrocytes and myelin sheaths of the CNS deteriorate and are replaced by hardened scar tissue, especially between the ages of 20 and 40. Nerve conduction is disrupted, with effects that depend on what part of the CNS is involved—double vision, blindness, speech defects, neurosis, tremors, or numbness, for example. Patients experience variable cycles of milder and worse symptoms until they eventually become bedridden. The cause of MS remains uncertain; most hypotheses suggest that it is an autoimmune disorder triggered by a virus in genetically susceptible individuals. There is no cure. There is conflicting evidence of how much it shortens a person's life expectancy, if at all. A few die within 1 year of diagnosis, but many people live with MS for 25 or 30 years.

*Tay–Sachs*[26] disease is a hereditary disorder seen mainly in infants of Eastern European Jewish ancestry. It results from the abnormal accumulation of a glycolipid called $GM_2$ (ganglioside) in the myelin sheath. $GM_2$ is normally decomposed by a lysosomal enzyme, but this enzyme is lacking from people who are homozygous recessive for the Tay–Sachs allele. As $GM_2$ accumulates, it disrupts the conduction of nerve signals and the victim typically suffers blindness, loss of coordination, and dementia. Signs begin to appear before the child is a year old, and most victims die by the age of 3 or 4 years. Asymptomatic adult carriers can be identified by a blood test and advised by genetic counselors on the risk of their children having the disease.

---

[23]*neuri* = nerve; *lemma* = husk, peel, sheath
[24]L. A. Ranvier (1835–1922), French histologist and pathologist

[25]*scler* = hard, tough; *osis* = condition
[26]Warren Tay (1843–1927), English physician; Bernard Sachs (1858–1944), American neurologist

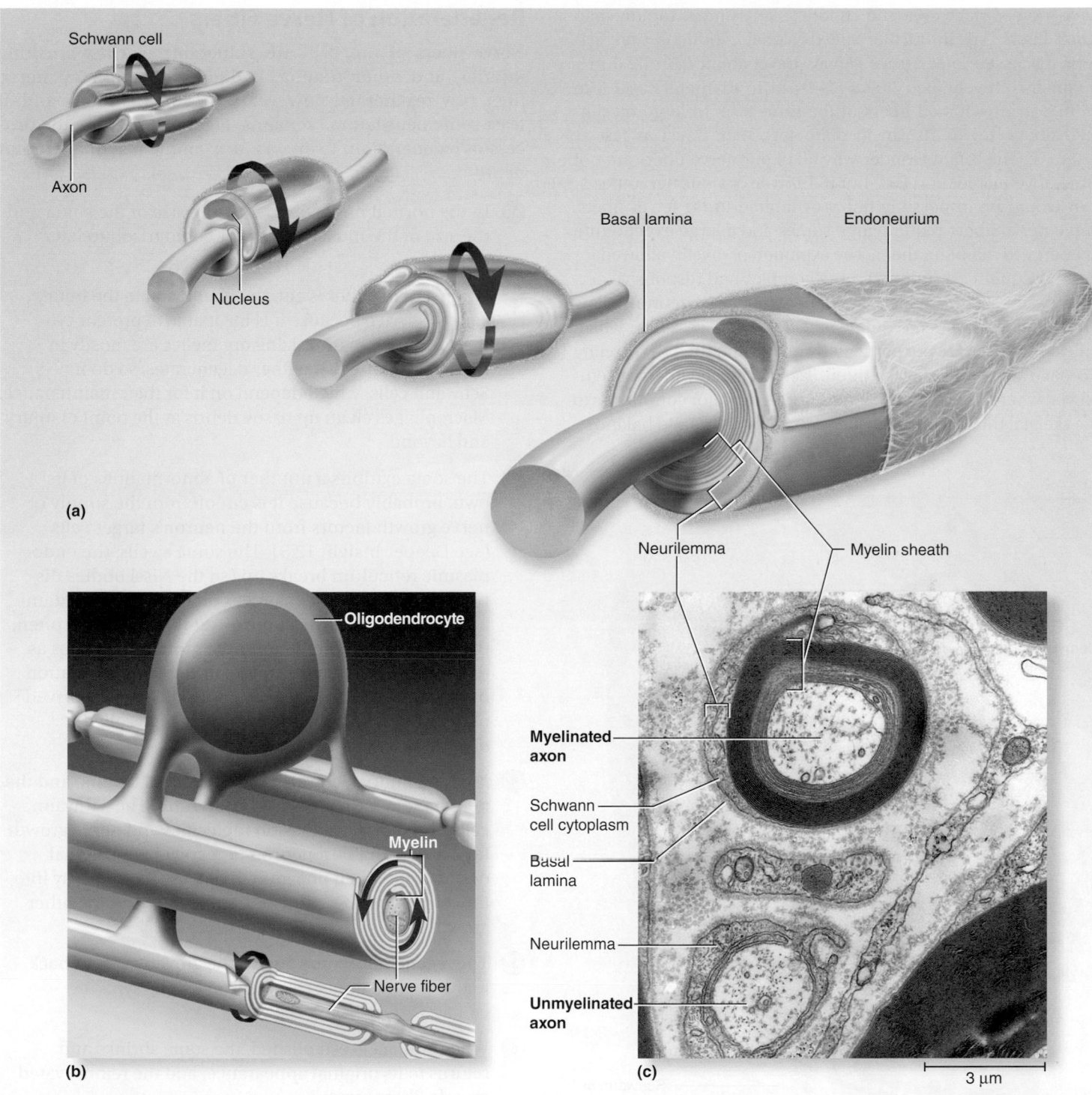

**FIGURE 12.7 Myelination.** (a) A Schwann cell of the PNS, wrapping repeatedly around an axon to form the multilayered myelin sheath. The myelin spirals outward away from the axon as it is laid down. The outermost coil of the Schwann cell constitutes the neurilemma. (b) An oligodendrocyte of the CNS wrapping around the axons of multiple neurons. Here, the myelin spirals inward toward the axon as it is laid down. These contrasting modes of myelination are called *centrifugal myelination* ("away from the center") in the PNS and *centripetal myelination* ("toward the center") in the CNS. (c) A myelinated axon (top) and unmyelinated axon (bottom) (TEM).

have more surface area and conduct signals more rapidly than small fibers. Myelin further speeds signal conduction for reasons discussed later. Nerve signals travel about 0.5 to 2.0 m/s in small unmyelinated fibers (2–4 μm in diameter) and 3 to 15 m/s in myelinated fibers of the same size. In large myelinated fibers (up to 20 μm in diameter), they travel as fast as 120 m/s. One might wonder why all of our nerve fibers are not large, myelinated, and fast; but if this were so, our nervous system would be impossibly bulky or limited to far fewer fibers. Large nerve fibers require large somas and a large expenditure of energy to maintain them. The evolution of myelin allowed for the subsequent evolution of more complex and responsive nervous systems with smaller, more energy-efficient neurons. Slow unmyelinated fibers are quite sufficient for processes in which quick responses are not particularly important, such as secreting stomach acid or dilating the pupil. Fast myelinated fibers are employed where speed is more important, as in motor commands to the skeletal muscles and sensory signals for vision and balance.

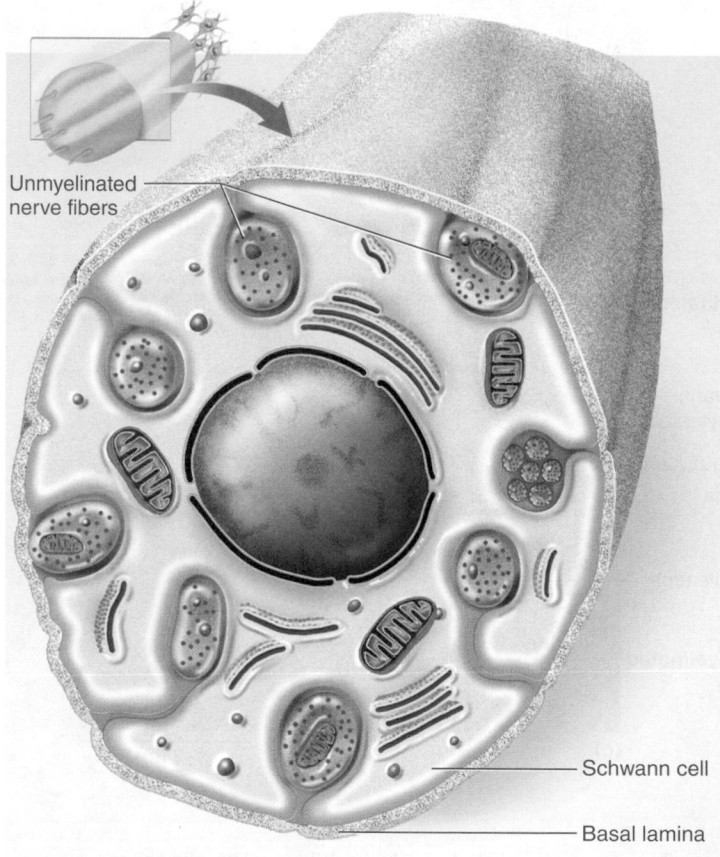

Unmyelinated nerve fibers

Schwann cell

Basal lamina

**FIGURE 12.8   Unmyelinated Nerve Fibers.**   Multiple unmyelinated fibers are enclosed in channels in the surface of a single Schwann cell. In some cases, smaller nerve fibers may be bundled together within a single Schwann cell channel.

❓ *What is the functional disadvantage of an unmyelinated nerve fiber? What is its anatomical advantage?*

## Regeneration of Nerve Fibers

Nerve fibers of the PNS are vulnerable to cuts, crushing injuries, and other trauma. A damaged peripheral nerve fiber may regenerate, however, if its soma is intact and at least some neurilemma remains. Figure 12.9 shows the process of regeneration, taking as its example a somatic motor neuron:

① In the normal nerve fiber, note the size of the soma and the size of the muscle fibers for comparison to later stages.

② When a nerve fiber is cut, the fiber distal to the injury cannot survive because it is incapable of protein synthesis. Protein-synthesizing organelles are mostly in the soma. As the distal fiber degenerates, so do its Schwann cells, which depend on it for their maintenance. Macrophages clean up tissue debris at the point of injury and beyond.

③ The soma exhibits a number of abnormalities of its own, probably because it is cut off from the supply of nerve growth factors from the neuron's target cells (see Deeper Insight 12.3). The soma swells, the endoplasmic reticulum breaks up (so the Nissl bodies disperse), and the nucleus moves off center. Not all damaged neurons survive; some die at this stage. But often, the axon stump sprouts multiple growth processes as the severed distal end shows continued degeneration of its axon and Schwann cells. Muscle fibers deprived of their nerve supply exhibit a shrinkage called *denervation atrophy.*

④ Near the injury, Schwann cells, the basal lamina, and the neurilemma form a **regeneration tube.** The Schwann cells produce cell-adhesion molecules and nerve growth factors that enable a neuron to regrow to its original destination. When one growth process finds its way into the tube, it grows rapidly (3–5 mm/day), and the other growth processes are retracted.

⑤ The regeneration tube guides the growing sprout back to the original target cells, reestablishing synaptic contact.

⑥ When contact is established, the soma shrinks and returns to its original appearance, and the reinnervated muscle fibers regrow.

Regeneration is not perfect. Some nerve fibers connect to the wrong muscle fibers or never find a muscle fiber at all, and some damaged motor neurons simply die. Nerve injury is therefore often followed by some degree of functional deficit. Even when regeneration is achieved, the slow rate of axon regrowth means that some nerve function may take as long as 2 years to be restored. Damaged nerve fibers in the CNS cannot regenerate at all, but since the CNS is enclosed in bone, it suffers less trauma than the PNS.

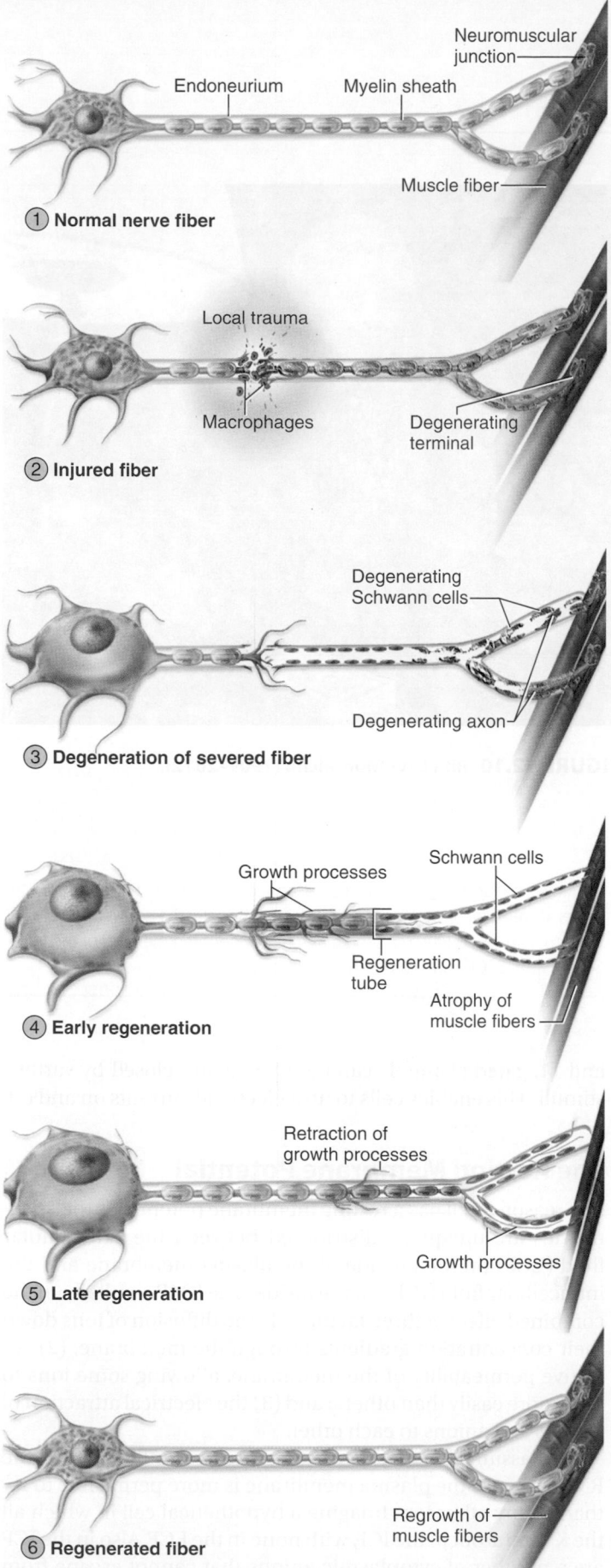

**FIGURE 12.9   Regeneration of a Damaged Nerve Fiber.** See text for explanation.

1. **Normal nerve fiber**

Neuromuscular junction
Endoneurium   Myelin sheath
Muscle fiber

2. **Injured fiber**

Local trauma
Macrophages   Degenerating terminal

3. **Degeneration of severed fiber**

Degenerating Schwann cells
Degenerating axon

4. **Early regeneration**

Growth processes   Schwann cells
Regeneration tube
Atrophy of muscle fibers

5. **Late regeneration**

Retraction of growth processes
Growth processes

6. **Regenerated fiber**

Regrowth of muscle fibers

**BEFORE YOU GO ON**

Answer the following questions to test your understanding of the preceding section:

8. How is a glial cell different from a neuron? List the six types of glial cells and discuss their functions.

9. How is myelin produced? How does myelin production in the CNS differ from that in the PNS?

10. How can a severed peripheral nerve fiber find its way back to the cells it originally innervated?

## 12.4   Electrophysiology of Neurons

### Expected Learning Outcomes

When you have completed this section, you should be able to

a. explain why a cell has an electrical charge difference (voltage) across its membrane;

b. explain how stimulation of a neuron causes a local electrical response in its membrane;

c. explain how local responses generate a nerve signal; and

d. explain how the nerve signal is conducted down an axon.

The nervous system has intrigued scientists and philosophers since ancient times. The Roman physician Galen thought that the brain pumped a vapor called *psychic pneuma* through hollow nerves and squirted it into the muscles to make them contract. The French philosopher René Descartes still argued for this theory in the seventeenth century. It finally fell out of favor in the eighteenth century, when Luigi Galvani discovered the role of electricity in muscle contraction (see p. 415). Further progress had to await improvements in microscope technology and histological staining methods. Italian histologist Camillo Golgi (1843–1926) developed an important method for staining neurons with silver. This enabled Spanish histologist Santiago Ramón y Cajal (1852–1934), with tremendous skill and patience, to trace the course of nerve fibers through tissue sections. He demonstrated that the nervous pathway was not a continuous "wire" or tube, but a series of cells separated by the gaps we now call synapses. Golgi and Ramón y Cajal shared the 1906 Nobel Prize for Physiology or Medicine for these important discoveries.

Cajal's theory suggested another direction for research—how do neurons communicate? Two key issues in neurophysiology are (1) How does a neuron generate an electrical signal? and (2) How does it transmit a meaningful message to the next cell? These are the questions to which this section and the next are addressed.

### Electrical Potentials and Currents

Neural communication, like muscle excitation, is based on electrophysiology—cellular mechanisms for producing electrical potentials and currents. An **electrical potential** is a

# DEEPER INSIGHT 12.3
## MEDICAL HISTORY

### Nerve Growth Factor—From Home Laboratory to Nobel Prize

It is remarkable what odds can be overcome by self-confident persistence. Neurobiologist Rita Levi-Montalcini (fig. 12.10) affords a striking example. Although born of a cultured and accomplished Italian Jewish family, she and her twin sister Paola were discouraged from considering a college education or career by their tradition-minded father. Determined to attend university anyway, they hired their own tutor to prepare them. Rita graduated *summa cum laude* in medicine and surgery in 1930 and embarked on advanced study in neurology. Paola became a renowned artist.

But then arose the sinister specter of anti-Semitism, as fascist dictator Mussolini barred Jews from professional careers. Rita despaired of pursuing medicine or research until a college friend reminded her of how much Cajal had achieved under very primitive conditions. That inspired her to set up a little laboratory in her bedroom, where she studied nervous system development in chick embryos. She had read of work by Viktor Hamburger in St. Louis, who believed that limb tissues secrete a chemical that attracts nerves to grow into them. Levi-Montalcini, however, believed that nerves grow into the limbs without such attractants, but die if they fail to receive a substance needed to sustain them.

Fleeing first from Allied bombing and then Hitler's invasion of Italy, Levi-Montalcini had to abandon her work and go into hiding until the end of the war. At war's end, Hamburger invited her to join him in America, where they found her hypothesis to be correct. She and Stanley Cohen isolated the nerve-sustaining substance and named it *nerve growth factor (NGF)*, for which they shared a 1986 Nobel Prize. NGF is a protein secreted by muscle and glial cells. It prevents apoptosis in growing neurons and thus enables them to establish connections with their target cells. It was the first of many cell growth factors discovered, and launched what is today a vibrant field of research in the use of growth factors to stimulate tissue development and repair (see Deeper Insight 12.4, for example).

**FIGURE 12.10  Rita Levi-Montalcini (1909–2012).**

Rita and Paola created the Levi-Montalcini Foundation to support the career development of young people and especially the scientific education of women in Africa. Rita also served as a highly honored member of the Italian Senate from 2001 until her death at the age of 103.

---

difference in the concentration of charged particles between one point and another. It is a form of potential energy that, under the right circumstances, can produce a current. An electrical **current** is a flow of charged particles from one point to another. A new flashlight battery, for example, typically has a potential, or charge, of 1.5 volts (V). If the two poles of the battery are connected by a wire, electrons flow through the wire from one pole of the battery to the other, creating a current that lights the bulb. As long as the battery has a potential (voltage), we say it is **polarized.**

Living cells are also polarized. As we saw in chapter 11, the charge difference across the plasma membrane is called the **resting membrane potential (RMP).** It is much less than the potential of a flashlight battery—typically about −70 millivolts (mV) in an unstimulated, "resting" neuron. The negative value means there are more negatively charged particles on the inside of the membrane than on the outside.

We do not have free electrons in the body as we do in an electrical circuit. Electrical currents in the body are created, instead, by the flow of ions such as $Na^+$ and $K^+$ through gated channels in the plasma membrane. As we saw in chapters 3

and 11, gated channels can be opened and closed by various stimuli. This enables cells to turn electrical currents on and off.

## The Resting Membrane Potential

The reason a cell has a resting membrane potential is that electrolytes are unequally distributed between the extracellular fluid (ECF) on the outside of the plasma membrane and the intracellular fluid (ICF) on the inside. The RMP results from the combined effect of three factors: (1) the diffusion of ions down their concentration gradients through the membrane; (2) selective permeability of the membrane, allowing some ions to pass more easily than others; and (3) the electrical attraction of cations and anions to each other.

Potassium ions ($K^+$) have the greatest influence on the RMP, because the plasma membrane is more permeable to $K^+$ than to any other ion. Imagine a hypothetical cell in which all the $K^+$ starts out in the ICF, with none in the ECF. Also in the ICF are a number of cytoplasmic anions that cannot escape from the cell because of their size or charge—phosphates, sulfates, small organic acids, proteins, ATP, and RNA. Assuming $K^+$ can

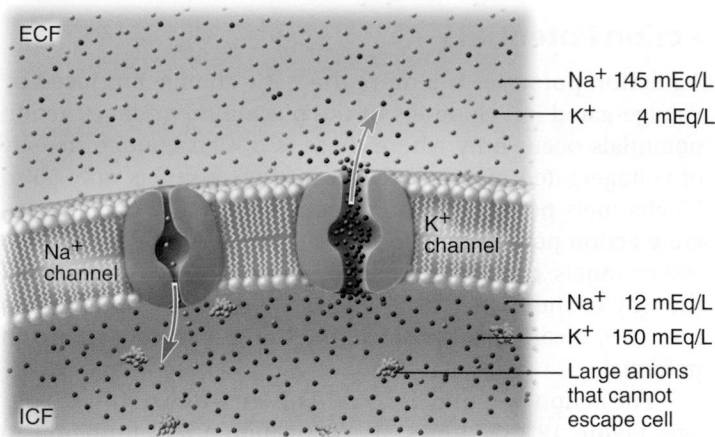

**FIGURE 12.11 Ionic Basis of the Resting Membrane Potential.** Note that sodium ions are much more concentrated in the extracellular fluid (ECF) than in the intracellular fluid (ICF), while potassium ions are more concentrated in the ICF. Large anions unable to penetrate the plasma membrane give the cytoplasm a negative charge relative to the ECF.

❓ *If we suddenly increased the concentration of Cl⁻ ions in the ICF, would the membrane potential become higher or lower?*

diffuse freely through channels in the plasma membrane, it diffuses down its concentration gradient and out of the cell, leaving these cytoplasmic anions behind (fig. 12.11). As a result, the ICF grows more and more negatively charged. But as the ICF becomes more negative, it exerts a stronger attraction for the positive potassium ions and attracts some of them back into the cell. Eventually an *equilibrium* is reached in which K⁺ is moving out of the cell (down its concentration gradient) and into the cell (by electrical attraction) at equal rates. There is no further *net* diffusion of K⁺. At the point of equilibrium, K⁺ is about 40 times as concentrated in the ICF as in the ECF.

If K⁺ were the only ion affecting the RMP, it would give the membrane a potential of about –90 mV. However, sodium ions (Na⁺) also enter the picture. Sodium is about 12 times as concentrated in the ECF as in the ICF. The resting plasma membrane is much less permeable to Na⁺ than to K⁺, but Na⁺ does diffuse down its concentration gradient into the cell, attracted by the negative charge in the ICF. This sodium leak is only a trickle, but it is enough to cancel some of the negative charge and reduce the voltage across the membrane.

Sodium leaks into the cell and potassium leaks out, but the sodium–potassium (Na⁺-K⁺) pump described in chapter 3 continually compensates for this leakage. It pumps 3 Na⁺ out of the cell for every 2 K⁺ it brings in, consuming 1 ATP for each exchange cycle. By removing more cations from the cell than it brings in, it contributes about –3 mV to the RMP. The net effect of all this—K⁺ diffusion out of the cell, Na⁺ diffusion inward, and the Na⁺-K⁺ pump—is the RMP of –70 mV.

The Na⁺-K⁺ pump accounts for about 70% of the energy (ATP) requirement of the nervous system. Every signal generated by a neuron slightly upsets the distribution of Na⁺ and K⁺, so the pump must work continually to restore equilibrium.

This is why nervous tissue has one of the highest rates of ATP consumption of any tissue in the body, and why it demands so much glucose and oxygen. Although a neuron is said to be resting when it is not producing signals, it is highly active maintaining its RMP and "waiting," as it were, for something to happen.

The uneven distribution of Na⁺ and K⁺ on the two sides of the plasma membrane pertains only to a very thin film of ions immediately adjacent to the membrane surfaces. The electrical events we are about to examine do not involve ions very far away from the membrane in either the ECF or the ICF.

## Local Potentials

Stimulation of a neuron causes local disturbances in membrane potential. Typically (but with exceptions), the response begins at a dendrite, spreads through the soma, travels down the axon, and ends at the synaptic knobs. We consider the process in that order.

Various neurons can be stimulated by chemicals, light, heat, or mechanical forces. We'll take as our example a neuron being chemically stimulated (fig. 12.12). The chemical—perhaps a pain signal from a damaged tissue or odor molecule in a breath of air—binds to receptors on the neuron. This opens sodium channels that allow Na⁺ to flow into the cell. The Na⁺ inflow cancels some of the internal negative charge, so the voltage across the membrane drifts toward zero. Any such case in which the voltage shifts to a less negative value is called **depolarization.**

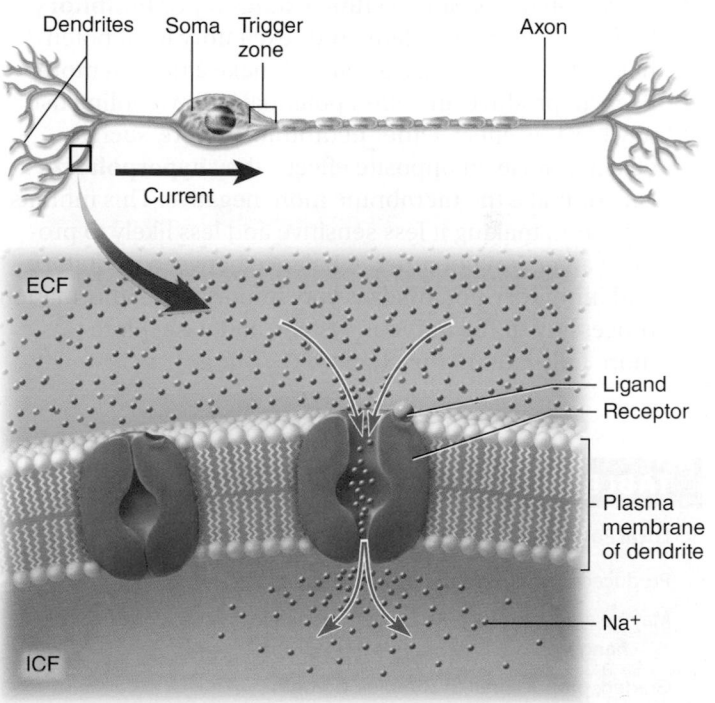

**FIGURE 12.12 Excitation of a Neuron by a Chemical Stimulus.** When the chemical (ligand) binds to a receptor on the neuron, the receptor acts as a ligand-gated channel that opens and allows Na⁺ to diffuse into the cell. This depolarizes the plasma membrane.

The incoming $Na^+$ diffuses for short distances along the inside of the plasma membrane and produces a current that travels from the point of stimulation toward the cell's trigger zone. This short-range change in voltage is called a **local potential.** There are four characteristics that distinguish local potentials from the action potentials we will study shortly (table 12.2). You will appreciate these distinctions more fully after you have studied action potentials.

1. Local potentials are **graded,** meaning they vary in magnitude (voltage) according to the strength of the stimulus. An intense or prolonged stimulus opens more ion channels than a weaker stimulus. Thus, more $Na^+$ enters the cell and the voltage changes more than it does with a weaker stimulus.

2. Local potentials are **decremental,** meaning they get weaker as they spread from the point of stimulation. The decline in strength occurs partly because as $Na^+$ spreads out under the plasma membrane and depolarizes it, $K^+$ flows out and reverses the effect of the $Na^+$ inflow, and partly because the $Na^+$ leaks back out of the cell through channels along its path. Therefore, the voltage shift caused by $Na^+$ diminishes rapidly with distance. This prevents local potentials from having long-distance effects.

3. Local potentials are **reversible,** meaning that if stimulation ceases, cation diffusion out of the cell quickly returns the membrane voltage to its resting potential.

4. Local potentials can be either **excitatory** or **inhibitory.** So far, we have considered only excitatory local potentials, which depolarize a cell and make a neuron more likely to produce an action potential. Acetylcholine usually has this effect. Other neurotransmitters, such as glycine, cause an opposite effect—they **hyperpolarize** a cell, or make the membrane more negative. This inhibits a neuron, making it less sensitive and less likely to produce an action potential. A balance between excitatory and inhibitory potentials is very important to information processing in the nervous system, and we explore this more fully later in the chapter.

## Action Potentials

An **action potential** is a more dramatic change produced by voltage-gated ion channels in the plasma membrane. Action potentials occur only where there is a high enough density of voltage-gated channels. Most of the soma has only 50 to 75 channels per square micrometer ($\mu m^2$) and cannot generate action potentials. The trigger zone, however, has 350 to 500 channels/$\mu m^2$. If an excitatory local potential spreads all the way to the trigger zone and is still strong enough when it arrives, it can open these channels and generate an action potential.

The action potential is a rapid up-and-down shift in voltage. Figure 12.13a shows an action potential numbered to correspond to the following description.

① When the local current arrives at the axon hillock, it depolarizes the membrane at that point. This appears as a steadily rising local potential.

② For anything more to happen, this local potential must rise to a critical voltage called the **threshold** (typically about −55 mV)—the minimum needed to open voltage-gated channels.

③ The neuron now "fires," or produces an action potential. At threshold, voltage-gated $Na^+$ channels open quickly, while $K^+$ channels open more slowly. The initial effect on membrane potential is therefore due to $Na^+$. Initially, only a few $Na^+$ channels open, but as $Na^+$ enters the cell, it further depolarizes the membrane. This stimulates still more voltage-gated $Na^+$ channels to open and admit even more $Na^+$, creating a positive feedback loop that makes the membrane voltage rise rapidly.

④ As the rising potential passes 0 mV, $Na^+$ channels are *inactivated* and begin closing. By the time they all close and $Na^+$ inflow ceases, the voltage peaks at approximately +35 mV. (The peak is as low as 0 mV in some neurons and as high as 50 mV in others.) The membrane is now positive on the inside and negative on the outside—its polarity is reversed compared to the RMP.

| **TABLE 12.2** | Comparison of Local Potentials and Action Potentials |
| --- | --- |
| **Local Potential** | **Action Potential** |
| Produced by gated channels on the dendrites and soma | Produced by voltage-gated channels on the trigger zone and axon |
| May be a positive (depolarizing) or negative (hyperpolarizing) voltage change | Always begins with depolarization |
| Graded; proportional to stimulus strength | All or none; either does not occur at all or exhibits the same peak voltage regardless of stimulus strength |
| Reversible; returns to RMP if stimulation ceases before threshold is reached | Irreversible; goes to completion once it begins |
| Local; has effects for only a short distance from point of origin | Self-propagating; has effects a great distance from point of origin |
| Decremental; signal grows weaker with distance | Nondecremental; signal maintains same strength regardless of distance |

⑤ By the time the voltage peaks, the slow K⁺ channels are fully open. Potassium ions, repelled by the positive ICF, now exit the cell. Their outflow **repolarizes** the membrane—that is, it shifts the voltage back into the negative numbers. The action potential consists of the up-and-down voltage shifts that occur from the time the threshold is reached to the time the voltage returns to the RMP.

⑥ Potassium channels stay open longer than Na⁺ channels, so slightly more K⁺ leaves the cell than the amount of Na⁺ that entered. Therefore, the membrane voltage drops to 1 or 2 mV more negative than the original RMP, producing a negative overshoot called *hyperpolarization* (or *afterpotential*).

⑦ As you can see, Na⁺ and K⁺ switch places across the membrane during an action potential. During hyperpolarization, Na⁺ diffusion into the cell and (in the CNS) the removal of extracellular K⁺ by astrocytes gradually restore the original RMP.

Figure 12.14 correlates these voltage changes with events in the plasma membrane. At the risk of being misleading, it is drawn as if most of the Na⁺ and K⁺ had traded places. In reality, only about one ion in a million crosses the membrane to produce an action potential, and an action potential involves only the thin layer of ions close to the membrane. If the illustration tried to represent these points accurately, the difference would be so slight you could not see it. Even after thousands of action potentials, the cytosol still has a higher concentration of K⁺ and a lower concentration of Na⁺ than the ECF does.

Figure 12.13a also is deliberately distorted. To demonstrate the different phases of the local potential and action potential, the magnitudes of the local potential and hyperpolarization are exaggerated, the local potential is stretched out to make it seem longer, and the duration of hyperpolarization is shrunken so the graph will fit the page. When these events are plotted on a more realistic timescale, they look more like figure 12.13b. The local potential is so brief it is unnoticeable, and hyperpolarization is very long but only slightly more negative than the RMP. An action potential is often called a *spike*; it is easy to see why from this figure.

Earlier we saw that local potentials are graded, decremental, and reversible. We can now contrast this with action potentials.

•   Action potentials follow an **all-or-none law.** If a stimulus depolarizes the neuron to threshold, the neuron fires at its maximum voltage (such as +35 mV); if threshold is not reached, the neuron does not fire at all. Above threshold, stronger stimuli do not produce stronger action potentials. Thus, action potentials are not graded (proportional to stimulus strength).

•   Action potentials are **nondecremental.** For reasons examined shortly, they do not get weaker with distance. The last action potential at the end of a nerve fiber is just as strong as the first one in the trigger zone, no matter how far.

•   Action potentials are **irreversible.** If a neuron reaches threshold, the action potential goes to completion; it cannot be stopped once it begins.

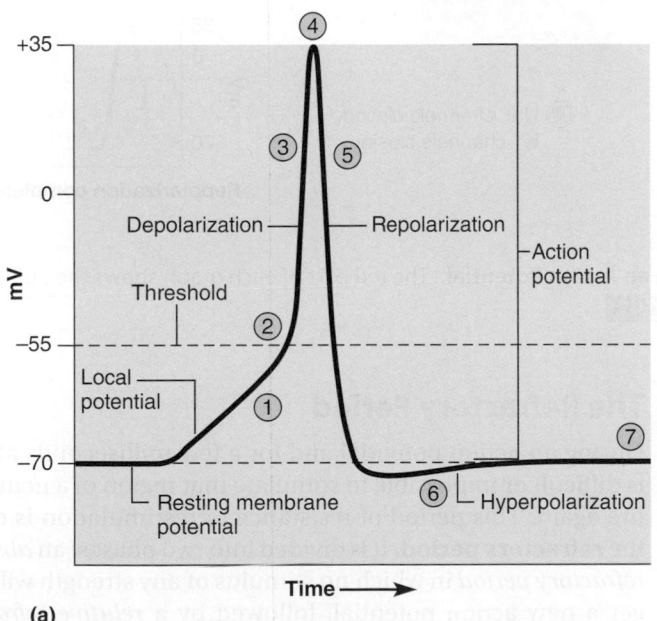

(a)

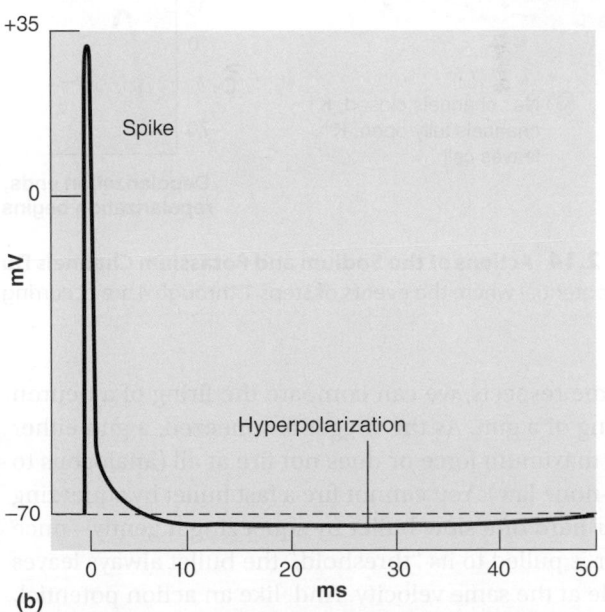

(b)

**FIGURE 12.13 An Action Potential.** (a) Diagrammed with a distorted timescale to make details of the action potential visible. Numbers correspond to stages discussed in the text. (b) On a more accurate timescale, the local potential is so brief it is imperceptible, the action potential appears as a spike, and the hyperpolarization is very prolonged.

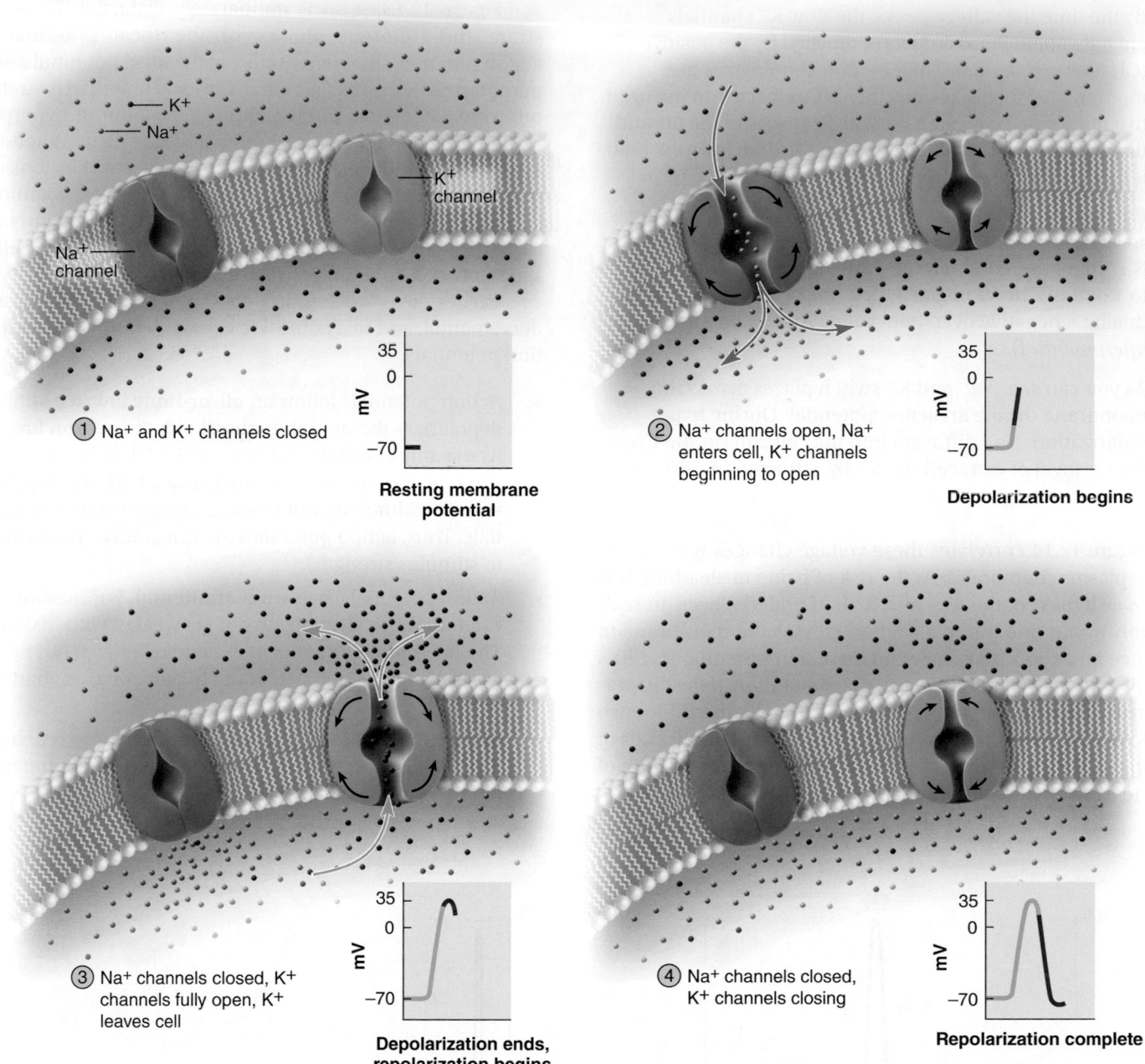

① Na⁺ and K⁺ channels closed

**Resting membrane potential**

② Na⁺ channels open, Na⁺ enters cell, K⁺ channels beginning to open

**Depolarization begins**

③ Na⁺ channels closed, K⁺ channels fully open, K⁺ leaves cell

**Depolarization ends, repolarization begins**

④ Na⁺ channels closed, K⁺ channels closing

**Repolarization complete**

**FIGURE 12.14 Actions of the Sodium and Potassium Channels During an Action Potential.** The red part of each graph shows the point in the action potential where the events of steps 1 through 4 are occurring. **AP|R**

In some respects, we can compare the firing of a neuron to the firing of a gun. As the trigger is squeezed, a gun either fires with maximum force or does not fire at all (analogous to the all-or-none law). You cannot fire a fast bullet by squeezing the trigger hard or a slow bullet by squeezing it gently—once the trigger is pulled to its "threshold," the bullet always leaves the muzzle at the same velocity. And, like an action potential, the firing of a gun is irreversible once the threshold is reached. Table 12.2 further contrasts a local potential with an action potential, including some characteristics of action potentials explained in the following sections.

## The Refractory Period

During an action potential and for a few milliseconds after, it is difficult or impossible to stimulate that region of a neuron to fire again. This period of resistance to restimulation is called the **refractory period.** It is divided into two phases: an *absolute refractory period* in which no stimulus of any strength will trigger a new action potential, followed by a *relative refractory period* in which it is possible to trigger a new action potential, but only with an unusually strong stimulus (fig. 12.15).

The absolute refractory period lasts from the start of the action potential until the membrane returns to the resting

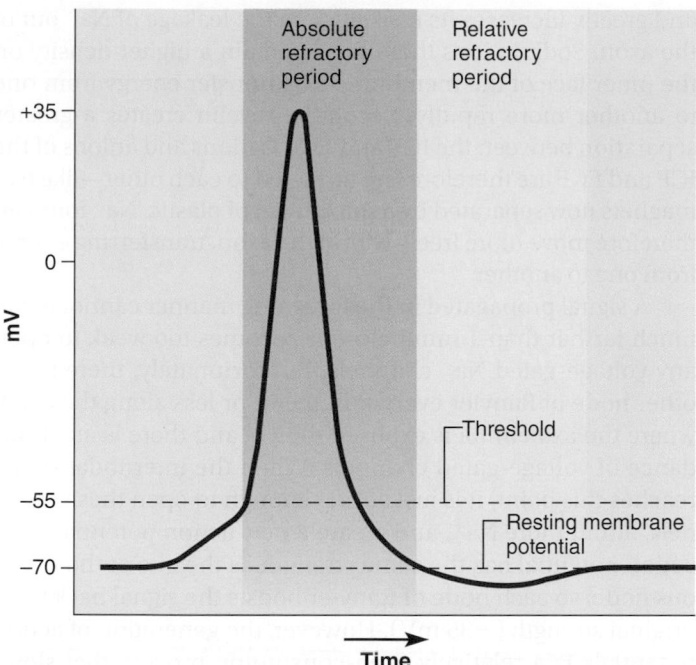

**FIGURE 12.15 The Absolute and Relative Refractory Periods in Relation to the Action Potential.**

short distance just beneath the plasma membrane. The resulting depolarization excites voltage-gated channels immediately distal to the action potential. Sodium and potassium channels open and close just as they did at the trigger zone, and a new action potential is produced. By repetition, this excites the membrane immediately distal to that. This chain reaction continues until the traveling signal reaches the end of the axon.

Note that an action potential itself does not travel along an axon; rather, it stimulates the production of a new action potential in the membrane just ahead of it. Thus, we can distinguish an *action potential* from a *nerve signal*. The nerve signal is a traveling wave of excitation produced by self-propagating action potentials. It is like a line of falling dominoes. No one domino travels to the end of the line, but each domino pushes over the next one and there is a transmission of energy from the first domino to the last. Similarly, no one action potential travels to the end of an axon; a nerve signal is a chain reaction of action potentials.

If one action potential stimulates the production of a new one next to it, you might think that the signal could also start traveling backward and return to the soma. This does not

potential—that is, for as long as the Na$^+$ channels are open and subsequently inactivated. The relative refractory period lasts until hyperpolarization ends. During this period, K$^+$ channels are still open. A new stimulus tends to admit Na$^+$ and depolarize the membrane, but K$^+$ diffuses out through the open channels as Na$^+$ comes in, and thus opposes the effect of the stimulus. It requires an especially strong stimulus to override the K$^+$ outflow and depolarize the cell enough to set off a new action potential. By the end of hyperpolarization, K$^+$ channels are closed and the cell is as responsive as ever.

The refractory period refers only to a small patch of membrane where an action potential has already begun, not to the entire neuron. Other parts of the neuron can still be stimulated while a small area of it is refractory, and even this area quickly recovers once the nerve signal has passed on.

## Signal Conduction in Nerve Fibers

If a neuron is to communicate with another cell, a signal has to travel to the end of the axon. We can now examine how this is achieved.

### Unmyelinated Fibers

Unmyelinated fibers present a relatively simple case of signal conduction, easy to understand based on what we have already covered (fig. 12.16). An unmyelinated fiber has voltage-gated channels along its entire length. When an action potential occurs at the trigger zone, Na$^+$ enters the axon and diffuses for a

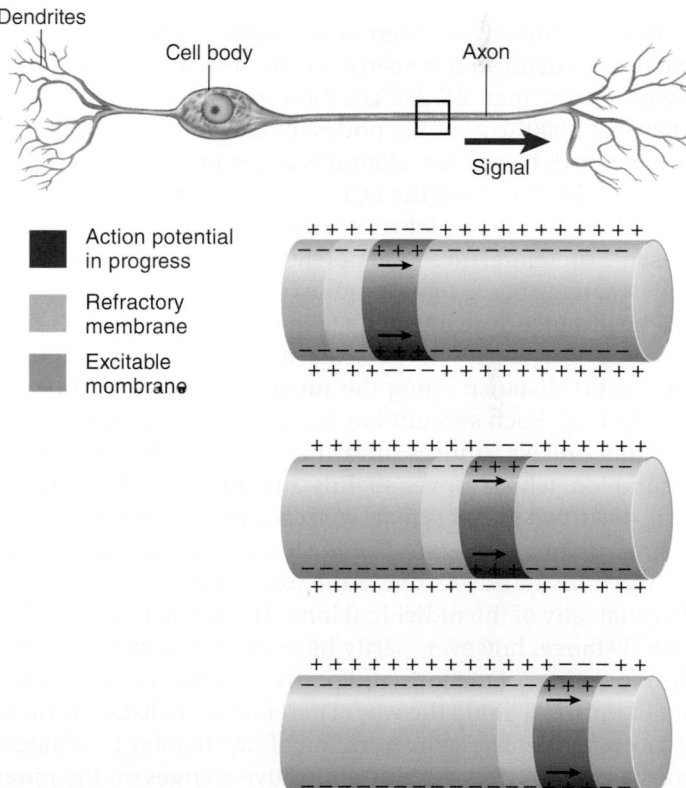

**FIGURE 12.16 Conduction of a Nerve Signal in an Unmyelinated Fiber.** Note that the membrane polarity is reversed in the region of the action potential (red). A region of membrane in its refractory period (yellow) trails the action potential and prevents the nerve signal from going backward toward the soma. The other membrane areas (green) are fully polarized and ready to respond. **AP|R**

occur, however, because the membrane behind the nerve signal is still in its refractory period and cannot be restimulated. Only the membrane ahead is sensitive to stimulation. The refractory period thus ensures that nerve signals are conducted in the proper direction, from the soma to the synaptic knobs.

A traveling nerve signal is an electrical current, but it is not the same as a current traveling through a wire. A current in a wire travels millions of meters per second and is decremental—it gets weaker with distance. A nerve signal is much slower (not more than 2 m/s in unmyelinated fibers), but as already noted, it is nondecremental. To clarify this concept, we can compare the nerve signal to a burning fuse. When a fuse is lit, the heat ignites powder immediately in front of this point, and this repeats itself in a self-propagating fashion until the end of the fuse is reached. At the end, the fuse burns just as hotly as it did at the beginning. In a fuse, the combustible powder is the source of potential energy that keeps the process going in a nondecremental fashion. In an axon, the potential energy comes from the ion gradient across the plasma membrane. Thus, the signal does not grow weaker with distance; it is self-propagating, like the burning of a fuse.

## Myelinated Fibers

Matters are somewhat different in myelinated fibers. Voltage-gated ion channels are scarce in the myelin-covered internodes—fewer than $25/\mu m^2$ in these regions compared with 2,000 to $12,000/\mu m^2$ at the nodes of Ranvier. There would be little point in having ion channels in the internodes—myelin insulates the fiber from the ECF at these points, and $Na^+$ from the ECF could not flow into the cell even if more channels were present. Therefore, no action potentials can occur in the internodes, and the nerve signal requires some other way of traversing the distance from one node to the next.

When $Na^+$ enters the axon at a node of Ranvier, it diffuses for a short distance along the inner face of the membrane (fig. 12.17a). Each sodium ion has an electrical field around it. When one $Na^+$ moves toward another, its field repels the second ion, which moves slightly and repels another, and so forth—like two magnets that repel each other if you try to push their north poles together. No one ion moves very far, but this energy transfer travels down the axon much faster and farther than any of the individual ions. The signal grows weaker with distance, however, partly because the axoplasm resists the movement of the ions and partly because $Na^+$ leaks back out of the axon along the way. Therefore with distance, there is a lower and lower concentration of $Na^+$ to relay the charge. Furthermore, with a surplus of positive charges on the inner face of the axolemma and a surplus of negative charges on the outer face, these cations and anions are attracted to each other through the membrane—like the opposite poles of two magnets attracting each other through a sheet of cardboard. This results in a "storage" (called *capacitance*) of unmoving or sluggishly moving charges on the membrane.

Myelin speeds up signal conduction in two ways. First of all, by wrapping tightly around the axon, it seals the nerve fiber and greatly increases its resistance to the leakage of $Na^+$ out of the axon. Sodium ions therefore maintain a higher density on the inner face of the membrane and transfer energy from one to another more rapidly. Secondly, myelin creates a greater separation between the ICF and ECF. Cations and anions of the ICF and ECF are therefore less attracted to each other—like two magnets now separated by a thick sheet of plastic. $Na^+$ ions can therefore move more freely within the axon, transferring energy from one to another.

A signal propagated in the foregoing manner cannot travel much farther than 1 mm before it becomes too weak to open any voltage-gated $Na^+$ channels. But fortunately, there is another node of Ranvier every millimeter or less along the axon, where the axolemma is exposed to ECF and there is an abundance of voltage-gated channels. When the internodal signal reaches this point, it is just strong enough to open these channels, admit more $Na^+$, and create a new action potential. This action potential has the same strength as the one at the previous node, so each node of Ranvier boosts the signal back to its original strength ($+35$ mV). However, the generation of action potentials is a relatively time-consuming process that slows down the nerve signal at the nodes.

Since action potentials occur only at the nodes, this mode of conduction creates a false impression that the nerve signal jumps from node to node. Conduction in myelinated fibers is therefore called **saltatory**[27] **conduction** (fig. 12.17b).

You might think of saltatory conduction by analogy to a crowded subway car. The doors open (like the $Na^+$ gates at a node), 20 more people get on (like $Na^+$ flowing into the axon), and everyone has to push to the rear of the car to make room for them. No one passenger moves from the door to the rear, but the crowding and transfer of energy from person to person forces even those at the rear to move a little, like the sodium ions at the next node. Events at one node thus create excitation at the next node some distance away.

In summary, saltatory conduction is based on a process that is very fast in the internodes (transfer of energy from ion to ion) but decremental. In the nodes, conduction is slower but nondecremental. Since most of the axon is covered with myelin, conduction occurs mainly by the fast internodal process. This is why myelinated fibers conduct signals much faster (up to 120 m/s) than unmyelinated ones (up to 2 m/s).

**BEFORE YOU GO ON**

Answer the following questions to test your understanding of the preceding section:

11. What causes $K^+$ to diffuse out of a resting cell? What attracts it into the cell?

12. What happens to $Na^+$ when a neuron is stimulated on its dendrite? Why does the movement of $Na^+$ raise the voltage on the plasma membrane?

---

[27]from *saltare* = to leap, to dance

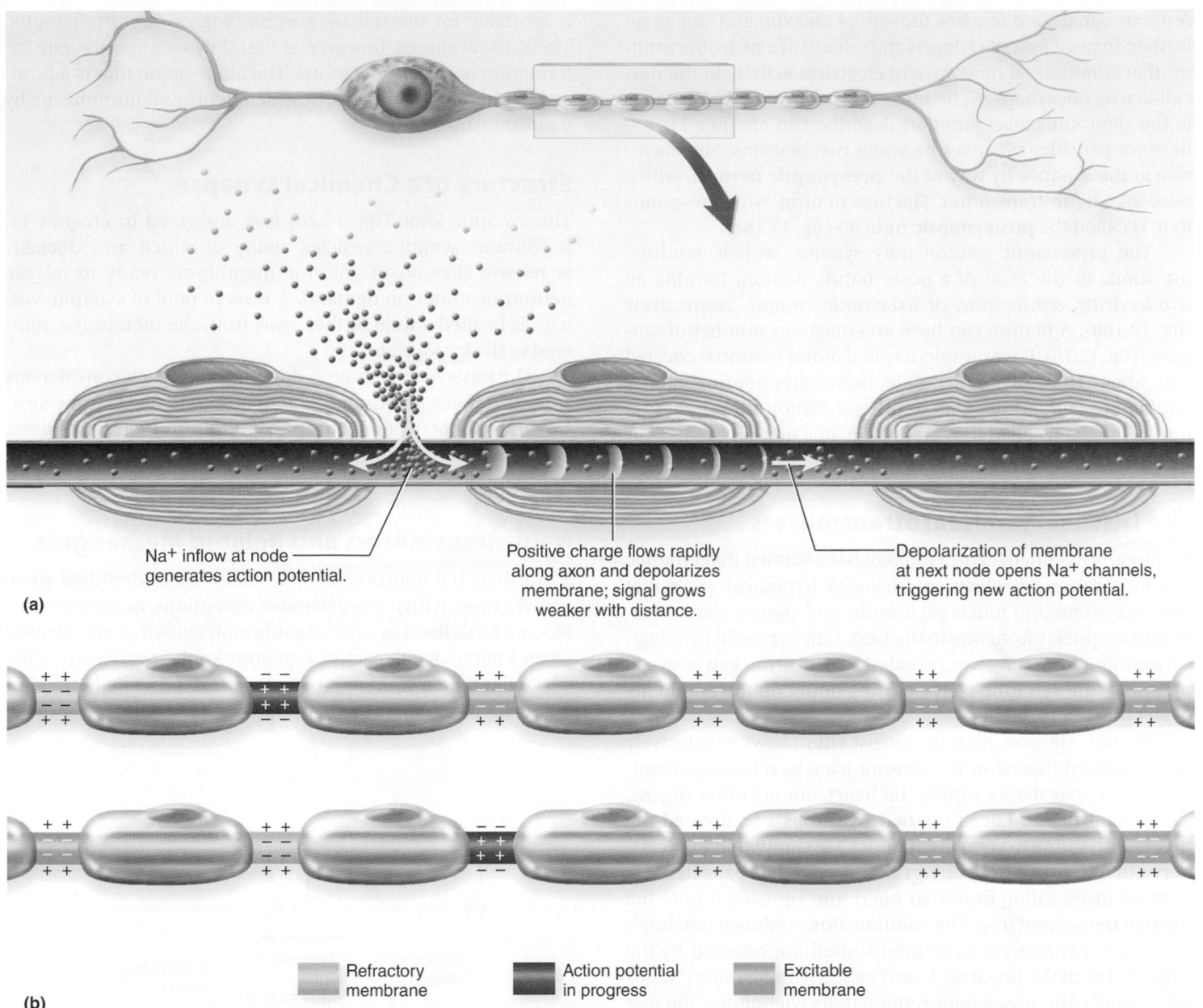

Na+ inflow at node generates action potential.

Positive charge flows rapidly along axon and depolarizes membrane; signal grows weaker with distance.

Depolarization of membrane at next node opens Na+ channels, triggering new action potential.

**(a)**

**(b)**

■ Refractory membrane   ■ Action potential in progress   ■ Excitable membrane

**FIGURE 12.17 Saltatory Conduction of a Nerve Signal in a Myelinated Fiber.** (a) Sodium inflow at a node of Ranvier initiates a force that is quickly transferred from ion to ion along the internode to the next node, getting weaker with distance. At the next node, Na+ ion (charge) density is just sufficient to open new voltage-gated channels and repeat the process. (b) Action potentials can occur only at nodes of Ranvier, so the nerve signal appears as if it were jumping from node to node.

13. What does it mean to say a local potential is graded, decremental, and reversible?

14. How does the plasma membrane at the trigger zone differ from that on the soma? How does it resemble the membrane at a node of Ranvier?

15. What makes an action potential rise to +35 mV? What makes it drop again after this peak?

16. List four ways in which an action potential is different from a local potential.

17. Explain why myelinated fibers conduct signals much faster than unmyelinated fibers.

## 12.5 Synapses

### Expected Learning Outcomes

When you have completed this section, you should be able to

a. explain how messages are transmitted from one neuron to another;

b. give examples of neurotransmitters and neuromodulators and describe their actions; and

c. explain how stimulation of a postsynaptic cell is stopped.

A nerve signal soon reaches the end of an axon and can go no farther. In most cases, it triggers the release of a neurotransmitter that stimulates a new wave of electrical activity in the next cell across the synapse. The most thoroughly studied synapse is the neuromuscular junction described in chapter 11, but here we consider synapses between two neurons. Signals arrive at the synapse by way of the **presynaptic neuron,** which releases a neurotransmitter. The next neuron, which responds to it, is called the **postsynaptic neuron** (fig. 12.18a).

The presynaptic neuron may synapse with a dendrite, the soma, or the axon of a postsynaptic neuron, forming an *axodendritic, axosomatic,* or *axoaxonic synapse,* respectively (fig. 12.18b). A neuron can have an enormous number of synapses (fig. 12.19). For example, a spinal motor neuron is covered with about 10,000 synaptic knobs from other neurons—8,000 ending on its dendrites and another 2,000 on the soma. In a part of the brain called the cerebellum, one neuron can have as many as 100,000 synapses.

## The Discovery of Neurotransmitters

In the early twentieth century, biologists assumed that synaptic communication was electrical—a logical hypothesis given that neurons seemed to touch each other and signals were transmitted so quickly from one to the next. Cajal's careful histological examinations, however, revealed a 20 to 40 nm gap between neurons—the **synaptic cleft**—casting doubt on the possibility of electrical transmission.

In 1921, German pharmacologist Otto Loewi conclusively demonstrated that neurons communicate by releasing chemicals. The *vagus nerves* supply the heart, among other organs, and slow it down. Loewi opened two frogs and flooded the hearts with saline to keep them moist. He stimulated the vagus nerve of one frog, and its heart rate dropped as expected. He then removed saline from that heart and squirted it onto the heart of the second frog. The solution alone reduced that frog's heart rate. Evidently it contained something released by the vagus nerve of the first frog. Loewi called it *Vagusstoffe* ("vagus substance") and it was later renamed acetylcholine—the first known neurotransmitter.

### ▶▶▶ APPLY WHAT YOU KNOW

*As described, does the previous experiment conclusively prove that the second frog's heart slowed as a result of something released by the vagus nerves? If you were Loewi, what control experiment would you do to rule out alternative explanations?*

Following Loewi's work, the idea of electrical communication between cells fell into disrepute. Now, however, we realize that some neurons, neuroglia, and cardiac and single-unit smooth muscle (see chapter 11) do indeed have **electrical synapses,** where adjacent cells are joined by gap junctions and ions diffuse directly from one cell into the next. These junctions have the advantage of quick transmission because there is no delay for the release and binding of neurotransmitter. Their disadvantage, however, is that they cannot integrate information and make decisions. The ability to do that is a property of **chemical synapses,** in which neurons communicate by neurotransmitters.

## Structure of a Chemical Synapse

The synaptic knob (fig. 12.20) was described in chapter 11. It contains synaptic vesicles, many of which are "docked" at release sites on the plasma membrane, ready to release neurotransmitter on demand. A reserve pool of synaptic vesicles is located a little farther away from the membrane, tethered to the cytoskeleton.

The postsynaptic neuron does not show such conspicuous specializations. At this end, the neuron has no synaptic vesicles and cannot release neurotransmitters. Its membrane does, however, have neurotransmitter receptors and ligand-gated ion channels.

## Neurotransmitters and Related Messengers

More than 100 neurotransmitters have been identified since Loewi's time. With a few debatable exceptions, neurotransmitters can be defined as small organic molecules that are released when a nerve signal reaches a synaptic knob or varicosity of the nerve fiber, then bind to a receptor on another cell and alter that cell's physiology. Some of the best-known ones are listed

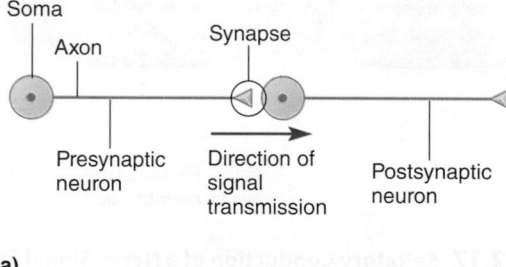

(a)

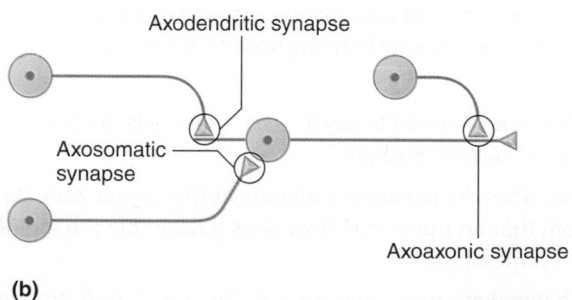

(b)

**FIGURE 12.18 Synaptic Relationships Between Neurons.**
(a) Pre- and postsynaptic neurons. (b) Types of synapses defined by the site of contact on the postsynaptic neuron.

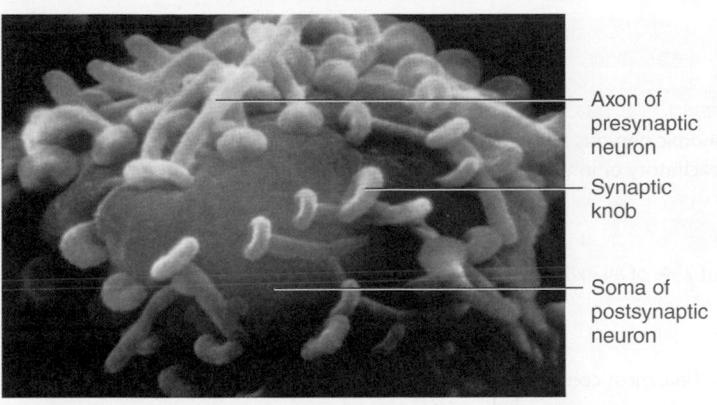

**FIGURE 12.19** **Synaptic Knobs on the Soma of a Neuron in a Marine Slug, *Aplysia* (SEM).**

Axon of presynaptic neuron

Synaptic knob

Soma of postsynaptic neuron

❓ *Are these synapses axodendritic, axosomatic, or axoaxonic?*

in table 12.3. Parts of the brain referred to in this table will become familiar as you study chapter 14, and you may wish to refer back to this table then to enhance your understanding of brain function. Most neurotransmitters fall into the following four categories (fig. 12.21).

1. **Acetylcholine** is in a class by itself. It is formed from acetic acid (acetate) and choline.

2. **Amino acid** neurotransmitters include glycine, glutamate, aspartate, and γ-aminobutyric acid (GABA).

3. **Monoamines (biogenic amines)** are synthesized from amino acids by removal of the –COOH group. They retain the $-NH_2$ (amino group), hence their name. Some monoamine neurotransmitters are epinephrine, norepinephrine, dopamine, histamine, ATP, and serotonin (5-hydroxytryptamine, or 5-HT). The first three of these are in a subclass called **catecholamines** (CAT-eh-COAL-uh-meens).

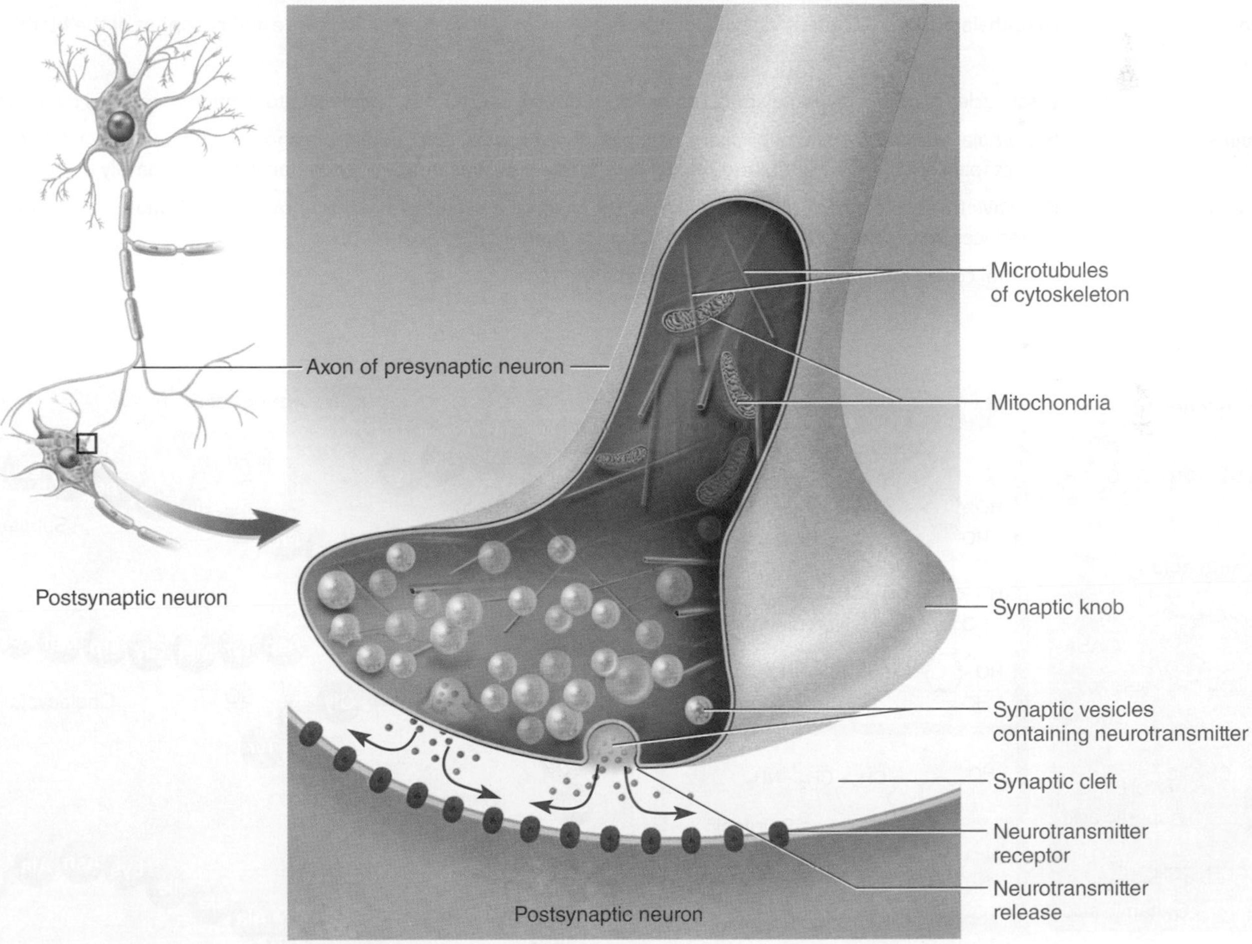

Axon of presynaptic neuron

Postsynaptic neuron

Microtubules of cytoskeleton

Mitochondria

Synaptic knob

Synaptic vesicles containing neurotransmitter

Synaptic cleft

Neurotransmitter receptor

Neurotransmitter release

Postsynaptic neuron

**FIGURE 12.20** **Structure of a Chemical Synapse.**

| TABLE 12.3 | Neurotransmitters (Selected Examples) |
|---|---|
| **Name** | **Locations and Actions** |
| *Acetylcholine (ACh)* | Neuromuscular junctions, most synapses of autonomic nervous system, retina, and many parts of the brain; excites skeletal muscle, inhibits cardiac muscle, and has excitatory or inhibitory effects on smooth muscle and glands depending on location |
| *Amino acids* | |
| Glutamate | Cerebral cortex and brainstem; accounts for about 75% of all excitatory synaptic transmission in the brain; involved in learning and memory |
| Aspartate | Spinal cord; effects similar to those of glutamate |
| Glycine | Inhibitory neurons of the brain, spinal cord, and retina; most common inhibitory neurotransmitter in the spinal cord |
| GABA | Thalamus, hypothalamus, cerebellum, occipital lobes of cerebrum, and retina; the most common inhibitory neurotransmitter in the brain |
| *Monoamines* | |
| Norepinephrine | Sympathetic nervous system, cerebral cortex, hypothalamus, brainstem, cerebellum, and spinal cord; involved in dreaming, waking, and mood; excites cardiac muscle; can excite or inhibit smooth muscle and glands depending on location |
| Epinephrine | Hypothalamus, thalamus, spinal cord, and adrenal medulla; effects similar to those of norepinephrine |
| Dopamine | Hypothalamus, limbic system, cerebral cortex, and retina; highly concentrated in substantia nigra of midbrain; involved in elevation of mood and control of skeletal muscles |
| Serotonin | Hypothalamus, limbic system, cerebellum, retina, and spinal cord; also secreted by blood platelets and intestinal cells; involved in sleepiness, alertness, thermoregulation, and mood |
| Histamine | Hypothalamus; also a potent vasodilator released by mast cells of connective tissue and basophils of the blood |
| *Neuropeptides* | |
| Substance P | Basal nuclei, midbrain, hypothalamus, cerebral cortex, small intestine, and pain-receptor neurons; mediates pain transmission |
| Enkephalins | Hypothalamus, limbic system, pituitary, pain pathways of spinal cord, and nerve endings of digestive tract; act as analgesics (pain relievers) by inhibiting substance P; inhibit intestinal motility; secretion increases sharply in women in labor |
| β-endorphin | Digestive tract, spinal cord, and many parts of the brain; also secreted as a hormone by the pituitary; suppresses pain; reduces perception of fatigue and may produce "runner's high" in athletes |
| Cholecystokinin | Cerebral cortex and small intestine; suppresses appetite |

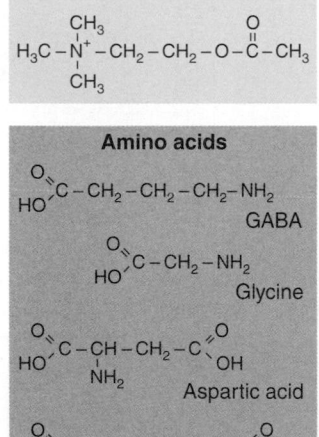

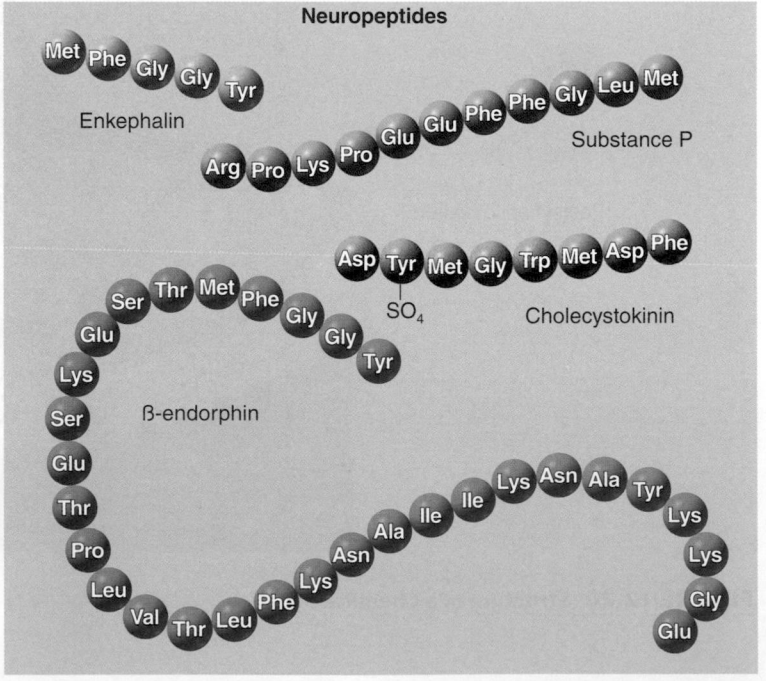

**FIGURE 12.21 Classification of Some Neurotransmitters.** The neuropeptides are chains of amino acids, each identified by its three-letter code. Table 2.8 (p. 66) explains the codes.

4. **Neuropeptides** are chains of 2 to 40 amino acids. Some examples are cholecystokinin (CCK) and substance P. Neuropeptides are stored in *secretory granules (dense-core vesicles)* that are about 100 nm in diameter, twice as large as typical synaptic vesicles. Some neuropeptides also function as hormones or as **neuromodulators,** whose action is discussed later in this chapter. Some are produced not only by neurons but also by the digestive tract; thus they are known as *gut–brain peptides.* Some of these cause cravings for specific nutrients such as fat, protein, or carbohydrates (see p. 998) and may be associated with certain eating disorders.

▶▶▶**APPLY WHAT YOU KNOW**

*Neuropeptides can be synthesized only in the soma and must be transported to the synaptic knobs. Why is their synthesis limited to the soma?*

We will see, especially in chapter 15, that a given neurotransmitter does not have the same effect everywhere in the body. There are multiple receptor types in the body for a particular neurotransmitter—over 14 receptor types for serotonin, for example—and it is the receptor that governs what effect a neurotransmitter has on its target cell. Most human and other mammalian neurons can secrete two or more neurotransmitters and can switch from one to another under different circumstances.

## Synaptic Transmission

Some neurotransmitters are excitatory, some are inhibitory, and for some the effect depends on what kind of receptor the postsynaptic cell has. Some open ligand-gated ion channels and others act through second messengers. Bearing this diversity in mind, we will examine three kinds of synapses with different modes of action.

## An Excitatory Cholinergic Synapse

A **cholinergic**[28] (CO-lin-UR-jic) synapse employs acetylcholine (ACh) as its neurotransmitter. ACh excites some postsynaptic cells (such as skeletal muscle) and inhibits others, but this discussion will describe an excitatory action. The steps in transmission at such a synapse are as follows (fig. 12.22):

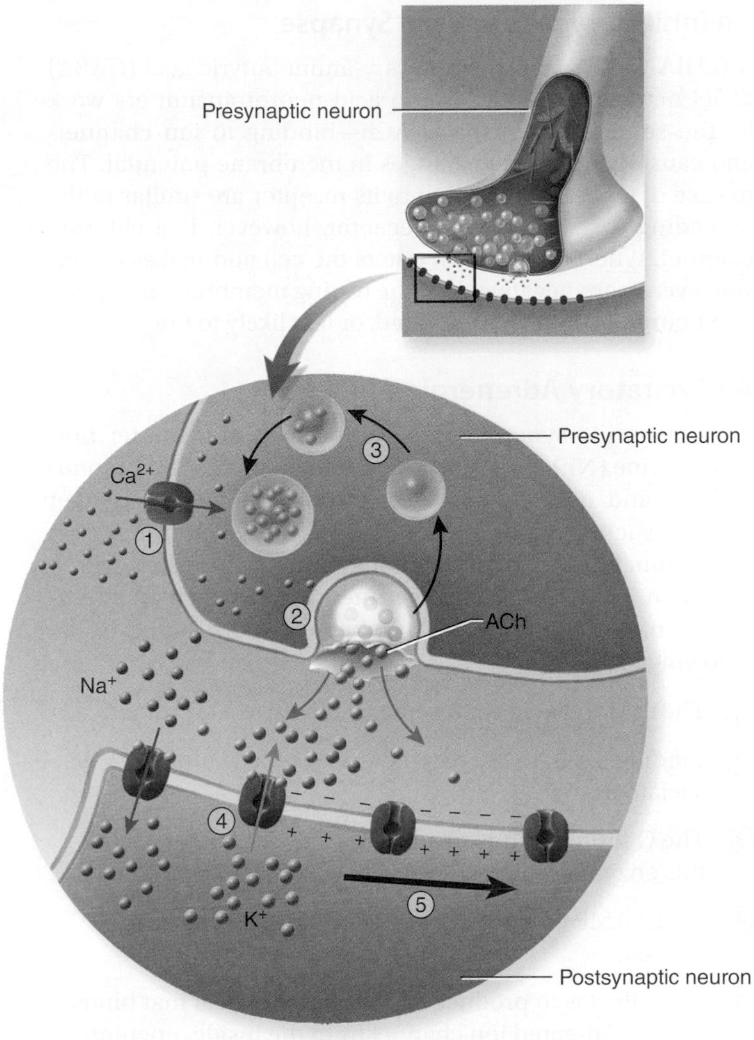

(1) The arrival of a nerve signal at the synaptic knob opens voltage-gated calcium channels.

(2) $Ca^{2+}$ enters the knob and triggers exocytosis of the synaptic vesicles, releasing ACh.

(3) Empty vesicles drop back into the cytoplasm to be refilled with ACh, while synaptic vesicles in the reserve pool move to the active sites and release their ACh—a bit like

[28]*cholin* = acetylcholine; *erg* = work, action

**FIGURE 12.22 Transmission at a Cholinergic Synapse.** Acetylcholine directly opens ion channels in the plasma membrane of the postsynaptic neuron. Numbered steps correspond to the description in the text. AP|R

a line of Revolutionary War soldiers firing their muskets and falling back to reload as another line moves to the fore.

(4) Meanwhile, ACh diffuses across the synaptic cleft and binds to ligand-gated channels on the postsynaptic neuron. These channels open, allowing $Na^+$ to enter the cell and $K^+$ to leave. Although illustrated separately, $Na^+$ and $K^+$ pass in opposite directions through the same gates.

(5) As $Na^+$ enters, it spreads out along the inside of the plasma membrane and depolarizes it, producing a local voltage shift called the **postsynaptic potential.** Like other local potentials, if this is strong and persistent enough (that is, if enough current makes it to the axon hillock), it opens voltage-gated ion channels in the trigger zone and causes the postsynaptic neuron to fire.

## An Inhibitory GABA-ergic Synapse

A **GABA-ergic** synapse employs γ-aminobutyric acid (GABA) as its neurotransmitter. Amino acid neurotransmitters work by the same mechanism as ACh—binding to ion channels and causing immediate changes in membrane potential. The release of GABA and binding to its receptor are similar to the preceding case. The GABA receptor, however, is a chloride channel. When it opens, Cl⁻ enters the cell and makes the inside even more negative than the resting membrane potential. The neuron is therefore inhibited, or less likely to fire.

## An Excitatory Adrenergic Synapse

An **adrenergic synapse** employs the neurotransmitter norepinephrine (NE), also called noradrenaline. NE, other monoamines, and neuropeptides act through second-messenger systems such as cyclic AMP (cAMP). The receptor is not an ion channel but a transmembrane protein associated with a G protein. Figure 12.23 shows some ways in which an adrenergic synapse can function, numbered to correspond to the following:

1. The unstimulated NE receptor is bound to a G protein.

2. Binding of NE to the receptor causes the G protein to dissociate from it.

3. The G protein binds to adenylate cyclase and activates this enzyme, which converts ATP to cAMP.

4. Cyclic AMP can induce several alternative effects in the cell.

5. One effect is to produce an internal chemical that binds to a ligand-gated ion channel from the inside, opening the channel and depolarizing the cell.

6. Another is to activate preexisting cytoplasmic enzymes, which can lead to diverse metabolic changes (for example, inducing a liver cell to break down glycogen and release glucose into the blood).

7. Yet another is for cAMP to induce genetic transcription, so that the cell produces new enzymes leading to diverse metabolic effects.

Although slower to respond than cholinergic and GABA-ergic synapses, adrenergic synapses do have an advantage— **enzyme amplification.** A single NE molecule binding to a receptor can induce the formation of many cAMPs, each of those can activate many enzyme molecules or induce the transcription of a gene to generate numerous mRNA molecules, and each of those can result in the production of a vast number of metabolic products such as glucose molecules.

As complex as synaptic events may seem, they typically require only 0.5 ms or so—an interval called **synaptic delay.** This is the time from the arrival of a signal at the axon terminal of a presynaptic cell to the beginning of an action potential in the postsynaptic cell.

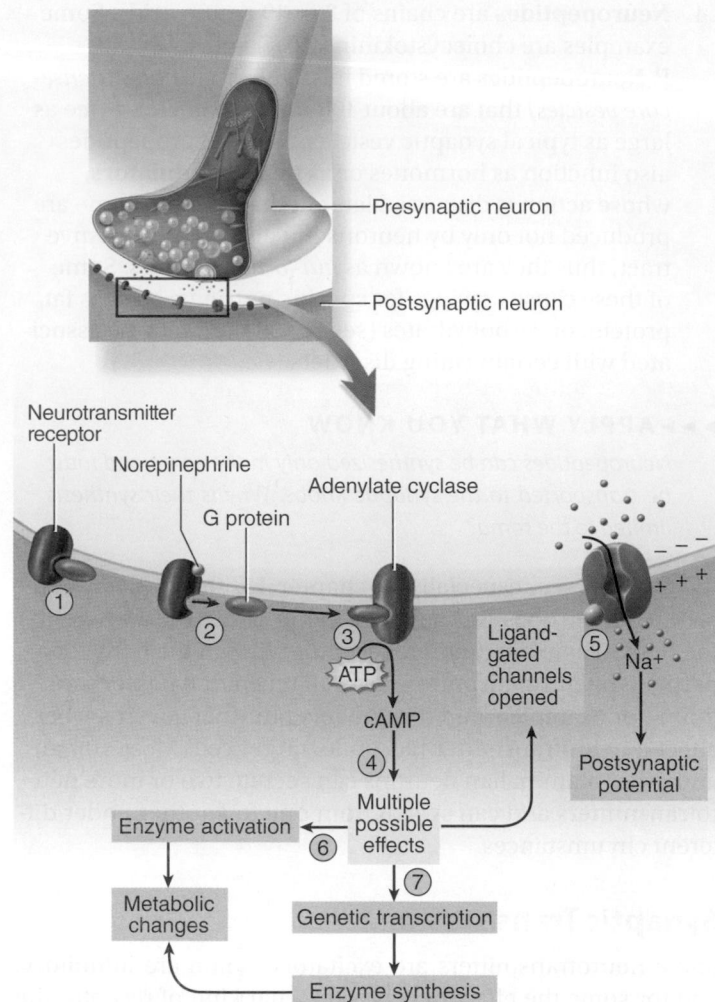

**FIGURE 12.23 Transmission at an Adrenergic Synapse.** The norepinephrine receptor is not an ion channel. It activates a second-messenger system with a variety of possible effects in the postsynaptic cell. Numbered steps correspond to the description in the text.

## Cessation of the Signal

It is important not only to stimulate a postsynaptic cell but also to turn off the stimulus in due time. Otherwise the postsynaptic neuron could continue firing indefinitely, causing a breakdown in physiological coordination. Here we examine some ways this is done.

A neurotransmitter molecule binds to its receptor for only 1 ms or so, then dissociates from it. If the presynaptic cell continues to release neurotransmitter, one molecule is quickly replaced by another and the postsynaptic cell is restimulated. This immediately suggests a way of stopping synaptic transmission—stop adding new neurotransmitter and get rid of that which is already there. The first step is achieved simply by the cessation of signals in the presynaptic nerve fiber. The second can be achieved in the following ways:

- **Diffusion.** Neurotransmitter escapes from the synapse into the nearby ECF. In the CNS, astrocytes absorb it and return it to the neurons.

- **Reuptake.** The synaptic knob reabsorbs amino acids and monoamines by endocytosis and breaks them down with an enzyme called **monoamine oxidase (MAO).** Some antidepressant drugs work by inhibiting MAO (see Deeper Insight 15.2, p. 574).

- **Degradation in the synaptic cleft.** The enzyme acetylcholinesterase (AChE), located in the synaptic cleft and on the postsynaptic membrane, breaks ACh down into acetate and choline. These breakdown products have no stimulatory effect on the postsynaptic cell. The synaptic knob reabsorbs the choline and uses it to synthesize more ACh.

## Neuromodulators

Physiologists have lately discovered that neurons sometimes secrete chemical signals that have long-term effects on entire groups of neurons instead of brief, quick effects at an individual synapse. Some call these **neuromodulators** to distinguish them from neurotransmitters; others use the term *neurotransmitter* broadly to include these. Neuromodulators adjust, or *modulate,* the activity of neuron groups in various ways—increasing the release of neurotransmitters by presynaptic neurons; adjusting the sensivity of postsynaptic neurons to neurotransmitters; or altering the rate of neurotransmitter reuptake or breakdown to prolong their effects.

The simplest neuromodulator is the gas **nitric oxide (NO).** NO diffuses readily into a postsynaptic cell and activates second-messenger pathways with such effects as relaxing smooth muscle. In small arteries, this has the effect of increasing blood flow to a tissue. One family of neuromodulators is called **neuropeptides,** as they are composed of chains of amino acids. Among these are the **enkephalins** and **endorphins,** which inhibit spinal neurons from transmitting pain signals to the brain (see p. 585). Some other neuromodulators include hormones and some neurotransmitters such as dopamine, serotonin, and histamine. The last point may seem confusing, but the terms *neurotransmitter, hormone,* and *neuromodulator* define not so much the chemical itself, but the role it plays in a given instance. One chemical can play two or more of these roles under different circumstances.

**BEFORE YOU GO ON**

Answer the following questions to test your understanding of the preceding section:

18. Concisely describe five steps that occur between the arrival of an action potential at the synaptic knob and the beginning of a new action potential in the postsynaptic neuron.

19. Contrast the actions of acetylcholine, GABA, and norepinephrine at their respective synapses.

20. Describe three mechanisms that stop synaptic transmission.

21. What is the function of neuromodulators? Compare and contrast neuromodulators and neurotransmitters.

## 12.6 Neural Integration

**Expected Learning Outcomes**

When you have completed this section, you should be able to

a. explain how a neuron "decides" whether or not to generate action potentials;

b. explain how the nervous system translates complex information into a simple code;

c. explain how neurons work together in groups to process information and produce effective output; and

d. describe how memory works at cellular and molecular levels.

Synaptic delay slows the transmission of nerve signals; the more synapses there are in a neural pathway, the longer it takes information to get from its origin to its destination. You might wonder, therefore, why we have synapses—why a nervous pathway is not, indeed, a continuous "wire" as biologists believed before the work of Cajal. The presence of synapses is not due to limitations on axon length—after all, one nerve fiber can reach from your toes to your brainstem; imagine how long some nerve fibers must be in a giraffe or whale! We also have seen that cells communicate much more quickly through gap junctions than through chemical synapses. So why have chemical synapses at all?

What we value most about our nervous system is its ability to process information, store it, and make decisions—and chemical synapses are the decision-making devices of the system. The more synapses a neuron has, the greater its information-processing capability. At this moment, you are using certain *pyramidal cells* of the cerebral cortex (see fig. 12.5) to read and comprehend this passage. Each pyramidal cell has about 40,000 synaptic contacts with other neurons. The cerebral cortex alone (the main information-processing tissue of your brain) is estimated to have 100 trillion ($10^{14}$) synapses. To get some impression of this number, imagine trying to count them. Even if you could count two synapses per second, day and night without stopping, and you were immortal, it would take you 1.6 million years. The ability of your neurons to process information, store and recall it, and make decisions is called **neural integration.**

## Postsynaptic Potentials

Neural integration is based on the postsynaptic potentials produced by neurotransmitters. Remember that a typical neuron has a resting membrane potential (RMP) of about −70 mV and a threshold of about −55 mV. A neuron has to be depolarized to this threshold in order to produce action potentials. Any voltage change in that direction makes a neuron more likely to fire and is therefore called an **excitatory postsynaptic potential (EPSP)** (fig. 12.24a). EPSPs usually result from Na⁺ flowing into the cell and neutralizing some of the negative charge on the inside of the membrane.

In other cases, a neurotransmitter hyperpolarizes the postsynaptic cell and makes it more negative than the RMP. Since this makes the postsynaptic cell less likely to fire, it is called an **inhibitory postsynaptic potential (IPSP)** (fig. 12.24b). Some IPSPs are produced by a neurotransmitter opening ligand-gated chloride channels, causing $Cl^-$ to flow into the cell and make the cytosol more negative. A less common way is to open selective $K^+$ channels, increasing $K^+$ diffusion out of the cell.

We must recognize that because of ion leakage through their membranes, all neurons fire at a certain background rate even when they are not being stimulated. EPSPs and IPSPs don't determine whether or not a neuron fires, but only change the rate of firing by stimulating or inhibiting the production of more action potentials.

Glutamate and aspartate are excitatory brain neurotransmitters that produce EPSPs. Glycine and GABA produce IPSPs and are therefore inhibitory. Acetylcholine (ACh) and norepinephrine are excitatory to some cells and inhibitory to others, depending on the type of receptors present. For example, ACh excites skeletal muscle but inhibits cardiac muscle because of different types of ACh receptors.

## Summation, Facilitation, and Inhibition

One neuron may receive input from thousands of other neurons. Some incoming nerve fibers may produce EPSPs while others produce IPSPs. The neuron's response depends on whether the *net* input is excitatory or inhibitory. If EPSPs override the IPSPs, threshold may be reached and set off an action potential; if IPSPs prevail, they inhibit the neuron from firing. **Summation** is the process of adding up postsynaptic potentials and responding to their net effect. It occurs in the trigger zone.

Suppose, for example, you are working in the kitchen and accidentally touch a hot pot. EPSPs in your motor neurons might cause you to jerk your hand back quickly and avoid being burned. Yet a moment later, you might nonchalantly pick up a cup of hot tea. Since you are expecting it to be hot, you don't jerk your hand away. You have learned that it won't injure you, so at some level of the nervous system, IPSPs prevail and inhibit the motor response.

It is fundamentally a balance between EPSPs and IPSPs that enables the nervous system to make decisions. A postsynaptic neuron is like a little cellular democracy acting on the "majority vote" of hundreds or thousands of presynaptic cells. In the teacup example, some presynaptic neurons send messages that signify "Hot! Danger!" in the form of EPSPs that may activate a hand-withdrawal reflex, while at the same time, others produce IPSPs that signify "Safe" and suppress the reflex. Whether you jerk your hand away depends on whether the EPSPs override the IPSPs or vice versa.

A single action potential in a synaptic knob does not produce enough activity to make a postsynaptic cell fire. An EPSP may be produced, but it fades before reaching threshold. A typical EPSP is a voltage change of only 0.5 mV and lasts only 15 to 20 ms. If a neuron has an RMP of −70 mV and a threshold of −55 mV, it needs at least 30 EPSPs to reach threshold and fire. There are two ways in which EPSPs can add up to do this, and both may occur simultaneously.

1. **Temporal summation** (fig. 12.25a). This occurs when a single synapse generates EPSPs so quickly that each is generated before the previous one fades. This allows the EPSPs to add up over time to a threshold voltage that triggers an action potential (fig. 12.26). Temporal summation can occur if even one presynaptic neuron stimulates the postsynaptic neuron at a fast enough rate.

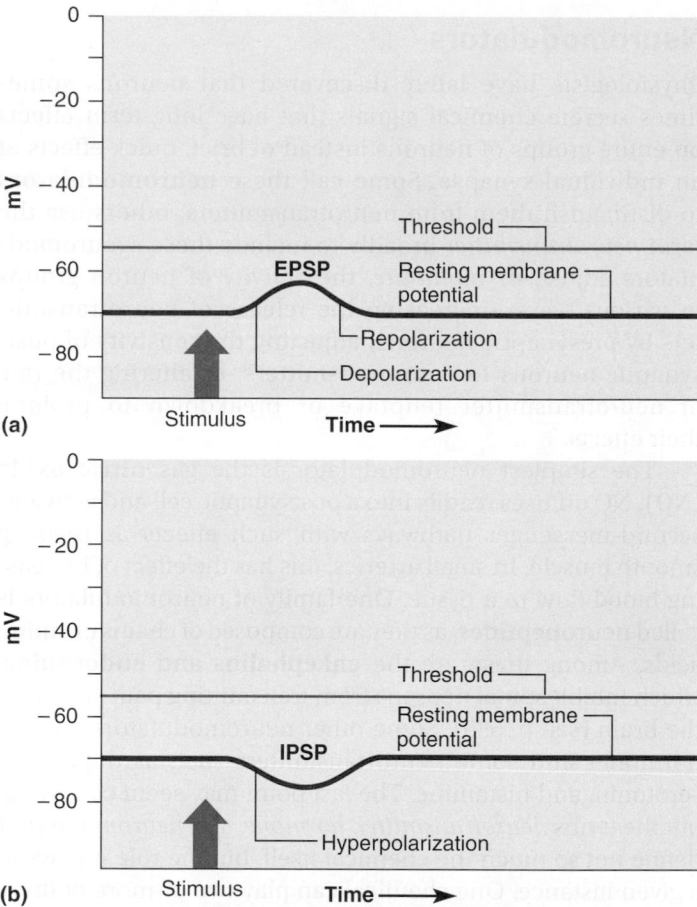

**FIGURE 12.24 Postsynaptic Potentials.** (a) An excitatory postsynaptic potential (EPSP), which shifts the membrane voltage closer to threshold and makes the cell more likely to fire. (b) An inhibitory postsynaptic potential (IPSP), which shifts the membrane voltage farther away from threshold and makes the cell less likely to fire. The sizes of these postsynaptic potentials are greatly exaggerated here for clarity; compare figure 12.26.

❓ *Why is a single EPSP insufficient to make a neuron fire?*

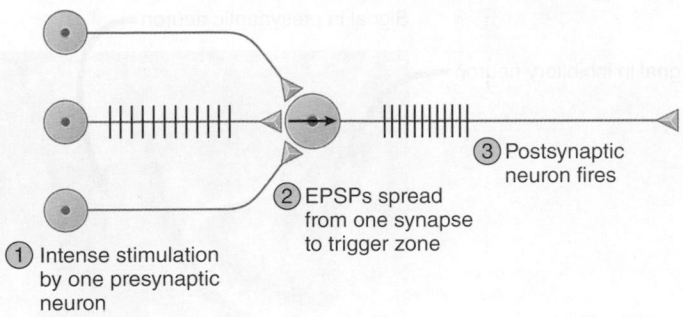

(1) Intense stimulation by one presynaptic neuron

(2) EPSPs spread from one synapse to trigger zone

(3) Postsynaptic neuron fires

**(a) Temporal summation**

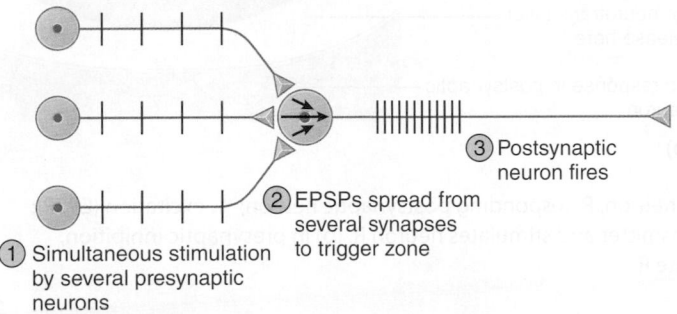

(1) Simultaneous stimulation by several presynaptic neurons

(2) EPSPs spread from several synapses to trigger zone

(3) Postsynaptic neuron fires

**(b) Spatial summation**

**FIGURE 12.25 Temporal and Spatial Summation.** Vertical lines on the nerve fibers indicate relative firing frequency. Arrows within the postsynaptic neuron indicate Na⁺ diffusion. (a) In temporal summation, a single presynaptic neuron stimulates the postsynaptic neuron so rapidly that its EPSPs add up to threshold and make it fire. (b) In spatial summation, multiple inputs to the postsynaptic cell each produce a moderate amount of stimulation, but collectively they produce enough EPSPs to add up to threshold at the trigger zone and make the cell fire.

2. **Spatial summation** (fig. 12.25b). This occurs when EPSPs from several synapses add up to threshold at the axon hillock. Any one synapse may admit only a moderate amount of Na⁺ into the cell, but several synapses acting together admit enough Na⁺ to reach threshold. The presynaptic neurons collaborate to induce the postsynaptic neuron to fire.

Neurons routinely work in groups to modify each other's actions. **Facilitation** is a process in which one neuron enhances the effect of another. In spatial summation, for example, one neuron acting alone may be unable to induce a postsynaptic neuron to fire. But when they collaborate, their combined "effort" does induce firing in the postsynaptic cell.

**Presynaptic inhibition** is the opposite of facilitation, a mechanism in which one presynaptic neuron suppresses another one. This mechanism is used to reduce or halt unwanted synaptic transmission. In figure 12.27, we see three neurons, which we will call neuron S for the stimulator, neuron I for the

inhibitor, and neuron R for the responder. Neuron I forms an axoaxonic synapse with S (synapses with the axon of S). When presynaptic inhibition is not occurring, neuron S releases its neurotransmitter and triggers a response in R. But when there is a need to block transmission across this pathway, neuron I releases the inhibitory neurotransmitter GABA. GABA prevents the voltage-gated calcium channels of neuron S from opening. Consequently, neuron S releases less neurotransmitter or none, and fails to stimulate neuron R.

## Neural Coding

The nervous system must interpret and pass along both quantitative and qualitative information about its environment—whether a light is dim or bright, red or green; whether a taste is mild or intense, salty or sour; whether a sound is loud or soft, high-pitched or low. Considering the complexity of information to be communicated about conditions in and around the body, it seems a marvel that it can be done in the form of something as simple as action potentials—particularly since all the action potentials of a given neuron are identical. Yet when we considered the genetic code in chapter 4, we saw that complex messages can indeed be expressed in simple codes. The way in which the nervous system converts information to a meaningful pattern of action potentials is called **neural coding** (or *sensory coding* when it occurs in the sense organs).

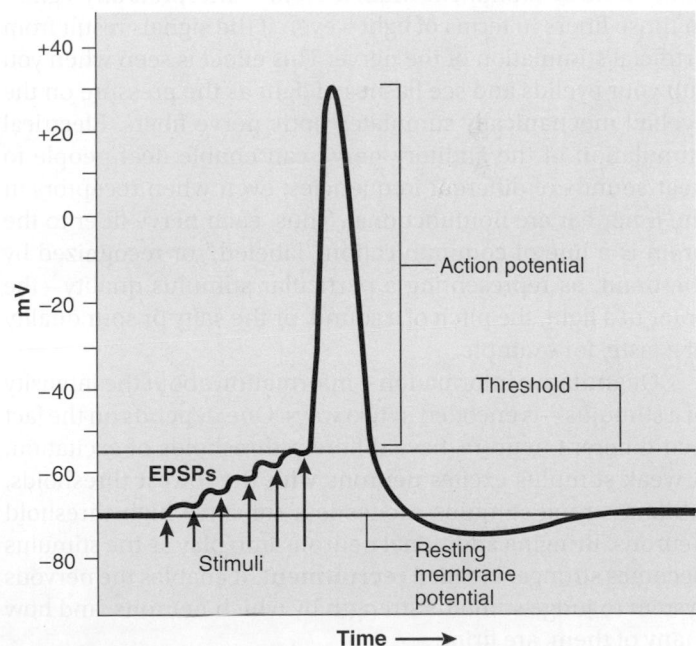

**FIGURE 12.26 Summation of EPSPs.** Each stimulus (arrows) produces one EPSP. If enough EPSPs arrive at the trigger zone faster than they fade, they can build on each other to bring the neuron to threshold and trigger an action potential.

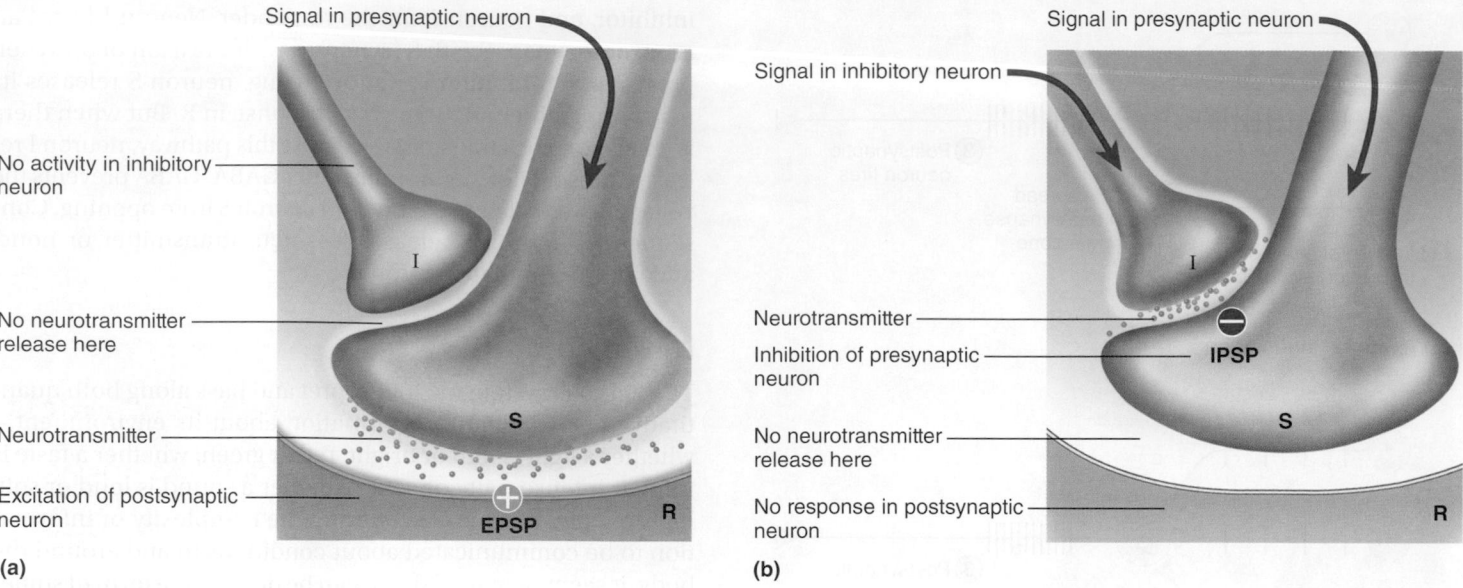

**FIGURE 12.27 Presynaptic Inhibition.**  S, stimulating neuron; I, inhibitory neuron; R, responding postsynaptic neuron; +, excitation (EPSP); –, inhibition (IPSP). (a) In the absence of inhibition, neuron S releases neurotransmitter and stimulates neuron R. (b) In presynaptic inhibition, neuron I suppresses the release of neurotransmitter by S, and S cannot stimulate R.

The most important mechanism for transmitting qualitative information is the **labeled line code.** This code is based on the fact that each nerve fiber to the brain leads from a receptor that specifically recognizes a particular stimulus type. Nerve fibers in the optic nerve, for example, carry signals only from light receptors in the eye; these fibers never carry information about taste or sound. The brain therefore interprets any signals in those fibers in terms of light—even if the signals result from artificial stimulation of the nerve. This effect is seen when you rub your eyelids and see flashes of light as the pressure on the eyeball mechanically stimulates optic nerve fibers. Electrical stimulation of the auditory nerve can enable deaf people to hear sounds of different frequencies, even when receptors in the inner ear are nonfunctional. Thus, each nerve fiber to the brain is a line of communication "labeled," or recognized by the brain, as representing a particular stimulus quality—the color of a light, the pitch of a sound, or the salty or sour quality of a taste, for example.

Quantitative information—information about the intensity of a stimulus—is encoded in two ways. One depends on the fact that different neurons have different thresholds of excitation. A weak stimulus excites neurons with the lowest thresholds, while a strong stimulus excites less sensitive high-threshold neurons. Bringing additional neurons into play as the stimulus becomes stronger is called **recruitment.** It enables the nervous system to judge stimulus strength by which neurons, and how many of them, are firing.

Another way of encoding stimulus strength depends on the fact that the more strongly a neuron is stimulated, the more frequently it fires. A weak stimulus may cause a neuron to generate 6 action potentials per second, and a strong stimulus,

600 per second. Thus, the central nervous system can judge stimulus strength from the firing frequency of afferent neurons (fig. 12.28).

There is a limit to how often a neuron can fire, set by its absolute refractory period. Think of an electronic camera flash by analogy. If you take a photograph and your flash unit takes 10 seconds to recharge, then you can't take more than six photos per minute. Similarly, if a nerve fiber takes 1 ms to repolarize after it has fired, then it can't fire more than 1,000 times per second. Refractory periods may be as short as 0.5 ms, which sets a theoretical limit to firing frequency of 2,000 action potentials per second. The highest frequencies actually observed, however, are 500 to 1,000 per second.

In summary, mild stimuli excite sensitive, low-threshold nerve fibers. As the stimulus intensity rises, these fibers fire at a higher and higher frequency, up to a certain maximum. If the stimulus intensity exceeds the capacity of these low-threshold fibers, it may recruit less sensitive, high-threshold fibers to begin firing. Still further increases in intensity cause these high-threshold fibers to fire at a higher and higher frequency.

#### ▶▶▶APPLY WHAT YOU KNOW

*How is neural recruitment related to the process of multiple motor unit summation described in chapter 11?*

## Neural Pools and Circuits

So far, we have dealt with interactions involving only two or three neurons at a time. Actually, neurons function in larger ensembles called **neural pools,** each of which consists of thousands to millions of interneurons concerned with a particular

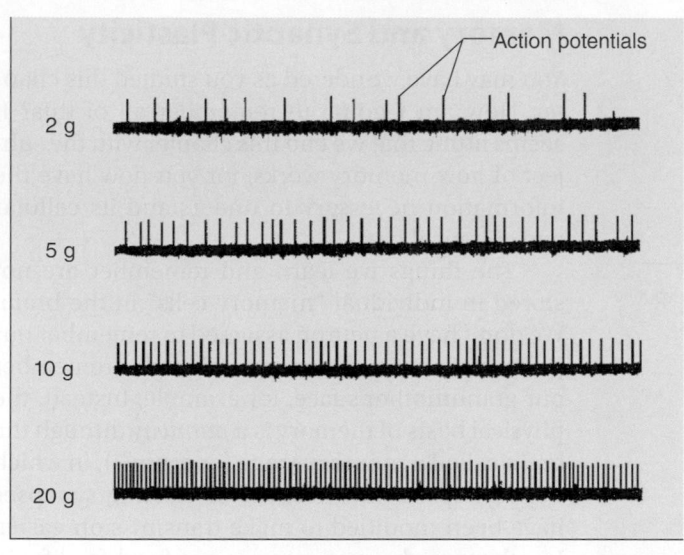

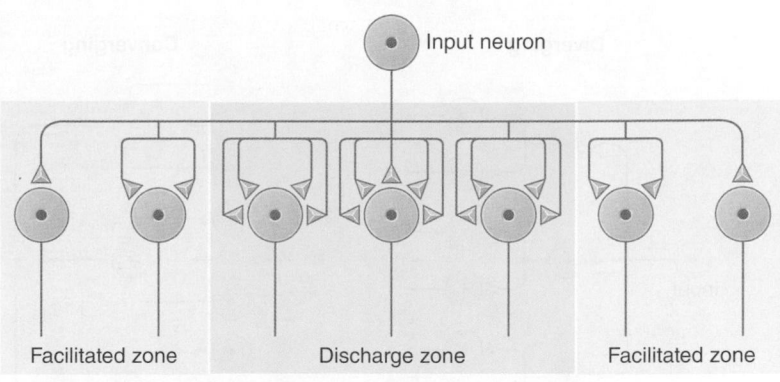

**FIGURE 12.29 Facilitated and Discharge Zones in a Neural Pool.** In the discharge zone, the presynaptic input neuron has so many synaptic contacts with each postsynaptic neuron that it alone can induce the postsynaptic cell to fire, employing spatial summation as in figure 12.25b. In a facilitated zone, the presynaptic neuron lacks enough synaptic contacts with a postsynaptic neuron to induce firing by itself. However, it can collaborate with other presynaptic neurons, facilitating each other in making the postsynaptic cell fire.

**FIGURE 12.28 An Example of Neural Coding.** This figure is based on recordings made from a sensory fiber of the frog sciatic nerve as the gastrocnemius muscle was stretched by suspending weights from it. As the stimulus strength (weight) and stretch increase, the firing frequency of the neuron increases. Firing frequency is a coded message that informs the CNS of stimulus intensity.

❓ *In what other way is the CNS informed of stimulus intensity?*

body function—one to control the rhythm of your breathing, one to move your limbs rhythmically as you walk, one to regulate your sense of hunger, and another to interpret smells, for example. At this point, we explore a few ways in which neural pools collectively process information.

Information arrives at a neural pool through one or more input neurons, which branch repeatedly and synapse with numerous interneurons in the pool. Some input neurons form multiple synapses with a single postsynaptic cell. They can produce EPSPs at all points of contact with that cell and, through spatial summation, make it fire more easily than if they synapsed with it at only one point. Within the **discharge zone** of an input neuron, that neuron acting alone can make the postsynaptic cells fire (fig. 12.29). But in a broader **facilitated zone,** it synapses with still other neurons in the pool, with fewer synapses on each of them. It can stimulate those neurons to fire only with the assistance of other input neurons; that is, it facilitates the others. It "has a vote" on what the postsynaptic cells in the facilitated zone will do, but it cannot determine the outcome alone. Such arrangements, repeated thousands of times throughout the central nervous system, give neural pools great flexibility in integrating input from several sources and "deciding" on an appropriate output.

The functioning of a radio can be understood from a circuit diagram showing its components and their connections. Similarly, the functions of a neural pool are partly determined by its **neural circuit**—the pathways among its neurons. Just as a wide variety of electronic devices are constructed from a

relatively limited number of circuit types, a wide variety of neural functions result from the operation of four principal kinds of neural circuits (fig. 12.30):

1. In a **diverging circuit,** one nerve fiber branches and synapses with several postsynaptic cells. Each of those may synapse with several more, so input from just one neuron may produce output through hundreds of neurons. Such a circuit allows one motor neuron of the brain, for example, to ultimately stimulate thousands of muscle fibers.

2. A **converging circuit** is the opposite of a diverging circuit—input from many nerve fibers is funneled to one neuron or neural pool. Such an arrangement allows input from your eyes, inner ears, and stretch receptors in your neck to be directed to an area of the brain concerned with the sense of balance. Also through neural convergence, a respiratory center in your brainstem receives input from other parts of your brain, from receptors for blood chemistry in your arteries, and from stretch receptors in your lungs. The respiratory center can then produce an output that takes all of these factors into account and sets an appropriate pattern of breathing.

3. In a **reverberating circuit,** neurons stimulate each other in a linear sequence from input to output neurons, but some of the neurons late in the path send axon collaterals back to neurons earlier in the path and restimulate them. As an exceedingly simplified model of such a circuit, consider a path such as A $\longrightarrow$ B $\longrightarrow$ C $\longrightarrow$ D, in which neuron C sends an axon collateral back to A. As a result, every time C fires it not only stimulates output neuron D, but also restimulates A and starts the process over. Such a circuit produces a prolonged or repetitive effect that lasts until one or

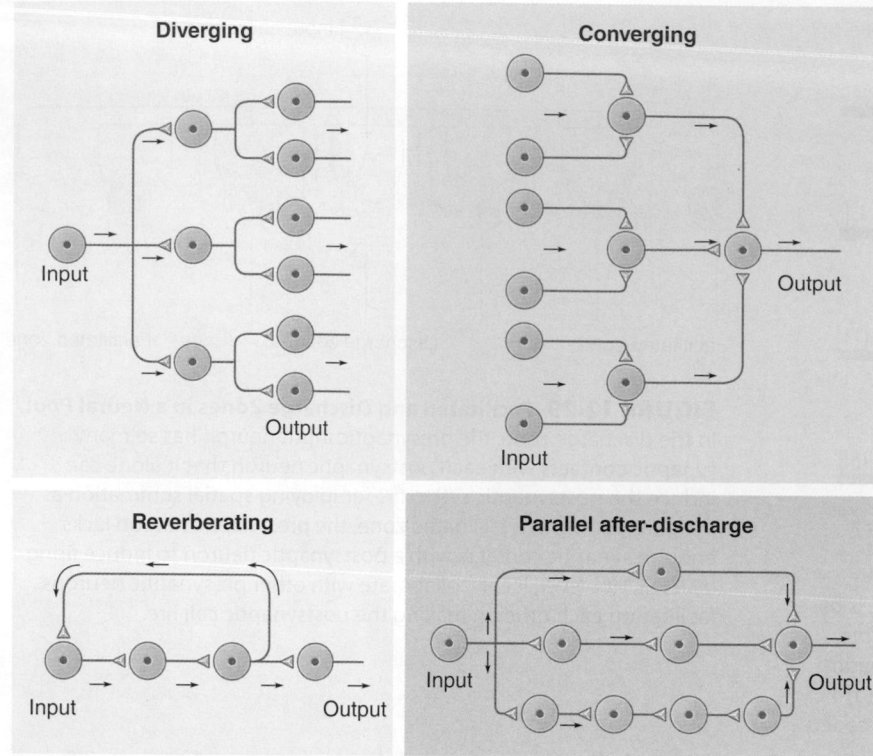

**FIGURE 12.30** **Four Types of Neural Circuits.** Arrows indicate the direction of the nerve signal.

**?** *Which of these four circuits is likely to fire the longest after a stimulus ceases? Why?*

more neurons in the circuit fail to fire, or an inhibitory signal from another source stops one of them from firing. A reverberating circuit sends repetitious signals to your diaphragm and intercostal muscles, for example, to make you inhale. When the circuit stops firing, you exhale; the next time it fires, you inhale again. Reverberating circuits may also be involved in short-term memory, as discussed in the next section, and they may play a role in the uncontrolled "storms" of neural activity that occur in epilepsy.

4. In a **parallel after-discharge circuit,** an input neuron diverges to stimulate several chains of neurons. Each chain has a different number of synapses, but eventually they all reconverge on one or few output neurons. Since the chains differ in total synaptic delay, their signals arrive at the output neurons at different times, and the output neurons may go on firing for some time after input has ceased. Unlike a reverberating circuit, this type has no feedback loop. Once all the neurons in the circuit have fired, the output ceases. Continued firing after the stimulus stops is called *after-discharge.* It explains why you can stare at a lamp, then close your eyes and continue to see an image of it for a while. Such a circuit is also important in withdrawal reflexes, in which a brief pain produces a longer-lasting output to the limb muscles and causes you to draw back your hand or foot from danger.

## Memory and Synaptic Plasticity

You may have wondered as you studied this chapter, How am I going to remember all of this? It seems fitting that we end this chapter with the subject of how memory works, for you now have the information necessary to understand its cellular and chemical basis.

The things we learn and remember are not stored in individual "memory cells" in the brain. We don't have a neuron assigned to remember our phone number and another assigned to remember our grandmother's face, for example. Instead, the physical basis of memory is a *pathway* through the brain called a **memory trace (engram[29]),** in which new synapses have formed or existing synapses have been modified to make transmission easier. In other words, synapses are not fixed for life; in response to experience, they can be added, taken away, or modified to make transmission easier or harder. Indeed, synapses can be created or deleted in as little as 1 or 2 hours. The ability of synapses to change is called **synaptic plasticity.**

Think about when you learned as a child to tie your shoes. The procedure was very slow, confusing, and laborious at first, but eventually it became so easy you could do it with little thought—like a motor program playing out in your brain without requiring your conscious attention. It became easier to do because the synapses in a certain pathway were modified to allow signals to travel more easily across them than across "untrained" synapses. The process of making transmission easier is called **synaptic potentiation** (one form of synaptic plasticity).

Neuroscientists still argue about how to classify the various forms of memory, but three kinds often recognized are *immediate memory, short-term memory,* and *long-term memory.* We also know of different modes of synaptic potentiation that last from just a few seconds to a lifetime, and we can correlate these at least tentatively with different forms of memory.

### Immediate Memory

**Immediate memory** is the ability to hold something in mind for just a few seconds. By remembering what just happened, we get a feeling for the flow of events and a sense of the present. Immediate memory is indispensable to the ability to read; you must remember the earliest words of a sentence until you get to its end in order to extract any meaning from it. You could not make any sense of what you read if you forgot each word as soon as you moved on to the next one. Immediate memory may be based on reverberating circuits. Our impression of what just happened can thus reecho in our minds for a few seconds as we experience the present moment and anticipate the next one.

---

[29]*en* = inner; *gram* = mark, trace, record

## Short-Term Memory

**Short-term memory (STM)** lasts from a few seconds to a few hours. Information stored in STM may be quickly forgotten if we stop mentally reciting it, we are distracted, or we have to remember something new. **Working memory** is a form of STM that allows us to hold an idea in mind long enough to carry out an action such as calling a telephone number we just looked up, working out the steps of a mathematics problem, or searching for a lost set of keys while remembering where we have already looked. It is limited to a few bits of information such as the digits of a telephone number. It has long been thought that working memory is based on persistent activity in a reverberating circuit of neurons, but recent evidence leans toward the storage of working memory in a circuit of facilitated synapses that can remain quiescent (consuming no energy) most of the time, but be reactivated by a new sensory input.

Such **synaptic facilitation,** as it is called (different from the facilitation of one neuron by another that we studied earlier in the chapter), can be induced by *tetanic stimulation,* the rapid arrival of repetitive signals at a synapse. Each signal causes a certain amount of $Ca^{2+}$ to enter the synaptic knob. If signals arrive very rapidly, the neuron cannot pump out all the $Ca^{2+}$ admitted by one action potential before the next action potential occurs. More and more $Ca^{2+}$ accumulates in the knob. Since $Ca^{2+}$ is what triggers the release of neurotransmitter, each new signal releases more neurotransmitter than the one before. With more neurotransmitter, the EPSPs in the postsynaptic cell become stronger and stronger, and that cell is more likely to fire.

Memories lasting for a few hours, such as remembering what someone said to you earlier in the day or remembering an upcoming appointment, may involve **posttetanic potentiation.** In this process, the $Ca^{2+}$ level in the synaptic knob stays elevated for so long that another signal, coming well after the tetanic stimulation has ceased, releases an exceptionally large burst of neurotransmitter. That is, if a synapse has been heavily used in the recent past, a new stimulus can excite the postsynaptic cell more easily. Thus, your memory may need only a slight jog to recall something from several hours earlier.

## Long-Term Memory

**Long-term memory (LTM)** lasts up to a lifetime and is less limited than STM in the amount of information it can store. LTM allows you to memorize the lines of a play, the words of a favorite song, or (one hopes!) textbook information for an exam. On a still longer timescale, it enables you to remember your name, the route to your home, and your childhood experiences.

There are two forms of long-term memory: declarative and procedural. **Declarative memory** is the retention of events and facts that you can put into words—numbers, names, dates, and so forth. **Procedural memory** is the retention of motor skills—how to tie your shoes, play a musical instrument, or type on a keyboard. These forms of memory involve different regions of the brain but are probably similar at the cellular level.

Some LTM involves the physical remodeling of synapses or the formation of new ones through the growth and branching of axon terminals and dendrites. In the pyramidal cells of the brain, the dendrites are studded with knoblike *dendritic spines* that increase the area of synaptic contact. Studies on fish and other experimental animals have shown that social and sensory deprivation causes these spines to decline in number, while a richly stimulatory environment causes them to proliferate—an intriguing clue to the importance of a stimulating environment to infant and child development. In some cases of LTM, a new synapse grows beside the original one, giving the presynaptic cell twice as much input into the postsynaptic cell.

LTM can also be grounded in molecular changes called **long-term potentiation.** This involves *NMDA*[30] *receptors,* which are glutamate-binding receptors found on the dendritic spines of pyramidal cells. NMDA receptors are usually blocked by magnesium ions $(Mg^{2+})$, but when they bind glutamate *and* are simultaneously subjected to tetanic stimulation, they expel the $Mg^{2+}$ and open to admit $Ca^{2+}$ into the dendrite. When $Ca^{2+}$ enters, it acts as a second messenger with multiple effects. It stimulates the neuron to produce even more NMDA receptors, making it more sensitive to glutamate in the future. The neuron also synthesizes proteins concerned with physically remodeling the synapse. It is thought, too, that the postsynaptic cell may send signals (perhaps nitric oxide, NO) back to the presynaptic cell that enhance its release of neurotransmitter.

You can see that in all of these ways, long-term potentiation can increase transmission across "experienced" synapses. Remodeling a synapse or installing more neurotransmitter receptors has longer-lasting effects than facilitation or posttetanic potentiation.

The anatomical sites of memory are discussed in chapter 14 in connection with brain anatomy. Regardless of the sites, however, the cellular mechanisms are as described here.

### BEFORE YOU GO ON

Answer the following questions to test your understanding of the preceding section:

22. Contrast the two types of summation at a synapse.

23. Describe how the nervous system communicates quantitative and qualitative information about stimuli.

24. List the four types of neural circuits and describe their similarities and differences. Discuss the unity of form and function in these four types—that is, explain why each type would not perform as it does if its neurons were connected differently.

25. How does long-term potentiation enhance the transmission of nerve signals along certain pathways?

---

[30]*N-methyl-D-aspartate,* a chemical similar to glutamate

# DEEPER INSIGHT 12.4
## CLINICAL APPLICATION
### Alzheimer and Parkinson Diseases

Alzheimer and Parkinson diseases are the two most common degenerative disorders of the brain. Both are associated with neurotransmitter deficiencies.

*Alzheimer*[31] *disease (AD)* may begin before the age of 50 with signs so slight and ambiguous that early diagnosis is difficult. One of its first signs is memory loss, especially for recent events. A person with AD may ask the same questions repeatedly, show a reduced attention span, and become disoriented and lost in previously familiar places. Family members often feel helpless and confused as they watch their loved one's personality gradually deteriorate beyond recognition. The AD patient may become moody, confused, paranoid, combative, or hallucinatory—he or she may ask irrational questions such as, Why is the room full of snakes? The patient may eventually lose even the ability to read, write, talk, walk, and eat. Death ensues from pneumonia or other complications of confinement and immobility.

AD affects about 11% of the U.S. population over the age of 65; the incidence rises to 47% by age 85. It accounts for nearly half of all nursing home admissions and is a leading cause of death among the elderly. AD claims about 100,000 lives per year in the United States.

Diagnosis of AD can be confirmed by autopsy. There is atrophy of some of the gyri (folds) of the cerebral cortex and the hippocampus, an important center of memory. Nerve cells exhibit *neurofibrillary tangles*—dense masses of broken and twisted cytoskeleton (fig. 12.31). Alois Alzheimer first observed these in 1907 in the brain of a patient who had died of senile dementia. The more severe the signs of disease, the more neurofibrillary tangles are seen at autopsy. In the intercellular spaces, there are *senile plaques* consisting of aggregations of cells, altered nerve fibers, and a core of *β-amyloid protein*—the breakdown product of a glycoprotein of plasma membranes. Amyloid protein is rarely seen in elderly people without AD. It is now widely believed to be the crucial factor that triggers all the other aspects of AD pathology.

Intense biomedical research efforts are currently geared toward identifying the causes of AD and developing treatment strategies. Three genes on chromosomes 1, 14, and 21 have been implicated in various forms of early- and late-onset AD. Interestingly, persons with Down syndrome (trisomy-21), who have three copies of chromosome 21 instead of the usual two, tend to show early-onset Alzheimer disease. Nongenetic (environmental) factors also seem to be involved.

As for treatment, considerable attention now focuses on trying to halt β-amyloid formation or stimulate the immune system to clear β-amyloid from the brain tissue, but clinical trials in both of these approaches have been suspended until certain serious side effects can be resolved. AD patients show deficiencies of acetylcholine (ACh) and nerve growth factor (NGF). Some patients show improvement when treated with NGF or cholinesterase inhibitors, but results so far have been modest.

*Parkinson*[32] *disease (PD),* also called *paralysis agitans* or *parkinsonism,* is a progressive loss of motor function beginning in a person's 50s or 60s. It is due to degeneration of dopamine-releasing neurons in a portion of the brain called the *substantia nigra.* A gene has recently been identified for a hereditary form of PD, but most cases are nonhereditary and of little-known cause; some authorities suspect environmental neurotoxins.

Dopamine (DA) is an inhibitory neurotransmitter that normally prevents excessive activity in motor centers of the brain called the *basal nuclei.* Degeneration of dopamine-releasing neurons leads to an excessive ratio of ACh to DA, causing hyperactivity of the basal nuclei. As a result, a person with PD suffers involuntary muscle contractions. These take such forms as shaking of the hands (tremor) and compulsive "pill-rolling" motions of the thumb and fingers. In addition, the facial muscles may become rigid and produce a staring, expressionless face with a slightly open mouth. The patient's range of motion

---

[31] Alois Alzheimer (1864–1915), German neurologist

[32] James Parkinson (1755–1824), British physician

diminishes. He or she takes smaller steps and develops a slow, shuffling gait with a forward-bent posture and a tendency to fall forward. Speech becomes slurred and handwriting becomes cramped and eventually illegible. Tasks such as buttoning clothes and preparing food become increasingly laborious.

Patients cannot be expected to recover from PD, but its effects can be alleviated with drugs and physical therapy. Treatment with dopamine is ineffective because it cannot cross the blood–brain barrier, but its precursor, levodopa (L-dopa), does cross the barrier and has been used to treat PD since the 1960s. L-dopa affords some relief, but it does not slow progression of the disease and it has undesirable side effects on the liver and heart. It is effective for only 5 to 10 years of treatment. A newer drug, deprenyl, is a monoamine oxidase (MAO) inhibitor that retards neural degeneration and slows the development of PD.

A surgical technique called *pallidotomy* has been used since the 1940s to quell severe tremors. It involves the destruction of a small portion of cerebral tissue in an area called the *globus pallidus*. Pallidotomy fell out of favor in the late 1960s when L-dopa came into common use. By the early 1990s, however, the limitations of L-dopa had become apparent, while MRI- and CT-guided methods had improved surgical precision and reduced the risks of brain surgery. Pallidotomy has thus made a comeback. Other surgical treatments for parkinsonism target brain areas called the *subthalamic nucleus* and the *ventral intermediate nucleus* of the thalamus, and involve either the destruction of tiny areas of tissue or the implantation of a stimulating electrode. Such procedures are generally used only in severe cases that are unresponsive to medication.

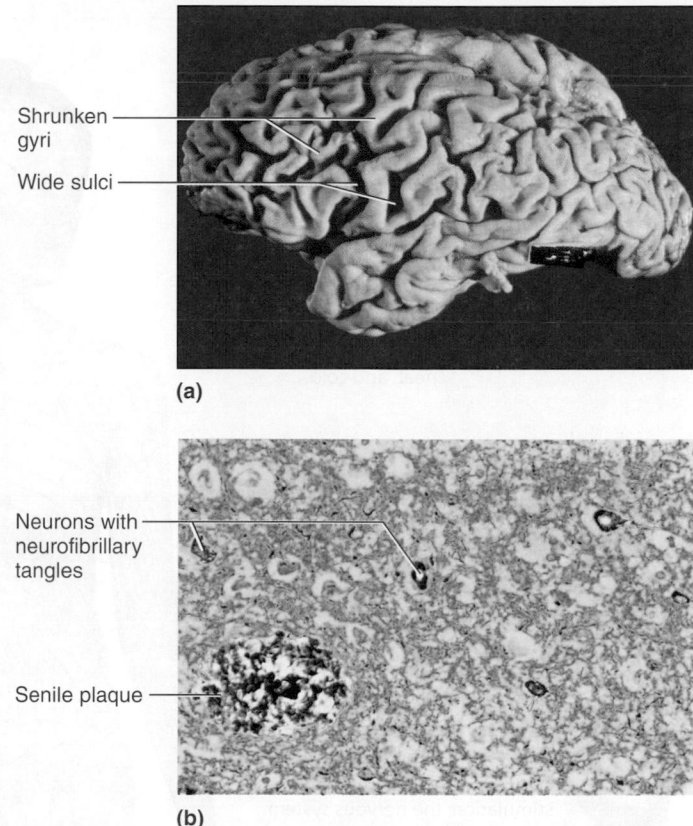

(a)

(b)

**FIGURE 12.31 Alzheimer Disease.** (a) Brain of a person who died of AD. Note the shrunken folds of cerebral tissue (gyri) and wide gaps (sulci) between them. (b) Cerebral tissue from a person with AD. Neurofibrillary tangles are present within the neurons, and a senile plaque is evident in the extracellular matrix.

## Effects of the NERVOUS SYSTEM
## On Other Organ Systems

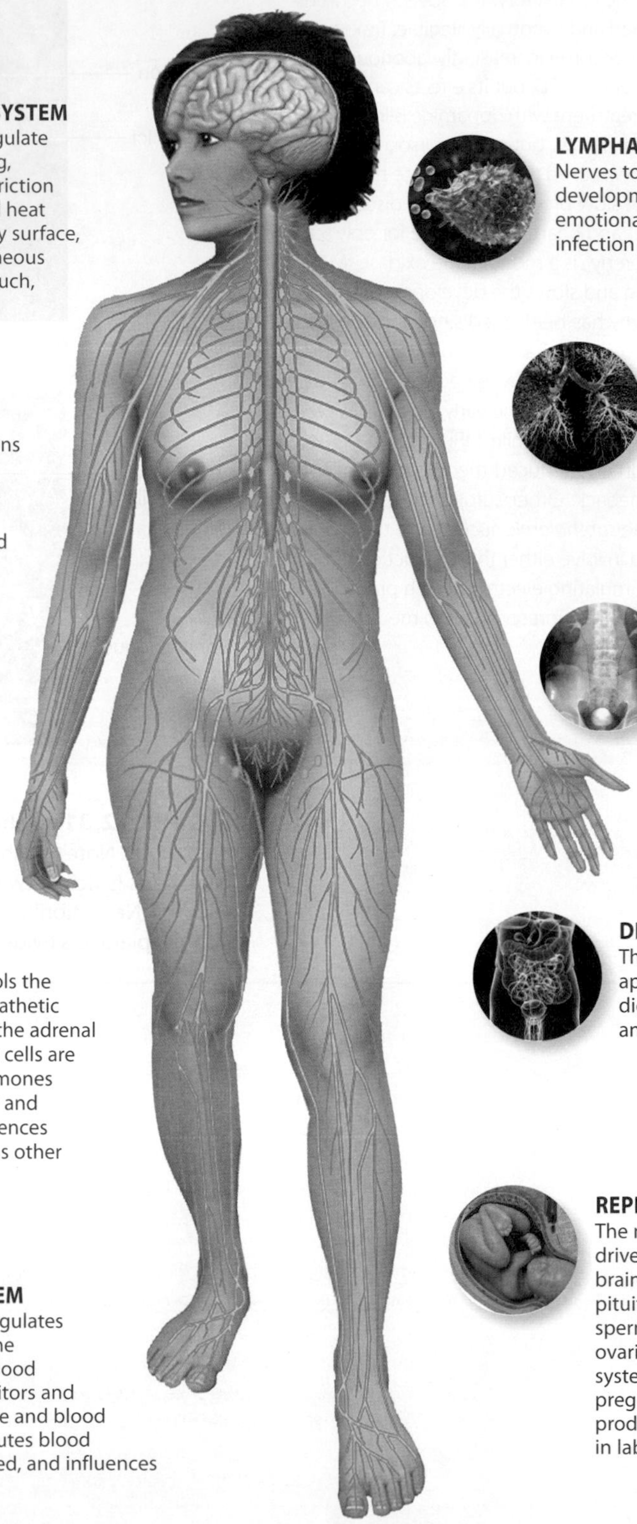

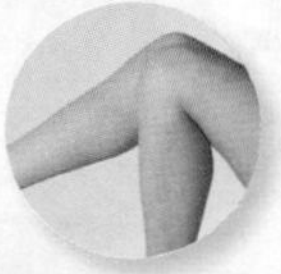

**INTEGUMENTARY SYSTEM**
Cutaneous nerves regulate piloerection, sweating, cutaneous vasoconstriction and vasodilation, and heat loss through the body surface, and provide for cutaneous sensations such as touch, itch, tickle, pressure, heat, and cold.

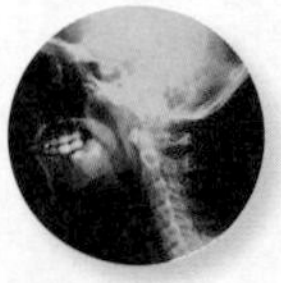

**SKELETAL SYSTEM**
Nervous stimulation maintains the muscle tension that stimulates bone growth and remodeling; nerves in the bones respond to strains and fractures.

**MUSCULAR SYSTEM**
Skeletal muscles cannot contract without nervous stimulation; the nervous system controls all body movements and muscle tone.

**ENDOCRINE SYSTEM**
The hypothalamus controls the pituitary gland; the sympathetic nervous system controls the adrenal medulla; neuroendocrine cells are neurons that secrete hormones such as oxytocin; sensory and other nervous input influences the secretion of numerous other hormones.

**CIRCULATORY SYSTEM**
The nervous system regulates the rate and force of the heartbeat, regulates blood vessel diameters, monitors and controls blood pressure and blood gas concentrations, routes blood to organs where needed, and influences blood clotting.

**LYMPHATIC/IMMUNE SYSTEM**
Nerves to lymphatic organs influence the development and activity of immune cells; emotional states influence susceptibility to infection and other failures of immunity.

**RESPIRATORY SYSTEM**
The brainstem regulates the rhythm of breathing, monitors blood pH and blood gases, and adjusts the respiratory rate and depth to control these within normal ranges.

**URINARY SYSTEM**
Sympathetic nerves modify the rate of urine production by the kidneys; nervous stimulation of urinary sphincters aids in urine retention in the bladder, and nervous reflexes control its emptying.

**DIGESTIVE SYSTEM**
The nervous system regulates appetite, feeding behavior, digestive secretion and motility, and defecation.

**REPRODUCTIVE SYSTEM**
The nervous system regulates sex drive, arousal, and orgasm; the brain regulates the secretion of pituitary hormones that control spermatogenesis in males and the ovarian cycle in females; the nervous system controls various aspects of pregnancy and childbirth; the brain produces oxytocin, which is involved in labor contractions and lactation.

# STUDY GUIDE

## ► Assess Your Learning Outcomes

*To test your knowledge, discuss the following topics with a study partner or in writing, ideally from memory.*

### 12.1 Overview of the Nervous System (p. 436)

1. What the nervous and endocrine systems have in common
2. Three fundamental functions of the nervous system; the roles of receptors and effectors in carrying out these functions
3. Difference between the central nervous system (CNS) and peripheral nervous system (PNS); between the sensory and motor divisions of the PNS; and between the somatic and visceral subdivisions of both the sensory and motor divisions
4. The autonomic nervous system and its two divisions

### 12.2 Properties of Neurons (p. 437)

1. Three fundamental physiological properties of neurons
2. Differences between sensory (afferent) neurons, interneurons (association neurons), and motor (efferent) neurons
3. The parts of a generalized multipolar neuron, and their functions
4. Differences between multipolar, bipolar, unipolar, and anaxonic neurons; an example of each
5. Ways in which neurons transport substances between the neurosoma and the distal ends of the axon

### 12.3 Supportive Cells (Neuroglia) (p. 442)

1. Six kinds of neuroglia; the structure and functions of each; and which kinds are found in the CNS and which ones in the PNS
2. Structure of the myelin sheath, and how CNS and PNS glial cells produce it
3. How fiber diameter and the presence or absence of myelin affect the conduction speed of a nerve fiber
4. The regeneration of a damaged nerve fiber; the role of Schwann cells, the basal lamina, and neurilemma in regeneration; and why CNS neurons cannot regenerate

### 12.4 Electrophysiology of Neurons (p. 447)

1. The meanings of *electrical potential* and *resting membrane potential (RMP);* the typical voltage of an RMP

2. What an electrical current is, and how sodium ions and gated membrane channels generate a current
3. How stimulation of a neuron generates a local potential; the physiological properties of a local potential
4. Special properties of the trigger zone and unmyelinated regions of a nerve fiber that enable these regions to generate action potentials
5. The mechanism of an action potential; how it relates to ion flows and the action of membrane channels; and what is meant by *depolarization* and *repolarization* of the plasma membrane during local and action potentials
6. The all-or-none law and how it applies to an action potential; other properties of action potentials in contrast to local potentials
7. The basis and significance of the refractory period that follows an action potential
8. How one action potential triggers another; how a chain reaction of action potentials constitutes a nerve signal in an unmyelinated nerve fiber; and what normally prevents the signal from traveling backward to the neurosoma
9. Saltatory conduction in a myelinated nerve fiber; differences in conduction mechanisms of the nodes of Ranvier and the internodes; and why signals travel faster in myelinated fibers than in unmyelinated fibers of comparable size

### 12.5 Synapses (p. 455)

1. The structure and locations of synapses
2. The role of neurotransmitters in synaptic transmission
3. Categories of neurotransmitters and common examples of each
4. Why the same neurotransmitter can have different effects on different cells
5. Excitatory synapses; how acetylcholine and norepinephrine excite a postsynaptic neuron
6. Inhibitory synapses; how γ-aminobutyric acid (GABA) inhibits a postsynaptic neuron
7. How second-messenger systems function at synapses
8. Three ways in which synaptic transmission is ended

9. Neuromodulators, their chemical nature, and how they affect synaptic transmission

### 12.6 Neural Integration (p. 461)

1. Why synapses slow down nervous communication; the overriding benefit of synapses
2. The meaning of *excitatory* and *inhibitory postsynaptic potentials (EPSPs* and *IPSPs)*
3. Why the production of an EPSP or IPSP may depend on both the neurotransmitter released by the presynaptic neuron and the type of receptor on the postsynaptic neuron
4. How a postsynaptic neuron's decision to fire depends on the ratio of EPSPs to IPSPs
5. Temporal and spatial summation, where they occur, and how they determine whether a neuron fires
6. Mechanisms of facilitation and presynaptic inhibition, and how communication between two neurons can be influenced by a third neuron employing one of these mechanisms
7. Mechanisms of neural coding; how a neuron communicates qualitative and quantitative information
8. Why the refractory period sets a limit to how frequently a neuron can fire
9. The meanings of *neural pool* and *neural circuit*
10. The difference between a neuron's discharge zone and facilitated zone, and how this relates to neurons working in groups
11. Diverging, converging, reverberating, and parallel after-discharge circuits of neurons; examples of their relevance to familiar body functions
12. The cellular basis of memory; what memory consists of in terms of neural pathways, and how it relates to synaptic plasticity and potentiation
13. Types of things remembered in immediate memory, short-term memory (STM), and long-term memory (LTM), and in the declarative and procedural forms of LTM
14. Neural mechanisms thought to be involved in these different forms of memory

# STUDY GUIDE

## ▶ Testing Your Recall

*Answers in Appendix B*

1. The integrative functions of the nervous system are performed mainly by
   a. afferent neurons.
   b. efferent neurons.
   c. neuroglia.
   d. sensory neurons.
   e. interneurons.

2. The highest density of voltage-gated ion channels is found on the _____ of a neuron.
   a. dendrites
   b. soma
   c. nodes of Ranvier
   d. internodes
   e. synaptic knobs

3. The soma of a mature neuron lacks
   a. a nucleus.
   b. endoplasmic reticulum.
   c. lipofuscin.
   d. centrioles.
   e. ribosomes.

4. The glial cells that fight infections in the CNS are
   a. microglia.
   b. satellite cells.
   c. ependymal cells.
   d. oligodendrocytes.
   e. astrocytes.

5. Posttetanic potentiation of a synapse increases the amount of _____ in the synaptic knob.
   a. neurotransmitter
   b. neurotransmitter receptors
   c. calcium
   d. sodium
   e. NMDA

6. An IPSP is _____ of the postsynaptic neuron.
   a. a refractory period
   b. an action potential
   c. a depolarization
   d. a repolarization
   e. a hyperpolarization

7. Saltatory conduction occurs only
   a. at chemical synapses.
   b. in the initial segment of an axon.
   c. in both the initial segment and axon hillock.
   d. in myelinated nerve fibers.
   e. in unmyelinated nerve fibers.

8. Some neurotransmitters can have either excitatory or inhibitory effects depending on the type of
   a. receptors on the postsynaptic cell.
   b. synaptic vesicles in the axon.
   c. synaptic potentiation that occurs.
   d. postsynaptic potentials on the synaptic knob.
   e. neuromodulator involved.

9. Differences in the volume of a sound are likely to be encoded by differences in _____ in nerve fibers from the inner ear.
   a. neurotransmitters
   b. signal conduction velocity
   c. types of postsynaptic potentials
   d. firing frequency
   e. voltage of the action potentials

10. Motor effects that depend on repetitive output from a neural pool are most likely to use
    a. parallel after-discharge circuits.
    b. reverberating circuits.
    c. facilitated circuits.
    d. diverging circuits.
    e. converging circuits.

11. Neurons that convey information to the CNS are called sensory, or _____, neurons.

12. To perform their role, neurons must have the properties of excitability, secretion, and _____.

13. The _____ is a period of time in which a neuron is producing an action potential and cannot respond to another stimulus of any strength.

14. Neurons receive incoming signals by way of specialized extensions of the cell called _____.

15. In the CNS, myelin is produced by glial cells called _____.

16. A myelinated nerve fiber can produce action potentials only in specialized regions called _____.

17. The trigger zone of a neuron consists of its _____ and _____.

18. The neurotransmitter secreted at an adrenergic synapse is _____.

19. A presynaptic nerve fiber cannot cause other neurons in its _____ to fire, but it can make them more sensitive to stimulation from other presynaptic fibers.

20. _____ are substances released along with a neurotransmitter that modify the neurotransmitter's effect.

## ▶ Building Your Medical Vocabulary

*Answers in Appendix B*

*State a medical meaning of each word element below, and give a term in which it or a slight variation of it is used.*

1. antero-

2. -aps

3. astro-

4. dendro-

5. -fer

6. gangli-

7. -grad

8. neuro-

9. sclero-

10. somato-

# STUDY GUIDE

## ► True or False

*Answers in Appendix B*

*Determine which five of the following statements are false, and briefly explain why.*

1. A neuron never has more than one axon.

2. Oligodendrocytes perform the same function in the brain as Schwann cells do in the peripheral nerves.

3. A resting neuron has a higher concentration of $K^+$ in its cytoplasm than in the extracellular fluid surrounding it.

4. During an action potential, most of the $Na^+$ and $K^+$ exchange places across the plasma membrane.

5. Excitatory postsynaptic potentials lower the threshold of a neuron and thus make it easier to stimulate.

6. The absolute refractory period sets an upper limit on how often a neuron can fire.

7. A given neurotransmitter has the same effect no matter where in the body it is secreted.

8. Myelinated nerve fibers conduct signals more rapidly than unmyelinated ones because they have nodes of Ranvier.

9. Learning occurs by increasing the number of neurons in the brain tissue.

10. Mature neurons are incapable of mitosis.

## ► Testing Your Comprehension

*Answers at www.mhhe.com/saladin7*

1. Schizophrenia is sometimes treated with drugs such as chlorpromazine that inhibit dopamine receptors. A side effect is that patients begin to develop muscle tremors, speech impairment, and other disorders similar to Parkinson disease. Explain.

2. Hyperkalemia is an excess of potassium in the extracellular fluid. What effect would this have on the resting membrane potentials of the nervous system and on neural excitability?

3. Suppose a poison were to slow down the $Na^+$–$K^+$ pumps of nerve cells. How would this affect the resting membrane potentials of neurons? Would it make neurons more excitable than normal, or make them more difficult to stimulate? Explain.

4. The unity of form and function is an important concept in understanding synapses. Give two structural reasons why nerve signals cannot travel backward across a chemical synapse. What might be the consequences if signals did travel freely in both directions?

5. The local anesthetics lidocaine (Xylocaine) and procaine (Novocaine) prevent voltage-gated $Na^+$ channels from opening. Explain why this would block the conduction of pain signals in a sensory nerve.

## ► Improve Your Grade at www.mhhe.com/saladln7

*Download animations to study when it fits your schedule. Practice quizzes, labeling activities, games, and flashcards offer fun ways to master the chapter concepts. Or, download image PowerPoint files for each chapter to create a study guide or for taking notes during lecture.*

CHAPTER

# 13

THE SPINAL CORD, SPINAL NERVES, AND SOMATIC REFLEXES

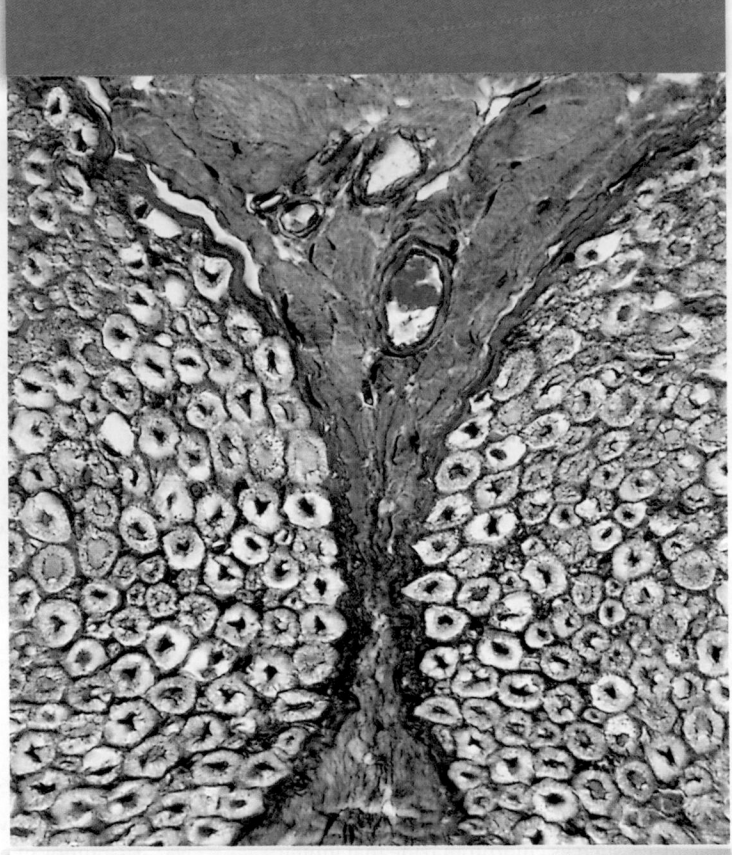

Cross section through two fascicles (bundles) of nerve fibers in a nerve

Anatomy & Physiology | REVEALED®
aprevealed.com

**Module 7: Nervous System**

## BRUSHING UP

- A knowledge of basic neuron structure (p. 438) is indispensable for understanding this chapter.

- In this chapter's discussion of spinal reflexes, it is necessary to be familiar with the ways muscles work in groups at a joint, especially antagonistic muscles. You can review that at page 314.

- An understanding of excitatory and inhibitory postsynaptic potentials (EPSPs and IPSPs) (p. 462) and the parallel after-discharge type of neural circuit (p. 466) are also important for understanding spinal reflexes.

Every year in the United States, thousands of people become paralyzed by spinal cord injuries, with devastating effects on their quality of life. The treatment of such injuries is one of the most lively areas of medical research today. Therapists in this specialty must know spinal cord anatomy and function to understand their patients' functional deficits and prospects for improvement and to plan a regimen of treatment. Such knowledge is necessary, as well, for understanding paralysis resulting from strokes and other brain injuries. The spinal cord is the "information highway" that connects the brain with the lower body; it contains the neural routes that explain why a lesion to a specific part of the brain results in a functional loss in a specific locality in the lower body.

In this chapter, we will study not only the spinal cord but also the spinal nerves that arise from it with ladderlike regularity at intervals along its length. Thus, we will examine components of both the central and peripheral nervous systems, but these are so closely related, structurally and functionally, that it is appropriate to consider them together. Similarly, the brain and cranial nerves will be considered together in the following chapter. Chapters 13 and 14 therefore elevate our study of the nervous system from the cellular level (chapter 12) to the organ and system levels.

## 13.1 The Spinal Cord

### Expected Learning Outcomes

When you have completed this section, you should be able to

a. state the three principal functions of the spinal cord;

b. describe its gross and microscopic structure; and

c. trace the pathways followed by nerve signals traveling up and down the spinal cord.

## Functions

The spinal cord serves four principal functions:

1. **Conduction.** It contains bundles of nerve fibers that conduct information up and down the cord, connecting different levels of the trunk with each other and with the brain. This enables sensory information to reach the brain, motor commands to reach the effectors, and input received at one level of the cord to affect output from another level.

2. **Neural integration.** Pools of spinal neurons receive input from multiple sources, integrate the information, and execute an appropriate output. For example, the spinal cord can integrate the stretch sensation from a full bladder with cerebral input concerning the appropriate time and place to urinate and execute control of the bladder accordingly.

3. **Locomotion.** Walking involves repetitive, coordinated contractions of several muscle groups in the limbs. Motor neurons in the brain initiate walking and determine its speed, distance, and direction, but the simple repetitive muscle contractions that put one foot in front of another, over and over, are coordinated by groups of neurons called **central pattern generators** in the cord. These neural circuits produce the sequence of outputs to the extensor and flexor muscles that cause alternating movements of the lower limbs.

4. **Reflexes.** Spinal reflexes play vital roles in posture, motor coordination, and protective responses to pain or injury.

## Surface Anatomy

The **spinal cord** (fig. 13.1) is a cylinder of nervous tissue that arises from the brainstem at the foramen magnum of the skull. It passes through the vertebral canal as far as the inferior margin of the first lumbar vertebra (L1) or slightly beyond. In adults, it averages about 45 cm long and 1.8 cm thick (about as thick as one's little finger). Early in fetal development, the cord extends for the full length of the vertebral column. However, the vertebral column grows faster than the spinal cord, so the cord extends only to L3 by the time of birth and to L1 in an adult. Thus, it occupies only the upper two-thirds of the vertebral canal; the lower one-third is described shortly.

The cord gives rise to 31 pairs of *spinal nerves.* Although the spinal cord is not visibly segmented, the part supplied by each pair of nerves is called a *segment.* The cord exhibits longitudinal grooves on its anterior and posterior sides—the *anterior median fissure* and *posterior median sulcus,* respectively (fig. 13.2b).

The spinal cord is divided into **cervical, thoracic, lumbar,** and **sacral regions.** It may seem odd that it has a sacral region when the cord itself ends well above the sacrum. These regions, however, are named for the level of the vertebral column from which the spinal nerves emerge, not for the vertebrae that contain the cord itself.

In two areas, the cord is a little thicker than elsewhere. In the inferior cervical region, a **cervical enlargement** gives rise to nerves of the upper limbs. In the lumbosacral region, there is a similar **lumbar enlargement** that issues nerves to the pelvic region and lower limbs. Inferior to the lumbar enlargement, the

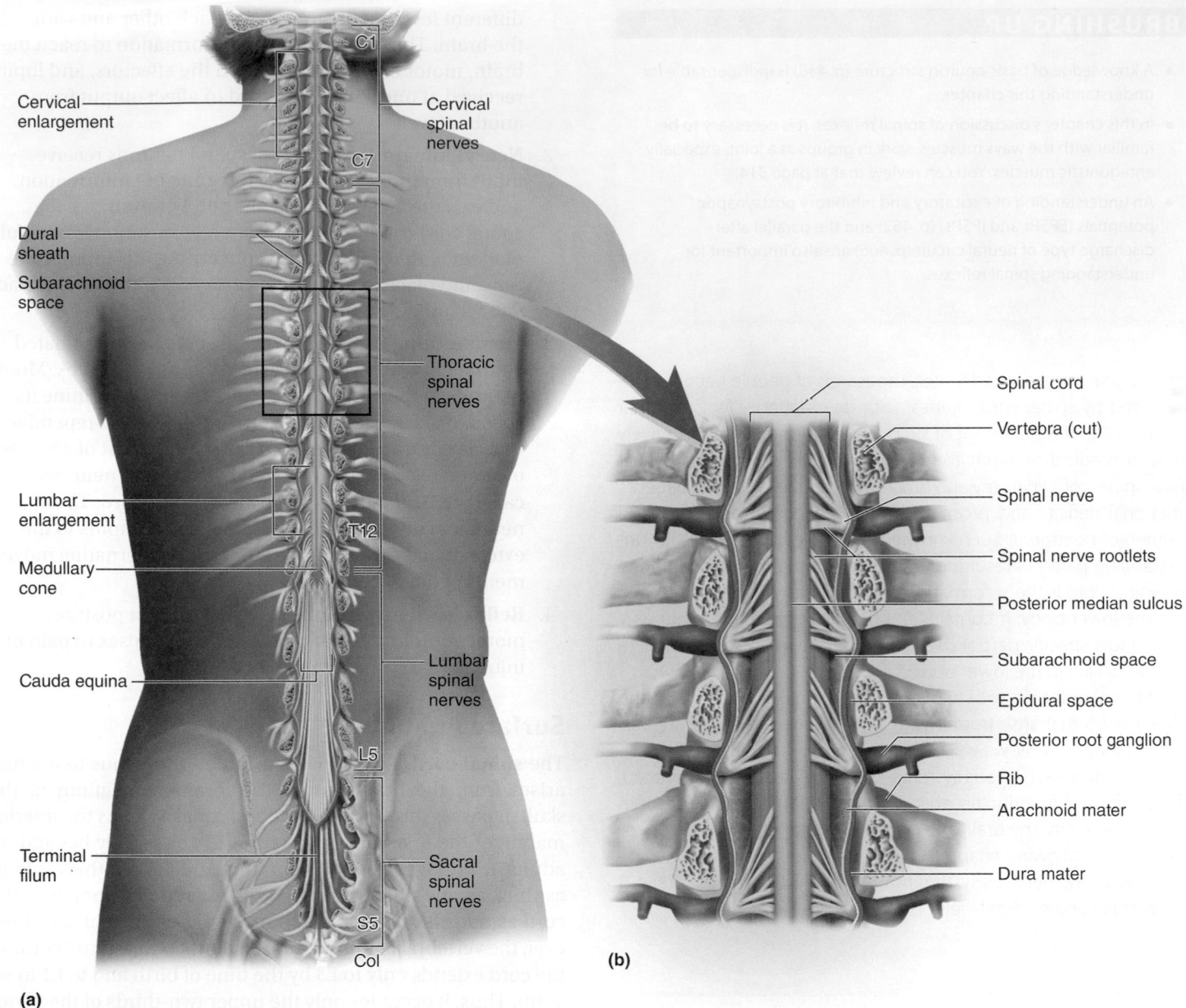

**(a)**

**(b)**

**FIGURE 13.1 The Spinal Cord, Posterior Aspect.** (a) Overview of spinal cord structure. (b) Detail of the spinal cord and associated structures. AP|R

cord tapers to a point called the **medullary cone (conus medullaris).** Arising from the lumbar enlargement and medullary cone is a bundle of nerve roots that occupy the vertebral canal from L2 to S5. This bundle, named the **cauda equina**[1] (CAW-duh ee-KWY-nah) for its resemblance to a horse's tail, innervates the pelvic organs and lower limbs.

▶▶▶ **APPLY WHAT YOU KNOW**

*Spinal cord injuries commonly result from fractures of vertebrae C5 to C6, but never from fractures of L3 to L5. Explain both observations.*

## Meninges of the Spinal Cord

The spinal cord and brain are enclosed in three fibrous membranes called **meninges**[2] (meh-NIN-jeez)—singular, *meninx* (MEN-inks) (fig. 13.2). These membranes separate the soft tissue of the central nervous system from the bones of the vertebrae and skull. From superficial to deep, they are the dura mater, arachnoid mater, and pia mater.

The **dura mater**[3] (DOO-ruh MAH-tur) forms a loose-fitting sleeve called the **dural sheath** around the spinal cord. It is a tough collagenous membrane about as thick as a rubber

---

[1]*cauda* = tail; *equin* = horse

[2]*menin* = membrane
[3]*dura* = tough; *mater* = mother, womb

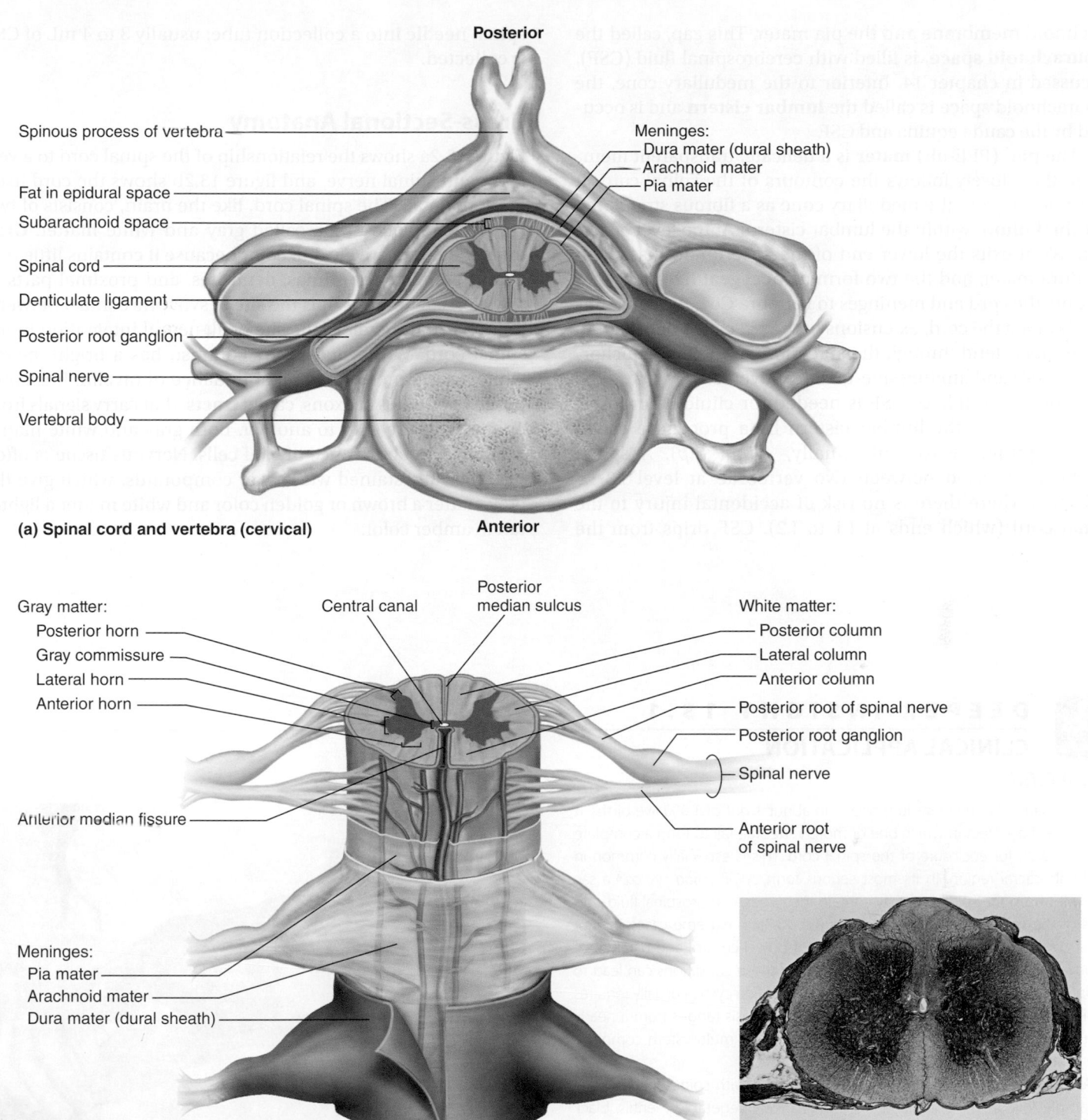

**Posterior**

Spinous process of vertebra

Meninges:
Dura mater (dural sheath)
Arachnoid mater
Pia mater

Fat in epidural space

Subarachnoid space

Spinal cord

Denticulate ligament

Posterior root ganglion

Spinal nerve

Vertebral body

**(a) Spinal cord and vertebra (cervical)**     **Anterior**

Gray matter:
Posterior horn
Gray commissure
Lateral horn
Anterior horn

Central canal

Posterior
median sulcus

White matter:
Posterior column
Lateral column
Anterior column
Posterior root of spinal nerve
Posterior root ganglion
Spinal nerve
Anterior root
of spinal nerve

Anterior median fissure

Meninges:
Pia mater
Arachnoid mater
Dura mater (dural sheath)

**(b) Spinal cord and meninges (thoracic)**

**(c) Lumbar spinal cord**

**FIGURE 13.2 Cross-Sectional Anatomy of the Spinal Cord.** (a) Relationship to the vertebra, meninges, and spinal nerve. (b) Detail of the spinal cord, meninges, and spinal nerves. (c) Cross section of the lumbar spinal cord with spinal nerves.

kitchen glove. The space between the sheath and vertebral bones, called the **epidural space,** is occupied by blood vessels, adipose tissue, and loose connective tissue. Anesthetics are sometimes introduced to this space to block pain signals during childbirth or surgery; this procedure is called *epidural anesthesia.*

The **arachnoid**[4] (ah-RACK-noyd) **mater** consists of a simple squamous epithelium, the *arachnoid membrane,* adhering to the inside of the dura, and a loose mesh of collagenous and elastic fibers spanning the gap between the

[4]*arachn* = spider, spider web; *oid* = resembling

arachnoid membrane and the pia mater. This gap, called the **subarachnoid space,** is filled with cerebrospinal fluid (CSF), discussed in chapter 14. Inferior to the medullary cone, the subarachnoid space is called the **lumbar cistern** and is occupied by the cauda equina and CSF.

The **pia**[5] (PEE-uh) **mater** is a delicate, transparent membrane that closely follows the contours of the spinal cord. It continues beyond the medullary cone as a fibrous strand, the **terminal filum,** within the lumbar cistern. At the level of vertebra S2, it exits the lower end of the cistern and fuses with the dura mater, and the two form a **coccygeal ligament** that anchors the cord and meninges to vertebra Co1. At regular intervals along the cord, extensions of the pia called **denticulate ligaments** extend through the arachnoid to the dura, anchoring the cord and limiting side-to-side movements.

When a sample of CSF is needed for clinical purposes, it is taken from the lumbar cistern by a procedure called **lumbar puncture** (or colloquially, *spinal tap*). A spinal needle is inserted between two vertebrae at level L3/L4 or L4/L5, where there is no risk of accidental injury to the spinal cord (which ends at L1 to L2). CSF drips from the spinal needle into a collection tube; usually 3 to 4 mL of CSF is collected.

## Cross-Sectional Anatomy

Figure 13.2a shows the relationship of the spinal cord to a vertebra and spinal nerve, and figure 13.2b shows the cord itself in more detail. The spinal cord, like the brain, consists of two kinds of nervous tissue called gray and white matter. **Gray matter** has a relatively dull color because it contains little myelin. It contains the somas, dendrites, and proximal parts of the axons of neurons. It is the site of synaptic contact between neurons, and therefore the site of all neural integration in the spinal cord. **White matter,** by contrast, has a bright, pearly white appearance due to an abundance of myelin. It is composed of bundles of axons, called **tracts,** that carry signals from one level of the CNS to another. Both gray and white matter also have an abundance of glial cells. Nervous tissue is often histologically stained with silver compounds, which give the gray matter a brown or golden color and white matter a lighter tan to amber color.

---

# DEEPER INSIGHT 13.1
## CLINICAL APPLICATION

### Spina Bifida

Spina bifida (SPY-nuh BIF-ih-duh) occurs in about 1 out of 1,000 live births. It is a congenital defect in which one or more vertebrae fail to form a complete vertebral arch for enclosure of the spinal cord. This is especially common in the lumbosacral region. In its most serious form, *spina bifida cystica,*[6] a sac protrudes from the spine and may contain meninges, cerebrospinal fluid, and parts of the spinal cord and nerve roots (fig. 13.3). In extreme cases, inferior spinal cord function is absent, causing lack of bowel control and paralysis of the lower limbs and urinary bladder. The last of these conditions can lead to chronic urinary infections and renal failure. Spina bifida cystica usually requires surgical closure within 72 hours of birth. The prognosis ranges from a nearly normal, productive life, to a lifetime of treatment for multisystem complications, or to infant death in extreme cases.

A woman can reduce the risk of having a child with spina bifida by taking ample folic acid—obtainable from green leafy vegetables, lentils, black beans, and enriched bread and pasta, or in tablet form. However, this must be done before pregnancy; spina bifida arises in the first 4 weeks of development, so by the time a woman knows she is pregnant, it is too late for folic acid supplements to prevent the disorder. This is the reason that the Food and Drug Administration decided in 1996 to add folic acid to flour and other grain products, a policy that has dramatically reduced the incidence of neural tube defects.

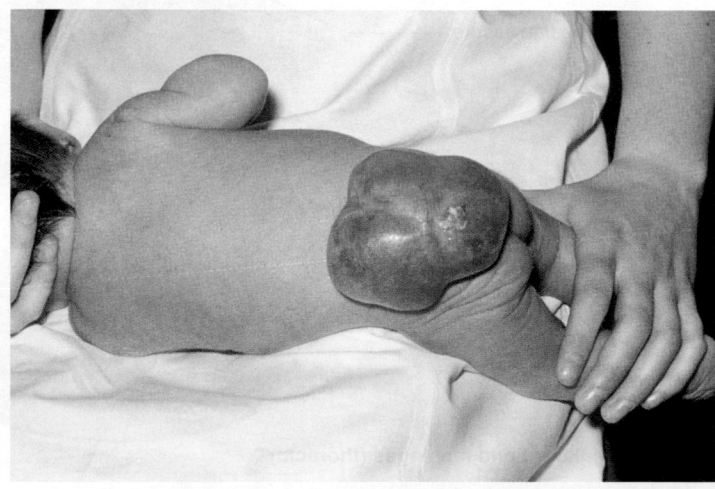

**FIGURE 13.3   Spina Bifida Cystica.** The sac in the lumbar region is called a myelomeningocele.

---

[5]*pia* = through mistranslation, now construed as tender, thin, or soft
[6]*bifid* = divided, forked; *cyst* = sac

## Gray Matter

The spinal cord has a central core of gray matter that looks somewhat butterfly- or H-shaped in cross sections. The core consists mainly of two **posterior (dorsal) horns,** which extend toward the posterolateral surfaces of the cord, and two thicker **anterior (ventral) horns,** which extend toward the anterolateral surfaces. The right and left sides are connected by a **gray commissure.** In the middle of the commissure is the **central canal,** which is collapsed in most areas of the adult spinal cord, but in some places (and in young children) remains open, lined with ependymal cells, and filled with CSF.

Near its attachment to the spinal cord, a spinal nerve branches into a **posterior (dorsal) root** and **anterior (ventral) root.** The posterior root carries sensory nerve fibers, which enter the posterior horn of the cord and sometimes synapse with an interneuron there. Such interneurons are especially numerous in the cervical and lumbar enlargements and are quite evident in histological sections at these levels. The anterior horns contain the large somas of the somatic motor neurons. Axons from these neurons exit by way of the anterior root of the spinal nerve and lead to the skeletal muscles. The spinal nerve roots are described more fully later in this chapter.

An additional **lateral horn** is visible on each side of the gray matter from segments T2 through L1 of the cord. It contains neurons of the sympathetic nervous system, which send their axons out of the cord by way of the anterior root along with the somatic efferent fibers.

## White Matter

The white matter of the spinal cord surrounds the gray matter. It consists of bundles of axons that course up and down the cord and provide avenues of communication between different levels of the CNS. These bundles are arranged in three pairs called **columns** or **funiculi**[7] (few-NIC-you-lie)—**a posterior (dorsal), lateral,** and **anterior (ventral) column** on each side. Each column consists of subdivisions called **tracts** or **fasciculi**[8] (fah-SIC-you-lye).

## Spinal Tracts

Knowledge of the locations and functions of the spinal tracts is essential in diagnosing and managing spinal cord injuries. **Ascending tracts** carry sensory information up the cord, and **descending tracts** conduct motor impulses down. All nerve fibers in a given tract have a similar origin, destination, and function. Many of these fibers have their origin or destination in a region called the *brainstem.* Described more fully in chapter 14 (see fig. 14.1), this is a vertical stalk that supports the large *cerebellum* at the rear of the head and, even larger, two *cerebral hemispheres* that dominate the brain. In the following discussion, you will find references to brainstem and other regions where spinal tracts begin and end. Spinal cord anatomy will grow in meaning as you study the brain.

Several of these tracts undergo **decussation**[9] (DEE-cuh-SAY-shun) as they pass up or down the brainstem and spinal cord—meaning that they cross over from the left side of the body to the right, or vice versa. As a result, the left side of the brain receives sensory information from the right side of the body and sends motor commands to that side, while the right side of the brain senses and controls the left side of the body. Therefore, a stroke that damages motor centers of the right side of the brain can cause paralysis of the left limbs and vice versa.

When the origin and destination of a tract are on opposite sides of the body, we say they are **contralateral**[10] to each other. When a tract does not decussate, its origin and destination are on the same side of the body and we say they are **ipsilateral.**[11]

The major spinal tracts are summarized in table 13.1 and figure 13.4. Bear in mind that each tract is repeated on the right and left sides of the spinal cord.

## Ascending Tracts

Ascending tracts carry sensory signals up the spinal cord. Sensory signals typically travel across three neurons from their origin in the receptors to their destination in the brain: a **first-order neuron** that detects a stimulus and transmits a signal to the spinal cord or brainstem; a **second-order neuron** that continues as far as a "gateway" called the *thalamus* at the upper end of the brainstem; and a **third-order neuron** that carries the signal the rest of the way to the cerebral cortex. The axons of these neurons are called the *first-* through *third-order nerve fibers* (fig. 13.5). Deviations from this pathway will be noted for some of the sensory systems to follow.

The major ascending tracts are as follows. The names of most of them consist of the prefix *spino-* followed by a root denoting the destination of its fibers in the brain, although this naming system does not apply to the first two.

- The **gracile**[12] **fasciculus** (GRAS-el fah-SIC-you-lus) (fig. 13.5a) carries signals from the midthoracic and lower parts of the body. Below vertebra T6, it composes the entire posterior column. At T6, it is joined by the cuneate fasciculus, discussed next. It consists of first-order nerve fibers that travel up the ipsilateral side of the spinal cord and terminate at the *gracile nucleus* in the medulla oblongata of the brainstem. These fibers carry signals for vibration, visceral pain, deep and discriminative touch (touch whose location one can precisely identify), and especially *proprioception*[13] from the lower limbs and lower trunk. Proprioception is the nonvisual sense of the position and movements of the body.

---

[7] *funicul* = little rope, cord
[8] *fascicul* = little bundle

[9] *decuss* = to cross, form an X
[10] *contra* = opposite; *later* = side
[11] *ipsi* = the same; *later* = side
[12] *gracil* = thin, slender
[13] *proprio* = one's own; *ception* = sensation

| TABLE 13.1 | Major Spinal Tracts | | |
|---|---|---|---|
| **Tract** | **Column** | **Decussation** | **Functions** |
| *Ascending (sensory) tracts* | | | |
| Gracile fasciculus | Posterior | In medulla | Sensations of limb and trunk position and movement, deep touch, visceral pain, and vibration, below level T6 |
| Cuneate fasciculus | Posterior | In medulla | Same as gracile fasciculus, from level T6 up |
| Spinothalamic | Lateral and anterior | In spinal cord | Sensations of light touch, tickle, itch, temperature, pain, and pressure |
| Spinoreticular | Lateral and anterior | In spinal cord (some fibers) | Sensation of pain from tissue injury |
| Posterior spinocerebellar | Lateral | None | Feedback from muscles (proprioception) |
| Anterior spinocerebellar | Lateral | In spinal cord | Same as posterior spinocerebellar |
| *Descending (motor) tracts* | | | |
| Lateral corticospinal | Lateral | In medulla | Fine control of limbs |
| Anterior corticospinal | Anterior | In spinal cord | Fine control of limbs |
| Tectospinal | Anterior | In midbrain | Reflexive head turning in response to visual and auditory stimuli |
| Lateral reticulospinal | Lateral | None | Balance and posture; regulation of awareness of pain |
| Medial reticulospinal | Anterior | None | Same as lateral reticulospinal |
| Lateral vestibulospinal | Anterior | None | Balance and posture |
| Medial vestibulospinal | Anterior | In medulla (some fibers) | Control of head position |

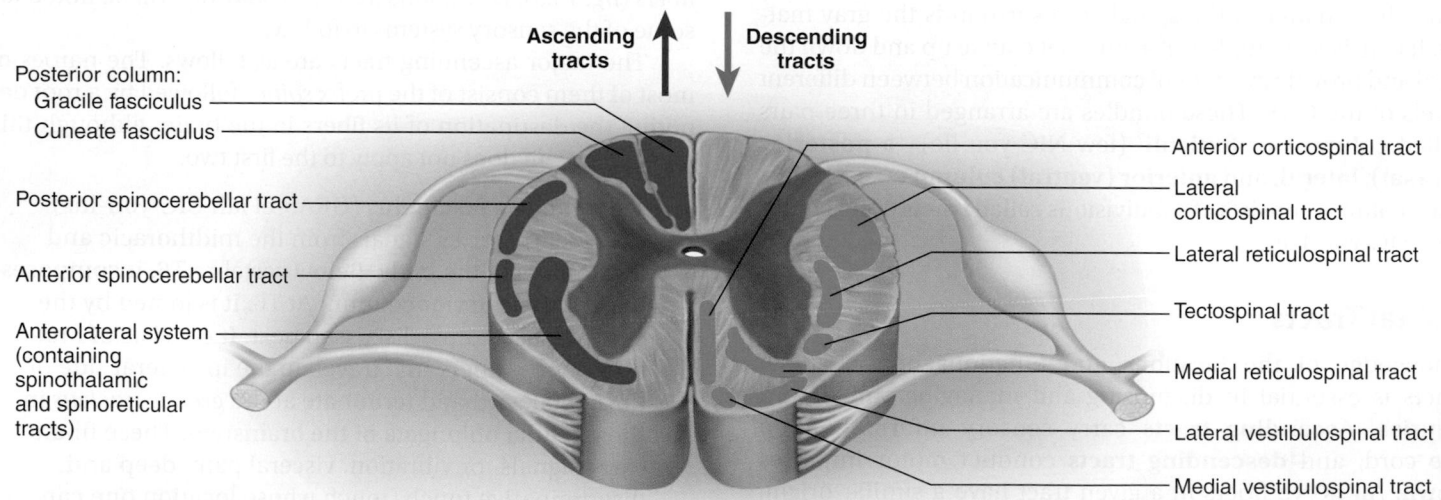

**FIGURE 13.4  Tracts of the Spinal Cord.**  All of the illustrated tracts occur on both sides of the cord, but only the ascending sensory tracts are shown on the left (red), and only the descending motor tracts on the right (green).

❓ *If you were told that this cross section is either at level T4 or T10, how could you determine which is correct?*

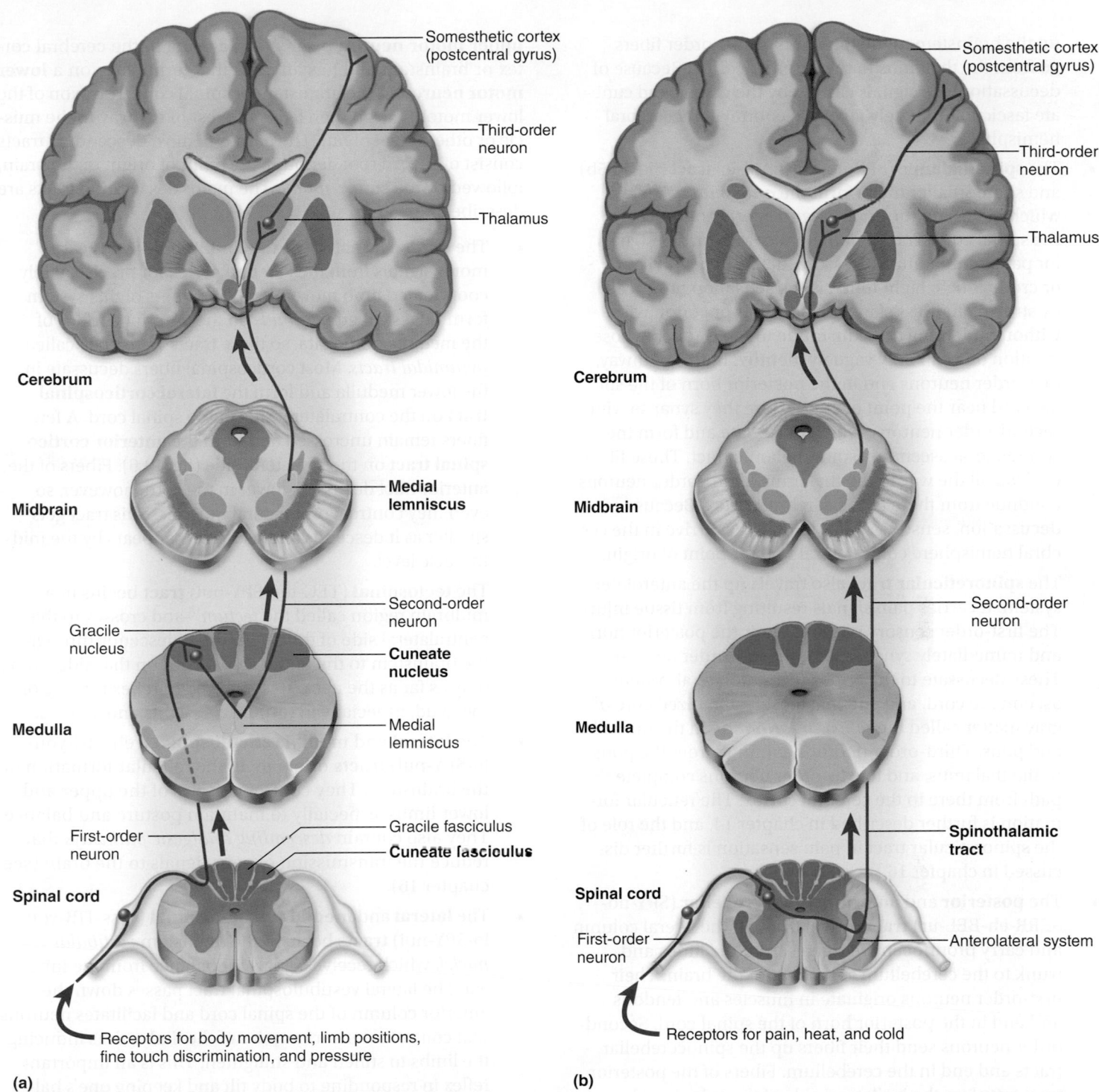

**FIGURE 13.5 Some Ascending Pathways of the CNS.** The spinal cord, medulla, and midbrain are shown in cross section and the cerebrum and thalamus (top) in frontal section. Nerve signals enter the spinal cord at the bottom of the figure and carry somatosensory information up to the cerebral cortex. (a) The cuneate fasciculus. (b) The spinothalamic tract.

- The **cuneate**[14] (CUE-nee-ate) **fasciculus** (fig. 13.5a) joins the gracile fasciculus at the T6 level. It occupies the lateral portion of the posterior column and forces the gracile fasciculus medially. It carries the same type of sensory signals, originating from T6 and up (from the upper limbs

and chest). Its fibers end in the *cuneate nucleus* on the ipsilateral side of the medulla oblongata. In the medulla, second-order fibers of the gracile and cuneate systems decussate and form the **medial lemniscus**[15] (lem-NIS-cus), a tract of nerve fibers that leads the rest of the way

---

[14]*cune* = wedge

[15]*lemniscus* = ribbon

up the brainstem to the thalamus. Third-order fibers go from the thalamus to the cerebral cortex. Because of decussation, the signals carried by the gracile and cuneate fasciculi ultimately go to the contralateral cerebral hemisphere.

- The **spinothalamic** (SPY-no-tha-LAM-ic) **tract** (fig. 13.5b) and some smaller tracts form the *anterolateral system,* which passes up the anterior and lateral columns of the spinal cord. The spinothalamic tract carries signals for pain, temperature, pressure, tickle, itch, and light or crude touch. Light touch is the sensation produced by stroking hairless skin with a feather or cotton wisp, without indenting the skin; crude touch is touch whose location one can only vaguely identify. In this pathway, first-order neurons end in the posterior horn of the spinal cord near the point of entry. Here they synapse with second-order neurons, which decussate and form the contralateral ascending spinothalamic tract. These fibers lead all the way to the thalamus. Third-order neurons continue from there to the cerebral cortex. Because of decussation, sensory signals in this tract arrive in the cerebral hemisphere contralateral to their point of origin.

- The **spinoreticular tract** also travels up the anterolateral system. It carries pain signals resulting from tissue injury. The first-order sensory neurons enter the posterior horn and immediately synapse with second-order neurons. These decussate to the opposite anterolateral system, ascend the cord, and end in a loosely organized core of gray matter called the *reticular formation* in the medulla and pons. Third-order neurons continue from the pons to the thalamus, and fourth-order neurons complete the path from there to the cerebral cortex. The reticular formation is further described in chapter 14, and the role of the spinoreticular tract in pain sensation is further discussed in chapter 16.

- The **posterior** and **anterior spinocerebellar** (SPY-no-SERR-eh-BEL-ur) **tracts** travel through the lateral column and carry proprioceptive signals from the limbs and trunk to the cerebellum at the rear of the brain. Their first-order neurons originate in muscles and tendons and end in the posterior horn of the spinal cord. Second-order neurons send their fibers up the spinocerebellar tracts and end in the cerebellum. Fibers of the posterior tract travel up the ipsilateral side of the spinal cord. Those of the anterior tract cross over and travel up the contralateral side but then cross back in the brainstem to enter the ipsilateral side of the cerebellum. Both tracts provide the cerebellum with feedback needed to coordinate muscle action, as discussed in chapter 14.

## Descending Tracts

Descending tracts carry motor signals down the brainstem and spinal cord. A descending motor pathway typically involves two neurons called the upper and lower motor neurons. The **upper motor neuron** begins with a soma in the cerebral cortex or brainstem and has an axon that terminates on a **lower motor neuron** in the brainstem or spinal cord. The axon of the lower motor neuron then leads the rest of the way to the muscle or other target organ. The names of most descending tracts consist of a word root denoting the point of origin in the brain, followed by the suffix *-spinal.* The major descending tracts are described here.

- The **corticospinal** (COR-tih-co-SPY-nul) **tracts** carry motor signals from the cerebral cortex for precise, finely coordinated limb movements. The fibers of this system form ridges called *pyramids* on the anterior surface of the medulla oblongata, so these tracts were once called *pyramidal tracts.* Most corticospinal fibers decussate in the lower medulla and form the **lateral corticospinal tract** on the contralateral side of the spinal cord. A few fibers remain uncrossed and form the **anterior corticospinal tract** on the ipsilateral side (fig. 13.6). Fibers of the anterior tract decussate lower in the cord, however, so even they control contralateral muscles. This tract gets smaller as it descends and usually disappears by the midthoracic level.

- The **tectospinal** (TEC-toe-SPY-nul) **tract** begins in a midbrain region called the *tectum*[16] and crosses to the contralateral side of the midbrain. It descends through the brainstem to the upper spinal cord on that side, going only as far as the neck. It is involved in reflex turning of the head, especially in response to sights and sounds.

- The **lateral** and **medial reticulospinal** (reh-TIC-you-lo-SPY-nul) **tracts** originate in the reticular formation of the brainstem. They control muscles of the upper and lower limbs, especially to maintain posture and balance. They also contain *descending analgesic pathways* that reduce the transmission of pain signals to the brain (see chapter 16).

- The **lateral** and **medial vestibulospinal** (vess-TIB-you-lo-SPY-nul) **tracts** begin in the brainstem *vestibular nuclei,* which receive signals for balance from the inner ear. The lateral vestibulospinal tract passes down the anterior column of the spinal cord and facilitates neurons that control extensor muscles of the limbs, thus inducing the limbs to stiffen and straighten. This is an important reflex in responding to body tilt and keeping one's balance. The medial vestibulospinal tract splits into ipsilateral and contralateral fibers that descend through the anterior column on both sides of the cord and terminate in the neck. It plays a role in the control of head position.

*Rubrospinal tracts* are prominent in other mammals, where they aid in muscle coordination. Although often pictured in illustrations of supposedly human anatomy, they are almost nonexistent in humans and have little functional importance.

---

[16]*tectum* = roof

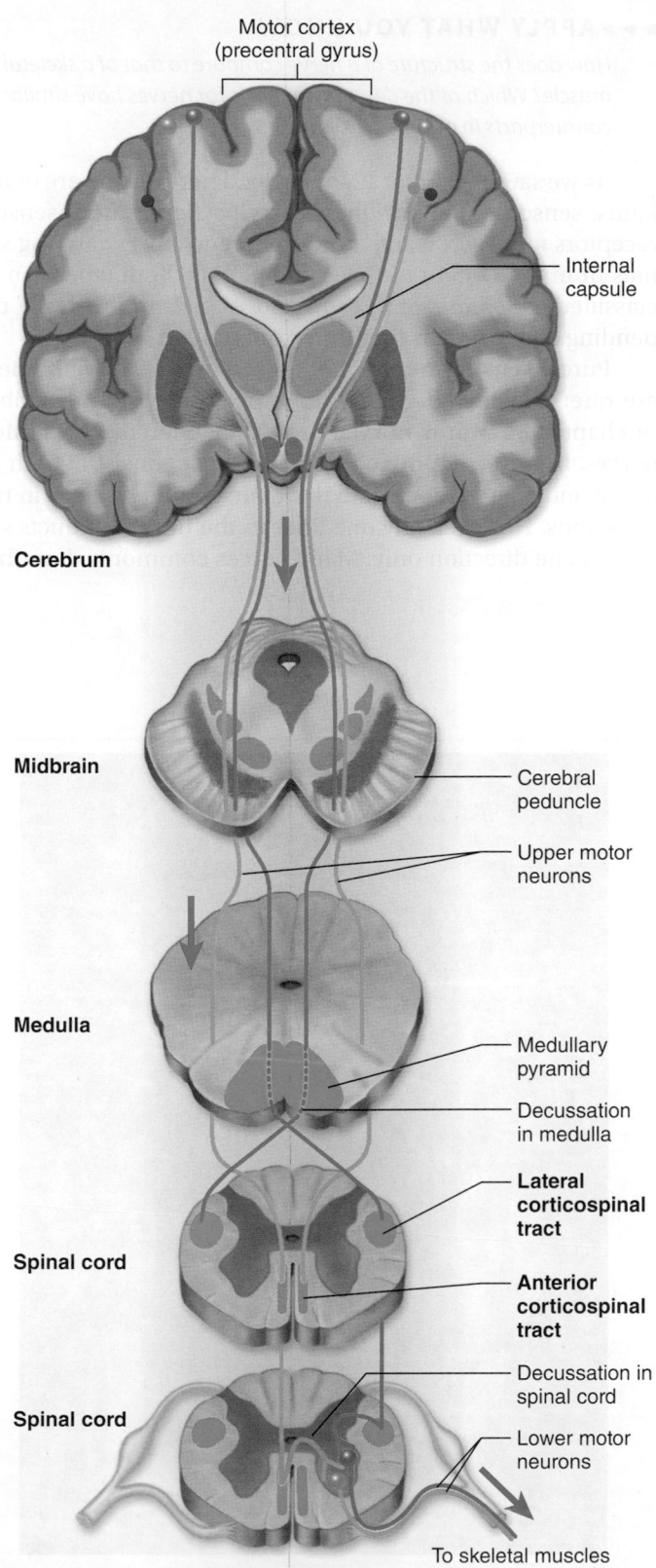

Motor cortex
(precentral gyrus)

Internal capsule

**Cerebrum**

**Midbrain**

Cerebral peduncle

Upper motor neurons

**Medulla**

Medullary pyramid

Decussation in medulla

**Lateral corticospinal tract**

**Spinal cord**

**Anterior corticospinal tract**

Decussation in spinal cord

**Spinal cord**

Lower motor neurons

To skeletal muscles

**FIGURE 13.6 Two Descending Pathways of the CNS.** The lateral and anterior corticospinal tracts, which carry signals for voluntary muscle contraction. Nerve signals originate in the cerebral cortex at the top of the figure and carry motor commands down the spinal cord.

▶▶▶**APPLY WHAT YOU KNOW**

*You are blindfolded and either a tennis ball or an iron ball is placed in your right hand. What spinal tract(s) would carry the signals that enable you to discriminate between these two objects?*

**BEFORE YOU GO ON**

Answer the following questions to test your understanding of the preceding section:

1. Name the four major regions and two enlargements of the spinal cord.

2. Describe the distal (inferior) end of the spinal cord and the contents of the vertebral canal from level L2 to S5.

3. Sketch a cross section of the spinal cord showing the anterior and posterior horns. Where are the gray and white matter? Where are the columns and tracts?

4. Give an anatomical explanation of why a stroke in the right cerebral hemisphere can paralyze the limbs on the left side of the body.

5. Identify each of the following spinal tracts—the gracile fasciculus and the lateral corticospinal, lateral reticulospinal, and spinothalamic tracts—with respect to whether it is ascending or descending; its origin and destination; and what sensory or motor purposes it serves.

## 13.2 The Spinal Nerves

### Expected Learning Outcomes

When you have completed this section, you should be able to

a. describe the anatomy of nerves and ganglia in general;

b. describe the attachments of a spinal nerve to the spinal cord;

c. trace the branches of a spinal nerve distal to its attachments;

d. name the five plexuses of spinal nerves and describe their general anatomy;

e. name some major nerves that arise from each plexus; and

f. explain the relationship of dermatomes to the spinal nerves.

### General Anatomy of Nerves and Ganglia

The spinal cord communicates with the rest of the body by way of the spinal nerves. Before we discuss those specific nerves, however, it is necessary to be familiar with the structure of nerves and ganglia in general.

A **nerve** is a cordlike organ composed of numerous nerve fibers (axons) bound together by connective tissue (fig. 13.8). If we compare a *nerve fiber* to a wire carrying an electrical current in one direction, a *nerve* would be comparable to an electrical cable composed of thousands of wires carrying currents in opposite directions. A nerve contains anywhere from a few nerve fibers to (in the optic nerve) a million. Nerves usually have a pearly white color and resemble frayed string as they divide into smaller and smaller branches.

Nerve fibers of the peripheral nervous system are ensheathed in Schwann cells, which form a neurilemma and often a myelin sheath around the axon (see p. 444). External to the neurilemma, each fiber is surrounded by a basal lamina and then a thin sleeve of loose connective tissue called the **endoneurium.** In most nerves, the fibers are gathered in bundles called **fascicles,** each wrapped in a sheath called the **perineurium.** The perineurium is composed of up to 20 layers of overlapping, squamous, epithelium-like cells. Several fascicles are then bundled together and wrapped in an outer epineurium to compose the nerve as a whole. The epineurium consists of dense irregular connective tissue and protects the nerve from stretching and injury. Nerves have a high metabolic rate and need a plentiful blood supply, which is furnished by blood vessels that penetrate these connective tissue coverings.

▶▶▶**APPLY WHAT YOU KNOW**

*How does the structure of a nerve compare to that of a skeletal muscle? Which of the descriptive terms for nerves have similar counterparts in muscle histology?*

As we saw in chapter 12, peripheral nerve fibers are of two kinds: sensory (afferent) fibers carrying signals from sensory receptors to the CNS, and motor (efferent) fibers carrying signals from the CNS to muscles and glands. Both types can be classified as *somatic* or *visceral* and as *general* or *special* depending on the organs they innervate (table 13.2).

Purely **sensory nerves,** composed only of afferent fibers, are rare; they include nerves for smell and vision described in chapter 14. **Motor nerves** carry only efferent fibers. Most nerves, however, are **mixed nerves,** which consist of both afferent and efferent fibers and therefore conduct signals in two directions. However, any one fiber in the nerve conducts signals in one direction only. Many nerves commonly described

# DEEPER INSIGHT 13.2

## CLINICAL APPLICATION

### *Poliomyelitis and Amyotrophic Lateral Sclerosis*

*Poliomyelitis*[17] and *amyotrophic lateral sclerosis*[18] *(ALS)* are two diseases that involve destruction of motor neurons. In both diseases, the skeletal muscles atrophy from lack of innervation.

Poliomyelitis is caused by the poliovirus, which destroys motor neurons in the brainstem and anterior horn of the spinal cord. Signs of polio include muscle pain, weakness, and loss of some reflexes, followed by paralysis, muscular atrophy, and sometimes respiratory arrest. The virus spreads by fecal contamination of water. Historically, polio afflicted many children who contracted the virus from swimming in contaminated pools. For a time, the polio vaccine nearly eliminated new cases, but the disease has lately begun to reemerge among children in some parts of the world because of political and religious opposition to vaccination.

ALS is also known as Lou Gehrig[19] disease after the baseball player who had to retire from the sport because of it. It is marked not only by the degeneration of motor neurons and atrophy of the muscles, but also sclerosis (scarring) of the lateral regions of the spinal cord—hence its name. Most cases occur when astrocytes fail to reabsorb the neurotransmitter glutamate from the tissue fluid, allowing it to accumulate to a neurotoxic level. The early signs of ALS include muscular weakness and difficulty in speaking, swallowing, and using the hands. Sensory and intellectual functions remain unaffected, as evidenced by the accomplishments of astrophysicist and best-selling author Stephen Hawking (fig. 13.7), who was stricken with ALS in college. Despite near-total paralysis, he remains highly productive and communicates with the aid of a speech synthesizer and computer. Tragically, many people are quick to assume that those who have lost most of their ability to communicate their ideas and feelings have few ideas and feelings to communicate. To a victim, this may be more unbearable than the loss of motor function itself.

**FIGURE 13.7 Stephen Hawking (1942–), Lucasian Professor of Mathematics at Cambridge University.**

---

[17]*polio* = gray matter; *myel* = spinal cord; *itis* = inflammation
[18]*a* = without; *myo* = muscle; *troph* = nourishment; *sclerosis* = hardening
[19]Lou Gehrig (1903–41), New York Yankees baseball player

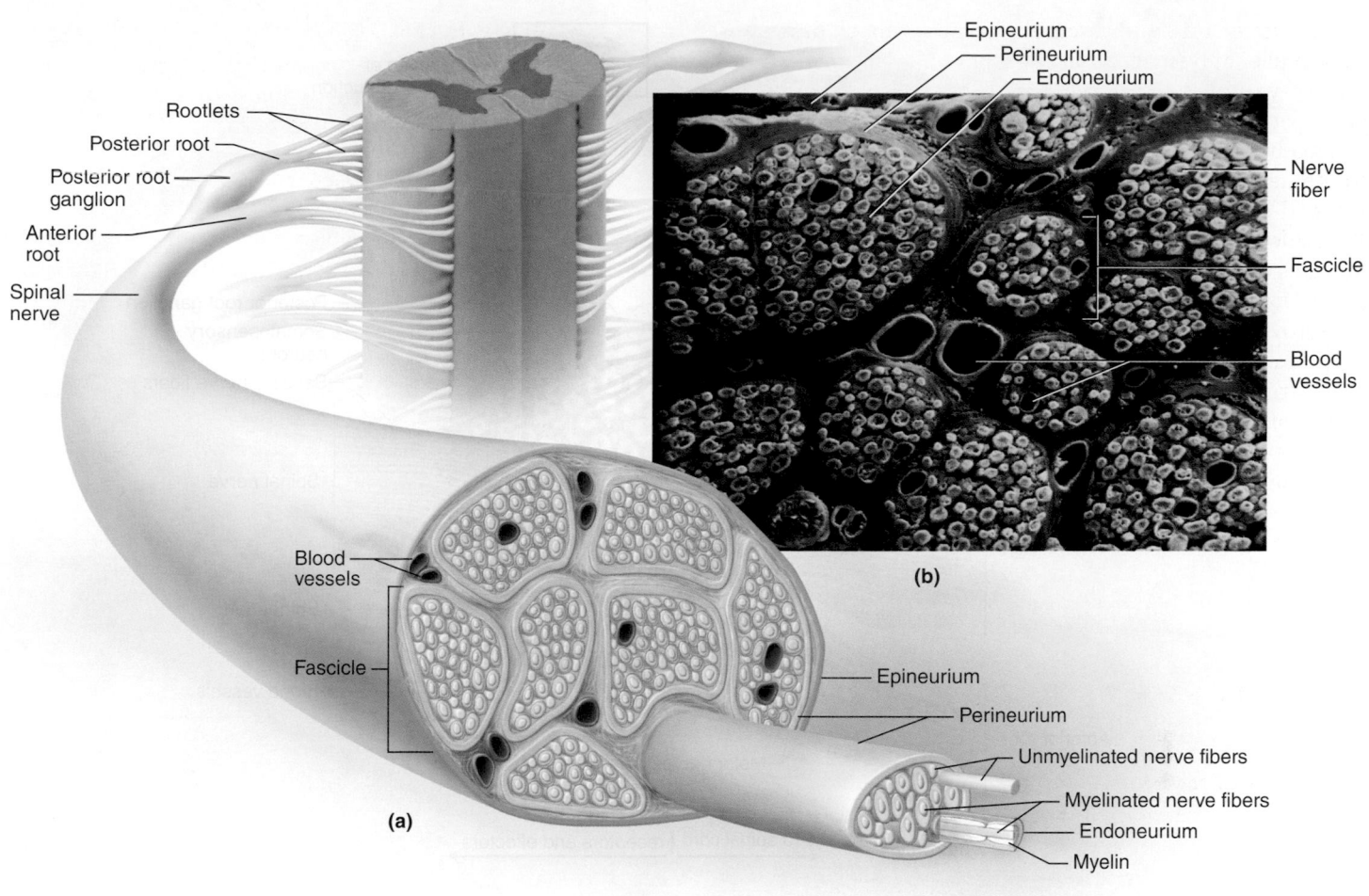

**FIGURE 13.8 Anatomy of a Nerve.** (a) A spinal nerve and its association with the spinal cord. (b) Cross section of a nerve (SEM). Myelinated nerve fibers appear in the photograph as white rings and unmyelinated fibers as solid gray. [Part (b) © Dr. Kessel & Dr. Kardon/Tissues and Organs/Visuals Unlimited, Inc.]

as motor are actually mixed because they carry sensory signals of proprioception from the muscle back to the CNS.

If a nerve resembles a thread, a **ganglion**[20] resembles a knot in the thread. A ganglion is a cluster of neurosomas outside the CNS. It is enveloped in an epineurium continuous with that of the nerve. Among the neurosomas are bundles of nerve fibers leading into and out of the ganglion. Figure 13.9 shows a type of ganglion associated with the spinal nerves.

## Spinal Nerves

There are 31 pairs of **spinal nerves:** 8 cervical (C1–C8), 12 thoracic (T1–T12), 5 lumbar (L1–L5), 5 sacral (S1–S5), and 1 coccygeal (Co1) (fig. 13.10). The first cervical nerve emerges between the skull and atlas, and the others emerge through intervertebral foramina, including the anterior and posterior foramina of the sacrum and the sacral hiatus. Thus, spinal nerves C1

| TABLE 13.2 | The Classification of Nerve Fibers |
|---|---|
| **Class** | **Description** |
| Afferent fibers | Carry sensory signals from receptors to the CNS |
| Efferent fibers | Carry motor signals from the CNS to effectors |
| Somatic fibers | Innervate skin, skeletal muscles, bones, and joints |
| Visceral fibers | Innervate blood vessels, glands, and viscera |
| General fibers | Innervate widespread organs such as muscles, skin, glands, viscera, and blood vessels |
| Special fibers | Innervate more localized organs in the head, including the eyes, ears, olfactory and taste receptors, and muscles of chewing, swallowing, and facial expression |

[20]*gangli* = knot

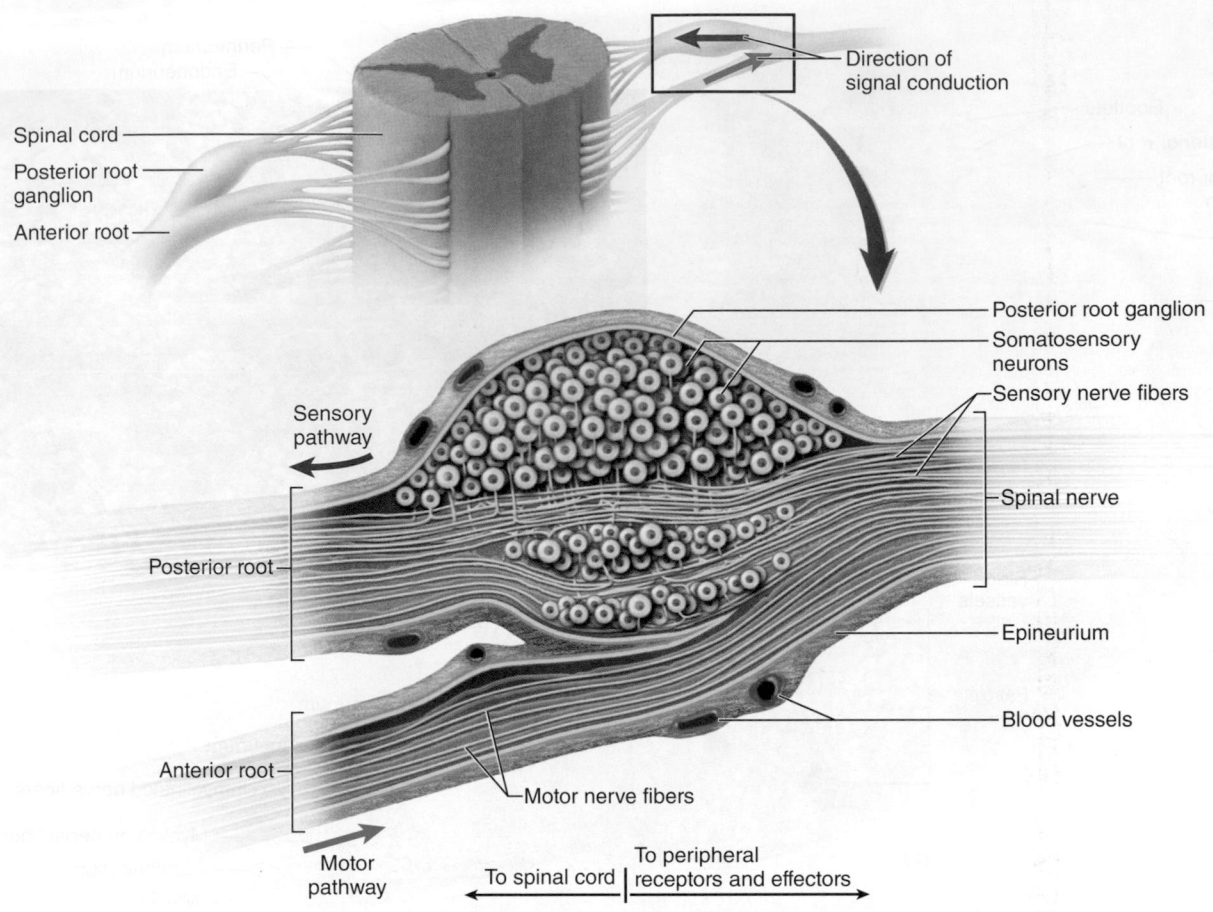

**FIGURE 13.9  Anatomy of a Ganglion (Longitudinal Section).** The posterior root ganglion contains the somas of unipolar sensory neurons conducting signals from peripheral sense organs toward the spinal cord. Below this is the anterior root of the spinal nerve, which conducts motor signals away from the spinal cord, toward peripheral effectors. (The anterior root is not part of the ganglion.)

❓ *Where are the somas of the motor neurons located?*

through C7 emerge superior to the correspondingly numbered vertebrae (nerve C5 above vertebra C5, for example); nerve C8 emerges inferior to vertebra C7; and below this, all the remaining nerves emerge inferior to the correspondingly numbered vertebrae (nerve L3 inferior to vertebra L3, for example).

## Proximal Branches

Each spinal nerve arises from two points of attachment to the spinal cord. In each segment of the cord, six to eight nerve **rootlets** emerge from the anterior surface and converge to form the **anterior (ventral) root** of the spinal nerve. Another six to eight rootlets emerge from the posterior surface and converge to form the **posterior (dorsal) root** (figs. 13.9, 13.11, and 13.12). A short distance away from the spinal cord, the

posterior root swells into a **posterior (dorsal) root ganglion,** which contains the somas of sensory neurons (fig. 13.9). There is no corresponding ganglion on the anterior root.

Slightly distal to the ganglion, the anterior and posterior roots merge, leave the dural sheath, and form the spinal nerve proper (fig. 13.11). The nerve then exits the vertebral canal through the intervertebral foramen. The spinal nerve is a mixed nerve, carrying sensory signals to the spinal cord by way of the posterior root and ganglion, and motor signals out to more distant parts of the body by way of the anterior root.

The anterior and posterior roots are shortest in the cervical region and become longer inferiorly. The roots that arise from segments L2 to Co1 of the cord form the cauda equina. Some viruses invade the CNS by way of the spinal nerve roots (see Deeper Insight 13.3).

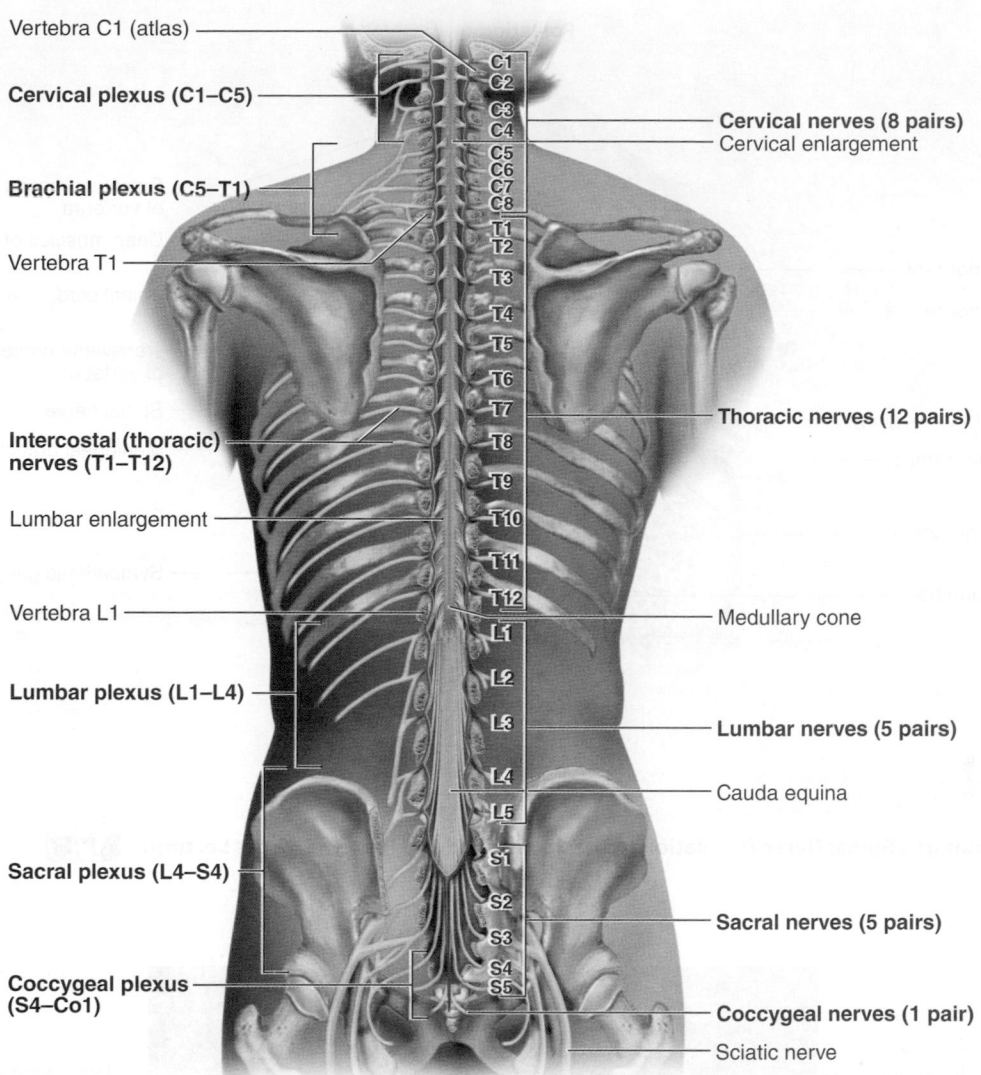

Vertebra C1 (atlas)

**Cervical plexus (C1–C5)**

**Brachial plexus (C5–T1)**

Vertebra T1

**Intercostal (thoracic) nerves (T1–T12)**

Lumbar enlargement

Vertebra L1

**Lumbar plexus (L1–L4)**

**Sacral plexus (L4–S4)**

**Coccygeal plexus (S4–Co1)**

**Cervical nerves (8 pairs)**
Cervical enlargement

**Thoracic nerves (12 pairs)**

Medullary cone

**Lumbar nerves (5 pairs)**

Cauda equina

**Sacral nerves (5 pairs)**

**Coccygeal nerves (1 pair)**
Sciatic nerve

**FIGURE 13.10** **The Spinal Nerve Roots and Plexuses, Posterior Aspect.**

## Distal Branches

Distal to the vertebrae, the branches of a spinal nerve are more complex (fig. 13.13). Immediately after emerging from the intervertebral foramen, the nerve divides into an **anterior ramus,**[21] a **posterior ramus,** and a small **meningeal branch.** Thus, each spinal nerve branches on both ends—into anterior and posterior *roots* approaching the spinal cord, and anterior and posterior *rami* leading away from the vertebral column.

The meningeal branch (see fig. 13.11) reenters the vertebral canal and innervates the meninges, vertebrae, and spinal ligaments with sensory and motor fibers. The posterior ramus innervates the muscles and joints in that region of the spine and

the skin of the back. The larger anterior ramus innervates the anterior and lateral skin and muscles of the trunk, and gives rise to nerves of the limbs.

The anterior ramus differs from one region of the trunk to another. In the thoracic region, it forms an **intercostal nerve,** which travels along the inferior margin of a rib and innervates the skin and intercostal muscles (thus contributing to breathing). It also innervates the internal oblique, external oblique, and transverse abdominal muscles. All other anterior rami form the *nerve plexuses* described next.

As shown in figure 13.13, the anterior ramus also gives off a pair of *communicating rami,* which connect with a string of *sympathetic chain ganglia* alongside the vertebral column. These are seen only in spinal nerves T1 through L2. They are components of the sympathetic nervous system and are discussed more fully in chapter 15.

[21]*ramus* = branch

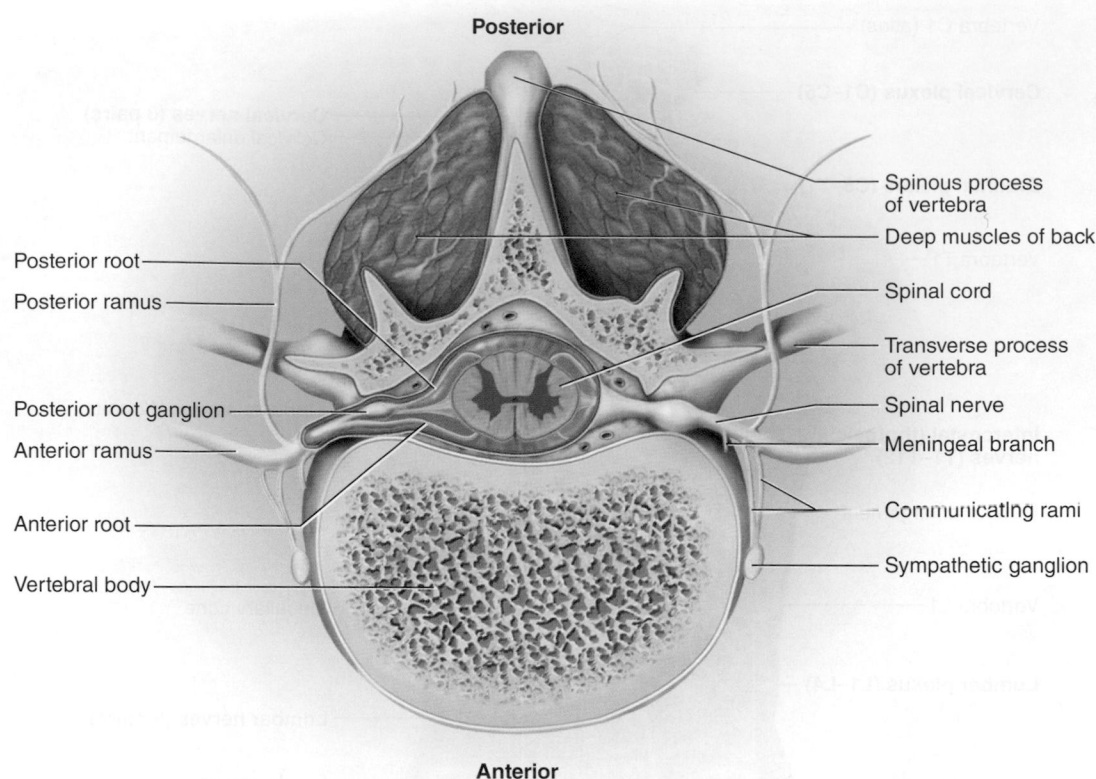

**Posterior**

Posterior root

Posterior ramus

Posterior root ganglion

Anterior ramus

Anterior root

Vertebral body

Spinous process of vertebra

Deep muscles of back

Spinal cord

Transverse process of vertebra

Spinal nerve

Meningeal branch

Communicating rami

Sympathetic ganglion

**Anterior**

**FIGURE 13.11  Branches of a Spinal Nerve in Relation to the Spinal Cord and Vertebra (Cross Section).** AP|R

Posterior median sulcus

Gracile fasciculus

Cuneate fasciculus

Lateral column

Segment C5

Cross section

Arachnoid mater

Dura mater

Neural arch of vertebra C3 (cut)

Spinal nerve C4

Vertebral artery

Spinal nerve C5:

Rootlets

Posterior root

Posterior root ganglion

Anterior root

**FIGURE 13.12  The Point of Entry of Two Spinal Nerves into the Spinal Cord.**  Posterior (dorsal) view with vertebrae cut away. Note that each posterior root divides into several rootlets that enter the spinal cord. A segment of the spinal cord is the portion receiving all the rootlets of one spinal nerve.

❓ *In the labeled rootlets of spinal nerve C5, are the nerve fibers afferent or efferent? How do you know?*

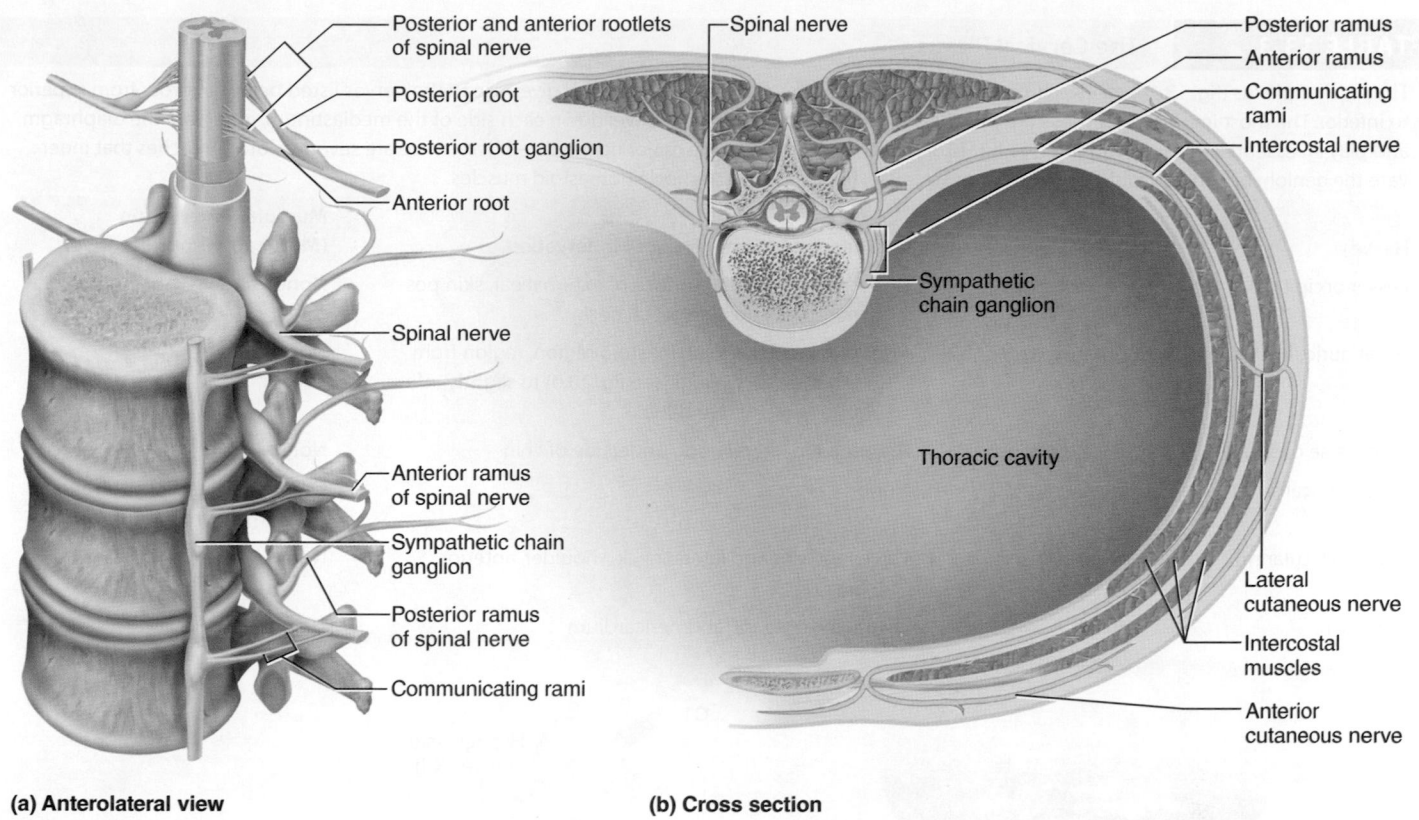

**(a) Anterolateral view**

**(b) Cross section**

**FIGURE 13.13  Rami of the Spinal Nerves.**  (a) Anterolateral view of the spinal nerves and their subdivisions in relation to the spinal cord and vertebrae. (b) Cross section of the thorax showing innervation of muscles and skin of the chest and back. This section is cut through the intercostal muscles between two ribs.

## Nerve Plexuses

Except in the thoracic region, the anterior rami branch and anastomose (merge) repeatedly to form five webs called nerve plexuses: the small **cervical plexus** in the neck, the **brachial plexus** near the shoulder, the **lumbar plexus** of the lower back, the **sacral plexus** immediately inferior to this, and finally, the tiny **coccygeal plexus** adjacent to the lower sacrum and coccyx. A general view of these plexuses is shown in figure 13.10; they are illustrated and described in tables 13.3 through 13.6. The spinal nerve roots that give rise to each plexus are indicated in violet in each table. Some of these roots give rise to smaller branches called *trunks, anterior divisions, posterior divisions,* and *cords,* which are color-coded and explained in the individual tables.

The nerves tabulated here have somatosensory and motor functions. *Somatosensory* means that they carry sensory signals from bones, joints, muscles, and the skin, in contrast to sensory input from the viscera or from special sense organs such as the eyes and ears. Somatosensory signals are for touch, heat, cold, stretch, pressure, pain, and other sensations. One of the most important sensory roles of these nerves is *proprioception,* in which the brain receives information about body position and movements from nerve endings in the muscles, tendons, and joints. The brain uses this information to adjust muscle actions and thereby maintain equilibrium (balance) and coordination.

The motor function of these nerves is primarily to stimulate the contraction of skeletal muscles. They also innervate the bones of the corresponding regions, and carry autonomic fibers to some viscera and blood vessels, thus adjusting blood flow to local needs.

The following tables identify the areas of skin innervated by the sensory fibers and the muscle groups innervated by the motor fibers of the individual nerves. The muscle tables in chapter 10 provide a more detailed breakdown of the muscles supplied by each nerve and the actions they perform. You may assume that for each muscle, these nerves also carry sensory fibers from its proprioceptors. Throughout these tables, *nerve* is abbreviated *n.* and *nerves* as *nn.*

## TABLE 13.3    The Cervical Plexus

The cervical plexus (fig. 13.14) receives fibers from the anterior rami of nerves C1 to C5 and gives rise to the nerves listed below, in order from superior to inferior. The most important of these are the *phrenic*[22] (FREN-ic) *nerves,* which travel down each side of the mediastinum, innervate the diaphragm, and play an essential role in breathing (see fig. 15.3, p. 561). In addition to the major nerves listed here, there are several motor branches that innervate the geniohyoid, thyrohyoid, scalene, levator scapulae, trapezius, and sternocleidomastoid muscles.

| Nerve | Composition | Cutaneous and Other Sensory Innervation | Muscular Innervation (Motor and Proprioceptive) |
|---|---|---|---|
| Lesser occipital n. | Somatosensory | Upper third of medial surface of external ear, skin posterior to ear, posterolateral neck | None |
| Great auricular n. | Somatosensory | Most of the external ear, mastoid region, region from parotid salivary gland (see fig. 10.6) to slightly inferior to angle of mandible | None |
| Transverse cervical n. | Somatosensory | Anterior and lateral neck, underside of chin | None |
| Ansa cervicalis | Mixed | None | Omohyoid, sternohyoid, and sternothyroid muscles |
| Supraclavicular nn. | Somatosensory | Lower anterior and lateral neck, shoulder, anterior chest | None |
| Phrenic n. | Mixed | Diaphragm, pleura, and pericardium | Diaphragm |

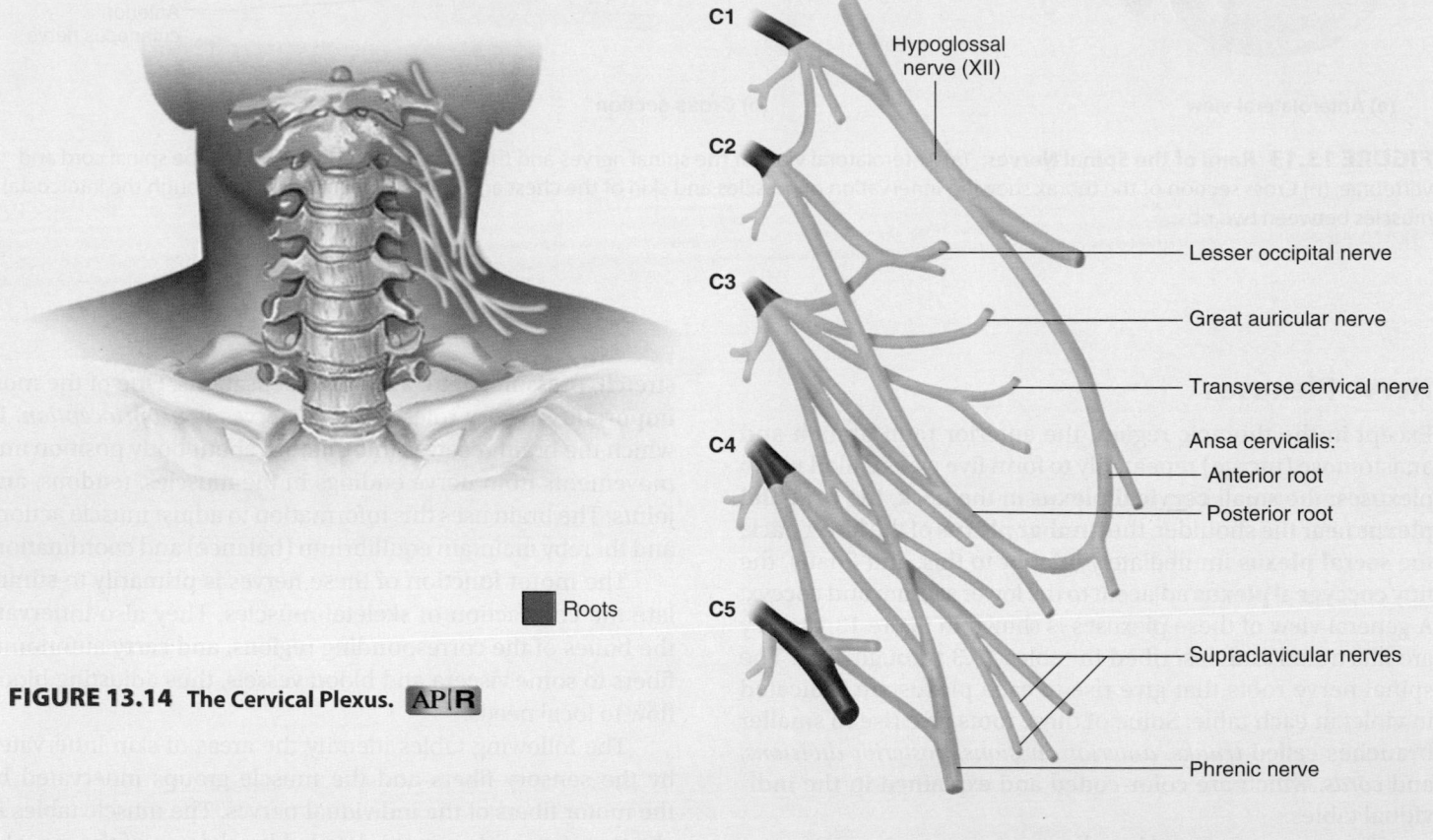

**FIGURE 13.14**  The Cervical Plexus. AP|R

Roots

C1

Hypoglossal nerve (XII)

C2

C3 — Lesser occipital nerve

— Great auricular nerve

— Transverse cervical nerve

Ansa cervicalis:
C4 — Anterior root
— Posterior root

C5 — Supraclavicular nerves

— Phrenic nerve

[22]*phren* = diaphragm

| **TABLE 13.4** | **The Brachial Plexus** |

The brachial plexus (figs. 13.15 and 13.16) is formed predominantly by the anterior rami of nerves C5 to T1 (C4 and T2 make small contributions). It passes over the first rib into the axilla and innervates the upper limb and some muscles of the neck and shoulder. This plexus is well known for its conspicuous M or W shape seen in cadaver dissections. The subdivisions of this plexus are called *roots, trunks, divisions,* and *cords* (color-coded in fig. 13.15). The five **roots** are the anterior rami of C5 through T1. Roots C5 and C6 converge to form the **upper trunk;** C7 continues as the **middle trunk;** and C8 and T1 converge to form the **lower trunk.** Each trunk divides into an **anterior** and **posterior division.** As the body is dissected from the anterior surface of the shoulder inward, the posterior divisions are found behind the anterior ones. Finally, the six divisions merge to form three large fiber bundles—the **lateral, posterior,** and **medial cords.** From these cords arise the following major nerves, listed in order of the illustration from superior to inferior.

| Nerve | Composition | Cord of Origin | Cutaneous and Joint Innervation (Sensory) | Muscular Innervation (Motor and Proprioceptive) |
|---|---|---|---|---|
| Musculocutaneous n. | Mixed | Lateral | Skin of anterolateral forearm; elbow joint | Brachialis, biceps brachii, and coracobrachialis muscles |
| Axillary n. | Mixed | Posterior | Skin of lateral shoulder and arm; shoulder joint | Deltoid and teres minor muscles |
| Radial n. | Mixed | Posterior | Skin of posterior arm; posterior and lateral forearm and wrist; joints of elbow, wrist, and hand | Mainly extensor muscles of posterior arm and forearm (see tables 10.10 and 10.11) |
| Median n. | Mixed | Lateral and medial | Skin of lateral two-thirds of hand; tips of digits I–IV; joints of hand | Mainly forearm flexors; thenar group and lumbricals I–II of hand (see tables 10.10 to 10.12) |
| Ulnar n. | Mixed | Medial | Skin of palmar and medial hand and digits III–V; joints of elbow and hand | Some forearm flexors; adductor pollicis; hypothenar group; interosseous muscles; lumbricals III–IV (see tables 10.11 and 10.12) |

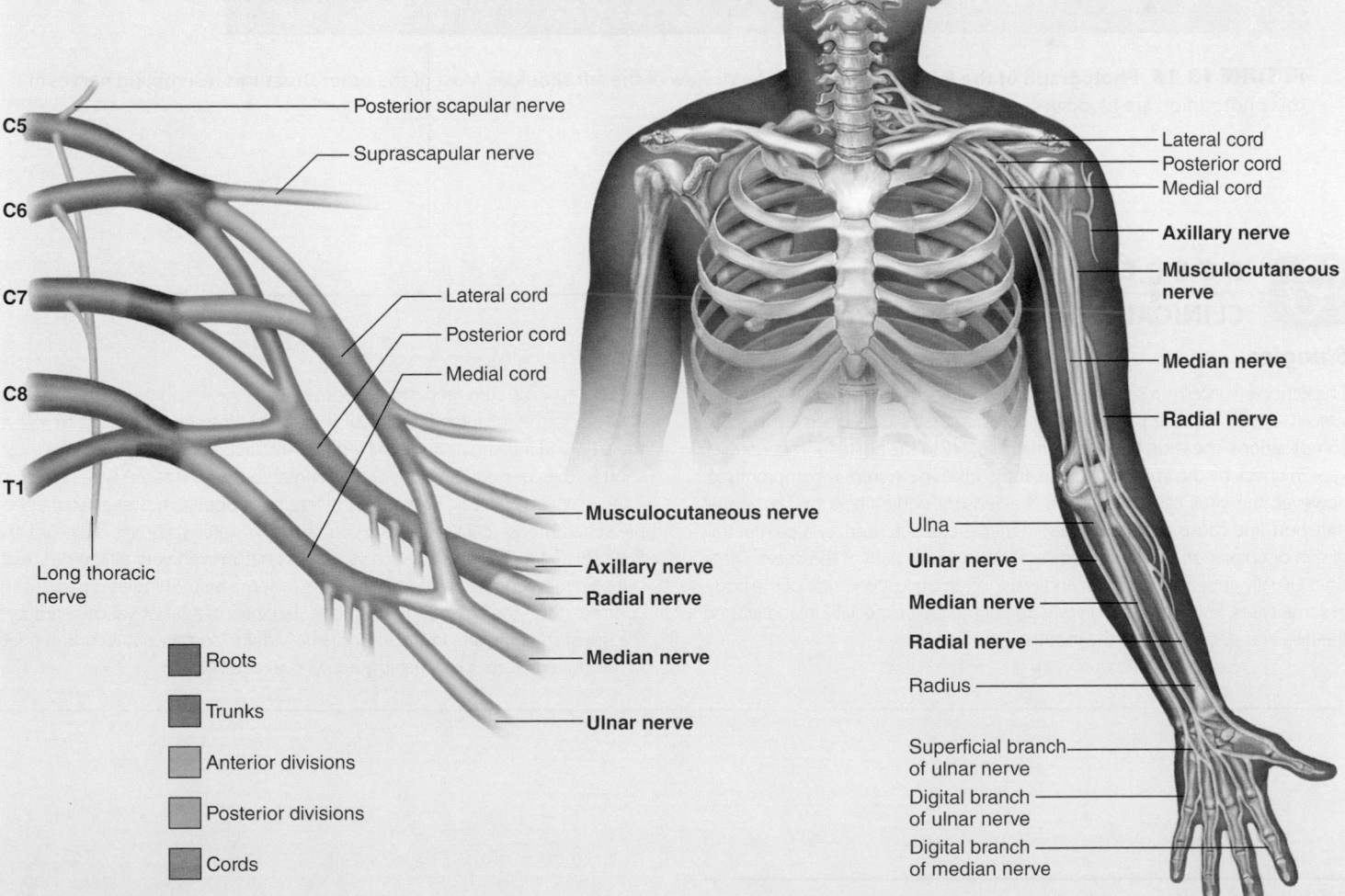

**FIGURE 13.15 The Brachial Plexus.** The labeled nerves innervate muscles tabulated in chapter 10, and those in boldface are further detailed in this table. AP|R

| TABLE 13.4 | The Brachial Plexus (continued) |
|---|---|

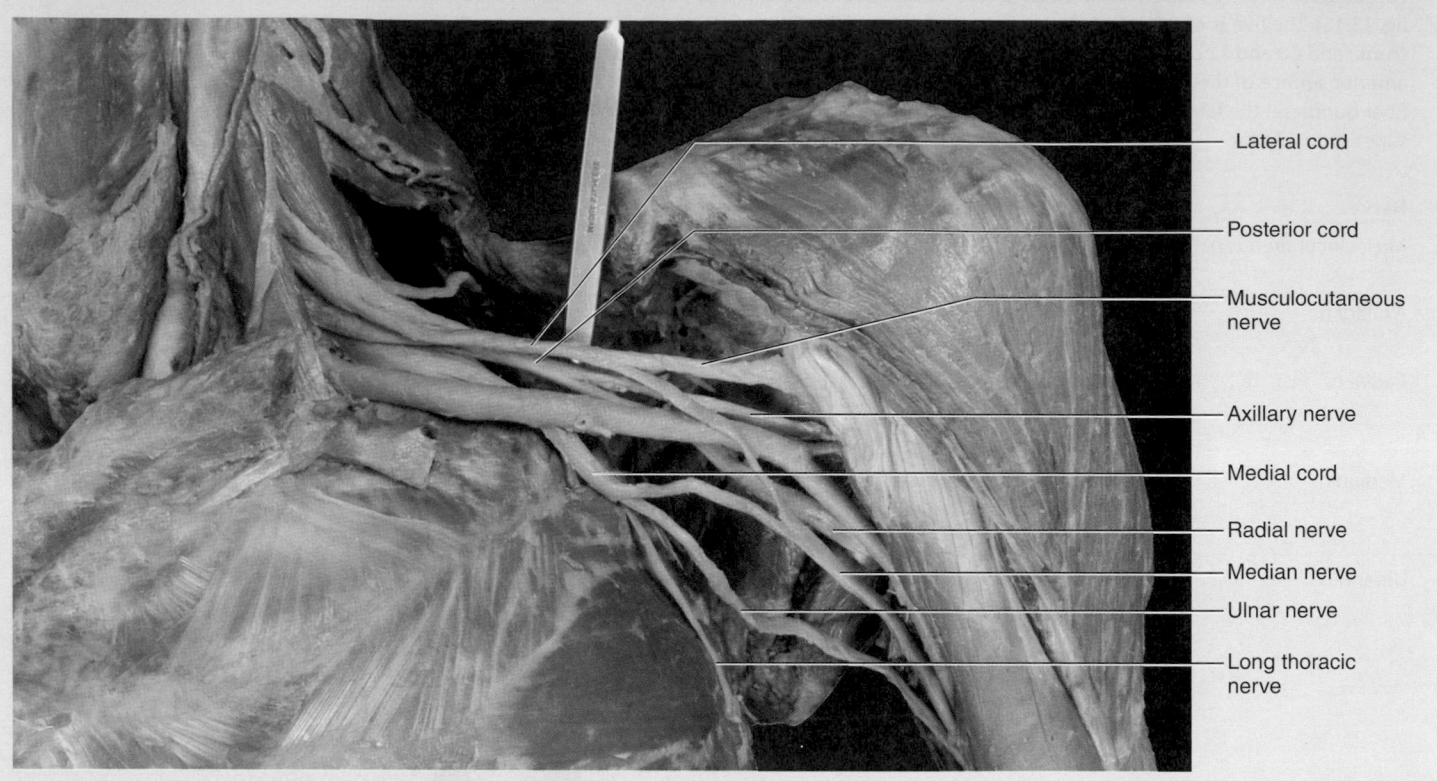

**FIGURE 13.16 Photograph of the Brachial Plexus.** Anterior view of the left shoulder. Most of the other structures resembling nerves in this photograph are blood vessels.

## DEEPER INSIGHT 13.3

### CLINICAL APPLICATION

#### Shingles

Chickenpox *(varicella)*, a common disease of early childhood, is caused by the *varicella-zoster*[23] virus. It produces an itchy rash that usually clears up without complications. The virus, however, remains for life in the posterior root ganglia, kept in check by the immune system. If the immune system is compromised, however, the virus can travel along the sensory nerve fibers by fast axonal transport and cause *shingles (herpes*[24] *zoster)*—characterized by a painful trail of skin discoloration and fluid-filled vesicles along the path of the nerve. These signs usually appear in the chest and waist, often on just one side of the body. In some cases, lesions appear on one side of the face, especially in and around the eye, and occasionally in the mouth.

There is no cure, and the vesicles usually heal spontaneously in 1 to 3 weeks. In the meantime, aspirin and steroidal ointments can help to relieve the pain and inflammation of the lesions. Antiviral drugs such as acyclovir can shorten the course of an episode of shingles, but only if taken within the first 2 to 3 days of outbreak. Even after the lesions disappear, however, some people suffer intense pain along the course of the nerve *(postherpetic neuralgia, PHN)*, lasting for months or even years. PHN has proven very difficult to treat, but pain relievers and antidepressants are of some help. Shingles is particularly common after the age of 50. Childhood vaccination against varicella reduces the risk of shingles later in life. A vaccine for adults (Zostavax) is recommended in the United States for all healthy adults over age 60.

---

[23]*varicella* = little spot; *zoster* = girdle
[24]*herpes* = creeping

| TABLE 13.5 | The Lumbar Plexus |
|---|---|

The lumbar plexus (fig. 13.17) is formed from the anterior rami of nerves L1 to L4 and some fibers from T12. With only five roots and two divisions, it is less complex than the brachial plexus. It gives rise to the following nerves.

| Nerve | Composition | Cutaneous and Joint Innervation (Sensory) | Muscular Innervation (Motor and Proprioceptive) |
|---|---|---|---|
| Iliohypogastric n. | Mixed | Skin of lower anterior abdominal and posterolateral gluteal regions | Internal and external abdominal oblique and transverse abdominal muscles |
| Ilioinguinal n. | Mixed | Skin of upper medial thigh; male scrotum and root of penis; female labia majora | Internal abdominal oblique |
| Genitofemoral n. | Mixed | Skin of middle anterior thigh; male scrotum; female labia majora | Male cremaster muscle (see p. 1036) |
| Lateral femoral cutaneous n. | Somatosensory | Skin of anterior and upper lateral thigh | None |
| Femoral n. | Mixed | Skin of anterior, medial, and lateral thigh and knee; skin of medial leg and foot; hip and knee joints | Iliacus, pectineus, quadriceps femoris, and sartorius muscles |
| Obturator n. | Mixed | Skin of medial thigh; hip and knee joints | Obturator externus; medial (adductor) thigh muscles (see table 10.13) |

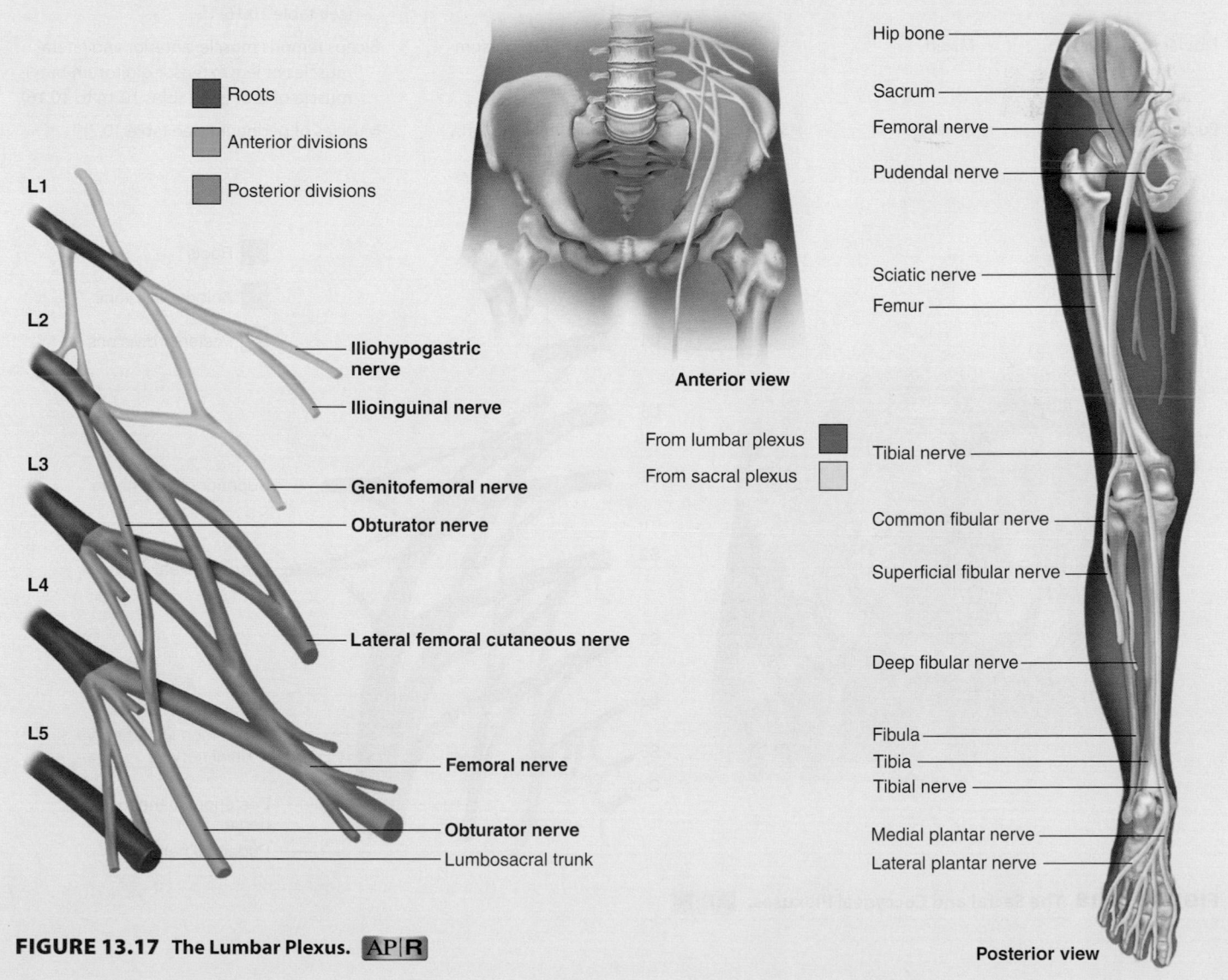

**FIGURE 13.17 The Lumbar Plexus.** AP|R

## TABLE 13.6    The Sacral and Coccygeal Plexuses

The sacral plexus is formed from the anterior rami of nerves L4, L5, and S1 through S4. It has six roots and anterior and posterior divisions. Since it is connected to the lumbar plexus by fibers that run through the *lumbosacral trunk,* the two plexuses are sometimes referred to collectively as the *lumbosacral plexus.* The coccygeal plexus is a tiny plexus formed from the anterior rami of S4, S5, and Co1 (fig. 13.18).

The *tibial* and *common fibular nerves* travel together through a connective tissue sheath; they are referred to collectively as the sciatic (sy-AT-ic) nerve. The sciatic nerve passes through the greater sciatic notch of the pelvis, extends for the length of the thigh, and ends at the popliteal fossa. Here, the tibial and common fibular nerves diverge and follow their separate paths into the leg. The tibial nerve descends through the leg and then gives rise to the medial and plantar nerves in the foot. The common fibular nerve divides into deep and superficial fibular nerves. The sciatic nerve is a common focus of injury and pain (see Deeper Insight 13.4).

| Nerve | Composition | Cutaneous and Joint Innervation (Sensory) | Muscular Innervation (Motor and Proprioceptive) |
|---|---|---|---|
| Superior gluteal n. | Mixed | None | Gluteus minimus, gluteus medius, and tensor fasciae latae muscles |
| Inferior gluteal n. | Mixed | None | Gluteus maximus muscle |
| Posterior cutaneous n. | Somatosensory | Skin of gluteal region, perineum, posterior and medial thigh, popliteal fossa, and upper posterior leg | None |
| Tibial n. | Mixed | Skin of posterior leg; plantar skin; knee and foot joints | Hamstring muscles; posterior muscles of leg (see tables 10.14 and 10.15); most intrinsic foot muscles (via plantar nerves) (see table 10.16) |
| Fibular (peroneal) nn. (common, deep, and superficial) | Mixed | Skin of anterior distal third of leg, dorsum of foot, and toes I–II; knee joint | Biceps femoris muscle; anterior and lateral muscles of leg; extensor digitorum brevis muscle of foot (see tables 10.14 to 10.16) |
| Pudendal n. | Mixed | Skin of penis and scrotum of male; clitoris, labia majora and minora, and lower vagina of female | Muscles of perineum (see table 10.7) |

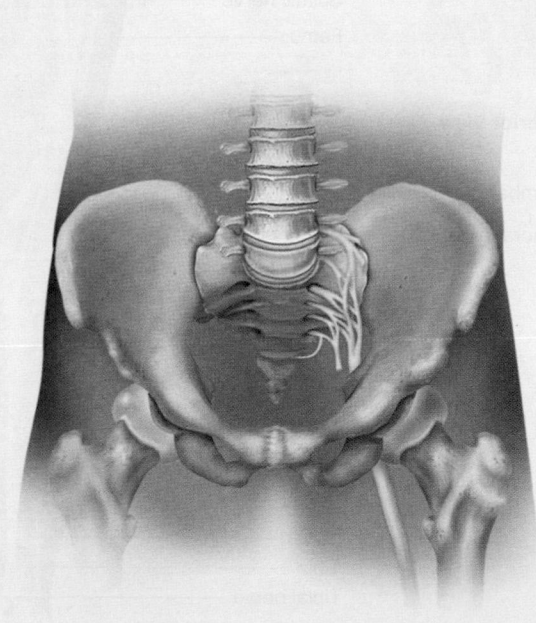
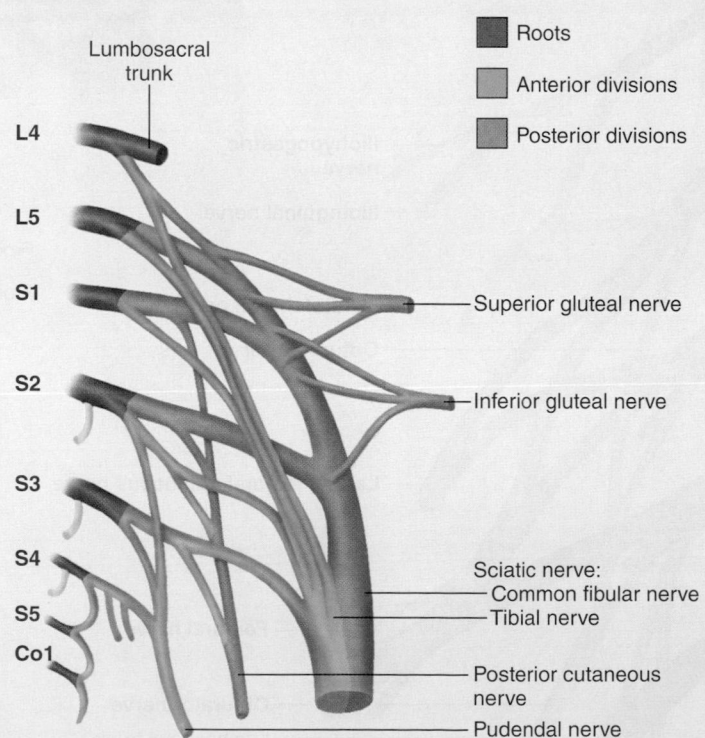

**FIGURE 13.18** The Sacral and Coccygeal Plexuses. AP|R

**DEEPER INSIGHT 13.4**

**CLINICAL APPLICATION**

*Nerve Injuries*

The radial and sciatic nerves are especially vulnerable to injury. The radial nerve, which passes through the axilla, may be compressed against the humerus by improperly adjusted crutches, causing *crutch paralysis*. A similar injury often resulted from the now-discredited practice of correcting a dislocated shoulder by putting a foot in a person's armpit and pulling on the arm. One consequence of radial nerve injury is *wrist drop*—the fingers, hand, and wrist are chronically flexed because the extensor muscles supplied by the radial nerve are paralyzed.

Because of its position and length, the sciatic nerve of the hip and thigh is the most vulnerable nerve in the body. Trauma to this nerve produces *sciatica*, a sharp pain that travels from the gluteal region along the posterior side of the thigh and leg as far as the ankle. Ninety percent of cases result from a herniated intervertebral disc or osteoarthritis of the lower spine, but sciatica can also be caused by pressure from a pregnant uterus, dislocation of the hip, injections in the wrong area of the buttock, or sitting for a long time on the edge of a hard chair. Men sometimes suffer sciatica because of the habit of sitting on a wallet carried in the hip pocket.

## Cutaneous Innervation and Dermatomes

Each spinal nerve except C1 receives sensory input from a specific area of skin called a **dermatome.**[25] A *dermatome map* (fig. 13.19) is a diagram of the cutaneous regions innervated by each spinal nerve. Such a map is oversimplified, however, because the dermatomes overlap at their edges by as much as 50%. Therefore, severance of one sensory nerve root does not entirely deaden sensation from a dermatome. It is necessary to sever or anesthetize three successive spinal nerves to produce a total loss of sensation from one dermatome. Spinal nerve damage is assessed by testing the dermatomes with pinpricks and noting areas in which the patient has no sensation.

**BEFORE YOU GO ON**

Answer the following questions to test your understanding of the preceding section:

6. What is meant by the anterior and posterior roots of a spinal nerve? Which of these Is sensory and which is motor?

7. Where are the neurosomas of the posterior root located? Where are the neurosomas of the anterior root?

8. List the five plexuses of spinal nerves and state where each one is located.

9. State which plexus gives rise to each of the following nerves: axillary, ilioinguinal, obturator, phrenic, pudendal, radial, and sciatic.

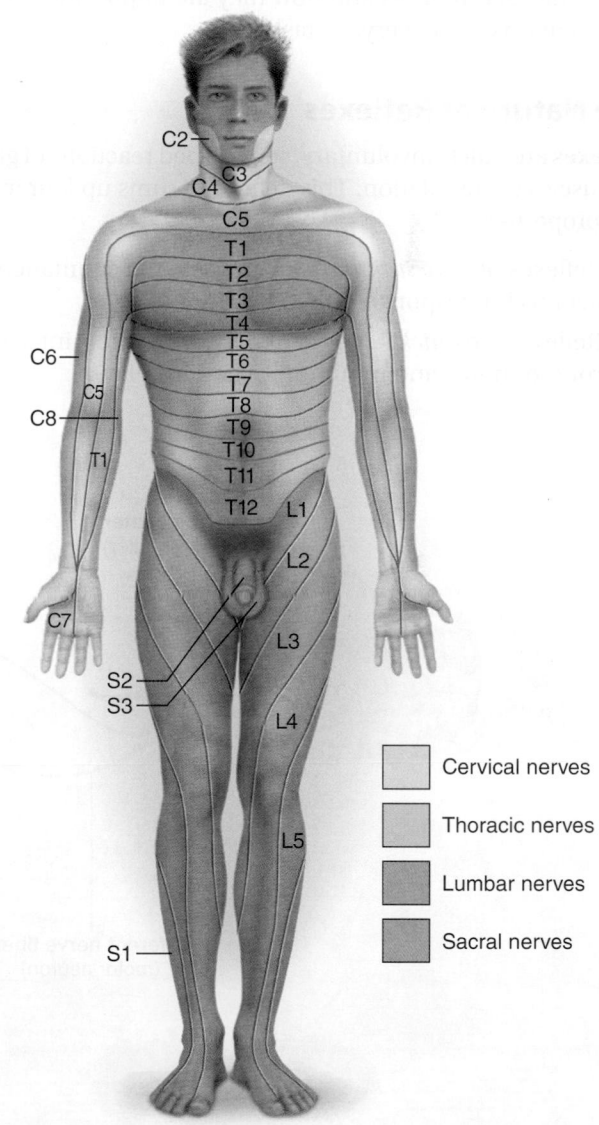

**FIGURE 13.19   A Dermatome Map of the Anterior Aspect of the Body.** Each zone of the skin is innervated by sensory branches of the spinal nerves indicated by the labels. Nerve C1 does not innervate the skin.

[25]*derma* = skin; *tome* = segment, part

## 13.3   Somatic Reflexes

### Expected Learning Outcomes

When you have completed this section, you should be able to

a. define *reflex* and explain how reflexes differ from other motor actions;

b. describe the general components of a typical reflex arc; and

c. explain how the basic types of somatic reflexes function.

Most of us have had our reflexes tested with a little rubber hammer; a tap below the knee produces an uncontrollable jerk of the leg, for example. In this section, we discuss what reflexes are and how they are produced by an assembly of receptors, neurons, and effectors. We also survey the different types of neuromuscular reflexes and how they are important in motor coordination of our everyday tasks.

### The Nature of Reflexes

**Reflexes** are quick, involuntary, stereotyped reactions of glands or muscles to stimulation. This definition sums up four important properties:

1. Reflexes *require stimulation*—they are not spontaneous actions but responses to sensory input.

2. Reflexes are *quick*—they generally involve few interneurons, or none, and minimum synaptic delay.

3. Reflexes are *involuntary*—they occur without intent, often without our awareness, and they are difficult to suppress. Given an adequate stimulus, the response is essentially automatic. You may become conscious of the stimulus that evoked a reflex, and this awareness may enable you to correct or avoid a potentially dangerous situation, but awareness is not a part of the reflex itself. It may come after the reflex action has been completed, and somatic reflexes can occur even if the spinal cord has been severed so that no stimuli reach the brain.

4. Reflexes are *stereotyped*—they occur in essentially the same way every time; the response is very predictable, unlike the variability of voluntary movement.

Reflexes include glandular secretion and contractions of all three types of muscle. The reflexes of skeletal muscle are called **somatic reflexes,** since they involve the somatic nervous system. Chapter 15 concerns the *visceral reflexes* of organs such as the heart and intestines. Somatic reflexes have traditionally been called *spinal reflexes,* but this is a misleading expression for two reasons: (1) Spinal reflexes are not exclusively somatic; visceral reflexes also involve the spinal cord. (2) Some somatic reflexes are mediated more by the brain than by the spinal cord.

A somatic reflex employs a **reflex arc,** in which signals travel along the following pathway (fig. 13.20):

1. *somatic receptors* in the skin, muscles, and tendons;

2. *afferent nerve fibers,* which carry information from these receptors to the posterior horn of the spinal cord or to the brainstem;

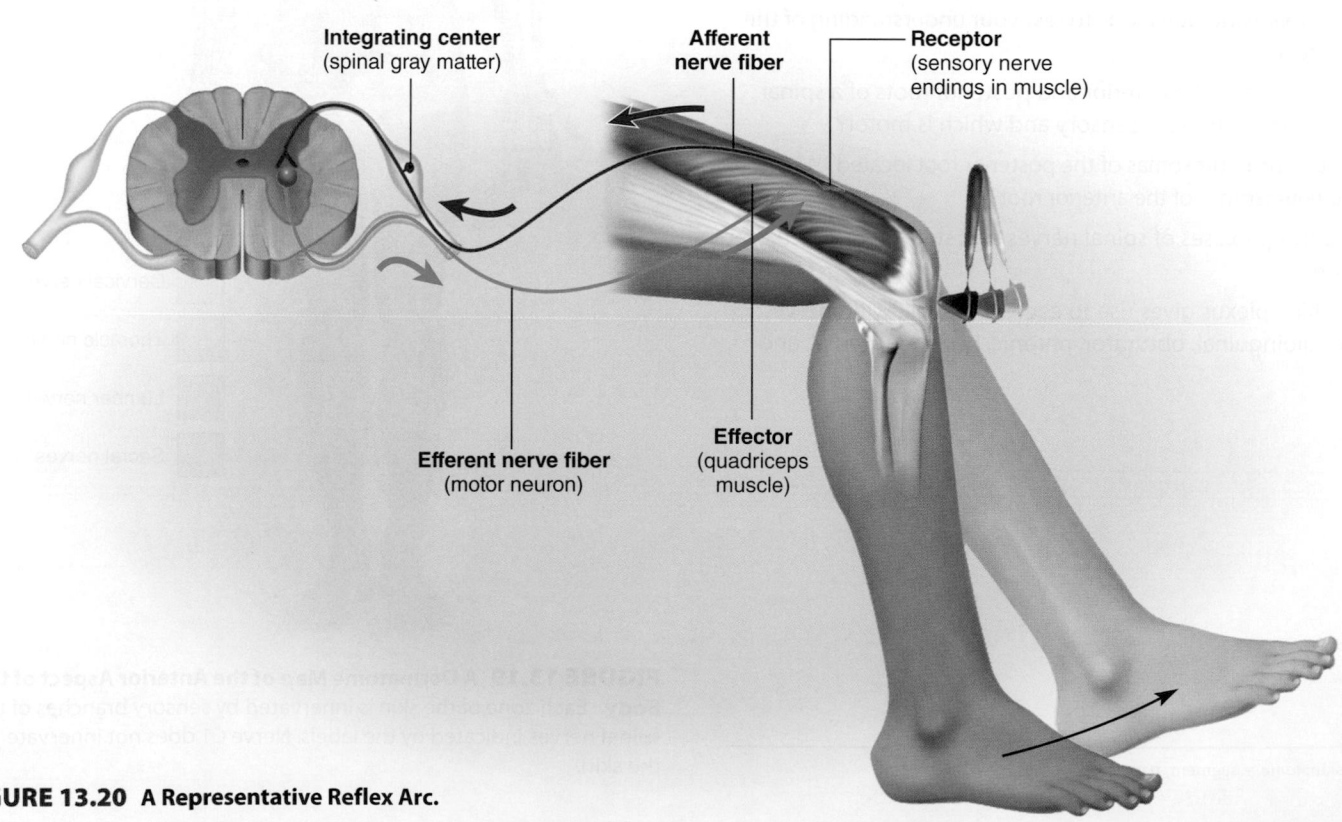

**Integrating center** (spinal gray matter)

**Afferent nerve fiber**

**Receptor** (sensory nerve endings in muscle)

**Efferent nerve fiber** (motor neuron)

**Effector** (quadriceps muscle)

**FIGURE 13.20  A Representative Reflex Arc.**

3. an *integrating center,* a point of synaptic contact between neurons in the gray matter of the cord or brainstem;

4. *efferent nerve fibers,* which carry motor impulses to the muscles; and

5. *effectors,* the muscles that carry out the response.

In most reflex arcs, the integrating center includes one or more interneurons. Synaptic events in the integrating center determine whether the efferent neurons issue signals to the muscles. The more interneurons there are, the more complex the information processing can be, but with more synapses, there is a longer delay between input and output.

## The Muscle Spindle

Many somatic reflexes involve stretch receptors called **muscle spindles** embedded in the muscles. These are among the body's **proprioceptors,** sense organs specialized to monitor the position and movement of body parts. The function of

muscle spindles is to inform the brain of muscle length and body movements. This enables the brain to send motor commands back to the muscles that control muscle tone, posture, coordinated movement, and corrective reflexes (for example, to keep one's balance). Spindles are especially abundant in muscles that require fine control. Hand and foot muscles have 100 or more spindles per gram of muscle, whereas there are relatively few in large muscles with coarse movements, and none at all in the middle-ear muscles.

A muscle spindle is a bundle of usually seven or eight small, modified muscle fibers enclosed in an elongated fibrous capsule about 5 to 10 mm long (fig. 13.21). Spindles are especially concentrated at the ends of a muscle, near its tendons. The modified muscle fibers within the spindle are called **intrafusal**[26] **fibers,** whereas those that make up the rest of the muscle and do its work are called **extrafusal fibers.**

---

[26]*intra* = within; *fus* = spindle

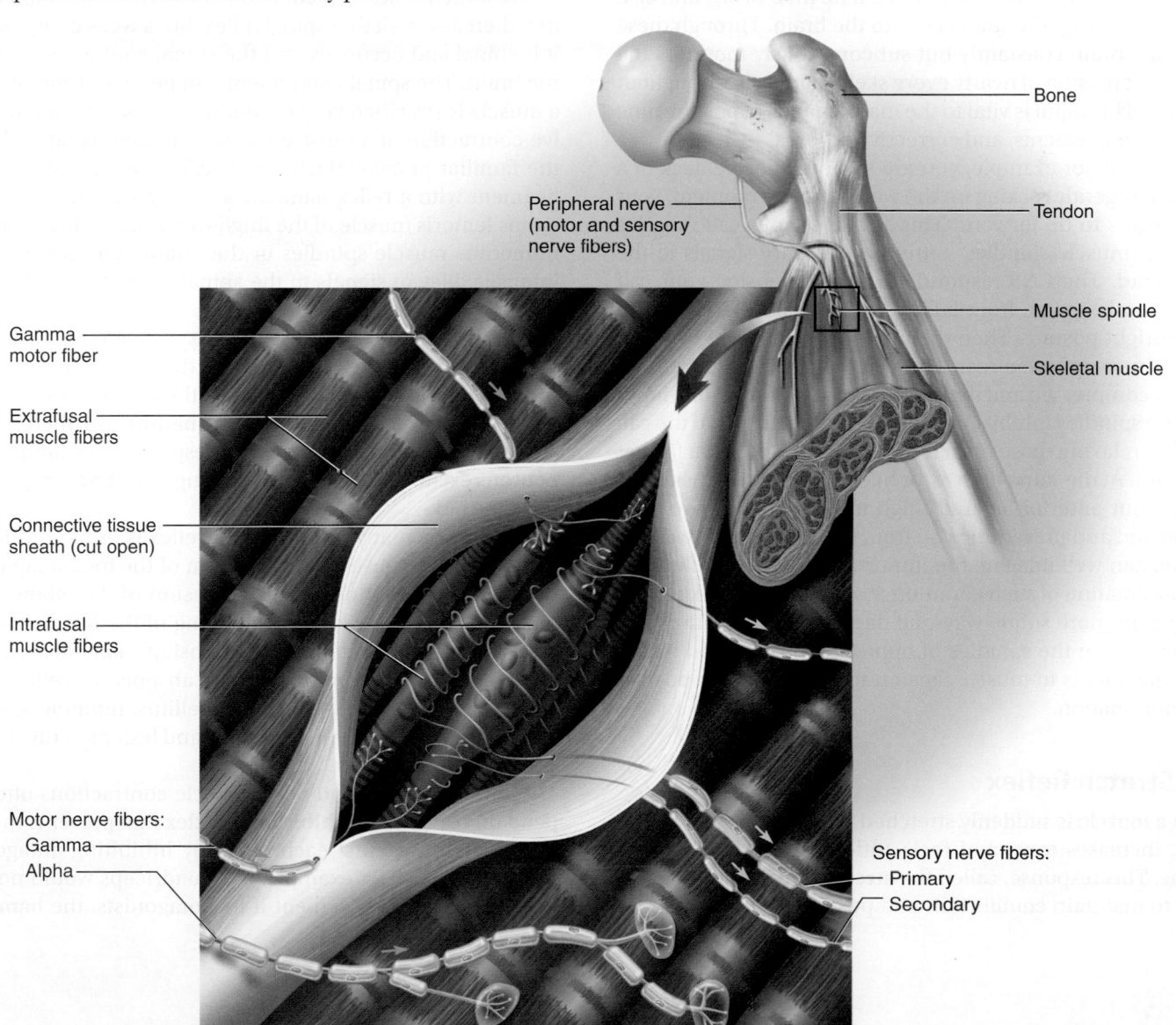

Bone

Peripheral nerve (motor and sensory nerve fibers)

Tendon

Muscle spindle

Skeletal muscle

Gamma motor fiber

Extrafusal muscle fibers

Connective tissue sheath (cut open)

Intrafusal muscle fibers

Motor nerve fibers:
Gamma
Alpha

Sensory nerve fibers:
Primary
Secondary

**FIGURE 13.21 A Muscle Spindle and Its Innervation.**

Each end of an intrafusal fiber has a few sarcomeres and is capable of contracting. A **gamma motor neuron** of the spinal cord innervates each end and stimulates its contraction. This maintains tension and sensitivity of the intrafusal fiber, preventing it from going slack like an unstretched rubber band when a muscle shortens. Spinal motor neurons that supply the extrafusal muscle fibers are called **alpha motor neurons.** Up to now, we have studied only that type and the neuromuscular junctions they form with muscle (chapter 11), but nearly one-third of all spinal motor neurons are the gamma type—evidence of the great importance of muscle spindles.

The long midportion of an intrafusal fiber lacks sarcomeres and cannot contract, but is supplied by two types of sensory nerve fibers: *primary afferent fibers* that monitor muscle length and how rapidly it changes, which are therefore very responsive to sudden body movements; and *secondary afferent fibers* that monitor length only, not rate of change. These sensory fibers enter the posterior horn of the spinal cord, synapse on the alpha motor neurons and regulate their firing, and also send branches up the spinal cord to the brain. Through these fibers, the brain constantly but subconsciously monitors the length and tension of nearly every skeletal muscle throughout the body. This input is vital to the maintenance of posture, fine control of movements, and corrective reflexes.

Suppose, for example, you are standing on the deck of a boat that is gently rocking on the waves. At one moment, your body begins to tip forward. This stretches your calf muscles and their muscle spindles, setting off sensory signals to the spinal cord. The CNS responds to this by tensing your calf muscles to keep you from falling and to restore or maintain your upright posture. Then the boat rocks the other way and you begin to tip to the rear. The spindles in the calf muscles are now compressed and their signaling rate drops. Such input from the spindles inhibits the alpha motor neurons of the calf muscles, relaxing those muscles so they don't pull you farther backward. At the same time, your backward tilt stretches spindles in your anterior leg and thigh muscles, leading to their contraction and preventing you from falling over backward.

You can well imagine the importance of these reflexes to the coordination of such common movements as walking and dancing. In more subtle ways, all day long, your brain monitors input from the spindles of opposing muscles and makes fine adjustments in muscle tension to maintain your posture and coordination.

## The Stretch Reflex

When a muscle is suddenly stretched, it "fights back"—it contracts, increases tone, and feels stiffer than an unstretched muscle. This response, called the **stretch (myotatic[27]) reflex,** helps to maintain equilibrium and posture, as we just saw in

the rocking boat example. To take another case, if your head starts to tip forward, it stretches muscles at the back of your neck. This stimulates their muscle spindles, which send signals to the cerebellum by way of the brainstem. The cerebellum integrates this information and relays it to the cerebral cortex, and the cortex sends signals back, via the brainstem, to the muscles. The muscles contract and raise your head.

Stretch reflexes often feed back not to a single muscle but to a set of synergists and antagonists. Since the contraction of a muscle on one side of a joint stretches the antagonist on the other side, the flexion of a joint creates a stretch reflex in the extensors, and extension creates a stretch reflex in the flexors. Consequently, stretch reflexes are valuable in stabilizing joints by balancing the tension of the extensors and flexors. They also dampen (smooth out) muscle action. Without stretch reflexes, a person's movements tend to be jerky. Stretch reflexes are especially important in coordinating vigorous and precise movements such as dance.

A stretch reflex is mediated primarily by the brain and is not, therefore, strictly a spinal reflex, but a weak component of it is spinal and occurs even if the spinal cord is severed from the brain. The spinal component can be more pronounced if a muscle is stretched very suddenly. This occurs in the reflexive contraction of a muscle when its tendon is tapped, as in the familiar *patellar* (knee-jerk) *reflex.* Tapping the patellar ligament with a reflex hammer abruptly stretches the quadriceps femoris muscle of the thigh (fig. 13.22). This stimulates numerous muscle spindles in the quadriceps and sends an intense volley of signals to the spinal cord, mainly by way of primary afferent fibers.

In the spinal cord, these fibers synapse directly with the alpha motor neurons that return to the muscle, thus forming **monosynaptic reflex arcs.** That is, there is only one synapse between the afferent and efferent neuron, so there is little synaptic delay and a very prompt response. The alpha motor neurons excite the quadriceps, making it contract and creating the knee jerk.

There are many other tendon reflexes. A tap on the calcaneal tendon causes plantar flexion of the foot, a tap on the triceps brachii tendon causes extension of the elbow, and a tap on the masseter causes clenching of the jaw. Testing somatic reflexes is valuable in diagnosing many diseases that cause exaggeration, inhibition, or absence of reflexes—for example, neurosyphilis, diabetes mellitus, multiple sclerosis, alcoholism, electrolyte imbalances, and lesions of the nervous system.

Stretch reflexes and other muscle contractions often depend on **reciprocal inhibition,** a reflex that prevents muscles from working against each other by inhibiting antagonists. In the knee jerk, for example, the quadriceps would not produce much joint movement if its antagonists, the hamstring

---

[27]*myo* = muscle; *tat* (from *tasis*) = stretch

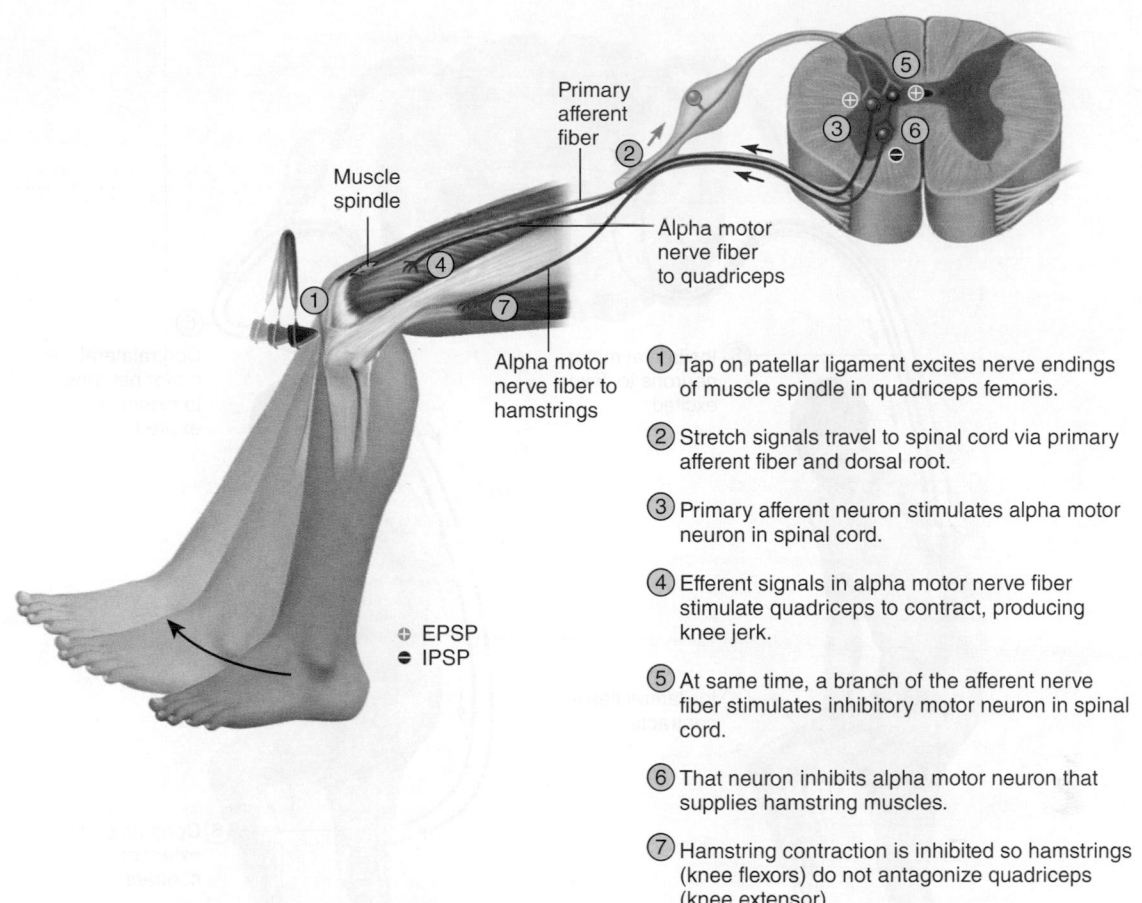

① Tap on patellar ligament excites nerve endings of muscle spindle in quadriceps femoris.

② Stretch signals travel to spinal cord via primary afferent fiber and dorsal root.

③ Primary afferent neuron stimulates alpha motor neuron in spinal cord.

④ Efferent signals in alpha motor nerve fiber stimulate quadriceps to contract, producing knee jerk.

⑤ At same time, a branch of the afferent nerve fiber stimulates inhibitory motor neuron in spinal cord.

⑥ That neuron inhibits alpha motor neuron that supplies hamstring muscles.

⑦ Hamstring contraction is inhibited so hamstrings (knee flexors) do not antagonize quadriceps (knee extensor).

⊕ EPSP
⊖ IPSP

**FIGURE 13.22 The Patellar Tendon Reflex Arc and Reciprocal Inhibition of the Antagonistic Muscle.** Plus signs indicate excitation of a postsynaptic cell (EPSPs), and the minus sign indicates inhibition (IPSPs). The tendon reflex occurs in the quadriceps femoris muscle (large red arrow), while the hamstring muscles exhibit reciprocal inhibition (large blue arrow) so they do not contract and oppose the quadriceps.

❓ *Why is no IPSP shown at point 7 if the contraction of this muscle is being inhibited?*

muscles, contracted at the same time. But reciprocal inhibition prevents that from happening. Some branches of the sensory fibers from the quadriceps muscle spindles stimulate spinal interneurons that, in turn, *inhibit* the alpha motor neurons of the hamstrings (fig. 13.22). The hamstring remain relaxed and allow the quadriceps to extend the knee.

## The Flexor (Withdrawal) Reflex

A **flexor reflex** is the quick contraction of flexor muscles resulting in the withdrawal of a limb from an injurious stimulus. For example, suppose you are wading in a lake and step on a broken bottle with your right foot (fig. 13.23). Even before you are consciously aware of the pain, you quickly pull your foot away before the glass penetrates any deeper. This action involves contraction of the flexors and relaxation of the extensors in that limb; the latter is another case of reciprocal inhibition.

The protective function of this reflex requires more than a quick jerk like a tendon reflex, so it involves more complex neural pathways. Sustained contraction of the flexors is produced by a parallel after-discharge circuit in the spinal cord (see fig. 12.30, p. 466). This circuit is part of a **polysynaptic reflex arc**—a pathway in which signals travel over many synapses on their way back to the muscle. Some signals follow routes with only a few synapses and return to the flexor muscles quickly. Others follow routes with more synapses, and therefore more delay, so they reach the flexor muscles a little later.

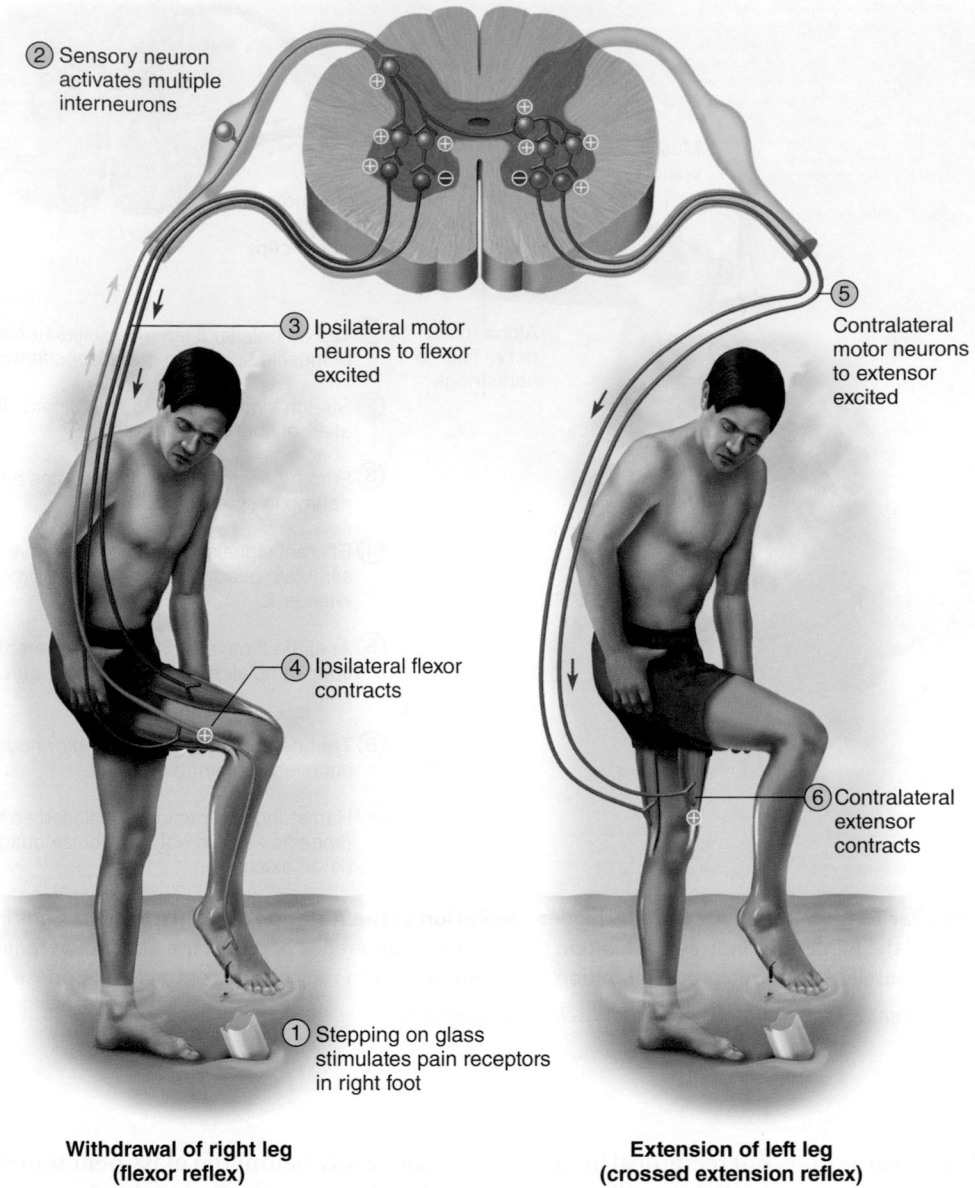

② Sensory neuron activates multiple interneurons

⑤ Contralateral motor neurons to extensor excited

③ Ipsilateral motor neurons to flexor excited

④ Ipsilateral flexor contracts

⑥ Contralateral extensor contracts

① Stepping on glass stimulates pain receptors in right foot

**Withdrawal of right leg (flexor reflex)**

**Extension of left leg (crossed extension reflex)**

**FIGURE 13.23  The Flexor and Crossed Extension Reflexes.**  A pain stimulus triggers a withdrawal reflex, which results in contraction of flexor muscles of the injured limb. At the same time, a crossed extension reflex results in contraction of extensor muscles of the opposite limb. The latter reflex aids in balance when the injured limb is raised. Note that for each limb, while the agonist contracts, the alpha motor neuron to its antagonist is inhibited, as indicated by the red minus signs in the spinal cord.  AP|R

❓ *Would you expect this reflex arc to show more synaptic delay, or less, than the ones in figure 13.22? Why?*

Consequently, the flexor muscles receive prolonged output from the spinal cord and not just one sudden stimulus as in a stretch reflex. By the time these efferent signals begin to die out, you will probably be consciously aware of the pain and begin taking voluntary action to prevent further harm.

## The Crossed Extension Reflex

In the preceding situation, if *all* you did was to quickly lift the injured leg from the lake bottom, you would fall over. To prevent this and maintain your balance, other reflexes shift your center of gravity over the leg that is still on the ground. The **crossed extension reflex** is the contraction of extensor muscles in the limb opposite from the one that is withdrawn (fig. 13.23). It extends that limb and enables you to keep your balance. To produce this reflex, branches of the afferent nerve fibers cross from the stimulated side of the body to the contralateral side of the spinal cord. There, they synapse with interneurons, which, in turn, excite or inhibit alpha motor neurons to the muscles of the contralateral limb.

In the ipsilateral leg (the side that was hurt), you would contract your flexors and relax your extensors to lift the leg from the ground. On the contralateral side, you would relax your flexors and contract the extensors to stiffen that leg, since it must suddenly support your entire body. At the same time, signals travel up the spinal cord and cause contraction of contralateral muscles of the hip and abdomen, such as your abdominal obliques, to shift your center of gravity over the extended leg. To a large extent, the coordination of all these muscles and maintenance of equilibrium are mediated by the cerebellum and cerebral cortex.

The flexor reflex employs an **ipsilateral reflex arc**—one in which the sensory input and motor output are on the same side of the spinal cord. The crossed extension reflex employs a **contralateral reflex arc,** in which the input and output are on opposite sides. An **intersegmental reflex arc** is one in which the input and output occur at different levels (segments) of the spinal cord—for example, when pain to the foot causes contractions of abdominal and hip muscles higher up the body. Note that all of these reflex arcs can function simultaneously to produce a coordinated protective response to pain.

## The Tendon Reflex

**Tendon organs** are proprioceptors located in a tendon near its junction with a muscle (fig. 13.24). A tendon organ is about 0.5 mm long. It consists of an encapsulated bundle of small, loose collagen fibers and one or more nerve fibers that penetrate the capsule and end in flattened leaflike processes between the collagen fibers. As long as the tendon is slack, its collagen fibers are slightly spread and put little pressure on the nerve endings. When muscle contraction pulls on the tendon, the collagen fibers come together like the two sides of a stretched rubber band and squeeze the nerve endings between them. The nerve fiber sends signals to the spinal cord that provide the CNS with feedback on the degree of muscle tension at the joint.

The **tendon reflex** is a response to excessive tension on the tendon. It inhibits alpha motor neurons to the muscle so the muscle does not contract as strongly. This serves to moderate muscle contraction before it tears a tendon or pulls it loose from the muscle or bone. Nevertheless, strong muscles and quick movements sometimes damage a tendon before the reflex can occur, causing such athletic injuries as a ruptured calcaneal tendon.

The tendon reflex also functions when some parts of a muscle contract more than others. It inhibits the muscle fibers connected with overstimulated tendon organs so their

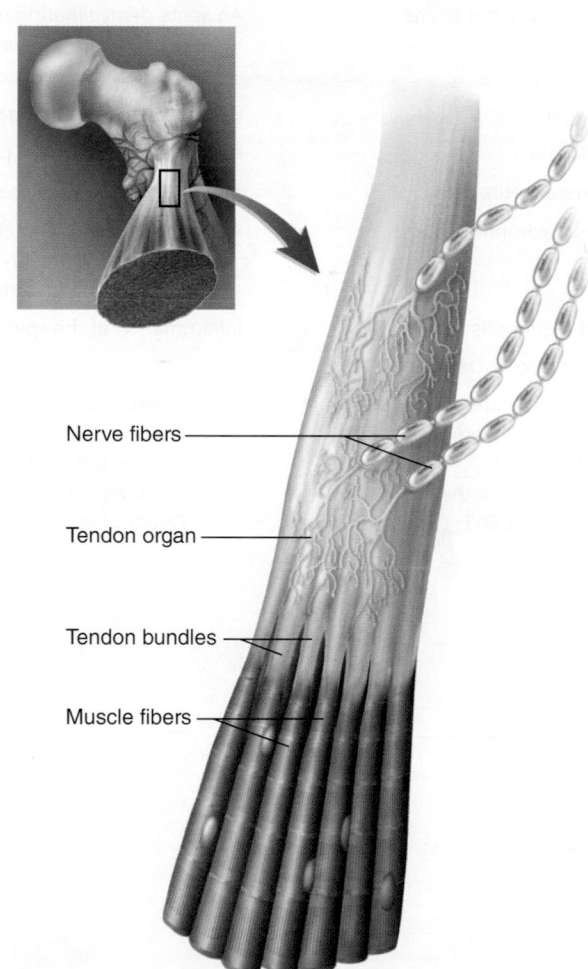

Nerve fibers

Tendon organ

Tendon bundles

Muscle fibers

**FIGURE 13.24  A Tendon Organ.**

contraction is more comparable to the contraction of the rest of the muscle. This spreads the workload more evenly over the entire muscle, which is beneficial in such actions as maintaining a steady grip on a tool.

Table 13.7 and Deeper Insight 13.5 describe some injuries and other disorders of the spinal cord and spinal nerves.

**BEFORE YOU GO ON**

Answer the following questions to test your understanding of the preceding section:

10. Name five structural components of a typical somatic reflex arc. Which of these is absent from a monosynaptic arc?

11. State the function of each of the following in a muscle spindle: intrafusal fibers, gamma motor neurons, and primary afferent fibers.

12. Explain how nerve fibers in a tendon sense the degree of tension in a muscle.

13. Why must the withdrawal reflex, but not the stretch reflex, involve a polysynaptic reflex arc?

14. Explain why the crossed extension reflex must accompany a withdrawal reflex of the leg.

| TABLE 13.7 | Some Disorders of the Spinal Cord and Spinal Nerves |
|---|---|
| Guillain–Barré syndrome | An acute demyelinating nerve disorder often triggered by viral infection, resulting in muscle weakness, elevated heart rate, unstable blood pressure, shortness of breath, and sometimes death from respiratory paralysis |
| Neuralgia | General term for nerve pain, often caused by pressure on spinal nerves from herniated intervertebral discs |
| Paresthesia | Abnormal sensations of prickling, burning, numbness, or tingling; a symptom of peripheral nerve disorders |
| Peripheral neuropathy | Any loss of sensory or motor function due to nerve injury; also called *nerve palsy* |
| Rabies (hydrophobia) | A disease usually contracted from animal bites, involving viral infection that spreads via somatic motor nerve fibers to the CNS and then autonomic nerve fibers; leads to seizures, coma, and death; invariably fatal if not treated before CNS symptoms appear |
| Spinal meningitis | Inflammation of the spinal meninges due to viral, bacterial, or other infection |

*Disorders described elsewhere*

| | | |
|---|---|---|
| Amyotrophic lateral sclerosis p. 484 | Leprosy p. 584 | Sciatica p. 495 |
| Carpal tunnel syndrome p. 352 | Multiple sclerosis p. 444 | Shingles p. 492 |
| Crutch paralysis p. 495 | Poliomyelitis p. 484 | Spina bifida p. 478 |
| Diabetic neuropathy pp. 584, 665 | Paraplegia p. 503 | Spinal cord trauma p. 503 |
| Hemiplegia p. 503 | Quadriplegia p. 503 | |

# DEEPER INSIGHT 13.5
## CLINICAL APPLICATION

### Spinal Cord Trauma

In the United States, 10,000 to 12,000 people become paralyzed annually by spinal cord trauma, usually as a result of vertebral fractures. The greatest incidence is among males from 16 to 30 years old, because of their high-risk behaviors. Fifty-five percent of their injuries are from automobile and motorcycle accidents, 18% from sports, and 15% from gunshot and stab wounds. Elderly people are also at above-average risk because of falls, and in times of war, battlefield injuries account for many cases.

### Effects of Injury

Complete *transection* (severance) of the spinal cord causes immediate loss of motor control at and below the level of the injury. Transection superior to segment C4 presents a threat of respiratory failure. Victims also lose sensation from the level of injury and below, although some patients temporarily feel burning pain within one or two dermatomes of the level of the lesion.

In the early stage, victims exhibit a syndrome called *spinal shock*. Muscles below the level of injury exhibit flaccid paralysis (inability to contract) and an absence of reflexes because of the lack of stimulation from higher levels of the CNS. For 8 days to 8 weeks after the accident, the patient typically lacks bladder and bowel reflexes and thus retains urine and feces. Lacking sympathetic stimulation to the blood vessels, a patient may exhibit *neurogenic shock* in which the vessels dilate and blood pressure drops dangerously low. Fever may occur because the hypothalamus cannot induce sweating to cool the body. Spinal shock can last from a few days to several weeks (usually 7 to 20 days).

As spinal shock subsides, somatic reflexes begin to reappear, at first in the toes and progressing to the feet and legs. Autonomic reflexes also reappear. Contrary to the earlier urinary and fecal retention, a patient now has the opposite problem, incontinence, as the rectum and bladder empty reflexively in response to stretch. Both the somatic and autonomic nervous systems typically exhibit exaggerated reflexes, a state called *hyperreflexia* or the *mass reflex reaction*. Stimuli such as a full bladder or cutaneous touch can trigger an extreme cardiovascular reaction. The systolic blood pressure, normally about 120 mm Hg, jumps to as high as 300 mm Hg. This causes intense headaches and sometimes a stroke. Pressure receptors in the major arteries sense this rise in blood pressure and activate a reflex that slows the heart, sometimes to a rate as low as 30 or 40 beats/minute *(bradycardia)*, compared with a normal rate of 70 to 80. The patient may also experience profuse sweating and blurred vision.

Men at first lose the capacity for erection and ejaculation. They may recover these functions later and become capable of ejaculating and fathering children, but without sexual sensation. In females, menstruation may become irregular or cease.

The most serious permanent effect of spinal cord trauma is paralysis. The flaccid paralysis of spinal shock later changes to spastic paralysis as spinal reflexes are regained but lack inhibitory control from the brain. Spastic paralysis typically starts with chronic flexion of the hips and knees (flexor spasms) and progresses to a state in which the limbs become straight and rigid (extensor spasms). Three forms of muscle paralysis are *paraplegia*, a paralysis of both lower limbs resulting from spinal cord lesions at levels T1 to L1; *quadriplegia*, the paralysis of all four limbs resulting from lesions above level C5; and *hemiplegia*, paralysis of one side of the body, usually resulting not from spinal cord injuries but from a stroke or other brain lesion. Spinal cord lesions from C5 to C7 can produce a state of partial quadriplegia—total paralysis of the lower limbs and partial paralysis (*paresis*, or weakness) of the upper limbs.

### Pathogenesis

Spinal cord trauma produces two stages of tissue destruction. The first is instantaneous—the destruction of cells by the traumatic event itself. The second stage, a wave of tissue death by necrosis and apoptosis, begins in minutes and lasts for days. It is far more destructive than the initial injury, typically converting a lesion in one spinal cord segment to a lesion that spans four or five segments, two above and two below the original site.

Microscopic hemorrhages appear in the gray matter and pia mater within minutes and grow larger over the next 2 hours. The white matter becomes edematous (swollen). Hemorrhaging and edema spread to adjacent segments of the cord and can fatally affect respiration or brainstem function when it occurs in the cervical region. *Ischemia* (iss-KEE-me-uh), the lack of blood, quickly leads to necrosis. The white matter regains circulation in about 24 hours, but the gray matter remains ischemic. Inflammatory cells (leukocytes and macrophages) infiltrate the lesion as the circulation recovers, and while they clean up necrotic tissue, they also contribute to the damage by releasing destructive free radicals and other toxic chemicals. The necrosis worsens, and is accompanied by another form of cell death, apoptosis (see p. 172). Apoptosis of the spinal oligodendrocytes, the myelinating glial cells of the CNS, results in demyelination of spinal nerve fibers, followed by death of the neurons.

In as little as 4 hours, this second wave of destruction, called *post-traumatic infarction*, consumes about 40% of the cross-sectional area of the spinal cord; within 24 hours, it destroys 70%. As many as five segments of the cord become transformed into a fluid-filled cavity, which is replaced with collagenous scar tissue over the next 3 to 4 weeks. This scar is one of the obstacles to the regeneration of lost nerve fibers.

### Treatment

The first priority in treating a spinal injury patient is to immobilize the spine to prevent further trauma. Respiratory or other life support may also be required. Methylprednisolone, a steroid, dramatically improves recovery. Given within 3 hours of the trauma, it reduces injury to cell membranes and inhibits inflammation and apoptosis.

After these immediate requirements are met, reduction (repair) of the fracture is important. If a CT or MRI scan indicates spinal cord compression by the vertebral canal, a *decompression laminectomy* may be performed, in which vertebral laminae are removed from the affected region. CT and MRI have helped a great deal in recent decades for assessing vertebral and spinal cord damage, guiding surgical treatment, and improving recovery. Physical therapy is important for maintaining muscle and joint function as well as promoting the patient's psychological recovery.

Treatment strategies for spinal cord injuries are a vibrant field of contemporary medical research. Some current interests are the use of antioxidants to reduce free radical damage, and the implantation of embryonic stem cells, which has produced significant (but not perfect) recovery from spinal cord lesions in rats. Public hopes have often been raised by promising studies reported in the scientific literature and news media, only to be dashed by the inability of other laboratories to repeat and confirm the results.

# STUDY GUIDE

## ► Assess Your Learning Outcomes

*To test your knowledge, discuss the following topics with a study partner or in writing, ideally from memory.*

### 13.1 The Spinal Cord (p. 475)

1. Functions of the spinal cord
2. Skeletal landmarks that mark the extent of the adult spinal cord, and what occupies the vertebral canal inferior to the spinal cord
3. The four regions of the spinal cord and the basis for their names
4. What defines one segment of the cord
5. Two enlargements of the cord and why the cord is wider at these points
6. Names and structures of the three spinal meninges, in order from superficial to deep, and the relationships of the epidural and subarachnoid spaces to the meninges
7. The two types of ligaments that arise from the pia mater; where they are found and what purpose they serve
8. Organization of spinal gray and white matter as seen in cross sections of the cord; how gray and white matter differ in composition; and why they are called *gray* and *white matter*
9. The position of the posterior and anterior horns of the gray matter; where lateral horns are also found, and the functions of all three
10. The anatomical basis for dividing white matter into three columns on each side of the cord, and for dividing each column into tracts
11. Names and functions of the ascending tracts of the spinal cord
12. The meaning of *first-* through *third-order* neurons in an ascending tract
13. Names and functions of the descending tracts

14. The locations and distinctions between upper and lower motor neurons in the descending tracts
15. Decussation and its implications for cerebral function in relation to sensation and motor control of the lower body, and for the effects of a stroke
16. What it means to say that the origin and destination of a tract, or any two body parts, are ipsilateral or contralateral

### 13.2 The Spinal Nerves (p. 483)

1. Structure of a nerve, especially the relationship of the endoneurium, perineurium, and epineurium to nerve fibers and fascicles
2. The basis for classifying nerve fibers as afferent or efferent, somatic or visceral, and special or general
3. The basis for classifying entire nerves as sensory, motor, or mixed
4. The definition and structure of a *ganglion*
5. The number of spinal nerves and their relationship to the spinal cord and intervertebral foramina
6. Anatomy of the posterior and anterior roots of a spinal nerve; the rootlets; and the posterior root ganglion
7. Anatomy of the anterior ramus, posterior ramus, and meningeal branch of a spinal nerve
8. What arises from the anterior ramus in the thoracic region as opposed to all other regions of the spinal cord
9. General structure of a spinal nerve plexus and the names and locations of the five plexuses
10. Distinctions between the roots, trunks, anterior and posterior divisions, and cords of a spinal nerve plexus; which of these five features occur in each of the five plexuses

11. Nerves that arise from each plexus and the body regions or structures to which each nerve provides sensory innervation, motor innervation, or both
12. Dermatomes and why they are relevant to the clinical diagnosis of nerve disorders

### 13.3 Somatic Reflexes (p. 496)

1. Four defining criteria of a reflex; how somatic reflexes differ from other types; and the flaw in calling somatic reflexes spinal reflexes
2. The pathway and constituents of a somatic reflex arc
3. The role of proprioceptors in somatic reflexes
4. Structure and function of muscle spindles
5. Stretch reflexes; one or more examples; the purpose they serve in everyday function; the mechanism of a stretch reflex; and an anatomical reason why stretch reflexes are often quicker than other types of somatic reflexes
6. Reciprocal inhibition and why it is important that it often accompany a stretch reflex
7. Flexor reflexes; a common purpose that they serve; and why it is beneficial for flexor reflexes to employ polysynaptic reflex arcs
8. Crossed extension reflexes and why it is important for this type of reflex to accompany a withdrawal reflex
9. The structure, location, and function of a tendon organ

# STUDY GUIDE

## ▶ Testing Your Recall

*Answers in Appendix B*

1. Below L2, the vertebral canal is occupied by a bundle of spinal nerve roots called
   a. the terminal filum.
   b. the descending tracts.
   c. the gracile fasciculus.
   d. the medullary cone.
   e. the cauda equina.

2. The brachial plexus gives rise to all of the following nerves *except*
   a. the axillary nerve.
   b. the radial nerve.
   c. the obturator nerve.
   d. the median nerve.
   e. the ulnar nerve.

3. Nerve fibers that adjust the tension in a muscle spindle are called
   a. intrafusal fibers.
   b. extrafusal fibers.
   c. alpha motor neurons.
   d. gamma motor neurons.
   e. primary afferent fibers.

4. A stretch reflex requires the action of _____ to prevent an antagonistic muscle from interfering with the agonist.
   a. gamma motor neurons
   b. a withdrawal reflex
   c. a crossed extension reflex
   d. reciprocal inhibition
   e. a contralateral reflex

5. A patient has a gunshot wound that caused a bone fragment to nick the spinal cord. The patient now feels no pain or temperature sensations from that level of the body down. Most likely, the _____ was damaged.
   a. gracile fasciculus
   b. medial lemniscus

   c. tectospinal tract
   d. lateral corticospinal tract
   e. spinothalamic tract

6. Which of these is *not* a region of the spinal cord?
   a. cervical
   b. thoracic
   c. pelvic
   d. lumbar
   e. sacral

7. In the spinal cord, the somas of the lower motor neurons are found in
   a. the cauda equina.
   b. the posterior horns.
   c. the anterior horns.
   d. the posterior root ganglia.
   e. the fasciculi.

8. The outermost connective tissue wrapping of a nerve is called the
   a. epineurium.
   b. perineurium.
   c. endoneurium.
   d. arachnoid mater.
   e. dura mater.

9. The intercostal nerves between the ribs arise from which spinal nerve plexus?
   a. cervical
   b. brachial
   c. lumbar
   d. sacral
   e. none of them

10. All somatic reflexes share all of the following properties *except*
    a. they are quick.
    b. they are monosynaptic.
    c. they require stimulation.

    d. they are involuntary.
    e. they are stereotyped.

11. Outside the CNS, the somas of neurons are clustered in swellings called _____.

12. Distal to the intervertebral foramen, a spinal nerve branches into an anterior and posterior _____.

13. The cerebellum receives feedback from the muscles and joints by way of the _____ tracts of the spinal cord.

14. In the _____ reflex, contraction of flexor muscles in one limb is accompanied by the contraction of extensor muscles in the contralateral limb.

15. Modified muscle fibers serving primarily to detect stretch are called _____.

16. The _____ nerves arise from the cervical plexus and innervate the diaphragm.

17. The crossing of a nerve fiber or tract from the right side of the CNS to the left, or vice versa, is called _____.

18. The nonvisual awareness of the body's position and movements is called _____.

19. The _____ ganglion contains the somas of neurons that carry sensory signals to the spinal cord.

20. The sciatic nerve is a composite of two nerves, the _____ and _____.

## ▶ Building Your Medical Vocabulary

*Answers in Appendix B*

*State a medical meaning of each word element below, and give a term in which it or a slight variation of it is used.*

1. arachno-

2. caudo-

3. contra-

4. cune-

5. ipsi-

6. phreno-

7. pia

8. proprio-

9. ram-

10. tecto-

# STUDY GUIDE

## ▶ True or False

*Answers in Appendix B*

*Determine which five of the following statements are false, and briefly explain why.*

1. The gracile fasciculus is a descending spinal tract.

2. At the inferior end, the adult spinal cord ends at a higher level than the vertebral column.

3. Each spinal cord segment has only one pair of spinal nerves.

4. Some spinal nerves are sensory and others are motor.

5. The dura mater adheres tightly to the bone of the vertebral canal.

6. The anterior and posterior horns of the spinal cord are composed of gray matter.

7. The corticospinal tracts carry motor signals down the spinal cord.

8. The dermatomes are nonoverlapping regions of skin innervated by different spinal nerves.

9. Somatic reflexes are those that do not involve the brain.

10. The tendon reflex acts to inhibit muscle contraction.

## ▶ Testing Your Comprehension

*Answers at* www.mhhe.com/saladin7

1. Jillian is thrown from a horse. She strikes the ground with her chin, causing severe hyperextension of the neck. Emergency medical technicians properly immobilize her neck and transport her to a hospital, but she dies 5 minutes after arrival. An autopsy shows multiple fractures of vertebrae C1, C6, and C7 and extensive damage to the spinal cord. Explain why she died rather than being left quadriplegic.

2. Wallace is the victim of a hunting accident. A bullet grazed his vertebral column, and bone fragments severed the left half of his spinal cord at segments T8 through T10. Since the accident, Wallace has had a condition called *dissociated sensory loss,* in which he feels no sensations of deep touch or limb position on the *left* side of his body below the injury, and no sensations of pain or heat from the *right* side. Explain what spinal tract(s) the injury has affected and why these sensory losses are on opposite sides of the body.

3. Anthony gets into a fight between rival gangs. As an attacker comes at him with a knife, he turns to flee, but stumbles. The attacker stabs him on the medial side of the right gluteal fold and Anthony collapses. He loses all use of his right limb, being unable to extend his hip, flex his knee, or move his foot. He never fully recovers these lost functions. Explain what nerve injury Anthony has most likely suffered.

4. Stand with your right shoulder, hip, and foot firmly against a wall. Raise your left foot from the floor without losing contact with the wall at any point. What happens? Why? What principle of this chapter does this demonstrate?

5. When a patient needs a tendon graft, surgeons sometimes use the tendon of the palmaris longus, a relatively dispensable muscle of the forearm. The median nerve lies nearby and looks very similar to this tendon. There have been cases in which a surgeon mistakenly removed a section of this nerve instead of the tendon. What effects do you think such a mistake would have on the patient?

## ▶ Improve Your Grade at www.mhhe.com/saladin7

*Download animations to study when it fits your schedule. Practice quizzes, labeling activities, games, and flashcards offer fun ways to master the chapter concepts. Or, download image PowerPoint files for each chapter to create a study guide or for taking notes during lecture.*

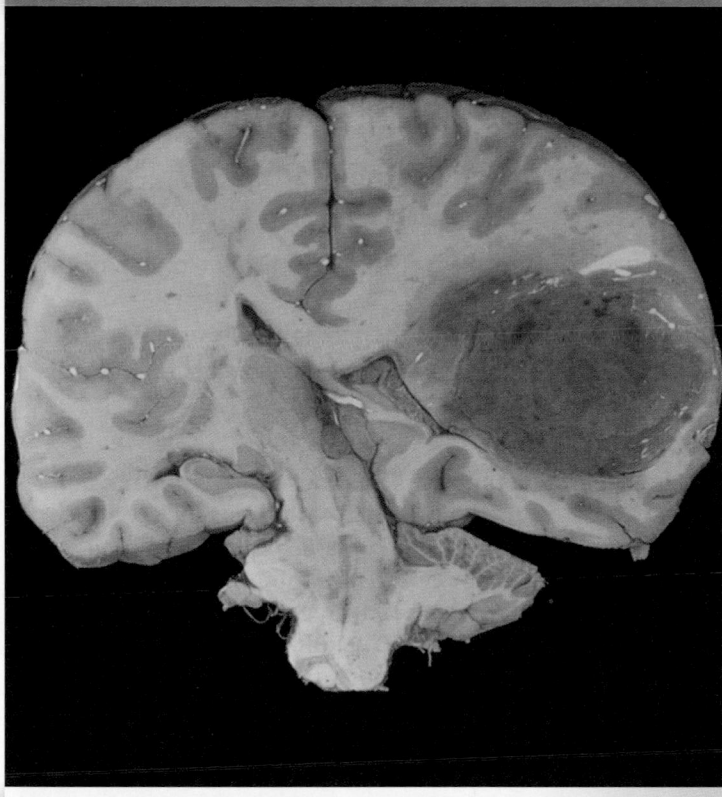

Frontal section of the brain with a large tumor (glioblastoma) in the left cerebral hemisphere

# THE BRAIN AND CRANIAL NERVES

apirevealed.com

**Module 7: Nervous System**

## BRUSHING UP

- The anatomy of the brain and cranial nerves is, in many respects, described in relation to the skull. Lobes of the cerebrum, for example, are named for adjacent cranial bones. Therefore, a review of pages 236 through 242 will be helpful.

- You must be familiar with general neuron anatomy (p. 438), glial cells and their functions (p. 442), and the gray and white matter of the CNS (p. 479).

- The brainstem contains extensions of the spinal cord tracts, so you will find it helpful to know those or refer back to table 13.1 (p. 480) as you study the brainstem.

- To best understand the cranial nerves, you should be familiar with the general structure of nerves and ganglia, afferent and efferent nerve fibers, and the distinction between sensory, motor, and mixed nerves (pp. 483–484).

The human brain has a high opinion of itself, often claiming to be the most complex object in the known universe. It would be hard to argue otherwise. It has a mystique that intrigues modern biologists and psychologists even as it did the philosophers of antiquity. Aristotle thought it was just a radiator for cooling the blood, but generations earlier, Hippocrates had expressed a more accurate view. "Men ought to know," he said, "that from the brain, and from the brain only, arise our pleasures, joy, laughter and jests, as well as our sorrows, pains, griefs and tears. Through it, in particular, we think, see, hear, and distinguish the ugly from the beautiful, the bad from the good, the pleasant from the unpleasant."

Brain function is so strongly associated with what it means to be alive and human that the cessation of brain activity is taken as a clinical criterion of death even when other organs of the body are still functioning. With its hundreds of neural pools and trillions of synapses, the brain performs sophisticated tasks beyond our present understanding. Still, all of our mental functions, no matter how complex, are ultimately based on the cellular activities described in chapter 12. The relationship of the mind and personality to the cellular function of the brain is a question that will provide fertile ground for scientific study and philosophical debate long into the future.

This chapter is a study of the brain and the cranial nerves directly connected to it. Here we will plumb some of the mysteries of motor control, sensation, emotion, thought, language, personality, memory, dreams, and plans. Brain circuitry and function easily fill many books the size of this one, and we can barely scratch the surface of this complex subject here. This coverage will, however, provide some intriguing insights and lay a foundation for further study in other courses.

## 14.1 Overview of the Brain

### Expected Learning Outcomes

When you have completed this section, you should be able to

a. describe the major subdivisions and anatomical landmarks of the brain;

b. describe the locations of its gray and white matter; and

c. describe the embryonic development of the CNS and relate this to adult brain anatomy.

In the evolution of the central nervous system from the simplest vertebrates to humans, the spinal cord changed very little while the brain changed a great deal. In fishes and amphibians, the brain weighs about the same as the spinal cord, but in humans, it weighs 55 times as much. It averages about 1,600 g (3.5 lb) in men and 1,450 g in women. The difference between the sexes is proportional to body size, not intelligence. The Neanderthal people had larger brains than modern humans do.

Ours is the most sophisticated brain when compared to others in awareness of the environment, adaptability to environmental variation and change, quick execution of complex decisions, fine motor control and mobility of the body, and behavioral complexity. Over the course of human evolution, it has shown its greatest growth in areas concerned with vision, memory, and motor control of the prehensile hand.

### Major Landmarks

We begin with a general overview of the major landmarks of the brain. These will provide important reference points as we progress through a more detailed study.

Two directional terms used in descriptions of CNS anatomy are *rostral* and *caudal*. **Rostral**[1] means "toward the nose" and **caudal**[2] means "toward the tail." These are apt descriptions for an animal such as a laboratory rat, on which so much brain research has been done. The terms are retained for human neuroanatomy as well, but in references to the brain, *rostral* means "toward the forehead" and *caudal* means "toward the spinal cord." In the spinal cord and brainstem, which are vertically oriented, *rostral* means "higher" and *caudal* means "lower."

We can conceptually divide the brain into three major portions—the *cerebrum, cerebellum,* and *brainstem*. The **cerebrum** (seh-REE-brum or SER-eh-brum) is about 83% of the brain's volume and consists of a pair of half globes called the **cerebral hemispheres** (fig. 14.1a). Each hemisphere is marked

---

[1] *rostr* = nose
[2] *caud* = tail

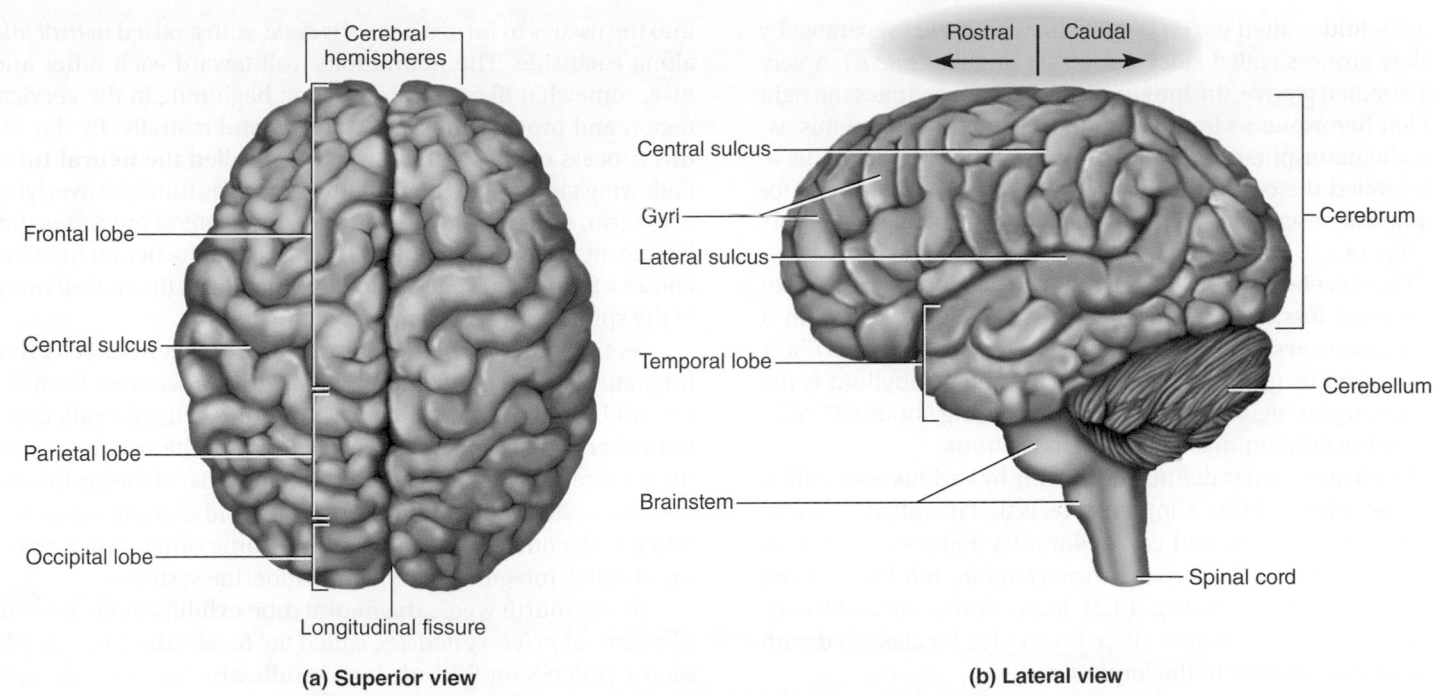

**(a) Superior view**

Cerebral hemispheres

Frontal lobe

Central sulcus

Parietal lobe

Occipital lobe

Longitudinal fissure

Rostral   Caudal

Central sulcus

Gyri

Lateral sulcus

Temporal lobe

Brainstem

Cerebrum

Cerebellum

Spinal cord

**(b) Lateral view**

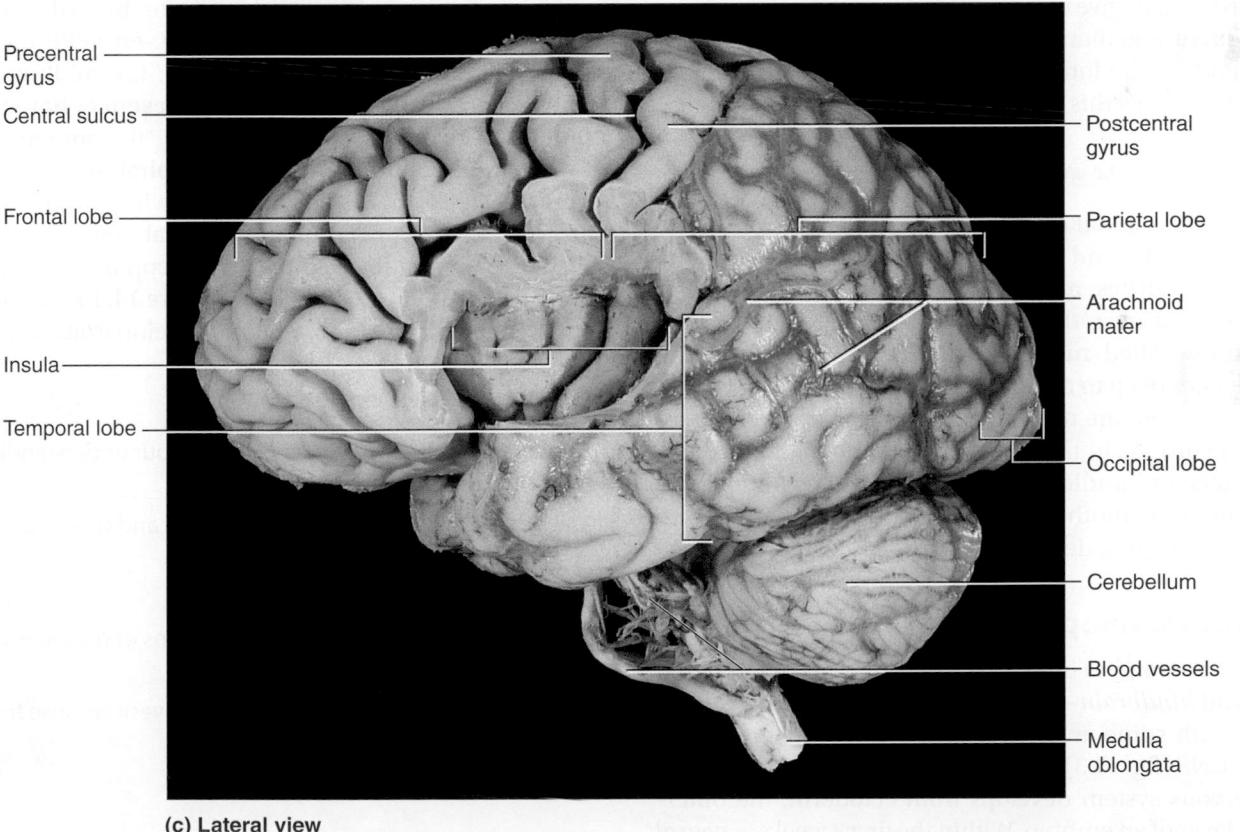

Precentral gyrus

Central sulcus

Frontal lobe

Insula

Temporal lobe

Postcentral gyrus

Parietal lobe

Arachnoid mater

Occipital lobe

Cerebellum

Blood vessels

Medulla oblongata

**(c) Lateral view**

**FIGURE 14.1  Surface Anatomy of the Brain.**  (a) Superior view of the cerebral hemispheres. (b) Left lateral view. (c) The partially dissected brain of a cadaver. Part of the left hemisphere is cut away to expose the insula. The arachnoid mater is removed from the anterior (rostral) half of the brain to expose the gyri and sulci. The arachnoid with its blood vessels is seen in the posterior (caudal) half. Blood vessels of the brainstem are left in place. **AP|R**

by thick folds called **gyri**[3] (JY-rye; singular, *gyrus*) separated by shallow grooves called **sulci**[4] (SUL-sye; singular, *sulcus*). A very deep median groove, the **longitudinal fissure,** separates the right and left hemispheres from each other. At the bottom of this fissure, the hemispheres are connected by a thick bundle of nerve fibers called the **corpus callosum**[5]—a prominent landmark for anatomical description with a distinctive C shape in sagittal section (fig. 14.2).

The **cerebellum**[6] (SER-eh-BEL-um) occupies the posterior cranial fossa inferior to the cerebrum, separated from it by the **transverse cerebral fissure** (fig. 14.1b, c; fig. 8.9). It is also marked by gyri, sulci, and fissures. The cerebellum is the second-largest region of the brain, constituting about 10% of its volume but containing over 50% of its neurons.

The **brainstem** is defined differently by various authorities. The original definition, adopted here, is that it is all of the brain except the cerebrum and cerebellum. Its major components, from rostral to caudal, are the *diencephalon, midbrain, pons,* and *medulla oblongata* (fig. 14.2). Many authorities, however, exclude the diencephalon, since it can also be classified with the cerebrum as part of the *forebrain.*

In a living person, the brainstem is oriented like a vertical stalk with the cerebrum perched on top like a mushroom cap. Postmortem changes give it a more oblique angle in the cadaver and, consequently, in many medical illustrations. Caudally, the brainstem ends at the foramen magnum of the skull, and the CNS continues below this as the spinal cord.

## Gray and White Matter

The brain, like the spinal cord, is composed of gray and white matter (see figs. 14.5 and 14.6c). Gray matter—the seat of the neurosomas, dendrites, and synapses—forms a surface layer called the **cortex** over the cerebrum and cerebellum, and deeper masses called **nuclei** surrounded by white matter. White matter lies deep to the cortical gray matter in most of the brain, opposite from the relationship of gray and white matter in the spinal cord. As in the spinal cord, white matter is composed of **tracts,** or bundles of axons, which here connect one part of the brain to another and to the spinal cord. These are described later in more detail.

## Embryonic Development

Mature brain anatomy is often described in terms of *forebrain, midbrain,* and *hindbrain*—three terms that can be fully appreciated only with some awareness of the embryonic development of the CNS (fig. 14.3).

The nervous system develops from ectoderm, the outermost tissue layer of an embryo. Within the first 3 weeks, a *neural plate* forms along the dorsal midline of the embryo and sinks into the tissues to form a *neural groove,* with a raised *neural fold* along each side. The neural folds roll toward each other and fuse, somewhat like a closing zipper, beginning in the cervical region and progressing both caudally and rostrally. By day 26, this process creates a hollow channel called the **neural tube.** Following closure, the neural tube separates from the overlying ectoderm, sinks a little deeper, and grows lateral processes that later form motor nerve fibers. The lumen of the neural tube becomes a fluid-filled space that later constitutes the *central canal* of the spinal cord and *ventricles* of the brain.

As the neural tube develops, some ectodermal cells that originally lay along the margin of the groove separate from the rest and form a longitudinal column on each side called the **neural crest.** Neural crest cells give rise to the two inner meninges (arachnoid mater and pia mater); most of the peripheral nervous system, including the sensory and autonomic nerves and ganglia and Schwann cells; and some other structures of the skeletal, integumentary, and endocrine systems.

By the fourth week, the neural tube exhibits three anterior dilations, or *primary vesicles,* called the **forebrain** *(prosencephalon*[7]*)* (PROSS-en-SEF-uh-lon), **midbrain** *(mesencephalon*[8]*)* (MEZ-en-SEF-uh-lon), and **hindbrain** *(rhombencephalon*[9]*)* (ROM-ben-SEF-uh-lon) (fig. 14.4). By the fifth week, it subdivides into five *secondary vesicles.* The forebrain divides into two of them, the **telencephalon**[10] (TEL-en-SEFF-uh-lon) and **diencephalon**[11] (DY-en-SEF-uh-lon); the midbrain remains undivided and retains the name **mesencephalon;** and the hindbrain divides into two vesicles, the **metencephalon**[12] (MET-en-SEF-uh-lon) and **myelencephalon**[13] (MY-el-en-SEF-uh-lon) (fig. 14.4b). The telencephalon has a pair of lateral outgrowths that later become the cerebral hemispheres, and the diencephalon exhibits a pair of small cuplike *optic vesicles* that become the retinas of the eyes. Figure 14.4 is color-coded to show which mature brain regions develop from each vesicle.

**BEFORE YOU GO ON**

Answer the following questions to test your understanding of the preceding section:

**1.** List the three major parts of the brain and describe their locations.

**2.** Define gyrus and sulcus.

**3.** Contrast the composition and locations of gray and white matter in the brain.

**4.** Explain how the five secondary brain vesicles arise from the neural tube.

---

[3]*gy* = turn, twist
[4]*sulc* = furrow, groove
[5]*corpus* = body; *call* = thick
[6]*cereb* = brain; *ellum* = little

[7]*pros* = before, in front; *encephal* = brain
[8]*mes* = middle
[9]*rhomb* = rhombus
[10]*tele* = end, remote
[11]*di* = through, between
[12]*met* = behind, beyond, distal to
[13]*myel* = spinal cord

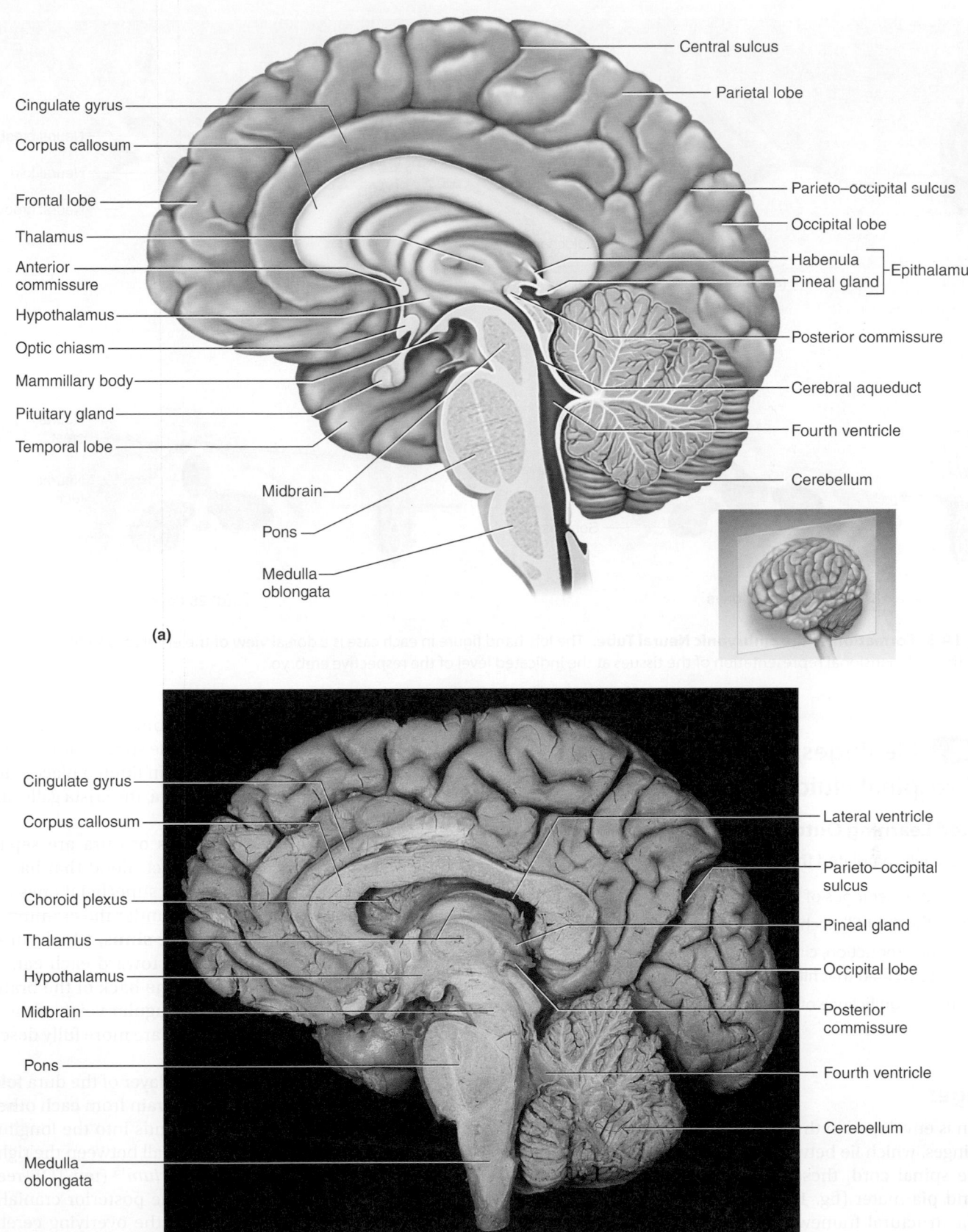

**(a)**

**(b)**

**FIGURE 14.2 Medial Aspect of the Brain.** (a) Major anatomical landmarks of the medial surface. (b) Median section of the cadaver brain. **AP|R**

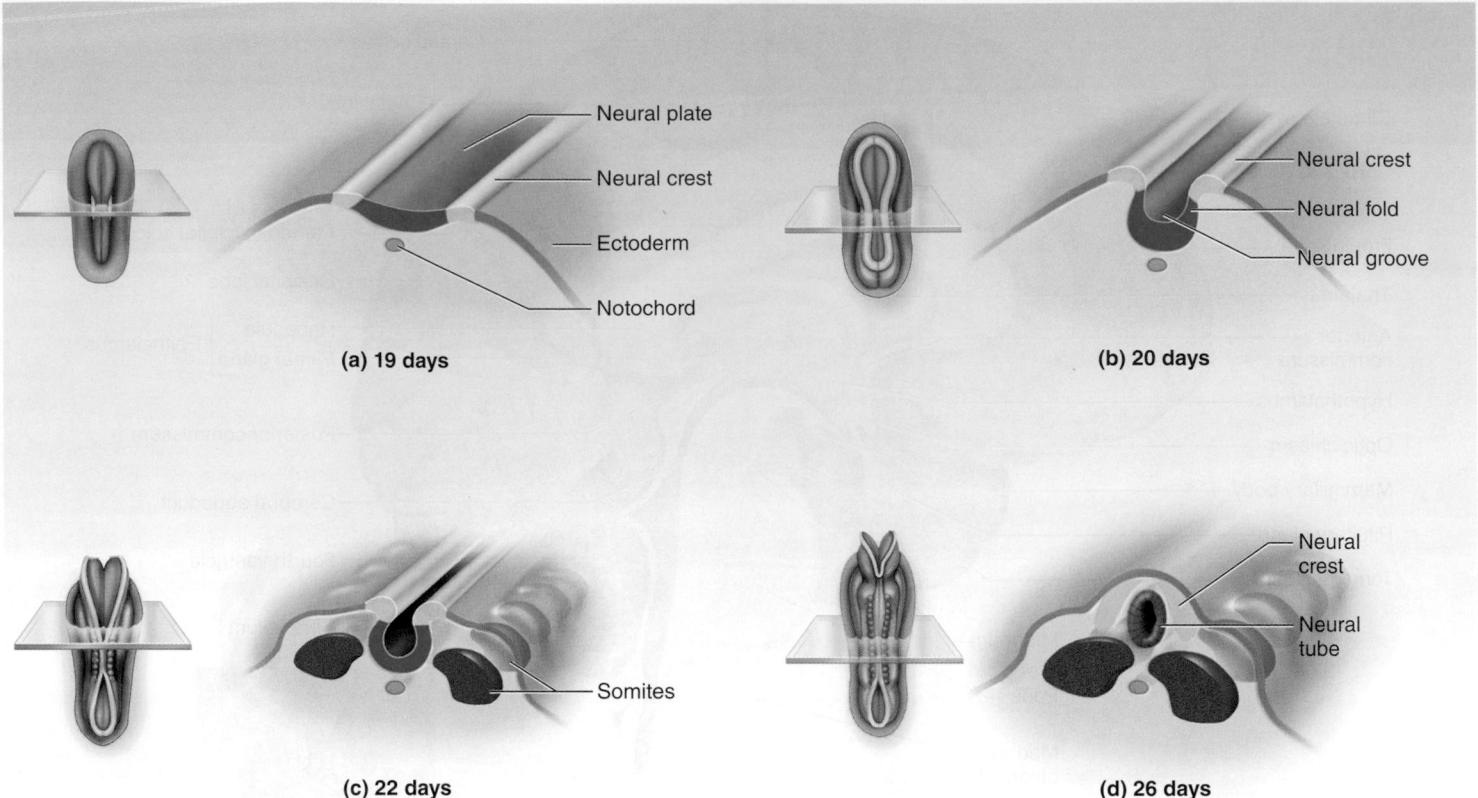

**(a) 19 days**
- Neural plate
- Neural crest
- Ectoderm
- Notochord

**(b) 20 days**
- Neural crest
- Neural fold
- Neural groove

**(c) 22 days**
- Somites

**(d) 26 days**
- Neural crest
- Neural tube

**FIGURE 14.3 Formation of the Embryonic Neural Tube.** The left-hand figure in each case is a dorsal view of the embryo, and the right-hand figure is a three-dimensional representation of the tissues at the indicated level of the respective embryo.

## 14.2 Meninges, Ventricles, Cerebrospinal Fluid, and Blood Supply

### Expected Learning Outcomes

When you have completed this section, you should be able to

a. describe the meninges of the brain;

b. describe the fluid-filled chambers within the brain;

c. discuss the production, circulation, and function of the cerebrospinal fluid that fills these chambers; and

d. explain the significance of the brain barrier system.

## Meninges

The brain is enveloped in three connective tissue membranes, the meninges, which lie between the nervous tissue and bone. As in the spinal cord, these are the dura mater, arachnoid mater, and pia mater (fig. 14.5). They protect the brain and provide a structural framework for its arteries and veins. In the cranial cavity, the dura mater consists of two layers—an outer *periosteal layer* equivalent to the periosteum of the cranial bones, and an inner *meningeal layer*. Only the meningeal layer continues into the vertebral canal, where it forms the dural sheath around the spinal cord. The cranial dura mater is pressed closely against the cranial bone, with no intervening epidural space like the one around the spinal cord. It is not attached to the bone, however, except in limited places: around the foramen magnum, the sella turcica, the crista galli, and the sutures of the skull.

In some places, the two layers of dura are separated by **dural sinuses,** spaces that collect blood that has circulated through the brain. Two major, superficial ones are the **superior sagittal sinus,** found just under the cranium along the median line, and the **transverse sinus,** which runs horizontally from the rear of the head toward each ear. These sinuses meet like an inverted T at the back of the brain and ultimately empty into the internal jugular veins of the neck. These and other sinuses of the brain are more fully described and pictured in chapter 20.

In certain places, the meningeal layer of the dura folds inward to separate major parts of the brain from each other: the *falx*[14] *cerebri* (falks SER-eh-bry) extends into the longitudinal fissure as a tough, crescent-shaped wall between the right and left cerebral hemispheres; the *tentorium*[15] (ten-TOE-ree-um) *cerebelli* stretches like a roof over the posterior cranial fossa and separates the cerebellum from the overlying cerebrum; and the *falx cerebelli* partially separates the right and left halves of the cerebellum on the inferior side.

---

[14]*falx* = sickle
[15]*tentorium* = tent

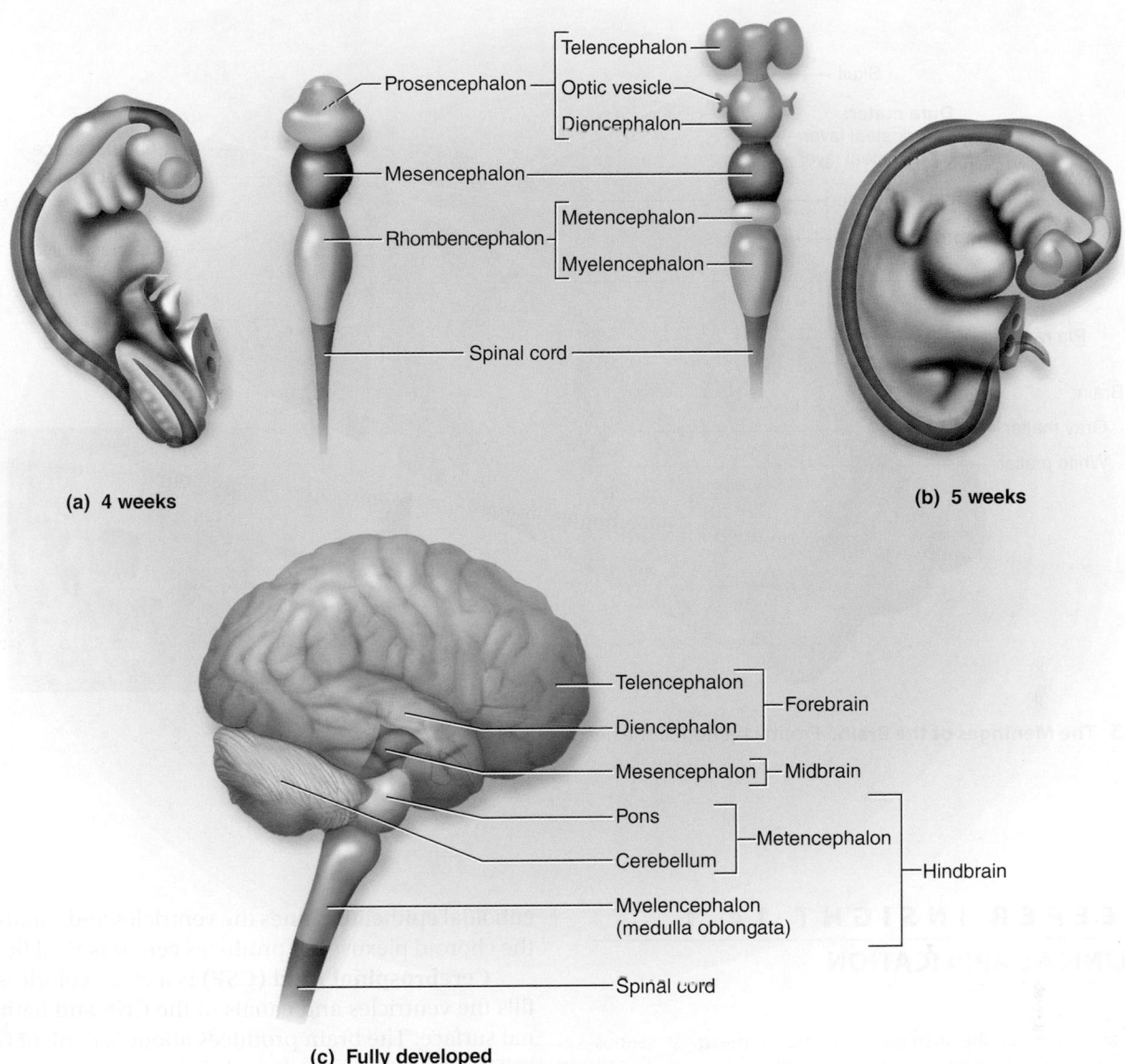

(a) 4 weeks

Prosencephalon

Telencephalon
Optic vesicle
Diencephalon

Mesencephalon

Rhombencephalon

Metencephalon
Myelencephalon

Spinal cord

(b) 5 weeks

Telencephalon
Diencephalon — Forebrain

Mesencephalon — Midbrain

Pons
Cerebellum — Metencephalon

Myelencephalon
(medulla oblongata) — Hindbrain

Spinal cord

(c) Fully developed

**FIGURE 14.4 Primary and Secondary Vesicles of the Embryonic Brain.** (a) The primary vesicles at 4 weeks. (b) The secondary vesicles at 5 weeks. (c) The fully developed brain, color-coded to relate its structures to the secondary embryonic vesicles.

The arachnoid mater and pia mater are similar to those of the spinal cord. The arachnoid mater is a transparent membrane over the brain surface, visible in the caudal half of the cerebrum in figure 14.1c. A *subarachnoid space* separates it from the pia below, and in some places, a *subdural space* separates it from the dura above. The pia mater is a very thin, delicate membrane that closely follows all the contours of the brain, even dipping into the sulci. It is not usually visible without a microscope.

## Ventricles and Cerebrospinal Fluid

The brain has four internal chambers called **ventricles** (fig. 14.6). The largest and most rostral ones are the two **lateral ventricles,**

which form an arc in each cerebral hemisphere. Through a tiny pore called the **interventricular foramen,** each lateral ventricle is connected to the **third ventricle,** a narrow median space inferior to the corpus callosum. From here, a canal called the **cerebral aqueduct** passes down the core of the midbrain and leads to the **fourth ventricle,** a small triangular chamber between the pons and cerebellum. Caudally, this space narrows and forms a **central canal** that extends through the medulla oblongata into the spinal cord.

On the floor or wall of each ventricle is a spongy mass of blood capillaries called a **choroid** (CO-royd) **plexus,** named for its histological resemblance to a fetal membrane called the chorion. *Ependyma,* a type of neuroglia that resembles a

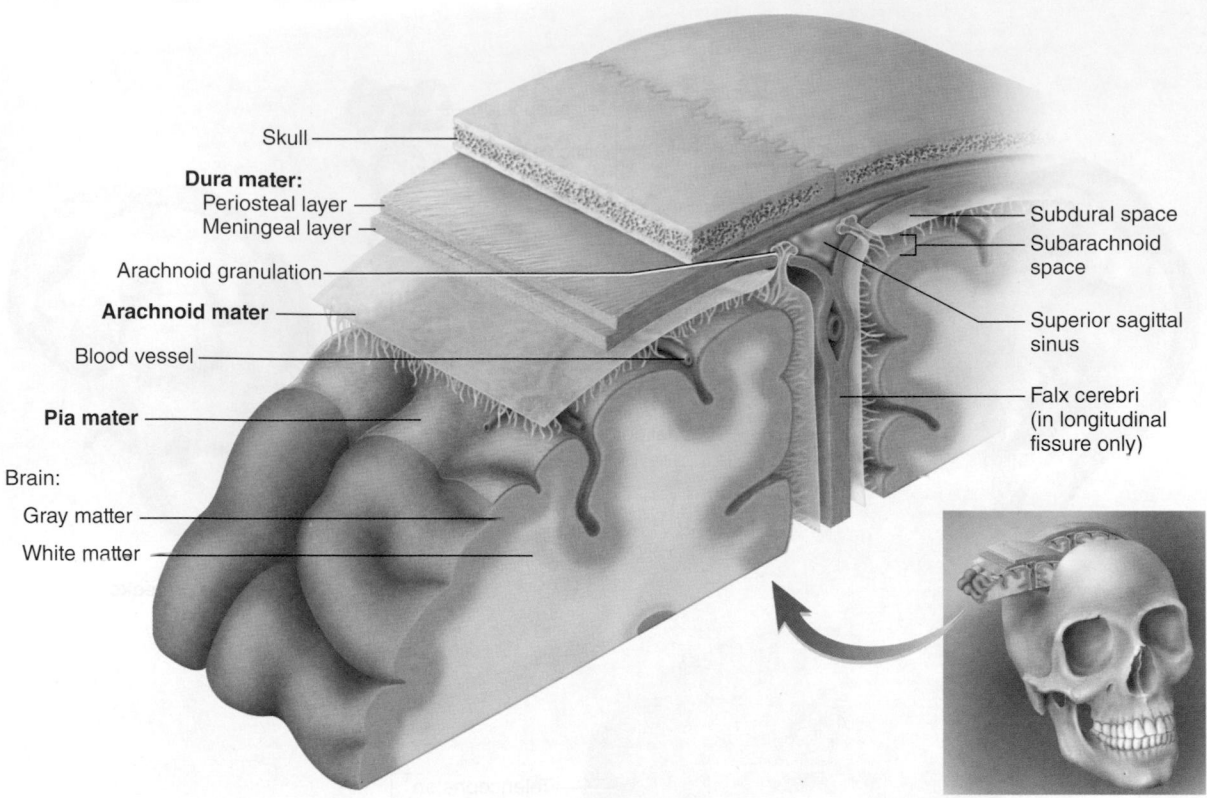

Skull

**Dura mater:**
  Periosteal layer
  Meningeal layer

Arachnoid granulation

**Arachnoid mater**

Blood vessel

**Pia mater**

Brain:
  Gray matter
  White matter

Subdural space
Subarachnoid space
Superior sagittal sinus
Falx cerebri (in longitudinal fissure only)

**FIGURE 14.5  The Meninges of the Brain.**  Frontal section of the head.

## DEEPER INSIGHT 14.1

### CLINICAL APPLICATION

#### Meningitis

*Meningitis*—inflammation of the meninges—is one of the most serious diseases of infancy and childhood. It occurs especially between 3 months and 2 years of age. Meningitis is caused by a variety of bacteria and viruses that invade the CNS by way of the nose and throat, often following respiratory, throat, or ear infections. The pia mater and arachnoid are most often affected, and from here the infection can spread to the adjacent nervous tissue. Meningitis can cause swelling of the brain, cerebral hemorrhaging, and sometimes death within mere hours of the onset of symptoms. Signs and symptoms include high fever, stiff neck, drowsiness, intense headache, and vomiting. Meningitis is diagnosed partly by examining the cerebrospinal fluid (CSF) for bacteria and white blood cells. The CSF is obtained by lumbar puncture (see p. 478).

Death from meningitis can occur so suddenly that infants and children with a high fever should therefore receive immediate medical attention. Freshman college students show a slightly elevated incidence of meningitis, especially those living in crowded dormitories rather than off campus.

cuboidal epithelium, lines the ventricles and canals and covers the choroid plexuses. It produces cerebrospinal fluid.

**Cerebrospinal fluid (CSF)** is a clear, colorless liquid that fills the ventricles and canals of the CNS and bathes its external surface. The brain produces about 500 mL of CSF per day, but the fluid is constantly reabsorbed at the same rate and only 100 to 160 mL is normally present at one time (but see Deeper Insight 14.2). About 40% of it is formed in the subarachnoid space external to the brain, 30% by the general ependymal lining of the brain ventricles, and 30% by the choroid plexuses. CSF production begins with the filtration of blood plasma through the capillaries of the brain. Ependymal cells modify the filtrate as it passes through them, so the CSF has more sodium and chloride than blood plasma, but less potassium, calcium, and glucose and very little protein.

CSF continually flows through and around the CNS, driven partly by its own pressure, partly by the beating of ependymal cilia, and partly by rhythmic pulsations of the brain produced by each heartbeat. The CSF of the lateral ventricles flows through the interventricular foramina into the third ventricle, then

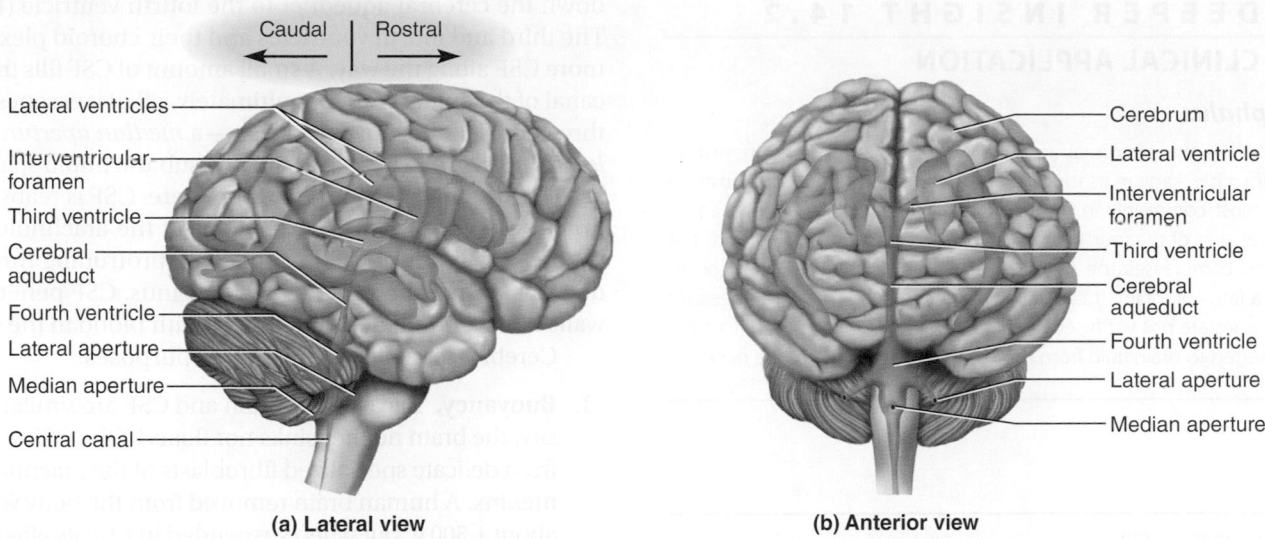

Caudal ← | → Rostral

**(a) Lateral view**

Lateral ventricles
Interventricular foramen
Third ventricle
Cerebral aqueduct
Fourth ventricle
Lateral aperture
Median aperture
Central canal

**(b) Anterior view**

Cerebrum
Lateral ventricle
Interventricular foramen
Third ventricle
Cerebral aqueduct
Fourth ventricle
Lateral aperture
Median aperture

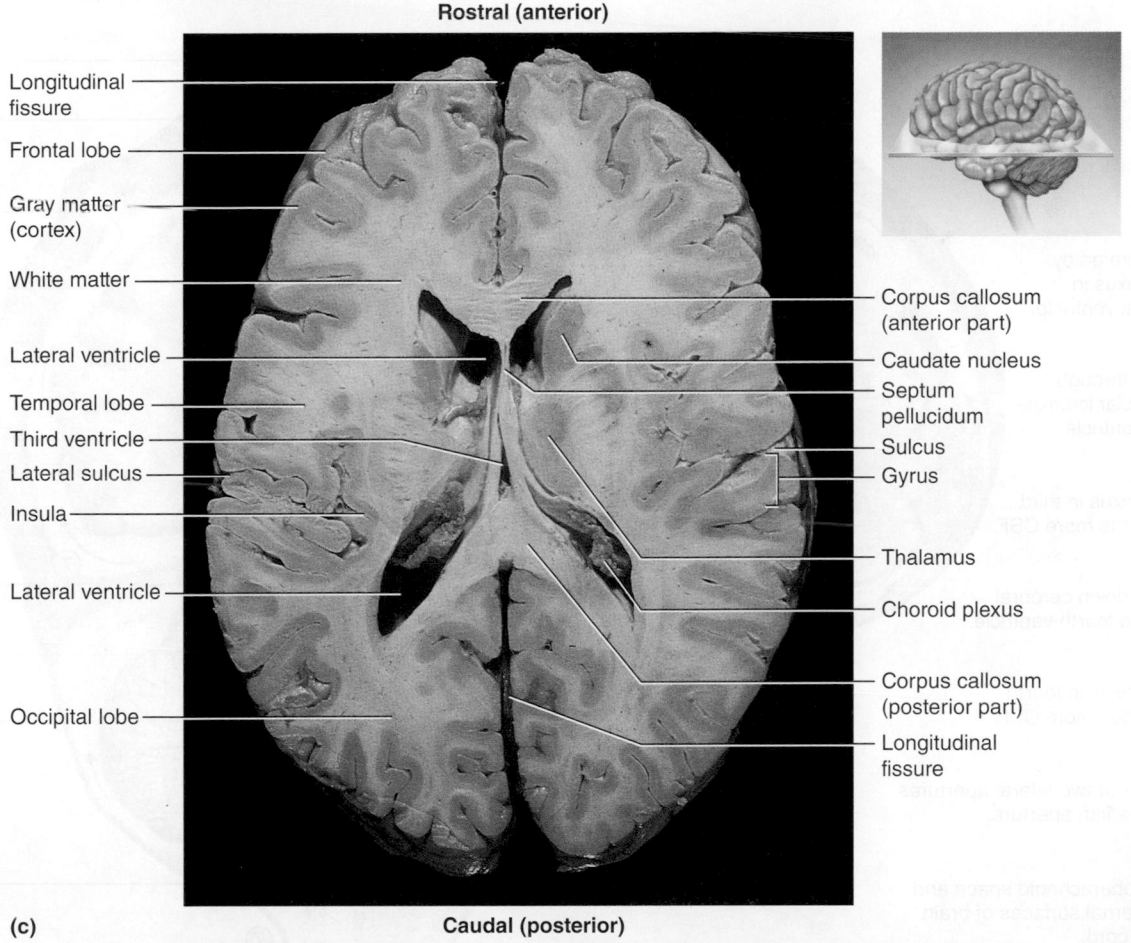

Rostral (anterior)

Longitudinal fissure
Frontal lobe
Gray matter (cortex)
White matter
Lateral ventricle
Temporal lobe
Third ventricle
Lateral sulcus
Insula
Lateral ventricle
Occipital lobe

Corpus callosum (anterior part)
Caudate nucleus
Septum pellucidum
Sulcus
Gyrus
Thalamus
Choroid plexus
Corpus callosum (posterior part)
Longitudinal fissure

**(c)**    Caudal (posterior)

**FIGURE 14.6  Ventricles of the Brain.** (a) Right lateral aspect. (b) Anterior aspect. (c) Superior view of a horizontal section of the cadaver brain, showing the lateral ventricles and some other features of the cerebrum. AP|R

# DEEPER INSIGHT 14.2

## CLINICAL APPLICATION

### Hydrocephalus

Hydrocephalus[16] is the abnormal accumulation of CSF in the brain, usually resulting from a blockage in its route of flow and reabsorption. Such obstructions occur most commonly in the interventricular foramen, cerebral aqueduct, and apertures of the fourth ventricle. The accumulated CSF expands the ventricles and compresses the nervous tissue, with potentially fatal consequences. In a fetus or infant, it can cause the entire head to enlarge because the cranial bones are not yet fused. Good recovery can be achieved if a tube (shunt) is inserted to drain fluid from the ventricles into a vein of the neck.

[16]hydro = water; cephal = head

down the cerebral aqueduct to the fourth ventricle (fig. 14.7). The third and fourth ventricles and their choroid plexuses add more CSF along the way. A small amount of CSF fills the central canal of the spinal cord, but ultimately, all of it escapes through three pores in the fourth ventricle—a *median aperture* and two *lateral apertures*. These lead into the subarachnoid space on the brain and spinal cord surface. From here, CSF is reabsorbed by **arachnoid granulations,** extensions of the arachnoid meninx shaped like little sprigs of cauliflower, protruding through the dura mater into the superior sagittal sinus. CSF penetrates the walls of the granulations and mixes with blood in the sinus.

Cerebrospinal fluid serves three purposes:

1. **Buoyancy.** Because the brain and CSF are similar in density, the brain neither sinks nor floats in the CSF. It hangs from delicate specialized fibroblasts of the arachnoid meninx. A human brain removed from the body weighs about 1,500 g, but when suspended in CSF its effective

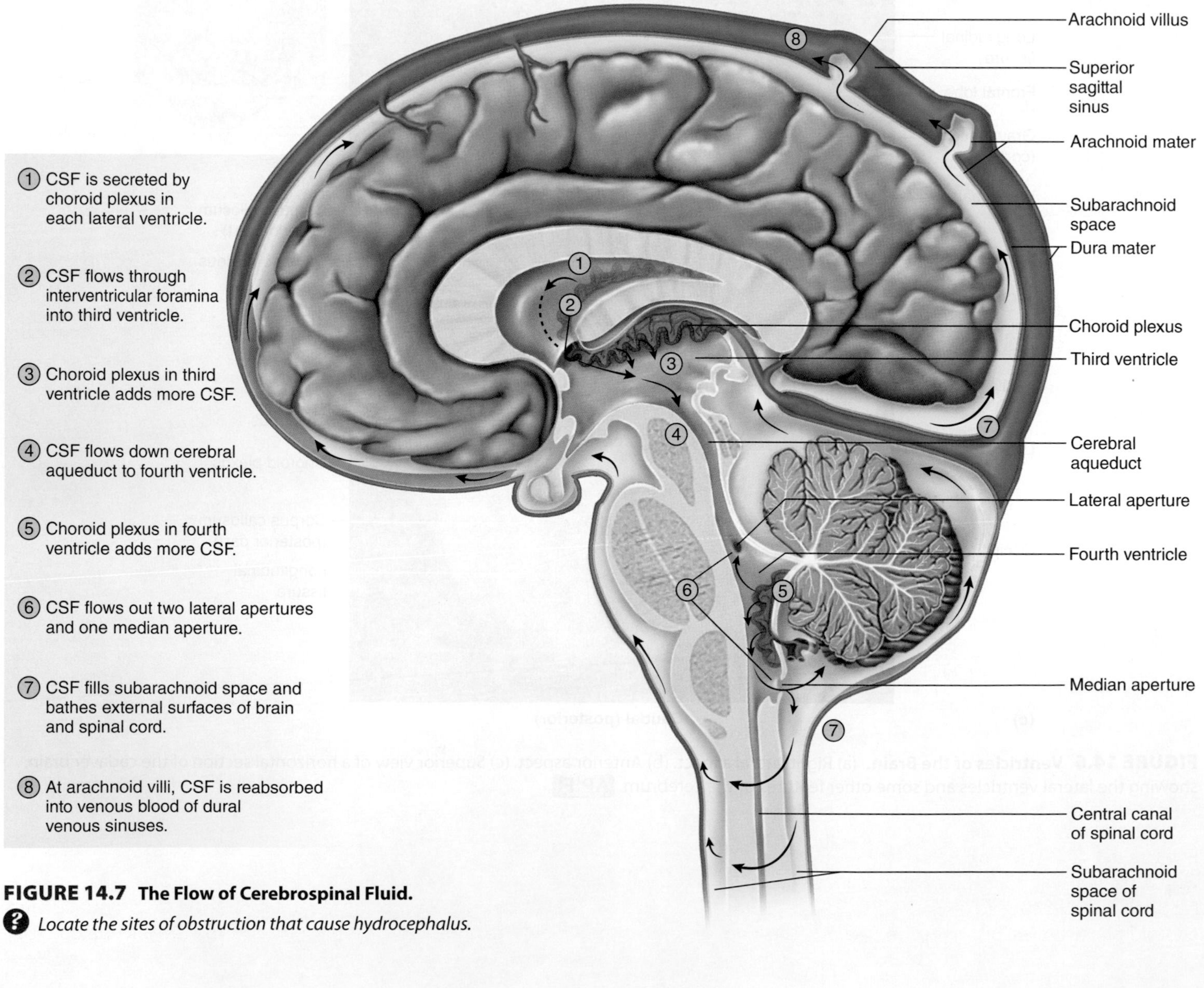

① CSF is secreted by choroid plexus in each lateral ventricle.

② CSF flows through interventricular foramina into third ventricle.

③ Choroid plexus in third ventricle adds more CSF.

④ CSF flows down cerebral aqueduct to fourth ventricle.

⑤ Choroid plexus in fourth ventricle adds more CSF.

⑥ CSF flows out two lateral apertures and one median aperture.

⑦ CSF fills subarachnoid space and bathes external surfaces of brain and spinal cord.

⑧ At arachnoid villi, CSF is reabsorbed into venous blood of dural venous sinuses.

Arachnoid villus
Superior sagittal sinus
Arachnoid mater
Subarachnoid space
Dura mater
Choroid plexus
Third ventricle
Cerebral aqueduct
Lateral aperture
Fourth ventricle
Median aperture
Central canal of spinal cord
Subarachnoid space of spinal cord

**FIGURE 14.7 The Flow of Cerebrospinal Fluid.**

Locate the sites of obstruction that cause hydrocephalus.

weight is only about 50 g. By analogy, consider how much easier it is to lift another person when you are standing in a lake than it is on land. This buoyancy allows the brain to attain considerable size without being impaired by its own weight. If the brain rested heavily on the floor of the cranium, the pressure would kill the nervous tissue.

2. **Protection.** CSF also protects the brain from striking the cranium when the head is jolted. If the jolt is severe, however, the brain still may strike the inside of the cranium or suffer shearing injury from contact with the angular surfaces of the cranial floor. This is one of the common findings in child abuse (shaken child syndrome) and in head injuries (concussions) from auto accidents, boxing, and the like.

3. **Chemical stability.** CSF rinses metabolic wastes from the nervous tissue and regulates its chemical environment. Slight changes in CSF composition can cause malfunctions of the nervous system. For example, a high glycine concentration disrupts the control of body temperature and blood pressure, and a high pH causes dizziness and fainting.

## Blood Supply and the Brain Barrier System

The blood vessels that serve the brain are detailed in chapter 20. Although the brain is only 2% of the adult body weight, it receives 15% of the blood (about 750 mL/min.) and consumes 20% of the oxygen and glucose. Because neurons have such a high demand for ATP, and therefore glucose and oxygen, the constancy of blood supply is especially critical to the nervous system. A mere 10-second interruption in blood flow can cause loss of consciousness; an interruption of 1 to 2 minutes can significantly impair neural function; and 4 minutes without blood usually causes irreversible brain damage. For more on cerebral blood flow and stroke, see page 765.

Despite its critical importance to the brain, blood is also a source of antibodies, macrophages, bacterial toxins, and other potentially harmful agents. Damaged brain tissue is essentially irreplaceable, and the brain therefore must be well protected. Consequently, there is a **brain barrier system** that strictly regulates what can get from the bloodstream into the tissue fluid of the brain.

There are two potential points of entry that must be guarded: the blood capillaries throughout the brain tissue and the capillaries of the choroid plexuses. At the former site, the brain is well protected by the **blood–brain barrier (BBB),** which consists of tight junctions between the endothelial cells that form the capillary walls. In the developing brain, astrocytes reach out and contact the capillaries with their perivascular feet, inducing endothelial cells to form tight junctions that completely seal off the gaps between them. This ensures that anything leaving the blood must pass through the cells and not between them. The endothelial cells are more selective than gaps between them would be, and can exclude harmful substances from the brain tissue while allowing necessary ones to pass through. At the choroid plexuses, the brain is protected by a similar **blood–CSF barrier** formed by tight junctions between the ependymal cells. Tight junctions are absent from ependymal cells elsewhere, because it is important to allow exchanges between the brain tissue and CSF. That is, there is no brain–CSF barrier.

The brain barrier system (BBS) is highly permeable to water, glucose, and lipid-soluble substances such as oxygen, carbon dioxide, alcohol, caffeine, nicotine, and anesthetics. It is slightly permeable to sodium, potassium, chloride, and the waste products urea and creatinine. While the BBS is an important protective device, it is an obstacle to the delivery of medications such as antibiotics and cancer drugs, and thus complicates the treatment of brain diseases.

Trauma and inflammation sometimes damage the BBS and allow pathogens to enter the brain tissue. Furthermore, there are places called **circumventricular organs (CVOs)** in the third and fourth ventricles where the barrier is absent and the blood has direct access to brain neurons. These enable the brain to monitor and respond to fluctuations in blood glucose, pH, osmolarity, and other variables. Unfortunately, CVOs also afford a route of invasion by the human immunodeficiency virus (HIV).

---

**BEFORE YOU GO ON**

Answer the following questions to test your understanding of the preceding section:

5. Name the three meninges from superficial to deep. How does the dura mater of the brain differ from that of the spinal cord?

6. Describe three functions of the cerebrospinal fluid.

7. Where does the CSF originate and what route does it take through and around the CNS?

8. Name the two components of the brain barrier system and explain the importance of this system.

---

## 14.3 The Hindbrain and Midbrain

### Expected Learning Outcomes
When you have completed this section, you should be able to

a. list the components of the hindbrain and midbrain and their functions; and

b. describe the location and functions of the reticular formation.

The study of the brain in the following pages will be organized around the five secondary vesicles of the embryonic brain and their mature derivatives. We will proceed in a caudal to rostral direction, beginning with the hindbrain and its relatively simple functions and progressing to the forebrain, the seat of such complex functions as thought, memory, and emotion.

### The Medulla Oblongata

As noted earlier, the embryonic hindbrain differentiates into two subdivisions: the myelencephalon and metencephalon (see fig. 14.4). The myelencephalon becomes just one adult structure, the **medulla oblongata** (meh-DULL-uh OB-long-GAH-ta).

The medulla (figs. 14.2 and 14.8) begins at the foramen magnum of the skull and extends for about 3 cm rostrally, ending at a groove between the medulla and pons. It looks superficially like an extension of the spinal cord, but slightly wider. Significant differences are apparent, however, on closer inspection of its gross and microscopic anatomy. Externally, the anterior surface features a pair of ridges called the **pyramids.** Resembling side-by-side baseball bats, these are wider at the rostral end, taper caudally, and are separated by a groove, the *anterior median fissure* continuous with that of the spinal cord. Lateral to each pyramid is a prominent bulge called the **olive.** Posteriorly, the **gracile** and **cuneate fasciculi** of the spinal cord continue as two pairs of ridges on the medulla.

All nerve fibers connecting the brain to the spinal cord pass through the medulla. As we saw in the cord, some of these are ascending (sensory) and some are descending (motor) fibers. The ascending fibers include first-order sensory fibers of the gracile and cuneate fasciculi, which end in the **gracile** and **cuneate nuclei** seen in figure 14.9c, a cross section of the medulla. Here, they synapse with second-order fibers that decussate and form the ribbonlike **medial lemniscus**[17] on each side. The second-order fibers rise to the thalamus, synapsing there with third-order fibers that complete the path to the cerebral cortex (compare fig. 13.5a, p. 481).

The largest group of descending fibers is the pair of **corticospinal tracts** filling the pyramids on the anterior surface. These carry motor signals from the cerebral cortex on the way to the spinal cord, ultimately to stimulate the skeletal muscles. Any time you carry out a body movement below the neck, the signals en route to your muscles pass through here. About 90% of these fibers cross over at the *pyramidal decussation,* an externally visible point near the caudal end of the pyramids (fig. 14.8a). As a result, muscles below the neck are controlled by the contralateral side of the brain. A smaller *tectospinal tract* controls the neck muscles.

The medulla contains neural networks involved in a multitude of fundamental sensory and motor functions. The former include the senses of hearing, equilibrium, touch, pressure, temperature, taste, and pain; the latter include chewing, salivation, swallowing, gagging, vomiting, respiration, speech, coughing, sneezing, sweating, cardiovascular and gastrointestinal control, and head, neck, and shoulder movements. Signals for these functions enter and leave the medulla not only by way of the spinal cord, but also by four pairs of cranial nerves that begin or end here: cranial nerves VIII (in part), IX, X, and XII. At the level of the section in figure 14.9c, we see the origins of two of them, the vagus and hypoglossal. The names and functions of these nerves are detailed in table 14.1 (pp. 543–550).

Another feature seen in cross section is the wavy **inferior olivary nucleus,** a major relay center for signals going from many levels of the brain and spinal cord to the cerebellum. The *reticular formation,* detailed later, is a loose network of nuclei extending throughout the medulla, pons, and midbrain. In the medulla, it includes a **cardiac center,** which regulates the rate and force of the heartbeat; a **vasomotor center,** which regulates blood pressure and flow by dilating and constricting blood vessels; two **respiratory centers,** which regulate the rhythm and depth of breathing; and other nuclei involved in the aforementioned motor functions.

## The Pons

The metencephalon develops into two structures, the pons and cerebellum. We will return to the cerebellum after finishing the brainstem. The **pons**[18] measures about 2.5 cm long. Most of it appears as a broad anterior bulge rostral to the medulla (figs. 14.2 and 14.8). Posteriorly, it consists mainly of two pairs of thick stalks called *cerebellar peduncles,* the cut edges in the upper half of figure 14.9b. They connect the cerebellum to the pons and midbrain (fig. 14.8b) and will be discussed with the cerebellum.

In cross section, the pons exhibits continuations of the previously mentioned reticular formation, medial lemniscus, tectospinal tract, and other spinal tracts. The anterior half of the pons (lower half of fig. 14.9b) is dominated by tracts of white matter, including transverse fascicles that cross between left and right and connect the two hemispheres of the cerebellum, and longitudinal fascicles that carry sensory and motor signals up and down the brainstem.

Cranial nerves V to VIII begin or end in the pons, although we see only the trigeminal nerve (V) at the level of figure 14.9b. The other three emerge from the groove between the pons and medulla. The functions of these four nerves, detailed in table 14.1, include sensory roles in hearing, equilibrium, and taste, and in facial sensations such as touch and pain; as well as motor roles in eye movement, facial expressions, chewing, swallowing, urination, and the secretion of saliva and tears. The reticular formation in the pons contains additional nuclei concerned with sleep, respiration, and posture.

## The Midbrain

The mesencephalon becomes just one mature brain structure, the **midbrain**—a short segment of brainstem that connects the hindbrain and forebrain (figs. 14.2 and 14.8). It contains the cerebral aqueduct, continuations of the medial lemniscus and reticular formation, and the motor nuclei of two cranial nerves that control eye movements: cranial nerves III (oculomotor) and IV (trochlear). Only the first of these is seen at the level of the cross section in figure 14.9a. (The *medial geniculate nucleus* seen in this figure is not part of the midbrain, but a part of the thalamus that happens to lie in the plane of this section.)

---

[17]*lemn* = ribbon; *iscus* = little

[18]*pons* = bridge

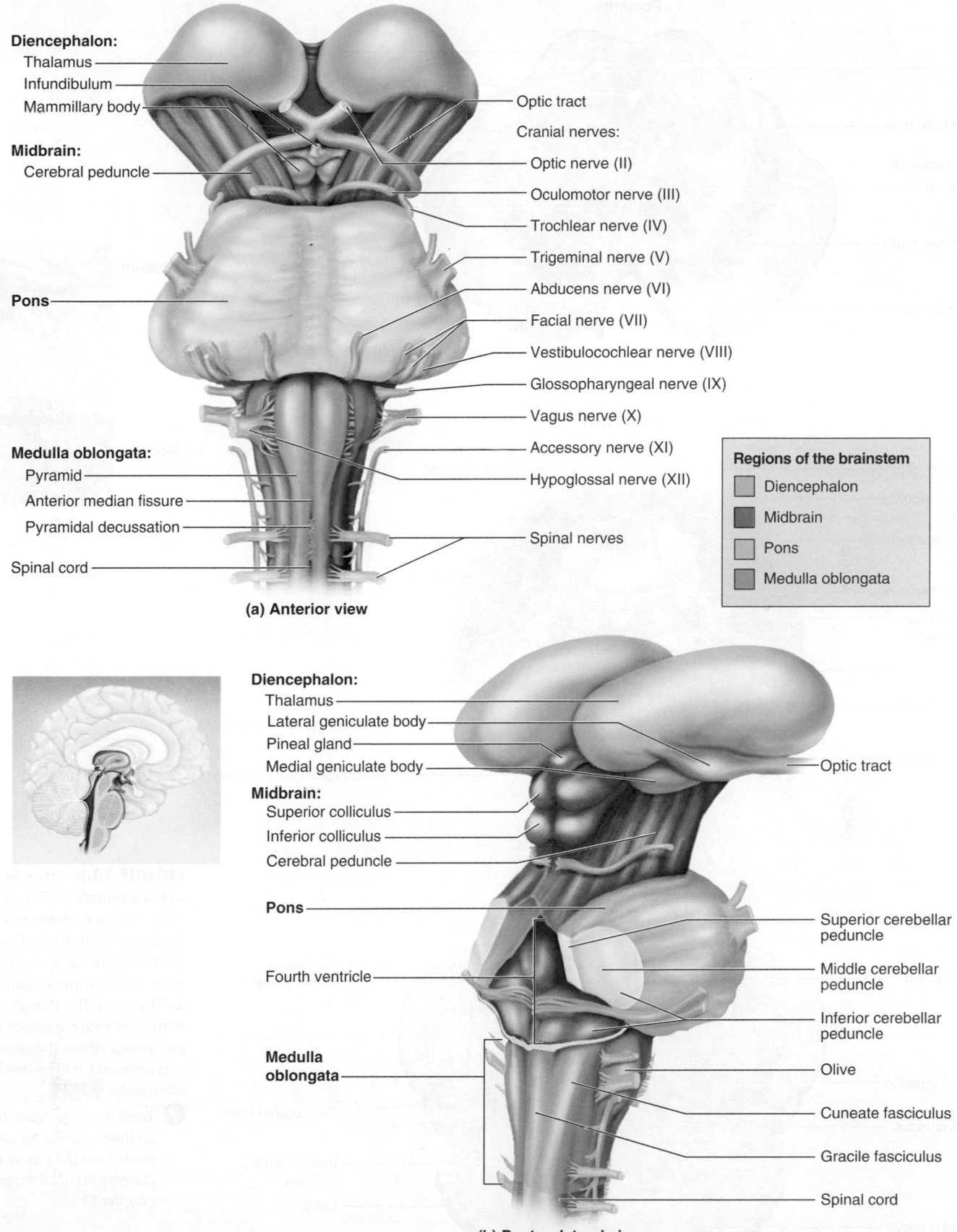

Diencephalon:
  Thalamus
  Infundibulum
  Mammillary body

Optic tract

Cranial nerves:

Midbrain:
  Cerebral peduncle

Optic nerve (II)

Oculomotor nerve (III)

Trochlear nerve (IV)

Trigeminal nerve (V)

Abducens nerve (VI)

Pons

Facial nerve (VII)

Vestibulocochlear nerve (VIII)

Glossopharyngeal nerve (IX)

Vagus nerve (X)

Medulla oblongata:
  Pyramid
  Anterior median fissure
  Pyramidal decussation

Accessory nerve (XI)

Hypoglossal nerve (XII)

Spinal nerves

Spinal cord

**(a) Anterior view**

**Regions of the brainstem**
  Diencephalon
  Midbrain
  Pons
  Medulla oblongata

Diencephalon:
  Thalamus
  Lateral geniculate body
  Pineal gland
  Medial geniculate body

Optic tract

Midbrain:
  Superior colliculus
  Inferior colliculus
  Cerebral peduncle

Pons

Superior cerebellar peduncle

Middle cerebellar peduncle

Fourth ventricle

Inferior cerebellar peduncle

Medulla oblongata

Olive

Cuneate fasciculus

Gracile fasciculus

Spinal cord

**(b) Posterolateral view**

**FIGURE 14.8  The Brainstem.**  (a) Anterior view. (b) Posterolateral view. Color-coded to match the embryonic origins in figure 14.4. The boundary between the middle and inferior cerebellar peduncles is indistinct. Authorities vary as to whether to include the diencephalon in the brainstem.

**AP|R**

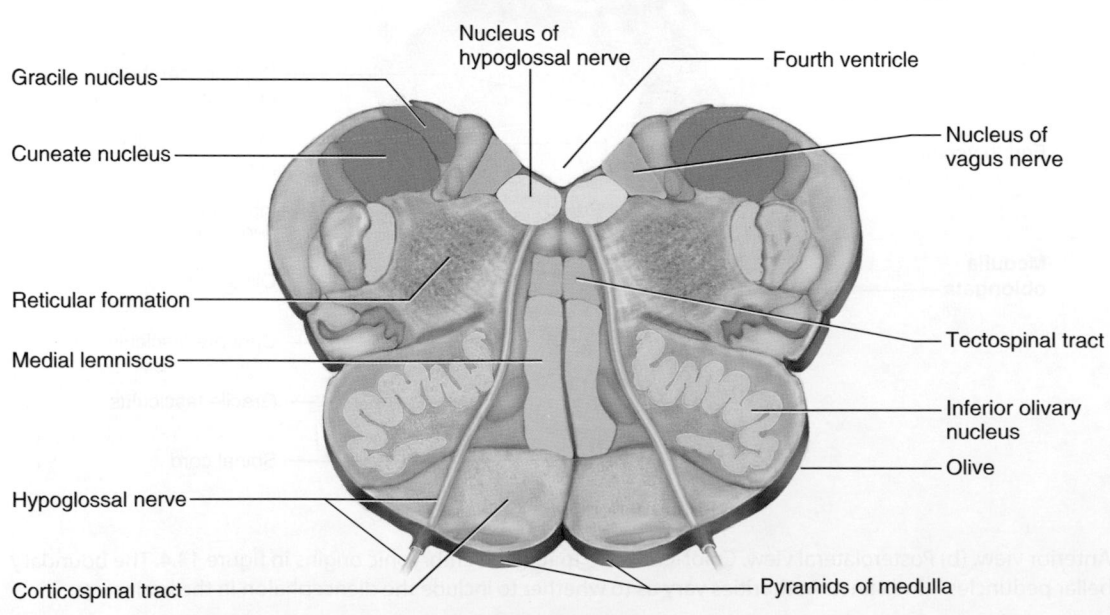

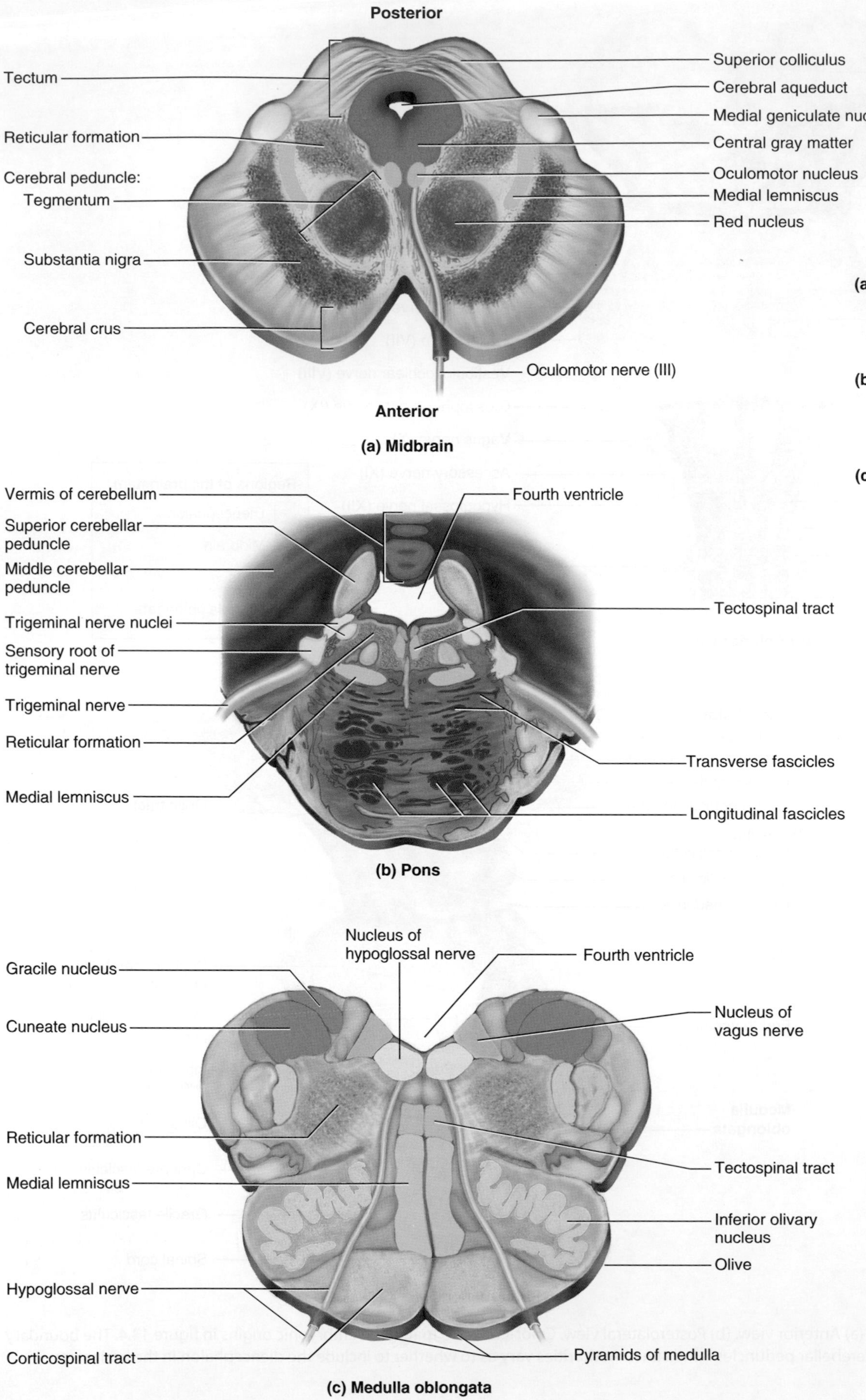

**Posterior**

Tectum

Reticular formation

Cerebral peduncle:
  Tegmentum

Substantia nigra

Cerebral crus

Superior colliculus

Cerebral aqueduct

Medial geniculate nucleus

Central gray matter

Oculomotor nucleus

Medial lemniscus

Red nucleus

Oculomotor nerve (III)

**Anterior**

**(a) Midbrain**

Vermis of cerebellum

Superior cerebellar peduncle

Middle cerebellar peduncle

Trigeminal nerve nuclei

Sensory root of trigeminal nerve

Trigeminal nerve

Reticular formation

Medial lemniscus

Fourth ventricle

Tectospinal tract

Transverse fascicles

Longitudinal fascicles

**(b) Pons**

Gracile nucleus

Cuneate nucleus

Reticular formation

Medial lemniscus

Hypoglossal nerve

Corticospinal tract

Nucleus of hypoglossal nerve

Fourth ventricle

Nucleus of vagus nerve

Tectospinal tract

Inferior olivary nucleus

Olive

Pyramids of medulla

**(c) Medulla oblongata**

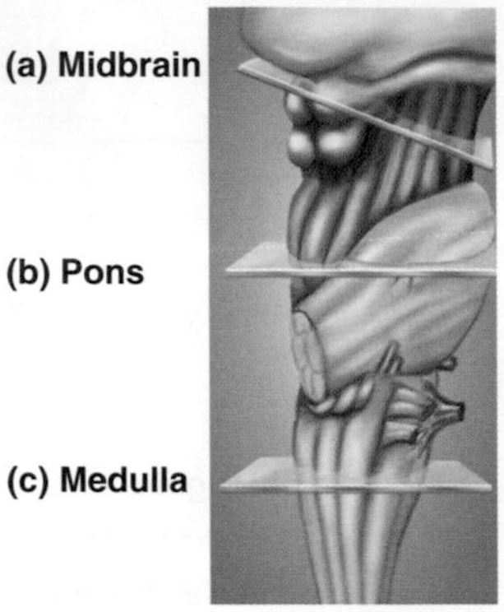

**(a) Midbrain**

**(b) Pons**

**(c) Medulla**

**FIGURE 14.9  Cross Sections of the Brainstem.** The level of each section is shown in the figure on the right. (a) The midbrain, cut obliquely to pass through the superior colliculi. (b) The pons. The straight edges indicate cut edges of the peduncles where the cerebellum was removed. (c) The medulla oblongata. **AP|R**

❓ *Trace the route taken through all three of these figures by fibers from the gracile and cuneate fasciculi described in chapter 13.*

The part of the midbrain posterior to the cerebral aqueduct is a rooflike **tectum.**[19] It exhibits four bulges, the **corpora quadrigemina.**[20] The upper pair, called the **superior colliculi**[21] (col-LIC-you-lye), functions in visual attention, visually tracking moving objects, and such reflexes as blinking, focusing, pupillary dilation and constriction, and turning the eyes and head in response to a visual stimulus (for example, to look at something that you catch sight of in your peripheral vision). The lower pair, called the **inferior colliculi,** receives signals from the inner ear and relays them to other parts of the brain, especially the thalamus. Among other functions, they mediate the reflexive turning of the head in response to a sound, and one's tendency to jump when startled by a sudden noise.

Anterior to the cerebral aqueduct, the midbrain consists mainly of the **cerebral peduncles**—two stalks that anchor the cerebrum to the brainstem. Each peduncle has three main components: tegmentum, substantia nigra, and cerebral crus. The **tegmentum**[22] is dominated by the **red nucleus,** named for a pink color imparted by its high density of blood vessels. Fibers from the red nucleus form the *rubrospinal tract* in most mammals, but in humans its connections go mainly to and from the cerebellum, with which it collaborates in fine motor control. The **substantia nigra**[23] (sub-STAN-she-uh NY-gruh) is a dark gray to black nucleus pigmented with melanin. It is a motor center that relays inhibitory signals to the thalamus and basal nuclei (both of which are discussed later), preventing unwanted body movement. Degeneration of the neurons in the substantia nigra leads to the muscle tremors of Parkinson disease (see Deeper Insight 12.4, p. 468). The **cerebral crus** (pronounced "cruss"; plural, *crura*) is a bundle of nerve fibers that connect the cerebrum to the pons and carry the corticospinal nerve tracts.

The cerebral aqueduct is encircled by the **central (periaqueductal) gray matter.** This is involved with the *reticulospinal tracts* in controlling awareness of pain, as further described in chapter 16.

#### ►►►APPLY WHAT YOU KNOW

*Why are the inferior colliculi shown in figure 14.8b but not in figure 14.9a? How are these two figures related?*

## The Reticular Formation

The **reticular**[24] **formation** is a loose web of gray matter that runs vertically through all levels of the brainstem, appearing at all three levels of figure 14.9. It occupies much of the space between the white fiber tracts and the more anatomically distinct brainstem nuclei, and has connections with many areas of the cerebrum (fig. 14.10). It consists of more than 100 small

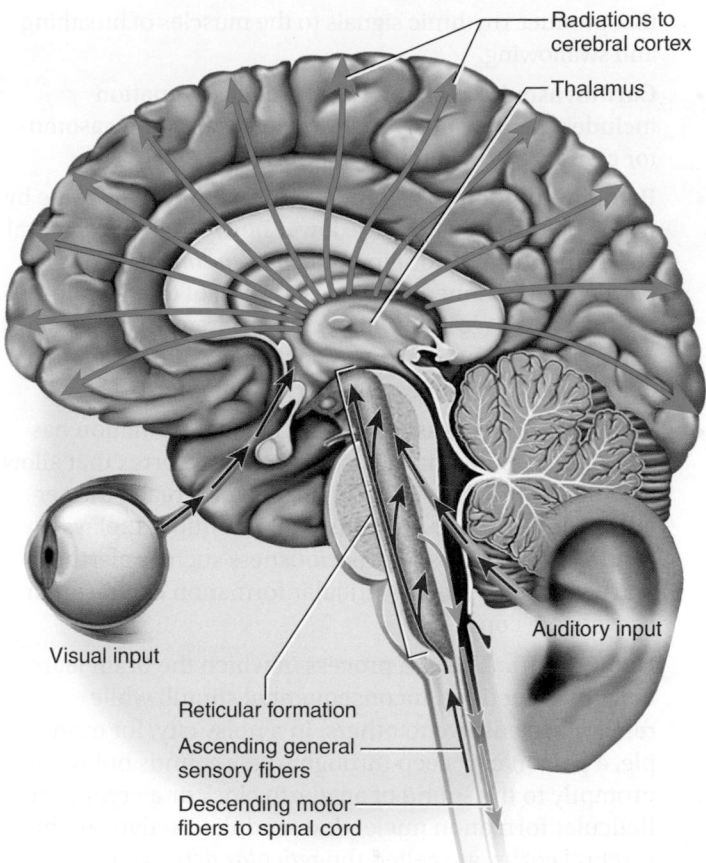

**FIGURE 14.10 The Reticular Formation.** The formation consists of over 100 nuclei scattered throughout the brainstem. Red arrows indicate routes of input to the reticular formation; blue arrows indicate the radiating relay of signals from the thalamus to the cerebral cortex; and green arrows indicate output from the reticular formation to the spinal cord.

**?** *Locate components of the reticular formation in all three parts of figure 14.9.*

neural networks defined less by anatomical boundaries than by their use of different neurotransmitters. The functions of these networks include the following:

- **Somatic motor control.** Some motor neurons of the cerebral cortex send their axons to reticular formation nuclei, which then give rise to the *reticulospinal tracts* of the spinal cord. These tracts adjust muscle tension to maintain tone, balance, and posture, especially during body movements. The reticular formation also relays signals from the eyes and ears to the cerebellum so the cerebellum can integrate visual, auditory, and vestibular (balance and motion) stimuli into its role in motor coordination. Other motor nuclei include *gaze centers,* which enable the eyes to track and fixate objects, and *central pattern generators*—neural pools

---

[19]*tectum* = roof, cover
[20]*corpora* = bodies; *quadrigemina* = quadruplets
[21]*colli* = hill; *cul* = little
[22]*tegmen* = cover
[23]*substantia* = substance; *nigra* = black
[24]*ret* = network; *icul* = little

that produce rhythmic signals to the muscles of breathing and swallowing.

- **Cardiovascular control.** The reticular formation includes the previously mentioned cardiac and vasomotor centers of the medulla oblongata.

- **Pain modulation.** The reticular formation is one route by which pain signals from the lower body reach the cerebral cortex. It is also the origin of the *descending analgesic pathways* mentioned in the description of the reticulospinal tracts on page 482. Under certain circumstances, the nerve fibers in these pathways act in the spinal cord to deaden one's awareness of pain (see chapter 16).

- **Sleep and consciousness.** The reticular formation has projections to the thalamus and cerebral cortex that allow it some control over what sensory signals reach the cerebrum and come to our conscious attention. It plays a central role in states of consciousness such as alertness and sleep. Injury to the reticular formation can result in irreversible coma.

- **Habituation.** This is a process in which the brain learns to ignore repetitive, inconsequential stimuli while remaining sensitive to others. In a noisy city, for example, a person can sleep through traffic sounds but wake promptly to the sound of an alarm clock or a crying baby. Reticular formation nuclei that modulate activity of the cerebral cortex are called the *reticular activating system* or *extrathalamic cortical modulatory system.*

## The Cerebellum

The **cerebellum** is the largest part of the hindbrain and second-largest part of the brain as a whole (fig. 14.11). It consists of right and left **cerebellar hemispheres** connected by a narrow wormlike bridge called the **vermis.**[25] Each hemisphere exhibits slender, transverse, parallel folds called **folia**[26] separated by shallow sulci. The cerebellum has a surface cortex of gray matter and a deeper layer of white matter. In a sagittal section, the white matter exhibits a branching, fernlike pattern called the **arbor vitae.**[27] Each hemisphere has four masses of gray matter called **deep nuclei** embedded in the white matter. All input to the cerebellum goes to the cortex and all of its output comes from the deep nuclei.

Although the cerebellum is only about 10% of the mass of the brain, it has about 60% as much surface area as the cerebral cortex and it contains more than half of all brain neurons—about 100 billion of them. Its tiny, densely spaced **granule cells** are the most abundant type of neuron in the entire brain. Its most distinctive neurons, however, are the unusually large, globose **Purkinje**[28] (pur-KIN-jee) **cells.** These have a tremendous profusion of dendrites compressed into a single plane like a flat tree (see fig. 12.5, p. 440, and the photograph on p. 435). The Purkinje cells are arranged in a single file, with these thick dendritic planes parallel to each other like books on a shelf. Their axons travel to the deep nuclei, where they synapse on output neurons that issue fibers to the brainstem.

The cerebellum is connected to the brainstem by three pairs of stalks called **cerebellar peduncles**[29] (peh-DUN-culs): a pair of *inferior peduncles* connected to the medulla oblongata, a pair of *middle peduncles* to the pons, and a pair of *superior peduncles* to the midbrain (see fig. 14.8b). These consist of thick bundles of nerve fibers that carry signals to and from the cerebellum. Connections between the cerebellum and brainstem are very complex, but overlooking some exceptions, we can draw a few generalizations. Most spinal input enters the cerebellum by way of the inferior peduncles; most input from the rest of the brain enters by way of the middle peduncles; and cerebellar output travels mainly by way of the superior peduncles.

The function of the cerebellum was unknown in the 1950s. By the 1970s, it had come to be regarded as a center for monitoring muscle contractions and aiding in motor coordination. People with cerebellar lesions exhibit serious deficits in coordination and locomotor ability; more will be said later in this chapter about the role of the cerebellum in movement. But cerebellar lesions also affect several sensory, linguistic, emotional, and other nonmotor functions. Recent studies by positron emission tomography (PET) and functional magnetic resonance imaging (fMRI) (both described on p. 23), and behavioral studies of people with cerebellar lesions, have created a much more expansive view of cerebellar function. It appears that its general role is the evaluation of certain kinds of sensory input, and monitoring muscle movement is only part of its broader function.

The cerebellum is highly active when a person explores objects with the fingertips, for example to compare the textures of two objects without looking at them. (Tactile nerve fibers from a rat's snout and a cat's forepaws also feed into the cerebellum.) Some spatial perception also resides here. The cerebellum is much more active when a person is required to solve a pegboard puzzle than when moving pegs randomly around the same puzzle board. People with cerebellar lesions also have difficulty identifying different views of a three-dimensional object as belonging to the same object.

The cerebellum is also a timekeeper. PET scans show increased cerebellar activity when a person is required to judge the elapsed time between two stimuli. People with cerebellar lesions have difficulty with rhythmic finger-tapping tasks and other tests of temporal judgment. An important aspect of

---

[25]*verm* = worm
[26]*foli* = leaf
[27]"tree of life"
[28]Johannes E. von Purkinje (1787–1869), Bohemian anatomist

[29]*ped* = foot; *uncle* = little

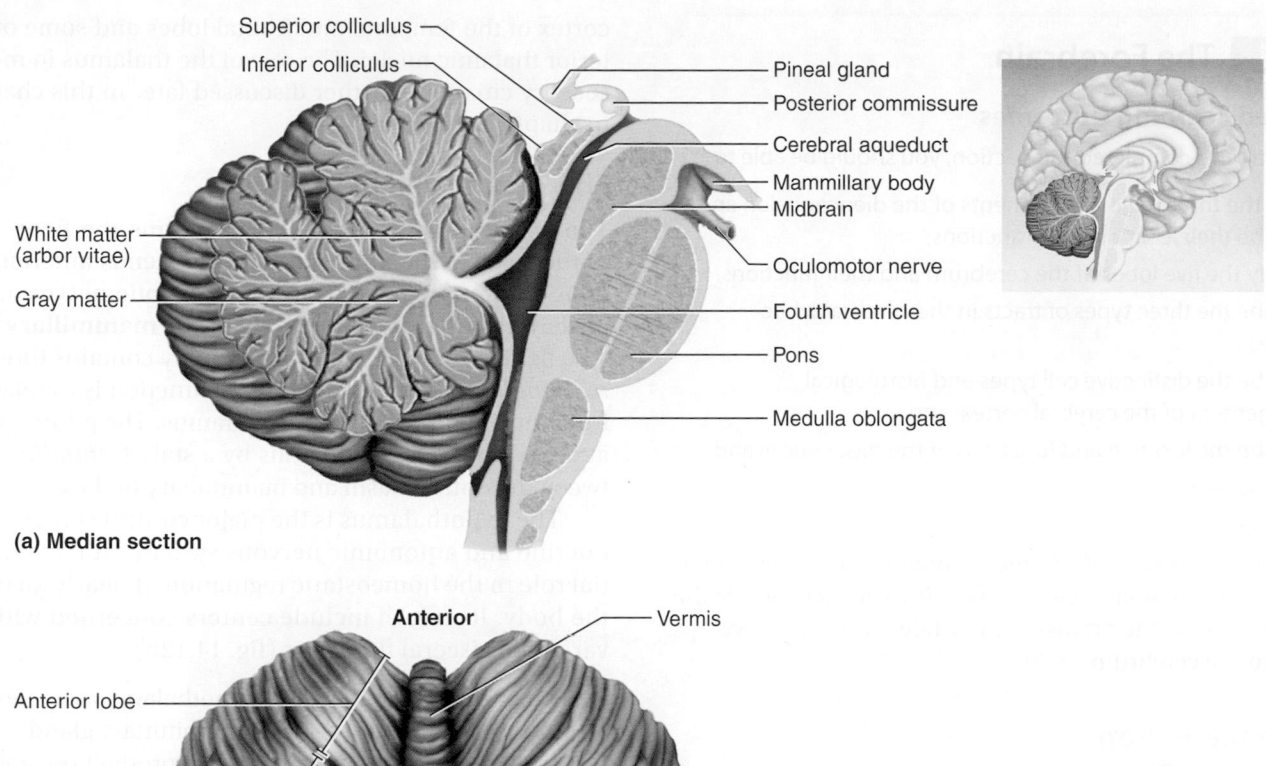

Superior colliculus
Inferior colliculus
Pineal gland
Posterior commissure
Cerebral aqueduct
Mammillary body
Midbrain
White matter (arbor vitae)
Gray matter
Oculomotor nerve
Fourth ventricle
Pons
Medulla oblongata

**(a) Median section**

**Anterior**

Vermis
Anterior lobe
Posterior lobe
Cerebellar hemisphere
Folia
**Posterior**

**(b) Superior view**

**FIGURE 14.11 The Cerebellum.** (a) Median section, showing relationship to the brainstem. (b) Superior aspect. AP|R

cerebellar timekeeping is the ability to predict where a moving object will be in the next second or so. You can imagine the importance of this to a predator chasing its prey, to a tennis player, or in driving a car in heavy traffic. The cerebellum also helps to predict how much the eyes must move in order to compensate for head movements and remain fixed on an object.

Even hearing has some newly discovered and surprising cerebellar components. Cerebellar lesions impair a person's ability to judge differences in pitch between two tones and to distinguish between similar-sounding words such as *rabbit* and *rapid*. Language output also involves the cerebellum. If a person is given a noun such as *apple* and told to think of a related verb such as *eat*, the cerebellum shows higher PET activity than if the person is told merely to repeat the word *apple*.

People with cerebellar lesions also have difficulty planning and scheduling tasks. They tend to overreact emotionally

and have difficulty with impulse control. Many children with attention-deficit/hyperactivity disorder (ADHD) have abnormally small cerebellums.

**BEFORE YOU GO ON**

Answer the following questions to test your understanding of the preceding section:

9. List several visceral functions controlled by nuclei of the medulla. What general function would be lost if the pyramids of the medulla were severed?

10. List several sensory and motor functions of the pons.

11. What functions are served by the superior and inferior colliculi? To what portion of the brainstem do they belong?

12. Where is the reticular formation found? Define it in a single sentence.

13. List several functions of the cerebellum.

## 14.4  The Forebrain

### Expected Learning Outcomes

When you have completed this section, you should be able to

a. name the three major components of the diencephalon and describe their locations and functions;

b. identify the five lobes of the cerebrum and their functions;

c. describe the three types of tracts in the cerebral white matter;

d. describe the distinctive cell types and histological arrangement of the cerebral cortex; and

e. describe the location and functions of the basal nuclei and limbic system.

The forebrain consists of the diencephalon and telencephalon. The diencephalon encloses the third ventricle and is the most rostral part of the brainstem. The telencephalon develops chiefly into the cerebrum.

## The Diencephalon

Three structures arise from the embryonic diencephalon: the *thalamus, hypothalamus,* and *epithalamus.*

### The Thalamus

Each side of the brain has a **thalamus,**[30] an ovoid mass perched at the superior end of the brainstem beneath the cerebral hemisphere (see figs. 14.6c, 14.8, and 14.17). The two thalami form about four-fifths of the diencephalon. Laterally, they protrude into the lateral ventricles. Medially, they protrude into the third ventricle and are joined to each other by a narrow *intermediate mass* in about 70% of people.

The thalamus consists of at least 23 nuclei, most of which fall into five groups: anterior, posterior, medial, lateral, and ventral. These groups and their functions are shown in figure 14.12a.

Broadly speaking, the thalamus is the "gateway to the cerebral cortex." Nearly all input to the cerebrum passes by way of synapses in the thalamic nuclei, including signals for taste, smell, hearing, equilibrium, vision, and such general senses as touch, pain, pressure, heat, and cold. (Some smell signals also get to the cerebrum by routes that bypass the thalamus.) The thalamic nuclei process this information and relay a small portion of it to the cerebral cortex.

The thalamus also serves in motor control by relaying signals from the cerebellum to the cerebrum and providing feedback loops between the cerebral cortex and the *basal nuclei* (deep cerebral motor centers). Finally, the thalamus is involved in the memory and emotional functions of the *limbic system,* a complex of structures that include some cerebral cortex of the temporal and frontal lobes and some of the anterior thalamic nuclei. The role of the thalamus in motor and sensory circuits is further discussed later in this chapter and in chapter 16.

### The Hypothalamus

The **hypothalamus** (see fig. 14.2) forms the floor and part of the walls of the third ventricle. It extends anteriorly to the *optic chiasm* (ky-AZ-um), where the optic nerves meet, and posteriorly to a pair of humps called the **mammillary**[31] **bodies** (see fig. 14.2a). Each mammillary body contains three to four *mammillary nuclei.* Their primary function is to relay signals from the limbic system to the thalamus. The pituitary gland is attached to the hypothalamus by a stalk *(infundibulum)* between the optic chiasm and mammillary bodies.

The hypothalamus is the major control center of the endocrine and autonomic nervous systems. It plays an essential role in the homeostatic regulation of nearly all organs of the body. Its nuclei include centers concerned with a wide variety of visceral functions (fig. 14.12b):

- **Hormone secretion.** The hypothalamus secretes hormones that control the anterior pituitary gland, thereby regulating growth, metabolism, reproduction, and stress responses. The hypothalamus also produces two hormones that are stored in the posterior pituitary gland, concerned with labor contractions, lactation, and water conservation. These relationships are explored especially in chapter 17.

- **Autonomic effects.** The hypothalamus is a major integrating center for the autonomic nervous system. It sends descending fibers to lower brainstem nuclei that influence heart rate, blood pressure, gastrointestinal secretion and motility, and pupillary diameter, among other functions.

- **Thermoregulation.** The *hypothalamic thermostat* consists of a collection of neurons, concentrated especially in the preoptic nucleus, that monitor body temperature. When the temperature deviates too much from its set point, this center activates mechanisms for lowering or raising the body temperature (see chapter 26).

- **Food and water intake.** The hypothalamus regulates sensations of hunger and satiety. One nucleus in particular, the *arcuate nucleus,* contains receptors for hormones that increase hunger and energy expenditure, other hormones that reduce both, and hormones that exert long-term control over body mass (see chapter 26). Hypothalamic neurons called *osmoreceptors* monitor blood osmolarity and stimulate water-seeking and drinking behavior when the body is dehydrated. Dehydration also stimulates the hypothalamus to produce *antidiuretic hormone,* which conserves water by reducing urine output.

---

[30] *thalamus* = chamber, inner room

[31] *mammill* = nipple

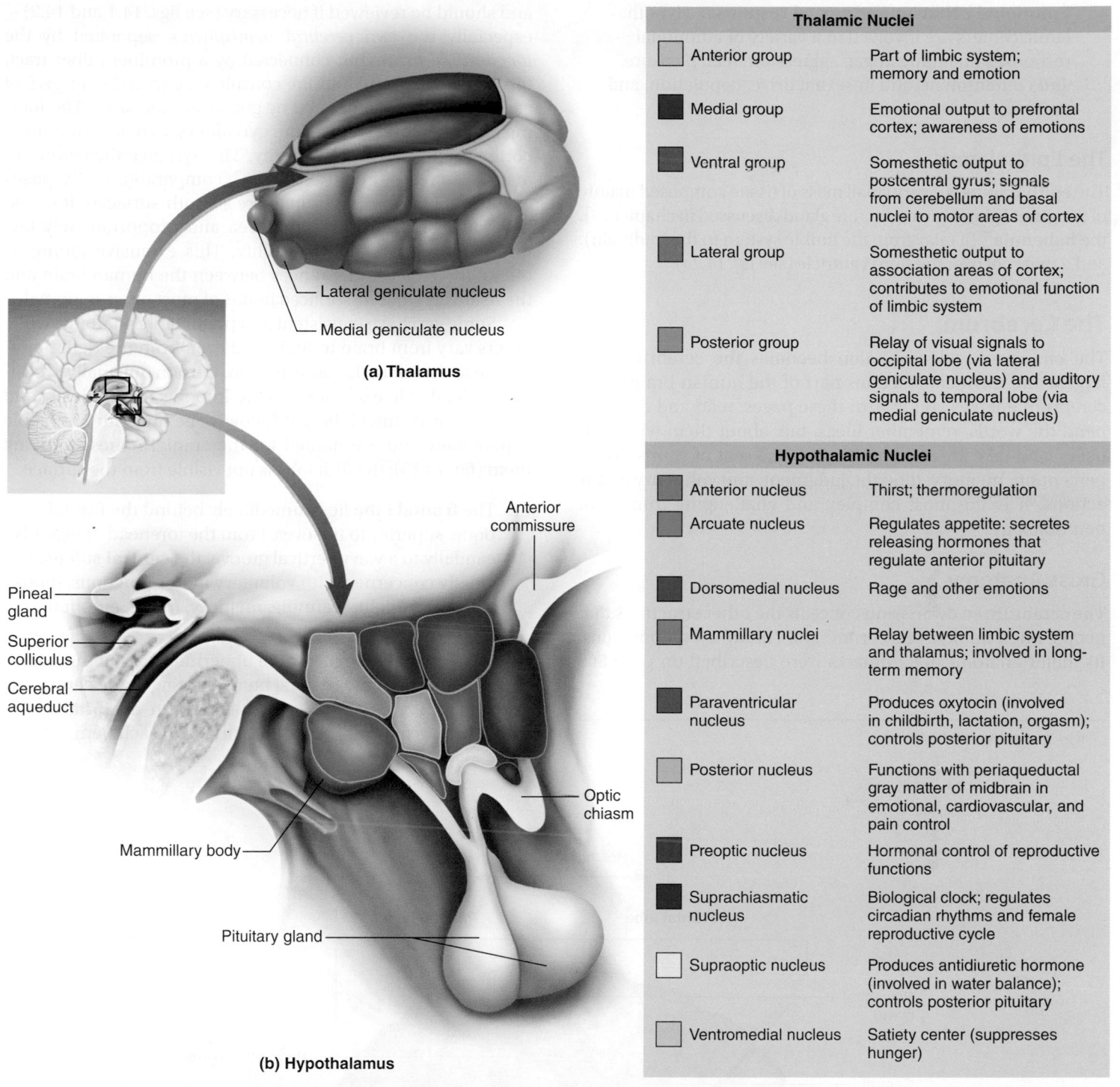

**Thalamic Nuclei**

| | | |
|---|---|---|
| Anterior group | | Part of limbic system; memory and emotion |
| Medial group | | Emotional output to prefrontal cortex; awareness of emotions |
| Ventral group | | Somesthetic output to postcentral gyrus; signals from cerebellum and basal nuclei to motor areas of cortex |
| Lateral group | | Somesthetic output to association areas of cortex; contributes to emotional function of limbic system |
| Posterior group | | Relay of visual signals to occipital lobe (via lateral geniculate nucleus) and auditory signals to temporal lobe (via medial geniculate nucleus) |

**Hypothalamic Nuclei**

| | | |
|---|---|---|
| Anterior nucleus | | Thirst; thermoregulation |
| Arcuate nucleus | | Regulates appetite: secretes releasing hormones that regulate anterior pituitary |
| Dorsomedial nucleus | | Rage and other emotions |
| Mammillary nuclei | | Relay between limbic system and thalamus; involved in long-term memory |
| Paraventricular nucleus | | Produces oxytocin (involved in childbirth, lactation, orgasm); controls posterior pituitary |
| Posterior nucleus | | Functions with periaqueductal gray matter of midbrain in emotional, cardiovascular, and pain control |
| Preoptic nucleus | | Hormonal control of reproductive functions |
| Suprachiasmatic nucleus | | Biological clock; regulates circadian rhythms and female reproductive cycle |
| Supraoptic nucleus | | Produces antidiuretic hormone (involved in water balance); controls posterior pituitary |
| Ventromedial nucleus | | Satiety center (suppresses hunger) |

Labels: Lateral geniculate nucleus; Medial geniculate nucleus; **(a) Thalamus**

Anterior commissure; Pineal gland; Superior colliculus; Cerebral aqueduct; Mammillary body; Pituitary gland; Optic chiasm; **(b) Hypothalamus**

**FIGURE 14.12  The Diencephalon.** Only some of the nuclei of the thalamus and hypothalamus are shown, and some of their functions are listed. These lists are by no means complete. AP|R

- **Sleep and circadian rhythms.** The caudal part of the hypothalamus is part of the reticular formation. It contains nuclei that regulate the rhythm of sleep and waking. Superior to the optic chiasm, the hypothalamus contains a *suprachiasmatic nucleus* that controls our 24-hour (circadian) rhythm of activity.

- **Memory.** The mammillary nuclei lie in the pathway of signals traveling from the hippocampus, an important memory center of the brain, to the thalamus. Thus, they are important in memory, and lesions to the mammillary nuclei cause memory deficits. (Memory is discussed more fully on p. 534.)

- **Emotional behavior and sexual response.** Hypothalamic centers are involved in a variety of emotional responses including anger, aggression, fear, pleasure, and contentment; and in sexual drive, copulation, and orgasm.

## The Epithalamus

The **epithalamus** is a very small mass of tissue composed mainly of the **pineal gland** (an endocrine gland discussed in chapter 17), the **habenula**[32] (a relay from the limbic system to the midbrain), and a thin roof over the third ventricle (see fig. 14.2a).

## The Cerebrum

The embryonic telencephalon becomes the cerebrum, the largest and most conspicuous part of the human brain. Your cerebrum enables you to turn these pages, read and comprehend the words, remember ideas, talk about them with your peers, and take an examination. It is the seat of your sensory perception, memory, thought, judgment, and voluntary motor actions. It is the most complex and challenging frontier of neurobiology.

## Gross Anatomy

The cerebrum so dwarfs and conceals the other structures that people often think of "cerebrum" and "brain" as synonymous. Its major anatomical landmarks were described on page 508

---

[32]*haben* = strap, rein; *ula* = little

and should be reviewed if necessary (see figs. 14.1 and 14.2)—especially the two *cerebral hemispheres,* separated by the *longitudinal fissure* but connected by a prominent fiber tract, the *corpus callosum;* and the conspicuous wrinkles, or *gyri,* of each hemisphere, separated by grooves called *sulci.* The folding of the cerebral surface into gyri allows a greater amount of cortex to fit in the cranial cavity. The gyri give the cerebrum a surface area of about 2,500 cm², comparable to 4½ pages of this book. If the cerebrum were smooth-surfaced, it would have only one-third as much area and proportionately less information-processing capability. This extensive folding is one of the greatest differences between the human brain and the relatively smooth-surfaced brains of most other mammals.

Some gyri have consistent and predictable anatomy, while others vary from brain to brain and even from the right hemisphere to the left in the same person. Certain unusually prominent sulci divide each hemisphere into five anatomically and functionally distinct lobes, as follows. The first four are visible superficially and are named for the cranial bones overlying them (fig. 14.13); the fifth lobe is not visible from the surface.

1. The **frontal lobe** lies immediately behind the frontal bone, superior to the eyes. From the forehead, it extends caudally to a wavy vertical groove, the **central sulcus.** It is chiefly concerned with voluntary motor functions, motivation, foresight, planning, memory, mood, emotion, social judgment, and aggression.

2. The **parietal lobe** forms the uppermost part of the brain and underlies the parietal bone. Starting at the central sulcus, it extends caudally to the **parieto–occipital sulcus,** visible on the medial surface of each hemisphere

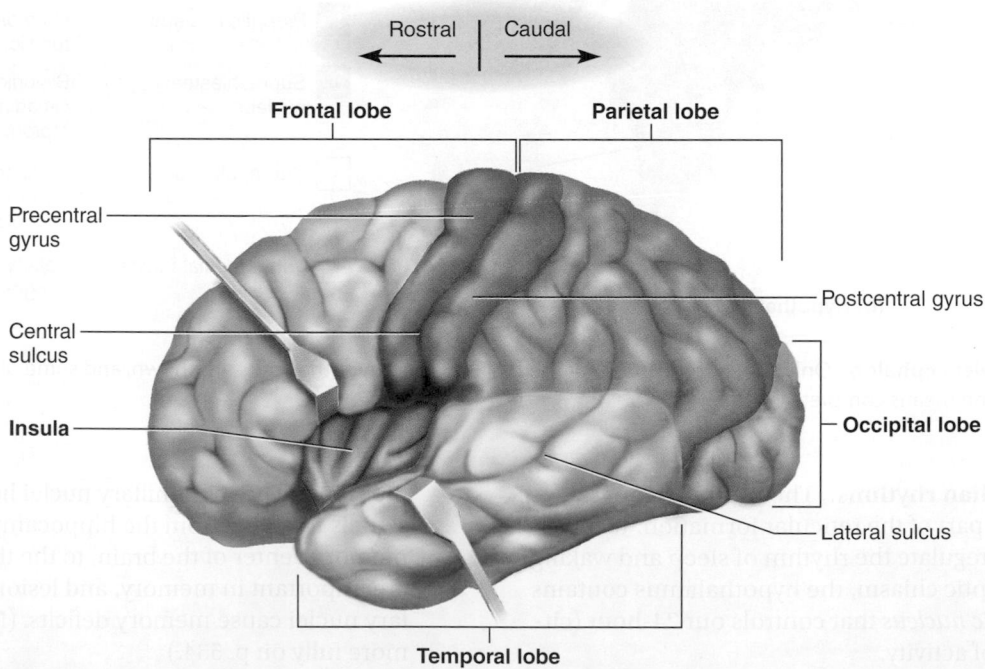

Rostral | Caudal

Frontal lobe | Parietal lobe

Precentral gyrus

Central sulcus

Insula

Postcentral gyrus

Occipital lobe

Lateral sulcus

Temporal lobe

**FIGURE 14.13  Lobes of the Cerebrum.** The frontal and temporal lobes are retracted slightly to reveal the insula. AP|R

(see fig. 14.2). It is the primary site for receiving and interpreting signals of the *general senses* described later; for taste (one of the *special senses*); and for some visual processing.

3. The **occipital lobe** is at the rear of the head, caudal to the parieto–occipital sulcus and underlying the occipital bone. It is the principal visual center of the brain.

4. The **temporal lobe** is a lateral, horizontal lobe deep to the temporal bone, separated from the parietal lobe above it by a deep **lateral sulcus.** It is concerned with hearing, smell, learning, memory, and some aspects of vision and emotion.

5. The **insula**[33] is a small mass of cortex deep to the lateral sulcus, made visible only by retracting or cutting away some of the overlying cerebrum (see figs. 14.1c, 14.6c, and

---

[33]*insula* = island

14.13). It is less understood than the other lobes because it is less accessible to testing in living subjects, but it apparently plays roles in understanding spoken language, in taste, and in integrating information from visceral receptors.

## The Cerebral White Matter

Most of the volume of the cerebrum is white matter. This is composed of glia and myelinated nerve fibers that transmit signals from one region of the cerebrum to another and between the cerebrum and lower brain centers. These fibers form bundles, or *tracts,* of three kinds (fig. 14.14):

1. **Projection tracts** extend vertically between higher and lower brain and spinal cord centers, and carry information between the cerebrum and the rest of the body. The

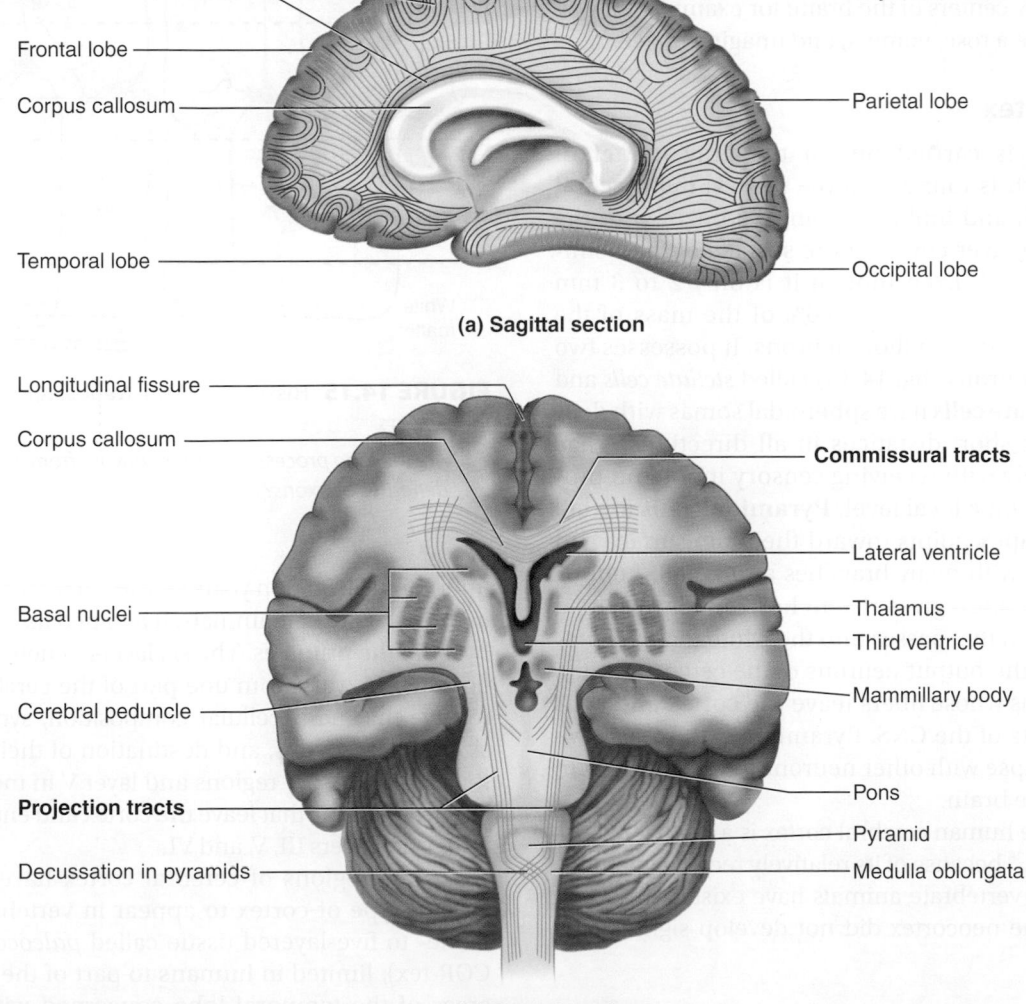

**FIGURE 14.14 Tracts of Cerebral White Matter.** (a) Left lateral aspect, showing association and projection tracts. (b) Frontal section, showing commissural and projection tracts.

 *What route can commissural tracts take other than the one shown here?*

corticospinal tracts, for example, carry motor signals from the cerebrum to the brainstem and spinal cord. Other projection tracts carry signals upward to the cerebral cortex. Superior to the brainstem, such tracts form a broad, dense sheet called the *internal capsule* between the thalamus and basal nuclei (described shortly), then radiate in a diverging, fanlike array (the *corona radiata*[34]) to specific areas of the cortex.

2. **Commissural tracts** cross from one cerebral hemisphere to the other through bridges called **commissures** (COM-ih-shurs). The great majority of commissural tracts pass through the large corpus callosum (see fig. 14.2). A few tracts pass through the much smaller **anterior** and **posterior commissures.** Commissural tracts enable the two sides of the cerebrum to communicate with each other.

3. **Association tracts** connect different regions within the same cerebral hemisphere. *Long association fibers* connect different lobes of a hemisphere to each other, whereas *short association fibers* connect different gyri within a single lobe. Among their roles, association tracts link perceptual and memory centers of the brain; for example, they enable you to see a rose, name it, and imagine its scent.

## The Cerebral Cortex

Neural integration is carried out in the gray matter of the cerebrum, which is found in three places: the cerebral cortex, basal nuclei, and limbic system. We begin with the **cerebral cortex,**[35] a layer covering the surface of the hemispheres (see fig. 14.6c). Even though it is only 2 to 3 mm thick, the cortex constitutes about 40% of the mass of the brain and contains 14 to 16 billion neurons. It possesses two principal types of neurons (fig. 14.15) called *stellate cells* and *pyramidal cells.* **Stellate cells** have spheroidal somas with dendrites projecting for short distances in all directions. They are concerned largely with receiving sensory input and processing information on a local level. **Pyramidal cells** are tall and conical. Their apex points toward the brain surface and has a thick dendrite with many branches and small, knobby *dendritic spines.* The base gives rise to horizontally oriented dendrites and an axon that passes into the white matter. Pyramidal cells include the output neurons of the cerebrum—the only cerebral neurons whose fibers leave the cortex and connect with other parts of the CNS. Pyramidal cell axons have collaterals that synapse with other neurons in the cortex or in deeper regions of the brain.

About 90% of the human cerebral cortex is a six-layered tissue called **neocortex**[36] because of its relatively recent evolutionary origin. Although vertebrate animals have existed for about 600 million years, the neocortex did not develop significantly

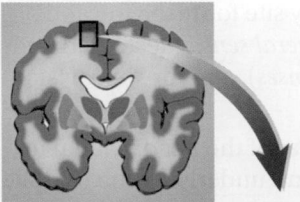

Cortical surface

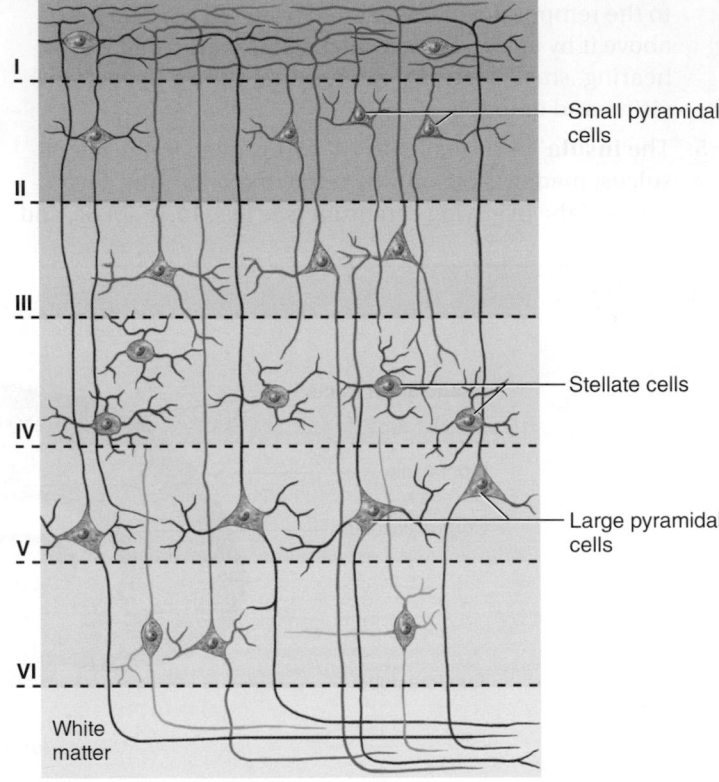

I

II — Small pyramidal cells

III

Stellate cells

IV

V — Large pyramidal cells

VI

White matter

**FIGURE 14.15  Histology of the Neocortex.** Neurons are arranged in six layers.

❓ *Are the long processes leading upward from each pyramidal cell body dendrites or axons?*

until about 60 million years ago, when there was a sharp increase in the diversity of mammals. It attained its highest development by far in the primates. The six layers of neocortex, numbered in figure 14.15, vary from one part of the cerebrum to another in relative thickness, cellular composition, synaptic connections, size of the neurons, and destination of their axons. Layer IV is thickest in sensory regions and layer V in motor regions, for example. All axons that leave the cortex and enter the white matter arise from layers III, V, and VI.

Some regions of cerebral cortex have fewer layers. The earliest type of cortex to appear in vertebrate evolution was a one- to five-layered tissue called *paleocortex* (PALE-ee-oh-COR-tex), limited in humans to part of the insula and certain areas of the temporal lobe concerned with smell. The next to evolve was a three-layered *archicortex* (AR-kee-COR-tex), found in the human hippocampus, a memory-forming center in the temporal lobe. The neocortex was the last to evolve.

---

[34]*corona* = crown; *radiata* = radiating
[35]*cortex* = bark, rind
[36]*neo* = new

## The Limbic System

The **limbic**[37] **system** is an important center of emotion and learning. It is a ring of structures on the medial side of the cerebral hemisphere, encircling the corpus callosum and thalamus. Its most anatomically prominent components are the **cingulate**[38] (SING-you-let) **gyrus,** which arches over the top of the corpus callosum in the frontal and parietal lobes; the **hippocampus**[39] in the medial temporal lobe (fig. 14.16); and the **amygdala**[40] (ah-MIG-da-luh) immediately rostral to the hippocampus, also in the temporal lobe. There are still differences of opinion on what structures to consider as parts of the limbic system, but these three are agreed upon. Other components include the mammillary bodies and other hypothalamic nuclei, some thalamic nuclei, parts of the basal nuclei, and parts of the frontal lobe called *prefrontal* and *orbitofrontal cortex.* Limbic system components are interconnected through a complex loop of fiber tracts allowing for somewhat circular patterns of feedback among its nuclei and cortical neurons. All of these structures are bilaterally paired; there is a limbic system in each cerebral hemisphere.

The limbic system was long thought to be associated with smell because of its close association with olfactory pathways, but beginning in the early 1900s and continuing even now, experiments have abundantly demonstrated more significant roles in emotion and memory. Most limbic system structures have centers for both gratification and aversion. Stimulation of a gratification center produces a sense of pleasure or reward; stimulation of an aversion center produces unpleasant sensations such as fear or sorrow. Gratification centers dominate some limbic structures, such as the *nucleus accumbens* (not illustrated), while aversion centers dominate others such as the amygdala. The roles of the amygdala in emotion and the hippocampus in memory are described in section 14.5.

## The Basal Nuclei

The basal nuclei are masses of cerebral gray matter buried deep in the white matter, lateral to the thalamus (fig. 14.17). They are often called *basal ganglia,* but the word *ganglion* is best restricted to clusters of neurons outside the CNS. Neuroanatomists disagree on how many brain centers to classify as basal nuclei, but agree on at least three: the **caudate**[41] **nucleus, putamen**[42] (pyu-TAY-men), and **globus pallidus.**[43] These three are collectively called the *corpus striatum*[44] because of their striped appearance. The putamen and globus pallidus together are also called the *lentiform*[45] *nucleus,* because they form a lens-shaped body. They are involved in motor control and are further discussed in a later section on that topic.

---

[37]*limbus* = border
[38]*cingul* = girdle
[39]*hippocampus* = sea horse, named for its shape
[40]*amygdala* = almond

[41]*caudate* = tailed, tail-like
[42]*putam* = pod, husk
[43]*glob* = globe, ball; *pall* = pale
[44]*corpus* = body; *striat* = stripe
[45]*lenti* = lens; *form* = shape

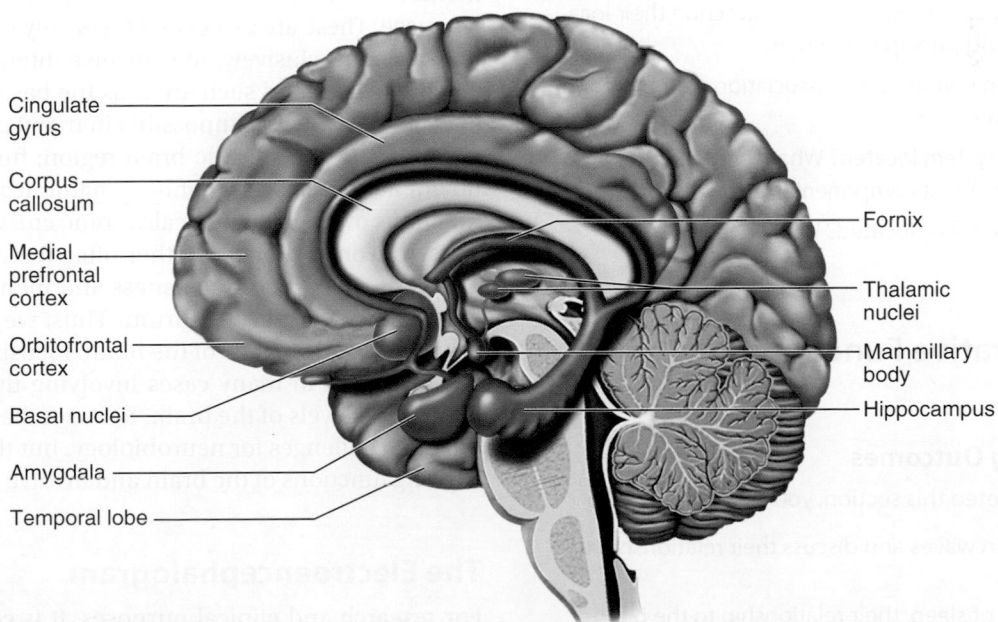

**FIGURE 14.16 The Limbic System.** This ring of structures (shown in violet) includes important centers of learning and emotion. In the frontal lobe, there is no sharp rostral boundary to limbic system components.

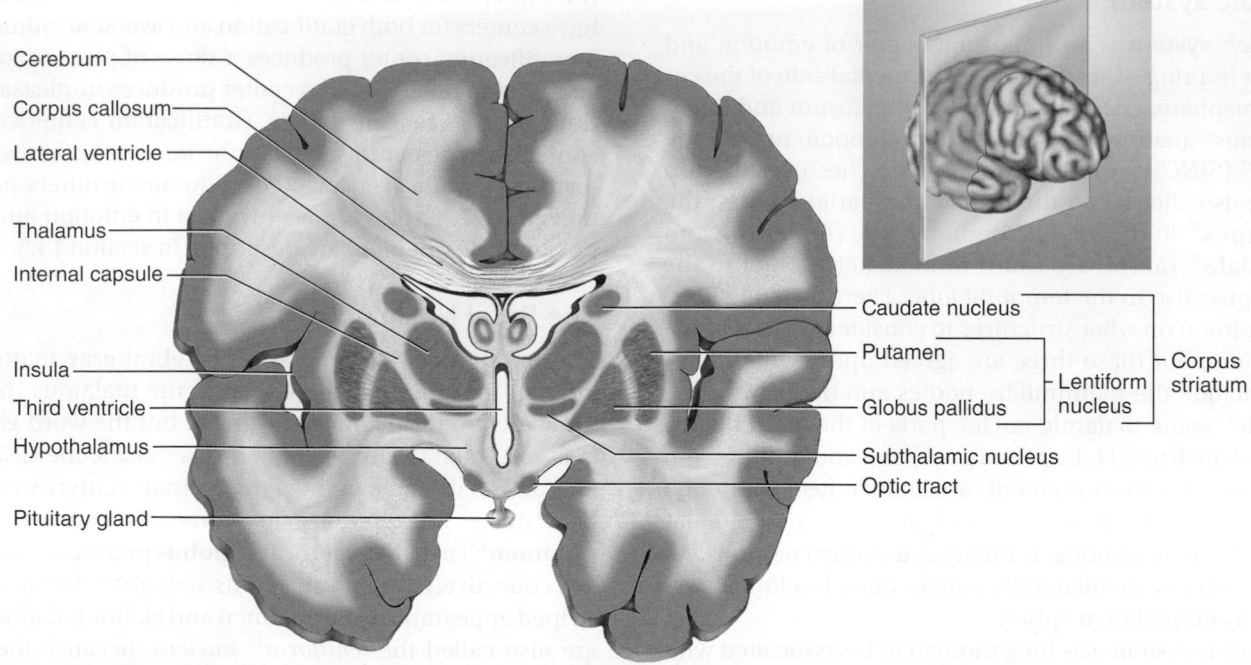

**FIGURE 14.17 The Basal Nuclei.** Frontal section of the brain. **AP|R**

**BEFORE YOU GO ON**

Answer the following questions to test your understanding of the preceding section:

14. What are the three major components of the diencephalon? Which ventricle does the diencephalon enclose?

15. What is the role of the thalamus in sensory function?

16. List at least six functions of the hypothalamus.

17. Name the five lobes of the cerebrum and describe their locations, boundaries, and principal functions.

18. Distinguish between commissural, association, and projection tracts of the cerebrum.

19. Where is the limbic system located? What component of it is involved in emotion? What component is involved in memory?

20. Where are the basal nuclei located? What is their general function?

## 14.5 Integrative Functions of the Brain

### Expected Learning Outcomes

When you have completed this section, you should be able to

a. list the types of brain waves and discuss their relationship to mental states;

b. describe the stages of sleep, their relationship to the brain waves, and the neural mechanisms of sleep;

c. identify the brain regions concerned with consciousness and thought, memory, emotion, sensation, motor control, and language; and

d. discuss the functional differences between the right and left cerebral hemispheres.

This section concerns such "higher" brain functions as sleep, memory, cognition, emotion, sensation, motor control, and language. These are associated especially with the cerebral cortex, but not exclusively; they involve interactions between the cerebral cortex and such areas as the basal nuclei, brainstem, and cerebellum. It is impossible in many cases to assign these functions to one specific brain region; functions of the brain do not have such easily defined anatomical boundaries. Some functions overlap anatomically; some cross anatomical boundaries from one region to another, often remote region; and some functions such as consciousness and memory are widely distributed through the cerebrum. Thus, we will consider these as *integrative* functions of the brain, focusing especially on the cerebrum but in many cases involving the combined action of multiple levels of the brain. Some of these present the most difficult challenges for neurobiology, but they are the most intriguing functions of the brain and involve its largest areas.

### The Electroencephalogram

For research and clinical purposes, it is common to monitor electrical activity called **brain waves.** Recorded with electrodes on the scalp (fig. 14.18a), these are rhythmic voltage changes resulting predominantly from synchronized postsynaptic potentials in the superficial layers of the cerebral cortex. The recording, called an **electroencephalogram**[46] **(EEG),** is useful in studying normal brain functions such as sleep and consciousness, and in diagnosing degenerative brain diseases, metabolic abnormalities, brain tumors, trauma, and so forth. States of consciousness ranging from high alertness to deep sleep are correlated with changes in the EEG. The complete and persistent absence of brain waves is a common clinical and legal criterion of brain death.

There are four types of brain waves (fig. 14.18b), distinguished by differences in amplitude (mV) and frequency. Frequency is expressed in hertz (Hz), or cycles per second:

1. **Alpha (α) waves** have a frequency of 8 to 13 Hz and are recorded especially in the parieto–occipital area. They dominate the EEG when a person is awake and resting, with the eyes closed and the mind wandering. They are suppressed when a person opens the eyes, receives specific sensory stimulation, or engages in a mental task such as performing mathematical calculations. They are absent during deep sleep.

2. **Beta (β) waves** have a frequency of 14 to 30 Hz and occur in the frontal to parietal region. They are accentuated during mental activity and sensory stimulation.

3. **Theta (θ) waves** have a frequency of 4 to 7 Hz. They are normal in children and in drowsy or sleeping adults, but a predominance of theta waves in awake adults suggests emotional stress or brain disorders.

4. **Delta (δ) waves** are high-amplitude "slow waves" with a frequency of less than 3.5 Hz. Infants exhibit delta waves when awake, and adults exhibit them in deep sleep. A predominance of delta waves in awake adults indicates serious brain damage.

## Sleep

**Sleep** can be defined as a temporary state of unconsciousness from which one can awaken when stimulated. It is one of many bodily functions that occur in cycles called **circadian**[47] (sur-CAY-dee-an) **rhythms,** so named because they are marked by events that reoccur at intervals of about 24 hours. Sleep is characterized by a stereotyped posture (usually lying down with the eyes closed) and inhibition of muscular activity *(sleep paralysis).* It superficially resembles other states of prolonged unconsciousness such as coma and animal hibernation, except that individuals cannot be aroused from those states by sensory stimulation.

Sleep occurs in distinct stages recognizable from changes in the EEG. In the first 30 to 45 minutes, the EEG waves drop in frequency but increase in amplitude as one passes through four sleep stages (fig. 14.19a):

- **Stage 1.** One feels drowsy, closes the eyes, and begins to relax. Thoughts come and go, often accompanied by a drifting sensation. One awakens easily if stimulated. The EEG is dominated by alpha waves.

---

[46] *electro* = electricity; *encephalo* = brain; *gram* = record

[47] *circa* = approximately; *dia* = a day, 24 hours

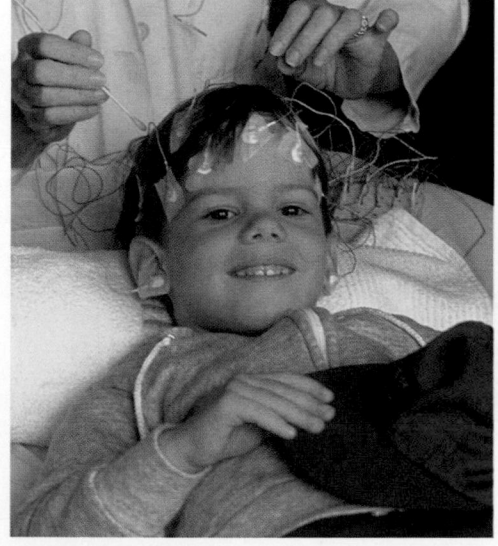

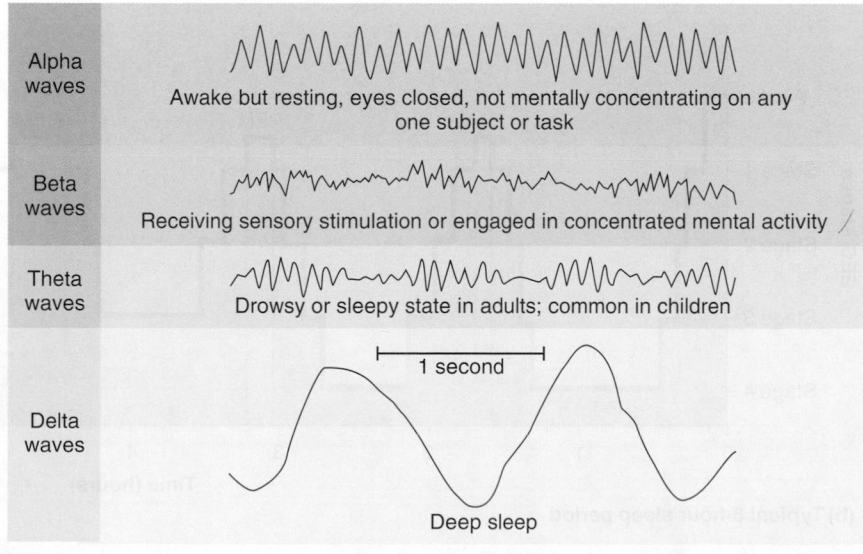

(a)　　　　　　　　　　　　　　　　(b)

**FIGURE 14.18** **The Electroencephalogram (EEG).** (a) An EEG is recorded from electrodes on the forehead and scalp. (b) Four classes of brain waves are seen in EEGs.

- **Stage 2.** One passes into light sleep. The EEG declines in frequency but increases in amplitude. Occasionally it exhibits 1 or 2 seconds of *sleep spindles,* high spikes resulting from interactions between neurons of the thalamus and cerebral cortex.

- **Stage 3.** This is moderate to deep sleep, typically beginning about 20 minutes after stage 1. Sleep spindles occur less often, and theta and delta waves appear. The muscles relax, and the *vital signs* (body temperature, blood pressure, and heart and respiratory rates) fall.

- **Stage 4.** This is also called *slow-wave sleep (SWS),* because the EEG is dominated by low-frequency, high-amplitude delta waves. The muscles are now very relaxed, vital signs are at their lowest levels, and one becomes difficult to awaken.

About five times a night, a sleeper backtracks from stage 3 or 4 to stage 2 and exhibits bouts of **rapid eye movement (REM) sleep** (fig. 14.19b). This is so named because the eyes oscillate back and forth as if watching a movie. It is also called *paradoxical sleep* because the EEG resembles the waking state, yet the sleeper is harder to arouse than in any other stage. The vital signs increase and the brain consumes even more oxygen than when awake. Sleep paralysis, other than in the muscles of eye movement, is especially strong during REM sleep. Paralysis may serve to prevent the sleeper from acting out his or her dreams and may have prevented our tree-dwelling ancestors from falling during their sleep.

Dreams occur in both REM and non-REM sleep, but REM dreams tend to be longer, more vivid, and more emotional than non-REM dreams. The parasympathetic nervous system is very active during REM sleep, causing constriction of the pupils and

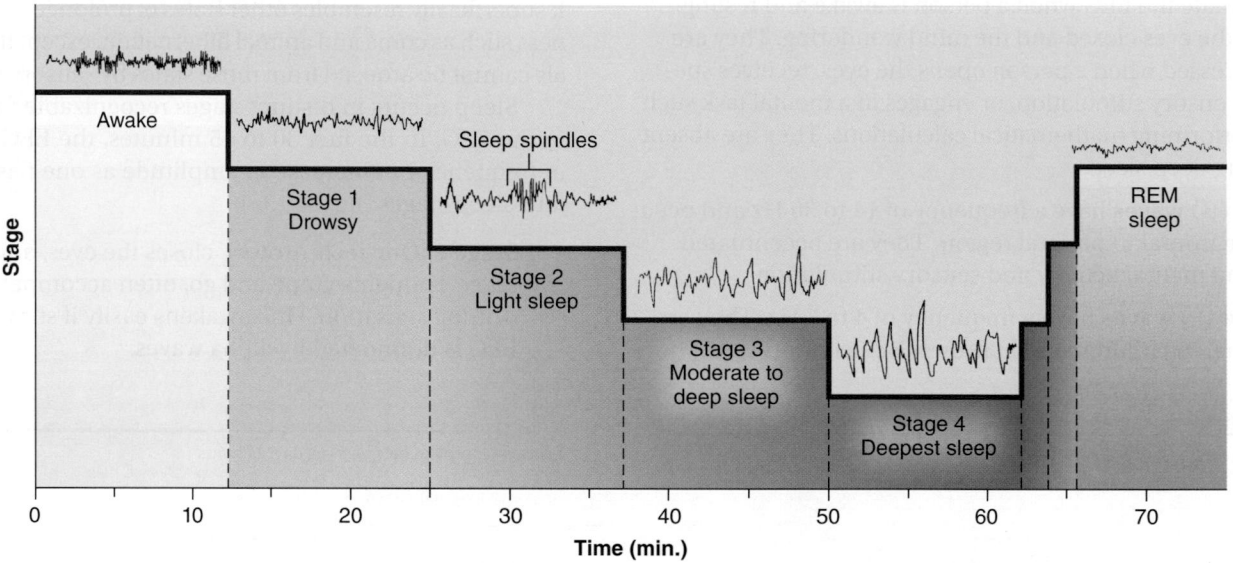

**(a) One sleep cycle**

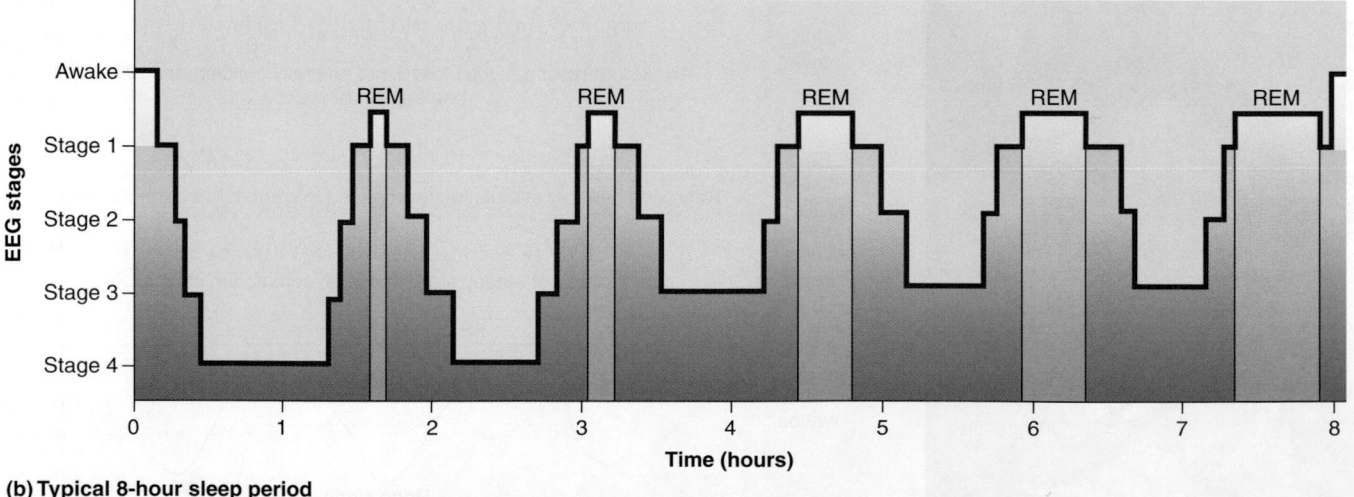

**(b) Typical 8-hour sleep period**

**FIGURE 14.19  Sleep Stages and Brain Activity.**  (a) A single sleep cycle, from waking to deep sleep, followed by 10 minutes of REM sleep. (b) Stages of sleep over an 8-hour night in a typical young adult. Stage 4 is attained only in the first two cycles. Periods of REM sleep increase from about 10 minutes in the first cycle to as long as 50 minutes in the last hour of sleep. Most dreaming occurs during REM sleep.

erection of the penis or clitoris. Erection, however, is seldom accompanied by sexual dream content.

The rhythm of sleep is well known, but its neurological mechanisms are still more of a mystery to science. The cycle of sleep and waking is controlled by a complex interaction between the cerebral cortex, thalamus, hypothalamus, and reticular formation. One of the control centers for sleep is the **suprachiasmatic** (SOO-pra-KY-az-MAT-ic) **nucleus (SCN),** located just above the optic chiasm in the anterior hypothalamus (see fig. 14.12b). Some nerve fibers from the eyes go to the SCN instead of the visual cortex of the cerebrum. The SCN uses this input to synchronize multiple body rhythms with the external rhythm of night and day—including not just sleep but also body temperature, urine production, hormone secretion, and other functions. It does not in itself induce sleep or waking, but regulates the time of day that a person sleeps.

Two related brain neuropeptides called **orexins** act as an important "sleep switch." Produced by the lateral and posterior hypothalamus, orexins strongly stimulate wakefulness and elevate the metabolic rate. Blocking orexin receptors induces sleep, and orexin levels are low or absent in a disorder called **narcolepsy,** in which a person experiences excessive daytime sleepiness and fatigue and may often fall asleep at work or school, with abnormally quick onset of REM sleep. Narcolepsy seems to be an autoimmune disease caused by antibody-mediated destruction of the orexin-producing neurons.

#### ▶▶▶ APPLY WHAT YOU KNOW

*Some animals have been shown to exhibit narcolepsy as the result of a mutation in the gene for an orexin receptor. Looking ahead in the book to type 2 diabetes mellitus (p. 665), can you identify a common thread in these two disorders?*

We still know little about the purposes of sleep and dreaming. Non-REM sleep seems to have a restorative effect on the body, and prolonged sleep deprivation is fatal to experimental animals. Yet it is unclear why quiet bed rest alone cannot serve the restorative purpose for humans—that is, why we must lose consciousness. One hypothesis is that sleep is a time to replenish such energy sources as glycogen and ATP. During waking hours, brain glycogen levels go down, while ATP consumption generates a sleep-inducing metabolite, adenosine (see p. 574). During sleep, ATP and glycogen levels rebound and the adenosine level goes down. Yet this is at best an incomplete explanation, because some animals with extraordinarily high metabolic rates (rapid ATP and glycogen consumption) sleep very little, and some with much lower metabolic rates (bats) sleep a great deal. From comparative zoology and evolutionary theory, there is evidence that sleep may have evolved to motivate animals to find a safe place and remain inactive during the most dangerous times of day, when such risks as predation may outweigh such benefits as food finding. Consistent with this, humans sleep more in infancy (when at their most vulnerable), as well as when they are sick or injured and must recover in safety.

Some researchers have suggested that REM sleep is a period in which the brain either "consolidates" and strengthens memories by reinforcing synaptic connections, or purges superfluous information from memory by weakening or eliminating other synapses. But while intriguing hypotheses for the purposes of sleep and dreaming abound, the evidence for any of them is still scanty.

## Cognition

**Cognition**[48] is the range of mental processes by which we acquire and use knowledge—sensory perception, thought, reasoning, judgment, memory, imagination, and intuition. Such functions are widely distributed over regions of cerebral cortex called **association areas,** which constitute about 75% of all brain tissue. This is the most difficult area of brain research and the most incompletely understood aspect of cerebral function. Much of what we know about it has come from studies of patients with brain lesions—areas of tissue destruction resulting from cancer, stroke, and trauma. The many brain injuries incurred in World Wars I and II yielded an abundance of insights into regional brain functions. More recently, imaging methods such as PET and fMRI have yielded much more sophisticated insights. These methods allow a researcher to scan a person's brain while that person performs various cognitive or motor tasks and to see which brain regions are most active in different mental and task states (see Deeper Insight 14.4, p. 553).

A few examples of the effects of cerebral lesions reveal some functions of the association areas:

- Parietal lobe lesions can cause people to become unaware of objects, or even their own limbs, on the other side of the body—a condition called *contralateral neglect syndrome.* In typical cases, men shave only half of the face, women apply makeup to only one side, patients dress only half of the body, and some people deny that one arm or leg belongs to them. Such patients are unable to find their way around—say, to describe the route from home to work or navigate within a familiar building.

- Temporal lobe lesions often result in *agnosia*[49] (ag-NO-zee-ah), the inability to recognize, identify, and name familiar objects. In *prosopagnosia,*[50] a person cannot remember familiar faces, even his or her own reflection in a mirror.

- Frontal lobe lesions are especially devastating to the qualities we think of as personality. The **prefrontal cortex (frontal association area),** the most rostral part of the frontal lobe, is well developed only in primates, especially humans. It integrates information from sensory and motor regions of the cortex and from other association areas. It gives us a sense of our relationship to the rest of the world, enabling us to think about it and to plan and execute

---

[48]*cognit* = to know
[49]*a* = without; *gnos* = knowledge
[50]*prosopo* = face, person

appropriate behavior. It is responsible for giving appropriate expression to our emotions. Lesions here may produce profound personality disorders and socially inappropriate behaviors.

As a broad generalization, we can conclude that the parietal association area is responsible for perceiving and attending to stimuli, the temporal association area for identifying them, and the frontal association area for planning our responses.

## Memory

**Memory** is one of the cognitive functions, but warrants special attention. We studied its forms and its neural and molecular mechanisms in chapter 12 (p. 466). Now that you have been introduced to the gross anatomy of the brain, we can consider where those processes occur anatomically.

Our subject is really a little broader than memory per se. Information management by the brain entails learning (acquiring new information), memory proper (information storage and retrieval), and forgetting (eliminating trivial information). Forgetting is as important as remembering. People with a pathological inability to forget trivial information have great difficulty in reading comprehension and other functions that require us to separate what is important from what is not. More often, though, brain-injured people are either unable to store new information **(anterograde amnesia)** or to recall things they knew before the injury **(retrograde amnesia).** *Amnesia* refers to defects in *declarative* memory (such as the ability to describe past events), not *procedural* memory (such as the ability to tie your shoes) (see definitions on p. 467).

The **hippocampus** of the limbic system is an important memory-forming center (see fig. 14.16). It does not store memories, but organizes sensory and cognitive experiences into a unified long-term memory. The hippocampus learns from sensory input while an experience is happening, but it has a short memory. Later, perhaps when one is sleeping, it plays this memory repeatedly to the cerebral cortex, which is a "slow learner" but forms longer-lasting memories through the processes described in chapter 12. This process of "teaching the cerebral cortex" until a long-term memory is established is called **memory consolidation.** Long-term memories are held in various areas of cortex. One's vocabulary and memory of faces and familiar objects, for example, reside in the superior temporal lobe, and memories of one's plans and social roles are in the prefrontal cortex.

Lesions of the hippocampus can cause profound anterograde amnesia. For example, in 1953, a famous patient known as H. M. (Henry Molaison, 1926–2008) underwent surgical removal of a large portion of both temporal lobes, including both hippocampi, to treat his severe epilepsy. The operation had no adverse effect on his intelligence, procedural memory, or declarative memory for things that had happened early in his life, but it left him with an inability to establish new memories. He could hold a conversation with his psychologist, but a few minutes later deny that it had taken place. He worked with the same psychologist for more than 40 years after his operation, but could not remember who she was from day to day.

Other parts of the brain involved in memory include the cerebellum, with a role in learning motor skills, and the amygdala, with a role in emotional memory.

## Emotion

Emotional feelings and memories are not exclusively cerebral functions, but result from an interaction between areas of the prefrontal cortex and diencephalon. Emotional control centers of the brain have been identified by studying people with brain lesions and by such techniques as surgical removal, ablation (destruction) of small regions with electrodes, and stimulation with electrodes and chemical implants, especially in experimental animals. Changes in behavior following such procedures give clues to the functions that a region performs. However, interpretation of the results is difficult and controversial because of the complex connections between the emotional brain and other regions.

The prefrontal cortex is the seat of judgment, intent, and control over the expression of our emotions. However we may feel, it is here that we decide the appropriate way to show those feelings. But the feelings themselves, and emotional memories, arise from deeper regions of the brain, especially the hypothalamus and amygdala. Here are the nuclei that stimulate us to recoil in fear from a rattlesnake or yearn for a lost love.

The amygdala is a major component of the limbic system described earlier (see fig. 14.16). It receives processed information from the general senses and from vision, hearing, taste, and smell. Such input enables it to mediate emotional responses to such stimuli as a disgusting odor, a foul taste, a beautiful sight, pleasant music, or a stomachache. It is especially important in the sense of fear, but also plays roles in food intake, sexual behavior, and drawing our attention to novel stimuli.

Output from the amygdala goes in two directions of special interest: (1) Some goes to the hypothalamus and lower brainstem and influences somatic and visceral motor systems. An emotional response to a stimulus may, through these connections, make one's heart race, raise the blood pressure, make the hair stand on end, or induce vomiting. (2) Other output goes to areas of the prefrontal cortex that mediate conscious control and expression of the emotions, such as the ability to express love, control anger, or overcome fear.

Many important aspects of personality depend on an intact, functional amygdala and hypothalamus. When specific regions are destroyed or artificially stimulated, humans and other animals exhibit blunted or exaggerated expressions of anger, fear, aggression, self-defense, pleasure, pain, love, sexuality, and parental affection, as well as abnormalities in learning, memory, and motivation. Lesions of the amygdala, for example, can completely abolish the sense of fear.

Much of our behavior is shaped by learned associations between stimuli, our responses to them, and the rewards or punishments that result. Nuclei involved in feelings of reward and punishment have been identified in the hypothalamus of cats, rats, monkeys, and other animals. In a representative experiment, an electrode is implanted in an area of an animal's hypothalamus called the **median forebrain bundle (MFB).** The animal is placed in a chamber with a foot pedal wired to that electrode. When the animal steps on the pedal, it receives a mild electrical stimulus to the MFB. Apparently the sensation is strongly rewarding, because the animal soon starts to press the pedal over and over and may spend most of its time doing so—even to the point of neglecting food and water. Rats have been known to bar-press 5,000 to 12,000 times an hour, and monkeys up to 17,000 times an hour, to stimulate their MFBs.

These animals cannot tell us what they are feeling, but electrode implants have also been used to treat people who suffer otherwise incurable schizophrenia, pain, or epilepsy. These patients also repeatedly press a button to stimulate the MFB, but they do not report feelings of joy or ecstasy. Some are unable to explain why they enjoy the stimulus, and others report "relief from tension" or "a quiet, relaxed feeling." With electrodes misplaced in other areas of the hypothalamus, subjects report feelings of fear or terror when stimulated.

### ▶▶▶ APPLY WHAT YOU KNOW

*MRI scans show that John has a pea-size tumor in his hippocampus and Allan has a tumor of the same size in his amygdala. One of these patients is in prison for violent crimes in which he seemed to show no fear or sense of self-preservation. The other patient cannot remember the name of his new granddaughter, no matter how many times he is told. Which behavioral outcome do you associate with each patient? Why?*

## Sensation

A great deal of the cerebrum is concerned with the senses—most cortex of the insula and of the parietal, occipital, and temporal lobes. Regions called **primary sensory cortex** are the sites where sensory input is first received and one becomes conscious of a stimulus. Adjacent to these are association areas where this sensory information is interpreted. For example, the primary visual cortex, which receives input from the eyes, is bordered by the visual association area, which interprets and makes cognitive sense of the visual stimuli, so we know what we are looking at. Some association areas are *multimodal*—instead of processing information from a single sensory source, they receive input from multiple senses and integrate this into our overall perception of our surroundings. For example, in the frontal lobe just above the eyes, there is a patch of multimodal cortex called the *orbitofrontal cortex*, which receives taste, smell, and visual input to form our overall impression of the desirability of a particular food.

The sense organs and their signaling pathways in the CNS are the subject of chapter 16. Here we will examine only the areas of cerebral cortex involved in sensory perception.

## The Special Senses

The **special senses** are limited to the head, and some employ relatively complex sense organs. They are vision, hearing, equilibrium, taste, and smell. Their primary cortices and association areas are located as follows (fig. 14.20).

- **Vision.** Visual signals are received by the **primary visual cortex** in the far posterior region of the occipital lobe. This is bordered anteriorly by the **visual association**

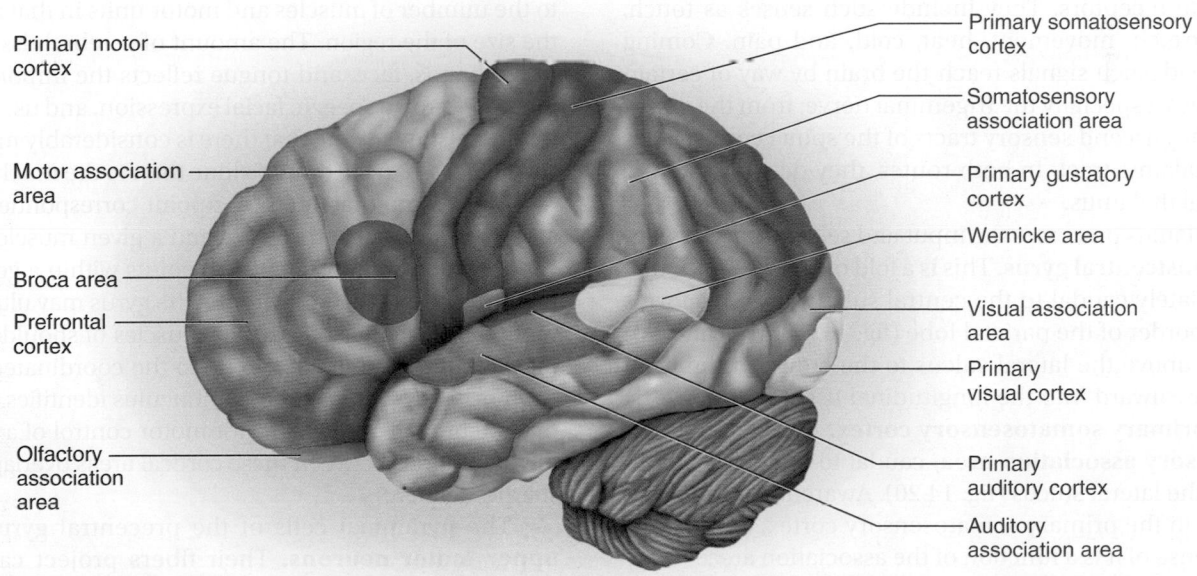

**FIGURE 14.20  Some Functional Regions of the Cerebral Cortex.** The Broca and Wernicke areas for language abilities are found in only one hemisphere, usually the left. The other regions shown here are mirrored in both hemispheres.

**area,** which includes all the remainder of the occipital lobe, some of the posterior parietal lobe (concerned with spatial perception), and much of the inferior temporal lobe, where we recognize faces and other familiar objects.

- **Hearing.** Auditory signals are received by the **primary auditory cortex** in the superior region of the temporal lobe and in the nearby insula. The **auditory association area** occupies areas of temporal lobe inferior to the primary auditory cortex and deep within the lateral sulcus. This is where we become capable of recognizing spoken words, a familiar piece of music, or a voice on the telephone.

- **Equilibrium.** Signals from the inner ear for equilibrium project mainly to the cerebellum and several brainstem nuclei concerned with head and eye movements and visceral functions. Some fibers of this system, however, are routed through the thalamus to areas of association cortex in the roof of the lateral sulcus and near the lower end of the central sulcus. This is the seat of consciousness of our body movements and orientation in space.

- **Taste and smell.** Gustatory (taste) signals are received by the **primary gustatory cortex** in the inferior end of the postcentral gyrus of the parietal lobe (discussed shortly) and an anterior region of the insula. Olfactory (smell) signals are received by the **primary olfactory cortex** in the medial surface of the temporal lobe and inferior surface of the frontal lobe. The *orbitofrontal cortex* mentioned earlier serves as a multimodal association area for both of these senses.

## The General Senses

The **general (somatosensory, somesthetic,**[51] or **somatic) senses** are distributed over the entire body and employ relatively simple receptors. They include such senses as touch, pressure, stretch, movement, heat, cold, and pain. Coming from the head, such signals reach the brain by way of certain cranial nerves, especially the trigeminal nerve; from the rest of the body, they ascend sensory tracts of the spinal cord such as the spinothalamic tract. In both routes, they decussate to the contralateral thalamus.

The thalamus processes the input and selectively relays signals to the **postcentral gyrus.** This is a fold of the cerebrum that lies immediately caudal to the central sulcus and thus forms the rostral border of the parietal lobe (fig. 14.21). We can trace it from just above the lateral sulcus to the crown of the head and then downward into the longitudinal fissure. Its cortex is called the **primary somatosensory cortex.** Adjacent to it is a **somatosensory association area,** caudal to the gyrus and in the roof of the lateral sulcus (fig. 14.20). Awareness of stimulation occurs in the primary somatosensory cortex, but making cognitive sense of it is a function of the association area.

Because of the aforementioned decussation in sensory pathways, the right postcentral gyrus receives input from the left side of the body and vice versa. The primary somatosensory cortex is like an upside-down sensory map of the contralateral side of the body, traditionally diagrammed as a *sensory homunculus*[52] (fig. 14.21). As the diagram shows, receptors in the lower limb project to superior and medial parts of the gyrus, and receptors in the face project to the inferior and lateral parts. Such point-for-point correspondence between an area of the body and an area of the CNS is called **somatotopy.**[53] The reason for the bizarre, distorted appearance of the homunculus is that the amount of cerebral tissue devoted to a given body region is proportional to how richly innervated and sensitive that region is, not to its size. Thus, the hands and face are represented by a much larger region of somatosensory cortex than the trunk is.

## Motor Control

The intention to contract a skeletal muscle begins in the **motor association (premotor) area** of the frontal lobes (fig. 14.20). This is where we plan our behavior—where neurons compile a program for the degree and sequence of muscle contractions required for an action such as dancing, typing, or speaking. The program is then transmitted to neurons of the **precentral gyrus (primary motor area),** which is the most posterior gyrus of the frontal lobe, immediately anterior to the central sulcus (fig. 14.22). Neurons here send signals to the brainstem and spinal cord, which ultimately results in muscle contractions.

The precentral gyrus, like the postcentral one, exhibits somatotopy. The neurons for toe movements, for example, are deep in the longitudinal fissure on the medial side of the gyrus. The summit of the gyrus controls the trunk, shoulder, and arm, and the inferolateral region controls the facial muscles. This map is diagrammed as a *motor homunculus* (fig. 14.22b). Like the sensory homunculus, it has a distorted look because the amount of cortex devoted to a given body region is proportional to the number of muscles and motor units in that region, not to the size of the region. The amount of cerebral tissue dedicated to the hands, face, and tongue reflects the importance of fine motor control in speech, facial expression, and use of the hands. It is interesting to note that there is considerably more brain tissue dedicated to the thumb alone than to the much larger thigh.

There is no exact point-for-point correspondence between an area of the precentral gyrus and a given muscle. A muscle is controlled by neurons at several points within a general area of the gyrus. Also, a given neuron in the gyrus may ultimately affect more than one muscle, such as muscles of shoulder and elbow movement that both contribute to the coordinated positioning of one's hand. Although the homunculus identifies cortical areas that are broadly responsible for motor control of a given region, the boundaries between these cortical areas overlap and are not sharply defined.

The pyramidal cells of the precentral gyrus are called **upper motor neurons.** Their fibers project caudally, with about 19 million fibers ending in nuclei of the brainstem and 1 million forming the corticospinal tracts. Most of these fibers

---

[51]*som* = body; *esthet* = feeling

[52]*hom* = man; *unculus* = little
[53]*somato* = body; *topy* = place

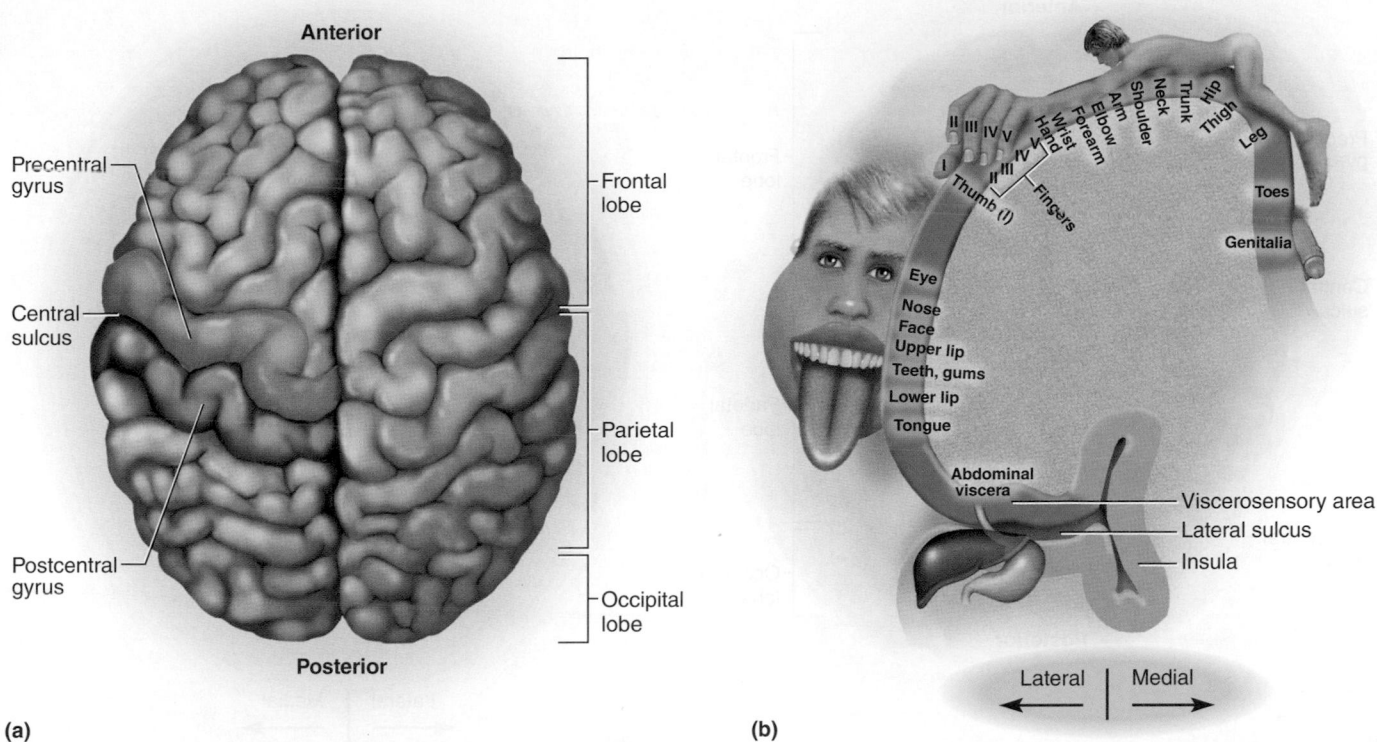

**FIGURE 14.21 The Primary Somatosensory Cortex (Postcentral Gyrus).** (a) Superior view of the brain showing the location of the postcentral gyrus (violet). (b) The sensory homunculus, drawn so that body parts are in proportion to the amount of cortex dedicated to their sensation. This gyrus also includes centers for visceral sensation from the intra-abdominal organs (the viscerosensory area). AP|R

decussate in the lower medulla oblongata (at the pyramidal decussation) and form the *lateral corticospinal tract* on each side of the spinal cord. A smaller number of fibers pass through the medulla without decussation and form the anterior corticospinal tracts, which cross over lower in the spinal cord. Each precentral gyrus thus controls muscles on the contralateral side of the body. In the brainstem or spinal cord, the fibers from the upper motor neurons synapse with **lower motor neurons** whose axons innervate the skeletal muscles (see fig. 13.6, p. 483).

Other areas of the brain important in muscle control are the basal nuclei (see fig. 14.17) and cerebellum. Among the roles of the basal nuclei are determining the onset and cessation of intentional movements; the repetitive hip and shoulder movements that occur in walking; and highly practiced, learned behaviors that one carries out with little thought—for example, writing, driving a car, and tying one's shoes. The basal nuclei lie in a feedback circuit from the cerebrum to the basal nuclei to the thalamus and back to the cerebrum. They receive signals from the substantia nigra of the midbrain and from all areas of cerebral cortex, except for the primary visual and auditory cortices. The basal nuclei process these and issue their output to the thalamus, which relays these signals back to the midbrain and cerebral cortex—especially to the prefrontal cortex, motor association area, and precentral gyrus.

Lesions of the basal nuclei cause movement disorders called **dyskinesias.**[54] These are sometimes characterized by abnormal difficulty initiating movement, such as rising from a chair or beginning to walk, and by a slow shuffling walk. Such motor dysfunctions are seen in Parkinson disease (see p. 468). Smooth, easy movements require the excitation of agonistic muscles and inhibition of their antagonists. In Parkinson disease, the antagonists are not inhibited. Therefore, opposing muscles at a joint fight each other, making it a struggle to move as one wishes. Other dyskinesias are characterized by exaggerated or unwanted movements, such as flailing of the limbs *(ballismus)* in Huntington disease.

The cerebellum is highly important in motor coordination. In addition to its cognitive functions described on page 522, it aids in learning motor skills, maintains muscle tone and posture, smooths muscle contractions, coordinates eye and body movements, and helps to coordinate the motions of different joints with each other (such as the shoulder and elbow in pitching a baseball). The cerebellum acts as a comparator in motor control. Through the middle peduncles, it receives information from the upper motor neurons of the cerebrum about the movements one intends to make, and information about body

---

[54]*dys* = bad, abnormal, difficult; *kines* = movement

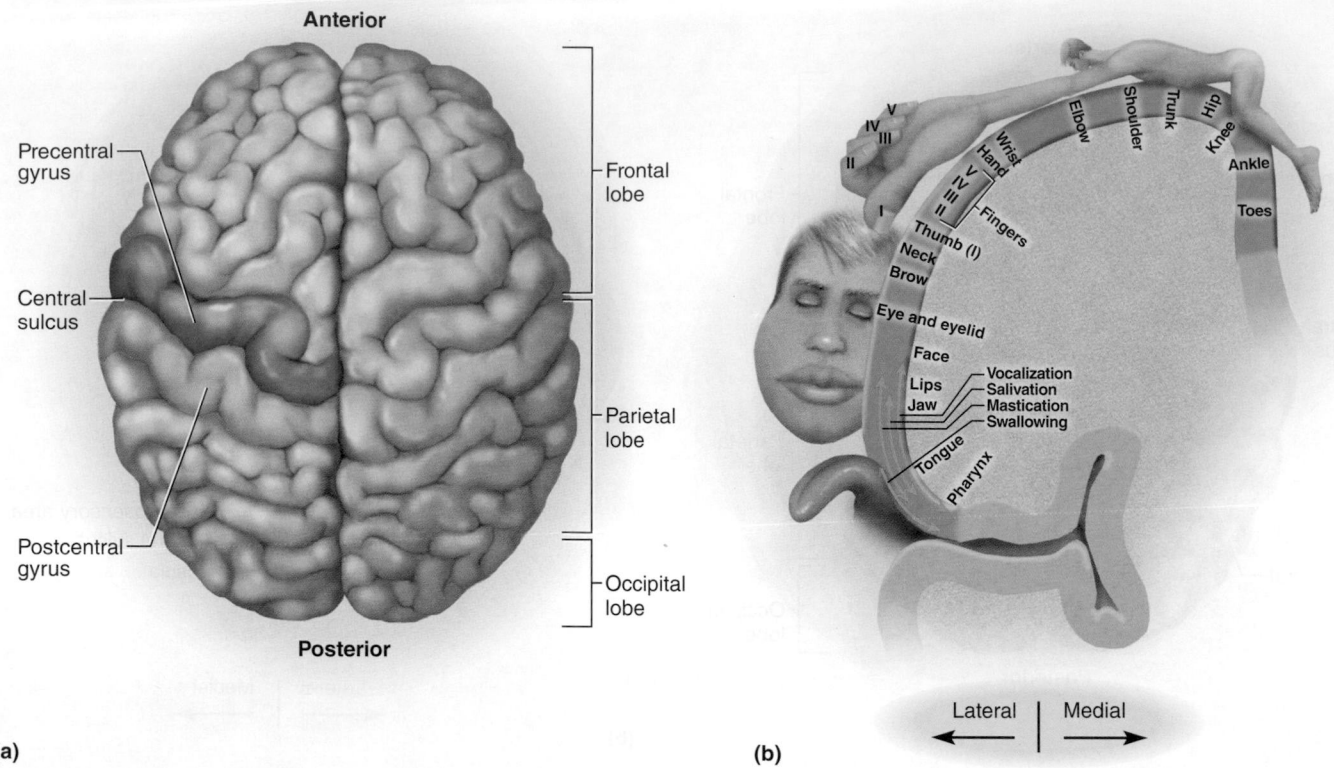

**(a)**

**(b)**

**FIGURE 14.22** **The Primary Motor Cortex (Precentral Gyrus).** (a) Superior view of the brain showing the location of the precental gyrus (blue). (b) Motor homunculus, drawn so that body parts are in proportion to the amount of primary motor cortex dedicated to their control. AP|R

❓ *Which body regions are controlled by the largest areas of motor cortex—regions with a few large muscles or regions with numerous small muscles?*

movement from the eyes and inner ears. Through the inferior peduncles, it receives information from proprioceptors in the muscles and joints about the actual performance of the movement (fig. 14.23, left). The Purkinje cells of the cerebellum compare the two. If there is a discrepancy between the intent and the performance, they signal the deep cerebellar nuclei. These, in turn, issue signals to the thalamus and lower brainstem, ultimately ascending to the motor association area of the cerebrum and the reticulospinal and vestibulospinal tracts of the spinal cord (fig. 14.23, right). Output from these areas corrects the muscle performance to match the intent. Lesions of the cerebellum can result in a clumsy, awkward gait *(ataxia)* and make some tasks such as climbing stairs virtually impossible.

## Language

Language includes several abilities—reading, writing, speaking, signing, and understanding words—assigned to different regions of cerebral cortex (fig. 14.24). The **Wernicke**[55] (WUR-ni-keh) **area** is responsible for the recognition of spoken and written language. It lies just posterior to the lateral sulcus, usually in the left hemisphere, at the crossroad between visual, auditory, and somatosensory areas of cortex, receiving input from all these neighboring regions. The *angular gyrus,* part of the parietal lobe

just caudal and superior to the Wernicke area, is important in the ability to read and write.

When we intend to speak, the Wernicke area formulates phrases according to learned rules of grammar and transmits a plan of speech to the **Broca**[56] **area,** located in the inferior prefrontal cortex of the same hemisphere. PET scans show a rise in the metabolic activity of the Broca area as one prepares to speak or sign (see fig. 14.40). This area generates a motor program for the muscles of the larynx, tongue, cheeks, and lips to produce speech, as well as for the hand motions of signing. It transmits this program to the primary motor cortex, which executes it— that is, it issues commands to the lower motor neurons that supply the relevant muscles.

The emotional aspect of language is controlled by regions in the opposite hemisphere that mirror the Wernicke and Broca areas. Opposite the Broca area is the *affective language area.* Lesions to this area result in *aprosody*—flat, emotionless speech. The cortex opposite Wernicke's area is concerned with recognizing the emotional content of another person's speech. Lesions here can result in such problems as inability to understand a joke.

▶▶▶ **APPLY WHAT YOU KNOW**

*Of all the language centers just described, which one best fits the concept of multimodal association cortex?*

---

[55]Karl Wernicke (1848–1905), German neurologist

[56]Pierre Paul Broca (1824–80), French surgeon and anthropologist

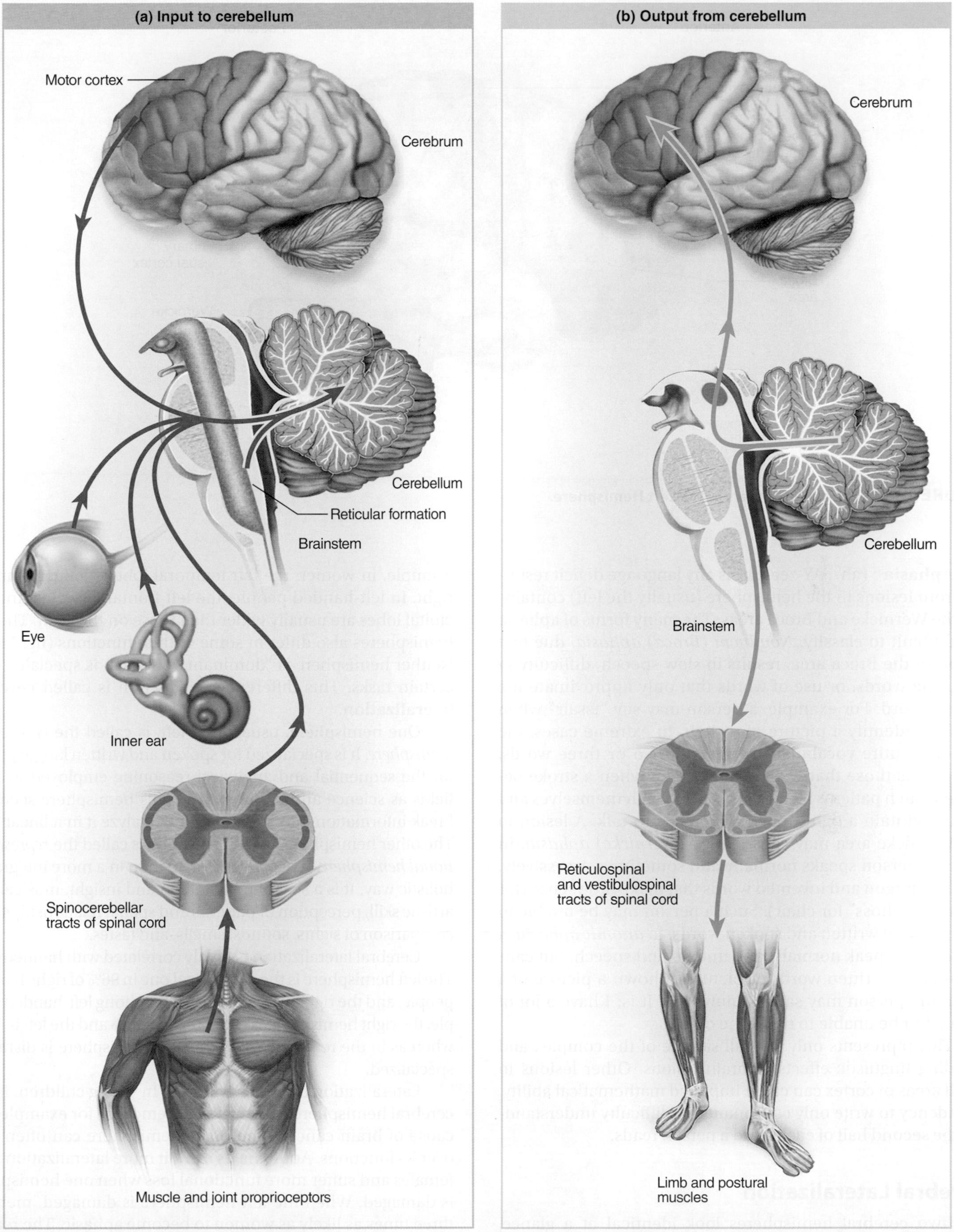

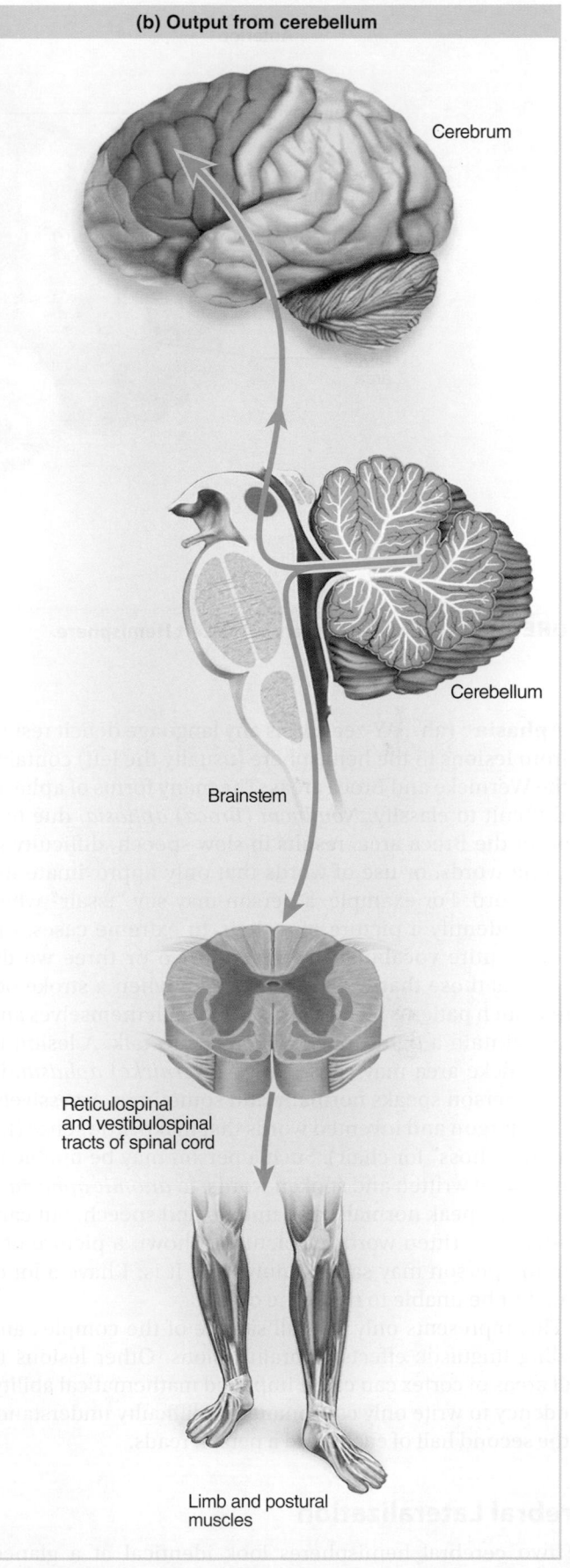

**(a) Input to cerebellum**

Motor cortex

Cerebrum

Cerebellum

Reticular formation

Brainstem

Eye

Inner ear

Spinocerebellar
tracts of spinal cord

Muscle and joint proprioceptors

**(b) Output from cerebellum**

Cerebrum

Cerebellum

Brainstem

Reticulospinal
and vestibulospinal
tracts of spinal cord

Limb and postural
muscles

**FIGURE 14.23  Motor Pathways Involving the Cerebellum.**  The cerebellum receives its input from the afferent pathways (red) on the left and sends its output through the efferent pathways (green) on the right.

**Anterior** **Posterior**

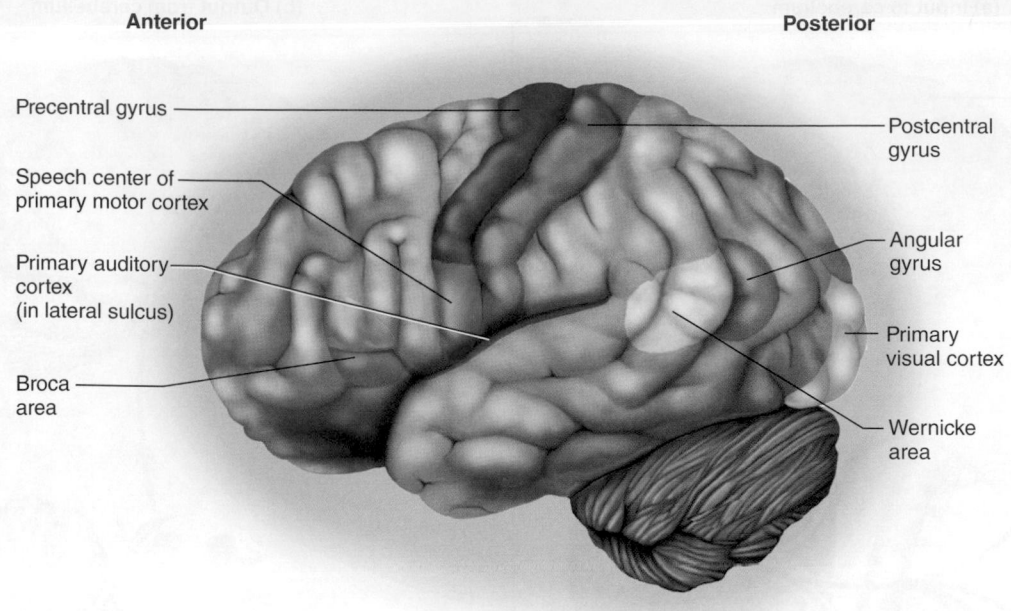

Precentral gyrus

Speech center of primary motor cortex

Primary auditory cortex (in lateral sulcus)

Broca area

Postcentral gyrus

Angular gyrus

Primary visual cortex

Wernicke area

**FIGURE 14.24 Language Centers of the Left Hemisphere.**

**Aphasia**[57] (ah-FAY-zee-uh) is any language deficit resulting from lesions in the hemisphere (usually the left) containing the Wernicke and Broca areas. The many forms of aphasia are difficult to classify. *Nonfluent (Broca) aphasia,* due to a lesion in the Broca area, results in slow speech, difficulty in choosing words, or use of words that only approximate the correct word. For example, a person may say "tssair" when asked to identify a picture of a chair. In extreme cases, the person's entire vocabulary consists of two or three words, sometimes those that were being spoken when a stroke occurred. Such patients feel very frustrated with themselves and often maintain a tight-lipped reluctance to talk. A lesion to the Wernicke area may cause *fluent (Wernicke) aphasia,* in which a person speaks normally and sometimes excessively, but uses jargon and invented words that make little sense (for example, "choss" for chair). Such a person may be unable to comprehend written and spoken words. In *anomic aphasia,* a person can speak normally and understand speech, but cannot identify written words or pictures. Shown a picture of a chair, the person may say, "I know what it is; I have a lot of them," but be unable to name the object.

This represents only a small sample of the complex and puzzling linguistic effects of brain lesions. Other lesions to small areas of cortex can cause impaired mathematical ability, a tendency to write only consonants, or difficulty understanding the second half of each word a person reads.

## Cerebral Lateralization

The two cerebral hemispheres look identical at a glance, but close examination reveals a number of differences. For example, in women the left temporal lobe is longer than the right. In left-handed people, the left frontal, parietal, and occipital lobes are usually wider than those on the right. The two hemispheres also differ in some of their functions (fig. 14.25). Neither hemisphere is "dominant," but each is specialized for certain tasks. This difference in function is called **cerebral lateralization.**

One hemisphere, usually the left, is called the *categorical hemisphere.* It is specialized for spoken and written language and for the sequential and analytical reasoning employed in such fields as science and mathematics. This hemisphere seems to break information into fragments and analyze it in a linear way. The other hemisphere, usually the right, is called the *representational hemisphere.* It perceives information in a more integrated, holistic way. It is a seat of imagination and insight, musical and artistic skill, perception of patterns and spatial relationships, and comparison of sights, sounds, smells, and tastes.

Cerebral lateralization is highly correlated with handedness. The left hemisphere is the categorical one in 96% of right-handed people, and the right hemisphere in 4%. Among left-handed people, the right hemisphere is categorical in 15% and the left in 70%, whereas in the remaining 15%, neither hemisphere is distinctly specialized.

Lateralization develops with age. In young children, if one cerebral hemisphere is damaged or removed (for example, because of brain cancer), the other hemisphere can often take over its functions. Adult males exhibit more lateralization than females and suffer more functional loss when one hemisphere is damaged. When the left hemisphere is damaged, men are three times as likely as women to become aphasic. The reason for this difference is not yet clear.

[57]*a* = without; *phas* = speech

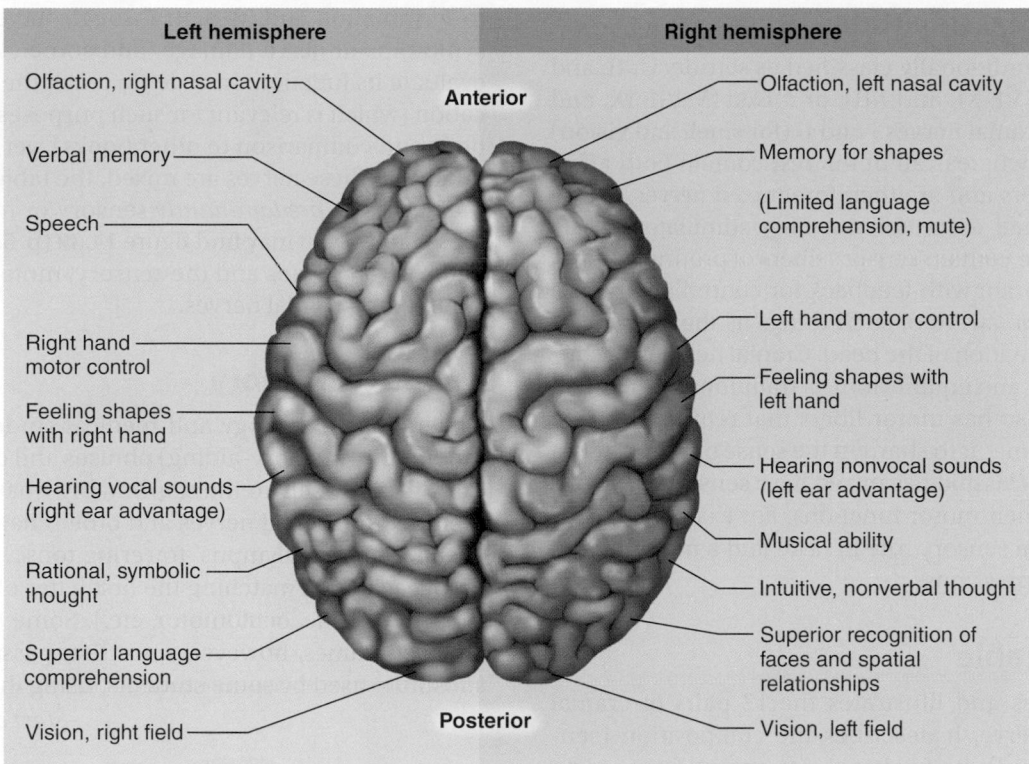

**FIGURE 14.25 Lateralization of Cerebral Functions.** The two cerebral hemispheres are not functionally identical.

Answer the following questions to test your understanding of the preceding section:

21. Suppose you are reading a novel and gradually fall asleep and begin to dream. How would your brain waves change during this sequence of events?

22. Describe the locations and functions of the somatosensory, visual, auditory, and frontal association areas.

23. Describe the somatotopy of the primary motor area and primary sensory area.

24. What are the roles of the Wernicke area, Broca area, and precentral gyrus in language?

## 14.6 The Cranial Nerves

### Expected Learning Outcomes

When you have completed this section, you should be able to

a. list the 12 cranial nerves by name and number;

b. identify where each cranial nerve originates and terminates; and

c. state the functions of each cranial nerve.

To be functional, the brain must communicate with the rest of the body. Most of its input and output travels by way of the spinal cord, but it also communicates by way of 12 pairs of **cranial nerves.** These arise primarily from the base of the brain, exit the cranium through its foramina, and lead to muscles and sense organs located mainly in the head and neck. The cranial nerves are numbered I to XII starting with the most rostral pair (fig. 14.26). Each nerve also has a descriptive name such as *optic nerve* and *vagus nerve.*

### Cranial Nerve Pathways

Most motor fibers of the cranial nerves begin in nuclei of the brainstem and lead to glands and muscles. The sensory fibers begin in receptors located mainly in the head and neck and lead mainly to the brainstem. These include the special senses such as vision and hearing, as well as general senses such as touch and proprioception. Pathways for the special senses are described in chapter 16. Sensory fibers for proprioception begin in the muscles innervated by motor fibers of the cranial nerves, but they often travel to the brain in a different nerve than the one that supplies the motor innervation.

Most cranial nerves carry fibers between the brainstem and ipsilateral receptors and effectors. Thus, a lesion in one side of the brainstem causes a sensory or motor deficit on the same side of the head. This contrasts with lesions of the motor and somatosensory cortex of the cerebrum, which, as we saw earlier, cause sensory and motor deficits on the *contralateral* side of the body. The exceptions are the optic nerve (II), where half the fibers decussate to the opposite side of the brain (see chapter 16), and the trochlear nerve (IV), in which all efferent fibers lead to a muscle of the contralateral eye.

## Cranial Nerve Classification

Cranial nerves are traditionally classified as sensory (I, II, and VIII), motor (III, IV, VI, XI, and XII), or mixed (V, VII, IX, and X). In reality, only cranial nerves I and II (for smell and vision) are purely sensory, whereas all of the rest contain both afferent and efferent fibers and are therefore mixed nerves. Those traditionally classified as motor not only stimulate muscle contractions but also contain sensory fibers of proprioception, which provide the brain with feedback for controlling muscle action and make one aware of such things as the position of the tongue and orientation of the head. Cranial nerve VIII, concerned with hearing and equilibrium, is traditionally classified as sensory, but it also has motor fibers that return signals to the inner ear and "tune" it to sharpen the sense of hearing. The nerves traditionally classified as mixed have sensory functions quite unrelated to their motor functions. For example, the facial nerve (VII) has a sensory role in taste and a motor role in controlling facial expressions.

## Cranial Nerve Table

Table 14.1 describes and illustrates the 12 pairs of cranial nerves. For each nerve, it describes the composition (sensory, motor, or mixed); its functions; its course from origin to termination and its path through the cranium; signs and symptoms of nerve damage; and some clinical tests used to evaluate its function. In order to teach the traditional classification (which is relevant for such purposes as board examinations and comparison to other books), yet remind you that all but two of these nerves are mixed, the table describes many of the nerves as *predominantly* sensory or motor. At the end of these tables, you may find figure 14.39 (p. 551) a helpful review of the target organs and the sensory, motor, or mixed composition of the cranial nerves.

## An Aid to Memory

Generations of biology and medical students have relied on mnemonic (memory-aiding) phrases and ditties, ranging from the sublimely silly to the unprintably ribald, to help them remember the cranial nerves and other anatomy. An old classic began, "On old Olympus' towering tops...," with the first letter of each word matching the first letter of each cranial nerve (olfactory, optic, oculomotor, etc.). Some cranial nerves have changed names, however, since that passage was devised. A substitute used by some students, using the first letter of each

*(Text continues on p. 545)*

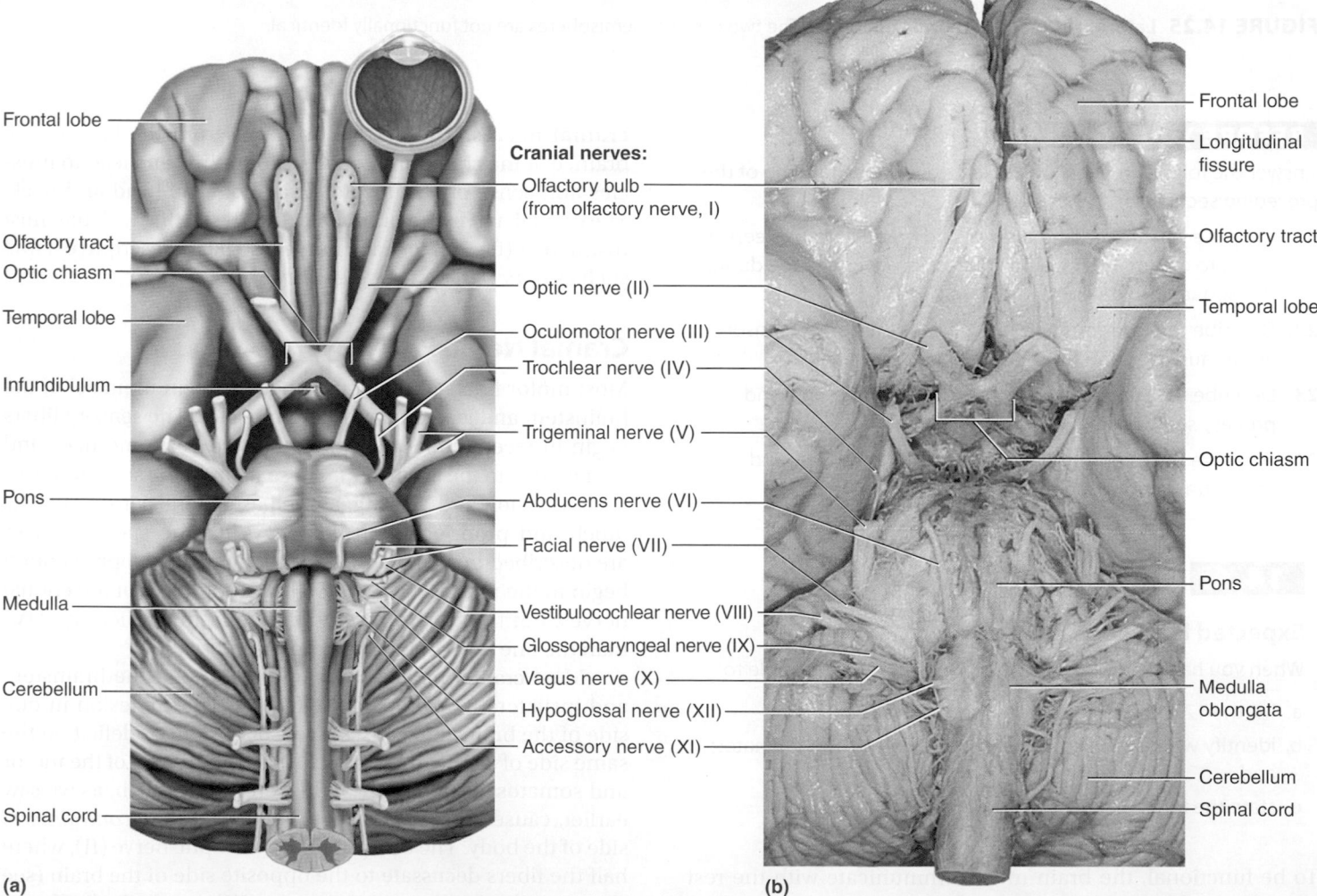

Cranial nerves:
- Frontal lobe
- Olfactory tract
- Optic chiasm
- Temporal lobe
- Infundibulum
- Pons
- Medulla
- Cerebellum
- Spinal cord
- Olfactory bulb (from olfactory nerve, I)
- Optic nerve (II)
- Oculomotor nerve (III)
- Trochlear nerve (IV)
- Trigeminal nerve (V)
- Abducens nerve (VI)
- Facial nerve (VII)
- Vestibulocochlear nerve (VIII)
- Glossopharyngeal nerve (IX)
- Vagus nerve (X)
- Hypoglossal nerve (XII)
- Accessory nerve (XI)
- Frontal lobe
- Longitudinal fissure
- Olfactory tract
- Temporal lobe
- Optic chiasm
- Pons
- Medulla oblongata
- Cerebellum
- Spinal cord

(a)    (b)

**FIGURE 14.26  The Cranial Nerves.**  (a) Base of the brain, showing the 12 cranial nerves. (b) Photograph of the cranial nerves.  **AP|R**

| TABLE 14.1 | The Cranial Nerves |
|---|---|

Origins of proprioceptive fibers are not tabulated. Nerves listed as mixed or sensory are agreed by all authorities to be either mixed or purely sensory nerves. Nerves classified as *predominantly* motor or sensory are traditionally classified that way but contain some fibers of the other type.

### I. Olfactory Nerve

This is the nerve for the sense of smell. It consists of several separate fascicles that pass independently through the cribriform plate in the roof of the nasal cavity. It is not visible on brains removed from the skull because these fascicles are severed by removal of the brain.

| Composition | Function | Origin | Termination | Cranial Passage | Effect of Damage | Clinical Test |
|---|---|---|---|---|---|---|
| Sensory | Smell | Olfactory mucosa in nasal cavity | Olfactory bulbs | Cribriform foramina of ethmoid bone | Impaired sense of smell | Determine whether subject can smell (not necessarily identify) aromatic substances such as coffee, vanilla, clove oil, or soap |

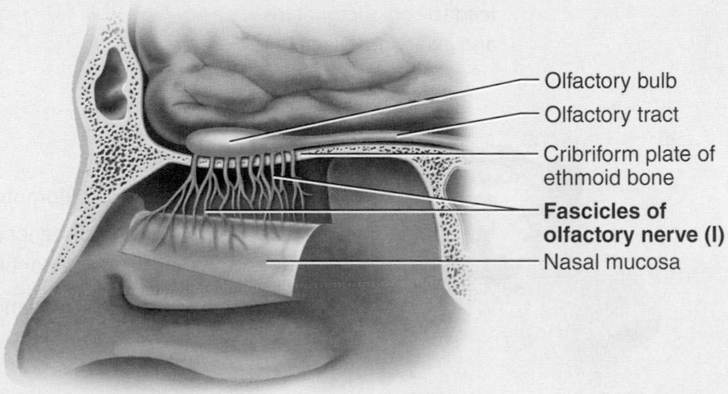

**FIGURE 14.27** The Olfactory Nerve (I). AP|R

### II. Optic Nerve

This is the nerve for vision.

| Composition | Function | Origin | Termination | Cranial Passage | Effect of Damage | Clinical Test |
|---|---|---|---|---|---|---|
| Sensory | Vision | Retina | Thalamus and midbrain | Optic foramen | Blindness in part or all of visual field | Inspect retina with ophthalmoscope; test peripheral vision and visual acuity |

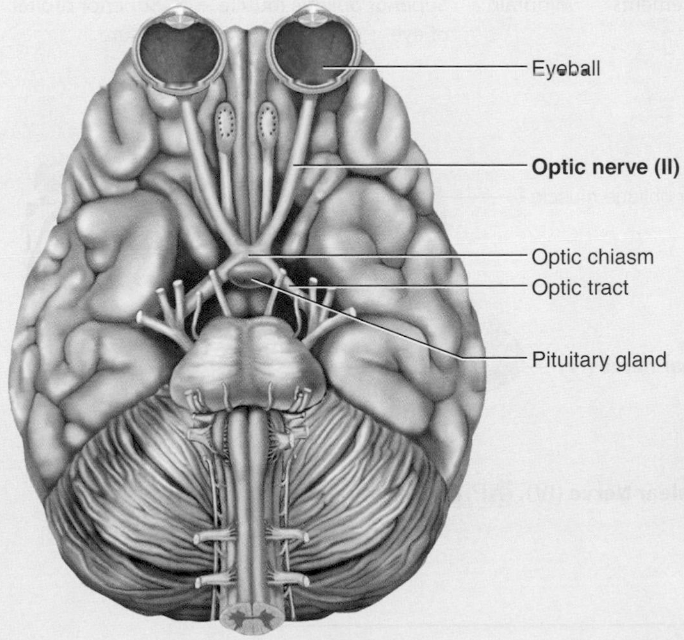

**FIGURE 14.28** The Optic Nerve (II). AP|R

| TABLE 14.1 | The Cranial Nerves (continued) |
|---|---|

### III. Oculomotor[58] Nerve (OC-you-lo-MO-tur)

This nerve controls muscles that turn the eyeball up, down, and medially, as well as controlling the iris, lens, and upper eyelid.

| Composition | Function | Origin | Termination | Cranial Passage | Effect of Damage | Clinical Test |
|---|---|---|---|---|---|---|
| Predominantly motor | Eye movements, opening of eyelid, pupillary constriction, focusing | Midbrain | Somatic fibers to levator palpebrae superioris; superior, medial, and inferior rectus muscles; and inferior oblique muscle of eye. Autonomic fibers enter eyeball and lead to constrictor of iris and ciliary muscle of lens. | Superior orbital fissure | Drooping eyelid; dilated pupil; inability to move eye in some directions; tendency of eye to rotate laterally at rest; double vision; difficulty focusing | Look for differences in size and shape of right and left pupils; test pupillary response to light; test ability to track moving objects |

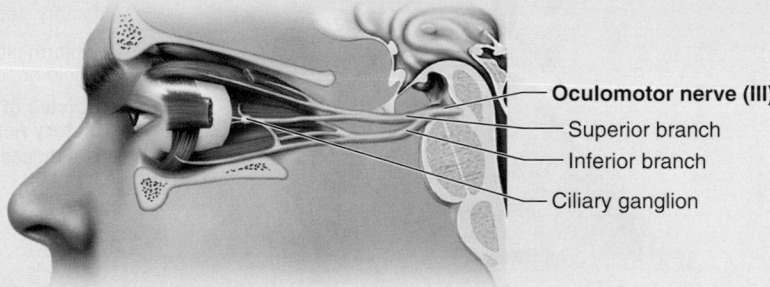

**Oculomotor nerve (III):**
Superior branch
Inferior branch
Ciliary ganglion

**FIGURE 14.29  The Oculomotor Nerve (III).** AP|R

### IV. Trochlear[59] Nerve (TROCK-lee-ur)

This nerve controls a muscle that directs the vision slightly downward and rotates the top of the eyeball toward the nose, especially in compensating for head movements. This is the only cranial nerve that arises from the posterior side of the brainstem. It is also unique in being the only one to completely decussate; the left trochlear nerve controls the right eye and vice versa..

| Composition | Function | Origin | Termination | Cranial Passage | Effect of Damage | Clinical Test |
|---|---|---|---|---|---|---|
| Predominantly motor | Eye movements | Midbrain | Superior oblique muscle of eye | Superior orbital fissure | Double vision and weakened ability to look downward; eye points superolaterally and subject tends to tuck the chin in and tilt the head downward to minimize the double vision | Ask subject to tilt the head toward one shoulder; affected eye shows upward deviation when head is tilted toward that side (*Bielschowsky's head tilt test*) |

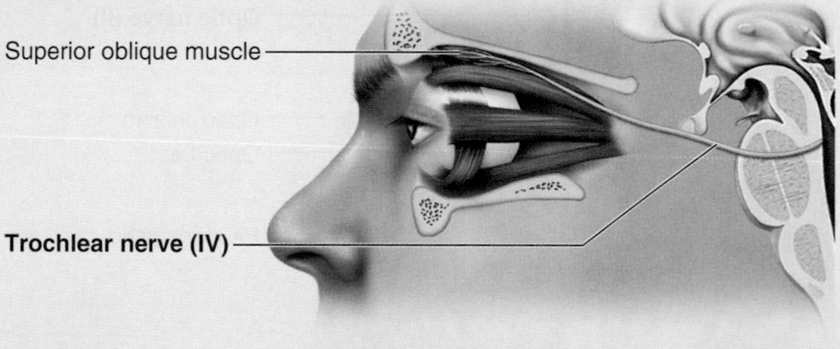

Superior oblique muscle

Trochlear nerve (IV)

**FIGURE 14.30  The Trochlear Nerve (IV).** AP|R

---

[58]*oculo* = eye; *motor* = mover
[59]*trochlea* = pulley (for a loop through which the muscle's tendon passes)

| TABLE 14.1 | The Cranial Nerves (continued) |
|---|---|

**V. Trigeminal[60] Nerve (tri-JEM-ih-nul)**

This is the largest of the cranial nerves and the most important sensory nerve of the face. It forks into three divisions: *ophthalmic (V₁)*, *maxillary (V₂)*, and *mandibular (V₃)* (fig. 14.31).

| Composition | Function | Origin | Termination | Cranial Passage | Effect of Damage | Clinical Test |
|---|---|---|---|---|---|---|
| **Ophthalmic division (V₁)** | | | | | | |
| Sensory | Touch, temperature, and pain sensations from upper face | Superior region of face as illustrated; surface of eyeball; lacrimal (tear) gland; superior nasal mucosa; frontal and ethmoid sinuses | Pons | Superior orbital fissure | Loss of sensation from upper face | Test corneal reflex (blinking in response to light touch to eyeball) |
| **Maxillary division (V₂)** | | | | | | |
| Sensory | Same as V₁, lower on face | Middle region of face as illustrated; nasal mucosa; maxillary sinus; palate; upper teeth and gums | Pons | Foramen rotundum and infraorbital foramen | Loss of sensation from middle face | Test sense of touch, pain, and temperature with light touch, pinpricks, and hot and cold objects |
| **Mandibular division (V₃)** | | | | | | |
| Mixed | *Sensory:* Same as V₁ and V₂, lower on face *Motor:* Mastication | *Sensory:* Inferior region of face as illustrated; anterior two-thirds of tongue (but not taste buds); lower teeth and gums; floor of mouth; dura mater *Motor:* Pons | *Sensory:* Pons *Motor:* Anterior belly of digastric; masseter, temporalis, mylohyoid, and pterygoid muscles; tensor tympani muscle of middle ear | Foramen ovale | Loss of sensation; impaired chewing | Assess motor functions by palpating masseter and temporalis while subject clenches teeth; test ability to move mandible from side to side and to open mouth against resistance |

nerve's name, is "Oh, once one takes the anatomy final, very good vacation ahead." The first two letters of *ahead* represent nerves XI and XII. One of the author's former students devised a mnemonic that can remind you of the first two to four letters of most cranial nerve names:

| | |
|---|---|
| **Ol**d | **ol**factory (I) |
| **Op**ie | **op**tic (II) |
| **oc**casionally | **oc**ulomotor (III) |
| **tr**ies | **tr**ochlear (IV) |
| **trig**onometry | **trig**eminal (V) |
| **a**nd | **a**bducens (VI) |
| **f**eels | **f**acial (VII) |
| **ve**ry | **ve**stibulocochlear (VIII) |
| **glo**omy, | **glo**ssopharyngeal (IX) |
| **vag**ue, | **vag**us (X) |
| **a**nd | **a**ccessory (XI) |
| **hypo**active. | **hypo**glossal (XII) |

## DEEPER INSIGHT 14.3

### CLINICAL APPLICATION

#### Some Cranial Nerve Disorders

*Trigeminal neuralgia*[61] *(tic douloureux*[62]*)* is a syndrome characterized by recurring episodes of intense stabbing pain in the trigeminal nerve. The cause is unknown; there is no visible change in the nerve. It usually occurs after the age of 50 and mostly in women. The pain lasts only a few seconds to a minute or two, but it strikes at unpredictable intervals and sometimes up to 100 times a day. The pain usually occurs in a specific zone of the face, such as around the mouth and nose. It may be triggered by touch, drinking, tooth brushing, or washing the face. Analgesics (pain relievers) give only limited relief. Severe cases are treated by cutting the nerve, but this also deadens most other sensation in that side of the face.

*Bell*[63] *palsy* is a degenerative disorder of the facial nerve, probably due to a virus. It is characterized by paralysis of the facial muscles on one side with resulting distortion of the facial features, such as sagging of the mouth or lower eyelid. The paralysis may interfere with speech, prevent closure of the eye, or sometimes inhibit tear secretion. There may also be a partial loss of the sense of taste. Bell palsy may appear abruptly, sometimes overnight, and often disappears spontaneously within 3 to 5 weeks.

[61]*neur* = nerve; *algia* = pain
[62](French) *douloureux* = painful; *tic* = twitch, spasm
[63] Sir Charles Bell (1774–1842), Scottish physician

[60]*tri* = three; *gem* = born (*trigem* = triplets)

**TABLE 14.1**      **The Cranial Nerves (continued)**

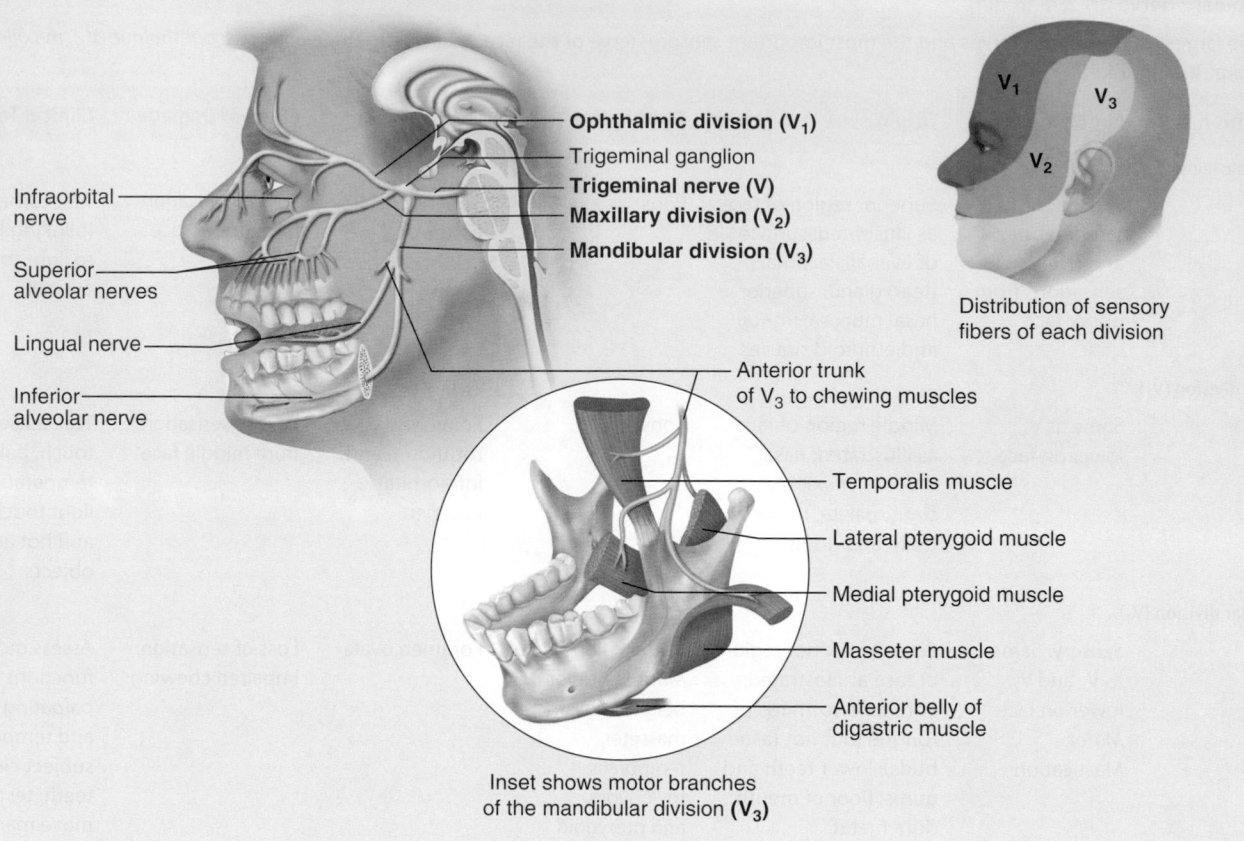

**FIGURE 14.31  The Trigeminal Nerve (V).** AP|R

## VI. Abducens[64] Nerve (ab-DOO-senz)

This nerve controls a muscle that turns the eyeball laterally.

| Composition | Function | Origin | Termination | Cranial Passage | Effect of Damage | Clinical Test |
|---|---|---|---|---|---|---|
| Predominantly motor | Lateral eye movement | Inferior pons | Lateral rectus muscle of eye | Superior orbital fissure | Inability to turn eye laterally; at rest, eye turns medially because of action of antagonistic muscles | Test lateral eye movement |

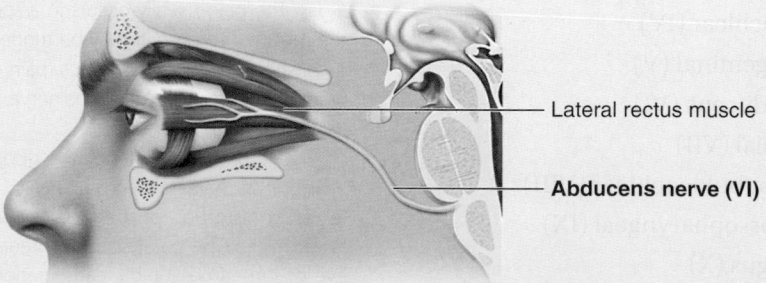

**FIGURE 14.32  The Abducens Nerve (VI).** AP|R

[64]*ab* = away; *duc* = to lead or turn

| TABLE 14.1 | The Cranial Nerves (continued) |
|---|---|

**VII. Facial Nerve**

This is the major motor nerve of the facial muscles. It divides into five prominent branches: *temporal, zygomatic, buccal, mandibular,* and *cervical*.

| Composition | Function | Origin | Termination | Cranial Passage | Effect of Damage | Clinical Test |
|---|---|---|---|---|---|---|
| Mixed | *Sensory:* Taste *Motor:* Facial expression; secretion of tears, saliva, nasal and oral mucus | *Sensory:* Taste buds of anterior two-thirds of tongue *Motor:* Pons | *Sensory:* Thalamus *Motor:* Somatic fibers to digastric muscle, stapedius muscle of middle ear, stylohyoid muscle, muscles of facial expression. Autonomic fibers to submandibular and sublingual salivary glands, tear glands, nasal and palatine glands. | Internal acoustic meatus and stylomastoid foramen | Inability to control facial muscles; sagging due to loss of muscle tone; distorted sense of taste, especially for sweets | Test anterior two-thirds of tongue with substances such as sugar, salt, vinegar, and quinine; test response of tear glands to ammonia fumes; test subject's ability to smile, frown, whistle, raise eyebrows, close eyes, etc. |

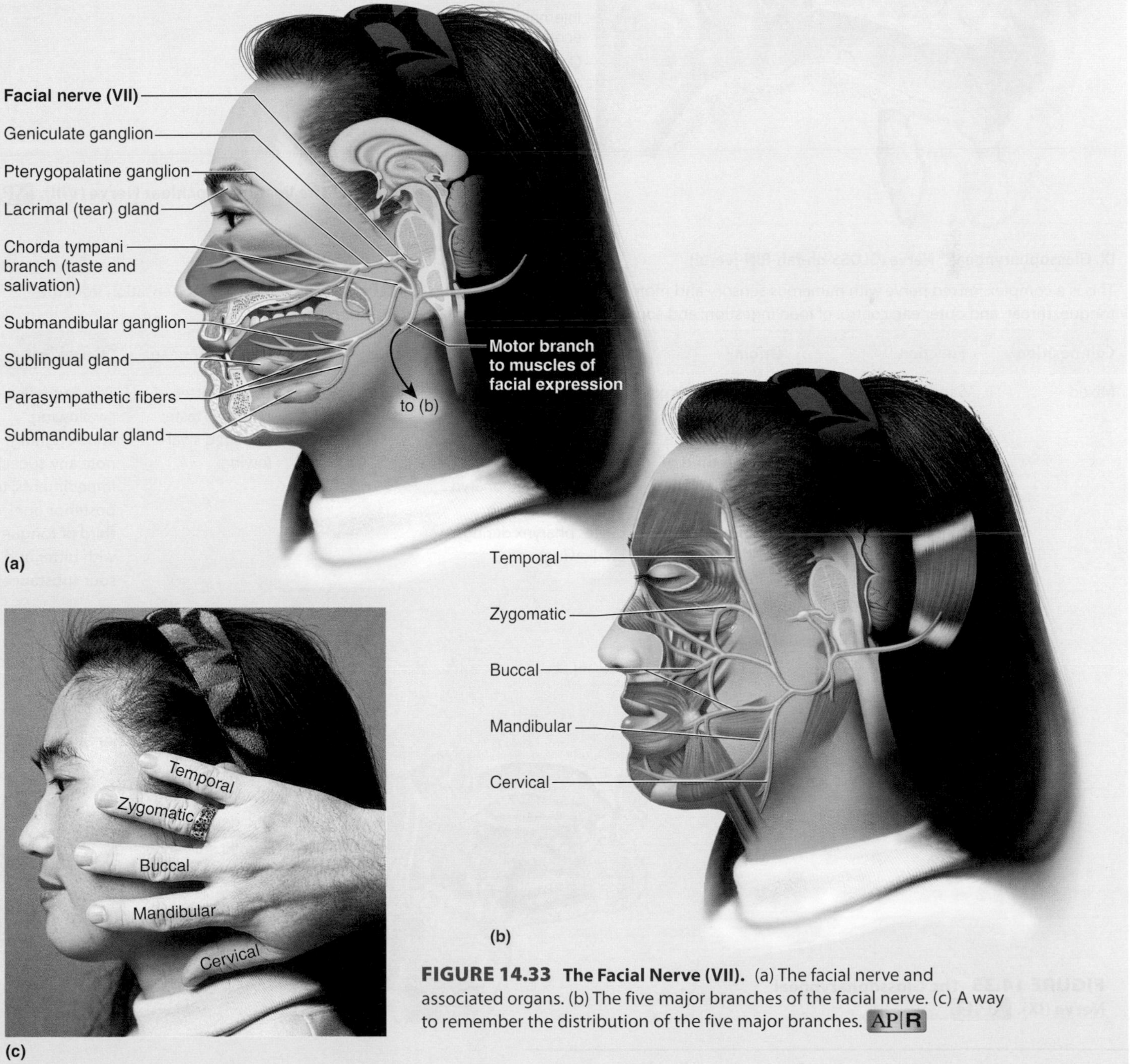

**FIGURE 14.33 The Facial Nerve (VII).** (a) The facial nerve and associated organs. (b) The five major branches of the facial nerve. (c) A way to remember the distribution of the five major branches. **AP|R**

**TABLE 14.1** The Cranial Nerves (continued)

### VIII. Vestibulocochlear[65] Nerve (vess-TIB-you-lo-COC-lee-ur)

This is the nerve of hearing and equilibrium, but it also has motor fibers that lead to cells of the cochlea that tune the sense of hearing (see chapter 16).

| Composition | Function | Origin | Termination | Cranial Passage | Effect of Damage | Clinical Test |
|---|---|---|---|---|---|---|
| Predominantly sensory | Hearing and equilibrium | *Sensory:* Cochlea, vestibule, and semicircular ducts of inner ear<br>*Motor:* Pons | *Sensory:* Fibers for hearing end in medulla; fibers for equilibrium end at junction of medulla and pons<br>*Motor:* Outer hair cells of cochlea of inner ear | Internal acoustic meatus | Nerve deafness, dizziness, nausea, loss of balance, and nystagmus (involuntary rhythmic oscillation of eyes from side to side) | Look for nystagmus; test hearing, balance, and ability to walk a straight line |

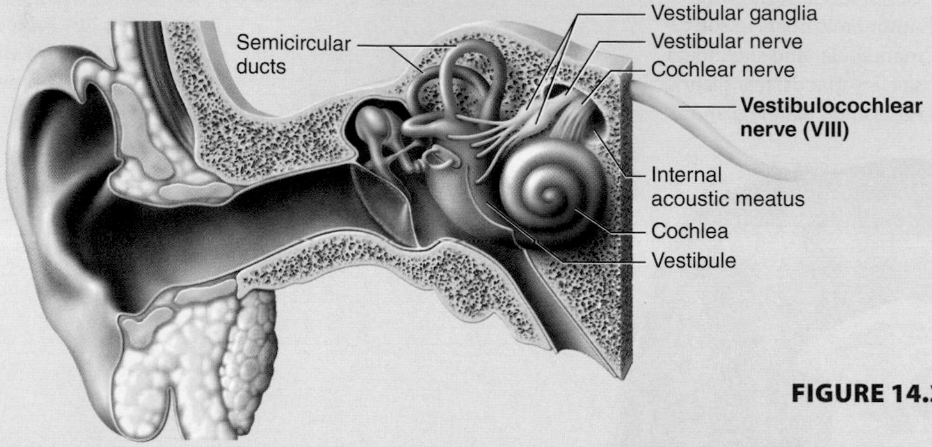

**FIGURE 14.34 The Vestibulocochlear Nerve (VIII).** AP|R

### IX. Glossopharyngeal[66] Nerve (GLOSS-oh-fah-RIN-jee-ul)

This is a complex, mixed nerve with numerous sensory and motor functions in the head, neck, and thoracic regions including sensation from the tongue, throat, and outer ear; control of food ingestion; and some aspects of cardiovascular and respiratory function.

| Composition | Function | Origin | Termination | Cranial Passage | Effect of Damage | Clinical Test |
|---|---|---|---|---|---|---|
| Mixed | *Sensory:* Taste; touch, pressure, pain, and temperature sensations from tongue and outer ear; regulation of blood pressure and respiration<br>*Motor:* Salivation, swallowing, gagging | *Sensory:* Pharynx; middle and outer ear; posterior one-third of tongue (including taste buds); internal carotid artery<br>*Motor:* Medulla oblongata | *Sensory:* Medulla oblongata<br>*Motor:* Parotid salivary gland; glands of posterior tongue; stylopharyngeal muscle (which dilates pharynx during swallowing) | Jugular foramen | Loss of bitter and sour taste; impaired swallowing | Test gag reflex, swallowing, and coughing; note any speech impediments; test posterior one-third of tongue with bitter and sour substances |

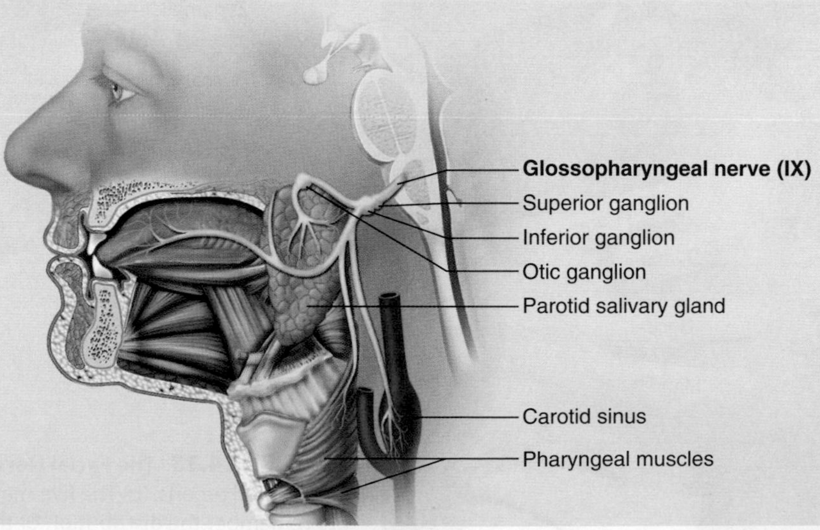

**FIGURE 14.35 The Glossopharyngeal Nerve (IX).** AP|R

---

[65]*vestibulo* = entryway (vestibule of inner ear); *cochlea* = conch, snail (cochlea of inner ear)
[66]*glosso* = tongue; *pharyng* = throat

| TABLE 14.1 | The Cranial Nerves (continued) |
| --- | --- |

### X. Vagus[67] Nerve (VAY-gus)

The vagus has the most extensive distribution of any cranial nerve, supplying not only organs in the head and neck but also most viscera of the thoracic and abdominopelvic cavities. It plays major roles in the control of cardiac, pulmonary, digestive, and urinary functions.

| Composition | Function | Origin | Termination | Cranial Passage | Effect of Damage | Clinical Test |
| --- | --- | --- | --- | --- | --- | --- |
| Mixed | *Sensory:* Taste; sensations of hunger, fullness, and gastrointestinal discomfort *Motor:* Swallowing, speech, deceleration of heart, bronchoconstriction, gastrointestinal secretion and motility | *Sensory:* Thoracic and abdominopelvic viscera, root of tongue, pharynx, larynx, epiglottis, outer ear, dura mater *Motor:* Medulla oblongata | *Sensory:* Medulla oblongata *Motor:* Tongue, palate, pharynx, larynx, lungs, heart, liver, spleen, digestive tract, kidney, ureter | Jugular foramen | Hoarseness or loss of voice; impaired swallowing and gastrointestinal motility; fatal if both vagus nerves are damaged | Examine palatal movements during speech; check for abnormalities of swallowing, absence of gag reflex, weak hoarse voice, inability to cough forcefully |

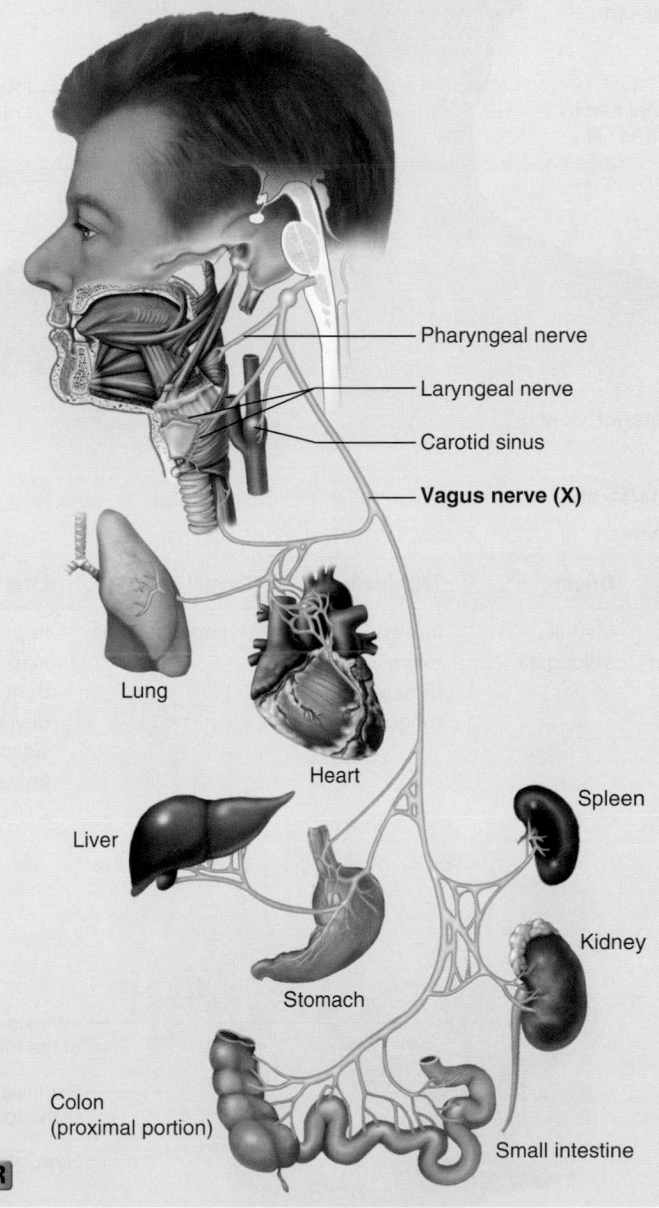

Pharyngeal nerve

Laryngeal nerve

Carotid sinus

**Vagus nerve (X)**

Lung

Heart

Liver

Spleen

Stomach

Kidney

Colon (proximal portion)

Small intestine

**FIGURE 14.36** The Vagus Nerve (X). AP|R

---

[67]*vag* = wandering

**TABLE 14.1**    The Cranial Nerves (continued)

### XI. Accessory Nerve

This nerve takes an unusual path. It arises from the upper spinal cord and not at all from the brain; therefore, strictly speaking, it is not a true cranial nerve. It ascends alongside the spinal cord, enters the cranial cavity through the foramen magnum, then exits the cranium through the jugular foramen, bundled with the vagus and glossopharyngeal nerves. The accessory nerve controls mainly swallowing and neck and shoulder muscles.

| Composition | Function | Origin | Termination | Cranial Passage | Effect of Damage | Clinical Test |
|---|---|---|---|---|---|---|
| Predominantly motor | Swallowing; head, neck, and shoulder movements | Spinal cord segments C1 to C6 | Palate, pharynx, trapezius and sterno-cleidomastoid muscles | Jugular foramen | Impaired movement of head, neck, and shoulders; difficulty shrugging shoulder on damaged side; paralysis of sternocleidomastoid, causing head to turn toward injured side | Test ability to rotate head and shrug shoulders against resistance |

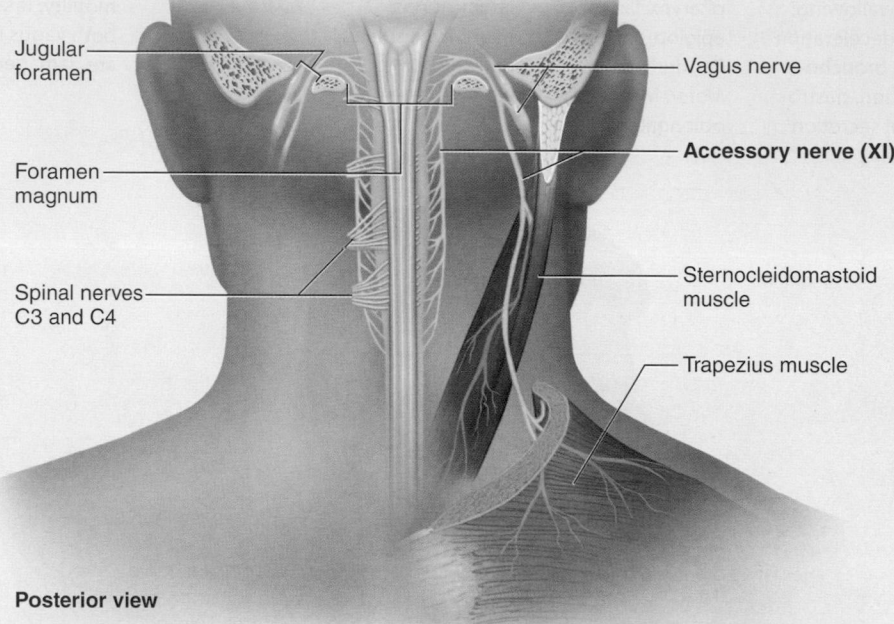

Jugular foramen

Foramen magnum

Spinal nerves C3 and C4

Vagus nerve

**Accessory nerve (XI)**

Sternocleidomastoid muscle

Trapezius muscle

**FIGURE 14.37**
**The Accessory Nerve (XI).** AP|R

Posterior view

### XII. Hypoglossal[68] Nerve (Hy-po-GLOSS-ul)

This nerve controls tongue movements.

| Composition | Function | Origin | Termination | Cranial Passage | Effect of Damage | Clinical Test |
|---|---|---|---|---|---|---|
| Predominantly motor | Tongue movements of speech, food manipulation, and swallowing | Medulla oblongata | Intrinsic and extrinsic muscles of tongue | Hypoglossal canal | Impaired speech and swallowing; inability to protrude tongue if both right and left nerves are damaged; deviation of tongue toward injured side, and atrophy on that side, if only one nerve is damaged | Note deviations of tongue as subject protrudes and retracts it; test ability to protrude tongue against resistance |

Intrinsic muscles of the tongue

Extrinsic muscles of the tongue

**Hypoglossal nerve (XII)**

**FIGURE 14.38**
**The Hypoglossal Nerve (XII).**
AP|R

[68]*hypo* = below; *gloss* = tongue

**FIGURE 14.39** **Review of Cranial Nerve Pathways.** Pathways concerned with sensory function are shown in green, and those concerned with motor function in red. All cranial nerves except I and II carry sensory fibers, but these functions are omitted in cases where the nerve is essentially motor and its sensory function is limited to proprioception from the muscles.

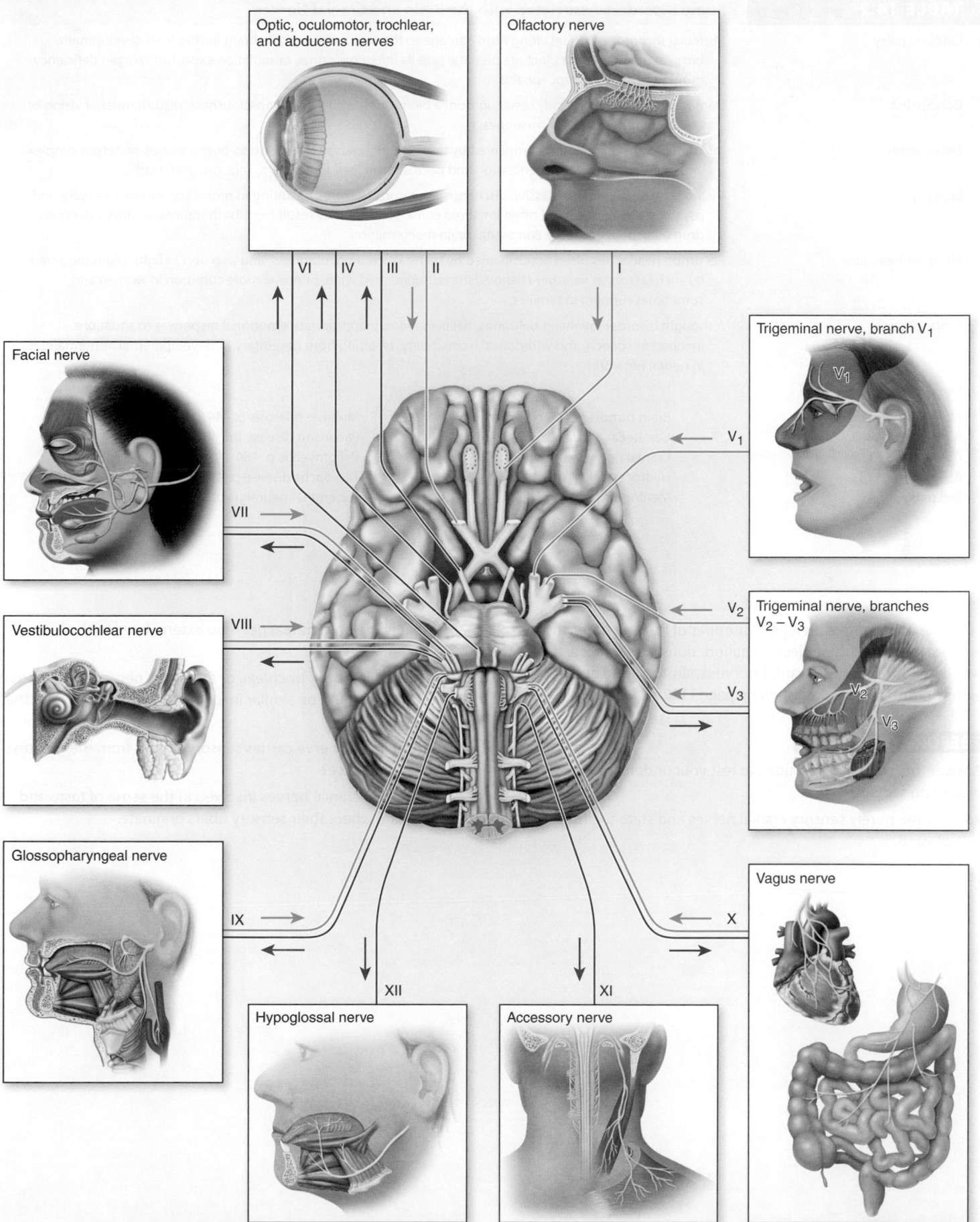

Optic, oculomotor, trochlear, and abducens nerves

Olfactory nerve

Facial nerve

Trigeminal nerve, branch V₁

Vestibulocochlear nerve

Trigeminal nerve, branches V₂ – V₃

Glossopharyngeal nerve

Vagus nerve

Hypoglossal nerve

Accessory nerve

| TABLE 14.2 | Some Disorders Associated with the Brain and Cranial Nerves |
|---|---|
| Cerebral palsy | Muscular incoordination resulting from damage to the motor areas of the brain during fetal development, birth, or infancy; causes include prenatal rubella infection, drugs, or radiation exposure; oxygen deficiency during birth; and hydrocephalus |
| Concussion | Damage to the brain typically resulting from a blow, often with loss of consciousness, disturbances of vision or equilibrium, and short-term amnesia |
| Encephalitis | Inflammation of the brain, accompanied by fever, usually caused by mosquito-borne viruses or herpes simplex virus; causes neuronal degeneration and necrosis; can lead to delirium, seizures, and death |
| Epilepsy | Disorder causing sudden, massive discharge of neurons (seizures) resulting in motor convulsions, sensory and psychic disturbances, and often impaired consciousness; may result from birth trauma, tumors, infections, drug or alcohol abuse, or congenital brain malformation |
| Migraine headache | Recurring headaches often accompanied by nausea, vomiting, dizziness, and aversion to light, often triggered by such factors as weather changes, stress, hunger, red wine, or noise; more common in women and sometimes running in families |
| Schizophrenia | A thought disorder involving delusions, hallucinations, inappropriate emotional responses to situations, incoherent speech, and withdrawal from society, resulting from hereditary or developmental abnormalities in neural networks |

*Disorders described elsewhere*

| | | |
|---|---|---|
| Alzheimer disease p. 468 | Brain tumors p. 443 | Multiple sclerosis p. 444 |
| Amnesia p. 534 | Cerebellar ataxia p. 538 | Parkinson disease pp. 468, 537 |
| Aphasia p. 540 | Cranial nerve injuries p. 552 | Poliomyelitis p. 484 |
| Aprosody p. 538 | Hydrocephalus pp. 245, 516 | Tay–Sachs disease p. 444 |
| Bell palsy p. 545 | Meningitis p. 514 | Trigeminal neuralgia p. 552 |

Like a machine with a great number of moving parts, the nervous system is highly subject to malfunctions. Table 14.2 lists a few well-known brain and cranial nerve dysfunctions. The effects of aging on the CNS are described on page 1120.

### BEFORE YOU GO ON

Answer the following questions to test your understanding of the preceding section:

25. List the purely sensory cranial nerves and state the function of each.

26. What is the only cranial nerve to extend beyond the head–neck region?

27. If the oculomotor, trochlear, or abducens nerve was damaged, the effect would be similar in all three cases. What would that effect be?

28. Which cranial nerve carries sensory signals from the greatest area of the face?

29. Name two cranial nerves involved in the sense of taste and describe where their sensory fibers originate.

## DEEPER INSIGHT 14.4
### CLINICAL APPLICATION

### Images of the Mind

Enclosed as it is in the cranium, there is no easy way to observe a living brain directly. This has long frustrated neurobiologists, who once had to content themselves with glimpses of brain function afforded by electroencephalograms, patients with brain lesions, and patients who remained awake and conversant during brain surgery and consented to experimentation while the brain was exposed. New imaging methods, however, are yielding dramatic perspectives on brain function. Two of these—positron emission tomography (PET) and magnetic resonance imaging (MRI)—were explained in Deeper Insight 1.5 (p. 22). Both techniques rely on transient increases in blood flow to parts of the brain called into action to perform specific tasks. By monitoring these changes, neuroscientists can identify which parts of the brain are involved in specific tasks.

To produce a PET scan of the brain, the subject is given an injection of radioactively labeled glucose and a scan is made in a *control state* before any specific mental task is begun. Then the subject is given a task. For example, the subject may be instructed to read the word *car* and speak a verb related to it, such as *drive*. New PET scans are made in the *task state* while the subject performs this task. Neither control- nor task-state images are very revealing by themselves, but the computer subtracts the control-state data from the task-state data and presents a color-coded image of the difference. To compensate for chance events and individual variation, the computer also produces an image that is either averaged from several trials with one person or from trials with several different people.

In such averaged images, the busiest areas of the brain seem to "light up" from moment to moment as the task is performed (fig. 14.40). This identifies the regions used for various stages of the task, such as reading the word, thinking of a verb to go with it, planning to say *drive,* and actually saying it. Among other things, such experiments demonstrate that the Broca and Wernicke areas are not involved in simply repeating words; they are active, however, when a subject must evaluate a word and choose an appropriate response—

that is, they function in formulating the new word the subject is going to say. PET scans also show that different neural pools take over a task as we practice and become more proficient at it.

*Functional magnetic resonance imaging (fMRI)* depends on the role of astrocytes in brain metabolism. The main excitatory neurotransmitter secreted by cerebral neurons is glutamate. After a neuron releases glutamate and glutamate stimulates the next neuron, astrocytes quickly remove it from the synapse and convert it to glutamine. Astrocytes acquire the energy for this from the anaerobic fermentation of glucose. High activity in an area of cortex thus requires an increased blood flow to supply this glucose, but it does not elevate oxygen consumption from that blood. Thus, the oxygen supply exceeds demand in that part of the brain, and blood leaving the region contains more oxygen than the blood leaving less active regions. Since the magnetic properties of hemoglobin depend on how much oxygen is bound to it, fMRI can detect changes in brain circulation.

fMRI is more precise than PET and pinpoints regions of brain activity with a precision of 1 to 2 mm. It also has the advantages of requiring no injected substances and no exposure to radioisotopes. While it takes about 1 minute to produce a PET scan, fMRI produces images much more quickly, which makes it more useful for determining how the brain responds immediately to sensory input or mental tasks.

PET and fMRI scanning have enhanced our knowledge of neurobiology by identifying shifting patterns of brain activity associated with attention and consciousness, sensory perception, memory, emotion, motor control, reading, speaking, musical judgment, planning a chess strategy, and so forth. In addition to their contribution to basic neuroscience, these techniques have proven very valuable to neurosurgery and psychopharmacology. They also are enhancing our understanding of brain dysfunctions such as depression, schizophrenia, and attention-deficit/hyperactivity disorder (ADHD). We have entered an exciting era in the safe visualization of normal brain function, producing pictures of the mind at work.

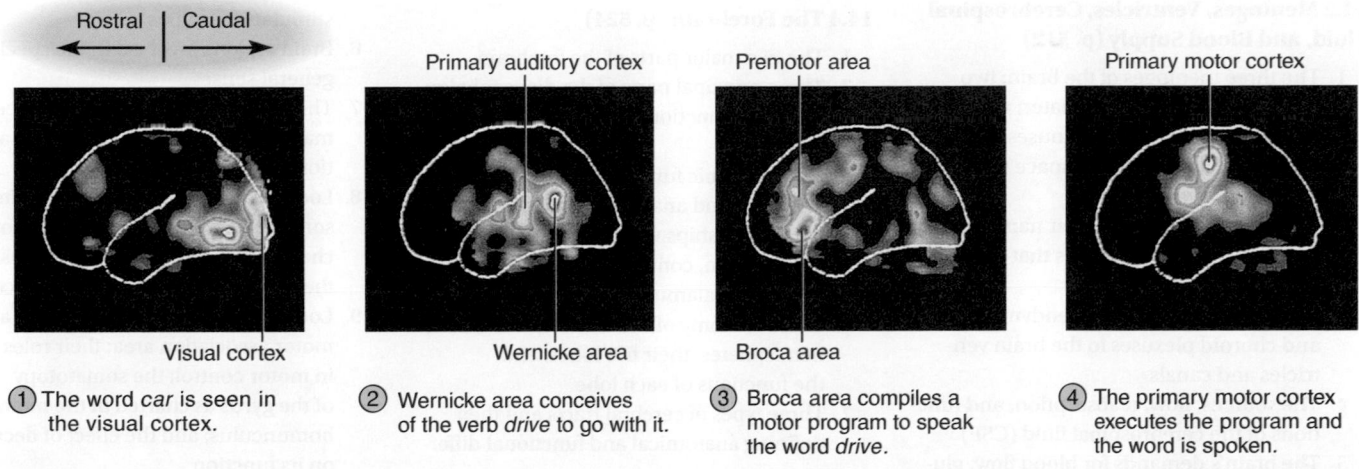

**FIGURE 14.40  PET Scans of the Brain Made During the Performance of a Language Task.** These images show the cortical regions that are active when a person reads words and then speaks them. The most active areas are shown in red and less active areas in blue.

# STUDY GUIDE

## ▶ Assess Your Learning Outcomes

*To test your knowledge, discuss the following topics with a study partner or in writing, ideally from memory.*

### 14.1 Overview of the Brain (p. 508)

1. Typical mass of the adult brain
2. Meanings of the directional terms *rostral* and *caudal* in brain anatomy
3. Three principal divisions of the brain, how definitions of *brainstem* vary, and the definition used in this book
4. Names of the tissue folds and shallow grooves in the cerebrum and cerebellum, and the names of the deep grooves between the two cerebral hemispheres and between the cerebrum and cerebellum
5. Locations of the gray and white matter in the cerebrum and cerebellum; the composition of gray and white matter and how this relates to their colors; and the meaning of *cortex, nucleus,* and *tract* in relation to brain gray and white matter
6. Embryonic development of the brain from neural plate to neural tube stage; differentiation into forebrain, midbrain, and hindbrain; development of the five embryonic brain vesicles; and the name and destiny of each vesicle

### 14.2 Meninges, Ventricles, Cerebrospinal Fluid, and Blood Supply (p. 512)

1. The three meninges of the brain; two subdivisions of the dura mater; and the relationship of the dural sinuses, subdural space, and subarachnoid space to the meninges
2. Ventricles of the brain, their names and locations, and the passages that connect them
3. The relationship of the ependymal cells and choroid plexuses to the brain ventricles and canals
4. The sources, flow, reabsorption, and functions of the cerebrospinal fluid (CSF)
5. The brain's demands for blood flow, glucose, and oxygen in relation to its percentage of the body weight
6. Structure and function of the blood–brain and blood–CSF barriers, and the clinical relevance of the brain barrier system
7. The locations, structural nature, function, and clinical relevance of circumventricular organs

### 14.3 The Hindbrain and Midbrain (p. 517)

1. The medulla oblongata: its location, gross anatomy, and general functions; which cranial nerves arise from it; and the specific functions of each of its tracts and nuclei
2. The pons: its location, gross anatomy, and functions; which cranial nerves arise from the pons and from the groove between pons and medulla; and its contributions to the anterolateral system and anterior spinocerebellar tract
3. The midbrain: its location, gross anatomy, and general functions; which cranial nerves arise from it; and the specific functions of its colliculi, substantia nigra, and central gray matter
4. How the reticular formation relates to the foregoing brainstem regions in location, structure, and functions
5. The cerebellum: its location, gross anatomy, unique neuron types, its three pairs of peduncles, and their relationship to cerebellar input and output
6. The classical view of cerebellar function and how this has lately expanded as a result of brain imaging studies and studies of people with cerebellar lesions

### 14.4 The Forebrain (p. 524)

1. The two major parts of the forebrain
2. Three principal parts of the diencephalon
3. Thalamic functions, location, and gross anatomy
4. Hypothalamic functions, location, gross anatomy, and anatomical and physiological relationships with the pituitary gland
5. The location, components, and functions of the epithalamus
6. Gross anatomy of the cerebral hemispheres, their five lobes, and the functions of each lobe
7. Three types of cerebral tracts and their defining anatomical and functional differences
8. Three locations of cerebral gray matter
9. The thickness, extent, and relative amount of tissue and number of neurons in the cerebral cortex
10. Two types of neurons in the cerebral cortex and their morphology
11. Evolutionary and structural distinctions between archicortex, paleocortex, and neo-

cortex, and where each occurs in the human brain
12. The location, constituents, and functions of the limbic system
13. Names, locations, and functions of the basal nuclei

### 14.5 Integrative Functions of the Brain (p. 530)

1. Four kinds of brain waves in the EEG, the states in which each type normally dominates the EEG, and the clinical relevance of the EEG
2. Stages of sleep; physiological characteristics of each; roles of the hypothalamus, orexins, and reticular formation in regulating the sleep cycle; and hypotheses on the functions of sleep
3. Association areas of the cerebral cortex; the involvement of some of these in cognitive function; and the contributions of brain injury patients and brain imaging methods to understanding the regional distribution of cognitive functions
4. Brain regions involved in memory; forms of amnesia
5. Brain regions involved in emotion, and insights into the neurobiology of emotion from brain trauma, ablation, and brain stimulation studies
6. Brain regions involved in the special and general senses
7. The functional relationship between primary sensory cortex and sensory association areas
8. Location of the postcentral gyrus; its somatosensory function; its somatotopy as charted in the sensory homunculus; and the effect of decussation on its function
9. Locations of the precentral gyrus and motor association area; their roles in motor control; the somatotopy of the gyrus as charted in the motor homunculus; and the effect of decussation on its function
10. Upper and lower motor neurons and the course of their axons
11. Roles of the basal nuclei and cerebellum in motor coordination and learned motor skills
12. Effects of Parkinson disease and basal nuclei lesions on motor control
13. Locations of the Wernicke and Broca areas; their roles in language; interaction

# STUDY GUIDE

of the Broca area and precentral gyrus in speech; and forms of aphasia and other language deficits resulting from damage to the language centers

14. Cerebral lateralization; functional differences between the categorical and representational hemispheres; and why it cannot be said that a particular function is necessarily "right-brained" or "left-brained"

## 14.6 The Cranial Nerves (p. 541)

1. Names and numbers of the 12 pairs of cranial nerves, and their relationships to the brainstem and skull foramina

2. Which cranial nerves are purely sensory, which are mixed, which have traditionally been regarded as motor, and why it is not entirely accurate to simply call them motor nerves

3. For each cranial nerve, its location, functions, origin, termination, passage through the skull

4. Effects of damage to each cranial nerve, and clinical methods of testing for damage

## ▶ Testing Your Recall

*Answers in Appendix B*

1. Which of these is caudal to the hypothalamus?
   a. the thalamus
   b. the optic chiasm
   c. the cerebral aqueduct
   d. the pituitary gland
   e. the corpus callosum

2. If the telencephalon was removed from a 5-week-old embryo, which of the following structures would fail to develop in the fetus?
   a. cerebral hemispheres
   b. the thalamus
   c. the midbrain
   d. the medulla oblongata
   e. the spinal cord

3. The blood–CSF barrier is formed by
   a. blood capillaries.
   b. endothelial cells.
   c. protoplasmic astrocytes.
   d. oligodendrocytes.
   e. ependymal cells.

4. The pyramids of the medulla oblongata contain
   a. descending corticospinal fibers.
   b. commissural fibers.
   c. ascending spinocerebellar fibers.
   d. fibers going to and from the cerebellum.
   e. ascending spinothalamic fibers.

5. Which of the following does *not* receive any input from the eyes?
   a. the hypothalamus
   b. the frontal lobe
   c. the thalamus
   d. the occipital lobe
   e. the midbrain

6. While studying in a noisy cafeteria, you get sleepy and doze off for a few minutes.

You awaken with a start and realize that all the cafeteria sounds have just "come back." While you were dozing, this auditory input was blocked from reaching your auditory cortex by
   a. the temporal lobe.
   b. the thalamus.
   c. the reticular activating system.
   d. the medulla oblongata.
   e. the vestibulocochlear nerve.

7. Because of a brain lesion, a certain patient never feels full, but eats so excessively that she now weighs nearly 270 kg (600 lb). The lesion is most likely in her
   a. hypothalamus.
   b. amygdala.
   c. hippocampus.
   d. basal nuclei.
   e. pons.

8. The _____ is most closely associated with the cerebellum in embryonic development and remains its primary source of input fibers throughout life.
   a. telencephalon
   b. thalamus
   c. midbrain
   d. pons
   e. medulla

9. Damage to the _____ nerve could result in defects of eye movement.
   a. optic
   b. vagus
   c. trigeminal
   d. facial
   e. abducens

10. All of the following *except* the _____ nerve begin or end in the orbit.
    a. optic
    b. oculomotor

    c. trochlear
    d. abducens
    e. accessory

11. The right and left cerebral hemispheres are connected to each other by a thick C-shaped bundle of fibers called the _____.

12. The brain has four chambers called _____ filled with _____ fluid.

13. On a sagittal plane, the cerebellar white matter exhibits a branching pattern called the _____.

14. Abnormal accumulation of cerebrospinal fluid in the ventricles can cause a condition called _____.

15. Cerebrospinal fluid is secreted partly by a mass of blood capillaries called the _____ in each ventricle.

16. The primary motor area of the cerebrum is the _____ gyrus of the frontal lobe.

17. A lesion in which lobe of the cerebrum is most likely to cause a radical alteration of the personality?

18. Areas of cerebral cortex that identify or interpret sensory information are called _____.

19. Linear, analytical, and verbal thinking occurs in the _____ hemisphere of the cerebrum, which is on the left in most people.

20. The motor pattern for speech is generated in an area of cortex called the _____ and then transmitted to the primary motor cortex to be carried out.

# STUDY GUIDE

## ▶ Building Your Medical Vocabulary

*Answers in Appendix B*

*State a medical meaning of each word element below, and give a term in which it or a slight variation of it is used.*

1. -algia
2. cephalo-
3. cerebro-
4. corpo-
5. encephalo-
6. -gram
7. insulo-
8. oculo-
9. trochle-
10. -uncle

## ▶ True or False

*Answers in Appendix B*

*Determine which five of the following statements are false, and briefly explain why.*

1. The two hemispheres of the cerebellum are separated by the longitudinal fissure.

2. The cerebral hemispheres would fail to develop if the neural crests of the embryo were destroyed.

3. The midbrain is caudal to the thalamus.

4. The Broca area is ipsilateral to the Wernicke area.

5. Most of the cerebrospinal fluid is produced by the choroid plexuses.

6. Hearing is a function of the occipital lobe.

7. Respiration is controlled by nuclei in both the pons and medulla oblongata.

8. The trigeminal nerve carries sensory signals from a larger area of the face than the facial nerve does.

9. Unlike other cranial nerves, the vagus nerve extends far beyond the head–neck region.

10. Most of the brain's neurons are found in the cerebral cortex.

## ▶ Testing Your Comprehension

*Answers at* www.mhhe.com/saladin7

1. Which cranial nerve conveys pain signals to the brain in each of the following situations: (a) sand blows into your eye; (b) you bite the back of your tongue; and (c) your stomach hurts from eating too much?

2. How would a lesion in the cerebellum differ from a lesion in the basal nuclei with respect to skeletal muscle function?

3. Suppose that a neuroanatomist performed two experiments on an animal with the same basic brainstem structure as a human's: In experiment 1, he selectively transected (cut across) the pyramids on the anterior side of the medulla oblongata, and in experiment 2, he selectively transected the gracile and cuneate fasciculi on the posterior side. How would the outcomes of the two experiments differ?

4. A person can survive destruction of an entire cerebral hemisphere but cannot survive destruction of the hypothalamus, which is a much smaller mass of brain tissue. Explain this difference and describe some ways that destruction of a cerebral hemisphere would affect one's quality of life.

5. What would be the most obvious effects of lesions that destroyed each of the following: (a) the hippocampus, (b) the amygdala, (c) the Broca area, (d) the occipital lobe, and (e) the hypoglossal nerve?

## ▶ Improve Your Grade at www.mhhe.com/saladin7

*Download animations to study when it fits your schedule. Practice quizzes, labeling activities, games, and flashcards offer fun ways to master the chapter concepts. Or, download image PowerPoint files for each chapter to create a study guide or for taking notes during lecture.*

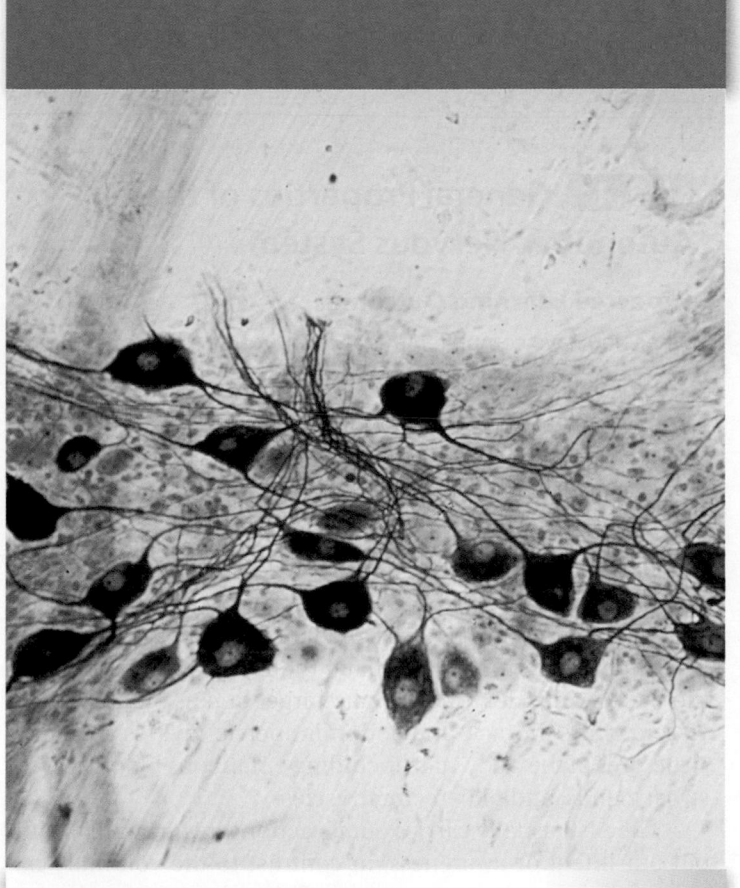

Autonomic neurons in the enteric nervous system of the digestive tract

CHAPTER

# 15

# THE AUTONOMIC NERVOUS SYSTEM AND VISCERAL REFLEXES

Module 7: Nervous System

## BRUSHING UP

- Figure 12.2 (p. 437) provides a helpful perspective on how the autonomic nervous system (ANS) fits into the scheme of the nervous system as a whole.

- It is important that you clearly understand neurotransmitters and receptors and why neurotransmitters can have either excitatory or inhitory effects on target cells (fig. 12.21 and table 12.3, p. 458).

- Many effects of the ANS are carried out by smooth muscle. The innervation of smooth muscle (p. 425) will help in understanding the effects described in this chapter.

- You should be familiar with the anatomy and functions of the hypothalamus (p. 524), a highly important ANS integrating center.

- Spinal cord and spinal nerve anatomy (pp. 475, 485) and cranial nerves III, VII, IX, and X (pp. 544–549) are especially important autonomic output pathways.

**W**e are consciously aware of many of the activities of our nervous system discussed in the preceding chapters—the general and special senses, our cognitive processes and emotions, and our voluntary movements. But there is another branch of the nervous system that operates in comparative secrecy, usually without our willing it, thinking about it, or even being able to consciously modify or suppress it.

This secretive agent is called the *autonomic nervous system (ANS)*. Its name means "self-governed," as it is almost fully independent of our will. Its job is to regulate such fundamental states and life processes as heart rate, blood pressure, body temperature, respiratory airflow, pupillary diameter, digestion, energy metabolism, defecation, and urination. In short, the ANS quietly manages a multitude of unconscious processes responsible for the body's homeostasis.

Walter Cannon (1871–1945), the American physiologist who coined such expressions as *"homeostasis"* and the *"fight-or-flight" reaction,* dedicated his career to the study of the autonomic nervous system. He found that an animal can live without a functional sympathetic nervous system (one of the two divisions of the ANS), but it must be kept warm and free of stress. It cannot regulate its body temperature, tolerate any strenuous exertion, or survive on its own. Indeed, the ANS is more necessary for survival than are many functions of the somatic nervous system; an absence of autonomic function is fatal because the body cannot maintain homeostasis without it. Thus, for an understanding of bodily function, the mode of action of many drugs, and other aspects of health care, we must be especially aware of how the ANS works.

## 15.1   General Properties of the Autonomic Nervous System

### Expected Learning Outcomes

When you have completed this section, you should be able to

a. explain how the autonomic and somatic nervous systems differ in form and function; and

b. explain how the two divisions of the autonomic nervous system differ in general function.

The **autonomic**[1] **nervous system (ANS)** can be defined as a motor nervous system that controls glands, cardiac muscle, and smooth muscle. It is also called the **visceral motor system** to distinguish it from the somatic motor system that controls the skeletal muscles. The primary target organs of the ANS are viscera of the thoracic and abdominopelvic cavities and some structures of the body wall, including cutaneous blood vessels, sweat glands, and piloerector muscles.

The ANS usually carries out its actions involuntarily, without our intent or awareness, in contrast to the voluntary nature of the somatic motor system. This voluntary–involuntary distinction is not, however, as clear-cut as it might seem. Some skeletal muscle responses are quite involuntary, such as the somatic reflexes, and some muscles are difficult or impossible to control, such as the middle-ear muscles. On the other hand, therapeutic uses of biofeedback have shown that some people can learn to voluntarily control such visceral functions as blood pressure.

Visceral effectors do not depend on the autonomic nervous system to function, but only to adjust their activity to the body's changing needs. The heart, for example, goes on beating even if all autonomic nerves to it are severed, but the ANS modulates the heart rate in conditions of rest or exercise. If the somatic nerves to a skeletal muscle are severed, the muscle exhibits flaccid paralysis—it no longer functions. But if the autonomic nerves to cardiac or smooth muscle are severed, the muscle exhibits exaggerated responses *(denervation hypersensitivity).*

### Visceral Reflexes

The ANS is responsible for the body's **visceral**[2] **reflexes**—unconscious, automatic, stereotyped responses to stimulation, much like the somatic reflexes discussed in chapter 13, but involving visceral receptors and effectors and slower responses. Some authorities regard the visceral afferent (sensory) pathways as part of the ANS, but most prefer to limit the term *ANS* to the efferent (motor) pathways. Regardless of this preference,

---

[1] *auto* = self; *nom* = rule (self-governed)
[2] *viscero* = internal organs

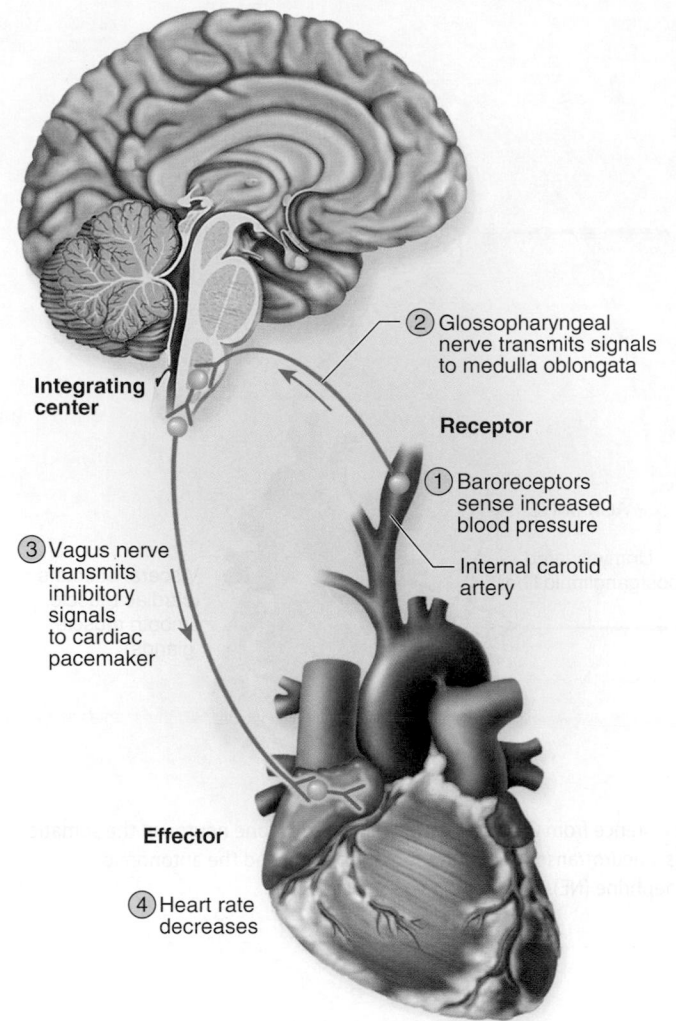

**Integrating center**

② Glossopharyngeal nerve transmits signals to medulla oblongata

**Receptor**

① Baroreceptors sense increased blood pressure

Internal carotid artery

③ Vagus nerve transmits inhibitory signals to cardiac pacemaker

**Effector**

④ Heart rate decreases

**FIGURE 15.1  An Autonomic Reflex Arc.** A rise in blood pressure is detected by baroreceptors in the carotid artery. The glossopharyngeal nerve transmits signals to the medulla oblongata, resulting in parasympathetic output from the vagus nerve that reduces the heart rate and lowers blood pressure. Both right and left carotid arteries contain baroreceptors; only the right one is shown here.

however, autonomic activity involves a visceral reflex arc that includes **receptors** (nerve endings that detect stretch, tissue damage, blood chemicals, body temperature, and other internal stimuli), afferent neurons leading to an **integrating center** in the CNS, interneurons in the CNS, efferent neurons carrying motor signals away from the CNS, and finally an **effector** that carries out the end response.

For example, high blood pressure activates a visceral *baroreflex.*[3] It stimulates stretch receptors called *baroreceptors* in the carotid arteries and aorta, and they transmit signals via

---

[3]*baro* = pressure

the glossopharyngeal nerves to the brainstem (fig. 15.1). The medulla integrates this with other information and transmits signals back to the heart by way of the vagus nerves. The vagus nerves slow down the heart and reduce blood pressure, thus completing a homeostatic negative feedback loop. A separate autonomic reflex accelerates the heart when blood pressure drops below normal—for example, when we move from a reclining to a standing position and gravity draws blood away from the upper body (see fig. 1.7, p. 16).

## Divisions of the Autonomic Nervous System

The ANS has two subsystems: sympathetic and parasympathetic. They differ in anatomy and function, but often innervate the same target organs and may have cooperative or contrasting effects on them. The **sympathetic division** adapts the body in many ways for physical activity—it increases alertness, heart rate, blood pressure, pulmonary airflow, blood glucose concentration, and blood flow to cardiac and skeletal muscle, but at the same time, it reduces blood flow to the skin and digestive tract. Cannon referred to extreme sympathetic responses as the "fight-or-flight" reaction because it comes into play when an animal must attack, defend itself, or flee from danger. In our own lives, this reaction occurs in situations involving arousal, exercise, competition, stress, danger, trauma, anger, or fear. Ordinarily, however, the sympathetic division has more subtle effects that we notice barely, if at all.

The **parasympathetic division,** by comparison, has a calming effect on many body functions. It is associated with reduced energy expenditure and normal bodily maintenance, including such functions as digestion and waste elimination. This is often called the "resting-and-digesting" state.

This does not mean that the body alternates between states where one system or the other is active. Normally both systems are active simultaneously. They exhibit a background rate of activity called **autonomic tone,** and the balance between *sympathetic tone* and *parasympathetic tone* shifts in accordance with the body's changing needs. Parasympathetic tone, for example, maintains smooth muscle tone in the intestines and holds the resting heart rate down to about 70 to 80 beats/min. If the parasympathetic vagus nerves to the heart are cut, the heart beats at its own intrinsic rate of about 100 beats/min. Sympathetic tone keeps most blood vessels partially constricted and thus maintains blood pressure. A loss of sympathetic tone can cause such a rapid drop in blood pressure that a person goes into shock and may faint.

Neither division has universally excitatory or inhibitory effects. The sympathetic division, for example, excites the heart but inhibits digestive and urinary functions, whereas the parasympathetic division has the opposite effects. We will later examine how differences in neurotransmitters and receptors account for these differences of effect.

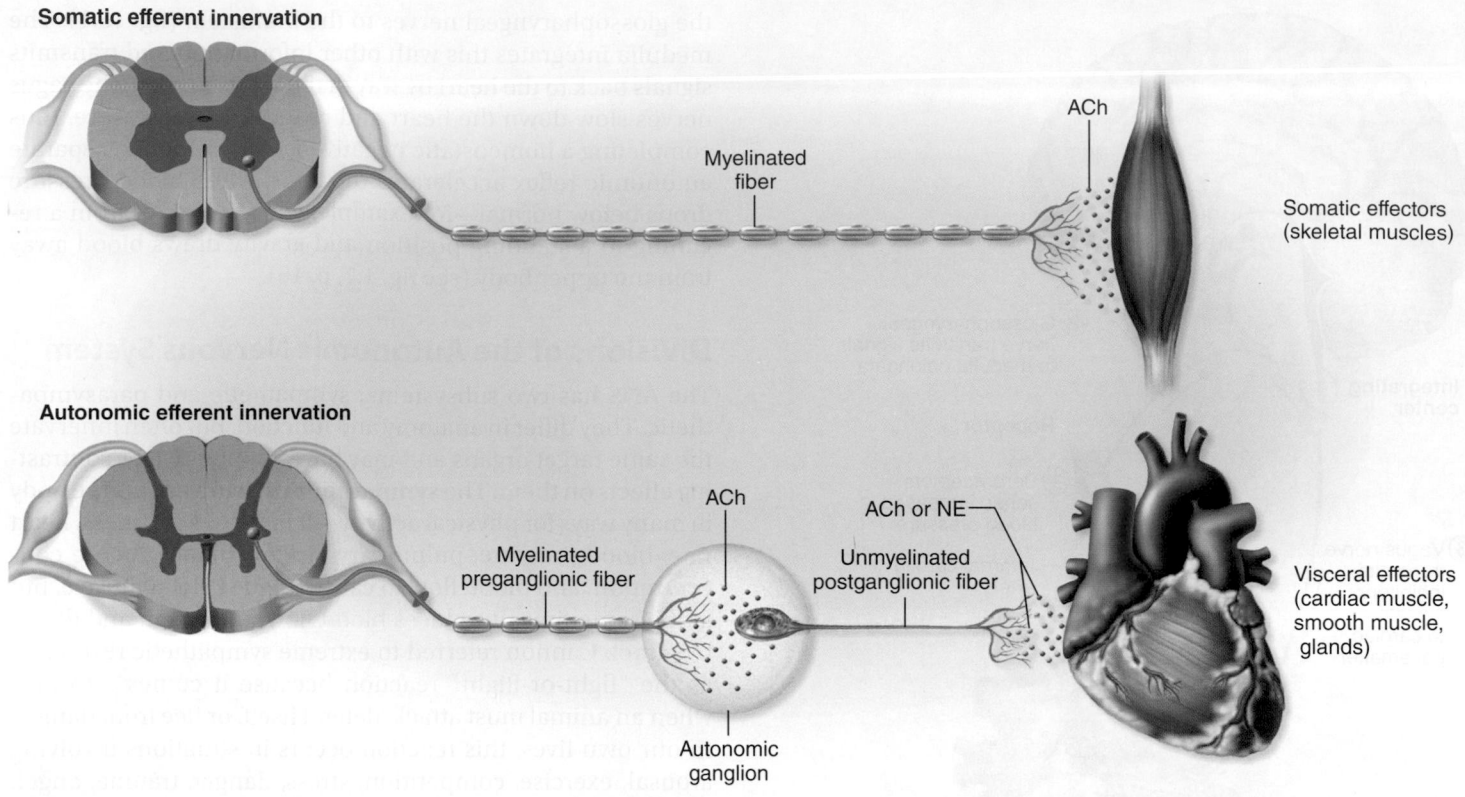

**Somatic efferent innervation**

Myelinated
fiber

ACh

Somatic effectors
(skeletal muscles)

**Autonomic efferent innervation**

ACh

Myelinated
preganglionic fiber

ACh or NE

Unmyelinated
postganglionic fiber

Visceral effectors
(cardiac muscle,
smooth muscle,
glands)

Autonomic
ganglion

**FIGURE 15.2 Comparison of Somatic and Autonomic Efferent Pathways.** The entire distance from CNS to effector is spanned by one neuron in the somatic system and two neurons in the autonomic system. Only acetylcholine (ACh) is employed as a neurotransmitter by the somatic neuron and the autonomic preganglionic fiber, but autonomic postganglionic fibers can employ either ACh or norepinephrine (NE).

## Autonomic Output Pathways

The ANS has components in both the central and peripheral nervous systems. It includes control nuclei in the hypothalamus and other regions of the brainstem, motor neurons in the spinal cord and peripheral ganglia, and nerve fibers that travel through the cranial and spinal nerves you have already studied.

The autonomic motor pathway to a target organ differs significantly from somatic motor pathways. In somatic pathways, a motor neuron in the brainstem or spinal cord issues a myelinated axon that reaches all the way to a skeletal muscle. In autonomic pathways, the signal must travel across two nerve fibers to get to the target organ, and it must cross a synapse where these two neurons meet in an autonomic ganglion (fig. 15.2). The first fiber, called the **preganglionic fiber** is myelinated and leads from a soma in the brainstem or spinal cord to the autonomic ganglion. It synapses there with a neuron that issues an unmyelinated **postganglionic fiber** to the target cells. In contrast to somatic motor neurons, postganglionic fibers of the ANS do not end by synapsing with a specific target cell, but with a chain of **varicosities** that diffusely release neurotransmitter into the tissue and may stimulate many cells simultaneously (see fig. 11.21, p. 425).

In summary, the autonomic nervous system is a division of the nervous system responsible for homeostasis, acting through the mostly unconscious and involuntary control of glands, smooth muscle, and cardiac muscle. Its target organs are mostly the thoracic and abdominopelvic viscera, but also include some cutaneous and other effectors. It acts through motor pathways that involve two nerve fibers, preganglionic and postganglionic, reaching from CNS to effector. The ANS has two divisions, sympathetic and parasympathetic, that often have cooperative or contrasting effects on the same target organ. Both divisions have excitatory effects on some target cells and inhibitory effects on others. These and other differences between the somatic and autonomic nervous systems are summarized in table 15.1.

**BEFORE YOU GO ON**

Answer the following questions to test your understanding of the preceding section:

1. How does the autonomic nervous system differ functionally and anatomically from the somatic motor system?

2. How do the general effects of the sympathetic division differ from those of the parasympathetic division?

| TABLE 15.1 | Comparison of the Somatic and Autonomic Nervous Systems | |
|---|---|---|
| **Feature** | **Somatic** | **Autonomic** |
| Effectors | Skeletal muscle | Glands, smooth muscle, cardiac muscle |
| Control | Usually voluntary | Usually involuntary |
| Distal nerve endings | Neuromuscular junctions | Varicosities |
| Efferent pathways | One nerve fiber from CNS to effector; no ganglia | Two nerve fibers from CNS to effector; synapse at a ganglion |
| Neurotransmitters | Acetylcholine (ACh) | ACh and norepinephrine (NE) |
| Effect on target cells | Always excitatory | Excitatory or inhibitory |
| Effect of denervation | Flaccid paralysis | Denervation hypersensitivity |

## 15.2 Anatomy of the Autonomic Nervous System

### Expected Learning Outcomes

When you have completed this section, you should be able to

a. identify the anatomical components and nerve pathways of the sympathetic and parasympathetic divisions;

b. discuss the relationship of the adrenal glands to the sympathetic nervous system; and

c. describe the enteric nervous system of the digestive tract and explain its significance.

### The Sympathetic Division

The sympathetic division is also called the *thoracolumbar division* because it arises from the thoracic and lumbar regions of the spinal cord. It has relatively short preganglionic and long postganglionic fibers. Neurosomas of the preganglionic fibers are in the lateral horns and nearby gray matter of the spinal cord. Their axons exit via spinal nerves T1 to L2 and lead to the nearby **sympathetic chain** of ganglia **(paravertebral[4] ganglia).** This is a longitudinal series of ganglia that lie adjacent to both sides of the vertebral column from the cervical to the coccygeal level. They are interconnected by longitudinal nerve cords (figs. 15.3 and 15.4). The number of

[4]*para* = next to; *vertebr* = vertebral column

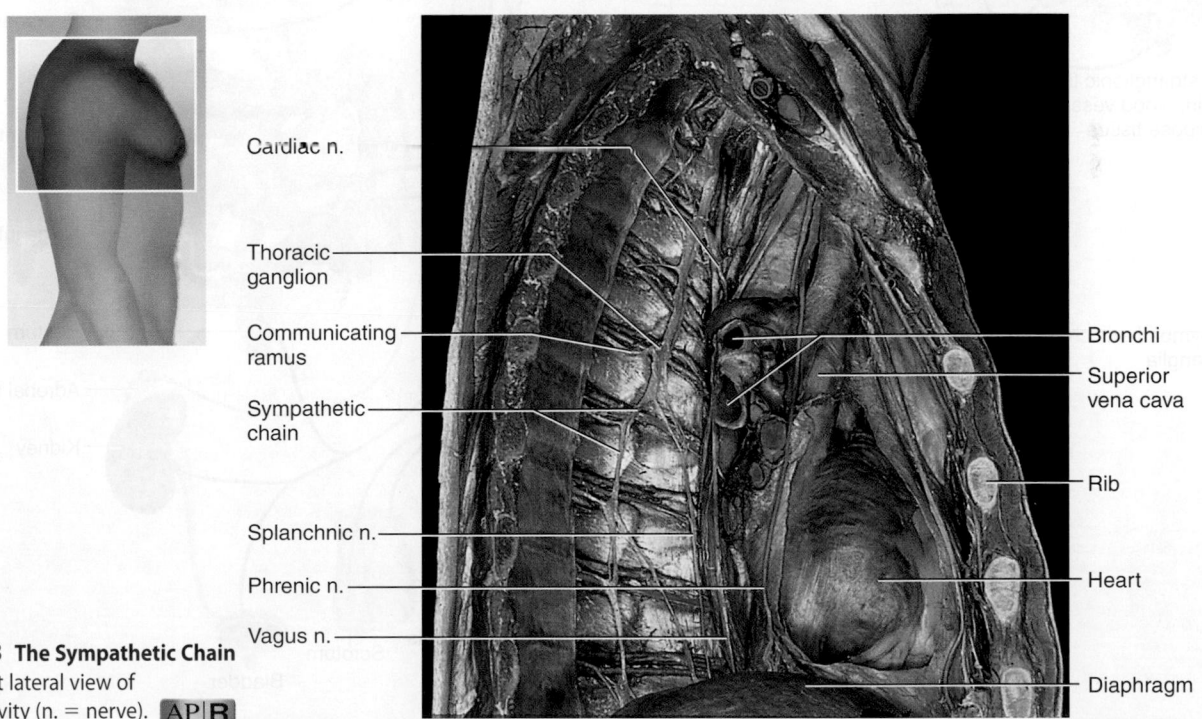

**FIGURE 15.3 The Sympathetic Chain Ganglia.** Right lateral view of the thoracic cavity (n. = nerve). **AP|R**

Cardiac n.
Thoracic ganglion
Communicating ramus
Sympathetic chain
Splanchnic n.
Phrenic n.
Vagus n.
Bronchi
Superior vena cava
Rib
Heart
Diaphragm

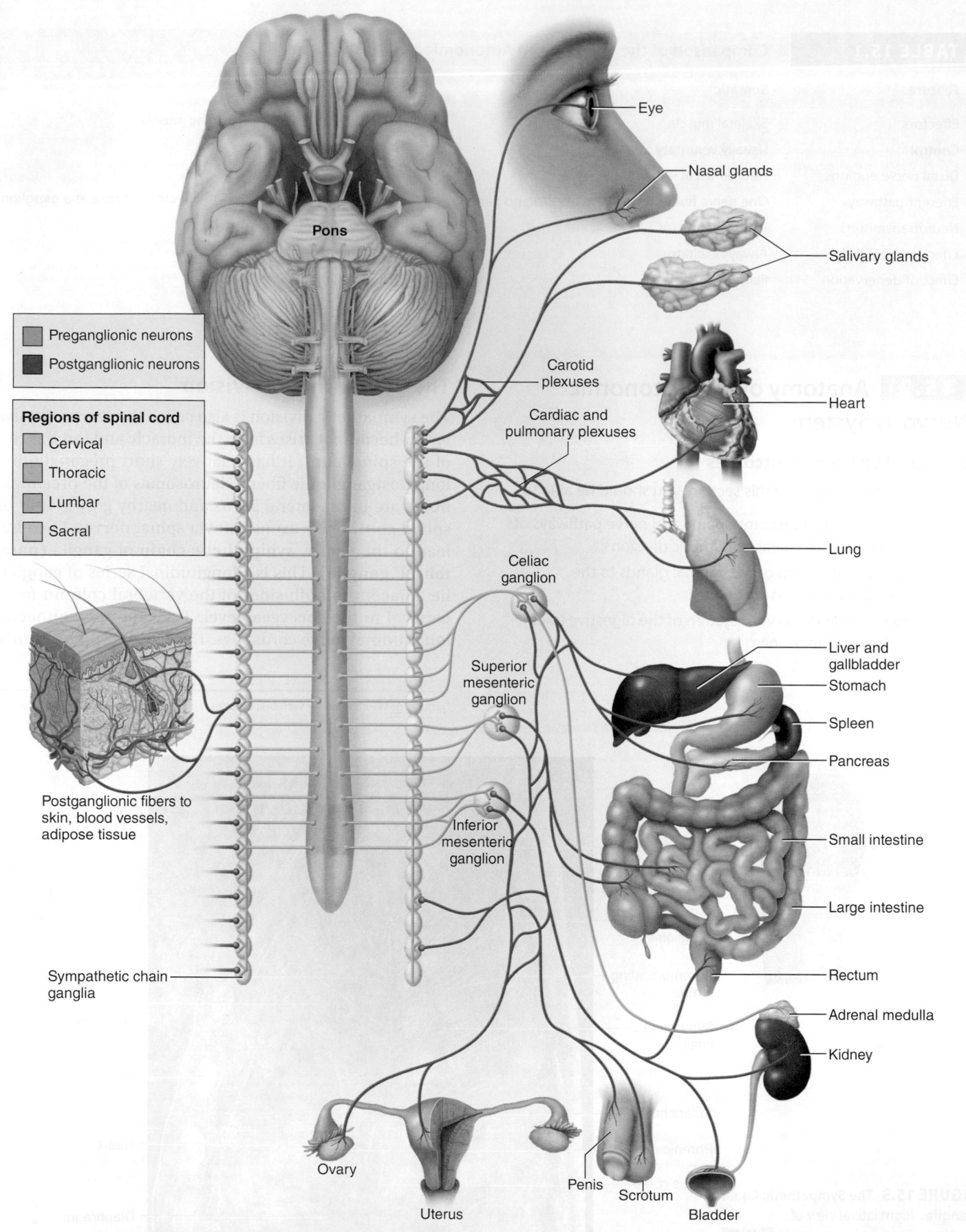

**Pons**

Preganglionic neurons
Postganglionic neurons

**Regions of spinal cord**
Cervical
Thoracic
Lumbar
Sacral

Postganglionic fibers to skin, blood vessels, adipose tissue

Sympathetic chain ganglia

Eye

Nasal glands

Salivary glands

Carotid plexuses

Cardiac and pulmonary plexuses

Heart

Lung

Celiac ganglion

Superior mesenteric ganglion

Inferior mesenteric ganglion

Liver and gallbladder

Stomach

Spleen

Pancreas

Small intestine

Large intestine

Rectum

Adrenal medulla

Kidney

Ovary

Uterus

Penis

Scrotum

Bladder

**FIGURE 15.4 Schematic of the Sympathetic Nervous System. AP|R**

❓ *Does the sympathetic innervation of the lungs make a person inhale and exhale? Explain.*

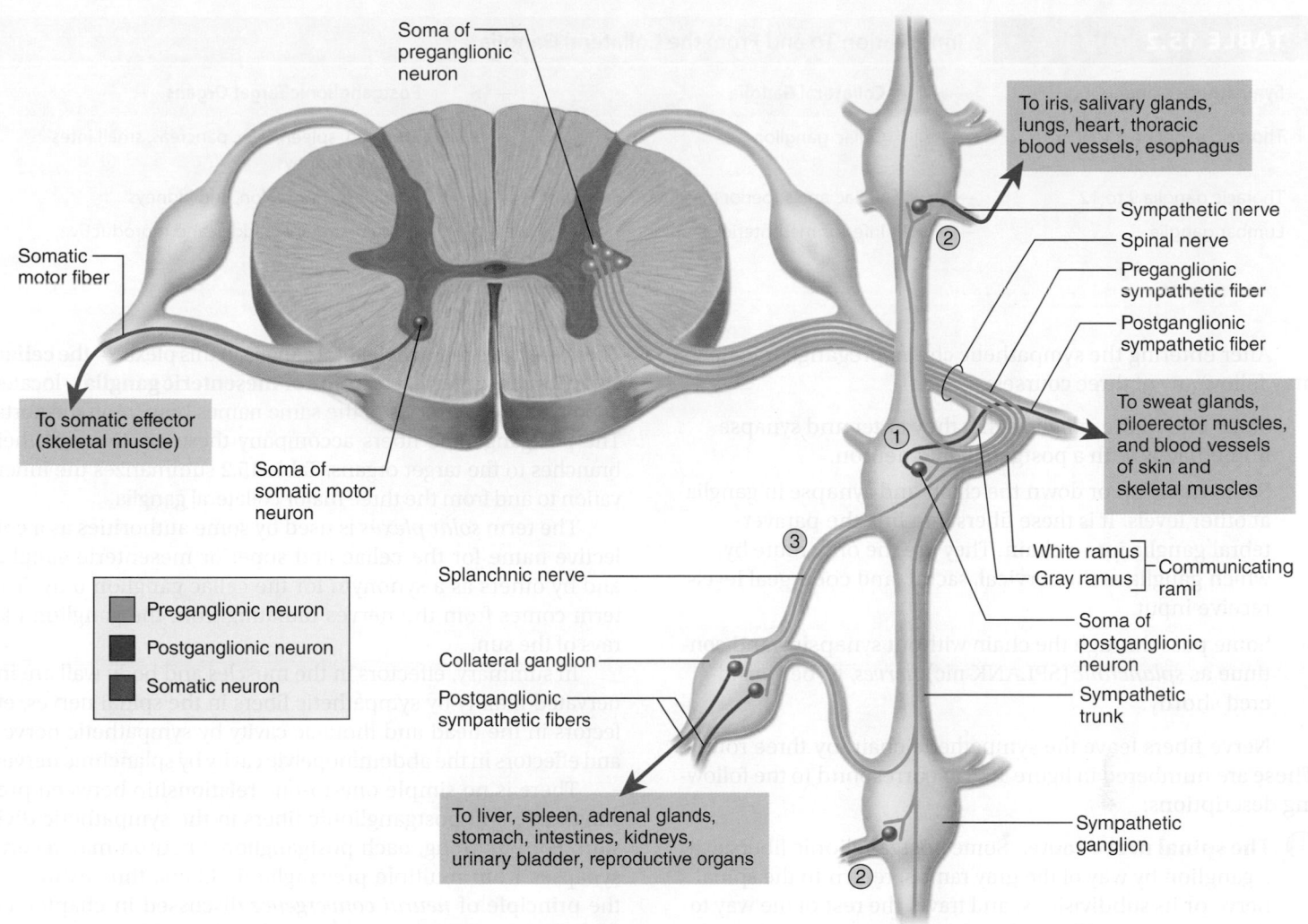

**FIGURE 15.5 Neural Pathways Through the Sympathetic Chain Ganglia.** Sympathetic fibers can follow any of the three numbered routes: (1) the spinal nerve route, (2) the sympathetic nerve route, or (3) the splanchnic nerve route. The somatic efferent pathway is shown on the left for comparison.

❓ *Name the parts of the spinal cord where the somas of the sympathetic and somatic efferent neurons are located.*

ganglia varies from person to person, but usually there are 3 cervical *(superior, middle,* and *inferior),* 11 thoracic, 4 lumbar, 4 sacral, and 1 coccygeal ganglion in each chain.

It may seem odd that sympathetic ganglia exist in the cervical, sacral, and coccygeal regions considering that sympathetic fibers arise only from the thoracic and lumbar regions of the spinal cord (levels T1 to L2). But as shown in figure 15.4, nerve cords from the thoracic region ascend to the ganglia in the neck, and cords from the lumbar region descend to the sacral and coccygeal ganglia. Consequently, sympathetic nerve fibers are distributed to every level of the body. As a general rule, the head receives sympathetic output arising from spinal cord segment T1, the neck from T2, the thorax and upper limbs from T3 to T6, the abdomen from T7 to T11, and the lower limbs from T12 to L2. There is considerable overlap and individual variation in this pattern, however.

In the thoracolumbar region, each paravertebral ganglion is connected to a spinal nerve by two branches called *communicating rami* (fig. 15.5). The preganglionic fibers are small myelinated fibers that travel from the spinal nerve to the ganglion by way of the **white communicating ramus,**[5] which gets its color and name from the myelin. Unmyelinated postganglionic fibers leave the ganglion by way of the **gray communicating ramus,** named for its lack of myelin and duller color, and by other routes. This ramus returns to the spinal nerve. Postganglionic fibers extend the rest of the way to the target organ.

#### ▶▶▶ APPLY WHAT YOU KNOW

*Would autonomic postganglionic fibers have faster or slower conduction speeds than somatic motor fibers? Why? (See hints in chapter 12.)*

---

[5]*ramus* = branch

| TABLE 15.2 | | Innervation To and From the Collateral Ganglia | | |
|---|---|---|---|---|
| **Sympathetic Ganglia** | $\longrightarrow$ | **Collateral Ganglia** | $\longrightarrow$ | **Postganglionic Target Organs** |
| Thoracic ganglion 5 to 9 or 10 | $\longrightarrow$ | Celiac ganglion | $\longrightarrow$ | Stomach, spleen, liver, pancreas, small intestine, and kidneys |
| Thoracic ganglia 9 to 12 | $\longrightarrow$ | Celiac and superior mesenteric ganglia | $\longrightarrow$ | Small intestine, colon, and kidneys |
| Lumbar ganglia | $\longrightarrow$ | Inferior mesenteric ganglion | $\longrightarrow$ | Rectum, urinary bladder, and reproductive organs |

After entering the sympathetic chain, preganglionic fibers may follow any of three courses:

- Some end in the ganglion that they enter and synapse immediately with a postganglionic neuron.

- Some travel up or down the chain and synapse in ganglia at other levels. It is these fibers that link the paravertebral ganglia into a chain. They are the only route by which ganglia at the cervical, sacral, and coccygeal levels receive input.

- Some pass through the chain without synapsing and continue as *splanchnic* (SPLANK-nic) *nerves,* to be considered shortly.

Nerve fibers leave the sympathetic chain by three routes. These are numbered in figure 15.5 to correspond to the following descriptions:

1. **The spinal nerve route.** Some postganglionic fibers exit a ganglion by way of the gray ramus, return to the spinal nerve or its subdivisions, and travel the rest of the way to the target organ. This is the route to most sweat glands, piloerector muscles, and blood vessels of the skin and skeletal muscles.

2. **The sympathetic nerve route.** Other postganglionic fibers leave by way of sympathetic nerves that extend to the heart, lungs, esophagus, and thoracic blood vessels. These nerves form a **carotid plexus** around each carotid artery of the neck and issue fibers from there to effectors in the head—including sweat, salivary, and nasal glands; piloerector muscles; blood vessels; and dilators of the iris. Some fibers from the superior and middle cervical ganglia form the *cardiac nerves* to the heart. (The cardiac nerves also contain parasympathetic fibers.)

3. **The splanchnic[6] nerve route.** Some of the fibers that arise from spinal nerves T5 to T12 pass through the sympathetic ganglia without synapsing. Beyond the ganglia, they continue as **splanchnic nerves,** which lead to a second set of ganglia called **collateral (prevertebral) ganglia.** Here the preganglionic fibers synapse with the postganglionics.

The collateral ganglia contribute to a network called the **abdominal aortic plexus** wrapped around the aorta (fig. 15.6).

There are three major collateral ganglia in this plexus—the **celiac, superior mesenteric,** and **inferior mesenteric ganglia**—located at points where arteries of the same names branch off the aorta. The postganglionic fibers accompany these arteries and their branches to the target organs. Table 15.2 summarizes the innervation to and from the three major collateral ganglia.

The term *solar plexus* is used by some authorities as a collective name for the celiac and superior mesenteric ganglia, and by others as a synonym for the celiac ganglion only. The term comes from the nerves radiating from the ganglion like rays of the sun.

In summary, effectors in the muscles and body wall are innervated mainly by sympathetic fibers in the spinal nerves, effectors in the head and thoracic cavity by sympathetic nerves, and effectors in the abdominopelvic cavity by splanchnic nerves.

There is no simple one-to-one relationship between preganglionic and postganglionic fibers in the sympathetic division. For one thing, each postganglionic neuron may receive synapses from multiple preganglionic fibers, thus exhibiting the principle of *neural convergence* discussed in chapter 12. Furthermore, each preganglionic fiber branches out to multiple postganglionic neurons, thus showing *neural divergence.* Most sympathetic preganglionic fibers synapse with 10 to 20 postganglionic neurons. Therefore, when one preganglionic neuron fires, it can excite multiple postganglionic fibers leading to different target organs. The sympathetic division thus tends to have relatively widespread effects—as suggested by the name *sympathetic.*[7]

## The Adrenal Glands

The paired **adrenal[8] (suprarenal) glands** rest like hats on the superior poles of the kidneys (fig. 15.6). Each adrenal is actually two glands with different functions and embryonic origins. The outer rind, the **adrenal cortex,** secretes steroid hormones discussed in chapter 17. The inner core, the **adrenal medulla,** is essentially a sympathetic ganglion. It consists of modified postganglionic neurons without dendrites or axons. Sympathetic preganglionic fibers penetrate through the cortex and terminate on these cells. The sympathetic nervous system and adrenal medulla are so closely related in development and function that they are referred to collectively as the *sympathoadrenal system.*

---

[6] *splanchn* = viscera

[7] *sym* = together; *path* = feeling
[8] *ad* = near; *ren* = kidney

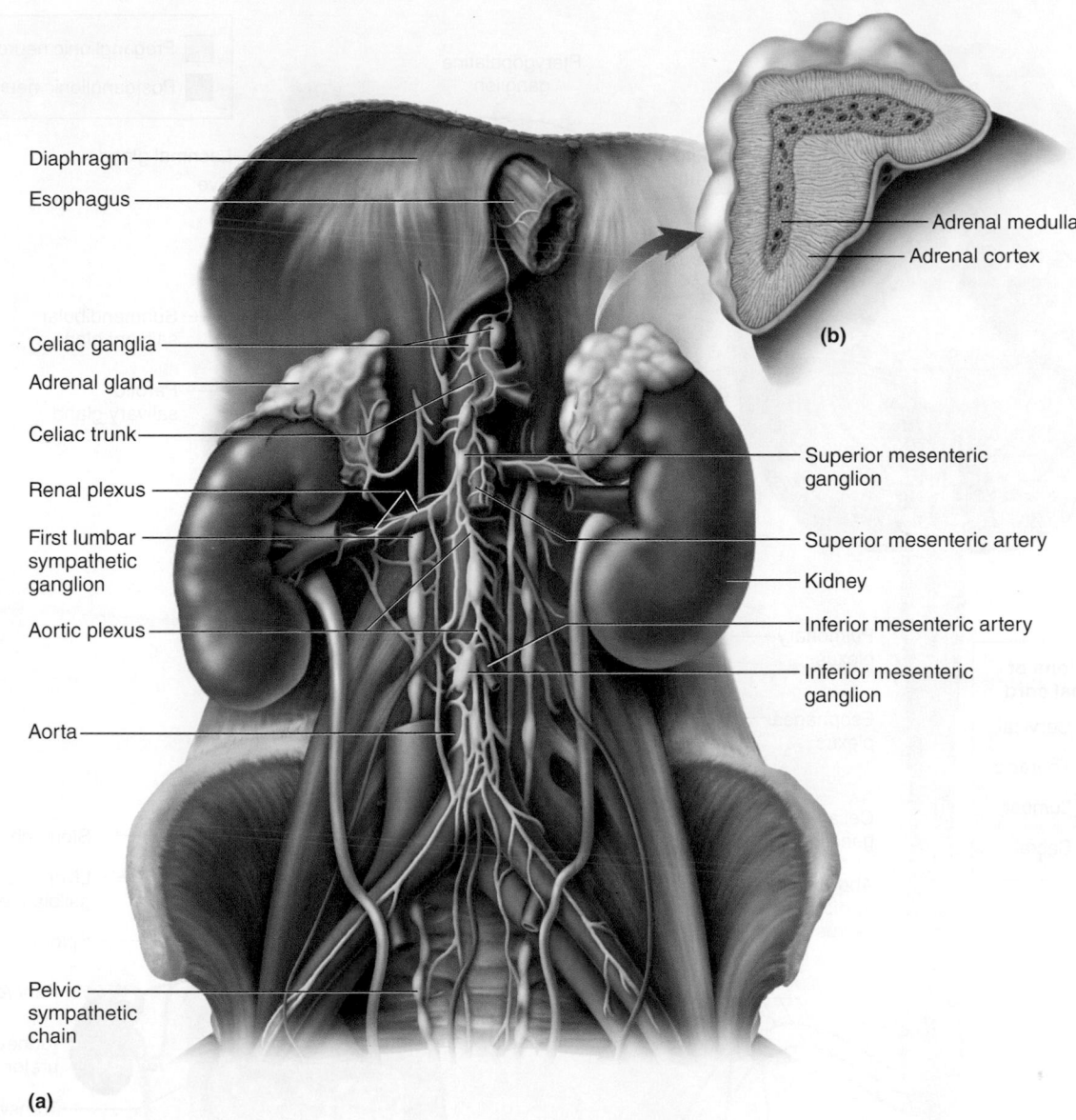

Diaphragm

Esophagus

Celiac ganglia

Adrenal gland

Celiac trunk

Renal plexus

First lumbar sympathetic ganglion

Aortic plexus

Aorta

Pelvic sympathetic chain

**(a)**

Adrenal medulla

Adrenal cortex

**(b)**

Superior mesenteric ganglion

Superior mesenteric artery

Kidney

Inferior mesenteric artery

Inferior mesenteric ganglion

**FIGURE 15.6 Abdominopelvic Components of the Sympathetic Nervous System.** (a) Collateral ganglia, abdominal aortic plexus, and adrenal glands. (b) The adrenal gland, frontal section. Only the adrenal medulla plays a role in the sympathetic nervous system; the adrenal cortex has unrelated roles described in chapter 17.

When stimulated, the adrenal medulla secretes a mixture of hormones into the bloodstream—about 85% epinephrine (adrenaline), 15% norepinephrine (noradrenaline), and a trace of dopamine. These hormones, the *catecholamines,* were briefly considered in chapter 12 because they also function as neurotransmitters.

## The Parasympathetic Division

The parasympathetic division is also called the *craniosacral division* because it arises from the brain and sacral region of the spinal cord; its fibers travel in certain cranial and sacral nerves. Somas of the preganglionic neurons are located in the midbrain, pons, medulla oblongata, and segments S2 to S4 of the

spinal cord (fig. 15.7). They issue long preganglionic fibers that end in **terminal ganglia** in or near the target organ. (If a terminal ganglion is embedded within the wall of a target organ, it is also called an *intramural*[9] *ganglion.*) Thus, the parasympathetic division has long preganglionic fibers, reaching almost all the way to the target cells, and short postganglionic fibers that cover the rest of the distance.

There is some neural divergence in the parasympathetic division, but much less than in the sympathetic. The parasympathetic division has a ratio of fewer than five postganglionic fibers to every preganglionic fiber. Furthermore, the preganglionic fiber reaches the target organ before even this slight divergence occurs. The parasympathetic division is therefore relatively selective in its stimulation of target organs.

[9]*intra* = within; *mur* = wall

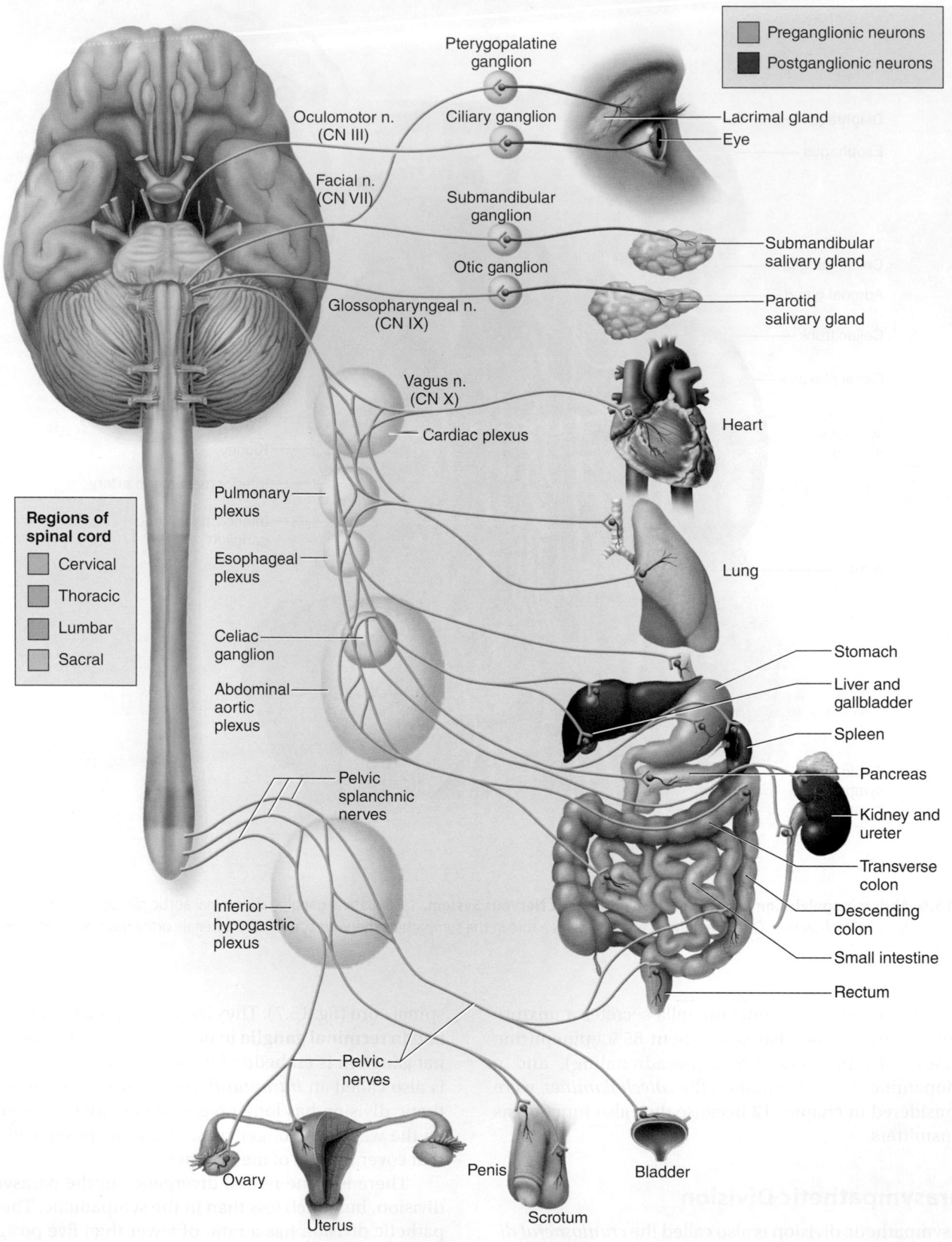

Pteryopalatine ganglion

Oculomotor n. (CN III)

Ciliary ganglion

Facial n. (CN VII)

Submandibular ganglion

Otic ganglion

Glossopharyngeal n. (CN IX)

Vagus n. (CN X)

Cardiac plexus

Pulmonary plexus

Esophageal plexus

Celiac ganglion

Abdominal aortic plexus

Pelvic splanchnic nerves

Inferior hypogastric plexus

Pelvic nerves

Preganglionic neurons
Postganglionic neurons

Lacrimal gland
Eye

Submandibular salivary gland

Parotid salivary gland

Heart

Lung

Stomach
Liver and gallbladder
Spleen
Pancreas
Kidney and ureter
Transverse colon
Descending colon
Small intestine
Rectum

**Regions of spinal cord**
Cervical
Thoracic
Lumbar
Sacral

Ovary
Uterus
Penis
Scrotum
Bladder

**FIGURE 15.7 Schematic of the Parasympathetic Nervous System.** AP|R

❓ *Which nerve carries the most parasympathetic nerve fibers?*

Parasympathetic fibers leave the brainstem in the following cranial nerves. The first three supply all parasympathetic innervation to the head, and the last one supplies viscera of the thoracic and abdominopelvic cavities.

1. **Oculomotor nerve (III).** The oculomotor nerve carries parasympathetic fibers that control the lens and pupil of the eye. The preganglionic fibers enter the orbit and terminate in the *ciliary ganglion* behind the eyeball. Postganglionic fibers enter the eyeball and innervate the *ciliary muscle,* which thickens the lens, and the *pupillary constrictor,* which narrows the pupil.

2. **Facial nerve (VII).** The facial nerve carries parasympathetic fibers that regulate the tear glands, salivary glands, and nasal glands. Soon after the facial nerve emerges from the pons, its parasympathetic fibers split away and form two smaller branches. The superior branch ends at the *pterygopalatine ganglion* near the junction of the maxilla and palatine bone. Postganglionic fibers then continue to the tear glands and glands of the nasal cavity, palate, and other areas of the oral cavity. The inferior branch crosses the middle-ear cavity and ends at the *submandibular ganglion* near the angle of the mandible. Postganglionic fibers from here supply salivary glands in the floor of the mouth.

3. **Glossopharyngeal nerve (IX).** The glossopharyngeal nerve carries parasympathetic fibers concerned with salivation. The preganglionic fibers leave this nerve soon after its origin and form the *tympanic nerve.* A continuation of this nerve crosses the middle-ear cavity and ends in the *otic*[10] *ganglion* near the foramen ovale. The postganglionic fibers then follow the trigeminal nerve to the *parotid salivary gland* just in front of the earlobe.

4. **Vagus nerve (X).** The vagus nerve carries about 90% of all parasympathetic preganglionic fibers. It travels down the neck and forms three networks in the mediastinum of the chest—the **cardiac plexus,** which supplies fibers to the heart; the **pulmonary plexus,** whose fibers accompany the bronchi and blood vessels into the lungs; and the **esophageal plexus,** whose fibers regulate swallowing.

At the lower end of the esophagus, these plexuses give off anterior and posterior **vagal trunks,** each of which contains fibers from both the right and left vagus nerves. These trunks penetrate the diaphragm, enter the abdominal cavity, and contribute to the extensive *abdominal aortic plexus* mentioned earlier. As we have seen, sympathetic fibers synapse here. The parasympathetic fibers, however, pass through the plexus without synapsing. They synapse farther along, in terminal ganglia in or near the liver, pancreas, stomach, small intestine, kidney, ureter, and proximal half of the colon.

The remaining parasympathetic fibers arise from levels S2 to S4 of the spinal cord. They travel a short distance in the anterior rami of the spinal nerves and then form **pelvic splanchnic nerves** that lead to the **inferior hypogastric plexus.** Some parasympathetic fibers synapse here, but most pass through this plexus and travel by way of **pelvic nerves** to the terminal ganglia in their target organs: the distal half of the colon, the rectum, urinary bladder, and reproductive organs. With few exceptions, the parasympathetic system does not innervate body wall structures (sweat glands, piloerector muscles, or cutaneous blood vessels).

The sympathetic and parasympathetic divisions of the ANS are compared in table 15.3.

▶▶▶ **APPLY WHAT YOU KNOW**

*Would autonomic functions be affected if the anterior roots of the cervical spinal nerves were damaged? Why or why not?*

## The Enteric Nervous System

The digestive tract has a nervous system of its own called the **enteric**[11] **nervous system.** Unlike the ANS proper, it does not arise from the brainstem or spinal cord, but like the ANS, it innervates smooth muscle and glands. Thus, opinions differ on whether it should be considered part of the ANS. It consists of about 100 million neurons embedded in the wall of the digestive tract (see photograph on p. 557)—perhaps more neurons than there are in the spinal cord—and it has its own reflex arcs. The enteric nervous system regulates the motility of the esophagus, stomach, and intestines and the secretion of digestive

---

[10]*ot* = ear; *ic* = pertaining to

[11]*enter* = intestines; *ic* = pertaining to

| **TABLE 15.3** | **Comparison of the Sympathetic and Parasympathetic Divisions** | |
|---|---|---|
| **Feature** | **Sympathetic** | **Parasympathetic** |
| Origin in CNS | Thoracolumbar | Craniosacral |
| Location of ganglia | Paravertebral ganglia adjacent to spinal column and prevertebral ganglia anterior to it | Terminal ganglia near or within target organs |
| Fiber lengths | Short preganglionic | Long preganglionic |
| | Long postganglionic | Short postganglionic |
| Neural divergence | Extensive | Minimal |
| Effects of system | Often widespread and general | More specific and local |

enzymes and acid. To function normally, however, these digestive activities also require regulation by the sympathetic and parasympathetic systems. The enteric nervous system is discussed in more detail in chapter 25. Its importance in intestinal motility becomes dramatically apparent when the system is absent (see Deeper Insight 15.1).

**BEFORE YOU GO ON**

Answer the following questions to test your understanding of the preceding section:

3. Explain why the sympathetic division is also called the thoracolumbar division even though its paravertebral ganglia extend all the way from the cervical to the sacral region.

4. Describe or diagram the structural relationships among the following: preganglionic fiber, postganglionic fiber, gray ramus, white ramus, and sympathetic ganglion.

5. Explain in anatomical terms why the parasympathetic division affects target organs more selectively than the sympathetic division does.

6. Trace the pathway of a parasympathetic fiber of the vagus nerve from the medulla oblongata to the small intestine.

**DEEPER INSIGHT 15.1**

**CLINICAL APPLICATION**

### Megacolon

The importance of the enteric nervous system becomes vividly clear when it is absent. Such is the case in a hereditary defect called *Hirschsprung disease*.[12] During normal embryonic development, neural crest cells migrate to the large intestine and establish the enteric nervous system. In Hirschsprung disease, however, they fail to supply the distal parts of the large intestine, leaving the sigmoid colon and rectum (see fig. 25.32, p. 985) without enteric ganglia. In the absence of these ganglia, the sigmoidorectal region lacks motility, constricts permanently, and obstructs the passage of feces. Feces accumulate and become impacted above the constriction, resulting in *megacolon*—a massive dilation of the bowel accompanied by abdominal distension and chronic constipation. The most life-threatening complications are colonic gangrene, perforation of the bowel, and bacterial infection of the peritoneum *(peritonitis)*. The treatment of choice is surgical removal of the affected segment and attachment of the healthy colon directly to the anal canal.

Hirschsprung disease is usually evident even in the newborn, which fails to have its first expected bowel movement. It affects four times as many infant boys as girls, and although its incidence in the general population is about 1 in 5,000 live births, it occurs in about 1 out of 10 infants with Down syndrome.

Hirschsprung disease is not the only cause of megacolon. In Central and South America, biting insects called *kissing bugs* transmit parasites called *trypanosomes* to humans. These parasites, similar to the ones that cause African sleeping sickness, cause *Chagas*[13] disease. Among other effects, they destroy the autonomic ganglia of the enteric nervous system, leading to a massively enlarged and often gangrenous colon.

---

[12] Harald Hirschsprung (1830–1916), Danish physician
[13] Carlos Chagas (1879–1934), Brazilian physician

## 15.3 Autonomic Effects on Target Organs

**Expected Learning Outcomes**

When you have completed this section, you should be able to

a. name the neurotransmitters employed at different synapses of the ANS;

b. name the receptors for these neurotransmitters and explain how they relate to autonomic effects;

c. explain how the ANS controls many target organs through dual innervation; and

d. explain how control is exerted in the absence of dual innervation.

### Neurotransmitters and Their Receptors

As noted earlier, the divisions of the ANS often have contrasting effects on an organ. The sympathetic division accelerates the heartbeat and the parasympathetic division slows it down, for example. But each division of the ANS also can have contrasting effects on different organs. For example, the parasympathetic division contracts the wall of the urinary bladder but relaxes the internal urethral sphincter; both actions are necessary for the expulsion of urine. It employs acetylcholine for both purposes. Similarly, the sympathetic division constricts most blood vessels but dilates the bronchioles of the lungs, and it achieves both effects with norepinephrine.

How can different autonomic neurons have such contrasting effects? There are two fundamental reasons: (1) Sympathetic and parasympathetic fibers secrete different neurotransmitters, and (2) target cells respond in different ways even to the same neurotransmitter depending on what type of receptors they have for it. All autonomic nerve fibers secrete either acetylcholine or norepinephrine, and each of these neurotransmitters has two major classes of receptors (fig. 15.8).

- **Acetylcholine (ACh).** ACh is secreted by the preganglionic fibers in both divisions and the postganglionic fibers of the parasympathetic division (table 15.4). A few sympathetic postganglionics also secrete ACh—those that innervate sweat glands and some blood vessels. Any nerve fiber that secretes ACh is called a **cholinergic** (CO-li-NUR-jic) **fiber,** and any receptor that binds it is called a **cholinergic receptor**. There are two categories of cholinergic receptors:

  - **Muscarinic** (MUSS-cuh-RIN-ic) **receptors.** These are named for muscarine, a mushroom toxin used in their discovery. All cardiac muscle, smooth muscle, and gland cells that receive cholinergic innervation have muscarinic receptors. There are different subclasses of muscarinic receptors with different

**(a) Parasympathetic fiber**

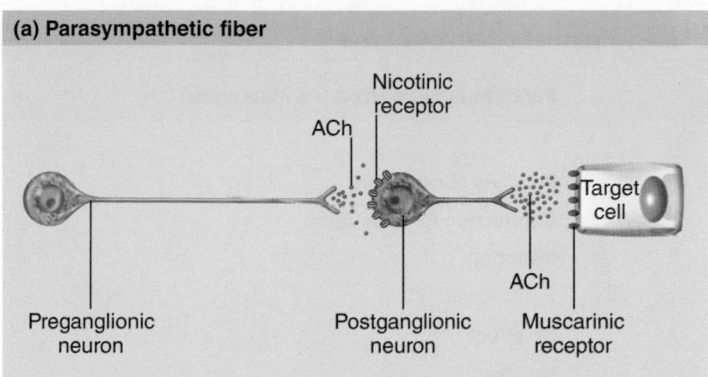

**(b) Sympathetic adrenergic fiber**

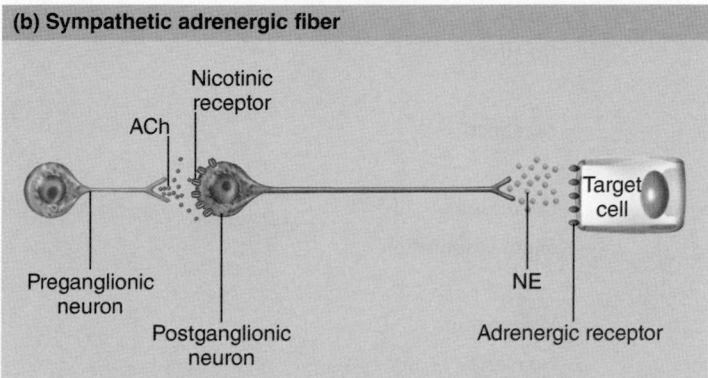

**(c) Sympathetic cholinergic fiber**

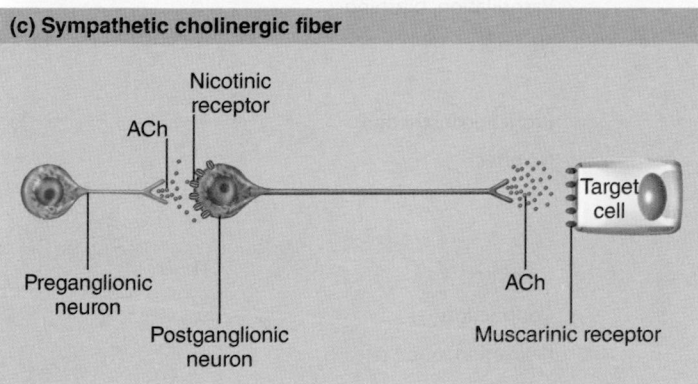

**FIGURE 15.8 Neurotransmitters and Receptors of the Autonomic Nervous System.** (a) All parasympathetic fibers are cholinergic. (b) Most sympathetic postganglionic fibers are adrenergic; they secrete norepinephrine (NE), and the target cell bears adrenergic receptors. (c) A few sympathetic postganglionic fibers are cholinergic; they secrete acetylcholine (ACh), and the target cell has cholinergic receptors of the muscarinic class.

| TABLE 15.4 | Locations of Cholinergic and Adrenergic Fibers in the ANS | |
| --- | --- | --- |
| **Division** | **Preganglionic Fibers** | **Postganglionic Fibers** |
| Sympathetic | Always cholinergic | Mostly adrenergic; a few cholinergic |
| Parasympathetic | Always cholinergic | Always cholinergic |

stimulate the postganglionic cells; on cells of the adrenal medulla; and at the neuromuscular junctions of skeletal muscle fibers. The binding of ACh to a nicotinic receptor is always excitatory. Nicotinic receptors work by opening ligand-gated ion channels and producing an excitatory postsynaptic potential in the target cell.

- **Norepinephrine (NE).** This neurotransmitter is secreted by nearly all sympathetic postganglionic fibers (table 15.4). Nerve fibers that secrete it are called **adrenergic fibers,** and the receptors for it are called **adrenergic receptors.** (NE is also called noradrenaline, the origin of the term *adrenergic.*) There are two principal categories of NE receptors:

  - **α-adrenergic receptors.** These usually have excitatory effects. For example, the binding of NE to α-adrenergic receptors promotes labor contractions, stimulates piloerection, and constricts dermal blood vessels, yet it inhibits intestinal motility. These contrasting effects result from the different actions of two subclasses of α-adrenergic receptors—$\alpha_1$ and $\alpha_2$. Receptors of the $\alpha_1$ type act through calcium ions as a second messenger, whereas $\alpha_2$ receptors inhibit the synthesis of cyclic AMP (cAMP).

  - **β-adrenergic receptors.** These are usually inhibitory. For example, NE relaxes and dilates the bronchioles (thus enhancing respiratory airflow) when it binds to β-adrenergic receptors of the smooth muscle. Yet when it binds to the β-adrenergic receptors of cardiac muscle, it has an excitatory effect. Such contrasting effects—increased pulmonary airflow and a stronger, faster heartbeat—are obviously appropriate to a state of exercise. Here again there are two receptor subclasses, $\beta_1$ and $\beta_2$, which mediate different effects. Both types, however, act through cAMP as a second messenger.

effects; thus ACh excites intestinal smooth muscle by binding to one type of muscarinic receptor, and inhibits cardiac muscle by binding to a different type. Muscarinic receptors work through a variety of second-messenger systems.

- **Nicotinic** (NIC-oh-TIN-ic) **receptors.** These are named for another botanical toxin helpful to their discovery—nicotine. They occur at all synapses in the autonomic ganglia, where the preganglionic fibers

Table 15.5 summarizes the effects of sympathetic and parasympathetic stimulation on different target organs and shows how some of these effects hinge on the receptor type. Knowledge of these receptor types is also vital to the field of

| TABLE 15.5 | Effects of the Sympathetic and Parasympathetic Nervous Systems | |
|---|---|---|
| **Target** | **Sympathetic Effect and Receptor Type** | **Parasympathetic Effect (All Muscarinic)** |
| **Eye** | | |
| Iris | Pupillary dilation (α) | Pupillary constriction |
| Ciliary muscle and lens | Relaxation for far vision (β) | Contraction for near vision |
| Lacrimal (tear) gland | None | Secretion |
| **Integumentary system** | | |
| Merocrine sweat glands (cooling) | Secretion (muscarinic) | No effect |
| Apocrine sweat glands (scent) | Secretion (α) | No effect |
| Piloerector muscles | Hair erection (α) | No effect |
| *Adipose tissue* | Decreased fat breakdown (α) | No effect |
| | Increased fat breakdown (α, β) | |
| *Adrenal medulla* | Hormone secretion (nicotinic) | No effect |
| **Circulatory system** | | |
| Heart rate and force | Increased (β) | Decreased |
| Deep coronary arteries | Vasodilation (β) | Slight vasodilation |
| | Vasoconstriction (α) | |
| Blood vessels of most viscera | Vasoconstriction (α) | Vasodilation |
| Blood vessels of skeletal muscles | Vasodilation (β) | No effect |
| Blood vessels of skin | Vasoconstriction (α) | Vasodilation, blushing |
| Platelets (blood clotting) | Increased clotting (α) | No effect |
| **Respiratory system** | | |
| Bronchi and bronchioles | Bronchodilation (β) | Bronchoconstriction |
| Mucous glands | Decreased secretion (α) | No effect |
| | Increased secretion (β) | |
| **Urinary system** | | |
| Kidneys | Reduced urine output (α) | No effect |
| Bladder wall | No effect | Contraction |
| Internal urinary sphincter | Contraction, urine retention (α) | Relaxation, urine release |
| **Digestive system** | | |
| Salivary glands | Thick mucous secretion (α) | Thin serous secretion |
| Gastrointestinal motility | Decreased (α, β) | Increased |
| Gastrointestinal secretion | Decreased (α) | Increased |
| Liver | Glycogen breakdown (α, β) | Glycogen synthesis |
| Pancreatic enzyme secretion | Decreased (α) | Increased |
| Pancreatic insulin secretion | Decreased (α) | No effect |
| | Increased (β) | |
| **Reproductive system** | | |
| Penile or clitoral erection | No effect | Stimulation |
| Glandular secretion | No effect | Stimulation |
| Orgasm, smooth muscle roles | Stimulation (α) | No effect |
| Uterus | Relaxation (β) | No effect |
| | Labor contractions (α) | |

neuropharmacology (see Deeper Insight 15.2). Many naturally occurring drugs bind selectively to one or another class or subclass of receptor. Atropine binds only to muscarinic receptors and curare only to nicotinic receptors, for example. Many synthetic drugs are designed to be similarly selective.

The autonomic effects on glandular secretion are often an indirect result of their effect on the blood vessels. Many glandular secretions begin as a filtrate of the blood, which is then modified by the gland cells. Increasing the blood flow through a gland (such as a salivary or sweat gland) tends to increase secretion, and reducing the blood flow reduces secretion.

Sympathetic effects tend to last longer than parasympathetic effects. After a parasympathetic fiber secretes ACh into a synapse, it is quickly broken down by acetylcholinesterase (AChE) and its effect lasts only a few seconds. The NE released by a sympathetic fiber, however, has various fates: (1) Some is reabsorbed by the nerve fiber, where it is either reused or broken down by an enzyme called *monoamine oxidase (MAO).* (2) Some diffuses into the adjacent tissues, where it is degraded by another enzyme, *catechol-O-methyltransferase (COMT).* (3) Much of it passes into the bloodstream, where MAO and COMT are absent. This NE and epinephrine from the adrenal gland circulate throughout the body and may exert their effects for several minutes before they are finally degraded by the liver.

ACh and NE are not the only neurotransmitters employed by the ANS. Although all autonomic fibers secrete one of these, many of them also secrete neuropeptides that modulate ACh or NE function. Sympathetic fibers may also secrete enkephalin, substance P, neuropeptide Y, somatostatin, neurotensin, or gonadotropin-releasing hormone. Some parasympathetic fibers relax blood vessels by stimulating the endothelial cells to release the gas nitric oxide (NO). NO inhibits smooth muscle tone in the vessel wall, thus allowing the vessel to dilate. This increases the blood flow through the vessel. Among other functions, this mechanism is crucial to penile erection (see Deeper Insight 27.4, p. 1051).

#### ▶▶▶ APPLY WHAT YOU KNOW

*Table 15.5 notes that the sympathetic nervous system has an α-adrenergic effect on blood platelets and promotes clotting. How can the sympathetic nervous system stimulate platelets, considering that platelets are drifting cell fragments in the bloodstream with no nerve fibers leading to them?*

## Dual Innervation

Most of the viscera receive nerve fibers from both the sympathetic and parasympathetic divisions and thus are said to have **dual innervation.** In such cases, the two divisions may have either *antagonistic* or *cooperative* effects on the same organ.

**Antagonistic effects** oppose each other. For example, the sympathetic division speeds up the heart and the parasympathetic division slows it down; the sympathetic division inhibits digestion and the parasympathetic division stimulates it; the sympathetic division dilates the pupil and the parasympathetic division constricts it. In some cases, these effects are exerted through dual innervation of the same effector cells, as in the heart, where nerve fibers of both divisions terminate on the same muscle cells. In other cases, antagonistic effects arise because each division innervates different effector cells with opposite effects on organ function. In the iris of the eye, for example, sympathetic fibers innervate the pupillary dilator and parasympathetic fibers innervate the constrictor (fig. 15.9).

**Cooperative effects** are seen when the two divisions act on different effectors to produce a unified overall effect. Salivation is a good example. The parasympathetic division stimulates serous cells of the salivary glands to secrete a watery, enzyme-rich secretion, while the sympathetic division stimulates mucous cells of the same glands to secrete mucus. The enzymes and mucus are both necessary components of the saliva.

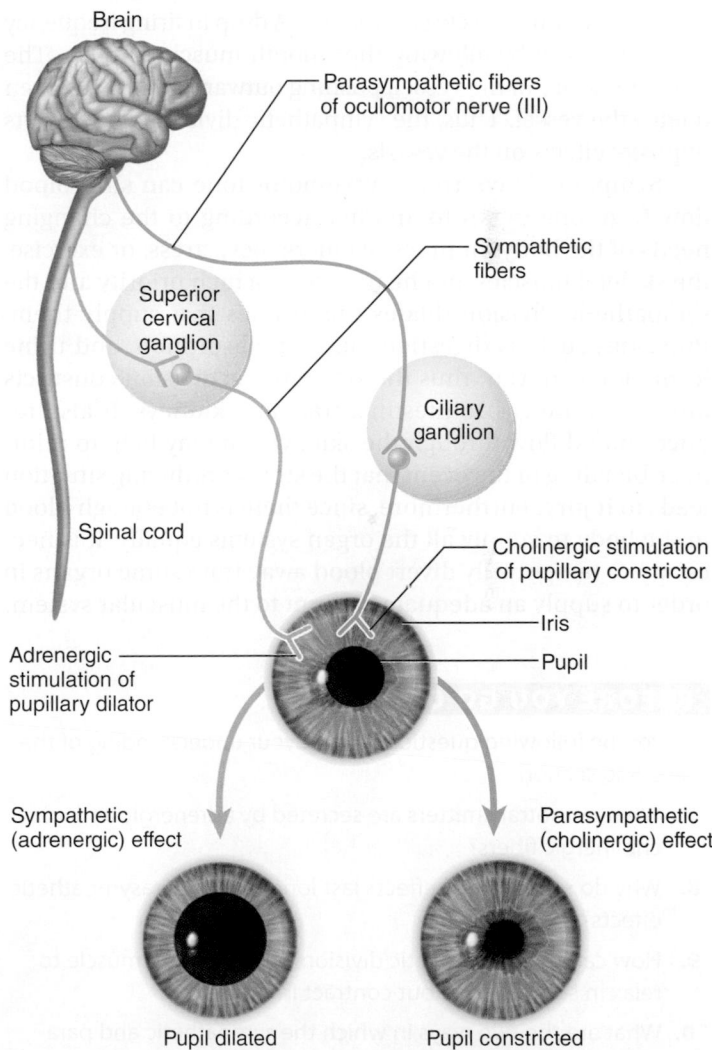

**FIGURE 15.9 Dual Innervation of the Iris.** Shows antagonistic effects of the sympathetic (yellow) and parasympathetic (blue) divisions on the iris.

❓ *If a person is in a state of fear, would you expect the pupils to be dilated or constricted? Why?*

Even when both divisions innervate a single organ, they do not always innervate it equally or exert equal influence. For example, the parasympathetic division forms an extensive plexus in the wall of the digestive tract and exerts much more influence over it than the sympathetic division does. In the ventricles of the heart, by contrast, there is much less parasympathetic than sympathetic innervation.

## Control Without Dual Innervation

Dual innervation is not always necessary for the ANS to produce opposite effects on an organ. The adrenal medulla, piloerector muscles, sweat glands, and many blood vessels receive only sympathetic fibers. The most significant example of control without dual innervation is regulation of blood pressure and routes of blood flow. The sympathetic fibers to a blood vessel have a baseline sympathetic tone, which keeps the vessels in a state of partial constriction called **vasomotor tone** (fig. 15.10). An increase in firing rate constricts a vessel by increasing smooth muscle contraction. A drop in firing frequency dilates a vessel by allowing the smooth muscle to relax. The blood pressure in the vessel, pushing outward on its wall, then dilates the vessel. Thus, the sympathetic division alone exerts opposite effects on the vessels.

Sympathetic control of vasomotor tone can shift blood flow from one organ to another according to the changing needs of the body. In times of emergency, stress, or exercise, the skeletal muscles and heart receive a high priority and the sympathetic division dilates the arteries that supply them. Processes such as digestion, nutrient absorption, and urine formation can wait; thus the sympathetic division constricts arteries to the gastrointestinal tract and kidneys. It also reduces blood flow through the skin, which may help to minimize bleeding in the event that the stress-producing situation leads to injury. Furthermore, since there is not enough blood in the body to supply all the organ systems equally, it is necessary to temporarily divert blood away from some organs in order to supply an adequate amount to the muscular system.

### BEFORE YOU GO ON

Answer the following questions to test your understanding of the preceding section:

7. What neurotransmitters are secreted by adrenergic and cholinergic fibers?

8. Why do sympathetic effects last longer than parasympathetic effects?

9. How can the sympathetic division cause smooth muscle to relax in some organs but contract in others?

10. What are the two ways in which the sympathetic and parasympathetic systems can affect each other when they both innervate the same target organ? Give examples.

11. How can the sympathetic nervous system have contrasting effects in a target organ without dual innervation?

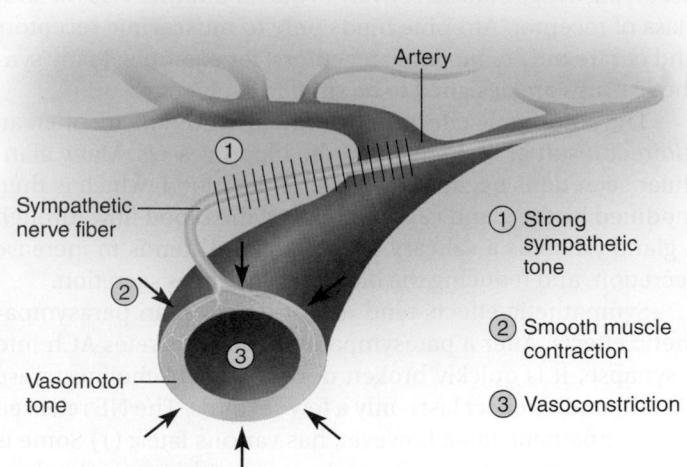

(a) **Vasoconstriction**

① Strong sympathetic tone

② Smooth muscle contraction

③ Vasoconstriction

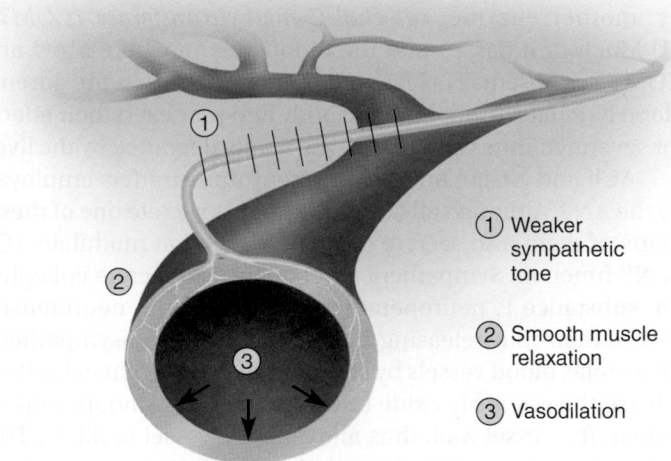

(b) **Vasodilation**

**FIGURE 15.10 Sympathetic and Vasomotor Tone.** (a) Vasoconstriction in response to a high rate of sympathetic nerve firing. (b) Vasodilation in response to a low rate of sympathetic nerve firing. Smooth muscle relaxation allows blood pressure within the vessel to push the vessel wall outward. Black lines crossing each nerve fiber represent action potentials, with a high firing frequency in part (a) and a lower frequency in part (b).

---

## 15.4 Central Control of Autonomic Function

### Expected Learning Outcome

When you have completed this section, you should be able to

a. describe how the autonomic nervous system is influenced by the central nervous system.

In spite of its name, the ANS is not an independent nervous system. In this section we briefly consider how it is influenced by various levels of the central nervous system.

- **Cerebral cortex.** Even if we usually cannot consciously control the ANS, it is clear that the mind does influence it. Anger raises the blood pressure, fear makes the heart race, thoughts of good food make the stomach rumble, sexual thoughts or images increase blood flow to the genitals, and anxiety inhibits sexual function. The limbic system (p. 529), an ancient part of the cerebral cortex, is involved in many emotional responses and has extensive connections with the hypothalamus, a site of several nuclei of autonomic control. Thus, the limbic system provides a pathway connecting sensory and mental experiences with the autonomic nervous system.

- **Hypothalamus.** Although the major site of CNS control over the somatic motor system is the primary motor cortex, the major control center of the visceral motor system is the hypothalamus. This small but vital region in the floor of the brain contains many nuclei for primitive functions, including hunger, thirst, thermoregulation, emotions, and sexuality. Artificial stimulation of different regions of the hypothalamus can activate the fight-or-flight response typical of the sympathetic nervous system or have the calming effects typical of the parasympathetic. Output from the hypothalamus travels largely to nuclei in more caudal regions of the brainstem, and from there to the cranial nerves and the sympathetic neurons in the spinal cord.

- **Midbrain, pons, and medulla oblongata.** These regions of the brainstem house numerous autonomic nuclei described in chapter 14: centers for cardiac and vasomotor control, salivation, swallowing, sweating, gastrointestinal secretion, bladder control, pupillary constriction and dilation, and other primitive functions. Many of these nuclei belong to the reticular formation, which extends from the medulla to the hypothalamus. Autonomic output from these nuclei travels by way of the spinal cord and the oculomotor, facial, glossopharyngeal, and vagus nerves.

- **Spinal cord.** Finally, the spinal cord integrates such autonomic reflexes as micturition (urination), defecation, erection, and ejaculation (details are in chapters 23, 25, and 27). Fortunately, the brain is able to inhibit defecation and urination consciously, but when injuries sever the spinal cord from the brain, the autonomic spinal reflexes alone control the elimination of urine and feces.

Table 15.6 describes some dysfunctions of the autonomic nervous system.

**BEFORE YOU GO ON**

Answer the following questions to test your understanding of the preceding section:

12. What system in the brain connects our conscious thoughts and feelings with the autonomic control centers of the hypothalamus?

13. List some autonomic responses that are controlled by nuclei in the hypothalamus.

14. What are the roles of the midbrain, pons, and medulla in autonomic control?

15. Name some visceral reflexes controlled by the spinal cord.

| **TABLE 15.6** | **Some Disorders of the Autonomic Nervous System** |
|---|---|
| Horner syndrome | Chronic unilateral pupillary constriction, sagging of the eyelid, withdrawal of the eye into the orbit, flushing of the skin, and lack of facial perspiration, resulting from lesions in the cervical ganglia, upper thoracic spinal cord, or brainstem that interrupt sympathetic innervation of the head. |
| Raynaud disease | Intermittent attacks of paleness, cyanosis, and pain in the fingers and toes, caused when cold or emotional stress triggers excessive vasoconstriction in the digits; most common in young women. In extreme cases, causes gangrene and may require amputation. Sometimes treated by severing sympathetic nerves to the affected regions. |
| *Disorders described elsewhere* | |
| Autonomic effects of cranial nerve injuries pp. 544, 549 | |
| Mass reflex reaction p. 503 | |
| Orthostatic hypotension p. 794 | |

# DEEPER INSIGHT 15.2
## CLINICAL APPLICATION

### Drugs and the Nervous System

*Neuropharmacology* is a branch of medicine that deals with the effects of drugs on the nervous system, especially drugs that mimic, enhance, or inhibit the action of neurotransmitters. A few examples will illustrate the clinical relevance of neurotransmitter and receptor functions.

A number of drugs work by stimulating adrenergic and cholinergic neurons or receptors. *Sympathomimetics*[14] are drugs that enhance sympathetic action. They stimulate adrenergic receptors or promote norepinephrine release. For example, phenylephrine, found in such cold medicines as Dimetapp and Sudafed PE, aids breathing by stimulating alpha receptors, dilating the bronchioles, and constricting nasal blood vessels, thus reducing swelling in the nasal mucosa. *Sympatholytics*[15] are drugs that suppress sympathetic action by inhibiting norepinephrine release or by binding to adrenergic receptors without stimulating them. Propranolol, for example, is a *beta-blocker*. It reduces hypertension partly by blocking β-adrenergic receptors. This interferes with the effects of epinephrine and norepinephrine on the heart and blood vessels. (It also reduces the production of *angiotensin II,* a hormone that stimulates vasoconstriction and raises blood pressure.)

*Parasympathomimetics* enhance parasympathetic effects. Pilocarpine, for example, relieves glaucoma (excessive pressure in the eyeball) by dilating a vessel that drains fluid from the eye. *Parasympatholytics* inhibit ACh release or block its receptors. Atropine, for example, blocks muscarinic receptors. It is sometimes used to dilate the pupils for eye examinations and to dry the mucous membranes of the respiratory tract before inhalation anesthesia. It is an extract of the deadly nightshade plant, *Atropa belladonna.*[16] Women of the Middle Ages used nightshade to dilate their pupils, which was regarded as a beauty enhancement.

The drugs we have mentioned so far act on the peripheral nervous system and its effectors. Many others act on the central nervous system. Strychnine, for example, blocks the inhibitory action of glycine on spinal motor neurons. The neurons then overstimulate the muscles, causing spastic paralysis and sometimes death by suffocation.

Sigmund Freud predicted that psychiatry would eventually draw upon biology and chemistry to deal with emotional problems once treated only by counseling and psychoanalysis. A branch of neuropharmacology called *psychopharmacology* has fulfilled his prediction. This field dates to the 1950s when chlorpromazine, an antihistamine, was incidentally found to relieve the symptoms of schizophrenia.

The management of clinical depression is one example of how contemporary psychopharmacology has supplemented counseling approaches. Some forms of depression result from deficiencies of the monoamine neurotransmitters. Thus, they yield to drugs that prolong the effects of the monoamines already present at the synapses. One of the earliest-discovered antidepressants was imipramine, which blocks the synaptic reuptake of serotonin and norepinephrine. However, it produces undesirable side effects such as dry mouth and irregular cardiac rhythms. It has been largely replaced by Prozac (fluoxetine), which blocks serotonin reuptake and prolongs its mood-elevating effect; thus Prozac is called a *selective serotonin reuptake inhibitor (SSRI).* It is also used to treat fear of rejection, excess sensitivity to criticism, lack of self-esteem, and inability to experience pleasure, all of which were long handled only through counseling, group therapy, or psychoanalysis. After monoamines are taken up from the synapse, they are degraded by monoamine oxidase (MAO). Drugs called *MAO inhibitors* interfere with the breakdown of monoamine neurotransmitters and provide another pharmacological approach to depression.

Our growing understanding of neurochemistry also gives us more insight into the action of addictive drugs of abuse such as amphetamines and cocaine. Amphetamines ("speed") chemically resemble norepinephrine and dopamine, two neurotransmitters associated with elevated mood. Dopamine is especially important in sensations of pleasure. Cocaine blocks dopamine reuptake and thus produces a brief rush of good feelings. But when dopamine is not reabsorbed by the neurons, it diffuses out of the synaptic cleft and is degraded elsewhere. Cocaine thus depletes the neurons of dopamine faster than they can synthesize it, so that finally there is no longer an adequate supply to maintain normal mood. The postsynaptic neurons make new dopamine receptors as if "searching" for the neurotransmitter—all of which leads ultimately to anxiety, depression, and the inability to experience pleasure without the drug.

Caffeine exerts its stimulatory effect by competing with adenosine. Adenosine, which you know as a component of DNA, RNA, and ATP, also functions as a neuromodulator in the brain, inhibiting ACh release by cholinergic neurons. One theory of sleepiness is that it results when prolonged metabolic activity breaks down so much ATP that the accumulated adenosine has a noticeably inhibitory effect. Caffeine has enough structural similarity to adenosine (fig. 15.11) to bind to its receptors, but it does not produce the inhibitory effect. Thus, it prevents adenosine from exerting its effect, more ACh is secreted, and a person feels more alert.

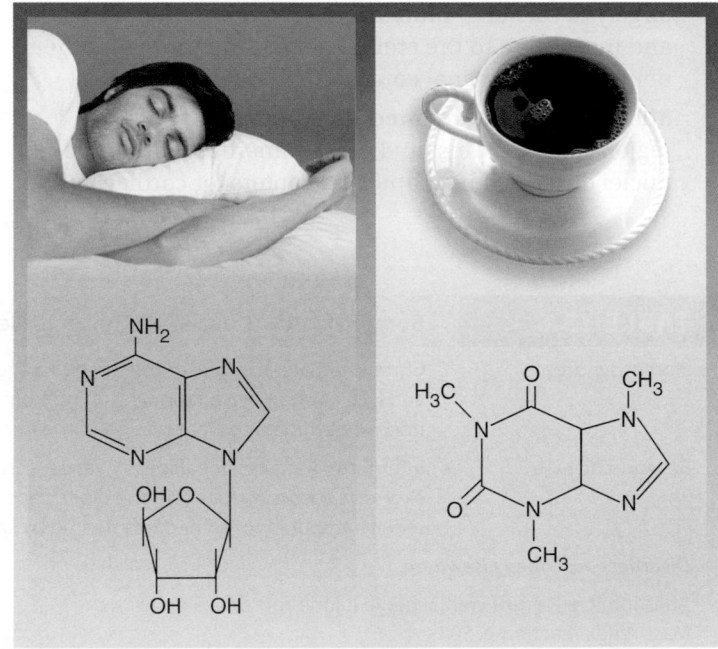

**FIGURE 15.11  Adenosine and Caffeine.** Adenosine is an inhibitory neuromodulator that inhibits ACh release and produces a sense of sleepiness. Caffeine is similar enough in structure to bind to adenosine receptors and block the action of adenosine. This results in increased ACh release and heightened arousal.

---

[14]*mimet* = imitate, mimic
[15]*lyt* = break down, destroy
[16]*bella* = beautiful, fine; *donna* = woman

# STUDY GUIDE

## ▶ Assess Your Learning Outcomes

*To test your knowledge, discuss the following topics with a study partner or in writing, ideally from memory.*

### 15.1 General Properties of the Autonomic Nervous System (p. 558)

1. The fundamental function and effectors of the autonomic nervous system (ANS)
2. Why this system is called *autonomic*; how it differs from the somatic motor system
3. The fundamental, contrasting functions of the sympathetic and parasympathetic divisions of the ANS
4. Why it cannot be said that at any given moment, either the sympathetic or the parasympathetic division is active; what is meant by autonomic tone
5. Anatomical components of the ANS
6. How autonomic efferent pathways differ from somatic efferent pathways; the meaning of *preganglionic* and *postganglionic* fibers

### 15.2 Anatomy of the Autonomic Nervous System (p. 561)

1. Origin of the sympathetic preganglionic fibers and the routes they take to the sympathetic chain ganglia
2. Anatomy of the sympathetic chain; the number of ganglia at its various levels; and the body regions supplied by nerve fibers issuing from each group of ganglia
3. The gray and white communicating rami that connect the sympathetic ganglia to the spinal nerves; the reason they are named *gray* and *white*; and the path that sympathetic nerve fibers take through these rami
4. Differences between the spinal nerve route, sympathetic nerve route, and splanchnic nerve route by which fibers leave the sympathetic chain
5. Various places in which a sympathetic preganglionic fiber may synapse with a postganglionic neuron
6. Locations of the celiac, superior mesenteric, and inferior mesenteric ganglia; the collective name for them; and the varied meanings of the expression *solar plexus*
7. The degree and significance of neural divergence in the sympathetic nervous system, and the effect of this on target organ stimulation
8. Why the adrenal medulla can be considered part of the sympathetic nervous system; what products it secretes when stimulated
9. Names and numbers of the cranial and spinal nerves that carry preganglionic fibers of the parasympathetic nervous system; which nerve carries the largest percentage of parasympathetic fibers
10. The path of the vagus nerve, and the names and locations of the plexuses and trunks to which it gives rise
11. Where, in general, terminal ganglia of the parasympathetic division are found, and therefore where the postganglionic fibers begin
12. The location and functions of the enteric nervous system

### 15.3 Autonomic Effects on Target Organs (p. 568)

1. Why the autonomic effect on a target cell depends on both the neurotransmitter released and the type of receptor on the target cell
2. The difference between cholinergic and adrenergic fibers and where each can be found in the ANS
3. What nicotinic and muscarinic receptors have in common, how they differ, and where they occur in the ANS
4. What α- and β-adrenergic receptors have in common, how they differ, and where they occur in the ANS
5. Neurotransmitters stability and how it relates to the duration of sympathetic versus parasympathetic effects
6. The variety of neurotransmitters and neuromodulators employed by them
7. Autonomic control of certain organs by dual innervation, and examples of antagonistic and cooperative effects on an organ
8. How the ANS can regulate organs that do not have dual innervation

### 15.4 Central Control of Autonomic Function (p. 572)

1. Examples of the influence of the cerebral cortex, hypothalamus, midbrain, pons, medulla oblongata, and spinal cord on the autonomic nervous system, and their involvement in autonomic effects

## ▶ Testing Your Recall

*Answers in Appendix B*

1. The autonomic nervous system innervates all of these *except*
   a. cardiac muscle.
   b. skeletal muscle.
   c. smooth muscle.
   d. salivary glands.
   e. blood vessels.

2. Muscarinic receptors bind
   a. epinephrine.
   b. norepinephrine.
   c. acetylcholine.
   d. cholinesterase.
   e. neuropeptides.

3. All of the following cranial nerves *except* the _____ carry parasympathetic fibers.
   a. vagus
   b. facial
   c. oculomotor
   d. glossopharyngeal
   e. hypoglossal

# STUDY GUIDE

4. Which of the following cranial nerves carries sympathetic fibers?
   a. oculomotor
   b. facial
   c. trigeminal
   d. vagus
   e. none of them

5. Which of these ganglia is *not* involved in the sympathetic division?
   a. intramural
   b. superior cervical
   c. paravertebral
   d. inferior mesenteric
   e. celiac

6. Epinephrine is secreted by
   a. sympathetic preganglionic fibers.
   b. sympathetic postganglionic fibers.
   c. parasympathetic preganglionic fibers.
   d. parasympathetic postganglionic fibers.
   e. the adrenal medulla.

7. The most significant autonomic control center within the CNS is
   a. the cerebral cortex.
   b. the limbic system.
   c. the midbrain.
   d. the hypothalamus.
   e. the sympathetic chain ganglia.

8. The gray communicating ramus contains
   a. visceral sensory fibers.
   b. parasympathetic motor fibers.
   c. sympathetic preganglionic fibers.
   d. sympathetic postganglionic fibers.
   e. somatic motor fibers.

9. Throughout the autonomic nervous system, the neurotransmitter released by the preganglionic fiber binds to _____ receptors on the postganglionic neuron.
   a. nicotinic
   b. muscarinic
   c. adrenergic
   d. alpha
   e. beta

10. Which of these does not result from sympathetic stimulation?
    a. dilation of the pupil
    b. acceleration of the heart
    c. digestive secretion
    d. enhanced blood clotting
    e. piloerection

11. Certain nerve fibers are called _____ fibers because they secrete norepinephrine.

12. _____ is a state in which a target organ receives both sympathetic and parasympathetic fibers.

13. _____ is a state of continual background activity of the sympathetic and parasympathetic divisions.

14. Most parasympathetic preganglionic fibers are found in the _____ nerve.

15. The digestive tract has a semi-independent nervous system called the _____ nervous system.

16. MAO and COMT are enzymes that break down _____ at certain ANS synapses.

17. The adrenal medulla consists of modified postganglionic neurons of the _____ nervous system.

18. The sympathetic nervous system has short _____ and long _____ nerve fibers.

19. Adrenergic receptors classified as $\alpha_1$, $\beta_1$, and $\beta_2$ act by changing the level of _____ in the target cell.

20. Sympathetic fibers to blood vessels maintain a state of partial vasoconstriction called _____.

## ▶ Building Your Medical Vocabulary

*Answers in Appendix B*

*State a medical meaning of each word element below, and give a term in which it or a slight variation of it is used.*

1. baro-
2. lyto-
3. muro-
4. nomo-
5. oto-
6. patho-
7. reno-
8. splanchno-
9. sym-
10. viscero-

## ▶ True or False

*Answers in Appendix B*

*Determine which five of the following statements are false, and briefly explain why.*

1. The parasympathetic nervous system shuts down when the sympathetic nervous system is active, and vice versa.

2. Blood vessels of the skin receive no parasympathetic innervation.

3. Voluntary control of the ANS is not possible.

4. The sympathetic nervous system stimulates digestion.

5. Some sympathetic postganglionic fibers are cholinergic.

6. Urination and defecation cannot occur without signals from the brain to the bladder and rectum.

7. Some parasympathetic nerve fibers are adrenergic.

8. Parasympathetic effects are more localized and specific than sympathetic effects.

9. The parasympathetic division shows less neural divergence than the sympathetic division does.

10. The two divisions of the ANS have antagonistic effects on the iris.

# STUDY GUIDE

## ▶ Testing Your Comprehension

*Answers at* www.mhhe.com/saladin7

1. You are dicing raw onions while preparing dinner, and the vapor makes your eyes water. Describe the afferent and efferent pathways involved in this response.

2. Suppose you are walking alone at night when you hear a dog growling close behind you. Describe the ways your sympathetic nervous system would prepare you to deal with this situation.

3. Suppose that the cardiac nerves were destroyed. How would this affect the heart and the body's ability to react to a stressful situation?

4. What would be the advantage to a wolf in having its sympathetic nervous system stimulate the piloerector muscles? What happens in a human when the sympathetic system stimulates these muscles?

5. Pediatric literature has reported many cases of poisoning in children with Lomotil, an antidiarrheal medicine. Lomotil works primarily by means of the morphinelike effects of its chief ingredient, diphenoxylate, but it also contains atropine. Considering the mode of action described for atropine in Deeper Insight 15.2, why might atropine contribute to the antidiarrheal effect of Lomotil? In atropine poisoning, would you expect the pupils to be dilated or constricted? The skin to be moist or dry? The heart rate to be elevated or depressed? The bladder to retain urine or void uncontrollably? Explain each answer. Atropine poisoning is treated with physostigmine, a cholinesterase inhibitor. Explain the rationale of this treatment.

## ▶ Improve Your Grade at www.mhhe.com/saladin7

*Download animations to study when it fits your schedule. Practice quizzes, labeling activities, games, and flashcards offer fun ways to master the chapter concepts. Or, download image PowerPoint files for each chapter to create a study guide or for taking notes during lecture.*

16

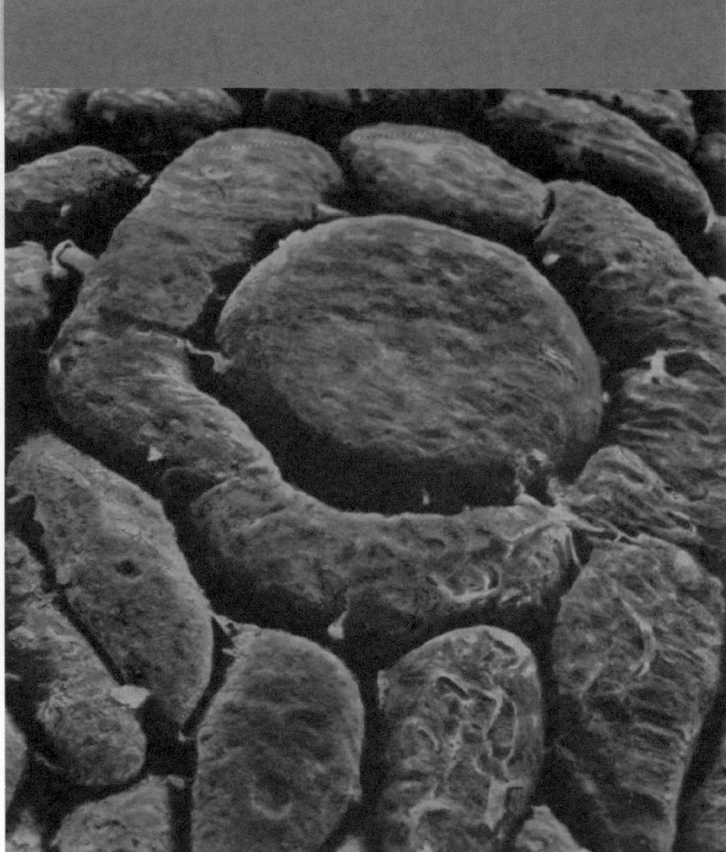

A vallate papilla of the tongue, where most taste buds are located (SEM)

Anatomy & Physiology | REVEALED®
aprevealed.com

**Module 7: Nervous System**

Anyone who enjoys music, art, fine food, or a good conversation appreciates the human senses. Yet their importance extends far beyond deriving pleasure from the environment. In the 1950s, behavioral scientists at Princeton University studied the methods used by Soviet Communists to extract confessions from political prisoners, including solitary confinement and sensory deprivation. Student volunteers were immobilized in dark soundproof rooms or suspended in dark chambers of water. In a short time, they experienced visual, auditory, and tactile hallucinations, incoherent thought patterns, deterioration of intellectual performance, and sometimes morbid fear or panic. Similar effects are seen in some burn patients who are immobilized and extensively bandaged (including the eyes) and thus suffer prolonged lack of sensory input. Patients connected to life-support equipment and confined under oxygen tents sometimes become delirious. Sensory input is vital to the integrity of personality and intellectual function.

Furthermore, much of the information communicated by the sense organs never comes to our conscious attention—blood pressure, body temperature, and muscle tension, for example. By monitoring such conditions, however, the sense organs initiate somatic and visceral reflexes that are indispensable to homeostasis and to our very survival in a ceaselessly changing and challenging environment.

## 16.1 Properties and Types of Sensory Receptors

### Expected Learning Outcomes

When you have completed this section, you should be able to

a. define *receptor* and *sense organ;*

b. list the four kinds of information obtained from sensory receptors, and describe how the nervous system encodes each type; and

c. outline three ways of classifying receptors.

A sensory **receptor** is any structure specialized to detect a stimulus. Some receptors are simple, bare nerve endings, such as the receptors for heat and pain, whereas others are true sense organs. A **sense organ** is a structure composed of nervous tissue along with other tissues that enhance its response to a certain type of stimulus. The accessory tissues may include epithelial, muscular, or connective tissue. Sense organs can be as complex as the eye and ear or as microscopic and simple as a dendrite wrapped in a little bit of connective tissue.

### General Properties of Receptors

The fundamental purpose of any sensory receptor is **transduction,** the conversion of one form of energy to another—light, sound, heat, touch, vibration, or other forms of stimulus energy into nerve signals. (Any device that converts one energy form to another is a *transducer*—whether a sense organ, gasoline engine, or lightbulb.)

The initial effect of a stimulus on a sensory cell is a small local electrical change called a **receptor potential.** In many cases, such as the senses of touch and smell, the sensory cell is a neuron. If the receptor potential is strong enough, the neuron fires off a volley of action potentials, generating a nerve signal to the CNS. In other cases, such as taste and hearing, the sensory cell is not a neuron but an epithelial cell. Nevertheless, it has synaptic vesicles at its base; it releases a neurotransmitter in response to a stimulus; and it stimulates an adjacent neuron. That neuron then generates signals to the CNS.

Not all sensory signals go to the brain, but when they do, we may experience a **sensation**—a subjective awareness of the stimulus. Yet most sensory signals delivered to the CNS produce no conscious sensation at all. Most of them are filtered out in the brainstem before reaching the cerebral cortex, a valuable function that keeps us from being driven mad by innumerable unimportant stimuli detected by the sense organs. Other nerve signals concern functions that do not require conscious awareness, such as monitoring blood pH and body temperature.

Sensory receptors transmit four kinds of information—*modality, location, intensity,* and *duration:*

1. **Modality** refers to the type of stimulus or the sensation it produces. Vision, hearing, and taste are examples of sensory modalities. The action potentials for vision are identical to the action potentials for taste or any other modality—so how can the brain tell a visual signal from a taste signal? It does this partly by "assuming" that if a signal comes from the retina, it must be a visual signal; if it comes from a taste bud, it must be a taste; and so on. It's as if each nerve pathway from sensory cells to the brain is "labeled" to identify its origin, and the brain employs these labels to interpret what modality the nerve signal represents—pain or tickle, sweet or bitter, green or red. This theory of sensory interpretation is called the **labeled line code.**

2. **Location** is also encoded by which nerve fibers issue signals to the brain. Any sensory neuron detects stimuli within an area called its **receptive field.** In the sense of touch, for example, a single sensory neuron may cover an area of skin as large as 7 cm in diameter. No matter where the skin is touched within that receptive field, it stimulates the same neuron. The brain may be unable to determine whether the skin was touched at "point A" or at some other point 1 or 2 cm away. If the skin is simultaneously touched at two places in the same receptive field, it can feel like a single touch (fig. 16.1a). In some areas of the body such as the back, however, it isn't necessary to make finer distinctions. On the other hand, we must be able to localize touch sensations much more precisely with the fingertips. Here, each sensory neuron may cover a receptive field 1 mm or less in diameter—there is a much higher density of tactile nerve fibers. Two points of contact just 2 mm apart will be felt as separate touches (fig. 16.1b). That is, the fingertips have finer *two-point touch discrimination* than the skin of the back. You can imagine the importance of this in reading braille, appreciating the texture of a fine fabric, or manipulating an object as small as a sesame seed. The receptive field concept pertains not only to touch, but also to other senses such as vision.

**Sensory projection** is the ability of the brain to identify the site of stimulation, including very small and specific areas within a receptor such as the retina. The pathways followed by sensory signals to their ultimate destinations in the CNS are called **projection pathways.**

3. **Intensity** refers to whether a sound is loud or soft, a light is bright or dim, a pain is mild or excruciating, and so forth. It is encoded in three ways: (a) as stimulus intensity rises, the firing frequencies of sensory nerve fibers rise (see fig. 12.28, p. 465); (b) intense stimuli recruit greater numbers of nerve fibers to fire; and (c) weak stimuli activate only the most sensitive nerve fibers, whereas strong stimuli can activate a less sensitive group of fibers with higher thresholds. Thus, the brain can distinguish intensities based on which fibers are firing, how many are doing so, and how fast they are firing.

4. **Duration,** or how long a stimulus lasts, is encoded by changes in firing frequency with the passage of time. All receptors exhibit the property of **sensory adaptation**—if the stimulus is prolonged, the firing of the neuron gets slower over time, and with it, we become less aware of the stimulus. Adapting to hot bathwater is an example of this. Receptors are classified according to how quickly they adapt. **Phasic receptors** generate a burst of action

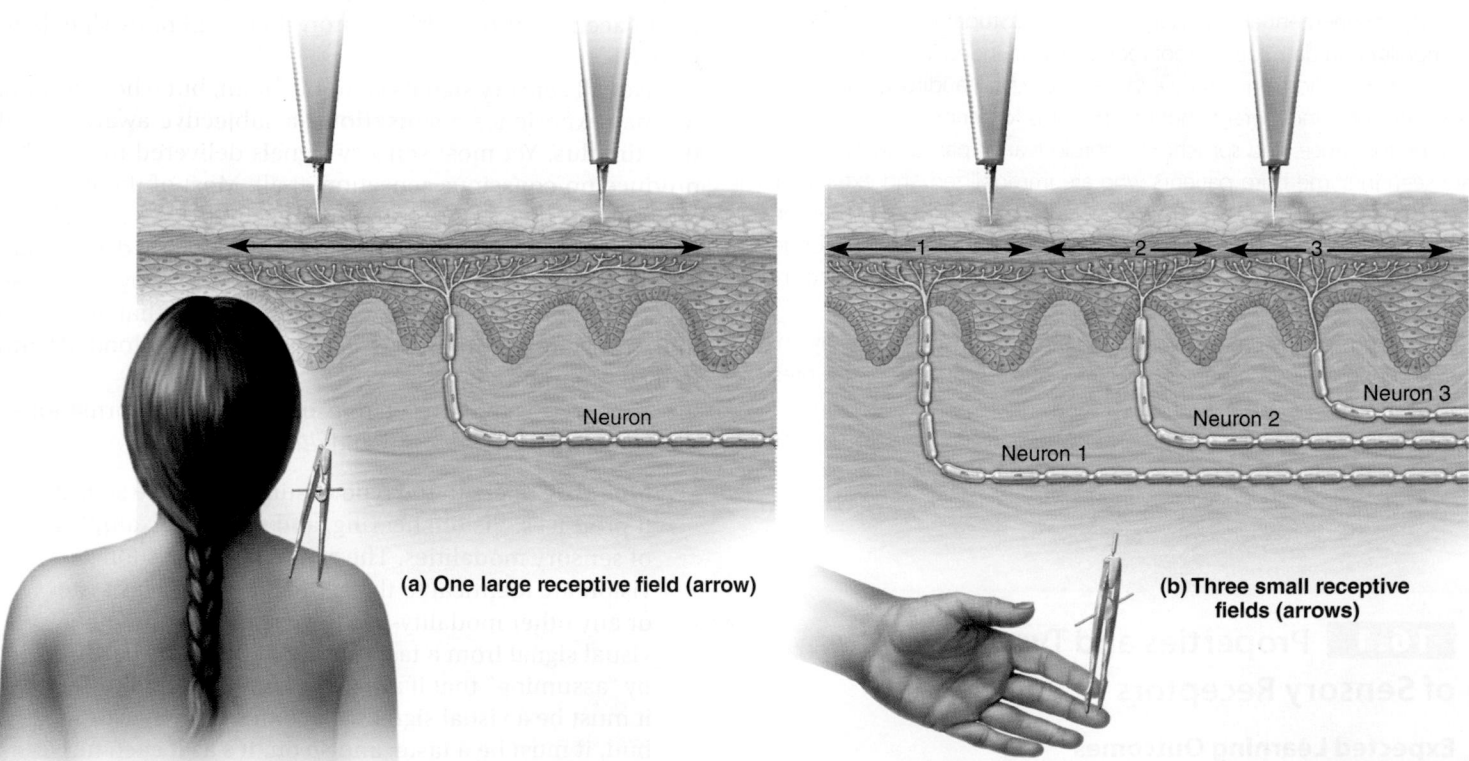

(a) **One large receptive field (arrow)**

(b) **Three small receptive fields (arrows)**

**FIGURE 16.1 Receptive Fields.** (a) A neuron with a large receptive field, as found in the skin of the back. If the skin is touched in two close-together places within this receptive field, the brain senses it as only one point of contact. (b) Neurons with small receptive fields, as found in the fingertips. Two close-together points of contact here are likely to stimulate two different neurons and be felt as separate touches.

❓ *If the receptive field in part (a) is 7 cm in diameter, is it possible for two touches 1 cm apart to be felt separately?*

potentials when first stimulated, then quickly adapt and sharply reduce or stop signaling even if the stimulus continues. Some of them fire again when the stimulus ceases. Phasic receptors are found in the senses of smell, hair movement, and cutaneous pressure. Their phasic nature explains why we may notice a suspicious odor such as a gas leak for several seconds, then the sensation fades in intensity even if the stimulus is still there. If we don't find the gas leak quickly, it becomes difficult to find it at all. **Tonic receptors** adapt more slowly and generate signals more steadily. Proprioceptors are among the most slowly adapting tonic receptors because the brain must always be aware of body position, muscle tension, and joint motion.

### ▶▶▶ APPLY WHAT YOU KNOW

*Although you may find it difficult to immerse yourself in a tub of hot water or a cold lake, you soon adapt and become more comfortable. In light of this, do you think cold and warm receptors are phasic or tonic? Explain.*

## Classification of Receptors

Receptors can be classified by several overlapping systems, some of which have been introduced in earlier chapters:

1. By stimulus modality:
   - **Thermoreceptors** respond to heat and cold.
   - **Photoreceptors,** the eyes, respond to light.
   - **Nociceptors**[1] (NO-sih-SEP-turs) are pain receptors; they respond to tissue injury or situations that threaten to damage a tissue.
   - **Chemoreceptors** respond to chemicals, including odors, tastes, and body fluid composition.
   - **Mechanoreceptors** respond to physical deformation of a cell or tissue caused by vibration, touch, pressure, stretch, or tension. They include the organs of hearing and balance and many receptors of the skin, viscera, and joints.

2. By the origin of the stimulus:
   - **Exteroceptors** sense stimuli external to the body. They include the receptors for vision, hearing, taste, smell, and cutaneous sensations such as touch, heat, cold, and pain.
   - **Interoceptors** detect stimuli in the internal organs such as the stomach, intestines, and bladder, and produce feelings of stretch, pressure, visceral pain, and nausea.
   - **Proprioceptors** sense the position and movements of the body or its parts. They occur in muscles, tendons, and joint capsules.

3. By the distribution of receptors in the body. There are two broad classes of senses:
   - **General (somatosensory, somesthetic) senses** employ widely distributed receptors in the skin, muscles, tendons, joints, and viscera. These include touch, pressure, stretch, heat, cold, and pain, as well as many stimuli that we do not perceive consciously, such as blood pressure and composition. Their receptors are relatively simple—sometimes nothing more than a bare dendrite.
   - **Special senses** are limited to the head, are innervated by the cranial nerves, and employ relatively complex sense organs. The special senses are vision, hearing, equilibrium, taste, and smell.

### BEFORE YOU GO ON

Answer the following questions to test your understanding of the preceding section:

1. Not every sensory receptor is a sense organ. Explain.
2. What does it mean to say sense organs are transducers? What form of energy do all receptors have as their output?
3. Not every sensory signal results in conscious awareness of a stimulus. Explain.
4. What is meant by the *modality* of a stimulus? Give some examples.
5. Three schemes of receptor classification were presented in this section. In each scheme, how would you classify the receptors for a full bladder? How would you classify taste receptors?
6. Nociceptors are tonic rather than phasic receptors. Speculate on why this is beneficial to homeostasis.

## 16.2 The General Senses

### Expected Learning Outcomes

When you have completed this section, you should be able to

a. list several types of somatosensory receptors;
b. describe the projection pathways for the general senses; and
c. explain the mechanisms of pain and the spinal blocking of pain signals.

Receptors for the general senses are relatively simple in structure and physiology. They consist of one or a few sensory nerve fibers and, usually, a sparse amount of connective tissue. These receptors are shown in figure 16.2 and described in table 16.1.

---

[1]*noci* = pain

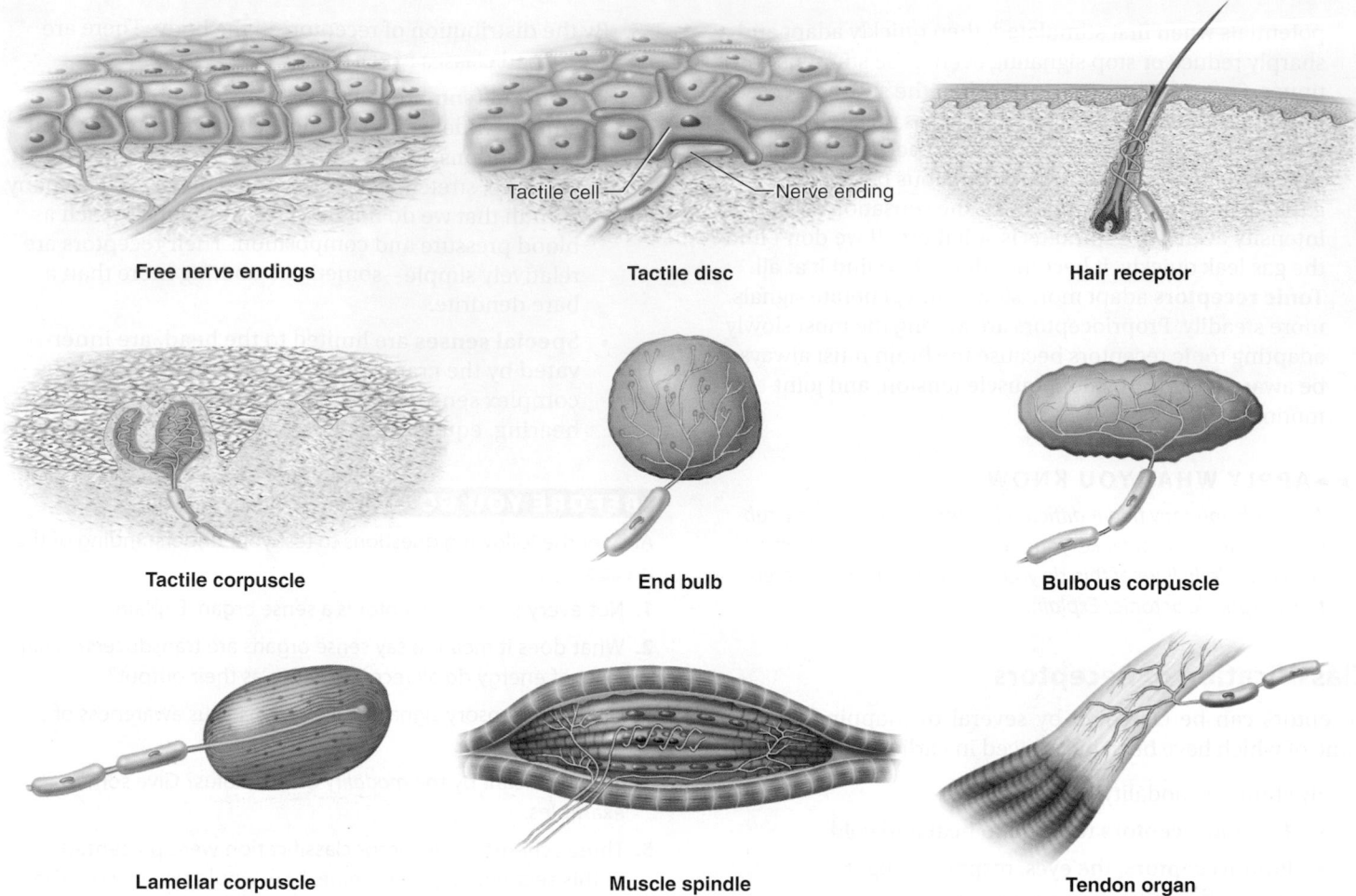

Free nerve endings

Tactile cell — Nerve ending

Tactile disc

Hair receptor

Tactile corpuscle

End bulb

Bulbous corpuscle

Lamellar corpuscle

Muscle spindle

Tendon organ

**FIGURE 16.2  Receptors of the General Senses.**  See table 16.1 for functions.

## Unencapsulated Nerve Endings

**Unencapsulated nerve endings** are dendrites with no connective tissue wrapping. They include the following:

- **Free nerve endings** include *warm receptors,* which respond to rising temperature; *cold receptors,* which respond to falling temperature; and *nociceptors* for pain. They are bare dendrites that have no special association with specific accessory cells or tissues. They are most abundant in the skin and mucous membranes. You can locate some of your cold receptors by gently touching your skin with the point of a graphite pencil, which conducts heat away from the skin. In spots where these receptors are located, this produces a sensation of cold.

- **Tactile (Merkel[2]) discs** are tonic receptors for light touch, thought to sense textures, edges, and shapes. They are flattened nerve endings that terminate adjacent to specialized *tactile (Merkel) cells* in the basal layer of the epidermis.

- **Hair receptors (peritrichial[3] endings)** are dendrites that coil around a hair follicle and respond to movements of the hair. They are stimulated when, for example, an ant walks across one's skin, bending one hair after another. However, they adapt quickly, so we are not constantly irritated by the feel of clothing against the skin. Hair receptors are particularly important in the eyelashes, where the slightest touch evokes a protective blink reflex.

## Encapsulated Nerve Endings

**Encapsulated nerve endings** are nerve fibers wrapped in glial cells or connective tissue. Most of them are mechanoreceptors for touch, pressure, and stretch. The connective tissue either enhances the sensitivity of the nerve fiber or makes it more selective with respect to which modality it responds to. We have already considered some encapsulated nerve

---

[2]Friedrich S. Merkel (1845–1919), German anatomist and physiologist

[3]*peri* = around; *trich* = hair

| TABLE 16.1 | Receptors of the General Senses | |
|---|---|---|
| Receptor Type | Locations | Modality |
| **Unencapsulated endings** | | |
| Free nerve endings | Widespread, especially in epithelia and connective tissues | Pain, heat, cold |
| Tactile discs | Stratum basale of epidermis | Light touch, pressure |
| Hair receptors | Around hair follicle | Light touch, movement of hairs |
| **Encapsulated nerve endings** | | |
| Tactile corpuscles | Dermal papillae of fingertips, palms, eyelids, lips, tongue, nipples, and genitals | Light touch, texture |
| End bulbs | Mucous membranes | Similar to tactile corpuscles |
| Bulbous corpuscles | Dermis, subcutaneous tissue, and joint capsules | Heavy continuous touch or pressure; joint movements |
| Lamellar corpuscles | Dermis, joint capsules, periosteum, breasts, genitals, and some viscera | Deep pressure, stretch, tickle, vibration |
| Muscle spindles | Skeletal muscles near tendon | Muscle stretch (proprioception) |
| Tendon organs | Tendons | Tension on tendons (proprioception) |

endings in chapter 13—muscle spindles and tendon organs. Others are as follows:

- **Tactile (Meissner[4]) corpuscles** are phasic receptors for light touch and texture. They are tall, ovoid to pear-shaped, and consist of two or three nerve fibers meandering upward through a mass of flattened Schwann cells. They occur in the dermal papillae of the skin and are especially concentrated in sensitive hairless areas such as the fingertips, palms, eyelids, lips, nipples, and genitals. Drag your fingernail lightly across the back of your hand, and then across your palm. The difference in sensation you feel is due to the high concentration of tactile corpuscles in the palmar skin. Tactile corpuscles enable you to tell the difference between silk and sandpaper, for example, by light strokes of your fingertips.

- **End bulbs (Krause[5] end bulbs)** are functionally similar to tactile corpuscles, but instead of occurring in the skin, they are found in the mucous membranes of the lips and tongue, in the conjunctiva on the anterior surface of the eye, and in the epineurium of large nerves. They are ovoid bodies composed of a connective tissue sheath around a sensory nerve fiber.

- **Lamellar (pacinian[6]) corpuscles** are phasic receptors for deep pressure, stretch, tickle, and vibration. They are large receptors, up to 1 or 2 mm long, and look like an onion slice in cross section. A single sensory dendrite travels through the center of the corpuscle. The innermost capsule layers around it are flattened Schwann cells, but the greater bulk of the capsule consists of concentric layers of fibroblasts with narrow fluid-filled spaces between them. These receptors occur in the periosteum of bone; in joint capsules; in the pancreas and some other viscera; and deep in the dermis, especially on the hands, feet, breasts, and genitals.

- **Bulbous (Ruffini[7]) corpuscles** are tonic receptors for heavy touch, pressure, stretching of the skin, and joint movements. They are flattened, elongated capsules containing a few nerve fibers and are located in the dermis, subcutaneous tissue, and joint capsules.

## Somatosensory Projection Pathways

From the receptor to the final destination in the brain, most somatosensory signals travel by way of three neurons called the **first-, second-,** and **third-order neurons.** Their axons are called first- through third-order nerve fibers. The first-order fibers for touch, pressure, and proprioception are large, myelinated, and fast; those for heat and cold are small, unmyelinated or lightly myelinated, and slower.

Somatosensory signals from the head, such as facial sensations, travel by way of several cranial nerves (especially V, the trigeminal nerve) to the pons and medulla oblongata. In the brainstem, the first-order fibers of these neurons synapse with second-order neurons that decussate and lead to the contralateral thalamus. Third-order neurons then complete the route to the cerebrum. Proprioceptive signals from the head are an exception, as the second-order fibers carry these signals to the cerebellum.

Below the head, the first-order fibers enter the posterior horn of the spinal cord. Signals ascend the cord in the spinothalamic and other pathways detailed in chapter 13 (see table 13.1 and figure 13.5, pp. 480–481). These pathways decussate either at or near the point of entry into the cord, or in the brainstem, so the primary somatosensory cortex in each cerebral hemisphere receives signals from the contralateral side of the body.

[4]Georg Meissner (1829–1905), German histologist
[5]Wilhelm J. F. Krause (1833–1910), German anatomist
[6]Filippo Pacini (1812–83), Italian anatomist

[7]Angelo Ruffini (1864–1929), Italian anatomist

Signals for proprioception below the head travel up the spinocerebellar tracts to the cerebellum. Signals from the thoracic and abdominal viscera travel to the medulla oblongata by way of sensory fibers in the vagus nerve (X). Visceral pain signals also ascend the spinal cord in the gracile fasciculus.

## Pain

Pain is a discomfort caused by tissue injury or noxious stimulation, typically leading to evasive action. Few of us enjoy pain and we may wish that no such thing existed, but it is one of our most important senses and we would be far worse off without it. We see evidence of its value in such diseases as leprosy and diabetes mellitus, where the sense of pain is lost because of nerve damage *(neuropathy)*. The absence of pain makes people unaware of minor injuries. They neglect them, so the injuries often become infected and worsen to the point that the victim may lose fingers, toes, or entire limbs.

In short, pain is an adaptive and necessary sensation. It is not simply an effect of the overstimulation of nerve fibers meant for other functions, but is mediated by its own specialized nerve fibers, the nociceptors. These are especially dense in the skin and mucous membranes, and occur in virtually all organs, although not in the brain or liver. In some brain surgery, the patient must remain conscious and able to talk with the surgeon; such patients need only a local anesthetic. Nociceptors occur in the meninges, however, and play an important role in headaches.

There are two types of nociceptors corresponding to different pain sensations. Myelinated pain fibers conduct at speeds of 12 to 30 m/s and produce the sensation of **fast (first) pain**—a feeling of sharp, localized, stabbing pain perceived at the time of injury. Unmyelinated pain fibers conduct at speeds of 0.5 to 2.0 m/s and produce the **slow (second) pain** that follows—a longer-lasting, dull, diffuse feeling. Pain from the skin, muscles, and joints is called **somatic pain,** and pain from the viscera is called **visceral pain.** The latter often results from stretch, chemical irritants, or ischemia, and is often accompanied by nausea.

Injured tissues release several chemicals that stimulate the nociceptors and trigger pain. **Bradykinin** is the most potent pain stimulus known; it hurts intensely when injected under the skin. It not only makes us aware of injuries but also activates a cascade of reactions that promote healing. Serotonin, prostaglandins, and histamine also stimulate nociceptors, as do potassium ions and ATP released from ruptured cells.

## Projection Pathways for Pain

It is notoriously difficult for clinicians to locate the origin of a patient's pain, because pain travels by such diverse and complex routes and the sensation can originate anywhere along any of the routes. Pain signals reach the brain by two main pathways, but there are multiple subroutes within each of them:

1.  Pain signals from the head travel to the brainstem by way of four cranial nerves: mainly the trigeminal (V), but also the facial (VII), glossopharyngeal (IX), and vagus (X) nerves. Trigeminal fibers enter the pons and descend to synapses in the medulla. Pain fibers of the other three

cranial nerves also end here. Second-order neurons arise in the medulla and ascend to the thalamus, which relays the message to the cerebral cortex. We will consider the relay from thalamus to cortex shortly.

2.  Pain signals from the neck down travel by way of three of the ascending spinal cord tracts: the spinothalamic tract, spinoreticular tract, and gracile fasciculus. These pathways are described in chapter 13 and need not be repeated here. The spinothalamic tract is the most significant pain pathway and carries most of the somatic pain signals that ultimately reach the cerebral cortex, making us conscious of pain. The spinoreticular tract carries pain signals to the reticular formation of the brainstem, and these are ultimately relayed to the hypothalamus and limbic system. These signals activate visceral, emotional, and behavioral reactions to pain, such as nausea, fear, and some reflexes. The gracile fasciculus carries signals to the thalamus for visceral pain, such as the pain of a stomachache or from passing a kidney stone. Figure 16.3 shows the spinothalamic and spinoreticular pain pathways.

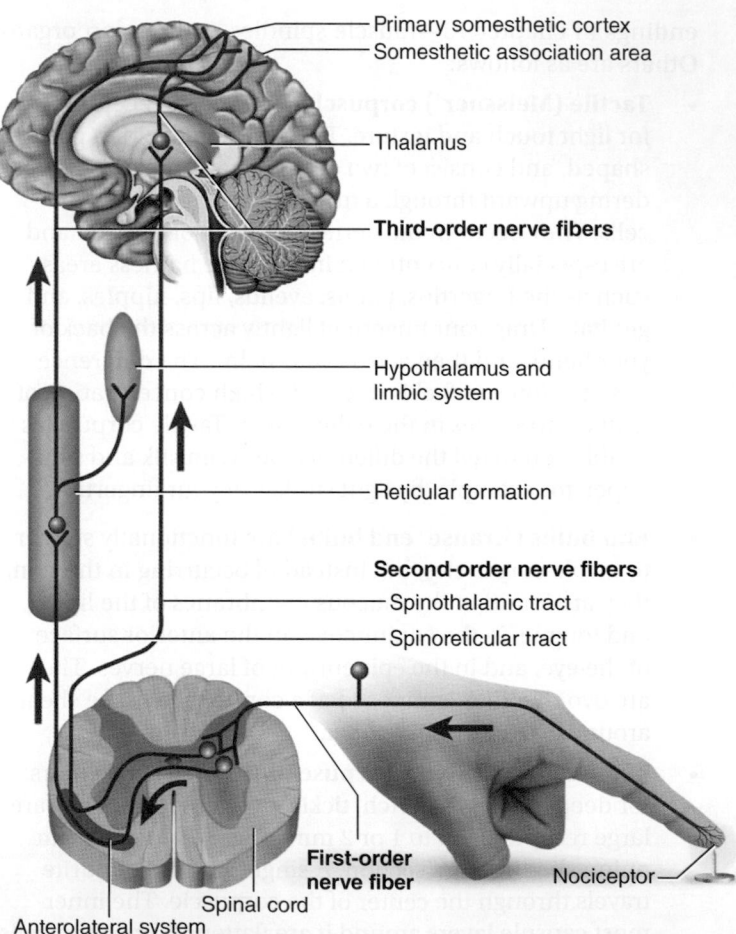

**FIGURE 16.3 Projection Pathways for Pain.** A first-order neuron conducts a pain signal to the posterior horn of the spinal cord, a second-order neuron conducts it to the thalamus, and a third-order neuron conducts it to the cerebral cortex. Signals from the spinothalamic tract pass through the thalamus. Signals from the spinoreticular tract bypass the thalamus on the way to the sensory cortex.

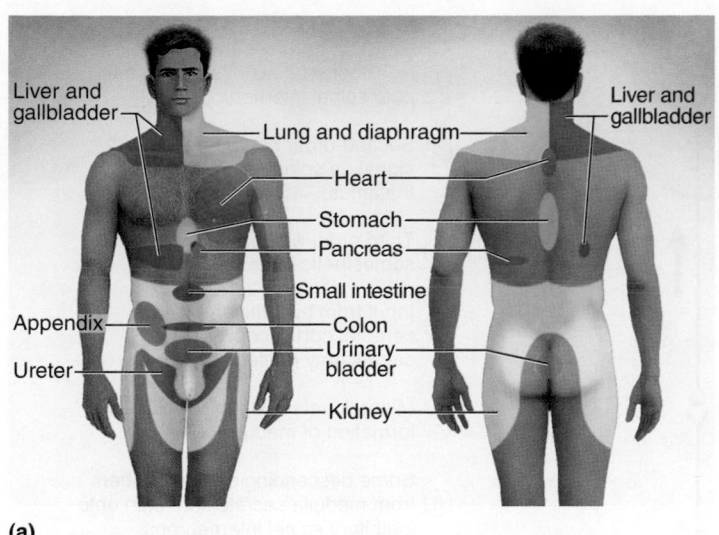

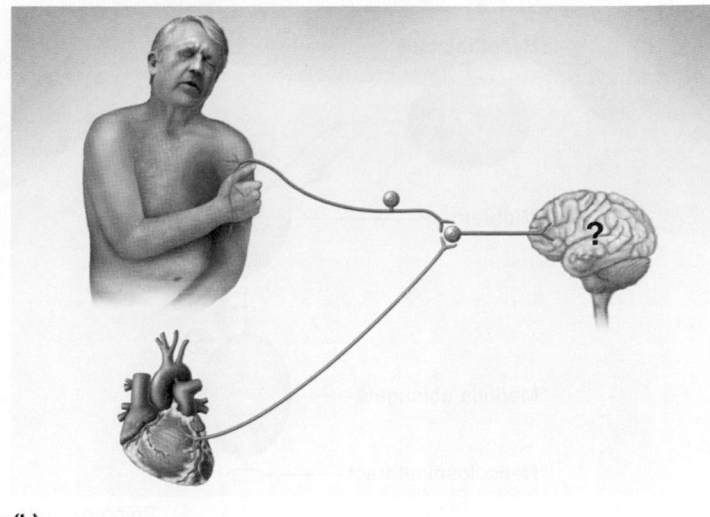

**FIGURE 16.4 Referred Pain.** (a) Areas of referred pain. Pain in these areas of the skin is often a symptom of problems in the indicated viscera. (b) The basis of referred pain in a heart attack. Pain signals from the heart and left arm converge on the same pathway to the sensory cortex, making the brain unable to correctly identify the source of pain.

When the thalamus receives pain signals from the foregoing sources, it relays most of them through third-order neurons to their final destination in the postcentral gyrus of the cerebrum. Exactly what part of this gyrus receives the signals depends on where the pain originated; recall the concept of somatotopy and the sensory homunculus in chapter 14. Most of this gyrus is somatosensory—that is, it receives signals for somatic pain and other senses. A region of the gyrus deep within the lateral sulcus of the brain, however, is a viscerosensory area, which receives the visceral signals conveyed by the gracile fasciculus (see fig. 14.21, p. 537).

Pain in the viscera is often mistakenly thought to come from the skin or other superficial sites—for example, the pain of a heart attack is felt "radiating" along the left shoulder and medial side of the arm. This phenomenon, called **referred pain,** results from the convergence of neural pathways in the CNS. In the case of cardiac pain, for example, spinal cord segments T1 to T5 receive input from the heart as well as from the chest and arm. Pain fibers from the heart and skin in this region converge on the same spinal interneurons, then follow the same pathway from there to the thalamus and cerebral cortex. The brain cannot distinguish which source the arriving signals are coming from. It acts as if it assumes that they are most likely from the skin, since skin has more pain receptors than the heart and suffers injury more often. Knowledge of the origins of referred pain is essential to the skillful diagnosis of organ dysfunctions (fig. 16.4).

## CNS Modulation of Pain

A person's physical and mental state can greatly affect his or her perception of pain. Many mortally wounded soldiers, for example, report little or no pain. The central nervous system (CNS) has **analgesic**[8] (pain-relieving) mechanisms that

are just beginning to be understood. Their discovery is tied to the long-known analgesic effects of opium, morphine, and heroin. In 1974, neurophysiologists discovered receptor sites in the brain for these drugs. Since these opiates do not occur naturally in the body, physiologists wondered what normally binds to these receptors. They soon found two analgesic oligopeptides with 200 times the potency of morphine, and named them **enkephalins.**[9] Larger analgesic neuropeptides, the **endorphins**[10] and **dynorphins,**[11] were discovered later. All three of these are known as **endogenous opioids** (which means "internally produced opium-like substances").

These opioids are secreted by the CNS, pituitary gland, digestive tract, and other organs in states of stress or exercise. In the CNS, they are found especially in the central gray matter of the midbrain (see fig. 14.9a, p. 520) and the posterior horn of the spinal cord. They are *neuromodulators* (see p. 461) that can block the transmission of pain signals and produce feelings of pleasure and euphoria. They may be responsible for the "second wind" ("runner's high") experienced by athletes and for the aforementioned battlefield reports. Their secretion rises sharply in women giving birth. Efforts to employ them in pain therapy have been disappointing, but exercise is an effective part of therapy for chronic pain and may help because it stimulates opioid secretion.

How do these opioids block pain? For pain to be perceived, signals from the nociceptors must get beyond the posterior horn of the spinal cord and travel to the brain. Through mechanisms called **spinal gating,** pain signals can be stopped at the posterior horn. Two of these mechanisms are described here.

---

[8]*an* = without; *alges* = pain

[9]*en* = within; *kephal* = head
[10]acronym from *en*dogenous mo*rphin*elike substance
[11]*dyn* = pain

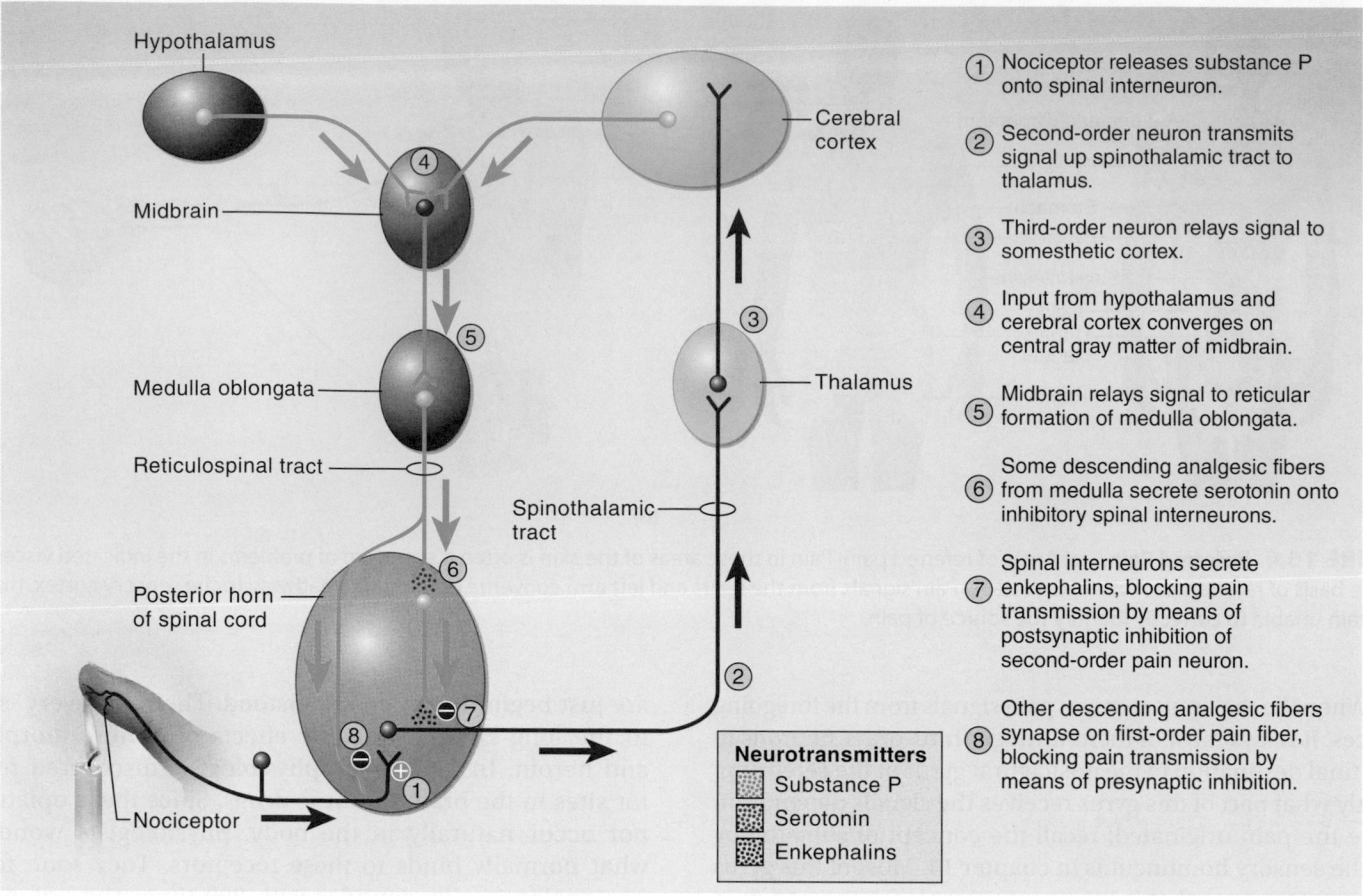

Hypothalamus

Midbrain

Medulla oblongata

Reticulospinal tract

Spinothalamic tract

Posterior horn of spinal cord

Nociceptor

Cerebral cortex

Thalamus

**Neurotransmitters**
- Substance P
- Serotonin
- Enkephalins

① Nociceptor releases substance P onto spinal interneuron.

② Second-order neuron transmits signal up spinothalamic tract to thalamus.

③ Third-order neuron relays signal to somesthetic cortex.

④ Input from hypothalamus and cerebral cortex converges on central gray matter of midbrain.

⑤ Midbrain relays signal to reticular formation of medulla oblongata.

⑥ Some descending analgesic fibers from medulla secrete serotonin onto inhibitory spinal interneurons.

⑦ Spinal interneurons secrete enkephalins, blocking pain transmission by means of postsynaptic inhibition of second-order pain neuron.

⑧ Other descending analgesic fibers synapse on first-order pain fiber, blocking pain transmission by means of presynaptic inhibition.

**FIGURE 16.5  Spinal Gating of Pain Signals.**

One mechanism involves **descending analgesic fibers**—nerve fibers that arise in the brainstem, travel down the spinal cord in the reticulospinal tract, and block pain signals from traveling up the cord to the brain. Figure 16.5 depicts one relatively simple mechanism of pain modulation in the posterior horn. The normal route of pain transmission is indicated by the red arrows and steps 1 through 3:

① A nociceptor stimulates a second-order nerve fiber. The neurotransmitter at this synapse is called **substance P** (think *P* for "pain"[12]).

② The second-order nerve fiber transmits signals up the spinothalamic tract to the thalamus.

③ The thalamus relays the signals through a third-order neuron to the cerebral cortex, where one becomes conscious of pain.

The pathway for pain blocking, or modulation, is indicated by descending green arrows and steps 4 through 8:

④ Signals from the hypothalamus and cerebral cortex feed into the central gray matter of the midbrain, allowing both autonomic and conscious influences on pain perception.

⑤ The midbrain relays signals to certain nuclei in the reticular formation of the medulla oblongata.

⑥ The medulla issues descending, serotonin-secreting analgesic fibers to the spinal cord. These fibers travel the reticulospinal tract and end in the posterior horn at all levels of the cord.

⑦ In the posterior horn, some of the descending analgesic fibers synapse on short spinal interneurons, which in turn synapse on the second-order pain fiber. These interneurons secrete enkephalins to inhibit the second-order neuron. This is an example of postsynaptic inhibition, working on the downstream side of the synapse between the first- and second-order pain neurons.

⑧ Some fibers from the medulla also exert presynaptic inhibition, synapsing on the axons of the nociceptors and blocking the release of substance P.

Another mechanism of spinal gating is one you may often have consciously employed without knowing why it worked.

---

[12]Named *substance P* because it was first discovered in a *powdered* extract of brain and intestine

Have you ever banged your elbow on the edge of a table or pinched your finger in a door, and found that you could ease the pain by rubbing or massaging the injured area? This works because pain-inhibiting interneurons of the posterior horn, like the one at step 7, also receive input from mechanoreceptors in the skin and deeper tissues. When you rub an injured area, you stimulate those mechanoreceptors; they stimulate the spinal interneurons; the interneurons secrete enkephalins; and enkephalins inhibit the second-order pain neurons.

The clinical control of pain has had a particularly interesting history, some of which is retold in Deeper Insight 16.5 (p. 624).

**BEFORE YOU GO ON**

Answer the following questions to test your understanding of the preceding section:

7. What stimulus modalities are detected by free nerve endings?

8. Name any four encapsulated nerve endings and identify their stimulus modalities.

9. Where do most second-order somatosensory neurons synapse with third-order neurons?

10. Explain the phenomenon of referred pain in terms of the neural pathways involved.

11. Explain the roles of bradykinin, substance P, and enkephalins in the perception of pain.

## 16.3 The Chemical Senses

### Expected Learning Outcomes

When you have completed this section, you should be able to

a. explain how taste and smell receptors are stimulated; and

b. describe the receptors and projection pathways for these two senses.

Taste and smell are the chemical senses. In both cases, environmental chemicals stimulate sensory cells. Other chemoreceptors, not discussed in this section, are located in the brain and blood vessels and monitor the chemistry of the body fluids.

## Taste

**Gustation (taste)** begins with the chemical stimulation of sensory cells clustered in about 4,000 **taste buds.** The chemical stimuli are called **tastants.** Most taste buds are on the tongue, but some occur inside the cheeks and on the soft palate, pharynx, and epiglottis, especially in infants and children.

## Anatomy

The visible bumps on the tongue are not taste buds but **lingual papillae.** There are four types of papillae—one without taste buds and three with taste buds just beneath the epithelial surface, where they are not visible to the eye (fig. 16.6a):

1. **Filiform**[13] **papillae** are tiny spikes without taste buds. They are responsible for the rough feel of a cat's tongue and are important to many mammals for grooming the fur. They are the most abundant papillae on the human tongue, but they are small and play no gustatory role. They serve, however, in one's sense of the texture of food.

2. **Foliate**[14] **papillae** are also weakly developed in humans. They form parallel ridges on the sides of the tongue about two-thirds of the way back from the tip, adjacent to the molar and premolar teeth, where most chewing occurs and most flavor chemicals are released from the food. Most of their taste buds degenerate by the age of 2 or 3 years. Perhaps this partially explains why children so often reject foods that are tolerated or enjoyed by adults.

3. **Fungiform**[15] (FUN-jih-form) **papillae** are shaped somewhat like mushrooms. Each has about three taste buds, located mainly on the apex. These papillae are widely distributed but especially concentrated at the tip and sides of the tongue.

4. **Vallate**[16] **(circumvallate) papillae** are large papillae arranged in a V at the rear of the tongue. Each is surrounded by a deep circular trench. There are only 7 to 12 of them, but they contain up to half of all taste buds—around 250 each, located on the wall of the papilla facing the trench (fig. 16.6b).

Regardless of location and sensory specialization, all taste buds look alike (fig. 16.6c, d). They are lemon-shaped groups of 50 to 150 *taste cells, supporting cells,* and *basal cells.* **Taste (gustatory) cells** are more or less banana-shaped and have a tuft of apical microvilli called **taste hairs,** which serve as receptor surfaces for tastants. The hairs project into a pit called a **taste pore** on the epithelial surface of the tongue. Taste cells are epithelial cells, not neurons, but they synapse with sensory nerve fibers at their base and have synaptic vesicles for the release of neurotransmitters. A taste cell lives for only 7 to 10 days. **Basal cells** are stem cells that multiply and replace taste cells that have died, but they also synapse with sensory nerve fibers of the taste bud and may play some role in the processing of sensory information before the signal goes to the brain. **Supporting cells** resemble taste cells but have no synaptic vesicles or sensory role.

---

[13]*fili* = thread; *form* = shaped
[14]*foli* = leaf; *ate* = like
[15]*fungi* = mushroom; *form* = shaped
[16]*vall* = wall; *ate* = like, possessing

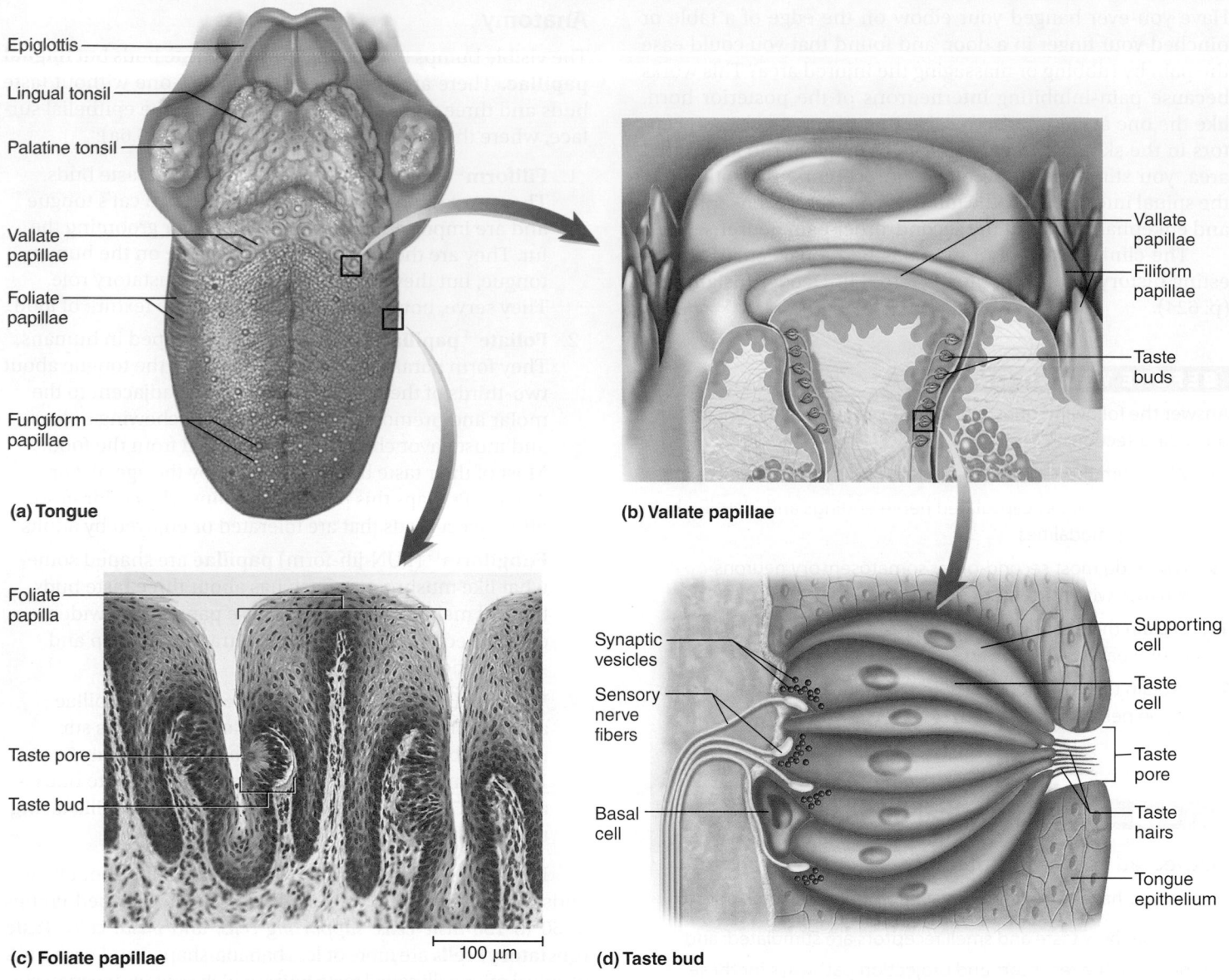

Epiglottis

Lingual tonsil

Palatine tonsil

Vallate papillae

Foliate papillae

Fungiform papillae

**(a) Tongue**

Vallate papillae

Filiform papillae

Taste buds

**(b) Vallate papillae**

Foliate papilla

Taste pore

Taste bud

**(c) Foliate papillae**

100 μm

Synaptic vesicles

Sensory nerve fibers

Basal cell

Supporting cell

Taste cell

Taste pore

Taste hairs

Tongue epithelium

**(d) Taste bud**

**FIGURE 16.6 Gustatory (Taste) Receptors.** (a) Superior aspect of the tongue and locations of its papillae. (b) Detail of the vallate papillae (compare photo on p. 578). (c) Taste buds on the walls of two adjacent foliate papillae. (d) Structure of a taste bud. AP|R

## Physiology

To be tasted, molecules must dissolve in the saliva and flood the taste pore. On a dry tongue, sugar or salt has as little taste as a sprinkle of sand. Physiologists currently recognize five primary taste sensations:

1. **Salty,** produced by metal ions such as sodium and potassium. Since these are vital electrolytes, there is obvious value in our ability to taste them and in having an appetite for salt. Electrolyte deficiencies can cause a craving for salt; many animals as diverse as insects, parrots, deer, and elephants seek salt deposits when necessary. Pregnancy can lower a woman's electrolyte concentrations and create a craving for salty food.

2. **Sweet,** produced by many organic compounds, especially sugars. Sweetness is associated with carbohydrates and foods of high caloric value. Many flowering plants have evolved sweet nectar and fruits that entice animals to eat them and disperse their pollen and seeds. Our fondness for fruit and sugar has coevolved with plant reproductive strategies.

3. **Sour,** usually associated with acids ($H^+$) in such foods as citrus fruits.

4. **Bitter,** associated with spoiled foods and alkaloids such as nicotine, caffeine, quinine, and morphine. Alkaloids are often poisonous, and the bitter taste sensation usually induces a human or animal to reject a food. While flowering plants make their fruits temptingly sweet, they often

load their leaves with bitter alkaloids to deter animals from eating them.

5. **Umami** is a "meaty" taste produced by amino acids such as aspartic and glutamic acids—the savory taste of beef or chicken broth. Pronounced "ooh-mommy," the word is Japanese slang for "delicious" or "yummy."

The many flavors we perceive are not simply a mixture of these five primary tastes, but are also influenced by food texture, aroma, temperature, appearance, and one's state of mind, among other things. Many flavors depend on smell; without their aromas, cinnamon merely has a faintly sweet taste, coffee and peppermint are bitter, and apples and onions taste almost identical. Some flavors such as pepper are due to stimulation of free endings of the trigeminal nerve rather than taste buds. Food technologists refer to the texture of food as *mouthfeel.* Filiform and fungiform papillae of the tongue are innervated by the *lingual nerve* (a branch of the trigeminal) and are sensitive to texture.

All primary tastes can be detected throughout the tongue, but some regions are more sensitive to one category than to others. The tip of the tongue is most sensitive to sweet tastes, which trigger such responses as licking, salivation, and swallowing. The lateral margins of the tongue are the most sensitive areas for salty and sour tastes. Taste buds in the vallate papillae at the rear of the tongue are especially sensitive to bitter compounds, which tend to trigger rejection responses such as gagging to protect against the ingestion of toxins. The threshold for bitter is lowest of all—that is, we can taste lower concentrations of alkaloids than of acids, salts, and sugars. The senses of sweet and salty are the least sensitive.

Sugars, alkaloids, and glutamate stimulate taste cells by binding to receptors on the membrane surface, which then activate G proteins and second-messenger systems within the cell. Sodium and acids penetrate into the cell and depolarize it directly. By either mechanism, taste cells then release neurotransmitters that stimulate the sensory dendrites at their base.

## Projection Pathways

The facial nerve (cranial nerve VII) collects sensory information from taste buds of the anterior two-thirds of the tongue, the glossopharyngeal nerve (IX) from the posterior one-third, and the vagus nerve (X) from taste buds of the palate, pharynx, and epiglottis. All taste fibers project to a site in the medulla oblongata called the *solitary nucleus.* Second-order neurons arise here and relay the signals to two destinations: (1) nuclei in the hypothalamus and amygdala that activate autonomic reflexes such as salivation, gagging, and vomiting; and (2) the thalamus, which relays signals to the insula and postcentral gyrus of the cerebrum, where we become conscious of the taste. Processed signals are further relayed to the orbitofrontal cortex (see fig. 14.16, p. 529), where they are integrated with signals from the nose and eyes and we form an overall impression of the flavor and palatability of food.

## Smell

**Olfaction,** the sense of smell, is a response to airborne chemicals called **odorants** in the nasal cavity. These are detected by receptor cells in a patch of epithelium, the **olfactory mucosa,** in the roof of the nasal cavity (fig. 16.7). This location places the olfactory cells close to the brain, but it is poorly ventilated; forcible sniffing is often needed to identify an odor or locate its source.

### Anatomy

The olfactory mucosa covers about 5 cm$^2$ of the superior concha, cribriform plate, and nasal septum of each nasal fossa. It consists of 10 to 20 million **olfactory cells** as well as epithelial supporting cells and basal stem cells. The rest of the nasal cavity is lined by a nonsensory *respiratory mucosa*.

Unlike taste cells, which are epithelial, olfactory cells are neurons. They are shaped a little like bowling pins (fig. 16.7c). The widest part, the neurosoma, contains the nucleus. The neck and head of the cell are a modified dendrite with a swollen tip. The head bears 10 to 20 cilia called **olfactory hairs.** These cilia are immobile, but they have binding sites for odor molecules. They lie in a tangled mass embedded in a thin layer of mucus. The basal end of each cell tapers to become an axon. These axons collect into small fascicles that leave the nasal cavity through pores *(cribriform foramina)* in the ethmoid bone. Collectively, the fascicles are regarded as cranial nerve I (the olfactory nerve).

Olfactory cells are the only neurons directly exposed to the external environment. Apparently this is hard on them, because they have a life span of only 60 days. Unlike most neurons, however, they are replaceable. The basal cells continually divide and differentiate into new olfactory cells.

### Physiology

Humans have a poorer sense of smell than most other mammals; it seems to have declined in primates as the visual system increased in significance. A 3 kg cat, for example, has a total of about 20 cm$^2$ of olfactory mucosa in its two nasal fossae, whereas a human with 20 times that body weight has only half as much olfactory mucosa. Nevertheless, our sense of smell is much more sensitive than our sense of taste; we can detect odor concentrations as low as a few parts per trillion. On average, women are more sensitive to odors than men are, and they are measurably more sensitive to some odors near the time of ovulation as opposed to other phases of the menstrual cycle. Olfaction is highly important in the social interactions of other animals and, in more subtle ways, to humans (see Deeper Insight 16.1).

Most people can distinguish 2,000 to 4,000 odors, and some can distinguish up to 10,000. Attempts to group these into odor classes have been inconclusive and controversial; suggested classes have included pungent, floral, musky, and earthy. Recent methods in molecular genetics have identified up to 1,200 olfactory receptor types and their genes in rats and mice, but in

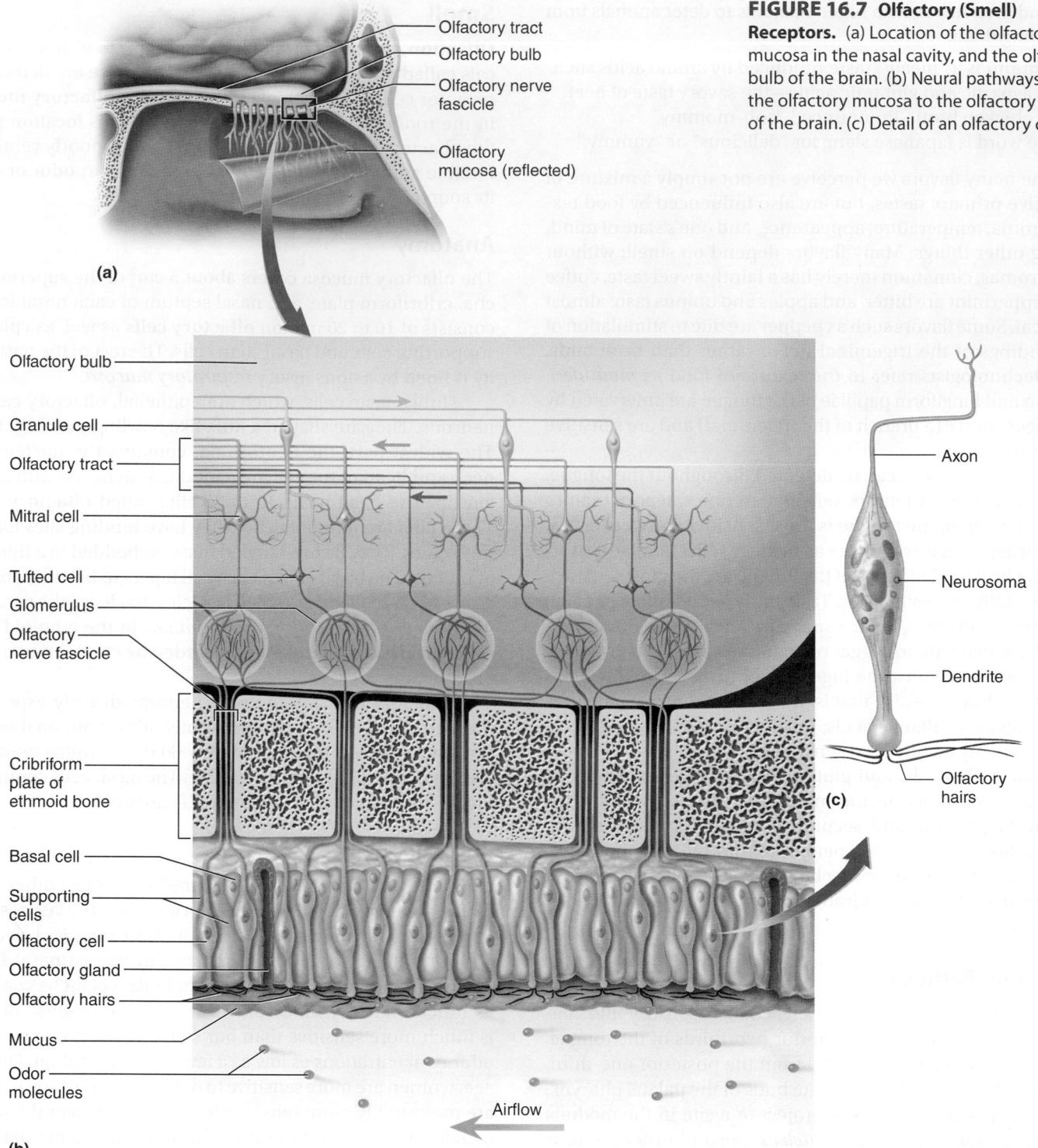

**(a)**

Olfactory tract
Olfactory bulb
Olfactory nerve
fascicle
Olfactory
mucosa (reflected)

**FIGURE 16.7 Olfactory (Smell) Receptors.** (a) Location of the olfactory mucosa in the nasal cavity, and the olfactory bulb of the brain. (b) Neural pathways from the olfactory mucosa to the olfactory tract of the brain. (c) Detail of an olfactory cell.

Olfactory bulb

Granule cell

Olfactory tract

Mitral cell

Tufted cell

Glomerulus

Olfactory
nerve fascicle

Cribriform
plate of
ethmoid bone

Basal cell

Supporting
cells

Olfactory cell

Olfactory gland

Olfactory hairs

Mucus

Odor
molecules

Airflow

**(b)**

Axon

Neurosoma

Dendrite

Olfactory
hairs

**(c)**

humans, two-thirds of the olfactory genes have mutated to the point of being inoperative *pseudogenes;* we have only about 350 kinds of olfactory receptors left. Each olfactory cell has only one receptor type and therefore binds only one odorant.

The first step in smell is that an odorant molecule must bind to a receptor on one of the olfactory hairs. Hydrophilic odorants diffuse freely through the mucus of the olfactory epithelium and bind directly to a receptor. Hydrophobic odorants are transported to the receptor by an *odorant-binding protein* in the mucus. When the receptor binds an odorant, it activates a G protein and through it, the cyclic adenosine monophosphate (cAMP) second-messenger system. The cAMP system

## DEEPER INSIGHT 16.1

### EVOLUTIONARY MEDICINE

#### Human Pheromones

There is an abundance of anecdote, but only equivocal evidence, that human body odors affect sexual behavior. There is more conclusive evidence, however, that a person's sweat and vaginal secretions affect other people's sexual physiology, even when the odors cannot be consciously smelled. Experimental data show that a woman's apocrine sweat can influence the timing of other women's menstrual cycles. This can produce a so-called *dormitory effect,* in which women who live together in the relative absence of men tend to have synchronous menstrual cycles. The presence of men seems to influence female ovulation. Conversely, when a woman is ovulating or close to it, and therefore fertile, her vaginal secretions contain pheromones called *copulines,* which have been shown to raise men's testosterone levels.

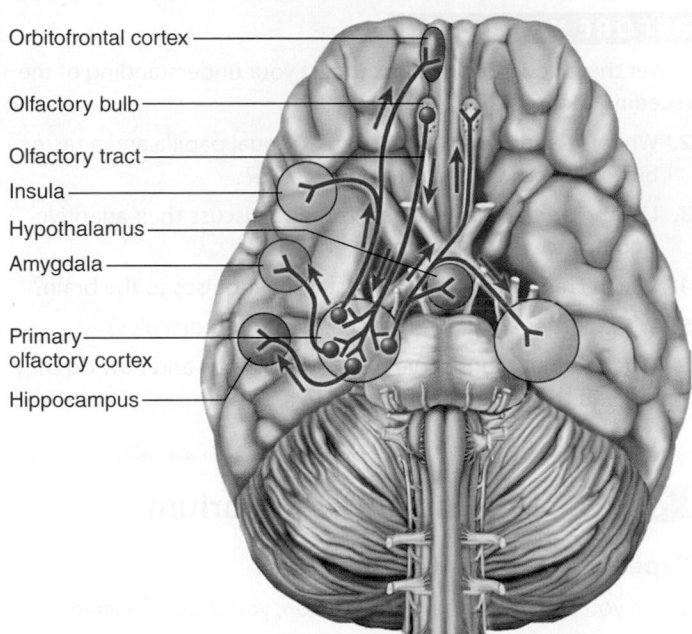

**FIGURE 16.8 Olfactory Projection Pathways in the Brain.**

ultimately opens ion channels in the plasma membrane, admitting cations ($Na^+$ or $Ca^{2+}$) into the cell and depolarizing it, creating a receptor potential. This triggers action potentials in the axon of the olfactory cell, and a signal is transmitted to the brain.

Some odorants, however, act on nociceptors of the trigeminal nerve rather than on olfactory cells. These include ammonia, menthol, chlorine, and the capsaicin of hot peppers. "Smelling salts" revive unconscious persons by stimulating the trigeminal nerve with ammonia fumes.

### Projection Pathways

When olfactory fibers pass through the roof of the nose, they enter a pair of **olfactory bulbs** beneath the frontal lobes of the brain (fig. 16.8). Here they synapse with the dendrites of neurons called *mitral cells* and *tufted cells.* Olfactory axons reach up and mitral and tufted cell dendrites reach down to meet each other in spherical clusters called *glomeruli* (fig. 16.7b). All olfactory fibers leading to any one glomerulus come from cells with the same receptor type; thus each glomerulus is dedicated to a particular type of odor. Higher brain centers interpret complex odors such as chocolate, wine, perfume, or coffee by decoding signals from a combination of odor-specific glomeruli. This is similar to the way our visual system decodes all the colors of the spectrum using input from just three color-specific receptor cells of the eye.

The tufted and mitral cells carry output from the glomeruli. Their axons form bundles called **olfactory tracts,** which course posteriorly along the underside of the frontal lobes. Most fibers of the olfactory tracts end in various regions of the inferior surface of the temporal lobe regarded as the **primary olfactory cortex.** It is noteworthy that olfactory signals can reach the cerebral cortex directly, without first passing through the thalamus; except for the spinoreticular pain pathway (see fig. 16.3), this is not true of any of our other senses. Even in

olfaction, however, some signals from the primary olfactory cortex continue to a relay in the thalamus on their way to olfactory association areas elsewhere.

From the primary olfactory cortex, signals travel to several other secondary destinations in the cerebrum and brainstem. Two important cerebral destinations are the insula and orbitofrontal cortex. The orbitofrontal cortex is where we identify and discriminate among odors. It receives inputs for both taste and smell and integrates these into our overall perception of flavor. Other secondary destinations include the hippocampus, amygdala, and hypothalamus. Considering the roles of these brain areas, it is not surprising that the odor of certain foods, a perfume, a hospital, or decaying flesh can evoke strong memories, emotional responses, and visceral reactions such as sneezing or coughing, the secretion of saliva and stomach acid, or vomiting.

Most areas of olfactory cortex also send fibers back to the olfactory bulbs by way of neurons called *granule cells.* Granule cells can inhibit the mitral and tufted cells. An effect of this feedback is that odors can change in quality and significance under different conditions. Food may smell more appetizing when you are hungry, for example, than when you have just eaten or when you are ill.

#### ▶▶▶ APPLY WHAT YOU KNOW

*Which taste sensations could be lost after damage to (1) the facial nerve or (2) the glossopharyngeal nerve? A fracture of which cranial bone would most likely eliminate the sense of smell?*

Answer the following questions to test your understanding of the preceding section:

12. What is the difference between a lingual papilla and a taste bud? Which is visible to the naked eye?

13. List the primary taste sensations and discuss their adaptive significance (survival value).

14. Which cranial nerves carry gustatory impulses to the brain?

15. What part of an olfactory cell binds odor molecules?

16. What brain regions serve the sense of smell and how do they differ in their olfactory functions?

## 16.4    Hearing and Equilibrium

### Expected Learning Outcomes

When you have completed this section, you should be able to

a. identify the properties of sound waves that account for pitch and loudness;

b. describe the gross and microscopic anatomy of the ear;

c. explain how the ear converts vibrations to nerve signals and discriminates between sounds of different intensity and pitch;

d. explain how the vestibular apparatus enables the brain to interpret the body's position and movements; and

e. describe the pathways taken by auditory and vestibular signals to the brain.

*Hearing* is a response to vibrating air molecules, and *equilibrium* is the sense of motion, body orientation, and balance. Both senses reside in the inner ear, a maze of fluid-filled passages and sensory cells. We will study how the fluid is set in motion and how the sensory cells convert this motion into an informative nerve signal.

### The Nature of Sound

To understand the physiology of hearing, it is necessary to know some basic properties of sound. **Sound** is any audible vibration of molecules. It can be transmitted through water, solids, or air, but not through a vacuum. Our discussion is limited to airborne sound.

Sound is produced by a vibrating object such as a tuning fork, a loudspeaker, or the vocal cords. Consider a loudspeaker. When the speaker cone moves forward, it pushes air molecules ahead of it. They collide with other molecules just ahead of them, and energy is transferred from molecule to molecule until it reaches the eardrum. No one molecule moves very far; they simply collide with each other like billiard balls until finally, some molecules collide with the eardrum and make it vibrate. The sensations we perceive as pitch and loudness are related to the physical properties of these vibrations.

### Pitch

**Pitch** is our sense of whether a sound is "high" (treble) or "low" (bass). It is determined by the frequency at which the sound source, eardrum, and other parts of the ear vibrate. One movement of a vibrating object back and forth is called a *cycle,* and the number of cycles per second (or hertz, Hz) is called **frequency.** The lowest note on a piano, for example, is 27.5 Hz, middle C is 261 Hz, and the highest note is 4,176 Hz. The most sensitive human ears can hear frequencies from 20 to 20,000 Hz. The *infrasonic* frequencies below 20 Hz are not detected by the ear, but we sense them through vibrations of the skull and skin and they play a significant role in our appreciation of music. The inaudible vibrations above 20,000 Hz are *ultrasonic.* Human ears are most sensitive to frequencies ranging from 1,500 to 5,000 Hz. In this range, we can hear sounds of relatively low energy (volume), whereas sounds above or below this range must be louder to be audible (fig. 16.9). Normal speech falls within this frequency range. Most of the hearing loss suffered with age is in the range of 250 to 2,050 Hz.

### Loudness

**Loudness** is the perception of sound energy, intensity, or the **amplitude** of vibration. In the speaker example, amplitude is a measure of how far forward and back the cone vibrates on each cycle and how much it compresses the air molecules in

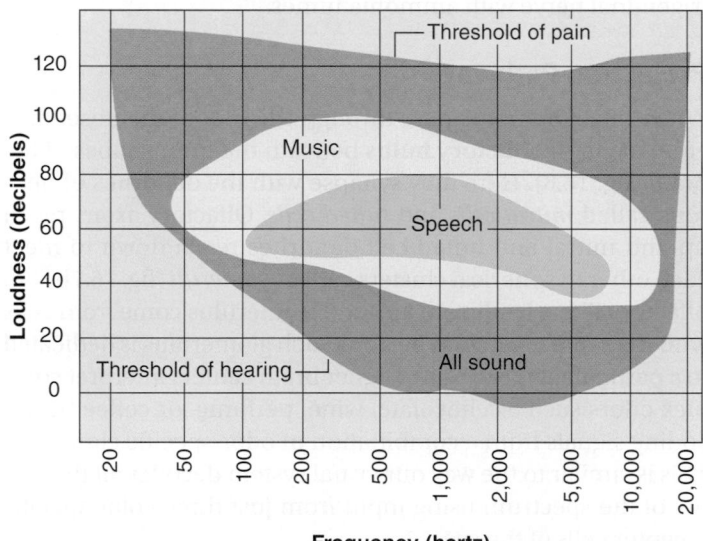

**FIGURE 16.9    The Range of Human Hearing.** People with very sensitive ears can hear sounds from 20 to 20,000 hertz (Hz), but to be heard, sounds at these extremes must be louder than those in the midrange. Our ears are most sensitive to frequencies of 1,500 to 5,000 Hz, where we can hear relatively soft sounds. Thus, the threshold of hearing varies with the frequency of the sound. Most sounds above 120 decibels (dB) are painful to the ear.

*?* *How would the shape of this graph change in a case of moderate hearing loss between 200 and 5,000 Hz?*

front of it. Loudness is expressed in decibels (dB), with 0 dB defined by a sound energy level that corresponds to the threshold of human hearing. Every 10 dB step up the scale represents a sound with 10 times greater intensity. Thus, 10 dB is 10 times threshold, 20 dB is 100 times threshold, 30 dB is 1,000 times threshold, and so forth. Normal conversation has a loudness of about 60 dB. At most frequencies, the threshold of pain is 120 to 140 dB, approximately the intensity of a loud thunderclap. Prolonged exposure to sounds greater than 90 dB can cause permanent loss of hearing.

## Anatomy of the Ear

The ear has three sections called the *outer, middle,* and *inner ear.* The first two are concerned only with transmitting sound to the inner ear, where vibration is converted to nerve signals.

## Outer Ear

The **outer (external)** ear is essentially a funnel for conducting airborne vibrations to the *tympanic membrane* (eardrum). It begins with the fleshy **auricle (pinna)** on the side of the head, shaped and supported by elastic cartilage except for the earlobe, which is mostly adipose tissue. The auricle has a predictable arrangement of named whorls and recesses that direct sound into the auditory canal (fig. 16.10).

The **auditory canal (external acoustic meatus)** is the passage leading through the temporal bone to the tympanic membrane. It follows a slightly S-shaped course for about 3 cm (fig. 16.11). It is lined with skin and supported by fibrocartilage at its opening and by the temporal bone for the rest of its length.

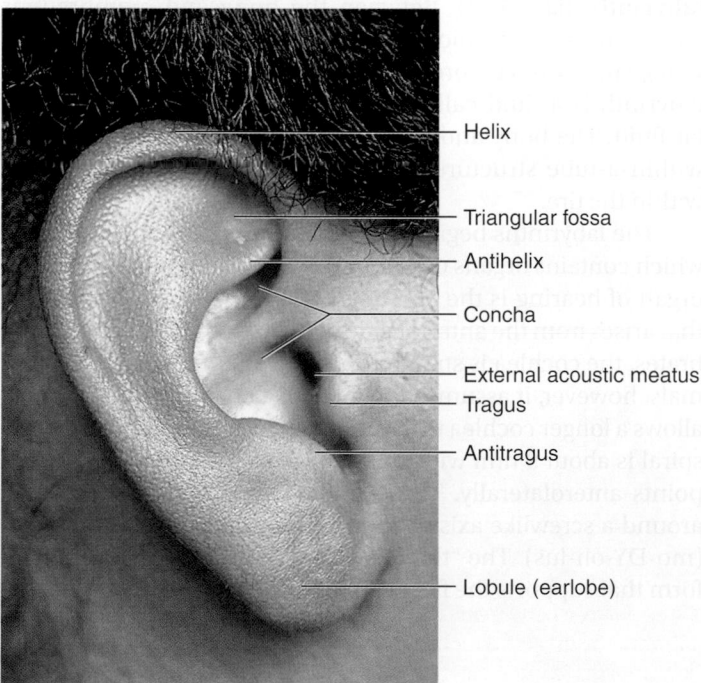

Helix

Triangular fossa

Antihelix

Concha

External acoustic meatus

Tragus

Antitragus

Lobule (earlobe)

**FIGURE 16.10** **Anatomy of the Auricle (Pinna) of the Ear.**

The outer end of the canal is protected by stiff **guard hairs.** The canal has ceruminous and sebaceous glands whose secretions mix with dead skin cells and form **cerumen** (earwax). Cerumen is sticky and coats the guard hairs, making them more effective in blocking foreign particles from the auditory canal. Its stickiness may also be a deterrent to insects, ticks, or other pests. In addition, it contains lysozyme and has a low pH, both of which inhibit bacterial growth; it waterproofs the canal and protects its skin and the tympanic membrane from absorbing water; and it keeps the tympanic membrane pliable. Cerumen normally dries and falls from the canal, but sometimes it becomes impacted and interferes with hearing.

## Middle Ear

The **middle ear** is located in the **tympanic cavity** of the temporal bone. What we colloquially call the eardrum is anatomically known as the **tympanic**[17] **membrane.** It closes the inner end of the auditory canal and separates it from the middle ear. The membrane is about 1 cm in diameter and slightly concave on its outer surface. It is suspended in a ring-shaped groove in the temporal bone and vibrates freely in response to sound. It is innervated by sensory branches of the vagus and trigeminal nerves and is highly sensitive to pain.

Posteriorly (behind the ear), the tympanic cavity is continuous with the mastoidal air cells in the mastoid process. It is filled with air that enters by way of the **auditory (eustachian**[18] **or pharyngotympanic) tube,** a passageway to the nasopharynx. (Be careful not to confuse *auditory tube* with *auditory canal.*) The auditory tube is normally flattened and closed, but swallowing or yawning opens it and allows air to enter or leave the tympanic cavity. This equalizes air pressure on both sides of the tympanic membrane, allowing it to vibrate freely. Excessive pressure on one side or the other muffles the sense of hearing and may cause pain, as one commonly experiences in airline flight. Unfortunately, the auditory tube also allows throat infections to spread to the middle ear (see Deeper Insight 16.2).

The tympanic cavity, a space only 2 to 3 mm wide between the outer and inner ears, contains the three smallest bones and the two smallest skeletal muscles of the body. The bones, called the **auditory ossicles,**[19] connect the tympanic membrane to the inner ear. Progressing inward, the first is the **malleus,**[20] which has an elongated *handle* attached to the inner surface of the tympanic membrane; a *head,* which is suspended by a ligament from the wall of the tympanic cavity; and a *short process,* which articulates with the next ossicle. The second bone, the **incus,**[21] has a roughly triangular *body* that articulates with the malleus; a *long limb* that articulates with the stapes; and a *short limb* (not illustrated) suspended by a ligament from the wall of

---

[17]*tympan* = drum
[18]Bartholomeo Eustachio (1524–74), Italian anatomist
[19]*oss* = bone; *icle* = little
[20]*malleus* = hammer
[21]*incus* = anvil

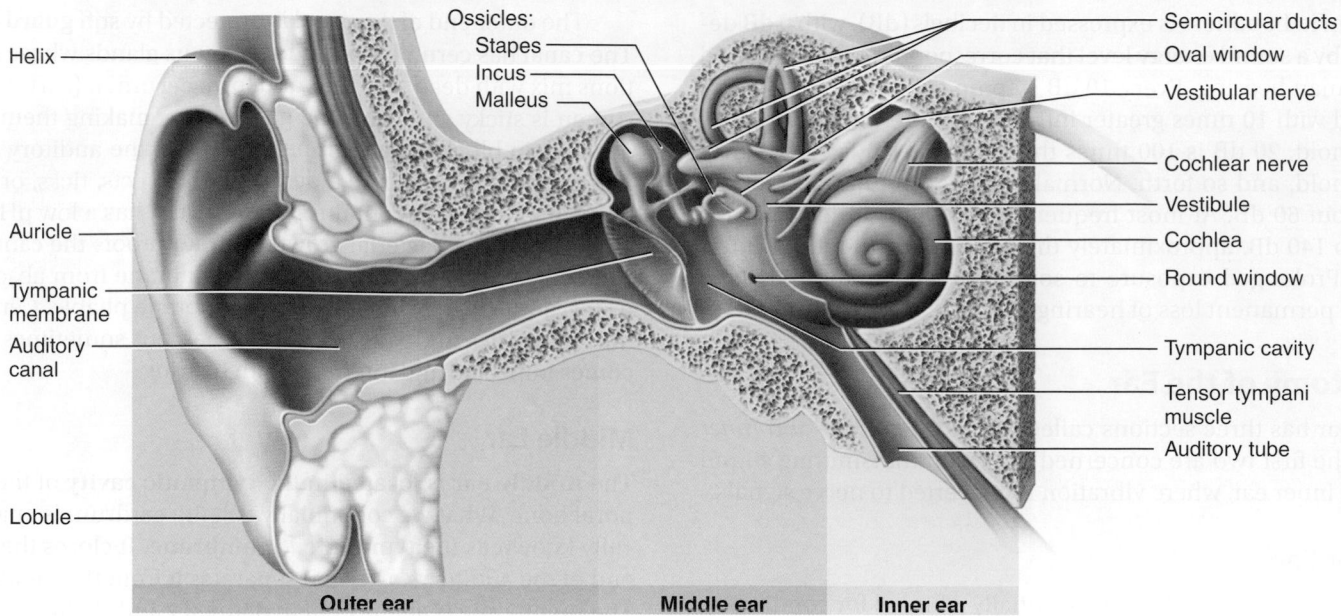

**FIGURE 16.11   Internal Anatomy of the Ear.**

the cavity. The **stapes**[22] (STAY-peez) is shaped like a stirrup. It has a *head* that articulates with the incus; two *limbs* that form an arch; and an elliptical *base (footplate)*. The base is held by a ringlike ligament in an opening called the **oval window,** where the inner ear begins. On the other side of the base is an inner-ear liquid called *perilymph.*

The muscles of the middle ear are the stapedius and tensor tympani. The **stapedius** (sta-PEE-dee-us) arises from the posterior wall of the cavity and inserts on the stapes. The **tensor tympani** (TEN-sor TIM-pan-eye) arises from the wall of the

auditory tube, travels alongside it, and inserts on the malleus. The function of these muscles is discussed under "The Physiology of Hearing."

## Inner Ear

The **inner (internal) ear** is housed in a maze of temporal bone passageways called the **bony (osseous) labyrinth,** which is lined by a system of fleshy tubes called the **membranous labyrinth** (fig. 16.12). Between the bony and membranous labyrinths is a cushion of fluid, the **perilymph** (PER-ih-limf), similar to cerebrospinal fluid. Within the membranous labyrinth is a fluid called **endolymph,** similar to intracellular fluid. The bony and membranous labyrinths form a tube-within-a-tube structure, somewhat like a bicycle inner tube within the tire.

The labyrinths begin with a chamber called the **vestibule,** which contains organs of equilibrium to be discussed later. The organ of hearing is the **cochlea**[25] (COC-lee-uh), a coiled tube that arises from the anterior side of the vestibule. In other vertebrates, the cochlea is straight or slightly curved. In most mammals, however, it assumes the form of a snail-like spiral, which allows a longer cochlea to fit in a compact space. In humans, the spiral is about 9 mm wide at the base and 5 mm high. Its apex points anterolaterally. The cochlea winds for about 2.5 coils around a screwlike axis of spongy bone called the **modiolus**[26] (mo-DY-oh-lus). The "threads of the screw" form a spiral platform that supports the fleshy tube of the cochlea.

## DEEPER INSIGHT 16.2
## CLINICAL APPLICATION

### Middle-Ear Infection

*Otitis*[23] *media* (middle-ear infection) is common in children because their auditory tubes are relatively short and horizontal. They allow upper respiratory infections to spread easily from the throat to the tympanic cavity and mastoidal air cells. Fluid accumulates in the cavity and produces pressure, pain, and impaired hearing. If otitis media goes untreated, it may spread from the mastoidal air cells and cause meningitis, a potentially deadly infection (see Deeper Insight 14.1, p. 514). Chronic otitis media can also cause fusion of the middle-ear bones and result in hearing loss. It is sometimes necessary to drain fluid from the tympanic cavity by lancing the tympanic membrane and inserting a tiny drainage tube—a procedure called *tympanostomy.*[24] The tube, which is eventually sloughed out of the ear, relieves the pressure and permits the infection to heal.

---

[22]*stapes* = stirrup
[23]*ot* = ear; *itis* = inflammation
[24]*tympano* = eardrum; *stomy* = making an opening

[25]*cochlea* = snail
[26]*modiolus* = hub

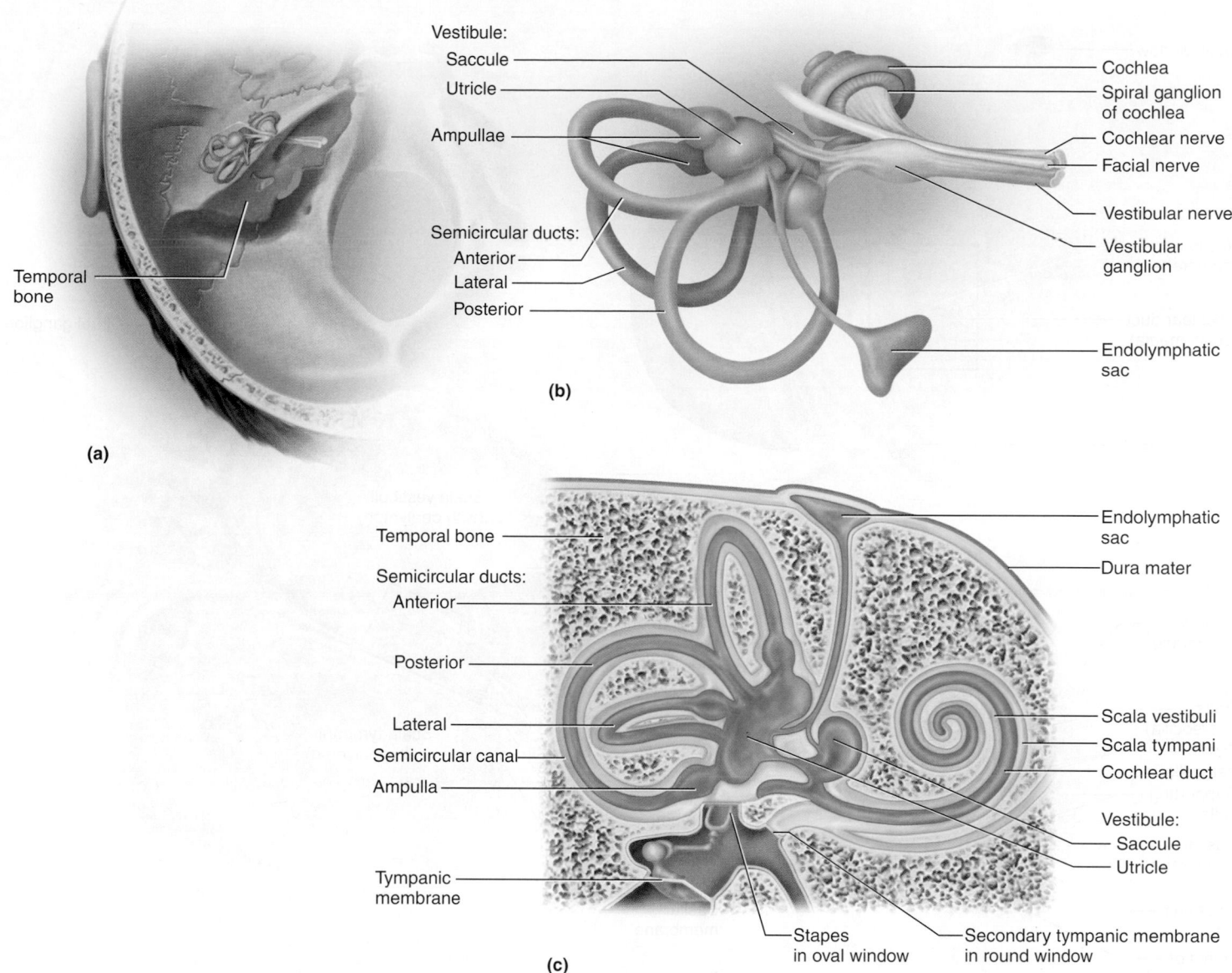

**FIGURE 16.12 Anatomy of the Membranous Labyrinth.** (a) Position and orientation within the petrous part of the temporal bone. (b) Structure of the membranous labyrinth and nerves. (c) Relationship of the perilymph (green) and endolymph (blue) to the labyrinth.

A vertical section cuts through the cochlea about five times (fig. 16.13a). A single cross section looks like figure 16.13b. It is important to realize that the structures seen in cross section actually have the form of spiral strips winding around the modiolus from base to apex.

The cochlea has three fluid-filled chambers separated by membranes. The superior chamber is called the **scala**[27] **vestibuli** (SCAY-la vess-TIB-you-lye) and the inferior one is the **scala tympani.** These are filled with perilymph and communicate with each other through a narrow channel at the apex of the cochlea. The scala vestibuli begins near the oval window and spirals to the apex; from there, the scala tympani

spirals back down to the base and ends at the **round window** (fig. 16.12). The round window is covered by a membrane called the *secondary tympanic membrane.*

The middle chamber is a triangular space, the **cochlear duct (scala media).** It is separated from the scala vestibuli above by a thin **vestibular membrane** and from the scala tympani below by a much thicker **basilar membrane.** Unlike those chambers, it is filled with endolymph rather than perilymph. The vestibular membrane separates the endolymph from the perilymph and helps to maintain a chemical difference between them. Within the cochlear duct, supported on the basilar membrane, is the **spiral organ,** also known as the

---

[27] *scala* = staircase

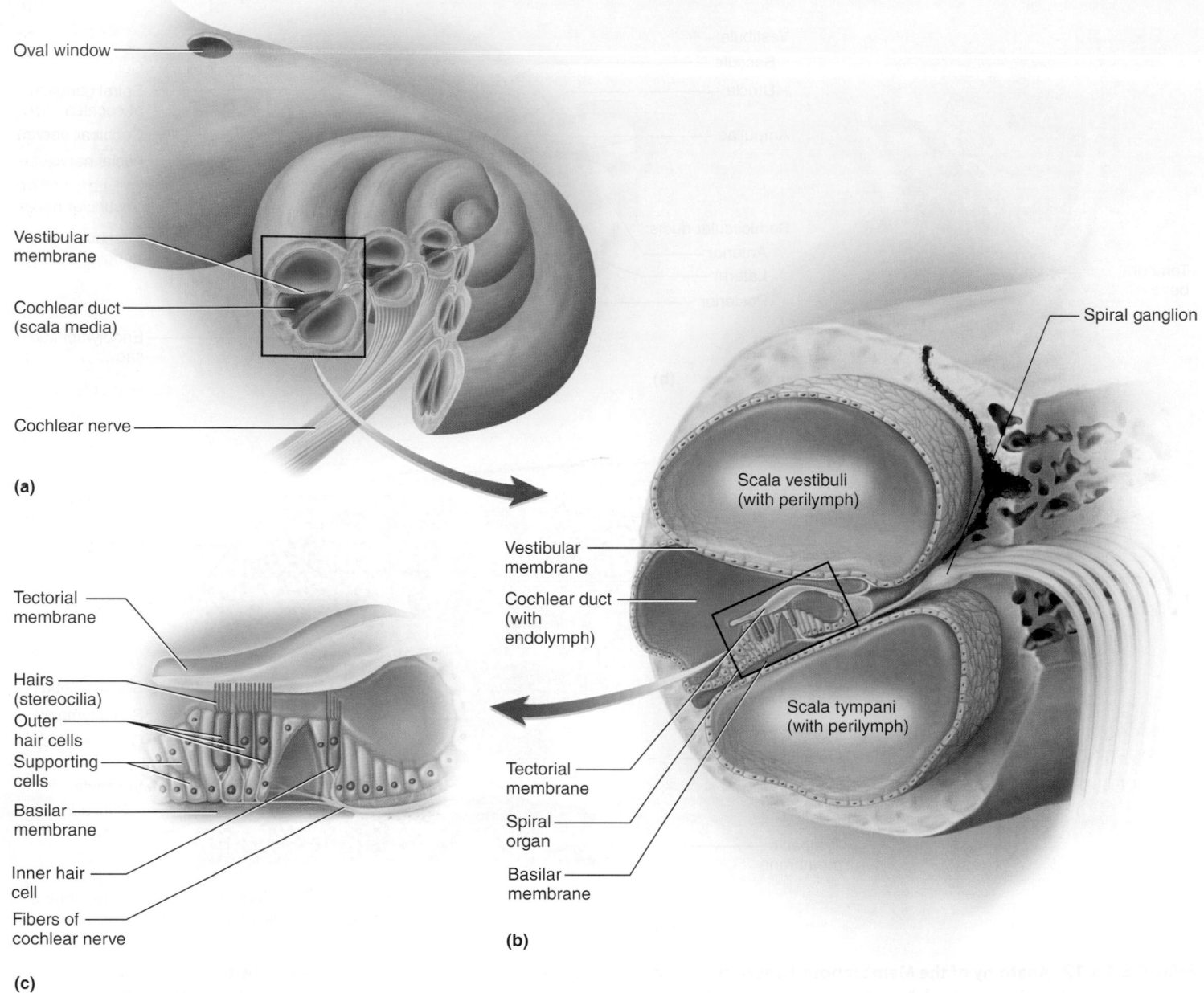

Oval window

Vestibular membrane

Cochlear duct (scala media)

Cochlear nerve

**(a)**

Spiral ganglion

Scala vestibuli (with perilymph)

Vestibular membrane

Cochlear duct (with endolymph)

Scala tympani (with perilymph)

Tectorial membrane

Spiral organ

Basilar membrane

**(b)**

Tectorial membrane

Hairs (stereocilia)

Outer hair cells

Supporting cells

Basilar membrane

Inner hair cell

Fibers of cochlear nerve

**(c)**

**FIGURE 16.13 Anatomy of the Cochlea.** (a) Vertical section. In anatomical position, the apex of the cochlea faces downward and anterolaterally. (b) Detail of one section through the cochlea. (c) Detail of the spiral organ.

*acoustic organ* or *organ of Corti*[28] (COR-tee)—a thick epithelium of sensory and supporting cells and associated membranes (fig. 16.13c). This is the device that converts vibrations into nerve impulses, so we must pay particular attention to its structural details.

The spiral organ has an epithelium composed of **hair cells** and **supporting cells.** Hair cells are named for the long, stiff microvilli called **stereocilia**[29] on their apical surfaces. (Stereocilia are not true cilia. They do not have an axoneme of microtubules as seen in cilia, and they do not move by themselves.)

Resting on top of the stereocilia is a gelatinous **tectorial**[30] **membrane.**

The spiral organ has four rows of hair cells spiraling along its length (fig. 16.14). About 3,500 of these, called **inner hair cells (IHCs),** are arranged in a row on the medial side of the basilar membrane (facing the modiolus). Each of these has a cluster of 50 to 60 stereocilia, graded from short to tall. Another 20,000 **outer hair cells (OHCs)** are neatly arranged in three rows across from the inner hair cells. Each OHC has about 100 stereocilia arranged in the form of a V, with their tips

---

[28]Alfonso Corti (1822–88), Italian anatomist
[29]*stereo* = solid

[30]*tect* = roof

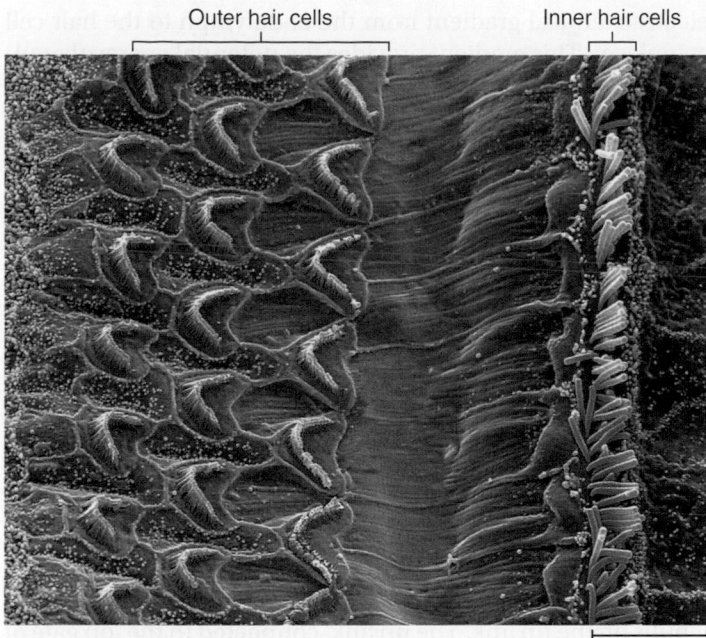

Outer hair cells          Inner hair cells

10 μm

**FIGURE 16.14  Apical Surfaces of the Cochlear Hair Cells.** All signals that we hear come from the inner hair cells on the right.

❓ *What is the function of the three rows of hair cells on the left?*

embedded in the tectorial membrane. All that we hear comes from the IHCs, which supply 90% to 95% of the sensory fibers of the cochlear nerve. The function of the OHCs is to adjust the response of the cochlea to different frequencies and enable the IHCs to work with greater precision. We will see later how this is done. Hair cells are not neurons, but synapse with nerve fibers at their base—the OHCs with both sensory and motor neurons and the IHCs with sensory neurons only.

## The Physiology of Hearing

We can now examine the way sound affects the ear and produces action potentials. Sound waves enter the auditory canal on one side and nerve signals exit the inner ear on the other. Connecting these is the middle ear, so we begin with an analysis of its contribution.

## The Middle Ear

One might wonder why we have a middle ear at all—why the tympanic membrane doesn't simply vibrate against the fluid-filled labyrinth of the inner ear. The reason is that the tympanic membrane, which moves in air, vibrates easily, whereas the stapes must push against the perilymph of the inner ear. This fluid resists motion much more than air does. If the tympanic membrane had air on one side and fluid on the other, the sound waves would not have enough energy to move the fluid adequately. The tympanic membrane, however, has 18 times the

area of the oval window. By concentrating the energy of the vibrating tympanic membrane on an area $^1/_{18}$ that size, the ossicles create a greater force per unit area at the oval window and overcome the inertia of the fluid.

The auditory ossicles do not, however, provide any mechanical advantage, any amplification of sound. Vibrations of the stapes against the inner ear normally have the same amplitude as vibrations of the tympanic membrane against the malleus. Why, then, have a lever system composed of three ossicles? Why not simply have one ossicle concentrating the mechanical energy of the tympanic membrane directly on the inner ear?

The answer is that the ossicles serve at times to *lessen* the transfer of energy to the inner ear. They and their muscles have a protective function. In response to a loud noise, the tensor tympani pulls the tympanic membrane inward and tenses it, while the stapedius reduces the motion of the stapes. This **tympanic reflex** muffles the transfer of vibrations from the tympanic membrane to the oval window. The reflex probably evolved in part for protection from loud but slowly building natural sounds such as thunder. The reflex has a latency of about 40 ms, which is not quick enough to protect the inner ear from sudden artificial noises such as gunshots. The tympanic reflex also does not adequately protect the ears from sustained loud noises such as factory noise or loud music. Such noises can irreversibly damage the hair cells of the inner ear. It is therefore imperative to wear ear protection when using firearms or working in noisy environments.

The middle-ear muscles also help to coordinate speech with hearing. Without them, the sound of your own speech would be so loud it could damage your inner ear, and it would drown out soft or high-pitched sounds from other sources. Just as you are about to speak, however, the brain signals these muscles to contract. This dampens the sense of hearing in phase with the inflections of your own voice and makes it easier to hear other people while you are speaking.

▶▶▶ **APPLY WHAT YOU KNOW**

*What type of muscle fibers—slow oxidative or fast glycolytic (see p. 421)—do you think constitute the stapedius and tensor tympani? That is, which type would best suit the purpose of these muscles?*

## Stimulation of Cochlear Hair Cells

The next step in hearing is based on movement of the cochlear hair cells relative to stationary structures nearby. In this section, we will see how movements of the inner-ear fluids and basilar membrane move the hair cells and why it is important that the tectorial membrane near the hair cells remains relatively still.

A simple mechanical model of the ear can help in visualizing how this happens (fig. 16.15). (The vestibular membrane is omitted from the model for simplicity.) As you listen to your favorite music, each inward movement of the tympanic

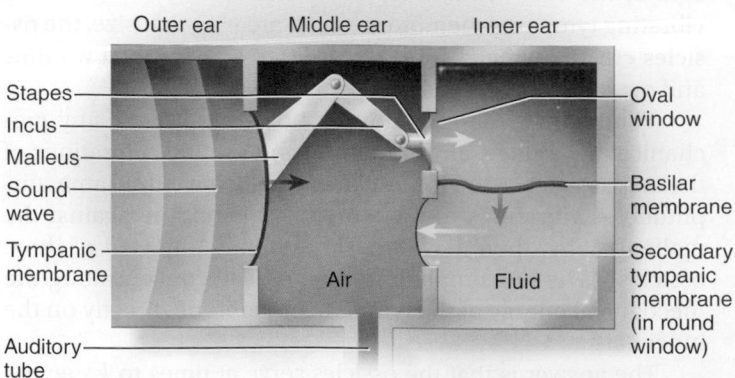

Outer ear Middle ear Inner ear

Stapes
Incus
Malleus
Sound wave
Tympanic membrane
Auditory tube

Oval window
Basilar membrane
Secondary tympanic membrane (in round window)

Air Fluid

**FIGURE 16.15 Mechanical Model of Auditory Function.** Each inward movement of the tympanic membrane pushes inward on the auditory ossicles of the middle ear and fluid of the inner ear. This pushes down on the basilar membrane, and pressure is relieved by an outward bulge of the secondary tympanic membrane. Thus the basilar membrane vibrates up and down in synchrony with the vibrations of the tympanic membrane.

❓ *Why would high air pressure in the middle ear reduce the movements of the basilar membrane of the inner ear?*

membrane pushes the middle-ear ossicles inward. The stapes, in turn, pushes on the perilymph in the scala vestibuli. Perilymph, like other liquids, cannot be compressed, so it flows away from the stapes footplate. The resulting pressure in the scala vestibuli pushes the vestibular membrane downward; this pushes on the endolymph in the cochlear duct; the endolymph pushes down on the basilar membrane; the basilar membrane pushes on the perilymph in the scala tympani; and finally, the secondary tympanic membrane bulges outward to relieve the pressure. As the cycle of vibration continues, the stapes pulls back from the oval window and all of this happens in reverse.

In short, as the stapes goes in-out-in, the secondary tympanic membrane goes out-in-out, and the basilar membrane goes down-up-down. It is not difficult to see how this happens—the only thing hard to imagine is that it can happen as often as 20,000 times per second! The important thing about all this is that the hair cells, affixed to the basilar membrane, go along for the ride, bobbing up and down as the basilar membrane moves.

To understand how all of this leads to electrical excitation of the hair cells, we must seemingly digress for a moment to examine the tips of the hair cells. These are bathed in a high-potassium fluid, the endolymph. Why is this fluid so rich in potassium, and why is that important?

Potassium is secreted into the endolymph by cells around the circumference of the cochlear duct (on the wall opposite from the modiolus). The vestibular membrane retains this fluid in the duct. Relative to the perilymph, the endolymph has an electrical potential of about +80 mV, and the interior of the hair cell about –40 mV. Thus there is an exceptionally strong

electrochemical gradient from the endolymph to the hair cell cytoplasm. This gradient provides the potential energy that ultimately enables the hair cell to work.

On the inner hair cells—the ones that generate all the signals we hear—each stereocilium has a single transmembrane protein at its tip that functions as a mechanically gated ion channel. A fine, stretchy protein filament called a **tip link** extends like a spring from the ion channel of one stereocilium to the sidewall of the stereocilium next to it (fig. 16.16). The stereocilia increase in height progressively, so that every stereocilium but the tallest one has a tip link leading to the next taller stereocilium beside it.

Now what of the tectorial membrane? This is a conspicuous structure of the spiral organ that is anchored to the core of the cochlea and remains relatively still as the hair cells dance up and down to the beat of the music. Each time the basilar membrane rises upward toward the tectorial membrane, the hair cell stereocilia are pushed against that membrane and tilt toward the tallest one. As each taller stereocilium bends over, it pulls on the tip link. The tip link, connected to the ion gate of the next shorter stereocilium, pulls the gate open and allows ions to flood into the cell. The gate is nonselective, but since the predominant ion of the endolymph is $K^+$, the primary effect of this gating is to allow a quick burst of $K^+$ to flow into the hair cell. This depolarizes the hair cell while the gate is open, and when the basilar membrane drops and the stereocilium bends the other way, the gate closes and the cell becomes briefly hyperpolarized. During each moment of depolarization, a hair cell releases a burst of neurotransmitter from its base, exciting the sensory dendrite with which the hair cell synapses. Each depolarization thus generates a signal in the cochlear nerve.

To summarize: Each upward movement of the basilar membrane pushes the inner hair cells closer to the stationary tectorial membrane. This forces the stereocilia to bend in the direction of the tallest one. Each stereocilium has a tip link connecting it to an ion channel at the top of the next shorter stereocilium. When the taller one bends over, it pulls the channel open. Potassium ions flow into the hair cell and depolarize it. The hair cell releases a burst of neurotransmitter, exciting the sensory processes of the cochlear nerve cells below it. Thus a signal is generated in the cochlear nerve and transmitted to the brain.

## Sensory Coding

For sounds to carry any meaning, we must distinguish differences in loudness and pitch. Our ability to do so stems from the fact that the cochlea responds differently to sounds of different amplitude and frequency. Variations in loudness (amplitude) cause variations in the intensity of cochlear vibration. A soft sound produces relatively slight up-and-down movements of the basilar membrane. Hair cells are stimulated only moderately, and a given sound frequency stimulates hair cells in a relatively limited, or focused, region of the cochlea. A louder

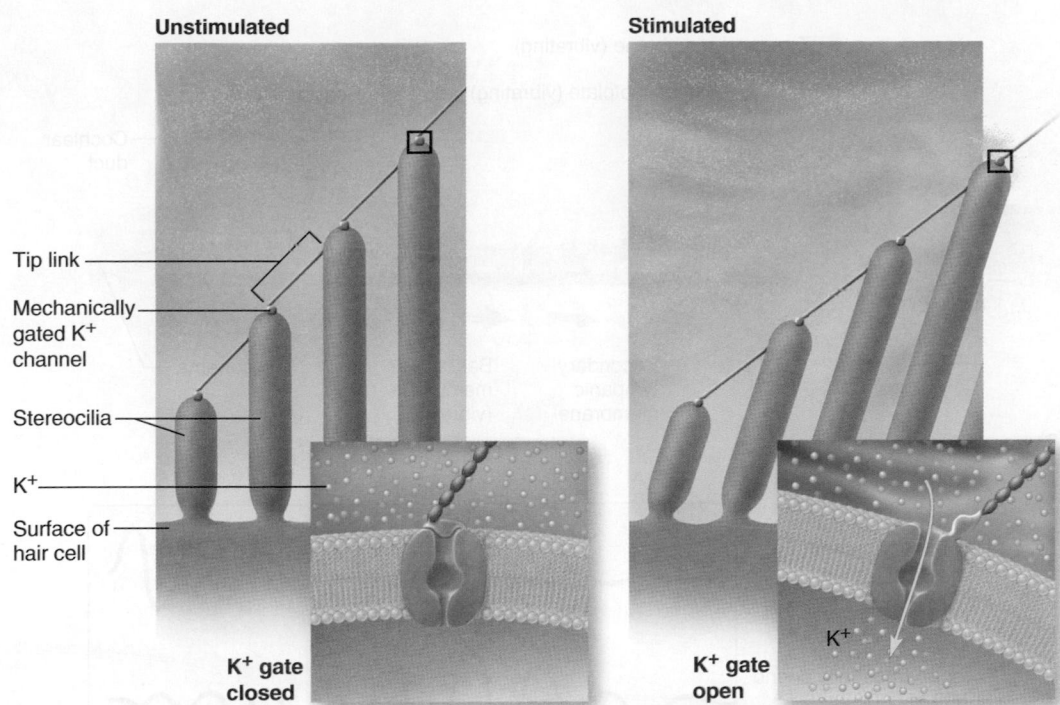

**Unstimulated**

**Stimulated**

Tip link

Mechanically
gated K⁺
channel

Stereocilia

K⁺

Surface of
hair cell

**K⁺ gate
closed**

**K⁺ gate
open**

K⁺

**FIGURE 16.16 Potassium Channels of the Cochlear Hair Cells.** Each stereocilium has a K⁺ channel at its tip. Vibrations of the cochlea cause each stereocilium to bend and, with its tip link, pull open the K⁺ channel of the adjacent stereocilium. The inflow of K⁺ depolarizes the hair cell.

sound makes the basilar membrane vibrate more vigorously. Hair cells respond more intensely, generating a higher firing frequency in the cochlear nerve. In addition, for a given frequency, a loud sound vibrates a longer segment of the basilar membrane and thus excites a greater number of hair cells. If the brain detects moderate firing rates associated with hair cells in relatively narrow bands of the cochlea, it interprets this as a soft sound. If it detects a high firing frequency in nerve fibers associated with broader bands, it interprets this as a loud sound.

Frequency discrimination is based on a structural gradient in the basilar membrane resembling a piano's gradient in sound frequency from the short strings to long strings. At its proximal end (the base of the cochlea), the membrane is attached, narrow, and stiff, like the piano's short strings. At its distal end (the apex of the cochlea), it is unattached, five times wider, and more flexible. Think of the basilar membrane as analogous to a wire stretched tightly between two posts. If you pluck the wire at one end, a wave of vibration travels down its length and back. This produces a standing wave, with some regions of the wire vertically displaced more than others. Similarly, a sound causes a standing wave in the basilar membrane. The peak amplitude of this wave is near the distal end in the case of low-frequency sounds and nearer the proximal end with sounds of higher frequencies. When the brain receives signals mainly from inner hair cells at the distal end, it interprets the sound as low-pitched; when signals come mainly from the proximal end, it interprets the sound as

high-pitched (fig. 16.17). Speech, music, and other everyday sounds, of course, are not pure tones—they create complex, ever-changing patterns of vibration in the basilar membrane that must be decoded by the brain.

## Cochlear Tuning

Just as we tune a radio to receive a certain frequency, we also tune our cochlea to receive some frequencies better than others. The outer hair cells (OHCs) are supplied with a few sensory fibers (5%–10% of those in the cochlear nerve), but more importantly, they receive motor fibers from the brain.

In response to sound, the OHCs send signals to the medulla oblongata by way of the sensory neurons, and the pons sends signals immediately back to the OHCs by way of the motor neurons. In response, the hair cells shorten up to 15%. Remember that an OHC is anchored to the basilar membrane below and its stereocilia to the tectorial membrane above. Therefore, tensing of an OHC reduces the basilar membrane's mobility. This results in some regions of the cochlea sending fewer signals to the brain than neighboring regions, so the brain can better distinguish between sound frequencies. When OHCs are experimentally incapacitated, the inner hair cells (IHCs) respond much less precisely to differences in pitch.

There is another mechanism of cochlear tuning involving the inner hair cells. The pons sends efferent fibers to the cochlea that synapse with the sensory nerve fibers near the base of the IHCs. The efferent fibers can inhibit the sensory fibers

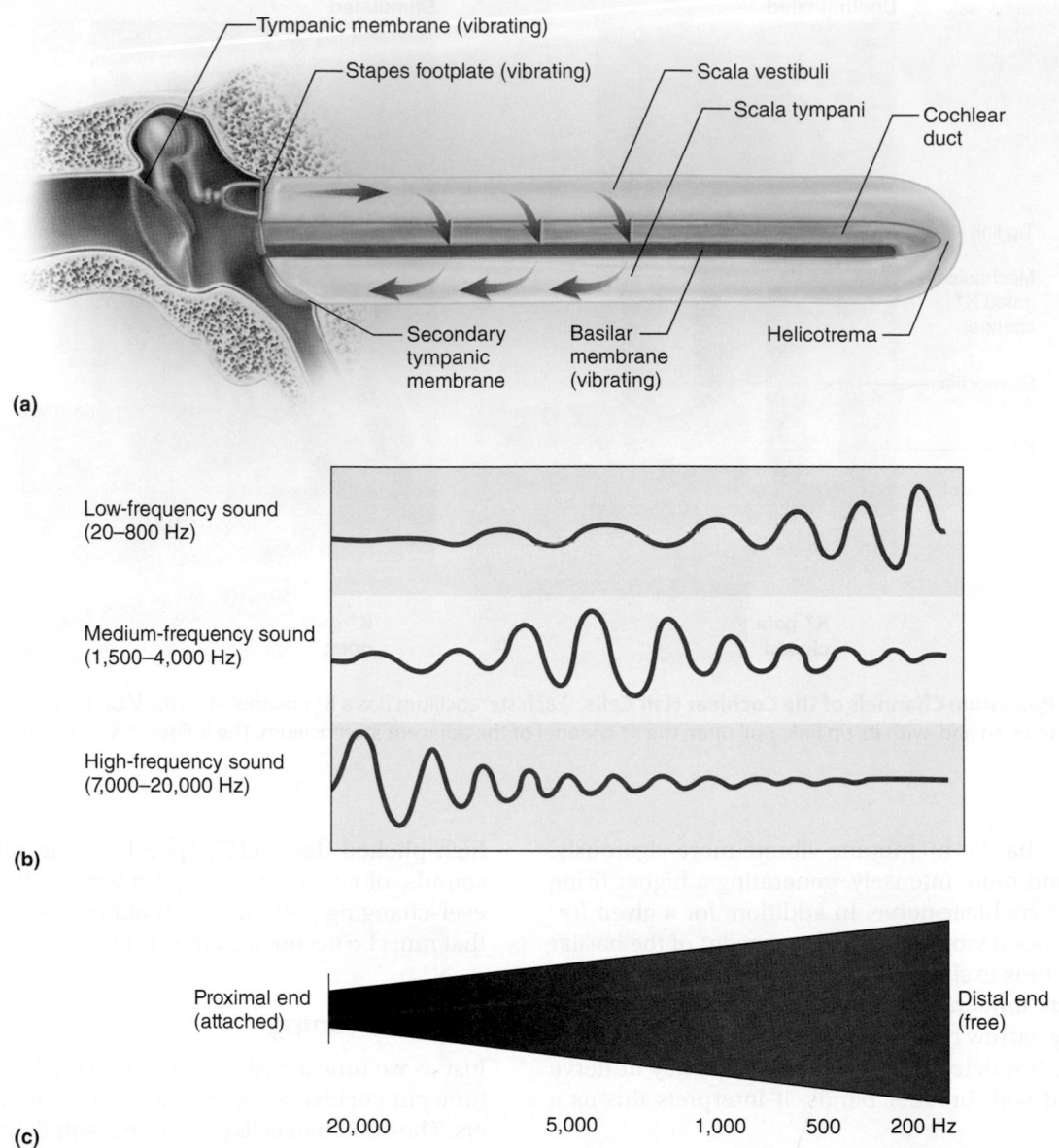

**FIGURE 16.17 Frequency Response of the Basilar Membrane of the Cochlea.** (a) The cochlea, uncoiled and laid out straight. (b) Sounds produce a standing wave of vibration along the basilar membrane. The peak amplitude of the wave varies with the frequency of the sound, as shown here. The amount of vibration is greatly exaggerated in this diagram to clarify the standing wave. (c) The taper of the basilar membrane and its correlation with sound frequencies. High frequencies (7,000–20,000 Hz) are best detected by hair cells near the narrow proximal end at the left, and low frequencies (20–800 Hz) by hair cells near the wider distal end at the right.

from firing in some areas of the cochlea, and thus enhance the contrast between signals from the more responsive and less responsive regions. Combined with the previously described role of the OHCs, this sharpens the tuning of the cochlea and our ability to discriminate pitch.

## The Auditory Projection Pathway

The sensory nerve fibers beginning at the bases of the hair cells belong to bipolar sensory neurons. Their somas form a coil, the **spiral ganglion,** around the modiolus (see fig. 16.13), and their

axons lead away from the cochlea as the **cochlear nerve.** This nerve joins the *vestibular nerve,* discussed later, and the two together become the *vestibulocochlear nerve* (cranial nerve VIII).

Each ear sends fibers to both sides of the medulla. There, they end in the *cochlear nuclei,* synapsing with second-order neurons that ascend to the *superior olivary nucleus* of the pons (fig. 16.18). By way of cranial nerve VIII, the superior olivary nucleus issues the efferent fibers back to the cochlea that are involved in cochlear tuning. By way of cranial nerves $V_3$ and VII, it also issues motor fibers to the tensor tympani and

## DEEPER INSIGHT 16.3
### CLINICAL APPLICATION

#### Deafness

*Deafness* means any hearing loss, from mild and temporary to complete and irreversible. *Conductive deafness* results from any condition that interferes with the transmission of vibrations to the inner ear. Such conditions include a damaged tympanic membrane, otitis media, blockage of the auditory canal, and otosclerosis. *Otosclerosis*[31] is fusion of the auditory ossicles to each other, or fusion of the stapes to the oval window. Either way, it prevents the bones from vibrating freely. *Sensorineural (nerve) deafness* results from the death of hair cells or any of the nervous elements concerned with hearing. It is a common occupational disease of factory and construction workers, musicians, and other people exposed to frequent or sustained loud sounds. Deafness leads some people to develop delusions of being talked about, disparaged, or cheated. Beethoven said his deafness drove him nearly to suicide.

---

stapedius muscles, respectively. The superior olivary nucleus also functions in **binaural**[32] **hearing**—comparing signals from the right and left ears to identify the direction from which a sound is coming.

Other fibers from the cochlear nuclei ascend to the inferior colliculi of the midbrain. The inferior colliculi help to locate the origin of a sound in space, process fluctuations in pitch that are important for such purposes as understanding another person's speech, and mediate the startle response and rapid head turning that occur in reaction to loud or sudden noises.

Third-order neurons begin in the inferior colliculi and lead to the thalamus. Fourth-order neurons complete the pathway from there to the primary auditory cortex; thus the auditory pathway, unlike most other sensory pathways, involves not three but four neurons from receptor to cerebral cortex. The primary auditory cortex lies in the superior margin of the temporal lobe deep within the lateral sulcus (see fig. 14.20, p. 535). The temporal lobe is the site of conscious perception of sound, and it completes the information processing essential to binaural hearing. Because of extensive decussation in the auditory pathway, damage to the right or left auditory cortex does not cause a unilateral loss of hearing.

## Equilibrium

The original function of the ear in vertebrate evolution was not hearing, but **equilibrium**—coordination, balance, and orientation in three-dimensional space. Only later did vertebrates evolve the cochlea, outer- and middle-ear structures, and auditory function of the ear. In humans, the receptors for equilibrium constitute the **vestibular apparatus,** which consists of three **semicircular ducts** and two chambers—an anterior

**saccule**[33] (SAC-yule) and a posterior **utricle**[34] (YOU-trih-cul) (see fig. 16.12b, c).

The sense of equilibrium is divided into **static equilibrium,** the perception of the orientation of the head when the body is stationary, and **dynamic equilibrium,** the perception of motion or acceleration. There are two kinds of acceleration: *linear acceleration,* a change in velocity in a straight line, as when riding in a car or elevator, and *angular acceleration,* a change in the rate of rotation, as when your car turns a corner or you swivel in a rotating chair. The saccule and utricle are responsible for static equilibrium and the sense of linear acceleration; the semicircular ducts detect only angular acceleration.

### The Saccule and Utricle

The saccule and utricle each contain a 2 × 3 mm patch of hair cells and supporting cells called a **macula.**[35] The **macula sacculi** lies nearly vertically on the wall of the saccule, and the **macula utriculi** lies nearly horizontally on the floor of the utricle (fig. 16.19a).

Each hair cell of a macula has 40 to 70 stereocilia and one true cilium called a **kinocilium.**[36] The tips of the stereocilia and kinocilium are embedded in a gelatinous **otolithic membrane.** This membrane is weighted with protein–calcium carbonate granules called **otoliths**[37] (fig. 16.19b), which add to the weight and inertia of the membrane and enhance the sense of gravity and motion.

Figure 16.19c shows how the macula utriculi detects tilt of the head. With the head erect, the otolithic membrane bears directly down on the hair cells, and stimulation is minimal. When the head is tilted, however, the heavy otolithic membrane sags and bends the stereocilia, stimulating the hair cells. Any orientation of the head causes a combination of stimulation to the utricules and saccules of the two ears. The brain interprets head orientation by comparing these inputs to each other and to other input from the eyes and stretch receptors in the neck, thereby detecting whether only the head is tilted or the entire body is tipping.

The inertia of the otolithic membranes is especially important in detecting linear acceleration. Suppose you are sitting in a car at a stoplight and then begin to move. The membrane of the macula utriculi briefly lags behind the rest of the tissues, bends the stereocilia backward, and stimulates the cells. When you stop at the next light, the macula stops but the otolithic membrane keeps going for a moment, bending the stereocilia forward. The hair cells convert this stimulation to nerve signals, and the brain is thus advised of changes in your linear velocity.

The macula sacculi is nearly vertical and its hair cells therefore respond to vertical acceleration and deceleration. If you are standing in an elevator and it begins to move up, the membrane of the macula sacculi lags behind briefly and pulls

---

[31]*oto* = ear; *scler* = hardening; *osis* = process, condition
[32]*bin* = two; *aur* = ears

[33]*saccule* = little sac
[34]*utricle* = little bag
[35]*macula* = spot
[36]*kino* = moving
[37]*oto* = ear; *lith* = stone

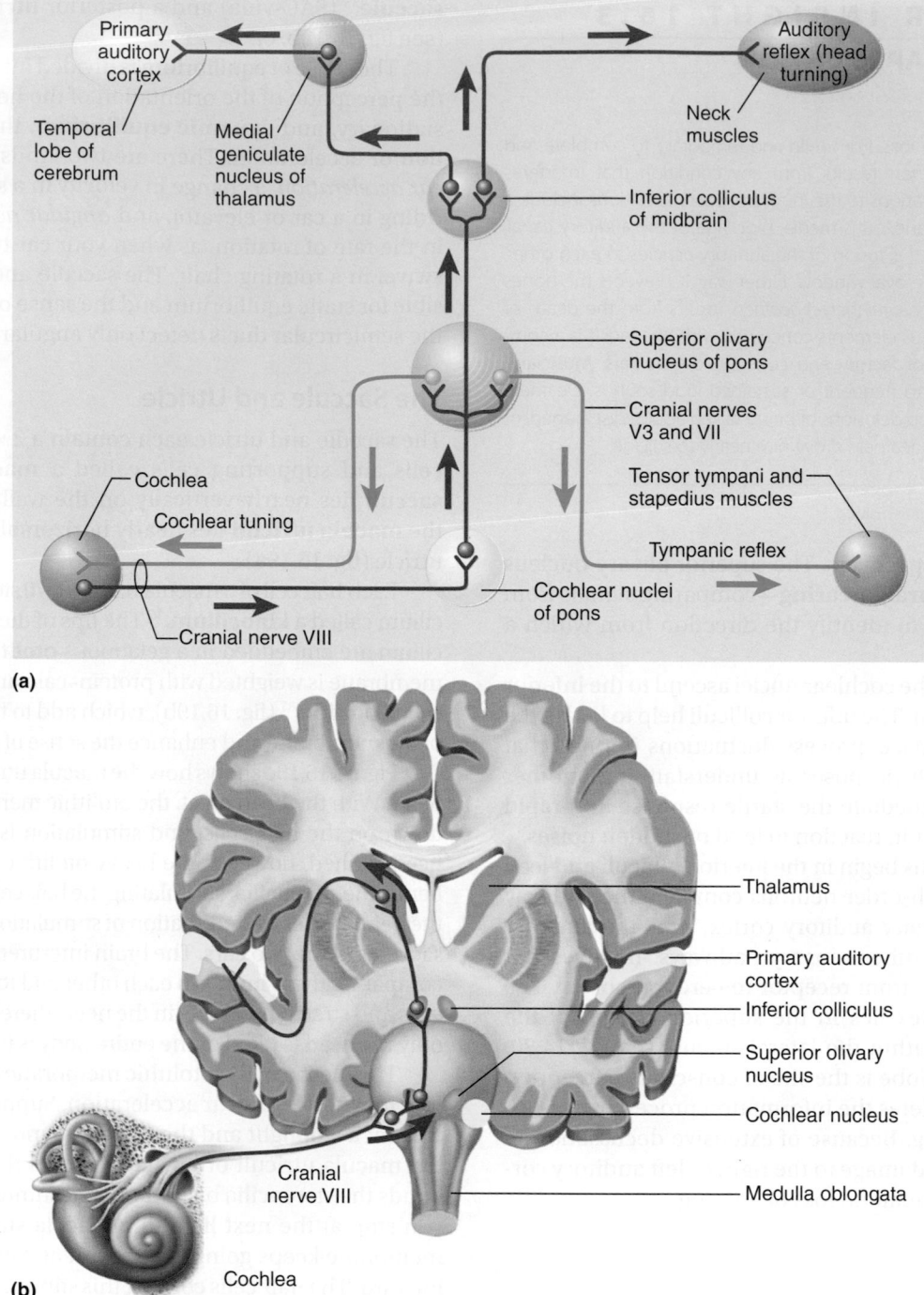

**FIGURE 16.18 Auditory Pathways in the Brain.** (a) Schematic. (b) Brainstem and frontal section of the cerebrum, showing the locations of auditory processing centers (cranial nerve V$_3$ = trigeminal nerve, mandibular division; CN VII = facial nerve; CN VIII = vestibulocochlear nerve).

down on the hairs. When the elevator stops, the membrane keeps going for a moment and bends the hairs upward. In both cases, the hair cells are stimulated and the brain is made aware of your vertical movements. These sensations are important in such ordinary actions as sitting down, and as your head bobs up and down during walking and running.

## The Semicircular Ducts

The head also experiences rotary movements, such as when you spin in a rotating chair, walk down a hall and turn a corner, or bend forward to pick something up from the floor. Such movements are detected by the three semicircular ducts

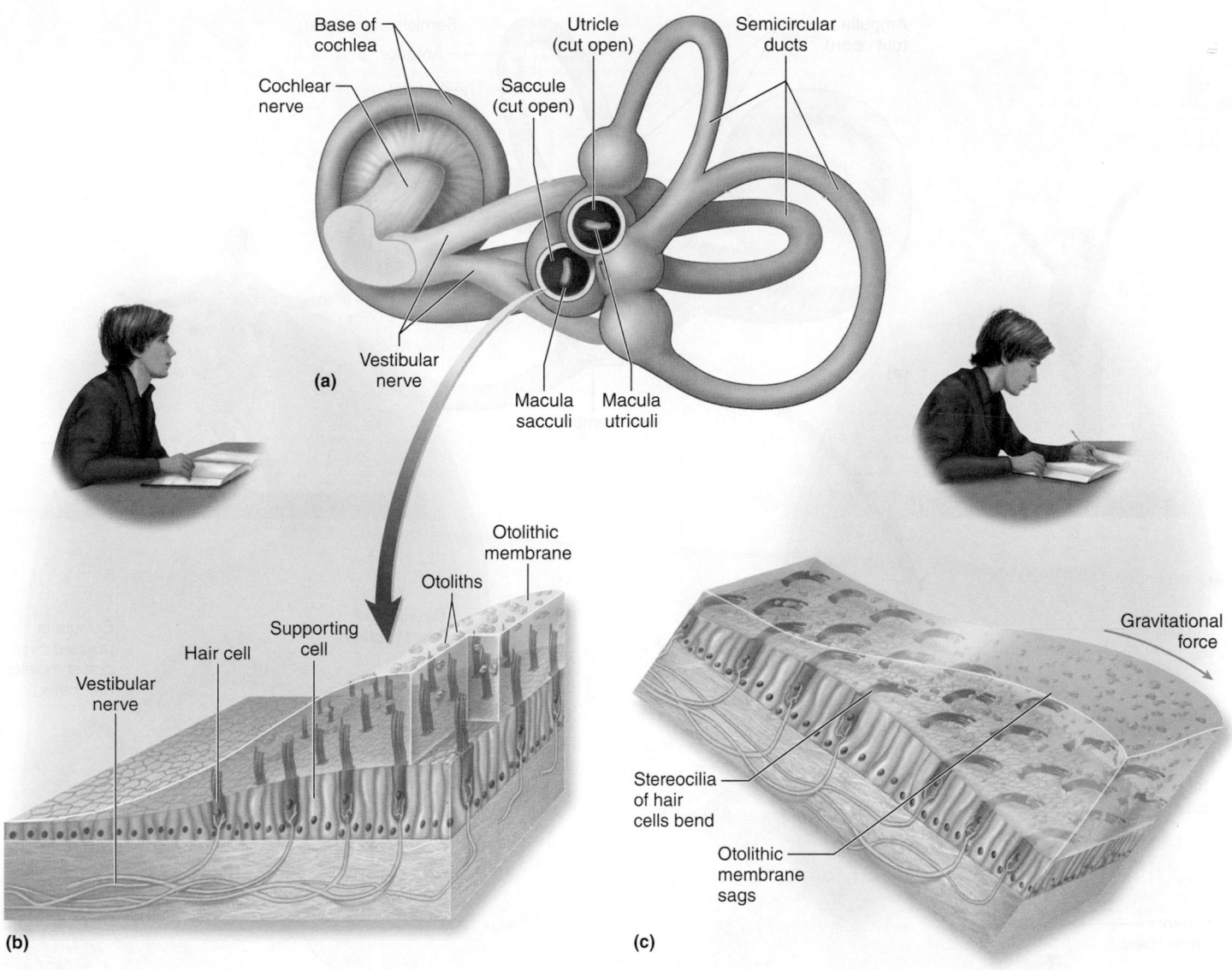

**FIGURE 16.19  The Saccule and Utricle.**  (a) Locations of the macula sacculi and macula utriculi. (b) Structure of a macula. (c) Action of the otolithic membrane on the hair cells when the head is tilted.

(fig. 16.20), housed in the bony *semicircular canals* of the temporal bone. The *anterior* and *posterior semicircular ducts* are oriented vertically at right angles to each other. The *lateral semicircular duct* is about 30° from the horizontal plane. The orientation of the ducts causes a different duct to be stimulated by rotation of the head in different planes.

The ducts are filled with endolymph. Each one opens into the utricle and has a dilated sac at one end called the **ampulla.**[38] Within the ampulla is a mound of hair cells and

supporting cells called the **crista**[39] **ampullaris.** The hair cells have stereocilia and a kinocilium embedded in the **cupula,**[40] a gelatinous cap that extends from the crista to the roof of the ampulla. When the head turns, the duct rotates but the endolymph lags behind. It pushes the cupula, bends the stereocilia, and stimulates the hair cells. After 25 to 30 seconds of continual rotation, however, the endolymph catches up with the movement of the duct and stimulation of the hair cells ceases.

---

[38]*ampulla* = little jar

[39]*crista* = crest, ridge
[40]*cupula* = little tub

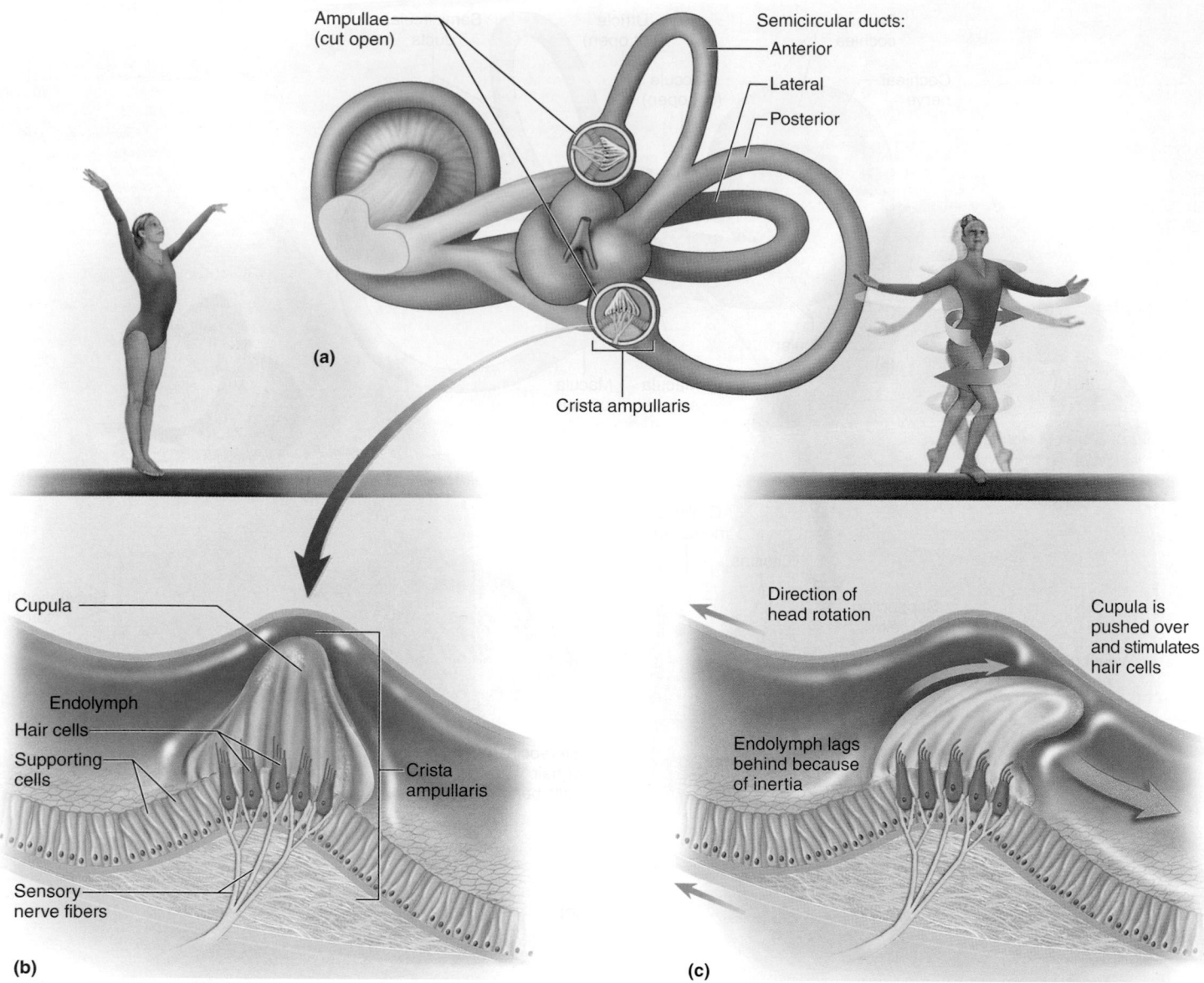

**FIGURE 16.20  The Semicircular Ducts.**  (a) Structure of the semicircular ducts, with each ampulla opened to show the crista ampullaris and cupula. (b) Detail of the crista ampullaris. (c) Action of the endolymph on the cupula and hair cells when the head is rotated.

## Projection Pathways

Hair cells of the macula sacculi, macula utriculi, and semicircular ducts synapse at their bases with sensory fibers of the **vestibular nerve.** This and the cochlear nerve merge to form the vestibulocochlear nerve (cranial nerve VIII). Fibers of the vestibular apparatus lead to a complex of four **vestibular nuclei** on each side of the pons and medulla. Nuclei on the right and left sides of the brainstem communicate extensively with each other, so each receives input from both the right and left ears. They process signals about the position and movement of the body and relay information to five targets (fig. 16.21):

1. The cerebellum, which integrates vestibular information into its control of head movements, eye movements, muscle tone, and posture.

2. Nuclei of the oculomotor, trochlear, and abducens nerves (cranial nerves III, IV, and VI). These nerves produce eye movements that compensate for movements of the head (the *vestibulo–ocular reflex*). To observe this effect, hold a book in front of you at a comfortable reading distance and fix your gaze on the middle of the page. Move the book left and right about once per second, and you will be unable to read it. Now hold the book still and shake

your head from side to side at the same rate. This time you will be able to read the page because the vestibulo-ocular reflex compensates for your head movements and keeps your eyes fixed on the target. This reflex enables you to keep your vision fixed on a distant object as you walk or run toward it.

3. The reticular formation, which adjusts breathing and blood circulation to changes in posture.

4. The spinal cord, where fibers descend the two vestibulospinal tracts on each side (see fig. 13.4, p. 480) and synapse on motor neurons that innervate the extensor (antigravity) muscles. This pathway allows you to make quick movements of the trunk and limbs to keep your balance.

5. The thalamus, which relays signals to two areas of the cerebral cortex. One is at the inferior end of the postcentral gyrus adjacent to sensory regions for the face. It is here that we become consciously aware of body position and movement. The other is slightly rostral to this, at the inferior end of the central sulcus in the transitional zone from primary sensory to motor cortex. This area is thought to be involved in motor control of the head and body.

**BEFORE YOU GO ON**

Answer the following questions to test your understanding of the preceding section:

17. What physical properties of sound waves correspond to the sensations of loudness and pitch?

18. What are the benefits of having auditory ossicles and muscles in the middle ear?

19. Explain how vibration of the tympanic membrane ultimately produces fluctuations of membrane voltage in a cochlear hair cell.

20. How does the brain recognize the difference between the musical notes high C and middle C? Between a loud sound and a soft one?

21. How does the function of the semicircular ducts differ from the function of the saccule and utricle?

22. How is sensory transduction in the semicircular ducts similar to that in the saccule and utricle?

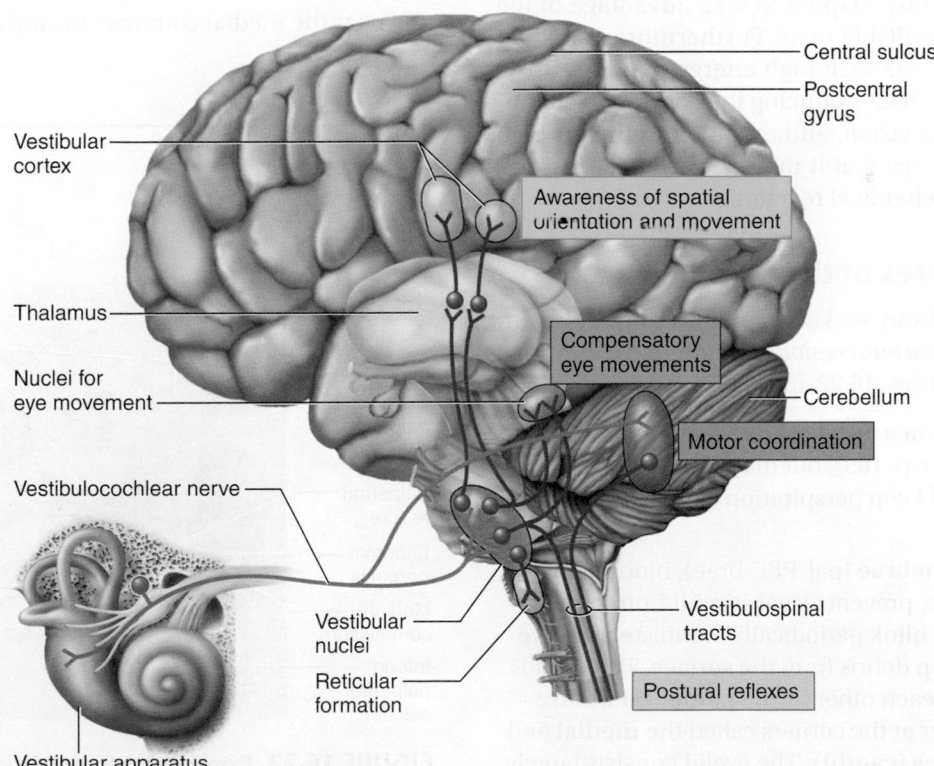

Vestibular cortex

Awareness of spatial orientation and movement

Thalamus

Nuclei for eye movement

Compensatory eye movements

Vestibulocochlear nerve

Cerebellum

Motor coordination

Vestibular nuclei

Vestibulospinal tracts

Reticular formation

Postural reflexes

Vestibular apparatus

Central sulcus

Postcentral gyrus

**FIGURE 16.21 Vestibular Projection Pathways in the Brain.**

## 16.5 Vision

### Expected Learning Outcomes
When you have completed this section, you should be able to

a. describe the anatomy of the eye and its accessory structures;

b. discuss the structure of the retina and its receptor cells;

c. explain how the optical system of the eye creates an image on the retina;

d. discuss how the retina converts this image to nerve signals;

e. explain why different types of receptor cells and neural circuits are required for day and night vision;

f. describe the mechanism of color vision; and

g. trace the visual projection pathways in the brain.

### Light and Vision

**Vision (sight)** is the perception of objects in the environment by means of the light they emit or reflect. *Light* is visible electromagnetic radiation. Human vision is limited to wavelengths ranging from about 400 to 700 nm. The *ultraviolet (UV)* radiation just below 400 nm and the *infrared (IR)* radiation just above 700 nm are invisible to us, although some animals can see a little farther into those ranges than we can. Most solar radiation that reaches the surface of the earth falls within this range; most radiation of shorter and longer wavelengths is filtered out by ozone, carbon dioxide, and water vapor in the atmosphere. Vision is thus adapted to take advantage of the radiation that is most available to us. Furthermore, radiation in the ultraviolet range has such high energy that it destroys macromolecules rather than producing the controlled chemical reactions needed for vision, and radiation in the infrared range has such low energy that it merely warms the tissues, also failing to energize chemical reactions.

### Accessory Structures of the Orbit

The eyeball occupies a bony socket called the **orbit.** This general area of the face, the *orbital region,* contains structures that protect and aid the eye (figs. 16.22 and 16.23):

- The **eyebrows** enhance facial expressions and nonverbal communication (see p. 189), but may also protect the eyes from glare and keep perspiration from running into the eye.

- The **eyelids,** or **palpebrae** (pal-PEE-bree), block foreign objects from the eye, prevent visual stimuli from disturbing one's sleep, and blink periodically to moisten the eye with tears and sweep debris from the surface. The eyelids are separated from each other by the **palpebral fissure** and meet each other at the corners called the **medial** and **lateral commissures (canthi).** The eyelid consists largely

of the orbicularis oculi muscle covered with skin (fig. 16.23a). It also contains a supportive fibrous **tarsal plate** that is thickened along the margin of the eyelid. Within the plate are 20 to 25 **tarsal glands** that open along the edge of the eyelid. They secrete an oil that coats the eye and reduces tear evaporation. The **eyelashes** are guard hairs that help to keep debris from the eye. Touching the eyelashes stimulates hair receptors and triggers the blink reflex.

- The **conjunctiva** (CON-junk-TY-vuh) is a transparent mucous membrane that covers the inner surface of the eyelid and anterior surface of the eyeball, except for the cornea. It secretes a thin mucous film that prevents the eyeball from drying. It is richly innervated and highly sensitive to pain. It is also very vascular, which is especially evident when the vessels are dilated and the eyes are "bloodshot." Because it is vascular and the cornea is not, the conjunctiva heals more readily than the cornea when injured.

- The **lacrimal**[41] **apparatus** (fig. 16.23b) consists of the lacrimal (tear) gland and a series of ducts that drain the tears into the nasal cavity. The **lacrimal gland,** about the size and shape of an almond, is nestled in a shallow fossa of the frontal bone in the superolateral corner of the orbit. About 12 short ducts lead from the gland to the surface of the conjunctiva. Tears cleanse and lubricate the eye surface, deliver oxygen and nutrients to the conjunctiva, and prevent infection by means of a bactericidal enzyme, *lysozyme.* After washing across the eye, tears collect near the medial commissure and flow into a tiny pore,

---

[41] *lacrim* = tear

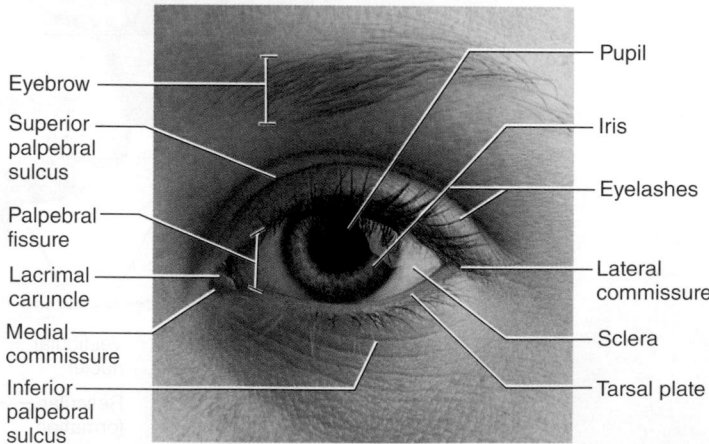

FIGURE 16.22 **External Anatomy of the Orbital Region.**

Eyebrow

Superior palpebral sulcus

Palpebral fissure

Lacrimal caruncle

Medial commissure

Inferior palpebral sulcus

Pupil

Iris

Eyelashes

Lateral commissure

Sclera

Tarsal plate

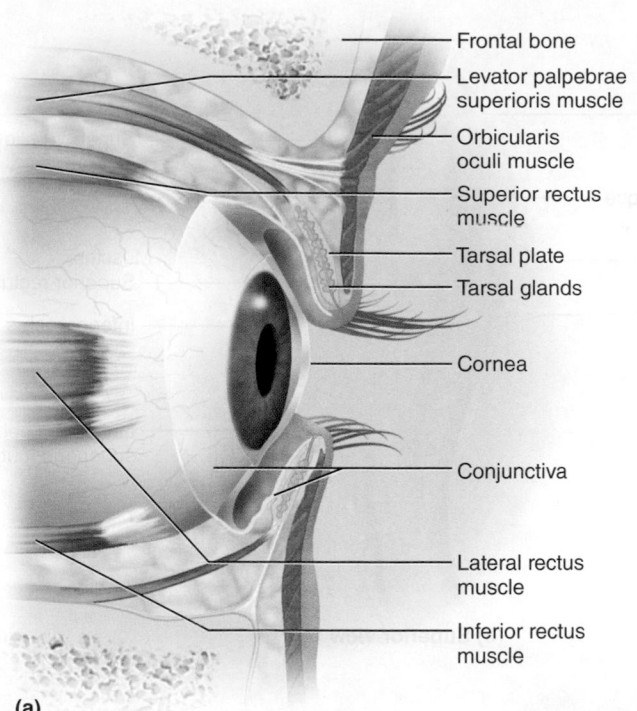

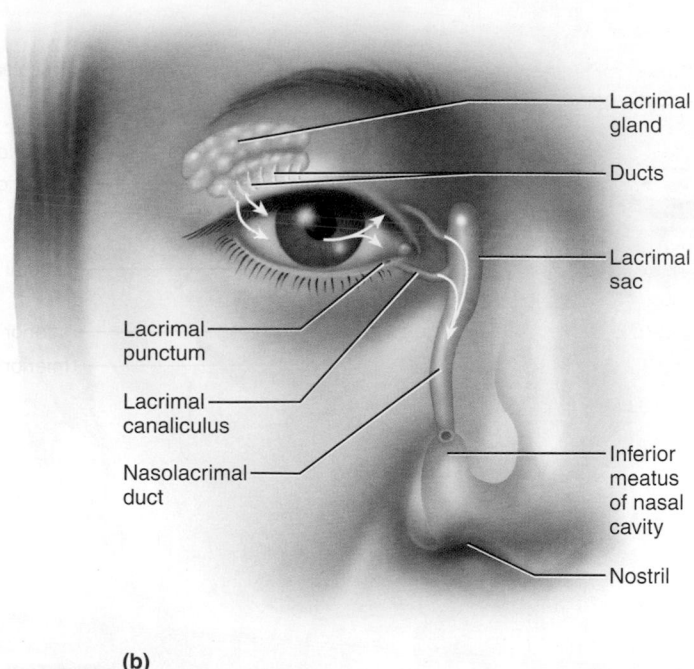

(a)

(b)

**FIGURE 16.23 Accessory Structures of the Orbit.** (a) Sagittal section of the eye and orbit. (b) The lacrimal apparatus. The arrows indicate the flow of tears from the lacrimal gland, across the front of the eye, into the lacrimal sac, and down the nasolacrimal duct.

❓ *What would be the effect of a blockage of the lacrimal puncta?*

the **lacrimal punctum,**[42] on the margin of each eyelid. The punctum opens into a short **lacrimal canaliculus,** which leads to the **lacrimal sac** in the medial wall of the orbit. From this sac, a **nasolacrimal duct** carries the tears to the inferior meatus of the nasal cavity—thus an abundance of tears from crying or watery eyes can result in a runny nose. Once the tears enter the nasal cavity, they normally flow back to the throat and are swallowed. When you have a cold, the nasolacrimal ducts become swollen and obstructed, the tears cannot drain, and they may overflow from the brim of the eye.

- Six **extrinsic eye muscles** attach to the walls of the orbit and the external surface of the eyeball. *Extrinsic* means arising externally; it distinguishes these from the *intrinsic* muscles inside the eye, to be considered later. The extrinsic muscles move the eye (fig. 16.24). They include four *rectus* ("straight") muscles and two *oblique* muscles. The **superior, inferior, medial,** and **lateral rectus** originate from a shared tendinous ring on the posterior wall of the orbit and insert on the anterior region of the eyeball, just beyond the visible "white of the eye."

They move the eye up, down, medially, and laterally. The **superior oblique** travels along the medial wall of the orbit. Its tendon passes through a fibrocartilage ring, the **trochlea**[43] (TROCK-lee-uh), and inserts on the superolateral aspect of the eyeball. The **inferior oblique** extends from the medial wall of the orbit to the inferolateral aspect of the eye. To visualize their function, suppose you turn your eyes to the right. The superior oblique muscle slightly depresses your right eye, while the inferior oblique slightly elevates the left eye. The opposite occurs when you look to the left. This is the primary function of the oblique muscles, but they also slightly rotate the eyes, turning the "twelve o'clock pole" of each eye toward or away from the nose. The superior oblique is innervated by the trochlear nerve (cranial nerve IV), the lateral rectus by the abducens (VI), and the rest of these muscles by the oculomotor nerve (III).

The eye is surrounded on the sides and back by **orbital fat.** It cushions the eye, allows it to move freely, and protects blood vessels and nerves in the rear of the orbit.

---

[42]*punct* = point

[43]*trochlea* = pulley

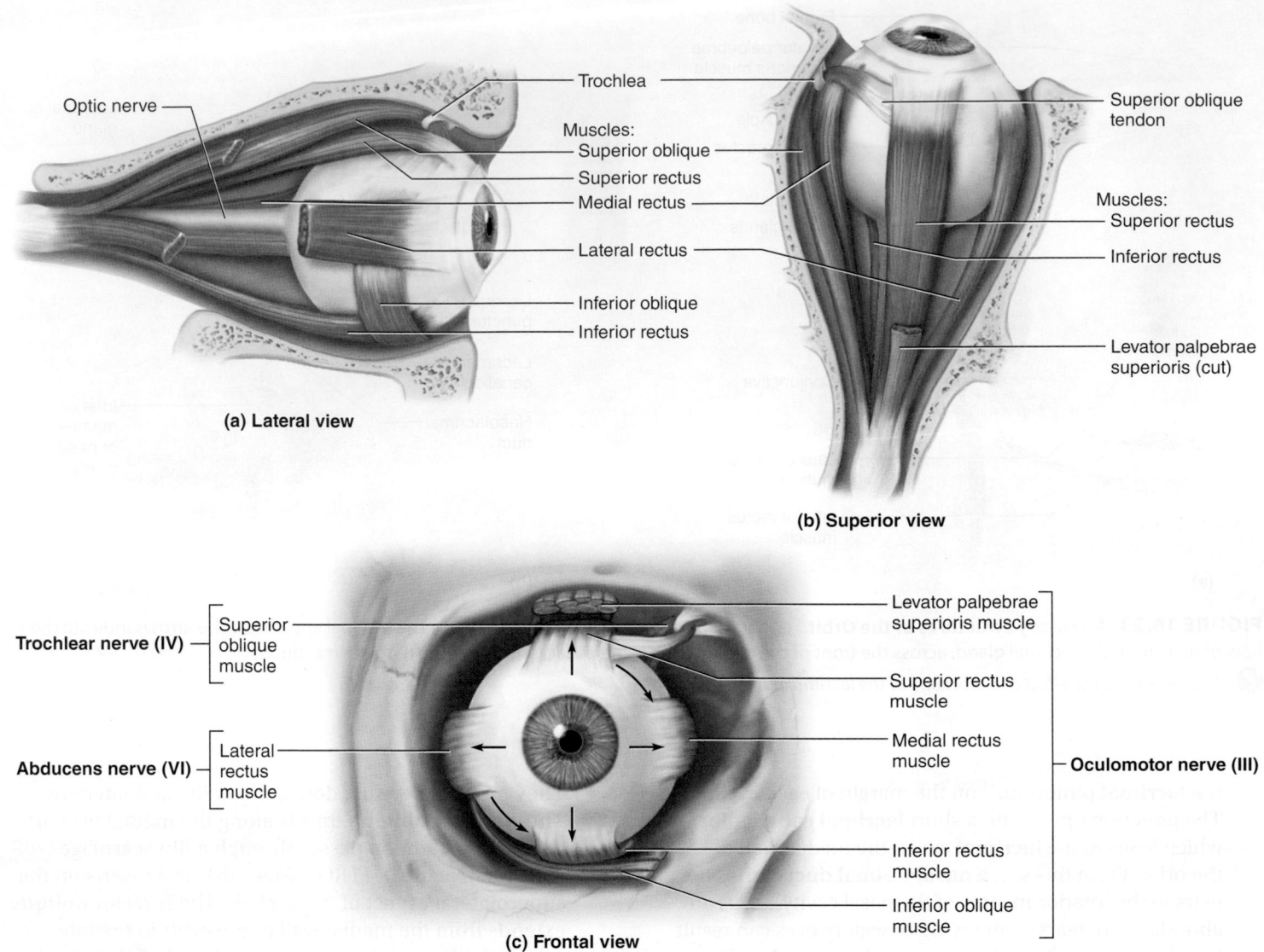

**FIGURE 16.24  Extrinsic Muscles of the Eye.** (a) Lateral view of the right eye. The lateral rectus muscle is cut to show a portion of the optic nerve. (b) Superior view of the right eye. (c) Innervation of the extrinsic muscles; arrows indicate the eye movement produced by each muscle.

❓ *What would cause the greatest loss of visual function—trauma to cranial nerve III, IV, or VI? Why?*

## Anatomy of the Eye

The eyeball is a sphere about 24 mm in diameter (fig. 16.25) with three principal components: (1) three layers (tunics) that form its wall; (2) optical components that admit and focus light; and (3) neural components, the retina and optic nerve. The retina is not only a neural component but also part of the inner tunic. The cornea is part of the outer tunic as well as one of the optical components.

## The Tunics

The three tunics of the eyeball are as follows:

- The outer **fibrous layer (tunica fibrosa).** This is divided into two regions: sclera and cornea. The **sclera**[44] (white of the eye) covers most of the eye surface and consists of dense collagenous connective tissue perforated by blood vessels and nerves. The **cornea** is the anterior transparent region of modified sclera that admits light into the

---

[44]*scler* = hard, tough

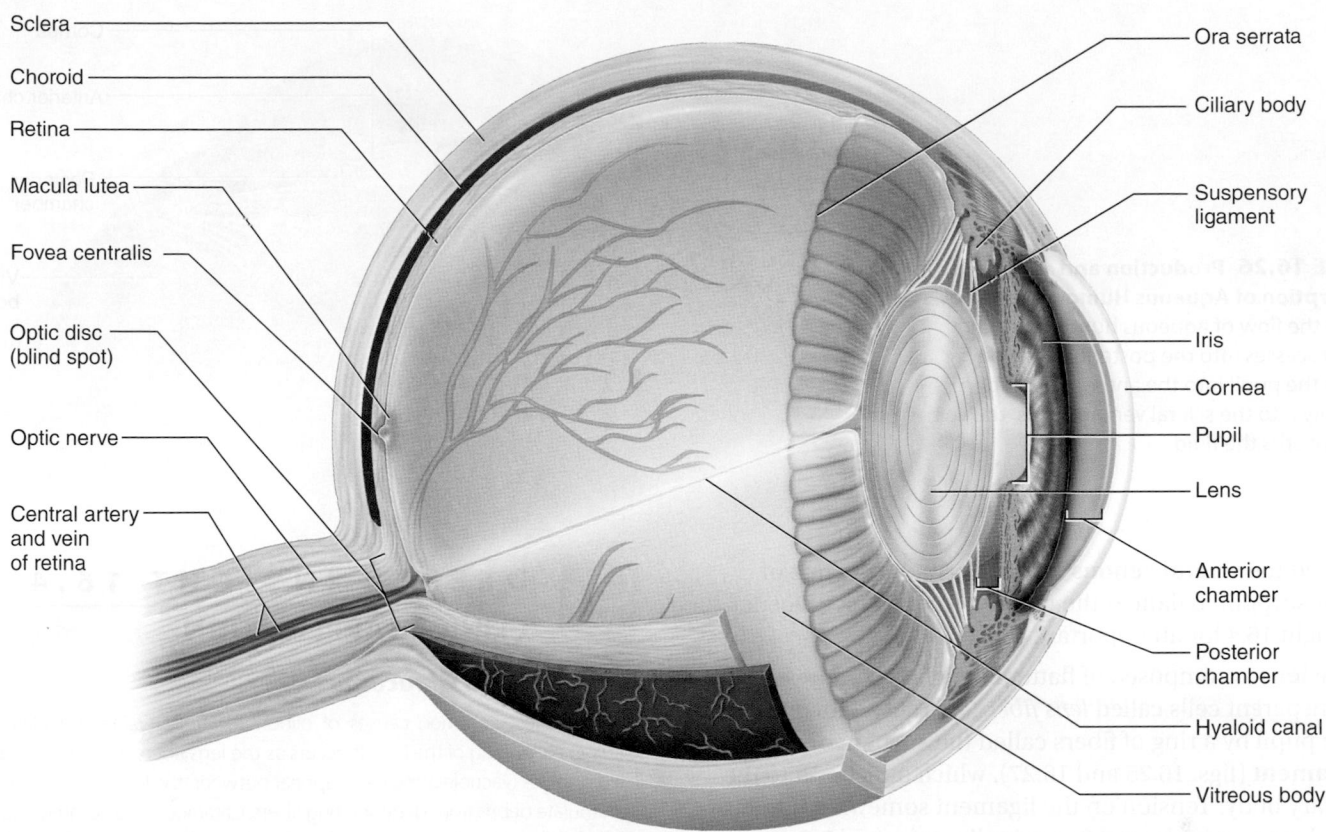

**FIGURE 16.25 The Eye.** Sagittal section.

eye. Most of it is composed of very compact layers of collagen fibrils and thin flat fibroblasts. It is covered by a thin stratified squamous epithelium anteriorly and a simple squamous epithelium posteriorly. Both epithelia pump sodium ions out of the corneal tissue. Water follows by osmosis, so this mechanism prevents the cornea from overhydrating, swelling, and losing transparency. The anterior epithelium also is a source of stem cells that give the cornea a great capacity for regeneration if it is injured.

- The middle **vascular layer (tunica vasculosa).** This is also called the **uvea**[45] (YOU-vee-uh) because it resembles a peeled grape in fresh dissection. It consists of three regions—the choroid, ciliary body, and iris. The **choroid** (CO-royd) is a highly vascular, deeply pigmented layer of tissue behind the retina. It gets its name from a histological resemblance to the chorion of a fetus. The **ciliary body,** a thickened extension of the choroid, forms a muscular ring around the lens. It supports the iris and lens and secretes a fluid called aqueous humor. The **iris** is an adjustable diaphragm that controls the diameter of the **pupil,** its central

opening. It has two pigmented layers: a posterior *pigment epithelium* that blocks stray light from reaching the retina, and the *anterior border layer,* which contains pigmented cells called **chromatophores.**[46] A high concentration of melanin in the chromatophores gives the iris a black, brown, or hazel color. If the melanin is scanty, light reflects from the posterior pigment epithelium and gives the iris a blue, green, or gray color.

- The **inner layer (tunica interna).** This consists of the retina and beginning of the optic nerve.

## The Optical Components

The optical components of the eye are transparent elements that admit light rays, bend (refract) them, and focus images on the retina. They include the *cornea, aqueous humor, lens,* and *vitreous body.* The cornea has been described already.

- The **aqueous humor** is a serous fluid secreted by the ciliary body into a space called the **posterior chamber** between the iris and lens (fig. 16.26). It flows through the pupil into the **anterior chamber** between the cornea and iris. From here, it is reabsorbed by a circular vein

[45]*uvea* = grape

[46]*chromato* = color; *phore* = bearer

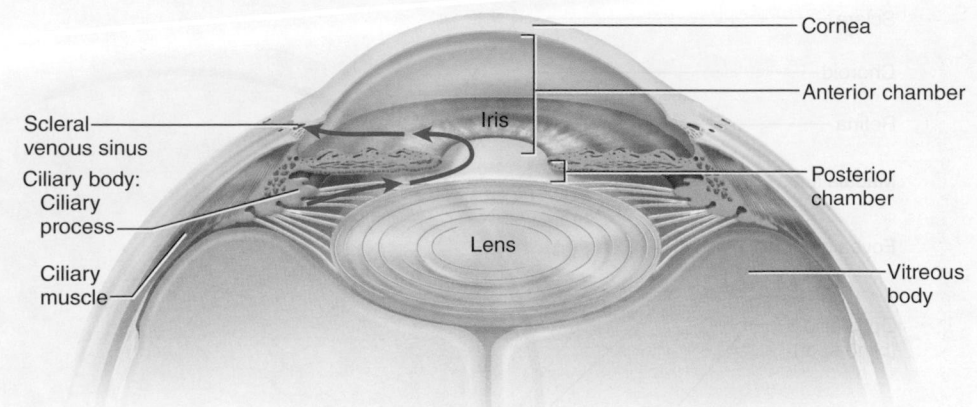

**FIGURE 16.26 Production and Reabsorption of Aqueous Humor.** Blue arrows indicate the flow of aqueous humor from the ciliary processes into the posterior chamber; through the pupil into the anterior chamber; and finally into the scleral venous sinus, the vein that reabsorbs the fluid.

called the **scleral venous sinus.** Normally the rate of reabsorption balances the rate of secretion (see Deeper Insight 16.4 for an important exception).

- The **lens** is composed of flattened, tightly compressed, transparent cells called *lens fibers.* It is suspended behind the pupil by a ring of fibers called the **suspensory ligament** (figs. 16.25 and 16.27), which attaches it to the ciliary body. Tension on the ligament somewhat flattens the lens so it is about 9.0 mm in diameter and 3.6 mm thick at the middle.

- The **vitreous**[47] **body (vitreous humor)** is a transparent jelly that fills a space called the *vitreous chamber* behind the lens. An oblique channel through this body called the *hyaloid canal* is the remnant of an artery present in the embryo (see fig. 16.25).

## The Neural Components

The neural components are the retina and optic nerve. The **retina** forms from a cup-shaped outgrowth of the diencephalon called the *optic vesicle* (see fig. 14.4, p. 513); it is actually a part of the brain—the only part that can be viewed without dissection. It is a thin transparent membrane attached to the rest of the eye at only two points: the **optic disc,** where the optic nerve leaves the rear *(fundus)* of the eye, and its scalloped anterior margin, the **ora serrata.**[48] The vitreous body presses the retina smoothly against the rear of the eyeball. The retina can separate from the wall of the eyeball because of blows to the head or insufficient pressure from the vitreous body. Such a *detached retina* may cause blurry areas in the field of vision. It leads to blindness if the retina remains separated for too long from the choroid, on which it depends for oxygen, nutrition, and waste removal.

The retina is examined with an illuminating and magnifying instrument called an *ophthalmoscope* (fig. 16.28). Directly posterior to the center of the lens, on the visual axis of

### DEEPER INSIGHT 16.4

#### CLINICAL APPLICATION

### Cataracts and Glaucoma

The two most common causes of blindness are cataracts and glaucoma. A *cataract* is clouding of the lens. It occurs as the lens fibers darken with age, fluid-filled bubbles (vacuoles) and clefts appear between the lens fibers, and the clefts accumulate debris from degenerating fibers. Cataracts are a common complication of diabetes mellitus, but can also be induced by heavy smoking, ultraviolet radiation, radiation therapy, certain viruses and drugs, and other causes. They cause the vision to appear milky or as if one were looking from behind a waterfall.[49] Cataracts can be treated by replacing the natural lens with a plastic one. The implanted lens improves vision almost immediately, but glasses still may be needed for near vision.

*Glaucoma* is a state of elevated pressure within the eye that occurs when the scleral venous sinus is obstructed so aqueous humor is not reabsorbed as fast as it is secreted. Pressure in the anterior and posterior chambers drives the lens back and puts pressure on the vitreous body. The vitreous body presses the retina against the choroid and compresses the blood vessels that nourish the retina. Without a good blood supply, retinal cells die and the optic nerve may atrophy, producing blindness. Symptoms often go unnoticed until the damage is irreversible. Illusory flashes of light are an early symptom of glaucoma. Late-stage symptoms include dimness of vision,[50] a narrowed visual field, and colored halos around artificial lights. Glaucoma can be halted with drugs or surgery, but lost vision cannot be restored. This disease can be detected at an early stage in the course of regular eye examinations. The field of vision is checked, the optic nerve is examined, and the intraocular pressure is measured with an instrument called a *tonometer.*

the eye, one sees a patch of cells called the **macula lutea**[51] about 3 mm in diameter. In the center of the macula is a tiny pit, the **fovea**[52] **centralis,** which produces the most finely detailed images for reasons explained later. About 3 mm medial to the macula lutea is the optic disc. Nerve fibers from all regions of the retina converge on this point and exit the eye to form the optic nerve. Blood vessels enter and leave the eye by

Suspensory ligament          Lens

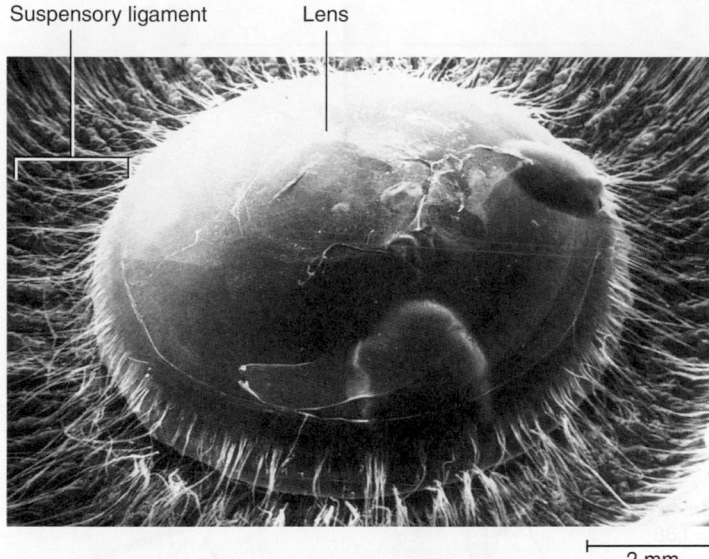

2 mm

**FIGURE 16.27  The Lens of the Eye (SEM).** Posterior view of the lens and the suspensory ligament that anchors it to the ciliary body.

way of the optic disc. Eye examinations serve for more than evaluating the visual system; they allow for a direct, noninvasive examination of blood vessels for signs of hypertension, diabetes mellitus, atherosclerosis, and other vascular diseases.

The optic disc contains no receptor cells, so it produces a **blind spot** in the visual field of each eye. You can detect your blind spot and observe an interesting visual phenomenon with the help of figure 16.29. Close or cover your right eye and hold the page about 30 cm (1 ft) from your face. Fixate on the X with your left eye. Without taking your gaze off the X, move the page slightly forward and back, or right and left, until the red dot disappears. This occurs because the image of the dot is falling on the blind spot of your left eye.

You should notice something else happen at the same time as the dot disappears—a phenomenon called **visual filling.** The green bar seems to fill in the space where the dot used to be. This occurs because the brain uses the image surrounding the blind spot to fill in the area with similar but imaginary information. The brain acts as if it is better to assume that the unseen area probably looks like its surroundings than to allow a dark blotch to disturb your vision. This is one reason why we don't see a blind patch in the visual field. Another is that the eyes continually undergo minute flickering movements **(saccades)** that ensure that the same area of the visual field doesn't always project onto the same area of retina. What an eye doesn't see at one moment, it will see just milliseconds later when the saccades redirect its visual axis. Eye saccades are the fastest muscular movements in the human body. Without them, you would be unable to read this page.

## Formation of an Image

The visual process begins when light rays enter the eye, focus on the retina, and produce a tiny inverted image. When fully

(a)

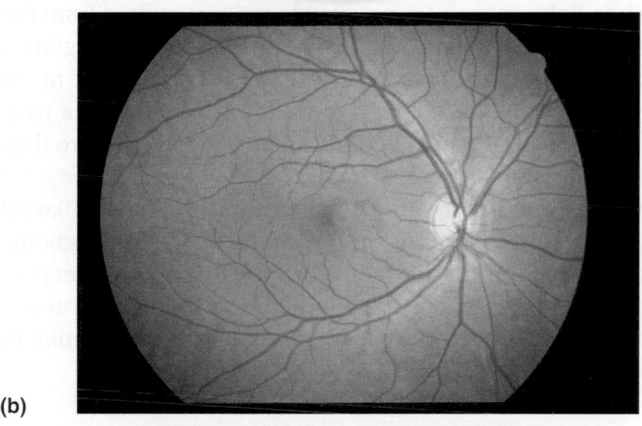

(b)

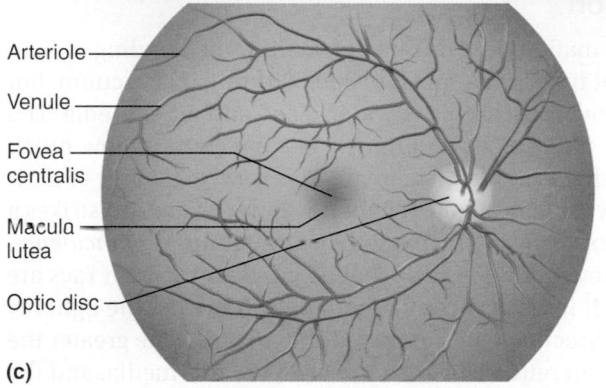

Arteriole

Venule

Fovea centralis

Macula lutea

Optic disc

(c)

**FIGURE 16.28  The Fundus (Rear) of the Eye.** (a) Use of the ophthalmoscope. (b) As seen with an ophthalmoscope. (c) Anatomical features of the fundus. Note the blood vessels diverging from the optic disc, where they enter the eye with the optic nerve. An eye examination also serves as a partial check on cardiovascular health.

❓ *Is part (b) the subject's right or left eye? How can you tell?*

**FIGURE 16.29  Demonstration of the Blind Spot and Visual Filling.** See text for explanation of how to conduct this demonstration.

dilated, the pupil admits five times as much light as it does when fully constricted. Its diameter is controlled by two sets of contractile elements in the iris: (1) The **pupillary constrictor** consists of smooth muscle cells that encircle the pupil. When stimulated by the parasympathetic nervous system, it narrows the pupil and admits less light to the eye. (2) The **pupillary dilator** consists of a spokelike arrangement of contractile *myo-epithelial cells.* When stimulated by the sympathetic nervous system, these cells contract, widen the pupil, and admit more light to the eye (see fig. 15.9, p. 571). Pupillary constriction and dilation occur in two situations: when light intensity changes and when we shift our gaze between distant and nearby objects. Constriction in response to a shift in gaze is part of the *near response* described later.

Pupillary constriction in response to light is called the **photopupillary reflex.** It is mediated by an autonomic reflex arc. When light intensity rises, signals are transmitted from the eye to the *pretectal region* of the upper midbrain. Preganglionic parasympathetic fibers travel by way of the oculomotor nerve from here to the *ciliary ganglion* in the orbit. From the ganglion, postganglionic fibers continue into the eye, where they stimulate the pupillary constrictor.

Sympathetic innervation to the pupil originates, like all other sympathetic efferents, in the spinal cord. Preganglionic fibers ascend from the thoracic cord to the superior cervical ganglion. From there, postganglionic fibers follow the carotid arteries into the head and lead ultimately to the pupillary dilator.

## Refraction

Image formation depends on **refraction,** the bending of light rays. Light travels at a speed of 300,000 km/s in a vacuum, but it slows down slightly in air, water, glass, and other media. The *refractive index* of a medium *(n)* is a measure of how much it retards light rays relative to air. The *refractive index* of air is arbitrarily set at $n = 1.00$. If light traveling through air strikes a medium of higher refractive index at a 90° *angle of incidence,* it slows down but does not change course—the light rays are not bent. If it strikes at any other angle, however, the light ray changes direction—it is refracted (fig. 16.30a). The greater the difference in refractive index between the two media, and the greater the angle of incidence, the stronger the refraction is.

As light enters the eye, it passes from a medium with $n = 1.00$ (air) to one with $n = 1.38$ (the cornea). Light rays striking the very center of the cornea pass straight through, but because of the curvature of the cornea, rays striking off center are bent toward the center (fig. 16.30b). The aqueous humor has a refractive index of 1.33 and doesn't greatly alter the path of the light. The lens has a refractive index of 1.40. As light passes from air to cornea, the refractive index changes by 0.38; but as it passes from aqueous humor to lens, the refractive index changes by only 0.07. Therefore, the cornea refracts light more than the lens does. The lens merely fine-tunes the image, especially as you shift your focus between near and distant objects.

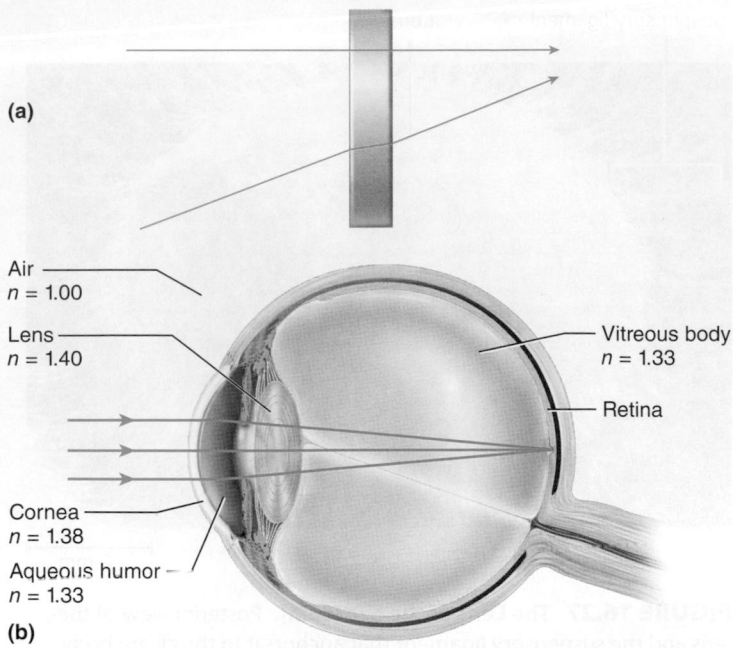

**(a)**

Air
$n = 1.00$

Lens
$n = 1.40$

Cornea
$n = 1.38$

Aqueous humor
$n = 1.33$

Vitreous body
$n = 1.33$

Retina

**(b)**

**FIGURE 16.30 Principles of Refraction.** (a) A refractive medium does not bend light rays that strike it at a 90° angle but does bend rays that enter or leave it at any other angle. (b) Refractive indices of the media from air to retina. The greater the difference between the refractive indices of two media, the more strongly light rays are refracted when passing from one to the next. In vision, most refraction (focusing) occurs as light passes from air to cornea. The lens makes only fine adjustments in the image.

## The Near Response

**Emmetropia**[53] (EM-eh-TRO-pee-uh) is a state in which the eye is relaxed and focused on an object more than 6 m (20 ft) away, the light rays coming from that object are essentially parallel, and the rays are focused on the retina without effort. If the gaze shifts to something closer, light rays from the source are too divergent to be focused without effort. In other words, the eye is automatically focused on things in the distance unless you make an effort to focus elsewhere. For a wild animal or our prehistoric ancestors, this arrangement would be adaptive because it allows for alertness to predators or prey at a distance.

The **near response** (fig. 16.31), or adjustment to close-range vision, involves three processes to focus an image on the retina:

1. **Convergence of the eyes.** Move your finger gradually closer to a baby's nose and the baby will look cross-eyed at it. This **convergence** of the eyes orients the visual axis of each eye toward the object in order to focus its image on each fovea. If the eyes cannot converge accurately—for example, when the extrinsic muscles are weaker in one eye than in the other—double vision, or *diplopia,*[54] results. The images fall on different parts of the two retinas and the brain sees

---

[53]*em* = in; *metr* = measure; *opia* = vision
[54]*dipl* = double; *opia* = vision

two images. You can simulate this effect by pressing gently on one eyelid as you look at this page; the image of the print will fall on noncorresponding regions of the two eyes and cause you to see double.

2. **Constriction of the pupil.** Lenses cannot refract light rays at their edges as well as they can closer to the center. The image produced by any lens is therefore somewhat blurry around the edges; this *spherical aberration* is quite evident in an inexpensive microscope. It can be minimized by screening out the peripheral light rays and looking only at the better-focused center. In the eye, the pupil serves this purpose by constricting as you focus on nearby objects. Like

the diaphragm setting (f-stop) of a camera, the pupil thus has a dual purpose: to adjust the eye to variations in brightness and to reduce spherical aberration.

3. **Accommodation of the lens.  Accommodation** is a change in the curvature of the lens that enables you to focus on a nearby object. When you look at something nearby, the ciliary muscle surrounding the lens contracts. This narrows the diameter of the ciliary body, relaxes the fibers of the suspensory ligament, and allows the lens to relax into a more convex shape (fig. 16.32). In emmetropia, the lens is about 3.6 mm thick at the center; in accommodation, it thickens to as much as 4.5 mm.

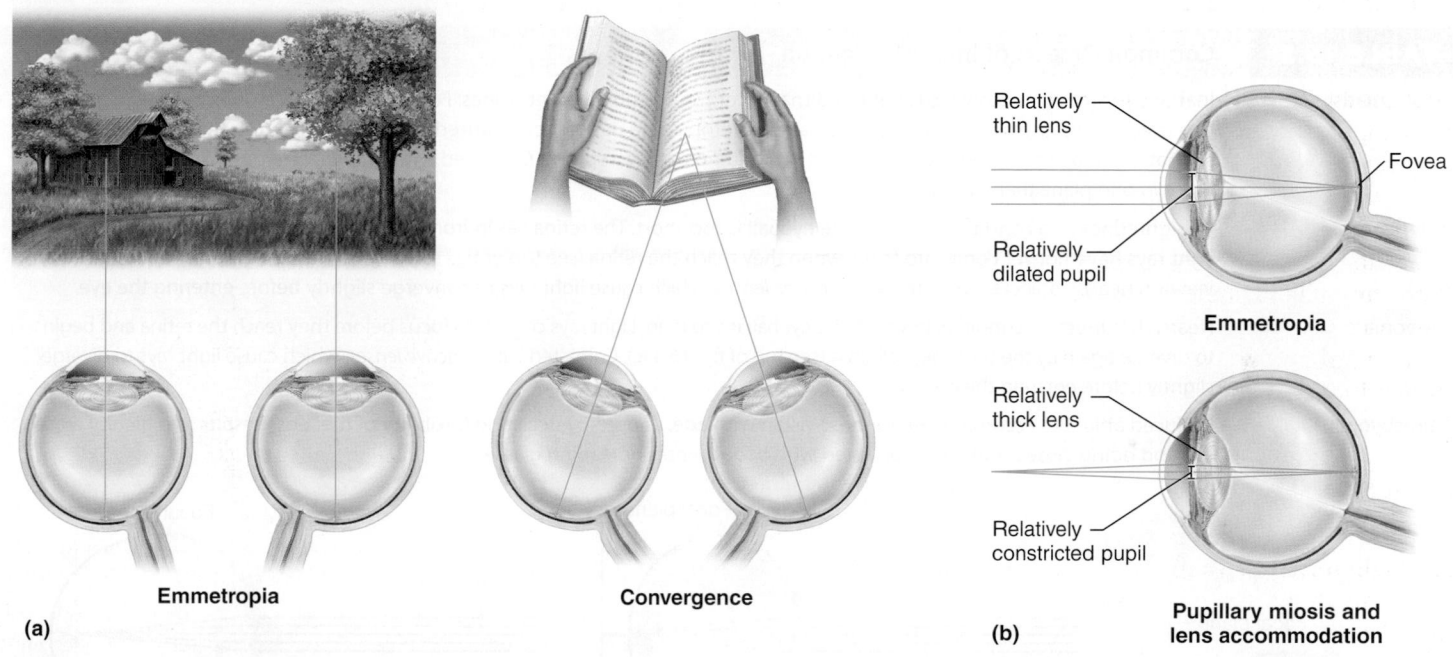

**FIGURE 16.31 Emmetropia and the Near Response.** (a) Superior view of both eyes fixed on an object more than 6 m away (left), and both eyes fixed on an object closer than 6 m (right). (b) Lateral view of the eye fixed on a distant object (top) and nearby object (bottom).

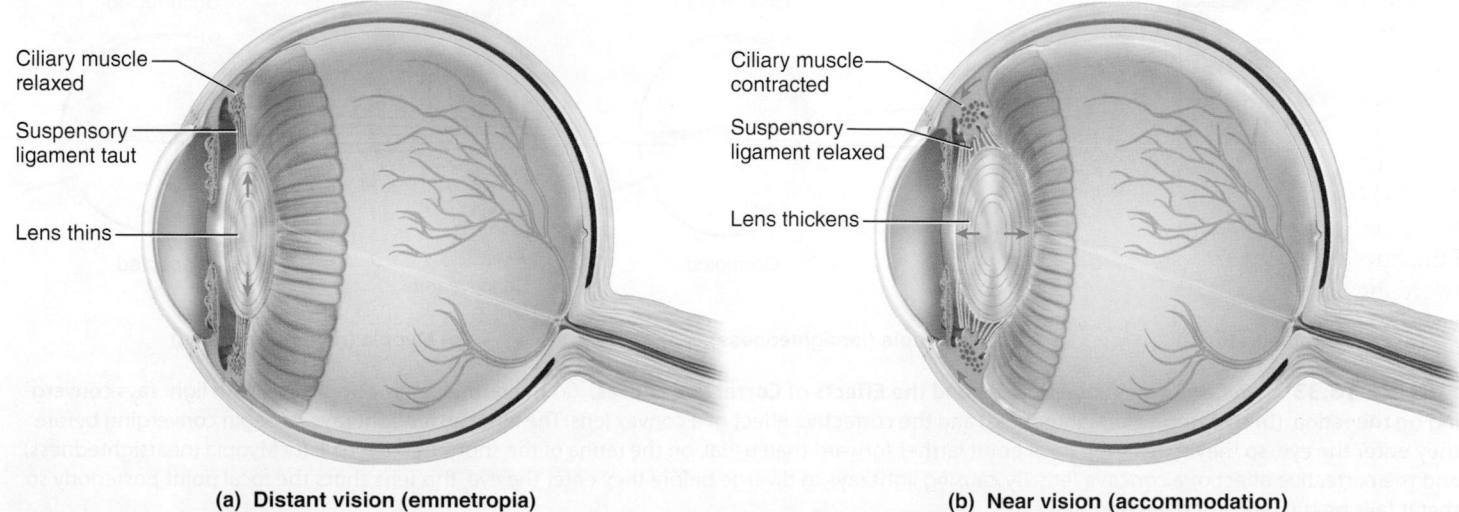

**FIGURE 16.32 Accommodation of the Lens.** (a) In the emmetropic eye, the ciliary muscle is relaxed and dilated. It puts tension on the suspensory ligament and flattens the lens. (b) In accommodation, the ciliary muscle contracts and narrows in diameter. This reduces tension on the suspensory ligament and allows the lens to relax into a more convex shape.

A more convex lens refracts light more strongly and focuses the divergent light rays onto the retina. The closest an object can be and still come into focus is called the **near point of vision.** It depends on the flexibility of the lens. The lens stiffens with age, so the near point averages about 9 cm at the age of 10 and 83 cm by the age of 60.

Some common defects in image formation are described in table 16.2.

▶▶▶ **APPLY WHAT YOU KNOW**

*Which extrinsic muscles of the eyes are the prime movers in convergence?*

## Sensory Transduction in the Retina

The conversion of light energy into action potentials occurs in the retina. We begin our exploration of this process with the cellular layout of the retina (fig. 16.34). From there we go to the pigments that absorb light and then to what happens when light is absorbed.

The most posterior part of the retina is the **pigment epithelium,** a darkly pigmented layer that serves, like the black inside of a film camera, to absorb stray light so it doesn't degrade the visual image.

The neural components of the retina consist of three principal cell layers. Progressing from the rear of the eye forward,

| TABLE 16.2 | Common Defects of Image Formation |
|---|---|
| Astigmatism[55] | Inability to simultaneously focus light rays that enter the eye on different planes. Focusing on vertical lines, such as the edge of a door, may cause horizontal lines, such as a tabletop, to go out of focus. Caused by a deviation in the shape of the cornea so that it is shaped like the back of a spoon rather than part of a sphere. Corrected with *cylindrical lenses,* which refract light more in one plane than another. |
| Hyperopia[56] | Farsightedness—a condition in which the eyeball is too short. The retina lies in front of the focal point of the lens, and the light rays have not yet come into focus when they reach the retina (see top of fig. 16.33b). Causes the greatest difficulty when viewing nearby objects. Corrected with *convex lenses,* which cause light rays to converge slightly before entering the eye. |
| Myopia[57] | Nearsightedness—a condition in which the eyeball is too long. Light rays come into focus before they reach the retina and begin to diverge again by the time they fall on it (see top of fig. 16.33c). Corrected with *concave lenses,* which cause light rays to diverge slightly before entering the eye. |
| Presbyopia[58] | Reduced ability to accommodate for near vision with age. Caused by declining flexibility of the lens. Results in difficulty reading and doing close handwork. Corrected with *bifocal lenses* or reading glasses. |

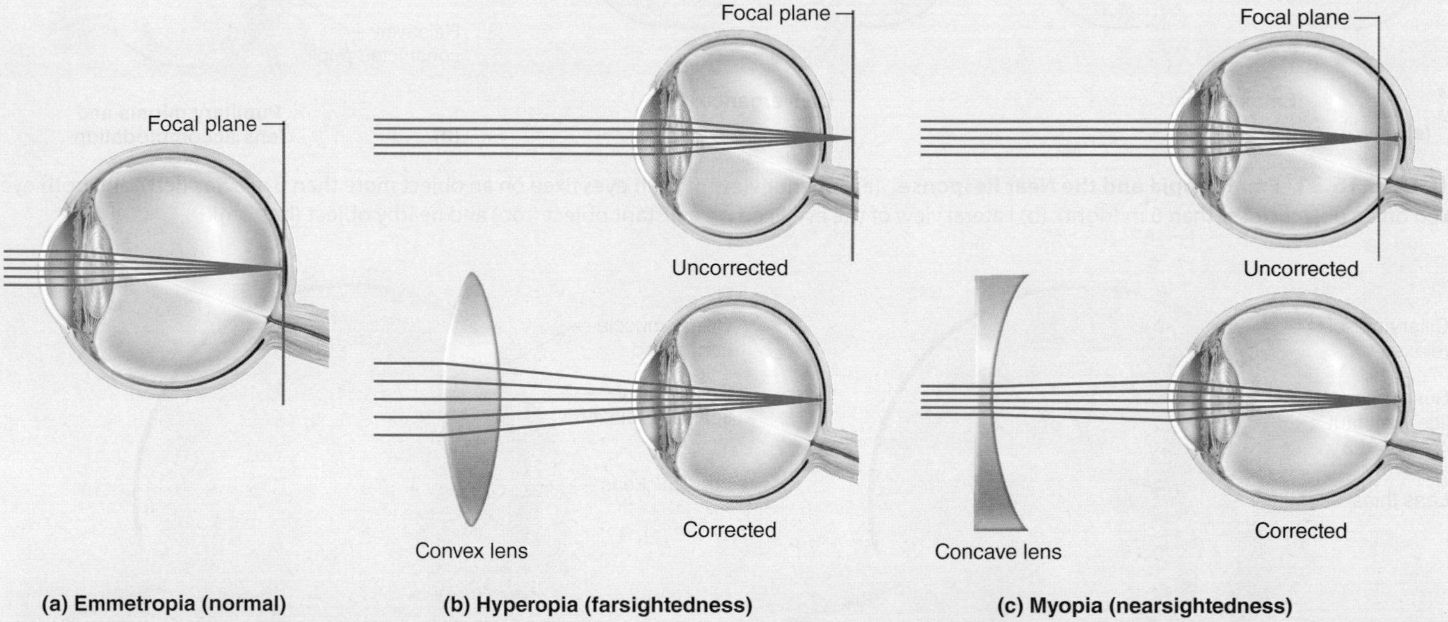

(a) Emmetropia (normal)    (b) Hyperopia (farsightedness)    (c) Myopia (nearsightedness)

**FIGURE 16.33 Two Common Visual Defects and the Effects of Corrective Lenses.** (a) The normal emmetropic eye, with light rays converging on the retina. (b) Hyperopia (farsightedness) and the corrective effect of a convex lens. The lens causes light rays to begin converging before they enter the eye, so they reach their focal point farther forward than usual, on the retina of the shortened eyeball. (c) Myopia (nearsightedness) and the corrective effect of a concave lens. By causing light rays to diverge before they enter the eye, this lens shifts the focal point posteriorly so that it falls on the retina of the elongated eye.

[55]*a* = not; *stigma* = point; *ism* = condition
[56]*hyper* = excessive; *op* = eye; *ia* = condition
[57]*my* = closed; *op* = eye; *ia* = condition
[58]*presby* = old; *op* = eye; *ia* = condition

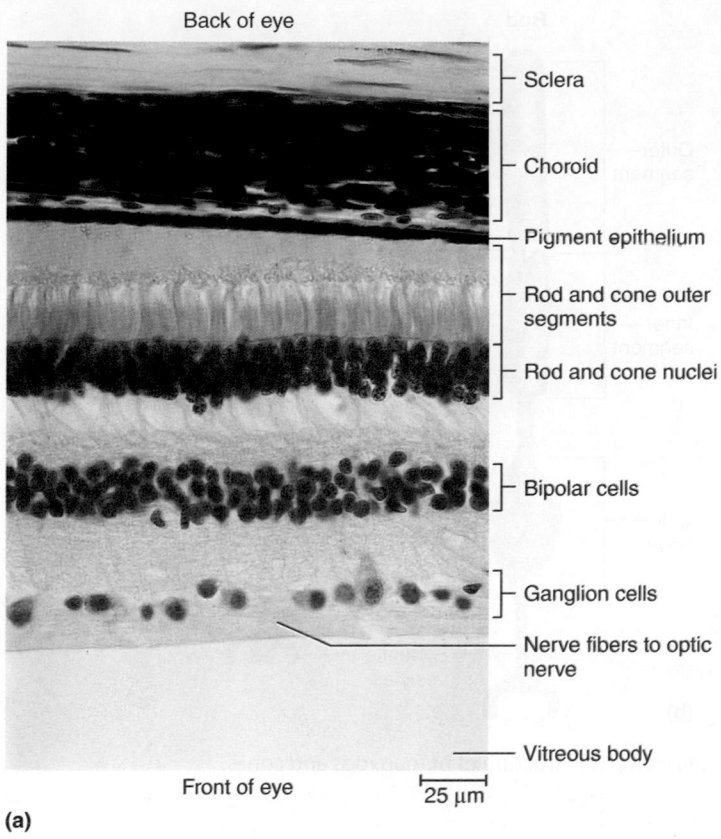

Back of eye

Sclera

Choroid

Pigment epithelium

Rod and cone outer segments

Rod and cone nuclei

Bipolar cells

Ganglion cells

Nerve fibers to optic nerve

Vitreous body

Front of eye

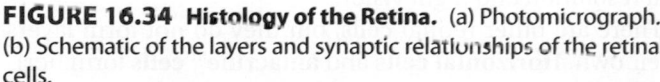

25 μm

**(a)**

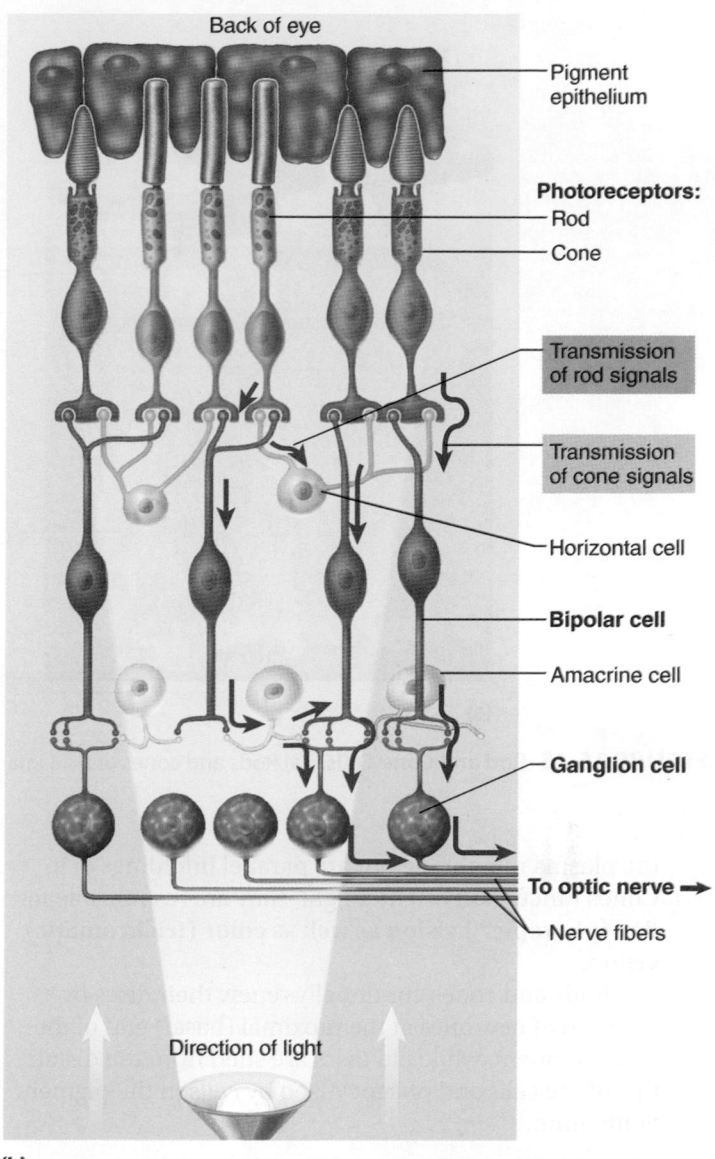

Back of eye

Pigment epithelium

**Photoreceptors:**
Rod
Cone

Transmission of rod signals

Transmission of cone signals

Horizontal cell

**Bipolar cell**

Amacrine cell

**Ganglion cell**

**To optic nerve** ➡

Nerve fibers

Direction of light

**(b)**

**FIGURE 16.34 Histology of the Retina.** (a) Photomicrograph. (b) Schematic of the layers and synaptic relationships of the retinal cells.

these are composed of *photoreceptor cells, bipolar cells,* and *ganglion cells:*

1. **Photoreceptor cells.** Photoreceptor cells absorb light and generate a chemical or electrical signal. There are three kinds: rods, cones, and certain ganglion cells. Only rods and cones produce visual images; the ganglion cells are discussed later. **Rods** and **cones** are not neurons, but are related to the ependymal cells of the brain. Each rod or cone has an **outer segment** that points toward the wall of the eye and an **inner segment** facing the interior (fig. 16.35). The two segments are separated by a constriction containing nine pairs of microtubules; the outer segment is actually a highly modified cilium specialized to absorb light. The inner segment contains mitochondria and other organelles. At its base, it gives rise to a cell body, which contains the nucleus, and

to processes that synapse with retinal neurons in the next layer.

In a rod, the outer segment is cylindrical and resembles a stack of coins in a wrapper—there is a plasma membrane around the outside and a neatly arrayed stack of about 1,000 membranous discs inside. Each disc is densely studded with globular proteins—the visual pigment *rhodopsin.* The membranes hold these pigment molecules in a position that results in the most efficient light absorption. Rod cells are responsible for **night (scotopic[59]) vision** and produce images only in shades of gray **(monochromatic vision).**

A cone cell is similar except that the outer segment tapers to a point, and the discs are not detached from

[59]*scot* = dark; *op* = vision

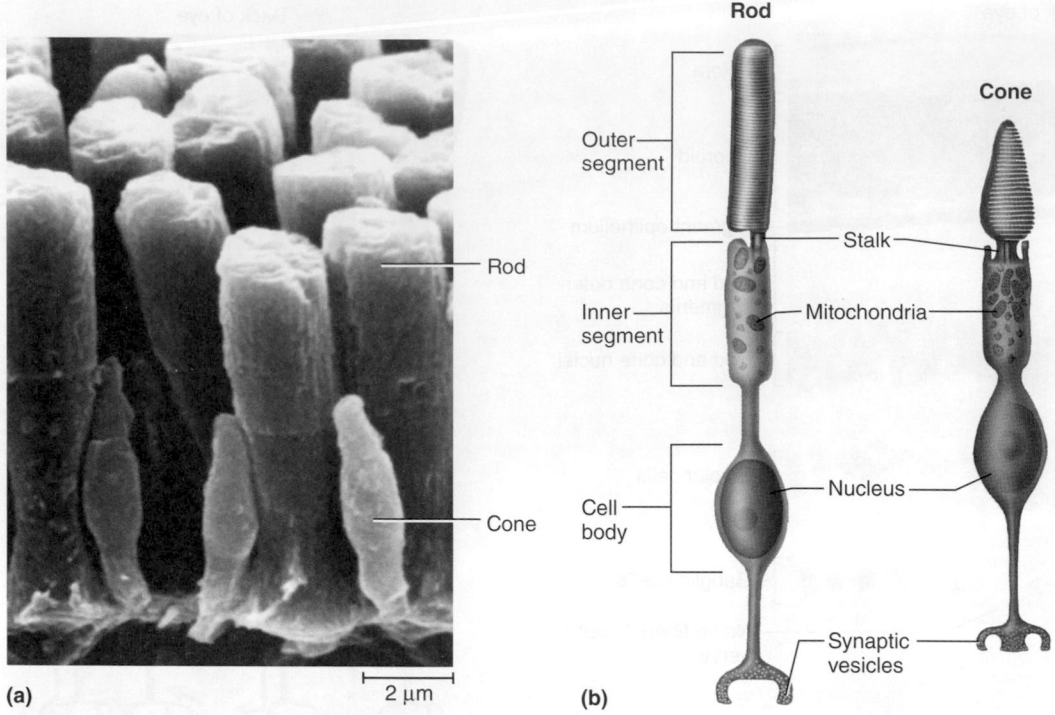

**FIGURE 16.35  Rod and Cone Cells.**  (a) Rods and cones of a salamander retina (SEM). (b) Structure of human rods and cones.

the plasma membrane but are parallel infoldings of it. Cones function in brighter light; they are responsible for **day (photopic**[60]**) vision** as well as **color (trichromatic) vision.**

Rods and cones continually renew their discs by addition of new ones at the proximal (basal) end of the outer segment, while old discs are shed from the distal tips of the cells and phagocytized by cells in the pigment epithelium.

2. **Bipolar cells.**  Rods and cones synapse with the dendrites of **bipolar cells,** the first-order neurons of the visual pathway. They in turn synapse with the ganglion cells described next (fig. 16.34b).

3. **Ganglion cells.  Ganglion cells** are the largest neurons of the retina, arranged in a single layer close to the vitreous body. They are the second-order neurons of the visual pathway. Most ganglion cells receive input from multiple bipolar cells. Their axons form the optic nerve. Some ganglion cells absorb light directly and transmit signals to brainstem nuclei that control pupillary diameter and the body's circadian rhythms. They do not contribute to visual images but detect only light intensity. Their sensory pigment is called **melanopsin.**

There are approximately 130 million rods and 6.5 million cones in one retina, but only 1 million nerve fibers in the optic

nerve. With a ratio of nearly 140 receptor cells to one optic nerve fiber, it is obvious that there must be substantial *neural convergence* and information processing in the retina itself before signals are conducted to the brain proper. Convergence occurs where multiple rods or cones synapse with one bipolar cell, and again where multiple bipolar cells feed into one ganglion cell. Later we will examine how convergence functions in visual resolution and night vision.

There are other retinal cells, but they do not form layers of their own. **Horizontal cells** and **amacrine**[61] **cells** form horizontal connections between rods, cones, and bipolar cells and intervene in the pathways from receptor cells to ganglion cells. They play diverse roles in enhancing the perception of contrast, the edges of objects, and changes in light intensity. In addition, much of the mass of the retina is composed of astrocytes and other types of glial cells.

## Visual Pigments

The visual pigment of the rods is called **rhodopsin** (ro-DOP-sin), or **visual purple.** It consists of two major parts (moieties): a protein called **opsin** and a vitamin A derivative called **retinal** (rhymes with "pal"), also known as **retinene** (fig. 16.36). Opsin is embedded in the disc membranes of the rod's outer segment. All rods contain a single kind of rhodopsin with an absorption peak at a wavelength of 500 nm. Rods are less sensitive to light of other wavelengths and cannot distinguish one color from another.

---

[60]*phot* = light; *op* = vision

[61]*a* = without; *macr* = long; *in* = fiber (lacking axons)

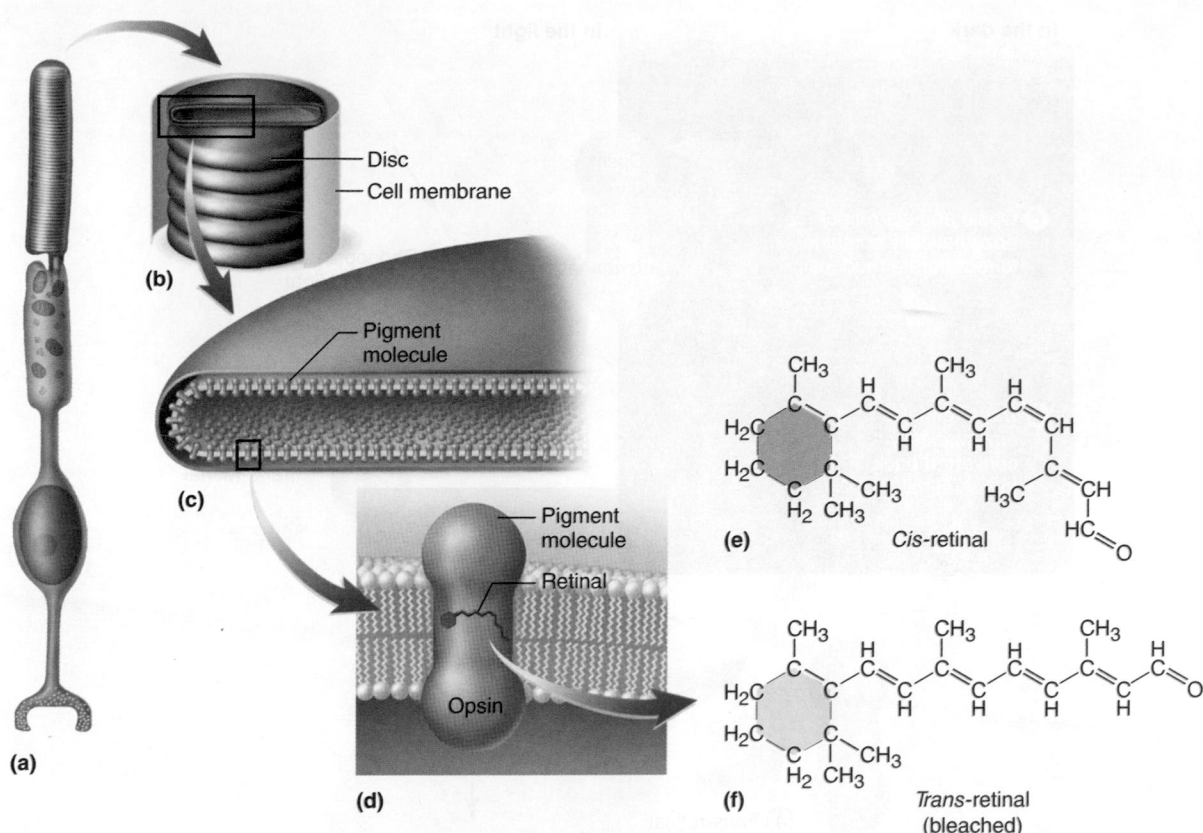

**FIGURE 16.36 Structure and Location of the Visual Pigments.** (a) A rod cell. (b) Detail of the rod outer segment. (c) One disc of the outer segment showing the membrane studded with pigment molecules. (d) A pigment molecule, embedded in the unit membrane of the disc, showing the protein moiety, opsin, and the vitamin A derivative, retinal. (e) *Cis*-retinal, the isomer present in the absence of light. (f) *Trans*-retinal, the isomer produced when the pigment absorbs a photon of light.

In cones, the pigment is called **photopsin (iodopsin).** Its retinal moiety is the same as that of rhodopsin, but the opsin moiety has a different amino acid sequence that determines which wavelengths of light the pigment absorbs. There are three kinds of cones, which are identical in appearance but optimally absorb different wavelengths of light. These differences enable us to perceive different colors.

## Generating the Optic Nerve Signal

The events of sensory transduction are probably the same in rods and cones, but are better known in rods. In the dark, their retinal has a bent shape called *cis*-**retinal.** When it absorbs light, it changes to a straight form called *trans*-retinal and breaks away from the opsin (fig. 16.37). This is called **bleaching,** because purified rhodopsin changes from violet to colorless in the light. For a rod to continue functioning, *trans*-retinal has to be converted back to the *cis* form and reunited with opsin, producing functional rhodopsin at a rate that keeps pace with bleaching. Fifty percent of the bleached rhodopsin is regenerated in about 5 minutes. Cones are faster and their photopsin is 50% regenerated in about 90 seconds.

In the dark, rods don't sit quietly doing nothing. They steadily release the neurotransmitter glutamate from the basal end of the cell (fig. 16.38a). When a rod absorbs light, glutamate secretion ceases (fig. 16.38b). We will not delve into the mechanistic details of why this occurs. The important point is that the bipolar cells, next in line, are sensitive to these on and off pulses of glutamate secretion. Some bipolar cells are inhibited by glutamate and excited when its secretion stops; these cells are therefore excited by rising light intensities. Other bipolar cells are excited by glutamate and therefore respond when light intensity drops. As your eye scans a scene, it passes areas of greater and lesser brightness. Their images on the retina cause a rapidly changing pattern of bipolar cell responses as the light intensity on a patch of retina rises and falls.

When bipolar cells detect fluctuations in light intensity, they stimulate ganglion cells either directly (by synapsing with them) or indirectly (via pathways that go through amacrine cells). Ganglion cells are the only retinal cells that produce action potentials; all other retinal cells produce only graded local potentials. Ganglion cells respond to the bipolar cells with rising and falling firing frequencies. Via the optic nerve, these changes provide visual signals to the brain.

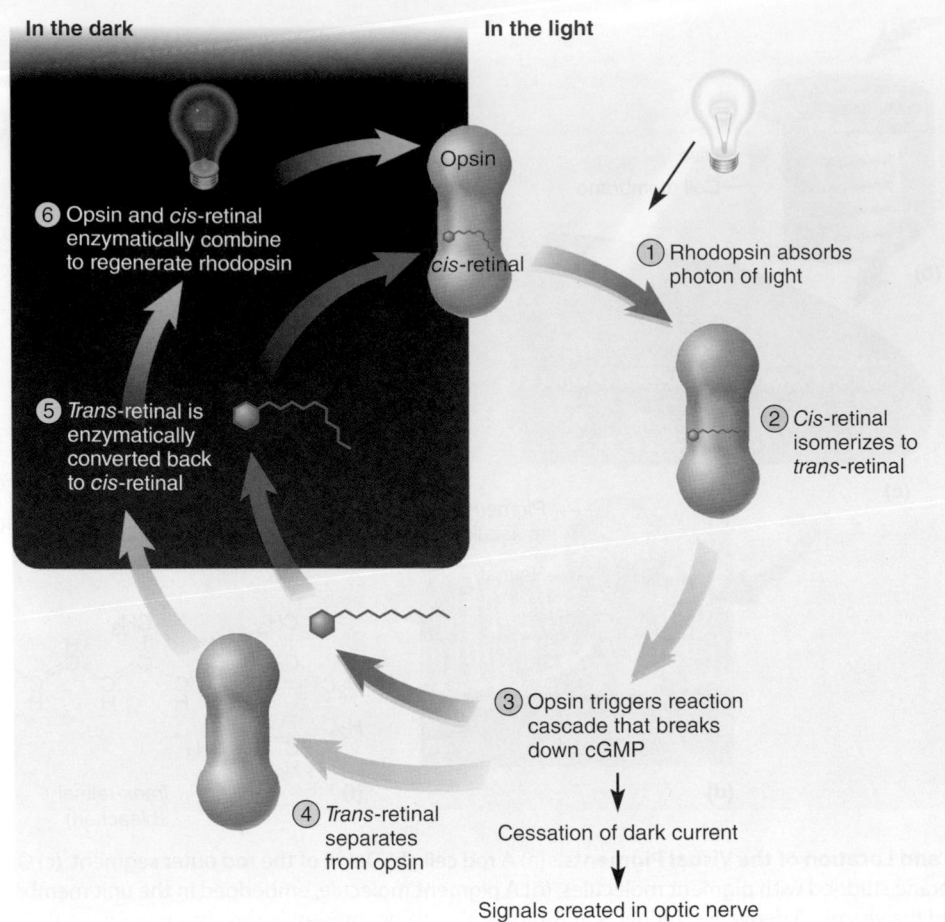

**In the dark**

⑥ Opsin and *cis*-retinal enzymatically combine to regenerate rhodopsin

⑤ *Trans*-retinal is enzymatically converted back to *cis*-retinal

Opsin

*cis*-retinal

④ *Trans*-retinal separates from opsin

**In the light**

① Rhodopsin absorbs photon of light

② *Cis*-retinal isomerizes to *trans*-retinal

③ Opsin triggers reaction cascade that breaks down cGMP

Cessation of dark current

Signals created in optic nerve

**FIGURE 16.37  The Bleaching and Regeneration of Rhodopsin.**  Numbers 1 through 4 indicate the bleaching events that occur in the light; numbers 5 and 6 indicate the regenerative events that are independent of light. The regenerative events occur in light and dark, but in the light, they are outpaced by bleaching.

## Light and Dark Adaptation

The human eye can detect light intensities ranging from a single photon at threshold (near-total darkness) to 10 billion times as bright (the intensity of bright sunlight reflected from snow). **Light adaptation** is an adjustment in vision that occurs when you go from a dark or dimly lit area into brighter light. If you wake up in the night and turn on a lamp, at first you see a harsh glare; you may experience discomfort from the overstimulated retinas. Your pupils quickly constrict to reduce the intensity of stimulation, but color vision and visual acuity (the ability to see fine detail) remain below normal for 5 to 10 minutes—the time needed for pigment bleaching to adjust retinal sensitivity to this light intensity. The rods bleach quickly in bright light, and cones take over. Even in typical indoor light, rod vision is nonfunctional.

On the other hand, suppose you are sitting in a bright room reading at night, and there is a power failure. Your eyes must undergo **dark adaptation** before you can see well enough to find your way in the dark. There is not enough light for cone

(photopic) vision, and it takes a little time for the rods (scotopic vision) to adjust to the dark. Your rod pigment was bleached by the lights in the room while the power was on, but now in the relative absence of light, rhodopsin regenerates faster than it bleaches. In a minute or two, scotopic vision begins to function, and after 20 to 30 minutes, the amount of regenerated rhodopsin is sufficient for your eyes to reach essentially maximum sensitivity. Dilation of the pupils also helps by admitting more light to the eye.

## The Dual Visual System

You may wonder why we have both rods and cones. Why can't we simply have one type of receptor cell that produces detailed color vision, both day and night? The **duplicity theory** of vision holds that a single receptor system cannot produce both high sensitivity and high resolution. It takes one type of cell and neural circuit to provide sensitive night vision and a different type to provide high-resolution daytime vision.

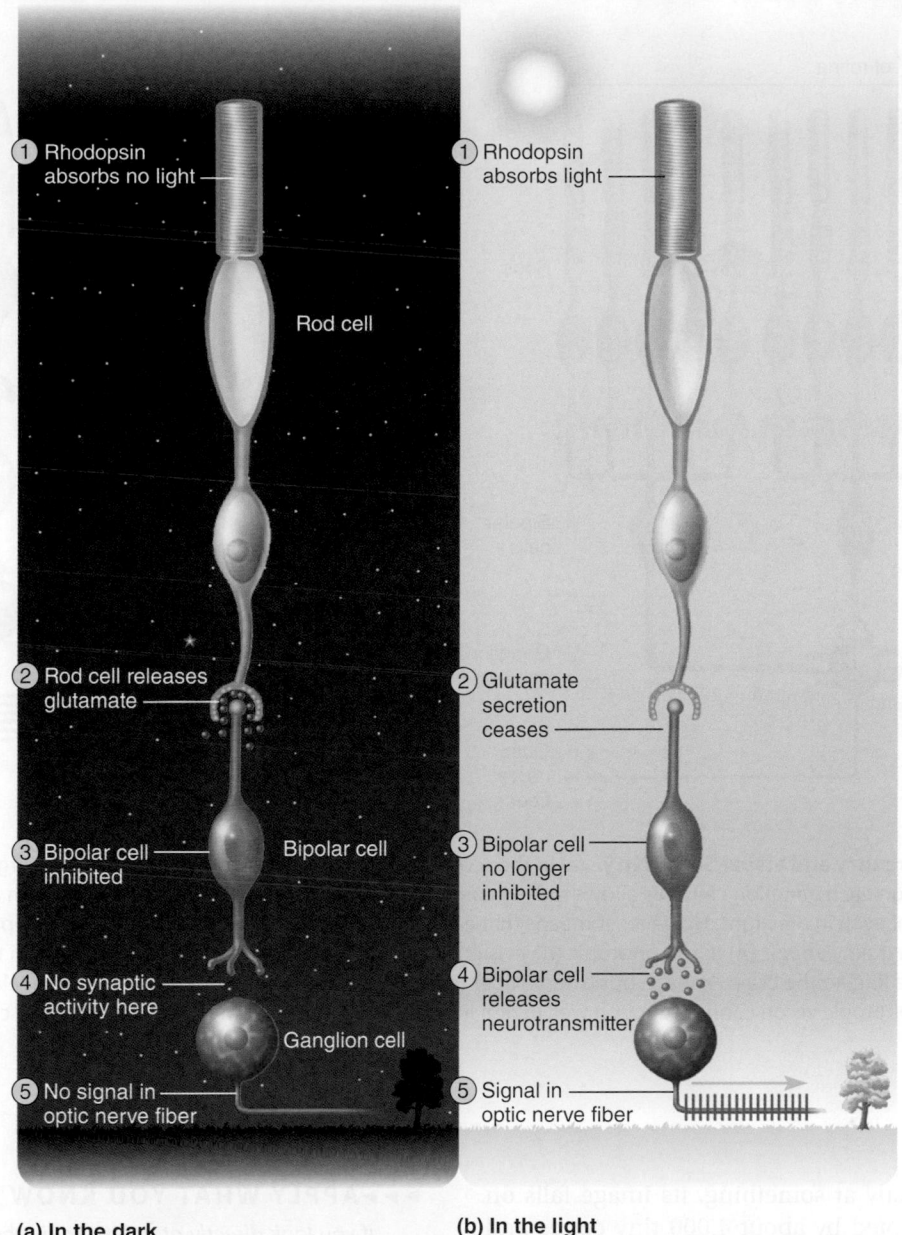

**(a) In the dark**

① Rhodopsin absorbs no light

Rod cell

② Rod cell releases glutamate

③ Bipolar cell inhibited

Bipolar cell

④ No synaptic activity here

Ganglion cell

⑤ No signal in optic nerve fiber

**(b) In the light**

① Rhodopsin absorbs light

② Glutamate secretion ceases

③ Bipolar cell no longer inhibited

④ Bipolar cell releases neurotransmitter

⑤ Signal in optic nerve fiber

**FIGURE 16.38  Mechanism of Generating Visual Signals.**  (a) In complete darkness, rod cells are active whereas certain bipolar cells and the ganglion cells are not. (b) As rod cells absorb light, they are inhibited and these bipolar cells and ganglion cells are activated. Cones are believed to function similarly. Some bipolar cells behave differently from the type used in this example.

The high sensitivity of rods in dim light stems partly from the extensive neural convergence that occurs between the rods and ganglion cells. Up to 600 rods converge on each bipolar cell, and many bipolar cells converge on each ganglion cell. This allows for a high degree of *spatial summation* (fig. 16.39a). Weak stimulation of many rods can produce an additive effect on one bipolar cell, and several bipolar cells can collaborate to excite one ganglion cell. Thus, a ganglion cell can respond in dim light that only weakly stimulates any individual rod. Scotopic vision is functional even at a light intensity less than starlight reflected from a sheet of white paper. A shortcoming of this system is that it cannot resolve finely detailed images.

One ganglion cell receives input from all the rods in about 1 mm² of retina—its receptive field. What the brain perceives is therefore a coarse, grainy image similar to an overenlarged newspaper photograph.

Around the edges of the retina, receptor cells are especially large and widely spaced. If you fixate on the middle of this page, you will notice that you cannot read the words near the margins. Visual acuity decreases rapidly as the image falls away from the fovea centralis. Our peripheral vision is a low-resolution system that serves mainly to alert us to motion in the periphery and to stimulate us to look that way to identify what is there.

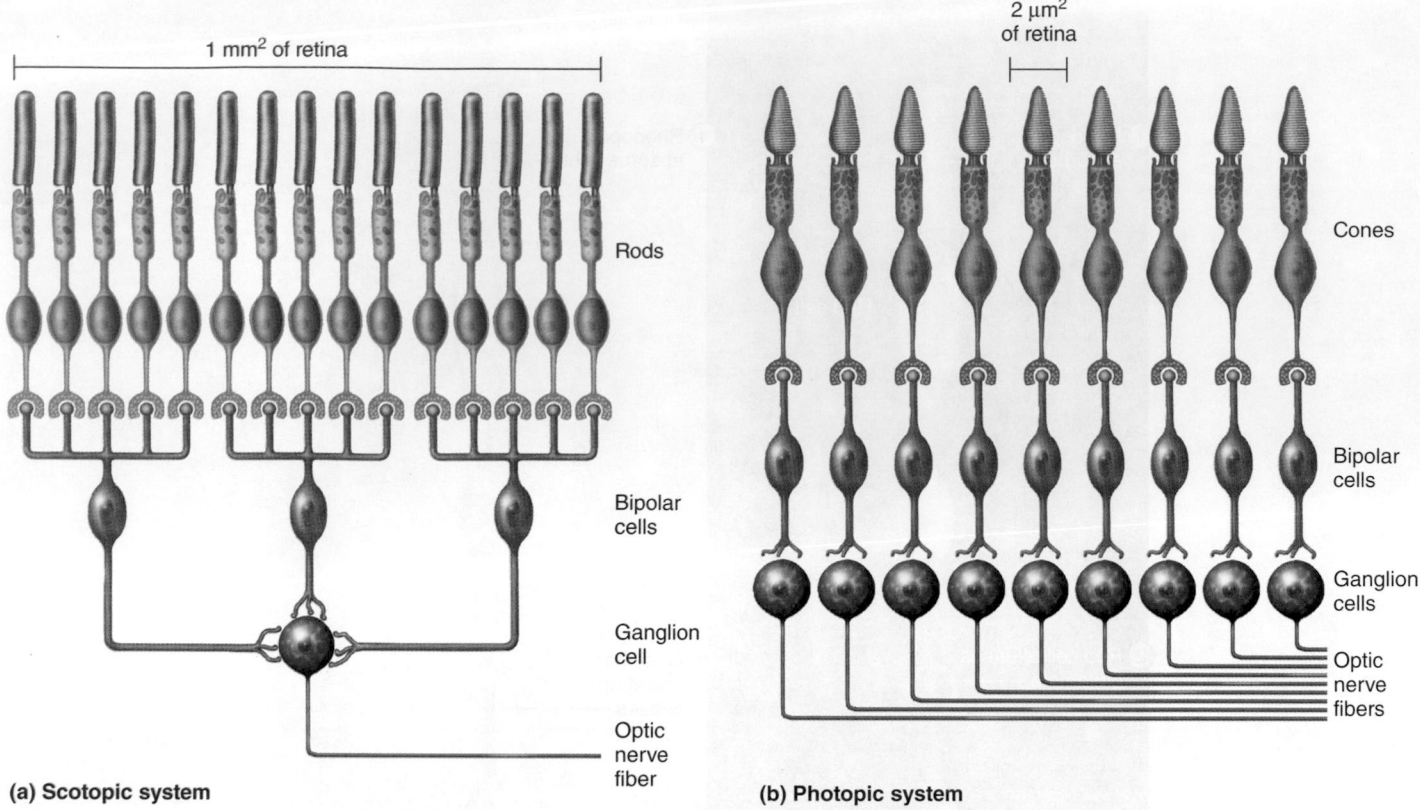

**FIGURE 16.39 Retinal Circuitry and Visual Sensitivity.** (a) In the scotopic (night vision) system, many rods converge on each bipolar cell and many bipolar cells converge on each ganglion cell. This allows rods to combine their effects through spatial summation and stimulate the ganglion cell, generating a nerve signal even in dim light. However, it means that each ganglion cell (and its optic nerve fiber) represents a relatively large area of retina and produces a grainy image. (b) In the photopic (day vision) system, there is little neural convergence. In the fovea, represented here, each cone has a "private line" to the brain, so each optic nerve fiber represents a tiny area of retina, and vision is relatively sharp. However, the lack of convergence means photopic vision cannot function well in dim light because weakly stimulated cones cannot collaborate to stimulate a ganglion cell.

When you look directly at something, its image falls on the fovea, which is occupied by about 4,000 tiny cones and no rods. The other neurons of the fovea are displaced to one side so they won't interfere with light falling on the cones. The smallness of these cones is like the smallness of the dots or pixels in a fine-grained (high-resolution) photograph; it is partially responsible for the high-resolution images formed at the fovea. In addition, the cones here show no neural convergence. Each cone synapses with only one bipolar cell and each bipolar cell with only one ganglion cell. This gives each foveal cone a "private line to the brain," and each ganglion cell of the fovea reports to the brain on a receptive field of just 2 μm² of retinal area (fig. 16.39b). Cones distant from the fovea exhibit some neural convergence but not nearly as much as rods do. The price of this lack of convergence at the fovea, however, is that cone cells are incapable of spatial summation, and the cone system is therefore less sensitive to light. The threshold of photopic (cone) vision lies between the intensity of starlight and moonlight reflected from white paper.

▶▶▶ **APPLY WHAT YOU KNOW**

*If you look directly at a dim star in the night sky, it disappears, and if you look slightly away, it reappears. Why?*

## Color Vision

Most nocturnal vertebrates have only rod cells, but many diurnal animals are endowed with cones and color vision. Color vision is especially well developed in primates for evolutionary reasons discussed in chapter 1. It is based on three kinds of cones named for the absorption peaks of their photopsins: **short-wavelength (S) cones,** with peak sensitivity at a wavelength of 420 nm; **medium-wavelength (M) cones,** which peak at 531 nm; and **long-wavelength (L) cones,** which peak at 558 nm. These were formerly called *blue, green,* and *red cones*—a less accurate terminology. "Red" cones, for example, do not peak in the red part of the spectrum (558 nm light is perceived as orange-yellow), but they are the only cones that respond at all to red light.

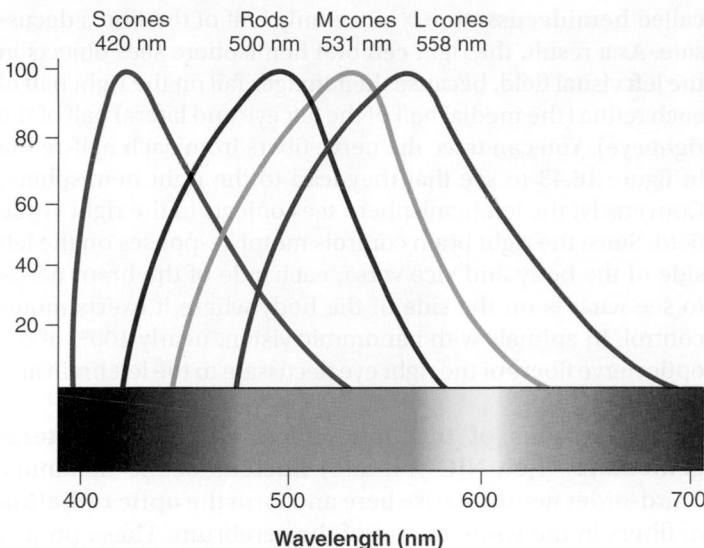

| Wavelength (nm) | Percentage of maximum cone response ( S : M : L ) | Perceived hue |
|---|---|---|
| 400 | 50 : 0 : 0 | Violet |
| 450 | 72 : 30 : 0 | Blue |
| 500 | 20 : 82 : 60 | Blue-green |
| 550 | 0 : 85 : 97 | Green |
| 625 | 0 : 3 : 35 | Orange |
| 675 | 0 : 0 : 5 | Red |

**FIGURE 16.40 Absorption Spectra of the Retinal Cells.** In the middle column of the table, each number indicates how strongly the respective cone cells respond as a percentage of their maximum capability. At 550 nm, for example, L cones respond at 97% of their maximum, M cones at 85%, and S cones not at all. The result is a perception of green light.

**?** *If you were to add another row to this table for 600 nm, what would you enter in the middle and right-hand columns?*

Our perception of colors is based on a mixture of nerve signals representing cones with different absorption peaks. In figure 16.40, note that light at 400 nm excites only S cones. At 500 nm, all three types of cones are stimulated; the L cones respond at 60% of their maximum capacity, M cones at 82% of their maximum, and S cones at 20%. The brain interprets this mixture as blue-green. The table in figure 16.40 shows how some other color sensations are generated by other response ratios.

Some individuals have a hereditary alteration or lack of one photopsin or another and thus exhibit **color blindness.** The most common form is *red–green color blindness,* which results from a lack of either L or M cones and causes difficulty distinguishing these and related shades from each other. For example, a person with normal *trichromatic* color vision sees figure 16.41 as showing the number 74, whereas a person with red–green color blindness sees no number. Red–green color blindness is a sex-linked recessive trait. It occurs in about 8% of males and 0.5% of females.

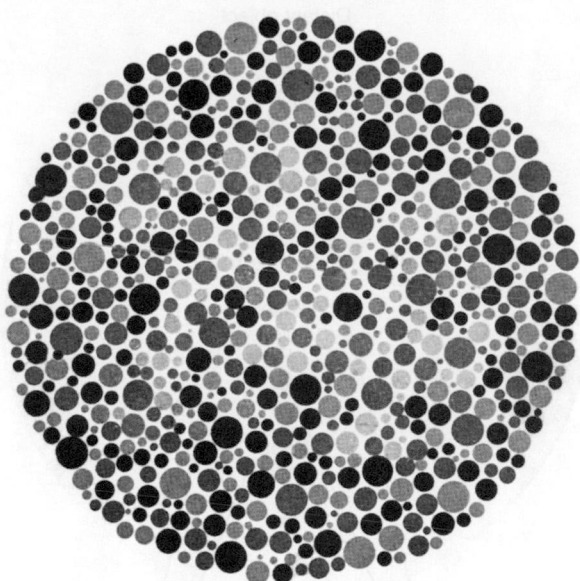

**FIGURE 16.41 A Test for Red–Green Color Blindness.** Persons with normal vision see the number 74. Persons with red–green color blindness see no discernible number.

## Stereoscopic Vision

**Stereoscopic vision (stereopsis)** is depth perception—the ability to judge how far away objects are. It depends on having two eyes with overlapping visual fields, which allows each eye to look at the same object from a different angle. Stereoscopic vision contrasts with the *panoramic vision* of mammals such as rodents and horses, in which the eyes are on opposite sides of the head. Although stereoscopic vision covers a smaller visual field than panoramic vision and provides less alertness to sneaky predators, it has the advantage of depth perception. The evolutionary basis of depth perception in primates was considered in chapter 1 (p. 10).

When you fixate on something within 30 m (100 ft), each eye views it from a slightly different angle and focuses its image on the fovea centralis. The point in space on which the eyes are focused is called the *fixation point.* Objects farther away than the fixation point cast an image somewhat medial to the foveas, and closer objects cast their images more laterally (fig. 16.42). The distance of an image from the two foveas provides the brain with information used to judge the position of other points relative to the fixation point.

## The Visual Projection Pathway

The first-order neurons in the visual pathway are the bipolar cells of the retina. The second-order neurons are the retinal ganglion cells, whose axons are the fibers of the optic nerve. The optic nerves leave each orbit through the optic canal and then converge to form an X, the **optic chiasm**[62] (ky-AZ-um), on

---

[62]*chiasm* = cross, X

Distant object
D

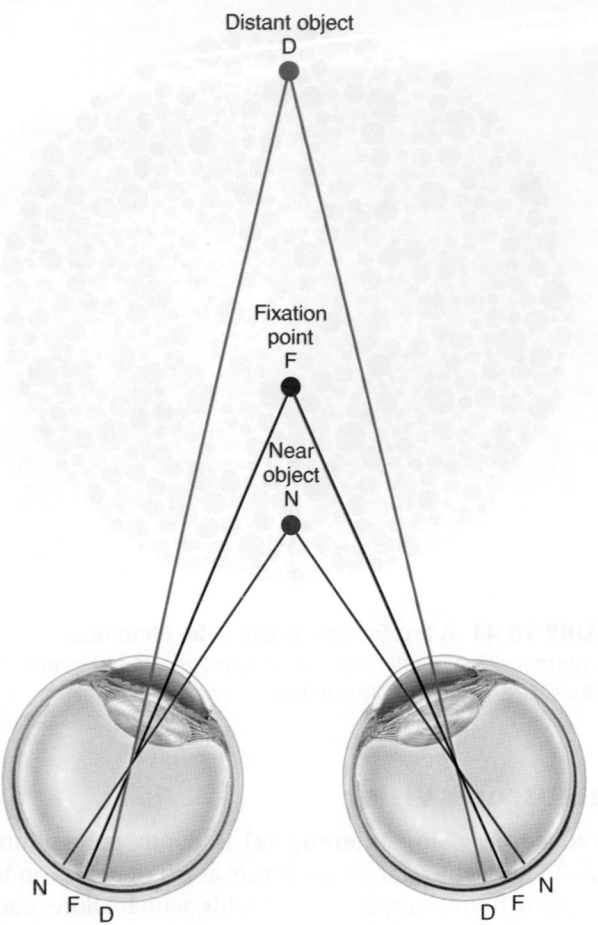

Fixation
point
F

Near
object
N

N
F
D

D F
N

**FIGURE 16.42  The Retinal Basis of Stereoscopic Vision (Depth Perception).** When the eyes converge on the fixation point (F), more distant objects (D) are focused on the retinas medial to the fovea and the brain interprets them as being farther away than the fixation point. Nearby objects (N) are focused lateral to the fovea and interpreted as being closer.

the base of the brain anterior to the pituitary. Beyond this, the fibers continue as a pair of **optic tracts.**

Within the chiasm, half of the fibers from each optic nerve cross over to the opposite side of the brain (fig. 16.43). This is called **hemidecussation,**[63] since only half of the fibers decussate. As a result, the right cerebral hemisphere sees objects in the left visual field, because their images fall on the right half of each retina (the medial half of the left eye and lateral half of the right eye). You can trace the nerve fibers from each half-retina in figure 16.43 to see that they lead to the right hemisphere. Conversely, the left hemisphere sees objects in the right visual field. Since the right brain controls motor responses on the left side of the body and vice versa, each side of the brain needs to see what is on the side of the body where it exerts motor control. In animals with panoramic vision, nearly 100% of the optic nerve fibers of the right eye decussate to the left brain and vice versa.

Most axons of the optic tracts end in the **lateral geniculate**[64] (jeh-NIC-you-late) **nucleus** of the thalamus. Third-order neurons arise here and form the **optic radiation** of fibers in the white matter of the cerebrum. These project to the primary visual cortex of the occipital lobe, where conscious visual sensation occurs. A lesion in the occipital lobe can cause blindness even if the eyes are fully functional.

A few optic nerve fibers come from the photosensitive, melanopsin-containing ganglion cells and take a different route, ending in the superior colliculi and pretectal nuclei of the midbrain. The superior colliculi control the visual reflexes of the extrinsic eye muscles, and the pretectal nuclei are involved in the photopupillary and accommodation reflexes.

Space does not allow us to consider much about the very complex processes of visual information processing in the brain. Some processing, such as contrast, brightness, motion, and stereopsis, begins in the retina. The primary visual cortex in the occipital lobe is connected by association tracts to nearby visual association areas in the posterior part of the parietal lobe and inferior part of the temporal lobe. These association areas process retinal data in ways that extract information about the location, motion, color, shape, boundaries, and other qualities of the objects we see. They also store visual memories and enable the brain to identify what we see—for example, to recognize printed words or name objects. What is yet to be learned about visual processing has important implications for biology, medicine, psychology, and even philosophy.

[63]*hemi* = half; *decuss* = to cross, form an X
[64]*geniculate* = bent like a knee

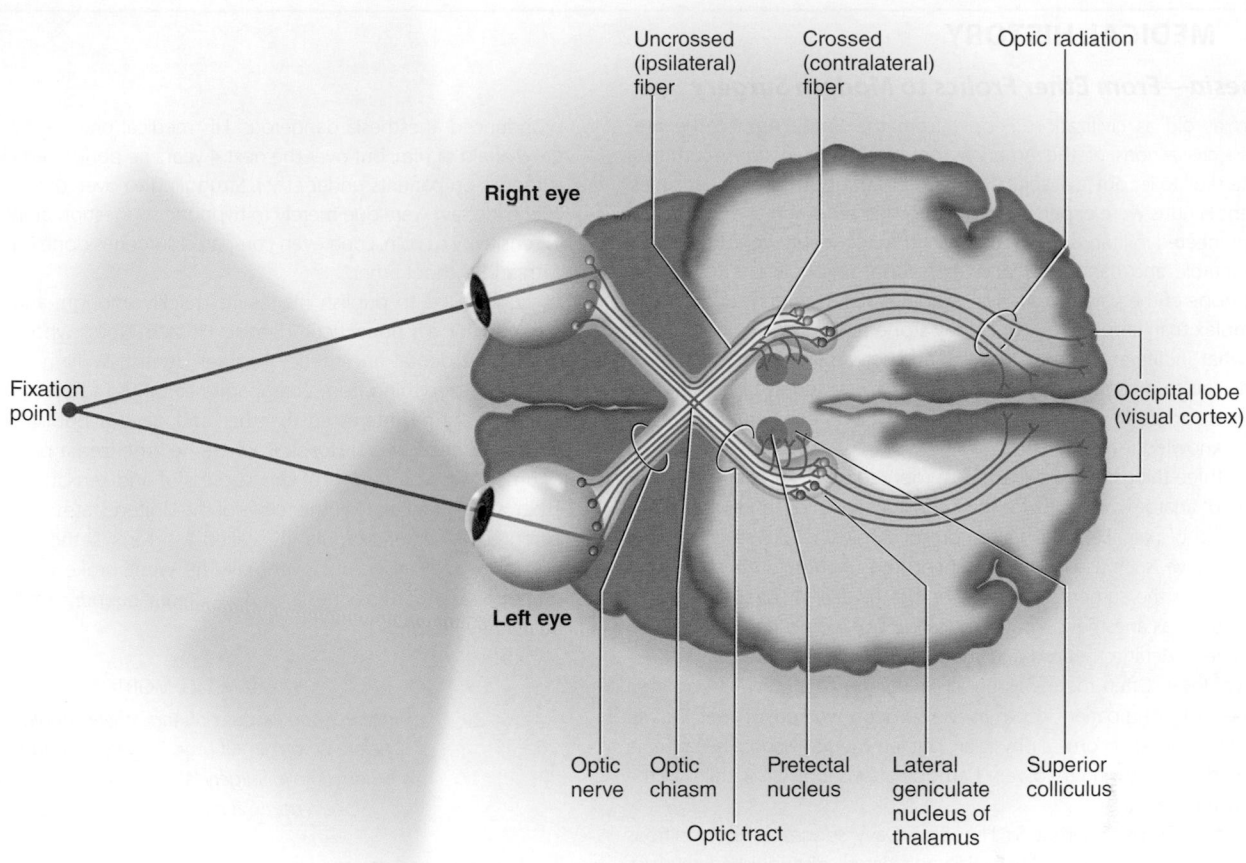

**Right eye**

Uncrossed (ipsilateral) fiber

Crossed (contralateral) fiber

Optic radiation

Fixation point

Occipital lobe (visual cortex)

**Left eye**

Optic nerve

Optic chiasm

Pretectal nucleus

Lateral geniculate nucleus of thalamus

Superior colliculus

Optic tract

**FIGURE 16.43 The Visual Projection Pathway.** Diagram of hemidecussation and projection to the primary visual cortex. Blue and yellow indicate the receptive fields of the left and right eyes; green indicates the area of overlap and stereoscopic vision. Nerve fibers from the medial side of the right eye (red) descussate to the left side of the brain, while fibers from the lateral side remain on the right side of the brain. The converse is true of the left eye. The right occipital lobe thus monitors the left side of the visual field and the left occipital lobe monitors the right side.

❓ *If a stroke destroyed the optic radiation of the right cerebral hemisphere, how would it affect a person's vision? Would it affect the person's visual reflexes?*

**BEFORE YOU GO ON**

Answer the following questions to test your understanding of the preceding section:

23. Why can't we see wavelengths below 350 nm or above 750 nm?

24. Why are light rays bent (refracted) more by the cornea than by the lens?

25. List as many structural and functional differences between rods and cones as you can.

26. Explain how the absorption of a photon of light leads to excitation of an optic nerve fiber.

27. Discuss the duplicity theory of vision, summarizing the advantage of having separate types of retinal photoreceptor cells and neural circuits for photopic and scotopic vision.

# DEEPER INSIGHT 16.5

## MEDICAL HISTORY

### Anesthesia—From Ether Frolics to Modern Surgery

Surgery is as old as civilization. People from the Stone Age to the pre-Columbian civilizations of the Americas practiced *trephination*—cutting a hole in the skull to let out "evil spirits" that were thought to cause headaches. The ancient Hindus were expert surgeons for their time, and the Greeks and Romans pioneered military surgery. But until the nineteenth century, surgery was a miserable and dangerous business, done only as a last resort and with little hope of the patient's survival. Surgeons rarely attempted anything more complex than amputations or kidney stone removal. A surgeon had to be somewhat indifferent to the struggles and screams of his patient. Most operations had to be completed in 3 minutes or less, and a strong arm and stomach were more important qualifications for a surgeon than extensive anatomical knowledge.

At least three things were needed for surgery to be more effective: better knowledge of anatomy, *asepsis*[65] for the control of infection, and *anesthesia*[66] for the control of pain. Early efforts to control surgical pain were crude and usually ineffective, such as choking a patient into unconsciousness and trying to complete the surgery before he or she awoke. Alcohol and opium were often used as anesthetics, but the dosage was poorly controlled; some patients were underanesthetized and suffered great pain anyway, and others died of overdoses. Often there was no alternative but for a few strong men to hold the struggling patient down as the surgeon worked. Charles Darwin originally intended to become a physician, but left medical school because he was sickened by observing "two very bad operations, one on a child," in the days before anesthesia.

In 1799, the English chemist Sir Humphry Davy suggested using nitrous oxide to relieve pain. His student, Michael Faraday, later a distinguished physicist, suggested ether. Neither of these ideas caught on for several decades, however. Nitrous oxide ("laughing gas") was a popular amusement in the 1800s, when traveling showmen went from town to town demonstrating its effects on volunteers from the audience. In 1841, at a medicine show in Georgia, some students were impressed with the volunteers' euphoric giggles and antics and asked a young local physician, Crawford W. Long, if he could make some nitrous oxide for them. Long lacked the equipment to synthesize it, but he recommended they try ether. Ether was commonly used in small oral doses for toothaches and "nervous ailments," but its main claim to popularity was its use as a party drug for so-called ether frolics. Long himself was a bit of a bon vivant who put on demonstrations for some of the young ladies, with the disclaimer that he could not be held responsible for whatever he might do under the influence of ether (such as stealing a kiss).

At these parties, Long noted that people sometimes fell and suffered considerable cuts and bruises without feeling pain. In 1842, he had a patient who was terrified of pain but needed a tumor removed from his neck. Long excised the tumor without difficulty as his patient sniffed ether from a towel. The operation created a sensation in town, but other physicians ridiculed Long and

pronounced anesthesia dangerous. His medical practice declined as people grew afraid of him, but over the next 4 years he performed eight more minor surgeries on patients under ether. Struggling to overcome criticisms that the effects he saw were due merely to hypnotic suggestion or individual variation in sensitivity to pain, Long even compared surgeries done on the same person with and without ether.

Long failed to publish his results quickly enough, and in 1844 he was scooped by a Connecticut dentist, Horace Wells, who had tried nitrous oxide as a dental anesthetic. Another dentist, William Morton of Boston, had tried everything from champagne to opium to kill pain in his patients. He too became interested in ether and gave a public demonstration at Massachusetts General Hospital, where he etherized a patient and removed a tumor. Within a month of this successful and sensational demonstration, ether was being used in other cities of the United States and England. Morton patented a "secret formula" he called Morton's Letheon,[67] which smelled suspiciously of ether, but eventually he went broke in a greedy effort to monopolize ether anesthesia, and he died a pauper. His grave near Boston bears the epitaph:

WILLIAM T. G. MORTON
*Inventor and Revealer of Anaesthetic Inhalation*
*Before Whom, in All Time, Surgery Was Agony.*
*By Whom Pain in Surgery Was Averted and Annulled.*
*Since Whom Science Has Control of Pain.*

Wells, who had engaged in a bitter feud to establish himself as the inventor of ether anesthesia, committed suicide at the age of 33. Crawford Long ran a successful medical practice in Athens, Georgia, but to his death he remained disappointed that he had not received credit as the first to perform surgery on etherized patients.

Ether and chloroform became obsolete when safer anesthetics were developed. *General anesthetics* such as isoflurane render a patient unconscious by crossing the blood–brain barrier and blocking nervous transmission through the brainstem. Certain general anesthetics deaden pain by activating GABA receptors and causing an inflow of $Cl^-$, which hyperpolarizes neurons and makes them less likely to fire. Diazepam (Valium) also employs this mechanism. *Local anesthetics* such as procaine (Novocaine) and tetracaine selectively deaden specific nerves. They decrease the permeability of membranes to $Na^+$, thereby reducing their ability to produce action potentials.

A sound knowledge of anatomy, control of infection and pain, and development of better tools converged to allow surgeons time to operate more carefully. As a result, surgery became more intellectually challenging and interesting. It attracted a more educated class of practitioner, which put it on the road to becoming the remarkable lifesaving approach that it is today.

---

[65]*a* = without; *sepsis* = infection
[66]*an* = without; *esthesia* = feeling, sensation

[67]*lethe* = oblivion, forgetfulness

# STUDY GUIDE

## ► Assess Your Learning Outcomes

*To test your knowledge, discuss the following topics with a study partner or in writing, ideally from memory.*

### 16.1 Properties and Types of Sensory Receptors (p. 579)

1. The definition of *receptor* and the range of complexity in sensory receptors
2. The definition of *sensory transduction* and the relationship of neural action potentials to that concept
3. The production and role of the receptor potential in sensory transduction
4. Four kinds of stimulus information transmitted by sensory receptors
5. Five categories of receptors classified by stimulus modality
6. Three categories of receptors classified by origin of their stimuli
7. Differences between general (somatosensory) and special senses

### 16.2 The General Senses (p. 581)

1. Three types of unencapsulated sensory nerve endings and what this term in general means
2. Six types of encapsulated nerve endings and how these differ from unencapsulated endings
3. The main routes for somatosensory signals from the head, and from the lower body, to the brainstem; the final destination of proprioceptive signals in the brain; and the final destination of most other sensory signals
4. Definitions of *pain* and *nociceptor*
5. Differences between fast pain and slow pain
6. Differences between somatic and visceral pain, and between superficial and deep somatic pain
7. Several chemicals released by injured tissues that stimulate nociceptors
8. The general three-neuron pathway typically taken by pain signals to the cerebral cortex
9. Pain pathways from receptors in the head to the cerebral cortex, including the cranial nerves that carry pain signals
10. Pain pathways from the lower body to the cerebral cortex, including the spinal cord tracts that carry pain signals

11. The special route and effect of pain signals that involve the reticular formation; how responses to these signals differ from other responses to pain
12. Referred pain and its anatomical basis
13. Spinal gating mechanisms that modify a person's sense of pain and awareness of body injury
14. Names of some analgesic neuropeptides and how they affect the sensation of pain

### 16.3 The Chemical Senses (p. 587)

1. Structure and locations of the taste buds
2. Types, locations, and functions of lingual papillae
3. Five primary taste sensations, and sensations other than taste that play a part in flavor
4. Mechanisms by which sugars, salts, alkaloids, acids, and glutamate excite taste cells
5. Which nerves carry taste signals, what routes they take to the brain, and what brain centers receive gustatory input
6. Structure and location of the olfactory mucosa and its receptor cells
7. How odor molecules excite olfactory cells
8. Which cranial nerve carries olfactory signals to the brain, and the route and point of termination of its nerve fibers
9. Sensory routes from the olfactory bulbs to the temporal lobes, insula, orbitofrontal cortex, hippocampus, amygdala, and hypothalamus
10. How the cerebral cortex influences olfactory bulb function and one's perception of smell

### 16.4 Hearing and Equilibrium (p. 592)

1. How sound is generated; what physical properties of a sound wave are measured in hertz and decibels; and what sensory qualities of sound correspond to those two physical properties
2. The total range of human hearing, in hertz, and the narrower range in which humans hear best
3. The decibel level of ordinary conversation and the thresholds of hearing and pain
4. Boundaries between the outer ear, middle ear, and inner ear

5. Anatomy of the outer ear and the function of its cerumen and guard hairs
6. Structure of the tympanic membrane and tympanic cavity; the names, structures, and anatomical arrangement of the auditory ossicles; the two middle-ear muscles; and anatomy of the auditory tube, its contribution to hearing, and its relevance to middle-ear infections
7. The bony and membranous labyrinths of the inner ear; the names and distribution of the two inner-ear fluids in relation to the labyrinths
8. Size and shape of the cochlea and its relationship to the modiolus
9. Cross-sectional anatomy of the cochlea
10. Structure of the spiral organ, especially the hair cells and tectorial membrane; differences between inner and outer hair cells
11. Function of the middle-ear ossicles and muscles; how the tympanic reflex works and how it protects one's hearing
12. How vibrations of the tympanic membrane lead to stimulation of the cochlear nerve
13. How the cochlea codes for differences in the pitch and loudness of sounds
14. How the outer hair cells tune the cochlea to improve its sensitivity to differences in pitch
15. The pathway from cochlear nerve to auditory centers of the brain; the feedback pathway from the pons back to the cochlea, and its purpose
16. Differences between static and dynamic equilibrium and between linear and angular acceleration
17. Structure of the saccule and utricle and the relevance of the spatial orientation of the macula in each one
18. How linear acceleration stimulates the hair cells of the saccule and utricle during linear acceleration; how the body senses the difference between vertical and horizontal acceleration
19. Structure of the semicircular ducts, especially the crista ampullaris and cupula
20. How acceleration stimulates hair cells of the crista ampullaris, and why the combined input of the six semicircular ducts enables

# STUDY GUIDE

the brain to sense tilting or rotation of the head in any direction

21. Why it cannot be said that the vestibular system senses motion of the head, but only changes in the rate of motion

22. The path taken by signals in the vestibular nerve to the cerebrum, cerebellum, reticular formation, spinal cord, and nuclei of the three cranial nerves for eye movements

23. Why it is important for eye movement to be coordinated with vestibular input

## 16.5 Vision (p. 606)

1. The definition of *vision* and the range of electromagnetic wavelengths over which human vision occurs

2. Six extrinsic eye muscles, their anatomy, the eye motions they produce, and the cranial nerves that control them

3. Components of the lacrimal apparatus and the route taken by tears as they wash over the eye and drain into the nasal cavity

4. Anatomy and functions of the eyebrow, eyelids, eyelashes, and conjunctiva

5. Three tunics of the eyeball and the structural components of each

6. Optical components of the eye

7. The secretion, flow, and reabsorption of aqueous humor

8. General structure of the retina; its two points of attachment to the wall of the eye; and the locations, structure, and functional significance of the optic disc, optic nerve, macula lutea, and fovea centralis

9. The cause of the blind spot and how the brain compensates for it

10. Structure and action of the pupillary constrictor and dilator; anatomy of their autonomic innervation; and the photopupillary reflex

11. Principles of refraction; points at which refraction occurs as light enters the eye; relative contributions of the cornea and lens to image formation, and the reason for the difference

12. The difference between emmetropia and the near response, and three mechanisms of the near response

13. Histological layers and cell types of the retina; three types of photoreceptor cells and their respective functions; functions of the retinal bipolar, ganglion, horizontal, and amacrine cells

14. The structures of rods and cones; where visual pigments are contained in these

cells; the general structure of rhodopsin and photopsin; and how these two pigments differ

15. Differences in rod and cone function

16. How light absorption generates an optic nerve signal

17. Mechanisms of light and dark adaptation

18. Why a single retinal receptor system cannot achieve both low-threshold night vision and high-resolution day vision; why rod vision works in very low light but sacrifices resolution to do so; and why cone vision gives high resolution but sacrifices light sensitivity

19. Difference between S, M, and L cones; how neural coding and three cone types produce sensitivity to innumerable colors; and what causes color blindness

20. The retinal basis of stereoscopic vision

21. Projection pathways from the eyes to the occipital lobe and to the superior colliculi and pretectal nuclei of the midbrain

22. Hemidecussation, where it occurs, and how it determines what areas of the visual field are seen by the right and left occipital lobes

## ▶ Testing Your Recall

*Answers in Appendix B*

1. Hot and cold stimuli are detected by
   a. free nerve endings.
   b. proprioceptors.
   c. end bulbs.
   d. lamellar corpuscles.
   e. tactile corpuscles.

2. _____ is a neurotransmitter that transmits pain sensations to second-order spinal neurons.
   a. Endorphin
   b. Enkephalin
   c. Substance P
   d. Acetylcholine
   e. Norepinephrine

3. _____ is a neuromodulator that blocks the conduction of pain signals by second-order spinal neurons.
   a. Endorphin
   b. Enkephalin

   c. Substance P
   d. Acetylcholine
   e. Norepinephrine

4. Most taste buds occur in
   a. the vallate papillae.
   b. the fungiform papillae.
   c. the filiform papillae.
   d. the palate.
   e. the lips.

5. The higher the frequency of a sound,
   a. the louder it sounds.
   b. the harder it is to hear.
   c. the more it stimulates the distal end of the spiral organ.
   d. the faster it travels through air.
   e. the higher its pitch.

6. Cochlear hair cells rest on
   a. the tympanic membrane.
   b. the secondary tympanic membrane.

   c. the tectorial membrane.
   d. the vestibular membrane.
   e. the basilar membrane.

7. The acceleration you feel when an elevator begins to rise is sensed by
   a. the anterior semicircular duct.
   b. the spiral organ.
   c. the crista ampullaris.
   d. the macula sacculi.
   e. the macula utriculi.

8. The color of light is determined by
   a. its velocity.
   b. its amplitude.
   c. its wavelength.
   d. refraction.
   e. how strongly it stimulates the rods.

# STUDY GUIDE

9. The retina receives its oxygen supply from
   a. the hyaloid artery.
   b. the vitreous body.
   c. the choroid.
   d. the pigment epithelium.
   e. the scleral venous sinus.

10. Which of the following statements about photopic vision is false?
    a. It is mediated by the cones.
    b. It has a low threshold.
    c. It produces fine resolution.
    d. It does not function in starlight.
    e. It does not employ rhodopsin.

11. The most finely detailed vision occurs when an image falls on a pit in the retina called the _____.

12. The only cells of the retina that generate action potentials are the _____ cells.

13. The visual pigment of a cone cell is _____.

14. The gelatinous membranes of the macula sacculi and macula utriculi are weighted by protein–calcium carbonate granules called _____.

15. Three rows of _____ in the cochlea have V-shaped arrays of stereocilia and tune the frequency sensitivity of the cochlea.

16. The _____ is a tiny bone that vibrates in the oval window and thereby transfers sound vibrations to the inner ear.

17. The _____ of the midbrain receives auditory input and triggers the head-turning auditory reflex.

18. The function of the _____ in a taste bud is to replace dead taste cells.

19. Olfactory neurons synapse with mitral cells and tufted cells in the _____, which lies inferior to the frontal lobe.

20. In the phenomenon of _____, pain from the viscera is perceived as coming from an area of the skin.

## ► Building Your Medical Vocabulary

*Answers in Appendix B*

*State a medical meaning of each word element below, and give a term in which it or a slight variation of it is used.*

1. bin-
2. decuss-
3. hemi-
4. lacrimo-
5. litho-
6. maculo-
7. noci-
8. scoto-
9. -sepsis
10. tricho-

## ► True or False

*Answers in Appendix B*

*Determine which five of the following statements are false, and briefly explain why*

1. The sensory (afferent) nerve fibers for touch end in the thalamus.

2. Things we touch with the left hand are perceived only in the right cerebral hemisphere.

3. Things we see with the left eye are perceived only in the right cerebral hemisphere.

4. Some chemoreceptors are interoceptors and some are exteroceptors.

5. The vitreous body occupies the posterior chamber of the eye.

6. Descending analgesic fibers prevent pain signals from reaching the spinal cord.

7. Cranial nerve VIII carries signals for both hearing and balance.

8. The tympanic cavity is filled with air, but the membranous labyrinth is filled with liquid.

9. Rods and cones release their neurotransmitter in the dark, not in the light.

10. All of the extrinsic muscles of the eye are controlled by the oculomotor nerve.

# STUDY GUIDE

## ► Testing Your Comprehension

*Answers at* www.mhhe.com/saladin7

1. The principle of neural convergence is explained on page 465. Discuss its relevance to referred pain and scotopic vision.

2. What type of cutaneous receptor enables you to feel an insect crawling through your hair? What type enables you to palpate a patient's pulse? What type enables a blind person to read braille?

3. Contraction of a muscle usually puts more tension on a structure, but contraction of the ciliary muscle puts less tension on the lens. Explain how.

4. Janet has terminal ovarian cancer and is in severe pelvic pain that has not yielded to any other treatment. A neurosurgeon performs an *anterolateral cordotomy*, cutting across the anterolateral region of her lumbar spinal cord. Explain the rationale of this treatment and its possible side effects.

5. What would be the benefit of a drug that blocks the receptors for substance P?

## ► Improve Your Grade at www.mhhe.com/saladin7

*Download animations to study when it fits your schedule. Practice quizzes, labeling activities, games, and flashcards offer fun ways to master the chapter concepts. Or, download image PowerPoint files for each chapter to create a study guide or for taking notes during lecture.*

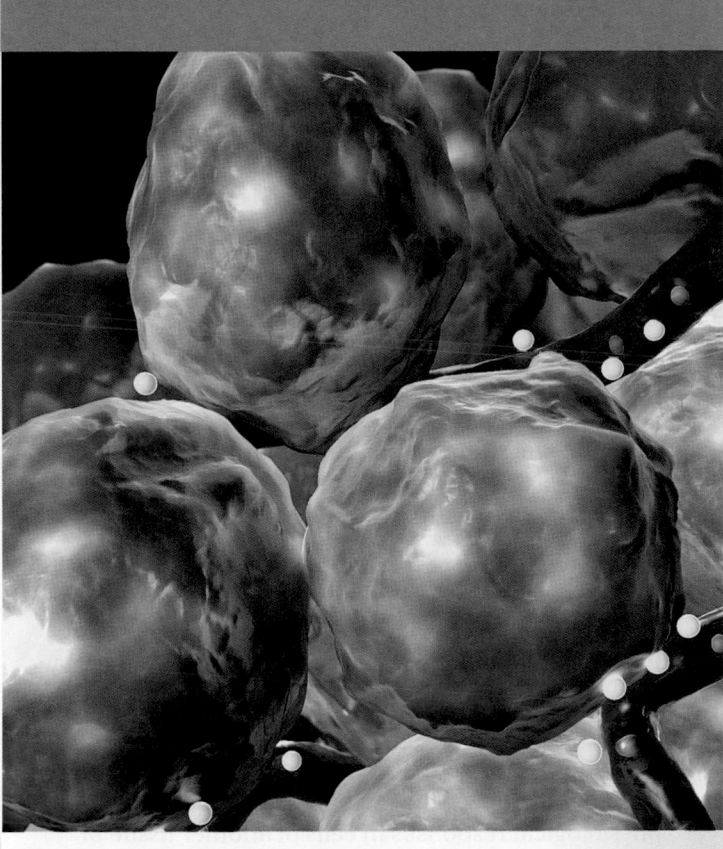

Artist's conception of beta cells of the pancreas releasing insulin molecules into the bloodstream

CHAPTER

# 17

# THE ENDOCRINE SYSTEM

## CHAPTER OUTLINE

## DEEPER INSIGHTS

**Module 8: Endocrine System**

If the body is to function as an integrated whole, its organs must communicate with each other and coordinate their activities. Even simple organisms composed of only a few cells have mechanisms for intercellular communication, suggesting that such mechanisms evolved very early in the history of life. In humans, two such systems are especially prominent—the nervous and endocrine systems, which communicate with neurotransmitters and hormones, respectively.

Nearly everyone has heard of at least some hormones—growth hormone, thyroid hormone, estrogen, and insulin, for example. At least passingly familiar, too, are some of the glands that secrete them (such as the pituitary and thyroid glands) and some disorders that result from hormone deficiency, excess, or dysfunction (such as diabetes, goiter, and dwarfism).

This chapter is primarily about the *endocrine* (hormonal) *system* of communication. We will start with the relatively familiar and large-scale aspects of this system—a survey of the endocrine glands, their hormones, and the principal effects of those hormones. We will then work our way down to the finer and less familiar details—the chemical identity of hormones, how they are made and transported, and how they produce their effects on their target cells. Shorter sections at the end of the chapter discuss the role of the endocrine system in adapting to stress, some hormonelike *paracrine* secretions, and the pathologies that result from endocrine dysfunction.

## 17.1 Overview of the Endocrine System

### Expected Learning Outcomes

When you have completed this section, you should be able to

a. define *hormone* and *endocrine system;*

b. name several organs of the endocrine system;

c. contrast endocrine with exocrine glands;

d. recognize the standard abbreviations for many hormones; and

e. compare and contrast the nervous and endocrine systems.

The body has four principal avenues of communication from cell to cell:

1. **Gap junctions** join single-unit smooth muscle, cardiac muscle, epithelial, and other cells to each other. They enable cells to pass nutrients, electrolytes, and signaling molecules directly from the cytoplasm of one cell to the cytoplasm of the next through pores in their plasma membranes (fig. 5.28, p. 163).

2. **Neurotransmitters** are released by neurons, diffuse across a narrow synaptic cleft, and bind to receptors on the surface of the next cell.

3. **Paracrines**[1] are secreted by one cell, diffuse to nearby cells in the same tissue, and stimulate their physiology. Some call them *local hormones*.

4. **Hormones,**[2] in the strict sense, are chemical messengers that are transported by the bloodstream and stimulate physiological responses in cells of another tissue or organ, often a considerable distance away. Certain hormones produced by the pituitary gland in the head, for example, act on organs as far away as the pelvic cavity.

This chapter is concerned mainly with hormones and, to some extent, paracrine secretions. The glands, tissues, and cells that secrete hormones constitute the **endocrine**[3] **system;** the study of this system and the diagnosis and treatment of its disorders is called **endocrinology.** The most familiar hormone sources are the organs traditionally recognized as **endocrine glands,** such as the pituitary, thyroid, and adrenal glands, among others (fig. 17.1). Growing knowledge of endocrinology has revealed, however, that hormones are also secreted by numerous organs and tissues not usually thought of as glands, such as the brain, heart, small intestine, bones, and adipose tissue.

## Comparison of Endocrine and Exocrine Glands

In chapter 5, we examined another category of glands, the *exocrine* glands. The classical distinction between exocrine and endocrine glands has been the presence or absence of ducts. Most exocrine glands secrete their products by way of a duct onto an epithelial surface such as the skin or the mucosa of the digestive tract. Endocrine glands, by contrast, are ductless and release their secretions into the bloodstream (see fig. 5.30, p. 165). For this reason, hormones were originally called the body's "internal secretions"; the word *endocrine* still alludes to this fact. Exocrine

---

[1]*para* = next to; *crin* = to secrete
[2]*hormone* = to excite, set in motion
[3]*endo* = into; *crin* = to secrete

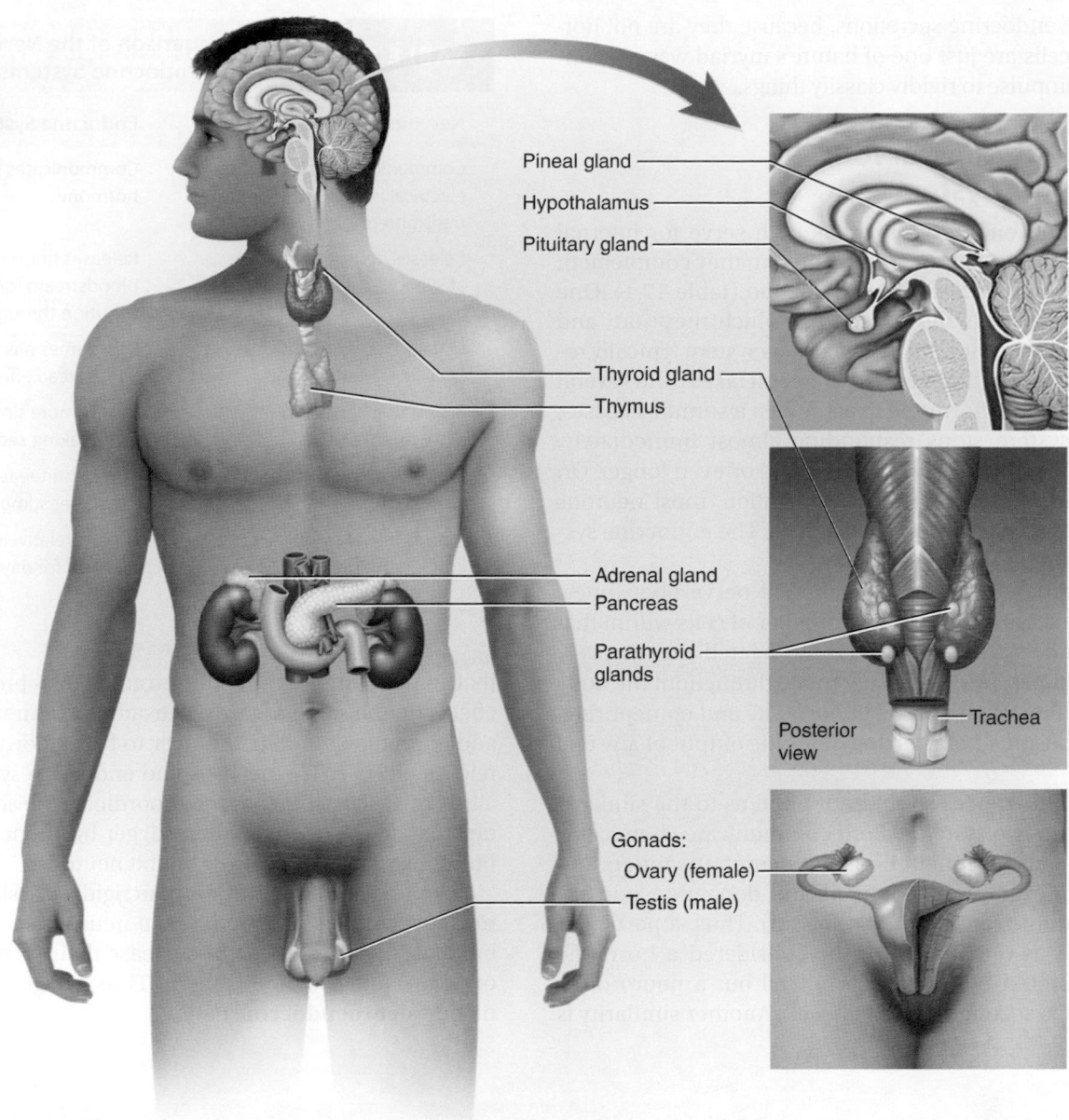

Pineal gland
Hypothalamus
Pituitary gland

Thyroid gland
Thymus

Adrenal gland
Pancreas

Parathyroid
glands

Posterior
view

Trachea

Gonads:
Ovary (female)
Testis (male)

**FIGURE 17.1  Major Organs of the Endocrine System.**  This system also includes gland cells in many other organs not shown here.  AP|R

❓ *After reading this chapter, name at least three hormone-secreting organs that are not shown in this illustration.*

secretions have extracellular effects such as the digestion of food, whereas endocrine secretions have intracellular effects—they alter cell metabolism.

Endocrine glands have an unusually high density of blood capillaries, which serve to pick up and carry away their hormones. These vessels are an especially permeable type called *fenestrated capillaries*, which have patches of large pores in their walls allowing for the easy uptake of matter from the gland tissue (see fig. 20.5, p. 749).

Some glands and secretory cells defy simple classification as endocrine or exocrine. Liver cells, for example, behave as exocrine cells in the traditional sense by secreting bile into ducts that lead ultimately to the small intestine. However, they also secrete hormones into the blood, and in this respect they act as endocrine cells. They secrete albumin and blood-clotting factors directly into the blood as well. These do not fit the traditional concept of exocrine secretions, because they are not released by way of ducts or onto epithelial surfaces; nor do they fit the concept of endocrine secretions, because they are not hormones. Liver cells are just one of nature's myriad ways of confounding our impulse to rigidly classify things.

## Comparison of the Nervous and Endocrine Systems

The nervous and endocrine systems both serve for internal communication, but they are not redundant; they complement rather than duplicate each other's function (table 17.1). One important difference is the speed with which they start and stop responding to a stimulus. The nervous system typically responds within a few milliseconds, whereas it takes from several seconds to days for a hormone to act. When a stimulus ceases, the nervous system stops responding almost immediately, whereas hormonal effects can last for days or even longer. On the other hand, under long-term stimulation, most neurons quickly adapt and their response declines. The endocrine system shows more persistent responses.

Another difference is that an efferent nerve fiber innervates only one organ and a limited number of cells within that organ, so its effects are usually precisely targeted and relatively specific. Hormones, by contrast, circulate throughout the body and some of them, such as growth hormone and epinephrine, have more widespread effects than does the output of any one nerve fiber.

But these differences should not blind us to the similarities between the two systems. Both communicate chemically, and several chemicals function as both neurotransmitters and hormones—for example, norepinephrine, dopamine, and antidiuretic hormone (arginine vasopressin). Thus, a particular chemical such as dopamine can be considered a hormone when it is secreted by an endocrine cell but a neurotransmitter when it is secreted by a nerve cell. Another similarity is that some hormones and neurotransmitters produce identical effects on the same organ. For example, both norepinephrine and glucagon stimulate the liver to break down glycogen and release glucose. The nervous and endocrine systems continually regulate each other as they coordinate the activities of other organ systems. Neurons often trigger hormone secretion, and hormones often stimulate or inhibit neurons.

Some cells defy any attempt to rigidly classify them as neurons or gland cells. They act like neurons in many respects, but like endocrine cells, they release their secretions (such as oxytocin) into the bloodstream. Thus, we give them a hybrid name—**neuroendocrine cells.**

| **TABLE 17.1** | Comparison of the Nervous and Endocrine Systems |
| --- | --- |
| **Nervous System** | **Endocrine System** |
| Communicates by means of electrical impulses and neurotransmitters | Communicates by means of hormones |
| Releases neurotransmitters at synapses at specific target cells | Releases hormones into bloodstream for general distribution throughout body |
| Usually has relatively local, specific effects | Sometimes has very general, widespread effects |
| Reacts quickly to stimuli, usually within 1–10 ms | Reacts more slowly to stimuli, often taking seconds to days |
| Stops quickly when stimulus stops | May continue responding long after stimulus stops |
| Adapts relatively quickly to continual stimulation | Adapts relatively slowly; may respond for days to weeks |

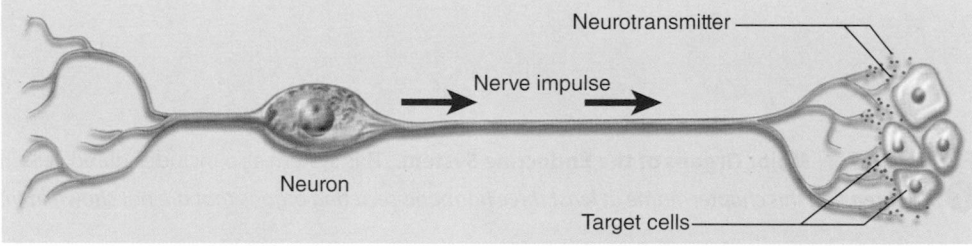

**(a) Nervous system**

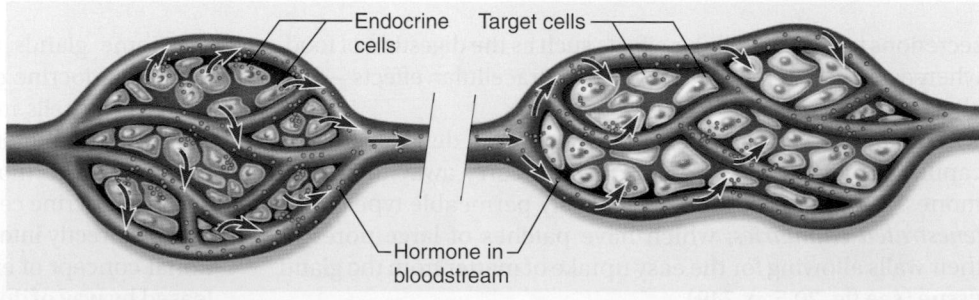

**(b) Endocrine system**

**FIGURE 17.2 Communication by the Nervous and Endocrine Systems.** (a) A neuron has a long fiber that delivers its neurotransmitter to the immediate vicinity of its target cells. (b) Endocrine cells secrete a hormone into the bloodstream (left). At a point often remote from its origin, the hormone leaves the bloodstream and enters or binds to its target cells (right).

We have seen that neurotransmitters depend on receptors in the receiving cell; they cannot exert any effect unless the receiving cell is equipped to bind and respond to them. This is true of hormones as well. When a hormone enters the bloodstream, it goes wherever the blood goes; there is no way to send it selectively to a particular organ. However, only certain **target organs** or **target cells** respond to it (fig. 17.2). Thyroid-stimulating hormone, for example, circulates everywhere the blood goes, but stimulates only the thyroid gland. In most cases, such selective responses are because only the target cells have receptors for a given hormone. They can also occur, however, because the circulating hormone is inactive and only the target cells have the enzyme needed to convert it to active form. Circulating testosterone, for example, is relatively inactive, but its target cells have an enzyme that converts it to dihydrotestosterone, which is much more potent.

## Hormone Nomenclature

Many hormones are denoted by standard abbreviations that are used repeatedly in this chapter. These are listed alphabetically in table 17.2 for use as a convenient reference while you work through the chapter. This list is by no means complete. It omits hormones that have no abbreviation, such as estrogen and insulin, and hormones that are not discussed much in this chapter. Synonyms used by many authors are indicated in parentheses, but the first name listed is the one used in this book.

| TABLE 17.2 | Names and Abbreviations for Hormones | |
|---|---|---|
| **Abbreviation** | **Name** | **Source** |
| ACTH | Adrenocorticotropic hormone (corticotropin) | Anterior pituitary |
| ADH | Antidiuretic hormone (arginine vasopressin) | Posterior pituitary |
| ANP | Atrial natriuretic peptide | Heart |
| CRH | Corticotropin-releasing hormone | Hypothalamus |
| DHEA | Dehydroepiandrosterone | Adrenal cortex |
| EPO | Erythropoietin | Kidney, liver |
| FSH | Follicle-stimulating hormone | Anterior pituitary |
| GH | Growth hormone (somatotropin) | Anterior pituitary |
| GHRH | Growth hormone–releasing hormone | Hypothalamus |
| GnRH | Gonadotropin-releasing hormone | Hypothalamus |
| IGFs | Insulin-like growth factors (somatomedins) | Liver, other tissues |
| LH | Luteinizing hormone | Anterior pituitary |
| NE | Norepinephrine | Adrenal medulla |
| OT | Oxytocin | Posterior pituitary |
| PIH | Prolactin-inhibiting hormone (dopamine) | Hypothalamus |
| PRL | Prolactin | Anterior pituitary |
| PTH | Parathyroid hormone (parathormone) | Parathyroids |
| $T_3$ | Triiodothyronine | Thyroid |
| $T_4$ | Thyroxine (tetraiodothyronine) | Thyroid |
| TH | Thyroid hormone ($T_3$ and $T_4$ collectively) | Thyroid |
| TRH | Thyrotropin-releasing hormone | Hypothalamus |
| TSH | Thyroid-stimulating hormone (thyrotropin) | Anterior pituitary |

### BEFORE YOU GO ON

Answer the following questions to test your understanding of the preceding section:

1. Define the word *hormone* and distinguish a hormone from a neurotransmitter. Why is this an imperfect distinction?

2. Name some sources of hormones other than purely endocrine glands.

3. Describe some distinctions between endocrine and exocrine glands.

4. List some similarities and differences between the endocrine and nervous systems.

5. Discuss why the target-cell concept is essential for understanding hormone function.

## 17.2 The Hypothalamus and Pituitary Gland

### Expected Learning Outcomes

When you have completed this section, you should be able to

a. describe the anatomical relationships between the hypothalamus and pituitary gland;

b. distinguish between the anterior and posterior lobes of the pituitary;

c. list the hormones produced by the hypothalamus and each lobe of the pituitary, and identify the functions of each hormone;

d. explain how the pituitary is controlled by the hypothalamus and its target organs; and

e. describe the effects of growth hormone.

There is no master control center that regulates the entire endocrine system, but the pituitary gland and hypothalamus have a more wide-ranging influence than any other part of the system, and several other endocrine glands cannot be adequately understood without first knowing how the pituitary influences them. This is therefore an appropriate place to begin a survey of the endocrine system.

## Anatomy

The hypothalamus, shaped like a flattened funnel, forms the floor and walls of the third ventricle of the brain; we studied its structure and function in chapter 14 (see figs. 14.2, p. 511; 14.12b, p. 525). It regulates primitive functions of the body ranging from water balance and thermoregulation to sex drive and childbirth. Many of its functions are carried out by way of the pituitary gland, which is closely associated with it both anatomically and physiologically.

The **pituitary gland (hypophysis[4])** is suspended from the floor of the hypothalamus by a stalk *(infundibulum[5])* and housed in a depression of the sphenoid bone, the sella turcica. The pituitary is usually about 1.3 cm wide and roughly the size and shape of a kidney bean; it grows about 50% larger in pregnancy. It is actually composed of two structures—the *anterior* and *posterior pituitary*—with independent origins and separate functions. The anterior pituitary arises from a pouch that grows upward from the embryonic pharynx, while the posterior pituitary arises as a bud growing downward from the brain (fig. 17.3). They come to lie side by side and are so closely joined that they look like a single gland.

The **anterior pituitary,** also called the *anterior lobe* or *adenohypophysis[6]* (AD-eh-no-hy-POFF-ih-sis) constitutes about three-quarters of the pituitary as a whole (figs. 17.4a, 17.5a). It has no nervous connection to the hypothalamus but is linked to it by a complex of blood vessels called the **hypophyseal portal system** (hy-POFF-ih-SEE-ul) (fig. 17.4b). This system consists of a network of *primary capillaries* in the hypothalamus, a group of small veins called *portal venules* that travel down the stalk, and a complex of *secondary capillaries* in the anterior pituitary. The hypothalamus controls the anterior pituitary by secreting hormones that enter the primary capillaries, travel down the venules, and diffuse out of the secondary capillaries into the pituitary tissue. The hypothalamic hormones regulate secretion by various types of pituitary cells that we will study later.

[4]*hypo* = below; *physis* = growth
[5]*infundibulum* = funnel

[6]*adeno* = gland

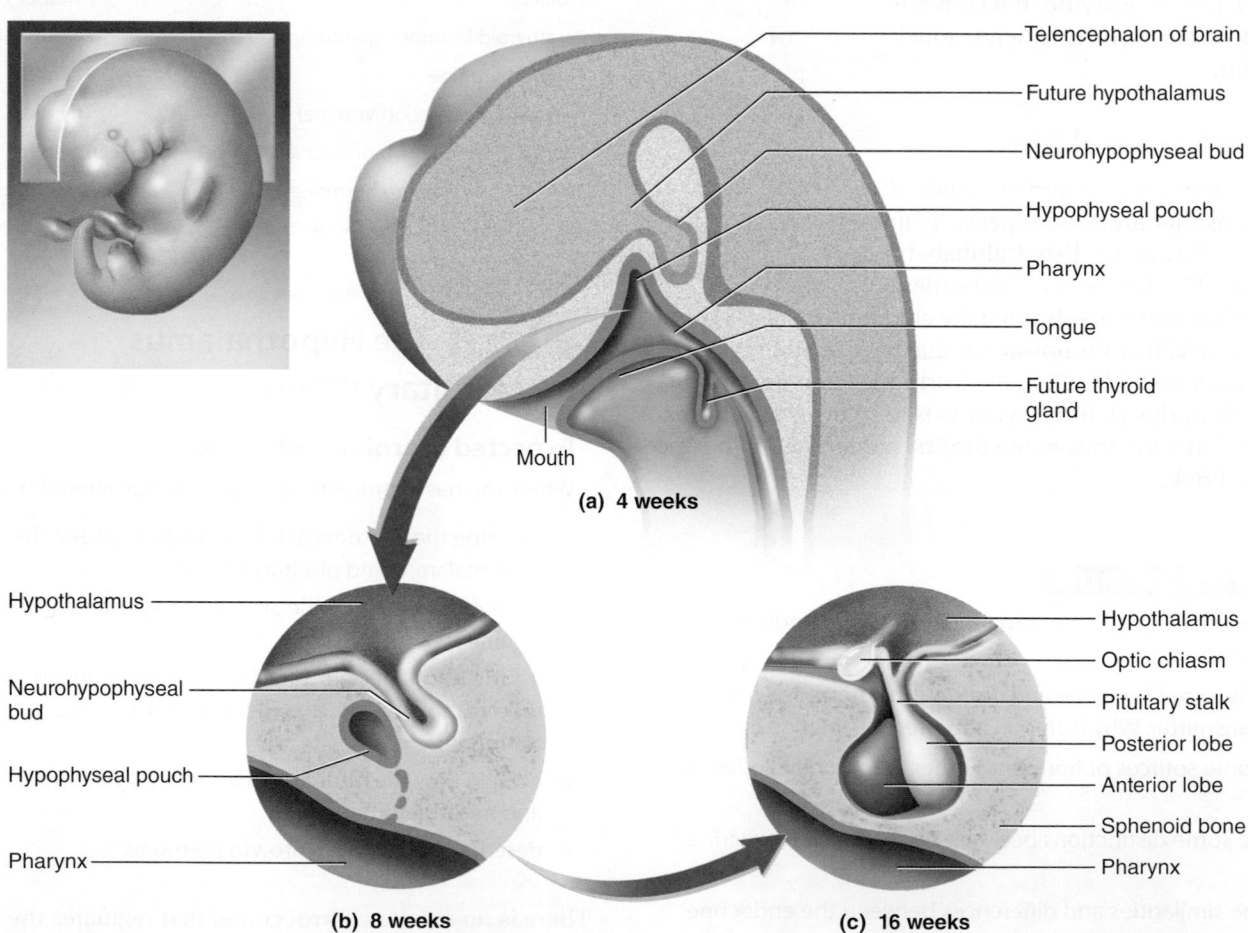

**FIGURE 17.3  Embryonic Development of the Pituitary Gland.**  (a) Sagittal section of the head showing the neural and pharyngeal origins of the pituitary. (b) Pituitary development at 8 weeks. The hypophyseal pouch, destined to become the anterior pituitary, has now separated from the pharynx. (c) Development at 16 weeks. The two lobes are now encased in bone and so closely associated they appear to be a single gland.

Anterior | Posterior

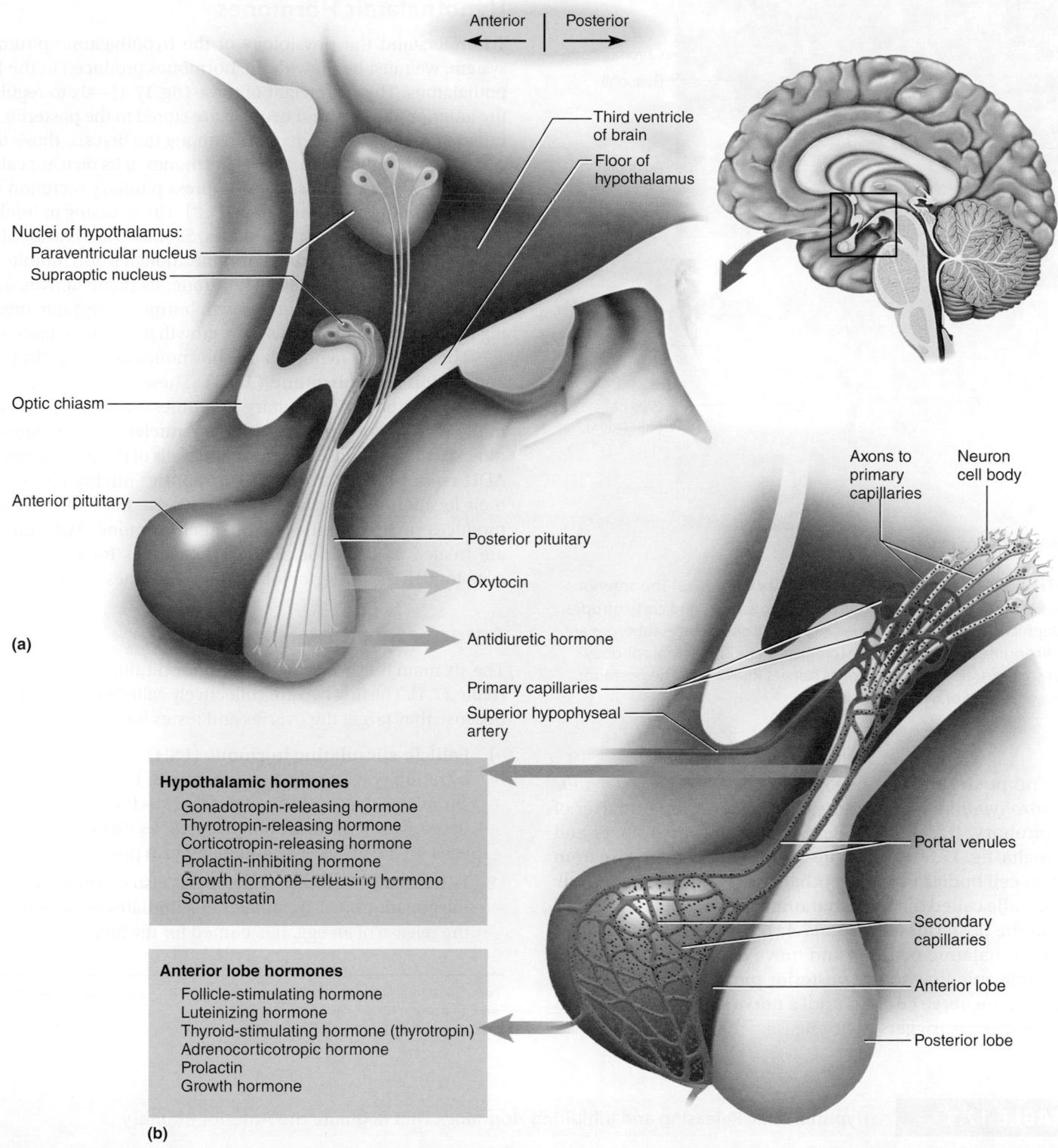

Third ventricle of brain

Floor of hypothalamus

Nuclei of hypothalamus:
Paraventricular nucleus
Supraoptic nucleus

Optic chiasm

Anterior pituitary

Posterior pituitary

Oxytocin

Antidiuretic hormone

Primary capillaries
Superior hypophyseal artery

Axons to primary capillaries

Neuron cell body

(a)

**Hypothalamic hormones**

Gonadotropin-releasing hormone
Thyrotropin-releasing hormone
Corticotropin-releasing hormone
Prolactin-inhibiting hormone
Growth hormone–releasing hormone
Somatostatin

**Anterior lobe hormones**

Follicle-stimulating hormone
Luteinizing hormone
Thyroid-stimulating hormone (thyrotropin)
Adrenocorticotropic hormone
Prolactin
Growth hormone

Portal venules

Secondary capillaries

Anterior lobe

Posterior lobe

(b)

**FIGURE 17.4 Anatomy of the Pituitary Gland.** (a) Major structures of the pituitary and hormones of the neurohypophysis. Note that these hormones are produced by two nuclei in the hypothalamus and later released from the posterior lobe of the pituitary. (b) The hypophyseal portal system, which regulates the anterior lobe of the pituitary. The hormones in the violet box are secreted by the hypothalamus and travel in the portal system to the anterior pituitary. The hormones in the pink box are secreted by the anterior pituitary under the control of the hypothalamic releasers and inhibitors. AP|R

? *Which lobe of the pituitary is essentially composed of brain tissue?*

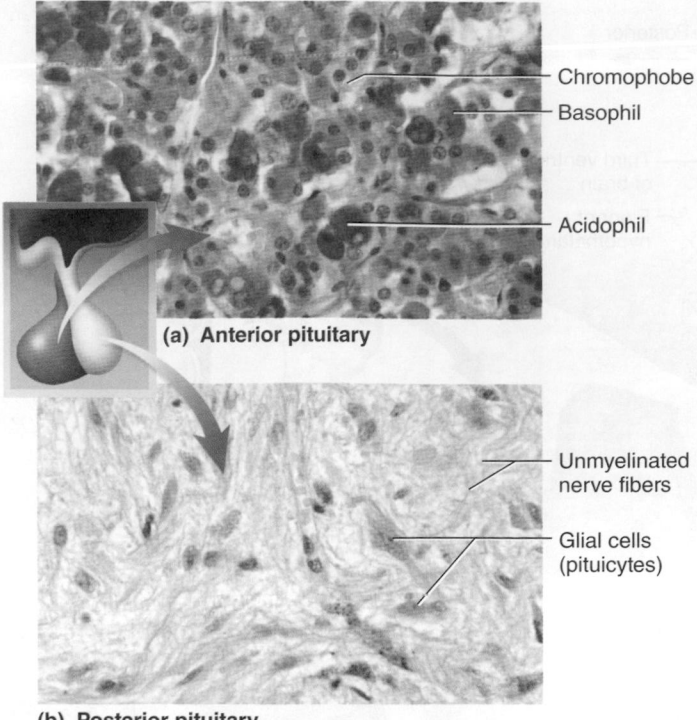

**(a) Anterior pituitary**

- Chromophobe
- Basophil
- Acidophil

**(b) Posterior pituitary**

- Unmyelinated nerve fibers
- Glial cells (pituicytes)

**FIGURE 17.5 Histology of the Pituitary Gland.** (a) The anterior lobe. Basophils include gonadotropes, thyrotropes, and corticotropes. Acidophils include somatotropes and lactotropes. These subtypes are not distinguishable with this histological stain. The chromophobes resist staining, and their function is not yet known. (b) The posterior lobe, composed of nervous tissue.

The **posterior pituitary,** also called the *posterior lobe* or *neurohypophysis,* constitutes the posterior one-quarter of the pituitary. It is actually nervous tissue (nerve fibers and neuroglia, fig. 17.5b), not a true gland. Nerve fibers arise from certain cell bodies in the hypothalamus, pass down the stalk as a bundle called the **hypothalamo–hypophyseal tract,** and end in the posterior lobe (fig. 17.4a). Hormones are made in the hypothalamic neurons and move down the nerve fibers by axoplasmic flow to the posterior pituitary. Here they are stored in the nerve endings until a nerve signal coming down the same axons triggers their release.

## Hypothalamic Hormones

To understand the physiology of the hypothalamic–pituitary system, we must begin with the hormones produced in the hypothalamus. There are eight of these (fig. 17.4)—six to regulate the anterior pituitary and two that are stored in the posterior pituitary and released on demand. Among the first six, those that stimulate the pituitary to release hormones of its own are called *releasing hormones;* those that suppress pituitary secretion are called *inhibiting hormones* (table 17.3). The releasing or inhibiting effect is identified in their names. *Somatostatin* is also called growth hormone–inhibiting hormone, though it also inhibits secretion of thyroid-stimulating hormone. Its name derives from *somatotropin,* a synonym for growth hormone, and *stat,* meaning to halt something (in this case, growth hormone secretion).

The other two hypothalamic hormones are **oxytocin (OT)** and **antidiuretic hormone (ADH).** These are stored and released by the posterior pituitary. OT comes mainly from neurons in the right and left **paraventricular**[7] **nuclei** of the hypothalamus, so called because they lie in the walls of the third ventricle. ADH comes mainly from the **supraoptic**[8] **nuclei,** named for their location just above the optic chiasm. Each nucleus also produces small quantities of the other hormone. ADH and OT are treated as posterior pituitary hormones for convenience even though the posterior lobe does not synthesize them.

## Anterior Pituitary Hormones

The six main hormones of the anterior pituitary are as follows (table 17.4). The first two are collectively called **gonadotropins**[9] because they target the ovaries and testes (gonads).

1. **Follicle-stimulating hormone (FSH).** FSH is secreted by pituitary cells called *gonadotropes.* In the ovaries, it stimulates the secretion of ovarian sex hormones and the development of the bubblelike *follicles* that contain the eggs. In the testes, it stimulates sperm production.

2. **Luteinizing hormone (LH).** LH is also secreted by the gonadotropes. In females, it stimulates *ovulation,* the release of an egg. It is named for the fact that after

---

[7] *para* = next to; *ventricular* = pertaining to the ventricle
[8] *supra* = above
[9] *gonado* = gonads; *trop* = to turn or change

| TABLE 17.3 | Hypothalamic Releasing and Inhibiting Hormones That Regulate the Anterior Pituitary |
|---|---|
| **Hormone** | **Principal Effects** |
| Thyrotropin-releasing hormone (TRH) | Promotes secretion of thyroid-stimulating hormone (TSH) and prolactin (PRL) |
| Corticotropin-releasing hormone (CRH) | Promotes secretion of adrenocorticotropic hormone (ACTH) |
| Gonadotropin-releasing hormone (GnRH) | Promotes secretion of follicle-stimulating hormone (FSH) and luteinizing hormone (LH) |
| Growth hormone–releasing hormone (GHRH) | Promotes secretion of growth hormone (GH) |
| Prolactin-inhibiting hormone (PIH) | Inhibits secretion of prolactin (PRL) |
| Somatostatin | Inhibits secretion of growth hormone (GH) and thyroid-stimulating hormone (TSH) |

ovulation, the follicle becomes a yellowish body called the *corpus luteum*.[10] LH also stimulates the corpus luteum to secrete progesterone, a hormone important in pregnancy. In males, LH stimulates the testes to secrete testosterone.

3. **Thyroid-stimulating hormone (TSH), or thyrotropin.** TSH is secreted by pituitary cells called *thyrotropes*. It stimulates growth of the thyroid gland and the secretion of thyroid hormone, which has widespread effects on metabolic rate, body temperature, and other functions detailed later.

4. **Adrenocorticotropic hormone (ACTH), or corticotropin.** ACTH is secreted by cells called *corticotropes*. Its target organ and the basis for its name is the adrenal cortex, studied later in this chapter. ACTH stimulates the cortex to secrete hormones called glucocorticoids (especially *cortisol*), which regulate glucose, protein, and fat metabolism and are important in the body's response to stress.

5. **Prolactin[11] (PRL).** PRL is secreted by pituitary cells called *lactotropes (mammotropes)*. The hormone and these cells are named for the role of PRL in lactation. During pregnancy, the lactotropes increase greatly in size and number, and PRL secretion rises proportionately, but it has no effect on the mammary glands until after a woman gives birth. Then, it stimulates them to synthesize milk.

6. **Growth hormone (GH), or somatotropin.** GH is secreted by *somatotropes,* the most numerous cells of the anterior pituitary. The pituitary produces at least a thousand times as much GH as any other hormone. The general effect of GH is to stimulate mitosis and cellular differentiation and thus to promote tissue growth throughout the body.

---

[10]*corpus* = body; *lute* = yellow
[11]*pro* = favoring, promoting; *lact* = milk

You can see that the anterior pituitary is involved in a chain of events linked by hormones: The hypothalamus secretes a releasing hormone; this induces a type of pituitary cell to secrete its hormone; that hormone is usually targeted to another endocrine gland elsewhere in the body; and finally that gland secretes a hormone with an effect of its own. For example, the hypothalamus secretes thyrotropin-releasing hormone (TRH); this induces the anterior pituitary to secrete thyroid-stimulating hormone (TSH); TSH, in turn, stimulates the thyroid gland to release thyroid hormone (TH); and finally, thyroid hormone exerts its metabolic effects throughout the body. Such a relationship between the hypothalamus, pituitary, and another downstream endocrine gland is called an **axis**—the *hypothalamo–pituitary–thyroid axis,* for example. Figure 17.6 summarizes these relationships between the hypothalamus, anterior pituitary, and more downstream target organs.

## Posterior Pituitary Hormones

The two posterior lobe hormones are ADH and OT (table 17.4). As we have already seen, they are synthesized in the hypothalamus, then transported to the posterior pituitary and stored until their release on command. Their functions are as follows:

1. **Antidiuretic hormone (ADH).** ADH increases water retention by the kidneys, reduces urine volume, and helps prevent dehydration. ADH also functions as a brain neurotransmitter and is usually called *arginine vasopressin (AVP)* in the neuroscience literature. This name refers to its ability to cause vasoconstriction, but this effect requires concentrations so unnaturally high for the human body that it is of doubtful significance except in pathological states.

2. **Oxytocin (OT).** OT has a variety of reproductive functions in situations ranging from intercourse to breast-feeding. It surges in both sexes during sexual arousal

| TABLE 17.4 | Pituitary Hormones | |
|---|---|---|
| **Hormone** | **Target Organ or Tissue** | **Principal Effects** |
| **Anterior Pituitary** | | |
| Follicle-stimulating hormone (FSH) | Ovaries, testes | *Female:* Growth of ovarian follicles and secretion of estrogen<br>*Male:* Sperm production |
| Luteinizing hormone (LH) | Ovaries, testes | *Female:* Ovulation, maintenance of corpus luteum<br>*Male:* Testosterone secretion |
| Thyroid-stimulating hormone (TSH) | Thyroid gland | Growth of thyroid, secretion of thyroid hormone |
| Adrenocorticotropic hormone (ACTH) | Adrenal cortex | Growth of adrenal cortex, secretion of glucocorticoids |
| Prolactin (PRL) | Mammary glands, testes | *Female:* Milk synthesis<br>*Male:* Increased LH sensitivity |
| Growth hormone (GH) | Liver, bone, cartilage, muscle, fat | Widespread tissue growth, especially in the stated tissues |
| **Posterior Pituitary** | | |
| Antidiuretic hormone (ADH) | Kidneys | Water retention |
| Oxytocin (OT) | Uterus, mammary glands | Labor contractions, milk release; possibly involved in ejaculation, sperm transport, sexual affection, and mother–infant bonding |

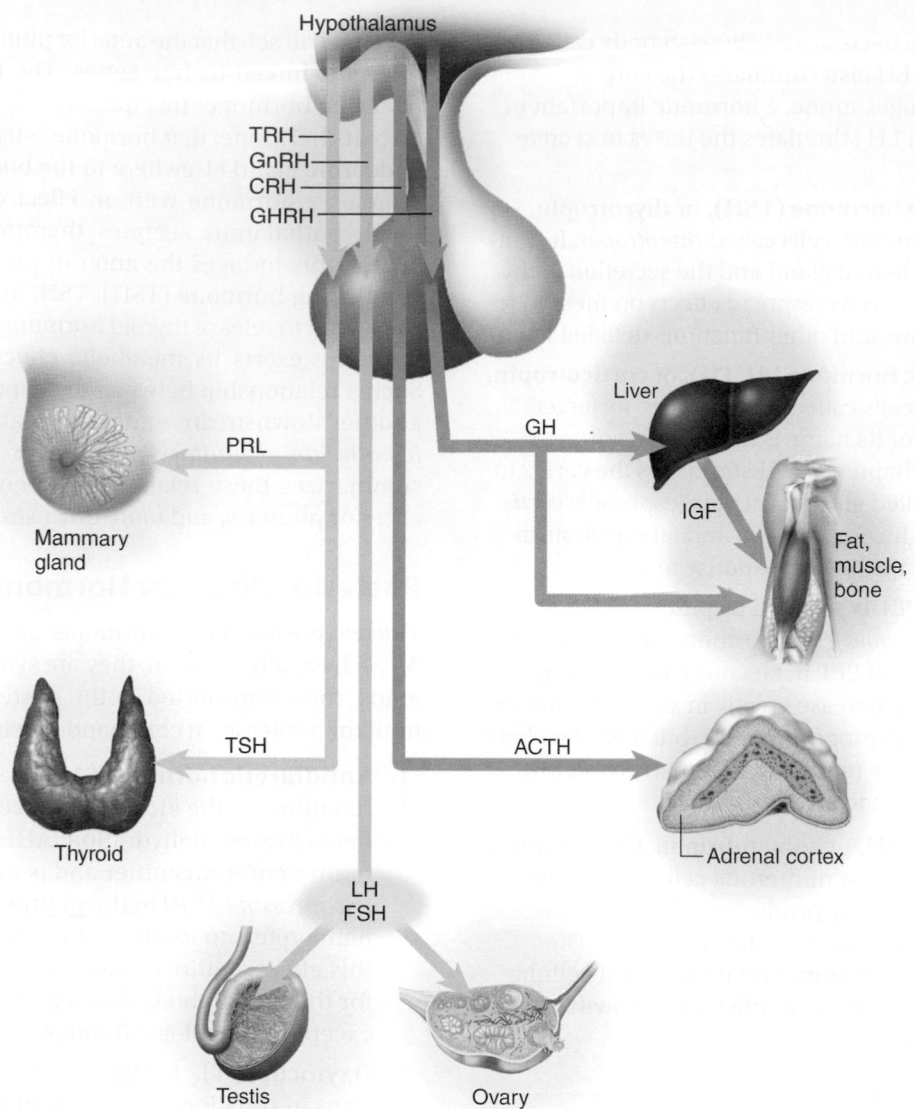

**FIGURE 17.6 Hypothalamo–Pituitary–Target Organ Relationships.** Hypothalamic releasing hormones, shown at the top, trigger secretion of all of the anterior pituitary hormones (bottom).

and orgasm, possibly aiding in the propulsion of semen through the male reproductive tract and stimulating uterine contractions that help transport sperm up the female tract. OT also functions in feelings of sexual satisfaction and emotional bonding between partners. In childbirth, it stimulates labor contractions, and in lactating mothers, it stimulates the flow of milk from deep in the mammary gland to the nipple, where it is accessible to the infant. It may also promote emotional bonding between the mother and infant. In the absence of oxytocin, other female mammals tend to neglect their helpless infants.

## Control of Pituitary Secretion

Pituitary hormones are not secreted at constant rates. GH is secreted mainly at night, LH peaks at the middle of the menstrual cycle, and OT surges during labor and nursing, for example. The timing and amount of pituitary secretion are regulated by the hypothalamus, other brain centers, and feedback from the target organs.

## Hypothalamic and Cerebral Control

Both lobes of the pituitary gland are strongly subject to control by the brain. Hypothalamic control enables the brain to monitor conditions within and outside the body and to stimulate or inhibit the release of anterior lobe hormones in response. For example, in times of stress, the hypothalamus triggers ACTH secretion, which leads to cortisol secretion and mobilization of materials needed for tissue repair. During pregnancy, the hypothalamus induces prolactin secretion so a woman will be prepared to lactate after the baby is born.

The posterior pituitary is controlled by **neuroendocrine reflexes**—the release of hormones in response to nerve signals. For example, dehydration raises the osmolarity of the blood, which is detected by hypothalamic neurons called *osmoreceptors*. They trigger ADH release, and ADH promotes water conservation. Excessive blood pressure, by contrast, stimulates stretch receptors in the heart and certain arteries. By another neuroendocrine reflex, this inhibits ADH release, increases urine output, and brings blood volume and pressure back to normal.

The suckling of an infant also triggers a neuroendocrine reflex mediated by oxytocin. Stimulation of the nipple sends nerve signals up the spinal cord and brainstem to the hypothalamus and from there to the posterior pituitary. This causes the release of oxytocin, which stimulates the release of milk. Neuroendocrine reflexes can also involve higher brain centers. For example, this lactation reflex can be triggered when the mother simply hears a baby cry—any baby—and emotional stress can affect gonadotropin secretion, thus disrupting ovulation, the menstrual rhythm, and fertility.

▶▶▶ **APPLY WHAT YOU KNOW**

*Which of the unifying themes at the end of chapter 1 (p. 21) is best exemplified by the neuroendocrine reflexes that govern ADH secretion?*

## Feedback from Target Organs

The regulation of other endocrine glands by the pituitary is not simply a system of "command from the top down." Those target organs also regulate the pituitary and hypothalamus through various feedback loops.

Most often, this takes the form of **negative feedback inhibition**—the pituitary stimulates another endocrine gland to secrete its hormone, and that hormone feeds back to the pituitary or hypothalamus and inhibits further secretion of the pituitary hormone. Figure 17.7 shows this in the hypothalamo–pituitary–thyroid axis as an example. The figure is numbered to correspond to the following description:

① The hypothalamus secretes thyrotropin-releasing hormone (TRH).

② TRH stimulates the anterior pituitary to secrete thyroid-stimulating hormone (TSH).

③ TSH stimulates the thyroid gland to secrete thyroid hormone (TH).

④ TH stimulates the metabolism of most cells throughout the body.

⑤ TH also *inhibits* the release of TSH by the pituitary.

⑥ To a lesser extent, TH also *inhibits* the release of TRH by the hypothalamus.

The negative feedback inhibition in this process consists of steps 5 and 6. It ensures that when the TH level is high, TSH secretion remains moderate. If thyroid hormone secretion drops, TSH secretion rises and stimulates the thyroid to secrete more hormone. This feedback keeps thyroid hormone levels oscillating around a set point in typical homeostatic fashion.

▶▶▶ **APPLY WHAT YOU KNOW**

*If the thyroid gland was removed from a cancer patient, would you expect the level of TSH to rise or fall? Why?*

Feedback from a target organ is not always inhibitory. As we saw in chapter 1, oxytocin triggers a positive feedback cycle during labor (see fig. 1.8, p. 17). Uterine stretching sends a

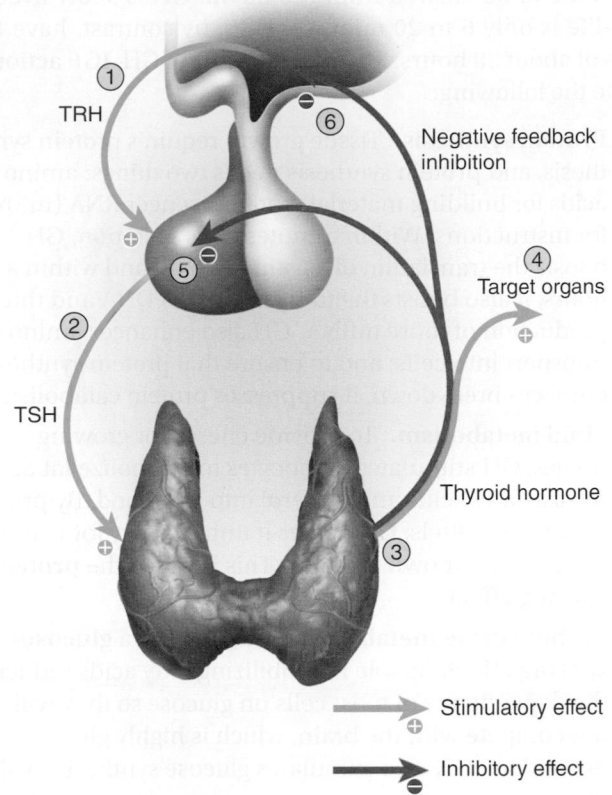

**FIGURE 17.7 Negative Feedback Inhibition of the Pituitary Gland by the Thyroid Gland.** See text for explanation of numbered steps.

nerve signal to the brain that stimulates OT release. OT stimulates uterine contractions, which push the infant downward. This stretches the lower end of the uterus some more, which results in a nerve signal that stimulates still more OT release. This positive feedback cycle continues until the infant is born.

## A Further Look at Growth Hormone

The ultimate effects of most pituitary hormones are exerted through endocrine glands farther downstream, such as the thyroid, adrenal cortex, and gonads. Thus, we will study those effects in the following section as we survey those glands. Growth hormone is a different matter and warrants further exploration at this point.

Unlike the other pituitary hormones, GH is not targeted to just one or a few organs but has widespread effects on the body, especially on cartilage, bone, muscle, and fat. It exerts its effects both directly and indirectly. GH itself directly stimulates these tissues, but it also induces the liver and other tissues to produce growth stimulants called **insulin-like growth factors (IGF-I and IGF-II),** or **somatomedins,**[12] which then stimulate target cells in diverse tissues (fig. 17.6). Most of these effects are caused by IGF-I, but IGF-II is important in fetal growth.

One effect of IGF is to prolong the action of GH. All hormones have a **half-life,** the time required for 50% of the

---

[12]Acronym for *somato*tropin *medi*ating prot*in*

hormone to be cleared from the blood. GH is short-lived; its half-life is only 6 to 20 minutes. IGFs, by contrast, have half-lives of about 20 hours. The mechanisms of GH–IGF action include the following:

- **Protein synthesis.** Tissue growth requires protein synthesis, and protein synthesis needs two things: amino acids for building material, and messenger RNA (mRNA) for instructions. Within minutes of its secretion, GH boosts the translation of existing mRNA, and within a few hours, it also boosts the transcription of DNA and thus the production of more mRNA. GH also enhances amino acid transport into cells; and to ensure that protein synthesis outpaces breakdown, it suppresses protein catabolism.

- **Lipid metabolism.** To provide energy for growing tissues, GH stimulates adipocytes to catabolize fat and release fatty acids and glycerol into the blood. By providing these fuels, GH makes it unnecessary for cells to consume their own proteins. This is called the **protein-sparing effect.**

- **Carbohydrate metabolism.** GH also has a **glucose-sparing effect.** Its role in mobilizing fatty acids reduces the dependence of most cells on glucose so they will not compete with the brain, which is highly glucose-dependent. GH also stimulates glucose synthesis by the liver.

- **Electrolyte balance.** GH promotes $Na^+$, $K^+$, and $Cl^-$ retention by the kidneys, enhances $Ca^{2+}$ absorption by the small intestine, and makes these electrolytes available to the growing tissues.

The most conspicuous effects of GH are on bone, cartilage, and muscle growth, especially during childhood and adolescence. IGF-I accelerates bone growth at the epiphyseal plates. It stimulates the multiplication of chondrocytes and osteogenic cells as well as protein deposition in the cartilage and bone matrix. In adulthood, it stimulates osteoblast activity and the appositional growth of bone; thus, it continues to influence bone thickening and remodeling.

GH secretion fluctuates greatly over the course of a day. It rises to 20 nanograms per milliliter of blood plasma (ng/mL) or higher during the first 2 hours of deep sleep, and may reach 30 ng/mL in response to vigorous exercise. GH secretion rises sharply in response to trauma, physical or emotional stress, hypoglycemia (low blood sugar), and other conditions. Small peaks occur after high-protein meals, but a high-carbohydrate meal tends to suppress GH secretion.

The GH level declines gradually with age. It averages about 6 ng/mL in adolescence and one-quarter of that in very old age. The resulting decline in protein synthesis may contribute to aging of the tissues, including wrinkling of the skin and reduced muscular mass and strength. At age 30, the average adult body is 10% bone, 30% muscle, and 20% fat; at age 75, it averages 8% bone, 15% muscle, and 40% fat.

---

**BEFORE YOU GO ON**

Answer the following questions to test your understanding of the preceding section:

6. What are two good reasons for considering the pituitary to be two separate glands?

7. Briefly contrast hypothalamic control of the anterior pituitary with its control of the posterior pituitary.

8. Name three anterior lobe hormones that have reproductive functions and three that have nonreproductive roles. What target organs are stimulated by each of these hormones?

9. In what sense does the pituitary "take orders" from the target organs under its command?

10. How does the liver promote GH function? How does GH affect the metabolism of proteins, fats, and carbohydrates?

---

### 17.3 Other Endocrine Glands

**Expected Learning Outcomes**

When you have completed this section, you should be able to

a. describe the structure and location of the remaining endocrine glands;

b. name the hormones these endocrine glands produce and state their functions; and

c. discuss the hormones produced by organs and tissues other than the classical endocrine glands.

## The Pineal Gland

The **pineal**[13] (PIN-ee-ul) **gland** is attached to the roof of the third ventricle of the brain, beneath the posterior end of the corpus callosum (see figs. 17.1 and 14.2, p. 511). Its name alludes to a shape like a pine cone. The philosopher René Descartes (1596–1650) thought it was the seat of the human soul. If so, children must have more soul than adults—a child's pineal gland is about 8 mm long and 5 mm wide, but after age 7 it regresses rapidly and is no more than a tiny shrunken mass of fibrous tissue in the adult. Such shrinkage of an organ is called **involution.**[14] Pineal secretion peaks between the ages of 1 and 5 years and declines 75% by the end of puberty.

We no longer look for the human soul in the pineal gland, but this little organ remains an intriguing mystery. It may play a role in establishing 24-hour *circadian rhythms* of physiological function synchronized with the cycle of daylight and darkness. At night, it synthesizes **melatonin,** a monoamine, from serotonin. Melatonin has been implicated in some human mood disorders, although the evidence is inconclusive (see Deeper Insight 17.1). Its secretion fluctuates seasonally with changes in day length; and in animals with seasonal breeding, it regulates the gonads and the annual breeding cycle. Melatonin may

---

[13]*pineal* = pine cone
[14]*in* = inward; *volution* = rolling or turning

## DEEPER INSIGHT 17.1

### CLINICAL APPLICATION

#### Melatonin, SAD, and PMS

There seems to be a relationship between melatonin and mood disorders, including depression and sleep disturbances. Some people experience a mood dysfunction called *seasonal affective disorder (SAD)*, especially in winter when the days are shorter and they get less exposure to sunlight, and in extreme northern and southern latitudes where sunlight may be dim to non-existent for months at a time. SAD thus affects about 20% of the population in Alaska but only 2.5% in Florida. The symptoms—which include depression, sleepiness, irritability, and carbohydrate craving—can be relieved by 2 or 3 hours of exposure to bright light each day *(phototherapy)*. Premenstrual syndrome (PMS) is similar to SAD and is also relieved by phototherapy. The melatonin level is elevated in both SAD and PMS and is reduced by phototherapy. However, there is also evidence that casts doubt on any causal link between melatonin and these mood disorders, so for now, "the jury is still out." Many people are taking melatonin for jet lag, and it is quite effective, but it is also risky to use when we know so little, as yet, about its potential effect on reproductive function.

---

suppress gonadotropin secretion; removal of the pineal from animals causes premature sexual maturation. Some physiologists think that the pineal gland may regulate the timing of puberty in humans, but a clear demonstration of its role has remained elusive. Pineal tumors cause premature onset of puberty in boys, but such tumors also damage the hypothalamus, so we cannot be sure the effect is due specifically to pineal damage.

## The Thymus

The **thymus** plays a role in three systems: endocrine, lymphatic, and immune. It is a bilobed gland in the mediastinum superior to the heart, behind the sternal manubrium. In the fetus and infant, it is enormous in comparison to adjacent organs, sometimes extending between the lungs from near the diaphragm to the base of the neck (fig. 17.8a). It continues to grow until the age of 5 or 6 years. In adults, the gland weighs about 20 g up to age 60, but then it too, like the pineal, undergoes involution and becomes increasingly fatty and less glandular. In the elderly, it is a small fibrous and fatty remnant barely distinguishable from the surrounding mediastinal tissues (fig. 17.8b).

The thymus is a site of maturation for certain white blood cells called T cells that are critically important for immune defense. It secretes several hormones **(thymopoietin, thymosin, and thymulin)** that stimulate the development of other lymphatic organs and regulate the development and activity of T cells. Its histology and immune function are discussed more fully in chapter 21.

## The Thyroid Gland

The **thyroid gland** is the largest adult gland to have a purely endocrine function, weighing about 25 g. It lies adjacent to the trachea immediately below the larynx, and is named for the nearby shieldlike thyroid[15] cartilage of the larynx. It is shaped like a butterfly wrapped around the trachea, with two wing-like lobes usually joined inferiorly by a narrow bridge of tissue, the **isthmus** (figs. 17.8, 17.9a). The thyroid receives one of the body's highest rates of blood flow per gram of tissue and consequently has a dark reddish brown color (fig. 17.9a).

Histologically, the thyroid is composed mostly of sacs called **thyroid follicles** (fig. 17.9b). Each is filled with a protein-rich colloid and lined by a simple cuboidal epithelium of **follicular cells.** These cells secrete mainly the hormone **thyroxine,** also known as **T₄** or **tetraiodothyronine** (TET-ra-EYE-oh-doe-THY-ro-neen) because of its four iodine atoms.

---

[15]*thyr* = shield; *oid* = resembling

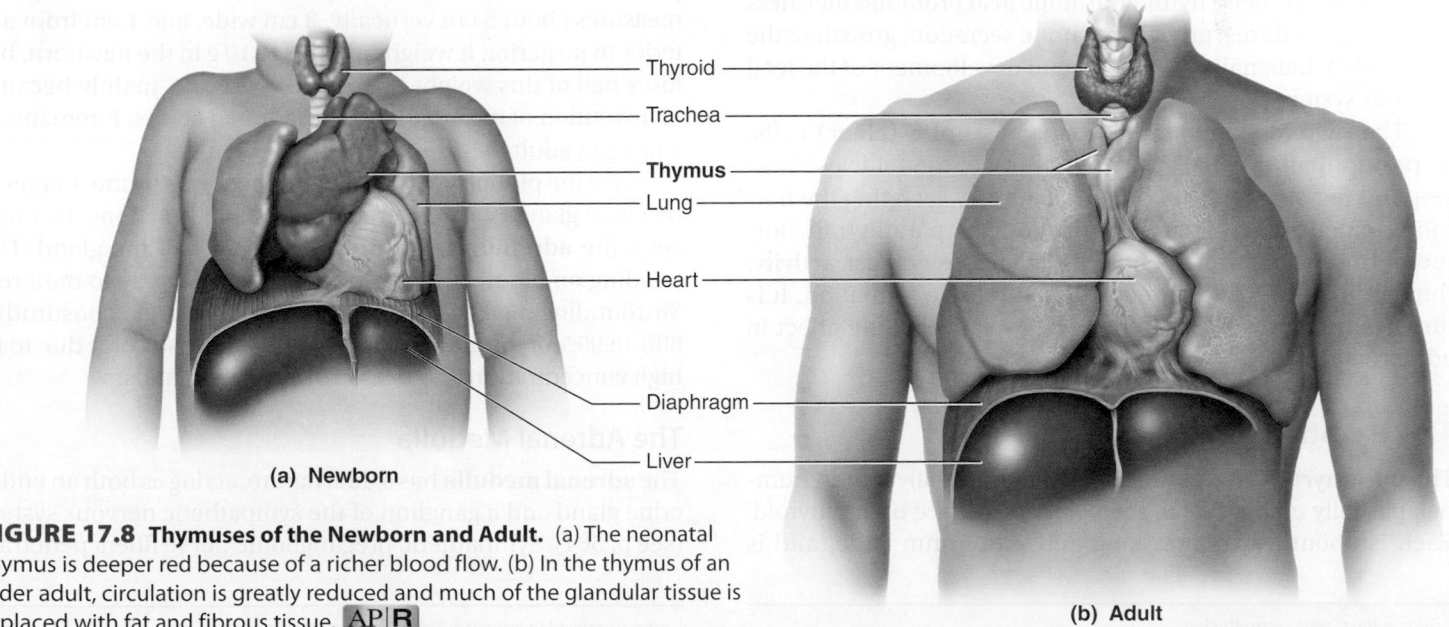

(a) **Newborn**        (b) **Adult**

**FIGURE 17.8  Thymuses of the Newborn and Adult.** (a) The neonatal thymus is deeper red because of a richer blood flow. (b) In the thymus of an older adult, circulation is greatly reduced and much of the glandular tissue is replaced with fat and fibrous tissue. **AP|R**

Labels: Thyroid, Trachea, **Thymus**, Lung, Heart, Diaphragm, Liver

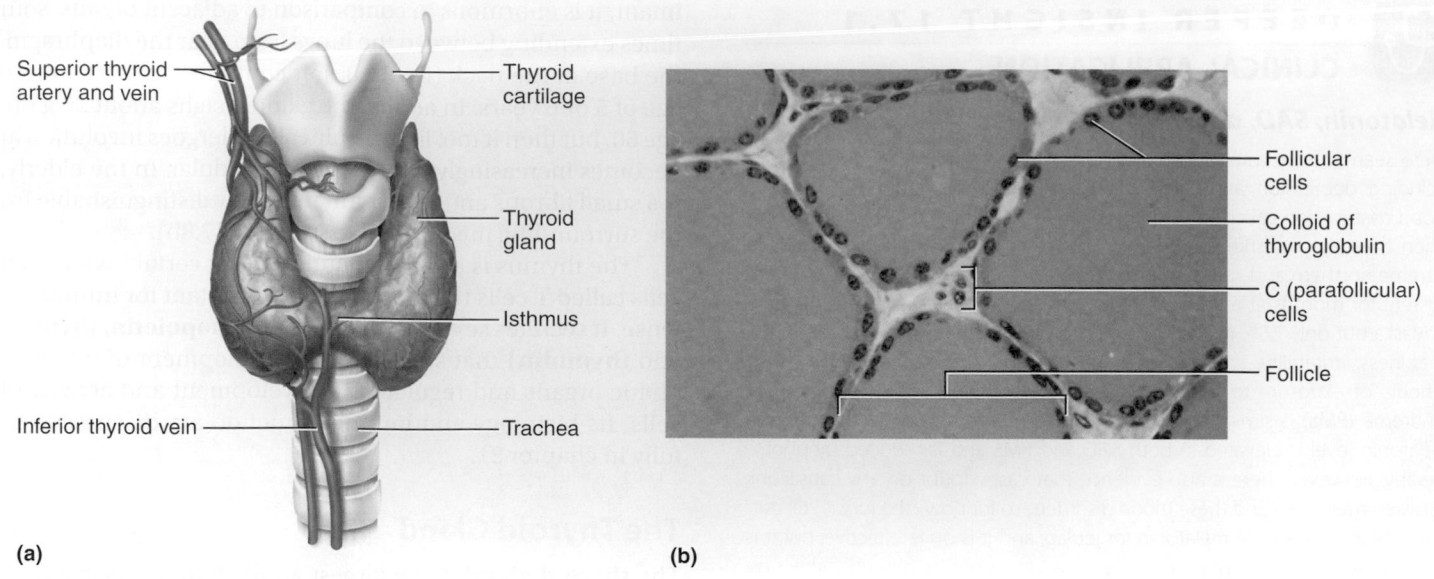

**FIGURE 17.9  Anatomy of the Thyroid Gland.** (a) Gross anatomy, anterior view. (b) Histology, showing the saccular thyroid follicles (source of thyroid hormone) and a nest of C cells (source of calcitonin). AP|R

It also produces **triiodothyronine (T$_3$),** with three iodine atoms. The expression **thyroid hormone (TH)** refers to T$_4$ and T$_3$ collectively. The average adult thyroid secretes about 80 micrograms (µg) of TH daily. About 90% of this is in the T$_4$ form.

Thyroid hormone is secreted in response to TSH from the pituitary. The primary effect of TH is to increase the body's metabolic rate. As a result, it raises oxygen consumption and has a **calorigenic**[16] **effect**—it increases heat production. To ensure an adequate blood and oxygen supply to meet this increased metabolic demand, thyroid hormone raises the respiratory rate, heart rate, and strength of the heartbeat. It stimulates the appetite and accelerates the breakdown of carbohydrates, fats, and protein for fuel. Thyroid hormone also promotes alertness and quicker reflexes; growth hormone secretion; growth of the bones, skin, hair, nails, and teeth; and development of the fetal nervous system.

The thyroid gland also contains nests of **C (clear) cells,** or **parafollicular cells,** at the periphery of the follicles. They respond to rising levels of blood calcium by secreting the hormone **calcitonin.** Calcitonin antagonizes parathyroid hormone (discussed shortly) and stimulates osteoblast activity, thus promoting calcium deposition and bone formation. It is important mainly in children, having relatively little effect in adults (see p. 218).

## The Parathyroid Glands

The **parathyroid glands** are ovoid glands, usually four in number, partially embedded in the posterior surface of the thyroid. Each is about 3 to 8 mm long and 2 to 5 mm wide, and is

separated from the thyroid follicles by a thin fibrous capsule (fig. 17.10). Often, they occur in other locations ranging from as high as the hyoid bone to as low as the aortic arch, and about 5% of people have more than four parathyroids. They secrete **parathyroid hormone (PTH)** in response to hypocalcemia. PTH raises blood calcium levels by mechanisms detailed in chapter 7.

## The Adrenal Glands

The **adrenal**[17] **(suprarenal) glands** sit like a cap on the superior pole of each kidney (fig. 17.11). Like the kidneys, they are retroperitoneal, lying outside the peritoneal cavity between the peritoneum and posterior body wall. The adult adrenal gland measures about 5 cm vertically, 3 cm wide, and 1 cm from anterior to posterior. It weighs about 8 to 10 g in the newborn, but loses half of this weight by the age of 2 years, mainly because of involution of its outer layer, the adrenal cortex. It remains at 4 to 5 g in adults.

Like the pituitary, the adrenal gland forms by the merger of two fetal glands with different origins and functions. Its inner core, the **adrenal medulla,** is 10% to 20% of the gland. Depending on blood flow, its color ranges from gray to dark red. Surrounding it is a much thicker **adrenal cortex,** constituting 80% to 90% of the gland and having a yellowish color due to its high concentration of cholesterol and other lipids.

## The Adrenal Medulla

The **adrenal medulla** has a dual nature, acting as both an endocrine gland and a ganglion of the sympathetic nervous system (see p. 564). Sympathetic preganglionic nerve fibers penetrate

---

[16]*calor* = heat; *genic* = producing

[17]*ad* = to, toward, near; *ren* = kidney; *al* = pertaining to

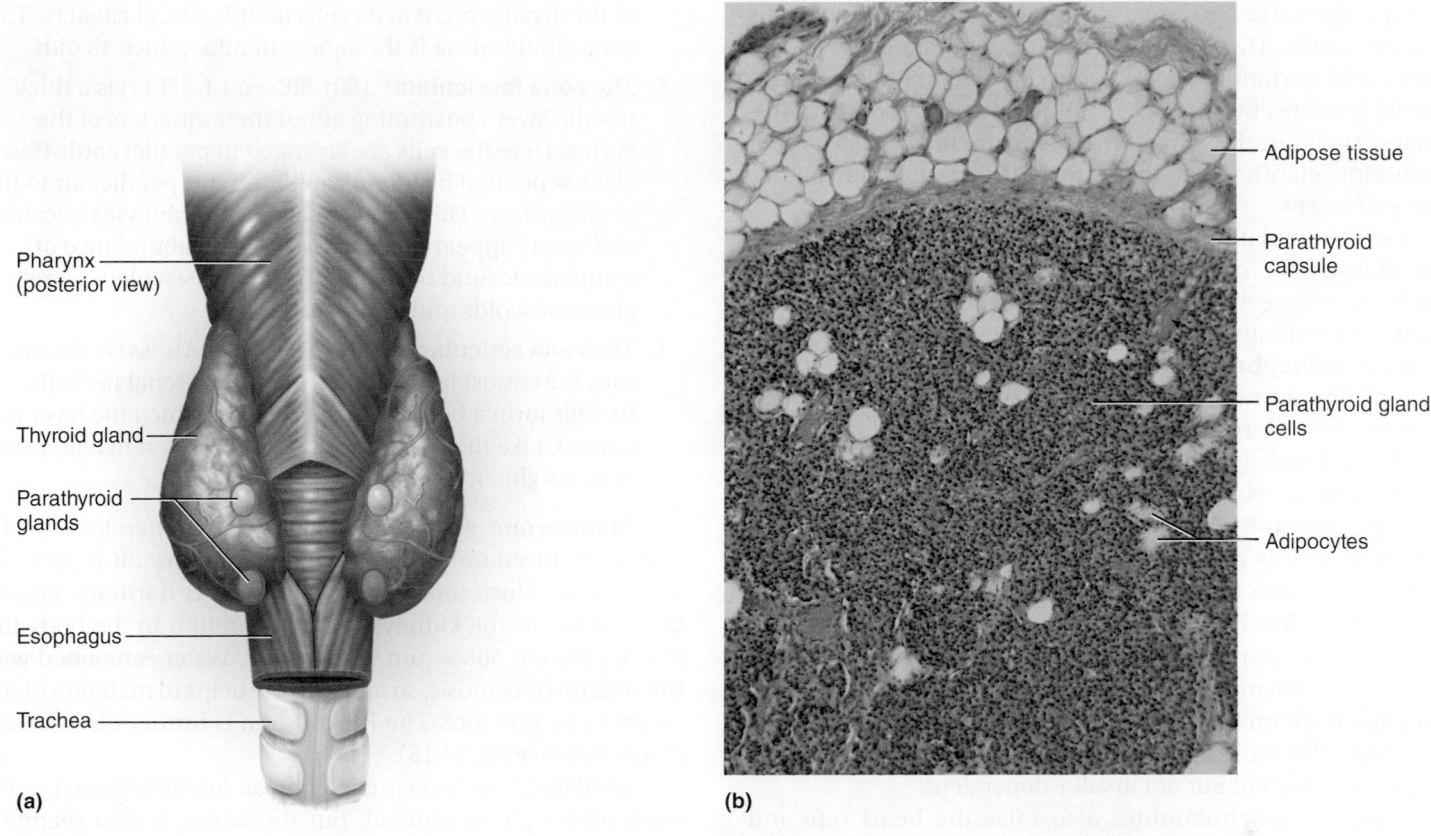

**FIGURE 17.10 The Parathyroid Glands.** (a) Location on the posterior surface of the thyroid. (b) Histology. AP|R

**FIGURE 17.11 The Adrenal Gland.** (a) Location and gross anatomy. (b) Drawing and photograph of adrenal histology. AP|R

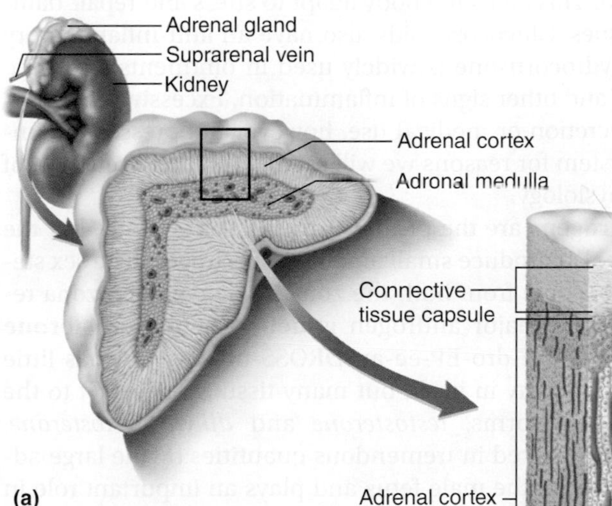

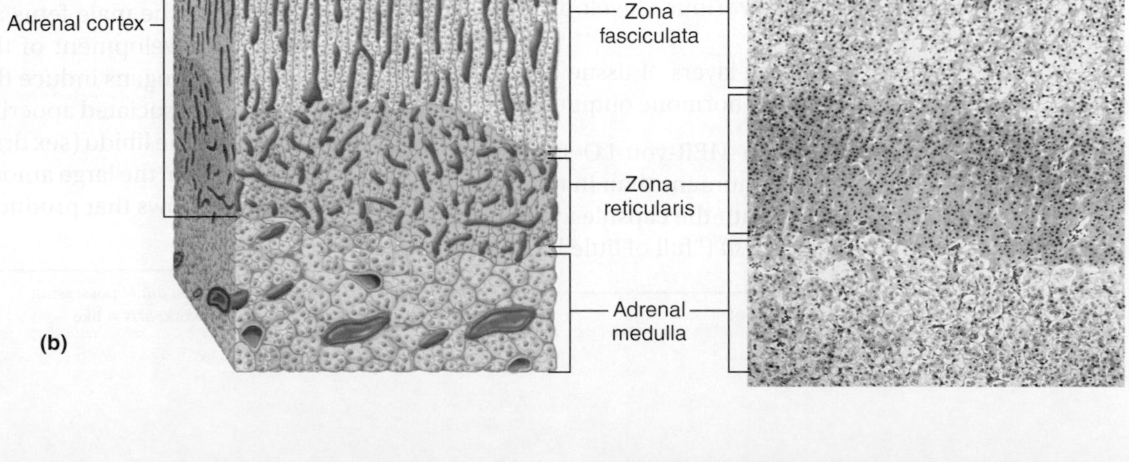

Labels for figure 17.10 (a):
- Pharynx (posterior view)
- Thyroid gland
- Parathyroid glands
- Esophagus
- Trachea

Labels for figure 17.10 (b):
- Adipose tissue
- Parathyroid capsule
- Parathyroid gland cells
- Adipocytes

Labels for figure 17.11 (a) and (b):
- Adrenal gland
- Suprarenal vein
- Kidney
- Adrenal cortex
- Adrenal medulla
- Connective tissue capsule
- Adrenal cortex
- Zona glomerulosa
- Zona fasciculata
- Zona reticularis
- Adrenal medulla

through the cortex to reach **chromaffin** (cro-MAFF-in) **cells** in the medulla. These cells, named for their tendency to stain brown with certain dyes, are essentially sympathetic postganglionic neurons, but they have no dendrites or axon and they release their products into the bloodstream like any other endocrine gland. Thus they are considered to be neuroendocrine cells.

Upon stimulation by the nerve fibers—usually in a situation of fear, pain, or other stress—the chromaffin cells release a mixture of catecholamines that we have previously encountered as neurotransmitters (see fig. 12.21, p. 458): about three-quarters **epinephrine,** one-quarter **norepinephrine,** and a trace of **dopamine.** Now we find them acting as hormones. They increase alertness and prepare the body in several ways for physical activity. They mobilize high-energy fuels such as lactate, fatty acids, and glucose. The liver boosts glucose levels by **glycogenolysis** (hydrolysis of glycogen to glucose) and **gluconeogenesis** (conversion of fats, amino acids, and other noncarbohydrates to glucose).

Epinephrine is said to have a **glucose-sparing effect.** It inhibits the secretion of insulin, so the muscles and other insulin-dependent organs absorb and consume less glucose. They fall back on alternative fuels such as fatty acids, while the blood glucose is left for use by the brain, which is more glucose-dependent but not insulin-dependent.

Adrenal catecholamines also raise the heart rate and blood pressure, stimulate circulation to the muscles, increase pulmonary airflow, and raise the metabolic rate. At the same time, they *inhibit* such temporarily inessential functions as digestion and urine production so that they do not compete for blood flow and energy.

## The Adrenal Cortex

The **adrenal cortex** surrounds the medulla on all sides. It produces more than 25 steroid hormones, known collectively as the **corticosteroids** or **corticoids.** All of them are synthesized from cholesterol; this and other lipids impart a yellow color to the cortex. Only five corticosteroids are secreted in physiologically significant amounts; the others are either negligible in quantity or, if more abundant, are in chemically less active forms. The five most important corticosteroids fall into three categories: **mineralocorticoids,** which regulate the body's electrolyte balance; **glucocorticoids,** which regulate the metabolism of glucose and other organic fuels; and **sex steroids,** with various developmental and reproductive functions.

The adrenal cortex has three layers of tissue (fig. 17.11b), which differ in their histology and hormone output.

1. The **zona glomerulosa**[18] (glo-MER-you-LO-suh) is a thin layer, less developed in humans than in many other mammals, located just beneath the capsule at the gland surface. The name *glomerulosa* ("full of little balls") refers

to the arrangement of its cells in little round clusters. The zona glomerulosa is the source of mineralocorticoids.

2. The **zona fasciculata**[19] (fah-SIC-you-LAH-ta) is a thick middle layer constituting about three-quarters of the cortex. Here the cells are arranged in parallel cords (fascicles), separated by blood capillaries, perpendicular to the gland surface. The cells are called **spongiocytes** because of a foamy appearance imparted by an abundance of cytoplasmic lipid droplets. The zona fasciculata secretes glucocorticoids and androgens.

3. The **zona reticularis**[20] (reh-TIC-you-LAR-iss) is the narrow, innermost layer, adjacent to the adrenal medulla. Its cells form a branching network for which the layer is named. Like the preceding layer, the zona reticularis also secretes glucocorticoids and androgens.

**Aldosterone** is the most significant mineralocorticoid, and is produced only by the zona glomerulosa. It is part of a vital renin–aldosterone–angiotensin (RAA) hormone system that stimulates the kidneys to retain sodium in the body fluids and excrete potassium in the urine. Water is retained with the sodium by osmosis, so aldosterone helps to maintain blood volume and pressure. The RAA system is further discussed in chapter 23 (see fig. 23.15).

**Cortisol** (also known clinically as *hydrocortisone*) is the most potent glucocorticoid, but the adrenals also secrete a weaker one called *corticosterone*. Glucocorticoids are secreted by the zona fasciculata and zona reticularis in response to ACTH from the pituitary. They stimulate fat and protein catabolism, gluconeogenesis, and the release of fatty acids and glucose into the blood. This helps the body adapt to stress and repair damaged tissues. Glucocorticoids also have an anti-inflammatory effect; hydrocortisone is widely used in ointments to relieve swelling and other signs of inflammation. Excessive glucocorticoid secretion or medical use, however, suppresses the immune system for reasons we will see later in the discussion of stress physiology.

**Androgens** are the primary adrenal sex steroids, but the adrenals also produce small amounts of estrogen. The sex steroids, too, come from both the zona fasciculata and zona reticularis. The major androgen is **dehydroepiandrosterone (DHEA)** (de-HY-dro-EP-ee-an-DROSS-tur-own). It has little biological activity in itself, but many tissues convert it to the more potent forms, *testosterone* and *dihydrotestosterone*. DHEA is produced in tremendous quantities by the large adrenal glands of the male fetus and plays an important role in the prenatal development of the male reproductive tract. In both sexes, androgens induce the growth of pubic and axillary hair and their associated apocrine scent glands in puberty, and they stimulate the libido (sex drive) throughout adolescent and adult life. In men, the large amount of androgen secreted by the testes overshadows that produced by the adrenals. In women,

---

[18]*zona* = zone; *glomerul* = little balls; *osa* = full of

[19]*fascicul* = little cords; *ata* = possessing
[20]*reticul* = little network; *aris* = like

however, the adrenal glands meet about 50% of the androgen requirement.

**Estradiol** is the main adrenal estrogen. It is normally of minor importance to women of reproductive age because its quantity is small compared with estrogen from the ovaries. After menopause, however, the ovaries no longer function and only the adrenals secrete estrogen. However, several other tissues, such as fat, convert androgens into additional estrogen. Both androgens and estrogens promote adolescent skeletal growth and help to sustain adult bone mass.

The medulla and cortex are not as functionally independent as once thought; each of them stimulates the other. Without stimulation by cortisol, the adrenal medulla atrophies significantly. Conversely, some chromaffin cells from the medulla extend into the cortex. When stress activates the sympathetic nervous system, these cells stimulate the cortex to secrete corticosterone and perhaps other corticosteroids.

#### ▶▶▶ APPLY WHAT YOU KNOW

*The zona fasciculata thickens significantly in pregnant women. What do you think would be the benefit of this phenomenon?*

## The Pancreatic Islets

The **pancreas** is an elongated, spongy gland located below and behind the stomach; most of it is retroperitoneal (fig. 17.12). It is primarily an exocrine digestive gland, and its gross anatomy is described in chapter 25. Scattered throughout the exocrine tissue, however, are 1 to 2 million endocrine cell clusters called **pancreatic islets (islets of Langerhans[21]).** Although they are less than 2% of the pancreatic tissue, the islets secrete hormones of vital importance, especially in the regulation of **glycemia,** the blood glucose concentration. A typical islet measures about 75 × 175 μm and contains from a few to 3,000 cells. Its main cell types are alpha cells (20%), beta cells (70%), and delta cells (5%). Their functions are as follows:

- **Alpha (α) cells,** or **A cells,** secrete **glucagon** between meals when the blood glucose concentration falls below 100 mg/dL. In the liver, glucagon stimulates gluconeogenesis, glycogenolysis, and the release of glucose into circulation, thus raising the blood glucose level.

---

[21]Paul Langerhans (1847–88), German anatomist

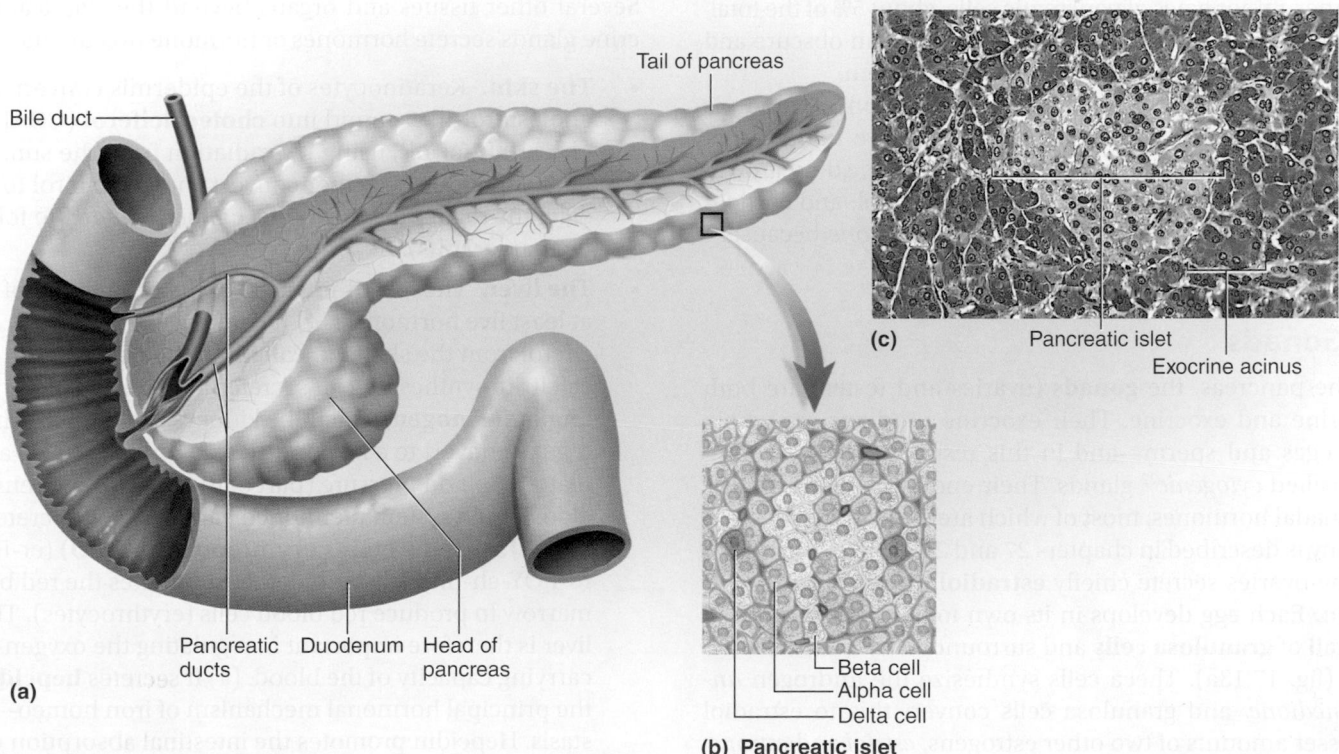

**FIGURE 17.12 The Pancreas.** (a) Gross anatomy and relationship to the duodenum. (b) Cells of a pancreatic islet. (PP and G cells are not shown; they are few in number and cannot be distinguished with ordinary histological staining.) (c) Light micrograph of a pancreatic islet amid the darker exocrine acini, which produce digestive enzymes AP|R.

Tail of pancreas

Bile duct

(c)

Pancreatic islet

Exocrine acinus

Pancreatic ducts   Duodenum   Head of pancreas

(a)

Beta cell
Alpha cell
Delta cell

**(b) Pancreatic islet**

In adipose tissue, it stimulates fat catabolism and the release of free fatty acids. Glucagon is also secreted in response to rising amino acid levels in the blood after a high-protein meal. It promotes amino acid absorption and thereby provides cells with the raw material for gluconeogenesis.

- **Beta (β) cells,** or **B cells,** secrete **insulin,** "the hormone of nutrient abundance," during and immediately following a meal when blood nutrient levels are rising. Its principal targets are the liver, skeletal muscles, and adipose tissue. In times of plenty, insulin stimulates cells to absorb glucose, fatty acids, and amino acids and to store or metabolize them; therefore, it lowers the level of blood glucose and other nutrients. It promotes the synthesis of glycogen, fat, and protein, thereby promoting the storage of excess nutrients for later use and enhancing cellular growth and differentiation. It also antagonizes glucagon, thus suppressing the use of already-stored fuels. The brain, liver, kidneys, and red blood cells absorb and use glucose without need of insulin, but insulin does promote glycogen synthesis in the liver. Insulin insufficiency or inaction is well known as the cause of diabetes mellitus, detailed later in this chapter.

- **Delta (δ) cells,** or **D cells,** secrete **somatostatin (growth hormone–inhibiting hormone)** concurrently with the release of insulin by the beta cells. Somatostatin inhibits the secretion of stomach acid.

Other, minor types of pancreatic cells, about 5% of the total, are called PP and G cells. Their functions remain obscure and controversial, so we will not further consider them.

Any hormone that raises blood glucose concentration is called a *hyperglycemic hormone.* You may have noticed that glucagon is not the only hormone that does so; so do growth hormone, epinephrine, norepinephrine, cortisol, and corticosterone. Insulin is called a *hypoglycemic hormone* because it lowers blood glucose levels.

## The Gonads

Like the pancreas, the **gonads** (ovaries and testes) are both endocrine and exocrine. Their exocrine products are whole cells—eggs and sperm—and in this respect they are sometimes called *cytogenic*[22] glands. Their endocrine products are the gonadal hormones, most of which are steroids. Their gross anatomy is described in chapters 27 and 28.

The ovaries secrete chiefly **estradiol, progesterone,** and **inhibin.** Each egg develops in its own follicle, which is lined by a wall of **granulosa cells** and surrounded by a capsule, the **theca** (fig. 17.13a). Theca cells synthesize the androgen *androstenedione,* and granulosa cells convert this to estradiol and lesser amounts of two other estrogens, *estriol* and *estrone.* In the middle of the monthly ovarian cycle, a mature follicle

ruptures and releases the egg. The remains of the follicle become the corpus luteum, which secretes progesterone for the next 12 days or so in a typical cycle (several weeks in the event of pregnancy).

The functions of estradiol and progesterone are detailed in chapter 28. In brief, they contribute to the development of the reproductive system and feminine physique, promote adolescent bone growth, regulate the menstrual cycle, sustain pregnancy, and prepare the mammary glands for lactation. Inhibin, which is also secreted by the follicle and corpus luteum, suppresses the FSH secretion by means of negative feedback inhibition of the anterior pituitary.

The testis consists mainly of minute *seminiferous tubules* that produce sperm. Its endocrine secretions are **testosterone,** lesser amounts of weaker androgens and estrogens, and inhibin. Inhibin comes from **sustentacular (Sertoli**[23]**) cells** that form the walls of the seminiferous tubules. By limiting FSH secretion, it regulates the rate of sperm production. Nestled between the tubules are clusters of **interstitial cells (cells of Leydig**[24]**),** the source of testosterone and the other sex steroids (fig. 17.13b). Testosterone stimulates development of the male reproductive system in the fetus and adolescent, the development of the masculine physique in adolescence, and the sex drive. It sustains sperm production and the sexual instinct throughout adult life.

## Endocrine Functions of Other Tissues and Organs

Several other tissues and organs beyond the classical endocrine glands secrete hormones or hormone precursors:

- **The skin.** Keratinocytes of the epidermis convert a cholesterol-like steroid into **cholecalciferol** (COAL-eh-cal-SIF-er-ol), using UV radiation from the sun. The liver and kidneys further convert cholecalciferol to a calcium-regulating hormone, **calcitriol** (see the following paragraphs).

- **The liver.** The liver is involved in the production of at least five hormones: (1) It converts the cholecalciferol from the skin into **calcidiol,** the next step in calcitriol synthesis. (2) It secretes a protein called **angiotensinogen,** which the kidneys, lungs, and other organs convert to a hormone called *angiotensin II,* a regulator of blood pressure (part of the renin–angiotensin–aldosterone system mentioned earlier). (3) It secretes about 15% of the body's **erythropoietin (EPO)** (er-RITH-ro-POY-eh-tin), a hormone that stimulates the red bone marrow to produce red blood cells (erythrocytes). The liver is therefore important in regulating the oxygen-carrying capacity of the blood. (4) It secretes **hepcidin,** the principal hormonal mechanism of iron homeostasis. Hepcidin promotes the intestinal absorption of dietary iron and the mobilization of iron for hemoglobin

---

[22]*cyto* = cell; *genic* = producing

[23]Enrico Sertoli (1824–1910), Italian histologist
[24]Franz von Leydig (1821–1908), German histologist

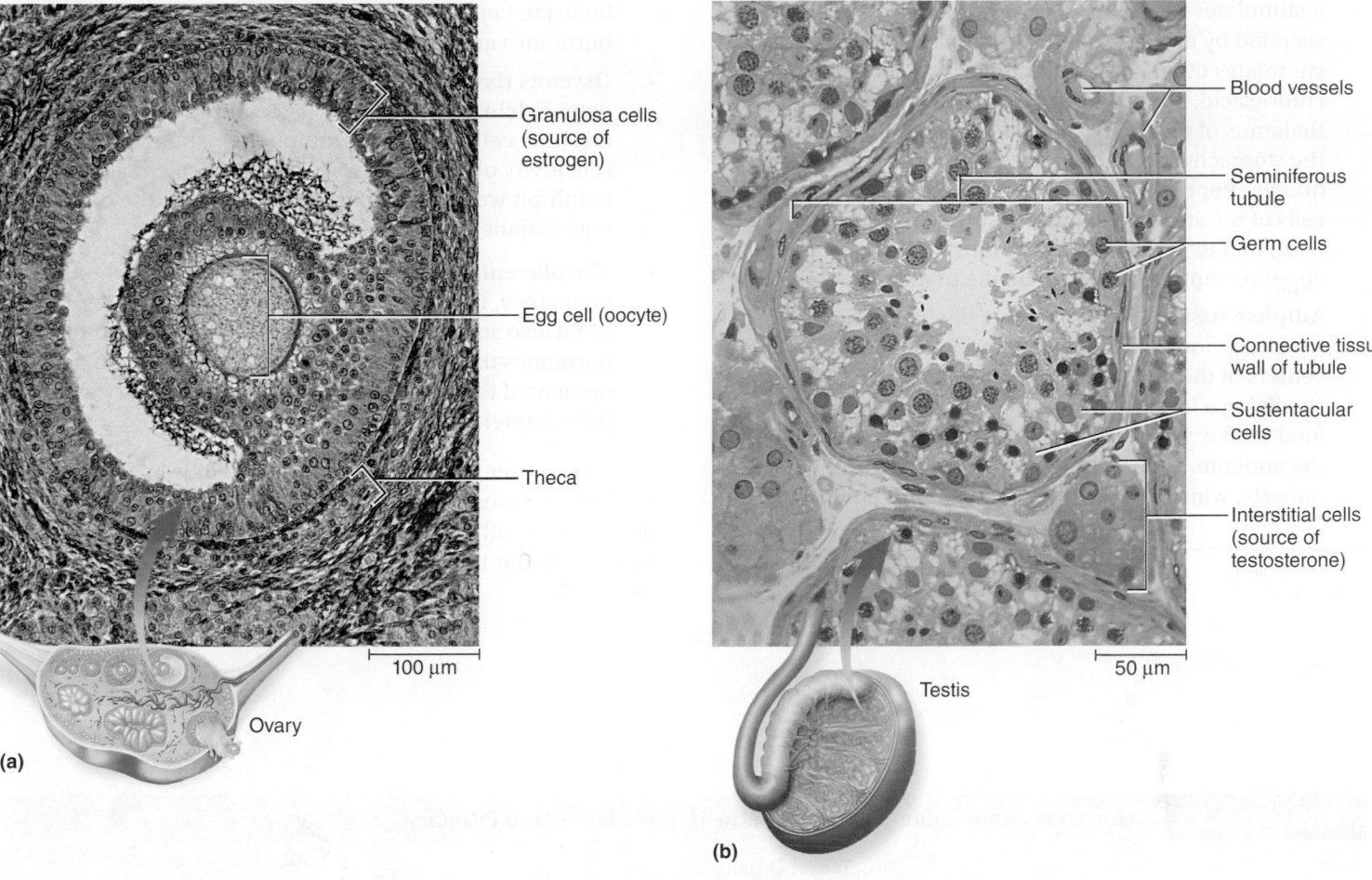

**FIGURE 17.13 The Gonads.**
(a) Histology of an ovarian follicle. (b) Histology of the testis. The granulosa cells of the ovary and interstitial cells of the testis are endocrine cells.
AP|R

synthesis and other uses. (5) It secretes **insulin-like growth factor I (IGF-I),** a hormone that mediates the action of growth hormone.

- **The kidneys.** The kidneys play endocrine roles in the production of three hormones—calcitriol, angiotensin II, and erythropoietin. (1) They convert calcidiol into **calcitriol (vitamin D₃),** thus completing the three-step process begun by the skin and liver. Calcitriol raises the blood concentration of calcium by promoting its intestinal absorption and slightly inhibiting its loss in the urine. This makes more calcium available for bone deposition and other uses. (2) They secrete an enzyme called **renin** (REE-nin), which converts angiotensinogen to angiotensin I. As angiotensin I circulates through various organs, especially the lungs, *angiotensin-converting enzyme (ACE)* on the linings of the blood capillaries converts it to **angiotensin II,** a small peptide. This is a very potent hormone that constricts blood vessels throughout the body and thereby raises blood pres-

sure. (3) The kidneys secrete about 85% of the body's erythropoietin.

- **The heart.** Rising blood pressure stretches the heart wall and stimulates cardiac muscle in the atria to secrete two similar **natriuretic**[25] **peptides.** These hormones increase sodium excretion and urine output and oppose the action of angiotensin II, described above. Together, these effects lower the blood pressure.

- **The stomach and small intestine.** These have various *enteroendocrine cells,*[26] which secrete at least 10 **enteric hormones.** One role of these hormones is to coordinate different regions and glands of the digestive system with each other. For example, **cholecystokinin**[27] **(CCK)** (COAL-eh-SIS-toe-KY-nin) is secreted by the small intestine in response to arriving fats. Among other effects,

[25]*natri* = sodium; *uretic* = pertaining to urine
[26]*entero* = intestine
[27]*chole* = bile; *cysto* = sac (gallbladder); *kin* = action, motion

it stimulates the gallbladder to release bile. **Gastrin**[28] is secreted by the stomach upon the arrival of food and stimulates other cells of the stomach to secrete hydrochloric acid. Some enteric hormones act on the hypothalamus of the brain. **Ghrelin**[29] (GRELL-in), secreted by the stomach when it is empty, produces the sensation of hunger. **Peptide YY (PYY),** secreted by enteroendocrine cells of the small and large intestines, signals satiety (fullness) and tends to terminate eating. CCK also has an appetite-suppressant effect on the brain.

- **Adipose tissue.** Fat cells secrete the hormone **leptin,** which has long-term effects on appetite-regulating centers of the hypothalamus. A low level of leptin, signifying a low level of body fat, increases appetite and food intake, whereas a high level of leptin tends to blunt the appetite. Leptin also serves as a signal for the onset of puberty, which is delayed in persons with abnormally low

[28]*gastr* = stomach
[29]*ghre* = growth

body fat. Leptin and some of the aforementioned enteric hormones are discussed more extensively in chapter 26.

- **Osseous tissue.** Osteoblasts secrete the hormone **osteocalcin,** which increases the number of pancreatic beta cells, pancreatic output of insulin, and insulin sensitivity of other body tissues. Osteocalcin also seems to inhibit weight gain (fat deposition) and the onset of type 2 diabetes mellitus.

- **The placenta.** This organ performs many functions in pregnancy, including fetal nutrition and waste removal. But it also secretes estrogen, progesterone, and other hormones that regulate pregnancy and stimulate development of the fetus and the mother's mammary glands (see chapter 28).

You can see that the endocrine system is extensive. It includes numerous discrete glands as well as individual cells in the tissues of other organs. The endocrine organs and tissues other than the hypothalamus and pituitary are reviewed in table 17.5.

| TABLE 17.5 | Hormones from Sources Other than the Hypothalamus and Pituitary | | |
|---|---|---|---|
| Source | Hormone | Target Organs and Tissues | Principal Effects |
| Pineal gland | Melatonin | Brain | Uncertain; may influence mood and sexual maturation |
| Thymus | Thymopoietin, thymosin, thymulin | Immune cells (T lymphocytes) | Stimulate T lymphocyte development and activity |
| Thyroid gland | Thyroxine ($T_4$) and triiodothyronine ($T_3$) | Most tissues | Elevate metabolic rate and heat production; increase respiratory rate, heart rate, and strength of heartbeat; stimulate appetite and accelerate breakdown of nutrients; promote alertness and quicken reflexes; stimulate growth hormone secretion and growth of skin, hair, nails, teeth, and fetal nervous system |
| | Calcitonin | Bone | Stimulates bone deposition, mainly in children |
| Parathyroid glands | Parathyroid hormone (PTH) | Bone, kidneys, small intestine | Raises blood $Ca^{2+}$ level by stimulating bone resorption and inhibiting deposition, reducing urinary $Ca^{2+}$ excretion, and enhancing calcitriol synthesis |
| Adrenal medulla | Epinephrine, norepinephrine, dopamine | Most tissues | Promote alertness; mobilize organic fuels; raise metabolic rate; stimulate circulation and respiration; increase blood glucose level; inhibit insulin secretion and glucose uptake by insulin-dependent organs (sparing glucose for brain) |
| Adrenal cortex | Aldosterone | Kidney | Promotes $Na^+$ and water retention and $K^+$ excretion; maintains blood pressure and volume |
| | Cortisol and corticosterone | Most tissues | Stimulate fat and protein catabolism, gluconeogenesis, stress resistance, and tissue repair |
| | Dehydroepiandrosterone | Bone, muscle, integument, brain, many other tissues | Precursor of testosterone; indirectly promotes growth of bones, pubic and axillary hair, apocrine glands, and fetal male reproductive tract; stimulates libido |

| TABLE 17.5 | Hormones from Sources Other than the Hypothalamus and Pituitary (continued) | | |
|---|---|---|---|
| **Source** | **Hormone** | **Target Organs and Tissues** | **Principal Effects** |
| Pancreatic islets | Glucagon | Primarily liver | Stimulates amino acid absorption, gluconeogenesis, glycogen and fat breakdown; raises blood glucose and fatty acid levels |
| | Insulin | Most tissues | Stimulates glucose and amino acid uptake; lowers blood glucose level; promotes glycogen, fat, and protein synthesis |
| | Somatostatin | Stomach, intestines, pancreatic islet cells | Modulates digestion, nutrient absorption, and glucagon and insulin secretion |
| | Pancreatic polypeptide | Pancreas, gallbladder | Inhibits release of bile and digestive enzymes |
| | Gastrin | Stomach | Stimulates acid secretion and gastric motility |
| Ovaries | Estradiol | Many tissues | Stimulates female reproductive development and adolescent growth; regulates menstrual cycle and pregnancy; prepares mammary glands for lactation |
| | Progesterone | Uterus, mammary glands | Regulates menstrual cycle and pregnancy; prepares mammary glands for lactation |
| | Inhibin | Anterior pituitary | Inhibits FSH secretion |
| Testes | Testosterone | Many tissues | Stimulates fetal and adolescent reproductive development, musculoskeletal growth, sperm production, and libido |
| | Inhibin | Anterior pituitary | Inhibits FSH secretion |
| Skin | Cholecalciferol | — | Precursor of calcitriol (see kidneys) |
| Liver | Calcidiol | — | Precursor of calcitriol (see kidneys) |
| | Angiotensinogen | — | Precursor of angiotensin II (see kidneys) |
| | Erythropoietin | Red bone marrow | Promotes red blood cell production, increases oxygen-carrying capacity of blood |
| | Hepcidin | Small intestine, liver | Promotes iron absorption and mobilization |
| | Insulin-like growth factor I | Many tissues | Prolongs and mediates action of growth hormone |
| Kidneys | Angiotensin I | — | Precursor of angiotensin II, a vasoconstrictor |
| | Calcitriol | Small intestine | Increases blood calcium level mainly by promoting intestinal absorption of $Ca^{2+}$ |
| | Erythropoietin | Red bone marrow | Promotes red blood cell production, increases oxygen-carrying capacity of blood |
| Heart | Natriuretic peptides | Kidney | Lower blood volume and pressure by promoting $Na^+$ and water loss |
| Stomach and small intestine | Cholecystokinin | Gallbladder, brain | Bile release; appetite suppression |
| | Gastrin | Stomach | Stimulates acid secretion |
| | Ghrelin | Brain | Stimulates hunger, initiates feeding |
| | Peptide YY | Brain | Produces sense of satiety, terminates feeding |
| | Other enteric hormones | Stomach, intestines | Coordinate secretion and motility in different regions of digestive tract |
| Adipose tissue | Leptin | Brain | Limits appetite over long term |
| Osseous tissue | Osteocalcin | Pancreas, adipose tissue | Stimulates pancreatic beta cells to multiply, increases insulin secretion, enhances insulin sensitivity of various tissues, and reduces fat deposition |
| Placenta | Estrogen, progesterone | Many tissues of mother and fetus | Stimulate fetal development and maternal bodily adaptations to pregnancy; prepare mammary glands for lactation |

Answer the following questions to test your understanding of the preceding section:

11. Identify three endocrine glands that are larger or more functional in infants or children than in adults. What is the term for the shrinkage of a gland with age?

12. Why does thyroid hormone have a calorigenic effect?

13. Name a glucocorticoid, a mineralocorticoid, and a catecholamine secreted by the adrenal gland.

14. Does the action of glucocorticoids more closely resemble that of glucagon or insulin? Explain.

15. Define *hypoglycemic* and *hyperglycemic hormone* and give an example of each.

16. What is the difference between a gonadal hormone and a gonadotropin?

## 17.4 Hormones and Their Actions

### Expected Learning Outcomes

When you have completed this section, you should be able to

a. identify the chemical classes to which various hormones belong;

b. describe how hormones are synthesized and transported to their target organs;

c. describe how hormones stimulate their target cells;

d. explain how target cells regulate their sensitivity to circulating hormones;

e. describe how hormones affect each other when two or more of them stimulate the same target cells; and

f. discuss how hormones are removed from circulation after they have performed their roles.

Having surveyed the body's major hormones and their effects, we have established a cast of characters for the endocrine story, but we're left with some deeper questions: Exactly what is a hormone? How are hormones synthesized and transported to their destinations? How does a hormone produce its effects on a target organ? Thus, we now address endocrinology at the molecular and cellular levels.

## Hormone Chemistry

Most hormones fall into three chemical classes: *steroids, monoamines,* and *peptides* (table 17.6, fig. 17.14).

1. **Steroid hormones** are derived from cholesterol. They include sex steroids produced by the testes and ovaries (such as estrogens, progesterone, and testosterone) and corticosteroids produced by the adrenal gland (such as cortisol, aldosterone, and DHEA). Calcitriol, the calcium-regulating hormone, is not a steroid but is derived from one and has the same hydrophobic

| TABLE 17.6 | Chemical Classification of Hormones |
|---|---|
| **Steroids and Steroid Derivatives** | |
| Aldosterone | |
| Androgens | |
| Calcitriol | |
| Corticosterone | |
| Cortisol | |
| Estrogens | |
| Progesterone | |
| **Monoamines** | |
| Dopamine | Norepinephrine |
| Epinephrine | Thyroid hormone |
| Melatonin | |
| **Peptides** | |
| **Oligopeptides (3–10 amino acids)** | |
| Angiotensin II | |
| Antidiuretic hormone | |
| Cholecystokinin | |
| Gonadotropin-releasing hormone | |
| Oxytocin | |
| Thyrotropin-releasing hormone | |
| **Polypeptides (more than 10 amino acids)** | |
| Adrenocorticotropic hormone | Hepcidin |
| Natriuretic peptides | Insulin |
| Calcitonin | Leptin |
| Corticotropin-releasing hormone | Pancreatic polypeptide |
| Gastrin | Parathyroid hormone |
| Ghrelin | Prolactin |
| Glucagon | Somatostatin |
| Growth hormone | Thymic hormones |
| Growth hormone–releasing hormone | |
| **Glycoproteins (protein–carbohydrate complexes)** | |
| Erythropoietin | Luteinizing hormone |
| Follicle-stimulating hormone | Thyroid-stimulating hormone |
| Inhibin | |

character and mode of action as the steroids, so it is commonly grouped with them.

2. **Monoamines (biogenic amines)** were introduced in chapter 12, since this class also includes several neurotransmitters (see fig. 12.21, p. 458). The monoamine hormones include dopamine, epinephrine, norepinephrine, melatonin, and thyroid hormone. The first three of these are also called *catecholamines*. Monoamines are made

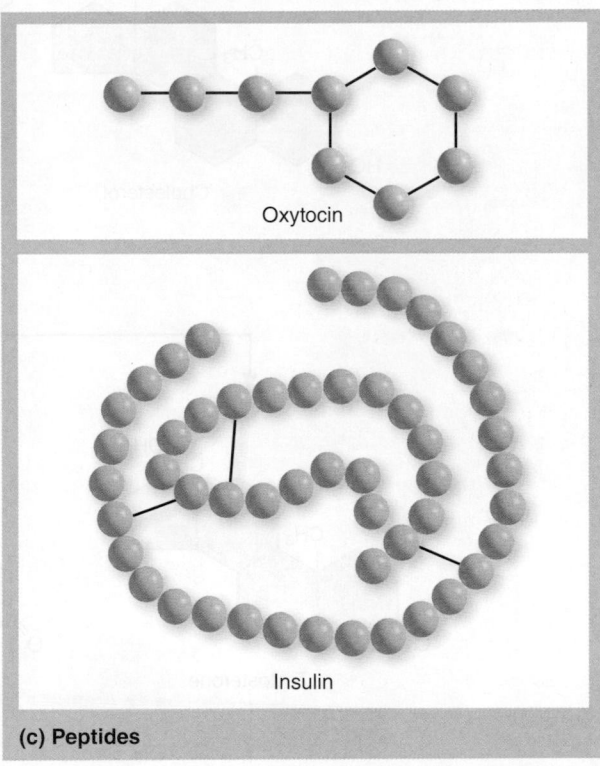

**FIGURE 17.14  The Chemical Classes of Hormones.** (a) Two steroid hormones, defined by their four-membered rings derived from cholesterol. (b) Two monoamines, derived from amino acids and defined by their –NH– or –NH$_2$ (amino) groups. (c) A small peptide hormone, oxytocin, and a protein hormone, insulin, defined by their chains of amino acids (the yellow circles).

from amino acids and retain an amino group, from which this hormone class gets its name.

3. **Peptide hormones** are chains of 3 to 200 or more amino acids. The two posterior pituitary hormones, oxytocin and antidiuretic hormone, are very similar oligopeptides of just nine amino acids. Probably the best-known large peptide (protein) hormone is insulin. Except for dopamine, the releasing and inhibiting hormones produced by the hypothalamus are polypeptides. Most hormones of the anterior pituitary are polypeptides or glycoproteins—polypeptides conjugated with short carbohydrate chains. Most glycoprotein hormones have an identical alpha chain of 92 amino acids and a variable beta chain that distinguishes them from each other.

## Hormone Synthesis

All hormones are made from either cholesterol or amino acids, with carbohydrate added in the case of glycoproteins.

## Steroids

Steroid hormones are synthesized from cholesterol and differ mainly in the functional groups attached to the four-ringed steroid backbone. Figure 17.15 shows the synthetic pathway for several steroid hormones. Notice that while estrogen and progesterone are typically thought of as "female" hormones and testosterone as a "male" hormone, these sex steroids are interrelated in their synthesis and have roles in both sexes.

## Peptides

Peptide hormones are synthesized the same way as any other protein. The gene for the hormone is transcribed to form a molecule of mRNA, and ribosomes translate the mRNA and assemble amino acids in the right order to make the peptide. After the basic amino acid sequence is assembled by the ribosomes, the rough endoplasmic reticulum and Golgi complex may further modify the peptide to form the mature hormone. Insulin, for example, begins as a single amino acid chain called *proinsulin*. A middle portion called the *connecting peptide* is removed to convert proinsulin to insulin, now composed of two polypeptide chains connected to each other by disulfide bridges (fig. 17.16).

▶▶▶ **APPLY WHAT YOU KNOW**

*During the synthesis of glycoprotein hormones, where in the cell would the carbohydrate be added? (See chapter 4.)*

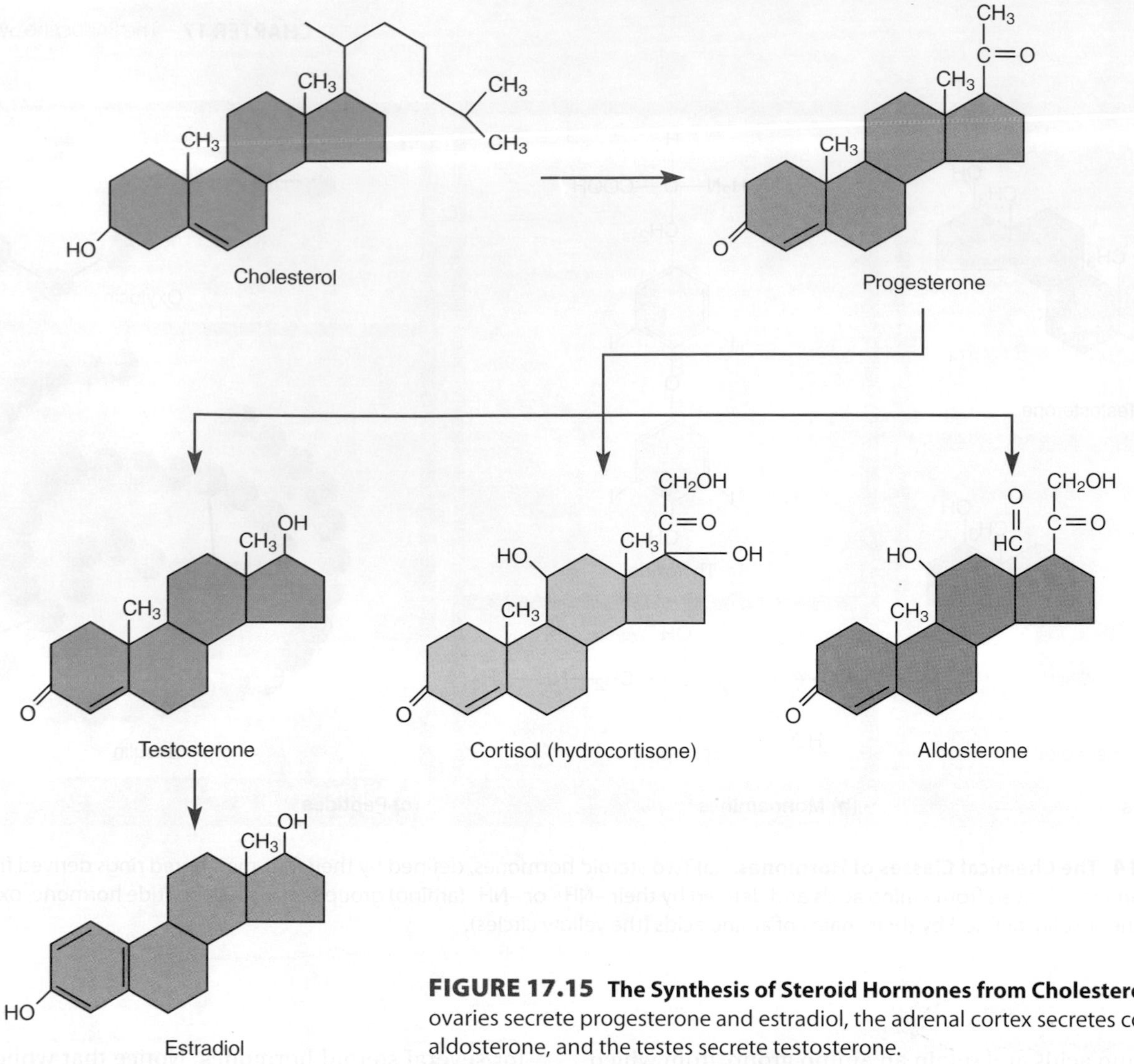

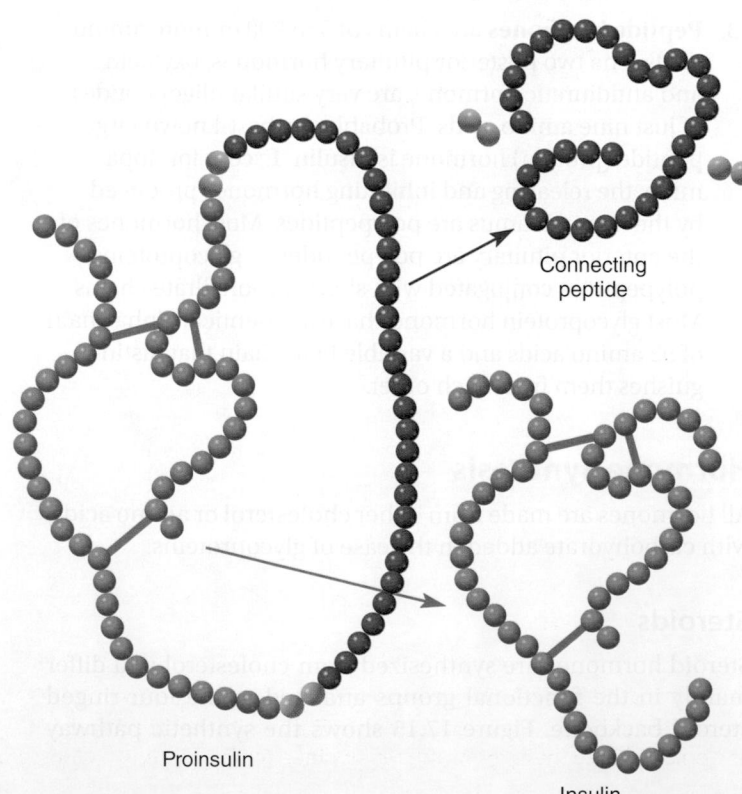

**FIGURE 17.15 The Synthesis of Steroid Hormones from Cholesterol.** The ovaries secrete progesterone and estradiol, the adrenal cortex secretes cortisol and aldosterone, and the testes secrete testosterone.

Cholesterol

Progesterone

Testosterone

Cortisol (hydrocortisone)

Aldosterone

Estradiol

**FIGURE 17.16 The Synthesis of Insulin, a Representative Polypeptide Hormone.** Proinsulin has a connecting (C) peptide, 31 amino acids long, that is removed to leave insulin. Insulin has two polypeptide chains, 30 and 21 amino acids long, joined by two disulfide (—S—S—) bridges represented by the green bars. A third disulfide bridge creates a loop in the short chain.

Connecting peptide

Proinsulin

Insulin

## Monoamines

Melatonin is synthesized from the amino acid tryptophan and the other monoamines from the amino acid tyrosine. Thyroid hormone (TH) is quite unusual in that it is made of *two* tyrosines, and is the only process in the human body that uses iodine; a lack of dietary iodine causes a thyroid disorder called *goiter* (see p. 663). Figure 17.17 shows the steps in TH synthesis.

(1) Cells of the thyroid follicle begin the process by absorbing iodide ($I^-$) ions from the blood of nearby capillaries. At the apical surface of the cells, facing the lumen of the follicle, they oxidize $I^-$ to a reactive form represented by $I^*$ in the figure.

(2) In the meantime, the follicle cells also synthesize a large protein called **thyroglobulin (Tg).** Each Tg has 123 tyrosines among its amino acids, but only 4 to 8 of them are used to make TH. The cells release thyroglobulin by exocytosis from their apical surfaces into the lumen.

(3) An enzyme at the cell surface adds iodine to some of the tyrosines. Some tyrosines receive one iodine and become *monoiodotyrosine (MIT)* (MON-oh-eye-OH-do-TY-ro-seen); some receive two and become *diiodotyrosine (DIT)* (fig. 17.17b). Where the Tg folds back on itself and two tyrosines meet, or where one tyrosine meets another on an adjacent Tg, the tyrosines link to each other through

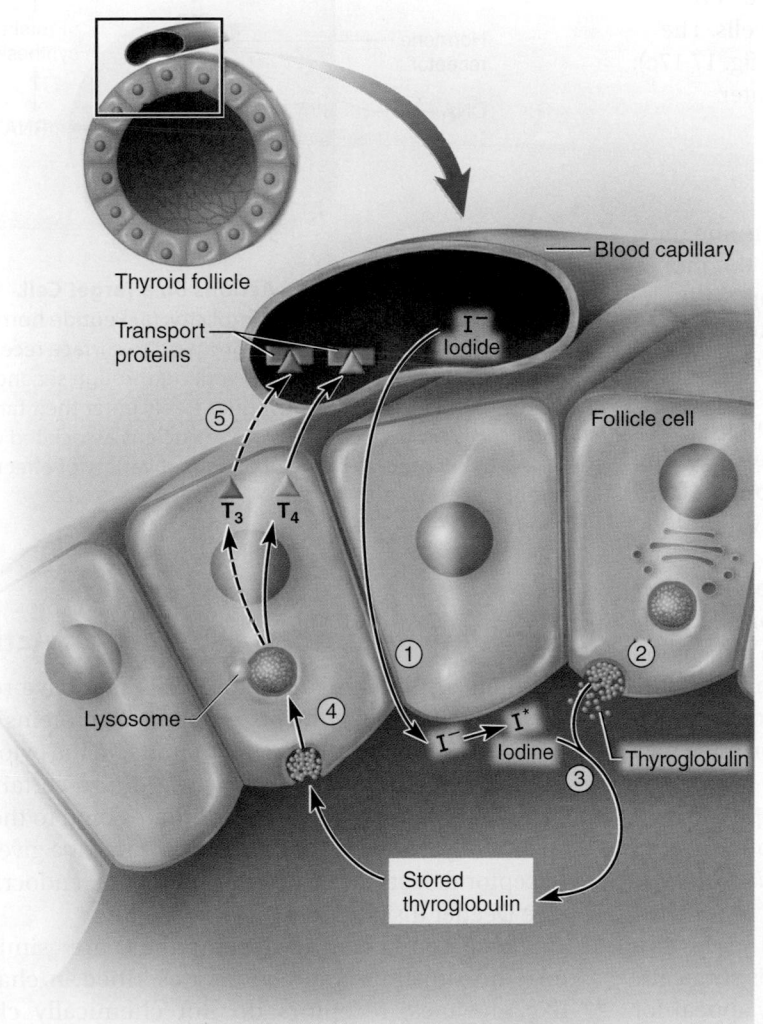

(a)

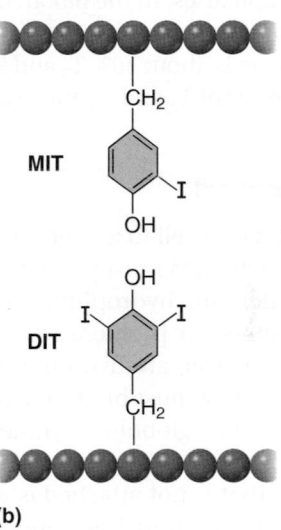

(b)

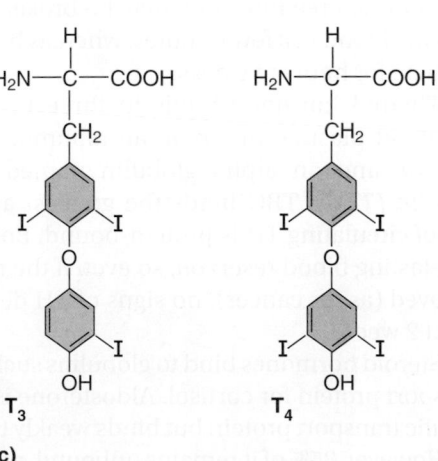

(c)

**FIGURE 17.17 Thyroid Hormone Synthesis and Secretion.** (a) TH synthesis in relation to the thyroid follicle. See text for explanation of numbered steps. (b) The meeting of an MIT and a DIT would produce the $T_3$ form of thyroid hormone, with three iodines. If both of these were DITs, it would produce the $T_4$ form with four iodines. (c) The two forms of mature thyroid hormone.

their side groups. If an MIT links up with a DIT, they form a complex with three iodines, destined to become the $T_3$ form of thyroid hormone; if two DITs unite, they form the forerunner of the $T_4$ form, with four iodines. One tyrosine then breaks away from its Tg, but for the time being, the hormone remains anchored to Tg through its other tyrosine. Tg is stored in the follicles until the thyroid gland receives a signal to release it. It is the pink-staining material in figure 17.9b.

④ When the follicle cells receive thyroid-stimulating hormone (TSH) from the anterior pituitary, they absorb droplets of Tg by pinocytosis. Within the cells, a lysosome contributes an enzyme that hydrolyzes the Tg chain, liberating thyroid hormone (TH).

⑤ TH is released from the basal side of the follicle cells into nearby blood capillaries. In the blood, it binds to various transport proteins that carry it to its target cells. The released hormone is about 10% $T_3$ and 90% $T_4$ (fig. 17.17c); the respective roles of $T_3$ and $T_4$ are discussed later.

## Hormone Transport

To get from an endocrine cell to a target cell, a hormone must travel in the blood, which is mostly water. Most of the monoamines and peptides are hydrophilic, so mixing with the blood plasma presents no problem for them. Steroids and thyroid hormone, however, are hydrophobic. To travel in the watery bloodstream, they must bind to hydrophilic **transport proteins**—albumins and globulins synthesized by the liver. A hormone attached to a transport protein is called a **bound hormone,** and one that is not attached is an **unbound (free) hormone.** Only the unbound hormone can leave a blood capillary and get to a target cell (fig. 17.18).

Transport proteins not only carry the hydrophobic hormones, but also prolong their half-lives. They protect hormones from being broken down by enzymes in the blood plasma and liver and from being filtered out of the blood by the kidneys. Free hormone may be broken down or removed from the blood in a few minutes, whereas bound hormone may circulate for hours to weeks.

Thyroid hormone binds to three transport proteins in the blood plasma: *albumin,* an albumin-like protein called *thyretin,* and an alpha globulin named *thyroxine-binding globulin (TBG).* TBG binds the greatest amount. More than 99% of circulating TH is protein-bound. Bound TH serves as a long-lasting blood reservoir, so even if the thyroid is surgically removed (as for cancer), no signs of TH deficiency appear for about 2 weeks.

Steroid hormones bind to globulins such as *transcortin,* the transport protein for cortisol. Aldosterone is unusual. It has no specific transport protein but binds weakly to albumin and others. However, 85% of it remains unbound, and correspondingly, it has a half-life of only 20 minutes.

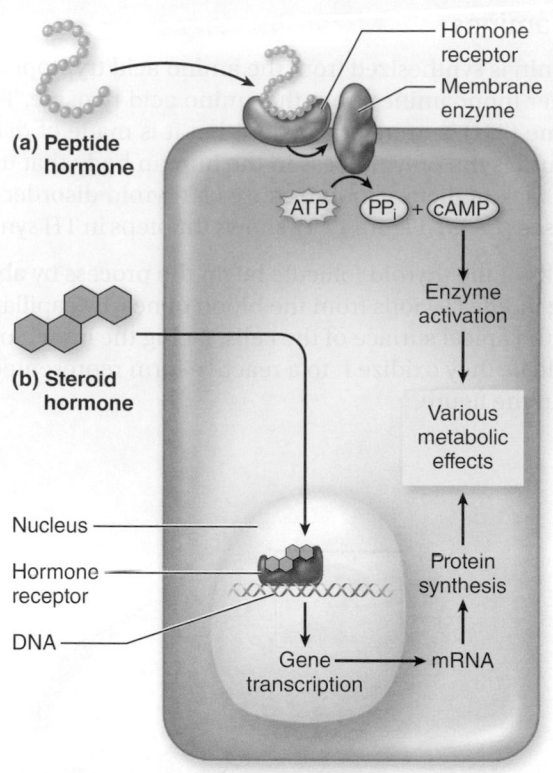

**FIGURE 17.18 Hormone Actions on a Target Cell.** Some process steps are omitted for simplicity. (a) Peptide hormones cannot enter their target cells. They bind to surface receptors and activate intracellular processes, working through second messengers such as cAMP. (b) Steroid hormones freely enter their target cells and usually bind to receptors in the nucleus associated with specific genes. Either process can lead to a great variety of effects on the target cell.

## Hormone Receptors and Mode of Action

Hormones stimulate only those cells that have receptors for them—their *target cells*. The receptors are proteins or glycoproteins located on the plasma membrane, in the cytoplasm, or in the nucleus. They act like switches to turn certain metabolic pathways on or off when the hormones bind to them. A target cell usually has a few thousand receptors for a given hormone. Receptor defects lie at the heart of several endocrine diseases (see Deeper Insight 17.2).

Receptor–hormone interactions are similar to the enzyme–substrate interactions described in chapter 2. Unlike enzymes, receptors do not chemically change their ligands, but they do exhibit enzymelike specificity and saturation. *Specificity* means that the receptor for one hormone will not bind other hormones. *Saturation* is the condition in which all the receptor molecules are occupied by hormone molecules. Adding more hormone cannot produce any greater effect.

# DEEPER INSIGHT 17.2

## CLINICAL APPLICATION

### Hormone Receptors and Therapy

In treating endocrine disorders, it is essential to understand the role of hormone receptors. For example, type 2 diabetes mellitus has long been thought to result from an insulin receptor defect or deficiency (among other possible causes). No amount of insulin replacement can correct this. And while growth hormone is now abundantly available thanks to genetic engineering, it is useless to children with *Laron dwarfism*, who have a hereditary defect in their GH receptors. *Androgen insensitivity syndrome* is due to an androgen receptor defect or deficiency; it causes genetic males to develop feminine genitalia and other features (see Deeper Insight 27.1, p. 1030). Estrogen stimulates the growth of some malignant tumors with estrogen receptors. For this reason, estrogen replacement therapy should not be used for women with estrogen-dependent cancer.

## Steroids and Thyroid Hormone

Some hormones enter the target-cell nucleus and act directly on the genes, changing target-cell physiology by either activating or inhibiting transcription of the gene for a metabolic enzyme or other protein. Such is the case with steroid hormones (fig. 17.18b). Being hydrophobic, they diffuse easily through the phospholipid regions of the plasma membrane. Most of them pass directly into the nucleus and bind to a receptor there; glucocorticoids, however, bind to a receptor in the cytosol, and the hormone–receptor complex is then transported into the nucleus. In either case, the receptor associates with the target gene in the nucleus, controlling its transcription.

Estrogen and progesterone afford a good example of this. In cells of the uterine mucosa, estrogen and its nuclear receptor activate the gene for a protein that functions as the progesterone receptor. Progesterone comes later in the menstrual cycle, binds to these receptors, and stimulates transcription of the gene for a glycogen-synthesizing enzyme. The uterine cell then synthesizes and accumulates glycogen for the nourishment of an embryo in the event of pregnancy. Progesterone has no effect on the uterine lining unless estrogen has been there earlier and prepared the way by inducing the synthesis of progesterone receptors.

Thyroid hormone also acts on nuclear receptors. It enters the target cell by means of an ATP-dependent transport protein. Surprisingly, although 90% of the TH secreted by the thyroid gland is thyroxine ($T_4$), $T_4$ has very little metabolic effect. Within the target-cell cytoplasm, an enzyme removes one iodine and converts it to the active form, $T_3$. This $T_3$, as well as a smaller amount of $T_3$ produced directly by the thyroid and absorbed from the blood, enters the nucleus and binds to receptors in the chromatin. One of the genes activated by $T_3$ is for the enzyme $Na^+$–$K^+$ ATPase (the $Na^+$–$K^+$ pump). As we saw in chapter 3 (p. 96), one effect of this pump is to generate heat, thus accounting for the calorigenic effect of thyroid hormone.

$T_3$ also activates the transcription of genes for a norepinephrine receptor and part of the muscle protein myosin, thus enhancing the responsiveness of cells such as cardiac muscle to sympathetic stimulation and increasing the strength of the heartbeat.

Steroid and thyroid hormones typically require several hours to days to show an effect. This lag is due to the time required for genetic transcription, translation, and accumulation of enough protein product to have a significant effect on target-cell metabolism.

## Peptides and Catecholamines

Peptides and catecholamines are hydrophilic and cannot penetrate a target cell, so they must stimulate its physiology indirectly. They bind to cell surface receptors, which are linked to second-messenger systems on the other side of the plasma membrane (fig. 17.18a). The best-known second messenger is cyclic adenosine monophosphate (cAMP). When glucagon binds to the surface of a liver cell, for example, its receptor activates a G protein, which in turn activates adenylate cyclase, the membrane enzyme that produces cAMP. cAMP leads ultimately to the activation of enzymes that hydrolyze glycogen stored in the cell (fig. 17.19). Somatostatin, by contrast, *inhibits* cAMP synthesis. Second messengers don't linger in the cell for long. cAMP, for example, is quickly broken down by an enzyme called **phosphodiesterase,** and the hormonal effect is therefore short-lived.

Two other second-messenger systems begin with one of the phospholipids in the plasma membrane. When activated by certain hormones (blue box in fig. 17.20), the receptor activates a G protein linked to a nearby enzyme, *phospholipase*, in the plasma membrane. Phospholipase splits a membrane phospholipid into two fragments—a small phosphate-containing piece called **inositol triphosphate ($IP_3$)** (eye-NOSS-ih-tol), and a larger piece, the triglyceride backbone with two fatty acids still attached, called **diacylglycerol (DAG)** (di-ACE-ul-GLISS-ur-ol). $IP_3$ and DAG are the second messengers that go on to activate a wide variety of metabolic changes in the target cells, depending on what cells are involved and what internal signaling pathways they use.

DAG activates a protein kinase (PK), much like cAMP does in figure 17.19. PK phosphorylates various other enzymes, turning them on or off and thereby activating or suppressing equally various metabolic processes in the target cell. For example, we saw earlier that thyroid-stimulating hormone (TSH) binds to thyroid follicle cells and stimulates them to release thyroid hormone into the bloodstream. TSH works through the DAG second-messenger system. In other cells, DAG stimulates mitosis and cell proliferation. Some cancer-causing agents (carcinogens) act by mimicking this *mitogenic* effect of DAG.

$IP_3$ works by increasing the calcium ($Ca^{2+}$) concentration in the target cell. It can open $Ca^{2+}$ channels in the plasma membrane, letting $Ca^{2+}$ into the cell from the extracellular fluid, or

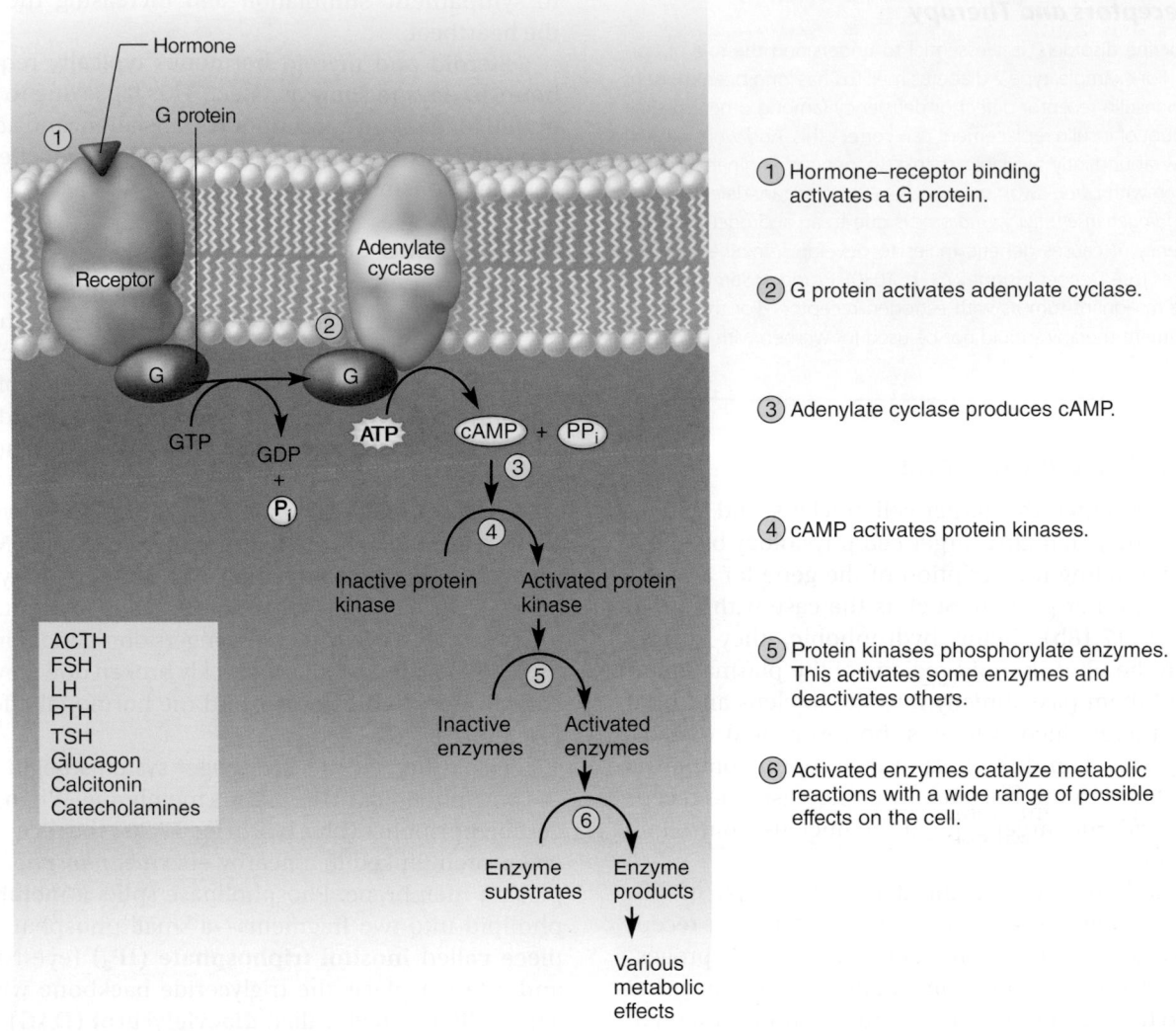

**FIGURE 17.19 Cyclic AMP as a Second Messenger.** The green box lists some hormones that act in this manner.

*Why are no steroid hormones listed in this figure?*

it can open channels in the endoplasmic reticulum, causing it to release a flood of $Ca^{2+}$ into the cytosol. Calcium then acts through several means to alter cell physiology:

1. It binds to certain calcium-dependent cytoplasmic enzymes that alter cell metabolism.

2. It can bind to a cytoplasmic calcium receptor, **calmodulin.** Not only is calcium-bound calmodulin the key to smooth muscle contraction, but it can also activate protein kinases with downstream effects just like cAMP or DAG.

3. It binds to membrane channels and changes their permeability to other solutes, in some cases altering the membrane potential (voltage) of the cell.

Childbirth affords one example of how an $IP_3$ second-messenger system works. Oxytocin (OT) from the pituitary gland binds to receptors on the smooth muscle cells of the uterus. It triggers the foregoing $IP_3$-releasing process, and $IP_3$, in turn, stimulates the sarcoplasmic reticulum to release $Ca^{2+}$. This initial burst of calcium then opens channels in the plasma membrane that admit still more calcium into the cell from the extracellular fluid. Calcium binds to calmodulin, as described in chapter 11 (p. 427), and stimulates contraction of the smooth muscle—that is, labor contractions.

The general point of all this is that hydrophilic hormones such as those listed in the blue box at the left side of figure 17.20 cannot enter the target cell. Yet by merely

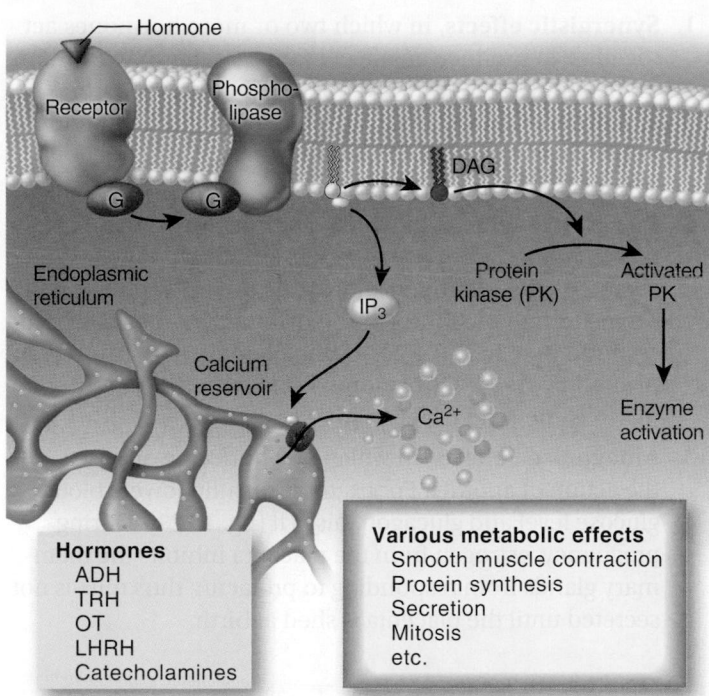

**FIGURE 17.20** **Diacylglycerol (DAG) and Inositol Triphosphate (IP₃) Second-Messenger System.** These are employed by hormones listed in the blue box.

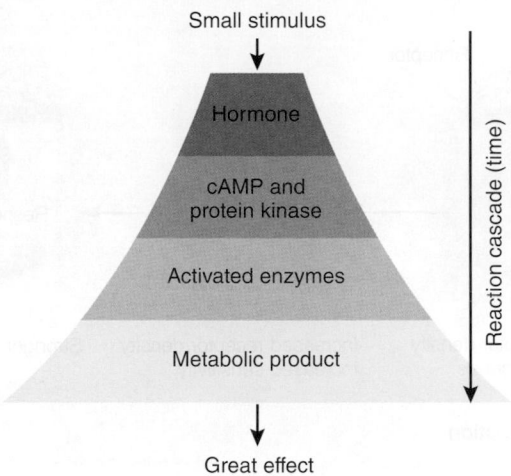

**FIGURE 17.21** **Signal Amplification.** A single hormone molecule can trigger the production of many cAMP molecules and activation of many molecules of protein kinase. Every protein kinase molecule can phosphorylate and activate many other enzymes. Each of those enzyme molecules can produce many molecules of a metabolic product. Amplification of the process at each step allows for a very small hormonal stimulus to cause a very large metabolic effect.

"knocking on the door" (binding to a surface receptor), they can initiate a flurry of metabolic activity within. The initial steps in this process are activation of a G protein and phospholipase. From there, divergent pathways are taken that involve DAG, IP₃, and $Ca^{2+}$ as second messengers. Ultimately these pathways lead to metabolic pathways being switched on or off within the cell.

Hormonal effects mediated through surface receptors tend to be relatively quick, because they do not depend on the cell synthesizing new proteins before anything else can happen. The oxytocin-induced labor contractions are a good example.

A given hormone doesn't always employ the same second messenger. ADH employs the IP₃–calcium system in smooth muscle but the cAMP system in kidney tubules. Insulin differs from all the foregoing mechanisms. Rather than using a second-messenger system, it binds to a plasma membrane enzyme, tyrosine kinase, which directly phosphorylates cytoplasmic proteins.

## Signal Amplification

Hormones are extraordinarily potent chemicals. Through a mechanism called **signal amplification** (or a **cascade effect**), one hormone molecule triggers the synthesis of not just one enzyme molecule but an enormous number (fig. 17.21). To put it in a simplistic but illustrative way, suppose 1 glucagon molecule triggered the formation of 1,000 molecules of

cAMP, cAMP activated a protein kinase, each protein kinase activated 1,000 other enzyme molecules, and each of those produced 1,000 molecules of a reaction product. These are modest numbers as chemical reactions go, and yet even at this low estimate, each glucagon molecule would trigger the production of 1 billion molecules of reaction product. Whatever the actual numbers may be, you can see how signal amplification enables a very small stimulus to produce a very large effect. Hormones are therefore powerfully effective in minute quantities. Their circulating concentrations are very low compared with other blood substances: on the order of nanograms per deciliter. Blood glucose, for example, is about 100 million times this concentrated. Because of signal amplification, target cells do not need a great number of hormone receptors.

## Modulation of Target-Cell Sensitivity

Target cells can adjust their sensitivity to a hormone by changing the number of receptors for it. In **up-regulation,** a cell increases the number of hormone receptors and becomes more sensitive to the hormone (fig. 17.22a). In late pregnancy, for example, the uterus produces oxytocin receptors, preparing itself for the surge of oxytocin that will occur during childbirth.

**Down-regulation** is the process in which a cell reduces its receptor population and thus becomes less sensitive to a hormone (fig. 17.22b). This sometimes happens in response to long-term exposure to a high hormone concentration. For example, adipocytes down-regulate when exposed to high concentrations of insulin, and cells of the testis down-regulate in response to high concentrations of luteinizing hormone.

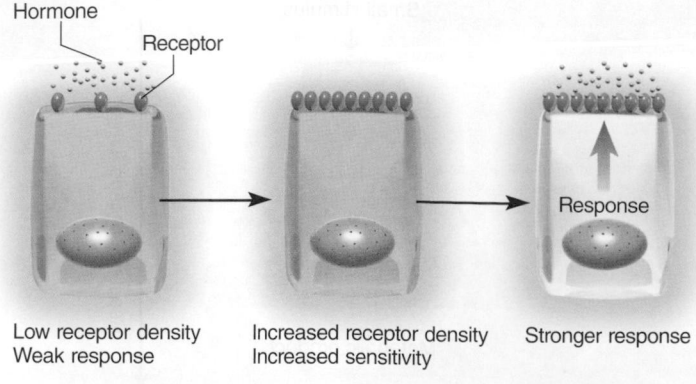

**(a) Up-regulation**

Hormone
Receptor

Low receptor density
Weak response

Increased receptor density
Increased sensitivity

Stronger response

Response

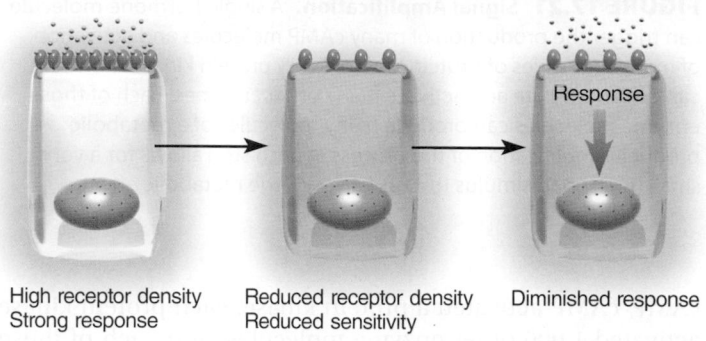

**(b) Down-regulation**

High receptor density
Strong response

Reduced receptor density
Reduced sensitivity

Diminished response

Response

**FIGURE 17.22** **Modulation of Target-Cell Sensitivity.**
(a) Up-regulation, in which a cell produces more receptors and increases its own sensitivity to a hormone. (b) Down-regulation, in which a cell reduces the density of its receptors and lessens its sensitivity to a hormone.

Hormone therapy often involves long-term use of abnormally high *pharmacological doses* of hormone, which may have undesirable side effects. Two ways in which adverse side effects can arise are (1) excess hormone may bind to receptor sites for other related hormones and mimic their effects; and (2) a target cell may convert one hormone into another, such as testosterone to estrogen. Thus, long-term high doses of testosterone can, paradoxically, have feminizing effects.

## Hormone Interactions

No hormone travels in the bloodstream alone, and no cell is exposed to only one hormone. Rather, there are many hormones in the blood and tissue fluid at once. Cells ignore the majority of them because they have no receptors for them, but most cells are sensitive to more than one. In these cases, the hormones may have three kinds of interactive effects:

1. **Synergistic effects,** in which two or more hormones act together to produce an effect that is greater than the sum of their separate effects. Neither FSH nor testosterone alone, for example, stimulates much sperm production. When they act together, however, the testes produce some 300,000 sperm per minute.

2. **Permissive effects,** in which one hormone enhances the target organ's response to a second hormone that is secreted later. Estrogen stimulates the up-regulation of progesterone receptors in the uterus. The uterus would respond poorly to progesterone, if at all, had it not been primed by the first hormone. Estrogen thus has a permissive effect on progesterone action.

3. **Antagonistic effects,** in which one hormone opposes the action of another. For example, insulin lowers blood glucose level and glucagon raises it (fig. 17.23). During pregnancy, estrogen from the placenta inhibits the mammary glands from responding to prolactin; thus milk is not secreted until the placenta is shed at birth.

## Hormone Clearance

Hormonal signals, like nervous signals, must be turned off when they have served their purpose. Most hormones are taken up and degraded by the liver and kidneys and then excreted in the bile or urine. Some are degraded by their target cells. As noted earlier, hormones that bind to transport proteins are removed from the blood much more slowly than hormones that do not employ transport proteins.

The rate of hormone removal is called the *metabolic clearance rate (MCR),* and the length of time required to clear 50% of the hormone from the blood is the half-life. The faster the MCR, the shorter is the half-life. Growth hormone, for example, uses no transport protein and has a half-life of only 6 to 20 minutes. Thyroxine, by contrast, is protected by transport proteins and maintains a physiologically effective level in the blood for up to 2 weeks after its secretion ceases.

**BEFORE YOU GO ON**

Answer the following questions to test your understanding of the preceding section:

17. What are the three chemical classes of hormones? Name at least one hormone in each class.

18. Why do corticosteroids and thyroid hormones require transport proteins to travel in the bloodstream?

19. Explain how MIT, DIT, $T_3$, and $T_4$ relate to each other structurally.

20. Where are hormone receptors located in target cells? Name one hormone that employs each receptor location.

21. Explain how one hormone molecule can activate millions of enzyme molecules.

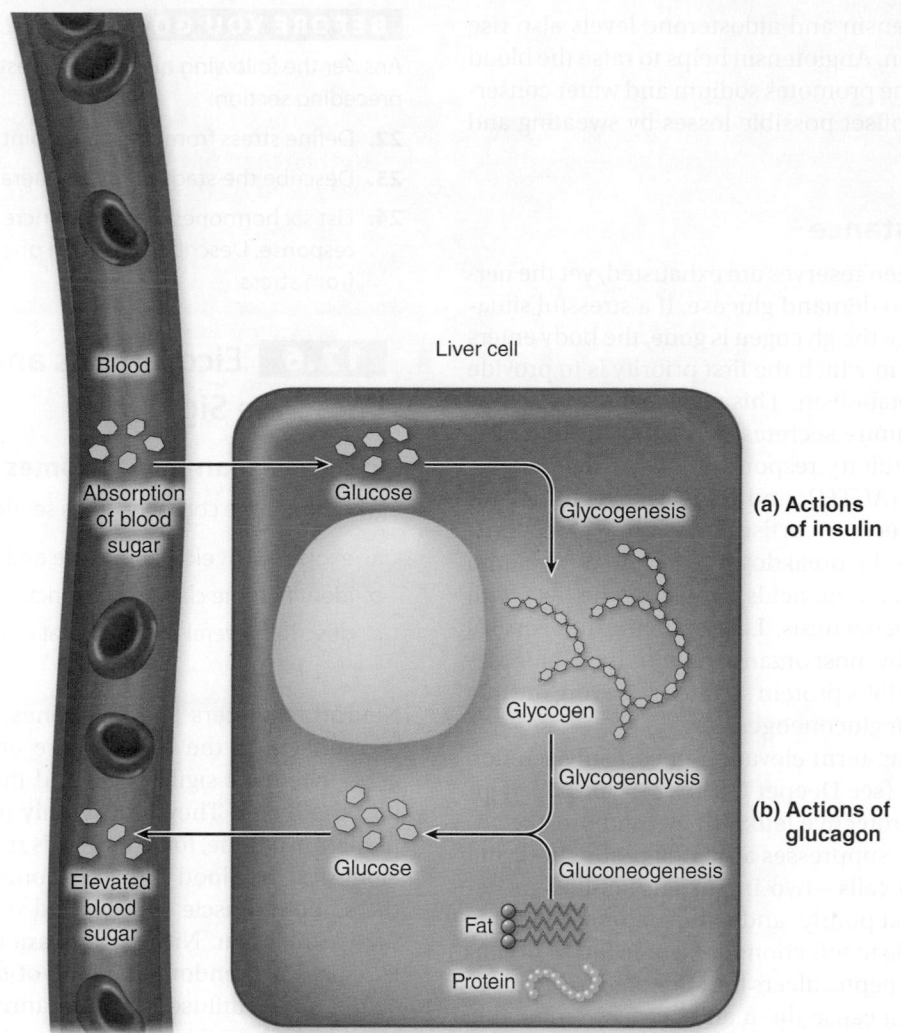

**FIGURE 17.23 Antagonistic Effects of Insulin and Glucagon on the Liver.** (a) Insulin promotes the uptake of blood glucose by cells and its polymerization to make glycogen, an energy-storage carbohydrate. In doing so, it lowers the blood sugar level. (b) Glucagon stimulates some cells, especially in the liver, to break down glycogen and release glucose to the bloodstream, thereby raising blood sugar level. Glycogenesis is the synthesis of glycogen; glycogenolysis is its hydrolysis (breakdown to glucose); and gluconeogenesis is the synthesis of glucose from noncarbohydrates, especially fats and proteins.

## 17.5 Stress and Adaptation

### Expected Learning Outcomes

When you have completed this section, you should be able to

a. give a physiological definition of *stress*; and
b. discuss how the body adapts to stress through its endocrine and sympathetic nervous systems.

**Stress** is defined as any situation that upsets homeostasis and threatens one's physical or emotional well-being. It affects us all from time to time, and we react to it in ways that are mediated mainly by the endocrine and sympathetic nervous systems. Physical causes of stress *(stressors)* include injury, surgery, hemorrhage, infection, intense exercise, temperature extremes, pain, and malnutrition. Emotional causes include anger, grief, depression, anxiety, and guilt.

Whatever the cause, the body reacts to stress in a fairly consistent way called the **stress response** or **general adaptation syndrome (GAS).** The response typically involves elevated levels of epinephrine and cortisol; some physiologists now define *stress* as any situation that raises the cortisol level. A pioneering researcher on stress physiology, Canadian biochemist Hans Selye, showed in 1936 that the GAS typically occurs in three stages, which he called the *alarm reaction,* the *stage of resistance,* and the *stage of exhaustion.*

### The Alarm Reaction

The initial response to stress is an **alarm reaction** mediated mainly by norepinephrine from the sympathetic nervous system and epinephrine from the adrenal medulla. These prepare the body to take action such as fighting or escaping danger. One of their effects, the consumption of stored glycogen, is particularly important in the transition to the next stage of the stress response. Angiotensin and aldosterone levels also rise during the alarm reaction. Angiotensin helps to raise the blood pressure, and aldosterone promotes sodium and water conservation, which helps to offset possible losses by sweating and bleeding.

## The Stage of Resistance

After a few hours, glycogen reserves are exhausted, yet the nervous system continues to demand glucose. If a stressful situation is not resolved before the glycogen is gone, the body enters the **stage of resistance**, in which the first priority is to provide alternative fuels for metabolism. This stage is dominated by cortisol. The hypothalamus secretes corticotropin-releasing hormone (CRH); the pituitary responds by secreting adrenocorticotropic hormone (ACTH); and this, in turn, stimulates the adrenal cortex to secrete cortisol and other glucocorticoids. Cortisol promotes the breakdown of fat and protein into glycerol, fatty acids, and amino acids, providing the liver with raw material for gluconeogenesis. Like epinephrine, cortisol inhibits glucose uptake by most organs and thus has a glucose-sparing effect. It also inhibits protein synthesis, leaving the free amino acids available for gluconeogenesis.

Unfortunately, a long-term elevation of cortisol secretion reduces one's immunity (see Deeper Insight 21.4, p. 842). It inhibits the synthesis of protective leukotrienes and prostaglandins (discussed shortly), suppresses antibody production, and kills immature T and B cells—two important families of immune cells. Wounds heal poorly, and a person under chronic stress is more susceptible to infections and some forms of cancer. Stress can aggravate peptic ulcers because of reduced resistance to the bacteria that cause them and because circulating epinephrine reduces secretion of the gastric mucus and pancreatic bicarbonate that normally protect the stomach lining.

Cortisol suppresses the secretion of sex hormones such as estrogen, testosterone, and luteinizing hormone, causing disturbances of fertility and sexual function.

## The Stage of Exhaustion

The body's fat reserves can carry it through months of stress, but when fat is depleted, stress overwhelms homeostasis. The **stage of exhaustion** sets in, often marked by rapid decline and death. With its fat stores gone, the body now relies primarily on protein breakdown to meet its energy needs. Thus, there is a progressive wasting away of the muscles and weakening of the body. After prolonged stimulation, the adrenal cortex may stop producing glucocorticoids, making it all the more difficult to maintain glucose homeostasis. Aldosterone sometimes promotes so much water retention that it creates a state of hypertension, and while it conserves sodium, it hastens the elimination of potassium and hydrogen ions. This creates a state of hypokalemia (potassium deficiency in the blood) and alkalosis (excessively high blood pH), resulting in nervous and muscular system dysfunctions. Death frequently results from heart failure, kidney failure, or overwhelming infection.

**BEFORE YOU GO ON**

Answer the following questions to test your understanding of the preceding section:

22. Define stress from the standpoint of endocrinology.

23. Describe the stages of the general adaptation syndrome.

24. List six hormones that show increased secretion in the stress response. Describe how each one contributes to recovery from stress.

## 17.6  Eicosanoids and Paracrine Signaling

### Expected Learning Outcomes

When you have completed this section, you should be able to

a. explain what eicosanoids are and how they are produced;

b. identify some classes and functions of eicosanoids; and

c. describe several physiological roles of prostaglandins.

Neurotransmitters and hormones are not the only chemical messengers in the body. There are also *paracrine* messengers—chemical signals released into the tissue fluid and not into the blood. They diffuse only to nearby cells in the same tissue. Histamine, for example, is released by mast cells that lie alongside the blood vessels in connective tissues. It diffuses to the smooth muscle of the blood vessel, relaxing it and allowing vasodilation. Nitric oxide, another paracrine vasodilator, is released by endothelial cells of the blood vessel itself. Catecholamines diffuse from the adrenal medulla to the cortex to stimulate corticosterone secretion. A single chemical can be considered a hormone, paracrine, or neurotransmitter depending on location and circumstances.

The **eicosanoids**[30] (eye-CO-sah-noyds) are an important family of paracrine secretions. They have 20-carbon backbones derived from a polyunsaturated fatty acid called **arachidonic** (ah-RACK-ih-DON-ic) **acid.** Some peptide hormones and other stimuli liberate arachidonic acid from one of the phospholipids of the plasma membrane, and the following two enzymes then convert it to various eicosanoids (fig. 17.24).

**Lipoxygenase** helps to convert arachidonic acid to **leukotrienes,** eicosanoids that mediate allergic and inflammatory reactions (see chapter 21). **Cyclooxygenase** converts arachidonic acid to three other eicosanoids:

1. **Prostacyclin** is produced by the walls of the blood vessels, where it inhibits blood clotting and vasoconstriction.

2. **Thromboxanes** are produced by blood platelets. In the event of injury, they override prostacyclin and stimulate vasoconstriction and clotting. Prostacyclin and thromboxanes are further discussed in chapter 18.

---

[30]*eicosa* (variation of *icosa*) = 20

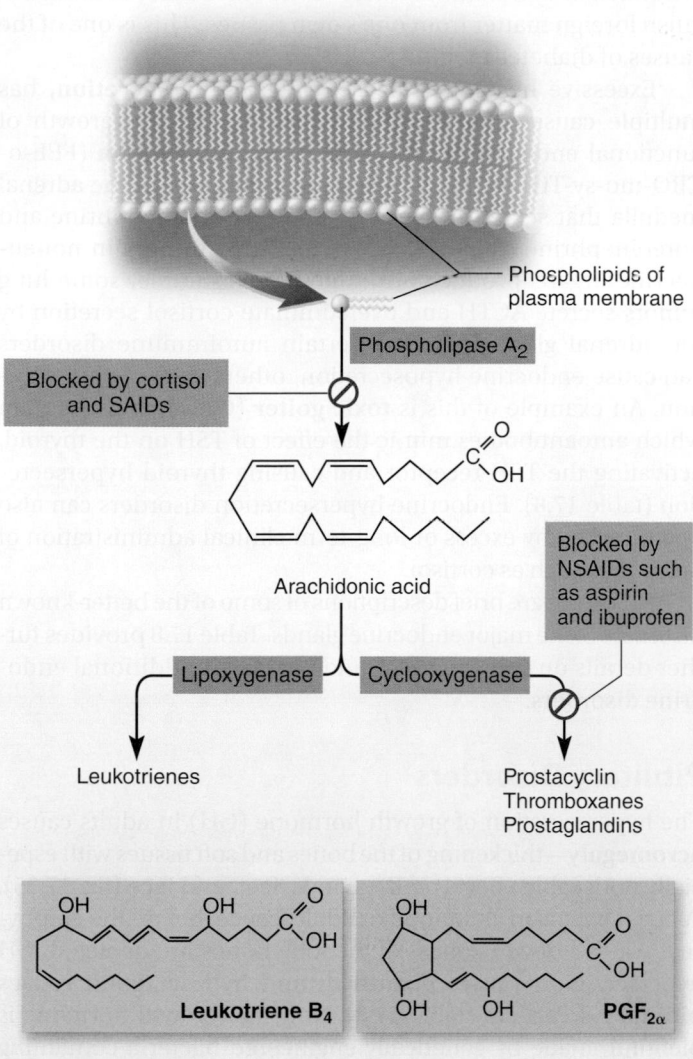

Understanding the pathways of eicosanoid synthesis makes it possible to understand the action of some familiar drugs (see Deeper Insight 17.3). The roles of prostaglandins and other eicosanoids are further explored in later chapters on blood, immunity, and reproduction.

**BEFORE YOU GO ON**

Answer the following questions to test your understanding of the preceding section:

25. What are eicosanoids and how do they differ from neurotransmitters and hormones?

26. Distinguish between a paracrine and endocrine effect.

27. State four functions of prostaglandins.

 **DEEPER INSIGHT 17.3**

**CLINICAL APPLICATION**

### Anti-Inflammatory Drugs

Cortisol and corticosterone are used as *steroidal anti-inflammatory drugs (SAIDs)*. They inhibit inflammation by blocking the release of arachidonic acid from the plasma membrane, thus inhibiting the synthesis of all eicosanoids. Their main disadvantage is that prolonged use causes side effects that mimic Cushing syndrome (see p. 663). Aspirin, ibuprofen (Motrin), and celecoxib (Celebrex) are *nonsteroidal anti-inflammatory drugs (NSAIDs)*, also called *COX inhibitors* because they block the action of cyclooxygenase (COX). Their advantage is that they do not affect lipoxygenase function or leukotriene production. One form of cyclooxygenase, COX-2, is specific to the cells of inflammation, so there has been a particular interest in developing safe COX-2 inhibitors that can treat inflammation without undesirable side effects such as gastrointestinal damage.

COX inhibitors are also useful in the treatment of fever and thrombosis (undesirable blood clotting). Fever is thought to result from the action of prostaglandins on the hypothalamus. Aspirin exerts its antipyretic (fever-reducing) effect by inhibiting prostaglandin synthesis and its antithrombotic effect by inhibiting thromboxane synthesis (see chapter 18).

**FIGURE 17.24 Eicosanoid Synthesis and Related Drug Actions.** SAIDs are steroidal anti-inflammatory drugs such as hydrocortisone; NSAIDs are nonsteroidal anti-inflammatory drugs such as aspirin and ibuprofen. A representative leukotriene and a prostaglandin are shown at the bottom.

❓ *How would the body be affected by a drug that selectively inhibited lipoxygenase?*

3. **Prostaglandins (PGs)** are the most diverse eicosanoids. They have a five-sided carbon ring in their backbone. They are named PG for *prostaglandin,* plus a third letter that indicates the type of ring structure (PGE, PGF, etc.) and a subscript that indicates the number of $C=C$ double bonds in the side chain—such as the $PGF_{2\alpha}$ shown in figure 17.24. They were first found in bull semen and the prostate gland, hence their name, but they are now thought to be produced in most organs of the body. The PGEs are usually antagonized by PGFs. For example, the PGE family relaxes smooth muscle in the bladder, intestines, bronchioles, and uterus and stimulates contraction of the smooth muscle of blood vessels. $PGF_{2\alpha}$ has precisely the opposite effects. Some other roles of prostaglandins are described in table 17.7.

| TABLE 17.7 | Some of the Roles of Prostaglandins |
|---|---|

*Inflammatory:* Promote fever and pain, two cardinal signs of inflammation

*Endocrine:* Mimic effects of TSH, ACTH, and other hormones; alter sensitivity of anterior pituitary to hypothalamic hormones; work with glucagon, catecholamines, and other hormones in regulation of fat mobilization

*Nervous:* Function as neuromodulators, altering the release or effects of neurotransmitters in the brain

*Reproductive:* Promote ovulation and formation of corpus luteum; induce labor contractions

*Gastrointestinal:* Inhibit gastric secretion

*Vascular:* Act as vasodilators and vasoconstrictors

*Respiratory:* Constrict or dilate bronchioles

*Renal:* Promote blood circulation through the kidney, increase water and electrolyte excretion

## 17.7  Endocrine Disorders

### Expected Learning Outcomes

When you have completed this section, you should be able to

a. explain some general causes and examples of hormone hyposecretion and hypersecretion;

b. briefly describe some common disorders of pituitary, thyroid, parathyroid, and adrenal function; and

c. in more detail, describe the causes and pathology of diabetes mellitus.

As we saw in the discussion of signal amplification, a little hormone can have a great effect. Thus it is necessary to tightly regulate hormone secretion and blood concentration. Variations in hormone concentration and target-cell sensitivity often have very noticeable effects on the body. This section deals with some of the better-known dysfunctions of the endocrine system. The effects of aging on the endocrine system are described on page 1120.

### Hyposecretion and Hypersecretion

Inadequate hormone release is called **hyposecretion.** It can result from tumors or lesions that destroy an endocrine gland or interfere with its ability to receive signals from another cell. For example, a fractured sphenoid bone can sever the hypothalamo–hypophyseal tract and prevent the transport of oxytocin and antidiuretic hormone (ADH) to the posterior pituitary. The resulting ADH hyposecretion disables the water-conserving capability of the kidneys and leads to **diabetes insipidus,** a condition of chronic polyuria without glucose in the urine. (*Insipidus* means "without taste" and refers to the lack of sweetness of the glucose-free urine, in contrast to the sugary urine of diabetes mellitus.) Autoimmune diseases can also lead to hormone hyposecretion when endocrine cells

are attacked by *autoantibodies*—antibodies that fail to distinguish foreign matter from one's own tissues. This is one of the causes of diabetes mellitus.

Excessive hormone release, called **hypersecretion,** has multiple causes. Some tumors result in the overgrowth of functional endocrine tissue. A **pheochromocytoma** (FEE-o-CRO-mo-sy-TOE-muh), for example, is a tumor of the adrenal medulla that secretes excessive amounts of epinephrine and norepinephrine (table 17.8, p. 666). Some tumors in nonendocrine organs produce hormones. For example, some lung tumors secrete ACTH and overstimulate cortisol secretion by the adrenal gland. Whereas certain autoimmune disorders can cause endocrine hyposecretion, others cause hypersecretion. An example of this is **toxic goiter** (Graves[31] disease), in which autoantibodies mimic the effect of TSH on the thyroid, activating the TSH receptor and causing thyroid hypersecretion (table 17.8). Endocrine hypersecretion disorders can also be mimicked by excess or long-term clinical administration of hormones such as cortisol.

Following are brief descriptions of some of the better-known disorders of the major endocrine glands. Table 17.8 provides further details on some of these and lists some additional endocrine disorders.

### Pituitary Disorders

The hypersecretion of growth hormone (GH) in adults causes **acromegaly**—thickening of the bones and soft tissues with especially noticeable effects on the hands, feet, and face (fig. 17.25). When it begins in childhood or adolescence, before the epiphyseal plates (growth zones) of the long bones are depleted, GH hypersecretion causes **gigantism** and hyposecretion causes **pituitary dwarfism** (table 17.8). Now that growth hormone is plentiful, made by genetically engineered bacteria containing the human GH gene, pituitary dwarfism has become rare.

---

[31]Robert James Graves (1796–1853), Irish physician

**Age 9**

**Age 16**

**Age 33**

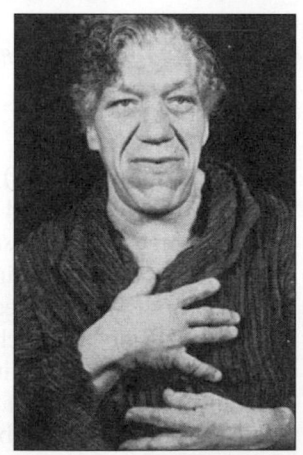

**Age 52**

**FIGURE 17.25  Acromegaly, a Condition Caused by Growth Hormone Hypersecretion in Adulthood.**  These four classic photographs show the same person taken at different ages. Note the characteristic thickening of the face and hands.

❓ *How would she have been affected if GH hypersecretion had begun at age 9?*

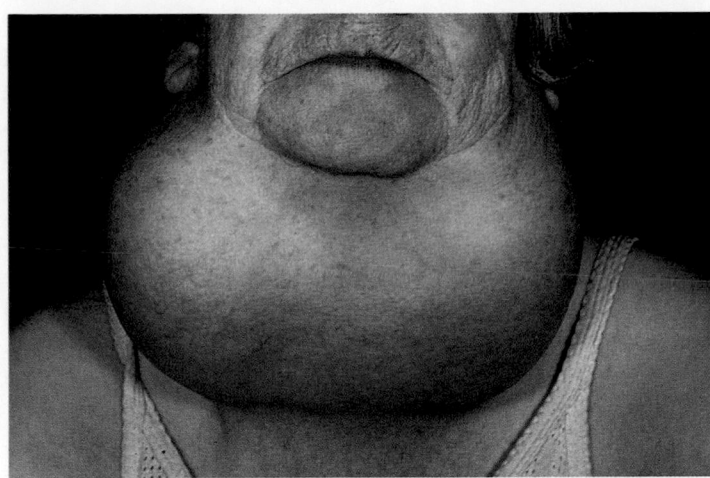

**FIGURE 17.26 Endemic Goiter.** An iodine deficiency in this person's diet resulted in a lack of thyroid hormone and extremely large goiter. For lack of negative feedback inhibition, the pituitary secreted elevated levels of thyroid-stimulating hormone (TSH). This resulted in hypertrophy of the thyroid gland.

## Thyroid and Parathyroid Disorders

**Congenital hypothyroidism** is thyroid hyposecretion present from birth. Severe or prolonged adult hypothyroidism can cause **myxedema** (MIX-eh-DEE-muh). Both syndromes are described in table 17.8, and both can be treated with oral thyroid hormone.

A more conspicuous, often striking abnormality of the thyroid is **endemic goiter** (fig. 17.26). It results from a deficiency of dietary iodine. Without iodine, the gland cannot synthesize TH. Without TH, the pituitary gland receives no feedback and acts as if the thyroid were understimulated. It produces extra TSH, which stimulates hypertrophy of the thyroid, visible as a swelling in the neck. Endemic goiter has become almost nonexistent in developed countries because of the addition of

iodine to table salt, animal feeds, and fertilizers. It occurs most often in localities that have neither these benefits nor access to iodine-rich seafood—notably central Africa and mountainous regions of South America, central Asia, and Indonesia. The word *endemic* refers to the occurrence of a disease in a defined geographic locality.

The parathyroids, because of their location and small size, are sometimes accidentally removed in thyroid surgery or degenerate when neck surgeries cut off their blood supply. Without hormone replacement therapy, the resulting **hypoparathyroidism** causes a rapid decline in blood calcium level; in as little as 2 or 3 days, this can lead to a fatal, suffocating spasm of the muscles of the larynx *(hypocalcemic tetany)*. **Hyperparathyroidism,** excess PTH secretion, is usually caused by a parathyroid tumor. It causes the bones to become soft, deformed, and fragile; it raises the blood levels of calcium and phosphate ions; and it promotes the formation of *renal calculi* (kidney stones) composed of calcium phosphate. Chapter 7 further describes the relationship among parathyroid function, blood calcium, and bone tissue.

## Adrenal Disorders

**Cushing**[32] **syndrome** is excess cortisol secretion owing to any of several causes: ACTH hypersecretion by the pituitary, ACTH-secreting tumors, or hyperactivity of the adrenal cortex independently of ACTH. Cushing syndrome disrupts carbohydrate and protein metabolism, leading to hyperglycemia, hypertension, muscular weakness, and edema. Muscle and bone mass are lost rapidly as protein is catabolized. Some patients exhibit abnormal fat deposition between the shoulders ("buffalo hump") or in the face ("moon face") (fig. 17.27). Long-term hydrocortisone therapy can have similar effects.

**Adrenogenital syndrome (AGS),** the hypersecretion of adrenal androgens, commonly accompanies Cushing

[32]Harvey Cushing (1869–1939), American physician

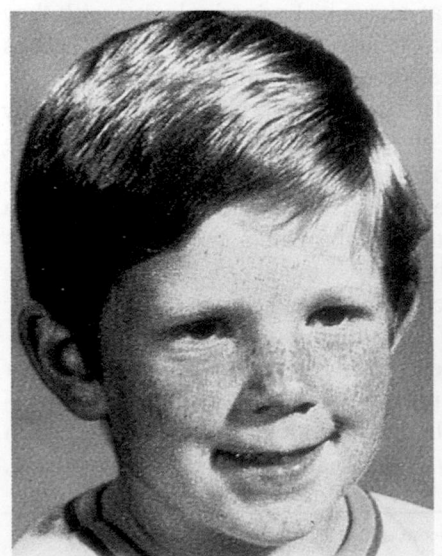

(a)

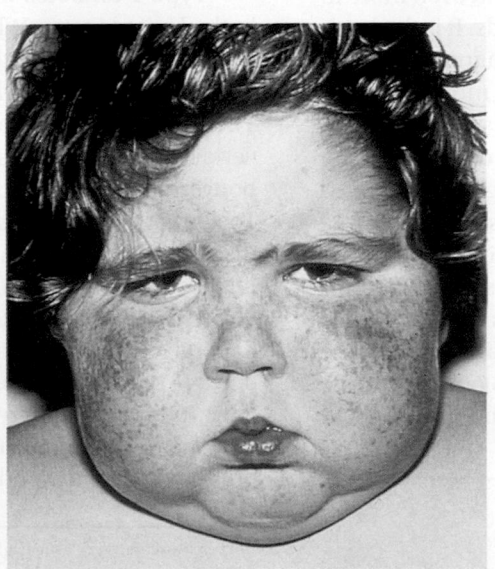

(b)

**FIGURE 17.27 Cushing Syndrome.**
(a) Patient before the onset of the syndrome.
(b) The same boy, only 4 months later, showing the "moon face" characteristic of Cushing syndrome.

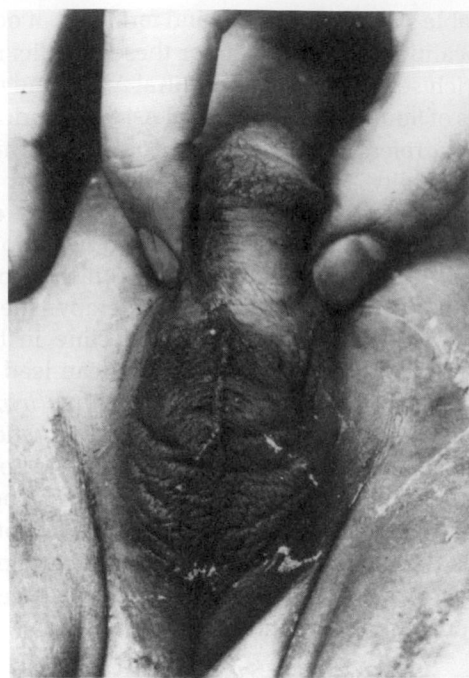

**FIGURE 17.28 Adrenogenital Syndrome (AGS).** These are the genitals of a baby girl with AGS, masculinized by prenatal hypersecretion of adrenal androgens. Note the fusion of the labia majora to resemble a scrotum and enlargement of the clitoris to resemble a penis. Such infants are easily mistaken for boys and raised as such.

syndrome. In children, AGS often causes enlargement of the penis or clitoris and the premature onset of puberty. Prenatal AGS can result in newborn girls exhibiting masculinized genitalia and being misidentified as boys (fig. 17.28). In women, AGS produces such masculinizing effects as increased body hair, deepening of the voice, and beard growth.

## Diabetes Mellitus

The world's most prevalent metabolic disease is diabetes mellitus, affecting about 7% of the U.S. population and even more in areas such as Scandinavia and the Pacific Islands. It is the leading cause of adult blindness, renal failure, gangrene, and the necessity for limb amputations, and warrants a more extended discussion than the less common endocrine diseases. The pathology of DM is described here, and the chapter ends with an essay on the history of insulin (see Deeper Insight 17.4).

**Diabetes mellitus**[33] **(DM)** can be defined as a disruption of carbohydrate, fat, and protein metabolism resulting from the hyposecretion or inaction of insulin. The classic signs and symptoms with which patients often first present to a physician are "the three polys": **polyuria**[34] (excessive urine output), **polydipsia**[35] (intense thirst), and **polyphagia**[36] (ravenous hunger). Blood and urine tests can confirm a diagnosis of DM by revealing three further signs: **hyperglycemia**[37] (elevated blood glucose), **glycosuria**[38] (glucose in the urine), and **ketonuria** (ketones in the urine). DM was originally named for the sweetness of the urine stemming from glycosuria. Before the advent of chemical tests for glucose, physicians tasted their patients' urine as part of their diagnostic process.

A little knowledge of kidney physiology is necessary to understand why glycosuria and polyuria occur. The kidneys filter blood plasma and convert the filtrate to urine. In a healthy person, the kidney tubules remove all glucose from the filtrate and return it to the blood, so there is little or no glucose in the urine. Water follows the glucose and other solutes by osmosis, so the tubules also reclaim most of the water in the filtrate.

But like any other carrier-mediated transport system, there is a limit to how fast the glucose transporters of the kidney can work. The maximum rate of reabsorption is called the *transport maximum,* $T_m$ (see p. 93). In diabetes mellitus, glucose enters the tubules so rapidly that it exceeds the $T_m$ and the tubules cannot reabsorb it fast enough. The excess passes through into the urine. Glucose and ketones in the tubules also raise the osmolarity of the tubular fluid and cause **osmotic diuresis**—water remains in the tubules with these solutes, so large amounts of water are passed in the urine. This accounts for the polyuria, dehydration, and thirst of diabetes. A person with untreated DM may pass 10 to 15 L of urine per day, compared with 1 or 2 L in a healthy person.

## Types and Treatment

There are two forms of diabetes mellitus: type 1 and type 2. These were formerly called juvenile-onset or insulin-dependent DM, and adult-onset or non-insulin-dependent DM, respectively. These terms have lately been abandoned because they are too misleading. Although insulin is always used to treat type 1, it is frequently used for type 2 diabetes as well, and either type can occur at any age. Indeed, with the burgeoning problem of childhood obesity, nearly half of all new cases of childhood diabetes are now type 2.

**Type 1 diabetes mellitus** accounts for 5% to 10% of cases in the United States. What causes it? Endocrinologists wish they had a better answer to that. It begins with heredity. Several genes have been identified that predispose a person to type 1 DM. Then, when a genetically susceptible individual is infected by certain viruses (rubella, cytomegalovirus, or a few others), the body produces autoantibodies that destroy pancreatic beta cells. To a great extent, this destruction is tolerated and produces no disease, but when 80% to 90% of the beta cells are gone, insulin falls to such a critically low level that it can no longer regulate glycemia, the blood glucose level. Now comes the problematic hyperglycemia and all of its insidious complications. Type 1 diabetes is usually diagnosed before the age of 30, but may occur later. Its victims require insulin to survive—usually periodic injections or continual subcutaneous delivery by a small insulin pump worn on the body. A dry insulin inhaler is now available,

---

[33]*diabet* = to flow through; *melli* = honey
[34]*poly* = much, excessive; *uri* = urine
[35]*dipsia* = drinking
[36]*phagia* = eating

[37]*hyper* = excess; *glyc* = sugar, glucose; *emia* = blood condition
[38]*glyco* = glucose, sugar; *uria* = urine condition

but is not suitable for all patients. Meal planning, exercise, and self-monitoring of blood glucose levels are also important aspects of the treatment regimen.

Some 90% to 95% of diabetics, however, have **type 2 DM.** Here, the chief problem is not lack of insulin, but *insulin resistance*—unresponsiveness of the target cells to the hormone. The level of insulin may actually be very high in the early stage of the disease, although it later tends to fall to normal or subnormal levels. Again, heredity is one of the causes, although no one gene, or even a mere few, can be blamed for the disease; more than 36 genes are known, so far, to contribute to the risk of type 2 DM. There are great differences in prevalence from one ethnic group to another; it is relatively high, for example, among people of Native American, Hispanic, and Asian descent. It also has a tendency to run in families, and shows very high concordance between genetically identical twins—if one twin develops type 2 DM, there is more than a 90% probability that the other will too. Other important risk factors are age, obesity, and a sedentary lifestyle. All of these are accompanied by the progressive replacement of muscular tissue with fat. Muscle plays a highly important role in absorbing blood glucose and buffering glycemia, so as muscle mass diminishes, a person becomes less and less able to regulate glycemia.

Type 2 DM develops slowly and is usually diagnosed after age 40, but it is becoming increasingly prevalent in young people because of early obesity. Aside from the loss of the glucose-buffering role of muscle, another apparent factor in type 2 DM is that adipose tissue secretes chemical signals that indirectly interfere with glucose transport into most cells—so the more body fat, the less efficient is glucose uptake. It is no surprise, then, that type 2 DM can often be successfully managed through a weight-loss program of diet and exercise, supplemented if necessary with glycemia-lowering oral medications. If these approaches prove inadequate, insulin therapy is also employed.

## Pathogenesis

When cells cannot absorb glucose, they must get their energy someplace else; they metabolize fat and protein. In time, this leads to muscular atrophy, emaciation, and weakness. Before insulin therapy was introduced in 1922, the victims of type 1 DM wasted away to an astonishing extent (see Deeper Insight 17.4). Diabetes was described in the first century as "a melting down of the flesh and limbs into urine." Adult patients weighed as little as 27 to 34 kg (60–75 lb) and looked like victims of severe famine. Their breath had a disagreeable sweet ketone smell, like rotten apples. One typical patient was described by medical historian Michael Bliss as "barely able to lift his head from his pillow, crying most of the time from pain, hunger, and despair." In the terminal stage, patients became increasingly drowsy, gasped for air, became comatose, and died within a few hours. Most diabetic children lived less than 1 year after diagnosis—a year of utmost misery at that. Such was the natural course of the disease in the centuries before insulin therapy.

Rapid fat catabolism elevates blood levels of free fatty acids and their breakdown products, the ketone bodies (acetoacetic acid, acetone, and β-hydroxybutyric acid). Ketonuria promotes

osmotic diuresis, flushes $Na^+$ and $K^+$ from the body, and creates electrolyte deficiencies that can lead to abdominal pain, vomiting, irregular heartbeat, and neurological dysfunction. As acids, ketones lower the pH of the blood and produce a condition called **ketoacidosis.** This causes a deep, gasping breathing called *Kussmaul*[39] *respiration*, typical of terminal diabetes. It also depresses the nervous system and produces diabetic coma.

DM also leads to long-term degenerative cardiovascular and neurological diseases—signs that were seldom seen before insulin therapy, when patients died too quickly to show the chronic effects. Through multiple, complex mechanisms, chronic hyperglycemia has devastating effects on small to medium blood vessels *(microvascular disease)*, including *atherosclerosis*, the obstruction of blood vessels by plaques of lipid and overgrown smooth muscle (see p. 753). Both types of DM also thicken the basement membrane of the blood vessels, interfering with the delivery of nutrients and hormones to the tissues and with the removal of their wastes. This leads to irreversible tissue degeneration in many organs. Two of the common complications of long-term DM are blindness and renal failure, brought on by arterial degeneration in the retinas and kidneys. Death from kidney failure is much more common in type 1 DM than in type 2. In type 2, the most common cause of death is heart failure stemming from coronary artery disease.

Another complication is *diabetic neuropathy*—nerve damage resulting from impoverished blood flow. This can lead to erectile dysfunction, incontinence, and loss of sensation from affected areas of the body. Microvascular disease in the skin results in poor healing of skin wounds, so even a minor break easily becomes ulcerated, infected, and gangrenous. This is especially common in the feet, because people take less notice of foot injuries; circulation is poorer in the feet (farthest from the heart) than anywhere else; pressure on the feet makes them especially susceptible to tissue injury; and neuropathy may make a person unaware of skin lesions or reluctant to consent to amputation of toes, or more, from which they feel no pain. DM outweighs all other reasons for the amputation of gangrenous appendages.

Diabetes mellitus is not the only kind of diabetes. Diabetes insipidus, a disease with no relation to insulin, has already been mentioned (p. 662), and other forms are discussed with the urinary system in chapter 23 (see p. 913).

---

**BEFORE YOU GO ON**

Answer the following questions to test your understanding of the preceding section:

28. Explain some causes of hormone hyposecretion, and give examples. Do the same for hypersecretion.

29. Why does a lack of dietary iodine lead to TSH hypersecretion? Why does the thyroid gland enlarge in endemic goiter?

30. In diabetes mellitus, explain the chain of events that lead to *(a)* osmotic diuresis, *(b)* ketoacidosis and coma, and *(c)* gangrene of the lower limbs.

---

[39]Adolph Kussmaul (1822–1902), German physician

| **TABLE 17.8** | **Some Disorders of the Endocrine System** |
|---|---|
| Addison[40] disease | Hyposecretion of adrenal glucocorticoids and mineralocorticoids, causing hypoglycemia, hypotension, weight loss, weakness, loss of stress resistance, darkening of the skin, and potentially fatal dehydration and electrolyte imbalances |
| Congenital hypothyroidism | Thyroid hormone hyposecretion present from birth, resulting in stunted physical development, thickened facial features, low body temperature, lethargy, and irreversible brain damage in infancy |
| Hyperinsulinism | Insulin excess caused by islet hypersecretion or injection of excess insulin, causing hypoglycemia, weakness, hunger, and sometimes *insulin shock*, which is characterized by disorientation, convulsions, or unconsciousness |
| Myxedema | A syndrome occurring in severe or prolonged adult hypothyroidism, characterized by low metabolic rate, sluggishness and sleepiness, weight gain, constipation, dry skin and hair, abnormal sensitivity to cold, hypertension, and tissue swelling |
| Pheochromocytoma | A tumor of the adrenal medulla that secretes excess epinephrine and norepinephrine. Causes hypertension, elevated metabolic rate, nervousness, indigestion, hyperglycemia, and glycosuria |
| Toxic goiter (Graves disease) | Thyroid hypertrophy and hypersecretion, occurring when autoantibodies mimic the effect of TSH and overstimulate the thyroid. Results in elevated metabolic rate and heart rate, nervousness, sleeplessness, weight loss, abnormal heat sensitivity and sweating, and bulging of the eyes (exophthalmos) resulting from eyelid retraction and edema of the orbital tissues |

**Disorders described elsewhere**

| | | |
|---|---|---|
| Acromegaly p. 662 | Diabetes insipidus p. 662 | Hyperparathyroidism p. 663 |
| Adrenogenital syndrome p. 663 | Diabetes mellitus p. 664 | Hypoparathyroidism p. 663 |
| Androgen-insensitivity syndrome p. 1030 | Endemic goiter p. 663 | Pituitary dwarfism p. 662 |
| Cushing syndrome p. 663 | Gigantism p. 662 | |

[40]Thomas Addison (1793–1860), English physician

# DEEPER INSIGHT 17.4

## MEDICAL HISTORY

### The Discovery of Insulin

At the start of the twentieth century, physicians felt nearly helpless in the face of diabetes mellitus. They put patients on useless diets—the oatmeal cure, the potato cure, and others—or on starvation diets as low as 750 cal per day so as not to "stress the system." They were resigned to the fact that their patients were doomed to die, and simple starvation seemed to produce the least suffering.

After the cause of diabetes was traced to the pancreatic islets in 1901, European researchers tried treating patients and experimental animals with extracts of pancreas, but became discouraged by the severe side effects of impurities in the extracts. They lacked the resources to pursue the problem to completion, and by 1913, the scientific community showed signs of giving up on diabetes.

But in 1920, Frederick Banting (1891–1941), a young Canadian physician with a failing medical practice, became intrigued with a possible method for isolating the islets from the pancreas and testing extracts of the islets alone. He returned to his alma mater, the University of Toronto, to present his idea to Professor J. J. R. Macleod (1876–1935), a leading authority on carbohydrate metabolism. Macleod was unimpressed with Banting, finding his knowledge of the diabetes literature and scientific method superficial. Nevertheless, he felt Banting's idea was worth pursuing and thought that with his military surgical training, Banting might be able to make some progress where others

had failed. He offered Banting laboratory space for the summer, giving him a marginal chance to test his idea. Banting was uncertain whether to accept, but when his fiancée broke off their engagement and an alternative job offer fell through, he closed his medical office, moved to Toronto, and began work. Little did either man realize that in 2 years' time, they would share a Nobel Prize and yet so thoroughly detest each other they would scarcely be on speaking terms.

#### A Modest Beginning

Macleod advised Banting on an experimental plan of attack and gave him an assistant, Charles Best (1899–1978). Best had just received his B.A. in physiology and looked forward to an interesting summer job with Banting before starting graduate school. Over the summer of 1921, they removed the pancreases from some dogs to render them diabetic and tied off the pancreatic ducts in other dogs to make most of the pancreas degenerate while leaving the islets intact. Their plan was to treat the diabetic dogs with extracts made from the degenerated pancreases of the others.

It was a difficult undertaking. Their laboratory was tiny, filthy, unbearably hot, and reeked of dog excrement. The pancreatic ducts were very small and difficult to tie, and it was hard to tell if all pancreatic tissue had been removed from the dogs intended to become diabetic. Several dogs died of overanesthesia, infection, and bleeding from Banting's clumsy surgical technique. Banting was also careless in reading his data and interpreting the results and had little interest in reading the literature to see what other researchers were doing. In Banting and Best's first publication, in early 1922, the data in their discussion disagreed with the data in the tables, and both disagreed with the data in their laboratory notebooks. These were not the signs of promising researchers.

In spite of themselves, Banting and Best achieved modest positive results over the summer. Crude extracts brought one dog back from a diabetic coma and reduced the hyperglycemia and glycosuria of others. Buoyed by these results, Banting demanded a salary, a better laboratory, and another assistant. Macleod grudgingly obtained salaries for the pair, but Banting began to loathe him for their disagreement over his demands, and he and Macleod grew in mutual contempt as the project progressed.

## Success and Conflict

Macleod brought biochemist J. B. Collip (1892–1965) into the project that fall in hopes that he could produce purer extracts. More competent in experimental science, Collip was the first to show that pancreatic extracts could eliminate ketosis and restore the liver's ability to store glycogen. He obtained better and better results in diabetic rabbits until, by January 1922, the group felt ready for human trials. Banting was happy to have Collip on the team initially, but grew intensely jealous of him as Collip not only achieved better results than he had, but also developed a closer relationship with Macleod. Banting, who had no qualifications to perform human experiments, feared he would be pushed aside as the project moved to its clinical phase. At one point, the tension between Banting and Collip nearly erupted into a fistfight in the laboratory.

Banting insisted that the first human trial be done with an extract he and Best prepared, not with Collip's. The patient was a 14-year-old boy who weighed only 29 kg (65 lb) and was on the verge of death. He was injected on January 11 with the Banting and Best extract, described by one observer as "a thick brown muck." The trial was an embarrassing failure, with only a slight lowering of his blood glucose and a severe reaction to the impurities in the extract. On January 23, the same boy was treated again, but with Collip's extract. This time, his ketonuria and glycosuria were almost completely eliminated and his blood sugar dropped 77%. This was the first successful clinical trial of insulin. Six more patients were treated in February 1922 and quickly became stronger, more alert, and in better spirits. In April, the Toronto group began calling the product *insulin,* and at a medical conference in May, they gave the first significant public report of their success.

Banting felt increasingly excluded from the project. He quit coming to the laboratory, stayed drunk much of the time, and day-dreamed of leaving diabetes research to work on cancer. He remained only because Best pleaded with him to stay. Banting briefly operated a private diabetes clinic, but fearful of embarrassment over alienating the discoverer of insulin, the university soon lured him back with a salaried appointment and hospital privileges.

Banting had a number of high-profile, successful cases in 1922, such as 14-year-old Elizabeth Hughes, who weighed only 20 kg (45 lb) before treatment. She began treatment in August and showed immediate, dramatic improvement. She was a spirited, optimistic, and articulate girl who kept enthusiastic diaries of being allowed to eat bread, potatoes, and macaroni and cheese for the first time since the onset of her illness. "Oh it is simply too wonderful for words this stuff," she exuberantly wrote to her mother—even though the still-impure extracts caused her considerable pain and swelling. The world quickly beat a path to Toronto begging for insulin. The pharmaceutical firm of Eli Lilly and Company entered into an agreement with the University of Toronto for the mass production of insulin, and by the fall of 1923, over 25,000 patients were being treated at more than 60 Canadian and U.S. clinics.

## The Bitter Fruits of Success

Banting's self-confidence was restored. He had become a public hero, and the Canadian Parliament awarded him an endowment generous enough to ensure him a life of comfort. Several distinguished physiologists nominated Banting and Macleod for the 1923 Nobel Prize, and they won. When the award was announced, Banting was furious about having to share it with Macleod. At first, he threatened to refuse it, but when he cooled down, he announced that he would split his share of the prize money with Best. Macleod quickly announced that half of his share would go to Collip.

Interestingly, Romanian physiologist Nicolae Paulescu (1869–1931) succeeded in isolating insulin (which he called pancreine) and treating diabetic dogs with it in 1916, years before Banting even conceived or began his work. Paulescu published four papers on it in April 1921, 8 months before Banting and Best published their first, and he patented his method of isolating insulin in April 1922. Paulescu's work never advanced to clinical trials on humans, however, and was overlooked by the Nobel Committee.

As a recipient of Canada's first Nobel Prize, Banting basked in his stature as a national hero. He made life at the university so unbearable to Macleod, however, that Macleod left in 1928 to accept a university post in Scotland. Banting stayed on at Toronto. Although now wealthy and surrounded by admiring students, he achieved nothing significant in science for the rest of his career. He was killed in a plane crash in 1941. Best replaced Macleod on the Toronto faculty, led a distinguished career, and developed the anticoagulant heparin. Collip went on to play a lead role in the isolation of PTH, ACTH, and other hormones.

Insulin made an industry giant of Eli Lilly and Company. It became the first protein whose amino acid sequence was determined, for which Frederick Sanger received a Nobel Prize in 1958. Diabetics today no longer depend on a limited supply of insulin extracted from beef and pork pancreas. Human insulin is now in plentiful supply, made by genetically engineered bacteria. Paradoxically, while insulin has dramatically reduced the suffering caused by diabetes mellitus, it has increased the number of people who have the disease—because thanks to insulin, diabetics are now able to live long enough to raise families and pass on the diabetes genes.

# CONNECTIVE ISSUES

## Effects of the ENDOCRINE SYSTEM
## On Other Organ Systems

**ALL SYSTEMS**
The development and metabolism of most tissues are affected by growth hormone, insulin, insulin-like growth factors, thyroid hormone, and glucocorticoids.

**INTEGUMENTARY SYSTEM**
Sex hormones affect skin pigmentation, development of body hair and apocrine glands, and subcutaneous fat deposition.

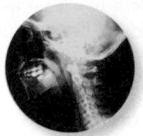

**SKELETAL SYSTEM**
Skeletal growth and maintenance are regulated by numerous hormones—calcitonin, calcitriol, parathyroid hormone, growth hormone, estrogen, testosterone, and others.

**MUSCULAR SYSTEM**
Growth hormone and testosterone stimulate muscular growth; insulin regulates glucose uptake by muscle; other hormones regulate the electrolyte balances that are important in muscular contraction.

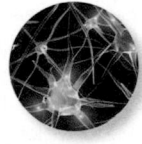

**NERVOUS SYSTEM**
Hormones exert negative feedback inhibition on the hypothalamus; several hormones affect nervous system development, mood, and behavior; hormones regulate the electrolyte balances that are important in neuron function.

**CIRCULATORY SYSTEM**
Angiotensin II, aldosterone, antidiuretic hormone, natriuretic peptides, and other hormones regulate blood volume and pressure; erythropoietin stimulates RBC production; thymic hormones stimulate WBC production; thrombopoietin stimulates platelet production; epinephrine, thyroid hormone, and other hormones affect the rate and force of the heartbeat.

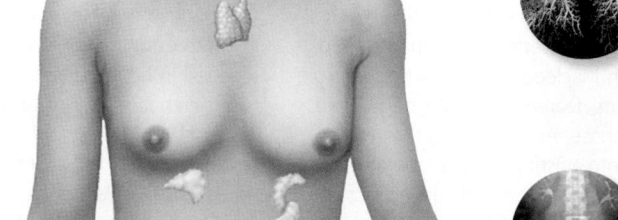
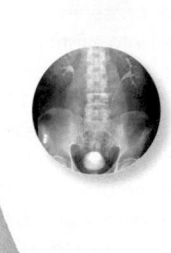
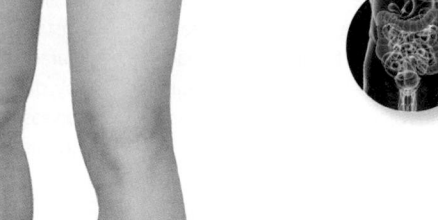
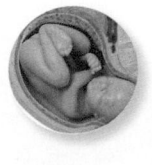

**LYMPHATIC/IMMUNE SYSTEM**
Thymic hormones activate immune cells; glucocorticoids suppress immunity and inflammation.

**RESPIRATORY SYSTEM**
Epinephrine and norepinephrine dilate the bronchioles and increase pulmonary airflow.

**URINARY SYSTEM**
Antidiuretic hormone regulates urine volume; calcitriol, parathyroid hormone, aldosterone, and natriuretic peptides regulate electrolyte absorption by the kidneys.

**DIGESTIVE SYSTEM**
Insulin and glucagon regulate nutrient storage and metabolism; enteric hormones control gastrointestinal secretion and motility; gut–brain peptides affect appetite and regulate food intake and body weight.

**REPRODUCTIVE SYSTEM**
Gonadotropins and sex steroids regulate sexual development, spermatogenesis and oogenesis, the ovarian and uterine cycles, sex drive, pregnancy, fetal development, and lactation.

# STUDY GUIDE

## ► Assess Your Learning Outcomes

*To test your knowledge, discuss the following topics with a study partner or in writing, ideally from memory.*

### 17.1 Overview of the Endocrine System (p. 630)

1. The importance of intercellular communication for survival, and the body's four mechanisms of intercellular communication
2. The general term for the cells and glands that secrete hormones, and the name of that branch of science and medicine that specializes in hormones
3. How endocrine glands differ from exocrine glands
4. Similarities, differences, and interactions between the nervous and endocrine systems
5. The term for organs or cells that are influenced by a given hormone, and why they are the only ones to respond to it even though the hormone travels throughout the body

### 17.2 The Hypothalamus and Pituitary Gland (p. 633)

1. Why the hypothalamus should be considered part of the endocrine system
2. The anatomical relationship of the hypothalamus to the pituitary gland; the two major parts of the pituitary; and how the hypothalamus communicates with each
3. Six hormones that are secreted by the hypothalamus to regulate the anterior pituitary, and their effects
4. Two hormones synthesized in the hypothalamus and stored in the posterior pituitary; how they get to the pituitary; and how their later release into the bloodstream is controlled
5. Six hormones secreted by the anterior pituitary, their abbreviations, and their target organs and functions
6. Two hormones secreted by the posterior pituitary, their abbreviations, their target organs and effects, and the role of neuroendocrine reflexes in their release

7. Examples and mechanisms of positive and negative feedback control of the hypothalamus and pituitary
8. The actions of growth hormone (GH) and the role of insulin-like growth factors in its effects

### 17.3 Other Endocrine Glands (p. 640)

1. Anatomy of the pineal gland; its involution; and its hormone and function
2. Anatomy of the thymus; its involution; and its hormones and functions
3. Anatomy of the thyroid gland; its hormones and functions; and the cells that produce each hormone
4. Anatomy of the parathyroid glands; their hormone and function
5. Anatomy of the adrenal glands, and structural differences between the cortex and medulla
6. Hormones and functions of the adrenal medulla
7. Three tissue zones of the adrenal cortex, the hormones of each zone, and their functions
8. Pancreatic islets and their cell types, hormones, and functions
9. Endocrine components of the ovaries and testes, and their hormones and functions
10. Hormones produced by the following tissues and organs, and their effects: the skin, liver, kidneys, heart, digestive tract, adipose tissue, osseous tissue, and placenta

### 17.4 Hormones and Their Actions (p. 650)

1. Three main chemical classes of hormones and examples of each
2. The synthesis of steroid hormones
3. The synthesis of peptide hormones such as insulin
4. Two amino acids that serve as hormone precursors and which hormones are produced from each
5. Thyroid hormone synthesis and secretion
6. The problem that must be overcome to transport thyroid hormone (TH) and steroid hormones in the blood, and how the transport mechanism affects their half-life

7. Where hormone receptors are located in the target cells, and differences between receptor systems for hydrophilic and hydrophobic hormones
8. Which hormones require second messengers to activate a target cell; how second messengers work, especially cAMP, DAG, and $IP_3$
9. How signal amplification enables small amounts of hormone to produce great physiological effects
10. How target cells modulate their hormone sensitivity
11. Three kinds of interactions that can occur when two or more of them act simultaneously on a target cell
12. How hormones are inactivated and cleared from the blood after completing their task

### 17.5 Stress and Adaptation (p. 659)

1. The physiological or medical definition of *stress*
2. Definition of the *stress response (general adaptation syndrome)*
3. The three stages of the stress response; the dominant hormones and physiological effects of each stage; and what marks the transition from one stage to the next

### 17.6 Eicosanoids and Paracrine Signaling (p. 660)

1. Paracrine secretions, examples, and how they compare and contrast with hormones
2. The general structure and metabolic precursor of eicosanoids
3. Synthesis and effects of leukotrienes
4. Synthesis and effects of the three cyclooxygenase (COX) products: prostacyclin, thromboxanes, and prostaglandins
5. How prostaglandins are named

### 17.7 Endocrine Disorders (p. 662)

1. Effects of growth hormone hyposecretion and hypersecretion, and how the effects differ between adult versus childhood onset
2. Myxedema, endemic goiter, and toxic goiter

# STUDY GUIDE

3. Effects of hypo- and hyperparathyroidism
4. Cushing syndrome and adrenogenital syndrome
5. The "three polys" of diabetes mellitus (DM), and three clinical findings that typically confirm DM
6. The mechanism of glycosuria and osmotic diuresis typical of DM; how this relates to the transport maximum ($T_m$) of carrier-mediated transport
7. Differences between the cause, pathology, and treatment of types 1 and 2 DM
8. Consequences of inadequately treated DM and why each of its many pathological effects occurs

## ► Testing Your Recall

*Answers in Appendix B*

1. CRH secretion would *not* raise the blood concentration of
   a. ACTH.
   b. thyroxine.
   c. cortisol.
   d. corticosterone.
   e. glucose.

2. Which of the following hormones has the least in common with the others?
   a. adrenocorticotropic hormone
   b. follicle-stimulating hormone
   c. thyrotropin
   d. thyroxine
   e. prolactin

3. Which hormone would no longer be secreted if the hypothalamo–hypophyseal tract were destroyed?
   a. oxytocin
   b. follicle-stimulating hormone
   c. growth hormone
   d. adrenocorticotropic hormone
   e. corticosterone

4. Which of the following is *not* a hormone?
   a. prolactin
   b. prolactin-inhibiting hormone
   c. thyroxine-binding globulin
   d. atrial natriuretic peptide
   e. cortisol

5. Where are the receptors for insulin located?
   a. in the pancreatic beta cells
   b. in the blood plasma
   c. on the target-cell membrane
   d. in the target-cell cytoplasm
   e. in the target-cell nucleus

6. What would be the consequence of defective ADH receptors?
   a. diabetes mellitus
   b. adrenogenital syndrome
   c. dehydration
   d. seasonal affective disorder
   e. none of these

7. Which of these has more exocrine than endocrine tissue?
   a. the pineal gland
   b. the adenohypophysis
   c. the thyroid gland
   d. the pancreas
   e. the adrenal gland

8. Which of these cells stimulate bone deposition?
   a. alpha cells
   b. beta cells
   c. C cells
   d. G cells
   e. T cells

9. Which of these hormones relies on cAMP as a second messenger?
   a. ACTH
   b. progesterone
   c. thyroxine
   d. testosterone
   e. estrogen

10. Prostaglandins are derived from
    a. phospholipase.
    b. cyclooxygenase.
    c. leukotriene.
    d. lipoxygenase.
    e. arachidonic acid.

11. The _____ develops from a pouch in the pharynx of the embryo.

12. Thyroxine ($T_4$) is synthesized by combining two iodinated molecules of the amino acid _____.

13. Growth hormone hypersecretion in adulthood causes a disease called _____.

14. The dominant hormone in the stage of resistance of the stress response is _____.

15. Adrenal steroids that regulate glucose metabolism are collectively called _____.

16. Testosterone is secreted by the _____ cells of the testis.

17. Target cells can reduce pituitary secretion by a process called _____.

18. Hypothalamic releasing factors are delivered to the anterior pituitary by way of a network of blood vessels called the _____.

19. A hormone is said to have a/an _____ effect when it stimulates the target cell to develop receptors for other hormones to follow.

20. _____ is a process in which a cell increases its number of receptors for a hormone, thus increasing its hormone sensitivity and response.

# STUDY GUIDE

## ► Building Your Medical Vocabulary

*Answers in Appendix B*

*State a medical meaning of each word element below, and give a term in which it or a slight variation of it is used.*

1. adeno-

2. chole-

3. diabet-

4. eicosa-

5. luteo-

6. -oid

7. -osa

8. pro-

9. tropo-

10. uri-

## ► True or False

*Answers in Appendix B*

*Determine which five of the following statements are false, and briefly explain why.*

1. Castration would raise a man's blood gonadotropin concentration.

2. Hormones in the glycoprotein class cannot have cytoplasmic or nuclear receptors in their target cells.

3. Epinephrine and thyroid hormone have the same effects on metabolic rate and blood pressure.

4. Tumors can lead to either hyposecretion or hypersecretion of various hormones.

5. All hormones are secreted by endocrine glands.

6. An atherosclerotic deposit that blocked blood flow in the hypophyseal portal system would cause the testes and ovaries to malfunction.

7. The pineal gland and thymus are larger in adults than in children.

8. A deficiency of dietary iodine would lead to negative feedback inhibition of TSH synthesis.

9. The tissue at the center of the adrenal gland is called the zona reticularis.

10. Of the endocrine organs covered in this chapter, only the adrenal glands are paired; the rest are single.

## ► Testing Your Comprehension

*Answers at* www.mhhe.com/saladin7

1. Propose a model of signal amplification for the effect of a steroid hormone and construct a diagram similar to figure 17.21 for your model. It may help to review protein synthesis in chapter 4.

2. Suppose you were browsing in a health-food store and saw a product advertised: "Put an end to heart disease. This herbal medicine will rid your body of cholesterol!" Would you buy it? Why or why not? If the product were as effective as claimed, what are some other effects it would produce?

3. A person with toxic goiter tends to sweat profusely. Explain this in terms of homeostasis.

4. How is the action of a peptide hormone similar to the action of the neurotransmitter norepinephrine?

5. A young man is involved in a motorcycle accident that fractures his sphenoid bone. Shortly thereafter, he begins to excrete enormous amounts of urine, up to 30 liters per day, and suffers intense thirst. His neurologist diagnoses the problem as diabetes insipidus. Explain how his head injury resulted in these effects on urinary function and thirst. Why would a sphenoid fracture be more likely than an occipital bone fracture to cause diabetes insipidus? What hormone imbalance resulted from this accident? Would you expect to find elevated glucose in the urine of this diabetic patient? Why or why not?

## ► Improve Your Grade at www.mhhe.com/saladin7

*Download animations to study when it fits your schedule. Practice quizzes, labeling activities, games, and flashcards offer fun ways to master the chapter concepts. Or, download image PowerPoint files for each chapter to create a study guide or for taking notes during lecture.*

# 18

# THE CIRCULATORY SYSTEM: BLOOD

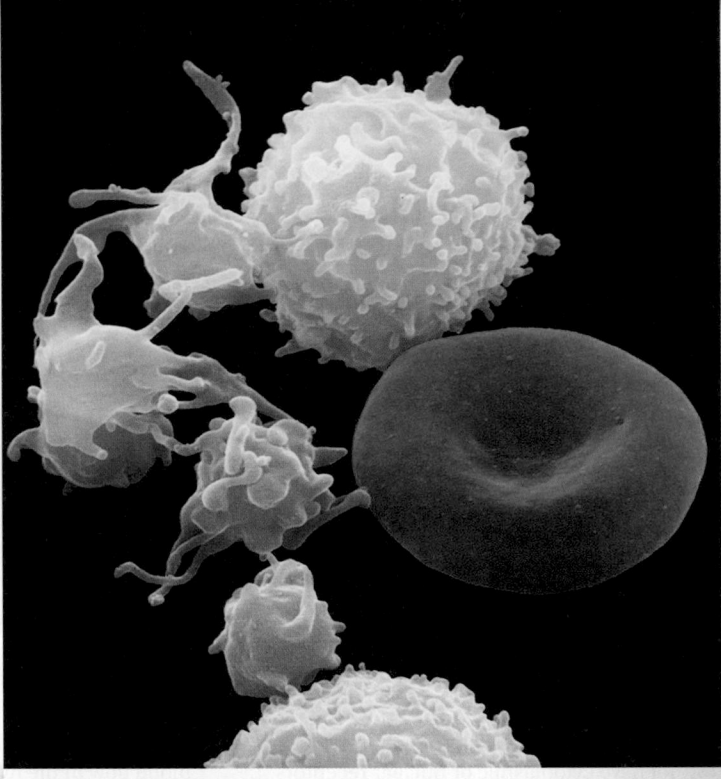

A red blood cell, white blood cells, and four platelets (SEM)

## CHAPTER OUTLINE

## DEEPER INSIGHTS

Anatomy & Physiology | REVEALED®
aprevealed.com

**Module 9: Cardiovascular System**

**B**lood has always had a special mystique. From time immemorial, people have seen blood flow from the body and, with it, the life of the individual. It is no wonder that blood was thought to carry a mysterious "vital force." Ancient Romans drank the blood of fallen gladiators in a belief that they could thus acquire the gladiator's vitality. Even today, we become especially alarmed when we find ourselves bleeding, and the emotional impact of blood is enough to make many people faint at the sight of it. From ancient Egypt to nineteenth-century America, physicians drained "bad blood" from their patients to treat everything from gout to headaches, from menstrual cramps to mental illness. It was long thought that hereditary traits were transmitted through the blood, and people still use such unfounded expressions as "I have one-quarter Cherokee blood."

Scarcely anything meaningful was known about blood until its cells were seen with the first microscopes. Even though blood is a uniquely accessible tissue, most of what we know about it dates only to the last 50 years. Recent developments in this field have empowered us to save and improve the lives of countless people who would otherwise suffer or die.

## 18.1 Introduction

### Expected Learning Outcomes
When you have completed this section, you should be able to

a. describe the functions and major components of the circulatory system;

b. describe the components and physical properties of blood;

c. describe the composition of blood plasma;

d. explain the significance of blood viscosity and osmolarity; and

e. describe in general terms how blood is produced.

## Functions of the Circulatory System

The **circulatory system** consists of the heart, blood vessels, and blood. The term **cardiovascular**[1] **system** refers only to the heart and vessels, which are the subject of chapters 19 and 20. The study of blood, specifically, is called **hematology.**[2]

The fundamental purpose of the circulatory system is to transport substances from place to place in the body. Blood is the liquid medium in which these materials travel, blood vessels ensure the proper routing of blood to its destinations, and the heart is the pump that keeps the blood flowing.

More specifically, the functions of the circulatory system are as follows:

*Transport*
- The blood carries oxygen from the lungs to all of the body's tissues, while it picks up carbon dioxide from those tissues and carries it to the lungs to be removed from the body.

- It picks up nutrients from the digestive tract and delivers them to all of the body's tissues.

- It carries metabolic wastes to the kidneys for removal.

- It carries hormones from endocrine cells to their target organs.

- It transports a variety of stem cells from the bone marrow and other origins to the tissues where they lodge and mature.

*Protection*
- The blood plays several roles in inflammation, a mechanism for limiting the spread of infection.

- White blood cells destroy microorganisms and cancer cells.

- Antibodies and other blood proteins neutralize toxins and help to destroy pathogens.

- Platelets secrete factors that initiate blood clotting and other processes for minimizing blood loss.

*Regulation*
- By absorbing or giving off fluid under different conditions, the blood capillaries help to stabilize fluid distribution in the body.

- By buffering acids and bases, blood proteins help to stabilize the pH of the extracellular fluids.

- Shifts in blood flow help to regulate body temperature by routing blood to the skin for heat loss or retaining it deeper in the body to conserve heat.

Considering the importance of efficiently transporting nutrients, wastes, hormones, and especially oxygen from place to place, it is easy to understand why an excessive loss of blood is quickly fatal, and why the circulatory system needs mechanisms for minimizing such losses.

---

[1]*cardio* = heart; *vas* = vessel
[2]*hem, hemato* = blood; *logy* = study of

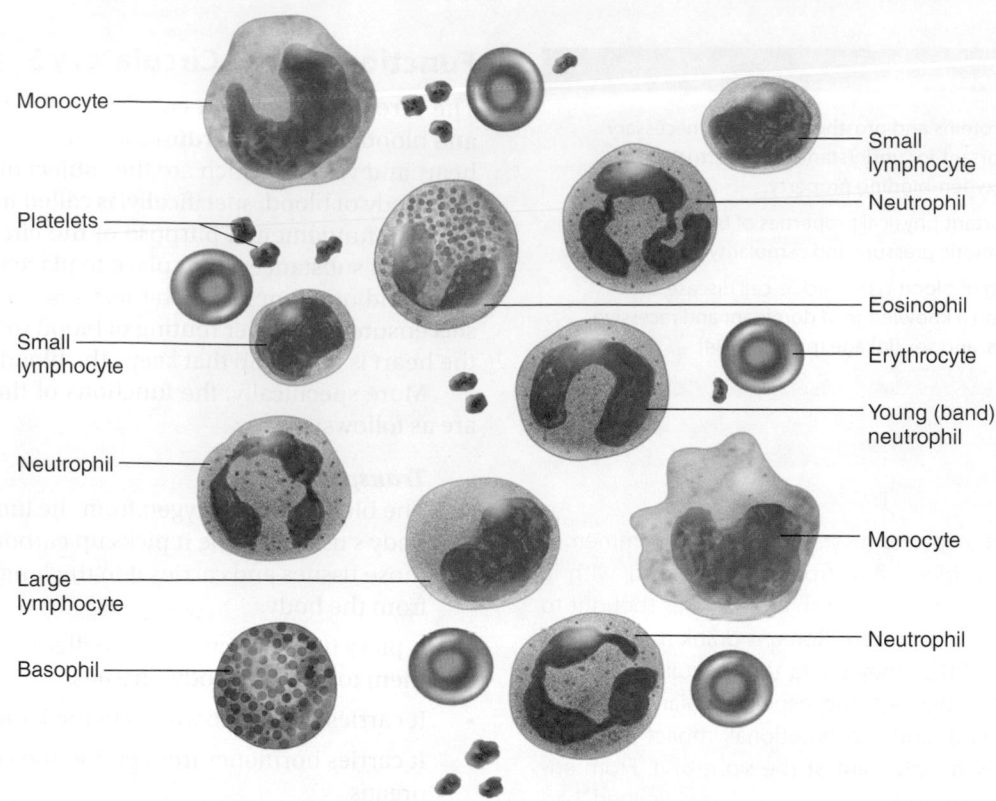

**FIGURE 18.1** **The Formed Elements of Blood.**

 *What do erythrocytes and platelets lack that the other formed elements have?*

## Components and General Properties of Blood

All of the foregoing functions depend, of course, on the characteristics of the blood. Adults generally have about 4 to 6 liters of blood. It is a liquid connective tissue composed, like other connective tissues, of cells and an extracellular matrix. The matrix is the blood **plasma,** a clear, light yellow fluid constituting a little over half of the blood volume. Suspended in the plasma are the **formed elements**—cells and cell fragments including the red blood cells, white blood cells, and platelets (fig. 18.1). The term *formed element* alludes to the fact that these are membrane-enclosed bodies with a definite structure visible with the microscope. Strictly speaking, they cannot all be called cells because the platelets, as explained later, are merely fragments torn from certain bone marrow cells.

The formed elements are classified as follows:

Erythrocytes[3] (red blood cells, RBCs)

Platelets

Leukocytes[4] (white blood cells, WBCs)

    Granulocytes

        Neutrophils

        Eosinophils

        Basophils

    Agranulocytes

        Lymphocytes

        Monocytes

Thus, there are seven kinds of formed elements: the erythrocytes, platelets, and five kinds of leukocytes. The five leukocyte types are divided into two categories, the *granulocytes* and *agranulocytes,* on grounds explained later.

The ratio of formed elements to plasma can be seen by taking a sample of blood in a tube and spinning it for a few minutes in a centrifuge (fig. 18.2). Erythrocytes, the densest elements, settle to the bottom of the tube and typically constitute about 37% to 52% of the total volume—a value called the *hematocrit.* WBCs and platelets settle into a narrow cream- or buff-colored zone called the *buffy coat* just above the RBCs; they total 1% or less of the blood volume. At the top of the tube is the plasma, which accounts for about 47% to 63% of the volume.

Table 18.1 lists several properties of the blood. Some of the terms in that table are defined later in the chapter.

**▶▶▶ APPLY WHAT YOU KNOW**

*Based on your body weight, estimate the volume (in liters) and weight (in kilograms) of your own blood, using the data in table 18.1.*

---

[3]*erythro* = red; *cyte* = cell
[4]*leuko* = white; *cyte* = cell

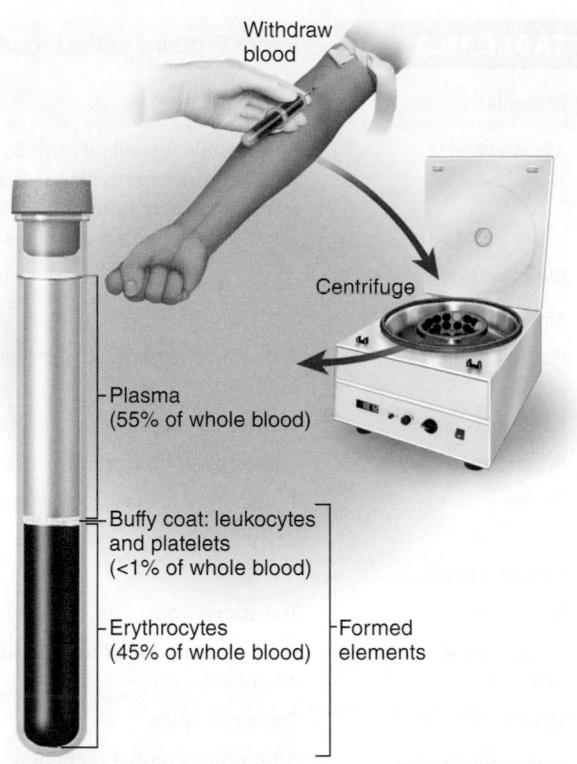

Withdraw
blood

Centrifuge

Plasma
(55% of whole blood)

Buffy coat: leukocytes
and platelets
(<1% of whole blood)

Erythrocytes — Formed
(45% of whole blood) elements

**FIGURE 18.2 The Components of Blood.** Centrifuging a sample of blood separates the erythrocytes from the white cells and platelets (buffy coat) and plasma. The hematocrit is the percentage of the volume composed of erythrocytes.

| TABLE 18.1 | General Properties of Blood |
|---|---|
| **Characteristic** | **Typical Values for Healthy Adults\*** |
| Mean fraction of body weight | 8% |
| Volume in adult body | Female: 4–5 L; male: 5–6 L |
| Volume/body weight | 80–85 mL/kg |
| Mean temperature | 38°C (100.4°F) |
| pH | 7.35–7.45 |
| Viscosity (relative to water) | Whole blood: 4.5–5.5; plasma: 2.0 |
| Osmolarity | 280–296 mOsm/L |
| Mean salinity (mainly NaCl) | 0.9% |
| Hematocrit (packed cell volume) | Female: 37% to 48%<br>Male: 45% to 52% |
| Hemoglobin | Female: 12–16 g/dL<br>Male: 13–18 g/dL |
| Mean RBC count | Female: 4.2–5.4 million/μL<br>Male: 4.6–6.2 million/μL |
| Platelet count | 130,000–360,000/μL |
| Total WBC count | 5,000–10,000/μL |

\*Values vary slightly depending on the testing methods used.

## Blood Plasma

Even though blood plasma has no anatomy that we can study visually, we cannot ignore its importance as the matrix of this liquid connective tissue we call blood. Plasma is a complex mixture of water, proteins, nutrients, electrolytes, nitrogenous wastes, hormones, and gases (table 18.2). When the blood clots and the solids are removed, the remaining fluid is the blood **serum.** Serum is essentially identical to plasma except for the absence of the clotting protein fibrinogen.

Protein is the most abundant plasma solute by weight, totaling 6 to 9 g/dL. Plasma proteins play a variety of roles including clotting, defense, and transport of other solutes such as iron, copper, lipids, and hydrophobic hormones. There are three major categories of plasma proteins: albumin, globulins, and fibrinogen (table 18.3). Many other plasma proteins are indispensable to survival, but they account for less than 1% of the total.

**Albumin** is the smallest and most abundant plasma protein. It serves to transport various solutes and buffer the pH of the plasma. It also makes a major contribution to two physical properties of blood: its *viscosity* and *osmolarity,* discussed shortly. Through its effects on these two variables, changes in albumin concentration can significantly affect blood volume, pressure, and flow. **Globulins** are divided into three subclasses; from smallest to largest in molecular weight, they are the alpha (α), beta (β), and gamma (γ) globulins. Globulins play various roles in solute transport, clotting, and immunity. **Fibrinogen** is a soluble precursor of *fibrin,* a sticky protein that forms the framework of a blood clot. Some of the other plasma proteins are enzymes involved in the clotting process.

The liver produces as much as 4 g of plasma protein per hour, contributing all of the major proteins except gamma globulins. The gamma globulins come from *plasma cells—* connective tissue cells that are descended from white blood cells called *B lymphocytes.*

### ▶▶▶ APPLY WHAT YOU KNOW

*How could a disease such as liver cancer or hepatitis result in impaired blood clotting?*

In addition to protein, the blood plasma contains such nitrogen-containing compounds as free amino acids and nitrogenous wastes. **Nitrogenous wastes** are toxic end products of catabolism. The most abundant is *urea,* a product of amino acid catabolism. These wastes are normally excreted by the kidneys at a rate that balances their production.

The plasma also transports nutrients absorbed by the digestive tract, including glucose, amino acids, fats, cholesterol, phospholipids, vitamins, and minerals. It transports dissolved oxygen, carbon dioxide, and nitrogen (see table 18.2). The dissolved nitrogen normally has no physiological role in the body (but see Deeper Insight 22.5, p. 884).

Electrolytes are another important component of the blood plasma. Sodium ions constitute about 90% of the plasma cations. Sodium is more important than any other solute for

| TABLE 18.2 | Composition of Blood Plasma |
|---|---|
| **Blood Component*** | **Typical Values for Healthy Adults** |
| *Water* | 92% by weight |
| *Proteins* | Total 6–9 g/dL |
| Albumin | 60% of total protein, 3.2–5.5 g/dL |
| Globulins | 36% of total protein, 2.3–3.5 g/dL |
| Fibrinogen | 4% of total protein, 0.2–0.3 g/dL |
| **Nutrients** | |
| Glucose (dextrose) | 70–110 mg/dL |
| Amino acids | 33–51 mg/dL |
| Lactic acid | 6–16 mg/dL |
| Total lipid | 450–850 mg/dL |
| Cholesterol | 120–220 mg/dL |
| Fatty acids | 190–420 mg/dL |
| High-density lipoprotein (HDL) | 30–80 mg/dL |
| Low-density lipoprotein (LDL) | 62–185 mg/dL |
| Triglycerides (neutral fats) | 40–150 mg/dL |
| Phospholipids | 6–12 mg/dL |
| Iron | 50–150 µg/dL |
| Trace elements | Traces |
| Vitamins | Traces |
| **Electrolytes** | |
| Sodium ($Na^+$) | 135–145 mEq/L |
| Calcium ($Ca^{2+}$) | 9.2–10.4 mEq/L |
| Potassium ($K^+$) | 3.5–5.0 mEq/L |
| Magnesium ($Mg^{2+}$) | 1.3–2.1 mEq/L |
| Chloride ($Cl^-$) | 100–106 mEq/L |
| Bicarbonate ($HCO_3^-$) | 23.1–26.7 mEq/L |
| Phosphate ($HPO_4^{2-}$) | 1.4–2.7 mEq/L |
| Sulfate ($SO_4^{2-}$) | 0.6–1.2 mEq/L |
| **Nitrogenous wastes** | |
| Urea | 10–20 mg/dL |
| Uric acid | 1.5–8.0 mg/dL |
| Creatinine | 0.6–1.5 mg/dL |
| Creatine | 0.2–0.8 mg/dL |
| Ammonia | 0.02–0.09 mg/dL |
| Bilirubin | 0–1.0 mg/dL |
| **Other components** | |
| Dissolved $CO_2$ | 2.62 mL/dL |
| Dissolved $O_2$ | 0.29 mL/dL |
| Dissolved $N_2$ | 0.98 mL/dL |
| Enzymes of diagnostic value | — |
| Hormones | — |

*This table is limited to substances of greatest relevance to this and later chapters. Concentrations refer to plasma only, not to whole blood.

| TABLE 18.3 | Major Proteins of the Blood Plasma |
|---|---|
| **Proteins** | **Functions** |
| *Albumin (60%)** | Responsible for colloid osmotic pressure; major contributor to blood viscosity; transports lipids, hormones, calcium, and other solutes; buffers blood pH |
| *Globulins (36%)** | |
| **Alpha (α) globulins** | |
| Haptoglobulin | Transports hemoglobin released by dead erythrocytes |
| Ceruloplasmin | Transports copper |
| Prothrombin | Promotes blood clotting |
| Others | Transport lipids, fat-soluble vitamins, and hormones |
| **Beta (β) globulins** | |
| Transferrin | Transports iron |
| Complement proteins | Aid in destruction of toxins and microorganisms |
| Others | Transport lipids |
| **Gamma (γ) globulins** | Antibodies; combat pathogens |
| *Fibrinogen (4%)** | Becomes fibrin, the major component of blood clots |

*Mean percentage of the total plasma protein by weight

the osmolarity of the blood. As such, it has a major influence on blood volume and pressure; people with high blood pressure are often advised to limit their sodium intake. Electrolyte concentrations are carefully regulated by the body and have rather stable concentrations in the plasma.

## Blood Viscosity and Osmolarity

Two important properties of blood—viscosity and osmolarity—arise from the formed elements and plasma composition. **Viscosity** is the resistance of a fluid to flow, resulting from the cohesion of its particles. Loosely speaking, it is the thickness or stickiness of a fluid. At a given temperature, mineral oil is more viscous than water, for example, and honey is more viscous than mineral oil. Whole blood is 4.5 to 5.5 times as viscous as water, mainly because of the RBCs; plasma alone is 2.0 times as viscous as water, mainly because of its protein. Viscosity is important in circulatory function because it partially governs the flow of blood through the vessels. An RBC or protein deficiency reduces viscosity and causes blood to flow too easily, whereas an excess causes blood to flow too sluggishly. Either of these conditions puts a strain on the heart that may lead to serious cardiovascular problems if not corrected.

The **osmolarity** of blood (total molarity of dissolved particles that cannot pass through the blood vessel wall) is another important factor in cardiovascular function. In order to nourish surrounding cells and remove their wastes, substances must pass between the bloodstream and tissue fluid through the capillary walls. This transfer of fluids depends on a balance between the filtration of fluid from the capillary and its reabsorption by osmosis (see chapter 20). The rate of reabsorption is governed by the relative osmolarity of the blood versus the tissue fluid. If the blood osmolarity is too high, the bloodstream absorbs too much water. This raises the blood volume, resulting in high blood pressure and a potentially dangerous strain on the heart and arteries. If its osmolarity drops too low, too much water remains in the tissues. They become edematous (swollen) and the blood pressure may drop to dangerously low levels because of the water lost from the bloodstream.

It is therefore important that the blood maintain an optimal osmolarity. The osmolarity of the blood is a product mainly of its sodium ions, protein, and erythrocytes. The contribution of protein to blood osmotic pressure—called the **colloid osmotic pressure (COP)**—is especially important, as we see from the effects of extremely low-protein diets (see Deeper Insight 18.1).

## How Blood Is Produced AP|R

We lose blood continually, not only from bleeding but also as blood cells grow old and die and plasma components are consumed or excreted from the body. Therefore, we must continually replace it. Every day, an adult typically produces 400 billion platelets, 200 billion RBCs, and 10 billion WBCs. The production of blood, especially its formed elements, is called **hemopoiesis**[5] (HE-mo-poy-EE-sis). A knowledge of this process provides an indispensable foundation for understanding leukemia, anemia, and other blood disorders.

The tissues that produce blood cells are called **hemopoietic tissues.** The first hemopoietic tissues of the human embryo form in the *yolk sac,* a membrane associated with all vertebrate embryos. In most vertebrates (fish, amphibians, reptiles, and birds), this sac encloses the egg yolk, transfers its nutrients to the growing embryo, and produces the forerunners of the first blood cells. Even animals that don't lay eggs, however, have a yolk sac that retains its hemopoietic function. (It is also the source of cells that later produce eggs and sperm.) Cell clusters called *blood islands* form here by the third week of human development. They produce primitive stem cells that migrate into the embryo proper and colonize the bone marrow, liver, spleen, and thymus. Here, the stem cells multiply and give rise to blood cells throughout fetal development. The liver stops producing blood cells around the time of birth. The spleen stops producing RBCs soon after, but it continues to produce lymphocytes for life.

From infancy onward, the red bone marrow produces all seven kinds of formed elements, while lymphocytes are also produced in the lymphatic tissues and organs—especially the thymus, tonsils, lymph nodes, spleen, and mucous membranes.

### Starvation and Plasma Protein Deficiency

Several conditions can lead to *hypoproteinemia,* a deficiency of plasma protein: extreme starvation or dietary protein deficiency, liver diseases that interfere with protein synthesis, and protein loss through the urine or body surface in the cases of kidney disease and severe burns, respectively. As the protein content of the blood plasma drops, so does its osmolarity. The bloodstream loses more fluid to the tissues than it reabsorbs by osmosis. Thus, the tissues become edematous and a pool of fluid may accumulate in the abdominal cavity—a condition called *ascites* (ah-SY-teez).

Children who suffer severe dietary protein deficiencies often exhibit a condition called *kwashiorkor* (KWASH-ee-OR-cor) (fig. 18.3). The arms and legs are emaciated for lack of muscle, the skin is shiny and tight with edema, and the abdomen is swollen by ascites. *Kwashiorkor* is an African word for a "deposed" or "displaced" child who is no longer breast-fed. Symptoms appear when a child is weaned and placed on a diet consisting mainly of rice or other cereals. Children with kwashiorkor often die of diarrhea and dehydration.

**FIGURE 18.3 Children of Angola with Kwashiorkor.** Note the thin limbs and fluid-distended abdomens.

Blood formation in the bone marrow and lymphatic organs is called, respectively, **myeloid**[6] and **lymphoid hemopoiesis.**

All formed elements trace their origins to a common type of **hemopoietic stem cell (HSC)** in the bone marrow. In the stem-cell terminology of chapter 5, HSCs would be classified as multipotent stem cells, destined to develop into multiple mature cell types. Hematologists, however, often call them *pluripotent stem cells (PPSCs).* (Stem-cell biology is a young science and specialists in different fields sometimes use different terminology.) HSCs multiply to maintain a small but persistent population in the bone marrow, but some of them go

---

[5]*hemo* = blood; *poiesis* = formation

[6]*myel* = bone marrow

on to become a variety of more specialized cells called **colony-forming units (CFUs).** Each CFU is destined to produce one or another class of formed elements. The specific processes leading from an HSC to RBCs, WBCs, and platelets are described at later points in this chapter.

Blood plasma also requires continual replacement. It is composed mainly of water, which it obtains primarily by absorption from the digestive tract. Its electrolytes and organic nutrients are also acquired there, and its gamma globulins come from connective tissue plasma cells and its other proteins mainly from the liver.

### BEFORE YOU GO ON

Answer the following questions to test your understanding of the preceding section:

1. Identify at least two each of the transport, protective, and regulatory functions of the circulatory system.

2. What are the two principal components of the blood?

3. List the three major classes of plasma proteins. Which one is absent from blood serum?

4. Define the *viscosity* and *osmolarity* of blood. Explain why each of these is important for human survival.

5. What does *hemopoiesis* mean? After birth, what one cell type is the starting point for all hemopoiesis?

## 18.2 Erythrocytes

### Expected Learning Outcomes

When you have completed this section, you should be able to

a. discuss the structure and function of erythrocytes (RBCs);

b. describe the structure and function of hemoglobin;

c. state and define some clinical measurements of RBC and hemoglobin quantities;

d. describe the life history of erythrocytes; and

e. name and describe the types, causes, and effects of RBC excesses and deficiencies.

**Erythrocytes,** or **red blood cells (RBCs),** have two principal functions: (1) to pick up oxygen from the lungs and deliver it to tissues elsewhere, and (2) to pick up carbon dioxide from the tissues and unload it in the lungs. RBCs are the most abundant formed elements of the blood and therefore the most obvious things one sees upon its microscopic examination. They are also the most critical to survival; a severe deficiency of leukocytes or platelets can be fatal within a few days, but a severe deficiency of RBCs can be fatal within mere minutes. It is the lack of life-giving oxygen, carried by erythrocytes, that leads rapidly to death in cases of major trauma or hemorrhage.

## Form and Function

An erythrocyte is a discoidal cell with a biconcave shape—a thick rim and a thin sunken center. It is about 7.5 μm in diameter and 2.0 μm thick at the rim (fig. 18.4). Although most cells, including white blood cells, have an abundance of organelles, RBCs lose their nucleus and other organelles during maturation and are thus remarkably devoid of internal structure. When viewed with the transmission electron microscope, the interior of an RBC appears uniformly gray. Lacking mitochondria, RBCs rely exclusively on anaerobic fermentation to produce ATP. The lack of aerobic respiration prevents them from consuming the oxygen that they must transport to other tissues. If they were aerobic and consumed oxygen, they would be like a bakery truck driver who was supposed to deliver doughnuts to the grocery store but ate half of them along the way. RBCs are made to deliver oxygen, not consume it. They are the only human cells that carry on anaerobic fermentation indefinitely.

The cytoplasm of an RBC consists mainly of a 33% solution of **hemoglobin** (about 280 million molecules per cell). This is the red pigment that gives an RBC its color and name. It is known especially for its role in oxygen transport, but it also aids in the transport of carbon dioxide and the buffering of blood pH. The cytoplasm also contains an enzyme, *carbonic anhydrase (CAH),* that catalyzes the reaction $CO_2 + H_2O \leftrightarrows H_2CO_3$. The role of CAH in gas transport and pH balance is discussed in chapters 22 and 24.

The plasma membrane of a mature RBC has glycolipids on the outer surface that determine a person's blood type. On its inner surface are two cytoskeletal proteins, *spectrin* and *actin,* that give the membrane resilience and durability. This is especially important when RBCs pass through small blood capillaries and sinusoids. Many of these passages are narrower than the diameter of an RBC, forcing the RBCs to stretch, bend, and fold as they squeeze through. When they enter larger vessels, RBCs spring back to their discoidal shape like an air-filled inner tube.

There has been appreciable, unresolved debate over whether the biconcave shape of the RBC has any functional advantage. Some suggest that it maximizes the ratio of cell surface area to volume and thereby promotes the quick diffusion of oxygen to all of the hemoglobin in the cell. This is hard to reconcile with the fact that the only place RBCs load oxygen is in the capillaries, and while squeezing through the tiny capillaries, they generally are not biconcave but compressed into ovoid or teardrop shapes. They spring back to the biconcave shape when reentering larger blood vessels, but no oxygen pickup occurs here. Another hypothesis is that the biconcave shape enables the dense slurry of RBCs to flow through the larger blood vessels with a smooth *laminar flow* (see p. 754) that minimizes turbulence. It has also been argued that it is simply the easiest, most stable shape for the cell and its cytoskeleton to relax into when the nucleus is removed, and it may have no physiological function at all.

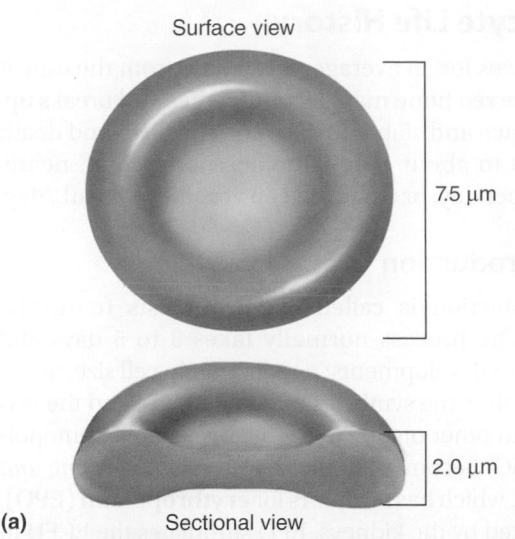

Surface view

7.5 μm

**(a)**

Sectional view

2.0 μm

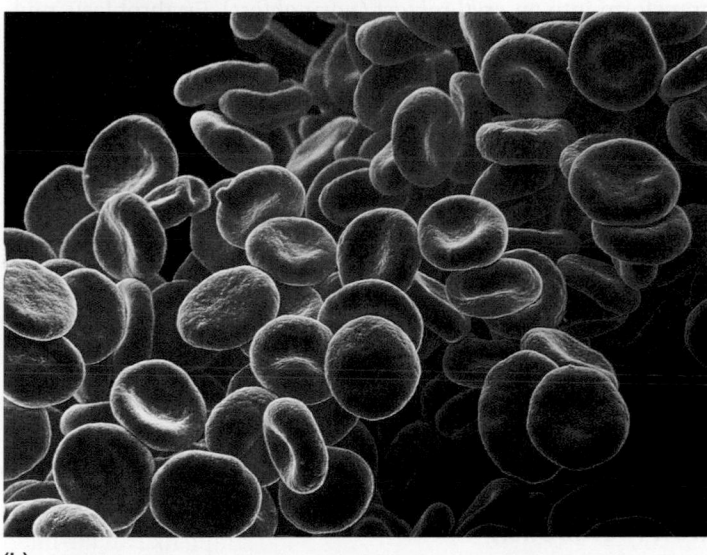

**(b)**

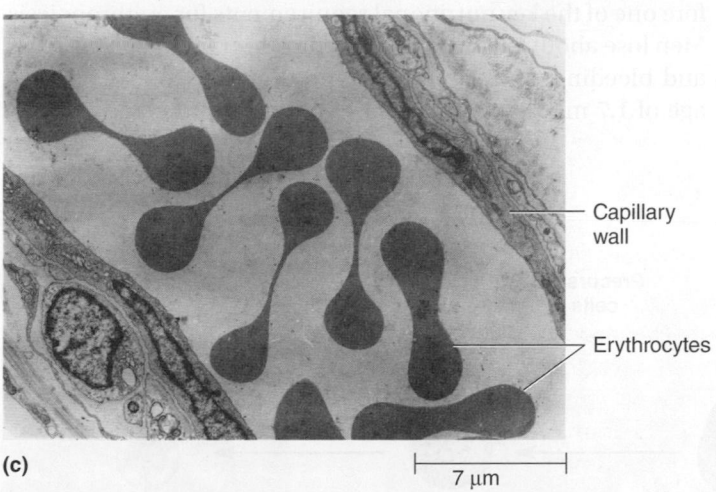

Capillary
wall

Erythrocytes

**(c)**

7 μm

**FIGURE 18.4 The Structure of Erythrocytes.** (a) Dimensions and shape of an erythrocyte. (b) Erythrocytes on the tip of a hypodermic needle (SEM). (c) Erythrocytes in a blood capillary (TEM). Note the absence of organelles or other internal features in the cells. **AP|R**

❓ *Why are erythrocytes caved in at the center?*

## Hemoglobin

Hemoglobin consists of four protein chains called **globins** (fig. 18.5). Two of these, the *alpha (α) chains,* are 141 amino acids long, and the other two, the *beta (β) chains,* are 146 amino acids long. Each chain is conjugated with a nonprotein moiety called the **heme** group, which binds oxygen to an iron atom (Fe) at its center. Each heme can carry one molecule of $O_2$; thus, the hemoglobin molecule as a whole can transport up to 4 $O_2$. About 5% of the $CO_2$ in the bloodstream is also transported by hemoglobin but is bound to the globin moiety rather than to the heme. Gas transport by hemoglobin is discussed in detail in chapter 22.

Hemoglobin exists in several forms with slight differences in the globin chains. The form just described is called *adult hemoglobin (HbA).* About 2.5% of an adult's hemoglobin, however, is of a form called HbA$_2$, which has two *delta (δ) chains* in place of the beta chains. The fetus produces a form called *fetal*

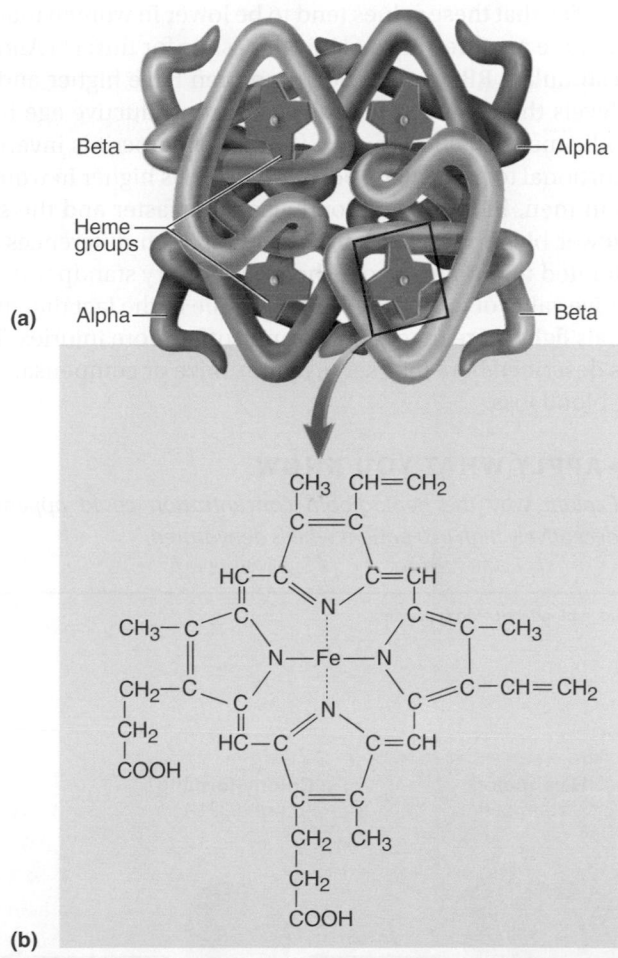

Beta — — Alpha

Heme
groups

**(a)** Alpha — — Beta

CH$_3$ CH=CH$_2$

**(b)**

**FIGURE 18.5 The Structure of Hemoglobin.** (a) The hemoglobin molecule consists of two alpha proteins and two beta proteins, each conjugated to a nonprotein heme group. (b) Structure of the heme group. Oxygen binds to Fe at the center of the heme.

❓ *In what way does this exemplify a quaternary protein structure? What is the prosthetic group of hemoglobin?*

*hemoglobin (HbF),* which has two *gamma* (γ) *chains* in place of the beta chains. The delta and gamma chains are the same length as the beta chains but differ in amino acid sequence. HbF binds oxygen more tightly than HbA does; thus, it enables the fetus to extract oxygen from the mother's bloodstream.

## Quantities of Erythrocytes and Hemoglobin

The RBC count and hemoglobin concentration are important clinical data because they determine the amount of oxygen the blood can carry. Three of the most common measurements are hematocrit, hemoglobin concentration, and RBC count. The **hematocrit[7] (packed cell volume, PCV)** is the percentage of whole blood volume composed of RBCs (see fig. 18.2). In men, it normally ranges between 42% and 52%; in women, between 37% and 48%. The **hemoglobin concentration** of whole blood is normally 13 to 18 g/dL in men and 12 to 16 g/dL in women. The RBC count is normally 4.6 to 6.2 million RBCs/μL in men and 4.2 to 5.4 million/μL in women. This is often expressed as cells per cubic millimeter (mm³); 1 μL = 1 mm³.

Notice that these values tend to be lower in women than in men. There are three physiological reasons for this: (1) Androgens stimulate RBC production, and men have higher androgen levels than women; (2) women of reproductive age have periodic menstrual losses; and (3) the hematocrit is inversely proportional to percentage body fat, which is higher in women than in men. In men, the blood also clots faster and the skin has fewer blood vessels than in women. Such differences are not limited to humans. From the evolutionary standpoint, the adaptive value of these differences may lie in the fact that male animals fight more than females and suffer more injuries. The traits described here may serve to minimize or compensate for their blood loss.

#### ▶▶▶APPLY WHAT YOU KNOW

*Explain why the hemoglobin concentration could appear deceptively high in a patient who is dehydrated.*

---

[7]*hemato* = blood; *crit* = to separate

## The Erythrocyte Life History

An erythrocyte lives for an average of 120 days from the time it is produced in the red bone marrow until it dies and breaks up. In a state of balance and stable RBC count, the birth and death of RBCs amount to about 1 million cells per second, nearly 100 billion cells per day, or a packed cell volume of 20 mL/day.

### Erythrocyte Production

Erythrocyte production is called **erythropoiesis** (eh-RITH-ro-poy-EE-sis). The process normally takes 3 to 5 days and involves four major developments: a reduction in cell size, an increase in cell number, the synthesis of hemoglobin, and the loss of the nucleus and other organelles. It begins when a hemopoietic stem cell (HSC) becomes an *erythrocyte colony-forming unit (ECFU)* (fig. 18.6), which has receptors for **erythropoietin (EPO)**, a hormone secreted by the kidneys. EPO stimulates the ECFU to transform into an *erythroblast (normoblast).* Erythroblasts multiply and synthesize hemoglobin. When this task is completed, the nucleus shrivels and is discharged from the cell. The cell is now called a *reticulocyte,* named for a temporary network (reticulum) composed of ribosome clusters (polyribosomes).

Reticulocytes leave the bone marrow and enter the circulating blood. In a day or two, the last of the polyribosomes disintegrate and disappear, and the cell is a mature erythrocyte. Normally, about 0.5% to 1.5% of the circulating RBCs are reticulocytes, but this percentage rises under certain circumstances. Blood loss, for example, stimulates accelerated erythropoiesis and leads to an increasing number of reticulocytes in circulation—as if the bone marrow is in such a hurry to replenish the lost RBCs that it releases many developing RBCs into circulation a little early.

### Iron Metabolism

Iron is a critical part of the hemoglobin molecule and therefore one of the key nutritional requirements for erythropoiesis. Men lose about 0.9 mg of iron per day through the urine, feces, and bleeding, and women of reproductive age lose an average of 1.7 mg/day through these routes and the added factor

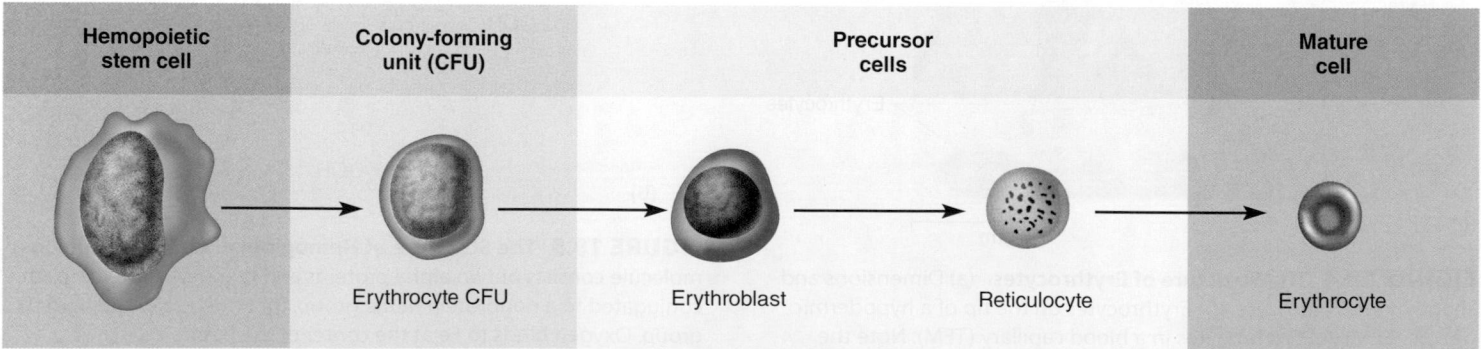

**FIGURE 18.6 Erythropoiesis.** Stages in the development of a red blood cell.

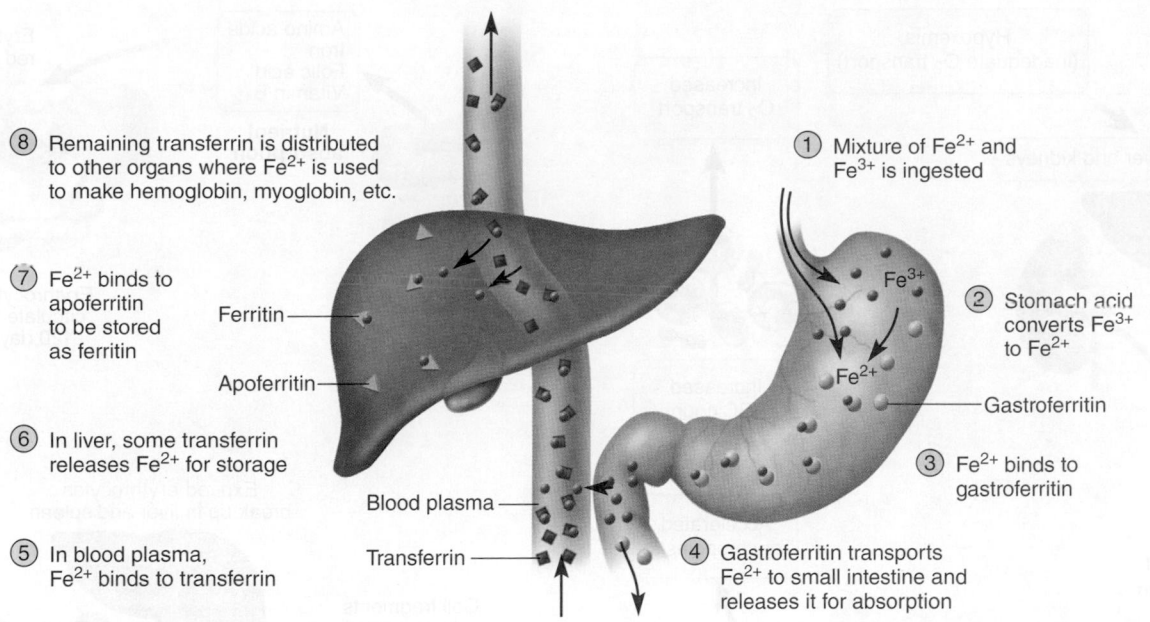

⑧ Remaining transferrin is distributed to other organs where $Fe^{2+}$ is used to make hemoglobin, myoglobin, etc.

⑦ $Fe^{2+}$ binds to apoferritin to be stored as ferritin

Ferritin

Apoferritin

⑥ In liver, some transferrin releases $Fe^{2+}$ for storage

Blood plasma

⑤ In blood plasma, $Fe^{2+}$ binds to transferrin

Transferrin

① Mixture of $Fe^{2+}$ and $Fe^{3+}$ is ingested

$Fe^{3+}$

② Stomach acid converts $Fe^{3+}$ to $Fe^{2+}$

$Fe^{2+}$

Gastroferritin

③ $Fe^{2+}$ binds to gastroferritin

④ Gastroferritin transports $Fe^{2+}$ to small intestine and releases it for absorption

**FIGURE 18.7**  Iron Metabolism.

of menstruation. Since we absorb only a fraction of the iron in our food, we must consume 5 to 20 mg/day to replace our losses. A pregnant woman needs 20 to 48 mg/day, especially in the last 3 months, to meet not only her own need but also that of the fetus.

Dietary iron exists in two forms: ferric ($Fe^{3+}$) and ferrous ($Fe^{2+}$) ions. Stomach acid converts most $Fe^{3+}$ to $Fe^{2+}$, the only form that can be absorbed by the small intestine (fig. 18.7). A protein called **gastroferritin,**[8] produced by the stomach, then binds $Fe^{2+}$ and transports it to the small intestine. Here, it is absorbed into the blood, binds to a plasma protein called **transferrin,** and travels to the bone marrow, liver, and other tissues. Bone marrow uses iron for hemoglobin synthesis; muscle uses it to make the oxygen-storage protein myoglobin; and nearly all cells use iron to make electron-transport molecules called cytochromes in their mitochondria. The liver binds surplus iron to a protein called **apoferritin,**[9] forming an iron-storage complex called **ferritin.** It releases $Fe^{2+}$ into circulation when needed.

Some other nutritional requirements for erythropoiesis are vitamin $B_{12}$ and folic acid, required for the rapid cell division and DNA synthesis that occurs in erythropoiesis, and vitamin C and copper, which are cofactors for some of the enzymes that synthesize hemoglobin. Copper is transported in the blood by an alpha globulin called *ceruloplasmin.*[10]

## Erythrocyte Homeostasis

The RBC count is maintained in a classic negative feedback manner (fig. 18.8). If the count should drop (for example,

because of hemorrhaging), it may result in a state of **hypoxemia**[11] (oxygen deficiency in the blood). The kidneys detect this and increase their EPO output. Three or four days later, the RBC count begins to rise and reverses the hypoxemia that started the process.

Hypoxemia has many causes other than blood loss. Another is a low level of oxygen in the atmosphere. If you were to move from Miami to Denver, for example, the lower $O_2$ level at the high elevation of Denver would produce temporary hypoxemia and stimulate EPO secretion and erythropoiesis. The blood of an average adult has about 5 million RBCs/$\mu$L, but people who live at high elevations may have counts of 7 to 8 million RBCs/$\mu$L. Another cause of hypoxemia is an abrupt increase in the body's oxygen consumption. If a formerly lethargic person takes up tennis or aerobics, for example, the muscles consume oxygen more rapidly and create a state of hypoxemia that stimulates erythropoiesis. Endurance-trained athletes commonly have RBC counts as high as 6.5 million RBCs/$\mu$L.

Not all hypoxemia can be corrected by increasing erythropoiesis. In emphysema, for example, less lung tissue is available to oxygenate the blood. Raising the RBC count cannot correct this, but the kidneys and bone marrow have no way of knowing it. The RBC count continues to rise in a futile attempt to restore homeostasis, resulting in a dangerous excess called *polycythemia,* discussed shortly.

## Erythrocyte Death and Disposal

The life of an RBC is summarized in figure 18.9. As an RBC ages and its membrane proteins (especially spectrin) deteriorate,

---

[8]*gastro* = stomach; *ferrit* = iron; *in* = protein
[9]*apo* = separated from; *ferrit* = iron; *in* = protein
[10]*cerulo* = blue-green, the color of oxidized copper; *plasm* = blood plasma; *in* = protein

[11]*hyp* = below normal; *ox* = oxygen; *emia* = blood condition

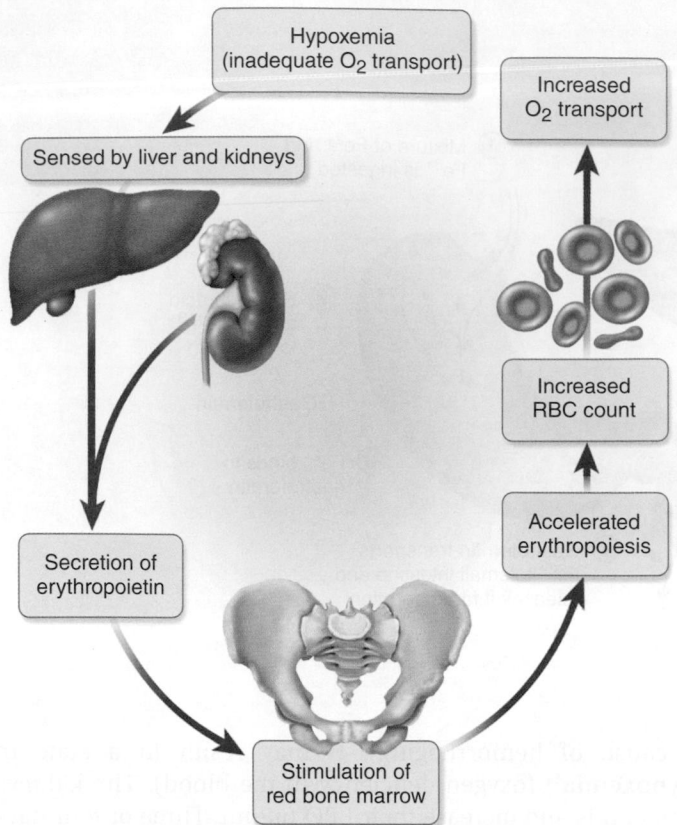

**FIGURE 18.8 Correction of Hypoxemia by a Negative Feedback Loop.**

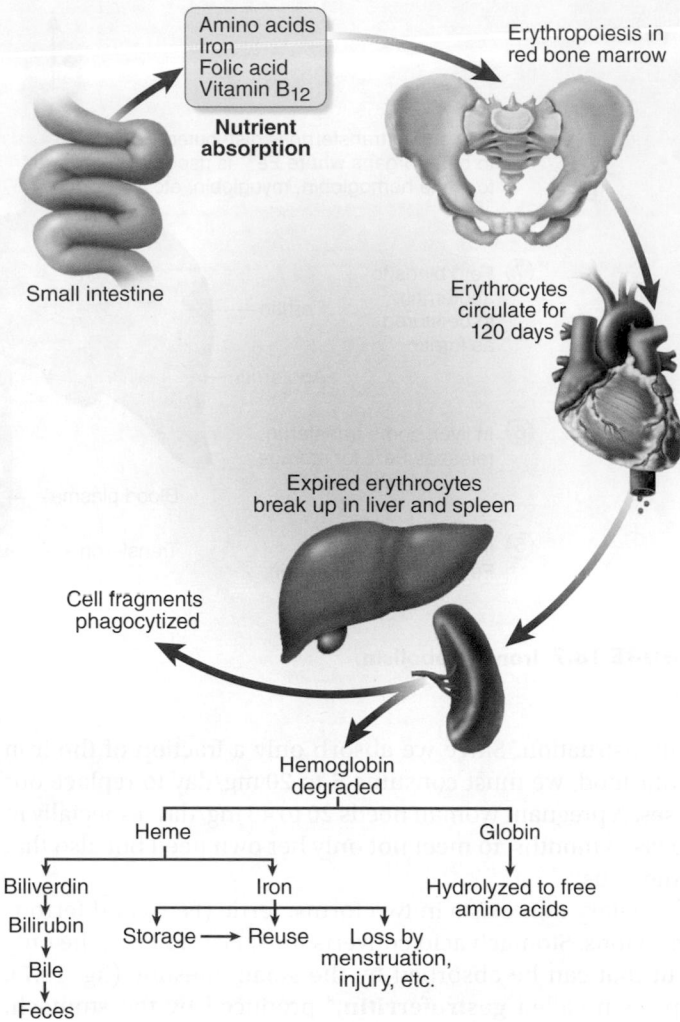

**FIGURE 18.9 The Life and Death of Erythrocytes.** Note especially the stages of hemoglobin breakdown and disposal.

the membrane grows increasingly fragile. Without a nucleus or ribosomes, an RBC cannot synthesize new spectrin. Many RBCs die in the spleen, which has been called the "erythrocyte graveyard." The spleen has channels as narrow as 3 μm that severely test the ability of old, fragile RBCs to squeeze through the organ. Old cells become trapped, broken up, and destroyed. An enlarged and tender spleen may indicate diseases in which RBCs are rapidly breaking down.

**Hemolysis**[12] (he-MOLL-ih-sis), the rupture of RBCs, releases hemoglobin and leaves empty plasma membranes. The membrane fragments are easily digested by macrophages in the liver and spleen, but hemoglobin disposal is a bit more complicated. It must be disposed of efficiently, however, or it can block kidney tubules and cause renal failure. Macrophages begin the disposal process by separating the heme from the globin. They hydrolyze the globin into free amino acids, which can be used for energy-releasing catabolism or recycled for protein synthesis.

Disposing of the heme is another matter. First, the macrophage removes the iron and releases it into the blood, where it combines with transferrin and is used or stored in the same way as dietary iron. The macrophage converts the rest of the heme

into a greenish pigment called **biliverdin**[13] (BIL-ih-VUR-din), then further converts most of this to a yellow-green pigment called **bilirubin.**[14] Bilirubin is released by the macrophages and binds to albumin in the blood plasma. The liver removes it from the albumin and secretes it into the bile, to which it imparts a dark green color as the bile becomes concentrated in the gallbladder. Biliverdin and bilirubin are collectively known as **bile pigments.** The gallbladder discharges the bile into the small intestine, where bacteria convert bilirubin to *urobilinogen,* responsible for the brown color of the feces. Another hemoglobin breakdown pigment, *urochrome,* produces the yellow color of urine. A high level of bilirubin in the blood causes *jaundice,* a yellowish cast in light-colored skin and the whites of eyes. Jaundice may be a sign of rapid hemolysis or a liver disease or bile duct obstruction that interferes with bilirubin disposal.

---

[12]*hemo* = blood; *lysis* = splitting, breakdown

[13]*bili* = bile; *verd* = green; *in* = substance
[14]*bili* = bile; *rub* = red; *in* = substance

## Erythrocyte Disorders

Any imbalance between the rates of erythropoiesis and RBC destruction may produce an excess or deficiency of red cells. An RBC excess is called *polycythemia*[15] (POL-ee-sy-THEE-me-uh), and a deficiency of either RBCs or hemoglobin is called *anemia.*[16]

## Polycythemia

**Primary polycythemia (polycythemia vera)** is due to cancer of the erythropoietic line of the red bone marrow. It can result in an RBC count as high as 11 million RBCs/μL and a hematocrit as high as 80%. Polycythemia from all other causes, called **secondary polycythemia,** is characterized by RBC counts as high as 6 to 8 million RBCs/μL. It can result from dehydration because water is lost from the bloodstream while erythrocytes remain and become abnormally concentrated. More often, it is caused by smoking, air pollution, emphysema, high altitude, excessive aerobic exercise, or other factors that create a state of hypoxemia and stimulate erythropoietin secretion.

The principal dangers of polycythemia are increased blood volume, pressure, and viscosity. Blood volume can double in primary polycythemia and cause the circulatory system to become tremendously engorged. Blood viscosity may rise to three times normal. Circulation is poor, the capillaries are clogged with viscous blood, and the heart is dangerously strained. Chronic (long-term) polycythemia can lead to embolism, stroke, or heart failure. The deadly consequences of emphysema and some other lung diseases are due in part to polycythemia.

## Anemia

The causes of **anemia** fall into three categories: (1) inadequate erythropoiesis or hemoglobin synthesis, (2) **hemorrhagic anemia** from bleeding, and (3) **hemolytic anemia** from RBC destruction. Table 18.4 gives specific examples and causes for each category.

Anemia often results from kidney failure, because RBC production depends on erythropoietin, which is produced mainly by the kidneys. Erythropoiesis also declines with age, simply because the kidneys atrophy and produce less and less EPO as we get older. Compounding this problem, elderly people tend to get less exercise and eat less well, and both of these factors reduce erythropoiesis.

Nutritional anemia results from a dietary deficiency of any of the requirements for erythropoiesis discussed earlier. Its most common form is **iron-deficiency anemia,** characterized by small pale erythrocytes. Iron-deficiency anemia is usually caused by blood loss without sufficiently compensatory iron ingestion. A deficiency of vitamin $B_{12}$ also causes anemia, but $B_{12}$ is so abundant in meat that a deficiency is rare except in strict vegetarians. More often, $B_{12}$ deficiency occurs when glands of the stomach fail to produce a substance called **intrinsic factor** that the small intestine needs to absorb the vitamin. Elderly people sometimes develop **pernicious anemia,** an autoimmune disease in which antibodies destroy stomach tissue. Pernicious anemia can also be hereditary. Without proper postsurgical management, gastric-bypass and gastrectomy patients can develop a similar anemia because of the removal of stomach tissue or surgical rearrangement of the stomach, disconnecting it from the small intestine where the intrinsic factor is needed. Such anemias are treatable with vitamin $B_{12}$ injections or oral $B_{12}$ and intrinsic factor.

**Hypoplastic**[17] **anemia** is caused by a decline in erythropoiesis, whereas the complete failure or destruction of the myeloid tissue produces **aplastic anemia,** a complete cessation of erythropoiesis. Aplastic anemia leads to grotesque tissue necrosis and blackening of the skin. Barring successful treatment, most victims die within a year. About half of all cases are of unknown or hereditary cause, especially in adolescents and young adults. Other causes are given in table 18.4.

[17]*hypo* = below normal; *plas* = formation; *tic* = pertaining to

| TABLE 18.4 | Causes of Anemia |
|---|---|
| **Categories of Anemia** | **Causes or Examples** |
| **Inadequate erythropoiesis** | |
| Iron-deficiency anemia | Dietary iron deficiency |
| Other nutritional anemias | Dietary folic acid, vitamin $B_{12}$, or vitamin C deficiency |
| Anemia due to renal insufficiency | Deficiency of EPO secretion |
| Pernicious anemia | Deficiency of intrinsic factor leading to inadequate vitamin $B_{12}$ absorption |
| Hypoplastic and aplastic anemia | Destruction of myeloid tissue by radiation, viruses, some drugs and poisons (arsenic, benzene, mustard gas), or autoimmune disease |
| Anemia of old age | Declining erythropoiesis due to nutritional deficiencies, reduced physical activity, gastric atrophy (reduced intrinsic factor secretion), or renal atrophy (depressed EPO secretion) |
| **Blood loss (hemorrhagic anemia)** | Trauma, hemophilia, menstruation, ulcer, ruptured aneurysm, etc. |
| **RBC destruction (hemolytic anemia)** | |
| Drug reactions | Penicillin allergy |
| Poisoning | Mushroom toxins, snake and spider venoms |
| Parasitic infection | RBC destruction by malaria parasites |
| Hereditary hemoglobin defects | Sickle-cell disease, thalassemia |
| Blood type incompatabilities | Hemolytic disease of the newborn, transfusion reactions |

[15]*poly* = many; *cyt* = cell; *hem* = blood; *ia* = condition
[16]*an* = without; *em* = blood; *ia* = condition

Anemia has three potential consequences:

1. The tissues suffer **hypoxia** (oxygen deprivation). The individual is lethargic and becomes short of breath upon physical exertion. The skin is pallid because of the deficiency of hemoglobin. Severe anemic hypoxia can cause life-threatening necrosis of brain, heart, and kidney tissues.

2. Blood osmolarity is reduced. More fluid thus transfers from the bloodstream to the intercellular spaces, resulting in edema.

3. Blood viscosity is reduced. Because the blood puts up so little resistance to flow, the heart beats faster than normal and cardiac failure may ensue. Blood pressure also drops because of the reduced volume and viscosity.

## Sickle-Cell Disease

Sickle-cell disease and thalassemia (see table 18.8) are hereditary hemoglobin defects that occur mostly among people of African and Mediterranean descent, respectively. **Sickle-cell disease** afflicts about 1.3% of people of African-American descent. It is caused by a recessive allele that modifies the hemoglobin. Sickle-cell hemoglobin (HbS) differs from normal HbA only in the sixth amino acid of the beta chain, where HbA has glutamic acid and HbS has valine. People who are homozygous for HbS exhibit sickle-cell disease. People who are heterozygous for it—about 8.3% of African-Americans—have *sickle-cell trait* but rarely have severe symptoms. However, if two carriers reproduce, their children each have a 25% chance of being homozygous and having the disease.

HbS does not bind oxygen very well. At low oxygen concentrations, it becomes deoxygenated, polymerizes, and forms a gel that causes the erythrocytes to become elongated and pointed at the ends (fig. 18.10), hence the name of the disease. Sickled erythrocytes are sticky; they **agglutinate**[18] (clump together) and block small blood vessels, causing intense pain in oxygen-starved tissues. Blockage of the circulation can also lead to kidney or heart failure, stroke, severe joint pain, or paralysis. Hemolysis of the fragile cells causes anemia and hypoxemia, which triggers further sickling in a deadly positive feedback loop. Chronic hypoxemia also causes fatigue, weakness, mental deficiency, and deterioration of the heart and other organs. In a futile effort to counteract the hypoxemia, the hemopoietic tissues become so active that bones of the cranium and elsewhere become enlarged and misshapen. The spleen reverts to a hemopoietic role, while also disposing of dead RBCs, and becomes enlarged and fibrous. Sickle-cell disease is a prime example of *pleiotropy*—the occurrence of multiple phenotypic effects from a change in a single gene (see p. 133).

Without treatment, a child with sickle-cell disease has little chance of living to age 2. Advances in treatment, however, have steadily raised life expectancy to a little beyond age 50.

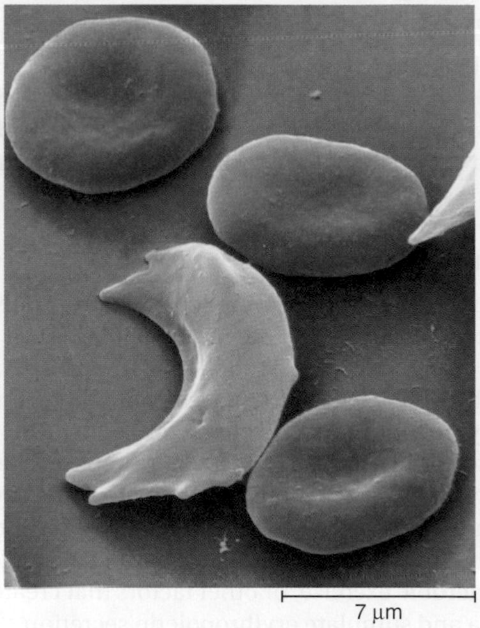

**FIGURE 18.10  Sickle-Cell Disease.**  Shows one deformed, pointed erythrocyte and three normal erythrocytes.

Why does sickle-cell disease exist? In Africa, where it originated, vast numbers of people die of malaria. Malaria is caused by a parasite that invades the RBCs and feeds on hemoglobin. Sickle-cell hemoglobin is detrimental to the parasites, and people heterozygous for sickle-cell disease are resistant to malaria. The lives saved by this gene far outnumber the deaths of homozygous individuals, so the gene persists in the population. The sickle-cell gene is less common in the United States and other essentially nonmalarious regions than it is in Africa.

### BEFORE YOU GO ON

Answer the following questions to test your understanding of the preceding section:

6. Describe the size, shape, and contents of an erythrocyte, and explain how it acquires its unusual shape.

7. What is the function of hemoglobin? What are its protein and nonprotein moieties called?

8. Define *hematocrit, hemoglobin concentration,* and *RBC count* and give the units of measurement in which each is expressed.

9. List the stages in the production of an RBC and describe how each stage differs from the previous one.

10. What is the role of erythropoietin in the regulation of RBC count? What is the role of gastroferritin?

11. What happens to each component of an RBC and its hemoglobin when it dies and disintegrates?

12. What are the three primary causes or categories of anemia? What are its three primary consequences?

---

[18]*ag* = together; *glutin* = glue

## 18.3 Blood Types

### Expected Learning Outcomes

When you have completed this section, you should be able to

a. explain what determines a person's ABO and Rh blood types and how this relates to transfusion compatibility;

b. describe the effect of an incompatibility between mother and fetus in Rh blood type; and

c. list some blood groups other than ABO and Rh and explain how they may be useful.

Blood types and transfusion compatibility are a matter of interactions between plasma proteins and erythrocytes. Ancient Greek physicians attempted to transfuse blood from one person to another by squeezing it from a pig's bladder through a porcupine quill into the recipient's vein. Although some patients benefited from the procedure, it was fatal to others. The reason some people have compatible blood and some don't remained obscure until 1900, when Karl Landsteiner discovered blood types A, B, and O—a discovery that won him a Nobel Prize in 1930; type AB was discovered later. World War II stimulated great improvements in transfusions, blood banking, and blood substitutes (see Deeper Insight 18.2).

Blood types are based on large molecules called *antigens* and *antibodies.* Explained more fully in chapter 21, these will be introduced only briefly here. **Antigens** are complex molecules such as proteins, glycoproteins, and glycolipids that are genetically unique to each individual (except identical twins). They occur on the surfaces of all cells and enable the body to distinguish its own cells from foreign matter. When the body detects an antigen of foreign origin, it activates an immune response. This response consists partly of the *plasma cells,* mentioned earlier, secreting proteins (gamma globulins) called **antibodies.**

Antibodies bind to antigens and mark them, or the cells bearing them, for destruction. One method of antibody action is **agglutination,** in which each antibody molecule binds to two or more antigen molecules and sticks them together. Repetition of this process produces large **antigen–antibody complexes** that immobilize the antigens until certain immune cells can break them down. Whole cells may become immobilized and clumped together when antibodies bind to their surface antigens.

Blood types result from reactions between antigens called **agglutinogens** (ah-glue-TIN-oh-jens) on the surfaces of the RBCs and antibodies called **agglutinins** (ah-GLUE-tih-nins) in the blood plasma. These names refer to the fact that they interact to agglutinate RBCs in the event of a mismatched transfusion.

 **DEEPER INSIGHT 18.2**

## MEDICAL HISTORY

### *Charles Drew—Blood-Banking Pioneer*

Charles Drew (fig. 18.11) was a scientist who lived and died in the grip of irony. After receiving his M.D. from McGill University of Montreal in 1933, Drew became the first black person to pursue the advanced degree of Doctor of Science in Medicine, for which he studied transfusion and blood banking at Columbia University. He became the director of a new blood bank at Columbia Presbyterian Hospital in 1939 and organized numerous blood banks during World War II.

Drew saved countless lives by convincing physicians to use plasma rather than whole blood for battlefield and other emergency transfusions. Whole blood could be stored for only a week and given only to recipients with compatible blood types. Plasma could be stored longer and was less likely to cause transfusion reactions.

When the U.S. War Department issued a directive forbidding the storage of Caucasian and Negro blood in the same military blood banks, Drew denounced the order and resigned his position. He became a professor of surgery at Howard University in Washington, D.C., and later chief of staff at Freedmen's Hospital. He was a mentor for numerous young black physicians and campaigned to get them accepted into the medical community. The American Medical Association, however, firmly refused to admit black members, even Drew himself.

Late one night in 1950, Drew and three colleagues set out to volunteer their medical services to an annual free clinic in Tuskegee, Alabama. Drew fell asleep at the wheel and was critically injured in the resulting accident. Doctors at the nearest hospital administered blood and attempted to revive him—yet for all the lives he saved through his pioneering work in transfusion, Drew himself bled to death at the age of 45.

**FIGURE 18.11** Charles Drew (1904–50).

| TABLE 18.5 | The ABO Blood Group | | | |
|---|---|---|---|---|
| **Characteristics** | **ABO Blood Type** | | | |
| | **Type O** | **Type A** | **Type B** | **Type AB** |
| Possible genotypes* | *ii* | *I^A I^A* or *I^A i* | *I^B I^B* or *I^B i* | *I^A I^B* |
| RBC antigen | None | A | B | A, B |
| Plasma antibody | Anti-A, anti-B | Anti-B | Anti-A | None |
| May safely receive RBCs of type | O | O, A | O, B | O, A, B, AB |
| May safely donate RBCs to | O, A, B, AB | A, AB | B, AB | AB |
| Frequency in U.S. population | | | | |
| White | 45% | 40% | 11% | 4% |
| Black | 49% | 27% | 20% | 4% |
| Hispanic | 63% | 14% | 20% | 3% |
| Japanese | 31% | 38% | 22% | 9% |
| Native American | 79% | 16% | 4% | < 1% |

*$I^A$ is the dominant allele for agglutinogen A; $I^B$ is the dominant allele for agglutinogen B; and allele *i* is recessive to both of these.

## The ABO Group

Blood types A, B, AB, and O form the **ABO blood group** (table 18.5). One's ABO blood type is determined by the hereditary presence or absence of antigens A and B on the RBCs. The genetic determination of blood types is explained on page 133. The antigens are glycolipids—membrane phospholipids with short carbohydrate chains bonded to them. Figure 18.12 shows how the carbohydrate moieties of RBC surface antigens determine the ABO blood types.

Antibodies of the ABO group begin to appear in the plasma 2 to 8 months after birth. They reach their maximum concentrations between 8 and 10 years of age and then slowly decline for the rest of one's life. They are produced mainly in response to bacteria that inhabit the intestines, but they cross-react with RBC antigens and are therefore best known for their significance in transfusions.

Antibodies of the ABO group react against any A or B antigen except one's own. The antibody that reacts against antigen A is called *alpha agglutinin,* or *anti-A;* it is present in the plasma of people with type O or type B blood—that is, anyone who does *not* possess antigen A. The antibody that reacts against antigen B is *beta agglutinin,* or *anti-B,* and is present in type O and type A individuals—those who do not possess antigen B. Each antibody molecule has 10 binding sites where it can attach to either an A or B antigen. An antibody can therefore attach to several RBCs at once and agglutinate them (fig. 18.13).

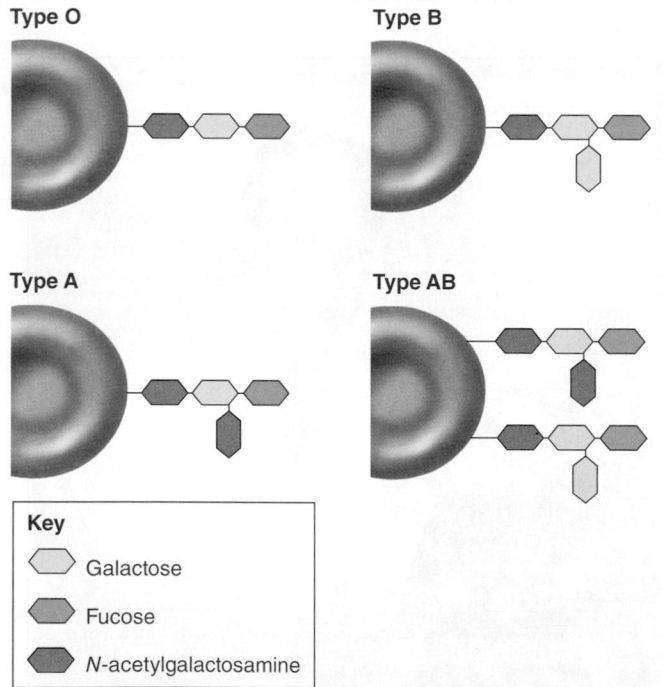

**Key**
- Galactose
- Fucose
- *N*-acetylgalactosamine

**FIGURE 18.12 Chemical Basis of the ABO Blood Types.** The terminal carbohydrates of the antigenic glycolipids are shown. All of them end with galactose and fucose (not to be confused with fructose). In type A, the galactose also has *N*-acetylgalactosamine added to it; in type B, it has another galactose; and in type AB, both of these chain types are present.

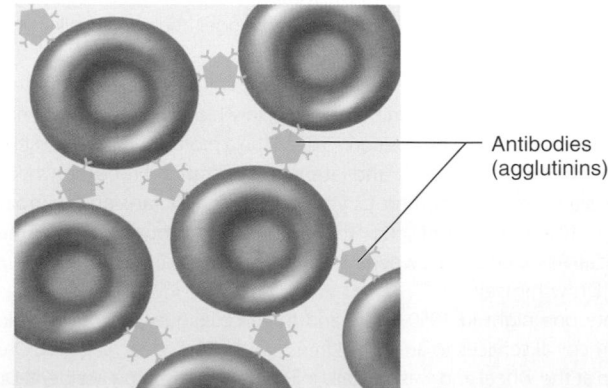

Antibodies (agglutinins)

**FIGURE 18.13 Agglutination of RBCs by an Antibody.** Anti-A and anti-B have 10 binding sites, located at the 2 tips of each of the 5 Ys, and can therefore bind multiple RBCs to each other.

A person's ABO blood type can be determined by placing one drop of blood in a pool of anti-A serum and another drop in a pool of anti-B serum. Blood type AB exhibits conspicuous agglutination in both antisera; type A or B agglutinates only in the corresponding antiserum; and type O does not agglutinate in either one (fig. 18.14).

Type O blood is the most common and AB is the rarest in the United States. Percentages differ from one region of the world to another and among ethnic groups because people tend to marry within their locality and ethnic group and perpetuate statistical variations particular to that group (see table 18.5).

In giving transfusions, it is imperative that the donor's RBCs not agglutinate as they enter the recipient's bloodstream. For example, if type B blood were transfused into a type A recipient, the recipient's anti-B would immediately agglutinate the donor's RBCs (fig. 18.15). A mismatch causes a **transfusion reaction**—the agglutinated RBCs block small blood vessels, hemolyze, and release their hemoglobin over the next few hours to days. Free hemoglobin can block the kidney tubules and cause death from acute renal failure within a week or so. For this reason, a person with type A (anti-B) blood must never be given a transfusion of type B or AB blood. A person with type B (anti-A) must never receive type A or AB blood. Type O (anti-A and anti-B) individuals cannot safely receive type A, B, or AB blood.

Type AB is sometimes called the *universal recipient* because this blood type lacks both anti-A and anti-B antibodies; thus, it will not agglutinate donor RBCs of any ABO type. However, this overlooks the fact that the *donor's* plasma can agglutinate the *recipient's* RBCs if it contains anti-A, anti-B, or both. For similar reasons, type O is sometimes called the *universal donor*. The plasma of a type O donor, however, can agglutinate the RBCs of a type A, B, or AB recipient. There are procedures for reducing the risk of a transfusion reaction in certain mismatches, however, such as giving packed RBCs with a minimum of plasma.

Contrary to some people's belief, blood type is not changed by transfusion. It is fixed at conception and remains the same for life.

### ▶▶▶ APPLY WHAT YOU KNOW

*Scientists have recently developed a method of enzymatically splitting N-acetylgalactosamine off the glycolipid of type A blood cells (fig. 18.12). What potential benefit do you think they saw as justifying their research effort?*

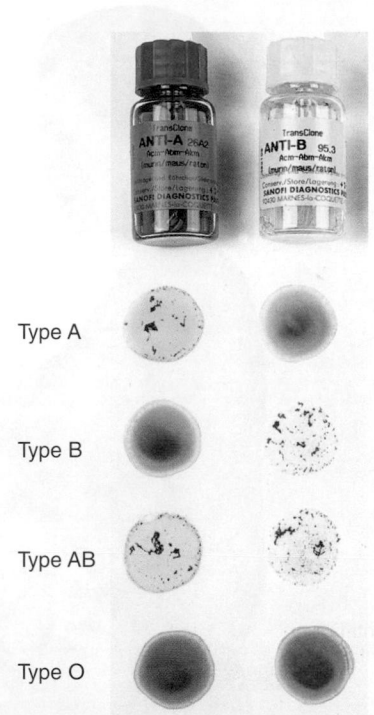

**FIGURE 18.14 ABO Blood Typing.** Each row shows the appearance of a drop of blood mixed with anti-A and anti-B antisera. Blood cells become clumped if they possess the antigens for the antiserum (top row left, second row right, third row both) but otherwise remain uniformly mixed. Thus, type A agglutinates only in anti-A; type B agglutinates only in anti-B; type AB agglutinates in both; and type O agglutinates in neither of them. The antisera in the vials at the top are artificially colored to make them more easily distinguishable.

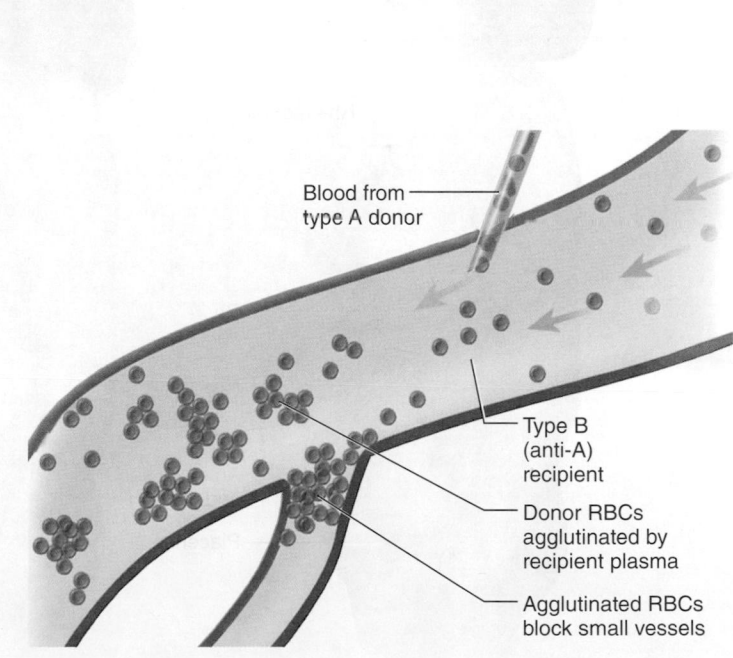

**FIGURE 18.15 Effects of a Mismatched Transfusion.** Donor RBCs become agglutinated in the recipient's blood plasma. The agglutinated RBCs lodge in smaller blood vessels downstream from this point and cut off the blood flow to vital tissues.

## The Rh Group

The **Rh blood group** is named for the rhesus monkey, in which the Rh antigens were discovered in 1940. This group includes numerous RBC antigens, of which the principal types are antigens C, D, and E. Antigen D is by far the most reactive of these, so a person is considered *Rh-positive (Rh+)* if he or she has the D antigen (genotype *DD* or *Dd*) and *Rh-negative (Rh−)* if it is lacking (genotype *dd*). The Rh blood type is tested by using an anti-D reagent. The Rh type is often combined with the ABO type in a single expression such as O+ for type O, Rh-positive, or AB− for type AB, Rh-negative. Rh frequencies vary among ethnic groups just as ABO frequencies do. About 85% of white Americans are Rh+ and 15% are Rh−, whereas about 99% of Asians are Rh+. ABO blood type has no influence on Rh type, or vice versa. If the frequency of type O whites in the United States is 45%, and 85% of these are also Rh+, then the frequency of O+ individuals is the product of these separate frequencies: $0.45 \times 0.85 = 0.38$, or 38%.

In contrast to the ABO group, anti-D antibodies are not normally present in the blood. They form only in Rh− individuals who are exposed to Rh+ blood. If an Rh− person receives an Rh+ transfusion, the recipient produces anti-D. Since anti-D does not appear instantaneously, this presents little danger in the first mismatched transfusion. But if that person should later receive another Rh+ transfusion, his or her anti-D could agglutinate the donor's RBCs.

A related condition sometimes occurs when an Rh− woman carries an Rh+ fetus. The first pregnancy is likely to be uneventful because the placenta normally prevents maternal and fetal blood from mixing. However, at the time of birth, or if a miscarriage occurs, placental tearing exposes the mother to Rh+ fetal blood. She then begins to produce anti-D antibodies (fig. 18.16). If she becomes pregnant again with an Rh+ fetus, her anti-D antibodies may pass through the placenta and agglutinate the fetal erythrocytes. Agglutinated RBCs hemolyze, and the baby is born with a severe anemia called **hemolytic disease of the newborn (HDN)**, or **erythroblastosis fetalis.** Not all HDN is due to Rh incompatibility, however. About 2% of cases result from incompatibility of ABO and other blood types. About 1 out of 10 cases of ABO incompatibility between mother and fetus results in HDN.

HDN, like so many other disorders, is easier to prevent than to treat. If an Rh− woman gives birth to (or miscarries) an Rh+ child, she can be given an *Rh immune globulin* (sold under

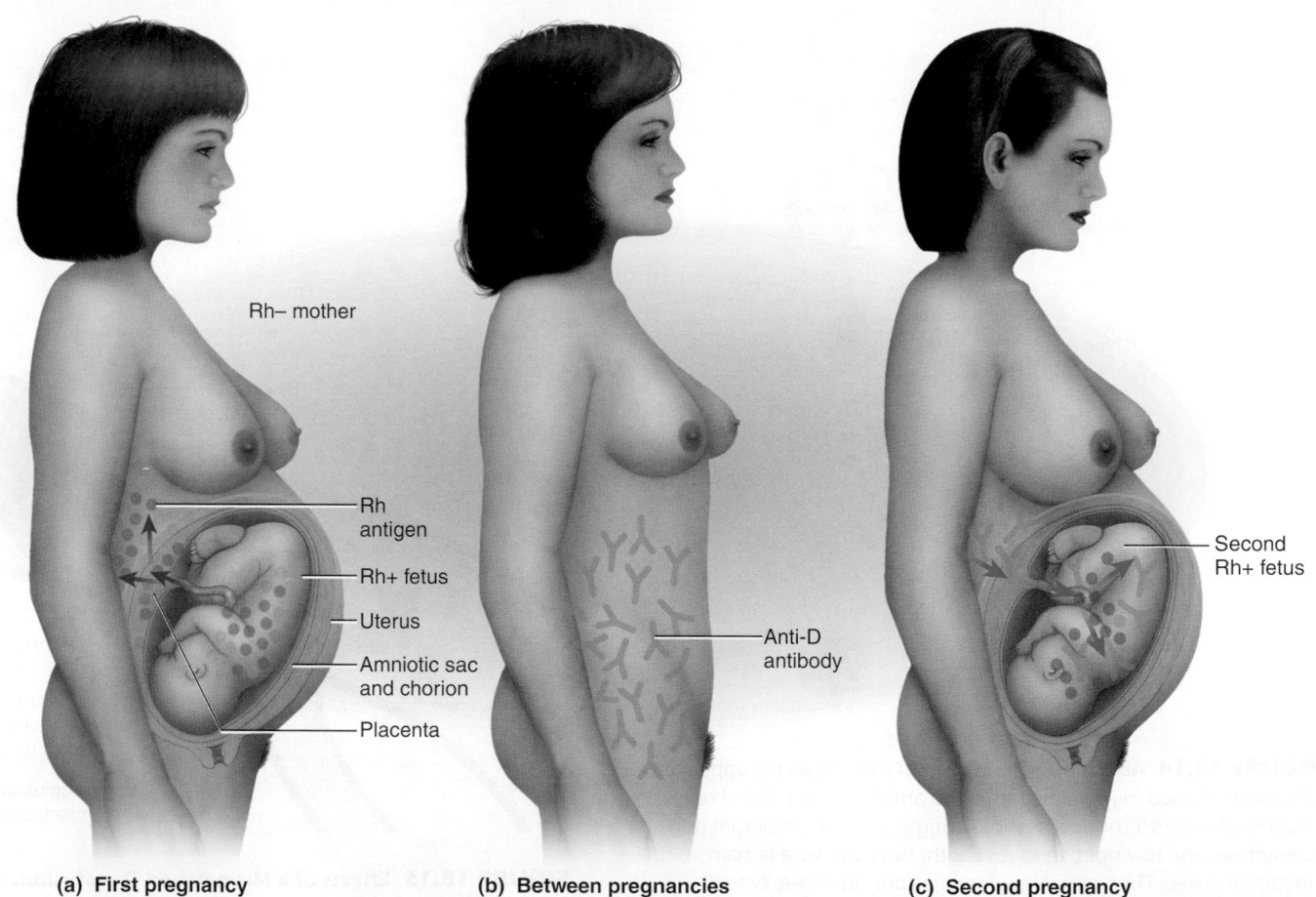

(a) **First pregnancy**  (b) **Between pregnancies**  (c) **Second pregnancy**

Rh− mother

Rh antigen
Rh+ fetus
Uterus
Amniotic sac and chorion
Placenta

Anti-D antibody

Second Rh+ fetus

**FIGURE 18.16 Hemolytic Disease of the Newborn (HDN).** (a) When an Rh− woman is pregnant with an Rh+ fetus, she is exposed to D (Rh) antigens, especially during childbirth. (b) Following that pregnancy, her immune system produces anti-D antibodies. (c) If she later becomes pregnant with another Rh+ fetus, her anti-D antibodies can cross the placenta and agglutinate the blood of that fetus, causing that child to be born with HDN.

such trade names as RhoGAM and Gamulin Rh). The immune globulin binds fetal RBC antigens so they cannot stimulate her immune system to produce anti-D. It is now common to give immune globulin at 28 to 32 weeks' gestation and at birth in any pregnancy in which the mother is Rh−.

If an Rh− woman has had one or more previous Rh+ pregnancies, her subsequent Rh+ children have about a 17% probability of being born with HDN. Infants with HDN are usually severely anemic. As the fetal hemopoietic tissues respond to the need for more RBCs, erythroblasts (immature RBCs) enter the circulation prematurely—hence the name *erythroblastosis fetalis.* Hemolyzed RBCs release hemoglobin, which is converted to bilirubin. High bilirubin levels can cause *kernicterus,* a syndrome of toxic brain damage that may kill the infant or leave it with motor, sensory, and mental deficiencies. HDN can be treated with *phototherapy*—exposing the infant to ultraviolet radiation, which degrades bilirubin as blood passes through capillaries of the skin. In more severe cases, an *exchange transfusion* may be given to completely replace the infant's Rh+ blood with Rh−. In time, the infant's hemopoietic tissues will replace the donor's RBCs with Rh+ cells, and by then the mother's antibody will have disappeared from the infant's blood.

▶▶▶ **APPLY WHAT YOU KNOW**

*A baby with HDN typically has jaundice and an enlarged spleen. Explain these effects.*

## Other Blood Groups

In addition to the ABO and Rh groups, there are at least 100 other known blood groups with a total of more than 500 antigens, including the MN, Duffy, Kell, Kidd, and Lewis groups. These rarely cause transfusion reactions, but they are useful for such legal purposes as paternity and criminal cases and for research in anthropology and population genetics. Reactions in the Kell, Kidd, and Duffy groups occasionally cause HDN.

**BEFORE YOU GO ON**

Answer the following questions to test your understanding of the preceding section:

13. What are antibodies and antigens? How do they interact to cause a transfusion reaction?

14. What antibodies and antigens are present in people with each of the four ABO blood types?

15. Describe the cause, prevention, and treatment of HDN.

16. Why might someone be interested in determining a person's blood type other than ABO/Rh?

# DEEPER INSIGHT 18.3
## CLINICAL APPLICATION

### *Bone Marrow and Cord Blood Transplants*

A bone marrow transplant is one treatment option for leukemia, sickle-cell disease, some forms of anemia, and other disorders. The principle is to replace cancerous or otherwise defective marrow with donor stem cells in hopes that they will rebuild a population of normal marrow and blood cells. The patient is first given chemotherapy or radiation to destroy the defective marrow and eliminate immune cells (T cells) that would attack the donated marrow. Bone marrow is drawn from the donor's sternum or hip bone and injected into the recipient's circulatory system. Donor stem cells colonize the patient's marrow cavities and, ideally, build healthy marrow.

There are, however, several drawbacks to bone marrow transplant. For one, it is difficult to find compatible donors. Surviving T cells in the patient may attack the donor marrow, and donor T cells may attack the patient's tissues (the *graft-versus-host response*). To inhibit graft rejection, the patient must take immunosuppressant drugs for life. These drugs leave a person vulnerable to infection and have many other adverse side effects. Infections are sometimes contracted from the donated marrow itself. In short, marrow

transplant is a high-risk procedure, up to one third of patients die from complications of treatment.

An alternative with several advantages is to use blood from placentas, which are normally discarded at every childbirth. Placental blood contains more stem cells than adult bone marrow, and is less likely to carry infectious microbes. With the parents' consent, it can be harvested from the umbilical cord with a syringe, and it can be stored almost indefinitely, frozen in liquid nitrogen at cord blood banks. The immature immune cells in cord blood have less tendency to attack the recipient's tissues; thus, cord blood transplants have lower rejection rates and do not require as close a match between donor and recipient, meaning that more donors are available to patients in need. Pioneered in the 1980s, cord blood transplants have successfully treated leukemia and a wide range of other blood diseases. Efforts are being made to further improve the procedure by stimulating fetal stem cells to multiply before the transplant, and by removing fetal T cells that may react against the recipient.

## 18.4   Leukocytes

### Expected Learning Outcomes

When you have completed this section, you should be able to

a. explain the function of leukocytes in general and the individual role of each leukocyte type;

b. describe the appearance and relative abundance of each type of leukocyte;

c. describe the formation and life history of leukocytes; and

d. discuss the types, causes, and effects of leukocyte excesses and deficiencies.

### Form and Function

**Leukocytes,** or **white blood cells (WBCs),** are the least abundant formed elements, totaling only 5,000 to 10,000 WBCs/$\mu$L. Yet we cannot live long without them, because they afford protection against infection and other diseases. WBCs are easily recognized in stained blood films because they have conspicuous nuclei that stain from light violet to dark purple with the most common blood stains. They are much more abundant in the body than their low number in blood films would suggest, because they spend only a few hours in the bloodstream, then migrate into the connective tissues and spend the rest of their lives there. It's as if the bloodstream were merely the subway that the WBCs take to work; in blood films, we see only the ones on their way to work, not the WBCs already at work in the tissues.

Leukocytes differ from erythrocytes in that they retain their organelles throughout life; thus, when viewed with the transmission electron microscope, they show a complex internal structure (fig. 18.17). Among their organelles are the usual instruments of protein synthesis—the nucleus, rough endoplasmic reticulum, ribosomes, and Golgi complex—for leukocytes must synthesize proteins in order to carry out their functions. Some of these proteins are packaged into lysosomes and other organelles, which appear as conspicuous cytoplasmic granules that distinguish one WBC type from another.

### Types of Leukocytes

As outlined at the beginning of this chapter, there are five kinds of leukocytes (table 18.6). They are distinguished from each other by their relative size and abundance, the size and shape of their nuclei, the presence or absence of certain cytoplasmic granules, the coarseness and staining properties of those granules, and most importantly by their functions.

All WBCs have lysosomes called **nonspecific (azurophilic[19]) granules** in the cytoplasm, so named because they absorb the blue or violet dyes of blood stains. Three of the five WBC types—neutrophils, eosinophils, and basophils—are

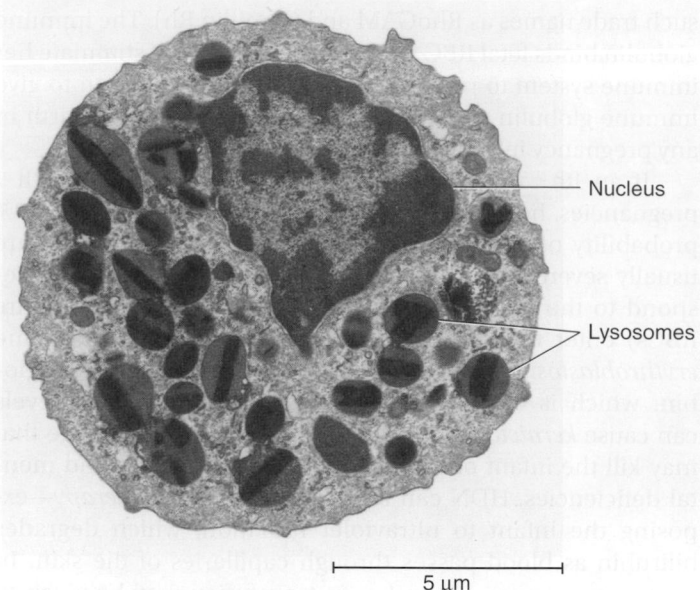

**FIGURE 18.17   The Structure of a Leukocyte (TEM).** This example is an eosinophil. The lysosomes seen here are the coarse pink granules seen in the cytoplasm of the eosinophil in table 18.6.

called **granulocytes** because they also have various types of **specific granules** that stain conspicuously and distinguish each cell type from the others. Basophils are named for the fact that their specific granules stain with methylene blue, a basic dye in a common blood-staining mixture called Wright's stain. Eosinophils are so named because they stain with eosin, an acidic dye in Wright's stain. The colors in the following descriptions are those typically seen with Wright's stain, but may differ on slides you examine because of the use of other stains. The specific granules of neutrophils do not stain intensely with either dye. Specific granules contain enzymes and other chemicals employed in defense against pathogens. The two remaining WBC types—monocytes and lymphocytes—are called **agranulocytes** because they lack specific granules. Nonspecific granules are inconspicuous to the light microscope, and these cells therefore have relatively clear-looking cytoplasm.

### *Granulocytes*

* **Neutrophils** (NEW-tro-fills) are the most abundant WBCs—generally about 4,150 cells/$\mu$L and constituting 60% to 70% of the circulating leukocytes. The nucleus is clearly visible and, in a mature neutrophil, typically consists of three to five lobes connected by slender nuclear strands. These strands are sometimes so delicate that they are scarcely visible, and the neutrophil may seem as if it had multiple nuclei. Young neutrophils have an undivided band-shaped nucleus and are called *band cells* or *stab[20] cells.* Neutrophils are also called *polymorphonuclear leukocytes (PMNs)* because of their varied nuclear shapes.

---

[19]*azuro* = blue; *philic* = loving

[20]*stab* = band, bar (German)

The cytoplasm contains fine reddish to violet specific granules, which contain lysozyme and other antimicrobial agents. The individual granules are barely visible with the light microscope, but their combined effect gives the cytoplasm a pale lilac color.

Neutrophils are aggressively antibacterial cells. Their numbers rise—a condition called *neutrophilia*—in response to bacterial infections. They destroy bacteria in ways detailed in chapter 21.

- **Eosinophils** (EE-oh-SIN-oh-fills) are harder to find in a blood film because they are only 2% to 4% of the WBC total, typically numbering about 170 cells/µL. The eosinophil count fluctuates greatly, however, from day to night, seasonally, and with the phase of the menstrual cycle. It rises *(eosinophilia)* in allergies, parasitic infections, collagen diseases, and diseases of the spleen and central nervous system. Although relatively scanty in the blood, eosinophils are abundant in the mucous membranes of the respiratory, digestive, and lower urinary tracts. The eosinophil nucleus usually has two large lobes connected by a thin strand, and the cytoplasm has an abundance of coarse rosy to orange-colored specific granules.

  Eosinophils secrete chemicals that weaken or destroy relatively large parasites such as hookworms and tapeworms, too big for any one WBC to phagocytize. They also phagocytize and dispose of inflammatory chemicals, antigen–antibody complexes, and allergens (foreign antigens that trigger allergies).

- **Basophils** are the rarest of the WBCs and, indeed, of all formed elements. They number about 40 cells/µL and usually constitute less than 0.5% of the WBC count. They can be recognized mainly by an abundance of very coarse, dark violet specific granules. The nucleus is largely hidden from view by these granules, but is large, pale, and typically S- or U-shaped.

  Basophils secrete two chemicals that aid in the body's defense processes: (1) **histamine,** a vasodilator that widens the blood vessels, speeds the flow of blood to an injured tissue, and makes the blood vessels more permeable so that blood components such as neutrophils and clotting proteins can get into the connective tissues more quickly; and (2) **heparin,** an anticoagulant that inhibits blood clotting and thus promotes the mobility of other WBCs in the area. They also release chemical signals that attract eosinophils and neutrophils to a site of infection.

### Agranulocytes

- **Lymphocytes** (LIM-fo-sites) are second to neutrophils in abundance and are thus quickly spotted when you examine a blood film. They number about 2,200 cells/µL and are 25% to 33% of the WBC count. There are several subclasses of lymphocytes with different immune functions (see chapter 21), but they look alike through the light microscope. They include the smallest WBCs; at 5 to 17 µm in diameter, they range from smaller than RBCs to two and a half times as large. They are sometimes classified into three size classes (table 18.6), but there are gradations between them. Medium and large lymphocytes are usually seen in fibrous connective tissues and only occasionally in the circulating blood (see fig. 18.1). The lymphocytes seen in blood films are mostly in the small size class. The lymphocyte nucleus is round, ovoid, or slightly dimpled on one side, and usually stains dark violet. In small lymphocytes, it fills nearly the entire cell and leaves only a narrow rim of light blue cytoplasm, often barely detectable, around the cell perimeter. The cytoplasm is more abundant in medium and large lymphocytes.

  Small lymphocytes are sometimes difficult to distinguish from basophils, but most basophils are conspicuously grainy, whereas the lymphocyte nucleus is uniform or merely mottled. Basophils also lack the rim of clear cytoplasm seen in most lymphocytes. Large lymphocytes are sometimes difficult to distinguish from monocytes.

- **Monocytes** (MON-oh-sites) are usually the largest WBCs seen on a blood slide, often two or three times the diameter of an RBC. They number about 460 cells/µL and about 3% to 8% of the WBC count. The nucleus is large and clearly visible, often a relatively light violet, and typically ovoid, kidney-shaped, or horseshoe-shaped. The cytoplasm is abundant and contains sparse, fine granules. In prepared blood films, monocytes often assume sharply angular to spiky shapes (see fig. 18.1).

  The monocyte count rises in inflammation and viral infections. Monocytes go to work only after leaving the bloodstream and transforming into large tissue cells called **macrophages** (MAC-ro-fay-jez). Macrophages are highly phagocytic cells that consume up to 25% of their own volume per hour. They destroy dead or dying host and foreign cells, pathogenic chemicals and microorganisms, and other foreign matter. They also chop up or "process" foreign antigens and "display" fragments of them on the cell surface to alert the immune system to the presence of a pathogen. Thus, they and a few other cells are called *antigen-presenting cells (APCs).* The functions of macrophages are further detailed in chapter 21.

| TABLE 18.6 | The White Blood Cells (Leukocytes) |
|---|---|

### Neutrophils

| Percentage of WBCs | 60% to 70% |
|---|---|
| Mean count | 4,150 cells/µL |
| Diameter | 9–12 µm |

*Appearance**

- Nucleus usually with 3–5 lobes in S- or C-shaped array
- Fine reddish to violet specific granules in cytoplasm

*Differential count*

- Increases in bacterial infections

*Functions*

- Phagocytize bacteria
- Release antimicrobial chemicals

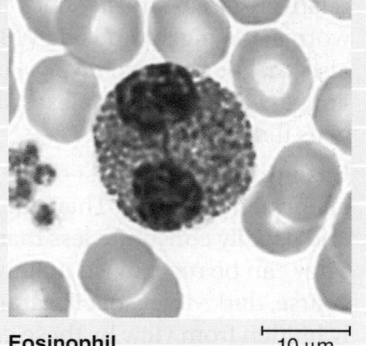

**Neutrophils**    10 µm

### Eosinophils

| Percentage of WBCs | 2% to 4% |
|---|---|
| Mean count | 165 cells/µL |
| Diameter | 10–14 µm |

*Appearance**

- Nucleus usually has two large lobes connected by thin strand
- Large orange-pink specific granules in cytoplasm

*Differential count*

- Fluctuates greatly from day to night, seasonally, and with phase of menstrual cycle
- Increases in parasitic infections, allergies, collagen diseases, and diseases of spleen and central nervous system

*Functions*

- Phagocytize antigen–antibody complexes, allergens, and inflammatory chemicals
- Release enzymes that weaken or destroy parasites such as worms

**Eosinophil**    10 µm

### Basophils

| Percentage of WBCs | < 0.5% |
|---|---|
| Mean count | 44 cells/µL |
| Diameter | 8–10 µm |

*Appearance**

- Nucleus large and U- to S-shaped, but typically pale and obscured from view
- Coarse, abundant, dark violet specific granules in cytoplasm

*Differential count*

- Relatively stable
- Increases in chickenpox, sinusitis, diabetes mellitus, myxedema, and polycythemia

*Functions*

- Secrete histamine (a vasodilator), which increases blood flow to a tissue
- Secrete heparin (an anticoagulant), which promotes mobility of other WBCs by preventing clotting

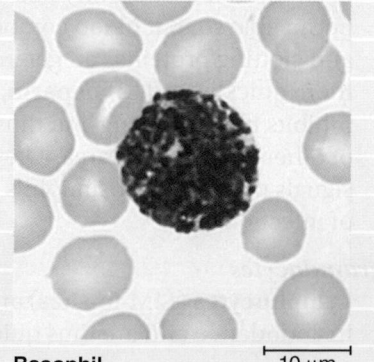

**Basophil**    10 µm

| TABLE 18.6 | The White Blood Cells (Leukocytes) (continued) |
|---|---|

**Lymphocytes**

| Percentage of WBCs | 25% to 33% |
|---|---|
| Mean count | 2,185 cells/µL |
| Diameter | |
| Small class | 5–8 µm |
| Medium class | 10–12 µm |
| Large class | 14–17 µm |

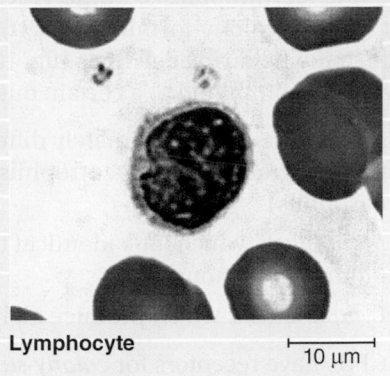

**Lymphocyte**    10 µm

*Appearance**

- Nucleus round, ovoid, or slightly dimpled on one side, of uniform or mottled dark violet color
- In small lymphocytes, nucleus fills nearly all of the cell and leaves only a scanty rim of clear, light blue cytoplasm.
- In larger lymphocytes, cytoplasm is more abundant; large lymphocytes may be hard to differentiate from monocytes.

*Differential count*

- Increases in diverse infections and immune responses

*Functions*

- Several functional classes usually indistinguishable by light microscopy
- Destroy cancer cells, cells infected with viruses, and foreign cells
- Present antigens to activate other cells of immune system
- Coordinate actions of other immune cells
- Secrete antibodies
- Serve in immune memory

**Monocytes**

| Percentage of WBCs | 3% to 8% |
|---|---|
| Mean count | 456 cells/µL |
| Diameter | 12–15 µm |

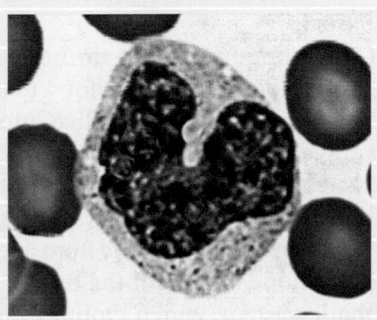

**Monocyte**    10 µm

*Appearance**

- Nucleus ovoid, kidney-shaped, or horseshoe-shaped; violet
- Abundant cytoplasm with sparse, fine nonspecific granules
- Sometimes very large with stellate or polygonal shapes

*Differential count*

- Increases in viral infections and inflammation

*Functions*

- Differentiate into macrophages (large phagocytic cells of the tissues)
- Phagocytize pathogens, dead neutrophils, and debris of dead cells
- Present antigens to activate other cells of immune system

*Appearance pertains to blood films dyed with Wright's stain.

## The Leukocyte Life History

**Leukopoiesis** (LOO-co-poy-EE-sis), the production of white blood cells, begins with the same hemopoietic stem cells (HSCs) as erythropoiesis. Some HSCs differentiate into distinct types of colony-forming units (CFUs) and then go on to produce the following cell lines (fig. 18.18), each of them now irreversibly committed to a certain outcome.

1. *Myeloblasts,* which ultimately differentiate into the three types of granulocytes (neutrophils, eosinophils, and basophils)

2. *Monoblasts,* which look identical to myeloblasts but lead ultimately to monocytes

3. *Lymphoblasts,* which produce all lymphocyte types

CFUs have receptors for *colony-stimulating factors (CSFs).* Mature lymphocytes and macrophages secrete several types of CSFs in response to infections and other immune challenges. Each CSF stimulates a different WBC type to develop in response to specific needs. Thus, a bacterial infection may trigger the production of neutrophils, whereas an allergy stimulates eosinophil production, each process working through its own CSF.

The red bone marrow stores granulocytes and monocytes until they are needed and contains 10 to 20 times more of these cells than the circulating blood does. Lymphocytes begin developing in the bone marrow but do not stay there. Some types mature there and others migrate to the thymus to complete their development. Mature lymphocytes from both locations then colonize the spleen, lymph nodes, and other lymphoid organs and tissues.

Circulating leukocytes do not stay in the blood for very long. Granulocytes circulate for 4 to 8 hours and then migrate into the tissues, where they live another 4 or 5 days. Monocytes travel in the blood for 10 to 20 hours, then migrate into the tissues and transform into a variety of macrophages, which can live as long as a few years. Lymphocytes, responsible for long-term immunity, survive from a few weeks to decades; they leave the bloodstream for the tissues and eventually enter the lymphatic system, which empties them back into the bloodstream. Thus, they are continually recycled from blood to tissue fluid to lymph and back to the blood.

When leukocytes die, they are generally phagocytized and digested by macrophages. Dead neutrophils, however, are responsible for the creamy color of pus, and are sometimes disposed of by the rupture of a blister onto the skin surface. The biology of leukocytes is discussed more extensively in chapter 21.

#### ▶▶▶ APPLY WHAT YOU KNOW

*It is sometimes written that RBCs do not live as long as WBCs because RBCs do not have a nucleus and therefore cannot repair and maintain themselves. Explain the flaw in this argument.*

## Leukocyte Disorders

The total WBC count is normally 5,000 to 10,000 WBCs/μL. A count below this range, called **leukopenia**[21] (LOO-co-PEE-nee-uh), is seen in lead, arsenic, and mercury poisoning; radiation sickness; and such infectious diseases as measles, mumps, chickenpox, polio, influenza, typhoid fever, and AIDS. It can also be produced by glucocorticoids, anticancer drugs, and immunosuppressant drugs given to organ-transplant patients. Since WBCs are protective cells, leukopenia presents an elevated risk of infection and cancer. A count above 10,000 WBCs/μL, called **leukocytosis,**[22] usually indicates infection, allergy, or other diseases but can also occur in response to dehydration or emotional disturbances. More useful than a total WBC count is a *differential WBC count,* which identifies what percentage of the total WBC count consists of each type of leukocyte (see Deeper Insight 18.4).

---

[21]*leuko* = white; *penia* = deficiency
[22]*leuko* = white; *cyt* = cell; *osis* = condition

### DEEPER INSIGHT 18.4

#### CLINICAL APPLICATION

### *The Complete Blood Count*

One of the most common clinical procedures in both routine physical examinations and the diagnosis of disease is a *complete blood count (CBC).* The CBC yields a highly informative profile of data on multiple blood values: the number of RBCs, WBCs, and platelets per microliter of blood; the relative numbers (percentages) of each WBC type, called a *differential WBC count;* hematocrit; hemoglobin concentration; and various *RBC indices* such as RBC size (*mean corpuscular volume, MCV*) and hemoglobin concentration per RBC (*mean corpuscular hemoglobin, MCH*).

RBC and WBC counts used to require the microscopic examination of films of diluted blood on a calibrated slide, and a differential WBC count required examination of stained blood films. Today, most laboratories use *electronic cell counters.* These devices draw a blood sample through a very narrow tube with sensors that identify cell types and measure cell sizes and hemoglobin content. These counters give faster and more accurate results based on much larger numbers of cells than the old visual methods. However, cell counters still misidentify some cells, and a medical technologist must review the results for suspicious abnormalities and identify cells that the instrument cannot.

The wealth of information gained from a CBC is too vast to give more than a few examples here. Various forms of anemia are indicated by low RBC counts or abnormalities of RBC size, shape, and hemoglobin content. A platelet deficiency can indicate an adverse drug reaction. A high neutrophil count suggests bacterial infection, and a high eosinophil count suggests an allergy or parasitic infection. Elevated numbers of specific WBC types or WBC stem cells can indicate various forms of leukemia. If a CBC does not provide enough information or if it suggests other disorders, additional tests may be done, such as coagulation time and bone marrow biopsy.

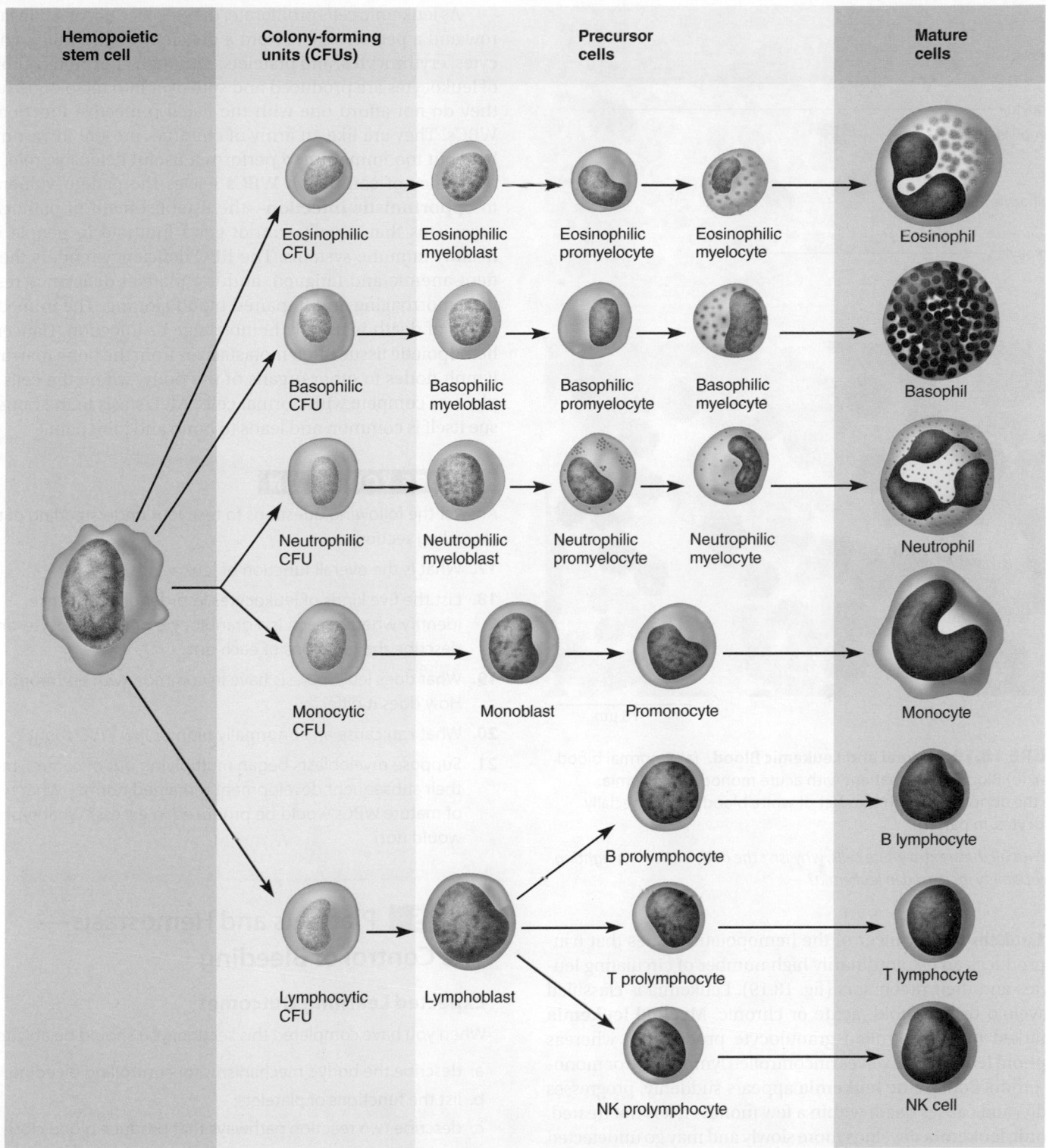

| Hemopoietic stem cell | Colony-forming units (CFUs) | Precursor cells | | | Mature cells |
|---|---|---|---|---|---|
| | Eosinophilic CFU | Eosinophilic myeloblast | Eosinophilic promyelocyte | Eosinophilic myelocyte | Eosinophil |
| | Basophilic CFU | Basophilic myeloblast | Basophilic promyelocyte | Basophilic myelocyte | Basophil |
| | Neutrophilic CFU | Neutrophilic myeloblast | Neutrophilic promyelocyte | Neutrophilic myelocyte | Neutrophil |
| | Monocytic CFU | Monoblast | Promonocyte | | Monocyte |
| | Lymphocytic CFU | Lymphoblast | B prolymphocyte | | B lymphocyte |
| | | | T prolymphocyte | | T lymphocyte |
| | | | NK prolymphocyte | | NK cell |

**FIGURE 18.18  Leukopoiesis.** Stages in the development of white blood cells. The hemopoietic stem cell at the left also is the ultimate source of red blood cells (see fig. 18.6) and platelet-producing cells.

❓ *Explain the meaning and relevance of the combining form* myelo- *seen in so many of these cell names.*

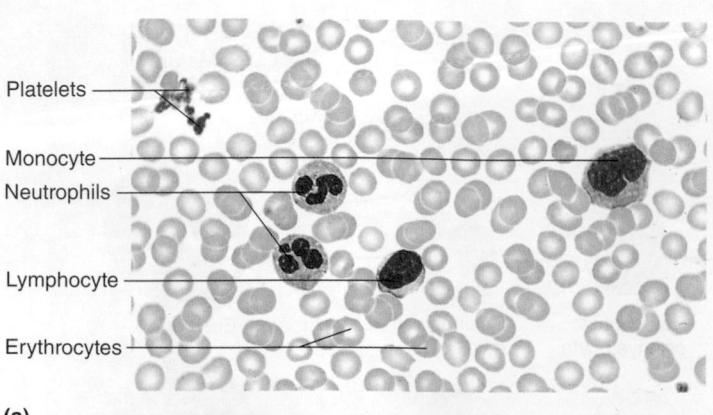

Platelets

Monocyte

Neutrophils

Lymphocyte

Erythrocytes

**(a)**

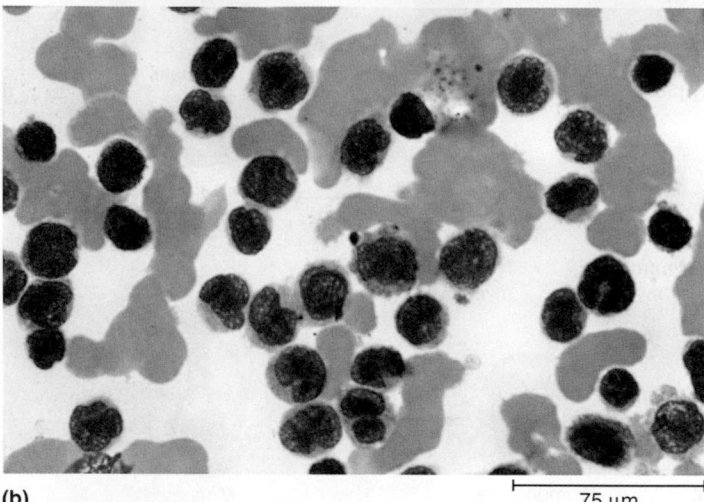

**(b)** 75 μm

**FIGURE 18.19 Normal and Leukemic Blood.** (a) A normal blood smear. (b) Blood from a patient with acute monocytic leukemia. Note the abnormally high number of white blood cells, especially monocytes, in part (b).

❓ *With all these extra white cells, why isn't the body's infection-fighting capability increased in leukemia?*

**Leukemia** is a cancer of the hemopoietic tissues that usually produces an extraordinarily high number of circulating leukocytes and their precursors (fig. 18.19). Leukemia is classified as myeloid or lymphoid, acute or chronic. **Myeloid leukemia** is marked by uncontrolled granulocyte production, whereas **lymphoid leukemia** involves uncontrolled lymphocyte or monocyte production. **Acute leukemia** appears suddenly, progresses rapidly, and causes death within a few months if it is not treated. **Chronic leukemia** develops more slowly and may go undetected for many months; if untreated, the typical survival time is about 3 years. Both myeloid and lymphoid leukemia occur in acute and chronic forms. The greatest success in treatment and cure has been with acute lymphoblastic leukemia, the most common type of childhood cancer. Treatment employs chemotherapy and marrow transplants along with the control of side effects such as anemia, hemorrhaging, and infection.

As leukemic cells proliferate, they replace normal bone marrow and a person suffers from a deficiency of normal granulocytes, erythrocytes, and platelets. Although enormous numbers of leukocytes are produced and spill over into the bloodstream, they do not afford one with the usual protective functions of WBCs. They are like an army of children, present in vast numbers but too immature to perform a useful defensive role. The deficiency of competent WBCs leaves the patient vulnerable to **opportunistic infection**—the establishment of pathogenic organisms that usually cannot get a foothold in people with healthy immune systems. The RBC deficiency renders the patient anemic and fatigued, and the platelet deficiency results in hemorrhaging and impaired blood clotting. The immediate cause of death is usually hemorrhage or infection. Cancerous hemopoietic tissue often metastasizes from the bone marrow or lymph nodes to other organs of the body, where the cells displace or compete with normal cells. Metastasis to the bone tissue itself is common and leads to bone and joint pain.

**BEFORE YOU GO ON**

Answer the following questions to test your understanding of the preceding section:

17. What is the overall function of leukocytes?

18. List the five kinds of leukocytes in order of abundance, identify whether each is a granulocyte or agranulocyte, and describe the functions of each one.

19. What does leukopoiesis have in common with erythropoiesis? How does it differ?

20. What can cause an abnormally high or low WBC count?

21. Suppose myeloblasts began multiplying out of control, but their subsequent development remained normal. What types of mature WBCs would be produced in excess? What types would not?

## 18.5 Platelets and Hemostasis— The Control of Bleeding

### Expected Learning Outcomes

When you have completed this section, you should be able to

a. describe the body's mechanisms for controlling bleeding;

b. list the functions of platelets;

c. describe two reaction pathways that produce blood clots;

d. explain what happens to blood clots when they are no longer needed;

e. explain what keeps blood from clotting in the absence of injury; and

f. describe some disorders of blood clotting.

Circulatory systems developed very early in animal evolution, and with them evolved mechanisms for stopping leaks, which are potentially fatal. **Hemostasis**[23] is the cessation of bleeding. Although hemostatic mechanisms may not stop a hemorrhage from a large blood vessel, they are quite effective at closing breaks in small ones. Platelets play multiple roles in hemostasis, so we begin with a consideration of their form and function.

## Platelet Form and Function

**Platelets** are not cells but small fragments of marrow cells called *megakaryocytes.* They are the second most abundant formed elements, after erythrocytes; a normal platelet count in blood from a fingerstick ranges from 130,000 to 400,000 platelets/μL (averaging about 250,000). The platelet count can vary greatly, however, under different physiological conditions and in blood samples taken from various places in the body. In spite of their numbers, platelets are so small (2 to 4 μm in diameter) that they contribute even less than WBCs to the blood volume.

Platelets have a complex internal structure that includes lysosomes, mitochondria, microtubules and microfilaments, **granules** filled with platelet secretions, and a system of channels called the **open canalicular system,** which open onto the platelet surface (fig. 18.20a). They have no nucleus. When activated, they form pseudopods and are capable of ameboid movement.

Despite their small size, platelets have a greater variety of functions than any of the true blood cells:

- They secrete *vasoconstrictors,* chemicals that stimulate spasmodic constriction of broken vessels and thereby help to reduce blood loss.

- They stick together to form temporary *platelet plugs* that seal small breaks in injured blood vessels.

- They secrete *procoagulants,* or clotting factors, which promote blood clotting.

- They initiate the formation of a clot-dissolving enzyme that dissolves blood clots that have outlasted their usefulness.

- They secrete chemicals that attract neutrophils and monocytes to sites of inflammation.

- They internalize and destroy bacteria.

- They secrete *growth factors* that stimulate mitosis in fibroblasts and smooth muscle and thereby help to maintain and repair blood vessels.

## Platelet Production

The production of platelets is a division of hemopoiesis called **thrombopoiesis.** (Platelets are occasionally called *thrombocytes.*[24]) Some hemopoietic stem cells produce receptors for the hormone *thrombopoietin,* from the liver and kidneys, thus

becoming megakaryoblasts—cells committed to the platelet-producing line. The megakaryoblast duplicates its DNA repeatedly without undergoing nuclear or cytoplasmic division. The result is a **megakaryocyte**[25] (MEG-ah-CAR-ee-oh-site), a gigantic cell up to 150 μm in diameter, visible to the naked eye, with a huge multilobed nucleus and multiple sets of chromosomes (fig. 18.20b). Most megakaryocytes live in the red bone marrow adjacent to blood-filled spaces called *sinusoids,* lined with a thin simple squamous epithelium called the *endothelium* (see fig. 21.9, p. 810).

A megakaryocyte sprouts long tendrils called *proplatelets* that protrude through the endothelium into the blood of the sinusoid. The blood flow shears off the proplatelets, which break up into platelets as they travel in the bloodstream. Much of this breakup is thought to occur when they pass through the small vessels of the lungs, because blood counts show more proplatelets entering the lungs than leaving and more platelets

---

[25]*mega* = giant; *karyo* = nucleus; *cyte* = cell

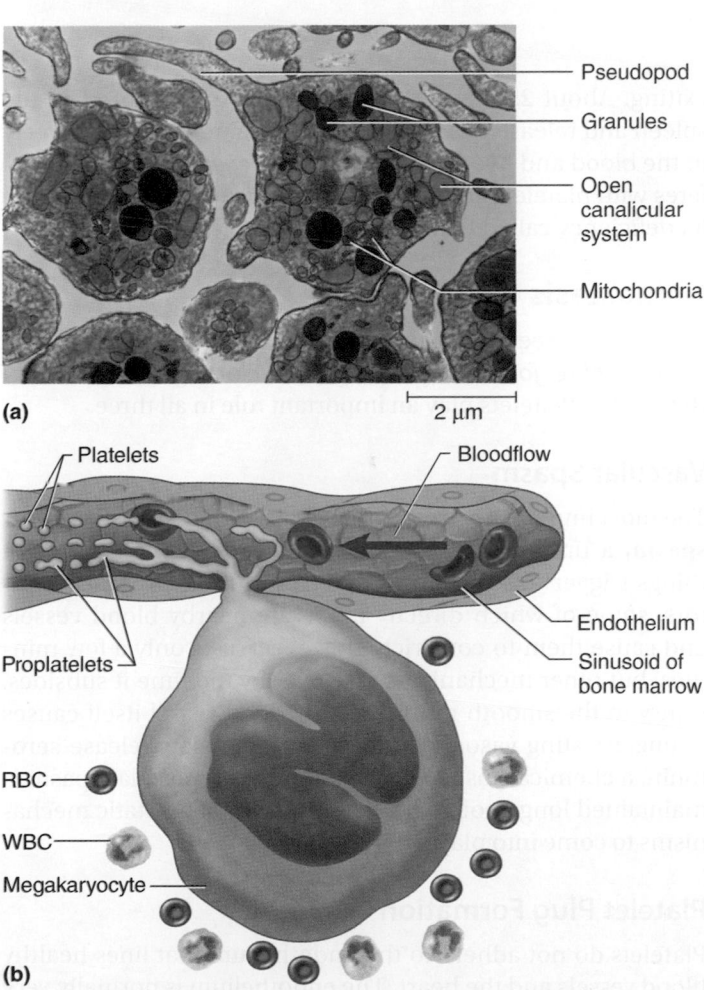

**(a)**

Pseudopod

Granules

Open canalicular system

Mitochondria

2 μm

**(b)**

Platelets

Bloodflow

Proplatelets

Endothelium

Sinusoid of bone marrow

RBC

WBC

Megakaryocyte

**FIGURE 18.20 Platelets.** (a) Structure of blood platelets (TEM). (b) Platelets being produced by the shearing of proplatelets from a megakaryocyte. Note the sizes of the megakaryocyte and platelets relative to RBCs and WBCs. **AP|R**

---

[23]*hemo* = blood; *stasis* = stability
[24]*thrombo* = clotting; *cyte* = cell

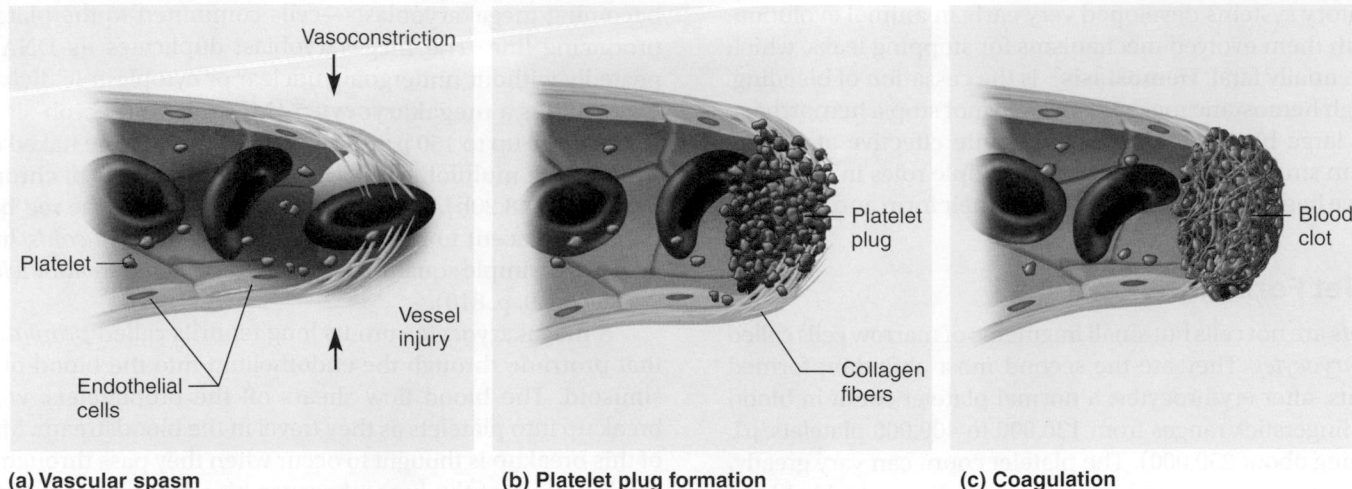

(a) Vascular spasm   (b) Platelet plug formation   (c) Coagulation

**FIGURE 18.21 Hemostasis.** (a) Vasoconstriction of a broken vessel reduces bleeding. (b) A platelet plug forms as platelets adhere to exposed collagen fibers of the vessel wall. The platelet plug temporarily seals the break. (c) A blood clot forms as platelets become enmeshed in fibrin threads. This forms a longer-lasting seal and gives the vessel a chance to repair itself.

? *How does a blood clot differ from a platelet plug?*

exiting. About 25% to 40% of the platelets are stored in the spleen and released as needed. The remainder circulate freely in the blood and live for about 5 to 6 days. Anything that interferes with platelet production can produce a dangerous platelet deficiency called **thrombocytopenia**[26] (see table 18.8).

## Hemostasis

There are three hemostatic mechanisms—*vascular spasm, platelet plug formation,* and *blood clotting* (coagulation) (fig. 18.21). Platelets play an important role in all three.

### Vascular Spasm

The most immediate protection against blood loss is **vascular spasm,** a prompt constriction of the broken vessel. Several things trigger this reaction. An injury stimulates pain receptors, some of which directly innervate nearby blood vessels and cause them to constrict. This effect lasts only a few minutes, but other mechanisms take over by the time it subsides. Injury to the smooth muscle of the blood vessel itself causes a longer-lasting vasoconstriction, and platelets release serotonin, a chemical vasoconstrictor. Thus, the vascular spasm is maintained long enough for the other two hemostatic mechanisms to come into play.

### Platelet Plug Formation

Platelets do not adhere to the endothelium that lines healthy blood vessels and the heart. The endothelium is normally very

smooth and coated with **prostacyclin,** a platelet repellent. When a vessel is broken, however, collagen fibers of its wall are exposed to the blood. Upon contact with collagen or other rough surfaces, platelets grow long spiny pseudopods that adhere to the vessel and to other platelets; the pseudopods then contract and draw the walls of the vessel together. The mass of platelets thus formed, called a **platelet plug,** may reduce or stop minor bleeding. The platelet plug is looser and more delicate than the blood clot to follow, which is why a bleeding injury should be blotted rather than wiped.

As platelets aggregate, they undergo **degranulation**—the exocytosis of their cytoplasmic granules and release of factors that promote hemostasis. Among these are serotonin, a vasoconstrictor; adenosine diphosphate (ADP), which attracts more platelets to the area and stimulates their degranulation; and **thromboxane A$_2$,** an eicosanoid that promotes platelet aggregation, degranulation, and vasoconstriction. Thus, a positive feedback cycle is activated that can quickly seal a small break in a blood vessel.

### Coagulation

**Coagulation (clotting)** of the blood is the last but most effective defense against bleeding. It is important for the blood to clot quickly when a vessel has broken, but equally important for it not to clot in the absence of vessel damage. Because of this delicate balance, coagulation is one of the most complex processes in the body, involving over 30 chemical reactions. It is presented here in a very simplified form.

---

[26]*thrombo* = clotting; *cyto* = cell; *penia* = deficiency

Perhaps clotting is best understood if we first consider its goal. The objective is to convert the plasma protein fibrinogen into **fibrin,** a sticky protein that adheres to the walls of a vessel. As blood cells and platelets arrive, they stick to the fibrin like insects in a spider web (fig. 18.22, inset). The resulting mass of fibrin, blood cells, and platelets ideally seals the break in the blood vessel. The complexity of clotting lies in how the fibrin is formed.

There are two reaction pathways to coagulation (fig. 18.22). One of them, the **extrinsic mechanism,** is initiated by clotting factors released by the damaged blood vessel and perivascular[27] tissues. The word *extrinsic* refers to the fact that these factors come from sources external to the blood itself. Blood may also clot, however, without these tissue factors—for example, when platelets adhere to a fatty plaque of atherosclerosis or to a test tube. The reaction pathway in this case is called the **intrinsic mechanism** because it uses only clotting factors found in the blood itself. In most cases of bleeding, both the extrinsic and intrinsic mechanisms work simultaneously and interact with each other to achieve hemostasis.

Clotting factors (table 18.7) are called **procoagulants,** in contrast to the **anticoagulants** discussed later (see Deeper Insight 18.6, p. 704). Most procoagulants are proteins produced by the liver. They are always present in the plasma in inactive form, but when one factor is activated, it functions as an enzyme that activates the next one in the pathway. That factor activates the next, and so on, in a sequence called a **reaction cascade**—a series of reactions, each of which depends on the product of the preceding one. Many of the clotting factors are identified by roman numerals, which indicate the order in which they were discovered, not the order of the reactions. Factors IV and VI are not included in table 18.7. These terms were abandoned when it was found that factor IV was calcium and factor VI was activated factor V. The last four procoagulants in the table are called *platelet factors* ($PF_1$ through $PF_4$) because they are produced by the platelets.

***Initiation of Coagulation***   The extrinsic mechanism is diagrammed on the top left side of figure 18.22. The damaged blood vessel and perivascular tissues release a lipoprotein mixture called **tissue thromboplastin**[28] **(factor III).** Factor III combines with factor VII to form a complex that, in the presence of $Ca^{2+}$, activates factor X. The extrinsic and intrinsic pathways differ only in how they arrive at active factor X. Therefore, before examining their common pathway from factor X to the end, let's consider how the intrinsic pathway reaches this step.

The intrinsic mechanism is diagrammed on the top right side of figure 18.22. Everything needed to initiate it is present in the plasma or platelets. When platelets degranulate, they release factor XII (Hageman factor, named for the patient in whom it was discovered). Through a cascade of reactions, this leads to activated factors XI, IX, and VIII, in that order—each serving as an enzyme that catalyzes the next step—and finally to factor X. This pathway also requires $Ca^{2+}$ and $PF_3$.

***Completion of Coagulation***   Once factor X is activated, the remaining events are identical in the intrinsic and extrinsic mechanisms (bottom half of figure). Factor X combines with factors III and V in the presence of $Ca^{2+}$ and $PF_3$ to produce *prothrombin activator.* This enzyme acts on a globulin called **prothrombin (factor II)** and converts it to the enzyme **thrombin.** Thrombin then converts fibrinogen into shorter strands of *fibrin monomer.* These monomers then covalently bond to each other end to end and form longer fibers of *fibrin polymer.* Factor XIII cross-links these strands to create a dense aggregation that forms the structural framework of the blood clot.

Once a clot begins to form, it launches a self-accelerating positive feedback process that seals off the damaged vessel more quickly. Thrombin works with factor V to accelerate the production of prothrombin activator, which in turn produces more thrombin.

The cascade of enzymatic reactions acts as an amplifying mechanism to ensure the rapid clotting of blood (fig. 18.23). Each activated enzyme in the pathway produces a larger number of enzyme molecules at the following step. One activated molecule of factor XII at the start of the intrinsic pathway, for example, very quickly produces thousands if not millions of fibrin molecules. Note the similarity of this process to the *signal amplification* that occurs in hormone action (see fig. 17.21, p. 657).

Notice that the extrinsic mechanism requires fewer steps to activate factor X than the intrinsic mechanism does; it is a "shortcut" to coagulation. It takes 3 to 6 minutes for a clot to form by the intrinsic pathway but only 15 seconds or so by the extrinsic pathway. For this reason, when a small wound bleeds, you can stop the bleeding sooner by massaging the site. This releases thromboplastin from the perivascular tissues and activates or speeds up the extrinsic pathway.

A number of laboratory tests are used to evaluate the efficiency of coagulation. Normally, the bleeding of a fingerstick should stop within 2 to 3 minutes, and a sample of blood in a clean test tube should clot within 15 minutes. **Bleeding time** is most precisely measured by the *Ivy method*—inflating a blood pressure cuff on the arm to 40 mm Hg, making a 1 mm deep incision in the forearm, and measuring the time for it to stop bleeding. Normally it should stop in 1 to 9 minutes. Other techniques are available that can separately assess the effectiveness of the intrinsic and extrinsic mechanisms.

---

[27]*peri* = around; *vas* = vessel; *cular* = pertaining to
[28]*thrombo* = clot; *plast* = forming; *in* = substance

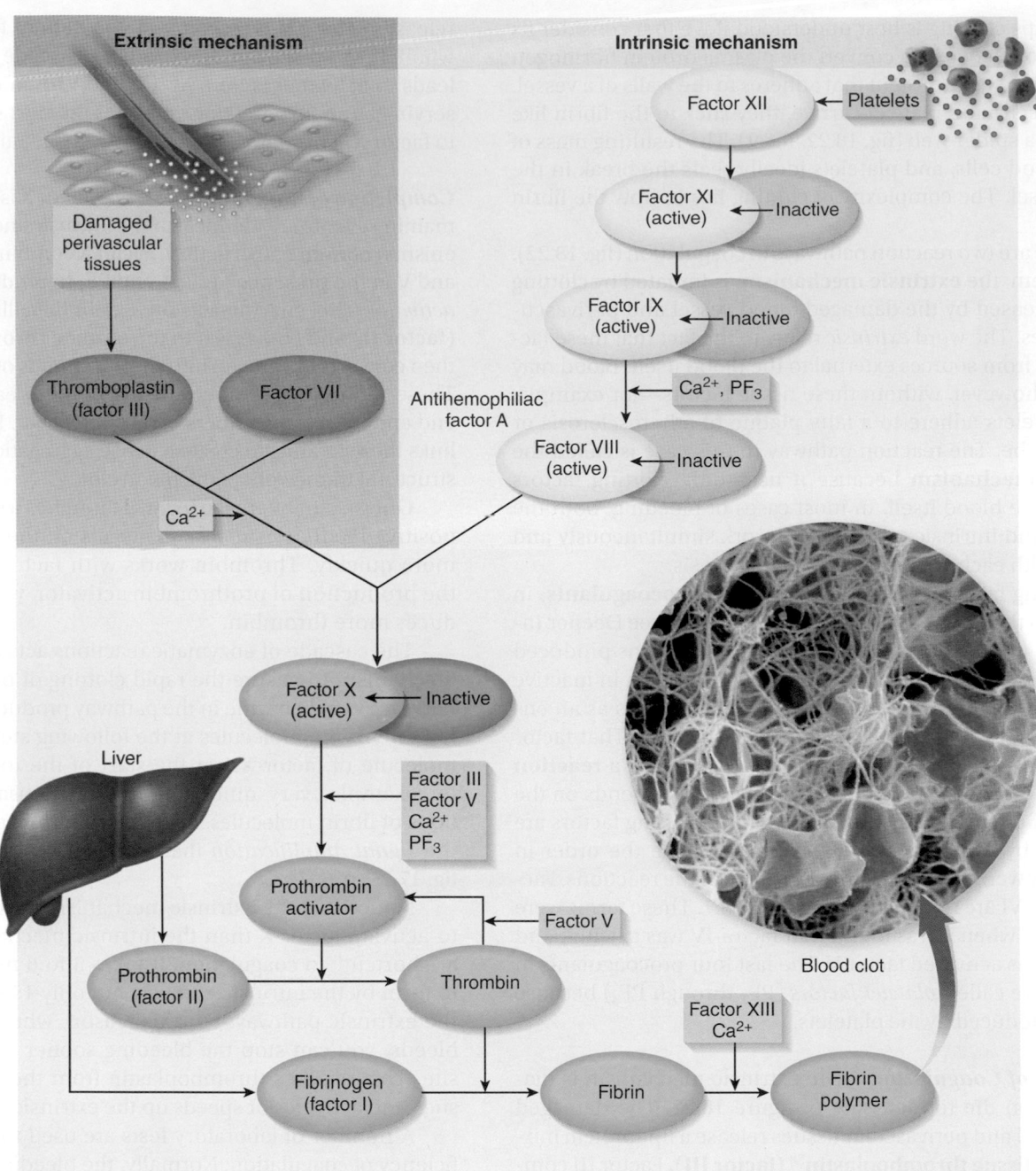

**FIGURE 18.22  The Pathways of Coagulation.**  Most clotting factors act as enzymes that convert the next factor from an inactive form to an active form. One enzyme molecule at any given level activates many enzyme molecules at the next level down, so the overall effect becomes amplified at each step. Inset: Platelets (orange) trapped in a mesh of sticky fibrin polymer (gray).

❓ *After you read about hemophilia C later in this chapter, explain whether it would affect the extrinsic mechanism, the intrinsic mechanism, or both.*

| TABLE 18.7 | | Clotting Factors (Procoagulants) | |
|---|---|---|---|
| **Number** | **Name** | **Origin** | **Function** |
| I | Fibrinogen | Liver | Precursor of fibrin |
| II | Prothrombin | Liver | Precursor of thrombin |
| III | Tissue thromboplastin | Perivascular tissue | Activates factor VII |
| V | Proaccelerin | Liver | Activates factor VII; combines with factor X to form prothrombin activator |
| VII | Proconvertin | Liver | Activates factor X in extrinsic pathway |
| VIII | Antihemophiliac factor A | Liver | Activates factor X in intrinsic pathway |
| IX | Antihemophiliac factor B | Liver | Activates factor VIII |
| X | Thrombokinase | Liver | Combines with factor V to form prothrombin activator |
| XI | Antihemophiliac factor C | Liver | Activates factor IX |
| XII | Hageman factor | Liver, platelets | Activates factor XI and plasmin; converts prekallikrein to kallikrein |
| XIII | Fibrin-stabilizing factor | Platelets, plasma | Cross-links fibrin filaments to make fibrin polymer and stabilize clot |
| $PF_1$ | Platelet factor 1 | Platelets | Same role as factor V; also accelerates platelet activation |
| $PF_2$ | Platelet factor 2 | Platelets | Accelerates thrombin formation |
| $PF_3$ | Platelet factor 3 | Platelets | Aids in activation of factor VIII and prothrombin activator |
| $PF_4$ | Platelet factor 4 | Platelets | Binds heparin during clotting to inhibit its anticoagulant effect |

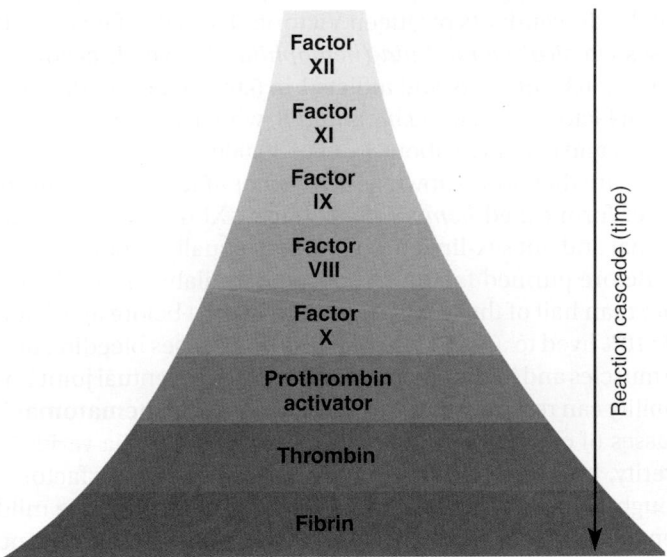

**FIGURE 18.23 The Reaction Cascade in Blood Clotting.** Each clotting factor produces many molecules of the next one, so the number of active clotting factors increases rapidly and a large amount of fibrin is quickly formed. The example shown here is for the intrinsic mechanism.

❓ *How does this compare with signal amplification in hormone action (compare fig. 17.21, p. 657)?*

## The Fate of Blood Clots

After a clot has formed, spinous pseudopods of the platelets adhere to strands of fibrin and contract. This pulls on the fibrin threads and draws the edges of the broken vessel together, like a drawstring closing a purse. Through this process of **clot retraction,** the clot becomes more compact within about 30 minutes.

Platelets and endothelial cells secrete a mitotic stimulant named **platelet-derived growth factor (PDGF).** PDGF stimulates fibroblasts and smooth muscle cells to multiply and repair the damaged blood vessel. Fibroblasts also invade the clot and produce fibrous connective tissue, which helps to strengthen and seal the vessel while the repairs take place.

Eventually, tissue repair is completed and the clot must be disposed of. **Fibrinolysis,** the dissolution of a clot, is achieved by a small cascade of reactions with a positive feedback component (fig. 18.24). In addition to promoting clotting, factor XII catalyzes the formation of a plasma enzyme called **kallikrein** (KAL-ih-KREE-in). Kallikrein, in turn, converts the inactive protein *plasminogen* into **plasmin,** a fibrin-dissolving enzyme that breaks up the clot. Thrombin also activates plasmin, and plasmin indirectly promotes the formation of more kallikrein, thus completing a positive feedback loop.

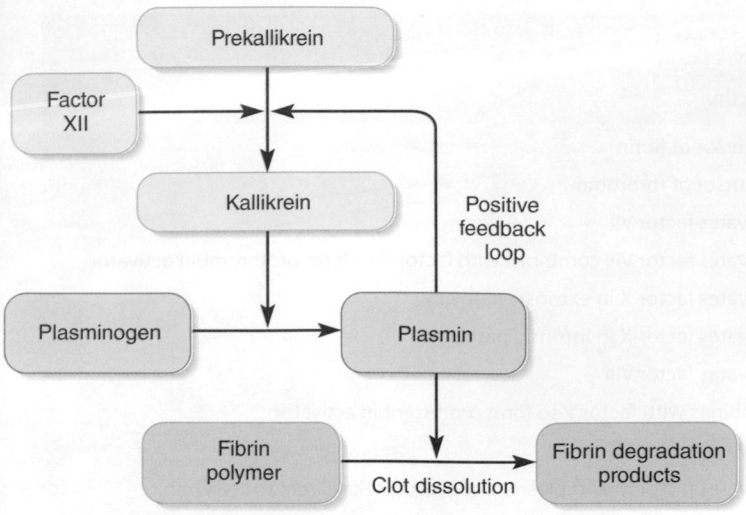

**FIGURE 18.24  The Mechanism for Dissolving Blood Clots.** Prekallikrein is converted to kallikrein. Kallikrein is an enzyme that catalyzes the formation of plasmin. Plasmin is an enzyme that dissolves the blood clot.

## Prevention of Inappropriate Clotting

Precise controls are required to prevent coagulation when it is not needed. These include the following:

- **Platelet repulsion.** As noted earlier, platelets do not adhere to the smooth prostacyclin-coated endothelium of healthy blood vessels.

- **Dilution.** Small amounts of thrombin form spontaneously in the plasma, but at normal rates of blood flow, the thrombin is diluted so quickly that a clot has little chance to form. If flow decreases, however, enough thrombin can accumulate to cause clotting. This can happen in circulatory shock, for example, when output from the heart is diminished and circulation slows down.

- **Anticoagulants.** Thrombin formation is suppressed by anticoagulants that are present in the plasma. **Antithrombin,** secreted by the liver, deactivates thrombin before it can act on fibrinogen. **Heparin,** secreted by basophils and mast cells, interferes with the formation of prothrombin activator, blocks the action of thrombin on fibrinogen, and promotes the action of antithrombin. Heparin is given by injection to patients with abnormal clotting tendencies.

## Clotting Disorders

In a process as complex as coagulation, it is not surprising that things can go wrong. Clotting deficiencies can result from causes as diverse as malnutrition, leukemia, and gallstones (see Deeper Insight 18.5).

## DEEPER INSIGHT 18.5

### CLINICAL APPLICATION

#### Liver Disease and Blood Clotting

Proper blood clotting depends on normal liver function for two reasons. First, the liver synthesizes most of the clotting factors. Therefore, diseases such as hepatitis, cirrhosis, and cancer that degrade liver function result in a deficiency of clotting factors. Second, the synthesis of clotting factors II, VII, IX, and X require vitamin K. The absorption of vitamin K from the diet requires bile, a liver secretion. Gallstones can lead to a clotting deficiency by obstructing the bile duct and thus interfering with bile secretion and vitamin K absorption. Efficient blood clotting is especially important in childbirth, since both the mother and infant bleed from the trauma of birth. Therefore, pregnant women should take vitamin K supplements to ensure fast clotting, and newborn infants may be given vitamin K injections.

A deficiency of any clotting factor can shut down the coagulation cascade. This happens in **hemophilia,** a family of hereditary diseases characterized by deficiencies of one factor or another. Because of the sex-linked recessive mechanism of heredity, hemophilia occurs predominantly in males. They can inherit it only from their mothers, however, as happened with the descendants of Queen Victoria. The lack of factor VIII causes *classical hemophilia (hemophilia A),* which accounts for about 83% of cases and afflicts 1 in 5,000 males worldwide. Lack of factor IX causes *hemophilia B,* which accounts for 15% of cases and occurs in about 1 out of 30,000 males. Factors VIII and IX are therefore known as *antihemophilic factors A and B.* A rarer form called *hemophilia C* (factor XI deficiency) is autosomal and not sex-linked, so it occurs equally in both sexes.

Before purified factor VIII became available in the 1960s, more than half of those with hemophilia died before age 5 and only 10% lived to age 21. Physical exertion causes bleeding into the muscles and joints. Excruciating pain and eventual joint immobility can result from intramuscular and joint **hematomas**[29] (masses of clotted blood in the tissues). Hemophilia varies in severity, however. Half of the normal level of clotting factor is enough to prevent the symptoms, and the symptoms are mild even in individuals with as little as 30% of the normal amount. Such cases may go undetected even into adulthood. Bleeding can be relieved for a few days by transfusion of plasma or purified clotting factors.

▶▶▶ **APPLY WHAT YOU KNOW**

*Why is it important for people with hemophilia not to use aspirin?* (Hint: *See fig. 17.24, p. 661, and do not say that aspirin "thins the blood," which is not true.*)

---

[29]*hemato* = blood; *oma* = mass

Failure of the blood to clot takes far fewer lives, however, than unwanted clotting. **Thrombosis,** the abnormal clotting of blood in an unbroken blood vessel, becomes increasingly problematic in old age. About 25% of people over age 50 experience venous blockage by thrombosis, especially people who do not exercise regularly or who are confined to a bed or wheelchair. The blood clots especially easily in the veins, where blood flow is slowest. A **thrombus** (clot) may grow large enough to obstruct a small vessel, or a piece of it may break loose and begin to travel in the bloodstream as an **embolus.**[30] An embolus can lodge in a small artery and block blood flow from that point on. If that vessel supplies vital tissue of the heart, brain, lung, or kidney, *infarction* (tissue death) may result. About 650,000 Americans die annually of *thromboembolism* (traveling blood clots) in the cerebral, coronary, and pulmonary arteries. Most strokes and heart attacks are due to thrombosis, and pulmonary failure often results from thromboembolism.

Thrombosis is more likely to occur in veins than in arteries because blood flows more slowly in the veins and does not dilute thrombin and fibrin as rapidly. It is especially common in the leg veins of inactive people and patients immobilized in a wheelchair or bed. Most venous blood flows directly to the heart and then to the lungs. Therefore, blood clots arising in the limbs commonly lodge in the lungs and cause *pulmonary embolism.* When blood cannot circulate freely through the lungs, it cannot receive oxygen and a person may die of hypoxia.

Table 18.8 describes some additional disorders of the blood. The effects of aging on the blood are described on page 1120.

### BEFORE YOU GO ON

Answer the following questions to test your understanding of the preceding section:

22. What are the three basic mechanisms of hemostasis?
23. How do the extrinsic and intrinsic mechanisms of coagulation differ? What do they have in common?
24. In what respect does blood clotting represent a negative feedback loop? What part of it is a positive feedback loop?
25. Describe some of the mechanisms that prevent clotting in undamaged vessels.
26. Describe a common source and effect of pulmonary embolism.

---

[30]*em* = in, within; *bolus* = ball, mass

| TABLE 18.8 | Some Disorders of the Blood |
|---|---|
| Disseminated intravascular coagulation (DIC) | Widespread clotting within unbroken vessels, limited to one organ or occurring throughout the body. Usually triggered by septicemia but also occurs when blood circulation slows markedly (as in cardiac arrest). Marked by widespread hemorrhaging, congestion of the vessels with clotted blood, and tissue necrosis in blood-deprived organs. |
| Infectious mononucleosis | Infection of B lymphocytes with Epstein–Barr virus, most commonly in adolescents and young adults. Usually transmitted by exchange of saliva, as in kissing. Causes fever, fatigue, sore throat, inflamed lymph nodes, and leukocytosis. Usually self-limiting and resolves within a few weeks. |
| Septicemia | Bacteremia (bacteria in the bloodstream) accompanying infection elsewhere in the body. Often causes fever, chills, and nausea, and may cause DIC or septic shock (see p. 764). |
| Thalassemia | A group of hereditary anemias most common in Greeks, Italians, and others of Mediterranean descent; shows a deficiency or absence of alpha or beta hemoglobin and RBC counts that may be less than 2 million/μL. |
| Thrombocytopenia | A platelet count below 100,000/mL. Causes include bone marrow destruction by radiation, drugs, poisons, or leukemia. Signs include small hemorrhagic spots in the skin or hematomas in response to minor trauma. |

*Disorders described elsewhere*

| | | |
|---|---|---|
| Anemia p. 683 | Hypoxemia p. 681 | Sickle-cell disease p. 684 |
| Hematoma p. 702 | Leukemia p. 696 | Thromboembolism p. 703 |
| Hemolytic disease of the newborn p. 688 | Leukocytosis p. 694 | Thrombosis p. 703 |
| Hemophilia p. 702 | Leukopenia p. 694 | Transfusion reaction p. 687 |
| Hypoproteinemia p. 677 | Polycythemia p. 683 | |

# DEEPER INSIGHT 18.6

## CLINICAL APPLICATION

### Clinical Management of Blood Clotting

For many cardiovascular patients, the goal of treatment is to prevent clotting or to dissolve clots that have already formed. Several strategies employ inorganic salts and products of bacteria, plants, and animals with anticoagulant and clot-dissolving effects.

#### Preventing Clots from Forming

Since calcium is an essential requirement for blood clotting, blood samples can be kept from clotting by adding a few crystals of sodium oxalate, sodium citrate, or EDTA[31]—salts that bind calcium ions and prevent them from participating in the coagulation reactions. Blood-collection equipment such as hematocrit tubes may also be coated with heparin, a natural anticoagulant whose action was explained earlier.

Since vitamin K is required for the synthesis of clotting factors, anything that antagonizes vitamin K usage makes the blood clot less readily. One vitamin K antagonist is *coumarin*[32] (COO-muh-rin), a sweet-smelling extract of tonka beans, sweet clover, and other plants, used in perfume. Taken orally by patients at risk for thrombosis, coumarin takes up to 2 days to act, but it has longer-lasting effects than heparin. A similar vitamin K antagonist is the pharmaceutical preparation *warfarin*[33] *(Coumadin),* which was originally developed as a pesticide—it makes rats bleed to death. Obviously, such anticoagulants must be used in humans with great care.

As explained in chapter 17, aspirin suppresses the formation of the eicosanoid thromboxane $A_2$, a factor in platelet aggregation. Low daily doses of aspirin can therefore suppress thrombosis and help to prevent heart attacks.

Many parasites feed on the blood of vertebrates and secrete anticoagulants to keep the blood flowing. Among these are aquatic worms known as leeches. Leeches secrete a local anesthetic that makes their bites painless; therefore, as early as 1567 BCE, physicians used them for bloodletting. This method was less painful and repugnant to their patients than *phlebotomy*[34]—cutting a vein—and indeed, leeching became very popular. In seventeenth-century France, it was quite the rage; tremendous numbers of leeches were used in ill-informed attempts to treat headaches, insomnia, whooping cough, obesity, tumors, menstrual cramps, mental illness, and almost anything else doctors or their patients imagined to be caused by "bad blood."

The first known anticoagulant was discovered in the saliva of the medicinal leech, *Hirudo medicinalis,* in 1884. Named *hirudin,* it is a polypeptide that prevents clotting by inhibiting thrombin. It causes the blood to flow freely while the leech feeds and for as long as an hour thereafter. While the doctrine of bad blood is now long discredited, leeches have lately reentered medical usage for other reasons (fig. 18.25). A major problem in reattaching a severed body part such as a finger or ear is that the tiny veins draining these organs are too small to reattach surgically. Since arterial blood flows into the reattached organ and cannot flow out as easily, it pools and clots there. This inhibits the regrowth of veins and the flow of fresh blood through the organ, and often leads to necrosis. Some vascular surgeons now place leeches on

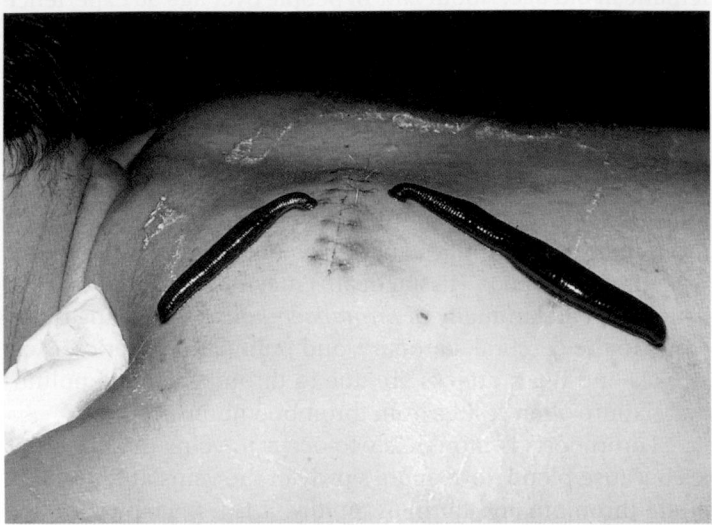

**FIGURE 18.25  A Modern Use of Leeching.** Two medicinal leeches are being used to remove clotted blood from a postsurgical hematoma. Despite their formidable size, the leeches secrete a natural anesthetic and produce a painless bite.

 *How does the modern theory behind leeching differ from the theory of leeching that was popular a few centuries ago?*

the reattached part. Their anticoagulant keeps the blood flowing freely and allows new veins to grow. After 5 to 7 days, venous drainage is restored and leeching can be stopped.

Anticoagulants also occur in the venom of some snakes. *Arvin,* for example, is obtained from the venom of the Malayan viper. It rapidly breaks down fibrinogen and may have potential as a clinical anticoagulant.

#### Dissolving Clots That Have Already Formed

When a clot has already formed, it can be treated with clot-dissolving drugs such as *streptokinase,* an enzyme made by certain bacteria (streptococci). Intravenous streptokinase is used to dissolve blood clots in coronary vessels, for example. It is nonspecific, however, and digests almost any protein. *Tissue plasminogen activator (TPA)* works faster, is more specific, and is now made by transgenic bacteria. TPA converts plasminogen into the clot-dissolving enzyme plasmin. Some anticoagulants of animal origin also work by dissolving fibrin. A giant Amazon leech, *Haementeria,* produces one such anticoagulant named *hementin.* This, too, has been successfully produced by genetically engineered bacteria and used to dissolve blood clots in cardiac patients.

---

[31]ethylenediaminetetraacetic acid
[32]*coumaru,* tonka bean tree
[33]acronym from *Wisconsin Alumni Research Foundation*
[34]*phlebo* = vein; *tomy* = cutting

# STUDY GUIDE

## ► Assess Your Learning Outcomes

*To test your knowledge, discuss the following topics with a study partner or in writing, ideally from memory.*

### 18.1 Introduction (p. 673)

1. Constituents of the circulatory system; the difference between circulatory system and cardiovascular system
2. The diverse functions of blood
3. Relative amounts of plasma and formed elements in the blood, and the three categories of formed elements
4. The composition of blood plasma
5. Importance of the viscosity and osmolarity of blood, what accounts for each, and the pathological effects of abnormal viscosity or osmolarity
6. The definition of *colloid osmotic pressure*
7. General aspects of hemopoiesis; where it occurs in the embryo, in the fetus, and after birth; and the stem cell with which all hemopoietic pathways begin

### 18.2 Erythrocytes (p. 678)

1. Erythrocyte (RBC) structure and function
2. The functions of hemoglobin and carbonic anhydrase
3. Hemoglobin structure and what parts of it bind $O_2$ and $CO_2$
4. Three ways of quantifying the RBCs and hemoglobin level of the blood; the definition and units of measurement of each; and reasons for the differences between male and female values
5. Stages of erythropoiesis and major transformations in each
6. Why iron is essential; how the stomach converts dietary iron to a usable form; and the roles of gastroferritin, transferrin, and ferritin in iron metabolism
7. Homeostatic regulation of erythropoiesis, including the origins and role of erythropoietin (EPO)
8. The life span of an RBC and how the body disposes of old RBCs
9. How the body disposes of the hemoglobin from expired RBCs and how this relates to the pigments of bile, feces, and urine
10. Excesses and deficiencies in RBC count and the forms, causes, and pathological consequences of each

11. Causes and effects of hemoglobin deficiencies and the pathology of sickle-cell disease and thalassemia

### 18.3 Blood Types (p. 685)

1. What determines a person's blood type; blood types of the ABO group and how they differ in genetics and RBC antigens
2. Why an individual does not have plasma antibodies against the ABO types at birth, but develops them during infancy; how these antibodies limit transfusion compatibility
3. The cause and mechanism of a transfusion reaction and why it can lead to renal failure and death; the meaning of *agglutination* and *hemolysis*
4. Blood types of the Rh group and how they differ in their genetics and RBC antigens
5. What can cause a person to develop antibodies against Rh-positive RBCs
6. Hemolytic disease of the newborn; why it seldom occurs in a woman's first susceptible child, but is more common in later pregnancies; and how it is treated
7. Blood groups other than ABO and Rh, and their usefulness for certain purposes

### 18.4 Leukocytes (p. 690)

1. The general function of all leukocytes (WBCs)
2. Three kinds of granulocytes, two kinds of agranulocytes, and what distinguishes granulocytes from agranulocytes as a class
3. The appearance, relative size and number, and functions of each WBC type, and the conditions under which each type increases in a differential WBC count
4. Three principal cell lines, the stages, and the anatomical sites of leukopoiesis
5. The relative length of time that WBCs travel in the bloodstream and spend in other tissues; which type recirculates into the blood and which types do not; and the relative life spans of WBCs
6. Causes and effects of leukopenia and leukocytosis
7. The naming and classification of various kinds of leukemia; why leukemia is typically accompanied by RBC and platelet deficiencies and elevated risk of opportunistic infection

### 18.5 Platelets and Hemostasis—The Control of Bleeding (p. 696)

1. Platelet structure and functions, a typical platelet count, and why platelets are not considered to be cells
2. The site and process of platelet production, and the hormone that stimulates it
3. Three mechanisms of hemostasis and their relative quickness and effectiveness
4. The general objective of coagulation; the end product of the coagulation reactions, and basic differences between the extrinsic and intrinsic mechanisms
5. Essentials of the extrinsic mechanism including the chemical that initiates it, other procoagulants involved, and the point at which it converges with the intrinsic mechanism at a common intermediate
6. Essentials of the intrinsic mechanism including the chemical that initiates it, other procoagulants involved, and the aforesaid point of convergence with the extrinsic mechanism
7. Steps in the continuation of coagulation from factor X to fibrin, including the procoagulants involved
8. The roles of positive feedback and enzyme amplification in coagulation
9. The processes of clot retraction, vessel repair, and fibrinolysis
10. Three mechanisms of preventing inappropriate coagulation in undamaged vessels
11. Causes of clotting deficiencies including the types, genetics, and pathology of hemophilia
12. Terms for unwanted or inappropriate clotting in a vessel, the clot itself, and a clot that breaks free and travels in the bloodstream
13. Why spontaneous clotting more often occurs in the veins than in the arteries; the danger presented by traveling blood clots; and why traveling clots so often lodge in the lungs even if they originate as far away as the lower limbs

# STUDY GUIDE

## ► Testing Your Recall

*Answers in Appendix B*

1. Antibodies belong to a class of plasma proteins called
   a. albumins.
   b. gamma globulins.
   c. alpha globulins.
   d. procoagulants.
   e. agglutinins.

2. Serum is blood plasma minus its
   a. sodium ions.
   b. calcium ions.
   c. clotting proteins.
   d. globulins.
   e. albumin.

3. Which of the following conditions is most likely to cause hemolytic anemia?
   a. folic acid deficiency
   b. iron deficiency
   c. mushroom poisoning
   d. alcoholism
   e. hypoxemia

4. It is impossible for a type O+ baby to have a type _____ mother.
   a. AB–
   b. O–
   c. O+
   d. A+
   e. B+

5. Which of the following is *not* a component of hemostasis?
   a. platelet plug formation
   b. agglutination
   c. clot retraction
   d. a vascular spasm
   e. degranulation

6. Which of the following contributes most to the viscosity of blood?
   a. albumin
   b. sodium
   c. globulins
   d. erythrocytes
   e. fibrin

7. Which of these is a granulocyte?
   a. a monocyte
   b. a lymphocyte
   c. a macrophage
   d. an eosinophil
   e. an erythrocyte

8. Excess iron is stored in the liver as a complex called
   a. gastroferritin.
   b. transferrin.
   c. ferritin.
   d. hepatoferritin.
   e. erythropoietin.

9. Pernicious anemia is a result of
   a. hypoxemia.
   b. iron deficiency.
   c. malaria.
   d. lack of intrinsic factor.
   e. Rh incompatibility.

10. The first clotting factor that the intrinsic and extrinsic pathways have in common is
    a. thromboplastin.
    b. Hageman factor.
    c. factor X.
    d. prothrombin activator.
    e. factor VIII.

11. Production of all the formed elements of blood is called _____.

12. The percentage of blood volume composed of RBCs is called the _____.

13. The extrinsic pathway of coagulation is activated by _____ from damaged perivascular tissues.

14. The RBC antigens that determine transfusion compatibility are called _____.

15. The hereditary lack of factor VIII causes a disease called _____.

16. The overall cessation of bleeding, involving several mechanisms, is called _____.

17. _____ results from a mutation that changes one amino acid in the hemoglobin molecule.

18. An excessively high RBC count is called _____.

19. Intrinsic factor enables the small intestine to absorb _____.

20. The kidney hormone _____ stimulates RBC production.

## ► Building Your Medical Vocabulary

*Answers in Appendix B*

*State a medical meaning of each word element below, and give a term in which it or a slight variation of it is used.*

1. an-

2. -blast

3. erythro-

4. glutino-

5. hemo-

6. leuko-

7. -penia

8. phlebo-

9. -poiesis

10. thrombo-

Failure of the blood to clot takes far fewer lives, however, than unwanted clotting. **Thrombosis,** the abnormal clotting of blood in an unbroken blood vessel, becomes increasingly problematic in old age. About 25% of people over age 50 experience venous blockage by thrombosis, especially people who do not exercise regularly or who are confined to a bed or wheelchair. The blood clots especially easily in the veins, where blood flow is slowest. A **thrombus** (clot) may grow large enough to obstruct a small vessel, or a piece of it may break loose and begin to travel in the bloodstream as an **embolus.**[30] An embolus can lodge in a small artery and block blood flow from that point on. If that vessel supplies vital tissue of the heart, brain, lung, or kidney, *infarction* (tissue death) may result. About 650,000 Americans die annually of *thromboembolism* (traveling blood clots) in the cerebral, coronary, and pulmonary arteries. Most strokes and heart attacks are due to thrombosis, and pulmonary failure often results from thromboembolism.

Thrombosis is more likely to occur in veins than in arteries because blood flows more slowly in the veins and does not dilute thrombin and fibrin as rapidly. It is especially common in the leg veins of inactive people and patients immobilized in a wheelchair or bed. Most venous blood flows directly to the heart and then to the lungs. Therefore, blood clots arising in the limbs commonly lodge in the lungs and cause *pulmonary embolism.* When blood cannot circulate freely through the lungs, it cannot receive oxygen and a person may die of hypoxia.

Table 18.8 describes some additional disorders of the blood. The effects of aging on the blood are described on page 1120.

### BEFORE YOU GO ON

Answer the following questions to test your understanding of the preceding section:

22. What are the three basic mechanisms of hemostasis?

23. How do the extrinsic and intrinsic mechanisms of coagulation differ? What do they have in common?

24. In what respect does blood clotting represent a negative feedback loop? What part of it is a positive feedback loop?

25. Describe some of the mechanisms that prevent clotting in undamaged vessels.

26. Describe a common source and effect of pulmonary embolism.

---

[30]*em* = in, within; *bolus* = ball, mass

| TABLE 18.8 | Some Disorders of the Blood |
|---|---|
| Disseminated intravascular coagulation (DIC) | Widespread clotting within unbroken vessels, limited to one organ or occurring throughout the body. Usually triggered by septicemia but also occurs when blood circulation slows markedly (as in cardiac arrest). Marked by widespread hemorrhaging, congestion of the vessels with clotted blood, and tissue necrosis in blood deprived organs. |
| Infectious mononucleosis | Infection of B lymphocytes with Epstein–Barr virus, most commonly in adolescents and young adults. Usually transmitted by exchange of saliva, as in kissing. Causes fever, fatigue, sore throat, inflamed lymph nodes, and leukocytosis. Usually self-limiting and resolves within a few weeks. |
| Septicemia | Bacteremia (bacteria in the bloodstream) accompanying infection elsewhere in the body. Often causes fever, chills, and nausea, and may cause DIC or septic shock (see p. 764). |
| Thalassemia | A group of hereditary anemias most common in Greeks, Italians, and others of Mediterranean descent; shows a deficiency or absence of alpha or beta hemoglobin and RBC counts that may be less than 2 million/$\mu$L. |
| Thrombocytopenia | A platelet count below 100,000/mL. Causes include bone marrow destruction by radiation, drugs, poisons, or leukemia. Signs include small hemorrhagic spots in the skin or hematomas in response to minor trauma. |

*Disorders described elsewhere*

# DEEPER INSIGHT 18.6

## CLINICAL APPLICATION

### Clinical Management of Blood Clotting

For many cardiovascular patients, the goal of treatment is to prevent clotting or to dissolve clots that have already formed. Several strategies employ inorganic salts and products of bacteria, plants, and animals with anticoagulant and clot-dissolving effects.

### Preventing Clots from Forming

Since calcium is an essential requirement for blood clotting, blood samples can be kept from clotting by adding a few crystals of sodium oxalate, sodium citrate, or EDTA[31]—salts that bind calcium ions and prevent them from participating in the coagulation reactions. Blood-collection equipment such as hematocrit tubes may also be coated with heparin, a natural anticoagulant whose action was explained earlier.

Since vitamin K is required for the synthesis of clotting factors, anything that antagonizes vitamin K usage makes the blood clot less readily. One vitamin K antagonist is *coumarin*[32] (COO-muh-rin), a sweet-smelling extract of tonka beans, sweet clover, and other plants, used in perfume. Taken orally by patients at risk for thrombosis, coumarin takes up to 2 days to act, but it has longer-lasting effects than heparin. A similar vitamin K antagonist is the pharmaceutical preparation *warfarin*[33] *(Coumadin),* which was originally developed as a pesticide—it makes rats bleed to death. Obviously, such anticoagulants must be used in humans with great care.

As explained in chapter 17, aspirin suppresses the formation of the eicosanoid thromboxane A$_2$, a factor in platelet aggregation. Low daily doses of aspirin can therefore suppress thrombosis and help to prevent heart attacks.

Many parasites feed on the blood of vertebrates and secrete anticoagulants to keep the blood flowing. Among these are aquatic worms known as leeches. Leeches secrete a local anesthetic that makes their bites painless; therefore, as early as 1567 BCE, physicians used them for bloodletting. This method was less painful and repugnant to their patients than *phlebotomy*[34]—cutting a vein—and indeed, leeching became very popular. In seventeenth-century France, it was quite the rage; tremendous numbers of leeches were used in ill-informed attempts to treat headaches, insomnia, whooping cough, obesity, tumors, menstrual cramps, mental illness, and almost anything else doctors or their patients imagined to be caused by "bad blood."

The first known anticoagulant was discovered in the saliva of the medicinal leech, *Hirudo medicinalis,* in 1884. Named *hirudin,* it is a polypeptide that prevents clotting by inhibiting thrombin. It causes the blood to flow freely while the leech feeds and for as long as an hour thereafter. While the doctrine of bad blood is now long discredited, leeches have lately reentered medical usage for other reasons (fig. 18.25). A major problem in reattaching a severed body part such as a finger or ear is that the tiny veins draining these organs are too small to reattach surgically. Since arterial blood flows into the reattached organ and cannot flow out as easily, it pools and clots there. This inhibits the regrowth of veins and the flow of fresh blood through the organ, and often leads to necrosis. Some vascular surgeons now place leeches on

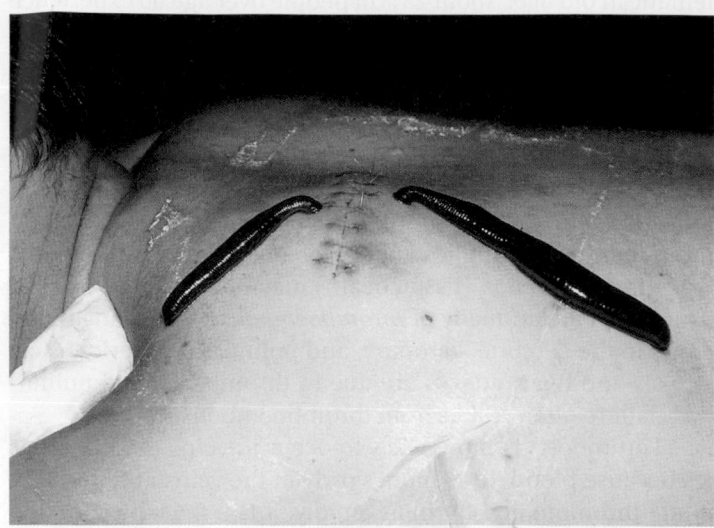

**FIGURE 18.25  A Modern Use of Leeching.** Two medicinal leeches are being used to remove clotted blood from a postsurgical hematoma. Despite their formidable size, the leeches secrete a natural anesthetic and produce a painless bite.

 *How does the modern theory behind leeching differ from the theory of leeching that was popular a few centuries ago?*

the reattached part. Their anticoagulant keeps the blood flowing freely and allows new veins to grow. After 5 to 7 days, venous drainage is restored and leeching can be stopped.

Anticoagulants also occur in the venom of some snakes. *Arvin,* for example, is obtained from the venom of the Malayan viper. It rapidly breaks down fibrinogen and may have potential as a clinical anticoagulant.

### Dissolving Clots That Have Already Formed

When a clot has already formed, it can be treated with clot-dissolving drugs such as *streptokinase,* an enzyme made by certain bacteria (streptococci). Intravenous streptokinase is used to dissolve blood clots in coronary vessels, for example. It is nonspecific, however, and digests almost any protein. *Tissue plasminogen activator (TPA)* works faster, is more specific, and is now made by transgenic bacteria. TPA converts plasminogen into the clot-dissolving enzyme plasmin. Some anticoagulants of animal origin also work by dissolving fibrin. A giant Amazon leech, *Haementeria,* produces one such anticoagulant named *hementin.* This, too, has been successfully produced by genetically engineered bacteria and used to dissolve blood clots in cardiac patients.

---

[31] ethylenediaminetetraacetic acid
[32] *coumaru,* tonka bean tree
[33] acronym from *Wisconsin Alumni Research Foundation*
[34] *phlebo* = vein; *tomy* = cutting

# STUDY GUIDE

## ▶ True or False

*Answers in Appendix B*

*Determine which five of the following statements are false, and briefly explain why.*

1. By volume, the blood usually contains more plasma than blood cells.

2. An increase in the albumin concentration of the blood would tend to increase blood pressure.

3. Anemia is caused by a low oxygen concentration in the blood.

4. *Hemostasis, coagulation,* and *clotting* are three terms for the same process.

5. A man with blood type A+ and a woman with blood type B+ could have a baby with type O–.

6. Lymphocytes are the most abundant WBCs in the blood.

7. Calcium ions are required for blood clotting.

8. All formed elements of the blood come ultimately from pluripotent stem cells.

9. When RBCs die and break down, the globin moiety of hemoglobin is excreted and the heme is recycled to new RBCs.

10. Leukemia is a severe deficiency of white blood cells.

## ▶ Testing Your Comprehension

*Answers at www.mhhe.com/saladin7*

1. Why would erythropoiesis not correct the hypoxemia resulting from lung cancer?

2. People with chronic kidney disease often have hematocrits of less than half the normal value. Explain why.

3. An elderly white woman is hit by a bus and severely injured. Accident investigators are informed that she lives in an abandoned warehouse, where her few personal effects include several empty wine bottles and an expired driver's license indicating she is 72 years old. At the hospital, she is found to be severely anemic. List all the factors you can think of that may contribute to her anemia.

4. How is coagulation different from agglutination?

5. Although fibrinogen and prothrombin are equally necessary for blood clotting, fibrinogen is about 4% of the plasma protein whereas prothrombin is present only in small traces. In light of the roles of these clotting factors and your knowledge of enzymes, explain this difference in abundance.

## ▶ Improve Your Grade at www.mhhe.com/saladin7

*Download animations to study when it fits your schedule. Practice quizzes, labeling activities, games, and flashcards offer fun ways to master the chapter concepts. Or, download image PowerPoint files for each chapter to create a study guide or for taking notes during lecture.*

# THE CIRCULATORY SYSTEM: HEART

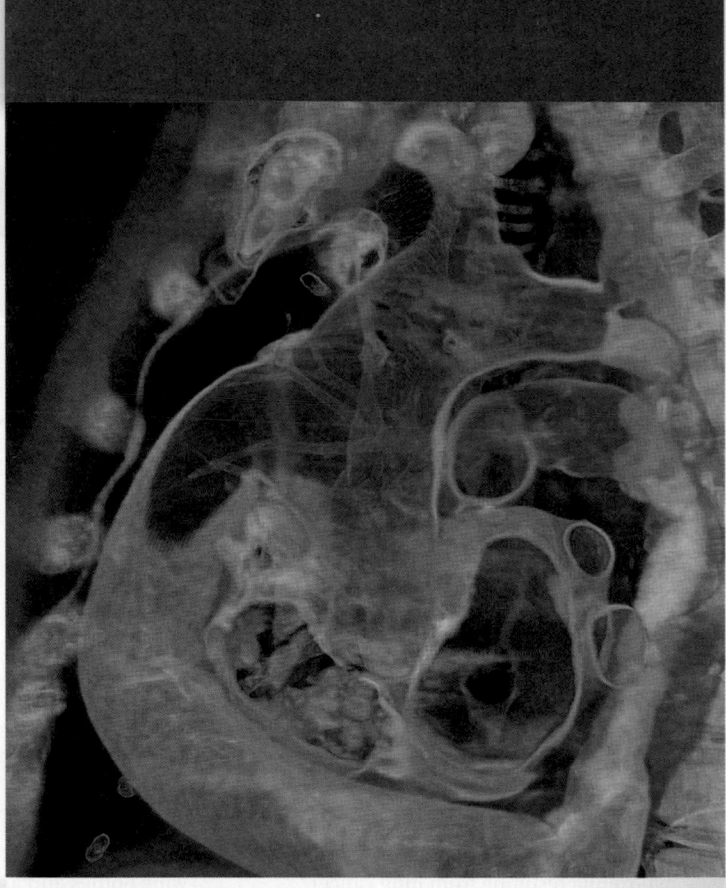

A three-dimensional CT scan of the heart; lateral view of a person facing left

Anatomy & Physiology **REVEALED®**
aprevealed.com

**Module 9: Cardiovascular System**

## BRUSHING UP

- Understanding the movement of blood through the heart chambers and valves calls on the principle of flow down gradients, especially pressure gradients (p. 17).

- For the best understanding of cardiac muscle, be sure you are familiar with desmosomes and gap junctions (pp. 163–164) and with the ultrastructure of striated muscle (p. 379).

- You must be familiar with membrane resting potentials (p. 447) and action potentials (p. 450) to understand cardiac pacemaker physiology and the excitation of cardiac muscle.

- Review excitation–contraction coupling in skeletal muscle (p. 408) for comparison with the process in cardiac muscle.

- The length–tension relationship of striated muscle (p. 412) helps to explain variation in the ejection of blood by the heart.

- Adjustment of cardiac output to states of rest and physical exertion hinges on understanding the mode of action of the sympathetic and parasympathetic nervous systems (pp. 561–567).

**W**e are more conscious of our heart than we are of most organs, and more wary of its failure. Speculation about the heart is at least as old as written history. Some ancient Chinese, Egyptian, Greek, and Roman scholars correctly surmised that the heart is a pump for filling the vessels with blood. Aristotle's views, however, were a step backward. Perhaps because the heart quickens its pace when we're emotionally aroused, and because grief causes "heartache," he regarded it primarily as the seat of emotion, as well as a source of heat to aid digestion. During the Middle Ages, Western medical schools clung dogmatically to the ideas of Aristotle. Perhaps the only significant advance came from Arabic medicine, when thirteenth-century physician Ibn an-Nafis described the role of the coronary blood vessels in nourishing the heart. The sixteenth-century dissections and anatomical charts of Vesalius, however, greatly improved knowledge of cardiovascular anatomy and set the stage for a more scientific study of the heart and treatment of its disorders—the science we now call **cardiology.**[1]

In the early decades of the twentieth century, little could be recommended for heart disease other than bed rest. Then nitroglycerin was found to improve coronary circulation and relieve the pain resulting from physical exertion, digitalis proved helpful for treating abnormal heart rhythms, and diuretics were first used to reduce hypertension. Coronary bypass surgery; replacement of diseased valves; clot-dissolving enzymes; heart transplants; and artificial pacemakers, valves, and hearts have made cardiology one of today's most dramatic and attention-getting fields of medicine.

## 19.1 Overview of the Cardiovascular System

### Expected Learning Outcomes
When you have completed this section, you should be able to

a. define and distinguish between the pulmonary and systemic circuits;

b. describe the general location, size, and shape of the heart; and

c. describe the pericardial sac that encloses the heart.

The **cardiovascular**[2] **system** consists of the heart and the blood vessels. The heart is a muscular pump that keeps blood flowing through the vessels. The vessels deliver the blood to all the body's organs and then return it to the heart (fig. 19.1). The broader term *circulatory system* also includes the blood, and some authorities use it to include the lymphatic system as well (described in chapter 21).

### The Pulmonary and Systemic Circuits

The cardiovascular system has two major divisions: a **pulmonary circuit,** which carries blood to the lungs for gas exchange and returns it to the heart, and a **systemic circuit,** which supplies blood to every organ of the body, including other parts of the lungs and the wall of the heart itself.

The right half of the heart supplies the pulmonary circuit. It receives blood that has circulated through the body, unloaded its oxygen and nutrients, and picked up a load of carbon dioxide and other wastes. It pumps this oxygen-poor blood into a large artery, the *pulmonary trunk,* which immediately divides into right and left *pulmonary arteries.* These transport blood to the air sacs *(alveoli)* of the lungs, where carbon dioxide is unloaded and oxygen is picked up. The oxygen-rich blood then flows by way of the *pulmonary veins* to the left side of the heart.

The left side supplies the systemic circuit. Blood leaves it by way of another large artery, the *aorta.* The aorta takes a sharp inverted U-turn, the *aortic arch,* and passes downward, posterior to the heart. The aortic arch gives off arteries that supply the head, neck, and upper limbs. The aorta then travels through the thoracic and abdominal cavities and issues smaller arteries to the other organs before branching into the lower limbs. After circulating through the body, the now deoxygenated systemic blood returns to the right side of the heart mainly by way of two large veins: the *superior vena cava* (draining the upper body) and *inferior vena cava* (draining everything below the diaphragm). The major arteries and veins entering and leaving the heart are called the *great vessels (great arteries* and *veins)* because of their relatively large diameters.

---

[1]*cardio* = heart; *logy* = study

[2]*cardio* = heart; *vas* = vessel

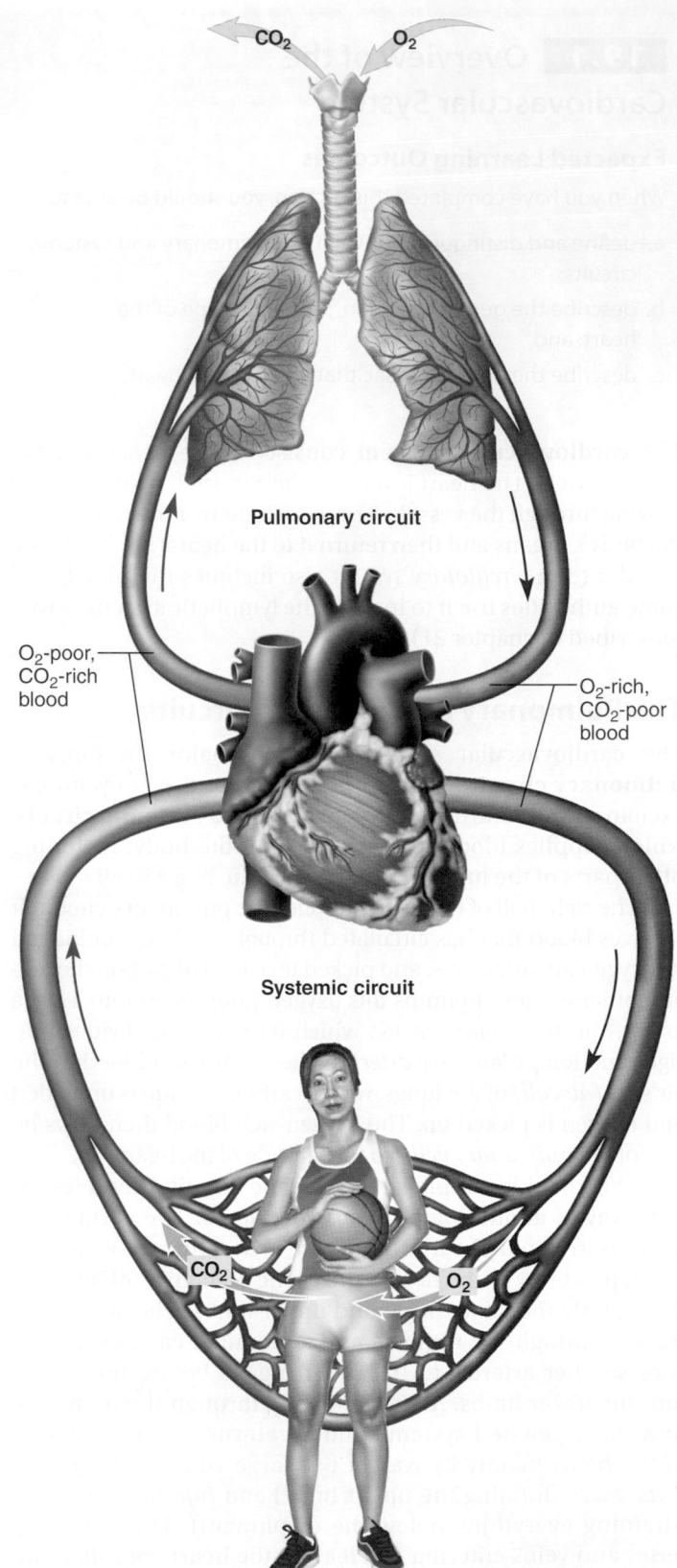

**FIGURE 19.1** **General Schematic of the Cardiovascular System.**

❓ *Are the lungs supplied by the pulmonary circuit, the systemic circuit, or both? Explain.* **AP|R**

## Position, Size, and Shape of the Heart

The heart lies within a thick partition called the **mediastinum** between the two lungs. It extends from a broad **base** at its uppermost end, where the great vessels are attached, to a bluntly pointed **apex** at the lower end, just above the diaphragm. It tilts toward the left from base to apex, so somewhat more than half the heart is to the left of the body's median plane. We can see this especially in a cross (horizontal) section through the thorax (fig. 19.2; see also figs. B.10–B.11, p. 386).

The adult heart is about 9 cm (3.5 in.) wide at the base, 13 cm (5 in.) from base to apex, and 6 cm (2.5 in.) from anterior to posterior at its thickest point. At any age, it is roughly the size of the same person's fist. It weighs about 300 g (10 ounces) in adults.

## The Pericardium

The heart is enclosed in a double-walled sac called the **pericardium.**[3] The outer wall, called the **pericardial sac (parietal pericardium),** has a tough, superficial *fibrous layer* of dense irregular connective tissue and a thin, deep *serous layer.* The serous layer turns inward at the base of the heart and forms the **visceral pericardium,** equivalent to the epicardium described shortly as part of the heart wall (fig. 19.3). The pericardial sac is anchored by ligaments to the diaphragm below and the sternum anterior to it, and more loosely anchored by fibrous connective tissue to mediastinal tissue posterior to the heart. The pericardium isolates the heart from other thoracic organs and allows it room to expand, yet resists excessive expansion. (See *cardiac tamponade* in table 19.2, p. 738.)

Between the parietal and visceral membranes is a space called the **pericardial cavity** (see figs. 19.2b and 19.3). The heart is not inside the pericardial cavity but enfolded by it. The relationship of the heart to the pericardium is often described by comparison to a fist pushed into an underinflated balloon (fig. 19.3c). The balloon surface in contact with the fist is like the epicardium; the outer balloon surface is like the pericardial sac, and the air space between them is like the pericardial cavity.

The pericardial cavity contains 5 to 30 mL of **pericardial fluid,** exuded by the serous layer of the pericardiac sac. The fluid lubricates the membranes and allows the heart to beat with minimal friction. In *pericarditis*—inflammation of the pericardium—the membranes may become roughened and produce a painful *friction rub* with each heartbeat.

### ▮ BEFORE YOU GO ON

Answer the following questions to test your understanding of the preceding section:

1. Distinguish between the pulmonary and systemic circuits and state which part of the heart supplies each one.

2. Predict the effect of a pericardial sac that fits too tightly around the heart. Predict the effect of a failure of the pericardial sac to secrete pericardial fluid.

---

[3]*peri* = around; *cardi* = heart

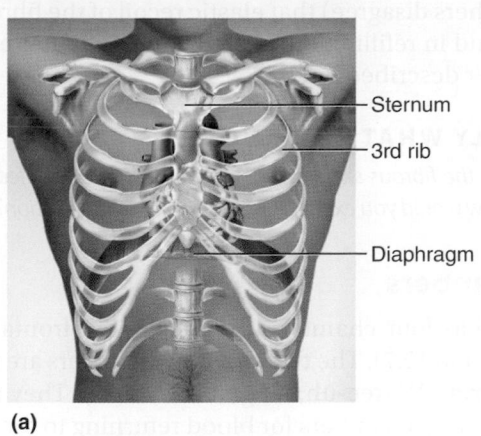

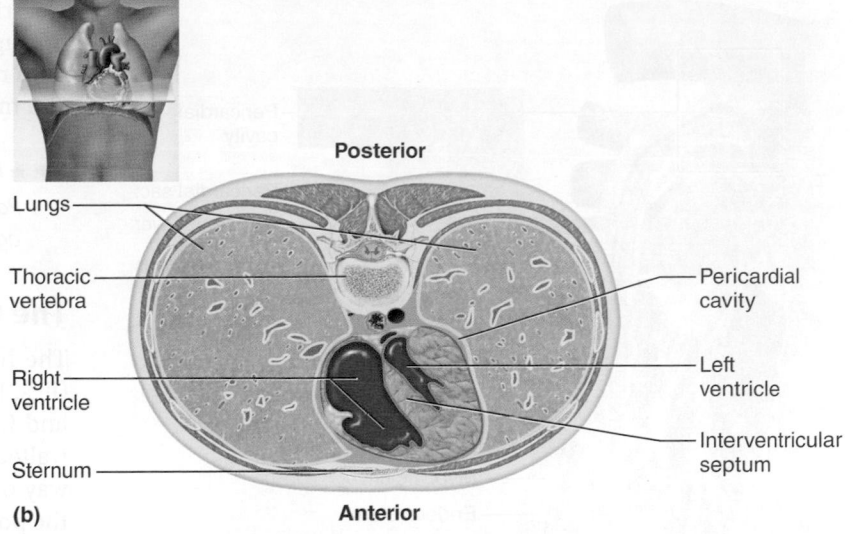

(a)

(b)

**Posterior**

Lungs

Thoracic vertebra

Right ventricle

Sternum

Pericardial cavity

Left ventricle

Interventricular septum

**Anterior**

**FIGURE 19.2  Position of the Heart in the Thoracic Cavity.**  (a) Relationship to the thoracic cage. (b) Cross section of the thorax at the level of the heart. (c) Frontal view with the lungs slightly retracted and the pericardial sac opened.  AP|R

❓ *Does most of the heart lie to the right or left of the median plane?*

## 19.2  Gross Anatomy of the Heart

### Expected Learning Outcomes

When you have completed this section, you should be able to

a.  describe the three layers of the heart wall;

b.  identify the four chambers of the heart;

c.  identify the surface features of the heart and correlate them with its internal four-chambered anatomy;

d.  identify the four valves of the heart;

e.  trace the flow of blood through the four chambers and valves of the heart and adjacent blood vessels; and

f.  describe the arteries that nourish the myocardium and the veins that drain it.

### The Heart Wall

The heart wall consists of three layers: *epicardium, myocardium,* and *endocardium.*

The **epicardium**[4] **(visceral pericardium)** is a serous membrane of the external heart surface. It consists mainly of a simple squamous epithelium overlying a thin layer of areolar tissue. In some places, it also includes a thick layer of adipose

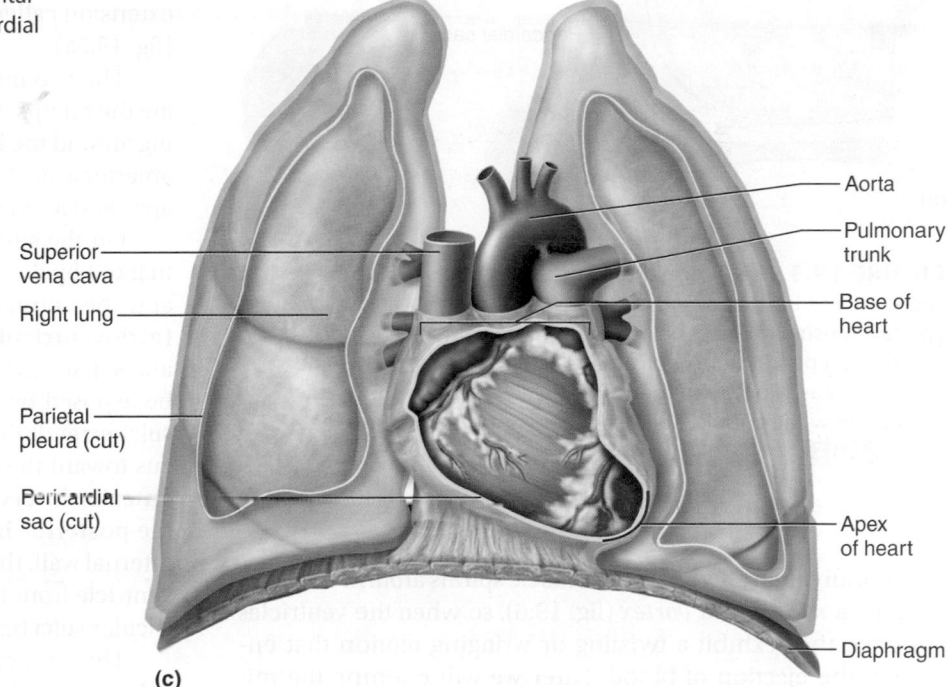

Superior vena cava

Right lung

Parietal pleura (cut)

Pericardial sac (cut)

Aorta

Pulmonary trunk

Base of heart

Apex of heart

Diaphragm

(c)

tissue, whereas in other areas it is fat-free and translucent, so the muscle of the underlying myocardium shows through (figs. 19.4a and 19.5). The largest branches of the coronary blood vessels travel through the epicardium.

The **endocardium,**[5] a similar layer, lines the interior of the heart chambers (figs. 19.3 and 19.4b). Like the epicardium, this is a simple squamous epithelium overlying a thin areolar tissue layer; however, it has no adipose tissue. The endocardium covers the valve surfaces and is continuous with the endothelium of the blood vessels.

The **myocardium**[6] between these two is composed of cardiac muscle. This is by far the thickest layer and performs the work of the heart. Its thickness is proportional to the workload

---

[4]*epi* = upon; *cardi* = heart

[5]*endo* = internal; *cardi* = heart
[6]*myo* = muscle; *cardi* = heart

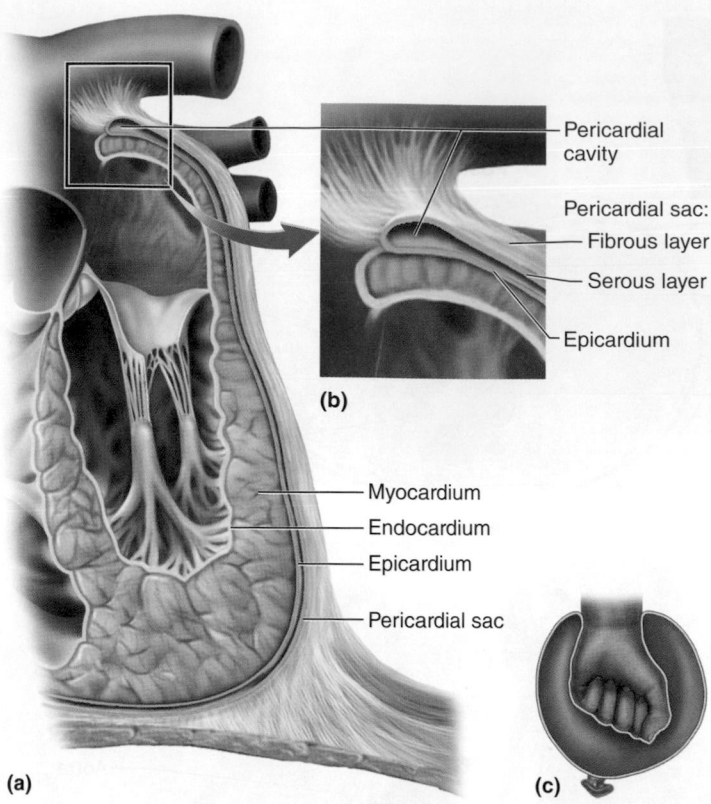

Pericardial
cavity

Pericardial sac:
- Fibrous layer
- Serous layer

Epicardium

**(b)**

Myocardium
Endocardium
Epicardium

Pericardial sac

**(a)**   **(c)**

**FIGURE 19.3   The Pericardium and Heart Wall.** (a) Frontal section of the heart showing the three layers of the heart wall and relationship to the pericardium. (b) Detail of the pericardial sac (parietal pericardium), epicardium (visceral pericardium), and pericardial cavity between them. (c) A fist in a balloon shows, by analogy, how the double-walled pericardium wraps around the heart. AP|R

on the individual chambers. Its muscle spirals around the heart forming a *myocardial vortex* (fig. 19.6), so when the ventricles contract, they exhibit a twisting or wringing motion that enhances the ejection of blood. Later we will examine the microscopic structure of the cardiac muscle cells, or *cardiocytes,* more closely.

The heart also has a framework of collagenous and elastic fibers that make up the **fibrous skeleton.** This tissue is especially concentrated in the walls between the heart chambers, in *fibrous rings (anuli fibrosi)* around the valves, and in sheets of tissue that interconnect these rings (see fig. 19.8). The fibrous skeleton has multiple functions: (1) It provides structural support for the heart, especially around the valves and the openings of the great vessels; it holds these orifices open and prevents them from excessively stretching when blood surges through them. (2) It anchors the cardiocytes and gives them something to pull against. (3) As a nonconductor of electricity, it serves as electrical insulation between the atria and the ventricles, so the atria cannot stimulate the ventricles directly. This insulation is important to the timing and coordination of

electrical and contractile activity. (4) Some authorities think (though others disagree) that elastic recoil of the fibrous skeleton may aid in refilling the heart with blood after each beat in a manner described later.

► ► ►**APPLY WHAT YOU KNOW**

*Parts of the fibrous skeleton sometimes become calcified in old age. How would you expect this to affect cardiac function?*

## The Chambers

The heart has four chambers, best seen in a frontal section (figs. 19.4b and 19.7). The two superior chambers are the **right** and **left atria** (AY-tree-uh; singular, *atrium*[7]). They are thin-walled receiving chambers for blood returning to the heart by way of the great veins. Most of the mass of each atrium is on the posterior side of the heart, so only a small portion is visible from an anterior view. Here, each atrium has a small earlike extension called an **auricle**[8] that slightly increases its volume (fig. 19.5a).

The two inferior chambers, the **right** and **left ventricles,**[9] are the pumps that eject blood into the arteries and keep it flowing around the body. The right ventricle constitutes most of the anterior aspect of the heart, whereas the left ventricle forms the apex and inferoposterior aspect.

On the surface, the boundaries of the four chambers are marked by three sulci (grooves), which are largely filled by fat and the coronary blood vessels (fig. 19.5a). The **coronary**[10] **(atrioventricular) sulcus** encircles the heart near the base and separates the atria above from the ventricles below. It can be exposed by lifting the margins of the atria. The other two sulci extend obliquely down the heart from the coronary sulcus toward the apex—one on the front of the heart called the **anterior interventricular sulcus** and one on the back called the **posterior interventricular sulcus.** These sulci overlie an internal wall, the *interventricular septum,* that divides the right ventricle from the left. The coronary sulcus and two interventricular sulci harbor the largest of the coronary blood vessels.

The atria exhibit thin flaccid walls corresponding to their light workload—all they do is pump blood into the ventricles immediately below (fig. 19.7). They are separated from each other by a wall called the **interatrial septum.** The right atrium and both auricles exhibit internal ridges of myocardium called **pectinate**[11] **muscles.**

The **interventricular septum** is a much more muscular, vertical wall between the ventricles. The right ventricle pumps blood only to the lungs and back to the left atrium, so its wall is only moderately muscular. The wall of the left ventricle is two to four times as thick because it bears the greatest workload of all four chambers, pumping blood through the entire body.

[7]*atrium* = entryway
[8]*auricle* = little ear
[9]*ventr* = belly, lower part; *icle* = little
[10]*coron* = crown; *ary* = pertaining to
[11]*pectin* = comb; *ate* = like

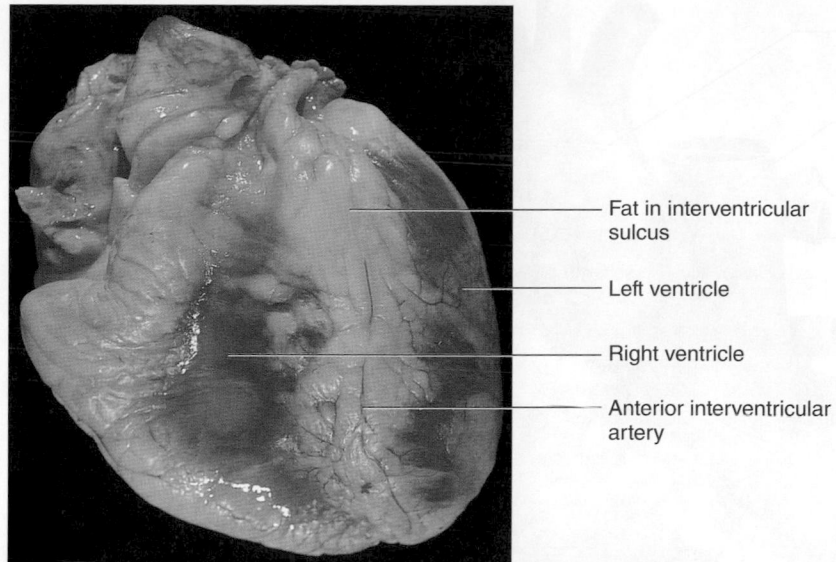

**(a) Anterior view, external anatomy**

Fat in interventricular sulcus

Left ventricle

Right ventricle

Anterior interventricular artery

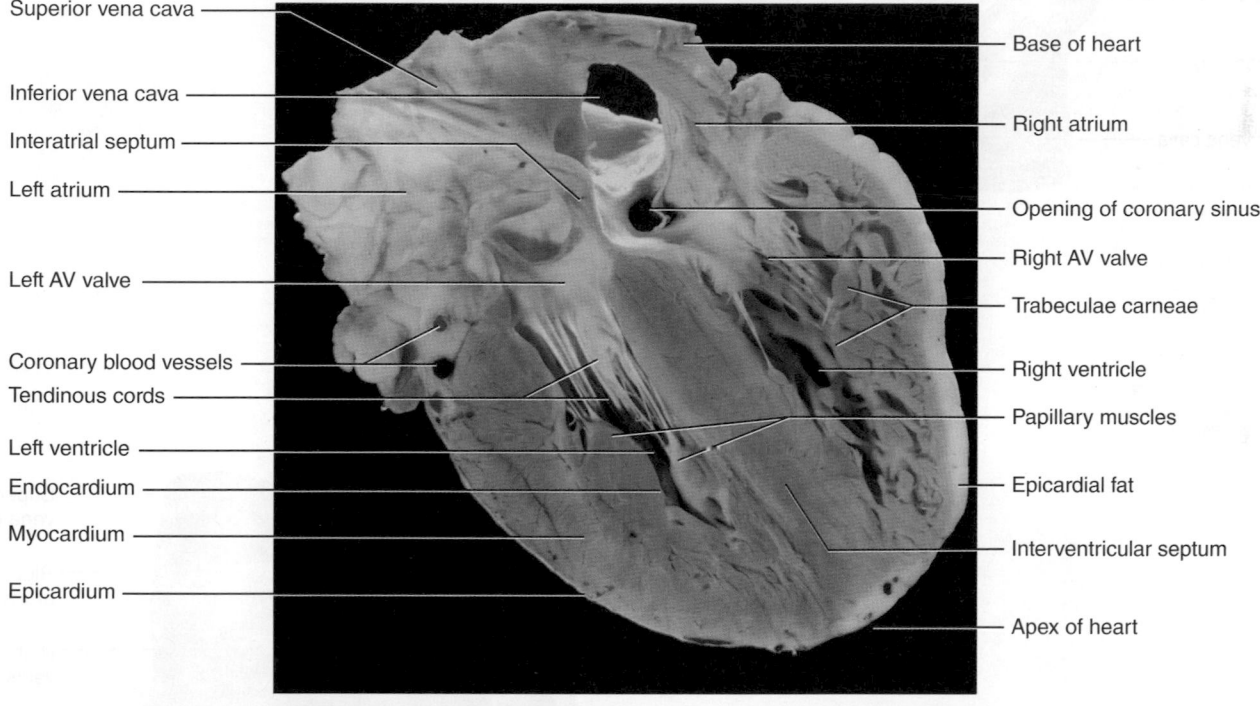

Superior vena cava

Inferior vena cava

Interatrial septum

Left atrium

Left AV valve

Coronary blood vessels

Tendinous cords

Left ventricle

Endocardium

Myocardium

Epicardium

Base of heart

Right atrium

Opening of coronary sinus

Right AV valve

Trabeculae carneae

Right ventricle

Papillary muscles

Epicardial fat

Interventricular septum

Apex of heart

**(b) Posterior view, internal anatomy**

**FIGURE 19.4 The Heart of a Human Cadaver.** AP|R

Both ventricles exhibit internal ridges called **trabeculae carneae**[12] (trah-BEC-you-lee CAR-nee-ee). It is thought that these ridges may serve to keep the ventricular walls from clinging to each other like suction cups when the heart contracts, allowing the chambers to expand more easily when they refill. If you wet your hands, press your palms firmly together, then pull them apart, you can appreciate how smooth wet surfaces cling to each other and how, without trabeculae carneae, the heart walls might also do so.

---

[12]*trabecula* = little beam; *carne* = flesh, meat

## The Valves

To pump blood effectively, the heart needs valves that ensure a one-way flow. There is a valve between each atrium and its ventricle and another at the exit from each ventricle into its great artery (fig. 19.7), but the heart has no valves where the great veins empty into the atria. Each valve consists of two or three fibrous flaps of tissue called **cusps** or **leaflets,** covered with endocardium.

The **atrioventricular (AV) valves** regulate the openings between the atria and ventricles. The **right AV (tricuspid) valve** has three cusps and the **left AV valve** has two (fig. 19.8).

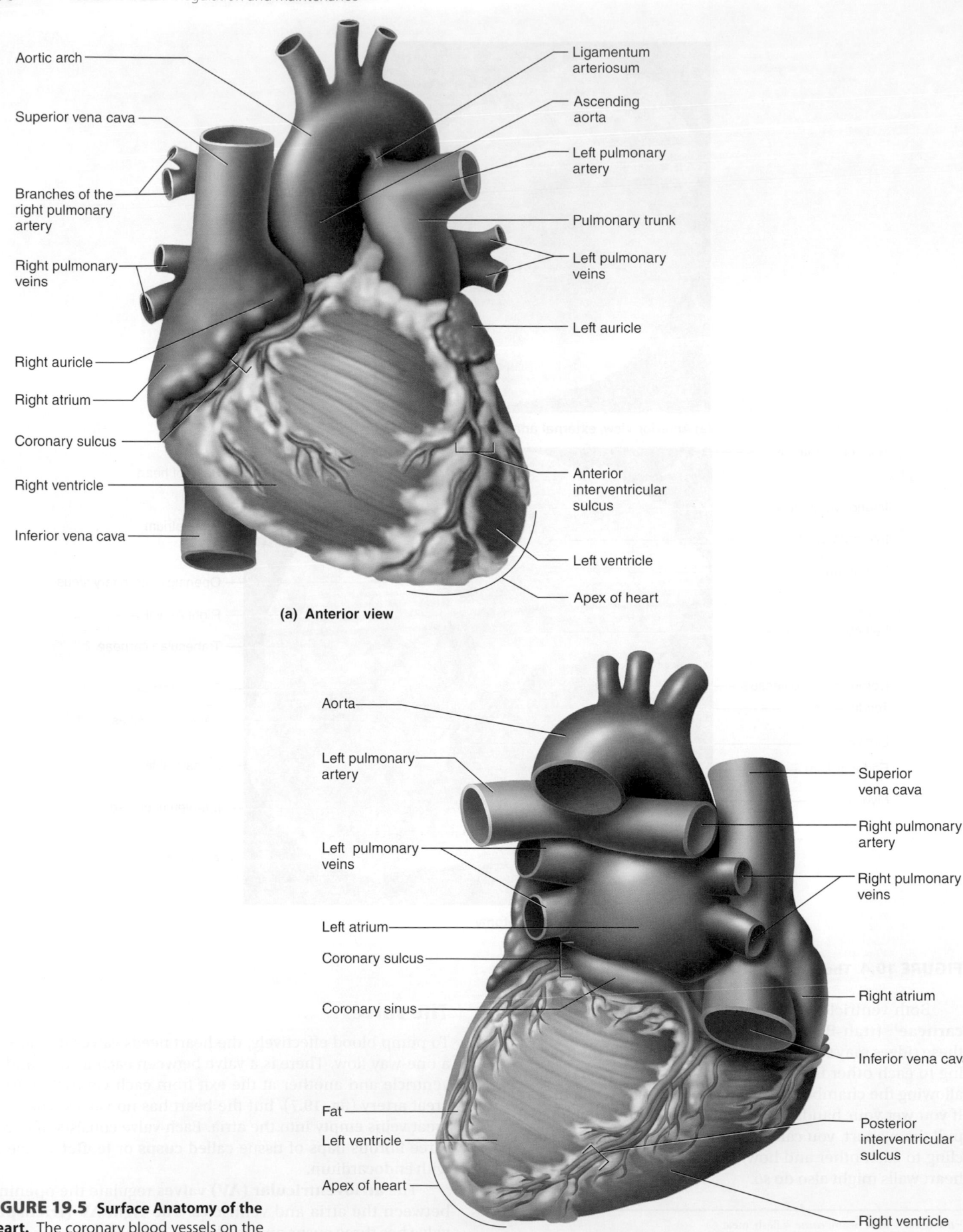

Aortic arch

Superior vena cava

Branches of the
right pulmonary
artery

Right pulmonary
veins

Right auricle

Right atrium

Coronary sulcus

Right ventricle

Inferior vena cava

Ligamentum
arteriosum

Ascending
aorta

Left pulmonary
artery

Pulmonary trunk

Left pulmonary
veins

Left auricle

Anterior
interventricular
sulcus

Left ventricle

Apex of heart

**(a) Anterior view**

Aorta

Left pulmonary
artery

Left pulmonary
veins

Left atrium

Coronary sulcus

Coronary sinus

Fat

Left ventricle

Apex of heart

Superior
vena cava

Right pulmonary
artery

Right pulmonary
veins

Right atrium

Inferior vena cava

Posterior
interventricular
sulcus

Right ventricle

**(b) Posterior view**

**FIGURE 19.5 Surface Anatomy of the
Heart.** The coronary blood vessels on the
heart surface are identified in figure 19.10.

AP|R

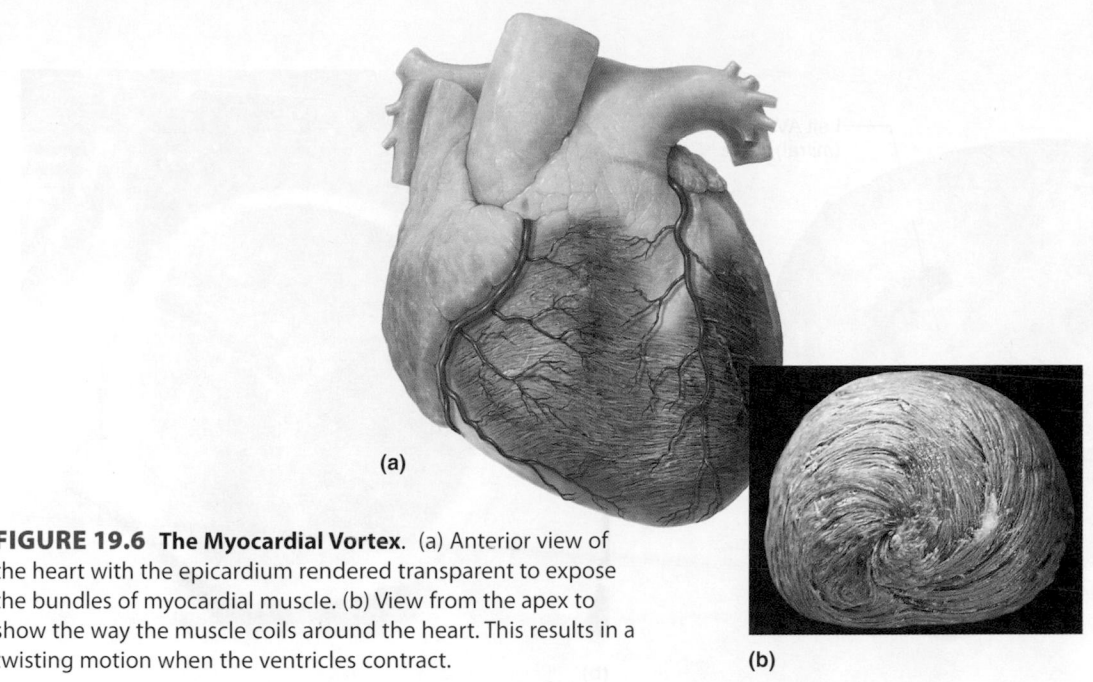

**(a)**

**FIGURE 19.6 The Myocardial Vortex.** (a) Anterior view of the heart with the epicardium rendered transparent to expose the bundles of myocardial muscle. (b) View from the apex to show the way the muscle coils around the heart. This results in a twisting motion when the ventricles contract.

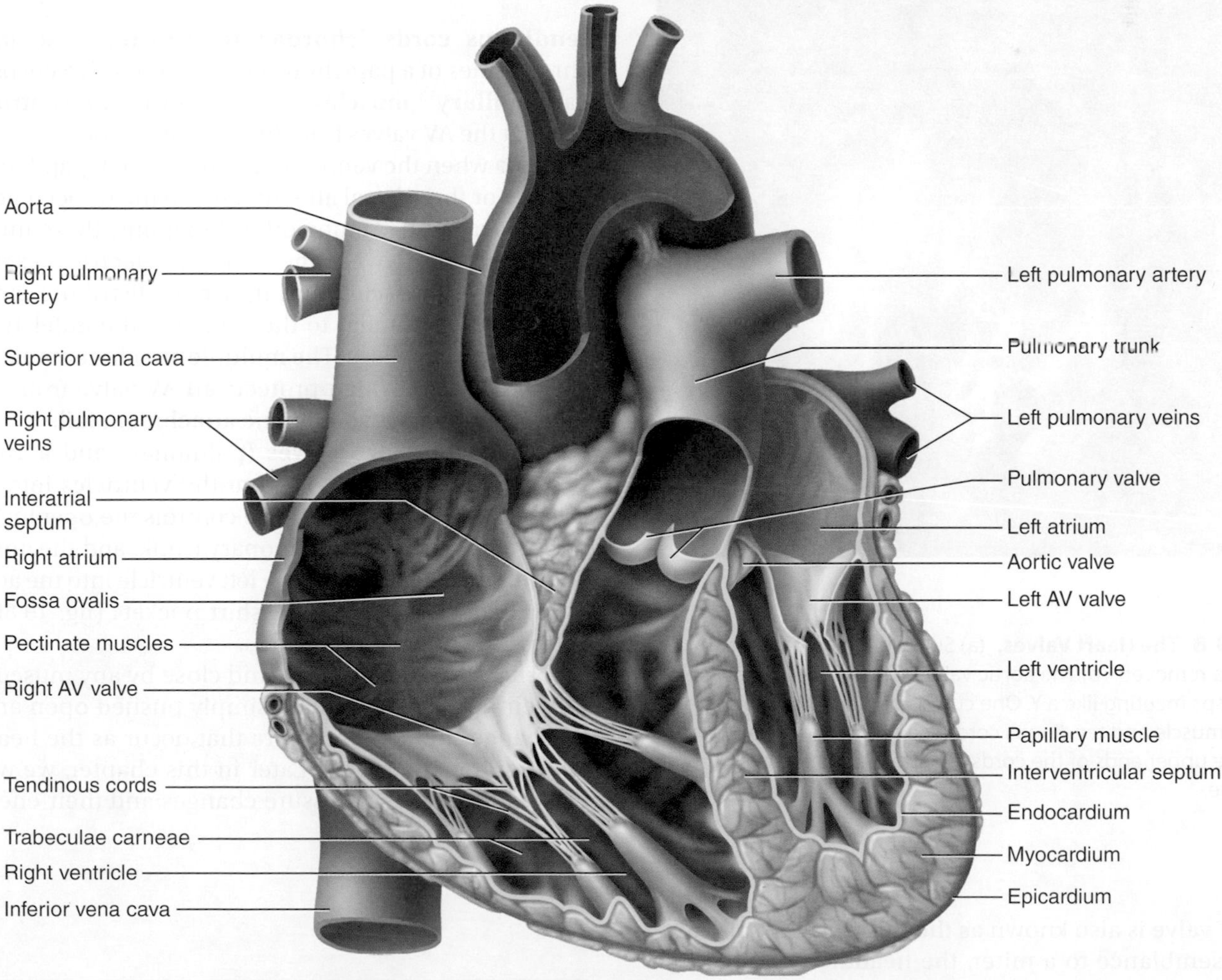

**(b)**

Aorta

Right pulmonary artery

Superior vena cava

Right pulmonary veins

Interatrial septum

Right atrium

Fossa ovalis

Pectinate muscles

Right AV valve

Tendinous cords

Trabeculae carneae

Right ventricle

Inferior vena cava

Left pulmonary artery

Pulmonary trunk

Left pulmonary veins

Pulmonary valve

Left atrium

Aortic valve

Left AV valve

Left ventricle

Papillary muscle

Interventricular septum

Endocardium

Myocardium

Epicardium

**FIGURE 19.7 Internal Anatomy of the Heart.** Anterior view. AP|R

❓ *Do the atrial pectinate muscles more nearly resemble the ventricular papillary muscles or the trabeculae carneae?*

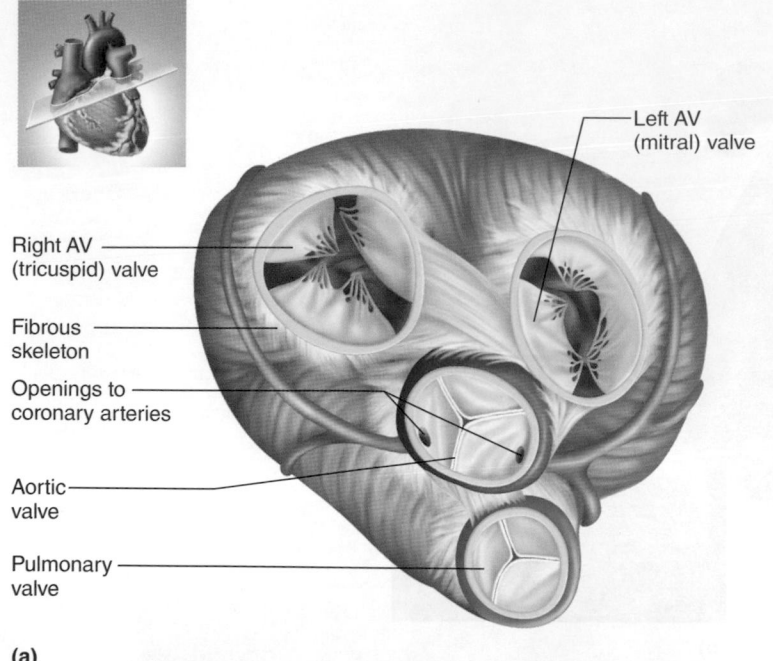

Right AV
(tricuspid) valve

Fibrous
skeleton

Openings to
coronary arteries

Aortic
valve

Pulmonary
valve

Left AV
(mitral) valve

**(a)**

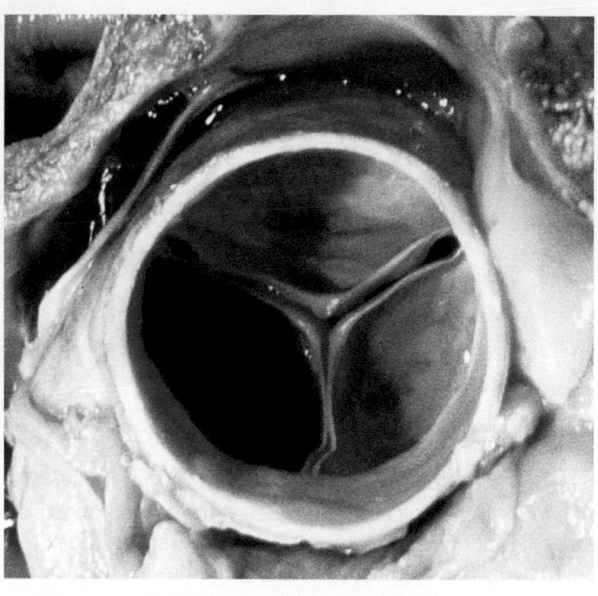

**(b)**

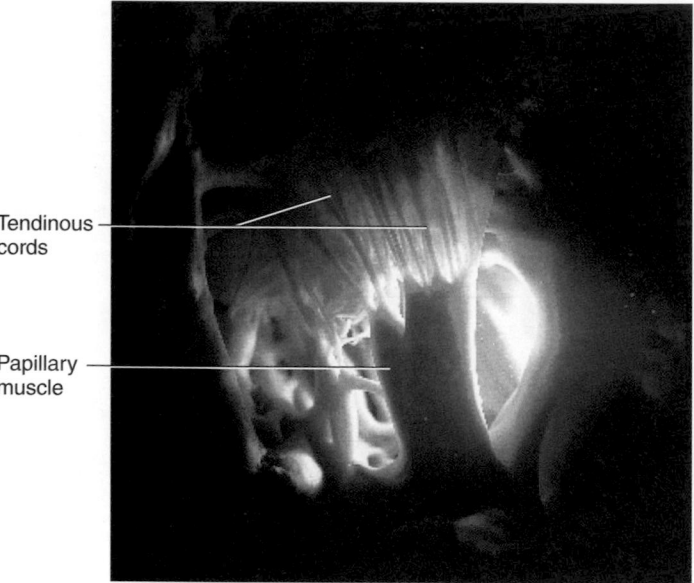

Tendinous
cords

Papillary
muscle

**(c)**

**FIGURE 19.8   The Heart Valves.**   (a) Superior view of the heart with the atria removed. (b) The aortic valve, superior view, showing the three cusps meeting like a Y. One cusp is darkened by a blood clot. (c) Papillary muscle and tendinous cords seen from within the right ventricle. The upper ends of the cords are attached to the cusps of the right AV valve.

**tendinous cords (chordae tendineae),** resembling the shroud lines of a parachute, connect the valve cusps to conical **papillary**[13] **muscles** on the floor of the ventricle. They prevent the AV valves from flipping inside out or bulging into the atria when the ventricles contract. Each papillary muscle has two or three basal attachments to the trabeculae carneae of the heart wall. Among other functions, these multiple attachments may govern the timing of electrical excitation of the papillary muscles, and they may distribute mechanical stress in a way similar to the weight of the Eiffel Tower supported on its four legs. The multiple attachments also provide some redundancy that protects an AV valve from complete mechanical failure should one attachment fail.

The **semilunar**[14] **valves** (pulmonary and aortic valves) regulate the flow of blood from the ventricles into the great arteries. The **pulmonary valve** controls the opening from the right ventricle into the pulmonary trunk, and the **aortic valve** controls the opening from the left ventricle into the aorta. Each has three cusps shaped like shirt pockets (fig. 19.8b). These valves have no tendinous cords.

The valves do not open and close by any muscular effort of their own. The cusps are simply pushed open and closed by changes in blood pressure that occur as the heart chambers contract and relax. Later in this chapter, we will take a closer look at these pressure changes and their effect on the valves.

The left AV valve is also known as the **mitral** (MY-trul) **valve** after its resemblance to a miter, the headdress of a church bishop; it has also formerly gone by the name of *bicuspid valve,* now considered inaccurate and obsolete. Stringy

---

[13]*papill* = nipple; *ary* = like, shaped
[14]*semi* = half; *lunar* = moonlike

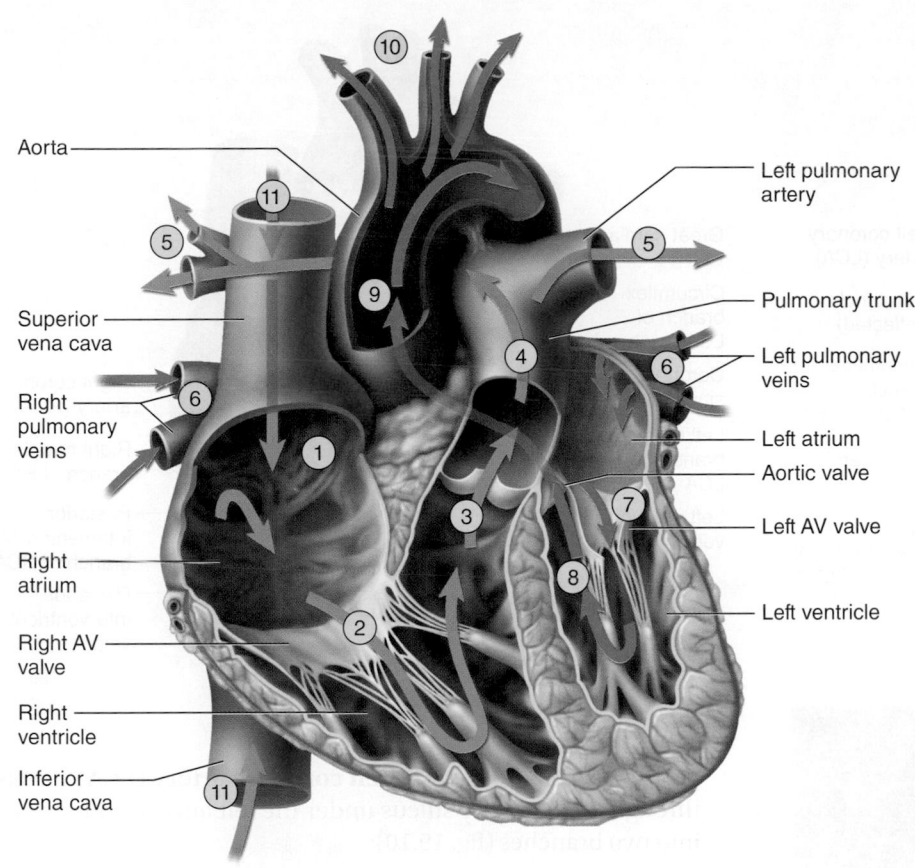

① Blood enters right atrium from superior and inferior venae cavae.

② Blood in right atrium flows through right AV valve into right ventricle.

③ Contraction of right ventricle forces pulmonary valve open.

④ Blood flows through pulmonary valve into pulmonary trunk.

⑤ Blood is distributed by right and left pulmonary arteries to the lungs, where it unloads $CO_2$ and loads $O_2$.

⑥ Blood returns from lungs via pulmonary veins to left atrium.

⑦ Blood in left atrium flows through left AV valve into left ventricle.

⑧ Contraction of left ventricle (simultaneous with step ③) forces aortic valve open.

⑨ Blood flows through aortic valve into ascending aorta.

⑩ Blood in aorta is distributed to every organ in the body, where it unloads $O_2$ and loads $CO_2$.

⑪ Blood returns to right atrium via venae cavae.

**FIGURE 19.9  The Pathway of Blood Flow Through the Heart.**  The pathway from 4 through 6 is the pulmonary circuit, and the pathway from 9 through 11 is the systemic circuit. Violet arrows indicate oxygen-poor blood; orange arrows indicate oxygen-rich blood. AP|R

## Blood Flow Through the Chambers

Until the sixteenth century, blood was thought to flow directly from the right ventricle into the left through invisible pores in the septum. This of course is not true. Blood is kept entirely separate on the right and left sides of the heart. Figure 19.9 shows the pathway of the blood as it travels from the right atrium through the body and back to the starting point.

Blood that has been through the systemic circuit returns by way of the superior and inferior venae cavae to the right atrium. It flows directly from the right atrium, through the right AV (tricuspid) valve, into the right ventricle. When the right ventricle contracts, it ejects blood through the pulmonary valve into the pulmonary trunk, on its way to the lungs to exchange carbon dioxide for oxygen.

Blood returns from the lungs by way of two pulmonary veins on the left and two on the right; all four of these empty into the left atrium. Blood flows through the left AV (mitral) valve into the left ventricle. Contraction of the left ventricle ejects this blood through the aortic valve into the ascending aorta, on its way to another trip around the systemic circuit.

## The Coronary Circulation

If your heart lasts for 80 years and beats an average of 75 times a minute, it will beat more than 3 billion times and pump more than 200 million liters of blood. It is, in short, a remarkably hardworking organ, and understandably, it needs an abundant supply of oxygen and nutrients. These needs are not met to any appreciable extent by the blood in the heart chambers, because the diffusion of substances from there through the myocardium would be too slow. Instead, the myocardium has its own supply of arteries and capillaries that deliver blood to every muscle cell. The blood vessels of the heart wall constitute the **coronary circulation.**

At rest, the coronary blood vessels supply the myocardium with about 250 mL of blood per minute. This constitutes about 5% of the circulating blood going to meet the metabolic needs of the heart, even though the heart is only 0.5% of the body's weight. It receives 10 times its "fair share" to sustain its strenuous workload.

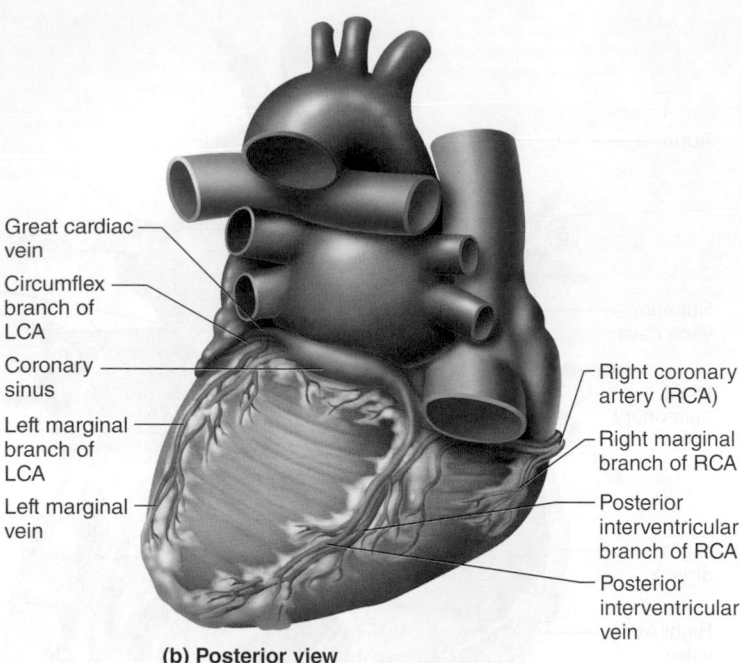

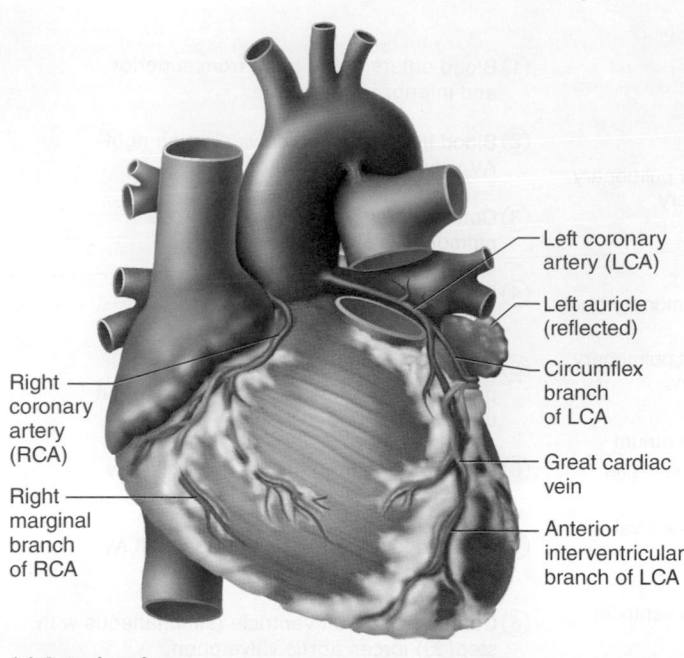

(a) Anterior view

(b) Posterior view

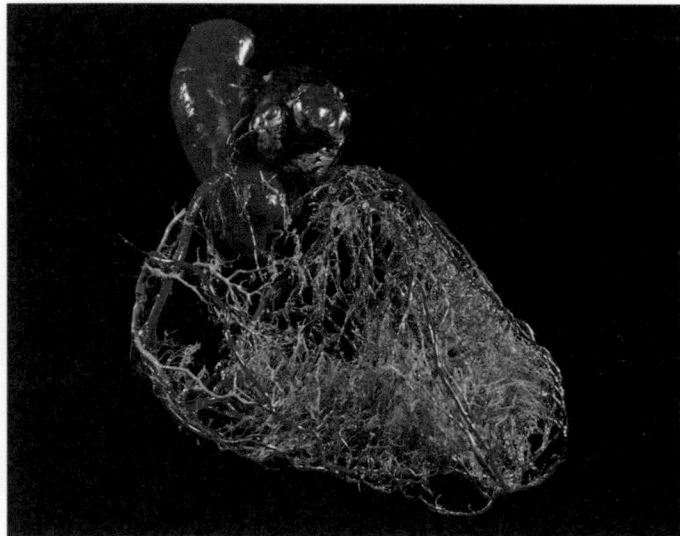

(c)

**FIGURE 19.10   The Principal Coronary Blood Vessels.**
(a) Anterior view. (b) Posterior view. (c) A polymer cast of the coronary circulation.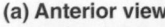

## Arterial Supply

The coronary circulation is the most variable aspect of cardiac anatomy. The following description covers only the pattern seen in about 70% to 85% of persons, and only the few largest vessels. (Compare the great density of coronary blood vessels seen in fig. 19.10c.)

Immediately after the aorta leaves the left ventricle, it gives off a right and left coronary artery. The orifices of these two arteries lie deep in the pockets formed by two of the aortic

valve cusps (fig. 19.8a). The **left coronary artery (LCA)** travels through the coronary sulcus under the left auricle and divides into two branches (fig. 19.10):

1. The **anterior interventricular branch** travels down the anterior interventricular sulcus to the apex, rounds the bend, and travels a short distance up the posterior side of the heart. There it joins the posterior interventricular branch described shortly. Clinically, it is also called the *left anterior descending (LAD) branch*. This artery supplies blood to both ventricles and the anterior two-thirds of the interventricular septum.

2. The **circumflex branch** continues around the left side of the heart in the coronary sulcus. It gives off a **left marginal branch** that passes down the left margin of the heart and furnishes blood to the left ventricle. The circumflex branch then ends on the posterior side of the heart. It supplies blood to the left atrium and posterior wall of the left ventricle.

The **right coronary artery (RCA)** supplies the right atrium and sinoatrial node (pacemaker), continues along the coronary sulcus under the right auricle, and gives off two branches of its own:

1. The **right marginal branch** runs toward the apex of the heart and supplies the lateral aspect of the right atrium and ventricle.

2. The RCA continues around the right margin of the heart to the posterior side, sends a small branch to the atrioventricular node, then gives off a large **posterior interventricular branch.** This branch travels down the corresponding sulcus and supplies the posterior walls of both ventricles

as well as the posterior portion of the interventricular septum. It ends by joining the anterior interventricular branch of the LCA.

The energy demand of the cardiac muscle is so critical that an interruption of the blood supply to any part of the myocardium can cause necrosis within minutes. A fatty deposit or blood clot in a coronary artery can cause a **myocardial infarction**[15] **(MI),** or heart attack (see Deeper Insight 19.1). Some protection from MI is provided by the aforementioned points where two arteries come together and combine their blood flow to points farther downstream. Such points of convergence, called *arterial anastomoses* (ah-NASS-tih-MO-seez), provide an alternative route, called **collateral circulation,** that can supply the heart tissue with blood if the primary route becomes obstructed.

In most organs, blood flow peaks when the ventricles contract and eject blood into the arteries, and diminishes when the ventricles relax and refill. The opposite is true in the coronary arteries: Flow peaks when the heart relaxes. There are three reasons for this. (1) Contraction of the myocardium compresses the coronary arteries and obstructs blood flow. (2) When the ventricles contract, the aortic valve is forced open and its cusps cover the openings to the coronary arteries, blocking blood from flowing into them. (3) When they relax, blood in the aorta briefly surges back toward the heart. It fills the aortic valve cusps and some of it flows into the coronary arteries, like water pouring into a bucket and flowing out through a

## DEEPER INSIGHT 19.1

### CLINICAL APPLICATION

#### Angina and Heart Attack

An obstruction of coronary blood flow can cause a chest pain known as *angina pectoris*[16] (an-JY-na PEC-toe-riss) or, more seriously, *myocardial infarction* (heart attack). Angina is a sense of heaviness or pain in the chest resulting from temporary and reversible *ischemia*[17] (iss-KEE-me-ah), or deficiency of blood flow to the cardiac muscle. It typically occurs when a partially blocked coronary artery constricts. The oxygen-deprived myocardium shifts to anaerobic fermentation, producing lactic acid, which stimulates pain receptors in the heart. The pain abates when the artery relaxes and normal blood flow resumes.

Myocardial infarction (MI), on the other hand, is the sudden death of a patch of myocardium resulting from long-term obstruction of the coronary circulation. Coronary arteries often become obstructed by a blood clot or fatty deposit called an atheroma (see Deeper Insight 19.4). As cardiac muscle downstream from the obstruction dies, the individual commonly feels a sense of heavy pressure or squeezing pain in the chest, often "radiating" to the shoulder and left arm. Some MIs are painless, "silent" heart attacks, especially in elderly or diabetic individuals. Infarctions weaken the heart wall and disrupt electrical conduction pathways, potentially leading to fibrillation and cardiac arrest (discussed later in this chapter). MI causes about 27% of deaths in the United States.

hole in the bottom. In the coronary blood vessels, therefore, blood flow increases during ventricular relaxation.

## Venous Drainage

**Venous drainage** refers to the route by which blood leaves an organ. After flowing through capillaries of the heart wall, about 5% to 10% of the coronary blood empties directly from multiple small *thebesian*[18] *veins (smallest cardiac veins)* into the heart chambers, especially the right ventricle. The rest returns to the right atrium by the following route (fig. 19.10):

- The **great cardiac vein** collects blood from the anterior aspect of the heart and travels alongside the anterior interventricular artery. It carries blood from the apex toward the coronary sulcus, then arcs around the left side of the heart and empties into the coronary sinus.

- The **posterior interventricular (middle cardiac) vein,** found in the posterior interventricular sulcus, collects blood from the posterior aspect of the heart. It, too, carries blood from the apex upward and drains into the same sinus.

- The **left marginal vein** travels from a point near the apex up the left margin, and also empties into the coronary sinus.

- The **coronary sinus,** a large transverse vein in the coronary sulcus on the posterior side of the heart, collects blood from all three of the aforementioned veins as well as some smaller ones. It empties blood into the right atrium.

### BEFORE YOU GO ON

Answer the following questions to test your understanding of the preceding section:

3. Name the three layers of the heart and describe their structural differences.

4. What are the functions of the fibrous skeleton?

5. Trace the flow of blood through the heart, naming each chamber and valve in order.

6. What are the three principal branches of the left coronary artery? Where are they located on the heart surface? What are the branches of the right coronary artery, and where are they located?

7. What is the medical significance of anastomoses in the coronary arterial system?

8. Why do the coronary arteries carry a greater blood flow during ventricular relaxation than they do during ventricular contraction?

9. What are the three major veins that empty into the coronary sinus?

---

[15]*infarct* = to stuff
[16]*angina* = to choke, strangle; *pectoris* = of the chest
[17]*isch* = holding back; *em* = blood; *ia* = condition

[18]Adam Christian Thebesius (1686–1732), German physician

## 19.3 Cardiac Muscle and the Cardiac Conduction System

### Expected Learning Outcomes

When you have completed this section, you should be able to

a. describe the unique structural and metabolic characteristics of cardiac muscle;

b. explain the nature and functional significance of the intercellular junctions between cardiac muscle cells;

c. describe the heart's pacemaker and internal electrical conduction system; and

d. describe the nerve supply to the heart and explain its role.

The most obvious physiological fact about the heart is its rhythmicity. It contracts at regular intervals, typically about 75 beats per minute (bpm) in a resting adult. Among invertebrates such as clams, crabs, and insects, each heartbeat is triggered by a pacemaker in the nervous system. The vertebrate heartbeat, however, is said to be *myogenic*[19] because the signal originates within the heart itself. The heart is described as being **autorhythmic**[20] because it doesn't depend on the nervous system for its rhythm. It has its own pacemaker and electrical system. We now turn our attention to the cardiac muscle, the pacemaker and internal electrical system, and the heart's nerve supply—the foundations for its electrical activity and rhythmic beat.

### Structure of Cardiac Muscle

The heart is mostly muscle. It is striated like skeletal muscle, but quite different from it in other structural and functional respects—and it has to be, if we want it to pump infallibly, more than once every second, for at least eight or nine decades.

**Cardiocytes (cardiomyocytes)** are relatively short, thick, branched cells, typically 50 to 100 μm long and 10 to 20 μm wide (fig. 19.11). The ends of the cell are slightly branched, like a log with deep notches in the end. Through these branches, each cardiocyte contacts several others, so collectively they form a network throughout each pair of heart chambers—one network in the atria and one in the ventricles. A cardiocyte usually has only one, centrally placed nucleus, often surrounded by a light-staining mass of glycogen. The sarcoplasmic reticulum is less developed than in skeletal muscle; it lacks terminal cisternae, although it does have footlike sacs associated with the T tubules. The T tubules are much larger than in skeletal muscle. During excitation of the cell, they admit calcium ions from the extracellular fluid to activate muscle contraction. Cardiocytes have especially large mitochondria for reasons noted shortly.

---

[19]*myo* = muscle; *genic* = arising from
[20]*auto* = self

Striations
Nucleus
Intercalated discs

**(a)**

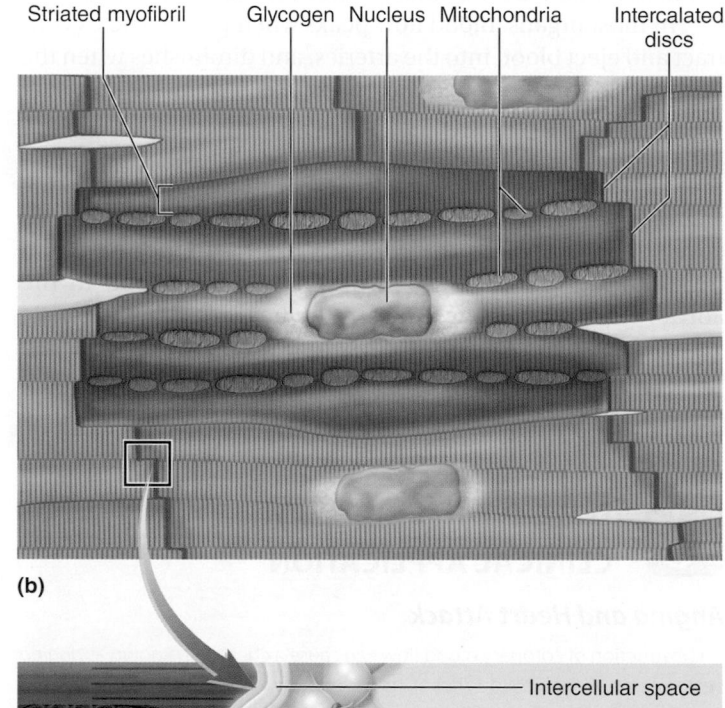

Striated myofibril    Glycogen  Nucleus  Mitochondria    Intercalated discs

**(b)**

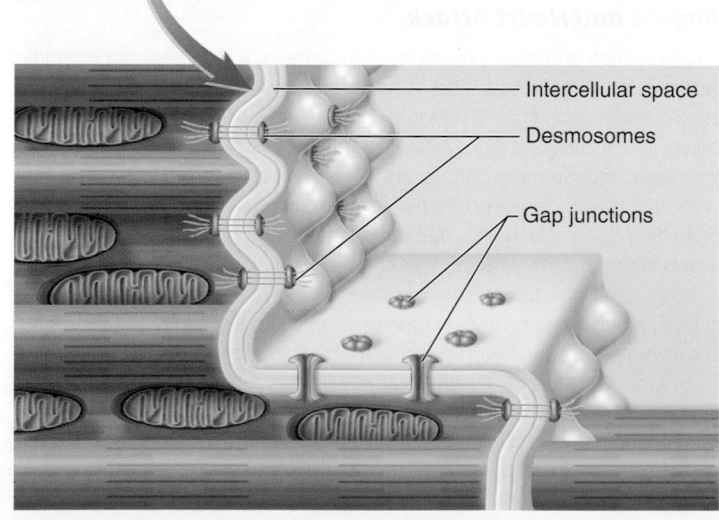

Intercellular space
Desmosomes
Gap junctions

**(c)**

**FIGURE 19.11  Cardiac Muscle.** (a) Light micrograph. (b) Structure of a cardiocyte and its relationship to adjacent cardiocytes. All of the colored area is a single cell. Note that it is notched at the ends and typically linked to two or more neighboring cardiocytes by the mechanical and electrical junctions of the intercalated discs. (c) Structure of an intercalated disc.

Cardiocytes are joined end to end by thick connections called **intercalated** (in-TUR-ka-LAY-ted) **discs.** With the right histological stain, these appear as dark lines thicker than the striations. An intercalated disc is a complex steplike structure with three distinctive features not found in skeletal muscle:

1. **Interdigitating folds.** The plasma membrane at the end of the cell is folded somewhat like the bottom of an egg carton. The folds of adjoining cells interlock with each other and increase the surface area of intercellular contact.

2. **Mechanical junctions.** The cells are tightly joined by two types of mechanical junctions: the fascia adherens and desmosomes. The *fascia adherens*[21] (FASH-ee-ah ad-HEER-enz) is the most extensive. It is a broad band in which the actin of the thin myofilaments is anchored to the plasma membrane and each cell is linked to the next via transmembrane proteins. The fascia adherens is interrupted here and there by *desmosomes*—patches of mechanical linkage that prevent the contracting cardiocytes from pulling apart. Both desmosomes and gap junctions (next) are described in greater detail on pages 163–164.

3. **Electrical junctions.** The intercalated discs also contain *gap junctions,* which form channels that allow ions to flow from the cytoplasm of one cardiocyte directly into the next. They enable each cardiocyte to electrically stimulate its neighbors. Thus, the entire myocardium of the two atria behaves almost like a single cell, as does the entire myocardium of the two ventricles. This unified action is essential for the effective pumping of a heart chamber.

Skeletal muscle contains satellite cells that can divide and replace dead muscle fibers to some extent. Cardiac muscle lacks these, however, so the repair of damaged cardiac muscle is almost entirely by fibrosis (scarring). The discovery of a limited capacity for myocardial mitosis and regeneration has raised hope that regeneration might one day be clinically enhanced to repair hearts damaged by myocardial infarction.

## Metabolism of Cardiac Muscle

Cardiac muscle depends almost exclusively on aerobic respiration to make ATP. It is very rich in myoglobin (a short-term source of oxygen for aerobic respiration) and glycogen (for stored energy). Its huge mitochondria fill about 25% of the cell; skeletal muscle fibers, by comparison, have much smaller mitochondria that occupy only 2% of the fiber. Cardiac muscle is relatively adaptable with respect to the organic fuels used. At rest, the heart gets about 60% of its energy from fatty acids, 35% from glucose, and 5% from other fuels such as ketones, lactic acid, and amino acids. Cardiac muscle is more vulnerable to an oxygen deficiency than it is to the lack of any specific fuel. Because it makes little use of anaerobic fermentation or the oxygen debt mechanism, it is not prone to fatigue. You can easily appreciate this fact by squeezing a rubber ball in your fist once every second for a minute or two. You will soon feel weakness and fatigue in your skeletal muscles and perhaps feel all the more grateful that cardiac muscle can maintain a rhythm like this, without fatigue, for a lifetime.

## The Conduction System

The heartbeat is coordinated by a **cardiac conduction system** composed of an internal pacemaker and nervelike conduction pathways through the myocardium. It generates and conducts rhythmic electrical signals in the following order (fig. 19.12):

[21]*fascia* = band; *adherens* = adhering

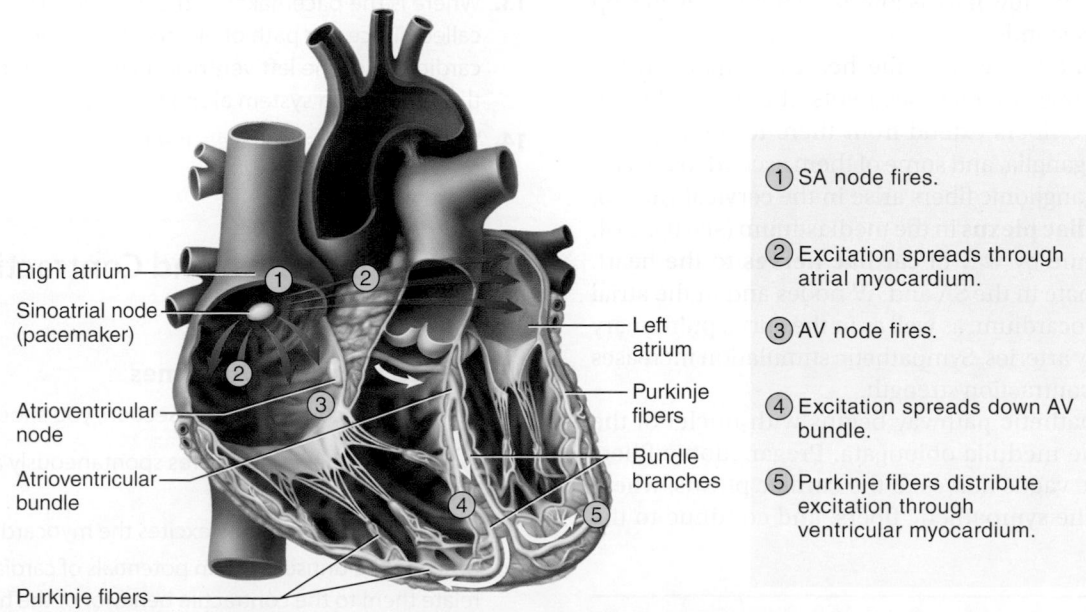

Right atrium

Sinoatrial node (pacemaker)

Atrioventricular node

Atrioventricular bundle

Purkinje fibers

Left atrium

Purkinje fibers

Bundle branches

(1) SA node fires.

(2) Excitation spreads through atrial myocardium.

(3) AV node fires.

(4) Excitation spreads down AV bundle.

(5) Purkinje fibers distribute excitation through ventricular myocardium.

**FIGURE 19.12  The Cardiac Conduction System.** Electrical signals travel along the pathways indicated by the arrows.

❓ *Which atrium is first to receive the signal that induces atrial contraction?*

①   The **sinoatrial (SA) node** is a patch of modified cardiocytes in the right atrium, just under the epicardium near the superior vena cava. This is the **pacemaker** that initiates each heartbeat and determines the heart rate.

②   Signals from the SA node spread throughout the atria, as shown by the red arrows in the figure.

③   The **atrioventricular (AV) node** is located at the lower end of the interatrial septum near the right AV valve. This node acts as an electrical gateway to the ventricles; the fibrous skeleton acts as an insulator to prevent currents from getting to the ventricles by any other route.

④   The **atrioventricular (AV) bundle,** or **bundle of His**[22] (pronounced "hiss"), is the pathway by which signals leave the AV node. The bundle soon forks into **right** and **left bundle branches,** which enter the interventricular septum and descend toward the apex.

⑤   **Purkinje**[23] (pur-KIN-jee) **fibers** are nervelike processes that arise from the lower end of the bundle branches and turn upward to spread throughout the ventricular myocardium. Purkinje fibers distribute the electrical excitation to the cardiocytes of the ventricles. They form a more elaborate network in the left ventricle than in the right.

Once the Purkinje fibers have delivered the electrical signal to their limits, the cardiocytes themselves perpetuate it by passing ions from cell to cell through their gap junctions.

## Nerve Supply to the Heart

Even though the heart has its own pacemaker, it does receive both sympathetic and parasympathetic nerves that modify the heart rate and contraction strength. Sympathetic stimulation can raise the heart rate to as high as 230 bpm, and parasympathetic stimulation can slow it to as low as 20 bpm or even stop the heart for a few seconds.

The sympathetic pathway to the heart originates in the lower cervical to upper thoracic segments of the spinal cord. Preganglionic nerve fibers extend from there to the adjacent sympathetic chain ganglia, and some of them ascend to the cervical ganglia. Postganglionic fibers arise in the cervical ganglia, pass through a **cardiac plexus** in the mediastinum (see fig. 15.4, p. 562), and continue by way of **cardiac nerves** to the heart. These fibers terminate in the SA and AV nodes and in the atrial and ventricular myocardium, as well as in the aorta, pulmonary trunk, and coronary arteries. Sympathetic stimulation increases the heart rate and contraction strength.

The parasympathetic pathway begins with nuclei of the vagus nerves in the medulla oblongata. Preganglionic fibers extend through the vagus nerves to the cardiac plexus, where they mingle with the sympathetic fibers, and continue to the heart by way of the cardiac nerves (see fig. 15.7, p. 566). They synapse with postganglionic neurons in the epicardial surface and within the heart wall. Postganglionic fibers from the right vagus nerve lead mainly to the SA node and those from the left vagus lead mainly to the AV node, but each has some fibers that cross over to the other target cells. Furthermore, some parasympathetic fibers terminate on sympathetic fibers and inhibit them from stimulating the heart and opposing their action. There is little or no parasympathetic innervation of the myocardium. Parasympathetic stimulation reduces the heart rate. In general, the parasympathetic nerves dominate the control of heart rate and the sympathetic nerves dominate the control of contraction strength. Autonomic effects on heart rate and contraction strength are described in more detail later in this chapter.

From the foregoing description, you can see that the cardiac nerves from the plexus to the heart contain both sympathetic and parasympathetic efferent fibers. They also carry sensory (afferent) fibers from the heart to the CNS. Those fibers are important in cardiovascular reflexes and the transmission of pain signals from the heart.

### BEFORE YOU GO ON

Answer the following questions to test your understanding of the preceding section:

10. What organelle(s) are less developed in cardiac muscle than in skeletal muscle? What one(s) are more developed? What is the functional significance of these differences between muscle types?

11. What exactly is an intercalated disc, and what function is served by each of its components?

12. Cardiac muscle rarely uses anaerobic fermentation to generate ATP. What benefit do we gain from this fact?

13. Where is the pacemaker of the heart located? What is it called? Trace the path of electrical excitation from there to a cardiocyte of the left ventricle, naming each component of the conduction system along the way.

14. Why does the heart have a nerve supply, since it continues to beat even without one?

## 19.4   Electrical and Contractile Activity of the Heart

### Expected Learning Outcomes

When you have completed this section, you should be able to

a. explain why the SA node fires spontaneously and rhythmically;

b. explain how the SA node excites the myocardium;

c. describe the unusual action potentials of cardiac muscle and relate them to the contractile behavior of the heart; and

d. interpret a normal electrocardiogram.

---

[22]Wilhelm His Jr. (1863–1934), German physiologist
[23]Johannes E. Purkinje (1787–1869), Bohemian physiologist

In this section, we examine how the electrical events in the heart produce its cycle of contraction and relaxation. Contraction is called **systole** (SIS-toe-lee) and relaxation is **diastole** (dy-ASS-toe-lee). These terms can refer to a specific part of the heart (for example, atrial systole), but if no particular chamber is specified, they usually refer to the more conspicuous and important ventricular action, which ejects blood from the heart.

## The Cardiac Rhythm

The normal heartbeat triggered by the SA node is called the **sinus rhythm.** At rest, the adult heart typically beats about 70 to 80 times per minute, although heart rates from 60 to 100 bpm are not unusual.

Any region of spontaneous firing other than the SA node is called an **ectopic**[24] **focus.** If the SA node is damaged, an ectopic focus may take over the governance of the heart rhythm. The most common ectopic focus is the AV node, which produces a slower heartbeat of 40 to 50 bpm called a **nodal rhythm.** If neither the SA nor AV node is functioning, other ectopic foci fire at rates of 20 to 40 bpm. The nodal rhythm is sufficient to sustain life, but a rate of 20 to 40 bpm provides too little flow to the brain to be survivable. This condition calls for an artificial pacemaker.

## Pacemaker Physiology

Why does the SA node spontaneously fire at regular intervals? Unlike skeletal muscle or neurons, cells of the SA node do not have a stable resting membrane potential. Their membrane potential starts at about −60 mV and drifts upward, showing a gradual depolarization called the **pacemaker potential** (fig. 19.13). This results primarily from a slow inflow of $Na^+$ without a compensating outflow of $K^+$.

When the pacemaker potential reaches a threshold of −40 mV, voltage-gated calcium channels open and $Ca^{2+}$ flows in from the extracellular fluid. This produces the rising (depolarizing) phase of the action potential, which peaks slightly above 0 mV. At that point, $K^+$ channels open and $K^+$ leaves the cell. This makes the cytosol increasingly negative and creates the falling (repolarizing) phase of the action potential. When repolarization is complete, the $K^+$ channels close and the pacemaker potential starts over, on its way to producing the next heartbeat. Each depolarization of the SA node sets off one heartbeat. When the SA node fires, it excites the other components in the conduction system; thus, the SA node serves as the system's pacemaker. At rest, it typically fires every 0.8 second or so, creating a heart rate of about 75 bpm.

## Impulse Conduction to the Myocardium

Firing of the SA node excites atrial cardiocytes and stimulates the two atria to contract almost simultaneously. The signal

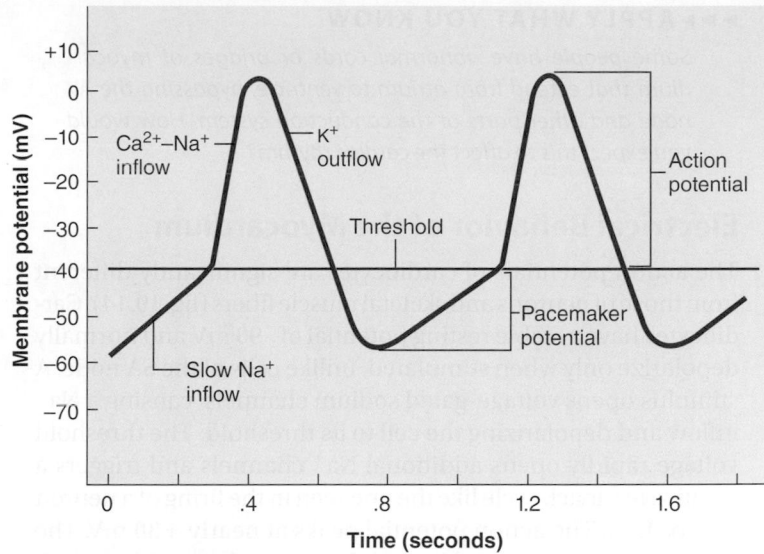

**FIGURE 19.13 Pacemaker Potentials and Action Potentials of the SA Node.**

travels at a speed of about 1 m/s through the atrial myocardium and reaches the AV node in about 50 ms. In the AV node, the signal slows down to about 0.05 m/s, partly because the cardiocytes here are thinner, but more importantly because they have fewer gap junctions over which the signal can be transmitted. This delays the signal at the AV node for about 100 ms—like highway traffic slowing down at a small town. This delay is essential because it gives the ventricles time to fill with blood before they begin to contract.

The ventricular myocardium has a conduction speed of only 0.3 to 0.5 m/s. If this were the only route of travel for the excitatory signal, some cardiocytes would be stimulated much sooner than others. Ventricular contraction would not be synchronized and the pumping effectiveness of the ventricles would be severely compromised. But signals travel through the AV bundle and Purkinje fibers at a speed of 4 m/s, the fastest in the conduction system, owing to their very high density of gap junctions. Consequently, the entire ventricular myocardium depolarizes within 200 ms after the SA node fires, causing the ventricles to contract in near unison.

Signals reach the papillary muscles slightly later than the rest of the myocardium. This gives the AV valves time to close before the papillary muscles contract and take up slack in the tendinous cords. Ventricular contraction causes blood to surge against the AV valves and press them closed more tightly. Ventricular systole begins at the apex of the heart, which is first to be stimulated, and progresses upward—pushing the blood upward toward the semilunar valves. Because of the spiral arrangement of the myocardial vortex, the ventricles twist slightly as they contract, like someone wringing out a towel.

---

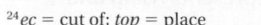
[24]*ec* = cut of; *top* = place

▶▶▶**APPLY WHAT YOU KNOW**

*Some people have abnormal cords or bridges of myocardium that extend from atrium to ventricle, bypassing the AV node and other parts of the conduction system. How would you expect this to affect the cardiac rhythm?*

## Electrical Behavior of the Myocardium

The action potentials of cardiocytes are significantly different from those of neurons and skeletal muscle fibers (fig. 19.14). Cardiocytes have a stable resting potential of −90 mV and normally depolarize only when stimulated, unlike cells of the SA node. A stimulus opens voltage-gated sodium channels, causing a Na⁺ inflow and depolarizing the cell to its threshold. The threshold voltage rapidly opens additional Na⁺ channels and triggers a positive feedback cycle like the one seen in the firing of a neuron (see p. 450). The action potential peaks at nearly +30 mV. The Na⁺ channels close quickly, and the rising phase of the action potential is very brief.

As action potentials spread over the plasma membrane, they open voltage-gated **slow calcium channels,** which admit a small amount of Ca²⁺ from the extracellular fluid into the cell. This Ca²⁺ binds to ligand-gated Ca²⁺ channels on the sarcoplasmic reticulum (SR), opening them and releasing a greater quantity of Ca²⁺ from the SR into the cytosol. This second wave of Ca²⁺ binds to troponin and triggers contraction in the same way as it does in skeletal muscle (see p. 480). The SR provides 90% to 98% of the Ca²⁺ needed for myocardial contraction.

In skeletal muscle and neurons, an action potential falls back to the resting potential within 2 ms. In cardiac muscle, however, the depolarization is prolonged for 200 to 250 ms (at a heart rate of 70–80 bpm), producing a long plateau in the action potential—perhaps because the Ca²⁺ channels of the SR are slow to close or because the SR is slow to remove Ca²⁺ from the cytosol.

As long as the action potential is in its plateau, the cardiocytes contract. Thus, in figure 19.14, you can see the development of muscle tension (myocardial contraction) following closely behind the depolarization and plateau. Rather than showing a brief twitch like skeletal muscle, cardiac muscle has a more sustained contraction necessary for expulsion of blood from the heart chambers. Both atrial and ventricular cardiocytes exhibit these plateaus, but they are more pronounced in the ventricles.

At the end of the plateau, Ca²⁺ channels close and K⁺ channels open. Potassium diffuses rapidly out of the cell and Ca²⁺ is transported back into the extracellular fluid and SR. Membrane voltage drops rapidly, and muscle tension declines soon afterward.

Cardiac muscle has an *absolute refractory period* of 250 ms, compared with 1 to 2 ms in skeletal muscle. This long refractory period prevents wave summation and tetanus, which would stop the pumping action of the heart.

▶▶▶**APPLY WHAT YOU KNOW**

*With regard to the ions involved, how does the falling (repolarization) phase of a myocardial action potential differ from that of a neuron's action potential? (See p. 450.)*

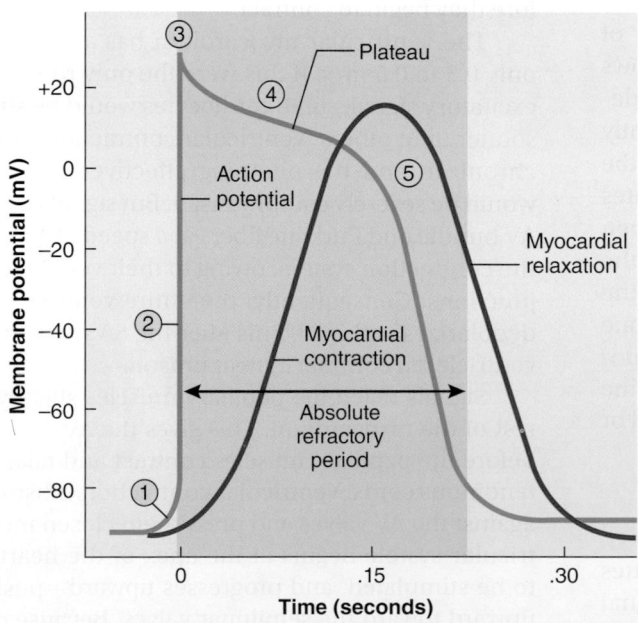

① Voltage-gated Na⁺ channels open.

② Na⁺ inflow depolarizes the membrane and triggers the opening of still more Na⁺ channels, creating a positive feedback cycle and a rapidly rising membrane voltage.

③ Na⁺ channels close when the cell depolarizes, and the voltage peaks at nearly +30 mV.

④ Ca²⁺ entering through slow Ca²⁺ channels prolongs depolarization of membrane, creating a plateau. Plateau falls slightly because of some K⁺ leakage, but most K⁺ channels remain closed until end of plateau.

⑤ Ca²⁺ channels close and Ca²⁺ is transported out of cell. K⁺ channels open, and rapid K⁺ outflow returns membrane to its resting potential.

**FIGURE 19.14 Action Potential of a Ventricular Cardiocyte.** The green curve is the action potential. The red curve represents rising and falling muscle tension as the myocardium contracts and relaxes.

 *What is the benefit of having such a long absolute refractory period in cardiac muscle?*

## The Electrocardiogram

We can detect electrical currents in the heart by means of electrodes (leads) applied to the skin. An instrument called the *electrocardiograph* amplifies these signals and produces a record, usually on a moving paper chart, called an **electrocardiogram**[25] (**ECG** or **EKG**[26]**).** To record an ECG, electrodes are typically attached to the wrists, ankles, and six locations on the chest. Simultaneous recordings can be made from electrodes at different distances from the heart; collectively, they provide a comprehensive image of the heart's electrical activity. An ECG is a composite recording of all action potentials produced by the nodal and myocardial cells—it should not be construed as a tracing of a single action potential.

Figure 19.15 shows a typical ECG. It shows three principal deflections above and below the baseline: the *P wave, QRS complex,* and *T wave.* (These letters were arbitrarily chosen; they do not stand for any words.) Figure 19.16 shows how these correspond to regions of the heart undergoing depolarization and repolarization.

---

[25]*graph* = recording instrument; *graphy* = recording procedure; *gram* = record of
[26]EKG is from the German spelling, Elektrokardiogramm

The **P wave** is produced when a signal from the SA node spreads through the atria and depolarizes them. Atrial systole begins about 100 ms after the P wave begins, during the *PQ segment.* This segment is about 160 ms long and represents the time required for impulses to travel from the SA node to the AV node.

The **QRS complex** consists of a small downward deflection (Q), a tall sharp peak (R), and a final downward deflection (S). It is produced when the signal from the AV node spreads through the ventricular myocardium and depolarizes the muscle. This is the most conspicuous part of the ECG because it is produced mainly by depolarization of the ventricles, which constitute the largest muscle mass of the heart and generate the greatest electrical current. Its complex shape is due to the different sizes of the two ventricles and the different times required for them to depolarize. Ventricular systole begins shortly after the QRS complex, in the *ST segment.* Atrial repolarization and diastole also occur during the QRS interval, but atrial repolarization sends a relatively weak signal that is obscured by the electrical activity of the more muscular ventricles. The ST segment corresponds to the plateau in the myocardial action potential and thus represents the time during which the ventricles contract and eject blood.

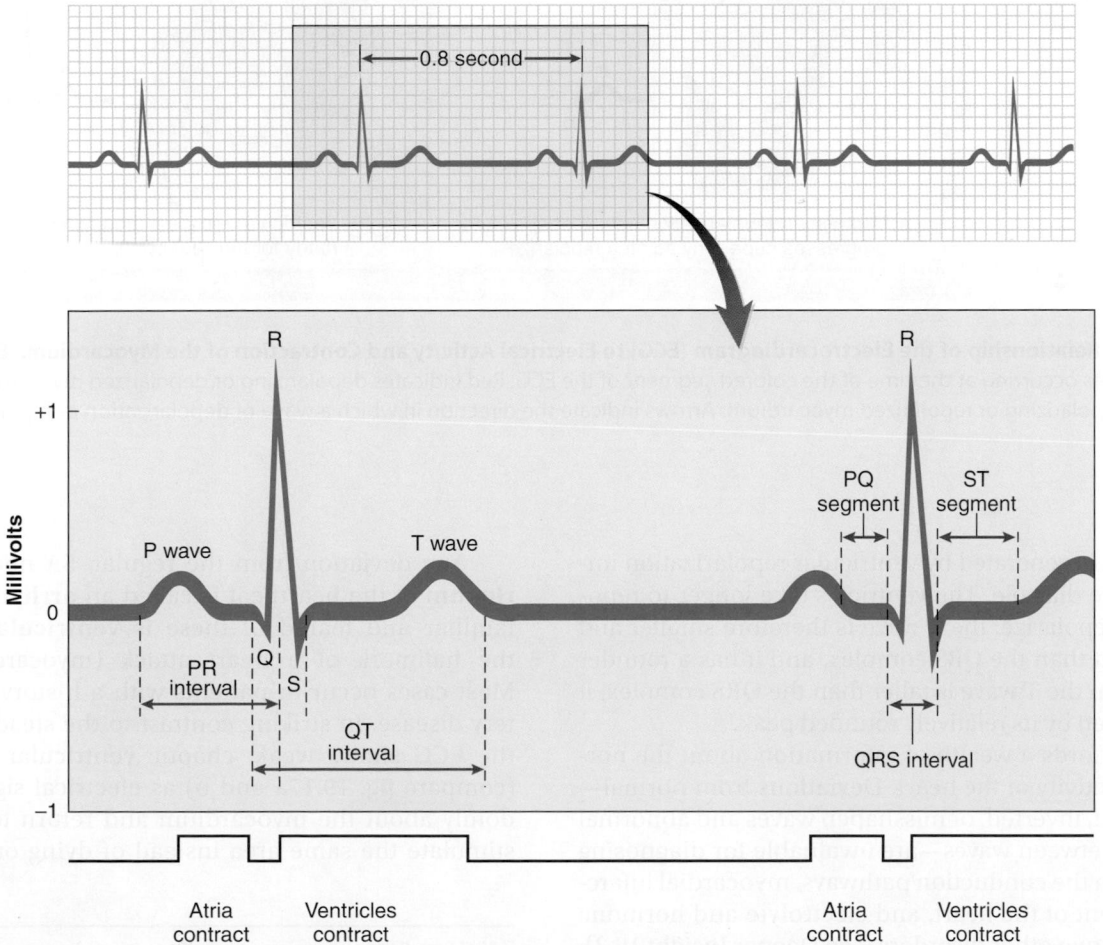

**FIGURE 19.15  The Normal Electrocardiogram.**

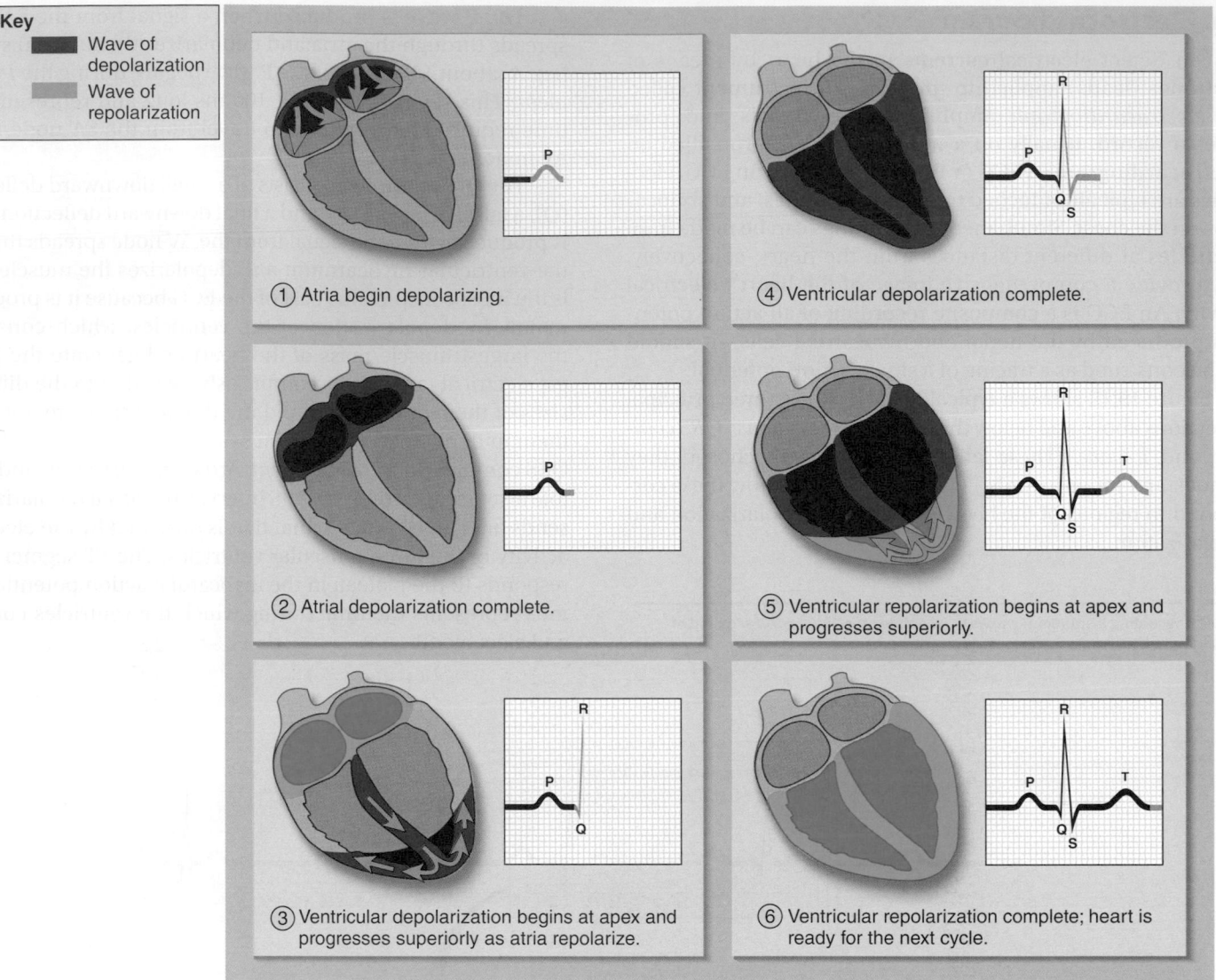

**Key**
- Wave of depolarization
- Wave of repolarization

① Atria begin depolarizing.

② Atrial depolarization complete.

③ Ventricular depolarization begins at apex and progresses superiorly as atria repolarize.

④ Ventricular depolarization complete.

⑤ Ventricular repolarization begins at apex and progresses superiorly.

⑥ Ventricular repolarization complete; heart is ready for the next cycle.

**FIGURE 19.16 Relationship of the Electrocardiogram (ECG) to Electrical Activity and Contraction of the Myocardium.** Each heart diagram indicates the events occurring at the time of the colored segment of the ECG. Red indicates depolarizing or depolarized myocardium, and green indicates repolarizing or repolarized myocardium. Arrows indicate the direction in which a wave of depolarization or repolarization is traveling. **AP|R**

The **T wave** is generated by ventricular repolarization immediately before diastole. The ventricles take longer to repolarize than to depolarize; the T wave is therefore smaller and more spread out than the QRS complex, and it has a rounder peak. Even when the T wave is taller than the QRS complex, it can be recognized by its relatively rounded peak.

The ECG affords a wealth of information about the normal electrical activity of the heart. Deviations from normal—such as enlarged, inverted, or misshapen waves and abnormal time intervals between waves—are invaluable for diagnosing abnormalities in the conduction pathways, myocardial infarction, enlargement of the heart, and electrolyte and hormone imbalances, among other disorders (see Deeper Insight 19.2).

Any deviation from the regular, SA node–driven **sinus rhythm** of the heartbeat is called an **arrhythmia.** The most familiar and feared of these is **ventricular fibrillation,**[27] the hallmark of a heart attack (myocardial infarction). Most cases occur in patients with a history of coronary artery disease. In striking contrast to the steady sinus rhythm, the ECG shows weak, chaotic ventricular depolarizations (compare fig. 19.17a and b) as electrical signals travel randomly about the myocardium and return to repeatedly restimulate the same area instead of dying out like a normal

[27] *fibril* = small fiber; *ation* = action, process

# DEEPER INSIGHT 19.2
## CLINICAL APPLICATION

### Cardiac Arrhythmias

Ventricular fibrillation is the most widely known arrhythmia, but others are not uncommon. *Atrial fibrillation* (fig. 19.17c) is a weak rippling contraction in the atria, manifested in the ECG by chaotic, high-frequency depolarizations (400–650/min.). Fibrillating atria fail to stimulate the ventricles, so we see a dissociation between the random atrial depolarizations and the ventricular QRS and T waves of the ECG. This is the most common atrial arrhythmia in the elderly. It can result from such causes as valvular disease, thyroid hormone excess, or myocardial inflammation, and is often seen in alcoholism.

*Heart block* is a failure of any part of the cardiac conduction system to conduct signals, usually as the result of disease and degeneration of conduction system fibers. In the ECG, one sees rhythmic atrial P waves, but the ventricles fail to receive the signal and no QRS wave follows the P (as in the second and third P waves of fig. 19.17d). A *bundle branch block* is a heart block resulting from damage to one or both branches of the AV bundle. Damage to the AV node causes *total heart block,* in which signals from the atria fail to reach the ventricles at all, and the ventricles beat at their own intrinsic rhythm of 20 to 40 bpm.

*Premature ventricular contraction (PVC)* is the result of a ventricular ectopic focus firing and setting off an extra beat *(extrasystole)* before the normal signal from the SA node arrives. The P wave is missing and the QRS wave is inverted and misshapen (see arrow in fig. 19.17e). PVCs can occur singly or in bursts. An occasional extra beat is not serious, and may result from emotional stress, lack of sleep, or irritation of the heart by stimulants (nicotine, caffeine). Persistent PVCs, however, can indicate more serious pathology and sometimes lead to ventricular fibrillation and sudden death.

---

ventricular depolarization. To the surgeon's eye and hand, a fibrillating ventricle exhibits squirming, uncoordinated contractions often described as feeling "like a bag of worms." A fibrillating ventricle pumps no blood, so there is no coronary blood flow and myocardial tissue rapidly dies of ischemia, as does cerebral tissue. *Cardiac arrest* is the cessation of cardiac output, with the ventricles either motionless or in fibrillation.

Fibrillation kills quickly if it is not stopped. *Defibrillation* is an emergency procedure in which the heart is given a strong electrical shock with a pair of electrodes. The purpose is to depolarize the entire myocardium and stop the fibrillation, with the hope that the SA node will resume its sinus rhythm. This doesn't correct the underlying cause of the fibrillation, but it may sustain a patient's life long enough to allow for other corrective action.

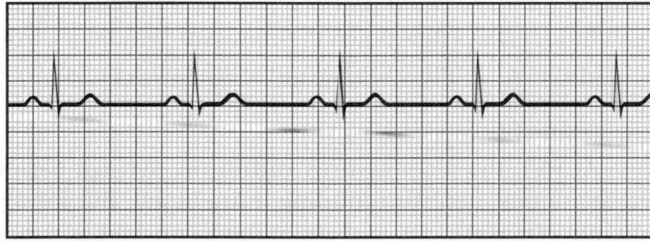

**(a) Sinus rhythm (normal)**

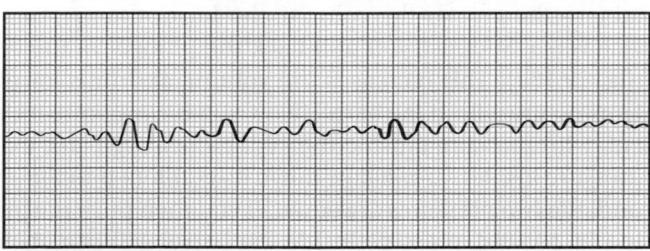

**(b) Ventricular fibrillation**

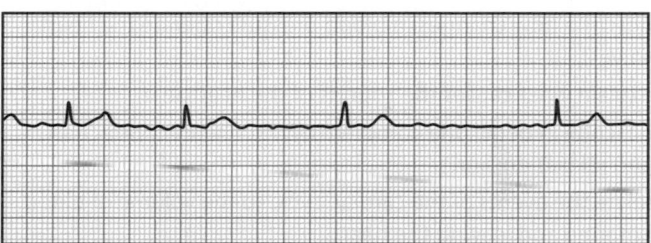

**(c) Atrial fibrillation**

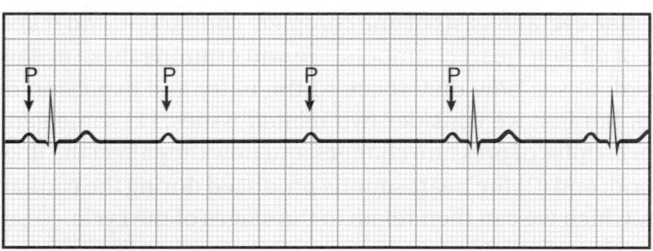

**(d) Heart block**

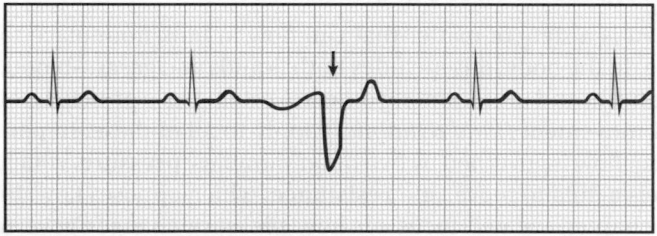

**(e) Premature ventricular contraction**

**FIGURE 19.17  Normal and Pathological Electrocardiograms.**  (a) Normal sinus rhythm. (b) Ventricular fibrillation, with grossly irregular waves of depolarization, as seen in a heart attack (myocardial infarction). (c) Atrial fibrillation; between heartbeats, the atria exhibit weak, chaotic, high-frequency depolarizations instead of normal P waves. (d) Heart block, in which some atrial depolarizations (P waves) are not conducted to the ventricles and not followed by ventricular QRS waves. (e) Premature ventricular contraction, or extrasystole (at arrow); note the absence of a P wave, the inverted QRS complex, and the misshapen QRS and elevated T.

**BEFORE YOU GO ON**

Answer the following questions to test your understanding of the preceding section:

15. Define systole and diastole.

16. How does the pacemaker potential of the SA node differ from the resting membrane potential of a neuron? Why is this important in creating the heart rhythm?

17. How does excitation–contraction coupling in cardiac muscle resemble that of skeletal muscle? How is it different?

18. What produces the plateau in the action potentials of cardiocytes? Why is this important to the pumping ability of the heart?

19. Identify the portion of the ECG that coincides with each of the following events: atrial depolarization, atrial systole, atrial repolarization, ventricular depolarization, ventricular systole, ventricular repolarization, ventricular diastole.

## 19.5    Blood Flow, Heart Sounds, and the Cardiac Cycle

### Expected Learning Outcomes

When you have completed this section, you should be able to

a. explain why blood pressure is expressed in millimeters of mercury;

b. describe how changes in blood pressure operate the heart valves;

c. explain what causes the sounds of the heartbeat;

d. describe in detail one complete cycle of heart contraction and relaxation; and

e. relate the events of the cardiac cycle to the volume of blood entering and leaving the heart.

A **cardiac cycle** consists of one complete contraction and relaxation of all four heart chambers. We will examine these events in detail to see how they relate to the entry and expulsion of blood, but first we consider two related issues: (1) some general principles of pressure changes and how they affect the flow of blood, and (2) the heart sounds produced during the cardiac cycle, which we can then relate to the stages of the cycle.

### Principles of Pressure and Flow

A fluid is a state of matter that can flow in bulk from place to place. In the body, this includes both liquids and gases—blood, lymph, air, and urine, among others. Certain basic principles of fluid movement (*fluid dynamics)* apply to all of these. In particular, flow is governed by two main variables: **pressure,** which impels a fluid to move, and **resistance,** which opposes flow. In this chapter, we will focus on how pressure changes govern the operation of the heart valves, the entry of blood into the heart chambers, and its expulsion into the arteries. In the next

chapter, we will examine the roles of pressure and resistance in the flow of blood through the blood vessels, and in chapter 22, we apply the same principles to respiratory airflow. The flow of blood and of air down their pressure gradients are two applications of the general principle of gradients and flow that we explored in chapter 1 (p. 17).

### Measurement of Pressure

Pressure is commonly measured by a device called a *manometer.* In simplest form, this is typically a J-shaped glass tube partially filled with mercury. The sealed upper end, above the mercury, contains a vacuum, whereas the lower end is open. Pressure applied at the lower end is measured in terms of how high it can push the mercury column up the evacuated end of the tube. In principle, any liquid would do, but mercury is used because it is so dense; it enables us to measure pressure with shorter columns than we would need with a less dense liquid such as water. Pressures are therefore commonly expressed in millimeters of mercury (mm Hg). Blood pressure, specifically, is usually measured with a **sphygmomanometer**[28] (SFIG-mo-ma-NOM-eh-tur)—a calibrated mercury manometer with its open lower end attached to an inflatable pressure cuff wrapped around the arm. Blood pressure and the method of measuring it are discussed in greater detail in chapter 20.

### Pressure Gradients and Flow

A fluid flows only if it is subjected to more pressure at one point than at another. The difference creates a **pressure gradient,** and fluids always flow down their pressure gradients, from the high-pressure point to the low-pressure point. Before we relate this to blood flow, it may be easier to begin with an analogy—an air-filled syringe (fig. 19.18).

At rest, the air pressures within the syringe barrel and in the atmosphere surrounding it are equal. But for a given quantity (mass) of air, and assuming a constant temperature, pressure is inversely proportional to the volume of the container—the greater the volume, the lower the pressure, and vice versa. Suppose you pull back the plunger of the syringe (fig. 19.18a). This increases the volume and thus lowers the air pressure within the barrel. Now you have a pressure gradient, with pressure outside the syringe being greater than the pressure inside. Air will flow down its gradient into the syringe until the two pressures are equal. If you then push the plunger in (fig. 19.18b), pressure inside the barrel will rise above the pressure outside, and air will flow out—again going down its pressure gradient but in the reverse direction.

The syringe barrel is analogous to a heart chamber such as the left ventricle. When the ventricle is expanding, its internal pressure falls. If the AV valve is open, blood flows into the ventricle from the atrium above. When the ventricle contracts, its internal pressure rises. When the aortic valve opens, blood is ejected from the ventricle into the aorta.

---

[28]*sphygmo* = pulse; *mano* = rare, sparse, roomy

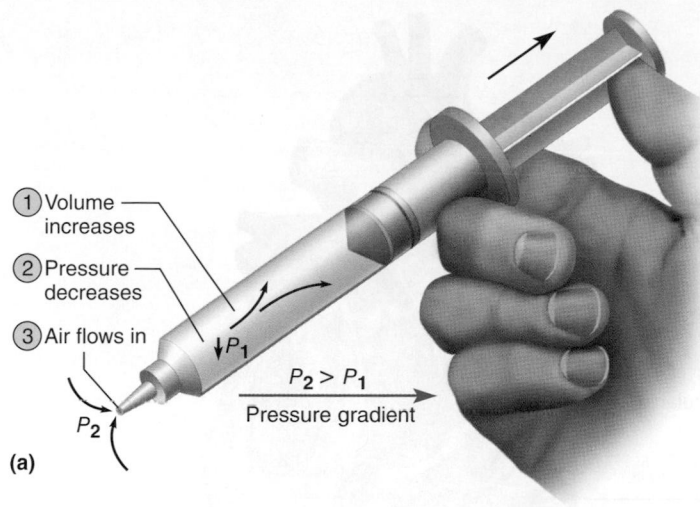

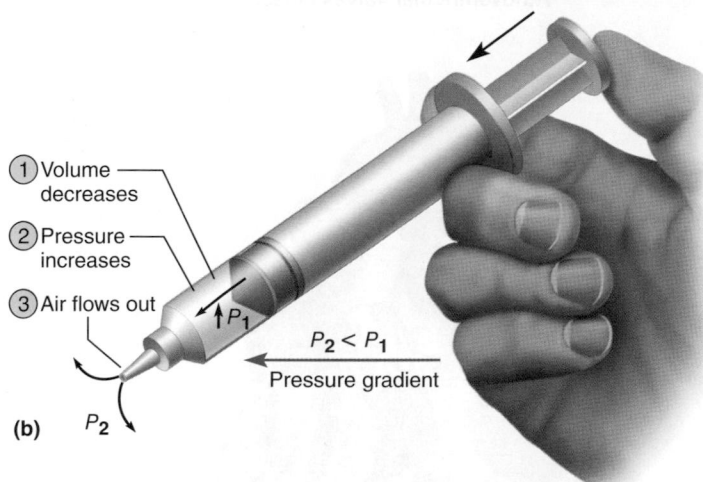

**FIGURE 19.18 Principles of Volume, Pressure, and Flow Illustrated with a Syringe.** (a) As the plunger is pulled back, the volume of the enclosed space increases, its pressure falls, and pressure inside the syringe ($P_1$) is lower than the pressure outside ($P_2$). The pressure gradient causes air to flow inward until the pressures are equal. This is analogous to the filling of an expanding heart chamber. (b) As the plunger is depressed, the volume of the enclosed space decreases, $P_1$ rises above $P_2$, and air flows out until the pressures are equal. This is analogous to the ejection of blood from a contracting heart chamber. In both cases, fluids flow down their pressure gradients.

The opening and closing of the heart valves are governed by these pressure changes. Remember that the valves are just soft flaps of connective tissue with no muscle. They do not exert any effort of their own, but are passively pushed open and closed by the changes in blood pressure on the upstream and downstream sides of the valve.

When the ventricles are relaxed and their pressure is low, the AV valve cusps hang down limply and both valves are open (fig. 19.19a). Blood flows freely from the atria into the ventricles

even before the atria contract. As the ventricles fill with blood, the cusps float upward toward the closed position. When the ventricles contract, their internal pressure rises sharply and blood surges against the AV valves from below. This pushes the cusps together, seals the openings, and prevents blood from flowing back into the atria. The papillary muscles contract slightly before the rest of the ventricular myocardium and tug on the tendinous cords, preventing the valves from bulging excessively (prolapsing) into the atria or turning inside out like windblown umbrellas. (See *mitral valve prolapse* in Deeper Insight 19.3.)

The rising pressure in the ventricles also acts on the aortic and pulmonary valves. Up to a point, pressure in the aorta and pulmonary trunk opposes their opening, but when the ventricular pressure rises above the arterial pressure, the valves open and blood is ejected from the heart. Then as the ventricles relax again and their pressure falls below that in the arteries, arterial blood briefly flows backward and fills the pocketlike cusps of the semilunar valves. The three cusps meet in the middle of the orifice and seal it, thereby preventing arterial blood from reentering the heart.

#### ▶▶▶ APPLY WHAT YOU KNOW

*How would aortic valvular stenosis (see Deeper Insight 19.3) affect the amount of blood pumped into the aorta? How might this affect a person's physical stamina? Explain your reasoning.*

 **DEEPER INSIGHT 19.3**

### CLINICAL APPLICATION

#### Valvular Insufficiency Disorders

*Valvular insufficiency* (incompetence) refers to any failure of a valve to prevent *reflux* (regurgitation)—the backward flow of blood. *Valvular stenosis*[29] is a form of insufficiency in which the cusps are stiffened and the opening is constricted by scar tissue. It frequently results from rheumatic fever, an autoimmune disease in which antibodies produced to fight a bacterial infection also attack the mitral and aortic valves. As the valves become scarred and constricted, the heart is overworked by the effort to force blood through the openings and may become enlarged. Regurgitation of blood through the incompetent valves creates turbulence that can be heard with a stethoscope as a *heart murmur*.

*Mitral valve prolapse (MVP)* is an insufficiency in which one or both mitral valve cusps bulge into the atrium during ventricular contraction. It is often hereditary and affects about 1 out of 40 people, especially young women. In many cases, it causes no serious dysfunction, but in some people it causes chest pain, fatigue, and shortness of breath.

In some cases, an incompetent valve can eventually lead to heart failure. A defective valve can be surgically repaired or replaced with an artificial valve or a valve transplanted from a pig heart.

---

[29]*steno* = narrow; *osis* = condition

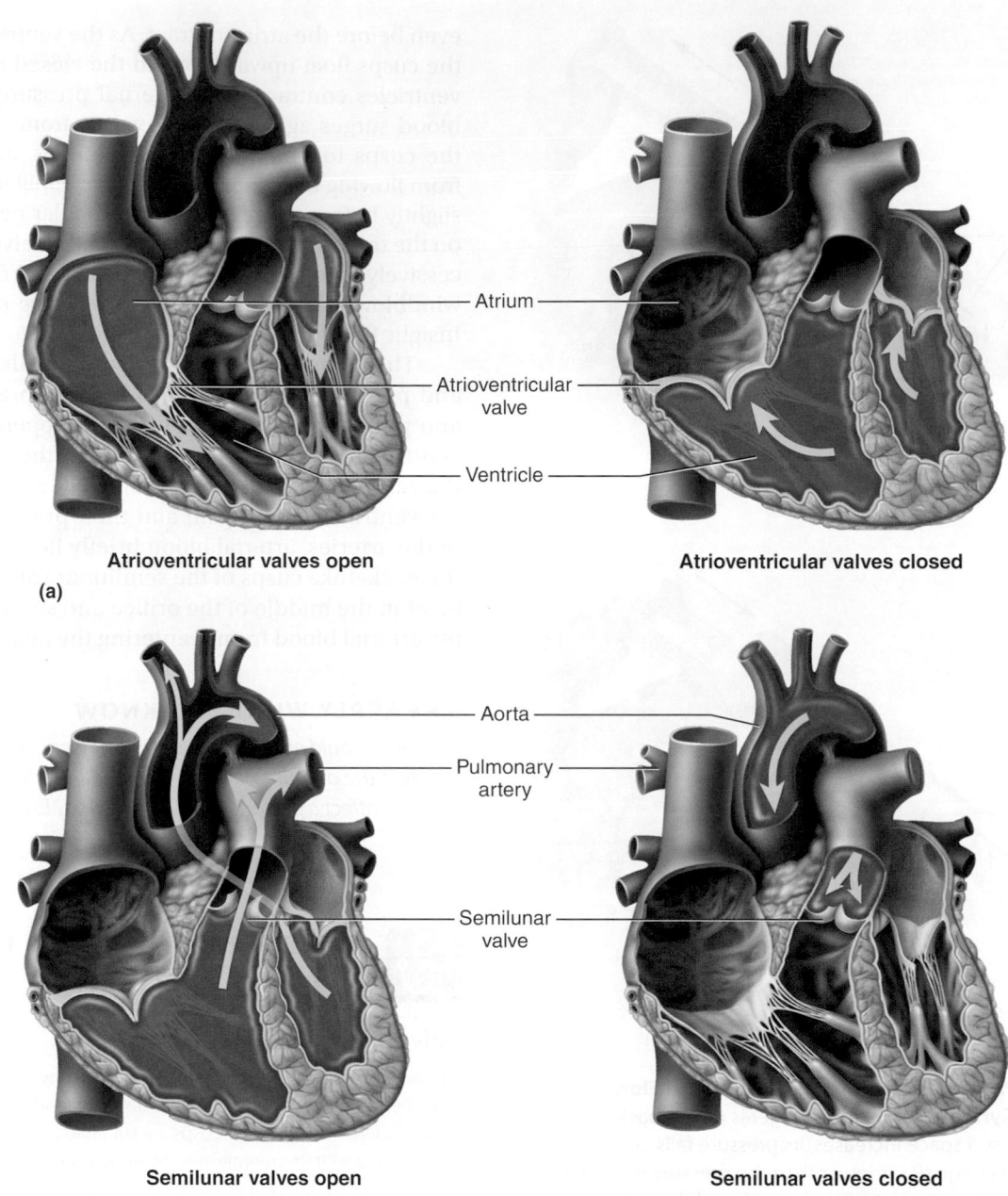

**Atrioventricular valves open**

**(a)**

**Atrioventricular valves closed**

- Atrium
- Atrioventricular valve
- Ventricle

**Semilunar valves open**

**(b)**

**Semilunar valves closed**

- Aorta
- Pulmonary artery
- Semilunar valve

**FIGURE 19.19 Operation of the Heart Valves.** (a) The atrioventricular valves. When atrial pressure is greater than ventricular pressure, the valve opens and blood flows through (green arrows). When ventricular pressure rises above atrial pressure, the blood in the ventricle pushes the valve cusps closed. (b) The semilunar valves. When the pressure in the ventricles is greater than the pressure in the great arteries, the semilunar valves are forced open and blood is ejected. When ventricular pressure is lower than arterial pressure, arterial blood holds these valves closed.

❓ *What role do the tendinous cords play?*

## Heart Sounds

As we follow events through the cardiac cycle, we will note the occurrence of two or three *heart sounds* audible with a stethoscope. Listening to sounds made by the body is called **auscultation** (AWS-cul-TAY-shun). The **first** and **second heart sounds,** symbolized $S_1$ and $S_2$, are often described as a "lubb-dupp"—$S_1$ is louder and longer and $S_2$ a little softer and sharper. In children and adolescents, it is normal to hear a **third heart**

**sound** ($S_3$). This is rarely audible in people older than 30, but when it is, the heartbeat is said to show a *triple rhythm* or *gallop,* which may indicate an enlarged and failing heart. If the normal sounds are roughly simulated by drumming two fingers on a table, a triple rhythm sounds a little like drumming with three fingers. The heart valves themselves operate silently, but $S_1$ and $S_2$ occur in conjunction with the closing of the valves as a result of turbulence in the bloodstream and movements of the heart wall.

## Phases of the Cardiac Cycle

We now examine the phases of the cardiac cycle, the pressure changes that occur, and how the pressure changes and valves govern the flow of blood. Cardiovascular physiologist Carl J. Wiggers (1883–1963) devised a chart, now known as the *Wiggers diagram* (fig. 19.20), for showing the major events that occur simultaneously at each moment throughout the cardiac cycle. Here, it is divided into colored and numbered bars to correspond to the following phases. Closely follow the figure as you study the text. Where to begin when describing a circular chain of events is somewhat arbitrary. However, in this presentation, we begin with the filling of the ventricles. Remember that all these events are completed in less than 1 second.

1. **Ventricular filling.** During diastole, the ventricles expand and their pressure drops below that of the atria. As a result, the AV valves open and blood flows into the ventricles, causing ventricular pressure to rise and atrial pressure to fall. Ventricular filling occurs in three phases: **(1a)** The first one-third is *rapid ventricular filling,* when blood enters especially quickly. **(1b)** The second one-third, called *diastasis* (di-ASS-tuh-sis), is marked by slower filling. The P wave of the electrocardiogram occurs at the end of diastasis, marking the depolarization of the atria. **(1c)** In the last one-third, *atrial systole* completes the filling process. The right atrium contracts slightly before the left because it is the first to receive the signal from the SA node. At the end of ventricular filling, each ventricle contains an **end-diastolic volume (EDV)** of about 130 mL of blood. Only 40 mL (31%) of this is contributed by atrial systole.

2. **Isovolumetric contraction.** The atria repolarize, relax, and remain in diastole for the rest of the cardiac cycle. The ventricles depolarize, generate the QRS complex, and begin to contract. Pressure in the ventricles rises sharply and reverses the pressure gradient between atria and ventricles. The AV valves close as ventricular blood surges back against the cusps. Heart sound $S_1$ occurs at the beginning of this phase and is produced mainly by the left ventricle; the right ventricle is thought to make little contribution. Causes of the sound are thought to include the tensing of ventricular tissues and tendinous cords (like the twang of a suddenly stretched rubber band), turbulence in the blood as it surges against the closed AV valves, and impact of the heart against the chest wall.

   This phase is called *isovolumetric*[30] because even though the ventricles contract, they do not eject blood yet and there is no change in their volume. This is because pressures in the aorta (80 mm Hg) and pulmonary trunk (10 mm Hg) are still greater than the pressures in the respective ventricles and thus oppose the opening of the

semilunar valves. The cardiocytes exert force, but with all four valves closed, the blood cannot go anywhere.

3. **Ventricular ejection.** The ejection of blood begins when ventricular pressure exceeds arterial pressure and forces the semilunar valves open. The pressure peaks at typically 120 mm Hg in the left ventricle and 25 mm Hg in the right. Blood spurts out of each ventricle rapidly at first *(rapid ejection),* then flows out more slowly under less pressure *(reduced ejection).* By analogy, suppose you were to shake up a bottle of soda pop and remove the cap. The soda would spurt out rapidly at high pressure and then more would dribble out at lower pressure, much like the blood leaving the ventricles. Ventricular ejection lasts about 200 to 250 ms, which corresponds to the plateau of the myocardial action potential but lags somewhat behind it (review the red tension curve in fig. 19.14). The T wave occurs late in this phase, beginning at the moment of peak ventricular pressure.

   The ventricles do not expel all their blood. In an average resting heart, each ventricle contains an EDV of 130 mL. The amount ejected, about 70 mL, is called the **stroke volume (SV).** This is about 54% of the EDV, a percentage called the **ejection fraction.** The blood remaining behind, about 60 mL in this case, is called the **end-systolic volume (ESV).** Note that EDV – SV = ESV. In vigorous exercise, the ejection fraction may be as high as 90%. Ejection fraction is an important measure of cardiac health. A diseased heart may eject much less than 50% of the blood it contains.

4. **Isovolumetric relaxation.** This is early ventricular diastole, when the T wave ends and the ventricles begin to expand. There are competing hypotheses as to how they expand. One is that the blood flowing into the ventricles inflates them. Another is that contraction of the ventricles deforms the fibrous skeleton, which subsequently springs back like the rubber bulb of a turkey baster that has been squeezed and released. This elastic recoil and expansion would cause pressure to drop rapidly and suck blood into the ventricles.

   At the beginning of ventricular diastole, blood from the aorta and pulmonary trunk briefly flows backward through the semilunar valves. The backflow, however, quickly fills the cusps and closes them, creating a slight pressure rebound that appears as the *dicrotic notch* of the aortic pressure curve (the top curve in fig. 19.20). Heart sound $S_2$ occurs as blood rebounds from the closed semilunar valves and the ventricles expand. This phase is called *isovolumetric* because the semilunar valves are closed, the AV valves have not yet opened, and the ventricles are therefore taking in no blood. When the AV valves open, ventricular filling (phase 1) begins again. Heart sound $S_3$, if it occurs, is thought to result from the transition from expansion of the empty ventricles to their sudden filling with blood.

---

[30]*iso* = same

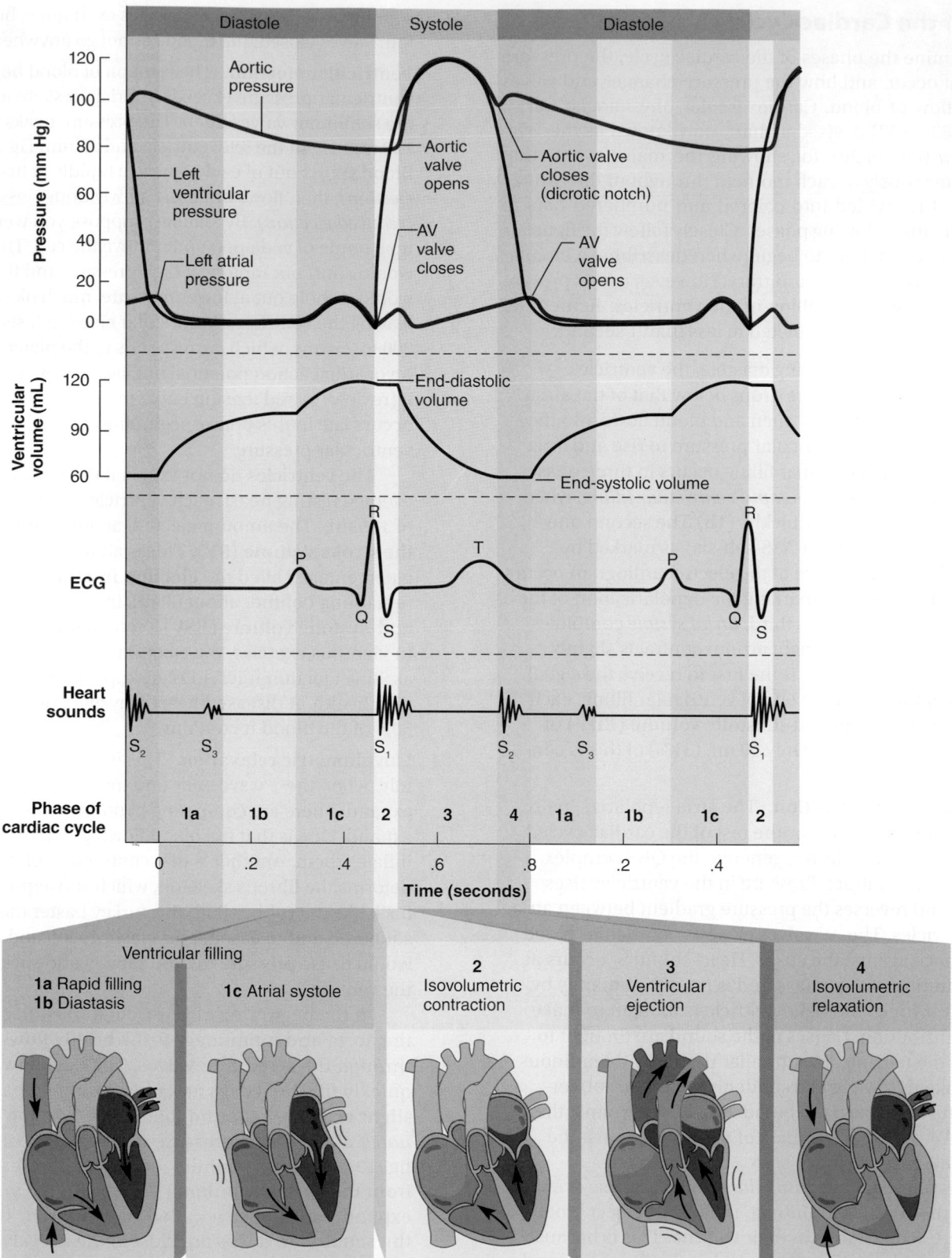

**FIGURE 19.20  Modified Wiggers Diagram, Illustrating Events of the Cardiac Cycle.**  Two cycles are shown. The phases are numbered across the bottom to correspond to the text description.

**?** *Explain why the aortic pressure curve begins to rise abruptly at about 0.5 second.*

In a resting person, atrial systole lasts about 0.1 second; ventricular systole, 0.3 second; and the *quiescent period* (when all four chambers are in diastole), 0.4 second. Total duration of the cardiac cycle is therefore 0.8 second (800 ms) in a heart beating at 75 bpm.

## Overview of Volume Changes

An additional perspective on the cardiac cycle can be gained if we review the volume changes that occur. This "balance sheet" is from the standpoint of one ventricle; both ventricles have equal volumes. The volumes vary somewhat from one person to another and depend on a person's state of activity.

| | |
|---|---|
| End-systolic volume (ESV) left from the previous heartbeat | 60 mL |
| Passively added to the ventricle during atrial diastole | + 30 mL |
| Added by atrial systole | + 40 mL |
| *Total:* End-diastolic volume (EDV) | 130 mL |
| Stroke volume (SV) ejected by ventricular systole | − 70 mL |
| *Leaves:* End-systolic volume (ESV) | 60 mL |

Notice that the ventricle pumps out as much blood as it received during diastole: 70 mL in this example. Both ventricles eject the same amount of blood even though pressure in the right ventricle is only about one-fifth the pressure in the left. Blood pressure in the pulmonary trunk is relatively low, so the right ventricle does not need to generate very much pressure to overcome it.

Equal output by the two ventricles is essential to homeostasis. If the right ventricle pumps more blood into the lungs than the left ventricle can handle on return, blood accumulates in the lungs, causing pulmonary hypertension, edema, and a risk of drowning in one's own body fluid (fig. 19.21a). One of the first signs of left ventricular failure is respiratory distress—shortness of breath and a sense of suffocation. Conversely, if the left ventricle pumps more blood than the right, blood accumulates in the systemic circuit, causing hypertension and widespread systemic edema (fig. 19.21b). Such systemic edema, once colloquially called *dropsy,* is marked by enlargement of the liver; **ascites** (ah-SITE-eez), a pooling of fluid in the abdominal cavity; distension of the jugular veins; and swelling of the fingers, ankles, and feet. It can lead to stroke or kidney failure. A failure of one ventricle increases the workload on the other, which stresses it and often leads to its eventual failure as well. In principle, if the output of the left ventricle were just 1% greater than output of the right, it would completely drain the lungs of blood in less than 10 minutes (although death would occur much sooner).

Fluid accumulation in either circuit due to insufficiency of ventricular pumping is called **congestive heart failure (CHF).** Common causes of CHF are myocardial infarction, chronic hypertension, valvular defects, and congenital defects in cardiac anatomy.

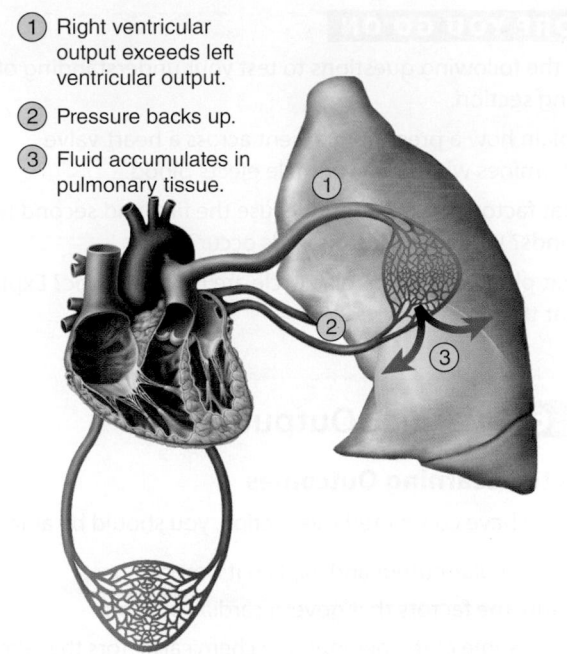

1. Right ventricular output exceeds left ventricular output.
2. Pressure backs up.
3. Fluid accumulates in pulmonary tissue.

**(a) Pulmonary edema**

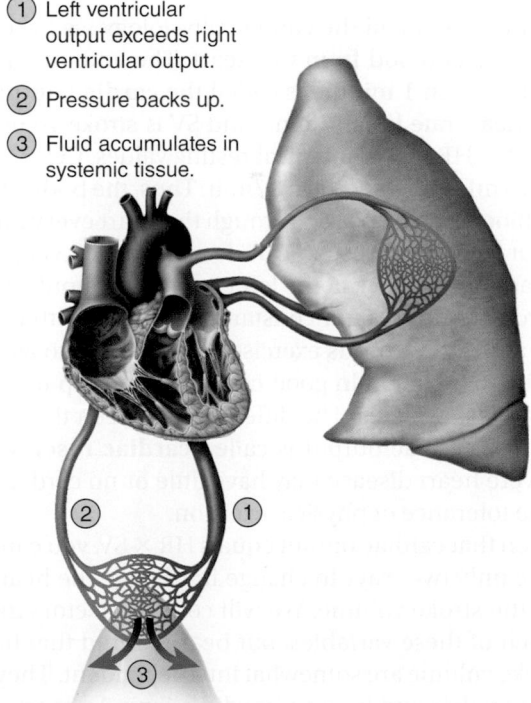

1. Left ventricular output exceeds right ventricular output.
2. Pressure backs up.
3. Fluid accumulates in systemic tissue.

**(b) Systemic edema**

**FIGURE 19.21 The Necessity of Balanced Ventricular Output.** (a) If the left ventricle pumps less blood than the right, blood pressure backs up into the lungs and causes pulmonary edema. (b) If the right ventricle pumps less blood than the left, pressure backs up in the systemic circulation and causes systemic edema. To maintain homeostasis, both ventricles must pump the same average amount of blood.

Answer the following questions to test your understanding of the preceding section:

20. Explain how a pressure gradient across a heart valve determines whether a ventricle ejects blood.

21. What factors are thought to cause the first and second heart sounds? When do these sounds occur?

22. What phases of the cardiac cycle are isovolumetric? Explain what this means.

## 19.6   Cardiac Output

### Expected Learning Outcomes

When you have completed this section, you should be able to

a. define *cardiac output* and explain its importance;

b. identify the factors that govern cardiac output;

c. discuss some of the nervous and chemical factors that alter heart rate, stroke volume, and cardiac output;

d. explain how the right and left ventricles achieve balanced output; and

e. describe some effects of exercise on cardiac output.

The entire point of all the cardiac physiology we have considered is to eject blood from the heart. The amount ejected by each ventricle in 1 minute is called the **cardiac output (CO).** If HR is heart rate (beats/min.) and SV is stroke volume (mL/beat), CO = HR × SV. At typical resting values, CO = 75 beats/min. × 70 mL/beat = 5,250 mL/min. Thus, the body's total volume of blood (4–6 L) passes through the heart every minute; or to look at it another way, an RBC leaving the left ventricle will, on average, arrive back at the left ventricle in about 1 minute.

Cardiac output is not constant, but varies with the body's state of activity. Vigorous exercise increases CO to as much as 21 L/min. in a person in good condition, and up to 35 L/min. in world-class athletes. The difference between the maximum and resting cardiac output is called **cardiac reserve.** People with severe heart disease may have little or no cardiac reserve and little tolerance of physical exertion.

Given that cardiac output equals HR × SV, you can see that there are only two ways to change it: Change the heart rate or change the stroke volume. We will consider factors that influence each of these variables, but bear in mind that heart rate and stroke volume are somewhat interdependent. They usually change together and in opposite directions. As heart rate goes up, stroke volume goes down, and vice versa.

### Heart Rate

Heart rate is most easily measured by taking a person's **pulse** at some point where an artery runs close to the body surface, such as the *radial artery* in the wrist or *common carotid artery*

in the neck. Each beat of the heart produces a surge of pressure that can be felt by palpating a superficial artery with the fingertips. Heart rate can be obtained by counting the number of pulses in 15 seconds and multiplying by 4 to get the beats per minute. In newborn infants, the resting heart rate is commonly 120 bpm or greater. It declines steadily with age, averaging 72 to 80 bpm in young adult females and 64 to 72 bpm in young adult males. It rises again in the elderly.

**Tachycardia**[31] is a persistent, resting adult heart rate above 100 bpm. It can be caused by stress, anxiety, stimulants, heart disease, or fever. Heart rate also rises to compensate to some extent for a drop in stroke volume. Thus, the heart races when the body has lost a significant quantity of blood or when there is damage to the myocardium.

**Bradycardia**[32] is a persistent, resting adult heart rate below 60 bpm. It is common during sleep and in endurance-trained athletes. Endurance training enlarges the heart and increases its stroke volume, enabling it to maintain the same output with fewer beats. Hypothermia (low body temperature) also slows the heart rate and may be deliberately induced in preparation for cardiac surgery. Diving mammals such as whales and seals exhibit bradycardia during the dive, as do humans to some extent when the face is immersed in cool water.

Factors that raise the heart rate are called **positive chronotropic**[33] **agents,** and factors that lower it are **negative chronotropic agents.** We next consider some chronotropic effects of the autonomic nervous system, hormones, electrolytes, and blood gases.

### Chronotropic Effects of the Autonomic Nervous System

Although the nervous system does not initiate the heartbeat, it does modulate its rhythm and force. The reticular formation of the medulla oblongata contains **cardiac centers,** with only vaguely defined anatomical boundaries, that initiate autonomic output to the heart. Some neurons in these centers have a *cardiostimulatory* effect and transmit signals to the heart by way of the sympathetic pathway described earlier; others have a *cardioinhibitory* effect communicated to the heart by way of the vagus nerves.

The sympathetic postganglionic fibers are adrenergic—they release norepinephrine, which binds to β-adrenergic fibers in the heart. This activates the cyclic adenosine monophosphate (cAMP) second-messenger system in the cardiocytes and nodal cells. Cyclic AMP activates an enzyme that opens a $Ca^{2+}$ channel in the plasma membrane. The $Ca^{2+}$ inflow accelerates depolarization of the SA node and contraction of the cardiocytes, so it speeds up the heart. In addition, cAMP accelerates the uptake of $Ca^{2+}$ by the sarcoplasmic reticulum and thereby enables cardiocytes to relax more quickly. By accelerating *both* contraction and relaxation, norepinephrine and cAMP increase the heart rate.

[31]*tachy* = speed, fast; *card* = heart; *ia* = condition
[32]*brady* = slow; *card* = heart; *ia* = condition
[33]*chrono* = time; *trop* = change, influence

Adrenergic stimulation can, in fact, raise the heart rate to as high as 230 bpm. This limit is set mainly by the refractory period of the SA node, which prevents it from firing any more frequently. Cardiac output peaks, however, at a heart rate of 160 to 180 bpm. At rates any higher than this, the ventricles have too little time to fill between beats. At a resting heart rate of 65 bpm, ventricular diastole lasts about 0.62 second, but at 200 bpm, it lasts only 0.14 second. Thus you can see that at excessively high heart rates, diastole is too brief to allow complete filling of the ventricles, and therefore stroke volume and cardiac output are reduced.

The parasympathetic vagus nerves, by contrast, have cholinergic, inhibitory effects on the SA and AV nodes. Acetylcholine (ACh) binds to muscarinic receptors and opens $K^+$ gates in the nodal cells. As $K^+$ exits the cells, they become hyperpolarized and fire less frequently, so the heart slows down. The vagus nerves have a faster-acting effect on the heart than the sympathetic nerves because ACh acts directly on ion channels in the plasma membrane; sympathetic effects are slower because of the time taken for the cAMP system to act on the ion channels.

If all sympathetic and parasympathetic stimulation of the heart is pharmacologically blocked, or if the cardiac nerves are severed, the heart beats at a rate of about 100 bpm. This is the intrinsic "natural" firing rate of the SA node free of autonomic influence. With intact, functional innervation, however, the resting heart rate is held down to about 70 to 80 bpm by **vagal tone,** a steady background firing rate of the vagus nerves. More extreme vagal stimulation can reduce the heart rate to as low as 20 bpm or even stop the heart briefly.

There is a benefit to placing heart rate under the influence of cardiac centers in the medulla—these centers can receive input from many other sources and integrate it into a "decision" as to whether the heart should beat more quickly or slowly. Sensory and emotional stimuli can act on the cardiac centers by way of the cerebral cortex, limbic system, and hypothalamus; therefore, heart rate can climb even as you anticipate taking the first plunge on a roller coaster or competing in an athletic event, and it is influenced by emotions such as love and anger. The medulla also receives input from receptors in the muscles, joints, arteries, and brainstem:

- **Proprioceptors** in the muscles and joints provide information on changes in physical activity. Thus, the heart can increase its output even before the metabolic demands of the muscles rise.

- **Baroreceptors (pressoreceptors)** are pressure sensors in the aorta and internal carotid arteries (see fig. 15.1, p. 559). They send a continual stream of signals to the medulla. When the heart rate rises, cardiac output increases and raises the blood pressure at the baroreceptors. The baroreceptors increase their signaling to the medulla and, depending on circumstances, the medulla may issue vagal output to lower the heart rate. Conversely, the baroreceptors also inform the medulla

of drops in blood pressure. The medulla can then issue sympathetic output to increase the heart rate, bringing cardiac output and blood pressure back up to normal (see fig. 1.7, p. 16). Either way, a negative feedback loop usually prevents the blood pressure from deviating too far from normal.

- **Chemoreceptors** occur in the aortic arch, carotid arteries, and the medulla oblongata itself, and are sensitive to blood pH, $CO_2$, and $O_2$ levels. They are more important in respiratory control than in cardiovascular control, but they do influence the heart rate. If circulation to the tissues is too slow to remove $CO_2$ as fast as the tissues produce it, then $CO_2$ accumulates in the blood and cerebrospinal fluid (CSF) and produces a state of *hypercapnia* ($CO_2$ excess). Furthermore, $CO_2$ generates hydrogen ions by reacting with water: $CO_2 + H_2O \rightarrow HCO_3^- + H^+$. The hydrogen ions lower the pH of the blood and CSF and may create a state of *acidosis* (pH < 7.35). Hypercapnia and acidosis stimulate the cardiac centers to increase the heart rate, thus improving perfusion of the tissues and restoring homeostasis. The chemoreceptors also respond to extreme *hypoxemia* (oxygen deficiency), as in suffocation, but the effect is usually to slow down the heart, perhaps so the heart does not compete with the brain for the limited oxygen supply.

Such responses to fluctuations in blood chemistry and blood pressure, called **chemoreflexes** and **baroreflexes,** are good examples of negative feedback loops. They are discussed more fully in the next chapter.

## Chronotropic Effects of Chemicals

Heart rate is influenced by many other chemicals besides the neurotransmitters of the cardiac nerves. Blood-borne epinephrine and norepinephrine from the adrenal medulla, for example, have the same effect as norepinephrine from the sympathetic nerves. The chronotropic action of some other chemicals can be understood from their relationships to this catecholamine–cAMP mechanism. Nicotine accelerates the heart by stimulating catecholamine secretion. Thyroid hormone stimulates the up-regulation of adrenergic receptors, making the heart more responsive to sympathetic stimulation; therefore, hyperthyroidism commonly produces tachycardia. Caffeine and the related stimulants in tea and chocolate accelerate the heart by inhibiting cAMP breakdown, prolonging the adrenergic effect.

The electrolyte with the greatest chronotropic effect is potassium ($K^+$). In *hyperkalemia,*[34] a potassium excess, $K^+$ diffuses into the cardiocytes and keeps the membrane voltage elevated, inhibiting cardiocyte repolarization. The myocardium becomes less excitable, the heart rate becomes slow and irregular, and the heart may arrest in diastole. In *hypokalemia,*

---

[34]*hyper* = excess; *kal* = potassium (Latin, *kalium*); *emia* = blood condition

a potassium deficiency, $K^+$ diffuses out of the cardiocytes and they become hyperpolarized—the membrane potential is more negative than normal. This makes them harder to stimulate. These potassium imbalances are very dangerous and require emergency medical treatment.

Calcium also affects heart rate. A calcium excess *(hypercalcemia)* causes a slow heartbeat, whereas a calcium deficiency *(hypocalcemia)* elevates the heart rate. Such calcium imbalances are relatively rare, however, and when they do occur, their primary effect is on contraction strength, which is considered in the coming section on contractility. Chapter 24 further explores the causes and effects of imbalances in potassium, calcium, and other electrolytes.

## Stroke Volume

The other factor in cardiac output is stroke volume. This, in turn, is governed by three variables called *preload, contractility,* and *afterload.* Increased preload or contractility increases stroke volume, whereas increased afterload opposes the emptying of the ventricles and reduces stroke volume.

## Preload

**Preload** is the amount of tension (stretch) in the ventricular myocardium immediately before it begins to contract. To understand how this influences stroke volume, imagine yourself engaged in heavy exercise. As active muscles massage your veins, they drive more blood back to the heart, increasing *venous return.* As more blood enters the heart, it stretches the myocardium. Because of the length–tension relationship of striated muscle, moderate stretch enables the cardiocytes to generate more tension when they contract—that is, stretch increases preload. When the ventricles contract more forcefully, they expel more blood, thus adjusting cardiac output to the increase in venous return.

This principle is summarized by the **Frank–Starling law of the heart.**[35] In a concise, symbolic way, it states that $SV \propto EDV$; that is, stroke volume is proportional to the end-diastolic volume. In other words, the ventricles tend to eject as much blood as they receive. Within limits, the more they are stretched, the harder they contract on the next beat.

Although relaxed skeletal muscle is normally at an optimum length for the most forceful contraction, relaxed cardiac muscle is at less than optimum length. Additional stretch therefore produces a significant increase in contraction force on the next beat. This helps balance the output of the two ventricles. For example, if the right ventricle begins to pump an increased amount of blood, this soon arrives at the left ventricle, stretches it more than before, and causes it to increase its stroke volume and match that of the right.

## Contractility

**Contractility** refers to how hard the myocardium contracts *for a given preload.* It does not describe the increase in tension produced by stretching the muscle, but rather an increase caused by factors that make the cardiocytes more responsive to stimulation. Factors that increase contractility are called **positive inotropic**[36] **agents,** and those that reduce it are **negative inotropic agents.**

Calcium has a strong, positive inotropic effect—it increases the strength of each contraction of the heart. This is not surprising, because $Ca^{2+}$ not only is essential to the excitation–contraction coupling of muscle, but also prolongs the plateau of the myocardial action potential. Calcium imbalances therefore affect not only heart rate, as we have already seen, but also contraction strength. In hypercalcemia, extra $Ca^{2+}$ diffuses into the cardiocytes and produces strong, prolonged contractions. In extreme cases, it can cause cardiac arrest in systole. In hypocalcemia, the cardiocytes lose $Ca^{2+}$ to the extracellular fluid, leading to a weak, irregular heartbeat and potentially to cardiac arrest in diastole. However, as explained in chapter 7, severe hypocalcemia is likely to kill through skeletal muscle paralysis and suffocation before the cardiac effects are felt.

Agents that affect calcium availability have not only the chronotropic effects already examined, but also inotropic effects. We have already seen that norepinephrine increases calcium levels in the sarcoplasm; consequently, it increases not only heart rate but also contraction strength (as does epinephrine, for the same reason). The pancreatic hormone glucagon exerts an inotropic effect by stimulating cAMP production; a solution of glucagon and calcium chloride is sometimes used for the emergency treatment of heart attacks. Digitalis, a cardiac stimulant from the foxglove plant, also raises the intracellular calcium level and contraction strength; it is used to treat congestive heart failure.

Hyperkalemia has a negative inotropic effect because it reduces the strength of myocardial action potentials and thus reduces the release of $Ca^{2+}$ into the sarcoplasm. The heart becomes dilated and flaccid. Hypokalemia, however, has little effect on contractility.

The vagus nerves have a negative inotropic effect on the atria, but they provide so little innervation to the ventricles that they have no significant effect on them.

There are other chronotropic and inotropic agents too numerous to mention here. The ones we have discussed are summarized in table 19.1.

### ▶▶▶ APPLY WHAT YOU KNOW

*Suppose a person has a heart rate of 70 bpm and a stroke volume of 70 mL. A negative inotropic agent then reduces the stroke volume to 50 mL. What would the new heart rate have to be to maintain the same cardiac output?*

---

[35] Otto Frank (1865–1944), German physiologist; Ernest Henry Starling (1866–1927), English physiologist

[36] *ino* = fiber; *trop* = change, influence

# Afterload

**Afterload** is the sum of all forces a ventricle must overcome before it can eject blood. The most significant contribution to afterload is the blood pressure in the aorta and pulmonary trunk immediately distal to the semilunar valves; it opposes the opening of these valves and thus limits stroke volume. For this reason, hypertension increases the afterload and opposes ventricular ejection. Anything that impedes arterial circulation can also increase the afterload. For example, in some lung diseases, scar tissue forms in the lungs and restricts pulmonary circulation. This increases the afterload in the pulmonary trunk. As the right ventricle works harder to overcome this resistance, it gets larger like any other muscle. Stress and hypertrophy of a ventricle can eventually cause it to weaken and fail. Right ventricular failure due to obstructed pulmonary circulation is called *cor pulmonale*[37] (CORE PUL-mo-NAY-lee). It is a common complication of emphysema, chronic bronchitis, and black lung disease (see chapter 22).

## Exercise and Cardiac Output

It is no secret that exercise makes the heart work harder, and it should come as no surprise that this increases cardiac output. The main reason the heart rate increases at the beginning of exercise is that proprioceptors in the muscles and joints transmit signals to the cardiac centers, signifying that the muscles are active and will quickly need an increased blood flow. Sympathetic output from the cardiac centers then increases cardiac output to meet the expected demand. As the exercise progresses, muscular activity increases venous return. This increases the preload on the right ventricle and is soon reflected in the left ventricle as more blood flows through the pulmonary circuit and reaches the left heart. As the heart rate and stroke volume rise, cardiac output rises, which compensates for the increased venous return.

A sustained program of exercise causes hypertrophy of the ventricles, which increases their stroke volume. As explained earlier, this allows the heart to beat more slowly and still maintain a normal resting cardiac output. Endurance athletes commonly have resting heart rates as low as 40 to 60 bpm, but because of the higher stroke volume, their resting cardiac output is about the same as that of an untrained person. Such athletes have greater cardiac reserve, so they can tolerate more exertion than a sedentary person can.

The effects of aging on the heart are discussed on page 1121, and some common heart diseases are listed in table 19.2. Disorders of the blood and blood vessels are described in chapters 18 and 20.

| TABLE 19.1 | Some Chronotropic and Inotropic Agents |
|---|---|
| **Chronotropic Agents (Influence Heart Rate)** | |
| **Positive** | **Negative** |
| Sympathetic stimulation | Parasympathetic stimulation |
| Epinephrine and norepinephrine | Acetylcholine |
| Thyroid hormone | Hyperkalemia |
| Hypocalcemia | Hypokalemia |
| Hypercapnia and acidosis | Hypercalcemia |
| | Hypoxia |
| **Inotropic Agents (Influence Contraction Strength)** | |
| **Positive** | **Negative** |
| Sympathetic stimulation | (Parasympathetic effect negligible) |
| Epinephrine and norepinephrine | Hyperkalemia |
| Hypercalcemia | Hypocalcemia |
| Digitalis | Myocardial hypoxia |
| Glucagon | Myocardial hypercapnia |
| Caffeine | Myocardial acidosis |

**BEFORE YOU GO ON**

Answer the following questions to test your understanding of the preceding section:

23. Define cardiac output in words and with a simple formula.

24. Describe the cardiac center and innervation of the heart.

25. Explain what is meant by positive and negative chronotropic and inotropic agents. Give two examples of each.

26. How do preload, contractility, and afterload influence stroke volume and cardiac output?

27. Explain the principle behind the Frank–Starling law of the heart. How does this mechanism normally prevent pulmonary or systemic congestion?

---

[37]*cor* = heart; *pulmo* = lung

| TABLE 19.2 | Some Disorders of the Heart |
|---|---|
| Acute pericarditis | Inflammation of the pericardium, sometimes due to infection, radiation therapy, or connective tissue disease, causing pain and friction rub |
| Cardiac tamponade | Compression of the heart by an abnormal accumulation of fluid or clotted blood in the pericardial cavity, interfering with ventricular filling and potentially leading to heart failure |
| Cardiomyopathy | Any disease of the myocardium not resulting from coronary artery disease, valvular dysfunction, or other cardiovascular disorders; can cause dilation and failure of the heart, thinning of the heart wall, or thickening of the interventricular septum |
| Infective endocarditis | Inflammation of the endocardium, usually due to bacterial infection, especially *Streptococcus* and *Staphylococcus* |
| Myocardial ischemia | Inadequate blood flow to the myocardium, usually because of coronary atherosclerosis; can lead to myocardial infarction |
| Pericardial effusion | Seepage of fluid from the pericardium into the pericardial cavity, often resulting from pericarditis and sometimes causing cardiac tamponade |
| Septal defects | Abnormal openings in the interatrial or interventricular septum, resulting in blood from the right atrium flowing directly into the left atrium, or blood from the left ventricle returning to the right ventricle; results in pulmonary hypertension, difficulty breathing, and fatigue; often fatal in childhood if uncorrected |

*Disorders described elsewhere*

| | | |
|---|---|---|
| Angina pectoris p. 719 | Cor pulmonale p. 737 | Premature ventricular contraction p. 727 |
| Atrial fibrillation p. 727 | Coronary artery disease p. 739 | Tachycardia p. 734 |
| Bradycardia p. 734 | Friction rub p. 710 | Total heart block p. 727 |
| Bundle branch block p. 727 | Heart murmur p. 729 | Valvular stenosis p. 729 |
| Cardiac arrest p. 727 | Mitral valve prolapse p. 729 | Ventricular fibrillation p. 726 |
| Congestive heart failure p. 733 | Myocardial infarction p. 719 | |

# DEEPER INSIGHT 19.4

## CLINICAL APPLICATION

### Coronary Artery Disease

*Coronary artery disease (CAD)* is a constriction of the coronary arteries usually resulting from *atherosclerosis*[38]—an accumulation of lipid deposits that degrade the arterial wall and obstruct the lumen. The most dangerous consequence of CAD is myocardial infarction.

### Pathogenesis

CAD begins when hypertension, diabetes, or other factors damage the arterial lining. Monocytes adhere to the lining, penetrate into the tissue, and become macrophages. Macrophages and smooth muscle cells absorb cholesterol and fat from the blood, which gives them a frothy appearance. They are then called *foam cells* and form visible *fatty streaks* on the arterial wall. Seen even in infants and children, these are harmless in themselves but have the potential to grow into atherosclerotic *plaques (atheromas*[39]).

Platelets adhere to these plaques and secrete a growth factor that stimulates local proliferation of smooth muscle and fibroblasts and deposition of collagen. The plaque grows into a bulging mass of lipid, fiber, and smooth muscle and other cells. When it obstructs 75% or more of the arterial lumen, it begins to cause such symptoms as angina pectoris. More seriously, inflammation of the plaque roughens its surface and creates a focal point for thrombosis. A blood clot can block what remains of the lumen, or break free and lodge in a smaller artery downstream. Sometimes a piece of plaque breaks free and travels as a *fatty embolus*. Furthermore, the plaque can contribute to spasms of the coronary artery, cutting off blood flow to the myocardium. If the lumen is already partially obstructed by a plaque and perhaps a blood clot, such a spasm can temporarily shut off the remaining flow and precipitate an attack of angina.

Over time, the resilient muscular and elastic tissue of an inflamed artery becomes increasingly replaced with scar tissue and calcium deposits, transforming an atheroma into a hard *complicated plaque* (fig. 19.22). Hardening of the arteries by calcified plaques is one cause of *arteriosclerosis*.[40] For reasons explained in chapter 20, this results in excessive surges of blood pressure that may weaken and rupture smaller arteries, leading to stroke and kidney failure.

### Risk, Prevention, and Treatment

A paramount risk factor for CAD is excess *low-density lipoproteins (LDLs)* in the blood combined with defective LDL receptors in the arterial walls. LDLs are protein-coated droplets of cholesterol, fats, free fatty acids, and phospholipids (see p. 1001). Most cells have LDL receptors that enable them to absorb these droplets from the blood so they can metabolize the cholesterol and other lipids. CAD can occur when the arterial cells have dysfunctional LDL receptors that "don't know when to quit," so the cells absorb and accumulate excess cholesterol.

Some risk factors for CAD are unavoidable—for example, heredity, aging, and being male. Most risk factors, however, are preventable—obesity, smoking,

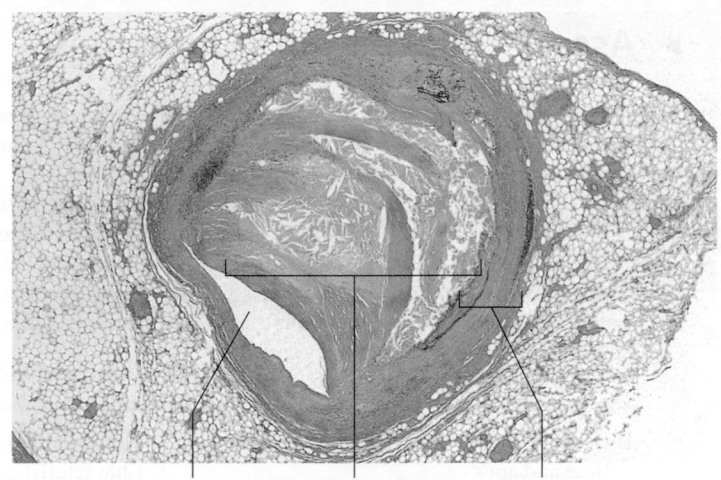

**Lumen   Complicated plaque   Artery wall**

**FIGURE 19.22 Diseased Coronary Artery (Cross Section).** A plaque has reduced the lumen to a very small space that can easily be blocked by thrombosis, embolism, or vasoconstriction.

lack of exercise, and a personality fraught with anxiety, stress, and aggression, all conducive to the hypertension that initiates arterial damage. Diet, of course, is very significant. Eating animal fat raises one's LDL level and reduces the number of LDL receptors. Foods high in soluble fiber (such as beans, apples, and oat bran) lower blood cholesterol by an interesting mechanism: The liver normally converts cholesterol to bile acids and secretes them into the small intestine to aid fat digestion. The bile acids are reabsorbed farther down the intestine and recycled to the liver for reuse. Soluble fiber, however, binds bile acids and carries them out in the feces. To replace them, the liver synthesizes more, thus consuming more cholesterol.

CAD is often treated with *coronary artery bypass surgery*. Sections of the great saphenous vein of the leg or small thoracic arteries are used to construct a detour around the obstruction in the coronary artery. In *balloon angioplasty*,[41] a slender catheter is threaded into the coronary artery and then a balloon at its tip is inflated to press the atheroma against the arterial wall, widening the lumen. In *laser angioplasty*, the surgeon views the interior of the diseased artery with an illuminated catheter and vaporizes the atheroma with a laser. Angioplasty is less risky and expensive than bypass surgery, but is often followed by *restenosis*—atheromas grow back and reobstruct the artery months later. Insertion of a tube called a *stent* into the artery can prevent restenosis.

---

[38]*athero* = fat, fatty; *sclerosis* = hardening
[39]*athero* = fat, fatty; *oma* = mass, tumor
[40]*arterio* = artery; *sclerosis* = hardening

[41]*angio* = vessel; *plasty* = surgical repair

# STUDY GUIDE

## ► Assess Your Learning Outcomes

*To test your knowledge, discuss the following topics with a study partner or in writing, ideally from memory.*

### 19.1 Overview of the Cardiovascular System (p. 709)

1. Two subdivisions of the cardiovascular system and their respective functions
2. Names of the great vessels directly connected to the heart, and their relations to the heart chambers
3. The exact location of the heart, its size, and its base and apex
4. Anatomy and function of the pericardium and pericardial fluid

### 19.2 Gross Anatomy of the Heart (p. 711)

1. Three layers of the heart wall and their histological differences
2. Relative thickness of the myocardium in different chambers; the functional significance of those differences; and significance of the myocardial vortex
3. Structure and function of the fibrous skeleton of the heart
4. Anatomy and functions of the atria and ventricles; the internal septa that separate the four chambers and the external sulci that mark the chamber boundaries
5. Names and synonyms for all four valves of the heart
6. Structural differences between the valves; anatomy and function of the papillary muscles and tendinous cords
7. The path of blood flow through the heart chambers and valves
8. Anatomy of the coronary arteries and their main branches
9. Causes of myocardial infarction (MI) and how the collateral circulation in the coronary arteries reduces the risk of MI
10. Why coronary artery blood flow is greater when the heart relaxes than when it contracts, in contrast to the arterial system almost everywhere else in the body
11. Anatomy of the major veins that drain the myocardium, where this blood goes, and how the major veins are supplemented by the thebesian veins

### 19.3 Cardiac Muscle and the Cardiac Conduction System (p. 720)

1. Structural properties of cardiac myocytes and how they relate to the unique function of cardiac muscle

2. Properties of cardiac muscle related to its nearly exclusive reliance on aerobic respiration
3. Components of the cardiac conduction system and the path traveled by electrical signals through the heart
4. Sympathetic and parasympathetic innervation of the heart and their effects

### 19.4 Electrical and Contractile Activity of the Heart (p. 722)

1. The meanings of *systole* and *diastole*
2. Characteristics of the sinus rhythm of the heart; some causes of premature ventricular contraction; why an ectopic focus may take over control of the rhythm; how a nodal rhythm differs from the sinus rhythm; and the general term for any abnormal cardiac rhythm
3. The mechanism that causes cells of the SA node to depolarize rhythmically; a graph of the time course and voltages of the pacemaker potentials; how often this repeats itself in a normal resting heart; and the role of gated ion channels and specific ion inflows and outflows in creating the nodal rhythm
4. The spread of excitation through the atria, AV node, AV bundle and bundle branches, and Purkinje fibers; changing conduction speeds at different points along this path, and why these changes are important; and the correlation of atrial and ventricular systole with the traveling wave of excitation
5. The twisting mode of ventricular contraction and the importance of the tendinous cords in preventing valvular prolapse
6. The cardiocyte resting potential; the actions of gated sodium, calcium, and potassium channels, and movements of these ions, in producing myocardial action potentials; how and why the shape of a myocardial action potential differs from that of a neuron; and how the plateau and unusually long refractory period of myocardial action potentials support the pumping effectiveness of the heart
7. Electrocardiograms and what happens in the heart during each ECG wave

### 19.5 Blood Flow, Heart Sounds, and the Cardiac Cycle (p. 728)

1. The principle on which a sphygmomanometer works, and why BP is expressed in millimeters of mercury (mm Hg)

2. The relationship of fluid volume, pressure, and flow, and how this relates to blood flow during the expansion and contraction of the heart chambers
3. Mechanisms that open and close the heart valves
4. The definition of *cardiac cycle,* and the names of its four phases
5. In each phase of the cardiac cycle, which chambers depolarize or repolarize, contract or relax; what each pair of valves does; what appears in the ECG; what accounts for the heart sounds; whether blood moves into or out of the atria or ventricles and where it is going when ejected; and blood volume and pressure changes in the left atrium, left ventricle, and aorta
6. The typical duration, in seconds, of atrial systole, ventricular systole, and the quiescent period, and how heart rate can be calculated from these values
7. The volume of blood typically found in each ventricle when it has finished filling; the volume ejected when a ventricle contracts; the percentage of ventricular blood that is ejected; the amount that remains behind when contraction is finished; and names of these four variables
8. Why it is necessary that each ventricle eject the same average amount of blood; what happens if either ventricle ejects more than the other over an extended period of time

### 19.6 Cardiac Output (p. 734)

1. The definition of *cardiac output (CO);* how it can be calculated from the heart rate and stroke volume; and a typical healthy value for cardiac output
2. A typical adult resting heart and how it changes over the life span
3. The meaning of *tachycardia* and *bradycardia*
4. General terms for chemicals that raise or lower the heart rate; examples of each
5. The brainstem center that modulates the heart rate; its connections to the heart; and inputs that it receives and integrates into its "decisions" on how to alter the heart rate
6. Mechanisms by which sympathetic and parasympathetic nerves raise and lower the heart rate, including the neurotransmitters, receptors, and ions involved

# STUDY GUIDE

7. Why cardiac output rises with heart rate up to a point, but then levels off; the heart rate associated with maximum cardiac output
8. The intrinsic rate of the SA node and the role of vagal tone in modifying this to produce a normal resting heart rate
9. Mechanisms by which epinephrine and norepinephrine, nicotine, thyroid hormone, and caffeine accelerate the heart; how potassium and calcium imbalances affect heart rate
10. Preload, contractility, and afterload and how these three variables affect stroke volume

11. How the Frank–Starling law of the heart matches stroke volume to venous return
12. General terms for chemicals that raise or lower the cardiac contractility; examples of each
13. Why calcium imbalances affect contractility; changes in cardiac function associated with hypocalcemia and hypercalcemia
14. Mechanisms by which epinephrine, norepinephrine, glucagon, digitalis, and hyperkalemia affect contractility
15. Conditions that increase afterload; the effect of afterload on cardiac output; and why certain lung diseases lead to cor pulmonale

16. Why cardiac output increases at the outset of exercise, and another factor that increases output as exercise continues
17. Why stroke volume may be unusually high and resting heart rate unusually low in people who have sustained programs of endurance exercise

## ► Testing Your Recall

*Answers in Appendix B*

1. The cardiac conduction system includes all of the following *except*
   a. the SA node.
   b. the AV node.
   c. the bundle branches.
   d. the tendinous cords.
   e. the Purkinje fibers.

2. To get from the right atrium to the right ventricle, blood flows through
   a. the pulmonary valve.
   b. the tricuspid valve.
   c. the bicuspid valve.
   d. the aortic valve.
   e. the mitral valve.

3. Assume that one ventricle of a child's heart has an EDV of 90 mL, an ESV of 60 mL, and a cardiac output of 2.55 L/min. What are the child's stroke volume (SV), ejection fraction (EF), and heart rate (HR)?
   a. SV = 60 mL; EF = 33%; HR = 85 bpm
   b. SV = 30 mL; EF = 60%; HR = 75 bpm
   c. SV = 150 mL; EF = 67%; HR = 42 bpm
   d. SV = 30 mL; EF = 33%; HR = 85 bpm
   e. Not enough information is given to calculate these.

4. A heart rate of 45 bpm and an absence of P waves suggest
   a. damage to the SA node.
   b. ventricular fibrillation.
   c. cor pulmonale.
   d. extrasystole.
   e. heart block.

5. There is/are _____ pulmonary vein(s) emptying into the right atrium of the heart.
   a. no
   b. one
   c. two
   d. four
   e. more than four

6. The coronary blood vessels are part of the _____ circuit of the circulatory system.
   a. cardiac
   b. pulmonary
   c. systematic
   d. systemic
   e. cardiovascular

7. The atria contract during
   a. the first heart sound.
   b. the second heart sound.
   c. the QRS complex.
   d. the PQ segment.
   e. the ST segment.

8. Cardiac muscle does not exhibit tetanus because it has
   a. fast $Ca^{2+}$ channels.
   b. scanty sarcoplasmic reticulum.
   c. a long absolute refractory period.
   d. electrical synapses.
   e. exclusively aerobic respiration.

9. The blood contained in a ventricle during isovolumetric relaxation is
   a. the end-systolic volume.
   b. the end-diastolic volume.
   c. the stroke volume.
   d. the ejection fraction.
   e. none of these; the ventricle is empty then.

10. Drugs that increase the heart rate have a _____ effect.
    a. myogenic
    b. negative inotropic
    c. positive inotropic
    d. negative chronotropic
    e. positive chronotropic

11. The contraction of any heart chamber is called _____ and its relaxation is called _____.

12. The circulatory route from aorta to the venae cavae is the _____ circuit.

13. The circumflex artery travels in a groove called the _____.

14. The pacemaker potential of the SA node cells results from the slow inflow of _____.

15. Electrical signals pass quickly from one cardiac myocyte to another through the _____ of the intercalated discs.

16. Repolarization of the ventricles produces the _____ of the electrocardiogram.

17. The _____ nerves innervate the heart and tend to reduce the heart rate.

18. The death of cardiac tissue from lack of blood flow is commonly known as a heart attack, but clinically called _____.

19. Blood in the heart chambers is separated from the myocardium by a thin membrane called the _____.

20. The Frank–Starling law of the heart explains why the _____ of the left ventricle is the same as that of the right ventricle.

# STUDY GUIDE

## ▶ Building Your Medical Vocabulary

*Answers in Appendix B*

*State a medical meaning of each word element below, and give a term in which it or a slight variation of it is used.*

1. atrio-
2. brady-
3. cardio-
4. corono-
5. lun-
6. papillo-
7. semi-
8. tachy-
9. vaso-
10. ventro-

## ▶ True or False

*Answers in Appendix B*

*Determine which five of the following statements are false, and briefly explain why.*

1. The blood supply to the myocardium is the coronary circulation; everything else is called the systemic circuit.

2. There are no valves at the point where venous blood flows into the atria.

3. No blood can enter the ventricles until the atria contract.

4. The vagus nerves reduce the heart rate but have little effect on the strength of ventricular contraction.

5. A high blood $CO_2$ level and low blood pH stimulate an increase in heart rate.

6. The first heart sound occurs at the time of the P wave of the electrocardiogram.

7. If all nerves to the heart were severed, the heart would instantly stop beating.

8. If the two pulmonary arteries were clamped shut, systemic edema would follow.

9. Ventricular cardiocytes have a stable resting membrane potential but myocytes of the SA node do not.

10. An electrocardiogram is a tracing of the action potential of a cardiocyte.

## ▶ Testing Your Comprehension

*Answers at* www.mhhe.com/saladin7

1. Verapamil is a calcium channel blocker used to treat hypertension. It selectively blocks slow calcium channels. Would you expect it to have a positive or negative inotropic effect? Explain. (See p. 85 to review calcium channel blockers.)

2. To temporarily treat tachycardia and restore the normal resting sinus rhythm, a physician may massage a patient's carotid artery near the angle of the mandible. Propose a mechanism by which this treatment would have the desired effect.

3. Becky, age 2, was born with a hole in her interventricular septum (*ventricular septal defect,* or *VSD*). Considering that the blood pressure in the left ventricle is significantly higher than blood pressure in the right ventricle, predict the effect of the VSD on Becky's pulmonary blood pressure, systemic blood pressure, and long-term changes in the ventricular walls.

4. In ventricular systole, the left ventricle is the first to begin contracting, but the right ventricle is the first to expel blood. Aside from the obvious fact that the pulmonary valve opens before the aortic valve, how can you explain this difference?

5. In dilated cardiomyopathy of the left ventricle, the ventricle can become enormously enlarged. Explain why this might lead to regurgitation of blood through the mitral valve (blood flowing from the ventricle back into the left atrium) during ventricular systole.

## ▶ Improve Your Grade at www.mhhe.com/saladin7

*Download animations to study when it fits your schedule. Practice quizzes, labeling activities, games, and flashcards offer fun ways to master the chapter concepts. Or, download image PowerPoint files for each chapter to create a study guide or for taking notes during lecture.*

Blood capillary beds

# THE CIRCULATORY SYSTEM: BLOOD VESSELS AND CIRCULATION

Anatomy & Physiology | REVEALED®
aprevealed.com

**Module 9: Cardiovascular System**

The route taken by the blood after it leaves the heart was a point of much confusion for many centuries. In Chinese medicine as early as 2650 BCE, blood was believed to flow in a complete circuit around the body and back to the heart, just as we know today. But in the second century CE, Roman physician Claudius Galen (129–c. 199) argued that it flowed back and forth in the veins, like air in the bronchial tubes. He believed that the liver received food directly from the esophagus and converted it to blood, the heart pumped the blood through the veins to all other organs, and those organs consumed it. The arteries were thought to contain only a mysterious "vital spirit."

The Chinese view was right, but the first experimental demonstration of this did not come for another 4,000 years. English physician William Harvey (1578–1657) (see p. 5) studied the filling and emptying of the heart in snakes, tied off the vessels above and below the heart to observe the effects on cardiac filling and output, and measured cardiac output in a variety of living animals and estimated it in humans. He concluded that (1) the heart pumps more blood in half an hour than there is in the entire body, (2) not enough food is consumed to account for the continual production of so much blood, and therefore (3) the blood returns to the heart rather than being consumed by the peripheral organs. He could not explain how, since the microscope had yet to be developed to the point that enabled Antony van Leeuwenhoek (1632–1723) and Marcello Malpighi (1628–94) to discover the blood capillaries.

Harvey's work was the first experimental study of animal physiology and a milestone in the history of biology and medicine. But so entrenched were the ideas of Aristotle and Galen in the medical community, and so strange was the idea of doing experiments on living animals, that Harvey's contemporaries rejected his ideas. Indeed, some of them regarded him as a crackpot because his conclusion flew in the face of common sense—if the blood was continually recirculated and not consumed by the tissues, they reasoned, then what purpose could it serve? We now know, of course, that he was

right. Harvey's case is one of the most interesting in biomedical history, for it shows how empirical science overthrows old theories and spawns better ones, and how common sense and blind allegiance to authority can interfere with the acceptance of truth. But most importantly, Harvey's contributions represent the birth of experimental physiology.

## 20.1 General Anatomy of the Blood Vessels

### Expected Learning Outcomes

When you have completed this section, you should be able to

a. describe the structure of a blood vessel;

b. describe the types of arteries, capillaries, and veins;

c. trace the general route usually taken by the blood from the heart and back again; and

d. describe some variations on this route.

There are three principal categories of blood vessels: arteries, veins, and capillaries (fig. 20.1). **Arteries** are the efferent vessels of the cardiovascular system—that is, vessels that carry blood away from the heart. **Veins** are the afferent vessels that carry blood back to the heart. **Capillaries** are microscopic, thin-walled vessels that connect the smallest arteries to the smallest veins. Aside from their general location and direction of blood flow, these three categories of vessels also differ in the histological structure of their walls.

### The Vessel Wall

The walls of arteries and veins are composed of three layers called *tunics* (fig. 20.2):

1. The **tunica interna (tunica intima)** lines the inside of the vessel and is exposed to the blood. It consists of a simple squamous epithelium called the **endothelium** overlying a basement membrane and a sparse layer of loose connective tissue; it is continuous with the endocardium of the heart. The endothelium acts as a selectively permeable barrier to materials entering or leaving the bloodstream; it secretes chemicals that stimulate dilation or constriction of the vessel; and it normally repels blood cells and platelets so that they flow freely without sticking to the vessel wall. When the endothelium is damaged, however, platelets may adhere to it and form a blood clot; and when the tissue around a vessel is inflamed, the endothelial cells produce *cell-adhesion molecules* that induce leukocytes to adhere to the surface. This causes leukocytes to congregate in tissues where their defensive actions are needed.

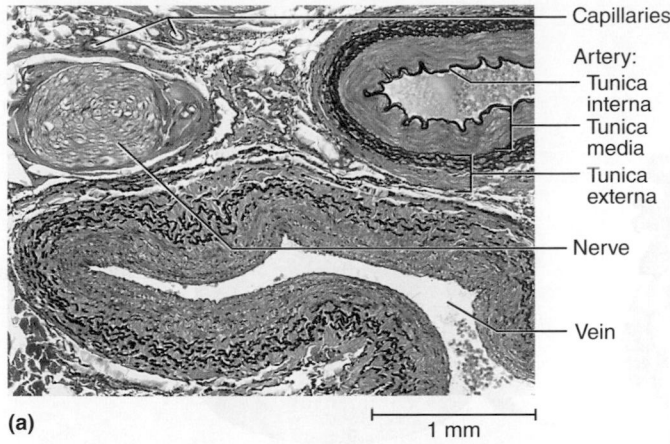

**(a)**

1 mm

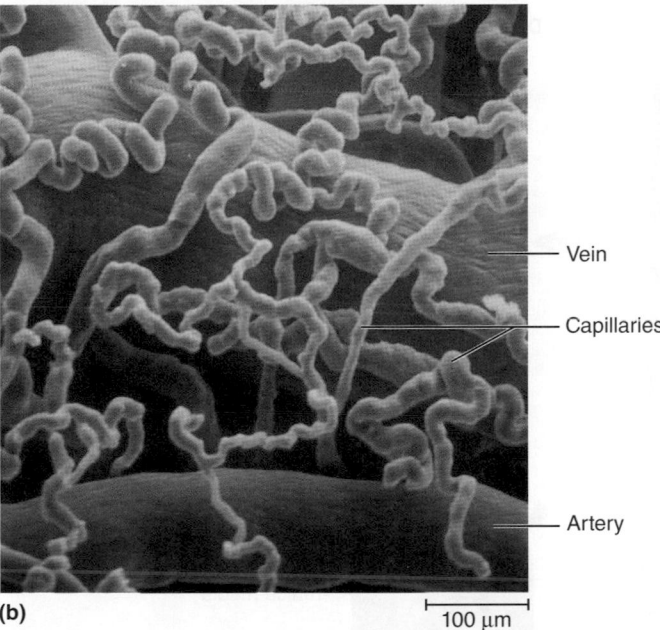

**(b)**

100 μm

**FIGURE 20.1** **Micrographs of Blood Vessels.** (a) A neurovascular bundle, composed of a small artery, vein, and nerve traveling together in a common sheath of connective tissue (LM). (b) A vascular cast of blood vessels of the eye (SEM). This was prepared by injecting the vessels with a polymer, digesting away all tissue to leave a replica of the vessels, and photographing the cast with the electron microscope.

2. The **tunica media,** the middle layer, is usually the thickest. It consists of smooth muscle, collagen, and in some cases, elastic tissue. The relative amounts of smooth muscle and elastic tissue vary greatly from one vessel to another and form a basis for classifying vessels as described in the next section. The tunica media strengthens the vessels and prevents blood pressure from rupturing them, and it regulates the diameter of a blood vessel.

3. The **tunica externa (tunica adventitia[1])** is the outermost layer. It consists of loose connective tissue that often merges with that of neighboring blood vessels, nerves, or other organs (fig. 20.1a). It anchors the vessel to adjacent tissues and provides passage for small nerves, lymphatic vessels, and smaller blood vessels that supply the tissues of the larger vessel. Small vessels called the **vasa vasorum[2]** (VAY-za vay-SO-rum) nourish at least the outer half of the wall of a larger vessel. Tissues of the inner half of the wall are thought to be nourished by diffusion from blood in the lumen.

## Arteries

Arteries are sometimes called the *resistance vessels* of the cardiovascular system because they have a relatively strong, resilient tissue structure. Each beat of the heart creates a surge of pressure in the arteries as blood is ejected into them, and arteries are built to withstand these surges. Being more muscular than veins, they retain their round shape even when empty, and they appear relatively circular in tissue sections. They are divided into three classes by size, but of course there is a gradual transition from one class to the next.

1. **Conducting (elastic or large) arteries** are the biggest. Examples include the aorta, common carotid and subclavian arteries, pulmonary trunk, and common iliac arteries. They have a layer of elastic tissue called the *internal elastic lamina* at the border between the interna and media, but microscopically, it is incomplete and difficult to distinguish from the elastic tissue of the tunica media. The tunica media consists of 40 to 70 layers of elastic sheets, perforated like slices of Swiss cheese, alternating with thin layers of smooth muscle, collagen, and elastic fibers. In histological sections, the view is dominated by this elastic tissue. The perforations allow for vasa vasorum and nerves to penetrate through all layers of the vessel and for smooth muscle cells to communicate with each other through gap junctions. There is an *external elastic lamina* at the border between the media and externa, but it, too, is difficult to distinguish from the elastic sheets of the tunica media. The tunica externa is quite sparse in the largest arteries but is well supplied with vasa vasorum.

   Conducting arteries expand as they receive blood during ventricular systole, and recoil during diastole. Their expansion takes some of the pressure off the blood so that smaller arteries downstream are subjected to less systolic stress. Their recoil between heartbeats prevents the blood pressure from dropping too low while the heart is relaxing and refilling. Further, their expansion stores potential energy and their recoil releases it to keep the blood flowing

---

[1]*advent* = added to
[2]*vasa* = vessels; *vasorum* = of the vessels

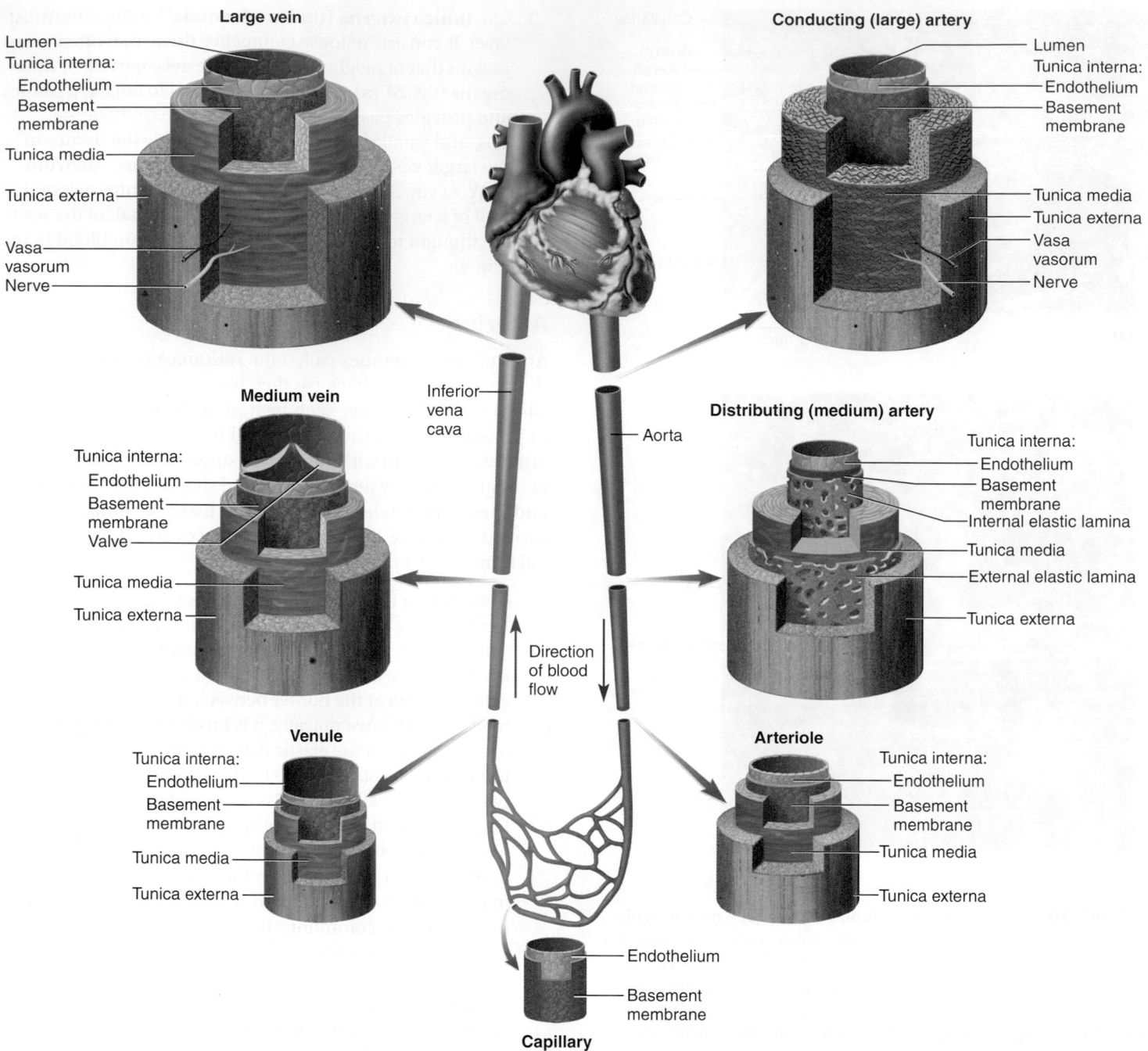

**FIGURE 20.2** Histology of the Blood Vessels.

❓ *Why do the arteries have so much more elastic tissue than the veins do?*

during diastole. These effects reduce the fluctuations in blood pressure that would otherwise occur. Arteries stiffen with age *(arteriosclerosis),* becoming less able to expand and recoil freely. Consequently, the downstream vessels are subjected to greater stress and are more likely to develop aneurysms and rupture (see Deeper Insight 20.1).

2. **Distributing (muscular** or **medium) arteries** are smaller branches that distribute blood to specific organs. You could compare a conducting artery to an interstate highway and distributing arteries to the exit ramps and state highways that serve individual towns. Most arteries that have specific anatomical names are in these first two size

## DEEPER INSIGHT 20.1
## CLINICAL APPLICATION

### Aneurysm

An *aneurysm*[3] is a weak point in an artery or in the heart wall. It forms a thin-walled, bulging sac that pulsates with each beat of the heart and may eventually rupture. In a *dissecting aneurysm*, blood accumulates between the tunics of an artery and separates them, usually because of degeneration of the tunica media. The most common sites of aneurysms are the abdominal aorta, renal arteries, and the arterial circle at the base of the brain. Even without hemorrhaging, aneurysms can cause pain or death by putting pressure on brain tissue, nerves, adjacent veins, pulmonary air passages, or the esophagus. Other consequences include neurological disorders, difficulty in breathing or swallowing, chronic cough, or congestion of the tissues with blood. Aneurysms sometimes result from congenital weakness of the blood vessels and sometimes from trauma or bacterial infections such as syphilis. The most common cause, however, is the combination of arteriosclerosis and hypertension.

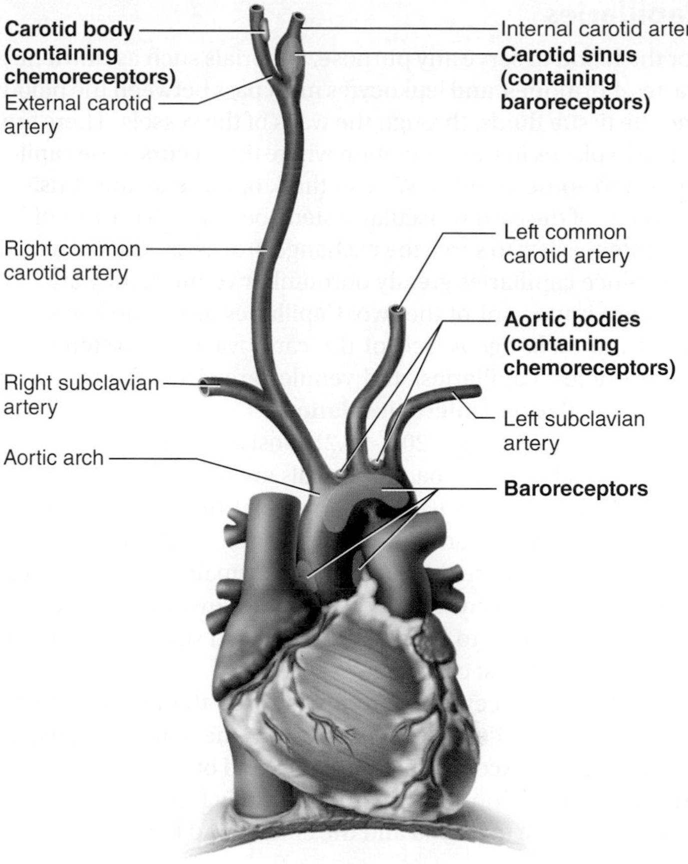

**FIGURE 20.3 Baroreceptors and Chemoreceptors in the Arteries Superior to the Heart.** The structures shown here in the right carotid arteries are repeated in the left carotids. AP|R

classes. Examples include the brachial, femoral, renal, and splenic arteries. Distributing arteries typically have up to 40 layers of smooth muscle constituting about three-quarters of the wall thickness. In histological sections, this smooth muscle is more conspicuous than the elastic tissue. Both the internal and external elastic laminae, however, are thick and often conspicuous.

3. **Resistance (small) arteries** are usually too variable in number and location to be given individual names. They exhibit up to 25 layers of smooth muscle and relatively little elastic tissue. Compared to large arteries, they have a thicker tunica media in proportion to the lumen. The smallest of these arteries, up to 200 μm in diameter and with only one to three layers of smooth muscle, are called **arterioles.** Arterioles have very little tunica externa. They are the major point of control over how much blood an organ or tissue receives, as we shall see later.

In some places, short vessels called **metarterioles**[4] link arterioles to capillaries or provide shortcuts through which blood can bypass the capillaries and flow directly to a venule. They will be further discussed with capillary beds.

## Arterial Sense Organs

Certain major arteries above the heart have sensory structures in their walls that monitor blood pressure and chemistry (fig. 20.3). These receptors transmit information to the brainstem that serves to regulate the heartbeat, blood vessel diameters, and respiration. They are of three kinds:

1. **Carotid sinuses.** These are *baroreceptors* (pressure sensors) that monitor blood pressure. Ascending the neck on each side is a *common carotid artery,* which branches

near the angle of the mandible, forming the *internal carotid artery* to the brain and *external carotid artery* to the face. The carotid sinuses are located in the wall of the internal carotid artery just above the branch point. The carotid sinus has a relatively thin tunica media and an abundance of glossopharyngeal nerve fibers in the tunica externa. The role of the baroreceptors in adjusting blood pressure, called the *baroreflex,* is described later in this chapter.

2. **Carotid bodies.** Also located near the branch of the common carotid arteries, these are oval receptors about $3 \times 5$ mm in size, innervated by sensory fibers of the glossopharyngeal nerves. They are *chemoreceptors* that monitor changes in blood composition. They primarily transmit signals to the brainstem respiratory centers, which adjust breathing to stabilize the blood pH and its $CO_2$ and $O_2$ levels.

3. **Aortic bodies.** These are one to three chemoreceptors located in the aortic arch near the arteries to the head and arms. They are structurally similar to the carotid bodies and have the same function, but transmit their signals to the brainstem via the vagus nerves.

[3]*aneurysm* = widening
[4]*meta* = beyond, next in a series

## Capillaries

For the blood to serve any purpose, materials such as nutrients, wastes, hormones, and leukocytes must pass between the blood and the tissue fluids, through the walls of the vessels. There are only two places in the circulation where this occurs—the capillaries and some venules. We can think of these as the "business end" of the cardiovascular system, because all the rest of the system exists to serve the exchange processes that occur here. Since capillaries greatly outnumber venules, they are the more important of the two. Capillaries are sometimes called the *exchange vessels* of the cardiovascular system; the arterioles, capillaries, and venules are also called the **microvasculature (microcirculation).**

Capillaries (see figs. 20.1–20.2) consist of only an endothelium and basal lamina. Their walls are as thin as 0.2 μm. They average about 5 μm in diameter at the proximal end (where they receive arterial blood), widen to about 9 μm at the distal end (where they empty into a small vein), and often branch along the way. Since erythrocytes are about 7.5 μm in diameter, they have to stretch into elongated shapes to squeeze through the smallest capillaries.

Scarcely any cell in the body is more than 60 to 80 μm (about four to six cell widths) away from the nearest capillary. There are a few exceptions: Capillaries and other blood vessels are scarce in tendons and ligaments, rarely found in cartilage, and absent from epithelia and the cornea and lens of the eye.

### Types of Capillaries

There are three types of capillaries, distinguished by the ease with which they allow substances to pass through their walls and by structural differences that account for their greater or lesser permeability.

1. **Continuous capillaries** (fig. 20.4) occur in most tissues, such as skeletal muscle. Their endothelial cells, held together by tight junctions, form a continuous tube. A thin protein–carbohydrate layer, the **basal lamina,** surrounds the endothelium and separates it from the adjacent connective tissues. The endothelial cells are separated by narrow **intercellular clefts** about 4 nm wide. Small solutes such as glucose can pass through these clefts, but most plasma protein, other large molecules, and platelets and blood cells are held back. The continuous capillaries of the brain lack intercellular clefts and have more complete tight junctions that form the blood–brain barrier discussed in chapter 14.

   Some continuous capillaries exhibit cells called **pericytes** that lie external to the endothelium. Pericytes have elongated tendrils that wrap around the capillary. They contain the same contractile proteins as muscle, and it is thought that they can contract and regulate blood flow through the capillaries. They also can differentiate into endothelial and smooth muscle cells and thus contribute to vessel growth and repair.

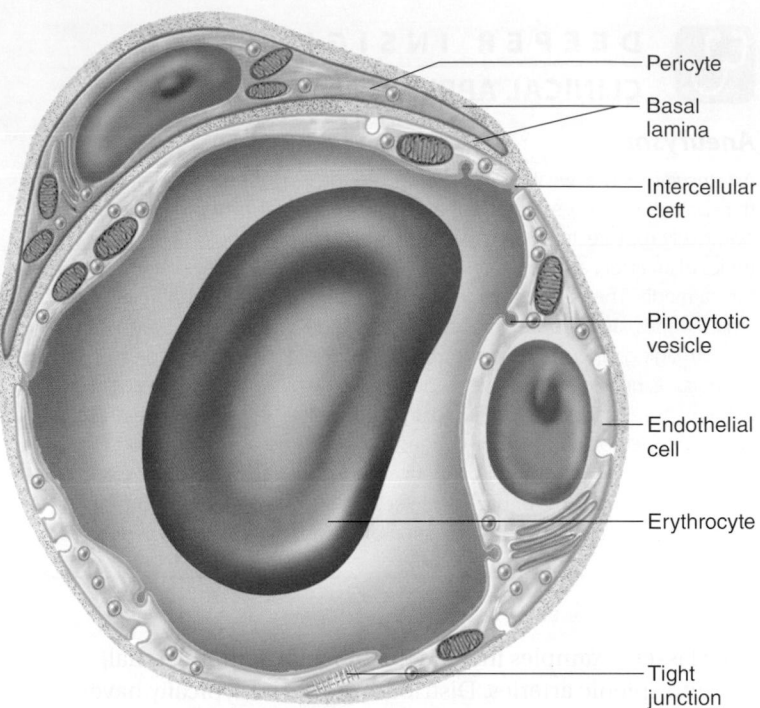

**FIGURE 20.4  Structure of a Continuous Capillary.** Cross section.

Labels: Pericyte; Basal lamina; Intercellular cleft; Pinocytotic vesicle; Endothelial cell; Erythrocyte; Tight junction

2. **Fenestrated capillaries** have endothelial cells riddled with patches of **filtration pores (fenestrations[5])** (fig. 20.5). These pores are about 20 to 100 nm in diameter, and are often spanned by a glycoprotein membrane that is much thinner than the cell's plasma membrane. They allow for the rapid passage of small molecules, but still retain most proteins and larger particles in the bloodstream. Fenestrated capillaries are important in organs that engage in rapid absorption or filtration—the kidneys, endocrine glands, small intestine, and choroid plexuses of the brain, for example.

3. **Sinusoids (discontinuous capillaries)** are irregular blood-filled spaces in the liver, bone marrow, spleen, and some other organs (fig. 20.6). They are twisted, tortuous passageways, typically 30 to 40 μm wide, that conform to the shape of the surrounding tissue. The endothelial cells are separated by wide gaps with no basal lamina, and the cells also frequently have especially large fenestrations through them. Even proteins and blood cells can pass through these pores; this is how albumin, clotting factors, and other proteins synthesized by the liver enter the blood, and how newly formed blood cells enter the circulation from the bone marrow and lymphatic organs. Some sinusoids contain macrophages or other specialized cells.

---

[5]*fenestra* = window

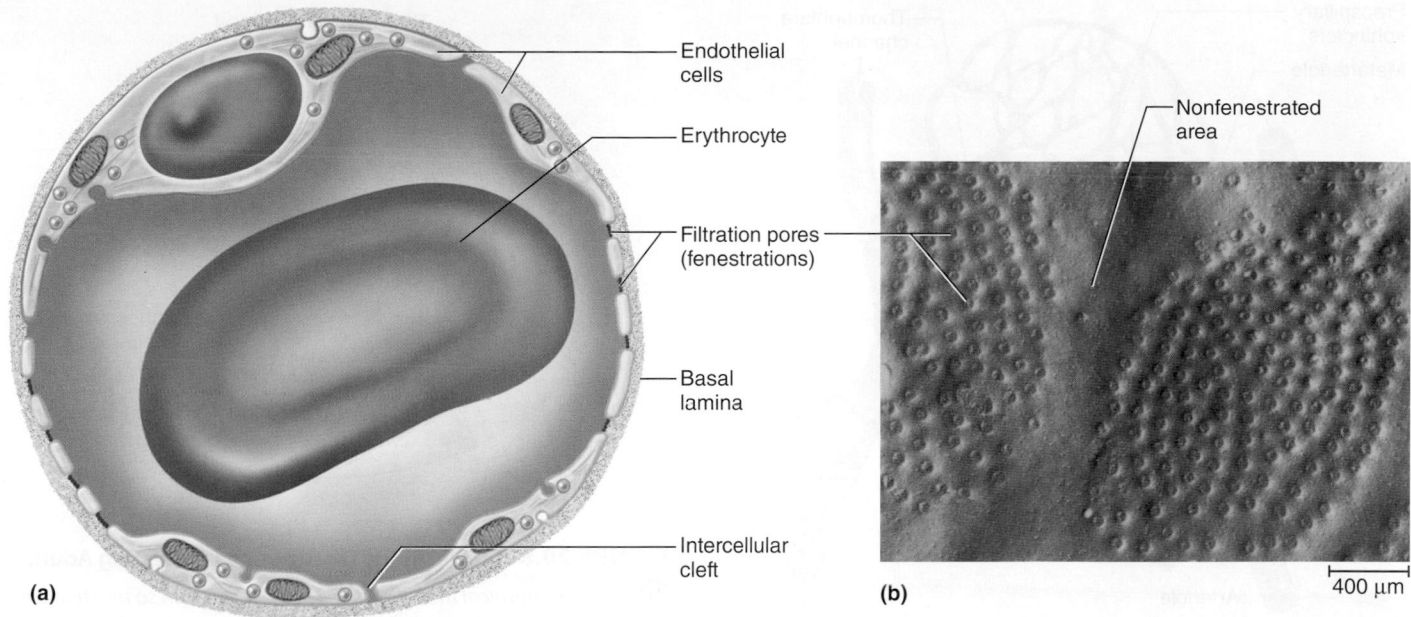

**FIGURE 20.5  Structure of a Fenestrated Capillary.**  (a) Cross section of the capillary wall. (b) Surface view of a fenestrated endothelial cell (SEM). The cell has patches of filtration pores (fenestrations) separated by nonfenestrated areas.

❓ *Identify some organs that have this type of capillary rather than continuous capillaries.*

## Capillary Beds

Capillaries are organized into weblike networks called **capillary beds**—typically 10 to 100 capillaries supplied by a single arteriole or metarteriole (fig. 20.7; see also photo on p. 743). At their distal end, capillaries transition to venules, gradually adding a thin tunica media. They may also drain into a distal continuation of the metarteriole called a *thoroughfare channel,* which then leads to a venule.

At any given time, about three-quarters of the body's capillaries are shut down because there isn't enough blood to supply all of them at once. In the skeletal muscles, for example, about 90% of the capillaries have little or no blood flow during periods of rest. During exercise, they receive an abundant flow while capillaries elsewhere—for example, in the skin and intestines—shut down to compensate. Capillary flow (perfusion) is usually regulated by the dilation or constriction of arterioles upstream from the capillary beds. In capillary beds supplied with metarterioles, there is often a single smooth muscle cell that wraps like a cuff around the opening to each capillary; it acts as a **precapillary sphincter** regulating blood flow. If the sphincters are relaxed, the capillaries are well perfused (fig. 20.7a). If many of the sphincters constrict, blood bypasses the capillaries, leaving them less perfused or even bloodless, and the blood takes a shortcut through the metarteriole and thoroughfare channel directly to a nearby venule (fig. 20.7b).

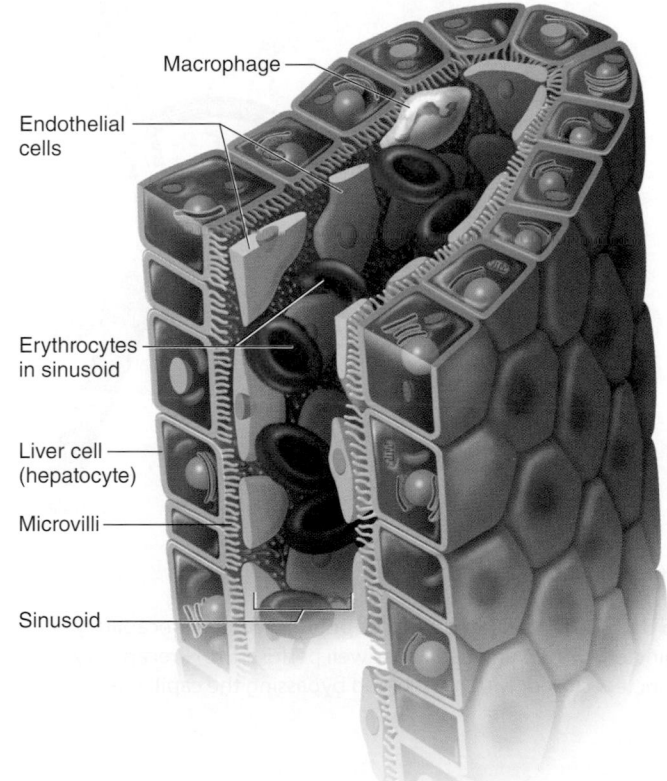

**FIGURE 20.6  A Sinusoid of the Liver.**  Large gaps between the endothelial cells allow blood plasma to directly contact the liver cells but retain blood cells in the lumen of the sinusoid.

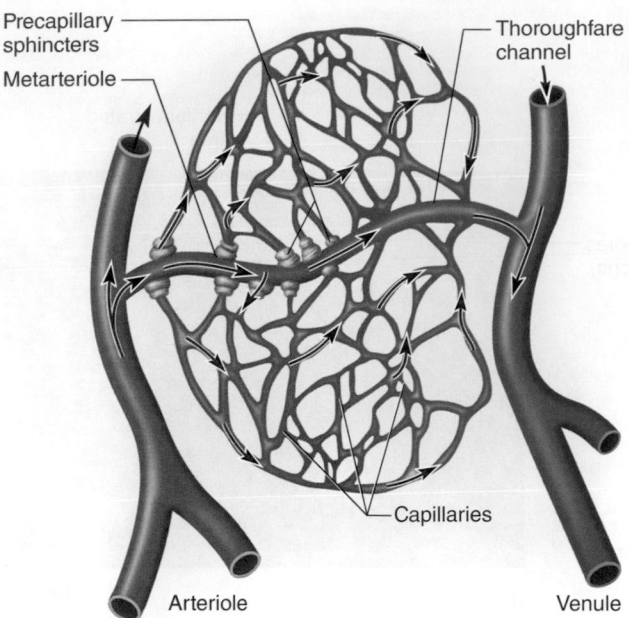

Precapillary sphincters

Metarteriole

Thoroughfare channel

Capillaries

Arteriole

Venule

**(a) Sphincters open**

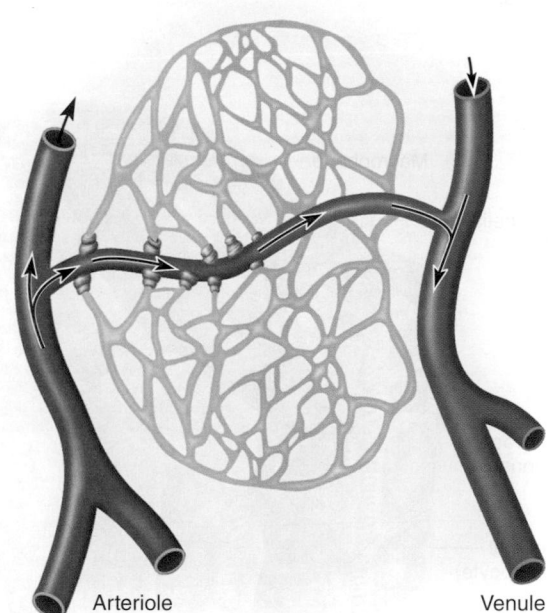

Arteriole

Venule

**(b) Sphincters closed**

**FIGURE 20.7   Perfusion of a Capillary Bed.** (a) Precapillary sphincters dilated and capillaries well perfused. (b) Precapillary sphincters closed, with most blood bypassing the capillaries.

## Veins

Veins are regarded as the *capacitance vessels* of the cardiovascular system because they are relatively thin-walled and flaccid, and expand easily to accommodate an increased volume

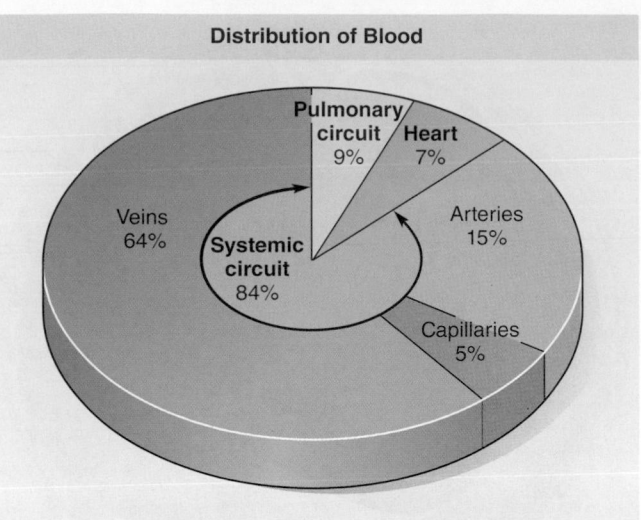

**Distribution of Blood**

Pulmonary circuit 9%

Heart 7%

Veins 64%

Systemic circuit 84%

Arteries 15%

Capillaries 5%

**FIGURE 20.8   Typical Blood Distribution in a Resting Adult.**

*What anatomical fact allows the veins to contain so much more blood than the arteries do?*

of blood; that is, they have a greater *capacity* for blood containment than arteries do. At rest, about 64% of the blood is found in the systemic veins as compared with only 13% in the systemic arteries (fig. 20.8). The reason that veins are so thin-walled and accommodating is that, being distant from the ventricles of the heart, they are subjected to relatively low blood pressure. In large arteries, blood pressure averages 90 to 100 mm Hg and surges to 120 mm Hg during systole, whereas in veins it averages about 10 mm Hg. Furthermore, the blood flow in the veins is steady, rather than pulsating with the heartbeat like the flow in the arteries. Veins therefore do not require thick, pressure-resistant walls. They collapse when empty and thus have relatively flattened, irregular shapes in histological sections (see fig. 20.1a).

As we trace blood flow in the arteries, we find it splitting off repeatedly into smaller and smaller *branches* of the arterial system. In the venous system, conversely, we find small veins merging to form larger and larger ones as they approach the heart. We refer to the smaller veins as *tributaries,* by analogy to the streams that converge and act as tributaries to rivers. In examining the types of veins, we will follow the direction of blood flow, working up from the smallest to the largest vessels.

1. **Postcapillary venules** are the smallest of the veins, beginning with diameters of about 10 to 20 μm. They receive blood from capillaries directly or by way of the distal ends of the thoroughfare channels. They have a tunica interna with only a few fibroblasts around it and no muscle. Like capillaries, they are often surrounded by pericytes. Postcapillary venules are even more porous than capillaries; therefore, venules also exchange fluid with the surrounding tissues. Most leukocytes emigrate from the bloodstream through the venule walls.

2. **Muscular venules** receive blood from the postcapillary venules. They are up to 1 mm in diameter. They have a tunica media of one or two layers of smooth muscle, and a thin tunica externa.

3. **Medium veins** range up to 10 mm in diameter. Most veins with individual names are in this category, such as the radial and ulnar veins of the forearm and the small and great saphenous veins of the leg. Medium veins have a tunica interna with an endothelium, basement membrane, loose connective tissue, and sometimes a thin internal elastic lamina. The tunica media is much thinner than it is in medium arteries; it exhibits bundles of smooth muscle, but not a continuous muscular layer as seen in arteries. The muscle is interrupted by regions of collagenous, reticular, and elastic tissue. The tunica externa is relatively thick.

   Many medium veins, especially in the limbs, exhibit infoldings of the tunica interna that meet in the middle of the lumen, forming **venous valves** directed toward the heart (see fig. 20.19). The pressure in the veins is not high enough to push all of the blood upward against the pull of gravity in a standing or sitting person. The upward flow of blood in these vessels depends partly on the massaging action of skeletal muscles and the ability of these valves to keep the blood from dropping down again when the muscles relax. When the muscles surrounding a vein contract, they force blood through these valves. The propulsion of venous blood by muscular massaging, aided by the venous valves, is a mechanism of blood flow called the *skeletal muscle pump.* Varicose veins result in part from the failure of the valves (see Deeper Insight 20.2).

4. **Venous sinuses** are veins with especially thin walls, large lumens, and no smooth muscle. Examples include the coronary sinus of the heart and the dural sinuses

of the brain. Unlike other veins, they are not capable of vasoconstriction.

5. **Large veins** have diameters greater than 10 mm. They have some smooth muscle in all three tunics. They have a relatively thin tunica media with only a moderate amount of smooth muscle; the tunica externa is the thickest layer and contains longitudinal bundles of smooth muscle. Large veins include the venae cavae, pulmonary veins, internal jugular veins, and renal veins.

## Circulatory Routes

The simplest and most common route of blood flow is heart → arteries → capillaries → veins → heart. Blood usually passes through only one network of capillaries from the time it leaves the heart until the time it returns (fig. 20.9a), but there are exceptions, notably portal systems and anastomoses.

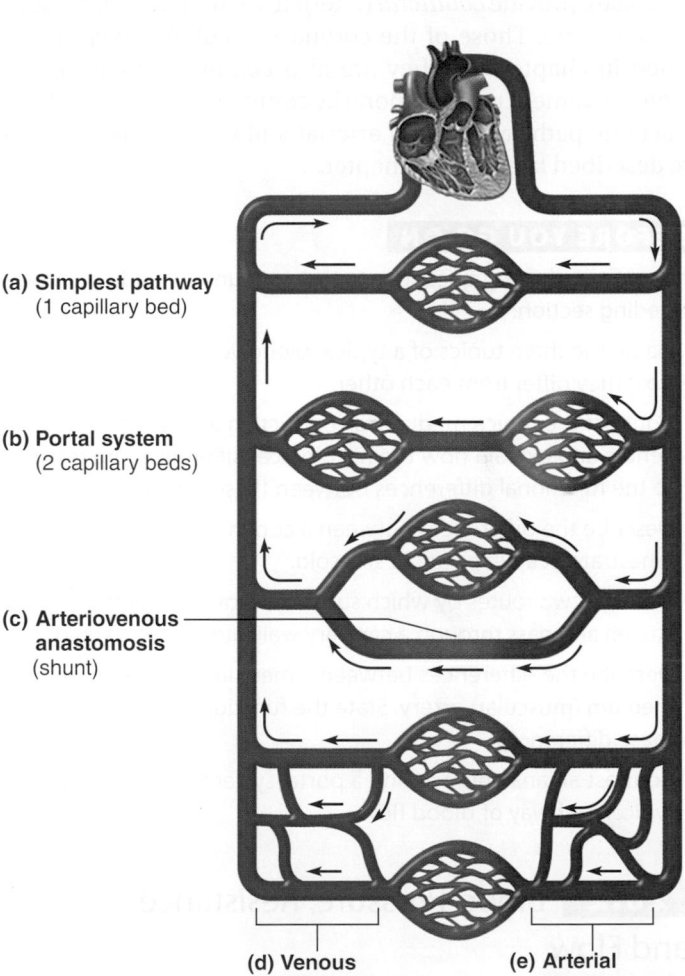

(a) **Simplest pathway**
(1 capillary bed)

(b) **Portal system**
(2 capillary beds)

(c) **Arteriovenous anastomosis**
(shunt)

(d) **Venous anastomoses**

(e) **Arterial anastomoses**

**FIGURE 20.9 Variations in Circulatory Pathways.**

❓ *After studying tables 20.2 through 20.12, identify specific sites in the body where one can find arterial anastomoses, venous anastomoses, and portal systems.*

In a **portal system** (fig. 20.9b), blood flows through two consecutive capillary networks before returning to the heart. Portal systems occur in the kidneys (chapter 23); connecting the hypothalamus and anterior pituitary (chapter 17); and connecting the intestines to the liver (table 20.8, part III).

An **anastomosis** is a point of convergence between two blood vessels other than capillaries. In an **arteriovenous anastomosis (shunt),** blood flows from an artery directly into a vein and bypasses the capillaries (fig. 20.9c). Shunts occur in the fingers, palms, toes, and ears, where they reduce heat loss in cold weather by allowing warm blood to bypass these exposed surfaces. Unfortunately, this makes these poorly perfused areas more susceptible to frostbite. The most common anastomoses are **venous anastomoses,** in which one vein empties directly into another (fig. 20.9d). These provide several alternative routes of drainage from an organ, so blockage of a vein is rarely as life-threatening as blockage of an artery. **Arterial anastomoses,** in which two arteries merge (fig. 20.9e), provide *collateral* (alternative) routes of blood supply to a tissue. Those of the coronary circulation were mentioned in chapter 19. They are also common around joints where movement may temporarily compress an artery and obstruct one pathway. Several arterial and venous anastomoses are described later in this chapter.

> **BEFORE YOU GO ON**
>
> Answer the following questions to test your understanding of the preceding section:
>
> 1. Name the three tunics of a typical blood vessel and explain how they differ from each other.
>
> 2. Contrast the tunica media of a conducting artery, arteriole, and venule and explain how the histological differences are related to the functional differences between these vessels.
>
> 3. Describe the differences between a continuous capillary, a fenestrated capillary, and a sinusoid.
>
> 4. Describe two routes by which substances can escape the bloodstream and pass through a capillary wall into the tissue fluid.
>
> 5. Describe the differences between a medium vein and a medium (muscular) artery. State the functional reasons for these differences.
>
> 6. Contrast an anastomosis and a portal system with the more typical pathway of blood flow.

## 20.2   Blood Pressure, Resistance, and Flow

### Expected Learning Outcomes

When you have completed this section, you should be able to

a. explain the relationship between blood pressure, resistance, and flow;

b. describe how blood pressure is expressed and how pulse pressure and mean arterial pressure are calculated;

c. describe three factors that determine resistance to blood flow;

d. explain how vessel diameter influences blood pressure and flow; and

e. describe some local, neural, and hormonal influences on vessel diameter.

To sustain life, the circulatory system must deliver oxygen and nutrients to the tissues, and remove their wastes, at a rate that keeps pace with tissue metabolism. Inadequate circulatory services to a tissue can lead within minutes to tissue necrosis and possibly death of the individual. Thus, it is crucial for the cardiovascular system to respond promptly to local needs and ensure that the tissues have an adequate blood supply at all times. This section of the chapter explores the mechanisms for achieving this.

The blood supply to a tissue can be expressed in terms of *flow* and *perfusion.* **Flow** is the amount of blood flowing through an organ, tissue, or blood vessel in a given time (such as mL/min.). **Perfusion** is the flow per given volume or mass of tissue (such as mL/min./g). Thus, a large organ such as the femur could have a *greater flow* but *less perfusion* than a small organ such as the ovary, because the ovary receives much more blood per gram of tissue.

In a resting individual, *total* flow is quite constant and is equal to cardiac output (typically 5.25 L/min.). Flow through individual organs, however, varies from minute to minute as blood is redirected from one organ to another. Digestion, for example, requires abundant flow to the intestines, and the cardiovascular system makes this available by reducing flow through other organs such as the kidneys. When digestion and nutrient absorption are over, blood flow to the intestines declines and a higher priority is given to the kidneys and other organs. Great variations in regional flow can occur with little or no change in total flow.

*Hemodynamics,* the physical principles of blood flow, are based mainly on pressure and resistance. These relationships can be concisely summarized by the formula $F \propto \Delta P/R$. In other words, the greater the pressure difference ($\Delta P$) between two points, the greater the flow *(F);* the greater the resistance *(R),* the less the flow. Therefore, to understand the flow of blood, we must consider the factors that affect pressure and resistance.

### Blood Pressure

**Blood pressure (BP)** is the force that the blood exerts against a vessel wall. It can be measured within a blood vessel or the heart by inserting a catheter or needle connected to an external manometer (pressure-measuring device). For routine clinical purposes, however, the measurement of greatest interest is the systemic arterial BP at a point close to the heart. We customarily measure it with a sphygmomanometer (see p. 728) connected to an inflatable cuff wrapped around the arm. The *brachial artery* passing through this region is sufficiently close to the level of the heart that the BP recorded here approximates the pressure at the exit from the left ventricle.

Two pressures are recorded: **Systolic pressure** is the peak arterial BP attained during ventricular contraction, and **diastolic pressure** is the minimum arterial BP occurring during the ventricular relaxation between heartbeats. For a healthy person age 20 to 30, these pressures are typically about 120 and 75 mm Hg, respectively. Arterial BP is written as a ratio of systolic over diastolic pressure: 120/75.

The difference between systolic and diastolic pressure is called **pulse pressure** (not to be confused with pulse *rate*). For the preceding BP, pulse pressure is 120 − 75 = 45 mm Hg. This is an important measure of the force that drives blood circulation and of the maximum stress exerted on small arteries by the pressure surges generated by the heart. Another measure of stress on the blood vessels is the **mean arterial pressure (MAP)**—the mean pressure you would obtain if you took measurements at several intervals (say every 0.1 second) throughout the cardiac cycle. MAP is not simply an arithmetic mean of systolic and diastolic pressures, however, because the low-pressure diastole lasts longer than the high-pressure systole. A close estimate of MAP is obtained by adding diastolic pressure and one-third of the pulse pressure. For a blood pressure of 120/75, MAP ≈ 75 + 45/3 = 90 mm Hg. This is typical for vessels at the level of the heart, but MAP varies with the influence of gravity. In a standing adult, it is about 62 mm Hg in the major arteries of the head and 180 mm Hg in major arteries of the ankle.

It is the mean arterial pressure that most influences the risk of disorders such as **syncope** (SIN-co-pee) (fainting), atherosclerosis, kidney failure, edema, and aneurysm. The importance of preventing excessive blood pressure is therefore clear. One of the body's chief means of doing so is the ability of the arteries to stretch and recoil during the cardiac cycle. If the arteries were rigid tubes, pressure would rise much higher in systole and drop to nearly zero in diastole. Blood throughout the circulatory system would flow and stop, flow and stop, and put great stress on the small vessels. But healthy conducting arteries expand with each systole, absorb some of the force of the ejected blood, and store potential energy. Then, when the heart is in diastole, their elastic recoil releases that energy, exerts pressure on the blood, and maintains blood flow throughout the cardiac cycle. The elastic arteries smooth out pressure fluctuations and reduce stress on the smaller arteries.

Nevertheless, blood flow in the arteries is *pulsatile*. In the aorta, blood rushes forward at 120 cm/s during systole and has an average speed of 40 cm/s over the cardiac cycle. When measured farther away from the heart, systolic and diastolic pressures are lower and there is less difference between them (fig. 20.10). In capillaries and veins, the blood flows at a steady speed with little if any pulsation because the pressure surges have been damped out by the distance traveled and the elasticity of the arteries. This is why an injured vein exhibits relatively slow, steady bleeding, whereas blood jets intermittently from a severed artery. In the inferior vena cava near the heart, however, venous flow fluctuates with the respiratory cycle for reasons explained later, and there is some fluctuation in the jugular veins of the neck.

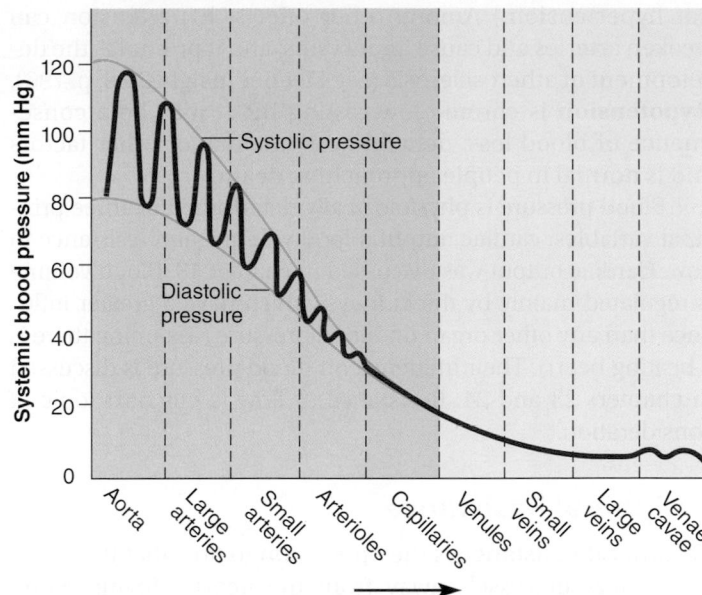

**FIGURE 20.10  Changes in Blood Pressure Relative to Distance from the Heart.** Because of arterial elasticity and the effect of friction against the vessel wall, all measures of blood pressure decline with distance—systolic pressure, diastolic pressure, pulse pressure, and mean arterial pressure. There is no pulse pressure beyond the arterioles, but there are slight pressure oscillations in the venae cavae caused by the respiratory pump described later in this chapter.

▶▶▶**APPLY WHAT YOU KNOW**

*Explain how the histological structure of large arteries relates to their ability to stretch during systole and recoil during diastole.*

As we get older, our arteries become less distensible and absorb less systolic force. This increasing stiffness of the arteries is called **arteriosclerosis**[6] ("hardening of the arteries"). The primary cause of it is cumulative damage by free radicals, which cause gradual deterioration of the elastic and other tissues of the arterial walls—much like old rubber bands that become less stretchy. Another contributing factor is **atherosclerosis,** the growth of lipid deposits in the arterial walls (see Deeper Insight 19.4, p. 739). These deposits can become calcified *complicated plaques,* giving the arteries a hard, bonelike consistency. As a result of these degenerative changes, blood pressure rises with age. Common blood pressures at the age of 20 are about 123/76 for males and 116/72 for females. For healthy persons at age 70, typical blood pressures are around 145/82 and 159/85 for the two sexes, respectively.

**Hypertension** (high BP) is commonly considered to be a chronic resting blood pressure higher than 140/90. (*Temporary* high BP resulting from emotion or exercise is

---

[6]*arterio* = artery; *sclerosis* = hardening

not hypertension.) Among other effects, hypertension can weaken arteries and cause aneurysms, and it promotes the development of atherosclerosis (see Deeper Insight 20.6, p. 796). **Hypotension** is chronic low resting BP. It may be a consequence of blood loss, dehydration, anemia, or other factors and is normal in people approaching death.

Blood pressure is physiologically determined by three principal variables: cardiac output, blood volume, and resistance to flow. Cardiac output was discussed in chapter 19. Blood volume is regulated mainly by the kidneys, which have a greater influence than any other organ on blood pressure (assuming there is a beating heart). Their influence on blood pressure is discussed in chapters 23 and 24. Resistance to flow is our next topic of consideration.

## Peripheral Resistance

**Peripheral resistance** is the opposition to flow that the blood encounters in vessels away from the heart. Moving blood would exert no pressure against a vessel wall unless it encountered at least some downstream resistance. Thus, pressure and resistance are not independent variables in blood flow—rather, pressure is affected by resistance and flow is affected by both. Resistance, in turn, hinges on three variables that we will now consider: blood viscosity, vessel length, and vessel radius.

### Blood Viscosity

Chapter 18 discusses the factors that affect the viscosity ("thickness") of the blood (p. 676). The most significant of these are the erythrocyte count and albumin concentration. A deficiency of erythrocytes (anemia) or albumin (hypoproteinemia) reduces viscosity and speeds up blood flow. On the other hand, viscosity increases and flow declines in such conditions as polycythemia and dehydration.

### Vessel Length

The farther a liquid travels through a tube, the more cumulative friction it encounters; pressure and flow therefore decline with distance. Partly for this reason, if you were to measure mean arterial pressure in a reclining person, you would obtain a higher value in the arm, for example, than in the ankle. In a reclining person, a strong pulse in the *dorsal pedal artery* of the foot is a good sign of adequate cardiac output. If perfusion is good at that distance from the heart, it is likely to be good elsewhere in the systemic circulation.

### Vessel Radius

Blood viscosity and vessel lengths do not change in the short term, of course. In a healthy individual, the only significant ways of controlling peripheral resistance from moment to moment are **vasoconstriction,** the narrowing of a vessel, and **vasodilation,** the widening of a vessel. Vasoconstriction occurs when the smooth muscle of the tunica media contracts. Vasodilation, however, is brought about not by any muscular

effort to widen a vessel, but rather by muscular passivity—relaxation of the smooth muscle, allowing blood pressure to expand the vessel. For lack of a better term, this chapter will refer to vasoconstriction and vasodilation collectively as **vasoreflexes,** whether they are controlled from afar by hormones or the autonomic nervous system or occur in direct response to the local metabolic state of a tissue.

The effect of vessel radius on blood flow stems from the friction of the moving blood against the vessel walls. Blood normally exhibits smooth, silent **laminar[7] flow.** That is, it flows in layers—faster near the center of a vessel, where it encounters less friction, and slower near the walls, where it drags against the vessel. You can observe a similar effect from the vantage point of a riverbank. The current may be very swift in the middle of a river but quite sluggish near shore, where the water encounters more friction against the riverbank and bottom. When a blood vessel dilates, a greater portion of the blood is in the middle of the stream and the average flow may be quite swift. When the vessel constricts, more of the blood is close to the wall and the average flow is slower (fig. 20.11).

Thus, the radius of a vessel markedly affects blood flow. Indeed, flow *(F)* is proportional not merely to vessel radius *(r)* but to the *fourth power* of radius—that is, $F \propto r^4$. This makes radius a very potent factor in the control of flow. For the sake of simplicity, consider a hypothetical blood vessel with a 1 mm radius when maximally constricted and a 3 mm radius when completely dilated. At a 1 mm radius, suppose the flow rate is 1 mL/min. By the formula $F \propto r^4$, consider how the flow would change as radius changed:

$$r = 1 \text{ mm } r^4 = 1^4 = \phantom{0}1 \; F = \phantom{0}1 \text{ mL/min. (given)}$$
$$r = 2 \text{ mm } r^4 = 2^4 = 16 \; F = 16 \text{ mL/min.}$$
$$r = 3 \text{ mm } r^4 = 3^4 = 81 \; F = 81 \text{ mL/min.}$$

These actual numbers do not matter; what matters is that a mere 3-fold increase in radius has produced an 81-fold increase in flow—a demonstration that vessel radius exerts a very powerful influence over flow.

Blood vessels are, indeed, capable of substantial changes in radius. The arteriole in figure 20.12, for example, has constricted to one-third of its relaxed diameter under the influence of a drop of epinephrine. Since blood viscosity and vessel length do not change from moment to moment, vessel radius is the most adjustable of all variables that govern peripheral resistance.

#### ▶▶▶ APPLY WHAT YOU KNOW

*Suppose a vessel with a radius of 1 mm had a flow of 3 mL/min., and then the vessel dilated to a radius of 5 mm. What would be the new flow rate?*

---

[7]*lamina* = layer

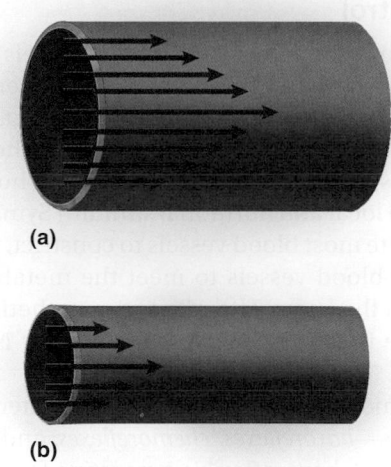

(a)

(b)

**FIGURE 20.11 Laminar Flow and the Effect of Vessel Radius.** Blood flows more slowly near the vessel wall, as indicated by shorter arrows, than it does near the center of the vessel. (a) When the vessel radius is large, the average velocity of flow is high. (b) When the radius is less, the average velocity is lower because a larger portion of the blood is slowed down by friction against the vessel wall.

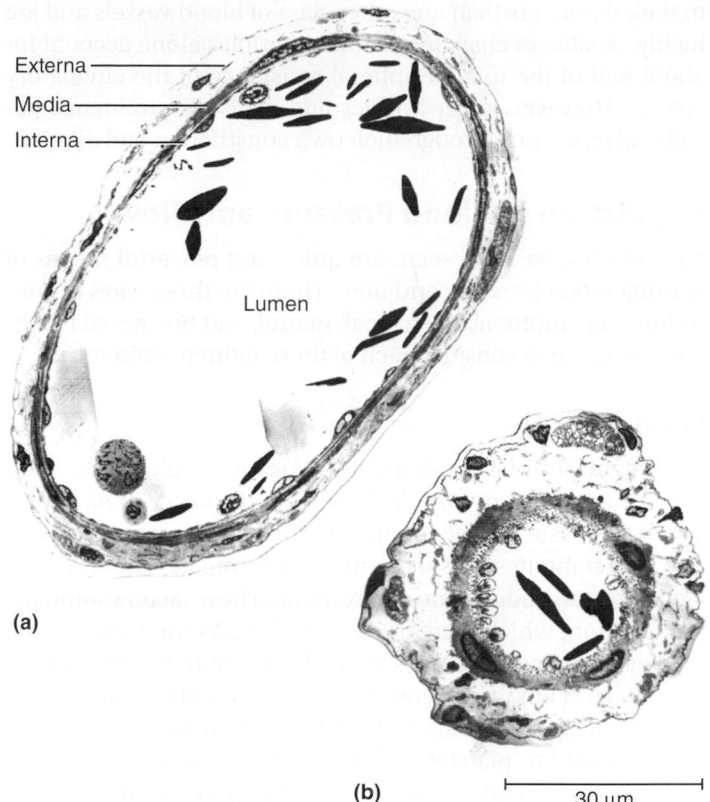

**FIGURE 20.12 The Capacity for Vasoconstriction in an Arteriole.** (a) A dilated arteriole (cross section, TEM). (b) The same arteriole, at a point just 1 mm from the area photographed in part (a). A single drop of epinephrine applied here has caused the arteriole to constrict to about one-third of its dilated diameter.

To integrate this information, consider how the velocity of blood flow differs from one part of the systemic circuit to another (table 20.1). Flow is fastest in the aorta because it is a large vessel close to the pressure source, the left ventricle. From aorta to capillaries, velocity diminishes for three reasons: (1) The blood has traveled a greater distance, so friction has slowed it down. (2) The arterioles and capillaries have smaller radii and therefore put up more resistance. (3) Even though the radii of individual vessels become smaller as we progress farther from the heart, the number of vessels and their *total* cross-sectional area become greater and greater. The aorta has a cross-sectional area of 3 to 5 cm², whereas the total cross-sectional area of all the capillaries is about 4,500 to 6,000 cm². Thus, a given volume of aortic blood is distributed over a greater total area in the capillaries, which *collectively* form a wider path in the bloodstream. Just as water slows down when a narrow mountain stream flows into a lake, blood slows down as it enters pathways with a greater total area or volume.

From capillaries to vena cava, velocity rises again. One reason for this is that the veins are larger than the capillaries, so they create less resistance. Furthermore, since many capillaries converge on one venule, and many venules on a larger vein, a large amount of blood is being forced into a progressively smaller channel—like water flowing from a lake into an outlet stream and thus flowing faster again. Note, however, that blood in the veins never regains the velocity it had in the large arteries. This is because the veins are farther from the pressure head (the heart) and because they are more compliant than arteries—they stretch to accommodate more blood, and this reduces pressure and flow.

| TABLE 20.1 | Blood Velocity in the Systemic Circuit | |
| --- | --- | --- |
| **Vessel** | **Typical Lumen Diameter** | **Velocity*** |
| Aorta | 2.5 cm | 1,200 mm/s |
| Arterioles | 20–50 μm | 15 mm/s |
| Capillaries | 5–9 μm | 0.4 mm/s |
| Venules | 20 μm | 5 mm/s |
| Inferior vena cava | 3 cm | 80 mm/s |

*Peak systolic velocity in the aorta; mean or steady velocity in other vessels, assuming no upstream vasoconstriction adding to resistance

Arterioles are the most significant point of control over peripheral resistance and blood flow because (1) they are on the proximal sides of the capillary beds, so they are best positioned to regulate flow into the capillaries; (2) they greatly outnumber any other class of arteries and thus provide the most numerous control points; and (3) they are more muscular in proportion to their diameters than any other class of blood vessels and are highly capable of changing radius. Arterioles alone account for about half of the total peripheral resistance of the circulatory system. However, larger arteries and veins also influence peripheral resistance through their own constriction and dilation.

## Regulation of Blood Pressure and Flow

Vasoreflexes, we have seen, are quick and powerful means of altering blood pressure and flow. There are three ways of controlling vasomotor activity: local, neural, and hormonal mechanisms. We now consider each of these influences in turn.

## Local Control

**Autoregulation** is the ability of tissues to regulate their own blood supply. According to the *metabolic theory of autoregulation,* if a tissue is inadequately perfused, it becomes hypoxic and its metabolites (waste products) accumulate—$CO_2$, $H^+$, $K^+$, lactic acid, and adenosine, for example. These factors stimulate vasodilation, which increases blood flow. As the bloodstream delivers oxygen and carries away the metabolites, the vessels reconstrict. Thus, a homeostatic dynamic equilibrium is established that adjusts perfusion to the tissue's metabolic needs.

In addition, platelets, endothelial cells, and the perivascular tissues secrete a variety of **vasoactive chemicals** that stimulate vasodilation under such conditions as trauma, inflammation, and exercise. These include histamine, bradykinin, and prostaglandins. The drag of blood flowing against the endothelial cells creates a *shear stress* (like rubbing your palms together) that stimulates them to secrete prostacyclin and nitric oxide, which are vasodilators.

If a tissue's blood supply is cut off for a time and then restored, it often exhibits **reactive hyperemia**—an increase above the normal level of flow. This may be due to the accumulation of metabolites during the period of ischemia. Reactive hyperemia is seen when the skin flushes after a person comes in from the cold. It also occurs in the forearm if a blood pressure cuff is inflated for too long and then loosened.

Over a longer time, a hypoxic tissue can increase its own perfusion by **angiogenesis**[8]—the growth of new blood vessels. (This term also refers to embryonic development of blood vessels.) Three situations in which this is important are the regrowth of the uterine lining after each menstrual period, the development of a higher density of blood capillaries in the muscles of well-conditioned athletes, and the growth of arterial bypasses around obstructions in the coronary circulation. Several growth factors and inhibitors control angiogenesis, but physiologists are not yet sure how it is regulated. There is great clinical importance in finding out. Malignant tumors secrete growth factors that stimulate a dense network of vessels to grow into them and provide nourishment to the cancer cells. Oncologists are interested in finding a way to block tumor angiogenesis, which would choke off a tumor's blood supply and perhaps shrink or kill it.

## Neural Control

In addition to local control, the blood vessels are under remote control by the central and autonomic nervous systems. The **vasomotor center** of the medulla oblongata exerts sympathetic control over blood vessels throughout the body. (Precapillary sphincters have no innervation, however, and respond only to local and hormonal stimuli.) Sympathetic nerve fibers stimulate most blood vessels to constrict, but they dilate the coronary blood vessels to meet the metabolic demands of exercise on the heart. The role of sympathetic tone and vasomotor tone in controlling vessel diameter is explained in chapter 15 (p. 572).

The vasomotor center is an integrating center for three autonomic reflexes—*baroreflexes, chemoreflexes,* and the *medullary ischemic reflex.* A **baroreflex**[9] is a negative feedback response to changes in blood pressure (see fig. 15.1, p. 559). The changes are detected by baroreceptors of the carotid sinuses (see p. 747). Glossopharyngeal nerve fibers from these sinuses transmit signals continually to the brainstem. When blood pressure rises, their signaling rate rises. This *inhibits* the sympathetic cardiac and vasomotor neurons and reduces sympathetic tone, and it *excites* the vagal fibers to the heart. Thus, it reduces the heart rate and cardiac output, dilates the arteries and veins, and reduces blood pressure (fig. 20.13). When blood pressure drops below normal, on the other hand, the opposite reactions occur and BP rises back to normal.

Baroreflexes are important chiefly in short-term regulation of BP, for example in adapting to changes in posture. Perhaps you have jumped quickly out of bed and felt a little dizzy for a moment. This occurs because gravity draws the blood into the large veins of the abdomen and lower limbs when you stand, which reduces venous return to the heart and cardiac output to the brain. Normally, the baroreceptors respond quickly to this drop in pressure and restore cerebral perfusion (see fig. 1.7, p. 16). Baroreflexes are not effective in correcting chronic hypertension, however. Within 2 days or less, they adjust their set point to the higher BP and maintain dynamic equilibrium at this new level.

A **chemoreflex** is an autonomic response to changes in blood chemistry, especially its pH and concentrations of $O_2$ and $CO_2$. It is initiated by the chemoreceptors called *aortic bodies* and *carotid bodies* (see p. 747). The primary role of chemoreflexes is to adjust respiration to changes in blood chemistry, but they have a secondary role in vasoreflexes. Hypoxemia (blood $O_2$ deficiency), hypercapnia ($CO_2$ excess), and acidosis (low blood pH) stimulate the chemoreceptors and act through the vasomotor center to induce widespread vasoconstriction. This increases overall BP, thus increasing perfusion of the lungs and the rate of gas exchange. Chemoreceptors also stimulate breathing, so increased ventilation of the lungs matches their increased perfusion. Increasing one without the other would be of little use.

---

[8]*angio* = vessels; *genesis* = production of

[9]*baro* = pressure

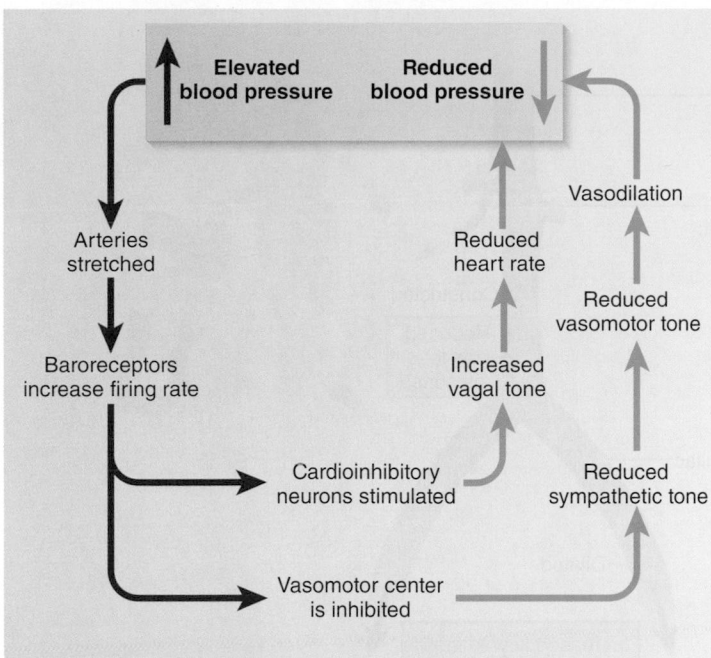

**FIGURE 20.13 Negative Feedback Control of Blood Pressure.** High blood pressure activates this cycle of reactions that return blood pressure to normal.

The **medullary ischemic** (iss-KEE-mic) **reflex** is an autonomic response to reduced perfusion of the brain; in other words, the medulla oblongata monitors its own blood supply and activates corrective reflexes when it senses a state of ischemia (insufficient perfusion). Within seconds of a drop in perfusion, the cardiac and vasomotor centers of the medulla send sympathetic signals to the heart and blood vessels that accelerate the heart and constrict the vessels. These actions raise the blood pressure and ideally restore normal cerebral perfusion. The cardiac and vasomotor centers also receive input from other brain centers, so stress, anger, and arousal can raise the blood pressure. The hypothalamus acts through the vasomotor center to redirect blood flow in response to exercise or changes in body temperature.

## Hormonal Control

All of the following hormones influence blood pressure, some through their vasoactive effects and some through means such as regulating water balance:

- **Angiotensin II.** This is a potent vasoconstrictor that raises the blood pressure. Its synthesis and action are detailed in chapter 23 (see fig. 23.15, p. 903). Its synthesis requires *angiotensin-converting enzyme (ACE).* Hypertension is often treated with drugs called *ACE inhibitors,* which block the action of this enzyme, thus lowering angiotensin II levels and blood pressure.

- **Aldosterone.** This "salt-retaining hormone" primarily promotes $Na^+$ retention by the kidneys. Since water follows sodium osmotically, $Na^+$ retention promotes water retention, thereby supporting blood pressure.

- **Natriuretic peptides.** These hormones, secreted by the heart, antagonize aldosterone. They increase $Na^+$ excretion by the kidneys, thus reducing blood volume and pressure. They also have a generalized vasodilator effect that helps to lower blood pressure.

- **Antidiuretic hormone.** ADH primarily promotes water retention, but at pathologically high concentrations it is also a vasoconstrictor—hence its alternate name, *arginine vasopressin.* Both of these effects raise blood pressure.

- **Epinephrine and norepinephrine.** These adrenal and sympathetic catecholamines bind to α-adrenergic receptors on the smooth muscle of most blood vessels. This stimulates vasoconstriction and raises the blood pressure. In the coronary blood vessels, however, they bind to β-adrenergic receptors and cause vasodilation, increasing blood flow to the myocardium during exercise.

▶▶▶ **APPLY WHAT YOU KNOW**

*Renin inhibitors are drugs used to treat hypertension. Explain how you think they would produce the desired effect.*

## Two Purposes of Vasoreflexes

Vasoreflexes (vasoconstriction and vasodilation) serve two physiological purposes: a generalized raising or lowering of blood pressure throughout the body, and selectively modifying the perfusion of a particular organ and rerouting blood from one region of the body to another.

A generalized increase in blood pressure requires centralized control—an action on the part of the medullary vasomotor center or by hormones that circulate throughout the system, such as angiotensin II or epinephrine. Widespread vasoconstriction raises the overall blood pressure because the whole "container" (the blood vessels) squeezes on a fixed amount of blood. This can be important in supporting cerebral perfusion in situations such as hemorrhaging or dehydration, in which blood volume has significantly fallen. Conversely, generalized vasodilation lowers BP throughout the system.

The rerouting of blood and changes in the perfusion of individual organs can be achieved by either central or local control. For example, during periods of exercise, the sympathetic nervous system can selectively reduce flow to the kidneys and digestive tract. Yet as we saw earlier, metabolite accumulation in a tissue can stimulate local vasodilation and increase perfusion of that tissue without affecting circulation elsewhere in the body.

If a specific artery constricts, pressure downstream from the constriction drops and pressure upstream from it rises. If blood can travel by either of two routes and one route puts up more resistance than the other, most blood follows the path of least resistance. This mechanism enables the body to redirect blood from one organ to another.

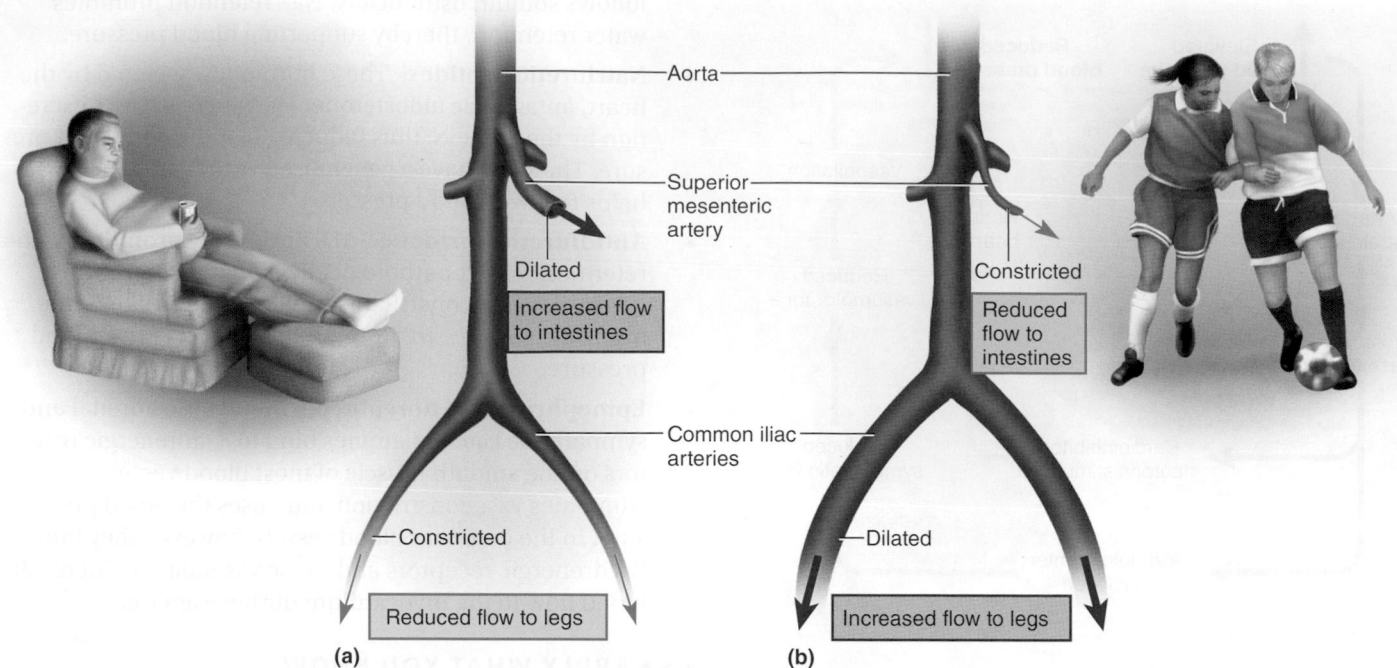

**FIGURE 20.14 Redirection of Blood Flow in Response to Changing Metabolic Needs.** (a) After a meal, the intestines receive priority and the skeletal muscles receive relatively little flow. (b) During exercise, the muscles receive higher priority. Although vasodilation and vasoconstriction are shown here in major arteries for illustration purposes, most control occurs at a microscopic level in the arterioles.

For example, if you are dozing in an armchair after a big meal (fig. 20.14a), vasoconstriction shuts down blood flow to 90% or more of the capillaries in the muscles of your lower limbs (and muscles elsewhere). This raises the BP above the limbs, where the aorta gives off a branch, the superior mesenteric artery, supplying the small intestine. High resistance in the circulation of the limbs and low resistance in the superior mesenteric artery route blood to the small intestine, where it is needed to absorb the nutrients you are digesting.

On the other hand, during vigorous exercise, the arteries in your lungs, coronary circulation, and muscles dilate. To increase the circulation in these routes, vasoconstriction must occur elsewhere, such as the kidneys and digestive tract (figs. 20.14b, 20.15). That reduces their perfusion for the time being, making more blood available to the organs important in sustaining exercise. Thus, local changes in peripheral resistance can shift blood flow from one organ system to another to meet the changing metabolic priorities of the body.

**BEFORE YOU GO ON**

Answer the following questions to test your understanding of the preceding section:

7. Explain why a drop in diastolic pressure would raise one's pulse pressure even if systolic pressure remained unchanged. How could this rise in pulse pressure adversely affect the blood vessels?

8. Explain why arterial blood flow is pulsatile and venous flow is not.

9. What three variables affect peripheral resistance to blood flow? Which of these is most able to change from one minute to the next?

10. What are the three primary mechanisms for controlling vessel radius? Briefly explain each.

11. Explain how the baroreflex serves as an example of homeostasis and negative feedback.

12. Explain how the body can shift the flow of blood from one organ system to another.

**20.3** Capillary Exchange

**Expected Learning Outcomes**

When you have completed this section, you should be able to

a. describe how materials get from the blood into the surrounding tissues;

b. describe and calculate the forces that enable capillaries to give off and reabsorb fluid; and

c. describe the causes and effects of edema.

Only 250 to 300 mL (5%) of the blood is in the capillaries at any given time. This is the most important blood in the body,

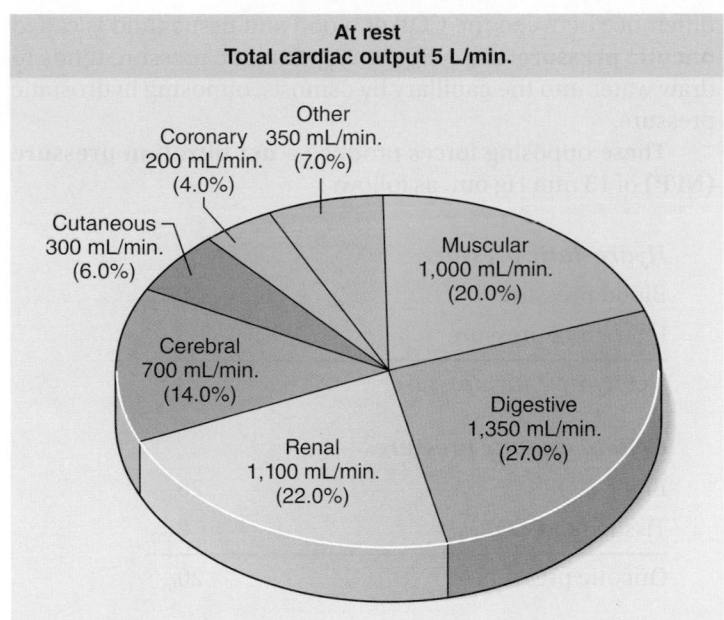

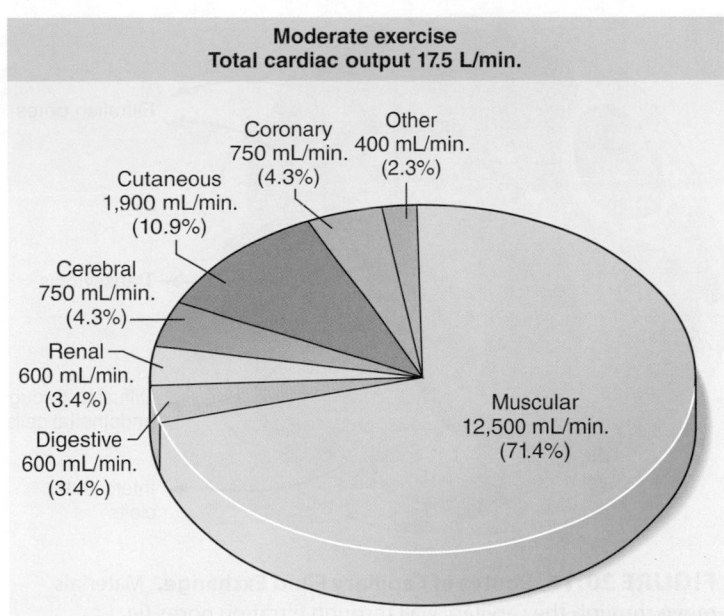

**FIGURE 20.15** Differences in Systemic Blood Flow During Rest and Exercise.

however, for it is mainly across capillary walls that exchanges occur between the blood and surrounding tissues. **Capillary exchange** refers to this two-way movement of fluid.

Chemicals given off by the capillary blood to the perivascular tissues include oxygen, glucose and other nutrients, antibodies, and hormones. Chemicals taken up by the capillaries include carbon dioxide and other wastes, and many of the same substances as they give off: glucose and fatty acids released from storage in the liver and adipose tissue; calcium and other minerals released from bone; antibodies secreted by immune cells; and hormones secreted by the endocrine glands. Thus, many of these chemicals have a two-way traffic between the blood and connective tissue, leaving the capillaries at one point and entering at another. Along with all these solutes, there is substantial movement of water into and out of the bloodstream across the capillary walls. Significant exchange also occurs across the walls of the venules, but capillaries are the more important exchange site because they so greatly outnumber the venules.

The mechanisms of capillary exchange are difficult to study quantitatively because it is hard to measure pressure and flow in such small vessels. For this reason, theories of capillary exchange remain in dispute. Few capillaries of the human body are accessible to direct, noninvasive observation, but those of the fingernail bed and eponychium (cuticle) at the base of the nails can be observed with a stereomicroscope and have been the basis for a number of studies. Their BP has been measured at 32 mm Hg at the arterial end and 15 mm Hg at the venous end, 1 mm away. Capillary BP drops rapidly because of the substantial friction the blood encounters in such narrow vessels. It takes 1 to 2 seconds for an RBC to pass through a nail bed capillary, traveling about 0.7 mm/s.

Chemicals pass through the capillary wall by three routes (fig. 20.16):

1. the endothelial cell cytoplasm;
2. intercellular clefts between the endothelial cells; and
3. filtration pores of the fenestrated capillaries.

The mechanisms of movement through the capillary wall are *diffusion, transcytosis, filtration,* and *reabsorption.*

## Diffusion

The most important mechanism of exchange is diffusion. Glucose and oxygen, being more concentrated in the systemic blood than in the tissue fluid, diffuse out of the blood. Carbon dioxide and other wastes, being more concentrated in the tissue fluid, diffuse into the blood. ($O_2$ and $CO_2$ diffuse in the opposite directions in the pulmonary circuit.) Such diffusion is possible only if the solute can either permeate the plasma membranes of the endothelial cells or find passages large enough to pass through—namely, the filtration pores and intercellular clefts. Such lipid-soluble substances as steroid hormones, $O_2$, and $CO_2$ diffuse easily through the plasma membranes. Substances insoluble in lipids, such as glucose and electrolytes, must pass through membrane channels, filtration pores, or intercellular clefts. Large molecules such as proteins are usually held back.

## Transcytosis

Transcytosis is a process in which endothelial cells pick up material on one side of the plasma membrane by pinocytosis or receptor-mediated endocytosis, transport the vesicles across the cell, and discharge the material on the other side by

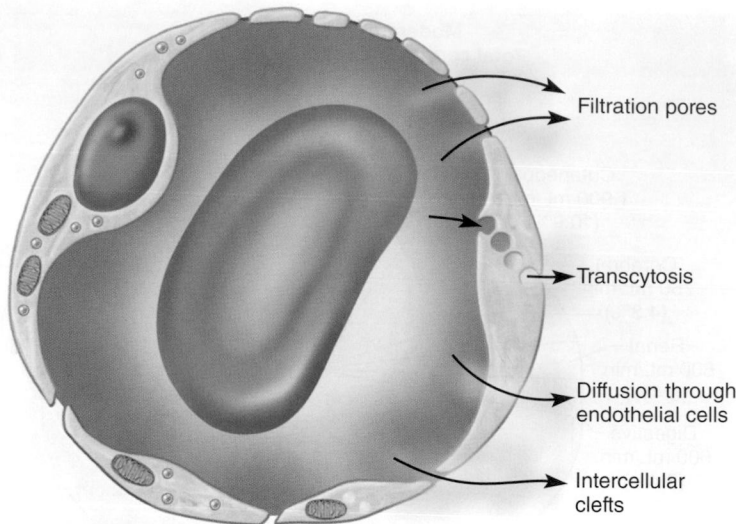

Filtration pores

Transcytosis

Diffusion through endothelial cells

Intercellular clefts

**FIGURE 20.16  Routes of Capillary Fluid Exchange.** Materials move through the capillary wall through filtration pores (in fenestrated capillaries only), by transcytosis, by diffusion through the endothelial cells, and through intercellular clefts.

exocytosis (see fig. 3.23, p. 98). This probably accounts for only a small fraction of solute exchange across the capillary wall, but fatty acids, albumin, and some hormones such as insulin move across the endothelium by this mechanism.

## Filtration and Reabsorption

The equilibrium between filtration and osmosis discussed in chapter 3 becomes particularly relevant when we consider capillary fluid exchange. Typically, fluid filters out of the arterial end of a capillary and osmotically reenters it at the venous end (fig. 20.17). This fluid delivers materials to the cells and removes their metabolic wastes. It may seem odd that a capillary could give off fluid at one point and reabsorb it at another. This comes about as the result of a shifting balance between hydrostatic and osmotic forces. **Hydrostatic pressure** is the physical force exerted by a liquid against a surface such as a capillary wall. Blood pressure is one example of hydrostatic pressure.

A typical capillary has a blood pressure of about 30 mm Hg at the arterial end. The hydrostatic pressure of the interstitial space has been difficult to measure, but a typical value accepted by many authorities is $-3$ mm Hg. The negative value indicates that this is a slight suction, which helps draw fluid out of the capillary. (This force will be represented hereafter as $3_{out}$.) In this case, the positive hydrostatic pressure within the capillary and the negative interstitial pressure work in the same direction, creating a total outward force of about 33 mm Hg.

These forces are opposed by **colloid osmotic pressure (COP),** the portion of the osmotic pressure due to protein. The blood has a COP of about 28 mm Hg, due mainly to albumin. Tissue fluid has less than one-third the protein concentration of blood plasma and has a COP of about 8 mm Hg. The

difference between the COP of blood and tissue fluid is called **oncotic pressure:** $28_{in} - 8_{out} = 20_{in}$. Oncotic pressure tends to draw water into the capillary by osmosis, opposing hydrostatic pressure.

These opposing forces produce a **net filtration pressure (NFP)** of 13 mm Hg out, as follows:

| *Hydrostatic pressure* | | |
|---|---|---|
| Blood pressure | | $30_{out}$ |
| Interstitial pressure | $+$ | $3_{out}$ |
| Net hydrostatic pressure | | $33_{out}$ |
| *Colloid osmotic pressure* | | |
| Blood COP | | $28_{in}$ |
| Tissue fluid COP | $-$ | $8_{out}$ |
| Oncotic pressure | | $20_{in}$ |
| *Net filtration pressure* | | |
| Net hydrostatic pressure | | $33_{out}$ |
| Oncotic pressure | $-$ | $20_{in}$ |
| Net filtration pressure | | $13_{out}$ |

The NFP of 13 mm Hg causes about 0.5% of the blood plasma to leave the capillaries at the arterial end.

At the venous end, however, capillary blood pressure is lower—about 10 mm Hg. All the other pressures are essentially unchanged. Thus, we get

| *Hydrostatic pressure* | | |
|---|---|---|
| Blood pressure | | $10_{out}$ |
| Interstitial pressure | $+$ | $3_{out}$ |
| Net hydrostatic pressure | | $13_{out}$ |
| *Net reabsorption pressure* | | |
| Oncotic pressure | | $20_{in}$ |
| Net hydrostatic pressure | $-$ | $13_{out}$ |
| Net reabsorption pressure | | $7_{in}$ |

The prevailing force is inward at the venous end because osmotic pressure overrides filtration pressure. The **net reabsorption pressure** of 7 mm Hg inward causes the capillary to reabsorb fluid at this end.

Now you can see why a capillary gives off fluid at one end and reabsorbs it at the other. The only pressure that changes significantly from the arterial end to the venous end is the capillary blood pressure, and this change is responsible for the shift from filtration to reabsorption. With a reabsorption pressure of 7 mm Hg and a net filtration pressure of 13 mm Hg, it might appear that far more fluid would leave the capillaries than reenter

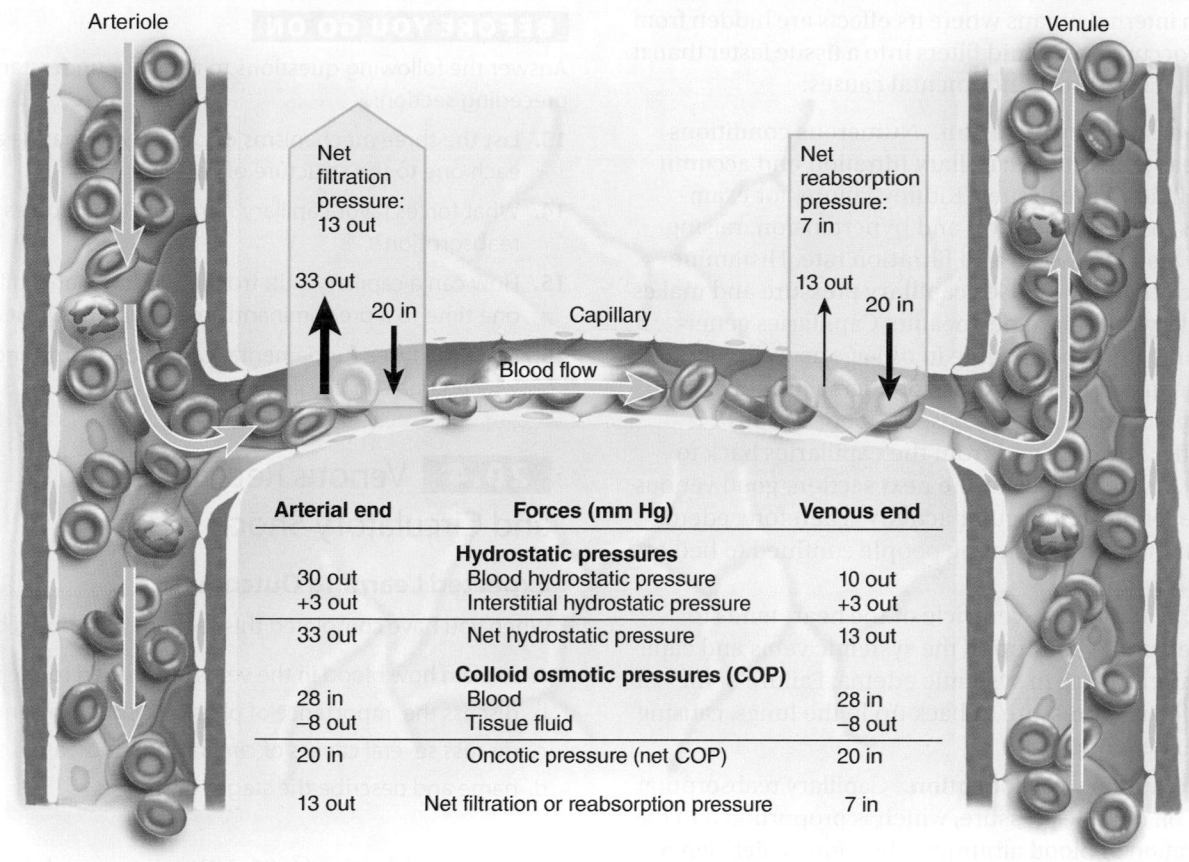

**FIGURE 20.17 The Forces of Capillary Filtration and Reabsorption.** Note the shift from net filtration at the arterial end (left) to net reabsorption at the venous end (right). **AP|R**

them. However, since capillaries branch along their length, there are more of them at the venous end than at the arterial end, which partially compensates for the difference between filtration and reabsorption pressures. They also typically have nearly twice the diameter at the venous end that they have at the arterial end, so there is more capillary surface area available to reabsorb fluid than to give it off. Consequently, capillaries reabsorb about 85% of the fluid they filter. The other 15% is absorbed and returned to the blood by way of the lymphatic system, as described in chapter 21.

Of course, water is not the only substance that crosses the capillary wall by filtration and reabsorption. Chemicals dissolved in the water are "dragged" along with it and pass through the capillary wall if they are not too large. This process, called **solvent drag,** will be important in our discussions of kidney and intestinal function in later chapters.

## Variations in Capillary Filtration and Reabsorption

The figures in the preceding discussion are only examples; circumstances differ from place to place in the body and from time to time in the same capillaries. Capillaries usually reabsorb most of the fluid they filter, but not always. The kidneys have capillary networks called *glomeruli* in which there is little or no reabsorption; they are entirely devoted to filtration. Alveolar capillaries of the lungs, by contrast, are almost entirely dedicated to absorption so fluid does not fill the air spaces.

Capillary activity also varies from moment to moment. In a resting tissue, most precapillary sphincters are constricted and the capillaries are collapsed. Capillary BP is very low (if there is any flow at all), and reabsorption predominates. When a tissue becomes metabolically active, its capillary flow increases. In active muscles, capillary pressure rises to the point that filtration overrides reabsorption along the entire length of the capillary. Fluid accumulates in the muscle and increases muscular bulk by as much as 25%. Capillary permeability is also subject to chemical influences. Traumatized tissue releases such chemicals as substance P, bradykinin, and histamine, which increase permeability and filtration.

## Edema

**Edema** is the accumulation of excess fluid in a tissue. It often shows as swelling of the face, fingers, abdomen, or ankles, but also occurs in internal organs where its effects are hidden from view. Edema occurs when fluid filters into a tissue faster than it is reabsorbed. It has three fundamental causes:

1. **Increased capillary filtration.** Numerous conditions can increase the rate of capillary filtration and accumulation of fluid in the tissues. Kidney failure, for example, leads to water retention and hypertension, raising capillary blood pressure and filtration rate. Histamine dilates arterioles and raises capillary pressure and makes the capillary wall more permeable. Capillaries generally become more permeable in old age as well, putting elderly people at increased risk of edema. Capillary blood pressure also rises in cases of poor venous return—the flow of blood from the capillaries back to the heart. As we will see in the next section, good venous return depends on muscular activity. Therefore, edema is a common problem among people confined to bed or a wheelchair.

   Failure of the right ventricle of the heart tends to cause pressure to back up in the systemic veins and capillaries, thus resulting in systemic edema. Failure of the left ventricle causes pressure to back up in the lungs, causing pulmonary edema.

2. **Reduced capillary reabsorption.** Capillary reabsorption depends on oncotic pressure, which is proportional to the concentration of blood albumin. Therefore, a deficiency of albumin (hypoproteinemia) produces edema by reducing the reabsorption of tissue fluid. Since albumin is produced by the liver, liver diseases such as cirrhosis tend to lead to hypoproteinemia and edema. Edema is commonly seen in regions of famine due to dietary protein deficiency (see kwashiorkor, p. 677). Hypoproteinemia and edema also commonly result from severe burns, owing to the loss of protein from body surfaces no longer covered with skin, and from kidney diseases that allow protein to escape in the urine.

3. **Obstructed lymphatic drainage.** The lymphatic system, described in detail in chapter 21, is a network of one-way vessels that collect fluid from the tissues and return it to the bloodstream. Obstruction of these vessels or the surgical removal of lymph nodes can interfere with fluid drainage and lead to the accumulation of tissue fluid distal to the obstruction (see fig. 21.2, p. 805).

Edema has multiple pathological consequences. As the tissues become congested with fluid, oxygen delivery and waste removal are impaired and the tissues may begin to die. Pulmonary edema presents a threat of suffocation as fluid replaces air in the lungs, and cerebral edema can produce headaches, nausea, and sometimes delirium, seizures, and coma. In severe edema, so much fluid may transfer from the blood vessels to the tissue spaces that blood volume and pressure drop low enough to cause circulatory shock (described in the next section).

---

**BEFORE YOU GO ON**

Answer the following questions to test your understanding of the preceding section:

13. List the three mechanisms of capillary exchange and relate each one to the structure of capillary walls.

14. What forces favor capillary filtration? What forces favor reabsorption?

15. How can a capillary shift from a predominantly filtering role at one time to a predominantly reabsorbing role at another?

16. State the three fundamental causes of edema and explain why edema can be dangerous.

## 20.4 Venous Return and Circulatory Shock

### Expected Learning Outcomes

When you have completed this section, you should be able to

a. explain how blood in the veins is returned to the heart;

b. discuss the importance of physical activity in venous return;

c. discuss several causes of circulatory shock; and

d. name and describe the stages of shock.

Hieronymus Fabricius (1537–1619) discovered the valves of the veins but did not understand their function. That was left to his student, William Harvey, who performed simple experiments on the valves that you can easily reproduce. In figure 20.18, by Harvey, the experimenter has pressed on a vein at point H to block flow from the wrist toward the elbow. With another finger, he has milked the blood out of it up to point O, the first valve proximal to H. When he tries to force blood downward, it stops at that valve. It can go no farther, and it causes the vein to swell at that point. Blood can flow from right to left through that valve but not from left to right. So as Harvey correctly surmised, the valves serve to ensure a one-way flow of blood toward the heart.

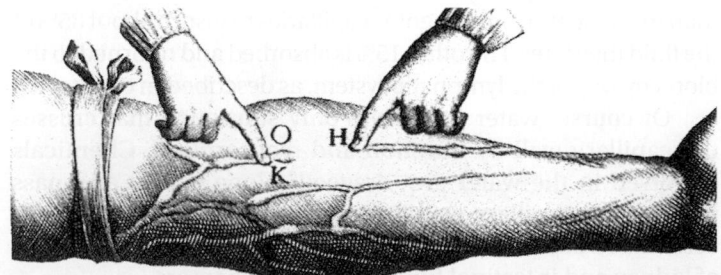

**FIGURE 20.18   An Illustration from William Harvey's *De Motu Cordis* (1628).** These experiments demonstrate the existence of one-way valves in veins of the arms. See text for explanation.

❓ *In the space between O and H, what (if anything) would happen if the experimenter lifted his finger from point O? What if he lifted his finger from point H? Why?*

You can easily demonstrate the action of these valves in your own hand. Hold your hand still, below waist level, until veins stand up on the back of it. (Do not apply a tourniquet!) Press on a vein close to your knuckles, and while holding it down, use another finger to milk that vein toward the wrist. It collapses as you force the blood out of it, and if you remove the second finger, it will not refill. The valves prevent blood from flowing back into it from above. When you remove the first finger, however, the vein fills from below.

## Mechanisms of Venous Return

The flow of blood back to the heart, called **venous return,** is achieved by five mechanisms:

1. **The pressure gradient.** Pressure generated by the heart is the most important force in venous flow, even though it is substantially weaker in the veins than in the arteries. Pressure in the venules ranges from 12 to 18 mm Hg, and pressure at the point where the venae cavae enter the heart, called **central venous pressure,** averages 4.6 mm Hg. Thus, there is a venous pressure gradient ($\Delta P$) of about 7 to 13 mm Hg favoring the flow of blood toward the heart. The pressure gradient and venous return increase when blood volume increases. Venous return also increases in the event of generalized, widespread vasoconstriction because this reduces the volume of the circulatory system and raises blood pressure and flow.

2. **Gravity.** When you are sitting or standing, blood from your head and neck returns to the heart simply by flowing "downhill" through the large veins above the heart. Thus, the large veins of the neck are normally collapsed or nearly so, and their venous pressure is close to zero. The dural sinuses of the brain, however, have more rigid walls and cannot collapse. Their pressure is as low as –10 mm Hg, creating a risk of *air embolism* if they are punctured (see Deeper Insight 20.3, p. 773).

3. **The skeletal muscle pump.** In the limbs, the veins are surrounded and massaged by the muscles. Contracting muscles squeeze the blood out of the compressed part of a vein, and the valves ensure that this blood can go only toward the heart (fig. 20.19).

4. **The thoracic (respiratory) pump.** This mechanism aids the flow of venous blood from the abdominal to the thoracic cavity. When you inhale, your thoracic cavity expands and its internal pressure drops, while downward movement of the diaphragm raises the pressure in your abdominal cavity. The *inferior vena cava (IVC),* your largest vein, is a flexible tube passing through both of these cavities. If abdominal pressure on the IVC rises while thoracic pressure on it drops, then blood is squeezed upward toward the heart. It is not forced back into the lower limbs because the venous valves there prevent this. Because of the thoracic pump, central venous pressure fluctuates

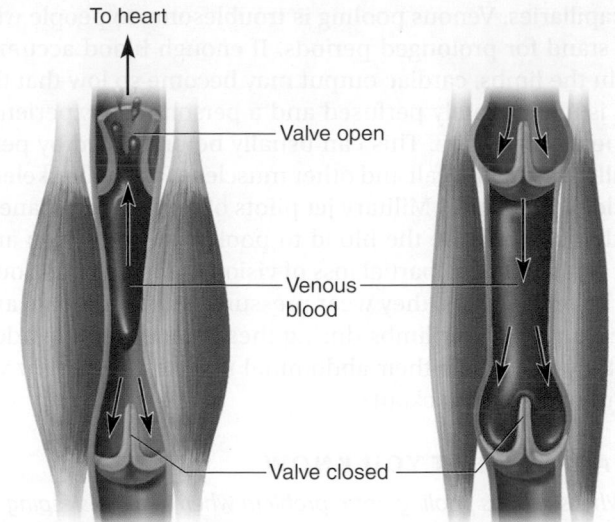

**(a) Contracted skeletal muscles**          **(b) Relaxed skeletal muscles**

**FIGURE 20.19  The Skeletal Muscle Pump.** (a) Muscle contraction squeezes the deep veins and forces blood through the next valve in the direction of the heart. Valves below the point of compression prevent backflow. (b) When the muscles relax, blood flows back downward under the pull of gravity but can flow only as far as the nearest valve.

from 2 mm Hg when you inhale to 6 mm Hg when you exhale, and blood flows faster when you inhale.

5. **Cardiac suction.** During ventricular systole, the tendinous cords pull the AV valve cusps downward, slightly expanding the atrial space. This creates a slight suction that draws blood into the atria from the venae cavae and pulmonary veins.

## Venous Return and Physical Activity

Exercise increases venous return for many reasons. The heart beats faster and harder, increasing cardiac output and blood pressure. Blood vessels of the skeletal muscles, lungs, and coronary circulation dilate, increasing flow. The increase in respiratory rate and depth enhances the action of the thoracic pump. Muscle contractions increase venous return by means of the skeletal muscle pump. Increased venous return then increases cardiac output, which is important in perfusion of the muscles just when they need it most.

Conversely, when a person is still, blood accumulates in the limbs because venous pressure is not high enough to override the weight of the blood and drive it upward. Such accumulation of blood is called **venous pooling.** To demonstrate this effect, hold one hand below your waist for about a minute and hold the other hand over your head. Then, quickly bring your two hands together and compare the palms. The hand held over head usually appears pale because its blood has drained out of it; the hand held below the waist appears redder than normal because of venous pooling in its veins and capillaries. Venous pooling is troublesome to people who must stand for prolonged periods. If enough blood accumulates in the limbs, cardiac output may become so low that the brain is inadequately perfused and a person may experience dizziness or syncope. This can usually be prevented by periodically tensing the calf and other muscles to keep the skeletal muscle pump active. Military jet pilots often perform maneuvers that could cause the blood to pool in the abdomen and lower limbs, causing partial loss of vision or loss of consciousness. To prevent this, they wear pressure suits that inflate and tighten on the lower limbs during these maneuvers; in addition, they may tense their abdominal muscles to prevent venous pooling and blackout.

►►►**APPLY WHAT YOU KNOW**

*Why is venous pooling not a problem when you are sleeping and the skeletal muscle pump is inactive?*

## Circulatory Shock

**Circulatory shock** (not to be confused with electrical or spinal shock) is any state in which cardiac output is insufficient to meet the body's metabolic needs. All forms of circulatory shock fall into two categories: (1) **cardiogenic shock,** caused by inadequate pumping by the heart, usually as a result of myocardial infarction; and (2) **low venous return (LVR) shock,** in which cardiac output is low because too little blood is returning to the heart.

There are three principal forms of LVR shock:

1. **Hypovolemic shock,** the most common form, is produced by a loss of blood volume as a result of hemorrhage, trauma, bleeding ulcers, burns, or dehydration. Dehydration is a major cause of death from heat exposure. In hot weather, the body excretes as much as 1.5 L of sweat per hour. Water transfers from the bloodstream to replace tissue fluid lost in the sweat, and blood volume may drop too low to maintain adequate circulation.

2. **Obstructed venous return shock** occurs when any object, such as a growing tumor or aneurysm, compresses a vein and impedes its blood flow.

3. **Venous pooling (vascular) shock** occurs when the body has a normal total blood volume, but too much of it accumulates in the lower body. This can result from long periods of standing or sitting or from widespread vasodilation. **Neurogenic shock** is a form of venous pooling shock that results from a sudden loss of vasomotor tone, allowing the vessels to dilate. This can result from causes as severe as brainstem trauma or as slight as an emotional shock.

Elements of both venous pooling and hypovolemic shock are present in certain cases, such as septic shock and anaphylactic shock, which involve both vasodilation and a loss of fluid through abnormally permeable capillaries. **Septic shock** occurs when bacterial toxins trigger vasodilation and increased capillary permeability. **Anaphylactic shock,** discussed more fully in chapter 21, results from exposure to an antigen to which a person is allergic, such as bee venom. Antigen–antibody complexes trigger the release of histamine, which causes generalized vasodilation and increased capillary permeability.

## Responses to Circulatory Shock

Shock is clinically described according to severity as compensated or decompensated. In **compensated shock,** several homeostatic mechanisms bring about spontaneous recovery. The hypotension resulting from low cardiac output triggers the sympathetic baroreflex and the production of angiotensin II, both of which counteract shock by stimulating vasoconstriction. Furthermore, if a person faints and falls to a horizontal position, gravity restores blood flow to the brain. Even quicker recovery is achieved if the person's feet are elevated to promote drainage of blood from the legs.

If these mechanisms prove inadequate, **decompensated shock** ensues and several life-threatening positive feedback loops occur. Poor cardiac output results in myocardial ischemia and infarction, which further weaken the heart and reduce output. Slow circulation of the blood can lead to disseminated intravascular coagulation (DIC) (see table 18.8, p. 703). As the vessels become congested with clotted blood, venous return grows even worse. Ischemia and acidosis of the brainstem depress the vasomotor and cardiac centers, causing loss of vasomotor tone, further vasodilation, and further drop in BP and cardiac output. Before long, damage to the cardiac and brain tissues may be too great to survive. About half of those who go into decompensated shock die from it.

**BEFORE YOU GO ON**

Answer the following questions to test your understanding of the preceding section:

17. Explain how respiration aids venous return.

18. Explain how muscular activity and venous valves aid venous return.

19. Define circulatory shock. What are some of the causes of low venous return shock?

## 20.5  Special Circulatory Routes

### Expected Learning Outcomes

When you have completed this section, you should be able to

a. explain how the brain maintains stable perfusion;

b. discuss the causes and effects of strokes and transient ischemic attacks;

c. explain the mechanisms that increase muscular perfusion during exercise; and

d. contrast the blood pressure of the pulmonary circuit with that of the systemic circuit, and explain why the difference is important in pulmonary function.

Certain circulatory pathways have special physiological properties adapted to the functions of their organs. Two of these are described in other chapters: the coronary circulation in chapter 19 and fetal and placental circulation in chapter 29. Here we take a closer look at the circulation to the brain, skeletal muscles, and lungs.

### Brain

Total blood flow to the brain fluctuates less than that of any other organ (about 700 mL/min. at rest). Such constancy is important because even a few seconds of oxygen deprivation causes loss of consciousness, and 4 or 5 minutes of anoxia is time enough to cause irreversible damage. Although total cerebral perfusion is fairly stable, blood flow can be shifted from one part of the brain to another in a matter of seconds as different parts engage in motor, sensory, or cognitive functions.

The brain regulates its own blood flow in response to changes in BP and chemistry. The cerebral arteries dilate when the systemic BP drops and constrict when it rises, thus minimizing fluctuations in cerebral BP. Cerebral blood flow thus remains quite stable even when mean arterial pressure (MAP) fluctuates from 60 to 140 mm Hg. However, an MAP below 60 mm Hg produces syncope and an MAP above 160 mm Hg causes cerebral edema.

The main chemical stimulus for cerebral autoregulation is pH. Poor perfusion allows $CO_2$ to accumulate in the brain. This lowers the pH of the tissue fluid and triggers local vasodilation, which improves perfusion. Extreme hypercapnia, however, depresses neural activity. The opposite condition, hypocapnia, raises the pH and stimulates vasoconstriction, thus reducing perfusion and giving $CO_2$ a chance to rise to a normal level. Hyperventilation (exhaling $CO_2$ faster than the body produces it) induces hypocapnia, which leads to cerebral vasoconstriction, ischemia, dizziness, and sometimes syncope.

Brief episodes of cerebral ischemia produce **transient ischemic attacks (TIAs),** characterized by temporary dizziness, loss of vision or other senses, weakness, paralysis, headache, or aphasia. A TIA may result from spasms of diseased cerebral arteries. It lasts from just a moment to a few hours and is often an early warning of an impending stroke. People with TIAs should receive prompt medical attention to identify the cause using brain imaging and other diagnostic means. Immediate treatment should be initiated to prevent a stroke.

A **stroke,** or **cerebrovascular accident (CVA),** is the sudden death (infarction) of brain tissue caused by ischemia. Cerebral ischemia can be produced by atherosclerosis, thrombosis, or a ruptured aneurysm. The effects of a CVA range from unnoticeable to fatal, depending on the extent of tissue damage and the function of the affected tissue. Blindness, paralysis, loss of sensation, and loss of speech are common. Some of the effects on speech are described on page 540. Recovery depends on the ability of neighboring neurons to take over the lost functions and on the extent of collateral circulation to regions surrounding the cerebral infarction.

### Skeletal Muscles

In contrast to the brain, the skeletal muscles receive a highly variable blood flow depending on their state of exertion. At rest, the arterioles are constricted, most of the capillary beds are shut down, and total flow through the muscular system is about 1 L/min. During exercise, the arterioles dilate in response to muscle metabolites such as lactic acid, nitric oxide (NO), adenosine, $CO_2$, and $H^+$. Blood flow through the muscles can increase more than 20-fold during strenuous exercise, which requires that blood be diverted from other organs such as the digestive tract and kidneys to meet the needs of the working muscles.

Muscular contraction compresses the blood vessels and impedes flow. For this reason, isometric contraction causes fatigue more quickly than intermittent isotonic contraction. If you squeeze a rubber ball as hard as you can without relaxing your grip, you feel the muscles fatigue more quickly than if you intermittently squeeze and relax.

### Lungs

After birth, the pulmonary circuit is the only route in which the arteries carry oxygen-poor blood and the veins carry oxygen-rich blood; the opposite situation prevails in the systemic circuit. The pulmonary arteries have thin distensible walls with less elastic tissue than the systemic arteries. Thus, they have a BP of only 25/10. Capillary hydrostatic pressure is about 10 mm Hg in the pulmonary circuit as compared with an average of 17 mm Hg in systemic capillaries. This lower pressure has two implications for pulmonary circulation: (1) Blood flows more slowly through the pulmonary capillaries, and therefore it has more time for gas exchange; and (2) oncotic pressure overrides hydrostatic pressure, so these capillaries are engaged almost entirely in absorption. This prevents fluid accumulation in the alveolar walls and lumens, which would compromise gas exchange. In a condition such as mitral valve stenosis, however, blood may back up in the pulmonary circuit, raising the capillary hydrostatic pressure and causing pulmonary edema, congestion, and hypoxemia.

#### ▶▶▶ APPLY WHAT YOU KNOW

*What abnormal skin coloration would result from pulmonary edema?*

Another unique characteristic of the pulmonary arteries is their response to hypoxia. Systemic arteries dilate in response to local hypoxia and improve tissue perfusion. By contrast, pulmonary arteries constrict. Pulmonary hypoxia indicates that part of the lung is not being ventilated well, perhaps because of mucous congestion of the airway or a degenerative lung disease. Vasoconstriction in poorly ventilated regions of the lung redirects blood flow to better ventilated regions.

#### BEFORE YOU GO ON

Answer the following questions to test your understanding of the preceding section:

20. In what conspicuous way does perfusion of the brain differ from perfusion of the skeletal muscles?

21. How does a stroke differ from a transient ischemic attack? Which of these bears closer resemblance to a myocardial infarction?

22. How does the low hydrostatic blood pressure in the pulmonary circuit affect the fluid dynamics of the capillaries there?

23. Contrast the vasomotor response of the lungs with that of skeletal muscles to hypoxia.

## 20.6 Anatomy of the Pulmonary Circuit

### Expected Learning Outcome

When you have completed this section, you should be able to

a. trace the route of blood through the pulmonary circuit.

The next three sections of this chapter center on the names and pathways of the principal arteries and veins. The pulmonary circuit is described here, and the systemic arteries and veins are described in the two sections that follow.

The pulmonary circuit (fig. 20.20) begins with the **pulmonary trunk,** a large vessel that ascends diagonally from the right ventricle and branches into the right and left **pulmonary arteries.** As it approaches the lung, the right pulmonary artery branches in two, and both branches enter the lung at a medial indentation called the *hilum* (see fig. 22.9, p. 857). The upper branch is the **superior lobar artery,** serving the superior lobe of the lung. The lower branch divides again within the lung to form the **middle lobar** and **inferior lobar arteries,** supplying the lower two lobes of that lung. The left pulmonary artery is much more variable. It gives off several superior lobar arteries to the superior lobe before entering the hilum, then enters the lung and gives off a variable number of inferior lobar arteries to the inferior lobe.

In both lungs, these arteries lead ultimately to small basketlike capillary beds that surround the pulmonary alveoli (air sacs). This is where the blood unloads $CO_2$ and picks up $O_2$. After leaving the alveolar capillaries, the pulmonary blood flows into venules and veins, ultimately leading to the main **pulmonary veins** that exit the lung at the hilum. The left atrium of the heart receives two pulmonary veins on each side (see fig. 19.5b, p. 714).

The purpose of the pulmonary circuit is primarily to exchange $CO_2$ for $O_2$. The lungs also receive a separate systemic blood supply by way of the *bronchial arteries* (see part I.1 in table 20.5).

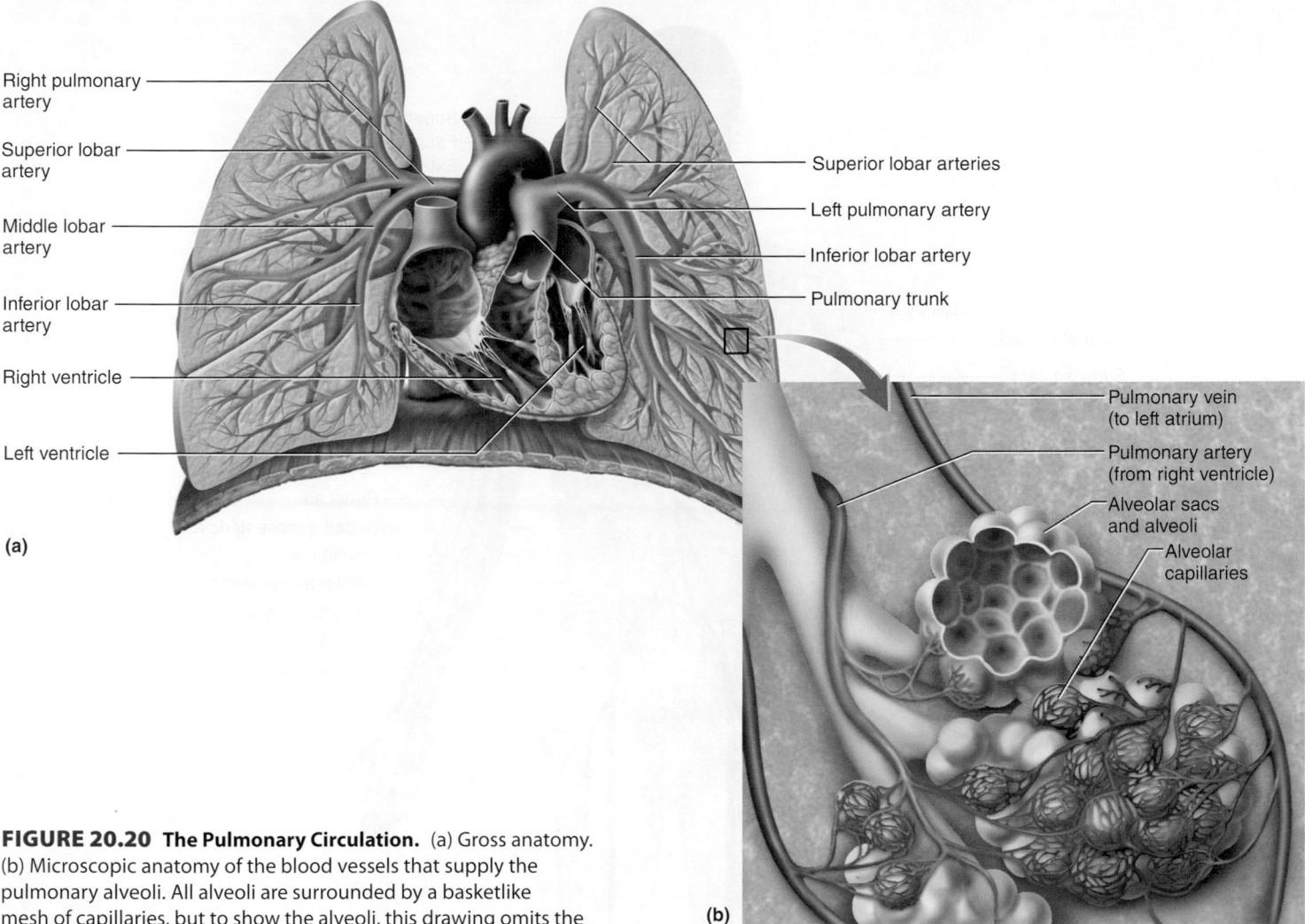

Right pulmonary artery

Superior lobar artery

Middle lobar artery

Inferior lobar artery

Right ventricle

Left ventricle

(a)

Superior lobar arteries

Left pulmonary artery

Inferior lobar artery

Pulmonary trunk

Pulmonary vein (to left atrium)

Pulmonary artery (from right ventricle)

Alveolar sacs and alveoli

Alveolar capillaries

(b)

**FIGURE 20.20  The Pulmonary Circulation.**  (a) Gross anatomy. (b) Microscopic anatomy of the blood vessels that supply the pulmonary alveoli. All alveoli are surrounded by a basketlike mesh of capillaries, but to show the alveoli, this drawing omits the capillaries from some of them.

**BEFORE YOU GO ON**

Answer the following questions to test your understanding of the preceding section:

**24.** Trace the flow of an RBC from right ventricle to left atrium and name the vessels along the way.

**25.** The lungs have two separate arterial supplies. Explain their functions.

## **20.7** Systemic Vessels of the Axial Region

### Expected Learning Outcomes

When you have completed this section, you should be able to

a. identify the principal systemic arteries and veins of the axial region; and

b. trace the flow of blood from the heart to any major organ of the axial region and back to the heart.

The systemic circuit (figs. 20.21 and 20.22) supplies oxygen and nutrients to all organs and removes their metabolic wastes. Part of it, the coronary circulation, was described in chapter 19. This section surveys the remaining arteries and veins of the axial region—the head, neck, and trunk. Tables 20.2 through 20.8 trace the arterial outflow and venous return, region by region. They outline only the most common circulatory pathways; there is a great deal of anatomical variation in the circulatory system from one person to another.

The names of the blood vessels often describe their location by indicating the body region traversed (as in the *axillary artery* and *brachial veins*), an adjacent bone (as in *temporal artery* and *ulnar vein*), or the organ supplied or drained by the vessel (as in *hepatic artery* and *renal vein*). In many cases, an artery and adjacent vein have similar names (*femoral artery* and *femoral vein,* for example).

As you trace blood flow in these tables, it is important to refer frequently to the illustrations. Verbal descriptions alone are likely to seem obscure if you do not make full use of the explanatory illustrations. Throughout these tables and figures, the abbreviations *a.* and *aa.* mean *artery* and *arteries,* and *v.* and *vv.* mean *vein* and *veins.*

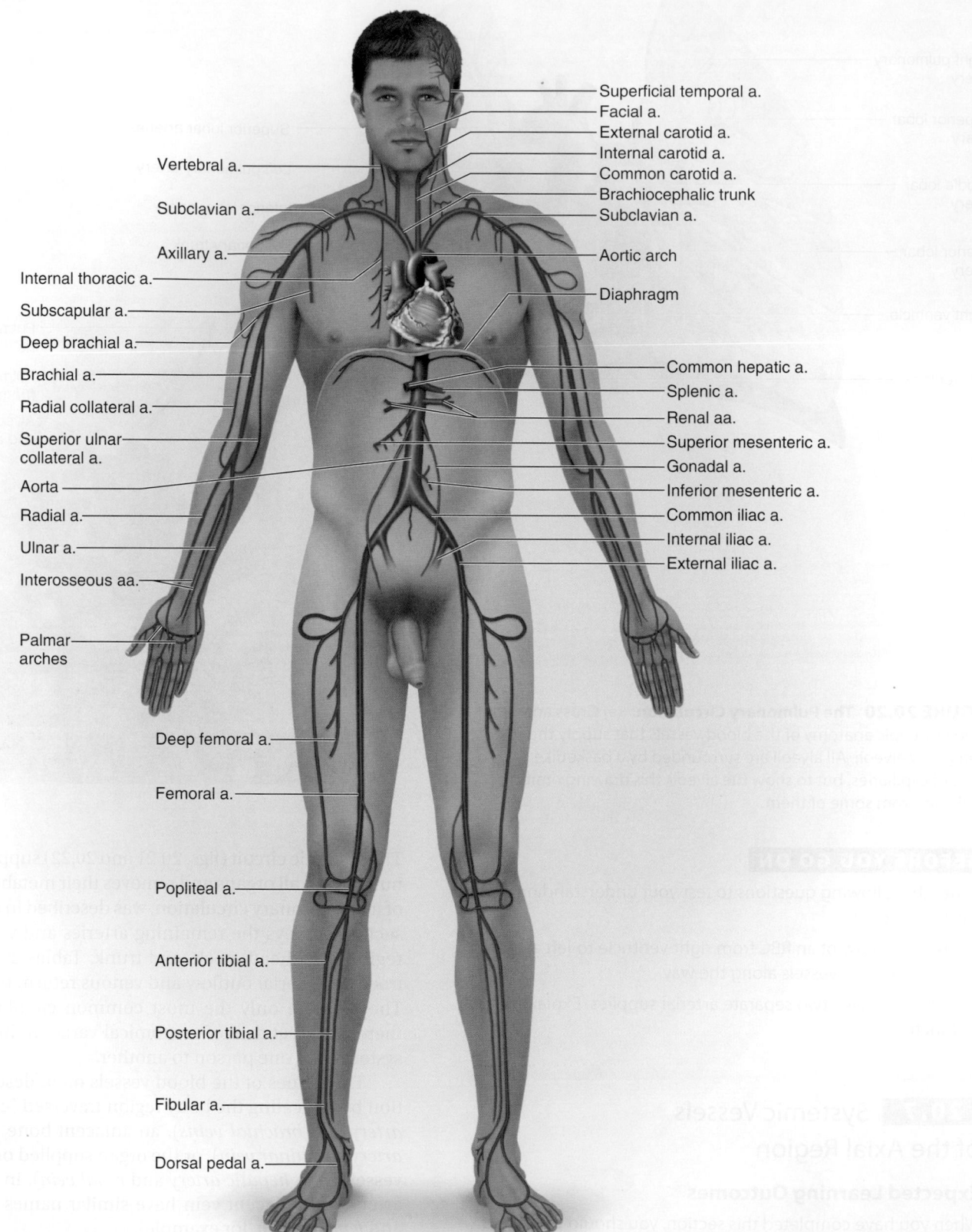

Superficial temporal a.
Facial a.
External carotid a.
Internal carotid a.
Common carotid a.
Brachiocephalic trunk
Subclavian a.
Aortic arch
Diaphragm
Common hepatic a.
Splenic a.
Renal aa.
Superior mesenteric a.
Gonadal a.
Inferior mesenteric a.
Common iliac a.
Internal iliac a.
External iliac a.

Vertebral a.
Subclavian a.
Axillary a.
Internal thoracic a.
Subscapular a.
Deep brachial a.
Brachial a.
Radial collateral a.
Superior ulnar collateral a.
Aorta
Radial a.
Ulnar a.
Interosseous aa.
Palmar arches
Deep femoral a.
Femoral a.
Popliteal a.
Anterior tibial a.
Posterior tibial a.
Fibular a.
Dorsal pedal a.

**FIGURE 20.21  The Major Systemic Arteries (Anterior View).** Different arteries are illustrated on the left than on the right for clarity, but nearly all of those shown occur on both sides.

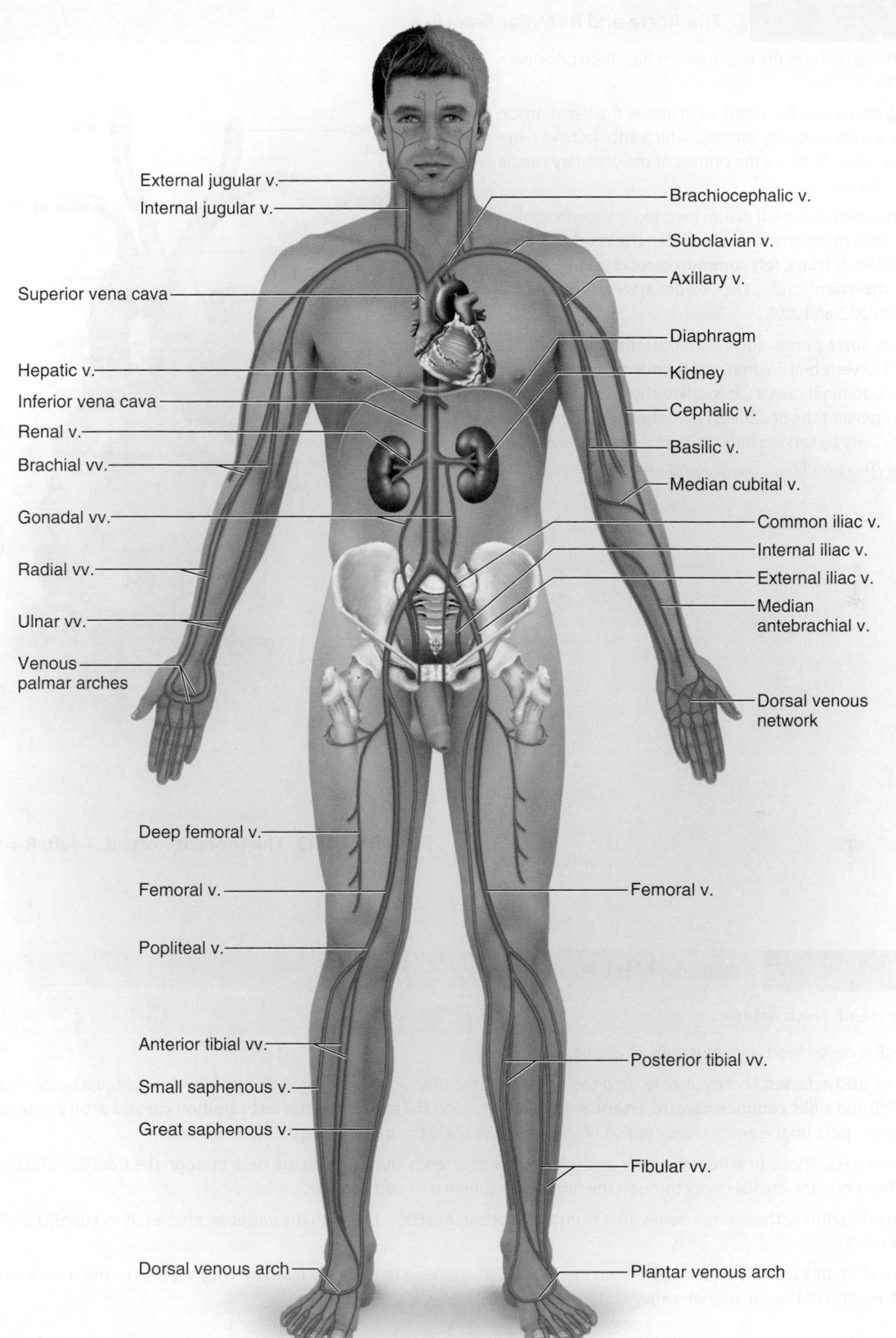

External jugular v.

Internal jugular v.

Superior vena cava

Hepatic v.

Inferior vena cava

Renal v.

Brachial vv.

Gonadal vv.

Radial vv.

Ulnar vv.

Venous palmar arches

Deep femoral v.

Femoral v.

Popliteal v.

Anterior tibial vv.

Small saphenous v.

Great saphenous v.

Dorsal venous arch

Brachiocephalic v.

Subclavian v.

Axillary v.

Diaphragm

Kidney

Cephalic v.

Basilic v.

Median cubital v.

Common iliac v.

Internal iliac v.

External iliac v.

Median antebrachial v.

Dorsal venous network

Femoral v.

Posterior tibial vv.

Fibular vv.

Plantar venous arch

**FIGURE 20.22 The Major Systemic Veins (Anterior View).** Different veins are illustrated on the left than on the right for clarity, but nearly all of those shown occur on both sides.

| **TABLE 20.2** | The Aorta and Its Major Branches |
| --- | --- |

All systemic arteries arise from the aorta, which has three principal regions (fig. 20.23):

1. The **ascending aorta** rises for about 5 cm above the left ventricle. Its only branches are the coronary arteries, which arise behind two cusps of the aortic valve. They are the origins of the coronary circulation described in chapter 19.

2. The **aortic arch** curves to the left like an inverted U superior to the heart. It gives off three major arteries in this order: the **brachiocephalic**[10] (BRAY-kee-oh-seh-FAL-ic) **trunk, left common carotid** (cah-ROT-id) **artery,** and **left subclavian**[11] (sub-CLAY-vee-un) **artery.** These are further traced in tables 20.3 and 20.9.

3. The **descending aorta** passes downward posterior to the heart, at first to the left of the vertebral column and then anterior to it, through the thoracic and abdominal cavities. It is called the *thoracic aorta* above the diaphragm and the *abdominal aorta* below it. It ends in the lower abdominal cavity by forking into the *right* and *left common iliac arteries* (see table 20.7, part IV).

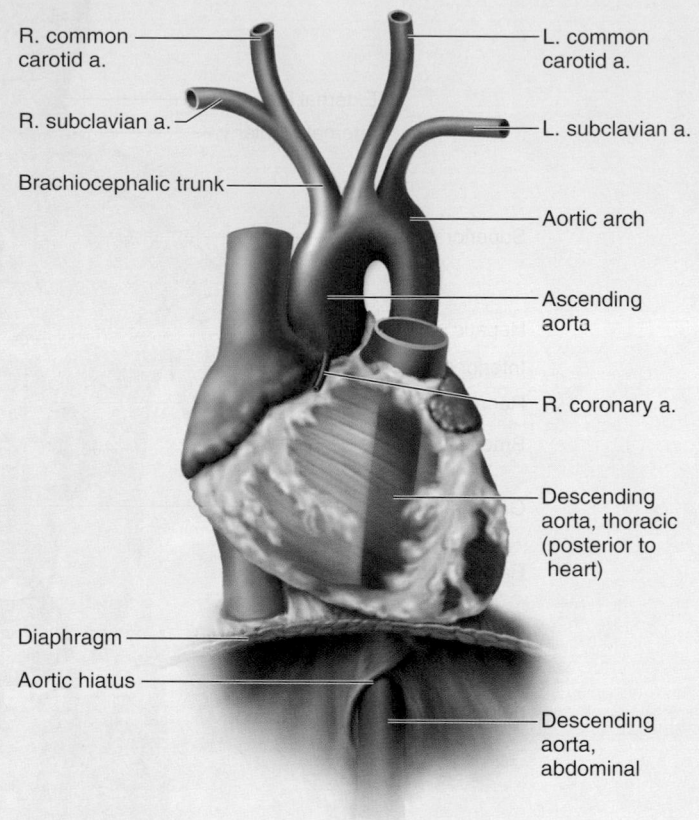

**FIGURE 20.23  The Thoracic Aorta (L. = left; R. = right).** AP|R

| **TABLE 20.3** | Arteries of the Head and Neck |
| --- | --- |

**I. Origins of the Head–Neck Arteries**

The head and neck receive blood from four pairs of arteries (fig. 20.24):

1. The **common carotid arteries.** Shortly after leaving the aortic arch, the brachiocephalic trunk divides into the *right subclavian artery* (further traced in table 20.5) and **right common carotid artery.** A little farther along the aortic arch, the **left common carotid artery** arises independently. The common carotids pass up the anterolateral region of the neck, alongside the trachea (see part II of this table).

2. The **vertebral arteries.** These arise from the right and left subclavian arteries and travel up the neck through the transverse foramina of vertebrae C1 through C6. They enter the cranial cavity through the foramen magnum (see part III of this table).

3. The **thyrocervical**[12] **trunks.** These tiny arteries arise from the subclavian arteries lateral to the vertebral arteries; they supply the thyroid gland and some scapular muscles.

4. The **costocervical**[13] **trunks.** These arteries arise from the subclavian arteries a little farther laterally. They supply the deep neck muscles and some of the intercostal muscles of the superior rib cage.

---

[10]*brachio* = arm; *cephal* = head
[11]*sub* = below; *clavi* = clavicle, collarbone
[12]*thyro* = thyroid gland; *cerv* = neck
[13]*costo* = rib; *cerv* = neck

| TABLE 20.3 | Arteries of the Head and Neck (continued) |
|---|---|

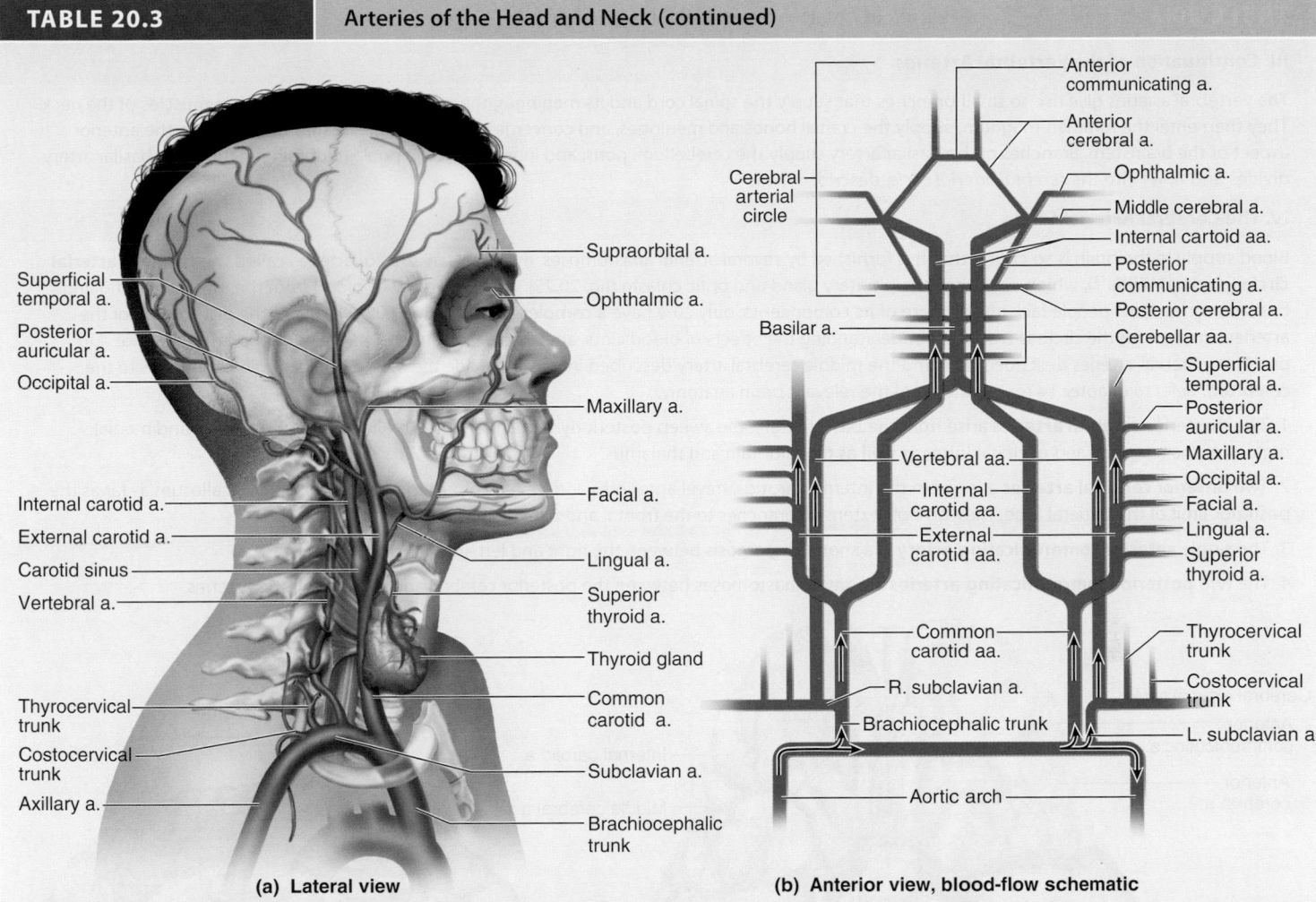

(a) Lateral view

(b) Anterior view, blood-flow schematic

**FIGURE 20.24 Superficial (Extracranial) Arteries of the Head and Neck.** The upper part of the schematic (b) depicts the cerebral circulation in figure 20.25. AP|R

❓ *List the arteries, in order, that an erythrocyte must travel to get from the left ventricle to the skin of the left side of the forehead.*

## II. Continuation of the Common Carotid Arteries

The common carotid arteries have the most extensive distribution of all the head–neck arteries. Near the laryngeal prominence ("Adam's apple"), each common carotid branches into an *external* and *internal carotid artery*.

1. The **external carotid artery** ascends the side of the head external to the cranium and supplies most external head structures except the orbits. It gives rise to the following arteries in ascending order:

   a. the **superior thyroid artery** to the thyroid gland and larynx;

   b. the **lingual artery** to the tongue;

   c. the **facial artery** to the skin and muscles of the face;

   d. the **occipital artery** to the posterior scalp;

   e. the **maxillary** to the teeth, maxilla, oral cavity, and external ear; and

   f. the **superficial temporal artery** to the chewing muscles, nasal cavity, lateral aspect of the face, most of the scalp, and the dura mater.

2. The **internal carotid artery** passes medial to the angle of the mandible and enters the cranial cavity through the carotid canal of the temporal bone. It supplies the orbits and about 80% of the cerebrum. Compressing the internal carotids near the mandible can therefore cause loss of consciousness. After entering the cranial cavity, each internal carotid gives rise to the following branches:

   a. the **ophthalmic artery** to the orbit, nose, and forehead;

   b. the **anterior cerebral artery** to the medial aspect of the cerebral hemisphere (see part IV of this table); and

   c. the **middle cerebral artery,** which travels in the lateral sulcus of the cerebrum, supplies the insula, and then issues numerous branches to the lateral region of the frontal, temporal, and parietal lobes of the brain.

| **TABLE 20.3** | Arteries of the Head and Neck (continued) |
|---|---|

### III. Continuation of the Vertebral Arteries

The vertebral arteries give rise to small branches that supply the spinal cord and its meninges, the cervical vertebrae, and deep muscles of the neck. They then enter the foramen magnum, supply the cranial bones and meninges, and converge to form a single **basilar artery** along the anterior aspect of the brainstem. Branches of the basilar artery supply the cerebellum, pons, and inner ear. At the pons–midbrain junction, the basilar artery divides and flows into the *cerebral arterial circle,* described next.

### IV. The Cerebral Arterial Circle

Blood supply to the brain is so critical that it is furnished by several arterial anastomoses, especially an array of arteries called the **cerebral arterial circle (circle of Willis**[14]**),** which surrounds the pituitary gland and optic chiasm (fig. 20.25). The circle receives blood from the internal carotid and basilar arteries. Most people lack one or more of its components; only 20% have a complete arterial circle. Knowledge of the distribution of the arteries arising from the circle is crucial for understanding the effects of blood clots, aneurysms, and strokes on brain function. The anterior and posterior cerebral arteries described here and the middle cerebral artery described in part II provide the most significant blood supplies to the cerebrum. Refer to chapter 14 for reminders of the relevant brain anatomy.

1. Two **posterior cerebral arteries** arise from the basilar artery and sweep posteriorly to the rear of the brain, serving the inferior and medial regions of the temporal and occipital lobes as well as the midbrain and thalamus.

2. Two **anterior cerebral arteries** arise from the internal carotids, travel anteriorly, and then arch posteriorly over the corpus callosum as far as the posterior limit of the parietal lobe. They give off extensive branches to the frontal and parietal lobes.

3. The single **anterior communicating artery** is a short anastomosis between the right and left anterior cerebral arteries.

4. The two **posterior communicating arteries** are small anastomoses between the posterior cerebral and internal carotid arteries.

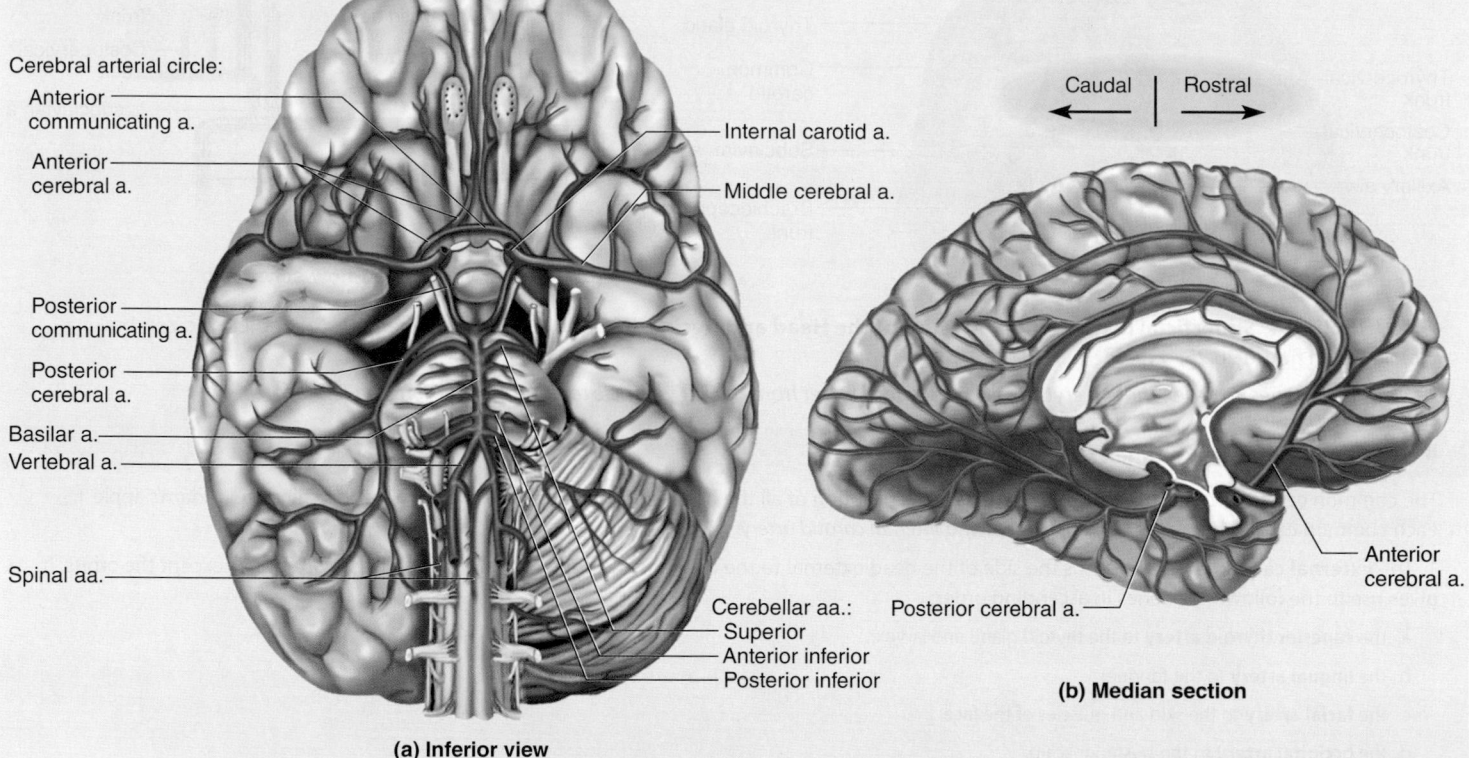

**FIGURE 20.25 The Cerebral Blood Supply.** (a) Inferior view of the brain showing the blood supply to the brainstem, cerebellum, and cerebral arterial circle. (b) Median section of the brain showing the more distal branches of the anterior and posterior cerebral arteries. Branches of the middle cerebral artery are distributed over the lateral surface of the cerebrum (not illustrated). AP|R

[14]Thomas Willis (1621–75), English anatomist

| TABLE 20.4 | Veins of the Head and Neck |
|---|---|

The head and neck are drained mainly by three pairs of veins—the *internal jugular, external jugular,* and *vertebral veins.* We will trace these from their origins to the *subclavian veins.*

### I. Dural Venous Sinuses

After blood circulates through the brain, it collects in large thin-walled veins called **dural venous sinuses**—blood-filled spaces between the layers of the dura mater (fig. 20.26a, b). A reminder of the structure of the dura mater will be helpful in understanding these sinuses. This tough membrane between the brain and cranial bone has a periosteal layer against the bone and a meningeal layer against the brain. In a few places, a space exists between these layers to accommodate a blood-collecting sinus. Between the two cerebral hemispheres is a vertical, sickle-shaped wall of dura called the *falx cerebri,* which contains two of the sinuses. There are about 13 dural venous sinuses in all; we survey only the few most prominent ones here.

1. The **superior sagittal sinus** is contained in the superior margin of the falx cerebri and overlies the longitudinal fissure of the brain (fig. 20.26a; see also figs. 14.5 and 14.7, pp. 514 and 516). It begins anteriorly near the crista galli of the skull and extends posteriorly to the very rear of the head, ending at the level of the posterior occipital protuberance of the skull. Here it bends, usually to the right, and drains into a *transverse sinus.*

2. The **inferior sagittal sinus** is contained in the inferior margin of the falx cerebri and arches over the corpus callosum, deep in the longitudinal fissure. Posteriorly, it joins the *great cerebral vein,* and their union forms the **straight sinus,** which continues to the rear of the head (see fig. 14.7). There, the superior sagittal and straight sinuses meet in a space called the **confluence of the sinuses.**

3. Right and left **transverse sinuses** lead away from the confluence and encircle the inside of the occipital bone, leading toward the ears (fig. 20.26b); their path is marked by grooves on the inner surface of the occipital bone (see fig. 8.5b, p. 235). The right transverse sinus receives blood mainly from the superior sagittal sinus, and the left one drains mainly the straight sinus. Laterally, each transverse sinus makes an S-shaped bend, the **sigmoid sinus,** then exits the cranium through the jugular foramen. From here, the blood flows down the internal jugular vein (see part II.1 of this table).

4. The **cavernous sinuses** are honeycombs of blood-filled spaces on each side of the body of the sphenoid bone (fig. 20.26b). They receive blood from the *superior ophthalmic vein* of the orbit and the *superficial middle cerebral vein* of the brain, among other sources. They drain through several outlets including the transverse sinus, internal jugular vein, and facial vein. They are clinically important because infections can pass from the face and other superficial sites into the cranial cavity by this route. Also, inflammation of a cavernous sinus can injure important structures that pass through it, including the internal carotid artery and cranial nerves III to VI.

### II. Major Veins of the Neck

Blood flows down the neck mainly through three veins on each side, all of which empty into the subclavian vein (fig. 20.26c).

1. The **internal jugular**[15] (JUG-you-lur) **vein** courses down the neck deep to the sternocleidomastoid muscle. It receives most of the blood from the brain; picks up blood from the **facial vein, superficial temporal vein,** and **superior thyroid vein** along the way; passes behind the clavicle; and joins the subclavian vein (which is further traced in table 20.6).

2. The **external jugular vein** courses down the side of the neck superficial to the sternocleidomastoid muscle and empties into the subclavian vein. It drains tributaries from the parotid salivary gland, facial muscles, scalp, and other superficial structures. Some of this blood also follows venous anastomoses to the internal jugular vein.

3. The **vertebral vein** travels with the vertebral artery in the transverse foramina of the cervical vertebrae. Although the companion artery leads to the brain, the vertebral vein does not come from there. It drains the cervical vertebrae, spinal cord, and some of the small deep muscles of the neck, and empties into the subclavian vein.

Table 20.6 traces this blood flow the rest of the way to the heart.

## DEEPER INSIGHT 20.3
### CLINICAL APPLICATION

#### Air Embolism

Injury to the dural sinuses or jugular veins presents less danger from loss of blood than from air sucked into the circulatory system. The presence of air in the bloodstream is called *air embolism.* This is an important concern to neurosurgeons, who sometimes operate with the patient in a sitting position. If a dural sinus is punctured, air can be sucked into the sinus and accumulate in the heart chambers, which blocks cardiac output and causes sudden death. Smaller air bubbles in the systemic circulation can cut off blood flow to the brain, lungs, myocardium, and other vital tissues.

---

[15]*jugul* = neck, throat

| **TABLE 20.4** | Veins of the Head and Neck (continued) |
|---|---|

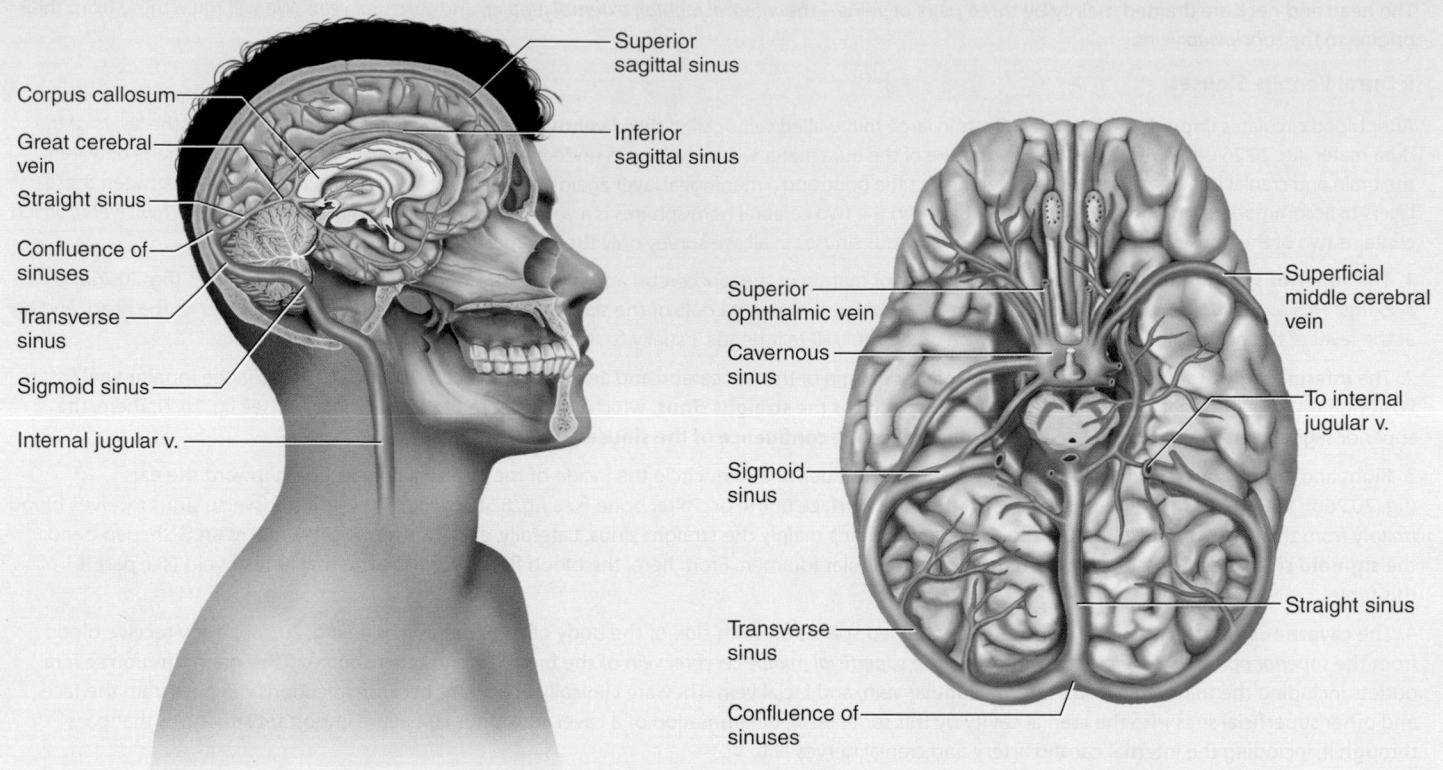

(a) **Dural venous sinuses, medial view**

Labels (a): Corpus callosum, Great cerebral vein, Straight sinus, Confluence of sinuses, Transverse sinus, Sigmoid sinus, Internal jugular v., Superior sagittal sinus, Inferior sagittal sinus

(b) **Dural venous sinuses, inferior view**

Labels (b): Superior ophthalmic vein, Cavernous sinus, Sigmoid sinus, Transverse sinus, Confluence of sinuses, Superficial middle cerebral vein, To internal jugular v., Straight sinus

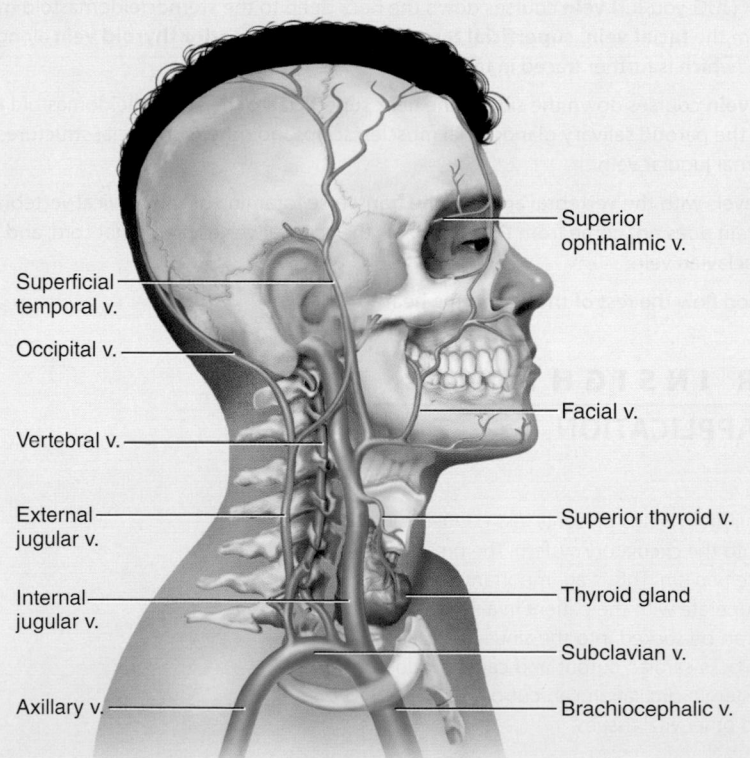

**FIGURE 20.26  Veins of the Head and Neck.** (a) Dural venous sinuses seen in a median section of the cerebrum. (b) Dural venous sinuses seen in an inferior view of the cerebrum. (c) Superficial (extracranial) veins of the head and neck. AP|R

(c) **Superficial veins of the head and neck**

Labels (c): Superficial temporal v., Occipital v., Vertebral v., External jugular v., Internal jugular v., Axillary v., Superior ophthalmic v., Facial v., Superior thyroid v., Thyroid gland, Subclavian v., Brachiocephalic v.

| **TABLE 20.5** | **Arteries of the Thorax** |
|---|---|

The thorax is supplied by several arteries arising directly from the aorta (parts I and II of this table) and from the subclavian and axillary arteries (part III). The thoracic aorta begins distal to the aortic arch and ends at the **aortic hiatus** (hy-AY-tus), a passage through the diaphragm. Along the way, it sends off numerous small branches to the thoracic viscera and the body wall (fig. 20.27), outlined on the next page.

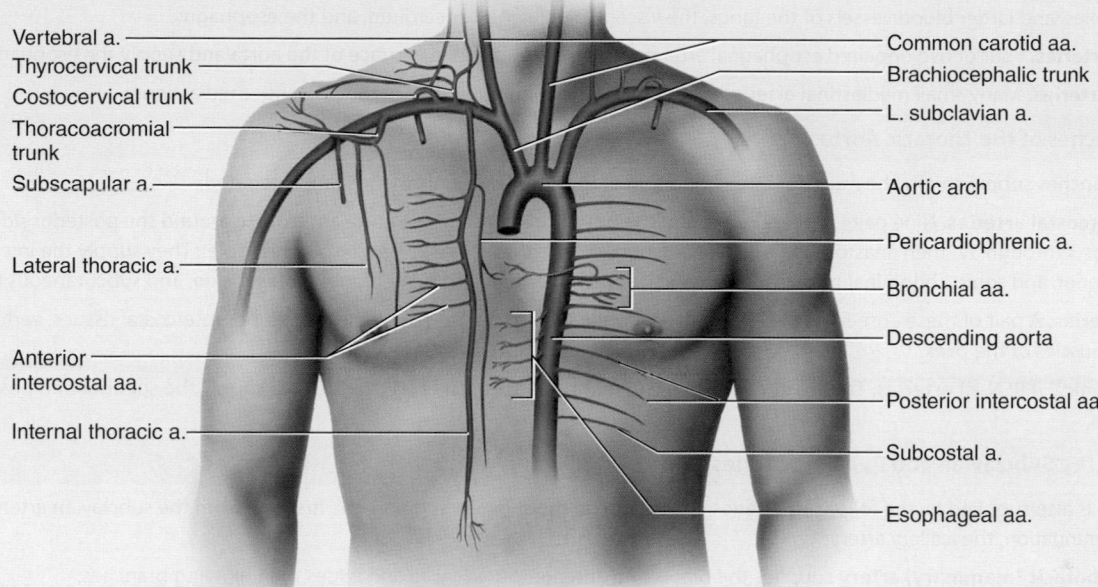

**(a) Major arteries**

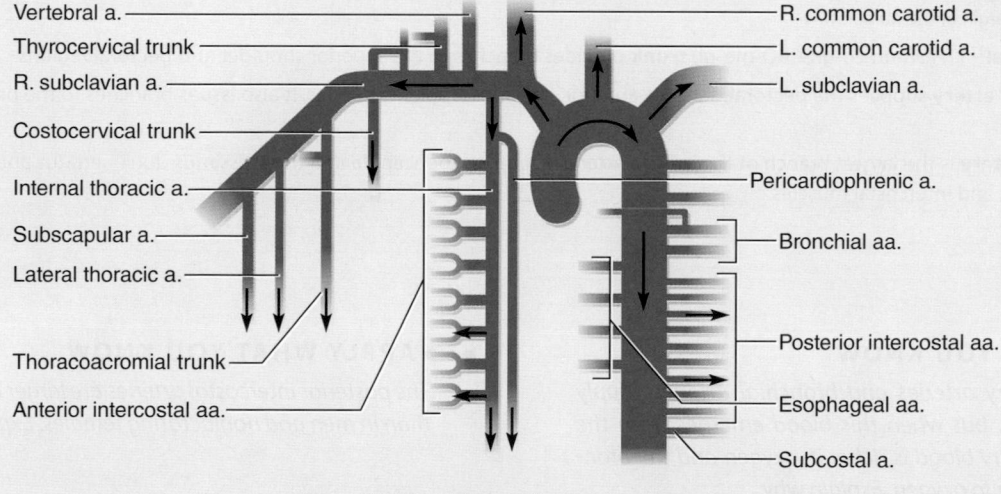

**(b) Blood-flow schematic**

**FIGURE 20.27  Arteries of the Thorax.** AP|R

| TABLE 20.5 | Arteries of the Thorax (continued) |
|---|---|

## I. Visceral Branches of the Thoracic Aorta

These supply the viscera of the thoracic cavity:

1. **Bronchial arteries.** Although variable in number and arrangement, there are usually two of these on the left and one on the right. The right bronchial artery usually arises from one of the left bronchial arteries or from a *posterior intercostal artery* (see part II.1). The bronchial arteries supply the bronchi, bronchioles, and larger blood vessels of the lungs, the visceral pleura, the pericardium, and the esophagus.

2. **Esophageal arteries.** Four or five unpaired esophageal arteries arise from the anterior surface of the aorta and supply the esophagus.

3. **Mediastinal arteries.** Many small mediastinal arteries (not illustrated) supply structures of the posterior mediastinum.

## II. Parietal Branches of the Thoracic Aorta

The following branches supply chiefly the muscles, bones, and skin of the chest wall; only the first are illustrated:

1. **Posterior intercostal arteries.** Nine pairs of these arise from the posterior surface of the aorta and course around the posterior side of the rib cage between ribs 3 through 12, then anastomose with the *anterior intercostal arteries* (see part III.1 in this table). They supply the intercostal, pectoralis, serratus anterior, and some abdominal muscles, as well as the vertebrae, spinal cord, meninges, breasts, skin, and subcutaneous tissue.

2. **Subcostal arteries.** A pair of these arise from the aorta inferior to the twelfth rib. They supply the posterior intercostal tissues, vertebrae, spinal cord, and deep muscles of the back.

3. **Superior phrenic**[16] (FREN-ic) **arteries.** These arteries, variable in number, arise at the aortic hiatus and supply the superior and posterior regions of the diaphragm.

## III. Branches of the Subclavian and Axillary Arteries

The thoracic wall is also supplied by the following arteries, which arise in the shoulder region—the first one from the subclavian artery and the other three from its continuation, the axillary artery:

1. The **internal thoracic (mammary) artery** supplies the breast and anterior thoracic wall and issues the following branches:

   a. The **pericardiophrenic artery** supplies the pericardium and diaphragm.

   b. The **anterior intercostal arteries** arise from the thoracic artery as it descends alongside the sternum. They travel between the ribs, supply the ribs and intercostal muscles, and anastomose with the posterior intercostal arteries. Each of these sends one branch along the lower margin of the rib above and another branch along the upper margin of the rib below.

2. The **thoracoacromial**[17] (THOR-uh-co-uh-CRO-me-ul) **trunk** provides branches to the superior shoulder and pectoral regions.

3. The **lateral thoracic artery** supplies the pectoral, serratus anterior, and subscapularis muscles. It also issues branches to the breast and is larger in females than in males.

4. The **subscapular artery** is the largest branch of the axillary artery. It supplies the scapula and the latissimus dorsi, serratus anterior, teres major, deltoid, triceps brachii, and intercostal muscles.

▶▶▶**APPLY WHAT YOU KNOW**

*Both the pulmonary arteries and bronchial arteries supply blood to the lungs, but when this blood emerges from the lungs, the pulmonary blood is richer in oxygen and the bronchial blood is poorer in oxygen. Explain why.*

▶▶▶**APPLY WHAT YOU KNOW**

*The posterior intercostal arteries are larger in lactating women than in men and nonlactating females. Explain why.*

---

[16]*phren* = diaphragm
[17]*thoraco* = chest; *acr* = tip; *om* = shoulder

| TABLE 20.6 | Veins of the Thorax |
|---|---|

### I. Tributaries of the Superior Vena Cava

The most prominent veins of the upper thorax are as follows. They carry blood from the shoulder region to the heart (fig. 20.28).

1. The **subclavian vein** drains the upper limb (see table 20.10). It begins at the lateral margin of the first rib and travels posterior to the clavicle. It receives the external jugular and vertebral veins, then ends (changes name) where it receives the internal jugular vein.

2. The **brachiocephalic vein** is formed by union of the subclavian and internal jugular veins. The right brachiocephalic is very short, about 2.5 cm, and the left is about 6 cm long. They receive tributaries from the vertebrae, thyroid gland, and upper thoracic wall and breast, then converge to form the next vein.

3. The **superior vena cava** is formed by the union of the right and left brachiocephalic veins. It travels inferiorly for about 7 cm and empties into the right atrium of the heart. Its main tributary is the *azygos vein*. It drains all structures superior to the diaphragm except the pulmonary circuit and coronary circulation. It also receives drainage from the abdominal cavity by way of the azygos system, described next.

### II. The Azygos System

The principal venous drainage of the thoracic organs is by way of the *azygos* (AZ-ih-goss) *system* (fig. 20.28). The most prominent vein of this system is the **azygos**[18] **vein,** which ascends the right side of the posterior thoracic wall and is named for the lack of a mate on the left. It receives the following tributaries, then empties into the superior vena cava at the level of vertebra T4.

1. The right **ascending lumbar vein** drains the right abdominal wall, then penetrates the diaphragm and enters the thoracic cavity. The azygos vein begins where the ascending lumbar vein meets the right **subcostal vein** beneath rib 12.

2. The right **posterior intercostal veins** drain the intercostal spaces. The first (superior) one empties into the right brachiocephalic vein; intercostals 2 and 3 join to form a *right superior intercostal vein* before emptying into the azygos; and intercostals 4 through 11 each enter the azygos vein separately.

3. The right **esophageal, mediastinal, pericardial,** and **bronchial veins** (not illustrated) drain their respective organs into the azygos.

4. The **hemiazygos**[19] **vein** ascends the posterior thoracic wall on the left. It begins where the left ascending lumbar vein, having just penetrated the diaphragm, joins the subcostal vein below rib 12. The hemiazygos then receives the lower three posterior intercostal veins, esophageal veins, and mediastinal veins. At the level of vertebra T9, it crosses to the right and empties into the azygos.

5. The **accessory hemiazygos vein** descends the posterior thoracic wall on the left. It receives drainage from posterior intercostal veins 4 through 8 and sometimes the left bronchial veins. It crosses to the right at the level of vertebra T8 and empties into the azygos vein.

The left posterior intercostal veins 1 to 3 are the only ones on this side that do not ultimately drain into the azygos vein. The first one usually drains directly into the left brachiocephalic vein. The second and third unite to form the *left superior intercostal vein,* which empties into the left brachiocephalic vein.

---

[18]unpaired; from *a* = without; *zygo* = union, mate
[19]*hemi* = half

| **TABLE 20.6** | **Veins of the Thorax (continued)** |
|---|---|

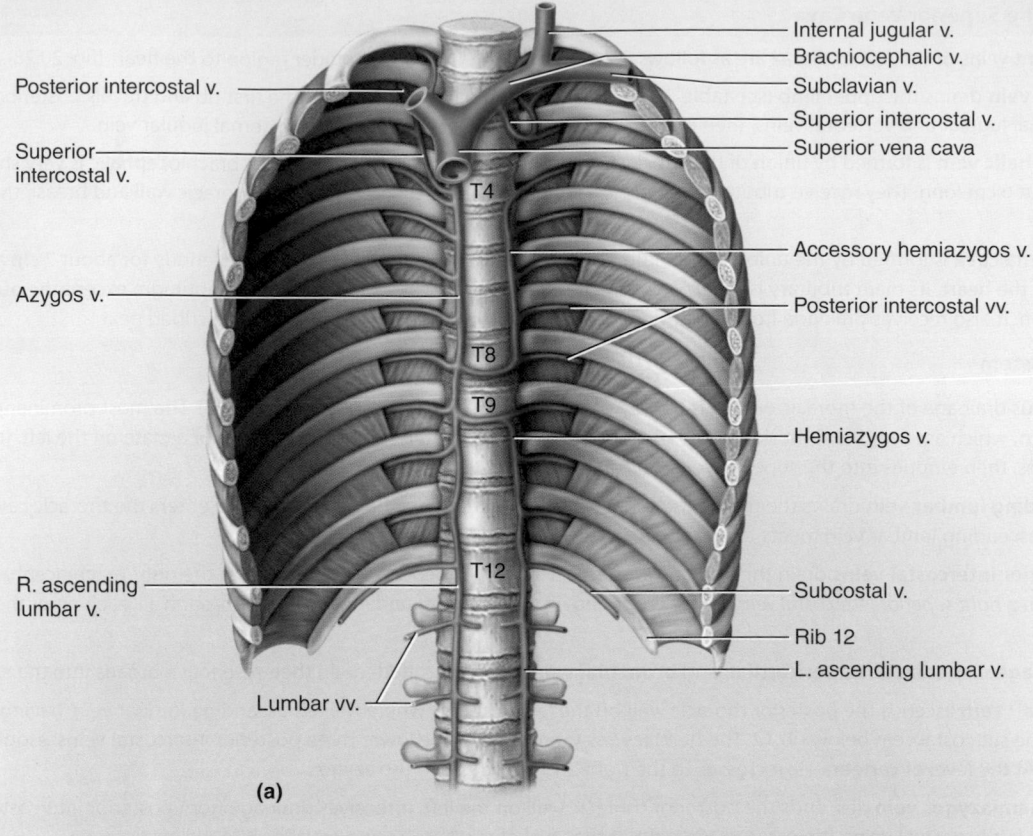

(a)

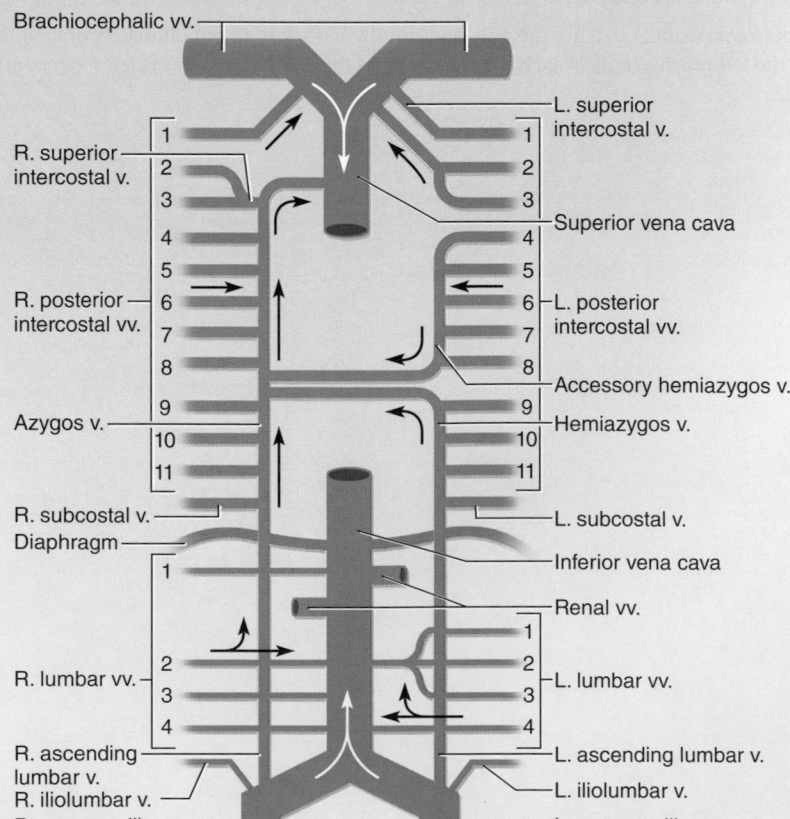

(b)

**FIGURE 20.28 Venous Drainage of the Posterior Wall of the Thorax and Abdomen.** (a) The azygos system of the thoracic wall. This system provides venous drainage from the wall and viscera of the thorax, but the visceral tributaries are not illustrated. (b) Blood-flow schematic of the thoracic and abdominal drainage. The components above the diaphragm constitute the azygos system. There is a great deal of individual variation in this anatomy. **AP|R**

After passing through the aortic hiatus, the aorta descends through the abdominal cavity and ends at the level of vertebra L4, where it branches into right and left *common iliac arteries*. The abdominal aorta is retroperitoneal.

## I. Major Branches of the Abdominal Aorta

The abdominal aorta gives off arteries in the order listed here (fig. 20.29). Those indicated in the plural are paired right and left, and those indicated in the singular are solitary median arteries.

1. The **inferior phrenic arteries** supply the inferior surface of the diaphragm. They may arise from the aorta, celiac trunk, or renal artery. Each issues two or three small **superior suprarenal arteries** to the ipsilateral adrenal (suprarenal) gland.

2. The **celiac**[20] (SEE-lee-ac) **trunk** supplies the upper abdominal viscera (see part II of this table).

3. The **superior mesenteric artery** supplies the intestines (see part III).

4. The **middle suprarenal arteries** arise laterally from the aorta, usually at the same level as the superior mesenteric artery; they supply the adrenal glands.

5. The **renal arteries** supply the kidneys and issue a small **inferior suprarenal artery** to each adrenal gland.

6. The gonadal arteries (**ovarian arteries** in the female and **testicular arteries** in the male) are long, slender arteries that arise from the midabdominal aorta and descend along the posterior body wall to the female pelvic cavity or male scrotum. The gonads begin their embryonic development near the kidneys, and the gonadal arteries are then quite short. As the gonads descend to the pelvic cavity, these arteries grow and acquire their peculiar length and course.

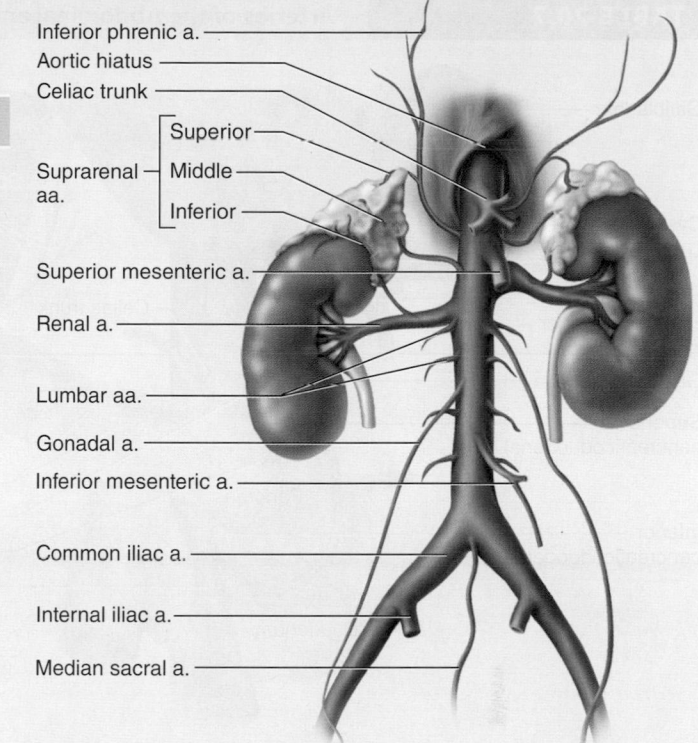

**FIGURE 20.29**   **The Abdominal Aorta and Its Major Branches.** AP|R

7. The **inferior mesenteric artery** supplies the distal end of the large intestine (see part III).

8. The **lumbar arteries** arise from the lower aorta in four pairs. They supply the posterior abdominal wall (muscles, joints, and skin) and the spinal cord and other tissues in the vertebral canal.

9. The **median sacral artery,** a tiny median artery at the inferior end of the aorta, supplies the sacrum and coccyx.

10. The **common iliac arteries** arise as the aorta forks at its inferior end. They are further traced in part IV of this table.

## II. Branches of the Celiac Trunk

The celiac circulation to the upper abdominal viscera is perhaps the most complex route off the abdominal aorta. Because it has numerous anastomoses, the bloodstream does not follow a simple linear path but divides and rejoins itself at several points (fig. 20.30). As you study the following description, locate these branches in the figure and identify the points of anastomosis.

    The short, stubby **celiac trunk,** barely more than 1 cm long, is a median branch of the aorta just below the diaphragm. It immediately gives rise to three branches—the *common hepatic, left gastric,* and *splenic arteries.*

1. The **common hepatic artery** passes to the right and issues two main branches—the gastroduodenal artery and the hepatic artery proper.

    a. The **gastroduodenal artery** gives off the **right gastro-omental (gastroepiploic**[21]**) artery** to the stomach. It then continues as the **pancreaticoduodenal** (PAN-cree-AT-ih-co-dew-ODD-eh-nul) **artery,** which splits into two branches that pass around the anterior and posterior sides of the head of the pancreas. These anastomose with the two branches of the *inferior pancreaticoduodenal artery,* discussed in part III.1.

    b. The **hepatic artery proper** ascends toward the liver. It gives off the **right gastric artery,** then branches into **right** and **left hepatic arteries.** The right hepatic artery issues a **cystic artery** to the gallbladder, then the two hepatic arteries enter the liver from below.

2. The **left gastric artery** supplies the stomach and lower esophagus, arcs around the *lesser curvature* (superomedial margin) of the stomach, and anastomoses with the right gastric artery (fig. 20.30b). Thus, the right and left gastric arteries approach from opposite directions and supply this margin of the stomach. The left gastric also has branches to the lower esophagus, and the right gastric also supplies the duodenum.

3. The **splenic artery** supplies blood to the spleen, but gives off the following branches on the way there:

    a. Several small **pancreatic arteries** supply the pancreas.

    b. The **left gastro-omental (gastroepiploic) artery** arcs around the *greater curvature* (inferolateral margin) of the stomach and anastomoses with the right gastro-omental artery. These two arteries stand off about 1 cm from the stomach itself and travel through the superior margin of the *greater omentum,* a fatty membrane suspended from the greater curvature (see figs. B.4, p. 380, and 25.3, p. 951). They furnish blood to both the stomach and omentum.

    c. The **short gastric arteries** supply the upper portion (fundus) of the stomach.

---

[20]*celi* = belly, abdomen          [21]*gastro* = stomach; *epi* = upon, above; *ploic* = pertaining to the greater omentum

**TABLE 20.7**            Arteries of the Abdominal and Pelvic Region (continued)

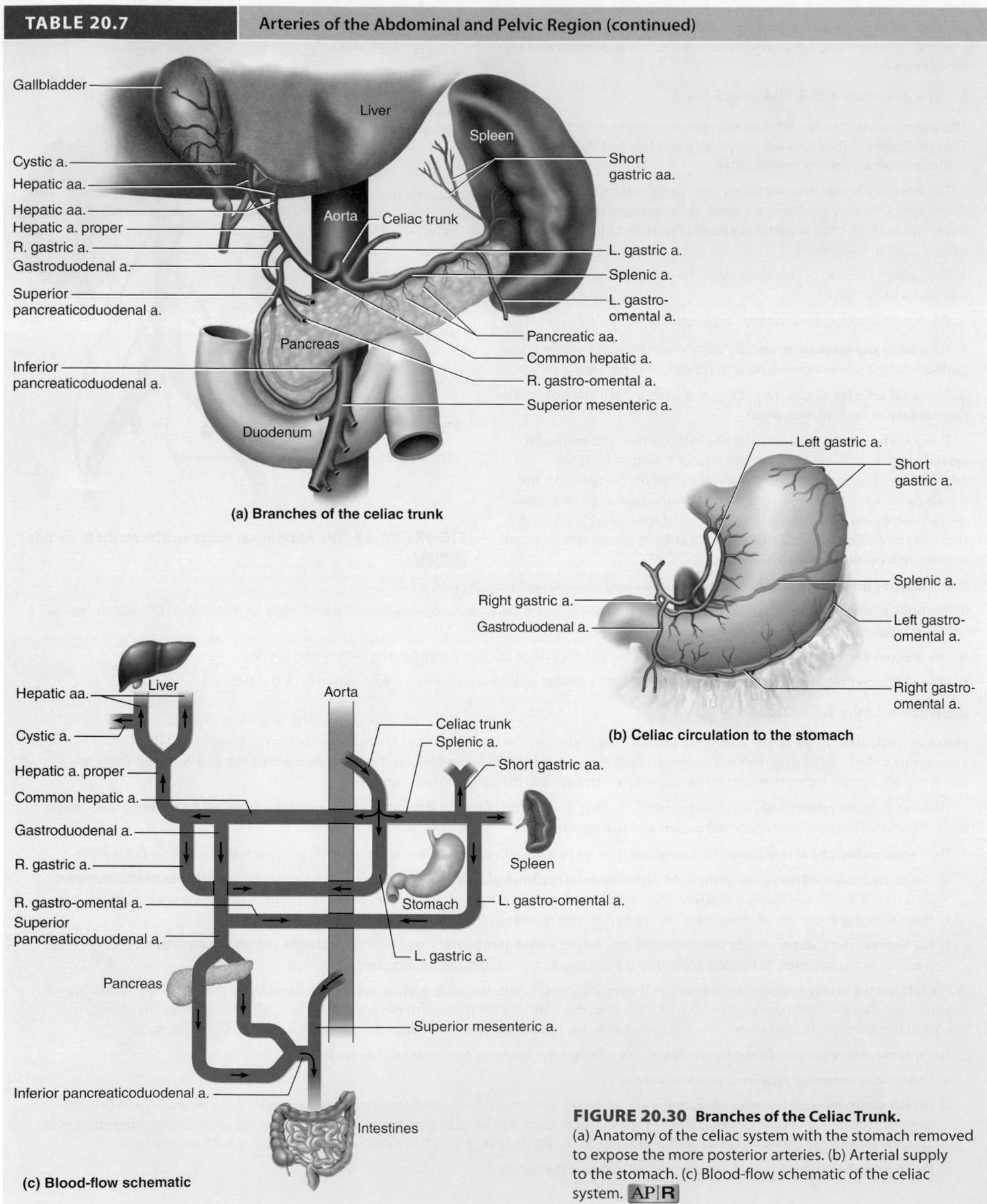

(a) Branches of the celiac trunk

- Gallbladder
- Cystic a.
- Hepatic aa.
- Hepatic aa.
- Hepatic a. proper
- R. gastric a.
- Gastroduodenal a.
- Superior pancreaticoduodenal a.
- Inferior pancreaticoduodenal a.
- Liver
- Spleen
- Aorta
- Celiac trunk
- Pancreas
- Duodenum
- Short gastric aa.
- L. gastric a.
- Splenic a.
- L. gastro-omental a.
- Pancreatic aa.
- Common hepatic a.
- R. gastro-omental a.
- Superior mesenteric a.

(b) Celiac circulation to the stomach

- Left gastric a.
- Short gastric a.
- Splenic a.
- Left gastro-omental a.
- Right gastro-omental a.
- Right gastric a.
- Gastroduodenal a.

(c) Blood-flow schematic

- Hepatic aa.
- Liver
- Aorta
- Cystic a.
- Hepatic a. proper
- Common hepatic a.
- Gastroduodenal a.
- R. gastric a.
- R. gastro-omental a.
- Superior pancreaticoduodenal a.
- Pancreas
- Inferior pancreaticoduodenal a.
- Celiac trunk
- Splenic a.
- Short gastric aa.
- Spleen
- Stomach
- L. gastro-omental a.
- L. gastric a.
- Superior mesenteric a.
- Intestines

**FIGURE 20.30  Branches of the Celiac Trunk.**
(a) Anatomy of the celiac system with the stomach removed to expose the more posterior arteries. (b) Arterial supply to the stomach. (c) Blood-flow schematic of the celiac system. **AP|R**

| **TABLE 20.7** | Arteries of the Abdominal and Pelvic Region (continued) |
|---|---|

### III. Mesenteric Circulation

The mesentery is a translucent sheet that suspends the intestines and other abdominal viscera from the posterior body wall (see figs. A.8, p. 35, and 25.3, p. 951). It contains numerous arteries, veins, and lymphatic vessels that supply and drain the intestines. The arterial supply arises from the *superior* and *inferior mesenteric arteries;* numerous anastomoses between these ensure adequate collateral circulation to the intestines even if one route is temporarily obstructed.

The **superior mesenteric artery** (fig. 20.31a) is the most significant intestinal blood supply, serving nearly all of the small intestine and the proximal half of the large intestine. It arises medially from the upper abdominal aorta and gives off the following branches:

1. The **inferior pancreaticoduodenal artery,** already mentioned, branches to pass around the anterior and posterior sides of the pancreas and anastomose with the two branches of the superior pancreaticoduodenal artery.

2. Twelve to 15 **jejunal** and **ileal arteries** form a fanlike array that supplies nearly all of the small intestine (portions called the jejunum and ileum).

3. The **ileocolic** (ILL-ee-oh-CO-lic) **artery** supplies the ileum, appendix, and parts of the large intestine (cecum and ascending colon).

4. The **right colic artery** also supplies the ascending colon.

5. The **middle colic artery** supplies most of the transverse colon.

The **inferior mesenteric artery** arises from the lower abdominal aorta and serves the distal part of the large intestine (fig. 20.31b); it gives off three main branches:

1. The **left colic artery** supplies the transverse and descending colon.

2. The **sigmoid arteries** supply the descending and sigmoid colon.

3. The **superior rectal artery** supplies the rectum.

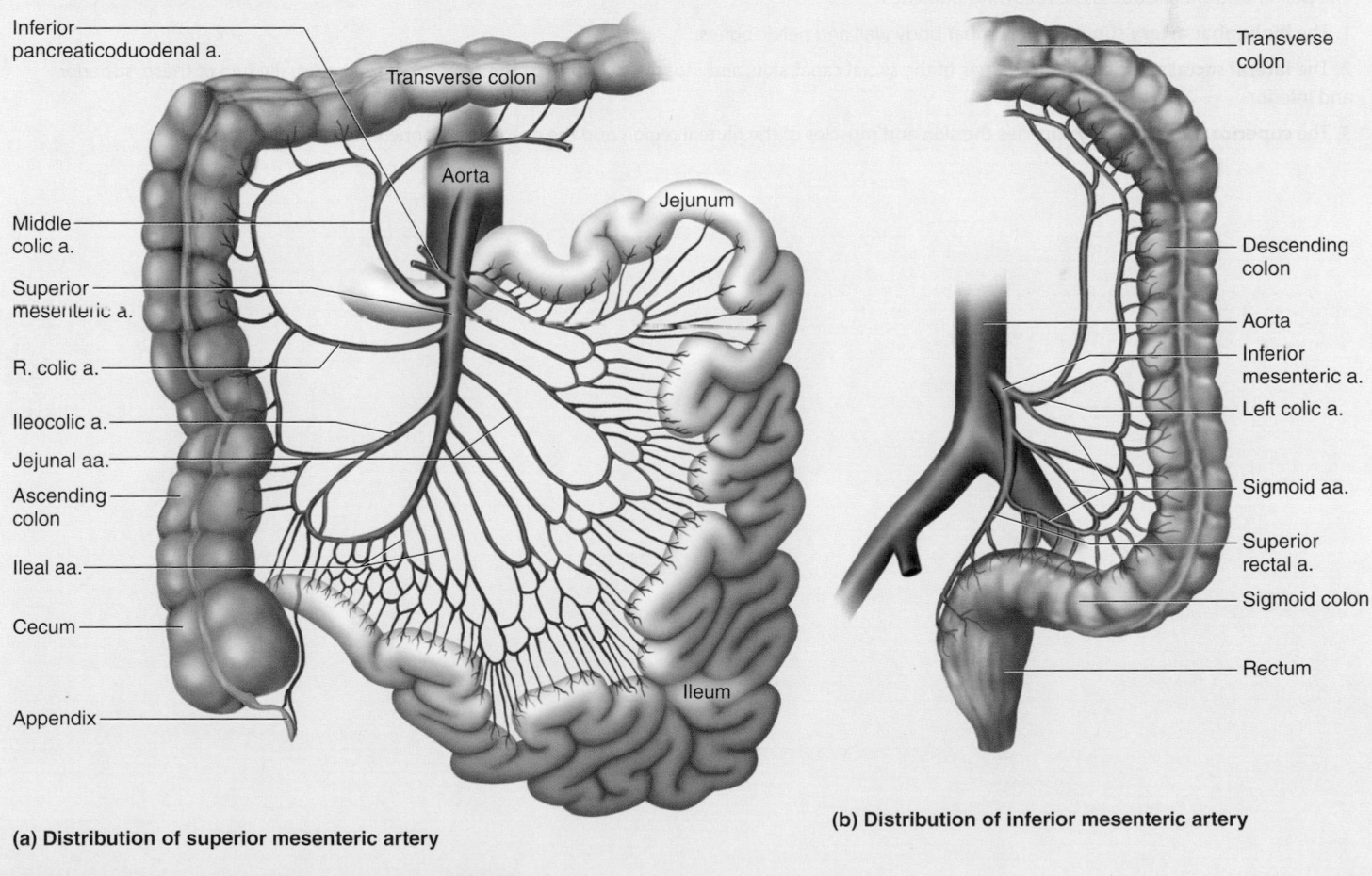

(a) Distribution of superior mesenteric artery

(b) Distribution of inferior mesenteric artery

**FIGURE 20.31  The Mesenteric Arteries.** AP|R

| TABLE 20.7 | Arteries of the Abdominal and Pelvic Region (continued) |
|---|---|

### IV. Arteries of the Pelvic Region   AP|R

The two common iliac arteries arise by branching of the aorta, descend for another 5 cm, and then, at the level of the sacroiliac joint, each divides into an external and internal iliac artery. The external iliac supplies mainly the lower limb (see table 20.11). The **internal iliac artery** supplies mainly the pelvic wall and viscera. Its branches are shown only in schematic form in figure 20.38.

Shortly after its origin, the internal iliac divides into anterior and posterior trunks. The anterior trunk produces the following branches:

1. The **superior vesical**[22] **artery** supplies the urinary bladder and distal end of the ureter. It arises indirectly from the anterior trunk by way of a short *umbilical artery,* a remnant of the artery that supplies the fetal umbilical cord. The rest of the umbilical artery becomes a closed fibrous cord after birth.

2. In men, the **inferior vesical artery** supplies the bladder, ureter, prostate gland, and seminal vesicle. In women, the corresponding vessel is the **vaginal artery,** which supplies the vagina and part of the bladder and rectum.

3. The **middle rectal artery** supplies the rectum.

4. The **obturator artery** exits the pelvic cavity through the obturator foramen and supplies the adductor muscles of the medial thigh.

5. The **internal pudendal**[23] (pyu-DEN-dul) **artery** serves the perineum and erectile tissues of the penis and clitoris; it supplies the blood for vascular engorgement during sexual arousal.

6. In women, the **uterine artery** is the main blood supply to the uterus and supplies some blood to the vagina. It enlarges substantially in pregnancy. It passes up the uterine margin, then turns laterally at the uterine tube and anastomoses with the *ovarian artery,* thus supplying blood to the ovary as well (see part I.6 of table 20.7, and fig. 28.7, p. 1065).

7. The **inferior gluteal artery** supplies the gluteal muscles and hip joint.

The posterior trunk produces the following branches:

1. The **iliolumbar artery** supplies the lumbar body wall and pelvic bones.

2. The **lateral sacral arteries** lead to tissues of the sacral canal, skin, and muscles posterior to the sacrum. There are usually two of these, superior and inferior.

3. The **superior gluteal artery** supplies the skin and muscles of the gluteal region and the muscle and bone tissues of the pelvic wall.

---

[22]*vesic* = bladder
[23]*pudend* = literally, "shameful parts"; the external genitals

| TABLE 20.8 | Veins of the Abdominal and Pelvic Region |
| --- | --- |

**I. Tributaries of the Inferior Vena Cava**

The **inferior vena cava (IVC)** is the body's largest blood vessel, having a diameter of about 3.5 cm. It forms by the union of the right and left common iliac veins at the level of vertebra L5 and drains many of the abdominal viscera as it ascends the posterior body wall. It is retroperitoneal and lies immediately to the right of the aorta. The IVC picks up blood from numerous tributaries in the following ascending order (fig. 20.32):

1. The **internal iliac veins** drain the gluteal muscles; the medial aspect of the thigh, the urinary bladder, rectum, prostate, and ductus deferens of the male; and the uterus and vagina of the female. They unite with the *external iliac veins,* which drain the lower limb and are described in table 20.12. Their union forms the **common iliac veins,** which then converge to form the IVC.

2. Four pairs of **lumbar veins** empty into the IVC, as well as into the ascending lumbar veins described in part II.

3. The **gonadal veins** (**ovarian veins** in the female and **testicular veins** in the male) drain the gonads. Like the gonadal arteries, and for the same reason (table 20.7, part I.6), these are long slender vessels that end far from their origins. The left gonadal vein empties into the left renal vein, whereas the right gonadal vein empties directly into the IVC.

4. The **renal veins** drain the kidneys into the IVC. The left renal vein also receives blood from the left gonadal and left suprarenal veins. It is up to three times as long as the right renal vein, since the IVC lies to the right of the median plane of the body.

5. The **suprarenal veins** drain the adrenal (suprarenal) glands. The right suprarenal empties directly into the IVC, and the left suprarenal empties into the left renal vein.

6. The **inferior phrenic veins** drain the inferior aspect of the diaphragm.

7. The **hepatic veins** drain the liver, extending a short distance from its superior surface to the IVC.

After receiving these inputs, the IVC penetrates the diaphragm and enters the right atrium of the heart from below. It does not receive any thoracic drainage.

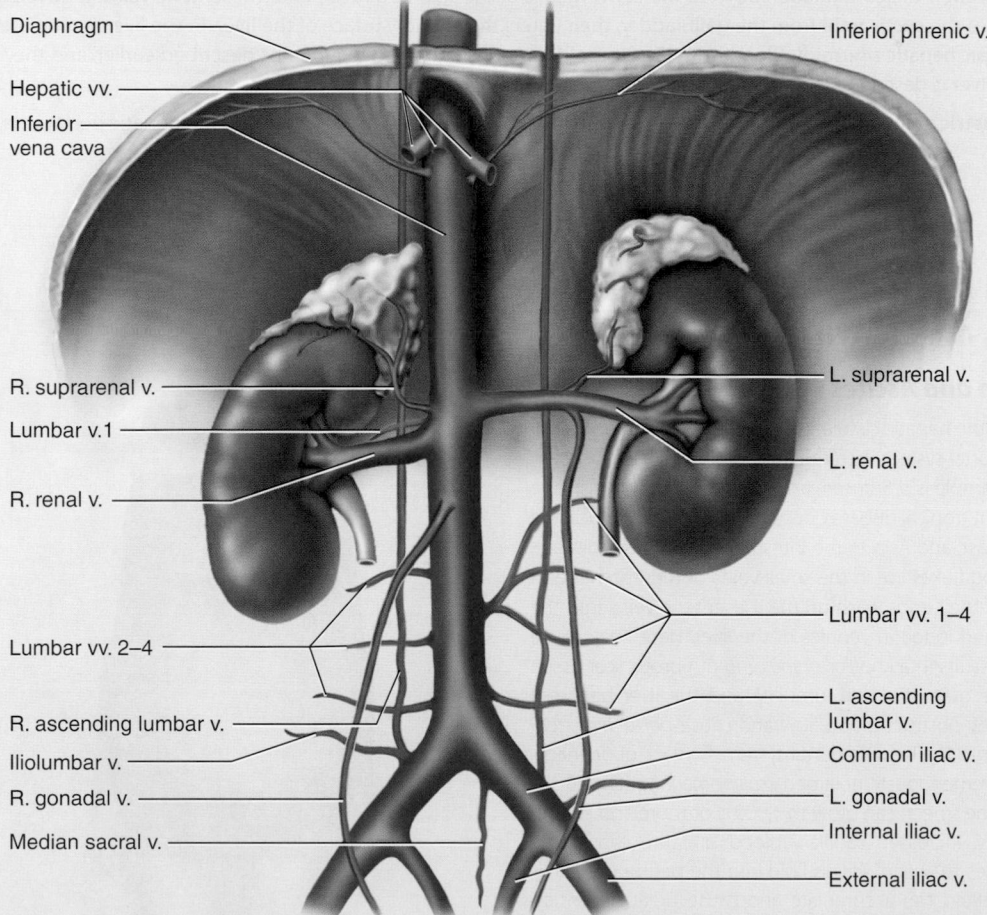

**FIGURE 20.32  The Inferior Vena Cava and Its Tributaries.**  Compare the blood-flow schematic in figure 20.28b. AP|R

❓ *Why do the veins that drain the ovaries and testes terminate so far away from the gonads?*

| **TABLE 20.8** | Veins of the Abdominal and Pelvic Region (continued) |
|---|---|

### II. Veins of the Abdominal Wall

A pair of **ascending lumbar veins** receives blood from the common iliac veins below and from the aforementioned lumbar veins of the posterior body wall (see fig. 20.28b). The ascending lumbar veins give off anastomoses with the inferior vena cava beside them as they ascend to the diaphragm. The left ascending lumbar vein passes through the diaphragm via the aortic hiatus and continues as the hemiazygos vein above. The right ascending lumbar vein passes through the diaphragm to the right of the vertebral column and continues as the azygos vein. The further paths of the azygos and hemiazygos veins are described in table 20.6.

### III. The Hepatic Portal System

The **hepatic portal system** receives all of the blood draining from the abdominal digestive tract, as well as from the pancreas, gallbladder, and spleen (fig. 20.34). It is called a portal system because it connects capillaries of the intestines and other digestive organs to modified capillaries (hepatic sinusoids) of the liver; thus, the blood passes through two capillary beds in series before it returns to the heart. Intestinal blood is richly laden with nutrients for a few hours following a meal. The hepatic portal system gives the liver first claim to these nutrients before the blood is distributed to the rest of the body. It also allows the blood to be cleansed of bacteria and toxins picked up from the intestines, an important function of the liver. Its principal veins are as follows:

1. The **inferior mesenteric vein** receives blood from the rectum and distal part of the colon. It converges in a fanlike array in the mesentery and empties into the splenic vein.

2. The **superior mesenteric vein** receives blood from the entire small intestine, ascending colon, transverse colon, and stomach. It, too, exhibits a fanlike arrangement in the mesentery and then joins the splenic vein to form the hepatic portal vein.

3. The **splenic vein** drains the spleen and travels across the abdominal cavity toward the liver. Along the way, it picks up **pancreatic veins** from the pancreas, then the inferior mesenteric vein, then ends where it meets the superior mesenteric vein.

4. The **hepatic portal vein** is the continuation beyond the convergence of the splenic and superior mesenteric veins. It travels about 8 cm upward and to the right, receives the **cystic vein** from the gallbladder, then enters the inferior surface of the liver. In the liver, it ultimately leads to the innumerable microscopic hepatic sinusoids. Blood from the sinusoids empties into the hepatic veins described earlier, and they empty into the IVC. Circulation within the liver is described in more detail in chapter 25 (p. 969).

5. The left and right **gastric veins** form an arc along the lesser curvature of the stomach and empty into the hepatic portal vein.

# DEEPER INSIGHT 20.4

## CLINICAL APPLICATION

### Portal Hypertension and Ascites

Liver diseases that obstruct the hepatic circulation can cause blood pressure to back up into the hepatic portal system with multiple effects on the upstream organs that it drains. An example is *schistosomiasis* (SHIS-to-so-MY-ah-sis), one of the world's most prevalent tropical diseases, occurring in South America, the Caribbean, Africa, the Mideast, and Asia. In the intestinal forms of the disease, parasitic worms called blood flukes live in the small veins of mesenteries and the intestinal wall. Some of their eggs wash up the mesenteric veins into the hepatic portal circulation and lodge in venules of the liver. Here, they cause severe inflammation that results in a knot or *granuloma* of fibrous scar tissue around each egg. As these granulomas accumulate and the liver becomes more and more fibrous, they obstruct blood flow and cause *portal hypertension*, a backup of pressure into the hepatic portal system. For lack of drainage, the spleen can become tremendously enlarged *(splenomegaly)* (fig. 20.33). Normally weighing 150 g, the spleen can grow to 1,000 g or more and extend even into the pelvic cavity. Increased capillary blood pressure causes the spleen, liver, and mesenteries to "weep" serous fluid into the peritoneal cavity. As much as 10 to 20 L of fluid can accumulate and cause great distension of the abdomen, a state called *ascites* (ah-SY-teez). Portal hypertension and ascites can also occur in many other obstructive liver diseases such as alcoholic cirrhosis.

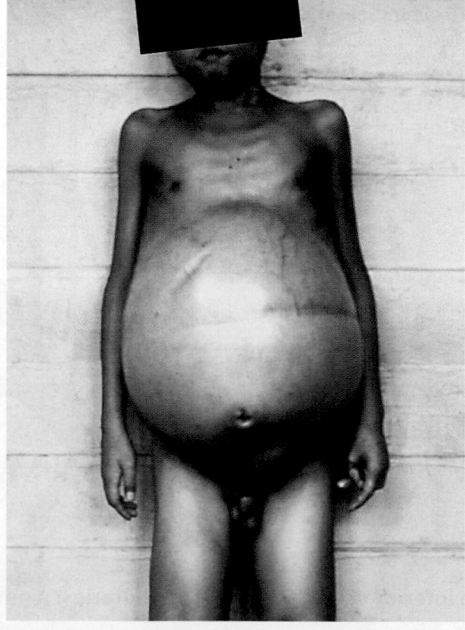

**FIGURE 20.33**
**Ascites.** The abdomen is distended with accumulated serous fluid that has filtered from the liver, spleen, and intestinal blood vessels.

| TABLE 20.8 | Veins of the Abdominal and Pelvic Region (continued) |
|---|---|

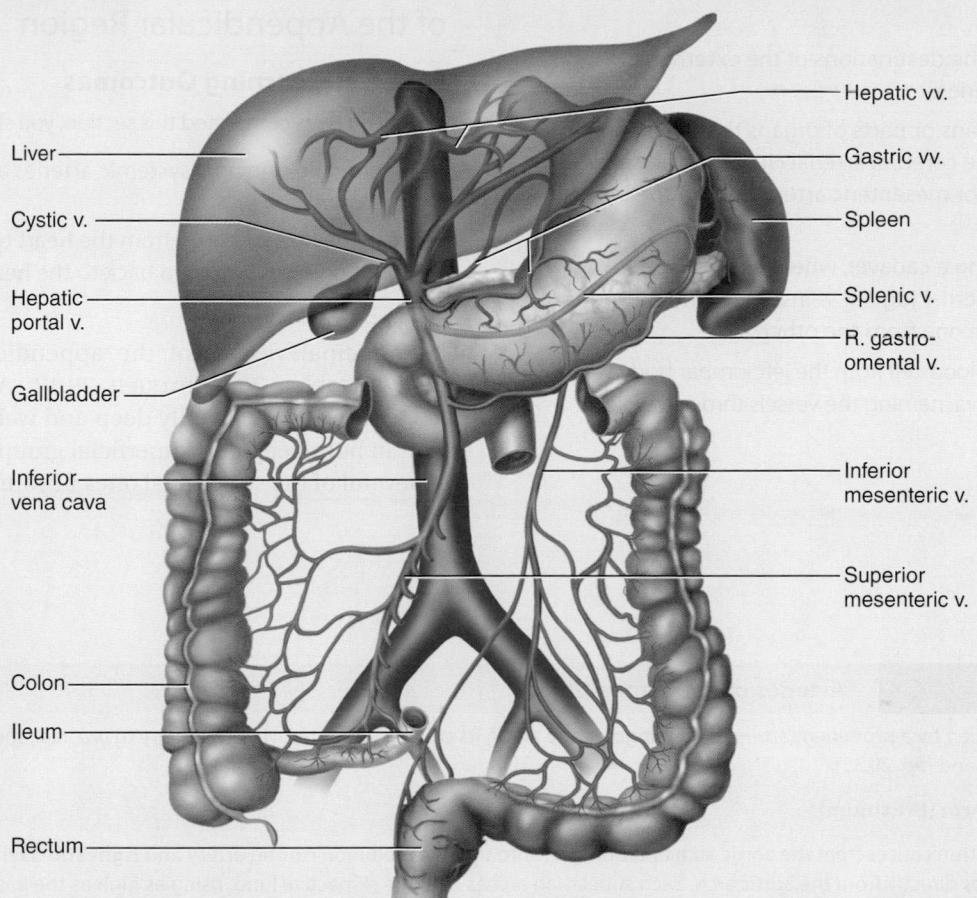

(a) Tributaries of the hepatic portal system

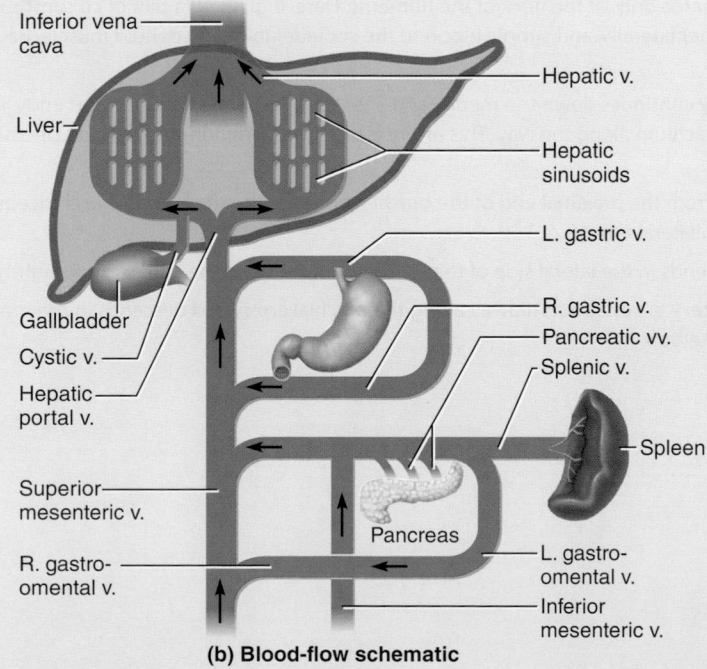

(b) Blood-flow schematic

**FIGURE 20.34  The Hepatic Portal System and Its Tributaries.** AP|R

## BEFORE YOU GO ON

Answer the following questions to test your understanding of the preceding section:

26. Concisely contrast the destinations of the external and internal carotid arteries.

27. Briefly state the organs or parts of organs that are supplied with blood by (a) the cerebral arterial circle, (b) the celiac trunk, (c) the superior mesenteric artery, and (d) the internal iliac artery.

28. If you were dissecting a cadaver, where would you look for the internal and external jugular veins? What muscle would help you distinguish one from the other?

29. Trace the path of a blood cell from the left lumbar body wall to the superior vena cava, naming the vessels through which it would travel.

## 20.8 Systemic Vessels of the Appendicular Region

### Expected Learning Outcomes

When you have completed this section, you should be able to

a. identify the principal systemic arteries and veins of the limbs; and

b. trace the flow of blood from the heart to any region of the upper or lower limb and back to the heart.

The principal vessels of the appendicular region are detailed in tables 20.9 through 20.12. Although the appendicular arteries are usually deep and well protected, the veins occur in both deep and superficial groups; you may be able to see several of the superficial ones in your forearms and hands.

| TABLE 20.9 | Arteries of the Upper Limb |
|---|---|

The upper limb is supplied by a prominent artery that changes name along its course from *subclavian* to *axillary* to *brachial*, then issues branches to the arm, forearm, and hand (fig. 20.35).

### I. The Shoulder and Arm (Brachium)

1. The brachiocephalic trunk arises from the aortic arch and branches into the right common carotid artery and **right subclavian artery;** the **left subclavian artery** arises directly from the aortic arch. Each subclavian arches over the respective lung, rising as high as the base of the neck slightly superior to the clavicle. It then passes posterior to the clavicle, downward over the first rib, and ends in name only at the rib's lateral margin. In the shoulder, it gives off several small branches to the thoracic wall and viscera, described in table 20.5.

2. As the artery continues past the first rib, it is named the **axillary artery.** It continues through the axillary region, gives off small thoracic branches (see table 20.5), and ends, again in name only, at the neck of the humerus. Here, it gives off a pair of **circumflex humeral arteries,** which encircle the humerus, anastomose with each other laterally, and supply blood to the shoulder joint and deltoid muscle. Beyond this loop, the vessel is called the brachial artery.

3. The **brachial** (BRAY-kee-ul) **artery** continues down the medial and anterior sides of the humerus and ends just distal to the elbow, supplying the anterior flexor muscles of the brachium along the way. This artery is the most common site of blood pressure measurement with the sphygmomanometer.

4. The **deep brachial artery** arises from the proximal end of the brachial and supplies the humerus and triceps brachii muscle. About midway down the arm, it continues as the radial collateral artery.

5. The **radial collateral artery** descends in the lateral side of the arm and empties into the radial artery slightly distal to the elbow.

6. The **superior ulnar collateral artery** arises about midway along the brachial artery and descends in the medial side of the arm. It empties into the ulnar artery slightly distal to the elbow.

| TABLE 20.9 | Arteries of the Upper Limb (continued) |
|---|---|

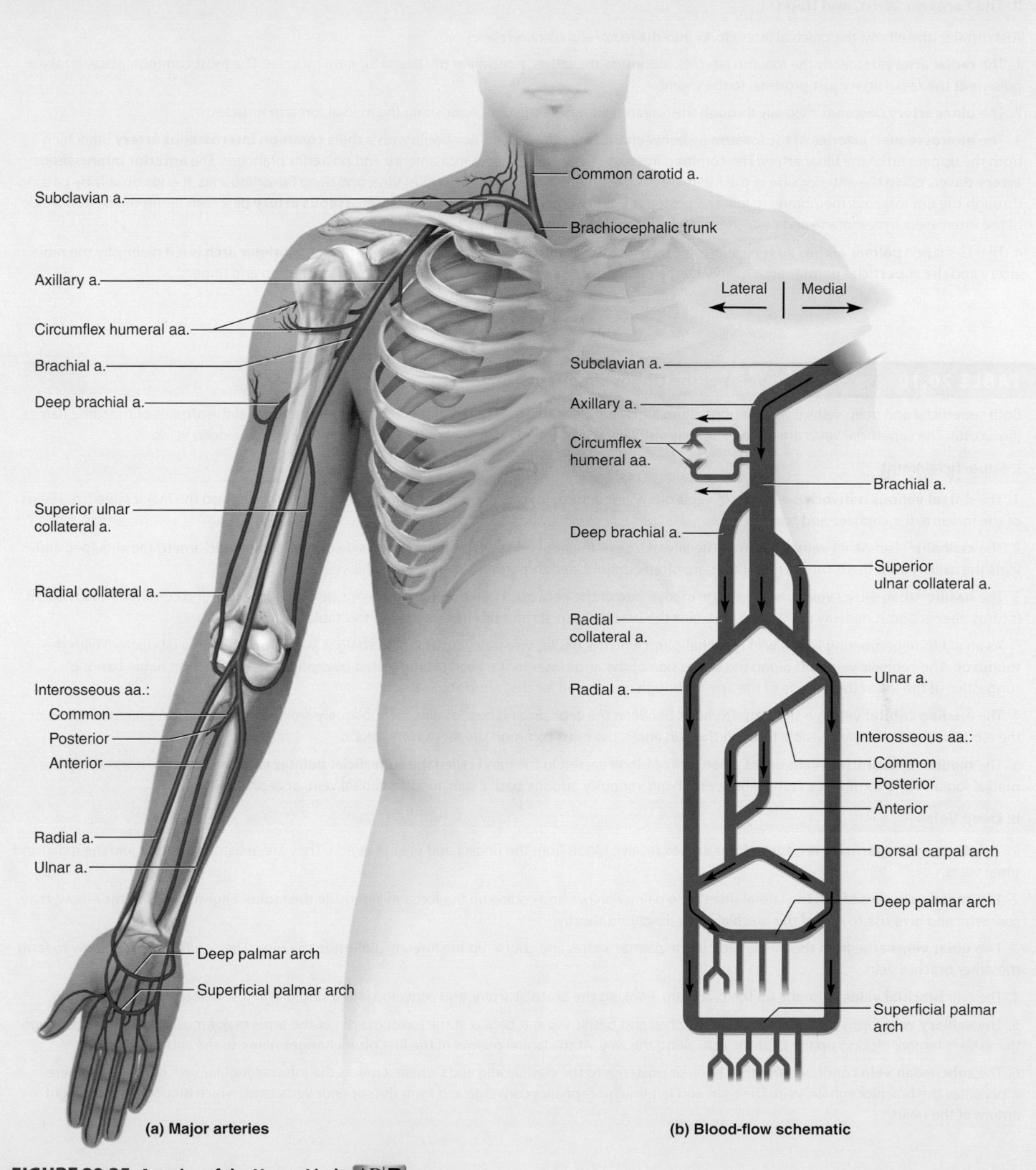

(a) Major arteries

(b) Blood-flow schematic

**FIGURE 20.35 Arteries of the Upper Limb.** AP|R

❓ *Why are arterial anastomoses especially common at joints such as the shoulder and elbow?*

| **TABLE 20.9** | Arteries of the Upper Limb (continued) |
| --- | --- |

**II. The Forearm, Wrist, and Hand**

Just distal to the elbow, the brachial artery forks into the *radial* and *ulnar arteries.*

1. The **radial artery** descends the forearm laterally, alongside the radius, nourishing the lateral forearm muscles. The most common place to take a pulse is at the radial artery just proximal to the thumb.

2. The **ulnar artery** descends medially through the forearm, alongside the ulna, nourishing the medial forearm muscles.

3. The **interosseous**[24] **arteries** of the forearm lie between the radius and ulna. They begin with a short **common interosseous artery** branching from the upper end of the ulnar artery. The common interosseous quickly divides into anterior and posterior branches. The **anterior interosseous artery** travels down the anterior side of the interosseous membrane, nourishing the radius, ulna, and deep flexor muscles. It ends distally by passing through the interosseous membrane to join the posterior interosseous artery. The **posterior interosseous artery** descends along the posterior side of the interosseous membrane and nourishes mainly the superficial extensor muscles.

4. Two U-shaped **palmar arches** arise by anastomosis of the radial and ulnar arteries at the wrist. The **deep palmar arch** is fed mainly by the radial artery and the **superficial palmar arch** mainly by the ulnar artery. The arches issue arteries to the palmar region and fingers.

| **TABLE 20.10** | Veins of the Upper Limb |
| --- | --- |

Both superficial and deep veins drain the upper limb, ultimately leading to axillary and subclavian veins that parallel the arteries of the same names (fig. 20.36). The superficial veins are often externally visible and are larger in diameter and carry more blood than the deep veins.

**I. Superficial Veins**

1. The **dorsal venous network** is a plexus of veins often visible through the skin on the back of the hand; it empties into the major superficial veins of the forearm, the cephalic and basilic.

2. The **cephalic**[25] (sef-AL-ic) **vein** arises from the lateral side of the network, travels up the lateral side of the forearm and arm to the shoulder, and joins the axillary vein there. Intravenous fluids are often administered through the distal end of this vein.

3. The **basilic**[26] (bah-SIL-ic) **vein** arises from the medial side of the network, travels up the posterior side of the forearm, and continues into the arm. It turns deeper about midway up the arm and joins the brachial vein at the axilla (see part II.4 of this table).

   As an aid to remembering which vein is cephalic and which is basilic, visualize your arm held straight away from the torso (abducted) with the thumb up. The cephalic vein runs along the upper side of the arm closer to the head (as suggested by *cephal,* "head"), and the name *basilic* is suggestive of the lower (basal) side of the arm (although not named for that reason).

4. The **median cubital vein** is a short anastomosis between the cephalic and basilic veins that obliquely crosses the cubital fossa (anterior bend of the elbow). It is often clearly visible through the skin and is the most common site for drawing blood.

5. The **median antebrachial vein** drains a network of blood vessels in the hand called the **superficial palmar venous network.** It travels up the medial forearm and terminates at the elbow, emptying variously into the basilic vein, median cubital vein, or cephalic vein.

**II. Deep Veins**

1. The **deep** and **superficial venous palmar arches** receive blood from the fingers and palmar region. They are anastomoses that join the radial and ulnar veins.

2. Two **radial veins** arise from the lateral side of the palmar arches and course up the forearm alongside the radius. Slightly distal to the elbow, they converge and give rise to one of the brachial veins described shortly.

3. Two **ulnar veins** arise from the medial side of the palmar arches and course up the forearm alongside the ulna. They unite near the elbow to form the other brachial vein.

4. The two **brachial veins** continue up the brachium, flanking the brachial artery, and converge into a single vein just before the axillary region.

5. The **axillary vein** forms by the union of the brachial and basilic veins. It begins at the lower margin of the teres major muscle and passes through the axillary region, picking up the cephalic vein along the way. At the lateral margin of the first rib, it changes name to the subclavian vein.

6. The **subclavian vein** continues into the shoulder posterior to the clavicle and ends where it meets the internal jugular vein of the neck. There it becomes the brachiocephalic vein. The right and left brachiocephalics converge and form the superior vena cava, which empties into the right atrium of the heart.

---

[24]*inter* = between; *osse* = bones
[25]*cephalic* = related to the head
[26]*basilic* = royal, prominent, important

| TABLE 20.10 | Veins of the Upper Limb (continued) |
|---|---|

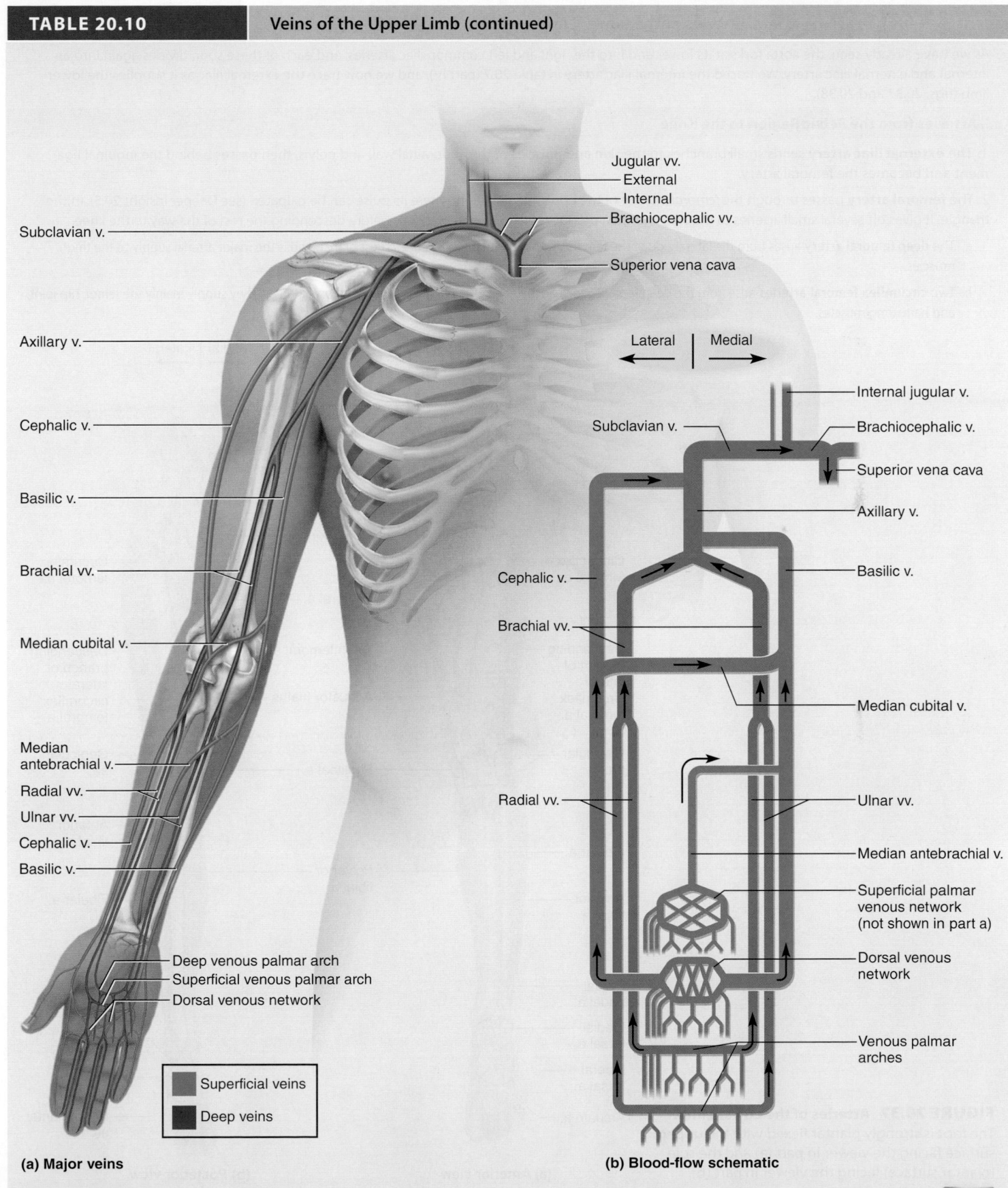

**(a) Major veins**

**(b) Blood-flow schematic**

**FIGURE 20.36 Veins of the Upper Limb.** Variations on this pattern are highly common. Many venous anastomoses are omitted for clarity. **AP|R**

❓ *Name three veins that are often visible through the skin of the upper limb.*

| TABLE 20.11 | Arteries of the Lower Limb |
| --- | --- |

As we have already seen, the aorta forks at its lower end into the right and left common iliac arteries, and each of these soon divides again into an internal and external iliac artery. We traced the internal iliac artery in table 20.7 (part IV), and we now trace the external iliac as it supplies the lower limb (figs. 20.37 and 20.38).

### I. Arteries from the Pelvic Region to the Knee

1. The **external iliac artery** sends small branches to the skin and muscles of the abdominal wall and pelvis, then passes behind the inguinal ligament and becomes the femoral artery.

2. The **femoral artery** passes through the *femoral triangle* of the upper medial thigh, where its pulse can be palpated (see Deeper Insight 20.5). In the triangle, it gives off several small arteries to the skin and then produces the following branches before descending the rest of the way to the knee.

 a. The **deep femoral artery** arises from the lateral side of the femoral, within the triangle. It is the largest branch and is the major arterial supply to the thigh muscles.

 b. Two **circumflex femoral arteries** arise from the deep femoral, encircle the head of the femur, and anastomose laterally. They supply mainly the femur, hip joint, and hamstring muscles.

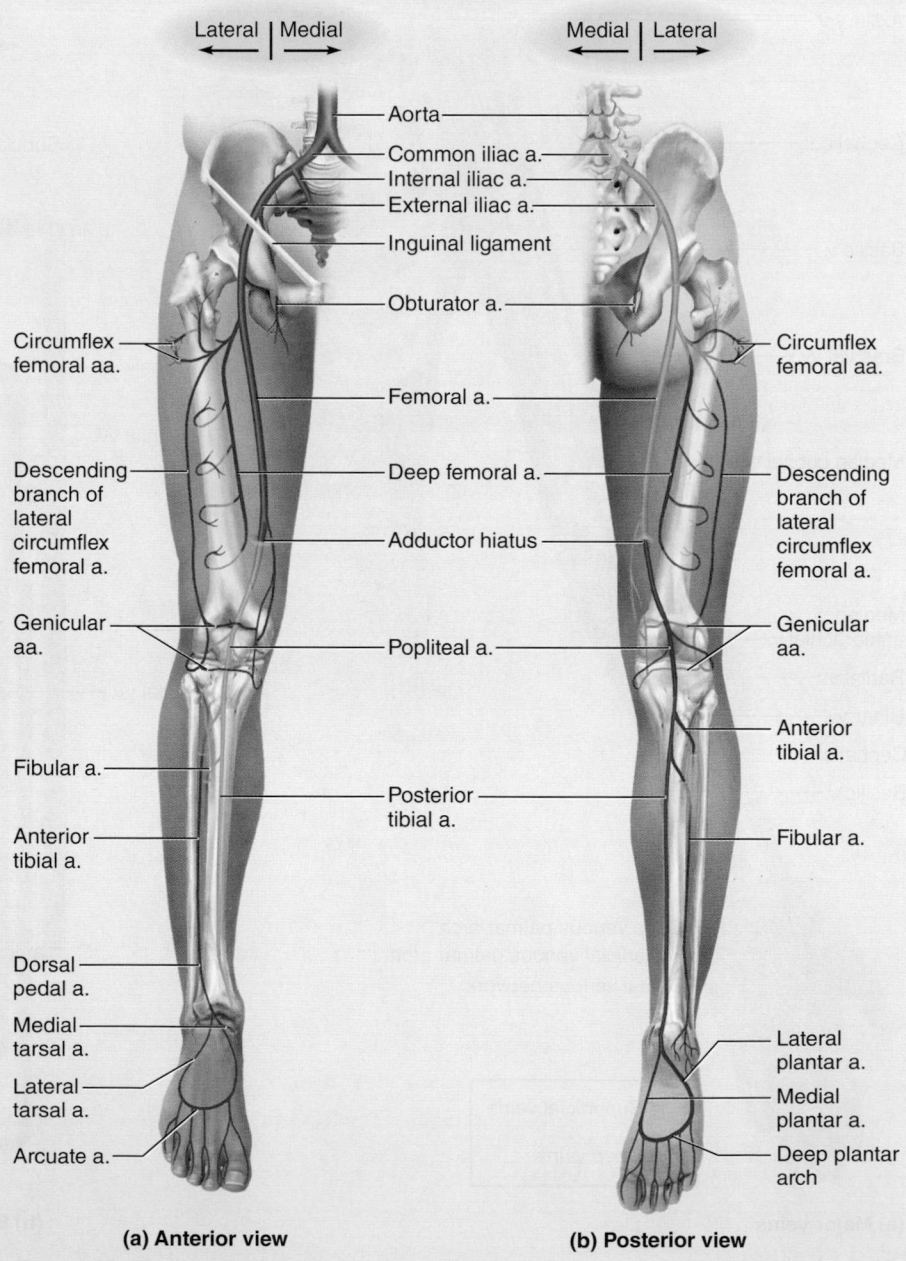

**FIGURE 20.37 Arteries of the Lower Limb.** The foot is strongly plantar flexed with the upper surface facing the viewer in part (a) and the sole (plantar surface) facing the viewer in part (b).

| TABLE 20.11 | Arteries of the Lower Limb (continued) |
|---|---|

3. The **popliteal artery** is a continuation of the femoral artery in the popliteal fossa at the rear of the knee. It begins where the femoral artery emerges from an opening *(adductor hiatus)* in the tendon of the adductor magnus muscle and ends where it splits into the *anterior* and *posterior tibial arteries*. As it passes through the popliteal fossa, it gives off anastomoses called **genicular**[27] **arteries** that supply the knee joint.

**II. Arteries of the Leg and Foot**

In the leg proper, the three most significant arteries are the anterior tibial, posterior tibial, and fibular arteries.

1. The **anterior tibial artery** arises from the popliteal artery and immediately penetrates through the interosseous membrane of the leg to the anterior compartment. There, it travels lateral to the tibia and supplies the extensor muscles. Upon reaching the ankle, it gives rise to the following dorsal arteries of the foot.

    a. The **dorsal pedal artery** traverses the ankle and upper medial surface of the foot and gives rise to the arcuate artery.

    b. The **arcuate artery** sweeps across the foot from medial to lateral and gives rise to vessels that supply the toes.

2. The **posterior tibial artery** is a continuation of the popliteal artery that passes down the leg, deep in the posterior compartment, supplying flexor muscles along the way. Inferiorly, it passes behind the medial malleolus of the ankle and into the plantar region of the foot. It gives rise to the following:

    a. The **medial** and **lateral plantar arteries** originate by branching of the posterior tibial artery at the ankle. The medial plantar artery supplies mainly the great toe. The lateral plantar artery sweeps across the sole of the foot and becomes the deep plantar arch.

    b. The **deep plantar arch** gives off another set of arteries to the toes.

3. The **fibular (peroneal) artery** arises from the proximal end of the posterior tibial artery near the knee. It descends through the lateral side of the posterior compartment, supplying lateral muscles of the leg along the way, and ends in a network of arteries in the heel.

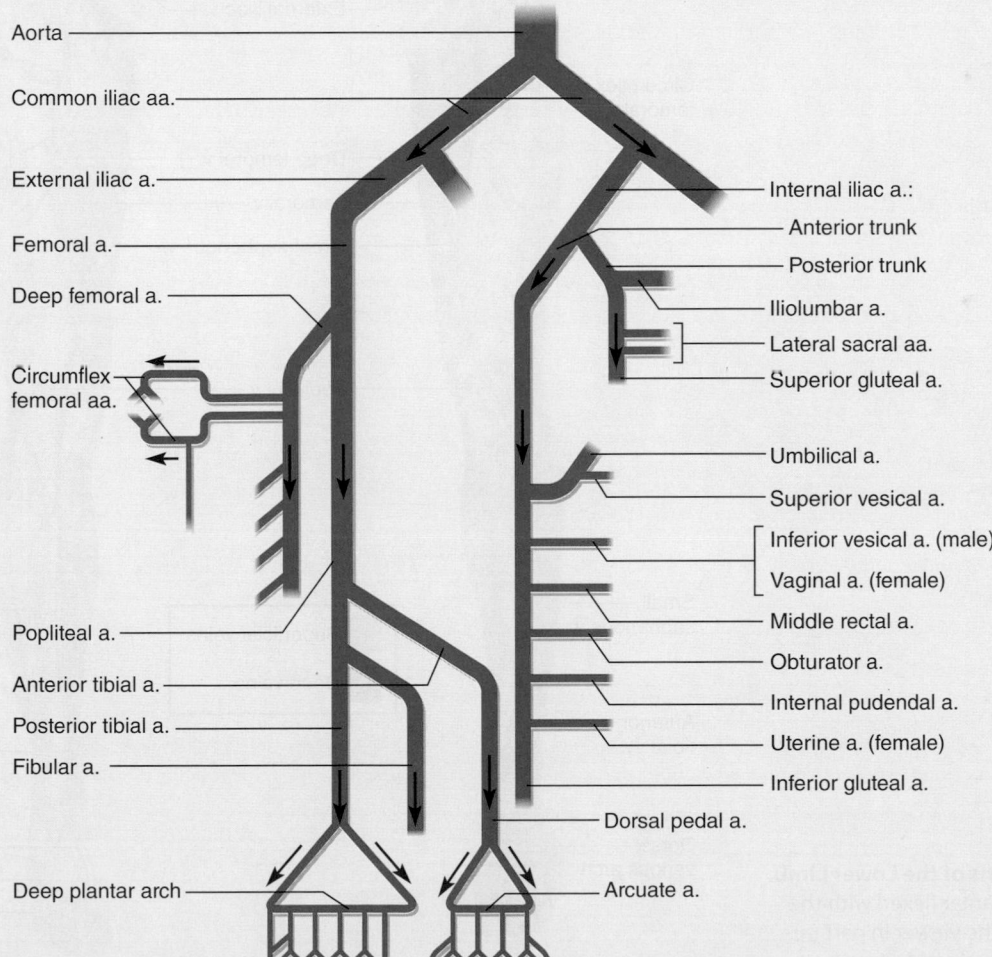

**FIGURE 20.38 Arterial Schematic of the Pelvic Region and Lower Limb (Anterior View).** The pelvic schematic on the right is stretched for clarity. These arteries are not located as far inferiorly as the arteries depicted adjacent to them on the left.

[27] *genic* = of the knee

| **TABLE 20.12** | **Veins of the Lower Limb** |
|---|---|

We will follow drainage of the lower limb from the toes to the inferior vena cava (figs. 20.39 and 20.40). As in the upper limb, there are deep and superficial veins with anastomoses between them. Most of the anastomoses are omitted from the illustrations.

### I. Superficial Veins

1. The **dorsal venous arch** (fig. 20.39a) is often visible through the skin on the dorsum of the foot. It collects blood from the toes and more proximal part of the foot, and has numerous anastomoses similar to the dorsal venous network of the hand. It gives rise to the following two veins.

2. The **small (short) saphenous**[28] (sah-FEE-nus) **vein** arises from the lateral side of the arch and passes up that side of the leg as far as the knee. There, it drains into the popliteal vein.

3. The **great (long) saphenous vein,** the longest vein in the body, arises from the medial side of the arch and travels all the way up the leg and thigh to the inguinal region. It empties into the femoral vein slightly inferior to the inguinal ligament. It is commonly used as a site for the long-term administration of intravenous fluids; it is a relatively accessible vein in infants and in patients in shock whose veins have collapsed. Portions of this vein are commonly used as grafts in coronary bypass surgery. The great and small saphenous veins are among the most common sites of varicose veins.

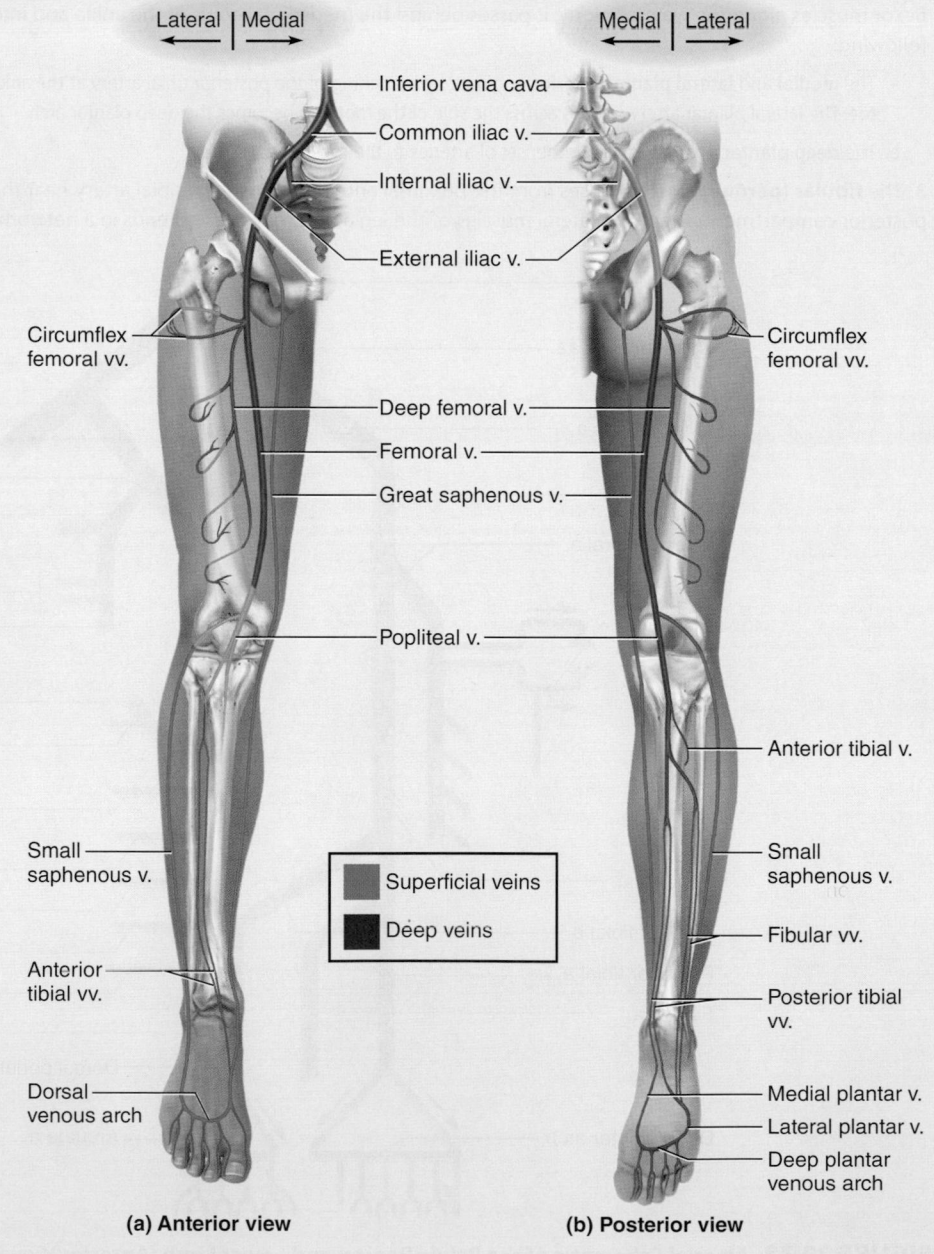

**FIGURE 20.39  Veins of the Lower Limb.** The foot is strongly plantar flexed with the upper surface facing the viewer in part (a) and the sole (plantar surface) facing the viewer in part (b). AP|R

(a) Anterior view    (b) Posterior view

[28]*saphen* = standing

| TABLE 20.12 | Veins of the Lower Limb (continued) |
|---|---|

**II. Deep Veins**

1. The **deep plantar venous arch** (fig. 20.39b) receives blood from the toes and gives rise to **lateral** and **medial plantar veins** on the respective sides. The lateral plantar vein gives off the *fibular veins,* then crosses over to the medial side and approaches the medial plantar vein. The two plantar veins pass behind the medial malleolus of the ankle and continue as a pair of *posterior tibial veins.*

2. The two **posterior tibial veins** pass up the leg embedded deep in the calf muscles. They converge like an inverted Y into a single vein about two-thirds of the way up the tibia.

3. The two **fibular (peroneal) veins** ascend the back of the leg and similarly converge like a Y.

4. The **popliteal vein** begins near the knee by convergence of these two inverted Ys. It passes through the popliteal fossa at the back of the knee.

5. The two **anterior tibial veins** travel up the anterior compartment of the leg between the tibia and fibula (fig. 20.39a). They arise from the medial side of the dorsal venous arch, converge just distal to the knee, and then flow into the popliteal vein.

6. The **femoral vein** is a continuation of the popliteal vein into the thigh. It drains blood from the deep thigh muscles and femur.

7. The **deep femoral vein** drains the femur and muscles of the thigh supplied by the deep femoral artery. It receives four principal tributaries along the shaft of the femur and then a pair of circumflex femoral veins that encircle the upper femur. It finally drains into the upper femoral vein.

8. The **external iliac vein** is formed by the union of the femoral and great saphenous veins near the inguinal ligament.

9. The **internal iliac vein** follows the course of the internal iliac artery and its distribution. Its tributaries drain the gluteal muscles; the medial aspect of the thigh; the urinary bladder, rectum, prostate, and ductus deferens in the male; and the uterus and vagina in the female.

10. The **common iliac vein** is formed by the union of the external and internal iliac veins. The right and left common iliacs then unite to form the inferior vena cava.

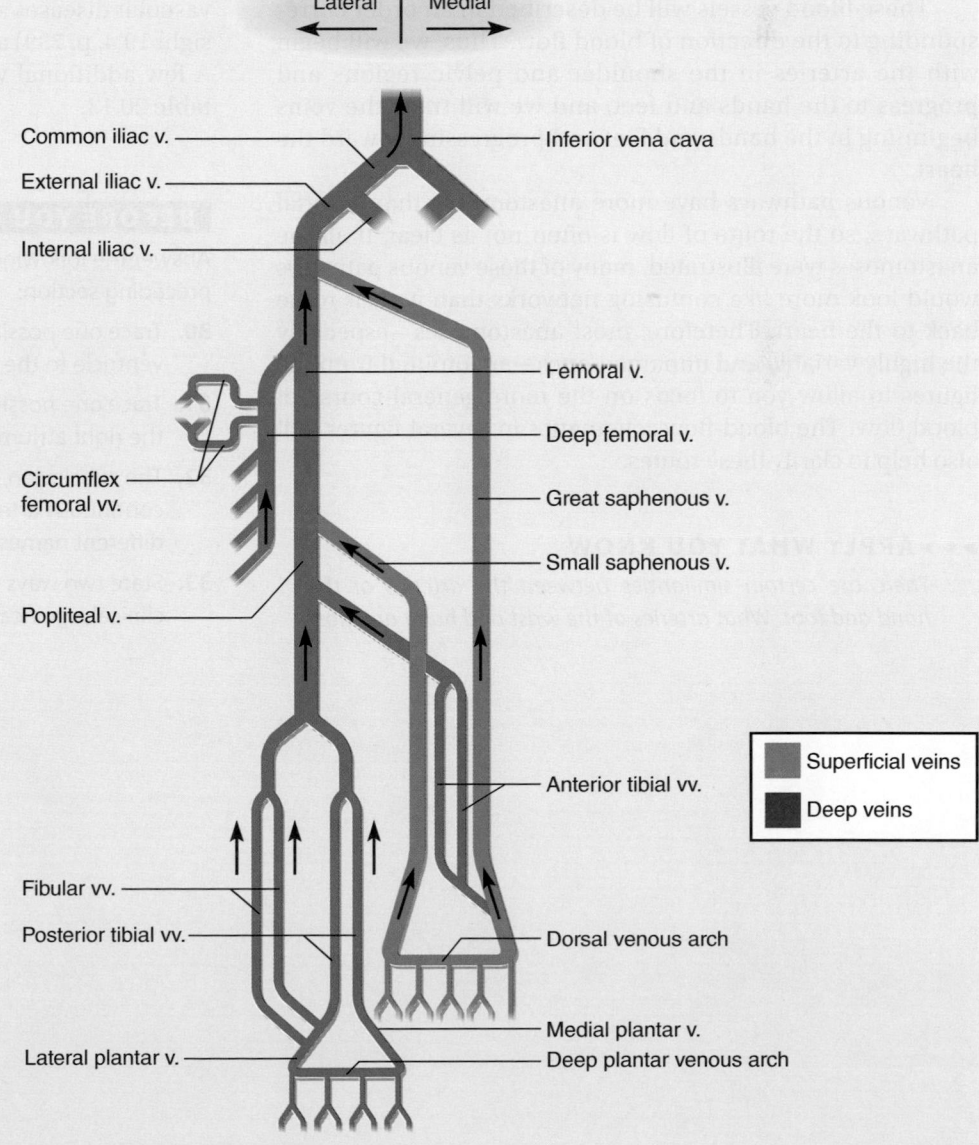

**FIGURE 20.40 Venous Schematic of the Lower Limb (Anterior View).**

| TABLE 20.13 | Some Disorders of the Arteries and Veins |
|---|---|
| Dissecting aneurysm | Splitting of the layers of an arterial wall from each other because of the accumulation of blood between layers. Results from either a tear in the tunica interna or rupture of the vasa vasorum. |
| Fat embolism | The presence of fat globules traveling in the bloodstream. Globules originate from bone fractures, fatty degeneration of the liver, and other causes and may block cerebral or pulmonary blood vessels. |
| Orthostatic hypotension | A decrease in blood pressure that occurs when one stands, often resulting in blurring of vision, dizziness, and syncope (fainting). Results from sluggish or inactive baroreflexes. |

*Disorders described elsewhere*

| | | |
|---|---|---|
| Aneurysm p. 747 | Edema p. 761 | Stroke p. 765 |
| Arteriosclerosis p. 753 | Embolism pp. 703, 773 | Thrombosis p. 703 |
| Atherosclerosis pp. 739, 753 | Hypertension pp. 753, 796 | Transient ischemic attack p. 765 |
| Circulatory shock p. 764 | Hypotension p. 754 | Varicose veins p. 751 |

Deep veins run parallel to the arteries and often have similar names (*femoral artery* and *femoral vein,* for example). In several cases, the deep veins occur in pairs flanking the corresponding artery (such as the two *radial veins* traveling alongside the *radial artery*).

These blood vessels will be described in an order corresponding to the direction of blood flow. Thus, we will begin with the arteries in the shoulder and pelvic regions and progress to the hands and feet, and we will trace the veins beginning in the hands and feet and progressing toward the heart.

Venous pathways have more anastomoses than arterial pathways, so the route of flow is often not as clear. If all the anastomoses were illustrated, many of these venous pathways would look more like confusing networks than a clear route back to the heart. Therefore, most anastomoses—especially the highly variable and unnamed ones—are omitted from the figures to allow you to focus on the more general course of blood flow. The blood-flow schematics in several figures will also help to clarify these routes.

▶▶▶ **APPLY WHAT YOU KNOW**

*There are certain similarities between the arteries of the hand and foot. What arteries of the wrist and hand are most comparable in arrangement and function to the arcuate artery and deep plantar arch of the foot?*

Deeper Insight 20.5 exemplifies the relevance of vascular anatomy to emergency first aid. The most common cardiovascular diseases are atherosclerosis (discussed in Deeper Insight 19.4, p. 739) and hypertension (see Deeper Insight 20.6). A few additional vascular disorders are briefly described in table 20.13.

**BEFORE YOU GO ON**

Answer the following questions to test your understanding of the preceding section:

30. Trace one possible path of a red blood cell from the left ventricle to the toes.

31. Trace one possible path of a red blood cell from the fingers to the right atrium.

32. The subclavian, axillary, and brachial arteries are really one continuous artery. What is the reason for giving it three different names along its course?

33. State two ways in which the great saphenous vein has special clinical significance. Where is this vein located?

**DEEPER INSIGHT 20.5**

## CLINICAL APPLICATION

### Arterial Pressure Points

In some places, major arteries come close enough to the body surface to be palpated. These places can be used to take a pulse, and they can serve as emergency *pressure points,* where firm pressure can be applied to temporarily reduce arterial bleeding (fig. 20.41). One of these points is the *femoral triangle* of the upper medial thigh. This is an important landmark for arterial supply, venous drainage, and innervation of the lower limb. Its boundaries are the sartorius muscle laterally, the inguinal ligament superiorly, and the adductor longus muscle medially. The femoral artery, vein, and nerve run close to the surface at this point.

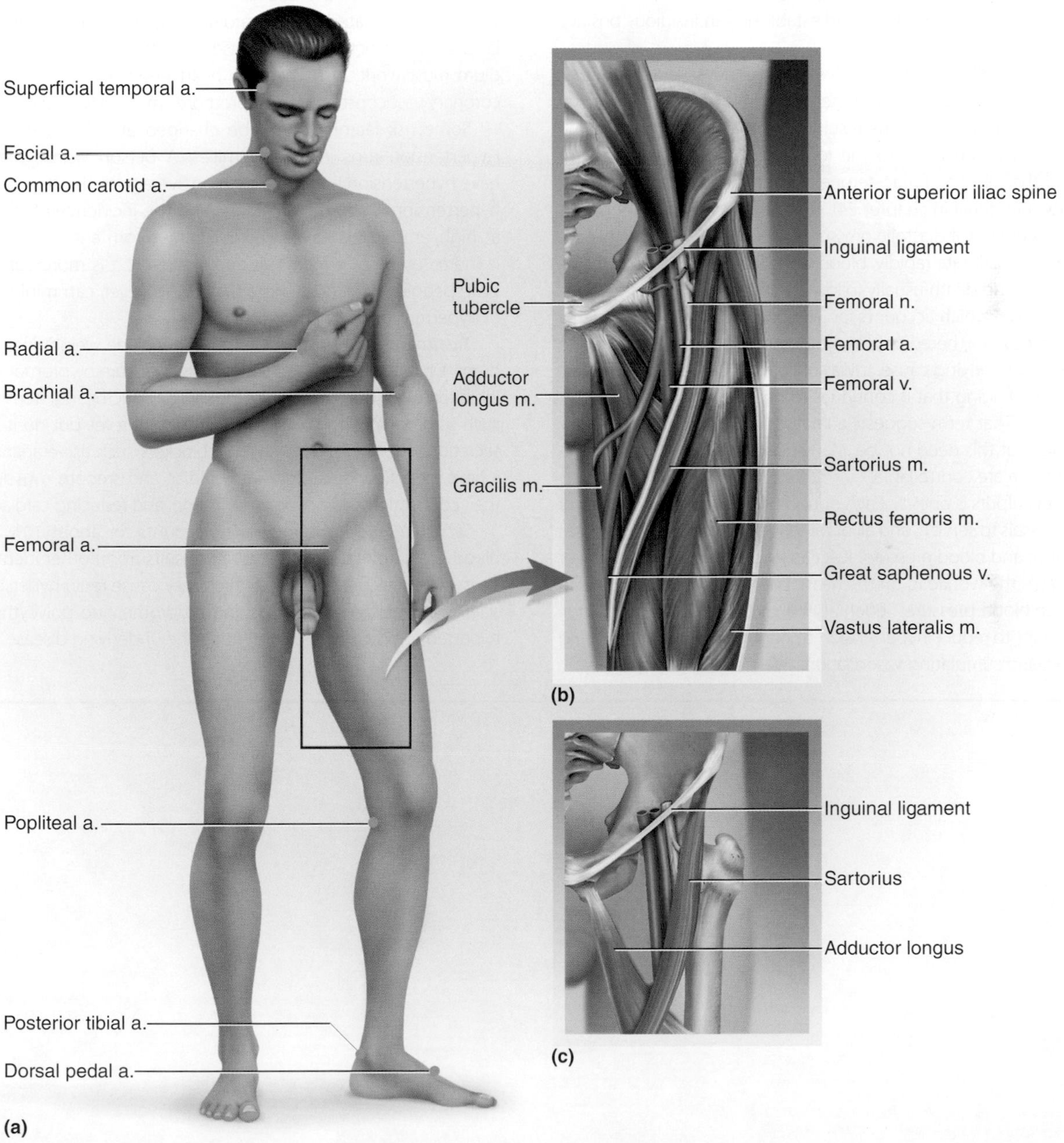

Superficial temporal a.

Facial a.

Common carotid a.

Radial a.

Brachial a.

Femoral a.

Popliteal a.

Posterior tibial a.

Dorsal pedal a.

(a)

Pubic tubercle

Adductor longus m.

Gracilis m.

Anterior superior iliac spine

Inguinal ligament

Femoral n.

Femoral a.

Femoral v.

Sartorius m.

Rectus femoris m.

Great saphenous v.

Vastus lateralis m.

(b)

Inguinal ligament

Sartorius

Adductor longus

(c)

**FIGURE 20.41 Arterial Pressure Points.** (a) Areas where arteries lie close enough to the surface that a pulse can be palpated or pressure can be applied to reduce arterial bleeding. (b) Structures in the femoral triangle. (c) The three boundaries that define the femoral triangle.

# DEEPER INSIGHT 20.6

## CLINICAL APPLICATION

### Hypertension—The "Silent Killer"

*Hypertension,* the most common cardiovascular disease, affects about 30% of Americans over age 50, and 50% by age 74. It is a "silent killer" that can wreak its destructive effects for 10 to 20 years before they are first noticed. Hypertension is the major cause of heart failure, stroke, and kidney failure. It damages the heart because it increases the afterload, which makes the ventricles work harder to expel blood. The myocardium enlarges up to a point (the *hypertrophic response*), but eventually it becomes excessively stretched and less efficient. Hypertension strains the blood vessels and tears the endothelium, thereby creating lesions that become focal points of atherosclerosis. Atherosclerosis then worsens the hypertension and establishes an insidious positive feedback cycle.

Another positive feedback cycle involves the kidneys. Their arterioles thicken in response to the stress, their lumens become narrower, and renal blood flow declines. In response to the resulting drop in blood pressure, the kidneys release renin, which leads to the formation of the vasoconstrictor angiotensin II and the release of aldosterone, a hormone that promotes salt retention (described in detail in chapter 24). These effects worsen the hypertension that already existed. If diastolic pressure exceeds 120 mm Hg, the kidneys and heart may deteriorate rapidly, blood vessels of the eye hemorrhage, blindness may ensue, and death usually follows within 2 years.

*Primary hypertension,* which accounts for 90% of cases, results from such a complex web of behavioral, hereditary, and other factors that it is difficult to sort out any specific underlying cause. It was once considered such a normal part of the "essence" of aging that it continues to be called by another name, *essential hypertension.* That term suggests a fatalistic resignation to hypertension as a fact of life, but this need not be. Many risk factors have been identified, and most of them are controllable.

One of the chief culprits is obesity. Each pound of extra fat requires miles of additional blood vessels to serve it, and all of this added vessel length increases peripheral resistance and blood pressure. Just carrying around extra weight, of course, also increases the workload on the heart. Even a small weight loss can significantly reduce blood pressure. Sedentary behavior is another risk factor. Aerobic exercise helps to reduce hypertension by controlling weight, reducing emotional tension, and stimulating vasodilation.

Dietary factors are also significant contributors to hypertension. Diets high in cholesterol and saturated fat contribute to atherosclerosis. Potassium and magnesium reduce blood pressure; thus, diets deficient in these minerals promote hypertension. The relationship of salt intake to hypertension has been a controversial subject. The kidneys compensate so effectively for excess salt intake that dietary salt has little effect on the blood pressure of most people. Reduced salt intake may, however, help to control hypertension in older people and in people with reduced renal function.

Nicotine makes a particularly devastating contribution to hypertension because it stimulates the myocardium to beat faster and harder, while it stimulates vasoconstriction and increases the afterload against which the myocardium must work. Just when the heart needs extra oxygen, nicotine causes coronary vasoconstriction and promotes myocardial ischemia.

Some risk factors cannot be changed at will—race, heredity, and sex. Hypertension runs in some families. A person whose parents or siblings have hypertension is more likely than average to develop it. The incidence of hypertension is about 30% higher, and the incidence of strokes about twice as high, among blacks as among whites. From ages 18 to 54, hypertension is more common in men, but above age 65, it is more common in women. Even people at risk from these factors, however, can minimize their chances of hypertension by changing risky behaviors.

Treatments for primary hypertension include weight loss, diet, and certain drugs. Diuretics lower blood volume and pressure by promoting urination. ACE inhibitors block the formation of the vasoconstrictor angiotensin II. Beta-blockers such as propranolol also lower angiotensin II level, but do it by inhibiting the secretion of renin. Calcium channel blockers such as verapamil and nifedipine inhibit the inflow of calcium into cardiac and smooth muscle, thus inhibiting their contraction, promoting vasodilation, and reducing cardiac workload.

*Secondary hypertension,* which accounts for about 10% of cases, is high blood pressure that is secondary to (results from) other identifiable disorders. These include kidney disease (which may cause renin hypersecretion), atherosclerosis, hyperthyroidism, Cushing syndrome, and polycythemia. Secondary hypertension is corrected by treating the underlying disease.

# CONNECTIVE ISSUES

## Effects of the CIRCULATORY SYSTEM
On Other Organ Systems

**ALL SYSTEMS**
Blood delivers $O_2$ to the tissues and removes $CO_2$ and other wastes from them; distributes nutrients and hormones throughout the body; and carries heat from deeper organs to the body surface for elimination.

**INTEGUMENTARY SYSTEM**
Dermal blood flow affects sweat production and skin temperature.

**SKELETAL SYSTEM**
Blood delivers the minerals needed for bone deposition; delivers hormones that regulate skeletal growth; and delivers hormones to the bone marrow that stimulate RBC, WBC, and platelet production.

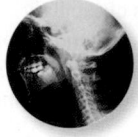

**MUSCULAR SYSTEM**
Blood removes the heat generated by exercise.

**NERVOUS SYSTEM**
Endothelial cells of the blood vessels maintain the blood–brain barrier and play a role in production of cerebrospinal fluid.

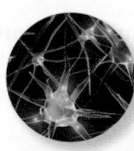

**ENDOCRINE SYSTEM**
Blood is the medium in which all hormones travel to their target organs.

**LYMPHATIC/IMMUNE SYSTEM**
Blood vessels produce tissue fluid, which becomes the lymph; blood contains the WBCs and plasma proteins employed in immunity.

**RESPIRATORY SYSTEM**
Blood picks up $O_2$ from the lungs and releases $CO_2$ to be exhaled; low capillary blood pressure and blood oncotic pressure keep alveoli from filling with fluid.

**URINARY SYSTEM**
Urine production begins with blood filtration; blood carries away the water and solutes reabsorbed by the kidneys; blood pressure maintains renal function.

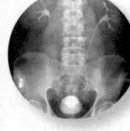

**DIGESTIVE SYSTEM**
Blood picks up absorbed nutrients and helps in reabsorption and recycling of bile salts and minerals from the intestines.

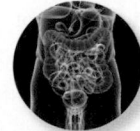

**REPRODUCTIVE SYSTEM**
Blood delivers the hormones that affect reproductive function; vasodilation produces erection in sexual response; blood provides nutrients, oxygen, and other metabolic needs to the fetus and removes its wastes.

# STUDY GUIDE

## ► Assess Your Learning Outcomes

*To test your knowledge, discuss the following topics with a study partner or in writing, ideally from memory.*

### 20.1 General Anatomy of the Blood Vessels (p. 744)

1. Definitions of *arteries, veins,* and *capillaries* with respect to the path of blood flow
2. Tunics of an artery or vein, and their general histological differences
3. Structure and functions of the endothelium
4. Location and function of the vasa vasorum
5. Three size classes of arteries; how and why they differ not just in diameter, but also histologically
6. The relationship of arterioles to metarterioles and capillaries, and the function of the precapillary sphincters of a metarteriole
7. Location, structure, and function of the carotid sinuses, carotid bodies, and aortic bodies
8. Histology of the three types of capillaries and how it relates to their functions
9. Organization of a capillary bed and how its perfusion is regulated
10. Why veins are called *capacitance vessels* and how this relates to the structural difference between veins and arteries
11. What capillaries and postcapillary venules have in common with respect to fluid exchange
12. Structural differences between muscular venules, medium veins, and large veins
13. Structure and purpose of the venous valves, where they occur, and the reason certain veins have valves but arteries of corresponding size do not
14. How venous sinuses differ from other veins, and where they occur
15. How portal systems and anastomoses differ from simpler routes of blood flow; types of anastomoses and their purposes

### 20.2 Blood Pressure, Resistance, and Flow (p. 752)

1. The difference between blood flow and perfusion
2. How blood flow is related to resistance and pressure differences; the mathematical expression of these relationships

3. How to determine systolic pressure, diastolic pressure, and pulse pressure; how to estimate mean arterial pressure (MAP), and why MAP differs from head to foot
4. The meanings of *hypertension* and *hypotension*
5. Why arterial expansion and recoil during the cardiac cycle reduce pulse pressure and ease the strain on small arteries
6. Why arterial flow is pulsatile but capillary and venous flow are not
7. Why blood pressure rises with age
8. Variables that determine blood pressure
9. Variables that determine peripheral resistance; whether each one is directly or inversely proportional to resistance; and which of them is most changeable from moment to moment
10. Terms for widening and narrowing of a blood vessel by muscular contraction and relaxation
11. The mathematical relationship between peripheral resistance and vessel radius; why this is related to the laminar flow of blood; and why it makes vasoreflexes such a powerful influence on blood flow
12. Why blood velocity declines from aorta to capillaries and rises again from capillaries to veins, but never rises as high in veins as it was in the aorta
13. Why arterioles exert a greater influence than any other category of blood vessels on tissue perfusion
14. Three levels of control over blood pressure and flow
15. Short- and long-term mechanisms of local control of blood flow; examples of vasoactive chemicals and how they can cause reactive hyperemia
16. Angiogenesis and its importance for cancer therapy
17. The role of the vasomotor center of the medulla oblongata in controlling blood flow; baroreflexes, chemoreflexes, and the medullary ischemic reflex
18. Mechanisms of action by angiotensin II, aldosterone, natriuretic peptides, antidiuretic hormone, epinephrine, and norepinephrine on blood pressure
19. How vasoreflexes can change systemwide blood pressure or redirect blood flow from one region to another; circumstances that call for redirection of blood flow

### 20.3 Capillary Exchange (p. 758)

1. The meaning of *capillary exchange*, and substances involved in the process
2. Three routes and four mechanisms by which materials pass through capillary walls
3. Substances exchanged by simple diffusion; factors that determine whether a substance can diffuse through a capillary wall
4. Capillary transcytosis and some substances exchanged this way
5. In capillary filtration, three forces that draw fluid out of the capillaries and one force that draws fluid into them
6. The values and net effects of capillary exchange forces at the arterial and venous ends of a capillary, and how they enable a capillary to give off fluid at one end and reabsorb it at the other
7. Relative amounts of fluid given off and reabsorbed by a model capillary, and what compensates for the difference between filtration and reabsorption
8. The role of solvent drag in capillary exchange
9. Why the dynamics of capillary absorption can change from moment to moment or differ in various places in the body; examples of places where the capillaries are engaged entirely in net filtration or reabsorption
10. Chemicals that affect capillary permeability and filtration
11. Three causes of edema, and its pathological consequences

### 20.4 Venous Return and Circulatory Shock (p. 762)

1. The meaning of *venous return,* and five mechanisms that drive it
2. How the skeletal muscle pump works and why it depends on venous valves
3. Why exercise increases venous return
4. Why physical inactivity can lead to venous pooling; consequences of venous pooling
5. Circulatory shock and how it differs from other forms of shock
6. Two basic categories of circulatory shock, three forms of low venous return (LVR) shock, and situations in which each form of shock may occur

# STUDY GUIDE

7. Why septic and anaphylactic shock cannot be strictly classified into any single category of LVR shock

8. Differences between compensated and decompensated shock

## 20.5 Special Circulatory Routes (p. 765)

1. A typical value for cerebral blood flow and why its constancy is important

2. How the brain regulates its blood flow and what chemical stimulus is the most potent in activating its regulatory mechanisms

3. The causes, effects, and difference between a transient ischemic attack (TIA) and cerebral vascular accident (stroke)

4. Variability of skeletal muscle perfusion; what stimuli increase perfusion to meet the demands of exercise; and why isometric contraction causes fatigue more quickly than isotonic contraction does

5. How pulmonary circulation differs from systemic circulation with respect to blood pressure, capillary exchange, relative oxygenation of arterial and venous blood, and the vasomotor response to hypoxia

## 20.6 Anatomy of the Pulmonary Circuit (p. 766)

1. The route of blood flow in the pulmonary circuit

2. Where the capillaries of the pulmonary circuit are found and the function they serve

3. How the function of the pulmonary circuit differs from that of the bronchial arteries, which also supply the lungs

## 20.7 Systemic Vessels of the Axial Region (p. 767)

1. For all named blood vessels in this outline, their anatomical location; the vessel from which they arise; the course they follow; and the organs, body regions, or other blood vessels they supply

2. The ascending aorta, aortic arch, and descending aorta, and the thoracic and abdominal segments of the descending aorta (table 20.2)

3. Branches that arise from the ascending aorta and aortic arch: the coronary arteries, brachiocephalic trunk, left common carotid artery, and left subclavian artery (table 20.2)

4. Four principal arteries of the neck: the common carotid, vertebral artery, thyro-

cervical trunk, and costocervical trunk (table 20.3, part I)

5. The external and internal carotid arteries; branches of the external carotid (superior thyroid, lingual, facial, occipital, maxillary, and superficial temporal arteries); and branches of the internal carotid (ophthalmic, anterior cerebral, and middle cerebral arteries) (table 20.3, part II)

6. Convergence of the vertebral arteries to form the basilar artery; the posterior cerebral arteries and arteries to the cerebellum, pons, and inner ear arising from the basilar artery (table 20.3, part III)

7. The location and constituents of the cerebral arterial circle (table 20.3, part IV)

8. Dural venous sinuses; the superior sagittal, inferior sagittal, transverse, and cavernous sinuses; outflow from the sinus system into the internal jugular veins (table 20.4, part I)

9. The internal jugular, external jugular, and vertebral veins of the neck (table 20.4, part II)

10. Visceral branches (bronchial, esophageal, and mediastinal arteries) and parietal branches (posterior intercostal, subcostal, and superior phrenic arteries) of the thoracic aorta (table 20.5, parts I–II)

11. Arteries of the thorax and shoulder that arise from the subclavian artery and its continuation, the axillary artery: the internal thoracic artery, thoracoacromial trunk, lateral thoracic artery, and subscapular artery (table 20.5, part III)

12. The subclavian vein, brachiocephalic vein, and superior vena cava; landmarks that define the transition from one to another (table 20.6, part I)

13. The azygos system of thoracic veins, especially the azygos, hemiazygos, and accessory hemiazygos veins; their tributaries, including the posterior intercostal, subcostal, esophageal, mediastinal, pericardial, bronchial, and ascending lumbar veins (table 20.6, part II)

14. Branches of the abdominal aorta: inferior phrenic arteries; celiac trunk; and superior mesenteric, middle suprarenal, renal, gonadal (ovarian or testicular), inferior mesenteric, lumbar, median sacral, and common iliac arteries (table 20.7, part I)

15. The general group of organs supplied by the celiac trunk; its three primary branches—the common hepatic, left

gastric, and splenic arteries—and smaller branches given off by each of these (table 20.7, part II)

16. Branches of the superior mesenteric artery: inferior pancreaticoduodenal, jejunal, ileal, and right and middle colic arteries (table 20.7, part III)

17. Branches of the inferior mesenteric artery: left colic, sigmoid, and superior rectal arteries (table 20.7, part III)

18. Two main branches of the common iliac artery, the posterior and anterior trunks of the internal iliac artery, and the organs supplied by those trunks (table 20.7, part IV)

19. Convergence of the internal and external iliac veins to form the common iliac vein; convergence of the right and left common iliac veins to form the inferior vena cava (IVC) (table 20.8, part I)

20. Abdominal tributaries of the IVC: lumbar, gonadal (ovarian or testicular), renal, suprarenal, hepatic, and inferior phrenic veins (table 20.8, part I)

21. The ascending lumbar vein, their drainage in the abdomen, and their continuation into the thorax (table 20.8, part II)

22. The hepatic portal system and its tributaries: the splenic vein; the pancreatic, inferior mesenteric, and superior mesenteric veins draining into it; continuation of the splenic vein as the hepatic portal vein; the cystic vein and gastric veins draining into the hepatic portal vein; hepatic sinusoids in the liver; and hepatic veins (table 20.8, part III)

## 20.8 Systemic Vessels of the Appendicular Region (p. 786)

1. The main artery to the upper limb, which changes name along its course from subclavian to axillary to brachial artery; branches of the brachial artery in the arm (deep brachial and superior ulnar collateral arteries); and the radial collateral artery (table 20.9, part I)

2. Brachial artery branches that supply the forearm: radial and ulnar arteries; anterior and posterior interosseous arteries; and deep and superficial palmar arches (table 20.9, part II)

3. The dorsal venous network of the hand; median antebrachial vein; and median cubital vein (table 20.10, part I)

# STUDY GUIDE

4. The venous palmar arches, and brachial, basilic, axillary, and subclavian veins (table 20.10, part II)

5. Continuation of the external iliac artery as the femoral artery; deep femoral and circumflex femoral branches of the femoral artery; popliteal artery; and anterior and posterior tibial arteries (table 20.11, part I)

6. The dorsal pedal and arcuate arteries that arise from the anterior tibial artery; fibular, medial plantar, and lateral plantar arteries; and deep plantar arch (table 20.11, part II)

7. The superficial dorsal venous arch, small and great saphenous veins, and popliteal vein (table 20.12, part I)

8. The deep plantar venous arch; the lateral and medial plantar veins; fibular and posterior tibial veins; and anterior tibial veins (table 20.12, part II)

9. The femoral, deep femoral, and external iliac veins (table 20.12)

## ► Testing Your Recall

*Answers in Appendix B*

1. Blood normally flows into a capillary bed from
   a. the distributing arteries.
   b. the conducting arteries.
   c. a metarteriole.
   d. a thoroughfare channel.
   e. the venules.

2. Plasma solutes enter the tissue fluid most easily from
   a. continuous capillaries.
   b. fenestrated capillaries.
   c. arteriovenous anastomoses.
   d. collateral vessels.
   e. venous anastomoses.

3. A blood vessel adapted to withstand a high pulse pressure would be expected to have
   a. an elastic tunica media.
   b. a thick tunica interna.
   c. one-way valves.
   d. a flexible endothelium.
   e. a rigid tunica media.

4. The substance most likely to cause a rapid drop in blood pressure is
   a. epinephrine.
   b. norepinephrine.
   c. angiotensin II.
   d. serotonin.
   e. histamine.

5. A person with a systolic blood pressure of 130 mm Hg and a diastolic pressure of 85 mm Hg would have a mean arterial pressure of about
   a. 85 mm Hg.
   b. 100 mm Hg.
   c. 108 mm Hg.
   d. 115 mm Hg.
   e. 130 mm Hg.

6. The velocity of blood flow decreases if
   a. vessel radius increases.
   b. blood pressure increases.
   c. viscosity increases.
   d. viscosity decreases.
   e. afterload increases.

7. Blood flows faster in a venule than in a capillary because venules
   a. have one-way valves.
   b. are more muscular.
   c. are closer to the heart.
   d. have higher blood pressures.
   e. have larger diameters.

8. In a case where interstitial hydrostatic pressure is negative, the only force causing capillaries to reabsorb fluid is
   a. colloid osmotic pressure of the blood.
   b. colloid osmotic pressure of the tissue fluid.
   c. capillary hydrostatic pressure.
   d. interstitial hydrostatic pressure.
   e. net filtration pressure.

9. Intestinal blood flows to the liver by way of
   a. the superior mesenteric artery.
   b. the celiac trunk.
   c. the inferior vena cava.
   d. the azygos system.
   e. the hepatic portal system.

10. The brain receives blood from all of the following vessels *except* the _____ artery or vein.
   a. basilar
   b. vertebral
   c. internal carotid
   d. internal jugular
   e. anterior communicating

11. The highest arterial blood pressure attained during ventricular contraction is called _____ pressure. The lowest attained during ventricular relaxation is called _____ pressure.

12. The capillaries of skeletal muscles are of the structural type called _____.

13. _____ shock occurs as a result of exposure to an antigen to which one is hypersensitive.

14. The role of breathing in venous return is called the _____.

15. The difference between the colloid osmotic pressure of blood and that of the tissue fluid is called _____.

16. Movement across the capillary endothelium by the uptake and release of fluid droplets is called _____.

17. All efferent fibers of the vasomotor center belong to the _____ division of the autonomic nervous system.

18. The pressure sensors in the major arteries near the head are called _____.

19. Most of the blood supply to the brain comes from a ring of arterial anastomoses called _____.

20. The major superficial veins of the arm are the _____ on the medial side and _____ on the lateral side.

# STUDY GUIDE

## ▶ Building Your Medical Vocabulary

*Answers in Appendix B*

*State a medical meaning of each word element below, and give a term in which it or a slight variation of it is used.*

1. angio-
2. brachio-
3. celi-
4. fenestra-
5. jugulo-
6. -orum
7. sapheno-
8. sub-
9. thoraco-
10. vesico-

## ▶ True or False

*Answers in Appendix B*

*Determine which five of the following statements are false, and briefly explain why.*

1. In some circulatory pathways, blood can get from an artery to a vein without going through capillaries.

2. In some cases, a blood cell may pass through two capillary beds in a single trip from left ventricle to right atrium.

3. The body's longest blood vessel is the great saphenous vein.

4. Arteries have a series of valves that ensure a one-way flow of blood.

5. If the radius of a blood vessel doubles and all other factors remain the same, blood flow through that vessel also doubles.

6. The femoral triangle is bordered by the inguinal ligament, sartorius muscle, and adductor longus muscle.

7. The lungs receive both pulmonary and systemic blood.

8. If blood capillaries fail to reabsorb all the fluid they emit, edema will occur.

9. An aneurysm is a ruptured blood vessel.

10. In the baroreflex, a drop in arterial blood pressure triggers a corrective vasodilation of the systemic blood vessels.

## ▶ Testing Your Comprehension

*Answers at* www.mhhe.com/saladin7

1. It is a common lay perception that systolic blood pressure should be 100 plus a person's age. Evaluate the validity of this statement.

2. Calculate the net filtration or reabsorption pressure at a point in a hypothetical capillary assuming a hydrostatic blood pressure of 28 mm Hg, an interstitial hydrostatic pressure of –2 mm Hg, a blood COP of 25 mm Hg, and an interstitial COP of 4 mm Hg. Give the magnitude (in mm Hg) and direction (in or out) of the net pressure.

3. Aldosterone secreted by the adrenal gland must be delivered to the kidney immediately below. Trace the route that an aldosterone molecule must take from the adrenal gland to the kidney, naming all major blood vessels in the order traveled.

4. People in shock commonly exhibit paleness, cool skin, tachycardia, and a weak pulse. Explain the physiological basis for each of these signs.

5. Discuss why it is advantageous to have baroreceptors in the aortic arch and carotid sinus rather than in some other location such as the common iliac arteries.

## ▶ Improve Your Grade at www.mhhe.com/saladin7

*Download animations to study when it fits your schedule. Practice quizzes, labeling activities, games, and flashcards offer fun ways to master the chapter concepts. Or, download image PowerPoint files for each chapter to create a study guide or for taking notes during lecture.*

# THE LYMPHATIC AND IMMUNE SYSTEMS

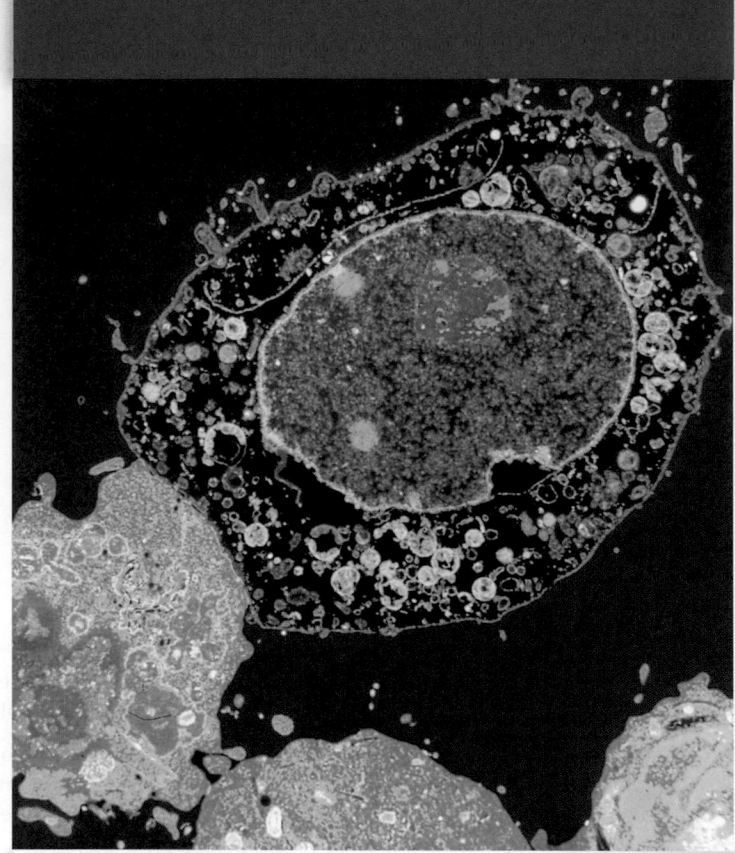

T lymphocytes (green) attacking a cancer cell (with blue nucleus) (TEM)

Anatomy & Physiology REVEALED®
aprevealed.com

**Module 10: Lymphatic System**

It may come as a surprise to know that the human body harbors at least 10 times as many bacterial cells as human cells. But it shouldn't be surprising. After all, human homeostasis works wonderfully not only to sustain our lives, but also to provide a predictable, warm, wet, nutritious habitat for our internal guests. It is a wonder that the body is not overrun and consumed by microbes—which indeed quickly happens when one dies and homeostasis ceases.

Many of these guest microbes are beneficial or even necessary to human health, but some have the potential to cause disease if they get out of hand. Furthermore, we are constantly exposed to new invaders through the food and water we consume, the air we breathe, and even the surfaces we touch. We must have a means of keeping such would-be colonists in check.

One of these defenses was discovered in 1882 by a moody, intense, Russian zoologist, Elie Metchnikoff (1845–1916). When studying the tiny transparent larvae of starfish, he observed mobile cells wandering throughout their bodies. He thought at first that they must be digestive cells, but when he saw similar cells in sea anemones ingest nonnutritive dye particles, he thought they must play a defensive role. Metchnikoff knew that mobile cells exist in human blood and pus and quickly surround a splinter introduced through the skin, so he decided to experiment to see if the starfish cells would do the same. He impaled a starfish larva on a rose thorn, and the next morning he found the thorn crawling with cells that seemed to be trying to devour it. He later saw similar cells devouring and digesting infectious yeast in tiny transparent crustaceans called water fleas. He coined the word *phagocytosis* for this reaction and termed the wandering cells *phagocytes*—terms we still use today.

Metchnikoff showed that animals from simple sea anemones and starfish to humans actively defend themselves against disease agents. His observations marked the founding of cellular and comparative immunology, and won him the scientific respect he had so long coveted. Indeed, he shared the 1908 Nobel Prize for Physiology or Medicine with Paul Ehrlich (1854–1915), who had developed the theory of humoral immunity, a process also discussed in this chapter.

This chapter focuses largely on the *immune system,* which is not an organ system but rather a population of cells that inhabit all of our organs and defend the body from agents of disease. But immune cells are especially concentrated in a true organ system, the *lymphatic system.* This is a network of organs and veinlike vessels that recover tissue fluid, inspect it for disease agents, activate immune responses, and return the fluid to the bloodstream. It is with the lymphatic system that we will begin this chapter's exploration.

## 21.1 The Lymphatic System AP|R

### Expected Learning Outcomes
When you have completed this section, you should be able to

a. list the functions of the lymphatic system;

b. explain how lymph forms and returns to the bloodstream;

c. name the major cells of the lymphatic system and state their functions;

d. name and describe the types of lymphatic tissue; and

e. describe the structure and function of the red bone marrow, thymus, lymph nodes, tonsils, and spleen.

The **lymphatic system** (fig. 21.1) consists of a network of vessels that penetrate nearly every tissue of the body, and a collection of tissues and organs that produce immune cells. It has three functions:

1. **Fluid recovery.** Fluid continually filters from blood capillaries into the tissue spaces. The capillaries reabsorb about 85% of it, but the 15% that they do not absorb would amount, over the course of a day, to 2 to 4 L of water and one-quarter to one-half of the plasma protein. One would die of circulatory failure within hours if this water and protein were not returned to the bloodstream. One task of the lymphatic system is to reabsorb this excess and return it to the blood. Even partial interference with lymphatic drainage can lead to severe edema (fig. 21.2).

2. **Immunity.** As the lymphatic system recovers tissue fluid, it also picks up foreign cells and chemicals from the tissues. On its way back to the bloodstream, the fluid passes through lymph nodes, where immune cells stand guard against foreign matter. When they detect anything potentially harmful, they activate a protective immune response.

3. **Lipid absorption.** In the small intestine, special lymphatic vessels called *lacteals* absorb dietary lipids that are not absorbed by the blood capillaries.

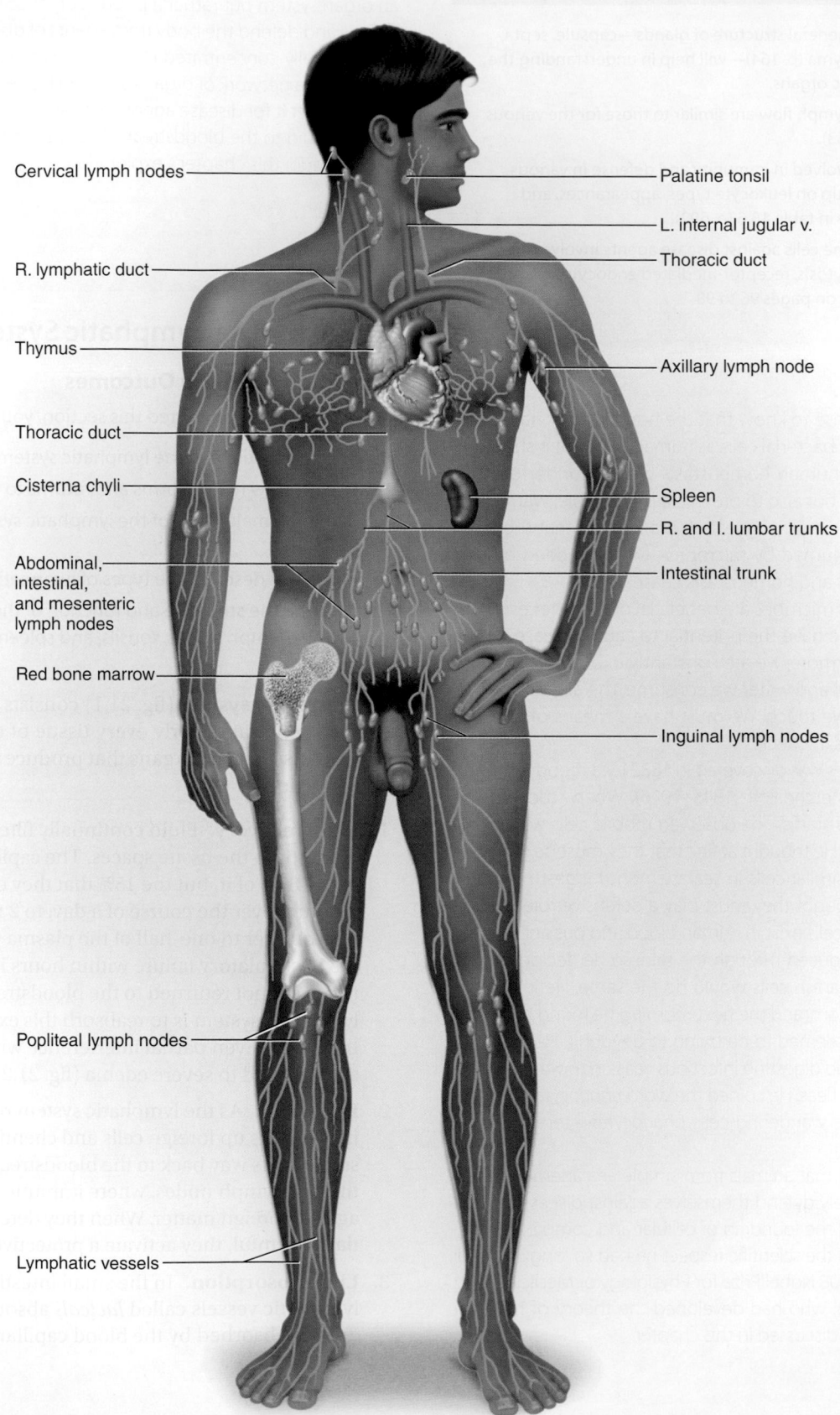

Cervical lymph nodes

Palatine tonsil

L. internal jugular v.

R. lymphatic duct

Thoracic duct

Thymus

Axillary lymph node

Thoracic duct

Cisterna chyli

Spleen

R. and l. lumbar trunks

Abdominal, intestinal, and mesenteric lymph nodes

Intestinal trunk

Red bone marrow

Inguinal lymph nodes

Popliteal lymph nodes

Lymphatic vessels

**FIGURE 21.1  The Lymphatic System.**

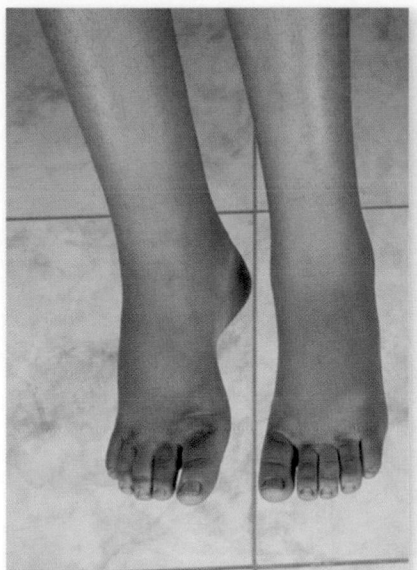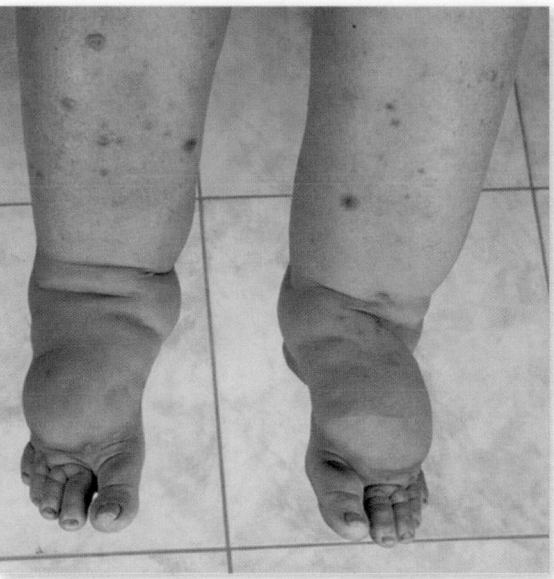

**FIGURE 21.2 Edema.** The person on the right shows severe edema of the legs and feet compared to a person without edema (left). Blockage of lymphatic vessels is one of several causes of edema.

The components of the lymphatic system are (1) *lymph,* the recovered fluid; (2) *lymphatic vessels,* which transport the lymph; (3) *lymphatic tissue,* composed of aggregates of lymphocytes and macrophages that populate many organs of the body; and (4) *lymphatic organs,* in which these cells are especially concentrated and which are set off from surrounding organs by connective tissue capsules.

## Lymph and the Lymphatic Vessels

**Lymph** is usually a clear, colorless fluid, similar to blood plasma but low in protein. It originates as tissue fluid that has been taken up by the lymphatic vessels. Its composition varies substantially from place to place. After a meal, for example, lymph draining from the small intestine has a milky appearance because of its lipid content. Lymph leaving the lymph nodes contains a large number of lymphocytes—indeed, this is the main supply of lymphocytes to the bloodstream. Lymph may also contain macrophages, hormones, bacteria, viruses, cellular debris, or even traveling cancer cells.

## Lymphatic Vessels

Lymph flows through a system of **lymphatic vessels (lymphatics)** similar to blood vessels. These begin with microscopic **lymphatic capillaries (terminal lymphatics),** which penetrate nearly every tissue of the body but are absent from the central nervous system, cartilage, cornea, bone, and bone marrow. They are closely associated with blood capillaries, but unlike them, they are closed at one end (fig. 21.3). A lymphatic capillary consists of a sac of thin endothelial cells that loosely overlap each other like the shingles of a roof. The cells are tethered to surrounding tissue by protein filaments that prevent the sac from collapsing.

Unlike the endothelial cells of blood capillaries, lymphatic endothelial cells are not joined by tight junctions, nor do they have a continuous basal lamina; indeed, the gaps between them are so large that bacteria, lymphocytes, and other cells and particles can enter along with the tissue fluid. Thus, the composition of lymph arriving at a lymph node is like a report on the state of the upstream tissues.

The overlapping edges of the endothelial cells act as valves that can open and close. When tissue fluid pressure is high, it pushes the flaps inward (open) and fluid flows into the capillary. When pressure is higher in the lymphatic capillary than in the tissue fluid, the flaps are pressed outward (closed).

### ▶▶▶ APPLY WHAT YOU KNOW

*Contrast the structure of a lymphatic capillary with that of a continuous blood capillary. Explain why their structural difference is related to their functional difference.*

Lymphatic vessels form in the embryo by budding from the veins, so it is not surprising that the larger ones have a similar histology. They have a *tunica interna* with an endothelium and valves (fig. 21.4), a *tunica media* with elastic fibers and smooth muscle, and a thin outer *tunica externa.* Their walls are thinner and their valves are closer together than those of the veins.

As the lymphatic vessels converge along their path, they become larger and larger vessels with changing names. The route from the tissue fluid back to the bloodstream is: lymphatic capillaries → collecting vessels → six lymphatic trunks → two collecting ducts → subclavian veins. Thus, there is a continual recycling of fluid from blood to tissue fluid to lymph and back to the blood (fig. 21.5).

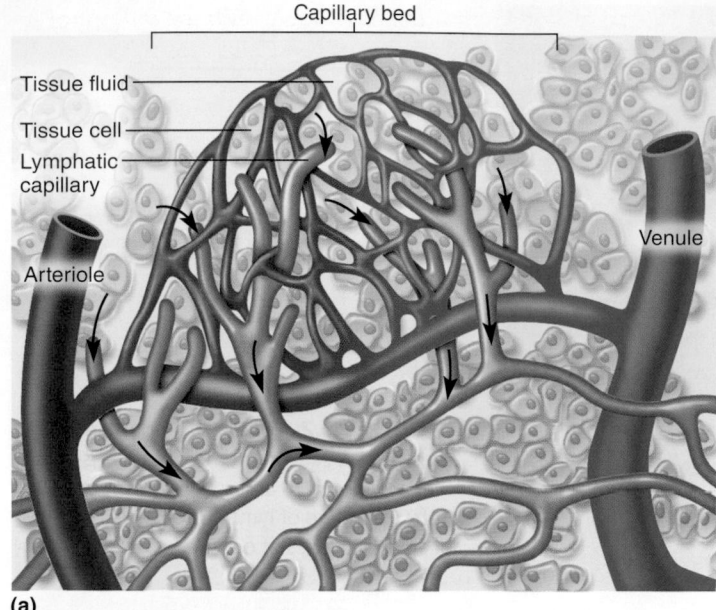

(a)

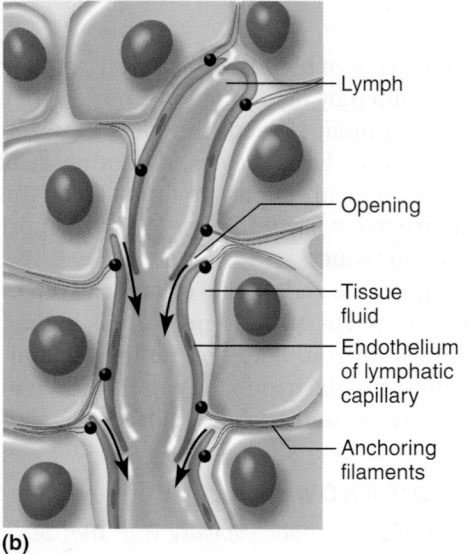

(b)

**FIGURE 21.3 Lymphatic Capillaries.** (a) Relationship of the lymphatic capillaries to a bed of blood capillaries. (b) Uptake of tissue fluid by a lymphatic capillary.

❓ *Why can metastasizing cancer cells get into the lymphatic system more easily than they can enter the bloodstream?*

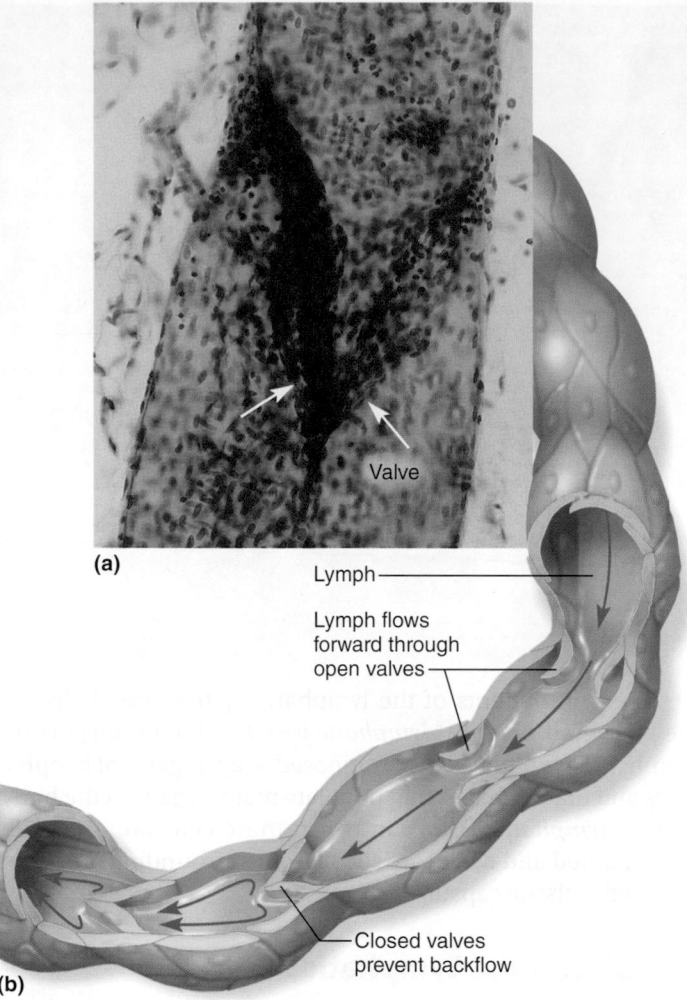

(a)

(b)

**FIGURE 21.4 Valves in the Lymphatic Vessels.** (a) Photograph of a lymphatic valve. (b) Operation of the valves to ensure a one-way flow of lymph.

❓ *What would be the consequence if these valves did not exist?*

The lymphatic capillaries converge to form **collecting vessels.** These often travel alongside veins and arteries and share a common connective tissue sheath with them. At irregular intervals, they empty into lymph nodes. The lymph trickles slowly through each node, where bacteria are phagocytized and immune cells monitor the fluid for foreign antigens. It leaves the other side of the node through another collecting vessel, traveling on and often encountering additional lymph nodes before it finally returns to the bloodstream.

Eventually, the collecting vessels converge to form larger **lymphatic trunks,** each of which drains a major portion of the body. There are six of these, whose names indicate their locations and parts of the body they drain: the *jugular, subclavian, bronchomediastinal, intercostal, intestinal,* and *lumbar trunks.* There is a single intestinal trunk, whereas all the others are paired. The lumbar trunk drains not only the lumbar region but also the lower limbs.

The lymphatic trunks converge to form two **collecting ducts,** the largest of the lymphatic vessels (fig. 21.6):

1. The **right lymphatic duct** is formed by the convergence of the right jugular, subclavian, and bronchomediastinal trunks in the right thoracic cavity. It receives lymphatic drainage from the right arm and right side of the thorax and head and empties into the right subclavian vein.

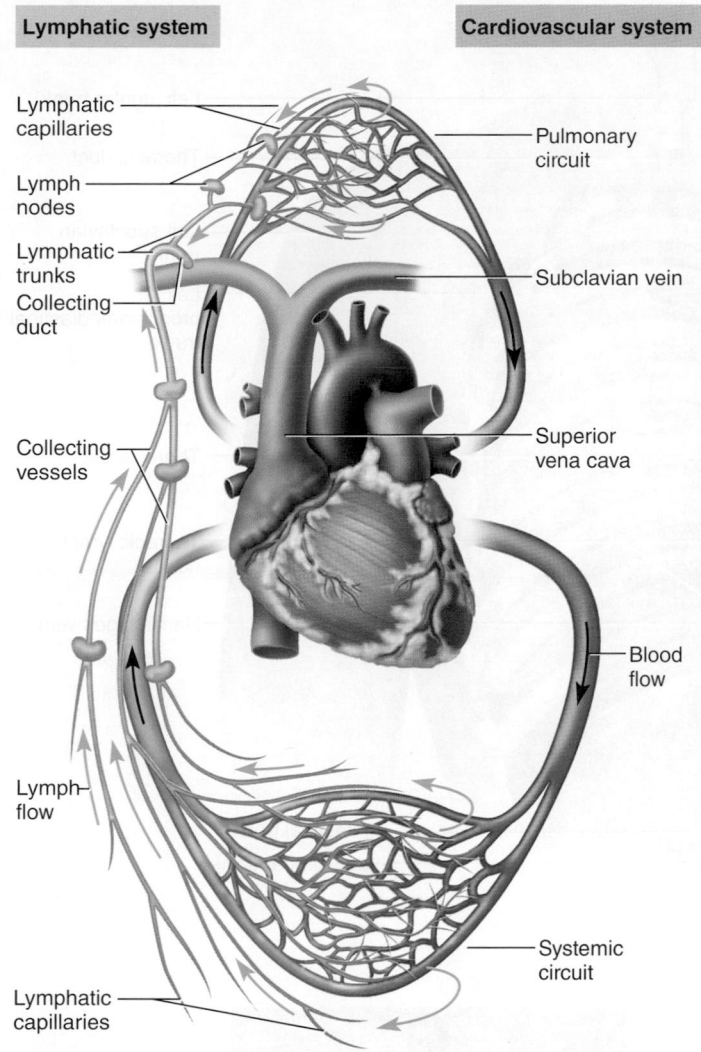

**Lymphatic system**

**Cardiovascular system**

- Lymphatic capillaries
- Lymph nodes
- Lymphatic trunks
- Collecting duct
- Collecting vessels
- Lymph flow
- Lymphatic capillaries

- Pulmonary circuit
- Subclavian vein
- Superior vena cava
- Blood flow
- Systemic circuit

**FIGURE 21.5** **Fluid Exchange Between the Circulatory and Lymphatic Systems.** Blood capillaries lose fluid to the tissue spaces. The lymphatic system picks up excess tissue fluid and returns it to the bloodstream.

❓ *Identify two benefits in having lymphatic capillaries pick up tissue fluid that is not reclaimed by the blood capillaries.*

2. The **thoracic duct,** on the left, is larger and longer. It begins just below the diaphragm anterior to the vertebral column at the level of the second lumbar vertebra. Here, the two lumbar trunks and the intestinal trunk join and form a prominent sac called the **cisterna chyli** (sis-TUR-nuh KY-lye), named for the large amount of *chyle* (fatty intestinal lymph) that it collects after a meal. The thoracic duct then passes through the diaphragm with the aorta and ascends the mediastinum, adjacent to the vertebral column. As it passes through the thorax, it receives additional lymph from the left bronchomediastinal, left subclavian, and left jugular trunks, then empties into the left subclavian vein. Collectively, this duct therefore drains

all of the body below the diaphragm, and the left upper limb and left side of the head, neck, and thorax.

## Flow of Lymph

Lymph flows under forces similar to those that govern venous return, except that the lymphatic system has no pump like the heart, and lymph flows at even lower pressure and speed than venous blood. The primary mechanism of flow is rhythmic contractions of the lymphatic vessels themselves, which contract when the fluid stretches them. The valves of lymphatic vessels, like those of veins, prevent the fluid from flowing backward. Lymph flow is also produced by skeletal muscles squeezing the lymphatic vessels, like the skeletal muscle pump that moves venous blood. Since lymphatic vessels are often wrapped with an artery in a common connective tissue sheath, arterial pulsation may also rhythmically squeeze the lymphatic vessels and contribute to lymph flow. A thoracic (respiratory) pump promotes the flow of lymph from the abdominal to the thoracic cavity as one inhales, just as it does in venous return. Finally, at the point where the collecting ducts empty into the subclavian veins, the rapidly flowing bloodstream draws the lymph into it. Considering these mechanisms of lymph flow, it should be apparent that physical exercise significantly increases the rate of lymphatic return.

### ▶▶▶APPLY WHAT YOU KNOW

*Why does it make more functional sense for the collecting ducts to connect to the subclavian veins than it would for them to connect to the subclavian arteries?*

## Lymphatic Cells AP|R

Another component of the lymphatic system is lymphatic tissue, which ranges from loosely scattered cells in the mucous membranes of the respiratory, digestive, urinary, and reproductive tracts to compact cell populations encapsulated in lymphatic organs. These tissues are composed of a variety of lymphocytes and other cells with various roles in defense and immunity:

1. **Natural killer (NK) cells** are large lymphocytes that attack and destroy bacteria, transplanted tissues, and *host cells* (cells of one's own body) that have either become infected with viruses or turned cancerous.

2. **T lymphocytes (T cells)** are lymphocytes that mature in the thymus and later depend on thymic hormones; the *T* stands for *thymus-dependent.* There are several subclasses of T cells that will be introduced later.

3. **B lymphocytes (B cells)** are lymphocytes that differentiate into *plasma cells*—connective tissue cells that secrete antibodies. They are named for an organ in chickens (the *bursa of Fabricius*[1]) in which they were first discovered. However, you may find it more helpful to think of *B* for *bone marrow,* the site where these cells mature in humans.

---

[1]Hieronymus Fabricius (Girolamo Fabrizzi) (1537–1619), Italian anatomist

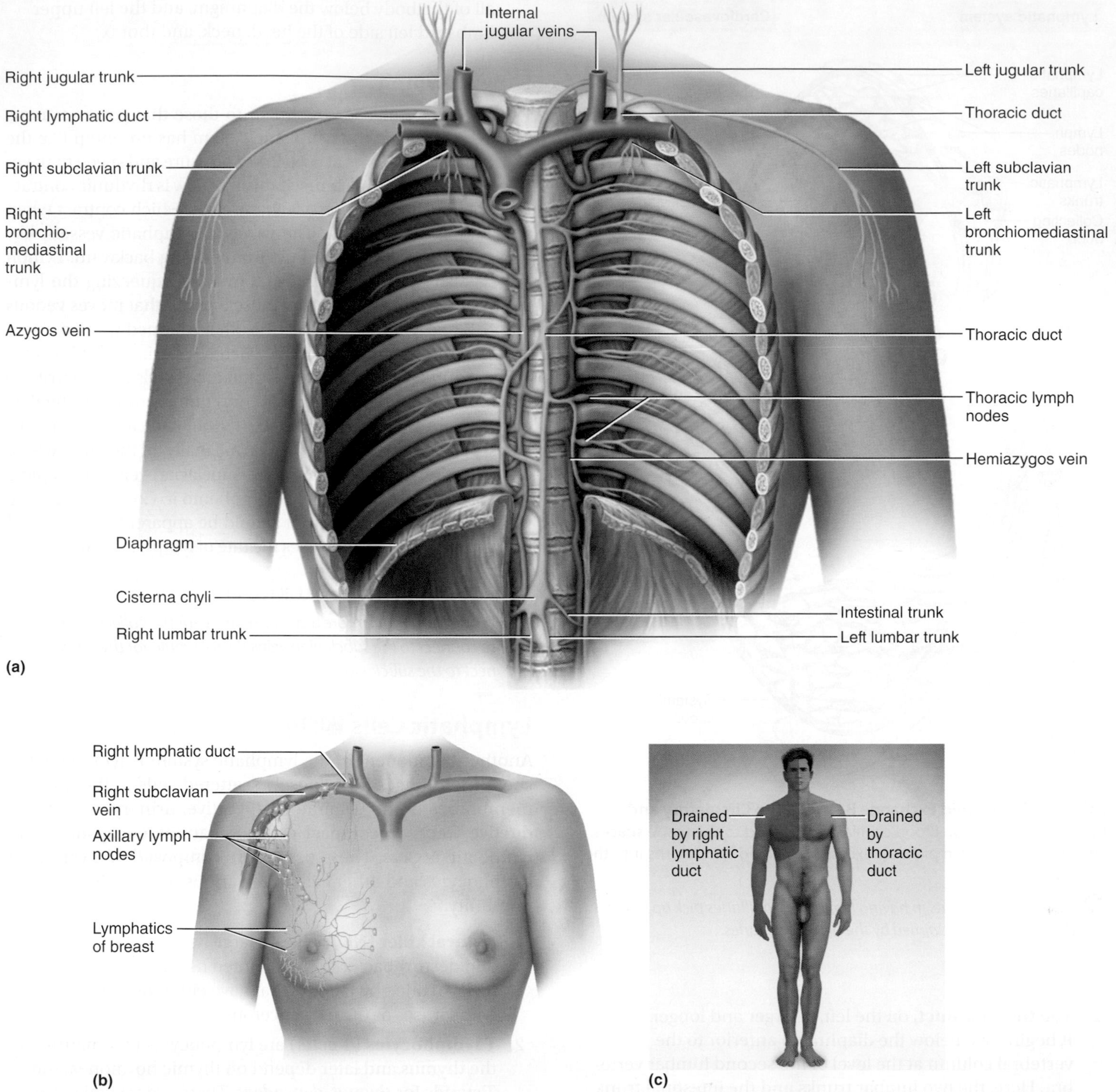

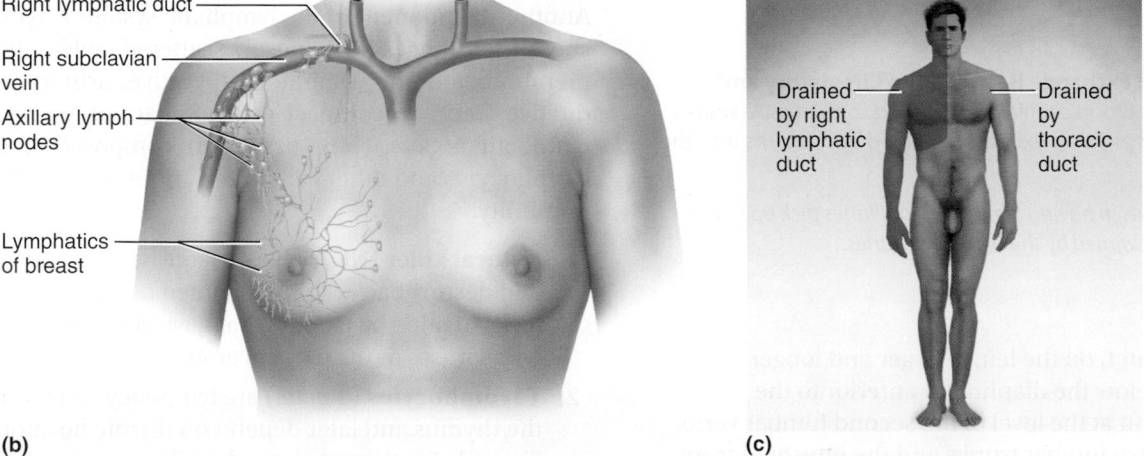

**FIGURE 21.6 Lymphatics of the Thoracic Region.** (a) Lymphatics of the thorax and upper abdomen and their relationship to the subclavian veins, where the lymph returns to the bloodstream. (b) Lymphatic drainage of the right mammary and axillary regions. (c) Regions of the body drained by the right lymphatic duct and thoracic duct.   AP|R

❓ *Why are the axillary lymph nodes often biopsied in cases of suspected breast cancer?*

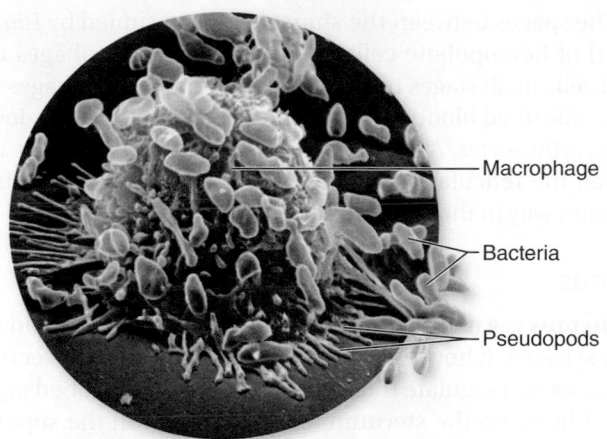

**FIGURE 21.7 Macrophages Phagocytizing Bacteria.** Filamentous pseudopods of the macrophages snare the rod-shaped bacteria and draw them to the cell surface, where they are phagocytized.

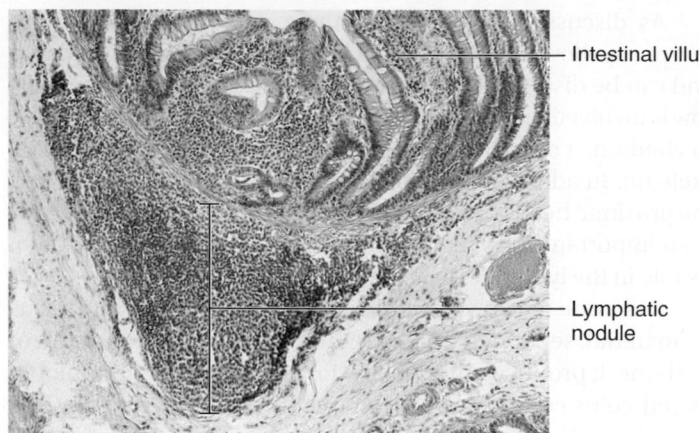

**FIGURE 21.8 Lymphatic Nodule in the Mucous Membrane of the Small Intestine.** This nodule is part of the mucosa-associated lymphatic tissue (MALT).

4. **Macrophages** are very large, avidly phagocytotic cells of the connective tissues. They develop from monocytes that have emigrated from the bloodstream. They phagocytize tissue debris, dead neutrophils, bacteria, and other foreign matter (fig. 21.7). They also process foreign matter and display antigenic fragments of it to certain T cells, thus alerting the immune system to the presence of an enemy. Macrophages and other cells that do this are collectively called *antigen-presenting cells (APCs).*

5. **Dendritic cells** are branched, mobile APCs found in the epidermis, mucous membranes, and lymphatic organs. (In the skin, they are often called *Langerhans*[2] *cells.*) They play an important role in alerting the immune system to pathogens that have breached the body surfaces. They engulf foreign matter by receptor-mediated endocytosis rather than phagocytosis, but otherwise function like macrophages. After internalizing an antigen, they migrate to a nearby lymph node and activate an immune reaction to it.

6. **Reticular cells** are branched, stationary APCs that contribute to the connective tissue framework (stroma) of the lymphatic organs (see fig. 21.10). (They should not be confused with reticular *fibers,* which are fine, branched collagen fibers common in lymphatic organs.)

## Lymphatic Tissues

**Lymphatic (lymphoid) tissues** are aggregations of lymphocytes in the connective tissues of mucous membranes and various organs. The simplest form is **diffuse lymphatic tissue,** in which the lymphocytes are scattered rather than densely clustered. It is particularly prevalent in body passages that are open to the exterior—the respiratory, digestive, urinary, and reproductive tracts—where it is called **mucosa-associated lymphatic tissue (MALT).**

In some places, lymphocytes and macrophages congregate in dense masses called **lymphatic nodules (follicles)** (fig. 21.8), which come and go as pathogens invade the tissues and the immune system answers the challenge. Abundant lymphatic nodules are, however, a relatively constant feature of the lymph nodes (see fig. 21.12), tonsils, and appendix. In the ileum, the distal portion of the small intestine, they form clusters called **Peyer**[3] **patches.**

## Lymphatic Organs

In contrast to the diffuse lymphatic tissue, **lymphatic (lymphoid) organs** have well-defined anatomical sites and at least partial connective tissue capsules that separate the lymphatic tissue from neighboring tissues. These organs include the red bone marrow, thymus, lymph nodes, tonsils, and spleen. The red bone marrow and thymus are regarded as *primary lymphatic organs* because they are the sites where B and T lymphocytes, respectively, become *immunocompetent*—that is, able to recognize and respond to antigens. The lymph nodes, tonsils, and spleen are called *secondary lymphatic organs* because immunocompetent lymphocytes migrate to these organs only after they mature in the primary lymphatic organs.

## Red Bone Marrow

Red bone marrow may not seem to be an organ; when aspirated from the bones for the purpose of biopsy or transfusion, it simply looks like extra-thick blood. Yet a careful microscopic examination of less disturbed marrow shows that it has a surprising degree of structure and is composed of multiple tissues, meeting the criteria of an organ, even if a very soft one.

---

[2]Paul Langerhans (1847–88), German anatomist

[3]Johann Conrad Peyer (1653–1712), Swiss anatomist

As discussed in chapter 7, there are two kinds of bone marrow: yellow and red. Yellow bone marrow is adipose tissue and can be disregarded for present purposes, but red bone marrow is involved in hemopoiesis (blood formation) and immunity. In children, it occupies the medullary spaces of nearly the entire skeleton. In adults, it is limited to parts of the axial skeleton and the proximal heads of the humerus and femur. Red bone marrow is an important supplier of lymphocytes to the immune system. Its role in the lymphocyte life history is described later.

Red bone marrow is a soft, loosely organized, highly vascular material, separated from osseous tissue by the endosteum of the bone. It produces all classes of formed elements of the blood; its red color comes from the abundance of erythrocytes. Numerous small arteries enter *nutrient foramina* on the bone surface, penetrate the bone, and empty into large *sinusoids* (45 to 80 μm wide) in the marrow (fig. 21.9). The sinusoids drain into a *central longitudinal vein* that exits the bone via the same route that the arteries entered. The sinusoids are lined by endothelial cells, like other blood vessels, and are surrounded by reticular cells and reticular fibers. The reticular cells secrete colony-stimulating factors that induce the formation of various leukocyte types. In the long bones of the limbs, aging reticular cells accumulate fat and transform into adipose cells, eventually replacing red bone marrow with yellow bone marrow.

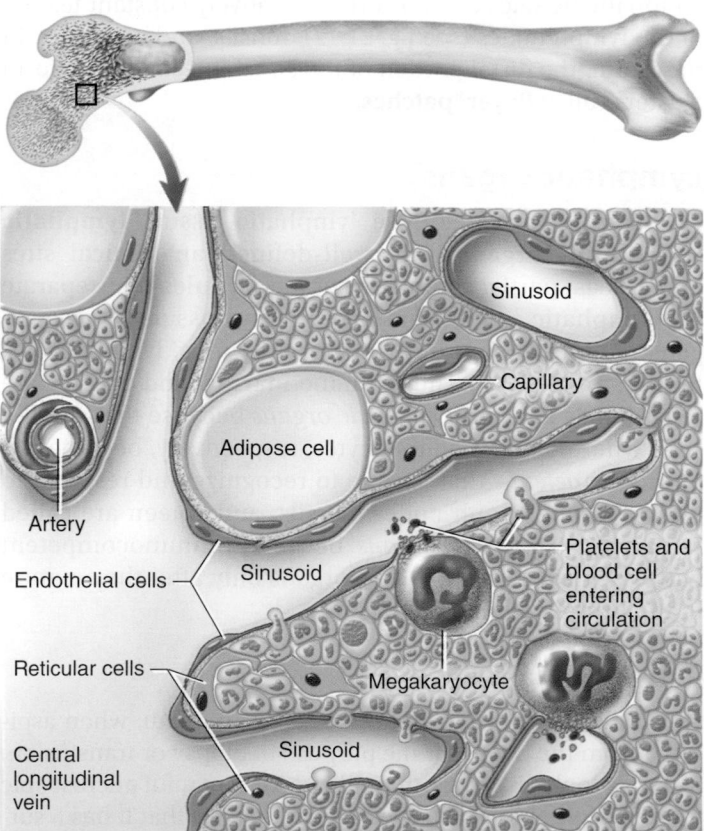

**FIGURE 21.9  Histology of the Red Bone Marrow.** The formed elements of blood squeeze through the endothelial cells into the sinuses, which converge on a central longitudinal vein at the lower left.

The spaces between the sinusoids are occupied by *islands (cords)* of hemopoietic cells, composed of macrophages and blood cells in all stages of development. The macrophages destroy malformed blood cells and the nuclei discarded by developing erythrocytes. As blood cells mature, they push their way through the reticular and endothelial cells to enter the sinus and flow away in the bloodstream.

## Thymus

The **thymus** is a member of the endocrine, lymphatic, and immune systems. It houses developing lymphocytes and secretes hormones that regulate their later activity. It is a bilobed organ located between the sternum and aortic arch in the superior mediastinum. The thymus shows a remarkable degree of degeneration (involution) with age, as described and illustrated earlier (fig. 17.8, p. 641).

The fibrous capsule of the thymus gives off trabeculae (septa) that divide the gland into several angular lobules. Each lobule has a light central *medulla* populated by T lymphocytes, surrounded by a dense, darker *cortex* (fig. 21.10). **Reticular epithelial cells** seal off the cortex from the medulla and surround blood vessels and lymphocyte clusters in the cortex. They thereby form a *blood–thymus barrier* that isolates developing lymphocytes from blood-borne antigens. After developing in the cortex, the T cells migrate to the medulla, where they spend another 3 weeks. There is no blood–thymus barrier in the medulla; mature T cells enter blood or lymphatic vessels here and leave the thymus. In the medulla, reticular epithelial cells form whorls called *thymic (Hassall[4]) corpuscles,* useful for identifying the thymus histologically.

Besides forming the blood–thymus barrier, reticular epithelial cells produce several signaling molecules that promote the development and action of T cells, including *thymosin, thymopoietin, thymulin, interleukins,* and *interferon.* If the thymus is removed from newborn mammals, they waste away and never develop immunity. Other lymphatic organs also seem to depend on thymosins or T cells and develop poorly in thymectomized animals. The role of the thymus in T cell development is discussed later.

## Lymph Nodes

**Lymph nodes** (fig. 21.11) are the most numerous lymphatic organs, numbering about 450 in a typical adult. They serve two functions: to cleanse the lymph and to act as a site of T and B cell activation. A lymph node is an elongated or bean-shaped structure, usually less than 3 cm long, often with an indentation called the *hilum* on one side (fig. 21.12). It is enclosed in a fibrous capsule with trabeculae that partially divide the interior of the node into compartments. Between the capsule and parenchyma is a narrow, relatively clear space called the *subcapsular sinus,* which contains reticular fibers, macrophages, and dendritic cells. Deep to this, the gland consists mainly of a

---

[4]Arthur H. Hassall (1817–94), British chemist and physician

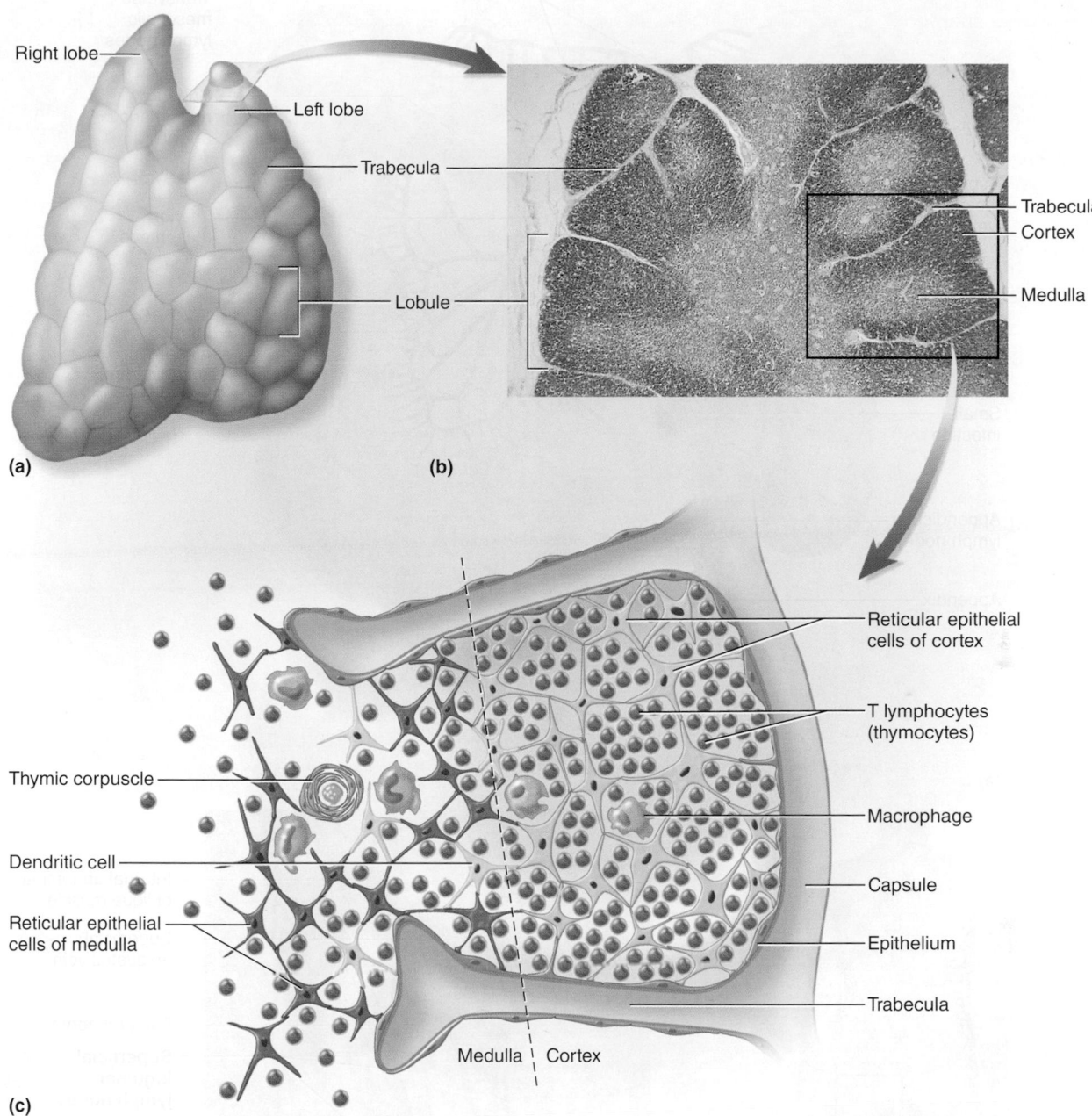

**FIGURE 21.10 The Thymus.** (a) Gross anatomy. (b) Histology. (c) Cellular architecture of a lobule. The reticular epithelial cells are interconnected to form a blood–thymus barrier. AP|R

stroma of reticular connective tissue and a parenchyma of lymphocytes and antigen-presenting cells.

The parenchyma is divided into an outer C-shaped **cortex** that encircles about four-fifths of the organ, and an inner **medulla** that extends to the surface at the hilum. The cortex consists mainly of ovoid to conical lymphatic nodules. When the lymph node is fighting a pathogen, these nodules acquire light-staining **germinal centers** where B cells multiply and differentiate into plasma cells. The medulla consists largely of a branching network of *medullary cords* composed of lymphocytes, plasma cells, macrophages, reticular cells, and reticular fibers. The cortex and medulla also contain lymph-filled sinuses continuous with the subcapsular sinus.

Several **afferent lymphatic vessels** lead into the node along its convex surface. Lymph flows from these vessels into the subcapsular sinus, percolates slowly through the sinuses of the cortex and medulla, and leaves the node through one to three **efferent lymphatic vessels** that emerge from the hilum. No other lymphatic organs have afferent lymphatic vessels; lymph nodes are the only organs that filter lymph as it flows along its course. The lymph node is a bottleneck that slows down lymph flow and allows time for cleansing it of foreign matter. Macrophages and reticular cells of the sinuses remove about 99% of the impurities before the lymph leaves the node. On its way to the bloodstream, lymph flows through one

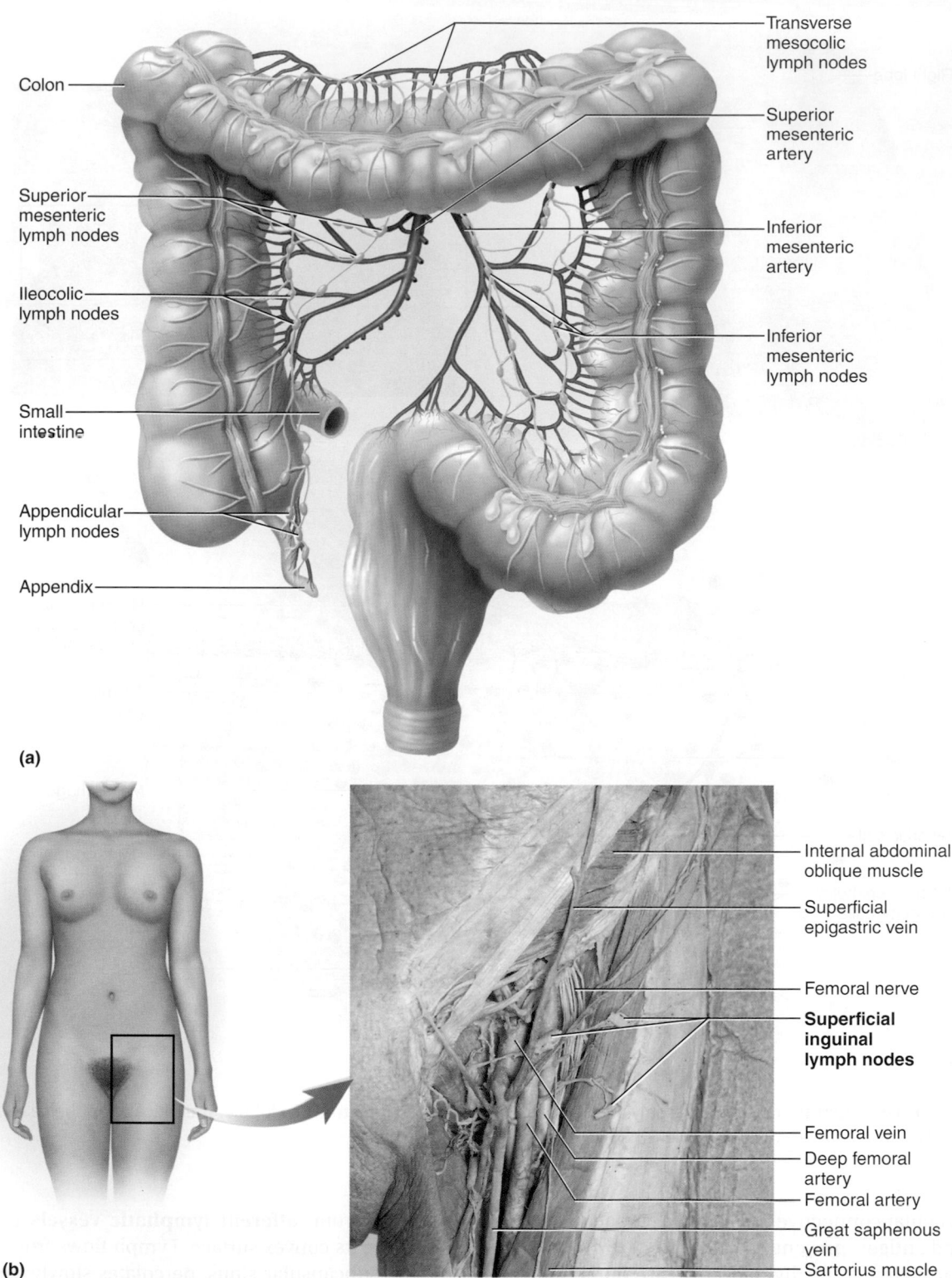

Colon

Superior mesenteric lymph nodes

Ileocolic lymph nodes

Small intestine

Appendicular lymph nodes

Appendix

Transverse mesocolic lymph nodes

Superior mesenteric artery

Inferior mesenteric artery

Inferior mesenteric lymph nodes

**(a)**

Internal abdominal oblique muscle

Superficial epigastric vein

Femoral nerve

**Superficial inguinal lymph nodes**

Femoral vein

Deep femoral artery

Femoral artery

Great saphenous vein

Sartorius muscle

**(b)**

**FIGURE 21.11  Some Areas of Lymph Node Concentration.**  (a) Mesenteric lymph nodes associated with the large intestine. (b) Inguinal lymph nodes in a female cadaver.  AP|R

lymph node after another and thus becomes quite thoroughly cleansed of impurities.

Blood vessels also penetrate the hilum. Arteries follow the medullary cords and give rise to capillary beds in the medulla and cortex. In the *deep cortex* near the junction with the medulla, lymphocytes emigrate from the bloodstream into the parenchyma of the node. Most lymphocytes in the deep cortex are T cells.

Lymph nodes are widespread but especially concentrated in the following locations (see fig. 21.1):

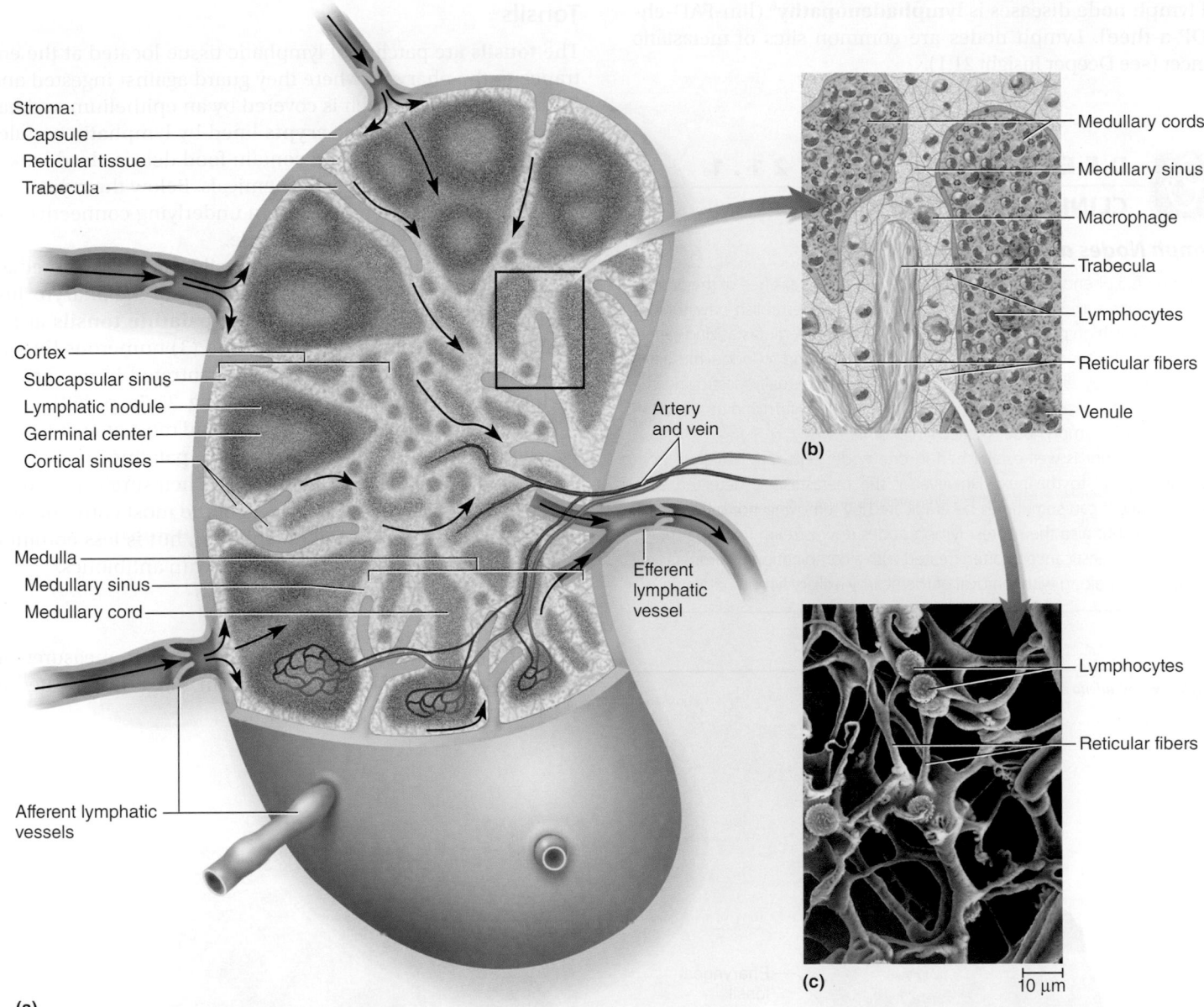

**(a)**

**FIGURE 21.12 Anatomy of a Lymph Node.** (a) Partially bisected lymph node showing pathway of lymph flow. (b) Detail of the boxed region in part (a). (c) Reticular fiber stroma and immune cells in a medullary sinus (SEM). AP|R

- *Cervical lymph nodes* occur in deep and superficial groups in the neck, and monitor lymph coming from the head and neck.

- *Axillary lymph nodes* are concentrated in the armpit (axilla) and receive lymph from the upper limb and breast (fig. 21.6b).

- *Thoracic lymph nodes* occur in the thoracic cavity, especially in the mediastinum, and receive lymph from the mediastinum, lungs, and airway.

- *Abdominal lymph nodes* occur in the posterior abdominopelvic wall and receive lymph from the urinary and reproductive systems.

- *Intestinal* and *mesenteric lymph nodes* are found in the mesenteries (fig. 21.11a) and adjacent to the appendix

and intestines; they receive lymph from the digestive tract.

- *Inguinal lymph nodes* occur in the groin (fig. 21.11b) and receive lymph from the entire lower limb.

- *Popliteal lymph nodes* occur at the back of the knee and receive lymph from the leg proper.

When a lymph node is under challenge by an antigen, it may become swollen and painful to the touch—a condition called **lymphadenitis**[5] (lim-FAD-en-EYE-tis). Physicians routinely palpate the accessible lymph nodes of the cervical, axillary, and inguinal regions for swelling. The collective term for

[5]*lymph* = water; *adeno* = gland; *itis* = inflammation

all lymph node diseases is **lymphadenopathy**[6] (lim-FAD-eh-NOP-a-thee). Lymph nodes are common sites of metastatic cancer (see Deeper Insight 21.1).

## DEEPER INSIGHT 21.1

### CLINICAL APPLICATION

#### Lymph Nodes and Metastatic Cancer

*Metastasis* is a phenomenon in which cancerous cells break free of the original *primary tumor*, travel to other sites in the body, and establish new tumors. Because of the high permeability of lymphatic capillaries, metastasizing cancer cells easily enter them and travel in the lymph. They tend to lodge in the first lymph node they encounter and multiply there, eventually destroying the node. Cancerous lymph nodes are swollen but relatively firm and usually painless. Cancer of a lymph node is called *lymphoma*.

Once a tumor is well established in one node, cells may emigrate from there and travel to the next. However, if the metastasis is detected early enough, cancer can sometimes be eradicated by removing not only the primary tumor, but also the nearest lymph nodes downstream from that point. For example, breast cancer is often treated with a combination of lumpectomy or mastectomy along with removal of the nearby axillary lymph nodes.

[6]*lymph* = water; *adeno* = gland; *pathy* = disease

## Tonsils

The **tonsils** are patches of lymphatic tissue located at the entrance to the pharynx, where they guard against ingested and inhaled pathogens. Each is covered by an epithelium and has deep pits called **tonsillar crypts** lined by lymphatic nodules (fig. 21.13). The crypts often contain food debris, dead leukocytes, bacteria, and antigenic chemicals. Below the crypts, the tonsils are partially separated from underlying connective tissue by an incomplete fibrous capsule.

There are three main sets of tonsils: (1) a single median **pharyngeal tonsil (adenoids)** on the wall of the pharynx just behind the nasal cavity; (2) a pair of **palatine tonsils** at the posterior margin of the oral cavity; and (3) numerous **lingual tonsils,** each with a single crypt, concentrated in patches on each side of the root of the tongue (see fig. 25.5a, p. 953).

The palatine tonsils are the largest and most often infected. *Tonsillitis* is an acute inflammation of the palatine tonsils, usually caused by a *Streptococcus* infection. Their surgical removal, called *tonsillectomy,* used to be one of the most common surgical procedures performed on children, but is less common today. Tonsillitis is now usually treated with antibiotics.

## Spleen

The **spleen,** the body's largest lymphatic organ, measures up to 12 cm long and usually weighs about 150 g. It is located in

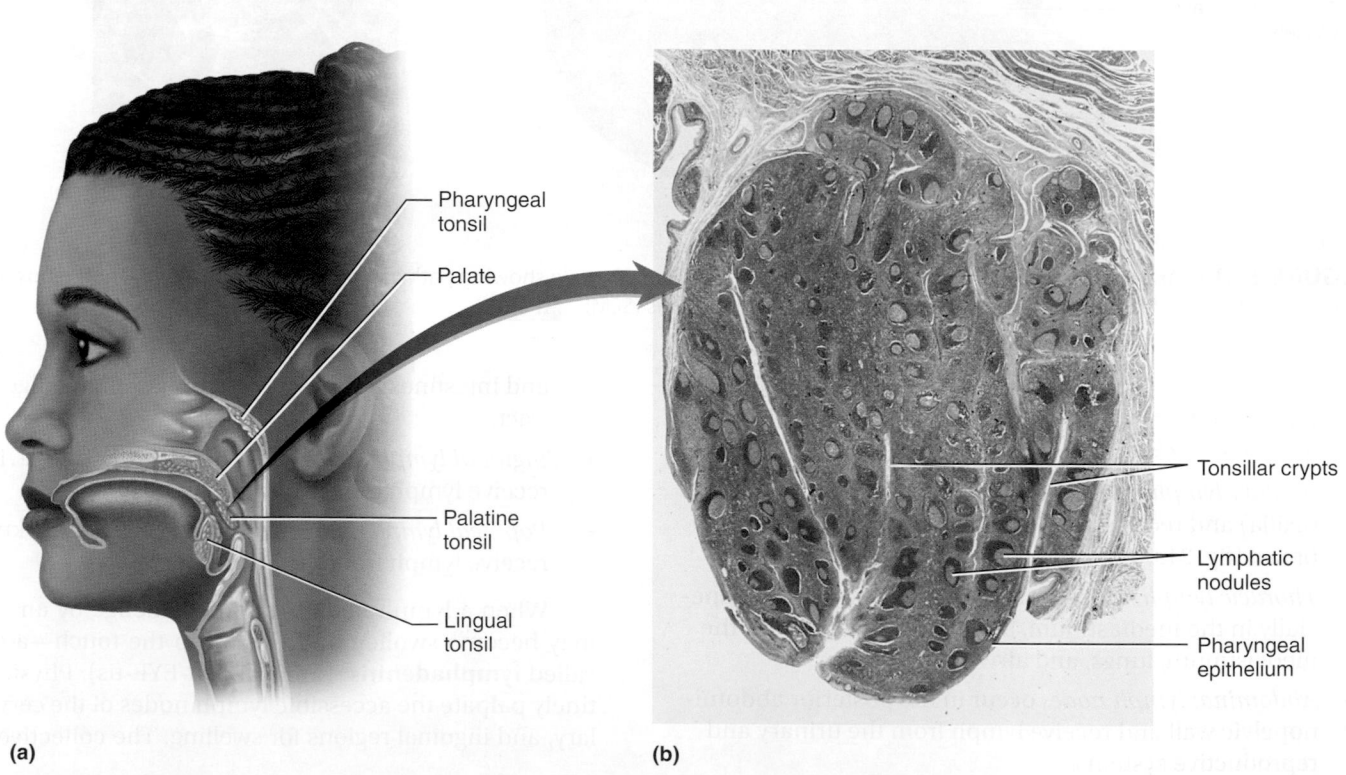

**FIGURE 21.13 The Tonsils.** (a) Locations of the tonsils. (b) Histology of the palatine tonsil. **AP|R**

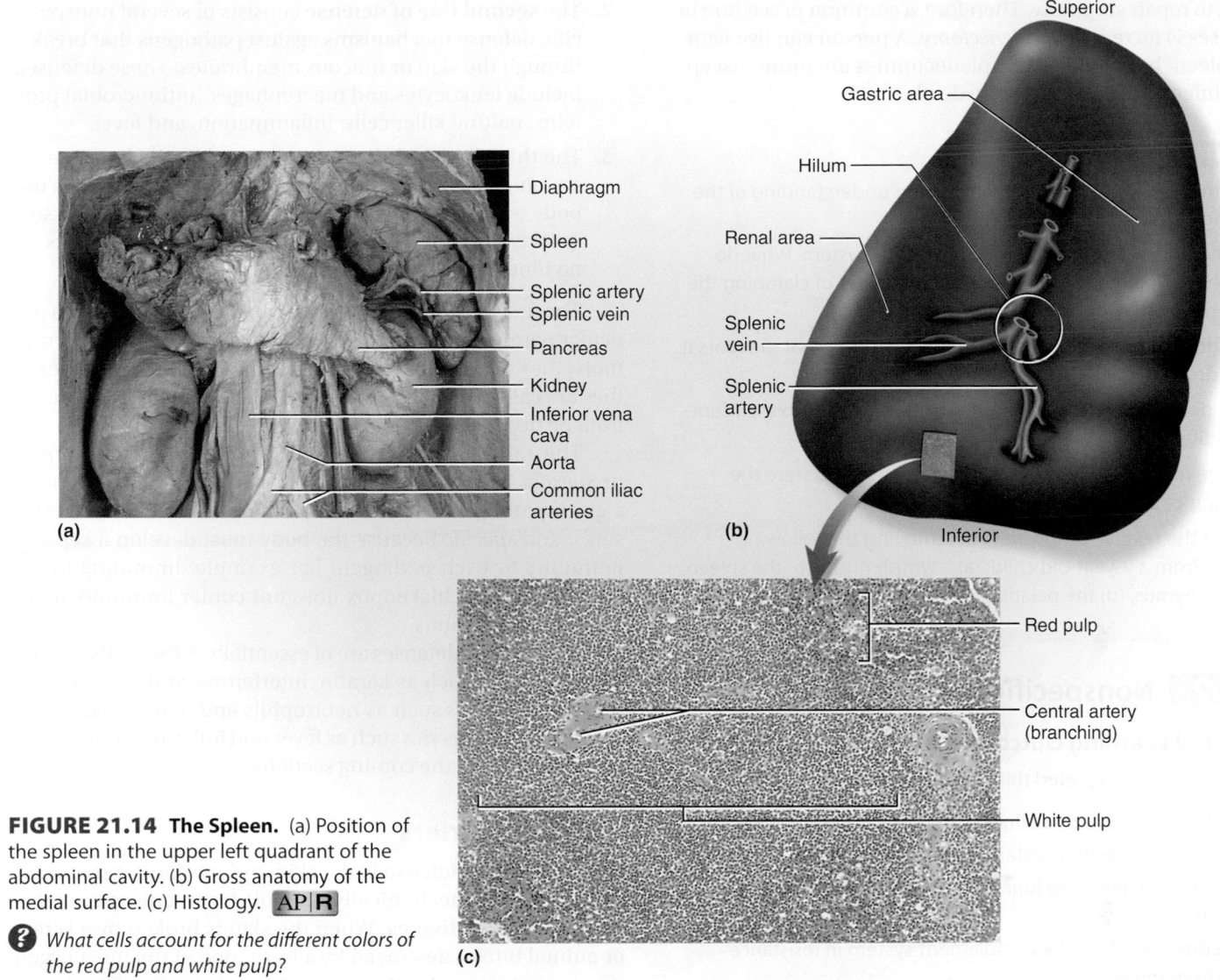

**FIGURE 21.14  The Spleen.** (a) Position of the spleen in the upper left quadrant of the abdominal cavity. (b) Gross anatomy of the medial surface. (c) Histology. AP|R

❓ *What cells account for the different colors of the red pulp and white pulp?*

the left hypochondriac region, just inferior to the diaphragm and posterolateral to the stomach (fig. 21.14; see also fig. B.6, p. 382). It is protected by ribs 10 through 12. The spleen fits snugly between the diaphragm, stomach, and kidney and has indentations called the *gastric area* and *renal area* where it presses against these adjacent viscera. It has a medial hilum penetrated by the splenic artery, splenic vein, and lymphatic vessels.

The parenchyma exhibits two types of tissue named for their appearance in fresh specimens (not in stained sections): **red pulp,** which consists of sinuses gorged with concentrated erythrocytes; and **white pulp,** which consists of lymphocytes and macrophages aggregated like sleeves along small branches of the splenic artery. In tissue sections, white pulp appears as an ovoid mass of lymphocytes with an arteriole passing through it. However, the three-dimensional shape is not egglike but cylindrical.

These two tissue types reflect the multiple functions of the spleen. Its blood capillaries are very permeable; they allow red blood cells (RBCs) to leave the bloodstream, accumulate in the

sinuses of the red pulp, and reenter the bloodstream later. The spleen is an "erythrocyte graveyard"—old, fragile RBCs rupture as they squeeze through the capillary walls into the sinuses. Macrophages phagocytize their remains, just as they dispose of blood-borne bacteria and other cellular debris. The spleen produces blood cells in the fetus and can resume this role in adults in the event of extreme anemia. Lymphocytes and macrophages of the white pulp monitor the blood for foreign antigens, much like the lymph nodes do the lymph. The spleen is a reservoir for a large "standing army" of monocytes, waiting in a state of emergency preparedness. In such events as microbial infection, myocardial infarction, or gaping wounds, angiotensin II stimulates the spleen to release great numbers of monocytes into the bloodstream. The monocytes help to combat pathogens and repair damaged tissues. The spleen also helps to stabilize blood volume by transferring excess plasma from the bloodstream into the lymphatic system.

The spleen is highly vascular and vulnerable to trauma and infection. A ruptured spleen can hemorrhage fatally, but is difficult to repair surgically. Therefore, a common procedure in such cases is its removal, *splenectomy.* A person can live without a spleen, but people with splenectomies are more susceptible to infections and premature death.

### BEFORE YOU GO ON

Answer the following questions to test your understanding of the preceding section:

1. List the primary functions of the lymphatic system. What do you think would be the most noticeable effect of clamping the right lymphatic duct closed?

2. How does fluid get into the lymphatic system? What prevents it from draining back out?

3. What do NK, T, and B cells have in common? How do their functions differ?

4. List five major cell types of lymphatic tissues and state the function of each.

5. Predict the relative seriousness of removing the following organs from a 2-year-old child: (a) a lymph node, (b) the spleen, (c) the thymus, (d) the palatine tonsils.

## 21.2  Nonspecific Resistance

### Expected Learning Outcomes

When you have completed this section, you should be able to

a. identify the body's three lines of defense against pathogens;

b. contrast nonspecific resistance with immunity;

c. describe the defensive functions of each kind of leukocyte;

d. describe the role of the complement system in resistance and immunity;

e. describe the process of inflammation and explain what accounts for its cardinal signs; and

f. describe the body's other nonspecific defenses.

For all living organisms, one of the greatest survival challenges is coping with microbes and other disease agents. **Pathogens**[7] are viruses, bacteria, fungi, and other microbes that cause disease. However, the body must also defend itself from nonliving disease agents such as toxins and radiation, which activate some of the same defenses. Our discussion will focus on infectious microbes. The human body has three lines of defense against disease agents:

1. The **first line of defense** consists of external barriers, notably the skin and mucous membranes, which are impenetrable to most of the pathogens that daily assault us.

2. The **second line of defense** consists of several nonspecific defense mechanisms against pathogens that break through the skin or mucous membranes. These defenses include leukocytes and macrophages, antimicrobial proteins, natural killer cells, inflammation, and fever.

3. The **third line of defense** is mediated by the immune system, which not only defeats a pathogen but leaves the body with a "memory" of it, enabling one to defeat it so quickly in future encounters that the pathogen causes no illness.

The first two defenses lack the capacity to remember a particular pathogen or react differently to it in the future. Furthermore, they defend equally against a broad range of diseases, so they are called *nonspecific defenses.* These defenses are present from birth.

The third defense confers a protection called *adaptive* or *specific immunity*—*adaptive* because the body adapts to a given pathogen and wards if off more easily in future exposures, and *specific* because the body must develop a separate immunity to each pathogen. For example, immunity to one disease such as chickenpox does not confer immunity to another such as tetanus.

Nonspecific defenses are of essentially three kinds: (1) *protective proteins* such as keratin, interferons, and complement; (2) *protective cells* such as neutrophils and macrophages; and (3) *protective processes* such as fever and inflammation. We will examine these in the coming sections.

### External Barriers

Our first line of defense is the skin and mucous membranes, which make it mechanically difficult for microorganisms to enter the body's tissues. When the skin is broken by a scrape or animal bite or destroyed by a burn, one of the most urgent treatment concerns is the prevention of infection. This attests to the importance of intact skin as a barrier. Its surface is composed mainly of keratin, a tough protein that few pathogens can penetrate. Furthermore, with exceptions such as the axillary and pubic areas, it is too dry and poor in nutrients to support much microbial growth. Even those microbes that do adhere to the epidermis are continually cast off by the exfoliation of dead surface keratinocytes (see p. 181). The skin also is coated with diverse antimicrobial chemicals. Sweat and sebum coat it with a protective **acid mantle**—a thin film of lactic and fatty acids that inhibit bacterial growth. Sweat also contains an antibacterial peptide called **dermicidin.** Keratinocytes, neutrophils, macrophages, and other cells also produce peptides called **defensins** and **cathelicidins**[8] (ca-THEL-ih-SY-dins) that destroy bacteria, viruses, and fungi. The effects of these defenses are enhanced by vitamin D (calcitriol), pointing to the benefits of a moderate amount of sunlight exposure for one's resistance to infection.

---

[7]*patho* = disease, suffering; *gen* = producing

[8]*catheli* = universal; *cid* = kill; *in* = substance, protein

The digestive, respiratory, urinary, and reproductive tracts are open to the exterior, making them vulnerable to invasion, but they are protected by mucous membranes. Mucus physically ensnares microbes. Organisms trapped in the respiratory mucus are moved by cilia to the pharynx, swallowed, and destroyed by stomach acid. Microbes are flushed from the upper digestive tract by saliva and from the lower urinary tract by urine. Mucus, tears, and saliva also contain **lysozyme,** an enzyme that destroys bacteria by dissolving their cell walls.

Beneath the epithelia of the skin and mucous membranes is a layer of areolar tissue. Its ground substance contains a giant glycosaminoglycan called **hyaluronic acid** (HI-ul-yur-ON-ic), which gives it a viscous consistency. It is normally difficult for microbes to migrate through this sticky tissue gel. Some organisms overcome this obstacle, however, by producing an enzyme called *hyaluronidase,* which breaks it down to a thinner consistency that is more easily penetrated. Hyaluronidase occurs in some snake venoms and bacterial toxins and is produced by some parasitic protozoans to facilitate their invasion of the connective tissues.

## Leukocytes and Macrophages

If microbes get past the physical barrier of the skin and mucous membranes, they are attacked by **phagocytes** (phagocytic cells) that have a voracious appetite for foreign matter. Leukocytes and macrophages play especially important roles in both nonspecific defense and adaptive immunity and, therefore, in both the second and third lines of defense.

## Leukocytes

The five types of leukocytes are illustrated and described in table 18.6 (p. 692). We will now examine their contributions to resistance and immunity in more detail.

1.  **Neutrophils** spend most of their lives wandering in the connective tissues killing bacteria. One of their methods is simple phagocytosis and digestion. The other is a more complex process that produces a cloud of bactericidal chemicals. When a neutrophil detects bacteria in the immediate area, its lysosomes migrate to the cell surface and *degranulate,* or discharge their enzymes into the tissue fluid. Here, the enzymes catalyze a reaction called the **respiratory burst:** The neutrophil rapidly absorbs oxygen and reduces it to *superoxide anions* ($O_2 \bullet^-$), which react with $H^+$ to form hydrogen peroxide ($H_2O_2$). Another lysosomal enzyme produces hypochlorite (HClO), the active ingredient in chlorine bleach, from chloride ions in the tissue fluid. Superoxide, hydrogen peroxide, and hypochlorite are highly toxic; they form a chemical **killing zone** around the neutrophil that destroys far more bacteria than the neutrophil can destroy by phagocytosis alone. Unfortunately for the neutrophil, it too is killed by these chemicals. These potent oxidizing agents can also damage connective tissues and sometimes contribute to rheumatoid arthritis.

2.  **Eosinophils** are found especially in the mucous membranes, standing guard against parasites, allergens (allergy-causing antigens), and other foes. They become especially concentrated at sites of allergy, inflammation, or parasitic infection. They help to kill parasites such as tapeworms and roundworms by producing superoxide, hydrogen peroxide, and various toxic proteins including a neurotoxin. They promote the action of basophils and mast cells (see next paragraph). They phagocytize and degrade antigen–antibody complexes. Finally, they secrete enzymes that degrade and limit the action of histamine and other inflammatory chemicals that, unchecked, could cause tissue damage.

3.  **Basophils** secrete chemicals that aid the mobility and action of other leukocytes: *leukotrienes* that activate and attract neutrophils and eosinophils; the vasodilator *histamine,* which increases blood flow and speeds the delivery of leukocytes to the area; and the anticoagulant *heparin,* which inhibits the formation of blood clots that would impede leukocyte mobility. These substances are also produced by **mast cells,** a type of connective tissue cell similar to basophils. Eosinophils promote basophil and mast cell action by stimulating them to release these secretions.

4.  **Lymphocytes** all look more or less alike in blood films, but there are several functional types. Three basic categories have already been mentioned: natural killer (NK) cells, T cells, and B cells. In the circulating blood, about 80% of the lymphocytes are T cells, 15% B cells, and 5% NK and stem cells. The roles of these lymphocyte types are too diverse for easy generalizations here, but are described in later sections on NK cells and adaptive immunity.

5.  **Monocytes** are leukocytes that emigrate from the blood into the connective tissues and transform into macrophages. All of the body's avidly phagocytotic cells except leukocytes are called the **macrophage system.** Dendritic cells are included even though they come from different stem cells than macrophages and employ receptor-mediated endocytosis instead of phagocytosis to internalize foreign matter. Some phagocytes are wandering cells that actively seek pathogens, whereas reticular cells and others are fixed in place and phagocytize only those pathogens that come to them—although they are strategically positioned for this to occur. Macrophages are widely distributed in the loose connective tissues, but there are also specialized forms with more specific localities: *microglia* in the central nervous system, *alveolar macrophages* in the lungs, and *hepatic macrophages* in the liver.

## Antimicrobial Proteins

Multiple types of proteins inhibit microbial reproduction and provide short-term, nonspecific resistance to pathogenic bacteria and viruses. We have already considered several at the skin surface, and now turn our attention to two families of blood-borne antimicrobial proteins.

## Interferons

When certain cells (especially leukocytes) are infected with viruses, they secrete proteins called **interferons.** These are of little benefit to the cell that secretes them, but are like its "dying words" that alert neighboring cells and protect them from becoming infected. They bind to surface receptors on those cells and activate second-messenger systems within. This induces the synthesis of dozens of antiviral proteins that defend a cell by such means as breaking down viral genes or preventing viral replication. Interferons also activate NK cells and macrophages, which destroy infected cells before they can liberate a swarm of newly replicated viruses. Interferons also confer resistance to cancer, since the activated NK cells destroy malignant cells.

## Complement System

The **complement system** is a group of 30 or more globulins that make powerful contributions to both nonspecific resistance and adaptive immunity. Immunology pioneer Paul Ehrlich named it *complement* because it "completes the action of antibody," and indeed this is the principal means of pathogen destruction in antibody-mediated immunity. Since Ehrlich's time, however, the complement system has also been found important in nonspecific defense.

Complement proteins are synthesized mainly by the liver. They circulate in the blood in inactive form and are activated in the presence of pathogens. The inactive proteins are named with the letter *C* and a number, such as C3. Activation splits them into fragments, which are further identified by lowercase letters (C3a and C3b, for example).

Activated complement contributes to pathogen destruction by four methods: inflammation, immune clearance, phagocytosis, and cytolysis. We will examine the pathways of activation with a view to understanding how each of these goals is achieved. There are three such routes (fig. 21.15): the classical, alternative, and lectin pathways.

The **classical pathway** requires an antibody to get it started; thus it is part of adaptive immunity. The antibody binds to an antigen on the surface of a microbe and changes shape, exposing a pair of *complement-binding sites* (see fig. 21.27). Complement C1 binds to these sites and sets off a reaction cascade. Like the cascade of blood-clotting reactions, each step generates an enzyme that catalyzes the production of many more molecules at the next step; each step is an amplifying process, so many molecules of product result from a small beginning. In the classical pathway, the cascade is called **complement fixation,** since it results in the attachment of a chain of complement proteins to the antibody.

The alternative and lectin pathways require no antibodies and thus belong to our nonspecific defenses. Complement C3 slowly and spontaneously breaks down in the blood into C3a

and C3b. In the **alternative pathway,** C3b binds directly to targets such as human tumor cells, viruses, bacteria, and yeasts. This, too, triggers a reaction cascade—this time with an *autocatalytic effect* in which C3b leads to the accelerated splitting of more C3 and production of even more C3b.

**Lectins** are plasma proteins that bind to carbohydrates. In the **lectin pathway,** a lectin binds to certain sugars of a microbial cell surface and sets off yet another reaction cascade leading to C3b production.

As we can see (fig. 21.15), the splitting of C3 into C3a and C3b is an intersection where all three pathways converge. These two C3 fragments then produce, directly or indirectly, the end results of the complement system:

1. **Inflammation.**  C3a stimulates mast cells and basophils to secrete histamine and other inflammatory chemicals. It also activates and attracts neutrophils and macrophages, the two key cellular agents of pathogen destruction in inflammation. The exact roles of these chemicals and cells are explained in the section on inflammation to follow.

2. **Immune clearance.**  C3b binds Ag–Ab complexes to red blood cells. As these RBCs circulate through the liver and spleen, the macrophages of those organs strip off and destroy the Ag–Ab complexes, leaving the RBCs unharmed. This is the principal means of clearing foreign antigens from the bloodstream.

3. **Phagocytosis.**  Bacteria, viruses, and other pathogens are phagocytized and digested by neutrophils and macrophages. However, those phagocytes cannot easily internalize "naked" microorganisms. C3b assists them by means of **opsonization:**[9] It coats microbial cells and serves as binding sites for phagocyte attachment. The way Elie Metchnikoff described this, opsonization "butters up" the foreign cells to make them more appetizing.

4. **Cytolysis.**[10]  C3b splits another complement protein, C5, into C5a and C5b. C5a joins C3a in its proinflammatory actions, but C5b plays a more important role in pathogen destruction. It binds to the enemy cell and then attracts complements C6, C7, and C8. This conglomeration of proteins (now called C5b678) goes on to bind up to 17 molecules of complement C9, which form a ring called the *membrane attack complex* (fig. 21.16). The complex forms a hole in the target cell up to 10 nm wide, about the diameter of a single protein molecule. The cell can no longer maintain homeostasis; electrolytes leak out, water flows rapidly in, and the cell ruptures.

---

[9]*opson* = to prepare food
[10]*cyto* = cell; *lysis* = split apart, break down

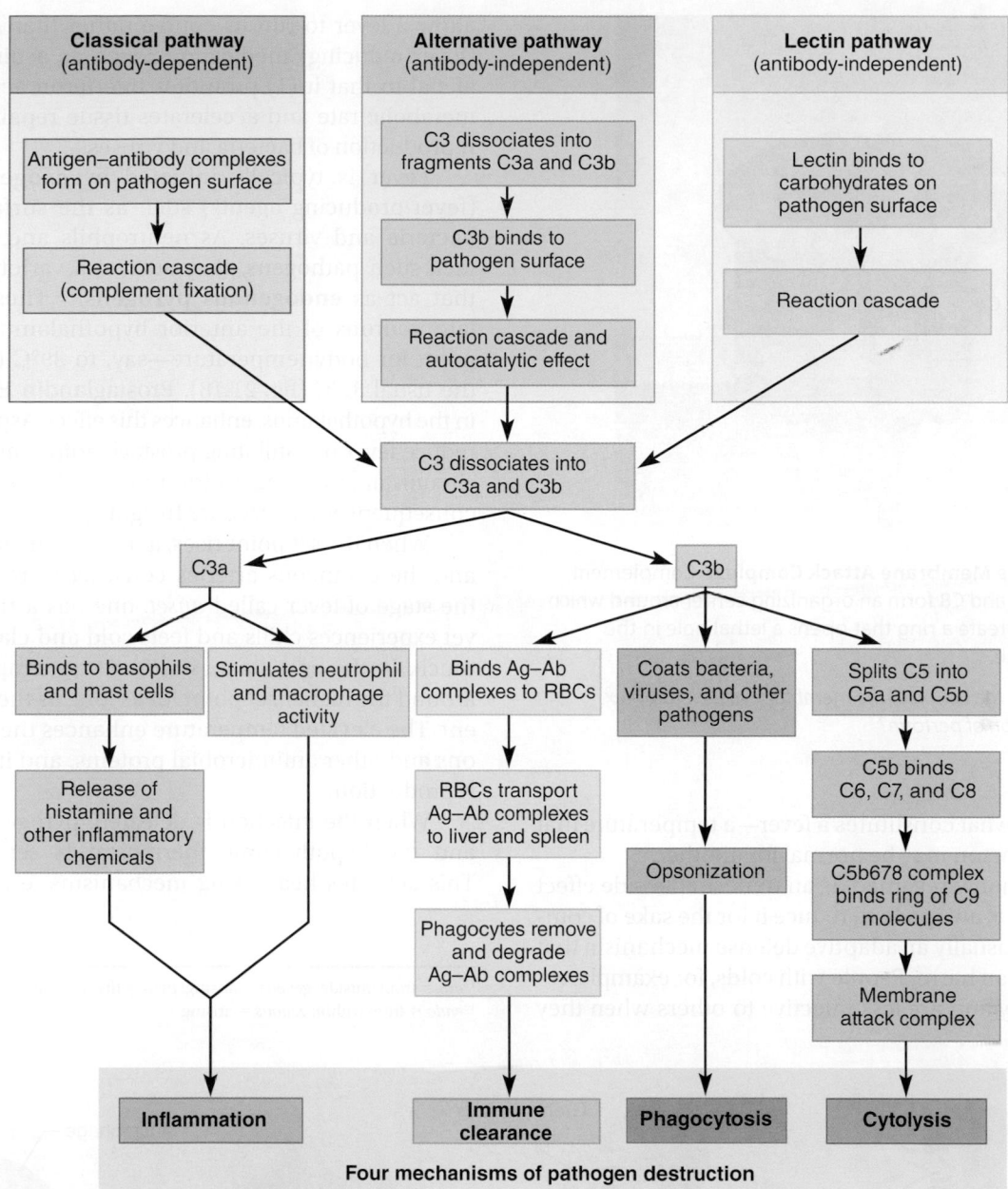

**FIGURE 21.15 Complement Activation.** The classical, alternative, and lectin pathways all lead to the cleavage of complement C3 into C3a and C3b. Those two fragments activate processes that lead to enhanced inflammation, immune clearance, phagocytosis, and cytolysis.

## Natural Killer Cells

Natural killer (NK) cells continually patrol the body "on the lookout" for pathogens or diseased host cells. They attack and destroy bacteria, cells of transplanted organs and tissues, cells infected with viruses, and cancer cells. Upon recognition of an enemy cell, the NK cell binds to it and releases proteins called **perforins,** which polymerize in a ring and create a hole in its plasma membrane (fig. 21.17). This is a kiss of death, for the hole allows a rapid inflow of water and salts. That alone may kill the target cell, but the NK cell also secretes a group of protein-degrading enzymes called **granzymes.** These enter the pore made by the perforins, destroy the target cell's enzymes, and induce apoptosis (programmed cell death).

## Fever

**Fever** is an abnormal elevation of body temperature. It is also known as **pyrexia,** and the term **febrile** means pertaining to fever (as in "febrile attack"). Fever results from trauma, infections, drug reactions, brain tumors, and several other causes. Because of variations in human body temperature, there is no

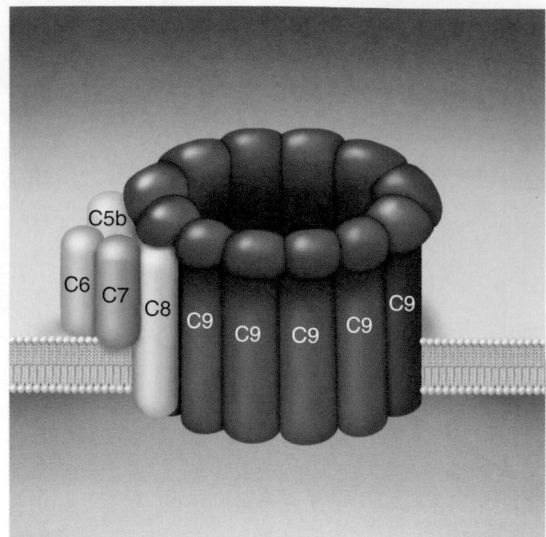

**FIGURE 21.16  The Membrane Attack Complex.** Complement proteins C5b, C6, C7, and C8 form an organizing center around which many C9 molecules create a ring that opens a lethal hole in the enemy cell membrane.

❓ *In what way does the action of the membrane attack complex resemble the action of perforin?*

exact criterion for what constitutes a fever—a temperature that is febrile for one person may be normal for another.

Fever is commonly regarded as an undesirable side effect of illness, and efforts are made to reduce it for the sake of comfort. However, it is usually an adaptive defense mechanism that does more good than harm. People with colds, for example, recover more quickly and are less infective to others when they

allow a fever to run its course rather than using **antipyretic** (fever-reducing) medications such as aspirin. Fever is beneficial in that it (1) promotes interferon activity, (2) elevates metabolic rate and accelerates tissue repair, and (3) inhibits reproduction of bacteria and viruses.

Fever is typically initiated by **exogenous pyrogens**[11] (fever-producing agents) such as the surface glycolipids of bacteria and viruses. As neutrophils and macrophages attack such pathogens, they secrete a variety of polypeptides that act as **endogenous pyrogens.**[12] These in turn stimulate neurons of the anterior hypothalamus to raise the set point for body temperature—say, to 39°C (102°F) instead of the usual 37°C (fig. 21.18). Prostaglandin $E_2$ ($PGE_2$), secreted in the hypothalamus, enhances this effect. Aspirin and ibuprofen reduce fever by inhibiting prostaglandin synthesis, but in some circumstances, using aspirin to control fever can have deadly consequences (see Deeper Insight 21.2).

When the set point rises, a person shivers to generate heat and the cutaneous arteries constrict to reduce heat loss. In the stage of fever called *onset,* one has a rising temperature, yet experiences chills and feels cold and clammy to another's touch. In the next stage, *stadium,* the temperature oscillates around the higher set point for as long as the pathogen is present. The elevated temperature enhances the action of interferons and other antimicrobial proteins, and it inhibits bacterial reproduction.

When the infection is defeated, pyrogen secretion ceases and the hypothalamic thermostat is set back to normal. This activates heat-losing mechanisms, especially cutaneous

---

[11]*exo* = from outside; *genous* = arising; *pyro* = fire, heat; *gen* = producing
[12]*endo* = from within; *genous* = arising

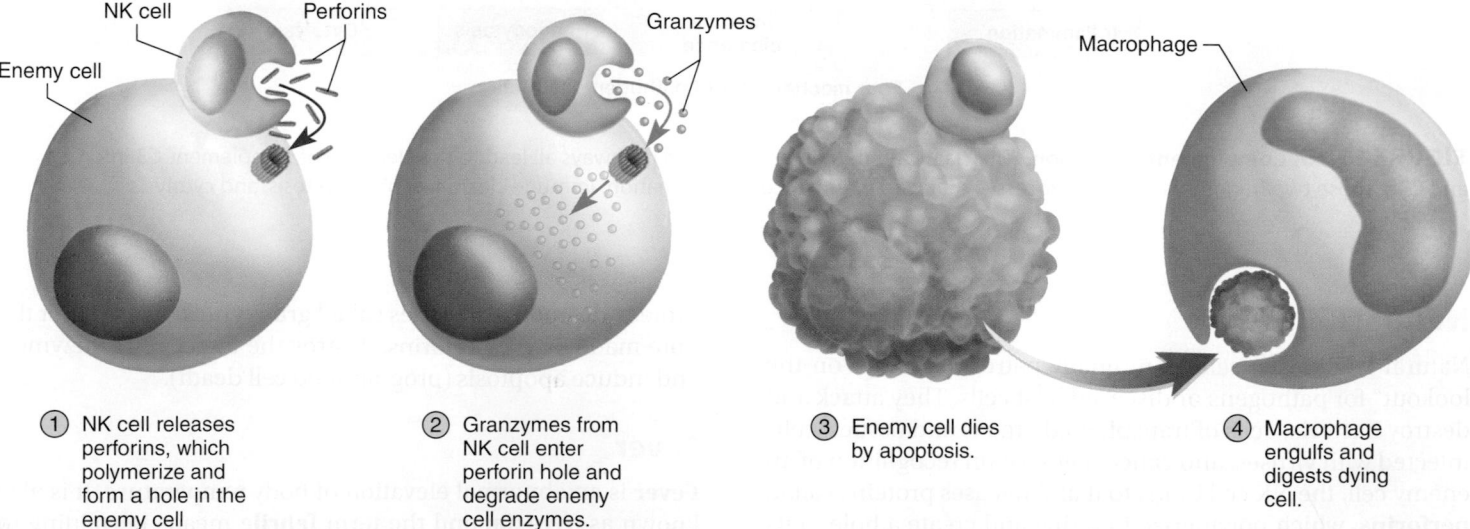

① NK cell releases perforins, which polymerize and form a hole in the enemy cell membrane.

② Granzymes from NK cell enter perforin hole and degrade enemy cell enzymes.

③ Enemy cell dies by apoptosis.

④ Macrophage engulfs and digests dying cell.

**FIGURE 21.17  The Action of a Natural Killer Cell.**

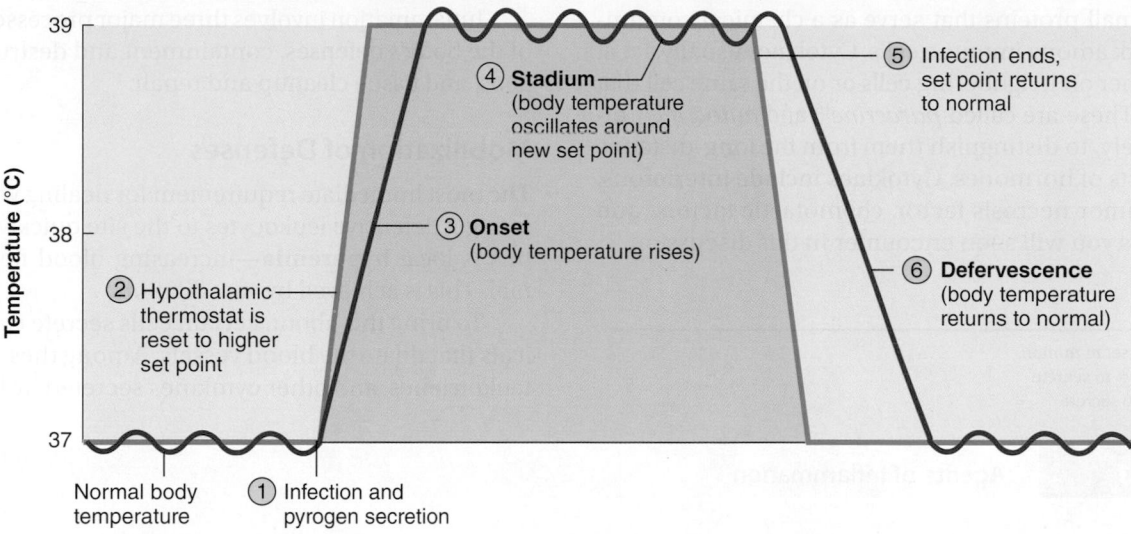

**FIGURE 21.18 The Course of a Fever.**

## DEEPER INSIGHT 21.2

### CLINICAL APPLICATION

#### Reye Syndrome

In children younger than 15, an acute viral infection such as chickenpox or influenza is sometimes followed by a serious disorder called *Reye*[13] *syndrome* (pronounced "rye"). First recognized in 1963, this disease is characterized by swelling of brain neurons and fatty infiltration of the liver and other viscera. Neurons die from hypoxia and the pressure of the swelling brain, which results in nausea, vomiting, disorientation, seizures, and coma. About 30% of victims die, and the survivors sometimes suffer mental retardation. Reye syndrome can be triggered by the use of aspirin to control fever; parents are now strictly advised never to give aspirin to children with chickenpox or flulike symptoms.

vasodilation and sweating. The skin is warm and flushed during this phase. The phase of falling temperature is called *defervescence* in general, *crisis (flush)* if the temperature drops abruptly, or *lysis* if it falls slowly.

Even though most fevers are beneficial, excessively high temperature can be dangerous because it speeds up different enzymatic pathways to different degrees, causing metabolic discoordination and cellular dysfunction. Fevers above 40.5°C (105°F) can make a person delirious. Convulsions and coma ensue at higher temperatures, and death or irreversible brain damage commonly results from fevers that range from 44° to 46°C (111° to 115°F).

[13]R. Douglas Reye (1912–77), Australian pathologist

## Inflammation

**Inflammation** is a local defensive response to tissue injury of any kind, including trauma and infection. Its general purposes are (1) to limit the spread of pathogens and ultimately destroy them, (2) to remove the debris of damaged tissue, and (3) to initiate tissue repair. Inflammation is characterized by four **cardinal signs:** redness, swelling, heat, and pain. Some authorities list impaired use as a fifth sign, but this may or may not occur and when it does, it is mostly because of the pain.

#### ▶▶▶ APPLY WHAT YOU KNOW

*In spite of the expression, only some of the four "cardinal signs" of inflammation are true signs and the others are symptoms. With the aid of the glossary, identify which are signs and which are symptoms.*

Words ending in the suffix *-itis* denote inflammation of specific organs and tissues: *arthritis, encephalitis, peritonitis, gingivitis,* and *dermatitis,* for example. Inflammation can occur anywhere in the body, but it is most common and observable in the skin, which is subject to more trauma than any other organ. Examples of cutaneous inflammation include an itchy mosquito bite, sunburn, a poison ivy rash, and the redness and blistering produced by manual labor, tight shoes, or a kitchen burn.

The following discussion will account for the four cardinal signs and explain how the three purposes of inflammation are achieved. Inflammation is mediated by several types of cells and chemicals summarized in table 21.1. Many of the chemicals that regulate inflammation and immunity are in a class called cytokines[14]—small proteins that serve as a chemical communication network among immune cells. Cytokines usually act at short range, either on neighboring cells or on the same cell that secretes them. These are called *paracrine*[15] and *autocrine*[16] effects, respectively, to distinguish them from the long-distance *endocrine* effects of hormones. Cytokines include interferons, interleukins, tumor necrosis factor, chemotactic factors, and other chemicals you will soon encounter in this discussion.

[14]*cyto* = cell; *kin* = to set in motion
[15]*para* = next to; *crin* = to secrete
[16]*auto* = self; *crin* = to secrete

Inflammation involves three major processes: mobilization of the body's defenses, containment and destruction of pathogens, and tissue cleanup and repair.

## Mobilization of Defenses

The most immediate requirement for dealing with tissue injury is to get defensive leukocytes to the site quickly. The way to do this is local **hyperemia**—increasing blood flow beyond normal. This is achieved by vasodilation.

To bring this about, certain cells secrete *vasoactive* chemicals that dilate the blood vessels. Among these are histamine, leukotrienes, and other cytokines secreted by basophils, mast

| TABLE 21.1 | Agents of Inflammation |
|---|---|
| **Cellular agents** | |
| Basophils | Secrete histamine, heparin, leukotrienes, and kinins |
| Endothelial cells | Produce cell-adhesion molecules to recruit leukocytes; secrete platelet-derived growth factor |
| Eosinophils* | Produce antiparasitic oxidizing agents and toxic proteins; stimulate basophils and mast cells; limit action of histamine and other inflammatory chemicals |
| Fibroblasts | Rebuild damaged tissue by secreting collagen, ground substance, and other tissue components |
| Helper T cells* | Secrete chemotactic factors and colony-stimulating factors |
| Macrophages* | Clean up tissue damage; phagocytize bacteria, tissue debris, dead and dying leukocytes and pathogens |
| Mast cells | Same actions as basophils |
| Monocytes | Emigrate into inflamed tissue and become macrophages |
| Neutrophils | Phagocytize bacteria; secrete bactericidal oxidizing agents; secrete cytokines that activate more leukocytes |
| Platelets | Secrete clotting factors and platelet-derived growth factor |
| **Chemical agents** | |
| Bradykinin | A kinin that stimulates vasodilation and capillary permeability, stimulates pain receptors, and is a neutrophil chemotactic factor |
| Chemotactic factors | Chemicals that provide a trail that neutrophils and other leukocytes can follow to specific sites of infection and tissue injury; include bradykinin, leukotrienes, and some complement proteins |
| Colony-stimulating factors | Hormones that raise the WBC count by stimulating leukopoiesis |
| Complement* | Proteins that promote phagocytosis and cytolysis of pathogens, activate and attract neutrophils and macrophages, and stimulate basophils and mast cells to secrete inflammatory chemicals |
| Cytokines | Small proteins produced mainly by WBCs; have autocrine, paracrine, and endocrine roles in cellular communication; act especially in inflammation and immune responses; include interleukins, interferons, colony-stimulating factors, chemotactic factors, and others |
| Fibrinogen | A plasma clotting protein that filters into inflamed tissue and forms fibrin, thus coagulating tissue fluid, sequestering pathogens, and forming a temporary scaffold for tissue rebuilding |
| Heparin | A polysaccharide secreted by basophils and mast cells that inhibits clotting of tissue fluid within the area walled off by fibrin, thus promoting free mobility of leukocytes that attack infectious microorganisms |
| Histamine | An amino acid derivative produced by basophils and mast cells that stimulates vasodilation and capillary permeability |
| Kinins | Plasma proteins that are activated by tissue injury and stimulate vasodilation, capillary permeability, and pain; include bradykinin |
| Leukotrienes | Eicosanoids that stimulate vasodilation, capillary permeability, and neutrophil chemotaxis, especially in inflammation and allergy |
| Prostaglandins | Eicosanoids that stimulate pain, fever, and vasodilation; promote neutrophil diapedesis; and enhance histamine and bradykinin action |
| Selectins | Cell-adhesion molecules of endothelial cells that adhere to circulating leukocytes, promoting margination and diapedesis |

\* These agents of inflammation have additional roles in adaptive immunity described in table 21.4.

cells, and cells damaged by the agents that triggered the inflammation. Hyperemia not only results in the more rapid delivery of leukocytes, but also washes toxins and metabolic wastes from the tissue more rapidly.

In addition to dilating local blood vessels, the vasoactive chemicals stimulate endothelial cells of the blood capillaries and venules to contract slightly, widening the gaps between them and increasing capillary permeability. This allows for the easier movement of fluid, leukocytes, and plasma proteins from the bloodstream into the surrounding tissue. Among the helpful proteins filtering from the blood are complement, antibodies, and clotting factors, all of which aid in combating pathogens.

Endothelial cells actively recruit leukocytes. In the area of injury, they produce cell-adhesion molecules called **selectins,** which make their membranes sticky and snag leukocytes arriving in the bloodstream. Leukocytes adhere loosely to the selectins and slowly tumble along the endothelium, sometimes coating it so thickly that they obstruct blood flow. This adhesion to the vessel wall is called **margination.** The leukocytes then crawl through the gaps between the endothelial cells—an action called **diapedesis**[17] or **emigration**—and enter the tissue fluid of the damaged tissue (fig. 21.19). Most diapedesis occurs across the walls of the postcapillary venules. Cells and chemicals that have left the bloodstream are said to be *extravasated.*[18]

In the events that have already transpired, we can see the basis for the four cardinal signs of inflammation: (1) The heat results from the hyperemia; (2) redness is also due to hyperemia and in some cases, such as sunburn, to extravasated erythrocytes in the tissue; (3) swelling (edema) is due to the increased fluid filtration from the capillaries; and (4) pain results from direct injury to the nerves, pressure on the nerves from the edema, and stimulation of pain receptors by prostaglandins, some bacterial toxins, and a kinin called **bradykinin.**

#### ►►►APPLY WHAT YOU KNOW

*Review eicosanoid synthesis (p. 660) and explain why aspirin eases the pain of inflammation.*

### Containment and Destruction of Pathogens

One priority in inflammation is to prevent pathogens from spreading through the body. The fibrinogen that filters into the tissue fluid clots in areas adjacent to the injury, forming a sticky mesh that sequesters (walls off and isolates) bacteria and other microbes. Heparin, the anticoagulant, prevents clotting in the immediate area of the injury, so bacteria or other pathogens are essentially trapped in a fluid pocket surrounded by a gelatinous capsule of clotted fluid. They are attacked by antibodies, phagocytes, and other defenses, while the surrounding areas of clotted tissue fluid prevent them from easily escaping this onslaught.

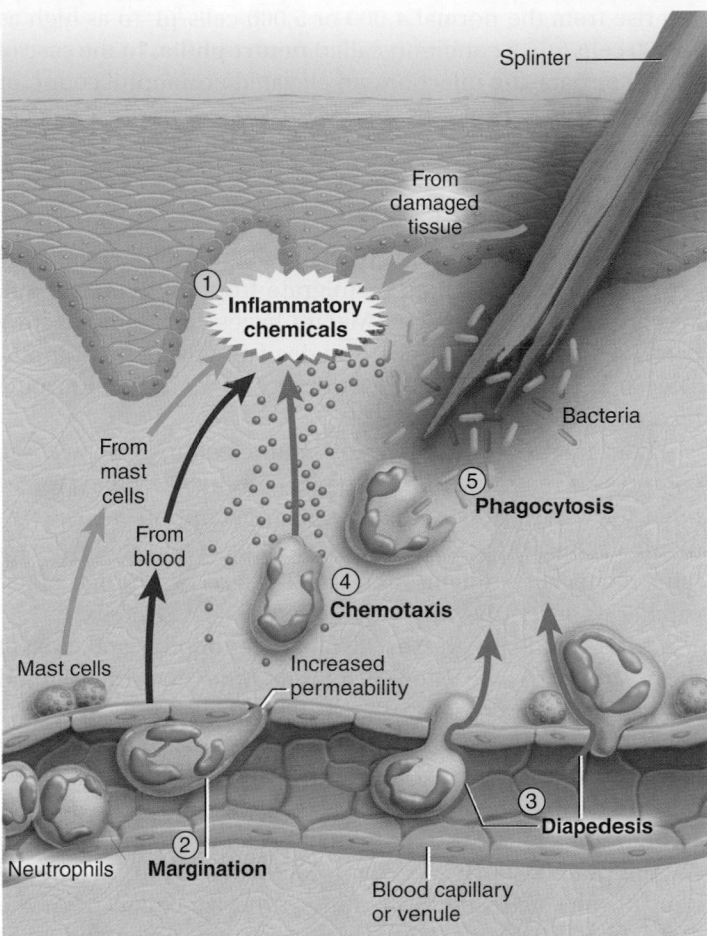

**FIGURE 21.19 Neutrophil Behavior in Inflammation.** Chemical messengers are released by basophils, mast cells, blood plasma, and damaged tissue. These inflammatory chemicals stimulate leukocyte margination (adhesion to the capillary wall), diapedesis (crawling through the capillary wall), chemotaxis (movement toward the source of the inflammatory chemicals), and phagocytosis (engulfing bacteria or other pathogens).

The chief enemies of bacteria are neutrophils, which accumulate in the inflamed tissue within an hour. After emigrating from the bloodstream, they exhibit **chemotaxis**—attraction to chemicals *(chemotactic factors)* such as bradykinin and leukotrienes that guide them to the site of injury or infection. As they encounter bacteria, neutrophils avidly phagocytize and digest them, and destroy many more by the respiratory burst described earlier. The major stages of neutrophil action are summarized in figure 21.19.

Neutrophils also recruit macrophages and additional neutrophils by secreting cytokines, like shouting "Over here!" to bring in reinforcements. Activated macrophages and T cells in the inflamed tissue secrete cytokines called *colony-stimulating factors,* which promote the production of more leukocytes (leukopoiesis) by the red bone marrow. Within a few hours of the onset of inflammation, the neutrophil count in the blood

---

[17]*dia* = through; *pedesis* = stepping
[18]*extra* = outside; *vas* = vessel

can rise from the normal 4,000 or 5,000 cells/μL to as high as 25,000 cells/μL, a condition called **neutrophilia.** In the case of allergy or parasitic infection, an elevated eosinophil count, or **eosinophilia,** may also occur. The task of eosinophils was described earlier.

## Tissue Cleanup and Repair

Monocytes are major agents of tissue cleanup and repair. They arrive within 8 to 12 hours, emigrate from the bloodstream, and turn into macrophages. Macrophages engulf and destroy bacteria, damaged host cells, and dead and dying neutrophils. They also act as antigen-presenting cells, activating immune responses described later.

Edema also contributes to tissue cleanup. The swelling compresses veins and reduces venous drainage, while it forces open the valves of lymphatic capillaries and promotes lymphatic drainage. The lymphatics can collect and remove bacteria, dead cells, proteins, and tissue debris better than blood capillaries or venules can.

As the battle progresses, all of the neutrophils and most of the macrophages die. These dead cells, other tissue debris, and tissue fluid form a pool of yellowish fluid called **pus,** which accumulates in a tissue cavity called an **abscess.**[19] Pus is usually absorbed, but sometimes it forms a blister between the epidermis and dermis and may be released by its rupture.

Blood platelets and endothelial cells in an area of injury secrete **platelet-derived growth factor,** an agent that stimulates fibroblasts to multiply and synthesize collagen. At the same time, hyperemia delivers oxygen, amino acids, and other necessities of protein synthesis, while the heat of inflamed tissue increases metabolic rate and the speed of mitosis and tissue repair. The fibrin clot in the tissue may provide a scaffold for reconstruction. Pain also contributes importantly to recovery. It is an important alarm signal that calls our attention to the injury and makes us limit the use of a body part so it has a chance to rest and heal.

### BEFORE YOU GO ON

Answer the following questions to test your understanding of the preceding section:

6. What are macrophages? Give four examples and state where they are found.

7. How do interferons and the complement system protect against disease?

8. Summarize the benefits of fever and the limits of these benefits.

9. List the cardinal signs of inflammation and state the cause of each.

---

## 21.3 General Aspects of Adaptive Immunity

### Expected Learning Outcomes

When you have completed this section, you should be able to

a. define *adaptive immunity;*

b. contrast cellular and humoral immunity, active and passive immunity, and natural and artificial immunity;

c. describe the chemical properties of antigens;

d. describe and contrast the development of T and B lymphocytes; and

e. describe the general roles played by lymphocytes, antigen-presenting cells, and interleukins in the immune response.

The remainder of this chapter is concerned with the **immune**[20] **system** and adaptive immunity, the third line of defense. The **immune system** consists of a large population of widely distributed cells that recognize foreign substances and act to neutralize or destroy them. Two characteristics distinguish adaptive (specific) immunity from nonspecific resistance:

1. **Specificity.** Immunity is directed against a particular pathogen. Immunity to one pathogen usually does not confer immunity to others.

2. **Memory.** When reexposed to the same pathogen, the body reacts so quickly that there is no noticeable illness. The reaction time for inflammation and other nonspecific defenses, by contrast, is just as long for later exposures as for the initial one.

These properties of immunity were recognized even in the fifth century BCE, when Greek historian Thucydides remarked that people who recover from a disease often become immune to that one but remain susceptible to others. A person might be immune to measles but still susceptible to polio, for example.

## Forms of Immunity

In the late 1800s, it was discovered that immunity can be transferred from one animal to another by way of the blood serum. In the mid-1900s, however, it was found that serum does not always confer immunity; sometimes only donor lymphocytes do so. Thus, we now recognize two types of immunity, called cellular and humoral immunity, although they interact extensively and often respond to the same pathogen.

**Cellular (cell-mediated) immunity** employs lymphocytes that directly attack and destroy foreign cells or diseased host

---

[19]*ab* = away; *scess* (from *cedere*) = to go

[20]*immuno* = free

cells. It is a means of ridding the body of pathogens that reside inside human cells, where they are inaccessible to antibodies: intracellular viruses, bacteria, yeasts, and protozoans, for example. Cellular immunity also acts against parasitic worms, cancer cells, and cells of transplanted tissues and organs.

**Humoral (antibody-mediated) immunity** employs antibodies, which do not directly destroy a pathogen, but tag them for destruction by mechanisms described later. The expression *humoral* comes from the fact that many of the antibodies are dissolved in the body fluids ("humors"). Humoral immunity is effective against extracellular viruses, bacteria, yeasts, protozoans, and molecular (noncellular) pathogens such as toxins, venoms, and allergens. In the unnatural event of a mismatched blood transfusion, it also destroys foreign erythrocytes.

Note that humoral immunity works only against the *extracellular* stages of infectious microorganisms. When such microorganisms invade host cells, antibodies cannot get at them. However, the *intracellular* stages are still vulnerable to cellular immunity, which destroys them by killing the cells that harbor them. Thus, humoral and cellular immunity sometimes attack the same microorganism at different points in its life cycle.

After our discussion of the details of the two processes, you will find cellular and humoral immunity summarized and compared in table 21.5 (p. 836).

Other ways of classifying immunity are active versus passive and natural versus artificial. In *active immunity,* the body makes its own antibodies or T cells against a pathogen, whereas in *passive immunity,* the body acquires them from another person or an animal that has developed immunity to the pathogen. Either type of immunity can occur naturally or, for treatment and prevention purposes, it can be induced artificially. Thus we can recognize four classes of immunity under this scheme:

1. **Natural active immunity.** This is the production of one's own antibodies or T cells as a result of natural exposure to an antigen.

2. **Artificial active immunity.** This is the production of one's own antibodies or T cells as a result of **vaccination** against diseases such as smallpox, tetanus, or influenza. A **vaccine** consists of either dead or *attenuated* (weakened) pathogens that can stimulate an immune response but cause little or no discomfort or disease. In some cases, periodic *booster shots* are given to restimulate immune memory and maintain a high level of protection (tetanus boosters, for example). Vaccination has eliminated smallpox worldwide and greatly reduced the incidence of life-threatening childhood diseases, but many people continue to die from influenza and other diseases that could be prevented by vaccination.

3. **Natural passive immunity.** This is a temporary immunity that results from acquiring antibodies produced by another person. The only natural way for this to happen is for a fetus to acquire antibodies from the mother through the placenta before birth, or for a baby to acquire them during breast-feeding.

4. **Artificial passive immunity.** This is a temporary immunity that results from the injection of an *immune serum* obtained from another person or from animals (such as horses) that have antibodies against a certain pathogen. Immune serum is used for emergency treatment of snakebites, botulism, tetanus, rabies, and other diseases.

Only the two forms of active immunity involve memory and thus provide future protection. Passive immunity typically lasts for only 2 or 3 weeks, until the acquired antibody is degraded. The remaining discussion is based on natural active immunity.

## Antigens

An **antigen**[21] **(Ag)** is any molecule that triggers an immune response. Some antigens are free molecules such as venoms and toxins; others are components of plasma membranes and bacterial cell walls. Small universal molecules such as glucose and amino acids are not antigenic; if they were, our immune systems would attack the nutrients and other molecules essential to our very survival. Most antigens have molecular weights over 10,000 amu and are complex molecules unique to each individual: proteins, polysaccharides, glycoproteins, and glycolipids. Their uniqueness enables the body to distinguish its own ("self") molecules from those of any other individual or organism ("nonself"). The immune system learns to distinguish self-antigens from nonself-antigens prior to birth; thereafter, it normally attacks only nonself-antigens.

Only certain regions of an antigen molecule, called **epitopes (antigenic determinants),** stimulate immune responses. One antigen molecule typically has several different epitopes, however, that can stimulate the production of different antibodies.

Some molecules, called **haptens,**[22] are too small to be antigenic in themselves, but they can stimulate an immune response by binding to a host macromolecule and creating a unique complex that the body recognizes as foreign. After the first exposure, the hapten alone may stimulate an immune response without needing to bind to a host molecule. Many people are allergic to haptens in cosmetics, detergents, industrial chemicals, poison ivy, and animal dander. The most common drug allergy is to penicillin—a hapten that binds to host proteins in allergic individuals, creating a complex that binds to mast cells and triggers massive release of histamine and other inflammatory chemicals. This can cause death from *anaphylactic shock.*

---

[21]acronym from *anti*body *gene*rating
[22]from *haptein* = to fasten

## Lymphocytes

The major cells of the immune system are lymphocytes, macrophages, and dendritic cells, which are especially concentrated at strategic places such as the lymphatic organs, skin, and mucous membranes. Lymphocytes fall into three classes: natural killer (NK) cells, T lymphocytes, and B lymphocytes. We have already discussed the nonspecific response of NK cells. Here, we must take a closer look at T and B lymphocytes.

## T Lymphocytes (T Cells)

The life history of a T cell basically involves three stages and three anatomical stations in the body. We can loosely think of these stages as their "birth," their "training" or maturation, and finally their "deployment" to locations where they will carry out their immune function.

T cells are "born" in the red bone marrow as descendants of the hemopoietic stem cells described in chapter 18. New T cells enter the bloodstream and travel to the thymus—the "school" where they mature into fully functional T cells (fig. 21.20). In the thymic cortex, reticular epithelial (RE) cells

secrete hormones that stimulate the T cells to develop surface antigen receptors. With receptors in place, the T cells are now **immunocompetent,** capable of recognizing antigens presented to them by APCs. The RE cells then test these cells by presenting self-antigens to them. There are two ways to fail the test: (1) inability to recognize the RE cells at all (specifically, their MHC proteins, described later), or (2) reacting to the self-antigens. Failure to recognize MHC proteins would mean the T cell would be incapable of recognizing a foreign attack on the body, but reacting to self-antigens would mean the T cells would attack one's own tissues. Either way, T cells that fail the test must be eliminated—a process called **negative selection.**

There are two forms of negative selection: (1) **clonal deletion,** in which self-reactive T cells die and macrophages phagocytize them, and (2) **anergy,**[23] in which they remain alive but unresponsive. Negative selection leaves the body in a state of **self-tolerance** in which the surviving, active T cells respond only to foreign antigens, not to one's own.

---

[23]*an* = without; *erg* = action, work

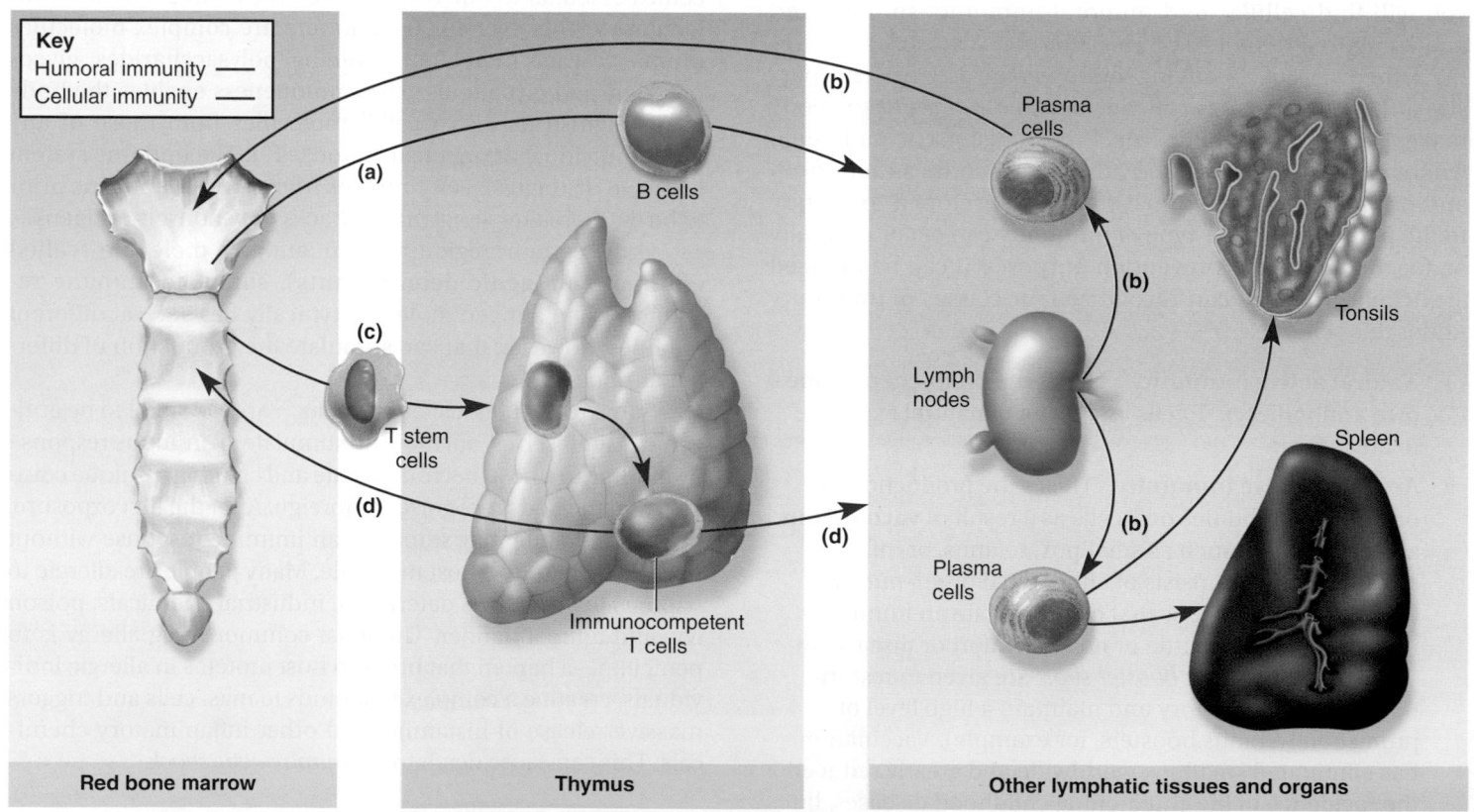

**FIGURE 21.20  The Life History and Migrations of B and T Cells.**  Humoral immunity is represented by the violet pathways and cellular immunity by the red. (a) B cells achieve immunocompetence in the red bone marrow (left), and many emigrate to lymphatic tissues and organs, including the lymph nodes, tonsils, and spleen (right). (b) Plasma cells develop in the lymph nodes (among other sites) and emigrate to the bone marrow and other lymphatic organs, where they spend a few days secreting antibodies. (c) T stem cells emigrate from the bone marrow and attain immunocompetence in the thymus. (d) Immunocompetent T cells emigrate from the thymus and recolonize the bone marrow or colonize various lymphatic organs (right).

T cells pass the test if they recognize the RE cells but do not react strongly to self-antigens. They now "graduate" to join the immune workforce. Only 2% of the T cells pass their graduation test. They move on to the medulla of the thymus, where they undergo **positive selection**—they multiply and form **clones** of identical T cells programmed to respond to a particular antigen. These cells, which are immunocompetent but have not yet encountered the "enemy" (foreign antigens), constitute the **naive lymphocyte pool.** Naive T cells leave the thymus and disperse, colonizing lymphatic tissues and organs everywhere (the bone marrow, lymph nodes, tonsils, and so forth). They are now ready to do battle.

▶▶▶ **APPLY WHAT YOU KNOW**

*Is clonal deletion a case of apoptosis or necrosis? Explain your answer. (Review these concepts in chapter 5 if necessary.)*

## B Lymphocytes (B Cells)

Another group of fetal stem cells remains in the bone marrow to differentiate into B cells. Those that respond to self-antigens undergo either anergy or clonal deletion, much like self-reactive T cells. Self-tolerant B cells, on the other hand, go on to produce surface receptors for antigens, divide, and produce immunocompetent B cell clones. These cells disperse throughout the body, colonizing the same organs as T cells (fig. 21.20). They are abundant in the lymph nodes, spleen, bone marrow, and mucous membranes.

## Antigen-Presenting Cells

Although the function of T cells is to recognize and attack foreign antigens, they usually cannot recognize such antigens on their own. They require the help of **antigen-presenting cells (APCs).** In addition to their other roles, dendritic cells, macrophages, reticular cells, and B cells function as APCs.

APC function hinges on a family of genes on chromosome 6 called the **major histocompatibility complex (MHC).** These genes code for **MHC proteins**—proteins on the APC surface that are shaped a little like hotdog buns, with an elongated groove for holding the "hotdog" of the foreign antigen. MHC proteins are structurally unique to every person except for identical twins. They act as "identification tags" that label every cell of your body as belonging to you.

When an APC encounters an antigen, it internalizes it by endocytosis, digests it into molecular fragments, and displays the relevant fragments (its epitopes) in the grooves of the MHC proteins (fig. 21.21a). These steps are called **antigen processing.** Wandering T cells regularly inspect APCs for displayed antigens (fig. 21.21b). If an APC displays a self-antigen, the T cells disregard it. If it displays a nonself-antigen, however, they initiate an attack. APCs thus alert the immune system to the presence of a foreign antigen. The key to a successful defense is then to quickly mobilize immune cells against it.

With so many cell types involved in immunity, it is not surprising that they require chemical messengers to coordinate their activities. Lymphocytes and APCs talk to each other with cytokines called **interleukins**[24]—chemical signals from one leukocyte (or leukocyte derivative) to another.

With this introduction to the main actors in immunity, we can now look at the more specific features of cellular and humoral immunity. Since the terminology of immune cells and chemicals is quite complex, you may find it helpful to refer often to table 21.4 (p. 836) as you read the following discussions.

**BEFORE YOU GO ON**

Answer the following questions to test your understanding of the preceding section:

10. How does adaptive immunity differ from nonspecific defense?
11. How does humoral immunity differ from cellular immunity?
12. Contrast active and passive immunity. Give natural and artificial examples of each.
13. What structural properties distinguish antigenic molecules from those that are not antigenic?
14. What is an immunocompetent lymphocyte? What does a lymphocyte have to produce in order to become immunocompetent?
15. What role does the thymus play in the life history of a T cell?
16. What role does an antigen-presenting cell play in the activation of a T cell?

## 21.4 Cellular Immunity AP|R

### Expected Learning Outcomes

When you have completed this section, you should be able to

a. list the types of lymphocytes involved in cellular immunity and describe the roles they play;
b. describe the process of antigen presentation and T cell activation;
c. describe how T cells destroy enemy cells; and
d. explain the role of memory cells in cellular immunity.

Cellular (cell-mediated) immunity is a form of adaptive immunity in which T lymphocytes directly attack and destroy diseased or foreign cells, and the immune system remembers

---

[24]*inter* = between; *leuk* = leukocytes

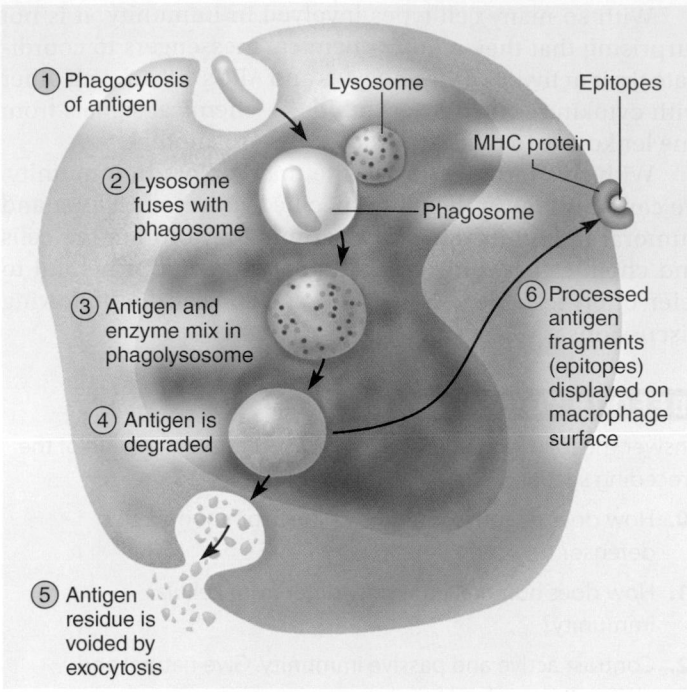

(a)

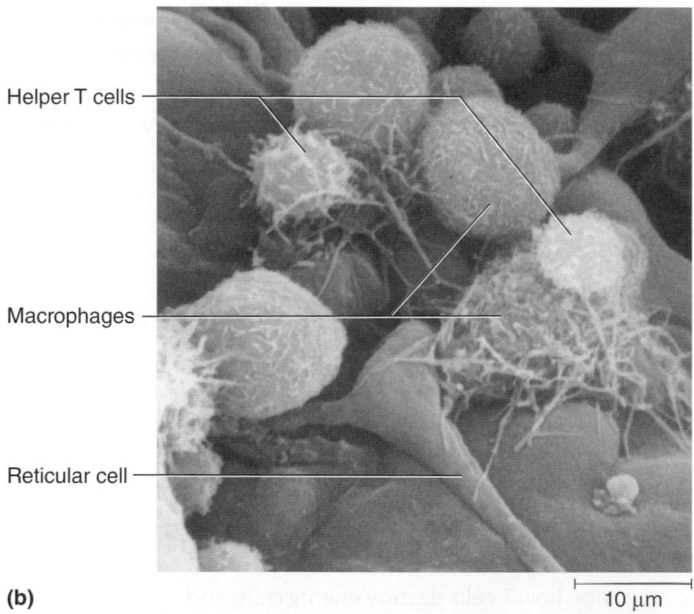

(b)

**FIGURE 21.21 The Action of Antigen-Presenting Cells (APCs).**
(a) Stages in the processing and presentation of an antigen by an APC such as a macrophage. (b) Macrophages and reticular cells presenting processed antigens to helper T cells. [© Dr. Kessel & Dr. Kardon/Tissues and Organs/Visuals Unlimited, Inc.]

the antigens of those invaders and prevents them from causing disease in the future. Cellular immunity employs four classes of T cells:

1. **Cytotoxic T ($T_C$) cells** are the "effectors" of cellular immunity that carry out the attack on foreign cells.

2. **Helper T ($T_H$) cells** promote the action of $T_C$ cells as well as playing key roles in humoral immunity and nonspecific resistance. All other T cells are involved in cellular immunity only.

3. **Regulatory T ($T_R$) cells,** or **T-regs,** limit the immune response by inhibiting multiplication and cytokine secretion by other T cells. They are still not well understood, but seem especially important in preventing autoimmune diseases, discussed later in this chapter.

4. **Memory T ($T_M$) cells** are descended from $T_C$ cells and are responsible for memory in cellular immunity.

$T_C$ cells are also known as T8, CD8, or CD8+ cells because they have a surface glycoprotein called CD8. $T_H$ and $T_R$ cells are also known as T4, CD4, or CD4+ cells, after their glycoprotein, CD4. (*CD* stands for *cluster of differentiation,* a classification system for many cell surface molecules.) These glycoproteins are cell-adhesion molecules that enable T cells to bind to other cells in the events to be described shortly.

Both cellular and humoral immunity occur in three stages that we can think of as recognition, attack, and memory (or "the three *R*s of immunity"—recognize, react, and remember). In cellular immunity, the events of each stage are as follows.

## Recognition

The recognition phase has two aspects: antigen presentation and T cell activation.

### Antigen Presentation

When an antigen-presenting cell (APC) encounters and processes an antigen, it typically migrates to the nearest lymph node and displays it to the T cells. Cytotoxic and helper T cells patrol the lymph nodes and other tissues as if looking for trouble. When they encounter a cell displaying an antigen on an MHC protein (MHCP), they initiate an immune response. T cells respond to two classes of MHCPs:

1. *MHC-I proteins* occur on every nucleated cell of the body (not erythrocytes). These proteins are constantly produced by the cell and transported to the plasma membrane. Along the way, they pick up small peptides in the cytoplasm and display these once they are installed in the membrane. If the peptides are normal self-antigens, they

do not elicit a T cell response. If they are viral proteins or abnormal antigens made by cancer cells, however, they do. In this case, the Ag–MHCP complex is like a tag on the host cell that says, "I'm diseased; kill me." Infected or malignant cells are then destroyed before they can do further harm to the body.

2. *MHC-II proteins* (also called *human leukocyte antigens, HLAs*) occur only on APCs and display only foreign antigens.

$T_C$ cells respond only to MHC-I proteins, and $T_H$ cells respond only to MHC-II (table 21.2).

## T Cell Activation

T cell activation is shown in figure 21.22. It begins when a $T_C$ or $T_H$ cell binds to an MHCP displaying an epitope that the T cell is programmed to recognize. Before the response can go any further, the T cell must bind to another APC protein, related to interleukins. In a sense, the T cell has to check twice to see if it really has bound to an APC displaying a suspicious antigen. This signaling process, called **costimulation,** helps to ensure that the immune system does not launch an attack in the absence of an enemy, which could turn against one's own body with injurious consequences.

Successful costimulation activates the process of **clonal selection:** The T cell undergoes repeated mitosis, giving rise to a clone of identical T cells programmed against the same epitope. Some cells in the clone become effector cells that carry out an immune attack, and some become memory T cells.

## Attack

Helper and cytotoxic T cells play different roles in the attack phase. Helper T ($T_H$) cells are necessary for most immune responses. They play a central coordinating role in both humoral and cellular immunity (fig. 21.23). When a $T_H$ cell recognizes an Ag–MHCP complex, it secretes interleukins that exert three effects: (1) They attract neutrophils and natural killer cells; (2) they attract macrophages, stimulate their phagocytic activity, and inhibit them from leaving the area; and (3) they stimulate T and B cell mitosis and maturation.

Cytotoxic T ($T_C$) cells are the only T lymphocytes that directly attack and kill other cells (fig. 21.24). When a $T_C$ cell recognizes a complex of antigen and MHC-I protein on a diseased or foreign cell, it "docks" on that cell, delivers a **lethal hit** of chemicals that will destroy it, and goes off in search of other enemy cells while the chemicals do their work. Among these chemicals are

- perforin and granzymes, which kill the target cell in the same manner as we saw earlier for NK cells (see fig. 21.17);

- interferons, which inhibit viral replication and recruit and activate macrophages, among other effects; and

- *tumor necrosis factor (TNF),* which aids in macrophage activation and kills cancer cells.

As more and more cells are recruited by helper T cells, the immune response exerts an overwhelming force against the pathogen. The primary response, seen on first exposure to a particular pathogen, peaks in about a week and then gradually declines.

#### ►►►APPLY WHAT YOU KNOW

*How is a cytotoxic T cell like a natural killer (NK) cell? How are they different?*

## Memory

The primary response is followed by immune memory. Following clonal selection, some T cells become memory cells. These cells are long-lived and much more numerous than naive T cells. Aside from their sheer numbers, they also require fewer steps to be activated, and therefore respond to antigens more rapidly. Upon reexposure to the same pathogen later in life, memory cells mount a quick attack called the **T cell recall response.** This time-saving response destroys a pathogen so quickly that no noticeable illness occurs—that is, the person is immune to the disease.

#### BEFORE YOU GO ON

Answer the following questions to test your understanding of the preceding section:

17. Name four types of lymphocytes involved in cellular immunity. Which of these is also essential to humoral immunity?

18. What are the three phases of an immune response?

19. Explain why cytotoxic T cells are activated by a broader range of host cells than are helper T cells.

20. Describe some ways in which cytotoxic T cells destroy target cells.

| **TABLE 21.2** | **Comparison of the Responses of Cytotoxic and Helper T Cells** | |
|---|---|---|
| Characteristic | $T_C$ Cells | $T_H$ Cells |
| Cells capable of stimulating a response | Any nucleated cell | Antigen-presenting cells |
| MHC protein | MHC-I | MHC-II |

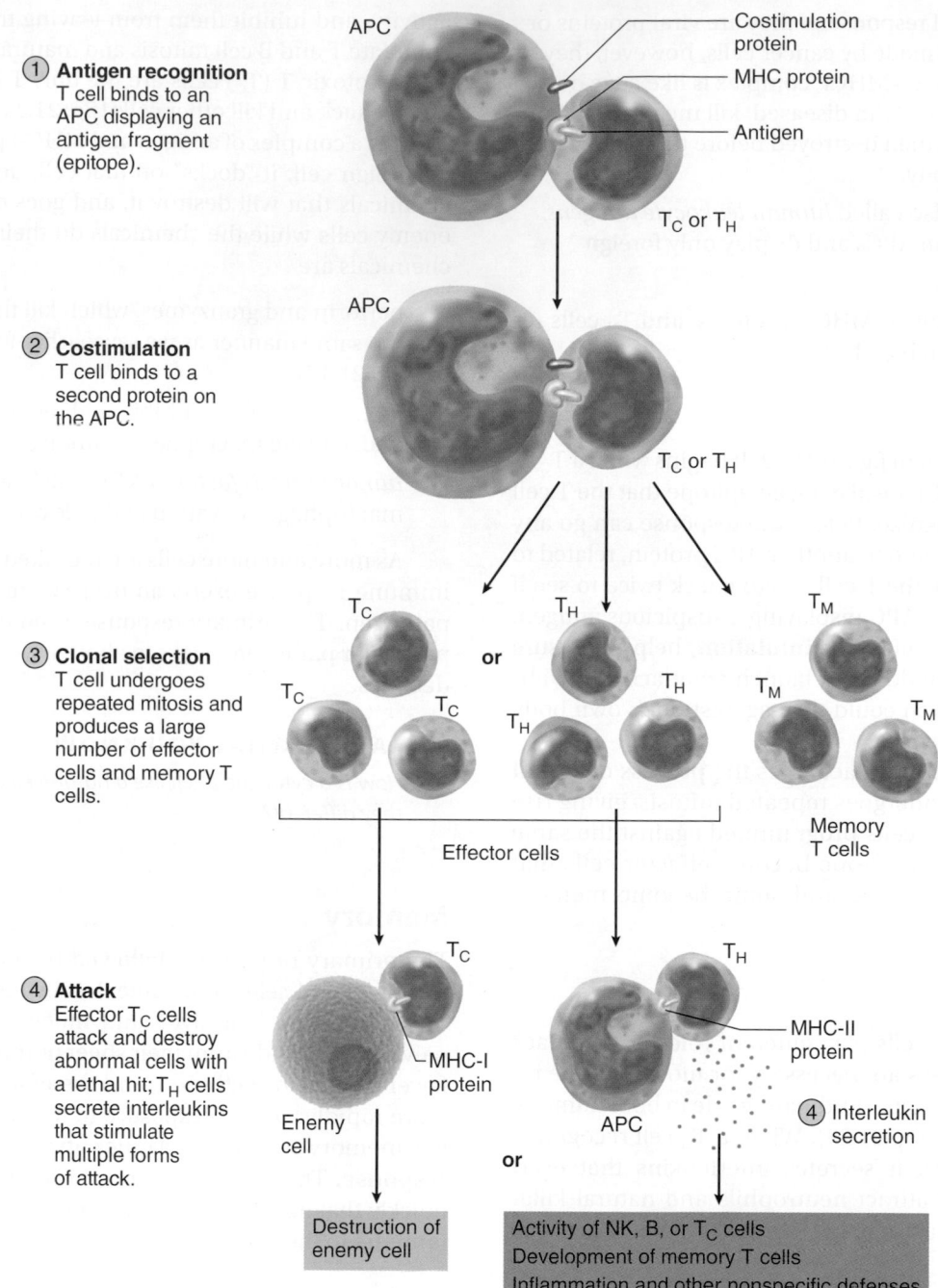

① **Antigen recognition**
T cell binds to an APC displaying an antigen fragment (epitope).

Costimulation protein
MHC protein
Antigen
APC
$T_C$ or $T_H$

② **Costimulation**
T cell binds to a second protein on the APC.
APC
$T_C$ or $T_H$

③ **Clonal selection**
T cell undergoes repeated mitosis and produces a large number of effector cells and memory T cells.
$T_C$    or    $T_H$    $T_M$
$T_C$    $T_C$    $T_H$    $T_H$    $T_M$    $T_M$
Effector cells
Memory T cells

④ **Attack**
Effector $T_C$ cells attack and destroy abnormal cells with a lethal hit; $T_H$ cells secrete interleukins that stimulate multiple forms of attack.
$T_C$
MHC-I protein
Enemy cell
$T_H$
MHC-II protein
APC
④ Interleukin secretion
or

Destruction of enemy cell

Activity of NK, B, or $T_C$ cells
Development of memory T cells
Inflammation and other nonspecific defenses

**FIGURE 21.22**   **T Cell Activation.**

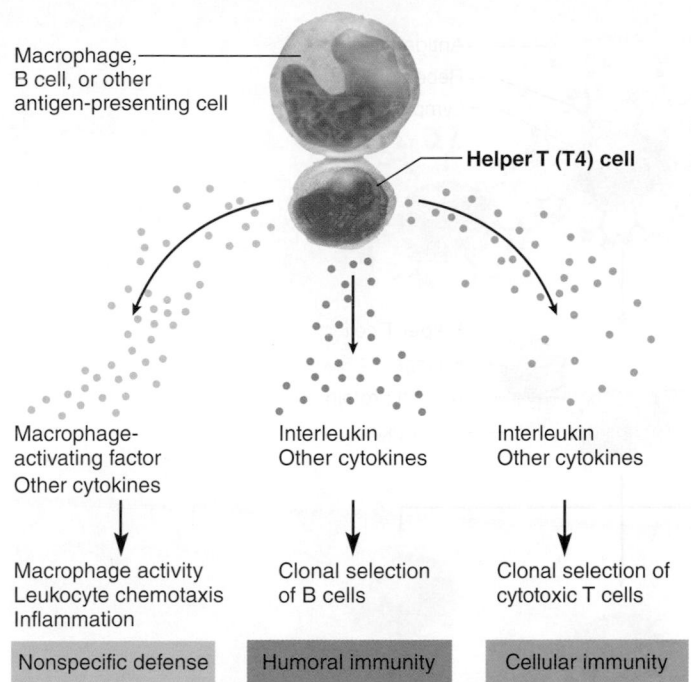

**FIGURE 21.23 The Role of Helper T Cells in Defense and Immunity.**

❓ *Why does AIDS reduce the effectiveness of all three defenses listed across the bottom of the figure?*

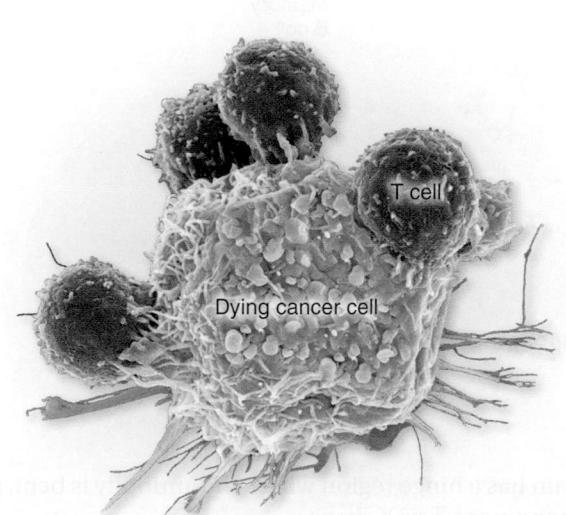

**FIGURE 21.24 Destruction of a Cancer Cell by Cytotoxic T Cells.**

## 21.5 Humoral Immunity

### Expected Learning Outcomes
When you have completed this section, you should be able to

a. explain how B cells recognize and respond to an antigen;

b. describe the structure, types, and actions of antibodies;

c. explain the mechanism of memory in humoral immunity; and

d. compare and contrast cellular and humoral immunity.

Humoral immunity is a more indirect method of defense than cellular immunity. Instead of directly attacking enemy cells, the B lymphocytes of humoral immunity produce antibodies that bind to antigens and tag them for destruction by other means. But like cellular immunity, humoral immunity works in three stages: recognition, attack, and memory.

### Recognition

An immunocompetent B cell has thousands of surface receptors for one antigen. B cell activation begins when an antigen binds to several of these receptors, links them together, and is taken into the cell by receptor-mediated endocytosis. One reason small molecules are not antigenic is that they are too small to link multiple receptors together. After endocytosis, the B cell processes (digests) the antigen, links some of the epitopes to its MHC-II proteins, and displays these on the cell surface.

Usually, the B cell response goes no further unless a helper T cell binds to this Ag–MHCP complex. (Some B cells are directly activated by antigens without the help of a $T_H$ cell.) When a $T_H$ cell binds to the complex, it secretes interleukins that activate the B cell. This triggers clonal selection—B cell mitosis giving rise to a battalion of identical B cells programmed against that antigen (fig. 21.25).

Most cells of the clone differentiate into **plasma cells.** These are larger than B cells and contain an abundance of rough endoplasmic reticulum (fig. 21.26). Plasma cells develop mainly in the germinal centers of the lymphatic nodules of the lymph nodes. About 10% of them remain in the lymph nodes, but the rest leave the nodes, take up residence in the bone marrow and elsewhere, and there produce antibodies until they die. A plasma cell secretes antibodies at the remarkable rate of 2,000 molecules per second over a life span of 4 to 5 days. These antibodies travel throughout the body in the blood and other body fluids. The first time you are exposed to a particular antigen, your plasma cells produce mainly an antibody class called IgM. In later exposures to the same antigen, they produce mainly IgG.

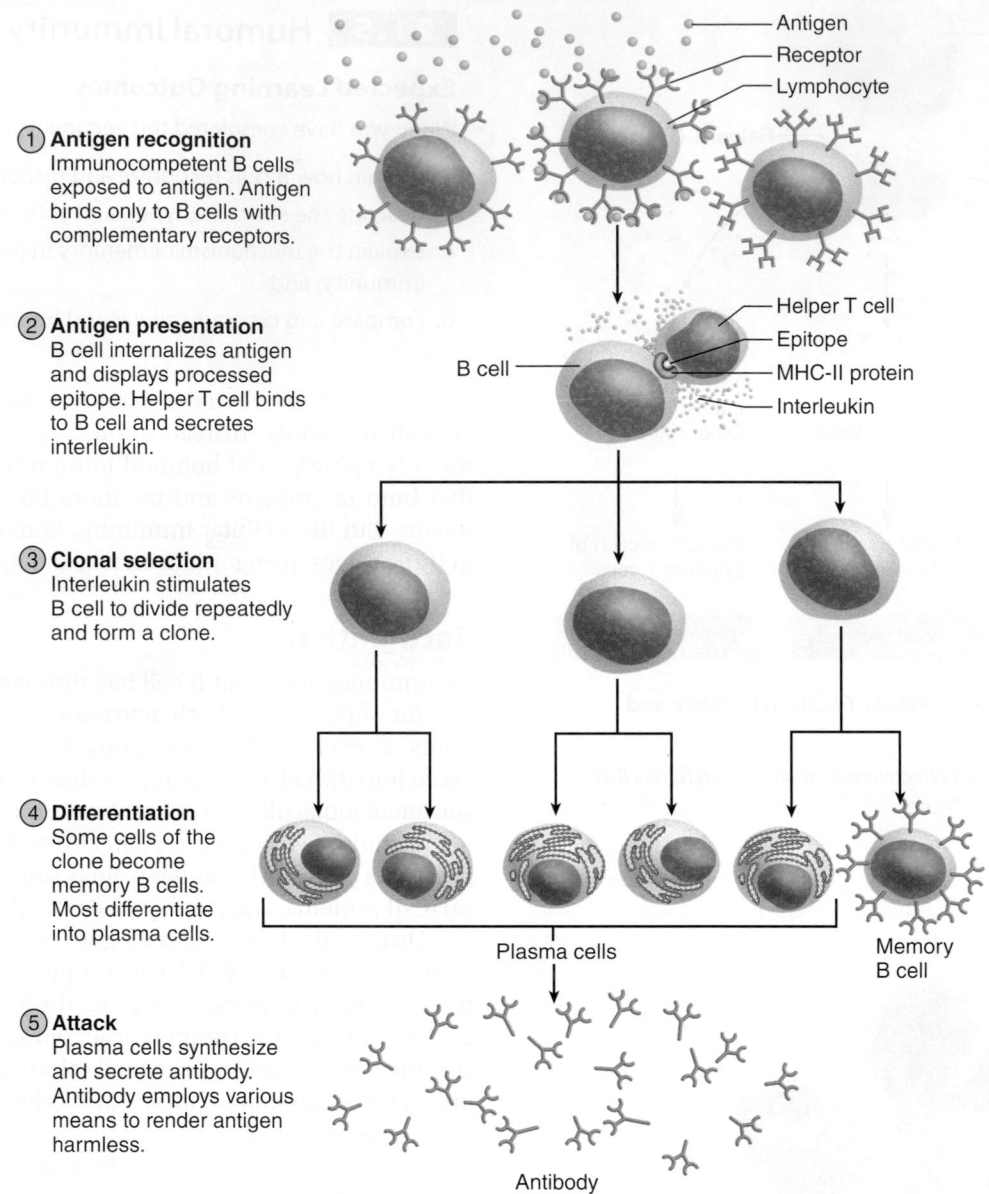

① **Antigen recognition**
Immunocompetent B cells exposed to antigen. Antigen binds only to B cells with complementary receptors.

② **Antigen presentation**
B cell internalizes antigen and displays processed epitope. Helper T cell binds to B cell and secretes interleukin.

③ **Clonal selection**
Interleukin stimulates B cell to divide repeatedly and form a clone.

④ **Differentiation**
Some cells of the clone become memory B cells. Most differentiate into plasma cells.

⑤ **Attack**
Plasma cells synthesize and secrete antibody. Antibody employs various means to render antigen harmless.

**FIGURE 21.25** Clonal Selection and Ensuing Events of the Humoral Immune Response.

## Attack

We have said much about antibodies already, and it is now time to take a closer look at what they are and how they work. Also called **immunoglobulins (Igs),** antibodies are defensive gamma globulins found in the blood plasma, tissue fluids, body secretions, and some leukocyte membranes. The basic structural unit of an antibody, an **antibody monomer,** is composed of four polypeptides linked by disulfide (—S—S—) bonds (fig. 21.27). The two **heavy chains** are about 400 amino acids long, and the two **light chains** are about half that long. Each

heavy chain has a hinge region where the antibody is bent, giving the monomer a T or Y shape.

All four chains have a **variable (V) region** that gives an antibody its uniqueness. The V regions of a heavy chain and light chain combine to form an **antigen-binding site** on each arm, which attaches to the epitope of an antigen molecule. The rest of each chain is a **constant (C) region,** which has the same amino acid sequence in all antibodies of a given class (within one person). The C region determines the mechanism of an antibody's action—for example, whether it can bind complement proteins.

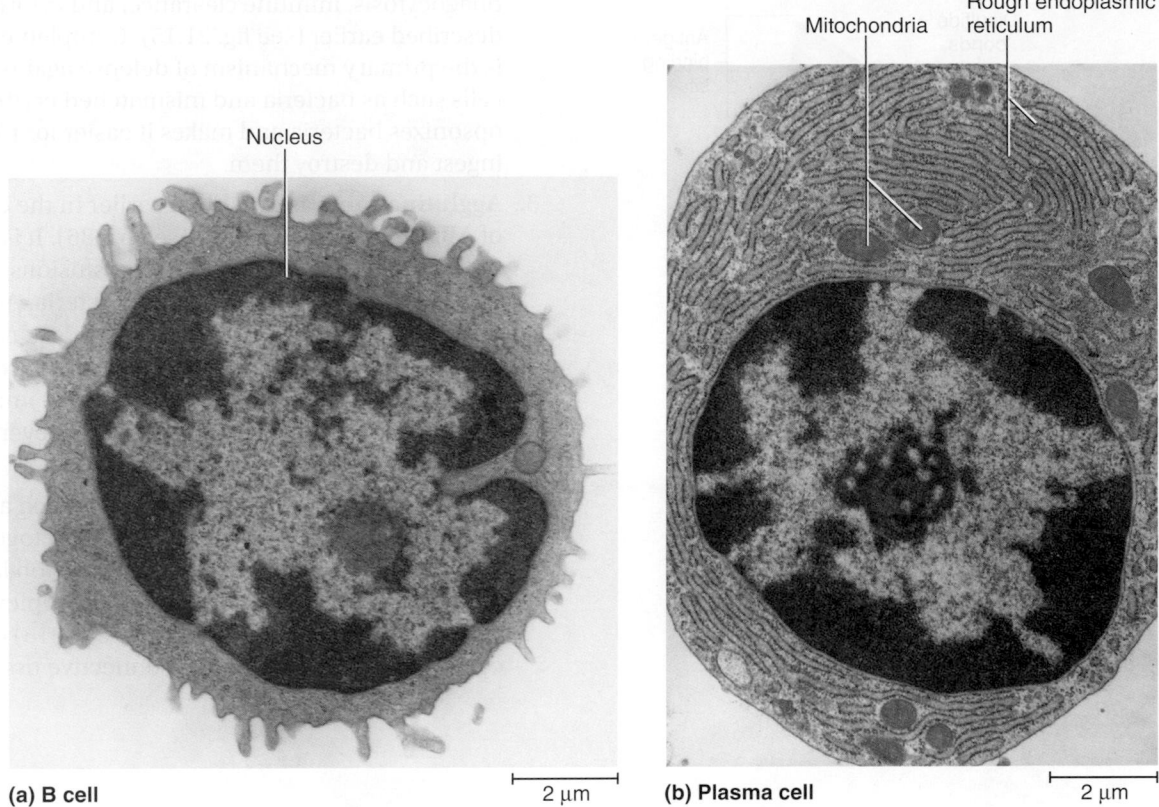

Nucleus

Mitochondria

Rough endoplasmic reticulum

**(a) B cell**     2 μm     **(b) Plasma cell**     2 μm

**FIGURE 21.26  B Cell and Plasma Cell.** (a) B cells have little cytoplasm and scanty organelles. (b) A plasma cell, which differentiates from a B cell, has an abundance of rough endoplasmic reticulum.

❓ *What does this endoplasmic reticulum do in the plasma cell?*

There are five classes of antibodies named **IgA, IgD, IgE, IgG,** and **IgM** (table 21.3) after the structures of their C regions *(alpha, delta, epsilon, gamma,* and *mu).* IgD, IgE, and IgG are monomers. IgA has a monomer form as well as a dimer composed of two cojoined monomers. IgM is a pentamer composed of five monomers. The surface antigen receptors synthesized by a developing B cell are IgD and IgM molecules. IgG is particularly important in the immunity of the newborn because it crosses the placenta with relative ease. Thus, it transfers immunity from the mother to her fetus. In addition, an infant acquires some maternal IgA through breast milk and colostrum (the fluid secreted for the first 2 or 3 days of breast-feeding).

The human immune system is believed capable of producing at least 10 billion and perhaps up to 1 trillion different antibodies. Any one person has a much smaller subset of these, but such an enormous potential helps to explain why we can deal with the tremendous diversity of antigens that exist in our environment. Yet such huge numbers are puzzling, because we are accustomed to thinking of each protein in the body being encoded by one gene, and we have as few as 20,000 genes, most of which have nothing to do with immunity. How can so few genes generate so many antibodies? Obviously there cannot be a different gene for each one. One means of generating

diversity is that the genome contains several hundred DNA segments that are shuffled and combined in various ways to produce antibody genes unique to each clone of B cells. This process is called **somatic recombination,** because it forms new combinations of DNA base sequences in somatic (nonreproductive) cells. Another mechanism of generating diversity is that B cells in the germinal centers of lymphatic nodules undergo exceptionally high rates of mutation, a process called **somatic hypermutation**—not just recombining preexisting DNA but creating wholly new DNA sequences. These and other mechanisms explain how we can produce such a tremendous variety of antibodies with a limited number of genes.

Once released by a plasma cell, antibodies use four mechanisms to render antigens harmless:

1. **Neutralization.** Only certain regions of an antigen are pathogenic—for example, the parts of a toxin molecule or virus that enable these agents to bind to human cells. Antibodies can neutralize an antigen by masking these active regions.

2. **Complement fixation.** IgM and IgG bind to enemy cells and change shape, exposing their complement-binding sites (fig. 21.27). This initiates the binding of complement to the enemy cell surface and leads to inflammation,

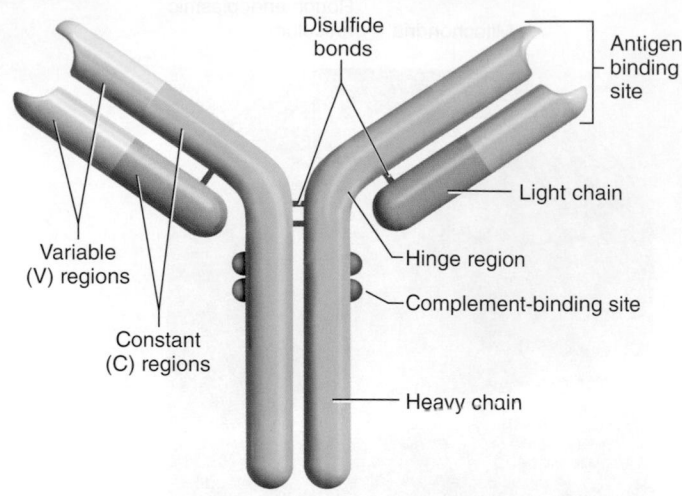

**FIGURE 21.27 Antibody Structure.** A molecule of IgG, a monomer.

phagocytosis, immune clearance, and cytolysis, as described earlier (see fig. 21.15). Complement fixation is the primary mechanism of defense against foreign cells such as bacteria and mismatched erythrocytes. It opsonizes bacteria and makes it easier for phagocytes to ingest and destroy them.

3. **Agglutination** was described earlier in the discussion of ABO and Rh blood types (see p. 686). It is effective not only in mismatched blood transfusions, but more importantly as a defense against bacteria. An antibody molecule has 2 to 10 binding sites; thus, it can bind to antigen molecules on two or more enemy cells at once and stick them together (fig. 21.28). This immobilizes microbes and antigen molecules and prevents them from spreading through the tissues.

4. **Precipitation** is a similar process in which antibodies link antigen molecules (not whole cells) together. This creates large Ag–Ab complexes that are too large to remain dissolved in solution. These complexes can be removed by immune clearance (see p. 818) or phagocytized by eosinophils in the connective tissues.

| TABLE 21.3 | The Five Classes of Antibodies | |
|---|---|---|
| **Class** | **Structure** | **Location and Function** |
| IgA | IgA    Monomer    Dimer | Found as a monomer in blood plasma and mainly as a dimer in mucus, tears, milk, saliva, and intestinal secretions. Sometimes also forms trimers and tetramers. Prevents pathogens from adhering to epithelia and penetrating underlying tissues. Provides passive immunity to the newborn. |
| IgD | IgD    Monomer | A transmembrane protein of B cells; functions in activation of B cells by antigens. |
| IgE | IgE    Monomer | A transmembrane protein of basophils and mast cells. Stimulates them to release histamine and other mediators of inflammation and allergy; important in immediate hypersensitivity reactions and in attracting eosinophils to sites of parasitic infection. |
| IgG | IgG    Monomer | Constitutes about 80% of circulating antibodies in blood plasma. The predominant antibody secreted in the secondary immune response. IgG and IgM are the only antibodies with significant complement-fixation activity. Crosses placenta and confers temporary immunity on the fetus. Includes the anti-D antibodies of the Rh blood group. |
| IgM | IgM    Monomer    Pentamer | Constitutes about 10% of circulating antibodies in plasma. Monomer is a transmembrane protein of B cells, where it functions as part of the antigen receptor. Pentamer occurs in blood plasma and lymph. The predominant antibody secreted in the primary immune response; very strong agglutinating and complement-fixation abilities; includes the anti-A and anti-B agglutinins of the ABO blood group. |

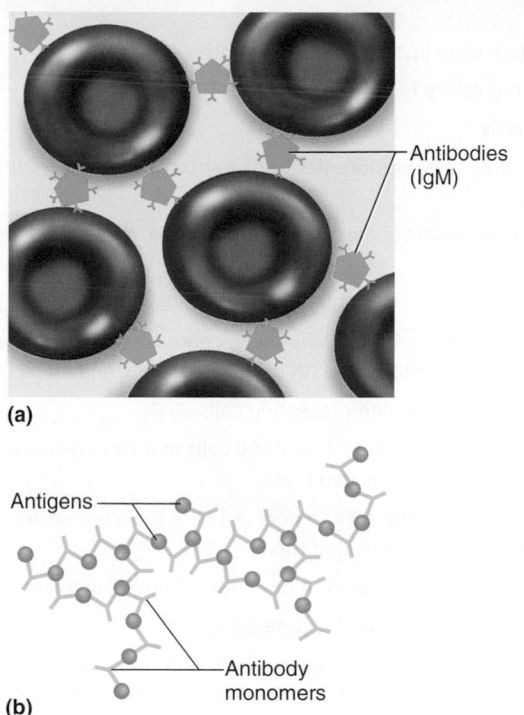

**(a)**

**(b)**

**FIGURE 21.28  Agglutination by Antibodies.**  (a) Agglutination of foreign erythrocytes by IgM, a pentamer. (b) An antigen–antibody complex involving a free molecular antigen and an antibody monomer such as IgG.

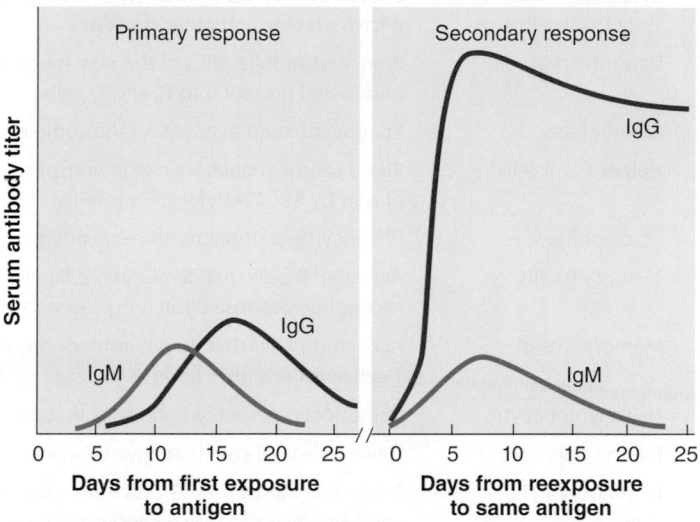

**FIGURE 21.29  The Primary and Secondary (Anamnestic) Responses in Humoral Immunity.**  The individual is exposed to antigen on day 0 in both cases. Note the differences in the speed of response, the height of the antibody titer, and the rate of decline in antibody titer.

You will note that antibodies do not directly destroy an antigen in any of these mechanisms. They render it harmless by masking its pathogenic sites or agglutinating it, and they mark it for destruction by other agents such as complement, macrophages, or eosinophils.

▶▶▶**APPLY WHAT YOU KNOW**

*Explain why IgM has a stronger power of agglutination than antibodies of any other class.*

## Memory

When a person is exposed to a particular antigen for the first time, the immune reaction is called the **primary response.** The appearance of protective antibodies is delayed for 3 to 6 days while naive B cells multiply and differentiate into plasma cells. As the plasma cells begin secreting antibody, the **antibody titer** (level in the blood plasma) begins to rise (fig. 21.29). IgM appears first, peaks in about 10 days, and soon declines. IgG levels rise as IgM declines, but even the IgG titer drops to a low level within a month.

The primary response, however, leaves one with an immune memory of the antigen. During clonal selection, some members of the clone become **memory B cells** rather than plasma cells (see fig. 21.25). Memory B cells, found mainly in the germinal centers of the lymph nodes, mount a very quick **secondary,** or **anamnestic**[25] (an-am-NESS-tic), **response** if reexposed to the same antigen. Plasma cells form within hours, so the IgG titer rises sharply and peaks within a few days. The response is so rapid that the antigen has little chance to exert a noticeable effect on the body, and no illness results. A low level of IgM is also secreted and quickly declines, but IgG remains elevated for weeks to years, conferring lasting protection. Memory does not last as long in humoral immunity, however, as it does in cellular immunity.

Table 21.4 summarizes many of the cellular and chemical agents involved in humoral and cellular immunity. Table 21.5 compares the main features of humoral and cellular immunity. Remember that these two processes often occur simultaneously, and in conjunction with inflammation as a three-pronged attack on the same pathogen.

[25]*ana* = back; *mnes* = remember

| TABLE 21.4 | Agents of Adaptive Immunity |
|---|---|
| **Cellular agents** | |
| B cells | Serve as antigen-presenting cells in humoral immunity; differentiate into antibody-secreting plasma cells |
| CD4 (T4) cells* | T lymphocytes with CD4 surface glycoproteins; helper and regulatory T cells |
| CD8 (T8) cells | T lymphocytes with CD8 surface glycoproteins; cytotoxic T cells |
| Cytotoxic T (killer T, $T_c$, or CD8) cells | Effectors of cellular immunity; directly attack and destroy enemy cells; produce perforin, granzymes, interferon, tumor necrosis factor, and other cytokines |
| Dendritic cells | Branched, mobile APCs of the skin, mucous membranes, and lymphatic tissues; internalize antigen, migrate to lymph nodes, and present it to $T_H$ and $T_C$ cells |
| Eosinophils* | Phagocytize and degrade Ag–Ab complexes |
| Helper T ($T_H$) cells* | Play a central regulatory role in nonspecific defense and humoral and cellular immunity; recognize antigen fragments displayed by APCs with MHC-II proteins; secrete interleukins that activate B, $T_C$, and NK cells, neutrophils, and macrophages |
| Macrophages* | Phagocytize pathogens and expended or damaged host cells; act as antigen-presenting cells (APCs) |
| Memory B cells | Activated B cells that do not immediately differentiate into plasma cells; act as a pool of B cells that can execute a quick secondary response upon reexposure to the same antigen that initially activated them |
| Memory T cells | Activated T cells that do not immediately differentiate into effector T cells; act as a pool of T cells that can execute a quick T cell recall response upon reexposure to the same antigen that initially activated them |
| Naive lymphocytes | Immunocompetent lymphocytes that are capable of responding to an antigen but have not yet encountered one |
| Plasma cells | Develop from B cells that have been activated by helper T cells; synthesize and secrete antibodies |
| Regulatory T ($T_R$) cells | T cells that inhibit other T cells from multiplying or secreting cytokines; important especially in preventing autoimmune diseases but may also play defensive roles against infection and cancer and protect the fetus from maternal immune responses |
| Reticular cells | Fixed cells in the stroma of the lymph nodes and other lymphatic organs; act as APCs in the thymus, with a role in maturation of T cells |
| **Chemical agents** | |
| Antibody (Ab) | A gamma globulin produced by plasma cells in response to an antigen; counteracts antigen by means of complement fixation, neutralization of toxins, agglutination, or precipitation |
| Antigen (Ag) | Molecule capable of triggering an immune response; usually a protein, polysaccharide, glycolipid, or glycoprotein |
| Complement* | Group of plasma proteins that help to destroy pathogens by cytolysis, phagocytosis, immune clearance, or nonspecific defense (inflammation) |
| Granzyme | Proteolytic enzyme produced by NK and $T_C$ cells; enters the pore made by perforins, degrades enzymes of the enemy cell, and induces apoptosis |
| Hapten | Small molecule initially unable to trigger an immune response by itself but able to bind to host molecules and produce a complex that is antigenic; may subsequently activate an immune response without binding to a host molecule |
| Interleukin | Cytokine produced by leukocytes and macrophages to stimulate other leukocytes |
| Perforin | A protein produced by NK and $T_C$ cells that binds to target cells, produces a hole, and admits granzymes into the cell |
| Tumor necrosis factor (TNF) | Cytokine secreted by $T_C$ cells that activates macrophages and kills cancer cells |

* These agents have additional roles in inflammation described in table 21.1.

| TABLE 21.5 | Some Comparisons Between Humoral and Cellular Immunity | |
|---|---|---|
| **Characteristic** | **Cellular Immunity** | **Humoral Immunity** |
| Disease agents | Intracellular viruses, bacteria, yeasts, and protozoans; parasitic worms; cancer cells; transplanted tissues and organs | Extracellular viruses, bacteria, yeasts, and protozoans; toxins, venoms, and allergens; mismatched RBCs |
| Effector cells | Cytotoxic T cells | Plasma cells (develop from B cells) |
| Other cells involved in attack | Helper T cells | Helper T cells |
| Antigen-presenting cells | B cells, macrophages, dendritic cells, nearly all cells | B cells |
| MHC proteins | MHC-I and MHC-II | MHC-II only |
| Chemical agents of attack | Perforins, granzymes, interferons, tumor necrosis factor | Antibodies, complement |
| Mechanisms of counteracting or destroying pathogens | Cytolysis, phagocytosis, apoptosis | Cytolysis, phagocytosis, immune clearance, inflammation, neutralization, agglutination, precipitation |
| Memory | T cell recall response | Secondary (anamnestic) response |

Answer the following questions to test your understanding of the preceding section:

21. What is the difference between a B cell and a plasma cell?

22. Describe four ways in which an antibody acts against an antigen.

23. Why does the secondary immune response prevent a pathogen from causing disease, while the primary immune response does not?

## 21.6  Immune System Disorders

### Expected Learning Outcomes

When you have completed this section, you should be able to

a. distinguish between the four classes of immune hypersensitivity and give an example of each;

b. explain the cause of anaphylaxis and distinguish local anaphylaxis from anaphylactic shock;

c. state some reasons immune self-tolerance may fail, and give examples of the resulting diseases; and

d. describe the pathology of immunodeficiency diseases, especially AIDS.

Because the immune system involves complex cellular interactions controlled by numerous chemical messengers, there are many points at which things can go wrong. The immune response may be too vigorous, too weak, or misdirected against the wrong targets. A few disorders are summarized here to illustrate the consequences.

### Hypersensitivity

**Hypersensitivity** is an excessive, harmful immune reaction to antigens. It includes reactions to tissues transplanted from another person *(alloimmunity)*, abnormal reactions to one's own tissues *(autoimmunity)*, and **allergies,**[26] which are reactions to environmental antigens. Such antigens, called **allergens,** occur in mold, dust, pollen, vaccines, bee and wasp venoms, animal dander, toxins from poison ivy and other plants, and foods such as nuts, milk, eggs, and shellfish. Drugs such as penicillin, tetracycline, and insulin are allergenic to some people.

One classification system recognizes four kinds of hypersensitivity, distinguished by the types of immune agents (antibodies or T cells) involved and their methods of attack on the antigen. In this system, type I is also characterized as *acute (immediate) hypersensitivity* because the response is very rapid, whereas types II and III are characterized as *subacute* because they exhibit a slower onset (1–3 hours after exposure) and last longer (10–15 hours). Type IV is a delayed cell-mediated response, whereas the other three are quicker antibody-mediated responses.

- **Type I (acute) hypersensitivity** includes the most common allergies. Some authorities use the word *allergy* for type I reactions only, and others use it for all four types. Type I is an IgE-mediated reaction that begins within seconds of exposure and usually subsides within 30 minutes, although it can be severe and even fatal. Allergens bind to IgE on the membranes of basophils and mast cells and stimulate them to secrete histamine and other inflammatory and vasoactive chemicals. These chemicals trigger glandular secretion, vasodilation, increased capillary permeability, smooth muscle spasms, and other effects. The clinical signs include local edema, mucus hypersecretion and congestion, watery eyes, a runny nose, hives (red itchy skin), and sometimes cramps, diarrhea, and vomiting. Some examples of type I hypersensitivity are food allergies and **asthma,**[27] a local inflammatory reaction to inhaled allergens (see Deeper Insight 21.3).

  **Anaphylaxis**[28] (AN-uh-fih-LAC-sis) is an immediate and severe type I reaction. Local anaphylaxis can be relieved with antihistamines. **Anaphylactic shock** is a severe, widespread acute hypersensitivity that occurs when an allergen such as bee venom or penicillin is introduced to the bloodstream of an allergic individual, or when a person ingests certain foods (such as peanuts) to which he or she is allergic. It is characterized by bronchoconstriction, dyspnea (labored breathing), widespread vasodilation, circulatory shock, and sometimes sudden death. Antihistamines are inadequate by themselves to counter anaphylactic shock, but epinephrine relieves the symptoms by dilating the bronchioles, increasing cardiac output, and restoring blood pressure. Fluid therapy and respiratory support are sometimes required.

- **Type II (antibody-dependent cytotoxic) hypersensitivity** occurs when IgG or IgM attacks antigens bound to cell surfaces. The reaction leads to complement activation and either lysis or opsonization of the target cell. Macrophages phagocytize and destroy opsonized platelets, erythrocytes, or other cells. Examples of cell destruction by type II reactions are blood transfusion reactions, pemphigus vulgaris (p. 164), and some drug reactions. In some other type II responses, an antibody binds to cell surface receptors and either interferes with their function (as in myasthenia gravis, p. 429) or overstimulates the cell (as in toxic goiter, p. 662).

- **Type III (immune complex) hypersensitivity** occurs when IgG or IgM forms antigen–antibody complexes that precipitate beneath the endothelium of the blood vessels or in other tissues. At the sites of deposition, these complexes activate complement and trigger intense inflammation, causing tissue destruction. Two examples of type III hypersensitivity are the autoimmune diseases acute

---

[26]*allo* = altered; *erg* = action, reaction

[27]*asthma* = panting

[28]*ana* = against; *phylax* = prevention

### Asthma

Asthma is the most common chronic illness of children, especially boys. It is the leading cause of school absenteeism and childhood hospitalization in the United States. About half of all cases develop before age 10 and only 15% after age 40. In the United States, it affects about 5% of adults and up to 10% of children, and takes about 5,000 lives per year. Moreover, asthma is on the rise; there are many more cases and deaths now than there were a few decades ago.

In *allergic (extrinsic) asthma,* the most common form, a respiratory crisis is triggered by allergens in pollen, mold, animal dander, food, dust mites, or cockroaches. The allergens stimulate plasma cells to secrete IgE, which binds to mast cells of the respiratory mucosa. Reexposure to the allergen causes the mast cells to release a complex mixture of inflammatory chemicals, which trigger intense airway inflammation. *Nonallergic (intrinsic) asthma* is not caused by allergens but can be triggered by infections, drugs, air pollutants, cold dry air, exercise, or emotions. This form is more common in adults than in children, but the effects are much the same.

Within minutes, the bronchioles constrict spasmodically *(bronchospasm),* causing severe coughing, wheezing, and sometimes fatal suffocation. A second respiratory crisis often occurs 6 to 8 hours later. Interleukins attract eosinophils to the bronchial tissue, where they secrete proteins that paralyze the respiratory cilia, severely damage the epithelium, and lead to scarring

and extensive long-term damage to the lungs. The bronchioles also become edematous and plugged with thick, sticky mucus. People who die of asthmatic suffocation typically show airways so plugged with gelatinous mucus that they could not exhale. The lungs remain hyperinflated even at autopsy.

Asthma is treated with epinephrine and other ß-adrenergic stimulants to dilate the airway and restore breathing, and with inhaled corticosteroids or nonsteroidal anti-inflammatory drugs to minimize airway inflammation and long-term damage. The treatment regimen can be very complicated, often requiring more than eight different medications daily, and compliance is therefore difficult for children and patients with low income or educational attainment.

Asthma runs in families and seems to result from a combination of heredity and environmental irritants. In the United States, asthma is most common, paradoxically, in two groups: (1) inner-city children who are exposed to crowding, poor sanitation, and poor ventilation, and who do not often go outside or get enough exercise; and (2) children from extremely clean homes, perhaps because they have had too little opportunity to develop normal immunities. Asthma is also more common in countries where vaccines and antibiotics are widely used. It is less common in developing countries and in farm children of the United States.

---

glomerulonephritis (p. 918) and systemic lupus erythematosus, a widespread inflammation of the connective tissues (see table 21.6).

- **Type IV (delayed) hypersensitivity** is a cell-mediated reaction in which the signs appear about 12 to 72 hours after exposure. It begins when APCs in the lymph nodes display antigens to helper T cells, and these T cells secrete interferon and other cytokines that activate cytotoxic T cells and macrophages. The result is a mixture of nonspecific and immune responses. Type IV reactions include allergies to haptens in cosmetics and poison ivy; graft rejection; the tuberculosis skin test; and the beta cell destruction that causes type 1 diabetes mellitus.

## Autoimmune Diseases

**Autoimmune diseases** are failures of self-tolerance—the immune system fails to distinguish self-antigens from foreign ones and produces **autoantibodies** that attack the body's own tissues. There are at least three reasons why self-tolerance may fail:

1. **Cross-reactivity.** Some antibodies against foreign antigens react to similar self-antigens. In rheumatic fever, for example, a streptococcus infection stimulates production of antibodies that react not only against the bacteria but also against antigens of the heart tissue. It often results in scarring and stenosis (narrowing) of the mitral and aortic valves.

2. **Abnormal exposure of self-antigens to the blood.** Some of our native antigens are normally not exposed to

the blood. For example, a blood–testis barrier (BTB) normally isolates sperm cells from the blood. Breakdown of the BTB can cause sterility when sperm first form in adolescence and activate the production of autoantibodies.

3. **Change in the structure of self-antigens.** Viruses and drugs may change the structure of self-antigens and cause the immune system to perceive them as foreign. One theory of type 1 diabetes mellitus is that a viral infection alters the antigens of the insulin-producing beta cells of the pancreatic islets, which leads to an autoimmune attack on the cells.

Evidence is emerging that not all self-reactive T cells are eliminated by clonal deletion in the thymus. We all apparently have at least some T cells poised to attack our own tissues, but regulatory T ($T_R$) cells keep them in check and normally prevent autoimmune disease.

## Immunodeficiency Diseases

In the foregoing diseases, the immune system reacts too vigorously or directs its attack against the wrong targets. In immunodeficiency diseases, by contrast, the immune system fails to respond vigorously enough.

### Severe Combined Immunodeficiency Disease

**Severe combined immunodeficiency disease (SCID)** is a group of disorders caused by recessive alleles that result in a scarcity or absence of both T and B cells. Children with SCID are highly vulnerable to opportunistic infections and must live in protective enclosures. The most publicized case was David Vetter, who spent his life in sterile plastic chambers and suits (fig. 21.30), finally succumbing at age 12 to cancer triggered by a viral infection. Children with SCID are sometimes helped by transplants of bone marrow or fetal thymus, but in some cases the transplanted cells fail to survive and multiply, or transplanted T cells attack the patient's tissues (the *graft-versus-host response*). David contracted the fatal virus from his sister through a bone marrow transplant.

## Acquired Immunodeficiency Syndrome

Other immunodeficiency diseases are nonhereditary and contracted after birth. The best-known example is **acquired immunodeficiency syndrome (AIDS),** a group of conditions in which infection with the **human immunodeficiency virus (HIV)** severely depresses the immune response.

HIV (fig. 21.31) has an inner core consisting of a protein *capsid* that encloses two molecules of RNA, two molecules of an enzyme called *reverse transcriptase,* and a few other enzyme molecules. The capsid is enclosed in another layer of viral protein, the *matrix*. External to this is a *viral envelope* composed of phospholipids and glycoproteins derived from the host cell. Like other viruses, HIV can be replicated only by a living host cell. It invades helper T (CD4) cells, dendritic cells, and macrophages. HIV adheres to a target cell by means of one of its envelope glycoproteins and "tricks" the target cell into internalizing it by receptor-mediated endocytosis. Within the host cell, reverse transcriptase uses the viral RNA as a template to synthesize DNA—the opposite of the usual process of genetic transcription. Viruses that carry out this RNA → DNA reverse transcription are called *retroviruses.*[29] The new DNA is inserted into the host cell's DNA, where it may lie dormant for months to years. When activated, however, it induces the host cell to produce new viral RNA, capsid proteins, and matrix proteins. As the new viruses emerge from the host cell (fig. 21.31b), they are coated with bits of the cell's plasma membrane, forming the new viral envelope. The new viruses then adhere to more host cells and repeat the process.

By destroying $T_H$ cells, HIV strikes at a central coordinating agent of nonspecific defense, humoral immunity, and cellular immunity (see figure 21.23). After an incubation period ranging from a few months to 12 years, the patient begins to experience flulike episodes of chills and fever as HIV attacks $T_H$ cells. At first, antibodies against HIV are produced and the $T_H$ count returns nearly to normal. As the virus destroys more and more cells, however, the signs and symptoms become more pronounced: night sweats, fatigue, headache, extreme weight loss, and lymphadenitis.

A normal $T_H$ count is 600 to 1,200 cells/μL, but a criterion of AIDS is a count less than 200/μL. With such severe depletion of $T_H$ cells, a person succumbs to opportunistic infections with such pathogens as *Toxoplasma* (a protozoan previously known mainly for causing birth defects), *Pneumocystis* (a group of respiratory fungi), herpes simplex virus, cytomegalovirus (which can cause blindness), and tuberculosis bacteria. White patches may appear in the mouth, caused by *Candida* (thrush) or Epstein–Barr[30] virus (leukoplakia). A cancer called Kaposi[31] sarcoma, common in AIDS patients, originates in the endothelial cells of the blood vessels and causes bruiselike purple lesions visible in the skin (fig. 21.32).

Patients with AIDS show no response to standard skin tests for delayed hypersensitivity. Slurred speech, loss of motor and cognitive functions, and dementia may occur as HIV invades the brain by way of infected phagocytes (microglia) and induces them to release toxins that destroy neurons and astrocytes. Death from cancer or infection is inevitable, usually within a few months but sometimes as long as 8 years after diagnosis. Some people, however, have been diagnosed as HIV-positive and yet have survived for 10 years or longer without developing AIDS.

HIV is transmitted through blood, semen, vaginal secretions, and breast milk. It can be transmitted from mother to fetus through the placenta or from mother to infant during childbirth or nursing. HIV occurs in saliva and tears, but is apparently not transmitted by those fluids. The most common means of transmission are sexual intercourse (vaginal, anal, or oral), contaminated blood products, and drug injections with contaminated needles. Worldwide, about 75% of HIV infections

**FIGURE 21.30 Boy with Severe Combined Immunodeficiency Disease.** David Vetter lived with SCID from birth (1971) to age 12 (1984). At the age of 6, he received a portable sterile suit designed by NASA that allowed him to leave the hospital for the first time.

[29]*retr* = an acronym from *reverse transcription*
[30]M. A. Epstein (1921–), British physician; Y. M. Barr (1932–), British virologist
[31]Moritz Kaposi (1837–1902), Austrian physician

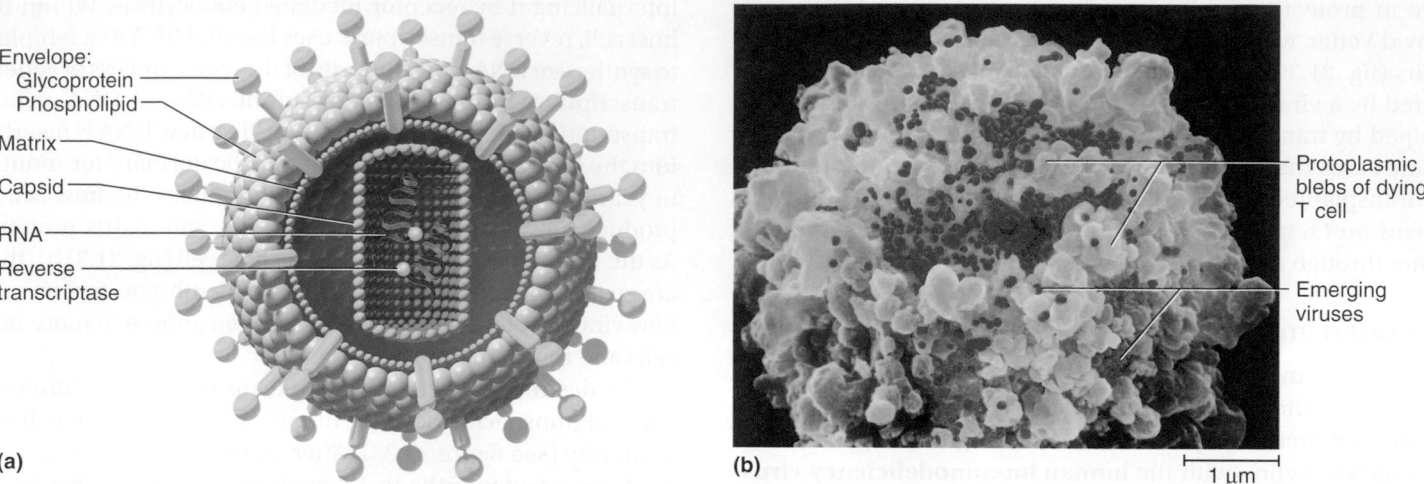

Envelope:
Glycoprotein
Phospholipid
Matrix
Capsid
RNA
Reverse transcriptase

(a)

Protoplasmic blebs of dying T cell

Emerging viruses

1 μm

(b)

**FIGURE 21.31  The Human Immunodeficiency Virus (HIV).**  (a) Structure of the virus. (b) Viruses emerging from a dying helper T cell. Each virus can now invade a new helper T cell and produce a similar number of descendants.

❓ *Which of the molecules in part (a) is the target of the drug azidothymidine (AZT)? How does AZT inhibit the spread of HIV?*

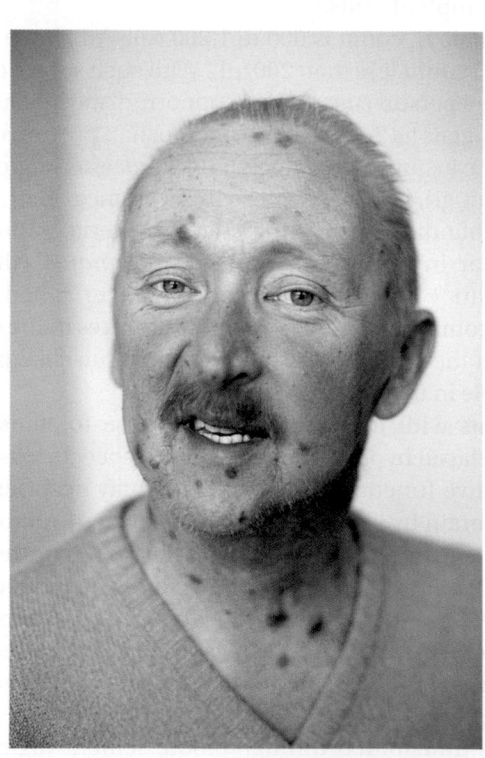

**FIGURE 21.32  Kaposi Sarcoma.**  Typical lesions on the face and neck of a person with AIDS.

are acquired through heterosexual, predominantly vaginal intercourse. In the United States, most cases occur in men who have sex with other men, but adolescents are the fastest-rising group of AIDS patients because of the exchange of unprotected sexual intercourse for drugs. The sharing of needles for drug use remains the chief means of transmission in urban ghettos.

Since 1984, all blood donated for transfusion has been tested for HIV, and the risk of infection from transfusions is now less than 1%. HIV cannot be contracted by donating blood, but irrational fear has resulted in an alarming drop in blood donors.

AIDS is not known to be transmitted through casual contact—for example, to family members, friends, coworkers, classmates, or medical personnel in charge of AIDS patients. It is not transmitted by kissing or by mosquitoes or other blood-sucking arthropods.

HIV survives poorly outside the human body. It is destroyed by laundering; dishwashing; exposure to heat (50°C [135°F] for at least 10 minutes); chlorination of swimming pools and hot tubs; and disinfectants such as bleach, Lysol, hydrogen peroxide, rubbing alcohol, and germicidal skin cleansers such as Betadine and Hibiclens. A properly used, undamaged latex condom is an effective barrier to HIV, but animal membrane condoms have gaps too large to block HIV transmission.

The AIDS epidemic has triggered an effort of unprecedented intensity to find a vaccine or cure. The strategies against HIV include efforts to prevent its binding to $T_H$ cells, disrupting the action of reverse transcriptase, or inhibiting the assembly of new viruses or their release from host cells. HIV is a difficult pathogen to attack. Since it hides within host cells, it usually escapes recognition by the immune system. In the brain, it is protected by the blood–brain barrier.

About 1% of HIV's genes mutate every year. Such rapid mutation is a barrier to both natural immunity and development of a vaccine. Even when immune cells do become sensitized to HIV, the virus soon evolves new surface antigens that escape recognition. The high mutation rate also would quickly make today's vaccine ineffective against tomorrow's strain of the virus. Another obstacle to treatment and prevention is the lack of animal models for vaccine and drug research and development. Most animals are not susceptible to HIV. The chimpanzee is an exception, but chimpanzees are difficult to maintain, and there are economic barriers and ethical controversies surrounding their use.

The first anti-HIV drug approved by the Food and Drug Administration (FDA) was azidothymidine (AZT, or Retrovir), which inhibited reverse transcriptase and prolonged the lives of some HIV-positive individuals. In 1996, a family of drugs called protease inhibitors became available, typically used in a "triple cocktail" combining these with two reverse transcriptase inhibitors. But by 1997, HIV had evolved resistance even to those drugs and they were failing in more than half of all patients. Today, at least 24 anti-HIV drugs are on the market, typically used in combinations of three or more. Such drug combinations have substantially reduced AIDS morbidity and mortality (disease and death) and reduced the hospital and hospice census, but none of them totally eliminate HIV from the body, and all of them have serious side effects that contraindicate their long-term use. The major unresolved questions in AIDS therapy today are when to start drug treatment, which drugs to use, when to switch drugs, how to improve patient compliance, and how to make therapy available in impoverished nations where AIDS is especially rampant.

There also remain a number of unanswered questions about the basic biology of HIV. It is still unknown, for example, why there are such strikingly different patterns of heterosexual versus homosexual transmission in different countries, and why some people succumb so rapidly to infection but others can be HIV-positive for years without developing AIDS. AIDS remains a stubborn problem sure to challenge virologists and epidemiologists for many years to come.

We have surveyed the major classes of immune system disorders and a few particularly notorious immune diseases. A few additional lymphatic and immune system disorders are described in table 21.6. The effects of aging on the immune system are described on page 1121.

### BEFORE YOU GO ON

Answer the following questions to test your understanding of the preceding section:

24. How does subacute hypersensitivity differ from acute hypersensitivity? Give an example of each.

25. Aside from the time required for a reaction to appear, how does delayed hypersensitivity differ from the acute and subacute types?

26. State some reasons why antibodies may begin attacking self-antigens to which they did not previously respond. What are these self-reactive antibodies called?

27. What is the distinction between a person who has an HIV infection and a person who has AIDS?

28. How does a reverse transcriptase inhibitor such as AZT slow the progress of AIDS?

| TABLE 21.6 | Some Disorders of the Lymphatic and Immune Systems |
|---|---|
| Contact dermatitis | A form of delayed hypersensitivity that produces skin lesions limited to the site of contact with an allergen or hapten; includes responses to poison ivy, cosmetics, latex, detergents, industrial chemicals, and some topical medicines. |
| Hives (urticaria[32]) | An allergic skin reaction characterized by a "wheal and flare" reaction—white blisters (wheals) surrounded by reddened areas (flares), usually with itching. Caused by local histamine release in response to allergens. Can be triggered by food or drugs, but sometimes by nonimmunological factors such as cold, friction, or emotional stress. |
| Hodgkin[33] disease | A lymph node malignancy, with early symptoms including enlarged painful nodes, especially in the neck, and fever; often progresses to neighboring lymph nodes. Radiation and chemotherapy cure about 75% of patients. |
| Splenomegaly[34] | Enlargement of the spleen, sometimes without underlying disease but often indicating infections, autoimmune diseases, heart failure, cirrhosis, Hodgkin disease, and other cancers. The enlarged spleen may "hoard" erythrocytes, causing anemia, and may become fragile and subject to rupture. |
| Systemic lupus erythematosus[35] | Formation of autoantibodies against DNA and other nuclear antigens, resulting in accumulation of antigen–antibody complexes in blood vessels and other organs, where they trigger widespread connective tissue inflammation. Named for skin lesions once likened to a wolf bite. Causes fever, fatigue, joint pain, weight loss, intolerance of bright light, and a "butterfly rash" across the nose and cheeks. Death may result from renal failure. |

*Disorders described elsewhere*

| | | |
|---|---|---|
| Acute glomerulonephritis p. 918 | Diabetes mellitus p. 664 | Rheumatic fever pp. 729, 838 |
| AIDS p. 839 | Edema p. 805 | Rheumatoid arthritis p. 303 |
| Allergy p. 837 | Lymphadenitis p. 813 | SCID p. 838 |
| Anaphylaxis p. 837 | Myasthenia gravis p. 429 | Toxic goiter p. 662 |
| Asthma p. 838 | Pemphigus vulgaris p. 164 | |

[32]*urtica* = nettle
[33]Thomas Hodgkin (1798–1866), British physician
[34]*megaly* = enlargement
[35]*lupus* = wolf; *erythema* = redness

**DEEPER INSIGHT 21.4**

## CLINICAL APPLICATION

### Neuroimmunology—The Mind–Body Connection

*Neuroimmunology* is a relatively new branch of medicine concerned with the relationship between mind and body in health and disease. It is attempting especially to understand how a person's state of mind influences health and illness through a three-way communication between the nervous, endocrine, and immune systems.

The sympathetic nervous system issues nerve fibers to the spleen, thymus, lymph nodes, and Peyer patches, where nerve fibers contact thymocytes, B cells, and macrophages. These immune cells have adrenergic receptors for norepinephrine and many other neurotransmitters such as neuropeptide Y, substance P, and vasoactive intestinal peptide (VIP). These neurotransmitters have been shown to influence immune cell activity in various ways. Epinephrine, for example, reduces the lymphocyte count and inhibits NK cell activity, thus suppressing both nonspecific defense and adaptive immunity. Cortisol, another stress hormone, inhibits T cell and macrophage activity, antibody production, and the secretion of inflammatory chemicals. It also induces atrophy of the thymus, spleen, and lymph nodes and reduces the number of circulating lymphocytes, macrophages, and eosinophils. Thus, it is not surprising that prolonged stress increases susceptibility to illnesses such as infections and cancer.

The immune system also sends messages to the nervous and endocrine systems. Immune cells synthesize numerous hormones and neurotransmitters that we normally associate with endocrine and nerve cells. B lymphocytes produce adrenocorticotropic hormone (ACTH) and enkephalins; T lymphocytes produce growth hormone, thyroid-stimulating hormone, luteinizing hormone, and follicle-stimulating hormone; and monocytes secrete prolactin, VIP, and somatostatin. The interleukins and tumor necrosis factor (TNF) secreted by immune cells produce feelings of fatigue and lethargy when we are sick, and stimulate the hypothalamus to secrete corticotropin-releasing hormone, leading to ACTH and cortisol secretion. It remains uncertain and controversial whether the quantities of some of these substances produced by immune cells are enough to have far-reaching effects on the body, but it seems increasingly possible that immune cells influence nervous and endocrine functions in ways that affect recovery from illness.

Although neuroimmunology has met with some skepticism among physicians, there is less and less room for doubt about the importance of a person's state of mind to immune function. People under stress, such as medical students during examination periods and people caring for relatives with Alzheimer disease, show more respiratory infections than other people and respond less effectively to hepatitis and flu vaccines. The attitudes, coping abilities, and social support systems of patients significantly influence survival time even in such serious diseases as AIDS and breast cancer. Women with breast cancer die at significantly higher rates if their husbands or partners cope poorly with stress. Attitudes such as optimism, cheer, depression, resignation, or despair in the face of disease significantly affect immune function. Religious beliefs can also influence the prospect of recovery. Indeed, ardent believers in voodoo sometimes die just from the belief that someone has cast a spell on them. The stress of hospitalization can counteract the treatment one gives to a patient, and neuroimmunology has obvious implications for treating patients in ways that minimize their stress and thereby promote recovery.

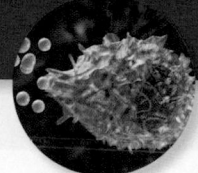

# CONNECTIVE ISSUES

## Effects of the LYMPHATIC AND IMMUNE SYSTEMS On Other Organ Systems

### ALL SYSTEMS
Lymphatic vessels drain excess tissue fluid, prevent edema, and remove cellular debris and pathogens. The immune system monitors all organs and provides defense against pathogens. Natural killer cells patrol the body and protect against cancer. All systems are subject to a variety of hypersensitivity and autoimmune disorders.

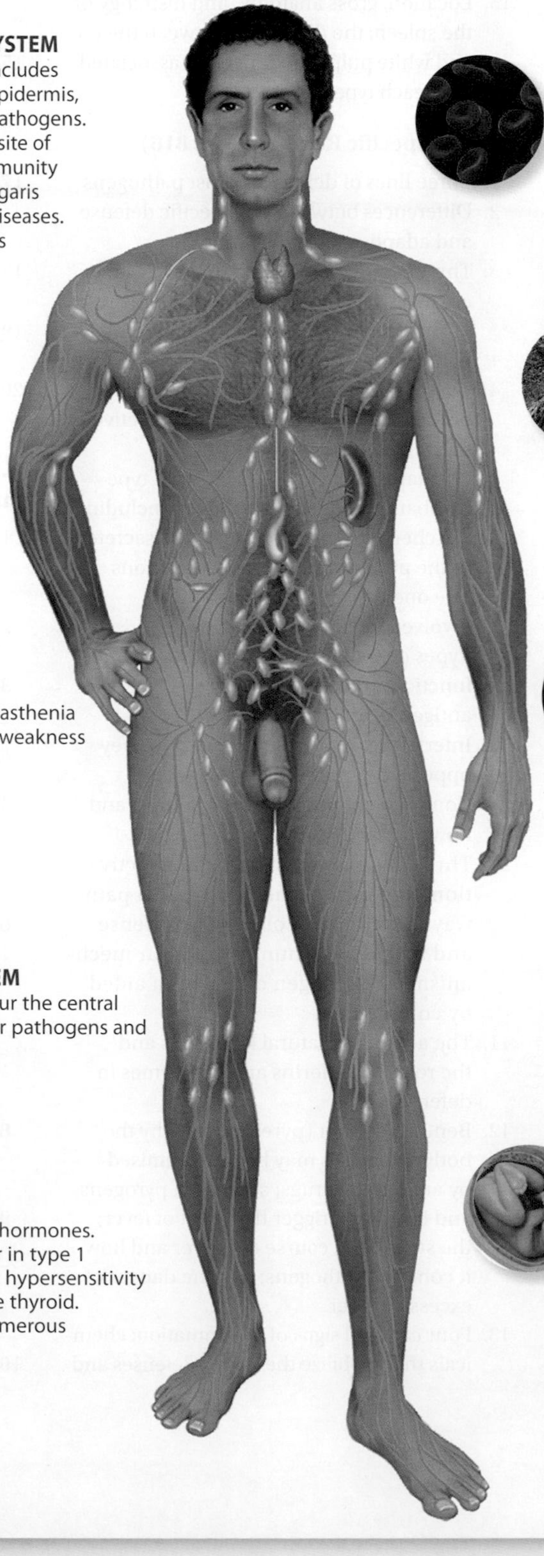

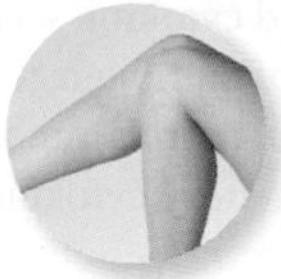

### INTEGUMENTARY SYSTEM
The immune system includes dendritic cells of the epidermis, which guard against pathogens. The skin is a common site of inflammation. Autoimmunity causes pemphigus vulgaris and some other skin diseases. Hypersensitivity causes skin eruptions such as hives and dermatitis.

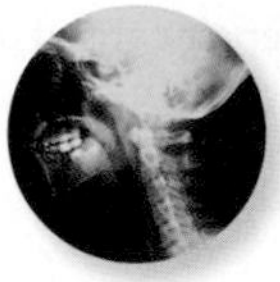

### SKELETAL SYSTEM
Autoimmunity causes rheumatoid arthritis.

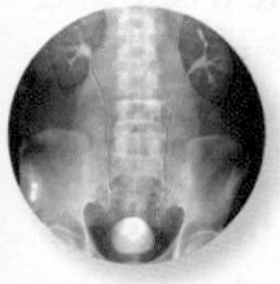

### MUSCULAR SYSTEM
Autoimmunity causes myasthenia gravis, leading to muscle weakness and paralysis.

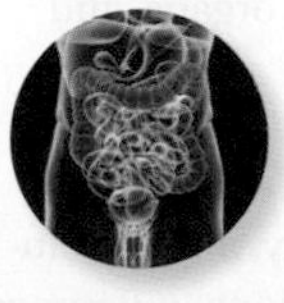

### NERVOUS SYSTEM
Microglia cells scour the central nervous system for pathogens and tissue debris.

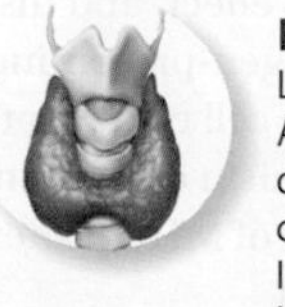

### ENDOCRINE SYSTEM
Lymph transports some hormones. Autoimmunity is a factor in type 1 diabetes mellitus. Type II hypersensitivity causes toxic goiter of the thyroid. Immune cells secrete numerous hormones.

### CIRCULATORY SYSTEM
Lymphatic vessels return fluid and lymphocytes to the bloodstream; the spleen disposes of expired RBCs; lymphatic organs filter microbes and debris from the blood. Autoimmunity damages heart valves in rheumatic fever, and immune hypersensitivity causes circulatory failure in anaphylactic shock.

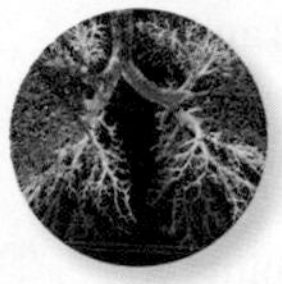

### RESPIRATORY SYSTEM
Alveolar macrophages remove debris from the lungs; pulmonary lymphatic vessels are especially abundant and are needed to prevent fluid accumulation in the lungs. Immune hypersensitivity has effects ranging from respiratory congestion to asthma.

### URINARY SYSTEM
Lymphatics absorb fluid and proteins in the kidneys, which is essential to enabling kidneys to concentrate the urine and conserve water. Autoimmunity causes acute glomerulonephritis.

### DIGESTIVE SYSTEM
Lymphatic vessels called lacteals in the small intestine absorb nearly all dietary lipids and fat-soluble vitamins.

### REPRODUCTIVE SYSTEM
Immunity to cells that are genetically different from other body cells requires the testes and ovaries to have barriers that protect sperm and eggs from immune destruction. A mismatch in Rh type can cause antibodies to attack fetal RBCs, causing hemolytic disease of the newborn.

# STUDY GUIDE

## ▶ Assess Your Learning Outcomes

*To test your knowledge, discuss the following topics with a study partner or in writing, ideally from memory.*

### 21.1 The Lymphatic System (p. 803)

1. Functions and basic constituents of the lymphatic system
2. The definition, appearance, and composition of lymph
3. How lymph is produced; characteristics of lymphatic capillaries that allow cells and other large particles to enter the lymph
4. Lymphatic collecting vessels, trunks, and collecting ducts; the similarity of lymphatic vessels to some blood vessels; and their relationship to the lymph nodes
5. Names of the six lymphatic trunks and two collecting ducts; the body regions drained by them; and the two points at which lymph empties into the bloodstream
6. Mechanisms that propel the flow of lymph
7. Six types of cells found in lymphatic tissue, and their functions
8. The nature of diffuse lymphatic tissue and where it is found; what *MALT* and *BALT* stand for and where they are found
9. How lymphatic nodules differ from diffuse lymphatic tissue; the name of the clusters of lymphatic nodules found in the distal small intestine
10. How lymphatic organs differ from diffuse lymphatic tissue and lymphatic nodules; the two primary lymphatic organs and three secondary lymphatic organs, and why they are called this
11. Structure and function of red bone marrow
12. The location, gross anatomy, and histology of the thymus; the functional difference between its cortex and medulla; the functions of its reticular endothelial cells; and the necessity of the thymus to immunity
13. Structure and function of lymph nodes; the significance of lymph nodes having both afferent and efferent lymphatic vessels, unlike any other lymphatic organs; the approximate number of lymph nodes and seven regions in which they are especially concentrated; and the meaning of *lymphadenitis* and *lymphadenopathy*

14. Types of tonsils, where they are located, and their structure and function; the most common cause of tonsillitis
15. Location, gross anatomy, and histology of the spleen; the difference between the red and white pulp; and functions associated with each type of pulp

### 21.2 Nonspecific Resistance (p. 816)

1. Three lines of defense against pathogens
2. Differences between nonspecific defense and adaptive immunity
3. Three properties of the skin that make it an effective barrier to pathogens; the roles of lactic acid and antimicrobial peptides in its barrier function
4. Three properties of the mucous membranes that also make them an effective barrier to pathogens
5. Mechanisms by which each WBC type combats pathogens and illness, including the chemicals that some of them secrete in the performance of these functions
6. The one type of lymphocyte that is involved in nonspecific defense
7. Types of macrophages; their origin, and functions; the action of macrophages as antigen-presenting cells
8. Interferons, their source, and how they oppose the spread of viruses
9. Complement proteins, their source, and how they are named
10. Three pathways of complement activation; how each is initiated; which pathways function in nonspecific defense and adaptive immunity; and four mechanisms of pathogen destruction aided by complement
11. The actions of natural killer cells and the roles of perforins and granzymes in defense
12. Benefits of fever (pyrexia) and why the body's defenses may be compromised by antipyretic drugs; sources of pyrogens and how they trigger the onset of fever; the stages and course of a fever and how it combats pathogens; and the danger of excessive fever
13. Four cardinal signs of inflammation; chemicals that mobilize the body's defenses and

initiate inflammation; and specific actions of these chemicals
14. The neutrophil actions of margination, diapedesis, chemotaxis, phagocytosis, the respiratory burst, and cytokine secretion
15. Other mechanisms of pathogen containment and destruction in inflammation
16. Examples of inflammatory cytokines and their roles
17. How hyperemia, bradykinin, and other factors account for the four cardinal signs of inflammation
18. The recruitment and action of macrophages
19. The formation, composition, and fate of pus
20. Mechanisms of tissue repair carried out after a pathogen is defeated

### 21.3 General Aspects of Adaptive Immunity (p. 824)

1. The definition of *immune system* and the two distinguishing characteristics of adaptive immunity
2. Two basic forms of adaptive immunity and the difference between them
3. How adaptive immunity is classified as active or passive and as natural or artificial; which types result in immune memory and lasting protection, and which do not
4. The definition of *antigen* and the chemical characteristics of antigens
5. The role of the epitope in the antigenicity of a molecule
6. Haptens and how they become antigenic
7. The life history of T lymphocytes including their origin, migration to the thymus, and attainment of immunocompetence; forms and effects of negative selection; positive selection and its effect; and dispersal of the naive lymphocyte pool
8. The life history of B lymphocytes including their origin; negative selection; positive selection and its effect; and dispersal
9. The necessity of antigen-presenting cells (APCs) to immunity; cell types that serve as APCs; the mechanism of antigen processing; and the role of MHC proteins in antigen presentation
10. Interleukins and their role in immunity

# STUDY GUIDE

## 21.4 Cellular Immunity (p. 827)

1. Four classes of T lymphocytes involved in cellular immunity, and the function of each
2. Three fundamental stages of cellular immunity
3. What an APC does when it detects a foreign antigen; functional differences between MHC-I and MHC-II proteins; how $T_H$ and $T_C$ cells respond to these MHC protein classes
4. Antigen recognition, costimulation, and clonal selection of a T cell; differentiation of selected T cells into effector cells and memory cells
5. How activated $T_H$ cells stimulate neutrophils, NK cells, and macrophages
6. How activated $T_C$ cells destroy target cells; the roles of interferons, perforin, granzymes, and tumor necrosis factor
7. Characteristics of immune memory and the T cell recall response in cellular immunity

## 21.5 Humoral Immunity (p. 831)

1. Similarities in and differences between humoral and cellular immunity
2. How an immunocompetent B cell responds when it encounters a foreign antigen; the roles of MHC-II proteins and a $T_H$ cell in its response
3. Clonal selection of activated B cell and differentiation of memory cells and plasma cells; the difference between a plasma cell and a B cell
4. Structure of antibody monomers, dimers, and pentamers; structural and functional differences between IgA, IgD, IgE, IgG, and IgM
5. Four mechanisms by which antibodies combat antigens
6. The secondary (anamnestic) response in humoral immune memory

## 21.6 Immune System Disorders (p. 837)

1. Three principal things that can go wrong with immune function

2. Hypersensitivity; names and characteristics of its four types, and examples of disorders of each type
3. The basic cause of autoimmune diseases; what normally prevents them; and three reasons why an autoimmune disease may appear, with an example of each
4. The basic cause of immunodeficiency diseases and the specific cause of severe combined immunodeficiency disease (SCID)
5. The pathology of acquired immunodeficiency disease (AIDS), including the structure of the human immunodeficiency virus (HIV); its mode of action; its effect on helper T cell count; the diseases to which AIDS makes a person more susceptible; how HIV is and is not transmitted; and treatment approaches to AIDS

## ▶ Testing Your Recall

*Answers in Appendix B*

1. The only lymphatic organ with both afferent and efferent lymphatic vessels is
   a. the spleen.
   b. a lymph node.
   c. a tonsil.
   d. a Peyer patch.
   e. the thymus.

2. Which of the following cells are involved in nonspecific resistance but not in adaptive immunity?
   a. helper T cells
   b. cytotoxic T cells
   c. natural killer cells
   d. B cells
   e. plasma cells

3. The respiratory burst is used by _____ to kill bacteria.
   a. neutrophils
   b. basophils
   c. mast cells
   d. NK cells
   e. cytotoxic T cells

4. Which of these is a macrophage?
   a. a microglial cell
   b. a plasma cell
   c. a reticular cell
   d. a helper T cell
   e. a mast cell

5. The cytolytic action of the complement system is most similar to the action of
   a. interleukin-1.
   b. platelet-derived growth factor.
   c. granzymes.
   d. perforin.
   e. IgE.

6. _____ become antigenic by binding to larger host molecules.
   a. Epitopes
   b. Haptens
   c. Interleukins
   d. Pyrogens
   e. Cell-adhesion molecules

7. Which of the following correctly states the order of events in humoral immunity? Let
   1 = antigen display,
   2 = antibody secretion,
   3 = secretion of interleukin,
   4 = clonal selection, and
   5 = endocytosis of an antigen.
   a. 3–4–1–5–2
   b. 5–3–1–2–4
   c. 3–5–1–4–2
   d. 5–3–1–4–2
   e. 5–1–3–4–2

8. The cardinal signs of inflammation include all of the following except
   a. redness.
   b. swelling.
   c. heat.
   d. fever.
   e. pain.

# STUDY GUIDE

9. A helper T cell can bind only to another cell that has
   a. MHC-II proteins.
   b. an epitope.
   c. an antigen-binding site.
   d. a complement-binding site.
   e. a CD4 protein.

10. Which of the following results from a lack of self-tolerance?
    a. SCID
    b. AIDS
    c. systemic lupus erythematosus
    d. anaphylaxis
    e. asthma

11. Any organism or substance capable of causing disease is called a/an _____.

12. Mucous membranes contain an antibacterial enzyme called _____.

13. _____ is a condition in which one or more lymph nodes are swollen and painful to the touch.

14. The movement of leukocytes through a capillary or venule wall is called _____.

15. In the process of _____, complement proteins coat bacteria and serve as binding sites for phagocytes.

16. Any substance that triggers a fever is called a/an _____.

17. The chemical signals produced by leukocytes to stimulate other leukocytes are called _____.

18. Part of an antibody called the _____ binds to part of an antigen called the _____.

19. Self-tolerance results from a process called _____, in which lymphocytes programmed to react against self-antigens die.

20. Any disease in which antibodies attack one's own tissues is called a/an _____ disease.

## ▶ Building Your Medical Vocabulary

*Answers in Appendix B*

*State a medical meaning of each word element below, and give a term in which it or a slight variation of it is used.*

1. ana-
2. crino-
3. extra-
4. -genous
5. immuno-
6. kino-
7. lympho-
8. -megaly
9. -pathy
10. pyro-

## ▶ True or False

*Answers in Appendix B*

*Determine which five of the following statements are false, and briefly explain why.*

1. Some bacteria employ lysozyme to liquefy the tissue gel and make it easier for them to get around.

2. T lymphocytes undergo clonal deletion and anergy in the thymus.

3. Interferons help to reduce inflammation.

4. T lymphocytes are involved only in cell-mediated immunity.

5. The white pulp of the spleen gets its color mainly from lymphocytes and macrophages.

6. Perforins are employed in both nonspecific resistance and cellular immunity.

7. Histamine and heparin are secreted by basophils and mast cells.

8. A person who is HIV-positive and has a $T_H$ (CD4) count of 1,000 cells/µL does not have AIDS.

9. Anergy is often a cause of autoimmune diseases.

10. Interferons kill pathogenic bacteria by making holes in their cell walls.

# STUDY GUIDE

## ▶ Testing Your Comprehension

*Answers at* www.mhhe.com/saladin7

1. Anti-D antibodies of an Rh– woman sometimes cross the placenta and hemolyze the RBCs of an Rh+ fetus (see p. 688). Yet the anti-B antibodies of a type A mother seldom affect the RBCs of a type B fetus. Explain this difference based on your knowledge of the five immunoglobulin classes.

2. In treating a woman for malignancy in the right breast, the surgeon removes some of her axillary lymph nodes. Following surgery, the patient experiences edema of her right arm. Explain why.

3. A girl with a defective heart receives a new heart transplanted from another child who was killed in an accident. The patient is given an antilymphocyte serum containing antibodies against her lymphocytes. The transplanted heart is not rejected, but the patient dies of an overwhelming bacterial infection. Explain why the antilymphocyte serum was given and why the patient was so vulnerable to infection.

4. A burn research center uses mice for studies of skin grafting. To prevent graft rejec-tion, the mice are thymectomized at birth. Even though B cells do not develop in the thymus, these mice show no humoral immune response and are very susceptible to infection. Explain why the removal of the thymus would improve the success of skin grafts but adversely affect humoral immunity.

5. Contrast the structure of a B cell with that of a plasma cell, and explain how their structural difference relates to their functional difference.

## ▶ Improve Your Grade at www.mhhe.com/saladin7

*Download animations to study when it fits your schedule. Practice quizzes, labeling activities, games, and flashcards offer fun ways to master the chapter concepts. Or, download image PowerPoint files for each chapter to create a study guide or for taking notes during lecture.*

# THE RESPIRATORY SYSTEM

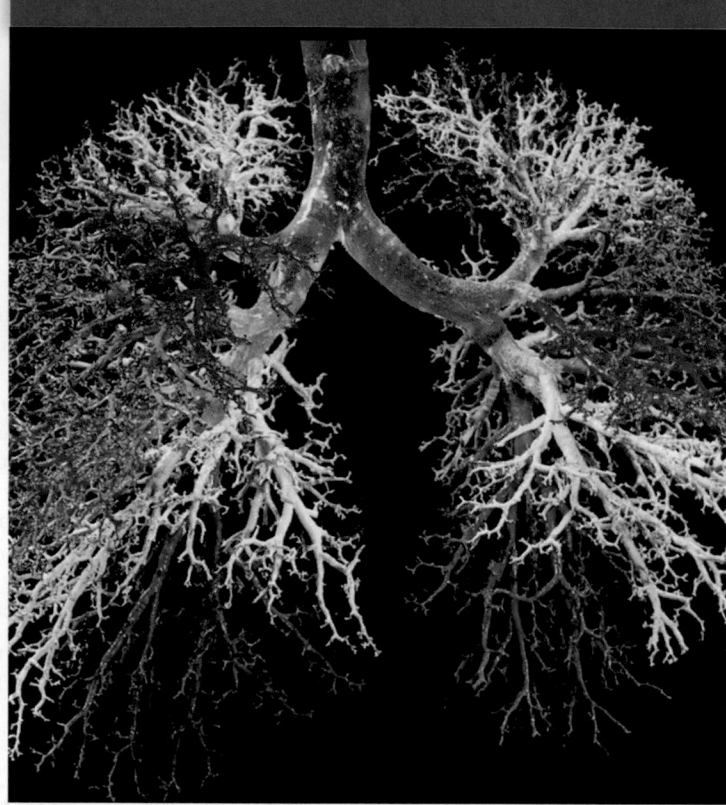

The bronchial trees

Anatomy & Physiology | REVEALED®
aprevealed.com

**Module 11: Respiratory System**

**B**reath represents life. The first breath of a baby and the last gasp of a dying person are two of the most dramatic moments of human experience. But why do we breathe? It comes down to the fact that most of our metabolism directly or indirectly requires ATP. Most ATP synthesis requires oxygen and generates carbon dioxide—thus driving the need to breathe in order to supply the former and eliminate the latter. The respiratory system consists essentially of tubes that deliver air to the lungs, where oxygen diffuses into the blood and carbon dioxide diffuses out.

The respiratory and cardiovascular systems collaborate to deliver oxygen to tissues throughout the body and to transport carbon dioxide to the lungs for elimination. Not only do these two systems have a close spatial relationship in the thoracic cavity, but they also have such a close functional relationship that they are often considered jointly as the *cardiopulmonary system*. A disorder that affects the lungs has direct and pronounced effects on the heart, and vice versa. As discussed in the next two chapters, the respiratory system also collaborates closely with the urinary system to regulate the body's acid–base balance, which is reason to consider these systems consecutively in this group of chapters.

## 22.1 Anatomy of the Respiratory System

### Expected Learning Outcomes

When you have completed this section, you should be able to

a. state the functions of the respiratory system;

b. name and describe the organs of this system;

c. trace the flow of air from the nose to the pulmonary alveoli; and

d. relate the function of any portion of the respiratory tract to its gross and microscopic anatomy.

The term *respiration* can mean ventilation of the lungs (breathing) or the use of oxygen in cellular metabolism. In this chapter, we are concerned with the first process. Cellular respiration was introduced in chapter 2 and is considered more fully in chapter 26.

The **respiratory system** is an organ system that rhythmically takes in air and expels it from the body, thereby supplying the body with oxygen and expelling the carbon dioxide that it generates. However, it has a broader range of functions than is commonly supposed:

1. **Gas exchange.** It provides for oxygen and carbon dioxide exchange between the blood and air.

2. **Communication.** It serves for speech and other vocalization (laughing, crying).

3. **Olfaction.** It provides the sense of smell, which is important in social interactions, food selection, and avoiding danger (such as a gas leak or spoiled food).

4. **Acid–base balance.** By eliminating $CO_2$, it helps to control the pH of the body fluids. Excess $CO_2$ reacts with water and generates acid; therefore, if respiration does not keep pace with $CO_2$ production, $H^+$ accumulates and the body fluids have an abnormally low pH *(acidosis)*.

5. **Blood pressure regulation.** The lungs carry out a step in synthesizing *angiotensin II,* which helps to regulate blood pressure.

6. **Blood and lymph flow.** Breathing creates pressure gradients between the thorax and abdomen that promote the flow of lymph and venous blood.

7. **Blood filtration.** The lungs filter small blood clots from the bloodstream and dissolve them, preventing clots from obstructing the more vital coronary, cerebral, and renal circulation.

8. **Expulsion of abdominal contents.** Breath-holding helps to expel abdominal contents during urination, defecation, and childbirth.

The principal organs of the respiratory system are the nose, pharynx, larynx, trachea, bronchi, and lungs (fig. 22.1). Within the lungs, air flows along a dead-end pathway consisting essentially of bronchi → bronchioles → alveoli (with some refinements to be introduced later). Incoming air stops in the *alveoli* (millions of tiny, thin-walled air sacs), exchanges gases with the bloodstream through the alveolar wall, and then flows back out.

The **conducting division** of the respiratory system consists of those passages that serve only for airflow, essentially from the nostrils through the major bronchioles. The walls of these passages are too thick for adequate diffusion of oxygen from the air into the blood. The **respiratory division** consists of the

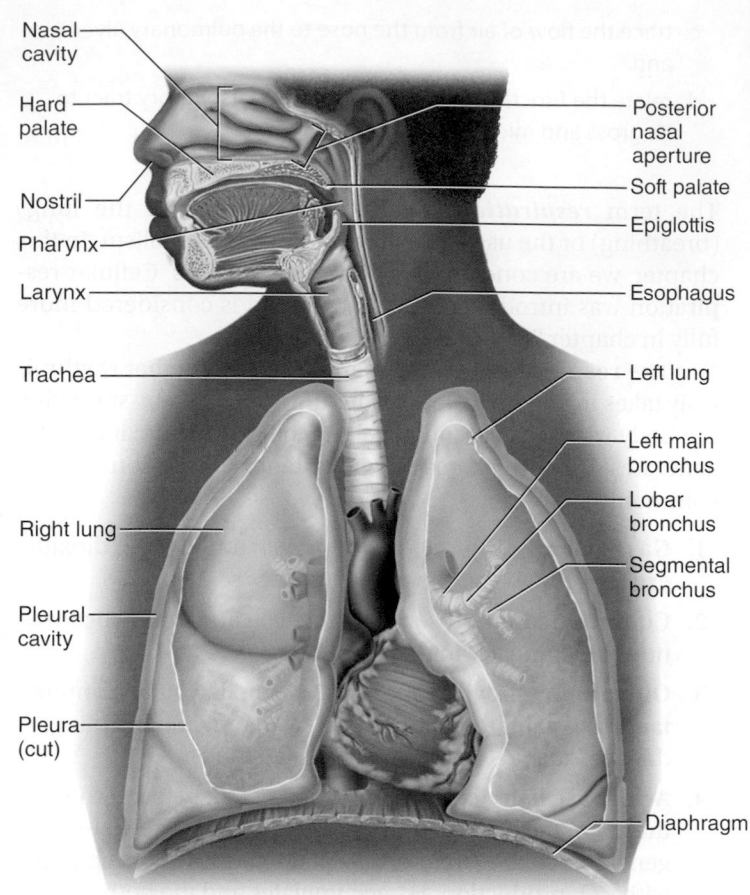

**FIGURE 22.1  The Respiratory System.** AP|R

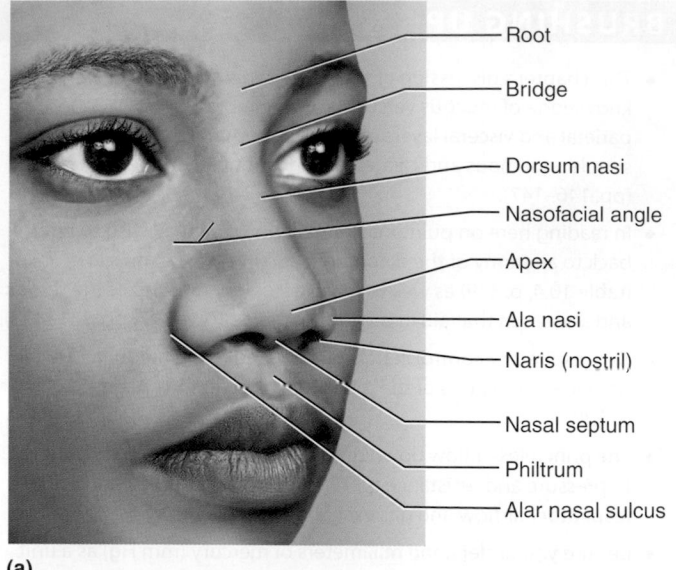

(a)

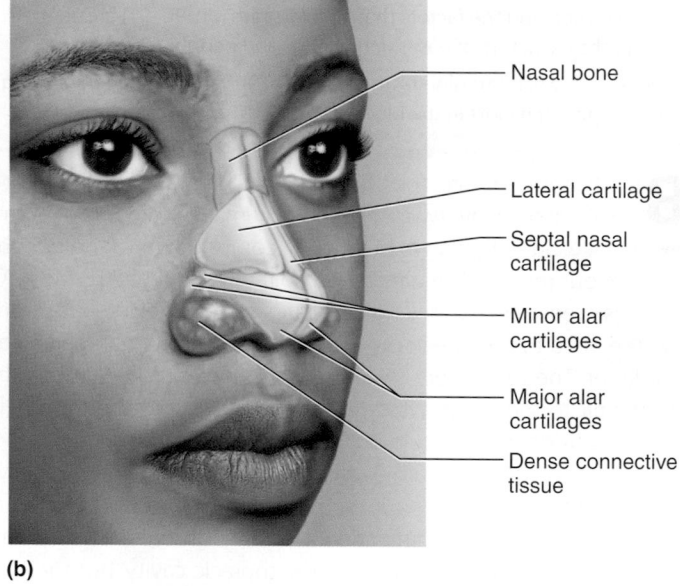

(b)

**FIGURE 22.2  Anatomy of the Nasal Region.** (a) External anatomy. (b) Connective tissues that shape the nose.

alveoli and other gas-exchange regions of the distal airway. The airway from the nose through the larynx is often called the **upper respiratory tract** (that is, the respiratory organs in the head and neck), and the regions from the trachea through the lungs compose the **lower respiratory tract** (the respiratory organs of the thorax). However, these are inexact terms and various authorities place the dividing line between the upper and lower tracts at different points.

## The Nose

The **nose** has several functions: It warms, cleanses, and humidifies inhaled air; it detects odors; and it serves as a resonating chamber that amplifies the voice. It extends from a pair of anterior openings called the **nostrils,** or **nares** (NAIR-eze) (singular, *naris*), to a pair of posterior openings called the **posterior nasal apertures, or choanae**[1] (co-AH-nee)**.**

The facial part of the nose is shaped by bone and hyaline cartilage. Its superior half is supported by a pair of small nasal bones medially and the maxillae laterally. The inferior half is supported by the **lateral** and **alar cartilages** (fig. 22.2). By palpating your own nose, you can easily find the boundary

between the bone and cartilage. The flared portion at the lower end of the nose, called the **ala nasi**[2] (AIL-ah NAZE-eye), is shaped by the alar cartilages and dense connective tissue.

The internal chamber of the nose, called the **nasal cavity,** is divided into right and left halves called **nasal fossae** (FAW-see). The dividing wall is a vertical plate, the **nasal septum,** composed of bone and hyaline cartilage. The vomer forms the inferior part of the septum, the perpendicular plate of the ethmoid bone forms its superior part, and the *septal cartilage*

---

[1]*choana* = funnel

[2]*ala* = wing; *nasi* = of the nose

forms its anterior part (fig. 22.3c). The ethmoid and sphenoid bones compose the roof of the nasal cavity, and the hard palate forms its floor. The palate separates the nasal cavity from the oral cavity and allows you to breathe while chewing food. The paranasal sinuses (see p. 236) and the nasolacrimal ducts of the orbits (see fig. 16.23, p. 607) drain into the nasal cavity.

The nasal cavity begins with a small dilated chamber called the **vestibule** just inside the nostril, bordered by the ala nasi. This space is lined with stratified squamous epithelium like the facial skin, and has stiff **guard hairs,** or **vibrissae** (vy-BRISS-ee), that block insects and debris from entering the nose. Posterior to the vestibule, the nasal cavity expands into a much larger chamber, but it does not have much open space. Most of it is occupied by three folds of tissue—the **superior, middle,** and **inferior nasal conchae**[3] (CON-kee), or **turbinates**—that project from the lateral walls toward the septum (fig. 22.3). Beneath each concha is a narrow air passage called a **meatus** (me-AY-tus). The narrowness of these passages and the turbulence caused by the conchae ensure that most air contacts the mucous membrane on its way through. As it does, most dust in the air sticks to the mucus, and the air picks up moisture and heat from the mucosa. The conchae thus enable the nose to cleanse, warm, and humidify the air more effectively than if the air had an unobstructed flow through a cavernous space.

Odors are detected by sensory cells in the **olfactory epithelium,** which covers a small area of the roof of the nasal fossa and adjacent parts of the septum and superior concha (see fig. 16.7, p. 590). The rest of the nasal cavity, except for the vestibule, is lined with **respiratory epithelium.** Both of these are ciliated pseudostratified columnar epithelia. However, in the olfactory epithelium, the cilia are immobile and serve to bind odor molecules. In the respiratory epithelium, they are mobile. The respiratory epithelium is similar to the one seen in figure 5.7 (p. 147). Its wineglass-shaped *goblet cells* secrete mucus, and its ciliated cells propel the mucus posteriorly toward the pharynx. The nasal mucosa also contains mucous glands, located in the lamina propria (the connective tissue layer beneath the epithelium). They supplement the mucus produced by the goblet cells. Inhaled dust, pollen, bacteria, and other foreign matter stick to the mucus and are swallowed; they are either digested or pass through the digestive tract rather than contaminating the lungs. The lamina propria is also well populated by lymphocytes and plasma cells that mount immune defenses against inhaled pathogens.

The lamina propria contains large blood vessels that help to warm the air. The inferior concha has an especially extensive venous plexus called the **erectile tissue (swell body).** Every 30 to 60 minutes, the erectile tissue on one side swells with blood and restricts airflow through that fossa. Most air is then directed through the other nostril, allowing the engorged side time to recover from drying. Thus, the preponderant flow of air shifts between the right and left nostrils once or twice each hour.

## The Pharynx

The **pharynx** (FAIR-inks, FAR-inks) is a muscular funnel extending about 13 cm (5 in.) from the posterior nasal apertures to the larynx. It has three regions: the *nasopharynx, oropharynx,* and *laryngopharynx* (fig. 22.3c).

The **nasopharynx** is distal to the posterior nasal apertures and above the soft palate. It receives the auditory (eustachian) tubes from the middle ears and houses the pharyngeal tonsil. Inhaled air turns 90° downward as it passes through the nasopharynx. Relatively large particles (>10 μm) generally cannot make the turn because of their inertia. They collide with the wall of the nasopharynx and stick to the mucosa near the tonsil, which is well positioned to respond to airborne pathogens.

The **oropharynx** is a space between the posterior margin of the soft palate and the epiglottis.

The **laryngopharynx** (la-RIN-go-FAIR-inks) lies mostly posterior to the larynx, extending from the superior margin of the epiglottis to the inferior margin of the cricoid cartilage. The esophagus begins at that point.

The nasopharynx passes only air and is lined by pseudostratified columnar epithelium, whereas the oropharynx and laryngopharynx pass air, food, and drink and are lined by stratified squamous epithelium. Muscles of the pharynx play necessary roles in swallowing and speech.

## The Larynx

The **larynx** (LAIR-inks), or voice box, is a cartilaginous chamber about 4 cm (1.5 in.) long (fig. 22.4). Its primary function is to keep food and drink out of the airway, but it evolved the additional role of sound production *(phonation)* in many animals.

The superior opening of the larynx is guarded by a flap of tissue called the **epiglottis.**[4] At rest, the epiglottis stands almost vertically. During swallowing, however, *extrinsic muscles* of the larynx pull the larynx upward toward the epiglottis, the tongue pushes the epiglottis downward to meet it, and the epiglottis closes the airway and directs food and drink into the esophagus behind it. The *vestibular folds* of the larynx, discussed shortly, play a greater role in keeping food and drink out of the airway, however.

In infants, the larynx is relatively high in the throat and the epiglottis touches the soft palate. This creates a more or less continuous airway from the nasal cavity to the larynx and allows an infant to breathe continually while swallowing. The epiglottis deflects milk away from the airstream, like rain running off a tent while it remains dry inside. By age 2, the root of the tongue becomes more muscular and forces the larynx to descend to a lower position. It then becomes impossible to breathe and swallow at the same time without choking.

The framework of the larynx consists of nine cartilages. The first three are solitary and relatively large. The most superior one, the **epiglottic cartilage,** is a spoon-shaped supportive plate

---

[3]*concha* = seashell

[4]*epi* = above, upon; *glottis* = back of the tongue

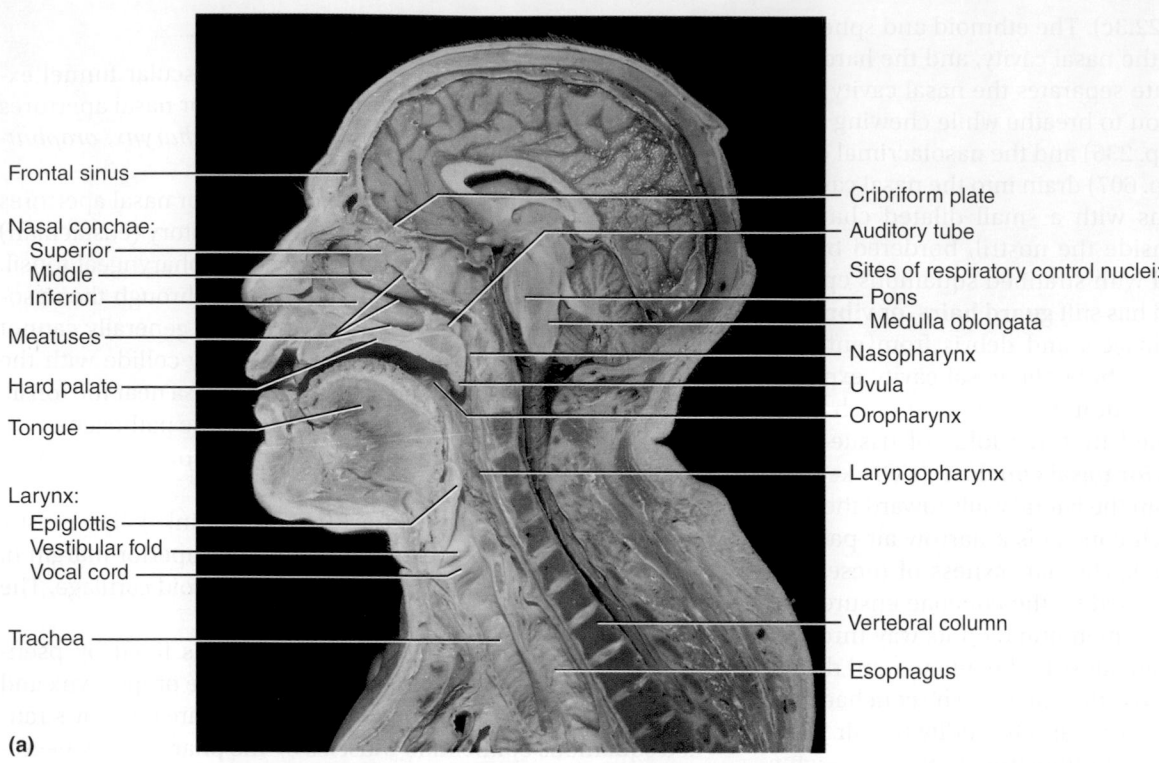

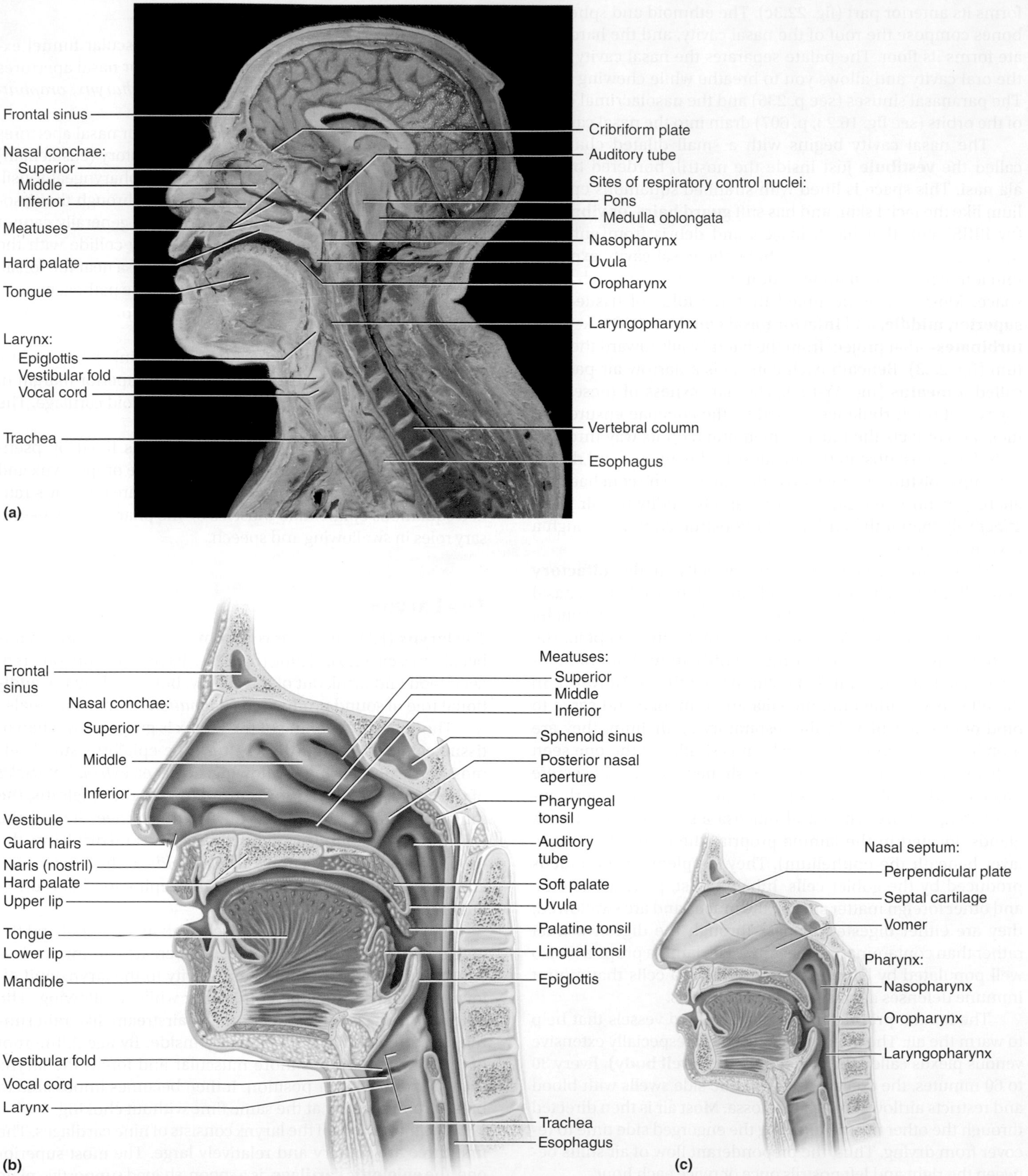

**(a)**

Frontal sinus

Nasal conchae:
- Superior
- Middle
- Inferior

Meatuses

Hard palate

Tongue

Larynx:
- Epiglottis
- Vestibular fold
- Vocal cord

Trachea

Cribriform plate

Auditory tube

Sites of respiratory control nuclei:
- Pons
- Medulla oblongata

Nasopharynx

Uvula

Oropharynx

Laryngopharynx

Vertebral column

Esophagus

**(b)**

Frontal sinus

Nasal conchae:
- Superior
- Middle
- Inferior

Vestibule

Guard hairs

Naris (nostril)

Hard palate

Upper lip

Tongue

Lower lip

Mandible

Vestibular fold

Vocal cord

Larynx

Meatuses:
- Superior
- Middle
- Inferior

Sphenoid sinus

Posterior nasal aperture

Pharyngeal tonsil

Auditory tube

Soft palate

Uvula

Palatine tonsil

Lingual tonsil

Epiglottis

Trachea

Esophagus

**(c)**

Nasal septum:
- Perpendicular plate
- Septal cartilage
- Vomer

Pharynx:
- Nasopharynx
- Oropharynx
- Laryngopharynx

**FIGURE 22.3 Anatomy of the Upper Respiratory Tract.** (a) Median section of the head. (b) Internal anatomy. (c) The nasal septum and regions of the pharynx.

❓ *Draw a line across part (b) of this figure to indicate the boundary between the upper and lower respiratory tract.*

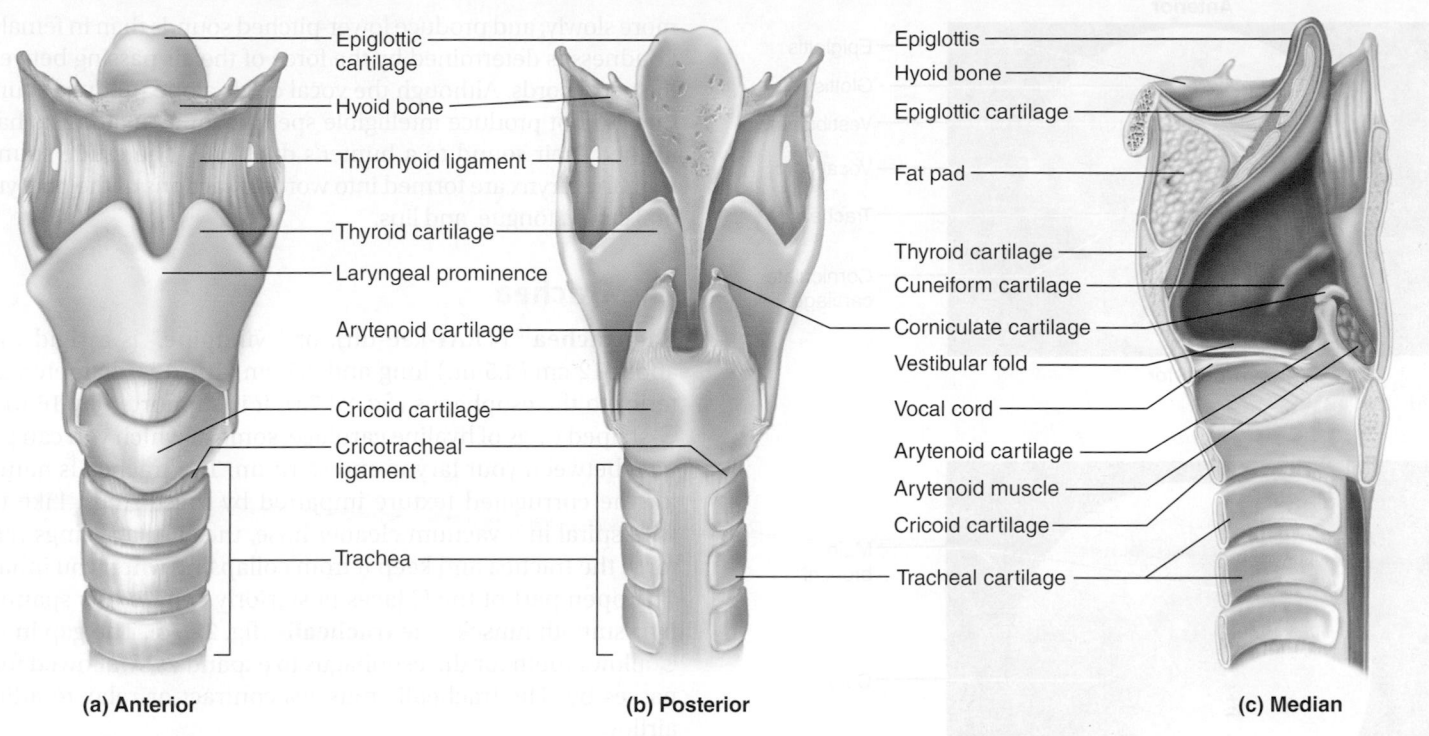

**(a) Anterior**

- Epiglottic cartilage
- Hyoid bone
- Thyrohyoid ligament
- Thyroid cartilage
- Laryngeal prominence
- Arytenoid cartilage
- Cricoid cartilage
- Cricotracheal ligament
- Trachea

**(b) Posterior**

**(c) Median**

- Epiglottis
- Hyoid bone
- Epiglottic cartilage
- Fat pad
- Thyroid cartilage
- Cuneiform cartilage
- Corniculate cartilage
- Vestibular fold
- Vocal cord
- Arytenoid cartilage
- Arytenoid muscle
- Cricoid cartilage
- Tracheal cartilage

**FIGURE 22.4 Anatomy of the Larynx.** Most muscles are removed in order to show the cartilages. AP|R

❓ *Which three cartilages in this figure are more mobile than any of the others?*

in the epiglottis. The largest, the **thyroid[5] cartilage,** is named for its shieldlike shape. It broadly covers the anterior and lateral aspects of the larynx. The "Adam's apple" is an anterior peak of the thyroid cartilage called the *laryngeal prominence.* Testosterone stimulates the growth of this prominence, which is therefore larger in males than in females. Inferior to the thyroid cartilage is a ringlike **cricoid[6]** (CRY-coyd) **cartilage.** The thyroid and cricoid cartilages essentially constitute the "box" of the voice box.

The remaining cartilages are smaller and occur in three pairs. Posterior to the thyroid cartilage are the two **arytenoid[7]** (AR-ih-TEE-noyd) **cartilages,** and attached to their upper ends is a pair of little horns, the **corniculate[8]** (cor-NICK-you-late) **cartilages.** The arytenoid and corniculate cartilages function in speech, as explained shortly. A pair of **cuneiform[9]** (cue-NEE-ih-form) **cartilages** supports the soft tissues between the arytenoids and the epiglottis.

A group of fibrous ligaments binds the cartilages of the larynx together and forms a suspension system for the upper airway. A broad sheet called the **thyrohyoid ligament** suspends the larynx from the hyoid bone above, and below, the **cricotracheal ligament** suspends the trachea from the cricoid

cartilage. These are collectively called the *extrinsic ligaments* because they link the larynx to other organs. The *intrinsic ligaments* are contained entirely within the larynx and link its nine cartilages to each other; they include ligaments of the vocal cords and vestibular folds described next.

The interior wall of the larynx has two folds on each side that stretch from the thyroid cartilage in front to the arytenoid cartilages in back. The superior **vestibular folds** (fig. 22.4c) play no role in speech but close the larynx during swallowing. They are supported by the **vestibular ligaments.** The inferior **vocal cords (vocal folds)** produce sound when air passes between them. They contain the **vocal ligaments** and are covered with stratified squamous epithelium, best suited to endure vibration and contact between the cords. The vocal cords and the opening between them are collectively called the **glottis** (fig. 22.5a).

The walls of the larynx are quite muscular. The superficial *extrinsic muscles* connect the larynx to the hyoid bone and elevate the larynx during swallowing. Also called the *infrahyoid group,* they are named and described in chapter 10 (table 10.2, p. 325).

The deeper *intrinsic muscles* control the vocal cords by pulling on the corniculate and arytenoid cartilages, causing the cartilages to pivot. Depending on their direction of rotation, the arytenoid cartilages abduct or adduct the vocal cords (fig. 22.6). Air forced between the adducted vocal cords vibrates them, producing a high-pitched sound when the cords are relatively taut and a lower-pitched sound when they are more slack. In adult males, the vocal cords are usually longer and thicker, vibrate

---

[5]*thyr* = shield; *oid* = resembling
[6]*crico* = ring; *oid* = resembling
[7]*aryten* = ladle; *oid* = resembling
[8]*corni* = horn; *cul* = little; *ate* = possessing
[9]*cune* = wedge; *form* = shape

**Anterior**

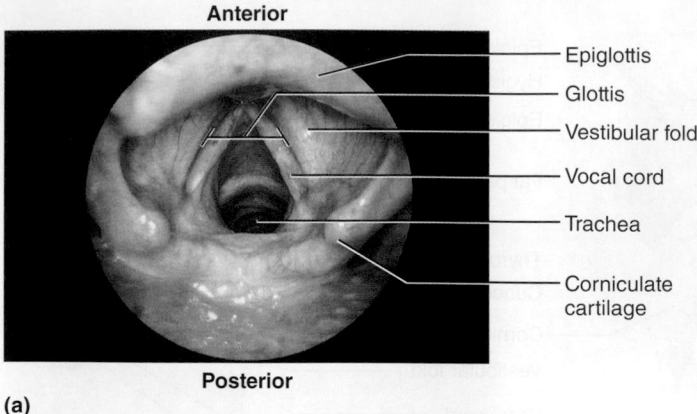

Epiglottis

Glottis

Vestibular fold

Vocal cord

Trachea

Corniculate cartilage

**Posterior**

**(a)**

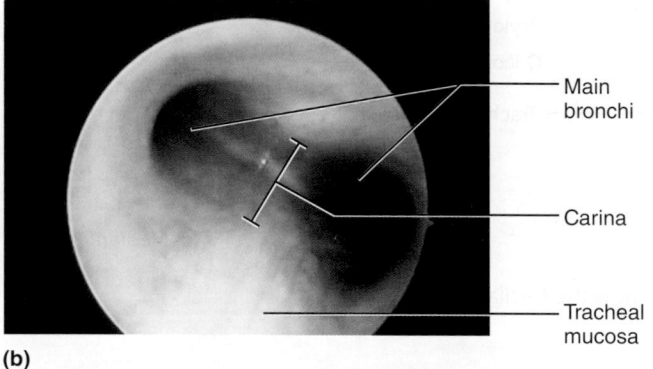

Main bronchi

Carina

Tracheal mucosa

**(b)**

**FIGURE 22.5  Endoscopic Views of the Respiratory Tract.**
(a) Superior view of the larynx, seen with a laryngoscope. (b) Lower end of the trachea, where it forks into the two primary bronchi, seen with a bronchoscope.

more slowly, and produce lower-pitched sounds than in females. Loudness is determined by the force of the air passing between the vocal cords. Although the vocal cords alone produce sound, they do not produce intelligible speech; some anatomists have likened their sound to a hunter's duck call. The crude sounds from the larynx are formed into words by actions of the pharynx, oral cavity, tongue, and lips.

## The Trachea

The **trachea**[10] (TRAY-kee-uh), or "windpipe," is a rigid tube about 12 cm (4.5 in.) long and 2.5 cm (1 in.) in diameter, anterior to the esophagus (fig. 22.7a). It is supported by 16 to 20 C-shaped rings of hyaline cartilage, some of which you can palpate between your larynx and sternum. The trachea is named for the corrugated texture imparted by these rings. Like the wire spiral in a vacuum cleaner hose, the cartilage rings reinforce the trachea and keep it from collapsing when you inhale. The open part of the C faces posteriorly, where it is spanned by a smooth muscle, the **trachealis** (fig. 22.7c). The gap in the C allows room for the esophagus to expand as swallowed food passes by. The trachealis muscles contract or relax to adjust airflow.

The inner lining of the trachea is a pseudostratified columnar epithelium composed mainly of mucus-secreting goblet cells, ciliated cells, and short basal stem cells (figs. 22.7b and 22.8). The mucus traps inhaled particles, and the upward beating of the cilia drives the debris-laden mucus

---

[10]*trache* = rough

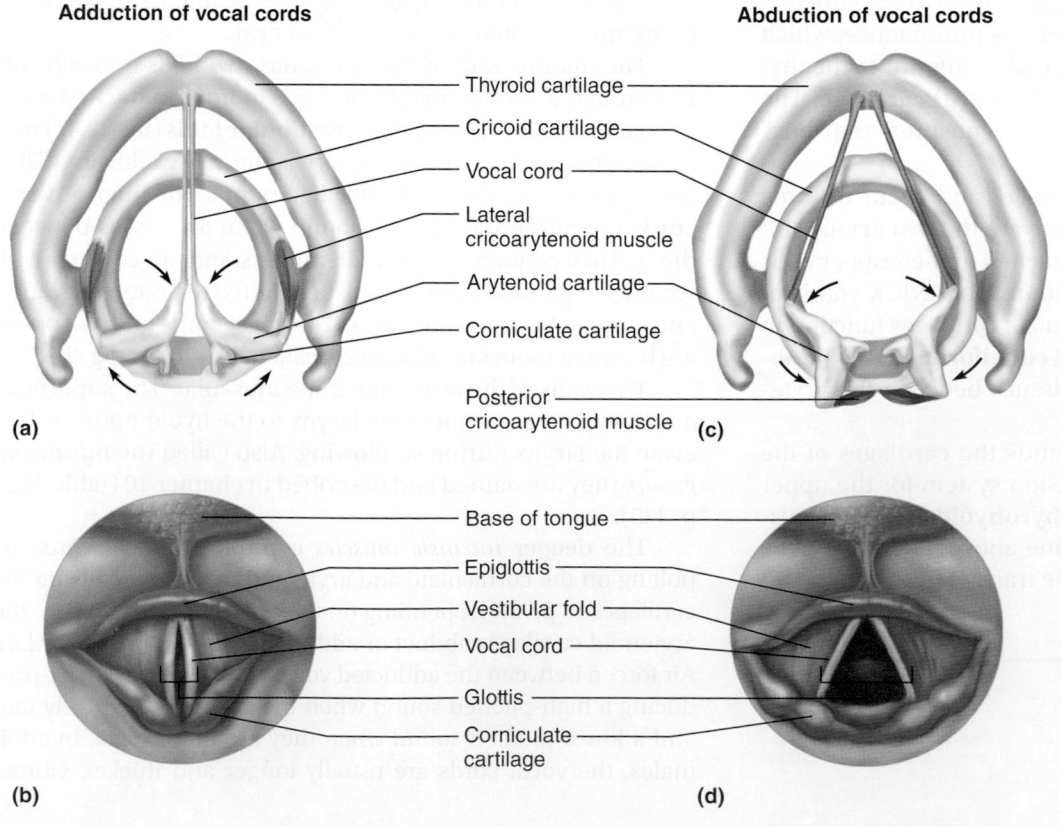

**Adduction of vocal cords**

**Abduction of vocal cords**

Thyroid cartilage

Cricoid cartilage

Vocal cord

Lateral cricoarytenoid muscle

Arytenoid cartilage

Corniculate cartilage

Posterior cricoarytenoid muscle

Anterior

Posterior

**(a)**

**(c)**

Base of tongue

Epiglottis

Vestibular fold

Vocal cord

Glottis

Corniculate cartilage

**(b)**

**(d)**

**FIGURE 22.6**
**Action of Some of the Intrinsic Laryngeal Muscles on the Vocal Cords.**  (a) Adduction of the vocal cords by the lateral cricoarytenoid muscles. (b) Adducted vocal cords seen with the laryngoscope. (c) Abduction of the vocal cords by the posterior cricoarytenoid muscles. (d) Abducted vocal cords seen with the laryngoscope.

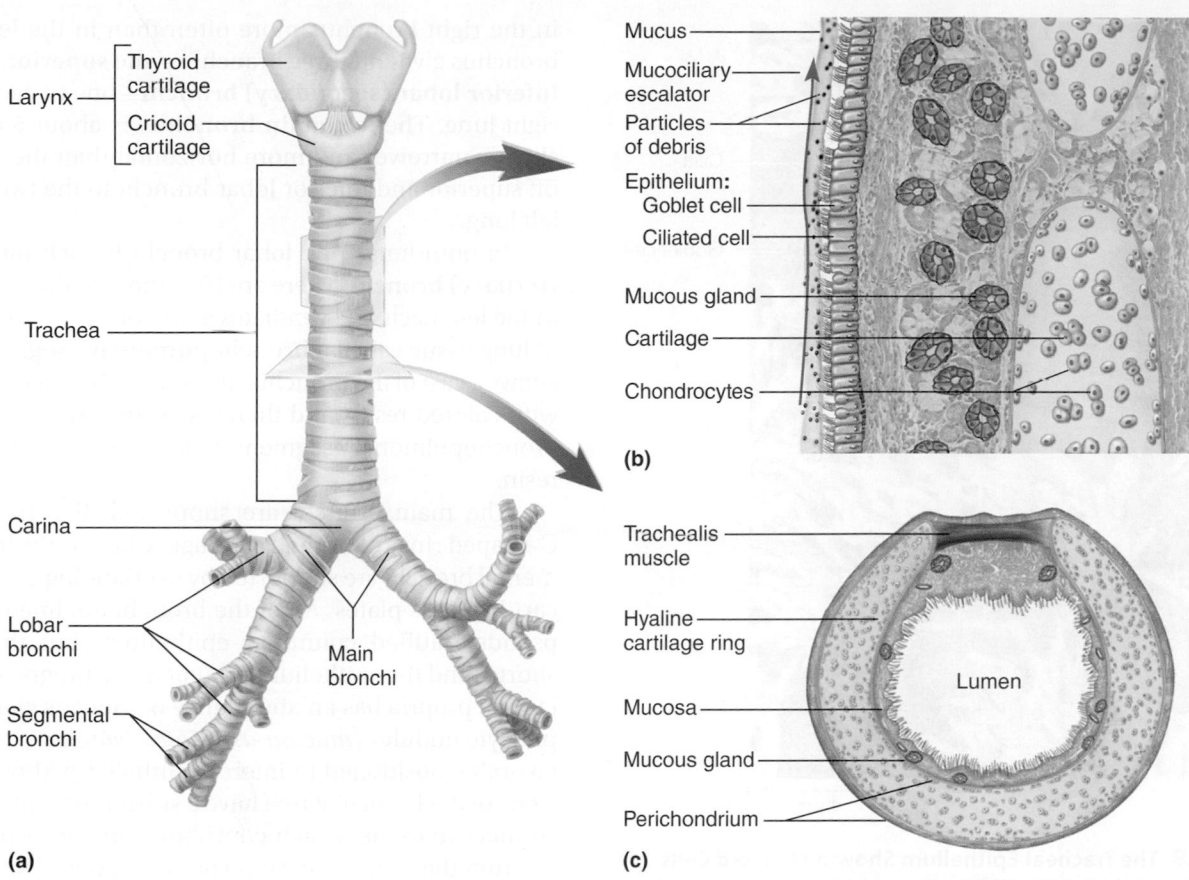

Larynx — Thyroid cartilage
Cricoid cartilage

Trachea —

Carina —

Lobar bronchi —

Segmental bronchi —

Main bronchi

**(a)**

Mucus —
Mucociliary escalator
Particles of debris
Epithelium:
  Goblet cell —
  Ciliated cell —

Mucous gland —
Cartilage —
Chondrocytes —

**(b)**

Trachealis muscle —

Hyaline cartilage ring —

Mucosa —
Mucous gland —
Perichondrium —

Lumen

**(c)**

**FIGURE 22.7  Anatomy of the Lower Respiratory Tract.** (a) Anterior view. (b) Longitudinal section of the trachea showing the action of the mucociliary escalator. (c) Cross section of the trachea showing the C-shaped tracheal cartilage.  AP|R

 *Why do inhaled objects more often go into the right main bronchus than into the left?*

toward the pharynx, where it is swallowed. This mechanism of debris removal is called the **mucociliary escalator.**

The connective tissue beneath the tracheal epithelium contains lymphatic nodules, mucous and serous glands, and the tracheal cartilages. The outermost layer of the trachea, called the **adventitia,** is fibrous connective tissue that blends into the adventitia of other organs of the mediastinum.

At the level of the sternal angle, the trachea forks into the right and left *main bronchi.* The lowermost tracheal cartilage has an internal median ridge called the **carina**[11] (ca-RY-na) that directs the airflow to the right and left (see fig. 22.5b). The bronchi are further traced in the discussion of the *bronchial tree* of the lungs.

## The Lungs and Bronchial Tree

Each **lung** is a somewhat conical organ with a broad, concave **base** resting on the diaphragm and a blunt peak called the **apex** projecting slightly above the clavicle (fig. 22.9). The broad **costal surface** is pressed against the rib cage, and the smaller

[11]*carina* = keel

### DEEPER INSIGHT 22.1
### CLINICAL APPLICATION

#### Tracheostomy

The functional importance of the nasal cavity becomes especially obvious when it is bypassed. If the upper airway is obstructed, it may be necessary to make a temporary opening in the trachea inferior to the larynx and insert a tube to allow airflow—a procedure called *tracheostomy*. This prevents asphyxiation, but the inhaled air bypasses the nasal cavity and thus is not humidified. If the opening is left for long, the mucous membranes of the respiratory tract dry out and become encrusted, interfering with the clearance of mucus from the tract and promoting infection. When a patient is on a ventilator and air is introduced directly into the trachea, the air must be filtered and humidified by the apparatus to prevent respiratory tract damage.

concave **mediastinal surface** faces medially. The mediastinal surface exhibits a slit called the **hilum** through which the lung receives the main bronchus, blood vessels, lymphatics, and nerves. These structures constitute the **root** of the lung.

The lungs are crowded by adjacent organs and neither fill the entire rib cage, nor are they symmetrical (fig. 22.10).

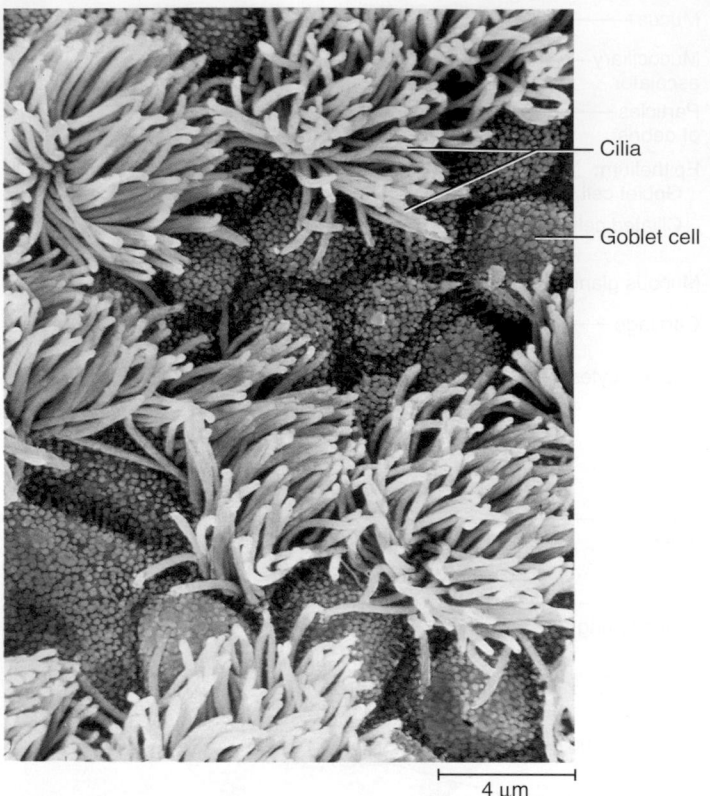

Cilia

Goblet cell

4 µm

**FIGURE 22.8  The Tracheal Epithelium Showing Ciliated Cells and Nonciliated Goblet Cells (SEM).**  The small bumps on the goblet cells are microvilli.

❓ *What is the function of the goblet cells?*

Inferior to the lungs and diaphragm, much of the space within the rib cage is occupied by the liver, spleen, and stomach (see fig. B.5, p. 381). The right lung is shorter than the left because the liver rises higher on the right. The left lung, although taller, is narrower than the right because the heart tilts toward the left and occupies more space on this side of the mediastinum. On the medial surface, the left lung has an indentation called the **cardiac impression** where the heart presses against it; part of this is visible anteriorly as a crescent-shaped **cardiac notch** in the margin of the lung. The right lung has three lobes— **superior, middle,** and **inferior.** A deep groove called the **horizontal fissure** separates the superior and middle lobes, and a similar **oblique fissure** separates the middle and inferior lobes. The left lung has only a **superior** and **inferior lobe** and a single oblique fissure.

## The Bronchial Tree

Each lung has a branching system of air tubes called the **bronchial tree,** extending from the main bronchus to about 65,000 *terminal bronchioles.* Arising from the fork in the trachea, the **right main (primary) bronchus** (BRON-cus) is about 2 to 3 cm long. It is slightly wider and more vertical than the left one; consequently, *aspirated* (inhaled) foreign objects lodge

in the right bronchus more often than in the left. The main bronchus gives off three branches—the **superior, middle,** and **inferior lobar (secondary) bronchi**—one to each lobe of the right lung. The **left main bronchus** is about 5 cm long and slightly narrower and more horizontal than the right. It gives off superior and inferior lobar bronchi to the two lobes of the left lung.

In both lungs, the lobar bronchi branch into **segmental (tertiary) bronchi.** There are 10 of these in the right lung and 8 in the left. Each one ventilates a functionally independent unit of lung tissue called a **bronchopulmonary segment.** Page 848 shows a cast of the bronchial trees made by injecting the airway with colored resins and then dissolving away the tissue. Each bronchopulmonary segment is identified by a different color of resin.

The main bronchi are supported, like the trachea, by C-shaped rings of hyaline cartilage, whereas the lobar and segmental bronchi are supported by overlapping crescent-shaped cartilaginous plates. All of the bronchi are lined with ciliated pseudostratified columnar epithelium, but the cells grow shorter and the epithelium thinner as we progress distally. The lamina propria has an abundance of mucous glands and lymphocyte nodules *(mucosa-associated lymphatic tissue, MALT),* favorably positioned to intercept inhaled pathogens. All divisions of the bronchial tree have a substantial amount of elastic connective tissue, which contributes to the recoil that expels air from the lungs in each respiratory cycle. The mucosa also has a well-developed layer of smooth muscle, the *muscularis mucosae,* which contracts or relaxes to constrict or dilate the airway, thus regulating airflow.

Branches of the *pulmonary artery* closely follow the bronchial tree on their way to the alveoli. The bronchial tree itself is serviced by the *bronchial artery,* which arises from the aorta and carries systemic blood.

**Bronchioles** (BRON-kee-olz) are continuations of the airway that lack supportive cartilage and are 1 mm or less in diameter. The portion of the lung ventilated by one bronchiole is called a **pulmonary lobule.** Bronchioles have a ciliated cuboidal epithelium and a well-developed layer of smooth muscle in their walls. Spasmodic contractions of this muscle at death cause the bronchioles to exhibit a wavy lumen in most histological sections.

Each bronchiole divides into 50 to 80 **terminal bronchioles,** the final branches of the conducting division. These measure 0.5 mm or less in diameter and have no mucous glands or goblet cells. They do have cilia, however, so that mucus draining into them from above can be driven back by the mucociliary escalator, preventing congestion of the terminal bronchioles and alveoli.

Each terminal bronchiole gives off two or more smaller **respiratory bronchioles,** which have alveoli budding from their walls. They are considered the beginning of the respiratory division because their alveoli participate in gas exchange. Their walls have scanty smooth muscle, and the smallest of them are nonciliated. Each respiratory bronchiole divides into 2 to 10 elongated, thin-walled passages called **alveolar ducts,**

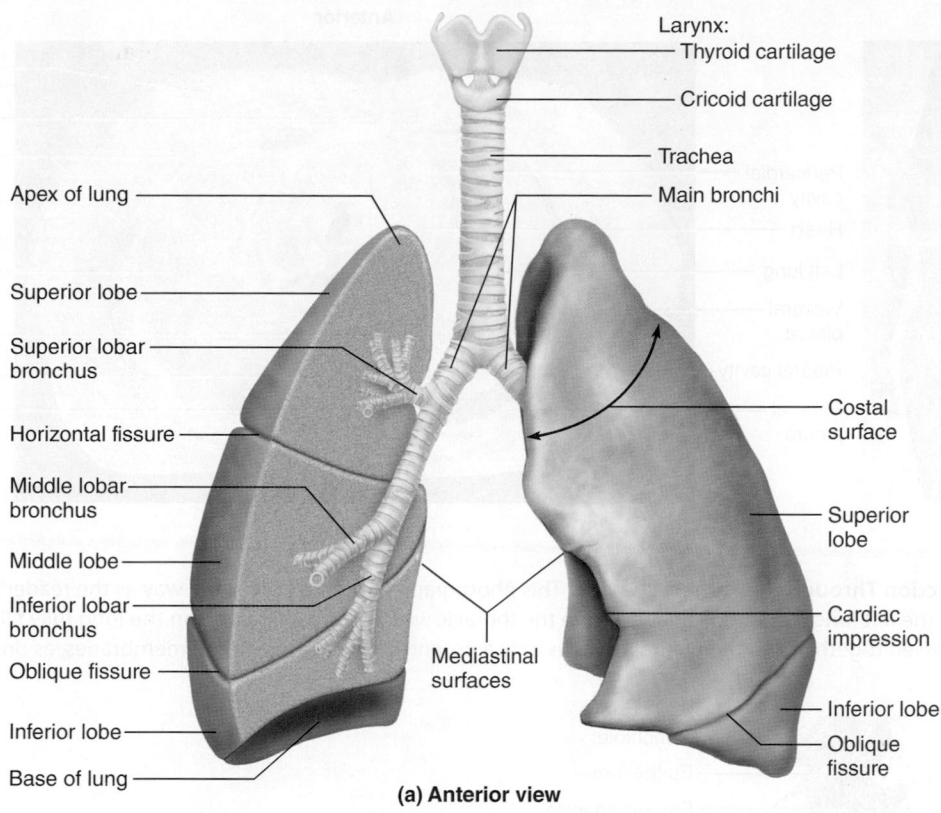

Larynx:
Thyroid cartilage
Cricoid cartilage
Trachea
Main bronchi

Apex of lung

Superior lobe

Superior lobar
bronchus

Horizontal fissure

Middle lobar
bronchus

Middle lobe

Inferior lobar
bronchus

Oblique fissure

Inferior lobe

Base of lung

Mediastinal
surfaces

Costal
surface

Superior
lobe

Cardiac
impression

Inferior lobe

Oblique
fissure

**(a) Anterior view**

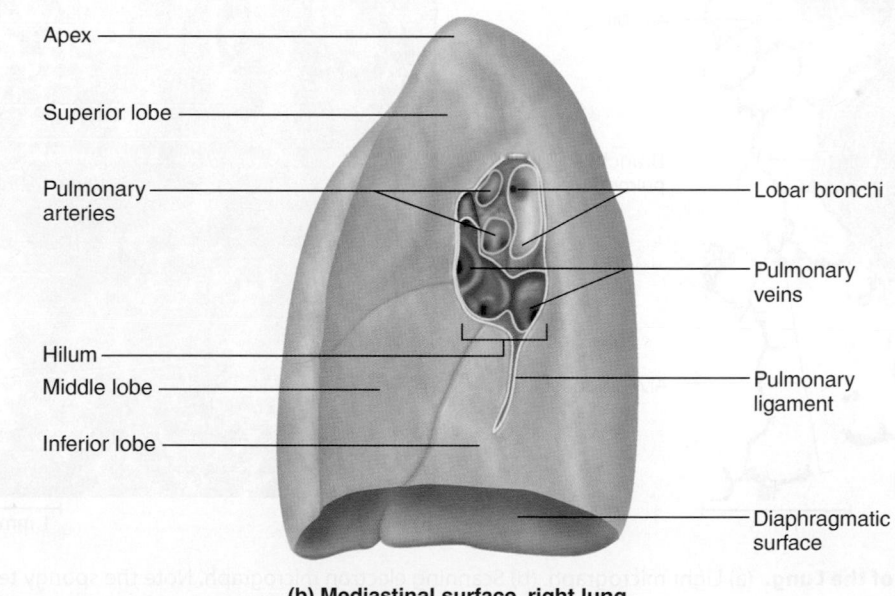

Apex

Superior lobe

Pulmonary
arteries

Hilum

Middle lobe

Inferior lobe

Lobar bronchi

Pulmonary
veins

Pulmonary
ligament

Diaphragmatic
surface

**(b) Mediastinal surface, right lung**

**FIGURE 22.9 Gross Anatomy of the Lungs.** AP|R

which also have alveoli along their walls (fig. 22.11). The alveolar ducts and smaller divisions have nonciliated simple squamous epithelia. The ducts end in **alveolar sacs,** which are clusters of alveoli arrayed around a central space called the **atrium.** The distinction between an alveolar duct and atrium is their shape—an elongated duct, or an atrium with about

equal length and width. It is sometimes a subjective judgment whether to regard a space as an alveolar duct or atrium.

In summary, the path of airflow is as follows. The first several passages belong to the conducting division, where there are no alveoli and the tissue walls are too thick for any significant exchange of oxygen or carbon dioxide with the blood: nasal

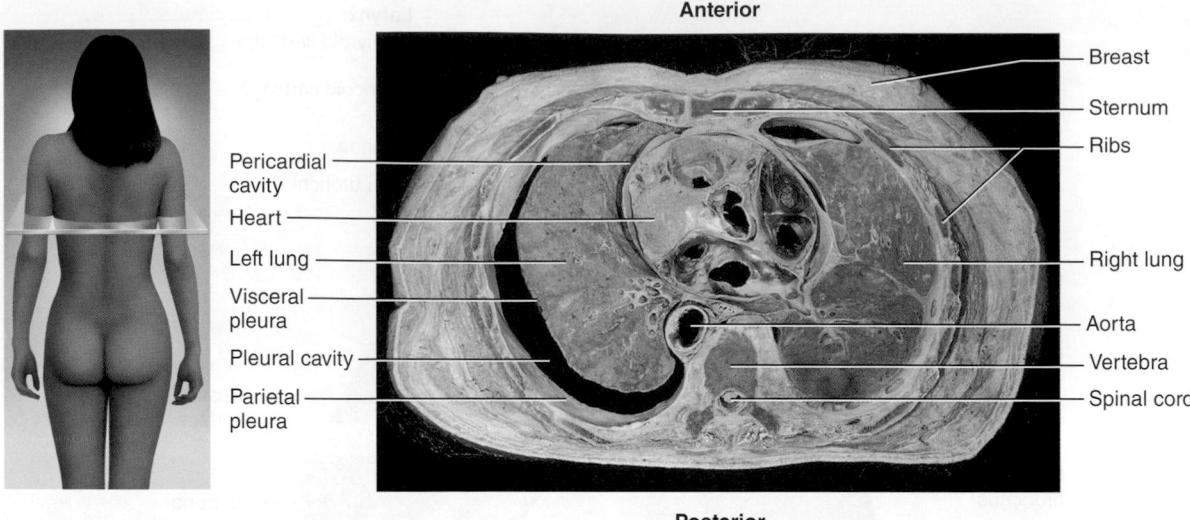

**Anterior**

Pericardial cavity

Heart

Left lung

Visceral pleura

Pleural cavity

Parietal pleura

Breast

Sternum

Ribs

Right lung

Aorta

Vertebra

Spinal cord

**Posterior**

**FIGURE 22.10  Cross Section Through the Thoracic Cavity.** This photograph is oriented the same way as the reader's body. The pleural cavity is especially evident where the left lung has shrunken away from the thoracic wall, but in a living person the lung fully fills this space, the parietal and visceral pleurae are pressed together, and the pleural cavity is only a potential space between the membranes, as on the right side of this photograph.

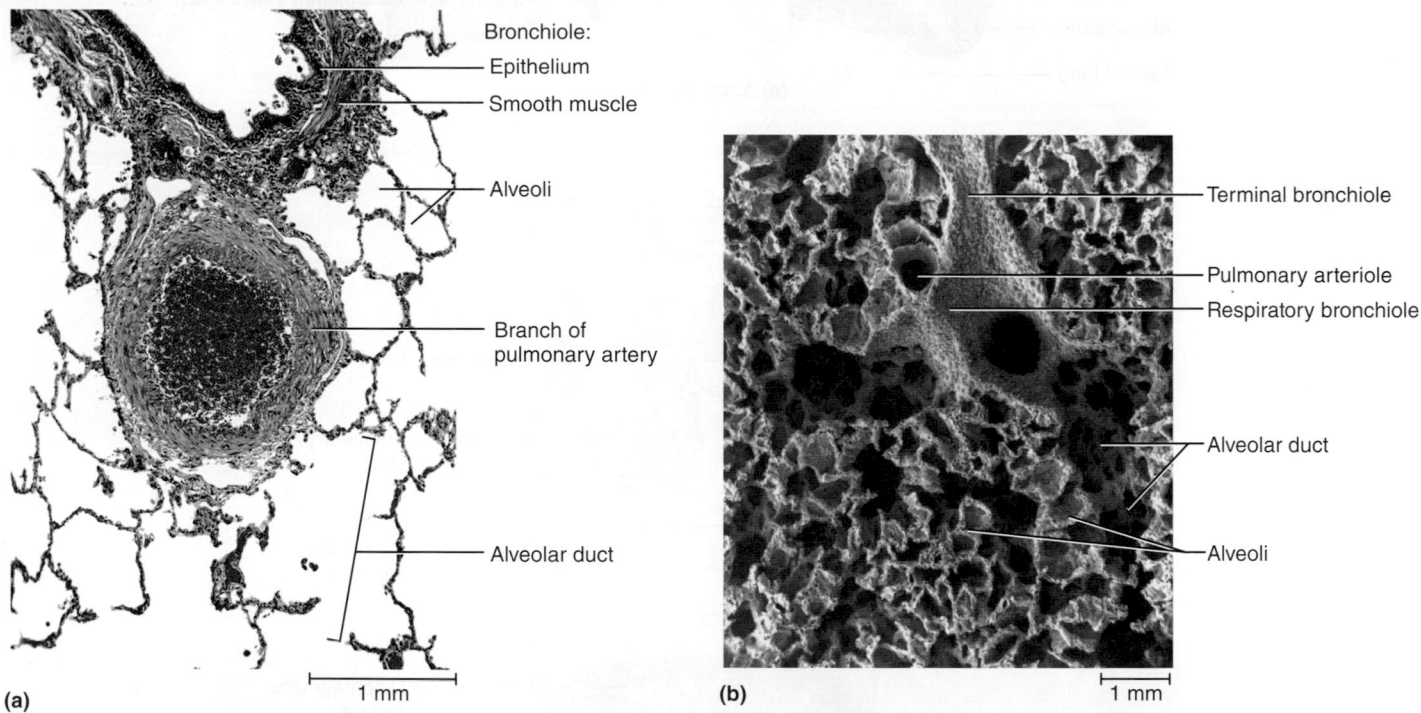

Bronchiole:

Epithelium

Smooth muscle

Alveoli

Branch of pulmonary artery

Alveolar duct

**(a)**    1 mm

Terminal bronchiole

Pulmonary arteriole

Respiratory bronchiole

Alveolar duct

Alveoli

**(b)**    1 mm

**FIGURE 22.11  Histology of the Lung.** (a) Light micrograph. (b) Scanning electron micrograph. Note the spongy texture of the lung.

❓ *Histologically, how can we tell that the passage at the top of part (a) is a bronchiole and not a bronchus?*

cavity → pharynx → trachea → main bronchus → lobar bronchus → segmental bronchus → bronchiole → terminal bronchiole. Then begins the respiratory division; all of the following passages have alveoli along their walls (or are themselves alveoli) and thus engage in gas exchange: respiratory bronchiole → alveolar duct → atrium → alveolus.

## Alveoli

The functional importance of human lung structure is best appreciated by comparison to the lungs of a few other animals. In frogs and other amphibians, the lung is a simple sac lined with blood vessels. This is sufficient to meet the oxygen needs of animals with relatively low metabolic rates. Mammals, with their high metabolic rates, could never have evolved with such a simple lung. Rather than consisting of one large sac, each human lung is a spongy mass composed of 150 million little sacs, the alveoli, which provide about 70 m² of surface for gas exchange—about equal to the floor area of a handball court or a room about 8.4 m (25 ft) square.

An **alveolus** (AL-vee-OH-lus) is a pouch about 0.2 to 0.5 mm in diameter (fig. 22.12). Thin, broad cells called **squamous (type I) alveolar cells** cover about 95% of the alveolar surface area. Their thinness allows for rapid gas diffusion between the air and blood. The other 5% is covered by round to cuboidal **great (type II) alveolar cells.** Even though they cover less surface area, these considerably outnumber the squamous alveolar cells. You might compare squamous and great alveolar cells to a given amount of dough rolled out into a pie crust or made into muffins, respectively, to understand why squamous cells cover more area yet great alveolar cells are more numerous. Great alveolar cells have two functions: (1) They repair the alveolar epithelium when the squamous cells are damaged, and (2) they secrete *pulmonary surfactant,* a mixture of phospholipids and protein that coats the alveoli and smallest bronchioles and prevents the bronchioles from collapsing when one exhales. This surfactant function is later explained in greater detail.

The most numerous of all cells in the lung are **alveolar macrophages (dust cells),** which wander the lumens of the alveoli and the connective tissue between them. These cells keep the alveoli free of debris by phagocytizing dust particles that escape entrapment by mucus in the higher parts of the respiratory tract. In lungs that are infected or bleeding, the macrophages also phagocytize bacteria and loose blood cells. As many as 100 million alveolar macrophages perish each day as they ride up the mucociliary escalator to be swallowed and digested, thus ridding the lungs of their load of debris.

Each alveolus is surrounded by a web of blood capillaries supplied by small branches of the pulmonary artery. The barrier between the alveolar air and blood, called the **respiratory membrane,** consists only of the squamous alveolar cell, the squamous endothelial cell of the capillary, and their shared basement membrane. These have a total thickness of only 0.5 μm, just 1/15 the diameter of a single erythrocyte.

It is very important to prevent fluid from accumulating in the alveoli, because gases diffuse too slowly through liquid to sufficiently aerate the blood. Except for a thin film of moisture on the alveolar wall, the alveoli are kept dry by the absorption of excess liquid by the blood capillaries. Their mean blood pressure is only 10 mm Hg and the oncotic pressure is

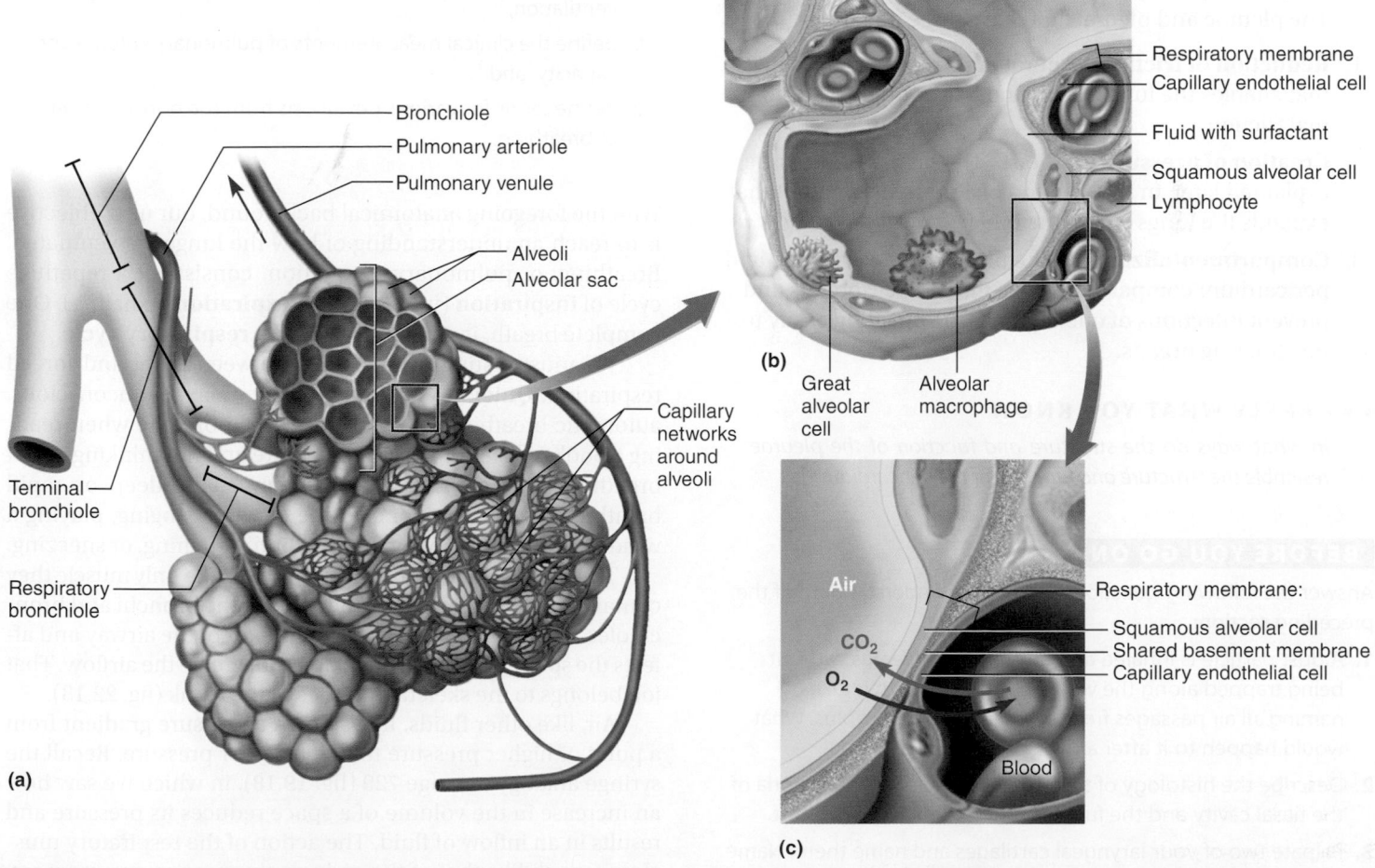

**FIGURE 22.12 Pulmonary Alveoli.** (a) Clusters of alveoli and their blood supply. (b) Structure of an alveolus. (c) Structure of the respiratory membrane.

28 mm Hg (compare p. 760), so the osmotic uptake of water overrides filtration and keeps the alveoli free of fluid. The lungs also have a more extensive lymphatic drainage than any other organ in the body. The low capillary blood pressure also prevents the rupture of the delicate respiratory membrane.

## The Pleurae

The surface of the lung consists of a serous membrane, the **visceral pleura** (PLOOR-uh), which extends into the fissures. At the hilum, the visceral pleura turns back on itself and forms the **parietal pleura,** which adheres to the mediastinum, inner surface of the rib cage, and superior surface of the diaphragm (see fig. 22.10). An extension of the parietal pleura, the *pulmonary ligament,* connects it to the diaphragm.

The space between the parietal and visceral pleurae is called the **pleural cavity.** The pleural cavity does not *contain* a lung, but rather, wraps around it, somewhat like wrapping a pillowcase (representing the two pleural membranes) around a watermelon, not putting the melon inside the pillowcase. The pleural cavity contains nothing but a film of slippery **pleural fluid;** the cavity is only a *potential space,* meaning there is normally no room between the membranes. However, under pathological conditions, the space can fill with air or liquid (see *pneumothorax,* p. 866).

The pleurae and pleural fluid have three functions:

1. **Reduction of friction.**  Pleural fluid acts as a lubricant that enables the lungs to expand and contract with minimal friction.
2. **Creation of pressure gradient.**  The pleurae play a role, explained later, in the creation of a pressure gradient that expands the lungs when one inhales.
3. **Compartmentalization.**  The pleurae, mediastinum, and pericardium compartmentalize the thoracic organs and prevent infections of one organ from spreading easily to neighboring organs.

#### ▶▶▶ APPLY WHAT YOU KNOW

*In what ways do the structure and function of the pleurae resemble the structure and function of the pericardium?*

#### BEFORE YOU GO ON

Answer the following questions to test your understanding of the preceding section:

1. A dust particle is inhaled and gets into an alveolus without being trapped along the way. Describe the path it takes, naming all air passages from external naris to alveolus. What would happen to it after arrival in the alveolus?
2. Describe the histology of the epithelium and lamina propria of the nasal cavity and the functions of the cell types present.
3. Palpate two of your laryngeal cartilages and name them. Name the ones that cannot be palpated on a living person.

4. Describe the roles of the intrinsic muscles, corniculate cartilages, and arytenoid cartilages in speech.
5. Contrast the epithelium of the bronchioles with that of the alveoli and explain how the structural difference is related to their functional difference.

### 22.2 Pulmonary Ventilation

#### Expected Learning Outcomes
When you have completed this section, you should be able to

a. name the muscles of respiration and describe their roles in breathing;
b. describe the brainstem centers that control breathing and the inputs they receive from other levels of the nervous system;
c. explain how pressure gradients account for the flow of air into and out of the lungs, and how those gradients are produced;
d. identify the sources of resistance to airflow and discuss their relevance to respiration;
e. explain the significance of anatomical dead space to alveolar ventilation;
f. define the clinical measurements of pulmonary volume and capacity; and
g. define terms for various deviations from the normal pattern of breathing.

With the foregoing anatomical background, our next objective is to reach an understanding of how the lungs are ventilated. Breathing, or pulmonary ventilation, consists of a repetitive cycle of **inspiration** (inhaling) and **expiration** (exhaling). One complete breath, in and out, is called a **respiratory cycle.**

We must distinguish at times between quiet and forced respiration. **Quiet respiration** refers to relaxed, unconscious, automatic breathing, the way one would breathe when reading a book or listening to a class lecture and not thinking about breathing. **Forced respiration** is unusually deep or rapid breathing, as in a state of exercise or when singing, playing a wind instrument, blowing up a balloon, coughing, or sneezing.

The lungs do not ventilate themselves. The only muscle they contain is smooth muscle in the walls of the bronchi and bronchioles. This muscle adjusts the diameter of the airway and affects the speed of airflow, but it does not create the airflow. That job belongs to the skeletal muscles of the trunk (fig. 22.13).

Air, like other fluids, flows down a pressure gradient from a point of higher pressure to one of lower pressure. Recall the syringe analogy on page 729 (fig. 19.18), in which we saw how an increase in the volume of a space reduces its pressure and results in an inflow of fluid. The action of the respiratory muscles is much like that of the syringe plunger—at one moment, to increase the volume and lower the pressure in the thoracic cavity, so that air flows in; at the next moment, to reduce thoracic volume and raise pressure, so that air flows out. We will next examine these muscular actions, how the nervous system controls them, and how the variables of pressure and resistance affect airflow and pulmonary ventilation.

## The Respiratory Muscles

The principal muscles of respiration are the diaphragm and intercostal muscles (fig. 22.13). The prime mover is the diaphragm (see table 10.4, p. 329); it alone produces about two-thirds of the pulmonary airflow. When relaxed, it bulges upward to its farthest extent, pressing against the base of the lungs. The lungs are at their minimum volume. When the diaphragm contracts, it tenses and flattens somewhat, dropping about 1.5 cm in relaxed inspiration and as much as 12 cm in deep breathing. Not only does its descent enlarge the superior-to-inferior dimension of the thoracic cage, but its flattening

also pushes outward on the sternum and ribs and enlarges the anterior-to-posterior dimension. Enlargement of the thoracic cavity lowers its internal pressure and produces an inflow of air. When the diaphragm relaxes, it bulges upward again, compresses the lungs, and expels air.

Several other muscles aid the diaphragm as synergists. Chief among these are the internal and external intercostal muscles between the ribs. Their primary function is to stiffen the thoracic cage during respiration and prevent it from caving inward when the diaphragm descends. However, they also contribute to enlargement and contraction of the thoracic cage and add about one-third of the air that ventilates the lungs. During quiet breathing, the scalene muscles of the neck fix (hold stationary) ribs 1 and 2, while the external intercostal muscles pull the other ribs upward. Since most ribs are anchored at both ends—by their attachment to the vertebral column at the proximal (posterior) end and their attachment through the costal cartilage to the sternum at the

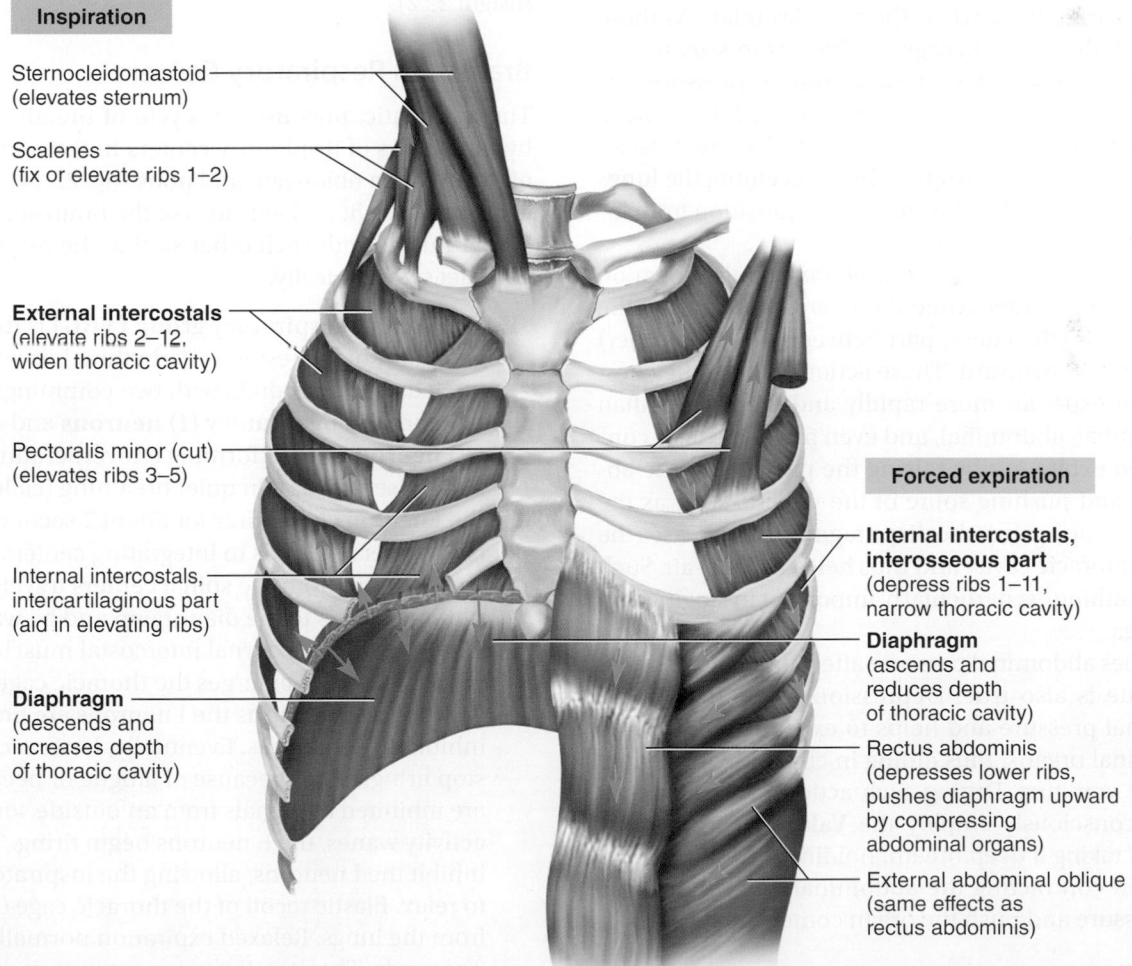

**FIGURE 22.13 The Respiratory Muscles.** Boldface indicates the principal respiratory muscles; the others are accessory. Arrows indicate the direction of muscle action. Muscles listed on the left are active during inspiration and those on the right are active during forced expiration. Note that the diaphragm is active in both phases, and different parts of the internal intercostal muscles serve for inspiration and expiration. Some other accessory muscles not shown here are discussed in the text.

distal (anterior) end—they swing upward like the handles on a bucket and thrust the sternum forward. These actions increase both the transverse (left to right) and anteroposterior diameters of the chest. In deep breathing, the anteroposterior dimension can increase as much as 20%.

Other muscles of the chest and abdomen also aid in breathing, especially during forced respiration; thus they are considered *accessory muscles* of respiration. Deep inspiration is aided by the erector spinae, which arches the back and increases chest diameter, and by several muscles that elevate the upper ribs: the sternocleidomastoids and scalenes of the neck; the pectoralis minor, pectoralis major, and serratus anterior of the chest; and the *intercartilaginous part* of the internal intercostals (the anterior part between the costal cartilages). Although the scalenes merely fix the upper ribs during quiet respiration, they elevate them during forced inspiration.

Normal expiration is an energy-saving passive process achieved by the elasticity of the lungs and thoracic cage. The bronchial tree, the attachments of the ribs to the spine and sternum, and the tendons of the diaphragm and other respiratory muscles spring back when the muscles relax. As these structures recoil, the thoracic cage diminishes in size, the air pressure in the lungs rises above the atmospheric pressure outside, and the air flows out. The only muscular effort involved in normal expiration is a braking action—that is, the muscles relax gradually rather than abruptly, thus preventing the lungs from recoiling too suddenly. This makes the transition from inspiration to expiration smoother.

In forced expiration, the *rectus abdominis* pulls down on the sternum and lower ribs, while the *interosseous part* of the internal intercostals (the lateral part between the ribs proper) pulls the other ribs downward. These actions reduce the chest dimensions and expel air more rapidly and thoroughly than usual. Other lumbar, abdominal, and even pelvic muscles contribute to forced expiration by raising the pressure in the abdominal cavity and pushing some of the viscera, such as the stomach and liver, up against the diaphragm. This increases the pressure in the thoracic cavity and thus helps to expel air. Such "abdominal breathing" is particularly important in singing and public speaking.

Not only does abdominal pressure affect thoracic pressure, but the opposite is also true. Depression of the diaphragm raises abdominal pressure and helps to expel the contents of certain abdominal organs, thus aiding in childbirth, urination, defecation, and vomiting. During such actions, we often consciously or unconsciously employ the **Valsalva**[12] **maneuver.** This consists of taking a deep breath, holding it by closing the glottis, and then contracting the abdominal muscles to raise abdominal pressure and push the organ contents out.

---

[12]Antonio Maria Valsalva (1666–1723), Italian anatomist

## Neural Control of Breathing

The heartbeat and breathing are the two most conspicuously rhythmic processes in the body. The heart, we have seen, has an internal pacemaker, but the lungs do not. No autorhythmic pacemaker cells for respiration have been found that are analogous to those of the heart, and the exact mechanism for setting the rhythm of respiration remains obscure. But we do know that breathing depends on repetitive stimuli from the brain. It ceases if the nerve connections to the thoracic muscles are severed or if the spinal cord is severed high on the neck. There are two reasons for this dependence on the brain: (1) Skeletal muscles, unlike cardiac muscle, cannot contract without nervous stimulation. (2) Breathing involves the well-orchestrated action of multiple muscles and thus requires a central coordinating mechanism.

Breathing is controlled at two levels of the brain. One is cerebral and conscious, enabling us to inhale or exhale at will. The other is unconscious and automatic. Most of the time, we breathe without thinking about it—fortunately, for we otherwise could not go to sleep without fear of respiratory arrest (see Deeper Insight 22.2).

### Brainstem Respiratory Centers

The automatic, unconscious cycle of breathing is controlled by three pairs of respiratory centers in the reticular formation of the medulla oblongata and pons (fig. 22.14). There is one of each on the right and left sides of the brainstem; the two sides communicate with each other so that the respiratory muscles contract symmetrically.

1. The **ventral respiratory group (VRG)** is the primary generator of the respiratory rhythm. It is an elongated nucleus in the medulla with two commingled webs of neurons—**inspiratory (I) neurons** and **expiratory (E) neurons**—each forming a reverberating neural circuit (see p. 465). In quiet breathing (called *eupnea*), the I neuron circuit fires for about 2 seconds at a time, issuing nerve signals to integrating centers in the spinal cord. Output from the spinal centers travels by way of the phrenic nerves to the diaphragm and by way of intercostal nerves to the external intercostal muscles. Contraction of these muscles enlarges the thoracic cage and causes inspiration. As long as the I neurons are firing, they also inhibit the E neurons. Eventually, however, the I neurons stop firing, either because of fatigue or because they are inhibited by signals from an outside source. As their activity wanes, the E neurons begin firing. They further inhibit the I neurons, allowing the inspiratory muscles to relax. Elastic recoil of the thoracic cage expels air from the lungs. Relaxed expiration normally lasts about 3 seconds. Then the E neuron activity wanes, the I neurons resume firing, and the cycle repeats itself. In eupnea, this oscillating pattern of neural activity, alternating between the I neuron and E neuron circuits, produces a respiratory rhythm of about 12 breaths per minute.

2. Obviously, one does not always breathe at that rate. Breathing can be faster or slower, shallower or deeper, because the VRG is subject to influence from external sources. Chief among these is the **dorsal respiratory group (DRG),** another web of neurons that extends for much of the length of the medulla between the VRG and the central canal of the brainstem. The DRG is apparently an integrating center that receives input from several sources detailed in the coming discussion: a respiratory center in the pons; a chemosensitive center of the anterior medulla oblongata; chemoreceptors in certain major arteries; and stretch and irritant receptors in the airway. The DRG issues output to the VRG that modifies the respiratory rhythm to adapt to varying conditions.

3. Furthermore, each side of the pons has a **pontine respiratory group (PRG)** (formerly called the *pneumotaxic center*) that modifies the rhythm of the VRG. The pontine group receives input from higher brain centers including the hypothalamus, limbic system, and cerebral cortex, and issues output to both the DRG and VRG. By acting on those centers in the medulla, it hastens or delays the transition from inspiration to expiration, making each breath shorter and shallower, or longer and deeper. The PRG adapts breathing to special circumstances such as sleep, exercise, vocalization, and emotional responses (for example, in crying, gasping, or laughing).

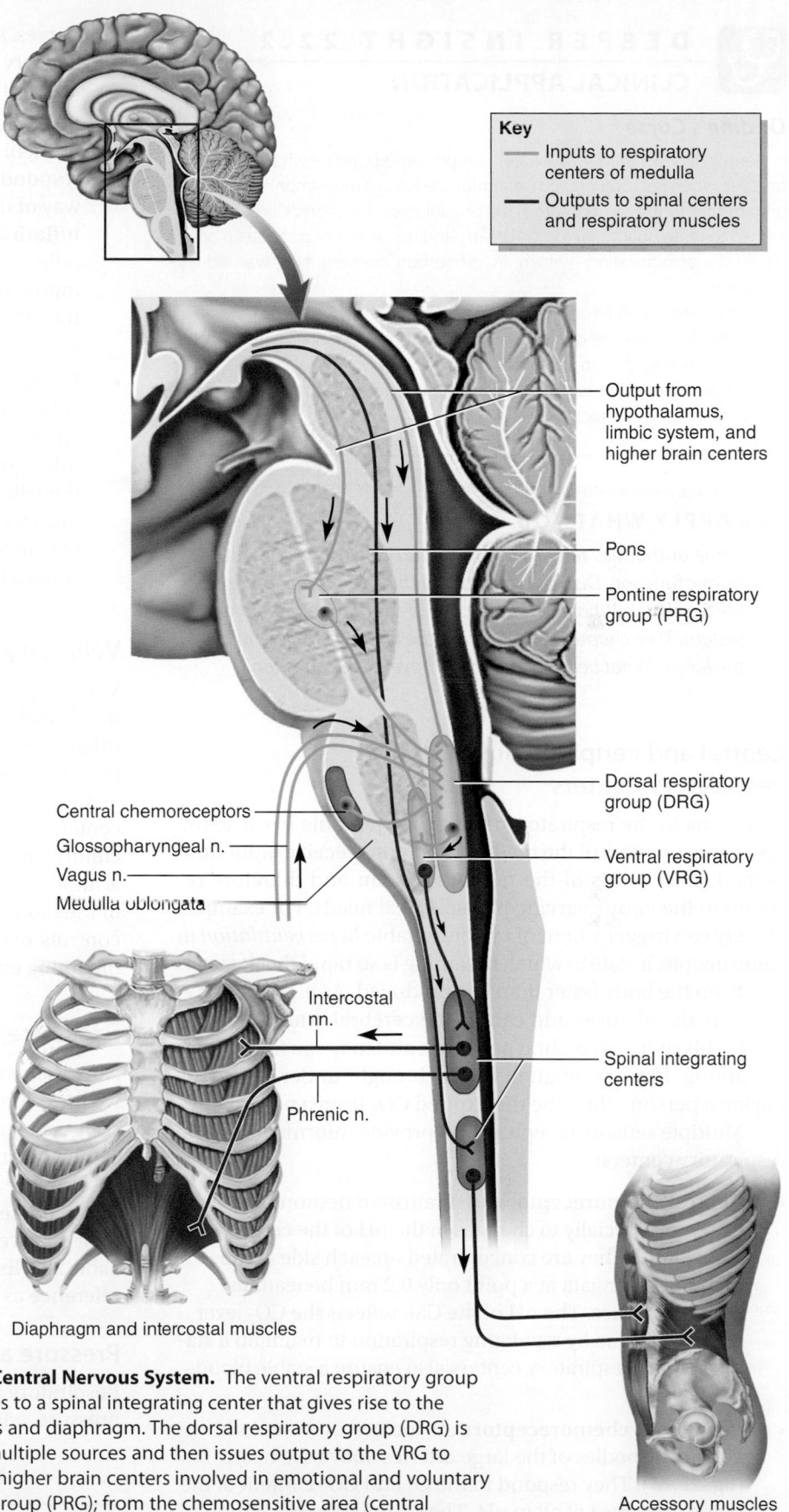

Key
— Inputs to respiratory centers of medulla
— Outputs to spinal centers and respiratory muscles

Output from hypothalamus, limbic system, and higher brain centers

Pons

Pontine respiratory group (PRG)

Dorsal respiratory group (DRG)

Ventral respiratory group (VRG)

Central chemoreceptors

Glossopharyngeal n.

Vagus n.

Medulla oblongata

Spinal integrating centers

Intercostal nn.

Phrenic n.

Diaphragm and intercostal muscles

Accessory muscles of respiration

**FIGURE 22.14 Respiratory Control Centers in the Central Nervous System.** The ventral respiratory group (VRG) generates the rhythm of breathing. Its output goes to a spinal integrating center that gives rise to the intercostal and phrenic nerves to the intercostal muscles and diaphragm. The dorsal respiratory group (DRG) is an integrating center that processes information from multiple sources and then issues output to the VRG to adjust the respiratory rhythm. The DRG gets input from higher brain centers involved in emotional and voluntary influences on respiration; from the pontine respiratory group (PRG); from the chemosensitive area (central chemoreceptors) of the medulla oblongata; and from peripheral chemoreceptors and stretch receptors by way of the vagus and glossopharyngeal nerves. Additional output from higher brain centers bypasses both the DRG and VRG and goes to spinal integrating centers that control the accessory muscles of respiration, such as the abdominal muscles.

## DEEPER INSIGHT 22.2
## CLINICAL APPLICATION

### Ondine's Curse

In German legend, there was a water nymph named Ondine who took a mortal lover. When her lover proved unfaithful, the king of the nymphs put a curse on him that took away his automatic physiological functions. Consequently, he had to remember to take each breath, and he could not go to sleep or he would die of suffocation—which, as exhaustion overtook him, was indeed his fate.

Some people suffer a disorder called *Ondine's curse,* in which the automatic respiratory functions are disabled—usually as a result of brainstem damage from poliomyelitis or as an accident of neurosurgery. Victims of Ondine's curse must remember to take each breath and cannot go to sleep without the aid of a mechanical ventilator.

### ►►► APPLY WHAT YOU KNOW

*Some authorities refer to the respiratory rhythm as an* auto-*nomic function. Discuss whether you think this is an appropriate word for it. What are the effectors of the autonomic nervous system? (See chapter 15.) What are the effectors that ventilate the lungs? What bearing might this have on the question?*

## Central and Peripheral Input to the Respiratory Centers

Variations in the respiratory rhythm are possible because the respiratory centers of the medulla and pons receive input from several other levels of the nervous system and therefore respond to the body's varying physiological needs. For example, anxiety can trigger a bout of uncontrollable *hyperventilation* in some people, a state in which breathing is so rapid that it expels $CO_2$ from the body faster than it is produced. As blood $CO_2$ levels drop, the pH rises and causes the cerebral arteries to constrict. This reduces cerebral perfusion and may cause dizziness or fainting. Hyperventilation can be brought under control by having a person rebreathe the expired $CO_2$ from a paper bag.

Multiple sensory receptors also provide information to the respiratory centers:

- **Central chemoreceptors** are brainstem neurons that respond especially to changes in the pH of the cerebrospinal fluid. They are concentrated on each side of the medulla oblongata at a point only 0.2 mm beneath its anterior surface. The pH of the CSF reflects the $CO_2$ level in the blood, so by regulating respiration to maintain a stable pH, the respiratory centers also ensure a stable blood $CO_2$ level.

- **Peripheral chemoreceptors** are located in the carotid and aortic bodies of the large arteries above the heart (fig. 22.15). They respond to the $O_2$ and $CO_2$ content of the blood, but most of all to pH. The carotid bodies communicate with the brainstem by way of the glossopharyngeal

nerves, and the aortic bodies by way of the vagus nerves. Sensory fibers in these nerves enter the medulla and synapse with neurons of the DRG.

- **Stretch receptors** are found in the smooth muscle of the bronchi and bronchioles and in the visceral pleura. They respond to inflation of the lungs and signal the DRG by way of the vagus nerves. Excessive inflation triggers the **inflation (Hering–Breuer)[13] reflex,** a protective somatic reflex that strongly inhibits the I neurons and stops inspiration. In infants, this may be a normal mechanism of transition from inspiration to expiration, but after infancy it is activated only by extreme stretching of the lungs.

- **Irritant receptors** are nerve endings amid the epithelial cells of the airway. They respond to smoke, dust, pollen, chemical fumes, cold air, and excess mucus. They transmit signals by way of the vagus nerves to the DRG, and the DRG returns signals to the respiratory and bronchial muscles, resulting in such protective reflexes as bronchoconstriction, shallower breathing, breath-holding *(apnea),* or coughing.

## Voluntary Control of Breathing

Voluntary control of breathing is important in singing, speaking, breath-holding, and other circumstances. Such control originates in the motor cortex of the cerebrum. The output neurons send impulses down the corticospinal tracts to the integrating centers in the spinal cord, bypassing the brainstem centers. There are limits to voluntary control. Temperamental children may threaten to hold their breath until they die, but it is impossible to do so. Holding one's breath raises the $CO_2$ level of the blood until a *breaking point* is reached when automatic controls override one's will. This forces a person to resume breathing even if he or she has lost consciousness.

## Pressure, Resistance, and Airflow

Now that we know the neuromuscular aspects of breathing, we'll explore how the expansion and contraction of the thoracic cage produce airflow into and out of the lungs.

Understanding pulmonary ventilation, the transport of gases in the blood, and the exchange of gases with the tissues draws on certain gas laws of physics. These are named after their discoverers and not intuitively easy to remember by name. Table 22.1 lists them for your convenience and may be a helpful reference as you progress through respiratory physiology.

## Pressure and Airflow

Respiratory airflow is governed by the same principles of flow, pressure, and resistance as blood flow. As we saw in chapter 20

---

[13]Heinrich Ewald Hering (1866–1948), German physiologist; Josef Breuer (1842–1925), Austrian physician

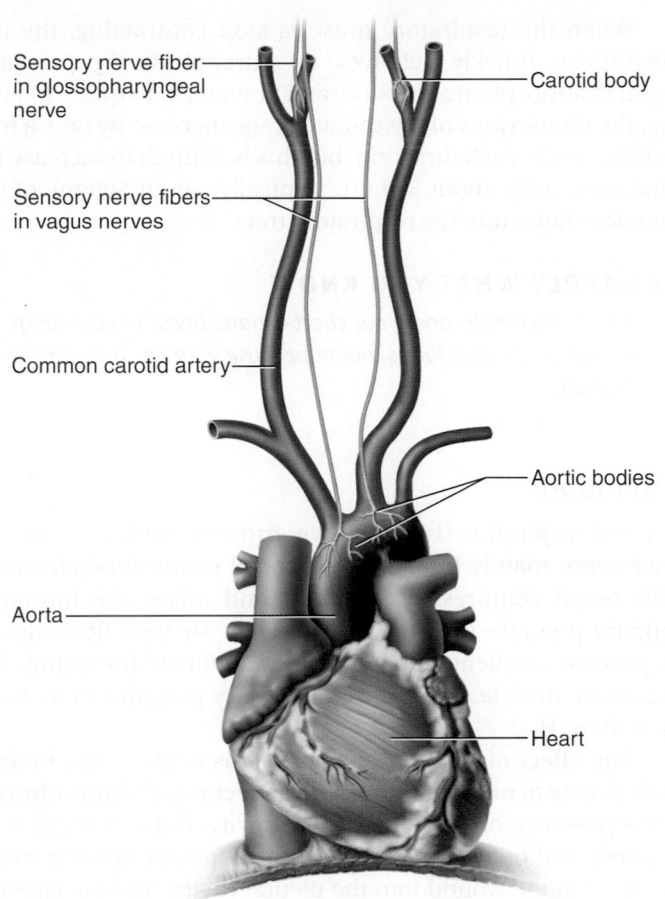

Sensory nerve fiber in glossopharyngeal nerve

Carotid body

Sensory nerve fibers in vagus nerves

Common carotid artery

Aortic bodies

Aorta

Heart

**FIGURE 22.15 The Peripheral Chemoreceptors of Respiration.** Peripheral chemoreceptors in the aortic arch and carotid bodies monitor blood gas concentrations and blood pH. They send signals about blood chemistry to the dorsal respiratory group of the medulla oblongata through sensory fibers in the vagus and glossopharyngeal nerves.

(p. 752), the flow *(F)* of a fluid is directly proportional to the pressure difference between two points *(ΔP)* and inversely proportional to resistance *(R)*:

$$F \propto \Delta P / R$$

For the moment, we will focus especially on *ΔP,* the pressure gradient that produces airflow. We will deal with resistance later.

The pressure that drives respiration is **atmospheric (barometric) pressure**—the weight of the air above us. At sea level, this averages 760 mm Hg, or by definition, *1 atmosphere (1 atm).* It fluctuates with the weather from day to day and is lower at higher elevations, but we will use this value as a reference point for discussion.

One way to change the pressure of an enclosed gas is to change the volume of its container. This fact is summarized by **Boyle's law,** which states that at a constant temperature, *the pressure of a given quantity of gas is inversely proportional to its volume.* If the lungs contain a quantity of gas and lung volume

increases, their internal pressure (**intrapulmonary pressure**) falls. Conversely, if lung volume decreases, intrapulmonary pressure rises. (Compare this to the syringe analogy.)

If the intrapulmonary pressure falls below the atmospheric pressure, then air tends to flow down its pressure gradient into the lungs. Conversely, if intrapulmonary pressure rises above atmospheric pressure, air flows out. Therefore, all we have to do to breathe is to cyclically raise and lower the intrapulmonary pressure, employing the neuromuscular mechanisms recently described.

Furthermore, when dealing with respiratory airflow and thoracic pressures generated by the foregoing muscle actions, we must use a new unit of measure, different from what you are accustomed to. In recent chapters, we used *millimeters of mercury (mm Hg)* as a measure of blood pressure, and we will use it again later in this chapter when we speak of atmospheric pressure and blood gases. Millimeters of mercury is a measure of how high up a vacuum tube a force such as blood pressure or the weight of the atmosphere can push a column of mercury. Mercury is a very heavy liquid, so we use it because pressures can be measured with a relatively short column of mercury, as in the sphygmomanometers of a doctor's office. But the pressures in respiratory airflow are so small that they could not move a mercury column much at all; mercury-based instruments are not sensitive enough. Respiratory physiologists therefore traditionally used water columns, which are more sensitive, and we measure these pressures in *centimeters of water (cm $H_2O$).* 1 mm Hg is about 13.6 mm $H_2O$ (1.4 cm $H_2O$). Small pressure changes will move a column of water more than a column of mercury; one can see them and measure them more accurately.

| TABLE 22.1 | The Gas Laws of Respiratory Physiology |
| --- | --- |
| Boyle's law[14] | The pressure of a given quantity of gas is inversely proportional to its volume (assuming a constant temperature). |
| Charles's law[15] | The volume of a given quantity of gas is directly proportional to its absolute temperature (assuming a constant pressure). |
| Dalton's law[16] | The total pressure of a gas mixture is equal to the sum of the partial pressures of its individual gases. |
| Henry's law[17] | At the air–water interface, the amount of gas that dissolves in water is determined by its solubility in water and its partial pressure in the air (assuming a constant temperature). |

[14]Robert Boyle (1627–91), Anglo-Irish physicist and chemist
[15]Jacques A. C. Charles (1746–1823), French physicist
[16]John Dalton (1766–1844), English physicist and chemist
[17]William Henry (1774–1836), English chemist

Since respiratory airflow is driven by a *difference* between surrounding air pressure and pressures in the chest, the following discussion is based on *relative* pressures. If it speaks of pressure in the pulmonary alveoli reaching –2 cm $H_2O$ during inspiration, we mean it falls 2 cm $H_2O$ below the atmospheric pressure external to the body; if it rises to +3 cm $H_2O$ during expiration, it is 3 cm $H_2O$ above atmospheric pressure.

## Inspiration

Now consider the flow of air into the lungs—inspiration. Figure 22.16 traces the events and pressure changes that occur throughout a respiratory cycle.

At the beginning, there is no movement of the thoracic cage, no difference between the air pressure within the lungs and external to the body, and no airflow. What happens when the thoracic cage expands? Why do the lungs not remain the same size and simply occupy less space in the chest? Consider the two layers of the pleura: the parietal pleura lining the rib cage and the visceral pleura constituting the lung surface. They are not anatomically attached to each other along their surfaces, but they are wet and cling together like sheets of wet paper. The space between them is only about 10 to 30 μm wide (about the width of one typical cell). At the end of a normal expiration, the chest wall (including the parietal pleura) tends to expand outward because of its elasticity while the lungs, because of their elasticity, tend to recoil inward. Thus, the lungs and chest wall are pulling in opposite directions. This creates a slightly negative **intrapleural pressure,** averaging about –5 cm $H_2O$, between the parietal and visceral pleurae.

When the ribs swing up and out during inspiration, the parietal pleura follows. If not for the cohesion of water, this might pull the chest wall away from the lungs; but because the two wet membranes cling to each other, it only reduces the intrapleural pressure a little more, to about –8 cm $H_2O$. As the visceral pleura (lung surface) is pulled outward, it stretches the alveoli just below the surface of the lung. Those alveoli are mechanically linked to deeper ones by their walls, and stretch the deeper ones as well. Thus the entire lung expands along with the thoracic cage. As in the syringe analogy, the alveoli increase in volume and decrease in pressure. Pressure within the alveoli, the **intrapulmonary (alveolar) pressure,** drops to an average of –1 cm $H_2O$. So now there is a pressure gradient from the atmosphere at the nose to the negative pressure in the alveoli. Air flows down its gradient and ventilates the alveoli. In short, we inhale.

Yet this is not the only force that expands the lungs. Another is warming of the inhaled air. As we see from **Charles's law** (see table 22.1), the volume of a given quantity of gas is directly proportional to its absolute temperature. Inhaled air is warmed to 37°C (99°F) by the time it reaches the alveoli. This means that on a cool day when the outdoor temperature is, say, 16°C (60°F), the air temperature will increase by 21°C (39°F) during inspiration. An inhaled volume of 500 mL will expand to 536 mL and this thermal expansion will contribute to inflation of the lungs.

When the respiratory muscles stop contracting, the inflowing air quickly achieves an intrapulmonary pressure equal to atmospheric pressure, and flow stops. In quiet breathing, the dimensions of the thoracic cage increase by only a few millimeters in each direction, but this is enough to increase its total volume by about 500 mL. Typically, about 500 mL of air therefore flows into the respiratory tract.

▶▶▶ **APPLY WHAT YOU KNOW**

*When you inhale, does your chest expand because your lungs inflate, or do your lungs inflate because your chest expands? Explain.*

## Expiration

Relaxed expiration is a passive process achieved, as we have seen, mainly by the elastic recoil of the thoracic cage. This recoil compresses the lungs and raises the intrapulmonary pressure to about +1 cm $H_2O$. Air thus flows down its pressure gradient, out of the lungs. In forced breathing, the accessory muscles raise intrapulmonary pressure to as high as +40 cm $H_2O$.

The effect of pulmonary elasticity is evident in a pathological state of pneumothorax and atelectasis. **Pneumothorax** is the presence of air in the pleural cavity. If the thoracic wall is punctured between the ribs, for example, inspiration sucks air through the wound into the pleural cavity, and the visceral and parietal pleurae separate; what was a *potential space* between them becomes an air-filled cavity. Without the negative intrapleural pressure to keep the lungs inflated, the lungs recoil and collapse. The collapse of part or all of a lung is called **atelectasis**[18] (AT-eh-LEC-ta-sis). Atelectasis can also result from airway obstruction—for example, by a lung tumor, aneurysm, swollen lymph node, or aspirated object. Blood absorbs gases from the alveoli distal to the obstruction, and that part of the lung collapses because it cannot be reinflated.

## Resistance to Airflow

Pressure is one determinant of airflow; the other is resistance. The greater the resistance, the slower the flow. But what governs resistance? Two factors are of particular importance: diameter of the bronchioles and pulmonary compliance.

Like arterioles, the large number of bronchioles, their small diameter, and their ability to change diameter make them the primary means of controlling resistance. The trachea and bronchi can also change diameter to a degree, but are more constrained by the supporting cartilages in their walls. An increase in the diameter of a bronchus or bronchiole is called **bronchodilation,** and a reduction in diameter is called **bronchoconstriction.** Epinephrine and the sympathetic nerves (norepinephrine) stimulate bronchodilation and

---

[18]*atel* = imperfect, incomplete; *ectasis* = expansion

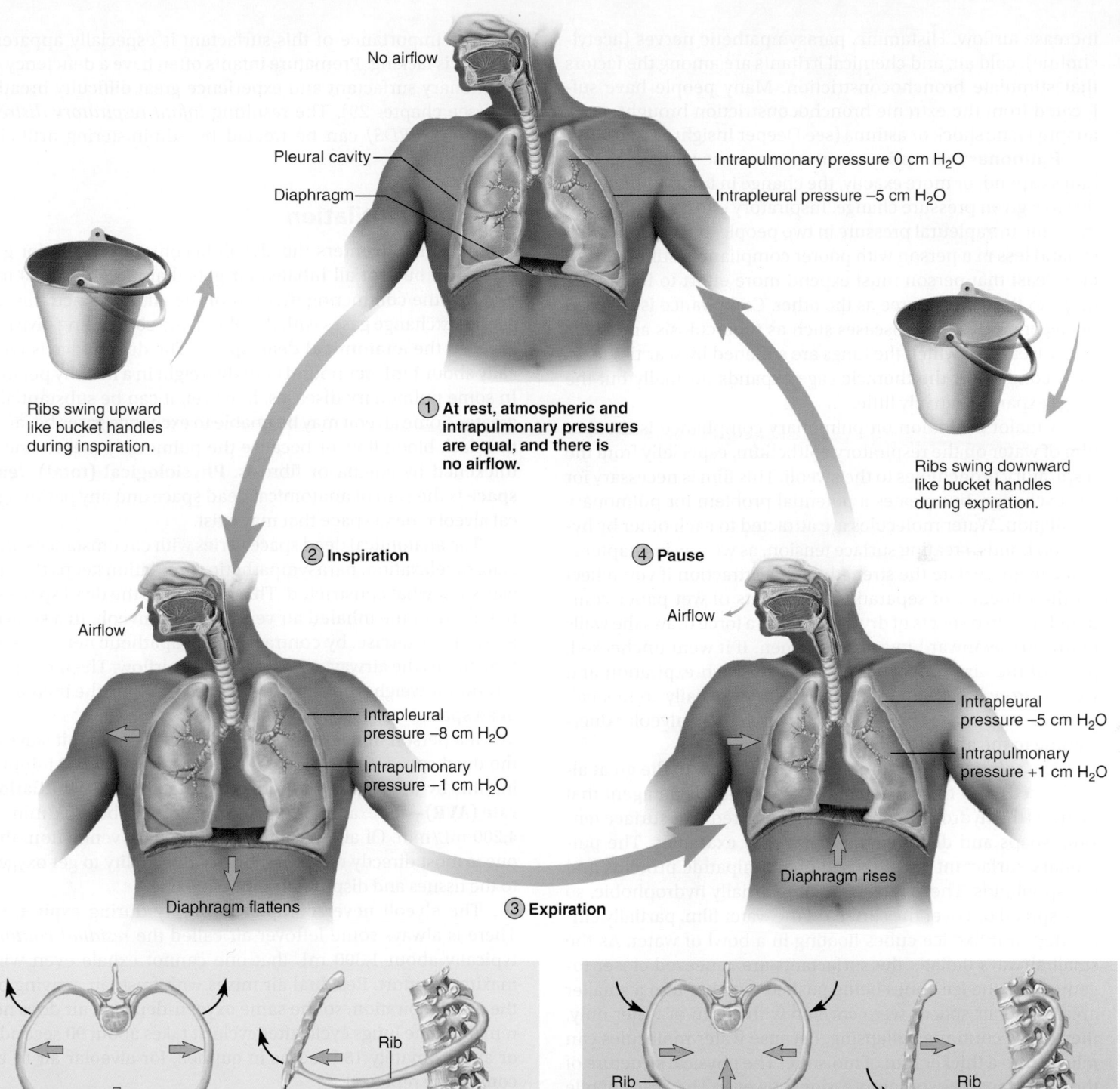

No airflow

Pleural cavity

Diaphragm

Intrapulmonary pressure 0 cm H₂O

Intrapleural pressure –5 cm H₂O

Ribs swing upward like bucket handles during inspiration.

Ribs swing downward like bucket handles during expiration.

① At rest, atmospheric and intrapulmonary pressures are equal, and there is no airflow.

② Inspiration

④ Pause

Airflow

Airflow

Intrapleural pressure –8 cm H₂O

Intrapulmonary pressure –1 cm H₂O

Intrapleural pressure –5 cm H₂O

Intrapulmonary pressure +1 cm H₂O

Diaphragm flattens

Diaphragm rises

③ Expiration

Rib

Sternum

Rib

Sternum

Rib

Sternum

Rib

Sternum

Ribs elevated, thoracic cavity expands laterally

Sternum swings up, thoracic cavity expands anteriorly

Ribs depressed, thoracic cavity narrows

Sternum swings down, thoracic cavity contracts posteriorly

② In inspiration, the thoracic cavity expands laterally, vertically, and anteriorly; intrapulmonary pressure drops 1 cm H₂O below atmospheric pressure, and air flows into the lungs.

③ In expiration, the thoracic cavity contracts in all three directions; intrapulmonary pressure rises 1 cm H₂O above atmospheric pressure, and air flows out of the lungs.

**FIGURE 22.16  The Respiratory Cycle.**  All pressures given here are relative to atmospheric pressure external to the body, which is considered to be zero as a point of reference. Note that pressures governing respiratory airflow are measured in cm H₂O (centimeters of water), not mm Hg. Like bucket handles, each rib is attached at both ends (to the spine and sternum) and swings up and down during inspiration and expiration. AP|R

increase airflow. Histamine, parasympathetic nerves (acetylcholine), cold air, and chemical irritants are among the factors that stimulate bronchoconstriction. Many people have suffocated from the extreme bronchoconstriction brought on by anaphylactic shock or asthma (see Deeper Insight 21.3, p. 838).

**Pulmonary compliance** means the ease with which the lungs expand, or more exactly, the change in lung volume relative to a given pressure change. Inspiratory effort may produce the same intrapleural pressure in two people, but the lungs will expand less in a person with poorer compliance (stiffer lungs), or at least that person must expend more effort to inflate the lungs to the same degree as the other. Compliance is reduced by degenerative lung diseases such as tuberculosis and black lung disease, in which the lungs are stiffened by scar tissue. In such conditions, the thoracic cage expands normally but the lungs expand relatively little.

A major limitation on pulmonary compliance is the thin film of water on the respiratory epithelium, especially from the respiratory bronchioles to the alveoli. This film is necessary for gas exchange, but creates a potential problem for pulmonary ventilation. Water molecules are attracted to each other by hydrogen bonds, creating surface tension, as we saw in chapter 2. You can appreciate the strength of this attraction if you reflect on the difficulty of separating two sheets of wet paper compared with two sheets of dry paper. Such a force draws the walls of the airway inward toward the lumen. If it went unchecked, parts of the airway would collapse with each expiration and would strongly resist reinflation. This is especially so in small airways such as the respiratory bronchioles and alveolar ducts leading to the alveoli.

The solution to this problem takes us back to the great alveolar cells and their surfactant. A surfactant is an agent that disrupts the hydrogen bonds of water and reduces surface tension; soaps and detergents are everyday examples. The pulmonary surfactant is composed of amphipathic proteins and phospholipids. These molecules are partially hydrophobic, so they spread out over the surface of the water film, partially embedded in it like ice cubes floating in a bowl of water. As the small airways deflate, the surfactants are squeezed closer together, like the ice cubes being pushed together into a smaller area. If the air spaces were covered with a film of water only, they could continue collapsing, because water molecules can pile up into a thicker film of moisture. The physical structure of the surfactants resists compression, however. They cannot pile up into a thicker layer, because their hydrophilic regions resist separation from the water below. As they become crowded into a small area and resist layering, they retard and then halt the collapse of the airway.

Deep breathing spreads pulmonary surfactant throughout the small airways. Patients recently out of surgery are encouraged to breathe deeply, even though it may hurt, in order to promote this spread of surfactant up the alveolar ducts and small bronchioles. Those who do not adhere to their breathing exercises can experience collapse of portions of the lung that are not adequately coated with surfactant.

The importance of this surfactant is especially apparent when it is lacking. Premature infants often have a deficiency of pulmonary surfactant and experience great difficulty breathing (see chapter 29). The resulting *infant respiratory distress syndrome (IRDS)* can be treated by administering artificial surfactant.

## Alveolar Ventilation

Air that actually enters the alveoli becomes available for gas exchange, but not all inhaled air gets that far. About 150 mL of it fills the conducting division of the airway. Since this air cannot exchange gases with the blood, the conducting division is called the **anatomical dead space.** The dead space is typically about 1 mL per pound of body weight in a healthy person. In some pulmonary diseases, however, it can be substantially greater. Some alveoli may be unable to exchange gases because they lack blood flow or because the pulmonary membrane is thickened by edema or fibrosis. **Physiological (total) dead space** is the sum of anatomical dead space and any pathological alveolar dead space that may exist.

The anatomical dead space varies with circumstances. In a state of relaxation, parasympathetic stimulation keeps the airway somewhat constricted. This minimizes the dead space so that more of the inhaled air ventilates the alveoli. In a state of arousal or exercise, by contrast, the sympathetic nervous system dilates the airway, which increases airflow. The increased airflow outweighs the air that is wasted by filling the increased dead space.

If a person inhales 500 mL of air and 150 mL of it stays in the dead space, then 350 mL ventilates the alveoli. Multiplying this by the respiratory rate gives the **alveolar ventilation rate (AVR)**—for example, 350 mL/breath × 12 breaths/min. = 4,200 mL/min. Of all measures of pulmonary ventilation, this one is most directly relevant to the body's ability to get oxygen to the tissues and dispose of carbon dioxide.

The alveoli never completely empty during expiration. There is always some leftover air called the *residual volume,* typically about 1,300 mL that one cannot exhale even with maximum effort. Residual air mixes with fresh air arriving on the next inspiration, so the same oxygen-depleted air does not remain in the lungs cycle after cycle. It takes about 90 seconds, or approximately 18 breaths in eupnea, for alveolar air to be completely replaced.

## Spirometry—The Measurement of Pulmonary Ventilation

Clinicians often measure a patient's pulmonary ventilation in order to assess the severity of a respiratory disease or monitor the patient's improvement or deterioration. The process of making such measurements is called **spirometry.**[19] It entails

---

[19]*spiro* = breath; *metry* = process of measuring

having the subject breathe into a device called a **spirometer,** which recaptures the expired breath and records such variables as the rate and depth of breathing, speed of expiration, and rate of oxygen consumption. Representative measurements for a healthy adult male are given in table 22.2 and explained in figure 22.17. Female values are somewhat lower because of smaller average body size.

Four of these values are called *respiratory volumes:* tidal volume, inspiratory reserve volume, expiratory reserve volume, and residual volume. **Tidal volume (TV)** is the amount of air inhaled and exhaled in one cycle; in quiet breathing, it averages about 500 mL. Beyond the amount normally inhaled, it is typically possible to inhale another 3,000 mL with maximum effort; this is the **inspiratory reserve volume (IRV).** Similarly, with maximum effort, one can normally exhale another 1,200 mL beyond the normal amount; this is the **expiratory reserve volume (ERV).** Even after a maximum voluntary expiration, there remains a **residual volume (RV)** of about 1,300 mL. This air allows gas exchange with the blood to continue even between the times one inhales fresh air.

Four other measurements, called *respiratory capacities,* are obtained by adding two or more of the respiratory volumes: **vital capacity (VC)** (ERV + TV + IRV), **inspiratory capacity** (TV + IRV), **functional residual capacity** (RV + ERV), and **total lung capacity** (RV + VC). Vital capacity, the maximum ability to ventilate the lungs in one breath, is an especially important measure of pulmonary health.

Spirometry helps to assess and distinguish between *restrictive* and *obstructive* lung disorders. **Restrictive disorders** are those that reduce pulmonary compliance, thus limiting the amount to which the lungs can be inflated. They show in spirometry as a reduced vital capacity. Any disease that produces pulmonary fibrosis has a restrictive effect: black lung disease and tuberculosis, for example. **Obstructive disorders** are those that interfere with airflow by narrowing or blocking the airway. They make it harder to inhale or exhale a given amount of air. Asthma and chronic bronchitis are the most common examples. Obstructive disorders can be measured by having the subject exhale as rapidly as possible into a spirometer and measuring **forced expiratory volume (FEV)**—the volume of air or the percentage of the vital capacity that can be exhaled in a given time interval. A healthy adult should be able to expel 75% to 85% of the vital capacity in 1.0 second (a value called the $FEV_{1.0}$). At home, asthma patients and others can monitor their respiratory function by blowing into a handheld meter that measures **peak flow,** the maximum speed of expiration.

The amount of air inhaled per minute is the **minute respiratory volume (MRV).** MRV largely determines the alveolar ventilation rate. It can be measured directly with a spirometer or obtained by multiplying tidal volume by respiratory rate. For example, if a person has a tidal volume of 500 mL per breath and a rate of 12 breaths per minute, the MRV would be $500 \times 12 = 6,000$ mL/min. During heavy exercise, MRV may be as high as 125 to 170 L/min. This is called **maximum voluntary ventilation (MVV),** formerly called *maximum breathing capacity.*

## Variations in the Respiratory Rhythm

Relaxed, quiet breathing is called **eupnea**[20] (YOOP-nee-uh). It is typically characterized by a tidal volume of about 500 mL and a respiratory rate of 12 to 15 breaths per minute. Conditions ranging from exercise or anxiety to various disease states can cause deviations such as abnormally fast, slow, or labored breathing. Table 22.3 defines the clinical terms for several such variations. You should familiarize yourself with these before proceeding, because later discussions in this chapter assume a working knowledge of some of these terms.

Other variations in pulmonary ventilation serve the purposes of speaking, expressing emotion (laughing, crying), yawning, hiccuping, expelling noxious fumes, coughing, sneezing, and expelling abdominal contents. Coughing is induced by irritants in the lower respiratory tract. To cough, we close the glottis and contract the muscles of expiration, producing high pressure in the lower respiratory tract. We then suddenly open the glottis and release an explosive burst of air at speeds over 900 km/h (600 mi./h). This drives mucus and foreign matter toward the pharynx and mouth. Sneezing is triggered by irritants in the nasal cavity. Its mechanism is similar to coughing except that the glottis is continually open, the soft palate and tongue block the flow of air while thoracic pressure builds, and then the soft palate is depressed to direct part of the airstream through the nose. These actions are coordinated by coughing and sneezing centers in the medulla oblongata.

### BEFORE YOU GO ON

Answer the following questions to test your understanding of the preceding section:

6. Explain why contraction of the diaphragm causes inspiration but contraction of the transverse abdominal muscle causes expiration.

7. Which brainstem respiratory nucleus is indispensable to respiration? What do the other nuclei do?

8. Explain why Boyle's law is relevant to the action of the respiratory muscles.

9. Explain why eupnea requires little or no action by the muscles of expiration.

10. Identify a benefit and a disadvantage of normal (nonpathological) bronchoconstriction.

11. Suppose a healthy person has a tidal volume of 650 mL, an anatomical dead space of 160 mL, and a respiratory rate of 14 breaths per minute. Calculate her alveolar ventilation rate.

12. Suppose a person has a total lung capacity of 5,800 mL, a residual volume of 1,200 mL, an inspiratory reserve volume of 2,400 mL, and an expiratory reserve volume of 1,400 mL. Calculate his tidal volume.

---

[20]*eu* = easy, normal; *pnea* = breathing

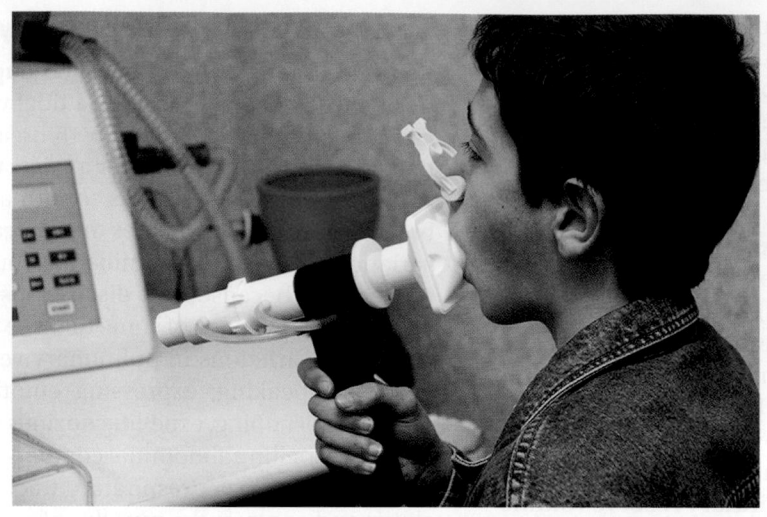

(a)

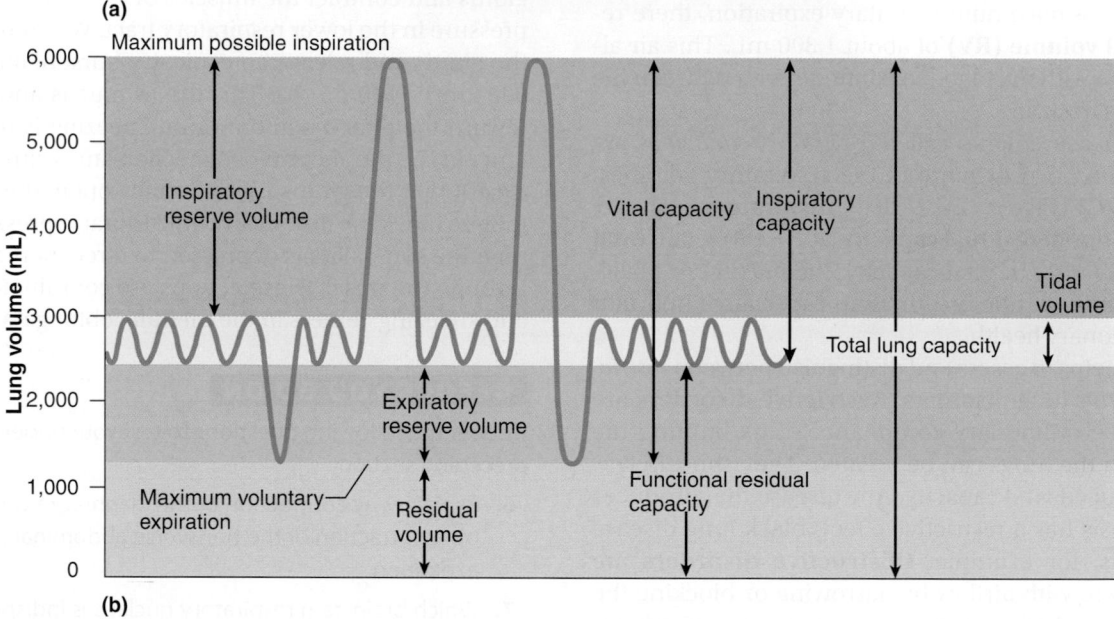

(b)

**FIGURE 22.17 Respiratory Volumes and Capacities.** (a) Boy breathing into a spirometer. (b) An idealized spirogram. The wavy line indicates inspiration when it rises and expiration when it falls. Compare with table 22.2.

| TABLE 22.2 | | Respiratory Volumes and Capacities for an Average Young Adult Male |
|---|---|---|
| **Measurement** | **Typical Value** | **Definition** |
| **Respiratory volumes** | | |
| Tidal volume (TV) | 500 mL | Amount of air inhaled and exhaled in one cycle during quiet breathing |
| Inspiratory reserve volume (IRV) | 3,000 mL | Amount of air in excess of tidal volume that can be inhaled with maximum effort |
| Expiratory reserve volume (ERV) | 1,200 mL | Amount of air in excess of tidal volume that can be exhaled with maximum effort |
| Residual volume (RV) | 1,300 mL | Amount of air remaining in the lungs after maximum expiration; the amount that can never voluntarily be exhaled |
| **Respiratory capacities** | | |
| Vital capacity (VC) | 4,700 mL | The amount of air that can be inhaled and then exhaled with maximum effort; the deepest possible breath (VC = ERV + TV + IRV) |
| Inspiratory capacity (IC) | 3,500 mL | Maximum amount of air that can be inhaled after a normal tidal expiration (IC = TV + IRV) |
| Functional residual capacity (FRC) | 2,500 mL | Amount of air remaining in the lungs after a normal tidal expiration (FRC = RV + ERV) |
| Total lung capacity (TLC) | 6,000 mL | Maximum amount of air the lungs can contain (TLC = RV + VC) |

| TABLE 22.3 | Variations in the Respiratory Rhythm |
|---|---|
| Apnea[21] (AP-nee-uh) | Temporary cessation of breathing (one or more skipped breaths) |
| Dyspnea[22] (DISP-nee-uh) | Labored, gasping breathing; shortness of breath |
| Hyperpnea[23] (HY-purp-NEE-uh) | Increased rate and depth of breathing in response to exercise, pain, or other conditions |
| Hyperventilation | Increased pulmonary ventilation in excess of metabolic demand, frequently associated with anxiety; expels $CO_2$ faster than it is produced, thus lowering the blood $CO_2$ concentration and raising the blood pH |
| Hypoventilation[24] | Reduced pulmonary ventilation; leads to an increase in blood $CO_2$ concentration if ventilation is insufficient to expel $CO_2$ as fast as it is produced |
| Kussmaul[25] respiration | Deep, rapid breathing often induced by acidosis; seen in diabetes mellitus |
| Orthopnea[26] (or-thop-NEE-uh) | Dyspnea that occurs when a person is lying down or in any position other than standing or sitting erect; seen in heart failure, asthma, emphysema, and other conditions |
| Respiratory arrest | Permanent cessation of breathing (unless there is medical intervention) |
| Tachypnea[27] (tack-ip-NEE-uh) | Accelerated respiration |

[21]$a$ = without; $pnea$ = breathing
[22]$dys$ = difficult, abnormal, painful; $pnea$ = breathing
[23]$hyper$ = above normal
[24]$hypo$ = below normal
[25]Adolph Kussmaul (1822–1902), German physician
[26]$ortho$ = straight, erect; $pnea$ = breathing
[27]$tachy$ = fast; $pnea$ = breathing

## 22.3   Gas Exchange and Transport

### Expected Learning Outcomes

When you have completed this section, you should be able to

a. define *partial pressure* and discuss its relationship to a gas mixture such as air;

b. contrast the composition of inspired and alveolar air;

c. discuss how partial pressure affects gas transport by the blood;

d. describe the mechanisms of transporting $O_2$ and $CO_2$;

e. describe the factors that govern gas exchange in the lungs and systemic capillaries;

f. explain how gas exchange is adjusted to the metabolic needs of different tissues; and

g. discuss the effect of blood gases and pH on the respiratory rhythm.

Ultimately, respiration is about gases, especially oxygen and carbon dioxide. We will turn our attention now to the behavior of these gases in the human body: how oxygen is obtained from inspired air and delivered to the tissues, and how carbon dioxide is removed from the tissues and released into the expired air. First, however, it is necessary to understand the composition of the air we inhale and how gases behave in contact with the water film that lines the alveoli.

### Composition of Air

Air consists of about 78.6% nitrogen; 20.9% oxygen; 0.04% carbon dioxide; several quantitatively minor gases such as argon, neon, helium, methane, and ozone; and a variable amount of water vapor. Water vapor constitutes from 0% to 4%, depending on temperature and humidity; we will use a value of 0.5%, typical of a cool clear day.

The total atmospheric pressure is a sum of the contributions of these individual gases—a principle known as **Dalton's law** (see table 22.1). The separate contribution of each gas in a mixture is called its **partial pressure** and is symbolized with a P followed by the formula of the gas, such as $P_{N_2}$. As we are now concerned with atmospheric pressures and how they influence the partial pressures of blood gases, we return to mm Hg as our unit of measurement (not cm $H_2O$ as when we were considering pulmonary ventilation). If we assume the average sea-level atmospheric pressure of 760 mm Hg and nitrogen is 78.6% of the atmosphere, then $P_{N_2}$ is simply

$$0.786 \times 760 \text{ mm Hg} = 597 \text{ mm Hg}$$

Applying Dalton's law to the mixture of gases in the preceding paragraph,

$$P_{N_2} + P_{O_2} + P_{H_2O} + P_{CO_2}$$
$$\approx 597 + 159 + 3.7 + 0.3$$
$$= 760.0 \text{ mm Hg}$$

These values change dramatically at higher altitude (see Deeper Insight 22.3).

This is the composition of the air we inhale, but it is not the composition of air in the alveoli. Alveolar air can be sampled with an apparatus that collects the last 10 mL of expired air. As we see in table 22.4, its composition differs from that of the atmosphere because of three influences: (1) It is humidified by contact with the mucous membranes, so its $P_{H_2O}$ is more than 10 times higher than that of the inhaled air. (2) Freshly inspired air mixes with residual air left from the previous respiratory cycle, so its oxygen is diluted and it is enriched with $CO_2$ from the residual air. (3) Alveolar air exchanges $O_2$ and $CO_2$ with the blood. Thus, the $P_{O_2}$ of alveolar air is about 65% that of inhaled air, and its $P_{CO_2}$ is more than 130 times higher.

| TABLE 22.4 | Composition of Inspired (Atmospheric) and Alveolar Air | | | |
|---|---|---|---|---|
| **Gas** | **Inspired Air*** | | **Alveolar Air** | |
| $N_2$ | 78.6% | 597 mm Hg | 74.9% | 569 mm Hg |
| $O_2$ | 20.9% | 159 mm Hg | 13.7% | 104 mm Hg |
| $H_2O$ | 0.5% | 3.7 mm Hg | 6.2% | 47 mm Hg |
| $CO_2$ | 0.04% | 0.3 mm Hg | 5.3% | 40 mm Hg |
| Total | 100% | 760 mm Hg | 100% | 760 mm Hg |

*Typical values for a cool clear day; values vary with temperature and humidity. Other gases present in small amounts are disregarded.

### ▶▶▶ APPLY WHAT YOU KNOW

*Expired air considered as a whole (not just the last 10 mL) is about 15.3% $O_2$ and 4.2% $CO_2$. Why would these values differ from the ones for alveolar air?*

## Alveolar Gas Exchange AP|R

Air in the alveolus is in contact with the film of water covering the alveolar epithelium. For oxygen to get into the blood, it must dissolve in this water and pass through the respiratory membrane separating the air from the bloodstream. For carbon dioxide to leave the blood, it must pass the other way and diffuse out of the water film into the alveolar air. This back-and-forth traffic of $O_2$ and $CO_2$ across the respiratory membrane is called **alveolar gas exchange.**

The reason that $O_2$ can diffuse in one direction and $CO_2$ in the other is that each gas diffuses down its own partial pressure

### DEEPER INSIGHT 22.3

### MEDICAL HISTORY

#### The Flight of the Zenith

As any aviator or mountain hiker knows, the composition of air changes dramatically from sea level to high altitude. This was all too sadly recognized by French physician Paul Bert (1833–86), who is commonly recognized as the founder of aerospace medicine. He invented the first pressure chamber capable of simulating the effects of high altitude, and undertook a variety of experiments on human and animal subjects to test the effects of variation in oxygen partial pressure. In 1875, balloonist Gaston Tissandier set out from Paris in a hot-air balloon with two of Bert's protégés in physiology. Their aim was to investigate the effects of the low oxygen pressures attainable only by ascending to a very high altitude. They ignored Bert's advice to breathe supplemental oxygen continually, rather than only when they felt the need for it. As they ascended, they observed each other and took notes. Seeing no ill effects, they continued to throw out ballast and go higher, eventually to 28,000 feet. But after passing 24,000 feet, they experienced stupefaction, muscular paralysis, euphoria, and finally unconsciousness. The balloon eventually descended on its own with two of the three men dead; only Tissandier lived to write about it.

gradient. Whenever air and water are in contact with each other, gases diffuse down their gradients until the partial pressure of each gas in the air is equal to its partial pressure in the water. If a gas has a greater partial pressure in the water than in the air, it diffuses into the air; the smell of chlorine near a swimming pool is evidence of this. If its partial pressure is greater in the air, it diffuses into the water.

**Henry's law** states that *at the air–water interface, for a given temperature, the amount of gas that dissolves in the water is determined by its solubility in water and its partial pressure in the air* (fig. 22.18). Thus, the greater the $P_{O_2}$ in the alveolar air, the more $O_2$ the blood picks up. And since blood arriving at an alveolus has a higher $P_{CO_2}$ than air, it releases $CO_2$ into the alveolar air. At the alveolus, the blood is said to *unload* $CO_2$ and *load* $O_2$. Each gas in a mixture behaves independently; the diffusion of one gas does not influence the diffusion of another.

Both $O_2$ loading and $CO_2$ unloading involve erythrocytes (RBCs). The efficiency of these processes therefore depends on how long an RBC spends in an alveolar capillary compared with how long it takes for each gas to be fully loaded or unloaded—that is, for them to reach equilibrium concentrations in the capillary blood. It takes about 0.25 second to reach equilibrium. At rest, when blood circulates at its slowest speed, an RBC passes through an alveolar capillary in about 0.75 second—plenty of time. Even in vigorous exercise, when the blood flows faster, an erythrocyte is in the alveolar capillary for about 0.3 second, which is still adequate.

Several variables affect the efficiency of alveolar gas exchange, and under abnormal conditions, some of these can prevent the complete loading and unloading of gases:

- **Pressure gradients of the gases.** The $P_{O_2}$ is about 104 mm Hg in alveolar air and 40 mm Hg in blood arriving at an alveolus. Oxygen therefore diffuses from the air into the blood, where it reaches a $P_{O_2}$ of 104 mm Hg. Before the blood leaves the lung, however, this drops to about 95 mm Hg. This oxygen dilution occurs because the pulmonary veins anastomose with the bronchial veins in the lungs, so there is some mixing of the oxygen-rich pulmonary blood with the oxygen-poor systemic blood.

  The $P_{CO_2}$ is about 46 mm Hg in blood arriving at the alveolus and 40 mm Hg in alveolar air. Carbon dioxide therefore diffuses from the blood into the alveoli. These changes are summarized here and at the middle of figure 22.19:

| ***Blood entering lungs*** | | ***Blood leaving lungs*** | |
|---|---|---|---|
| $P_{O_2}$ | 40 mm Hg | $P_{O_2}$ | 95 mm Hg |
| $P_{CO_2}$ | 46 mm Hg | $P_{CO_2}$ | 40 mm Hg |

These gradients differ under special circumstances such as high elevation and *hyperbaric oxygen therapy* (treatment with oxygen at greater than 1 atm of pressure) (fig. 22.20). At high elevations, the partial pressures of all atmospheric gases are lower. Atmospheric $P_{O_2}$, for

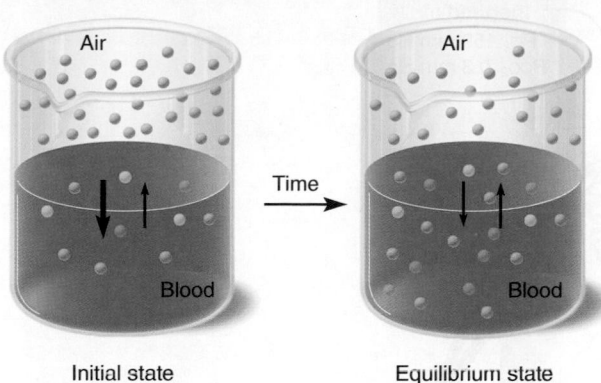

**(a) Oxygen**

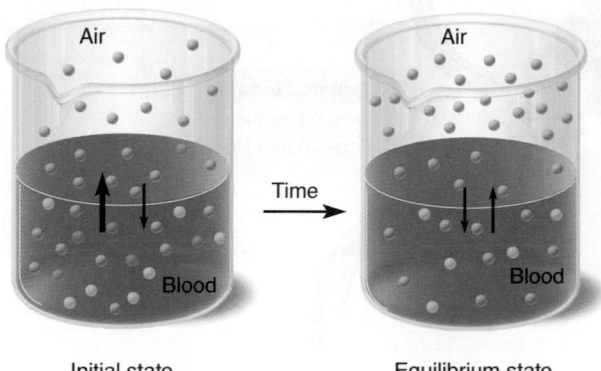

**(b) Carbon dioxide**

**FIGURE 22.18 Henry's Law and Its Relationship to Alveolar Gas Exchange.** (a) The $P_{O_2}$ of alveolar air is initially higher than the $P_{O_2}$ of the blood arriving at an alveolus. Oxygen diffuses into the blood until the two are in equilibrium. (b) The $P_{CO_2}$ of the arriving blood is initially higher than the $P_{CO_2}$ of alveolar air. Carbon dioxide diffuses into the alveolus until the two are in equilibrium. It takes about 0.25 second for both gases to reach equilibrium.

example, is 159 mm Hg at sea level but only 110 mm Hg at 3,000 m (10,000 feet). The $O_2$ gradient from air to blood is proportionately less, so as we can predict from Henry's law, less $O_2$ diffuses into the blood. In a hyperbaric oxygen chamber, by contrast, a patient is exposed to 3 to 4 atm of oxygen to treat such conditions as gangrene (to kill anaerobic bacteria) and carbon monoxide poisoning (to displace the carbon monoxide from hemoglobin). The $P_{O_2}$ ranges from 2,300 to 3,000 mm Hg. Thus, there is a very steep gradient of $P_{O_2}$ from alveolus to blood and diffusion into the blood is accelerated.

- **Solubility of the gases.** Gases differ in their ability to dissolve in water. Carbon dioxide is about 20 times as

soluble as oxygen, and oxygen is about twice as soluble as nitrogen. Even though the pressure gradient of $O_2$ is much greater than that of $CO_2$ across the respiratory membrane, equal amounts of the two gases are exchanged because $CO_2$ is so much more soluble and diffuses more rapidly.

- **Membrane thickness.** The respiratory membrane between the blood and alveolar air is only 0.5 μm thick in most places—much less than the 7 to 8 μm diameter of a single RBC. Thus, it presents little obstacle to diffusion (fig. 22.21a). In such heart conditions as left ventricular failure, however, blood pressure builds up in the lungs and promotes capillary filtration into the connective tissues, causing the respiratory membranes to become edematous and thickened (similar to their condition in pneumonia; fig. 22.21b). The gases have farther to travel between blood and air and cannot equilibrate fast enough to keep pace with blood flow. Under these circumstances, blood leaving the lungs has an unusually high $P_{CO_2}$ and low $P_{O_2}$.

- **Membrane area.** In good health, each lung has about 70 m² of respiratory membrane available for gas exchange. Since the alveolar capillaries contain a total of only 100 mL of blood at any one time, this blood is spread very thinly. Several pulmonary diseases, however, decrease the alveolar surface area and thus lead to low blood $P_{O_2}$—for example, emphysema (fig. 22.21c) and lung cancer.

- **Ventilation–perfusion coupling.** Gas exchange requires not only good ventilation of the alveoli but also good perfusion of their capillaries. *Ventilation–perfusion coupling* refers to physiological responses that match airflow to blood flow and vice versa. For example, if part of a lung were poorly ventilated because of tissue destruction or an airway obstruction, it would be pointless to direct much blood to that tissue. Poor ventilation leads to a low $P_{O_2}$ in that region of the lung. This stimulates local vasoconstriction, rerouting the blood to better-ventilated areas of the lung where it can pick up more oxygen (fig. 22.22a, left). In contrast, increased ventilation raises the local blood $P_{O_2}$ and this stimulates vasodilation, increasing blood flow to that region to take advantage of the oxygen availability (fig. 22.22a, right). These reactions of the pulmonary arteries are opposite from the reactions of systemic arteries, which dilate in response to hypoxia. Furthermore, changes in the blood flow to a region of a lung stimulate bronchoconstriction or dilation, adjusting ventilation so that air is directed to the best-perfused parts of the lung (fig. 22.22b).

## Gas Transport

**Gas transport** is the process of carrying gases from the alveoli to the systemic tissues and vice versa. This section explains how the blood loads and transports $O_2$ and $CO_2$.

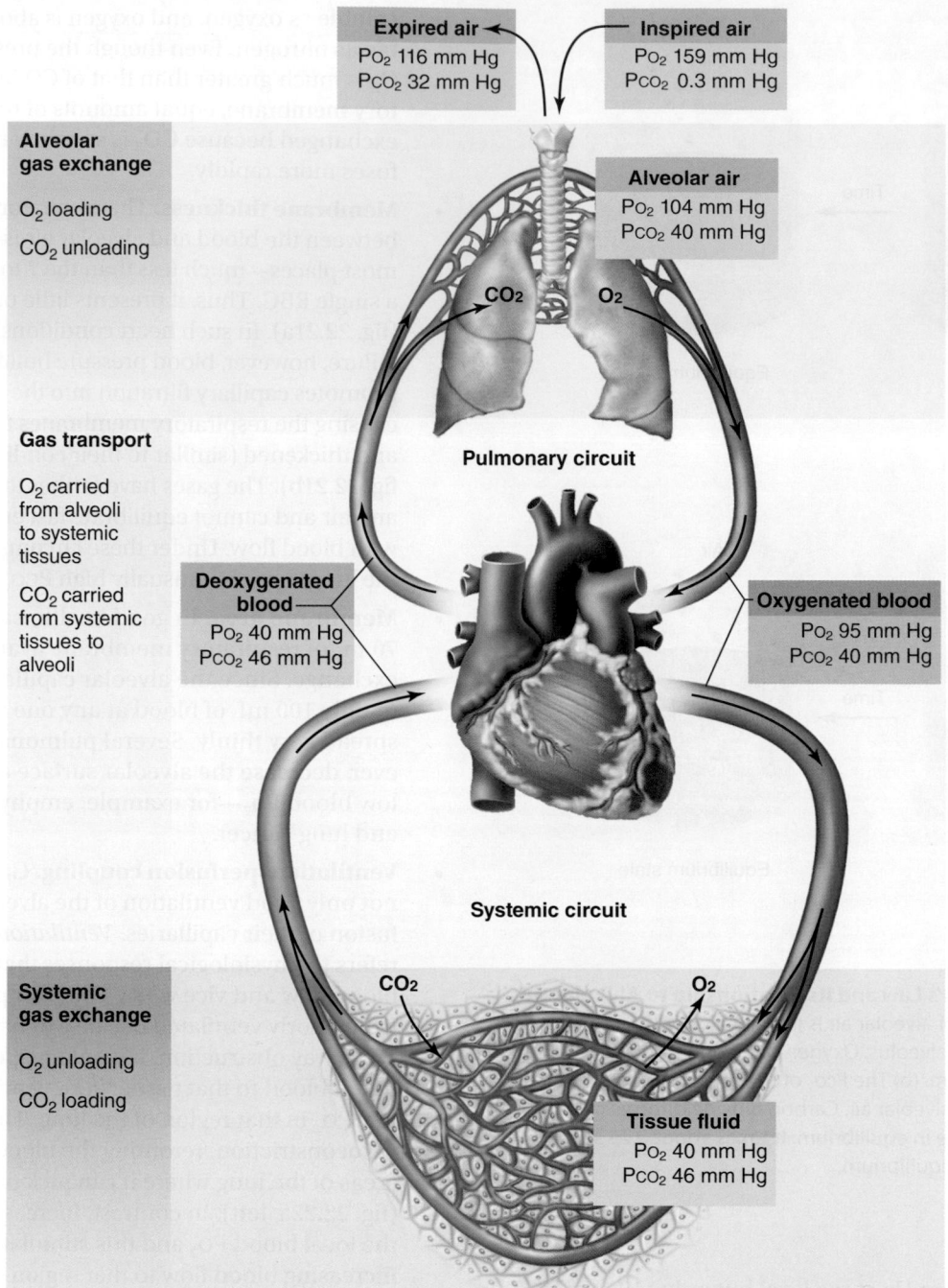

**Expired air**
Po₂ 116 mm Hg
Pco₂ 32 mm Hg

**Inspired air**
Po₂ 159 mm Hg
Pco₂ 0.3 mm Hg

**Alveolar
gas exchange**

O₂ loading

CO₂ unloading

**Alveolar air**
Po₂ 104 mm Hg
Pco₂ 40 mm Hg

CO₂          O₂

**Pulmonary circuit**

**Gas transport**

O₂ carried
from alveoli
to systemic
tissues

CO₂ carried
from systemic
tissues to
alveoli

**Deoxygenated
blood**
Po₂ 40 mm Hg
Pco₂ 46 mm Hg

**Oxygenated blood**
Po₂ 95 mm Hg
Pco₂ 40 mm Hg

**Systemic circuit**

**Systemic
gas exchange**

O₂ unloading

CO₂ loading

CO₂          O₂

**Tissue fluid**
Po₂ 40 mm Hg
Pco₂ 46 mm Hg

**FIGURE 22.19  Changes in Po₂ and Pco₂ Along the Circulatory Route.**  AP|R

❓ *Trace the partial pressure of oxygen from inspired air to expired air and explain each change in Po₂ along the way. Do the same for Pco₂.*

## Oxygen

Arterial blood carries about 20 mL of oxygen per deciliter. About 98.5% of it is bound to hemoglobin in the RBCs and 1.5% is dissolved in the blood plasma. Hemoglobin is specialized for oxygen transport. It consists of four protein (globin) chains, each with one heme group (see fig. 18.5, p. 679). Each heme can bind 1 $O_2$ to the iron atom at its center; thus, one hemoglobin molecule can carry up to 4 $O_2$. If one or more molecules of $O_2$ are bound to hemoglobin, the

compound is called **oxyhemoglobin (HbO₂),** whereas hemoglobin with no oxygen bound to it is **deoxyhemoglobin (HHb).** When hemoglobin is 100% saturated, every molecule of it carries 4 $O_2$; if it is 75% saturated, there is an average of 3 $O_2$ per hemoglobin molecule; if it is 50% saturated, there is an average of 2 $O_2$ per hemoglobin; and so forth. The poisonous effect of carbon monoxide stems from its competition for the $O_2$ binding site (see Deeper Insight 22.4).

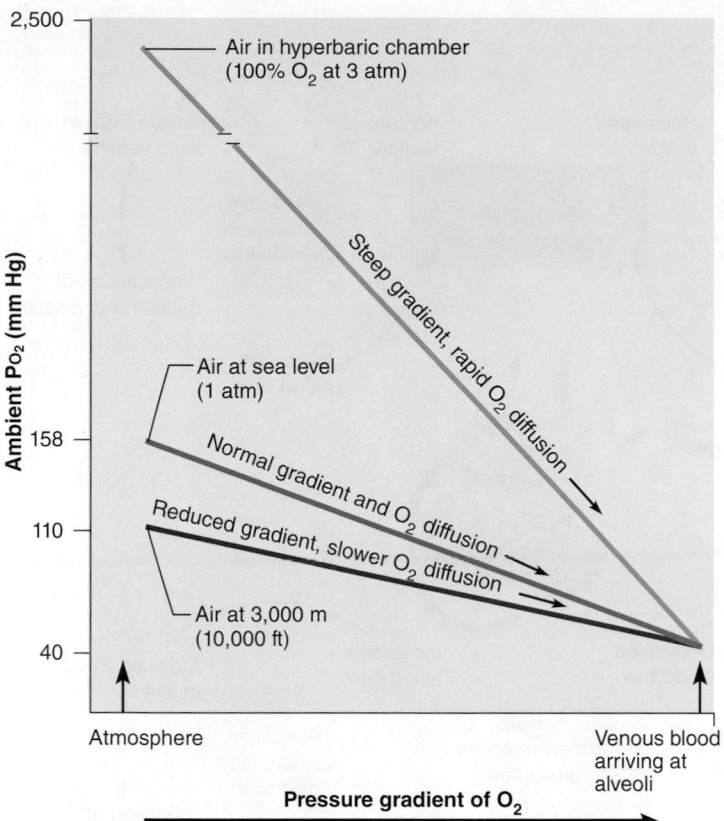

**FIGURE 22.20 Oxygen Loading in Relation to Partial Pressure Gradient.** The rate of loading depends on the steepness of the gradient from alveolar air to the venous blood arriving at the alveolar capillaries. Compared with the oxygen gradient at sea level (blue line), the gradient is less steep at high elevation (red line) because the $P_{O_2}$ of the atmosphere is lower. Thus, oxygen loading of the pulmonary blood is slower. In a hyperbaric chamber with 100% oxygen, the gradient from air to blood is very steep (green line), and oxygen loading is correspondingly rapid. This is an illustration of Henry's law and has important effects in diving, aviation, mountain climbing, and oxygen therapy.

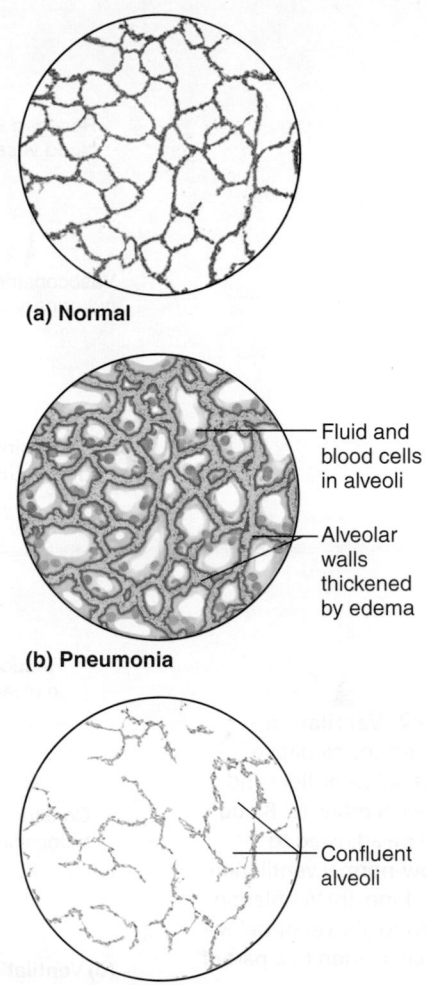

**FIGURE 22.21 Pulmonary Alveoli in Health and Disease.** (a) In a healthy lung, the alveoli are small and have thin respiratory membranes. (b) In pneumonia, the respiratory membranes (alveolar walls) are thick with edema, and the alveoli contain fluid and blood cells. (c) In emphysema, alveolar membranes break down and neighboring alveoli join to form larger, fewer alveoli with less total surface area.

The relationship between hemoglobin saturation and $P_{O_2}$ is shown by the **oxyhemoglobin dissociation curve** (fig. 22.23). As you can see, it is not a simple linear relationship. At low $P_{O_2}$, the curve rises slowly; then there is a rapid increase in oxygen loading as $P_{O_2}$ rises further. This reflects the way hemoglobin loads oxygen. When the first heme group binds $O_2$, hemoglobin changes shape in a way that facilitates uptake of the second $O_2$ by another heme group. This, in turn, promotes the uptake of the third and then the fourth $O_2$—hence the rapidly rising midportion of the curve. At high $P_{O_2}$ levels, the curve levels off because the hemoglobin approaches 100% saturation and cannot load much more oxygen.

#### ▶▶▶ APPLY WHAT YOU KNOW

*Is oxygen loading a positive or negative feedback process? Explain.*

 **DEEPER INSIGHT 22.4**

## CLINICAL APPLICATION

### Carbon Monoxide Poisoning

The lethal effect of carbon monoxide (CO) is well known. This colorless, odorless gas occurs in cigarette smoke, engine exhaust, and fumes from furnaces and space heaters. It binds to the iron of hemoglobin to form *carboxyhemoglobin (HbCO)*. Thus, it competes with oxygen for the same binding site. Not only that, but it binds 210 times as tightly as oxygen. Thus, CO tends to tie up hemoglobin for a long time. Less than 1.5% of the hemoglobin is occupied by carbon monoxide in most nonsmokers, but this figure rises to as much as 3% in residents of heavily polluted cities and 10% in heavy smokers. An atmospheric concentration of 0.1% CO is enough to bind 50% of a person's hemoglobin, and a concentration of 0.2% is quickly lethal.

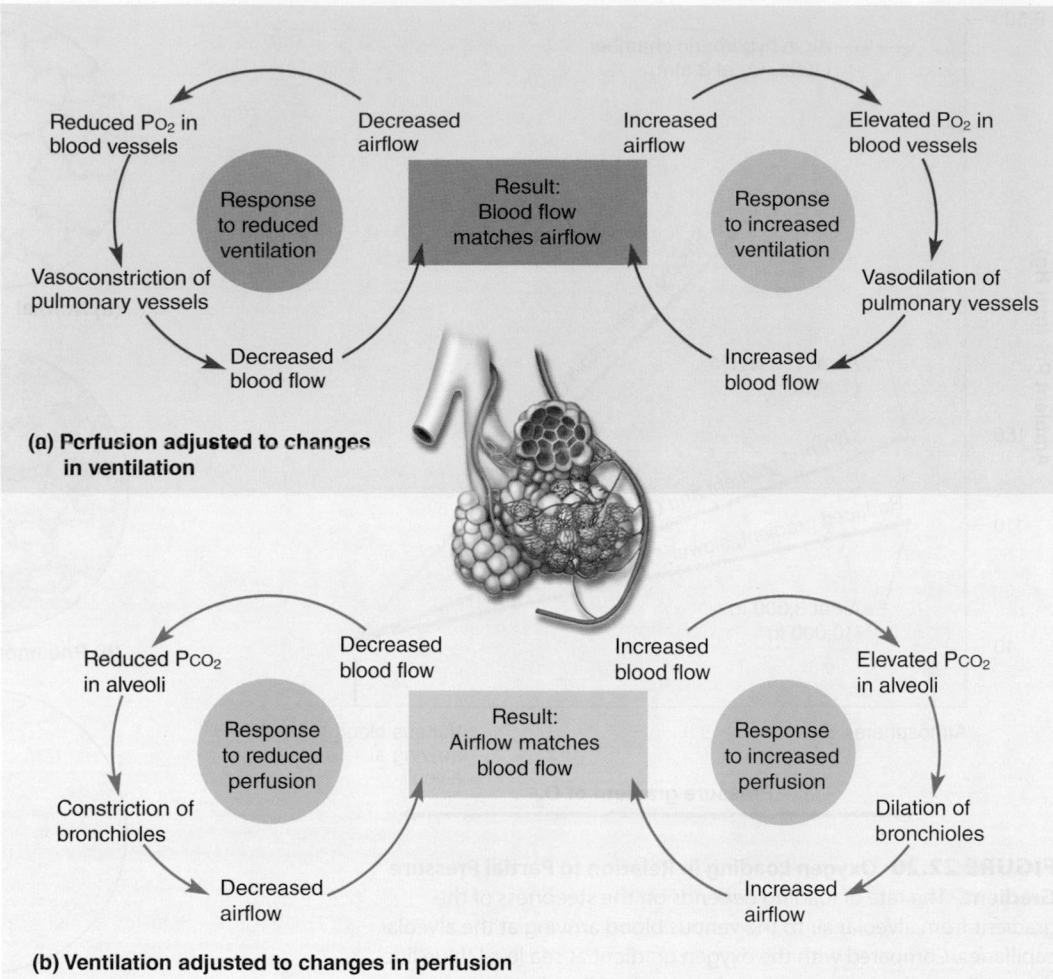

**FIGURE 22.22 Ventilation–Perfusion Coupling.** Negative feedback loops adjust airflow and blood flow to each other. (a) Blood circulation can be adjusted to above- or below-normal ventilation of a part of the lung. (b) Ventilation can be adjusted to above- or below-normal blood circulation to a part of the lung.

## Carbon Dioxide

Carbon dioxide is transported in three forms: carbonic acid, carbamino compounds, and dissolved gas.

1. About 90% of the $CO_2$ is hydrated (reacts with water) to form **carbonic acid,** which then dissociates into bicarbonate and hydrogen ions:

$$CO_2 + H_2O \rightarrow H_2CO_3 \rightarrow HCO_3^- + H^+$$

   More will be said about this reaction shortly.

2. About 5% binds to the amino groups of plasma proteins and hemoglobin to form **carbamino compounds**—chiefly **carbaminohemoglobin (HbCO$_2$).** The reaction with hemoglobin can be symbolized $Hb + CO_2 \rightarrow HbCO_2$. Carbon dioxide does not compete with oxygen because $CO_2$ and $O_2$ bind to different sites on the hemoglobin molecule—oxygen to the heme moiety and $CO_2$ to the polypeptide chains. Hemoglobin can therefore transport both $O_2$ and $CO_2$ simultaneously. As we will see, however, each gas somewhat inhibits transport of the other.

3. The remaining 5% of the $CO_2$ is carried in the blood as dissolved gas, like the $CO_2$ in carbonated soft drinks and sparkling wines.

The relative amounts of $CO_2$ exchanged between the blood and alveolar air differ from the percentages just given. About 70% of the *exchanged* $CO_2$ comes from carbonic acid, 23% from carbamino compounds, and 7% from the dissolved gas. That is, blood gives up the dissolved $CO_2$ gas and $CO_2$ from the carbamino compounds more easily than it gives up the $CO_2$ in bicarbonate.

## Systemic Gas Exchange

Systemic gas exchange is the unloading of $O_2$ and loading of $CO_2$ at the systemic capillaries (see fig. 22.19, bottom, and fig. 22.24).

## Carbon Dioxide Loading

Aerobic respiration produces a molecule of $CO_2$ for every molecule of $O_2$ it consumes. The tissue fluid therefore contains a relatively high $P_{CO_2}$ and there is typically a $CO_2$ gradient of $46 \rightarrow 40$ mm Hg from tissue fluid to blood. Consequently, $CO_2$ diffuses into the bloodstream, where it is carried in the three forms already noted. Most of it reacts with water to produce bicarbonate ($HCO_3^-$) and hydrogen ($H^+$) ions. This reaction occurs slowly in the blood plasma but much faster in the RBCs, where it is catalyzed by the enzyme *carbonic anhydrase.* An antiport called the *chloride–bicarbonate exchanger* then pumps

**FIGURE 22.23 The Oxyhemoglobin Dissociation Curve.** This curve shows the relative amount of hemoglobin that is saturated with oxygen (*y*-axis) as a function of ambient (surrounding) oxygen partial pressure (*x*-axis). As it passes through the alveolar capillaries where the $P_{O_2}$ is high, hemoglobin becomes saturated with oxygen. As it passes through the systemic capillaries where the $P_{O_2}$ is low, it typically gives up about 22% of its oxygen (color bar at top of graph).

❓ *What would be the approximate utilization coefficient if the systemic tissues had a $P_{O_2}$ of 20 mm Hg?*

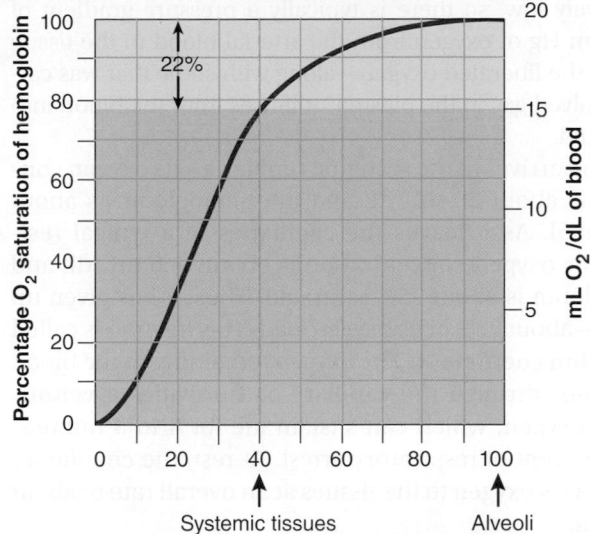

most of the $HCO_3^-$ out of the RBC in exchange for $Cl^-$ from the blood plasma. This exchange is called the **chloride shift.** Most of the $H^+$ binds to hemoglobin or oxyhemoglobin, which thus buffers the intracellular pH.

## Oxygen Unloading

When $H^+$ binds to oxyhemoglobin ($HbO_2$), it reduces the affinity of hemoglobin for $O_2$ and tends to make hemoglobin release it. Oxygen consumption by respiring tissues keeps the $P_{O_2}$ of tissue

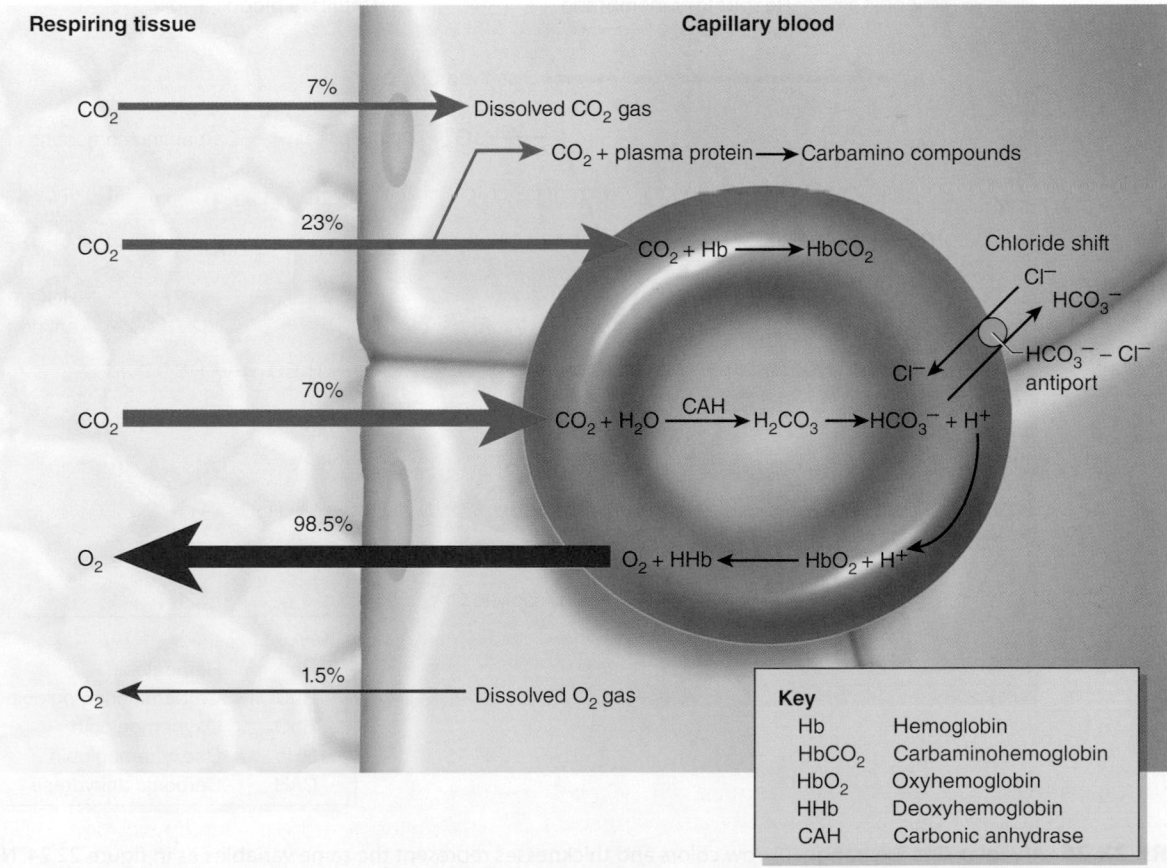

**FIGURE 22.24 Systemic Gas Exchange.** Blue arrows show the three mechanisms of $CO_2$ loading and transport; their thickness represents the relative amounts of $CO_2$ loaded in each of the three forms. Red arrows show the two mechanisms of $O_2$ unloading; their thickness indicates the relative amounts unloaded by each mechanism. Note that $CO_2$ loading releases hydrogen ions in the erythrocyte, and hydrogen ions promote $O_2$ unloading.

fluid relatively low, so there is typically a pressure gradient of $95 \rightarrow 40$ mm Hg of oxygen from the arterial blood to the tissue fluid. Thus, the liberated oxygen—along with some that was carried as dissolved gas in the plasma—diffuses from the blood into the tissue fluid.

As blood arrives at the systemic capillaries, its oxygen concentration is about 20 mL/dL and the hemoglobin is about 97% saturated. As it leaves the capillaries of a typical resting tissue, its oxygen concentration is about 15.6 mL/dL and the hemoglobin is about 75% saturated. Thus, it has given up 4.4 mL/dL—about 22% of its oxygen load. This fraction is called the **utilization coefficient.** The oxygen remaining in the blood after it passes through the capillary bed provides a **venous reserve** of oxygen, which can sustain life for 4 to 5 minutes even in the event of respiratory arrest. At rest, the circulatory system releases oxygen to the tissues at an overall rate of about 250 mL/min.

## Alveolar Gas Exchange Revisited

The processes illustrated in figure 22.24 make it easier to understand alveolar gas exchange more fully. As shown in figure 22.25, the reactions that occur in the lungs are essentially the reverse of systemic gas exchange. As hemoglobin loads oxygen, its affinity for $H^+$ declines. Hydrogen ions dissociate from the hemoglobin and bind with bicarbonate ($HCO_3^-$) ions transported from the plasma into the RBCs. Chloride ions are transported back out of the RBC (a reverse chloride shift). The reaction of $H^+$ and $HCO_3^-$ reverses the hydration reaction and generates free $CO_2$. This diffuses into the alveolus to be exhaled—as does the $CO_2$ released from carbaminohemoglobin and $CO_2$ gas that was dissolved in the plasma.

## Adjustment to the Metabolic Needs of Individual Tissues

Hemoglobin does not unload the same amount of oxygen to all tissues. Some tissues need more and some less, depending on their state of activity. Hemoglobin responds to such variations and unloads more oxygen to the tissues that need it most. In exercising skeletal muscles, for example, the utilization coefficient may be as high as 80%. Four factors adjust the rate of oxygen unloading to the metabolic rates of different tissues:

1. **Ambient $Po_2$.** Since an active tissue consumes oxygen rapidly, the $Po_2$ of its tissue fluid remains low. From the

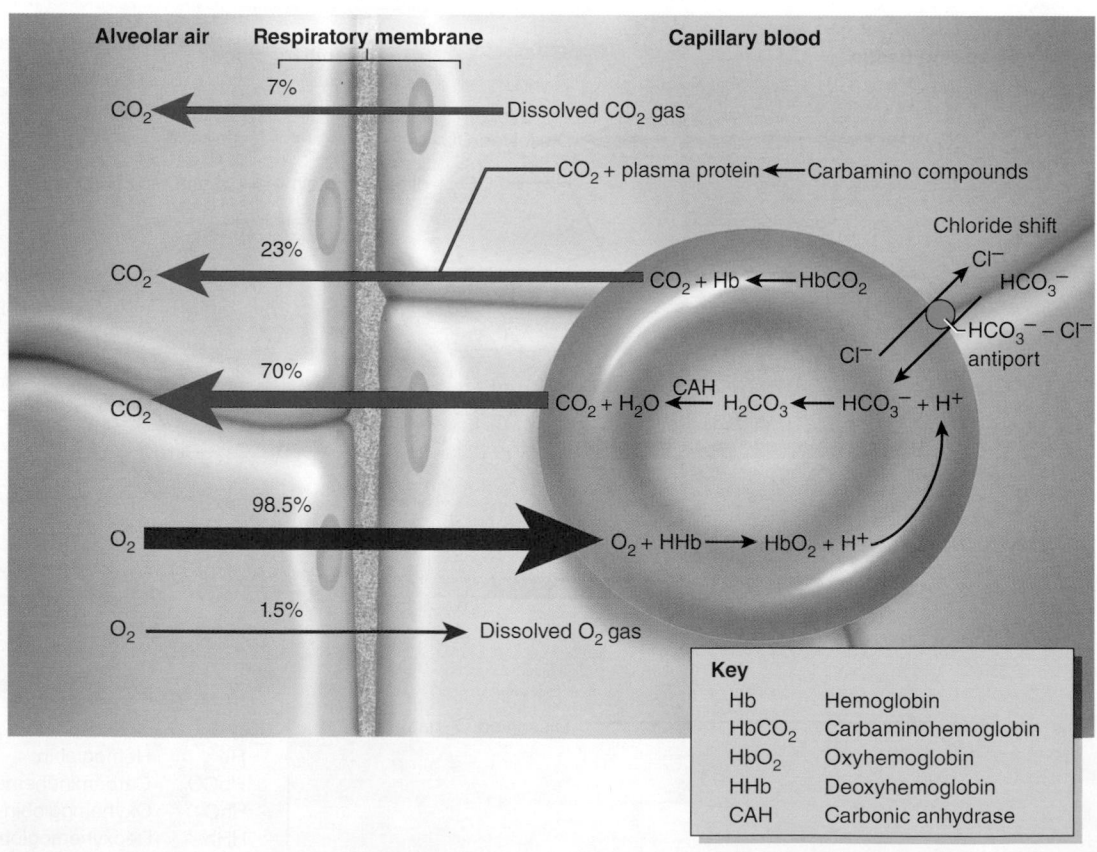

**FIGURE 22.25  Alveolar Gas Exchange.**  Arrow colors and thicknesses represent the same variables as in figure 22.24. Note that $O_2$ loading promotes the decomposition of carbonic acid into $H_2O$ and $CO_2$, and most exhaled $CO_2$ comes from the erythrocytes.

❓ *In what fundamental way does this differ from the preceding figure? Following alveolar gas exchange, will the blood contain a higher or lower concentration of bicarbonate ions than it did before?*

oxyhemoglobin dissociation curve (see fig. 22.23), you can see that at a low $P_{O_2}$, $HbO_2$ releases more oxygen.

2. **Temperature.** When temperature rises, the oxyhemoglobin dissociation curve shifts to the right (fig. 22.26a); in other words, elevated temperature promotes oxygen unloading. Active tissues are warmer than less active ones and thus extract more oxygen from the blood passing through them.

3. **The Bohr effect.** Active tissues also generate extra $CO_2$, which raises the $H^+$ concentration and lowers the pH of the blood. Hydrogen ions weaken the bond between hemoglobin and oxygen and thereby promote oxygen unloading—a phenomenon called the **Bohr**[28] **effect.** This can be seen in the oxyhemoglobin dissociation curve, where a drop in pH shifts the curve to the right (fig. 22.26b). The effect is less pronounced at the high $P_{O_2}$ present in the lungs, so pH has relatively little effect on pulmonary oxygen loading. In the systemic capillaries, however, $P_{O_2}$ is lower and the Bohr effect is more pronounced.

4. **BPG.** Erythrocytes have no mitochondria and meet their energy needs solely by anaerobic fermentation. One of their metabolic intermediates is **bisphosphoglycerate (BPG),** which binds to hemoglobin and promotes oxygen unloading. An elevated body temperature (as in fever) stimulates BPG synthesis, as do thyroxine, growth hormone, testosterone, and epinephrine. All of these hormones thus promote oxygen unloading to the tissues.

The rate of $CO_2$ loading is also adjusted to varying needs of the tissues. A low level of oxyhemoglobin ($HbO_2$) enables the blood to transport more $CO_2$, a phenomenon known as the **Haldane**[29] **effect.** It occurs for two reasons: (1) $HbO_2$ does not bind $CO_2$ as well as deoxyhemoglobin (HHb) does. (2) HHb binds more hydrogen ions than $HbO_2$ does, and by removing $H^+$ from solution, HHb shifts the carbonic acid reaction ($H_2O + CO_2 \rightarrow HCO_3^- + H^+$) to the right. A high metabolic rate keeps oxyhemoglobin levels relatively low and thus allows more $CO_2$ to be transported by these two mechanisms.

## Blood Gases and the Respiratory Rhythm

Normally the systemic arterial blood has a $P_{O_2}$ of 95 mm Hg, a $P_{CO_2}$ of 40 mm Hg, and a pH of $7.40 \pm 0.05$. The rate and depth of breathing are adjusted to maintain these values. This is possible only because the brainstem respiratory centers receive input from central and peripheral chemoreceptors that monitor the composition of the blood and CSF, as described on page 864. Of these three chemical stimuli, the most potent stimulus for breathing is pH, followed by $CO_2$; perhaps surprisingly, the least significant is $O_2$.

[28]Christian Bohr (1855–1911), Danish physiologist
[29]John Scott Haldane (1860–1936), Scottish physiologist

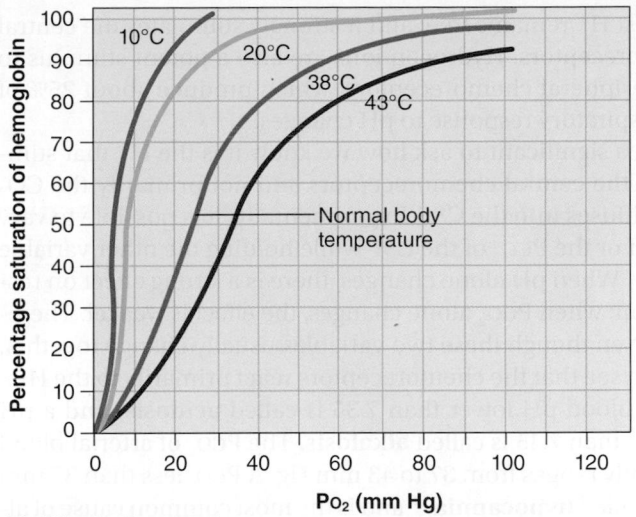

**(a) Effect of temperature**

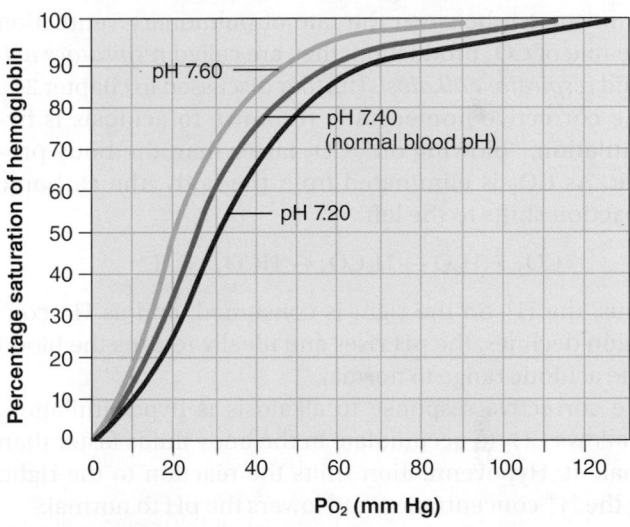

**(b) Effect of pH**

**FIGURE 22.26 Effects of Temperature and pH on Oxyhemoglobin Dissociation.** (a) For a given $P_{O_2}$, hemoglobin unloads more oxygen at higher temperatures. (b) For a given $P_{O_2}$, hemoglobin unloads more oxygen at lower pH (the Bohr effect). Both mechanisms cause hemoglobin to release more oxygen to tissues with higher metabolic rates.

❓ *Why is it physiologically beneficial to the body that the curves in part (a) shift to the right as temperature increases?*

## Hydrogen Ions

Ultimately, pulmonary ventilation is adjusted to maintain the pH of the brain. The central chemoreceptors in the medulla oblongata produce about 75% of the change in respiration induced by pH shifts, and yet $H^+$ does not cross the blood–brain barrier very easily. However, $CO_2$ does, and once it is in the CSF, it reacts with water to produce carbonic acid, and the carbonic acid dissociates into bicarbonate and hydrogen ions. The CSF contains relatively little protein to buffer the hydrogen ions, so most $H^+$ remains free, and it strongly stimulates the central chemoreceptors. Hydrogen ions are also a potent stimulus to the peripheral chemoreceptors, which produce about 25% of the respiratory response to pH changes.

It is significant to ask how we know it is the $H^+$ that stimulates the central chemoreceptors and not primarily the $CO_2$ that diffuses into the CSF. Experimentally, it is possible to vary the pH or the $P_{CO_2}$ of the CSF while holding the other variable steady. When pH alone changes, there is a strong effect on respiration; when $P_{CO_2}$ alone changes, the effect is weaker. Therefore, even though these two variables usually change together, we can see that the chemoreceptors react primarily to the $H^+$.

A blood pH lower than 7.35 is called **acidosis,** and a pH greater than 7.45 is called **alkalosis.** The $P_{CO_2}$ of arterial blood normally ranges from 37 to 43 mm Hg. A $P_{CO_2}$ less than 37 mm Hg is called **hypocapnia,**[30] and is the most common cause of alkalosis. The most common cause of acidosis is **hypercapnia,** a $P_{CO_2}$ greater than 43 mm Hg. When these pH imbalances result from a mismatch between the rate of pulmonary ventilation and the rate of $CO_2$ production, they are called *respiratory acidosis* and *respiratory alkalosis* (further discussed in chapter 24).

The corrective homeostatic response to acidosis is hyperventilation, "blowing off" $CO_2$ faster than the body produces it. As $CO_2$ is eliminated from the body, the carbonic acid reaction shifts to the left:

$$CO_2 + H_2O \leftarrow H_2CO_3 \leftarrow HCO_3^- + H^+$$

Thus, the $H^+$ on the right is consumed, and as $H^+$ concentration declines, the pH rises and ideally returns the blood from the acidotic range to normal.

The corrective response to alkalosis is hypoventilation, which allows $CO_2$ to accumulate in the body fluids faster than we exhale it. Hypoventilation shifts the reaction to the right, raising the $H^+$ concentration and lowers the pH to normal:

$$CO_2 + H_2O \rightarrow H_2CO_3 \rightarrow HCO_3^- + H^+$$

Although pH changes usually result from $P_{CO_2}$ changes, they can have other causes. In diabetes mellitus, for example, rapid fat oxidation releases acidic ketone bodies, causing an abnormally low pH called *ketoacidosis.* Ketoacidosis tends to induce a form of dyspnea called *Kussmaul respiration* (see table 22.3). Hyperventilation cannot reduce the level of ketone bodies in the blood, but by blowing off $CO_2$, it reduces the concentration of $CO_2$-generated $H^+$ and compensates to some degree for the $H^+$ released by the ketone bodies.

## Carbon Dioxide

Although the arterial $P_{CO_2}$ has a strong influence on respiration, we have seen that it is mostly an indirect one, mediated through its effects on the pH of the CSF. Yet the experimental evidence described earlier shows that $CO_2$ has some effect even when pH remains stable. At the beginning of exercise, the rising

blood $CO_2$ level may directly stimulate the peripheral chemoreceptors and trigger an increase in ventilation more quickly than the central chemoreceptors do.

## Oxygen

The partial pressure of oxygen usually has little effect on respiration. Even in eupnea, the hemoglobin is 97% saturated with $O_2$, so little can be added by increasing pulmonary ventilation. Arterial $P_{O_2}$ significantly affects respiration only if it drops below 60 mm Hg. At low elevations, such a low $P_{O_2}$ seldom occurs even in prolonged holding of the breath. A moderate drop in $P_{O_2}$ does stimulate the peripheral chemoreceptors, but another effect overrides this: As the level of $HbO_2$ falls, hemoglobin binds more $H^+$ (see fig. 22.24). This raises the blood pH, which inhibits respiration and counteracts the effect of low $P_{O_2}$.

At about 10,800 feet (3,300 m), $P_{O_2}$ falls to 60 mm Hg and the stimulatory effect of hypoxemia on the carotid bodies overrides the inhibitory effect of the pH increase. This produces heavy breathing in people who are not acclimated to high elevation. Long-term hypoxemia can lead to a condition called **hypoxic drive,** in which respiration is driven more by the low $P_{O_2}$ than by $CO_2$ or pH. This occurs in situations such as emphysema and pneumonia, which interfere with alveolar gas exchange, and in mountain climbing of at least 2 or 3 days' duration.

## Respiration and Exercise

It is common knowledge that we breathe more heavily during exercise, and it's tempting to think this occurs because exercise raises $CO_2$ levels, lowers the blood pH, and lowers blood $O_2$ levels. However, this is not true; all these values remain essentially the same in exercise as they do at rest. It appears that the increased respiration has other causes: (1) When the brain sends motor commands to the muscles (via the lower motor neurons of the spinal cord), it also sends this information to the respiratory centers, so they increase pulmonary ventilation in anticipation of the needs of the exercising muscles. (2) Exercise stimulates proprioceptors of the muscles and joints, and they transmit excitatory signals to the brainstem respiratory centers. Thus, the respiratory centers increase breathing because they are informed that the muscles have been told to move or are actually moving. The increase in pulmonary ventilation keeps blood gas values at their normal levels in spite of the elevated $O_2$ consumption and $CO_2$ generation by the muscles.

In summary, the main chemical stimulus to pulmonary ventilation is the $H^+$ in the CSF and tissue fluid of the brain. These hydrogen ions arise mainly from $CO_2$ diffusing into the CSF and brain and generating $H^+$ through the carbonic acid reaction. Therefore, the $P_{CO_2}$ of the arterial blood is an important driving force in respiration, even though its action on the chemoreceptors is indirect. Ventilation is adjusted to maintain arterial pH at about 7.40 and arterial $P_{CO_2}$ at about 40 mm Hg. This automatically ensures that the blood is at least 97% saturated with $O_2$ as well. Under ordinary circumstances, arterial

---

[30]*capn* = smoke

$Po_2$ has relatively little effect on respiration. When it drops below 60 mm Hg, however, it excites the peripheral chemoreceptors and stimulates an increase in ventilation. This can be significant at high elevations and in certain lung diseases. The increase in respiration during exercise results from the expected or actual activity of the muscles, not from any change in blood gas pressures or pH.

**BEFORE YOU GO ON**

Answer the following questions to test your understanding of the preceding section:

13. Why is the composition of alveolar air different from that of the atmosphere?

14. What four factors affect the efficiency of alveolar gas exchange?

15. Explain how perfusion of a pulmonary lobule changes if it is poorly ventilated.

16. How is most oxygen transported in the blood, and why does carbon monoxide interfere with this?

17. What are the three ways in which blood transports $CO_2$?

18. Give two reasons why highly active tissues can extract more oxygen from the blood than less active tissues do.

19. Define *hypocapnia* and *hypercapnia*. Name the pH imbalances that result from these conditions and explain the relationship between $Pco_2$ and pH.

20. What is the most potent chemical stimulus to respiration, and where are the most effective chemoreceptors for it located?

21. Explain how changes in pulmonary ventilation can correct pH imbalances.

## 22.4 Respiratory Disorders

### Expected Learning Outcomes

When you have completed this section, you should be able to

a. describe the forms and effects of oxygen deficiency and oxygen excess;

b. describe the chronic obstructive pulmonary diseases and their consequences; and

c. explain how lung cancer begins, progresses, and exerts its lethal effects.

The delicate lungs are exposed to a wide variety of inhaled pathogens and debris, so it is not surprising that they are prone to a host of diseases. Several already have been mentioned in this chapter and some others are briefly described in table 22.5. The effects of aging on the respiratory system are discussed on page 1121.

## Oxygen Imbalances

*Hypoxia* is a deficiency of oxygen in a tissue or the inability to use oxygen. It is not a respiratory disease in itself but is often a consequence of respiratory diseases. Hypoxia is classified according to cause:

- **Hypoxemic hypoxia,** a state of low arterial $Po_2$, is usually due to inadequate pulmonary gas exchange. Some of its root causes include atmospheric deficiency of oxygen at high elevations; impaired ventilation, as in drowning or aspiration of foreign matter; respiratory arrest; and the degenerative lung diseases discussed shortly. It also occurs in carbon monoxide poisoning, which prevents hemoglobin from transporting oxygen.

- **Ischemic hypoxia** results from inadequate circulation of the blood, as in congestive heart failure.

- **Anemic hypoxia** is due to anemia and the resulting inability of the blood to carry adequate oxygen.

- **Histotoxic hypoxia** occurs when a metabolic poison such as cyanide prevents the tissues from using the oxygen delivered to them.

Hypoxia is often marked by **cyanosis,** blueness of the skin. Whatever the cause, its primary effect is the necrosis of oxygen-starved tissues. This is especially critical in organs with the highest metabolic demands, such as the brain, heart, and kidneys.

An oxygen excess is also dangerous. It is safe to breathe 100% oxygen at 1 atm for a few hours, but **oxygen toxicity** rapidly develops when pure oxygen is breathed at 2.5 atm or greater. Excess oxygen generates hydrogen peroxide and free radicals that destroy enzymes and damage nervous tissue; thus, it can lead to seizures, coma, and death. This is why scuba divers breathe a mixture of oxygen and nitrogen rather than pure compressed oxygen (see Deeper Insight 22.5). Hyperbaric oxygen was formerly used to treat premature infants for respiratory distress syndrome, but it caused retinal deterioration and blinded many infants before the practice was discontinued.

## Chronic Obstructive Pulmonary Diseases

**Chronic obstructive pulmonary diseases (COPDs)** are defined by a long-term obstruction of airflow and substantial reduction of pulmonary ventilation. The major COPDs are *chronic bronchitis* and *emphysema*. COPDs are the fourth highest cause of adult mortality in the United States. They are almost always caused by cigarette smoking, but occasionally result from air pollution, occupational exposure to airborne irritants, or a hereditary defect. Most COPD patients exhibit mixed chronic bronchitis and emphysema, but one form or the other often predominates.

**Chronic bronchitis** is severe, persistent inflammation of the lower respiratory tract. Goblet cells of the bronchial mucosa enlarge and secrete excess mucus, while at the same time, the cilia are immobilized and unable to discharge it.

Thick, stagnant mucus accumulates in the lungs and furnishes a growth medium for bacteria. Furthermore, tobacco smoke incapacitates the alveolar macrophages and reduces one's defense against respiratory infection. Smokers with chronic bronchitis develop a chronic cough, bringing up sputum (SPEW-tum), a thick mixture of mucus and cellular debris. Since blood flowing through congested areas of the lung cannot load a normal amount of oxygen, the ventilation–perfusion ratio is reduced and such patients commonly exhibit hypoxemia and cyanosis.

In **emphysema**[31] (EM-fih-SEE-muh), alveolar walls break down and alveoli converge into fewer and larger spaces (see fig. 22.21c). Thus, there is much less respiratory membrane available for gas exchange. In severe cases, the lungs are flabby and cavitated with spaces as big as grapes or even ping-pong balls. The severity of the disease may not be fully appreciated by looking only at histological specimens since such large spaces are not seen on microscope slides. The lungs also become fibrotic and less elastic. The air passages open adequately during inspiration, but they tend to collapse and obstruct the outflow of air. Air becomes trapped in the lungs,

---

[31]*emphys* = inflamed

and over a period of time a person becomes barrel-chested. The overly stretched thoracic muscles contract weakly, which further contributes to the difficulty of expiration. Since proportionate amounts of alveolar wall and capillaries are both destroyed, the ventilation–perfusion ratio of the lung is relatively normal, and persons with emphysema do not necessarily show the cyanosis that typifies chronic bronchitis. People with emphysema can become exhausted and emaciated because they expend three to four times the normal amount of energy just to breathe. Even slight physical exertion, such as walking across a room, can cause severe shortness of breath.

### ►►► APPLY WHAT YOU KNOW

*Explain how the length–tension relationship of skeletal muscle (see p. 412) accounts for the weakness of the respiratory muscles in emphysema.*

COPD tends to reduce vital capacity and causes hypoxemia, hypercapnia, and respiratory acidosis. Hypoxemia stimulates the kidneys to secrete erythropoietin, which leads to accelerated erythrocyte production and polycythemia, as discussed in chapter 18. COPD also leads to **cor pulmonale**—hypertrophy and potential failure of the right heart due to obstruction of the pulmonary circulation (see chapter 19).

| TABLE 22.5 | Some Disorders of the Respiratory System |
|---|---|
| Acute rhinitis | The common cold. Caused by many types of viruses that infect the upper respiratory tract. Symptoms include congestion, increased nasal secretion, sneezing, and dry cough. Transmitted especially by contact of contaminated hands with mucous membranes; not transmitted orally. |
| Adult respiratory distress syndrome | Acute lung inflammation and alveolar injury stemming from trauma, infection, burns, aspiration of vomit, inhalation of noxious gases, drug overdoses, and other causes. Alveolar injury is accompanied by severe pulmonary edema and hemorrhaging, followed by fibrosis that progressively destroys lung tissue. Fatal in about 40% of cases under age 60 and in 60% of cases over age 65. |
| Pneumonia | A lower respiratory infection caused by any of several viruses, fungi, or protozoans, but most often the bacterium *Streptococcus pneumoniae*. Causes filling of alveoli with fluid and dead leukocytes and thickening of the respiratory membrane, which interferes with gas exchange and causes hypoxemia. Especially dangerous to infants, the elderly, and people with compromised immune systems, such as AIDS and leukemia patients. |
| Sleep apnea | Cessation of breathing for 10 seconds or longer during sleep; sometimes occurs hundreds of times per night, often accompanied by restlessness and alternating with snoring. Can result from altered function of CNS respiratory centers, airway obstruction, or both. Over time, may lead to daytime drowsiness, hypoxemia, polycythemia, pulmonary hypertension, congestive heart failure, and cardiac arrhythmia. Most common in obese people and in men. |
| Tuberculosis (TB) | Pulmonary infection with the bacterium *Mycobacterium tuberculosis*, which invades the lungs by way of air, blood, or lymph. Stimulates the lung to form fibrous nodules called tubercles around the bacteria. Progressive fibrosis compromises the elastic recoil and ventilation of the lungs. Especially common among impoverished and homeless people and becoming increasingly common among people with AIDS. |

*Disorders described elsewhere*

## Smoking and Lung Cancer

Lung cancer (fig. 22.27) accounts for more deaths than any other form of cancer. The most important cause of lung cancer is cigarette smoking, distantly followed by air pollution. Cigarette smoke contains at least 60 carcinogenic compounds. Lung cancer commonly follows or accompanies COPD.

There are three forms of lung cancer, the most common of which is **squamous-cell carcinoma.** In its early stage, basal cells of the bronchial epithelium multiply and the ciliated pseudostratified epithelium transforms into the stratified squamous type. As the dividing epithelial cells invade the underlying tissues of the bronchial wall, the bronchus develops bleeding lesions. Dense swirled masses of keratin appear in the lung parenchyma and replace functional respiratory tissue.

A second form of lung cancer, nearly as common, is **adenocarcinoma,**[32] which originates in the mucous glands of the lamina propria. The least common (10%–20% of malignancies) but most dangerous form is **small-cell (oat-cell) carcinoma,** named for clusters of cells that resemble oat grains. This originates in the main bronchi but invades the mediastinum and metastasizes quickly to other organs.

Over 90% of lung tumors originate in the mucous membranes of the large bronchi. As a tumor invades the bronchial wall and grows around it, it compresses the airway and may cause atelectasis (collapse) of more distal parts of the lung. Growth of the tumor produces a cough, but coughing is such an everyday occurrence among smokers it seldom causes much alarm. Often, the first sign of serious trouble is coughing up blood. Lung cancer metastasizes so rapidly that it has usually spread to other organs by the time it is diagnosed. Common sites of metastasis are the pericardium, heart, bones, liver, lymph nodes, and brain. The chance of recovery is poor, with only 7% of patients surviving for 5 years after diagnosis.

### BEFORE YOU GO ON

Answer the following questions to test your understanding of the preceding section.

22. Describe the four classes of hypoxia.

23. Name and compare the two COPDs and describe some pathological effects that they have in common.

24. In what lung tissue does lung cancer originate? How does it kill?

---

[32]*adeno* = gland; *carcino* = cancer; *oma* = tumor

**(a) Healthy lung, mediastinal surface**

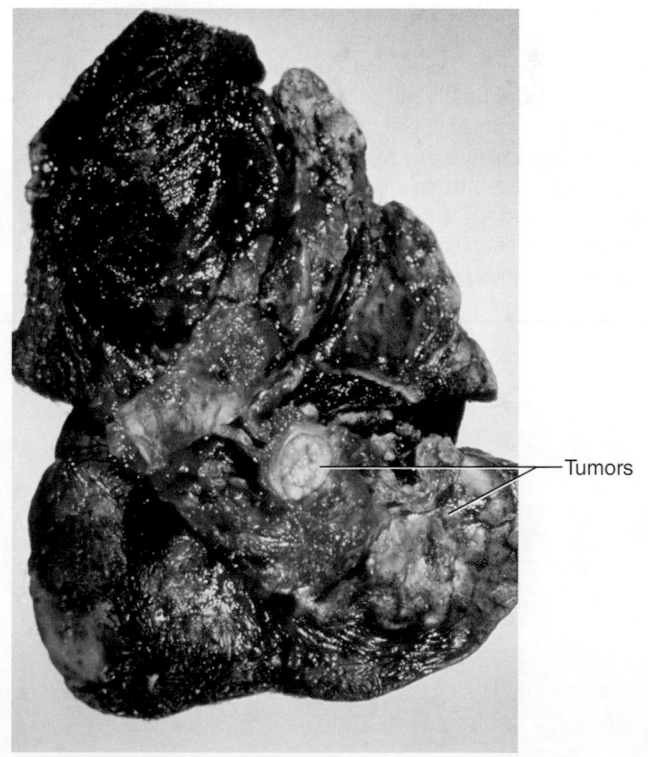

**(b) Smoker's lung with carcinoma**

Tumors

**FIGURE 22.27 Effect of Smoking.**

# DEEPER INSIGHT 22.5

## CLINICAL APPLICATION

### Diving Physiology and Decompression Sickness

Because of the popularity of scuba diving, many people know something about the scientific aspects of breathing under high pressure. But diving is by no means a new fascination. As early as the fifth century BCE, Aristotle described divers using snorkels and taking containers of air underwater in order to stay down longer. Some Renaissance artists depicted divers many meters deep breathing from tubes to the water surface. In reality, this would be physically impossible. For one thing, such tubes would have so much dead space that fresh air from the surface would not reach the diver. The short snorkels used today are about the maximum length that will work for surface breathing. Another reason snorkels cannot be used at greater depths is that water pressure increases by 1 atm for every 10 m of depth, and even at 1 m the pressure is so great that a diver cannot expand the chest muscles without help. This is one reason why scuba divers use pressurized air tanks. The tanks create a positive intrapulmonary pressure and enable the diver to inhale with only slight assistance from the thoracic muscles. Scuba tanks also have regulators that adjust the outflow pressure to the diver's depth and the opposing pressure of the surrounding water.

But breathing pressurized (hyperbaric) gas presents its own problems. Divers cannot use pure oxygen because of the problem of oxygen toxicity. Instead, they use compressed air—a mixture of 21% oxygen and 79% nitrogen. On land, nitrogen presents no physiological problems; it dissolves poorly in blood and it is physiologically inert. But under hyperbaric conditions, larger amounts of nitrogen dissolve in the blood. (Which of the gas laws applies here?) Even more dissolves in adipose tissue and the myelin of the brain, since nitrogen is more soluble in lipids. In the brain, it causes *nitrogen narcosis,* or what scuba inventor Jacques Cousteau (1910–1997) termed "rapture of the deep." A diver can become dizzy, euphoric, and dangerously disoriented; for every 15 to 20 m of depth, the effect is said to be equivalent to that of one martini on an empty stomach.

Strong currents, equipment failure, and other hazards sometimes make scuba divers panic, hold their breath, and quickly swim to the surface (a *breath-hold ascent*). Ambient (surrounding) pressure falls rapidly as a diver ascends, and the air in the lungs expands just as rapidly. (Which gas law is demonstrated here?) It is imperative that an ascending diver keep his or her airway open to exhale the expanding gas; otherwise it is likely to cause *pulmonary barotrauma*—ruptured alveoli. Then, when the diver takes a breath of air at the surface, alveolar air goes directly into the bloodstream and causes air embolism. After passing through the heart, the emboli enter the cerebral circulation because the diver is head-up and air bubbles rise in liquid. The resulting cerebral embolism can cause motor and sensory dysfunction, seizures, unconsciousness, and drowning.

Barotrauma can be fatal even at the depths of a backyard swimming pool. In one case, children trapped air in a bucket 1 m underwater and then swam under the bucket to breathe from the air space. Because the bucket was under water, the air in it was compressed. One child filled his lungs under the bucket, did a "mere" 1 m breath-hold ascent, and his alveoli ruptured. He died in the hospital, partly because the case was mistaken for drowning and not treated for what it really was. This would not have happened to a person who inhaled at the surface, did a breath-hold dive, and then resurfaced—nor is barotrauma a problem for those who do breath-hold dives to several meters. (Why? What is the difference?)

Even when not holding the breath, but letting the expanding air escape from the mouth, a diver must ascend slowly and carefully to allow for decompression of the nitrogen that has dissolved in the tissues. *Decompression tables* prescribe safe rates of ascent based on the depth and the length of time a diver has been down. When pressure drops, nitrogen dissolved in the tissues can go either of two places—it can diffuse into the alveoli and be exhaled, or it can form bubbles like the $CO_2$ in a bottle of soda when the cap is removed. The diver's objective is to ascend slowly, allowing for the former and preventing the latter. If a diver ascends too rapidly, nitrogen "boils" from the tissues—especially in the 3 m just below the surface, where the relative pressure change is greatest. A diver may double over in pain from bubbles in the joints, bones, and muscles—a disease called the *bends* or *decompression sickness (DCS)*. Nitrogen bubbles in the pulmonary capillaries cause *chokes*—substernal pain, coughing, and dyspnea. DCS is sometimes accompanied by mood changes, seizures, numbness, and itching. These symptoms usually occur within an hour of surfacing, but they are sometimes delayed for up to 36 hours. DCS is treated by putting the individual in a hyperbaric chamber to be recompressed and then *slowly* decompressed.

DCS is also called *caisson disease.* A caisson is a watertight underwater chamber filled with pressurized air. Caissons are used in underwater construction work on bridges, tunnels, ships' hulls, and so forth. Caisson disease was first reported in the late 1800s among workmen building the foundations of the Brooklyn Bridge.

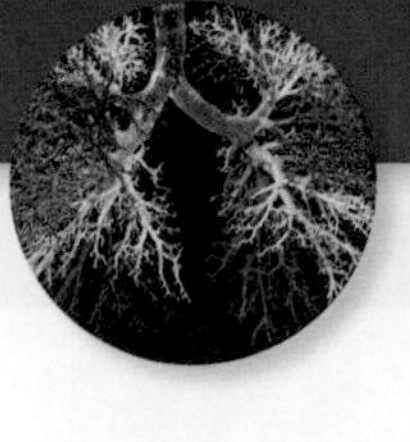

# CONNECTIVE ISSUES

## Effects of the **RESPIRATORY SYSTEM** On Other Organ Systems

**ALL SYSTEMS**
Delivers oxygen to the tissues and removes their carbon dioxide; maintains proper acid–base balance in the tissues

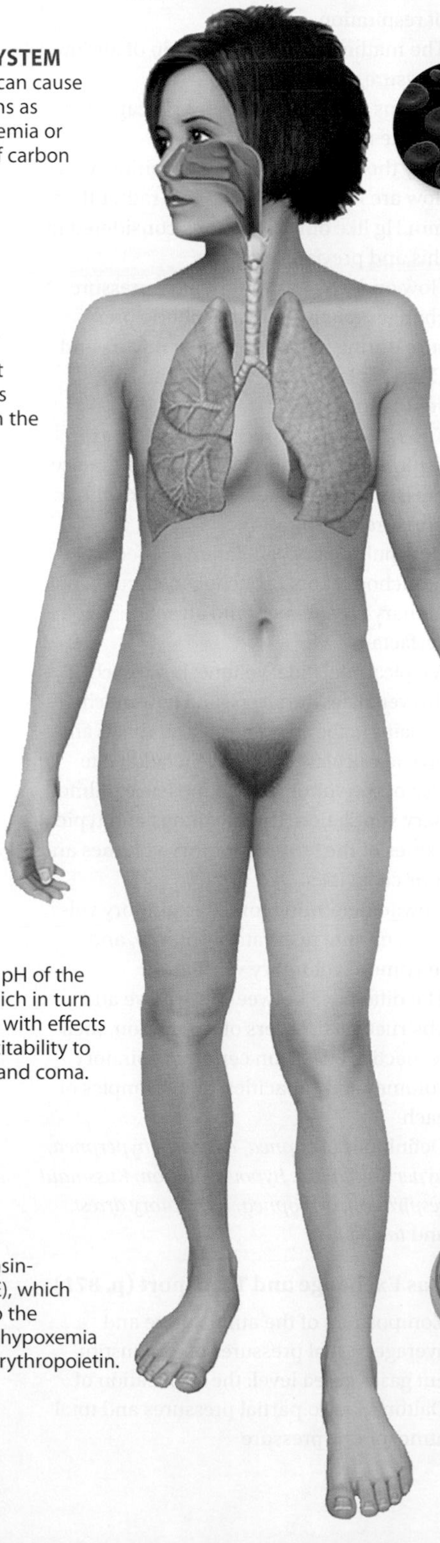

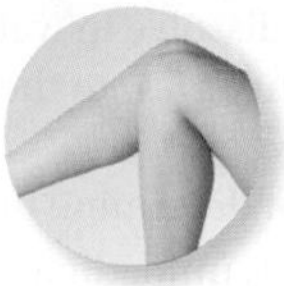

**INTEGUMENTARY SYSTEM**
Respiratory disorders can cause such skin discolorations as the cyanosis of hypoxemia or the cherry-red color of carbon monoxide poisoning.

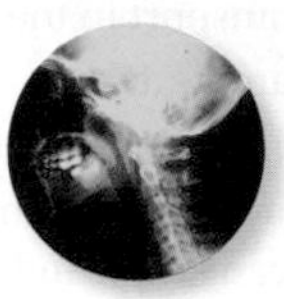

**SKELETAL SYSTEM**
Any respiratory disorder that causes hypoxemia stimulates accelerated erythropoiesis in the red bone marrow.

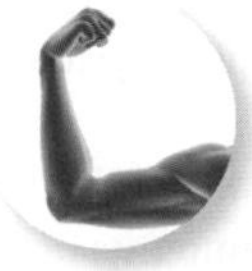

**MUSCULAR SYSTEM**
Acid–base imbalances of respiratory origin can affect neuromuscular function.

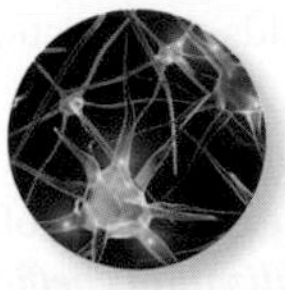

**NERVOUS SYSTEM**
Respiration affects the pH of the cerebrospinal fluid, which in turn affects neural function with effects ranging from hyperexcitability to depressed excitability and coma.

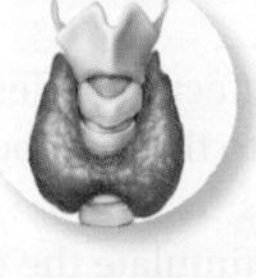

**ENDOCRINE SYSTEM**
Lungs produce angiotensin-converting enzyme (ACE), which converts angiotensin I to the hormone angiotensin II; hypoxemia stimulates secretion of erythropoietin.

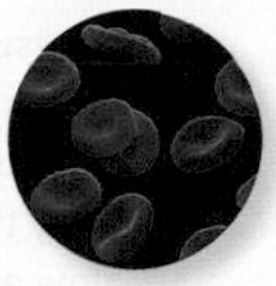

**CIRCULATORY SYSTEM**
Thoracic pump aids venous return of blood; proplatelets break up into platelets in the lungs; angiotensin II, produced in the lungs, stimulates vasoconstriction and helps regulate blood volume and pressure; respiration strongly influences blood pH; obstruction of pulmonary circulation can lead to right-sided heart failure; lungs filter blood clots and emboli and prevent them from obstructing vital arteries elsewhere.

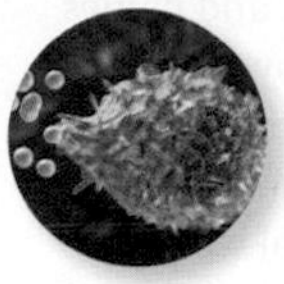

**LYMPHATIC/ IMMUNE SYSTEM**
Thoracic pump promotes lymph flow and its return to the bloodstream.

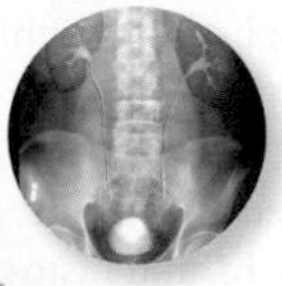

**URINARY SYSTEM**
Valsalva maneuver aids in urination; urinary and respiratory systems collaborate in acid–base balance and compensate for each other's deficiencies in maintaining normal pH; hypoxemia stimulates kidneys to secrete erythropoietin.

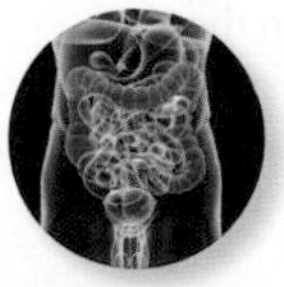

**DIGESTIVE SYSTEM**
Valsalva maneuver aids in defecation.

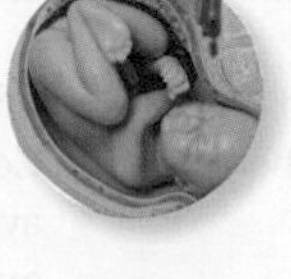

**REPRODUCTIVE SYSTEM**
Valsalva maneuver aids in childbirth.

# STUDY GUIDE

## ▶ Assess Your Learning Outcomes

*To test your knowledge, discuss the following topics with a study partner or in writing, ideally from memory.*

### 22.1 Anatomy of the Respiratory System (p. 849)

1. Two meanings of the word *respiration*
2. Functions of the respiratory system
3. The distinction between the respiratory and conducting divisions of this system, and constituents of each division
4. The distinction between the upper and lower respiratory tract and the dividing line between them
5. The extent of the nasal cavity, names of its anterior and posterior openings, and names of the two chambers separated by the nasal septum, and histology of its mucosa
6. Names and functions of the scroll-like folds that arise from each lateral wall of the nasal cavity
7. Anatomy and functions of the pharynx, larynx, and trachea
8. Gross anatomy of the lungs; how the right and left lungs differ; and the structures that enter or leave through the hilum
9. Divisions of the bronchial tree from main bronchus to segmental bronchi, and histological changes along the way
10. How bronchioles differ from bronchi; two types of bronchioles; and how the two differ in histology and function
11. Alveolar ducts and alveoli; cell types of the alveoli and their functions; the relationship of pulmonary blood vessels to the alveoli; and the structure of the respiratory membrane in relation to its function
12. The parietal and visceral pleurae, pleural cavity, and pleural fluid

### 22.2 Pulmonary Ventilation (p. 860)

1. The two phases of the respiratory cycle
2. Actions of the respiratory muscles; the prime mover and synergists of respiration
3. Locations and functions of the brainstem respiratory centers; their connections with each other and with other levels of the CNS; and routes of CNS output to the respiratory muscles
4. Locations and roles of the central and peripheral chemoreceptors, stretch receptors, and irritant receptors in modulating the respiratory rhythm
5. The neural pathway for voluntary control of respiration
6. The mathematical relationship of airflow, pressure, and resistance
7. Actions of the sternum and rib cage during the respiratory cycle
8. Why the pressures driving respiratory airflow are measured in cm $H_2O$ rather than mm Hg like other pressures considered in this and previous chapters
9. How and why intrapulmonary pressure changes relative to atmospheric pressure during inspiration; how Boyle's and Charles's laws relate to pulmonary ventilation
10. The role of elastic recoil of the thorax in expiration; how and why intrapulmonary pressure changes relative to atmospheric pressure in expiration
11. How pulmonary ventilation is affected by bronchodilation, bronchoconstriction, pulmonary compliance, and alveolar surfactant
12. A typical adult tidal volume; how much of this ventilates the alveoli and how much remains in the anatomical dead space; and how to calculate alveolar ventilation rate
13. Use of the spirometer to measure pulmonary ventilation; the meanings and typical values of the four respiratory volumes and four capacities
14. How to determine forced expiratory volume, minute respiratory volume, and maximum voluntary ventilation
15. The difference between restrictive and obstructive disorders of respiration, their respective effects on certain respiratory volumes and capacities, and examples of each
16. Definitions of *eupnea, dyspnea, hyperpnea, hyperventilation, hypoventilation, Kussmaul respiration, orthopnea, respiratory arrest,* and *tachypnea*

### 22.3 Gas Exchange and Transport (p. 871)

1. Composition of the atmosphere and average partial pressures of its constituent gases at sea level; the application of Dalton's law to partial pressures and total atmospheric pressure
2. Differences between the composition of atmospheric air and alveolar air, and reasons for the differences
3. Why gas exchange depends on the ability of the gases to dissolve in water; the application of Henry's law to the air–water interface in the alveoli
4. Four variables that determine the rate of $O_2$ loading and $CO_2$ unloading by blood passing through the alveolar capillaries
5. How ventilation–perfusion coupling matches pulmonary airflow to blood flow for optimal gas exchange
6. The two modes of $O_2$ transport in the blood and the relative amounts of $O_2$ transported by each; where $O_2$ binds to the hemoglobin molecule; how much $O_2$ a hemoglobin molecule can carry; and what hemoglobin is called when $O_2$ is bound to it
7. Interpretation of the oxyhemoglobin dissociation curve, including the reason for its shape and how it can be used to show the amount of $O_2$ unloading as hemoglobin passes through a typical systemic tissue
8. Three modes of $CO_2$ transport in the blood and the relative amounts of $CO_2$ transported by each; where $CO_2$ binds to hemoglobin; and what hemoglobin is called when $CO_2$ is bound to it
9. How carbonic anhydrase (CAH) and the chloride shift aid in the loading of $CO_2$ from the tissue fluids; the reaction catalyzed by CAH
10. How the loading of $CO_2$ in systemic tissues influences the unloading of $O_2$; the meaning of the *utilization coefficient* and a typical resting value
11. How the loading of $O_2$ in the lungs influences the unloading of $CO_2$
12. Four mechanisms that adjust the amount of $O_2$ unloaded by hemoglobin to the needs of individual tissues
13. How the Haldane effect modifies $CO_2$ loading in relation to the metabolic rates of individual tissues
14. Three factors that stimulate the central and peripheral chemoreceptors, and their relative influences on breathing

# STUDY GUIDE

15. The normal pH range of the blood; terms for deviations above and below this range; terms for the $CO_2$ imbalances that cause these pH deviations; and how the body homeostatically regulates blood pH

16. The mechanism by which exercise increases respiration

**22.4 Respiratory Disorders (p. 881)**

1. The definition of *hypoxia;* its four varieties and the cause of each; and the consequences of uncorrected hypoxia

2. The mechanism and effects of oxygen toxicity

3. The names, most common cause, and pathology of the two chronic obstructive pulmonary diseases (COPDs)

4. The most common cause of lung cancer, and the names and pathological differences between the three forms of lung cancer

## ▶ Testing Your Recall

*Answers in Appendix B*

1. The nasal cavity is divided by the nasal septum into right and left
   a. nares.
   b. vestibules.
   c. fossae.
   d. choanae.
   e. conchae.

2. The intrinsic laryngeal muscles regulate speech by rotating
   a. the extrinsic laryngeal muscles.
   b. the corniculate cartilages.
   c. the arytenoid cartilages.
   d. the hyoid bone.
   e. the vocal cords.

3. The largest air passages that engage in gas exchange with the blood are
   a. the respiratory bronchioles.
   b. the terminal bronchioles.
   c. the primary bronchi.
   d. the alveolar ducts.
   e. the alveoli.

4. Respiratory arrest would most likely result from a tumor of the
   a. pons.
   b. midbrain.
   c. thalamus.
   d. cerebellum.
   e. medulla oblongata.

5. Which of these values would normally be the highest?
   a. tidal volume
   b. inspiratory reserve volume
   c. expiratory reserve volume
   d. residual volume
   e. vital capacity

6. The _____ protects the lungs from injury by excessive inspiration.
   a. pleura
   b. rib cage
   c. inflation reflex
   d. Haldane effect
   e. Bohr effect

7. According to _____, the warming of air as it is inhaled helps to inflate the lungs.
   a. Boyle's law
   b. Charles's law
   c. Dalton's law
   d. the Bohr effect
   e. the Haldane effect

8. Poor blood circulation causes _____ hypoxia.
   a. ischemic
   b. histotoxic
   c. hemolytic
   d. anemic
   e. hypoxemic

9. Most of the $CO_2$ that diffuses from the blood into an alveolus comes from
   a. dissolved gas.
   b. carbaminohemoglobin.
   c. carboxyhemoglobin.
   d. carbonic acid.
   e. expired air.

10. The duration of an inspiration is set by
    a. the pneumotaxic center.
    b. the phrenic nerves.
    c. the vagus nerves.
    d. the I neurons.
    e. the E neurons.

11. The superior opening into the larynx is guarded by a tissue flap called the _____.

12. Within each lung, the airway forms a branching complex called the _____.

13. The great alveolar cells secrete a phospholipid–protein mixture called _____.

14. Intrapulmonary pressure must be lower than _____ pressure for inspiration to occur.

15. _____ disorders reduce the speed of airflow through the airway.

16. Some inhaled air does not participate in gas exchange because it fills the _____ of the respiratory tract.

17. Inspiration depends on the ease of pulmonary inflation, called _____, whereas expiration depends on _____, which causes pulmonary recoil.

18. Inspiration is caused by the firing of I neurons in the _____ of the medulla oblongata.

19. The matching of airflow to blood flow in any region of the lung is called _____.

20. A blood pH > 7.45 is called _____ and can be caused by a $CO_2$ deficiency called _____.

# STUDY GUIDE

## ▶ Building Your Medical Vocabulary

*Answers in Appendix B*

*State a medical meaning of each word element below, and give a term in which it or a slight variation of it is used.*

1. atel-

2. capni-

3. carcino-

4. corni-

5. eu-

6. -meter

7. naso-

8. -pnea

9. spiro-

10. thyro-

## ▶ True or False

*Answers in Appendix B*

*Determine which five of the following statements are false, and briefly explain why.*

1. The phrenic nerves fire during inspiration only.

2. The lungs contain more respiratory bronchioles than terminal bronchioles.

3. In alveolar capillaries, oncotic pressure is greater than the mean blood pressure.

4. If you increase the volume of a given quantity of gas, its pressure increases.

5. Pneumothorax is the only cause of atelectasis.

6. Obstruction of the bronchial tree results in a reduced FEV.

7. At a given $P_{O_2}$ and pH, hemoglobin carries less oxygen at warmer temperatures than it does at cooler temperatures.

8. Most of the air one inhales never makes it to the alveoli.

9. The greater the $P_{CO_2}$ of the blood is, the lower its pH is.

10. Most of the $CO_2$ transported by the blood is in the form of dissolved gas.

## ▶ Testing Your Comprehension

*Answers at www.mhhe.com/saladin7*

1. Discuss how the different functions of the conducting division and respiratory division relate to differences in their histology.

2. State whether hyperventilation would raise or lower each of the following—the blood $P_{O_2}$, $P_{CO_2}$, and pH—and explain why. Do the same for emphysema.

3. Some competitive swimmers hyperventilate before a race, thinking they can "load up extra oxygen" and hold their breaths longer underwater. While they can indeed hold their breaths longer, it is not for the reason they think. Furthermore, some have lost consciousness and drowned because of this practice. What is wrong with this thinking, and what accounts for the loss of consciousness?

4. Consider a man in good health with a 650 mL tidal volume and a respiratory rate of 11 breaths per minute. Report his minute respiratory volume in liters per minute. Assuming his anatomical dead space is 185 mL, calculate his alveolar ventilation rate in liters per minute.

5. An 83-year-old woman is admitted to the hospital, where a critical care nurse attempts to insert a nasoenteric tube ("stomach tube") for feeding. The patient begins to exhibit dyspnea, and a chest X-ray reveals air in the right pleural cavity and a collapsed right lung. The patient dies 5 days later from respiratory complications. Name the conditions revealed by the X-ray and explain how they could have resulted from the nurse's procedure.

## ▶ Improve Your Grade at www.mhhe.com/saladin7

*Download animations to study when it fits your schedule. Practice quizzes, labeling activities, games, and flashcards offer fun ways to master the chapter concepts. Or, download image PowerPoint files for each chapter to create a study guide or for taking notes during lecture.*

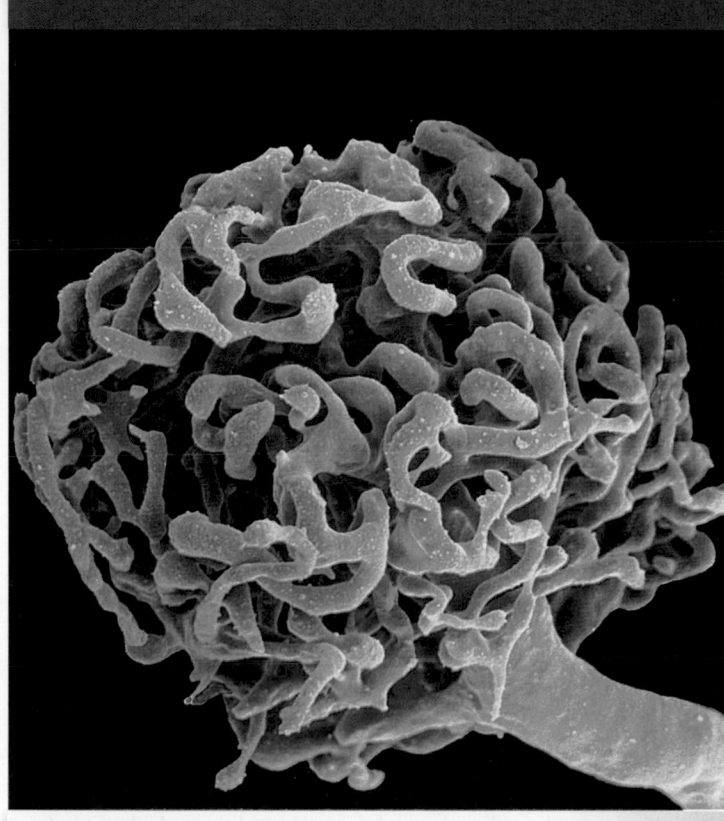

The renal glomerulus, a mass of capillaries where the kidney filters the blood (SEM of a resin cast)

# THE URINARY SYSTEM

Anatomy & Physiology | REVEALED
aprevealed.com

**Modue 13: Urinary System**

- Many aspects of kidney physiology are governed by the principle of flow down pressure, osmotic, and electrochemical gradients (p. 17).

- You must be familiar with the forces of capillary fluid exchange (p. 760) to understand glomerular filtration, the first step in urine production.

- The concepts of osmosis, tonicity, and osmolarity (p. 91) are central to understanding the stages of urine production and the role of the kidneys in regulating fluid balance and urine concentration.

- Refresh your memory of carrier-mediated transport, especially symports and antiports (p. 94), in order to best understand how the kidney processes a blood plasma filtrate to produce urine.

To live is to metabolize, and metabolism unavoidably produces a variety of waste products that are not merely unneeded by the body, but indeed toxic if allowed to accumulate. We rid the body of some of these wastes through the respiratory and digestive tracts and the sweat glands, but the urinary system is the principal means of waste excretion. The kidneys are glands that separate metabolic wastes from the blood. The rest of the urinary system serves only for the transport, storage, and elimination of urine. Most of our focus in this chapter is therefore on the kidneys.

Their task goes far beyond waste excretion. As we will see, the kidneys also play indispensable roles in regulating blood volume, pressure, and composition. In performing these tasks, they have a very close physiological relationship with the endocrine, circulatory, and respiratory systems, covered in recent chapters.

Anatomically, the urinary system is closely associated with the reproductive system. In many animals, the eggs and sperm are emitted through the urinary tract, and the two systems have a shared embryonic development and adult anatomical relationship. This is reflected in humans, where the systems develop together in the embryo and, in the male, the urethra continues to serve as a passage for both urine and sperm. Thus, the urinary and reproductive systems are often collectively called the *urogenital (U–G) system,* and *urologists* treat both urinary and reproductive disorders. We examine the anatomical relationship between the urinary and reproductive systems in chapter 27, but the physiological link to the circulatory and respiratory systems is more important to consider at this time.

## 23.1   Functions of the Urinary System

### Expected Learning Outcomes

When you have completed this section, you should be able to

a. name and locate the organs of the urinary system;

b. list several functions of the kidneys in addition to urine formation;

c. name the major nitrogenous wastes and identify their sources; and

d. define *excretion* and identify the systems that excrete wastes.

The **urinary system** consists of six principal organs: two **kidneys,** two **ureters,** the **urinary bladder,** and the **urethra.** Figure 23.1 shows these organs in anterior and posterior views. The urinary tract has important spatial relationships with the vagina and uterus in females and the prostate gland in males. These relationships are best appreciated from the sagittal views of figures 27.10 and 28.1 (pp. 1039 and 1060) and can be seen in the cadaver in figure B.14 (p. 388).

## Functions of the Kidneys  AP|R

Although the primary role of the kidneys is excretion, they play more roles than are commonly realized:

- They filter the blood plasma and excrete the toxic metabolic wastes.

- They regulate blood volume, pressure, and osmolarity by regulating water output.

- They regulate the electrolyte and acid–base balance of the body fluids.

- They secrete the hormone *erythropoietin,* which stimulates the production of red blood cells and thus supports the oxygen-carrying capacity of the blood.

- They help to regulate calcium homeostasis and bone metabolism by participating in the synthesis of calcitriol.

- They clear hormones and drugs from the blood and thereby limit their action.

- They detoxify free radicals.

- In conditions of extreme starvation, they help to support the blood glucose level by synthesizing glucose from amino acids.

In view of such diverse roles, it is easy to see why renal failure can lead to the collapse of many other physiological functions as well.

## Nitrogenous Wastes

A **waste** is any substance that is useless to the body or present in excess of the body's needs. A **metabolic waste,** more specifically, is a waste substance produced by the body. The food residue in feces, for example, is a waste but not a metabolic waste, since it was not produced by the body and, indeed, never entered the body's tissues.

Among the most toxic of our metabolic wastes are small nitrogen-containing compounds called **nitrogenous wastes** (fig. 23.2). About 50% of the nitrogenous waste is urea, a byproduct of protein catabolism. Proteins are hydrolyzed to amino acids, and then the $-NH_2$ group is removed from each

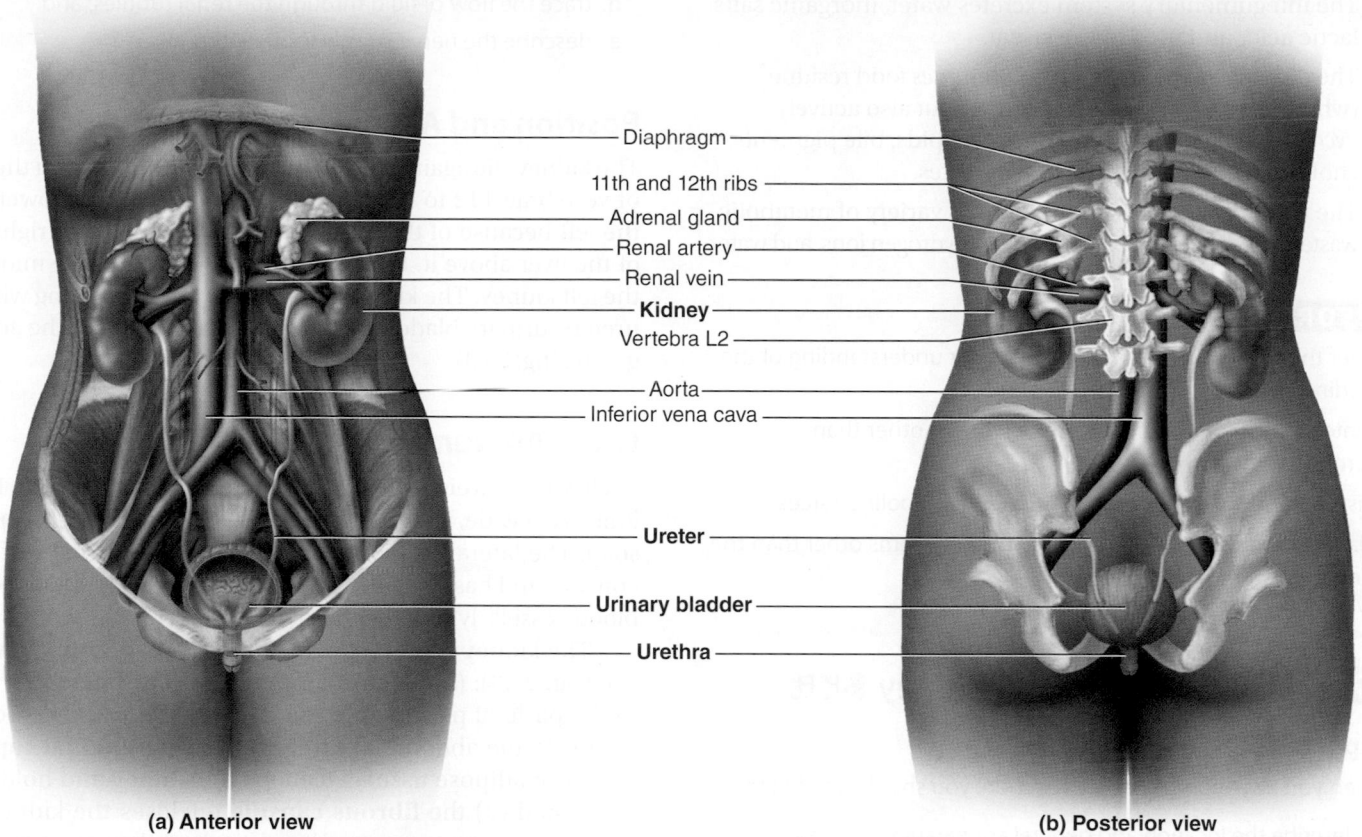

Diaphragm

11th and 12th ribs

Adrenal gland

Renal artery

Renal vein

**Kidney**

Vertebra L2

Aorta

Inferior vena cava

**Ureter**

**Urinary bladder**

**Urethra**

**(a) Anterior view**

**(b) Posterior view**

**FIGURE 23.1 The Urinary System.** Organs of the urinary system are indicated in boldface.   AP|R

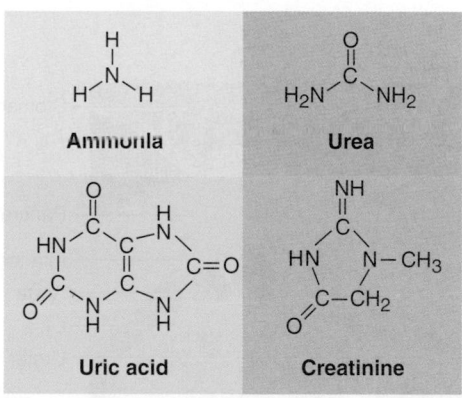

Ammonia

Urea

Uric acid

Creatinine

**FIGURE 23.2 The Major Nitrogenous Wastes.**

❓ *How is each of these wastes produced in the body?*

amino acid. The $-NH_2$ forms ammonia, which is exceedingly toxic but which the liver quickly converts to urea, $CO(NH_2)_2$, a somewhat less toxic waste:

$$2\ NH_3 + CO_2 \rightarrow H_2N\overset{\overset{\displaystyle O}{\|}}{-C-}NH_2 + H_2O$$

Other nitrogenous wastes in the urine include **uric acid** and **creatinine** (cree-AT-ih-neen), produced by the catabolism of nucleic acids and creatine phosphate, respectively. Although less toxic than ammonia and less abundant than urea, these too are potentially harmful.

The level of nitrogenous waste in the blood is typically expressed as **blood urea nitrogen (BUN).** The normal concentration of blood urea is 10 to 20 mg/dL. An elevated BUN is called **azotemia**[1] (AZ-oh-TEE-me-uh) and may indicate renal insufficiency. It can progress to **uremia** (you-REE-me-uh), a syndrome of diarrhea, vomiting, dyspnea, and cardiac arrhythmia stemming from the toxicity of the nitrogenous wastes. Convulsions, coma, and death can follow within a few days. Deeper Insight 23.5 discusses treatments for renal insufficiency.

## Excretion

**Excretion** is the process of separating wastes from the body fluids and eliminating them from the body. It is carried out by four organ systems:

1. The respiratory system excretes carbon dioxide, small amounts of other gases, and water.

---

[1] *azot* = nitrogen; *emia* = blood condition

2. The integumentary system excretes water, inorganic salts, lactic acid, and urea in the sweat.

3. The digestive system not only *eliminates* food residue (which is not a process of excretion) but also actively *excretes* water, salts, carbon dioxide, lipids, bile pigments, cholesterol, and other metabolic wastes.

4. The urinary system excretes a broad variety of metabolic wastes, toxins, drugs, hormones, salts, hydrogen ions, and water.

### BEFORE YOU GO ON

Answer the following questions to test your understanding of the preceding section:

1. State at least four functions of the kidneys other than forming urine.

2. List four nitrogenous wastes and their metabolic sources.

3. Name some wastes eliminated by three systems other than the urinary system.

## 23.2   Anatomy of the Kidney  AP|R

### Expected Learning Outcomes

When you have completed this section, you should be able to

a. describe the location and general appearance of the kidney;

b. identify the external and internal features of the kidney;

c. trace the flow of blood through the kidney;

d. trace the flow of fluid through the renal tubules; and

e. describe the nerve supply to the kidney.

### Position and Associated Structures

The kidneys lie against the posterior abdominal wall at the level of vertebrae T12 to L3. The right kidney is slightly lower than the left because of the space occupied by the large right lobe of the liver above it. Rib 12 crosses the approximate middle of the left kidney. The kidneys are retroperitoneal, along with the ureters, urinary bladder, renal artery and vein, and the adrenal glands (fig. 23.3).

### Gross Anatomy

Each kidney weighs about 150 g and measures about 11 cm long, 6 cm wide, and 3 cm thick—about the size of a bar of bath soap. The lateral surface is convex, and the medial surface is concave and has a slit, the **hilum,** that admits the renal nerves, blood vessels, lymphatics, and ureter.

The kidney is protected by three layers of connective tissue (fig. 23.3): (1) A fibrous **renal fascia,** immediately deep to the parietal peritoneum, binds the kidney and associated organs to the abdominal wall; (2) the **perirenal fat capsule,** a layer of adipose tissue, cushions the kidney and holds it in place; and (3) the **fibrous capsule** encloses the kidney like a cellophane wrapper anchored at the hilum, and protects it from trauma and infection. The kidneys are suspended by collagen fibers that extend from the fibrous capsule, through

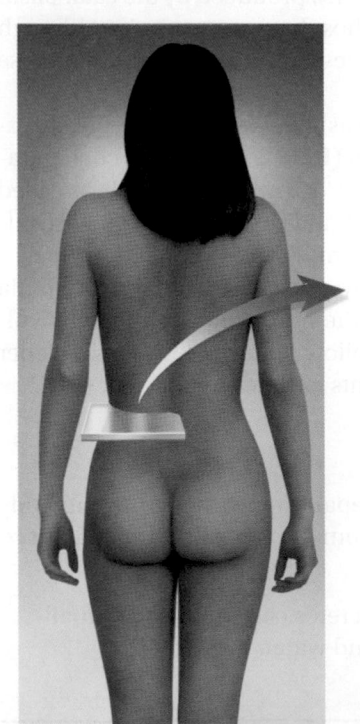
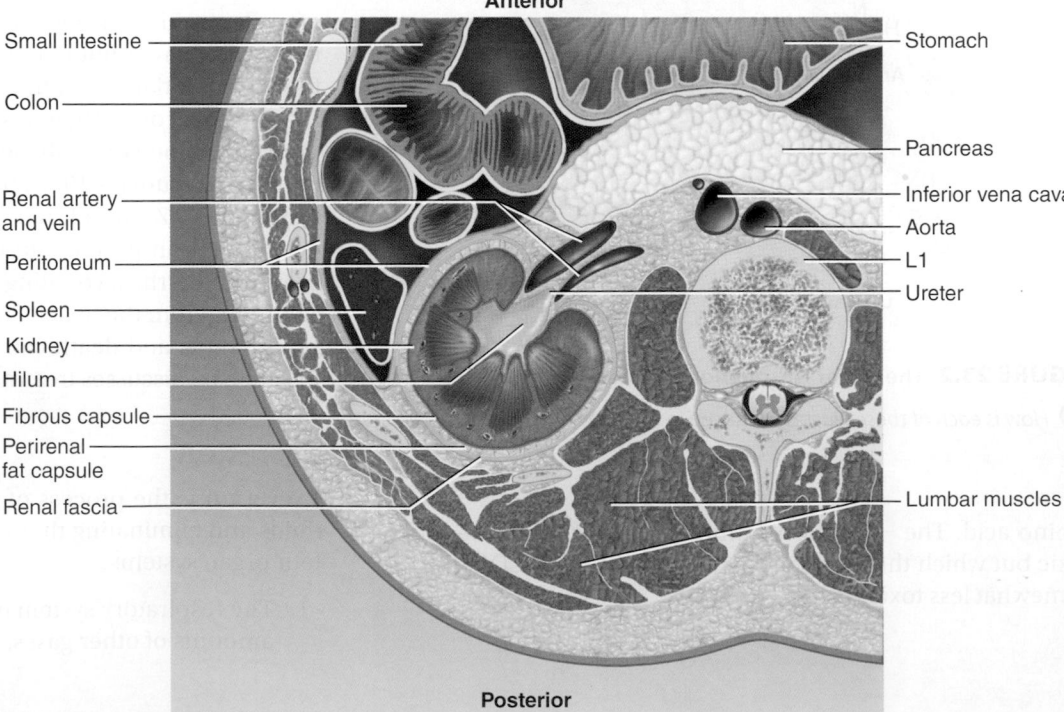

**FIGURE 23.3  Retroperitoneal Position of the Kidney.** Cross section of the abdomen at the level of vertebra L1.

❓ *If the kidney were not retroperitoneal, where on this figure would you have to relocate it?*

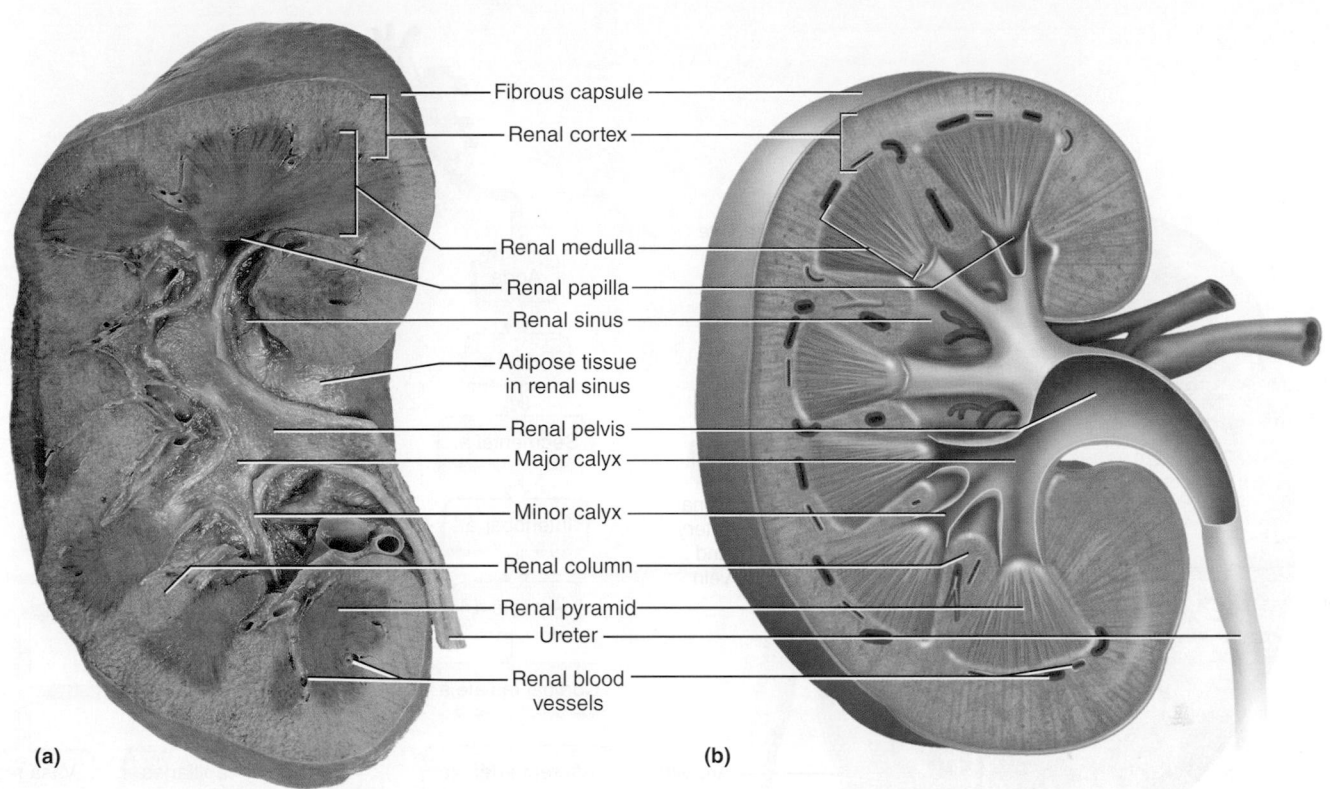

**FIGURE 23.4  Gross Anatomy of the Kidney.**  Posterior views. (a) Photograph of frontal section. (b) Major anatomical features.  AP|R

the fat, to the renal fascia. The renal fascia is fused with the peritoneum anteriorly and with the fascia of the lumbar muscles posteriorly. In spite of all this, the kidneys drop about 3 cm when one goes from lying down to standing, as when getting out of bed in the morning. Under some circumstances, they become detached and drift even lower, with pathological results (see nephroptosis, or "floating kidney," in table 23.3, p. 918).

The renal parenchyma—the glandular tissue that forms the urine—appears C-shaped in frontal section (fig. 23.4). It encircles a medial cavity, the **renal sinus,** occupied by blood and lymphatic vessels, nerves, and urine-collecting structures. Adipose tissue fills the remaining space in the sinus and holds these structures in place.

The parenchyma is divided into two zones: an outer **renal cortex** about 1 cm thick and an inner **renal medulla** facing the sinus. Extensions of the cortex called **renal columns** project toward the sinus and divide the medulla into 6 to 10 **renal pyramids.** Each pyramid is conical, with a broad base facing the cortex and a blunt point called the **renal papilla** facing the sinus. One pyramid and the overlying cortex constitute one *lobe* of the kidney.

The papilla of each renal pyramid is nestled in a cup called a **minor calyx**[2] (CAY-lix), which collects its urine. Two or three minor calyces (CAY-lih-seez) converge to form a **major calyx,** and two or three major calyces converge in the sinus to form the

funnel-like **renal pelvis.**[3] The ureter is a tubular continuation of the renal pelvis that drains the urine down to the urinary bladder.

## Renal Circulation

The kidneys account for only 0.4% of the body weight, but receive about 1.2 liters of blood per minute, or 21% of the cardiac output (the *renal fraction*)—more for the purpose of waste removal than to meet the metabolic demands of the kidney tissue. This is a hint of how important the kidneys are in regulating blood volume and composition.

The larger divisions of the renal circulation are shown in figure 23.5. Each kidney is supplied by a **renal artery** arising from the aorta. Just before or after entering the hilum, the renal artery divides into a few **segmental arteries,** and each of these further divides into a few **interlobar arteries.** An interlobar artery penetrates each renal column and travels between the pyramids toward the *corticomedullary junction,* the boundary between the cortex and medulla. Along the way, it branches again to form **arcuate arteries,** which make a sharp 90° bend and travel along the base of the pyramid. Each arcuate artery gives rise to several **cortical radiate arteries,** which pass upward into the cortex.

The finer branches of the renal circulation are shown in figure 23.6. As a cortical radiate artery ascends through the cortex, a series of **afferent arterioles** arise from it at

---

[2] *calyx* = cup

[3] *pelvis* = basin

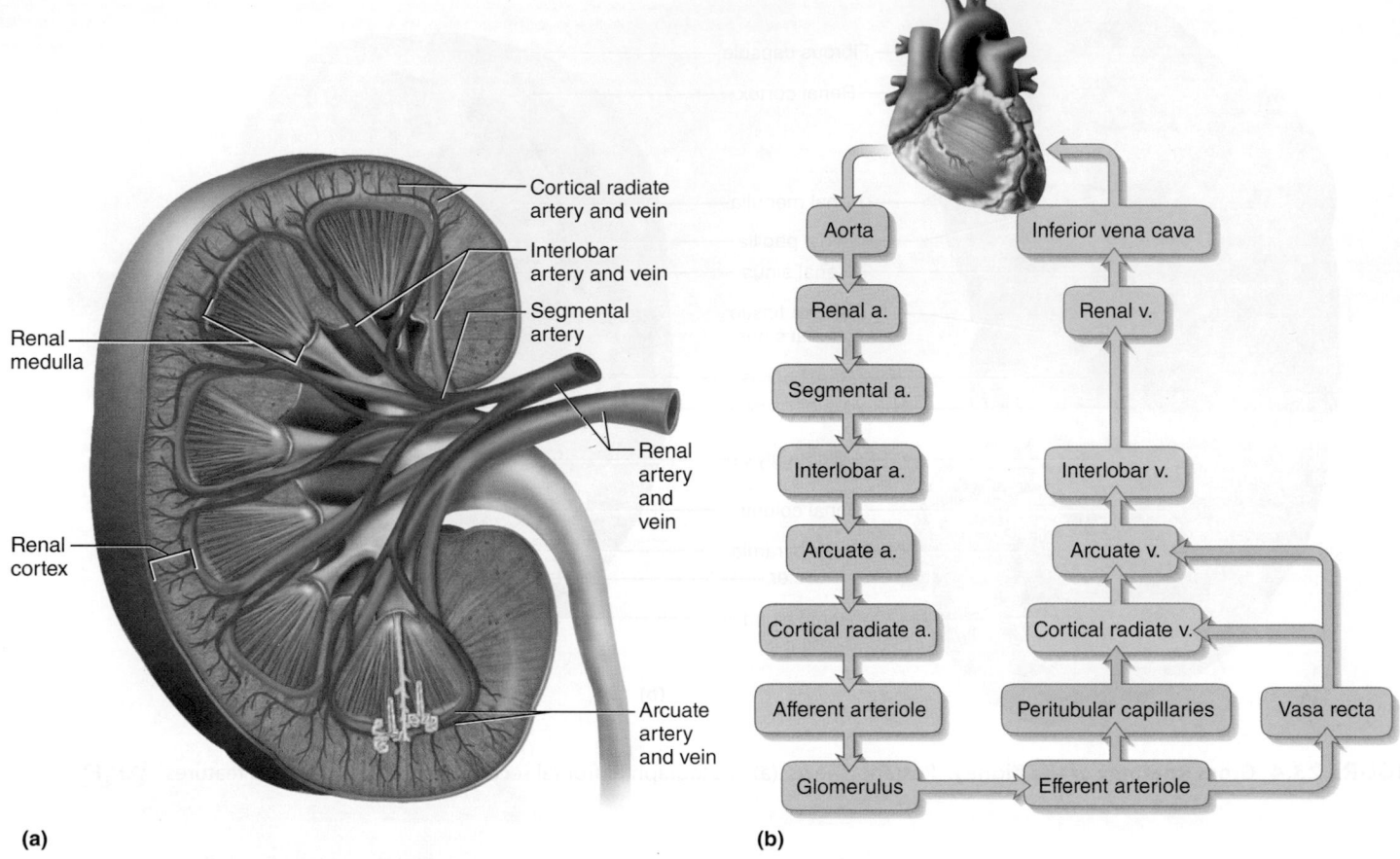

**(a)**

**(b)**

**FIGURE 23.5  Renal Circulation.**  (a) The larger blood vessels of the kidney. (b) Flowchart of renal circulation. The pathway through the vasa recta (instead of peritubular capillaries) applies only to the juxtamedullary nephrons.

nearly right angles like limbs arising from the trunk of a pine tree. Each afferent arteriole supplies one functional unit of the kidney called a **nephron**[4] (NEF-ron). The afferent arteriole leads to a ball of capillaries called a **glomerulus**[5] (glo-MERR-you-lus), enclosed in a sphere called the *glomerular capsule.* Blood leaves the glomerulus by way of an **efferent arteriole.**

The efferent arteriole usually leads to a plexus of **peritubular capillaries,** named for the fact that they form a network around another part of the nephron, the *renal tubule.* The renal tubule reabsorbs most of the water and solutes that filtered out of the blood at the glomerulus and returns these to the bloodstream by way of these peritubular capillaries. The peritubular capillaries carry it away to the **cortical radiate veins, arcuate veins, interlobar veins,** and the **renal vein,** in that order. These veins travel parallel to the arteries of the same names. (There are, however, no segmental veins corresponding to the segmental arteries.) The renal vein leaves the hilum and drains into the inferior vena cava.

The renal medulla receives only 1% to 2% of the total renal blood flow, supplied by a network of vessels called the **vasa recta**[6] (VAH-za REC-ta). These arise from nephrons in the deep cortex close to the medulla. Here, the efferent arterioles descend immediately into the medulla and give rise to the vasa recta instead of peritubular capillaries. Capillaries of the vasa recta lead into venules that ascend and empty into the arcuate and cortical radiate veins. Capillaries of the vasa recta are wedged into the tight spaces between the medullary parts of the renal tubule, and carry away water and solutes reabsorbed by those sections of the tubule. Figure 23.5b summarizes the route of renal blood flow.

#### ▶▶▶ APPLY WHAT YOU KNOW

*Can you identify a portal system in the renal circulation?*

### The Nephron  AP|R

Each kidney has about 1.2 million nephrons. To understand how just one of these works is to understand nearly everything

---

[4] *nephro* = kidney
[5] *glomer* = ball; *ulus* = little

[6] *vasa* = vessels; *recta* = straight

Cortical nephron

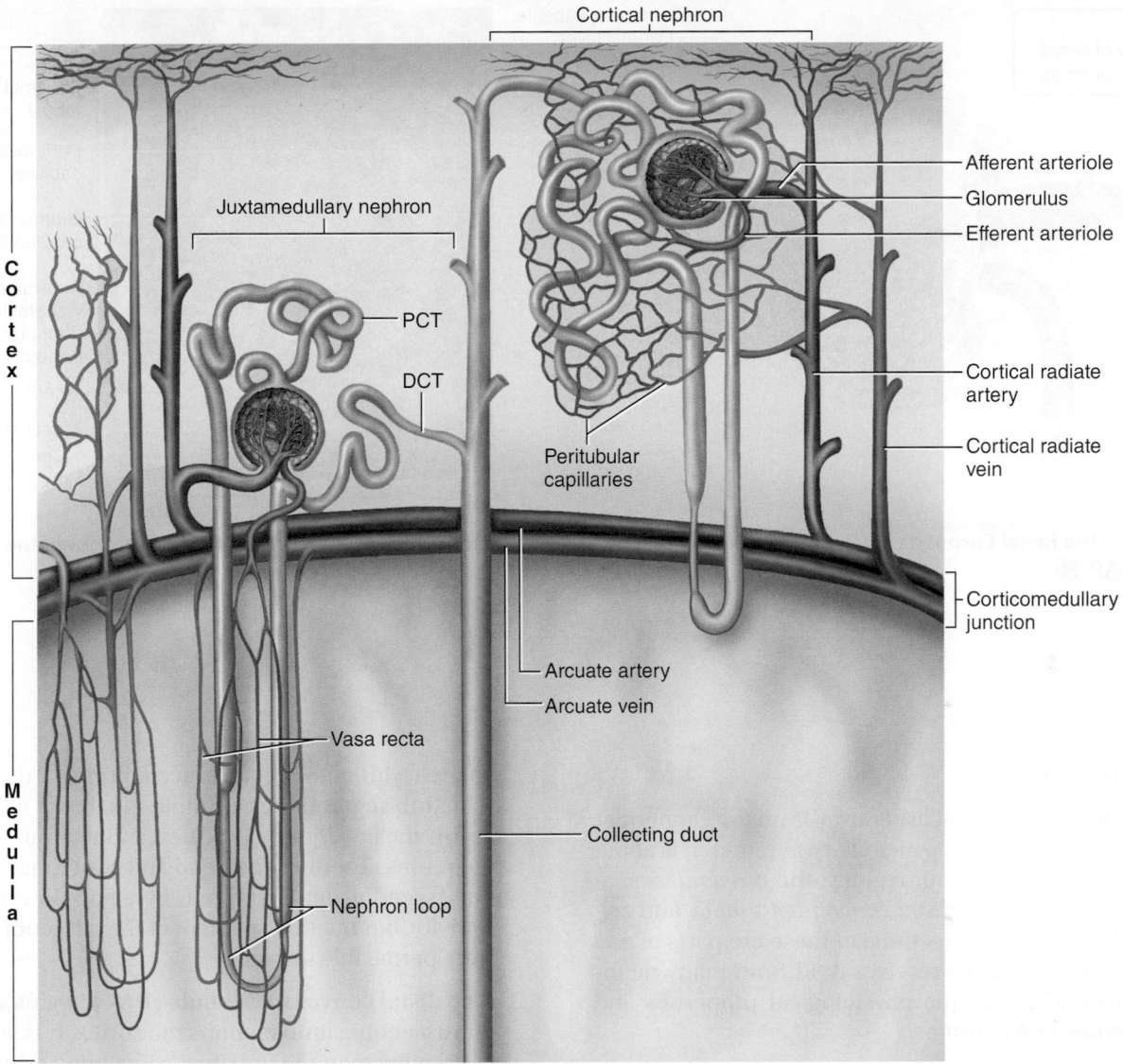

Juxtamedullary nephron

Cortex

PCT

DCT

Peritubular capillaries

Medulla

Vasa recta

Nephron loop

Afferent arteriole

Glomerulus

Efferent arteriole

Cortical radiate artery

Cortical radiate vein

Corticomedullary junction

Arcuate artery

Arcuate vein

Collecting duct

**FIGURE 23.6 Microcirculation of the Kidney.** For clarity, vasa recta are shown only on the left and peritubular capillaries only on the right. In the juxtamedullary nephron (left), the efferent arteriole gives rise to the vasa recta of the medulla. In the cortical nephron (right), the nephron loop barely dips into the renal medulla and the efferent arteriole gives rise to peritubular capillaries (DCT = distal convoluted tubule; PCT = proximal convoluted tubule).

about how the whole kidney works. Each nephron is composed of two principal parts: a *renal corpuscle,* which filters the blood plasma, and a long coiled *renal tubule,* which converts the filtrate to urine.

## The Renal Corpuscle

The **renal corpuscle** (fig. 23.7) consists of the glomerulus described earlier and a two-layered **glomerular capsule** that encloses it. The parietal (outer) layer is a simple squamous epithelium, and the visceral (inner) layer consists of elaborate cells called **podocytes**[7] wrapped around the capillaries of the

glomerulus. The two layers are separated by a filtrate-collecting **capsular space.** In tissue sections, this space appears as an empty circular or C-shaped space around the glomerulus.

Opposite sides of the renal corpuscle are called the vascular and urinary poles. At the **vascular pole,** the afferent arteriole enters the capsule, bringing blood to the glomerulus, and the efferent arteriole leaves the capsule and carries blood away. The afferent arteriole is significantly larger than the efferent arteriole. Thus, the glomerulus has a large inlet and a small outlet—a point whose functional significance will become apparent later. At the **urinary pole,** the parietal wall of the capsule turns away from the corpuscle and gives rise to the renal tubule. The simple squamous epithelium of the capsule becomes simple cuboidal in the tubule.

[7] *podo* = foot; *cyte* = cell

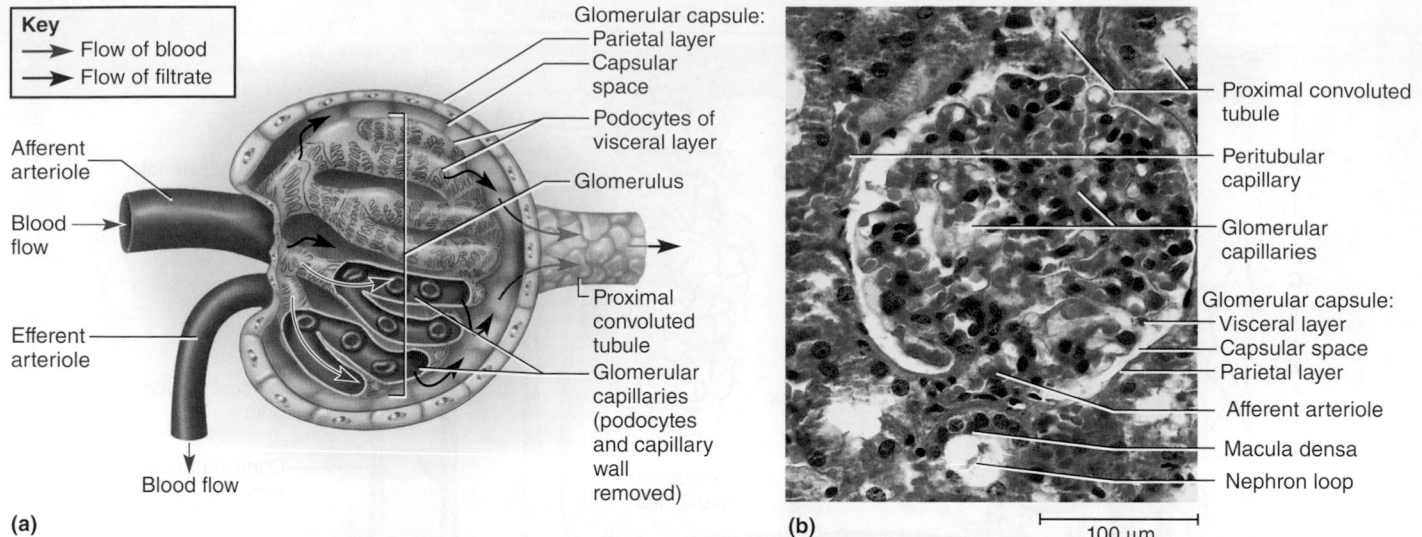

**FIGURE 23.7** **The Renal Corpuscle.** (a) Anatomy of the corpuscle. (b) Light micrograph of the renal corpuscle and sections of the surrounding renal tubule. AP|R

## The Renal Tubule

The **renal tubule** is a duct that leads away from the glomerular capsule and ends at the tip of a medullary pyramid. It is about 3 cm long and divided into four regions: the *proximal convoluted tubule, nephron loop, distal convoluted tubule,* and *collecting duct* (fig. 23.8). The first three of these are parts of one nephron; the collecting duct receives fluid from many nephrons. Each region has unique physiological properties and roles in the production of urine.

1. The **proximal convoluted tubule (PCT)** arises from the glomerular capsule. It is the longest and most coiled of the four regions and therefore dominates histological sections of renal cortex. It has a simple cuboidal epithelium with prominent microvilli (a brush border), which attests to the great deal of absorption that occurs here. The microvilli give the epithelium a distinctive shaggy look.

2. The **nephron loop** is a long U-shaped portion of the renal tubule found mostly in the medulla. It begins where the PCT straightens out and dips toward or into the medulla, forming the **descending limb** of the loop. At its deep end, the loop turns 180° and forms the **ascending limb,** which returns to the cortex, traveling parallel and close to the descending limb. The loop is divided into thick and thin segments. The **thick segments** have a simple cuboidal epithelium. They form the initial part of the descending limb and part or all of the ascending limb. The cells here are heavily engaged in active transport of salts, so they have very high metabolic activity and are

loaded with mitochondria, accounting for their thickness. The **thin segment** has a simple squamous epithelium. It forms the lower part of the descending limb, and in some nephrons, it rounds the bend and continues partway up the ascending limb. The cells here have low metabolic activity, but the thin segment of the descending limb is very permeable to water.

3. The **distal convoluted tubule (DCT)** begins shortly after the ascending limb reenters the cortex. It is shorter and less coiled than the proximal convoluted tubule, so fewer sections of it are seen in histological sections. It has a cuboidal epithelium with smooth-surfaced cells nearly devoid of microvilli. The DCT is the end of the nephron.

4. The **collecting duct** receives fluid from the DCTs of several nephrons as it passes back into the medulla. Numerous collecting ducts converge toward the tip of a medullary pyramid, and near the papilla, they merge to form a larger **papillary duct.** About 30 papillary ducts end in pores at the conical tip of each papilla. Urine drains from these pores into the minor calyx that encloses the papilla. The collecting and papillary ducts are lined with simple cuboidal epithelium.

The flow of fluid from the point where the glomerular filtrate is formed to the point where urine leaves the body is: glomerular capsule → proximal convoluted tubule → nephron loop → distal convoluted tubule → collecting duct → papillary duct → minor calyx → major calyx → renal pelvis → ureter → urinary bladder → urethra.

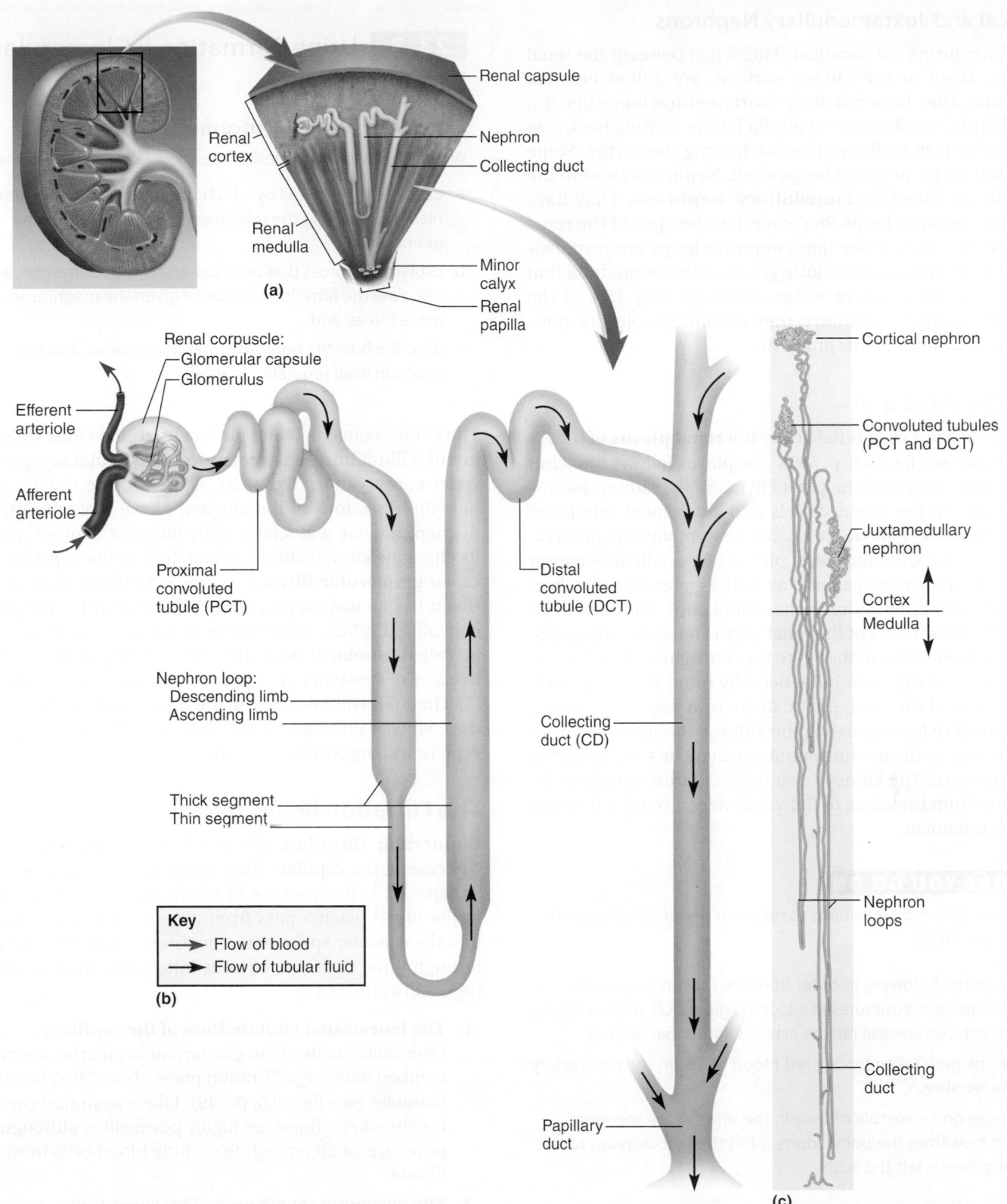

- Renal capsule
- Renal cortex
- Nephron
- Collecting duct
- Renal medulla
- Minor calyx
- Renal papilla

**(a)**

Renal corpuscle:
- Glomerular capsule
- Glomerulus

- Efferent arteriole
- Afferent arteriole

- Proximal convoluted tubule (PCT)

Nephron loop:
- Descending limb
- Ascending limb

- Thick segment
- Thin segment

**Key**
→ Flow of blood
→ Flow of tubular fluid

**(b)**

- Distal convoluted tubule (DCT)

- Collecting duct (CD)

- Papillary duct

- Cortical nephron
- Convoluted tubules (PCT and DCT)
- Juxtamedullary nephron
- Cortex
- Medulla
- Nephron loops
- Collecting duct

**(c)**

**FIGURE 23.8 Microscopic Anatomy of the Nephron.** (a) Location of the nephrons in one wedge-shaped lobe of the kidney. (b) Structure of a nephron. For clarity, the nephron is stretched out to separate the convoluted tubules. The nephron loop is greatly shortened for the purpose of illustration. (c) The true proportions of the nephron loops relative to the convoluted tubules. Three nephrons are shown. Their proximal and distal convoluted tubules are commingled in a single tangled mass in each nephron. Note the extreme lengths of the nephron loops. **AP|R**

## Cortical and Juxtamedullary Nephrons

Not all nephrons are identical. Those just beneath the renal capsule, close to the kidney surface, are called **cortical nephrons.** They have relatively short nephron loops that dip only slightly into the outer medulla before turning back (see fig. 23.6) or turn back even before leaving the cortex. Some of them have no nephron loops at all. Nephrons close to the medulla are called **juxtamedullary**[8] **nephrons.** They have very long nephron loops that extend to the apex of the renal pyramid. As you will see later, nephron loops are responsible for maintaining an osmotic gradient in the medulla that helps the body conserve water. Although only 15% of the nephrons are juxtamedullary, they are almost solely responsible for maintaining this gradient.

## Renal Innervation

Wrapped around each renal artery is a **renal plexus** of nerves and ganglia (see fig. 15.6, p. 565). The plexus follows branches of the renal artery into the parenchyma of the kidney, issuing nerve fibers to the blood vessels and convoluted tubules of the nephrons. The renal plexus carries sympathetic innervation from the abdominal aortic plexus (especially its superior mesenteric and celiac ganglia) as well as afferent pain fibers from the kidneys en route to the spinal cord. Stimulation by the sympathetic fibers of the renal plexus tends to reduce glomerular blood flow and therefore the rate of urine production, although these rates are influenced by other factors as well. Another role of the sympathetic fibers is to respond to falling blood pressure by stimulating the kidneys to secrete *renin,* an enzyme that activates hormonal mechanisms for restoring blood pressure. The kidneys also receive parasympathetic innervation from branches of the vagus nerve, but the function of this is unknown.

### BEFORE YOU GO ON

Answer the following questions to test your understanding of the preceding section:

4. Arrange the following in order from the most numerous to the least numerous structures in a kidney: glomeruli, major calyces, minor calyces, cortical radiate arteries, interlobar arteries.

5. Trace the path taken by one red blood cell from the renal artery to the renal vein.

6. Consider one molecule of urea in the urine. Trace the route that it took from the point where it left the bloodstream to the point where it left the body.

---

[8]*juxta* = next to

### 23.3 Urine Formation I: Glomerular Filtration AP|R

#### Expected Learning Outcomes
When you have completed this section, you should be able to

a. describe the process by which the kidney filters the blood plasma, including the relevant cellular structure of the glomerulus;

b. Explain the forces that promote and oppose filtration, and calculate the filtration pressure if given the magnitude of these forces; and

c. describe how the nervous system, hormones, and the nephron itself regulate filtration.

The kidney converts blood plasma to urine in four stages: glomerular filtration, tubular reabsorption, tubular secretion, and water conservation (fig. 23.9). These are the themes for the next three sections of the chapter. As we trace fluid through the nephron, we will refer to it by different names that reflect its changing composition: (1) The fluid in the capsular space, called **glomerular filtrate,** is similar to blood plasma except that it has almost no protein. (2) The fluid from the proximal convoluted tubule through the distal convoluted tubule will be called **tubular fluid.** It differs from the glomerular filtrate because of substances removed and added by the tubule cells. (3) The fluid will be called **urine** once it enters the collecting duct, since it undergoes little alteration beyond that point except for a change in water content.

### The Filtration Membrane

**Glomerular filtration,** discussed in this section, is a special case of the capillary fluid exchange process described in chapter 20. It is a process in which water and some solutes in the blood plasma pass from capillaries of the glomerulus into the capsular space of the nephron. To do so, fluid passes through three barriers that constitute a **filtration membrane** (figs. 23.10 and 23.11):

1. **The fenestrated endothelium of the capillary.** Endothelial cells of the glomerular capillaries are honeycombed with large filtration pores about 70 to 90 nm in diameter (see fig. 20.5, p. 749). Like fenestrated capillaries elsewhere, these are highly permeable, although their pores are small enough to exclude blood cells from the filtrate.

2. **The basement membrane.** This consists of a proteoglycan gel. Passing large molecules through it would be like trying to grind sand through a kitchen sponge. A few particles may penetrate its small spaces, but most would be held back. On the basis of size alone, the basement membrane excludes molecules larger than

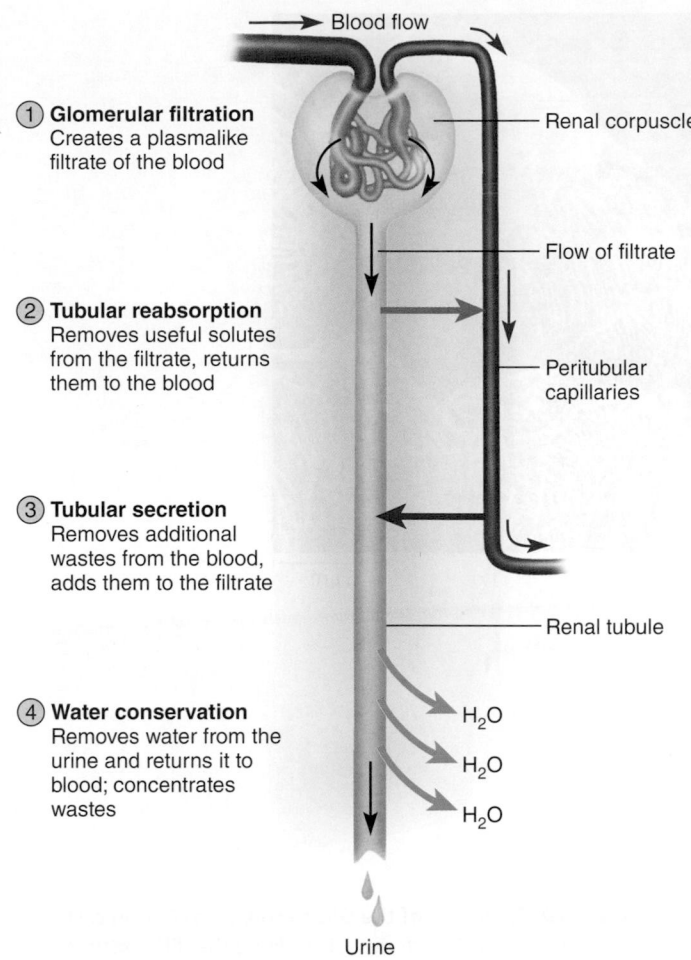

**① Glomerular filtration**
Creates a plasmalike
filtrate of the blood

Blood flow

Renal corpuscle

Flow of filtrate

**② Tubular reabsorption**
Removes useful solutes
from the filtrate, returns
them to the blood

Peritubular
capillaries

**③ Tubular secretion**
Removes additional
wastes from the blood,
adds them to the filtrate

Renal tubule

**④ Water conservation**
Removes water from the
urine and returns it to
blood; concentrates
wastes

$H_2O$

$H_2O$

$H_2O$

Urine

**FIGURE 23.9 Basic Stages of Urine Formation.**

8 nm. Even some smaller molecules, however, are held back by a negative charge on the proteoglycans. Blood albumin is slightly smaller than 7 nm, but it is also negatively charged and thus repelled by the basement membrane. Although the blood plasma is 7% protein, the glomerular filtrate is only 0.03% protein. It has traces of albumin and smaller polypeptides, including some hormones.

3. **Filtration slits.** A podocyte of the glomerular capsule is shaped somewhat like an octopus, with a bulbous cell body and several thick arms. Each arm has numerous little extensions called **foot processes (pedicels[9])** that wrap around the capillaries and interdigitate with each other, like wrapping your hands around a pipe and lacing your fingers together. The foot processes have negatively charged **filtration slits** about 30 nm wide between them, which are an additional obstacle to large anions.

Almost any molecule smaller than 3 nm can pass freely through the filtration membrane into the capsular space. This includes water, electrolytes, glucose, fatty acids, amino acids, nitrogenous wastes, and vitamins. Such solutes have about the same concentration in the glomerular filtrate as in the blood plasma. Some solutes of low molecular weight are retained in the bloodstream, however, because they are bound to plasma proteins that cannot get through the membrane. For example, most calcium, iron, and thyroid hormone in the blood are bound to plasma proteins that retard their filtration by the kidneys. The small fraction that is unbound, however, passes freely through the membrane and appears in the urine.

Kidney infections and trauma can damage the filtration membrane and allow albumin or blood cells to filter through. Kidney disease is sometimes marked by the presence of protein (especially albumin) or blood in the urine—conditions called **proteinuria (albuminuria)** and **hematuria,**[10] respectively. Distance runners and swimmers often experience temporary proteinuria and hematuria. Strenuous exercise greatly reduces perfusion of the kidneys, and the glomerulus deteriorates under the prolonged hypoxia, thus leaking protein and sometimes blood into the filtrate.

## Filtration Pressure

Glomerular filtration follows the same principles that govern filtration in other blood capillaries (see pp. 760–761), but there are significant differences in the magnitude of the forces involved:

- The blood hydrostatic pressure (BHP) is much higher here than elsewhere—about 60 mm Hg compared with 10 to 15 mm Hg in most other capillaries. This results from the fact that the afferent arteriole is substantially larger than the efferent arteriole, giving the glomerulus a large inlet and small outlet (fig. 23.10a).

- The hydrostatic pressure in the capsular space is about 18 mm Hg, compared with the slightly negative interstitial pressures elsewhere. This results from the high rate of filtration and continual accumulation of fluid in the capsule.

- The colloid osmotic pressure (COP) of the blood is about the same here as anywhere else, 32 mm Hg.

- The glomerular filtrate is almost protein-free and has no significant COP. (This can change markedly in kidney diseases that allow protein to filter into the capsular space.)

On balance, then, we have a high outward pressure of 60 mm Hg, opposed by two inward pressures of 18 and 32 mm Hg (fig. 23.12), giving a **net filtration pressure (NFP)** of

$$60_{out} - 18_{in} - 32_{in} = 10 \text{ mm Hg}_{out}$$

---

[9] *pedi* = foot; *cel* = little

[10] *hemat* = blood; *ur* = urine; *ia* = condition

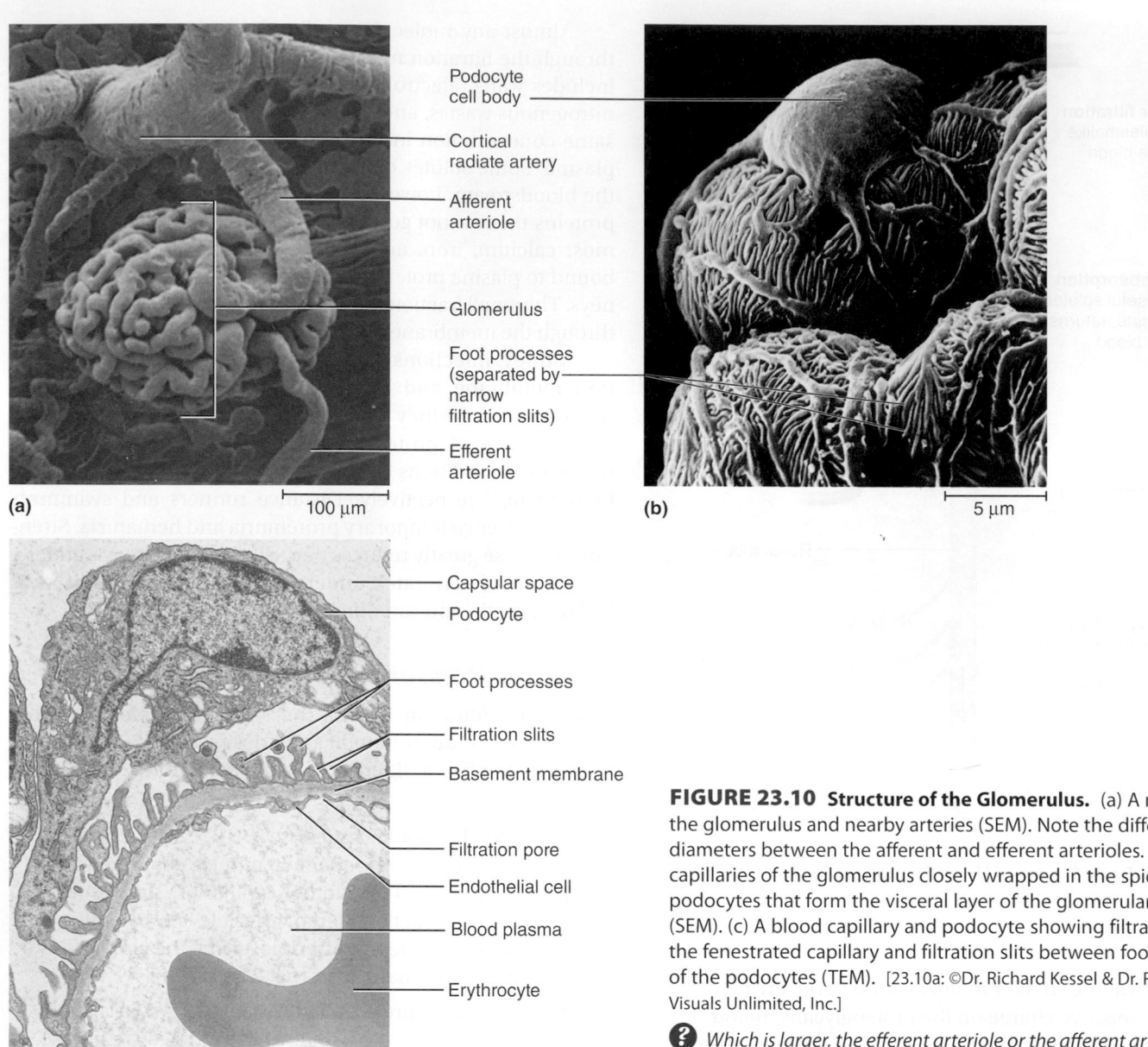

**(a)**    100 µm

**(b)**    5 µm

- Podocyte cell body
- Cortical radiate artery
- Afferent arteriole
- Glomerulus
- Foot processes (separated by narrow filtration slits)
- Efferent arteriole

- Capsular space
- Podocyte
- Foot processes
- Filtration slits
- Basement membrane
- Filtration pore
- Endothelial cell
- Blood plasma
- Erythrocyte

**(c)**    0.5 µm

**FIGURE 23.10  Structure of the Glomerulus.** (a) A resin cast of the glomerulus and nearby arteries (SEM). Note the difference in diameters between the afferent and efferent arterioles. (b) Blood capillaries of the glomerulus closely wrapped in the spidery podocytes that form the visceral layer of the glomerular capsule (SEM). (c) A blood capillary and podocyte showing filtration pores of the fenestrated capillary and filtration slits between foot processes of the podocytes (TEM). [23.10a: ©Dr. Richard Kessel & Dr. Randy Kardon/ Visuals Unlimited, Inc.]

❓ *Which is larger, the efferent arteriole or the afferent arteriole? How does the difference affect the function of the glomerulus?*

In most blood capillaries, the BHP drops low enough at the venous end that osmosis overrides filtration and the capillaries reabsorb fluid. Although BHP also drops along the course of the glomerular capillaries, it remains high enough that these capillaries are engaged solely in filtration. They reabsorb little or no fluid.

The high blood pressure in the glomeruli makes the kidneys especially vulnerable to hypertension, which can have devastating effects on renal function. Hypertension ruptures glomerular capillaries and leads to scarring of the kidneys *(nephrosclerosis).* It promotes atherosclerosis of the renal blood vessels just as it does elsewhere in the body and thus diminishes renal blood supply. Over time, hypertension often leads to renal failure and renal failure leads to worsening hypertension in an insidious positive feedback loop.

## Glomerular Filtration Rate

**Glomerular filtration rate (GFR)** is the amount of filtrate formed per minute by the two kidneys combined. For every 1 mm Hg of net filtration pressure, the kidneys of a young adult male produce about 12.5 mL of filtrate per minute. This value, called the *filtration coefficient ($K_f$),* depends on the permeability and surface area of the filtration barrier. $K_f$ is about 10% lower in women than in men. For males,

$$\text{GFR} = \text{NFP} \times K_f = 10 \times 12.5 = 125 \text{ mL/min.}$$

In young adult females, the GFR is about 105 mL/min.

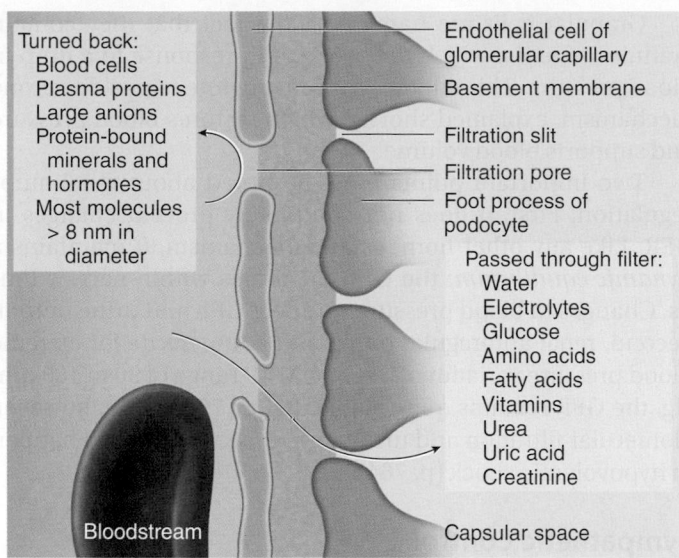

**FIGURE 23.11  The Glomerular Filtration Membrane.**

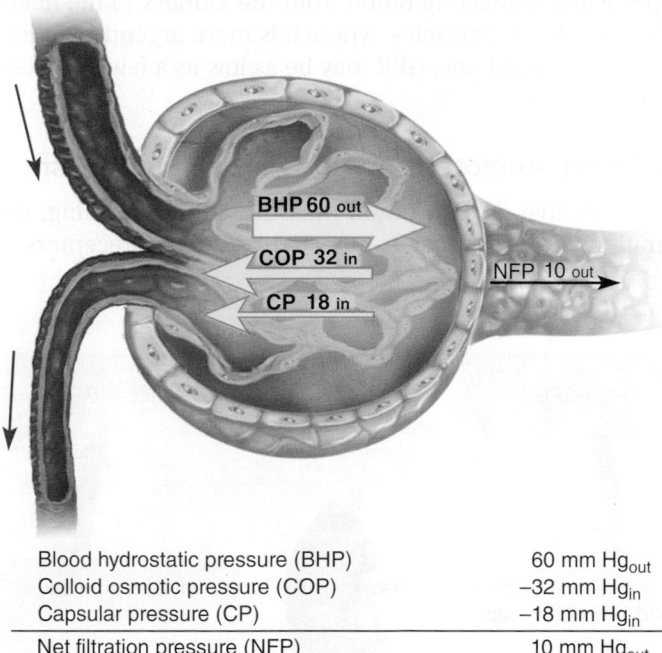

| Blood hydrostatic pressure (BHP) | 60 mm Hg$_{out}$ |
|---|---|
| Colloid osmotic pressure (COP) | −32 mm Hg$_{in}$ |
| Capsular pressure (CP) | −18 mm Hg$_{in}$ |
| Net filtration pressure (NFP) | 10 mm Hg$_{out}$ |

**FIGURE 23.12  The Forces Involved in Glomerular Filtration.**

These rates are equivalent to 180 L/day in males and 150 L/day in females—impressive numbers considering that this is about 60 times the amount of blood in the body and 50 to 60 times the amount of filtrate produced by all other capillaries combined. Obviously, only a small portion of this is eliminated as urine. An average adult reabsorbs 99% of the filtrate and excretes 1 to 2 L of urine per day.

## Regulation of Glomerular Filtration

GFR must be precisely controlled. If it is too high, fluid flows through the renal tubules too rapidly for them to reabsorb the usual amount of water and solutes. Urine output rises and creates a threat of dehydration and electrolyte depletion. If GFR is too low, fluid flows sluggishly through the tubules, they reabsorb wastes that should be eliminated in the urine, and azotemia may occur.

The only way to adjust GFR from moment to moment is to change glomerular blood pressure. This is achieved by three homeostatic mechanisms: renal autoregulation, sympathetic control, and hormonal control.

### Renal Autoregulation

**Renal autoregulation** is the ability of the nephrons to adjust their own blood flow and GFR without external (nervous or hormonal) control. It enables them to maintain a relatively stable GFR in spite of changes in arterial blood pressure. If the mean arterial pressure (MAP) rose from 100 to 125 mm Hg and there were no renal autoregulation, urine output would increase from the normal 1 to 2 L/day to more than 45 L/day. Because of renal autoregulation, however, urine output increases only a few percent even if MAP rises as high as 160 mm Hg. Renal autoregulation thus helps to ensure stable fluid and electrolyte balance in spite of the many circumstances that substantially alter one's blood pressure. There are two mechanisms of autoregulation: the myogenic mechanism and tubuloglomerular feedback.

*The Myogenic*[11] *Mechanism*  This mechanism of stabilizing the GFR is based on the tendency of smooth muscle to contract when stretched. When arterial blood pressure rises, it stretches the afferent arteriole. The arteriole constricts and prevents blood flow into the glomerulus from changing very much. Conversely, when blood pressure falls, the afferent arteriole relaxes and allows blood to flow more easily into the glomerulus. Either way, glomerular blood flow and filtration remain fairly stable.

*Tubuloglomerular Feedback*  This is a mechanism by which the glomerulus receives feedback on the status of the downstream tubular fluid and adjusts filtration to regulate its composition, stabilize nephron performance, and compensate for fluctuations in blood pressure. It involves a structure, the **juxtaglomerular** (JUX-tuh-glo-MER-you-lur) **apparatus,** found at the very end of the nephron loop where it has just reentered the renal cortex. Here, the loop contacts the afferent and efferent arterioles at the vascular pole of the renal corpuscle (fig. 23.13).

---

[11] *myo* = muscle; *genic* = produced by

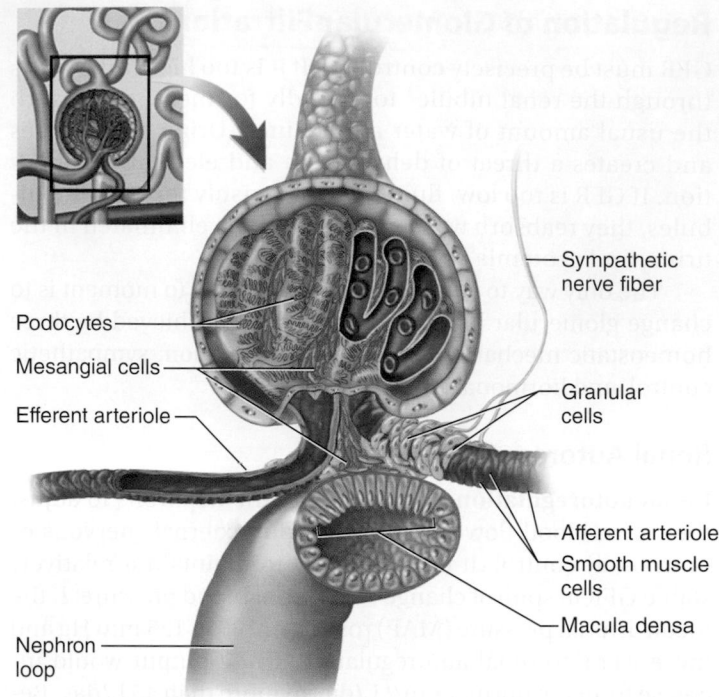

Sympathetic
nerve fiber

Podocytes

Mesangial cells

Efferent arteriole

Granular
cells

Afferent arteriole

Smooth muscle
cells

Macula densa

Nephron
loop

**FIGURE 23.13  The Juxtaglomerular Apparatus.**

Tubuloglomerular feedback begins with the **macula densa,**[12] a patch of slender, closely spaced sensory cells on one side of the loop. When the GFR is very high, the sodium chloride level in the nephron loop is also sharply elevated. Cells of the macula densa absorb $Na^+$, $K^+$, and $Cl^-$; swell; and secrete ATP from their basal surfaces. The ATP is metabolized to adenosine by nearby **mesangial**[13] (mez-AN-jee-ul) **cells,** which pack the spaces between the arterioles and within the glomerulus.

Adenosine acts as a paracrine messenger that stimulates nearby **granular (juxtaglomerular) cells.** These are modified smooth muscle cells wrapped around the afferent arteriole and to a lesser extent the efferent arteriole. They respond to a rising adenosine level by constricting the afferent arteriole. That reduces blood flow into the glomerulus, thereby reducing GFR and completing the negative feedback loop (fig. 23.14). The mesangial cells of the glomerulus may also contract, constricting the glomerular capillaries and reducing filtration. Mesangial cells also form a supportive matrix for the glomerulus and phagocytize tissue debris.

Just beyond the juxtaglomerular apparatus, the distal convoluted tubule has a limited capacity for NaCl reabsorption. The tubuloglomerular feedback process may help to prevent overloading the DCT with NaCl, and thus prevent excessive NaCl and water loss in the urine.

---

[12] *macula* = spot, patch; *densa* = dense
[13] *mes* = in the middle; *angi* = vessel

Granular cells are named for the fact that they contain granules of *renin,* which they secrete in response to a drop in blood pressure. This initiates a renin–angiotensin–aldosterone mechanism, explained shortly, which restores blood pressure and supports blood volume.

Two important points must be noted about renal autoregulation. First, it does not completely prevent changes in GFR. Like any other homeostatic mechanism, it maintains a *dynamic equilibrium;* the GFR fluctuates within narrow limits. Changes in blood pressure *do* affect GFR and urine output. Second, renal autoregulation cannot compensate for extreme blood pressure variations. Over an MAP range of 90 to 180 mm Hg, the GFR remains quite stable. Below 70 mm Hg, however, glomerular filtration and urine output cease. This can happen in hypovolemic shock (p. 764).

## Sympathetic Control

Sympathetic nerve fibers richly innervate the renal blood vessels. In strenuous exercise or acute conditions such as circulatory shock, sympathetic stimulation and adrenal epinephrine constrict the afferent arterioles. This reduces GFR and urine output while redirecting blood from the kidneys to the heart, brain, and skeletal muscles, where it is more urgently needed. Under such conditions, GFR may be as low as a few milliliters per minute.

## The Renin–Angiotensin–Aldosterone Mechanism

Any substantial drop in blood pressure due to bleeding, dehydration, or other cause is detected by the baroreceptors in

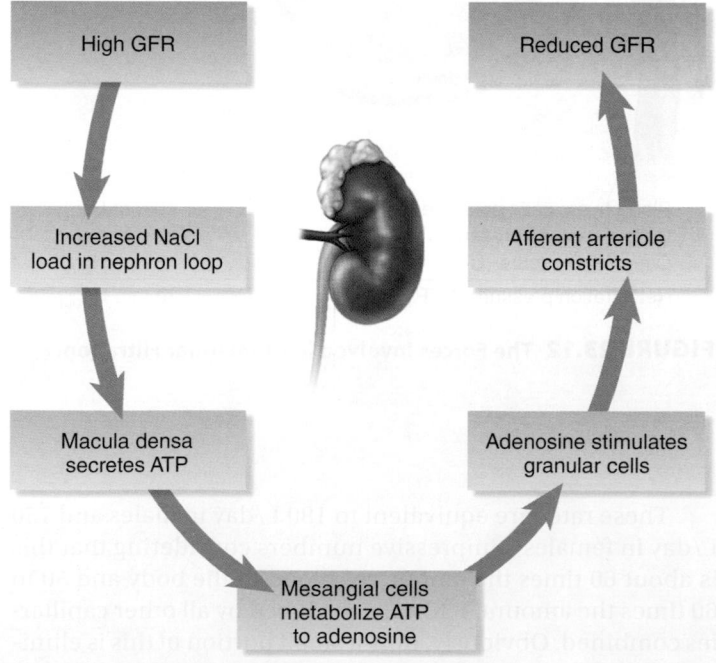

High GFR

Reduced GFR

Increased NaCl
load in nephron loop

Afferent arteriole
constricts

Macula densa
secretes ATP

Adenosine stimulates
granular cells

Mesangial cells
metabolize ATP
to adenosine

**FIGURE 23.14  Tubuloglomerular Feedback.**

the aorta and carotid arteries. They transmit a signal to the brainstem, leading to various corrective sympathetic reflexes (p. 756). One such response is that sympathetic fibers stimulate the granular cells to secrete the enzyme **renin** (REE-nin). Renin acts on *angiotensinogen,* a protein in the blood plasma, to split off a 10-amino-acid peptide called angiotensin I. In the lungs and kidneys, **angiotensin-converting enzyme (ACE)** removes two more amino acids, converting it to **angiotensin II,** a hormone that acts in several ways to restore fluid volume and blood pressure (fig. 23.15):

- It is a potent vasoconstrictor. Widespread vasoconstriction raises the mean arterial blood pressure throughout the body.

- In the kidneys, it constricts the efferent arterioles and to a somewhat lesser degree, the afferent arterioles. By constricting the glomerular outlet more than the inlet, it raises glomerular blood pressure and GFR, or at least prevents a drastic reduction in GFR, thus ensuring continued filtration of wastes from the blood even when blood pressure has fallen.

- Constriction of the efferent arteriole lowers the blood pressure (BP) in the peritubular capillaries. Since capillary BP normally opposes fluid reabsorption, this reduction in BP strongly enhances the reabsorption of NaCl and water from the nephron. More water is returned to the bloodstream instead of being lost in the urine.

- Angiotensin II stimulates the adrenal cortex to secrete aldosterone, which promotes sodium and water reabsorption by the distal convoluted tubule and collecting duct. Angiotensin II also directly stimulates sodium and water reabsorption in the proximal convoluted tubule.

- It stimulates the posterior pituitary gland to secrete antidiuretic hormone, which promotes water reabsorption by the collecting duct.

- It stimulates the sense of thirst and encourages water intake.

Some of these effects are explained more fully later in this chapter and in chapter 24. Collectively, they raise blood pressure by reducing water loss, encouraging water intake, and constricting blood vessels.

▶▶▶**APPLY WHAT YOU KNOW**

*Would you expect ACE inhibitors (see p. 757) to increase or reduce urine output? Why?*

To summarize the events thus far: Glomerular filtration occurs because the high blood pressure of the glomerular capillaries overrides the colloid osmotic pressure. The filtration membrane allows most plasma solutes into the capsular space while retaining formed elements and protein in the bloodstream. Glomerular filtration is maintained at a fairly steady rate of about 105 to 125 mL/min. (female and male, respectively) in spite of variations in systemic blood pressure.

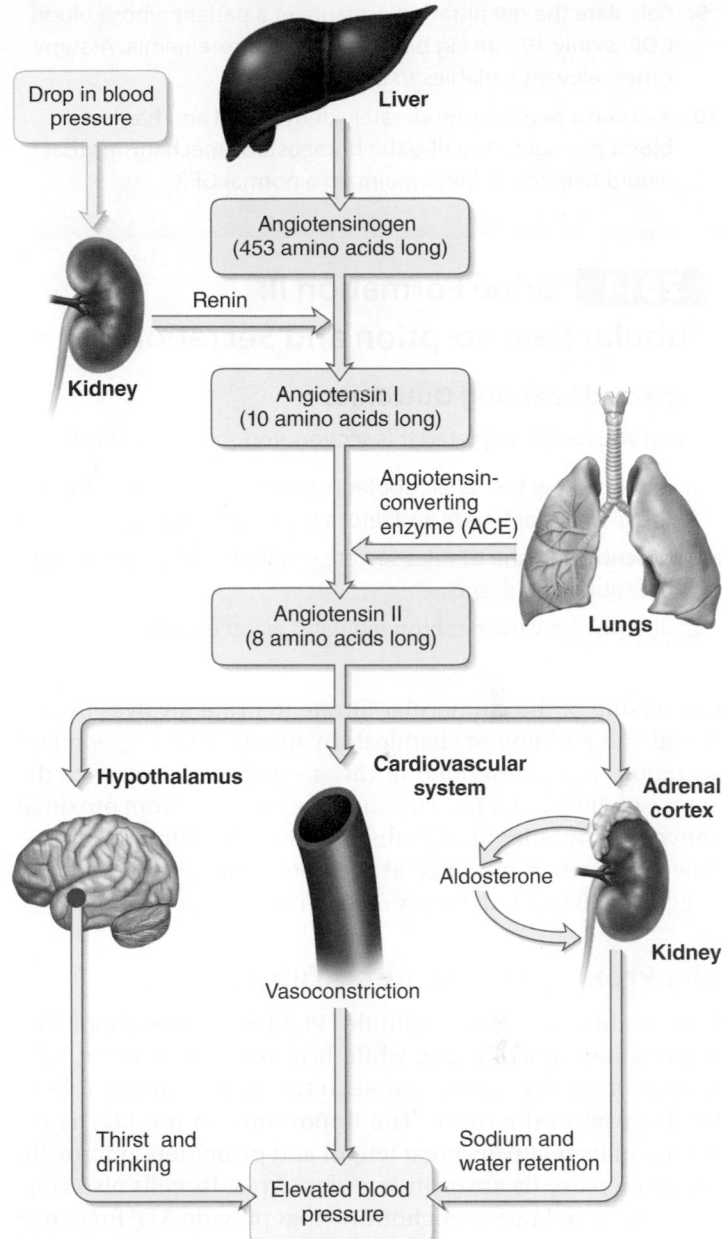

**FIGURE 23.15  The Renin–Angiotensin–Aldosterone Mechanism.** This chain of events is activated by a drop in blood pressure and acts to raise it again.

Stability is achieved by renal autoregulation, sympathetic control, and hormonal control.

**BEFORE YOU GO ON**

Answer the following questions to test your understanding of the preceding section:

7. Name the four major processes in urine production.

8. Trace the movement of a urea molecule from the blood to the capsular space, and name the barriers it passes through.

9. Calculate the net filtration pressure in a patient whose blood COP is only 10 mm Hg because of hypoproteinemia. Assume other relevant variables to be normal.

10. Assume a person is moderately dehydrated and has low blood pressure. Describe the homeostatic mechanisms that would help the kidneys maintain a normal GFR.

## 23.4 Urine Formation II: Tubular Reabsorption and Secretion

### Expected Learning Outcomes

When you have completed this section, you should be able to

a. describe how the renal tubules reabsorb useful solutes from the glomerular filtrate and return them to the blood;

b. describe how the tubules secrete solutes from the blood into the tubular fluid; and

c. describe how the nephron regulates water excretion.

Conversion of the glomerular filtrate to urine involves the removal and addition of chemicals by tubular reabsorption and secretion, to be described in this section. Here we trace the course of the tubular fluid through the nephron, from proximal convoluted tubule through distal convoluted tubule, and see how the filtrate is modified at each point along the way. Refer to figure 23.9 to put these processes into perspective.

### The Proximal Convoluted Tubule

The proximal convoluted tubule (PCT) reabsorbs about 65% of the glomerular filtrate, while it also removes some substances from the blood and secretes them into the tubule for disposal in the urine. The importance of the PCT is reflected in its relatively great length and prominent microvilli, which increase its absorptive surface area. Its cells also contain abundant large mitochondria that provide ATP for active transport. The PCTs alone account for about 6% of one's resting ATP and calorie consumption.

### Tubular Reabsorption

**Tubular reabsorption** is the process of reclaiming water and solutes from the tubular fluid and returning them to the blood. The PCT reabsorbs a greater variety of chemicals than any other part of the nephron.

There are two routes of reabsorption: (1) the **transcellular**[14] **route,** in which substances pass through the cytoplasm and out the base of the epithelial cells; and (2) the **paracellular**[15] **route,**

in which substances pass through gaps between the cells. The "tight" junctions between the epithelial cells are quite leaky and allow significant amounts of water to pass through. As it travels through the epithelium, water carries with it a variety of dissolved solutes—a process called **solvent drag.** Whether by the transcellular or paracellular route, water and solutes enter the tissue fluid at the base of the epithelium, and from there they are taken up by the peritubular capillaries. In the following discussion and figure 23.16, we examine mechanisms of reabsorption.

*Sodium Chloride*   Sodium reabsorption is the key to almost everything else, because it creates an osmotic and electrical gradient that drives the reabsorption of water and the other solutes. It is the most abundant cation in the glomerular filtrate, with a concentration of 140 mEq/L in the fluid entering the proximal convoluted tubule and only 12 mEq/L in the cytoplasm of the epithelial cells. This is a very steep concentration gradient, favoring its diffusion into the epithelial cells.

Two types of transport proteins in the apical cell surface are responsible for sodium uptake: (1) various *symports* that simultaneously bind $Na^+$ and another solute such as glucose, amino acids, or lactate; and (2) an $Na^+–H^+$ *antiport* that pulls $Na^+$ into the cell while pumping $H^+$ out of the cell into the tubular fluid. This antiport is a means not only of reabsorbing sodium, but also of eliminating acid from the body fluids. Angiotensin II activates the $Na^+–H^+$ antiport and thereby exerts a strong influence on sodium reabsorption.

Sodium is prevented from accumulating in the epithelial cells by $Na^+–K^+$ pumps in the basal domain of the plasma membrane, which pump $Na^+$ out into the extracellular fluid. From there, it is picked up by the peritubular capillaries and returns to the bloodstream. These $Na^+–K^+$ pumps, like those anywhere, are ATP-consuming active transport pumps. Although the sodium-transporting symports in the apical membrane do not consume ATP, they are considered an example of **secondary active transport** because of their dependence on the $Na^+–K^+$ pumps at the base of the cell.

Chloride ions, being negatively charged, follow $Na^+$ because they are electrically attracted to it. There also are various antiports in the apical cell membrane that absorb $Cl^-$ in exchange for other anions that they eject into the tubular fluid. Chloride and potassium ions are driven out through the basal cell surface by a $K^+–Cl^-$ symport. Both $Na^+$ and $Cl^-$ also pass through the tubule epithelium by the paracellular route.

*Other Electrolytes*   Potassium, magnesium, and phosphate ions pass through the paracellular route with water. Phosphate is also cotransported into the epithelial cells with $Na^+$. Roughly 52% of the filtered calcium is reabsorbed by the paracellular route and 14% by the transcellular route in the PCT. Calcium absorption here is independent of hormonal influence, but another 33% of the calcium is reabsorbed later in the nephron under the influence of parathyroid hormone, to be discussed later. The remaining 1%, normally, is excreted in the urine.

---

[14] *trans* = across
[15] *para* = next to

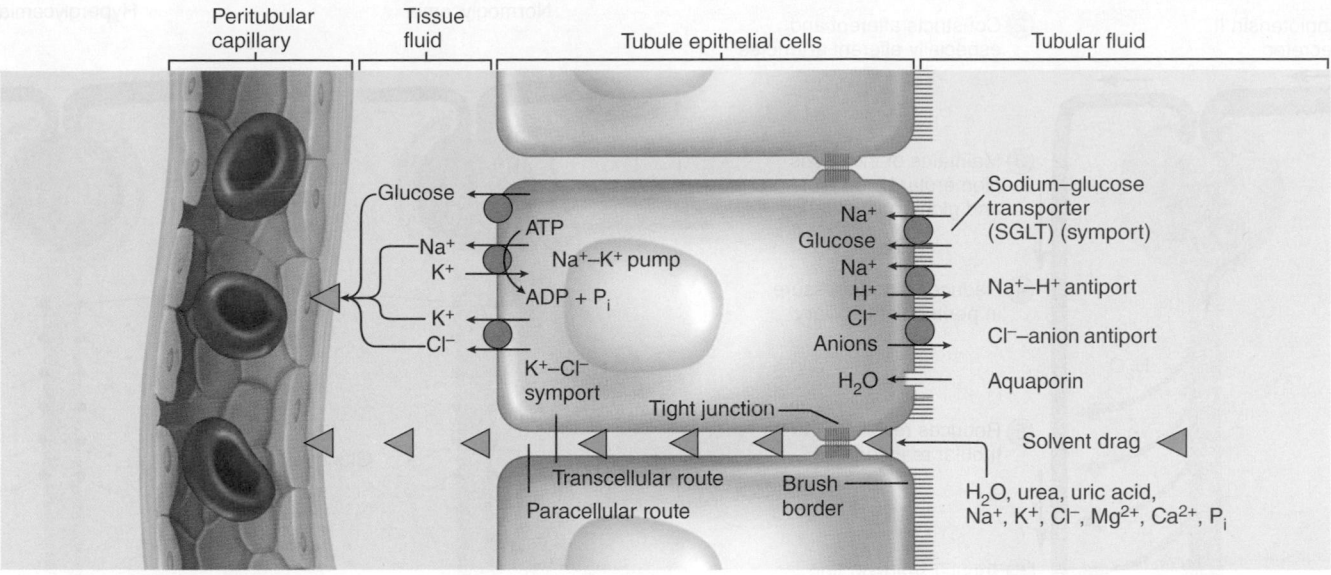

**FIGURE 23.16 Reabsorption in the Proximal Convoluted Tubule.** Water and solutes in the tubular fluid (right) are carried through the tubule epithelium by various means including symports, antiports, aquaporins, and the paracellular route between cells. They enter the tissue fluid at the base of the epithelium and are picked up by the peritubular capillaries (left). Many other solutes not shown here are reabsorbed by similar means.

*How would increased $Na^+$ reabsorption affect the pH of the urine? Why?*

***Glucose***   Glucose is cotransported with $Na^+$ by symports called **sodium–glucose transporters (SGLTs).** It is then removed from the basolateral surface of the cell by facilitated diffusion. Normally, all glucose in the tubular fluid is reabsorbed and there is none in the urine.

***Nitrogenous Wastes***   Urea passes through the epithelium with water. The nephron as a whole reabsorbs 40% to 60% of the urea in the tubular fluid, but since it reabsorbs 99% of the water, urine has a substantially higher urea concentration than blood or glomerular filtrate. When blood enters the kidney, its urea concentration is about 20 mg/dL; when it leaves the kidney, it is typically down to 10.4 mg/dL. Thus, the kidney removes about half of the urea, keeping its concentration down to a safe level but not completely clearing the blood of it.

The PCT reabsorbs nearly all the uric acid entering it, but later parts of the nephron secrete it back into the tubular fluid. Creatinine is not reabsorbed at all, but stays in the tubule and is all passed in the urine.

***Water***   The kidneys reduce about 180 L of glomerular filtrate to 1 or 2 L of urine each day, so obviously water reabsorption is a significant function. About two-thirds of the water is reabsorbed by the PCT. The reabsorption of all the salt and organic solutes as just described makes the tubule cells and tissue fluid hypertonic to the tubular fluid. Water follows the solutes by osmosis through both the paracellular and transcellular routes. Transcellular absorption occurs by way of water channels called **aquaporins** in the apical and basolateral domains of the plasma membrane, enabling water to enter the tubule cells at

the apical surface and leave them (to return to the blood) via the basolateral surface.

Because the PCT reabsorbs proportionate amounts of solutes and water, the osmolarity of the tubular fluid remains unchanged here. Elsewhere in the nephron, water reabsorption is continually modulated by hormones according to the body's state of hydration. In the PCT, however, water is reabsorbed at a constant rate called **obligatory water reabsorption.**

## Uptake by the Peritubular Capillaries

After water and solutes leave the basal surface of the tubule epithelium, they are reabsorbed by the peritubular capillaries. The mechanisms of capillary absorption are osmosis and solvent drag. Three factors promote osmosis into these capillaries: (1) The accumulation of reabsorbed fluid on the basal side of the epithelium cells creates a high tissue fluid pressure that physically drives water into the capillaries. (2) The narrowness of the efferent arteriole lowers the blood hydrostatic pressure (BHP) from 60 mm Hg in the glomerulus to only 8 mm Hg in the peritubular capillaries, so there is less resistance to reabsorption here than in most systemic capillaries. (3) As blood passes through the glomerulus, a lot of water is filtered out but nearly all of the protein remains in the blood. Therefore, the blood has an elevated colloid osmotic pressure (COP) by the time it leaves the glomerulus. With a high COP and low BHP in the capillaries and a high hydrostatic pressure in the tissue fluid, the balance of forces in the peritubular capillaries strongly favors reabsorption. This tendency is even further accentuated by angiotensin II. By constricting the afferent and efferent arterioles, this hormone reduces blood pressure in the

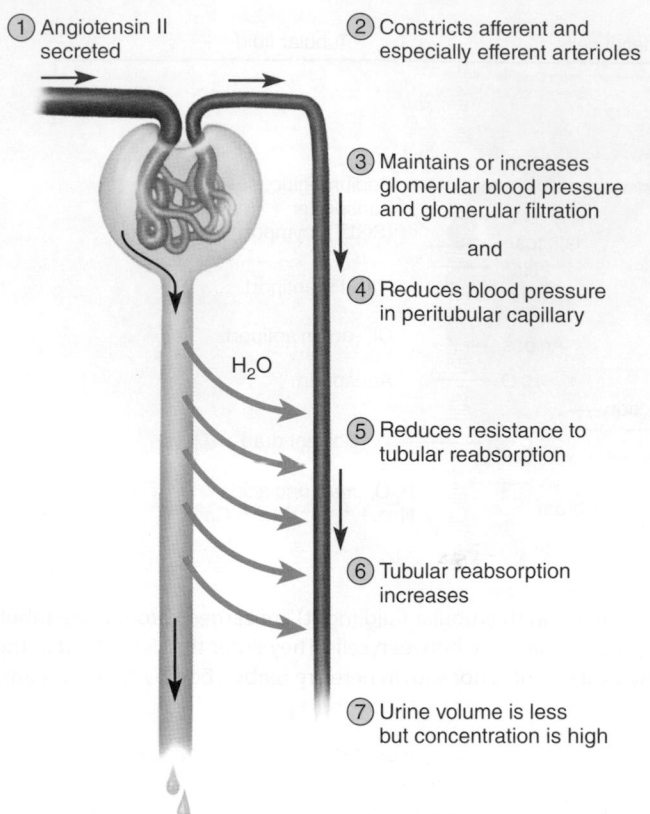

① Angiotensin II secreted

② Constricts afferent and especially efferent arterioles

③ Maintains or increases glomerular blood pressure and glomerular filtration

and

④ Reduces blood pressure in peritubular capillary

H$_2$O

⑤ Reduces resistance to tubular reabsorption

⑥ Tubular reabsorption increases

⑦ Urine volume is less but concentration is high

**FIGURE 23.17** **The Effect of Angiotensin II on Tubular Reabsorption.**

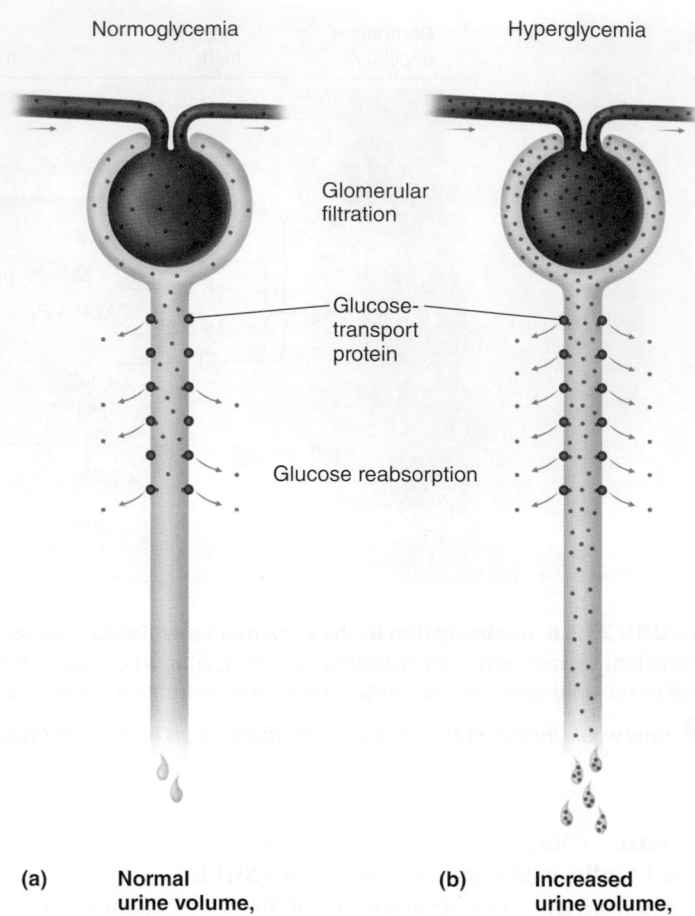

Normoglycemia                    Hyperglycemia

Glomerular filtration

Glucose-transport protein

Glucose reabsorption

(a)    **Normal urine volume, glucose-free**          (b)    **Increased urine volume, with glycosuria**

**FIGURE 23.18** **The Transport Maximum.** (a) At a normal blood glucose concentration (normoglycemia), all glucose filtered by the glomerulus is reabsorbed by glucose-transport proteins in the proximal convoluted tubule, and the urine is glucose-free. (b) At a high blood glucose concentration (hyperglycemia), more glucose is filtered than the transport proteins, now saturated, are able to absorb. The glomerular filtration of glucose now exceeds the transport maximum $T_m$ of the renal tubule. Excess glucose escapes reabsorption and appears in the urine (glycosuria).

peritubular capillaries and thereby reduces their resistance to fluid reabsorption (fig. 23.17).

## The Transport Maximum

There is a limit to the amount of solute that the renal tubule can reabsorb because there are limited numbers of transport proteins in the plasma membranes. If all the transporters are occupied as solute molecules pass through, some solute will escape reabsorption and appear in the urine. The maximum rate of reabsorption is the **transport maximum ($T_m$),** which is reached when the transporters are saturated (see p. 93).

Each solute reabsorbed by the renal tubule has its own $T_m$. Glucose, for example, has a $T_m$ of 320 mg/min. It normally enters the renal tubule at a rate of 125 mg/min., well within the $T_m$; thus all of it is reabsorbed. But at blood glucose levels above 220 mg/dL, glucose is filtered faster than the renal tubule can reabsorb it, and the excess passes in the urine—a condition called **glycosuria**[16] (GLY-co-SOOR-ee-uh) (fig. 23.18). In untreated diabetes mellitus, the plasma glucose concentration may exceed 400 mg/dL, so glycosuria is one of the classic signs of this disease.

[16] *glycos* = sugar; *uria* = urine condition

## Tubular Secretion

**Tubular secretion** is a process in which the renal tubule extracts chemicals from the capillary blood and secretes them into the tubular fluid (see fig. 23.9). In the proximal convoluted tubule and nephron loop, it serves three purposes: (1) It contributes to acid–base balance by secreting varying proportions of hydrogen to bicarbonate ions, as detailed in chapter 24. (2) It extracts wastes from the blood, including urea, uric acid, bile acids, ammonia, and a little creatinine. Uric acid secretion compensates for its reabsorption earlier in the PCT and accounts for all of the uric acid in the urine. (3) It clears drugs and contaminants from the blood, such as morphine, penicillin, and aspirin. One reason why so many drugs must be taken three or four times a day is to keep pace with this clearance and maintain a therapeutically effective drug concentration in the blood.

## The Nephron Loop

The primary function of the nephron loop is to generate an osmotic gradient that enables the collecting duct to concentrate the urine and conserve water, as discussed later. But in addition, the loop reabsorbs about 25% of the Na$^+$, K$^+$, and Cl$^-$ and 15% of the water in the glomerular filtrate. Cells in the thick segment of the ascending limb of the loop have proteins in the apical membranes that simultaneously bind 1 Na$^+$, 1 K$^+$, and 2 Cl$^-$ from the tubular fluid and cotransport them into the cytoplasm. These ions leave the basolateral cell surfaces by active transport of Na$^+$ and diffusion of K$^+$ and Cl$^-$. Potassium reenters the cell by means of the Na$^+$–K$^+$ pump and then reenters the tubular fluid, but NaCl remains in the tissue fluid of the renal medulla. The thick segment is impermeable to water; thus water cannot follow the reabsorbed electrolytes, and tubular fluid becomes very dilute (200 mOsm/L) by the time it passes from the nephron loop into the distal convoluted tubule.

## The Distal Convoluted Tubule and Collecting Duct

Fluid arriving in the DCT still contains about 20% of the water and 7% of the salts from the glomerular filtrate. If this were all passed as urine, it would amount to 36 L/day, so obviously a great deal of fluid reabsorption is still to come. The DCT and collecting duct reabsorb variable amounts of water and salts and are regulated by several hormones—particularly aldosterone, natriuretic peptides, antidiuretic hormone, and parathyroid hormone.

There are two kinds of cells in the DCT and collecting duct. The **principal cells** are the more abundant; they have receptors for the foregoing hormones and are involved chiefly in salt and water balance. **Intercalated cells** are fewer in number. They reabsorb K$^+$ and secrete H$^+$ into the tubule and are involved mainly in acid–base balance, as discussed in chapter 24. The major hormonal influences on the DCT and collecting duct are as follows.

## Aldosterone

Aldosterone, the "salt-retaining hormone," is a steroid secreted by the adrenal cortex when the blood Na$^+$ concentration falls or its K$^+$ concentration rises. A drop in blood pressure also induces aldosterone secretion, but indirectly—it stimulates the kidney to secrete renin; this leads to the production of angiotensin II; and angiotensin II stimulates aldosterone secretion (see fig. 23.15).

Aldosterone acts on the thick segment of the ascending limb of the nephron loop, on the DCT, and on the cortical portion of the collecting duct. It stimulates these segments of the nephron to reabsorb Na$^+$ and secrete K$^+$. Water and Cl$^-$ follow the Na$^+$, so the net effect is that the body retains NaCl and

water, urine volume is reduced, and the urine has an elevated K$^+$ concentration. Water retention helps to maintain blood volume and pressure. Chapter 24 deals further with the action of aldosterone.

## Natriuretic Peptides

The heart secretes natriuretic peptides in response to high blood pressure. These hormones exert four actions that result in the excretion of more salt and water in the urine, thereby reducing blood volume and pressure:

1. They dilate the afferent arteriole and constrict the efferent arteriole, which increases the GFR.
2. They antagonize the renin–angiotensin–aldosterone mechanism by inhibiting renin and aldosterone secretion.
3. They inhibit the secretion of antidiuretic hormone and its action on the kidney.
4. They inhibit NaCl reabsorption by the collecting duct.

## Antidiuretic Hormone (ADH)

Dehydration, loss of blood volume, and rising blood osmolarity stimulate arterial baroreceptors and hypothalamic osmoreceptors. In response, the posterior pituitary gland secretes ADH. ADH makes the collecting duct more permeable to water, so water in the tubular fluid reenters the tissue fluid and blood rather than being lost in the urine. Its mechanisms of doing so are described later.

## Parathyroid Hormone (PTH)

A calcium deficiency (hypocalcemia) stimulates the parathyroid glands to secrete PTH, which acts in several ways to restore calcium homeostasis. Its effect on bone metabolism was described in chapter 7. In the kidney, it acts on the PCT to inhibit phosphate reabsorption and acts on the DCT and thick segment of the nephron loop to increase calcium reabsorption. On average, about 25% of the filtered calcium is reabsorbed by the thick segment and 8% by the DCT. PTH therefore increases the phosphate content and lowers the calcium content of the urine. This helps to minimize further decline in the blood calcium level. Because phosphate is not retained along with the calcium, the calcium ions stay in circulation rather than precipitating into the bone tissue as calcium phosphate. Calcitriol and calcitonin have similar but weaker effects on the DCT. PTH also stimulates calcitriol synthesis by the epithelial cells of the PCT.

In summary, the PCT reabsorbs about 65% of the glomerular filtrate and returns it to the blood of the peritubular capillaries. Much of this reabsorption occurs by osmotic and cotransport mechanisms linked to the active transport of sodium ions. The nephron loop reabsorbs another 25% of the filtrate, although its primary role, detailed later, is to aid the function of the collecting duct. The DCT reabsorbs more sodium chloride and water, but its rates of reabsorption are subject to control by hormones, especially aldosterone and ANP. These tubules also extract drugs, wastes, and some other solutes from the blood and secrete them into the tubular fluid. The DCT essentially completes the process of determining the chemical composition of the urine. The principal function left to the collecting duct is to conserve water.

### BEFORE YOU GO ON

Answer the following questions to test your understanding of the preceding section:

11. The reabsorption of water, Cl⁻, and glucose by the PCT is linked to the reabsorption of Na⁺, but in three very different ways. Contrast these three mechanisms.

12. Explain why a substance appears in the urine if its rate of glomerular filtration exceeds the $T_m$ of the renal tubule.

13. Contrast the effects of aldosterone and natriuretic peptides on the renal tubule.

## 23.5 Urine Formation III: Water Conservation

### Expected Learning Outcomes

When you have completed this section, you should be able to

a. explain how the collecting duct and antidiuretic hormone regulate the volume and concentration of urine; and

b. explain how the kidney maintains an osmotic gradient in the renal medulla that enables the collecting duct to function.

The kidney serves not just to eliminate metabolic waste from the body but also to prevent excessive water loss in doing so, and thus to support the body's fluid balance. As the kidney returns water to the tissue fluid and bloodstream, the fluid remaining in the renal tubule, and ultimately passed as urine, becomes more and more concentrated. In this section, we examine the kidney's mechanisms for conserving water and concentrating the urine.

### The Collecting Duct

The collecting duct (CD) begins in the cortex, where it receives tubular fluid from numerous nephrons. As it passes through the medulla, it usually reabsorbs water and concentrates the urine. When urine enters the upper end of the CD, it is isotonic with blood plasma (300 mOsm/L), but by the time it leaves the lower end, it can be up to four times as concentrated—that is, highly hypertonic to the plasma. This ability to concentrate wastes and control water loss was crucial to the evolution of terrestrial animals such as ourselves (see Deeper Insight 23.1).

Two facts enable the collecting duct to produce such hypertonic urine: (1) The osmolarity of the extracellular fluid is four times as high in the lower medulla as it is in the cortex, and (2) the medullary portion of the CD is more permeable to water than to solutes. Therefore, as urine passes down the CD through the increasingly hypertonic medulla, water leaves the tubule by osmosis, most NaCl and other wastes remain behind, and the urine becomes more and more concentrated (fig. 23.19).

### Control of Water Loss

Just how concentrated the urine becomes depends on the body's state of hydration. For example, if you drink a large volume of water, you soon produce a large volume of hypotonic urine—a response called *water diuresis*[17] (DY-you-REE-sis). Under such conditions, the cortical portion of the CD reabsorbs NaCl but is impermeable to water. Salt is removed from the urine, water stays in it, and urine osmolarity may be as low as 50 mOsm/L.

Dehydration, on the other hand, causes the urine to be scanty and more concentrated. The high blood osmolarity of a dehydrated person stimulates the pituitary to release ADH. ADH increases water reabsorption (reduces urine output) by two mechanisms. (1) Within seconds, cells of the collecting duct transfer aquaporins from storage vesicles in the cytoplasm to the apical cell surface and the cells begin taking up more water from the tubular fluid. (2) If the ADH level remains elevated for ≥24 hours, it induces the cell to transcribe the aquaporin gene and manufacture more aquaporins, further raising the water permeability of the collecting duct.

[17] *diuresis* = passing urine

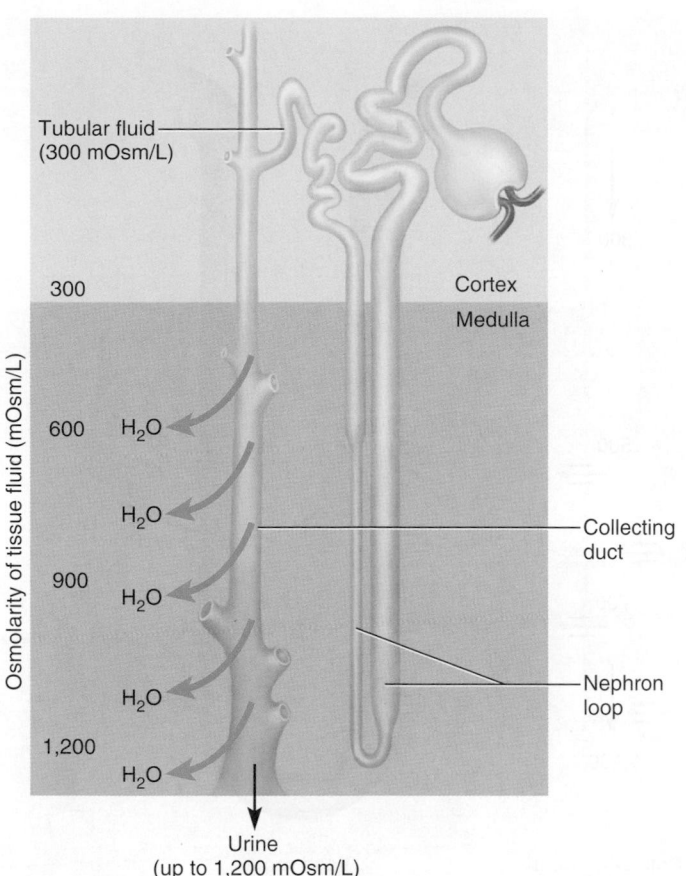

Tubular fluid
(300 mOsm/L)

Osmolarity of tissue fluid (mOsm/L)

300 — Cortex / Medulla

600 — $H_2O$

$H_2O$

900 — $H_2O$ — Collecting duct

$H_2O$ — Nephron loop

1,200 — $H_2O$

Urine
(up to 1,200 mOsm/L)

**FIGURE 23.19  Water Reabsorption by the Collecting Duct.** Note that the osmolarity of the tissue fluid increases fourfold from 300 mOsm/L in the cortex to 1,200 mOsm/L deep in the medulla. Urine concentration increases proportionately as water leaves the duct through its aquaporins.

By contrast, when you are well hydrated, ADH secretion falls; the tubule cells remove aquaporins from the plasma membrane and store them in cytoplasmic vesicles. The duct is then less permeable to water, so more water remains in the duct and you produce abundant, dilute urine.

In extreme cases, the blood pressure of a dehydrated person is low enough to significantly reduce the glomerular filtration rate. When the GFR is low, fluid flows more slowly through the renal tubules and there is more time for tubular reabsorption. Less salt remains in the urine as it enters the CD, so there is less opposition to the osmosis of water out of the duct and into the ECF. More water is reabsorbed and less urine is produced.

## The Countercurrent Multiplier

The ability of the CD to concentrate urine depends on the osmotic gradient of the renal

medulla. It may seem surprising that the ECF osmolarity is four times as great deep in the medulla as in the cortex. We would expect salt to diffuse toward the cortex until it was evenly distributed through the kidney. However, there is a mechanism that overrides this—the nephron loop acts as a **countercurrent multiplier,** which continually recaptures salt and returns it to the deep medullary tissue. It is called a *multiplier* because it multiplies the osmolarity deep in the medulla, and a *countercurrent* mechanism because it is based on fluid flowing in opposite directions in two adjacent tubules—downward in the descending limb and upward in the ascending limb.

Figure 23.20 shows how this works. Steps 2 through 5 form a positive feedback loop. As fluid flows down the descending limb of the nephron loop, it passes through an environment of increasing osmolarity. Most of the descending limb is very permeable to water but not to NaCl; therefore, water passes by osmosis from the tubule into the ECF, leaving NaCl behind. The tubule contents increase in osmolarity, reaching about 1,200 mOsm/L by the time the fluid rounds the bend at the lower end of the loop.

Most or all of the ascending limb (its thick segment), by contrast, is impermeable to water, but has pumps that cotransport $Na^+$, $K^+$, and $Cl^-$ into the ECF. This keeps the osmolarity of the renal medulla high. Since water remains in the tubule, the tubular fluid becomes more and more dilute as it approaches the cortex and is only about 100 mOsm/L at the top of the loop.

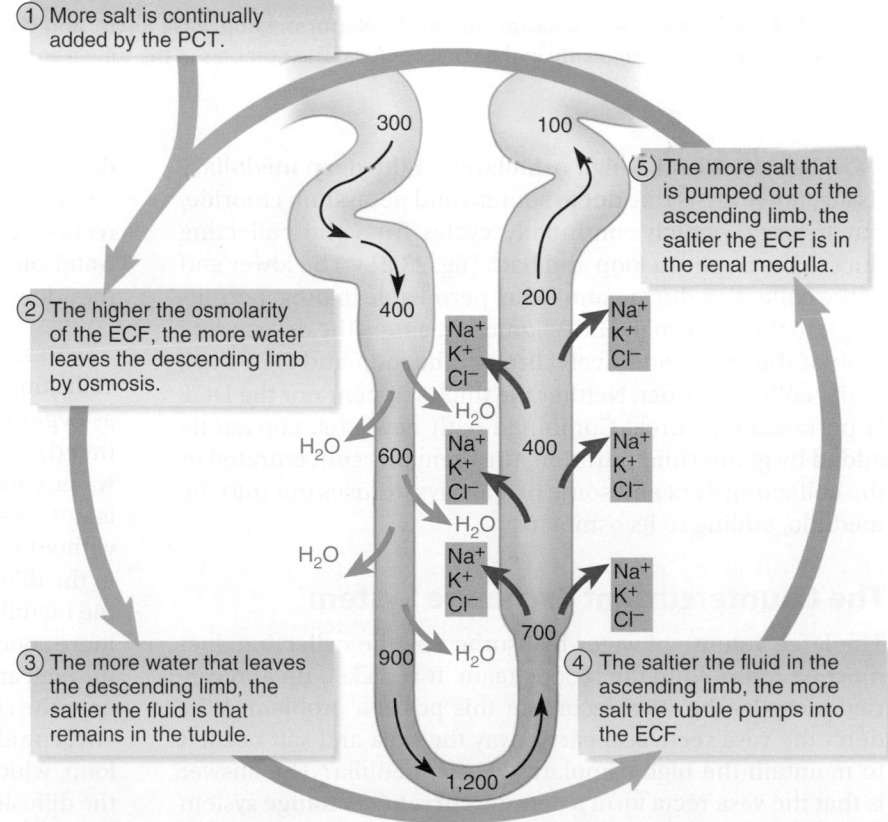

① More salt is continually added by the PCT.

300 · 100

⑤ The more salt that is pumped out of the ascending limb, the saltier the ECF is in the renal medulla.

② The higher the osmolarity of the ECF, the more water leaves the descending limb by osmosis.

400 · 200 · $Na^+$ $K^+$ $Cl^-$

$H_2O$

$H_2O$ · 600 · $Na^+$ $K^+$ $Cl^-$ · 400 · $Na^+$ $K^+$ $Cl^-$

$H_2O$

$H_2O$ · $Na^+$ $K^+$ $Cl^-$ · $Na^+$ $K^+$ $Cl^-$

③ The more water that leaves the descending limb, the saltier the fluid is that remains in the tubule.

900 · $H_2O$ · 700

④ The saltier the fluid in the ascending limb, the more salt the tubule pumps into the ECF.

1,200

**FIGURE 23.20  The Countercurrent Multiplier of the Nephron Loop.** The numbers in the tubule are in mOsm/L.

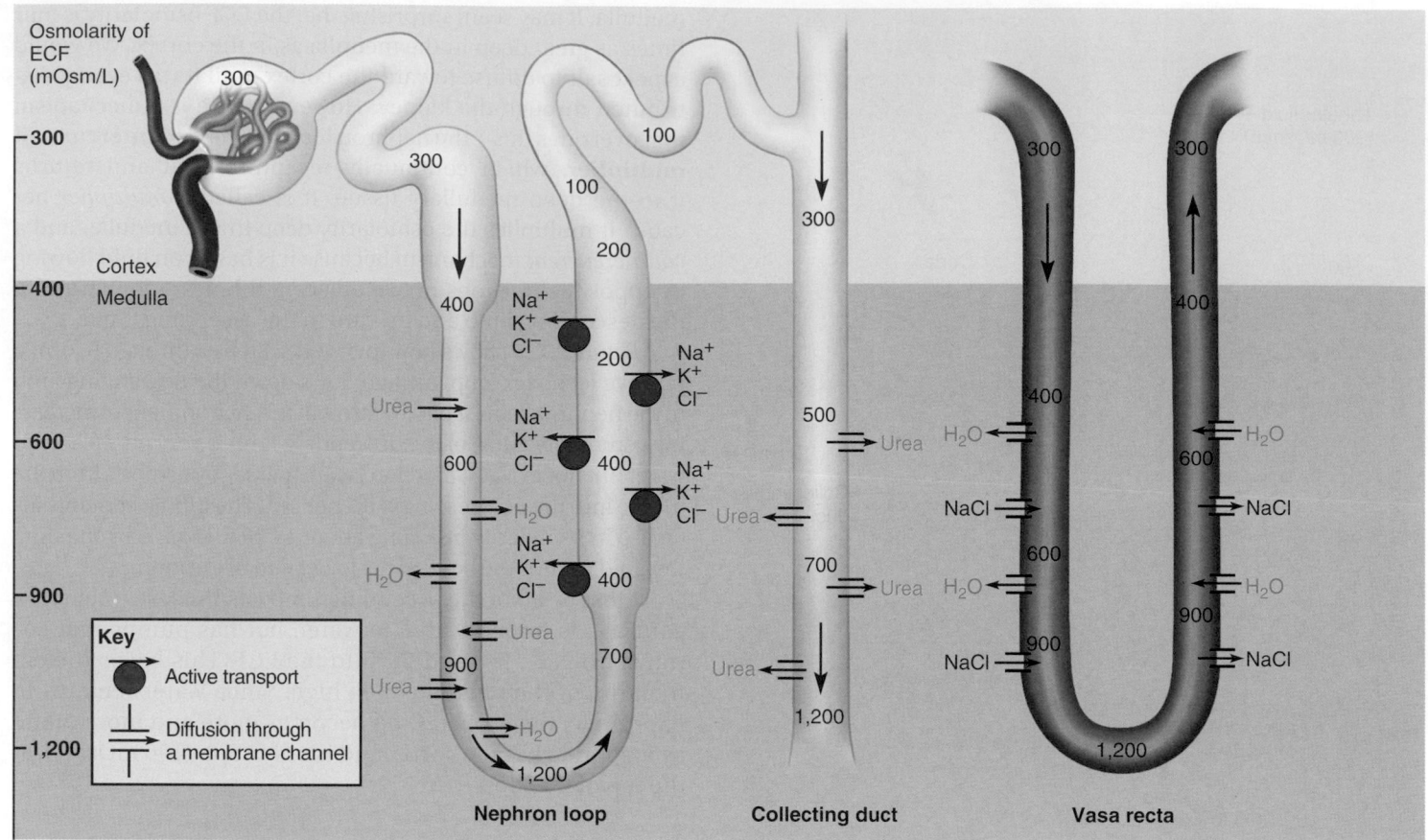

**FIGURE 23.21  Functional Relationship of the Nephron Loop, Vasa Recta, and Collecting Duct.** These three structures work together to maintain a gradient of osmolarity in the renal medulla. The numbers in the tubule and vasa recta are in mOsm/L.

About 40% of the high osmolarity in the deep medullary tissue, however, is due not to sodium and potassium chloride, but to urea—which continually cycles from the collecting duct to the nephron loop and back (fig. 23.21). The lower end of the collecting duct is somewhat permeable to urea, permitting it to diffuse into the ECF. Some of it enters the descending limb of the loop and travels through the loop and DCT back to the collecting duct. Neither the thick segment nor the DCT is permeable to urea. Combined with new urea constantly added by glomerular filtration, urea remains concentrated in the collecting duct and some of it always diffuses out into the medulla, adding to its osmolarity.

## The Countercurrent Exchange System

The large volume of water reabsorbed by the collecting duct must be returned to the bloodstream. It is picked up and carried away by the vasa recta, but this poses a problem: Why don't the vasa recta also carry away the urea and salt needed to maintain the high osmolarity of the medulla? The answer is that the vasa recta form a **countercurrent exchange** system that prevents this from happening. Blood flows in opposite directions in adjacent parallel capillaries. As it flows downward into the medulla, the vessels exchange water for salt—water

diffuses out of the capillaries and salt diffuses in. But as the blood flows back up to the cortex, the opposite occurs; the vasa recta give up salt and absorb water. Indeed, they absorb more water on the way out than they give up on the way in. Thus, they do not subtract from the osmolarity of the medulla.

To summarize what we have studied in this section, the collecting duct can adjust water reabsorption to produce urine as hypotonic (dilute) as 50 mOsm/L or as hypertonic (concentrated) as 1,200 mOsm/L, depending on the body's need for water conservation or removal. In a state of hydration, ADH is not secreted and the cortical part of the CD reabsorbs salt without reabsorbing water; the water remains to be excreted in the dilute urine. In a state of dehydration, ADH is secreted, the medullary part of the CD reabsorbs water, and the urine is more concentrated. The CD is able to do this because it passes through an osmotic gradient in the medulla from 300 mOsm/L near the cortex to 1,200 mOsm/L near the papilla. This gradient is produced by a countercurrent multiplier of the nephron loop, which concentrates NaCl in the lower medulla, and by the diffusion of urea from the collecting duct into the medulla. The vasa recta are arranged as a countercurrent exchange system that enables them to remove water from the medulla without subtracting from its osmotic gradient.

| TABLE 23.1 | Hormones Affecting Renal Function | |
|---|---|---|
| **Hormone** | **Renal Targets** | **Effects** |
| Aldosterone | Nephron loop, DCT, CD | Promotes $Na^+$ reabsorption and $K^+$ secretion; indirectly promotes $Cl^-$ and $H_2O$ reabsorption; maintains blood volume and reduces urine volume |
| Angiotensin II | Afferent and efferent arterioles, PCT | Reduces water loss, stimulates thirst and encourages water intake, and constricts blood vessels, thus raising blood pressure. Reduces GFR; stimulates PCT to reabsorb $NaCl$ and $H_2O$; stimulates aldosterone and ADH secretion |
| Antidiuretic hormone | Collecting duct | Promotes $H_2O$ reabsorption; reduces urine volume, increases concentration |
| Natriuretic peptides | Afferent and efferent arterioles, collecting duct | Dilate afferent arteriole, constrict efferent arteriole, increase GFR; inhibit secretion of renin, ADH, and aldosterone; inhibit $NaCl$ reabsorption by collecting duct; increase urine volume and lower blood pressure |
| Calcitonin | DCT | Weak effects similar to those of parathyroid hormone |
| Calcitriol | DCT | Weak effects similar to those of parathyroid hormone |
| Epinephrine and norepinephrine | Juxtaglomerular apparatus, afferent arteriole | Induce renin secretion; constrict afferent arteriole; reduce GFR and urine volume |
| Parathyroid hormone | PCT, DCT, nephron loop | Promotes $Ca^{2+}$ reabsorption by loop and DCT; increases phosphate excretion by PCT; promotes calcitriol synthesis |

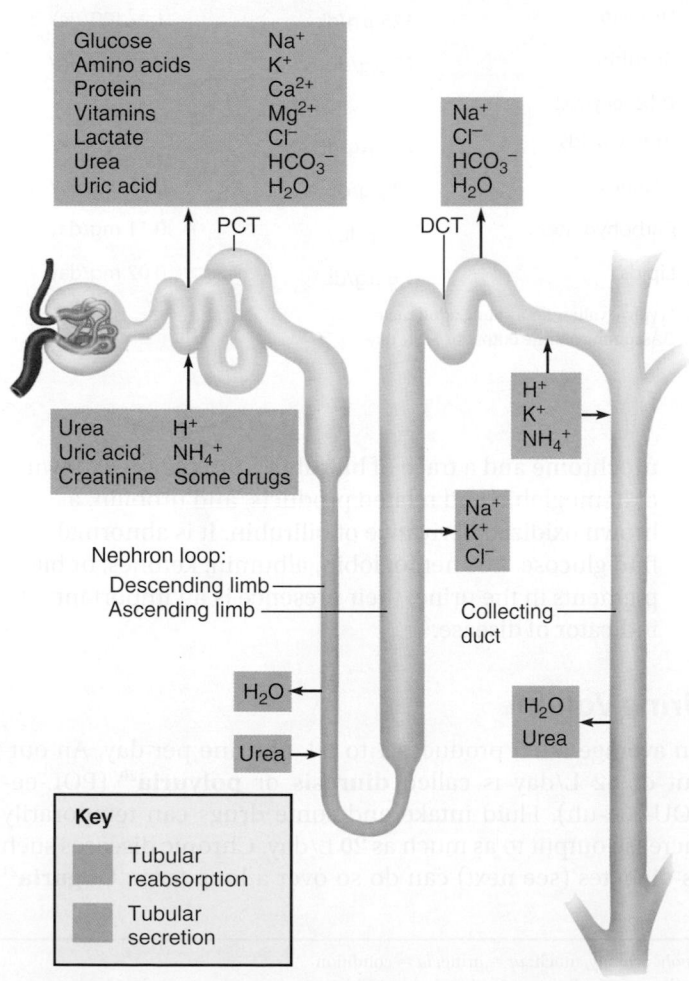

**FIGURE 23.22  Solutes Reabsorbed and Secreted in Each Portion of the Renal Tubule.**

Figure 23.22 summarizes the major solutes reabsorbed and secreted in each part of the renal tubule. Table 23.1 summarizes the hormones that affect renal function.

### BEFORE YOU GO ON

Answer the following questions to test your understanding of the preceding section:

14. Predict how ADH hypersecretion would affect the sodium concentration of the urine, and explain why.

15. Concisely contrast the role of the countercurrent multiplier with that of the countercurrent exchanger.

16. How would the function of the collecting duct change if the nephron loop did not exist?

### 23.6  Urine and Renal Function Tests

#### Expected Learning Outcomes

When you have completed this section, you should be able to

a. describe the composition and properties of urine; and

b. carry out some calculations to evaluate renal function.

Medical diagnosis often rests on determining the current and recent physiological state of the tissues. No two fluids are as valuable for this purpose as blood and urine. **Urinalysis,** the examination of the physical and chemical properties of urine, is therefore one of the most routine procedures in medical examinations. The principal characteristics of urine and certain tests used to evaluate renal function are described here.

## Composition and Properties of Urine

The basic composition and properties of urine are as follows:

- **Appearance.** Urine varies from almost colorless to deep amber, depending on the body's state of hydration. The yellow color of urine is due to **urochrome,**[18] a pigment produced by the breakdown of hemoglobin from expired erythrocytes. Pink, green, brown, black, and other colors result from certain foods, vitamins, drugs, and metabolic diseases. Urine is normally clear but turns cloudy upon standing because of bacterial growth. Pus in the urine (**pyuria**[19]) makes it cloudy and suggests kidney infection. Blood in the urine (**hematuria**) may be due to a urinary tract infection, trauma, or kidney stones. Cloudiness or blood in a urine specimen sometimes, however, simply indicates contamination with semen or menstrual fluid.

- **Odor.** Fresh urine has a distinctive but not repellent odor. As it stands, however, bacteria multiply, degrade urea to ammonia, and produce the pungent odor typical of stale wet diapers. Asparagus and other foods can impart distinctive aromas to the urine. Diabetes mellitus gives it a sweet, fruity odor of acetone. A mousy odor suggests phenylketonuria (PKU), and a rotten odor may indicate urinary tract infection.

- **Specific gravity.** This is a ratio of the density (g/mL) of a substance to the density of distilled water. Distilled water has a specific gravity of 1.000, and urine ranges from 1.001 when it is very dilute to 1.028 when it is very concentrated. Multiplying the last two digits of the specific gravity by a proportionality constant of 2.6 gives an estimate of the grams of solid matter per liter of urine. For example, a specific gravity of 1.025 indicates a solute concentration of $25 \times 2.6 = 65$ g/L.

- **Osmolarity.** Urine can have an osmolarity as low as 50 mOsm/L in a very hydrated person or as high as 1,200 mOsm/L in a dehydrated person. Compared with the osmolarity of blood (300 mOsm/L), then, urine can be either hypotonic or hypertonic.

- **pH.** The body constantly generates metabolic acids and gets rid of them by excreting mildly acidic urine, usually with a pH of about 6.0 (but ranging from 4.5 to 8.2). The regulation of urine pH is discussed extensively in chapter 24.

- **Chemical composition.** Urine averages 95% water and 5% solutes by volume (table 23.2). Normally, the most abundant solute is urea, followed by sodium chloride, potassium chloride, and lesser amounts of creatinine, uric acid, phosphates, sulfates, and traces of calcium, magnesium, and sometimes bicarbonate. Urine contains

| TABLE 23.2 | Properties and Composition of Urine | |
|---|---|---|
| **Physical Properties** | | |
| Specific gravity | 1.001–1.028 | |
| Osmolarity | 50–1,200 mOsm/L | |
| pH | 6.0 (range 4.5–8.2) | |
| **Solute** | **Concentration*** | **Output** ** |
| **Inorganic ions** | | |
| Chloride | 533 mg/dL | 6.4 g/day |
| Sodium | 333 mg/dL | 4.0 g/day |
| Potassium | 166 mg/dL | 2.0 g/day |
| Phosphate | 83 mg/dL | 1 g/day |
| Ammonia | 60 mg/dL | 0.68 g/day |
| Calcium | 17 mg/dL | 0.2 g/day |
| Magnesium | 13 mg/dL | 0.16 g/day |
| **Nitrogenous wastes** | | |
| Urea | 1.8 g/dL | 21 g/day |
| Creatinine | 150 mg/dL | 1.8 g/day |
| Uric acid | 40 mg/dL | 0.5 g/day |
| Urobilin | 125 µg/dL | 1.52 mg/day |
| Bilirubin | 20 µg/dL | 0.24 mg/day |
| **Other organics** | | |
| Amino acids | 288 µg/dL | 3.5 mg/day |
| Ketones | 17 µg/dL | 0.21 mg/day |
| Carbohydrates | 9 µg/dL | 0.11 mg/day |
| Lipids | 1.6 µg/dL | 0.02 mg/day |

*Typical values for a young adult male
**Assuming a urine output of 1.2 L/day

urochrome and a trace of bilirubin from the breakdown of hemoglobin and related products, and urobilin, a brown oxidized derivative of bilirubin. It is abnormal to find glucose, free hemoglobin, albumin, ketones, or bile pigments in the urine; their presence is an important indicator of disease.

## Urine Volume

An average adult produces 1 to 2 L of urine per day. An output of >2 L/day is called **diuresis** or **polyuria**[20] (POL-ee-YOU-ree-uh). Fluid intake and some drugs can temporarily increase output to as much as 20 L/day. Chronic diseases such as diabetes (see next) can do so over a long term. **Oliguria**[21]

---

[18] *uro* = urine; *chrom* = color
[19] *py* = pus; *ur* = urine; *ia* = condition

[20] *poly* = many, much; *ur* = urine; *ia* = condition
[21] *oligo* = few, a little; *ur* = urine; *ia* = condition

(oll-ih-GURE-ee-uh) is an output of <500 mL/day, and **anuria**[22] is an output of 0 to 100 mL/day. Low output can result from kidney disease, dehydration, circulatory shock, prostate enlargement, and other causes. If urine output drops to <400 mL/day, the body cannot maintain a safe, low concentration of wastes in the blood plasma. The result is azotemia.

## Diabetes

**Diabetes**[23] is any metabolic disorder resulting in chronic polyuria. There are at least four forms of diabetes: *diabetes mellitus type 1* and *type 2, gestational diabetes,* and *diabetes insipidus.* In most cases, the polyuria results from a high concentration of glucose in the renal tubule. Glucose osmotically retains water in the tubule, so more water passes in the urine *(osmotic diuresis)* and a person may become severely dehydrated. In diabetes mellitus and gestational diabetes, the high glucose level in the tubular fluid is a result of hyperglycemia, a high level in the blood. About 1% to 3% of pregnant women experience gestational diabetes, in which pregnancy reduces the mother's insulin sensitivity, resulting in hyperglycemia and glycosuria. Diabetes insipidus results from ADH hyposecretion. Without ADH, the collecting duct does not reabsorb much water, so more water passes in the urine.

Diabetes mellitus and gestational diabetes are characterized by glycosuria. Before chemical tests for urine glucose were developed, physicians diagnosed diabetes mellitus[24] by tasting the patient's urine for sweetness. Tests for glycosuria are now as simple as dipping a chemical test strip into the urine specimen—an advance in medical technology for which urologists are no doubt grateful. In diabetes insipidus,[25] the urine contains no glucose and, by the old diagnostic method, would not taste sweet.

## Diuretics

A **diuretic** is any chemical that increases urine volume. Some diuretics act by increasing glomerular filtration rate, such as caffeine, which dilates the afferent arteriole. Others act by reducing tubular reabsorption of water. Alcohol, for example, inhibits ADH secretion and thereby reduces reabsorption in the collecting duct. **Loop diuretics** such as furosemide (Lasix) act on the nephron loop to inhibit the $Na^+$–$K^+$–$Cl^-$ symport. This impairs the countercurrent multiplier, thus reducing the osmotic gradient in the renal medulla and making the collecting duct unable to reabsorb as much water as usual. Diuretics are commonly administered to treat hypertension and congestive heart failure by reducing the body's fluid volume and blood pressure.

## Renal Function Tests

There are several tests for diagnosing kidney diseases, evaluating their severity, and monitoring their progress. Here we examine two methods used to determine renal clearance and glomerular filtration rate.

## Renal Clearance

**Renal clearance** is the volume of blood plasma from which a particular waste is completely removed in 1 minute. It represents the net effect of three processes:

> Glomerular filtration of the waste
> + Amount added by tubular secretion
> – Amount removed by tubular reabsorption
> ———————————————————————
> Renal clearance

In principle, we could determine renal clearance by sampling blood entering and leaving the kidney and comparing their waste concentrations. In practice, it is not practical to draw blood samples from the renal vessels, but clearance can be assessed indirectly by collecting samples of blood and urine, measuring the waste concentration in each, and measuring the rate of urine output.

Suppose the following values were obtained for urea:

| | |
|---|---|
| U (urea concentration in urine) | = 6.0 mg/mL |
| V (rate of urine output) | = 2 mL/min. |
| P (urea concentration in plasma) | = 0.2 mg/mL |

Renal clearance (C) is

$$C = UV/P$$
$$= (6.0 \text{ mg/mL})(2 \text{ mL/min.})/0.2 \text{ mg/mL}$$
$$= 60 \text{ mL/min.}$$

This means the equivalent of 60 mL of blood plasma is completely cleared of urea per minute. If this person has a normal GFR of 125 mL/min., then the kidneys have cleared urea from 60/125 = 48% of the glomerular filtrate. This is a normal rate of urea clearance and is sufficient to maintain safe levels of urea in the blood.

### ▶▶▶ APPLY WHAT YOU KNOW

*What would you expect the value of renal clearance of glucose to be in a healthy individual? Why?*

---

[22] *an* = without; *ur* = urine; *ia* = condition
[23] *diabetes* = passing through
[24] *melli* = honey, sweet
[25] *insipid* = tasteless

## Glomerular Filtration Rate

Assessment of kidney disease often calls for a measurement of GFR. We cannot determine GFR from urea excretion for two reasons: (1) Some of the urea in the urine is secreted by the renal tubule, not filtered by the glomerulus, and (2) much of the urea filtered by the glomerulus is reabsorbed by the tubule. To measure GFR ideally requires a substance that is not secreted or reabsorbed at all, so that all of it in the urine gets there by glomerular filtration.

There doesn't appear to be a single urine solute produced by the body that is not secreted or reabsorbed to some degree. However, several plants, including garlic and artichoke, produce a polysaccharide called inulin (not to be confused with insulin) that is useful for GFR measurement. All inulin filtered by the glomerulus remains in the renal tubule and appears in the urine; none is reabsorbed, nor does the tubule secrete it. GFR can be measured by injecting inulin and subsequently measuring the rate of urine output and the concentrations of inulin in the blood and urine.

For inulin, GFR is equal to the renal clearance. Suppose, for example, that a patient's plasma concentration of inulin is P = 0.5 mg/mL, the urine concentration is U = 30 mg/mL, and urine output is V = 2 mL/min. This person has a normal GFR:

$$GFR = UV/P$$
$$= (30 \text{ mg/mL})(2 \text{ mL/min.})/0.5 \text{mg/mL}$$
$$= 120 \text{ mL/min.}$$

In clinical practice, GFR is more often estimated from creatinine excretion. This has a small but acceptable error of measurement, and is an easier procedure than injecting inulin.

A solute that is reabsorbed by the renal tubules will have a renal clearance *less* than the GFR (provided its tubular secretion is less than its rate of reabsorption). This is why the renal clearance of urea is about 60 mL/min. A solute that is secreted by the renal tubules will have a renal clearance *greater* than the GFR (provided its reabsorption does not exceed its secretion). Creatinine, for example, has a renal clearance of 140 mL/min.

### BEFORE YOU GO ON

Answer the following questions to test your understanding of the preceding section:

17. Define *oliguria* and *polyuria*. Which of these is characteristic of diabetes?

18. Identify a cause of glycosuria other than diabetes mellitus.

19. How is the diuresis produced by furosemide like the diuresis produced by diabetes mellitus? How are they different?

20. Explain why GFR cannot be determined by measuring the amount of NaCl in the urine.

## 23.7 Urine Storage and Elimination

### Expected Learning Outcomes

When you have completed this section, you should be able to

a. describe the functional anatomy of the ureters, urinary bladder, and male and female urethra; and

b. explain how the nervous system and urethral sphincters control the voiding of urine.

Urine is produced continually, but fortunately it does not drain continually from the body. Urination is episodic—occurring when we allow it. This is made possible by an apparatus for storing urine and by neural controls for its timely release.

### The Ureters

The renal pelvis funnels urine into the ureter, a retroperitoneal, muscular tube that extends to the urinary bladder. The ureter is about 25 cm long and reaches a maximum diameter of about 1.7 cm near the bladder. The ureters pass posterior to the bladder and enter it from below, passing obliquely through its muscular wall and opening onto its floor. A small flap of mucosa acts as a valve at the opening of each ureter into the bladder, preventing urine from backing up into the ureter when the bladder contracts.

The ureter has three layers: an adventitia, muscularis, and mucosa. The adventitia is a connective tissue layer that binds it to the surrounding tissues. The muscularis consists of two layers of smooth muscle over most of its length, but a third layer appears in the lower ureter. The mucosa has a transitional epithelium that begins in the minor calyces of the kidney and extends from there through the bladder.

When urine enters the ureter and stretches it, the muscularis contracts and initiates a peristaltic wave that milks the urine from the renal pelvis down to the bladder. These contractions occur every few seconds to few minutes, proportional to the rate at which urine enters the ureter. The lumen of the ureter is very narrow and is easily obstructed or injured by kidney stones (see Deeper Insight 23.2).

### The Urinary Bladder

The urinary bladder (fig. 23.23) is a muscular sac on the floor of the pelvic cavity, inferior to the peritoneum and posterior to the pubic symphysis. It is covered by parietal peritoneum on its flattened superior surface and by a fibrous adventitia elsewhere. Its muscularis, called the **detrusor**[26] (deh-TROO-zur), consists of three layers of smooth muscle. The openings of the two ureters and the urethra mark a smooth-surfaced triangular

---

[26] *de* = down; *trus* = push

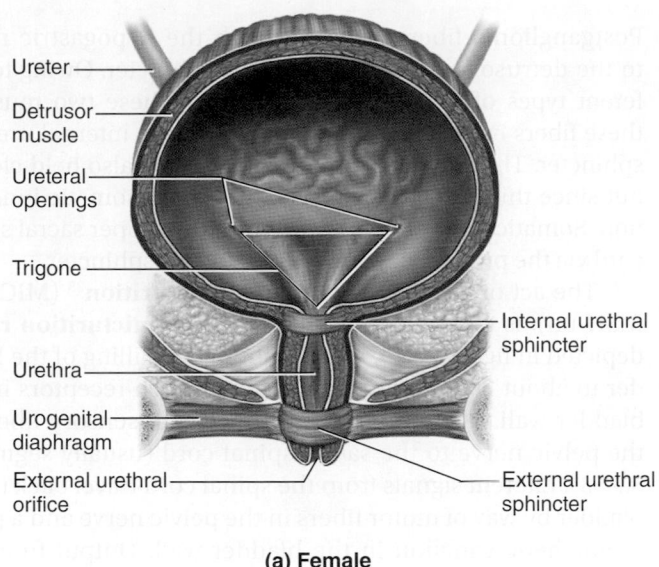

Ureter
Detrusor muscle
Ureteral openings
Trigone
Internal urethral sphincter
Urethra
Urogenital diaphragm
External urethral orifice
External urethral sphincter

**(a) Female**

**FIGURE 23.23 The Urinary Bladder and Urethra (Frontal Sections).** AP|R

 *Why are women more susceptible than men to bladder infections?*

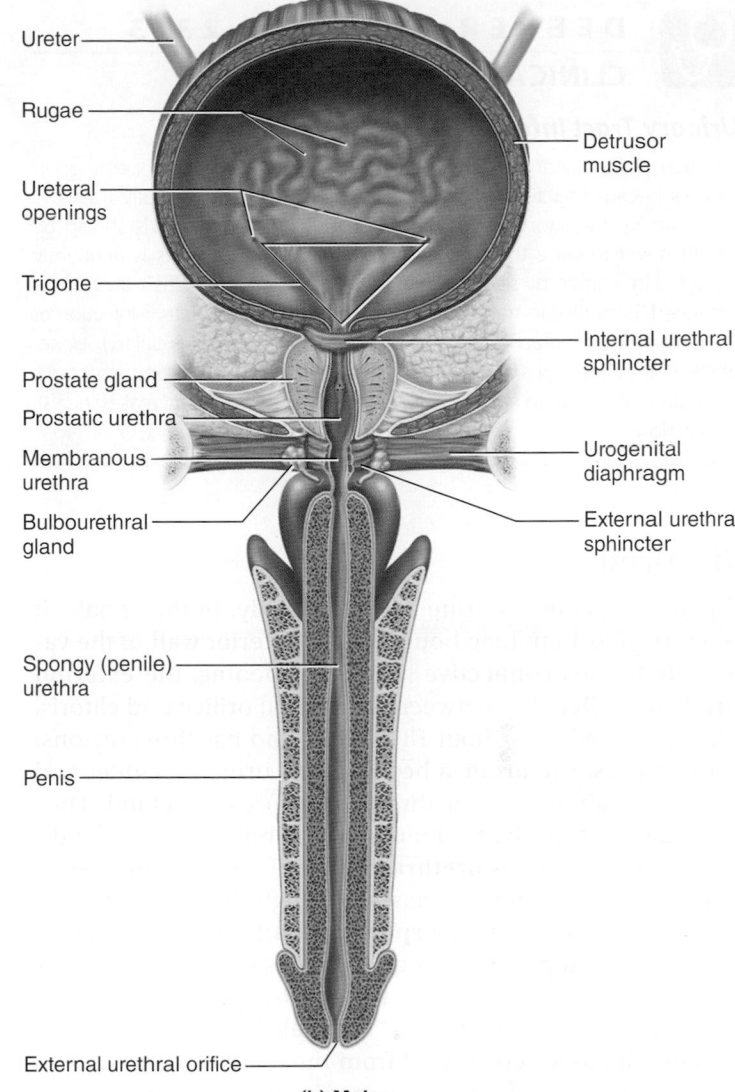

Ureter
Rugae
Detrusor muscle
Ureteral openings
Trigone
Internal urethral sphincter
Prostate gland
Prostatic urethra
Membranous urethra
Urogenital diaphragm
External urethral sphincter
Bulbourethral gland
Spongy (penile) urethra
Penis
External urethral orifice

**(b) Male**

## DEEPER INSIGHT 23.2

### CLINICAL APPLICATION

#### Kidney Stones

A *renal calculus*[27] (kidney stone) is a hard granule composed usually of calcium phosphate or calcium oxalate, but sometimes uric acid or a magnesium salt called struvite. Renal calculi form in the renal pelvis and are usually small enough to pass unnoticed in the urine flow. Some, however, grow as large as several centimeters and block the renal pelvis or ureter, which can lead to the destruction of nephrons as pressure builds in the kidney. A large, jagged calculus passing down the ureter stimulates strong contractions that can be excruciatingly painful. It can also damage the ureter and cause hematuria. Causes of renal calculi include hypercalcemia, dehydration, pH imbalances, frequent urinary tract infections, or an enlarged prostate gland causing urine retention. Calculi are sometimes treated with stone-dissolving drugs, but often they require surgical removal. A nonsurgical technique called *lithotripsy*[28] uses ultrasound to pulverize the calculi into fine granules easily passed in the urine.

---

[27] *calc* = calcium, stone; *ul* = little
[28] *litho* = stone; *tripsy* = crushing

area called the **trigone**[29] (TRY-goan) on the bladder floor. This is a common site of bladder infection (see Deeper Insight 23.3).

The mucosa is lined with transitional epithelium, whose unique surface *umbrella cells* protect it from the hypertonic and acidic urine as described in chapter 5 (p. 150). The epithelium is thicker here than anywhere else in the urinary tract, since it is subject to the most prolonged exposure to urine. When the bladder is empty, it is typically five or six cells thick and exhibits wrinkles called **rugae**[30] (ROO-jee). As the bladder fills, the rugae flatten and the epithelium thins to two or three cells thick. A moderately full bladder contains about 500 mL of urine and extends about 12.5 cm from top to bottom. The maximum capacity is 700 to 800 mL.

---

[29] *tri* = three; *gon* = angle
[30] *rugae* = folds, wrinkles

## DEEPER INSIGHT 23.3
### CLINICAL APPLICATION

#### Urinary Tract Infection (UTI)

Infection of the urinary bladder is called *cystitis*.[31] It is especially common in females because bacteria such as *Escherichia coli* can travel easily from the perineum up the short urethra. Because of this risk, young girls should be taught never to wipe the anus in a forward direction. Cystitis is frequently triggered in women by sexual intercourse ("honeymoon cystitis"). If cystitis is untreated, bacteria can spread up the ureters and cause *pyelitis*,[32] infection of the renal pelvis. If it reaches the renal cortex and nephrons, it is called *pyelonephritis*. Kidney infections can also result from invasion by blood-borne bacteria. Urine stagnation due to renal calculi or prostate enlargement increases the risk of infection.

## The Urethra

The urethra conveys urine out of the body. In the female, it is a tube 3 to 4 cm long bound to the anterior wall of the vagina by fibrous connective tissue. Its opening, the **external urethral orifice,** lies between the vaginal orifice and clitoris. The male urethra is about 18 cm long and has three regions: (1) The **prostatic urethra** begins at the urinary bladder and passes for about 2.5 cm through the prostate gland. During orgasm, it receives semen from the reproductive glands. (2) The **membranous urethra** is a short (0.5 cm), thin-walled portion where the urethra passes through the muscular floor of the pelvic cavity. (3) The **spongy (penile) urethra** is about 15 cm long and passes through the penis to the external urethral orifice. It is named for the *corpus spongiosum* of the penis, through which it passes. The male urethra assumes an S shape: It passes downward from the bladder, turns anteriorly as it enters the root of the penis, and then turns about 90° downward again as it enters the external, pendant part of the penis. The mucosa has a transitional epithelium near the bladder, a pseudostratified epithelium for most of its length, and finally a stratified squamous epithelium near the external urethral orifice. There are mucous **urethral glands** in its wall.

In both sexes, the detrusor is thickened near the urethra to form an **internal urethral sphincter,** which compresses the urethra and retains urine in the bladder. Since this sphincter is composed of smooth muscle, it is under involuntary control. Where the urethra passes through the pelvic floor, it is encircled by an **external urethral sphincter** of skeletal muscle, which provides voluntary control over the voiding of urine.

## Voiding Urine

Between acts of urination, when the bladder is filling, it is important that the detrusor relaxes and the urethral sphincters remain tightly closed. This is ensured by sympathetic pathways that originate in the upper lumbar spinal cord.

Postganglionic fibers travel through the hypogastric nerve to the detrusor and internal urethral sphincter. Owing to different types of adrenergic receptors on these two muscles, these fibers *relax* the detrusor and *excite* the internal urethral sphincter. The external urethral sphincter is also held closed, but since this is skeletal muscle, it receives somatic innervation. Somatic motor fibers course from the upper sacral spinal cord via the pudendal nerve to the external sphincter.

The act of urinating, also called **micturition**[33] (MIC-too-RISH-un), is controlled partly by a spinal **micturition reflex** depicted in figure 23.24, steps 1 through 4. Filling of the bladder to about 200 mL or more excites stretch receptors in the bladder wall. They issue signals by way of sensory fibers in the pelvic nerve to the sacral spinal cord (usually segments S2–S3). Efferent signals from the spinal cord travel back to the bladder by way of motor fibers in the pelvic nerve and a parasympathetic ganglion in the bladder wall. Output from the ganglion excites the detrusor and relaxes the internal urethral sphincter. This results in emptying of the bladder. If there was no voluntary control over urination, this reflex would be the only means of control—which is the case in very young children and in people with spinal cord injuries that disconnect the brain from the lower spinal cord.

Normally, however, one also has voluntary control over urination (steps 5–8). Some input from the stretch receptors ascends the spinal cord to a nucleus in the pons called the **micturition center.** This nucleus integrates information about bladder tension with information from other brain centers such as the amygdala and cerebrum. Thus, urination can be prompted by fear or inhibited by knowledge that the circumstances are inappropriate for urination.

Fibers from the micturition center descend the spinal cord through the reticulospinal tracts. Some of these fibers inhibit sympathetic neurons that normally keep the internal urethral sphincter contracted, thus allowing for its relaxation. Others descend farther to the sacral spinal cord and excite the parasympathetic neurons that stimulate the detrusor and relax the internal urethral sphincter. The initial contraction of the detrusor raises pressure in the bladder, further exciting the stretch receptors that started the process; thus, a positive feedback loop is established and intensifies bladder contraction as urination proceeds.

For the bladder to begin emptying, however, one final obstacle must be overcome—the external urethral sphincter. Nerve fibers from the cerebral cortex descend by way of the corticospinal tracts and inhibit sacral somatic motor neurons that normally keep that sphincter constricted. It is this voluntary component of micturition that gives a person conscious control of when to urinate and the ability to stop urination in midstream. Voluntary control develops as the nervous system matures in early childhood. Males expel the last few milliliters of urine by voluntarily contracting the bulbocavernosus muscle that ensheaths the root of the penis. This helps to reduce the retention of urine in the longer male urethra.

---

[31] *cyst* = bladder; *itis* = inflammation
[32] *pyel* = pelvis; *itis* = inflammation

[33] *mictur* = to urinate

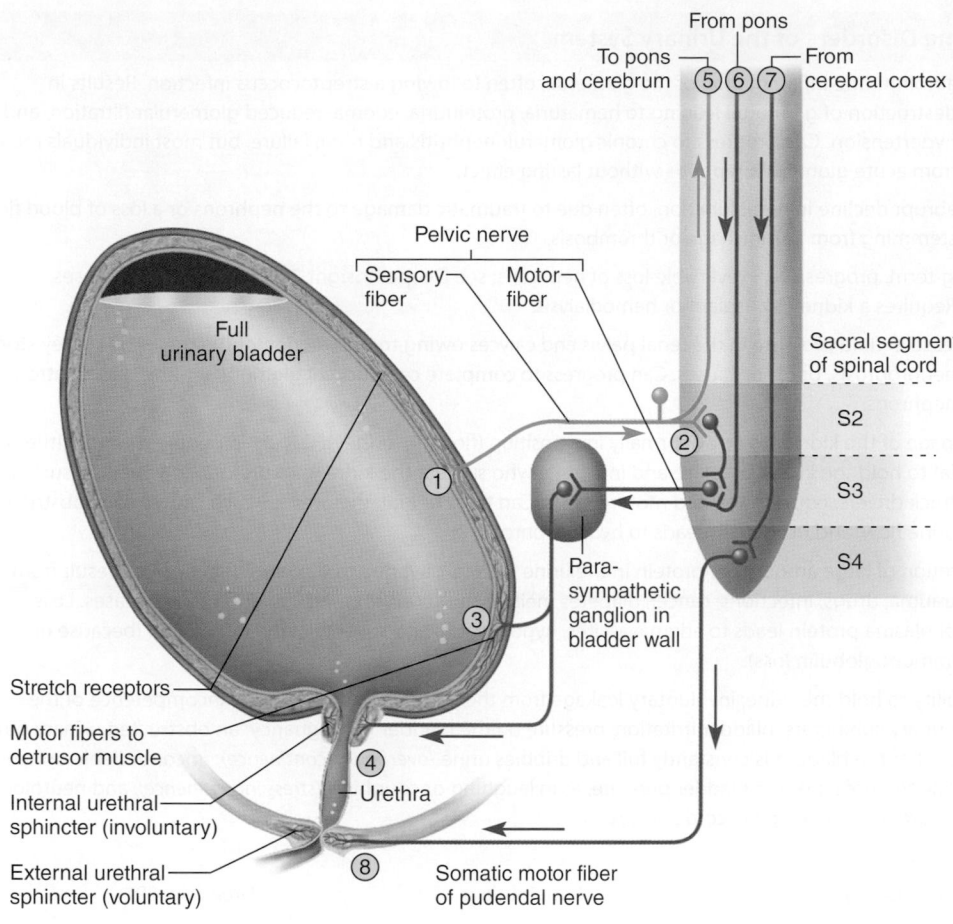

From pons

To pons — and cerebrum ⑤⑥⑦ — From cerebral cortex

Pelvic nerve

Sensory fiber    Motor fiber

Full urinary bladder

Sacral segments of spinal cord

S2

② S3

S4

① 

Para-sympathetic ganglion in bladder wall

③ 

Stretch receptors

Motor fibers to detrusor muscle

Internal urethral sphincter (involuntary)

④ — Urethra

External urethral sphincter (voluntary)

⑧    Somatic motor fiber of pudendal nerve

**FIGURE 23.24  Neural Control of Micturition.**

**Involuntary micturition reflex**

① Stretch receptors detect filling of bladder, transmit afferent signals to spinal cord.

② Signals return to bladder from spinal cord segments S2 and S3 via parasympathetic fibers in pelvic nerve.

③ Efferent signals excite detrusor muscle.

④ Efferent signals relax internal urethral sphincter. Urine is involuntarily voided if not inhibited by brain.

**Voluntary control**

⑤ For voluntary control, micturition center in pons receives signals from stretch receptors.

⑥ If it is timely to urinate, pons returns signals to spinal interneurons that excite detrusor and relax internal urethral sphincter. Urine is voided.

⑦ If it is untimely to urinate, signals from cerebrum excite spinal interneurons that keep external urethral sphincter contracted. Urine is retained in bladder.

⑧ If it is timely to urinate, signals from cerebrum inhibit sacral neurons that keep external sphincter closed. External urethral sphincter relaxes and urine is voided.

If the urge to urinate arises at an inconvenient time and one must suppress it, the stretch receptors fatigue and stop firing. As bladder tension increases, however, the signals return with increasing frequency and persistence. Conversely, there are times when the bladder is not full enough to trigger the micturition reflex, but one wishes to "go" anyway because of a long drive or lecture coming up. In this case, the Valsalva maneuver (p. 862) is used to compress the bladder and excite the stretch receptors early, thereby getting the reflex started. The Valsalva maneuver also aids in emptying the bladder.

The effects of aging on the urinary system are discussed on page 1121. Some disorders of this system are briefly described in table 23.3.

**DEEPER INSIGHT 23.4**

**CLINICAL APPLICATION**

**Urination and Spinal Cord Injuries**

Knowledge of the neural control of micturition is particularly important for understanding and treating persons with spinal cord injuries. Transection of the spinal cord, as in many cervical fractures, disconnects the *supraspinal* control centers (cerebrum and pons) from the spinal cord circuits that control urination. During the period of spinal shock (see p. 503), a person is generally incontinent—lacking any control over urination. Bladder control returns as the spinal cord recovers, but is limited to the involuntary micturition reflex. The bladder often cannot empty completely, and there is consequently an increased incidence of cystitis.

| TABLE 23.3 | Some Disorders of the Urinary System |
|---|---|
| Acute glomerulonephritis | An autoimmune inflammation of the glomeruli, often following a streptococcus infection. Results in destruction of glomeruli leading to hematuria, proteinuria, edema, reduced glomerular filtration, and hypertension. Can progress to chronic glomerulonephritis and renal failure, but most individuals recover from acute glomerulonephritis without lasting effect. |
| Acute renal failure | An abrupt decline in renal function, often due to traumatic damage to the nephrons or a loss of blood flow stemming from hemorrhage or thrombosis. |
| Chronic renal failure | Long-term, progressive, irreversible loss of nephrons; see Deeper Insight 23.5 for a variety of causes. Requires a kidney transplant or hemodialysis. |
| Hydronephrosis[34] | Increase in fluid pressure in the renal pelvis and calyces owing to obstruction of the ureter by kidney stones, nephroptosis, or other causes. Can progress to complete cessation of glomerular filtration and atrophy of nephrons. |
| Nephroptosis[35] (NEFF-rop-TOE-sis) | Slippage of the kidney to an abnormally low position (floating kidney). Occurs in people with too little body fat to hold the kidney in place and in people who subject the kidneys to prolonged vibration, such as truck drivers, equestrians, and motorcyclists. Can twist or kink the ureter, which causes pain, obstructs urine flow, and potentially leads to hydronephrosis. |
| Nephrotic syndrome | Excretion of large amounts of protein in the urine (≥3.5 g/day) due to glomerular injury. Can result from trauma, drugs, infections, cancer, diabetes mellitus, lupus erythematosus, and other diseases. Loss of plasma protein leads to edema, ascites, hypotension, and susceptibility to infection (because of immunoglobulin loss). |
| Urinary incontinence | Inability to hold the urine; involuntary leakage from the bladder. Can result from incompetence of the urinary sphincters; bladder irritation; pressure on the bladder in pregnancy; an obstructed urinary outlet so that the bladder is constantly full and dribbles urine (overflow incontinence); uncontrollable urination due to brief surges in bladder pressure, as in laughing or coughing (stress incontinence); and neurological disorders such as spinal cord injuries. |

*Disorders described elsewhere*

| | | | |
|---|---|---|---|
| Anuria p. 913 | Kidney stones p. 915 | Proteinuria p. 899 | Uremia p. 891 |
| Azotemia p. 891 | Nephrosclerosis p. 900 | Pyuria p. 912 | Urinary tract infection p. 916 |
| Hematuria p. 899 | Oliguria p. 912 | | |

[34] *hydro* = water; *nephr* = kidney; *osis* = medical condition
[35] *nephro* = kidney; *ptosis* = sagging, falling

**BEFORE YOU GO ON**

Answer the following questions to test your understanding of the preceding section:

21. Describe the location and function of the detrusor.

22. Compare and contrast the functions of the internal and external urethral sphincters.

23. In males, the sympathetic nervous system triggers ejaculation and, at the same time, stimulates constriction of the internal urethral sphincter. What purpose is served by the latter action?

## DEEPER INSIGHT 23.5

### CLINICAL APPLICATION

#### *Renal Insufficiency and Hemodialysis*

*Renal insufficiency* is a state in which the kidneys cannot maintain homeostasis due to extensive destruction of their nephrons. Some causes of nephron destruction include

- Hypertension.
- Chronic or repetitive kidney infections.
- Trauma from such causes as blows to the lower back or continual vibration from machinery.
- Prolonged ischemia and hypoxia, as in long-distance runners and swimmers.
- Poisoning by heavy metals such as mercury and lead and solvents such as carbon tetrachloride, acetone, and paint thinners. These are absorbed into the blood from inhaled fumes or by skin contact and then filtered by the glomeruli. They kill renal tubule cells.
- Blockage of renal tubules with proteins small enough to be filtered by the glomerulus—for example, myoglobin released by skeletal muscle damage and hemoglobin released by a transfusion reaction.
- Atherosclerosis, which reduces blood flow to the kidney.
- Glomerulonephritis, an autoimmune disease of the glomerular capillaries.

Nephrons can regenerate and restore kidney function after short-term injuries. Even when some of the nephrons are irreversibly destroyed, others hypertrophy and compensate for their lost function. Indeed, a person can survive on as little as one-third of one kidney. When 75% of the nephrons

are lost, however, urine output may be as low as 30 mL/h compared with the normal rate of 50 to 60 mL/h. This is insufficient to maintain homeostasis and is accompanied by azotemia and acidosis. Uremia develops when there is 90% loss of renal function. Renal insufficiency also tends to cause anemia because the diseased kidneys produce too little erythropoietin (EPO), the hormone that stimulates red blood cell formation.

*Hemodialysis* is a procedure for artificially clearing wastes from the blood when the kidneys are not adequately doing so (fig. 23.25). Blood is pumped from the radial artery to a *dialysis machine* (artificial kidney) and returned to the patient by way of a vein. In the dialysis machine, the blood flows through a semipermeable cellophane tube surrounded by dialysis fluid. Urea, potassium, and other solutes that are more concentrated in the blood than in the dialysis fluid diffuse through the membrane into the fluid, which is discarded. Glucose, electrolytes, and drugs can be administered by adding them to the dialysis fluid so they will diffuse through the membrane into the blood. People with renal insufficiency accumulate substantial amounts of excess body water between treatments, and dialysis serves also to remove it. Patients are typically given erythropoietin (EPO) to compensate for the lack of EPO from the failing kidneys.

Hemodialysis patients typically have three sessions per week for 4 to 8 hours per session. In addition to inconvenience, hemodialysis carries risks of infection and thrombosis. Blood tends to clot when exposed to foreign surfaces, so an anticoagulant such as heparin is added during dialysis. Unfortunately, this inhibits clotting in the patient's body as well, and dialysis patients sometimes suffer internal bleeding.

A procedure called *continuous ambulatory peritoneal dialysis (CAPD)* is more convenient. It can be carried out at home by the patient, who is provided with plastic bags of dialysis fluid. Fluid is introduced into the abdominal cavity through an indwelling catheter. Here, the peritoneum provides over 2 $m^2$ of blood-rich semipermeable membrane. The fluid is left in the body cavity for 15 to 60 minutes to allow the blood to equilibrate with it; then it is drained, discarded, and replaced with fresh dialysis fluid. The patient is not limited by a stationary dialysis machine and can go about most normal activities. CAPD is less expensive and promotes better morale than conventional hemodialysis, but it is less efficient in removing wastes and it is more often complicated by infection.

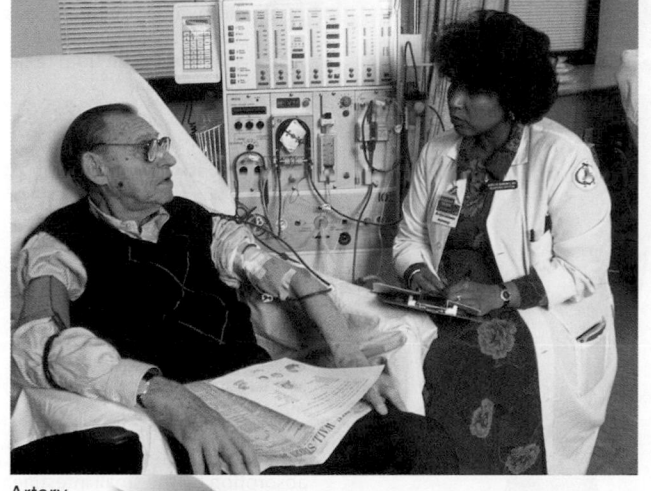

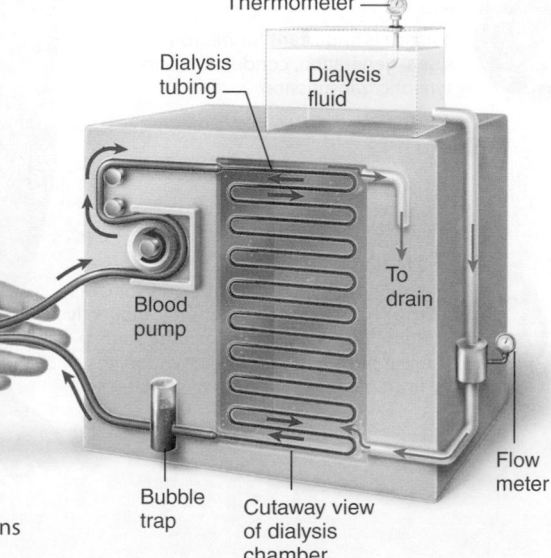

**FIGURE 23.25 Hemodialysis.** Blood is pumped into a dialysis chamber, where it flows through a selectively permeable membrane surrounded by dialysis fluid. Blood leaving the chamber passes through a bubble trap to remove air before it is returned to the patient's body. The fluid picks up excess water and metabolic wastes from the patient's blood and may contain medications that diffuse into the blood.

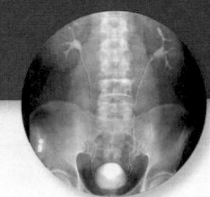

# CONNECTIVE ISSUES

## Effects of the URINARY SYSTEM
## On Other Organ Systems

**ALL SYSTEMS**
Excretes metabolic wastes to prevent poisoning of the tissues; maintains fluid, electrolyte, and acid–base balance necessary for homeostasis

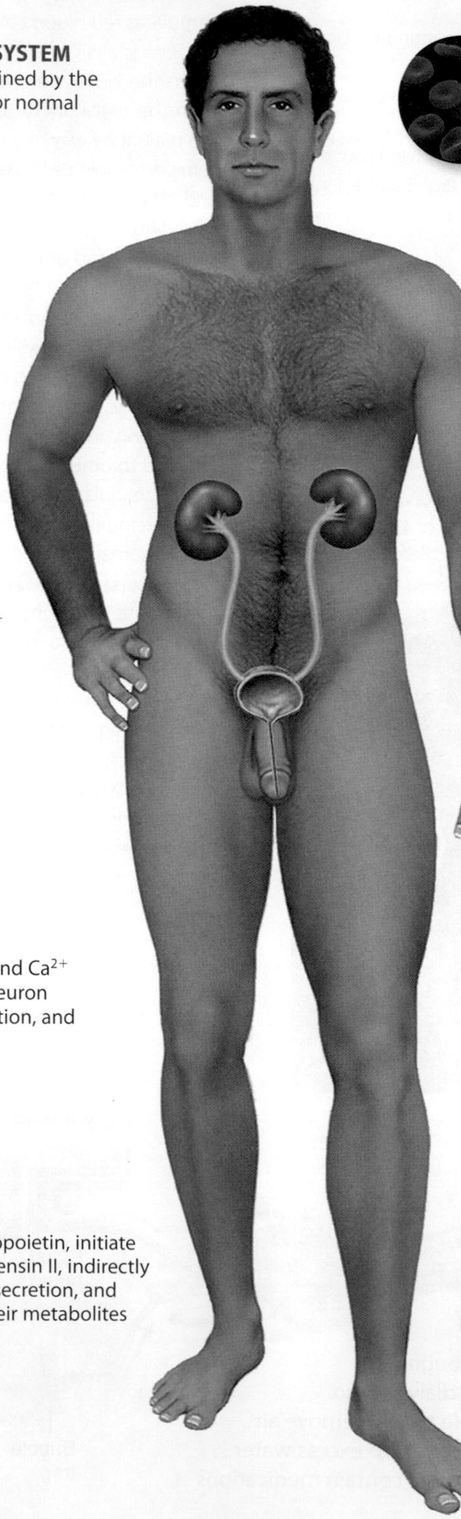

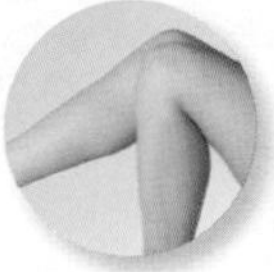

**INTEGUMENTARY SYSTEM**
Fluid balance maintained by the kidneys is essential for normal secretion of sweat.

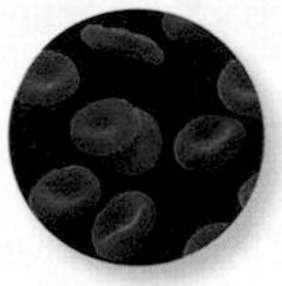

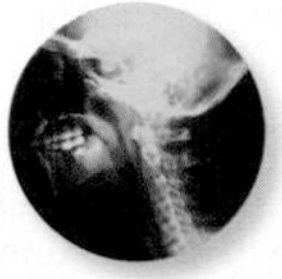

**SKELETAL SYSTEM**
Calcitriol synthesis and other roles of the kidneys in calcium and phosphate homeostasis are necessary for normal bone deposition and maintenance.

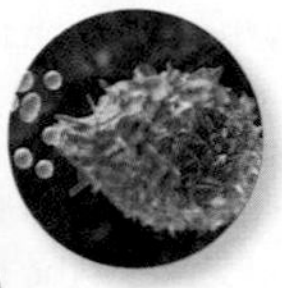

**MUSCULAR SYSTEM**
Renal control of $Na^+$, $K^+$, and $Ca^{2+}$ balance is important for muscle excitability and contractility.

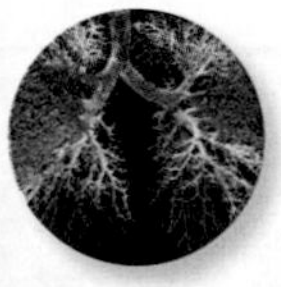

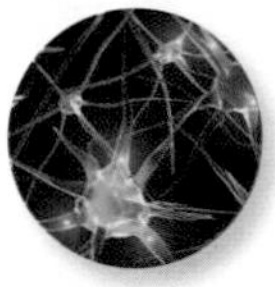

**NERVOUS SYSTEM**
Renal control of $Na^+$, $K^+$, and $Ca^{2+}$ balance is important for neuron signal generation, conduction, and synaptic transmission.

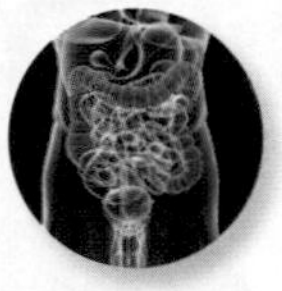

**ENDOCRINE SYSTEM**
Kidneys secrete erythropoietin, initiate the synthesis of angiotensin II, indirectly stimulate aldosterone secretion, and clear hormones and their metabolites from the body.

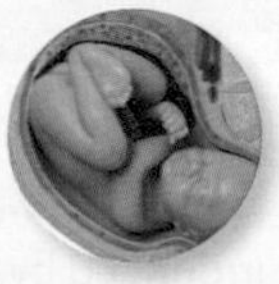

**CIRCULATORY SYSTEM**
Kidneys affect blood pressure more than any other organ but the heart and regulate blood composition; renal dysfunction can cause electrolyte imbalances that affect the cardiac rhythm.

**LYMPHATIC/ IMMUNE SYSTEM**
Acidity of urine provides nonspecific defense against urinary tract infections; renal failure burdens lymphatic system by creating fluid retention and edema.

**RESPIRATORY SYSTEM**
Respiratory rhythm is sensitive to acid–base imbalances that may result from renal dysfunction.

**DIGESTIVE SYSTEM**
Kidneys excrete toxins absorbed by intestines; kidneys excrete metabolites generated by the liver; calcitriol secreted by the kidneys stimulates calcium absorption by the small intestine.

**REPRODUCTIVE SYSTEM**
Male urethra serves as common passage for urine and semen; maternal urinary system excretes fetal wastes.

# STUDY GUIDE

## ▶ Assess Your Learning Outcomes

*To test your knowledge, discuss the following topics with a study partner or in writing, ideally from memory.*

### 23.1 Functions of the Urinary System (p. 890)

1. The six main organs of the urinary system
2. Six or more functions of the kidneys
3. Four main nitrogenous wastes and their metabolic sources; how metabolic wastes differ from other wastes
4. Blood urea nitrogen, azotemia, and uremia, and the pathological effects of uremia
5. The meaning of *excretion,* and four human organ systems that perform it

### 23.2 Anatomy of the Kidney (p. 892)

1. Location of the kidneys in relation to adjacent tissues and organs
2. Structures that enter and leave the kidney through the hilum
3. Three layers of tissue that surround and encapsulate the kidney
4. Arrangement of the renal parenchyma around the renal sinus; organization of the cortex, medulla, and renal pyramids
5. The relationship of a renal pyramid to a minor calyx, and of the minor calyces to the major calyces, renal pelvis, and ureter
6. The term for the microscopic functional units of the kidney, and their approximate number per kidney
7. Blood flow from the renal artery through the kidney to the renal vein, including circulation through both the cortex and the medulla
8. Structure of the renal corpuscle
9. Fluid flow from the point where it filters from the blood to the point where it leaves the kidney; those parts of the renal tubule that belong to a single nephron and the part that is shared by multiple nephrons
10. The distinction between glomerular filtrate, tubular fluid, and urine in relation to the progress of the fluid through the renal tubule
11. Differences in the structure and function of cortical nephrons and juxtamedullary nephrons
12. Innervation of the kidney and two effects of sympathetic stimulation on renal function

### 23.3 Urine Formation I: Glomerular Filtration (p. 898)

1. Four basic stages of urine formation
2. Structure of the glomerular filtration membrane; roles of the capillary fenestrations, basement membrane, and filtration slits in producing glomerular filtrate; the difference between the filtrate and blood plasma
3. Definitions of *proteinuria* and *hematuria,* and their causes
4. Forces of capillary filtration that account for the net filtration pressure (NFP) in the glomerulus; the magnitude of each and of the NFP
5. How glomerular filtration rate can be calculated from the NFP and filtration coefficient; typical GFR values
6. The meaning of *renal autoregulation*
7. How the myogenic mechanism of renal autoregulation works
8. Structure and function of the juxtaglomerular apparatus and how tubuloglomerular feedback works
9. How the sympathetic nervous system regulates GFR
10. The renin–angiotensin–aldosterone mechanism of regulating GFR; multiple effects of angiotensin II on the kidney and on the rest of the body
11. Effects of antidiuretic hormone and aldosterone on the kidney and how they reduce water loss from the body

### 23.4 Urine Formation II: Tubular Reabsorption and Secretion (p. 904)

1. The percentage of the glomerular filtrate that is eliminated as urine, and percentages of the water in the filtrate that are reabsorbed by PCT and nephron loop
2. Transcellular and paracellular routes of reabsorption, and the role of solvent drag
3. How the PCT reabsorbs NaCl, other electrolytes, glucose, urea, and water; why the reabsorption of water and solutes depends directly or indirectly on the sodium–glucose transporter (SGLT) and sodium reabsorption
4. How tubular reabsorption is limited by the transport maximum ($T_m$); how this relate to glycosuria in diabetes mellitus

5. Substances added to the tubular fluid by tubular secretion in the PCT
6. The primary functions of the nephron loop and DCT, and what they reabsorb
7. Effects of aldosterone, natriuretic peptides, parathyroid hormone, calcitonin, and calcitriol on nephron function

### 23.5 Urine Formation III: Water Conservation (p. 908)

1. Function of the collecting duct (CD) and the range of urine osmolarities it can produce
2. How the osmotic gradient of the renal medulla and selective permeability of the CD act to concentrate the urine
3. The effect of antidiuretic hormone (ADH) on the CD and the role of aquaporins in this effect
4. Function of the countercurrent multiplier and how it performs that role
5. Function of the countercurrent exchanger and how it performs that role

### 23.6 Urine and Renal Function Tests (p. 911)

1. Why urine is yellow; why the shade of yellow varies; and some causes of other, unusual colors
2. Normal ranges of urine specific gravity, osmolarity, and pH
3. Normal and abnormal odors of urine and some reasons for the latter
4. The three most abundant solutes in urine, and some causes of other, unusual solutes
5. Typical daily output of urine, and terms for abnormally low and high outputs
6. The defining sign of diabetes in general; four forms of diabetes and their causes
7. The general effect of diuretics; modes of diuretic action of caffeine, alcohol, and loop diuretics
8. Methods of measuring a person's GFR and renal clearance; ability to calculate these if given the necessary data

### 23.7 Urine Storage and Elimination (p. 914)

1. The route and mechanism of urine transport from the kidney to the urinary bladder; anatomy and histology of the ureters and their relationship to the bladder

# STUDY GUIDE

2. Anatomy and histology of the urinary bladder; the detrusor, mucosal epithelium, rugae, trigone, and internal urethral sphincter

3. Anatomy of the female urethra and external urethral sphincter

4. Anatomy of the male urethra, its three segments, and the external urethral sphincter

5. The mechanism of the spinal micturition reflex and the neural anatomy involved

6. Mechanisms of brainstem and cerebral control of micturition

## ▶ Testing Your Recall

*Answers in Appendix B*

1. Micturition occurs when the _____ contracts.
   a. detrusor
   b. internal urethral sphincter
   c. external urethral sphincter
   d. muscularis of the ureter
   e. all of the above

2. The compact ball of capillaries in a nephron is called
   a. the nephron loop.
   b. the peritubular plexus.
   c. the renal corpuscle.
   d. the glomerulus.
   e. the vasa recta.

3. Which of these is the most abundant nitrogenous waste in the blood?
   a. uric acid
   b. urea
   c. ammonia
   d. creatinine
   e. albumin

4. Which of these lies closest to the renal cortex?
   a. the parietal peritoneum
   b. the renal fascia
   c. the fibrous capsule
   d. the perirenal fat capsule
   e. the renal pelvis

5. Most sodium is reabsorbed from the glomerular filtrate by
   a. the vasa recta.
   b. the proximal convoluted tubule.
   c. the distal convoluted tubule.
   d. the nephron loop.
   e. the collecting duct.

6. A glomerulus and glomerular capsule make up one
   a. renal capsule.
   b. renal corpuscle.
   c. kidney lobule.
   d. kidney lobe.
   e. nephron.

7. The kidney has more _____ than any of the other structures listed.
   a. arcuate arteries
   b. minor calyces
   c. medullary pyramids
   d. afferent arterioles
   e. collecting ducts

8. The renal clearance of _____ is normally zero.
   a. sodium
   b. potassium
   c. uric acid
   d. urea
   e. amino acids

9. Beavers have relatively little need to conserve water and could therefore be expected to have _____ than humans do.
   a. fewer nephrons
   b. longer nephron loops
   c. shorter nephron loops
   d. longer collecting ducts
   e. longer convoluted tubules

10. Increased ADH secretion should cause the urine to have
    a. a higher specific gravity.
    b. a lighter color.
    c. a higher pH.
    d. a lower urea concentration.
    e. a lower potassium concentration.

11. The _____ reflex is an autonomic reflex activated by pressure in the urinary bladder.

12. _____ is the ability of a nephron to adjust its GFR independently of nervous or hormonal influences.

13. The two ureters and the urethra form the boundaries of a smooth area called the _____ on the floor of the urinary bladder.

14. The _____ is a group of epithelial cells of the nephron loop that monitors the composition of the tubular fluid.

15. To enter the capsular space, filtrate must pass between foot processes of the _____, cells that form the visceral layer of the glomerular capsule.

16. Glycosuria occurs if the rate of glomerular filtration of glucose exceeds the _____ of the proximal convoluted tubule.

17. _____ is a hormone that regulates the amount of water reabsorbed by the collecting duct.

18. The _____ sphincter is under involuntary control and relaxes during the micturition reflex.

19. Very little _____ is found in the glomerular filtrate because it is negatively charged and is repelled by the basement membrane of the glomerulus.

20. Blood flows through the _____ arteries just before entering the cortical radiate arteries.

# STUDY GUIDE

## ▶ Building Your Medical Vocabulary

*Answers in Appendix B*

*State a medical meaning of each word element below, and give a term in which it or a slight variation of it is used.*

1. azoto-

2. cysto-

3. glomer-

4. juxta-

5. meso-

6. nephro-

7. podo-

8. -ptosis

9. pyelo-

10. recto-

## ▶ True or False

*Answers in Appendix B*

*Determine which five of the following statements are false, and briefly explain why.*

1. The proximal convoluted tubule is not subject to hormonal influence.

2. Sodium is the most abundant solute in the urine.

3. The kidney has more distal convoluted tubules than collecting ducts.

4. Tight junctions prevent material from leaking between the epithelial cells of the renal tubule.

5. All forms of diabetes are characterized by glucose in the urine.

6. If all other conditions remain the same, constriction of the afferent arteriole reduces the glomerular filtration rate.

7. Angiotensin II reduces urine output.

8. The minimum osmolarity of urine is 300 mOsm/L, equal to the osmolarity of the blood.

9. A sodium deficiency (hyponatremia) could cause glycosuria.

10. Micturition depends on contraction of the detrusor.

## ▶ Testing Your Comprehension

*Answers at* www.mhhe.com/saladin7

1. How would the glomerular filtration rate be affected by kwashiorkor (see p. 677)?

2. A patient produces 55 mL of urine per hour. Urea concentration is 0.25 mg/mL in her blood plasma and 8.6 mg/mL in her urine. (a) What is her rate of renal clearance for urea? (b) About 95% of adults excrete urea at a rate of 12.6 to 28.6 g/day. Is this patient above, within, or below this range? Show how you calculated your answers.

3. A patient with poor renal perfusion is treated with an ACE inhibitor and goes into renal failure. Explain the reason for the renal failure.

4. Drugs called *renin inhibitors* are used to treat hypertension. Explain how they would have this effect.

5. Discuss how the unity of form and function is exemplified by each of the following comparisons: (a) the thin and thick segments of the nephron loop; (b) the proximal and distal convoluted tubules; and (c) the afferent and efferent arterioles.

## ▶ Improve Your Grade at www.mhhe.com/saladin7

*Download animations to study when it fits your schedule. Practice quizzes, labeling activities, games, and flashcards offer fun ways to master the chapter concepts. Or, download image PowerPoint files for each chapter to create a study guide or for taking notes during lecture.*

# WATER, ELECTROLYTE, AND ACID–BASE BALANCE

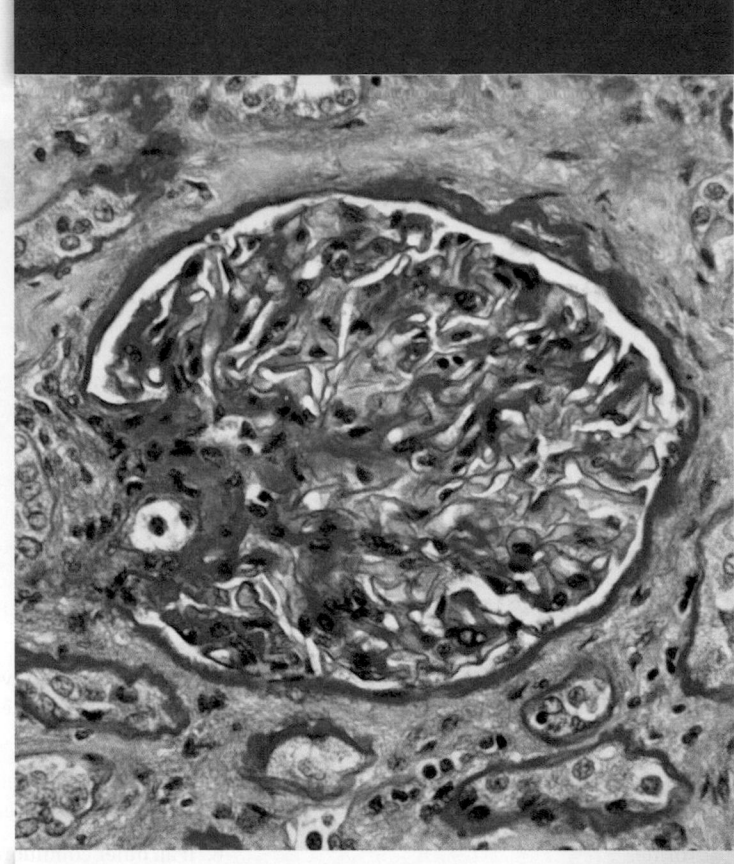

Water balance begins at the renal glomeruli like the one shown here

**Module 13: Urinary System**

Cellular function requires a fluid medium with a carefully controlled composition. If the quantity, osmolarity, electrolyte concentration, or pH of this medium is altered, life-threatening disorders of cellular function may result. Consequently, the body has several mechanisms for keeping these variables within narrow limits and maintaining three types of homeostatic balance:

1. *water balance,* in which average daily water intake and loss are equal;

2. *electrolyte balance,* in which the amount of electrolytes absorbed by the small intestine balance the amount lost from the body, chiefly through the urine; and

3. *acid–base balance,* in which the body rids itself of acid (hydrogen ions) at a rate that balances its metabolic production, thus maintaining a stable pH.

These balances are maintained by the collective action of the urinary, respiratory, digestive, integumentary, endocrine, nervous, cardiovascular, and lymphatic systems. This chapter describes the homeostatic regulation of water, electrolyte, and acid–base balance and shows the close relationship of these variables to each other. These balances are so crucial that fluid therapy, aimed at restoring one or more of these, is often a critical aspect of patient care. Problems of fluid and electrolyte balance also arise frequently in sports and recreational medicine, in situations ranging from summer football practice to back-country hiking.

## 24.1 Water Balance

### Expected Learning Outcomes

When you have completed this section, you should be able to

a. name the major fluid compartments and explain how water moves from one to another;

b. list the body's sources of water and routes of water loss;

c. describe the mechanisms of regulating water intake and output; and

d. describe some conditions in which the body has a deficiency or excess of water or an improper distribution of water among the fluid compartments.

We enter the world in a rather soggy condition, having swallowed, excreted, and floated in amniotic fluid for months. At birth, a baby's weight is as much as 75% water; infants normally lose a little weight in the first day or two as they excrete the excess. Young adult men average 55% to 60% water; women average slightly less because they have more adipose tissue, which is nearly free of water. Obese and elderly people are as little as 45% water by weight. The **total body water (TBW)** content of a 70 kg (150 lb) young male is about 40 L.

## Fluid Compartments

Body water is distributed among certain **fluid compartments,** areas separated by selectively permeable membranes and differing from each other in chemical composition. The major fluid compartments are:

65% *intracellular fluid (ICF)* and

35% *extracellular fluid (ECF),* subdivided into

    25% *tissue (interstitial) fluid,*

    8% *blood plasma* and *lymph,* and

    2% *transcellular fluid,* a catch-all category for cerebrospinal, synovial, peritoneal, pleural, and pericardial fluids; vitreous and aqueous humors of the eye; bile; and fluid in the digestive, urinary, and respiratory tracts.

Fluid is continually exchanged between compartments by way of capillary walls and plasma membranes (fig. 24.1). Water moves by osmosis from the digestive tract to the bloodstream and by capillary filtration from the blood to the tissue fluid. From the tissue fluid, it may be reabsorbed by the capillaries, osmotically absorbed into cells, or taken up by the lymphatic system, which returns it to the bloodstream.

Because water moves so easily through plasma membranes, osmotic gradients between the ICF and ECF never last for very long. If a local imbalance arises, osmosis usually restores the balance within seconds so that intracellular and extracellular osmolarity are equal. If the osmolarity of the tissue fluid rises, water moves out of the cells; if it falls, water moves into the cells.

Osmosis from one fluid compartment to another is determined by the relative concentration of solutes in each compartment. The most abundant solute particles by far are the electrolytes—especially sodium salts in the ECF and potassium salts in the ICF. Electrolytes play the principal role in governing the body's water distribution and total water content; the subjects of water and electrolyte balance are therefore inseparable.

## Water Gain and Loss

A person is in a state of **fluid balance** when daily gains and losses are equal. We typically gain and lose about 2,500 mL/day (fig. 24.2). The gains come from two sources. One of these is **metabolic water** (about 200 mL/day), which is produced as a by-product of dehydration synthesis reactions and aerobic respiration:

$$C_6H_{12}O_6 + 6\,O_2 \rightarrow 6\,CO_2 + 6\,H_2O$$

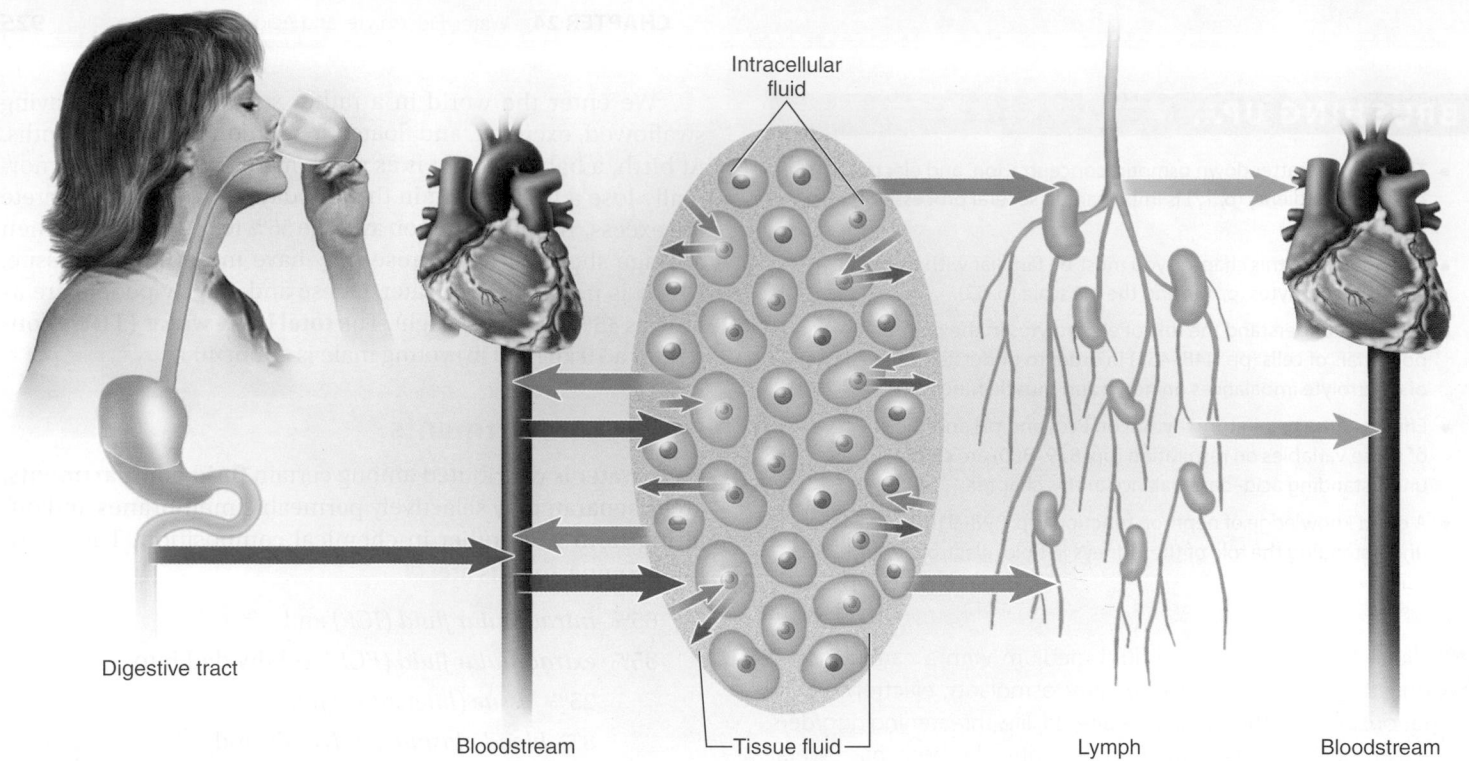

**FIGURE 24.1 The Movement of Water Between the Major Fluid Compartments.** Ingested water is absorbed by the bloodstream. There is a two-way exchange of water between the blood and tissue fluid and between the tissue fluid and intracellular fluids. Excess tissue fluid is picked up by the lymphatic system, which returns it to the bloodstream.

❓ *In which of these compartments would fluid accumulate in edema?*

The other source is **preformed water,** which is ingested in food (700 mL/day) and drink (1,600 mL/day).

The routes of water loss are more varied:

- 1,500 mL/day is excreted as urine.
- 200 mL/day is eliminated in the feces.
- 300 mL/day is lost in the expired breath. You can easily visualize this by breathing onto a cool surface such as a mirror.
- 100 mL/day of sweat is secreted by a resting adult at an ambient air temperature of 20°C (68°F).
- 400 mL/day is lost as **cutaneous transpiration,**[1] water that diffuses through the epidermis and evaporates. This is not the same as sweat; it is not a glandular secretion. A simple way to observe it is to cup the palm of your hand for a minute against a cool nonporous surface such as a laboratory benchtop or mirror. When you take your hand away, you will notice the water that transpired through the skin and condensed on that surface, even in places that were not in contact with your skin.

Water loss varies greatly with physical activity and environmental conditions. Respiratory loss increases in cold

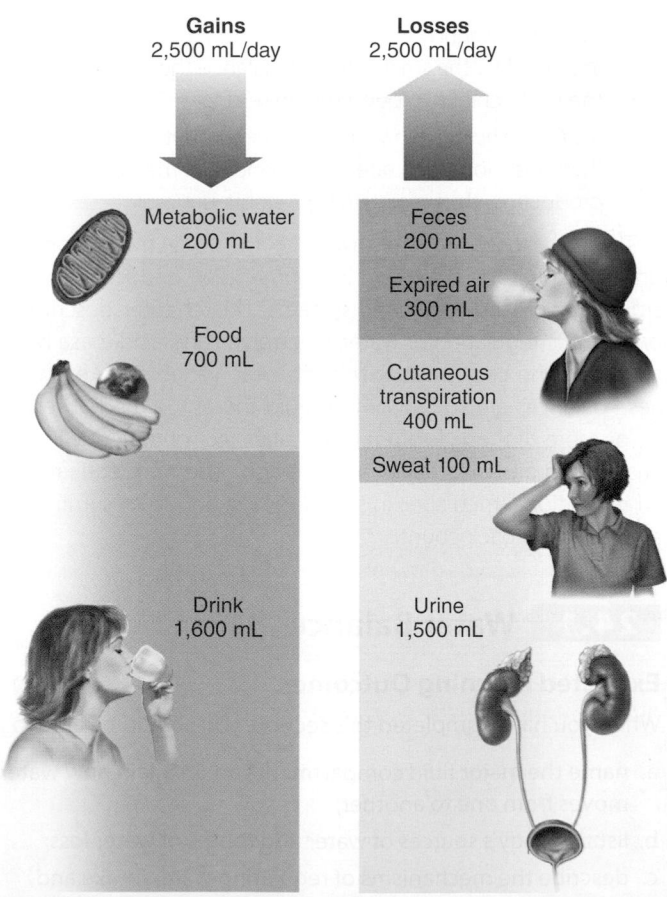

**FIGURE 24.2 Typical Daily Water Gains and Losses in a State of Fluid Balance.**

---

[1] *trans* = through; *spiration* = breathing

weather, for example, because cold air is drier and absorbs more body water from the respiratory tract. Hot, humid weather slightly reduces the respiratory loss but increases perspiration to as much as 1,200 mL/day. Prolonged, heavy work can raise the respiratory loss to 650 mL/day and perspiration to as much as 5 L/day, though it reduces urine output by nearly two-thirds.

Output through the breath and cutaneous transpiration is called **insensible water loss** because we are not usually conscious of it. **Sensible water loss** is noticeable output, particularly through the urine and in case of sufficient sweating to produce obvious wetness of the skin. **Obligatory water loss** is output that is relatively unavoidable: expired air, cutaneous transpiration, sweat, fecal moisture, and the minimum urine output, about 400 mL/day, needed to prevent azotemia. Even dehydrated individuals cannot prevent such losses; thus, they become further dehydrated.

## Regulation of Intake

Fluid intake is governed mainly by thirst, which is controlled by the mechanisms shown in figure 24.3. Dehydration reduces blood volume and pressure and raises blood osmolarity. The hypothalamus has at least three groups of neurons called **osmoreceptors** that respond to angiotensin II and to rising osmolarity of the ECF—both of which are signs that the body has a water deficit. The osmoreceptors communicate with other hypothalamic neurons that produce antidiuretic hormone (ADH), thus promoting water conservation; and they apparently communicate with the cerebral cortex to produce a conscious sense of thirst. A 2% to 3% increase in plasma osmolarity makes a person intensely thirsty, as does a 10% to 15% blood loss.

When we are thirsty, we salivate less. There are two reasons for this: (1) The osmoreceptor response leads to sympathetic output from the hypothalamus that inhibits the salivary glands. (2) Saliva is produced primarily by capillary filtration, but in a dehydrated person, this is opposed by the lower capillary blood pressure and higher osmolarity of the blood. Reduced salivation produces a dry, sticky-feeling mouth and a desire to drink, but it is by no means certain that this is the primary motivation to drink. Some people do not secrete saliva, yet they do not drink more than normal individuals except when eating, when they need water to moisten the food. The same is true of experimental animals that have the salivary ducts tied off.

Long-term satiation of thirst depends on absorbing water from the small intestine and lowering the osmolarity of the blood. Reduced osmolarity stops the osmoreceptor response, promotes capillary filtration, and makes the saliva more abundant and watery. However, these changes require 30 minutes or longer to take effect, and it would be rather impractical if we had to drink that long while waiting to feel satisfied. Water intake would be grossly excessive. Fortunately, there are mechanisms that act more quickly to temporarily quench the thirst and allow time for the change in blood osmolarity to occur.

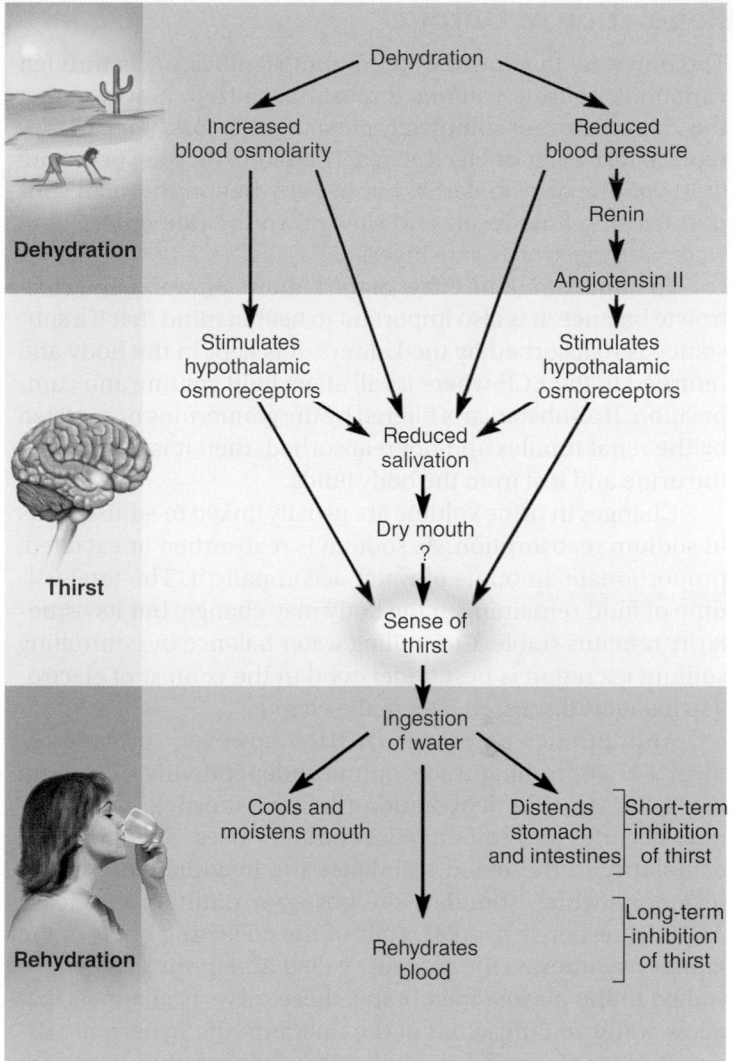

**FIGURE 24.3 Dehydration, Thirst, and Rehydration.**

Experiments with rats and dogs have isolated the stimuli that quench the thirst. One of these is cooling and moistening the mouth; rats drink less if their water is cool than if it is warm, and simply moistening the mouth temporarily satisfies an animal even if the water is drained from its esophagus before it reaches the stomach. Distension of the stomach and small intestine is another inhibitor of thirst. If a dog is allowed to drink while the water is drained from its esophagus but its stomach is inflated with a balloon, its thirst is satisfied for a time. If the water is drained away but the stomach is not inflated, satiation does not last as long. Such fast-acting stimuli as coolness, moisture, and filling of the stomach stop an animal (and presumably a human) from drinking an excessive amount of liquid, but they are effective for only 30 to 45 minutes. If they are not soon followed by absorption of water into the bloodstream, the thirst soon returns. Only a drop in blood osmolarity produces a lasting effect.

## Regulation of Output

The only way to control water output significantly is through variations in urine volume. It must be realized, however, that the kidneys cannot completely prevent water loss, nor can they replace lost water or electrolytes. Therefore, they never restore fluid volume or osmolarity, but in dehydration they can support existing fluid levels and slow down the rate of loss until water and electrolytes are ingested.

To understand the effect of the kidneys on water and electrolyte balance, it is also important to bear in mind that if a substance is reabsorbed by the kidneys, it is kept in the body and returned to the ECF, where it will affect fluid volume and composition. If a substance is filtered by the glomerulus or secreted by the renal tubules and not reabsorbed, then it is excreted in the urine and lost from the body fluids.

Changes in urine volume are usually linked to adjustments in sodium reabsorption. As sodium is reabsorbed or excreted, proportionate amounts of water accompany it. The total volume of fluid remaining in the body may change, but its osmolarity remains stable. Controlling water balance by controlling sodium excretion is best understood in the context of electrolyte balance, discussed later in the chapter.

Antidiuretic hormone (ADH), however, provides a means of controlling water output independently of sodium (fig. 24.4). In true dehydration (defined shortly), blood volume declines and sodium concentration rises. The increased osmolarity of the blood stimulates the hypothalamic osmoreceptors, which stimulate the posterior pituitary to release ADH. In response to ADH, cells of the collecting ducts of the kidneys synthesize the proteins called aquaporins. When installed in the plasma membrane, these serve as channels that allow water to diffuse out of the duct into the hypertonic tissue fluid of the renal medulla. The kidneys then reabsorb more water and produce less urine. Sodium continues to be excreted, so the *ratio* of sodium to water in the urine increases (the urine becomes more concentrated). By helping the kidneys retain water, ADH slows down the decline in blood volume and the rise in its osmolarity. Thus, the ADH mechanism forms a negative feedback loop.

Conversely, if blood volume and pressure are too high or blood osmolarity is too low, ADH release is inhibited. The renal tubules reabsorb less water, urine output increases, and total body water declines. This is an effective way of compensating for hypertension. Since the lack of ADH increases the ratio of water to sodium in the urine, it raises the sodium concentration and osmolarity of the blood.

## Disorders of Water Balance

The body is in a state of fluid imbalance if there is an abnormality of total *volume, concentration,* or *distribution* of fluid among the compartments.

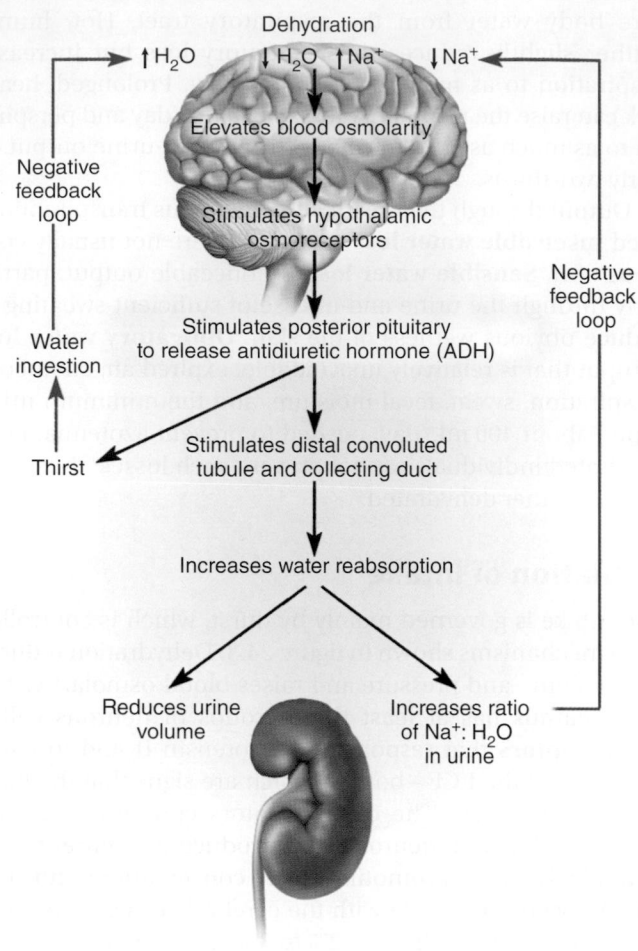

**FIGURE 24.4  The Action of Antidiuretic Hormone.** Pathways shown in red represent negative feedback.

## Fluid Deficiency

Fluid deficiency arises when output exceeds intake over a long enough period of time. The two kinds of deficiency—volume depletion and dehydration—differ in the relative loss of water and electrolytes and the resulting osmolarity of the ECF. This important distinction calls for different strategies of fluid replacement therapy (see Deeper Insight 24.2 at the end of the chapter).

**Volume depletion (hypovolemia[2])** occurs when proportionate amounts of water *and* sodium are lost without replacement. Total body water declines but osmolarity remains normal. Volume depletion occurs in cases of hemorrhage, severe burns, and chronic vomiting or diarrhea. A less common

---

[2]*hypo* = below normal; *vol* = volume; *emia* = blood condition

cause is aldosterone hyposecretion (Addison disease), which results in inadequate sodium and water reabsorption by the kidneys.

**Dehydration (negative water balance)** occurs when the body eliminates significantly more water than sodium, so the ECF osmolarity rises. The simplest cause of dehydration is a lack of drinking water—for example, when stranded in a desert or at sea. It can be a serious problem for elderly and bedridden people who depend on others to provide them with water, especially for those who cannot express their need or whose caretakers are insensitive to it. Diabetes mellitus, ADH hyposecretion (diabetes insipidus), profuse sweating, and overuse of diuretics are additional causes of dehydration. Cold weather can dehydrate a person just as much as hot weather (see Deeper Insight 24.1).

For three reasons, infants are more vulnerable to dehydration than adults: (1) Their high metabolic rate produces toxic metabolites faster, and they must excrete more water to eliminate them. (2) Their kidneys are not fully mature and cannot concentrate urine as effectively. (3) They have a greater ratio of body surface to volume; consequently, compared with adults, they lose twice as much water per kilogram of body weight by evaporation.

Dehydration affects all fluid compartments. Suppose, for example, that you play a strenuous tennis match on a hot summer day and lose a liter of sweat. Where does this fluid come from? Most of it filters out of the bloodstream through the capillaries of the sweat glands. In principle, 1 L of sweat would amount to about one-third of the blood plasma. However, as the blood loses water, its osmolarity rises and water from the tissue fluid enters the bloodstream to balance the loss. This raises the osmolarity of the tissue fluid, so water moves out of the cells to balance that (fig. 24.5). Ultimately, all three fluid compartments (the ICF, blood, and tissue fluid) lose water. To excrete 1 L of sweat, about 300 mL of water would come from the tissue fluid and 700 mL from the ICF. Immoderate exercise without fluid replacement can lead to losses greater than 1 L per hour.

The most serious effects of fluid deficiency are circulatory shock due to loss of blood volume and neurological dysfunction due to dehydration of brain cells. Volume depletion by diarrhea is a major cause of infant mortality, especially under unsanitary conditions that lead to intestinal infections such as cholera.

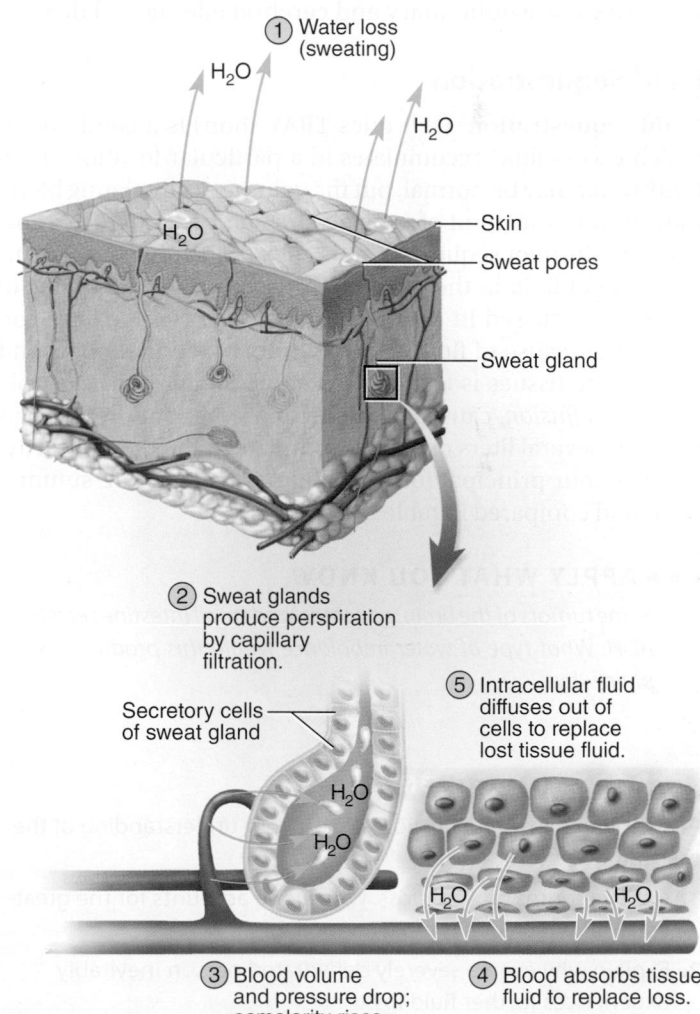

**FIGURE 24.5 Effects of Profuse Sweating on the Fluid Compartments.** In extreme dehydration, the loss of intracellular fluid can cause cellular shrinkage and dysfunction.

---

## Fluid Excess

Fluid excess is less common than fluid deficiency because the kidneys are highly effective at compensating for excessive intake by excreting more urine (fig. 24.6). Renal failure and other causes, however, can lead to excess fluid retention.

Fluid excesses are of two types called volume excess and hypotonic hydration. In **volume excess,** both sodium and water are retained and the ECF remains isotonic. This can result from aldosterone hypersecretion or renal failure. In **hypotonic hydration** (also called **water intoxication** or **positive water balance**), more water than sodium is retained or ingested and the ECF becomes hypotonic. This can occur if you lose a large amount of water *and* salt through urine and sweat and you replace it by drinking plain water. Without a proportionate intake of electrolytes, water dilutes the ECF, makes it hypotonic, and causes cellular swelling. ADH hypersecretion can cause hypotonic hydration by stimulating excessive water retention as sodium continues to be excreted. Among the most serious effects of either type of fluid excess are pulmonary and cerebral edema and death.

## Fluid Sequestration

**Fluid sequestration**[3] (seh-ques-TRAY-shun) is a condition in which excess fluid accumulates in a particular location. Total body water may be normal, but the volume of circulating blood may drop to the point of causing circulatory shock. The most common form of sequestration is *edema,* the abnormal accumulation of fluid in the interstitial spaces, causing swelling of a tissue (discussed in detail in chapter 20). Hemorrhage can be another cause of fluid sequestration; blood that pools and clots in the tissues is lost to circulation. Yet another example is *pleural effusion,* caused by some lung infections, in which as much as several liters of fluid accumulate in the pleural cavity.

The four principal forms of fluid imbalance are summarized and compared in table 24.1.

#### ►►► APPLY WHAT YOU KNOW

*Some tumors of the brain, pancreas, and small intestine secrete ADH. What type of water imbalance would this produce? Explain why.*

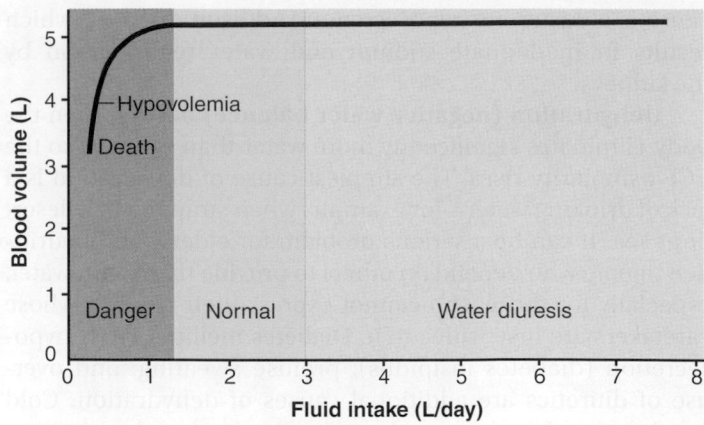

**FIGURE 24.6  The Relationship of Blood Volume to Fluid Intake.** The kidneys cannot compensate very well for inadequate fluid intake. Below an intake of about 1 L/day, blood volume drops significantly and there may be a threat of death from hypovolemic shock. The kidneys compensate very well, on the other hand, for abnormally high fluid intake; they eliminate the excess by water diuresis and maintain a stable blood volume.

| TABLE 24.1 | Forms of Fluid Imbalance | |
|---|---|---|
| **Form** | **Total Body Water** | **Osmolarity** |
| **Fluid deficiency** | | |
| Volume depletion (hypovolemia) | Reduced | Isotonic (normal) |
| Dehydration (negative water balance) | Reduced | Hypertonic (elevated) |
| **Fluid excess** | | |
| Volume excess | Elevated | Isotonic (normal) |
| Hypotonic hydration (positive water balance, water intoxication) | Elevated | Hypotonic (reduced) |

#### BEFORE YOU GO ON

Answer the following questions to test your understanding of the preceding section:

1. List five routes of water loss. Which one accounts for the greatest loss? Which one is most controllable?

2. Explain why even a severely dehydrated person inevitably experiences further fluid loss.

3. Suppose there were no mechanisms to stop the sense of thirst until the blood became sufficiently hydrated. Explain why we would routinely suffer hypotonic hydration.

4. Summarize the effect of ADH on total body water and blood osmolarity.

5. Name and define the four types of fluid imbalance, and give an example of a situation that could produce each type.

[3] *sequestr* = to isolate

## 24.2 Electrolyte Balance

### Expected Learning Outcomes

When you have completed this section, you should be able to

a. describe the physiological roles of sodium, potassium, chloride, calcium, magnesium, and phosphates;

b. describe the hormonal and renal mechanisms that regulate the concentrations of these electrolytes; and

c. state the term for an excess or deficiency of each electrolyte and describe the consequences of these imbalances.

Electrolytes are physiologically important for multiple reasons: They are chemically reactive and participate in metabolism, they determine the electrical potential (charge difference) across cell membranes, and they strongly affect the osmolarity of the body fluids and the body's water content and distribution. Strictly speaking, electrolytes are salts such as sodium chloride, not just sodium or chloride ions. In common usage, however, the individual ions are often referred to as electrolytes. The major cations are sodium ($Na^+$), potassium ($K^+$), calcium ($Ca^{2+}$), magnesium ($Mg^{2+}$), and hydrogen ($H^+$); the major anions are chloride ($Cl^-$), bicarbonate ($HCO_3^-$), and phosphates ($P_i$). Hydrogen and bicarbonate regulation are discussed later under acid–base balance. Here we focus on the other six.

The typical concentrations of these ions in the blood plasma and intracellular fluid are compared in figure 24.7. Notice that in spite of great differences in electrolyte concentrations, the two fluid compartments have the same osmolarity (300 mOsm/L). Blood plasma is the most accessible fluid for measurements of electrolyte concentration, so excesses and deficiencies are defined with reference to normal plasma concentrations. Concentrations in the tissue fluid differ only slightly from those in the plasma. The prefix *normo-* denotes a normal electrolyte concentration (for example, *normokalemia*), and *hyper-* and *hypo-* denote concentrations that are, respectively, sufficiently above or below normal to cause physiological disorders.

## Sodium

### Functions

Sodium is one of the principal ions responsible for the resting membrane potentials of cells, and the inflow of sodium through membrane channels is an essential event in the depolarization that underlies nerve and muscle function. Sodium is the principal cation of the ECF; sodium salts account for 90% to 95% of its osmolarity. Sodium is therefore the most significant solute in determining total body water and the distribution of water among fluid compartments. Sodium gradients across the plasma membrane provide the potential energy that is tapped

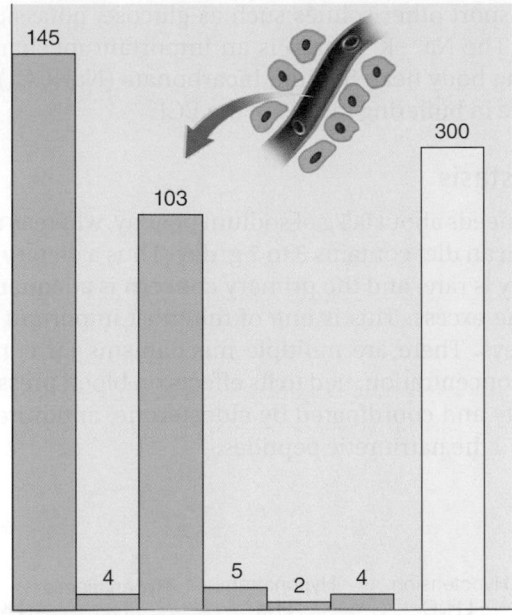

**(a) Blood plasma**

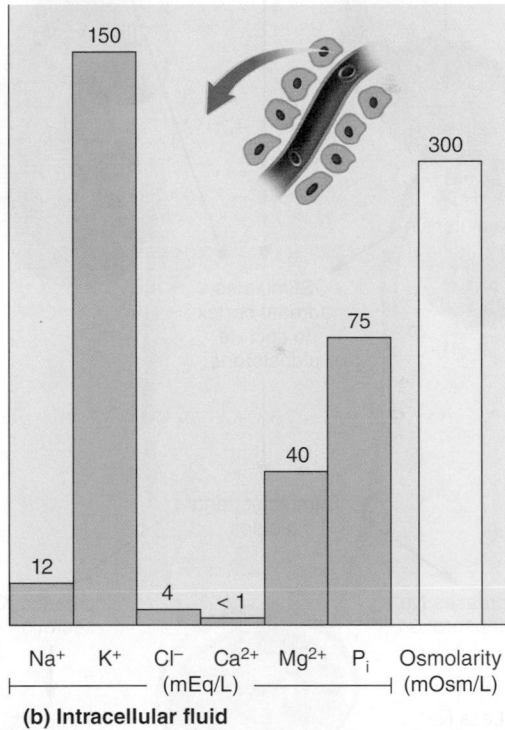

Na⁺  K⁺  Cl⁻  Ca²⁺  Mg²⁺  Pᵢ  Osmolarity
├────────── (mEq/L) ──────────┤  (mOsm/L)

**(b) Intracellular fluid**

**FIGURE 24.7 Electrolyte Concentrations in the Blood Plasma and Intracellular Fluid.** Concentrations in the blood plasma (top) are shown directly above concentrations of the same ions in the intracellular fluid for contrast. $P_i$ collectively represents inorganic phosphates in the forms of $PO_4^{3-}$, $HPO_4^{2-}$, and $H_2PO_4^-$. Electrolyte concentrations (green bars) are given in mEq/L. Concentrations in mmol/L are the same for $Na^+$, $K^+$, and $Cl^-$; one-half of the illustrated values for $Ca^{2+}$ and $Mg^{2+}$; and one-third for $PO_4^{3-}$. Osmolarity (yellow bars) is given in mOsm/L.

to cotransport other solutes such as glucose, potassium, and calcium. The $Na^+-K^+$ pump is an important mechanism for generating body heat. Sodium bicarbonate ($NaHCO_3$) plays a major role in buffering the pH of the ECF.

## Homeostasis

An adult needs about 0.5 g of sodium per day, whereas the typical American diet contains 3 to 7 g/day. Thus a dietary sodium deficiency is rare, and the primary concern is adequate excretion of the excess. This is one of the most important roles of the kidneys. There are multiple mechanisms for controlling sodium concentration, tied to its effects on blood pressure and osmolarity and coordinated by aldosterone, antidiuretic hormone, and the natriuretic peptides.

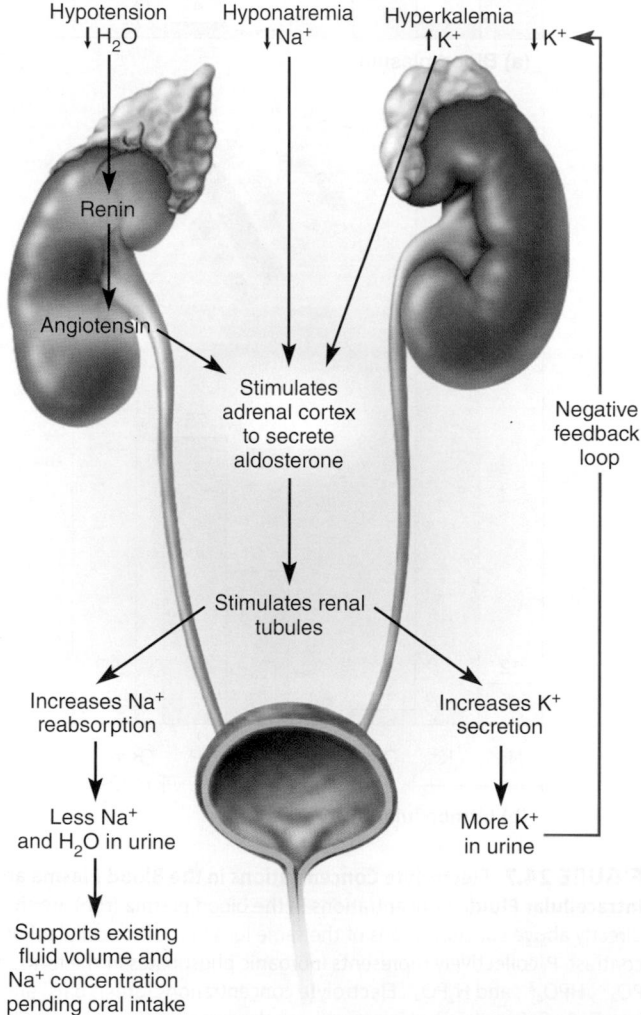

**FIGURE 24.8 The Action of Aldosterone.** The pathway shown in red represents negative feedback.

❓ *What is required, in addition to aldosterone, to increase blood volume?*

Aldosterone, the "salt-retaining hormone," plays the primary role in adjustment of sodium excretion. Hyponatremia and hyperkalemia directly stimulate the adrenal cortex to secrete aldosterone, and hypotension stimulates its secretion by way of the renin–angiotensin–aldosterone mechanism (fig. 24.8).

Only cells in the ascending limb of the nephron loop, the distal convoluted tubule, and the cortical part of the collecting duct have aldosterone receptors. Aldosterone, a steroid, binds to nuclear receptors and activates transcription of a gene for the $Na^+-K^+$ pump. In 10 to 30 minutes, enough $Na^+-K^+$ pumps are synthesized and installed in the plasma membrane to produce a noticeable effect—sodium concentration in the urine begins to fall and potassium concentration rises as the tubules reabsorb more sodium and secrete more hydrogen and potassium ions. Water and chloride passively follow sodium. Thus, the primary effects of aldosterone are that the urine contains less NaCl and more potassium and has a lower pH. An average adult male excretes 5 g of sodium per day, but the urine can be virtually sodium-free when aldosterone level is high. Although aldosterone strongly influences sodium reabsorption, it has little effect on plasma sodium *concentration* because reabsorbed sodium is accompanied by a proportionate amount of water.

Elevated blood pressure inhibits the renin–angiotensin–aldosterone mechanism. The kidneys then reabsorb almost no sodium beyond the proximal convoluted tubule (PCT), and the urine contains up to 30 g of sodium per day.

Aldosterone has only slight effects on urine volume, blood volume, and blood pressure in spite of the tendency of water to follow sodium osmotically. Even in aldosterone hypersecretion, blood volume is rarely more than 5% to 10% above normal. An increase in blood volume increases blood pressure and glomerular filtration rate (GFR). Even though aldosterone increases the tubular reabsorption of sodium and water, this is offset by the rise in GFR and there is only a small drop in urine output.

Antidiuretic hormone modifies water excretion independently of sodium excretion. Thus, unlike aldosterone, it can change sodium *concentration*. A high concentration of sodium in the blood stimulates the posterior lobe of the pituitary gland to release ADH. The kidneys then reabsorb more water, which slows down any further increase in blood sodium concentration. ADH alone cannot lower the concentration; this requires water ingestion to dilute the existing sodium. A drop in sodium concentration, by contrast, inhibits ADH release. More water is excreted, thereby raising the relative amount of sodium that remains in the blood.

The natriuretic peptides inhibit sodium and water reabsorption and the secretion of renin and ADH. The kidneys then eliminate more sodium and water and lower the blood pressure. Angiotensin II, by contrast, activates the $Na^+-H^+$ antiport in the PCT and thereby increases sodium reabsorption and reduces urinary sodium output.

Several other hormones also affect sodium homeostasis. Estrogen mimics the effect of aldosterone and causes women to retain water during pregnancy and part of the menstrual cycle. Progesterone reduces sodium reabsorption and has a diuretic effect. High levels of glucocorticoids promote sodium reabsorption and edema.

In some cases, sodium homeostasis is achieved by regulation of salt intake. A craving for salt occurs in people who are depleted of sodium—for example, by blood loss or Addison disease. Pregnant women sometimes develop a craving for salty foods. Salt craving is not limited to humans; many animals ranging from elephants to butterflies seek out wet salty soil where they can obtain this vital mineral.

## Imbalances

True imbalances in sodium concentration are relatively rare because sodium excess or depletion is almost always accompanied by proportionate changes in water volume. **Hypernatremia**[4] is a plasma sodium concentration in excess of 145 mEq/L. It can result from the administration of intravenous saline (see Deeper Insight 24.2). Its major consequences are water retention, hypertension, and edema. **Hyponatremia** (less than 130 mEq/L) is usually the result of excess body water rather than excess sodium excretion, as in the case mentioned

[4]*natr* = sodium; *emia* = blood condition

earlier of a person who loses large volumes of sweat or urine and replaces it by drinking plain water. Usually, hyponatremia is quickly corrected by excretion of the excess water, but if uncorrected it produces the symptoms of hypotonic hydration described earlier.

## Potassium

### Functions

Potassium is the most abundant cation of the ICF and is the greatest determinant of intracellular osmolarity and cell volume. Along with sodium, it produces the resting membrane potentials and action potentials of nerve and muscle cells (fig. 24.9a). Potassium is as important as sodium to the $Na^+–K^+$ pump and its functions of cotransport and thermogenesis (heat production). It is an essential cofactor for protein synthesis and some other metabolic processes.

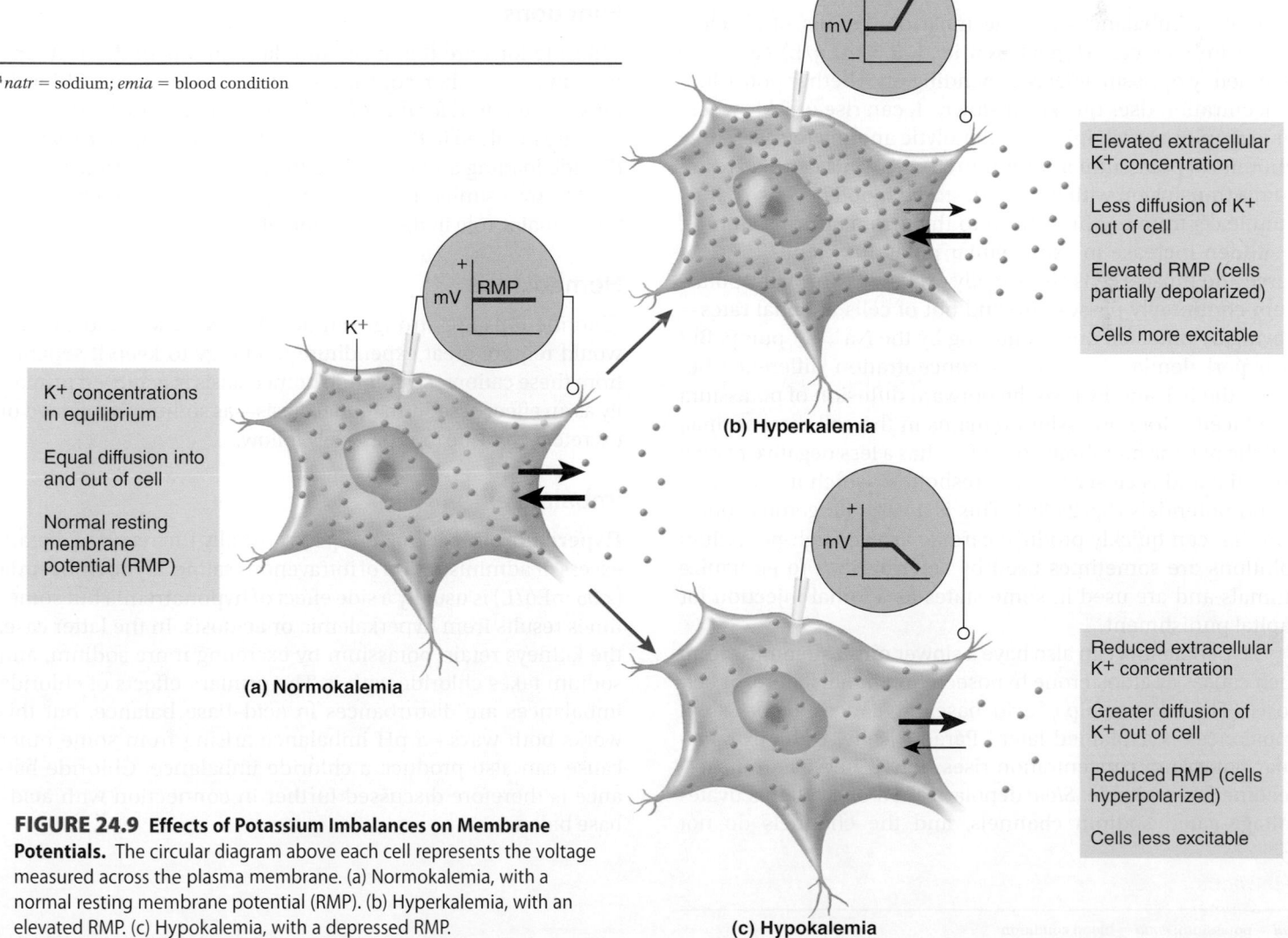

K⁺ concentrations in equilibrium

Equal diffusion into and out of cell

Normal resting membrane potential (RMP)

**(a) Normokalemia**

Elevated extracellular K⁺ concentration

Less diffusion of K⁺ out of cell

Elevated RMP (cells partially depolarized)

Cells more excitable

**(b) Hyperkalemia**

Reduced extracellular K⁺ concentration

Greater diffusion of K⁺ out of cell

Reduced RMP (cells hyperpolarized)

Cells less excitable

**(c) Hypokalemia**

**FIGURE 24.9 Effects of Potassium Imbalances on Membrane Potentials.** The circular diagram above each cell represents the voltage measured across the plasma membrane. (a) Normokalemia, with a normal resting membrane potential (RMP). (b) Hyperkalemia, with an elevated RMP. (c) Hypokalemia, with a depressed RMP.

## Homeostasis

Potassium homeostasis is closely linked to that of sodium. Regardless of the body's state of potassium balance, about 90% of the potassium filtered by the glomerulus is reabsorbed by the PCT and the rest is excreted in the urine. Variations in potassium excretion are controlled later in the nephron by changing the amount of potassium returned to the tubular fluid by the distal convoluted tubule and cortical portion of the collecting duct (CD). When potassium concentration is high, they secrete more potassium into the filtrate and the urine may contain more potassium than the glomerulus filters from the blood. When blood potassium level is low, the CD secretes less. The distal convoluted tubule and collecting duct reabsorb potassium through their intercalated cells.

Aldosterone regulates potassium balance along with sodium (fig. 24.8). A rise in potassium concentration stimulates the adrenal cortex to secrete aldosterone. Aldosterone stimulates renal secretion of potassium at the same time that it stimulates reabsorption of sodium. The more sodium there is in the urine, the less potassium, and vice versa.

## Imbalances

Potassium imbalances are the most dangerous of all electrolyte imbalances. **Hyperkalemia**[5] (>5.5 mEq/L) can have completely opposite effects depending on whether potassium concentration rises quickly or slowly. It can rise quickly when, for example, a crush injury or hemolytic anemia releases large amounts of potassium from ruptured cells. This can also result from a transfusion with outdated, stored blood because potassium leaks from erythrocytes into the plasma during storage. A sudden increase in extracellular potassium tends to make nerve and muscle cells abnormally excitable. Normally, potassium continually passes into and out of cells at equal rates—leaving by diffusion and reentering by the $Na^+$–$K^+$ pump. But in hyperkalemia, there is less concentration difference between the ICF and ECF, so the outward diffusion of potassium is reduced. More potassium remains in the cell than normal, and the plasma membrane therefore has a less negative resting potential and is closer to the threshold at which it will set off action potentials (fig. 24.9b). This is a very dangerous condition that can quickly produce cardiac arrest. High-potassium solutions are sometimes used by veterinarians to euthanize animals and are used in some states as a lethal injection for capital punishment.

Hyperkalemia can also have a slower onset stemming from such causes as aldosterone hyposecretion, renal failure, or acidosis. (The relationship of acid–base imbalances to potassium imbalances is explained later.) Paradoxically, if the extracellular potassium concentration rises slowly, nerve and muscle become *less* excitable. *Slow* depolarization of a cell inactivates voltage-gated sodium channels, and the channels do not

become excitable again until the membrane repolarizes. Inactivated sodium channels cannot produce action potentials.

**Hypokalemia** (<3.5 mEq/L) rarely results from a dietary deficiency, because most diets contain ample amounts of potassium; it can occur, however, in people with depressed appetites. Hypokalemia more often results from heavy sweating, chronic vomiting or diarrhea, excessive use of laxatives, aldosterone hypersecretion, or alkalosis. As ECF potassium concentration falls, more potassium moves from the ICF to the ECF. With the loss of these cations from the cytoplasm, cells become hyperpolarized and nerve and muscle cells are less excitable (fig. 24.9c). This is reflected in muscle weakness, loss of muscle tone, depressed reflexes, and irregular electrical activity of the heart.

### ▶▶▶ APPLY WHAT YOU KNOW

*Some tumors of the adrenal cortex secrete excess aldosterone and may cause paralysis. Explain this effect and identify the electrolyte and fluid imbalances you would expect to observe in such a case.*

## Chloride
### Functions

Chloride ions are the most abundant anions of the ECF and thus make a major contribution to its osmolarity. Chloride ions are required for the formation of stomach acid (HCl), and they are involved in the chloride shift that accompanies carbon dioxide loading and unloading by the erythrocytes (see chapter 22). By a similar mechanism explained later, chloride ions play a major role in the regulation of body pH.

### Homeostasis

Chloride ions are strongly attracted to $Na^+$, $K^+$, and $Ca^{2+}$. It would require great expenditure of energy to keep it separate from these cations, so chloride homeostasis is achieved primarily as an effect of sodium homeostasis—as sodium is retained or excreted, chloride ions passively follow.

### Imbalances

**Hyperchloremia** (>105 mEq/L) is usually the result of dietary excess or administration of intravenous saline. **Hypochloremia** (<95 mEq/L) is usually a side effect of hyponatremia but sometimes results from hyperkalemia or acidosis. In the latter case, the kidneys retain potassium by excreting more sodium, and sodium takes chloride with it. The primary effects of chloride imbalances are disturbances in acid–base balance, but this works both ways—a pH imbalance arising from some other cause can also produce a chloride imbalance. Chloride balance is therefore discussed further in connection with acid–base balance.

---

[5]*kal* = potassium; *emia* = blood condition

# Calcium

## Functions

Calcium lends strength to the skeleton, activates the sliding filament mechanism of muscle contraction, serves as a second messenger for some hormones and neurotransmitters, activates exocytosis of neurotransmitters and other cellular secretions, and is an essential factor in blood clotting. Cells maintain a very low intracellular calcium concentration because they require a high concentration of phosphate ions (for reasons discussed shortly). If calcium and phosphate were both very concentrated in a cell, calcium phosphate crystals would precipitate in the cytoplasm. To maintain a high phosphate concentration but avoid crystallization of calcium phosphate, cells must pump out $Ca^{2+}$ and keep it at a low intracellular concentration, or else sequester $Ca^{2+}$ in the smooth ER and release it only when needed. Cells that store $Ca^{2+}$ often have a protein called *calsequestrin,* which binds the stored $Ca^{2+}$ and keeps it chemically unreactive.

## Homeostasis

Calcium concentration is regulated chiefly by parathyroid hormone, calcitriol, and in children, calcitonin. These hormones work through their effects on bone deposition and resorption, intestinal absorption of calcium, and urinary excretion. For flowcharts of calcium homeostasis, see figure 7.16 (p. 219); the associated discussion gives further details on the mode of action of these hormones.

## Imbalances

**Hypercalcemia** (>5.8 mEq/L) can result from alkalosis, hyperparathyroidism, or hypothyroidism. It reduces the sodium permeability of plasma membranes and inhibits the depolarization of nerve and muscle cells. At concentrations $\geq 12$ mEq/dL, hypercalcemia causes muscular weakness, depressed reflexes, and cardiac arrhythmia.

**Hypocalcemia** (<4.5 mEq/L) can result from vitamin D deficiency, diarrhea, pregnancy, lactation, acidosis, hypoparathyroidism, or hyperthyroidism. It increases the sodium permeability of plasma membranes, causing the nervous and muscular systems to be overly excitable. Tetany occurs when calcium concentration drops to 6 mg/dL and may be lethal at 4 mg/dL (2 mEq/L) due to laryngospasm and suffocation.

# Magnesium

## Functions

About 54% of the body's magnesium ($Mg^{2+}$) is in the bone tissue and 45% in the intracellular fluid, especially in the skeletal muscles. Magnesium is the second most abundant intracellular cation after calcium. Most ICF $Mg^{2+}$ is complexed with ATP, but $Mg^{2+}$ is also a necessary cofactor for many enzymes, membrane transport proteins, and nucleic acids. Magnesium imbalances can therefore have wide-ranging effects on membrane transport, membrane electrical potentials, cell metabolism, and DNA replication.

## Homeostasis

Magnesium levels in the blood plasma normally range from 1.5 to 2.0 mEq/L, whereas intracellular concentrations are quite variable from one tissue to another, but range up to 40 mEq/L in skeletal muscle. Dietary intake of magnesium is typically 140 to 360 mg/day, but only 30% to 40% of it is absorbed by the small intestine and the rest passes through unused. Its intestinal absorption is regulated mainly by vitamin D.

About two-thirds of the body's $Mg^{2+}$ loss is via the feces and one-third via the urine. Retention or loss of plasma $Mg^{2+}$ is regulated mainly by the thick segment of the ascending limb of the nephron loop, where about 70% of the filtered $Mg^{2+}$ is reabsorbed; smaller amounts are reabsorbed in other segments of the nephron. Reabsorption is mainly by the paracellular route (between tubule epithelial cells), driven by the positive electrical potential of the tubular fluid repelling the positive magnesium ions. Parathyroid hormone governs the rate of reabsorption and is the primary regulator of plasma $Mg^{2+}$ level.

## Imbalances

Magnesium imbalances are usually due to excessive loss from the body rather than dietary deficiency. **Hypomagnesemia,** a plasma $Mg^{2+}$ deficiency (< 1.5 mEq/L), can result from intestinal malabsorption, vomiting, diarrhea, or renal disease. It results in hyperirritability of the nervous and muscular systems; muscle tremors, spasms, or tetanus; hypertension resulting from excessive vasoconstriction; and tachycardia and ventricular arrhythmia.

**Hypermagnesemia,** an excess (> 2.0 mEq/L), is rare except in renal insufficiency. It tends to have a sedative effect, with lethargy, muscle weakness, and weak reflexes; and it can cause respiratory depression or failure, hypotension due to lack of vasomotor tone, and flaccid, diastolic cardiac arrest.

# Phosphates

## Functions

The inorganic phosphates ($P_i$) of the body fluids are an equilibrium mixture of phosphate ($PO_4^{3-}$), monohydrogen phosphate ($HPO_4^{2-}$), and dihydrogen phosphate ($H_2PO_4^{-}$) ions. Phosphates are relatively concentrated in the ICF, where they are generated by the hydrolysis of ATP and other phosphate compounds. They are a component of nucleic acids, phospholipids, ATP, GTP, cAMP, creatine phosphate, and related compounds. Every process that depends on ATP depends on phosphate ions. Phosphates activate many metabolic pathways by phosphorylating enzymes and substrates such as glucose. They are also important as buffers that help stabilize the pH of the body fluids.

## Homeostasis

The average diet provides ample amounts of phosphate ions, which are readily absorbed by the small intestine. Plasma phosphate concentration is usually maintained at about 4 mEq/L, with continual loss of excess phosphate by glomerular filtration. If plasma phosphate concentration drops much below this level, however, the renal tubules reabsorb all filtered phosphate.

Parathyroid hormone increases the excretion of phosphate as part of the mechanism for increasing the concentration of free calcium ions in the ECF. Lowering the ECF phosphate concentration minimizes the formation of calcium phosphate and thus helps support plasma calcium concentration. Rates of phosphate excretion are also strongly affected by the pH of the urine, as discussed shortly.

## Imbalances

Phosphate homeostasis is not as critical as that of other electrolytes. The body can tolerate broad variations several times above or below the normal concentration with little immediate effect on physiology.

---

**BEFORE YOU GO ON**

Answer the following questions to test your understanding of the preceding section:

6. Which do you think would have the most serious effect, and why—a 5 mEq/L increase in the plasma concentration of sodium, potassium, chloride, or calcium?

7. Answer the same question for a 5 mEq/L.

8. Explain why ADH is more likely than aldosterone to change the osmolarity of the blood plasma.

9. Explain why aldosterone hyposecretion could cause hypochloremia.

10. Magnesium sulfate, commonly sold as Epsom salt, can be used as a bath salt to relax tight, aching muscles. Explain this effect in view of what you have learned in this chapter section.

11. Why are more phosphate ions required in the ICF than in the ECF? How does this affect the distribution of calcium ions between these fluid compartments?

---

## 24.3 Acid–Base Balance

### Expected Learning Outcomes

When you have completed this section, you should be able to

a. define *buffer* and write chemical equations for the bicarbonate, phosphate, and protein buffer systems;

b. discuss the relationship between pulmonary ventilation, pH of the extracellular fluids, and the bicarbonate buffer system;

c. explain how the kidneys secrete hydrogen ions and how these ions are buffered in the tubular fluid;

d. identify some types and causes of acidosis and alkalosis, and describe the effects of these pH imbalances; and

e. explain how the respiratory and urinary systems correct acidosis and alkalosis, and compare their effectiveness and limitations.

As we saw in chapter 2, metabolism depends on the functioning of enzymes, and enzymes are very sensitive to pH. Slight deviations from the normal pH can shut down metabolic pathways as well as alter the structure and function of other macromolecules. Consequently, acid–base balance is one of the most important aspects of homeostasis.

The blood and tissue fluid normally have a pH of 7.35 to 7.45. Such a narrow range of variation is remarkable considering that our metabolism constantly produces acid: lactic acid from anaerobic fermentation, phosphoric acids from nucleic acid catabolism, fatty acids and ketones from fat catabolism, and carbonic acid from carbon dioxide. Here we examine mechanisms for resisting these challenges and maintaining acid–base balance.

### Acids, Bases, and Buffers

The pH of a solution is determined solely by its hydrogen ions ($H^+$). An acid is any chemical that releases $H^+$ in solution. A **strong acid** such as hydrochloric acid (HCl) ionizes freely, gives up most of its hydrogen ions, and can markedly lower the pH of a solution. A **weak acid** such as carbonic acid ($H_2CO_3$) ionizes only slightly and keeps most hydrogen in a chemically bound form that does not affect pH. A base is any chemical that accepts $H^+$. A **strong base** such as the hydroxide ion ($OH^-$) has a strong tendency to bind $H^+$ and raise the pH, whereas a **weak base** such as the bicarbonate ion ($HCO_3^-$) binds less of the available $H^+$ and has less effect on pH.

A **buffer,** broadly speaking, is any mechanism that resists changes in pH by converting a strong acid or base to a weak one. The body has both physiological and chemical buffers. A **physiological buffer** is a system—namely, the respiratory or urinary system—that stabilizes pH by controlling the body's output of acids, bases, or $CO_2$. Of all buffer systems, the urinary system buffers the greatest quantity of acid or base, but it requires several hours to days to exert an effect. The respiratory system exerts an effect within a few minutes but cannot alter the pH as much as the urinary system can.

A **chemical buffer** is a substance that binds $H^+$ and removes it from solution as its concentration begins to rise, or releases $H^+$ into solution as its concentration falls. Chemical buffers can restore normal pH within a fraction of a second. They function as mixtures called **buffer systems** composed of a weak acid and a weak base. The three major chemical buffer systems of the body are the bicarbonate, phosphate, and protein systems.

The amount of acid or base that can be neutralized by a chemical buffer system depends on two factors: the concentration of the buffers and the pH of their working environment. Each system has an optimum pH at which it functions best; its effectiveness is greatly reduced if the pH of its environment deviates too far from this. The relevance of these factors will become apparent as you study the following buffer systems.

## The Bicarbonate Buffer System

The **bicarbonate buffer system** is a solution of carbonic acid and bicarbonate ions. Carbonic acid ($H_2CO_3$) forms by the hydration of carbon dioxide and then dissociates into bicarbonate ($HCO_3^-$) and $H^+$:

$$CO_2 + H_2O \leftrightarrows H_2CO_3 \leftrightarrows HCO_3^- + H^+$$

This is a reversible reaction. When it proceeds to the right, carbonic acid acts as a weak acid by releasing $H^+$ and lowering pH. When the reaction proceeds to the left, bicarbonate acts as a weak base by binding $H^+$, removing the ions from solution, and raising pH.

At a pH of 7.4, the bicarbonate system would not ordinarily have a particularly strong buffering capacity outside of the body. This is too far from its optimum pH of 6.1. If a strong acid was added to a beaker of carbonic acid–bicarbonate solution at pH 7.4, the preceding reaction would shift only slightly to the left. Much surplus $H^+$ would remain and the pH would be substantially lower. In the body, by contrast, the bicarbonate system works quite well because the lungs and kidneys constantly remove $CO_2$ and prevent an equilibrium from being reached. This keeps the reaction moving to the left, and more $H^+$ is neutralized. Conversely, if there is a need to lower the pH, the kidneys excrete $HCO_3^-$, keep this reaction moving to the right, and elevate the $H^+$ concentration of the ECF. Thus, you can see that the physiological and chemical buffers of the body function together in maintaining acid–base balance.

▶▶▶**APPLY WHAT YOU KNOW**

*In the systemic circulation, arterial blood has a mean pH of 7.40 and venous blood has a mean of 7.35. What do you think causes this difference?*

## The Phosphate Buffer System

The **phosphate buffer system** is a solution of $HPO_4^{2-}$ and $H_2PO_4^-$. It works in much the same way as the bicarbonate system. The following reaction can proceed to the right to liberate $H^+$ and lower pH, or it can proceed to the left to bind $H^+$ and raise pH:

$$H_2PO_4^- \leftrightarrows HPO_4^{2-} + H^+$$

The optimal pH for this system is 6.8, closer to the actual pH of the ECF (7.4). Thus, the phosphate buffer system has a stronger buffering effect than an equal amount of bicarbonate buffer. However, phosphates are much less concentrated in the ECF than bicarbonate, so they are less important in buffering the ECF. They are more important in the renal tubules and ICF, where not only are they more concentrated, but the pH is lower

and closer to their functional optimum. In the ICF, the constant production of metabolic acids creates pH values ranging from 4.5 to 7.4, probably averaging 7.0. The reason for the low pH in the renal tubules is discussed later.

## The Protein Buffer System

Proteins are more concentrated than either bicarbonate or phosphate buffers, especially in the ICF. The **protein buffer system** accounts for about three-quarters of all chemical buffering in the body fluids. The buffering ability of proteins is due to certain side groups of their amino acid residues. Some have carboxyl (−COOH) side groups, which release $H^+$ when pH begins to rise and thus lower pH:

$$-COOH \rightarrow -COO^- + H^+$$

Others have amino (−$NH_2$) side groups, which bind $H^+$ when pH falls too low, thus raising pH toward normal:

$$-NH_2 + H^+ \rightarrow -NH_3^+$$

▶▶▶**APPLY WHAT YOU KNOW**

*What protein do you think is the most important buffer in blood plasma? In erythrocytes?*

## Respiratory Control of pH

The equation for the bicarbonate buffer system shows that the addition of $CO_2$ to the body fluids raises $H^+$ concentration and lowers pH, while the removal of $CO_2$ has the opposite effects. This is the basis for the strong buffering capacity of the respiratory system. Indeed, this system can neutralize two or three times as much acid as the chemical buffers can.

Carbon dioxide is constantly produced by aerobic metabolism and is normally eliminated by the lungs at an equivalent rate. As explained in chapter 22, rising $CO_2$ concentration and falling pH stimulate peripheral and central chemoreceptors, which stimulate an increase in pulmonary ventilation. This expels excess $CO_2$ and thus reduces $H^+$ concentration. The free $H^+$ becomes part of the water molecules produced by this reaction:

$$HCO_3^- + H^+ \rightarrow H_2CO_3 \rightarrow CO_2 \text{ (expired)} + H_2O$$

Conversely, a drop in $H^+$ concentration raises pH and reduces pulmonary ventilation. This allows metabolic $CO_2$ to accumulate in the ECF faster than it is expelled, thus lowering pH to normal.

These are classic negative feedback mechanisms that result in acid–base homeostasis. Respiratory control of pH has some limitations, however, that are discussed later under acid–base imbalances.

## Renal Control of pH

The kidneys can neutralize more acid or base than either the respiratory system or the chemical buffers. The essence of this mechanism is that the renal tubules secrete $H^+$ into the

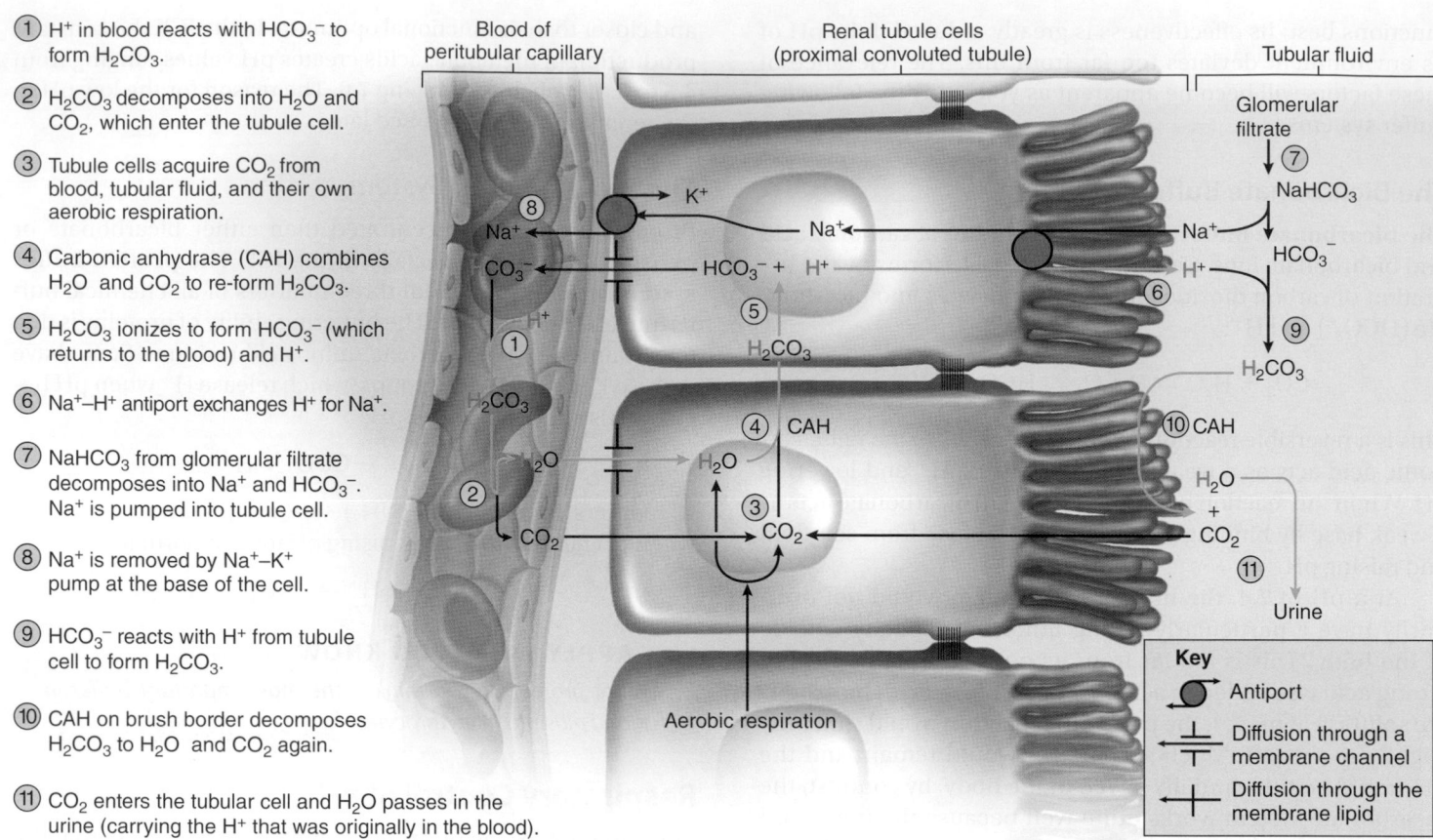

① H⁺ in blood reacts with $HCO_3^-$ to form $H_2CO_3$.

② $H_2CO_3$ decomposes into $H_2O$ and $CO_2$, which enter the tubule cell.

③ Tubule cells acquire $CO_2$ from blood, tubular fluid, and their own aerobic respiration.

④ Carbonic anhydrase (CAH) combines $H_2O$ and $CO_2$ to re-form $H_2CO_3$.

⑤ $H_2CO_3$ ionizes to form $HCO_3^-$ (which returns to the blood) and H⁺.

⑥ Na⁺–H⁺ antiport exchanges H⁺ for Na⁺.

⑦ $NaHCO_3$ from glomerular filtrate decomposes into Na⁺ and $HCO_3^-$. Na⁺ is pumped into tubule cell.

⑧ Na⁺ is removed by Na⁺–K⁺ pump at the base of the cell.

⑨ $HCO_3^-$ reacts with H⁺ from tubule cell to form $H_2CO_3$.

⑩ CAH on brush border decomposes $H_2CO_3$ to $H_2O$ and $CO_2$ again.

⑪ $CO_2$ enters the tubular cell and $H_2O$ passes in the urine (carrying the H⁺ that was originally in the blood).

**FIGURE 24.10 Secretion and Neutralization of Hydrogen Ions in the Kidneys.** The colored hydrogen symbols allow you to trace hydrogen from H⁺ in the blood to $H_2O$ in the urine.

❓ *If the pH of the tubular fluid went down, how would its Na⁺ concentration change?*

tubular fluid, where most of it binds to bicarbonate, ammonia, and phosphate buffers. Bound and free H⁺ are then excreted in the urine. Thus the kidneys, in contrast to the lungs, actually expel H⁺ from the body. The other buffer systems only reduce its concentration by binding it to another chemical.

Figure 24.10 shows how the renal tubule secretes and neutralizes H⁺. The hydrogen ions are colored so you can trace them from the blood (step 1) to the tubular fluid (step 6). Notice that it is not a simple matter of transporting free H⁺ across the tubule cells; rather, the H⁺ travels in the form of carbonic acid and water molecules.

The tubular secretion of H⁺ takes place at step 6, where the ion is pumped out of the tubule cell into the tubular fluid. This can happen only if there is a steep enough concentration gradient between a high H⁺ concentration within the cell and a lower concentration in the tubular fluid. If the pH of the tubular fluid drops any lower than 4.5, H⁺ concentration in the fluid is so high that tubular secretion ceases. Thus, pH 4.5 is the **limiting pH** for tubular secretion. This has added significance later in our discussion.

In a person with normal acid–base balance, all bicarbonate ions ($HCO_3^-$) in the tubular fluid are consumed by neutralizing H⁺; thus there is no $HCO_3^-$ in the urine. Bicarbonate ions are filtered by the glomerulus, gradually disappear from the tubular fluid, and appear in the peritubular capillary blood. It *appears* as if $HCO_3^-$ is reabsorbed by the renal tubules, but this is not the case; indeed, the renal tubules are incapable of reabsorbing $HCO_3^-$ directly. The cells of the proximal convoluted tubule, however, have carbonic anhydrase (CAH) on their brush borders facing the lumen. This breaks down the $H_2CO_3$ in the tubular fluid to $CO_2 + H_2O$ (step 10). It is the $CO_2$ that is reabsorbed, not the bicarbonate. For every $CO_2$ reabsorbed, however, a *new* bicarbonate ion is formed in the tubule cell and released into the blood (step 5). The effect is the same as if the tubule cells had reabsorbed bicarbonate itself.

Note that for every bicarbonate ion that enters the peritubular capillaries, a sodium ion does too. Thus, Na⁺ reabsorption by the renal tubules is part of the process of neutralizing acid. The more acid the kidneys excrete, the less sodium the urine contains.

The tubules secrete somewhat more H⁺ than the available bicarbonate can neutralize. The urine therefore contains a slight excess of free H⁺, which gives it a pH of about 5 to 6. Yet if all of the excess H⁺ secreted by the tubules remained in this free ionic form, the pH of the tubular fluid would drop far below the limiting pH of 4.5, and H⁺ secretion would stop. This must

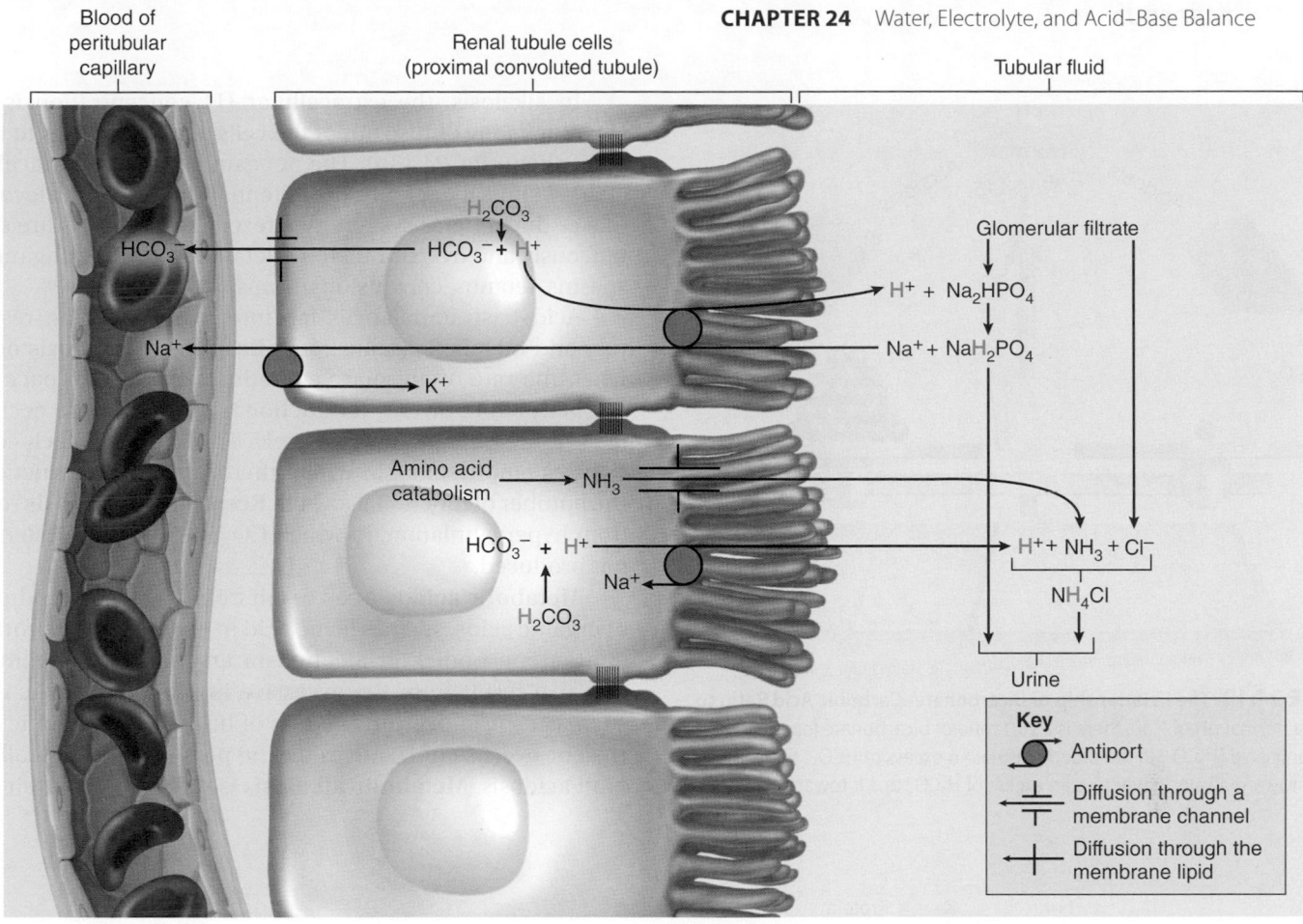

**FIGURE 24.11  Acid Buffering in the Urine.**  Reactions in the tubule cells are the same as in figure 24.10 but are simplified in this diagram. The essential differences are the buffering mechanisms shown in the tubular fluid. Hydrogen symbols are colored to allow tracing them from carbonic acid to the urine.

be prevented, and there are additional buffers in the tubular fluid to do so.

The glomerular filtrate contains $Na_2HPO_4$ (dibasic sodium phosphate), which reacts with some of the $H^+$ (fig. 24.11). A hydrogen ion replaces one of the sodium ions in the buffer, forming $NaH_2PO_4$ (monobasic sodium phosphate). This is passed in the urine, and the displaced $Na^+$ is transported into the tubule cell and from there to the bloodstream.

In addition, tubule cells catabolize certain amino acids and release ammonia ($NH_3$) as a product (fig. 24.11). Ammonia diffuses into the tubular fluid, where it acts as a base to neutralize acid. It reacts with $H^+$ and $Cl^-$ (the most abundant anion in the glomerular filtrate) to form ammonium chloride ($NH_4Cl$), which is passed in the urine.

Since there is so much chloride in the tubular fluid, you might ask why $H^+$ is not simply excreted as hydrochloric acid (HCl). Why involve ammonia? The reason is that HCl is a strong acid—it dissociates almost completely, so most of its hydrogen would be in the form of free $H^+$. The pH of the tubular fluid would drop below the limiting pH and prevent excretion of more acid. Ammonium chloride, by contrast, is a weak acid—most of its hydrogen remains bound to it and does not lower the pH of the tubular fluid.

## Disorders of Acid–Base Balance

Figure 24.12 represents acid–base balance with an instructive metaphor to show its dependence on the bicarbonate buffer system. At a normal pH of 7.4, the ECF has a 20:1 ratio of $HCO_3^-$ to $H_2CO_3$. Excess hydrogen ions convert $HCO_3^-$ to $H_2CO_3$ and tip the balance to a lower pH. A pH below 7.35 is considered to be a state of **acidosis.** On the other hand, a $H^+$ deficiency causes $H_2CO_3$ to dissociate into $H^+$ and $HCO_3^-$, thus tipping the balance to a higher pH. A pH above 7.45 is a state of **alkalosis.** Either of these imbalances has potentially fatal effects. A person cannot live more than a few hours if the blood pH is below 7.0 or above 7.7, and a pH below 6.8 or above 8.0 is quickly fatal.

In acidosis, $H^+$ diffuses down its concentration gradient into cells, and to maintain electrical balance, $K^+$ diffuses out (fig. 24.13a). The $H^+$ is buffered by intracellular proteins, so this exchange results in a net loss of cations from the cell. This makes the resting membrane potential more negative than usual (hyperpolarized) and makes nerve and muscle cells more difficult to stimulate. This is why acidosis depresses the central nervous system and causes such symptoms as confusion, disorientation, and coma.

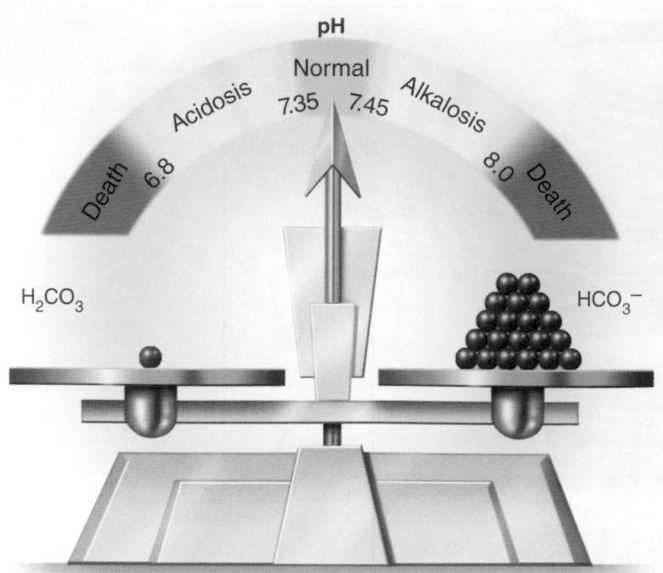

**FIGURE 24.12  The Relationship of Bicarbonate–Carbonic Acid Ratio to pH.** At a normal pH of 7.40, there is a 20:1 ratio of bicarbonate ions ($HCO_3^-$) to carbonic acid ($H_2CO_3$) in the blood plasma. An excess of $HCO_3^-$ tips the balance toward alkalosis, whereas an excess of $H_2CO_3$ tips it toward acidosis.

In alkalosis, the extracellular $H^+$ concentration is low. Hydrogen ions diffuse out of the cells and $K^+$ diffuses in to replace them (fig. 24.13b). The net gain in positive intracellular charges shifts the membrane potential closer to firing level and makes the nervous system hyperexcitable. Neurons fire spontaneously and overstimulate skeletal muscles, causing muscle spasms, tetanus, convulsions, or respiratory paralysis.

Acid–base imbalances fall into two categories, respiratory and metabolic (table 24.2). **Respiratory acidosis** occurs when the rate of alveolar ventilation fails to keep pace with the body's rate of $CO_2$ production. Carbon dioxide accumulates in the ECF and lowers its pH. This occurs in such conditions as emphysema, in which there is a severe reduction in the number of functional alveoli. **Respiratory alkalosis** results from hyperventilation, in which $CO_2$ is eliminated faster than it is produced.

**Metabolic acidosis** can result from increased production of organic acids, such as lactic acid in anaerobic fermentation and ketone bodies in alcoholism and diabetes mellitus. It can also result from the excessive ingestion of acidic drugs such as aspirin or from the loss of base due to chronic diarrhea or overuse of laxatives. Dying persons also typically exhibit acidosis. **Metabolic alkalosis** is rare but can result from

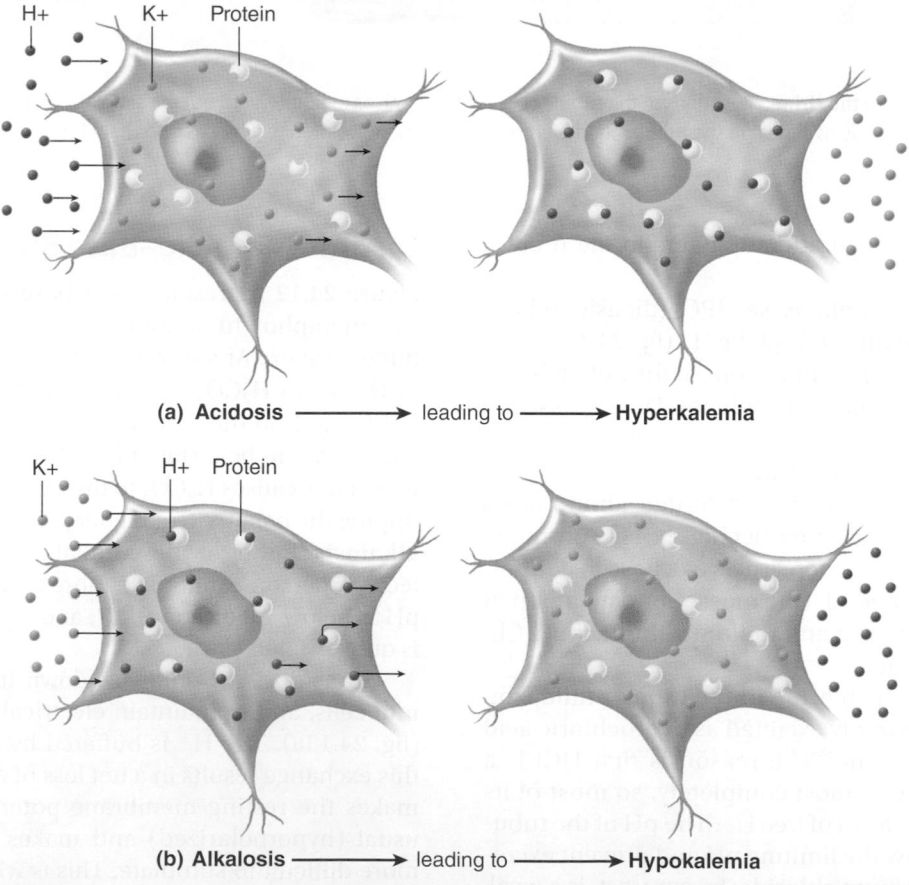

**(a) Acidosis ⟶ leading to ⟶ Hyperkalemia**

**(b) Alkalosis ⟶ leading to ⟶ Hypokalemia**

**FIGURE 24.13  The Relationship Between Acid–Base Imbalances and Potassium Imbalances.** (a) In acidosis, $H^+$ diffuses into the cells and drives out $K^+$, elevating the $K^+$ concentration of the ECF. (b) In alkalosis, $H^+$ diffuses out of the cells and $K^+$ diffuses in to replace it, lowering the $K^+$ concentration of the ECF.

❓ *How would you change part (a) to show the effect of hyperkalemia on the pH of the ECF?*

| TABLE 24.2 | Some Causes of Acidosis and Alkalosis | |
|---|---|---|
| | **Acidosis** | **Alkalosis** |
| *Respiratory* | Hypoventilation, apnea, or respiratory arrest; asthma; emphysema; cystic fibrosis; chronic bronchitis; narcotic overdose | Hyperventilation due to pain or emotions such as anxiety; oxygen deficiency (as at high elevation) |
| *Metabolic* | Excess production of organic acids as in diabetes mellitus and starvation; long-term anaerobic fermentation; hyperkalemia; chronic diarrhea; excessive alcohol consumption; drugs such as aspirin and laxatives | Rare but can result from chronic vomiting; overuse of bicarbonates (antacids); aldosterone hypersecretion |

overuse of bicarbonates (such as oral antacids and intravenous bicarbonate solutions) or from the loss of stomach acid by chronic vomiting.

## Compensation for Acid–Base Imbalances

In **compensated** acidosis or alkalosis, either the kidneys compensate for pH imbalances of respiratory origin, or the respiratory system compensates for pH imbalances of metabolic origin. **Uncompensated** acidosis or alkalosis is a pH imbalance that the body cannot correct without clinical intervention.

In **respiratory compensation,** changes in pulmonary ventilation correct the pH of the body fluids by expelling or retaining $CO_2$. If there is a $CO_2$ excess (hypercapnia), pulmonary ventilation increases to expel $CO_2$ and bring the blood pH back up to normal. If there is a $CO_2$ deficiency (hypocapnia), ventilation is reduced to allow $CO_2$ to accumulate in the blood and lower the pH to normal.

This is very effective in correcting pH imbalances due to abnormal $P_{CO_2}$ but not very effective in correcting other causes of acidosis and alkalosis. In diabetic acidosis, for example, the lungs cannot reduce the concentration of ketone bodies in the blood, although one can somewhat compensate for the $H^+$ that ketones release by increasing pulmonary ventilation and exhausting extra $CO_2$. The respiratory system can adjust a blood pH of 7.0 back to 7.2 or 7.3 but not all the way back to the normal 7.4. Although the respiratory system has a very powerful buffering effect, its ability to stabilize pH is therefore limited.

**Renal compensation** is an adjustment of pH by changing the rate of $H^+$ secretion by the renal tubules. The kidneys are slower to respond to pH imbalances but better at restoring a fully normal pH. Urine usually has a pH of 5 to 6, but in acidosis it may fall as low as 4.5 because of excess $H^+$, whereas in alkalosis it may rise as high as 8.2 because of excess $HCO_3^-$. The kidneys cannot act quickly enough to compensate for short-term pH imbalances, such as the acidosis that might result from an asthmatic attack lasting an hour or two, or the alkalosis resulting from a brief episode of emotional hyperventilation. They are effective, however, at compensating for pH imbalances that last for a few days or longer.

In acidosis, the renal tubules increase the rate of $H^+$ secretion. The extra $H^+$ in the tubular fluid must be buffered; otherwise, the fluid pH could exceed the limiting pH and $H^+$ secretion would stop. Therefore, in acidosis, the renal tubules

secrete more ammonia to buffer the added $H^+$, and the amount of ammonium chloride in the urine may rise to 7 to 10 times normal.

### ▶▶▶ APPLY WHAT YOU KNOW

*Suppose you measured the pH and ammonium chloride concentration of urine from a person with emphysema and urine from a healthy individual. How would you expect the two to differ, and why?*

In alkalosis, the bicarbonate concentration and pH of the urine are elevated. This is partly because there is more $HCO_3^-$ in the blood and glomerular filtrate and partly because there is not enough $H^+$ in the tubular fluid to neutralize all the $HCO_3^-$ in the filtrate.

## Acid–Base Imbalances in Relation to Electrolyte and Water Imbalances

The foregoing discussion once again stresses a point made early in this chapter—we cannot understand or treat imbalances of water, electrolyte, or acid–base balance in isolation from each other, because each of these frequently affects the other two. Table 24.3 itemizes and explains a few of these interactions. This is by no means a complete list of how fluid, electrolytes, and pH affect each other, but it does demonstrate their interdependence. Note that many of these relationships are reciprocal—for example, acidosis can cause hyperkalemia, and conversely, hyperkalemia can cause acidosis.

### BEFORE YOU GO ON

Answer the following questions to test your understanding of the preceding section:

12. Write two chemical equations that show how the bicarbonate buffer system compensates for acidosis and alkalosis and two equations that show how the phosphate buffer system compensates for these imbalances.

13. Why are phosphate buffers more effective in the cytoplasm than in the blood plasma?

14. Renal tubules cannot reabsorb $HCO_3^-$; yet $HCO_3^-$ concentration in the tubular fluid falls while in the blood plasma it rises. Explain this apparent contradiction.

15. In acidosis, the renal tubules secrete more ammonia. Why?

| TABLE 24.3 | | Some Relationships Among Fluid, Electrolyte, and Acid–Base Imbalances | |
|---|---|---|---|
| **Cause** | | **Potential Effect** | **Reason** |
| Acidosis | → | Hyperkalemia | $H^+$ diffuses into cells and displaces $K^+$ (see fig. 24.13a). As $K^+$ leaves the ICF, $K^+$ concentration in the ECF rises. |
| Hyperkalemia | → | Acidosis | Opposite from the above; high $K^+$ concentration in the ECF causes less $K^+$ to diffuse out of the cells than normally. $H^+$ diffuses out to compensate, and this lowers the extracellular pH. |
| Alkalosis | → | Hypokalemia | $H^+$ diffuses from ICF to ECF. More $K^+$ remains in the ICF to compensate for the $H^+$ loss, causing a drop in ECF $K^+$ concentration (see fig. 24.13b). |
| Hypokalemia | → | Alkalosis | Opposite from the above; low $K^+$ concentration in the ECF causes $K^+$ to diffuse out of cells. $H^+$ diffuses in to replace $K^+$, lowering the $H^+$ concentration of the ECF and raising its pH. |
| Acidosis | → | Hypochloremia | More $Cl^-$ is excreted as $NH_4Cl$ to buffer the excess acid in the renal tubules, leaving less $Cl^-$ in the ECF. |
| Alkalosis | → | Hyperchloremia | More $Cl^-$ is reabsorbed from the renal tubules, so ingested $Cl^-$ accumulates in the ECF rather than being excreted. |
| Hyperchloremia | → | Acidosis | More $H^+$ is retained in the blood to balance the excess $Cl^-$, causing hyperchloremic acidosis. |
| Hypovolemia | → | Alkalosis | More $Na^+$ is reabsorbed by the kidney. $Na^+$ reabsorption is coupled to $H^+$ secretion (see fig. 24.10), so more $H^+$ is secreted and pH of the ECF rises. |
| Hypervolemia | → | Acidosis | Less $Na^+$ is reabsorbed, so less $H^+$ is secreted into the renal tubules. $H^+$ retained in the ECF causes acidosis. |
| Acidosis | → | Hypocalcemia | Acidosis causes more $Ca^{2+}$ to bind to plasma protein and citrate ions, lowering the concentration of free, ionized $Ca^{2+}$ and causing symptoms of hypocalcemia. |
| Alkalosis | → | Hypercalcemia | Alkalosis causes more $Ca^{2+}$ to dissociate from plasma protein and citrate ions, raising the concentration of free $Ca^{2+}$. |

# DEEPER INSIGHT 24.2

## CLINICAL APPLICATION

### Fluid Replacement Therapy

One of the most significant problems in the treatment of seriously ill patients is the restoration and maintenance of proper fluid volume, composition, and distribution among the fluid compartments. Fluids may be administered to replenish total body water, restore blood volume and pressure, shift water from one fluid compartment to another, or restore and maintain electrolyte and acid–base balance.

Drinking water is the simplest method of fluid replacement, but it does not replace electrolytes. Heat exhaustion can occur when you lose water and salt in the sweat and replace the fluid by drinking plain water. Broths, juices, and sports drinks replace water, carbohydrates, and electrolytes.

Patients who cannot take fluids by mouth must be treated by alternative routes. Some fluids can be given by enema and absorbed through the colon. All routes of fluid administration other than the digestive tract are called *parenteral*[6] routes. The most common of these is the intravenous (I.V.) route, but for various reasons, including inability to find a suitable vein, fluids are sometimes given by subcutaneous (sub-Q), intramuscular (I.M.), or other parenteral routes. Many kinds of sterile solutions are available to meet the fluid replacement needs of different patients.

In cases of extensive blood loss, there may not be time to type and cross-match blood for a transfusion. The more urgent need is to replenish blood volume and pressure. *Normal saline* (isotonic, 0.9% NaCl) is a relatively quick and simple way to raise blood volume while maintaining normal osmolarity, but it has significant shortcomings. It takes three to five times as much saline as whole blood to rebuild normal volume because much of the saline escapes the circulation into the interstitial fluid compartment or is excreted by the kidneys. In addition, normal saline can induce hypernatremia and hyperchloremia, because the body excretes the water but retains much of the NaCl. Hyperchloremia can, in turn, produce acidosis. Normal saline also lacks potassium, magnesium, and calcium. Indeed, it dilutes those electrolytes that are already present and creates a risk of cardiac arrest from hypocalcemia. Saline also dilutes plasma albumin and RBCs, creating still greater risks for patients who have suffered extensive blood loss. Nevertheless, the emergency maintenance of blood volume sometimes takes temporary precedence over these other considerations.

Fluid therapy is also used to correct pH imbalances. Acidosis may be treated with *Ringer's lactate solution,* which includes sodium to rebuild ECF volume, potassium to rebuild ICF volume, lactate to balance the cations, and enough glucose to make the solution isotonic. Alkalosis can be treated with potassium chloride. This must be administered very carefully, because

potassium ions can cause painful venous spasms, and even a small potassium excess can cause cardiac arrest. High-potassium solutions should never be given to patients in renal failure or whose renal status is unknown, because in the absence of renal excretion of potassium, they can bring on lethal hyperkalemia. Ringer's lactate or potassium chloride also must be administered very cautiously, with close monitoring of blood pH, to avoid causing a pH imbalance opposite the one that was meant to be corrected. Too much Ringer's lactate causes alkalosis and too much KCl causes acidosis.

*Plasma volume expanders* are hypertonic solutions or colloids that are retained in the bloodstream and draw interstitial water into it by osmosis. They include albumin, sucrose, mannitol, and dextran. Plasma expanders are also used to combat hypotonic hydration by drawing water out of swollen cells, averting such problems as seizures and coma. A plasma expander can draw several liters of water out of the intracellular compartment within a few minutes.

Patients who cannot eat are often given isotonic 5% dextrose (glucose). A fasting patient loses as much as 70 to 85 g of protein per day from the tissues as protein is broken down to fuel the metabolism. Giving 100 to 150 g of I.V. glucose per day reduces this by half and is said to have a *protein-sparing effect.* More than glucose is needed in some cases—for example, if a patient has not eaten for several days and cannot be fed by nasogastric tube (due to lesions of the digestive tract, for example) or if large amounts of nutrients are needed for tissue repair following severe trauma, burns, or infections. In *total parenteral nutrition (TPN),* or *hyperalimentation,*[7] a patient is provided with complete I.V. nutritional support, including a protein hydrolysate (amino acid mixture), vitamins, electrolytes, 20% to 25% glucose, and on alternate days, a fat emulsion.

The water from parenteral solutions is normally excreted by the kidneys. If the patient has renal insufficiency, however, excretion may not keep pace with intake, and there is a risk of hypotonic hydration. Intravenous fluids are usually given slowly, by *I.V. drip,* to avoid abrupt changes or overcompensation for the patient's condition. In addition to pH, the patient's heart rate, blood pressure, hematocrit, and plasma electrolyte concentrations are monitored, and the patient is examined periodically for respiratory sounds indicating pulmonary edema.

The delicacy of fluid replacement therapy underscores the close relationships among fluids, electrolytes, and pH. It is dangerous to manipulate any one of these variables without close attention to the others. Parenteral fluid therapy is usually used for persons who are seriously ill. Their homeostatic mechanisms are already compromised and leave less room for error than in a healthy person.

---

[6]*para* = beside; *enter* = intestine

[7]*hyper* = above normal; *aliment* = nourishment

# STUDY GUIDE

## ▶ Assess Your Learning Outcomes

*To test your knowledge, discuss the following topics with a study partner or in writing, ideally from memory.*

### 24.1 Water Balance (p. 925)

1. Fluid compartments; total body water content of a typical young adult; and what percentages of this are in the intracellular and extracellular fluid compartments
2. How water moves from one fluid compartment to another; which differs more from one compartment to another—osmolarity or chemical composition—and why
3. What it means to be in a state of fluid balance
4. The meanings of *sensible, insensible,* and *obligatory water loss*
5. Typical daily water gain and loss; sources of water gain and avenues of water loss, and the typical amounts of each
6. How the hypothalamus senses the body's state of hydration and how it promotes fluid intake and water conservation when necessary
7. Short- and long-term mechanisms by which thirst is satiated; why it is important to have the short-term mechanisms
8. The role of antidiuretic hormone in regulating the body's rate of fluid loss
9. The meaning of *fluid deficiency;* two forms of fluid deficiency and how they differ; and potential consequences of fluid deficiency
10. The meaning of *fluid excess;* two forms of fluid excess and how they differ; and potential consequences of fluid excess
11. The meaning of *fluid sequestration;* examples and potential consequences of fluid sequestration

### 24.2 Electrolyte Balance (p. 931)

1. Functions of electrolytes in general; the body's common electrolytes; and the relative ECF and ICF concentrations of $Na^+$, $K^+$, $Cl^-$, $Ca^{2+}$, $Mg^{2+}$, and phosphate ions
2. Physiological functions of sodium; how it is regulated by aldosterone, antidiuretic hormone, and natriuretic peptides; and causes and effects of hyper- and hyponatremia
3. Physiological functions of potassium; how it is regulated by the kidneys and aldosterone; and causes and effects of hyper- and hypokalemia
4. Physiological functions of chloride; why chloride levels are determined mainly by the sodium-regulating mechanisms; and causes and effects of hyper- and hypochloremia
5. Physiological functions of calcium; why cells usually must maintain a low intracellular calcium level; how calcium homeostasis is regulated by parathyroid hormone, calcitriol, and calcitonin; and causes and effects of hyper- and hypocalcemia
6. Physiological functions of magnesium; how vitamin D and parathyroid hormone regulate magnesium level; and causes and effects of magnesium deficiency and excess
7. Three forms of inorganic phosphate ions ($P_i$); the physiological functions of phosphate; and how phosphate homeostasis is regulated

### 24.3 Acid–Base Balance (p. 936)

1. The normal pH range of the blood and most other ECF, and why acid–base balance is so crucial to homeostasis
2. Strong and weak acids and bases, and examples
3. How a buffer resists pH changes; the body's principal physiological and chemical buffer systems
4. How the bicarbonate, phosphate, and protein buffer systems neutralize excess acid or base; their effectiveness relative to each other and to the physiological buffer systems
5. How the respiratory system buffers pH
6. How the renal tubule secretes acid; why urine is normally bicarbonate-free; why acid secretion is linked to sodium reabsorption; and the roles of $Na_2HPO_4$ and $NH_3$ in buffering urinary acid
7. The ratio of $H_2CO_3$ to $HCO_3^-$ at normal blood pH, and how this ratio changes in acidosis and alkalosis; the pH levels at which acidosis and alkalosis are soon fatal
8. Mechanisms behind the neuromuscular effects of acidosis and alkalosis
9. Common causes of respiratory and metabolic acidosis and alkalosis
10. The difference between compensated and uncompensated acidosis and alkalosis, and how the respiratory and urinary systems compensate for pH imbalances
11. Examples of how water, electrolyte, and acid–base imbalances can each cause, or be caused by, imbalances in the other two categories

## ▶ Testing Your Recall

*Answers in Appendix B*

1. The greatest percentage of the body's water is in
   a. the blood plasma.
   b. the lymph.
   c. the intracellular fluid.
   d. the interstitial fluid.
   e. the extracellular fluid.

2. Hypertension is likely to increase the secretion of
   a. natriuretic peptide.
   b. antidiuretic hormone.
   c. bicarbonate ions.
   d. aldosterone.
   e. ammonia.

3. _____ increases water reabsorption without increasing sodium reabsorption.
   a. Antidiuretic hormone
   b. Aldosterone
   c. Natriuretic peptide
   d. Parathyroid hormone
   e. Calcitonin

# STUDY GUIDE

4. Hypotonic hydration can result from
   a. ADH hypersecretion.
   b. ADH hyposecretion.
   c. aldosterone hypersecretion.
   d. aldosterone hyposecretion.
   e. *a* and *d* only.

5. Tetanus is most likely to result from
   a. hypernatremia.
   b. hypokalemia.
   c. hyperkalemia.
   d. hypocalcemia.
   e. *c* and *d* only.

6. The principal determinant of intracellular osmolarity and cellular volume is
   a. protein.
   b. phosphate.
   c. potassium.
   d. sodium.
   e. chloride.

7. Increased excretion of ammonium chloride in the urine most likely indicates
   a. hypercalcemia.
   b. hyponatremia.
   c. hypochloremia.
   d. alkalosis.
   e. acidosis.

8. The most effective buffer in the intracellular fluid is
   a. phosphate.
   b. protein.
   c. bicarbonate.
   d. carbonic acid.
   e. ammonia.

9. Tubular secretion of hydrogen is directly linked to
   a. tubular secretion of potassium.
   b. tubular secretion of sodium.
   c. tubular reabsorption of potassium.
   d. tubular reabsorption of sodium.
   e. tubular secretion of chloride.

10. Hyperchloremia is most likely to result in
    a. alkalosis.
    b. acidosis.
    c. hypernatremia.
    d. hyperkalemia.
    e. hypovolemia.

11. The most abundant cation in the ECF is _____.

12. The two most abundant cations in the ICF are _____ and _____.

13. Water produced by the body's chemical reactions is called _____.

14. The skin loses water by two processes, sweating and _____.

15. Any abnormal accumulation of fluid in a particular place in the body is called _____.

16. An excessive concentration of potassium ions in the blood is called _____.

17. A deficiency of sodium ions in the blood is called _____.

18. A blood pH of 7.2 caused by inadequate pulmonary ventilation would be classified as _____.

19. Tubular secretion of hydrogen ions ceases if the acidity of the tubular fluid falls below a value called the _____.

20. Long-term satiation of thirst depends on a reduction of the _____ of the blood.

## ► Building Your Medical Vocabulary

*Answers in Appendix B*

*State a medical meaning of each word element below, and give a term in which it or a slight variation of it is used.*

1. aliment-

2. -emia

3. entero-

4. kali-

5. natri-

6. para-

7. sequestr-

8. spiro-

9. trans-

10. vol-

## ► True or False

*Answers in Appendix B*

*Determine which five of the following statements are false, and briefly explain why.*

1. Hypokalemia lowers the resting membrane potentials of nerve and muscle cells and makes them less excitable.

2. Aldosterone promotes sodium and water retention and can therefore greatly increase blood pressure.

3. Injuries that rupture a lot of cells tend to elevate the K+ concentration of the ECF.

4. It is possible for a person to suffer circulatory shock even without losing a significant amount of fluid from the body.

5. Parathyroid hormone promotes calcium and phosphate reabsorption by the kidneys.

6. The bicarbonate system buffers more acid than any other chemical buffer.

7. The more sodium the renal tubules reabsorb, the more hydrogen ion they secrete into the tubular fluid.

8. The body compensates for respiratory acidosis by increasing the respiratory rate.

9. In true dehydration, the body fluids remain isotonic although total body water is reduced.

10. Long-term quenching of thirst results primarily from wetting and cooling of the mouth when one drinks.

# STUDY GUIDE

## ▶ Testing Your Comprehension

*Answers at* www.mhhe.com/saladin7

1. A duck hunter is admitted to the hospital with a shotgun injury to the abdomen. He has suffered extensive blood loss but is conscious. He complains of being intensely thirsty. Explain the physiological mechanism connecting his injury to his thirst.

2. A woman living at poverty level finds bottled water at the grocery store next to the infant formula. The label on the water states that it is made especially for infants, and she construes this to mean that it can be used as a nutritional supplement. The water is much cheaper than formula, so she gives her baby several ounces of bottled water a day as a substitute for formula. After several days the baby has seizures and is taken to the hospital, where it is found to have edema, acidosis, and a plasma sodium concentration of 116 mEq/L. The baby is treated with anticonvulsants followed by normal saline and recovers. Explain each of the signs.

3. Explain why the respiratory and urinary systems are both necessary for the bicarbonate buffer system to work effectively in the blood plasma.

4. A 4-year-old child is caught up in tribal warfare in Africa. In a refugee camp, the only drinking water is from a sewage-contaminated pond. The child soon develops severe diarrhea and dies 10 days later of cardiac arrest. Explain the possible physiological cause(s) of his death.

5. The left column indicates some increases or decreases in blood plasma values. In the right column, replace the question mark with an up or down arrow to indicate the expected effect. Explain each effect.

| Cause | Effect |
|---|---|
| a. $\uparrow H_2O$ | ? $Na^+$ |
| b. $\uparrow Na^+$ | ? $Cl^-$ |
| c. $\downarrow K^+$ | ? $H^+$ |
| d. $\uparrow H^+$ | ? $K^+$ |
| e. $\downarrow Ca^{2+}$ | ? $PO_4^{3-}$ |

## ▶ Improve Your Grade at www.mhhe.com/saladin7

*Download animations to study when it fits your schedule. Practice quizzes, labeling activities, games, and flashcards offer fun ways to master the chapter concepts. Or, download image PowerPoint files for each chapter to create a study guide or for taking notes during lecture.*

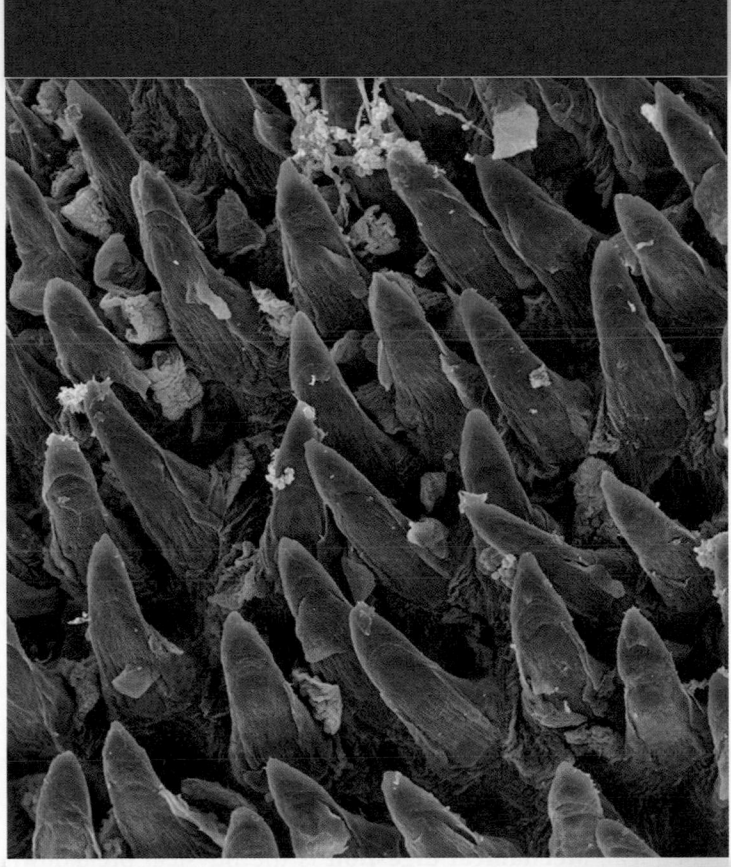

Filiform papillae of the human tongue (SEM)

Anatomy &
Physiology | **REVEALED®**
aprevealed.com

**Module 12: Digestive System**

- All chemical digestion consists of hydrolysis reactions, so you should understand this process (p. 59) and its bearing on the relationship between polymers and monomers.

- You must also know the basic structure of polysaccharides, disaccharides, triglycerides, and proteins (pp. 59–67) to understand how these are digested.

- The effect of pH on enzyme action (p. 70) helps to explain activation and deactivation of various digestive enzymes in different regions of the digestive tract.

- An understanding of the modes of membrane transport (pp. 90–98) is essential for understanding intestinal absorption of nutrients.

- You will find close parallels between the neuromuscular control of urination (p. 916) and defecation.

Most of the nutrients we eat cannot be used in their existing form. They must be broken down into smaller components, such as amino acids and monosaccharides, that are universal to all species. Consider what happens if you eat a piece of beef, for example. The myosin of beef differs very little from that of your own muscles, but the two are not identical, and even if they were, beef myosin could not be absorbed, transported in the blood, and properly installed in your muscle cells. Like any other dietary protein, it must be broken down into amino acids before it can be used. Since beef and human proteins are made of the same 20 amino acids, those of beef proteins might indeed become part of your own myosin but could equally well end up in your insulin, fibrinogen, collagen, or any other protein.

The digestive system is essentially a disassembly line—its primary purpose is to break nutrients down into forms that can be used by the body and to absorb them so they can be distributed to the tissues. The study of the digestive tract and the diagnosis and treatment of its disorders is called **gastroenterology.**[1]

## 25.1 General Anatomy and Digestive Processes

### Expected Learning Outcomes

When you have completed this section, you should be able to

a. list the functions and major physiological processes of the digestive system;

b. distinguish between mechanical and chemical digestion;

c. describe the basic chemical process underlying all chemical digestion, and name the major substrates and products of this process;

d. list the regions of the digestive tract and the accessory organs of the digestive system;

e. identify the layers of the digestive tract and describe its relationship to the peritoneum; and

f. describe the general neural and chemical controls over digestive function.

### Digestive Function AP|R

The **digestive system** is the organ system that processes food, extracts nutrients from it, and eliminates the residue. It does this in five stages:

1. **ingestion,** the selective intake of food;

2. **digestion,** the mechanical and chemical breakdown of food into a form usable by the body;

3. **absorption,** the uptake of nutrient molecules into the epithelial cells of the digestive tract and then into the blood or lymph;

4. **compaction,** absorbing water and consolidating the indigestible residue into feces; and finally,

5. **defecation,** the elimination of feces.

The digestion stage itself has two facets, mechanical and chemical. **Mechanical digestion** is the physical breakdown of food into smaller particles. It is achieved by the cutting and grinding action of the teeth and the churning contractions of the stomach and small intestine. Mechanical digestion exposes more food surface to the action of digestive enzymes. **Chemical digestion** is a series of hydrolysis reactions that break dietary macromolecules into their monomers (*residues*): polysaccharides into monosaccharides, proteins into amino acids, fats into monoglycerides and fatty acids, and nucleic acids into nucleotides. It is carried out by digestive enzymes produced by the salivary glands, stomach, pancreas, and small intestine. Some nutrients are already present in usable form in the ingested food and are absorbed without being digested: vitamins, free amino acids, minerals, cholesterol, and water.

### General Anatomy

The digestive system has two anatomical subdivisions, the digestive tract and the accessory organs (fig. 25.1). The **digestive tract (alimentary**[2] **canal)** is a muscular tube extending from mouth to anus. It measures about 5 m (16 ft) long in a living person, but about 9 m (30 ft) in the cadaver due to the loss of muscle tone at death. It includes the mouth, pharynx, esophagus, stomach, small intestine, and large intestine. Part of this, the stomach and intestines, constitute the *gastrointestinal (GI) tract*. The **accessory organs** are the teeth, tongue, salivary glands, liver, gallbladder, and pancreas.

The digestive tract is open to the environment at both ends. Most of the material in it has not entered any body tissues and is considered to be external to the body until it is absorbed

---

[1]*gastro* = stomach; *entero* = intestines; *logy* = study of

[2]*aliment* = food

by epithelial cells of the alimentary canal. In the strict sense, defecated food residue was never in the body.

Most of the digestive tract follows the basic structural plan shown in figure 25.2, with a wall composed of the following tissue layers, in order from the inner to the outer surface:

Mucosa
> Epithelium
> Lamina propria
> Muscularis mucosae

Submucosa

Muscularis externa
> Inner circular layer
> Outer longitudinal layer

Serosa
> Areolar tissue
> Mesothelium

Slight variations on this theme are found in different regions of the tract.

The inner lining of the digestive tract, called the **mucosa** or **mucous membrane,** consists of an inner epithelium, a loose connective tissue layer called the **lamina propria,** and a thin layer of smooth muscle called the **muscularis mucosae** (MUSS-cue-LERR-is mew-CO-see). The epithelium is simple columnar in most of the digestive tract, but stratified squamous from the oral cavity through the esophagus and in the lower anal canal, where the tract is subject to more abrasion. The muscularis mucosae tenses the mucosa, creating grooves and ridges that enhance its surface area and contact with food. This improves the efficiency of digestion and nutrient absorption. The mucosa exhibits an abundance of lymphocytes and lymphatic nodules—the *mucosa-associated lymphatic tissue (MALT)* (see p. 809).

The **submucosa** is a thicker layer of loose connective tissue containing blood vessels and lymphatics, a nerve plexus, and in some places, glands that secrete lubricating mucus into the lumen. The MALT extends into the submucosa in some parts of the GI tract.

The **muscularis externa** consists of usually two layers of muscle near the outer surface. Cells of the inner layer encircle the tract while those of the outer layer run longitudinally. In some places, the circular layer is thickened to form valves (sphincters) that regulate the passage of material through the tract. The muscularis externa is responsible for the motility that propels food and residue through the digestive tract.

The **serosa** is composed of a thin layer of areolar tissue topped by a simple squamous mesothelium. The serosa begins in the lower 3 to 4 cm of the esophagus and ends just before the rectum. The pharynx, most of the esophagus, and the rectum have no serosa but are surrounded by a fibrous connective tissue layer called the **adventitia,** which blends into the adjacent connective tissue of other organs.

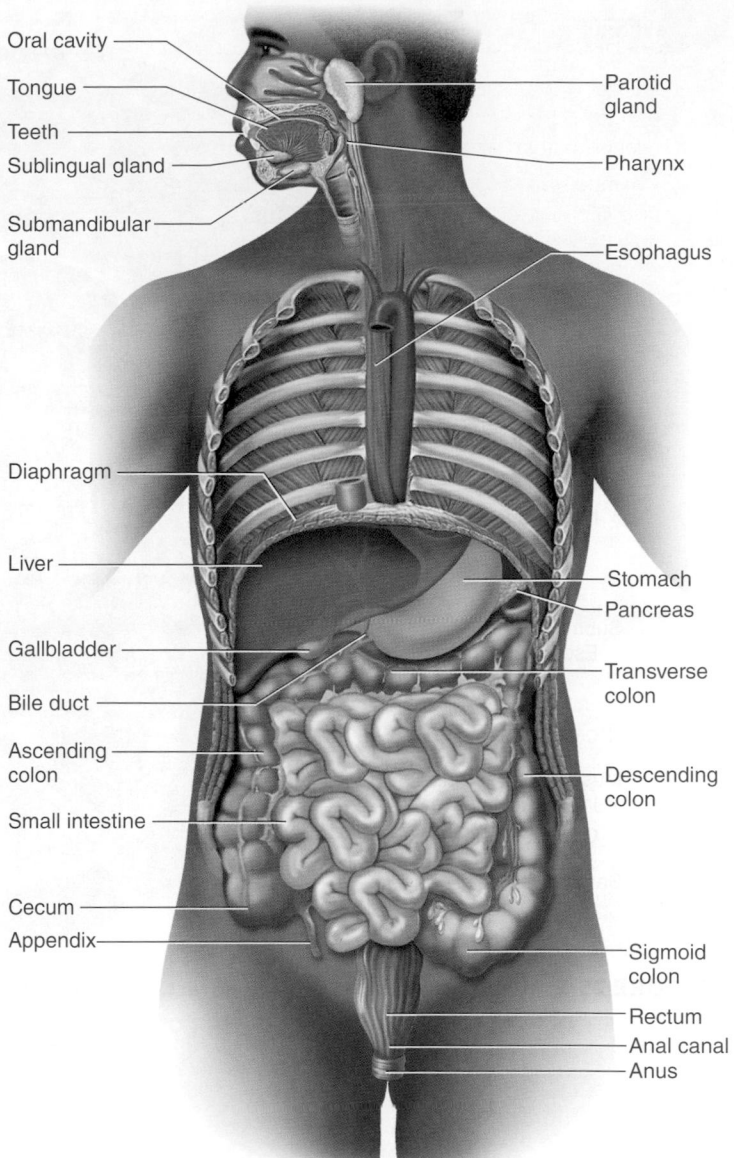

**FIGURE 25.1  The Digestive System.**

The esophagus, stomach, and intestines have a nervous network called the **enteric[3] nervous system,** which regulates digestive tract motility, secretion, and blood flow. This system is thought to have over 100 million neurons—more than the spinal cord! It can function independently of the central nervous system, although the CNS usually exerts a significant influence on its action. It is usually regarded as part of the autonomic nervous system, but opinions on this vary.

---

[3]*enter* = intestine

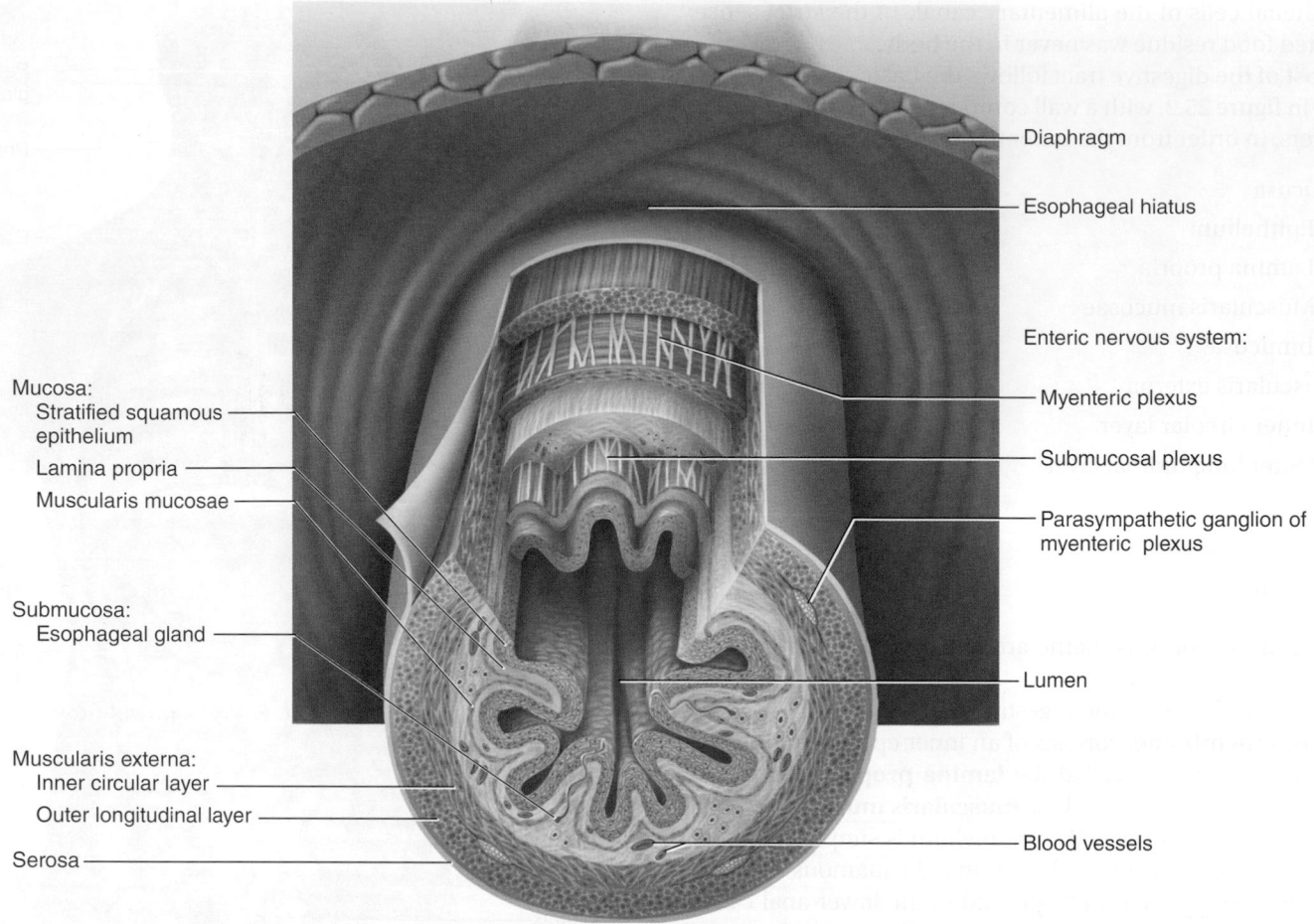

Mucosa:
  Stratified squamous epithelium
  Lamina propria
  Muscularis mucosae

Submucosa:
  Esophageal gland

Muscularis externa:
  Inner circular layer
  Outer longitudinal layer

Serosa

Diaphragm

Esophageal hiatus

Enteric nervous system:
  Myenteric plexus
  Submucosal plexus

Parasympathetic ganglion of myenteric plexus

Lumen

Blood vessels

**FIGURE 25.2 Tissue Layers of the Digestive Tract.** Cross section of the esophagus just below the diaphragm where it meets the stomach.

The enteric nervous system is composed of two networks of neurons: the **submucosal (Meissner[4]) plexus** in the submucosa and the **myenteric (Auerbach[5]) plexus** of parasympathetic ganglia and nerve fibers between layers of the muscularis externa. Parasympathetic preganglionic fibers of the vagus nerves terminate in the ganglia of the myenteric plexus. Postganglionic fibers arising in this plexus not only innervate the muscularis externa, but also pass through its inner circular layer and contribute to the submucosal plexus. The myenteric plexus controls peristalsis and other contractions of the muscularis externa, and the submucosal plexus controls movements of the muscularis mucosae and glandular secretion of the mucosa. The enteric nervous system also includes sensory neurons that monitor tension in the gut wall and conditions in the lumen.

## Relationship to the Peritoneum

In processing food, the stomach and intestines undergo such strenuous contractions that they need freedom to move in the abdominal cavity. They are not tightly bound to the abdominal wall, but over most of their length, they are loosely suspended from it by connective tissue sheets called **mesenteries.** Mesenteries hold the abdominal viscera in their proper relationship to each other and prevent the small intestine, especially, from becoming twisted and tangled by its own contractions or changes in body position. Furthermore, the mesenteries provide passage for the blood vessels and nerves that supply the digestive tract, and contain many lymph nodes and lymphatic vessels.

The parietal peritoneum is a serous membrane that lines the wall of the abdominal cavity. Along the posterior (dorsal) midline of the body, it turns inward and forms the **posterior (dorsal) mesentery,** a translucent two-layered membrane extending to the digestive tract. Upon reaching an organ such as the stomach or small intestine, the two layers of the mesentery separate and pass around opposite sides of the

[4]Georg Meissner (1829–1905), German histologist
[5]Leopold Auerbach (1828–97), German anatomist

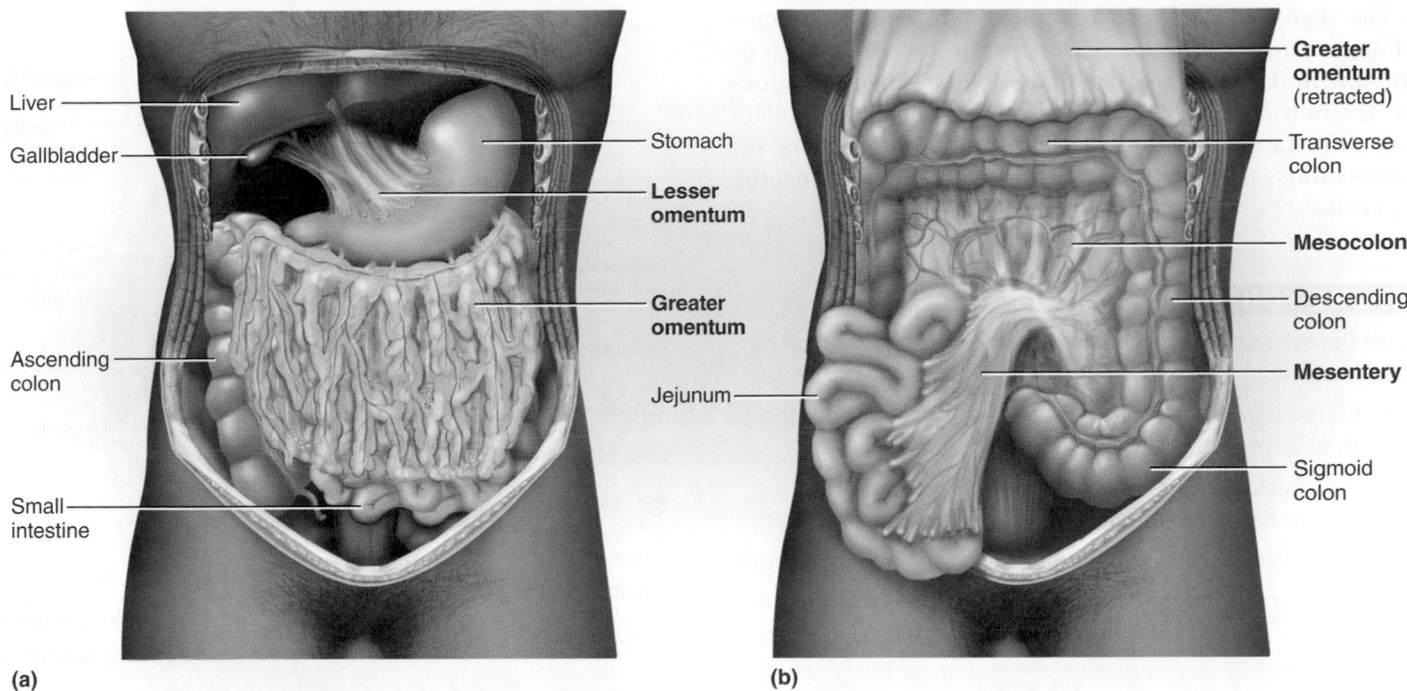

**FIGURE 25.3 Serous Membranes Associated with the Digestive Tract.** (a) The greater and lesser omenta. (b) Greater omentum and small intestine retracted to show the mesocolon and mesentery. These membranes contain the mesenteric arteries and veins. **AP|R**

organ, forming the serosa. In some places, the two layers come together again on the far side of that organ and continue as another sheet of tissue, the **anterior (ventral) mesentery.** This mesentery may hang freely in the abdominal cavity or attach to the anterior abdominal wall or other organs. The relationship between the mesenteries and the serosa is shown in figures A.7 and A.8 (pp. 34–35).

Two anterior mesenteries called *omenta* are associated with the stomach (see fig. 25.3). The **lesser omentum** extends the short distance from the liver to the right superior margin *(lesser curvature)* of the stomach. A much larger and fatty **greater omentum** hangs like an apron from the left inferior margin *(greater curvature)* of the stomach, loosely covering the small intestine. At its lower end, the greater omentum turns back on itself and passes upward, behind the superficial layer, forming a deep, empty pouch between the layers like an apron turned up at the hem. At the superior margin, the upturned layer continues as a serosa that encloses the spleen and transverse colon. From the transverse colon, it continues to the posterior abdominal wall and anchors the colon. This mesentery of the colon is called the **mesocolon.**

The omenta have a loosely organized, lacy appearance due partly to many holes or gaps in the membranes and partly to an irregular distribution of adipose tissue. They also contain many lymph nodes, lymphatic vessels, blood vessels, and nerves. The omenta adhere to perforations or inflamed areas of the stomach or intestines, contribute immune cells to the site, and isolate infections that might otherwise give rise to peritonitis. When a person gains abdominal weight, a great deal of it is fat deposited in these mesenteries.

When an organ is enclosed by mesentery (serosa) on both sides, it is considered to be within the peritoneal cavity, or **intraperitoneal.** When an organ lies against the posterior body wall and is covered by peritoneum on the anterior side only, it is said to be outside the peritoneal cavity, or **retroperitoneal.** The duodenum, most of the pancreas, and parts of the large intestine are retroperitoneal. The stomach, liver, and other parts of the small and large intestines are intraperitoneal.

## Regulation of the Digestive Tract

The motility and secretion of the digestive tract are controlled by neural, hormonal, and paracrine mechanisms. The neural controls include short and long autonomic reflexes. In **short (myenteric) reflexes,** stretching or chemical stimulation of the digestive tract acts through the myenteric plexus to stimulate contractions in nearby regions of the muscularis externa, such as the *peristaltic* contractions of swallowing. **Long (vagovagal) reflexes** act through autonomic nerve fibers that carry sensory signals from the digestive tract to the brainstem and motor commands back to the digestive tract. Parasympathetic fibers of the vagus nerves are especially important in stimulating digestive motility and secretion by way of these long reflexes.

The digestive tract also produces numerous hormones such as *gastrin* and *secretin,* and paracrine secretions such as *histamine* and *prostaglandins,* that stimulate digestive function. The hormones are secreted into the blood and stimulate relatively distant parts of the digestive tract. The paracrine secretions diffuse through the tissue fluids and stimulate nearby target cells.

### BEFORE YOU GO ON

Answer the following questions to test your understanding of the preceding section:

1. What is the term for the serous membrane that suspends the intestines from the abdominal wall?

2. Which physiological process of the digestive system truly moves a nutrient from the outside to the inside of the body?

3. What one type of reaction is the basis of all chemical digestion?

4. Name some nutrients that are absorbed without being digested.

## 25.2    The Mouth Through Esophagus

### Expected Learning Outcomes

When you have completed this section, you should be able to

a. describe the gross anatomy of the digestive tract from the mouth through the esophagus;

b. describe the composition and functions of saliva; and

c. describe the neural control of salivation and swallowing.

### The Mouth

The mouth is also known as the **oral,** or **buccal** (BUCK-ul), **cavity.** Its functions include ingestion (food intake), taste and other sensory responses to food, mastication (chewing), chemical digestion (starch is partially digested in the mouth), swallowing, speech, and respiration. The mouth is enclosed by the cheeks, lips, palate, and tongue (fig. 25.4). Its anterior opening between the lips is the **oral fissure** and its posterior opening into the throat is the **fauces**[6] (FAW-seez). The mouth is lined with stratified squamous epithelium. This epithelium is keratinized in areas subject to the greatest food abrasion, such as the gums and hard palate, and nonkeratinized in other areas such as the floor of the mouth, the soft palate, and the inside of the cheeks and lips.

### The Cheeks and Lips

The cheeks and lips retain food and push it between the teeth for chewing. They are essential for articulate speech and for sucking and blowing actions, including suckling by infants.

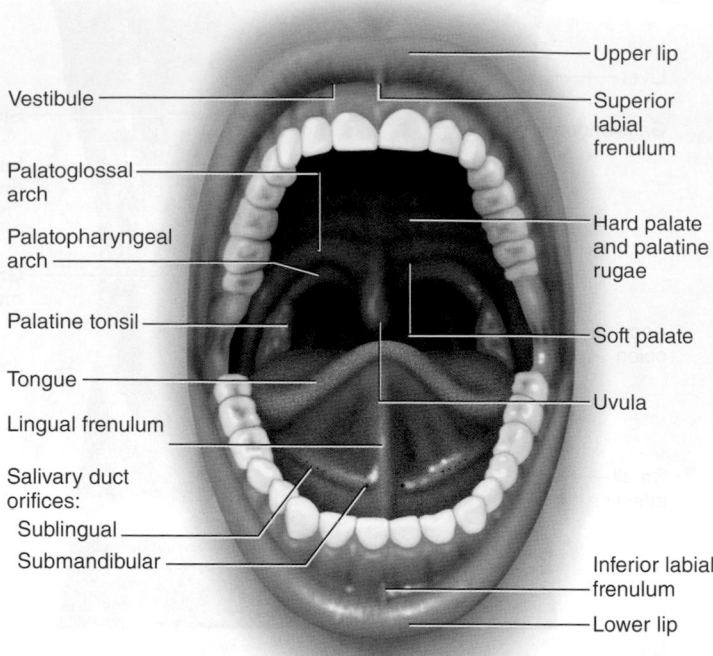

**FIGURE 25.4   The Oral Cavity.** For a photographic medial view, see figure B.2 (p. 378). **AP|R**

Their fleshiness is due mainly to subcutaneous fat, the buccinator muscles of the cheeks, and the orbicularis oris muscle of the lips. A median fold called the **labial frenulum**[7] attaches each lip to the gum, between the anterior incisors. The **vestibule** is the space between the cheeks or lips and the teeth—the space where you insert your toothbrush when brushing the outer surfaces of the teeth.

The lips are divided into three areas: (1) The *cutaneous area* is colored like the rest of the face and has hair follicles and sebaceous glands; on the upper lip, this is where a mustache grows. (2) The *red area (vermilion)* is the hairless region where the lips meet (where one might apply lipstick). It has unusually tall dermal papillae, which allow blood capillaries and nerve endings to come closer to the epidermal surface. Thus, this area is redder and more sensitive than the cutaneous area. (3) The *labial mucosa* is the inner surface of the lip, facing the gums and teeth.

### The Tongue

The tongue, although muscular and bulky, is a remarkably agile and sensitive organ. It manipulates food between the teeth while it avoids being bitten, it can extract food particles from the teeth after a meal, and it is sensitive enough to feel a stray

---

[6]*fauces* = throat

[7]*labi* = lip; *frenulum* = little bridle

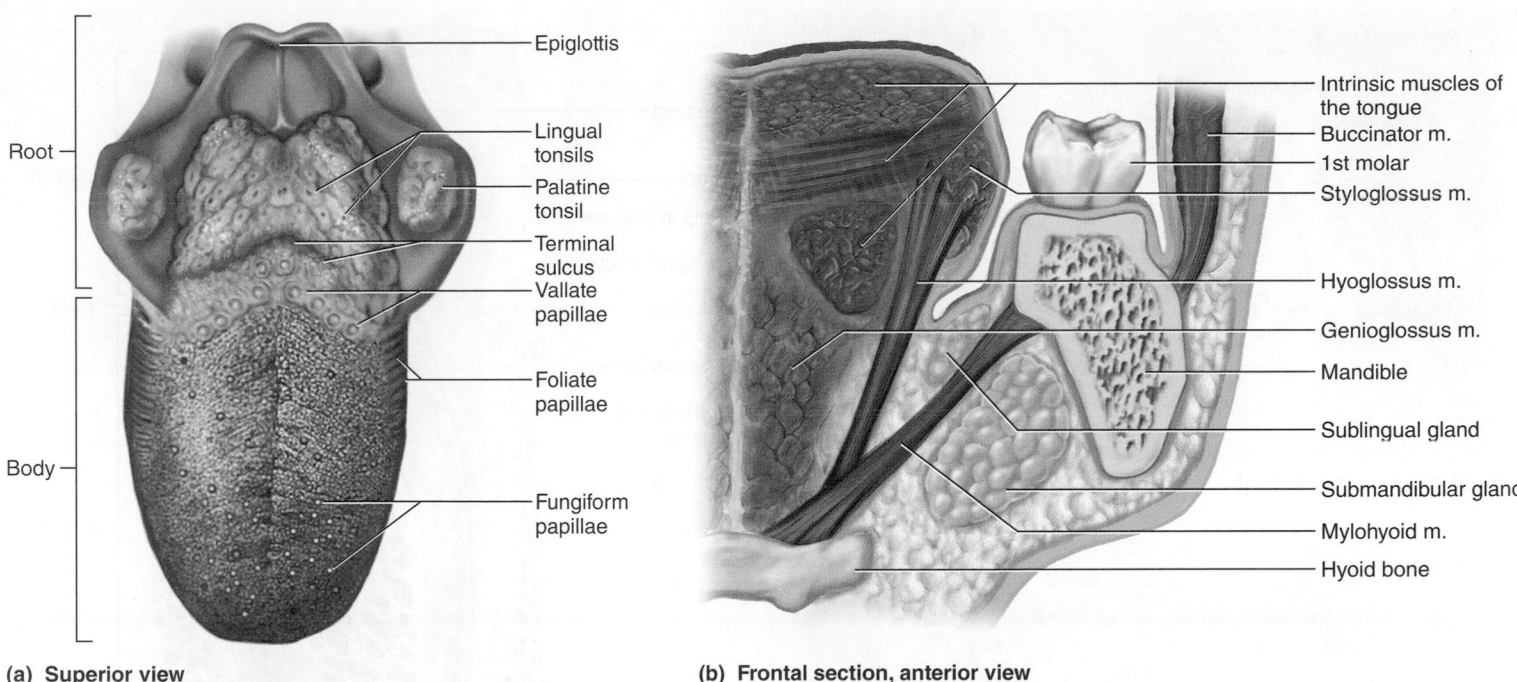

**(a) Superior view**

**(b) Frontal section, anterior view**

**FIGURE 25.5 The Tongue.** For a sagittal section, see figure B.2 (p. 378).

hair in a bite of food. Its surface is covered with nonkeratinized stratified squamous epithelium and exhibits bumps and projections called **lingual papillae,** the site of most taste buds. The types of papillae and sense of taste are discussed in chapter 16, and the general anatomy of the tongue is shown in figure 25.5.

#### ►►►APPLY WHAT YOU KNOW

*How does proprioception protect the tongue from being bitten?*

The anterior two-thirds of the tongue, called the **body,** occupies the oral cavity; and the posterior one-third, the **root,** occupies the oropharynx. The boundary between them is marked by a V-shaped row of **vallate papillae** and, behind these, a groove called the **terminal sulcus.** The body is attached to the floor of the mouth by a median fold called the **lingual frenulum.**

The muscles of the tongue, which compose most of its mass, are described in chapter 10. The **intrinsic muscles,** contained entirely within the tongue, produce the relatively subtle tongue movements of speech. The **extrinsic muscles,** with origins elsewhere and insertions in the tongue, produce most of the stronger movements of food manipulation. The extrinsic muscles include the **genioglossus, hyoglossus, palatoglossus, and styloglossus** (fig 25.5b; see also fig. 10.9, p. 323). Amid the muscles are serous and mucous **lingual glands,** which secrete a portion of the saliva. The lingual tonsils are contained in the root.

### The Palate

The palate, separating the oral cavity from the nasal cavity, makes it possible to breathe while chewing food. Its anterior portion, the **hard (bony) palate,** is supported by the palatine processes of the maxillae and by the smaller palatine bones. It has transverse ridges called *palatine rugae* that aid the tongue in holding and manipulating food. Posterior to this is the **soft palate,** which has a more spongy texture and is composed mainly of skeletal muscle and glandular tissue, but no bone. It has a conical medial projection, the **uvula,**[8] visible at the rear of the mouth. The uvula helps to retain food in the mouth until one is ready to swallow.

At the rear of the mouth, two muscular arches on each side begin at the roof near the uvula and descend to the floor. The anterior one is the **palatoglossal arch** and the posterior one is the **palatopharyngeal arch.** The latter arch marks the beginning of the pharynx. The palatine tonsils are located on the wall between the arches.

### The Teeth

The teeth are collectively called the **dentition.** They serve to *masticate* the food, breaking it into smaller pieces. This not

[8]*uvula* = little grape

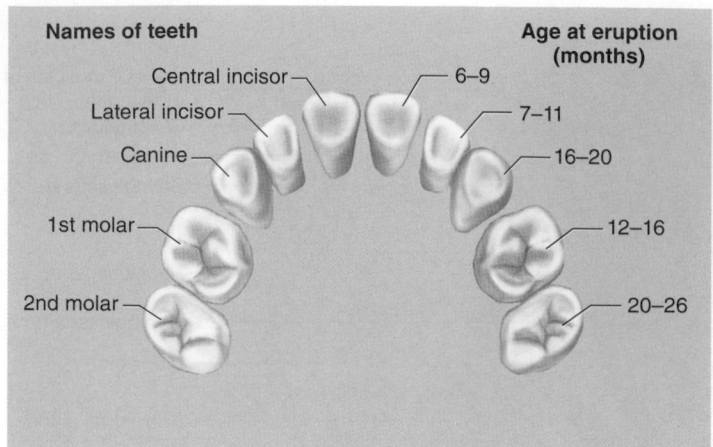

**Names of teeth**

**Age at eruption (months)**

Central incisor — 6–9
Lateral incisor — 7–11
Canine — 16–20
1st molar — 12–16
2nd molar — 20–26

**(a) Deciduous (baby) teeth**

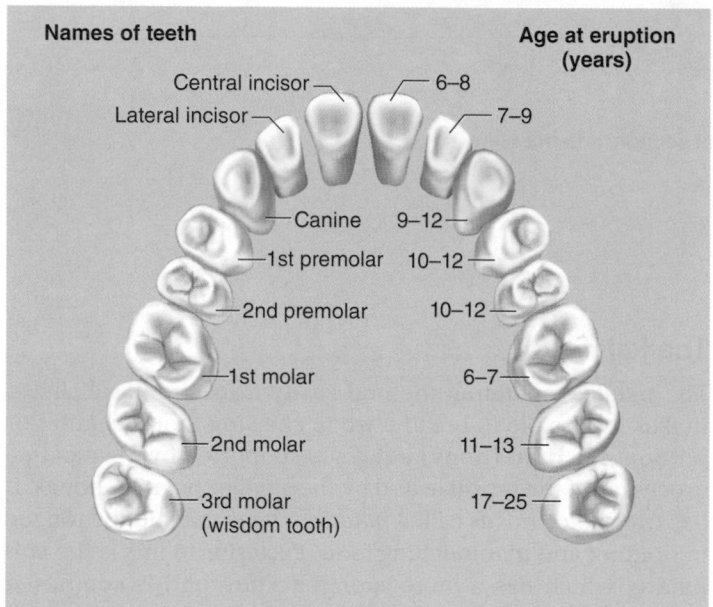

**Names of teeth**

**Age at eruption (years)**

Central incisor — 6–8
Lateral incisor — 7–9
Canine    9–12
1st premolar    10–12
2nd premolar    10–12
1st molar    6–7
2nd molar    11–13
3rd molar (wisdom tooth)    17–25

**(b) Permanent teeth**

**FIGURE 25.6  The Dentition.** Each figure shows only the upper teeth. The ages at eruption are composite ages for the corresponding upper and lower teeth. Generally, the lower (mandibular) teeth erupt somewhat earlier than their upper (maxillary) counterparts. **AP|R**

 *Which teeth are absent from a 3-year-old child?*

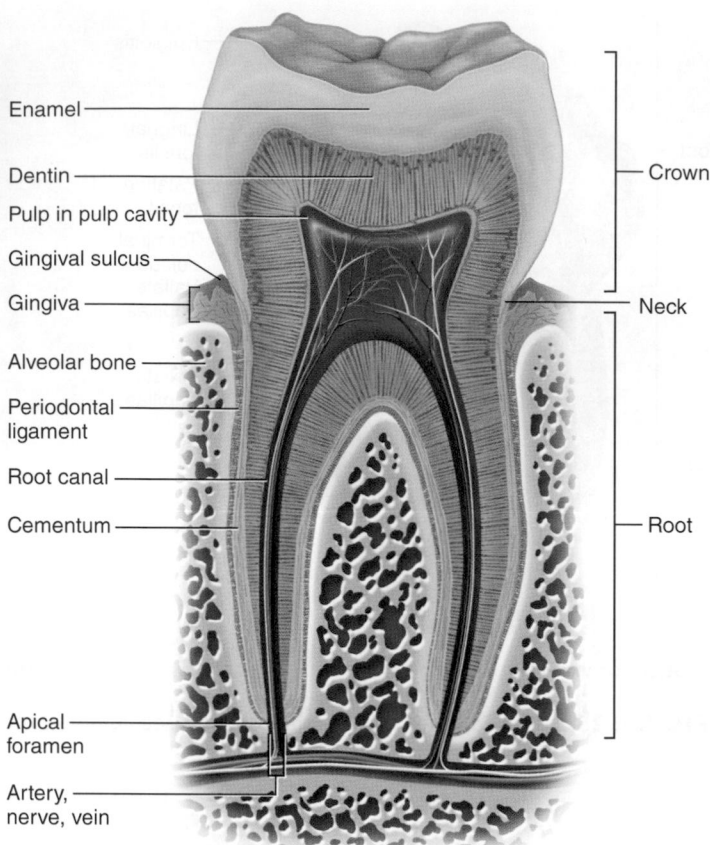

Enamel
Dentin
Pulp in pulp cavity
Gingival sulcus
Gingiva
Alveolar bone
Periodontal ligament
Root canal
Cementum
Crown
Neck
Root
Apical foramen
Artery, nerve, vein

**FIGURE 25.7  Structure of a Tooth and Its Alveolus.** Shows typical anatomy of the tooth and periodontal tissues. This particular example is a molar.

 *Of all the components shown here, which is or are not living tissue(s)?*

only makes the food easier to swallow, but also exposes more surface area to the action of digestive enzymes and thus speeds up chemical digestion. Adults normally have 16 teeth in the mandible and 16 in the maxilla. From the midline to the rear of each jaw, there are two incisors, a canine, two premolars, and up to three molars on each side (fig. 25.6b). The **incisors** are chisel-like cutting teeth used to bite off a piece of food. The **canines** are more pointed and act to puncture and shred it. They serve as weapons in many mammals but became reduced

in the course of human evolution until they now project barely beyond the other teeth. The **premolars** and **molars** have relatively broad surfaces adapted for crushing, shredding, and grinding.

Each tooth is embedded in a socket called an **alveolus,** forming a joint called a *gomphosis* between the tooth and bone (fig. 25.7). The alveolus is lined by a **periodontal** (PERR-ee-oh-DON-tul) **ligament,** a modified periosteum whose collagen fibers penetrate into the bone on one side and into the tooth on the other. This anchors the tooth firmly in the alveolus but allows for slight movement under the stress of chewing. The gum, or **gingiva** (JIN-jih-vuh), covers the alveolar bone. Regions of a tooth are defined by their relationship to the gingiva: The **crown** is the portion above the gum; the **root** is the portion below the gum, embedded in alveolar bone; and the **neck** is the point where the crown, root, and gum meet. The space between the tooth and gum is the **gingival sulcus.** The hygiene of this sulcus is especially important to dental health (see Deeper Insight 25.1).

## CLINICAL APPLICATION

### Tooth and Gum Disease

The human mouth is home to more than 700 species of microorganisms, especially bacteria. Bacteria and sugars form a sticky residue on the teeth called *plaque*. If plaque is not thoroughly removed by brushing and flossing, bacteria multiply, metabolize the sugars, and release lactic acid and other acids. These acids dissolve the minerals of enamel and dentin, and the bacteria enzymatically digest the collagen and other organic components. The eroded "cavities" of the tooth are known as *dental caries*.[9] If not repaired, caries may fully penetrate the dentin and spread to the pulp cavity. This requires either extraction of the tooth or *root canal therapy*, in which the pulp is removed and replaced with inert material.

When plaque calcifies on the tooth surface, it is called *calculus (tartar)*. Calculus in the gingival sulcus wedges the tooth and gum apart and allows bacterial invasion of the sulcus. This leads to *gingivitis*, or gum inflammation. Nearly everyone has gingivitis at some time. In some cases, bacteria spread from the sulcus into the alveolar bone and begin to dissolve it, producing *periodontal disease*. About 86% of people over age 70 have periodontal disease and many suffer tooth loss as a result. This accounts for 80% to 90% of adult tooth loss.

five cusps. Cusps of the upper and lower premolars and molars mesh when the jaws are closed and slide over each other as the jaw makes lateral chewing motions. This grinds and tears food more effectively than if the occlusal surfaces were flat.

Teeth develop beneath the gums and **erupt** (emerge) in predictable order. Twenty **deciduous teeth** (*milk teeth* or *baby teeth*) erupt from the ages of 6 to 30 months, beginning with the incisors (fig. 25.6a). Between 6 and 25 years of age, these are replaced by the 32 **permanent teeth.** As a permanent tooth grows below a deciduous tooth (fig. 25.8), the root of the deciduous tooth dissolves and leaves little more than the crown by the time it falls out. The third molars (*wisdom teeth*) erupt around ages 17 to 25, if at all. Over the course of human evolution, the face became flatter and the jaws shorter, leaving little room for the third molars. Thus, they often remain below the gum and become *impacted*—so crowded against neighboring teeth and bone that they cannot erupt.

## Mastication

**Mastication** (chewing) breaks food into pieces small enough to be swallowed and exposes more surface to the action of digestive enzymes. It is the first step in mechanical digestion. Mastication requires little thought because food stimulates oral receptors that trigger an involuntary chewing reflex. The tongue, buccinator, and orbicularis oris muscles manipulate food and push it between the teeth. The masseter and

Most of a tooth consists of hard yellowish tissue called **dentin,** covered with **enamel** in the crown and **cementum** in the root. Dentin and cementum are living connective tissues with cells or cell processes embedded in a calcified matrix. Cells of the cementum *(cementocytes)* are scattered more or less randomly and occupy tiny cavities similar to the lacunae of bone. Cells of the dentin *(odontoblasts)* line the pulp cavity and have slender processes that travel through tiny parallel tunnels in the dentin. Enamel is not a tissue but a cell-free secretion produced before the tooth erupts above the gum. Damaged dentin and cementum can regenerate, but damaged enamel cannot—it must be artificially repaired.

Internally, a tooth has a dilated **pulp cavity** in the crown and a narrow **root canal** in the lower root. These spaces are occupied by **pulp**—a mass of loose connective tissue, blood and lymphatic vessels, and nerves. These nerves and vessels enter the tooth through a pore, the **apical foramen,** at the basal end of each root canal.

The meeting of the teeth when the mouth closes is called **occlusion** and the surfaces where they meet are called the **occlusal** (ah-CLUE-zul) **surfaces.** The occlusal surface of a premolar has two rounded bumps called **cusps;** thus, the premolars are also known as **bicuspids.** The molars have four to

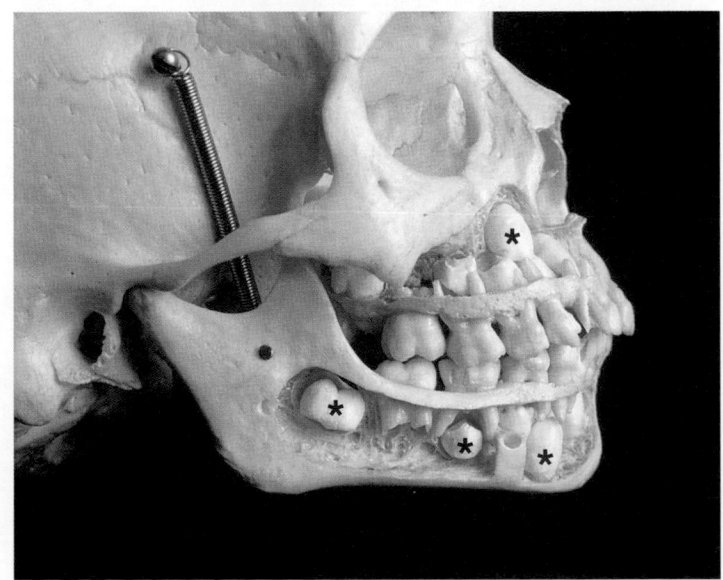

**FIGURE 25.8  Permanent and Deciduous Teeth in a Child's Skull.** This dissection shows erupted deciduous teeth and, deep to them and marked with asterisks, the permanent teeth waiting to erupt.

---

[9]*caries* = rottenness

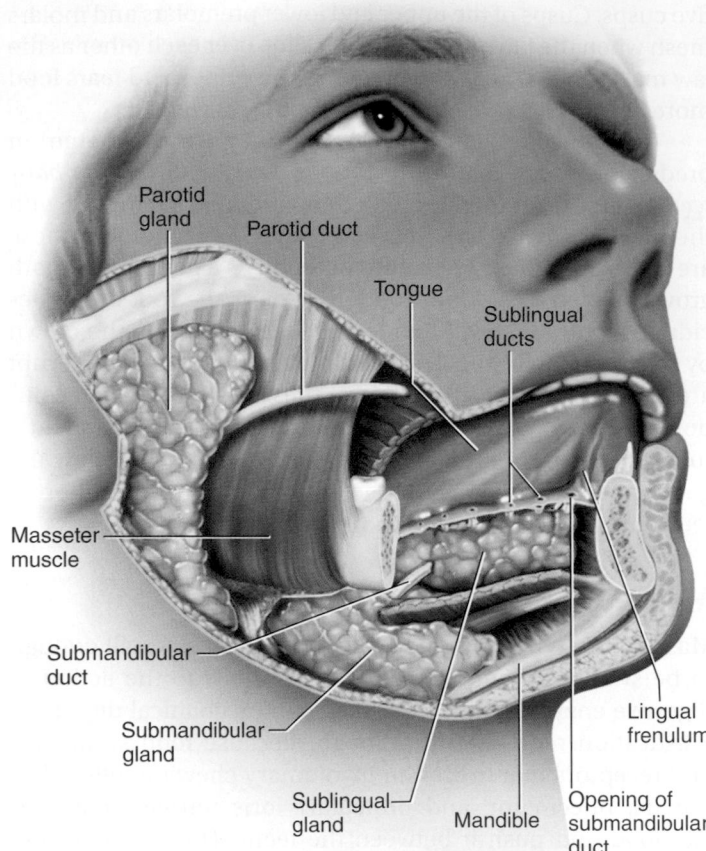

Parotid gland

Parotid duct

Tongue

Sublingual ducts

Masseter muscle

Submandibular duct

Submandibular gland

Sublingual gland

Mandible

Lingual frenulum

Opening of submandibular duct

**FIGURE 25.9 The Extrinsic Salivary Glands.** Part of the mandible has been removed to expose the sublingual gland medial to it. **AP|R**

temporalis muscles produce the up-and-down crushing action of the teeth, and the lateral and medial pterygoid muscles and masseters produce side-to-side grinding action.

## Saliva and the Salivary Glands

Saliva moistens and cleanses the mouth, inhibits bacterial growth, dissolves molecules so they can stimulate the taste buds, digests a little starch and fat, and makes swallowing easier by binding the food particles into a soft mass (bolus) and lubricating it with mucus. It is a hypotonic solution of 97.0% to 99.5% water, a pH of 6.8 to 7.0, and the following solutes:

- **mucus,** which binds and lubricates the food bolus;
- **electrolytes,** salts of $Na^+$, $K^+$, $Cl^-$, phosphate, and bicarbonate;
- **lysozyme,** an enzyme that kills bacteria;
- **immunoglobulin A (IgA),** an antibacterial antibody;
- **salivary amylase,** an enzyme that begins starch digestion in the mouth; and

- **lingual lipase,** an enzyme that begins fat digestion in the mouth (but mainly after the food is swallowed).

## The Salivary Glands

There are two kinds of salivary glands, intrinsic and extrinsic. The **intrinsic salivary glands** are an indefinite number of small glands dispersed amid the other oral tissues—*lingual glands* in the tongue, *labial glands* on the inside of the lips, *palatine glands* of the palate, and *buccal glands* on the inside of the cheeks. They secrete saliva at a fairly constant rate whether we are eating or not, but in relatively small amounts. This saliva contains lingual lipase and lysozyme.

The **extrinsic salivary glands** are three pairs of larger, more discrete organs located outside of the oral mucosa. They are compound tubuloacinar glands (see p. 165) with a treelike duct system leading to the oral cavity (fig. 25.9). The secretory acini at the twig ends of the tree are in some cases purely mucous, in others purely serous, and in mixed acini, composed of both mucous and serous cells (fig. 25.10). The serous cells secrete a watery fluid rich in enzymes and electrolytes. The three extrinsic salivary glands are

1. The **parotid**[10] **glands,** located just beneath the skin anterior to the earlobes. The parotid duct passes superficially over the masseter muscle, pierces the buccinator muscle, and opens into the mouth opposite the second upper molar tooth. *Mumps* is a parotid inflammation and swelling caused by a virus.

2. The **submandibular glands,** located halfway along the body of the mandible, medial to its margin, just deep to the mylohyoid muscle. The submandibular duct empties into the mouth at a papilla on the side of the lingual frenulum, near the lower central incisors.

3. The **sublingual glands,** located in the floor of the mouth. They have multiple ducts that empty into the mouth posterior to the papillae of the submandibular ducts.

## Salivation

The extrinsic salivary glands secrete 1.0 to 1.5 L of saliva per day, mainly in response to food in the mouth. Food stimulates the oral taste, tactile, and pressure receptors, which transmit signals to a group of **salivatory nuclei** in the medulla oblongata and pons. These nuclei integrate this information with input from higher brain centers, so even the aroma, sight, or thought of food stimulates salivation. Irritation of the stomach and esophagus by spicy foods, stomach acid, or toxins also stimulates salivation, perhaps serving to dilute and rinse away the irritants.

The salivatory nuclei issue signals to the glands by way of autonomic fibers in the facial and glossopharyngeal nerves.

---

[10]*par* = next to; *ot* = ear

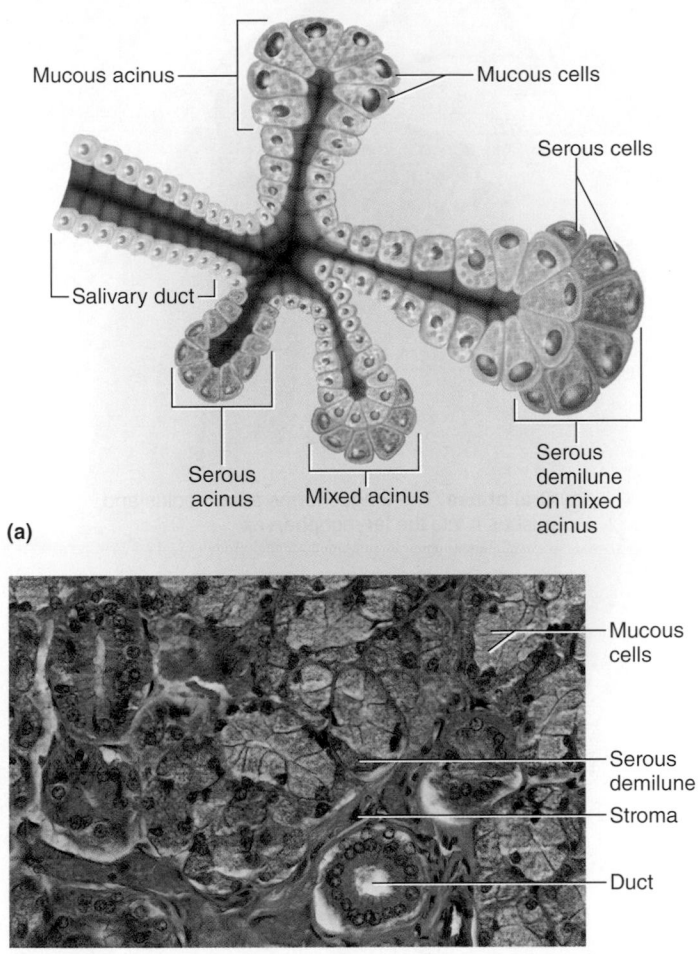

(a)

(b)

**FIGURE 25.10 Microscopic Anatomy of the Salivary Glands.**
(a) Duct and acini of a generalized salivary gland with a mixture of mucous and serous cells. Serous cells often form crescent-shaped caps called serous demilunes over the ends of mucous acini. (b) Histology of the sublingual salivary gland.

Parasympathetic fibers stimulate the glands to produce abundant, thin saliva rich in enzymes. Sympathetic stimulation, by contrast, briefly enhances salivation but its primary effect is to produce less abundant, thicker saliva with more mucus. This is why the mouth can feel sticky or dry under conditions of stress. Sympathetic stimulation constricts the blood vessels of the salivary glands. Since saliva begins as a filtrate from the blood capillaries, this vasoconstriction reduces saliva output. Dehydration similarly reduces salivation by reducing capillary filtration.

Saliva production begins with the filtration of water and electrolytes from the blood capillaries of the glands. Acinar cells add mucin, amylase, and lysozyme to the filtrate and the ducts modify its electrolyte composition. Salivary amylase begins to digest starch as the food is chewed; lingual lipase begins fat digestion to a slight extent; and the mucus binds the masticated food particles into an easily swallowed bolus. Without mucus, one must drink much more liquid to swallow food.

## The Pharynx

The pharynx, described in chapter 22, is a muscular funnel that connects the oral cavity to the esophagus and the nasal cavity to the larynx; thus, it is a point where the digestive and respiratory tracts intersect. It has a deep layer of longitudinally oriented skeletal muscle and a superficial layer of circular skeletal muscle. The circular muscle is divided into superior, middle, and inferior **pharyngeal constrictors,** which force food downward during swallowing. When food is not being swallowed, the inferior constrictor remains contracted to exclude air from the esophagus. This constriction is regarded as the **upper esophageal sphincter,** although it is not an anatomical feature of the esophagus. It disappears at the time of death when the muscle relaxes. Thus, it is regarded as a *physiological sphincter* rather than a constant anatomical structure.

## The Esophagus

The **esophagus** is a straight muscular tube 25 to 30 cm long (see figs. 25.1 and 25.2). It begins at a level between vertebra C6 and the cricoid cartilage, inferior to the larynx and posterior to the trachea. After passing downward through the mediastinum, it penetrates the diaphragm at an opening called the *esophageal hiatus,* continues another 3 to 4 cm, and meets the stomach at the level of vertebra T7. Its opening into the stomach is called the **cardiac orifice** (named for its proximity to the heart). Food pauses briefly at this point before entering the stomach because of a constriction called the **lower esophageal sphincter (LES).** The LES prevents stomach contents from regurgitating into the esophagus, thus protecting the esophageal mucosa from the erosive effect of stomach acid. "Heartburn" has nothing to do with the heart, but is the burning sensation produced by acid reflux into the esophagus.

The wall of the esophagus is organized into the tissue layers described earlier, with some regional specializations. The mucosa has a nonkeratinized stratified squamous epithelium. The submucosa contains **esophageal glands** that secrete lubricating mucus into the lumen. When the esophagus is empty, the mucosa and submucosa are deeply folded into longitudinal ridges, giving the lumen a starlike shape in cross section.

The muscularis externa is composed of skeletal muscle in the upper one-third of the esophagus, a mixture of skeletal and smooth muscle in the middle one-third, and only smooth muscle in the lower one-third.

Most of the esophagus is in the mediastinum. Here, it is covered with a connective tissue adventitia that merges into the adventitias of the trachea and thoracic aorta. The short segment below the diaphragm is partially covered by a serosa.

# Swallowing

Swallowing, or **deglutition** (DEE-glu-TISH-un), is a complex action involving over 22 muscles in the mouth, pharynx, and esophagus, coordinated by the **swallowing center,** a pair of nuclei in the medulla oblongata. This center communicates with muscles of the pharynx and esophagus by way of the trigeminal, facial, glossopharyngeal, and hypoglossal nerves (cranial nerves V, VII, IX, and XII). Swallowing occurs in three phases (fig. 25.11).

1. The **oral phase** is under voluntary control. During chewing, the tongue collects food, presses it against the palate to form a bolus, and pushes it posteriorly. Food thus accumulates in the oropharynx in front of the blade of the epiglottis. When the bolus reaches a critical size, the epiglottis tips posteriorly and the bolus slides around it, through a space on each side, into the laryngopharynx.

2. The **pharyngeal phase** is involuntary. The soft palate and root of the tongue block food and drink from entering the nasal cavity or reentering the mouth. To prevent choking, breathing is automatically suspended, the infrahyoid muscles pull the larynx up to meet the epiglottis and cover the laryngeal opening, and the vocal cords adduct to close the airway. These actions also widen the upper esophagus to receive the food. The pharyngeal constrictors contract in order from superior to middle and inferior, driving the bolus downward into the esophagus.

3. The **esophageal phase** is a wave of involuntary contractions called **peristalsis,** controlled jointly by the brainstem swallowing center and the myenteric plexus in the esophageal wall. The bolus stimulates stretch receptors that feed into the plexus, which transmits signals to the muscularis externa above and below the bolus. The circular muscle layer above the bolus constricts and pushes the food downward. Below the bolus, the circular muscle relaxes while the longitudinal muscle contracts. The latter action pulls the wall of the esophagus slightly upward, making it a little shorter and dilating it to receive the descending food.

When one is standing or sitting upright, most food and liquid drop through the esophagus by gravity faster than the peristaltic wave can keep up with it. Peristalsis, however, propels more solid food pieces and ensures that you can swallow regardless of the body's position—even standing on your head! Liquid normally reaches the stomach in 1 to 2 seconds and a food bolus in 4 to 8 seconds. As a bolus reaches the lower end of the esophagus, the lower esophageal sphincter relaxes to let it pass into the stomach.

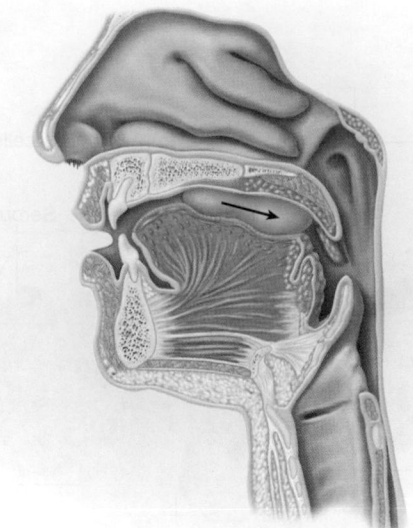

1 **Oral phase.** The tongue forms a food bolus and pushes it into the laryngopharynx.

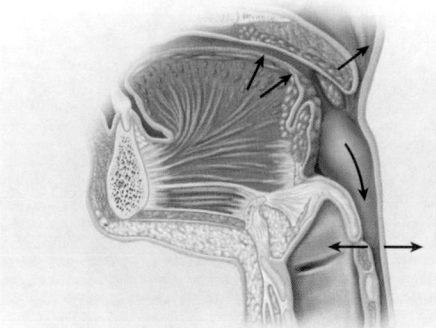

2 **Pharyngeal phase.** The palate, tongue, vocal cords, and epiglottis block the oral and nasal cavities and airway while pharyngeal constrictors push the bolus into the esophagus.

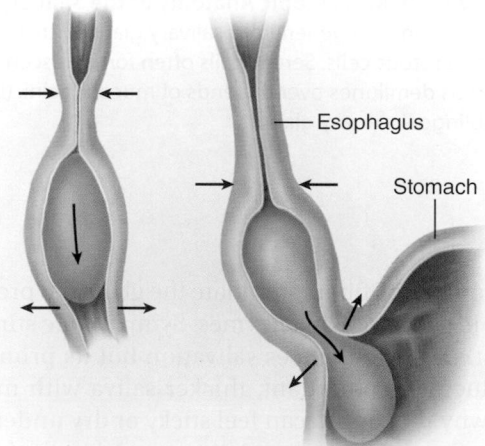

— Esophagus

Stomach

3 **Esophageal phase.** Peristalsis drives the bolus downward, and relaxation of the lower esophageal sphincter admits it into the stomach.

**FIGURE 25.11  Swallowing.** See numbered steps in the text for further explanation. AP|R

❓ *What actions prevent the pharynx from forcing food back into the mouth or nose?*

Answer the following questions to test your understanding of the preceding section:

5. List as many functions of the tongue as you can.

6. Imagine a line from the mandibular bone to the root canal of a tooth. Name the tissues, in order, through which this line would pass.

7. What is the difference in function and location between intrinsic and extrinsic salivary glands? Name the extrinsic salivary glands and describe their locations.

8. Identify at least two histological features of the esophagus that are especially tied to its role in swallowing.

9. Describe the mechanisms that prevent food from entering the nasal cavity and larynx during swallowing.

## 25.3 The Stomach AP|R

### Expected Learning Outcomes

When you have completed this section, you should be able to

a. describe the gross and microscopic anatomy of the stomach;

b. state the function of each type of epithelial cell in the gastric mucosa;

c. identify the secretions of the stomach and state their functions;

d. explain how the stomach produces hydrochloric acid and pepsin;

e. describe the contractile responses of the stomach to food; and

f. describe the three phases of gastric function and how gastric activity is activated and inhibited.

The stomach is a muscular sac in the upper left abdominal cavity immediately inferior to the diaphragm. It functions primarily as a food storage organ, with an internal volume of about 50 mL when empty and 1.0 to 1.5 L after a typical meal. When extremely full, it may hold up to 4 L and extend nearly as far as the pelvis.

Well into the nineteenth century, authorities regarded the stomach as essentially a grinding chamber, fermentation vat, or cooking pot. Some even attributed digestion to a supernatural spirit in the stomach. We now know that it mechanically breaks up food particles, liquefies the food, and begins the chemical digestion of proteins and fat. This produces a soupy or pasty mixture of semidigested food called **chyme**[11] (pronounced "kime"). Most digestion occurs after the chyme passes on to the small intestine.

## Gross Anatomy

The stomach is J-shaped (fig. 25.12), relatively vertical in tall people and more nearly horizontal in short people. It is divided into four regions: (1) The **cardiac region (cardia)** is a small area within about 3 cm of the cardiac orifice. (2) The **fundic region (fundus)** is the dome superior to the esophageal attachment. (3) The **body (corpus)** is the greatest part distal to the cardiac orifice. (4) The **pyloric region** is a slightly narrower pouch at the inferior end; it is subdivided into a funnel-like **antrum**[12] and a narrower **pyloric canal.** The latter terminates at the **pylorus,**[13] a narrow passage into the duodenum. The pylorus is surrounded by a thick ring of smooth muscle, the **pyloric (gastroduodenal) sphincter,** which regulates the passage of chyme into the duodenum.

Between the esophagus and duodenum, the stomach has two margins called the **greater** and **lesser curvatures.** The greater curvature is the long way around, about 40 cm, along the inferolateral surface. The greater omentum, overhanging the small intestine, is suspended from the greater curvature. The lesser curvature is the shorter distance, about 10 cm, along the superomedial margin facing the liver. The lesser omentum spans the space between the liver and lesser curvature (see fig. 25.3).

## Innervation and Circulation

The stomach receives parasympathetic nerve fibers from the vagus nerves and sympathetic fibers from the celiac ganglia (see p. 564). It is supplied with blood by branches of the celiac trunk (see p. 779). All blood drained from the stomach and intestines enters the hepatic portal circulation and filters through the liver before returning to the heart.

## Microscopic Anatomy

The stomach wall has tissue layers similar to those of the esophagus, with some variations. The mucosa is covered with a simple columnar glandular epithelium (fig. 25.13). The apical regions of its cells are filled with mucin, which swells with water and becomes mucus after it is secreted. The mucosa and submucosa are flat and smooth when the stomach is full, but as it empties, these layers form conspicuous longitudinal wrinkles called **gastric rugae**[14] (ROO-jee). The lamina propria is almost entirely occupied by tubular glands, to be described shortly. The muscularis externa has three layers, rather than two: outer longitudinal, middle circular, and inner oblique layers (fig. 25.12).

---

[11]*chyme* = juice

[12]*antrum* = cavity
[13]*pylorus* = gatekeeper
[14]*rugae* = folds, creases

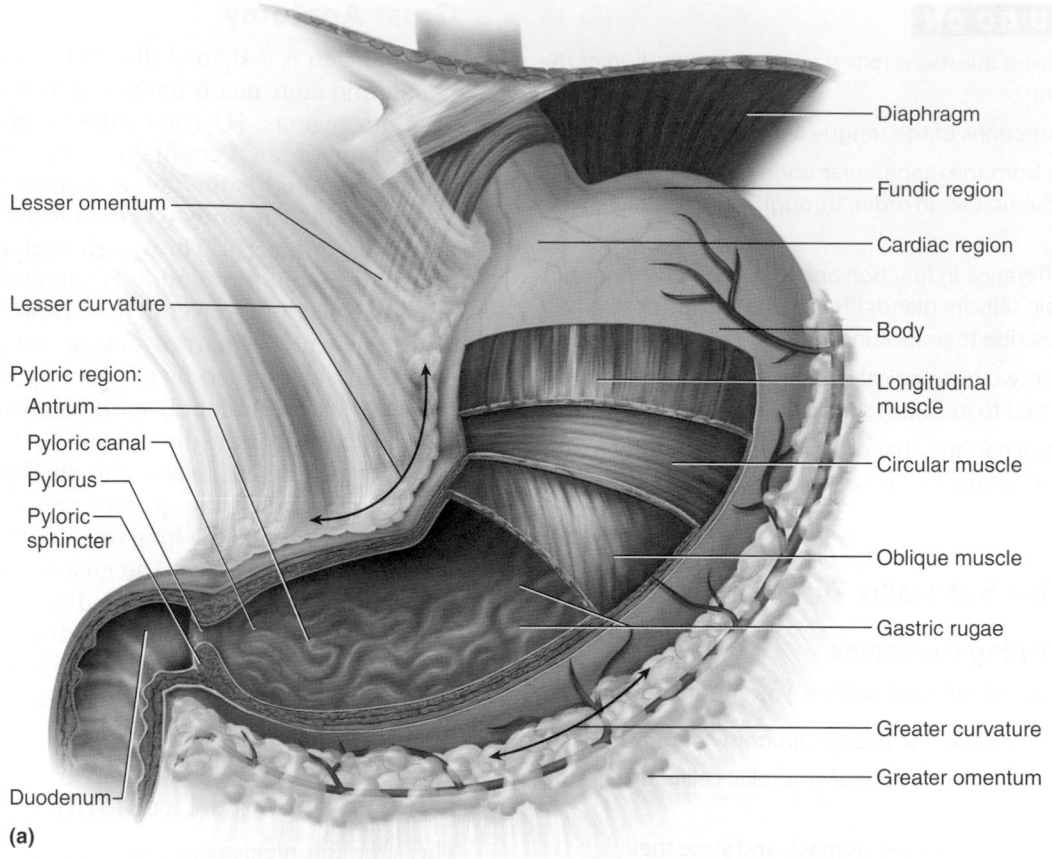

Diaphragm

Fundic region

Cardiac region

Body

Longitudinal muscle

Circular muscle

Oblique muscle

Gastric rugae

Greater curvature

Greater omentum

Lesser omentum

Lesser curvature

Pyloric region:
Antrum
Pyloric canal
Pylorus
Pyloric sphincter

Duodenum

(a)

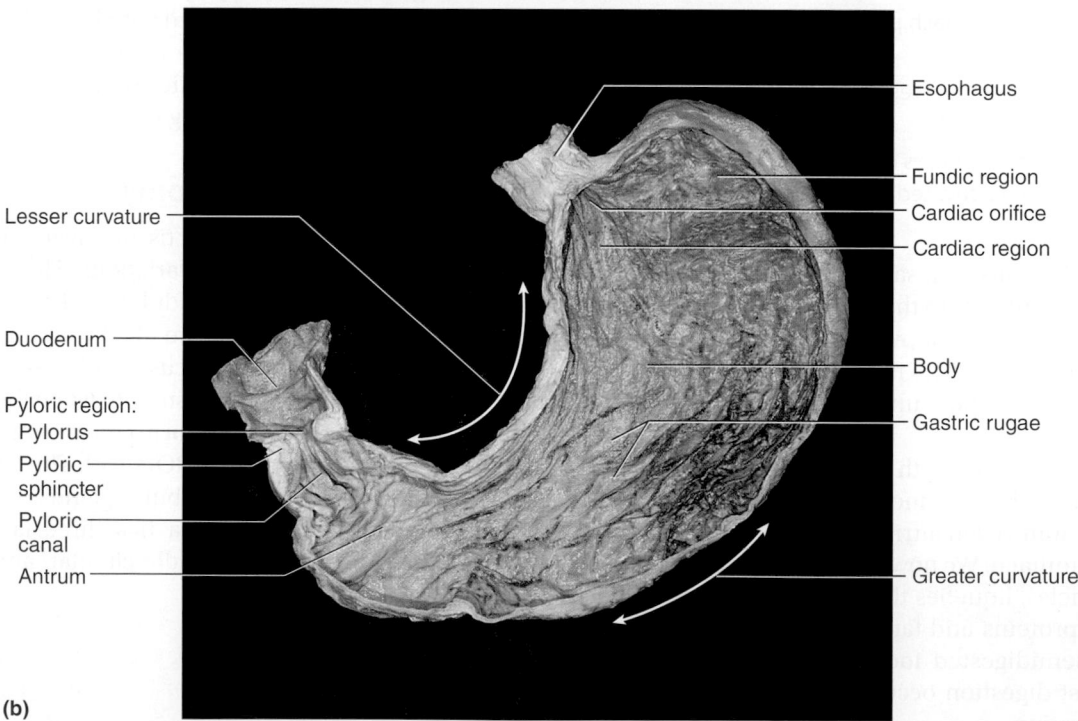

Esophagus

Fundic region
Cardiac orifice
Cardiac region

Body

Gastric rugae

Greater curvature

Lesser curvature

Duodenum

Pyloric region:
Pylorus
Pyloric sphincter
Pyloric canal
Antrum

(b)

**FIGURE 25.12  The Stomach.**  (a) Gross anatomy. (b) Photograph of the internal surface.  **AP|R**

❓ *How does the muscularis externa of the stomach differ from that of the esophagus?*

▶▶▶**APPLY WHAT YOU KNOW**

*Contrast the epithelium of the esophagus with that of the stomach. Why is each epithelial type best suited to the function of its respective organ?*

The gastric mucosa is pocked with depressions called **gastric pits** lined with the same columnar epithelium as the surface (fig. 25.13). Two or three tubular glands open into the bottom of each gastric pit and span the rest of the lamina propria. In the cardiac and pyloric regions, they are called **cardiac glands** and **pyloric glands,** respectively. In the rest of the stomach, they are called **gastric glands.** These three glands differ in cellular composition, but collectively have the following cell types:

- **Mucous cells,** which secrete mucus, predominate in the cardiac and pyloric glands. In gastric glands, they are called *mucous neck cells* and are concentrated in the narrow *neck* of the gland, where it opens into the gastric pit.

- **Regenerative (stem) cells,** found in the base of the pit and neck of the gland, divide rapidly and produce a continual supply of new cells. Newly generated cells migrate upward to the gastric surface as well as downward into the glands to replace cells that die.

- **Parietal cells,** found mostly in the upper half of the gland, secrete *hydrochloric acid, intrinsic factor,* and an appetite-regulating hormone called *ghrelin* (see p. 995). They are found mostly in the gastric glands, but a few occur in the pyloric glands.

- **Chief cells,** so named because they are the most numerous, secrete the enzymes *gastric lipase* and *pepsinogen.* They dominate the lower half of the gastric glands but are absent from cardiac and pyloric glands.

- **Enteroendocrine cells,** concentrated especially in the lower end of a gland, secrete hormones and paracrine messengers that regulate digestion. They occur in all regions of the stomach, but are most abundant in the gastric and pyloric glands. There are at least eight kinds of enteroendocrine cells in the stomach, each of which produces a different chemical messenger.

In general, the cardiac and pyloric glands secrete mainly mucus; acid and enzyme secretion occur predominantly in the gastric glands; and hormones are secreted throughout the stomach.

## Gastric Secretions

The gastric glands produce 2 to 3 L of **gastric juice** per day, composed mainly of water, hydrochloric acid, and pepsin.

### Hydrochloric Acid

Gastric juice has a high concentration of hydrochloric acid (HCl) and a pH as low as 0.8. Such concentrated acid could cause a serious chemical burn to the skin. How, then, does the stomach produce and tolerate such acidity?

The reactions that produce HCl (fig. 25.14) may seem familiar by now because they have been discussed in previous chapters—most recently in connection with renal excretion of $H^+$ in chapter 24. Parietal cells contain carbonic anhydrase (CAH), which catalyzes the first step in the following reaction:

$$CO_2 + H_2O \xrightarrow{\text{CAH}} H_2CO_3 \longrightarrow HCO_3^- + H^+$$

Parietal cells pump the $H^+$ from this reaction into the lumen of a gastric gland by an active-transport protein similar to the $Na^+$–$K^+$ pump, called $\mathbf{H^+}$–$\mathbf{K^+}$ **ATPase.** This is an antiport that uses the energy of ATP to pump $H^+$ out of the cell and $K^+$ into it. HCl secretion does not affect the pH within the parietal cell because $H^+$ is pumped out as fast as it is generated. The bicarbonate ions ($HCO_3^-$) are exchanged for chloride ions ($Cl^-$) from the blood plasma—the same *chloride-shift* process that occurs in the renal tubules and red blood cells—and the $Cl^-$ is pumped into the lumen of the gastric gland to join the $H^+$.

Thus, HCl accumulates in the stomach while bicarbonate ions accumulate in the blood. Because of the bicarbonate, blood leaving the stomach has a higher pH when digestion is occurring than when the stomach is empty. This high-pH blood is called the *alkaline tide.*

Stomach acid has several functions: (1) It activates the enzymes pepsin and lingual lipase, as discussed shortly. (2) It breaks up connective tissues and plant cell walls, helping to liquefy food and form chyme. (3) It converts ingested ferric ions ($Fe^{3+}$) to ferrous ions ($Fe^{2+}$), a form of iron that can be absorbed and used for hemoglobin synthesis. (4) It contributes to nonspecific disease resistance by destroying most ingested pathogens.

### Pepsin

Several digestive enzymes are secreted as inactive proteins called **zymogens** and then converted to active enzymes by the removal of some of their amino acids. In the stomach, chief cells secrete a zymogen called **pepsinogen.** Hydrochloric acid removes some of its amino acids and converts it to **pepsin.** Since pepsin digests protein, and pepsinogen itself is a protein, pepsin has an *autocatalytic* effect—as some pepsin is formed, it converts pepsinogen into more pepsin (fig. 25.15). The ultimate function of pepsin, however, is to digest dietary proteins to shorter peptide chains, which then pass to the small intestine, where their digestion is completed.

### Gastric Lipase

The chief cells also secrete **gastric lipase.** This enzyme and lingual lipase, which plays a minor role, digest 10% to 15% of the dietary fat in the stomach. The remainder is digested in the small intestine.

### Intrinsic Factor

Parietal cells also secrete a glycoprotein called **intrinsic factor** that is essential to the absorption of vitamin $B_{12}$ by the small intestine. Intrinsic factor binds vitamin $B_{12}$ and the intestinal cells

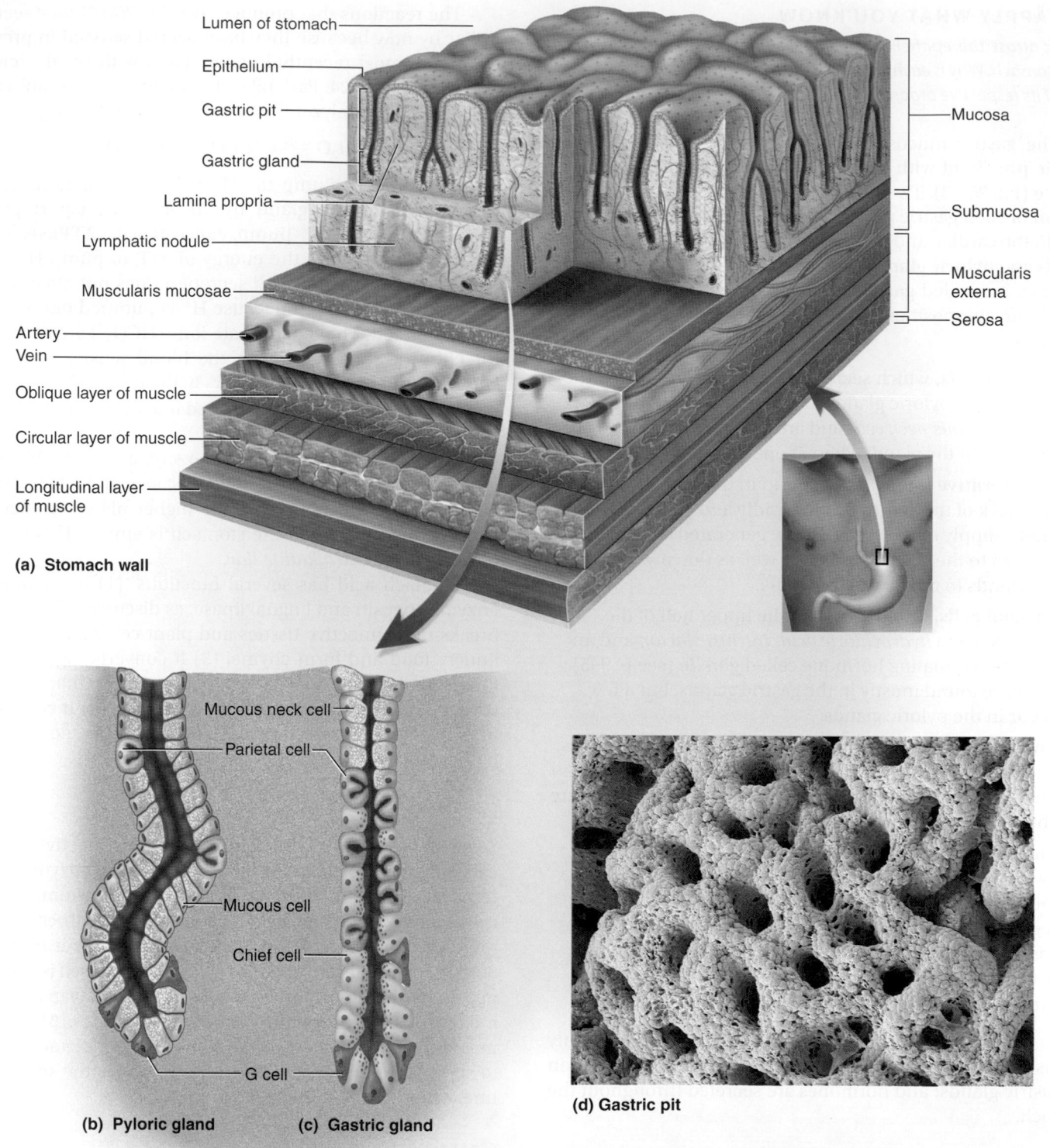

Lumen of stomach

Epithelium

Gastric pit

Gastric gland

Lamina propria

Lymphatic nodule

Muscularis mucosae

Artery

Vein

Oblique layer of muscle

Circular layer of muscle

Longitudinal layer of muscle

Mucosa

Submucosa

Muscularis externa

Serosa

**(a) Stomach wall**

Mucous neck cell

Parietal cell

Mucous cell

Chief cell

G cell

**(b) Pyloric gland**    **(c) Gastric gland**

**(d) Gastric pit**

**FIGURE 25.13  Microscopic Anatomy of the Stomach Wall.** (a) A block of tissue showing all layers from the mucosa (top) to the serosa (bottom). (b) A pyloric gland from the inferior end of the stomach. Note the absence of chief cells and relatively few parietal cells. (c) A gastric gland, the most widespread type in the stomach. (d) The opening of a gastric pit into the stomach, surrounded by the rounded apical surfaces of the columnar epithelial cells of the mucosa (SEM). **AP|R**

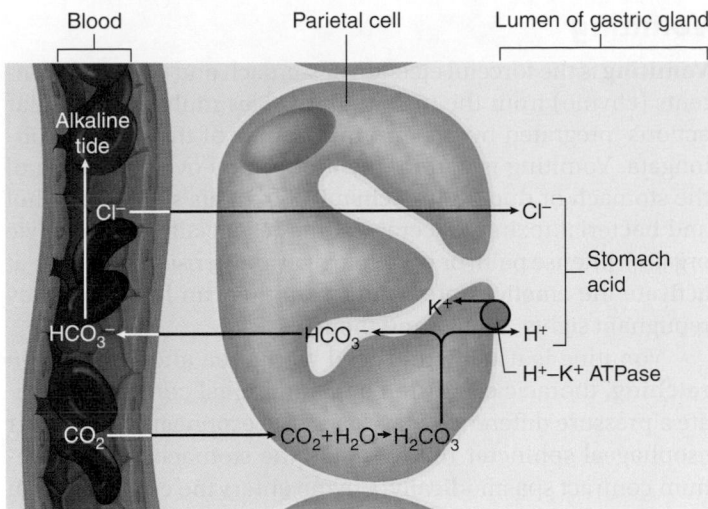

**FIGURE 25.14 The Mode of Hydrochloric Acid Secretion.** The parietal cell combines water with $CO_2$ from the blood to form carbonic acid (bottom line of figure). Carbonic acid breaks down into bicarbonate ion ($HCO_3^-$) and hydrogen ion ($H^+$). $HCO_3^-$ returns to the blood. In exchange, $Cl^-$ enters the lumen with $H^+$ and the two form hydrochloric acid.

❓ *What role does active transport play in this process?*

then absorb this complex by receptor-mediated endocytosis. Without vitamin $B_{12}$, hemoglobin cannot be synthesized and anemia develops (see p. 683). The secretion of intrinsic factor is the only indispensable function of the stomach. Digestion can continue following removal of the stomach *(gastrectomy),* but

a person usually must then take vitamin $B_{12}$ by injection, or vitamin $B_{12}$ and intrinsic factor orally. As we age, the gastric mucosa atrophies and less intrinsic factor is secreted, increasing the risk of anemia. Some people, especially in old age, develop *pernicious anemia,* the result of an autoimmune disease that destroys gastric mucosa and reduces intrinsic factor secretion.

## Chemical Messengers

The gastric and pyloric glands have various kinds of enteroendocrine cells that collectively produce as many as 20 chemical messengers. Most of these are hormones—they travel in the bloodstream and stimulate distant target cells. Some also behave as paracrine secretions, diffusing a short distance away and stimulating other cells in the gastric mucosa. Several of them are peptides produced in both the digestive tract and the central nervous system; thus, they are called **gut–brain peptides.** These include substance P, vasoactive intestinal peptide (VIP), secretin, gastric inhibitory peptide (GIP), cholecystokinin, and neuropeptide Y (NPY). The functions of some of these peptides in digestion will be explained in the following sections, and their roles in appetite regulation are discussed in chapter 26.

Several of the gastric secretions are summarized in table 25.1. Some of the functions listed there are explained later in the chapter.

## Gastric Motility

As you begin to swallow, food stimulates mechanoreceptors in the pharynx and they transmit signals to the medulla

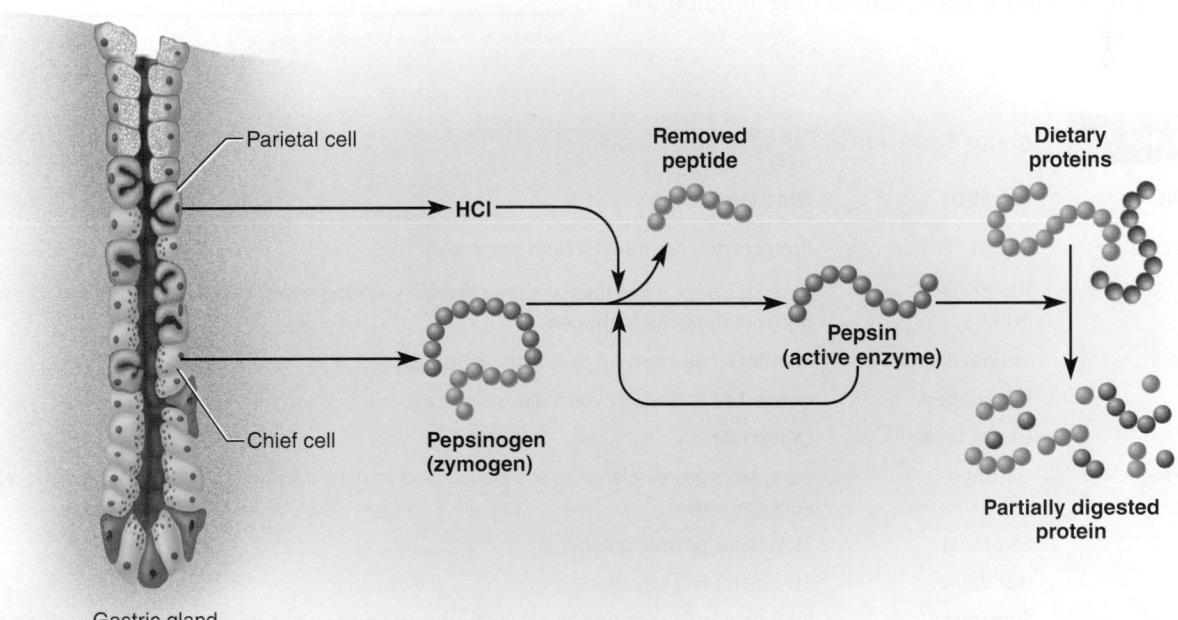

**FIGURE 25.15 The Production and Action of Pepsin.** The chief cells secrete pepsinogen and the parietal cells secrete HCl. HCl removes some of the amino acids from pepsinogen and converts it to pepsin. Pepsin catalyzes the production of more pepsin (autocatalytic effect), as well as partially digesting dietary protein.

oblongata. The medulla relays signals to the stomach by way of the vagus nerves. The stomach reacts with a **receptive-relaxation response,** in which it briefly resists stretching but then relaxes to accommodate the arriving food.

Soon, the stomach shows a rhythm of peristaltic contractions governed by pacemaker cells in the muscularis externa. The upper stomach (fundus) does not participate in these, but below the fundus, around midbody, a tight ring of constriction appears about every 20 seconds and progresses downward toward the antrum, becoming stronger as it goes. After food has been in the stomach for 30 minutes or so, these contractions become intense. They churn the food, mix it with gastric juice, and promote its physical breakup and chemical digestion.

The antrum holds about 30 mL of chyme at a time. The muscularis is thickest here, and acts as a strong *antral pump* that breaks up semidigested food into smaller particles and prepares it for the small intestine. A leading wave of antral contraction proceeds to the pyloric valve and closes it tightly. A trailing wave then comes along, churning and breaking up the chyme. Chyme jets backward through the trailing constriction into the gastric body, where it awaits the next wave of contraction to drive it down again. The repetitive downward propulsion and reverse jetting of chyme break the food into smaller and smaller particles. Food particles are not allowed to pass into the duodenum until they are reduced to 1 to 7 mm in size, and only about 3 mL of chyme is squirted into the duodenum at a time. Allowing only small amounts into the duodenum enables the duodenum to neutralize the stomach acid and digest nutrients little by little. If the duodenum becomes overfilled, it inhibits gastric motility and postpones receiving more chyme; the mechanism for this is discussed later. A typical meal is emptied from the stomach in about 4 hours, but it takes less time if the meal is more liquid, longer if the stomach contents are more acidic, and as long as 6 hours if the meal is high in fat.

## Vomiting

**Vomiting** is the forceful ejection of stomach and intestinal contents (chyme) from the mouth. It involves multiple muscular actions integrated by the **emetic**[15] **center** of the medulla oblongata. Vomiting is commonly induced by overstretching of the stomach or duodenum; chemical irritants such as alcohol and bacterial toxins; visceral trauma (especially to the pelvic organs); intense pain; or psychological and sensory stimuli that activate the emetic center (thus, vomiting can be induced by repugnant sights, smells, and thoughts).

Vomiting is usually preceded by nausea and retching. In **retching,** thoracic expansion and abdominal contraction create a pressure difference that dilates the esophagus. The lower esophageal sphincter relaxes while the stomach and duodenum contract spasmodically. Chyme enters the esophagus but then drops back into the stomach as the muscles relax; it does not get past the upper esophageal sphincter. Retching is often accompanied by tachycardia, profuse salivation, and sweating. Vomiting occurs when abdominal contraction and rising thoracic pressure force the upper esophageal sphincter open, the esophagus and body of the stomach relax, and chyme is driven out of the stomach and mouth by strong abdominal contraction combined with reverse peristalsis of the gastric antrum and duodenum. **Projectile vomiting** is sudden vomiting with no prior nausea or retching. It may be caused by neurological lesions but is also common in infants after feeding.

Chronic vomiting can cause dangerous fluid, electrolyte, and acid–base imbalances. In cases of frequent vomiting, as in the eating disorder *bulimia,* the tooth enamel becomes

[15]*emet* = vomiting

| TABLE 25.1 | Major Secretions of the Gastric Glands | |
|---|---|---|
| **Secretory Cells** | **Secretion** | **Function** |
| Mucous neck cells | Mucus | Protects mucosa from HCl and enzymes |
| Parietal cells | Hydrochloric acid | Activates pepsin and lingual lipase; helps liquefy food; reduces dietary iron to usable form ($Fe^{2+}$); destroys ingested pathogens |
| | Intrinsic factor | Enables small intestine to absorb vitamin $B_{12}$ |
| Chief cells | Pepsinogen | Converted to pepsin, which digests protein |
| | Gastric lipase | Digests fat |
| Enteroendocrine cells | Gastrin | Stimulates gastric glands to secrete HCl and enzymes; stimulates intestinal motility; relaxes ileocecal valve |
| | Serotonin | Stimulates gastric motility |
| | Histamine | Stimulates HCl secretion |
| | Somatostatin | Inhibits gastric secretion and motility; delays emptying of stomach; inhibits secretion by pancreas; inhibits gallbladder contraction and bile secretion; reduces blood circulation and nutrient absorption in small intestine |
| | Gut–brain peptides | Various roles in short- and long-term appetite regulation and energy balance |

severely eroded by the hydrochloric acid in the chyme. Aspiration (inhalation) of this acid is very destructive to the respiratory tract. Many have died from aspiration of vomit when they were unconscious or semiconscious. This is the reason that surgical anesthesia, which may induce nausea, must be preceded by fasting until the stomach and small intestine are empty.

## Digestion and Absorption

Salivary and gastric enzymes partially digest protein and lesser amounts of starch and fat in the stomach, but most digestion and nearly all nutrient absorption occur after the chyme passes into the small intestine. The stomach does not absorb any significant amount of nutrients but does absorb aspirin and some lipid-soluble drugs. Alcohol is absorbed mainly by the small intestine, so its intoxicating effect depends partly on how rapidly the stomach is emptied.

## Protection of the Stomach

One might think that the stomach would be its own worst enemy; it is, after all, made of meat. Some people enjoy haggis and tripe, dishes made from animal stomachs, and have no difficulty digesting those. Why, then, doesn't the human stomach digest itself? The answer is that the living stomach is protected in three ways from the harsh acidic and enzymatic environment it creates:

1. **Mucous coat.** The thick, highly alkaline mucus resists the action of acid and enzymes.

2. **Tight junctions.** The epithelial cells are joined by tight junctions that prevent gastric juice from seeping between them and digesting the connective tissue below.

3. **Epithelial cell replacement.** In spite of these other protections, the stomach's epithelial cells live only 3 to 6 days and are then sloughed off into the chyme and digested with the food. They are replaced just as rapidly, however, by cell division in the gastric pits.

The breakdown of these protective mechanisms can result in inflammation and peptic ulcer (see Deeper Insight 25.2).

## Regulation of Gastric Function

The nervous and endocrine systems collaborate to increase gastric secretion and motility when food is eaten and to suppress them as the stomach empties. Gastric activity is divided into three stages called the *cephalic, gastric,* and *intestinal phases,* based on whether the stomach is being controlled by the brain, by itself, or by the small intestine, respectively (fig. 25.17). These phases overlap and all three can occur simultaneously.

### The Cephalic Phase

The **cephalic phase** is the stage in which the stomach responds to the mere sight, smell, taste, or thought of food. These sensory and mental inputs converge on the hypothalamus, which relays signals to the medulla oblongata. Vagus nerve fibers from the medulla stimulate the enteric nervous system of the stomach, which, in turn, stimulates secretion by the parietal cells (acid) and G cells (gastrin). About 40% of the stomach's acid secretion occurs in the cephalic phase.

### The Gastric Phase

The **gastric phase** is a period in which swallowed food and semidigested protein (peptides and amino acids) activate gastric activity. About one-half of acid secretion and two-thirds of total gastric secretion occur during this phase. Ingested food stimulates gastric activity in two ways: by stretching the stomach and by raising the pH of its contents. Stretch activates two reflexes: a short reflex mediated through the myenteric plexus and a long reflex mediated through the vagus nerves and brainstem.

Gastric secretion is stimulated chiefly by three chemicals: acetylcholine (ACh), histamine, and gastrin. ACh is secreted by parasympathetic nerve fibers of both the short and long reflex pathways. Histamine is a paracrine secretion from enteroendocrine cells in the gastric glands. **Gastrin** is a hormone produced by enteroendocrine **G cells** in the pyloric glands.

All three of these stimulate parietal cells to secrete hydrochloric acid and intrinsic factor. The chief cells secrete pepsinogen in response to gastrin and especially ACh, and ACh also stimulates mucus secretion.

As dietary protein is digested, it breaks down into smaller peptides and amino acids, which directly stimulate the G cells to secrete even more gastrin—a positive feedback loop that accelerates protein digestion (fig. 25.18). Small peptides also buffer stomach acid so the pH does not fall excessively low. But as digestion continues and these peptides are emptied from the stomach, the pH drops lower and lower. Below pH of 2, stomach acid inhibits the parietal cells and G cells—a negative feedback loop that winds down the gastric phase as the need for pepsin and HCl declines.

### The Intestinal Phase

The **intestinal phase** is a stage in which the duodenum responds to arriving chyme and moderates gastric activity through hormones and nervous reflexes. The duodenum initially enhances gastric secretion, but soon inhibits it. Stretching of the duodenum accentuates vagovagal reflexes that stimulate the stomach, and peptides and amino acids in the chyme stimulate G cells of the duodenum to secrete more gastrin, which further stimulates the stomach.

Soon, however, the acid and semidigested fats in the duodenum trigger the **enterogastric reflex**—the duodenum sends inhibitory signals to the stomach by way of the enteric nervous system, and sends signals to the medulla that (1) inhibit the vagal nuclei, thus reducing vagal stimulation of the stomach; and (2) stimulate sympathetic neurons, which send inhibitory signals to the stomach. Chyme also stimulates duodenal enteroendocrine cells to release **secretin** and **cholecystokinin** (CO-leh-SIS-toe-KY-nin) **(CCK).** They primarily stimulate the pancreas and gallbladder, as discussed later, but also suppress

# DEEPER INSIGHT 25.2

## CLINICAL APPLICATION

### Peptic Ulcer

Inflammation of the stomach, called *gastritis*, can lead to a *peptic ulcer* as pepsin and hydrochloric acid erode the stomach wall (fig. 25.16). Peptic ulcers occur even more commonly in the duodenum and occasionally in the esophagus. If untreated, they can perforate the organ and cause fatal hemorrhaging or peritonitis. Most such fatalities occur in people over age 65.

There is no evidence to support the popular belief that peptic ulcers result from psychological stress. Hypersecretion of acid and pepsin is sometimes involved, but even normal secretion can cause ulceration if the mucosal defense is compromised by other causes. Most ulcers involve an acid-resistant bacterium, *Helicobacter pylori*, that invades the mucosa of the stomach and duodenum and opens the way to chemical damage to the tissue. Other risk factors include smoking and the use of aspirin and

other nonsteroidal anti-inflammatory drugs (NSAIDs). NSAIDs suppress the synthesis of prostaglandins, which normally stimulate the secretion of protective mucus and acid-neutralizing bicarbonate. Aspirin itself is an acid that directly irritates the gastric mucosa.

At one time, the most widely prescribed drug in the United States was cimetidine (Tagamet), which was designed to treat peptic ulcers by reducing acid secretion. Histamine stimulates acid secretion by binding to sites on the parietal cells called $H_2$ receptors; cimetidine, an $H_2$ blocker, prevents this binding. Lately, however, ulcers have been treated more successfully with antibiotics against *Helicobacter* combined with bismuth suspensions such as Pepto-Bismol. This is a much shorter and less expensive course of treatment and permanently cures about 90% of peptic ulcers, as compared with a cure rate of only 20% to 30% for $H_2$ blockers.

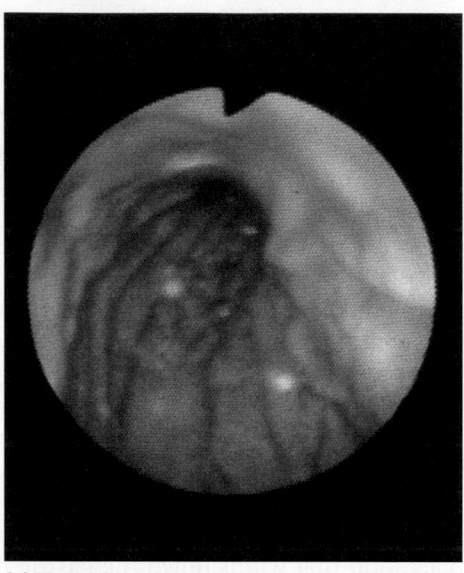

**(a) Normal**

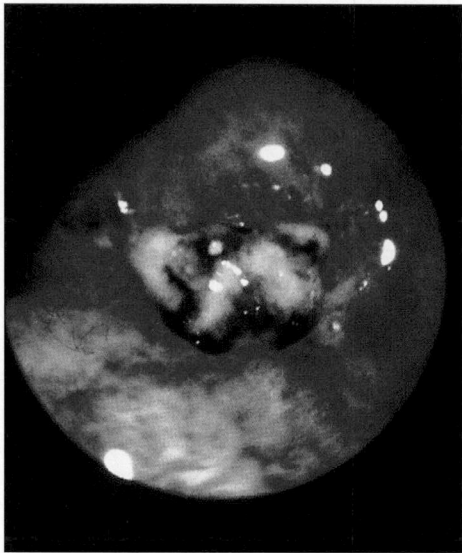

**(b) Peptic ulcer**

**FIGURE 25.16  Endoscopic Views of the Gastroesophageal Junction.**  The esophagus can be seen opening into the cardiac stomach. (a) A view of the cardiac orifice from above, showing a healthy esophageal mucosa. The small white spots are reflections of light from the endoscope. (b) A bleeding peptic ulcer. A peptic ulcer typically has an oval shape and yellow-white color. Here the yellowish floor of the ulcer is partially obscured by black blood clots, and fresh blood is visible around the margin of the ulcer.

gastric secretion and motility. The effect of this is that gastrin secretion declines and the pyloric sphincter contracts tightly to limit the admission of more chyme into the duodenum. This gives the duodenum time to work on the chyme it has already received before being loaded with more.

The enteroendocrine cells also secrete **glucose-dependent insulinotropic peptide (GIP)**. Originally called *gastric-inhibitory peptide* (the original source of the GIP abbreviation), it is no longer thought to have a significant effect on the stomach, but seems more concerned with stimulating insulin secretion in preparation for processing the nutrients about to be absorbed by the small intestine.

**BEFORE YOU GO ON**

Answer the following questions to test your understanding of the preceding section:

10. Name four types of epithelial cells of the gastric and pyloric glands and state what each one secretes.

11. Explain how the gastric glands produce hydrochloric acid and why this produces an alkaline tide.

12. What positive feedback cycle can you identify in the formation and action of pepsin?

13. How does food in the duodenum inhibit motility and secretion in the stomach?

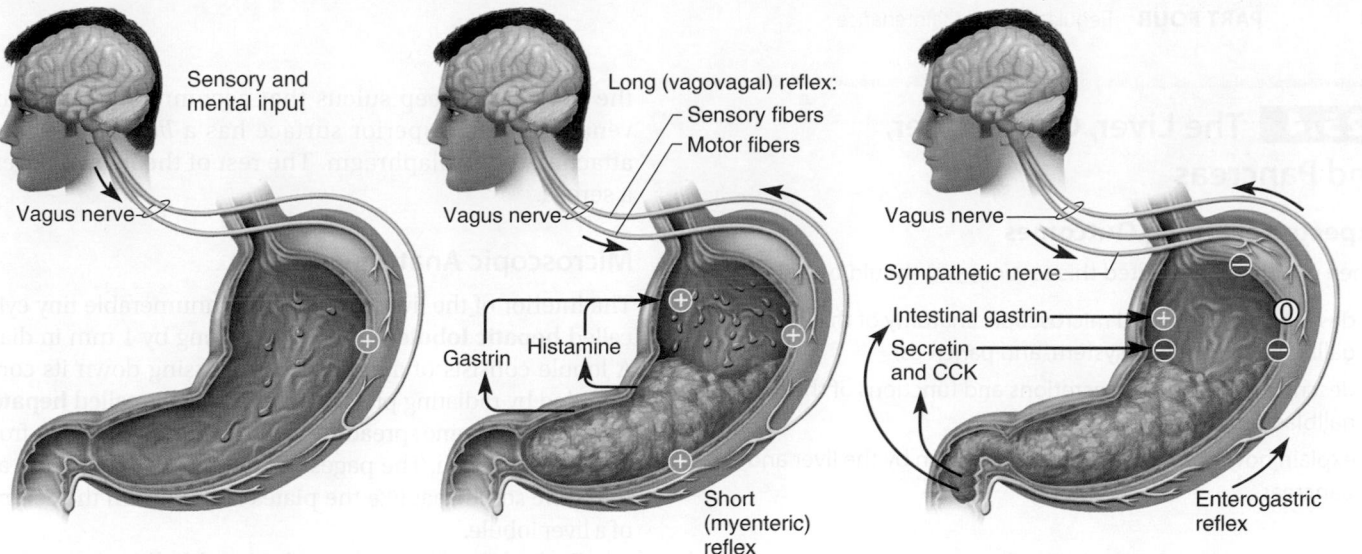

① **Cephalic phase**
Vagus nerve stimulates gastric secretion even before food is swallowed.

**Key**
⊕ Stimulation
⊖ Inhibition
⓪ Reduced or no effect

② **Gastric phase**
Food stretches the stomach and activates myenteric and vagovagal reflexes. These reflexes stimulate gastric secretion. Histamine and gastrin also stimulate acid and enzyme secretion.

③ **Intestinal phase**
Intestinal gastrin briefly stimulates the stomach, but then secretin, CCK, and the enterogastric reflex inhibit gastric secretion and motility while the duodenum processes the chyme already in it. Sympathetic nerve fibers suppress gastric activity, while vagal (parasympathetic) stimulation of the stomach is now inhibited.

**FIGURE 25.17** Neural and Hormonal Control of the Stomach.

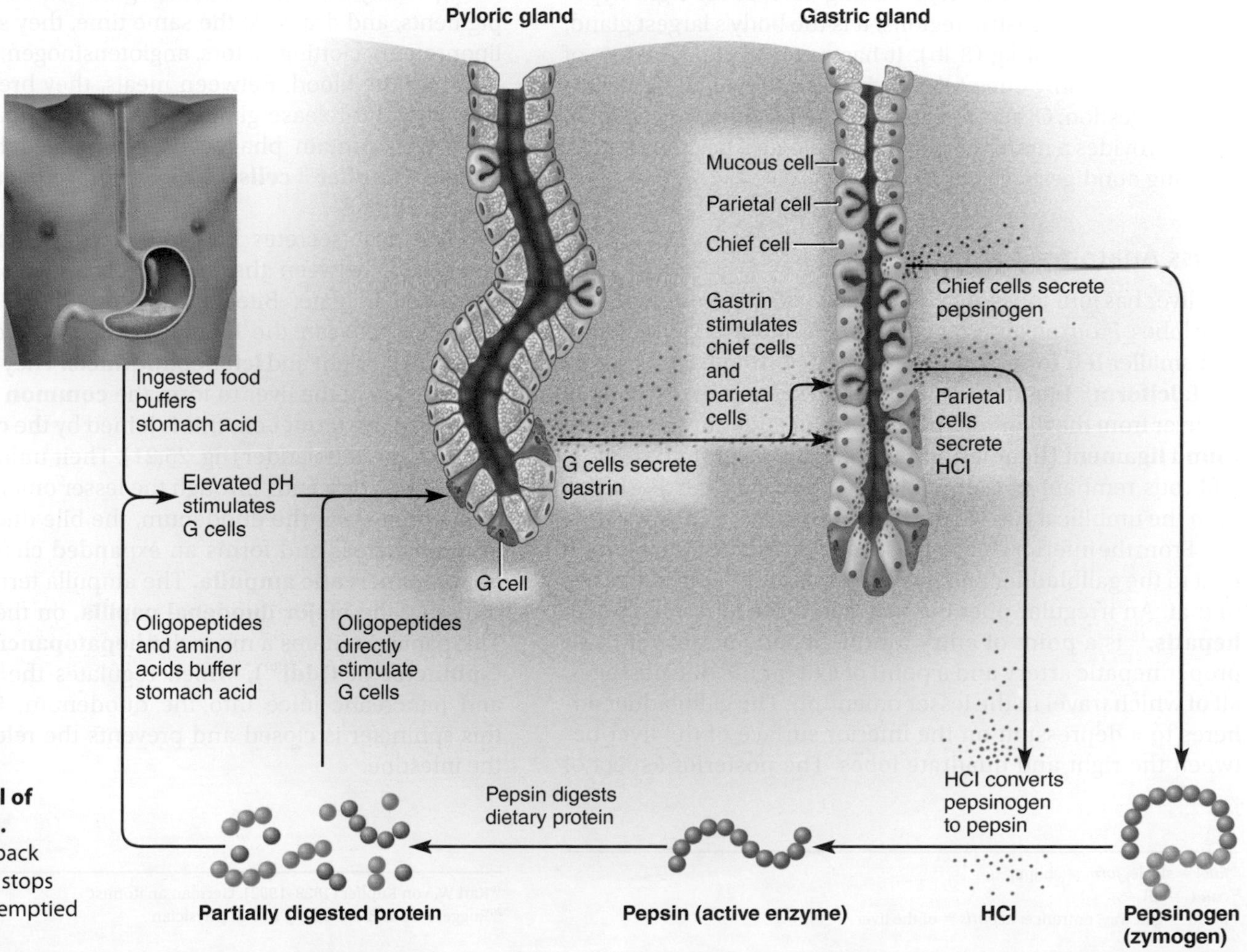

**FIGURE 25.18 Feedback Control of Gastric Secretion.**
This positive feedback loop declines and stops as the stomach is emptied and the pH drops.

Pyloric gland

Gastric gland

Ingested food buffers stomach acid

Elevated pH stimulates G cells

Oligopeptides and amino acids buffer stomach acid

Oligopeptides directly stimulate G cells

G cells secrete gastrin

G cell

Mucous cell
Parietal cell
Chief cell

Gastrin stimulates chief cells and parietal cells

Chief cells secrete pepsinogen

Parietal cells secrete HCl

HCl converts pepsinogen to pepsin

Pepsin digests dietary protein

**Partially digested protein**

**Pepsin (active enzyme)**

**HCl**

**Pepsinogen (zymogen)**

## 25.4 The Liver, Gallbladder, and Pancreas

### Expected Learning Outcomes

When you have completed this section, you should be able to

a. describe the gross and microscopic anatomy of the liver, gallbladder, bile duct system, and pancreas;

b. describe the digestive secretions and functions of the liver, gallbladder, and pancreas; and

c. explain how hormones regulate secretion by the liver and pancreas.

The small intestine receives not only chyme from the stomach but also secretions from the liver and pancreas, which enter the digestive tract near the junction of the stomach and small intestine. These secretions are so important to the digestive processes of the small intestine that it is necessary to understand them before continuing with intestinal physiology.

## The Liver AP|R

The liver (fig. 25.19) is a reddish brown gland located immediately inferior to the diaphragm, filling most of the right hypochondriac and epigastric regions. It is the body's largest gland, weighing about 1.4 kg (3 lb). It has a tremendous variety of functions, but only one of them, the secretion of bile, contributes to digestion. Others are discussed in the following chapter, which provides a more thorough physiological basis for understanding nondigestive liver functions.

### Gross Anatomy

The liver has four lobes called the right, left, quadrate, and caudate lobes. From an anterior view, we see only a large **right lobe** and smaller **left lobe.** They are separated from each other by the **falciform**[16] **ligament,** a sheet of mesentery that suspends the liver from the diaphragm and anterior abdominal wall. The **round ligament (ligamentum teres),** also visible anteriorly, is a fibrous remnant of the umbilical vein, which carries blood from the umbilical cord to the liver of a fetus.

From the inferior view, we also see a squarish **quadrate lobe** next to the gallbladder and a tail-like **caudate**[17] **lobe** posterior to that. An irregular opening between these lobes, the **porta hepatis,**[18] is a point of entry for the hepatic portal vein and proper hepatic artery and a point of exit for the bile passages, all of which travel in the lesser omentum. The gallbladder adheres to a depression on the inferior surface of the liver between the right and quadrate lobes. The posterior aspect of the liver has a deep sulcus that accommodates the inferior vena cava. The superior surface has a *bare area* where it is attached to the diaphragm. The rest of the liver is covered by a serosa.

### Microscopic Anatomy

The interior of the liver is filled with innumerable tiny cylinders called **hepatic lobules,** about 2 mm long by 1 mm in diameter. A lobule consists of a **central vein** passing down its core, surrounded by radiating plates of cuboidal cells called **hepatocytes** (fig. 25.20). Imagine spreading a book wide open until its front and back covers touch. The pages of the book would fan out around the spine somewhat like the plates fan out from the central vein of a liver lobule.

Each plate of hepatocytes is an epithelium one or two cells thick. The spaces between the plates are blood-filled channels called **hepatic sinusoids.** These are lined by a fenestrated endothelium that separates the hepatocytes from the blood cells, but allows blood plasma into the space between the hepatocytes and endothelium. The hepatocytes have a brush border of microvilli that project into this space. Blood filtering through the sinusoids comes directly from the stomach and intestines. After a meal, the hepatocytes absorb glucose, amino acids, iron, vitamins, and other nutrients from it for metabolism or storage. They also remove and degrade hormones, toxins, bile pigments, and drugs. At the same time, they secrete albumin, lipoproteins, clotting factors, angiotensinogen, and other products into the blood. Between meals, they break down stored glycogen and release glucose into the circulation. The sinusoids also contain phagocytic cells called **hepatic macrophages (Kupffer**[19] **cells),** which remove bacteria and debris from the blood.

The liver secretes bile into narrow channels, the **bile canaliculi,** between the back-to-back layers of hepatocytes within each plate. Bile passes from there into small **bile ductules** between the lobules, and these converge to ultimately form **right** and **left hepatic ducts.** They converge on the inferior side of the liver to form the **common hepatic duct.** A short distance farther on, this is joined by the **cystic duct** coming from the gallbladder (fig. 25.21). Their union forms the **bile duct,** which descends through the lesser omentum toward the duodenum. Near the duodenum, the bile duct joins the duct of the pancreas and forms an expanded chamber called the **hepatopancreatic ampulla.** The ampulla terminates at a fold of tissue, the **major duodenal papilla,** on the duodenal wall. This papilla contains a muscular **hepatopancreatic sphincter (sphincter of Oddi**[20]**),** which regulates the passage of bile and pancreatic juice into the duodenum. Between meals, this sphincter is closed and prevents the release of bile into the intestine.

---

[16]*falci* = sickle; *form* = shape
[17]*caud* = tail
[18]*porta* = gateway, entrance; *hepatis* = of the liver

[19]Karl W. von Kupffer (1829–1902), German anatomist
[20]Ruggero Oddi (1864–1913), Italian physician

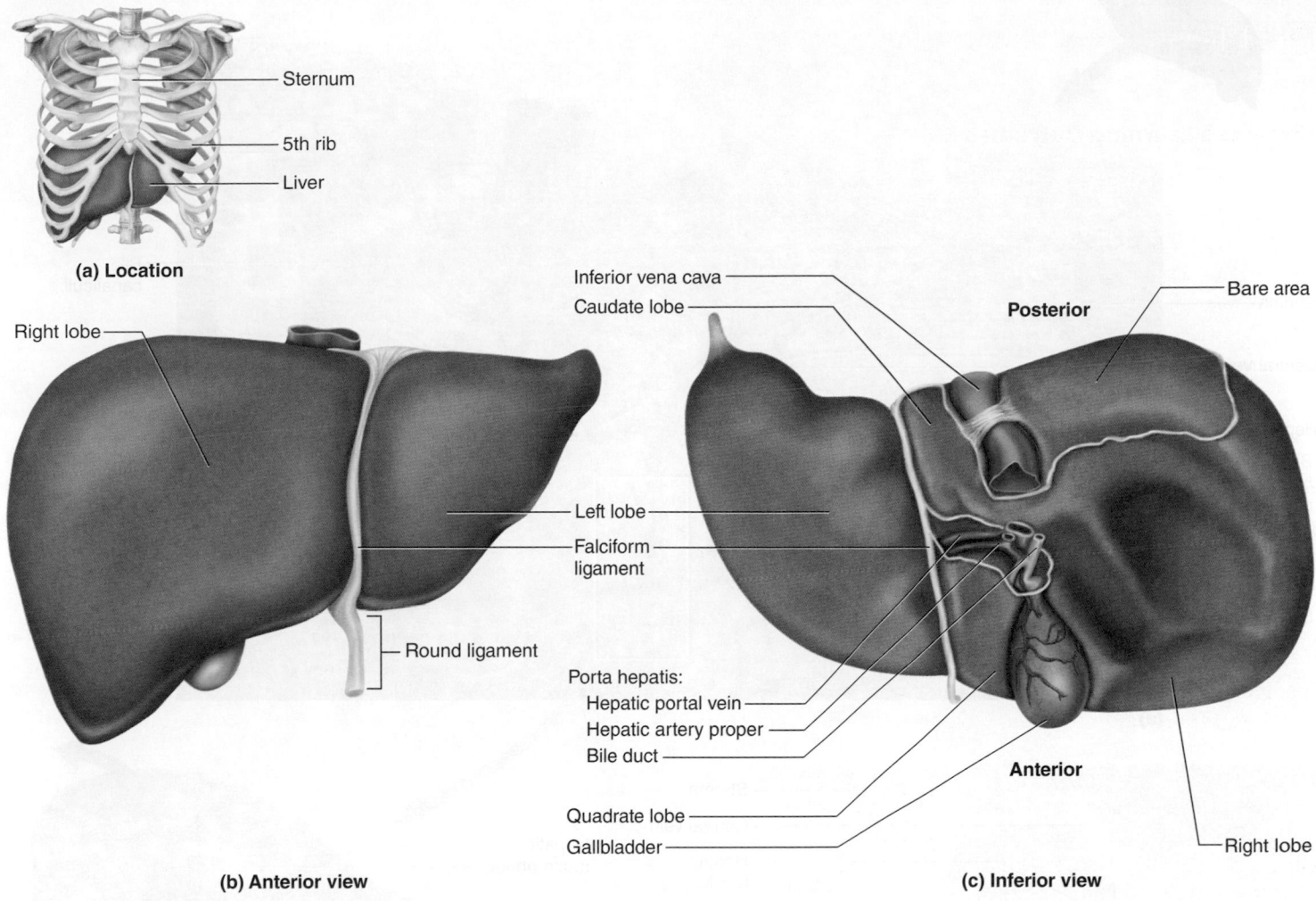

**(a) Location**
- Sternum
- 5th rib
- Liver

**(b) Anterior view**
- Right lobe
- Left lobe
- Falciform ligament
- Round ligament

Inferior vena cava
Caudate lobe
**Posterior**
Bare area
Left lobe
**Porta hepatis:**
Hepatic portal vein
Hepatic artery proper
Bile duct
**Anterior**
Quadrate lobe
Gallbladder
Right lobe

**(c) Inferior view**

**FIGURE 25.19  Gross Anatomy of the Liver.  AP|R**

The hepatic lobules are separated by a sparse connective tissue stroma. In cross sections, the stroma is especially visible in the triangular areas where three or more lobules meet. Here there is often a **hepatic triad** consisting of a bile ductule and two blood vessels—branches of the hepatic artery proper and the hepatic portal vein.

## Circulation

The liver receives blood from two sources: about 70% from the **hepatic portal vein** and 30% from the **hepatic arteries.** The hepatic portal vein receives blood from the stomach, intestines, pancreas, and spleen, and carries it into the liver at the porta hepatis; see the *hepatic portal system* in table 20.8 (p. 784). All nutrients absorbed by the small intestine reach the liver by this route except for lipids (transported in the lymphatic system). Arterial blood bound for the liver exits the aorta at the celiac trunk and follows the route shown in figure 20.30 (p. 780): celiac trunk → common hepatic artery → hepatic artery proper → right and left hepatic arteries, which enter the liver at the porta hepatis. These arteries deliver oxygen and other materials to the liver.

Branches of the hepatic portal vein and hepatic arteries meet each other in the spaces between the liver lobules, and both drain into the liver sinusoids. Hence, there is an unusual mixing of venous and arterial blood in the sinusoids. After processing by the hepatocytes, the blood collects in the central vein at the core of the lobule. Blood from the central veins ultimately converges in the right and left hepatic veins, which exit the superior surface of the liver and empty into the nearby inferior vena cava.

## The Gallbladder and Bile

The **gallbladder** is a pear-shaped sac on the underside of the liver that serves to store and concentrate bile. It is about 10 cm long and internally lined by a highly folded mucosa with a simple columnar epithelium. Its head *(fundus)* usually projects slightly beyond the inferior margin of the liver. Its neck *(cervix)* leads into the cystic duct, which leads in turn to the bile duct.

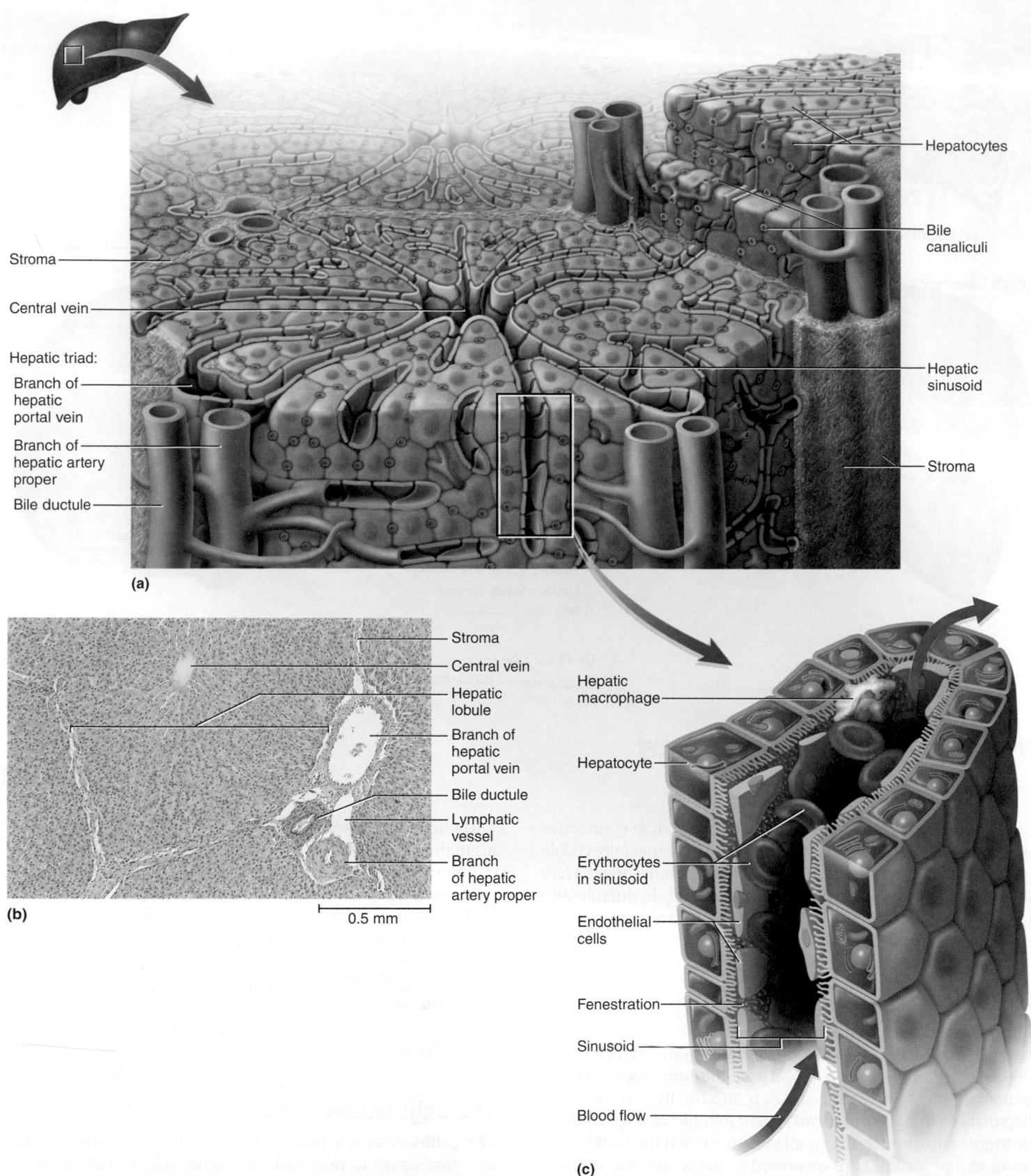

**FIGURE 25.20  Microscopic Anatomy of the Liver.**  (a) The hepatic lobules and their relationship to the blood vessels and bile tributaries. (b) Histological section of the liver. (c) A hepatic sinusoid.   AP|R

❓ *Identify two blood vessels in chapter 20 that supply blood to the hepatic sinusoids.*

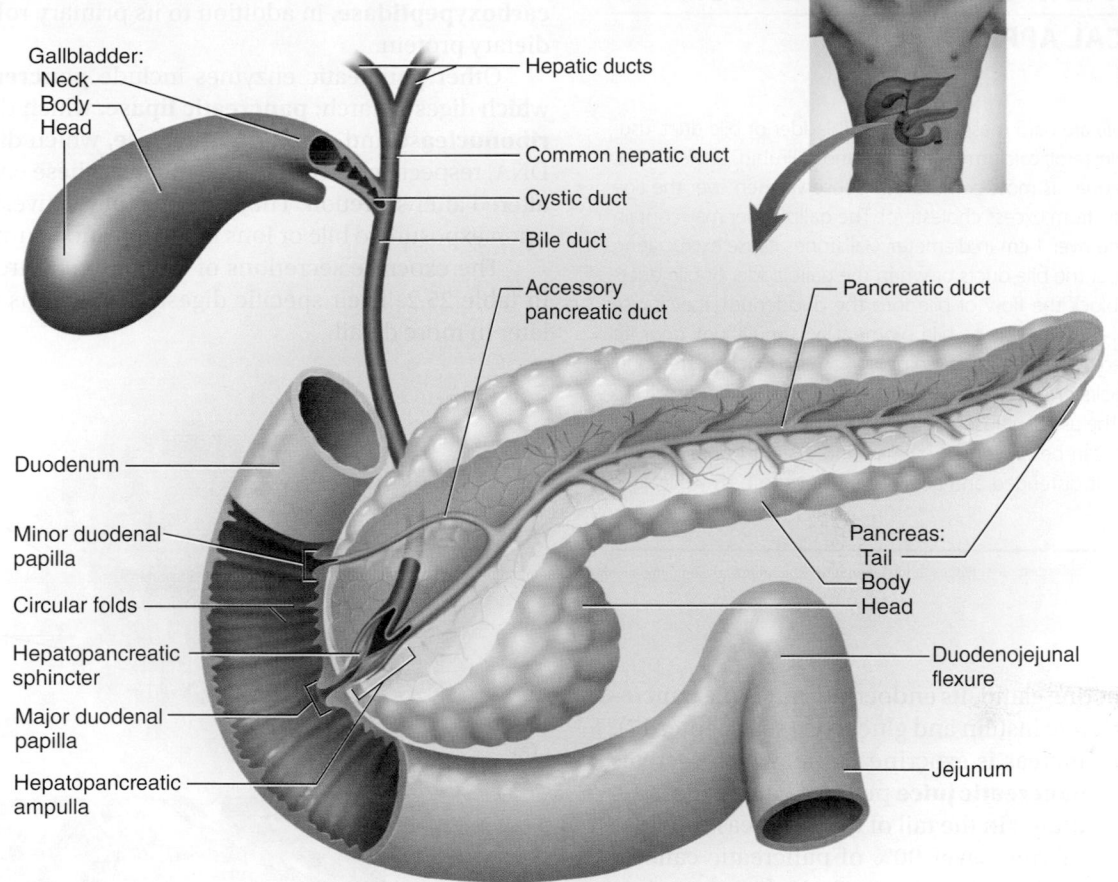

**FIGURE 25.21 Gross Anatomy of the Gallbladder, Pancreas, and Bile Passages.** The liver is omitted to show more clearly the gallbladder, which adheres to its inferior surface, and the hepatic ducts, which emerge from the liver tissue. **AP|R**

**Bile** is a green fluid containing minerals, cholesterol, neutral fats, phospholipids, bile pigments, and bile acids. The principal pigment is **bilirubin,** derived from the decomposition of hemoglobin. Bacteria of the large intestine metabolize bilirubin to **urobilinogen,** which is responsible for the brown color of feces. In the absence of bile secretion, the feces are grayish white and marked with streaks of undigested fat *(acholic feces).* **Bile acids (bile salts)** are steroids synthesized from cholesterol. Bile acids and **lecithin,** a phospholipid, aid in fat digestion and absorption, as discussed later. All other components of the bile are wastes destined for excretion in the feces. When these waste products become excessively concentrated, they may form gallstones (see Deeper Insight 25.3).

Bile gets into the gallbladder by first filling the bile duct, then overflowing into the gallbladder. Between meals, the gallbladder absorbs water and electrolytes from the bile and concentrates it by a factor of 5 to 20 times. The liver secretes about 500 to 1,000 mL of bile per day.

About 80% of the bile acids are reabsorbed in the ileum, the last portion of the small intestine, and returned to the liver,

where the hepatocytes absorb and resecrete them. This route of secretion, reabsorption, and resecretion, called the *enterohepatic circulation,* reuses the bile acids two or more times during the digestion of an average meal. The 20% of the bile that is not reabsorbed is excreted in the feces. This is the body's only way of eliminating excess cholesterol. The liver synthesizes new bile acids from cholesterol to replace the quantity lost in the feces.

▶▶▶**APPLY WHAT YOU KNOW**

*Certain drugs designed to reduce blood cholesterol work by blocking the reabsorption of bile acids in the ileum. Explain why they would have this cholesterol-lowering effect.*

## The Pancreas

The pancreas (fig. 25.21) is a spongy retroperitoneal gland posterior to the greater curvature of the stomach. It measures 12 to 15 cm long and about 2.5 cm thick. It has a globose *head* encircled by the duodenum, a midportion called the *body,* and a blunt, tapered *tail* on the left. The pancreas is both an

## DEEPER INSIGHT 25.3

### CLINICAL APPLICATION

#### Gallstones

*Gallstones (biliary calculi)* are hard masses in the gallbladder or bile duct, usually composed of cholesterol, calcium carbonate, and bilirubin. *Cholelithiasis,* the formation of gallstones, is most common in obese women over the age of 40 and usually results from excess cholesterol. The gallbladder may contain several gallstones, some over 1 cm in diameter. Gallstones cause excruciating pain when they obstruct the bile ducts or when the gallbladder or bile ducts contract. When they block the flow of bile into the duodenum, they cause jaundice (yellowing of the skin due to bile pigment accumulation), poor fat digestion, and impaired absorption of fat-soluble vitamins. Once treated only by surgical removal, gallstones are now often treated with stone-dissolving drugs or by *lithotripsy,* the use of ultrasonic vibration to pulverize them without surgery. Reobstruction can be prevented by inserting a stent (tube) into the bile duct, which keeps it distended and allows gallstones to pass while they are still small.

endocrine and exocrine gland. Its endocrine part is the pancreatic islets, which secrete insulin and glucagon (see chapter 17). About 99% of the pancreas is exocrine tissue, which secretes 1,200 to 1,500 mL of **pancreatic juice** per day. Pancreatic islets are relatively concentrated in the tail of the pancreas, whereas the head is more exocrine. Over 90% of pancreatic cancers arise from the ducts of the exocrine portion *(ductal carcinomas),* so cancer is most common in the head of the gland.

The cells of the secretory acini exhibit a high density of rough ER and secretory vesicles *(zymogen granules)* (fig. 25.22). The acini open into a system of branched ducts that eventually converge on the main **pancreatic duct.** This duct runs lengthwise through the middle of the gland and joins the bile duct at the hepatopancreatic ampulla. The hepatopancreatic sphincter thus controls the release of both bile and pancreatic juice into the duodenum. Usually, however, there is a smaller **accessory pancreatic duct** that branches from the main pancreatic duct and opens independently into the duodenum at the **minor duodenal papilla,** proximal to the major papilla. The accessory duct bypasses the sphincter and allows pancreatic juice to be released into the duodenum even when bile is not.

Pancreatic juice is an alkaline mixture of water, enzymes, zymogens, sodium bicarbonate, and other electrolytes. The acini secrete the enzymes and zymogens, whereas the ducts secrete the sodium bicarbonate. The bicarbonate buffers HCl arriving from the stomach.

The pancreatic zymogens are **trypsinogen** (trip-SIN-oh-jen), **chymotrypsinogen** (KY-mo-trip-SIN-o-jen), and **procarboxypeptidase** (PRO-car-BOC-see-PEP-tih-dase). When trypsinogen is secreted into the intestinal lumen, it is converted to **trypsin** by **enterokinase,** an enzyme secreted by the mucosa of the small intestine (fig. 25.23). Trypsin is autocatalytic—it converts trypsinogen into still more trypsin. It

also converts the other two zymogens into **chymotrypsin** and **carboxypeptidase,** in addition to its primary role of digesting dietary protein.

Other pancreatic enzymes include **pancreatic amylase,** which digests starch; **pancreatic lipase,** which digests fat; and **ribonuclease** and **deoxyribonuclease,** which digest RNA and DNA, respectively. Unlike the zymogens, these enzymes are not altered after secretion. They become fully active, however, only upon exposure to bile or ions in the intestinal lumen.

The exocrine secretions of the pancreas are summarized in table 25.2. Their specific digestive functions are explained later in more detail.

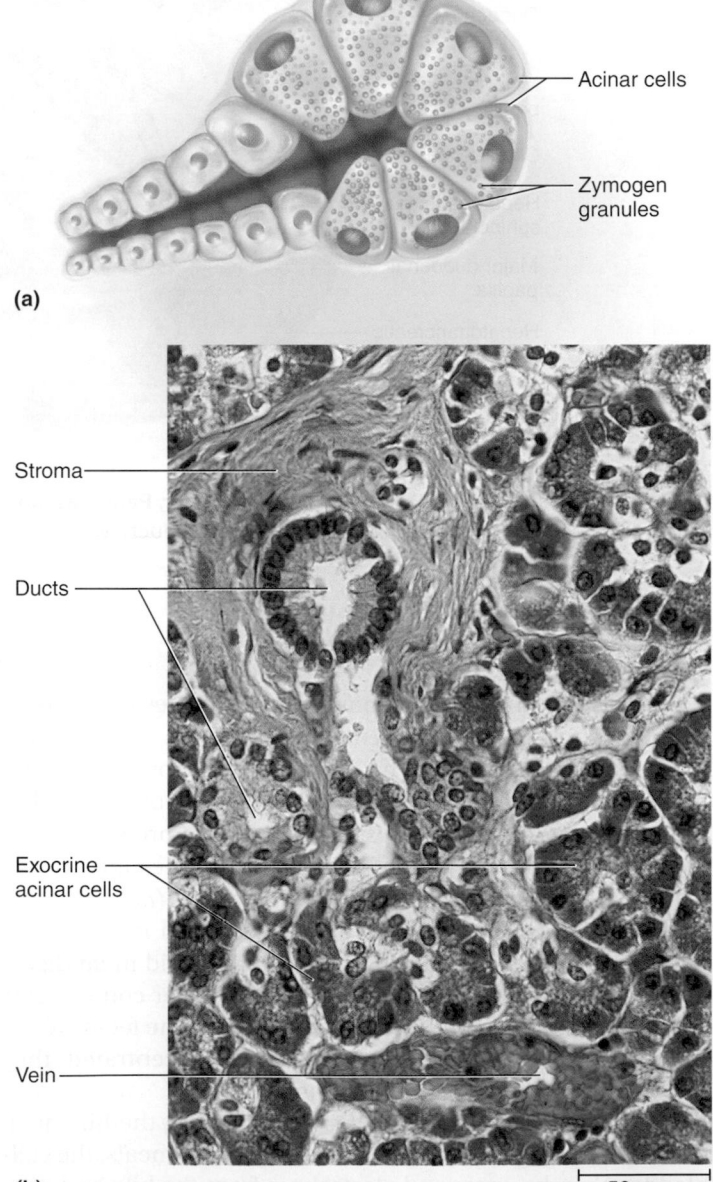

(a)

(b)

Acinar cells

Zymogen granules

Stroma

Ducts

Exocrine acinar cells

Vein

50 μm

**FIGURE 25.22  Microscopic Anatomy of the Pancreas.** (a) An acinus. (b) Histological section of the exocrine tissue and some of the connective tissue stroma.

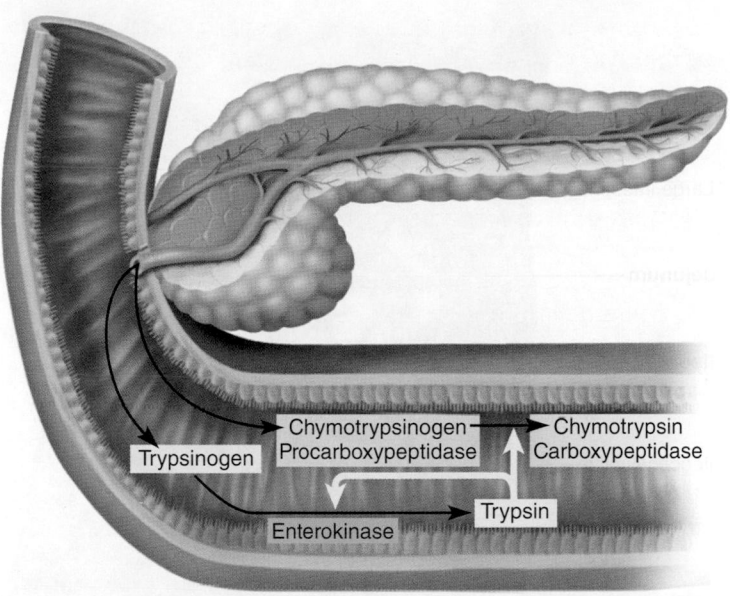

**FIGURE 25.23 The Activation of Pancreatic Enzymes in the Small Intestine.** The pancreas secretes trypsinogen, and enterokinase secreted by the duodenum converts it to trypsin. Trypsin not only digests dietary protein but also catalyzes the production of more trypsin and activates two other pancreatic zymogens—chymotrypsinogen and procarboxypeptidase.

| TABLE 25.2 | Exocrine Secretions of the Pancreas |
|---|---|
| **Secretion** | **Function** |
| *Sodium bicarbonate* | Neutralizes HCl |
| *Zymogens* | Converted to active digestive enzymes after secretion |
| Trypsinogen | Becomes trypsin, which digests protein |
| Chymotrypsinogen | Becomes chymotrypsin, which digests protein |
| Procarboxypeptidase | Becomes carboxypeptidase, which hydrolyzes the terminal amino acid from the carboxyl (–COOH) end of small peptides |
| *Enzymes* | |
| Pancreatic amylase | Digests starch |
| Pancreatic lipase | Digests fat |
| Ribonuclease | Digests RNA |
| Deoxyribonuclease | Digests DNA |

## Regulation of Secretion

Three stimuli are chiefly responsible for the release of pancreatic juice and bile.

1. **Acetylcholine (ACh),** coming from the vagus nerves and enteric neurons. ACh stimulates the pancreatic acini to secrete their enzymes even during the cephalic phase of gastric control, before food is swallowed. The enzymes remain stored in the pancreatic acini and ducts, however, in preparation for release later when chyme enters the duodenum.

2. **Cholecystokinin[21] (CCK),** secreted by the mucosa of the duodenum and proximal jejunum (the next segment of the small intestine), primarily in response to fats in the small intestine. CCK also stimulates the pancreatic acini to secrete enzymes, but it is named for its strongly stimulatory effect on the gallbladder. It induces contractions of the gallbladder and relaxation of the hepatopancreatic sphincter, discharging bile into the duodenum.

3. **Secretin,** produced by the same regions of the small intestine, mainly in response to the acidity of chyme from the stomach. Secretin stimulates the ducts of both the liver and pancreas to secrete an abundant sodium bicarbonate solution. In the pancreas, this flushes the enzymes into the duodenum. Sodium bicarbonate buff-

ers the hydrochloric acid arriving from the stomach, with the reaction

$$HCl + NaHCO_3 \longrightarrow NaCl + H_2CO_3 \text{ (carbonic acid)}$$

The carbonic acid then breaks down to carbon dioxide and water. $CO_2$ is absorbed into the blood and ultimately exhaled. What is left in the small intestine, therefore, is salt water—NaCl and $H_2O$. Sodium bicarbonate is therefore important in protecting the intestinal mucosa from HCl as well as raising the intestinal pH to the level needed for activity of the pancreatic and intestinal digestive enzymes.

▶▶▶ **APPLY WHAT YOU KNOW**

*Draw a negative feedback loop showing how secretin influences duodenal pH.*

**BEFORE YOU GO ON**

Answer the following questions to test your understanding of the preceding section:

14. What does the liver contribute to digestion?

15. Trace the pathway taken by bile acids from the liver and back. What is this pathway called?

16. Name two hormones, four enzymes, and one buffer secreted by the pancreas, and state the function of each.

17. What stimulates cholecystokinin (CCK) secretion, and how does CCK affect other parts of the digestive system?

[21]*chole* = bile; *cysto* = bladder (gallbladder); *kin* = action

## 25.5    The Small Intestine

### Expected Learning Outcomes

When you have completed this section, you should be able to

a. describe the gross and microscopic anatomy of the small intestine;

b. state how the mucosa of the small intestine differs from that of the stomach, and explain the functional significance of the differences;

c. define *contact digestion* and describe where it occurs; and

d. describe the types of movement that occur in the small intestine.

Nearly all chemical digestion and nutrient absorption occur in the small intestine. To perform these roles efficiently and thoroughly, the small intestine is the longest part of the digestive tract—about 5 m long (range 3 to 7 m) in a living person; in the cadaver, where there is no muscle tone, it is up to 8 m long. The term *small* intestine refers not to its length but to its diameter—about 2.5 cm (1 in.).

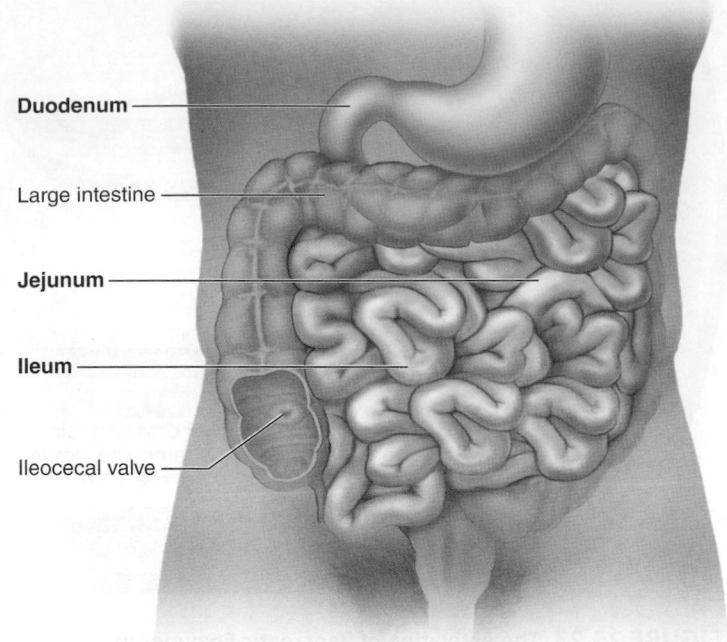

**FIGURE 25.24  Gross Anatomy of the Small Intestine.** AP|R

### Gross Anatomy

The small intestine is a coiled mass filling most of the abdominal cavity inferior to the stomach and liver. It is divided into three regions (fig. 25.24): the *duodenum, jejunum,* and *ileum.*

The **duodenum** (dew-ODD-eh-num, DEW-oh-DEE-num) constitutes the first 25 cm (10 in.). Its name refers to its length, about equal to the width of 12 fingers.[22] It begins at the pyloric valve, arcs around the head of the pancreas and passes to the left, and ends at a sharp bend called the **duodenojejunal flexure.** Slightly distal to the pyloric valve, it exhibits the previously described wrinkles called the major and minor duodenal papillae, where it receives the pancreatic duct and accessory pancreatic duct, respectively. Along with the pancreas, most of the duodenum is retroperitoneal. It receives the stomach contents, pancreatic juice, and bile. Stomach acid is neutralized here, fats are physically broken up (emulsified) by the bile acids, pepsin is inactivated by the elevated pH, and pancreatic enzymes take over the job of chemical digestion.

The **jejunum** (jeh-JOO-num), by definition, is the first 40% of the small intestine beyond the duodenum—about 1.0 to 1.7 m in a living person. Its name refers to the fact that early anatomists typically found it to be empty.[23] The jejunum begins in the upper left quadrant of the abdomen but lies mostly within the umbilical region (see fig. A.4, p. 31). It has large, tall, closely spaced circular folds. Its wall is thick and muscular, and it has an especially rich blood supply, which gives it a relatively red color. Most digestion and nutrient absorption occur here.

The **ileum**[24] forms the last 60% of the postduodenal small intestine (about 1.6 to 2.7 m). It occupies mainly the hypogastric region and part of the pelvic cavity. Compared with the jejunum, its wall is thinner, less muscular, less vascular, and has a paler pink color. On the side opposite from its mesenteric attachment, the ileum has prominent lymphatic nodules in clusters called **Peyer**[25] **patches,** which are readily visible to the naked eye and become progressively larger approaching the large intestine.

The end of the small intestine is the **ileocecal** (ILL-ee-oh-SEE-cul) **junction,** where the ileum joins the *cecum* of the large intestine. The muscularis of the ileum is thickened at this point to form a sphincter, the **ileocecal** (ILL-ee-oh-SEE-cul) **valve,** which protrudes into the cecum. It regulates the passage of food residue into the large intestine and prevents feces from backing up into the ileum.

### Microscopic Anatomy

The tissue layers of the small intestine are reminiscent of those in the esophagus and stomach with modifications appropriate for nutrient digestion and absorption. The lumen is lined with simple columnar epithelium. The muscularis externa is notable for a thick inner circular layer and a thinner outer longitudinal layer. The jejunum and ileum are intraperitoneal and thus covered on all sides with a serosa, which is continuous with the complex, folded mesentery that suspends the small

---

[22]*duoden* = 12
[23]*jejun* = empty, dry

[24]*from eilos* = twisted
[25]Johann K. Peyer (1653–1712), Swiss anatomist

intestine from the posterior abdominal wall. Most of the duodenum is retroperitoneal and has a serosa only on its anterior surface; its other surfaces are covered by adventitia.

Effective digestion and absorption require the small intestine to have a large internal surface area. This is provided by its relatively great length and by three kinds of internal folds or projections: the *circular folds, villi,* and *microvilli.* If the mucosa were smooth, like the inside of a hose, it would have a surface area of about 0.3 to 0.5 m², but with these surface elaborations, its actual surface area is about 200 m²—clearly a great advantage for nutrient absorption. The circular folds increase the surface area by a factor of 2 to 3; the villi by a factor of 10; and the microvilli by a factor of 20.

#### ▶▶▶ APPLY WHAT YOU KNOW

*The small intestine exhibits some of the same structural adaptations as the proximal convoluted tubule of the kidney, and for the same reasons. Discuss what they have in common, the reasons for it, and how this relates to this book's theme of the unity of form and function.*

**Circular folds,** the largest of these elaborations, are transverse to spiral ridges up to 1 cm high (see fig. 25.21). These involve only the mucosa and submucosa; they are not visible on the external surface, which is smooth. They slow the progress of the chyme and make it flow on a somewhat spiral path, which increases its contact with the mucosa and promotes more thorough mixing and nutrient absorption. Circular folds begin in the duodenum, attain their greatest height in the jejunum, and become smaller and more sparse in the ileum. They are absent from the distal half of the ileum, but most nutrient absorption is completed by that point.

**Villi** (VIL-eye; singular, *villus*) are tiny projections that give the inner lining of the intestine a fuzzy texture, like a terry cloth towel. They are about 0.5 to 1.0 mm high, with tongue- to fingerlike shapes (fig. 25.25). Villi are largest in the duodenum and become progressively smaller in more distal regions of the intestine. Villi are covered with two kinds of epithelial cells: columnar **enterocytes (absorptive cells)** and mucus-secreting **goblet cells.** Like epithelial cells of the stomach, those of the small intestine are joined by tight junctions that prevent digestive enzymes from seeping between them and eroding the underlying tissue.

The core of a villus is filled with areolar tissue of the lamina propria and contains an arteriole, blood capillaries, a venule, and a lymphatic capillary called a **lacteal** (LAC-tee-ul). The blood capillaries absorb most nutrients, but the lacteal absorbs most lipids. Lipids give its contents a milky appearance for which the lacteal is named.[26] The core of the villus also has a few smooth muscle cells that contract periodically. This enhances mixing of the chyme in the intestinal lumen and milks lymph down the lacteal to the larger lymphatics of the submucosa.

**Microvilli** are much smaller plasma membrane extensions, about 1 μm high, that form a fuzzy **brush border** on the surface of each enterocyte. In addition to increasing surface area, they contain **brush border enzymes** in the plasma membrane. These enzymes carry out some of the final stages of chemical digestion. They are not secreted into the lumen; instead, the chyme must contact the brush border for digestion to occur. This process, called **contact digestion,** is one reason why it is so important that intestinal contractions churn the chyme and ensure that it all contacts the mucosa.

On the floor of the small intestine, between the bases of the villi, there are numerous pores that open into tubular glands called **intestinal crypts (crypts of Lieberkühn;**[27] LEE-berkoohn). These crypts, similar to the gastric glands, extend as far as the muscularis mucosae. In the upper half, they consist of enterocytes and goblet cells like those of the villi. The lower half is dominated by dividing stem cells. In its life span of 3 to 6 days, an epithelial cell migrates up the crypt to the tip of the villus, where it is sloughed off and digested. A few **Paneth**[28] **cells** are clustered at the base of each crypt. They secrete lysozyme, phospholipase, and defensins—defensive proteins that resist bacterial invasion of the mucosa.

The duodenum has prominent **duodenal (Brunner**[29]**) glands** in the submucosa. They secrete an abundance of bicarbonate-rich mucus, which neutralizes stomach acid and shields the mucosa from its erosive effects. Throughout the small intestine, the lamina propria and submucosa have a large population of lymphocytes that intercept pathogens before they can invade the bloodstream. In some places, these are aggregated into conspicuous lymphatic nodules such as the Peyer patches of the ileum.

### Intestinal Secretion

The intestinal crypts secrete 1 to 2 L of **intestinal juice** per day, especially in response to acid, hypertonic chyme, and distension of the intestine. This fluid has a pH of 7.4 to 7.8. It contains water and mucus but relatively little enzyme. Most enzymes that function in the small intestine are found in the brush border and pancreatic juice.

### Intestinal Motility

Contractions of the small intestine serve three functions: (1) to mix chyme with intestinal juice, bile, and pancreatic juice, allowing these fluids to neutralize acid and digest nutrients more effectively; (2) to churn chyme and bring it into contact with the mucosa for contact digestion and nutrient absorption; and (3) to move residue toward the large intestine.

**Segmentation** is a movement in which stationary ringlike constrictions appear at several places along the intestine and then relax as new constrictions form elsewhere (fig. 25.26a).

---

[26]*lact* = milk

[27]Johann N. Lieberkühn (1711–56), German anatomist
[28]Josef Paneth (1857–90), Austrian physician
[29]Johann C. Brunner (1653–1727), Swiss anatomist

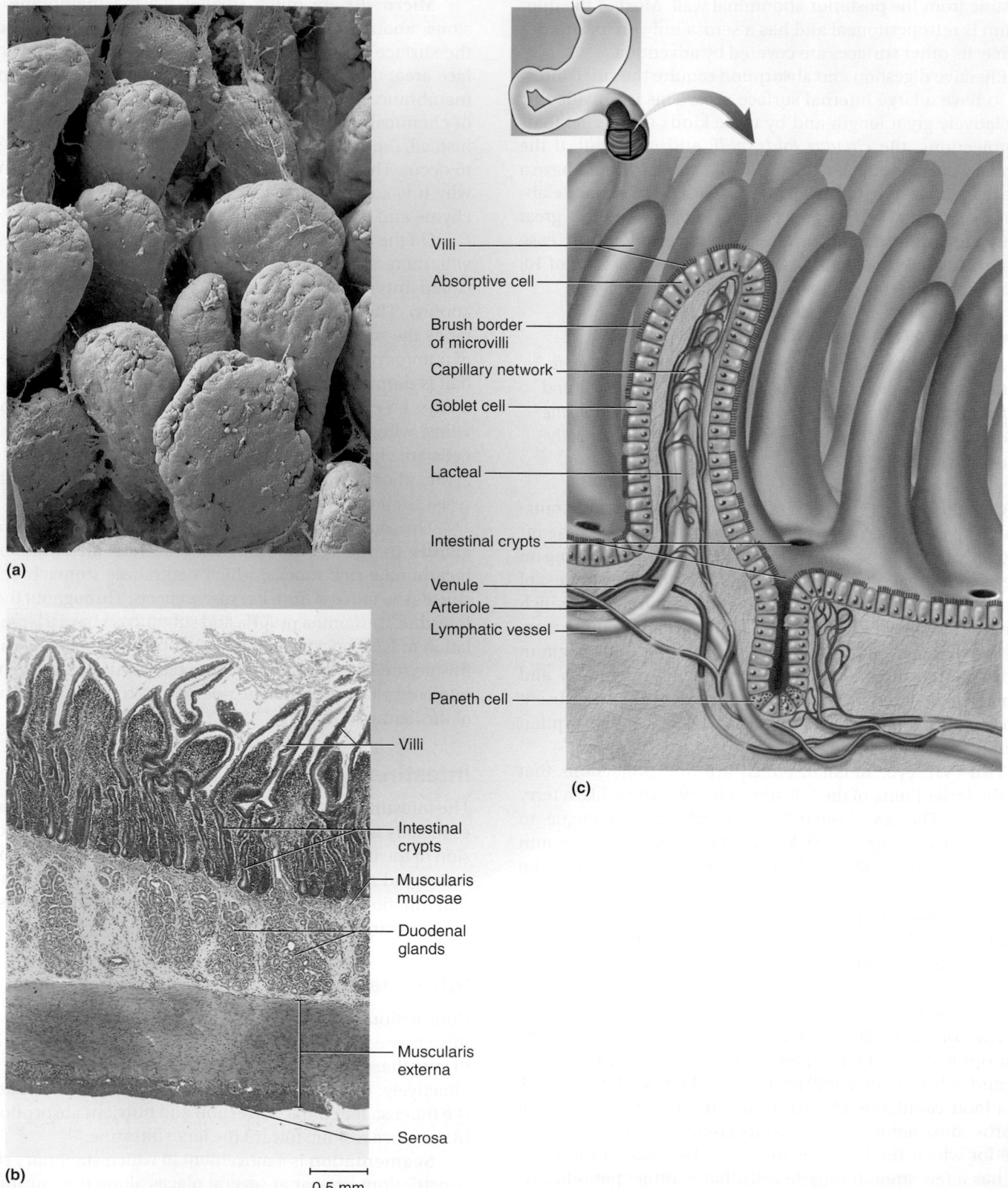

**FIGURE 25.25 Intestinal Villi.** (a) Villi (SEM). Each villus is about 1 mm high. (b) Histological section of the duodenum showing villi, intestinal crypts, and duodenal glands. (c) Structure of a villus.

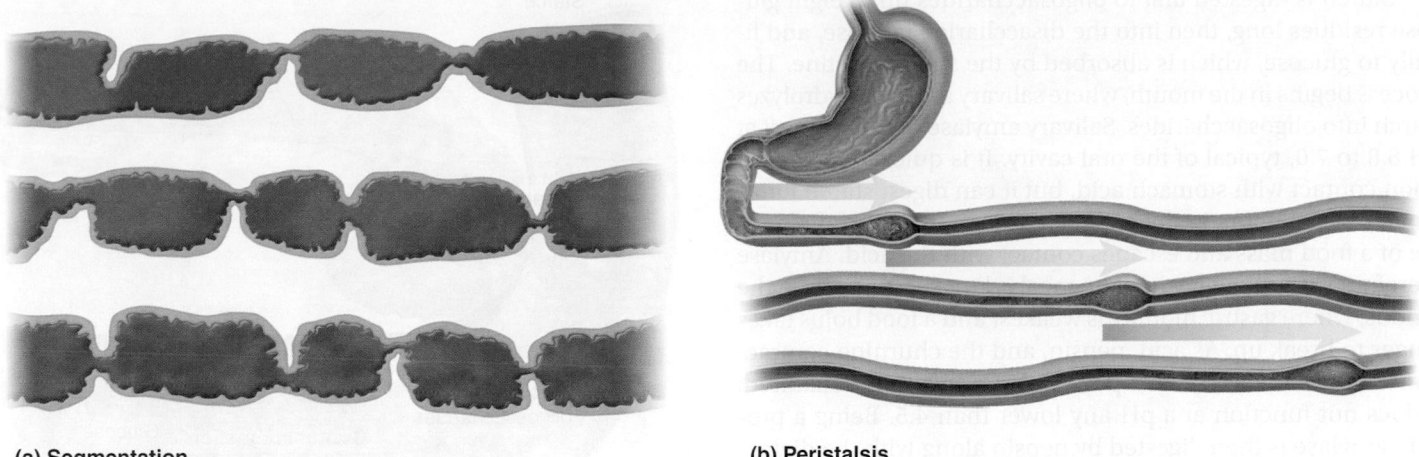

**(a) Segmentation**

**(b) Peristalsis**

**FIGURE 25.26 Contractions of the Small Intestine.** (a) Segmentation, in which circular constrictions of the intestine cut into the contents, churning and mixing them. (b) The migrating motor complex of peristalsis, in which successive waves of peristalsis overlap each other. Each wave travels partway down the intestine and milks the contents toward the colon.

This is the most common type of intestinal contraction. Its effect is to knead or churn the contents. Pacemaker cells of the muscularis externa set the rhythm of segmentation, with contractions about 12 times per minute in the duodenum and 8 to 9 times per minute in the ileum. Since the contractions are less frequent distally, segmentation causes slow progression of the chyme toward the colon. The intensity (but not frequency) of contractions is modified by nervous and hormonal influences.

When most nutrients have been absorbed and little remains but undigested residue, segmentation declines and peristalsis begins. A peristaltic wave begins in the duodenum, travels 10 to 70 cm, and dies out, only to be followed by another wave that begins a little farther down the tract than the first one did (fig. 25.26b). These successive, overlapping waves of contraction are called a **migrating motor complex.** They milk the chyme toward the colon over a period of about 2 hours. A second complex then expels residue and bacteria from the small intestine, thereby helping to limit bacterial colonization. Refilling of the stomach at the next meal suppresses peristalsis and reactivates segmentation.

The ileocecal valve is usually closed. Food in the stomach, however, triggers both the release of gastrin and the **gastroileal reflex,** both of which enhance segmentation in the ileum and relax the valve. As the cecum fills with residue, the pressure pinches the valve shut and prevents the reflux of cecal contents into the ileum.

**BEFORE YOU GO ON**

Answer the following questions to test your understanding of the preceding section:

18. What three structures increase the absorptive surface area of the small intestine?

19. Sketch a villus and label its epithelium, brush border, lamina propria, blood capillaries, and lacteal.

20. Distinguish between segmentation and the migrating motor complex of the small intestine. How do these differ in function?

## 25.6 Chemical Digestion and Absorption

### Expected Learning Outcomes

When you have completed this section, you should be able to

a. describe how each major class of nutrients is chemically digested, name the enzymes involved, and discuss the functional differences among these enzymes; and

b. describe how each type of nutrient is absorbed by the small intestine.

Chemical digestion and nutrient absorption are essentially finished by the time food residue leaves the small intestine and enters the cecum. But before going on to the functions of the large intestine, we trace each major class of nutrients—especially carbohydrates, proteins, and fats—from the mouth through the small intestine to see how it is chemically degraded and absorbed. Figures 25.27 through 25.30 depict details of the digestion and absorption of the major classes of organic nutrients, and then figure 25.31 sums all of this up for you in the "big picture."

### Carbohydrates

Most digestible dietary carbohydrate is starch. Cellulose is indigestible and is not considered here, although its importance as dietary fiber is discussed in chapter 26. The amount of glycogen in the diet is negligible, but it is digested in the same manner as starch.

Starch is digested first to oligosaccharides up to eight glucose residues long, then into the disaccharide maltose, and finally to glucose, which is absorbed by the small intestine. The process begins in the mouth, where salivary amylase hydrolyzes starch into oligosaccharides. Salivary amylase functions best at pH 6.8 to 7.0, typical of the oral cavity. It is quickly denatured upon contact with stomach acid, but it can digest starch for as long as 1 to 2 hours in the stomach as long as it is in the middle of a food mass and escapes contact with the acid. Amylase therefore works longer when the meal is larger, especially in the fundus, where gastric motility is weakest and a food bolus takes longer to break up. As acid, pepsin, and the churning contractions of the stomach break up the bolus, amylase is denatured; it does not function at a pH any lower than 4.5. Being a protein, amylase is then digested by pepsin along with the dietary proteins.

About 50% of the dietary starch is digested before it reaches the small intestine. Its digestion resumes in the small intestine when the chyme mixes with pancreatic amylase (fig. 25.27). Starch is entirely converted to oligosaccharides and maltose within 10 minutes. Its digestion is completed as the chyme contacts the brush border of the enterocytes. Two brush border enzymes, **dextrinase** and **glucoamylase,** hydrolyze oligosaccharides that are three or more residues long. The third, **maltase,** hydrolyzes maltose to glucose.

Maltose is also present in some foods, but the major dietary disaccharides are sucrose (cane sugar) and lactose (milk sugar). They are digested by the brush border enzymes **sucrase** and **lactase,** respectively, and the resulting monosaccharides are immediately absorbed (glucose and fructose from the former; glucose and galactose from the latter). In most of the world population, however, lactase production ceases or declines to a low level after age 4 and lactose becomes indigestible (see Deeper Insight 25.4).

The plasma membrane of the enterocytes has transport proteins that absorb monosaccharides as soon as the brush border enzymes release them (fig. 25.28). About 80% of the absorbed sugar is glucose, which is taken up by a sodium–glucose transporter (SGLT) like that of the kidney tubules (see p. 905). The glucose is subsequently transported out the base of the cell into the extracellular fluid (ECF). Sugar entering the ECF increases its osmolarity, and this draws water osmotically from the lumen of the intestine, through the now leaky tight junctions between the epithelial cells. Water carries more glucose and other nutrients with it by *solvent drag,* much as it does in the kidney. After a high-carbohydrate meal, solvent drag absorbs two to three times as much glucose as the SGLT.

The SGLT also absorbs galactose, whereas fructose is absorbed by facilitated diffusion using a separate carrier that does not depend on Na⁺. Inside the enterocyte, most fructose is converted to glucose. Glucose, galactose, and the small amount of remaining fructose are then transported out the base of the cell by facilitated diffusion and absorbed by the blood capillaries of the villus. The hepatic portal system delivers them to the liver; chapter 26 follows the fate of these sugars from there.

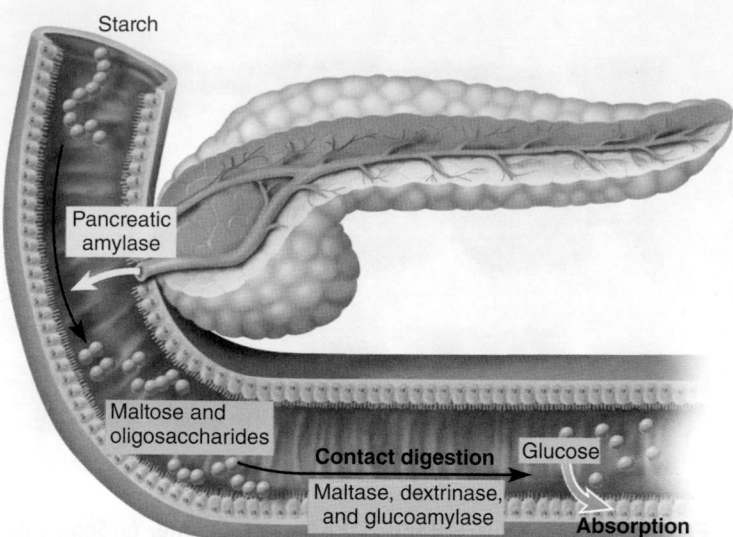

**FIGURE 25.27  Starch Digestion in the Small Intestine.** Pancreatic amylase digests starch into maltose and small oligosaccharides. Brush border enzymes (maltase, dextrinase, and glucoamylase) digest these to glucose, which is absorbed by the epithelial cells.

## Proteins

The amino acids absorbed by the small intestine come from three sources: (1) dietary proteins, (2) digestive enzymes digested by each other, and (3) sloughed epithelial cells digested by these enzymes. The endogenous amino acids from the last two sources total about 30 g/day, compared with about 44 to 60 g/day from the diet.

Enzymes that digest proteins are called **proteases (peptidases).** They are absent from the saliva but begin work in the stomach. Here, pepsin hydrolyzes any peptide bond between tyrosine and phenylalanine, thereby digesting 10% to 15% of the dietary protein into shorter polypeptides and a small amount of free amino acids (fig. 25.29). Pepsin has an optimal pH of 1.5 to 3.5, so it is inactivated when it passes into the duodenum and mixes with the alkaline pancreatic juice (pH 8).

In the small intestine, the pancreatic enzymes trypsin and chymotrypsin take over protein digestion by hydrolyzing polypeptides into even shorter oligopeptides. Finally, these are taken apart one amino acid at a time by three more enzymes: (1) **carboxypeptidase** removes amino acids from the −COOH end of the chain; (2) **aminopeptidase** removes them from the −NH₂ end; and (3) **dipeptidase** splits dipeptides in the middle and releases the last two free amino acids. The last two of these are brush border enzymes, whereas carboxypeptidase is a pancreatic secretion.

Amino acid absorption is similar to that of monosaccharides. Enterocytes have several sodium-dependent amino acid cotransporters for different classes of amino acids. Dipeptides and tripeptides can also be absorbed, but they are hydrolyzed within the enterocytes before their amino acids

# DEEPER INSIGHT 25.4

## CLINICAL APPLICATION

### Lactose Intolerance

Humans are a strange species. Unique among mammals, we go on drinking milk in adulthood, and moreover, the milk of other species! This odd habit is largely limited, however, to people of western and northern Europe, Mongolia, a few pastoral tribes of Africa, and their descendants in the Americas and elsewhere. They have an ancestral history of milking domestic animals, a practice that goes back about 10,000 years and has coincided with the continued production of lactase into adulthood.

People without lactase have *lactose intolerance*. If they consume milk, lactose passes undigested into the large intestine, increases the osmolarity of the intestinal contents, and causes colonic water retention and diarrhea. In addition, lactose fermentation by intestinal bacteria produces gas, resulting in painful cramps and flatulence.

Lactose intolerance occurs in about 15% of American whites; 90% of American blacks, who are predominantly descended from nonpastoral African tribes; 70% or more of Mediterraneans; and nearly all people of Asian descent, including those of us descended from the native migrants into North, Central, and South America. People with lactose intolerance can consume products such as yogurt and cheese, in which bacteria have broken down the lactose, and they can digest milk and ice cream with the aid of lactase drops or tablets.

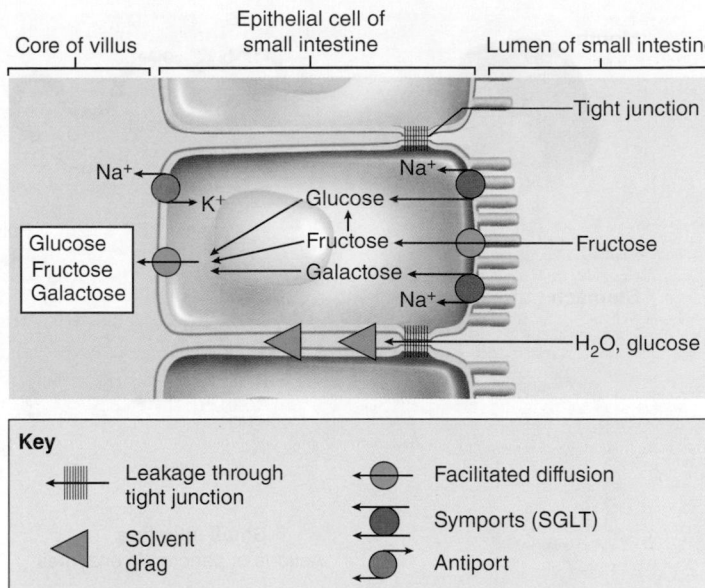

**FIGURE 25.28 Monosaccharide Absorption by the Small Intestine.** Glucose and galactose are absorbed by the SGLT symport in the apical membrane of the absorptive cell (right). Glucose is also absorbed along with water through the paracellular route (between cells) by solvent drag. Fructose is absorbed separately by facilitated diffusion. Most fructose is converted to glucose within the epithelial cell. The monosaccharides pass through the basal membrane of the cell by facilitated diffusion (left) and are then absorbed by the blood capillaries of the villus.

are released to the bloodstream. At the basal surfaces of the cells, amino acids behave like the monosaccharides discussed previously—they leave the cell by facilitated diffusion, enter the capillaries of the villus, and are carried away in the hepatic portal circulation.

The absorptive cells of infants can take up intact proteins by pinocytosis and release them to the blood by exocytosis. This allows IgA from breast milk to pass into an infant's bloodstream and confer passive immunity from mother to infant. It has the disadvantage, however, that intact proteins entering the infant's blood are detected as foreign antigens and sometimes trigger food allergies. As the intestine matures, its ability to pinocytose protein declines but never completely ceases.

## Lipids

The hydrophobic quality of lipids makes their digestion and absorption more complicated than that of carbohydrates and proteins (fig. 25.30). Fats are digested by enzymes called **lipases.** *Lingual lipase,* secreted by the intrinsic salivary glands of the tongue, digests a small amount of fat while food is still in the mouth, but becomes more active at the acidic pH of the stomach. Here it is joined by *gastric lipase,* which makes a much larger contribution to preduodenal fat digestion. About 10% to 15% of dietary fat is digested before the chyme passes on to the duodenum.

Being hydrophobic, ingested fat takes the form of large globules that, without further physical processing, could be attacked by these lipases only at their surface. This would result in rather slow, inefficient digestion. The vigorous *antral*

*pumping* described earlier, however, breaks the fat up into small droplets dispersed through the watery chyme—that is, it *emulsifies* the fat before passing it on to the duodenum, exposing much more fat surface to enzymatic action. In the duodenum, the little **emulsification droplets** received from the stomach are promptly coated by certain components of the bile—lecithin and bile acids. These agents have hydrophobic regions attracted to the surface of a fat globule and hydrophilic regions attracted to the surrounding water. The agitation produced by intestinal segmentation breaks the fat up further into droplets as small as 1 μm, and the coating of lecithin and bile acids keeps it broken up, preventing the droplets from coalescing into larger globules.

There is enough *pancreatic lipase* in the small intestine after a meal to digest the average daily fat intake in as little as 1 or 2 minutes. When lipase acts on a triglyceride, it removes the first and third fatty acids from the glycerol backbone and usually leaves the middle one. The products of lipase action are therefore two free fatty acids (FFAs) and a monoglyceride.

The absorption of these products and other lipids depends on minute droplets in the bile called **micelles**[30] (my-SELLS). Micelles, made in the liver, consist of 20 to 40 bile

---

[30]*mic* = grain, crumb; *elle* = little

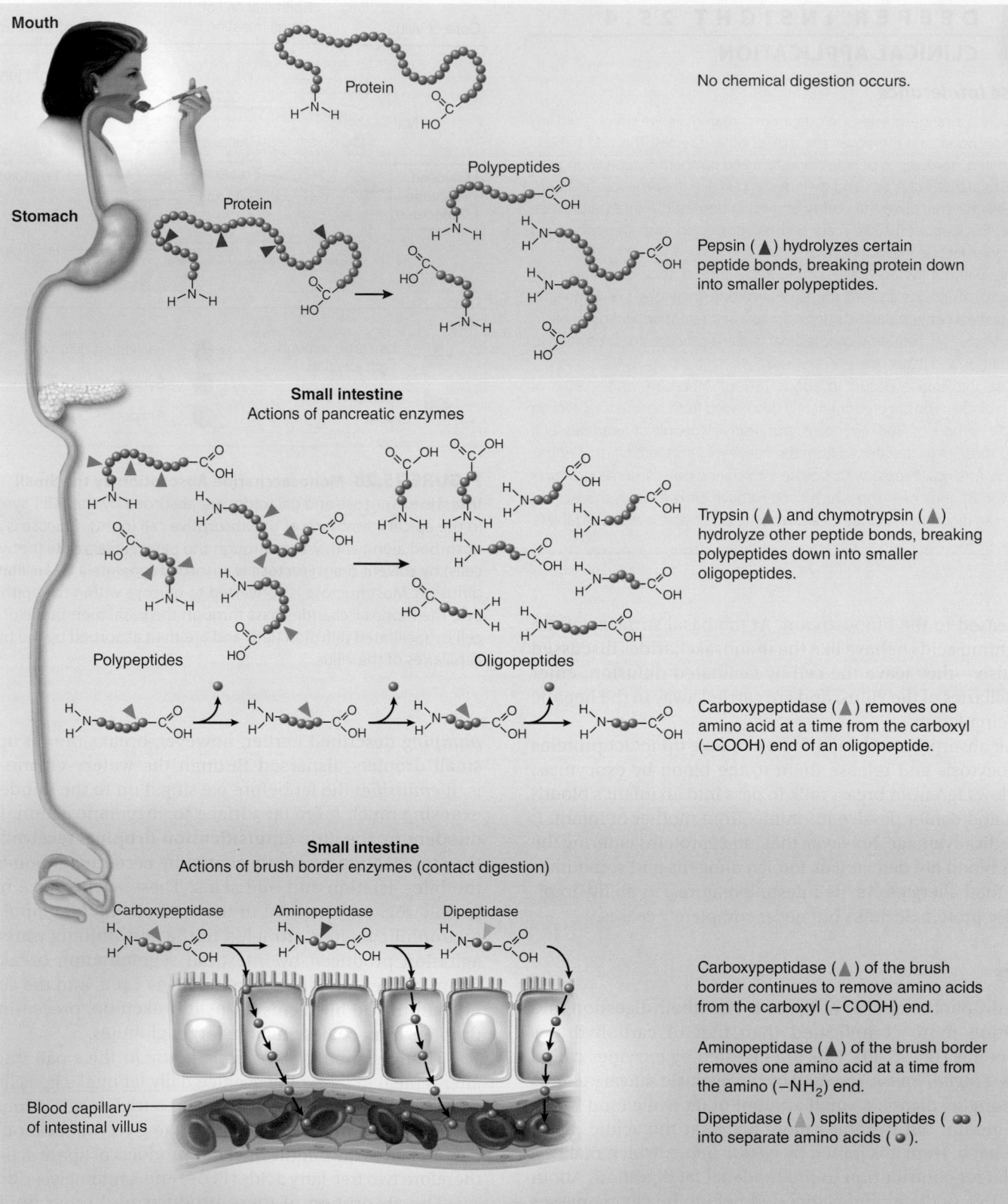

**Mouth**

Protein

No chemical digestion occurs.

**Stomach**

Protein

Polypeptides

Pepsin (▲) hydrolyzes certain peptide bonds, breaking protein down into smaller polypeptides.

**Small intestine**
Actions of pancreatic enzymes

Polypeptides

Oligopeptides

Trypsin (▲) and chymotrypsin (▲) hydrolyze other peptide bonds, breaking polypeptides down into smaller oligopeptides.

Carboxypeptidase (▲) removes one amino acid at a time from the carboxyl (–COOH) end of an oligopeptide.

**Small intestine**
Actions of brush border enzymes (contact digestion)

Carboxypeptidase    Aminopeptidase    Dipeptidase

Blood capillary of intestinal villus

Carboxypeptidase (▲) of the brush border continues to remove amino acids from the carboxyl (–COOH) end.

Aminopeptidase (▲) of the brush border removes one amino acid at a time from the amino (–NH₂) end.

Dipeptidase (▲) splits dipeptides (●●) into separate amino acids (●).

**FIGURE 25.29 Protein Digestion and Absorption.**

acid molecules aggregated with their hydrophilic side groups facing outward and their hydrophobic steroid rings facing inward. Bile phospholipids and cholesterol diffuse into the center of the micelle to form its core. The micelles pass down the bile duct into the duodenum, where they absorb fat-soluble vitamins, more cholesterol, and the FFAs and monoglycerides produced by fat digestion. Because of their charged, hydrophilic surfaces, micelles remain suspended in water more easily than free lipids do. They transport lipids to the surfaces of the enterocytes, where the lipids leave the micelles and diffuse through the plasma membrane into the cells. The micelles are reused, picking up another cargo of lipids and ferrying them to the enterocytes. Without micelles, the small intestine absorbs only about 40% to 50% of the dietary fat and almost no cholesterol.

Within the enterocytes, FFAs and monoglycerides are transported into the smooth endoplasmic reticulum and re-synthesized into triglycerides. The Golgi complex combines these with a small amount of cholesterol and coats the complex with a film of phospholipids and protein, forming droplets 75 to 1,200 nm in diameter called **chylomicrons**[31] (KY-lo-MY-crons). It packages chylomicrons into secretory vesicles that migrate to the basal surface of the cell and release their contents into the core of the villus. Although some FFAs enter the blood capillaries, chylomicrons are too large to penetrate the endothelium. They are taken up by the more porous lacteal into the lymph. The white, fatty intestinal lymph *(chyle)* flows through larger and larger lymphatic vessels of the mesenteries, eventually passing through the cisterna chyli (see fig. 21.1, p. 804) to the thoracic duct, then entering the bloodstream at the left subclavian vein. The further fate of dietary fat is described in chapter 26.

For a summary of the major aspects of carbohydrate, protein, and fat digestion and absorption, see figure 25.31.

▶▶▶ **APPLY WHAT YOU KNOW**

*Explain why the right lymphatic duct does not contribute any dietary fat to the bloodstream.*

## Nucleic Acids

The nucleic acids, DNA and RNA, are present in much smaller quantities than the polymers discussed previously. The **nucleases** (ribonuclease and deoxyribonuclease) of pancreatic juice hydrolyze these to their constituent nucleotides. **Nucleosidases** and **phosphatases** of the brush border then decompose the nucleotides into phosphate ions, ribose (from RNA) or deoxyribose (from DNA), and nitrogenous bases. These products are transported across the intestinal epithelium by membrane carriers and enter the capillary blood of the villus.

## Vitamins

Vitamins are absorbed unchanged. The fat-soluble vitamins A, D, E, and K are absorbed with other lipids as just described. Therefore, if they are ingested without fat-containing food, they are not absorbed at all but are passed in the feces and wasted. Water-soluble vitamins (the B complex and vitamin C) are absorbed by simple diffusion. An exception is vitamin $B_{12}$, an unusually large molecule that is absorbed poorly unless bound to intrinsic factor (IF) from the stomach. The $B_{12}$–IF complex then binds to receptors on absorptive cells of the distal ileum, where it is taken up by receptor-mediated endocytosis.

## Minerals

Minerals (electrolytes) are absorbed along the entire length of the small intestine. Sodium ions are cotransported with sugars and amino acids. Chloride ions are actively transported in the distal ileum by a pump that exchanges them for bicarbonate ions, reversing the chloride–bicarbonate exchange that occurs in the stomach. Potassium ions are absorbed by simple diffusion. The $K^+$ concentration of the chyme rises as water is absorbed from it, creating a gradient favorable to $K^+$ absorption. In diarrhea, when water absorption is hindered, potassium ions remain in the intestine and pass with the feces; therefore, chronic diarrhea can lead to hypokalemia.

Iron and calcium are unusual in that they are absorbed in proportion to the body's need, whereas other minerals are absorbed at fairly constant rates regardless of need, leaving it to the kidneys to excrete any excess. Iron absorption is stimulated by the liver hormone hepcidin. The enterocytes bind ferrous ions ($Fe^{2+}$) and internalize them by active transport; they are unable to absorb ferric ions ($Fe^{3+}$), but stomach acid (HCl) reduces most $Fe^{3+}$ to absorbable $Fe^{2+}$. $Fe^{2+}$ is transported to the basal surface of the cell and there taken up by the extracellular protein *transferrin*. The transferrin–iron complex diffuses into the blood and is carried to such places as the bone marrow for hemoglobin synthesis, muscular tissue for myoglobin synthesis, and the liver for storage (see fig. 18.7, p. 681). Excess dietary iron, if absorbed, binds irreversibly to ferritin in the enterocyte and is held there until that cell sloughs off and passes in the feces.

▶▶▶ **APPLY WHAT YOU KNOW**

*Young adult women have four times as many iron transport proteins in the intestinal mucosa as men have. Can you explain this in terms of functional significance?*

The small intestine absorbs nearly all dietary phosphate, predominantly by active transport. By contrast, it absorbs only about 40% of the dietary calcium, leaving the rest to pass in the feces. In the duodenum, calcium is absorbed by the transcellular route. It enters the enterocytes through calcium channels in the apical plasma membrane and binds to a cytoplasmic protein called *calbindin*. This keeps the intracellular concentration of free calcium low, maintaining a gradient that favors calcium uptake. What free calcium there is in the cytoplasm is

---

[31]*chyl* = juice; *micr* = small

## Emulsification

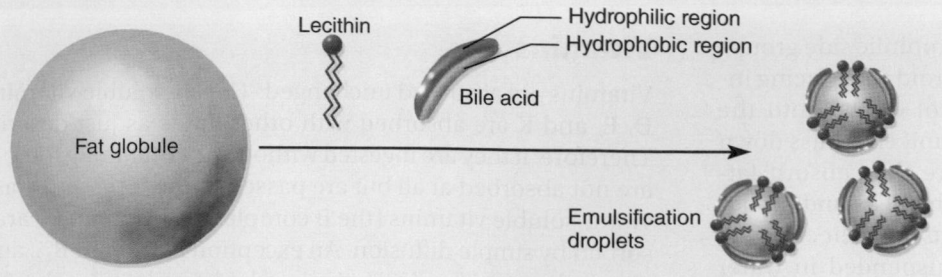

Hydrophilic region

Hydrophobic region

Lecithin

Bile acid

Fat globule

Emulsification droplets

Fat globule is broken up and coated by lecithin and bile acids.

## Fat hydrolysis

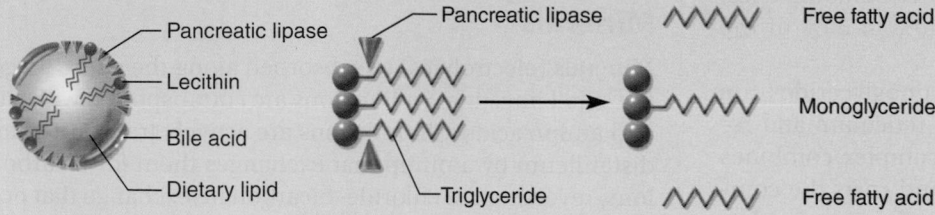

Pancreatic lipase

Lecithin

Bile acid

Dietary lipid

Pancreatic lipase

Triglyceride

Free fatty acid

Monoglyceride

Free fatty acid

Emulsification droplets are acted upon by pancreatic lipase, which hydrolyzes the first and third fatty acids from triglycerides, usually leaving the middle fatty acid.

## Lipid uptake by micelles

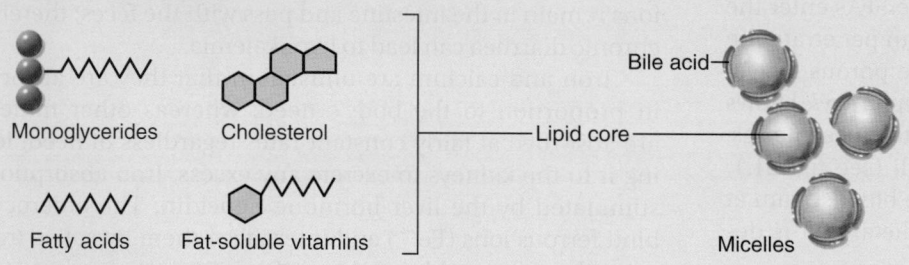

Monoglycerides

Cholesterol

Fatty acids

Fat-soluble vitamins

Lipid core

Bile acid

Micelles

Micelles in the bile pass to the small intestine and pick up several types of dietary and semidigested lipids.

## Chylomicron formation

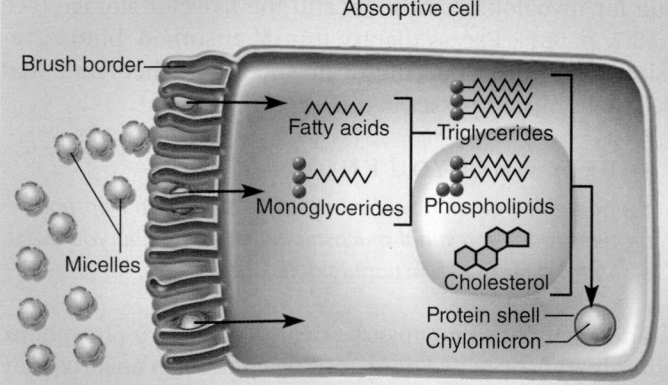

Absorptive cell

Brush border

Fatty acids

Monoglycerides

Triglycerides

Phospholipids

Cholesterol

Protein shell

Chylomicron

Micelles

Intestinal cells absorb lipids from micelles, resynthesize triglycerides, and package triglycerides, cholesterol, and phospholipids into protein-coated chylomicrons.

## Chylomicron exocytosis and lymphatic uptake

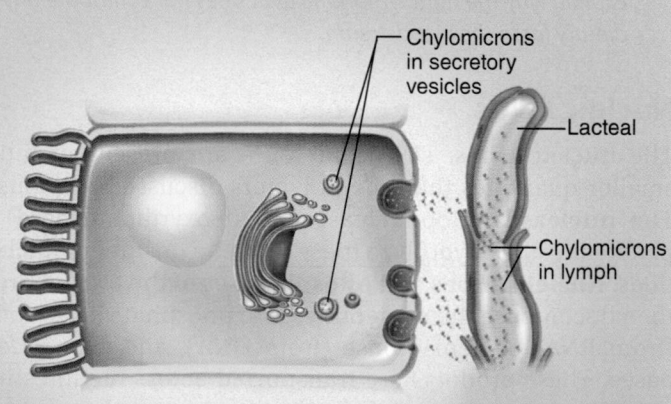

Chylomicrons in secretory vesicles

Lacteal

Chylomicrons in lymph

Golgi complex packages chylomicrons into secretory vesicles; chylomicrons are released from basal cell membrane by exocytosis and enter the lacteal (lymphatic capillary) of the villus.

**FIGURE 25.30  Fat Digestion and Absorption.**

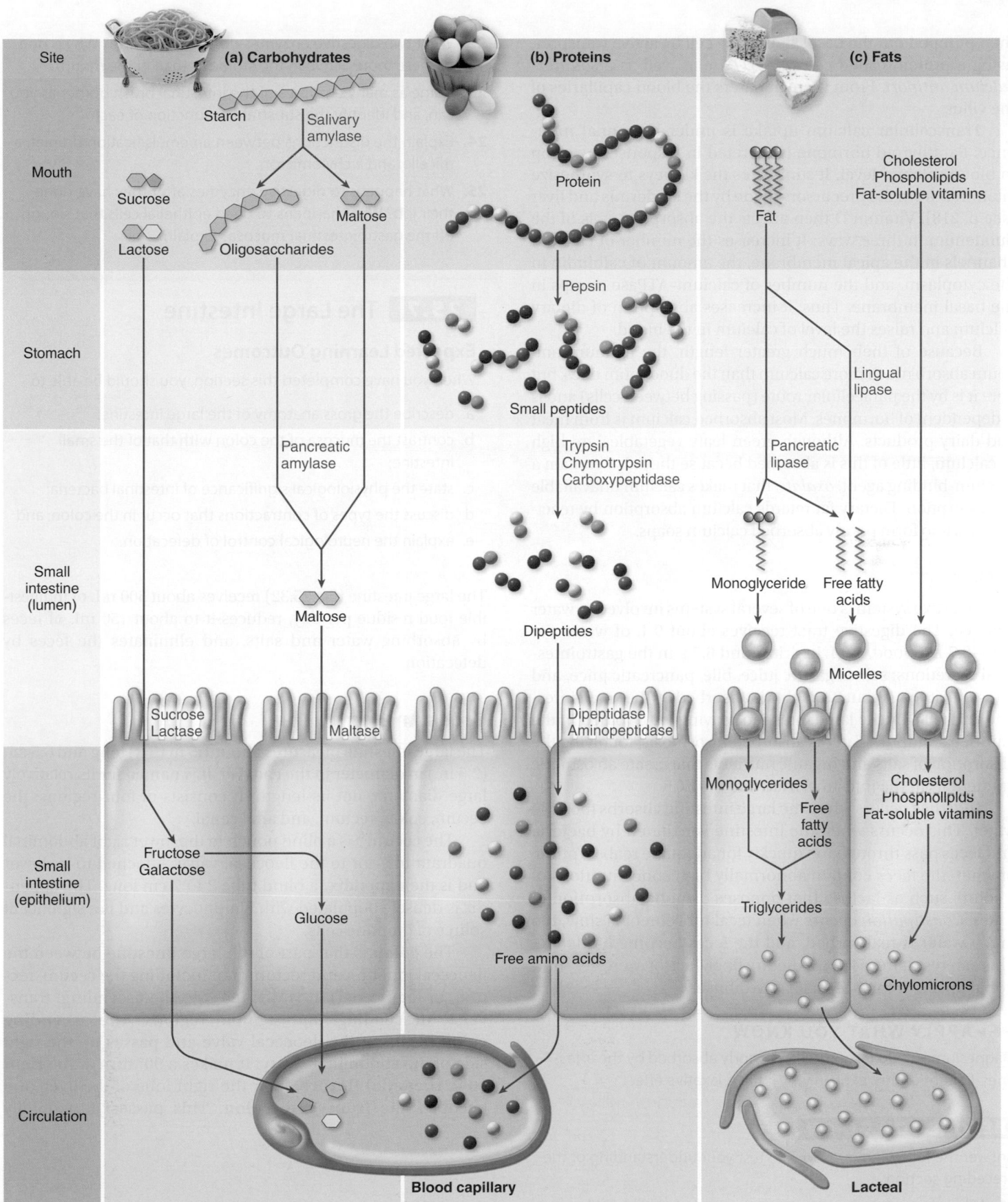

**FIGURE 25.31 Summary of the Digestion and Absorption of the Major Organic Nutrients.** Enzyme names are indicated in red.

❓ *As they leave the small intestine, sugars and amino acids arrive at the liver before any other organ, but lipids do not. Why not? See chapter 21 and trace the route that a dietary fat would have to take to reach the liver.*

then pumped out the basal side of the cell by active transport, using a protein called *calcium–ATPase* as well as a *sodium–calcium antiport.* From there, it enters the blood capillaries of the villus.

Transcellular calcium uptake is under hormonal influence. Parathyroid hormone is secreted in response to a drop in blood calcium level. It stimulates the kidneys to synthesize vitamin D from the precursors made by the epidermis and liver (see p. 218). Vitamin D then affects the absorptive cells of the duodenum in three ways: It increases the number of calcium channels in the apical membrane, the amount of calbindin in the cytoplasm, and the number of calcium–ATPase pumps in the basal membrane. Thus, it increases absorption of dietary calcium and raises the level of calcium in the blood.

Because of their much greater length, the jejunum and ileum absorb much more calcium than the duodenum does, but here it is by the paracellular route (passing between cells) and is independent of hormones. Most absorbed calcium is from meat and dairy products. Although green leafy vegetables are high in calcium, little of this is absorbed because they also contain a calcium-binding agent, *oxalate,* that makes calcium unavailable for absorption. Dietary fat retards calcium absorption by reacting with it to form poorly absorbed calcium soaps.

## Water

The digestive system is one of several systems involved in water balance. The digestive tract receives about 9 L of water per day—0.7 L in food, 1.6 L in drink, and 6.7 L in the gastrointestinal secretions: saliva, gastric juice, bile, pancreatic juice, and intestinal juice. About 8 L of this is absorbed by the small intestine and 0.8 L by the large intestine, leaving 0.2 L voided in the daily fecal output. Water is absorbed by osmosis, following the absorption of salts and organic nutrients that create an osmotic gradient from the intestinal lumen to the ECF.

*Diarrhea* occurs when the large intestine absorbs too little water. This occurs when the intestine is irritated by bacteria and feces pass through too quickly for adequate reabsorption, or when the feces contain abnormally high concentrations of a solute such as lactose that opposes osmotic absorption of water. *Constipation* occurs when fecal movement is slow, too much water is reabsorbed, and the feces become hardened. This can result from lack of dietary fiber, lack of exercise, emotional upset, or long-term laxative abuse.

### ▶▶▶ APPLY WHAT YOU KNOW

Magnesium sulfate (Epsom salt) is poorly absorbed by the intestines. In light of this, explain why it has a laxative effect.

### BEFORE YOU GO ON

Answer the following questions to test your understanding of the preceding section:

21. What three classes of nutrients are most abundant? What are the end products of enzymatic digestion of each?

22. What two digestive enzymes occur in the saliva? Why is one of these more active in the stomach than in the mouth?

23. Name as many enzymes of the intestinal brush border as you can, and identify the substrate or function of each.

24. Explain the distinctions between an emulsification droplet, a micelle, and a chylomicron.

25. What happens to digestive enzymes after they have done their job? What happens to dead epithelial cells that slough off the gastrointestinal mucosa? Explain.

## 25.7   The Large Intestine

### Expected Learning Outcomes

When you have completed this section, you should be able to

a. describe the gross anatomy of the large intestine;

b. contrast the mucosa of the colon with that of the small intestine;

c. state the physiological significance of intestinal bacteria;

d. discuss the types of contractions that occur in the colon; and

e. explain the neurological control of defecation.

The large intestine (fig. 25.32) receives about 500 mL of indigestible food residue per day, reduces it to about 150 mL of feces by absorbing water and salts, and eliminates the feces by defecation.

## Gross Anatomy

The large intestine measures about 1.5 m (5 ft) long and 6.5 cm (2.5 in.) in diameter in the cadaver. It is named for its relatively large diameter, not its length. It consists of four regions: the cecum, colon, rectum, and anal canal.

The **cecum**[32] is a blind pouch in the lower right abdominal quadrant inferior to the ileocecal valve. Attached to its lower end is the **appendix,** a blind tube 2 to 7 cm long. The appendix is densely populated with lymphocytes and is a significant source of immune cells.

The **colon** is that part of the large intestine between the ileocecal junction and rectum (not including the cecum, rectum, or anal canal). It is divided into the ascending, transverse, descending, and sigmoid regions. The **ascending colon** begins at the ileocecal valve and passes up the right side of the abdominal cavity. It makes a 90° turn at the **right colic (hepatic) flexure,** near the right lobe of the liver, and becomes the **transverse colon.** This passes horizontally

---

[32]*cec* = blind

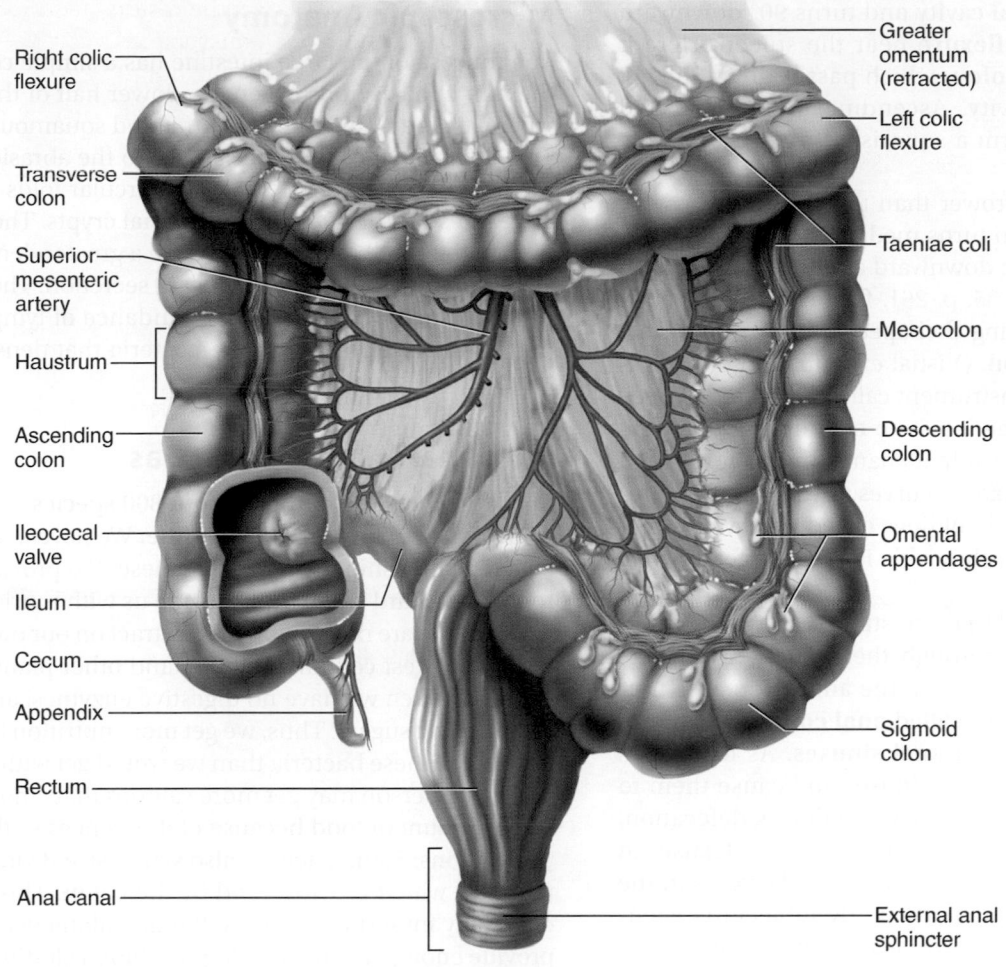

Right colic flexure

Transverse colon

Superior mesenteric artery

Haustrum

Ascending colon

Ileocecal valve

Ileum

Cecum

Appendix

Rectum

Anal canal

Greater omentum (retracted)

Left colic flexure

Taeniae coli

Mesocolon

Descending colon

Omental appendages

Sigmoid colon

External anal sphincter

**(a) Gross anatomy**

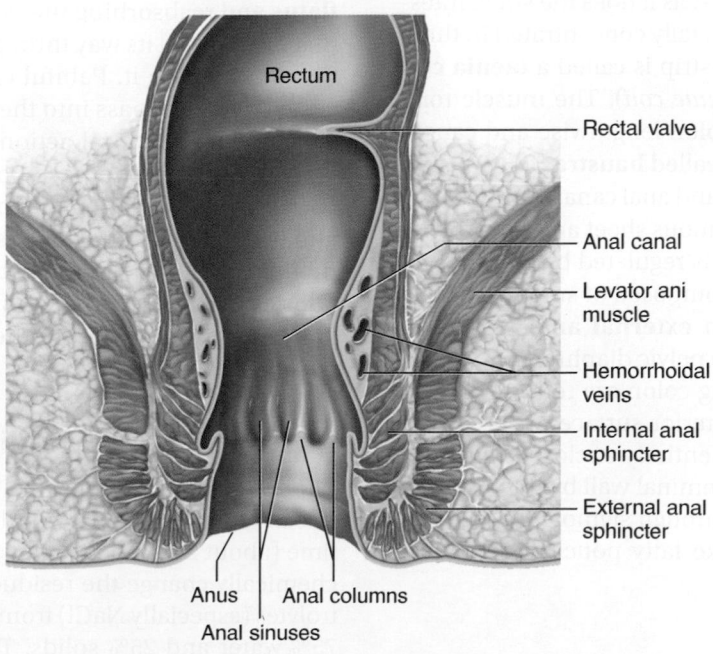

Rectum

Rectal valve

Anal canal

Levator ani muscle

Hemorrhoidal veins

Internal anal sphincter

External anal sphincter

Anus   Anal columns

Anal sinuses

**(b) Anal canal**

**FIGURE 25.32**
**The Large Intestine.** **AP|R**

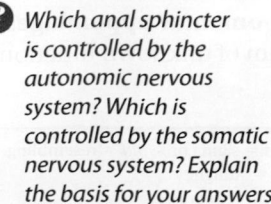

*Which anal sphincter is controlled by the autonomic nervous system? Which is controlled by the somatic nervous system? Explain the basis for your answers.*

across the upper abdominal cavity and turns 90° downward at the **left colic (splenic) flexure** near the spleen. Here it becomes the **descending colon,** which passes down the left side of the abdominal cavity. Ascending, transverse, and descending colons thus form a squarish, three-sided frame around the small intestine.

The pelvic cavity is narrower than the abdominal cavity, so at the hip bone, the colon turns medially and travels along the iliac fossa before turning downward at the pelvic inlet into the pelvic cavity. (See fig. 8.35, p. 261, for review of the skeletal landmarks.) The resulting S-shaped portion of the tract is called the **sigmoid**[33] **colon.** (Visual examination of this region is performed with an instrument called a *sigmoidoscope.*) In the pelvic cavity, the large intestine continues as the **rectum,**[34] about 15 cm long. Despite its name, the rectum is not quite straight but has three lateral curves as well as an anteroposterior curve. It has three infoldings called **transverse rectal folds (rectal valves),** which enable it to retain feces while passing gas.

The final 3 cm of the large intestine is the **anal canal** (fig. 25.32b), which passes through the levator ani muscle of the pelvic floor and terminates at the anus. Here, the mucosa forms longitudinal ridges called **anal columns** with depressions between them called **anal sinuses.** As feces pass through the canal, they press the sinuses and cause them to exude extra mucus and lubricate the canal during defecation. Prominent **hemorrhoidal veins** form superficial plexuses in the anal columns and around the orifice. Unlike veins in the limbs, they lack valves and are particularly subject to distension and venous pooling. *Hemorrhoids* are permanently distended veins that protrude into the anal canal or form bulges external to the anus.

The muscularis externa of the colon is unusual. Although it completely encircles the colon just as it does the small intestine, its longitudinal fibers are especially concentrated in three thickened, ribbonlike strips. Each strip is called a **taenia coli** (TEE-nee-ah CO-lye) (plural, *taeniae coli*). The muscle tone of the taeniae coli contracts the colon lengthwise and causes its wall to bulge, forming pouches called **haustra**[35] (HAW-stra; singular, *haustrum*). In the rectum and anal canal, however, the longitudinal muscle forms a continuous sheet and haustra are absent. The anus, like the urethra, is regulated by two sphincters: an **internal anal sphincter** composed of smooth muscle of the muscularis externa and an **external anal sphincter** composed of skeletal muscle of the pelvic diaphragm.

The ascending and descending colon are retroperitoneal and have a serosa only on the anterior surface, whereas the transverse and sigmoid colon are entirely enclosed in serosa and anchored to the posterior abdominal wall by the mesocolon. The serosa of the transverse through sigmoid colon often has **omental appendages,** clublike fatty pouches of peritoneum of unknown function.

## Microscopic Anatomy

The mucosa of the large intestine has a simple columnar epithelium in all regions except the lower half of the anal canal, where it has a nonkeratinized stratified squamous epithelium. The latter provides more resistance to the abrasion caused by the passage of feces. There are no circular folds or villi in the large intestine, but there are intestinal crypts. They are deeper than in the small intestine and have a greater density of goblet cells; mucus is their only significant secretion. The lamina propria and submucosa have an abundance of lymphatic tissue, providing protection from the bacteria that densely populate the large intestine.

## Intestinal Microbes and Gas

The large intestine harbors about 800 species of bacteria collectively called the **gut microbiome.** We have a mutually beneficial relationship with many of these. We provide them with room and board while they provide us with nutrients from our food that we are not equipped to extract on our own. For example, they digest cellulose, pectin, and other plant polysaccharides for which we have no digestive enzymes, and we absorb the resulting sugars. Thus, we get more nutrition from our food because of these bacteria than we would get without them. Indeed, one person may get more calories than another from the same amount of food because of differences in their bacterial populations. Some bacteria also synthesize B vitamins and vitamin K, which are absorbed by the colon. This vitamin K is especially important because the diet alone usually does not provide enough to ensure adequate blood clotting.

One of the less desirable and sometimes embarrassing products of these bacteria is intestinal gas. The large intestine contains about 7 to 10 L of gas, expelling about 500 mL/day as **flatus** and reabsorbing the rest. Much of this is swallowed air that has worked its way through the digestive tract, but the gut microbes add to it. Painful cramping can result when undigested nutrients pass into the colon and furnish an abnormal substrate for bacterial action—for example, in lactose intolerance. Flatus is composed of nitrogen ($N_2$), carbon dioxide ($CO_2$), hydrogen ($H_2$), methane ($CH_4$), hydrogen sulfide ($H_2S$), and two amines: indole and skatole. Indole, skatole, and $H_2S$ produce the odor of flatus and feces, whereas the others are odorless. The hydrogen gas is combustible and has been known to explode during the use of electrical cauterization in surgery.

## Absorption and Motility

The large intestine takes 36 to 48 hours to reduce the residue of a meal to feces, with the residue spending the greatest time (about 24 h) in the transverse colon. The colon does not chemically change the residue, but reabsorbs water and electrolytes (especially NaCl) from it. Feces usually consist of about 75% water and 25% solids. The solids are about 30% bacteria, 30% undigested dietary fiber, 10% to 20% fat, and smaller

---

[33]*sigm* = sigma or S; *oid* = resembling
[34]*rect* = straight
[35]*haustr* = to draw

stimulates it to contract. This churns and mixes the residue, promotes water and salt absorption, and passes the residue distally to another haustrum. Stronger contractions called **mass movements** occur one to three times a day. They last about 15 minutes and move residue several centimeters at a time. They are often triggered by the **gastrocolic** and **duodenocolic reflexes,** in which filling of the stomach and duodenum stimulates motility of the colon. Mass movements occur especially in the transverse to sigmoid colon, often within an hour after breakfast, moving the feces that accumulated and stretched the colon overnight.

## Defecation

Stretching of the rectum stimulates the defecation reflexes, which account for the urge to defecate that is often felt soon after a meal. The predictability of this response is useful in house training pets and toilet training children. The process involves two reflexes:

1. The **intrinsic defecation reflex.** This reflex is mediated entirely by the myenteric plexus. Stretch signals travel through the plexus to the muscularis of the descending and sigmoid colon and the rectum. This activates a peristaltic wave that drives feces downward, and it relaxes the internal anal sphincter. This reflex is relatively weak, however, and usually requires the cooperative action of the following reflex.

2. The **parasympathetic defecation reflex.** This is a spinal reflex. Its principal events (fig. 25.33) are that stretch signals are transmitted to the spinal cord, and motor signals return by way of the pelvic nerves to intensify peristalsis in the descending and sigmoid colon and rectum and to relax the internal anal sphincter.

These reflexes are involuntary and are the sole means of controlling defecation in infants and some people with transecting spinal cord injuries. However, the external anal sphincter, like the external urethral sphincter controlling urination, is under voluntary control, enabling one to limit defecation to appropriate circumstances. Voluntary retention of feces is also aided by the **puborectalis muscle,** which loops around the rectum like a sling and creates a sharp anorectal angle that blocks the passage of feces. Defecation normally occurs only when the external anal sphincter and puborectalis muscle are voluntarily relaxed. The kink in the rectum then straightens out and the sphincter opens to allow the feces to fall away. Defecation is also aided by the voluntary Valsalva maneuver, in which a breath hold and contraction of the abdominal muscles increase abdominal pressure, compress the rectum, and squeeze the feces from it. This maneuver can also initiate the defecation reflex by forcing feces from the descending colon into the rectum. The external anal sphincter and external urethral sphincter are controlled together by inhibitory signals from the brainstem, so as this inhibition is released, defecation is usually accompanied by urination.

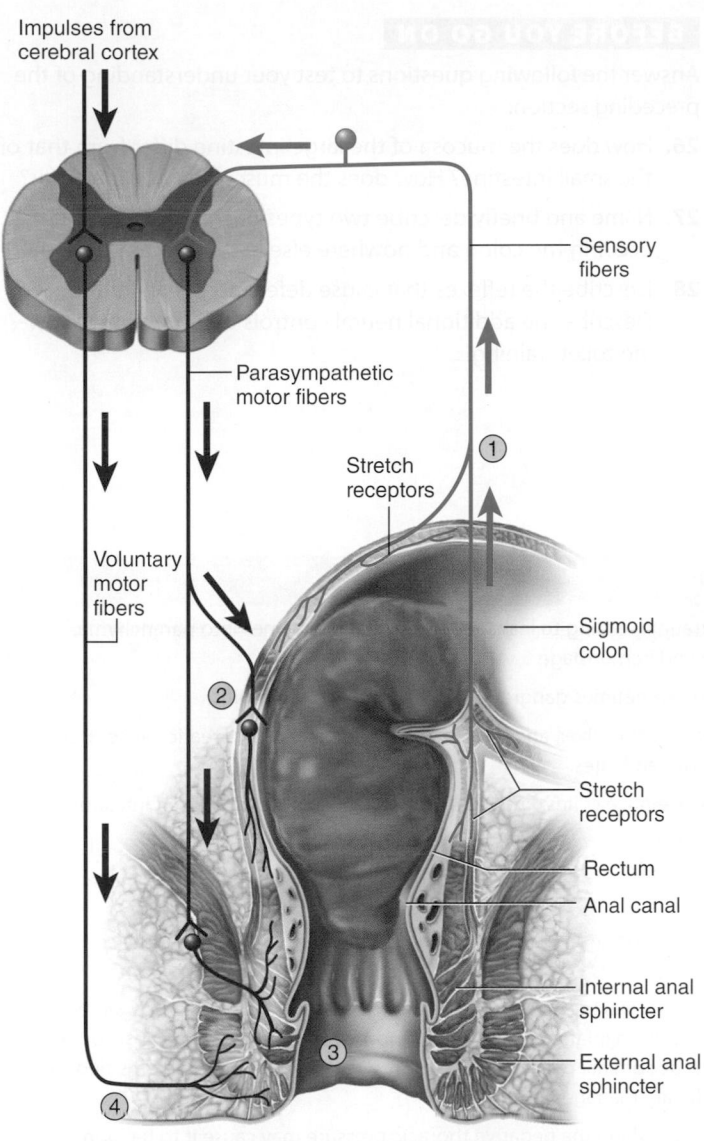

Impulses from cerebral cortex

Sensory fibers

Parasympathetic motor fibers

Stretch receptors

Voluntary motor fibers

Sigmoid colon

Stretch receptors

Rectum

Anal canal

Internal anal sphincter

External anal sphincter

① Feces stretch the rectum and stimulate stretch receptors, which transmit signals to the spinal cord.

② A spinal reflex stimulates contraction of the rectum.

③ The spinal reflex also relaxes the internal anal sphincter.

④ Impulses from the brain prevent untimely defecation by keeping the external anal sphincter contracted. Defecation occurs only if this sphincter also relaxes.

**FIGURE 25.33 Neural Control of Defecation.**

amounts of protein, sloughed epithelial cells, salts, mucus, and other digestive secretions. The fat is not from the diet but from bacteria and broken-down epithelial cells.

The most common type of colonic motility is a type of segmentation called **haustral contractions,** which occur about every 30 minutes. Distension of a haustrum with feces

If the defecation urge is suppressed, contractions cease in a few minutes and the rectum relaxes. The defecation reflexes reoccur a few hours later or when another mass movement propels more feces into the rectum.

The effects of aging on the digestive system are discussed on page 1122. Table 25.3 lists and describes some common digestive disorders.

Answer the following questions to test your understanding of the preceding section:

26. How does the mucosa of the large intestine differ from that of the small intestine? How does the muscularis externa differ?

27. Name and briefly describe two types of contractions that occur in the colon and nowhere else in the alimentary canal.

28. Describe the reflexes that cause defecation in an infant. Describe the additional neural controls that function following toilet training.

| TABLE 25.3 | Some Digestive System Diseases |
|---|---|
| Acute pancreatitis | Severe pancreatic inflammation perhaps caused by trauma leading to leakage of pancreatic enzymes into parenchyma, where they digest tissue and cause inflammation and hemorrhage. |
| Appendicitis | Inflammation of the appendix, with swelling, pain, and sometimes gangrene, perforation, and peritonitis. |
| Cancers | Malignant tumors especially of the esophagus, stomach, colon, liver, and pancreas, with colon and pancreatic cancer being among the leading causes of cancer death in the United States. |
| Crohn disease | Inflammation of small and large intestines, similar to ulcerative colitis. Produces granular lesions and fibrosis of intestine, diarrhea, and lower abdominal pain. Often hereditary. |
| Diverticulitis | Presence of inflamed herniations (outpocketings, diverticula) of the colon, associated especially with low-fiber diets. Diverticula may rupture, leading to peritonitis. |
| Dysphagia | Difficulty swallowing. Can result from esophageal obstructions (tumors, constrictions) or impaired peristalsis (due to neuromuscular disorders). |
| Gluten-sensitive enteropathy | Formerly called *sprue* or *celiac disease.* Atrophy of intestinal villi triggered in genetically susceptible individuals by *gluten,* the protein component of cereal grains. Onset is usually in infancy or early childhood. Results in severe malabsorption of most nutrients, causing watery or fatty diarrhea, abdominal pain, diminished growth, and multiple problems tied to nutritional deficiencies. Treatable with intensive dietary management. |
| Hiatal hernia | Protrusion of part of the stomach into the thoracic cavity, where the negative thoracic pressure may cause it to balloon. Often causes gastroesophageal reflux (especially when a person is supine) and esophagitis (inflammation of the esophagus). |
| Ulcerative colitis | Chronic inflammation resulting in ulceration of the large intestine, especially the sigmoid colon and rectum. Tends to be hereditary but exact causes are not well known. |

*Disorders described elsewhere*

| | | |
|---|---|---|
| Ascites pp. 677, 784 | Gingivitis p. 955 | Impacted molars p. 955 |
| Constipation p. 984 | Heartburn p. 957 | Lactose intolerance p. 979 |
| Dental caries p. 955 | Hemorrhoids p. 986 | Peptic ulcer p. 966 |
| Diarrhea p. 984 | Hepatic cirrhosis p. 1016 | Periodontal disease p. 955 |
| Gallstones p. 972 | Hepatitis p. 1016 | |

# DEEPER INSIGHT 25.5
## MEDICAL HISTORY

### The Man with a Hole in His Stomach

Perhaps the most famous episode in the history of digestive physiology began with a grave accident in 1822 on Mackinac Island in northern Michigan. Alexis St. Martin, a 28-year-old Canadian voyageur (fig. 25.34), was standing outside a trading post when he was accidentally hit by a shotgun blast from 3 feet away. A frontier Army doctor stationed at Fort Mackinac, William Beaumont, was summoned to examine St. Martin. As Beaumont later wrote, "a portion of the lung as large as a turkey's egg" protruded through St. Martin's lacerated and burnt flesh. Below that was a portion of the stomach with a puncture in it "large enough to receive my forefinger." Beaumont did his best to pick out bone fragments and dress the wound, though he did not expect St. Martin to survive.

Surprisingly, he lived. Over a period of months the wound extruded pieces of bone, cartilage, gunshot, and gun wadding. As the wound healed, a fistula (hole) remained in the stomach, so large that Beaumont had to cover it with a compress to prevent food from coming out. The opening remained, covered only by a loose flap of skin, for the rest of St. Martin's life. A fold of tissue later grew over the fistula, but it was easily opened. A year later, St. Martin was still feeble. Town authorities decided they could no longer support him on public funds and wanted to ship him 1,500 miles to his home. Beaumont, however, was imbued with a passionate sense of destiny. Very little was known about digestion, and he saw the accident as a unique opportunity to learn. He took St. Martin in at his personal expense and performed 238 experiments on him over several years. Beaumont had never attended medical school and had little idea how scientists work, yet he proved to be an astute experimenter. Under crude frontier conditions and with almost no equipment, he discovered many of the basic facts of gastric physiology discussed in this chapter.

"I can look directly into the cavity of the stomach, observe its motion, and almost see the process of digestion," Beaumont wrote. "I can pour in water with a funnel and put in food with a spoon, and draw them out again with a siphon." He put pieces of meat on a string into the stomach and removed them hourly for examination. He sent vials of gastric juice to the leading chemists of America and Europe, who could do little but report that it contained hydrochloric acid. He proved that digestion required HCl and could even occur outside the stomach, but he found that HCl alone did not digest meat; gastric juice must contain some other digestive ingredient. Theodor Schwann, one of the founders of the cell theory, identified that ingredient as pepsin. Beaumont also demonstrated that gastric juice is secreted only in response to food; it did not accumulate between meals as previously thought. He disproved the idea that hunger is caused by the walls of the empty stomach rubbing against each other.

Now disabled from wilderness travel, St. Martin agreed to participate in Beaumont's experiments in exchange for room and board—though he felt helpless and humiliated by it all. The fur trappers taunted him as "the man with a hole in his stomach," and he longed to return to his work in the wilderness. He had a wife and daughter in Canada whom he rarely got to see, and

he ran away repeatedly to join them. He was once gone for 4 years before poverty made him yield to Beaumont's financial enticement to come back. Beaumont despised St. Martin's drunkenness and profanity and was quite insensitive to his embarrassment and discomfort over the experiments. Yet St. Martin's temper enabled Beaumont to make the first direct observations of the relationship between emotion and digestion. When St. Martin was particularly distressed, Beaumont noted little digestion occurring—as we now know, the sympathetic nervous system inhibits digestive activity.

Beaumont published a book in 1833 that laid the foundation for modern gastric physiology and dietetics. It was enthusiastically received by the medical community and had no equal until Russian physiologist Ivan Pavlov (1849–1936) performed his celebrated experiments on digestion in animals. Building on the methods pioneered by Beaumont, Pavlov received the 1904 Nobel Prize for Physiology or Medicine.

In 1853, Beaumont slipped on some ice, suffered a blow to the base of his skull, and died a few weeks later. St. Martin continued to tour medical schools and submit to experiments by other physiologists, whose conclusions were often less correct than Beaumont's. Some, for example, attributed chemical digestion to lactic acid instead of hydrochloric acid. St. Martin lived in wretched poverty in a tiny shack with his wife and several children, and died 28 years after Beaumont. By then he was senile and believed he had been to Paris, where Beaumont had often promised to take him.

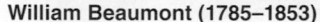
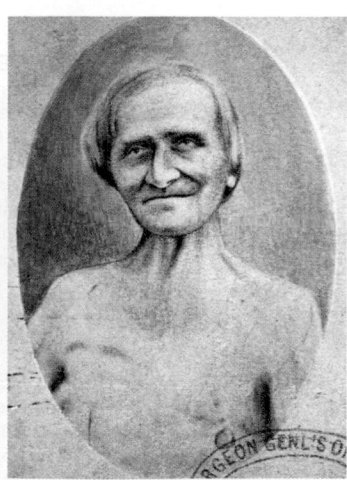

**William Beaumont (1785–1853)**    **Alexis St. Martin (1794–1880)**

**FIGURE 25.34  Doctor and Patient in a Pioneering Study of Digestion.**

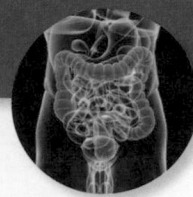

# CONNECTIVE ISSUES

## Effects of the DIGESTIVE SYSTEM
## On Other Organ Systems

**ALL SYSTEMS**
Digestion provides all systems with nutrients in a form usable for cellular metabolism and tissue growth and maintenance.

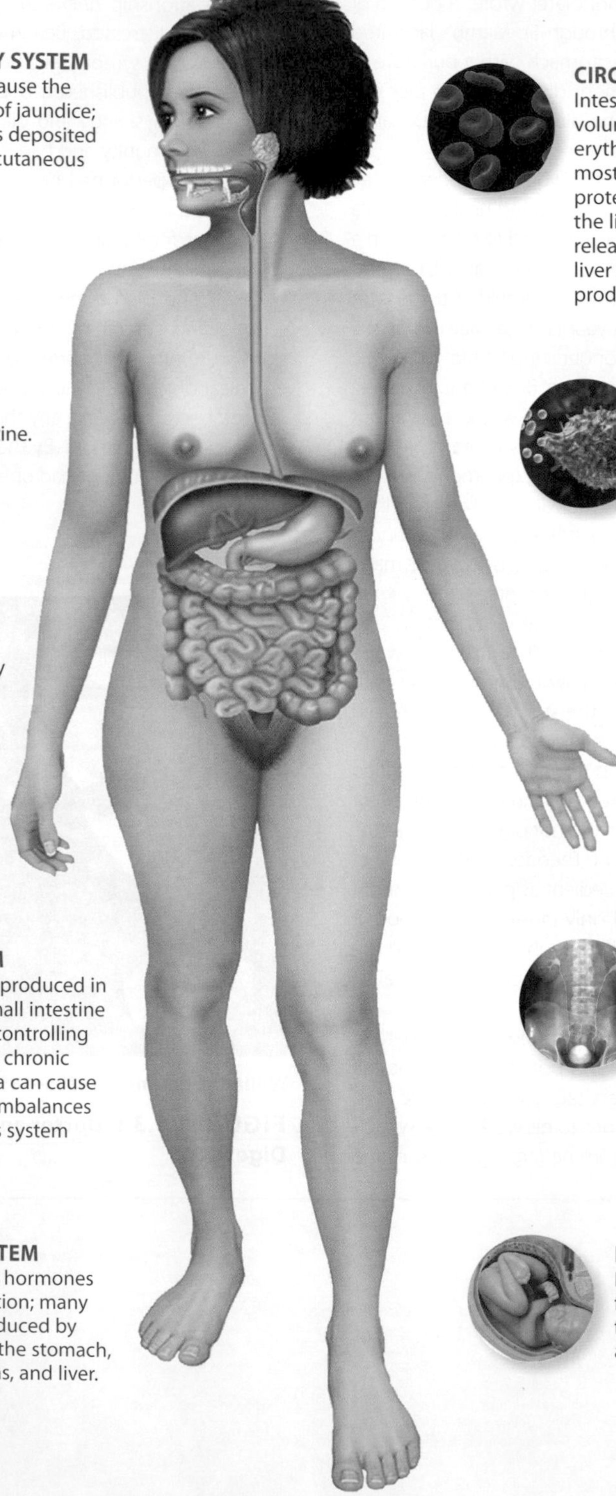

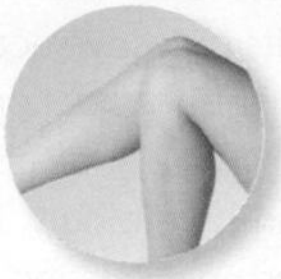

**INTEGUMENTARY SYSTEM**
Liver disease can cause the skin discoloration of jaundice; excess dietary fat is deposited in dermal and subcutaneous adipose tissue.

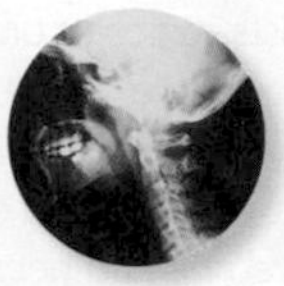

**SKELETAL SYSTEM**
Bone deposition and maintenance depend on calcium and phosphate absorption by small intestine.

**MUSCULAR SYSTEM**
The liver promotes recovery from muscle fatigue by metabolizing lactic acid generated by the muscles; the liver and intestinal epithelium store iron and release it as needed for myoglobin synthesis.

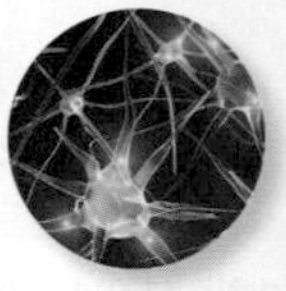

**NERVOUS SYSTEM**
Gut–brain peptides produced in the stomach and small intestine stimulate appetite-controlling centers in the brain; chronic vomiting or diarrhea can cause electrolyte and pH imbalances that lead to nervous system dysfunction.

**ENDOCRINE SYSTEM**
The liver degrades hormones and limits their action; many hormones are produced by endocrine cells of the stomach, intestines, pancreas, and liver.

**CIRCULATORY SYSTEM**
Intestinal fluid absorption supports blood volume; the liver degrades the heme from dead erythrocytes; the liver synthesizes the albumin, most blood-clotting factors, and other plasma proteins; the liver stabilizes blood glucose level; the liver and intestinal epithelium store iron and release it as needed for hemoglobin synthesis; the liver secretes erythropoietin, which stimulates RBC production.

**LYMPHATIC/IMMUNE SYSTEM**
The small intestine is the source of lipids transported by lymphatic vessels; the intestinal mucosa is a major source of lymphocytes; acid, enzymes, and lysozyme provide nonspecific defense against ingested pathogens; infants acquire passive immunity by intestinal absorption of IgA from breast milk.

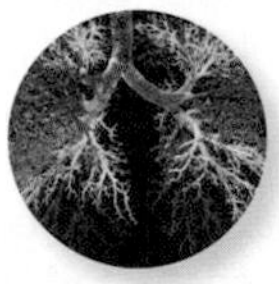

**RESPIRATORY SYSTEM**
Contraction of the abdominal muscles pushes the stomach against the diaphragm and aids in forced expiration.

**URINARY SYSTEM**
The liver synthesizes the urea excreted by the kidneys; this urea also contributes to the osmotic gradient that enables the kidneys to concentrate urine; intestines complement the kidneys in water and electrolyte reabsorption.

**REPRODUCTIVE SYSTEM**
Digestion provides nutrients for fetal growth; certain aspects of egg fertilization depend on calcium absorbed by the small intestine.

# STUDY GUIDE

## ► Assess Your Learning Outcomes

*To test your knowledge, discuss the following topics with a study partner or in writing, ideally from memory.*

### 25.1 General Anatomy and Digestive Processes (p. 948)

1. Three functions of the digestive system and the five stages in which it carries these out
2. The difference between the digestive tract and the accessory organs of the digestive system; the organs that belong in each category
3. Tissue layers typical of most regions of the digestive tract
4. Functions of the enteric nervous system; its two subdivisions, their locations, and their respective functions
5. Functions of the mesenteries and their relationship to the abdominal digestive organs
6. Hormones, paracrines, and visceral reflexes that regulate motility and secretion in the digestive tract

### 25.2 The Mouth Through Esophagus (p. 952)

1. Seven functions of the oral cavity
2. Anatomical boundaries of the mouth; variations in its mucosal epithelium; and all digestive system organs contained in it
3. Anatomy of the cheeks and lips; the three regions of the lip
4. Anatomy and functions of the tongue; what forms the border between the body and root of the tongue; its intrinsic and extrinsic muscles; the lingual glands and lingual tonsils
5. Anatomy of the hard and soft palates; the two arches that mark the border between the oral cavity and pharynx
6. The structure of a typical tooth and periodontal tissues; the four kinds of teeth and number and position of each; the mode of replacement of deciduous teeth by permanent teeth; and the functions of mastication
7. Six functions of saliva; its composition and pH; general histology of salivary glands; names and function of the intrinsic salivary glands; the three pairs of extrinsic salivary glands; and how the nervous system regulates salivation

8. Anatomy of the pharynx; the pharyngeal constrictor muscles and upper esophageal sphincter, and the functions of these muscles
9. Gross anatomy and histology of the esophagus; the distribution of skeletal and smooth muscle in the esophageal wall; the location and function of the esophageal glands; the cardiac orifice and lower esophageal sphincter
10. The physiology of swallowing; the swallowing center and the cranial nerves involved in the process; what occurs in the buccal and pharyngoesophageal phases of swallowing; how peristalsis is controlled; and to what extent swallowing depends on peristalsis or occurs independently of it

### 25.3 The Stomach (p. 959)

1. Anatomy and functions of the stomach; features that mark its beginning and end; and the volume of the empty stomach and its full capacity
2. Innervation of the stomach; its blood supply and relation to the hepatic portal system
3. Structure of the gastric mucosa including the gastric pits; the glands that open into them; and differences in the spatial distribution and functions of gastric, cardiac, and pyloric glands; and five cell types of these glands and their respective functions
4. The composition of gastric juice
5. The cells that secrete hydrochloric acid, how they do so, and the functions of the acid
6. The difference between a zymogen and an active enzyme
7. The source of pepsinogen; how it is converted to pepsin; and the function of pepsin
8. The source and function of gastric lipase
9. The source and function of intrinsic factor; the effect of hyposecretion of intrinsic factor; and how a deficiency of intrinsic factor is treated
10. Hormones and paracrine secretions that regulate gastrointestinal function; why some of these are called gut–brain peptides

11. The nature and functions of the receptive-relaxation response and peristalsis in the stomach; the reason very little chyme is passed into the duodenum at a time
12. The degree of digestion that occurs in the stomach; what is absorbed by the stomach and what is not
13. How the stomach is protected from its own acid and enzymes
14. How gastric activity is controlled; the regulatory mechanisms of the cephalic, gastric, and intestinal phases

### 25.4 The Liver, Gallbladder, and Pancreas (p. 968)

1. The location, gross anatomy, and digestive function of the liver
2. Structure and spatial arrangement of the hepatic lobules
3. The route of blood flow through the liver
4. Composition and functions of the bile, the route of bile flow from the hepatocytes to the duodenum; and the recycling of bile acids and its relationship to the elimination of cholesterol from the body
5. Structure and function of the gallbladder, and its connection to the bile duct
6. Location, gross anatomy, and digestive functions of the pancreas
7. Structure of the pancreatic acini, the duct system, and its connection to the duodenum
8. Composition and digestive functions of pancreatic juice; the names and functions of its digestive zymogens and enzymes
9. Hormones that regulate the secretion of bile and pancreatic juice

### 25.5 The Small Intestine (p. 974)

1. Structures that mark the beginning and end of the small intestine; the three regions of the small intestine and their respective lengths and histological differences
2. How the small intestine is protected from the erosive effect of stomach acid
3. The importance of surface area for the function of the small intestine, and four features that give it a large surface area
4. Histology of the intestinal villi and crypts; the cell types found in each, and

# STUDY GUIDE

their respective functions; the relationship of the lacteal to the lymphatic system, and its function

5. Brush border enzymes of the small intestine and their functions
6. Two types of intestinal motility and their functional difference

## 25.6 Chemical Digestion and Absorption (p. 977)

1. Steps in carbohydrate digestion from the mouth to small intestine; the enzymes involved at each step and their respective contributions to carbohydrate hydrolysis
2. Mechanisms of monosaccharide absorption by the intestinal mucosa
3. Steps in protein digestion from the stomach to small intestine; the enzymes involved at each step and their respective contributions to peptide hydrolysis
4. Mechanisms of amino acid absorption by the intestinal mucosa

5. Steps in fat digestion from the stomach to small intestine; the enzymes involved at each step and their respective contributions to peptide hydrolysis
6. The necessity of emulsification for efficient fat digestion; the role of bile acids and lecithin in this process
7. Mechanisms of lipid absorption by the intestinal mucosa; the role of the lacteals; and how and why lipid absorption and transport differ from the absorption and transport of sugars and amino acids
8. Differences between emulsification droplets, micelles, and chylomicrons in lipid processing
9. Digestion of DNA and RNA in the small intestine, and the mode of absorption of their decomposition products
10. Nutrients that are not digested but simply absorbed; special aspects of the absorption of fat-soluble vitamins and vitamin $B_{12}$

11. Modes of absorption of water and minerals, particularly $Na^+$, $K^+$, $Cl^-$, $Fe^{2+}$, and $Ca^{2+}$; the roles of hepcidin, calcitriol, and parathyroid hormone in iron and calcium absorption

## 25.7 The Large Intestine (p. 984)

1. Gross anatomy, histology, and functions of the large intestine, including its six segments and total length
2. The gut microbiome of the colon; the composition and sources of flatus
3. Control of colonic motility; differences between its haustral contractions and mass movements; and the role of gastrocolic and duodenocolic reflexes
4. Mechanisms of the intrinsic and parasympathetic defecation reflexes, and of voluntary control over defecation; the neuroanatomy involved in these control mechanisms; and how the Valsalva maneuver can trigger and aid defecation

## ▶ Testing Your Recall

*Answers in Appendix B*

1. Which of the following enzymes acts in the stomach?
   a. chymotrypsin
   b. lingual lipase
   c. carboxypeptidase
   d. enterokinase
   e. dextrinase

2. Which of the following enzymes does *not* digest any nutrients?
   a. chymotrypsin
   b. lingual lipase
   c. carboxypeptidase
   d. enterokinase
   e. dextrinase

3. Which of the following is *not* an enzyme?
   a. chymotrypsin
   b. enterokinase
   c. secretin
   d. pepsin
   e. nucleosidase

4. The substance in question 3 that is *not* an enzyme is
   a. a zymogen.
   b. a nutrient.
   c. an emulsifier.
   d. a neurotransmitter.
   e. a hormone.

5. The lacteals absorb
   a. chylomicrons.
   b. micelles.
   c. emulsification droplets.
   d. amino acids.
   e. monosaccharides.

6. All of the following contribute to the absorptive surface area of the small intestine *except*
   a. its length.
   b. the brush border.
   c. haustra.
   d. circular folds.
   e. villi.

7. Which of the following is a periodontal tissue?
   a. the gingiva
   b. the enamel
   c. the cementum
   d. the pulp
   e. the dentin

8. Anatomically, the _____ of the stomach most closely resemble the _____ of the small intestine.
   a. gastric pits, intestinal crypts
   b. pyloric glands, intestinal crypts
   c. rugae, Peyer patches

   d. parietal cells, goblet cells
   e. gastric glands, duodenal glands

9. Which of the following cells secrete digestive enzymes?
   a. chief cells
   b. mucous neck cells
   c. parietal cells
   d. goblet cells
   e. enteroendocrine cells

10. What phase of gastric regulation includes inhibition by the enterogastric reflex?
    a. the intestinal phase
    b. the gastric phase
    c. the buccal phase
    d. the cephalic phase
    e. the pharyngoesophageal phase

11. Cusps are a feature of the _____ surfaces of the molars and premolars.

12. The acidity of the stomach halts the action of _____ but promotes the action of _____, both of which are salivary enzymes.

13. The _____ salivary gland is named for its proximity to the ear.

# STUDY GUIDE

14. The submucosal and myenteric plexuses collectively constitute the _____ nervous system.

15. Nervous stimulation of gastrointestinal activity is mediated mainly through the parasympathetic fibers of the _____ nerves.

16. Food in the stomach causes G cells to secrete _____, which in turn stimulates the secretion of HCl and pepsinogen.

17. Hepatic macrophages occur in blood-filled spaces of the liver called _____.

18. The brush border enzyme that finishes the job of starch digestion, producing glucose, is called _____. Its substrate is _____.

19. Fats are taken into the lymphatic capillaries as droplets called _____.

20. Within the absorptive cells of the small intestine, ferritin binds the nutrient _____.

## ► Building Your Medical Vocabulary

*Answers in Appendix B*

*State a medical meaning of each word element below, and give a term in which it or a slight variation of it is used.*

1. antro-

2. chylo-

3. -elle

4. emet-

5. freno-

6. hepato-

7. jejuno-

8. porto-

9. pyloro-

10. sigmo-

## ► True or False

*Answers in Appendix B*

*Determine which five of the following statements are false, and briefly explain why.*

1. Fat is not digested until it reaches the duodenum.

2. A tooth is composed mostly of enamel.

3. Hepatocytes secrete bile into the hepatic sinusoids.

4. Cholecystokinin stimulates the release of bile into the duodenum.

5. Peristalsis is controlled by the myenteric nerve plexus.

6. Pepsinogen, trypsinogen, and procarboxypeptidase are enzymatically inactive zymogens.

7. The absorption of dietary iron depends on intrinsic factor.

8. Filling of the stomach stimulates contractions of the colon.

9. The duodenum secretes a hormone that inhibits contractions of the stomach.

10. Tight junctions of the small intestine prevent anything from leaking between the epithelial cells.

## ► Testing Your Comprehension

*Answers at www.mhhe.com/saladin7*

1. On Monday, David has a waffle with syrup for breakfast at 7:00, and by 10:30 he feels hungry again. On Tuesday, he has bacon and eggs and feels satisfied until his noon lunch hour approaches. Assuming an equal quantity of food on both days and all other factors to be equal, explain this difference.

2. Which of these do you think would have the most severe effect on digestion: surgical removal of the stomach, gallbladder, or pancreas? Explain.

3. What do carboxypeptidase and aminopeptidase have in common? Identify as many differences between them as you can.

4. What do micelles and chylomicrons have in common? Identify as many differences between them as you can.

5. Explain why most dietary lipids must be absorbed by the lacteals rather than by the blood capillaries of a villus.

## ► Improve Your Grade at www.mhhe.com/saladin7

*Download animations to study when it fits your schedule. Practice quizzes, labeling activities, games, and flashcards offer fun ways to master the chapter concepts. Or, download image PowerPoint files for each chapter to create a study guide or for taking notes during lecture.*

# CHAPTER 26

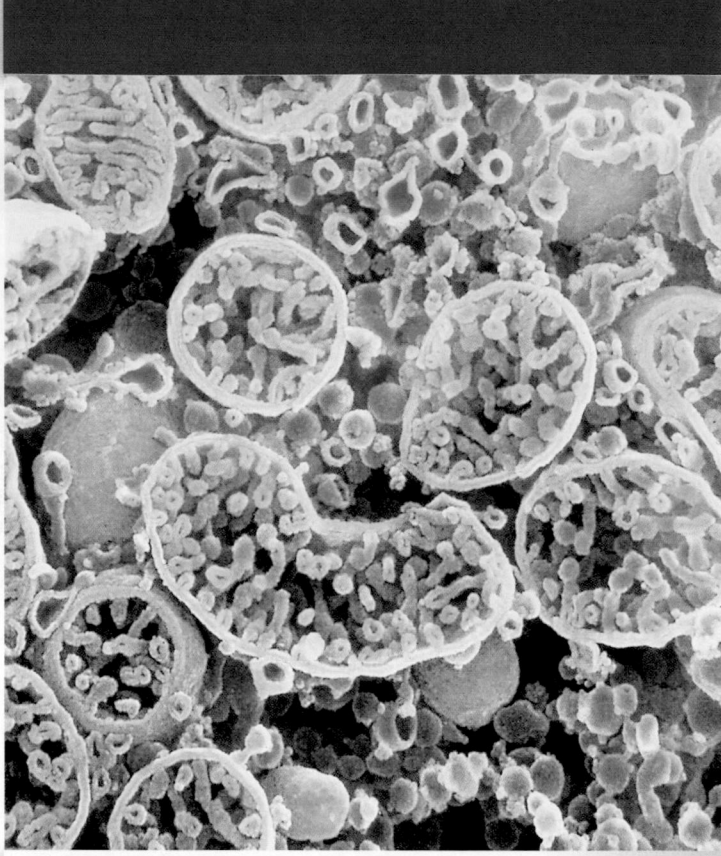

Mitochondria (green) and smooth endoplasmic reticulum in a cell of the ovary (SEM)

CHAPTER

# 26

# NUTRITION AND METABOLISM

Anatomy & Physiology | REVEALED®

aprevealed.com

**Module 12: Digestive System**

## BRUSHING UP

- Understanding metabolism requires prior familiarity with oxidation, reduction, catabolism, and anabolism (pp. 56–57), and metabolic pathways (p. 70).

- You should be familiar with the chemistry of cholesterol and saturated and unsaturated fats (pp. 61–64) before reading on LDLs, HDLs, and other blood lipoproteins in this chapter.

- Lipoprotein processing involves receptor-mediated endocytosis, introduced at page 96.

- The catabolic reactions discussed in this chapter are predominantly ways of making ATP. You must understand the structure and functions of ATP (p. 70).

- Insulin and glucagon are particularly important in regulating metabolism; brush up on these at pages 645–646 if necessary.

- In the regulation of body temperature, heat transfer follows the principle of flow down thermal gradients introduced on page 18.

Nutrition is the starting point and basis for all human form and function. From the time a single-celled, fertilized egg divides in two, nutrition provides the matter needed for cell division, growth, and development. It is the source of fuel that provides the energy for all biological work and of the raw materials for replacement of worn-out biomolecules and cells. The fact that it provides only the raw materials means, further, that chemical change—metabolism—lies at the foundation of form and function. In chapter 25, we saw how the digestive system breaks nutrients down into usable form and absorbs them into the blood and lymph. We now consider these nutrients in more depth, follow their fate after absorption, and explore related issues of metabolism and body heat.

## 26.1 Nutrition

### Expected Learning Outcomes

When you have completed this section, you should be able to

a. describe some factors that regulate hunger and satiety;

b. define *nutrient* and list the six major categories of nutrients;

c. state the function of each class of macronutrients, the approximate amounts required in the diet, and some major dietary sources of each;

d. name the blood lipoproteins, state their functions, and describe how they differ from each other; and

e. name the major vitamins and minerals required by the body and the general functions they serve.

## Body Weight and Energy Balance

The subject of nutrition quickly brings to mind the subject of body weight and the popular desire to control it. Weight is determined by one's energy balance—if energy intake and output are equal, body weight is stable. We gain weight if intake exceeds output and lose weight if output exceeds intake. Body weight usually remains quite stable over many years' time and seems to have a homeostatic set point. This has been experimentally demonstrated in animals. If an animal is force-fed until it becomes obese and then allowed to feed at will, it voluntarily reduces intake and soon stabilizes at its former weight. Similarly, if an animal is undernourished until it loses much of its weight and then allowed to feed at will, it increases its intake and again quickly stabilizes at its former weight.

In humans, the set point varies greatly from person to person, and body weight results from a combination of hereditary and environmental influences. From studies of identical twins and other people, it appears that about 30% to 50% of the variation in human weight is due to heredity, and the rest to environmental factors such as eating and exercise habits.

## Appetite

The struggle for weight control often seems to be a struggle against the appetite. Since the early 1990s, physiologists have discovered a still-growing list of peptide hormones and regulatory pathways that control short- and long-term appetite and body weight. Some of the hormones have been called *gut-brain peptides* because they act as chemical signals from the gastrointestinal tract to the brain. A few will be described here to give some idea of regulatory mechanisms known to date and where a great deal of research is currently focused.

### Short-Term Regulators of Appetite

The following three peptides work over periods of minutes to hours, making one feel hungry and begin eating, then feel satiated and end a meal:

- **Ghrelin.**[1] This is secreted by parietal cells in the gastric fundus, especially when the stomach is empty. It produces the sensation of hunger and stimulates the hypothalamus to secrete growth hormone–releasing hormone, priming the body to take advantage of the nutrients about to be absorbed. Within an hour after eating, ghrelin secretion ceases.

- **Peptide YY (PYY).** This is a member of a family of hormones related to *neuropeptide Y (NPY)*. It is secreted by enteroendocrine cells in the ileum and colon, but they sense that food has arrived even as it enters the stomach. They secrete PYY long before the chyme reaches the ileum, and in quantities proportional to the calories consumed. The primary effect of PYY is to signal satiety and terminate eating. Thus, ghrelin is one of the signals that begins a meal, and PYY is one of the signals that ends it. PYY remains elevated well after a meal. It acts as an *ileal brake* that prevents the stomach from emptying too quickly, and therefore prolongs the sense of satiety.

---

[1]named partly from *ghre* = growth, and partly as an acronym derived from growth hormone–releasing hormone

- **Cholecystokinin (CCK).**  As we saw in chapter 25, CCK is secreted by enteroendocrine cells in the duodenum and jejunum. It stimulates the secretion of bile and pancreatic enzymes, but also stimulates the brain and sensory fibers of the vagus nerves, producing an appetite-suppressing effect. Thus, it joins PYY as a signal to stop eating.

## Long-Term Regulators of Appetite

Other peptides regulate appetite, metabolic rate, and body weight over the longer term, thus governing one's average rate of caloric intake and energy expenditure over periods of weeks to years. The following two members of this group work as "adiposity signals," informing the brain of how much adipose tissue the body has and activating mechanisms for adding or reducing fat.

- **Leptin.**[2]  Leptin is secreted by adipocytes throughout the body. Its level is proportional to one's fat stores, so this is the brain's primary way of knowing how much body fat we have. Animals with a leptin deficiency or a defect in leptin receptors exhibit *hyperphagia* (overeating) and extreme obesity. With few exceptions, however, obese humans are not leptin-deficient or aided by leptin injections. More commonly, it seems that obesity is linked to unresponsiveness to leptin—a receptor defect rather than a hormone deficiency. Adipose tissue is increasingly seen as an important source of multiple hormones that influence the body's energy balance.

- **Insulin.**  As we saw in chapter 17, insulin is secreted by the pancreatic beta cells. It stimulates glucose and amino acid uptake and promotes glycogen and fat synthesis. But it also has receptors in the brain and functions, like leptin, as an index of the body's fat stores. It has a weaker effect on appetite than leptin does, however.

An important brain center for appetite regulation is the **arcuate nucleus** of the hypothalamus. All five of the aforementioned peptides have receptors in the arcuate nucleus, although they act on other target cells in the body as well. The arcuate nucleus has two neural networks involved in hunger. One group secretes **neuropeptide Y (NPY),** itself a potent appetite stimulant. The other secretes **melanocortin,** which inhibits eating. Ghrelin stimulates NPY secretion, whereas insulin, PYY, and leptin inhibit it. Leptin also stimulates melanocortin secretion (fig. 26.1) and inhibits the secretion of appetite stimulants called *endocannabinoids,* named for their resemblance to the tetrahydrocannabinol (THC) of marijuana.

▶▶▶**APPLY WHAT YOU KNOW**

*A friend encourages you to invest in a company that proposes to produce tablets of leptin and CCK to be taken as oral diet pills. Would you consider this a wise investment? Why or why not?*

---

[2]*lept* = thin

## DEEPER INSIGHT 26.1
## CLINICAL APPLICATION

### Obesity

*Obesity* is clinically defined as a weight more than 20% above the recommended norm for one's age, sex, and height. In the United States, about 30% of the population is obese and another 35% is overweight; there has lately been an alarming increase in the number of children who are morbidly obese by the age of 10. You can judge whether you are overweight or obese by calculating your *body mass index (BMI).* If W is your weight in kilograms and H is your height in meters, BMI = $W/H^2$. (Or if using weight in pounds and height in inches, BMI = $703W/H^2$.) A BMI of 20 to 25 $kg/m^2$ is considered to be optimal for most people. A BMI over 27 $kg/m^2$ is considered overweight, and above 30 $kg/m^2$ is considered obese.

Excess weight shortens life expectancy and increases a person's risk of atherosclerosis, hypertension, diabetes mellitus, joint pain and degeneration, kidney stones, and gallstones; cancer of the breast, uterus, and liver in women; and cancer of the colon, rectum, and prostate gland in men. Excess thoracic fat impairs breathing and results in increased blood $P_{CO_2}$, sleepiness, and reduced vitality. Obesity is also a significant impediment to successful surgery.

Heredity plays as much role in obesity as in height, and even more than in many other disorders generally acknowledged to be hereditary. However, a predisposition to obesity is often greatly worsened by overfeeding in infancy and childhood. Consumption of excess calories in childhood causes adipocytes to increase in size and number. In adulthood, adipocytes do not multiply except in some extreme weight gains; their number remains constant while weight gains and losses result from changes in cell size (cellular hypertrophy).

As so many dieters learn, it is very difficult to substantially reduce one's adult weight. Most diets are unsuccessful over the long run as dieters lose and regain the same weight over and over. From an evolutionary standpoint, this is not surprising. The body's appetite- and weight-regulating mechanisms have evolved more to limit weight loss than weight gain, for a scarcity of food was surely a more common problem than a food surplus for our prehistoric ancestors. Were it not for the mechanisms that thwart weight loss, our ancestors might not have made it through the lean eons and we might not be here; but now that we are surrounded with a glut of tempting food, these survival mechanisms have become mechanisms of pathology.

Understandably, pharmaceutical companies are keenly interested in developing effective weight-control drugs. There could be an enormous profit, for example, in a drug that inhibits ghrelin signaling or enhances or mimics leptin or melanocortin signaling. Such efforts have so far met with little success, but clearly a prerequisite to drug development is a better understanding of appetite-regulating peptides and their receptors. This subject is generating an abundant literature, and undoubtedly much more will be known about it before this chapter even gets to the printing press.

---

Gut–brain peptides certainly are not the whole story behind appetite regulation. Hunger is also stimulated partly by gastric peristalsis. Mild **hunger contractions** begin soon after the stomach is emptied and increase in intensity over a period of hours. They can become quite a painful and powerful incentive to eat, yet they do not affect the amount of food

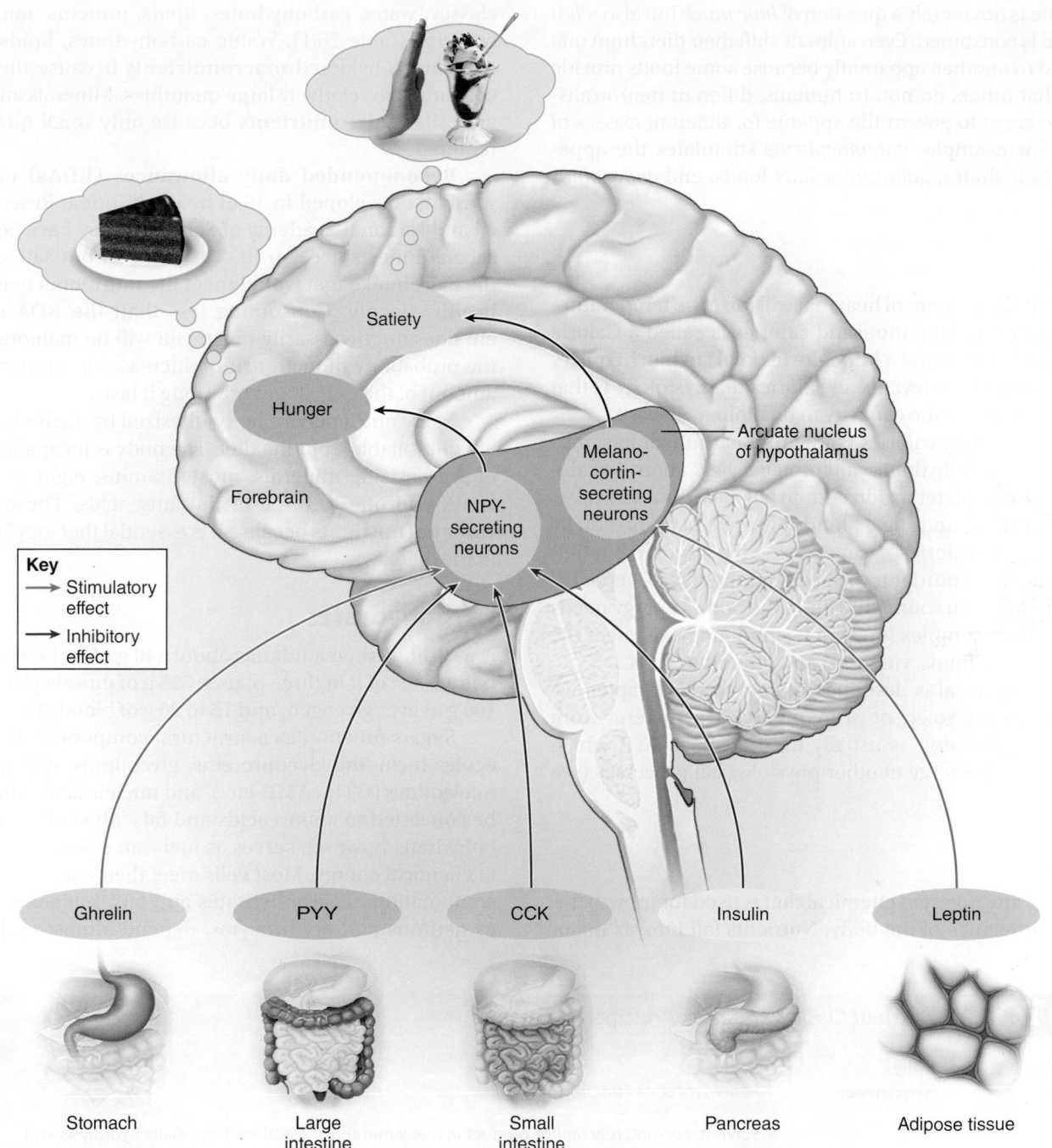

**FIGURE 26.1 Principal Pathways of Appetite Regulation by Gut–Brain Peptides.** Tissues and organs at the bottom of the figure are sources of peptides that stimulate or inhibit appetite-regulating neurons in the arcuate nucleus of the hypothalamus. Depending on the balance of stimulation and inhibition, those neurons secrete NPY or melanocortin to create a conscious sensation of hunger or satiety, respectively (PYY = peptide YY; CCK = cholecystokinin; NPY = neuropeptide Y).

consumed—this remains much the same even when nervous connections to the stomach and intestines are severed to cut off all conscious perception of hunger contractions. Food intake is terminated not only by PYY and CCK, but also in ways similar to the way that water intake slakes the thirst (see p. 927). Merely chewing and swallowing food briefly satisfy the appetite, even

if the food is removed through an esophageal fistula (opening) before reaching the stomach. Inflating the stomach with a balloon inhibits hunger even in an animal that has not actually swallowed any food. Satiation produced by these mechanisms, however, is very short-lived. Lasting satiation depends on the absorption of nutrients into the blood.

Appetite is not merely a question of *how much* but also *what kind* of food is consumed. Even animals shift their diets from one kind of food to another, apparently because some foods provide nutrients that others do not. In humans, different neurotransmitters also seem to govern the appetite for different classes of nutrients. For example, *norepinephrine* stimulates the appetite for carbohydrates, *galanin* for fatty foods, and *endorphins* for protein.

## Calories

One calorie is the amount of heat that will raise the temperature of 1 g of water 1°C. One thousand calories is called a Calorie (capital *C*) in dietetics and a **kilocalorie (kcal)** in biochemistry and physiology. The relevance of calories to physiology is that they are a measure of the capacity to do biological work.

Nearly all dietary calories come from carbohydrates, proteins, and fats. Carbohydrates and proteins yield about 4 kcal/g when they are completely oxidized, and fats yield about 9 kcal/g. Alcohol (7.1 kcal/g) and sugary foods promote malnutrition by providing "empty calories"—suppressing the appetite but failing to provide other nutrients the body requires (see Deeper Insight 26.4, p. 1022). In sound nutrition, the body's energy needs are met by more complex foods that simultaneously meet the need for proteins, lipids, vitamins, and other nutrients.

When a chemical is described as **fuel** in this chapter, we mean it is oxidized solely or primarily to extract energy from it. The extracted energy is usually used to make ATP, which then transfers the energy to other physiological processes (see fig. 2.31, p. 72).

## Nutrients

A **nutrient** is any ingested chemical that is used for growth, repair, or maintenance of the body. Nutrients fall into six major classes: water, carbohydrates, lipids, proteins, minerals, and vitamins (table 26.1). Water, carbohydrates, lipids, and proteins are considered **macronutrients** because they must be consumed in relatively large quantities. Minerals and vitamins are called **micronutrients** because only small quantities are required.

**Recommended daily allowances (RDAs)** of nutrients were first developed in 1943 by the National Research Council and National Academy of Sciences; they have been revised several times since. An RDA is a liberal but safe estimate of the daily intake that would meet the nutritional needs of most healthy people. Consuming less than the RDA of a nutrient does not necessarily mean you will be malnourished, but the probability of malnutrition increases in proportion to the amount of the deficit and how long it lasts.

Many nutrients can be synthesized by the body when they are unavailable from the diet. The body is incapable, however, of synthesizing minerals, most vitamins, eight of the amino acids, and one to three of the fatty acids. These are called **essential nutrients** because it is essential that they be included in the diet.

## Carbohydrates

A well-nourished adult has about 440 g of carbohydrate in the body, most of it in three places: 325 g of muscle glycogen, 90 to 100 g of liver glycogen, and 15 to 20 g of blood glucose.

Sugars function as a structural component of other molecules including glycoproteins, glycolipids, ATP and related nucleotides (GTP, cAMP, etc.), and nucleic acids; and they can be converted to amino acids and fats. Most of the body's carbohydrate, however, serves as fuel—an easily oxidized source of chemical energy. Most cells meet their energy needs from a combination of carbohydrates and fats, but some cells, such as neurons and erythrocytes, depend almost exclusively on

| TABLE 26.1 | Nutrient Classes and Their Principal Functions | |
|---|---|---|
| **Nutrient** | **Daily Requirement** | **Representative Functions** |
| Water | 2.5 L | Solvent; coolant; reactant or product in many metabolic reactions (especially hydrolysis and condensation); dilutes and eliminates metabolic wastes; supports blood volume and pressure |
| Carbohydrates | 125–175 g | Fuel; a component of nucleic acids, ATP and other nucleotides, glycoproteins, and glycolipids |
| Lipids | 80–100 g | Fuel; plasma membrane structure; myelin sheaths of nerve fibers; hormones; eicosanoids; bile acids; insulation; protective padding around organs; absorption of fat-soluble vitamins; vitamin D synthesis; some blood-clotting factors |
| Proteins | 44–60 g | Muscle contraction; ciliary and flagellar motility; structure of cellular membranes and extracellular material; enzymes; major component of connective tissues; transport of plasma lipids; some hormones; oxygen binding and transport pigments; blood-clotting factors; blood viscosity and osmolarity; antibodies; immune recognition; neuromodulators; buffers; emergency fuel |
| Minerals | 0.05–3,300 mg | Structure of bones and teeth; component of some structural proteins, hormones, ATP, phospholipids, and other chemicals; cofactors for many enzymes; electrolytes; oxygen transport by hemoglobin and myoglobin; buffers; stomach acid; osmolarity of body fluids |
| Vitamins | 0.002–60 mg | Coenzymes for many metabolic pathways; antioxidants; component of visual pigment; one hormone (vitamin D) |

carbohydrates. Even a brief period of **hypoglycemia**[3] (deficiency of blood glucose) causes nervous system disturbances felt as weakness or dizziness.

Blood glucose concentration is therefore carefully regulated, mainly through the interplay of insulin and glucagon (see chapter 17 and later in this chapter). Among other effects, these hormones regulate the balance between glycogen and free blood glucose. If blood glucose concentration drops too low, the body draws on its stores of glycogen to meet its energy needs. If glycogen stores are depleted, physical endurance is greatly reduced. Thus, it is important to consume enough carbohydrate to ensure that the body maintains adequate stores of glycogen for periods of exercise and fasting (including sleep).

Carbohydrate intake also influences the metabolism of other nutrients. When glucose and glycogen levels are too low to meet our energy needs, we oxidize fat as fuel; conversely, excess carbohydrate is converted to fat. This is why the consumption of starchy and sugary foods has a pronounced effect on body weight. It is unwise, however, to try to "burn off fat" by excessively reducing carbohydrate intake. As shown later in this chapter, the complete and efficient oxidation of fats depends on adequate carbohydrate intake and the presence of certain intermediates of carbohydrate metabolism. If these are lacking, fats are incompletely oxidized to ketone bodies, which may cause metabolic acidosis.

## Requirements

Because carbohydrates are rapidly oxidized, they are required in greater amounts than any other nutrient. The RDA is 125 to 175 g. The brain alone consumes about 120 g of glucose per day. Most Americans get about 40% to 50% of their calories from carbohydrates, but highly active people should get up to 60%.

Carbohydrate consumption in the United States has become excessive over the past century because of a combination of fondness for sweets, increased use of sugar in processed foods, and reduced physical activity (see Deeper Insight 26.2). A century ago, Americans consumed an average of 1.8 kg (4 lb) of sugar per year. Now, with sucrose and high-fructose corn syrup so widely used in foods and beverages, the average American ingests 200 to 300 g of carbohydrate per day and the equivalent of 27 kg (60 lb) of table sugar and 21 kg (46 lb) of corn syrup per year. A single 355 mL (12 oz) serving of nondiet soft drink contains 38 to 43 g of sugar (about 8 teaspoons).

Dietary carbohydrates come in three principal forms: monosaccharides, disaccharides, and polysaccharides (complex carbohydrates). The only nutritionally significant polysaccharide is starch. Although glycogen is a polysaccharide, only trivial amounts of it are present in cooked meats. Cellulose, another polysaccharide, is not considered a nutrient because it is not digested and never enters the human tissues. Its importance as dietary fiber, however, is discussed shortly.

---

[3]*hypo* = below normal; *glyc* = sugar; *emia* = blood condition

## DEEPER INSIGHT 26.2
### EVOLUTIONARY MEDICINE

#### *Evolution of the Sweet Tooth*

Our craving for sugar doubtlessly originated in our prehistoric ancestors. Not only did they have to work much harder to survive than we do, but high-calorie foods were scarce and people were at constant risk of starvation. Those who were highly motivated to seek and consume sugary, high-calorie foods passed their "sweet tooth" on to us, their descendants—along with a similarly adaptive appetite for other rare but vital nutrients, namely fat and salt. The tastes that were essential to our ancestors' survival can now be a disadvantage in a culture in which salty, fatty, and sugary foods are all too easy to obtain; the food industry eagerly capitalizes on these tastes.

The three major dietary disaccharides are sucrose, lactose, and maltose. The monosaccharides—glucose, galactose, and fructose—arise mainly from the digestion of starch and disaccharides. The small intestine and liver convert fructose and galactose to glucose, so ultimately all carbohydrate digestion generates glucose. Outside of the hepatic portal system, glucose is the only monosaccharide present in the blood in significant quantity; thus, it is known as *blood sugar.* Its concentration is normally maintained at 70 to 110 mg/dL in peripheral venous blood.

### ►►►APPLY WHAT YOU KNOW

*Glucose concentration is about 15 to 30 mg/dL higher in arterial blood than in most venous blood. Explain why.*

The effect of a dietary carbohydrate on one's blood glucose level can be expressed as the **glycemic index (GI).** The effect of ingesting 50 g of glucose on blood glucose level over the next 2 hours is set at 100, and the effects of other carbohydrates are expressed in relation to this. A carbohydrate with a GI of 50, for example, would produce half the effect of pure glucose. Carbohydrates with a high GI (≥70) are quickly digested and absorbed and rapidly raise blood glucose. Such carbohydrates are found, for example, in white bread, white rice, baked white potatoes, and many processed breakfast cereals. High-GI carbohydrates stimulate a high insulin demand and raise the risk of obesity and type 2 diabetes mellitus. Carbohydrates with a low GI (≤55) are digested more slowly and raise blood glucose more gradually. These include the carbohydrates found in most fruits and vegetables, legumes, milk, and grainy bread and pasta. The glycemic index of a given food varies, however, from person to person and even in the same person from day to day, and depends as well on how the food is cooked.

Ideally, most carbohydrate intake should be in the form of starch. This is partly because foods that provide starch also usually provide other nutrients. Simple sugars not only provide empty calories but also promote tooth decay. A typical

American, however, now obtains only 50% of his or her carbohydrates from starch and the other 50% from sucrose and corn syrup.

## Dietary Sources

Nearly all dietary carbohydrates come from plants—particularly grains, legumes, fruits, and root vegetables. Sucrose is refined from sugarcane and sugar beets. Fructose is present in fruits and corn syrup. Maltose is present in some foods such as germinating cereal grains. Lactose is the most abundant solute in cow's milk (about 4.6% lactose by weight).

## Fiber

*Dietary fiber* refers to all fibrous materials of plant and animal origin that resist digestion. Most is plant matter—the carbohydrates cellulose and pectin and such noncarbohydrates as gums and lignin. Although it is not a nutrient, fiber is an essential component of the diet. The recommended daily allowance is about 30 g, but average intake varies greatly from country to country—from 40 to 150 g/day in India and Africa to only 12 g/day in the United States.

**Water-soluble fiber** includes pectin and certain other carbohydrates found in oats, beans, peas, carrots, brown rice, and fruits. It reduces blood cholesterol and low-density lipoprotein (LDL) levels (see Deeper Insight 19.4, p. 739). **Water-insoluble fiber** includes cellulose, hemicellulose, and lignin. It apparently has no effect on cholesterol or LDL levels, but it absorbs water and swells, thereby softening the stool and increasing its bulk by 40% to 100%. This effect stretches the colon and stimulates peristalsis, thereby quickening the passage of feces. In doing so, water-insoluble fiber reduces the risk of constipation and diverticulitis (see table 25.3, p. 988).

Contrary to previous medical opinion, dietary fiber has no clear effect on the incidence of colorectal cancer. Excess fiber can actually have a deleterious effect on health by interfering with the absorption of iron, calcium, magnesium, phosphorus, and some trace elements.

## Lipids

Healthy young men and women average, respectively, about 15% and 25% fat by weight. Fat accounts for most of the body's stored energy. Lesser amounts of phospholipid, cholesterol, and other lipids also play vital structural and physiological roles.

A well-nourished adult meets 80% to 90% of his or her resting energy needs from fat. Fat is superior to carbohydrates for energy storage for two reasons: (1) Carbohydrates are hydrophilic, absorb water, and thus expand and occupy more space in the tissues. Fat, however, is hydrophobic, contains almost no water, and is a more compact energy storage substance. (2) Fat is less oxidized than carbohydrate and contains over twice as much energy (9 kcal/g of fat compared with 4 kcal/g of carbohydrate). A man's typical fat reserves contain enough energy for 119 hours of running, whereas his carbohydrate stores would suffice for only 1.6 hours.

Fat has **glucose-sparing** and **protein-sparing effects**—as long as enough fat is available to meet the energy needs of the tissues, protein is not catabolized for fuel and glucose is spared for consumption by cells that cannot use fat, such as neurons.

Vitamins A, D, E, and K are fat-soluble vitamins, which depend on dietary fat for their absorption by the intestine. People who ingest less than 20 g of fat per day are at risk of vitamin deficiency because there is not enough fat in the intestine to transport these vitamins into the body tissues.

Phospholipids and cholesterol are major structural components of plasma membranes and myelin. Cholesterol is also important as a precursor of steroid hormones, bile acids, and vitamin D. Thromboplastin, an essential blood-clotting factor, is a lipoprotein. Two fatty acids—arachidonic acid and linoleic acid—are precursors of prostaglandins and other eicosanoids.

In addition to its metabolic and structural roles, fat has important protective and insulating functions described under adipose tissue in chapter 5.

## Requirements

Fat should account for no more than 30% of one's daily caloric intake; no more than 10% of fat intake should be saturated fat; and average cholesterol intake should not exceed 300 mg/day (one egg yolk contains about 240 mg). A typical American consumes 30 to 150 g of fat per day, obtains 40% to 50% of his or her calories from fat, and ingests twice as much cholesterol as the recommended limit.

Most fatty acids can be synthesized by the body. **Essential fatty acids** are those we cannot synthesize and therefore must obtain from the diet. These include linoleic acid and possibly linolenic and arachidonic acids; there are differences of opinion about the body's ability to synthesize the last two. As long as 1% to 2% of the total energy intake comes from linoleic acid, people show no signs of essential fatty acid deficiency. In the typical Western diet, linoleic acid provides about 6% of the calories.

## Sources

Saturated fats are predominantly of animal origin. They occur in meat, egg yolks, and dairy products but also in some plant products such as coconut and palm oils (common in nondairy coffee creamers and other products). Processed foods such as hydrogenated oils and vegetable shortening are also high in saturated fat, which is therefore abundant in many baked goods. Unsaturated fats predominate in nuts, seeds, and most vegetable oils. The essential fatty acids are amply provided by the vegetable oils in mayonnaise, salad dressings, and margarine and by whole grains and vegetables. Excessive consumption of saturated and unsaturated fats is a risk factor for diabetes mellitus, cardiovascular disease, and breast and colon cancer.

The richest source of cholesterol is egg yolks, but it is also prevalent in milk products; shellfish (especially shrimp); organ meats such as kidneys, liver, and brains; and other mammalian meat. Foods of plant origin contain only insignificant traces of cholesterol. The serum cholesterol level is strongly influenced by the types and quantity of fatty acids in the diet. This relationship is explained in the next section.

## Cholesterol and Serum Lipoproteins

Lipids are an important part of the diet and must be transported to all cells of the body, yet they are hydrophobic and do not dissolve in the aqueous blood plasma. This problem is overcome by complexes called **lipoproteins**—tiny droplets with a core of cholesterol and triglycerides and a coating of proteins and phospholipids. The coating not only enables the lipids to remain suspended in the blood, but also serves as a recognition marker for cells that absorb them. The complexes are often referred to as *serum lipoproteins* because their concentrations are expressed in terms of a volume of blood serum, not whole blood.

Lipoproteins are classified into four major categories (and some lesser ones) by their density: **chylomicrons, high-density lipoproteins (HDLs), low-density lipoproteins (LDLs),** and **very low–density lipoproteins (VLDLs)** (fig. 26.2a). The higher the proportion of protein to lipid, the higher the density. These particles also differ considerably in size: Chylomicrons range widely from 75 to 1,200 nm in diameter, but the others diminish in size from VLDLs (30–80 nm), to LDLs (18–25 nm), to HDLs (5–12 nm). Their most important differences, however, are in composition and function. Figure 26.2b shows the three primary pathways by which they are made and processed.

Chylomicrons form in the absorptive cells of the small intestine and then pass into the lymphatic system and ultimately the bloodstream (see p. 981). Endothelial cells of the blood capillaries have a surface enzyme called **lipoprotein lipase** that hydrolyzes chylomicron triglycerides into monoglycerides and free fatty acids (FFAs). These products can then pass through the capillary walls into adipocytes, where they are resynthesized into storage triglycerides. Some FFAs, however, remain in the blood plasma bound to albumin. The remainder of a chylomicron after the triglycerides have been extracted, called a *chylomicron remnant,* is removed and degraded by the liver.

VLDLs, produced by the liver, transport lipids to the adipose tissue for storage. When their triglycerides are removed in the adipose tissue, the VLDLs become LDLs and contain mostly cholesterol. Cells that need cholesterol (usually for membrane structure or steroid hormone synthesis) absorb LDLs by receptor-mediated endocytosis, digest them with lysosomal enzymes, and release the cholesterol for intracellular use.

HDL production begins in the liver, which produces an empty, collapsed protein shell. This shell travels in the blood and picks up cholesterol and phospholipids from other organs. The next time it circulates through the liver, the liver removes the cholesterol and eliminates it in the bile as either cholesterol

or bile acids. HDLs are therefore a vehicle for removing excess cholesterol from the body.

It is desirable to maintain a total cholesterol concentration of 200 mg/dL or less in the blood plasma. From 200 to 239 mg/dL is considered borderline high, and levels over 240 mg/dL are pathological.

Most of the body's cholesterol is endogenous (internally synthesized) rather than dietary, and the body compensates for variations in dietary intake. High intake steps down cholesterol synthesis by the liver, whereas a low dietary intake steps it up. Thus, lowering dietary cholesterol reduces the serum cholesterol level by no more than 5%. More important is the fact that certain saturated fatty acids (SFAs) raise serum cholesterol level. For example, palmitic acid, a 16-carbon SFA, raises serum cholesterol by blocking its uptake by the tissues (yet stearic acid, an 18-carbon SFA, does not raise it). Some food advertising is deceptive on this point. It may truthfully advertise a food as being cholesterol-free, but neglect to mention that it contains SFAs that may raise the consumer's cholesterol level anyway. A moderate reduction of saturated fatty acid intake can lower blood cholesterol by 15% to 20%—considerably more effective than reducing dietary cholesterol per se.

Vigorous exercise also lowers blood cholesterol levels. The mechanism is somewhat roundabout: Exercise reduces the sensitivity of the right atrium of the heart to blood pressure, so the heart secretes less natriuretic peptide. Consequently, the kidneys excrete less sodium and water, and the blood volume rises. This dilutes the lipoproteins in the blood, and the adipocytes compensate by producing more lipoprotein lipase. The adipocytes therefore consume more blood triglycerides. This shrinks the VLDL particles, which shed some of their cholesterol in the process, and HDLs pick up this free cholesterol for removal by the liver.

Blood cholesterol is not the only important measure of healthy lipid concentrations, however. A high LDL concentration is a warning sign because, as you can see from the function of LDLs described previously, it signifies a high rate of cholesterol deposition in the arteries. LDLs are elevated not only by saturated fats but also by cigarette smoking, coffee, and stress. A high proportion of HDL, on the other hand, is beneficial because it indicates that cholesterol is being removed from the arteries and transported to the liver for disposal. Thus it is desirable to increase your ratio of HDL to LDL. This is best done with a diet low in calories and saturated fats and is promoted by regular aerobic exercise.

## Proteins

Protein constitutes about 12% to 15% of the body mass; 65% of it is in the skeletal muscles. Proteins are responsible for muscle contraction and the motility of cilia and flagella. They are a major structural component of all cellular membranes, with multiple roles such as membrane receptors, pumps, ion channels, and cell-identity markers. Fibrous proteins such as collagen, elastin, and keratin make up much of the structure

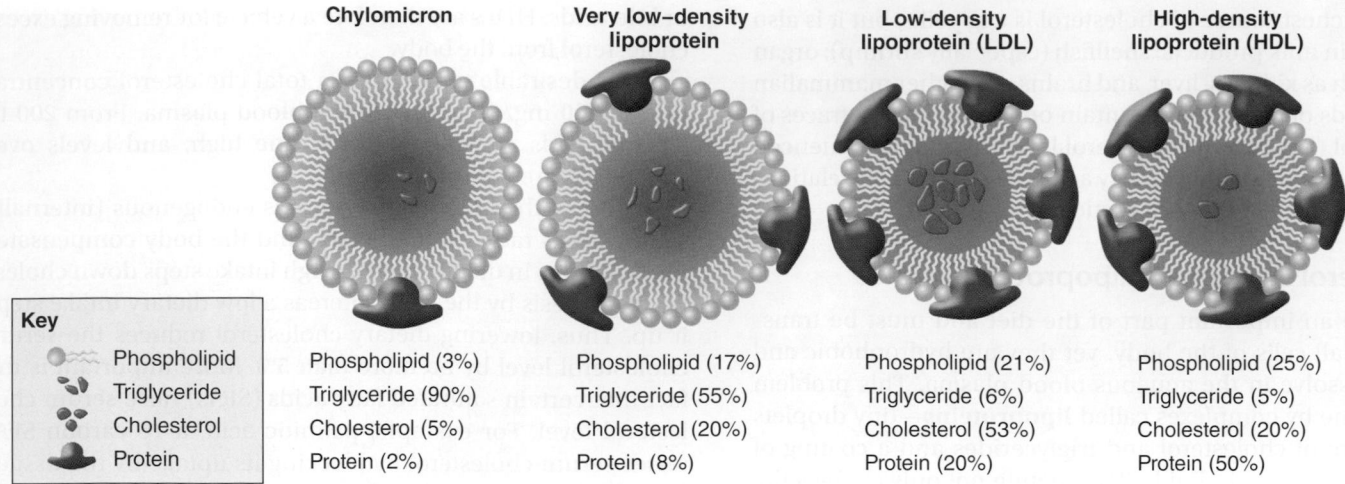

**Chylomicron**

Phospholipid (3%)
Triglyceride (90%)
Cholesterol (5%)
Protein (2%)

**Very low–density lipoprotein**

Phospholipid (17%)
Triglyceride (55%)
Cholesterol (20%)
Protein (8%)

**Low-density lipoprotein (LDL)**

Phospholipid (21%)
Triglyceride (6%)
Cholesterol (53%)
Protein (20%)

**High-density lipoprotein (HDL)**

Phospholipid (25%)
Triglyceride (5%)
Cholesterol (20%)
Protein (50%)

**Key**
- Phospholipid
- Triglyceride
- Cholesterol
- Protein

**(a) Lipoprotein types**

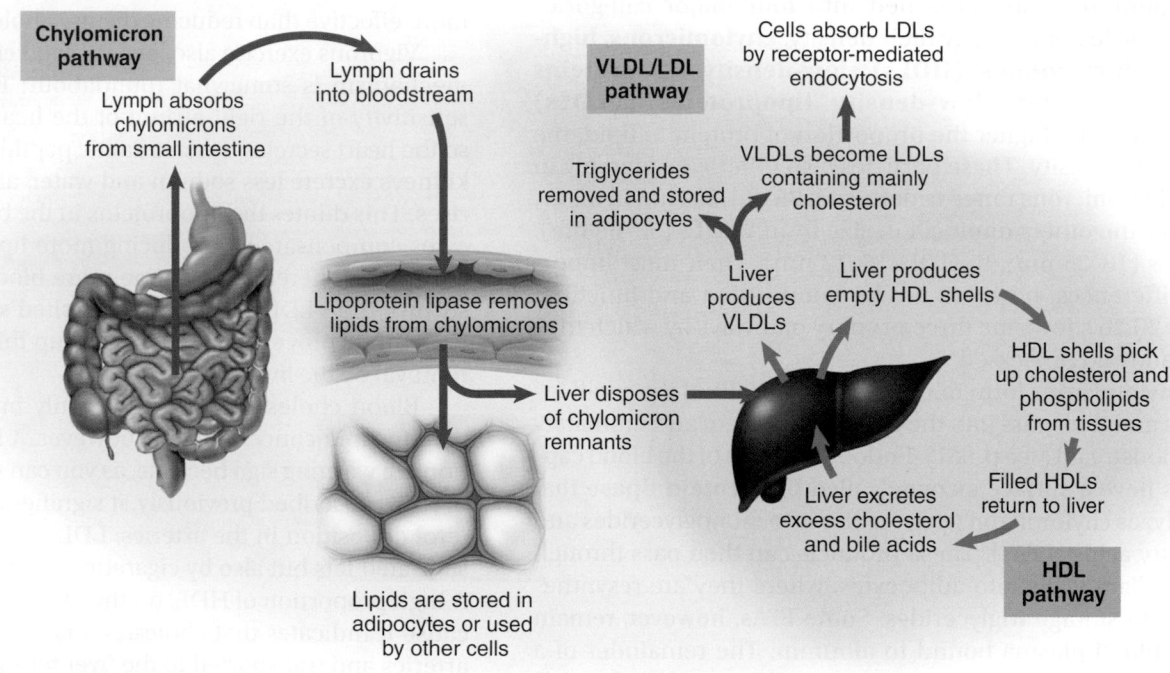

**Chylomicron pathway**

Lymph absorbs chylomicrons from small intestine

Lymph drains into bloodstream

Lipoprotein lipase removes lipids from chylomicrons

Lipids are stored in adipocytes or used by other cells

**VLDL/LDL pathway**

Cells absorb LDLs by receptor-mediated endocytosis

VLDLs become LDLs containing mainly cholesterol

Triglycerides removed and stored in adipocytes

Liver produces VLDLs

Liver disposes of chylomicron remnants

Liver produces empty HDL shells

HDL shells pick up cholesterol and phospholipids from tissues

Filled HDLs return to liver

Liver excretes excess cholesterol and bile acids

**HDL pathway**

**(b) Lipoprotein-processing pathways**

**FIGURE 26.2 Lipoprotein Processing.** (a) The four types of serum lipoproteins. (b) The three pathways of lipoprotein processing.

❓ *Why is a high HDL:LDL ratio healthier than a high LDL:HDL ratio?*

of bone, cartilage, tendons, ligaments, skin, hair, and nails. Globular proteins include antibodies, hormones, neuromodulators, hemoglobin, myoglobin, and about 2,000 enzymes that control nearly every aspect of cellular metabolism. They also include the albumin and other plasma proteins that maintain blood viscosity and osmolarity and transport lipids and some other plasma solutes. Proteins buffer the pH of body fluids

and contribute to the resting membrane potentials of all cells. No other class of biomolecules has such a broad variety of functions.

## Requirements

For persons of average weight, the RDA of protein is 44 to 60 g, depending on age and sex. Multiplying your weight in pounds by 0.37 or your weight in kilograms by 0.8 gives an estimate of your protein RDA in grams. A higher intake is recommended, however, under conditions of stress, infection, injury, and pregnancy. Infants and children require more protein than adults relative to body weight. Excessive protein intake, however, overloads the kidneys with nitrogenous waste and can cause renal damage. This is a risk in certain high-protein fad diets.

Total protein intake is not the only significant measure of dietary adequacy. The nutritional value of a protein depends on whether it supplies the right amino acids in the proportions needed for building human proteins. Adults can synthesize 12 of the 20 amino acids from other organic compounds when they are not available from the diet, but there are 8 **essential amino acids** that we cannot synthesize: isoleucine, leucine, lysine, methionine, phenylalanine, threonine, tryptophan, and valine. (Infants also require histidine.) In addition, 2 amino acids can be synthesized only from essential amino acids: cysteine from methionine and tyrosine from phenylalanine. The other 10 (9 in infants) are called **inessential amino acids**—not because the body does not require them but because it can synthesize its own when the diet does not supply them.

Cells do not store surplus amino acids for later use. When a protein is to be synthesized, all of the amino acids necessary must be present at once, and if even one is missing, the protein cannot be made. High-quality **complete proteins** are those that provide all of the essential amino acids in the necessary proportions for human tissue growth, maintenance, and nitrogen balance. Lower-quality **incomplete proteins** lack one or more essential amino acids. For example, cereals are low in lysine and legumes are low in methionine.

Protein quality is also determined by **net protein utilization**—the percentage of the amino acids in a protein that the human body uses. We typically use 70% to 90% of animal protein but only 40% to 70% of plant protein. It therefore takes a larger serving of plant protein than animal protein to meet our needs—for example, we need 400 g (about 14 oz) of rice and beans to provide as much usable protein as 115 g (about 4 oz) of hamburger. However, reducing meat intake and increasing plant intake have advantages. Among other considerations, plant foods provide more vitamins, minerals, and fiber; less saturated fat; no cholesterol; and less pesticide. In an increasingly crowded world, it must also be borne in mind that it requires far more land to produce meat than to produce food crops.

## Dietary Sources

The animal proteins of meat, eggs, and dairy products closely match human proteins in amino acid composition. Thus, animal products provide high-quality complete protein, whereas plant proteins are incomplete. Nevertheless, this does not mean that your dietary protein *must* come from meat; indeed, about two-thirds of the world's population receives adequate protein nutrition from diets containing very little meat. We can combine plant foods so that one provides what another lacks—beans and rice, for example, are a complementary combination of legume and cereal. Beans provide the isoleucine and lysine lacking in grains, while rice provides the tryptophan and cysteine lacking in beans.

## Nitrogen Balance

Proteins are our chief dietary source of nitrogen. **Nitrogen balance** is a state in which the rate of nitrogen ingestion equals the rate of excretion (chiefly as nitrogenous wastes). Growing children exhibit a state of **positive nitrogen balance** because they ingest more than they excrete, thus retaining protein for tissue growth. Pregnant women and athletes in resistance training also show positive nitrogen balance. When excretion exceeds ingestion, a person is in a state of **negative nitrogen balance.** This indicates that body proteins are being broken down and used as fuel. Proteins of the muscles and liver are more easily broken down than others; therefore, negative nitrogen balance tends to be associated with muscle atrophy. Negative nitrogen balance may occur if carbohydrate and fat intake are insufficient to meet the need for energy. Carbohydrates and fats are said to have a protein-sparing effect because they prevent protein catabolism when present in sufficient amounts to meet one's energy needs.

Nitrogen balance is affected by some hormones. Growth hormone and sex steroids promote protein synthesis and positive nitrogen balance during childhood, adolescence, and pregnancy. Glucocorticoids, on the other hand, promote protein catabolism and negative nitrogen balance in states of stress.

▶▶▶ **APPLY WHAT YOU KNOW**

*Would you expect a person recovering from a long infectious disease to be in a state of positive or negative nitrogen balance? Why?*

## Minerals and Vitamins

Minerals are inorganic elements that plants extract from soil or water and introduce into the food web. Vitamins are small dietary organic compounds that are necessary to metabolism. Neither is used as fuel, but both are essential to our ability to use other nutrients. With the exception of a few vitamins, these nutrients cannot be synthesized by the body and must be included in the diet. They are, however, required in relatively small quantities. Mineral RDAs range from 0.05 mg of chromium and selenium to 1,200 mg of calcium and phosphorus. Vitamin RDAs range from about 0.002 mg of vitamin $B_{12}$ to 60 mg of vitamin C. Despite the small quantities involved, minerals and vitamins have very potent effects on physiology. Indeed, excessive amounts are toxic and potentially lethal.

## Minerals

Minerals constitute about 4% of the body mass, with three-quarters of this being the calcium and phosphorus in the bones and teeth. Phosphorus is also a key structural component of phospholipids, DNA, RNA, ATP, cAMP, GTP, and creatine phosphate, and is the basis of the phosphate buffer system (see chapter 24). Calcium, iron, magnesium, and manganese function as cofactors for enzymes. Iron is essential to the oxygen-carrying capacity of hemoglobin and myoglobin. Chlorine is a component of stomach acid (HCl). Many mineral salts function as electrolytes and govern the function of nerve and muscle cells, osmotically regulate the content and distribution of water in the body, and maintain blood volume.

Table 26.2 summarizes adult mineral requirements and dietary sources. Broadly speaking, the best sources of minerals are vegetables, legumes, milk, eggs, fish, shellfish, and some other meats. Cereal grains are a relatively poor source, but processed cereals may be mineral-fortified.

Sodium chloride has been both a prized commodity and a curse. Animal tissues contain relatively large amounts of salt, and carnivores rarely lack ample salt in their diets. Plants, however, are relatively poor in salt, so herbivores often must supplement their diet by ingesting salt from the soil. As humans developed agriculture and became more dependent on plants, they also became increasingly dependent on supplemental salt. Salt has often been used as a form of payment for goods and services—the word *salary* comes from *sal* (salt). Our fondness for salt and high sensitivity to it undoubtedly stem from its physiological importance and its scarcity in a largely vegetarian diet.

Now, however, this fondness has become a bane. The recommended sodium intake is 1.1 g/day, but a typical American diet contains about 4.5 g/day. This is due not just to the use of table salt but more significantly to the large amounts of salt in processed foods, much of it "disguised" in soy sauce, MSG (monosodium glutamate), baking soda, and baking powder. In some areas of Japan, salt intake averages 27 g/day and most people die before age 70 of stroke and other complications of hypertension.

Hypertension is a leading cause of death among American blacks, who have twice the risk of hypertension and 10 times the risk of dying from it that American whites have. The reason for this is not excessive salt intake, but rather that people of West African descent have kidneys with an especially strong tendency to retain salt.

## Vitamins

Vitamins were originally named with letters in the order of their discovery, but they also have chemically descriptive names such as ascorbic acid (vitamin C) and riboflavin (vitamin $B_2$). Most vitamins must be obtained from the diet (table 26.3), but the body synthesizes some of them from precursors called *provitamins*—niacin from the amino acid tryptophan; vitamin D from cholesterol; and vitamin A from carotene, which is abundantly present in carrots, squash, and other yellow vegetables and fruits. Vitamin K, pantothenic acid, biotin, and folic acid are produced by the bacteria of the large intestine. The feces contain more biotin than food does.

Vitamins are classified as water-soluble or fat-soluble. **Water-soluble vitamins** are absorbed with water from the small

| **TABLE 26.2** | **Mineral Requirements and Some Dietary Sources** | |
|---|---|---|
| **Mineral** | **RDA (mg)** | **Some Dietary Sources*** |
| **Major minerals** | | |
| Calcium | 1,200 | Milk, fish, shellfish, greens, tofu, orange juice |
| Phosphorus | 1,200 | Red meat, poultry, fish, eggs, milk, legumes, whole grains, nuts |
| Sodium | 1,500 | Table salt, processed foods; usually present in excess |
| Chloride | 2,300 | Table salt, some vegetables; usually present in excess |
| Magnesium | 280–350 | Milk, greens, whole grains, nuts, legumes, chocolate |
| Potassium | 4,700 | Red meat, poultry, fish, cereals, spinach, squash, bananas, apricots |
| Sulfur | Unknown | Meats, milk, eggs, legumes; almost any proteins |
| **Trace minerals** | | |
| Zinc | 12–15 | Red meat, seafood, cereals, wheat germ, legumes, nuts, yeast |
| Iron | 10–15 | Red meat, liver, shellfish, eggs, dried fruits, nuts, legumes, molasses |
| Manganese | 2.5–5.0 | Greens, fruits, legumes, whole grains, nuts |
| Copper | 1.5–3.0 | Red meat, liver, shellfish, legumes, whole grains, nuts, cocoa |
| Fluoride | 1.5–4.0 | Fluoridated water and toothpaste, tea, seafood, seaweed |
| Iodine | 0.15 | Marine fish, fish oils, shellfish, iodized salt |
| Molybdenum | 0.07–0.25 | Beans, whole grains, nuts |
| Chromium | 0.05–0.25 | Meats, liver, cheese, eggs, whole grains, yeast, wine |
| Selenium | 0.05–0.07 | Red meats, organ meats, fish, shellfish, eggs, cereals |
| Cobalt | Unknown | Red meat, poultry, fish, liver, milk |

*"Red meat" refers to mammalian muscle such as beef and pork. "Organ meat" refers to brain, pancreas, heart, kidney, etc. Liver is specified separately and refers to beef, pork, and chicken livers, which are similar for most nutrients.

intestine, dissolve freely in the body fluids, and are quickly excreted by the kidneys. They cannot be stored in the body and therefore seldom accumulate to excess. The water-soluble vitamins are ascorbic acid (vitamin C) and the B vitamins. Ascorbic acid promotes hemoglobin synthesis, collagen synthesis, and sound connective tissue structure; and it is an antioxidant that scavenges free radicals and possibly reduces the risk of cancer. The B vitamins function as coenzymes or parts of coenzyme molecules; they assist enzymes by transferring electrons from one metabolic reaction to another, making it possible for enzymes to catalyze these reactions. Some of their functions arise later in this chapter as we consider carbohydrate metabolism.

**Fat-soluble vitamins** are incorporated into lipid micelles in the small intestine and absorbed with dietary lipids. They are more varied in function than water-soluble vitamins. Vitamin A is a component of the visual pigments and promotes proteoglycan synthesis and epithelial maintenance. Vitamin D promotes calcium absorption and bone mineralization. Vitamin K is essential to prothrombin synthesis and blood clotting. Vitamins A and E are antioxidants, like ascorbic acid.

It is common knowledge that various diseases result from vitamin deficiencies, but it is less commonly known that **hypervitaminosis** (vitamin excess) also causes disease. A *deficiency* of vitamin A, for example, can result in night blindness, dry skin and hair, a dry conjunctiva and cloudy cornea, and increased incidence of urinary, digestive, and respiratory infections. This is the world's most common vitamin deficiency. An *excess* of vitamin A, however, may cause anorexia, nausea and vomiting, headache, pain and fragility of the bones, hair loss, an enlarged liver and spleen, and birth defects. Vitamins $B_6$, C, D, and E have also been implicated in toxic hypervitaminosis.

Some people take *megavitamins*—doses 10 to 1,000 times the RDA—thinking that they will improve athletic performance. Since vitamins are not burned as fuel, and small amounts fully meet the body's metabolic needs, there is no evidence that vitamin supplements improve performance except when used to correct a dietary deficiency. Megadoses of fat-soluble vitamins can be especially harmful.

**BEFORE YOU GO ON**

Answer the following questions to test your understanding of the preceding section:

1. Name two hormones that regulate short-term hunger and satiety. How does leptin differ from these in its effects?

2. Explain the following statement: Cellulose is an important part of a healthy diet but it is not a nutrient.

3. What class of nutrients provides most of the calories in the diet? What class of nutrients provides the body's major reserves of stored energy?

4. Contrast the functions of VLDLs, LDLs, and HDLs. Explain how this is related to the fact that a high blood HDL level is desirable, but a high VLDL–LDL level is undesirable.

5. Why do some proteins have more nutritional value than others?

| TABLE 26.3 | | Vitamin Requirements and Some Dietary Sources |
|---|---|---|
| **Vitamin** | **RDA (mg)** | **Some Dietary Sources*** |
| **Water-soluble vitamins** | | |
| Ascorbic acid (C) | 60 | Citrus fruits, strawberries, tomatoes, greens, cabbage, cauliflower, broccoli, brussels sprouts |
| **B complex** | | |
| Thiamine ($B_1$) | 1.5 | Red meat, liver, other organ meats, eggs, greens, asparagus, legumes, whole grains, seeds, yeast |
| Riboflavin ($B_2$) | 1.7 | Widely distributed, and deficiencies are rare; all types of meat, milk, eggs, greens, whole grains, apricots, legumes, mushrooms, yeast |
| Pyridoxine ($B_6$) | 2.0 | Red meat, fish, liver, other organ meats, greens, apricots, legumes, whole grains, seeds |
| Cobalamin ($B_{12}$) | 0.002 | Red meat, liver, other organ meats, shellfish, eggs, milk; absent from food plants |
| Niacin (nicotinic acid) | 19 | Readily synthesized from tryptophan, which is present in any diet with adequate protein; red meat, liver, other organ meats, poultry, fish, apricots, legumes, whole grains, mushrooms |
| Panthothenic acid | 4–7 | Widely distributed, and deficiencies are rare; red meat, liver, other organ meats, eggs, green and yellow vegetables, legumes, whole grains, mushrooms, yeast |
| Folic acid (folacin) | 0.2 | Eggs, liver, greens, citrus fruits, legumes, whole grains, seeds |
| Biotin | 0.03–0.10 | Red meat, liver, other organ meats, eggs, cheese, cabbage, cauliflower, bananas, legumes, nuts |
| **Fat-soluble vitamins** | | |
| Retinol (A) | 1.0 | Fish oils, eggs, cheese, milk, greens, other green and yellow vegetables and fruits, margarine |
| Calcitriol (D) | 0.01 | Formed by exposure of skin to sunlight; fish, fish oils, milk |
| α-tocopherol (E) | 10 | Fish oils, greens, seeds, wheat germ, vegetable oils, margarine, nuts |
| Phylloquinone (K) | 0.08 | Most of the RDA is met by synthesis by intestinal bacteria; liver, greens, cabbage, cauliflower |

*See footnote in table 26.2.

## 26.2  Carbohydrate Metabolism

### Expected Learning Outcomes

When you have completed this section, you should be able to

a. describe the principal reactants and products of each major step of glucose oxidation;

b. contrast the functions and products of anaerobic fermentation and aerobic respiration;

c. explain where and how cells produce ATP; and

d. describe the production, function, and use of glycogen.

Most dietary carbohydrate is burned as fuel within a few hours of absorption. Although three monosaccharides are absorbed from digested food—glucose, galactose, and fructose—the last two are quickly converted to glucose, and all oxidative carbohydrate consumption is essentially a matter of glucose catabolism. The overall reaction for this is

$$C_6H_{12}O_6 + 6\ O_2 \rightarrow 6\ CO_2 + 6\ H_2O$$

The function of this reaction is not to produce carbon dioxide and water but to transfer energy from glucose to ATP.

Along the pathway of glucose oxidation are several links through which other nutrients—especially fats and amino acids—can also be oxidized as fuel. Carbohydrate catabolism therefore provides a central vantage point from which we can view the catabolism of all fuels and the generation of ATP.

### Glucose Catabolism

If the preceding reaction were carried out in a single step, it would generate a short, intense burst of heat—like the burning of paper, which has the same chemical equation. Not only would this be useless to the body's metabolism, it would kill the cells. In the body, however, the process is carried out in a series of small steps, each controlled by a separate enzyme. Energy is released in small manageable amounts, and as much as possible is transferred to ATP. The rest is released as heat.

There are three major pathways of glucose catabolism:

1. **glycolysis,** which splits a glucose molecule into two molecules of pyruvic acid;

2. **anaerobic fermentation,** which reduces pyruvic acid to lactic acid without using oxygen; and

3. **aerobic respiration,** which requires oxygen and oxidizes pyruvic acid to carbon dioxide and water.

You may find it helpful to review figure 2.32 (see p. 72) for a broad overview of these processes and their relationship to ATP production. Figures 26.3 to 26.6 examine these processes in closer detail; the first two of these are labeled with numbers that correspond to the reaction steps to follow.

Coenzymes are vitally important to these reactions. Enzymes remove electrons (as hydrogen atoms) from the intermediate compounds of these pathways, but they do not bind them. Instead, they transfer the hydrogen atoms to coenzymes, and the coenzymes donate them to other compounds later in one of the reaction pathways. Thus, the enzymes of glucose catabolism cannot function without their coenzymes.

The two coenzymes of special importance to glucose catabolism are **NAD**$^+$ (nicotinamide adenine dinucleotide) and **FAD** (flavin adenine dinucleotide). Both are derived from B vitamins: NAD$^+$ from niacin and FAD from riboflavin. Hydrogen atoms are removed from metabolic intermediates in pairs—that is, two protons and two electrons ($2\ H^+$ and $2\ e^-$) at a time—and transferred to a coenzyme. This produces a reduced coenzyme with a higher free energy content than it had before the reaction. Coenzymes thus become the temporary carriers of the energy extracted from glucose metabolites. The reactions for this are

$$FAD + 2\ H \rightarrow FADH_2$$

and

$$NAD^+ + 2\ H \rightarrow NADH + H^+$$

FAD binds two protons and two electrons to become FADH$_2$. NAD$^+$, however, binds the two electrons but only one of the protons to become NADH. The other proton remains a free hydrogen ion, H$^+$ (or H$_3$O$^+$, but it is represented in this chapter as H$^+$).

### Glycolysis

Upon entering a cell, glucose begins a series of conversions called glycolysis[4] (fig. 26.3):

①  **Phosphorylation.**  The enzyme *hexokinase* transfers an inorganic phosphate (P$_i$) group from ATP to glucose, producing glucose 6-phosphate (G6P). This has two effects:

- It keeps the intracellular concentration of glucose low, maintaining a concentration gradient that favors the continued diffusion of more glucose into the cell.

- Phosphorylated compounds cannot pass through the plasma membrane, so this prevents the sugar from leaving the cell. In most cells, step 1 is irreversible because the cells lack the enzyme to convert G6P back to glucose. The few exceptions are cells that must be able to release free glucose to the blood: absorptive cells of the small intestine, proximal convoluted tubule cells in the kidney, and liver cells.

G6P is a versatile molecule that can be converted to fat or amino acids, polymerized to form glycogen for storage, or further oxidized to extract its energy. For now, we are mainly concerned with its further oxidation (glycolysis), the general

---

[4]*glyco* = sugar; *lysis* = splitting

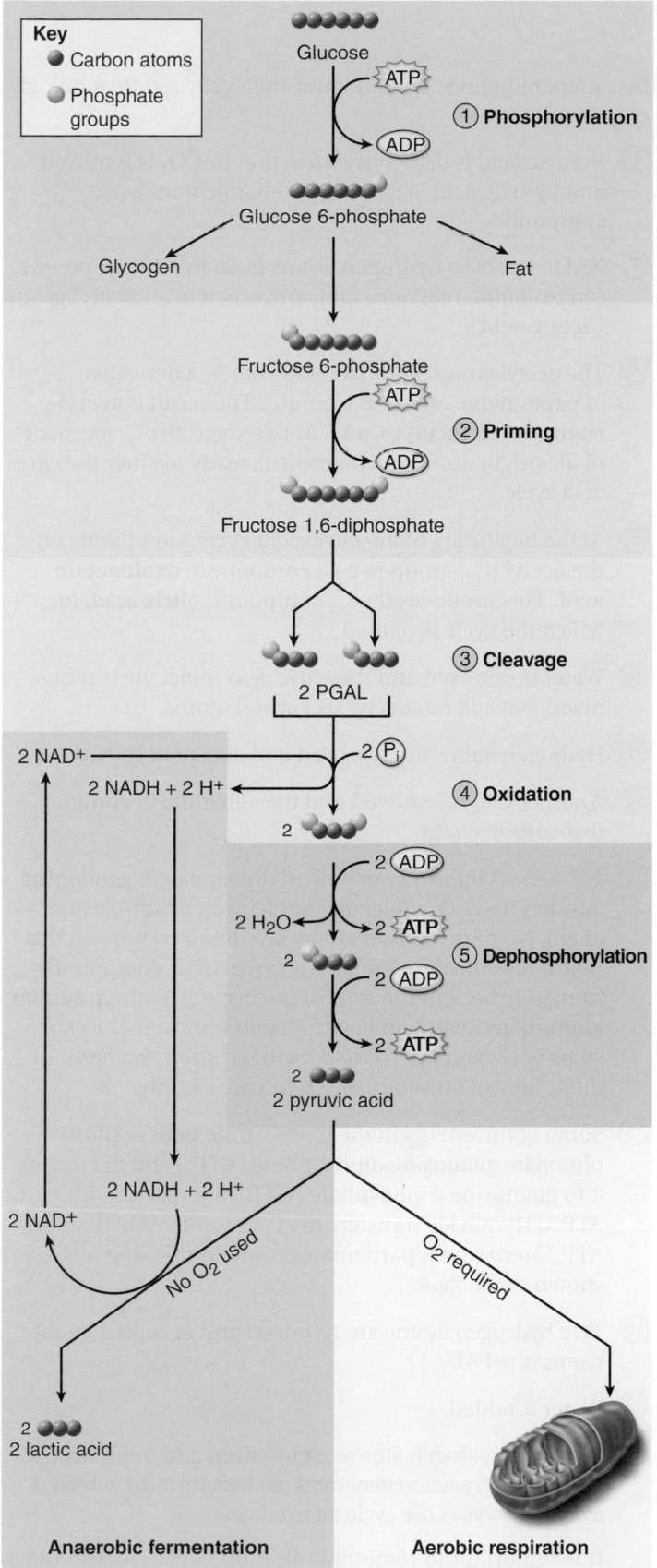

**FIGURE 26.3 Glycolysis and Anaerobic Fermentation.**
Numbered reaction steps are explained in the text.

❓ *At what point would this reaction stop, and what reaction intermediate would accumulate, if NAD⁺ was unavailable to the cell? What process replenishes the NAD⁺ supply?*

effect of which is to split G6P (a six-carbon sugar, $C_6$) into two three-carbon ($C_3$) molecules of **pyruvic acid (pyruvate).** Continue tracing these steps in figure 26.3 as you read.

② **Priming.** G6P is rearranged (isomerized) to form fructose 6-phosphate, which is phosphorylated again to form fructose 1,6-diphosphate. This "primes" the process by providing activation energy, somewhat like the heat of a match used to light a fireplace. Two molecules of ATP have already been consumed, but just as a fire gives back more heat than it takes to start it, aerobic respiration eventually gives back far more ATP than it takes to prime glycolysis.

③ **Cleavage.** The "lysis" part of glycolysis occurs when fructose 1,6-diphosphate splits into two three-carbon ($C_3$) molecules. Through a slight rearrangement of one of them (not shown in the figure), this generates two molecules of **PGAL (phosphoglyceraldehyde,** also called **glyceraldehyde 3-phosphate).**

④ **Oxidation.** Each PGAL is then oxidized by removing a pair of hydrogen atoms. The electrons and one proton are picked up by NAD⁺ and the other proton is released into the cytosol, yielding NADH + H⁺. At this step, a phosphate ($P_i$) group is also added to each of the $C_3$ fragments. Unlike the earlier steps, this $P_i$ is not supplied by ATP but comes from the cell's pool of free phosphate ions.

⑤ **Dephosphorylation.** In the next two steps, phosphate groups are taken from the glycolysis intermediates and transferred to ADP, converting it to ATP. The $C_3$ compound becomes pyruvic acid.

The net end products of glycolysis are therefore

$$2 \text{ pyruvic acid} + 2 \text{ NADH} + 2 \text{ H}^+ + 2 \text{ ATP}$$

Note that 4 ATP are actually produced (step 5), but 2 ATP were consumed to initiate glycolysis (steps 1 and 2), so the net gain is 2 ATP per glucose. Some of the energy originally in the glucose is contained in this ATP, some is in the NADH, and some is lost as heat. Most of the energy, however, remains in the pyruvic acid.

## Anaerobic Fermentation

The fate of pyruvic acid depends on whether or not oxygen is available. In an exercising muscle, the demand for ATP may exceed the supply of oxygen, in which case the muscle cells must rely on the 2 ATP produced by glycolysis. The only ATP the cells can make under these circumstances is the 2 ATP produced by glycolysis. Cells without mitochondria, such as erythrocytes, are also restricted to making ATP by this method.

But glycolysis would quickly come to a halt if the reaction stopped at pyruvic acid. Why? Because it would use up the supply of NAD⁺, which is needed to accept electrons at step 4 and keep glycolysis going. NAD⁺ must be replenished.

In the absence of oxygen, a cell resorts to a one-step reaction called anaerobic fermentation. (This is often inaccurately called *anaerobic respiration,* but strictly speaking, human cells do not carry out anaerobic respiration; that is a process found only in certain bacteria.) In this pathway, NADH donates a pair of electrons to pyruvic acid, thus reducing it to **lactic acid** and regenerating $NAD^+$.

▶▶▶**APPLY WHAT YOU KNOW**

*Does lactic acid have more free energy than pyruvic acid or less? Explain.*

Lactic acid leaves the cells that generate it and travels by way of the bloodstream to the liver. When oxygen becomes available again, the liver oxidizes lactic acid back to pyruvic acid, which can then enter the aerobic pathway described shortly. The oxygen required to do this is part of the *excess postexercise oxygen consumption* created by exercising skeletal muscles (see p. 421). The liver can also convert lactic acid back to G6P and can do either of two things with that: (1) polymerize it to form glycogen for storage or (2) remove the phosphate group and release free glucose into the blood.

Although anaerobic fermentation keeps glycolysis running a little longer, it has some drawbacks. One is that it is wasteful, because most of the energy of glucose is still in the lactic acid and has contributed no useful work. The other is that lactic acid is toxic.

Skeletal muscle is relatively tolerant of anaerobic fermentation, and cardiac muscle is less so. The brain employs almost no anaerobic fermentation. During birth, when the infant's blood supply is cut off, almost every organ of its body switches to anaerobic fermentation so they do not compete with the brain for the limited supply of oxygen.

## Aerobic Respiration

Most ATP is generated in the mitochondria, which require oxygen as the final electron acceptor. In the presence of oxygen, pyruvic acid enters the mitochondria and is oxidized by aerobic respiration. This occurs in two principal steps:

- a group of reactions we will call the **matrix reactions,** because their controlling enzymes are in the fluid of the mitochondrial matrix; and

- reactions we will call the **membrane reactions,** because their controlling enzymes are bound to the membranes of the mitochondrial cristae.

## The Matrix Reactions

The matrix reactions are shown in figure 26.4, where the reaction steps are numbered to resume where figure 26.3 ended. Most of the matrix reactions constitute a series called the **citric acid (Krebs[5]) cycle.** Preceding this, however, are three steps

that prepare pyruvic acid to enter the cycle and thus link glycolysis to it.

⑥ Pyruvic acid is *decarboxylated;* that is, $CO_2$ is removed and pyruvic acid, a $C_3$ compound, becomes a $C_2$ compound.

⑦ $NAD^+$ removes hydrogen atoms from the $C_2$ compound (an oxidation reaction) and converts it to an **acetyl group (acetic acid).**

⑧ The acetyl group binds to coenzyme A, a derivative of pantothenic acid (a B vitamin). The result is **acetyl-coenzyme A (acetyl-CoA).** At this stage, the $C_2$ remnant of the original glucose molecule is ready to enter the citric acid cycle.

⑨ At the beginning of the citric acid cycle, CoA hands off the acetyl ($C_2$) group to a $C_4$ compound, **oxaloacetic acid.** This produces the $C_6$ compound **citric acid,** for which the cycle is named.

⑩ Water is removed and the citric acid molecule is reorganized, but still retains its six carbon atoms.

⑪ Hydrogen atoms are removed and accepted by $NAD^+$.

⑫ Another $CO_2$ is removed and the substrate becomes a five-carbon chain.

⑬ - ⑭ Steps 11 and 12 are essentially repeated, generating another free $CO_2$ molecule and leaving a four-carbon chain. No more carbon atoms are removed beyond this point; the substrate remains a series of $C_4$ compounds from here back to the start of the cycle. The three carbon atoms of pyruvic acid have all been removed as $CO_2$ at steps 6, 12, and 14. These *decarboxylation reactions* are the source of most of the $CO_2$ in your breath.

⑮ Some of the energy in the $C_4$ substrate goes to phosphorylate guanosine diphosphate (GDP) and to convert it to guanosine triphosphate (GTP), a molecule similar to ATP. GTP quickly transfers the $P_i$ group to ADP to make ATP. Coenzyme A participates again in this step but is not shown in the figure.

⑯ Two hydrogen atoms are removed and accepted by the coenzyme FAD.

⑰ Water is added.

⑱ Two final hydrogen atoms are removed and transferred to $NAD^+$. This reaction generates oxaloacetic acid, which is available to start the cycle all over again.

It is important to remember that for every glucose molecule that entered glycolysis, all of these matrix reactions occur

---

[5]Sir Hans Krebs (1900–1981), German biochemist

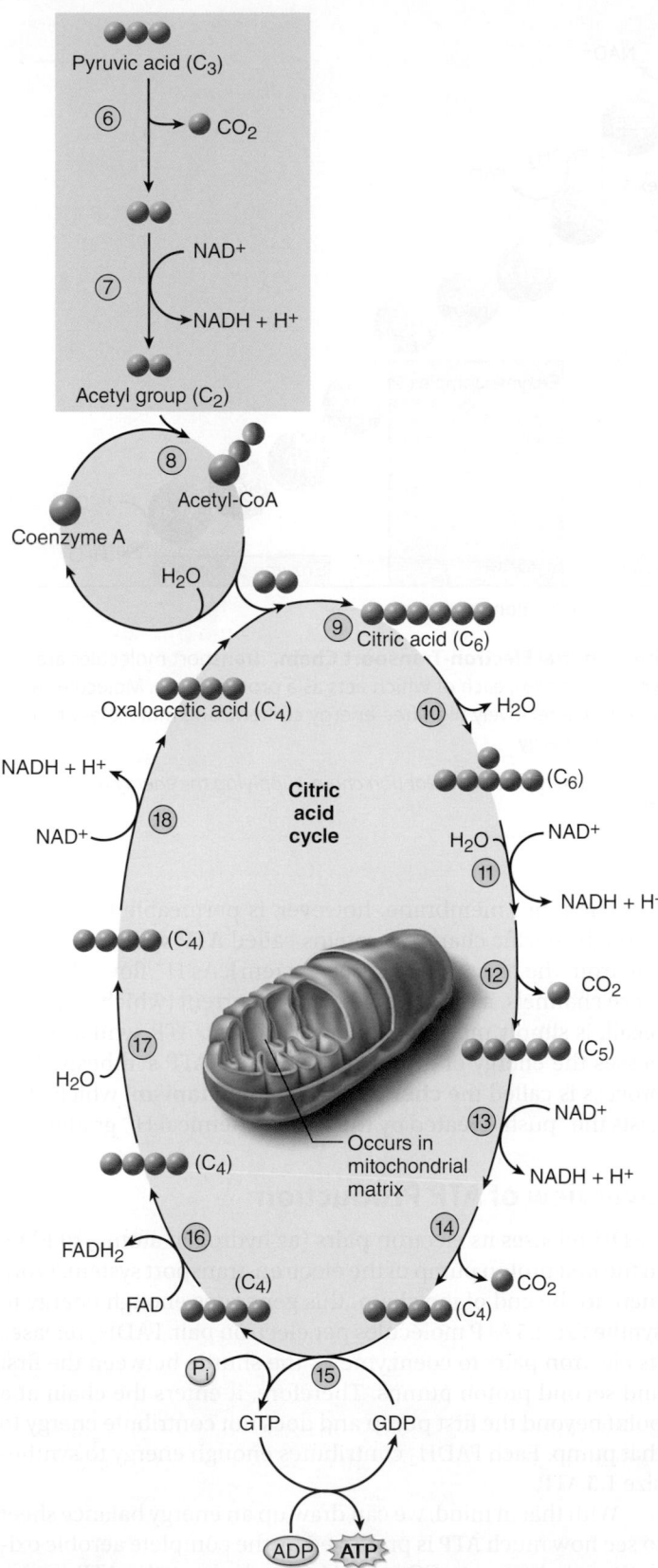

twice (once for each pyruvic acid). The matrix reactions can be summarized:

$$2 \text{ pyruvate} + 6\,H_2O \rightarrow 6\,CO_2$$
$$+ 2\,ADP + 2\,P_i \rightarrow 2\,ATP$$
$$+ 8\,NAD^+ + 8\,H_2 \rightarrow 8\,NADH + 8\,H^+$$
$$+ 2\,FAD + 2\,H_2 \rightarrow 2\,FADH_2$$

There is nothing left of the organic matter of the glucose; its carbon atoms have all been carried away as $CO_2$. Although still more of its energy is lost as heat along the way, some is stored in the additional 2 ATP, and most of it, by far, is in the reduced coenzymes—8 NADH and 2 $FADH_2$ molecules generated by the matrix reactions and 2 NADH generated by glycolysis. These must be oxidized to extract the energy from them.

The citric acid cycle not only oxidizes glucose metabolites but is also a pathway and a source of intermediates for the synthesis of fats and nonessential amino acids. The connections between the citric acid cycle and the metabolism of other nutrients are discussed later.

## The Membrane Reactions

The membrane reactions have two purposes: (1) to further oxidize NADH and $FADH_2$ and transfer their energy to ATP and (2) to regenerate $NAD^+$ and FAD and make them available again to earlier reaction steps. The membrane reactions are carried out by a series of compounds called the **mitochondrial electron-transport chain** (fig. 26.5). Most members of the chain are bound to the inner mitochondrial membrane. They are arranged in a precise order that enables each one to receive a pair of electrons from the member on one side of it (or, in two cases, from NADH and $FADH_2$) and pass these electrons along to the member on the other side—like a row of people passing along a hot potato. By the time the "potato" reaches the last member in the chain, it is relatively "cool"—its energy has been used to make ATP.

The members of this transport chain are as follows:

- **Flavin mononucleotide (FMN),** a derivative of riboflavin similar to FAD, bound to a membrane protein. FMN accepts electrons from NADH.

- **Iron–sulfur (Fe-S) centers,** complexes of iron and sulfur atoms bound to membrane proteins.

- **Coenzyme Q (CoQ),** which accepts electrons from $FADH_2$. Unlike the other members, this is a relatively small, mobile molecule that moves about in the membrane.

- **Copper (Cu) ions** bound to two membrane proteins.

- **Cytochromes,**[6] five enzymes with iron cofactors, so named because they are brightly colored in pure form. In order of participation in the chain, they are cytochromes $b$, $c_1$, $c$, $a$, and $a_3$.

**FIGURE 26.4 The Mitochondrial Matrix Reactions.** Numbered reaction steps are explained in the text.

---

[6]*cyto* = cell; *chrom* = color

***Electron Transport***   Figure 26.5 shows the order in which electrons are passed along the chain. Hydrogen atoms are split apart as they transfer from coenzymes to the chain. The protons are pumped into the intermembrane space (fig. 26.6), and the electrons travel in pairs ($2 \, e^-$) along the transport chain. Each electron carrier in the chain becomes reduced when it receives an electron pair and oxidized again when it passes the electrons along to the next carrier. Energy is liberated at each transfer.

The final electron acceptor in the chain is oxygen. Each oxygen atom (half of an $O_2$ molecule) accepts two electrons ($2 \, e^-$) from cytochrome $a_3$ and two protons ($2 \, H^+$) from the mitochondrial matrix. The result is a molecule of water:

$$1/2 \, O_2 + 2 \, e^- + 2 \, H^+ \rightarrow H_2O$$

This is the body's primary source of *metabolic water*—water synthesized in the body rather than ingested in food and drink. This reaction also explains why the body requires oxygen. Without it, this reaction stops and, like a traffic jam, stops all the other processes leading to it. As a result, a cell produces too little ATP to sustain life, and death ensues within a few minutes.

***The Chemiosmotic Mechanism***   Of primary importance is what happens to the energy liberated by the electrons as they pass along the chain. Some of it is unavoidably lost as heat, but some of it drives the **respiratory enzyme complexes.** The first complex includes FMN and five or more Fe–S centers; the second complex includes cytochromes $b$ and $c_1$ and an Fe–S center; and the third complex includes two copper centers and cytochromes $a$ and $a_3$. Each complex collectively acts as a **proton pump** that removes $H^+$ from the mitochondrial matrix and pumps it into the space between the inner and outer mitochondrial membranes (fig. 26.6). Coenzyme Q is a shuttle that transfers electrons from the first pump to the second, and cytochrome $c$ shuttles electrons from the second pump to the third.

These pumps create a very high $H^+$ concentration (low pH) and positive charge between the membranes compared with a low $H^+$ concentration and negative charge in the mitochondrial matrix. That is, they create a steep electrochemical gradient across the inner mitochondrial membrane. If the inner membrane were freely permeable to $H^+$, these ions would have a strong tendency to diffuse down this gradient and back into the matrix.

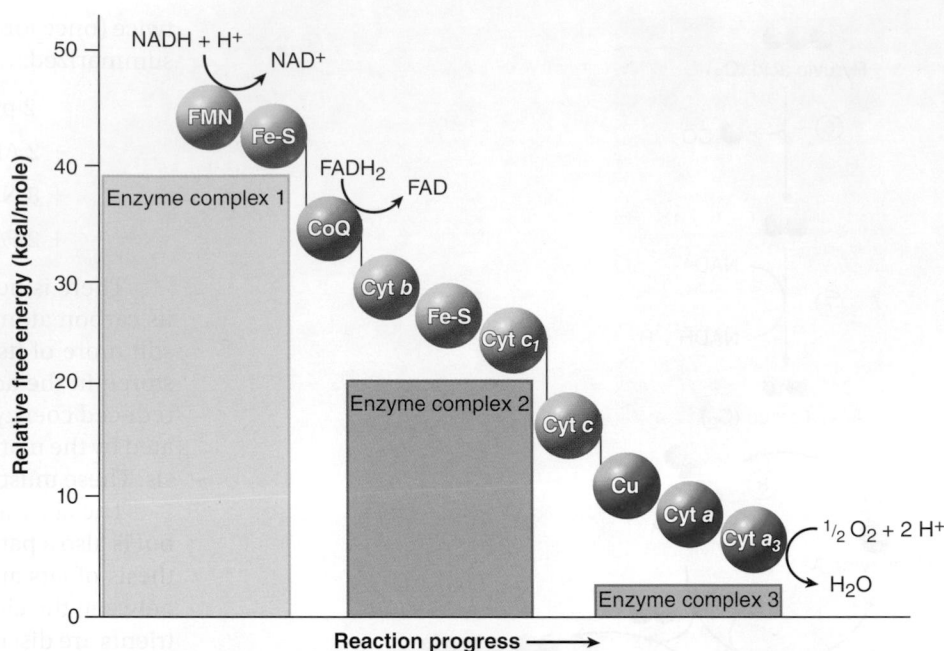

**FIGURE 26.5  The Mitochondrial Electron-Transport Chain.**  Transport molecules are grouped into three enzyme complexes, each of which acts as a proton pump. Molecules at the upper left of the figure have a relatively high free energy content, and molecules at the lower right are relatively low in energy.

❓ *What two molecules import energy into this reaction chain, supplying the energy that becomes stored in ATP?*

The inner membrane, however, is permeable to $H^+$ only through specific channel proteins called **ATP synthase** (separate from the electron-transport system). As $H^+$ flows through these channels, it creates an electrical current (which, you may recall, is simply moving charged particles). ATP synthase harnesses the energy of this current to drive ATP synthesis. This process is called the **chemiosmotic[7] mechanism,** which suggests the "push" created by the electrochemical $H^+$ gradient.

## Overview of ATP Production

NADH releases its electron pairs (as hydrogen atoms) to FMN in the first proton pump of the electron-transport system. From there to the end of the chain, this generates enough energy to synthesize 2.5 ATP molecules per electron pair. $FADH_2$ releases its electron pairs to coenzyme Q, the shuttle between the first and second proton pumps. Therefore, it enters the chain at a point beyond the first pump and does not contribute energy to that pump. Each $FADH_2$ contributes enough energy to synthesize 1.5 ATP.

With that in mind, we can draw up an energy balance sheet to see how much ATP is produced by the complete aerobic oxidation of glucose to $CO_2$ and $H_2O$ and where the ATP comes

---

[7]*chemi* = chemical; *osmo* = push

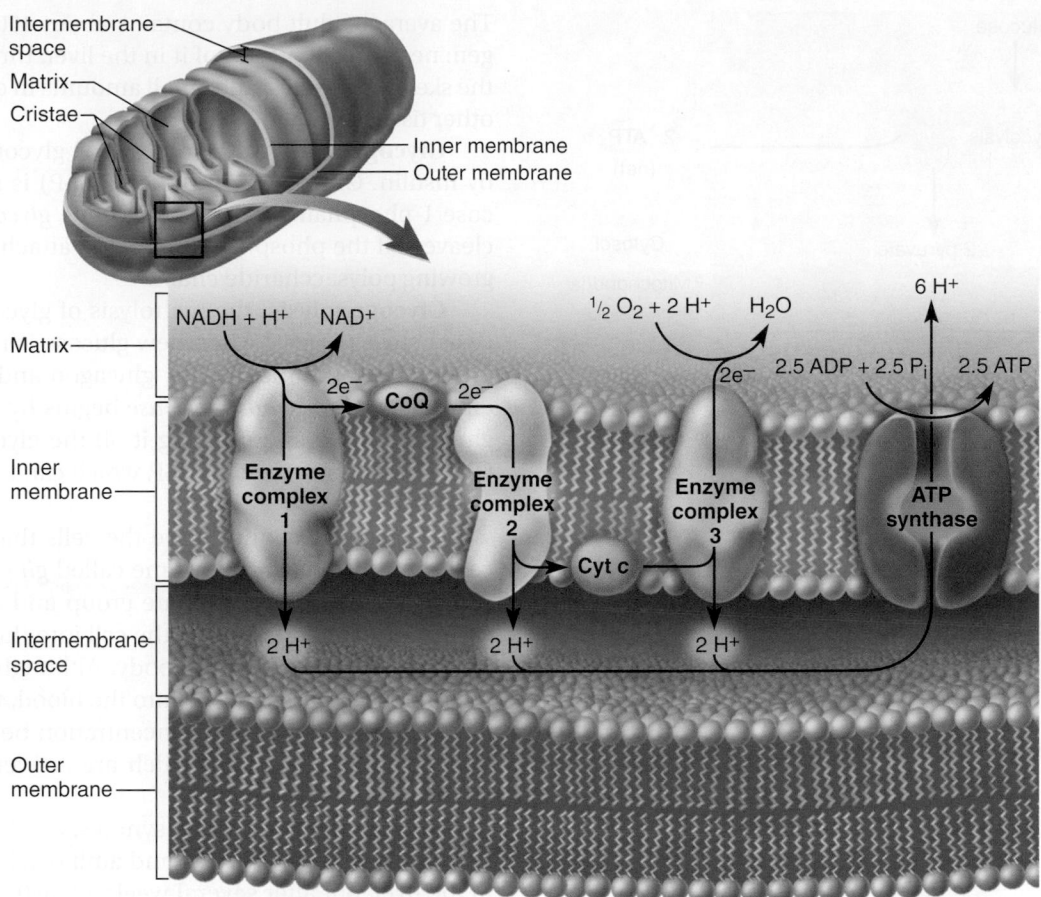

**FIGURE 26.6 The Chemiosmotic Mechanism of ATP Synthesis.** Each enzyme complex pumps hydrogen ions into the space between the mitochondrial membranes. These hydrogen ions diffuse back into the matrix by way of ATP synthase, which taps their energy to synthesize ATP.

from; see also figure 26.7. This summary refers back to the re-action steps 1 to 18 in figures 26.3 and 26.4. For each glucose molecule, there are

|  | 10 | NADH produced at steps 4, 7, 11, 13, and 18 |
|  | × 2.5 | ATP per NADH produced by the electron-transport chain |
|  | **25** | **ATP** generated by NADH |
| Plus: | 2 | FADH$_2$ produced at step 16 |
|  | × 1.5 | ATP per FADH$_2$ produced by the electron-transport chain |
|  | = **3** | **ATP** generated by FADH$_2$ |
| Plus: | 2 | **ATP** net amount generated by glycolysis (step 5 offset by step 2) |
|  | 2 | **ATP** generated by the matrix reactions (step 15) |
| Total: | **32** | **ATP** per glucose |

This should be viewed as a theoretical maximum. There is some uncertainty about how much H$^+$ must be pumped between the mitochondrial membranes to generate 1 ATP, and some of the energy from the H$^+$ current is consumed by pumping ATP from the mitochondrial matrix into the cytosol and exchanging it for more raw materials (ADP and P$_i$) pumped from the cytosol into the mitochondria.

Furthermore, the NADH generated by glycolysis can-not enter the mitochondria and donate its electrons directly to the electron-transport chain. In liver, kidney, and heart cells, NADH passes its electrons to *malate,* a shuttle molecule that delivers the electrons to the beginning of the electron-transport chain and generates the maximum amount of ATP. In skeletal muscle and brain cells, however, the glycolytic NADH transfers its electrons to *glycerol phosphate,* a different shuttle that donates the electrons farther down the electron-transport chain and generates less ATP. Therefore, the amount of ATP produced per NADH differs from one cell type to another and is still unknown for others.

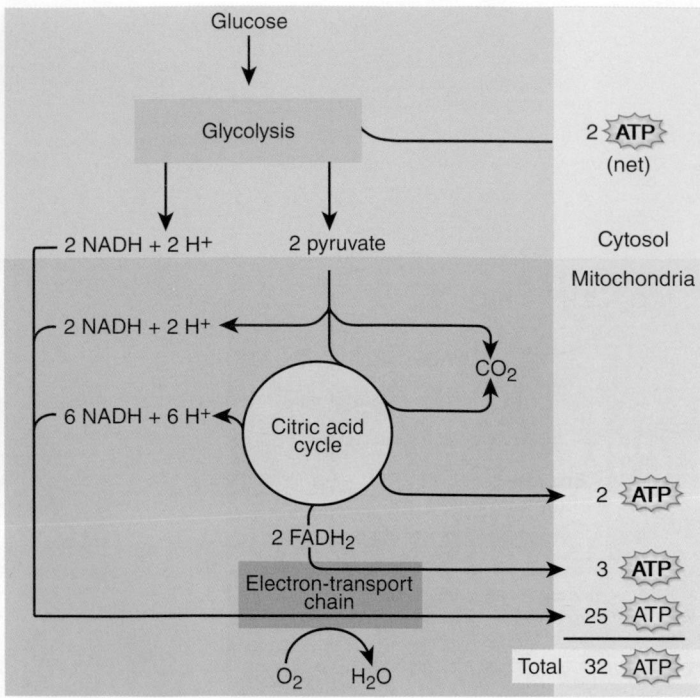

**FIGURE 26.7   Summary of the Sources of ATP Generated by the Complete Oxidation of Glucose.**

But if we assume the maximum ATP yield, every mole (180 g) of glucose releases enough energy to synthesize up to 32 moles of ATP. Glucose has an energy content of 686 kcal/mole and ATP has 7.3 kcal/mole (233.6 kcal in 32 moles). This means that aerobic respiration has an **efficiency** (a ratio of energy output to input) of up to 233.6/686 kcal = 34%. The other 66% is lost as body heat.

The pathways of glucose catabolism are summarized in table 26.4. The aerobic respiration of glucose can be represented in the summary equation:

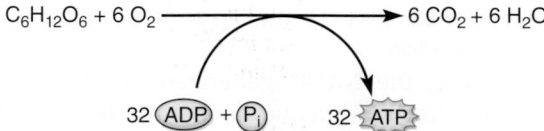

## Glycogen Metabolism

ATP is quickly used after it is synthesized—it is an *energy-transfer* molecule, not an *energy-storage* molecule. Therefore, if the body has an ample amount of ATP and there is still more glucose in the blood, it does not produce and store excess ATP but converts the glucose to other compounds better suited for energy storage—namely, glycogen and fat. Fat synthesis is considered later. Here we consider the synthesis and use of glycogen, as well as the generation of glucose from other sources.

The average adult body contains about 400 to 450 g of glycogen: nearly one-quarter of it in the liver, three-quarters of it in the skeletal muscles, and small amounts in cardiac muscle and other tissues.

**Glycogenesis,** the synthesis of glycogen, is stimulated by insulin. Glucose 6-phosphate (G6P) is isomerized to glucose 1-phosphate (G1P). The enzyme *glycogen synthase* then cleaves off the phosphate group and attaches the glucose to a growing polysaccharide chain.

**Glycogenolysis,** the hydrolysis of glycogen, releases glucose between meals when new glucose is not being ingested. The process is stimulated by glucagon and epinephrine. The enzyme *glycogen phosphorylase* begins by phosphorylating a glucose residue and splitting it off the glycogen molecule as G1P. This is isomerized to G6P, which can then enter the pathway of glycolysis.

G6P usually cannot leave the cells that produce it. Liver cells, however, have an enzyme called *glucose 6-phosphatase,* which removes the phosphate group and produces free glucose. This can diffuse out of the cell into the blood, where it is available to any cells in the body. Although muscle cells cannot directly release glucose into the blood, they contribute indirectly to blood glucose concentration because they release pyruvic and lactic acids, which are converted to glucose by the liver.

**Gluconeogenesis**[8] is the synthesis of glucose from noncarbohydrates such as glycerol and amino acids. It occurs chiefly in the liver, but after several weeks of fasting, the kidneys also undertake this process and eventually produce just as much glucose as the liver does.

The processes described here are summarized in figure 26.8, and the distinctions among those similar terms are summarized in table 26.5.

**BEFORE YOU GO ON**

Answer the following questions to test your understanding of the preceding section:

6. Identify the reaction steps in figures 26.3 and 26.4 at which vitamins are needed for glucose catabolism.

7. In the laboratory, glucose can be oxidized in a single step to $CO_2$ and $H_2O$. Why is it done in so many little steps in cells?

8. Explain the origin of the word *glycolysis* and why this is an appropriate name for the function of that reaction pathway.

9. What are two advantages of aerobic respiration over anaerobic fermentation?

10. What important enzyme is found in the inner mitochondrial membrane other than those of the electron-transport chain? Explain how its function depends on the electron-transport chain.

11. Describe how the liver responds to (a) an excess and (b) a deficiency of blood glucose.

---

[8]*gluco* = sugar, glucose; *neo* = new; *genesis* = production of

| TABLE 26.4 | Pathways of Glucose Catabolism | | |
|---|---|---|---|
| **Stage** | **Principal Reactants** | **Principal Products** | **Purpose** |
| Glycolysis | Glucose, 2 ADP, 2 $P_i$, 2 $NAD^+$ | 2 pyruvic acid, 2 ATP, 2 NADH, 2 $H_2O$ | Reorganizes glucose and splits it in two in preparation for further oxidation by the mitochondria; sole source of ATP in anaerobic conditions |
| Anaerobic fermentation | 2 pyruvic acid, 2 NADH | 2 lactic acid, 2 $NAD^+$ | Regenerates $NAD^+$ so glycolysis can continue to function (and generate ATP) in the absence of oxygen |
| Aerobic respiration | | | |
| Matrix reactions | 2 pyruvic acid, 8 $NAD^+$, 2 FAD, 2 ADP, 2 $P_i$, 8 $H_2O$ | 6 $CO_2$, 8 NADH, 2 $FADH_2$, 2 ATP, 2 $H_2O$ | Remove electrons from pyruvic acid and transfer them to coenzymes $NAD^+$ and FAD; produce some ATP |
| Membrane reactions | 10 NADH, 2 $FADH_2$, 6 $O_2$ | Up to 28 ATP, 12 $H_2O$ | Finish oxidation and produce most of the ATP of cellular respiration |

## 26.3 Lipid and Protein Metabolism

### Expected Learning Outcomes

When you have completed this section, you should be able to

a. describe the processes of lipid catabolism and anabolism;

b. describe the processes of protein catabolism and anabolism; and

c. explain the metabolic source of ammonia and how the body disposes of it.

In the foregoing discussion, glycolysis and the mitochondrial reactions were treated from the standpoint of carbohydrate oxidation. These pathways also serve for the oxidation of proteins and lipids as fuel and as a source of metabolic intermediates that can be used for protein and lipid synthesis. Here we examine these related metabolic pathways.

### Lipids

Triglycerides are stored primarily in the body's adipocytes, where a given molecule remains for about 2 to 3 weeks. Although the total amount of stored triglyceride remains quite constant, there is a continual turnover as lipids are released, transported in the blood, and either oxidized for energy or redeposited in other adipocytes. Synthesizing fats from other types of molecules is called **lipogenesis,** and breaking down fat for fuel is called **lipolysis** (lih-POL-ih-sis).

### Lipogenesis

It is common knowledge that a diet high in sugars causes us to put on fat. Lipogenesis employs compounds such as sugars and amino acids to synthesize glycerol and fatty acids, the triglyceride precursors. PGAL, one of the intermediates of glycolysis, can be converted to glycerol. As glucose and amino

| TABLE 26.5 | Some Terminology Related to Glucose and Glycogen Metabolism |
|---|---|
| **Anabolic (synthesis) reactions** | |
| Glycogenesis | The synthesis of glycogen by polymerizing glucose |
| Gluconeogenesis | The synthesis of glucose from noncarbohydrates such as glycerol and amino acids |
| **Catabolic (breakdown) reactions** | |
| Glycolysis | The splitting of glucose into two molecules of pyruvic acid in preparation for anaerobic fermentation or aerobic respiration |
| Glycogenolysis | The hydrolysis of glycogen to release free glucose or glucose 1-phosphate |

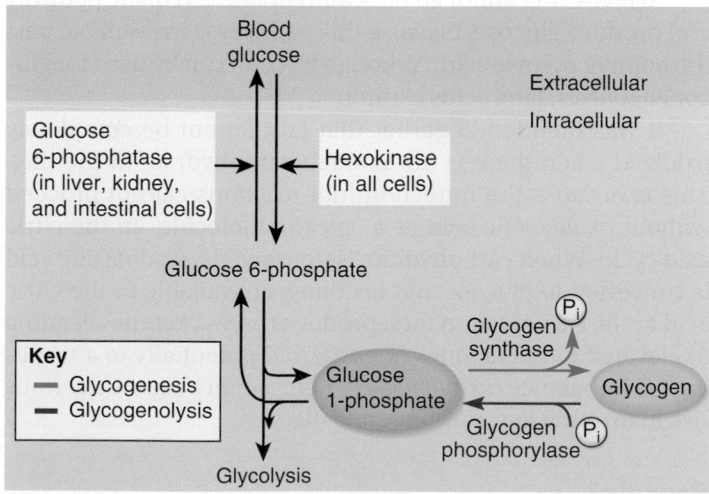

**FIGURE 26.8 Major Pathways of Glucose Storage and Use.**
In most cells, the glucose 1-phosphate generated by glycogenolysis can undergo only glycolysis. In liver, kidney, and intestinal cells, it can be converted back to free glucose and released into circulation.

acids enter the citric acid cycle by way of acetyl-CoA, the acetyl-CoA can also be diverted to make fatty acids. The glycerol and fatty acids can then be condensed to form a triglyceride, which can be stored in the adipose tissue or converted to other lipids. These pathways are summarized in figure 26.9.

## Lipolysis

Lipolysis, also shown in figure 26.9, begins with the hydrolysis of a triglyceride into glycerol and fatty acids—a process stimulated by epinephrine, norepinephrine, glucocorticoids, thyroid hormone, and growth hormone. The glycerol and fatty acids are further oxidized by separate pathways. Glycerol is easily converted to PGAL and thus enters the pathway of glycolysis. It generates only half as much ATP as glucose, however, because it is a $C_3$ compound compared with glucose ($C_6$); thus, it leads to the production of only half as much pyruvic acid.

The fatty acid component is catabolized in the mitochondrial matrix by a process called **beta oxidation,** which removes 2 carbon atoms at a time. The resulting acetyl ($C_2$) groups are bonded to coenzyme A to make acetyl-CoA—the entry point into the citric acid cycle. A fatty acid of 16 carbon atoms can yield 129 molecules of ATP—obviously a much richer source of energy than a glucose molecule; remember that each fat molecule generates 3 fatty acids.

Excess acetyl groups can be metabolized by the liver in a process called **ketogenesis.** Two acetyl groups are condensed to form acetoacetic acid, and some of this is further converted to β-hydroxybutyric acid and acetone. These three products are the **ketone bodies.** Some cells convert acetoacetic acid back to acetyl-CoA and thus feed the $C_2$ fragments into the citric acid cycle to extract their energy. When the body is rapidly oxidizing fats, however, excess ketone bodies accumulate. This causes the ketoacidosis typical of type 1 diabetes mellitus, in which cells must oxidize fats because they cannot absorb glucose.

Acetyl-CoA cannot go backward up the glycolytic pathway and produce glucose, because this pathway is irreversible past the point of pyruvic acid. Although glycerol can be used for gluconeogenesis, fatty acids cannot.

It was mentioned earlier that fats cannot be completely oxidized when there is not enough carbohydrate in the diet. This is because the mitochondrial reactions cannot proceed without oxaloacetic acid as a "pickup molecule" in the citric acid cycle. When carbohydrate is unavailable, oxaloacetic acid is converted to glucose and becomes unavailable to the citric acid cycle. Fat oxidation then produces excess ketones, leading to elevated blood ketones *(ketosis)* and potentially to a resulting pH imbalance *(ketoacidosis)*. Ketosis can become a serious risk in extreme low-carbohydrate diets.

## Proteins

About 100 g of tissue protein breaks down each day into free amino acids. These combine with the amino acids from the diet to form an **amino acid pool** that cells can draw upon to

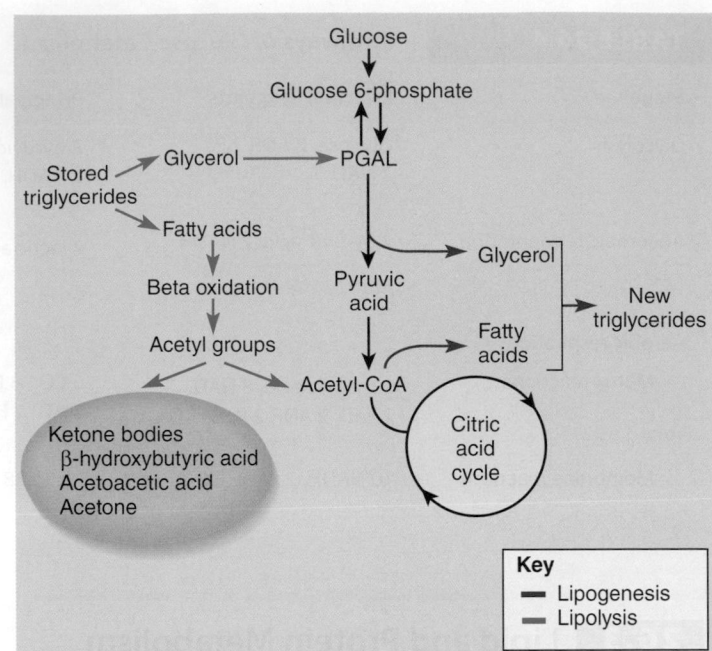

**FIGURE 26.9** **Pathways of Lipolysis and Lipogenesis in Relation to Glycolysis and the Citric Acid Cycle.**

*Name the acid–base imbalance that results from the accumulation of the ketone bodies shown in the oval.*

make new proteins. The fastest rate of tissue protein turnover is in the intestinal mucosa, where epithelial cells are replaced at a very high rate. Dead cells are digested along with the food and thus contribute to the amino acid pool. Of all the amino acid absorbed by the small intestine, about 50% is from the diet, 25% from dead epithelial cells, and 25% from enzymes that have digested each other.

Some amino acids in the pool can be converted to others. Free amino acids also can be converted to glucose and fat or directly used as fuel. Such conversions involve three processes: (1) **deamination,** the removal of an amino group (–$NH_2$); (2) **amination,** the addition of –$NH_2$; or (3) **transamination,** the transfer of –$NH_2$ from one molecule to another. The following discussion shows how these processes are involved in amino acid metabolism.

## Use as Fuel

The first step in using amino acids as fuel is to deaminate them. After the –$NH_2$ group is removed, the remainder of the molecule is called a *keto acid.* Depending on which amino acid is involved, the resulting keto acid may be converted to pyruvic acid, acetyl-CoA, or one of the acids of the citric acid cycle (fig. 26.10). It is important to note that some of these reactions are reversible. When there is a deficiency of amino acids in the body, citric acid cycle intermediates can be aminated and converted to amino acids, which are then available for protein synthesis. In gluconeogenesis, keto acids are used to synthesize glucose, essentially through a reversal of the glycolysis reactions.

## Transamination, Ammonia, and Urea

When an amino acid is deaminated, its amino group is transferred to a citric acid cycle intermediate, α-ketoglutaric acid, converting it to glutamic acid. Such transamination reactions are the route by which several amino acids enter the citric acid cycle.

Glutamic acid can travel from any of the body's cells to the liver. Here, its –NH$_2$ group is removed, converting it back to α-ketoglutaric acid. The –NH$_2$ becomes ammonia (NH$_3$), which is extremely toxic to cells and cannot be allowed to accumulate. In a pathway called the **urea cycle,** the liver quickly combines ammonia with carbon dioxide to produce a less toxic waste, urea. Urea is then excreted in the urine as one of the body's nitrogenous wastes. Other nitrogenous wastes and their sources are described in chapter 23 (see p. 890). When a diseased liver cannot carry out the urea cycle, NH$_3$ accumulates in the blood and death from *hepatic coma* may ensue within a few days.

## Protein Synthesis

Protein synthesis, described in detail in chapter 4, is a complex process involving DNA, mRNA, tRNA, ribosomes, and often the rough ER. It is stimulated by growth hormone, thyroid hormones, and insulin; and it requires a supply of all the amino acids necessary for a particular protein. The liver can make many of these amino acids from other amino acids or from citric acid cycle intermediates by transamination reactions. The essential amino acids, however, must be obtained from the diet.

## Liver Functions in Metabolism

You may notice that the liver plays a wide variety of roles in the processes discussed in this chapter—especially carbohydrate, lipid, and protein metabolism. Although it is connected to the digestive tract and regarded as a digestive gland, most of its functions are nondigestive (table 26.6). Except for phagocytosis, all of them are performed by the hepatocytes described in chapter 25. Such functional diversity is remarkable in light of the uniform structure of these cells. Because of the numerous critical functions performed by the liver, degenerative liver diseases such as hepatitis, cirrhosis, and liver cancer are especially life-threatening (see Deeper Insight 26.3).

### BEFORE YOU GO ON

Answer the following questions to test your understanding of the preceding section:

12. Which of the processes in table 26.5 is most comparable to lipogenesis? Which is most comparable to lipolysis? Explain.

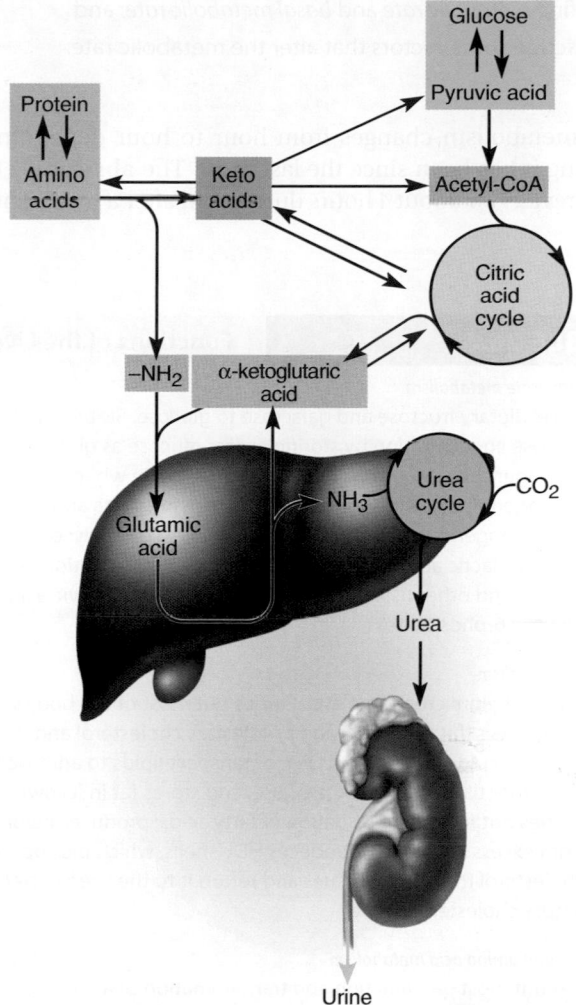

**FIGURE 26.10 Pathways of Amino Acid Metabolism in Relation to Glycolysis and the Citric Acid Cycle.**

❓ *Find a pathway for gluconeogenesis in this diagram.*

13. When fats are converted to glucose, only the glycerol component is used in this way, not the fatty acid. Explain why, and state what happens to the fatty acids.

14. What metabolic process produces ammonia? How does the body dispose of ammonia?

## 26.4 Metabolic States and Metabolic Rate

### Expected Learning Outcomes

When you have completed this section, you should be able to

a. define the *absorptive* and *postabsorptive states;*

b. explain what happens to carbohydrates, fats, and proteins in each of these states;

c. describe the hormonal and nervous regulation of each state;

d. define *metabolic rate* and *basal metabolic rate;* and

e. describe some factors that alter the metabolic rate.

One's metabolism changes from hour to hour depending on how long it has been since the last meal. The **absorptive (fed) state** prevails for about 4 hours during and after a meal. This is a time in which nutrients are being absorbed and may be used immediately to meet energy and other needs. The **postabsorptive (fasting) state** prevails in the late morning, late afternoon, and overnight. During this time, the stomach and small intestine are empty and the body's energy needs are met from stored fuels. The two states are compared in table 26.7 and explained in the following discussion.

| **TABLE 26.6** | **Functions of the Liver** |
|---|---|

*Carbohydrate metabolism*
Converts dietary fructose and galactose to glucose. Stabilizes blood glucose concentration by storing excess glucose as glycogen (glycogenesis), releasing glucose from glycogen when needed (glycogenolysis), and synthesizing glucose from fats and amino acids (gluconeogenesis) when glucose demand exceeds glycogen reserves. Receives lactic acid generated by anaerobic fermentation in skeletal muscle and other tissues and converts it back to pyruvic acid or glucose 6-phosphate.

*Lipid metabolism*
Degrades chylomicron remnants. Carries out most of the body's lipogenesis (fat synthesis) and synthesizes cholesterol and phospholipids; produces VLDLs to transport lipids to adipose tissue and other tissues for storage or use; and stores fat in its own cells. Carries out most beta oxidation of fatty acids; produces ketone bodies from excess acetyl-CoA. Produces HDL shells, which pick up excess cholesterol from other tissues and return it to the liver; excretes the excess cholesterol in bile.

*Protein and amino acid metabolism*
Carries out most deamination and transamination of amino acids. Removes $-NH_2$ from glutamic acid and converts the resulting ammonia to urea by means of the urea cycle. Synthesizes nonessential amino acids by transamination reactions.

*Synthesis of plasma proteins*
Synthesizes nearly all the proteins of blood plasma, including albumin, alpha and beta globulins, fibrinogen, prothrombin, and several other clotting factors. (Does not synthesize plasma enzymes or gamma globulins.)

*Vitamin and mineral metabolism*
Converts vitamin $D_3$ to calcidiol, a step in the synthesis of calcitriol; stores a 3- to 4-month supply of vitamin D. Stores a 10-month supply of vitamin A and enough vitamin $B_{12}$ to last from one to several years. Secretes hepcidin to regulate iron absorption; stores iron in ferritin and releases it as needed. Excretes excess calcium by way of the bile.

*Digestion*
Synthesizes bile acids and lecithin, which emulsify fat and promote its digestion; produces micelles, which aid in absorption of dietary lipids.

*Disposal of drugs, toxins, and hormones*
Detoxifies alcohol, antibiotics, and many other drugs. Metabolizes bilirubin from RBC breakdown and excretes it as bile pigments. Deactivates thyroxine and steroid hormones and excretes them or converts them to a form more easily excreted by the kidneys.

*Phagocytosis*
Macrophages cleanse blood of bacteria and other foreign matter.

# DEEPER INSIGHT 26.3

## CLINICAL APPLICATION

### Hepatitis and Cirrhosis

*Hepatitis,* inflammation of the liver, is usually caused by one of the five strains of hepatitis viruses. They differ in mode of transmission, severity of the resulting illness, affected age groups, and the best strategies for prevention. Hepatitis A is common and mild. Over 45% of people in urban areas of the United States have had it. It spreads rapidly in such settings as day-care centers and residential institutions for psychiatric patients, and it can be acquired by eating uncooked seafood such as oysters. Hepatitis B and C are far more serious. Both are transmitted sexually and through blood and other body fluids; the incidence of hepatitis C has surpassed AIDS as a sexually transmitted disease. Initial signs and symptoms of hepatitis include fatigue, malaise, nausea, vomiting, and weight loss. The liver becomes enlarged and tender. Jaundice, or yellowing of the skin, tends to follow as hepatocytes are destroyed, bile passages are blocked, and bile pigments accumulate in the blood. Hepatitis A causes up to 6 months of illness, but most people recover and then have permanent immunity to it. Hepatitis B and C, however, often lead to chronic hepatitis,

which can progress to cirrhosis or liver cancer. More liver transplants are necessitated by hepatitis C than by any other cause.

*Cirrhosis* is an irreversible inflammatory liver disease. It develops slowly over a period of years, but has a high mortality rate and is one of the leading causes of death in the United States. It is characterized by a disorganized liver histology in which regions of scar tissue alternate with nodules of regenerating cells, giving the liver a lumpy or knobby appearance and hardened texture. As in hepatitis, blockage of the bile passages results in jaundice. Protein synthesis declines as the liver deteriorates, leading to ascites, impaired blood clotting, and other cardiovascular effects (see Deeper Insight 26.4). Obstruction of the hepatic circulation by scar tissue leads to *angiogenesis,* the growth of new blood vessels to bypass the liver. Deprived of blood, the condition of the liver worsens, with increasing necrosis and, often, liver failure. Most cases of cirrhosis result from alcohol abuse, but hepatitis, gallstones, pancreatic inflammation, and other conditions can also bring it about. The prognosis for recovery is often poor.

## The Absorptive State

In the absorptive state, blood glucose is readily available for ATP synthesis. It serves as the primary fuel and spares the body from having to draw on stored fuels. The status of major nutrient classes during this phase is as follows:

- **Carbohydrates.** Absorbed sugars are transported by the hepatic portal system to the liver. Most glucose passes through the liver and becomes available to cells everywhere in the body. Glucose in excess of immediate need, however, is absorbed by the liver and may be converted to glycogen or fat. Most fat synthesized in the liver is released into the circulation; its further fate is comparable to that of dietary fats, discussed next.

- **Fats.** Fats enter the lymphatic system as chylomicrons and initially bypass the liver. As described earlier, lipoprotein lipase removes fats from the chylomicrons for uptake by the tissues, especially adipose and muscular tissue. The liver disposes of the chylomicron remnants. Fats are the primary energy substrate for hepatocytes, adipocytes, and muscle.

- **Amino acids.** Amino acids, like sugars, circulate first to the liver. Most pass through and become available to other cells for protein synthesis. Some, however, are removed by the liver and have one of the following fates: (1) to be used for protein synthesis; (2) to be deaminated and used as fuel for ATP synthesis; or (3) to be deaminated and used for fatty acid synthesis.

### Regulation of the Absorptive State

The absorptive state is regulated largely by insulin, which is secreted in response to elevated blood glucose and amino acid levels and to the intestinal hormones gastrin, secretin, cholecystokinin, and glucose-dependent insulinotropic peptide (GIP). Insulin regulates the rate of glucose uptake by nearly all cells except neurons, kidney cells, and erythrocytes, which have independent rates of uptake. With those exceptions, insulin has the following effects on its target cells:

- Within minutes, it increases the cellular uptake of glucose by as much as 20-fold. As cells absorb glucose, the blood glucose concentration falls.

- It stimulates glucose oxidation, glycogenesis, and lipogenesis.

- It inhibits gluconeogenesis, which makes sense because blood glucose concentration is already high and there is no immediate need for more.

- It stimulates the active transport of amino acids into cells and promotes protein synthesis.

- It acts on the brain as an adiposity signal, an index of the body's fat stores.

Following a high-protein, low-carbohydrate meal, it may seem that the amino acids would stimulate insulin secretion; insulin would accelerate both amino acid and glucose uptake; and since there was relatively little glucose in the ingested food, this would create a risk of hypoglycemia. In actuality, this is prevented by the fact that a high amino acid level stimulates the secretion of *both* insulin and glucagon. Glucagon, you may recall, is an insulin antagonist. It supports an adequate level of blood glucose to meet the needs of the brain.

## The Postabsorptive State

The essence of the postabsorptive state is to homeostatically regulate blood glucose concentration within about 90 to 100 mg/dL. This is especially critical to the brain, which cannot

| TABLE 26.7 | Major Aspects of the Absorptive and Postabsorptive States | |
|---|---|---|
| | **Absorptive** | **Postabsorptive** |
| *Regulatory hormones* | Principally insulin | Principally glucagon |
| | Also gastrin, secretin, CCK, GIP | Also epinephrine, growth hormone |
| *Carbohydrate metabolism* | Blood glucose rising | Blood glucose falling |
| | Glucose uptake | |
| | Glucose stored by glycogenesis | Glucose released by glycogenolysis |
| | Gluconeogenesis suppressed | Gluconeogenesis stimulated |
| *Lipid metabolism* | Lipogenesis occurring | Lipolysis occurring |
| | Lipid uptake from chylomicrons | Fatty acids oxidized for fuel |
| | Lipid storage in fat and muscle | Glycerol used for gluconeogenesis |
| *Protein metabolism* | Amino acid uptake, protein synthesis | Amino acids oxidized if glycogen and fat stores are inadequate for energy needs |
| | Excess amino acids burned as fuel | |

use alternative energy substrates except in cases of prolonged fasting. The postabsorptive status of major nutrients is as follows:

- **Carbohydrates.** Glucose is drawn from the body's glycogen reserves (glycogenolysis) or synthesized from other compounds (gluconeogenesis). The liver usually stores enough glycogen after a meal to support 4 hours of postabsorptive metabolism before significant gluconeogenesis occurs.

- **Fats.** Adipocytes and hepatocytes hydrolyze fats and convert the glycerol to glucose. Free fatty acids (FFAs) cannot be converted to glucose, but they can favorably affect blood glucose concentration. As the liver oxidizes them to ketone bodies, other cells absorb and use these, or use FFAs directly, as their source of energy. By switching from glucose to fatty acid catabolism, they leave glucose for use by the brain (the glucose-sparing effect). After 4 to 5 days of fasting, the brain begins to use ketone bodies as supplemental fuel.

- **Proteins.** If glycogen and fat reserves are depleted, the body begins to use proteins as fuel. Some proteins are more resistant to catabolism than others. Collagen is almost never broken down for fuel, but muscle protein goes quickly. People with cancer and certain other chronic diseases sometimes exhibit **cachexia**[9] (ka-KEX-ee-ah), an extreme wasting away due to altered metabolism and loss of appetite (anorexia).

## Regulation of the Postabsorptive State

Postabsorptive metabolism is more complex than the absorptive state. It is regulated mainly by the sympathetic nervous system and glucagon, but several other hormones are involved. As blood glucose level drops, insulin secretion declines and the pancreatic alpha cells secrete glucagon. Glucagon promotes glycogenolysis and gluconeogenesis, raising the blood glucose level, and it promotes lipolysis and a rise in FFA levels. Thus, it makes both glucose and lipids available for fuel.

The sympathoadrenal system also promotes glycogenolysis and lipolysis, especially under conditions of injury, fear, anger, and other forms of stress. Adipose tissue is richly innervated by the sympathetic nervous system, while adipocytes, hepatocytes, and muscle cells also respond to epinephrine from the adrenal medulla. In circumstances in which there is likely to be tissue injury and a need for repair, the sympathoadrenal system therefore mobilizes stored energy reserves and makes them available to meet the demands of tissue repair. Stress also stimulates the release of cortisol, which promotes fat and protein catabolism and gluconeogenesis (see p. 660).

Growth hormone is secreted in response to a rapid drop in blood glucose level and in states of prolonged fasting. It opposes insulin and raises the blood glucose concentration.

---

[9]*cac* = bad; *exia* = body condition

## Metabolic Rate

**Metabolic rate** means the amount of energy liberated in the body per unit of time, expressed in such terms as kcal/h or kcal/day. Metabolic rate can be measured directly by putting a person in a **calorimeter,** a closed chamber with water-filled walls that absorb the heat given off by the body. The rate of energy release is measured from the temperature change of the water. Metabolic rate can also be measured indirectly with a spirometer, an apparatus described in chapter 22 that can be used to measure the amount of oxygen a person consumes. For every liter of oxygen, approximately 4.82 kcal of energy is released from organic nutrients. This is only an estimate, because the number of kilocalories per liter of oxygen varies slightly with the type of nutrients the person is oxidizing at the time of measurement.

Metabolic rate depends on physical activity, mental state, absorptive or postabsorptive status, thyroid hormone and other hormones, and other factors. The **basal metabolic rate (BMR)** is a baseline or standard of comparison that minimizes the effects of such variables. It is the metabolic rate when one is awake but relaxed, in a room at comfortable temperature, in a postabsorptive state 12 to 14 hours after the last meal. It is not the minimum metabolic rate needed to sustain life, however. When one is asleep, the metabolic rate is slightly lower than the BMR. **Total metabolic rate (TMR)** is the sum of BMR and energy expenditure for voluntary activities, especially muscular contractions.

The BMR of an average adult is about 2,000 kcal/day for a male and slightly less for a female. Roughly speaking, one must therefore consume at least 2,000 kcal/day to fuel essential metabolic tasks—active transport, muscle tone, brain activity, cardiac and respiratory rhythms, renal function, and other essential processes. Even a relatively sedentary lifestyle requires another 500 kcal/day to support a low level of physical activity, and someone who does hard physical labor may require as much as 5,000 kcal/day.

Aside from physical activity, some factors that raise the TMR and caloric requirements include pregnancy, anxiety (which stimulates epinephrine release and muscle tension), fever (TMR rises about 14% for each 1°C of body temperature), eating (TMR rises after a meal), and the catecholamine and thyroid hormones. TMR is relatively high in children and declines with age. Therefore, as we reach middle age we often find ourselves gaining weight with no apparent change in food intake.

Some factors that lower TMR include apathy, depression, and prolonged starvation. In weight-loss diets, loss is often rapid at first and then goes more slowly. This is partly because the initial loss is largely water and partly because the TMR drops over time, fewer dietary calories are "burned off," and there is more lipogenesis even with the same caloric intake. As one reduces food intake, the body reduces its metabolic rate to conserve body mass—thus making weight loss all the more difficult.

Answer the following questions to test your understanding of the preceding section:

15. Define *absorptive* and *postabsorptive states.* In which state is the body storing excess fuel? In which state is it drawing from these stored fuel reserves?

16. What hormone primarily regulates the absorptive state, and what are the major effects of this hormone?

17. Explain why triglycerides have a glucose-sparing effect.

18. List a variety of factors and conditions that raise a person's total metabolic rate above basal metabolic rate.

## 26.5 Body Heat and Thermoregulation

### Expected Learning Outcomes

When you have completed this section, you should be able to

a. identify the principal sources of body heat;

b. describe some factors that cause variations in body temperature;

c. define and contrast the different forms of heat loss;

d. describe how the hypothalamus monitors and controls body temperature; and

e. describe conditions in which the body temperature is excessively high or low.

The enzymes that control our metabolism depend on an optimal, stable working temperature. In order to maintain this, heat loss from the body must be matched by heat generation. An excessively low body temperature, called **hypothermia,** can slow down the metabolism to the point that it cannot sustain life. Conversely, an excessively high body temperature, called **hyperthermia,** can make some metabolic pathways race ahead of others and disrupt their coordination to the point that this, too, can lead to death. **Thermoregulation,** the balance between heat production and loss, is therefore a critically important aspect of homeostasis.

### Body Temperature

"Normal" body temperature depends on when, where, and in whom it is measured. Body temperature fluctuates about 1°C (1.8°F) in a 24-hour cycle. It tends to be lowest in the early morning and highest in the late afternoon. Temperature also varies from place to place in one body.

The most important body temperature is the **core temperature**—the temperature of organs in the cranial, thoracic, and abdominal cavities. Rectal temperature is relatively easy to measure and gives an estimate of core temperature: usually 37.2° to 37.6°C (99.0°–99.7°F), but as high as 38.5°C (101°F) in active children and some adults.

**Shell temperature** is the temperature closer to the surface, especially skin and oral temperature. Here, heat is lost from the body and temperatures are slightly lower than rectal temperature. Adult oral temperature is typically 36.6° to 37.0°C (97.9°–98.6°F) but may be as high as 40°C (104°F) during hard exercise. Shell temperature fluctuates as a result of processes that serve to maintain a stable core temperature.

### Heat Production and Loss

Most body heat comes from exergonic (energy-releasing) chemical reactions such as nutrient oxidation and ATP use. A little heat is generated by joint friction, blood flow, and other movements. At rest, most heat is generated by the brain, heart, liver, and endocrine glands; the skeletal muscles contribute about 20% to 30% of the total resting heat. Increased muscle tone or exercise greatly increases heat generation in the muscles, however; in vigorous exercise, they produce 30 to 40 times as much heat as the rest of the body.

The body loses heat in four ways: radiation, conduction, convection, and evaporation:

1. **Radiation** is the emission of infrared (IR) rays by moving molecules. In essence, heat means molecular motion, and all molecular motion produces IR rays. When an object absorbs IR rays, its molecular motion and temperature increase. Therefore, IR radiation removes heat from its source and adds heat to anything that absorbs it. The heat lamps in bathrooms and restaurants work on this principle. Our bodies continually receive IR from the objects around us and give off IR to our surroundings. Since we are usually warmer than the objects around us, we usually lose more heat this way than we gain.

2. **Conduction** is the transfer of kinetic energy from molecule to molecule as they collide with one another. Heat generated in the body core is conducted to the body surface through the tissues, then lost from the body by conduction from the skin to any cooler objects or medium in contact with it. The warmth of your body adds to the molecular motion of your clothes, the chair you sit in, the air around you, or water if you go swimming or sit in a cool bath. With this transfer of kinetic energy to your surroundings, you lose body heat. You can also gain heat by conduction, as on a very hot day when the air temperature is greater than your shell temperature, or when you use a heating pad for sore muscles, bask in a hot tub, or lie on hot sand at the beach.

3. **Convection** is the transfer of heat to a moving fluid—blood, air, or water. Most of the heat generated by metabolism in the body core is carried by convection in the bloodstream to the body surface. At the skin surface, body heat warms the adjacent air. Warm air is less dense than cool air, so it rises from the body and is replaced by cooler air from below. This can be visualized by a

technique called *schlieren photography* (fig. 26.11). Water does this as well if one stands still enough in a cool lake. Such movement of a fluid caused entirely by its change in temperature and density is called **natural convection.** When air movement is forced by a fan or the wind, even if the air itself is no cooler, it carries heat away from the body more rapidly. This effect, called **forced convection,** is the reason why, even at the same temperature, we feel cooler on a windy day than on a day when the air is still. It is the basis for the *windchill factor* of a cold, windy day. Forced convection increases heat loss by both conduction and evaporation (discussed next), but has no effect on radiation.

4. **Evaporation** is the change from a liquid to a gaseous state. The cohesion of water molecules hampers their vibratory movement in response to heat input. If the temperature of water is raised sufficiently, however, its molecular motion becomes great enough for molecules to break free and evaporate. Evaporation of water thus carries a substantial amount of heat with it (0.58 kcal/g). This is the significance of perspiration. Sweat wets the skin surface and its evaporation carries heat away. In extreme conditions, the body can lose 2 L or more of sweat per hour and dissipate up to 600 kcal of heat per hour by evaporative loss. Evaporative heat loss is increased by forced convection, as you can readily feel when you are sweaty and stand in front of a fan.

The relative amounts of heat lost by different methods depend on prevailing conditions. A nude body at an air temperature of 21°C (70°F) loses about 60% of its heat by radiation, 18% by conduction, and 22% by evaporation. If air temperature is higher than skin temperature, evaporation becomes the only means of heat loss because radiation and conduction add more heat to the body than they remove from it. Hot, humid weather hinders even evaporative cooling because there is less of a humidity gradient from skin to air. Such conditions increase the risk of heatstroke (discussed shortly).

## Thermoregulation

Thermoregulation is achieved through several negative feedback loops. The preoptic area of the hypothalamus (anterior to the optic chiasm) functions as a **hypothalamic thermostat.** It monitors blood temperature and receives signals also from **peripheral thermoreceptors** located mainly in the skin. In turn, it sends signals either to the **heat-loss center,** a nucleus still farther anterior in the hypothalamus, or to the **heat-promoting center,** a more posterior nucleus near the mammillary bodies.

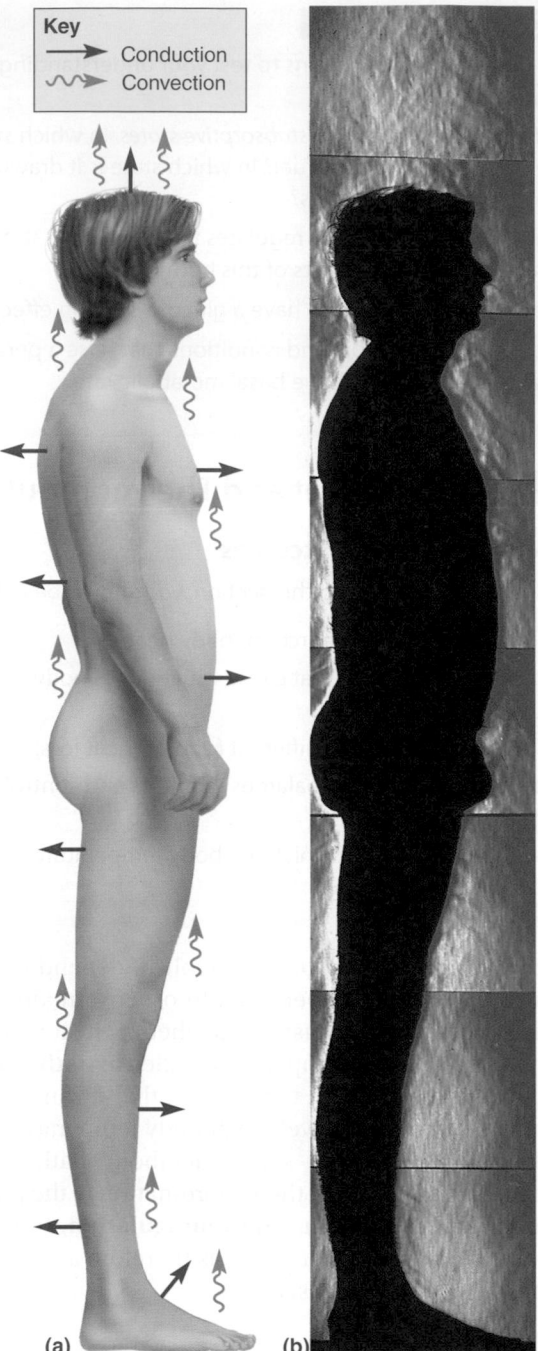

**Key**
→ Conduction
∿→ Convection

(a) (b)

**FIGURE 26.11  Heat Loss by Conduction and Convection.**
(a) Heat transfers from the body to surrounding air molecules by conduction. Warm air then rises from the body by convection, carrying heat away. Cool air replaces this warm air from below. (b) Schlieren photograph of the column of warm air rising from the body.

When the heat-loss center senses that the blood temperature is too high, it activates heat-losing mechanisms. The first and simplest of these is cutaneous vasodilation, which increases blood flow close to the body surface and thus promotes heat loss. If this fails to restore normal temperature, the heat-loss center triggers sweating. It also inhibits the heat-promoting center.

When the heat-promoting center senses that the blood temperature is too low, it activates mechanisms to conserve body heat or generate more. By way of the sympathetic nervous system, it causes cutaneous vasoconstriction. Warm blood is then retained deeper in the body and less heat is lost through the skin. The sympathetic nervous system in other mammals also stimulates the piloerector muscles, which make the hair stand on end. This traps an insulating blanket of still air near the skin. The human sympathetic nervous system attempts to do this as well, but since our body hair is so scanty, the only noticeable effect of this is goose bumps.

If dermal vasoconstriction cannot restore or maintain normal core temperature, the body resorts to **shivering thermogenesis.** If you leave a warm house on a cold day, you may notice that your muscles become tense, sometimes even painfully taut, and you begin to shiver. Shivering involves a spinal reflex that causes tiny alternating contractions in antagonistic muscle pairs. Every muscle contraction releases heat from ATP, and shivering can increase the body's heat production as much as fourfold.

**Nonshivering thermogenesis** is a more long-term mechanism for generating heat, occurring especially in the colder seasons of the year. The sympathetic nervous system raises the metabolic rate as much as 30% after several weeks of cold weather. More nutrients are burned as fuel; we consume more calories to "stoke the furnace"; and consequently, we have greater appetites in the winter than in the summer. Infants can generate heat by breaking down *brown fat,* a tissue in which lipolysis is not linked to ATP synthesis, so all the energy released from the fat is in the form of heat.

In addition to these physiological mechanisms, and of even greater importance, humans and other animals practice **behavioral thermoregulation**—behaviors that raise or lower the body's heat gains and losses. Just getting out of the sun greatly cuts down heat gain by radiation, for example, and shedding heavy clothing or kicking off a blanket at night helps to cool the body.

In summary, you can see that thermoregulation is a function of multiple organs: the brain, autonomic nerves, thyroid gland, skin, blood vessels, and skeletal muscles.

## Disturbances of Thermoregulation

Chapter 21 described the mechanism of fever and its importance in combating infection. Fever is a normal protective mechanism that should be allowed to run its course if it is not excessively high. A body temperature above 42° to 43°C (108°–110°F), however, can be very dangerous. The high temperature elevates the metabolic rate and the body generates heat faster than its heat-losing mechanisms can disperse it. Therefore, the elevated metabolic rate increases the fever and the fever increases the metabolic rate in a dangerous positive feedback loop. At a core body temperature of 44° to 45°C (111°–113°F), metabolic dysfunction and neurological damage can be fatal.

Exposure to excessive heat causes heat cramps, heat exhaustion, and heatstroke. **Heat cramps** are painful muscle spasms that result from excessive electrolyte loss in the sweat. They occur especially when one begins to relax after strenuous exertion and heavy sweating. **Heat exhaustion** results from more severe water and electrolyte loss and is characterized by hypotension, dizziness, vomiting, and sometimes fainting. Prolonged heat waves, especially if accompanied by high humidity, bring on many deaths from **heatstroke (sunstroke).** The body gains heat by radiation and conduction, but the humidity retards evaporative cooling. Heatstroke is clinically defined as a state in which the core body temperature is over 40°C (104°F); the skin is hot and dry; and the subject exhibits nervous system dysfunctions such as delirium, convulsions, or coma. It is also accompanied by tachycardia, hyperventilation, inflammation, and multiorgan dysfunction; it is often fatal.

Hypothermia can result from exposure to cold weather or immersion in icy water. It, too, entails life-threatening positive feedback loops. If the core temperature falls below 33°C (91°F), the metabolic rate drops so low that heat production cannot keep pace with heat loss, and the temperature falls even more. Death from cardiac fibrillation may occur below 32°C (90°F), but some people survive body temperatures as low as 29°C (84°F) in a state of suspended animation. A body temperature below 24°C (75°F) is usually fatal. It is dangerous to give alcohol to someone in a state of hypothermia; the chemical "burn" produces an illusion of warmth, but alcohol actually accelerates heat loss by dilating cutaneous blood vessels.

**BEFORE YOU GO ON**

Answer the following questions to test your understanding of the preceding section:

19. What is the primary source of body heat? What are some lesser sources?

20. What mechanisms of heat loss are aided by convection?

21. Describe the major heat-promoting and heat-losing mechanisms of the body.

22. Describe the positive feedback loops that can cause death from hyperthermia and hypothermia.

# DEEPER INSIGHT 26.4

## CLINICAL APPLICATION

### Alcohol and Alcoholism

Alcohol not only is a popular mind-altering drug but also is regarded in many cultures as a food staple. As a source of empty calories, an addictive drug, and a toxin, it can have a broad spectrum of adverse effects on the body.

### Absorption and Metabolism

Alcohol is rapidly absorbed from the digestive tract—about 10% of it in the stomach and 90% in the proximal small intestine. Carbonation, as in beer and sparkling wines, increases its rate of absorption by moving it more quickly to the small intestine, whereas food reduces its absorption by delaying gastric emptying. Alcohol is soluble in both water and fat, so it is rapidly distributed to all body tissues and easily crosses the blood–brain barrier to exert its intoxicating effects on the brain.

Alcohol is detoxified by the hepatic enzyme *alcohol dehydrogenase*, which oxidizes it to acetaldehyde. This enters the citric acid cycle and is oxidized to $CO_2$ and $H_2O$. The average adult male can clear the blood of about 10 mL of 100% (200 proof) alcohol per hour—the amount in about 30 mL (1 oz) of whiskey or 355 mL (12 oz) of beer. Women have less alcohol dehydrogenase and clear alcohol from the bloodstream more slowly. They are also more vulnerable to alcohol-related illnesses such as cirrhosis of the liver.

Tolerance to alcohol, the ability to "hold your liquor," results from two factors: behavioral modification, such as giving in less readily to lowered inhibitions, and increased levels of alcohol dehydrogenase in response to routine alcohol consumption. Alcohol dehydrogenase also deactivates other drugs, and drug dosages must be adjusted to compensate for this when treating alcoholics for other diseases.

### Physiological Effects

#### Nervous System
Alcohol is a depressant that inhibits the release of norepinephrine and disrupts the function of GABA receptors. In low doses, it depresses inhibitory synapses and creates sensations of confidence, euphoria,

and giddiness. As the dosage rises, however, the breakdown products of ethanol enhance the diffusion of $K^+$ out of neurons, hyperpolarizing them and making them less responsive to neurotransmitters. Thus, the timing and coordination of communication between neurons are impaired, resulting in such symptoms of intoxication as slurred speech, poor coordination, and slower reaction time. These symptoms begin to become significant at a blood alcohol level of 80 to 100 mg/dL—the legal criterion of intoxication in many states. Above 400 mg/dL, alcohol can so disrupt the electrophysiology of neurons as to induce coma and death.

#### Liver
The liver's role in metabolizing alcohol makes it especially susceptible to long-term toxic effects. Heavy drinking stresses the liver with a high load of acetaldehyde and acetate; this depletes its oxidizing agents and reduces its ability to catabolize these intermediates as well as fatty acids. Alcoholism often produces a greatly enlarged and fatty liver for multiple reasons: The calories provided by alcohol make it unnecessary to burn fat as fuel, fatty acids are poorly oxidized, and acetaldehyde is converted to new fatty acids. Acetaldehyde also causes inflammation of the liver and pancreas (*hepatitis* and *pancreatitis*), leading to disruption of digestive function. Acetaldehyde and other toxic intermediates destroy hepatocytes faster than they can be regenerated, leading to cirrhosis (see Deeper Insight 26.3). Many symptoms of alcoholism stem from deterioration of liver functions. Hepatic coma may occur as the liver becomes unable to produce urea, thus allowing ammonia to accumulate in the blood. Jaundice results from the liver's inability to excrete bilirubin.

#### Circulatory System
Deteriorating liver functions exert several effects on the blood and cardiovascular system. Blood clotting is impaired because the liver cannot synthesize clotting factors adequately. Edema results from inadequate synthesis of blood albumin. Cirrhosis obstructs the hepatic portal blood

circulation. Portal hypertension results, and combined with hypoproteinemia, this causes the liver and other organs to "weep" serous fluid into the peritoneal cavity. This leads to *ascites*[10] (ah-SY-teez)—swelling of the abdomen with as much as several liters of serous fluid. The combination of hypertension and impaired clotting often leads to hemorrhaging. *Hematemesis,*[11] the vomiting of blood, may occur as enlarged veins of the esophagus hemorrhage. Alcohol abuse also destroys myocardial tissue, reduces contractility of the heart, and causes cardiac arrhythmia.

### Digestive System and Nutrition

Alcohol breaks down the protective mucous barrier of the stomach and the tight junctions between its epithelial cells. Thus, it may cause gastritis and bleeding. Alcohol is commonly believed to be a factor in peptic ulcers, but there is little concrete evidence of this. Heavy drinking, especially in combination with smoking, increases the incidence of esophageal cancer. Malnutrition is a typical complication of alcoholism, partly because the empty calories of alcohol suppress the appetite for more nutritious foods. The average American gets about 4.5% of his or her calories from alcohol (more when nondrinkers are excluded), but heavy drinkers may obtain half or more of their calories from alcohol and have less appetite for foods that would meet their other nutritional requirements. In addition, acetaldehyde interferes with vitamin absorption and use. Thiamine deficiency is common in alcoholism, and thiamine is routinely given to alcoholics in treatment.

### Addiction

Alcohol is the most widely available addictive drug in America. In many respects it is almost identical to barbiturates in its toxic effects, its potential for tolerance and dependence, and the risk of overdose. The difference is that obtaining barbiturates usually requires a prescription, while obtaining alcohol requires, at most, proof of age.

*Alcoholism* is defined by a combination of criteria, including the pathological changes just described; physiological tolerance of high concentrations; impaired physiological, psychological, and social functionality; and withdrawal symptoms occurring when intake is reduced or stopped. Heavy drinking followed by a period of abstinence—for example, when a patient is admitted to the hospital and cannot get access to alcohol—may trigger *delirium tremens (DT),* characterized by restlessness, insomnia, confusion, irritability, tremors, incoherent speech, hallucinations, convulsions, and coma. DT has a 5% to 15% mortality rate.

Most alcoholism (type I) sets in after age 25 and is usually associated with stress or peer pressure. These influences lead to increased drinking, which can start a vicious cycle of illness, reduced job performance, family and social problems, arrest, and other stresses leading to still more drinking. A smaller number of alcoholics have type II alcoholism, which is at least partially hereditary. Most people with type II alcoholism are men who become addicted before age 25, especially the sons of other type II alcoholics. Type II alcoholics show abnormally rapid increases in blood acetaldehyde levels when they drink, and they have unusual brain waves (EEGs) even when not drinking. Children of alcoholics have a higher than average incidence of becoming alcoholic even when raised by nonalcoholic foster parents. It is by no means inevitable that such people will become alcoholics, but stress or peer pressure can trigger alcoholism more easily in those who are genetically predisposed to it.

Alcoholism is treated primarily through behavior modification—abstinence, peer support, avoidance or correction of the stresses that encourage drinking, and sometimes psychotherapy. Drugs such as disulfiram (Antabuse) have been used to support behavior modification programs by producing unpleasant effects from alcohol consumption, but drug treatment has been fraught with potentially dangerous side effects and little evidence of effectiveness.

---

[10]*asc* = bag; *ites* = like, resembling
[11]*hemat* = blood; *emesis* = vomiting

# STUDY GUIDE

## ► Assess Your Learning Outcomes

*To test your knowledge, discuss the following topics with a study partner or in writing, ideally from memory.*

### 26.1 Nutrition (p. 995)

1. Evidence of a homeostatic set point for body weight; relative contributions of heredity and behavior to differences in body weight

2. The meaning of *gut–brain peptides*

3. Sources and actions of ghrelin, peptide YY, and cholecystokinin as short-term regulators of appetite

4. Sources and actions of leptin and insulin as adiposity signals and long-term regulators of appetite

5. Roles of the arcuate nucleus, neuropeptide YY, and melanocortin in appetite regulation

6. The role of hunger contractions in the onset of feeding; factors that satiate hunger over the short and long terms, and how these resemble the satiation of thirst

7. Hormones that stimulate an appetite for specific classes of nutrients such as carbohydrates, fats, and proteins

8. The definition of *calorie* and how this relates to dietary Calories (kilocalories)

9. Principal dietary sources of calories; the relative yield from fats as compared to carbohydrates and proteins; and the meaning of *empty calories*

10. The definition of *nutrient;* why some nutrients are not digested and yield no calories; and why some things are not considered nutrients even though they are important components of a healthy diet

11. The difference between macronutrients and micronutrients; the nutrients in each category

12. Why some substances are considered to be essential nutrients

13. Forms and amounts of stored and mobile carbohydrates in the body; how the body uses carbohydrates; how dietary carbohydrates influence the metabolism of fats

14. Hormones that regulate the balance between blood glucose and stored glycogen; the normal range of blood glucose concentration

15. The recommended daily allowance (RDA) of carbohydrates; the percentage of calories that come from carbohydrates

in a typical U.S. diet; the forms in which carbohydrates exist in the diet, and their relative amounts; dietary sources of carbohydrates

16. The meaning of *dietary fiber;* the RDA of fiber and how actual consumption varies around the world; the forms of dietary fiber; which forms are classified as soluble and insoluble fiber, and differences in the health benefits of these two classes; the detrimental effects of too much dietary fiber

17. Two reasons why the body stores more energy as fat than as carbohydrate; the caloric yield from fat compared to carbohydrate; the typical percentages of body fat in the reference male and female

18. Why fat is said to have glucose-sparing and protein-sparing effects

19. Metabolically important lipids other than fat, and their uses in the body

20. The RDA of fat and recommended upper limit for cholesterol, and how the typical U.S. diet compares to these recommendations

21. Why linoleic acid is called an essential fatty acid; two other fatty acids that might be essential

22. Dietary sources of saturated and unsaturated fats, essential fatty acids, and cholesterol; the health risks of excessive saturated and unsaturated fats

23. Types of lipoproteins found in the bloodstream—chylomicrons, VLDLs, LDLs, and HDLs—and the differences in their source, composition, and functions; the relevance of LDLs and HDLs to cardiovascular health and how these relate to the colloquial expressions "good cholesterol" and "bad cholesterol"

24. Functions of proteins in the body, and a typical percentage of the body mass composed of protein

25. The RDA of protein; how that can be estimated from body weight; good dietary sources of protein; conditions that call for a protein intake greater than the normal RDA; and the risks from excessive dietary protein

26. Why nutritional value of a protein depends on its amino acid composition; the eight essential amino acids; the difference between complete and incomplete proteins; the meaning of net protein utilization and why it is different for dietary

proteins of plant and animal origin; and some dietary and ecological advantages of plant protein

27. The meaning of positive and negative nitrogen balance and the conditions under which each of them occurs

28. The definition and ultimate source of *dietary minerals;* the functions of minerals in the body; the most abundant minerals in the body; and, in general, good and poor dietary sources of minerals

29. The general history of human salt consumption; the RDA of sodium and some reasons why most U.S. diets greatly exceed this; and the consequences of excessive sodium intake

30. The definition of *vitamins;* dietary and nondietary sources; the functions of vitamins in the body; the difference between water-soluble and fat-soluble vitamins

### 26.2 Carbohydrate Metabolism (p. 1006)

1. The summary equation for the complete aerobic oxidation of glucose

2. Function of the coenzymes $NAD^+$ and FAD in glucose oxidation

3. The general process and outcome of glycolysis, and its net ATP yield

4. Anaerobic fermentation and its primary purpose

5. The cellular site of aerobic respiration, its end products, and its advantages

6. The citric acid cycle, where it occurs, the fate of the carbon atoms that originated in the glucose, and the cycle's yield of ATP, NADH, and $FADH_2$

7. Mitochondrial membrane reactions; where they occur in the organelle; the components of the mitochondrial electron-transport chain; how mitochondria transport electrons from NADH and $FADH_2$ to oxygen; and how the reactions produce metabolic water

8. Mitochondrial proton pumps, the chemiosmotic mechanism, and ATP synthase in producing ATP

9. The net ATP yield of glycolysis and aerobic respiration; the amount of ATP produced at each step from glucose to $H_2O$; and why the yield varies slightly between different cell types

10. The efficiency of aerobic respiration and how to calculate this

# STUDY GUIDE

11. How excess glucose is converted to glycogen; the body's typical glycogen store and where these reserves are located; and the processes and purposes of glycogenesis, glycogenolysis, and gluconeogenesis

## 26.3 Lipid and Protein Metabolism (p. 1013)

1. What cells are primarily responsible for storing and releasing triglycerides; the meaning of the lipogenesis and lipolysis carried out by these and other cells

2. The process of lipolysis including the hydrolysis of triglycerides and the beta oxidation of fatty acids; the ATP yield from complete oxidation of a typical fatty acid

3. The meaning of *ketogenesis;* the metabolic use of ketone bodies and the pathological effects of excessive ketone levels; common circumstances in which excessive ketogenesis occurs

4. A typical daily rate of protein turnover in the body; where and why the fastest rate of protein turnover occurs; and the dietary and nondietary sources of the amino acids absorbed by the small intestine

5. The uses of free amino acids in the amino acid pool

6. What occurs in the deamination, amination, and transamination processes in amino acid metabolism; the body's uses of deaminated amino acids

7. How amino acids are shuttled into the citric acid cycle for oxidation as fuel

8. How the liver produces urea

9. Other nondigestive functions of the liver

## 26.4 Metabolic States and Metabolic Rate (p. 1015)

1. When the body is in its absorptive state; what things occur in this state with respect to carbohydrate, fat, and protein metabolism

2. The main hormone that regulates the absorptive state; its primary metabolic effects in this state; and what antagonist modulates its effects

3. When the body is in its postabsorptive state; what things occur in this state with respect to carbohydrate, fat, and sometimes protein metabolism

4. Hormones that regulate the postabsorptive state, their effects, and the role of the sympathetic nervous system in its regulation

5. The meaning of *metabolic rate;* how it is measured and in what units of measurement it is expressed; what factors cause it to vary; and how the basal metabolic rate (BMR) differs from total metabolic rate (TMR)

6. Typical values for the BMR and TMR under different conditions of physical exertion

## 26.5 Body Heat and Thermoregulation (p. 1019)

1. The meaning of *thermoregulation;* terms for abnormally low and high body temperatures; and reasons why those two conditions can be fatal

2. Typical core and shell body temperatures and how the two are measured

3. How most body heat is produced, and which organs are the most important sources of body heat at rest and in exercise

4. Four mechanisms by which body heat is lost, and the percentages of total heat loss attributable to each of them at rest and at a comfortable ambient temperature (21°C)

5. How the hypothalamic thermostat monitors core and shell body temperature

6. Two mechanisms for lowering body temperature and two mechanisms of raising it; regions of the hypothalamus involved in each

7. The meanings of *nonshivering thermogenesis* and *behavioral thermoregulation,* and examples of the latter

8. Differences between heat cramps, heat exhaustion, and heatstroke; the role of positive feedback loops in hyperthermia; and how and at what temperature range heatstroke can lead to death

9. The role of positive feedback loops in hypothermia, and how and at what temperature range hypothermia can lead to death

## ▶ Testing Your Recall

*Answers in Appendix B*

1. _____ are not used as fuel and are required in relatively small quantities.
   a. Micronutrients
   b. Macronutrients
   c. Essential nutrients
   d. Proteins
   e. Lipids

2. The only significant digestible polysaccharide in the diet is
   a. glycogen.
   b. cellulose.
   c. starch.
   d. maltose.
   e. fiber.

3. Which of the following store(s) the greatest amount of energy for the smallest amount of space in the body?
   a. glucose
   b. triglycerides
   c. glycogen
   d. proteins
   e. vitamins

4. The lipoproteins that remove cholesterol from the tissues are
   a. chylomicrons.
   b. lipoprotein lipases.
   c. VLDLs.
   d. LDLs.
   e. HDLs.

5. Which of the following is most likely to make you hungry?
   a. leptin
   b. ghrelin
   c. cholecystokinin
   d. peptide YY
   e. melanocortin

6. The primary function of B-complex vitamins is to act as
   a. structural components of cells.
   b. sources of energy.
   c. components of pigments.
   d. antioxidants.
   e. coenzymes.

# STUDY GUIDE

7. FAD is reduced to $FADH_2$ in
   a. glycolysis.
   b. anaerobic fermentation.
   c. the citric acid cycle.
   d. the electron-transport chain.
   e. beta oxidation of lipids.

8. The primary, direct benefit of anaerobic fermentation is to
   a. regenerate $NAD^+$.
   b. produce $FADH_2$.
   c. produce lactic acid.
   d. dispose of pyruvic acid.
   e. produce more ATP than glycolysis does.

9. Which of these occurs in the mitochondrial matrix?
   a. glycolysis
   b. chemiosmosis
   c. the cytochrome reactions
   d. the citric acid cycle
   e. anaerobic fermentation

10. When the body emits more infrared energy than it absorbs, it is losing heat by
    a. convection.
    b. forced convection.
    c. conduction.
    d. radiation.
    e. evaporation.

11. A/an _____ protein lacks one or more essential amino acids.

12. In the postabsorptive state, glycogen is hydrolyzed to liberate glucose. This process is called _____.

13. Synthesis of glucose from amino acids or triglycerides is called _____.

14. The major nitrogenous waste resulting from protein catabolism is _____.

15. The organ that synthesizes the nitrogenous waste in question 14 is the _____.

16. The absorptive state is regulated mainly by the hormone _____.

17. The temperature of organs in the body cavities is called _____.

18. The appetite hormones ghrelin, leptin, CCK, and others act on part of the hypothalamus called the _____ nucleus.

19. The brightly colored, iron-containing, electron-transfer molecules of the inner mitochondrial membrane are called _____.

20. The flow of $H^+$ from the intermembrane space to the mitochondrial matrix creates an electrical current used by the enzyme _____ to make _____.

## ▶ Building Your Medical Vocabulary

*Answers in Appendix B*

*State a medical meaning of each word element below, and give a term in which it or a slight variation of it is used.*

1. asco-

2. cac-

3. chromo-

4. -genesis

5. glyco-

6. -ites

7. lepto-

8. -lysis

9. neo-

10. osmo-

## ▶ True or False

*Answers in Appendix B*

*Determine which five of the following statements are false, and briefly explain why.*

1. Ghrelin and leptin are two hormones that stimulate the appetite.

2. Water is a nutrient, but oxygen and cellulose are not.

3. An extremely low-fat diet can cause vitamin-deficiency diseases.

4. Most of the body's cholesterol comes from the diet.

5. There is no harm in maximizing one's daily protein intake.

6. Aerobic respiration produces more ATP than anaerobic fermentation.

7. Reactions occurring on the mitochondrial inner membrane produce more ATP than glycolysis and the matrix reactions combined.

8. Gluconeogenesis occurs especially in the absorptive state during and shortly after a meal.

9. Brown fat generates more ATP than white fat and is therefore especially important for thermoregulation.

10. At a comfortable air temperature, the body loses more heat as infrared radiation than by any other means.

# STUDY GUIDE

## ▶ Testing Your Comprehension

*Answers at* www.mhhe.com/saladin7

1. Cyanide blocks the transfer of electrons from cytochrome $a_3$ to oxygen. In light of this, explain why it is so lethal. Also explain whether cyanide poisoning could be treated by giving a patient supplemental oxygen, and justify your answer.

2. Chapter 17 defines and describes some hormone actions that are synergistic and antagonistic. Identify some synergistic and antagonistic hormone interactions in the postabsorptive state of metabolism.

3. Mrs. Jones, a 42-year-old, complains that, "Everything I eat goes to fat. But my husband and my son eat twice as much as I do, and they're both as skinny as can be." How would you explain this to her?

4. A television advertisement proclaims, "Feeling tired? Need more energy? Order your supply of Zippy Megavitamins and feel better fast!" Your friend Cathy is about to send in her order, and you try to talk her out of wasting her money. Summarize the argument you would use.

5. Explain why a patient whose liver has been extensively damaged by hepatitis could show elevated concentrations of thyroid hormone and bilirubin in the blood.

## ▶ Improve Your Grade at www.mhhe.com/saladin7

*Download animations to study when it fits your schedule. Practice quizzes, labeling activities, games, and flashcards offer fun ways to master the chapter concepts. Or, download image PowerPoint files for each chapter to create a study guide or for taking notes during lecture.*

# 27

# THE MALE REPRODUCTIVE SYSTEM

Seminiferous tubules, where sperm are produced. Sperm tails are seen as hairlike masses in the center of the tubules (SEM).

Anatomy & Physiology | REVEALED®
aprevealed.com

**Module 14: Reproductive System**

- The flow of heat down thermal gradients (p. 18) is important in the temperature control of the testes.

- Your understanding of male reproductive anatomy may benefit if you refresh your memory of the pelvic girdle (p. 259) and muscles of the pelvic floor (p. 337).

- Chromosome structure (p. 115), the human karyotype (p. 131), and mitosis (p. 129) are important for understanding sperm production.

- Sexual development and adult function depend on the gonadotropins and sex steroids introduced on pages 636–637 and 646. You should know the anatomy of the pituitary gland (p. 634) and the mechanism of negative feedback inhibition (p. 639).

From all we have learned of the structural and functional complexity of the human body, it seems a wonder that it works at all! The fact is, however, that even with modern medicine we cannot keep it working forever. The body suffers various degenerative changes as we age, and eventually our time is up and we must say good-bye. Yet our genes live on in new containers—our offspring. The production of offspring is the subject of these last three chapters. In this chapter, we examine some general aspects of human reproductive biology and then focus on the role of the male in reproduction. Chapter 28 focuses on the female and chapter 29 on the embryonic development of humans and on changes at the other end of the life span—the changes of old age.

## 27.1 Sexual Reproduction and Development

### Expected Learning Outcomes

When you have completed this section, you should be able to

a. identify the most fundamental biological distinction between male and female;

b. define *primary sex organs, secondary sex organs,* and *secondary sex characteristics;*

c. explain the role of the sex chromosomes in determining sex;

d. explain how the Y chromosome determines the response of the fetal gonad to prenatal hormones;

e. identify which of the male and female external genitalia are homologous to each other; and

f. describe the descent of the gonads and explain why it is important.

### The Two Sexes

The essence of sexual reproduction is that it is biparental—the offspring receive genes from two parents and therefore are not genetically identical to either one. To achieve this, the parents must produce **gametes**[1] (sex cells) that meet and combine their genes in a **zygote**[2] (fertilized egg). The gametes must have two properties for reproduction to be successful: motility so they can achieve contact, and nutrients for the developing embryo. A single cell cannot perform both of these roles optimally, because to contain ample nutrients means to be relatively large and heavy, and this is inconsistent with the need for motility. Therefore, these tasks are usually apportioned to two kinds of gametes. The small motile one—little more than DNA with a propeller—is the **sperm (spermatozoon),** and the large nutrient-laden one is the **egg (ovum).**

In any sexually reproducing species, by definition, an individual that produces eggs is female and one that produces sperm is male. These criteria are not always that simple, as we see in certain abnormalities in sexual development. Genetically, however, any human with a Y sex chromosome is classified as male and anyone lacking a Y is classified as female.

In mammals, the female is also the parent that provides a sheltered internal environment for the development and nutrition of the embryo. For fertilization and development to occur in the female, the male must have a copulatory organ, the penis, for introducing his gametes into the female reproductive tract, and the female must have a copulatory organ, the vagina, for receiving the sperm. This is the most obvious difference between the sexes, but appearances can be deceiving (see fig. 17.28, p. 664, and Deeper Insight 27.1).

### Overview of the Reproductive System

The **reproductive system** in the male serves to produce sperm and introduce them into the female body. The female reproductive system produces eggs, receives the sperm, provides for the union of these gametes, harbors the fetus, gives birth, and nourishes the offspring.

The reproductive system consists of primary and secondary sex organs. The **primary sex organs,** or **gonads,**[3] are organs that produce the gametes—**testes** of the male and **ovaries** of the female. The **secondary sex organs** are organs other than gonads that are necessary for reproduction. In the male, they constitute a system of ducts, glands, and the penis, concerned with the storage, survival, and conveyance of sperm. In the female, they include the uterine tubes, uterus, and vagina, concerned with uniting the sperm and egg and harboring the fetus.

According to location, the reproductive organs are classified as **external** and **internal genitalia** (table 27.1). The external genitalia are located in the perineum (see fig. 27.6). Most of them are externally visible, except for the accessory glands of the female perineum. The internal genitalia are located mainly in the pelvic cavity, except for the male testes and some associated ducts contained in the scrotum.

**Secondary sex characteristics** are features that further distinguish the sexes and play a role in mate attraction. They

---

[1]*gam* = marriage, union
[2]*zygo* = yoke, union
[3]*gon* = seed

## DEEPER INSIGHT 27.1
## CLINICAL APPLICATION

### Androgen-Insensitivity Syndrome

Occasionally, a girl shows all the usual changes of puberty except that she fails to menstruate. Physical examination shows the presence of testes in the abdomen and a karyotype reveals that she has the XY chromosomes of a male. The testes produce normal male levels of testosterone, but the target cells lack receptors for it. This is called *androgen-insensitivity syndrome (AIS)*, or *testicular feminization*.

The external genitals develop female anatomy as if no testosterone were present. At puberty, breasts and other feminine secondary sex characteristics develop (fig. 27.1) because the testes secrete small amounts of estrogen and there is no overriding influence of testosterone. However, there is no uterus or menstruation. If the abdominal testes are not removed, the person has a high risk of testicular cancer.

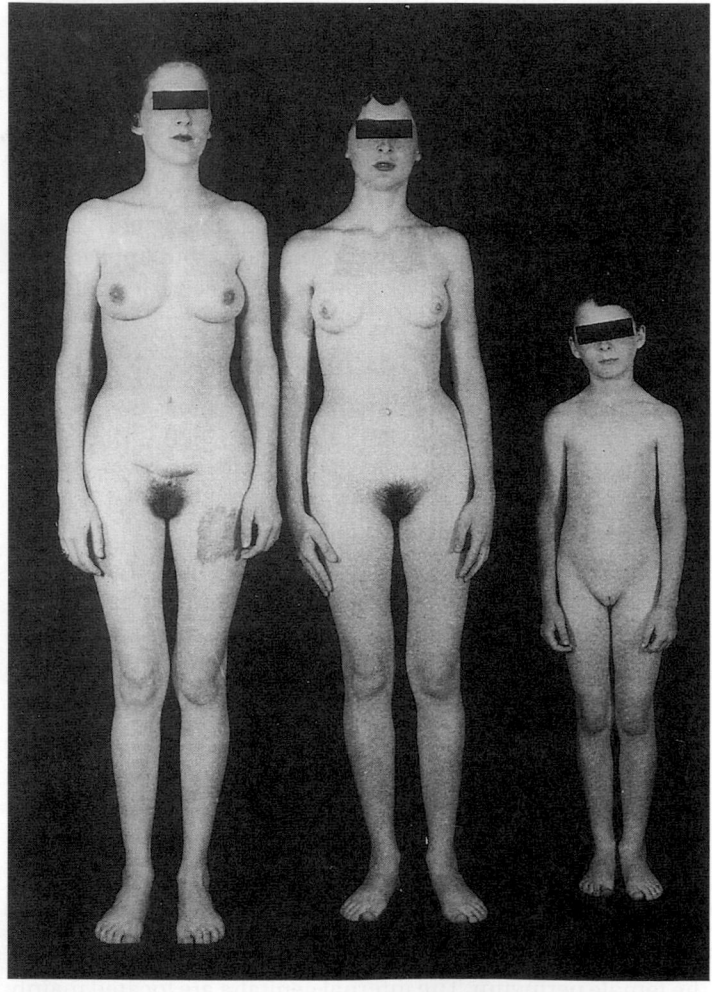

**FIGURE 27.1 Androgen-Insensitivity Syndrome.** These siblings are genetically male (XY). Testes are present and secrete testosterone, but the target cells lack receptors for it, so testosterone cannot exert its masculinizing effects. The external genitalia and secondary sex characteristics are feminine, but there are no ovaries, uterus, or vagina.

*In what way is androgen-insensitivity syndrome similar to type 2 diabetes mellitus?*

| TABLE 27.1 | The External and Internal Genitalia |
|---|---|
| **External Genitalia** | **Internal Genitalia** |
| **Male** | |
| Penis | Testes (s., *testis*) |
| Scrotum | Epididymides (s., *epididymis*) |
| | Ductus deferentes (s., *ductus deferens*) |
| | Seminal vesicles |
| | Prostate |
| | Bulbourethral glands |
| **Female** | |
| Mons pubis | Ovaries |
| Labia majora (s., *labium majus*) | Uterine tubes |
| Labia minora (s., *labium minus*) | Uterus |
| Clitoris | Vagina |
| Vaginal orifice | |
| Vestibular bulbs | |
| Vestibular glands | |
| Paraurethral glands | |

typically appear only as an animal approaches sexual maturity (during adolescence in humans). From the call of a bullfrog to the tail of a peacock, these are well known in the animal kingdom. In humans, the physical attributes that contribute to mate attraction are so culturally conditioned that it is more difficult to identify what secondary sex characteristics are biologically fundamental. Generally accepted as such in both sexes are the pubic and axillary hair and their associated scent glands, and the pitch of the voice. Other traits commonly regarded as male secondary sex characteristics are the facial hair, relatively coarse and visible hair on the torso and limbs, and the relatively muscular physique. In females, they include the distribution of body fat, enlargement of the breasts (independently of lactation), and relatively hairless appearance of the skin.

## Chromosomal Sex Determination

What determines whether a zygote develops into a male or female? The distinction begins with the combination of sex chromosomes bequeathed to the zygote. Most of our cells have 23 pairs of chromosomes: 22 pairs of *autosomes* and 1 pair of *sex chromosomes* (see fig. 4.17, p. 131). A sex chromosome can be either a large X chromosome or a small Y chromosome. Every egg contains an X chromosome, but half of the sperm carry an X and the other half carry a Y. If an egg is fertilized with an X-bearing sperm, it produces an XX zygote that is destined to become a female. If it is fertilized with a Y-bearing sperm,

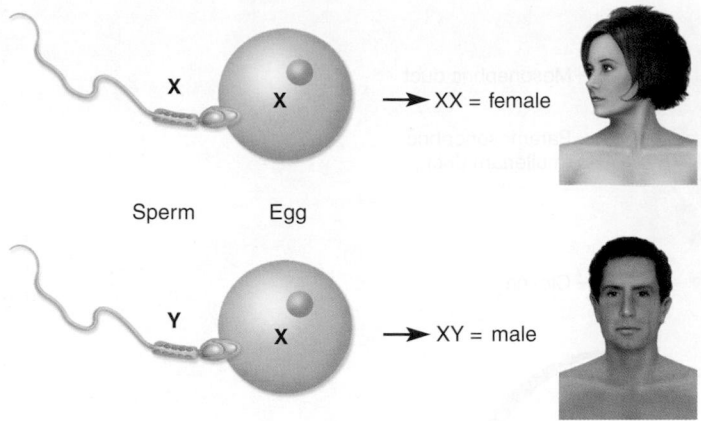

**FIGURE 27.2  Chromosomal Sex Determination.** All eggs carry the X chromosome. The sex of a child is determined by whether the egg is fertilized by an X-bearing sperm or a Y-bearing sperm.

it produces an XY zygote destined to become a male. Thus, the sex of a child is determined at conception (fertilization), and not by the mother's egg but by the sperm that fertilizes it (fig. 27.2).

## Prenatal Hormones and Sexual Differentiation

Sex determination does not end with fertilization, however. It requires an interaction between genetics and hormones produced by the mother and fetus. Just as we have seen with other hormones (see chapter 17), those involved here require specific receptors on their target cells to exert an effect.

Up to a point, a fetus is sexually undifferentiated, or "noncommittal" as to which sex it will become. Its gonads begin to develop at 5 to 6 weeks as *gonadal ridges,* each lying alongside a primitive kidney, the *mesonephros,* which later degenerates. Adjacent to each gonadal ridge are two ducts: the **mesonephric**[4] (MEZ-oh-NEF-ric) (**wolffian**[5]) **duct,** which originally serves the mesonephros, and the **paramesonephric**[6] (**müllerian**[7]) **duct.** In males, the mesonephric ducts develop into the reproductive tract and the paramesonephric ducts degenerate. In females, the opposite occurs (fig. 27.3).

But why? The Y chromosome has a gene called **SRY** (sex-determining region of the Y) that codes for a protein called **testis-determining factor (TDF).** TDF then interacts with genes on some of the other chromosomes, including a gene on the X chromosome for androgen receptors and genes that initiate the development of male anatomy. By 8 to 9 weeks, the

male gonadal ridge has become a rudimentary testis that begins to secrete testosterone. Testosterone stimulates the mesonephric duct on its own side to develop into the system of male reproductive ducts. By this time, the testis also secretes a hormone called **müllerian-inhibiting factor (MIF)** that causes atrophy of the paramesonephric (müllerian) duct on that side. Even an adult male, however, retains a tiny Y-shaped vestige of the paramesonephric ducts, like a vestigial uterus and uterine tubes, in the area of the prostatic urethra.

It may seem as if androgens should induce the formation of a male reproductive tract and estrogens induce a female reproductive tract. However, the estrogen level is always high during pregnancy, so if this mechanism were the case, it would feminize all fetuses. Thus, the development of a female results from the absence of androgens, not the presence of estrogens.

## Development of the External Genitalia

You perhaps regard the external genitals as the most definitive characteristics of a male or female, yet there is more similarity between the sexes than most people realize. In the embryo, the genitals begin developing from identical structures in both sexes. By 6 weeks, the embryo has the following (fig. 27.4):

- a **genital tubercle,** an anterior median bud;
- **urogenital folds,** a pair of medial tissue folds slightly posterior to the genital tubercle; and
- **labioscrotal folds,** a larger pair of tissue folds lateral to the urogenital folds.

By the end of week 9, the fetus begins to show sexual differentiation, and either male or female genitalia are distinctly formed by the end of week 12. In the female, the three structures just listed become the clitoral glans, labia minora, and labia majora, respectively; all of these are more fully described in chapter 28. In the male, the genital tubercle elongates to form the *phallus;* the urogenital folds fuse to enclose the urethra, joining the phallus to form the penis; and the labioscrotal folds fuse to form the scrotum, a sac that will later contain the testes.

Male and female organs that develop from the same embryonic structure are said to be **homologous.** Thus, the penis is homologous to the clitoris and the scrotum is homologous to the labia majora. This becomes strikingly evident in some abnormalities of sexual development. In the presence of excess androgen, the clitoris may become greatly enlarged and resemble a small penis. In other cases, the ovaries descend into the labia majora as if they were testes descending into a scrotum. Such abnormalities sometimes result in mistaken identification of the sex of an infant at birth.

## Descent of the Gonads

Both male and female gonads initially develop high in the abdominal cavity, near the kidneys, and migrate into the pelvic cavity (ovaries) or scrotum (testes). In the embryo,

---

[4]*meso* = middle; *nephr* = kidney
[5]Kaspar F. Wolff (1733–94), German anatomist
[6]*para* = next to
[7]Johannes P. Müller (1801–58), German physician

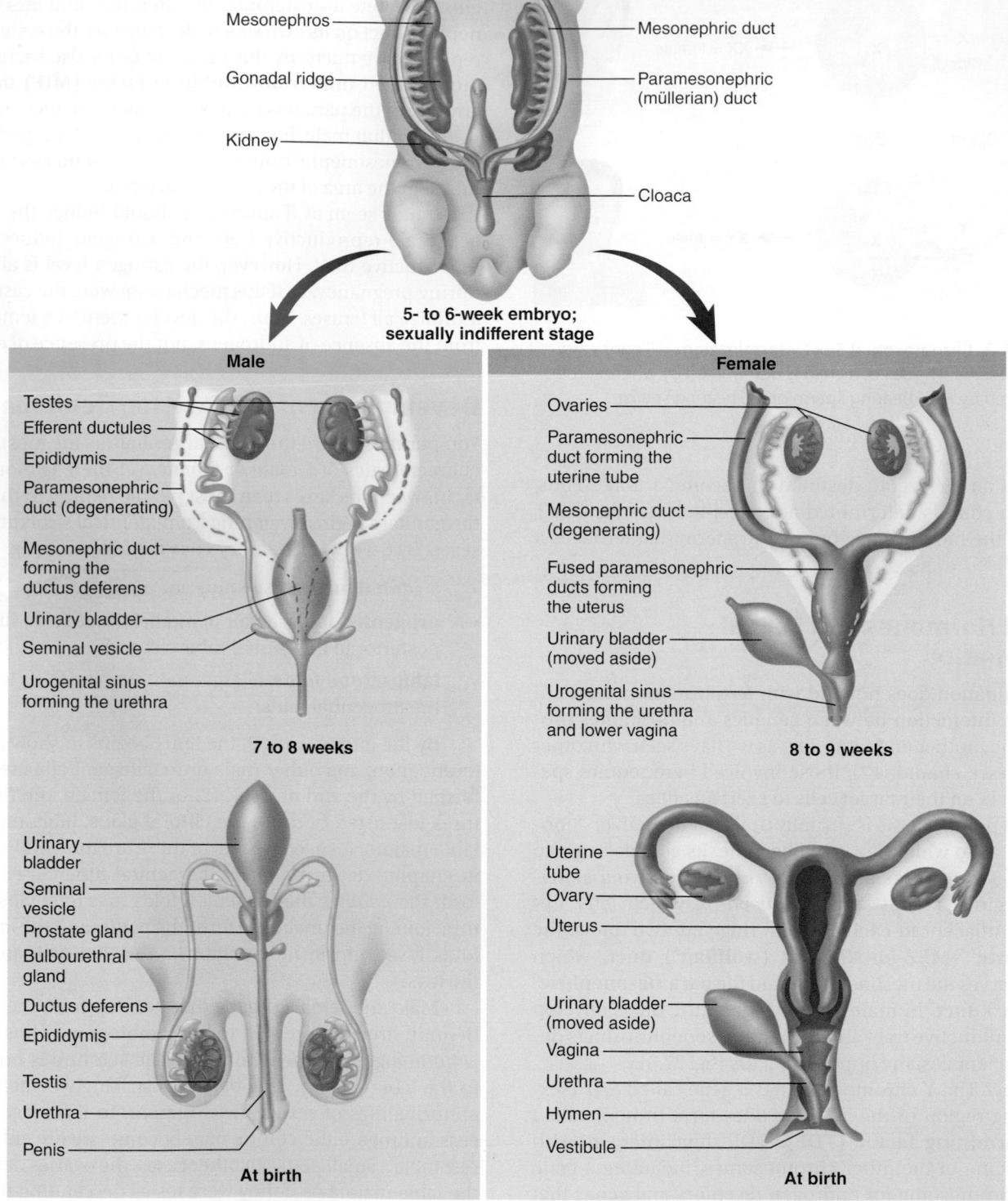

**FIGURE 27.3  Embryonic Development of the Male and Female Reproductive Tracts.** Note that the male tract develops from the mesonephric duct and the female tract from the paramesonephric duct; the other duct in each sex degenerates.

a connective tissue cord called the *gubernaculum*[8] (GOO-bur-NACK-you-lum) extends from the gonad to the floor of the abdominopelvic cavity. As the male gubernaculum continues to grow, it passes between the internal and external abdominal oblique muscles and into the scrotal swelling. Independently of migration of the testis, the peritoneum also

---

[8]*gubern* = rudder, to guide

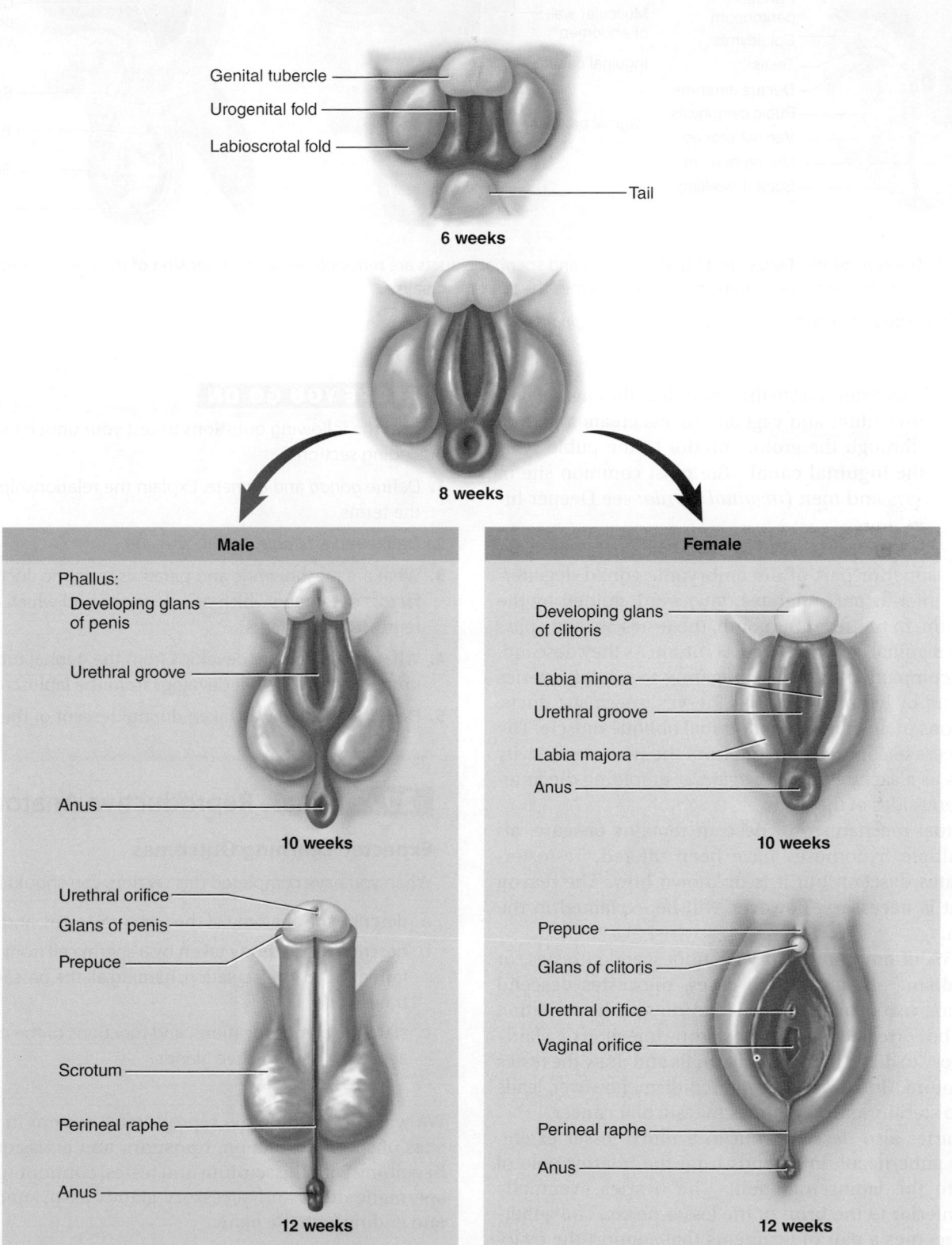

**FIGURE 27.4  Development of the External Genitalia.**  By 6 weeks, the embryo has three primordial structures—the genital tubercle, urogenital folds, and labioscrotal folds—which will become the male or female genitalia. At 8 weeks, these structures have grown but the sexes are still indistinguishable. Slight sexual differentiation is noticeable at 10 weeks, and the sexes are fully distinguishable by 12 weeks. Matching colors identify homologous structures of the male and female.

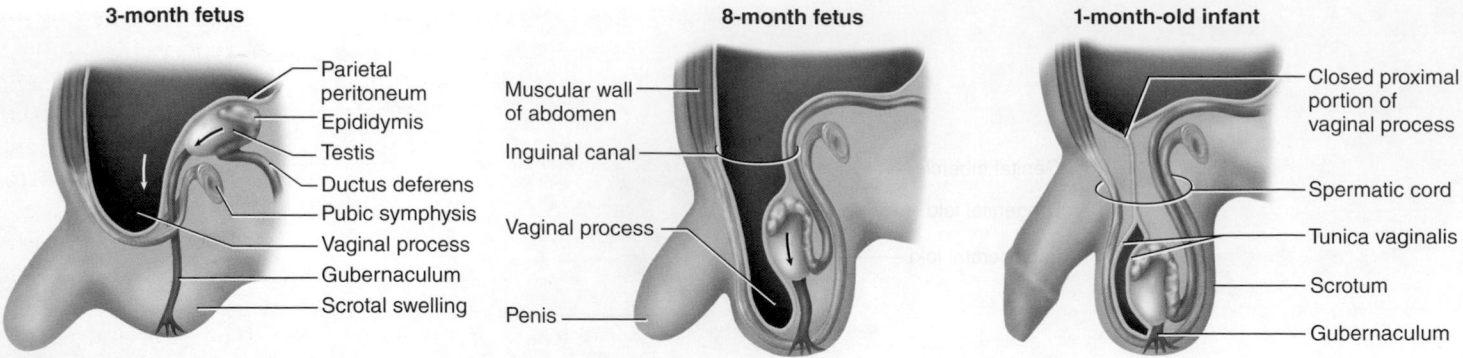

**FIGURE 27.5 Descent of the Testis.** Note that the testis and spermatic ducts are retroperitoneal. An extension of the peritoneum called the *vaginal process* passes through the inguinal canal and becomes the tunica vaginalis.

❓ *Why is this structure of male anatomy called the tunica vaginalis?*

develops a fold that extends into the scrotum as the *vaginal process.*[9] The gubernaculum and vaginal process create a path of low resistance through the groin, anterior to the pubic symphysis, called the **inguinal canal**—the most common site of herniation in boys and men *(inguinal hernia;* see Deeper Insight 10.3, p. 339).

The **descent of the testes** (fig. 27.5) begins as early as week 6. The superior part of the embryonic gonad degenerates and its inferior part migrates downward, guided by the gubernaculum. In the seventh month, the testes abruptly pass through the inguinal canals into the scrotum. As they descend, they are accompanied by ever-elongating testicular arteries and veins and by lymphatic vessels, nerves, spermatic ducts, and extensions of the internal abdominal oblique muscle. The vaginal process becomes separated from the peritoneal cavity and persists as a sac, the *tunica vaginalis,* enfolding the anterior and lateral sides of the testis.

The actual mechanism of descent remains obscure, although multiple hypotheses have been offered. Testosterone stimulates descent, but it is unknown how. The reason this descent is necessary, however, will be explained in the next section.

About 3% of boys are born with undescended testes, or **cryptorchidism.**[10] In most such cases, the testes descend within the first year of infancy, but if they do not, the condition can usually be corrected with a testosterone injection or a fairly simple surgery to dilate the inguinal canals and draw the testes into the scrotum. Uncorrected cryptorchidism, however, leads inevitably to sterility and sometimes to testicular cancer.

The ovaries also descend, but to a much lesser extent. The female gubernaculum extends from the inferior pole of the ovary to the labioscrotal fold. The ovaries eventually lodge just inferior to the brim of the lesser pelvis. The gubernaculum becomes a pair of ligaments that support the ovary and uterus.

---

[9]*vagin* = sheath
[10]*crypto* = hidden; *orchid* = testis; *ism* = condition

**BEFORE YOU GO ON**

Answer the following questions to test your understanding of the preceding section:

1. Define *gonad* and *gamete.* Explain the relationship between the terms.
2. Define *male, female, sperm,* and *egg.*
3. What are mesonephric and paramesonephric ducts? What factors determine which one develops and which one regresses in the fetus?
4. What male structure develops from the genital tubercle and urogenital folds? What develops from the labioscrotal folds?
5. Describe the pathway taken during descent of the male gonad.

## 27.2 Male Reproductive Anatomy AP|R

### Expected Learning Outcomes
When you have completed this section, you should be able to

a. describe the anatomy of the scrotum, testes, and penis;
b. describe the pathway taken by a sperm cell from its formation to its ejaculation, naming all the passages it travels; and
c. state the names, locations, and functions of the male accessory reproductive glands.

We will survey the male reproductive system in order of the sites of sperm formation, transport, and emission—therefore beginning with the scrotum and testes, continuing through the spermatic ducts and accessory glands associated with them, and ending with the penis.

### The Scrotum

The scrotum and penis constitute the external genitalia of the male and occupy the **perineum** (PERR-ih-NEE-um). This is

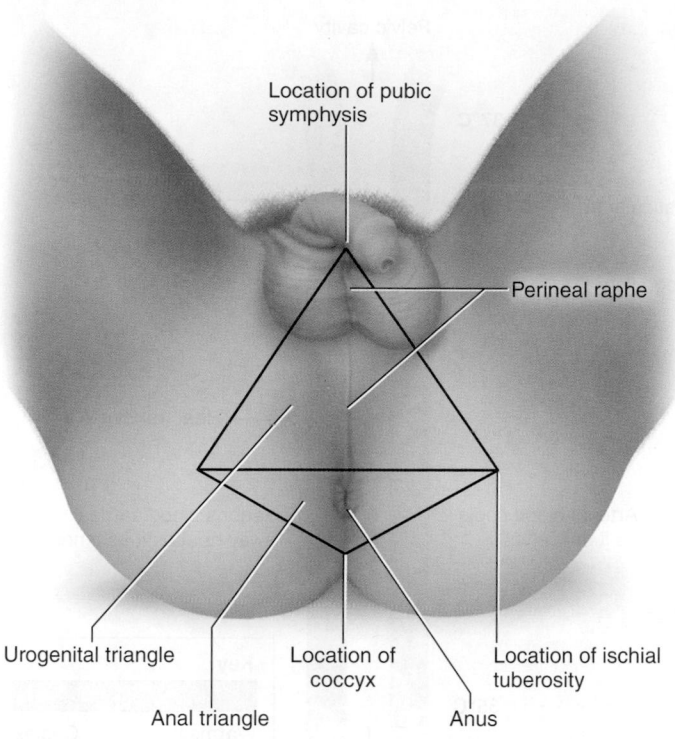

Location of pubic symphysis

Perineal raphe

Urogenital triangle

Location of coccyx

Location of ischial tuberosity

Anal triangle

Anus

**FIGURE 27.6 The Male Perineum.** AP|R

a diamond-shaped area between the thighs bordered by the pubic symphysis, ischial tuberosities, and coccyx (fig. 27.6).

The **scrotum**[11] is a pendulous pouch of skin, muscle, and fibrous connective tissue, containing the testes (fig. 27.7). The skin has sebaceous glands, sparse hair, rich sensory innervation, and somewhat darker pigmentation than skin elsewhere. The scrotum is divided into right and left compartments by an internal **median septum,** which protects each testis from infections of the other one. The location of the septum is externally marked by a seam called the **perineal raphe**[12] (RAY-fee), which also extends anteriorly along the ventral side of the penis and posteriorly as far as the margin of the anus (fig. 27.6). The left testis is usually suspended lower than the right so the two are not compressed against each other between the thighs.

Posteriorly, the scrotum contains the **spermatic cord,** a bundle of fibrous connective tissue containing the *ductus deferens* (a sperm duct), blood and lymphatic vessels, and testicular nerves. It passes upward behind and superior to the testis, where it is easily palpated through the skin of the scrotum. It continues across the anterior side of the pubis and into

---

[11]*scrotum* = bag
[12]*raphe* = seam

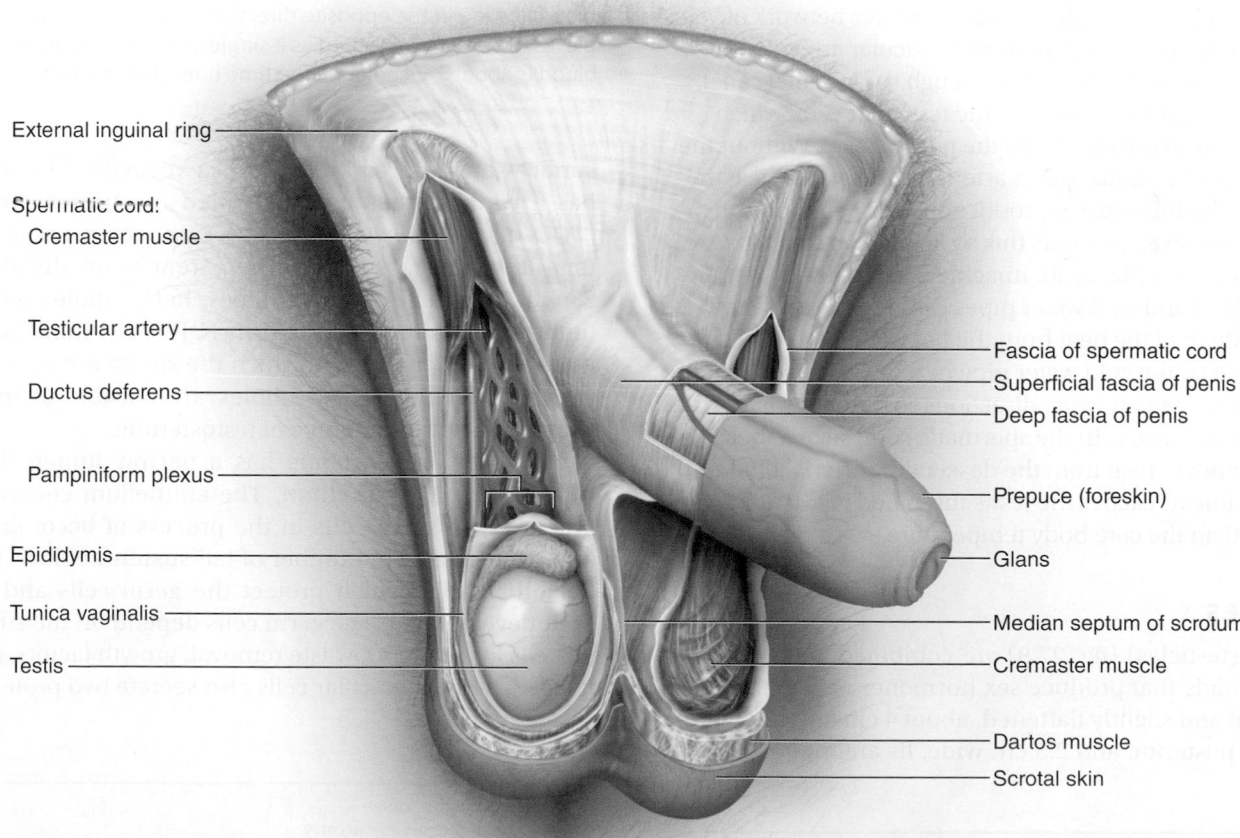

External inguinal ring

Spermatic cord:
Cremaster muscle

Testicular artery

Ductus deferens

Pampiniform plexus

Epididymis

Tunica vaginalis

Testis

Fascia of spermatic cord
Superficial fascia of penis
Deep fascia of penis

Prepuce (foreskin)

Glans

Median septum of scrotum

Cremaster muscle

Dartos muscle
Scrotal skin

**FIGURE 27.7 The Scrotum and Spermatic Cord.** AP|R

a 4-cm-long **inguinal canal,** which leads through the muscles of the groin and emerges into the pelvic cavity. The inferior entrance into the inguinal canal is called the *external inguinal ring,* and its superior exit into the pelvic cavity is the *internal inguinal ring.*

The original reason that a scrotum evolved is a subject of debate among reproductive biologists; it is absent from some mammals, such as elephants, sea cows, and whales, so it is not universally necessary for mammalian sperm production. For whatever reason human testes reside in the scrotum, however, they are now adapted to this cooler environment and cannot produce sperm at the core body temperature of 37°C; they must be held at about 35°C. The scrotum has three mechanisms for regulating the temperature of the testes:

1. The **cremaster**[13] **muscle**—strips of the internal abdominal oblique muscle that enmesh the spermatic cord. When it is cold, the cremaster contracts and draws the testes closer to the body to keep them warm. When it is warm, the cremaster relaxes and the testes are suspended farther from the body.

2. The **dartos**[14] **muscle**—a subcutaneous layer of smooth muscle. It, too, contracts when it is cold, and the scrotum becomes taut and wrinkled. The tautness of the scrotum helps to hold the testes snugly against the warm body and it reduces the surface area of the scrotum, thus reducing heat loss.

3. The **pampiniform**[15] **plexus**—an extensive network of veins from the testis that surround the testicular artery in the spermatic cord. As they pass through the inguinal canal, these veins converge to form the testicular vein, which emerges from the canal into the pelvic cavity. Without the pampiniform plexus, warm arterial blood would heat the testis and inhibit sperm production. The pampiniform plexus, however, prevents this by acting as a *countercurrent heat exchanger* (fig. 27.8). Imagine that a house had uninsulated hot and cold water pipes running close to each other. Much of the heat from the hot water pipe would be absorbed by the cold water pipe next to it—especially if the water flowed in opposite directions so the cold water carried the heat away. In the spermatic cord, such a mechanism removes heat from the descending arterial blood, so by the time it reaches the testis this blood is 1.5° to 2.5°C cooler than the core body temperature.

## The Testes

The testes (testicles) (fig. 27.9) are combined endocrine and exocrine glands that produce sex hormones and sperm. Each testis is oval and slightly flattened, about 4 cm long, 3 cm from anterior to posterior, and 2.5 cm wide. Its anterior and lateral

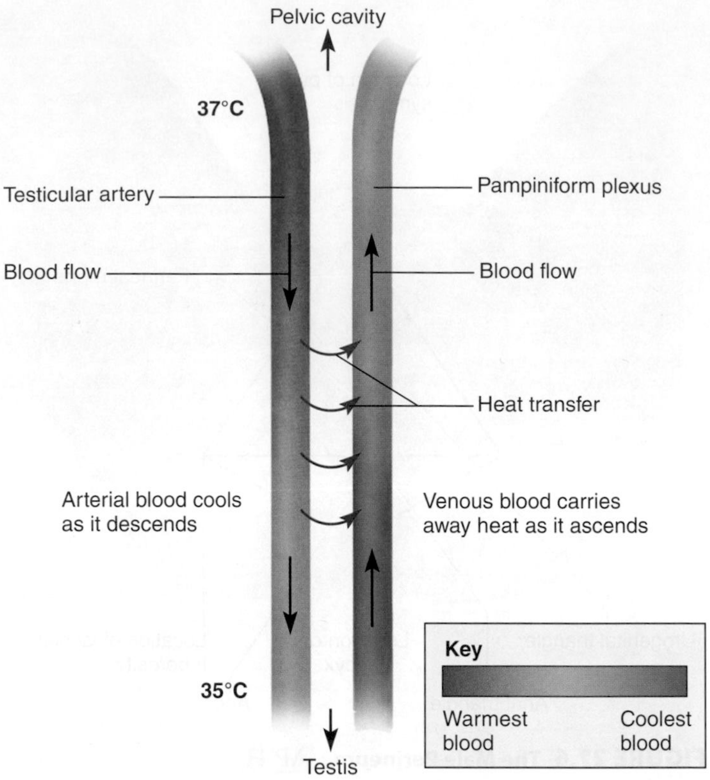

**FIGURE 27.8  The Countercurrent Heat Exchanger.** Warm blood flowing down the testicular artery loses some of its heat to the cooler blood flowing in the opposite direction through the pampiniform plexus of veins (represented as a single vessel for simplicity). Arterial blood is about 2°C cooler by the time it reaches the testis.

surfaces are covered by the tunica vaginalis. The testis itself has a white fibrous capsule called the **tunica albuginea**[16] (TOO-nih-ca AL-byu-JIN-ee-uh). Connective tissue septa extend from the capsule into the parenchyma, dividing it into 250 to 300 wedge-shaped lobules. Each lobule contains one to three **seminiferous**[17] (SEM-ih-NIF-er-us) **tubules**—slender ducts up to 70 cm long in which the sperm are produced. Between the seminiferous tubules are clusters of **interstitial (Leydig**[18]**) cells,** the source of testosterone.

A seminiferous tubule has a narrow lumen lined by a thick **germinal epithelium**. The epithelium consists of several layers of **germ cells** in the process of becoming sperm, and a much smaller number of tall **sustentacular**[19] **(nurse or Sertoli**[20]**) cells,** which protect the germ cells and promote their development. The germ cells depend on the sustentacular cells for nutrients, waste removal, growth factors, and other needs. The sustentacular cells also secrete two proteins called

---

[13]*cremaster* = suspender
[14]*dartos* = skinned
[15]*pampin* = tendril; *form* = shape

[16]*tunica* = coat; *alb* = white
[17]*semin* = seed, sperm; *fer* = to carry
[18]Franz von Leydig (1821–1908), German anatomist
[19]*sustentacul* = support
[20]Enrico Sertoli (1842–1910), Italian histologist

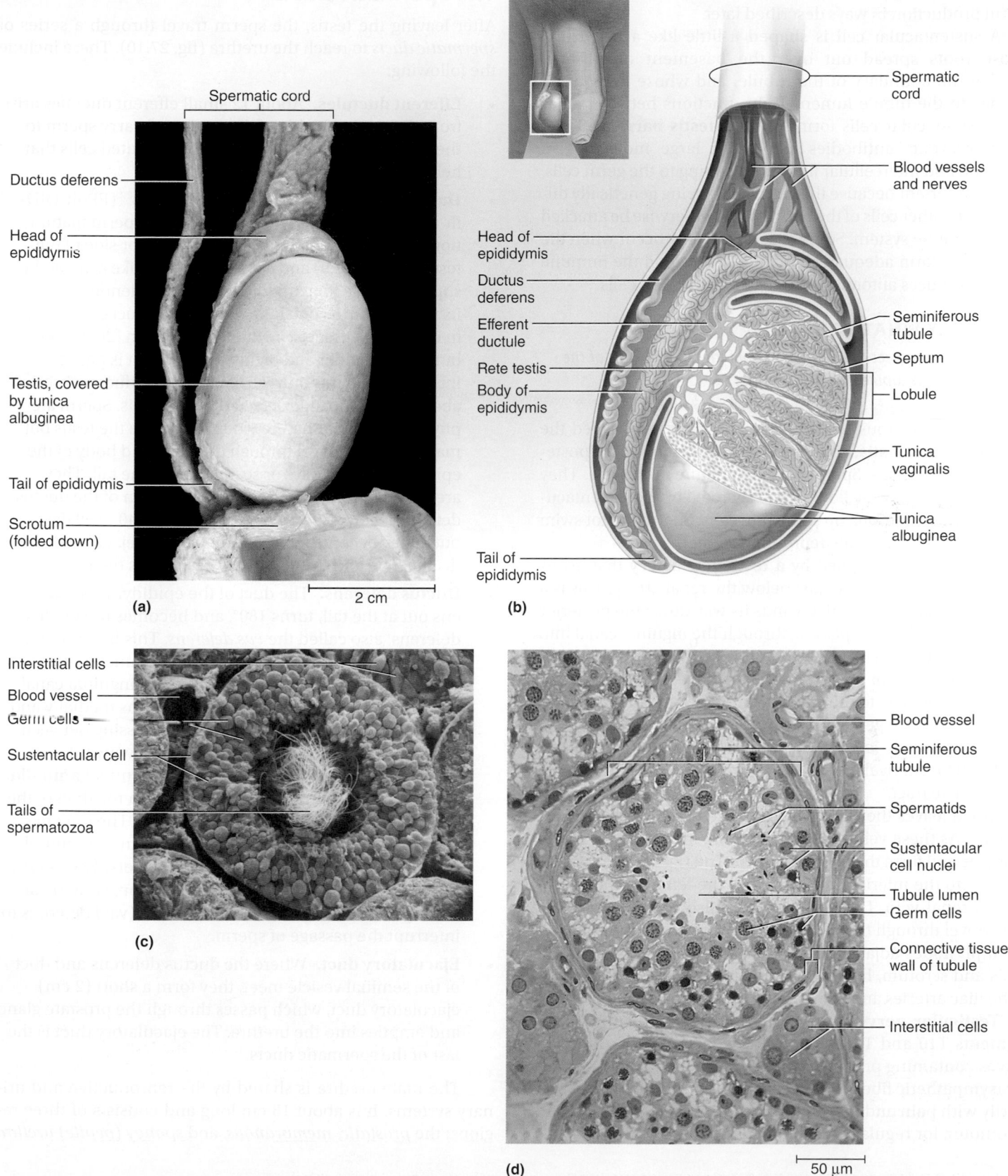

**(a)**

Spermatic cord

Ductus deferens

Head of epididymis

Testis, covered by tunica albuginea

Tail of epididymis

Scrotum (folded down)

2 cm

**(b)**

Spermatic cord

Blood vessels and nerves

Head of epididymis

Ductus deferens

Efferent ductule

Rete testis

Body of epididymis

Tail of epididymis

Seminiferous tubule

Septum

Lobule

Tunica vaginalis

Tunica albuginea

**(c)**

Interstitial cells

Blood vessel

Germ cells

Sustentacular cell

Tails of spermatozoa

**(d)**

Blood vessel

Seminiferous tubule

Spermatids

Sustentacular cell nuclei

Tubule lumen

Germ cells

Connective tissue wall of tubule

Interstitial cells

50 μm

**FIGURE 27.9  The Testis and Associated Structures.**  (a) The scrotum is opened and folded downward to reveal the testis and associated organs. (b) Anatomy of the testis, epididymis, and spermatic cord. (c) Scanning electron micrograph of the seminiferous tubules. (d) Light micrograph. This region of the tubule did not have mature sperm in it at the time.  AP|R

*androgen-binding protein* and *inhibin,* both of which regulate sperm production in ways described later.

A sustentacular cell is shaped a little like a tree trunk whose roots spread out over the basement membrane, forming the boundary of the tubule, and whose thick trunk reaches to the tubule lumen. Tight junctions between adjacent sustentacular cells form a **blood–testis barrier (BTB),** which prevents antibodies and other large molecules in the blood and intercellular fluid from getting to the germ cells. This is important because the germ cells, being genetically different from other cells of the body, would otherwise be attacked by the immune system. Some cases of sterility occur when the BTB fails to form adequately in adolescence and the immune system produces autoantibodies against the germ cells.

#### ►►►APPLY WHAT YOU KNOW

*Would you expect to find blood capillaries in the walls of the seminiferous tubules? Why or why not?*

The seminiferous tubules lead into a network called the **rete**[21] (REE-tee) **testis,** embedded in the capsule on the posterior side of the testis. Sperm partially mature in the rete. They are moved along by the flow of fluid secreted by the sustentacular cells and by the cilia on some rete cells. Sperm do not swim while they are in the male reproductive tract.

Each testis is supplied by a **testicular artery** that arises from the abdominal aorta just below the renal artery. This is a very long, slender artery that winds its way down the posterior abdominal wall before passing through the inguinal canal into the scrotum (see fig. 27.7). Its blood pressure is very low, and indeed this is one of the few arteries to have no pulse. Consequently, blood flow to the testes is quite meager and the testes receive a poor oxygen supply. In response to this, the sperm develop unusually large mitochondria, which may precondition them for survival in the hypoxic environment of the female reproductive tract.

Blood leaves the testis by way of the pampiniform plexus of veins. As these veins pass through the inguinal canal, they converge and form the **testicular vein.** The right testicular vein drains into the inferior vena cava and the left one drains into the left renal vein. Lymphatic vessels also drain each testis. They travel through the inguinal canal with the veins and lead to lymph nodes adjacent to the lower aorta. Lymph from the penis and scrotum, however, travels to lymph nodes adjacent to the iliac arteries and veins and in the inguinal region.

**Testicular nerves** lead to the gonads from spinal cord segments T10 and T11. They are mixed sensory and motor nerves containing predominantly sympathetic but also some parasympathetic fibers. The sensory fibers are concerned primarily with pain and the autonomic fibers are predominantly vasomotor, for regulation of blood flow.

### The Spermatic Ducts

After leaving the testis, the sperm travel through a series of *spermatic ducts* to reach the urethra (fig. 27.10). These include the following:

- **Efferent ductules.** About 12 small efferent ductules arise from the posterior side of the testis and carry sperm to the epididymis. They have clusters of ciliated cells that help drive the sperm along.

- **Duct of the epididymis.** The **epididymis**[22] (EP-ih-DID-ih-miss; plural, *epididymides*) is a site of sperm maturation and storage. It adheres to the posterior side of the testis (see fig. 27.9) and consists of a clublike *head* at the superior end, a long middle *body,* and a slender *tail* at its inferior end. It contains a single coiled duct embedded in connective tissue. The duct is about 6 m (20 ft) long, but it is so slender and highly coiled that it is packed into an epididymis only 7.5 cm long. The duct reabsorbs about 90% of the fluid secreted by the testis. Sperm are physiologically immature when they leave the testis but mature as they travel through the head and body of the epididymis. In 20 days or so, they reach the tail. They are stored here and in the adjacent portion of the ductus deferens. Stored sperm remain fertile for 40 to 60 days, but if they become too old without being ejaculated, they disintegrate and the epididymis reabsorbs them.

- **Ductus deferens.** The duct of the epididymis straightens out at the tail, turns 180°, and becomes the ductus deferens, also called the *vas deferens.* This is a muscular tube about 45 cm long and 2.5 mm in diameter. It passes upward through the spermatic cord and inguinal canal and enters the pelvic cavity. There, it turns medially and approaches the urinary bladder. After passing between the bladder and ureter, the duct turns downward behind the bladder and widens into a terminal **ampulla.** The ductus deferens ends by uniting with the duct of the seminal vesicle, a gland considered later. The duct has a very narrow lumen and a thick wall of smooth muscle well innervated by sympathetic nerve fibers. *Vasectomy,* the surgical method of male contraception, consists of cutting out a short portion of the ductus (vas) deferens to interrupt the passage of sperm.

- **Ejaculatory duct.** Where the ductus deferens and duct of the seminal vesicle meet, they form a short (2 cm) ejaculatory duct, which passes through the prostate gland and empties into the urethra. The ejaculatory duct is the last of the spermatic ducts.

The male urethra is shared by the reproductive and urinary systems. It is about 18 cm long and consists of three regions: the *prostatic, membranous,* and *spongy (penile) urethra*

---

[21]*rete* = network

[22]*epi* = upon; *didym* = twins, testes

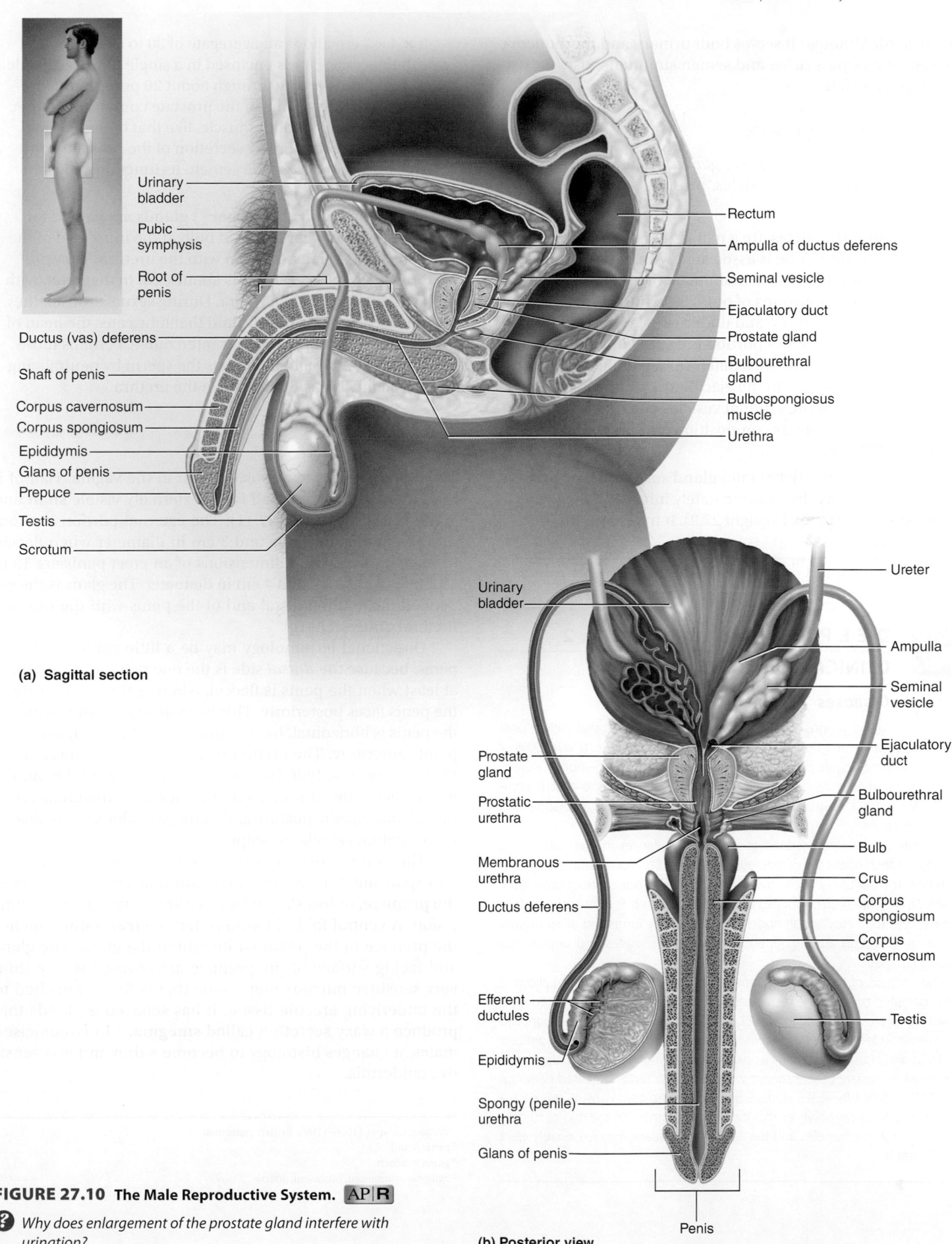

**(a) Sagittal section**

Urinary bladder
Pubic symphysis
Root of penis
Ductus (vas) deferens
Shaft of penis
Corpus cavernosum
Corpus spongiosum
Epididymis
Glans of penis
Prepuce
Testis
Scrotum

Rectum
Ampulla of ductus deferens
Seminal vesicle
Ejaculatory duct
Prostate gland
Bulbourethral gland
Bulbospongiosus muscle
Urethra

**(b) Posterior view**

Urinary bladder
Prostate gland
Prostatic urethra
Membranous urethra
Ductus deferens
Efferent ductules
Epididymis
Spongy (penile) urethra
Glans of penis

Ureter
Ampulla
Seminal vesicle
Ejaculatory duct
Bulbourethral gland
Bulb
Crus
Corpus spongiosum
Corpus cavernosum
Testis
Penis

**FIGURE 27.10  The Male Reproductive System.** AP|R

❓ *Why does enlargement of the prostate gland interfere with urination?*

(see p. 916). Although it serves both urinary and reproductive roles, it cannot pass urine and semen simultaneously for reasons explained later.

## The Accessory Glands

There are three sets of *accessory glands* in the male reproductive system: the seminal vesicles, prostate gland, and bulbourethral glands:

1. The **seminal vesicles** are a pair of glands posterior to the urinary bladder; one is associated with each ductus deferens. A seminal vesicle is about 5 cm long, with approximately the dimensions of one's little finger. It has a connective tissue capsule and underlying layer of smooth muscle. The secretory portion is a very convoluted duct with numerous branches that form a complex labyrinth. The duct empties into the ejaculatory duct. The yellowish secretion of the seminal vesicles constitutes about 60% of the semen; its composition and functions are discussed later.

2. The **prostate**[23] (PROSS-tate) **gland** surrounds the urethra and ejaculatory ducts immediately inferior to the urinary bladder (see Deeper Insight 27.2). It measures about

---

[23]*pro* = before; *stat* = to stand; commonly misspelled and mispronounced "prostrate"

## DEEPER INSIGHT 27.2
### CLINICAL APPLICATION

#### Prostate Diseases

The prostate gland weighs about 20 g by age 20, remains at that weight until age 45 or so, and then begins to grow slowly again. By age 70, over 90% of men show some degree of *benign prostatic hyperplasia (BPH)*—noncancerous enlargement of the gland. The major complication of this is that it compresses the urethra, obstructs the flow of urine, and may promote bladder and kidney infections.

*Prostate cancer* is the second most common cancer in men (after lung cancer); it affects about 9% of men over the age of 50. Prostate tumors tend to form near the periphery of the gland, where they do not obstruct urine flow; therefore, they often go unnoticed until they cause pain. Prostate cancer often metastasizes to nearby lymph nodes and then to the lungs and other organs. It is more common among American blacks than whites and uncommon among Japanese.

The position of the prostate immediately anterior to the rectum allows it to be palpated through the rectal wall to check for tumors. This procedure is called *digital rectal examination (DRE)*. Prostate cancer can also be diagnosed from elevated levels of *serine protease* (also known as *prostate-specific antigen, PSA*) and *acid phosphatase* (another prostatic enzyme) in the blood. Up to 80% of men with prostate cancer survive when it is detected and treated early, but only 10% to 50% survive if it spreads beyond the prostatic capsule. It is such a slow-growing cancer, however, that if discovered late in life, the risks of surgery may outweigh the benefits and the doctor and patient may reasonably elect not to treat it.

---

$2 \times 4 \times 3$ cm and is an aggregate of 30 to 50 compound tubuloacinar glands enclosed in a single fibrous capsule. These glands empty through about 20 pores in the urethral wall. The stroma of the prostate consists of connective tissue and smooth muscle, like that of the seminal vesicles. The thin, milky secretion of the prostate constitutes about 30% of the semen. Its functions, too, are considered later.

3. The **bulbourethral (Cowper**[24]**) glands** are named for their position near a dilated bulb at the inner end of the penis and their association with the urethra. They are brownish, spherical glands about 1 cm in diameter, with a 2.5 cm duct to the urethra. During sexual arousal, they produce a clear slippery fluid that lubricates the head of the penis in preparation for intercourse. Perhaps more important, though, it protects the sperm by neutralizing the acidity of residual urine in the urethra.

## The Penis

The **penis**[25] serves to deposit semen in the vagina. Half of it is an internal **root** and half is the externally visible **shaft** and **glans**[26] (figs. 27.10 and 27.11). The external portion is about 8 to 10 cm (3–4 in.) long and 3 cm in diameter when flaccid (nonerect); the typical dimensions of an erect penis are 13 to 18 cm (5–7 in.) long and 4 cm in diameter. The glans is the expanded head at the distal end of the penis with the external urethral orifice at its tip.

Directional terminology may be a little confusing in the penis, because the *dorsal* side is the one that faces anteriorly, at least when the penis is flaccid, whereas the *ventral* side of the penis faces posteriorly. This is because in most mammals, the penis is horizontal, held against the abdomen by skin, and points anteriorly. The urethra passes through its lower, more obviously ventral, half. Directional terminology in the human penis follows the same convention as for other mammals, even though our bipedal posture and more pendulous penis change these anatomical relationships.

The skin is loosely attached to the penile shaft, allowing for expansion during erection. It continues over the glans as the **prepuce,** or foreskin, which is often removed by circumcision. A ventral fold of tissue called the **frenulum** attaches the prepuce to the proximal margin of the glans. The glans and facing surface of the prepuce are covered with a thin, very sensitive mucous membrane that is firmly attached to the underlying erectile tissue. It has sebaceous glands that produce a waxy secretion called **smegma.**[27] In circumcised males, it changes histology to become a thin and less sensitive epidermis.

---

[24]William Cowper (1666–1709), British anatomist
[25]*penis* = tail
[26]*glans* = acorn
[27]*smegma* = unguent, ointment, soap

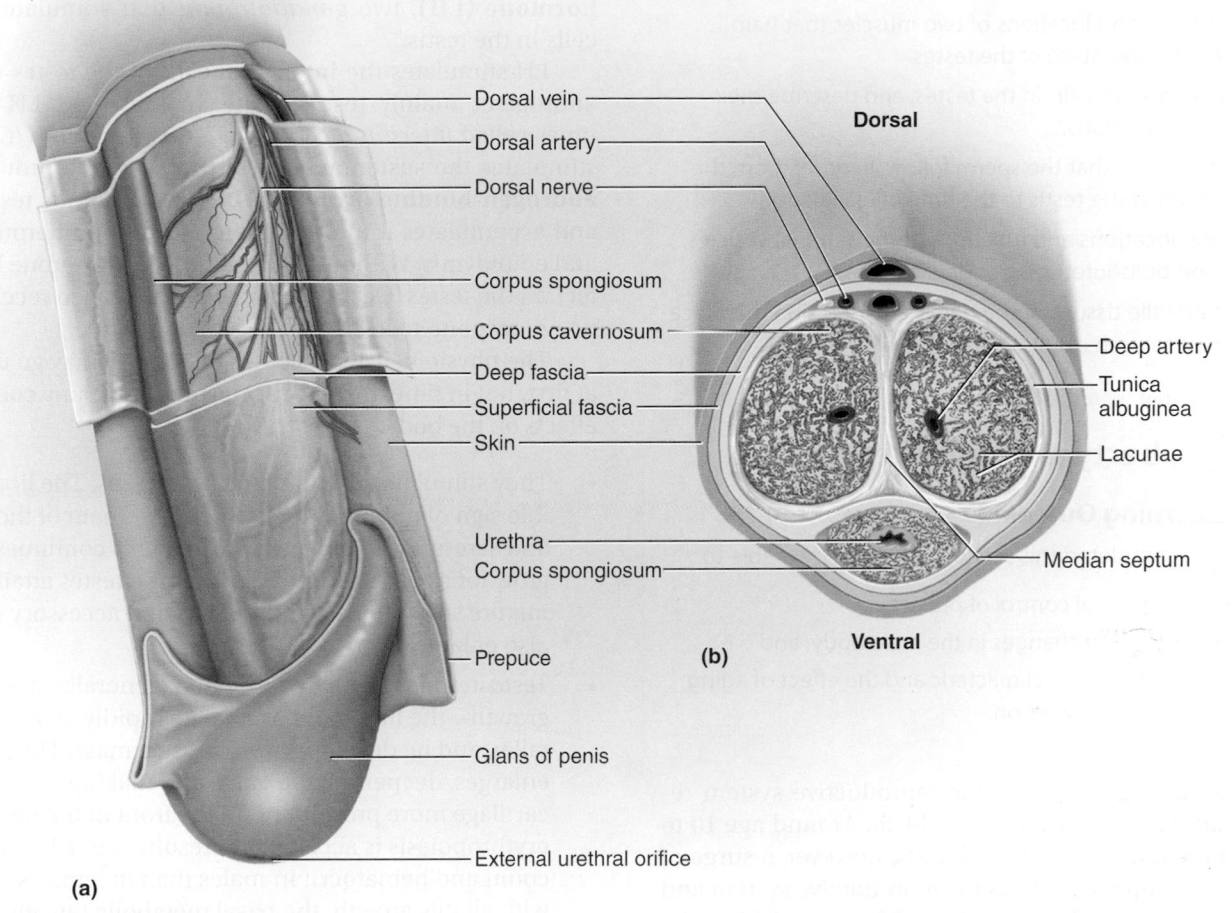

**FIGURE 27.11 Anatomy of the Penis.** (a) Superficial dissection of shaft, lateral view. (b) Cross section at midshaft. **AP|R**

❓ *What is the functional benefit of the corpus spongiosum not having a tunica albuginea?*

The penis consists mainly of three cylindrical bodies called **erectile tissues,** which fill with blood during sexual arousal and account for its enlargement and erection. A single erectile body, the **corpus spongiosum,** passes along the ventral side of the penis and encloses the penile urethra. It expands at the distal end to fill the entire glans. Proximal to the glans, the dorsal side of the penis has a **corpus cavernosum** (plural, *corpora cavernosa*) on each side. Each is ensheathed in a fibrous **tunica albuginea,** and they are separated from each other by a **median septum.** (Note that the testes also have a tunica albuginea and the scrotum also has a median septum.)

All three cylinders of erectile tissue are spongy in appearance and contain numerous tiny blood sinuses called **lacunae.** The partitions between lacunae, called **trabeculae,** are composed of connective tissue and smooth **trabecular muscle.** In the flaccid penis, trabecular muscle tone collapses the lacunae, which appear as tiny slits in the tissue.

At the body surface, the penis turns 90° posteriorly and continues inward as the root. The corpus spongiosum terminates internally as a dilated **bulb,** which is ensheathed in the bulbospongiosus muscle and attached to the lower surface of the muscular pelvic floor within the urogenital triangle (see p. 337). The corpora cavernosa diverge like the arms of a Y. Each arm, called a **crus** (pronounced "cruss"; plural, *crura*), attaches the penis to the pubic arch (ischiopubic ramus) and perineal membrane on its respective side. Each crus is enveloped by an ischiocavernosus muscle. The innervation and blood supply to the penis are discussed later in connection with the mechanism of erection.

Answer the following questions to test your understanding of the preceding section:

6. State the names and locations of two muscles that help regulate the temperature of the testes.

7. Name three types of cells in the testes, and describe their locations and functions.

8. Name all the ducts that the sperm follow, in order, from the time they form in the testis to the time of ejaculation.

9. Describe the locations and functions of the seminal vesicles, prostate, and bulbourethral glands.

10. Name the erectile tissues of the penis, and describe their locations relative to each other.

## 27.3   Puberty and Climacteric

### Expected Learning Outcomes

When you have completed this section, you should be able to

a. describe the hormonal control of puberty;

b. describe the resulting changes in the male body; and

c. define and describe male climacteric and the effect of aging on male reproductive function.

Unlike any other organ system, the reproductive system remains dormant for several years after birth. Around age 10 to 12 in most boys and 8 to 10 in most girls, however, a surge of pituitary gonadotropins awakens the reproductive system and begins preparing it for adult reproductive function. This is the onset of puberty.

Definitions of *adolescence* and *puberty* vary. This book uses **adolescence**[28] to mean the period from the onset of gonadotropin secretion and reproductive development until a person attains full adult height. **Puberty**[29] is the first few years of adolescence, until the first menstrual period in girls or the first ejaculation of viable sperm in boys. In North America, this is typically attained around age 12 in girls and age 14 in boys.

### Endocrine Control of Puberty

The testes secrete substantial amounts of testosterone in the first trimester (3 months) of fetal development. Even in the first few months of infancy, testosterone levels are about as high as they are in midpuberty, but then the testes become dormant for the rest of infancy and childhood. From puberty through adulthood, reproductive function is regulated by hormonal links between the hypothalamus, pituitary gland, and gonads (fig. 27.12).

As the hypothalamus matures, it begins producing **gonadotropin-releasing hormone (GnRH),** which travels by way of the hypophyseal portal system to the anterior lobe of the pituitary. Here it stimulates cells called *gonadotropes* to secrete **follicle-stimulating hormone (FSH)** and **luteinizing hormone (LH),** two *gonadotropins* that stimulate different cells in the testis.

LH stimulates the interstitial cells of the testes to secrete androgens, mainly testosterone. In the male, LH is sometimes called *interstitial cell–stimulating hormone (ICSH)*. FSH stimulates the sustentacular cells to secrete a protein called **androgen-binding protein (ABP),** which binds testosterone and accumulates it in the lumen of the seminiferous tubules and epididymis. Without FSH and ABP, testosterone has no effect on the testes. Germ cells have no androgen receptors and do not respond to it.

The physiological processes of puberty may go unnoticed at first, but in time, the androgens produce many conspicuous effects on the body:

- They stimulate growth of the sex organs. The first visible sign of puberty is usually enlargement of the testes and scrotum around age 13. The penis continues to grow for about 2 more years after the testes attain their mature size. Internally, the ducts and accessory glands also enlarge.

- Testosterone stimulates a burst of generalized body growth—the limb bones elongate rapidly, a boy grows taller, and he develops more muscle mass. The larynx enlarges, deepening the voice and making the thyroid cartilage more prominent on the front of the neck. Even erythropoiesis is accelerated, resulting in a higher RBC count and hematocrit in males than in females. Along with all this growth, the basal metabolic rate increases, accompanied by an increase in appetite.

- One of the androgens, dihydrotestosterone (DHT), stimulates development of the pubic hair, axillary hair, and later the facial hair. The skin becomes darker and thicker and secretes more sebum, which often leads to acne; acne patients have 2 to 20 times the normal level of DHT in their skin. The apocrine scent glands of the perineal, axillary, and beard areas develop in conjunction with the hair in those regions.

- The accumulation of testosterone by ABP leads to the onset and rising rate of sperm production. If testosterone secretion ceases, the sperm count and semen volume decline rapidly and a male becomes sterile.

- Testosterone inhibits GnRH secretion by the hypothalamus and GnRH sensitivity of the pituitary. Consequently, FSH and LH secretion is held in check. Over the course of puberty, however, the pituitary becomes less sensitive to this negative feedback and secretes increasing amounts of gonadotropins, resulting in a rising level of androgens.

- Testosterone also stimulates the brain and awakens the **libido** (sex drive)—although, perhaps surprisingly, the neurons convert it to estrogen, which is what directly affects the behavior. With increasing libido and sensitivity

---

[28]*adolesc* = to grow up
[29]*puber* = grown up

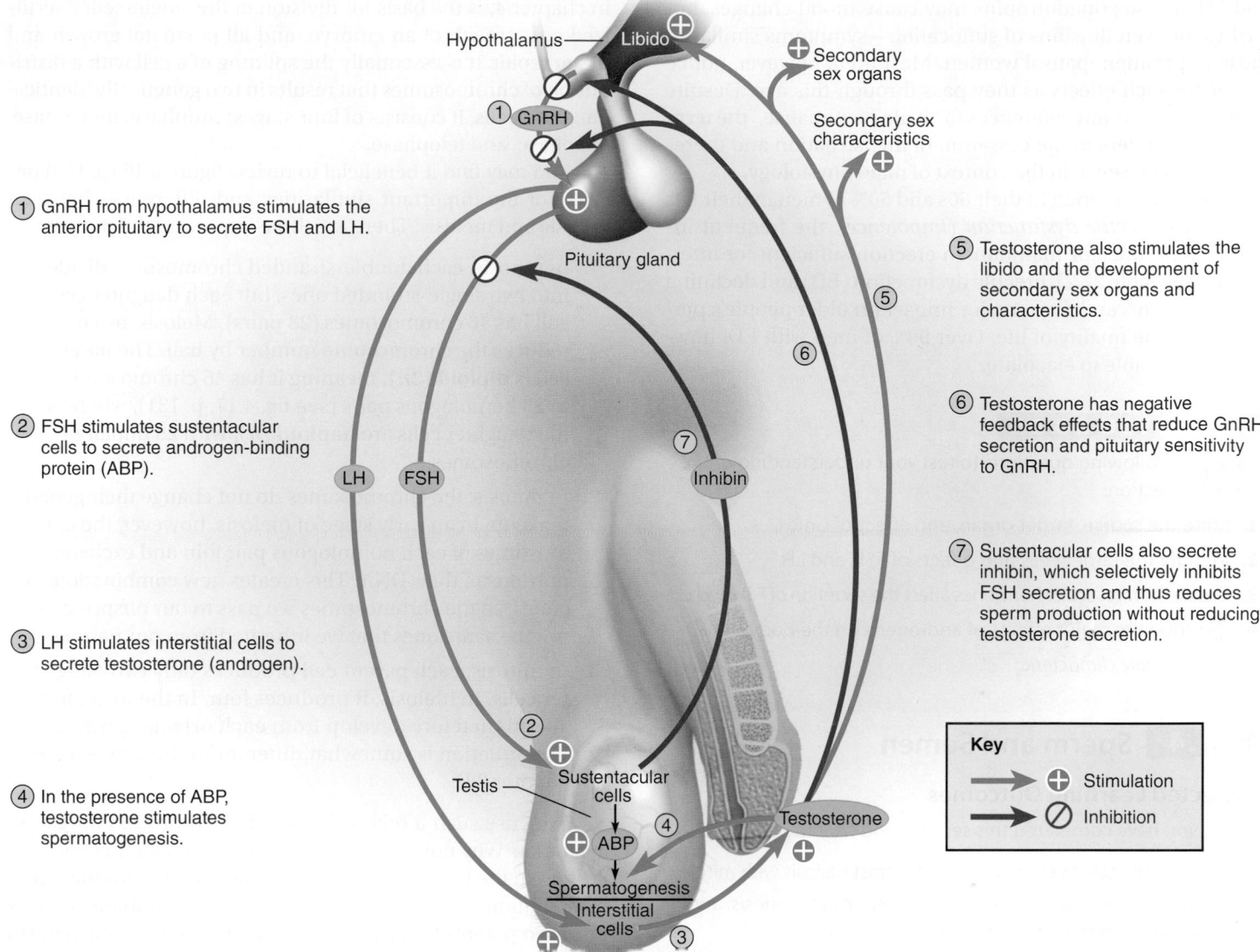

① GnRH from hypothalamus stimulates the anterior pituitary to secrete FSH and LH.

② FSH stimulates sustentacular cells to secrete androgen-binding protein (ABP).

③ LH stimulates interstitial cells to secrete testosterone (androgen).

④ In the presence of ABP, testosterone stimulates spermatogenesis.

⑤ Testosterone also stimulates the libido and the development of secondary sex organs and characteristics.

⑥ Testosterone has negative feedback effects that reduce GnRH secretion and pituitary sensitivity to GnRH.

⑦ Sustentacular cells also secrete inhibin, which selectively inhibits FSH secretion and thus reduces sperm production without reducing testosterone secretion.

**Key**
→⊕ Stimulation
→⊘ Inhibition

**FIGURE 27.12  Hormonal Relationships Between the Hypothalamus, Pituitary Gland, and Testis.**

to stimulation, erections occur frequently and ejaculation often occurs during sleep (nocturnal emissions, or "wet dreams").

Throughout adulthood, testosterone sustains the male reproductive tract, sperm production, and libido.

The body can modulate FSH secretion and sperm production without having to reduce LH and testosterone secretion. The mechanism for this is the hormone **inhibin,** secreted by the sustentacular cells of the seminiferous tubules. Inhibin selectively suppresses FSH output from the pituitary. When the sperm count drops below 20 million sperm/mL, however, inhibin secretion drops and FSH secretion rises.

▶▶▶**APPLY WHAT YOU KNOW**

*If a male animal is castrated, would you expect FSH and LH levels to rise, fall, or be unaffected. Why?*

## Aging and Sexual Function

**Male climacteric,** or **andropause,** is a controversial concept, not accepted by all authorities. It is a period of declining reproductive function that becomes noticeable, if at all, typically in the early 50s. It is brought on by falling levels of testosterone and inhibin. Testosterone secretion peaks at about 7 mg/day at age 20, then declines steadily to as little as one-fifth of this level by age 80. There is a corresponding decline in the number and secretory activity of the interstitial cells (the source of testosterone) and sustentacular cells (the source of inhibin). As testosterone and inhibin levels decline, so does feedback inhibition of the pituitary. Along with the declining testosterone level, the sperm count and libido diminish. By age 65, sperm count is typically about one-third of what it was in a man's 20s. Nevertheless, men remain capable of fathering a child throughout old age.

With reduced levels of testosterone and inhibin, the pituitary is less inhibited and it secretes elevated levels of FSH and LH. These gonadotropins may cause mood changes, hot flashes, or even illusions of suffocation—symptoms similar to those in perimenopausal women. Most men, however, notice few or no such effects as they pass through this age. Despite comedic or sardonic references to "male menopause," the term *menopause* refers to the cessation of menstruation and therefore makes no sense in the context of male physiology.

About 20% of men in their 60s and 50% of men in their 80s experience *erectile dysfunction (impotence),* the frequent inability to produce or maintain an erection sufficient for intercourse (see table 27.2). Erectile dysfunction (ED) and declining sexual activity can have a major impact on older people's perception of the quality of life. Over 90% of men with ED, however, remain able to ejaculate.

**BEFORE YOU GO ON**

Answer the following questions to test your understanding of the preceding section:

11. State the source, target organ, and effect of GnRH.

12. Identify the target cells and effects of FSH and LH.

13. Explain how testicular hormones affect the secretion of FSH and LH.

14. Describe the major effects of androgens on the body.

15. Define *male climacteric.*

## 27.4    Sperm and Semen

### Expected Learning Outcomes

When you have completed this section, you should be able to

a. describe the stages of meiosis and contrast meiosis with mitosis;

b. describe the sequence of cell types in spermatogenesis, and relate these to the stages of meiosis;

c. describe the role of the sustentacular cell in spermatogenesis;

d. describe or draw and label a sperm cell; and

e. describe the composition of semen and functions of its components.

**Spermatogenesis** is the process of sperm production. It occurs in the seminiferous tubules and involves three principal events: (1) division and remodeling of a relatively large germ cell into four small, mobile cells with flagella; (2) reduction of the chromosome number by one-half; and (3) a shuffling of the genes so that each chromosome of the sperm carries new gene combinations that did not exist in the chromosomes of the parents. This ensures genetic variety in the offspring. The genetic recombination and reduction in chromosome number are achieved through a form of cell division called **meiosis,** which produces four daughter cells that subsequently differentiate into sperm.

## Meiosis

In nearly all living organisms except bacteria, there are two forms of cell division: mitosis and meiosis. Mitosis, described

in chapter 4, is the basis for division of the single-celled fertilized egg, growth of an embryo, and all postnatal growth and tissue repair. It is essentially the splitting of a cell with a distribution of chromosomes that results in two genetically identical daughter cells. It consists of four stages: prophase, metaphase, anaphase, and telophase.

You may find it beneficial to review figure 4.16 (p. 130) because of the important similarities and differences between mitosis and meiosis. There are three important differences:

1. In mitosis, each double-stranded chromosome divides into two single-stranded ones, but each daughter cell still has 46 chromosomes (23 pairs). Meiosis, by contrast, reduces the chromosome number by half. The parent cell is **diploid (2*n*),** meaning it has 46 chromosomes in 23 homologous pairs (see fig. 4.17, p. 131), whereas the daughter cells are **haploid (*n*),** with 23 unpaired chromosomes.

2. In mitosis, the chromosomes do not change their genetic makeup. In an early stage of meiosis, however, the chromosomes of each homologous pair join and exchange portions of their DNA. This creates new combinations of genes, so the chromosomes we pass to our offspring are not the same ones that we inherited from our parents.

3. In mitosis, each parent cell produces only two daughter cells. In meiosis, it produces four. In the male, four sperm therefore develop from each original germ cell. The situation is somewhat different in the female (see chapter 28).

Why use such a relatively complicated process for gametogenesis? Why not use mitosis, as we do for all other cell replication in the body? The answer is that sexual reproduction is, by definition, biparental. If we are going to combine gametes from two parents to make a child, there must be a mechanism for keeping the chromosome number constant from generation to generation. Mitosis would produce eggs and sperm with 46 chromosomes each. If these gametes combined, the zygote and the next generation would have 92 chromosomes per cell, the generation after that would have 184, and so forth. To prevent the chromosome number from doubling in every generation, the number is reduced by half during gametogenesis. Meiosis[30] is sometimes called *reduction division* for this reason.

The stages of meiosis are fundamentally the same in both sexes. Briefly, it consists of two cell divisions in succession and occurs in the following phases: prophase I, metaphase I, anaphase I, telophase I, interkinesis, prophase II, metaphase II, anaphase II, and telophase II. These events are detailed in figure 27.13, but let us note some of the unique and important aspects of meiosis.

In prophase I, pairs of homologous chromosomes line up side by side and form **tetrads** (*tetra* denoting the four chromatids). One chromosome of each tetrad is from the individual's father (the paternal chromosome) and the other is from the

---

[30]*meio* = less, fewer

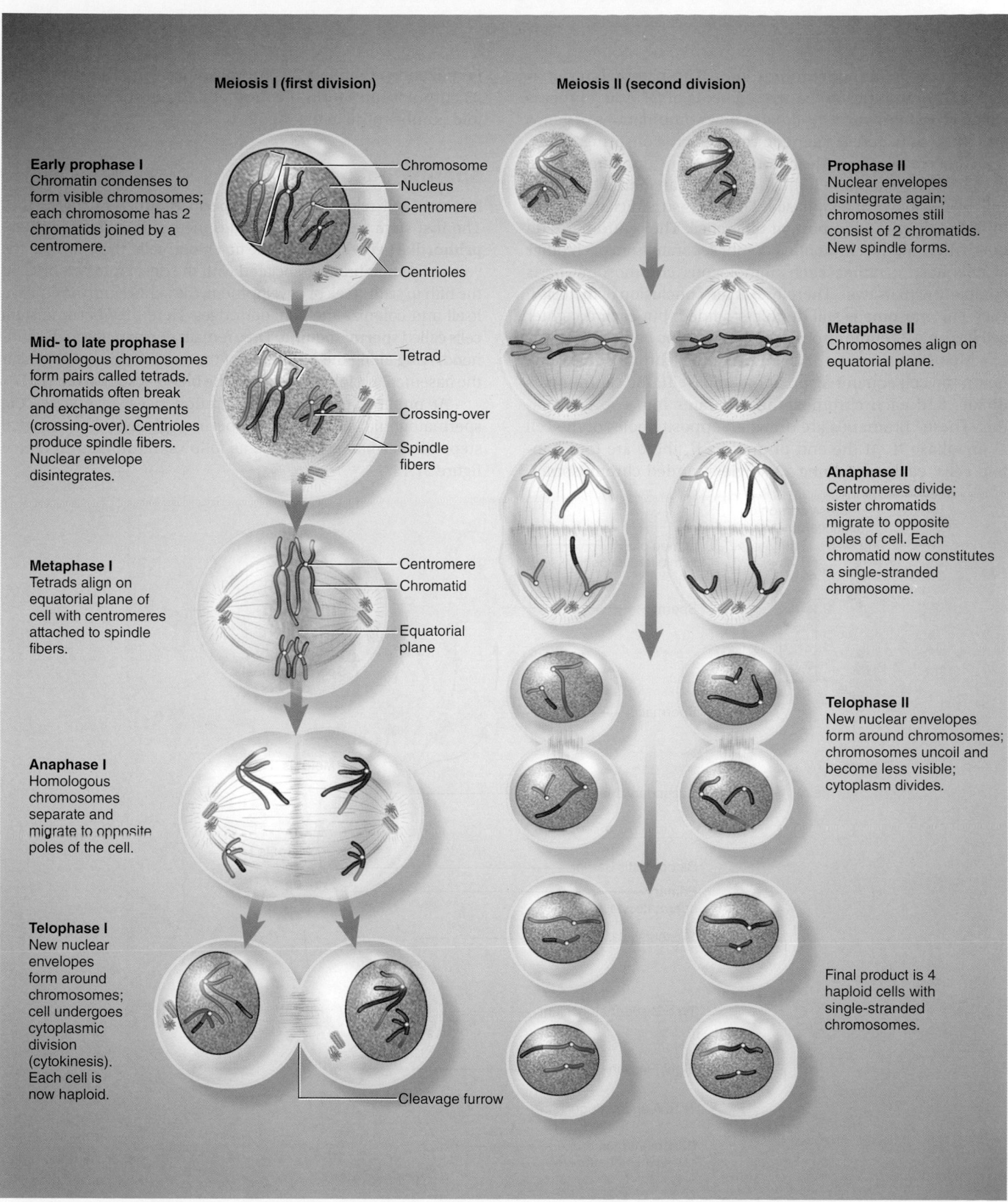

**Meiosis I (first division)**

**Early prophase I**
Chromatin condenses to form visible chromosomes; each chromosome has 2 chromatids joined by a centromere.

Chromosome
Nucleus
Centromere
Centrioles

**Mid- to late prophase I**
Homologous chromosomes form pairs called tetrads. Chromatids often break and exchange segments (crossing-over). Centrioles produce spindle fibers. Nuclear envelope disintegrates.

Tetrad
Crossing-over
Spindle fibers

**Metaphase I**
Tetrads align on equatorial plane of cell with centromeres attached to spindle fibers.

Centromere
Chromatid
Equatorial plane

**Anaphase I**
Homologous chromosomes separate and migrate to opposite poles of the cell.

**Telophase I**
New nuclear envelopes form around chromosomes; cell undergoes cytoplasmic division (cytokinesis). Each cell is now haploid.

Cleavage furrow

**Meiosis II (second division)**

**Prophase II**
Nuclear envelopes disintegrate again; chromosomes still consist of 2 chromatids. New spindle forms.

**Metaphase II**
Chromosomes align on equatorial plane.

**Anaphase II**
Centromeres divide; sister chromatids migrate to opposite poles of cell. Each chromatid now constitutes a single-stranded chromosome.

**Telophase II**
New nuclear envelopes form around chromosomes; chromosomes uncoil and become less visible; cytoplasm divides.

Final product is 4 haploid cells with single-stranded chromosomes.

**FIGURE 27.13 Meiosis.** For simplicity, the cell is shown with only two pairs of homologous chromosomes. Human cells begin meiosis with 23 pairs.

❓ *Although we pass the same genes to our offspring as we inherit from our parents, we do not pass on the same chromosomes. What process in this figure accounts for the latter fact?*

mother (the maternal chromosome). The paternal and maternal chromosomes exchange segments of DNA in a process called **crossing-over.** This creates new combinations of genes and thus contributes to genetic variety in the offspring.

After crossing-over, the chromosomes line up at the midline of the cell in metaphase I, they separate at anaphase I, and the cell divides in two at telophase I. This looks superficially like mitosis, but there is an important difference: The centromeres do not divide and the chromatids do not separate from each other at anaphase I; rather, each homologous chromosome parts company with its twin. Therefore, at the conclusion of meiosis I, each chromosome is still double-stranded, but each daughter cell has only 23 chromosomes—it has become haploid.

Meiosis II is more like mitosis—the chromosomes line up on the cell equator again at metaphase II, the centromeres divide, and each chromosome separates into two chromatids. These chromatids are drawn to opposite poles of the cell at anaphase II. At the end of meiosis II, there are four haploid cells, each containing 23 single-stranded chromosomes.

Fertilization combines 23 chromosomes from the father with 23 chromosomes from the mother and reestablishes the diploid number of 46 in the zygote.

## Spermatogenesis

Now we will relate meiosis to sperm production (fig. 27.14). The first stem cells specifically destined to become sperm are **primordial germ cells.** Like the first blood cells, these form in the yolk sac, a membrane associated with the developing embryo. In the fifth to sixth week of development, they crawl into the embryo itself and colonize the gonadal ridges. Here they become stem cells called **spermatogonia.** They remain dormant through childhood, lying along the periphery of the seminiferous tubule near the basement membrane, outside the blood–testis barrier (BTB).

At puberty, testosterone secretion rises, reactivates the spermatogonia, and brings on spermatogenesis. The essential steps of spermatogenesis are as follows, numbered to match figure 27.14.

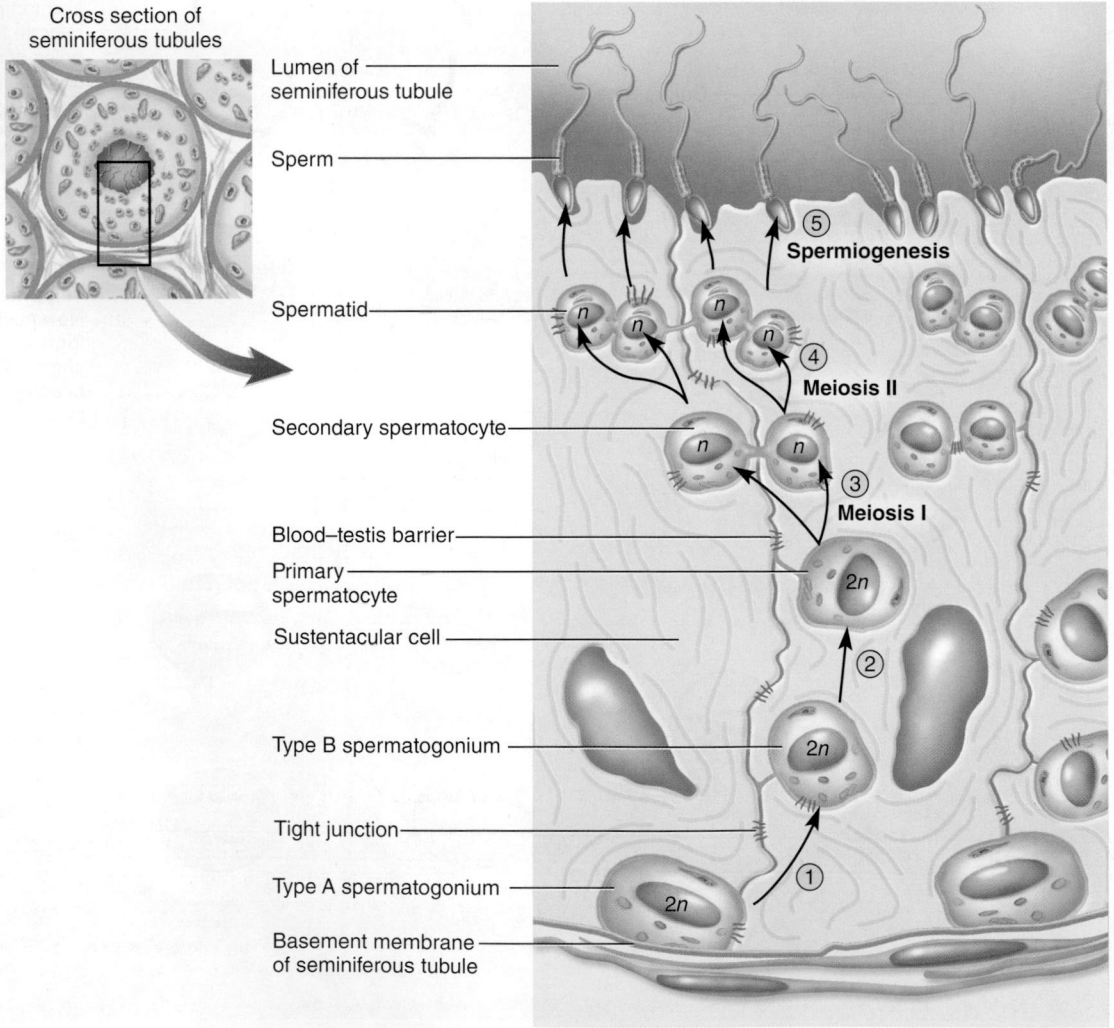

Cross section of seminiferous tubules

Lumen of seminiferous tubule

Sperm

⑤ **Spermiogenesis**

Spermatid — $n$ $n$ $n$ $n$

④ **Meiosis II**

Secondary spermatocyte — $n$ $n$

③ **Meiosis I**

Blood–testis barrier

Primary spermatocyte — $2n$

Sustentacular cell

②

Type B spermatogonium — $2n$

Tight junction

①

Type A spermatogonium — $2n$

Basement membrane of seminiferous tubule

**FIGURE 27.14 Spermatogenesis.** *2n* indicates diploid cells and *n* indicates haploid cells. The process proceeds from the bottom of the figure to the top. The daughter cells from secondary spermatocytes through spermatids remain connected by slender cytoplasmic processes until spermiogenesis is complete and individual spermatozoa are released. See text for explanation of steps 1 to 5.

❓ *Why must the primary spermatocyte move through the blood–testis barrier before undergoing meiosis?*

1. Spermatogonia divide by mitosis. One daughter cell from each division remains near the tubule wall as a stem cell called a *type A spermatogonium.* Type A spermatogonia serve as a lifetime supply of stem cells, so men normally remain fertile throughout old age. The other daughter cell, called a *type B spermatogonium,* migrates slightly away from the wall on its way to producing sperm.

2. The type B spermatogonium enlarges and becomes a **primary spermatocyte.** Since this cell is about to undergo meiosis and become genetically different from other cells of the body, it must be protected from the immune system. Ahead of the primary spermatocyte, the tight junction between two sustentacular cells is dismantled, opening a door for the movement of the spermatocyte toward the lumen. Behind it, a new tight junction forms like a door closing between the spermatocyte and the blood supply in the periphery of the tubule. The spermatocyte moves forward, like an astronaut passing through the double-door airlock of a spaceship, and is now protected by the BTB closing behind it.

3. Now safely isolated from blood-borne antibodies, the primary spermatocyte undergoes meiosis I, which gives rise to two equal-size, haploid and genetically unique **secondary spermatocytes.**

4. Each secondary spermatocyte undergoes meiosis II, dividing into two **spermatids**—a total of four for each spermatogonium.

5. A spermatid divides no further, but undergoes a transformation called **spermiogenesis,** in which it differentiates into a single spermatozoon (fig. 27.15). The spermatid sprouts a tail (flagellum) and discards most of its cytoplasm, making the sperm a lightweight, self-propelled cell. It will not move under its own power, however, until ejaculation.

Each stage of this process is a little closer to the lumen than the earlier stages. All stages on the lumenal side of the BTB are bound to the sustentacular cells by tight junctions and gap junctions, and are closely enveloped in tendrils of the sustentacular cells. Throughout their meiotic divisions, the daughter cells never completely separate, but remain connected to each other by cytoplasmic bridges. Eventually, however, the mature sperm separate from each other, depart from their supportive sustentacular cells, and are washed down the tubule by a slow flow of fluid. Spermatogenesis occurs in cycles, progressing down the tubule in *spermatogenic waves.* At any particular time, therefore, cells in a given portion of the tubule are in the same stage of spermatogenesis, not evenly distributed among all stages.

It takes about 70 days for a type B spermatogonium to become mature spermatozoa. A young man produces about 300,000 sperm per minute, or 400 million per day.

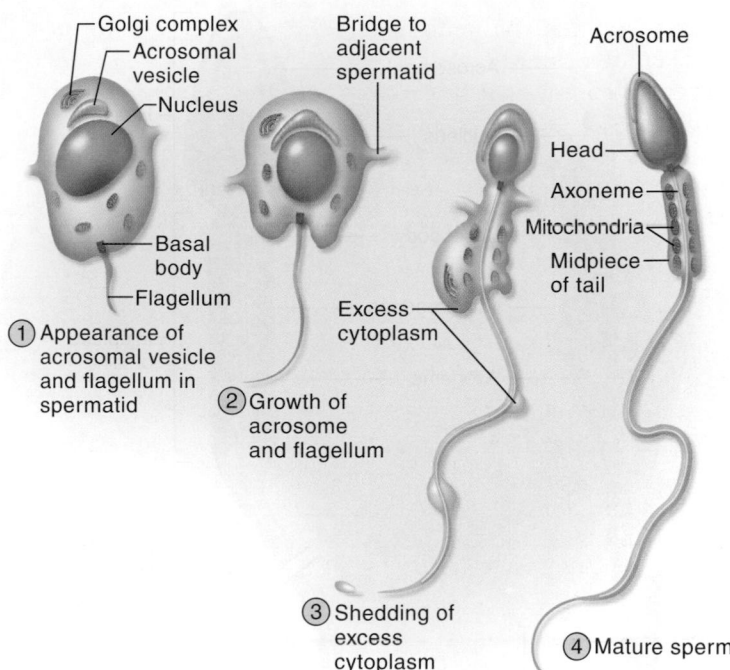

**FIGURE 27.15 Spermiogenesis.** In this process, the spermatids discard excess cytoplasm, grow tails, and become spermatozoa.

## The Spermatozoon

The spermatozoon is an example, par excellence, of the unity of form and function—shaped by evolution for lightness, streamlining, motility, and the effective delivery of its cargo of DNA. It has two parts: a pear-shaped head and a long tail (fig. 27.16). The **head,** about 4 to 5 μm long and 3 μm wide at its broadest part, contains three structures: a nucleus, acrosome, and flagellar basal body. The most important of these is the nucleus, which fills most of the head and contains a haploid set of condensed, genetically inactive chromosomes. The **acrosome**[31] is a lysosome in the form of a thin cap covering the apical half of the nucleus. It contains enzymes that are later used to penetrate the egg if the sperm is successful. The basal body of the tail flagellum is nestled in an indentation at the basal end of the nucleus.

The **tail** is divided into three regions called the midpiece, principal piece, and endpiece. The **midpiece,** a cylinder about 5 to 9 μm long and half as wide as the head, is the thickest part. It contains numerous large mitochondria that coil tightly around the axoneme of the flagellum. They produce the ATP needed for the beating of the tail when the sperm migrates up the female reproductive tract. The **principal piece,** 40 to 45 μm long, constitutes most of the tail and consists of the axoneme surrounded by a sheath of supportive fibers. The **endpiece,** 4 to 5 μm long, consists of the axoneme only and is the narrowest part of the sperm.

---

[31]*acro* = tip, peak; *some* = body

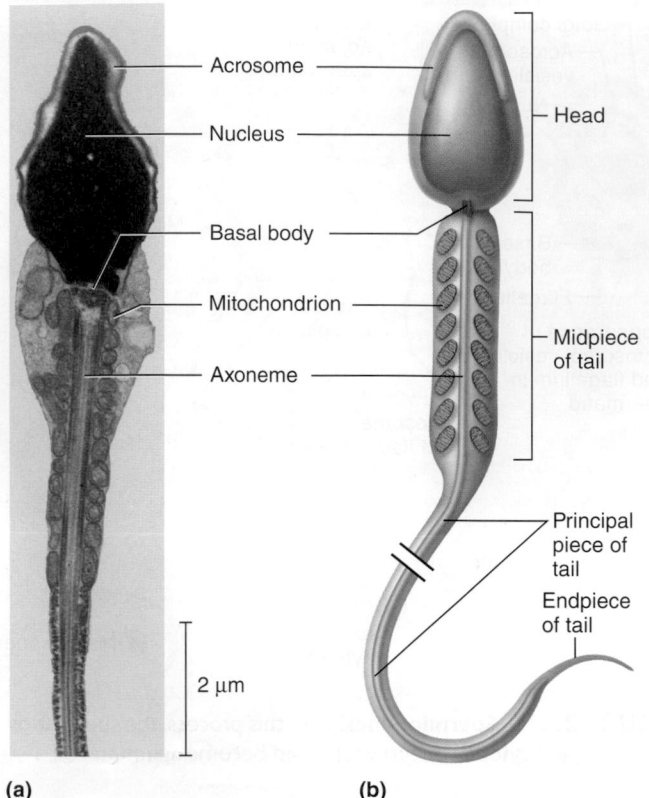

Acrosome — — Head

Nucleus

Basal body

Mitochondrion — — Midpiece of tail

Axoneme

Principal piece of tail

Endpiece of tail

2 μm

(a)                    (b)

**FIGURE 27.16  The Mature Spermatozoon.** (a) Head and part of the tail of a spermatozoon (TEM). (b) Sperm structure.

## Semen

The fluid expelled during orgasm is called **semen**[32] (seminal fluid). A typical ejaculation discharges 2 to 5 mL of semen, composed of about 10% sperm and spermatic duct secretions, 30% prostatic fluid, 60% seminal vesicle fluid, and a trace of bulbourethral fluid. Most of the sperm emerge in the first one or two jets of semen. The semen usually has a **sperm count** of 50 to 120 million sperm/mL. A sperm count any lower than 20 to 25 million sperm/mL is usually associated with **infertility** (sterility), the inability to fertilize an egg (see table 27.2). The prostate and seminal vesicles contribute the following constituents to the semen:

1. The prostate produces a thin, milky white fluid containing calcium, citrate, and phosphate ions; a clotting enzyme; and a protein-hydrolyzing enzyme called *serine protease (prostate-specific antigen, PSA)* (see Deeper Insight 27.2).

2. The seminal vesicles contribute a viscous yellowish fluid, the last component of the semen to emerge. It contains fructose and other carbohydrates, citrate, prostaglandins (discovered in and named for the bovine prostate, but more abundant in the seminal vesicle fluid), and a protein called *prosemenogelin*.

[32]*semen* = seed

A well-known property of semen is its stickiness, an adaptation that promotes fertilization. It arises when the clotting enzyme from the prostate activates prosemenogelin, converting it to a sticky fibrinlike protein, **semenogelin.** Semenogelin entangles the sperm, sticks to the walls of the inner vagina and cervix, and ensures that the semen does not simply drain back out of the vagina. It may also promote the uptake of sperm-laden clots of semen into the uterus. Twenty to 30 minutes after ejaculation, the **serine protease** of the prostatic fluid breaks down semenogelin and liquifies the semen.

Two requirements must be met for sperm motility: an elevated pH and an energy source. The pH of the vagina is about 3.5 to 4.0, and the male spermatic ducts are also quite

### DEEPER INSIGHT 27.3
## CLINICAL APPLICATION

### Reproductive Effects of Pollution

In recent decades, there has been a great deal of interest within the scientific community and popular media concerning *endocrine disrupting chemicals (EDCs)*—environmental agents that interfere with our natural hormones and may disrupt reproduction, development, and homeostasis.

Proven or suspected EDCs occur in industrial solvents and lubricants, pesticides, plastics (including the plastic lining of food cans), prescription drugs and medical devices, and even infant soy milk formulas. We're exposed to EDCs through food, drinking water, contaminated air and soil, and household chemicals, and in certain occupations such as farming and manufacturing that use such chemicals. Some EDCs may affect not only the exposed person, but also that person's descendants through the DNA methylation and epigenetic effects explained in chapter 4 (p. 135). Your grandchildren could be genetically affected by your chemical exposures today.

EDCs act by several mechanisms. They or their metabolites can mimic the effects of estrogen by activating its receptors; antagonize the effects of androgens; alter gene expression; or disrupt positive and negative feedback loops that regulate the body's secretion of its own estrogens and androgens.

EDCs are suspected or implicated, at least tentatively, in a broad spectrum of reproductive abnormalities—in males, cryptorchidism, hypospadias (a urethra opening ventrally on the penis instead of at the tip), low sperm count and reduced motility, and testicular cancer; in females, premature breast development, premature menopause, breast cancer, uterine fibroids, endometriosis, and disrupted ovulation and lactation. Some authorities hypothesize that EDCs may underlie historical trends in declining sperm count and increasing breast cancer over the last 50 to 70 years.

But in many such cases, the data are contradictory, weak, or in dispute. The evidence, however disturbing, is often vague and indirect. It is enormously difficult to prove a link between a particular reproductive disorder and a suspected EDC. For one thing, we cannot experiment on this with humans, and the results of animal experiments often do not translate to humans. For another, the effects may be delayed by years, decades, or perhaps even generations; and there are so many variables in human life that they cover up the trail of causation—prenatal or infant exposure with no effects visible until adulthood; changes in occupation, residence, and environmental exposure; migration and international adoption; and exposure to complex mixtures of environmental chemicals—making it impossible to single out any one cause or to know how they interact to produce effects that no one of them would produce alone. Despite such difficulties, EDCs pose a compelling problem that demands continuing investigation.

acidic. The sperm remain still at such low pHs. But the prostatic fluid buffers the vaginal and seminal acidity, raising the pH to about 7.5 and activating the sperm at the time of ejaculation. The sperm must synthesize a lot of ATP to power their movements. They get the energy for this from the fructose and other sugars contributed by the seminal vesicles.

The activated sperm now thrash with their tails and crawl up the mucosa of the vagina and uterus. The prostaglandins of the semen may thin the mucus in the *cervical canal* of the female (see chapter 28), making it easier for sperm to migrate from the vagina into the uterus, and they may stimulate peristaltic waves in the uterus and uterine tubes that help to spread semen through the female reproductive tract.

### BEFORE YOU GO ON

Answer the following questions to test your understanding of the preceding section:

16. Name the stages of spermatogenesis from spermatogonium to spermatozoon. How do they differ in the number of chromosomes per cell and chromatids per chromosome?

17. Describe the two major parts of a spermatozoon and state what organelles or cytoskeletal components are contained in each.

18. List the major contributions of the seminal vesicles and prostate gland to the semen, and state the functions of these components.

## 27.5 Male Sexual Response

### Expected Learning Outcomes
When you have completed this section, you should be able to

a. describe the blood and nerve supply to the penis; and

b. explain how these govern erection and ejaculation.

The physiology of sexual intercourse was unexplored territory before the 1950s because of repressive attitudes toward the subject. British psychologist Havelock Ellis published the groundbreaking six-volume *Studies in the Psychology of Sex* (1897–1910), only to find it banned for several years as too controversial for victorian tastes. In the 1950s, William Masters and Virginia Johnson daringly launched the first physiological studies of sexual response in the laboratory. In 1966, they published *Human Sexual Response,* which detailed measurements and observations on more than 10,000 sexual acts by nearly 700 volunteer men and women. Masters and Johnson then turned their attention to disorders of sexual function and pioneered modern therapy for sexual dysfunctions. They divided intercourse into four recognizable phases, which they called *excitement, plateau, orgasm,* and *resolution.* The following discussion is organized around this model, although other authorities have modified it or proposed alternatives. Sexual intercourse is also known as **coitus, coition,**[33] or **copulation.**[34]

### Anatomical Foundations

To understand male sexual function, we must give closer attention to the blood circulation and nerve supply to the penis.

Each internal iliac artery gives rise to an **internal pudendal (penile) artery,** which enters the root of the penis and divides in two. One branch, the **dorsal artery,** travels dorsally along the penis not far beneath the skin (see fig. 27.11), supplying blood to the skin, fascia, and corpus spongiosum. The other branch, the **deep artery,** travels through the core of the corpus cavernosum and gives off smaller **helicine**[35] **arteries,** which penetrate the trabeculae and empty into the lacunae. When the penis is flaccid, most of its blood supply comes from the dorsal arteries. When the deep artery dilates, the lacunae fill with blood and the penis becomes erect. There are numerous anastomoses between the dorsal and deep arteries, so neither of them is the exclusive source of blood to any one erectile tissue. A median **deep dorsal vein** drains blood from the penis. It runs between the two dorsal arteries beneath the deep fascia and empties into a plexus of prostatic veins.

The penis is richly innervated by sensory and motor nerve fibers. The glans has an abundance of tactile, pressure, and temperature receptors, especially on its proximal margin and frenulum. They lead by way of a pair of prominent **dorsal nerves** to the **internal pudendal nerves,** then via the sacral plexus to segments S2 to S4 of the spinal cord. Sensory fibers of the shaft, scrotum, perineum, and elsewhere are also highly important to sexual stimulation.

Both autonomic and somatic motor fibers carry impulses from integrating centers in the spinal cord to the penis and other pelvic organs. Sympathetic fibers arise from levels T12 to L2, pass through the hypogastric and pelvic plexuses, and innervate the penile arteries, trabecular muscle, spermatic ducts, and accessory glands. They dilate the penile arteries and can induce erection even when the sacral region of the spinal cord is damaged. They also initiate erection in response to input to the special senses and to sexual thoughts.

Parasympathetic fibers extend from segments S2 to S4 of the spinal cord through the pudendal nerves to the arteries of the penis. They are involved in an autonomic reflex arc that causes erection in response to direct stimulation of the penis and perineal region.

---

[33]*coit* = to come together
[34]*copul* = link, bond
[35]*helic* = coil, helix

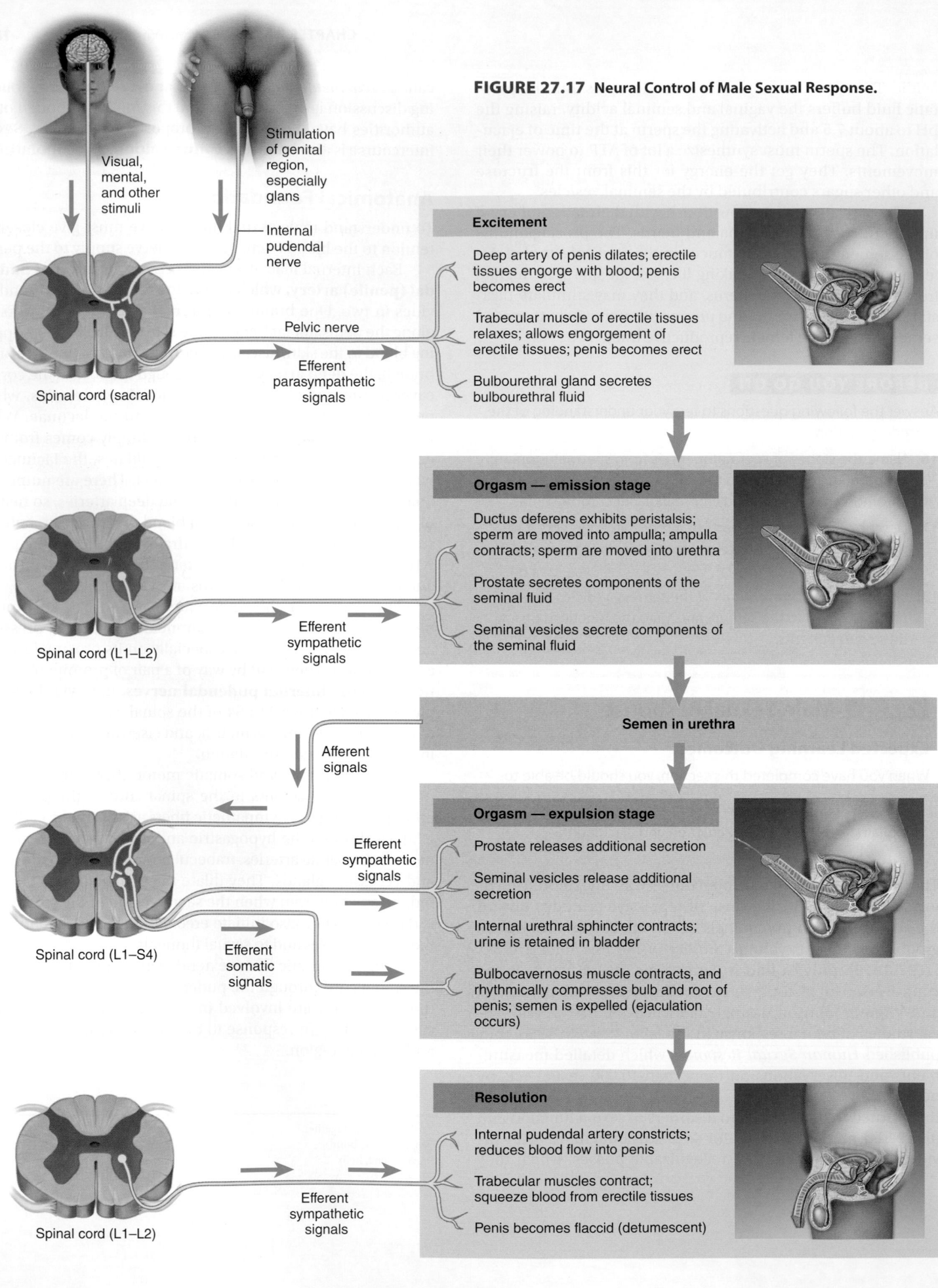

**FIGURE 27.17** Neural Control of Male Sexual Response.

Visual, mental, and other stimuli

Stimulation of genital region, especially glans

Internal pudendal nerve

Spinal cord (sacral)

Pelvic nerve

Efferent parasympathetic signals

**Excitement**

Deep artery of penis dilates; erectile tissues engorge with blood; penis becomes erect

Trabecular muscle of erectile tissues relaxes; allows engorgement of erectile tissues; penis becomes erect

Bulbourethral gland secretes bulbourethral fluid

Spinal cord (L1–L2)

Efferent sympathetic signals

**Orgasm — emission stage**

Ductus deferens exhibits peristalsis; sperm are moved into ampulla; ampulla contracts; sperm are moved into urethra

Prostate secretes components of the seminal fluid

Seminal vesicles secrete components of the seminal fluid

**Semen in urethra**

Afferent signals

Spinal cord (L1–S4)

Efferent sympathetic signals

Efferent somatic signals

**Orgasm — expulsion stage**

Prostate releases additional secretion

Seminal vesicles release additional secretion

Internal urethral sphincter contracts; urine is retained in bladder

Bulbocavernosus muscle contracts, and rhythmically compresses bulb and root of penis; semen is expelled (ejaculation occurs)

Spinal cord (L1–L2)

Efferent sympathetic signals

**Resolution**

Internal pudendal artery constricts; reduces blood flow into penis

Trabecular muscles contract; squeeze blood from erectile tissues

Penis becomes flaccid (detumescent)

## Excitement and Plateau

The **excitement phase** is characterized by **vasocongestion** (swelling of the genitals with blood); **myotonia** (muscle tension); and increases in heart rate, blood pressure, and pulmonary ventilation (fig. 27.17). The bulbourethral glands secrete their fluid during this phase. The excitement phase can be initiated by a broad spectrum of erotic stimuli—sights, sounds, aromas, touch—and even by dreams or thoughts. Conversely, emotions can inhibit sexual response and make it difficult to function when a person is anxious, stressed, or preoccupied with other thoughts.

The most obvious manifestation of male sexual arousal is **erection** of the penis, which makes entry of the vagina possible. Erection is an autonomic reflex mediated predominantly by parasympathetic nerve fibers that travel alongside the deep and helicine arteries of the penis. These fibers trigger the secretion of nitric oxide (NO), which leads to the relaxation of the deep arteries and lacunae (see Deeper Insight 27.4). Whether this is enough to cause erection, or whether it is also necessary to block the outflow of blood from the penis, is still debated. According to one hypothesis, as lacunae near the deep arteries fill with blood, they compress lacunae closer to the periphery of the erectile tissue. This is where blood leaves the erectile tissues, so the compression of the peripheral lacunae helps retain blood in the penis. Their compression is aided by the fact that each corpus

cavernosum is wrapped in a tunica albuginea, which fits over the erectile tissue like a tight fibrous sleeve and contributes to its tension and firmness. In addition, the bulbospongiosus and ischiocavernosus muscles aid in erection by compressing the root of the penis and forcing blood forward into the shaft.

As the corpora cavernosa expand, the penis becomes enlarged, rigid, and elevated to an angle conducive to entry of the vagina. Once **intromission** (entry) is achieved, the tactile and pressure sensations produced by vaginal massaging of the penis further accentuate the erection reflex.

The corpus spongiosum has neither a central artery nor a tunica albuginea. It swells and becomes more visible as a cordlike ridge along the ventral surface of the penis, but it does not become nearly as engorged and hardened as the corpora cavernosa. Vasocongestion is not limited to the penis; the testes also become as much as 50% larger during excitement.

In the **plateau phase,** variables such as respiratory rate, heart rate, and blood pressure are sustained at a high level, or rise slightly, for a few seconds to a few minutes before orgasm. This phase may be marked by increased vasocongestion and myotonia.

**▶▶▶APPLY WHAT YOU KNOW**

*Why is it important that the corpus spongiosum not become as engorged and rigid as the corpora cavernosa?*

---

 # DEEPER INSIGHT 27.4

## CLINICAL APPLICATION

### Treating Erectile Dysfunction

Among the most lucrative drugs developed in the 1990s were the popular treatments for erectile dysfunction: sildenafil (Viagra), vardenafil (Levitra), and tadalafil (Cialis). The basis for developing them was a seemingly unrelated discovery: the role of nitric oxide in cell signaling. When sexual stimulation triggers NO secretion, NO activates the enzyme guanylate cyclase. Guanylate cyclase produces cyclic guanosine monophosphate (cGMP). cGMP then relaxes the smooth muscle of the deep arteries and lacunae of the corpora cavernosa, increasing blood flow into these erectile tissues and bringing about an erection (fig. 27.18).

The erection subsides when cGMP is broken down by the enzyme *phosphodiesterase type 5 (PDE5)*. The problem for many men and their partners is that it subsides too soon, and the solution is to prevent cGMP from breaking down so fast. The aforementioned drugs are in a family called *phosphodiesterase inhibitors.* By slowing down the action of PDE5, they prolong the life of cGMP and the duration of the erection.

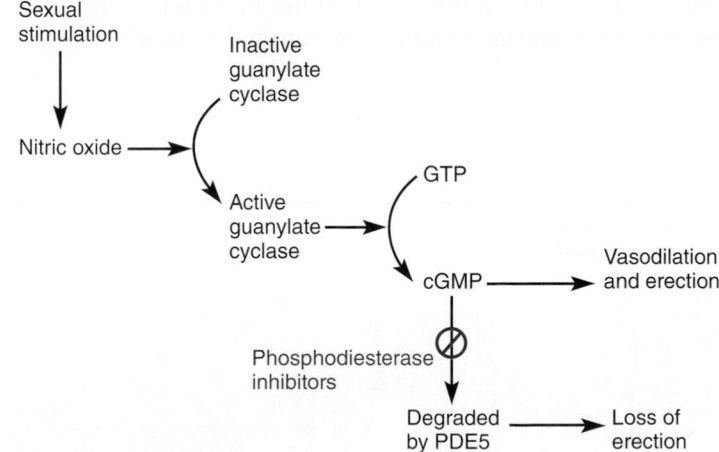

**FIGURE 27.18 The Action of Viagra and Other Phosphodiesterase Inhibitors.**

## Orgasm and Ejaculation

The **orgasm,**[36] or **climax,** is a short but intense reaction that lasts 3 to 15 seconds and usually is marked by the discharge of semen. The heart rate increases to as high as 180 beats/min., blood pressure rises proportionately, and the respiratory rate becomes as high as 40 breaths/min. From the standpoint of producing offspring, the most significant aspect of male orgasm is the **ejaculation**[37] of semen into the vagina.

Ejaculation occurs in two stages called emission and expulsion. In **emission,** the sympathetic nervous system stimulates peristalsis in the smooth muscle of the ductus deferens, which propels sperm from the tail of the epididymis, along the ductus, and into the ampulla. Contractions of the ampulla propel the sperm into the prostatic urethra, and contractions of smooth muscle in the prostate gland force prostatic fluid into the urethra. Secretions of the seminal vesicles join the semen soon after the prostatic secretion. The contractions and seminal flow of this phase create an urgent sensation that ejaculation is inevitable.

Semen in the urethra activates somatic and sympathetic reflexes that result in its **expulsion.** Sensory signals travel to the spinal cord via the internal pudendal nerve and reach an integrating center in the upper lumbar region. Sympathetic nerve fibers carry motor signals from here out to the prostate gland and seminal vesicles, causing the smooth muscle in their walls to express more fluid into the urethra. The sympathetic reflex also constricts the internal urethral sphincter so urine cannot enter the urethra and semen cannot enter the bladder.

Somatic motor signals leave the third and fourth sacral segments of the cord and travel to the bulbospongiosus, ischiocavernosus, and levator ani muscles. The bulbospongiosus, which envelops the root of the penis (see fig. 10.20a, p. 338), undergoes five or six strong, spasmodic contractions that compress the urethra and forcibly expel the semen. Most sperm are ejected in the first milliliter of semen, mixed primarily with prostatic fluid. The seminal vesicle secretion follows and flushes most remaining sperm from the ejaculatory ducts and urethra. Some sperm may seep from the penis prior to ejaculation, and pregnancy can therefore result from genital contact even without orgasm.

Orgasm is accompanied by an intense feeling of release from tension. Ejaculation and orgasm are not the same. Although they usually occur together, it is possible to have all of the sensations of orgasm without ejaculating, and ejaculation occasionally occurs with little or no sensation of orgasm.

## Resolution

Immediately following orgasm comes the **resolution** phase. Discharge of the sympathetic nervous system constricts the internal pudendal artery and reduces the flow of blood into the penis. It also causes contraction of the trabecular muscles, which squeeze blood from the lacunae of the erectile tissues. The penis may remain semierect long enough to continue intercourse, which may be important to the female's attainment of climax, but gradually the penis undergoes **detumescence**—it becomes soft and flaccid again. The resolution phase is also a time in which cardiovascular and respiratory functions return to normal. Many people break out in sweat during the resolution phase. In men, resolution is followed by a **refractory period,** lasting anywhere from 10 minutes to a few hours, in which it is usually impossible to attain another erection and orgasm.

Men and women have many similarities and a few significant differences in sexual response. The response cycle of women is described in chapter 28.

Two of the most common concerns related to sex are sexually transmitted diseases (STDs) and contraception. Understanding most contraceptive methods requires a prior understanding of female anatomy and physiology, so contraceptive techniques for both sexes are discussed at the end of the next chapter, while STDs are discussed in this chapter (see Deeper Insight 27.5). This is not to imply, of course, that STDs are only a male concern and contraception only a female concern.

Reproductive disorders specific to males and females are briefly summarized in tables 27.2 and 28.5, respectively. The effects of aging on the reproductive system are described in chapter 29 (p. 1122).

---

[36]*orgasm* = swelling
[37]*e* = ex = out; *jacul* = to throw

| **TABLE 27.2** | **Some Male Reproductive Disorders** |
|---|---|
| Breast cancer | Accounts for 0.2% of male cancers in the United States; usually seen after age 60 but sometimes in children and adolescents. (For every male who gets breast cancer, about 175 females do so.) Usually felt as a lump near the nipple, often with crusting and discharge from nipple. Often quite advanced by the time of diagnosis, with poor prospects for recovery, because of denial and delay in seeking treatment. |
| Erectile dysfunction (impotence) | Inability to maintain an erection adequate for vaginal entry in half or more of one's attempts. Can stem from aging and declining testosterone level as well as cardiovascular and neurological diseases, diabetes mellitus, medications, fear of failure, depression, and other causes. |
| Hypospadias[38] (HY-po-SPAY-dee-us) | A congenital defect in which the urethra opens on the ventral side or base of the penis rather than at the tip; usually corrected surgically at about 1 year of age. |
| Infertility | Inability to fertilize an egg because of a low sperm count (lower than 20 to 25 million/mL), poor sperm motility, or a high percentage of deformed sperm (two heads, defective tails, etc.). May result from malnutrition, gonorrhea and other infections, toxins, or testosterone deficiency. |
| Penile cancer | Accounts for 1% of male cancers in the United States; most common in black males ages 50 to 70 and of low income. Most often seen in men with nonretractable foreskins (phimosis) combined with poor penile hygiene; least common in men circumcised at birth. |
| Testicular cancer | The most common solid tumor in men 15 to 34 years old, especially white males of middle to upper economic classes. Typically begins as a painless lump or enlargement of the testis. Highly curable if detected early. Men should routinely palpate the testes for normal size and smooth texture. |
| Varicocele (VAIR-ih-co-seal) | Abnormal dilation of veins of the spermatic cord, so that they resemble a "bag of worms." Occurs in 10% of males in the United States. Caused by absence or incompetence of venous valves. Reduces testicular blood flow and often causes infertility. |

*Disorders described elsewhere*

| | | |
|---|---|---|
| Androgen-insensitivity syndrome p. 1030 | Genital herpes p. 1054 | Prostate cancer p. 1040 |
| Benign prostatic hyperplasia p. 1040 | Genital warts p. 1054 | Syphilis p. 1054 |
| Chlamydia p. 1054 | Gonorrhea p. 1054 | |

[38]*hypo* = below; *spad* = to draw off (the urine)

**BEFORE YOU GO ON**

Answer the following questions to test your understanding of the preceding section:

19. Explain how penile blood circulation changes during sexual arousal and why the penis becomes enlarged and stiffened.

20. State the roles of the sympathetic, parasympathetic, and somatic nervous systems in male sexual response.

## DEEPER INSIGHT 27.5

### CLINICAL APPLICATION

#### Sexually Transmitted Diseases

*Sexually transmitted diseases (STDs)* have been well known since the writings of Hippocrates. Here we discuss three bacterial STDs—gonorrhea, chlamydia, and syphilis—and three viral STDs—genital herpes, genital warts, and hepatitis. AIDS is discussed in chapter 21.

All of these STDs have an *incubation period* in which the pathogen multiplies without symptoms, and a *communicable period* in which one can transmit the disease to others, even in the absence of symptoms. STDs often cause fetal deformity, stillbirth, and neonatal death.

*Gonorrhea* (GON-oh-REE-uh) is caused by the bacterium *Neisseria gonorrhoeae.* Galen, thinking the pus discharged from the penis was semen, named the disease *gonorrhea* ("flow of seed"). Gonorrhea causes abdominal discomfort, genital pain and discharge, painful urination, and abnormal uterine bleeding, but most infected women are asymptomatic. It can cause scarring of the uterine tubes, resulting in infertility. Gonorrhea is treated with antibiotics.

*Nongonococcal urethritis (NGU)* is any urethral inflammation caused by agents other than the gonorrhea bacterium. NGU often produces pain or discomfort on urination. The most common bacterial NGU is *chlamydia,* caused by *Chlamydia trachomatis.* Most chlamydia infections are asymptomatic, but they may cause urethral discharge and pain in the testes or pelvic region. Gonorrhea and chlamydia frequently occur together.

*Pelvic inflammatory disease (PID)* is bacterial infection of the female pelvic organs, usually with *Chlamydia* or *Neisseria.* It often results in sterility and may require surgical removal of infected uterine tubes or other organs. The incidence of PID in the United States has increased from 17,800 cases in 1970 to more than a million cases per year currently. PID is responsible for many cases of ectopic pregnancy (see p. 1102).

*Syphilis* (SIFF-ih-liss) is caused by a spiral bacterium named *Treponema pallidum.* After an incubation period of 2 to 6 weeks, a small, hard lesion called a *chancre* (SHAN-kur) appears at the site of infection—usually on the penis of a male but sometimes out of sight in the vagina of a female. It disappears in 4 to 6 weeks, ending the first stage of syphilis and often creating an illusion of recovery. A second stage ensues, however, with a widespread pink rash, other skin eruptions, fever, joint pain, and hair loss. This subsides in 3 to 12 weeks, but symptoms can come and go for up to 5 years. A person is contagious even when symptoms are not present. The disease may progress to a third stage, *tertiary syphilis (neurosyphilis),* with cardiovascular damage and brain lesions that can cause paralysis and dementia. Syphilis is treatable with antibiotics.

*Genital herpes* is the most common STD in the United States, with 20 to 40 million infected people. It is usually caused by the *herpes simplex virus type 2 (HSV-2).* After an incubation period of 4 to 10 days, the virus causes blisters on the penis of the male; on the labia, vagina, or cervix of the female; and sometimes on the thighs and buttocks of either sex. Over 2 to 10 days, these blisters rupture, seep fluid, and begin to form scabs. The initial infection may be painless or may cause intense pain, urethritis, and watery discharge from the penis or vagina. The lesions heal in 2 to 3 weeks and leave no scars.

During this time, however, HSV colonizes sensory nerves and ganglia. Here it can lie dormant for years, later migrating along the nerves and causing epithelial lesions anywhere on the body. The movement from place to place is the basis of the name *herpes.*[39] Most patients have five to seven recurrences, ranging from several years apart to several times a year. An infected person is contagious to a sexual partner when the lesions are present and sometimes even when they are not. HSV may increase the risk of cervical cancer and AIDS.

*Genital warts (condylomas)* are one of the most rapidly increasing STDs today, with about 6.2 million new cases per year. They are caused by various strains of *human papillomavirus (HPV).* In the male, lesions usually appear on the penis, perineum, or anus; and in the female, they are usually on the cervix, vaginal wall, perineum, or anus. Lesions are sometimes small and almost invisible. HPV has been implicated in cancer of the penis, vagina, cervix, and anus; it is found in about 90% of cervical cancers. About 90% of genital warts, however, involve strains that have not been linked to cancer. Genital warts are sometimes treated with cryosurgery (freezing and excision), laser surgery, or interferon. A vaccine against HPV, Gardasil, is now available. Given the alarming incidence of sexual activity and HPV among young adolescents, the U.S. Food and Drug Administration recommends vaccination of girls and boys at the age of puberty in hopes of immunizing most before the onset of sexual activity.

*Hepatitis B* and *C* are inflammatory liver diseases caused by the hepatitis B and C viruses (HBV, HCV), introduced in Deeper Insight 26.3 (p. 1016). Although they can be transmitted by means other than sex, they are becoming increasingly common as STDs. Hepatitis C threatens to become a major epidemic of the twenty-first century. It already far surpasses the prevalence of AIDS and is the leading reason for liver transplants in the United States.

---

[39]*herp* = to creep

# STUDY GUIDE

## ► Assess Your Learning Outcomes

*To test your knowledge, discuss the following topics with a study partner or in writing, ideally from memory.*

### 27.1 Sexual Reproduction and Development (p. 1029)

1. Essential characteristic of sexual reproduction; what defines *male* and *female* in any sexual species; and the names and defining characteristics of their respective gametes and gonads

2. Multiple functions of the male and female reproductive systems

3. Which organs of each sex are considered to be primary and secondary sex organs; which are considered to be internal and external genitalia; and how secondary sex characteristics differ from secondary sex organs

4. The difference between sex chromosomes and autosomes; the number of each; names of the sex chromosomes; and how males and females differ chromosomally

5. The male-determining gene on the Y chromosome; the name of the protein encoded by it; and the effect of that protein on embryonic development

6. Fates of the mesonephric and paramesonephric ducts in the male and female embryos, and why this is indirectly determined by the *SRY* gene

7. Structures of the embryonic genital tubercle, urogenital folds, and labioscrotal folds, and mature structures of the male and female that arise from each

8. Descent of the gonads; similarities and differences in this process in the male and female fetus; and consequences for the male if it fails to occur to completion

### 27.2 Male Reproductive Anatomy (p. 1034)

1. Anatomy and functions of the scrotum and spermatic cord

2. Why the testes must be kept cooler than the core body temperature, and three mechanisms for achieving this

3. Anatomy of the testis and functions of its seminiferous tubules and interstitial cells

4. The germinal epithelium of the seminiferous tubule; its cell types and their functions; the necessity and structure of the blood–testis barrier

5. The arterial supply and venous drainage of the testis; testicular nerves and lymphatic vessels

6. Gross anatomy of the epididymis; the series of spermatic ducts from the efferent ductules to the ejaculatory duct; and differences in their anatomy and relationships to adjacent organs

7. Three sets of male accessory glands; their anatomy, functions, and relationships to adjacent organs

8. The root, shaft, and glans of the penis; the prepuce and frenulum distally; the bulb and the crura proximally

9. Erectile tissues of the penis; their histological structure; the lacunae, trabeculae, and trabecular muscle, differences between the corpus spongiosum and the corpora cavernosa

### 27.3 Puberty and Climacteric (p. 1042)

1. Definitions of *puberty* and *adolescence;* the typical age range of each

2. The hormonal trigger for the onset of puberty; the roles of GnRH, FSH, LH, androgens, and androgen-binding protein in male adolescence

3. Bodily changes of male adolescence, onset of the libido, and their respective hormonal causes

4. The source of inhibin and its role in male sexual physiology

5. Changes in the levels of testosterone, inhibin, FSH, and LH over the male life span; how these relate to negative feedback inhibition of the pituitary; and the effects experienced by some men in male climacteric

6. Effects of old age on male sexual function

### 27.4 Sperm and Semen (p. 1044)

1. The meaning of *spermatogenesis* and distinction between spermatogenesis and spermiogenesis

2. Why meiosis is necessary in sexually reproducing species, how it affects the chromosome number of gametes, and why chromosome number remains constant from one generation to the next

3. The nine stages of meiosis—four in meiosis I, interkinesis, and four in meiosis II; events of each stage; the final number of functional daughter cells, their chromo-

some number, and final chromosome structure in the male

4. Multiple reasons that meiosis results in genetic diversity of the gametes

5. The origin of primordial germ cells, their migration to the embryonic gonads, and the stage they have reached by the time of birth

6. The mode and location of production of type A and type B spermatogonia and the difference between them

7. Stages in the transformation of spermatogonia to sperm; how the germ cells migrate through the germinal epithelium as this is occurring; when they must pass through the blood–testis barrier, why, and how they do so

8. The process and effects of spermiogenesis

9. Morphology of a mature sperm and the functions of its parts

10. Composition of semen and percentages of it that come from its three main sources; functions of the fructose, semenogelin, clotting enzyme, prostaglandins, and serine protease; and the identity and clinical relevance of prostate-specific antigen (PSA)

11. The typical sperm count and the minimum necessary for fertility

### 27.5 Male Sexual Response (p. 1049)

1. Blood vessels of the penis and how they function in the flaccid penis and in erection

2. The nerve supply of the penis and its relationship to the sacral plexus and spinal cord; three types of nerve fibers that the penis receives

3. Physiological changes that the male undergoes during the excitement phase of the sexual response

4. The mechanism of erection including the roles of parasympathetic stimulation, nitric oxide, and vasodilation; why the corpora cavernosa become more engorged than the corpus spongiosum

5. The physiological state of the male in the plateau phase of sexual response

6. The mechanism of male orgasm, including the stages of ejaculation

7. Characteristics of the male resolution phase; the mechanism of penile detumescence; and the refractory period

# STUDY GUIDE

## ▶ Testing Your Recall

*Answers in Appendix B*

1. The ductus deferens develops from the
   ___ of the embryo.
   a. mesonephric duct
   b. paramesonephric duct
   c. phallus
   d. labioscrotal folds
   e. urogenital folds

2. The protein that clots and causes the
   stickiness of the semen is
   a. semenogelin.
   b. prostaglandin.
   c. fibrin.
   d. phosphodiesterase.
   e. serine protease.

3. The expulsion of semen occurs when the
   bulbospongiosus muscle is stimulated by
   a. somatic efferent neurons.
   b. somatic afferent neurons.
   c. sympathetic efferent neurons.
   d. parasympathetic efferent neurons.
   e. prostaglandins.

4. Prior to ejaculation, sperm are stored
   primarily in
   a. the seminiferous tubules.
   b. the rete testis.
   c. the epididymis.
   d. the seminal vesicles.
   e. the ejaculatory ducts.

5. The penis is attached to the pubic arch
   by crura of
   a. the corpora cavernosa.
   b. the corpus spongiosum.
   c. the perineal membrane.
   d. the bulbospongiosus.
   e. the ischiocavernosus.

6. The first hormone secreted at the onset of
   puberty is
   a. follicle-stimulating hormone.
   b. interstitial cell–stimulating hormone.
   c. human chorionic gonadotropin.
   d. gonadotropin-releasing hormone.
   e. testosterone.

7. When it is necessary to reduce sperm
   production without reducing testosterone
   secretion, the sustentacular cells secrete
   a. dihydrotestosterone.
   b. androgen-binding protein.
   c. LH.
   d. FSH.
   e. inhibin.

8. Four spermatozoa arise from each
   a. primordial germ cell.
   b. type A spermatogonium.
   c. type B spermatogonium.
   d. secondary spermatocyte.
   e. spermatid.

9. The point in meiosis at which sister
   chromatids separate from each other is
   a. prophase I.
   b. metaphase I.
   c. anaphase I.
   d. anaphase II.
   e. telophase II.

10. Blood is forced out of the penile lacunae
    by contraction of the _____ muscles.
    a. bulbospongiosus
    b. ischiocavernosus
    c. cremaster
    d. trabecular
    e. dartos

11. Under the influence of androgens, the
    embryonic _____ duct develops into the
    male reproductive tract.

12. Spermatozoa obtain energy for locomotion
    from _____ in the semen.

13. The _____, a network of veins in the
    spermatic cord, helps keep the testes
    cooler than the core body temperature.

14. All germ cells beginning with the _____
    are genetically different from the rest
    of the body cells and therefore must be
    protected by the blood–testis barrier.

15. The corpora cavernosa as well as the
    testes have a fibrous capsule called
    the _____.

16. Over half of the semen consists of
    secretions from a pair of glands called
    the _____.

17. The blood–testis barrier is formed by tight
    junctions between the _____ cells.

18. The earliest haploid stage of spermato-
    genesis is the _____.

19. Erection of the penis occurs when nitric
    oxide causes the _____ arteries to
    dilate.

20. A sperm penetrates the egg by means of
    enzymes in its _____.

## ▶ Building Your Medical Vocabulary

*Answers in Appendix B*

*State a medical meaning of each word element
below, and give a term in which it or a slight
variation of it is used.*

1. crypto-

2. didymo-

3. e-

4. gamo-

5. -ism

6. meio-

7. orchido-

8. rete

9. -some

10. vagino-

# STUDY GUIDE

## ▶ True or False

*Answers in Appendix B*

*Determine which five of the following statements are false, and briefly explain why.*

1. The male scrotum is homologous to the female labia majora.

2. Sperm cannot develop at the core body temperature.

3. Luteinizing hormone stimulates the testes to secrete testosterone.

4. The testes and penis are the primary sex organs of the male.

5. A high testosterone level makes a fetus develop a male reproductive system, and a high estrogen level makes it develop a female reproductive system.

6. Most of the semen is produced by the seminal vesicles.

7. The pampiniform plexus serves to keep the testes warm.

8. Prior to ejaculation, sperm are stored mainly in the seminal vesicles.

9. Male menopause is the cessation of sperm production around the age of 55.

10. Erection is caused by parasympathetic stimulation of the penile arteries.

## ▶ Testing Your Comprehension

*Answers at* www.mhhe.com/saladin7

1. Explain why testosterone may be considered both an endocrine and a paracrine secretion of the testes. (Review paracrines in chapter 17 if necessary.)

2. A young man is in a motorcycle accident that severs his spinal cord at the neck and leaves him paralyzed from the neck down. When informed of the situation, his wife asks the physician if her husband will be able to have erections and father any children. What should the doctor tell her? Explain your answer.

3. Considering the temperature in the scrotum, would you expect hemoglobin to unload more oxygen to the testes, or less, than it unloads in the warmer internal organs? Why? (*Hint:* See fig. 22.26, p. 879.) How would you expect this fact to influence sperm development?

4. Why is it possible for spermatogonia to be outside the blood–testis barrier, yet necessary for primary spermatocytes and later stages to be within the barrier, isolated from the blood?

5. A 68-year-old man taking medication for hypertension complains to his physician that it has made him impotent. Explain why this could be an effect of antihypertension drugs.

## ▶ Improve Your Grade at www.mhhe.com/saladin7

*Download animations to study when it fits your schedule. Practice quizzes, labeling activities, games, and flashcards offer fun ways to master the chapter concepts. Or, download image PowerPoint files for each chapter to create a study guide or for taking notes during lecture.*

# THE FEMALE REPRODUCTIVE SYSTEM

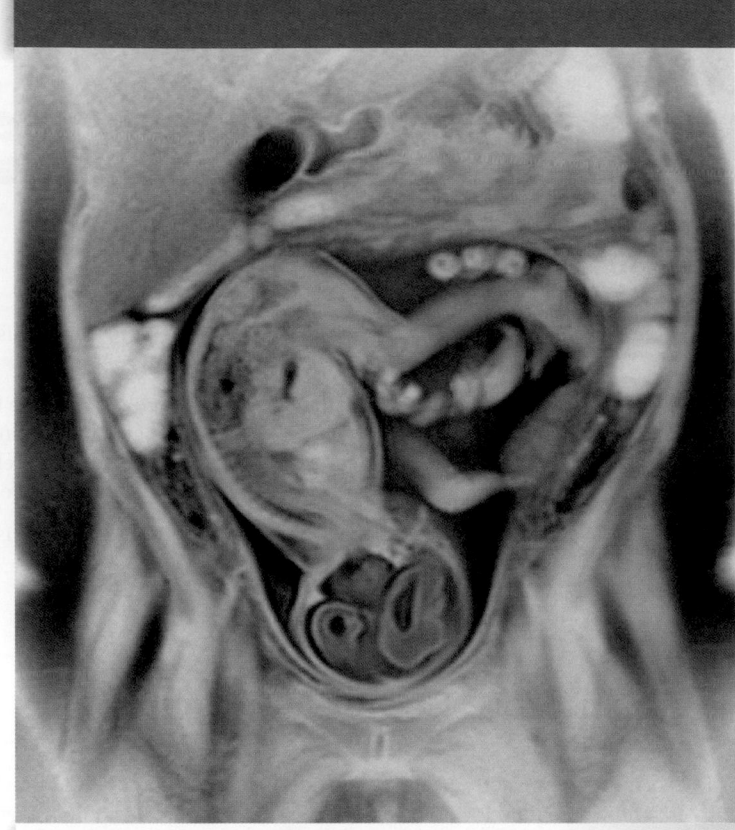

MRI scan of a 36-week-old fetus in the uterus

aprevealed.com

**Module 14: Reproductive System**

- Female reproductive anatomy is best understood from the perspective of the embryonic development of the reproductive system described on page 1031.

- You will need a good understanding of gonadotropin-releasing hormone, follicle-stimulating hormone, and luteinizing hormone; the hypothalamo–pituitary–ovarian axis; and negative feedback inhibition of the hypothalamus and pituitary (pp. 636–639) to understand several aspects of ovarian development, pregnancy, and lactation.

- Hormonal interactions in the reproductive cycle and pregnancy involve the synergistic, permissive, and antagonistic effects described on page 658.

- Oogenesis (egg production) involves the same stages of meiosis as spermatogenesis in the male (p. 1046); only the differences between oogenesis and spermatogenesis are presented in this chapter.

- Female sexual response is much like that of the male (p. 1049); only the differences between female and male are presented in this chapter.

The female reproductive system is more complex than the male's because it serves more purposes. Whereas the male needs only to produce and deliver gametes, the female must do this as well as provide nutrition and safe harbor for fetal development, then give birth and nourish the infant. Furthermore, female reproductive physiology is more conspicuously cyclic, and female hormones are secreted in a more complex sequence compared with the relatively steady, simultaneous secretion of regulatory hormones in the male.

This chapter discusses the anatomy of the female reproductive system; the production of gametes and how it relates to the ovarian and menstrual cycles; the female sexual response; and the physiology of pregnancy, birth, and lactation. Embryonic and fetal development are treated in chapter 29.

## 28.1 Reproductive Anatomy AP|R

### Expected Learning Outcomes

When you have completed this section, you should be able to

a. describe the structure of the ovary;

b. trace the female reproductive tract and describe the gross anatomy and histology of each organ;

c. identify the ligaments that support the female reproductive organs;

d. describe the blood supply to the female reproductive tract;

e. identify the external genitalia of the female; and

f. describe the structure of the nonlactating breast.

## Sexual Differentiation

The female reproductive system (fig. 28.1) is obviously different from that of the male, but as we saw earlier, the two sexes are indistinguishable for the first 8 to 10 weeks of development (see fig. 27.4, p. 1033). The female reproductive tract develops from the paramesonephric duct not because of the positive action of any hormone, but because of the absence of testosterone and müllerian-inhibiting factor (MIF). Without testosterone, the mesonephric duct degenerates while the genital tubercle becomes the glans of the clitoris, the urogenital folds develop into labia minora, and the labioscrotal folds develop into labia majora. Without MIF, the paramesonephric ducts develop into the uterine tubes, uterus, and vagina (see fig. 27.3, p. 1032). This developmental pattern can be disrupted, however, by abnormal hormonal exposure before birth, as happens in adrenogenital syndrome (see fig. 17.28, p. 664).

## The Genitalia

The internal genitalia include the ovaries and a duct system that runs from the vicinity of each ovary to the outside of the body—the uterine tubes, uterus, and vagina. The external genitalia include principally the clitoris, labia minora, and labia majora. These occupy the perineum, which is defined by the same skeletal landmarks as in the male (see fig. 27.6, p. 1035). The ovaries are the primary sex organs, and the other internal and external genitalia are the secondary sex organs.

## The Ovaries

The female gonads are the **ovaries,**[1] which produce egg cells (ova) and sex hormones. The ovary is an almond-shaped organ nestled in the *ovarian fossa,* a depression in the posterior pelvic wall. It measures about 3 cm long, 1.5 cm wide, and 1 cm thick. Its capsule, like that of the testis, is called the **tunica albuginea.** The interior of the ovary is indistinctly divided into a central **medulla** and an outer **cortex** (fig. 28.2). The medulla is a core of fibrous connective tissue occupied by the principal arteries and veins of the ovary. The cortex is the site of the ovarian **follicles,** each of which consists of one developing ovum surrounded by numerous small follicular cells. The ovary does not have a system of tubules like the testis; eggs are released one at a time by the bursting of the follicles *(ovulation).* In childhood, the ovaries are smooth-surfaced. During the reproductive years, they become more corrugated because growing follicles of various ages produce bulges in the surface. After menopause, the ovaries are shrunken and composed mostly of scar tissue. Figure 28.2 shows several types of follicles that coincide with different stages of egg maturation, as discussed later.

---

[1]*ov* = egg; *ary* = place for

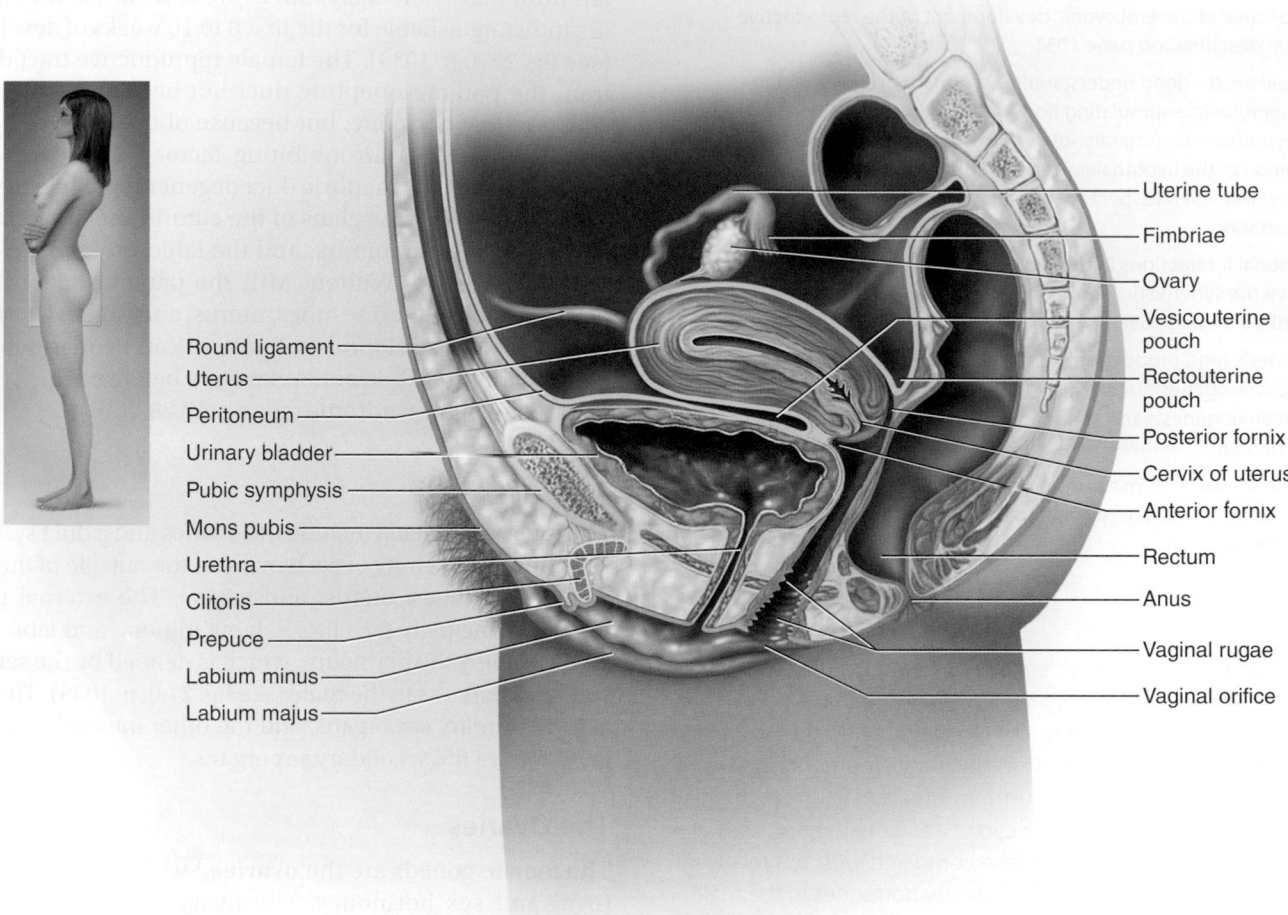

Round ligament
Uterus
Peritoneum
Urinary bladder
Pubic symphysis
Mons pubis
Urethra
Clitoris
Prepuce
Labium minus
Labium majus

Uterine tube
Fimbriae
Ovary
Vesicouterine pouch
Rectouterine pouch
Posterior fornix
Cervix of uterus
Anterior fornix
Rectum
Anus
Vaginal rugae
Vaginal orifice

**FIGURE 28.1  The Female Reproductive System.**  AP|R

Several connective tissue ligaments hold the ovaries and other internal genitalia in place (fig. 28.3). The medial pole of the ovary is attached to the uterus by the **ovarian ligament** and its lateral pole is attached to the pelvic wall by the **suspensory ligament.** The anterior margin of the ovary is anchored by a peritoneal fold called the **mesovarium.**[2] This ligament extends to a sheet of peritoneum called the *broad ligament,* which flanks the uterus and encloses the uterine tube in its superior margin. If you picture these ligaments as a sideways T, the vertical bar would represent the broad ligament, flanking the uterus and enfolding the uterine tube at its superior end, and the horizontal bar would represent the mesovarium enfolding the ovary at its free end.

The ovary receives blood from two arteries: the **ovarian branch of the uterine artery,** which passes through the mesovarium and approaches the medial pole of the ovary, and the

**ovarian artery,** which passes through the suspensory ligament and approaches the lateral pole. The ovarian artery is the female equivalent of the male testicular artery described in chapter 27, arising high on the aorta and traveling down to the gonad along the posterior body wall. The ovarian and uterine arteries anastomose along the margin of the ovary and give off multiple small arteries that enter the ovary on that side. Ovarian veins, lymphatics, and nerves also travel through the suspensory ligament. The veins and lymphatics follow courses similar to those described for the testes (p. 1038).

### The Uterine Tubes

The **uterine tubes,** also called **oviducts** or **fallopian**[3] **tubes,** are canals about 10 cm long leading from each ovary to the uterus. At the distal (ovarian) end, the tube flares into a

---

[2]*mes* = middle; *ovari* = ovary

[3]Gabriele Fallopio (1523–62), Italian anatomist and physician

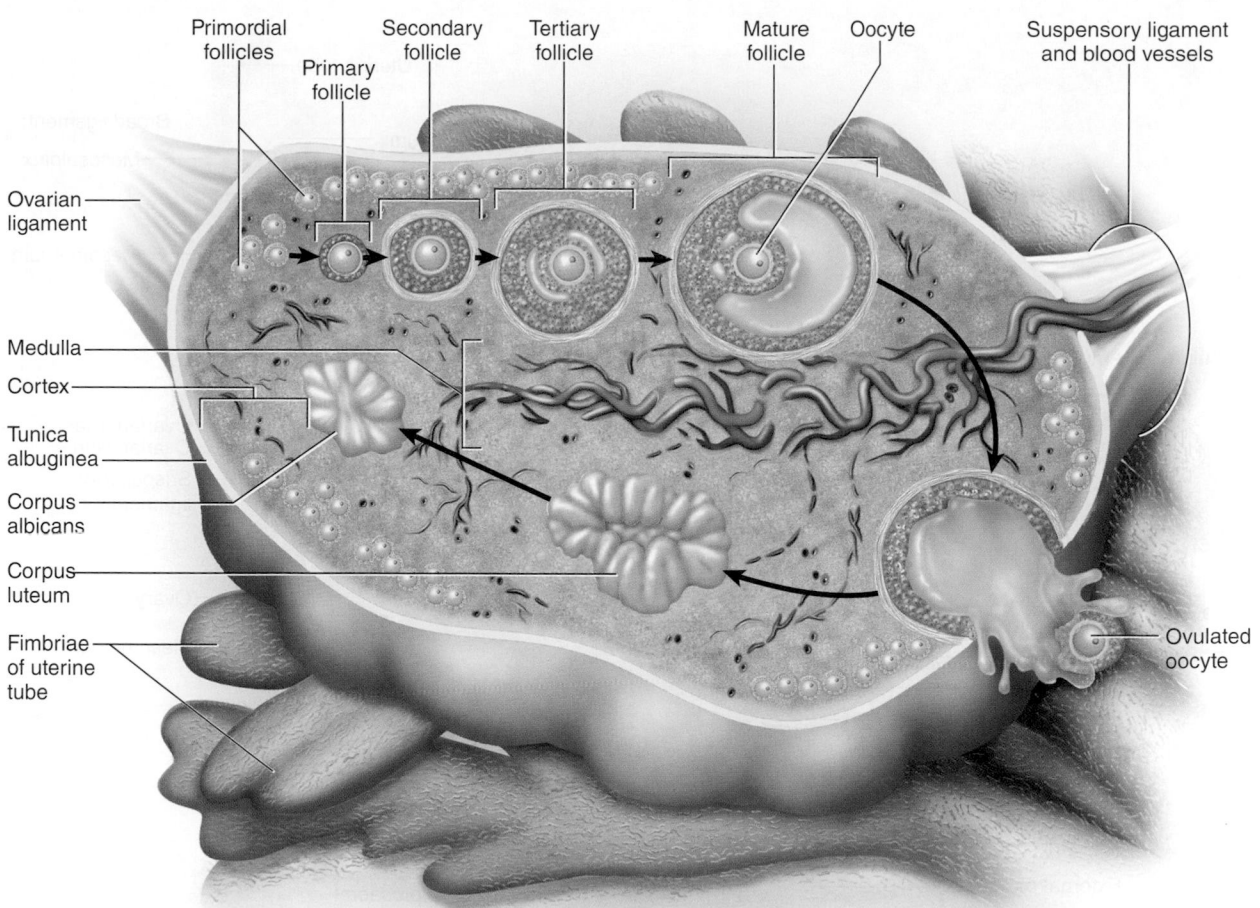

**FIGURE 28.2 Structure of the Ovary.** Arrows indicate the developmental sequence of an ovarian follicle and are not to imply that follicles migrate around the ovary. AP|R

trumpet-shaped **infundibulum**[4] with feathery projections called **fimbriae**[5] (FIM-bree-ee); the middle and longest part of the tube is the **ampulla;** and near the uterus it forms a narrower **isthmus.** The uterine tube is enfolded in the **mesosalpinx**[6] (MEZ-oh-SAL-pinks), which is the superior margin of the broad ligament.

The wall of the uterine tube is well endowed with smooth muscle. Its mucosa is highly folded into longitudinal ridges and has an epithelium of ciliated cells and a smaller number of secretory cells (fig. 28.4). The cilia beat toward the uterus and, with the help of muscular contractions of the tube, convey the egg in that direction.

## The Uterus

The **uterus**[7] is a thick muscular chamber that opens into the roof of the vagina and usually tilts forward over the urinary bladder (see fig. 28.1). Its function is to harbor the fetus, provide a source of nutrition, and expel the fetus at the end of its development. It is somewhat pear-shaped, with a broad superior curvature called the **fundus,** a midportion called the **body (corpus),** and a cylindrical inferior end called the **cervix** (fig. 28.3). The uterus measures about 7 cm from cervix to fundus, 4 cm wide at its broadest point on the fundus, and 2.5 cm thick, but it is somewhat larger in women who have been pregnant.

The lumen of the uterus is roughly triangular, with its two upper corners opening into the uterine tubes. In the non-pregnant uterus, the lumen is not a hollow cavity but rather a *potential space* (see p. 34); the mucous membranes of the opposite walls are pressed against each other with little room between them. The lumen communicates with the vagina by way of a narrow passage through the cervix called the **cervical canal.** The superior opening of this canal into the body of the uterus is the *internal os*[8] (pronounced "oz" or "ose") and its opening into the vagina is the *external os*. The canal contains **cervical glands** that secrete mucus, thought to prevent the spread of

---

[4]*infundibulum* = funnel
[5]*fimbria* = fringe
[6]*meso* = mesentery; *salpin* = trumpet
[7]*uterus* = womb

[8]*os* = mouth

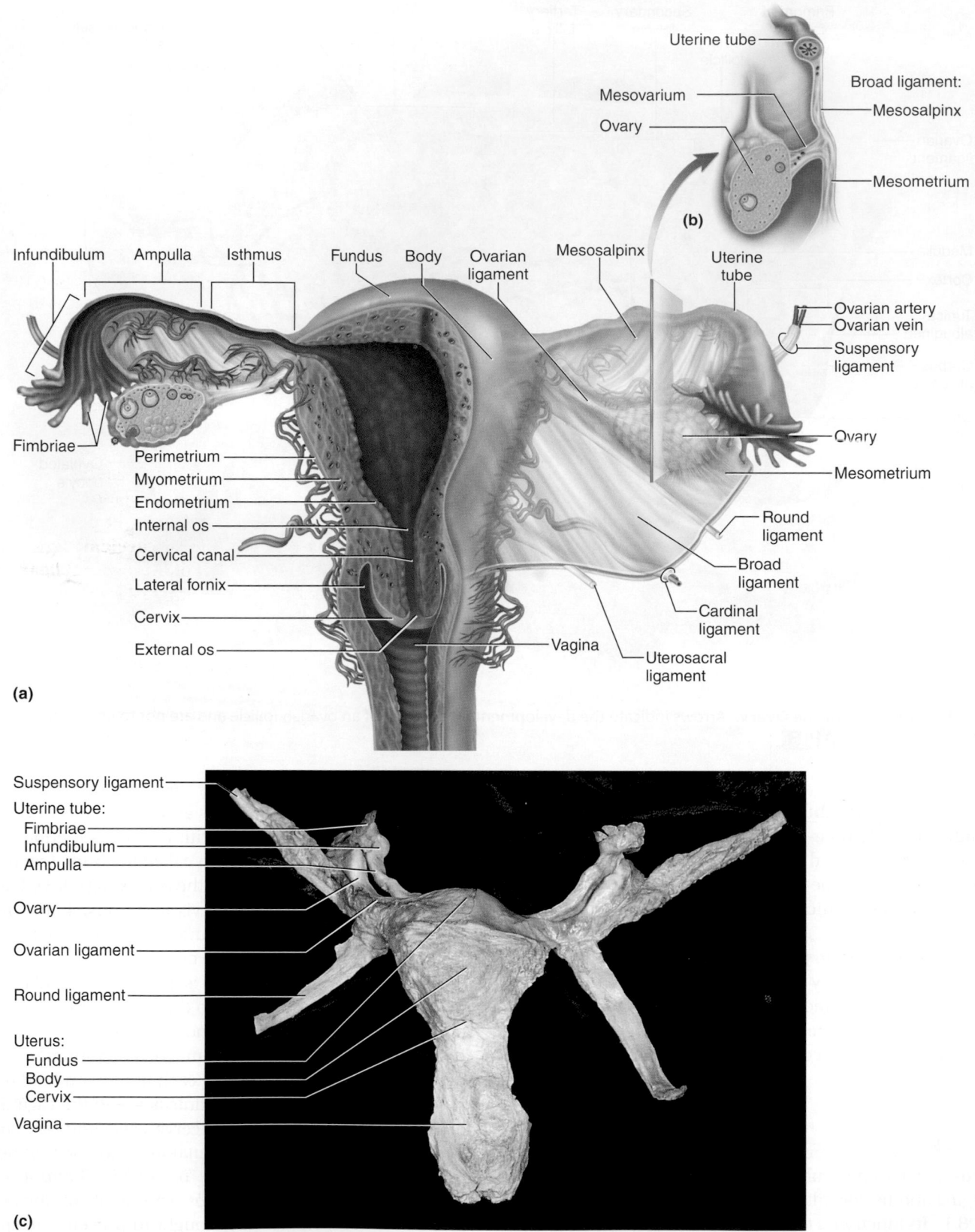

**FIGURE 28.3  The Female Reproductive Tract and Supportive Ligaments.**  (a) Posterior view of the reproductive tract. (b) Relationship of the uterine tube and ovary to the supporting ligaments. (c) Anterior view of the major female reproductive organs from a cadaver.  AP|R

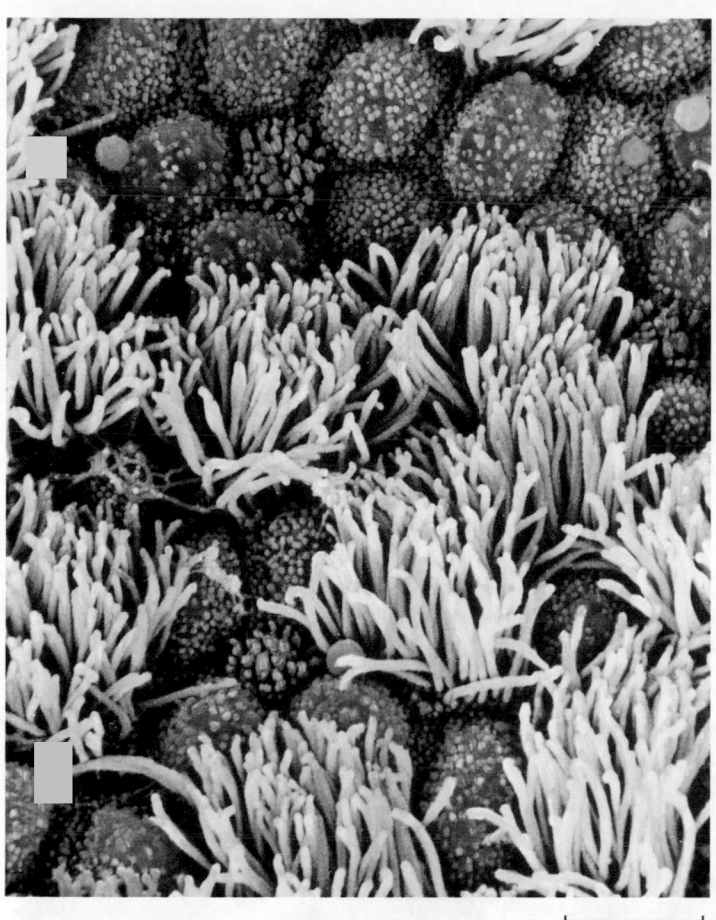

4 μm

**FIGURE 28.4** Epithelial Lining of the Uterine Tube. Secretory cells are shown in red and green, and cilia of the ciliated cells in yellow (SEM). AP|R

❓ *What purpose do these cilia serve?*

microorganisms from the vagina into the uterus. Near the time of ovulation, the mucus becomes thinner than usual and allows easier passage for sperm.

***Uterine Wall*** The uterine wall consists of an external serosa called the *perimetrium,* a middle muscular layer called the *myometrium,* and an inner mucosa called the *endometrium.* The **myometrium**[9] constitutes most of the wall; it is about 1.25 cm thick in the nonpregnant uterus. It is composed mainly of bundles of smooth muscle that sweep downward from the fundus and spiral around the body of the uterus. The myometrium is less muscular and more fibrous near the cervix; the cervix itself is almost entirely collagenous. The muscle cells

of the myometrium are about 40 μm long immediately after menstruation, but they are twice this long at the middle of the menstrual cycle and 10 times as long in pregnancy. The function of the myometrium is to produce the labor contractions that help to expel the fetus.

The **endometrium**[10] has a simple columnar epithelium, compound tubular glands, and a lamina propria populated by leukocytes, macrophages, and other cells (see fig. 28.6). The superficial half to two-thirds of it, called the **functional layer (stratum functionalis),** is shed in each menstrual period. The deeper layer, called the **basal layer (stratum basalis),** stays behind and regenerates a new functional layer in the next cycle (see fig. 28.16). When pregnancy occurs, the endometrium is the site of attachment of the embryo and forms the maternal part of the *placenta* from which the fetus is nourished.

***Ligaments*** The uterus is supported by the muscular floor of the pelvic outlet and folds of peritoneum that form supportive ligaments around the organ, as they do for the ovary and uterine tube (see fig. 28.3a). The **broad ligament** has two parts: the *mesosalpinx* mentioned earlier and the *mesometrium* on each side of the uterus. The cervix and superior part of the vagina are supported by **cardinal (lateral cervical) ligaments** extending to the pelvic wall. A pair of **uterosacral ligaments** attach the posterior side of the uterus to the sacrum, and a pair of **round ligaments** arise from the anterior surface of the uterus, pass through the inguinal canals, and terminate in the labia majora, much like the gubernaculum of the male terminating in the scrotum.

As the peritoneum folds around the various pelvic organs, it creates several dead-end recesses and pouches (extensions of the peritoneal cavity). Two major ones are the *vesicouterine*[11] *pouch,* which forms the space between the uterus and urinary bladder, and *rectouterine pouch* between the uterus and rectum (see fig. 28.1).

***Blood Supply*** The uterine blood supply is particularly important to the menstrual cycle and pregnancy. A **uterine artery** arises from each internal iliac artery and travels through the broad ligament to the uterus (see fig. 28.7). It gives off several branches that penetrate into the myometrium and lead to **arcuate arteries.** Each arcuate artery travels in a circle around the uterus and anastomoses with the arcuate artery on the other side. Along its course, it gives rise to smaller arteries that penetrate the rest of the way through the myometrium, into the endometrium, and produce the **spiral arteries.** The spiral arteries wind tortuously between the endometrial glands toward the surface of the mucosa (see fig. 28.16). They rhythmically constrict and dilate, making the mucosa alternately blanch and flush with blood.

---

[9]*myo* = muscle; *metr* = uterus

[10]*endo* = inside; *metr* = uterus
[11]*vesico* = bladder

## DEEPER INSIGHT 28.1

## CLINICAL APPLICATION

### Pap Smears and Cervical Cancer

Cervical cancer occurs most often between the ages of 30 and 50, especially in women who smoke, who began sexual activity at an early age, and who have histories of frequent sexually transmitted diseases or cervical inflammation. It is almost always caused by the human papillomavirus (HPV), a sexually transmitted pathogen (see p. 1054). Cervical cancer usually begins in the epithelial cells of the lower cervix, develops slowly, and remains a local, easily removed lesion for several years. If the cancerous cells spread to the subepithelial connective tissue, however, the cancer is said to be *invasive* and is much more dangerous and potentially fatal.

The best protection against cervical cancer is early detection by means of a *Pap*[12] smear—a procedure in which loose cells are removed from the cervix and vagina with a small flat stick and cervical brush, then microscopically examined. The pathologist looks for cells with signs of *dysplasia* (abnormal development) or carcinoma (fig. 28.5). One system of grading Pap smears classifies abnormal results into three grades of *cervical intraepithelial neoplasia (CIN)*. Findings are rated on the following scale, and further vigilance or treatment planned accordingly:

ASCUS—atypical squamous cells of undetermined significance
CIN I—mild dysplasia with cellular changes typically associated with HPV
CIN II—moderate dysplasia with precancerous lesions
CIN III—severe dysplasia, *carcinoma in situ* (preinvasive carcinoma of surface cells)

A rating of ASCUS or CIN I calls for a repeat Pap smear and visual examination of the cervix *(colposcopy)* in 3 to 6 months. CIN II calls for a biopsy, often done with an "electric scalpel" in a procedure called LEEP (loop electrosurgical excision procedure). A cone of tissue is removed to evaluate the depth of invasion by the malignant or premalignant cells. This in itself may be curative if all margins of the specimen are normal, indicating all abnormal cells were removed. CIN III may be cause for radiation therapy or *hysterectomy*[13] (removal of the uterus).

An average woman is typically advised to have annual Pap smears for 3 years and may then have them less often at the discretion of her physician. Women with any of the risk factors listed may be advised to have more frequent examinations.

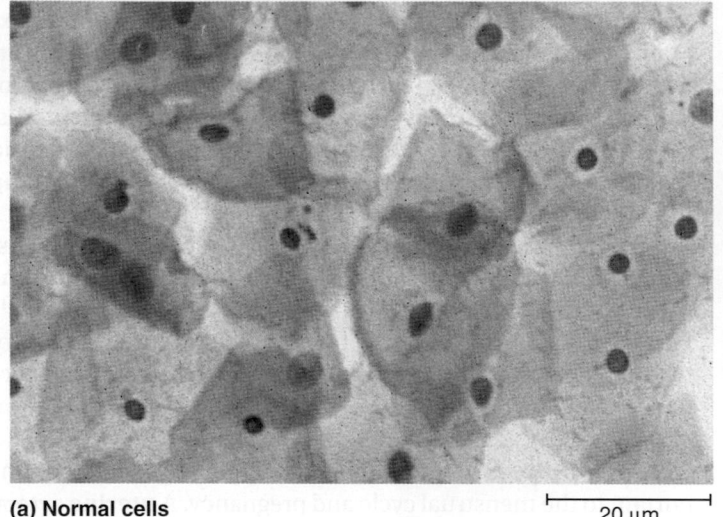

**(a) Normal cells**                    20 μm

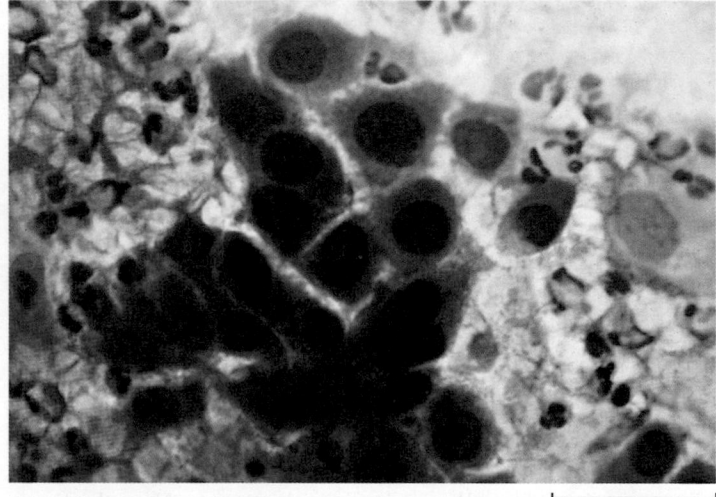

**(b) Malignant (CIN III) cells**                    20 μm

**FIGURE 28.5  Pap Smears.**  These are smears of squamous epithelial cells scraped from the cervix. In the malignant (cancerous) cells, note the loss of cell volume and the greatly enlarged nuclei.

## The Vagina

The **vagina**[14] is a tube about 8 to 10 cm long that allows for the discharge of menstrual fluid, receipt of the penis and semen, and birth of a baby. The vaginal wall is thin but very distensible. It consists of an outer adventitia, a middle muscularis, and an inner mucosa. The vagina tilts posteriorly between the urethra and rectum; the urethra is bound to its anterior wall. The vagina has no glands, but it is lubricated by the seepage of serous fluid through its walls *(transudation)* and by mucus from the cervical glands above it. The vagina extends slightly beyond the cervix and forms blind-ended spaces called *fornices*[15] (FOR-nih-sees; singular, *fornix*) surrounding it (see figs. 28.1 and 28.3a).

---

[12]George N. Papanicolaou (1883–1962), Greek–American physician and cytologist
[13]*hyster* = uterus; *ectomy* = cutting out
[14]*vagina* = sheath

[15]*fornix* = arch, vault

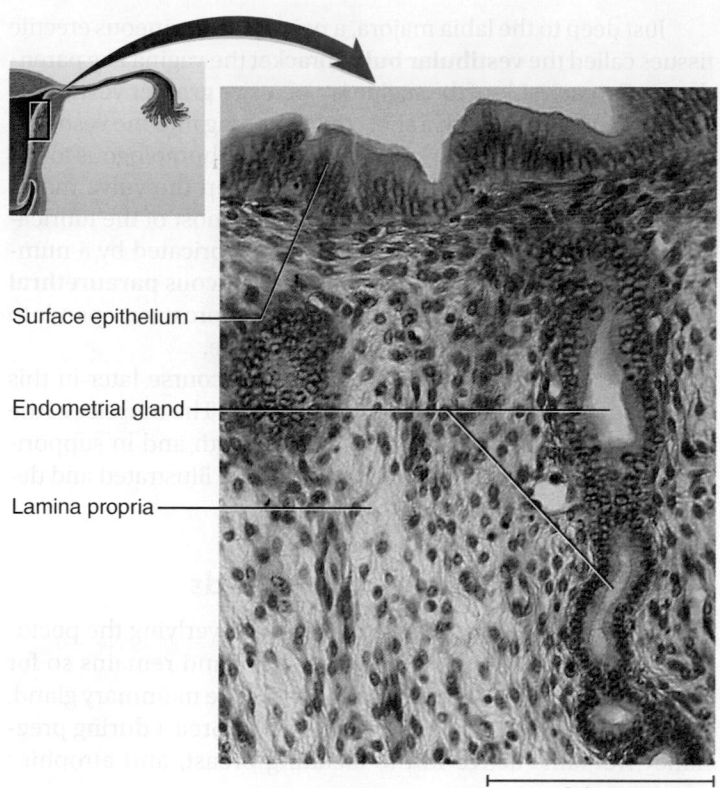

Surface epithelium

Endometrial gland

Lamina propria

0.1 mm

**FIGURE 28.6** **Functional Layer of the Endometrium.** AP|R

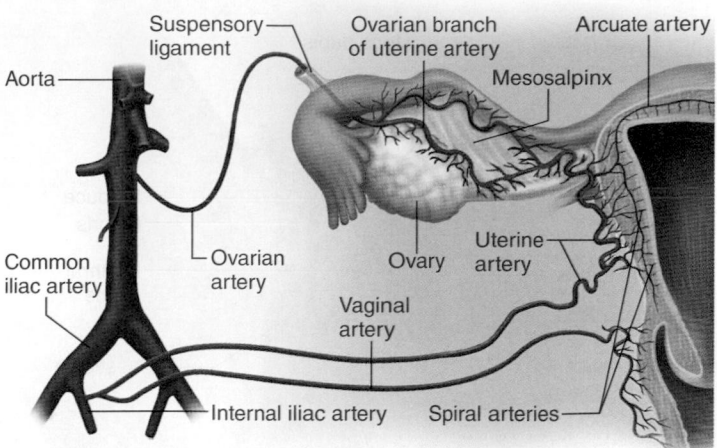

Suspensory ligament — Ovarian branch of uterine artery — Arcuate artery

Aorta — Mesosalpinx

Common iliac artery — Ovarian artery — Ovary — Uterine artery

Vaginal artery

Internal iliac artery — Spiral arteries

**FIGURE 28.7** **Blood Supply to the Female Reproductive Tract.** The vaginal, uterine, and ovarian arteries are exaggerated in length by the perspective of the drawing, moving the aorta away from the uterus for clarity.

The lower end of the vagina has transverse friction ridges, or **vaginal rugae,** which contribute to both male and female stimulation during intercourse. At the vaginal orifice, the mucosa folds inward and forms a membrane, the **hymen,** which stretches across the opening. The hymen has one or more openings to allow discharge of menstrual fluid, but it usually must rupture to allow for intercourse. A little bleeding often accompanies the first act of intercourse; however, the hymen is commonly ruptured before then by tampons, medical examinations, or strenuous exercise.

The vaginal epithelium is simple cuboidal in childhood, but the estrogens of puberty transform it into a stratified squamous epithelium. This is an example of *metaplasia,* the transformation of one tissue type to another. The epithelial cells are rich in glycogen. Bacteria ferment this to lactic acid, which produces a low vaginal pH (about 3.5–4.0) that inhibits the growth of pathogens. Recall from chapter 27 that this acidity is neutralized by the semen so it does not harm the sperm. The mucosa also has antigen-presenting **dendritic cells,** which are a route by which HIV from infected semen invades the female body.

#### ▶▶▶ APPLY WHAT YOU KNOW

*For what functional reason do you think the vaginal epithelium changes type at puberty? Of all types of epithelium it might become, why stratified squamous?*

### The External Genitalia

The external genitalia occupy most of the perineum and are collectively called the **vulva**[16] (**pudendum**[17]); they include the mons pubis, labia majora and minora, clitoris, vaginal orifice, and accessory glands and erectile tissues (fig. 28.8).

The **mons**[18] **pubis** consists mainly of an anterior mound of adipose tissue overlying the pubic symphysis, bearing most of the pubic hair (see fig. A.3a, p. 30). The **labia majora**[19] (singular, *labium majus*) are a pair of thick folds of skin and adipose tissue inferior to the mons; the fissure between them is the *pudendal cleft.* Pubic hair grows on the lateral surfaces of the labia majora at puberty, but the medial surfaces remain hairless. Medial to the labia majora are the much thinner, entirely hairless **labia minora**[20] (singular, *labium minus*). The area enclosed by them, called the **vestibule,** contains the urinary and vaginal orifices. At the anterior margin of the vestibule, the labia minora meet and form a hoodlike **prepuce** over the clitoris.

The **clitoris**[21] (CLIT-er-is, cli-TOR-is) is structured like the penis in many respects, but has no urinary role. Its function is entirely sensory, serving as the primary center of sexual stimulation. Unlike the penis, it is almost entirely internal, it has no corpus spongiosum, and it does not enclose the urethra. Essentially, it is a pair of corpora cavernosa enclosed in connective tissue. Its head, the **glans,** protrudes slightly from the prepuce.

[16]*vulva* = covering
[17]*pudend* = shameful
[18]*mons* = mountain
[19]*labi* = lip; major = larger, greater
[20]*minor* = smaller, lesser
[21] origin uncertain; possibly from *kleis* = door key, or *klei* + *ein* = to close

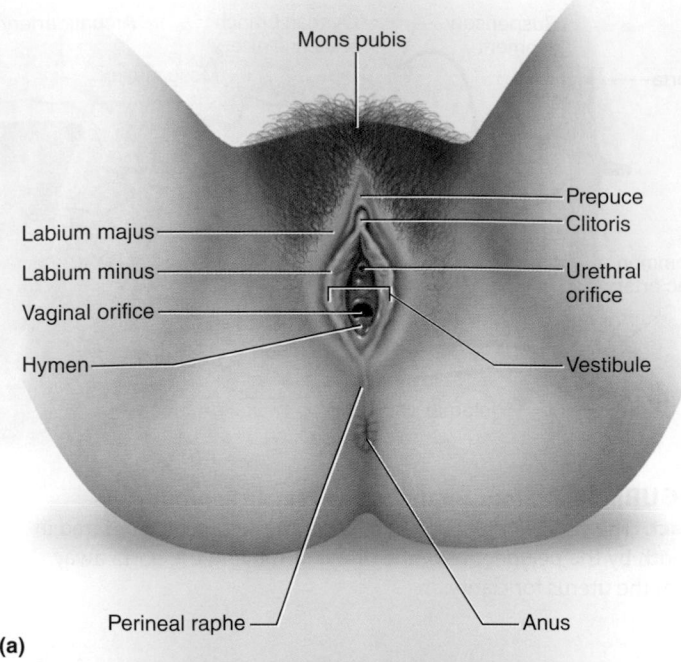

**(a)**

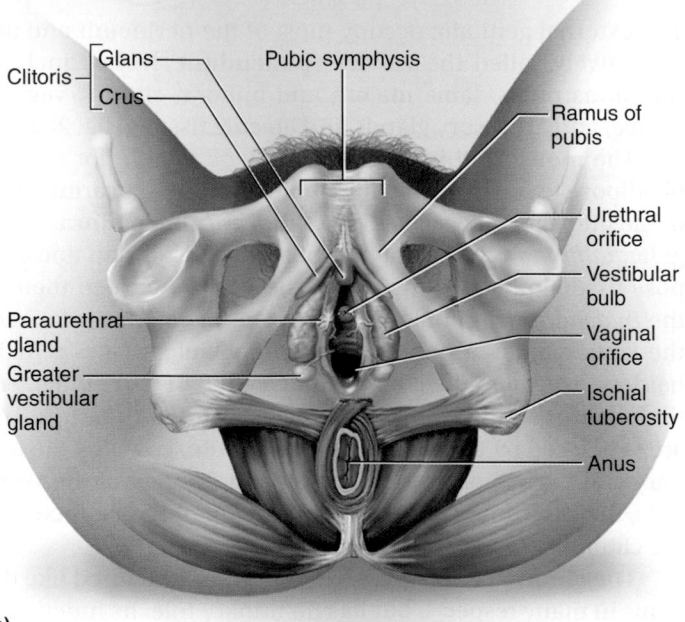

**(b)**

**FIGURE 28.8 The Female Perineum.** (a) Surface anatomy. (b) Subcutaneous structures. **AP|R**

❓ *The male prostate gland is homologous with the female _____ gland(s).*

The **body (corpus)** passes internally, inferior to the pubic symphysis (see fig. 28.1). At its internal end, the corpora cavernosa diverge like a Y as a pair of **crura,** which, like those of the penis, attach the clitoris to each side of the pubic arch. The circulation and innervation of the clitoris are much like those of the penis (see p. 1049).

Just deep to the labia majora, a pair of subcutaneous erectile tissues called the **vestibular bulbs** bracket the vagina like parentheses. On each side of the vagina is a pea-size **greater vestibular (Bartholin[22]) gland** with a short duct opening into the vestibule or lower vagina (fig. 28.8b). These glands are homologous to the bulbourethral glands of the male. They keep the vulva moist, and during sexual excitement they provide most of the lubrication for intercourse. The vestibule is also lubricated by a number of **lesser vestibular glands.** A pair of mucous **paraurethral (Skene[23]) glands,** homologous to the male prostate, open into the vestibule near the external urethral orifice.

A discussion of the physiology of intercourse later in this chapter details the functions of these organs. The muscles of the pelvic floor are highly important in childbirth and in supporting the female reproductive organs. They are illustrated and described in chapter 10 (p. 337).

## The Breasts and Mammary Glands

The **breast** (fig. 28.9) is a mound of tissue overlying the pectoralis major muscle. It enlarges at puberty and remains so for life, but most of this time it contains very little mammary gland. The **mammary gland** develops within the breast during pregnancy, remains active in the lactating breast, and atrophies when a woman ceases to nurse.

The breast has two principal regions: the conical to pendulous **body,** with the nipple at its apex, and an extension toward the armpit called the **axillary tail.** Lymphatics of the axillary tail are especially important as a route of breast cancer metastasis.

The nipple is surrounded by a circular zone, the **areola,** usually darker than the rest of the breast. Dermal blood capillaries and nerves come closer to the surface here than in the surrounding skin, accentuating the color and sensitivity of the areola. Pregnancy increases melanin deposition in the areola and nipple, making them more visible to the indistinct vision of a nursing infant. Sensory nerve fibers of the areola are important in triggering a *milk ejection reflex* when an infant nurses. The areola has sparse hairs and **areolar glands,** visible as small bumps on the surface. These glands are intermediate between sweat glands and mammary glands in their degree of development. When a woman is nursing, secretions of the areolar glands and sebaceous glands protect the areola and nipple from chapping and cracking. The dermis of the areola has smooth muscle fibers that contract in response to cold, touch, and sexual arousal, wrinkling the skin and erecting the nipple.

Internally, the nonlactating breast consists mostly of adipose and collagenous tissue (fig. 28.9b). Breast size is determined by the amount of adipose tissue and has no relationship to the amount of milk the mammary gland can produce. **Suspensory ligaments** attach the breast to the dermis of the overlying skin and to the fascia of the pectoralis major.

[22]Caspar Bartholin (1655–1738), Danish anatomist
[23]Alexander J. C. Skene (1838–1900), American gynecologist

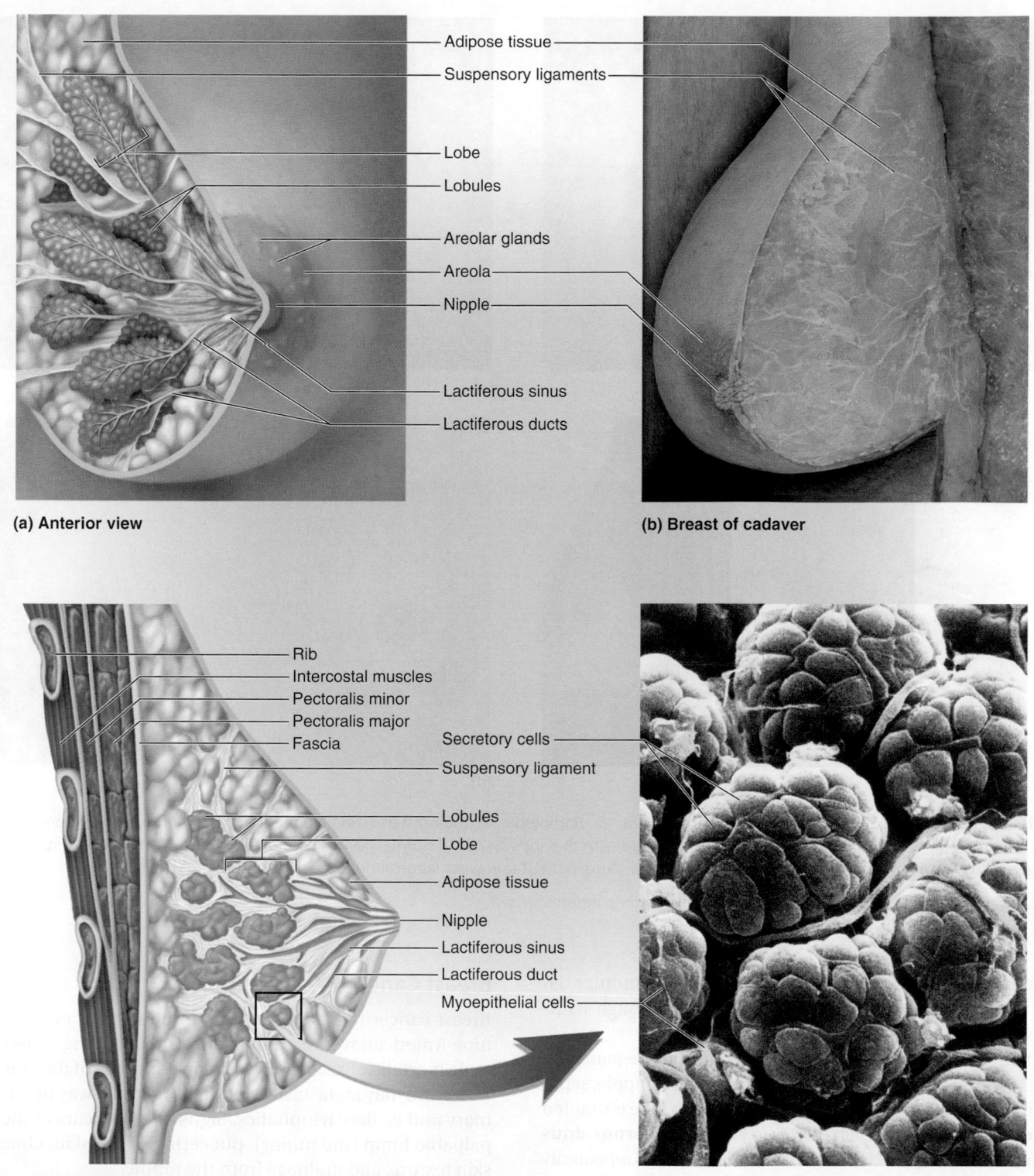

**(a) Anterior view**

Adipose tissue
Suspensory ligaments
Lobe
Lobules
Areolar glands
Areola
Nipple
Lactiferous sinus
Lactiferous ducts

**(b) Breast of cadaver**

**(c) Sagittal section**

Rib
Intercostal muscles
Pectoralis minor
Pectoralis major
Fascia
Suspensory ligament
Lobules
Lobe
Adipose tissue
Nipple
Lactiferous sinus
Lactiferous duct
Myoepithelial cells

Secretory cells

**(d) Mammary acinus**

**FIGURE 28.9 The Breast.** Parts (a), (c), and (d) depict the breast in a lactating state. Some of the features in parts (a) and (c) are absent from the nonlactating breast in part (b). The cluster of lobules boxed in part (c) would contain numerous microscopic acini like the ones in part (d). AP|R

? *What is the function of the myoepithelial cells in part (d)?*

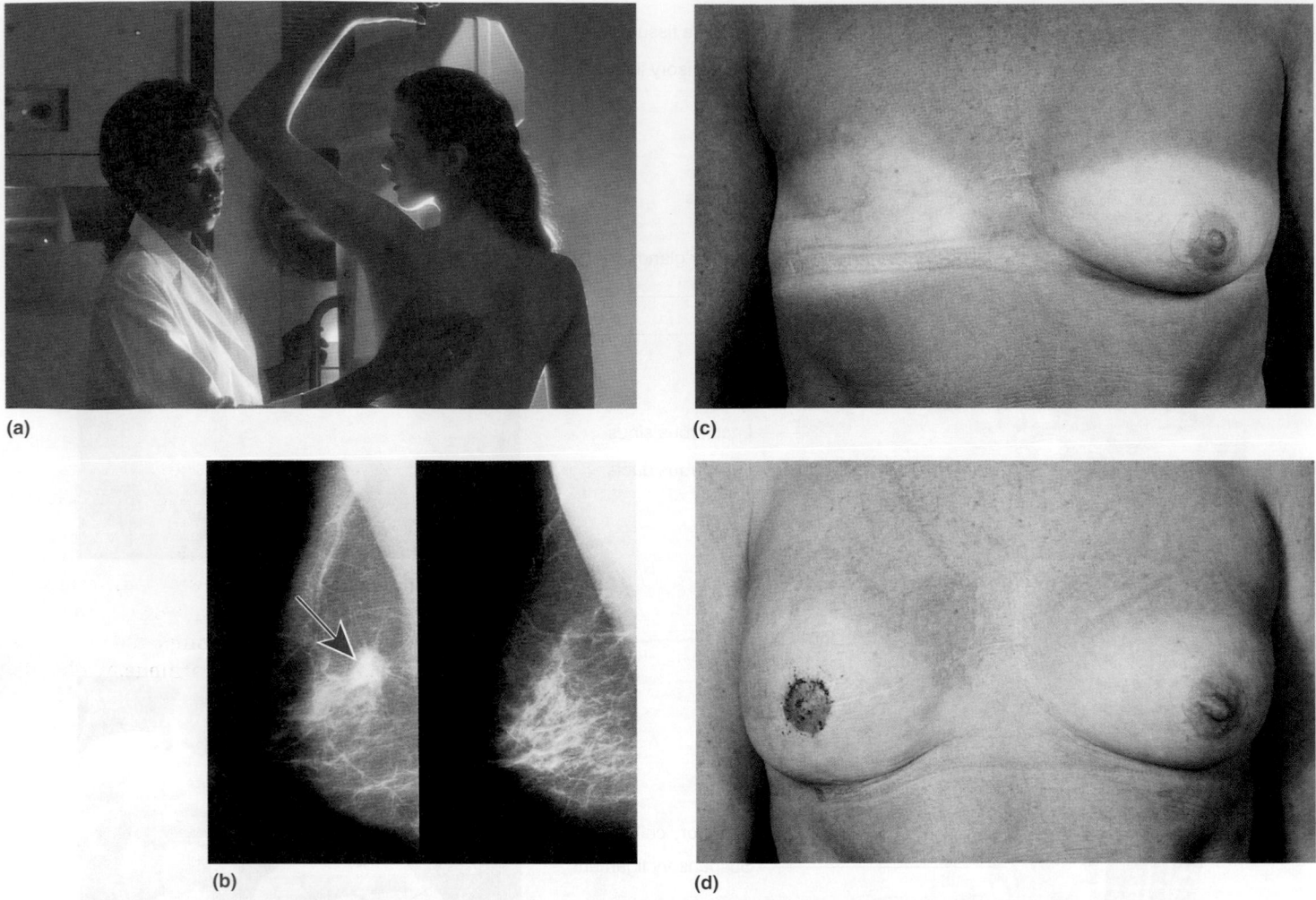

(a)

(b)

(c)

(d)

**FIGURE 28.10  Breast Cancer Screening and Treatment.**  (a) Radiologic technologist assisting a patient in mammography. (b) Mammogram of a breast with a tumor visible at the arrow (left), compared with the appearance of normal fibrous connective tissue of the breast (right). (c) Patient following mastectomy of the right breast. (d) The same patient following surgical breast reconstruction.  **AP│R**

❓ *What is the diagnostic benefit of compressing the breast for a mammogram?*

Although the nonlactating breast contains little glandular tissue, it does have a system of ducts branching through its fibrous stroma and converging on the nipple.

When the mammary gland develops during pregnancy, it exhibits 15 to 20 lobes arranged radially around the nipple, separated from each other by stroma (fig. 28.9a). Each lobe is drained by a **lactiferous**[24] **duct,** which dilates to form a **lactiferous sinus** opening onto the nipple. Distally, each duct branches repeatedly with the finest branches ending in sacs called acini. The acini are organized into grapelike clusters (lobules) within each lobe of the breast. Each acinus consists of a sac of pyramidal secretory cells arranged around a central lumen (see fig. 5.30, p. 165). Like an orange in a mesh bag, the acinus is surrounded by a network of contractile **myoepithelial cells** (fig. 28.9d). Their role in milk release, and other aspects of the lactating breast, are described later in this chapter.

## Breast Cancer

Breast cancer (fig. 28.10) occurs in one out of every eight or nine American women and is one of the leading causes of female mortality. Breast tumors begin with cells of the mammary ducts and may metastasize to other organs by way of the mammary and axillary lymphatics. Signs of breast cancer include a palpable lump (the tumor), puckering of the skin, changes in skin texture, and drainage from the nipple.

Two breast cancer genes were discovered in the 1990s, named *BRCA1* and *BRCA2,* but most breast cancer is nonhereditary. Some breast tumors are stimulated by estrogen. Consequently, breast cancer is more common among women who begin menstruating early in life and who reach menopause relatively late—that is, women who have a long period of fertility and estrogen exposure. Other risk factors include aging, exposure to ionizing radiation and carcinogenic chemicals, excessive alcohol and fat intake, and smoking. Over 70% of cases, however, lack any identifiable risk factors.

---

[24]*lact* = milk; *fer* = to carry

The majority of tumors are discovered during breast self-examination (BSE), which should be a monthly routine for all women. *Mammograms* (breast X-rays), however, can detect tumors too small to be noticed by BSE. Although opinions vary, a schedule commonly recommended is to have a baseline mammogram in the late 30s and then have one every 2 years from ages 40 to 49 and every year beginning at age 50.

Treatment of breast cancer is usually by *lumpectomy* (removal of the tumor only) or *simple mastectomy* (removal of the breast tissue only or breast tissue and some axillary lymph nodes). *Radical mastectomy,* rarely done since the 1970s, involves the removal of not only the breast but also the underlying muscle, fascia, and lymph nodes. Although very disfiguring, it proved to be no more effective than simple mastectomy or lumpectomy. Surgery is generally followed by radiation or chemotherapy, and estrogen-sensitive tumors may also be treated with an estrogen blocker such as tamoxifen. A natural-looking breast can often be reconstructed from skin, fat, and muscle from other parts of the body.

**BEFORE YOU GO ON**

Answer the following questions to test your understanding of the preceding section:

1. How do the site of female gamete production and mode of release from the gonad differ from those in the male?

2. How is the structure of the uterine tube mucosa related to its function?

3. Contrast the function of the endometrium with that of the myometrium.

4. List the subcutaneous erectile tissues and glands of the female perineum and state the function of each.

## 28.2 Puberty and Menopause

### Expected Learning Outcomes

When you have completed this section, you should be able to

a. name the hormones that regulate female reproductive function, and state their roles;

b. describe the principal signs of puberty;

c. describe the hormonal changes of female climacteric and their effects; and

d. define and describe menopause, and distinguish menopause from climacteric.

Puberty and menopause are physiological transitions at the beginning and end of a female's reproductive life.

## Puberty

Puberty begins at ages 8 to 10 for most girls in the United States and Europe, but significantly later in many countries. However, a 1997 study of more than 17,000 girls in the United States showed 3% of black girls and 1% of white girls beginning puberty by age 3, and 27% and 7%, respectively, by age 7. The youngest mother known to history was Lina Medina (1933– ) of Peru, who began menstruating at 8 months of age, was impregnated at age 4, and gave birth at age 5, in 1939. You can find her story online.

Puberty is triggered by the same hypothalamic and pituitary hormones in girls as it is in boys. Rising levels of gonadotropin-releasing hormone (GnRH) stimulate the anterior lobe of the pituitary to secrete follicle-stimulating hormone (FSH) and luteinizing hormone (LH). FSH, especially, stimulates development of the ovarian follicles, which, in turn, secrete estrogens, progesterone, inhibin, and a small amount of androgen. These hormone levels rise gradually from ages 8 to 12 and then more sharply in the early teens. The **estrogens**[25] are feminizing hormones with widespread effects on the body. They include *estradiol* (the most abundant), *estriol,* and *estrone.* Most of the visible changes at puberty result from estradiol and androgens.

The earliest noticeable sign of puberty is **thelarche**[26] (theel-AR-kee), the onset of breast development. Estrogen, progesterone, and prolactin initially induce the formation of lobules and ducts in the breast. Duct development is completed under the influence of glucocorticoids and growth hormone, while adipose and fibrous tissue enlarge the breast. Breast development is complete around age 20, but minor changes occur in each menstrual cycle and major changes occur in pregnancy.

Thelarche is soon followed by **pubarche** (pyu-BAR-kee), the appearance of pubic and axillary hair, sebaceous glands, and axillary glands. Androgens from the ovaries and adrenal cortex stimulate pubarche as well as the libido. Women secrete about 0.5 mg of androgens per day, compared with 6 to 8 mg/day in men.

Next comes **menarche**[27] (men-AR-kee), the first menstrual period. In Europe and America, the average age at menarche declined from age 16.5 in 1860 to age 12 in 1997, mostly because of improved nutrition. Menarche cannot occur until a girl has attained about 17% body fat, and adult menstruation generally ceases if a woman drops below 22% fat. This is about the minimum needed to sustain pregnancy and lactation; thus, the body reacts as if to prevent a futile pregnancy when it is too lean.

As explained in chapter 26, the brain monitors the amount of body fat from the amount of leptin in the blood. In addition to its role in regulating appetite, leptin stimulates gonadotropin secretion. Therefore, if body fat and leptin levels drop too low, gonadotropin secretion declines and a girl's or woman's menstrual cycle may cease. Adolescent girls with very low body fat, such as avid dancers and gymnasts, tend to begin menstruating at a later age than average.

Menarche does not necessarily signify fertility. A girl's first few menstrual cycles are typically *anovulatory* (no egg is

---

[25]*estro* = desire, frenzy; *gen* = to produce
[26]*thel* = breast, nipple; *arche* = beginning
[27]*men* = monthly; *arche* = beginning

ovulated). Most girls begin ovulating regularly about a year after they begin menstruating.

Estradiol stimulates many other changes of puberty. It causes the vaginal metaplasia described earlier. It stimulates growth of the ovaries and secondary sex organs. It stimulates growth hormone secretion and causes a rapid increase in height and widening of the pelvis. Estradiol is largely responsible for the feminine physique because it stimulates fat deposition in the mons pubis, labia majora, hips, thighs, buttocks, and breasts. It makes a girl's skin thicken, but the skin remains thinner, softer, and warmer than in males of corresponding age.

**Progesterone**[28] acts primarily on the uterus, preparing it for possible pregnancy in the second half of each menstrual cycle and playing roles in pregnancy discussed later. Estrogens and progesterone also suppress FSH and LH secretion through negative feedback inhibition of the anterior pituitary. **Inhibin** selectively suppresses FSH secretion.

Thus, we see many hormonal similarities in males and females from puberty onward. The sexes differ less in the identity of the hormones that are present than in their relative amounts—high levels of androgens and low levels of estrogens in males and the opposite in females. Another difference is that these hormones are secreted more or less continually and simultaneously in males, whereas in females, secretion is distinctly cyclic and the hormones are secreted in sequence. This will be very apparent as you read about the ovarian and menstrual cycles.

## Climacteric and Menopause

Women, like men, go through a midlife change in hormone secretion called the **climacteric.** In women, it is accompanied by **menopause,** the cessation of menstruation (see Deeper Insight 28.2).

A female is born with about 2 million eggs in her ovaries, each in its own follicle. The older she gets, the fewer follicles remain. Climacteric begins not at any specific age, but when she has only about 1,000 follicles left. Even the remaining follicles are less responsive to gonadotropins, so they secrete less estrogen and progesterone. Without these steroids, the uterus, vagina, and breasts atrophy. Intercourse may become uncomfortable, and vaginal infections more common, as the vagina becomes thinner, less distensible, and drier. The skin becomes thinner, cholesterol levels rise (increasing the risk of cardiovascular disease), and bone mass declines (increasing the risk of osteoporosis). Blood vessels constrict and dilate in response to shifting hormone balances, and the sudden dilation of cutaneous arteries may cause **hot flashes**—a spreading sense of heat from the abdomen to the thorax, neck, and face. Hot flashes may occur several times a day, sometimes accompanied by headaches resulting from the sudden vasodilation of arteries in the head.

In some people, the changing hormonal profile also causes mood changes. Many physicians prescribe hormone replacement therapy (HRT)—low doses of estrogen and progesterone usually taken orally or by a skin patch—to relieve some of these symptoms. The risks and benefits of HRT are still being debated.

### ▶▶▶ APPLY WHAT YOU KNOW

*FSH and LH secretion rise at climacteric and these hormones attain high concentrations in the blood. Explain this using the preceding information and what you know about the pituitary–gonadal relationship.*

Menopause is the cessation of menstrual cycles, usually occurring between the ages of 45 and 55. The average age has increased steadily in the last century and is now about 52. It is difficult to precisely establish the time of menopause because the menstrual periods can stop for several months and then begin again. Menopause is generally considered to have occurred when there has been no menstruation for a year or more.

### BEFORE YOU GO ON

Answer the following questions to test your understanding of the preceding section:

5. Describe the similarities and differences between male and female puberty.

6. Describe the major changes that occur in female climacteric and the principal cause of these changes.

7. What is the difference between climacteric and menopause?

## DEEPER INSIGHT 28.2

### EVOLUTIONARY MEDICINE

#### The Evolution of Menopause

There has been considerable speculation about why women do not remain fertile to the end of their lives, as men do. Some theorists argue that menopause served a biological purpose for our prehistoric foremothers. Human offspring take a long time to rear. Beyond a certain point, the frailties of age make it unlikely that a woman could rear another infant to maturity or even survive the stress of pregnancy. She might do better in the long run to become infertile and finish rearing her last child, or help to rear her grandchildren, instead of having more. In this view, menopause was biologically advantageous for our ancestors—in other words, an evolutionary adaptation.

Others argue against this "grandmother hypothesis" on the grounds that Pleistocene (Ice Age) skeletons indicate that early hominids rarely lived past age 40. If this is true, menopause setting in at 45 to 55 years of age could have served little purpose. In this view, Pleistocene women may indeed have been fertile to the end of their lives; menopause now may be just an artifact of modern nutrition and medicine, which have made it possible for us to live much longer than our ancestors did.

[28]*pro* = favoring; *gest* = pregnancy; *sterone* = steroid hormone

## 28.3 Oogenesis and the Sexual Cycle AP|R

### Expected Learning Outcomes

When you have completed this section, you should be able to

a. describe the process of egg production (oogenesis);

b. describe changes in the ovarian follicles (folliculogenesis) in relation to oogenesis;

c. describe the hormonal events that regulate the ovarian cycle;

d. describe how the uterus changes during the menstrual cycle; and

e. construct a chart of the phases of the monthly sexual cycle showing the hormonal, ovarian, and uterine events of each phase.

The reproductive lives of women are conspicuously cyclic. They include the **reproductive cycle,** which encompasses the sequence of events from fertilization to giving birth and returning to a state of fertility, and the **sexual cycle,** which encompasses the events that recur every month when pregnancy does not intervene. The sexual cycle, in turn, consists of two interrelated cycles controlled by shifting patterns of hormone secretion: the **ovarian cycle,** consisting of events in the ovaries, and the **menstrual cycle,** consisting of parallel changes in the uterus.

Before we delve into the familiar 28-day sexual cycle, let us look at the developmental stages that the eggs and their follicles go through. Then we can integrate that with the controlling hormones and the monthly rhythm of ovulation and menstruation.

## Oogenesis

Egg production is called **oogenesis**[29] (OH-oh-JEN-eh-sis) (fig. 28.11). Like spermatogenesis, it produces a haploid gamete by means of meiosis. There are, however, numerous differences between the two. The most obvious, perhaps, is that males produce sperm continually, whereas oogenesis is a distinctly cyclic event that normally releases only one egg per month. Oogenesis is accompanied by cyclic changes in hormone secretion and in the histological structure of the ovaries and uterus; the uterine changes result in the monthly menstrual flow.

Oogenesis begins before a girl is born. The first germ cells arise, like those of the male, from the embryonic yolk sac. They migrate to the gonadal ridges in the first 5 to 6 weeks of development and then differentiate into **oogonia** (OH-oh-GO-nee-uh). Oogonia multiply until the fifth month, reach 6 to 7 million in number, then go into a state of arrested development until shortly before birth. At that time, some of them transform into **primary oocytes** and go as far as early meiosis I. All of them reach this stage by birth; no oogonia remain after that. Any stage from the primary oocyte to the time of fertilization can be called an **egg,** or **ovum.**

Most primary oocytes undergo a process of degeneration called **atresia** (ah-TREE-zhee-uh) before a girl is born. Two million remain at the time of birth; most of those undergo atresia during childhood; and by puberty, only about 200,000 remain. This is the female's lifetime supply of gametes, but it is more than ample; even if she ovulated every 28 days from the age of 14 to 50, she would ovulate only 480 times. All of the others undergo atresia between puberty and menopause.

Egg development resumes in adolescence. Each month for about 30 years, a cohort of about two dozen arrested follicles is recruited to resume development. (After 30 years, the process becomes more irregular, and at menopause, it ceases.) It takes about 290 days for one member of a cohort to fully mature and ovulate. Meiosis I is completed on the day of ovulation and produces two daughter cells—a large one called the **secondary oocyte** and a much smaller one called the **first polar body.** The polar body ultimately disintegrates. It is merely a means of discarding the extra haploid set of chromosomes. The secondary oocyte proceeds as far as metaphase II, then stops. If it is not fertilized, it dies and never finishes meiosis. If fertilized, it completes meiosis II and casts off a second polar body, which disposes of one chromatid from each chromosome. The chromosomes of the large remaining egg then unite with those of the sperm. Further development of the fertilized egg is discussed in chapter 29.

Note the contrast between spermatogenesis and oogenesis. In spermatogenesis, a primary spermatocyte gives rise to four equal-sized sperm. But in oogenesis, a primary oocyte gives rise to only one mature egg; the other three daughter cells (at most) are tiny polar bodies that die. In oogenesis it is important to produce an egg with as much cytoplasm as possible, because if fertilized it must divide repeatedly and produce numerous daughter cells. Splitting each oocyte into four equal but small parts would run counter to this purpose.

## Folliculogenesis

As an egg undergoes oogenesis, the follicle around it undergoes **folliculogenesis,** passing through the following stages.

1. **Primordial follicles.** A primordial follicle is about 40 μm in diameter. It consists of a primary oocyte in early meiosis, surrounded by a single layer of squamous follicular cells and a basement membrane separating the follicle from the surrounding stroma of the ovary (fig. 28.12a). Primordial follicles are concentrated in the ovarian cortex, close to the capsule. All of them form in the sixth through ninth month of gestation, but they persist into adulthood. In the adult ovary, 90% to 95% of the follicles are primordial. Those that survive childhood atresia must wait at least 13 years—and some as long as 50 years—before they are reactivated, resume their development, and have a chance to ovulate. **Recruitment (primordial follicle activation)**

---

[29]*oo* = egg; *genesis* = production

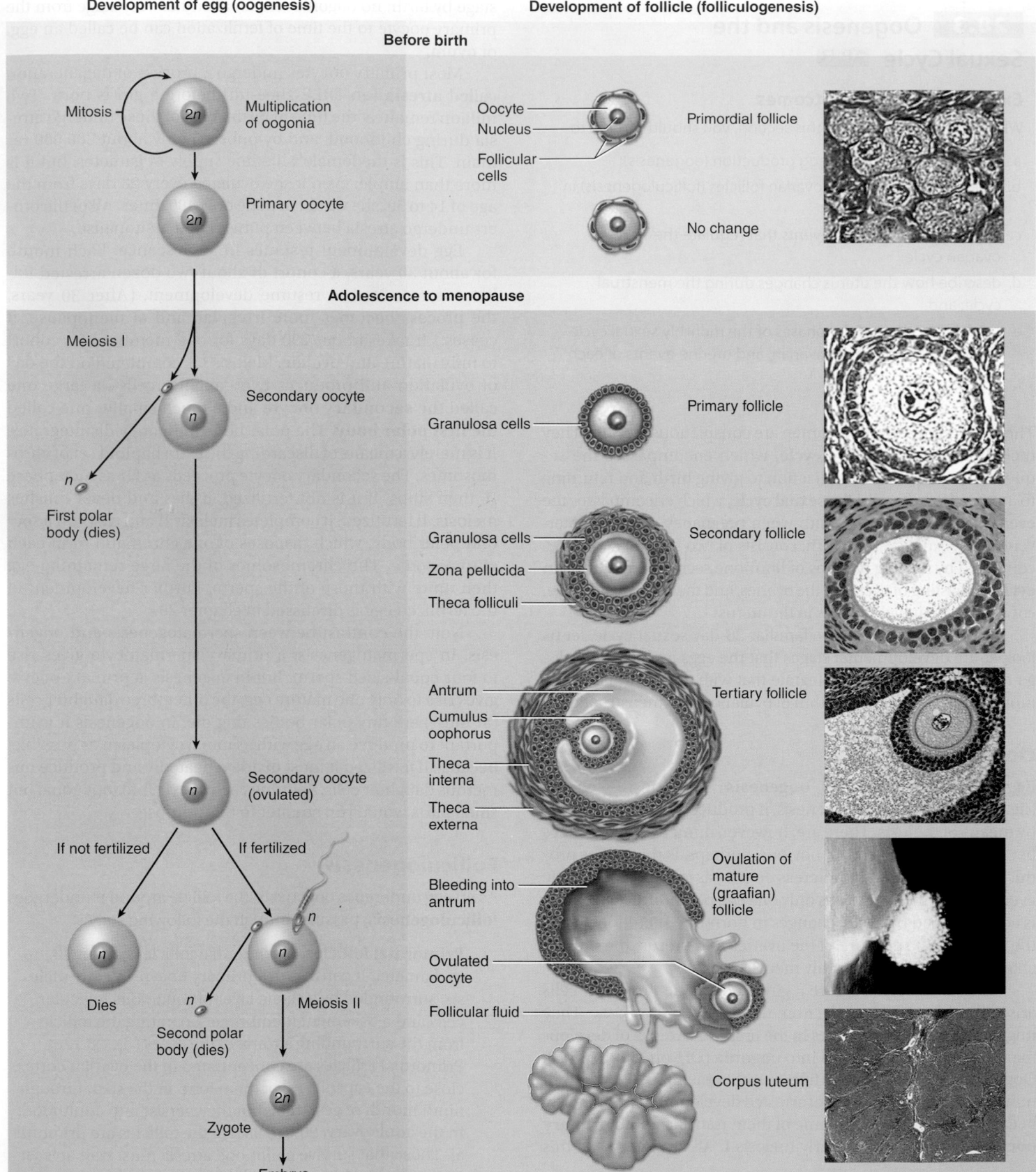

**Development of egg (oogenesis)**

**Development of follicle (folliculogenesis)**

**Before birth**

Mitosis — Multiplication of oogonia (2n)

Primary oocyte (2n)

Oocyte
Nucleus
Follicular cells — Primordial follicle

No change

**Adolescence to menopause**

Meiosis I

Secondary oocyte (n)

First polar body (dies) (n)

Granulosa cells — Primary follicle

Granulosa cells
Zona pellucida
Theca folliculi — Secondary follicle

Secondary oocyte (ovulated) (n)

Antrum
Cumulus oophorus
Theca interna
Theca externa — Tertiary follicle

If not fertilized / If fertilized

Dies (n)

Second polar body (dies) (n)

Meiosis II

Bleeding into antrum — Ovulation of mature (graafian) follicle

Ovulated oocyte
Follicular fluid

Zygote (2n)

Corpus luteum

Embryo

**FIGURE 28.11 Oogenesis (Left) and Corresponding Development of the Follicle (Right).** AP|R

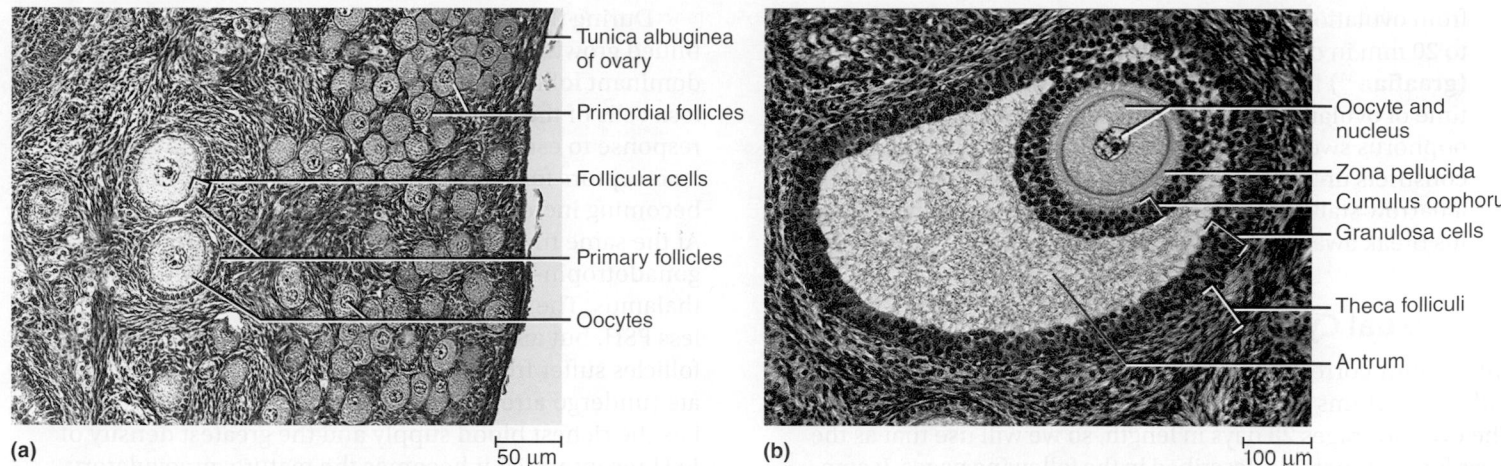

**(a)**  |—50 μm—|

**(b)**  |—100 μm—|

Labels (a): Tunica albuginea of ovary; Primordial follicles; Follicular cells; Primary follicles; Oocytes

Labels (b): Oocyte and nucleus; Zona pellucida; Cumulus oophorus; Granulosa cells; Theca folliculi; Antrum

**FIGURE 28.12 Ovarian Follicles.** (a) Note the very thin layer of squamous cells around the oocyte in a primordial follicle, and the single layer of cuboidal cells in a primary follicle. (b) A mature (graafian) follicle. Just before ovulation, this follicle will grow to as much as 25 mm in diameter. AP|R

awakens about two dozen primordial follicles to begin a 290-day march to maturity; usually, only one of these will eventually ovulate and the rest will die.

2. **Primary follicles.** About 140 days into the cycle, the recruited primordial follicles will have become primary follicles about 100 μm in diameter. These have a larger, now secondary oocyte and their follicular cells are now cuboidal, but still form only a single layer. Development from primordial to primary follicle is controlled by local growth factors, not by gonadotropins (FSH or LH), for which they still have no receptors. The granulosa cells, however, soon develop FSH receptors and the follicle depends on FSH to progress to the next stage.

3. **Secondary follicles.** These appear about 170 days into the cycle, still about six menstrual cycles before ovulation. They are distinguished by follicular cells that have multiplied and piled atop each other to form two or more layers and are now called **granulosa cells.** Secondary follicles are about 200 μm in diameter and the oocyte alone is about 120 μm—five times its original size in preparation for nourishment and development of the early embryo. The oocyte is surrounded by a layer of glycoprotein gel, the **zona pellucida.**[30]

The follicle is now enclosed in a tough husk called the **theca**[31] **folliculi** (THEE-ca fol-IC-you-lye). The theca is richly supplied with blood vessels, which bring it nutrients, hormones, and cholesterol. It divides into an outer *theca externa* of innervated smooth muscle and an inner *theca interna* of steroid-synthesizing cells. LH and insulin stimulate the theca interna to absorb cholesterol from the blood and convert it to androgens—androstenedione and a lesser amount of testosterone. These androgens diffuse inward to the granulosa cells, where FSH stimulates them to convert it to estrogens, especially estradiol.

4. **Tertiary follicles.** About 60 days (2 menstrual cycles) before ovulation, the granulosa cells begin secreting **follicular fluid,** which pools in the follicle wall (see fig. 28.14). The presence of these pools defines the tertiary follicle. As they enlarge, the pools merge and become a single fluid-filled cavity, the **antrum.** Because of this, tertiary and mature follicles are called antral follicles, whereas the earlier stages are called *preantral.* On one side of the antrum, a mound of granulosa cells called the **cumulus oophorus**[32] covers the oocyte and secures it to the follicle wall.

The innermost layer of cells in the cumulus, surrounding the zona pellucida and oocyte, is called the **corona radiata.**[33] These cells and the oocyte sprout microvilli that reach out and interdigitate like the fingers of two people laced together. Gap junctions form between them, enabling the cells to pass nutrients, wastes, and chemical signals to each other. Some granulosa microvilli penetrate deeply into the oocyte and almost reach its nucleus. Nothing can get to the oocyte except by going through the corona radiata, which forms a protective barrier around the egg functionally similar to the blood–testis barrier described in chapter 27.

5. **Mature follicles.** Normally only one follicle in each month's cohort becomes a mature follicle (fig. 28.12b), destined to ovulate while the rest degenerate. About 20 days before ovulation (that is, late in the previous menstrual cycle), one follicle in the cohort is selected to be the dominant follicle, the one destined to ovulate. It somehow (no one yet knows how) captures and sequesters FSH, while FSH does not accumulate in the other follicles and their development slows down. By the time a woman's menstrual period begins—day 1 of the cycle in which this follicle will ovulate—the dominant follicle is about 5 mm in diameter and two weeks away

---

[30]*zona* = zone; *pellucid* = clear, transparent
[31]*theca* = box, case

[32]*cumulus* = little mound; *oo* = egg; *phor* = to carry
[33]*corona* = crown; *radiata* = radiating

from ovulation. About 5 days before ovulation, it is up to 20 mm in diameter and is considered a **preovulatory (graafian**[34]**)** follicle; it attains a size up to 30 mm by the time of ovulation. As ovulation approaches, the cumulus oophorus swells and its attachment to the follicle wall constricts until the oocyte is attached to the wall by only a narrow stalk. In the last day or so, the oocyte and cumulus break away and float freely in the antrum.

## The Sexual Cycle

We can now correlate these changes in the egg cell and follicle with the rhythms of the ovaries and uterus—the sexual cycle. The cycle averages 28 days in length, so we will use that as the basis for the timetable described in the following pages. It commonly varies from 20 to 45 days, however, so be aware that the timetable given in this discussion may differ from person to person and from month to month. As you study this cycle, bear in mind that it is regulated by the **hypothalamo–pituitary-ovarian axis:** Hormones of the hypothalamus regulate the pituitary gland; pituitary hormones regulate the ovaries; and the ovaries, in turn, secrete hormones that regulate the uterus. That is, the basic hierarchy of control can be represented: hypothalamus → pituitary → ovaries → uterus. However, the ovaries also exert positive and negative feedback controls over the hypothalamus and pituitary.

Let us start with a brief preview of the sexual cycle as a whole. It begins with a 2-week *follicular phase.* The first 3 to 5 days of this are marked by menstruation, the vaginal discharge of blood and endometrial tissue. The uterus then replaces the lost tissue by mitosis. While this is going on, a cohort of tertiary follicles grows until the dominant follicle ovulates around day 14. After ovulation, the remainder of that follicle becomes a body called the *corpus luteum.* Over the next 2 weeks, called the *luteal phase,* the corpus luteum stimulates endometrial secretion, making the endometrium thicken still more, up to about day 26. If pregnancy does not occur, the endometrium breaks down again in the last 2 days. As loose tissue and blood accumulate, menstruation begins and the cycle starts over.

## The Ovarian Cycle

We will now see, in three principal steps, what happens in the ovaries and in their relationship to the hypothalamus and pituitary gland.

1. **Follicular phase.** The follicular phase of the cycle extends from the beginning of menstruation until ovulation—that is, from day 1 to day 14 in an average cycle. The portion from the *end* of menstruation until ovulation is also called the *preovulatory phase.* The follicular phase is the most variable part of the cycle, so unfortunately for family planning or pregnancy avoidance, it is seldom possible to reliably predict the date of ovulation.

During the follicular phase, FSH stimulates continued growth of all follicles in the cohort, but of the dominant follicle above all. FSH stimulates the granulosa cells of the antral follicles to secrete estradiol. In response to estradiol, the dominant follicle up-regulates its receptors for FSH, LH, and estradiol itself, thereby becoming increasingly sensitive to these hormones. At the same time, estradiol inhibits the secretion of gonadotropin-releasing hormone (GnRH) by the hypothalamus. The anterior pituitary gland secretes less and less FSH, but an increasing amount of LH. Most antral follicles suffer from the reduced FSH level and degenerate (undergo atresia). The dominant follicle, however, has the richest blood supply and the greatest density of FSH receptors, so it becomes the mature, preovulatory follicle. The ovary, at this stage, also exhibits follicles in many other stages, belonging to other cohorts trailing behind the lead cohort like freshmen to juniors trailing behind the senior class.

2. **Ovulation.** Ovulation is the rupture of the mature follicle and the release of its egg and attendant cells, typically around day 14. Dramatic changes over the preceding day signify its imminence. Estradiol stimulates a surge of LH and a lesser spike in FSH secretion by the pituitary (fig. 28.13 and the midpoint of fig. 28.14). LH induces several momentous events. The primary oocyte completes meiosis I, producing a haploid secondary oocyte and the first polar body.

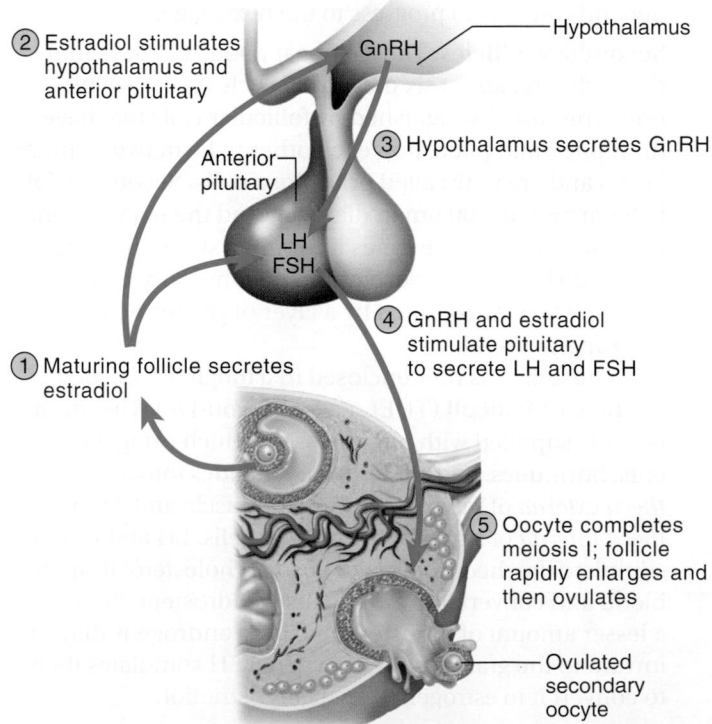

**FIGURE 28.13   Control of Ovulation by Pituitary and Ovarian Hormones.**

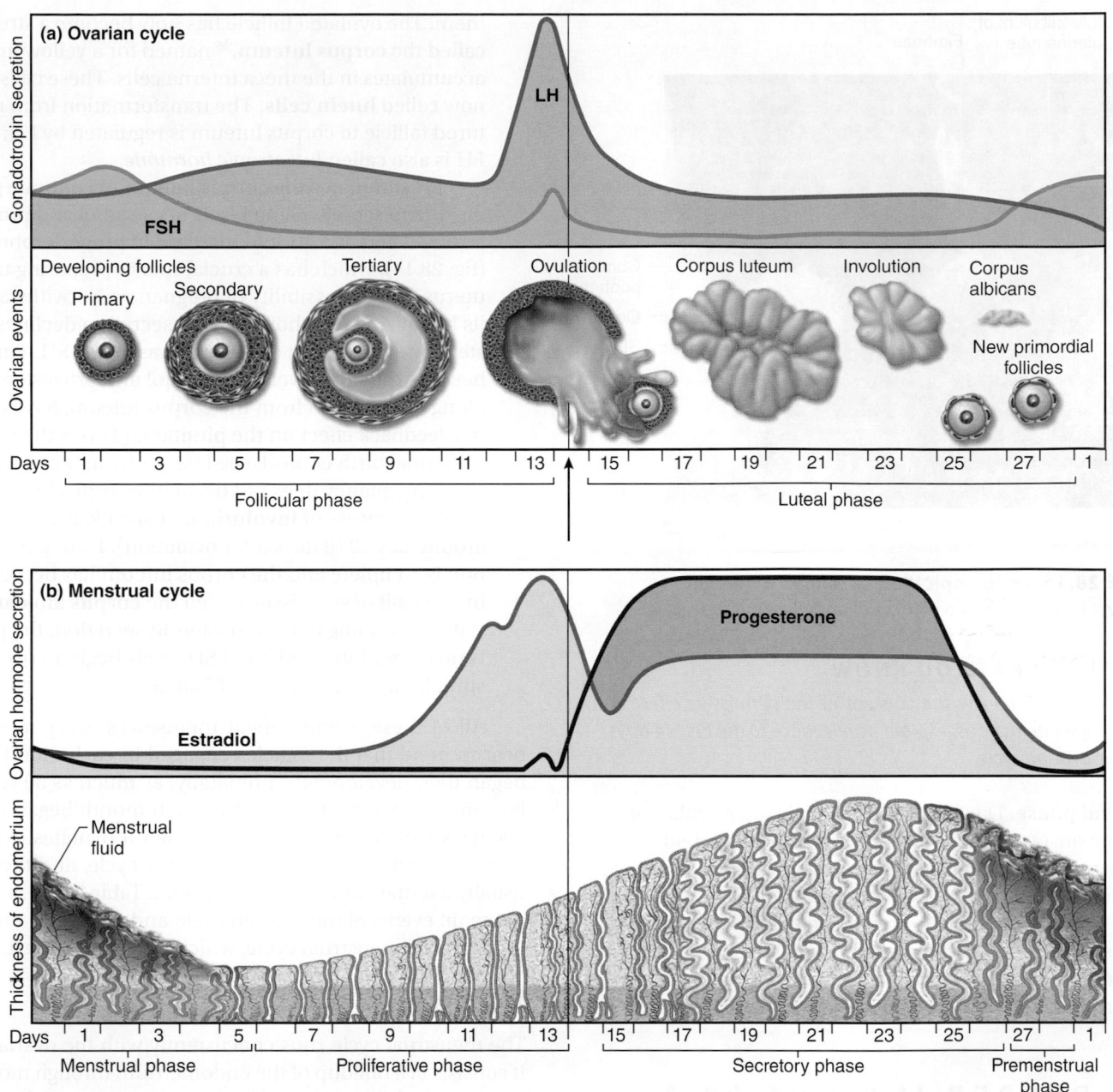

**FIGURE 28.14 The Female Sexual Cycle.** (a) The ovarian cycle (events in the ovary). (b) The menstrual cycle (events in the uterus). The two hormone levels in part (a) are drawn to the same scale, but those in part (b) are not. The peak progesterone concentration is about 17 times as high as the peak estradiol concentration. **AP|R**

Follicular fluid builds rapidly; the follicle swells to as much as 25 or 30 mm in diameter and contains up to 7 mL of fluid. Its size is astounding, considering that the entire ovary is only 10 mm thick. The preovulatory follicle bulges from the ovary like a blister. Macrophages and leukocytes are attracted to the area and secrete enzymes that weaken the follicular wall and adjacent ovarian tissue. A nipplelike **stigma** appears on the ovarian surface over the follicle. With mounting internal pressure and a weakening wall, the mature follicle approaches rupture.

Meanwhile, the uterine tube prepares to catch the oocyte when it emerges. It swells with edema; its fimbriae envelop and caress the ovary in synchrony with the woman's heartbeat; and its cilia create a gentle current in the nearby peritoneal fluid.

Ovulation itself takes only 2 or 3 min. The stigma seeps follicular fluid for 1 or 2 min., and then the follicle bursts. The remaining fluid oozes out, carrying the oocyte and cumulus oophorus (fig. 28.15). These are normally swept up by the ciliary current and taken into the uterine tube, although many oocytes fall into the pelvic cavity and die.

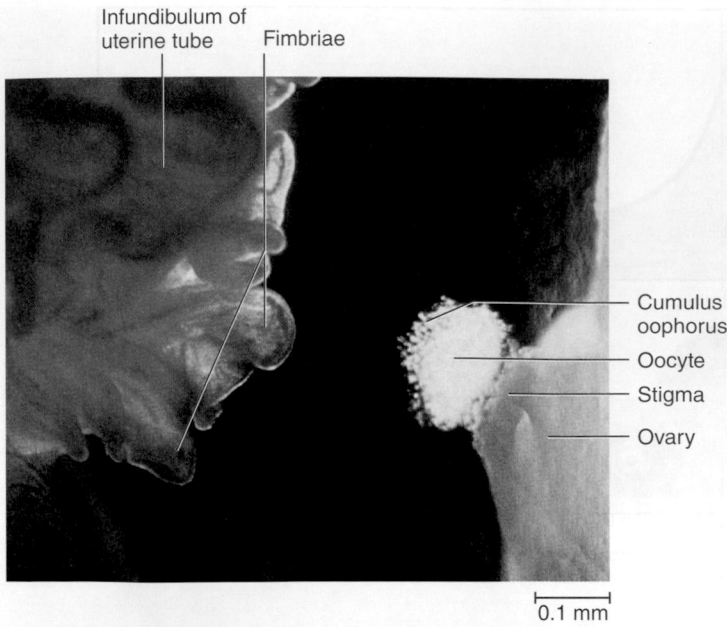

**FIGURE 28.15 Endoscopic View of Human Ovulation.**
The oocyte is shrouded by a blanket of cumulus oophorus cells.

▶▶▶ **APPLY WHAT YOU KNOW**

*In chapter 17, review the concept of the permissive effect in hormone interactions; explain its relevance to the first 14 days of the ovarian cycle.*

3. **Luteal phase.** Days 15 to 28, from just after ovulation to the onset of menstruation, are called the **luteal (postovulatory) phase.** Assuming pregnancy does not occur, the major events of this phase are as follows.

  When the follicle ruptures, it collapses and bleeds into the antrum. As the clotted blood is slowly absorbed, granulosa and theca interna cells multiply and fill the antrum, and a dense bed of blood capillaries grows amid

# DEEPER INSIGHT 28.3

## CLINICAL APPLICATION

### Signs of Ovulation

If a couple is attempting to conceive a child or to avoid pregnancy, it is important to be able to tell when ovulation occurs. The signs are subtle but detectable. For one, the cervical mucus becomes thinner and more stretchy. Also, the resting body temperature *(basal temperature)* rises 0.2° to 0.3°C (0.4°–0.6°F). This is best measured first thing in the morning, before rising from bed; the change can be detected if basal temperatures are recorded for several days before ovulation in order to see the difference. The LH surge that occurs about 24 hours before ovulation can be detected with a home testing kit. Finally, some women experience twinges of ovarian pain known by the German name, *mittelschmerz,*[35] which lasts from a few hours to a day or so at the time of ovulation. The most likely time to become pregnant is within 24 hours after the cervical mucus changes consistency and the basal temperature rises.

[35]*mittel* = in the middle; *schmerz* = pain

them. The ovulated follicle has now become a structure called the **corpus luteum,**[36] named for a yellow lipid that accumulates in the theca interna cells. These cells are now called **lutein cells.** The transformation from ruptured follicle to corpus luteum is regulated by LH; hence, LH is also called *luteotropic hormone.*

  LH stimulates the corpus luteum to continue growing and to secrete rising levels of estradiol and progesterone. There is a 10-fold increase in progesterone level (fig. 28.14), which has a crucial role in preparing the uterus for the possibility of pregnancy. Notwithstanding its luteinizing role, however, LH secretion declines steadily over the rest of the cycle, as does FSH. This is because the high levels of estradiol and progesterone, along with inhibin from the corpus luteum, have a negative feedback effect on the pituitary. (This is the basis for hormonal birth control; see Deeper Insight 28.5.)

  If pregnancy does not occur, the corpus luteum begins a process of **involution,** or shrinkage, beginning around day 22 (8 days after ovulation). By day 26, involution is complete and the corpus luteum has become an inactive bit of scar tissue called the **corpus albicans.**[37] With the waning of ovarian steroid secretion, the pituitary is no longer inhibited and FSH levels begin to rise again, stimulating a new cohort of follicles.

  All of these events repeat themselves every month, but bear in mind that the follicles engaged in each monthly cycle began their development prenatally, as much as 50 years earlier, and the oocyte that ovulates each month began ripening 290 days earlier, not in the cycle when it ovulates. Ovulation normally occurs in only one ovary per cycle, and the ovaries usually alternate from month to month. Table 28.1 summarizes the main events of the ovarian cycle and correlates them with events of the menstrual cycle, which we examine next.

## The Menstrual Cycle

The menstrual cycle runs concurrently with the ovarian cycle. It consists of a buildup of the endometrium through most of the sexual cycle, followed by its breakdown and vaginal discharge. It is divided into a *proliferative phase, secretory phase, premenstrual phase,* and *menstrual phase* (fig. 28.14). The menstrual phase averages 5 days long, and the first day of noticeable vaginal discharge is defined as day 1 of the sexual cycle. But even though it begins our artificial timetable for the cycle, the reason for menstruation is best understood after you become acquainted with the buildup of endometrial tissue that precedes it. Thus, we begin our survey of the cycle with the proliferative phase.

1. **Proliferative phase.** The functional layer of endometrial tissue lost in the last menstruation is rebuilt during the **proliferative phase.** At the end of menstruation, around day 5, the endometrium is about 0.5 mm thick and consists only of the basal layer. But as a new cohort of

[36]*corpus* = body; *lute* = yellow
[37]*corpus* = body; *alb* = white

| TABLE 28.1 | | Phases of the Ovarian Cycle |
|---|---|---|
| **Days** | **Phase** | **Major Features** |
| *1–14* | *Follicular phase* | Development of ovarian follicles and secretion primarily of estradiol. Coincides with menstrual and proliferative phases of the menstrual cycle. |
| | Primordial follicle | Formed prenatally and many persist into adulthood. Consists of an oocyte surrounded by a single layer of squamous follicular cells. |
| | Primary follicle | Consists of an oocyte surrounded by one layer of cuboidal follicular cells. |
| | Secondary follicle | Follicular cells stratify, become granulosa cells, and secrete a zona pellucida. Theca folliculi forms around follicle. |
| | Tertiary follicle | Develops from a secondary follicle in each cycle. Forms an antrum filled with follicular fluid and exhibits a cumulus oophorus, corona radiata, zona pellucida, and bilayered theca. |
| | Dominant follicle | The tertiary follicle that is destined to ovulate. Present by the end of the menstrual phase. Hormonally dominates the rest of the cycle, while other follicles in the cohort undergo atresia. Secretes mainly estradiol. Coincides with the proliferative phase of the menstrual cycle, in which the uterine endometrium thickens by mitosis. |
| | Mature (graafian) follicle | The dominant follicle just prior to ovulation. Attains a diameter of 20 to 30 mm and builds to high internal fluid pressure as adjacent ovarian wall weakens. |
| *14* | *Ovulation* | Rupture of mature follicle and release of oocyte. |
| *15–28* | *Luteal (postovulatory) phase* | Dominated by corpus luteum. Coincides with secretory and premenstrual phases of the menstrual cycle. |
| | Corpus luteum | Develops from ovulated follicle by proliferation of granulosa and theca interna cells. Progesterone stimulates thickening of endometrium by secretion (secretory phase of the menstrual cycle). Begins to involute by day 22 in the absence of pregnancy; involution complete by day 26. |
| | Corpus albicans | Scar tissue left by involution of corpus luteum; not hormonally active. In the absence of progesterone, endometrium exhibits ischemia, necrosis, and sloughing of tissue. Necrotic endometrial tissue mixes with blood and forms menstrual fluid. |

follicles develops, they secrete more and more estrogen. Estrogen stimulates mitosis in the basal layer and the prolific regrowth of blood vessels, thus regenerating the functional layer (fig. 28.16a). By day 14, the endometrium is 2 to 3 mm thick. Estrogen also stimulates endometrial cells to produce progesterone receptors, priming them for the progesterone-dominated secretory phase to follow.

2. **Secretory phase.** The endometrium thickens still more during the **secretory phase,** but as a result of secretion and fluid accumulation rather than mitosis. This phase extends from day 15 (after ovulation) to day 26 of a typical cycle. After ovulation, the corpus luteum secretes mainly progesterone. This hormone stimulates the endometrial glands to secrete glycogen. The glands grow wider, longer, and more coiled, and the lamina propria swells with tissue fluid (fig. 28.16b). By the end of this phase, the endometrium is 5 to 6 mm thick—a soft, wet, nutritious bed available for embryonic development in the event of pregnancy.

3. **Premenstrual phase.** The last 2 days or so of the cycle are the **premenstrual phase,** a period of endometrial degeneration. As we have already seen, when there is no pregnancy, the corpus luteum atrophies and the progesterone level falls sharply. The drop in progesterone triggers spasmodic contractions of the spiral arteries of the endometrium, causing endometrial ischemia (interrupted blood

flow). The premenstrual phase is therefore also called the **ischemic** (iss-KEE-mic) **phase.** Ischemia brings on tissue necrosis (and menstrual cramps). As the endometrial glands, stroma, and blood vessels degenerate, pools of blood accumulate in the functional layer. Necrotic endometrium falls away from the uterine wall, mixes with blood and serous fluid in the lumen, and forms the **menstrual fluid** (fig. 28.16c).

4. **Menstrual phase.** When enough menstrual fluid accumulates in the uterus, it begins to be discharged from the vagina for a period called the **menstrual phase (menses).** The first day of discharge marks day 1 of a new cycle. The average woman expels about 40 mL of blood and 35 mL of serous fluid over a 5-day period. Menstrual fluid contains fibrinolysin, so it does not clot. The vaginal discharge of clotted blood may indicate uterine pathology rather than normal menstruation.

To summarize the female sexual cycle, the ovaries go through a follicular phase characterized by growing follicles, then ovulation, and then a postovulatory (mostly luteal) phase dominated by the corpus luteum. The uterus, in the meantime, goes through a menstrual phase in which it discharges its stratum functionalis; then a proliferative phase in which it replaces that tissue by mitosis; then a secretory phase in which the endometrium thickens by the accumulation of secretions;

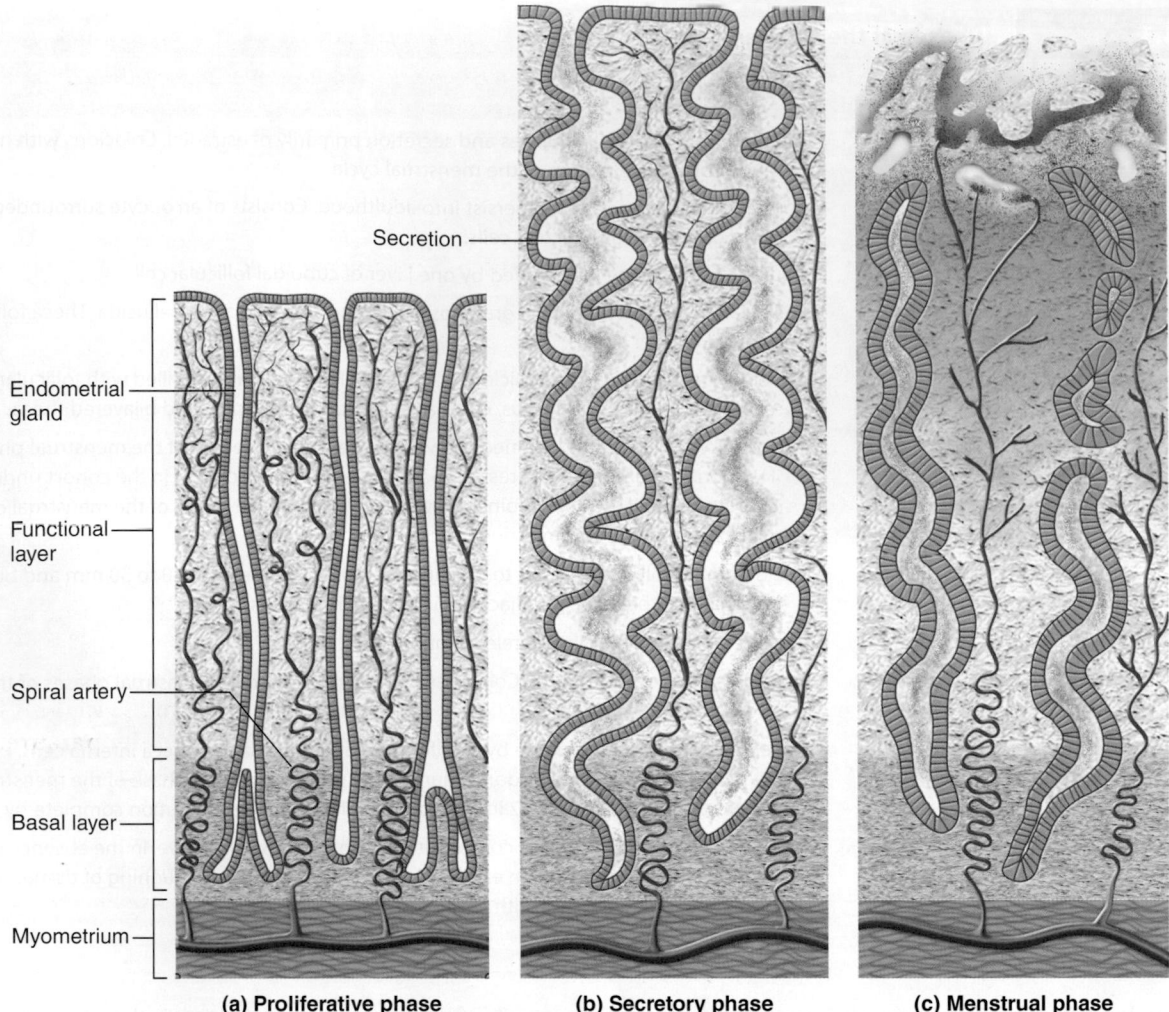

Secretion

Endometrial
gland

Functional
layer

Spiral artery

Basal layer

Myometrium

**(a) Proliferative phase**       **(b) Secretory phase**       **(c) Menstrual phase**

**FIGURE 28.16 Endometrial Changes Through the Menstrual Cycle.** (a) Late proliferative phase. The endometrium is 2 to 3 mm thick and has relatively straight, narrow endometrial glands. Spiral arteries penetrate upward between the endometrial glands. (b) Secretory phase. The endometrium has thickened to 5 to 6 mm by accumulating glycogen and mucus. The endometrial glands are much wider and more distinctly coiled, showing a zigzag or "sawtooth" appearance in histological sections. (c) Menstrual phase. Ischemic tissue has begun to die and fall away from the uterine wall, with bleeding from broken blood vessels and pooling of blood in the tissue and uterine lumen.

and finally, a premenstrual (ischemic) phase in which the stratum functionalis breaks down again. The first half of the cycle is governed largely by follicle-stimulating hormone (FSH) from the pituitary gland and estrogen from the ovaries. Ovulation is triggered by luteinizing hormone (LH) from the pituitary, and the second half of the cycle is governed mainly by LH and progesterone, the latter secreted by the ovaries.

**BEFORE YOU GO ON**

Answer the following questions to test your understanding of the preceding section:

8. Name the sequence of cell types in oogenesis and identify the ways in which oogenesis differs from spermatogenesis.

9. Distinguish between primordial, primary, secondary, and tertiary follicles. Describe the major structures of a mature follicle.

10. Describe what happens in the ovary during the follicular and postovulatory phases.

11. Describe what happens in the uterus during the menstrual, proliferative, secretory, and premenstrual phases.

12. Describe the effects of FSH and LH on the ovary.

13. Describe the effects of estrogen and progesterone on the uterus, hypothalamus, and anterior pituitary.

 **DEEPER INSIGHT 28.4**

**CLINICAL APPLICATION**

### Endometriosis

In some women, endometrial tissue lodges and grows in the pelvic cavity, often on the peritoneum or external surface of the ovary. This condition, called *endometriosis,* affects 6% to 10% of women, often causing pelvic pain and sometimes infertility. The most widely accepted theory of endometriosis is *retrograde menstruation*—menstrual fluid flowing backward and exiting through the uterine tube instead of the vagina. Management options include hormone therapy and surgery, but no cure is known at present.

## 28.4 Female Sexual Response

### Expected Learning Outcomes

When you have completed this section, you should be able to

a. describe the female sexual response at each phase of intercourse; and

b. compare and contrast the female and male sexual responses.

Female sexual response, the physiological changes that occur during intercourse, may be viewed in terms of the four phases identified by Masters and Johnson and discussed in chapter 27: excitement, plateau, orgasm, and resolution (fig. 28.17). The neurological and vascular controls of the female response are essentially the same as in the male (pp. 1049–1052) and need not be repeated here. The emphasis here is on ways the female response differs from that of the male.

### Excitement and Plateau

Excitement is marked by myotonia; vasocongestion; and increased heart rate, blood pressure, and respiratory rate. Although vasocongestion works by the same mechanism in both sexes, its effects are quite different in females. The labia minora become congested and often protrude beyond the labia majora. The labia majora redden and enlarge, then flatten and spread away from the vaginal orifice. The lower one-third of the vagina constricts to form a narrow passage called the **orgasmic platform,** owing partly to vasocongestion of the vestibular bulbs. The narrower canal and the vaginal rugae (friction ridges) enhance stimulation and help induce orgasm in both partners. The upper end of the vagina, in contrast, dilates and becomes cavernous.

Increased blood flow in the vaginal wall turns it purple and produces a serous fluid, the **vaginal transudate,** that seeps through the wall into the canal. Along with secretions of the greater vestibular glands, this moistens the vestibule and provides lubrication.

The uterus, which normally tilts forward over the urinary bladder, stands more erect during excitement and the cervix withdraws from the vagina. In plateau, the uterus is nearly vertical and extends into the greater pelvis. This is called the **tenting effect.**

Although the vagina is the female copulatory organ, the clitoris is more comparable to the penis in structure, physiology, and importance as the primary focus of sexual stimulation. It has a high concentration of sensory nerve endings, which, by contrast, are relatively scanty in the vagina. Recall that the penis and clitoris are homologous structures. Both have a pair of corpora cavernosa with *deep arteries* and become engorged by the same mechanism. The glans and shaft of the clitoris swell to two or three times their unstimulated size, but

since the clitoris cannot swing upward away from the body like the penis, it tends to withdraw beneath the prepuce. Thrusting of the penis in the vagina tugs on the labia minora and, by extension, pulls on the prepuce and stimulates the clitoris. The clitoris may also be stimulated by pressure between the pubic symphyses of the partners.

The breasts also become congested and swollen during the excitement phase, and the nipples become erect. Stimulation of the breasts also enhances sexual arousal.

### Orgasm

Late in plateau, many women experience involuntary pelvic thrusting, followed by 1 to 2 seconds of "suspension" or "stillness" preceding orgasm. Orgasm is commonly described as an intense sensation spreading from the clitoris through the pelvis, sometimes with pelvic throbbing and a spreading sense of warmth. The orgasmic platform gives three to five strong contractions about 0.8 seconds apart, while the cervix plunges spasmodically into the vagina and pool of semen, should this be present. The uterus exhibits peristaltic waves of contraction; it is still debated whether or not this helps to draw semen from the vagina. The anal and urethral sphincters constrict, and the paraurethral glands, homologous to the prostate, sometimes expel copious fluid similar to prostatic fluid ("female ejaculation"). Tachycardia and hyperventilation occur; the breasts enlarge still more and the areolae often become engorged; and in many women, a reddish, rashlike flush appears on the lower abdomen, chest, neck, and face.

### Resolution

During resolution, the uterus drops forward to its resting position. The orgasmic platform quickly relaxes, while the inner end of the vagina returns more slowly to its normal dimensions. The flush disappears quickly and the areolae and nipples undergo rapid detumescence, but it may take 5 to 10 minutes for the breasts to return to their normal size. In many women (and men), there is a postorgasmic outbreak of perspiration. Unlike men, women do not have a refractory period and may quickly experience additional orgasms.

#### BEFORE YOU GO ON

Answer the following questions to test your understanding of the preceding section:

14. What are the female sources of lubrication in coitus?

15. What female tissues and organs become vasocongested?

16. Describe the actions of the uterus throughout the sexual response cycle.

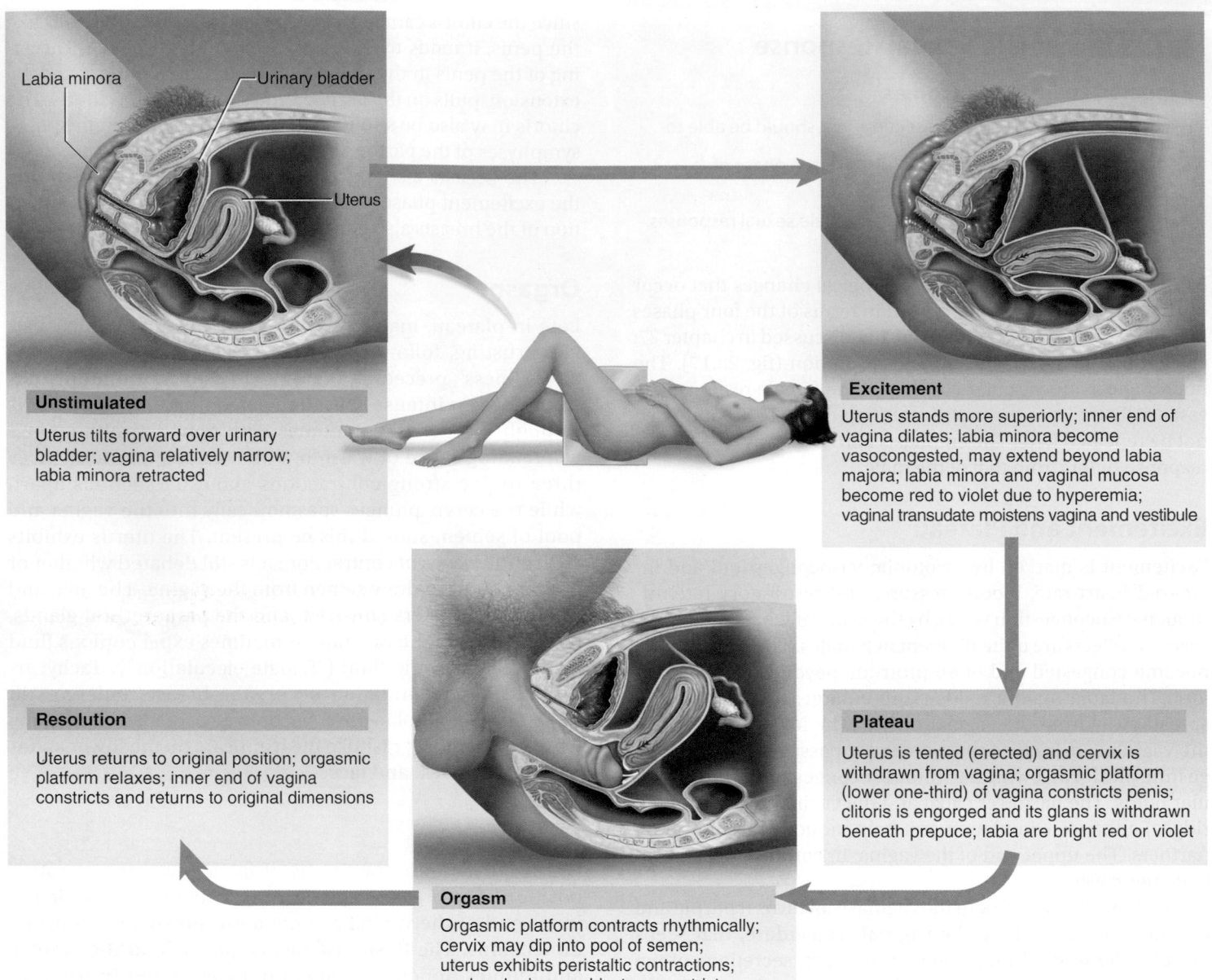

**Unstimulated**

Uterus tilts forward over urinary bladder; vagina relatively narrow; labia minora retracted

Labia minora    Urinary bladder

Uterus

**Excitement**

Uterus stands more superiorly; inner end of vagina dilates; labia minora become vasocongested, may extend beyond labia majora; labia minora and vaginal mucosa become red to violet due to hyperemia; vaginal transudate moistens vagina and vestibule

**Resolution**

Uterus returns to original position; orgasmic platform relaxes; inner end of vagina constricts and returns to original dimensions

**Plateau**

Uterus is tented (erected) and cervix is withdrawn from vagina; orgasmic platform (lower one-third) of vagina constricts penis; clitoris is engorged and its glans is withdrawn beneath prepuce; labia are bright red or violet

**Orgasm**

Orgasmic platform contracts rhythmically; cervix may dip into pool of semen; uterus exhibits peristaltic contractions; anal and urinary sphincters constrict

**FIGURE 28.17  Stages of the Female Sexual Response.**  Anatomy is shown in the supine position.

## 28.5    Pregnancy and Childbirth

### Expected Learning Outcomes

When you have completed this section, you should be able to

a.  list the major hormones that regulate pregnancy and explain their roles;

b.  describe a woman's bodily adaptations to pregnancy;

c.  identify the physical and chemical stimuli that increase uterine contractility in late pregnancy;

d.  describe the mechanism of labor contractions;

e.  name and describe the three stages of labor; and

f.  describe the physiological changes that occur in a woman during the weeks following childbirth.

This section treats pregnancy from the maternal standpoint—that is, adjustments of the woman's body to pregnancy and the mechanism of childbirth. Development of the fetus is described in chapter 29.

**Gestation** (pregnancy) lasts an average of 266 days from conception to childbirth, but the gestational calendar is usually measured from the first day of the woman's last menstrual period (LMP). Thus, the birth is predicted to occur 280 days (40 weeks) from the LMP. The duration of pregnancy, called its *term,* is commonly described in 3-month intervals called **trimesters.**

## Prenatal Development

A few fundamental facts of fetal development must be introduced as a foundation for understanding maternal physiology. All the products of conception—the embryo or fetus as well as the placenta and membranes associated with it—are collectively called the **conceptus.** The developing individual is a hollow ball called a *blastocyst* for much of the first 2 weeks, an *embryo* from day 16 through week 8, and a *fetus* from the beginning of week 9 until birth. The fetus is attached by way of an *umbilical cord* to a disc-shaped *placenta* on the uterine wall. The placenta provides fetal nutrition and waste disposal, and secretes hormones that regulate pregnancy, mammary development, and fetal development. For the first 6 weeks after birth, the infant is called a *neonate.*[38]

## Hormones of Pregnancy

The hormones with the strongest influences on pregnancy are estrogens, progesterone, human chorionic gonadotropin, and human chorionic somatomammotropin. The levels of these hormones in the maternal blood over the course of the pregnancy provide a good indicator of the well-being of the fetus. They are secreted primarily by the placenta, but the corpus luteum is an important source of hormones in the first several weeks. If the corpus luteum is removed before week 7, abortion almost always occurs. From weeks 7 to 17, the corpus luteum degenerates and the placenta takes over its endocrine functions.

## Human Chorionic Gonadotropin

**Human chorionic gonadotropin (HCG)** is secreted by the blastocyst and placenta. Its presence in the urine is the basis of pregnancy tests and can be detected with home testing kits as early as 8 or 9 days after conception. HCG secretion peaks around 10 to 12 weeks and then falls to a relatively low level for the rest of gestation (fig. 28.18). Like LH, it stimulates growth of the corpus luteum, which doubles in size and secretes increasing amounts of progesterone and estrogen. Without HCG, the corpus luteum would atrophy and the uterus would expel the conceptus.

## Estrogens

Estrogen secretion increases to about 30 times the usual amount by the end of gestation. The corpus luteum is an important source of estrogen for the first 12 weeks; after that, it comes mainly from the placenta. The adrenal glands of the mother and fetus secrete androgens, which the placenta converts to

[38]*neo* = new; *nate* = born, birth

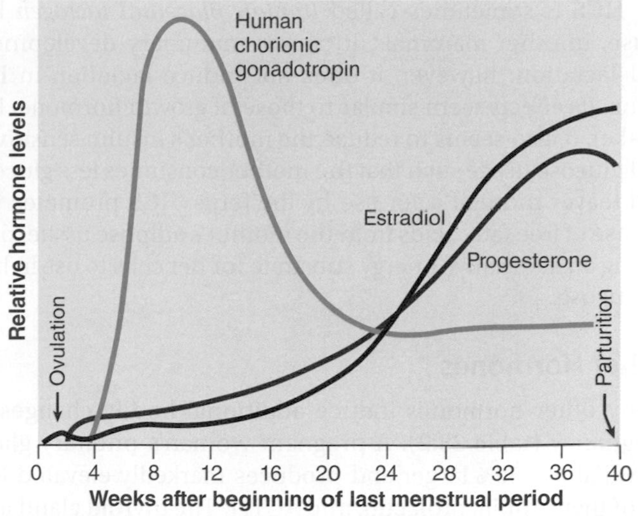

**FIGURE 28.18**  Hormone Levels over the Course of Pregnancy.

*How does the changing ratio of estradiol to progesterone relate to labor contractions?*

estrogens. The most abundant estrogen of pregnancy is estriol, but its effects are relatively weak; estradiol is less abundant but 100 times as potent.

Estrogen stimulates tissue growth in the fetus and mother. It causes the mother's uterus and external genitalia to enlarge, the mammary ducts to grow, and the breasts to increase to nearly twice their former size. It makes the pubic symphysis more elastic and the sacroiliac joints more limber, so the pelvis widens during pregnancy and the pelvic outlet expands during childbirth.

## Progesterone

The placenta secretes a great deal of progesterone, and early in the pregnancy, so does the corpus luteum. Progesterone and estrogen suppress pituitary secretion of FSH and LH, thereby preventing more follicles from developing during pregnancy. (This is the basis for contraceptive pills and implants; see Deeper Insight 28.5.) Progesterone also suppresses uterine contractions so the conceptus is not prematurely expelled. It prevents menstruation and promotes the proliferation of *decidual cells* of the endometrium, on which the blastocyst feeds. Once estrogen has stimulated growth of the mammary ducts, progesterone stimulates development of the secretory acini—another step toward lactation.

## Human Chorionic Somatomammotropin

**Human chorionic somatomammotropin (HCS)** is secreted in amounts several times that of all the other hormones combined, yet its function is the least understood. The placenta begins secreting HCS around the fifth week and output increases steadily from then until term, in proportion to the size of the placenta.

HCS is sometimes called *human placental lactogen* because, in other mammals, it causes mammary development and lactation; however, it does not induce lactation in humans. Its effects seem similar to those of growth hormone, but weaker. It also seems to reduce the mother's insulin sensitivity and glucose usage such that the mother consumes less glucose and leaves more of it for use by the fetus. HCS promotes the release of free fatty acids from the mother's adipose tissue, providing an alternative energy substrate for her cells to use in lieu of glucose.

## Other Hormones

Many other hormones induce additional bodily changes in pregnancy (table 28.2). A pregnant woman's pituitary gland grows about 50% larger and produces markedly elevated levels of thyrotropin, prolactin, and ACTH. The thyroid gland also enlarges about 50% under the influence of HCG, pituitary thyrotropin, and *human chorionic thyrotropin* from the placenta. Elevated thyroid hormone secretion increases the metabolic rate of the mother and fetus. The parathyroid glands enlarge and stimulate osteoclast activity, liberating calcium from the mother's bones for fetal use. ACTH stimulates glucocorticoid secretion, which may serve primarily to mobilize amino acids for fetal protein synthesis. Aldosterone secretion rises and promotes fluid retention, contributing to the mother's increased blood volume. The corpus luteum and placenta secrete *relaxin,* which relaxes the pubic symphysis in other animals but does not seem to have this effect in humans. In humans, it synergizes progesterone in stimulating the multiplication of decidual cells in early pregnancy and promotes the growth of blood vessels in the pregnant uterus.

## Adjustments to Pregnancy

Pregnancy places considerable stress on a woman's body and requires adjustments in nearly all the organ systems. A few of the major adjustments and effects of pregnancy are described here.

## Digestive System, Nutrition, and Metabolism

For many women, one of the first signs of pregnancy is **morning sickness**—nausea, especially after rising from bed—in the first few months of gestation. The cause is unknown. One hypothesis is that it stems from the reduced intestinal motility caused by the steroids of pregnancy. Another is that it is an evolutionary adaptation to protect the fetus from toxins. The fetus is most vulnerable to toxins at the same time that morning sickness peaks. Women with morning sickness tend to prefer bland foods and avoid spicy and pungent foods, which are highest in compounds that might be toxic to the fetus. In some women, the nausea progresses to vomiting. Occasionally, this is severe enough to require hospitalization (see *hyperemesis gravidarum* in table 28.5).

Constipation and heartburn are common in pregnancy. The former is another result of reduced intestinal motility. The latter is due to the enlarging uterus pressing upward on the stomach, causing the reflux of gastric contents into the esophagus.

The basal metabolic rate rises about 15% in the second half of gestation. Pregnant women often feel overheated because of this and the effort of carrying the extra weight. The appetite may be strongly stimulated, but a pregnant woman needs only 300 extra kcal/day even in the last trimester. Some

| TABLE 28.2 | The Hormones of Pregnancy |
|---|---|
| **Hormone** | **Effects** |
| Human chorionic gonadotropin (HCG) | Prevents involution of corpus luteum and stimulates its growth and secretory activity; basis of pregnancy tests |
| Estrogens | Stimulate maternal and fetal tissue growth, including enlargement of uterus and maternal genitalia; stimulate development of mammary ducts; soften pubic symphysis and sacroiliac joints, facilitating pelvic expansion in pregnancy and childbirth; suppress FSH and LH secretion |
| Progesterone | Suppresses premature uterine contractions; prevents menstruation; stimulates proliferation of decidual cells, which nourish embryo; stimulates development of mammary acini; suppresses FSH and LH secretion |
| Human chorionic somatomammotropin (HCS) | Has weak growth-stimulating effects similar to growth hormone and glucose-sparing effect on mother, making glucose more available to fetus; mobilizes fatty acids for use as maternal fuel |
| Pituitary thyrotropin | Stimulates thyroid activity and metabolic rate |
| Human chorionic thyrotropin | Same effect as pituitary thyrotropin |
| Parathyroid hormone | Stimulates osteoclasts and mobilizes maternal calcium for fetal use |
| Adrenocorticotropic hormone | Stimulates glucocorticoid secretion; thought to mobilize amino acids for fetal protein synthesis |
| Aldosterone | Causes fluid retention, contributing to increased maternal blood volume |
| Relaxin | Promotes development of decidual cells and blood vessels in the pregnant uterus |

women overeat, however, and gain as much as 34 kg (75 lb) of weight compared with a healthy average of 11 kg (24 lb). Maternal nutrition should emphasize the quality of food eaten, not quantity.

During the last trimester, the fetus needs more nutrients than the mother's digestive tract can absorb. In preparation for this, the placenta stores nutrients early in gestation and releases them in the final trimester. The demand is especially high for protein, iron, calcium, and phosphates. A pregnant woman needs an extra 600 mg of iron for her own hemopoiesis and 375 mg for the fetus. She is likely to become anemic if she does not ingest enough iron during late pregnancy. Supplemental vitamin K is often given late in pregnancy to promote prothrombin synthesis in the fetus. In the United States, newborns are routinely given an injection of vitamin K to minimize the risk of neonatal hemorrhage, especially in the brain, caused by the stresses of birth. A vitamin D supplement helps to ensure adequate calcium absorption to meet fetal demands. Supplemental folic acid reduces the risk of neurological disorders in the fetus, such as spina bifida and anencephaly (failure of the cerebrum, cerebellum, and calvaria to develop), but it is effective only if taken habitually prior to conception (see Deeper Insight 13.1, p. 478).

## Circulatory System

By full term, the placenta requires about 625 mL/min. of blood from the mother. The mother's blood volume rises about 30% during pregnancy because of fluid retention and hemopoiesis; she eventually has about 1 to 2 L of extra blood. Cardiac output rises about 30% to 40% above normal by 27 weeks, but for unknown reasons, it falls almost to normal in the last 8 weeks. As the pregnant uterus puts pressure on the large pelvic blood vessels, it interferes with venous return from the legs and pelvic region. This can result in hemorrhoids, varicose veins, and edema of the ankles and feet.

## Respiratory System

Over the course of pregnancy, respiratory rate remains constant but the tidal volume and minute ventilation increase about 40%. There are two reasons for this: (1) The oxygen demand rises in proportion to the woman's increased metabolic rate and the increasing needs of the fetus. (2) Progesterone increases the sensitivity of the woman's respiratory chemoreceptors to carbon dioxide, and ventilation is adjusted to keep her arterial $P_{CO_2}$ lower than normal. Low maternal $P_{CO_2}$ promotes the diffusion of $CO_2$ from the fetal bloodstream through the placenta and into the maternal blood. As pregnancy progresses, many women feel an increasing "air hunger" (dyspnea) and make more conscious efforts to breathe. This sensation apparently stems from increased $CO_2$ sensitivity and, late in pregnancy, pressure on the diaphragm from the growing uterus. In the last month, however, the pelvis usually expands enough for the fetus to drop lower in the abdominopelvic cavity, taking some pressure off the diaphragm and allowing one to breathe more easily.

## Urinary System

Aldosterone and the steroids of pregnancy promote water and salt retention by the kidneys. Nevertheless, the glomerular filtration rate increases by 50% and urine output is slightly elevated. This enables a woman to dispose of both her own and the fetus's metabolic wastes. As the pregnant uterus compresses the bladder and reduces its capacity, urination becomes more frequent and some women experience uncontrollable leakage of urine (incontinence).

## Integumentary System

The skin grows to accommodate expansion of the abdomen and breasts and the added fat deposition in the hips and thighs. Stretching of the dermis often tears the connective tissue and causes *striae*, or *stretch marks*. These appear reddish at first but fade after pregnancy. Melanocyte activity increases in some areas and darkens the areolae and linea alba. The latter often becomes a dark line, the **linea nigra**[39] (LIN-ee-uh NY-gruh), from the umbilical to the pubic region. Some women also acquire a temporary blotchy darkening of the skin over the nose and cheeks called the "mask of pregnancy," or **chloasma**[40] (clo-AZ-muh), which usually disappears when the pregnancy is over.

## Uterine Growth and Weight Gain

The uterus weighs about 50 g when a woman is not pregnant and about 900 g by the end of pregnancy. Its growth is monitored by palpating the fundus, which eventually reaches almost to the xiphoid process (fig. 28.19). Table 28.3 shows the distribution of weight gain in a typical healthy pregnancy.

## Childbirth

In the seventh month of gestation, the fetus normally turns into a head-down *vertex position*. Consequently, most babies are born head first, the head acting as a wedge that widens the mother's cervix, vagina, and vulva during birth. The ancients thought that the fetus kicked against the uterus and pushed itself out head first. The fetus, however, is a rather passive player in its own birth; its expulsion is achieved only by the contractions of the mother's uterine and abdominal muscles. Yet there is evidence that the fetus may play some role in its birth by chemically stimulating labor contractions and perhaps even sending chemical signals that signify when it is developed enough to be born.

---

[39]*linea* = line; *nigra* = black
[40]*chlo* = green; *asma* = to be

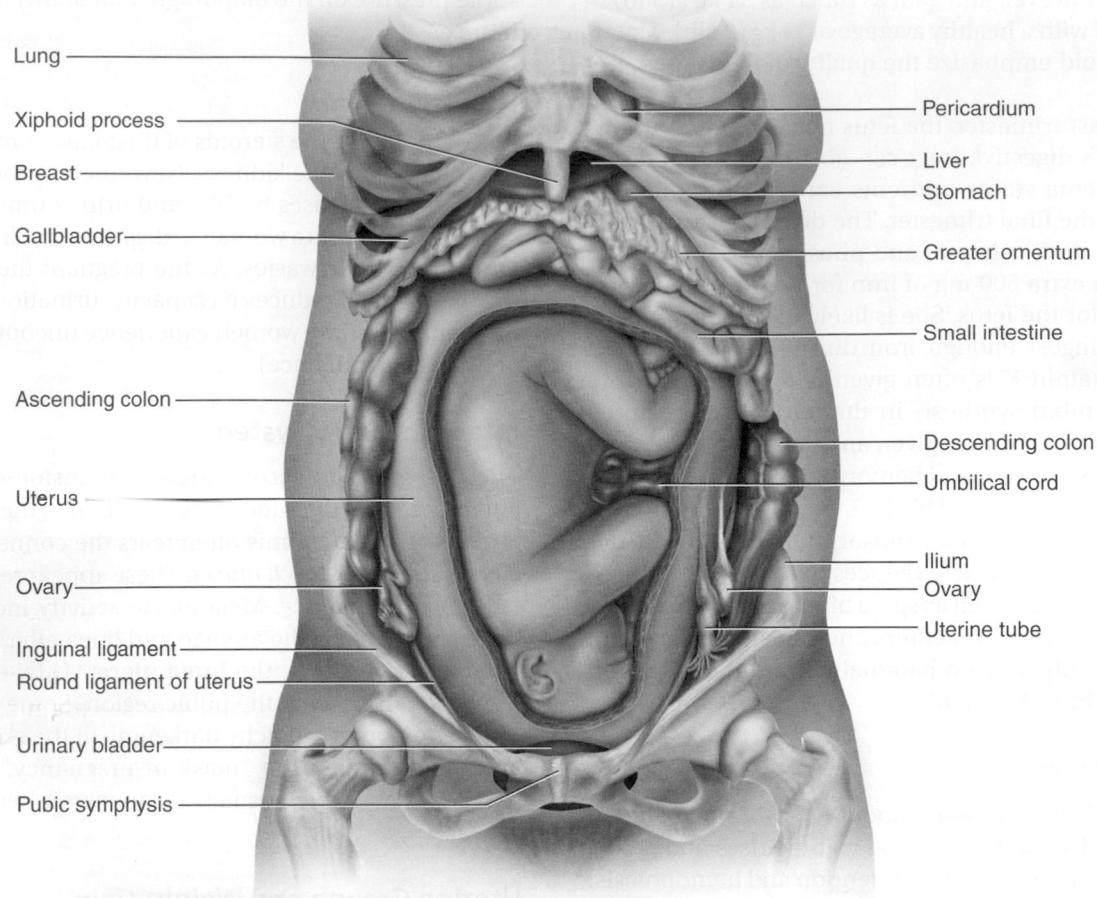

Lung

Xiphoid process

Breast

Gallbladder

Ascending colon

Uterus

Ovary

Inguinal ligament

Round ligament of uterus

Urinary bladder

Pubic symphysis

Pericardium

Liver

Stomach

Greater omentum

Small intestine

Descending colon

Umbilical cord

Ilium

Ovary

Uterine tube

**FIGURE 28.19  The Full-Term Fetus in Vertex Position.**  Note the displacement and compression of the abdominal viscera.

| TABLE 28.3 | Distribution of Weight Gain in Pregnancy |
|---|---|
| Fetus | 3   kg (7 lb) |
| Placenta, fetal membranes, and amniotic fluid | 1.8 kg (4 lb) |
| Blood and tissue fluid | 2.7 kg (6 lb) |
| Fat | 1.4 kg (3 lb) |
| Uterus | 0.9 kg (2 lb) |
| Breasts | 0.9 kg (2 lb) |
| **Total** | **11  kg (24 lb)** |

## Uterine Contractility

Over the course of gestation, the uterus exhibits relatively weak **Braxton Hicks**[41] **contractions.** These become stronger in late pregnancy and often send women rushing to the hospital with "false labor." At term, however, these contractions transform suddenly into the more powerful **labor contractions.** True

---

[41]John Braxton Hicks (1823–97), British gynecologist

labor contractions mark the onset of **parturition** (PAR-too-RISH-un), the process of giving birth.

Progesterone and estrogen (estradiol) balance may be one factor in this pattern of increasing contractility. Both hormone levels increase over the course of gestation. Progesterone inhibits uterine contractions, but its secretion levels off or declines slightly after 6 months, while estradiol secretion continues to rise (fig. 28.18). Estradiol stimulates uterine contractions and may be a factor in the irritability of the uterus in late pregnancy.

Also, as the pregnancy nears full term, the posterior pituitary releases more oxytocin (OT) and the uterine muscle equips itself with more OT receptors. Oxytocin promotes labor in two ways: (1) It directly stimulates muscle of the myometrium; and (2) it stimulates the fetal membranes to secrete prostaglandins, which are synergists of OT in producing labor contractions. Labor is prolonged if OT or prostaglandins are lacking, and it may be induced or accelerated by giving intravenous OT or a vaginal prostaglandin suppository.

The conceptus itself may produce chemical stimuli promoting its own birth. Fetal cortisol secretion rises in late pregnancy and may enhance estrogen secretion by the placenta. The fetal pituitary gland also produces oxytocin, which does not enter the maternal circulation but may promote prostaglandin secretion as noted above.

Uterine stretching is also thought to play a role in initiating labor. Stretching any smooth muscle increases its contractility, and movements of the fetus produce the sort of intermittent stretch that is especially stimulatory to the myometrium. Twins are born an average of 19 days earlier than solitary infants, probably because of the greater stretching of the uterus. When a fetus is in the vertex position, its head pushes against the cervix, which is especially sensitive to stretch.

## Labor Contractions

Labor contractions begin about 30 minutes apart. As labor progresses, they become more intense and eventually occur every 1 to 3 minutes. It is important that they be intermittent rather than one long, continual contraction. Each contraction sharply reduces maternal blood flow to the placenta, so the uterus must periodically relax to restore flow and oxygen delivery to the fetus. Contractions are strongest in the fundus and body of the uterus and weaker near the cervix, thus pushing the fetus downward.

According to the **positive feedback theory of labor,** labor contractions are induced by stretching of the cervix. This triggers a reflex contraction of the uterine body that pushes the fetus downward and stretches the cervix still more. Thus, there is a self-amplifying cycle of stretch and contraction. In addition, cervical stretching induces a neuroendocrine reflex through the spinal cord, hypothalamus, and posterior pituitary. The posterior pituitary releases oxytocin, which is carried in the blood and stimulates the uterine muscle both directly and through the action of prostaglandins. This, too, is a positive feedback cycle: cervical stretching → oxytocin secretion → uterine contraction → cervical stretching (see fig. 1.8, p. 17).

As labor progresses, a woman feels a growing urge to "bear down." A reflex arc extends from the uterus to the spinal cord and back to the skeletal muscles of the abdomen. Contraction of these muscles—partly reflexive and partly voluntary—aids in expelling the fetus, especially when combined with the Valsalva maneuver for increasing intra-abdominal pressure.

The pain of labor is due at first mainly to ischemia of the myometrium—muscle hurts when deprived of blood, and each labor contraction temporarily restricts uterine circulation. As the fetus enters the vaginal canal, the pain becomes stronger because of stretching of the cervix, vagina, and perineum and sometimes the tearing of tissue. At this stage, the obstetrician may perform an *episiotomy*—an incision in the vulva to widen the vaginal orifice and prevent random tearing. The pain of human childbirth, compared with the relative ease with which other mammals give birth, is an evolutionary product of two factors: the unusually large brain and head of the human infant, and the narrowing of the pelvic outlet, which helped to adapt hominids to bipedal locomotion (see p. 269).

## Stages of Labor

Labor occurs in three stages (fig. 28.20). The duration of each tends to be longer in a **primipara**[42] (a woman giving birth for the first time) than in a **multipara**[43] (a woman who has previously given birth).

1. **Dilation (first) stage.** This is the longest stage, lasting 8 to 24 hours in a primipara but as little as a few minutes in a multipara. It is marked by the **dilation** (widening) of the cervical canal and **effacement** (thinning) of the cervix. The cervix reaches a maximum diameter of about 10 cm (the diameter of the baby's head). During dilation, the fetal membranes usually rupture and the *amniotic fluid* is discharged (the "breaking of the waters").

2. **Expulsion (second) stage.** This stage typically lasts about 30 to 60 minutes in a primipara and as little as 1 minute in a multipara. It begins when the baby's head enters the vagina and lasts until the baby is entirely expelled. The baby is said to be **crowning** when the top of its head is visible, stretching the vulva. Delivery of the head is the most difficult part, with the rest of the body following more easily. An episiotomy may be performed during this stage. An attendant often uses a suction bulb to remove mucus from the baby's mouth and nose even before it is fully delivered. When the baby is fully expelled, an attendant drains the blood of the umbilical vein into the baby, clamps the umbilical cord in two places, and cuts the cord between the clamps.

3. **Placental (third) stage.** The uterus continues to contract after expulsion of the baby. The placenta, however, is a nonmuscular organ that cannot contract, so it buckles away from the uterine wall. About 350 mL of blood is typically lost at this stage, but contractions of the myometrium compress the blood vessels and prevent more extensive bleeding. The placenta, amnion, and other fetal membranes are expelled by uterine contractions, which may be aided by a gentle pull on the umbilical cord. The membranes *(afterbirth)* must be carefully inspected to be sure everything has been expelled. If any of these structures remain in the uterus, they can cause postpartum hemorrhaging. The umbilical blood vessels are counted because an abnormal number in the cord may indicate cardiovascular abnormalities in the infant.

---

[42]*primi* = first; *para* = birth
[43]*multi* = many; *para* = births

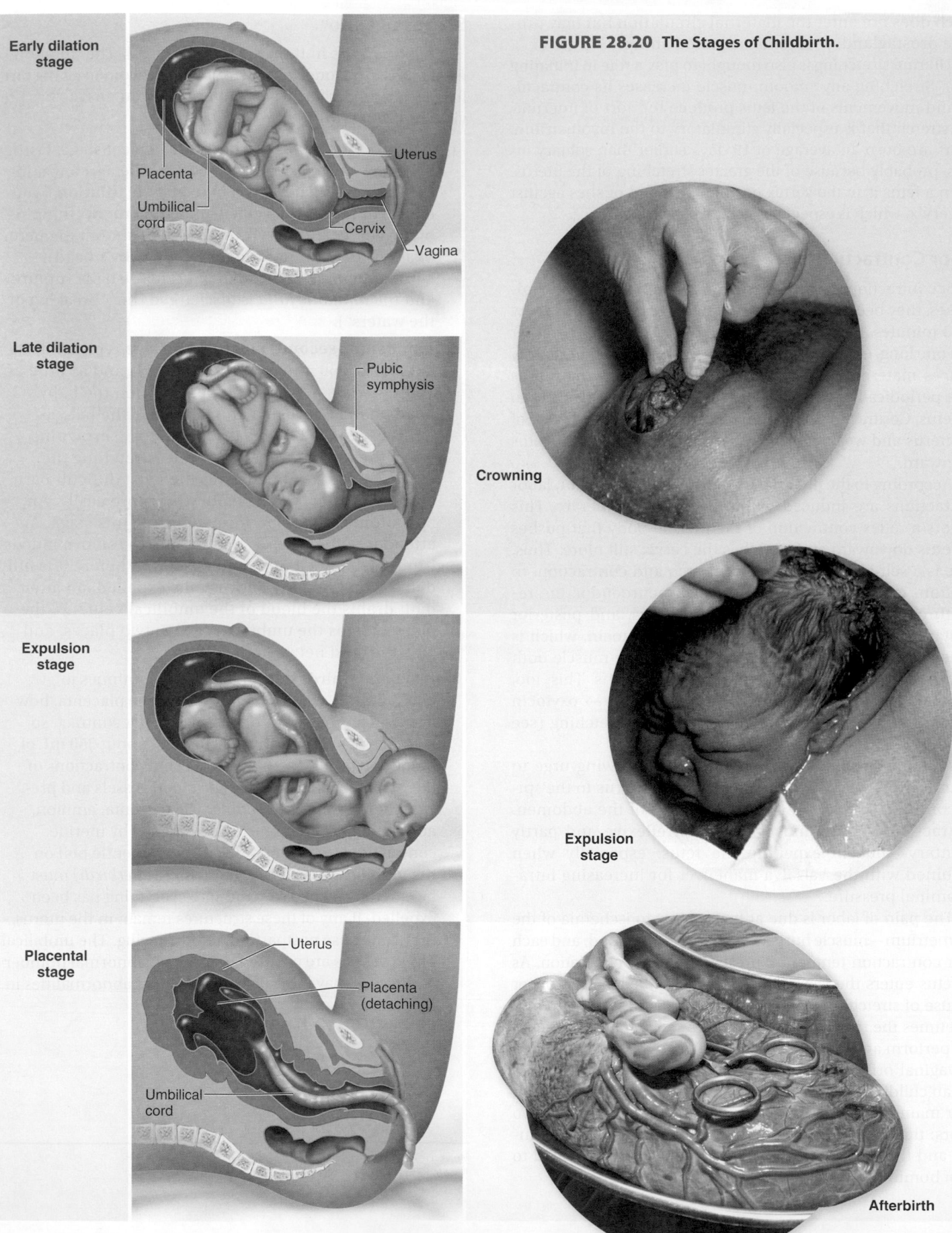

**Early dilation stage**

Placenta
Umbilical cord
Uterus
Cervix
Vagina

**Late dilation stage**

Pubic symphysis

**Expulsion stage**

**Placental stage**

Uterus
Placenta (detaching)
Umbilical cord

**FIGURE 28.20 The Stages of Childbirth.**

Crowning

Expulsion stage

Afterbirth

## The Puerperium

The first 6 weeks **postpartum** (after birth) are called the **puerperium**[44] (PYU-er-PEER-ee-um), a period in which the mother's anatomy and physiology stabilize and the reproductive organs return nearly to the pregravid state (their condition prior to pregnancy). Shrinkage of the uterus during this period is called **involution.** In a lactating woman, it loses about 50% of its weight in the first week and is nearly at its pregravid weight in 4 weeks. Involution is achieved by **autolysis** (self-digestion) of uterine cells by their own lysosomal enzymes. For about 10 days, this produces a vaginal discharge called **lochia,** which is bloody at first and then turns clear and serous. Breast-feeding promotes involution because (1) it suppresses estrogen secretion, which would otherwise cause the uterus to remain more flaccid; and (2) it stimulates oxytocin secretion, which causes the myometrium to contract and firm up the uterus sooner. It is important for the puerperium to be undisturbed, as emotional upset can inhibit lactation in some women.

### BEFORE YOU GO ON

Answer the following questions to test your understanding of the preceding section:

17. List the roles of HCG, estrogen, progesterone, and HCS in pregnancy.

18. What is the role of the corpus luteum in pregnancy? What eventually takes over this role?

19. List and briefly explain the special nutritional requirements of pregnancy.

20. How much weight does the average woman gain in pregnancy? What contributes to this weight gain other than the fetus?

21. Describe the positive feedback theory of labor.

22. What major events define the three stages of labor?

## 28.6 Lactation

### Expected Learning Outcomes

When you have completed this section, you should be able to

a. describe development of the breasts in pregnancy;

b. describe the shifting hormonal balance that regulates the onset and continuation of lactation;

c. describe the mechanism of milk ejection;

d. contrast colostrum with breast milk; and

e. discuss the benefits of breast-feeding.

**Lactation** is the synthesis and ejection of milk from the mammary glands. It lasts for as little as 1 week in women who do not breast-feed, but it can continue for many years as long as the breast is stimulated by a nursing infant or mechanical device (breast pump). Numerous studies conducted before the widespread marketing of artificial infant formulas suggest that worldwide, women traditionally nursed their infants until a median age of about 2.8 years.

## Development of the Mammary Glands in Pregnancy

The high estrogen level in pregnancy causes the ducts of the mammary glands to grow and branch extensively. Growth hormone, insulin, glucocorticoids, and prolactin also contribute to this development. Once the ducts are complete, progesterone stimulates the budding and development of acini at the ends of the ducts.

## Colostrum and Milk Synthesis

In late pregnancy, the mammary acini and ducts are distended with a secretion called **colostrum.** This is similar to breast milk in protein and lactose content but contains about one-third less fat. It is the infant's only natural source of nutrition for the first 1 to 3 days postpartum. Colostrum has a thin watery consistency and a cloudy yellowish color. The amount of colostrum secreted per day is at most 1% of the amount of milk secreted later, but since infants are born with excess body water and ample fat, they do not require high calorie or fluid intake at first. A major benefit of colostrum is that it contains immunoglobulins, especially IgA. IgA resists digestion and may protect the infant from gastroenteritis. It is also thought to be pinocytosed by the small intestine and to confer wider, systemic immunity to the neonate.

Milk synthesis is promoted by prolactin, a hormone of the anterior pituitary gland. In the nonpregnant state, dopamine (prolactin-inhibiting hormone) from the hypothalamus inhibits prolactin secretion. Prolactin secretion begins 5 weeks into the pregnancy, and by full term it is 10 to 20 times its normal level. Even so, prolactin has little effect on the mammary glands until after birth. While the steroids of pregnancy prepare the mammary glands for lactation, they antagonize prolactin and suppress milk synthesis. When the placenta is discharged at birth, the steroid levels abruptly drop and allow prolactin to have a stronger effect. Milk is synthesized in increasing quantity over the following week. Milk synthesis also requires the action of growth hormone, cortisol, insulin, and parathyroid hormone to mobilize the necessary amino acids, fatty acids, glucose, and calcium.

At the time of birth, baseline prolactin secretion drops to the nonpregnant level. Every time the infant nurses, however, it jumps to 10 to 20 times this level for the next hour and stimulates the synthesis of milk for the next feeding (fig. 28.21). These prolactin surges are accompanied by smaller increases in estrogen and progesterone secretion. If the mother does not nurse or these hormone surges are absent (due to pituitary damage, for

---

[44]*puer* = child; *per* (from *par*) = birth

example), the mammary glands stop producing milk in about a week. Even if she does nurse, milk production declines after 7 to 9 months.

Only 5% to 10% of women become pregnant again while breast-feeding. Apparently, either prolactin or nerve signals from the breast inhibit GnRH secretion, which, in turn, results in reduced gonadotropin secretion and ovarian cycling. This mechanism may have evolved as a natural means of spacing births, but breast-feeding is not a reliable means of contraception. Even in women who breast-feed, the ovarian cycle sometimes resumes several months postpartum. In those who do not breast-feed, the cycles resume in a few weeks, but for the first 6 months they are usually anovulatory.

## Milk Ejection

Milk is continually secreted into the mammary acini, but it does not easily flow into the ducts. Its flow, called **milk ejection (letdown),** is controlled by a neuroendocrine reflex. The infant's suckling stimulates nerve endings of the nipple and areola, which in turn signal the hypothalamus and posterior pituitary to release oxytocin. Oxytocin stimulates the myoepithelial cells that enmesh each gland acinus (see fig. 28.9d). These cells are of epithelial origin, but are packed with actin and contract like smooth muscle to squeeze milk from the acinus into the duct. The infant does not get any milk for the first 30 to 60 seconds of suckling, but milk soon fills the ducts and lactiferous sinuses and is then easily sucked out.

▶▶▶ **APPLY WHAT YOU KNOW**

*When a woman is nursing her baby at one breast, would you expect only that breast, or both breasts, to eject milk? Explain why.*

## Breast Milk

Table 28.4 compares the composition of colostrum, human milk, and cow's milk. Breast milk changes composition over the first 2 weeks, varies from one time of day to another, and changes even during the course of a single feeding. For example, at the end of a feeding there is less lactose and protein in the milk, but six times as much fat, as there is at the beginning.

Cow's milk is not a good substitute for human milk. It has one-third less lactose but three to five times as much protein and minerals. The excess protein forms a harder curd in the infant's stomach, so cow's milk is not digested and absorbed as efficiently as mother's milk. It also increases the infant's nitrogenous waste excretion, which increases the incidence and severity of diaper rash, particularly as bacteria in the diaper break urea down to ammonia, a skin irritant.

Colostrum and milk have a laxative effect that helps to clear the neonatal intestine of *meconium,* a greenish black, sticky fecal matter composed of bile, epithelial cells, and other wastes that accumulated during fetal development. By clearing bile and bilirubin from the body, breast-feeding also reduces the incidence and degree of jaundice in neonates. Breast milk promotes colonization of the neonatal intestine

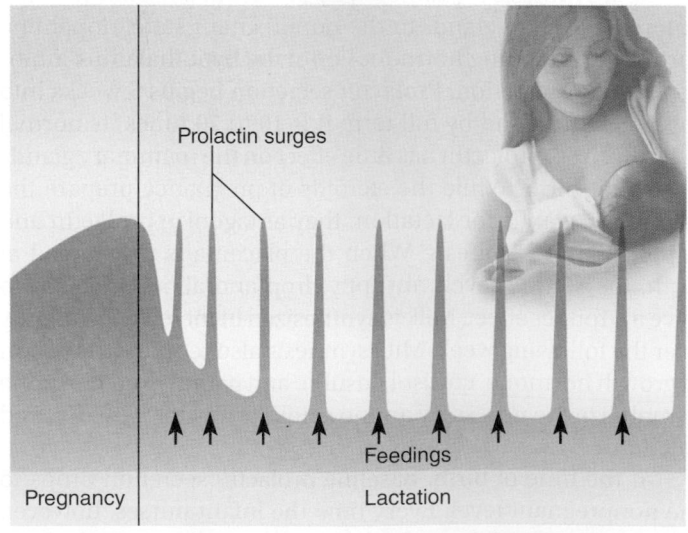

**FIGURE 28.21   Prolactin Secretion in the Lactating Female.**
Each time the infant nurses, maternal prolactin secretion surges. This prolactin stimulates synthesis of the milk that will be available at the next feeding.

| **TABLE 28.4** | A Comparison of Colostrum, Human Milk, and Cow's Milk* | | |
|---|---|---|---|
| **Nutrient** | **Human Colostrum** | **Human Milk** | **Cow's Milk** |
| Total protein (g/L) | 22.9 | 10.6 | 30.9 |
| Lactalbumin (g/L) | — | 3.7 | 25.0 |
| Casein (g/L) | — | 3.6 | 2.3 |
| Immunoglobulins (g/L) | 19.4 | 0.09 | 0.8 |
| Fat (g/L) | 29.5 | 45.4 | 38.0 |
| Lactose (g/L) | 57 | 71 | 47 |
| Calcium (mg/L) | 481 | 344 | 1,370 |
| Phosphorus (mg/L) | 157 | 141 | 910 |

*Colostrum data are for the first day postpartum, and human milk data are for "mature milk" at about 15 days postpartum.

with beneficial bacteria and supplies antibodies that lend protection against infection by pathogenic bacteria. Breast-feeding also tends to promote a closer bond between mother and infant.

A woman nursing one baby eventually produces about 1.5 L of milk per day; women with twins produce more. Lactation places a great metabolic demand on the mother. It is equivalent to losing 50 g of fat, 100 g of lactose (made from her blood glucose), and 2 to 3 g of calcium phosphate per day. A woman is at greater risk of bone loss when breast-feeding than when she is pregnant, because much of the infant's skeleton is still cartilage at birth and becomes mineralized at her expense in the first year postpartum. If a nursing mother does not have enough calcium and vitamin D in her own diet, lactation stimulates parathyroid hormone secretion and osteoclast activity, taking calcium from her bones to supply her baby.

To conclude this chapter, table 28.5 briefly describes some of the common disorders of pregnancy. Other reproductive disorders are discussed elsewhere: cervical cancer in Deeper Insight 28.1, breast cancer on page 1068, and sexually transmitted diseases at the end of chapter 27.

**BEFORE YOU GO ON**

Answer the following questions to test your understanding of the preceding section:

23. Why is little or no milk secreted while a woman is pregnant?

24. How does a lactating breast differ from a nonlactating breast in structure? What stimulates these differences to develop during pregnancy?

25. What is colostrum and what is its significance?

26. How does suckling stimulate milk ejection?

27. Why is breast milk superior to cow's milk for an infant?

| TABLE 28.5 | Some Disorders of Pregnancy |
|---|---|
| Abruptio placentae[45] | Premature separation of the placenta from the uterine wall, often associated with preeclampsia or cocaine use. May require birth by cesarean section. |
| Ectopic[46] pregnancy | Implantation of the conceptus anywhere other than the uterus; usually in the uterine tube (tubal pregnancy) but occasionally in the abdominopelvic cavity. See Deeper Insight 29.2 (p. 1102) for further details. |
| Gestational diabetes | A form of diabetes mellitus that develops in about 1% to 3% of pregnant women, characterized by insulin insensitivity, hyperglycemia, glycosuria, and a risk of excessive fetal size and birth trauma. Glucose metabolism often returns to normal after delivery of the infant, but 40% to 60% of women with gestational diabetes develop diabetes mellitus within 15 years after the pregnancy. |
| Hyperemesis gravidarum[47] | Severe vomiting, dehydration, alkalosis, and weight loss in early pregnancy, often requiring hospitalization to stabilize fluid, electrolyte, and acid–base balance; sometimes associated with liver damage |
| Placenta previa[48] | Blockage of the cervical canal by the placenta, preventing birth of the infant before the placenta separates from the uterus. Requires birth by cesarean section. |
| Preeclampsia[49] | Gestational hypertension and proteinuria, often with edema of the face and hands, occurring especially in the third trimester in primiparas. Correlated with abnormal development of placental arteries, potentially leading to widespread thrombosis and organ dysfunction in the mother. Occurs in 5% to 8% of pregnancies. Sometimes progresses to *eclampsia* (seizures), which can be fatal to the mother, fetus, or both. Eclampsia may occur *postpartum*. |
| Spontaneous abortion | Occurs in 10% to 15% of pregnancies, usually because of fetal deformities or chromosomal abnormalities incompatible with survival, but may also result from maternal abnormalities, infectious disease, and drug abuse. |

[45]*ab* = away; *rupt* = to tear; *placentae* = of the placenta
[46]*ec* = out of; *top* = place
[47]*hyper* = excessive; *emesis* = vomiting; *gravida* = pregnant woman
[48]*pre* = before; *via* = the way (obstructing the way)
[49]*ec* = forth; *lampsia* = shining

# DEEPER INSIGHT 28.5

## CLINICAL APPLICATION

### Methods of Contraception

The term *contraception* is used here to mean any procedure or device intended to prevent pregnancy (the presence of an implanted conceptus in the uterus). This essay describes the most common methods of contraception, some issues involved in choosing among them, and the relative reliability of the various methods. Several of those options are shown in figure 28.22.

### Behavioral Methods

*Abstinence* (refraining from intercourse) is, obviously, a completely reliable method if used consistently. The *rhythm method* (periodic abstinence) is based on avoiding intercourse near the time of expected ovulation. Among typical users, it has a 25% failure rate, partly due to lack of restraint and partly because it is difficult to predict the exact date of ovulation. Intercourse must be avoided for at least 7 days before ovulation so there will be no surviving sperm in the reproductive tract when the egg is ovulated, and for at least 2 days after ovulation so there will be no fertile egg present when sperm are introduced. The rhythm method is valuable, however, for couples who are trying to conceive a child by having intercourse at the time of apparent ovulation.

*Withdrawal (coitus interruptus)* requires the male to withdraw the penis before ejaculation. This often fails because of lack of willpower, because some sperm are present in the preejaculatory fluid, and because sperm ejaculated anywhere in the vulva can potentially get into the reproductive tract.

### Barrier and Spermicidal Methods

Barrier methods are designed to prevent sperm from getting into or beyond the vagina. They are most effective when used with chemical *spermicides,* available as nonprescription foams, creams, and jellies.

The *male condom* is a sheath of latex, rubber, or animal membrane (lamb intestine) that is unrolled over the erect penis and collects the semen. It is inexpensive, convenient, and very reliable when used carefully. About 25% of American couples who use contraceptives use only condoms, which rank second to birth-control pills in popularity.

The *female condom* is less used. It is a polyurethane sheath with a flexible ring at each end. The inner ring fits over the cervix and the outer ring covers the external genitalia. Male and female condoms are the only contraceptives that also protect against disease transmission. Animal membrane condoms, however, are porous to HIV and hepatitis B viruses and do not afford dependable protection from disease.

The *diaphragm* is a latex or rubber dome that is placed over the cervix to block sperm migration. It requires a physical examination and prescription to ensure proper fit, but is otherwise comparable to the condom in convenience and reliability, provided it is used with a spermicide. Without a spermicide, it is not very effective.

The *sponge* is a foam disc inserted before intercourse to cover the cervix. It is impregnated with a spermicide and acts by trapping and killing the sperm.

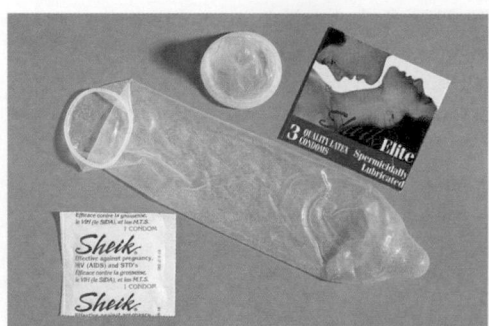

Male condom

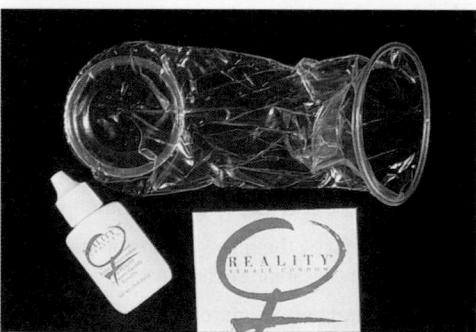

Female condom

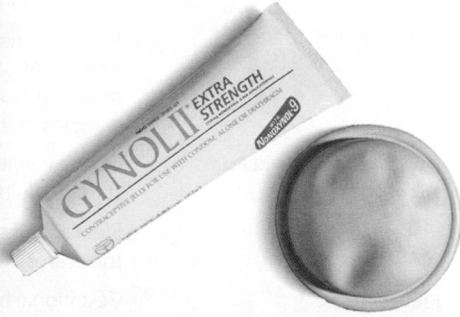

Diaphragm with contraceptive jelly

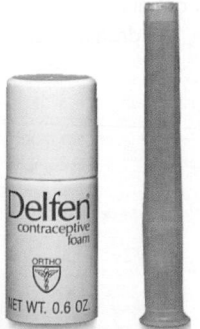

Contraceptive foam with vaginal applicator

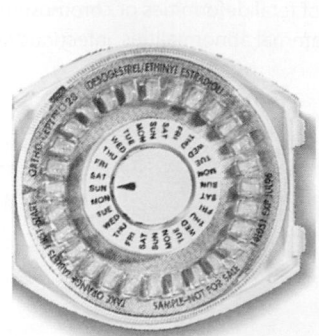

Birth-control pills

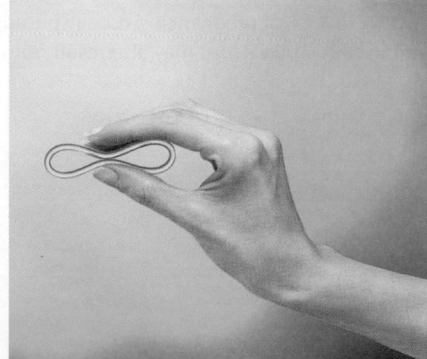

NuvaRing

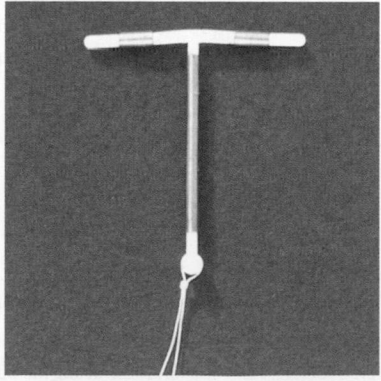

Intrauterine device (IUD)

**FIGURE 28.22  Contraceptive Devices.**

It requires no prescription or fitting. The sponge provides protection for up to 24 hours, and must be left in place for at least 6 hours after the last act of intercourse.

Contraceptive sponges and other barrier methods date to antiquity. Ancient Egyptians and Greeks used vaginal sponges soaked in lemon juice, which had a mild spermicidal effect. Some Egyptian women used vaginal pessaries made of crocodile dung and honey, but crocodile dung is difficult to find in pharmacies these days, limiting the modern usefulness of this idea.

### Hormonal Methods

Most hormonal methods of contraception are aimed at preventing ovulation. Efforts to develop a hormonal contraceptive for men have so far been unsuccessful, but are continuing. In women, hormonal contraceptives mimic the negative feedback effect of ovarian hormones, inhibiting FSH and LH secretion so follicles do not mature. For most women, they are highly effective and present minimal complications. Their differences lie largely in method of application and convenience of use and, to some extent, in reliability and risk of complications.

The oldest and still the most widely used hormonal method in the United States is the *combined oral contraceptive,* or *birth-control pill,* first approved for use in 1960. "The pill" is composed of estrogen and progestin (a synthetic progesterone-like hormone). It must be taken daily, at the same time of day, for 21 days each cycle. It comes in 28-day packets, marked day by day, with the last 7 pills being plain sugar just to keep the user in the habit of taking one every day. The 7-day withdrawal from hormones allows for menstruation. Side effects include an elevated risk of heart attack or stroke in smokers and in women with a history of diabetes, hypertension, or clotting disorders.

Other hormonal methods avoid the need to remember a daily pill. One option is a skin patch marketed as Ortho Evra in the United States, which releases estrogen and progestin transdermally. It is changed at 7-day intervals (three patches per month and 1 week without). The NuvaRing is a soft flexible vaginal ring that releases estrogen and progestin for absorption through the vaginal mucosa. It must be worn continually for 3 weeks and removed for the fourth week of each cycle. Reportedly, it cannot be felt even during intercourse, and there is hope that when more data are in, it will prove even more reliable than the pill. *Medroxyprogesterone* (trade name Depo-Provera) is a progestin administered by injection two to four times per year. It provides highly reliable, long-term contraception, although in some women it causes headaches, nausea, or weight gain, and fertility may not return immediately when its use is discontinued.

Some drugs can be taken orally after intercourse to prevent implantation of a conceptus. These are called emergency contraceptive pills (ECPs), or "morning-after pills" (trade names Plan B, Levonelle). An ECP is a high dose of estrogen and progestin or a progestin alone. It can be taken within 72 hours after intercourse and induces menstruation within 2 weeks. ECPs work on several fronts: inhibiting ovulation; inhibiting sperm or egg transport in the uterine tube; and preventing implantation. They do not work if a blastocyst is already implanted. ECPs are available without a prescription in some states, but availability has been limited or delayed elsewhere by political controversy.

*Mifepristone* (Mifeprex, also known as RU-486) is a progesterone antagonist. It is used less as a contraceptive than as an *abortifacient;* in high doses, it induces abortion up to 2 months into pregnancy. But at a dose of 2 mg/day, it prevents ovulation like other steroidal contraceptives, and a single 10 mg dose can also be used as an emergency "morning-after" contraceptive if taken after intercourse but before ovulation.

### The Intrauterine Device

*Intrauterine devices (IUDs)* are springy, often T-shaped devices inserted through the cervical canal into the uterus. Some IUDs act by releasing a synthetic progesterone, but most have a copper wire wrapping or copper sleeve. IUDs act by irritating the uterine lining and interfering with blastocyst implantation, and copper IUDs also inhibit sperm motility. An IUD can be left in place for 5 to 12 years.

### Surgical Sterilization

People who are confident that they do not want more children (or any) often elect to be surgically sterilized. This entails the cutting and tying or clamping of the genital ducts, thus blocking the passage of sperm or eggs. Surgical sterilization has the advantage of convenience, since it requires no further attention. Its initial cost is higher, however, and for people who later change their minds, surgical reversal is much more expensive than the original procedure and is often unsuccessful. *Vasectomy* is the severing of the ductus (vas) deferens, done through a small incision in the scrotum. In *tubal ligation,*[50] the uterine tubes are cut. This can be done through small abdominal incisions to admit a cutting instrument and laparoscope (viewing device).

### Issues in Choosing a Contraceptive

Many issues enter into the appropriate choice of a contraceptive, including personal preference, pattern of sexual activity, medical history, religious views, convenience, initial and ongoing costs, and disease prevention. For most people, however, the two primary issues are safety and reliability.

The following table shows the expected rates of failure for several types of contraception as reported by the World Health Organization (WHO). Each column shows the number of sexually active women who typically become pregnant within 1 year while they or their partners are using the indicated contraceptives. The lowest rate (perfect use) is for those who use the method correctly and consistently, whereas the higher rate (typical use) is based on random surveys of users and takes human error (lapses and incorrect usage) into account.

We have not considered all the currently available methods of contraception or all the issues important to the choice of a contraceptive. No one contraceptive method can be recommended as best for all people. Further information necessary to a sound choice and proper use of contraceptives should be sought from a health department, college health service, physician, or other such sources.

### Failure Rates of Contraceptive Methods

| Method | Rate of Failure (Pregnancies per 100 Users) | |
| --- | --- | --- |
| | Perfect Use | Typical Use |
| No protection | 85 | 85 |
| Rhythm method | 3–5 | 25 |
| Withdrawal | 4 | 27 |
| Spermicide alone | 18 | 26 |
| Condom alone (male or female) | 2–5 | 15–21 |
| Diaphragm with spermicide | 6 | 16 |
| Vaginal sponge | 9–20 | 16–32 |
| Birth-control pill, patch, or NuvaRing | 0.3–0.5 | 8 |
| Medroxyprogesterone | 0.3 | 3 |
| Intrauterine device | 0.2–0.6 | 0.2–0.8 |
| Vasectomy | 0.10 | 0.15 |
| Tubal ligation | 0.5 | 0.5 |

[50]*ligat* = to tie

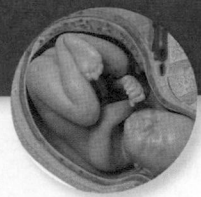

# CONNECTIVE ISSUES

## Effects of the **REPRODUCTIVE SYSTEM**
## On Other Organ Systems

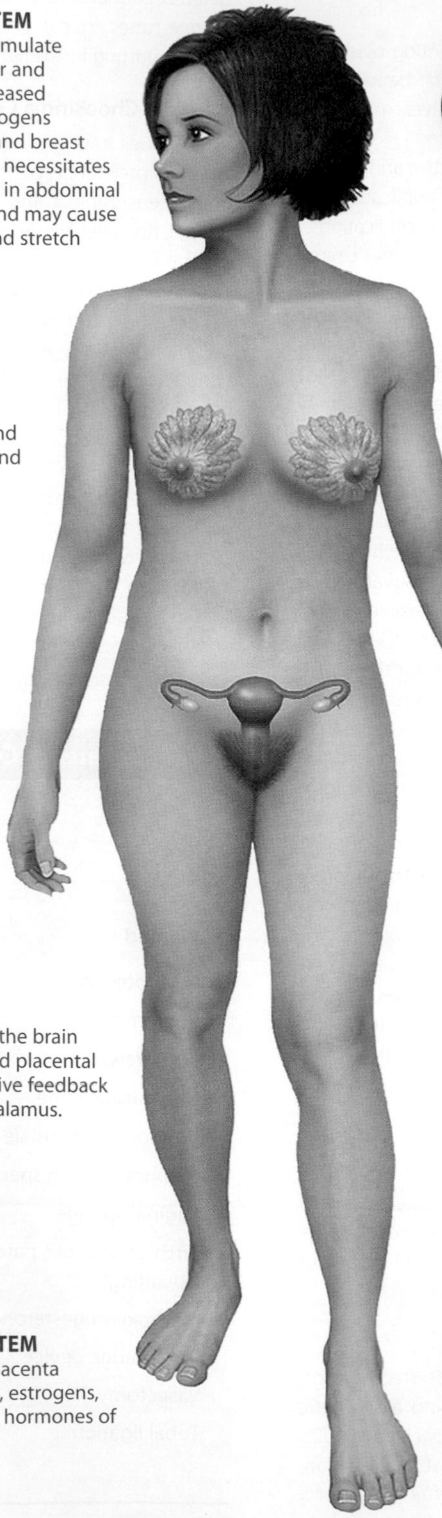

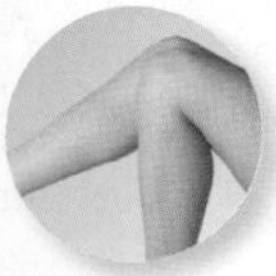

**INTEGUMENTARY SYSTEM**
At puberty, androgens stimulate development of body hair and apocrine glands and increased sebaceous secretion; estrogens stimulate fat deposition and breast development; pregnancy necessitates growth of skin, especially in abdominal and mammary regions, and may cause pigmentation changes and stretch marks.

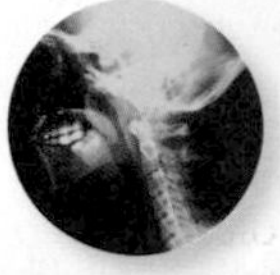

**SKELETAL SYSTEM**
Androgens and estrogens stimulate bone deposition and adolescent skeletal growth and maintain adult bone mass.

**MUSCULAR SYSTEM**
Androgens stimulate muscle growth; sexual climax and childbirth involve contractions of specific skeletal muscles.

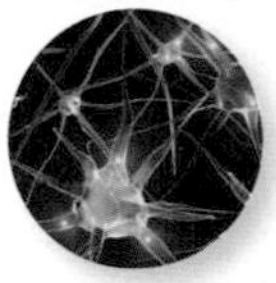

**NERVOUS SYSTEM**
Sex steroids stimulate the brain and libido; gonadal and placental hormones exert negative feedback control on the hypothalamus.

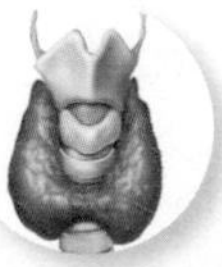

**ENDOCRINE SYSTEM**
The gonads and placenta secrete androgens, estrogens, progesterone, and hormones of pregnancy.

**CIRCULATORY SYSTEM**
Androgens stimulate erythropoiesis; estrogens inhibit atherosclerosis in females; pregnancy increases blood volume and cardiac output and may cause varicose veins.

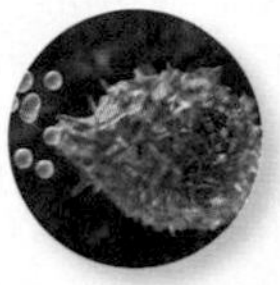

**LYMPHATIC/IMMUNE SYSTEM**
Barriers in the testis and ovary protect germ cells from antibodies; androgens somewhat inhibit immunity and increase susceptibility to infectious diseases.

**RESPIRATORY SYSTEM**
Sexual arousal increases pulmonary ventilation; pregnancy increases $CO_2$ sensitivity of respiratory chemoreceptors and increases tidal volume and minute ventilation.

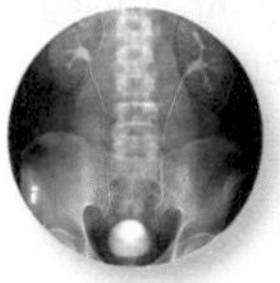

**URINARY SYSTEM**
Sexual arousal constricts the internal urinary sphincter, which prevents reflux of semen into the male urinary bladder; prostatic hyperplasia can impede urine flow; pregnancy crowds the bladder, reduces its capacity, and may cause incontinence; pregnancy promotes salt and water reabsorption by the kidneys and increases glomerular filtration rate and urine output.

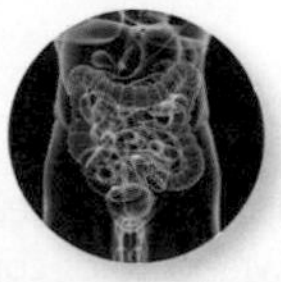

**DIGESTIVE SYSTEM**
A growing fetus crowds the stomach and intestines and may cause heartburn; pregnancy is often associated with constipation and nausea.

# STUDY GUIDE

## ▶ Assess Your Learning Outcomes

*To test your knowledge, discuss the following topics with a study partner or in writing, ideally from memory.*

### 28.1 Reproductive Anatomy (p. 1059)

1. Why the female paramesonephric duct develops into a reproductive tract whereas the male's paramesonephric duct does not; what mature female structures arise from the duct
2. What mature female structures arise from the embryonic genital tubercle, urogenital folds, and labioscrotal folds
3. Internal structure of the ovary; its supportive ligaments; and its blood and nerve supplies
4. General structure of ovarian follicles; their function and how they compare and contrast with the male's seminiferous tubules; and their location in the ovary
5. Gross anatomy of the uterine (fallopian) tube; its three segments; its supportive ligament; and the structure of its mucosa and relationship of that structure to its function
6. Gross anatomy of the uterus; its supportive ligaments; its relationship to the vesicouterine and rectouterine pouches; its blood supply; and function of the cervical glands
7. Tissue layers of the uterine wall; histology of the endometrium; functions of the endometrial sublayers
8. The tilt of the vagina and its relationship to adjacent organs; histology of its mucosa in childhood and adulthood; significance of its dendritic cells; sources of its lubrication; and anatomy of the hymen
9. Anatomy of the vulva including the mons pubis, labia majora and minora, clitoris and prepuce, vaginal and urethral orifices, accessory glands, and erectile tissues
10. Anatomy of the mature breast in the resting and lactating states
11. The prevalence of breast cancer; its diagnostic signs; genetic and other risk factors for breast cancer; preventive breast care; and treatment options

### 28.2 Puberty and Menopause (p. 1069)

1. The typical age of onset of female puberty in the United States and Europe

2. The hormonal trigger for the onset of female puberty; roles of GnRH, FSH, LH, inhibin, androgens, and estrogens; and three types of estrogen
3. Thelarche, pubarche, and menarche as signs of female puberty; their hormonal causes; and the hormonal basis of the libido
4. The typical age of menarche and its relationship to the onset of ovulation
5. Bodily effects of estradiol, progesterone, and inhibin in female puberty
6. Ovarian and hormonal changes that bring on female climacteric and menopause; effects commonly experienced by perimenopausal women; and differences between female and male climacteric
7. The criterion for determining that a woman has passed through menopause, and the reason one cannot identify the exact time of menopause

### 28.3 Oogenesis and the Sexual Cycle (p. 1071)

1. Meanings of and distinctions between the female *sexual cycle, ovarian cycle,* and *menstrual cycle*
2. The meaning of *oogenesis* and ways in which it differs from spermatogenesis
3. Prenatal development of oogonia and primary oocytes; the peak number of oogonia typically attained in the fetus; why this number is so much less at the time of birth and again by the onset of puberty; and the name for the prenatal and childhood degeneration of female germ cells
4. Ways in which meiosis in the female differs from that in the male; why male gametogenesis produces four functional gametes per stem cell and female gametogenesis produces only one; what happens to the other three meiotic daughter cells in the female
5. How far meiosis has progressed by the time the egg is ovulated, and what must happen thereafter for meiosis to be completed
6. Development of a follicle from primary to mature (graafian) types; the structural differences between the stages; and how folliculogenesis is correlated with oogenesis
7. Structural details of a mature follicle

8. The timetable of oogenesis and folliculogenesis; why the ovarian cycle is considered to average 28 days long whereas any given egg and follicle take much longer to mature; and what event marks day 1 of a cycle
9. How many follicles begin to develop in each cycle of folliculogenesis; how many of them normally ovulate; what happens to the rest, and what that fate is called
10. Roles of FSH and LH in regulating the ovarian cycle
11. The process of ovulation, the day on which it occurs in a typical cycle, and how the egg gets into the uterine tube
12. Production, structure, function, and eventual involution of the corpus luteum; why days 15 through 28 of a typical cycle are called the luteal phase; what remains after a corpus luteum has fully involuted
13. Four phases of the menstrual cycle; what happens histologically to the endometrium in each phase; what days of the cycle are spanned by each phase; and what hormones regulate these changes

### 28.4 Female Sexual Response (p. 1079)

1. Ways in which the female sexual response differs from the male's in the excitement phase; vasocongestion of the labia, clitoris, and breasts; sources of lubrication of the vagina and vulva; anatomical changes in the vagina and uterus during sexual arousal
2. Physiological responses of orgasm and resolution in the female and how they differ from those of the male; absence of a refractory period and potential for multiple orgasms

### 28.5 Pregnancy and Childbirth (p. 1080)

1. The timetable of gestation and how the date of birth is predicted
2. What is included in the conceptus
3. Human chorionic gonadotropin (HCG), its source and effects, the time course of its rising and falling secretion during pregnancy, and its usefulness in pregnancy tests
4. Sources and effects of estrogen, progesterone, and human chorionic gonadotropin in pregnancy

# STUDY GUIDE

5. Effects of thyroid hormone, parathyroid hormone, glucocorticoids, aldosterone, and relaxin in pregnancy
6. Causes of morning sickness, constipation, and heartburn in pregnancy; the change in basal metabolic rate and the related nutritional needs of pregnancy
7. Effects of pregnancy on blood volume and cardiac output; how pregnancy can cause edema, hemorrhoids, and varicose veins
8. Effects of pregnancy on respiratory function; the mechanism for enhancing diffusion of carbon dioxide from the fetal blood into the maternal blood of the placenta
9. Effects of pregnancy on glomerular filtration, urine output, and the capacity of the bladder
10. Effects of pregnancy on the skin; causes of striae (stretch marks), the linea nigra, and chloasma

11. The vertex position and the developmental age at which the fetus typically assumes it
12. The nature and possible cause of Braxton Hicks contractions, when they occur, and how they differ from true labor contractions
13. Factors that stimulate the onset of labor contractions; the roles of oxytocin, positive feedback, and the voluntary abdominal muscles in labor
14. Events that mark each stage of labor, and the names of the stages
15. The puerperium, its time course, the postpartum changes in a woman's body during this time

## 28.6 Lactation (p. 1087)

1. Influences of estrogen, growth hormone, insulin, glucocorticoids, and prolactin on mammary gland development during pregnancy

2. The fluid secreted by the mammary glands for the first few days postpartum, how it differs from breast milk, and its benefits to the neonate
3. Why prolactin stimulates milk synthesis after birth but not during pregnancy
4. The neuroendocrine reflex stimulated by the suckling of an infant, and the roles of oxytocin and prolactin in breast-feeding
5. Composition of breast milk in comparison to colostrum and cow's milk; reasons why cow's milk is less healthy than breast milk for an infant
6. How breast milk varies in composition from one time to another; which components of the milk are released early, and which are released nearer the end, of a single feeding
7. The daily quantity of breast milk typically produced (eventually) and its nutritional demands on the mother

## ▶ Testing Your Recall

*Answers in Appendix B*

1. Of the following organs, the one(s) most comparable to the penis in structure is/are
   a. the clitoris.
   b. the vagina.
   c. the vestibular bulbs.
   d. the labia minora.
   e. the prepuce.

2. The ovaries secrete all of the following *except*
   a. estrogens.
   b. progesterone.
   c. androgens.
   d. follicle-stimulating hormone.
   e. inhibin.

3. The first haploid stage in oogenesis is
   a. the oogonium.
   b. the primary oocyte.
   c. the secondary oocyte.
   d. the second polar body.
   e. the zygote.

4. The haploid secondary oocyte is shielded from immune attack by
   a. the zona pellucida.
   b. the theca folliculi.
   c. the cumulus oophorus.
   d. follicular fluid.
   e. inhibin.

5. The hormone that most directly influences the secretory phase of the menstrual cycle is
   a. HCG.
   b. FSH.
   c. LH.
   d. estrogen.
   e. progesterone.

6. The ischemic phase of the uterus results from
   a. rising progesterone levels.
   b. falling progesterone levels.
   c. stimulation by oxytocin.
   d. stimulation by prostaglandins.
   e. stimulation by estrogens.

7. Before secreting milk, the mammary glands secrete
   a. prolactin.
   b. colostrum.
   c. lochia.
   d. meconium.
   e. chloasma.

8. Few women become pregnant while nursing because _____ inhibit(s) GnRH secretion.
   a. FSH.
   b. prolactin.
   c. prostaglandins.
   d. oxytocin.
   e. HCG.

9. Smooth muscle cells of the myometrium and myoepithelial cells of the mammary glands are the target cells for
   a. prostaglandins.
   b. LH.
   c. oxytocin.
   d. progesterone.
   e. FSH.

10. Which of these is *not* true of the luteal phase of the sexual cycle?
    a. Progesterone level is high.
    b. The endometrium stores glycogen.
    c. Ovulation occurs.
    d. Fertilization may occur.
    e. The endometrial glands enlarge.

11. Each egg cell develops in its own fluid-filled space called a/an _____.

12. The mucosa of the uterus is called the _____.

13. A girl's first menstrual period is called _____.

14. A yellowish structure called the _____ secretes progesterone during the secretory phase of the menstrual cycle.

15. The layer of cells closest to a mature secondary oocyte is the _____.

# STUDY GUIDE

16. A tertiary follicle differs from a primary follicle in having a cavity called the _____.

17. Menopause occurs during a midlife period of changing hormone secretion called _____.

18. All the products of fertilization, including the embryo or fetus, the placenta, and the embryonic membranes, are collectively called the _____.

19. The funnel-like distal end of the uterine tube is called the _____ and has feathery processes called _____.

20. Postpartum uterine involution produces a vaginal discharge called _____.

## ▶ Building Your Medical Vocabulary

*Answers in Appendix B*

*State a medical meaning of each word element below, and give a term in which it or a slight variation of it is used.*

1. -arche

2. -arum

3. cumulo-

4. gesto-

5. hystero-

6. lacto-

7. ligat-

8. metri-

9. oo-

10. primi-

## ▶ True or False

*Answers in Appendix B*

*Determine which five of the following statements are false, and briefly explain why.*

1. After ovulation, a follicle begins to move down the uterine tube to the uterus.

2. Human chorionic gonadotropin is secreted by the granulosa cells of the follicle.

3. An oocyte never completes meiosis II unless it is fertilized.

4. A slim girl who is active in dance and gymnastics is likely to begin menstruating at a later age than an overweight inactive girl.

5. There are more future egg cells in the ovary at puberty than there are at birth.

6. Women do not lactate while they are pregnant because prolactin is not secreted until after birth.

7. Colostrum contains more protein than milk, but less fat.

8. Several follicles develop in each ovarian cycle even though only one of them usually ovulates.

9. Progesterone inhibits uterine contractions.

10. The entire endometrium is shed in each menstrual period.

## ▶ Testing Your Comprehension

*Answers at* www.mhhe.com/saladin7

1. Would you expect puberty to create a state of positive or negative nitrogen balance? Explain. (See chapter 26 for nitrogen balance.)

2. Aspirin and ibuprofen can inhibit the onset of labor and are sometimes used to prevent premature birth. Review your knowledge of these drugs and the mechanism of labor, and explain this effect.

3. At 6 months postpartum, a nursing mother is in an automobile accident that fractures her skull and severs the hypophyseal portal vessels. How would you expect this to affect her milk production? How would you expect it to affect her future ovarian cycles? Explain the difference.

4. If the ovaries are removed in the first 6 weeks of pregnancy, the embryo will be aborted. If they are removed later in pregnancy, the pregnancy can go to a normal full term. Explain the difference.

5. A breast-feeding woman leaves her baby at home and goes shopping. There, she hears another woman's baby crying and notices her blouse becoming wet with a little exuded milk. Explain the physiological link between hearing that sound and the ejection of milk.

## ▶ Improve Your Grade at www.mhhe.com/saladin7

*Download animations to study when it fits your schedule. Practice quizzes, labeling activities, games, and flashcards offer fun ways to master the chapter concepts. Or, download image PowerPoint files for each chapter to create a study guide or for taking notes during lecture.*

# CHAPTER 29

## HUMAN DEVELOPMENT AND AGING

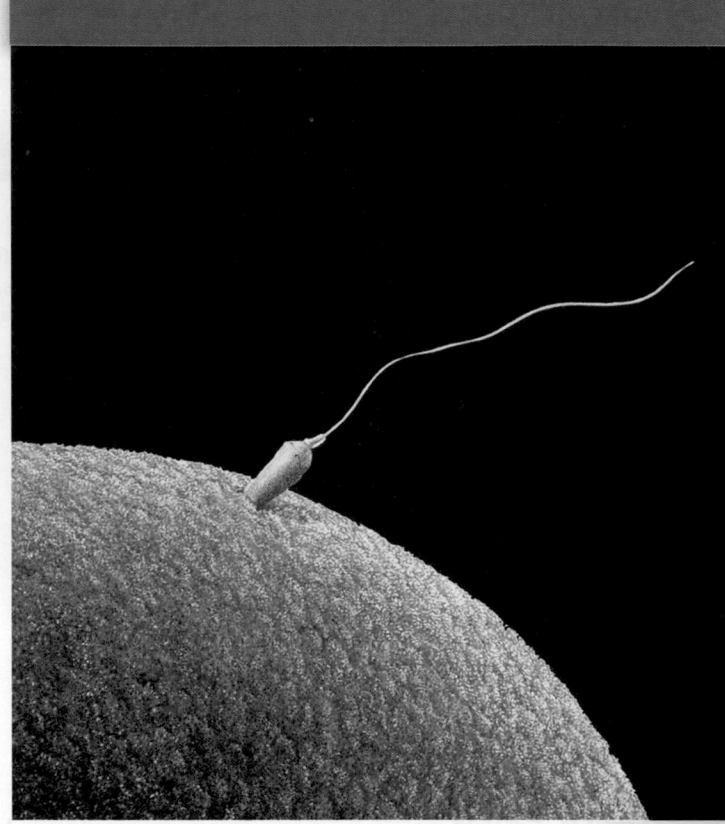

Boy meets girl: the union of sperm and egg (SEM)

### CHAPTER OUTLINE

### DEEPER INSIGHTS

Anatomy & Physiology | REVEALED®
aprevealed.com

**Module 14: Reproductive System**

## BRUSHING UP...

- Understanding the fertilization process requires familiarity with sperm structure, described on page 1047.
- Understanding fertilization and the events immediately following it also depends on a knowledge of oogenesis (p. 1071) and the structure of the mature ovarian follicle (p. 1073).
- Understanding embryonic implantation in the uterus and the development of the placenta requires a knowledge of the histology of the uterine endometrium, described on page 1063.

Perhaps the most dramatic, miraculous aspect of human life is the transformation of a one-celled fertilized egg into an independent, fully developed individual. From the beginning of recorded thought, people have pondered how a baby forms in the mother's body and how two parents can produce another human being who, although unique, possesses characteristics of each. Aristotle, in his quest to understand prenatal development, dissected bird embryos and established the sequence in which their organs appeared and took shape. He also speculated that the hereditary traits of a child resulted from the mixing of the male's semen with the female's menstrual blood. Such misconceptions about human development persisted for many centuries. In the seventeenth century, scientists thought that the features of the infant existed in a preformed state in the egg or the sperm, and simply unfolded and expanded as the embryo developed. Some thought that the head of the sperm had a miniature human curled up in it, while others thought that the miniature person existed in the egg and the sperm were parasites in the semen.

The modern science of **embryology**—the study of prenatal development—was not born until the nineteenth century, largely because darwinism at last gave biologists a systematic framework for asking the right questions and discovering unifying themes in the development of diverse species of animals, including humans. It was in that era, too, that the human egg was first observed. Embryology is now a part of **developmental biology,** a broader science that embraces changes in form and function from fertilized egg through old age. A rapidly expanding area of developmental biology today is the genetic regulation of development.

In this book's closing chapter, it is fitting that we reflect on the closing chapter of life as well. Why do our bodies wear out? Isn't there something we can do about that? Is there any cure for old age on the horizon? Our scope of discussion here embraces the human life span from conception to death.

## 29.1 Fertilization and the Preembryonic Stage

### Expected Learning Outcomes

When you have completed this section, you should be able to

a. describe the process of sperm migration and fertilization;

b. explain how an egg prevents fertilization by more than one sperm;

c. describe the major events that transform a fertilized egg into an embryo; and

d. describe the implantation of the preembryo in the uterine wall.

Authorities attach different meanings to the word **embryo.** Some use it to denote stages beginning with the fertilized egg or at least with the two-celled stage produced by its first division. Others first apply the word *embryo* to an individual 16 days old, when it consists of three **primary germ layers** called the *ectoderm, mesoderm,* and *endoderm.* The events leading up to that stage are called *embryogenesis,* so the first 16 days after fertilization are called the *preembryonic stage.* This is the sense in which we will use such terms in this book. We begin with the process in which a sperm locates and fertilizes the egg.

### Sperm Migration

If it is to survive, an egg must be fertilized within 12 to 24 hours of ovulation; yet it takes about 72 hours for an egg to reach the uterus. Therefore, in order to fertilize an egg before it dies, sperm must encounter it somewhere in the distal one-third of the uterine tube. The vast majority of sperm never make it that far. Many are destroyed by vaginal acid or drain out of the vagina. Others fail to penetrate the mucus of the cervical canal, and those that do are often destroyed by leukocytes in the uterus. Of those that get past the uterus, probably half go up the wrong uterine tube. Finally, about 200 spermatozoa reach the vicinity of the egg—not many of the 300 million that were ejaculated.

Sperm migrate mainly by means of the snakelike lashing of their tails as they crawl along the female mucosa, but they are assisted by certain aspects of female physiology. Strands of mucus guide them through the cervical canal. Although female orgasm is not required for fertilization, orgasm does involve uterine contractions that may suck semen from the vagina and spread it throughout the uterus, like hand lotion pressed between your palms. The egg itself may release a chemical that attracts sperm from a short distance; this has been demonstrated for some animals but remains unproven for humans.

## Sperm Capacitation

Sperm can reach the distal uterine tube in half an hour or less after ejaculation, but they cannot fertilize an egg for about 10 hours. While migrating, they must undergo a process called **capacitation** that makes them capable of penetrating an egg. In fresh sperm, the plasma membrane is toughened by cholesterol. This prevents the premature release of acrosomal enzymes while sperm are still in the male, and thus avoids wastage of sperm. It also prevents enzymatic damage to the spermatic ducts. After ejaculation, however, fluids of the female reproductive tract leach cholesterol from the plasma membrane and dilute other inhibitory factors in the semen. The membrane of the sperm head becomes more fragile and more permeable to calcium ions, which diffuse into the sperm and stimulate more powerful lashing of the tail.

Sperm remain viable for up to 6 days after ejaculation, so there is little chance of pregnancy from intercourse occurring more than a week before ovulation. Fertilization also is unlikely if intercourse takes place more than 14 hours after ovulation, because the egg would no longer be viable by the time the sperm became capacitated. For those wishing to conceive

a child, the optimal "window of opportunity" is therefore from a few days before ovulation to 14 hours after. Those wishing to avoid pregnancy, however, should allow a wider margin of safety for variations in sperm and egg longevity, capacitation time, and time of ovulation—variations that make the rhythm method of contraception so unreliable.

## Fertilization

When the sperm encounters an egg, it undergoes an **acrosomal reaction**—exocytosis of the acrosome, releasing the penetration enzymes. But the first sperm to reach an egg is not the one to fertilize it. Sperm must first penetrate the granulosa cells and zona pellucida that surround it (fig. 29.1). It may require numerous sperm to clear a path for the one that penetrates the egg proper.

Two of the acrosomal enzymes are **hyaluronidase,** which digests the hyaluronic acid that binds granulosa cells together, and **acrosin,** a protease similar to the trypsin of pancreatic juice. When a path has been cleared through the granulosa cells, a sperm binds to the zona pellucida and releases its enzymes, digesting a pathway through the zona until it contacts

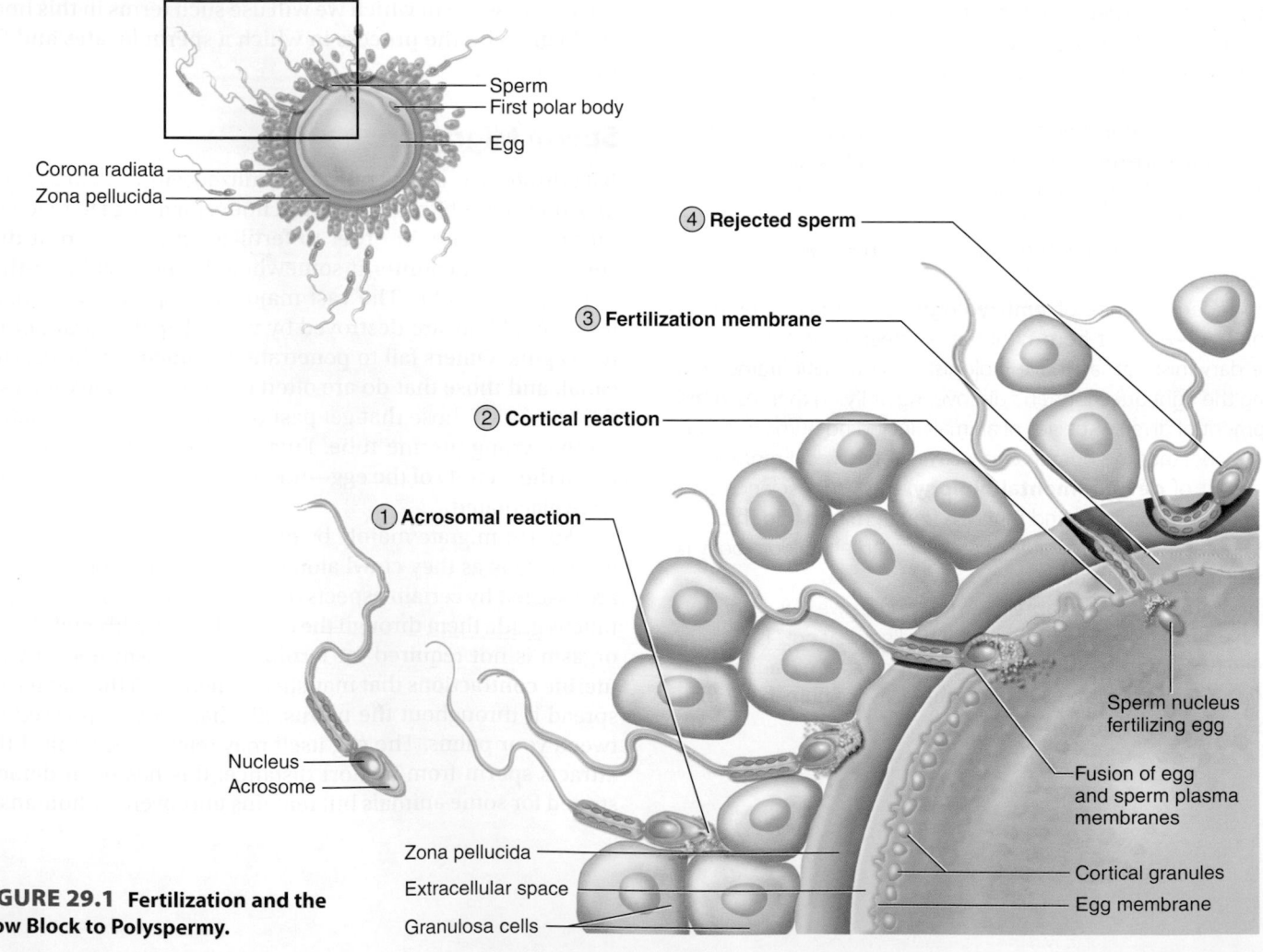

**FIGURE 29.1  Fertilization and the Slow Block to Polyspermy.**

the egg itself. The sperm head and midpiece enter the egg, but the egg destroys the sperm mitochondria and passes only maternal mitochondria on to the offspring.

Fertilization combines the haploid (*n*) set of sperm chromosomes with the haploid set of egg chromosomes, producing a diploid (*2n*) set. Fertilization by two or more sperm, called **polyspermy,** would produce a triploid (*3n*) or larger set of chromosomes and the egg would die. Thus it is important for the egg to prevent this, and it has two mechanisms for doing so: a fast block and slow block to polyspermy. In the **fast block,** binding of the sperm to the egg opens $Na^+$ channels in the egg membrane. The rapid inflow of $Na^+$ depolarizes the membrane and inhibits the attachment of any more sperm. The **slow block** involves secretory vesicles called **cortical granules** just beneath the membrane. Sperm penetration releases an inflow of $Ca^{2+}$; this, in turn, stimulates a **cortical reaction** in which the cortical granules release their secretion beneath the zona pellucida. The secretion swells with water, pushes any remaining sperm away from the egg, and creates an impenetrable **fertilization membrane** between the egg and zona pellucida.

▶ ▶ ▶ **APPLY WHAT YOU KNOW**

*What similarity can you see between the slow block to polyspermy and the release of acetylcholine from the synaptic vesicles of a neuron? (Compare p. 459.)*

## Meiosis II

A secondary oocyte begins meiosis II before ovulation (see p. 1071) and completes it only if fertilized. Through the formation of a second polar body, the fertilized egg discards one chromatid from each chromosome. The sperm and egg nuclei then swell and become **pronuclei.** A mitotic spindle forms between them, each pronucleus ruptures, and the chromosomes of the two gametes mix into a single diploid set (fig. 29.2). The fertilized egg, now called a **zygote,** is ready for its first mitotic division.

As in chapter 28, we will use the term *conceptus* for everything that arises from this zygote—not only the developing individual but also the placenta, umbilical cord, and membranes associated with the embryo and fetus.

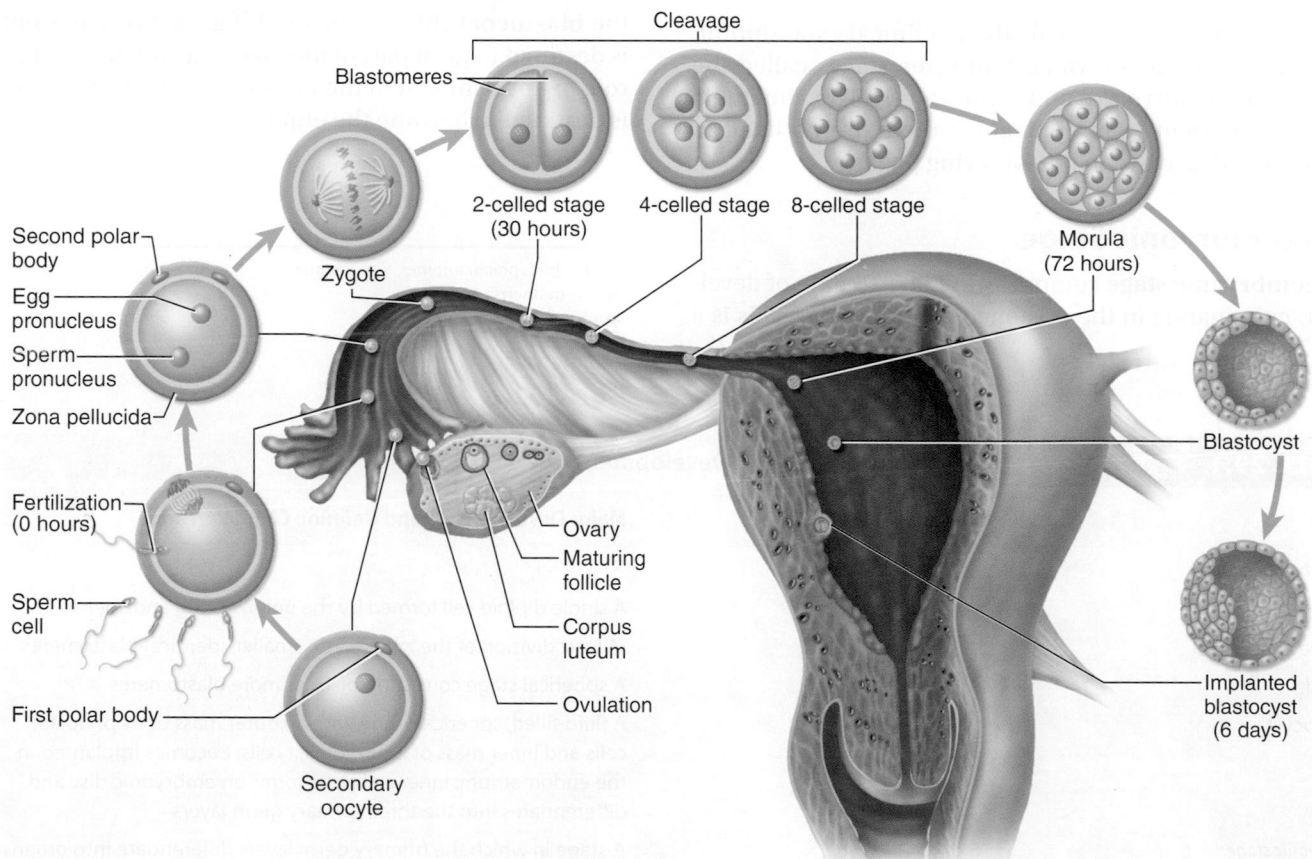

**Cleavage**

Blastomeres

2-celled stage (30 hours)

4-celled stage

8-celled stage

Morula (72 hours)

Zygote

Second polar body

Egg pronucleus

Sperm pronucleus

Zona pellucida

Fertilization (0 hours)

Sperm cell

First polar body

Secondary oocyte

Ovary

Maturing follicle

Corpus luteum

Ovulation

Blastocyst

Implanted blastocyst (6 days)

**FIGURE 29.2 Migration of the Conceptus.** The egg is fertilized in the distal end of the uterine tube, and the preembryo begins cleavage as it migrates to the uterus.

 *Why can't the egg be fertilized in the uterus?*

## Major Stages of Prenatal Development

Clinically, the course of a pregnancy is divided into 3-month intervals called **trimesters:**

1. The **first trimester** extends from fertilization through the first 12 weeks. This is the most precarious stage, for more than half of all embryos die then. The conceptus is most vulnerable to stress, drugs, and nutritional deficiencies during this time.

2. The **second trimester** (weeks 13 through 24) is a period in which the organs complete most of their development. It becomes possible with sonography to see good anatomical detail in the fetus. By the end of this trimester, the fetus looks distinctly human, and with intensive care, infants born at the end of the second trimester have a chance of survival.

3. In the **third trimester** (week 25 to birth), the fetus grows rapidly and the organs achieve enough cellular differentiation to support life outside the womb. Some organs, such as the brain, liver, and kidneys, however, require further differentiation after birth to become fully functional. At 35 weeks from fertilization, the fetus typically weighs about 2.5 kg (5.5 lb). It is considered mature at this weight, and usually survives if born early. Most twins are born at about 35 weeks' gestation and solitary infants at 40 weeks.

From a more biological than clinical standpoint, human development is divided into three stages called the *preembryonic, embryonic,* and *fetal stages.* The timetable and landmark events that distinguish them are outlined in table 29.1 and described in the following pages.

### The Preembryonic Stage

The **preembryonic stage** comprises the first 16 days of development, culminating in the existence of an embryo. This is a period in which the zygote divides into hundreds of cells, the cells organize themselves into the primary germ layers, and the conceptus becomes firmly attached to the uterine wall. It can be summarized in three words: *cleavage, implantation,* and *embryogenesis.*

### Cleavage

**Cleavage** refers to mitotic divisions that occur in the first 3 days, while the conceptus migrates down the uterine tube (fig. 29.2). The first cleavage occurs about 30 hours after fertilization and produces the first two daughter cells, or **blastomeres.**[1] These divide simultaneously at shorter and shorter time intervals, doubling the number of blastomeres each time. By the time the conceptus arrives in the uterus, about 72 hours after ovulation, it consists of 16 or more cells and somewhat resembles a mulberry—hence, it is called a **morula.**[2] The morula is no larger than the zygote; cleavage merely produces smaller and smaller blastomeres and a larger number of cells from which to form different embryonic tissues.

The morula lies free in the uterine cavity for 4 to 5 days and divides into 100 cells or so. Meanwhile, the zona pellucida disintegrates and releases the conceptus, which is now at a stage called the **blastocyst**—a hollow sphere with an outer layer of squamous cells called the **trophoblast,**[3] an inner cell mass called the **embryoblast,** and an internal cavity called the **blastocoel** (BLAST-oh-seal) (fig. 29.4a). The trophoblast is destined to form part of the placenta and play an important role in nourishment of the embryo, whereas the embryoblast is destined to become the embryo itself.

---

[1]*blast* = bud, precursor; *mer* = segment, part
[2]*mor* = mulberry; *ula* = little
[3]*troph* = food, nourishment

| TABLE 29.1 | The Stages of Prenatal Development | |
|---|---|---|
| **Stage** | **Age\*** | **Major Developments and Defining Characteristics** |
| *Preembryonic stage* | | |
| Zygote | 0–30 hours | A single diploid cell formed by the union of egg and sperm |
| Cleavage | 30–72 hours | Mitotic division of the zygote into smaller, identical blastomeres |
| Morula | 3–4 days | A spherical stage consisting of 16 or more blastomeres |
| Blastocyst | 4–16 days | A fluid-filled, spherical stage with an outer mass of trophoblast cells and inner mass of embryoblast cells; becomes implanted in the endometrium; inner cell mass forms an embryonic disc and differentiates into the three primary germ layers |
| *Embryonic stage* | 16 days–8 weeks | A stage in which the primary germ layers differentiate into organs and organ systems; ends when all organ systems are present |
| *Fetal stage* | 8–38 weeks | A stage in which organs grow and mature at a cellular level to the point of being capable of supporting life independently of the mother |

\*From the time of fertilization

# DEEPER INSIGHT 29.1
## CLINICAL APPLICATION

### Twins

There are two ways in which twins are produced (and, by extension, other multiple births). About two-thirds of twins are *dizygotic (DZ)*—produced when two eggs are ovulated and fertilized by separate sperm. They are no more or less genetically similar than any other siblings and may be of different sexes. Multiple ovulation can also result in triplets, quadruplets, or even greater numbers of offspring. DZ twins implant separately on the uterine wall and each forms its own placenta, although their placentas may fuse if they implant close together (fig. 29.3).

*Monozygotic (MZ)* twins are produced when a single egg is fertilized and the cell mass (embryoblast) later divides into two. MZ twins are genetically identical, or nearly so, and are therefore of the same sex and nearly identical appearance. In most cases, they share the same placenta. Identical triplets and quadruplets occasionally result from the splitting of a single embryoblast.

Reproductive biologists are beginning to question whether MZ twins are truly genetically identical. They have suggested that blastomeres may undergo mutation in the course of DNA replication, and the splitting of the embryoblast may represent an attempt of each cell mass to reject the other one as genetically different and seemingly foreign.

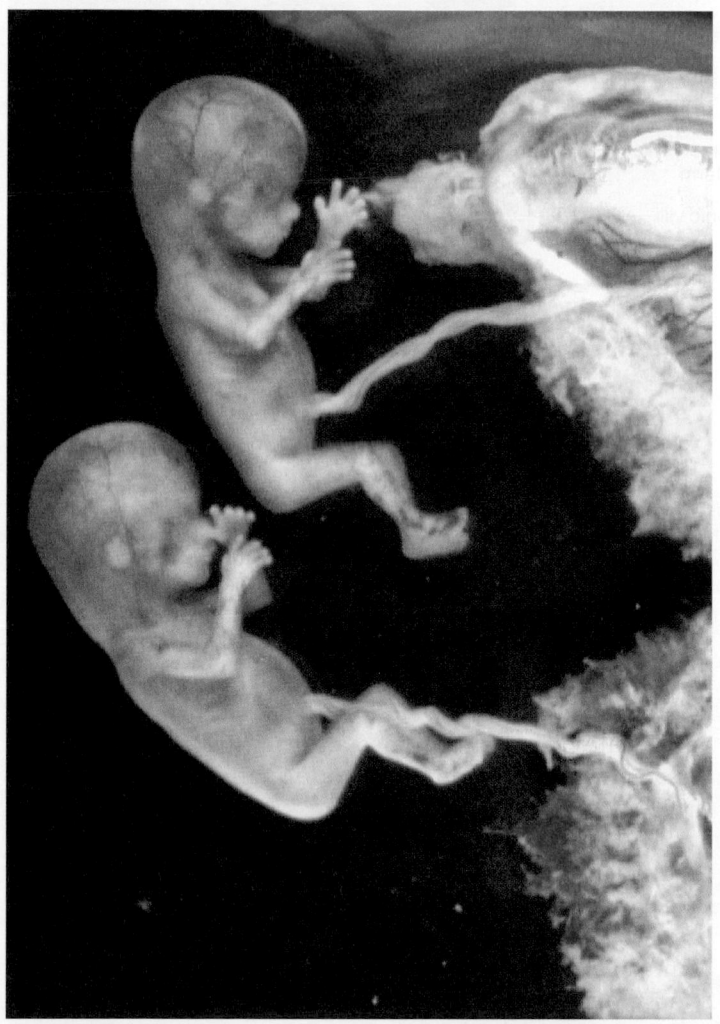

**FIGURE 29.3 Dizygotic Twins with Separate Placentas.**

## Implantation

About 6 days after ovulation, the blastocyst attaches to the endometrium, usually on the fundus or posterior wall of the uterus. The process of attachment, called **implantation,** begins when the blastocyst adheres to the endometrium. The trophoblast cells on this side separate into two layers. In the superficial layer, in contact with the endometrium, the plasma membranes break down and the trophoblast cells fuse into a multinucleate mass called the **syncytiotrophoblast**[4] (sin-SISH-ee-oh-TRO-fo-blast). (A *syncytium* is any body of protoplasm containing multiple nuclei.) The deep layer, close to the embryoblast, is called the **cytotrophoblast** because it retains individual cells divided by membranes (fig. 29.4b).

The syncytiotrophoblast grows into the uterus like little roots, digesting endometrial cells along the way. The endometrium reacts to this injury by growing over the blastocyst and eventually covering it, so the conceptus becomes completely buried in endometrial tissue (fig. 29.4c). Implantation takes about a week and is completed about the time the next menstrual period would have occurred if the woman had not become pregnant.

Another role of the trophoblast is to secrete human chorionic gonadotropin (HCG). HCG stimulates the corpus luteum to secrete estrogen and progesterone, and progesterone suppresses menstruation. The level of HCG in the mother's blood rises until the end of the second month. During this time, the trophoblast develops into a membrane called the *chorion,* which takes over the role of the corpus luteum and makes HCG unnecessary. The ovaries then become inactive for the rest of the pregnancy, but estrogen and progesterone levels rise dramatically as they are secreted by the ever-growing chorion (see fig. 28.18, p. 1081).

## Embryogenesis

During implantation, the embryoblast undergoes **embryogenesis**—arrangement of the blastomeres into the three primary germ layers: *ectoderm, mesoderm,* and *endoderm.* At the beginning of this phase, the embryoblast separates slightly from the trophoblast, creating a narrow space between them called the **amniotic cavity.** The embryoblast flattens into an **embryonic disc** composed initially of two layers: the *epiblast* facing the amniotic cavity and the *hypoblast* facing away. Some hypoblast cells multiply and form a membrane called the *yolk sac* enclosing the blastocoel. Now the embryonic disc is flanked by two spaces: the amniotic cavity on one side and the yolk sac on the other (fig. 29.4c).

Meanwhile, the embryonic disc elongates and, around day 15, a thickened cell layer called the **primitive streak** forms along the midline of the epiblast, with a **primitive groove** running down its middle (fig. 29.5). These events make the embryo bilaterally symmetric and define its future right and left sides, dorsal and ventral surfaces, and cephalic and caudal ends.

[4]*syn* = together; *cyt* = cell

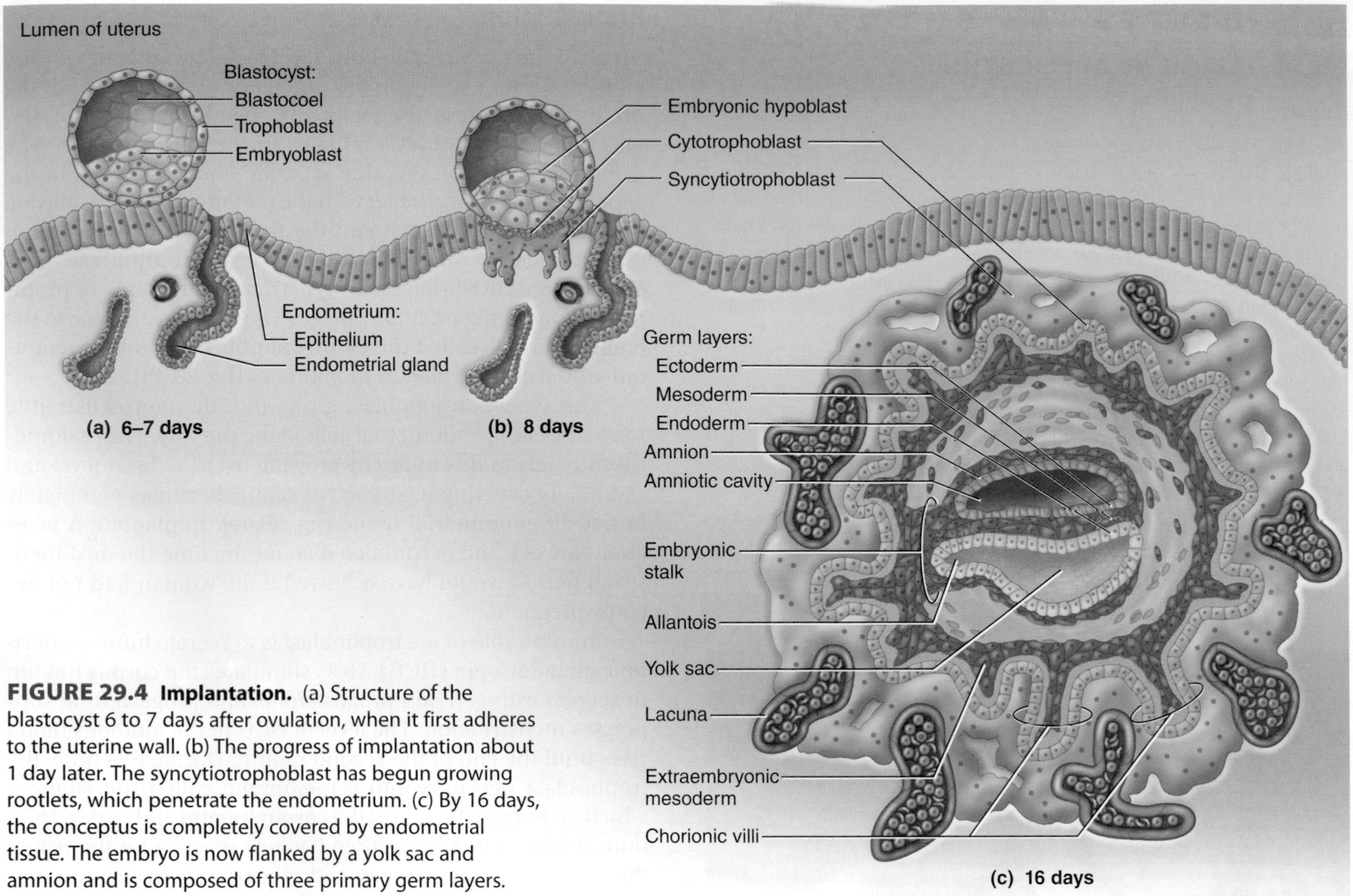

Lumen of uterus

Blastocyst:
— Blastocoel
— Trophoblast
— Embryoblast

Embryonic hypoblast
Cytotrophoblast
Syncytiotrophoblast

Endometrium:
— Epithelium
— Endometrial gland

**(a) 6–7 days**

**(b) 8 days**

Germ layers:
— Ectoderm
— Mesoderm
— Endoderm
Amnion
Amniotic cavity

Embryonic
stalk

Allantois

Yolk sac

Lacuna

Extraembryonic
mesoderm

Chorionic villi

**(c) 16 days**

**FIGURE 29.4 Implantation.** (a) Structure of the blastocyst 6 to 7 days after ovulation, when it first adheres to the uterine wall. (b) The progress of implantation about 1 day later. The syncytiotrophoblast has begun growing rootlets, which penetrate the endometrium. (c) By 16 days, the conceptus is completely covered by endometrial tissue. The embryo is now flanked by a yolk sac and amnion and is composed of three primary germ layers.

## DEEPER INSIGHT 29.2

### CLINICAL APPLICATION

#### Ectopic Pregnancy

In about 1 out of 300 pregnancies, the blastocyst implants somewhere other than the uterus, producing an *ectopic*[5] *pregnancy*. Most cases are *tubal pregnancies,* implantation in the uterine tube. This usually occurs because the conceptus encounters a constriction resulting from such causes as earlier pelvic inflammatory disease, tubal surgery, previous ectopic pregnancies, or repeated miscarriages. The uterine tube cannot expand enough to accommodate the growing conceptus for long; if the situation is not detected and treated early, the tube usually ruptures within 12 weeks, potentially with fatal hemorrhaging. Occasionally, a conceptus implants in the abdominopelvic cavity, producing an *abdominal pregnancy*. It can grow anywhere it finds an adequate blood supply—for example, on the broad ligament or the outside of the uterus, colon, or bladder. About 1 pregnancy in 7,000 is abdominal. This is a serious threat to the mother's life and usually requires abortion, but about 9% of abdominal pregnancies result in live birth by cesarean section.

The next step is **gastrulation**—multiplying epiblast cells migrate medially toward the primitive groove and down into it. They replace the original hypoblast with a layer now called **endoderm,** which will become the inner lining of the digestive tract among other things. A day later, migrating epiblast cells form a third layer between the first two, called **mesoderm.** Once this is formed, the remaining epiblast is called **ectoderm.** Thus, all three primary germ layers arise from the original epiblast. Some mesoderm overflows the embryonic disc and becomes an extensive *extraembryonic mesoderm,* which contributes to formation of the placenta (fig. 29.4c).

The ectoderm and endoderm are epithelia composed of tightly joined cells, but the mesoderm is a more loosely organized tissue. It later differentiates into a loose fetal connective tissue called **mesenchyme,** which gives rise to such tissues as muscle, bone, and blood. Mesenchyme is composed of a loose network of wispy *mesenchymal cells* embedded in a gelatinous ground substance.

[5]*ec* = outside; *top* = place

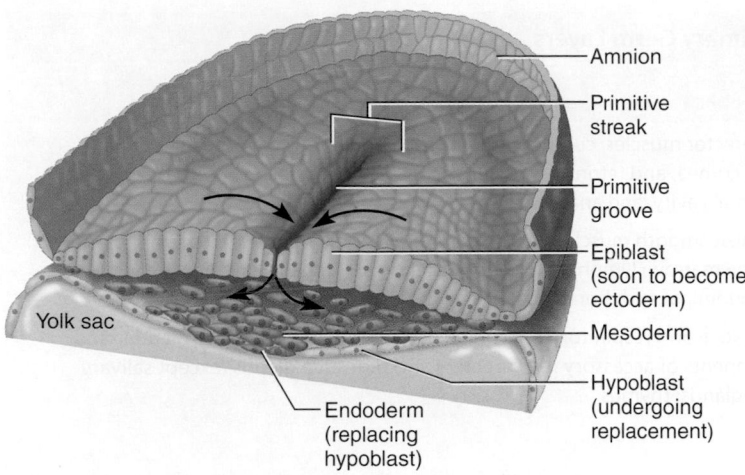

**FIGURE 29.5 Formation of the Primary Germ Layers (Gastrulation).** Composite view of the embryonic disc at 15 to 16 days. Epiblast cells migrate over the surface and down into the primitive groove, first replacing the hypoblast cells with endoderm, then filling the space with mesoderm. Upon completion of this process, the uppermost layer is considered ectoderm.

Once the three primary germ layers are formed, embryogenesis is complete and the individual is considered an embryo. It is about 2 mm long and 16 days old at this point.

> **BEFORE YOU GO ON**
>
> Answer the following questions to test your understanding of the preceding section:
>
> 1. How soon can a sperm reach an egg after ejaculation? How soon can it fertilize an egg? What accounts for the difference?
>
> 2. Describe two ways a fertilized egg prevents the entry of excess sperm.
>
> 3. In the blastocyst, what are the cells called that eventually give rise to the embryo? What are the cells that carry out implantation?
>
> 4. What major characteristic distinguishes an embryo from a preembryo?

## 29.2 The Embryonic and Fetal Stages

### Expected Learning Outcomes

When you have completed this section, you should be able to

a. describe the formation and functions of the placenta;

b. explain how the conceptus is nourished before the placenta takes over this function;

c. describe the embryonic membranes and their functions;

d. identify the major tissues derived from the primary germ layers;

e. describe the major events of fetal development; and

f. describe the fetal circulatory system.

Sixteen days after conception, the germ layers are present and the **embryonic stage** of development begins. Over the next 6 weeks, a placenta forms on the uterine wall and becomes the embryo's primary means of nutrition, while the germ layers differentiate into organs and organ systems—a process called **organogenesis** (table 29.2). Although these organs are still far from functional, it is their presence at 8 weeks that marks the transition from the embryonic stage to the fetal stage. In the following pages, we will examine the transformation from embryo to fetus, how the membranes collectively known as the "afterbirth" develop around the fetus, and how the conceptus is nourished throughout its gestation.

### Embryonic Folding and Organogenesis

One of the major transformations to occur in the embryonic stage is conversion of the flat embryonic disc of figure 29.4c into a somewhat cylindrical form. This occurs during week 4 as the embryo rapidly grows and folds around the yolk sac (fig. 29.6). As the cephalic and caudal ends curve around the ends of the yolk sac, the embryo becomes C-shaped, with the head and tail almost touching. At the same time, the lateral margins of the disc fold around the sides of the yolk sac to form the ventral surface of the embryo. This lateral folding encloses a longitudinal channel, the *primitive gut,* which later becomes the digestive tract.

As a result of embryonic folding, the entire surface is covered with ectoderm, which later produces the epidermis of the skin. In the meantime, the mesoderm splits into two layers. One of them adheres to the ectoderm and the other to the endoderm, opening a body cavity between them called the **coelom** (SEE-loam) (fig. 29.6c). The coelom divides into the thoracic cavity and peritoneal cavity separated by a wall, the diaphragm. By the end of week 5, the thoracic cavity further subdivides into pleural and pericardial cavities.

Two more especially significant events in organogenesis are the appearance of a **neural tube** (fig. 29.6b), which will later become the brain and spinal cord, and segmentation of the mesoderm into blocks of tissue called **somites** (see fig. 29.7a, b), which will give rise to the vertebral column, trunk muscles, and dermis of the skin.

We cannot delve at greater length into development of all the organ systems, but this description is at least enough to see how some of them begin to form. Some have also been described in earlier chapters. Some of the highlights of prenatal development through the end of gestation are summarized

| TABLE 29.2 | Derivatives of the Three Primary Germ Layers |
|---|---|
| **Germ Layer** | **Major Derivatives** |
| Ectoderm | Epidermis; hair follicles and piloerector muscles; cutaneous glands; nervous system; adrenal medulla; pineal and pituitary glands; lens, cornea, and intrinsic muscles of the eye; internal and external ear; salivary glands; epithelia of nasal cavity, oral cavity, and anal canal |
| Mesoderm | Skeleton; skeletal, cardiac, and most smooth muscle; cartilage; adrenal cortex; middle ear; dermis; blood; blood and lymphatic vessels; bone marrow; lymphoid tissue; epithelium of kidneys, ureters, gonads, and genital ducts; mesothelium of abdominal and thoracic cavities |
| Endoderm | Most mucosal epithelium of digestive and respiratory tracts; mucosal epithelium of urinary bladder and parts of urethra; epithelial components of accessory reproductive and digestive glands (except salivary glands); thyroid and parathyroid glands; thymus |

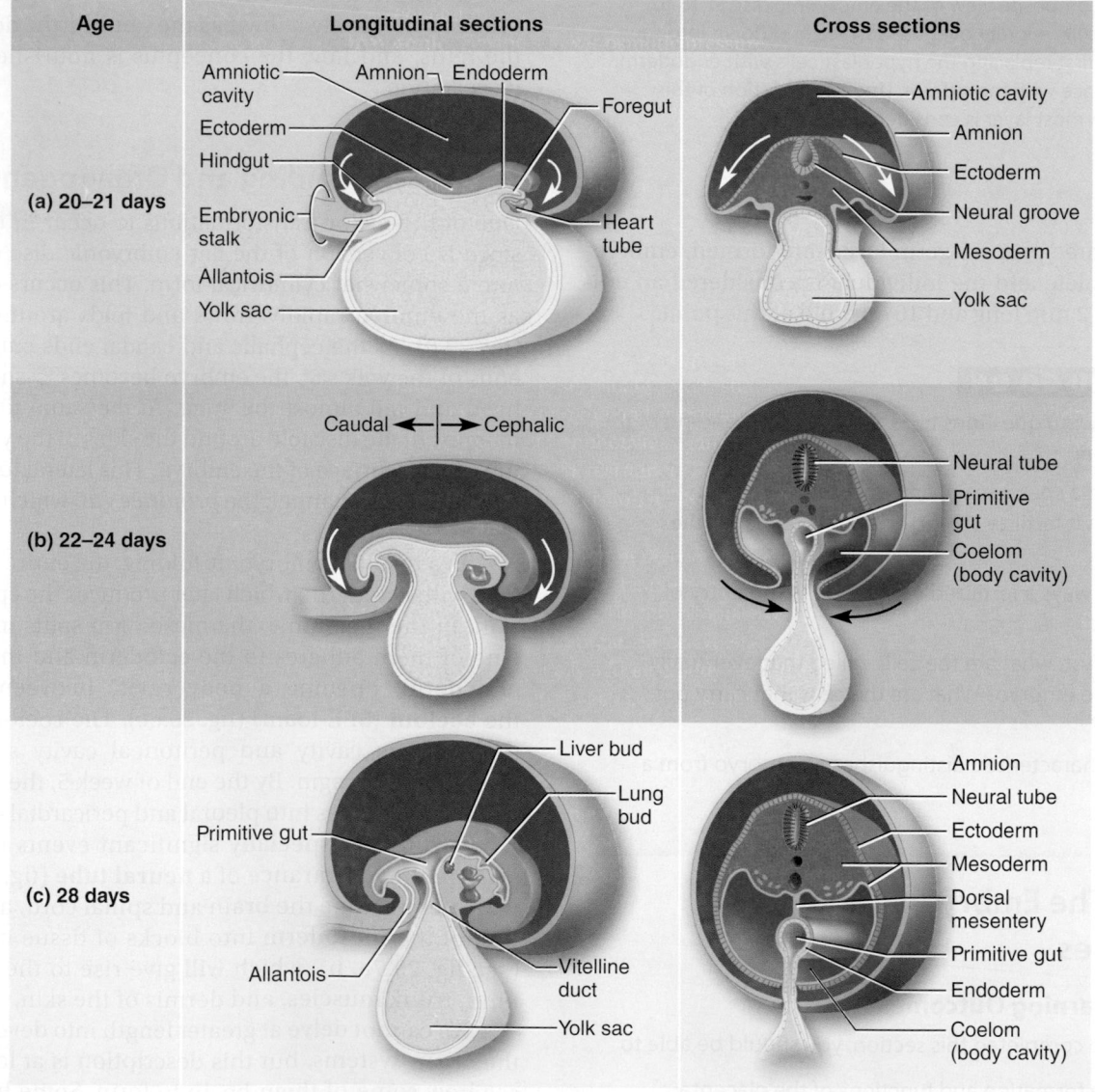

**FIGURE 29.6 Embryonic Folding.** The right-hand figures are cross sections cut about midway along the figures on the left. Part (a) corresponds to figure 29.4c at a slightly later stage of development. Note the general trend for the cephalic and caudal (head and tail) ends of the embryo to curl toward each other (left-hand figures) until the embryo assumes a C shape, and for the flanks of the embryo to fold laterally (right-hand figures), converting the flat embryonic disc into a more cylindrical body and eventually enclosing a body cavity (c).

in table 29.3. Prenatal growth is charted by weight and body length. Body length is customarily measured from the crown of the head to the curve of the buttocks *(crown-to-rump length, CRL),* thus excluding the lower limbs. Figure 29.7 illustrates some key features of embryonic and fetal development. In the following pages, we will see how these features unfold.

▶▶▶**APPLY WHAT YOU KNOW**

*List the four primary tissue types of the adult body (see chapter 5) and identify which of the three primary germ layers of the embryo predominantly gives rise to each.*

## Embryonic Membranes

Several accessory organs develop alongside the embryo: a *placenta; umbilical cord;* and four embryonic membranes called the *amnion, yolk sac, allantois,* and *chorion* (figs. 29.6 and 29.8).

To understand these membranes, it helps to realize that all mammals evolved from egg-laying reptiles. Within the shelled, self-contained egg of a reptile, the embryo rests atop a yolk, which is enclosed in the yolk sac; it floats in a little sea of liquid contained in the amnion; it stores its toxic wastes in the allantois; and to breathe, it has a chorion permeable to gases. All of these membranes persist in mammals, including humans, but are modified in their functions.

The **amnion** is a transparent sac that develops from cells of the epiblast (see figs. 29.4c and 29.5). It grows to completely enclose the embryo and is penetrated only by the umbilical cord. The amnion fills with **amniotic fluid** (see fig. 29.7d), which protects the embryo from trauma, infection, and temperature fluctuations; allows the freedom of movement important to muscle development; enables the embryo to develop symmetrically; prevents body parts from adhering to each other, such as an arm to the trunk; and stimulates lung development as the fetus "breathes" the fluid. At first, the amniotic

| TABLE 29.3 | | Major Events of Prenatal Development, with Emphasis on the Fetal Stage |
|---|---|---|
| **End of Week** | **Crown-to-Rump Length; Weight** | **Developmental Events** |
| 4 | 0.6 cm; <1 g | Vertebral column and central nervous system begin to form; limbs represented by small limb buds; heart begins beating around day 22; no visible eyes, nose, or ears |
| 8 | 3 cm; 1 g | Eyes form, eyelids fused shut; nose flat, nostrils evident but plugged with mucus; head nearly as large as the rest of the body; brain waves detectable; bone calcification begins; limb buds form paddlelike hands and feet with ridges called **digital rays,** which then separate into distinct fingers and toes; blood cells and major blood vessels form; genitals present but sexes not yet distinguishable |
| 12 | 9 cm; 45 g | Eyes well developed, facing laterally; eyelids still fused; nose develops bridge; external ears present; limbs well formed, digits exhibit nails; fetus swallows amniotic fluid and produces urine; fetus moves, but too weakly for mother to feel it; liver is prominent and produces bile; palate is fusing; sexes can be distinguished |
| 16 | 14 cm; 200 g | Eyes face anteriorly, external ears stand out from head, face looks more distinctly human; body larger in proportion to head; skin is bright pink, scalp has hair; joints forming; lips exhibit sucking movements; kidneys well formed; digestive glands forming and **meconium**[6] (fetal feces) accumulating in intestine; heartbeat can be heard with a stethoscope |
| 20 | 19 cm; 460 g | Body covered with fine hair called **lanugo**[7] and cheeselike sebaceous secretion called **vernix caseosa,**[8] which protects it from amniotic fluid; skin bright pink; brown fat forms and will be used for postpartum heat production; fetus is now bent forward into "fetal position" because of crowding; **quickening** occurs—mother can feel fetal movements |
| 24 | 23 cm; 820 g | Eyes partially open; skin wrinkled, pink, and translucent; lungs begin producing surfactant; rapid weight gain |
| 28 | 27 cm; 1,300 g | Eyes fully open; skin wrinkled and red; full head of hair present; eyelashes formed; fetus turns into upside-down **vertex position;** testes begin to descend into scrotum; marginally viable if born at 28 weeks |
| 32 | 30 cm; 2,100 g | Subcutaneous fat deposition gives fetus a more plump, babyish appearance, with lighter, less wrinkled skin; testes descending; twins usually born at this stage |
| 36 | 34 cm; 2,900 g | More subcutaneous fat deposited, body plump; lanugo is shed; nails extend to fingertips; limbs flexed; firm hand grip |
| 38 | 36 cm; 3,400 g | Prominent chest, protruding breasts; testes in inguinal canal or scrotum; fingernails extend beyond fingertips |

[6]*mecon* = poppy juice, opium; refers to an appearance similar to black tar opium
[7]*lan* = down, wool
[8]*vernix* = varnish; *caseo* = cheese

**FIGURE 29.7  The Developing Human.**  Parts (a) through (d) show development through the end of the embryonic stage. Parts (e) and (f) represent the fetal stage of development.

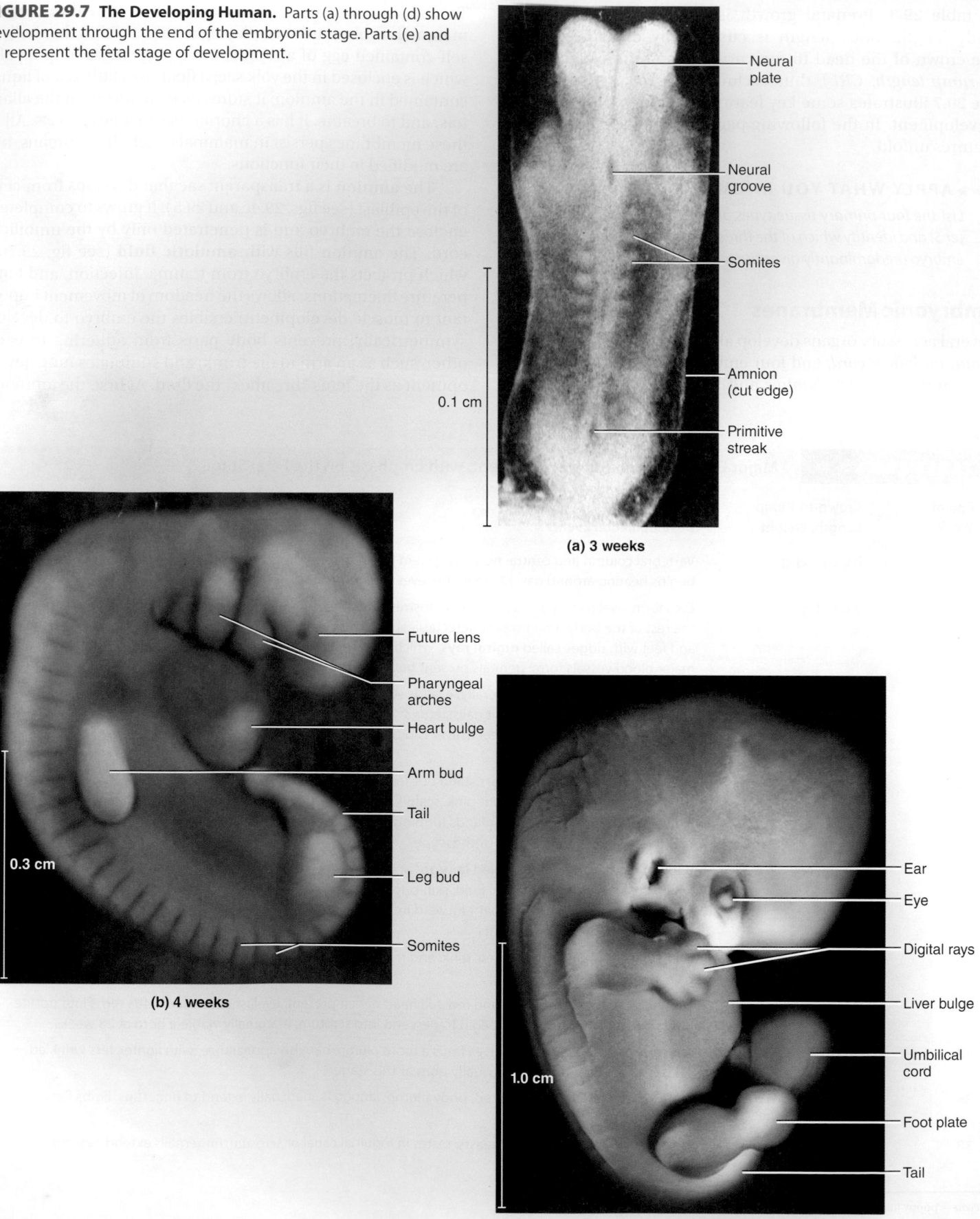

(a) 3 weeks

Neural plate

Neural groove

Somites

Amnion (cut edge)

Primitive streak

0.1 cm

(b) 4 weeks

Future lens

Pharyngeal arches

Heart bulge

Arm bud

Tail

Leg bud

Somites

0.3 cm

(c) 7 weeks

Ear

Eye

Digital rays

Liver bulge

Umbilical cord

Foot plate

Tail

1.0 cm

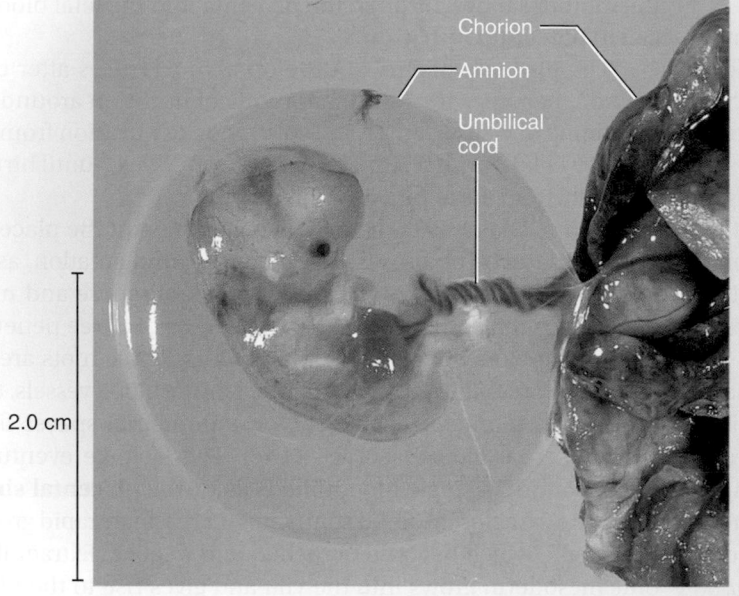

Chorion
Amnion
Umbilical cord

2.0 cm

**(d) 8 weeks**

2.0 cm

**(e) 12 weeks**

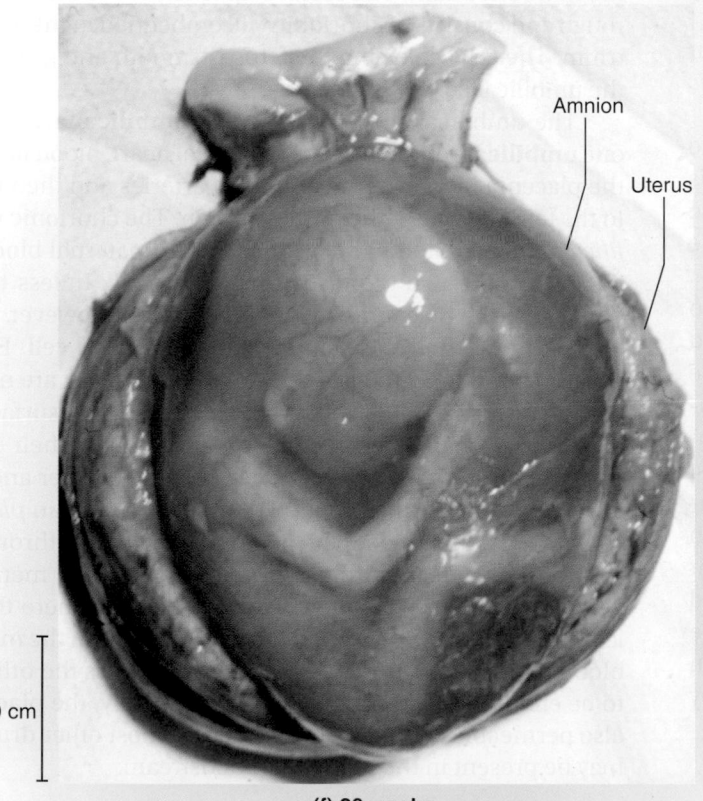

Amnion
Uterus

5.0 cm

**(f) 20 weeks**

fluid forms by filtration of the mother's blood plasma, but beginning at 8 to 9 weeks, the fetus urinates into the amniotic cavity about once an hour and contributes substantially to the fluid volume. The volume grows slowly, however, because the fetus swallows amniotic fluid at a comparable rate. At term, the amnion contains 700 to 1,000 mL of fluid.

The **yolk sac** arises from hypoblast cells opposite the amnion. It is a small sac suspended from the ventral side of the embryo. It contributes to the formation of the digestive tract and produces the first blood cells and forerunners of the future egg or sperm cells.

The **allantois** (ah-LON-toe-iss) begins as an outpocketing of the yolk sac (see fig. 29.4c), but eventually becomes an outgrowth of the caudal end of the gut. It forms the foundation for the umbilical cord and becomes part of the urinary bladder. It can be seen in cross sections cut near the fetal end of a mature umbilical cord.

The **chorion** is the outermost membrane, enclosing all the rest of the membranes and the embryo (fig. 29.8). Initially, it has shaggy outgrowths called **chorionic villi** around its entire surface, but as the pregnancy advances, the villi of the placental region grow and branch while the rest of them degenerate. At the placental attachment, the chorion is then called the *villous chorion,* and the rest is called the *smooth chorion.* The villous chorion forms the fetal portion of the placenta, discussed shortly.

## Prenatal Nutrition

Over the course of gestation, the conceptus is nourished in three different, overlapping ways: by *uterine milk, trophoblastic nutrition,* and *placental nutrition.*

**Uterine milk** is a glycogen-rich secretion of the uterine tubes and endometrial glands. The conceptus absorbs this fluid as it travels down the tube and lies free in the uterine cavity before implantation. The accumulating fluid forms the blastocoel in figure 29.4a.

As it implants, the conceptus makes a transition to **trophoblastic nutrition,** in which it consumes so-called **decidual**[9] **cells** of the endometrium. Progesterone from the corpus luteum stimulates these cells to proliferate and accumulate a store of glycogen, proteins, and lipids. As the conceptus burrows into the endometrium, the syncytiotrophoblast digests the decidual cells and supplies the nutrients to the embryoblast. Trophoblastic nutrition is the only mode of nutrition for the first week after implantation. It remains the dominant source of nutrients through the end of week 8; the period from implantation through week 8 is therefore called the **trophoblastic phase** of the pregnancy. Trophoblastic nutrition wanes as placental nutrition takes over, and ceases entirely by the end of week 12 (fig. 29.9).

The **placenta**[10] is the fetus's life-support system—a disc-shaped organ attached to the uterine wall on one side, and on the other, attached to the fetus by way of the **umbilical cord** (fig. 29.8). It is the fetus's source of oxygen and nutrients, and its means of waste disposal. The diffusion of nutrients from the mother's blood through the placenta into the fetal blood is called **placental nutrition.**

The placenta begins to develop about 11 days after conception, becomes the dominant mode of nutrition around the beginning of week 9, and is the sole mode of nutrition from the end of week 12 until birth. The period from week 9 until birth is called the **placental phase** of the pregnancy.

Figure 29.4 depicts the early development of the placenta, or *placentation.* The process begins during implantation, as extensions of the syncytiotrophoblast penetrate more and more deeply into the endometrium, like the roots of a tree penetrating into the nourishing "soil" of the uterus. These roots are the early chorionic villi. As they penetrate uterine blood vessels, they become surrounded by *lacunae,* or endometrial spaces filled with maternal blood (see fig. 29.4c). The lacunae eventually merge to form a single blood-filled cavity, the **placental sinus.** Exposure to maternal blood stimulates increasingly rapid growth of the villi, which become branched and treelike. Extraembryonic mesoderm grows into the villi and gives rise to the blood vessels that connect to the embryo by way of the umbilical cord.

When fully developed, the placenta is about 20 cm in diameter, 3 cm thick, and weighs about one-sixth as much as the newborn infant. The surface attached to the uterine wall is rough and consists of chorionic villi embedded in the endometrium. The surface facing the fetus is smooth and gives rise to the umbilical cord (fig. 29.8c, d).

The umbilical cord contains two **umbilical arteries** and one **umbilical vein.** Pumped by the fetal heart, blood flows into the placenta by way of the umbilical arteries and then returns to the fetus by way of the umbilical vein. The chorionic villi are *filled with* fetal blood and *surrounded by* maternal blood (see fig. 29.8b); the two bloodstreams do not mix unless there is damage to the placental barrier. The barrier, however, is only 3.5 μm thick—half the diameter of a red blood cell. Early in development, the villi have thick membranes that are not very permeable to nutrients and wastes, and their total surface area is relatively small. As the villi grow and branch, their surface area increases and the membranes become thinner and more permeable. This brings about a dramatic increase in *placental conductivity,* the rate at which substances diffuse through the membrane. Materials diffuse from the side of the membrane where they are more concentrated to the side where they are less so. Therefore, oxygen and nutrients pass from the maternal blood to the fetal blood, while fetal wastes pass the other way to be eliminated by the mother. Unfortunately, the placenta is also permeable to nicotine, alcohol, and most other drugs that may be present in the maternal bloodstream.

---

[9]*decid* = falling off

[10]*placenta* = flat cake

**FIGURE 29.8 The Placenta and Embryonic Membranes.** (a) A 12-week fetus. The placenta is completely formed. (b) A portion of the mature placenta and umbilical cord, showing the relationship between fetal and maternal circulation. (c) The fetal side of the placenta, showing blood vessels, the umbilical cord, and some of the amniotic sac attached to the lower left margin. (d) The maternal (uterine) side, where chorionic villi give the placenta a rougher texture.

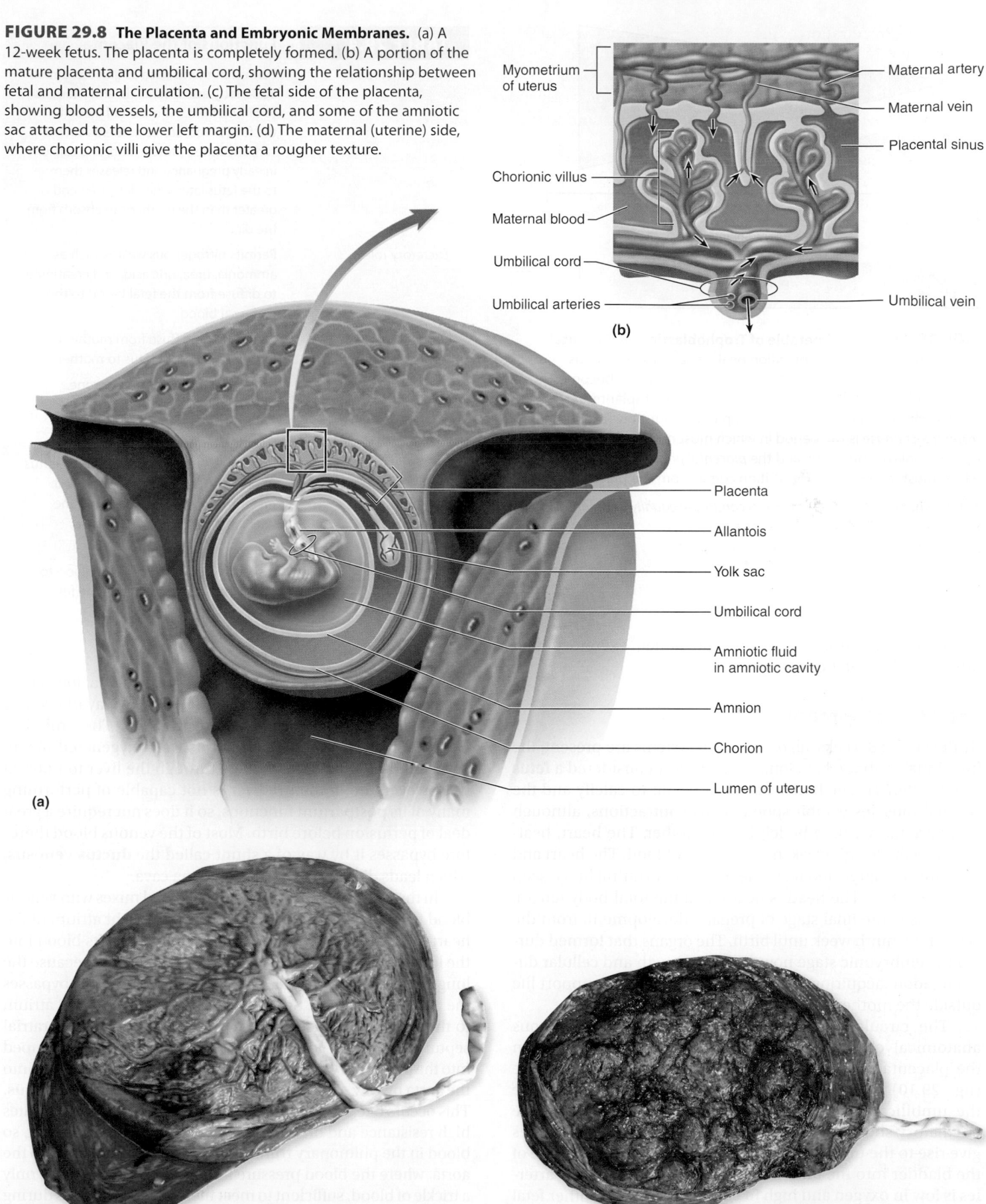

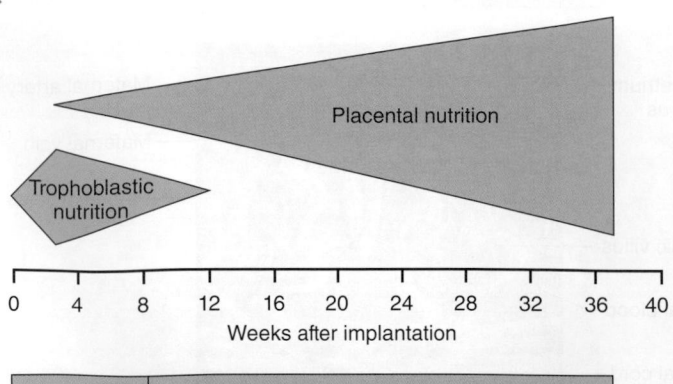

**FIGURE 29.9 The Timetable of Trophoblastic and Placental Nutrition.** Trophoblastic nutrition peaks at 2 weeks and ends by 12 weeks. Placental nutrition begins at 2 weeks and becomes increasingly important until birth, 37 weeks after implantation. The two modes of nutrition overlap up to the eighth week, but the *trophoblast phase* is the period in which most nutrients are supplied by trophoblastic nutrition, and the *placental phase* is the period in which most (eventually all) nutrition comes from the placenta.

❓ *At what point do the two modes contribute equally to prenatal nutrition?*

| TABLE 29.4 | Functions of the Placenta |
|---|---|
| Nutritional roles | Permits nutrients such as glucose, amino acids, fatty acids, minerals, and vitamins to diffuse from the maternal blood to the fetal blood; stores nutrients such as carbohydrates, protein, iron, and calcium in early pregnancy and releases them to the fetus later, when fetal demand is greater than the mother can absorb from the diet |
| Excretory roles | Permits nitrogenous wastes such as ammonia, urea, uric acid, and creatinine to diffuse from the fetal blood to the maternal blood |
| Respiratory roles | Permits $O_2$ to diffuse from mother to fetus and $CO_2$ from fetus to mother |
| Endocrine roles | Secretes estrogens, progesterone, relaxin, human chorionic gonadotropin, and human chorionic somatomammotropin; allows other hormones synthesized by the conceptus to pass into the mother's blood and maternal hormones to pass into the fetal blood |
| Immune roles | Transfers maternal antibodies (especially IgG) into the fetal blood to confer passive immunity on the fetus |

Table 29.4 summarizes the nutritional, excretory, and other functions of the placenta.

## Fetal Development

By the end of 8 weeks, all of the organ systems are present, the individual is about 3 cm long, and it is now considered a **fetus** (see fig. 29.7d). The bones have just begun to calcify and the skeletal muscles exhibit spontaneous contractions, although these are too weak to be felt by the mother. The heart, beating since the fourth week, now circulates blood. The heart and liver are very large and form the prominent ventral bulge seen in figure 29.7c. The head is nearly half the total body length. The fetus is the final stage of prenatal development, from the start of the ninth week until birth. The organs that formed during the embryonic stage now undergo growth and cellular differentiation, acquiring the functional capability to support life outside the mother.

The circulatory system shows the most conspicuous anatomical changes from a prenatal state, dependent on the placenta, to the independent neonatal (newborn) state (fig. 29.10). The unique aspects of fetal circulation are the umbilical–placental circuit and the presence of three circulatory shortcuts called *shunts.* The internal iliac arteries give rise to the umbilical arteries, which pass on either side of the bladder into the umbilical cord. The blood in these arteries is low in oxygen and high in carbon dioxide and other fetal wastes; thus, they are depicted in blue in figure 29.10a. The arterial blood discharges its wastes in the placenta, loads oxygen and nutrients, and returns to the fetus by way of a single umbilical vein, which leads toward the liver. The umbilical vein is depicted in red because of its well-oxygenated blood. Some of this venous blood filters through the liver to nourish it. However, the immature liver is not capable of performing many of its postpartum functions, so it does not require a great deal of perfusion before birth. Most of the venous blood therefore bypasses it by way of a shunt called the **ductus venosus,** which leads directly to the inferior vena cava.

In the inferior vena cava, placental blood mixes with venous blood from the fetus's body and flows to the right atrium of the heart. After birth, the right ventricle pumps all of its blood into the lungs, but there is little need for this in the fetus because the lungs are not yet functional. Therefore, most fetal blood bypasses the pulmonary circuit. Some goes directly from the right atrium to the left through the **foramen ovale,** a hole in the interatrial septum. Some also goes into the right ventricle and is pumped into the pulmonary trunk, but most of this is shunted directly into the aorta by way of a short passage called the **ductus arteriosus.** This occurs because the collapsed state of the fetal lungs creates high resistance and blood pressure in the pulmonary circuit, so blood in the pulmonary trunk flows through the ductus into the aorta, where the blood pressure is lower. The lungs receive only a trickle of blood, sufficient to meet their metabolic needs during development. Blood leaving the left ventricle enters the general

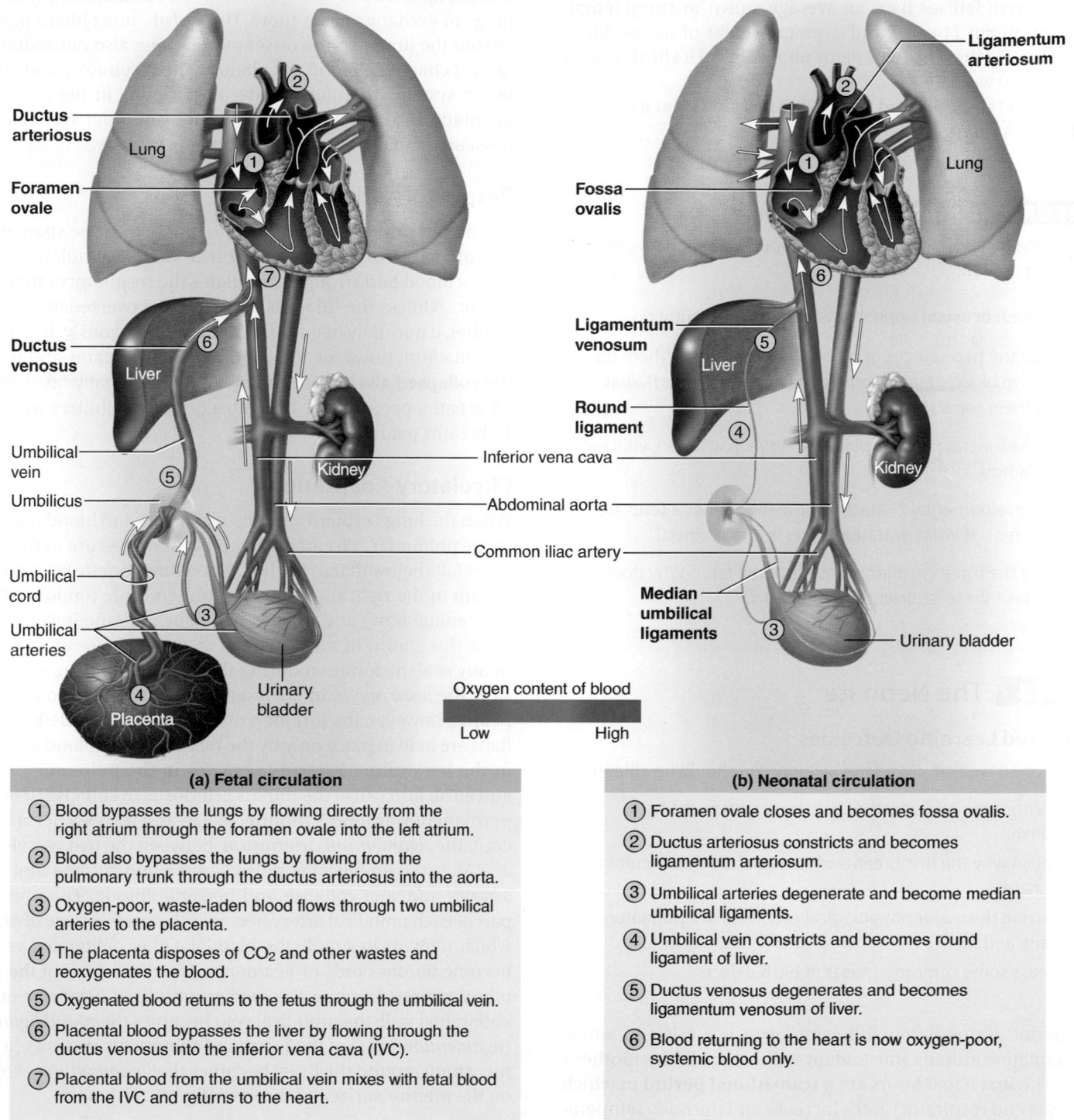

Oxygen content of blood

Low          High

**(a) Fetal circulation**

1. Blood bypasses the lungs by flowing directly from the right atrium through the foramen ovale into the left atrium.

2. Blood also bypasses the lungs by flowing from the pulmonary trunk through the ductus arteriosus into the aorta.

3. Oxygen-poor, waste-laden blood flows through two umbilical arteries to the placenta.

4. The placenta disposes of $CO_2$ and other wastes and reoxygenates the blood.

5. Oxygenated blood returns to the fetus through the umbilical vein.

6. Placental blood bypasses the liver by flowing through the ductus venosus into the inferior vena cava (IVC).

7. Placental blood from the umbilical vein mixes with fetal blood from the IVC and returns to the heart.

**(b) Neonatal circulation**

1. Foramen ovale closes and becomes fossa ovalis.

2. Ductus arteriosus constricts and becomes ligamentum arteriosum.

3. Umbilical arteries degenerate and become median umbilical ligaments.

4. Umbilical vein constricts and becomes round ligament of liver.

5. Ductus venosus degenerates and becomes ligamentum venosum of liver.

6. Blood returning to the heart is now oxygen-poor, systemic blood only.

**FIGURE 29.10 Blood Circulation in the Fetus and Newborn.** Boldface terms in part (a) indicate the three shunts in the fetal circulation, which allow most blood to bypass the liver and lungs. Boldface terms in part (b) indicate the postpartum vestiges of fetal structures.

systemic circulation, and some of this returns to the placenta. This circulatory pattern changes dramatically at birth, when the neonate is cut off from the placenta and the lungs expand with air. Those changes will be described later.

Full-term fetuses have an average crown-to-rump length of about 36 cm (14 in.) and average weight of about 3.0 to 3.4 kg (6.6–7.5 lb). The fetus gains about 50% of its birth weight in the last 10 weeks.

Review table 29.3 and figure 29.7 for additional aspects of fetal development.

<hr>

**BEFORE YOU GO ON**

Answer the following questions to test your understanding of the preceding section:

5. Distinguish between trophoblastic and placental nutrition.

6. Identify the two sources of blood to the placenta. Where do these two bloodstreams come closest to each other? What keeps them separated?

7. State the functions of the placenta, amnion, chorion, yolk sac, and allantois.

8. What developmental characteristic distinguishes a fetus from an embryo? At what gestational age is this attained?

9. Identify the three circulatory shunts of the fetus. Why does the blood take these "shortcuts" before birth?

## **29.3**  The Neonate

### Expected Learning Outcomes

When you have completed this section, you should be able to

a. describe how and why the circulatory system changes at birth;

b. explain why the first breaths of air are relatively difficult for a neonate;

c. describe the major physiological problems of a premature infant; and

d. discuss some common causes of birth defects.

The period immediately following birth is a crisis in which the neonate suddenly must adapt to life outside the mother's body. The first 6 to 8 hours are a **transitional period** in which the heart and respiratory rates increase and the body temperature falls. Physical activity then declines and the baby sleeps for about 3 hours. In its second period of activity, the baby often gags on mucus and debris in the pharynx. The baby then sleeps again, becomes more stable, and begins a cycle of waking every 3 to 4 hours to feed. The first 6 weeks of life constitute the **neonatal period.**

## Adapting to Life Outside the Uterus

The most dramatic and suspenseful event at birth is for the neonate to begin breathing on its own. Breathing, of course, is useless unless most or all of the blood circulates through the lungs to exchange gases there. Until birth, most blood has bypassed the lungs, so the onset of breathing also necessitates a radical change in blood circulation. The respiratory and circulatory systems therefore figure prominently in the neonate's adaptations to life outside the uterus, but other systems also must adapt to this new and challenging lifestyle.

### Respiratory Adaptations

It is an old misconception that a neonate must be spanked to stimulate it to breathe. During birth, $CO_2$ accumulates in the baby's blood and strongly stimulates the respiratory chemoreceptors. Unless the infant is depressed by oversedation of the mother, it normally begins breathing spontaneously. It requires a great effort, however, to take the first few breaths and inflate the collapsed alveoli. For the first 2 weeks, a baby takes about 45 breaths per minute, but subsequently stabilizes at about 12 breaths per minute.

### Circulatory Adaptations

When the lungs expand with air, resistance and blood pressure in the pulmonary circuit drop rapidly and pressure in the right heart falls below that in the left. Blood flows briefly from the left atrium to the right through the foramen ovale (opposite from its prenatal flow) and pushes two flaps of tissue into place to close this shunt. In most people, these flaps fuse and permanently seal the foramen during the first year, leaving a depression, the *fossa ovalis,* in the interatrial septum. In about 25% of people, however, the foramen ovale remains unsealed and the flaps are held in place only by the relatively high blood pressure in the left atrium. Pressure changes in the pulmonary trunk and aorta also cause the ductus arteriosus to collapse. It closes permanently around 3 months of age and leaves a permanent cord, the *ligamentum arteriosum,* between the two vessels.

After the umbilical cord is clamped and cut, the umbilical arteries and vein collapse and become fibrotic. The proximal part of each umbilical artery becomes the *superior vesical artery,* which remains to supply the bladder. Other obliterated vessels become fibrous cords or ligaments: The distal parts of the umbilical arteries become the *median umbilical ligaments* of the abdominal wall; the umbilical vein becomes the *round ligament (ligamentum teres)* of the liver; and the ductus venosus (a former shunt around the liver) becomes the *ligamentum venosum* on the inferior surface of the liver (fig. 29.10b).

### Immunological Adaptations

Cellular immunity begins to appear early in fetal development, but the immune responses of the neonate are still weak. Fortunately, an infant is born with a near-adult level of IgG acquired from the mother through the placenta. This maternal IgG breaks down rapidly after birth, declining to about half the initial level in the first month and to essentially none by 10 months. Nevertheless, maternal IgG levels remain high enough for 6 months to protect the infant from measles, diphtheria, polio, and most other infectious diseases (but not whooping cough). By 6 months, the infant's own IgG reaches about half the typical adult level. The lowest total (maternal + infant) level of IgG exists around 5 to 6 months of age, and respiratory infections are especially common at that age. A breast-fed neonate also acquires protection from gastroenteritis from the IgA present in the colostrum.

## Other Adaptations

Thermoregulation and fluid balance are also critical aspects of neonatal physiology. An infant has a larger ratio of surface area to volume than an adult does, so it loses heat more easily. One of its defenses against hypothermia is brown fat, a special adipose tissue deposited from weeks 17 to 20 of fetal development. The mitochondria of brown fat release all the energy of pyruvic acid as heat rather than using it to make ATP; thus, this is a heat-generating tissue. As a baby grows, its metabolic rate increases and it accumulates even more subcutaneous fat, thus producing and retaining more heat. Nevertheless, body temperature is more variable in infants and children than in adults.

The kidneys are not fully developed at birth and cannot concentrate the urine as much as a mature kidney can. Consequently, infants have a relatively high rate of water loss and require more fluid intake, relative to body weight, than adults do.

In addition, the liver is still not fully functional at birth, most joints are not yet ossified, and myelination of the nervous system is not completed until adolescence. Indeed, humans are born in a very immature state compared with other mammals—a fact necessitated by the narrow outlet of the female pelvis, which in turn was a product of the evolution of bipedal locomotion.

## Premature Infants

Neonates weighing under 2.5 kg (5.5 lb) are generally considered **premature.** They have multiple difficulties in respiration, thermoregulation, excretion, digestion, and liver function. Most neonates weighing 1.5 to 2.5 kg are viable, but with difficulty. Those weighing under 500 g rarely survive.

The respiratory system is adequately developed by 7 months of gestation to support independent life. Infants born before this have a deficiency of pulmonary surfactant, causing **infant respiratory distress syndrome (IRDS),** also called *hyaline membrane disease.* The alveoli collapse each time the infant exhales, and a great effort is needed to reinflate them. The infant becomes very fatigued by the high energy demand of breathing. IRDS may be treated by ventilating the lungs with oxygen-enriched air at a positive pressure to keep the lungs inflated between breaths, and by administering surfactant as an inhalant. Nevertheless, IRDS remains the most common cause of neonatal death.

## DEEPER INSIGHT 29.3

### CLINICAL APPLICATION

#### Neonatal Assessment

A newborn infant is immediately evaluated for general appearance, vital signs (temperature, pulse, and respiratory rate), weight, length, and head circumference and other dimensions; and it is screened for congenital disorders such as phenylketonuria (PKU). At 1 minute and 5 minutes after birth, the heart rate, respiratory effort, muscle tone, reflexes, and skin color are noted and given a score of 0 (poor), 1, or 2 (excellent). The total (0–10), called the *Apgar*[11] *score,* is a good predictor of infant survival. Infants with low Apgar scores may have neurological damage and need immediate attention if they are to survive. A low score at 1 minute suggests asphyxiation and may demand assisted ventilation. A low score at 5 minutes indicates a high probability of death.

A premature infant has an incompletely developed hypothalamus and therefore cannot thermoregulate effectively. Body temperature must be controlled by placing the infant in a warmer.

It is difficult for premature infants to ingest milk because of their small stomach volume and undeveloped sucking and swallowing reflexes. Some must be fed by nasogastric or nasoduodenal tubes. Most of them, however, can tolerate human milk or formulas. Infants under 1.5 kg (3.3 lb) require nutritional supplements of calcium, phosphorus, and protein.

The liver is also poorly developed, and bearing in mind its very diverse functions (see table 26.6, p. 1016), you can probably understand why this would have several serious consequences. The liver synthesizes inadequate amounts of albumin, so the baby suffers hypoproteinemia. This upsets the balance between capillary filtration and reabsorption and leads to edema. The infant bleeds easily because of a deficiency of the clotting factors synthesized by the liver. This is true to some degree even in full-term infants, however, because the baby's intestines are not yet colonized by the bacteria that synthesize vitamin K, which is essential for the synthesis of clotting factors. Vitamin K injections are now routine (but somewhat controversial) for newborns in the United States. Jaundice is common in neonates, especially premature babies, because the liver cannot dispose of bile pigments such as bilirubin efficiently. A moderately elevated bilirubin level *(neonatal hyperbilirubinemia)* is normal and desirable in all infants, however. Bilirubin plays a valuable role as an antioxidant until the infant can develop its other antioxidant systems.

## Birth Defects

A birth defect, or **congenital anomaly,**[12] is the abnormal structure or position of an organ at birth, resulting from a defect in prenatal development. The study of birth defects is called

[11]Virginia Apgar (1909–74), American anesthesiologist
[12]*con* = with; *gen* = born; *a* = without; *nomaly* = evenness, regularity

teratology.[13] Birth defects are the most common cause of infant mortality in North America. Not all of them are noticeable at birth; some are detected months to years later. Thus, by the age of 2 years, 6% of children are diagnosed with congenital anomalies, and by age 5 the incidence is 8%. The following sections discuss some known causes of congenital anomalies, but in 50% to 60% of cases, the cause is unknown.

## Teratogens

**Teratogens**[14] are agents that cause anatomical deformities in the fetus. They fall into three major classes: drugs and other chemicals, infectious diseases, and radiation such as X-rays. The effect of a teratogen depends on the genetic susceptibility of the embryo, the dosage of the teratogen, and the time of exposure. Teratogen exposure during the first 2 weeks usually does not cause birth defects, but may cause spontaneous abortion. Teratogens can exert destructive effects at any stage of development, but the period of greatest vulnerability is weeks 3 through 8. Different organs have different critical periods. For example, limb abnormalities are most likely to result from teratogen exposure at 24 to 36 days, and brain abnormalities from exposure at 3 to 16 weeks.

Perhaps the most notorious teratogenic drug is thalidomide, a sedative first marketed in West Germany in 1957. Thalidomide was taken by many women in early pregnancy to relieve morning sickness, and by others as a sleeping aid even before they knew they were pregnant. By the time it was removed from the market in 1961, it had affected an estimated 10,000 to 20,000 babies worldwide, many of them born with unformed arms or legs (fig. 29.11) and often with defects of the ears, heart, and intestines. The U.S. Food and Drug Administration never approved thalidomide for market, but many American women obtained it by participation in clinical drug trials or from foreign sources. Thalidomide has recently been reintroduced and used under more tightly controlled conditions for more limited purposes such as treating leprosy. People in some Third World countries still take thalidomide in a misguided attempt to treat AIDS and other diseases, resulting in an upswing in severe birth defects. A general lesson to be learned from the thalidomide tragedy and other cases is that pregnant women should avoid all sedatives, barbiturates, and opiates. Even the acne medicine isotretinoin (Accutane) has caused severe birth defects. Many teratogens produce less obvious or delayed effects, including physical or mental retardation, inattention, hyperirritability, strokes, seizures, respiratory arrest, crib death, and cancer.

Alcohol causes more birth defects than any other drug. Even one drink a day can have noticeable effects on fetal and childhood development, some of which are not noticed until a child begins school. Alcohol abuse during pregnancy can cause **fetal alcohol syndrome (FAS),** characterized by a small head, malformed facial features, cardiac and central nervous

**FIGURE 29.11  Schoolboy Showing the Effect of Thalidomide on Upper Limb Development.**

system defects, stunted growth, and behavioral signs such as hyperactivity, nervousness, and a poor attention span. Cigarette smoking also contributes to fetal and infant mortality, ectopic pregnancy, anencephaly (failure of the cerebrum to develop), cleft palate and lip, and cardiac abnormalities. Diagnostic X-rays should be avoided during pregnancy because radiation can have teratogenic effects.

Infectious diseases are largely beyond the scope of this book, but it must be noted at least briefly that several microorganisms can cross the placenta and cause serious congenital anomalies, stillbirth, or neonatal death. Common viral infections of the fetus and newborn include herpes simplex, rubella, cytomegalovirus, and human immunodeficiency virus (HIV). Congenital bacterial infections include gonorrhea and syphilis. *Toxoplasma,* a protozoan contracted from meat, unpasteurized milk, and house cats, is another common cause of fetal deformity. Some of these pathogens have relatively mild effects on adults, but because of its immature immune system, the fetus is vulnerable to devastating effects such as blindness, hydrocephalus, cerebral palsy, seizures, and profound physical and mental retardation. These diseases are treated in greater detail in microbiology textbooks.

## Mutagens and Genetic Anomalies

Genetic anomalies are the most common known cause of birth defects, accounting for an estimated one-third of all cases and 85% of those with an identifiable cause. One cause of genetic defects is **mutations,** or changes in DNA structure. Among other disorders, mutations cause achondroplastic dwarfism (see Deeper Insight 7.2, p. 215), microcephaly (abnormal

---

[13]*terato* = monster; *logy* = study of
[14]*terato* = monster; *gen* = producing

smallness of the head), stillbirth, and childhood cancer. Mutations can occur through errors in DNA replication during the cell cycle or under the influence of environmental agents called **mutagens,** including some chemicals, viruses, and radiation.

Some of the most common genetic disorders result not from mutagens, however, but from the failure of homologous chromosomes to separate during meiosis. Recall that homologous chromosomes pair up during prophase I and normally separate from each other at anaphase I (see p. 1044). This separation, called *disjunction,* produces daughter cells with 23 chromosomes each.

In **nondisjunction,** a pair of chromosomes fails to separate. Both chromosomes go to the same daughter cell, which receives 24 chromosomes while the other cell receives 22.

**Aneuploidy**[15] (AN-you-PLOY-dee), the presence of an extra chromosome or lack of one, accounts for about 50% of spontaneous abortions. The lack of a chromosome, leaving one chromosome without a match, is called **monosomy,** whereas the presence of one extra chromosome, producing a triple set, is called **trisomy.** Aneuploidy can be detected prior to birth by **amniocentesis**, the examination of cells in a sample of amniotic fluid, or by **chorionic villus sampling (CVS),** the biopsy of cells from the chorion.

Figure 29.12 compares normal disjunction of the X chromosomes with some effects of nondisjunction. In nondisjunction, an egg may receive both X chromosomes. If it is fertilized

---

[15]*an* = not, without; *eu* = true, normal; *ploid,* from *diplo* = double, paired

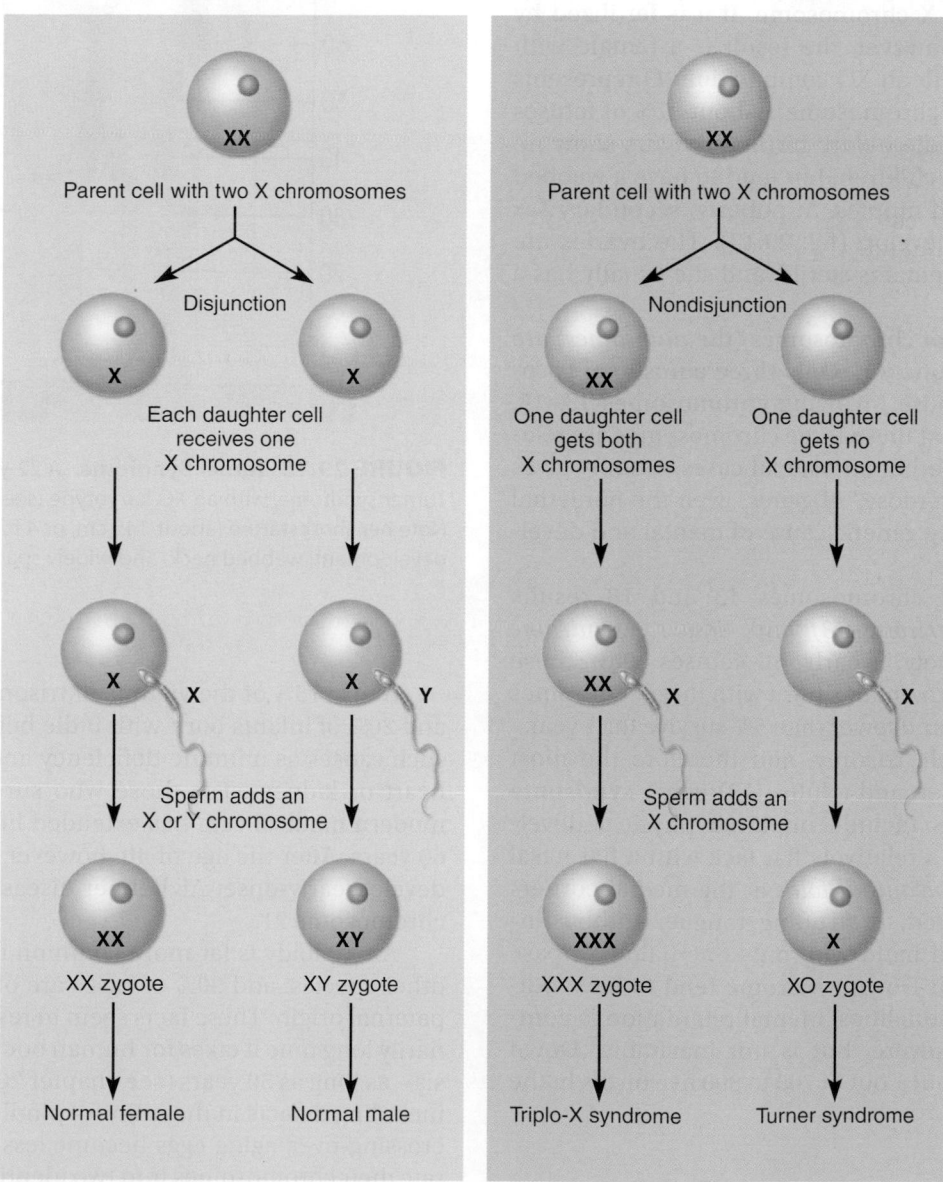

(a) **Normal disjunction of X chromosomes**

(b) **Nondisjunction of X chromosomes**

**FIGURE 29.12 Disjunction and Nondisjunction.** (a) The outcome of normal disjunction and fertilization by X- or Y-bearing sperm. (b) Two of the possible outcomes of nondisjunction followed by fertilization with an X-bearing sperm.

❓ *In the right half of the figure, what would the two outcomes be if the sperm carried a Y chromosome instead of an X?*

by an X-bearing sperm, the result is an XXX zygote and a set of anomalies called the **triplo-X syndrome.** Triplo-X females are sometimes infertile and sometimes have mild intellectual impairments. If an XX egg is fertilized by a Y-bearing sperm, the result is an XXY combination and **Klinefelter**[16] **syndrome.** People with Klinefelter syndrome are sterile males, usually of average intelligence, but with undeveloped testes, sparse body hair, unusually long arms and legs, and enlarged breasts (*gynecomastia*[17]). This syndrome often goes undetected until puberty, when failure to develop the secondary sex characteristics may prompt genetic testing.

The other possible outcome of X chromosome nondisjunction is that an egg may receive no X chromosome (both X chromosomes are discarded in the first polar body). If fertilized by a Y-bearing sperm, it dies for lack of the indispensable genes on the X chromosome. If it is fertilized by an X-bearing sperm, however, the result is a female with **Turner**[18] **syndrome,** with an XO combination (O represents the absence of one sex chromosome). About 97% of fetuses with Turner syndrome die before birth. Survivors show no serious impairments as children, but tend to have a webbed neck and widely spaced nipples. At puberty, secondary sex characteristics fail to develop (fig. 29.13). The ovaries are nearly absent, the girl remains sterile, and she usually has a short stature.

The other 22 pairs of chromosomes (the *autosomes*) are also subject to nondisjunction. Only three autosomal trisomies are survivable: those involving chromosomes 13, 18, and 21. The reason is that these three chromosomes are relatively gene-poor. In all other autosomal cases, trisomy gives the embryo a lethal "overdose" of genes. Even the nonlethal trisomies are the leading genetic cause of mental and developmental abnormalities.

Nondisjunction of chromosomes 13 and 18 results in *Patau syndrome (trisomy-13)* and *Edward syndrome (trisomy-18)*, respectively. Nearly all fetuses with these trisomies die before birth. Infants born with these syndromes are severely deformed, and fewer than 5% survive for 1 year.

The most survivable trisomy, and therefore the most common among children and adults, is **Down**[19] **syndrome (trisomy-21).** Its signs include impaired physical development; short stature; a relatively flat face with a flat nasal bridge; low-set ears; *epicanthal folds* at the medial corners of the eyes; an enlarged, protruding tongue; stubby fingers; and a short broad hand with only one palmar crease (fig. 29.14). People with Down syndrome tend to have outgoing, affectionate personalities. Mental retardation is common and sometimes severe, but is not inevitable. Down syndrome occurs in about 1 out of 700 to 800 live births in the United States.

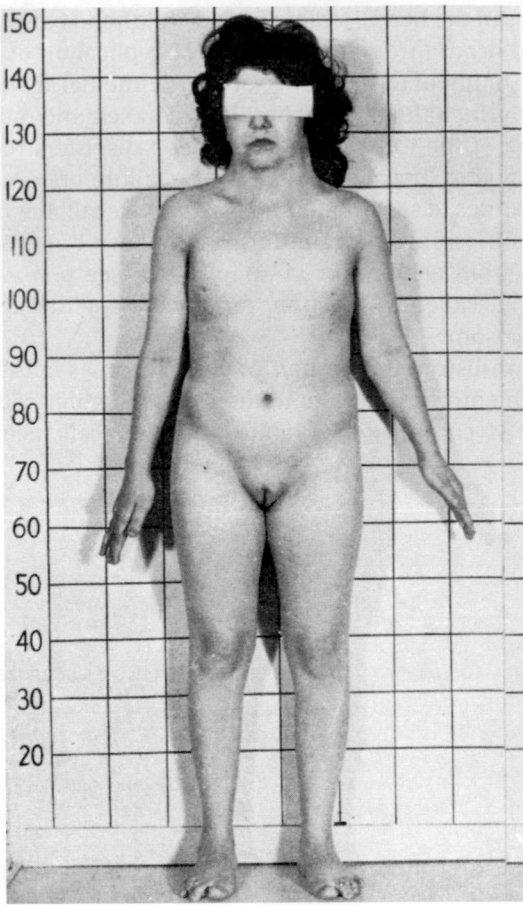

**FIGURE 29.13  Turner Syndrome.** A 22-year-old woman with Turner syndrome, with an XO karyotype (see figure 29.12, far right). Note her short stature (about 145 cm, or 4 ft 9 in.), lack of sexual development, webbed neck, and widely spaced nipples.

About 75% of the victims of trisomy-21 die before birth, and 20% of infants born with it die before the age of 10 from such causes as immune deficiency and abnormalities of the heart or kidneys. For those who survive beyond that age, modern medical care has extended life expectancy to about 60 years. After the age of 40, however, many of these people develop early-onset Alzheimer disease, linked to a gene on chromosome 21.

Aneuploidy is far more common in humans than in any other species, and 90% of cases are of maternal rather than paternal origin. These facts seem to result from the extraordinarily long time it takes for human oocytes to complete meiosis—as long as 50 years (see chapter 28). For various reasons, including defects in the mitotic spindle and in chromosomal crossing-over, aging eggs become less and less able to separate their chromosomes into two identical sets. This is evident in the statistics of Down syndrome: The chance of having a child with Down syndrome is about 1 in 3,000 for a woman under 30, 1 in 365 by age 35, and 1 in 9 by age 48.

---

[16]Harry F. Klinefelter Jr. (1912–90), American physician
[17]*gyneco* = female; *mast* = breast; *ia* = condition
[18]Henry H. Turner (1892–1970), American endocrinologist
[19]John Langdon H. Down (1828–96), British physician

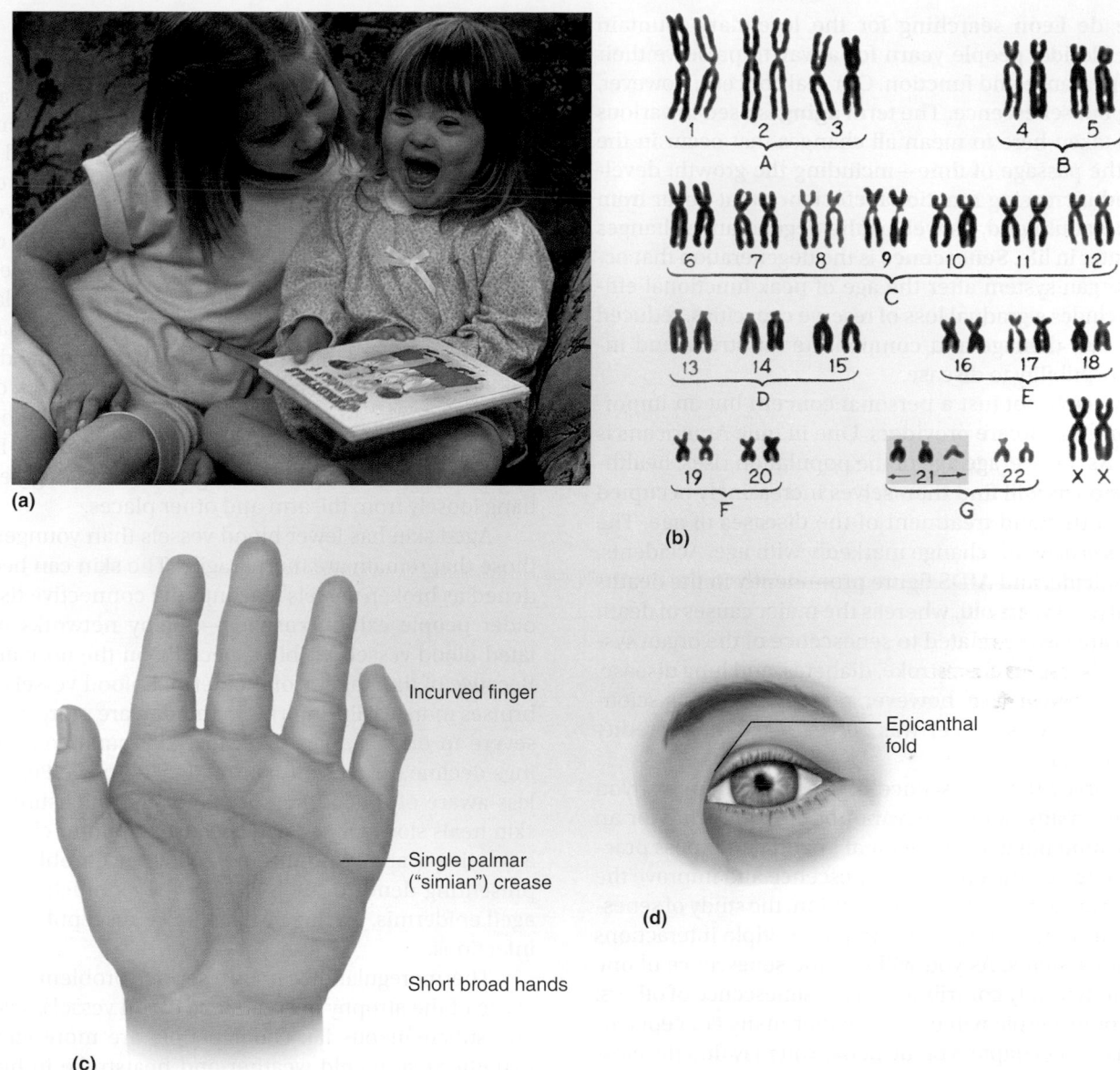

**FIGURE 29.14 Down Syndrome.** (a) A child with Down syndrome (right) and her sister. (b) The karyotype of Down syndrome, showing the trisomy of chromosome 21. (c) Characteristics of the hand in Down syndrome. (d) The epicanthal fold over the medial commissure (canthus) of the left eye.

❓ *What was the sex of the person from whom the karyotype in part (b) was obtained?*

**BEFORE YOU GO ON**

Answer the following questions to test your understanding of the preceding section.

10. How does inflation of the lungs at birth affect the route of blood flow through the heart?

11. Why is respiratory distress syndrome common in premature infants?

12. Define *nondisjunction* and explain how it causes aneuploidy. Name two syndromes resulting from aneuploidy.

## **29.4** Aging and Senescence

### Expected Learning Outcomes

When you have completed this section, you should be able to

a. define *senescence* and distinguish it from aging;

b. describe some major changes that occur with aging in each organ system;

c. summarize some current theories of senescence; and

d. be able to explain how exercise and other factors can slow the rate of senescence.

Like Ponce de León searching for the legendary fountain of youth in Florida, people yearn for a way to preserve their youthful appearance and function. Our real concern, however, is not aging but senescence. The term **aging** is used in various ways but is taken here to mean all changes that occur in the body with the passage of time—including the growth, development, and increasing functional efficiency that occur from childhood to adulthood, as well as the degenerative changes that occur later in life. **Senescence** is the degeneration that occurs in an organ system after the age of peak functional efficiency. It includes a gradual loss of reserve capacities, reduced ability to repair damage and compensate for stress, and increased susceptibility to disease.

Senescence is not just a personal concern but an important issue for health-care providers. One in nine Americans is 65 or older. As the average age of the population rises, health-care professionals will find themselves increasingly occupied by the prevention and treatment of the diseases of age. The leading causes of death change markedly with age. Accidents, homicide, suicide, and AIDS figure prominently in the deaths of people 18 to 34 years old, whereas the major causes of death after age 55 are clearly related to senescence of the organ systems: heart disease, cancer, stroke, diabetes, and lung disease. The causes of senescence, however, remain as much a scientific mystery today as cancer was 50 years ago and heredity was 100 years ago.

As we survey the senescence of the organ systems, you should notice many points relevant not only to caring for an aging population but also to personal health and fitness practices that can lessen the effects of senescence and improve the quality of life in your later years. In addition, the study of senescence calls for renewed attention to the multiple interactions among organ systems. As you will see, the senescence of one organ system typically contributes to the senescence of others. Your study of this topic will bring together many concepts introduced in earlier chapters of the book. You may find the glossary helpful in refreshing your memory of concepts revisited in the following discussion.

## Senescence of the Organ Systems

Organ systems do not all degenerate at the same rate. For example, from ages 30 to 80, the speed of nerve conduction declines only 10% to 15%, but the number of functional glomeruli in the kidneys declines about 60%. Some physiological functions show only moderate changes at rest but more pronounced differences when tested under exercise conditions. The organ systems also vary widely in the age at which senescence becomes noticeable. There are forerunners of atherosclerosis, for example, even in infants; visual and auditory sensitivities begin to decline soon after puberty. By contrast, the female reproductive system does not show significant senescence until menopause and then its decline is relatively abrupt. Aside from these unusual examples, most physiological measures of performance peak between the late teens and age 30, then decline at a rate influenced by the level of use of the organs.

## Integumentary System

Two-thirds of people age 50 and over, and nearly all people over age 70, have medical concerns or complaints about their skin. Senescence of the integumentary system often becomes noticeable by the late 40s. The hair turns grayer and thinner as melanocytes die out, mitosis slows down, and dead hairs are not replaced. The atrophy of sebaceous glands leaves the skin and hair drier. As epidermal mitosis declines and collagen is lost from the dermis, the skin becomes almost paper-thin and translucent. It becomes looser because of a loss of elastic fibers and flattening of the dermal papillae, which normally form a stress-resistant corrugated boundary between the dermis and epidermis. If you pinch a fold of skin on the back of a child's hand, it quickly springs back when you let go; do the same on an older person and the skinfold remains longer. Because of its loss of elasticity, aged skin sags to various degrees and may hang loosely from the arm and other places.

Aged skin has fewer blood vessels than younger skin, and those that remain are more fragile. The skin can become reddened as broken vessels leak into the connective tissue. Many older people exhibit **rosacea**—patchy networks of tiny, dilated blood vessels visible especially on the nose and cheeks. Because of the fragility of the dermal blood vessels, aged skin bruises more easily. Injuries to the skin are more common and severe in old age, partly because the cutaneous nerve endings decline by two-thirds from age 20 to age 80, leaving one less aware of touch, pressure, and injurious stimuli. Injured skin heals slowly in old age because of poorer circulation and a relative scarcity of immune cells and fibroblasts. Antigen-presenting dendritic cells decline by as much as 40% in the aged epidermis, leaving the skin more susceptible to recurring infections.

Thermoregulation can be a serious problem in old age because of the atrophy of cutaneous blood vessels, sweat glands, and subcutaneous fat. Older people are more vulnerable to hypothermia in cold weather and heatstroke in hot weather. Heat waves and cold spells take a disproportionate toll among elderly poor who suffer from a combination of reduced homeostasis and inadequate housing.

These are all "normal" changes in the skin, or **intrinsic aging**—changes that occur more or less inevitably with the passage of time. In addition, there is **photoaging**—degenerative changes in proportion to a person's lifetime exposure to ultraviolet radiation. UV radiation accounts for more than 90% of the integumentary changes that people find medically troubling or cosmetically disagreeable: skin cancer; yellowing and mottling of the skin; age spots, which resemble enlarged freckles on the back of the hand and other sun-exposed areas; and wrinkling, which affects the face, hands, and arms more than areas of the body that receive less exposure. A lifetime of outdoor activity can give the skin a leathery, deeply wrinkled, "outdoorsy" appearance (fig. 29.15), but beneath this rugged exterior is a less happy histological appearance. The sun-damaged skin shows many malignant and premalignant cells, extensive damage to the dermal blood

**FIGURE 29.15 Senescence of the Skin.** The skin exhibits both intrinsic aging and photoaging. The deep creases seen in the old woman's face result mainly from photoaging.

vessels, and dense masses of coarse, frayed elastic fibers underlying the surface wrinkles and creases.

Senescence of the skin has far-reaching effects on other organ systems. Cutaneous vitamin D production declines as much as 75% in old age. This is all the more significant because the elderly spend less time outdoors; and because of increasing lactose intolerance, they often avoid dairy products, the only dietary source of vitamin D. Consequently, the elderly are at high risk of calcium deficiency, which, in turn, contributes to bone loss, muscle weakness, and impaired glandular secretion and synaptic transmission.

## Skeletal System

After age 30, osteoblasts become less active than osteoclasts. This imbalance results in **osteopenia,** the loss of bone; when the loss is severe enough to compromise a person's physical activity and health, it is called *osteoporosis* (see p. 224). After age 40, women lose about 8% of their bone mass per decade and men about 3%. Bone loss from the jaws is a contributing factor in tooth loss.

Not only does bone density decline with age, but the bones become more brittle as the cells synthesize less protein. Fractures occur more easily and heal more slowly. A fracture may impose a long period of immobility, which makes a person more vulnerable to pneumonia and other infectious diseases.

People notice more stiffness and pain in the synovial joints as they age, and degenerative joint diseases affect the lifestyle of 85% of people over age 75. Synovial fluid is less abundant and the articular cartilage is thinner or absent. Exposed bone surfaces abrade each other and cause friction, pain, and reduced mobility. *Osteoarthritis* is the most common joint disease of older people and one of the most common causes of physical disability (p. 303). Even breathing becomes more difficult and tiring in old age because expansion of the thorax is restricted by calcification of the sternocostal joints. Degeneration of the intervertebral discs causes back pain and stiffness, but herniated discs are less common in old age than in youth because the discs become more fibrous and stronger, with less nucleus pulposus.

## Muscular System

One of the most noticeable changes we experience with age is the replacement of lean body mass (muscle) with fat. The change is dramatically exemplified by CT scans of the thigh. In a young well-conditioned male, muscle accounts for 90% of the cross-sectional area of the midthigh, whereas in a frail 90-year-old woman, it is only 30%. Muscular strength and mass peak in the 20s; by the age of 80, most people have only half as much strength and endurance. Many people over age 75 cannot lift a 4.5 kg (10 lb) weight with their arms; such simple tasks as carrying a sack of groceries into the house may become impossible. The loss of strength is a major contributor to falls, fractures, and dependence on others for the routine activities of daily living. Fast-twitch fibers exhibit the earliest and most severe atrophy, thus increasing reaction time and reducing coordination.

There are multiple reasons for the loss of strength. Aged muscle fibers have fewer myofibrils, so they are smaller and weaker. The sarcomeres are increasingly disorganized, and muscle mitochondria are smaller and have reduced quantities of oxidative enzymes. Aged muscle has less ATP, creatine phosphate, glycogen, and myoglobin; consequently, it fatigues quickly. Muscles also exhibit more fat and fibrosis with age, which limits their movement and blood circulation. With reduced circulation, muscle injuries heal more slowly and with more scar tissue.

But the weakness and easy fatigue of aged muscle also stem from the senescence of other organ systems. There are fewer motor neurons in the spinal cord, and some muscle shrinkage may represent denervation atrophy. The remaining neurons produce less acetylcholine and show less efficient synaptic transmission, which makes the muscles slower to respond to stimulation. As muscle atrophies, motor units have fewer muscle fibers per motor neuron, and more motor units must be recruited to perform a given task. Tasks that used to be easy, such as buttoning the clothes or eating a meal, take more time and effort. The sympathetic nervous system is also less efficient in old age; consequently, blood flow to the muscles does not respond efficiently to exercise and this contributes to their rapid fatigue.

## Nervous System

The nervous system reaches its peak development around age 30. The average brain weighs 56% less at age 75 than at age 30. The cerebral gyri are narrower, the sulci are wider, the cortex is thinner, and there is more space between the brain and meninges. The remaining cortical neurons have fewer synapses, and for multiple reasons, synaptic transmission is less efficient: The neurons produce less neurotransmitter, they have fewer receptors, and the neuroglia around the synapses is more leaky and allows neurotransmitter to diffuse away. The degeneration of myelin sheaths with age also slows down nerve conduction.

Neurons exhibit less rough ER and Golgi complex with age, which indicates that their metabolism is slowing down. Old neurons accumulate lipofuscin pigment and show more neurofibrillary tangles—dense mats of cytoskeletal elements in their cytoplasm. In the extracellular material, plaques of fibrillar protein (amyloid) appear, especially in people with Down syndrome and Alzheimer disease (AD). AD is the most common nervous disability of old age (p. 468).

Not all functions of the central nervous system are equally affected by senescence. Motor coordination, intellectual function, and short-term memory decline more than language skills and long-term memory. Elderly people are often better at remembering things in the distant past than remembering recent events.

The sympathetic nervous system loses adrenergic receptors with age and becomes less sensitive to norepinephrine. This contributes to a decline in homeostatic control of such variables as body temperature and blood pressure. Many elderly people experience *orthostatic hypotension*—a drop in blood pressure when they stand, which sometimes results in dizziness, loss of balance, or fainting.

## Sense Organs

Some sensory functions decline shortly after adolescence. Presbyopia (loss of flexibility in the lenses) makes it more difficult for the eyes to focus on nearby objects. Visual acuity declines and often requires corrective lenses at a young age. Cataracts (cloudiness of the lenses) are more common in old age. Night vision is impaired as more and more light is needed to stimulate the retina. This has several causes: There are fewer receptor cells in the retina, the vitreous body becomes less transparent, and the pupil becomes narrower as the pupillary dilators atrophy. Dark adaptation takes longer as the enzymatic reactions of the photoreceptor cells become slower. Changes in the structure of the iris, ciliary body, or lens can block the reabsorption of aqueous humor, thereby increasing the risk of glaucoma. Having to give up reading and driving can be among the most difficult changes of lifestyle in old age.

Auditory sensitivity peaks in adolescence and declines afterward. The tympanic membrane and the joints between the auditory ossicles become stiffer, so vibrations are transferred less effectively to the inner ear, creating a degree of conductive deafness. Nerve deafness occurs as the number of cochlear hair cells and auditory nerve fibers declines. The greatest hearing loss occurs at high frequencies and in the frequency range of most conversation. The death of receptor cells in the semicircular ducts, utricle, and saccule, and of nerve fibers in the vestibular nerve and neurons in the cerebellum, results in poor balance and dizziness—another factor in falls and bone fractures.

The senses of taste and smell are blunted as the taste buds, olfactory cells, and second-order neurons in the olfactory bulbs decline in number. Food may lose its appeal, and declining sensory function can therefore be a factor in malnutrition.

## Endocrine System

The endocrine system degenerates less than any other organ system. The reproductive hormones drop sharply and growth hormone and thyroid hormone secretion declines steadily after adolescence, but other hormones continue to be secreted at fairly stable levels even into old age. Target-cell sensitivity declines, however, so some hormones have less effect. For example, the pituitary gland is less sensitive to negative feedback inhibition by adrenal glucocorticoids; consequently, the response to stress is more prolonged than usual. Diabetes mellitus is more common in old age, largely because target cells have fewer insulin receptors. In part, this is an effect of the greater percentage of body fat in the elderly. The more fat at any age, the less sensitive other cells are to insulin. Body fat increases as the muscles atrophy, and muscle is one of the body's most significant glucose-buffering tissues. Because of the blunted insulin response, glucose levels remain elevated longer than normal after a meal.

## Circulatory System

Cardiovascular disease is a leading cause of death in old age. Senescence has multiple effects on the blood, heart, arteries, and veins. Anemia may result from nutritional deficiencies, inadequate exercise, disease, and other causes. The factors that cause anemia in older people are so complicated it is almost impossible to control them enough to determine whether aging alone causes it. Evidence suggests that there is no change in the baseline rate of erythropoiesis in old age. Hemoglobin concentration, cell counts, and other variables are about the same among healthy people in their 70s as in the 30s. However, older people do not adapt well to stress on the hemopoietic system, perhaps because of the senescence of other organ systems. As the gastric mucosa atrophies, for example, it produces less of the intrinsic factor needed for vitamin $B_{12}$ absorption. This increases the risk of anemia. As the kidneys age and the number of nephrons declines, less erythropoietin is secreted. There may also be a limit to how many times the hemopoietic stem cells can divide and continue giving rise to new blood cells. Whatever its cause, anemia limits the amount of oxygen that can be transported and thus contributes to the atrophy of tissues everywhere in the body.

▶▶▶**APPLY WHAT YOU KNOW**

*Draw a positive feedback loop showing how anemia and senescence of the kidneys could affect each other.*

Everyone is affected to some degree in old age by arteriosclerosis and atherosclerosis (see p. 753 for the distinction). Coronary atherosclerosis leads to the degeneration of myocardial tissue; angina pectoris and myocardial infarction become more common; the heart wall becomes thinner and weaker; and stroke volume, cardiac output, and cardiac reserve decline. Like other connective tissues, the fibrous skeleton of the heart becomes less elastic. This limits cardiac distension and reduces the force of systole. Degenerative changes in the SA and AV nodes and conduction pathways of the heart lead to a higher incidence of cardiac arrhythmia and heart block. Physical endurance is compromised by the drop in cardiac output.

Arteries stiffened by arteriosclerosis cannot expand as effectively to accommodate the pressure surges of cardiac systole. Blood pressure therefore rises steadily with age. Atherosclerosis also narrows the arteries and reduces the perfusion of most organs. The effects of reduced circulation on the skin, skeletal muscles, and brain have already been noted. The combination of atherosclerosis and hypertension also weakens the arteries and increases the risk of aneurysm and stroke.

Atherosclerotic plaques trigger thrombosis, especially in the lower limbs, where flow is relatively slow and the blood clots more easily. About 25% of people over age 50 experience venous blockage by thrombosis—especially people who do not exercise regularly.

Degenerative changes in the veins are most evident in the limbs. The valves become weaker and less able to stop the backflow of blood. Blood pools in the legs and feet, raises capillary blood pressure, and causes edema. Chronic stretching of the vessels often produces varicose veins and hemorrhoids. Support hose can reduce edema by compressing the tissues and forcing tissue fluid to return to the bloodstream, but physical activity is even more important in promoting venous return.

## Immune System

The amounts of lymphatic tissue and red bone marrow decline with age; consequently, there are fewer hemopoietic stem cells, disease-fighting leukocytes, and antigen-presenting cells (APCs). Also, the lymphocytes produced in these tissues often fail to mature and become immunocompetent. Both humoral and cellular immunity depend on APCs and helper T cells,

and therefore both types of immune response are blunted. As a result, an older person is less protected against cancer and infectious diseases. It becomes especially important in old age to be vaccinated against influenza and other acute seasonal infections.

## Respiratory System

Pulmonary ventilation declines steadily after the 20s and is one of several factors in the gradual loss of stamina. The costal cartilages and joints of the thoracic cage become less flexible, the lungs have less elastic tissue, and the lungs have fewer alveoli. Vital capacity, minute respiratory volume, and forced expiratory volume fall. The elderly are also less capable of clearing the lungs of irritants and pathogens and are therefore increasingly vulnerable to respiratory infections. Pneumonia causes more deaths than any other infectious disease and is often contracted in hospitals and nursing homes.

The chronic obstructive pulmonary diseases (COPDs)—emphysema and chronic bronchitis—are more common in old age since they represent the cumulative effects of a lifetime of degenerative change. They are among the leading causes of death in old age. Pulmonary obstruction also contributes to cardiovascular disease, hypoxemia, and hypoxic degeneration in all the organ systems. Respiratory health is therefore a major concern in aging.

## Urinary System

The kidneys exhibit a striking degree of atrophy with age. From ages 25 to 85, the number of nephrons declines 30% to 40% and up to a third of the remaining glomeruli become atherosclerotic, bloodless, and nonfunctional. The kidneys of a 90-year-old are 20% to 40% smaller than those of a 30-year-old and receive only half as much blood. The glomerular filtration rate is proportionately lower and the kidneys are less efficient at clearing wastes from the blood. Although baseline renal function is adequate even in old age, there is little reserve capacity; thus, other diseases can lead to surprisingly rapid renal failure. Drug doses often need to be reduced in old age because the kidneys cannot clear drugs from the blood as rapidly; this is a contributing factor in overmedication among the aged.

Water balance becomes more precarious in old age because the kidneys are less responsive to antidiuretic hormone and because the sense of thirst is sharply reduced. Even when given free access to water, elderly people may not drink enough to maintain normal blood osmolarity. Dehydration is therefore common.

Voiding and bladder control become problematic for both men and women. About 80% of men over the age of 80 are affected by benign prostatic hyperplasia. The enlarged prostate compresses the urethra and interferes with emptying of the bladder. Urine retention may cause pressure to back up in the kidneys, aggravating the failure of the nephrons. Older women are subject to incontinence (leakage of urine), especially if their history of pregnancy and childbearing has weakened the pelvic muscles and urethral sphincters. Senescence of the sympathetic nervous system and nervous disorders such as stroke and Alzheimer disease can also cause incontinence.

## Digestive System and Nutrition

Restaurants commonly offer a "seniors' menu" and reduced prices to customers over 60 or 65 years old, knowing that they tend to eat less. There are multiple reasons for this reduced appetite. Older people have lower metabolic rates and tend to be less active than younger people and, hence, need fewer calories. The stomach atrophies with age, and it takes less to fill it up. For many, food has less aesthetic appeal in old age because of losses in the senses of smell, taste, and even vision. In addition, older people secrete less saliva, making food less flavorful and swallowing more difficult. Reduced salivation also makes the teeth more prone to caries. Dentures are an unpleasant fact of life for many people over 65 who have lost their teeth to caries and periodontitis. Atrophy of the epithelium of the oral cavity and esophagus makes these surfaces more subject to abrasion and may further detract from the ease of chewing and swallowing.

The reduced mobility of old age makes shopping and meal preparation more troublesome, and with food losing its sensory appeal, some decide that it simply isn't worth the trouble. However, one's protein, vitamin, and mineral requirements remain essentially unchanged, so vitamin and mineral supplements may be needed to compensate for reduced food intake and poorer intestinal absorption. Malnutrition is common among older people and is an important factor in anemia and reduced immunity.

As the gastric mucosa atrophies, it secretes less acid and intrinsic factor. Acid deficiency reduces the absorption of calcium, iron, zinc, and folic acid. Heartburn becomes more common as the weakening lower esophageal sphincter fails to prevent the reflux of stomach contents into the esophagus. The most common digestive complaint of older people is constipation, which results from reduced muscle tone and weaker peristalsis of the colon. This seems to stem from a combination of factors: atrophy of the muscularis externa, reduced sensitivity to neurotransmitters that promote motility, less fiber and water in the diet, and less exercise. The liver, gallbladder, and pancreas show only slightly reduced function, but the drop in liver function reduces the rate of drug deactivation and can contribute to overmedication.

## Reproductive System

In men, the senescent changes in the reproductive system are relatively gradual; they include declining testosterone secretion, sperm count, and libido. By age 65, sperm count is about one-third of what it was in a man's 20s. Men remain fertile (capable of fathering a child) well into old age, but impotence (inability to maintain an erection) can occur because of atherosclerosis, hypertension, medication, and psychological reasons.

In women, the changes are more pronounced and develop more rapidly. Over the course of menopause, the ovarian follicles are used up, gametogenesis ceases, and the ovaries stop producing sex steroids. This may result in vaginal dryness, genital atrophy, and reduced libido and may make sex less enjoyable. With the loss of ovarian steroids, a postmenopausal woman has an elevated risk of osteoporosis and atherosclerosis.

## Exercise and Senescence

Other than the mere passage of time, senescence results from obesity and insufficient exercise more than from any other causes. Conversely, good nutrition and exercise are the best ways to slow its progress.

There is no clear evidence that exercise will prolong your life, but there is little doubt that it improves the quality of life in old age. It maintains endurance, strength, and joint mobility, and reduces the incidence and severity of hypertension, osteoporosis, obesity, and diabetes mellitus. This is especially true if you begin a program of regular physical exercise early in life and make a lasting habit of it. If you stop exercising regularly after middle age, the body rapidly becomes deconditioned, although appreciable reconditioning can be achieved even when an exercise program is begun late in life. A person in his or her 90s can increase muscle strength two- or threefold in 6 months with as little as 40 minutes of isometric exercise a week. The improvement results from a combination of muscle hypertrophy and neural efficiency.

Resistance exercises may be the most effective way of reducing accidental injuries such as bone fractures, whereas endurance exercises reduce body fat and increase cardiac output and maximum oxygen uptake. A general guideline for ideal endurance training is to have three to five periods of aerobic exercise per week, each 20 to 60 minutes long and vigorous enough to reach 60% to 90% of your maximum heart rate. The maximum is best determined by a stress test but averages about 220 beats per minute minus one's age in years.

An exercise program should ideally be preceded by a complete physical examination and stress test. Warm-up and cool-down periods are especially important in avoiding soft tissue injuries and undue cardiovascular stress. Because of their lower capacity for thermoregulation, older people must be careful not to overdo exercise, especially in hot weather. At the outset of a new exercise program, it is best to "start low and go slow."

## Theories of Senescence

Why do our organs wear out? Why must we die? There still is no general theory on this. The question actually comes down to two issues: (1) What are the mechanisms that cause the organs to deteriorate with age? (2) Why has natural selection not eliminated these and produced bodies capable of longer life?

## Mechanisms of Senescence

Numerous hypotheses have been proposed and discarded to explain why organ function degenerates with age. Some authorities maintain that senescence is an intrinsic process governed by inevitable or even preprogrammed changes in cellular function. Others attribute senescence to extrinsic (environmental) factors that progressively damage our cells over the course of a lifetime.

There is good evidence of a hereditary component to longevity. Unusually long and short lives tend to run in families. Monozygotic (identical) twins are more likely than dizygotic twins to die at a similar age. One striking genetic defect called *progeria*[20] is characterized by greatly accelerated senescence (fig. 29.16). Symptoms begin to appear by age 2. The child's growth rate declines, the muscles and skin become flaccid, most victims lose their hair, and most die in early adolescence from advanced atherosclerosis. In Werner syndrome, caused by a defective gene on chromosome 8, people show marked senescence beginning in their 20s and usually die by age 50. There is some controversy over the relevance or similarity of these syndromes to normal senescence, but they do demonstrate that many of the changes associated with old age can be brought on by a genetic anomaly.

Knowing that senescence is partially hereditary, however, does not answer the question about why tissues degenerate. Quite likely, no one theory explains all forms of senescence, but let's briefly examine some of them.

***Replicative Senescence***   Normal organ function usually depends on a rate of cell renewal that keeps pace with cell death. There is a limit, however, to how many times cells can divide. Human cells cultured in the laboratory divide 80 to 90 times if taken from a fetus, but only 20 to 30 times if taken from older people. After reaching their maximum number of divisions, cultured cells degenerate and die. This decline in mitotic potential with age is called **replicative senescence.**

Why this occurs is a subject of lively research. Much of the evidence points to the **telomere,**[21] a "cap" on each end of a chromosome analogous to the plastic tip of a shoelace. In humans, it consists of a noncoding nucleotide sequence CCCTAA repeated 1,000 times or more. One of its functions may be to stabilize the chromosome and prevent it from unraveling or sticking to other chromosomes. Also, during DNA replication, DNA polymerase cannot reproduce the very ends of the DNA molecule. If there were functional genes at the end, they would not get duplicated. The telomere may therefore provide a bit of "disposable" DNA at the end, so that DNA polymerase doesn't fail to replicate genes that would otherwise be there. Every time DNA is replicated, 50 to 100 bases are lost from the telomere. In old age, the telomere may be exhausted and the polymerase may then indeed fail to replicate some of the terminal genes. Old chromosomes may therefore be more vulnerable

**FIGURE 29.16  Progeria.** This is a genetic disorder in which senescence appears to be greatly accelerated. The individuals here, from left to right, are 15, 12, and 26 years old. Few people with progeria live as long as the woman on the right.

to damage, replication errors, or both, causing old cells to be increasingly dysfunctional. The "immortality" of cancer cells results from an enzyme called *telomerase,* lacking from healthy cells, which enables cancer cells to repair telomere damage and escape the limit on number of cell divisions.

***DNA Damage Theory***   Telomere damage and replicative senescence are clearly not the entire answer to why organs degenerate. Skeletal muscle fibers and brain neurons exhibit pronounced senescence, yet these cells are nonmitotic. Another leading theory of senescence is unrepaired DNA damage. DNA suffers 10,000 to 100,000 damaging events per day in the average mammalian cell, especially oxidative stress from free radicals generated by the cell's own metabolism. Most of it is fixed by DNA repair enzymes, but this is not 100% efficient. Some damage persists and accumulates as cells age, especially nondividing cells like neurons, cardiocytes, and skeletal muscle fibers. Such cumulative damage has been shown to cause age-related decline in liver and kidney function; cardiac and muscular strength; and brain functions related to neuronal survival, synaptic plasticity, learning, and memory.

***Cross-Linking Theory***   About one-fourth of the body's protein is collagen. With age, collagen molecules become cross-linked by more and more disulfide bridges, thus making the

---

[20]*pro* = before; *ger* = old age
[21]*telo* = end; *mer* = piece

fibers less soluble and more stiff. This is thought to be a factor in several of the most noticeable changes of the aging body, including stiffening of the joints, lenses, and arteries. Similar cross-linking of DNA and enzyme molecules could progressively impair their functions as well.

***Other Protein Abnormalities***   Not only collagen but also many other proteins exhibit increasingly abnormal structure in older tissues and cells. The changes are not in amino acid sequence—therefore not attributable to DNA mutations—but lie in the way the proteins are folded and other moieties such as carbohydrates are attached to them. This is another reason that cells accumulate more dysfunctional proteins as they age.

***Autoimmune Theory***   Some of the altered macromolecules described previously may be recognized as foreign antigens

# DEEPER INSIGHT 29.4
## CLINICAL APPLICATION

### Reproductive Technology—Making Babies in the Laboratory

Fertile heterosexual couples who have frequent intercourse and use no contraception have a 90% chance of conceiving within 1 year. About one in six American couples, however, is *infertile*—unable to conceive. Infertility can sometimes be corrected by hormone therapy or surgery, but when this fails, parenthood may still be possible through other reproductive technologies discussed here. These techniques, called Assisted Reproduction Technology (ART), are performed under strict guidelines regulated by the U.S. Food and Drug Administration.

### Artificial Insemination

If only the male is infertile, the oldest and simplest solution is *artificial insemination (AI)*, in which a physician introduces donor semen into or near the cervix. This was first done in the 1890s, when the donor was often the physician himself or a medical student who donated semen for payment. In 1953, a technique was developed for storing semen in glass ampules frozen in liquid nitrogen; the first commercial sperm banks opened in 1970. Most women undergoing AI use sperm from anonymous donors but are able to select from a catalog that specifies the donors' physical and intellectual traits. A man with a low sperm count can donate semen at intervals over a course of several weeks and have it pooled, concentrated, and used to artificially inseminate his partner. Men planning vasectomies sometimes donate sperm for storage as insurance against the death of a child, divorce and remarriage, or a change in family planning. Some cases of infertility are due to sperm destruction by the woman's immune system. This can sometimes be resolved by *sperm washing*—a technique in which the sperm are collected, washed to remove antigenic proteins from their surfaces, and then introduced by AI.

### Oocyte Donation

The counterpart to sperm donation is *oocyte donation*, in which fresh oocytes are obtained from a donor, fertilized, and transplanted to the uterus of the client. A woman may choose this procedure for a variety of reasons: being past menopause, having had her ovaries removed, or having a hereditary disorder she does not want to pass to her children, for example. The donated oocytes are sometimes provided by a relative or may be left over from another woman's in vitro fertilization (see next paragraph). The first baby conceived by oocyte donation was born in 1984. This procedure has a success rate of 20% to 50%.

### In Vitro Fertilization

In some women the uterus is normal but the uterine tubes are scarred by pelvic inflammatory disease or other causes. *In vitro fertilization (IVF)* can be an option for these and other reasons. The woman is given gonadotropins to induce the "superovulation" of multiple eggs. The physician views the ovary with a laparoscope and removes eggs by suction. These are placed in a solution that mimics the chemical environment of her reproductive tract, and sperm are added to the dish. The term *in vitro*[22] *fertilization* refers to the fact that fertilization occurs in laboratory glassware; children conceived by IVF are often misleadingly called "test-tube babies." In some cases, fertilization is assisted by piercing the zona pellucida before the sperm are added (*zona drilling*) or by injecting sperm directly into the egg through a micropipette. By the day after fertilization, some of the preembryos reach the 8- to 16-celled stage. Several of these are transferred to the mother's uterus through the cervix and her blood HCG level is monitored to determine whether implantation has occurred. Excess IVF preembryos may be donated to other infertile couples or frozen and used in later attempts. In cases where a woman has lost her ovaries to disease, the oocytes may be provided by another donor, often a relative.

In the United States, IVF costs about $10,000 to $15,000 per attempt. It results in live birth in 30% to 40% of women up to age 34, using their own eggs, and declines to about 10% by age 43. Using donor eggs, the success rate is about 50% in recipients from the late 20s to mid-40s. Some attempts are "too successful." Multiple preembryos are usually introduced to the uterus as insurance against the low probability that any one of them will implant and survive. Sometimes, however, this results in multiple births—in rare cases, up to seven babies (septuplets). One advantage of IVF is that when the preembryo reaches the eight-celled stage, one or two cells can be removed and tested for genetic defects before the preembryo is introduced to the uterus.

IVF has been used in animal breeding since the 1950s, but the first child conceived this way was Louise Joy Brown (fig. 29.17), born in England in 1978. The practice grew rapidly, and by 2009, more than 1 million babies had been born through IVF. The inventor of IVF, Robert Edwards, received the 2010 Nobel Prize for Physiology or Medicine for this achievement.

### Surrogate Mothers

IVF is an option only for women who have a functional uterus. A woman who has had a hysterectomy or is otherwise unable to become pregnant or maintain a pregnancy may contract with a *surrogate mother* who provides a "uterus for hire." Some surrogates are both genetic and gestational mothers, and others gestational only. In the former case, the surrogate is artificially inseminated by a man's sperm and agrees to give the baby to the man and his partner at birth. In the latter case, oocytes are collected from one woman's ovaries, fertilized in vitro, and the preembryos are placed in the surrogate's uterus. This is typical of cases in which a woman has functional ovaries but no functional uterus. A surrogate typically receives a fee of about $10,000 plus medical and legal costs. Several hundred babies have been produced this way in the United States. In at least one case, a woman carried the child of her infertile daughter, thus giving birth to her own granddaughter.

---

[22]*in vitro* = in glass

and stimulate lymphocytes to mount an immune response against the body's own tissues. Autoimmune diseases do, in fact, become more common in old age.

## Evolution and Senescence

If certain genes contribute to senescence, it raises an evolutionary question—Why does natural selection not eliminate them? In an attempt to answer this, biologists once postulated that senescence and death were for the good of the species—a way for older, worn-out individuals to make way for younger, healthier ones. We can see the importance of death for the human population by imagining that science had put an end to senescence and people died at a rate of only 1 per 1,000 per year regardless of age (the rate at which American 18-year-olds now die). If so, the median age of the population would be 163,

### Gamete Intrafallopian Transfer

The limited success rate of IVF led to a search for more reliable and cost-effective techniques. *Gamete intrafallopian transfer (GIFT)* was developed in the mid-1980s on the conjecture that pregnancy would be more successful if the oocyte was fertilized and began cleavage in a more natural environment. Eggs are obtained from a woman after a weeklong course of ovulation-inducing drug treatment. The most active sperm cells are isolated from the semen, and the eggs and sperm are introduced into her uterine tube proximal to any existing obstruction. GIFT is about half as expensive as IVF and succeeds about 25% to 30% of the time. In a modification called *zygote intrafallopian transfer (ZIFT)*, fertilization occurs in vitro and the preembryo is introduced into the uterine tube. Traveling down the uterine tube seems to improve the chance of implantation when the conceptus reaches the uterus.

### Embryo Adoption

*Embryo adoption* is used when a woman has malfunctioning ovaries but a normal uterus. A man's sperm are used to artificially inseminate another woman. A few days later, the preembryo is flushed from the donor's uterus before it implants and is transferred to the uterus of the woman who wishes to have a child.

### Ethical and Legal Issues

Like many other advances in medicine, reproductive technology has created its own ethical and legal dilemmas, some of which are especially confounding. Perhaps the most common problem is the surrogate mother who changes her mind. Surrogates enter into a contract to surrender the baby to a couple at birth, but after carrying a baby for 9 months and giving birth, they sometimes feel differently. This raises questions about the definition of motherhood, especially if she is the gestational but not the genetic mother.

The converse problem is illustrated by a case in which the child had hydrocephalus and neither the contracting couple nor the surrogate mother wanted it. In this case, genetic testing showed that the child actually had been fathered by the surrogate's husband, not the man who had contracted for her service. The surrogate then accepted the baby as her own. Nevertheless, the case raised the question of whether the birth of a genetically defective child constituted fulfillment of the contract and obligated the contracting couple to accept the child, or whether such a contract implied that the surrogate mother must produce a healthy child.

In another case, a wealthy couple was killed in an accident and left frozen preembryos in an IVF clinic. A lawsuit was filed on behalf of the preembryos on the grounds that they were heirs to the couple's estate and should be carried to birth by a surrogate mother so they could inherit it. The court ruled against the suit and the preembryos were allowed to die. In still another widely publicized case, a man sued his wife for custody of their frozen preembryos as part of a divorce settlement.

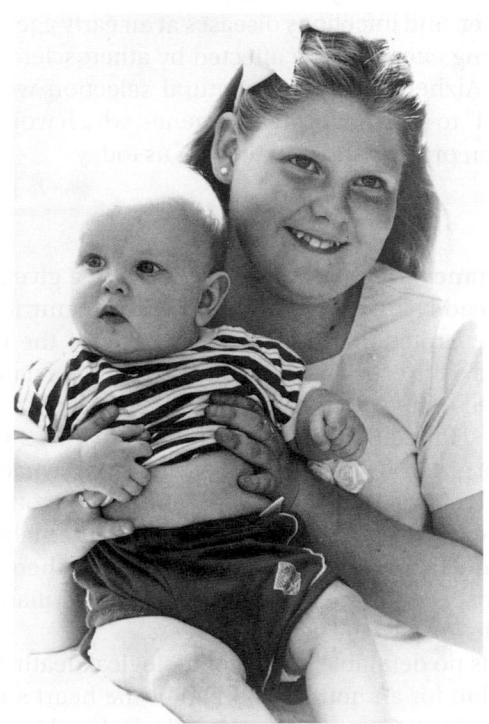

**FIGURE 29.17 New Beginnings Through Reproductive Technology.** Louise Joy Brown, shown here at age 10, was the first child ever conceived by in vitro fertilization (IVF). She is holding Andrew Macheta, another IVF baby, at a 1998 10-year anniversary celebration at the clinic near London where both were conceived.

IVF also creates a question of what to do with the excess preembryos. Some people view their disposal as a form of abortion, even if the preembryo is only a mass of 8 to 16 undifferentiated cells. On the other hand, there are those who see such excess preembryos as a research opportunity to obtain information that could not be obtained in any other way. In 1996, an IVF clinic in England was allowed to destroy 3,300 unclaimed preembryos, but only after heated public controversy.

It is common for scientific advances to require new advances in law and ethics. The parallel development of these disciplines is necessary if we are to benefit from the developments of science and ensure that knowledge is applied in an ethical and humane manner.

and 13% of us would live to be 2,000 years old. The implications for world population, social order, and competition for resources would be staggering. Thus, it is easy to understand why death was once interpreted as a self-sacrificing phenomenon for the good of the species.

But this hypothesis has several weaknesses. One of them is the fact that natural selection works exclusively through the effects of genes on the reproductive rates of individuals. A species evolves only because some members reproduce more than others. A gene that does not affect reproductive rate can be neither eliminated nor favored by natural selection. Genes for disorders such as Alzheimer disease have little or no effect until a person is past reproductive age. Our prehistoric and even fairly recent ancestors usually died of accidents, predation, starvation, weather, and infectious diseases at an early age. Few people lived long enough to be affected by atherosclerosis, colon cancer, or Alzheimer disease. Natural selection would have been "blind" to such death-dealing genes, which would escape the selection process and remain with us today.

## Death

**Life expectancy,** the average length of life in a given population, has steadily increased in industrialized countries. People born in the United States at the beginning of the twentieth century had a life expectancy of only 45 to 50 years; nearly half of them died of infectious disease. As of 2010, the average boy born in the United States could expect to live 76.2 years and the average girl 81.0 years. This is due mostly to victories over infant and child mortality, not to advances at the other end of the life span. **Life span,** the maximum age attainable by humans, has not increased for many centuries and there seems to be little prospect that it ever will. There is no verifiable record of anyone living past the age of 122 years.

There is no definable instant of biological death. Some organs function for an hour or more after the heart stops beating. During this time, even if a person is declared legally dead, living organs may be removed for transplantation. For legal purposes, death was once defined as the loss of a spontaneous heartbeat and respiration. Now that cardiopulmonary functions can be artificially maintained for years, this criterion is less distinct. Clinical death is now widely defined in terms of **brain death**—a lack of cerebral activity indicated by a flat electroencephalogram for 30 minutes to 24 hours (depending on state laws), accompanied by a lack of reflexes or lack of spontaneous respiration and heartbeat.

Death usually results from the failure of a particular organ, which then has a cascading effect on other organs. Kidney failure, for example, leads to the accumulation of toxic wastes in the blood, which in turn leads to loss of consciousness, brain function, respiration, and heartbeat.

Ninety-nine percent of us will die before age 100, and there is little chance that this outlook will change within our lifetimes. We cannot presently foresee any "cure for old age" or significant extension of the human life span. The real issue is to maintain the best possible quality of life, and when the time comes to die, to do so in comfort and dignity.

**BEFORE YOU GO ON**

Answer the following questions to test your understanding of the preceding section:

13. Define aging and senescence.

14. List some tissues or organs in which changes in collagenous and elastic connective tissues lead to senescence.

15. Many older people have difficulty with mobility and simple self-maintenance tasks such as dressing and cooking. Name some organ systems whose senescence is most relevant to these limitations.

16. Explain why both endurance and resistance exercises are important in old age.

17. Summarize five mechanisms that may be responsible for senescence.

# STUDY GUIDE

## ► Assess Your Learning Outcomes

*To test your knowledge, discuss the following topics with a study partner or in writing, ideally from memory.*

### 29.1 Fertilization and the Preembryonic Stages (p. 1097)

1. Why sperm must meet an egg near the distal end of the uterine tube; factors that may aid in sperm migration
2. Why freshly ejaculated sperm cannot fertilize an egg; the process in which they acquire that ability
3. The acrosomal reaction, barriers to fertilization, and why multiple sperm must collaborate to fertilize an egg
4. The term for fertilization of an egg by two or more sperm, and how the egg normally prevents this
5. Events that occur between penetration by a sperm and the mingling of sperm and egg chromosomes; the term for a fertilized egg
6. The division of pregnancy into three trimesters and into preembryonic, embryonic, and fetal stages
7. Duration of the preembryonic stage; the three major events that occur in this stage; and the end product of this stage
8. The meaning of *cleavage;* the term for the resulting cells; development and characteristics of the morula and blastocyst; and structural and functional distinctions between the trophoblast and embryoblast
9. The process of implantation; the cytotrophoblast and syncytiotrophoblast; and the role of the latter in implantation and nourishment of the conceptus
10. The source and function of human chorionic gonadotropin
11. The process of gastrulation; the formation and names of the three primary germ layers; and why the process is called embryogenesis

### 29.2 The Embryonic and Fetal Stages (p. 1103)

1. Major events that occur in the embryonic stage and the developmental ages at which it begins and ends
2. Longitudinal and lateral folding of the embryo; how this gives rise to the body cavity and gut; and the term for the differentiation of the embryonic organs

3. Structure and function of the four extraembryonic membranes associated with the embryo and fetus
4. The source and functions of amniotic fluid
5. How the conceptus is nourished prior to implantation
6. The mode of trophoblastic nutrition; what the trophoblast digests for its nourishment at this stage; the developmental age at which this yields to the placenta as the dominant mode of nutrition; and the age at which it ends entirely
7. When and how the placenta begins to form; the structure of the placenta; the method of transfer of nutrients and wastes between the maternal and fetal blood
8. Structure of the umbilical cord and the origin and termination of the umbilical blood vessels in the fetal body
9. When the individual is considered to be a fetus, and by what criterion
10. Developments that occur in the fetal stage; the meaning of *digital rays, meconium, lanugo, vernix caseosa, quickening,* and *vertex position*
11. Why the fetal circulatory system differs from the neonatal system, and names of the circulatory shunts in the fetus

### 29.3 The Neonate (p. 1112)

1. What occurs in the neonate in the transitional period of 6 to 8 hours after birth; the duration of the neonatal period
2. What happens to the umbilical arteries and fetal circulatory shunts after birth; names of the adult remnants of the shunts
3. Why breathing is so difficult for the neonate
4. Sources of neonatal immunity; the age at which an infant produces ample amounts of antibody on its own
5. Why neonatal thermoregulation is so critical; the functional importance of brown fat in the neonate
6. Why the neonate needs more fluid intake relative to body weight than adults do
7. Why premature infants are at risk of infant respiratory distress syndrome, hypothermia, hypoproteinemia, edema, clotting deficiency, and jaundice
8. Four principal causes of congenital anomalies (birth defects)

9. Three classes of teratogens, with examples of each; why organ systems vary in age of maximum susceptibility to teratogenesis
10. Some of the more common and serious infectious diseases of the newborn
11. Nondisjunction and how it gives rise to triplo-X, Klinefelter, Turner, and Down syndromes; the signs of these syndromes; the term for any condition in which there are more or fewer than two sex chromosomes or two copies of each autosome

### 29.4 Aging and Senescence (p. 1117)

1. The difference between aging and senescence
2. Differences in the leading causes of death in old age compared to those of young adulthood
3. Senescent changes in the integumentary system; differences between intrinsic aging and photoaging of the skin; and why integumentary senescence raises the risk of lactose intolerance, bone loss, muscle weakness, and poorer glandular secretion and synaptic transmission
4. Senescent changes in the skeletal system and how males and females differ in this respect
5. Senescent changes in the muscular system, and how they arise partially from nervous system senescence
6. Senescent changes in the nervous system and why this has especially widespread effects on homeostasis
7. Forms of declining sensory function in old age
8. The relatively slight senescence of most endocrine functions except for reproductive hormones; reasons for increased risk of diabetes mellitus as one ages
9. Senescent changes in the circulatory system including the blood, heart, and blood vessels; how circulatory system senescence contributes to atrophy in many other organs
10. Why senescent losses of immune function raise the risk and severity of infectious diseases and cancer
11. Senescent changes in the respiratory system, including the thoracic cage; why pneumonia and chronic obstructive pulmonary diseases become such prevalent causes of death

# STUDY GUIDE

12. Senescent changes in the urinary system; how this affects fluid needs and drug therapy in old age; and reasons for increasing urine retention in elderly men and incontinence in elderly women

13. Senescent changes in the digestive system and their effects on nutrition

14. Senescent changes in the male and female reproductive systems and the hormonal effects of this senescence on other organ systems

15. The relative contributions of regular exercise to longevity and quality of life in old age, and disorders for which the risk is reduced by regular exercise

16. Theories of senescence including replicative senescence and the telomere theory; protein and DNA cross-linking; protein misfolding and other structural defects; the free radical theory; and the autoimmune theory

17. The hereditary contribution to life expectancy; the reason natural selection cannot eliminate genes that cause the degenerative diseases of old age

18. The difference between life span and life expectancy, and why medical science has been able to extend one but not the other

19. Clinical and legal criteria of death and the issue of how clinical death differs from complete biological death

## ▶ Testing Your Recall

*Answers in Appendix B*

1. When a conceptus arrives in the uterus, it is at what stage of development?
   a. zygote
   b. morula
   c. blastomere
   d. blastocyst
   e. embryo

2. The entry of a sperm nucleus into an egg must be preceded by
   a. the cortical reaction.
   b. the acrosomal reaction.
   c. the fast block.
   d. implantation.
   e. cleavage.

3. The stage of the conceptus that implants in the uterine wall is
   a. a blastomere.
   b. a morula.
   c. a blastocyst.
   d. an embryo.
   e. a zygote.

4. Chorionic villi develop from
   a. the zona pellucida.
   b. the endometrium.
   c. the syncytiotrophoblast.
   d. the embryoblast.
   e. the corona radiata.

5. Which of these results from aneuploidy?
   a. Turner syndrome
   b. fetal alcohol syndrome
   c. nondisjunction
   d. progeria
   e. rubella

6. Fetal urine accumulates in the _____ and contributes to the fluid there.
   a. placental sinus
   b. yolk sac
   c. allantois
   d. chorion
   e. amnion

7. One theory of senescence is that it results from a lifetime of damage by
   a. teratogens.
   b. aneuploidy.
   c. free radicals.
   d. cytomegalovirus.
   e. nondisjunction.

8. Photoaging is a major factor in the senescence of
   a. the integumentary system.
   b. the eyes.
   c. the nervous system.
   d. the skeletal system.
   e. the cardiovascular system.

9. Which of these is *not* a common effect of senescence?
   a. reduced synthesis of vitamin D
   b. atrophy of the kidneys
   c. atrophy of the cerebral gyri
   d. increased herniation of the intervertebral discs
   e. reduced pulmonary vital capacity

10. For the first 8 weeks of gestation, a conceptus is nourished mainly by
   a. the placenta.
   b. amniotic fluid.
   c. colostrum.
   d. decidual cells.
   e. yolk cytoplasm.

11. Viruses and chemicals that cause congenital anatomical deformities are called _____.

12. Aneuploidy is caused by _____, the failure of two homologous chromosomes to separate in meiosis.

13. The maximum age attainable by a member of the human species is called the _____.

14. The average age attained by humans in a given population is called the _____.

15. Fetal blood flows through growths called _____, which project into the placental sinus.

16. The enzymes with which a sperm penetrates an egg are contained in an organelle called the _____.

17. Stiffening of the arteries, joints, and lenses in old age may be a result of cross-linking between _____ molecules.

18. An enlarged tongue, epicanthal folds of the eyes, and mental retardation are characteristic of a genetic anomaly called _____.

19. The fossa ovalis is a remnant of a fetal shunt called the _____.

20. A developing individual is first classified as a/an _____ when the three primary germ layers have formed.

# STUDY GUIDE

## ► Building Your Medical Vocabulary

*Answers in Appendix B*

*State a medical meaning of each word element below, and give a term in which it or a slight variation of it is used.*

1. con-
2. decidu-
3. gero-
4. mor-
5. placento-
6. -ploid
7. senesc-
8. syn-
9. telo-
10. terato-

## ► True or False

*Answers in Appendix B*

*Determine which five of the following statements are false, and briefly explain why.*

1. Freshly ejaculated sperm are more capable of fertilizing an egg than are sperm several hours old.

2. Fertilization normally occurs in the lumen of the uterus.

3. An egg is usually fertilized by the first sperm that contacts it.

4. By the time a conceptus reaches the uterus, it has already undergone several cell divisions and consists of 16 cells or more.

5. The conceptus is first considered a fetus when all of the organ systems are present.

6. The placenta becomes increasingly permeable as it develops.

7. The endocrine system shows less senescence in old age than most other organ systems.

8. Fetal blood bypasses the nonfunctional liver by passing through the foramen ovale.

9. Blood in the umbilical vein has a higher $P_{O_2}$ than blood in the umbilical arteries.

10. It is well established that people who exercise regularly live longer than those who do not.

## ► Testing Your Comprehension

*Answers at www.mhhe.com/saladin7*

1. Suppose a woman had a mutation resulting in a tough zona pellucida that did not disintegrate after the egg was fertilized. How would this affect her fertility? Why?

2. Suppose a drug were developed that could slow down the rate of collagen cross-linking with age. What diseases of old age could be made less severe with such a drug?

3. Some health-food stores market the enzyme superoxide dismutase (SOD) as an oral antioxidant to retard senescence. Explain why it would be a waste of your money to buy it.

4. In some children, the ductus arteriosus fails to close after birth—a condition that eventually requires surgery. Predict how this condition would affect (a) pulmonary blood pressure, (b) systemic diastolic pressure, and (c) the right ventricle of the heart.

5. Only one sperm is needed to fertilize an egg, yet a man who ejaculates fewer than 10 million sperm is usually infertile. Explain this apparent contradiction. Supposing 10 million sperm were ejaculated, predict how many would come within close range of the egg. How likely is it that any one of these sperm would fertilize it?

## ► Improve Your Grade at www.mhhe.com/saladin7

*Download animations to study when it fits your schedule. Practice quizzes, labeling activities, games, and flashcards offer fun ways to master the chapter concepts. Or, download image PowerPoint files for each chapter to create a study guide or for taking notes during lecture.*

# APPENDIX A

## PERIODIC TABLE OF THE ELEMENTS

Nineteenth-century chemists discovered that when they arranged the known elements by atomic weight, certain properties reappeared periodically. In 1869, Russian chemist Dmitri Mendeleev published the first modern periodic table of the elements, leaving gaps for those that had not yet been discovered. He accurately predicted properties of the missing elements, which helped other chemists discover and isolate them.

Each row in the table is a *period* and each column is a *group (family)*. Each period has one electron shell more than the period above it, and as we progress from left to right within a period, each element has one more proton and electron than the one before. The dark steplike line from boron (5) to astatine (85) separates the metals to the left of it (except hydrogen) from the nonmetals to the right. Each period begins with a soft, light, highly reactive *alkali metal,* with one valence electron, in family IA. Progressing from left to right, the metallic properties of the elements become less and less pronounced. Elements in family VIIA are highly reactive gases called *halogens,* with seven valence electrons. Elements in family VIIIA, called *noble (inert) gases,* have a full valence shell of eight electrons, which makes them chemically unreactive.

Ninety-one of the elements occur naturally on earth. Physicists have created elements up to atomic number 118 in the laboratory, but the International Union of Pure and Applied Chemistry has established formal names only through element 109 to date.

The 24 elements with normal roles in human physiology are color-coded according to their relative abundance in the body (see chapter 2). Others, however, may be present as contaminants with very destructive effects (such as arsenic, lead, and radiation poisoning).

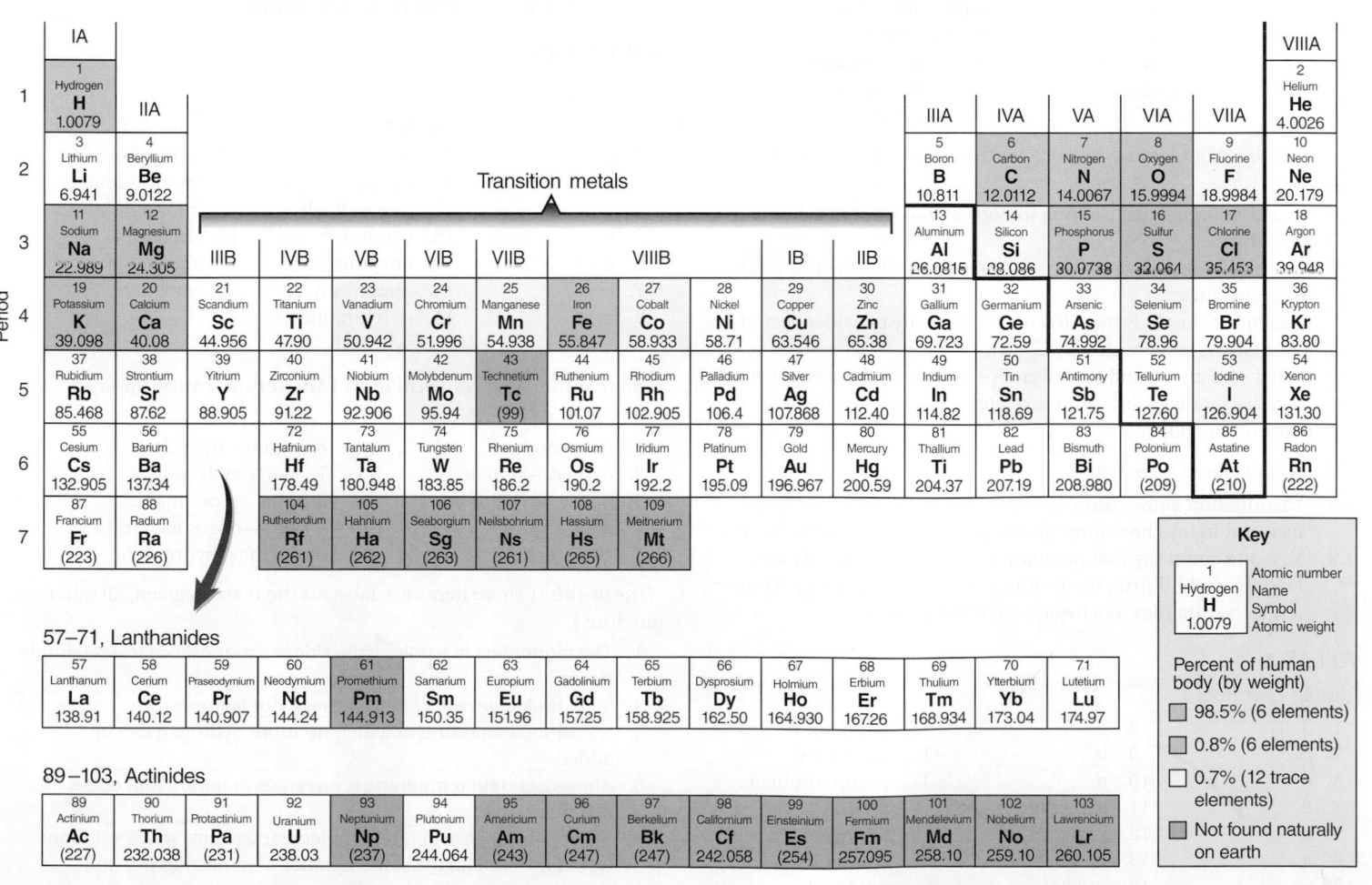

**Key**

| 1 Hydrogen H 1.0079 | Atomic number / Name / Symbol / Atomic weight |

**Percent of human body (by weight)**

- 98.5% (6 elements)
- 0.8% (6 elements)
- 0.7% (12 trace elements)
- Not found naturally on earth

# APPENDIX B

## ANSWER KEYS

This appendix provides answers to the end-of-chapter Testing Your Recall, Building Your Medical Vocabulary, and True or False questions, and questions in the figure legends. Answers to Apply What You Know and Testing Your Comprehension questions are available on the Saladin website at www.mhhe.com/saladin7.

### CHAPTER 1

*Testing Your Recall*

| | | |
|---|---|---|
| 1. a | 8. c | 15. homeostasis |
| 2. e | 9. d | 16. set point |
| 3. d | 10. b | 17. negative feedback |
| 4. a | 11. dissection | 18. organ |
| 5. c | 12. gradient | 19. stereoscopic |
| 6. c | 13. deduction | 20. prehensile, |
| 7. a | 14. psychosomatic | opposable |

*Building Your Medical Vocabulary* (Answers may vary; these are acceptable examples.)

| | |
|---|---|
| 1. listen—auscultation | 6. nature—physiology |
| 2. apart—dissection | 7. cut—dissection |
| 3. the same—homeostasis | 8. to stay—homeostasis |
| 4. change—metabolism | 9. solid—stereoscopic |
| 5. touch—palpation | 10. to cut—tomography |

*True or False* (These items are false for the reasons given; all others are true.)

3. Auscultation means listening to body sounds, not inspecting its appearance.
6. Leeuwenhoek was a textile merchant who built microscopes to examine fabric.
7. A scientific theory is founded on a large body of evidence and summarizes what is already known.
8. Both the treatment and control groups consist of volunteer patients.
10. Negative feedback is a self-corrective process with a beneficial effect on the body.

*Answers to Figure Legend Questions*

1.6 Vasodilation allows more blood to flow close to the body surface and to lose heat through the skin; thus, it cools the body.
1.8 Yes; one could say that pregnancy activates a series of events leading to childbirth, the termination of the pregnancy. Thus, it has the qualities of a negative feedback loop.

### ATLAS A

*Testing Your Recall*

| | | |
|---|---|---|
| 1. d | 8. d | 15. hand, foot |
| 2. c | 9. b | 16. meninges |
| 3. e | 10. d | 17. retroperitoneal |
| 4. d | 11. mesenteries | 18. medial |
| 5. d | 12. parietal | 19. inferior |
| 6. a | 13. mediastinum | 20. cubital, |
| 7. a | 14. nuchal | popliteal |

*Building Your Medical Vocabulary* (Answers may vary; these are acceptable examples.)

| | |
|---|---|
| 1. before—antebrachium | 6. within—intraperitoneal |
| 2. neck—cervical | 7. wall—parietal |
| 3. above—epigastric | 8. around—peritoneum |
| 4. below—hypochondriac | 9. behind—retroperitoneal |
| 5. groin—inguinal | 10. arrow—sagittal |

*True or False* (These items are false for the reasons given; all others are true.)

4. The diaphragm is inferior to the lungs.
5. The esophagus is superior to the stomach.
6. The liver extends from the hypochondriac to the epigastric region, superior to the lumbar region.
9. The peritoneum lines the outside of the stomach and intestines.
10. The sigmoid colon is in the lower left quadrant.

*Answers to Figure Legend Questions*

A.4 Right lower quadrant (RLQ)
A.8 No, it is inferior to the peritoneal cavity.

### CHAPTER 2

*Testing Your Recall*

| | | |
|---|---|---|
| 1. a | 9. b | 16. -ose, -ase |
| 2. c | 10. d | 17. phospholipids |
| 3. a | 11. cation | 18. cyclic adenosine |
| 4. c | 12. free radicals | monophosphate |
| 5. a | 13. catalyst, | 19. anaerobic |
| 6. e | enzymes | fermentation |
| 7. b | 14. anabolism | 20. substrate |
| 8. c | 15. dehydration | |
| | synthesis | |

*Building Your Medical Vocabulary* (Answers may vary; these are acceptable examples.)

| | |
|---|---|
| 1. not—atom | 6. water—hydrolysis |
| 2. oxygen—aerobic | 7. part—polymer |
| 3. both—amphiphilic | 8. one—monomer |
| 4. heat—calorie | 9. few—oligosaccharide |
| 5. glue—colloid | 10. loving—hydrophilic |

*True or False* (These items are false for the reasons given; all others are true.)

1. The monomers of a polysaccharide are monosaccharides (simple sugars).
3. Such molecules are called isomers, not isotopes.
6. A saturated fat is one to which no more hydrogen can be added.
8. Above a certain temperature, enzymes denature and cease working.
9. These solutes have different molecular weights, so 2% solutions would not contain the same number of molecules per unit volume.

2.1 Sodium gives up an electron.

2.8 Because water molecules are attracted to each other, it requires more thermal energy for any one of them to break free and evaporate.

2.13 Decomposition

2.27 No, the amount of energy released is the same with or without an enzyme.

## CHAPTER 3

*Testing Your Recall*

| | | |
|---|---|---|
| 1. e | 9. d | 16. exocytosis |
| 2. b | 10. b | 17. Ribosomes, proteasomes |
| 3. d | 11. micrometers | |
| 4. b | 12. second messenger | 18. smooth ER, peroxisomes |
| 5. e | 13. Voltage-gated | |
| 6. e | 14. hydrostatic pressure | 19. ligand-regulated gate |
| 7. a | | |
| 8. c | 15. hypertonic | 20. cisterna |

*Building Your Medical Vocabulary* (Answers may vary; these are acceptable examples.)

1. opposite—antiport
2. color—chromatin
3. together—cotransport
4. cell—cytoplasm
5. into—endocytosis
6. easy—facilitated
7. spindle—fusiform
8. study of—cytology
9. process—pinocytosis
10. eat—phagocytosis

*True or False* (These items are false for the reasons given; all others are true.)

1. Osmosis does not require ATP.
3. Second messengers activate enzymes in the cell; they are not transport proteins.
5. A channel could not move material from the outside of a cell to the inside unless it extended all the way across the membrane; it must be a transmembrane protein.
6. The plasma membrane consists primarily of phospholipid molecules.
7. The brush border is composed of microvilli.

*Answers to Figure Legend Questions*

3.9 Adenylate cyclase is a transmembrane protein. The G protein is peripheral.

3.20 The Na⁺-K⁺ pump requires ATP, whereas osmosis does not. ATP is quickly depleted after a cell dies.

3.23 Transcytosis is a combination of endocytosis and exocytosis.

3.27 Large molecules such as enzymes and RNA must pass through the nuclear pores, but pores in the plasma membrane must be small enough to prevent such large molecules from escaping the cell.

3.33 A centriole is composed of a cylinder of nine groups of microtubules, but in a centriole, there are three microtubules in each group and in an axoneme there are only two. Also, an axoneme usually has a central pair of microtubules, whereas a centriole does not.

## CHAPTER 4

*Testing Your Recall*

| | | |
|---|---|---|
| 1. a | 8. d | 15. RNA polymerase |
| 2. e | 9. d | 16. genome |
| 3. c | 10. a | 17. 46, 92, 92 |
| 4. c | 11. cytokinesis | 18. ribosome |
| 5. e | 12. alleles | 19. growth factors |
| 6. b | 13. genetic code | 20. autosomes |
| 7. a | 14. polyribosome | |

*Building Your Medical Vocabulary* (Answers may vary; these are acceptable examples.)

1. different—allele
2. finger—polydactyly
3. double—diploid
4. half—haploid
5. different—heterozygous
6. nucleus—karyotype
7. next in a series—metaphase
8. shape—polymorphism
9. change—mutation
10. many—polydactyly

*True or False* (These items are false for the reasons given; all others are true.)

1. There are no ribosomes on the Golgi complex; they are on the rough ER.
2. There are no genes for steroids, carbohydrates, or phospholipids, but only for proteins.
6. This law describes the pairing of bases between the two strands of DNA, not between mRNA and tRNA.
9. Males have only one X chromosome, but have two sex chromosomes (the X and Y).
10. Several RNA polymerase molecules at once can transcribe a gene.

*Answers to Figure Legend Questions*

4.1 The helix would bulge where two purines were paired and would be constricted where two pyrimidines were paired.

4.8 The ribosome would have no way of holding the partially completed peptide in place while adding the next amino acid.

## CHAPTER 5

*Testing Your Recall*

| | | |
|---|---|---|
| 1. a | 9. b | 17. basement membrane |
| 2. b | 10. b | |
| 3. c | 11. necrosis | 18. matrix (extracellular material) |
| 4. e | 12. mesothelium | |
| 5. c | 13. lacunae | 19. multipotent |
| 6. a | 14. fibers | 20. simple |
| 7. b | 15. collagen | |
| 8. e | 16. skeletal muscle | |

*Building Your Medical Vocabulary* (Answers may vary; these are acceptable examples.)

1. away—apoptosis
2. cartilage—chondrocyte
3. outer—ectoderm
4. producing—collagen
5. tissue—histology
6. whole—holocrine
7. glassy—hyaline
8. dead—necrosis
9. formed—neoplasia
10. scale—squamous

*True or False* (These items are false for the reasons given; all others are true.)

1. The esophageal epithelium is nonkeratinized.
5. Adipose tissue is an exception; cells constitute most of its volume.
6. Adipocytes are also found in areolar tissue, either singly or in small clusters.
7. Tight junctions serve mainly to restrict the passage of material between cells.
10. Perichondrium is lacking from fibrocartilage and from hyaline articular cartilage.

*Answers to Figure Legend Questions*

5.2 They are longitudinal sections. In a cross section, both the egg white and yolk would look circular. In an oblique section, the white would look elliptical but the yolk would still look circular.
5.12 The epithelia of the tongue, oral cavity, esophagus, and anal canal would look similar to this.
5.28 Gap junctions
5.30 Exocytosis
5.31 The sketch would look like one of the purple sacs in part (b), budding directly from the epithelial surface with no duct.
5.32 Holocrine glands, because entire cells break down to become the secretion, and these must be continually replaced.

## CHAPTER 6

*Testing Your Recall*

| | | |
|---|---|---|
| 1. d | 8. a | 14. cyanosis |
| 2. c | 9. a | 15. dermal papillae |
| 3. d | 10. d | 16. earwax |
| 4. b | 11. Insensible | 17. sebaceous glands |
| 5. a | perspiration | 18. anagen |
| 6. e | 12. piloerector | 19. dermal papilla |
| 7. c | 13. debridement | 20. third-degree |

*Building Your Medical Vocabulary* (Answers may vary; these are acceptable examples.)

1. substance—melanin
2. white—albinism
3. skin—dermatology
4. through—diaphoresis
5. same—homograft
6. injure—lesion
7. black—melanoma
8. tumor—carcinoma
9. nail—eponychium
10. hair—piloerector

*True or False* (These items are false for the reasons given; all others are true.)

3. Keratin is the protein of the epidermis; the dermis is composed mainly of collagen.
4. Vitamin D synthesis begins in the keratinocytes.
7. The hypodermis is not considered to be a layer of the skin.
8. Different races have about the same density of melanocytes but different amounts of melanin.
9. A genetic lack of melanin causes albinism, not pallor. Pallor is a temporary, nonhereditary paleness of the skin.

*Answers to Figure Legend Questions*

6.6 Keratinocytes
6.8 Cuticle
6.11 Holocrine, because cells die to become the secretion and thus require rapid replacement
6.12 Asymmetry (A), irregular border (B), and color (C). The photo does not provide enough information to judge the diameter of the lesion (D).

## CHAPTER 7

*Testing Your Recall*

| | | |
|---|---|---|
| 1. e | 8. e | 15. hypocalcemia |
| 2. a | 9. b | 16. Osteoblasts |
| 3. d | 10. d | 17. calcitriol |
| 4. c | 11. hydroxyapatite | 18. osteoporosis |
| 5. d | 12. canaliculi | 19. metaphysis |
| 6. a | 13. appositional | 20. osteomalacia |
| 7. d | 14. solubility product | |

*Building Your Medical Vocabulary* (Answers may vary; these are acceptable examples.)

1. calcium—hypocalcemia
2. destroy—osteoclast
3. softening—osteomalacia
4. marrow—osteomyelitis
5. straight—orthopedics
6. bone—osseous
7. bone—osteocyte
8. growth—diaphysis
9. dart—spicule
10. place—ectopic

*True or False* (These items are false for the reasons given; all others are true.)

3. The most common bone disease is osteoporosis, not fractures.
4. Bones elongate at the epiphyseal plate, not the articular cartilage.
5. Osteoclasts develop from stem cells in the bone marrow, not from osteoblasts.
7. Hydroxyapatite is the major mineral of bone; the major protein is collagen.
9. The major effect of vitamin D is bone resorption, though it also promotes deposition.

*Answers to Figure Legend Questions*

7.1 The wider epiphyses provide surface area for muscle attachment and bone articulation, while the narrowness of the diaphysis minimizes weight.
7.4 Spongy bone
7.6 Places where bone comes close to the skin, such as the sternum and hips
7.7 Temporal bone, parietal bone, and several others
7.9 Humerus, radius, ulna, femur, tibia, fibula
7.10 An infant's joints are still cartilaginous.
7.12 The zones of cell proliferation and cell hypertrophy

## CHAPTER 8

*Testing Your Recall*

| | | |
|---|---|---|
| 1. b | 8. b | 15. anulus fibrosus |
| 2. e | 9. e | 16. dens |
| 3. a | 10. b | 17. auricular |
| 4. d | 11. fontanels | 18. styloid |
| 5. a | 12. temporal | 19. pollex, hallux |
| 6. e | 13. sutures | 20. medial |
| 7. c | 14. sphenoid | longitudinal |

*Building Your Medical Vocabulary* (Answers may vary; these are acceptable examples.)

1. rib—intercostal
2. helmet—cranium
3. tough—dura mater
4. tongue—hypoglossal
5. little—ossicle
6. breast—mastoid
7. foot—bipedal
8. wing—pterygoid
9. above—supraorbital
10. ankle—metatarsal

*True or False* (explanation of the false statements only)

2. Each hand and foot has 14 phalanges.
3. The female pelvis is wider and shallower than the male's.
7. The lumbar vertebrae have transverse processes but no transverse costal facets.
8. The most frequently broken bone is the clavicle.
9. *Arm* refers to the region containing only the humerus; *leg* refers to the region containing the tibia and fibula.

*Answers to Figure Legend Questions*

8.10 Any five of these: the occipital, parietal, sphenoid, zygomatic, and palatine bones, and the mandible and maxilla

8.12 Any five of these: the frontal, lacrimal, and sphenoid bones, and the vomer, maxilla, and inferior concha

8.24 Rupture of this ligament allows the atlas to slip anteriorly and the dens of the axis to tear into the spinal cord.

8.34 The adult hand lacks epiphyseal plates, the growth zones of a child's long bones.

8.40 The three cuneiforms and the cuboid bone of the tarsus are arranged in a row similar to the distal carpal bones (trapezium, trapezoid, capitate, and hamate), with the trapezium corresponding to the median cuneiform (proximal to digit I in each case). In the proximal row, the navicular bone of the tarsus is somewhat similar to the scaphoid of the carpus, being proximal to digit I and articulating with three bones of the distal row, but the calcaneus and talus are very different, being adapted to their load-bearing role.

## CHAPTER 9

*Testing Your Recall*

| | | | | | |
|---|---|---|---|---|---|
| 1. | c | 8. | d | 15. | gomphosis |
| 2. | b | 9. | b | 16. | serrate |
| 3. | a | 10. | d | 17. | extension |
| 4. | e | 11. | synovial fluid | 18. | range of motion |
| 5. | c | 12. | bursa | 19. | articular disc |
| 6. | c | 13. | pivot | 20. | talus |
| 7. | a | 14. | Kinesiology | | |

*Building Your Medical Vocabulary* (Answers may vary; these are acceptable examples.)

1. away—abduction
2. joint—arthritis
3. characterized by—cruciate
4. letter X—cruciate
5. leg—talocrural
6. to lead—adduction
7. movement—kinesiology
8. moon—meniscus
9. to lay back—supination
10. to pull—protraction

*True or False* (These items are false for the reasons given; all others are true.)

1. Osteoarthritis occurs in almost everyone after a certain age; rheumatoid arthritis is less common.
2. A kinesiologist studies joint movements; a rheumatologist treats arthritis.
3. Synovial joints are diarthroses and amphiarthroses, but never synarthroses.
7. The round ligament is somewhat slack and probably does not secure the femoral head.
9. Synovial fluid is secreted by the synovial membrane of the joint capsule and fills the bursae.

*Answers to Figure Legend Questions*

9.2 The gomphosis, because a tooth is not a bone

9.4 The pubic symphysis consists of the cartilaginous interpubic disc and the adjacent parts of the two pubic bones.

9.5 Interphalangeal joints are not subjected to a great deal of compression.

9.7 MA = 1.0. Shifting the fulcrum to the left would increase the MA of this lever, while the lever would remain first-class.

9.19 The atlas (C1)

## CHAPTER 10

*Testing Your Recall*

| | | | | | |
|---|---|---|---|---|---|
| 1. | b | 8. | a | 14. | hamstring |
| 2. | b | 9. | d | 15. | flexor retinacula |
| 3. | a | 10. | c | 16. | urogenital triangle |
| 4. | c | 11. | origin | 17. | linea alba |
| 5. | e | 12. | fascicle | 18. | synergist |
| 6. | e | 13. | prime mover (agonist) | 19. | bipennate |
| 7. | b | | | 20. | sphincter |

*Building Your Medical Vocabulary* (Answers may vary; these are acceptable examples.)

1. head—splenius capitis
2. work—synergist
3. bundle—fascicle
4. lip—levator labii superioris
5. lower back—quadratus lumborum
6. mouse—muscle
7. muscle—perimysium
8. shoulder—omohyoid
9. feather—bipennate
10. third—peroneus tertius

*True or False* (These items are false for the reasons given; all others are true.)

3. The mastoid process is its insertion.
7. The trapezius is superficial to the scalenes.
8. Normal exhalation does not employ these muscles.
9. They result from rapid extension of the knee, not flexion.
10. They are on opposite sides of the tibia and act as antagonists.

*Answers to Figure Legend Questions*

10.4 The brachialis and lateral head of the triceps brachii have direct attachments, the biceps brachii and long head of the triceps brachii have indirect attachments.

10.7 The zygomaticus major, levator palpebrae superioris, and orbicularis oris

10.15 Pectoralis minor, serratus anterior, and all three layers of the upper intercostal muscles

10.26 *Teres* refers to the round or cordlike shape of the first muscle, and *quadratus* refers to the four-sided shape of the second.

10.27 Part (c) represents a cross section cut too high on the forearm to include these muscles.

10.33 Climbing stairs, walking, running, or riding a bicycle

10.37 The soleus

## ATLAS B

*Answers to Figure Legend Questions*

B.1 Orbicularis oris; trapezius

B.5 The lungs, heart, liver, stomach, gallbladder, and spleen, among others

B.8 Sternocleidomastoids

B.11 Posterior

B.13 Fat (adipose tissue)

B.18 One tendon proximally (at the humerus) and four distally (in the hand)

B.19 The mark would belong close to where the leader for the styloid process of the radius ends in figure B.19b.

B.20 The mark would belong close to where the leader for the rectus femoris ends; the vastus intermedius is deep to this.

B.21 The fibula

B.24 There is no such bone; digit I (the hallux) has only a proximal and distal phalanx.

B.25 Answers to the muscle test are as follows:

| | | |
|---|---|---|
| 1. f | 11. x | 21. k |
| 2. b | 12. m | 22. d |
| 3. k | 13. n | 23. f |
| 4. p | 14. e | 24. b |
| 5. h | 15. g | 25. a |
| 6. y | 16. v | 26. u |
| 7. z | 17. f | 27. j |
| 8. w | 18. c | 28. i |
| 9. c | 19. x | 29. g |
| 10. a | 20. w | 30. q |

## CHAPTER 11

*Testing Your Recall*

| | | |
|---|---|---|
| 1. a | 8. c | 15. acetylcholine |
| 2. c | 9. e | 16. myoglobin |
| 3. b | 10. b | 17. Z discs |
| 4. d | 11. threshold | 18. varicosities |
| 5. c | 12. oxidative | 19. muscle tone |
| 6. c | 13. terminal | 20. Isometric |
| 7. e | cisternae | contraction |
| | 14. myosin | |

*Building Your Medical Vocabulary* (Answers may vary; these are acceptable examples.)

| | |
|---|---|
| 1. weak—myasthenia | 6. muscle—myocyte |
| 2. self—autorhythmic | 7. flesh—sarcolemma |
| 3. abnormal—dystrophy | 8. time—temporal |
| 4. same—isometric | 9. tension—isotonic |
| 5. length—isometric | 10. growth—dystrophy |

*True or False* (These items are false for the reasons given; all others are true.)

1. A motor neuron may supply 1,000 or more muscle fibers; a motor unit consists of one motor neuron and all the muscle fibers it innervates.

2. Calcium binds to troponin, not to myosin.

6. Thick and thin filaments are present but not arranged in a way that produces striations.

7. Under natural conditions, a muscle seldom or never attains complete tetanus.

9. A muscle produces most of its ATP during this time by anaerobic fermentation, which generates lactic acid; it does not consume lactic acid.

*Answers to Figure Legend Questions*

11.1 The striations distinguish it from smooth muscle; the multiple nuclei adjacent to the plasma membrane and the parallel fibers distinguish it from both cardiac and smooth muscle.

11.2 The electrical excitation spreading down the T tubule must excite the opening of calcium gates in the terminal cisternae.

11.13 ATP is needed to pump $Ca^{2+}$ back into the sarcoplasmic reticulum by active transport and to induce each myosin head to release actin so the sarcomere can relax.

11.16 The gluteus maximus and quadriceps femoris

11.17 The muscle tension curve would drop gradually while the muscle length curve would rise.

## CHAPTER 12

*Testing Your Recall*

| | | |
|---|---|---|
| 1. e | 9. d | 16. nodes of Ranvier |
| 2. c | 10. b | 17. axon hillock, |
| 3. d | 11. afferent | initial segment |
| 4. a | 12. conductivity | 18. norepinephrine |
| 5. c | 13. absolute | 19. facilitated zone |
| 6. e | refractory period | 20. Neuromodulators |
| 7. d | 14. dendrites | |
| 8. a | 15. oligodendrocytes | |

*Building Your Medical Vocabulary* (Answers may vary; these are acceptable examples.)

| | |
|---|---|
| 1. forward—anterograde | 6. knot—ganglion |
| 2. to touch—synapse | 7. to walk—retrograde |
| 3. star—astrocyte | 8. nerve—neuroglia |
| 4. tree—dendrite | 9. hard—sclerosis |
| 5. carry—afferent | 10. body—somatic |

*True or False* (These items are false for the reasons given; all others are true.)

4. Only a small fraction of the neuron's $Na^+$ and $K^+$ exchange places across the plasma membrane.

5. The threshold stays the same but an EPSP brings the membrane potential closer to the threshold.

7. The effect of a neurotransmitter varies from place to place depending on the type of receptor present.

8. The signals travel rapidly through the internodes and slow down at each node of Ranvier.

9. Learning involves modification of the synapses of existing neurons, not an increase in the neuron population.

*Answers to Figure Legend Questions*

12.8 Its conduction speed is relatively slow, but it has a small diameter and contributes relatively little bulk to the nervous tissue.

12.11 The membrane potential will become lower (more negative).

12.19 Axosomatic

12.24 One EPSP is a voltage change of only 0.5 mV or so. A change of about 15 mV is required to reach threshold and make a neuron fire.

12.28 The CNS interprets a stimulus as more intense if it receives signals from high-threshold sensory neurons than if it receives signals only from low-threshold neurons.

12.30 A reverberating circuit, because a neuron early in the circuit is continually restimulated

## CHAPTER 13

*Testing Your Recall*

| | | |
|---|---|---|
| 1. e | 8. a | 15. intrafusal fibers |
| 2. c | 9. e | 16. phrenic |
| 3. d | 10. b | 17. decussation |
| 4. d | 11. ganglia | 18. proprioception |
| 5. e | 12. ramus | 19. posterior root |
| 6. c | 13. spinocerebellar | 20. tibial, common |
| 7. c | 14. crossed | fibular |
| | extension | |

*Building Your Medical Vocabulary* (Answers may vary; these are acceptable examples.)

1. spider—arachnoid
2. tail—cauda equina
3. opposite—contralateral
4. wedge—cuneate
5. same—ipsilateral
6. diaphragm—phrenic
7. tender—pia mater
8. oneself—proprioception
9. branch—ramus
10. roof—tectospinal

*True or False* (These items are false for the reasons given; all others are true.)

1. The gracile fasciculus is an ascending (sensory) tract.
4. All spinal nerves are mixed nerves; none are purely sensory or motor.
5. The dura is separated from the bone by a fat-filled epidural space.
8. Dermatomes overlap each other by as much as 50%.
9. Some somatic reflexes are mediated primarily through the brainstem and cerebellum.

*Answers to Figure Legend Questions*

13.4 If it were T10, there would be no cuneate fasciculus; that exists only from T6 up.
13.9 They are in the anterior horn of the spinal cord.
13.12 They are afferent, because they arise from the posterior root of the spinal nerve.
13.22 Motor neurons are capable only of exciting skeletal muscle (end-plate potentials are always excitatory). To inhibit muscle contraction, it is necessary to inhibit the motor neuron at the CNS level (point 7).
13.23 They would show more synaptic delay, because there are more synapses in the pathway.

## CHAPTER 14

*Testing Your Recall*

| | | |
|---|---|---|
| 1. c | 8. d | 14. hydrocephalus |
| 2. a | 9. e | 15. choroid plexus |
| 3. e | 10. e | 16. precentral |
| 4. a | 11. corpus callosum | 17. frontal |
| 5. b | 12. ventricles, | 18. association areas |
| 6. c | cerebrospinal | 19. categorical |
| 7. a | 13. arbor vitae | 20. Broca area |

*Building Your Medical Vocabulary* (Answers may vary; these are acceptable examples.)

1. pain—neuralgia
2. head—hydrocephalus
3. brain—cerebrospinal
4. body—corpus callosum
5. brain—electroencephalogram
6. record of—electroencephalogram
7. island—insula
8. eye—oculomotor
9. pulley—trochlea
10. little—peduncle

*True or False* (These items are false for the reasons given; all others are true.)

1. This fissure separates the cerebral hemispheres, not the cerebellar hemispheres.
2. The cerebral hemispheres do not develop from neural crest tissue.
5. The choroid plexuses produce only 30% of the CSF.

6. Hearing is a temporal lobe function; vision resides in the occipital lobe.
10. The cerebellum contains more than half of all brain neurons.

*Answers to Figure Legend Questions*

14.7 The most common sites of obstruction are the interventricular foramen at label 2, the cerebral aqueduct at label 4, and the lateral and median apertures indicated by label 6.
14.9 Signals in the cuneate fasciculus ascend to the cuneate nucleus in part (c), and signals in the gracile fasciculus ascend to the nearby gracile nucleus. Both of them decussate together to the contralateral medial lemniscus in parts (b) and (a) and travel this route to the thalamus.
14.10 The reticular formation is labeled on all three parts of the figure.
14.14 Commissural tracts also cross through the anterior and posterior commissures shown in figure 14.2
14.15 Dendrites
14.22 Regions with numerous small muscles

## CHAPTER 15

*Testing Your Recall*

| | | |
|---|---|---|
| 1. b | 8. d | 15. enteric |
| 2. c | 9. a | 16. norepinephrine |
| 3. e | 10. c | 17. sympathetic |
| 4. e | 11. adrenergic | 18. preganglionic, |
| 5. a | 12. Dual innervation | postganglionic |
| 6. e | 13. Autonomic tone | 19. cAMP |
| 7. d | 14. vagus | 20. vasomotor tone |

*Building Your Medical Vocabulary* (Answers may vary; these are acceptable examples.)

1. pressure—baroreflex
2. dissolve—sympatholytic
3. wall—intramural
4. rule—autonomic
5. ear—otic
6. feeling—parasympathetic
7. kidney—adrenal
8. internal organs—splanchnic
9. together—sympathetic
10. internal organs—visceral

*True or False* (These items are false for the reasons given; all others are true.)

1. Both systems are always simultaneously active.
3. In biofeedback and other circumstances, limited voluntary control of the ANS is possible.
4. The sympathetic division inhibits digestion.
6. Waste elimination can occur by autonomic spinal reflexes without necessarily involving the brain.
7. All parasympathetic fibers are cholinergic.

*Answers to Figure Legend Questions*

15.4 No; inhaling and exhaling are controlled by the somatic motor system and skeletal muscles.
15.5 The soma of the somatic efferent neuron is in the anterior horn and the soma of the sympathetic preganglionic neuron is in the lateral horn.
15.7 The vagus nerve
15.9 The pupils dilate because fear increases sympathetic output, which induces dilation.

## CHAPTER 16

*Testing Your Recall*

1. a
2. c
3. b
4. a
5. e
6. e
7. d
8. c
9. c
10. b
11. fovea centralis
12. ganglion
13. photopsin
14. otoliths
15. outer hair cells
16. stapes
17. inferior colliculi
18. basal cells
19. olfactory bulb
20. referred pain

*Building Your Medical Vocabulary* (Answers may vary; these are acceptable examples.)

1. two—binaural
2. cross over—hemidecussation
3. half—hemidecussation
4. tears—lacrimal
5. stone—otolithic
6. spot—macula sacculi
7. pain—nociceptor
8. dark—scotopic
9. infection—asepsis
10. hair—peritrichial

*True or False* (These items are false for the reasons given; all others are true.)

1. These fibers end in the medulla oblongata.
3. Because of hemidecussation, each hemisphere receives signals from both eyes.
5. The posterior chamber, the space between iris and lens, is filled with aqueous humor.
6. Descending analgesic fibers block signals that have reached the dorsal horn of the spinal cord.
10. The trochlear and abducens nerves control the superior oblique and lateral rectus, respectively.

*Answers to Figure Legend Questions*

16.1 Yes; two touches are felt separately if they straddle the boundary between two separate receptive fields.
16.9 The lower margin of the blue zone ("all sound") would be higher in that frequency range.
16.14 They are the outer hair cells, which function to "tune the cochlea" and improve discrimination between sounds of different pitches.
16.15 It would oppose the inward movement of the tympanic membrane and, thus, reduce the amount of vibration transferred to the inner ear.
16.23 This would prevent tears from draining into the lacrimal canals, resulting in more watery eyes.
16.24 Cranial nerve III, because it controls more eye movements than IV or VI
16.28 It is the right eye. The optic disc is always medial to the fovea, so this has to be a view of the observer's left and the subject's right.
16.40 Approximately 68:20:0, and yellow
16.43 It would cause blindness in the left half of the visual field. It would not affect the visual reflexes.

## CHAPTER 17

*Testing Your Recall*

1. b
2. d
3. a
4. c
5. c
6. c
7. d
8. c
9. a
10. e
11. adenohypophysis
12. tyrosine
13. acromegaly
14. cortisol
15. glucocorticoids
16. interstitial
17. negative feedback inhibition
18. hypophyseal portal system
19. permissive
20. Up-regulation

*Building Your Medical Vocabulary* (Answers may vary; these are acceptable examples.)

1. gland—adenohypophysis
2. bile—cholecystokinin
3. flow through—diabetes
4. twenty—eicosanoid
5. yellow—luteinizing
6. resembling—thyroid
7. full of—glomerulosa
8. favoring—progesterone
9. turn—gonadotropin
10. urine—glycosuria

*True or False* (These items are false for the reasons given; all others are true.)

5. Hormones are also secreted by the heart, liver, kidneys, and other organs not generally regarded as glands.
7. The pineal gland and thymus undergo involution with age.
8. Without iodine, there is no thyroid hormone (TH); without TH, there can be no negative feedback inhibition.
9. The tissue at the center is the adrenal medulla.
10. There are also two testes, two ovaries, and four parathyroid glands.

*Answers to Figure Legend Questions*

17.1 Heart, liver, stomach, small intestine, placenta (any three)
17.4 The neurohypophysis
17.19 Steroids enter the target cell; they do not bind to membrane receptors or activate second messengers.
17.24 Such a drug would block leukotriene synthesis and thus inhibit allergic and inflammatory responses.
17.25 She would have been a pituitary giant.

## CHAPTER 18

*Testing Your Recall*

1. b
2. c
3. c
4. a
5. b
6. d
7. d
8. c
9. d
10. c
11. hemopoiesis
12. hematocrit (packed cell volume)
13. thromboplastin
14. agglutinogens
15. hemophilia
16. hemostasis
17. Sickle-cell disease
18. polycythemia
19. vitamin $B_{12}$
20. erythropoietin

*Building Your Medical Vocabulary* (Answers may vary; these are acceptable examples.)

1. without—anemia
2. producing—erythroblast
3. red—erythrocyte
4. aggregate—agglutination
5. blood—hemostasis
6. white—leukocyte
7. deficiency—leukopenia
8. vein—phlebotomy
9. formation—hemopoiesis
10. clot—thrombosis

*True or False* (These items are false for the reasons given; all others are true.)

3. Oxygen deficiency is the result of anemia, not its cause.
4. Clotting (coagulation) is one mechanism of hemostasis, but hemostasis includes other mechanisms (vascular spasm and platelet plug), so it is not synonymous with the other two terms.
6. The most abundant WBCs are neutrophils.
9. The heme is excreted; the globin is broken down into amino acids that can be reused.
10. In leukemia, there is an excess of WBCs. A WBC deficiency is leukopenia.

18.1 A nucleus

18.4 The sunken center represents the former location of the nucleus.

18.5 Hemoglobin consists of a noncovalent association of four protein chains. The prosthetic group is the heme moiety of each of the four chains.

18.18 *Myelo-* refers to the bone marrow, where these cells develop.

18.19 Although numerous, these WBCs are immature and incapable of performing their defensive roles.

18.21 A platelet plug lacks the fibrin mesh that a blood clot has.

18.22 It would affect only the intrinsic mechanism.

18.23 In both blood clotting and the signal amplification mechanism of hormone action, the product of one reaction step is an enzyme that catalyzes the production of many more molecules of product at the next step. Thus, there is a geometric increase in the number of product molecules at each step and ultimately, a large final result from a small beginning.

18.25 The older theory of leeching was that many disorders are caused by "bad blood," which could be removed painlessly by medicinal leeches. The modern practice is to take advantage of the anticoagulants in the leech saliva to promote blood flow to a tissue or to dissolve and remove clots that have already formed.

## CHAPTER 19

*Testing Your Recall*

| | | |
|---|---|---|
| 1. d | 9. a | 16. T wave |
| 2. b | 10. e | 17. vagus |
| 3. d | 11. systole, diastole | 18. myocardial |
| 4. a | 12. systemic | infarction |
| 5. a | 13. atrioventricular | 19. endocardium |
| 6. d | (coronary) | 20. cardiac output |
| 7. d | sulcus | |
| 8. c | 14. Na⁺ | |
| | 15. gap junctions | |

*Building Your Medical Vocabulary* (Answers may vary; these are acceptable examples.)

1. entryway—atrium
2. slow—bradycardia
3. heart—cardiology
4. crown—coronary
5. moon—semilunar
6. nipple—papillary
7. semi—semilunar
8. fast—tachycardia
9. vessel—vasomotor
10. belly—ventricle

*True or False* (These items are false for the reasons given; all others are true.)

1. The coronary circulation is part of the systemic circuit; the other division is the pulmonary circuit.

3. The first two-thirds of ventricular filling occurs before the atria contract. The atria add only about 31% of the blood that fills the ventricles.

6. The first heart sound occurs at the time of the QRS complex.

7. The heart has its own internal pacemaker and would continue beating; the nerves only alter the heart rate.

10. The ECG is a composite record of the electrical activity of the entire myocardium, not a record from a single myocyte. It looks much different from an action potential.

19.1 Both; they receive pulmonary arteries from the pulmonary circuit and bronchial arteries from the systemic circuit.

19.2 To the left

19.7 The trabeculae carneae

19.12 The right atrium

19.14 It ensures that wave summation and tetanus will not occur, thus ensuring relaxation and refilling of the heart chambers.

19.19 They prevent prolapse of the AV valves during ventricular systole.

19.20 This is the point at which the aortic valve opens and blood is ejected into the aorta, raising its blood pressure.

## CHAPTER 20

*Testing Your Recall*

| | | |
|---|---|---|
| 1. c | 8. a | 14. thoracic pump |
| 2. b | 9. e | 15. oncotic pressure |
| 3. a | 10. d | 16. transcytosis |
| 4. e | 11. systolic, diastolic | 17. sympathetic |
| 5. b | 12. continuous | 18. baroreceptors |
| 6. c | capillaries | 19. the arterial circle |
| 7. e | 13. Anaphylactic | 20. basilic, cephalic |

*Building Your Medical Vocabulary* (Answers may vary; these are acceptable examples.)

1. vessel—angiogenesis
2. arm—brachiocephalic
3. abdomen—celiac
4. window—fenestrations
5. neck—jugular
6. belonging to—vasa vasorum
7. standing—saphenous
8. below—subclavian
9. chest—thoracoacromial
10. bladder—vesical

*True or False* (These items are false for the reasons given; all others are true.)

4. Some veins have valves, but arteries do not.

5. By the formula $F \propto r^4$, the flow increases 16-fold.

8. The capillaries normally reabsorb about 85% of the fluid they filter; the rest is absorbed by the lymphatic system.

9. An aneurysm is a weak, bulging vessel that *may* rupture.

10. The response to falling blood pressure is a corrective vasoconstriction.

20.2 Veins are subjected to less pressure than arteries and have less need of elasticity.

20.5 Endocrine glands, kidneys, the small intestine, and choroid plexuses of the brain

20.8 Veins have less muscular and elastic tissue, so they expand more easily than arteries.

20.9 Arterial anastomoses: the arterial circle of the brain, the celiac circulation, encircling the heads of the humerus and femur, and the arterial arches of the hand and foot. Venous anastomoses: the jugular veins, the azygos system, the mesenteric veins, and venous networks of the hand and foot. Portal systems: the hepatic portal system and (outside of this chapter) the hypophyseal portal system. Answers may vary.

20.18 Nothing would happen if he lifted his finger from point O because the valve at that point would prevent blood from flowing downward and filling the vein. If he lifted his finger from point H, blood would flow upward, fill the vein, and the vein between O and H would stand out.

20.24 Aorta → left common carotid a. → external carotid a. → superficial temporal a.

20.32 The ovaries and testes begin their embryonic development near the kidneys. The gonadal veins elongate as the gonads descend to the pelvic cavity and scrotum.

20.35 Joint movements may temporarily compress an artery. Anastomoses allow for continued blood flow through alternative routes to more distal regions.

20.36 The cephalic, basilic, and median cubital vv.

## CHAPTER 21

*Testing Your Recall*

1. b
2. c
3. a
4. a
5. d
6. b
7. e
8. d
9. a
10. c
11. pathogen
12. lysozyme
13. Lymphadenitis
14. diapedesis (emigration)
15. opsonization
16. pyrogen
17. interleukins
18. antigen-binding site, epitope
19. clonal deletion
20. autoimmune

*Building Your Medical Vocabulary* (Answers may vary; these are acceptable examples.)

1. apart—anaphylactic
2. secrete—paracrine
3. outside—extravasated
4. arising—endogenous
5. freedom—immunology
6. set in motion—cytokine
7. water—lymphatic
8. enlargement—splenomegaly
9. disease—lymphadenopathy
10. fire—pyrogen

*True or False* (These items are false for the reasons given; all others are true.)

1. Lysozyme is a bacteria-killing enzyme.
3. Interferons promote inflammation.
4. Helper T cells are also necessary to humoral immunity.
9. Anergy is a loss of lymphocyte activity, whereas autoimmune diseases result from misdirected activity.
10. Interferons inhibit viral replication; perforins lyse bacteria.

*Answers to Figure Legend Questions*

21.3 There are much larger gaps between the endothelial cells of lymphatic capillaries than between those of blood capillaries.

21.4 There would be no consistent one-way flow of lymph. Lymph and tissue fluid would accumulate, especially in the lower regions of the body.

21.5 (1) Prevention of excess tissue fluid accumulation and (2) monitoring the tissue fluids for pathogens

21.6 Lymph flows from the breast to the axillary lymph nodes. Therefore, metastatic cancer cells tend to lodge first in those nodes.

21.14 Erythrocytes in the red pulp; lymphocytes and macrophages in the white pulp

21.16 Both of these produce a ring of proteins in the target cell plasma membrane, opening a hole in the membrane through which the cell contents escape.

21.23 All three defenses depend on the action of helper T cells, which are destroyed by HIV.

21.26 The ER is the site of antibody synthesis.

21.31 AZT targets reverse transcriptase. If this enzyme is unable to function, HIV cannot produce viral DNA and insert it into the host cell DNA, and the virus therefore cannot be replicated.

## CHAPTER 22

*Testing Your Recall*

1. c
2. c
3. a
4. e
5. e
6. c
7. b
8. a
9. d
10. a
11. epiglottis
12. bronchial tree
13. pulmonary surfactant
14. atmospheric
15. Obstructive
16. anatomical dead space
17. compliance, elasticity
18. ventral respiratory group
19. ventilation–perfusion coupling
20. alkalosis, hypocapnia

*Building Your Medical Vocabulary* (Answers may vary; these are acceptable examples.)

1. imperfect—atelectasis
2. smoke—hypercapnia
3. cancer—carcinoma
4. horn—corniculate
5. true—eupnea
6. measuring device—spirometer
7. nose—nasofacial
8. breathing—dyspnea
9. breath—spirometry
10. shield—thyroid

*True or False* (These items are false for the reasons given; all others are true.)

1. They also fire during expiration (although at a lower rate) to exert a braking action on the diaphragm.
4. When volume increases, pressure decreases.
5. Atelectasis can have other causes such as airway obstruction.
8. In an average 500 mL tidal volume, 350 mL reaches the alveoli.
10. Most $CO_2$ is transported as bicarbonate ions.

*Answers to Figure Legend Questions*

22.3 The line would cross the figure just slightly above the trachea label.

22.4 Epiglottic, corniculate, and arytenoid

22.7 The right main bronchus is slightly wider and more vertical than the left, making it easier for aspirated objects to fall into the right.

22.8 To secrete mucus

22.11 It lacks cartilage.

22.19 $P_{O_2}$ drops from 104 to 95 mm Hg on its way out of the lungs because of some mixing with systemic blood. It drops farther to 40 mm Hg when the blood gives up $O_2$ to respiring tissues, and remains at this level until the blood is reoxygenated back in the lungs. $P_{CO_2}$ is 40 mm Hg leaving the lungs and rises to 46 mm Hg when $CO_2$ is picked up from respiring tissues. It remains at that level until the blood returns to the lungs and unloads $CO_2$.

22.23 About 70%

22.25 In the alveoli, $CO_2$ leaves the blood, $O_2$ enters, and all the chemical reactions are the reverse of those in figure 22.23. The blood bicarbonate concentration will be reduced following alveolar gas exchange.

22.26 A higher temperature suggests a relatively high metabolic rate and, thus, an elevated demand for oxygen. Comparison of these curves shows that for a given $P_{O_2}$, hemoglobin gives up more oxygen at warmer temperatures.

# CHAPTER 23

## Testing Your Recall

| | | |
|---|---|---|
| 1. a | 9. c | 16. transport |
| 2. d | 10. a | maximum |
| 3. b | 11. micturition | 17. Antidiuretic |
| 4. c | 12. Renal | hormone |
| 5. b | autoregulation | 18. internal urethral |
| 6. b | 13. trigone | 19. protein |
| 7. d | 14. macula densa | 20. arcuate |
| 8. e | 15. podocytes | |

## Building Your Medical Vocabulary (Answers may vary; these are acceptable examples.)

1. nitrogen—azotemia
2. bladder—cystitis
3. ball—glomerulus
4. next to—juxtaglomerular
5. middle—mesangial
6. kidney—nephron
7. foot—podocyte
8. sagging—nephroptosis
9. pus—pyelonephritis
10. straight—vasa recta

## True or False (These items are false for the reasons given; all others are true.)

1. Calcium and sodium reabsorption by the PCT are influenced by parathyroid hormone and angiotensin II.
2. Urine contains more urea and chloride than sodium.
4. A substantial amount of tubular fluid is reabsorbed by the paracellular route, passing through leaky tight junctions.
5. Glycosuria does not occur in diabetes insipidus.
8. Urine can be as dilute as 50 mOsm/L.

## Answers to Figure Legend Questions

23.2 Ammonia is produced by the deamination of amino acids; urea is produced from ammonia and carbon dioxide; uric acid is produced from nucleic acids; and creatinine is produced from creatine phosphate.

23.3 It would be in the dark space at the top of the figure, where the jejunum, colon, and spleen are shown.

23.10 The afferent arteriole is larger. The relatively large inlet to the glomerulus and its small outlet result in high blood pressure in the glomerulus. This is the force that drives glomerular filtration.

23.16 It lowers the urine pH because of the $Na^+$-$H^+$ antiport. The more $Na^+$ that is reabsorbed, the more $H^+$ is secreted into the tubular fluid.

23.23 The relatively short female urethra is less of an obstacle for bacteria traveling from the perineum to the urinary bladder.

# CHAPTER 24

## Testing Your Recall

| | | |
|---|---|---|
| 1. c | 9. d | 16. hyperkalemia |
| 2. a | 10. b | 17. hyponatremia |
| 3. a | 11. $Na^+$ | 18. respiratory |
| 4. a | 12. $K^+$ and $Mg^{2+}$ | acidosis |
| 5. d | 13. metabolic water | 19. limiting pH |
| 6. c | 14. cutaneous | 20. osmolarity |
| 7. e | transpiration | |
| 8. b | 15. fluid | |
| | sequestration | |

## Building Your Medical Vocabulary (Answers may vary; these are acceptable examples.)

1. food—hyperalimentation
2. blood—hypoxemia
3. intestine—parenteral
4. potassium—hyperkalemia
5. sodium—hyponatremia
6. next to—parenteral
7. isolate—sequestration
8. breathing—transpiration
9. across—transpiration
10. volume—hypovolemia

## True or False (These items are false for the reasons given; all others are true.)

2. Aldosterone has only a small influence on blood pressure.
5. PTH promotes calcium absorption but phosphate excretion.
6. Protein buffers more acid than bicarbonate or phosphates do.
9. More water than salt is lost, so the body fluids become hypertonic.
10. Oral wetting and cooling have only a short-term effect on thirst.

## Answers to Figure Legend Questions

24.1 The tissue fluid compartment
24.8 Ingestion of water
24.10 It would decrease.
24.13 Reverse both arrows to point to the left.

# CHAPTER 25

## Testing Your Recall

| | | |
|---|---|---|
| 1. b | 8. a | 15. vagus |
| 2. d | 9. a | 16. gastrin |
| 3. c | 10. a | 17. sinusoids |
| 4. e | 11. occlusal | 18. maltase, maltose |
| 5. a | 12. amylase, lipase | 19. chylomicrons |
| 6. c | 13. parotid | 20. iron |
| 7. a | 14. enteric | |

## Building Your Medical Vocabulary (Answers may vary; these are acceptable examples.)

1. cavity—antrum
2. juice—chylomicron
3. little—micelle
4. vomiting—emetic
5. bridle—frenulum
6. liver—hepatocyte
7. dry—jejunum
8. gateway—portal
9. gateway—pyloric
10. S-shaped—sigmoid

## True or False (These items are false for the reasons given; all others are true.)

1. Fat digestion begins in the stomach.
2. Most of the tooth is dentin.
3. Hepatocytes secrete bile into the bile canaliculi.
7. Intrinsic factor is involved in the absorption of vitamin $B_{12}$.
10. Water, glucose, and other nutrients pass between cells, through the tight junctions.

## Answers to Figure Legend Questions

25.6 The first and second premolars and the third molar
25.7 The enamel is a secretion, not a tissue; all the rest are living tissues.
25.11 Blockage of the mouth by the root of the tongue and blockage of the nose by the soft palate
25.12 The muscularis externa of the esophagus has two layers of muscle, with skeletal muscle in the upper to middle regions and smooth muscle in the middle to lower regions. In the stomach, it has three layers of muscle, all of which is smooth muscle.

25.14 It exchanges H⁺ for K⁺ (H⁺-K⁺ ATPase is an active transport pump).

25.20 The hepatic artery and the hepatic portal vein

25.31 Lipids do not enter the hepatic portal system that leads directly to the liver. Dietary fats are absorbed into the lacteals of the small intestine, then would have to travel the following route, at a minimum, to reach the liver: intestinal trunk → thoracic duct → left subclavian vein → heart → aorta → celiac trunk → common hepatic artery → hepatic artery proper → hepatic arteries → liver.

25.32 The internal anal sphincter is composed of smooth muscle and therefore controlled by the autonomic nervous system. The external anal sphincter is composed of skeletal muscle and therefore controlled by the somatic nervous system.

## CHAPTER 26

*Testing Your Recall*

| | | |
|---|---|---|
| 1. a | 8. a | 15. liver |
| 2. c | 9. d | 16. insulin |
| 3. b | 10. d | 17. core temperature |
| 4. e | 11. incomplete | 18. arcuate |
| 5. b | 12. glycogenolysis | 19. cytochromes |
| 6. e | 13. gluconeogenesis | 20. ATP synthase, |
| 7. c | 14. urea | ATP |

*Building Your Medical Vocabulary* (Answers may vary; these are acceptable examples.)

1. bag—ascites
2. bad—cachexia
3. color—cytochrome
4. producing—lipogenesis
5. sugar—hypoglycemia
6. like—ascites
7. thin—leptin
8. splitting—glycolysis
9. new—gluconeogenesis
10. push—chemiosmotic

*True or False* (These items are false for the reasons given; all others are true.)

1. Leptin suppresses the appetite.
4. Most of the cholesterol is endogenous, not dietary.
5. Excessive protein intake can cause renal damage.
8. Gluconeogenesis is a postabsorptive phenomenon.
9. Brown fat does not generate ATP.

*Answers to Figure Legend Questions*

26.2 A high HDL:LDL ratio indicates that excess cholesterol is being transported to the liver for removal from the body. A high LDL:HDL ratio indicates a high rate of cholesterol deposition in the walls of the arteries.

26.3 It would stop at step 3 and PGAL would accumulate. Anaerobic fermentation replenishes NAD⁺.

26.5 NADH and FADH₂

26.9 Acidosis (or ketoacidosis or metabolic acidosis)

26.10 From amino acids to keto acids to pyruvic acid to glucose

## CHAPTER 27

*Testing Your Recall*

| | | |
|---|---|---|
| 1. a | 9. d | 15. tunica albuginea |
| 2. a | 10. d | 16. seminal vesicles |
| 3. a | 11. mesonephric | 17. sustentacular |
| 4. c | 12. fructose | 18. secondary |
| 5. a | 13. pampiniform | spermatocyte |
| 6. d | plexus | 19. deep |
| 7. e | 14. secondary | 20. acrosome |
| 8. c | spermatocytes | |

*Building Your Medical Vocabulary* (Answers may vary; these are acceptable examples.)

1. hidden—cryptorchidism
2. twins—epididymis
3. out—ejaculation
4. union—gamete
5. condition—cryptorchidism
6. reduction—meiosis
7. testis—cryptorchidism
8. network—rete testis
9. body—acrosome
10. sheath—tunica vaginalis

*True or False* (These items are false for the reasons given; all others are true.)

4. Only the testes are primary sex organs.
5. Female development results from a low testosterone level, not from estrogen.
7. The pampiniform plexus prevents the testes from overheating.
8. Sperm are stored in the epididymis.
9. There is no such phenomenon as male menopause, and sperm production normally continues throughout old age.

*Answers to Figure Legend Questions*

27.1 Both disorders result from defects in hormone receptors rather than from a lack of the respective hormone.

27.5 The word *vagina* means "sheath." The tunica vaginalis partially ensheaths the testis.

27.10 An enlarged prostate gland compresses the urethra and interferes with emptying of the bladder.

27.11 An overly engorged corpus spongiosum would compress the urethra and obstruct the expulsion of semen.

27.13 The crossing-over in prophase I

27.14 The next cell stage in meiosis, the secondary spermatocyte, is genetically different from the other cells of the body and would be subject to immune attack if not isolated from the antibodies in the blood.

## CHAPTER 28

*Testing Your Recall*

| | | |
|---|---|---|
| 1. a | 8. b | 15. corona radiata |
| 2. d | 9. c | 16. antrum |
| 3. c | 10. c | 17. climacteric |
| 4. a | 11. follicle | 18. conceptus |
| 5. e | 12. endometrium | 19. infundibulum, |
| 6. b | 13. menarche | fimbriae |
| 7. b | 14. corpus luteum | 20. lochia |

*Building Your Medical Vocabulary* (Answers may vary; these are acceptable examples.)

1. beginning—menarche
2. of—gravidarum
3. mound—cumulus
4. pregnancy—progesterone
5. uterus—hysterectomy
6. milk—lactation
7. to tie—tubal ligation
8. uterus—endometrium
9. egg—oogenesis
10. first—primipara

*True or False* (These items are false for the reasons given; all others are true.)

1. Only the ovum and corona radiata enter the uterine tube, not the whole follicle.
2. HCG is secreted by the placenta.
5. Many eggs and follicles undergo atresia during childhood, so their number is greatly reduced by the age of puberty.
6. Prolactin is secreted during pregnancy but does not induce lactation then.
10. Only the superficial layer (functionalis) is shed.

*Answers to Figure Legend Questions*

28.4 To move the egg or conceptus toward the uterus

28.8 paraurethral

28.9 They cause milk to flow from the acinus into the ducts of the mammary gland.

28.10 This results in a clearer image since the X-rays do not have to penetrate such a thick mass of tissue.

28.18 The rising ratio of estradiol to progesterone makes the uterus more irritable.

## CHAPTER 29

*Testing Your Recall*

| | | |
|---|---|---|
| 1. b | 8. a | 15. chorionic villi |
| 2. b | 9. d | 16. acrosome |
| 3. c | 10. d | 17. collagen |
| 4. c | 11. teratogens | 18. Down syndrome |
| 5. a | 12. nondisjunction | (trisomy-21) |
| 6. e | 13. life span | 19. foramen ovale |
| 7. c | 14. life expectancy | 20. embryo |

*Building Your Medical Vocabulary* (Answers may vary; these are acceptable examples.)

1. with—congenital
2. falling off—decidual
3. old age—progeria
4. mulberry—morula
5. flat cake—placenta
6. double—aneuploidy
7. aging—senescence
8. together—syncytiotrophoblast
9. end—telomere
10. monster—teratogen

*True or False* (These items are false for the reasons given; all others are true.)

1. Sperm require about 10 hours to become capacitated and able to fertilize an egg.
2. Fertilization occurs in the uterine tube.
3. Several early-arriving sperm clear a path for the one that fertilizes the egg.
8. Blood bypasses the lungs via the foramen ovale.
10. Exercise improves the quality of life in old age, but has not been shown to increase life expectancy significantly.

*Answers to Figure Legend Questions*

29.2 An unfertilized egg dies long before it reaches the uterus.

29.9 Eight weeks

29.12 XXY (Klinefelter syndrome) and YO (a zygote that would not survive)

29.14 Female, as seen from the two X chromosomes at the lower right

# APPENDIX C

## SYMBOLS, WEIGHTS, AND MEASURES

### Units of Length

| | |
|---|---|
| m | meter |
| km | kilometer ($10^3$ m) |
| cm | centimeter ($10^{-2}$ m) |
| mm | millimeter ($10^{-3}$ m) |
| μm | micrometer ($10^{-6}$ m) |
| nm | nanometer ($10^{-9}$ m) |

### Units of Mass and Weight

| | |
|---|---|
| amu | atomic mass unit |
| MW | molecular weight |
| mole | MW in grams |
| g | gram |
| kg | kilograms ($10^3$ g) |
| mg | milligrams ($10^{-3}$ g) |

### Units of Pressure

| | |
|---|---|
| atm | atmospheres (1 atm = 760 mm Hg) |
| mm Hg | millimeters of mercury |
| $P_X$ | partial pressure of gas × (as in $P_{O_2}$) |

### Conversion Factors

| | |
|---|---|
| 1 in. = 2.54 cm | 1 cm = 0.394 in. |
| 1 fl oz = 29.6 mL | 1 mL = 0.034 fl oz |
| 1 qt = 0.946 L | 1 L = 1.057 qt |
| 1 g = 0.0035 oz | 1 oz = 28.38 g |
| 1 lb = 0.45 kg | 1 kg = 2.2 lb |
| °C = (5/9)(°F – 32) | °F = (9/5)(°C) + 32 |

### Units of Volume

| | |
|---|---|
| L | liter |
| dL | deciliter (= 100 mL) ($10^{-1}$ L) |
| mL | milliliter ($10^{-3}$ L) |
| μL | microliter (= 1 mm³) ($10^{-6}$ L) |

### Units of Concentration

| | |
|---|---|
| mEq/L | milliequivalents per liter |
| Osm/L | osmoles per liter |
| mOsm/L | milliosmoles per liter |
| M | molar |
| mM | millimolar |
| pH | negative log of $H^+$ molarity |

### Units of Heat

| | |
|---|---|
| cal | "small" calories |
| kcal | kilocalories (Calories; 1 kcal = 1,000 cal) |
| Cal | "large" (dietary) calories (1 Cal = 1,000 cal) |

### Greek Letters

| | |
|---|---|
| α | alpha |
| β | beta |
| γ | gamma |
| Δ | delta (uppercase) |
| δ | delta (lowercase) |
| η | eta |
| θ | theta |
| μ | mu |

# APPENDIX D

## BIOMEDICAL ABBREVIATIONS

| | |
|---|---|
| $\delta+, \delta-$ | slight positive or negative charge |
| $2n$ | diploid |
| a., aa. | artery, arteries |
| A | adenine |
| Ab | antibody |
| ACh | acetylcholine |
| AChE | acetylcholinesterase |
| ACTH | adrenocorticotropic hormone |
| AD | Alzheimer disease |
| ADH | antidiuretic hormone |
| ADP | adenosine diphosphate |
| Ag | antigen |
| AIDS | acquired immunodeficiency syndrome |
| amu | atomic mass unit |
| ANF | atrial natriuretic factor |
| ANS | autonomic nervous system |
| APC | antigen-presenting cell |
| ATP | adenosine triphosphate |
| AV | atrioventricular |
| BBB | blood–brain barrier |
| BMR | basal metabolic rate |
| BP | blood pressure |
| bpm | beats per minute |
| C | cytosine, carbon |
| $Ca^{2+}$ | calcium ion |
| CAH | carbonic anhydrase |
| cAMP | cyclic adenosine monophosphate |
| CCK | cholecystokinin |
| CHF | congestive heart failure |
| $Cl^-$ | chloride ion |
| CNS | central nervous system |
| COP | colloid osmotic pressure |
| COPD | chronic obstructive pulmonary disease |
| CP | creatine phosphate |
| CRH | corticotropin-releasing hormone |
| c.s. | cross section |
| CSF | cerebrospinal fluid |
| DNA | deoxyribonucleic acid |
| ECF | extracellular fluid |
| ECG | electrocardiogram |
| EPSP | excitatory postsynaptic potential |
| ER | endoplasmic reticulum |
| FAD | flavin adenine dinucleotide |
| $Fe^{2+}$ | ferrous ion |
| $Fe^{3+}$ | ferric ion |
| FSH | follicle-stimulating hormone |
| G | guanine |
| GABA | gamma-aminobutyric acid |
| GAG | glycosaminoglycan |
| GFR | glomerular filtration rate |
| GH | growth hormone |
| GHRH | growth hormone–releasing hormone |
| GnRH | gonadotropin-releasing hormone |
| $H^+$ | hydrogen ion |
| Hb | hemoglobin |
| HCG | human chorionic gonadotropin |
| $HCO_3^-$ | bicarbonate ion |
| HDL | high-density lipoprotein |
| HIV | human immunodeficiency virus |
| Hz | hertz (cycles per second) |
| $I^-$ | iodide ion |
| ICF | intracellular fluid |
| Ig | immunoglobulin |
| IL | interleukin |
| I.M. | intramuscular |
| IPSP | inhibitory postsynaptic potential |
| IR | infrared |
| I.V. | intravenous |
| $K^+$ | potassium ion |
| LDL | low-density lipoprotein |
| LH | luteinizing hormone |
| LM | light microscope |
| l.s. | longitudinal section |
| m., mm. | muscle, muscles |
| MHC | major histocompatibility complex |
| MI | myocardial infarction |
| mRNA | messenger ribonucleic acid |
| MS | multiple sclerosis |
| MSH | melanocyte-stimulating hormone |
| mV | millivolts |
| MW | molecular weight |
| $n$ | haploid |
| n., nn. | nerve, nerves |
| $Na^+$ | sodium ion |
| $NAD^+$ | nicotinamide adenine dinucleotide |
| NE | norepinephrine |
| NFP | net filtration pressure |
| $OH^-$ | hydroxyl ion |
| OT | oxytocin |
| $P_i$ | inorganic phosphate group |
| PG | prostaglandin |
| PIH | prolactin-inhibiting hormone |
| PNS | peripheral nervous system |
| PRL | prolactin (luteotropin) |
| PTH | parathyroid hormone (parathormone) |
| RBC | red blood cell |
| Rh | rhesus factor |
| RMP | resting membrane potential |
| RNA | ribonucleic acid |
| rRNA | ribosomal ribonucleic acid |
| SA | sinoatrial |
| SEM | scanning electron microscope |
| T | thymine |
| $T_{1/2}$ | half-life |
| $T_3$ | triiodothyronine |
| $T_4$ | thyroxine (tetraiodothyronine) |
| $T_m$ | transport maximum |
| TEM | transmission electron microscope |
| TRH | thyrotropin-releasing hormone |
| tRNA | transfer ribonucleic acid |
| TSH | thyroid-stimulating hormone |
| UV | ultraviolet |
| v., vv. | vein, veins |
| VLDL | very low–density lipoprotein |
| $Vo_{2max}$ | maximum oxygen uptake |
| WBC | white blood cell |

## THE GENETIC CODE

To determine what amino acid is encoded by a given mRNA codon, choose the first letter of the codon from the column on the left, the second letter from the row across the top, and the third letter from the column on the right. The amino acid symbol is at the intersection of these three. For example, codon GCU codes for alanine (Ala) and CAC codes for histidine (His). Codons UAA, UAG, and UGA are *stop codons*, which do not code for any amino acid but indicate the end of the message to be translated, like the period at the end of a sentence.

| First Position | Second Position | | | | Third Position |
|---|---|---|---|---|---|
| | **A** | **G** | **C** | **U** | |
| **A** | Lys | Arg | Thr | Ile | **A** |
| | Lys | Arg | Thr | Met | **G** |
| | Asn | Ser | Thr | Ile | **C** |
| | Asn | Ser | Thr | Ile | **U** |
| **G** | Glu | Gly | Ala | Val | **A** |
| | Glu | Gly | Ala | Val | **G** |
| | Asp | Gly | Ala | Val | **C** |
| | Asp | Gly | Ala | Val | **U** |
| **C** | Gln | Arg | Pro | Leu | **A** |
| | Gln | Arg | Pro | Leu | **G** |
| | His | Arg | Pro | Leu | **C** |
| | His | Arg | Pro | Leu | **U** |
| **U** | STOP | STOP | Ser | Leu | **A** |
| | STOP | Trp | Ser | Leu | **G** |
| | Tyr | Cys | Ser | Phe | **C** |
| | Tyr | Cys | Ser | Phe | **U** |

**❓** *The following genetic code represents a well-known hormone, angiotensin II: UUU-CCC-CAC-AUA-UAU-GUA-AGG-GAC. The amino acid sequence represented by this code begins Phe-Pro.... Write the rest of the sequence.*

# GLOSSARY

Terms defined here are not necessarily the most important ones in the book, but are terms that are reintroduced most often and not redefined each time they arise. The index indicates where you can find definitions or explanations of additional terms. Terms are defined only in the sense that they are used in this book. Some have broader meanings, even within biology and medicine, that are beyond its scope. Terms that are commonly abbreviated, such as ATP and PET, are defined under the full spelling.

## A

**abdominal cavity** The body cavity between the diaphragm and pelvic brim. fig. A.5

**abduction** (ab-DUC-shun) Movement of a body part away from the median plane, as in raising an arm away from the side of the body. fig. 9.13

**absorption** 1. Process in which a chemical passes through a membrane or tissue surface and becomes incorporated into a body fluid or tissue. 2. Any process in which one substance passes into another and becomes a part of it. *Compare* adsorption.

**acetylcholine (ACh)** (ASS-eh-till-CO-leen) A neurotransmitter released by somatic motor fibers, parasympathetic fibers, and some other neurons, composed of choline and an acetyl group. fig. 12.21

**acetylcholinesterase (AChE)** (ASS-eh-till-CO-lin-ESS-ter-ase) An enzyme that hydrolyzes acetylcholine, thus halting signal transmission at a cholinergic synapse.

**acid** A proton (H⁺) donor; a chemical that releases protons into solution.

**acidosis** An acid–base imbalance in which the blood pH is lower than 7.35.

**acinus** (ASS-ih-nus) A sac of secretory cells at the inner end of a gland duct. fig. 5.30

**actin** A filamentous intracellular protein that provides cytoskeletal support and interacts with other proteins, especially myosin, to cause cellular movement; important in muscle contraction and membrane actions such as phagocytosis, ameboid movement, and cytokinesis.

**action** The movement produced by the contraction of a particular muscle.

**action potential** A rapid voltage change in which a plasma membrane briefly reverses electrical polarity; has a self-propagating effect that produces a traveling wave of excitation in nerve and muscle cells.

**active site** The region of a protein that binds to a ligand, such as the substrate-binding site of an enzyme or the hormone-binding site of a receptor.

**active transport** Transport of particles through a selectively permeable membrane, up their concentration gradient, with the aid of a carrier that consumes ATP.

**acute** Pertaining to a disease with abrupt onset, intense symptoms, and short duration. *Compare* chronic.

**adaptation** 1. An evolutionary process leading to the establishment of species characteristics that favor survival and reproduction. 2. Any characteristic of anatomy, physiology, or behavior that promotes survival and reproduction. 3. A sensory process in which a receptor adjusts its sensitivity or response to the prevailing level of stimulation, such as dark adaptation of the eye.

**adaptive immunity** Mechanisms of defense against a pathogen or other antigen, characterized by specificity, immune memory, and rapid response to the same antigen upon reexposure.

**adduction** (ah-DUC-shun) Movement of a body part toward the median plane, such as bringing the feet together from a spread-legged position. fig. 9.13

**adenosine triphosphate (ATP)** (ah-DEN-oh-seen tri-FOSS-fate) A molecule composed of adenine, ribose, and three phosphate groups that functions as a universal energy-transfer molecule; yields adenosine diphosphate (ADP) and an inorganic phosphate group (Pᵢ) upon hydrolysis. fig. 2.30a

**adenylate cyclase** (ah-DEN-ih-late SY-clase) An enzyme of the plasma membrane that removes two phosphate molecules from ATP and makes cyclic adenosine monophosphate (cAMP); important in the activation of the cAMP second-messenger system.

**adipocyte** (AD-ih-po-site) A fat cell.

**adipose tissue** A connective tissue composed predominantly of adipocytes; fat.

**adsorption** The binding of one substance to the surface of another without becoming a part of the latter. *Compare* absorption.

**aerobic exercise** (air-OH-bic) Exercise in which oxygen is used to produce ATP; endurance exercise.

**aerobic respiration** Oxidation of organic compounds in a reaction series that requires oxygen and produces ATP.

**afferent** (AFF-uh-rent) Carrying toward, as in *afferent neurons,* which carry signals toward the central nervous system, and *afferent arterioles,* which carry blood toward a tissue.

**agglutination** (ah-GLUE-tih-NAY-shun) Clumping of cells or molecules by antibodies. fig. 18.13

**albumin** (al-BYU-min) A class of small proteins constituting about 60% of the protein fraction of the blood plasma; plays roles in blood viscosity, colloid osmotic pressure, and solute transport.

**aldosterone** (AL-doe-steh-RONE, al-DOSS-teh-rone) A steroid hormone secreted by the adrenal cortex that acts on the kidneys to promote sodium retention and potassium excretion.

**alkalosis** An acid–base imbalance in which the blood pH is higher than 7.45.

**allele** (ah-LEEL) Any of the alternative forms that one gene can take, such as dominant and recessive alleles.

**alveolus** (AL-vee-OH-lus) 1. A microscopic air sac of the lung. 2. A gland acinus. 3. A tooth socket. 4. Any small anatomical space.

**amino acids** Small organic molecules with an amino group and a carboxyl group; the monomers of which proteins are composed.

**amphipathic** (AM-fih-PATH-ic) Pertaining to a molecule that has both hydrophilic and hydrophobic regions, such as phospholipids, bile acids, and some proteins.

**ampulla** (am-PULL-uh) A wide or saclike portion of a tubular organ such as a semicircular duct or uterine tube.

**anabolism** (ah-NAB-oh-lizm) Any metabolic reactions that consume energy and construct more complex molecules with higher free energy from less complex molecules with lower free energy; for example, the synthesis of proteins from amino acids. *Compare* catabolism.

**anaerobic fermentation** (AN-err-OH-bic) A reduction reaction independent of oxygen that converts pyruvic acid to lactic acid and enables glycolysis to continue under anaerobic conditions.

**anastomosis** (ah-NASS-tih-MO-sis) An anatomical convergence, the opposite of a branch; a point where two blood vessels merge and combine their bloodstreams or where two nerves or ducts converge. fig. 20.9

**anatomical position** A reference posture that allows for standardized anatomical terminology. A subject in anatomical position is standing with the feet flat on the floor, arms down to the sides, and the palms and eyes directed forward. fig. A.1

**androgen** (AN-dro-jen) Testosterone or a related steroid hormone. Stimulates bodily changes at puberty in both sexes, adult libido in both sexes, development of male anatomy in the fetus and adolescent, and spermatogenesis.

**aneurysm** (AN-you-rizm) A weak, bulging point in the wall of a heart chamber or blood vessel that presents a threat of hemorrhage.

**angiogenesis** (AN-jee-oh-GEN-eh-sis) The growth of new blood vessels.

**angiotensin II** (AN-jee-oh-TEN-sin) A hormone produced from angiotensinogen (a plasma protein) by the kidneys and lungs; raises blood pressure by stimulating vasoconstriction and stimulating the adrenal cortex to secrete aldosterone.

**anion** (AN-eye-on) An ion with more electrons than protons and consequently a net negative charge.

**antagonist** 1. A muscle that opposes the agonist at a joint. 2. Any agent, such as a hormone or drug, that opposes another.

**antebrachium** (AN-teh-BRAY-kee-um) The region from elbow to wrist; the forearm.

**anterior** Pertaining to the front (facial-abdominal aspect) of the body; ventral.

**anterior root** The branch of a spinal nerve that emerges from the anterior side of the spinal cord and carries efferent (motor) nerve fibers; often called *ventral root*. fig. 13.2b

**antibody** A protein of the gamma globulin class that reacts with an antigen and aids in protecting the body from its harmful effects; found in the blood plasma, in other body fluids, and on the surfaces of certain leukocytes and their derivatives.

**antidiuretic hormone (ADH)** (AN-tee-DYE-you-RET-ic) A hormone released by the posterior lobe of the pituitary gland in response to low blood pressure; promotes water retention by the kidneys. Also known as *arginine vasopressin*.

**antigen** (AN-tih-jen) Any large molecule capable of binding to an antibody or immune cells and triggering an immune response.

**antigen-presenting cell (APC)** A cell that phagocytizes an antigen and displays fragments of it on its surface for recognition by other cells of the immune system; chiefly macrophages and B lymphocytes.

**antiport** A cotransport protein that moves two or more solutes in opposite directions through a cellular membrane; for example, the Na$^+$–K$^+$ pump.

**apical surface** The uppermost surface of an epithelial cell, usually exposed to the lumen of an organ. fig. 3.5

**apocrine** Pertaining to certain sweat glands with large lumens and relatively thick, aromatic secretions and to similar glands such as the mammary gland; formerly thought to form secretions by pinching off bits of apical cytoplasm.

**apoptosis** (AP-op-TOE-sis) Programmed cell death; the normal death of cells that have completed their function. *Compare* necrosis.

**appendicular** (AP-en-DIC-you-lur) Pertaining to the limbs and their supporting skeletal girdles. fig. 8.1

**areolar tissue** (AIR-ee-OH-lur) A fibrous connective tissue with loosely organized, widely spaced fibers and cells and an abundance of fluid-filled space; found under nearly every epithelium, among other places. fig. 5.14

**arteriole** (ar-TEER-ee-ole) A small artery that empties into a metarteriole or capillary.

**arteriosclerosis** (ar-TEER-ee-o-sclair-O-sis) Stiffening of the arteries correlated with age or disease processes, caused primarily by cumulative free radical damage and tissue deterioration. *Compare* atherosclerosis.

**articular cartilage** A thin layer of hyaline cartilage covering the articular surface of a bone at a synovial joint, serving to reduce friction and ease joint movement. fig. 9.5

**articulation** A skeletal joint; any point at which two bones meet; may or may not be movable.

**aspect** A particular view of the body or one of its structures, or a part that faces in a particular direction, such as the anterior aspect.

**atherosclerosis** (ATH-ur-oh-skleh-ROE-sis) A degenerative disease of the blood vessels characterized by the presence of lipid deposits; often leading to calcification of the vessel wall and obstruction of coronary, cerebral, or other vital arteries. *Compare* arteriosclerosis.

**atrioventricular (AV) node** (AY-tree-oh-ven-TRIC-you-lur) A group of autorhythmic cells in the interatrial septum of the heart that relays excitation from the atria to the ventricles.

**atrioventricular (AV) valves** The bicuspid (left) and tricuspid (right) valves between the atria and ventricles of the heart.

**atrophy** (AT-ro-fee) Shrinkage of a tissue due to age, disuse, or disease.

**autoantibody** An antibody that fails to distinguish the body's own molecules from foreign molecules and thus attacks host tissues, causing autoimmune diseases.

**autoimmune disease** Any disease in which antibodies fail to distinguish between foreign and self-antigens and attack the body's own tissues; for example, systemic lupus erythematosus and rheumatic fever.

**autolysis** (aw-TOLL-ih-sis) Digestion of cells by their own internal enzymes.

**autonomic nervous system (ANS)** (AW-toe-NOM-ic) A motor division of the nervous system that innervates glands, smooth muscle, and cardiac muscle; consists of sympathetic and parasympathetic divisions and functions largely without voluntary control. *Compare* somatic nervous system.

**autoregulation** The ability of a tissue to adjust its own blood supply through vasomotion or angiogenesis.

**autosome** (AW-toe-some) Any chromosome except the sex chromosomes. Genes on the autosomes are inherited without regard to the sex of the individual.

**axial** (AC-see-ul) Pertaining to the head, neck, and trunk; the part of the body excluding the appendicular portion. fig. 8.1

**axillary** (ACK-sih-LERR-ee) Pertaining to the armpit.

**axon** A fibrous extension of a neuron that transmits action potentials; also called a *nerve fiber*. There is only one axon to a neuron, and it is usually much longer and less branched than the dendrites. fig. 12.4

# B

**baroreceptors** (BAR-oh-re-SEP-turz) Pressure sensors located in the heart, aortic arch, and carotid sinuses that trigger autonomic reflexes in response to fluctuations in blood pressure.

**base** 1. A chemical that binds protons from solution; a proton acceptor. 2. Any of the purines or pyrimidines of a nucleic acid (adenine, thymine, guanine, cytosine, or uracil) serving in part to code for protein structure. 3. The broadest part of a tapered organ such as the uterus or heart or the inferior aspect of an organ such as the brain.

**basement membrane** A thin layer of glyco-proteins, collagen, and glycosaminoglycans beneath the deepest cells of an epithelium, serving to bind the epithelium to the underlying tissue. fig. 5.33

**basophil** (BAY-so-fill) A granulocyte with coarse cytoplasmic granules that produces heparin, histamine, and other chemicals involved in inflammation. table 18.6

**belly** The thick part of a skeletal muscle between its origin and insertion. fig. 10.4

**bicarbonate buffer system** An equilibrium mixture of carbonic acid, bicarbonate ions, and hydrogen ions ($H_2CO_3 \leftrightarrow HCO_3^- + H^+$) that stabilizes the pH of the body fluids.

**bicarbonate ion** An anion, $HCO_3^-$, that functions as a base in the buffering of body fluids.

**biogenic amines** A class of chemical messengers with neurotransmitter and hormonal functions, synthesized from amino acids and retaining an amino group; also called *monoamines*. Examples include epinephrine and thyroxine.

**bipedalism** The habit of walking on two legs; a defining characteristic of the family Hominidae that underlies many skeletal and other characteristics of humans.

**blood–brain barrier (BBB)** A barrier between the bloodstream and nervous tissue of the brain that is impermeable to many blood solutes and thus prevents them from affecting the brain tissue; formed by the tight junctions between capillary endothelial cells, the basement membrane of the endothelium, and the perivascular feet of astrocytes.

**B lymphocyte** A lymphocyte that functions as an antigen-presenting cell and, in humoral immunity, differentiates into an antibody-producing plasma cell; also called a *B cell*.

**body** 1. The entire organism. 2. Part of a cell, such as a neuron, containing the nucleus and most other organelles. 3. The largest or principal part of an organ such as the stomach or uterus; also called the *corpus*.

**brachial** (BRAY-kee-ul) Pertaining to the arm proper, the region from shoulder to elbow.

**bradykinin** (BRAD-ee-KY-nin) An oligopeptide produced in inflammation that stimulates vasodilation, increases capillary permeability, and stimulates pain receptors.

**brainstem** The stalklike lower portion of the brain, composed of all of the brain except the cerebrum and cerebellum. (Many authorities also exclude the diencephalon and regard only the medulla oblongata, pons, and midbrain as the brainstem.) fig. 14.8

**bronchiole** (BRONK-ee-ole) A pulmonary air passage that is usually 1 mm or less in diameter and lacks cartilage but has relatively abundant smooth muscle, elastic tissue, and a simple cuboidal, usually ciliated epithelium.

**bronchus** (BRONK-us) A relatively large pulmonary air passage with supportive cartilage in the wall; any passage beginning with the main bronchus at the fork in the trachea and ending with segmental bronchi, from which air continues into the bronchioles.

**brush border** A fringe of microvilli on the apical surface of an epithelial cell, serving to enhance surface area and promote absorption. fig. 5.6

**bursa** A sac filled with synovial fluid at a synovial joint, serving to facilitate muscle or joint action. fig. 9.24

# C

**calcification** The hardening of a tissue due to the deposition of calcium salts.

**calorie** The amount of thermal energy that will raise the temperature of 1 g of water by 1°C. Also called a *small calorie*.

**Calorie** *See* kilocalorie.

**canaliculus** (CAN-uh-LIC-you-lus) A microscopic canal, as in osseous tissue. fig. 7.4

**capillary** (CAP-ih-LERR-ee) The narrowest type of vessel in the cardiovascular and lymphatic systems; engages in fluid exchanges with surrounding tissues.

**capillary exchange** The process of fluid transfer between the bloodstream and tissue fluid.

**capsule** The fibrous covering of a structure such as the spleen or a synovial joint.

**carbohydrate** A hydrophilic organic compound composed of carbon and a 2:1 ratio of hydrogen to oxygen; includes sugars, starches, glycogen, and cellulose.

**carbonic anhydrase** An enzyme found in erythrocytes and kidney tubule cells that catalyzes the decomposition of carbonic acid into carbon dioxide and water or the reverse reaction ($H_2CO_3 \leftrightarrows CO_2 + H_2O$).

**carcinogen** (car-SIN-oh-jen) An agent capable of causing cancer, including certain chemicals, viruses, and ionizing radiation.

**cardiac output (CO)** The amount of blood pumped by each ventricle of the heart in one minute.

**cardiovascular system** An organ system consisting of the heart and blood vessels, serving for the transport of blood. *Compare* circulatory system.

**carpal** Pertaining to the wrist (carpus).

**carrier** 1. A protein in a cellular membrane that performs carrier-mediated transport. 2. A person who is heterozygous for a recessive allele and does not exhibit the associated phenotype, but may transmit this allele to his or her children; for example, a carrier for sickle-cell disease.

**carrier-mediated transport** Any process of transporting materials through a cellular membrane that involves reversible binding to a transport protein.

**catabolism** (ca-TAB-oh-lizm) Any metabolic reactions that release energy and break relatively complex molecules with high free energy into less complex molecules with lower free energy; for example, digestion and glycolysis. *Compare* anabolism.

**catecholamine** (CAT-eh-COAL-uh-meen) A subclass of biogenic amines that includes epinephrine, norepinephrine, and dopamine. fig. 12.21

**cation** (CAT-eye-on) An ion with more protons than electrons and consequently a net positive charge.

**caudal** (CAW-dul) 1. Pertaining to a tail or narrow tail-like part of an organ. 2. Pertaining to the inferior part of the trunk of the body, where the tail of other animals arises. *Compare* cranial. 3. Relatively distant from the forehead, especially in reference to structures of the brain and spinal cord; for example, the medulla oblongata is caudal to the pons. *Compare* rostral.

**celiac** (SEE-lee-ac) Pertaining to the abdomen.

**central** Located relatively close to the median axis of the body, as in the central nervous system; opposite of peripheral.

**central nervous system (CNS)** The brain and spinal cord.

**centriole** (SEN-tree-ole) An organelle composed of a short cylinder of nine triplets of microtubules, usually paired with another centriole perpendicular to it; origin of the mitotic spindle; identical to the basal body of a cilium or flagellum. fig. 3.33

**cephalic** (seh-FAL-ic) Pertaining to the head.

**cerebellum** (SERR-eh-BEL-um) A large portion of the brain posterior to the brainstem and inferior to the cerebrum, responsible for equilibrium, motor coordination, and memory of learned motor skills. fig. 14.11

**cerebrospinal fluid (CSF)** (SERR-eh-bro-SPY-nul, seh-REE-bro-SPY-nul) A liquid that fills the ventricles of the brain, the central canal of the spinal cord, and the space between the CNS and dura mater.

**cerebrum** (SERR-eh-brum, seh-REE-brum) The largest and most superior part of the brain, divided into two convoluted cerebral hemispheres separated by a deep longitudinal fissure.

**cervical** (SUR-vih-cul) Pertaining to the neck or any cervix.

**cervix** (SUR-vix) 1. The neck. 2. A narrow or necklike part of an organ such as the uterus and gallbladder. fig. 28.3

**channel protein** A protein in the plasma membrane that has a pore through it for the passage of materials between the cytoplasm and extracellular fluid. fig. 3.8

**chemical bond** A force that attracts one atom to another, such as their opposite charges or the sharing of electrons.

**chemical synapse** A meeting of a nerve fiber and another cell with which the neuron communicates by releasing neurotransmitters. fig. 12.20

**chemoreceptor** An organ or cell specialized to detect chemicals, as in the carotid bodies and taste buds.

**chief cells** The majority type of cell in an organ or tissue such as the parathyroid glands or gastric glands.

**cholecystokinin (CCK)** (CO-lee-SIS-toe-KY-nin) A polypeptide employed as a hormone and neurotransmitter, secreted by some brain neurons and cells of the digestive tract. fig. 12.21

**cholesterol** (co-LESS-tur-ol) A steroid that functions as part of the plasma membrane and as a precursor for all other steroids in the body.

**cholinergic** (CO-lin-UR-jic) Pertaining to acetylcholine (ACh), as in cholinergic nerve fibers that secrete ACh, cholinergic receptors that bind it, or cholinergic effects on a target organ.

**chondrocyte** (CON-dro-site) A cartilage cell; a former chondroblast that has become enclosed in a lacuna in the cartilage matrix. fig. 5.21

**chorion** (CO-ree-on) A fetal membrane external to the amnion; forms part of the placenta and has diverse functions including fetal nutrition, waste removal, and hormone secretion. fig. 29.8

**chromatin** (CRO-muh-tin) Filamentous material in the interphase nucleus, composed of DNA and associated proteins.

**chromosome** A complex of DNA and protein carrying the genetic material of a cell's nucleus. Normally there are 46 chromosomes in the nucleus of each cell except germ cells. fig. 4.5

**chronic** 1. Long-lasting. 2. Pertaining to a disease that progresses slowly and has a long duration. *Compare* acute.

**chronic bronchitis** A chronic obstructive pulmonary disease characterized by damaged and immobilized respiratory cilia, excessive mucus secretion, infection of the lower respiratory tract, and bronchial inflammation; caused especially by cigarette smoking. *See also* chronic obstructive pulmonary disease.

**chronic obstructive pulmonary disease (COPD)** Certain lung diseases (chronic bronchitis and emphysema) that result in long-term obstruction of airflow and substantially reduced pulmonary ventilation; one of the leading causes of death in old age.

**cilium** (SIL-ee-um) A hairlike process, with an axoneme, projecting from the apical surface of an epithelial cell; often motile and serving to propel matter across the surface of an epithelium, but sometimes nonmotile and serving sensory roles. fig. 3.11

**circulatory shock** A state of cardiac output inadequate to meet the metabolic needs of the body.

**circulatory system** An organ system consisting of the heart, blood vessels, and blood. *Compare* cardiovascular system.

**circumduction** A joint movement in which one end of an appendage remains relatively stationary and the other end is moved in a circle. fig. 9.16

**cirrhosis** (sih-RO-sis) A degenerative liver disease characterized by replacement of functional parenchyma with fibrous and adipose tissue; causes include alcohol, other poisons, and viral and bacterial inflammation.

**cisterna** (sis-TUR-nuh) A fluid-filled space or sac, such as the cisterna chyli of the lymphatic system and a cisterna of the endoplasmic reticulum or Golgi complex. fig. 3.28

**climacteric** A period in the lives of men and women, usually in the early 50s, marked by changes in the level of reproductive hormones, a variety of somatic and psychological effects, and in women, cessation of ovulation and menstruation (menopause).

**clone** A population of cells that are mitotically descended from the same parent cell and are identical to each other genetically or in other respects.

**coagulation** (co-AG-you-LAY-shun) The clotting of blood, lymph, tissue fluid, or semen.

**coenzyme** (co-EN-zime) A small organic molecule, usually derived from a vitamin, that is needed to make an enzyme catalytically active; acts by accepting electrons from an enzymatic reaction and transferring them to a different reaction chain.

**cofactor** A nonprotein such as a metal ion or coenzyme needed for an enzyme to function.

**cohesion** The clinging of identical molecules such as water to each other.

**collagen** (COLL-uh-jen) The most abundant protein in the body, forming the fibers of many connective tissues in places such as the dermis, tendons, and bones.

**colloid** An aqueous mixture of particles that are too large to pass through most selectively permeable membranes but small enough to remain evenly dispersed through the solvent by the thermal motion of solvent particles; for example, the proteins in blood plasma.

**colloid osmotic pressure (COP)** A portion of the osmotic pressure of a body fluid that is due to its protein. *Compare* oncotic pressure.

**columnar** A cellular shape that is significantly taller than it is wide. fig. 5.6

**commissure** (COM-ih-shur) 1. A bundle of nerve fibers that crosses from one side of the brain or spinal cord to the other. fig. 14.2 2. A corner or angle at which the eyelids, lips, or genital labia meet; in the eye, also called the *canthus*. fig. 16.22

**complement** 1. To complete or enhance the structure or function of something else, as in the coordinated action of two hormones. 2. A system of plasma proteins involved in defense against pathogens.

**computerized tomography (CT)** A method of medical imaging that uses X-rays and a computer to create an image of a thin section of the body; also called a *CT scan*.

**concentration gradient** A difference in chemical concentration from one point to another, as on two sides of a plasma membrane.

**conception** The fertilization of an egg, producing a zygote.

**conceptus** All products of conception, ranging from a fertilized egg to the full-term fetus with its embryonic membranes, placenta, and umbilical cord. *Compare* embryo, fetus, preembryo.

**condyle** (CON-dile) A rounded knob on a bone serving to smooth the movement of a joint. fig. 8.2

**conformation** The three-dimensional structure of a protein that results from interaction among its amino acid side groups, its interactions with water, and the formation of disulfide bonds.

**congenital** Present at birth; for example, an anatomical defect, a syphilis infection, or a hereditary disease.

**conjugated** A state in which one organic compound is bound to another compound of a different class, such as a protein conjugated with a carbohydrate to form a glycoprotein.

**connective tissue** A tissue usually composed of more extracellular than cellular volume and usually with a substantial amount of extracellular fiber; forms supportive frameworks and capsules for organs, binds structures together, holds them in place, stores energy (as in adipose tissue), or transports materials (as in blood).

**contractility** 1. The ability to shorten. 2. The amount of force that a contracting muscle fiber generates for a given stimulus; may be increased by epinephrine, for example, while stimulus strength remains constant.

**contralateral** On opposite sides of the body, as in reflex arcs in which the stimulus comes from one side of the body and a response is given by muscles on the other side. *Compare* ipsilateral.

**cooperative effect** Effect in which two hormones, or both divisions of the autonomic nervous system, work together to produce a single overall result.

**corona** A halo- or crownlike structure, as in the corona radiata or the coronal suture of the skull.

**corona radiata** 1. An array of nerve tracts in the brain that arise mainly from the thalamus and fan out to different regions of the cerebral cortex. 2. The first layer of cuboidal cells immediately external to the zona pellucida around an egg cell.

**coronal plane** *See* frontal plane.

**coronary circulation** A system of blood vessels that serve the wall of the heart. fig. 19.10

**corpus** Body or mass; the main part of an organ, as opposed to such regions as a head, tail, or cervix.

**cortex** (plural, *cortices*) The outer layer of some organs such as the adrenal gland, cerebrum, lymph node, and ovary; usually covers or encloses tissue called the medulla.

**corticosteroid** (COR-tih-co-STERR-oyd) Any steroid hormone secreted by the adrenal cortex, such as aldosterone, cortisol, and sex steroids.

**costal** (COSS-tul) Pertaining to the ribs.

**cotransport** A form of carrier-mediated transport in which a membrane protein transports two solutes simultaneously or within the same cycle of action by either facilitated diffusion or active transport; for example, the sodium–glucose transporter and the Na$^+$–K$^+$ pump.

**countercurrent** A situation in which two fluids flow side by side in opposite directions, as in the countercurrent multiplier of the kidney and the countercurrent heat exchanger of the scrotum.

**cranial** (CRAY-nee-ul) 1. Pertaining to the cranium of the skull. 2. In a position relatively close to the head or a direction toward the head. *Compare* caudal.

**cranial nerve** Any of 12 pairs of nerves connected to the base of the brain and passing through foramina of the cranium.

**crista** An anatomical crest, such as the crista galli of the ethmoid bone or the crista of a mitochondrion.

**cross section** A cut perpendicular to the long axis of the body or an organ.

**crural** (CROO-rul) Pertaining to the leg proper or to the crus of a organ. *See* crus.

**crus (cruss)** (plural, *crura*) 1. The leg proper; the region from the knee to the ankle. 2. A leglike extension of an organ such as the penis and clitoris. figs. 27.10b, 28.8

**cuboidal** (cue-BOY-dul) A cellular shape that is roughly like a cube or in which the height and width are about equal; typically looks squarish in tissue sections. fig. 5.5

**current** A moving stream of charged particles such as ions or electrons.

**cusp** 1. One of the flaps of a valve of the heart, veins, and lymphatic vessels. 2. A conical projection on the occlusal surface of a premolar or molar tooth.

**cutaneous** (cue-TAY-nee-us) Pertaining to the skin.

**cyanosis** (SY-uh-NO-sis) A bluish color of the skin and mucous membranes due to ischemia or hypoxemia.

**cyclic adenosine monophosphate (cAMP)** A cyclic molecule produced from ATP by the enzymatic removal of two phosphate groups; serves as a second messenger in many hormone and neurotransmitter actions. fig. 2.30b

**cytolysis** (sy-TOL-ih-sis) The rupture and destruction of a cell by such agents as complement proteins and hypotonic solutions.

**cytoplasm** The contents of a cell between its plasma membrane and its nuclear envelope, consisting of cytosol, organelles, inclusions, and the cytoskeleton.

**cytoskeleton** A system of protein microfilaments, intermediate filaments, and microtubules in a cell, serving in physical support, cellular movement, and the routing of molecules and organelles to their destinations within the cell. fig. 3.25

**cytosol** A clear, featureless, gelatinous colloid in which the organelles and other internal structures of a cell are embedded.

**cytotoxic T cell** A T lymphocyte that directly attacks and destroys infected body cells, cancerous cells, and the cells of transplanted tissues.

# D

**daughter cells** Cells that arise from a parent cell by mitosis or meiosis.

**deamination** (dee-AM-ih-NAY-shun) Removal of an amino group from an organic molecule; a step in the catabolism of amino acids.

**decomposition reaction** A chemical reaction in which a larger molecule is broken down into smaller ones. *Compare* synthesis reaction.

**decussation** (DEE-cuh-SAY-shun) The crossing of nerve fibers from the right side of the central nervous system to the left or vice versa, especially in the spinal cord, medulla oblongata, and optic chiasma.

**deep** Relatively far from the body surface; opposite of *superficial*. For example, most bones are deep to the skeletal muscles.

**degranulation** Exocytosis and disappearance of cytoplasmic granules, especially in platelets and granulocytes.

**denaturation** A change in the three-dimensional conformation of a protein that destroys its enzymatic or other functional properties, usually caused by extremes of temperature or pH.

**dendrite** Extension of a neuron that receives information from other cells or from environmental stimuli and conducts signals to the soma. Dendrites are usually shorter, more branched, and more numerous than the axon and are incapable of producing action potentials. fig. 12.4

**dendritic cell** An antigen-presenting cell of the epidermis and mucous membranes. fig. 6.3

**denervation atrophy** The shrinkage of skeletal muscle that occurs when the motor neuron dies or is severed from the muscle.

**dense connective tissue** A connective tissue with a high density of fiber, relatively little ground substance, and scanty cells; seen in tendons and the dermis, for example.

**depolarization** A shift in the electrical potential across a plasma membrane to a value less negative than the resting membrane potential; associated with excitation of a nerve or muscle cell. *Compare* hyperpolarization.

**dermal papilla** A bump or ridge of dermis that extends upward to interdigitate with the epidermis and create a wavy boundary that resists stress and slippage of the epidermis.

**dermis** The deeper of the two layers of the skin, underlying the epidermis and composed of fibrous connective tissue.

**desmosome** (DEZ-mo-some) A patchlike intercellular junction that mechanically links two cells together. fig. 5.28

**diabetes** (DY-uh-BEE-teez) Any disease characterized by chronic polyuria of metabolic origin; diabetes mellitus unless otherwise specified.

**diabetes insipidus** (in-SIP-ih-dus) A form of diabetes that results from hyposecretion of antidiuretic hormone; unlike other forms, it is not characterized by hyperglycemia or glycosuria.

**diabetes mellitus (DM)** (mel-EYE-tus) A form of diabetes that results from hyposecretion of insulin or from a deficient target cell response to it; signs include hyperglycemia and glycosuria.

**diaphysis** (dy-AFF-ih-sis) The shaft of a long bone. fig. 7.1

**diarthrosis** (DY-ar-THRO-sis) *See* synovial joint.

**diastole** (dy-ASS-tuh-lee) A period in which a heart chamber relaxes and fills with blood; especially ventricular relaxation.

**diencephalon** (DY-en-SEFF-uh-lon) A portion of the brain between the midbrain and corpus callosum; composed of the thalamus, epithalamus, and hypothalamus. fig. 14.12

**differentiation** Development of a relatively unspecialized cell or tissue into one with a more specific structure and function.

**diffusion** Spontaneous net movement of particles from a place of high concentration to a place of low concentration (down a concentration gradient).

**disaccharide** (dy-SAC-uh-ride) A carbohydrate composed of two simple sugars (monosaccharides) joined by a glycosidic bond; for example, lactose, sucrose, and maltose. fig. 2.17

**disseminated intravascular coagulation (DIC)** Widespread clotting of the blood within unbroken vessels, leading to hemorrhaging, congestion of the vessels with clotted blood, and ischemia and necrosis of organs.

**distal** Relatively distant from a point of origin or attachment; for example, the wrist is distal to the elbow. *Compare* proximal.

**disulfide bond** A covalent bond that links two cysteine residues through their sulfur atoms, serving to join one peptide chain to another or to hold a single chain in its three-dimensional conformation.

**diuretic** (DY-you-RET-ic) A chemical that increases urine output.

**dominant 1.** Pertaining to a genetic allele that is phenotypically expressed in the presence of any other allele. **2.** Pertaining to a trait that results from a dominant allele.

**dopamine** (DOE-puh-meen) An inhibitory catecholamine neurotransmitter of the central nervous system, especially of the basal nuclei, where it acts to suppress unwanted motor activity. fig. 12.21

**dorsal** Toward the back (spinal) side of the body; in humans, usually synonymous with *posterior*.

**dorsal root** *See* posterior root.

**dorsiflexion** (DOR-sih-FLEC-shun) A movement of the ankle that reduces the joint angle and raises the toes. fig. 9.22

**duodenum** (DEW-oh-DEE-num, dew-ODD-eh-num) The first portion of the small intestine extending for about 25 cm from the pyloric valve of the stomach to a sharp bend called the duodenojejunal flexure; receives chyme from the stomach and secretions from the liver and pancreas. fig. 25.24

**dynein** (DINE-een) A motor protein involved in the beating of cilia and flagella and in the movement of molecules and organelles within cells, as in retrograde transport in a nerve fiber.

# E

**ectopic** (ec-TOP-ic) In an abnormal location; for example, ectopic pregnancy and ectopic pacemakers of the heart.

**edema** (eh-DEE-muh) Abnormal accumulation of tissue fluid resulting in swelling of the tissue.

**effector** A molecule, cell, or organ that carries out a response to a stimulus.

**efferent** (EFF-ur-unt) Carrying away or out, such as a blood vessel that carries blood away from a tissue or a nerve fiber that conducts signals away from the central nervous system.

**elastic fiber** A connective tissue fiber, composed of the protein elastin, that stretches under tension and returns to its original length when released; responsible for the resilience of organs such as the skin and lungs.

**elasticity** The tendency of a stretched structure to return to its original dimensions when tension is released.

**electrical synapse** A gap junction that enables one cell to stimulate another directly, without the intermediary action of a neurotransmitter; such synapses connect the cells of cardiac muscle and single-unit smooth muscle.

**electrolyte** A salt that ionizes in water and produces a solution that conducts electricity; loosely speaking, any ion that results from the dissociation of such salts, such as sodium, potassium, calcium, chloride, and bicarbonate ions.

**elevation** A joint movement that raises a body part, as in hunching the shoulders or closing the mouth.

**embolism** (EM-bo-lizm) The obstruction of a blood vessel by an embolus.

**embolus** (EM-bo-lus) Any abnormal traveling object in the bloodstream, such as agglutinated bacteria or blood cells, a blood clot, or an air bubble.

**embryo** A developing individual from the sixteenth day of gestation when the three primary germ layers have formed, through the end of the eighth week when all of the organ systems are present. *Compare* conceptus, fetus, preembryo.

**emphysema** (EM-fih-SEE-muh) A degenerative lung disease characterized by a breakdown of alveoli and diminishing surface area available for gas exchange; occurs with aging of the lungs but is greatly accelerated by smoking or air pollution.

**endocrine gland** (EN-doe-crin) A ductless gland that secretes hormones into the bloodstream; for example, the thyroid and adrenal glands. *Compare* exocrine gland.

**endocytosis** (EN-doe-sy-TOE-sis) Any process in which a cell forms vesicles from its plasma membrane and takes in large particles, molecules, or droplets of extracellular fluid; for example, phagocytosis and pinocytosis.

**endoderm** The innermost of the three primary germ layers of an embryo; gives rise to the mucosae of the digestive and respiratory tracts and to their associated glands.

**endogenous** (en-DODJ-eh-nus) Originating internally, such as the endogenous cholesterol synthesized in the body in contrast to the exogenous cholesterol coming from the diet. *Compare* exogenous.

**endometrium** (EN-doe-MEE-tree-um) The mucosa of the uterus; the site of implantation and source of menstrual discharge.

**endoplasmic reticulum (ER)** (EN-doe-PLAZ-mic reh-TIC-you-lum) An extensive system of interconnected cytoplasmic tubules or channels; classified as rough ER or smooth ER depending on the presence or absence of ribosomes on its membrane. fig. 3.28

**endothelium** (EN-doe-THEEL-ee-um) A simple squamous epithelium that lines the lumens of the blood vessels, heart, and lymphatic vessels.

**enteric** (en-TERR-ic) Pertaining to the small intestine, as in enteric hormones.

**eosinophil** (EE-oh-SIN-oh-fill) A granulocyte with a large, often bilobed nucleus and coarse cytoplasmic granules that stain with eosin; phagocytizes antigen–antibody complexes, allergens, and inflammatory chemicals and secretes enzymes that combat parasitic infections. table 18.6

**epidermis** A stratified squamous epithelium that constitutes the superficial layer of the skin, overlying the dermis. fig. 6.3

**epinephrine** (EP-ih-NEFF-rin) A catecholamine that functions as a neurotransmitter in the sympathetic nervous system and as a hormone secreted by the adrenal medulla; also called *adrenaline.* fig. 12.21

**epiphyseal plate** (EP-ih-FIZZ-ee-ul) A plate of hyaline cartilage between the epiphysis and diaphysis of a long bone in a child or adolescent, serving as a growth zone for bone elongation. figs. 7.9, 7.11

**epiphysis** (eh-PIF-ih-sis) **1.** The head of a long bone. fig. 7.1 **2.** The pineal gland (epiphysis cerebri).

**epithelium** A type of tissue consisting of one or more layers of closely adhering cells with little intercellular material and no blood vessels; forms the coverings and linings of many organs and the parenchyma of the glands.

**erectile tissue** A tissue that functions by swelling with blood, as in the penis and clitoris and inferior concha of the nasal cavity.

**erythrocyte** (eh-RITH-ro-site) A red blood cell.

**erythropoiesis** (eh-RITH-ro-poy-EE-sis) The production of erythrocytes.

**erythropoietin** (eh-RITH-ro-POY-eh-tin) A hormone that is secreted by the kidneys and liver in response to hypoxemia and stimulates erythropoiesis.

**estrogens** (ESS-tro-jenz) A family of steroid hormones known especially for producing female secondary sex characteristics and regulating various aspects of the menstrual cycle and pregnancy; major forms are estradiol, estriol, and estrone.

**evolution** A change in the relative frequencies of alleles in a population over a period of time; the mechanism that produces adaptations in human form and function. *See also* adaptation.

**excitability** The ability of a cell to respond to a stimulus, especially the ability of nerve and muscle cells to produce membrane voltage changes in response to stimuli; also called *irritability.*

**excitation–contraction coupling** Events that link the synaptic stimulation of a muscle cell to the onset of contraction.

**excitatory postsynaptic potential (EPSP)** A partial depolarization of a postsynaptic neuron or muscle cell in response to a neurotransmitter, making it more likely to reach threshold and produce an action potential.

**excretion** The process of eliminating metabolic waste products from a cell or from the body. *Compare* secretion.

**exocrine gland** (EC-so-crin) A gland that secretes its products into another organ or onto the body surface, usually by way of a duct; for example, salivary and gastric glands. *Compare* endocrine gland.

**exocytosis** (EC-so-sy-TOE-sis) A process in which a vesicle in the cytoplasm of a cell

fuses with the plasma membrane and releases its contents from the cell; used in the elimination of cellular wastes and in the release of gland products and neurotransmitters.

**exogenous** (ec-SODJ-eh-nus) Originating externally, such as exogenous (dietary) cholesterol; extrinsic. *Compare* endogenous.

**expiration 1.** Exhaling. **2.** Dying.

**extension** Movement of a joint that increases the angle between articulating bones (straightens the joint). fig. 9.12 *Compare* flexion.

**extracellular fluid (ECF)** Any body fluid that is not contained in the cells; for example, blood, lymph, and tissue fluid.

**extrinsic** (ec-STRIN-sic) **1.** Originating externally, such as extrinsic blood-clotting factors; exogenous. **2.** Not fully contained within an organ but acting on it, such as the extrinsic muscles of the hand and eye. *Compare* intrinsic.

# F

**facilitated diffusion** The process of transporting a chemical through a cellular membrane, down its concentration gradient, with the aid of a carrier that does not consume ATP; enables substances to diffuse through the membrane that would do so poorly, or not at all, without a carrier.

**facilitation** Making a process more likely to occur, such as the firing of a neuron, or making it occur more easily or rapidly, as in facilitated diffusion.

**fascia** (FASH-ee-uh) A layer of connective tissue between the muscles or separating the muscles from the skin. fig. 10.1

**fascicle** (FASS-ih-cul) A bundle of muscle or nerve fibers ensheathed in connective tissue; multiple fascicles bound together constitute a muscle or nerve as a whole. figs. 10.1, 13.8

**fat 1.** A triglyceride molecule. **2.** Adipose tissue.

**fatty acid** An organic molecule composed of a chain of an even number of carbon atoms with a carboxyl group at one end and a methyl group at the other; one of the structural subunits of triglycerides and phospholipids.

**fenestrated** (FEN-eh-stray-ted) Perforated with holes or slits, as in fenestrated blood capillaries and the elastic sheets of large arteries. fig. 20.5

**fetus** In human development, an individual from the beginning of the ninth week when all of the organ systems are present, through the time of birth. *Compare* conceptus, embryo, preembryo.

**fibrin** (FY-brin) A sticky fibrous protein formed from fibrinogen in blood, tissue fluid, and lymph; forms the matrix of a blood clot.

**fibroblast** A connective tissue cell that produces collagen fibers and ground substance; the only type of cell in tendons and ligaments.

**fibrosis** Replacement of damaged tissue with fibrous scar tissue rather than by the original tissue type; scarring. *Compare* regeneration.

**fibrous connective tissue** Any connective tissue with a preponderance of fiber, such as areolar, reticular, dense regular, and dense irregular connective tissues.

**filtrate** A fluid formed by filtration, as at the renal glomerulus and other capillaries.

**filtration** A process in which hydrostatic pressure forces a fluid through a selectively permeable membrane (especially a capillary wall).

**fire** To produce an action potential, as in nerve and muscle cells.

**fix 1.** To hold a structure in place; for example, by fixator muscles that prevent unwanted joint movements. **2.** To preserve a tissue by means of a fixative such as formalin.

**flexion** A joint movement that, in most cases, decreases the angle between two bones. fig. 9.12 *Compare* extension.

**fluid balance** *See* water balance.

**fluid compartment** Any of the major categories of fluid in the body, separated by selectively permeable membranes and differing from each other in chemical composition. Primary examples are the intracellular fluid, tissue fluid, blood, and lymph.

**follicle** (FOLL-ih-cul) **1.** A small space, such as a hair follicle, thyroid follicle, or ovarian follicle. **2.** An aggregation of lymphocytes in a lymphatic organ or mucous membrane.

**follicle-stimulating hormone (FSH)** A hormone secreted by the anterior pituitary gland that stimulates development of the ovarian follicles and egg cells.

**foramen** (fo-RAY-men) A hole through a bone or other organ, in many cases providing passage for blood vessels and nerves.

**formed element** An erythrocyte, leukocyte, or platelet; any cellular component of blood or lymph as opposed to the extracellular fluid component.

**fossa** (FOSS-uh) A depression in an organ or tissue, such as the fossa ovalis of the heart or a cranial fossa of the skull.

**fovea** (FOE-vee-uh) A small pit, such as the fovea capitis of the femur or fovea centralis of the retina.

**free energy** The potential energy in a chemical that is available to do work.

**free radical** A particle derived from an atom or molecule, having an unpaired electron that makes it highly reactive and destructive to cells; produced by intrinsic processes such as aerobic respiration and by extrinsic agents such as chemicals and ionizing radiation.

**frenulum** (FREN-you-lum) A fold of tissue that attaches a movable structure to a relatively immovable one, such as the lip to the gum or the tongue to the floor of the mouth. fig. 25.4

**frontal plane** An anatomical plane that passes through the body or an organ from right to left and superior to inferior; also called a *coronal plane*. fig. A.1

**fundus** The base, the broadest part, or the part farthest from the opening of certain viscera such as the stomach and uterus.

**fusiform** (FEW-zih-form) Spindle-shaped; elongated, thick in the middle, and tapered at both ends, such as the shape of a smooth muscle cell or a muscle spindle.

# G

**gamete** (GAM-eet) An egg or sperm cell.

**gametogenesis** (GAM-eh-toe-JEN-eh-sis) The production of eggs or sperm.

**ganglion** (GANG-glee-un) A cluster of nerve cell bodies in the peripheral nervous system, often resembling a knot in a string.

**gangrene** Tissue necrosis resulting from ischemia.

**gap junction** A junction between two cells consisting of a pore surrounded by a ring of proteins in the plasma membrane of each cell; allows solutes to diffuse from the cytoplasm of one cell to the next; functions include cell-to-cell nutrient transfer in the developing embryo and electrical communication between cells of cardiac and smooth muscle. *See also* electrical synapse. fig. 5.28

**gastric** Pertaining to the stomach.

**gate** A protein channel in a cellular membrane that can open or close in response to chemical, electrical, or mechanical stimuli, thus controlling when substances are allowed to pass through the membrane.

**gene** An information-containing segment of DNA that codes for the production of a molecule of RNA, which in most cases goes on to play a role in the synthesis of one or more proteins.

**gene locus** The site on a chromosome where a given gene is located.

**genome** (JEE-nome) All the genes of one individual, estimated at about 20,000 genes in humans.

**genotype** (JEE-no-type) The pair of alleles possessed by an individual at one gene locus on a pair of homologous chromosomes; strongly influences the individual's phenotype for a given trait.

**germ cell** A gamete or any precursor cell destined to become a gamete.

**germ layer** Any of first three tissue layers of an embryo: ectoderm, mesoderm, or endoderm.

**gestation** (jess-TAY-shun) Pregnancy.

**globulin** (GLOB-you-lin) A globular protein such as an enzyme, antibody, or albumin; especially a family of proteins in the blood plasma that includes albumin, antibodies, fibrinogen, and prothrombin.

**glomerular capsule** (glo-MERR-you-lur) A double-walled capsule around each glomerulus of the kidney; receives glomerular filtrate and empties into the proximal convoluted tubule. Also called *Bowman's capsule*. fig. 23.7

**glomerulus 1.** A spheroid mass of blood capillaries in the kidney that filters plasma and produces glomerular filtrate, which is further processed to form the urine. fig. 23.7 **2.** A spheroidal mass of nerve endings in the olfactory bulb where olfactory neurons from the nose synapse with mitral and dendritic cells of the bulb. fig. 16.7

**glucagon** (GLUE-ca-gon) A hormone secreted by alpha cells of the pancreatic islets in response to hypoglycemia; promotes glycogenolysis and other effects that raise blood glucose concentration.

**glucocorticoid** (GLUE-co-COR-tih-coyd) Any hormone of the adrenal cortex that affects carbohydrate, fat, and protein metabolism; chiefly cortisol and corticosterone.

**gluconeogenesis** (GLUE-co-NEE-oh-JEN-eh-sis) The synthesis of glucose from non-carbohydrates such as fats and amino acids.

**glucose** A monosaccharide ($C_6H_{12}O_6$) also known as blood sugar; glycogen, starch, cellulose, and maltose are made entirely of glucose, and glucose constitutes half of a sucrose or lactose molecule. The isomer involved in human physiology is also called *dextrose*.

**glucose-sparing effect** An effect of fats or other energy substrates in which they are used as fuel by most cells, so that those cells do not consume glucose; this makes more glucose available to cells such as neurons that cannot use alternative energy substrates.

**glycocalyx** (GLY-co-CAY-licks) A layer of carbohydrate molecules covalently bonded to the phospholipids and proteins of a plasma membrane; forms a surface coat on all human cells.

**glycogen** (GLY-co-jen) A glucose polymer synthesized by liver, muscle, uterine, and vaginal cells that serves as an energy-storage polysaccharide.

**glycogenesis** (GLY-co-JEN-eh-sis) The synthesis of glycogen.

**glycogenolysis** (GLY-co-jeh-NOLL-ih-sis) The hydrolysis of glycogen, releasing glucose.

**glycolipid** (GLY-co-LIP-id) A phospholipid molecule with a carbohydrate covalently bonded to it, found in the plasma membranes of cells.

**glycolysis** (gly-COLL-ih-sis) A series of anaerobic oxidation reactions that break a glucose molecule into two molecules of pyruvic acid and produce a small amount of ATP.

**glycoprotein** (GLY-co-PRO-teen) A protein molecule with a smaller carbohydrate covalently bonded to it; found in mucus and the glycocalyx of cells, for example.

**glycosaminoglycan (GAG)** (GLY-cose-ah-ME-no-GLY-can) A polysaccharide composed of modified sugars with amino groups; the major component of a proteoglycan. GAGs are largely responsible for the viscous consistency of tissue gel and the stiffness of cartilage.

**glycosuria** (GLY-co-SOOR-ee-uh) The presence of glucose in the urine, typically indicative of a kidney disease, diabetes mellitus, or other endocrine disorder.

**goblet cell** A mucus-secreting gland cell, shaped somewhat like a wineglass, found in the epithelia of many mucous membranes. fig. 5.33a

**Golgi complex** (GOAL-jee) An organelle composed of several parallel cisternae, somewhat like a stack of saucers, that modifies and packages newly synthesized proteins and synthesizes carbohydrates. fig. 3.29

**Golgi vesicle** A membrane-bounded vesicle pinched from the Golgi complex, containing its chemical product; may be retained in the cell as a lysosome or become a secretory vesicle that releases the product by exocytosis.

**gonad** The ovary or testis.

**gonadotropin** (go-NAD-oh-TRO-pin) A pituitary hormone that stimulates the gonads; specifically FSH and LH.

**G protein** A protein of the plasma membrane that is activated by a membrane receptor and, in turn, opens an ion channel or activates an intracellular physiological response; important in linking ligand–receptor binding to second-messenger systems.

**gradient** A difference or change in any variable, such as pressure or chemical concentration, from one point in space to another; provides a basis for molecular movements such as gas exchange, osmosis, and facilitated diffusion, and for bulk movements such as the flow of blood, air, and heat.

**gray matter** A zone or layer of tissue in the central nervous system where the neuron cell bodies, dendrites, and synapses are found; forms the cerebral cortex and basal nuclei; cerebellar cortex and deep nuclei; nuclei of the brainstem; and core of the spinal cord. fig. 14.6c

**gross anatomy** Bodily structure that can be observed without magnification.

**growth factor** A chemical messenger that stimulates mitosis and differentiation of target cells that have receptors for it; important in such processes as fetal development, tissue maintenance and repair, and hemopoiesis; sometimes a contributing factor in cancer.

**growth hormone (GH)** A hormone of the anterior pituitary gland with multiple effects on many tissues, generally promoting tissue growth.

**gyrus** (JY-rus) A wrinkle or fold in the cortex of the cerebrum or cerebellum.

# H

**hair cell** Sensory cell of the cochlea, semicircular ducts, utricle, and saccule, with a fringe of surface microvilli that respond to the relative motion of a gelatinous membrane at their tips; responsible for the senses of hearing, body position, and motion.

**hair follicle** An epidermal pit that contains a hair and extends into the dermis or hypodermis.

**half-life** ($T_{1/2}$) **1.** The time required for one-half of a quantity of a radioactive element to decay to a stable isotope (*physical half-life*) or to be cleared from the body through a combination of radioactive decay and physiological excretion (*biological half-life*). **2.** The time required for one-half of a quantity of hormone to be cleared from the bloodstream.

**haploid** (*n*) In humans, having 23 unpaired chromosomes instead of the usual 46 chromosomes in homologous pairs; in any organism or cell, having half the normal diploid number of chromosomes for that species.

**helper T cell** A type of lymphocyte that performs a central coordinating role in humoral and cellular immunity; target of the human immunodeficiency virus (HIV).

**hematocrit** (he-MAT-oh-crit) The percentage of blood volume that is composed of erythrocytes; also called *packed cell volume.*

**hematoma** (HE-muh-TOE-muh) A mass of clotted blood in the tissues; forms a bruise when visible through the skin.

**heme** (heem) The nonprotein, iron-containing prosthetic group of hemoglobin or myoglobin; oxygen binds to its iron atom. fig. 18.5

**hemoglobin** (HE-mo-GLO-bin) The red gas transport pigment of an erythrocyte.

**hemopoiesis** (HE-mo-poy-EE-sis) Production of any of the formed elements of blood.

**hemopoietic stem cell (HSC)** A cell of the red bone marrow that can give rise, through a series of intermediate cells, to erythrocytes, platelets, various kinds of macrophages, and any type of leukocyte.

**heparin** (HEP-uh-rin) A polysaccharide secreted by basophils and mast cells that inhibits blood clotting.

**hepatic** (heh-PAT-ic) Pertaining to the liver.

**hepatic portal system** A network of blood vessels that connect capillaries of the intestines to capillaries (sinusoids) of the liver, thus delivering newly absorbed nutrients directly to the liver.

**heterozygous** (HET-er-oh-ZY-gus) Having nonidentical alleles at the same gene locus of two homologous chromosomes.

**high-density lipoprotein (HDL)** A lipoprotein of the blood plasma that is about 50% lipid and 50% protein; functions to transport phospholipids and cholesterol from other organs to the liver for disposal. A high proportion of HDL to low-density lipoprotein (LDL) is desirable for cardiovascular health.

**hilum** (HY-lum) A point on the surface of an organ where blood vessels, lymphatic vessels, or nerves enter and leave, usually marked by a depression and slit; the midpoint of the concave surface of any organ that is roughly bean-shaped, such as the lymph nodes, kidneys, and lungs. Also called the *hilus*. fig. 22.9b

**histamine** (HISS-ta-meen) An amino acid derivative secreted by basophils, mast cells, and some neurons; functions as a paracrine secretion and neurotransmitter to stimulate effects such as gastric secretion, bronchoconstriction, and vasodilation. fig. 12.21

**histology** **1.** The microscopic structure of tissues and organs. **2.** The study of such structure.

**homeostasis** (HO-me-oh-STAY-sis) The tendency of a living body to maintain relatively stable internal conditions in spite of greater changes in its external environment.

**homologous** (ho-MOLL-uh-gus) **1.** Having the same embryonic or evolutionary origin but not necessarily the same function, such as the scrotum and labia majora. **2.** Pertaining to two chromosomes with identical structures and gene loci but not necessarily identical alleles; each member of the pair is inherited from a different parent.

**homozygous** (HO-mo-ZY-gus) Having identical alleles at the same gene locus of two homologous chromosomes.

**hormone** A chemical messenger that is secreted by an endocrine gland or isolated gland cell, travels in the bloodstream, and triggers a physiological response in distant cells with receptors for it.

**host cell** Any cell belonging to the human body, as opposed to foreign cells introduced to it by such causes as infections and tissue transplants.

**human chorionic gonadotropin (HCG)** (COR-ee-ON-ic) A hormone of pregnancy secreted by the chorion that stimulates continued growth of the corpus luteum and secretion of its hormones. HCG in urine is the basis for pregnancy testing.

**human immunodeficiency virus (HIV)** A virus that infects human helper T cells and other cells, suppresses immunity, and causes AIDS.

**hyaline cartilage** (HY-uh-lin) A form of cartilage with a relatively clear matrix and fine collagen fibers but no conspicuous elastic fibers or coarse collagen bundles as in other types of cartilage.

**hyaluronic acid** (HY-uh-loo-RON-ic) A glycosaminoglycan that is particularly abundant in connective tissues, where it becomes hydrated and forms the tissue gel.

**hydrogen bond** A weak attraction between a slightly positive hydrogen atom on one molecule and a slightly negative oxygen or nitrogen atom on another molecule, or between such atoms on different parts of the same molecule; responsible for the cohesion of water and the coiling of protein and DNA molecules, for example.

**hydrolysis** (hy-DROL-ih-sis) A chemical reaction that breaks a covalent bond in a molecule by adding an –OH group to one side of the bond and –H to the other side, thus consuming a water molecule.

**hydrophilic** (HY-dro-FILL-ic) Pertaining to molecules that attract water or dissolve in it because of their polar nature.

**hydrophobic** (HY-dro-FOE-bic) Pertaining to molecules that do not attract water or dissolve in it because of their nonpolar nature; such molecules tend to dissolve in lipids and other nonpolar solvents.

**hydrostatic pressure** The physical force generated by a liquid such as blood or tissue fluid, as opposed to osmotic and atmospheric pressures.

**hypercapnia** (HY-pur-CAP-nee-uh) An excess of carbon dioxide in the blood.

**hyperextension** A joint movement that increases the angle between two bones beyond 180°. fig. 9.12

**hyperglycemia** (HY-pur-gly-SEE-me-uh) An excess of glucose in the blood.

**hyperkalemia** (HY-pur-ka-LEE-me-uh) An excess of potassium ions in the blood.

**hypernatremia** (HY-pur-na-TREE-me-uh) An excess of sodium ions in the blood.

**hyperplasia** (HY-pur-PLAY-zhuh) The growth of a tissue through cellular multiplication, not cellular enlargement. *Compare* hypertrophy.

**hyperpolarization** A shift in the electrical potential across a plasma membrane to a value more negative than the resting membrane potential, tending to inhibit a nerve or muscle cell. *Compare* depolarization.

**hypersecretion** Excessive secretion of a hormone or other gland product; can lead to endocrine disorders such as Addison disease or gigantism, for example.

**hypertension** Excessively high blood pressure; criteria vary but it is often considered to be a condition in which systolic pressure exceeds 140 mm Hg or diastolic pressure exceeds 90 mm Hg at rest.

**hypertonic** Having a higher osmotic pressure than human cells or some other reference solution and tending to cause osmotic shrinkage of cells.

**hypertrophy** (hy-PUR-tro-fee) The growth of a tissue through cellular enlargement, not cellular multiplication; for example, the growth of muscle under the influence of exercise. *Compare* hyperplasia.

**hypocalcemia** (HY-po-cal-SEE-me-uh) A deficiency of calcium ions in the blood.

**hypocapnia** (HY-po-CAP-nee-uh) A deficiency of carbon dioxide in the blood.

**hypodermis** (HY-po-DUR-miss) A layer of connective tissue deep to the skin; also called *superficial fascia, subcutaneous tissue,* or when it is predominantly adipose, *subcutaneous fat.*

**hypoglycemia** (HY-po-gly-SEE-me-uh) A deficiency of glucose in the blood.

**hypokalemia** (HY-po-ka-LEE-me-uh) A deficiency of potassium ions in the blood.

**hyponatremia** (HY-po-na-TREE-me-uh) A deficiency of sodium ions in the blood.

**hyposecretion** Inadequate secretion of a hormone or other gland product; can lead to endocrine disorders such as diabetes mellitus or pituitary dwarfism, for example.

**hypothalamic thermostat** (HY-po-thuh-LAM-ic) A nucleus in the hypothalamus that monitors body temperature and sends afferent signals to hypothalamic heat-promoting or heat-losing centers to maintain thermal homeostasis.

**hypothalamus** (HY-po-THAL-uh-mus) The inferior portion of the diencephalon of the brain, forming the walls and floor of the third ventricle and giving rise to the posterior pituitary gland; controls many fundamental physiological functions such as appetite, thirst, and body temperature and exerts many of its effects through the endocrine and autonomic nervous systems. fig. 14.12b

**hypothesis** An informed conjecture that is capable of being tested and potentially falsified by experimentation or data collection.

**hypotonic** Having a lower osmotic pressure than human cells or some other reference solution and tending to cause osmotic swelling and lysis of cells.

**hypoxemia** (HY-pock-SEE-me-uh) A deficiency of oxygen in the bloodstream.

**hypoxia** (hy-POCK-see-uh) A deficiency of oxygen in any tissue.

# I

**immunity** The ability to ward off a specific infection or disease, usually as a result of prior exposure and the body's production of antibodies or lymphocytes against a pathogen. *Compare* resistance.

**immunoglobulin** (IM-you-no-GLOB-you-lin) *See* antibody.

**implantation** The attachment of a conceptus to the endometrium of the uterus.

**inclusion** Any visible object in the cytoplasm of a cell other than an organelle or cytoskeletal element; usually a foreign body or a stored cell product, such as a virus, dust particle, lipid droplet, glycogen granule, or pigment.

**infarction** (in-FARK-shun) **1.** The sudden death of tissue from a lack of blood perfusion. **2.** An area of necrotic tissue produced by this process; also called an *infarct.*

**inferior** Lower than another structure or point of reference from the perspective of anatomical position; for example, the stomach is inferior to the diaphragm.

**inflammation** (IN-fluh-MAY-shun) A complex of tissue responses to trauma or infection serving to ward off a pathogen and promote tissue repair; recognized by the cardinal signs of redness, heat, swelling, and pain.

**inguinal** (IN-gwih-nul) Pertaining to the groin.

**inhibitory postsynaptic potential (IPSP)** Hyperpolarization of a postsynaptic neuron in response to a neurotransmitter, making it less likely to reach threshold and fire.

**innervation** (IN-ur-VAY-shun) The nerve supply to an organ.

**insertion** The point at which a muscle attaches to another tissue (usually a bone) and produces movement, opposite from its stationary origin; the origin and insertion of a given muscle sometimes depend on what muscle action is being considered. *Compare* origin. fig. 10.4

**inspiration 1.** Inhaling. **2.** The stimulus that resulted in this book.

**integral protein** A protein of the plasma membrane that penetrates into or all the way through the phospholipid bilayer. fig. 3.7

**integration** A process in which a neuron receives input from multiple sources and their combined effects determine its output; the cellular basis of information processing by the nervous system.

**intercalated disc** (in-TUR-kuh-LAY-ted) A complex of fascia adherens, gap junctions, and desmosomes that join two cardiac muscle cells end to end, microscopically visible as a dark line that helps to histologically distinguish this muscle type; functions as a mechanical and electrical link between cells. fig. 19.11

**intercellular** Between cells.

**intercostal** (IN-tur-COSS-tul) Between the ribs, as in the intercostal muscles, arteries, veins, and nerves.

**interdigitate** (IN-tur-DIDJ-ih-tate) To fit together like the fingers of two folded hands; for example, at the dermal–epidermal boundary, intercalated discs of the heart, and foot processes of the podocytes in the kidney. fig. 23.10b

**interleukin** (IN-tur-LOO-kin) A hormonelike chemical messenger from one leukocyte to another, serving as a means of communication and coordination during immune responses.

**interneuron** (IN-tur-NEW-ron) A neuron that is contained entirely in the central nervous system and, in the path of signal conduction, lies anywhere between an afferent pathway and an efferent pathway.

**interosseous membrane** (IN-tur-OSS-ee-us) A fibrous membrane that connects the radius to the ulna and the tibia to the fibula along most of the shaft of each bone. fig. 8.33

**interphase** That part of the cell cycle between one mitotic phase and the next, from the end of cytokinesis to the beginning of the next prophase.

**interstitial** (IN-tur-STISH-ul) **1.** Pertaining to the extracellular spaces in a tissue. **2.** Located between other structures, as in the interstitial cells of the testis.

**intervertebral disc** A cartilaginous pad between the bodies of two adjacent vertebrae.

**intracellular** Within a cell.

**intracellular fluid (ICF)** The fluid contained in the cells; one of the major fluid compartments.

**intravenous (I.V.) 1.** Present or occurring within a vein, such as an intravenous blood clot. **2.** Introduced directly into a vein, such as an intravenous injection or I.V. drip.

**intrinsic** (in-TRIN-sic) **1.** Arising from within, such as intrinsic blood-clotting factors; endogenous. **2.** Fully contained within an organ, such as the intrinsic muscles of the hand and eye. *Compare* extrinsic.

**involuntary** Not under conscious control, including tissues such as smooth and cardiac muscle and events such as reflexes.

**involution** (IN-vo-LOO-shun) Shrinkage of a tissue or organ by autolysis, such as involution of the thymus after childhood and of the uterus after pregnancy.

**ion** A chemical particle with unequal numbers of electrons or protons and consequently a net negative or positive charge; it may have a single atomic nucleus as in a sodium ion or a few atoms as in a bicarbonate ion, or it may be a large molecule such as a protein.

**ionic bond** The force that binds a cation to an anion.

**ionizing radiation** High-energy electromagnetic rays that eject electrons from atoms or molecules and convert them to ions, frequently causing cellular damage; for example, X-rays and gamma rays.

**ipsilateral** (IP-sih-LAT-ur-ul) On the same side of the body, as in reflex arcs in which a muscular response occurs on the same side of the body as the stimulus. *Compare* contralateral.

**ischemia** (iss-KEE-me-uh) Insufficient blood flow to a tissue, typically resulting in metabolite accumulation and sometimes tissue death.

**isotonic** Having the same osmotic pressure as human cells or some other reference solution.

# J

**jaundice** (JAWN-diss) A yellowish color of the skin, corneas, mucous membranes, and body fluids due to an excessive concentration of bilirubin; usually indicative of a liver disease, obstructed bile secretion, or hemolytic disease.

# K

**ketone** (KEE-tone) Any organic compound with a carbonyl (C=O) group covalently bonded to a two-carbon backbone.

**ketone bodies** Certain ketones (acetone, acetoacetic acid, and β-hydroxybutyric acid) produced by the incomplete oxidation of fats, especially when fats are being rapidly catabolized. *See also* ketosis.

**ketonuria** (KEE-toe-NEW-ree-uh) The abnormal presence of ketones in the urine; a sign of diabetes mellitus but also occurring in other conditions that entail rapid fat oxidation.

**ketosis** (kee-TOE-sis) An abnormally high concentration of ketone bodies in the blood, occurring in pregnancy, starvation, diabetes mellitus, and other conditions; tends to cause acidosis and to depress the nervous system.

**kilocalorie** The amount of heat energy needed to raise the temperature of 1 kg of water by 1°C; 1,000 calories. Also called a *Calorie* or *large calorie. See also* calorie.

**kinase** Any enzyme that adds an inorganic phosphate ($P_i$) group to another organic molecule. Also called a *phosphokinase.*

# L

**labium** (LAY-bee-um) A lip, such as those of the mouth and the labia majora and minora of the vulva.

**lactic acid** A small organic acid produced as an end product of the anaerobic fermentation of pyruvic acid.

**lacuna** (la-CUE-nuh) A small cavity or depression in a tissue such as bone, cartilage, and the erectile tissues.

**lamella** (la-MELL-uh) A little plate, such as a lamella of bone. fig. 7.4

**lamina** (LAM-ih-nuh) A thin layer, such as the lamina of a vertebra or the lamina propria of a mucous membrane. fig. 8.22

**lamina propria** (PRO-pree-uh) A thin layer of areolar tissue immediately deep to the epithelium of a mucous membrane. fig. 5.33a

**larynx** (LAIR-inks) A cartilaginous chamber in the neck containing the vocal cords; colloquially called the voicebox.

**latent period** The interval between a stimulus and response, especially in the action of nerve and muscle cells.

**lateral** Away from the midline of an organ or median plane of the body; toward the side. *Compare* medial.

**law** A verbal or mathematical description of a predictable natural phenomenon or of the relationship between variables; for example, Boyle's law of gases and the law of complementary base pairing in DNA.

**lesion** A circumscribed zone of tissue injury, such as a skin abrasion or myocardial infarction.

**leukocyte** (LOO-co-site) A white blood cell.

**leukotriene** (LOO-co-TRY-een) An eicosanoid that promotes allergic and inflammatory responses such as vasodilation and neutrophil chemotaxis; secreted by basophils, mast cells, and damaged tissues.

**libido** (lih-BEE-do) Sex drive.

**ligament** A cord or band of tough collagenous tissue binding one organ to another, especially one bone to another, and serving to hold organs in place; for example, the cruciate ligaments of the knee, broad ligament of the uterus, and falciform ligament of the liver.

**ligand** (LIG-and, LY-gand) A chemical that binds reversibly to a receptor site on a protein, such as a neurotransmitter that binds to a membrane receptor or a substrate that binds to an enzyme.

**ligand-gated channel** A channel protein in a plasma membrane that opens or closes when a ligand binds to it, enabling the ligand to determine when substances can enter or leave the cell.

**light microscope (LM)** A microscope that produces images with visible light.

**linea** (LIN-ee-uh) An anatomical line, such as the linea alba.

**lingual** (LING-gwul) Pertaining to the tongue, as in lingual papillae.

**lipase** (LY-pace) An enzyme that hydrolyzes a triglyceride into fatty acids and glycerol.

**lipid** A hydrophobic organic compound composed mainly of carbon and a high ratio of hydrogen to oxygen; includes fatty acids, fats, phospholipids, steroids, and prostaglandins.

**lipoprotein** (LIP-oh-PRO-teen) A protein-coated lipid droplet in the blood plasma or lymph, serving as a means of lipid transport; for example, chylomicrons and high- and low-density lipoproteins.

**load 1.** To pick up a gas for transport in the bloodstream. **2.** The resistance acted upon by a muscle.

**lobule** (LOB-yool) A small subdivision of an organ or of a lobe of an organ, especially of a gland.

**locus** *See* gene locus.

**long bone** A bone such as the femur or humerus that is markedly longer than wide and that generally serves as a lever.

**longitudinal** Oriented along the longest dimension of the body or of an organ.

**loose connective tissue** *See* areolar tissue.

**low-density lipoprotein (LDL)** A blood-borne droplet of about 20% protein and 80% lipid (mainly cholesterol) that transports cholesterol from the liver to other tissues for their use.

**lumbar** Pertaining to the lower back and sides, between the thoracic cage and pelvis.

**lumen** (LOO-men) The internal space of a hollow organ such as a blood vessel or the esophagus, or a space surrounded by secretory cells as in a gland acinus.

**lymph** The fluid contained in lymphatic vessels and lymph nodes, produced by the absorption of tissue fluid.

**lymphatic system** (lim-FAT-ic) An organ system consisting of lymphatic vessels, lymph nodes, the tonsils, spleen, and thymus; functions include tissue fluid recovery and immunity.

**lymph node** A small organ found along the course of a lymphatic vessel that filters the lymph and contains lymphocytes and macrophages, which respond to antigens in the lymph. fig. 21.12

**lymphocyte** (LIM-foe-site) A relatively small agranulocyte with numerous types and roles in nonspecific defense, humoral immunity, and cellular immunity. table 18.6

**lysosome** (LY-so-some) A membrane-bounded organelle containing a mixture of enzymes with a variety of intracellular and extracellular roles in digesting foreign matter, pathogens, and expired organelles.

**lysozyme** (LY-so-zime) An enzyme found in tears, milk, saliva, mucus, and other body fluids that destroys bacteria by digesting their cell walls. Also called *muramidase.*

# M

**macromolecule** Any molecule of large size and high molecular weight, such as a protein, nucleic acid, polysaccharide, or triglyceride.

**macrophage** (MAC-ro-faje) Any cell of the body, other than a leukocyte, that is specialized for phagocytosis; usually derived from a blood monocyte and often functioning as an antigen-presenting cell.

**macula** (MAC-you-luh) A patch or spot, such as the macula lutea of the retina.

**malignant** (muh-LIG-nent) Pertaining to a cell or tumor that is cancerous; capable of metastasis.

**mast cell** A connective tissue cell, similar to a basophil, that secretes histamine, heparin, and other chemicals involved in inflammation; often concentrated along the course of blood capillaries.

**matrix** **1.** The extracellular material of a tissue. **2.** The fluid within a mitochondrion containing enzymes of the citric acid cycle. **3.** The substance or framework within which other structures are embedded, such as the fibrous matrix of a blood clot. **4.** A mass of epidermal cells from which a hair root or nail root develops.

**mechanoreceptor** A sensory nerve ending or organ specialized to detect mechanical stimuli such as touch, pressure, stretch, or vibration.

**medial** Toward the midline of an organ or median plane of the body. *Compare* lateral.

**median plane** The sagittal plane that divides the body or an organ into equal right and left halves; also called *midsagittal plane.* fig. A.1

**mediastinum** (ME-dee-ass-TY-num) The thick median partition of the thoracic cavity that separates one pleural cavity from the other and contains the heart, great blood vessels, esophagus, trachea, and thymus. fig. A.5

**medulla** (meh-DULE-uh, meh-DULL-uh) Tissue deep to the cortex of certain two-layered organs such as the lymph nodes, adrenal glands, hairs, and kidneys.

**medulla oblongata** (OB-long-GAH-ta) The most caudal part of the brainstem, immediately superior to the foramen magnum of the skull, connecting the spinal cord to the rest of the brain. figs. 14.2, 14.8

**meiosis** (my-OH-sis) A form of cell division in which a diploid cell divides twice and produces four haploid daughter cells; occurs only in gametogenesis.

**melanocyte** A cell of the stratum basale of the epidermis that synthesizes melanin and transfers it to the keratinocytes.

**meninges** (meh-NIN-jeez) (singular, *meninx*) Three fibrous membranes between the central nervous system and surrounding bone: the dura mater, arachnoid mater, and pia mater. fig. 14.5

**merocrine** (MERR-oh-crin) Pertaining to gland cells that release their product by exocytosis; also called *eccrine.*

**mesenchyme** (MEZ-en-kime) A gelatinous embryonic connective tissue derived from the mesoderm; differentiates into all permanent connective tissues and most muscle.

**mesentery** (MESS-en-tare-ee) A serous membrane that binds the intestines together and suspends them from the abdominal wall; the visceral continuation of the peritoneum. fig. 25.3

**mesoderm** (MEZ-oh-durm) The middle layer of the three primary germ layers of an embryo; gives rise to muscle and connective tissue.

**metabolism** (meh-TAB-oh-lizm) The sum of all chemical reactions in the body.

**metabolite** (meh-TAB-oh-lite) Any chemical produced by metabolism.

**metaplasia** Transformation of one mature tissue type into another; for example, a change from pseudostratified to stratified squamous epithelium in an overventilated nasal cavity.

**metastasis** (meh-TASS-tuh-sis) The spread of cancer cells from the original tumor to a new location, where they seed the development of a new tumor.

**microtubule** An intracellular cylinder composed of the protein tubulin, forming centrioles, the axonemes of cilia and flagella, and part of the cytoskeleton.

**microvillus** An outgrowth of the plasma membrane that increases the surface area of a cell and functions in absorption and some sensory processes; distinguished from cilia and flagella by its smaller size and lack of an axoneme.

**milliequivalent** One-thousandth of an equivalent, which is the amount of an electrolyte that would neutralize 1 mole of $H^+$ or $OH^-$. Electrolyte concentrations are commonly expressed in milliequivalents per liter (mEq/L).

**mitochondrion** (MY-toe-CON-dree-un) An organelle specialized to synthesize ATP, enclosed in a double unit membrane with infoldings of the inner membrane called cristae.

**mitosis** (my-TOE-sis) A form of cell division in which a cell divides once and produces two genetically identical daughter cells; sometimes used to refer only to the division of the genetic material or nucleus and not to include cytokinesis, the subsequent division of the cytoplasm.

**moiety** (MOY-eh-tee) A chemically distinct subunit of a macromolecule, such as the heme and globin moieties of hemoglobin or the lipid and carbohydrate moieties of a glycolipid.

**molarity** A measure of chemical concentration expressed as moles of solute per liter of solution.

**mole** The mass of a chemical equal to its molecular weight in grams, containing $6.023 \times 10^{23}$ molecules.

**monocyte** An agranulocyte specialized to migrate into the tissues and transform into a macrophage. table 18.6

**monomer** (MON-oh-mur) **1.** One of the identical or similar subunits of a larger molecule in the dimer to polymer range; for example, the glucose monomers of starch, the amino acids of a protein, or the nucleotides of DNA. **2.** One subunit of an antibody molecule, composed of four polypeptides.

**monosaccharide** (MON-oh-SAC-uh-ride) A simple sugar, or sugar monomer; chiefly glucose, fructose, and galactose.

**motor end plate** *See* neuromuscular junction.

**motor neuron** A neuron that transmits signals from the central nervous system to any effector (muscle or gland cell); its axon is an efferent nerve fiber.

**motor protein** Any protein that produces movements of a cell or its components owing to its ability to undergo quick repetitive changes in conformation and to bind reversibly to other molecules; for example, myosin, dynein, and kinesin.

**motor unit** One motor neuron and all the skeletal muscle fibers innervated by it.

**mucosa** (mew-CO-suh) A tissue layer that forms the inner lining of an anatomical tract that is open to the exterior (the respiratory, digestive, urinary, and reproductive tracts). Composed of epithelium, connective tissue (lamina propria), and often smooth muscle (muscularis mucosae). fig. 5.33a

**mucous membrane** A mucosa.

**multipotent** Pertaining to a stem cell that has the potential to develop into two or more types of fully differentiated, functional cells, but not into an unlimited variety of cell types.

**muscle fiber** One skeletal muscle cell.

**muscle tone** A state of continual, partial contraction of resting skeletal or smooth muscle.

**muscularis externa** The external muscular wall of certain viscera such as the esophagus and small intestine. fig. 25.2

**muscularis mucosae** (MUSS-cue-LERR-iss mew-CO-see) A layer of smooth muscle immediately deep to the lamina propria of a mucosa. fig. 5.33a

**muscular system** An organ system composed of the skeletal muscles, specialized mainly for maintaining postural support and producing movements of the bones.

**mutagen** (MEW-tuh-jen) Any agent that causes a mutation, including viruses, chemicals, and ionizing radiation.

**mutation** Any change in the structure of a chromosome or a DNA molecule, often resulting in a change of organismal structure or function.

**myelin** (MY-eh-lin) A lipid sheath around a nerve fiber, formed from closely spaced spiral layers of the plasma membrane of

a Schwann cell or oligodendrocyte. fig. 12.7

**myocardium** (MY-oh-CAR-dee-um) The middle, muscular layer of the heart.

**myocyte** A muscle cell, especially a cell of cardiac or smooth muscle.

**myoepithelial cell** An epithelial cell that has become specialized to contract like a muscle cell; important in dilation of the pupil and ejection of secretions from gland acini.

**myofilament** A protein microfilament responsible for the contraction of a muscle cell, composed mainly of myosin or actin. fig. 11.3

**myoglobin** (MY-oh-GLO-bin) A red oxygen-storage pigment of muscle; supplements hemoglobin in providing oxygen for aerobic muscle metabolism.

**myosin** A motor protein that constitutes the thick myofilaments of muscle and has globular, mobile heads of ATPase that bind to actin molecules.

# N

**necrosis** (neh-CRO-sis) Pathological tissue death due to such causes as infection, trauma, or hypoxia. *Compare* apoptosis.

**negative feedback** A self-corrective mechanism that underlies most homeostasis, in which a bodily change is detected and responses are activated that reverse the change and restore stability and preserve normal body function.

**negative feedback inhibition** A mechanism for limiting the secretion of a pituitary tropic hormone. The tropic hormone stimulates another endocrine gland to secrete its own hormone, and that hormone inhibits further release of the tropic hormone.

**neonate** (NEE-oh-nate) An infant up to 6 weeks old.

**neoplasia** (NEE-oh-PLAY-zee-uh) Abnormal growth of new tissue, such as a tumor, with no useful function.

**nephron** One of approximately 1 million blood-filtering, urine-producing units in each kidney; consists of a glomerulus, glomerular capsule, proximal convoluted tubule, nephron loop, and distal convoluted tubule. fig. 23.8

**nerve** A cordlike organ of the peripheral nervous system composed of multiple nerve fibers ensheathed in connective tissue.

**nerve fiber** The axon of a single neuron.

**nerve impulse** A wave of self-propagating action potentials traveling along a nerve fiber.

**nervous tissue** A tissue composed of neurons and neuroglia.

**net filtration pressure** A net force favoring filtration of fluid from a capillary or venule when all the hydrostatic and osmotic pressures of the blood and tissue fluids are taken into account.

**neural pool** A group of interconnected neurons of the central nervous system that perform a single collective function; for example, the vasomotor center of the brainstem and speech centers of the cerebral cortex.

**neural tube** A dorsal hollow tube in the embryo that develops into the central nervous system. fig. 14.3

**neuroglia** (noo-ROG-lee-uh) All cells of nervous tissue except neurons; cells that perform various supportive and protective roles for the neurons.

**neuromuscular junction** A synapse between a nerve fiber and a muscle fiber; also called a *motor end plate*. fig. 11.7

**neuron** (NOOR-on) A nerve cell; an electrically excitable cell specialized for producing and transmitting action potentials and secreting chemicals that stimulate adjacent cells.

**neuropeptide** A peptide secreted by a neuron, often serving to modify the action of a neurotransmitter; for example, endorphins, enkephalin, and cholecystokinin. fig. 12.21

**neurotransmitter** A chemical released at the distal end of an axon that stimulates an adjacent cell; for example, acetylcholine, norepinephrine, or serotonin.

**neutrophil** (NOO-tro-fill) A granulocyte, usually with a multilobed nucleus, that serves especially to destroy bacteria by means of phagocytosis, intracellular digestion, and secretion of bactericidal chemicals. table 18.6

**nitrogenous base** (ny-TRODJ-eh-nus) An organic molecule with a single or double carbon–nitrogen ring that forms one of the building blocks of ATP, other nucleotides, and nucleic acids; the basis of the genetic code. fig. 4.1

**nitrogenous waste** Any nitrogen-containing substance produced as a metabolic waste and excreted in the urine; chiefly ammonia, urea, uric acid, and creatinine.

**nociceptor** (NO-sih-SEP-tur) A nerve ending specialized to detect tissue damage and produce a sensation of pain; pain receptor.

**norepinephrine** (nor-EP-ih-NEF-rin) A catecholamine that functions as a neurotransmitter and adrenal hormone, especially in the sympathetic nervous system. fig. 12.21

**nuclear envelope** (NEW-clee-ur) A pair of membranes enclosing the nucleus of a cell, with prominent pores allowing traffic of molecules between the nucleoplasm and cytoplasm. fig. 3.27

**nucleic acid** (new-CLAY-ic) An acidic polymer of nucleotides found or produced in the nucleus, functioning in heredity and protein synthesis; of two types, DNA and RNA.

**nucleotide** (NEW-clee-oh-tide) An organic molecule composed of a nitrogenous base, a monosaccharide, and a phosphate group; the monomer of a nucleic acid.

**nucleus** (NEW-clee-us) 1. A cell organelle containing DNA and surrounded by a double membrane. 2. A mass of neurons (gray matter) surrounded by white matter of the brain, including the basal nuclei and brainstem nuclei. 3. The positively charged core of an atom, consisting of protons and neutrons. 4. A central structure, such as the nucleus pulposus of an intervertebral disc.

## O

**olfaction** (ole-FAC-shun) The sense of smell.

**oncotic pressure** (ong-COT-ic) The difference between the colloid osmotic pressure of the blood and that of the tissue fluid, usually favoring fluid absorption by the blood capillaries. *Compare* colloid osmotic pressure.

**oocyte** (OH-oh-site) In the development of an egg cell, any haploid stage between meiosis I and fertilization.

**oogenesis** (OH-oh-JEN-eh-sis) The production of a fertilizable egg cell through a series of mitotic and meiotic cell divisions; female gametogenesis.

**opposition** A movement of the thumb in which it approaches or touches any fingertip of the same hand.

**orbit** The eye socket of the skull.

**organ** Any anatomical structure that is composed of at least two different tissue types, has recognizable structural boundaries, and has a discrete function different from the structures around it. Many organs are microscopic and many organs contain smaller organs, such as the skin containing numerous microscopic sense organs.

**organelle** Any structure within a cell that carries out one of its metabolic roles, such as mitochondria, centrioles, endoplasmic reticulum, and the nucleus; an intracellular structure other than the cytoskeleton and inclusions.

**origin** The relatively stationary attachment of a skeletal muscle. fig. 10.4 *Compare* insertion.

**osmolality** (OZ-mo-LAL-ih-tee) The molar concentration of dissolved particles in 1 kg of water.

**osmolarity** (OZ-mo-LERR-ih-tee) The molar concentration of dissolved particles in 1 L of solution.

**osmoreceptor** (OZ-mo-re-SEP-tur) A neuron of the hypothalamus that responds to changes in the osmolarity of the extracellular fluid.

**osmosis** (oz-MO-sis) The net flow of water through a selectively permeable membrane, resulting from either a chemical concentration difference or a mechanical force across the membrane.

**osmotic pressure** The amount of pressure that would have to be applied to one side of a selectively permeable membrane to stop osmosis; proportional to the concentration of nonpermeating solutes on that side and therefore serving as an indicator of solute concentration.

**osseous** (OSS-ee-us) Pertaining to bone.

**ossification** (OSS-ih-fih-CAY-shun) Bone formation.

**osteoblast** Bone-forming cell that arises from an osteogenic cell, deposits bone matrix, and eventually becomes an osteocyte.

**osteoclast** Macrophage of the bone surface that dissolves the matrix and returns minerals to the extracellular fluid.

**osteocyte** A mature bone cell formed when an osteoblast becomes surrounded by its own matrix and entrapped in a lacuna.

**osteon** A structural unit of compact bone consisting of a central canal surrounded by concentric cylindrical lamellae of matrix. fig. 7.4

**osteoporosis** (OSS-tee-oh-pore-OH-sis) A degenerative bone disease characterized by a loss of bone mass, increasing susceptibility to spontaneous fractures, and sometimes deformity of the vertebral column; causes include aging, estrogen hyposecretion, and insufficient resistance exercise.

**ovulation** (OV-you-LAY-shun) The release of a mature oocyte by the bursting of an ovarian follicle.

**ovum** Any stage of the female gamete from the conclusion of meiosis I until fertilization; a primary or secondary oocyte; an egg.

**oxidation** A chemical reaction in which one or more electrons are removed from a molecule, lowering its free energy content; opposite of reduction and always linked to a reduction reaction.

## P

**pancreatic islet** (PAN-cree-AT-ic EYE-let) A small cluster of endocrine cells in the pancreas that secretes insulin, glucagon, somatostatin, and other intercellular messengers; also called *islet of Langerhans*. fig. 17.12

**papilla** (pa-PILL-uh) A conical or nipplelike structure, such as a lingual papilla of the tongue or the papilla of a hair bulb.

**papillary** (PAP-ih-lerr-ee) 1. Pertaining to or shaped like a nipple, such as the papillary muscles of the heart. 2. Having papillae, such as the papillary layer of the dermis.

**paracrine** (PERR-uh-crin) 1. A chemical messenger similar to a hormone whose effects are restricted to the immediate vicinity of the cells that secrete it; sometimes called a local hormone. 2. Pertaining to such a secretion, as opposed to *endocrine*.

**parasympathetic nervous system** (PERR-uh-SIM-pa-THET-ic) A division of the autonomic nervous system that issues efferent fibers through the cranial and sacral nerves and exerts cholinergic effects on its target organs.

**parathyroid hormone (PTH)** A hormone secreted by the parathyroid glands that raises blood calcium concentration by stimulating bone resorption by osteoclasts, promoting intestinal absorption of calcium, and inhibiting urinary excretion of calcium.

**parenchyma** (pa-REN-kih-muh) The tissue that performs the main physiological functions of an organ, especially a gland, as opposed to the tissues (stroma) that mainly provide structural support.

**parietal** (pa-RY-eh-tul) 1. Pertaining to a wall, as in the parietal cells of the gastric glands and parietal bone of the skull. 2. The outer or more superficial layer of a two-layered membrane such as the pleura, pericardium, or glomerular capsule. *Compare* visceral. fig. A.6

**pathogen** Any disease-causing microorganism.

**pedicle** (PED-ih-cul) A small footlike process, as in the vertebrae and the renal podocytes; also called a *pedicel*.

**pelvic cavity** The space enclosed by the true (lesser) pelvis, containing the urinary bladder, rectum, and internal reproductive organs.

**pelvis** A basinlike structure such as the pelvic girdle of the skeleton or the urine-collecting space near the hilum of the kidney. figs. 8.35, 23.4

**peptide** Any chain of two or more amino acids. *See also* polypeptide, protein.

**peptide bond** A group of four covalently bonded atoms (a –C$=$O group bonded to an –NH group) that links two amino acids in a protein or other peptide. fig. 2.24b

**perfusion** The amount of blood supplied to a given mass of tissue in a given period of time (such as mL/g/min.)

**perichondrium** (PERR-ih-CON-dree-um) A layer of fibrous connective tissue covering the surface of hyaline or elastic cartilage.

**perineum** (PERR-ih-NEE-um) The region between the thighs bordered by the coccyx, pubic symphysis, and ischial tuberosities; contains the orifices of the urinary, reproductive, and digestive systems. figs. 27.6, 28.8

**periosteum** (PERR-ee-OSS-tee-um) A layer of fibrous connective tissue covering the surface of a bone. fig. 7.1

**peripheral** (peh-RIF-eh-rul) Away from the center of the body or of an organ, as in peripheral vision and peripheral blood vessels.

**peripheral nervous system (PNS)** A subdivision of the nervous system composed of all nerves and ganglia; all of the nervous system except the central nervous system.

**peristalsis** (PERR-ih-STAL-sis) A wave of constriction traveling along a tubular organ such as the esophagus or ureter, serving to propel its contents.

**peritoneum** (PERR-ih-toe-NEE-um) A serous membrane that lines the peritoneal cavity of the abdomen and covers the mesenteries and viscera.

**perivascular** (PERR-ih-VASS-cue-lur) Pertaining to the region surrounding a blood vessel.

**phagocytosis** (FAG-oh-sy-TOE-sis) A form of endocytosis in which a cell surrounds a foreign particle with pseudopods and engulfs it, enclosing it in a cytoplasmic vesicle called a phagosome.

**pharynx** (FAIR-inks) A muscular passage in the throat at which the respiratory and digestive tracts cross.

**phospholipid** An amphipathic molecule composed of two fatty acids and a phosphate-containing group bonded to the three carbons of a glycerol molecule; composes most of the molecules of the plasma membrane and other cellular membranes.

**phosphorylation** Addition of an inorganic phosphate ($P_i$) group to an organic molecule.

**piloerector** A bundle of smooth muscle cells associated with a hair follicle, responsible for erection of the hair; also called a *pilomotor muscle* or *arrector pili*. fig. 6.7

**pinocytosis** (PIN-oh-sy-TOE-sis) A form of endocytosis in which the plasma membrane sinks inward and imbibes droplets of extracellular fluid.

**plantar** (PLAN-tur) Pertaining to the sole of the foot.

**plaque** A small scale or plate of matter, such as dental plaque, the fatty plaques of atherosclerosis, and the amyloid plaques of Alzheimer disease.

**plasma** The noncellular portion of the blood.

**plasma membrane** The membrane that encloses a cell and controls the traffic of molecules in and out of the cell. fig. 3.6

**platelet** A formed element of the blood derived from a megakaryocyte, known especially for its role in stopping bleeding, but with additional roles in dissolving blood clots, stimulating inflammation, promoting tissue growth and maintenance, and destroying bacteria.

**pleura** (PLOOR-uh) A double-walled serous membrane that encloses each lung.

**plexus** A network of blood vessels, lymphatic vessels, or nerves, such as a choroid plexus of the brain or brachial plexus of nerves.

**pluripotent stem cell (PPSC)** A cell of the inner cell mass of a blastocyst that is capable of developing into any type of embryonic cell, but not into cells of the accessory organs of pregnancy.

**polymer** A molecule that consists of a long chain of identical or similar subunits, such as protein, DNA, and starch.

**polypeptide** Any chain of more than 10 or 15 amino acids. *See also* protein.

**polysaccharide** (POL-ee-SAC-uh-ride) A polymer of simple sugars; for example, glycogen, starch, and cellulose.

**polyuria** (POL-ee-YOU-ree-uh) Excessive output of urine.

**popliteal** (po-LIT-ee-ul) Pertaining to the posterior aspect of the knee.

**positron emission tomography (PET)** A method of producing a computerized image of the physiological state of a tissue using injected radioisotopes that emit positrons.

**posterior** Near or pertaining to the back or spinal side of the body; dorsal.

**posterior root** The branch of a spinal nerve that enters the posterior side of the spinal cord and carries afferent (sensory) nerve fibers; often called *dorsal root*. fig. 13.2b

**postganglionic** (POST-gang-glee-ON-ic) Pertaining to a neuron that conducts signals from a ganglion to a more distal target organ.

**postsynaptic** (POST-sih-NAP-tic) Pertaining to a neuron or other cell that receives signals from the presynaptic neuron at a synapse. fig. 12.18

**potential** A difference in electrical charge from one point to another, especially on opposite sides of a plasma membrane; usually measured in millivolts.

**preembryo** A developing individual from the time of fertilization to the time, at 16 days, when the three primary germ layers have formed. *Compare* conceptus, embryo, fetus.

**preganglionic** (PRE-gang-glee-ON-ic) Pertaining to a neuron that conducts signals from the central nervous system to a ganglion.

**presynaptic** (PRE-sih-NAP-tic) Pertaining to a neuron that conducts signals to a synapse. fig. 12.18

**primary germ layers** The ectoderm, mesoderm, and endoderm; the three tissue layers of an early embryo from which all later tissues and organs arise.

**prime mover** The muscle that produces the most force in a given joint action; agonist.

**programmed cell death** *See* apoptosis.

**pronation** (pro-NAY-shun) A rotational movement of the forearm that turns the palm downward or posteriorly. fig. 9.18

**proprioception** (PRO-pree-oh-SEP-shun) The nonvisual perception, usually subconscious, of the position and movements of the body, resulting from input from proprioceptors and the vestibular apparatus of the inner ear.

**proprioceptor** (PRO-pree-oh-SEP-tur) A sensory receptor of the muscles, tendons, and joint capsules that detects muscle contractions and joint movements.

**prostaglandin** (PROSS-ta-GLAN-din) An eicosanoid with a five-sided carbon ring in the middle of a hydrocarbon chain, playing a variety of roles in inflammation, neurotransmission, vasomotion, reproduction, and metabolism. fig. 2.22

**prostate gland** (PROSS-tate) A male reproductive gland that encircles the urethra immediately inferior to the bladder and contributes to the semen. fig. 27.10

**protein** A large polypeptide; while criteria for a protein are somewhat subjective and variable, polypeptides over 50 amino acids long are generally classified as proteins.

**proteoglycan** (PRO-tee-oh-GLY-can) A large molecule composed of a bristlelike arrangement of glycosaminoglycans surrounding a protein core in a shape resembling a bottle brush. Binds cells to extracellular materials and gives the tissue fluid a gelatinous consistency.

**proximal** Relatively near a point of origin or attachment; for example, the shoulder is proximal to the elbow. *Compare* distal.

**pseudopod** (SOO-doe-pod) A temporary cytoplasmic extension of a cell used for locomotion (ameboid movement) and phagocytosis.

**pseudostratified columnar** A type of epithelium with tall columnar cells reaching the free surface and shorter basal cells that do not reach the surface, but with all cells resting on the basement membrane; creates a false appearance of stratification. fig. 5.7

**pulmonary circuit** A route of blood flow that supplies blood to the pulmonary alveoli for gas exchange and then returns it to the heart; all blood vessels between the right ventricle and the left atrium of the heart.

**pyrogen** (PY-ro-jen) A fever-producing agent.

# R

**ramus** (RAY-mus) An anatomical branch, as in a nerve or in the pubis.

**receptor** 1. A cell or organ specialized to detect a stimulus, such as a taste cell or the eye. 2. A protein molecule that binds and responds to a chemical such as a hormone, neurotransmitter, or odor molecule.

**receptor-mediated endocytosis** A process in which certain molecules in the extracellular fluid bind to receptors in the plasma membrane, these receptors gather together, the membrane sinks inward at that point, and the molecules are incorporated into vesicles in the cytoplasm. fig. 3.22

**receptor potential** A variable change in membrane voltage produced by a stimulus acting on a receptor cell; generates an action potential if it reaches threshold.

**reduction** 1. A chemical reaction in which one or more electrons are added to a molecule, raising its free energy content; opposite of *oxidation* and always linked to an oxidation reaction. 2. Treatment of a fracture by restoring the broken parts of a bone to their proper alignment.

**reflex** A stereotyped, automatic, involuntary response to a stimulus; includes somatic reflexes, in which the effectors are skeletal muscles, and visceral (autonomic) reflexes, in which the effectors are usually visceral muscle, cardiac muscle, or glands.

**reflex arc** A simple neural pathway that mediates a reflex; involves a receptor, an afferent nerve fiber, sometimes one or more interneurons, an efferent nerve fiber, and an effector.

**refractory period** 1. A period of time after a nerve or muscle cell has responded to a stimulus in which it cannot be reexcited by a threshold stimulus. 2. A period of time after male orgasm when it is not possible to reattain erection or ejaculation.

**regeneration** Replacement of damaged tissue with new tissue of the original type. *Compare* fibrosis.

**renin** (REE-nin) An enzyme secreted by the kidneys in response to hypotension; converts the plasma protein angiotensinogen to angiotensin I, leading indirectly to a rise in blood pressure.

**repolarization** Reattainment of the resting membrane potential after a nerve or muscle cell has depolarized.

**residue** Any one of the amino acids in a protein or other peptide.

**resistance** 1. A nonspecific ability to ward off infection or disease regardless of whether the body has been previously exposed to it. *Compare* immunity. 2. A force that opposes the flow of a fluid such as air or blood. 3. A force, or load, that opposes the action of a muscle or lever.

**resting membrane potential (RMP)** A stable voltage across the plasma membrane of an unstimulated nerve or muscle cell.

**reticular cell** (reh-TIC-you-lur) A delicate, branching phagocytic cell found in the reticular connective tissue of the lymphatic organs.

**reticular fiber** A fine, branching collagen fiber coated with glycoprotein, found in the stroma of lymphatic organs and some other tissues and organs.

**reticular tissue** A connective tissue composed of reticular cells and reticular fibers, found in bone marrow, lymphatic organs, and in lesser amounts elsewhere.

**ribosome** A granule found free in the cytoplasm or attached to the rough endoplasmic reticulum or nuclear envelope composed of ribosomal RNA and enzymes; specialized to read the nucleotide sequence of messenger RNA and assemble a corresponding sequence of amino acids to make a protein.

**risk factor** Any environmental factor or characteristic of an individual that increases one's chance of developing a particular disease; includes such intrinsic factors as age, sex, and race and such extrinsic factors as diet, smoking, and occupation.

**rostral** Relatively close to the forehead, especially in reference to structures of the brain and spinal cord; for example, the frontal lobe is rostral to the parietal lobe. *Compare* caudal.

**ruga** (ROO-ga) 1. An internal fold or wrinkle in the mucosa of a hollow organ such as the stomach and urinary bladder; typically present when the organ is empty and relaxed but not when the organ is full and stretched. 2. Tissue ridges in such locations as the hard palate and vagina. fig. 25.12

# S

**saccule** (SAC-yule) A saclike receptor in the inner ear with a vertical patch of hair cells, the macula sacculi; senses the orientation of the head and responds to vertical acceleration, as when riding in an elevator or standing up. fig. 16.19

**sagittal plane** (SADJ-ih-tul) Any plane that extends from anterior to posterior and cephalic to caudal and that divides the body into right and left portions. *Compare* median plane.

**sarcomere** (SAR-co-meer) In skeletal and cardiac muscle, the portion of a myofibril from one Z disc to the next, constituting one contractile unit. fig. 11.5

**sarcoplasmic reticulum (SR)** The smooth endoplasmic reticulum of a muscle cell, serving as a calcium reservoir. fig. 11.2

**scanning electron microscope (SEM)** A microscope that uses an electron beam in place of light to form high-resolution, three-dimensional images of the surfaces of objects; capable of much higher magnifications than a light microscope.

**sclerosis** (scleh-RO-sis) Hardening or stiffening of a tissue, as in multiple sclerosis of the central nervous system or atherosclerosis of the blood vessels.

**sebum** (SEE-bum) An oily secretion of the sebaceous glands that keeps the skin and hair pliable.

**second messenger** A chemical that is produced within a cell (such as cAMP) or that enters a cell (such as calcium ions) in response to the binding of a messenger to a membrane receptor, and that triggers a metabolic reaction in the cell.

**secondary active transport** A mechanism in which solutes are moved through a plasma membrane by a carrier that does not itself use ATP but depends on a concentration gradient established by an active transport pump elsewhere in the cell.

**secretion** 1. A chemical released by a cell to serve a physiological function, such as a hormone or digestive enzyme. 2. The process of releasing such a chemical, often by exocytosis. *Compare* excretion.

**selectively permeable membrane** A membrane that allows some substances to pass through while excluding others; for example, the plasma membrane and dialysis membranes.

**semicircular ducts** Three ring-shaped, fluid-filled tubes of the inner ear that detect angular accelerations of the head; each is enclosed in a bony passage called the semicircular canal. fig. 16.20

**semilunar valve** A valve that consists of crescent-shaped cusps, including the aortic and pulmonary valves of the heart and valves of the veins and lymphatic vessels. fig. 19.8

**semipermeable membrane** *See* selectively permeable membrane.

**senescence** (seh-NESS-ense) Degenerative changes that occur with age.

**sensation** Conscious perception of a stimulus; pain, taste, and color, for example, are not stimuli but sensations resulting from stimuli.

**sensory nerve fiber** An axon that conducts information from a receptor to the central nervous system; an afferent nerve fiber.

**serosa** (seer-OH-sa) *See* serous membrane.

**serous fluid** (SEER-us) A watery, low-protein fluid similar to blood serum, formed as a filtrate of the blood or tissue fluid or as a secretion of serous gland cells; moistens the serous membranes.

**serous membrane** A membrane such as the peritoneum, pleura, or pericardium that lines a body cavity or covers the external surfaces of the viscera; composed of a simple squamous mesothelium and a thin layer of areolar connective tissue. Also called *serosa*. fig. 5.33b

**sex chromosomes** The X and Y chromosomes, which determine the sex of an individual.

**shock 1.** Circulatory shock, a state of cardiac output that is insufficient to meet the body's physiological needs, with consequences ranging from fainting to death. **2.** Insulin shock, a state of severe hypoglycemia caused by administration of insulin. **3.** Spinal shock, a state of depressed or lost reflex activity inferior to a point of spinal cord injury. **4.** Electrical shock, the effect of a current of electricity passing through the body, often causing muscular spasm and cardiac arrhythmia or arrest.

**sign** An objective manifestation of illness that any observer can see, such as cyanosis or edema. *Compare* symptom.

**simple epithelium** An epithelium in which all cells rest directly on the basement membrane; includes simple squamous, cuboidal, and columnar types, and pseudostratified columnar. fig. 5.3

**sinus 1.** An air-filled space in the cranium. **2.** A modified, relatively dilated vein that lacks smooth muscle and is incapable of vasomotion, such as the dural sinuses of the cerebral circulation and coronary sinus of the heart. **3.** A small fluid-filled space in an organ such as the spleen and lymph nodes. **4.** Pertaining to the sinoatrial node of the heart, as in *sinus rhythm*.

**sodium–glucose transporter (SGLT)** A symport that simultaneously transports Na⁺ and glucose into a cell.

**somatic 1.** Pertaining to the body as a whole. **2.** Pertaining to the skin, bones, and skeletal muscles as opposed to the viscera. **3.** Pertaining to cells other than germ cells.

**somatic nervous system** A division of the nervous system that includes efferent fibers mainly from the skin, muscles, and skeleton and afferent fibers to the skeletal muscles. *Compare* autonomic nervous system.

**somatosensory 1.** Pertaining to widely distributed *general senses* in the skin, muscles, tendons, joint capsules, and viscera, as opposed to the *special senses* found in the head only; also called *somesthetic*. **2.** Pertaining to the cerebral cortex of the postcentral gyrus, which receives input from such receptors.

**somite** One segment in a linear series of mesodermal masses that form on each side of the neural tube and give rise to trunk muscles, vertebrae, and dermis. fig. 29.7

**spermatogenesis** (SPUR-ma-toe-JEN-eh-sis) The production of sperm cells through a series of mitotic and meiotic cell divisions; male gametogenesis.

**spermatozoon** (SPUR-ma-toe-ZO-on) A sperm cell.

**sphincter** (SFINK-tur) A ring of muscle that opens or closes an opening or passageway; found, for example, in the eyelids, around the urinary orifice, and at the beginning of a blood capillary.

**spinal nerve** Any of the 31 pairs of nerves that arise from the spinal cord and pass through the intervertebral foramina.

**spindle 1.** An elongated structure that is thick in the middle and tapered at the ends (fusiform). **2.** A football-shaped complex of microtubules that guide the movement of chromosomes in mitosis and meiosis. fig. 4.16 **3.** A stretch receptor in the skeletal muscles. fig. 13.21

**spine 1.** The vertebral column. **2.** A pointed process or sharp ridge on a bone, such as the styloid process of the cranium and spine of the scapula.

**splanchnic** (SPLANK-nic) Pertaining to the digestive tract.

**squamous** (SKWAY-mus) Having a flat, scaly shape; pertains especially to a class of epithelial cells. figs. 5.4, 5.12

**stem cell** Any undifferentiated cell that can divide and differentiate into more functionally specific cell types such as blood cells and germ cells.

**stenosis** (steh-NO-sis) The narrowing of a passageway such as a heart valve or

uterine tube; a permanent, pathological constriction as opposed to physiological constriction of a passageway.

**steroid** (STERR-oyd, STEER-oyd) A lipid molecule that consists of four interconnected carbon rings; cholesterol and several of its derivatives.

**stimulus** A chemical or physical agent in a cell's surroundings that is capable of creating a physiological response in the cell; especially agents detected by sensory cells, such as chemicals, light, and pressure.

**strain** The extent to which a body, such as a bone, is deformed when subjected to stress. *Compare* stress.

**stratified epithelium** A type of epithelium in which some cells rest on top of others instead of on the basement membrane; includes stratified squamous, cuboidal, and columnar types, and transitional epithelium. fig. 5.3

**stress 1.** A mechanical force applied to any part of the body; important in stimulating bone growth, for example. *Compare* strain. **2.** A condition in which any environmental influence disturbs the homeostatic equilibrium of the body and stimulates a physiological response, especially involving the increased secretion of certain adrenal hormones.

**stroke volume** The volume of blood ejected by one ventricle of the heart in one contraction.

**stroma** The connective tissue framework of a gland, lymphatic organ, or certain other viscera, as opposed to the tissue (parenchyma) that performs the physiological functions of the organ.

**subcutaneous** (SUB-cue-TAY-nee-us) Beneath the skin.

**substrate 1.** A chemical that is acted upon and changed by an enzyme. **2.** A chemical used as a source of energy, such as glucose and fatty acids.

**substrate specificity** The ability of an enzyme to bind only one substrate or a limited range of related substrates.

**sulcus** (SUL-cuss) A groove in the surface of an organ, as in the cerebrum or heart.

**summation 1.** A phenomenon in which multiple stimuli combine their effects on a cell to produce a response; seen especially in nerve and muscle cells. **2.** A phenomenon in which multiple muscle twitches occur so closely together that a muscle fiber cannot fully relax between twitches but develops more tension than a single twitch produces. fig. 11.15

**superficial** Relatively close to the surface; opposite of deep. For example, the ribs are superficial to the lungs.

**superior** Higher than another structure or point of reference from the perspective of anatomical position; for example, the lungs are superior to the diaphragm.

**supination** (SOO-pih-NAY-shun) A rotational movement of the forearm that turns the palm so that it faces upward or forward. fig. 9.18

**surfactant** (sur-FAC-tent) A chemical that reduces the surface tension of water and enables it to penetrate other substances more effectively. Examples include pulmonary surfactant and bile acids.

**sympathetic nervous system** A division of the autonomic nervous system that issues efferent fibers through the thoracic and lumbar nerves and usually exerts adrenergic effects on its target organs; includes a chain of paravertebral ganglia adjacent to the vertebral column, and the adrenal medulla.

**symphysis** (SIM-fih-sis) A joint in which two bones are held together by fibrocartilage; for example, between bodies of the vertebrae and between the right and left pubic bones.

**symport** A cotransport protein that moves two solutes simultaneously through a plasma membrane in the same direction, such as the sodium-glucose transporter.

**symptom** A subjective manifestation of illness that only the ill person can sense, such as dizziness or nausea. *Compare* sign.

**synapse** (SIN-aps) **1.** A junction at the end of an axon where it stimulates another cell. **2.** A gap junction between two cardiac or smooth muscle cells at which one cell electrically stimulates the other; called an *electrical synapse.*

**synaptic cleft** (sih-NAP-tic) A narrow space between the synaptic knob of an axon and the adjacent cell, across which a neurotransmitter diffuses. fig. 12.20

**synaptic knob** The swollen tip at the distal end of an axon; the site of synaptic vesicles and neurotransmitter release. fig. 12.20

**synaptic vesicle** A spheroidal organelle in a synaptic knob containing neurotransmitter. fig. 12.20

**syndrome** A suite of related signs and symptoms stemming from a specific pathological cause.

**synergist** (SIN-ur-jist) A muscle that works with the prime mover (agonist) to contribute to the same overall action at a joint.

**synergistic** An effect in which two agents working together (such as two hormones) exert an effect that is greater than the sum of their separate effects. For example, neither follicle-stimulating hormone nor testosterone alone stimulates significant sperm production, but the two of them together stimulate production of vast numbers of sperm.

**synovial fluid** (sih-NO-vee-ul) A lubricating fluid similar to egg white in consistency, found in the synovial joint cavities and bursae.

**synovial joint** A point where two bones are separated by a narrow, encapsulated space filled with lubricating synovial fluid; most such joints are relatively mobile. Also called *diarthrosis.*

**synthesis reaction** A chemical reaction in which relatively small molecules are combined to form a larger one. *Compare* decomposition reaction.

**systemic** (sis-TEM-ic) Widespread or pertaining to the body as a whole, as in the systemic circulation.

**systemic circuit** All blood vessels that convey blood from the left ventricle to all organs of the body and back to the right atrium of the heart; all of the cardiovascular system except the heart and pulmonary circuit.

**systole** (SIS-toe-lee) The contraction of any heart chamber; ventricular contraction unless otherwise specified.

**systolic pressure** (sis-TOLL-ic) The peak arterial blood pressure measured during ventricular systole.

# T

**target cell** A cell acted upon by a nerve fiber, hormone, or other chemical messenger.

**tarsal** Pertaining to the ankle (tarsus).

**T cell** A type of lymphocyte involved in nonspecific defense, humoral immunity, and cellular immunity; occurs in several forms including helper, cytotoxic, and suppressor T cells and natural killer cells.

**tendon** A collagenous band or cord associated with a muscle, usually attaching it to a bone and transferring muscular tension to it.

**tetanus** **1.** A state of sustained muscle contraction produced by temporal summation as a normal part of contraction; also called *tetany.* **2.** Spastic muscle paralysis produced by the toxin of the bacterium *Clostridium tetani.*

**thalamus** (THAL-uh-muss) The largest part of the diencephalon, located immediately inferior to the corpus callosum and bulging into each lateral ventricle; a point of synaptic relay of nearly all signals passing from lower levels of the CNS to the cerebrum. figs. 14.8, 14.12a

**theory** An explanatory statement, or set of statements, that concisely summarizes the state of knowledge on a phenomenon and provides direction for further study; for example, the fluid mosaic theory of the plasma membrane and the sliding filament theory of muscle contraction.

**thermogenesis** The production of heat, for example, by shivering or by the action of thyroid hormones.

**thermoreceptor** A neuron specialized to respond to heat or cold, found in the skin and mucous membranes, for example.

**thermoregulation** Homeostatic regulation of the body temperature within a narrow range by adjustments of heat-promoting and heat-losing mechanisms.

**thorax** A region of the trunk between the neck and the diaphragm; the chest.

**threshold** **1.** The minimum voltage to which the plasma membrane of a nerve or muscle cell must be depolarized before it produces an action potential. **2.** The minimum combination of stimulus intensity and duration needed to generate an afferent signal from a sensory receptor.

**thrombosis** (throm-BO-sis) The formation or presence of a thrombus. *Compare* embolism.

**thrombus** A clot that forms in a blood vessel or heart chamber; may break free and travel in the bloodstream as a thromboembolus. *Compare* embolus.

**thyroid hormone** Either of two similar hormones, thyroxine and triiodothyronine, synthesized from iodine and tyrosine.

**thyroid-stimulating hormone (TSH)** A hormone of the anterior pituitary gland that stimulates the thyroid gland; also called *thyrotropin.*

**thyroxine (T₄)** (thy-ROCK-seen) The thyroid hormone secreted in greatest quantity, with four iodine atoms; also called *tetraiodothyronine.* fig. 17.17

**tight junction** A region in which adjacent cells are bound together by fusion of the outer phospholipid layer of their plasma membranes; forms a zone that encircles each cell near its apical pole and reduces or prevents flow of material between cells. fig. 5.28

**tissue** An aggregation of cells and extracellular materials, usually forming part of an organ and performing some discrete function for it; the four primary classes are epithelial, connective, muscular, and nervous tissue.

**totipotent** Pertaining to a stem cell of the early preembryo, prior to development of a blastocyst, that has the potential to develop into any type of embryonic, extraembryonic, or adult cell.

**trabecula** (tra-BEC-you-la) A thin plate or layer of tissue, such as the calcified trabeculae of spongy bone or the fibrous trabeculae that subdivide a gland. fig. 7.4

**trachea** (TRAY-kee-uh) A cartilage-supported tube from the inferior end of the larynx to the origin of the main bronchi; conveys air to and from the lungs; colloquially called the *windpipe.*

**translation** The process in which a ribosome reads an mRNA molecule and synthesizes the protein specified by its genetic code.

**transmembrane protein** An integral protein that extends through a plasma membrane and contacts both the extracellular and intracellular fluid. fig. 3.7

**transmission electron microscope (TEM)** A microscope that uses an electron beam in place of light to form high-resolution, two-dimensional images of ultrathin slices of cells or tissues; capable of extremely high magnification.

**triglyceride** (try-GLISS-ur-ide) A lipid composed of three fatty acids joined to a glycerol; also called a *triacylglycerol* or *neutral fat.* fig. 2.19

**trunk** **1.** That part of the body excluding the head, neck, and appendages. **2.** A major blood vessel, lymphatic vessel, or nerve that gives rise to smaller branches; for example, the pulmonary trunk and spinal nerve trunks.

**T tubule** Transverse tubule; a tubular extension of the plasma membrane of a muscle cell that conducts action potentials into the sarcoplasm and excites the sarcoplasmic reticulum. fig. 11.2

**tunic** (TOO-nic) A layer that encircles or encloses an organ, such as the tunics of a blood vessel or eyeball.

**tympanic membrane** The eardrum.

# U

**ultraviolet radiation** Invisible, ionizing, electromagnetic radiation with shorter wavelength and higher energy than violet light; causes skin cancer and photoaging of the skin but is required in moderate amounts for the synthesis of vitamin D.

**unipotent** Pertaining to a stem cell that has the potential to develop into only one type of fully differentiated, functional cell, such as an epidermal cell that can become only a keratinocyte.

**unmyelinated** (un-MY-eh-lih-nay-ted) Lacking a myelin sheath. fig. 12.8

**urea** (you-REE-uh) A nitrogenous waste produced from two ammonia molecules and carbon dioxide; the most abundant nitrogenous waste in the blood and urine. fig. 23.2

**uterine tube** A duct that extends from the ovary to the uterus and conveys an egg or conceptus to the uterus; also called *fallopian tube* or *oviduct.*

# V

**van der Waals force** A weak attraction between two atoms occurring when a brief fluctuation in the electron cloud density of one atom induces polarization of an adjacent atom; important in association of lipids with each other, protein folding, and protein–ligand binding.

**varicose vein** A vein that has become permanently distended and convoluted due to a loss of competence of the venous valves; especially common in the lower limb, esophagus, and anal canal (where they are called hemorrhoids).

**vas (vass)** (plural, *vasa*) A vessel or duct.

**vascular** Pertaining to blood vessels.

**vasoconstriction** (VAY-zo-con-STRIC-shun) The narrowing of a blood vessel due to muscular contraction of its tunica media.

**vasodilation** (VAY-zo-dy-LAY-shun) The widening of a blood vessel due to relaxation of the muscle of its tunica media and the outward pressure of the blood exerted against the wall.

**vasomotor center** A nucleus in the medulla oblongata that transmits efferent signals to the blood vessels and regulates vessel diameter.

**ventral** Pertaining to the front of the body, the regions of the chest and abdomen; anterior.

**ventral root** *See* anterior root.

**ventricle** (VEN-trih-cul) A fluid-filled chamber of the brain or heart.

**venule** (VEN-yool) The smallest type of vein, receiving drainage from capillaries.

**vertebra** (VUR-teh-bra) One of the bones of the vertebral column.

**vertebral column** (VUR-teh-brul) A posterior series of usually 33 vertebrae; encloses the spinal cord, supports the skull and thoracic cage, and provides attachment for the limbs and postural muscles. Also called *spine* or *spinal column.*

**vesicle** (VESS-ih-cul) A fluid-filled tissue sac or an organelle such as a synaptic or secretory vesicle.

**vesicular transport** The movement of particles or fluid droplets through the plasma membrane by the process of endocytosis or exocytosis.

**viscera** (VISS-er-uh) (singular, *viscus*) The organs contained in the body cavities, such as the brain, heart, lungs, stomach, intestines, and kidneys.

**visceral** (VISS-er-ul) **1.** Pertaining to the viscera. **2.** The inner or deeper layer of a two-layered membrane such as the pleura, pericardium, or glomerular capsule. *Compare* parietal. fig. A.6

**visceral muscle** Single-unit smooth muscle found in the walls of blood vessels and the digestive, respiratory, urinary, and reproductive tracts.

**viscosity** The resistance of a fluid to flow; the thickness or stickiness of a fluid.

**voluntary muscle** Muscle that is usually under conscious control; skeletal muscle.

**vulva** The female external genitalia; the mons, labia majora, and all superficial structures between the labia majora.

# W

**water balance** An equilibrium between fluid intake and output or between the amounts of fluid contained in the body's different fluid compartments.

**white matter** White myelinated nervous tissue deep to the cortex of the cerebrum and cerebellum and superficial to the gray matter of the spinal cord. fig. 14.6

# X

**X chromosome** The larger of the two sex chromosomes; males have one X chromosome and females have two in each somatic cell.

**X-ray** **1.** A high-energy, penetrating electromagnetic ray with wavelengths in the range of 0.1 to 10 nm; used in diagnosis and therapy. **2.** A photograph made with X-rays; radiograph.

# Y

**Y chromosome** Smaller of the two sex chromosomes, found only in males and having little if any genetic function except development of the testis.

**yolk sac** An embryonic membrane that encloses the yolk in vertebrates that lay eggs and serves in humans as the origin of the first blood and germ cells.

# Z

**zygomatic arch** An arch of bone anterior to the ear, formed by the zygomatic processes of the temporal, frontal, and zygomatic bones; origin of the masseter muscle.

**zygote** A single-celled, fertilized egg.

**Connective Issues**
**Circulatory System:** ©Andrew Syred/Getty Images; **Lymphatic and Immune Systems:** © Russell Kightley/Science Source; **Respiratory System:** © Ralph Hutchings/Visuals Unlimited/Getty Images; **Urinary System:** © Medical Body Scans/Science Source; **Reproductive System, Digestive System, Nervous System, Muscular System, Integumentary System:** iStockphoto; **Skeletal System:** © U.H.B. Trust/Tony Stone Images/Getty Images.

**Chapter 1**
**Opener:** ©Science Photo Library/Getty Images; **1.1a:** Neurosurg Focus © 2004 American Association of Neurological Surgeons; **1.1b:** ©SPL/Science Source; **1.2a:** Courtesy of the Armed Forces Institute of Pathology; **1.2b:** ©Corbis-Bettmann; **1.3b:** ©Tim Davis/Science Source; **1.10a:** ©U.H.B. Trust/Tony Stone Images/Getty Images; **1.10b:** ©Custom Medical Stock Photo, Inc.; **1.10c:** ©CNR/Phototake; **1.10d:** ©Tony Stone Images/Getty Images; **1.10e:** ©Monte S. Buchsbaum, University of California, San Diego; **1.11a:** ©Alexander Tsiaras/Science Source; **1.11b:** ©Ken Saladin.

**Atlas A**
**Opener:** ©SPL/Science Source; **A.1–A.3d:** ©McGraw-Hill Education/Joe DeGrandis; **A.8b:** ©MedicImage/Getty Image.

**Chapter 2**
**Opener:** ©Alfred Pasieka/Science Photo Library; **2.3:** ©Science Source; **2.10:** ©Ken Saladin.

**Chapter 3**
**Opener:** ©P.M. Motta & T. Naguro/Science Source; **3.3:** ©K.G. Murti/Visuals Unlimited; **3.4a,b:** From *Cell Ultrastructure* by William A. Jensen and Roderick B. Park. ©1967 by Wadsworth Publishing Co., Inc. Reprinted by permission of the publisher; **3.6a:** © Don Fawcett/Science Source; **3.10a:** Courtesy of Dr. S. Ito, Harvard Medical School; **3.10b:** ©Biophoto Associates/Science Source; **3.11a:** ©Custom Medical Stock Photo, Inc.; **3.11c:** © Don Fawcett/Science Source; **3.16a–c:** ©Dr. David M. Phillips/Visuals Unlimited; **3.22(all):** Company of Biologists, Ltd.; **3.23:** ©Don Fawcett/Science Source; **3.24b:** Courtesy of Dr. Birgit Satir, Albert Einstein College of Medicine; **3.25b:** ©K.G. Murti/Visuals Unlimited; **3.27a:** ©Richard Chao; **3.27b:** ©E.G. Pollock; **3.28a,b:** ©Don Fawcett/Science Source; **3.29:** ©Visuals Unlimited; **3.30a,b:** ©Don Fawcett/Science Source; **3.32:** ©Dr. Donald Fawcett & Dr. Porter/Visuals Unlimited; **3.33:** From Manley McGill, D.P. Highfield, T.M. Monahan, and Brinkley, B.R. "Effects of Nucleic Acid Specific Dyes on Centrioles of Mammalian Cells," published in the *Journal of Ultrastructure Research* 57, 43–53 (1976), pg. 48, fig. 6, with permission from Elsevier.

**Chapter 4**
**Opener:** ©Gopal Murti/Phototake NYC; **4.3a:** From *The Double Helix* by James D. Watson, 1968, Atheneum Press, NY. Courtesy of Cold Springs Harbor Laboratory; **4.3b:** Courtesy of King's College, London; **4.3c:** © A. Barrington Brown/Science Source.; **4.4a:** ©P. Motta & T. Naguro/SPL/Science Source; **4.5b:** © Biophoto Associates/Science Source; **4.9:** ©E. V. Kiseleva, February 5 Letter 257; 251-253, 1989; **4.16(1–4):** ©Ed Reschke; **4.17:** ©Science Source; **4.18a(left):** © Brad Barket/Getty Images; **(right):** ©Kurt Krieger/Corbis; **4.19(top):** © Comstock/Getty Images RF; **(middle, bottom):** ©Photodisc Red/Getty Images RF; **4.20(Phenotype 1):** From G. Pierrard, A. Nikkels. April 5, 2001, "A Medical Mystery" *The New England Journal of Medicine*, 344: p. 1057. ©2001 Massachusetts Medical Society. All rights reserved; **(Phenotype 2):** *British Journal of Ophthalmology* 1999; 83:680 ©by BMJ Publishing Group Ltd, http://www.bjophthalmol.com; **(Phenotype 3):** From G. Pierrard, A. Nikkels. April 5, 2001, "A Medical Mystery" *The New England Journal of Medicine*, 344: p. 1057. ©2001 Massachusetts Medical Society. All rights reserved; **4.24:** From the University of Alabama at Birmingham, Department of Pathology PEIR Digital Library ©http://peir.net.

**Chapter 5**
**Opener:** ©Don Fawcett/Science Source; **5.4a, 5.5a:** ©McGraw-Hill Education/Dennis Strete; **5.6a:** ©Lester V. Bergman; **5.7a:** ©McGraw-Hill Education/Dennis Strete; **5.8a, 5.9a:** © Ed Reschke; **5.10a:** ©McGraw-Hill Education/Dennis Strete; **5.11a:** ©Johnny R. Howze; **5.12:** ©Dr. Kessel & Dr. Kardon/Tissues and Organs/Visuals Unlimited, Inc; **5.13:** ©McGraw-Hill Education/Rebecca Gray; **5.14a:** ©McGraw-Hill Education/Dennis Strete; **5.15a:** ©McGraw-Hill Education/Al Telser; **5.16a, 5.17a, 5.18a:** ©McGraw-Hill Education/Dennis Strete; **5.19a, 5.20a:** ©Ed Reschke; **5.21a:** © Dr. Alvin Telser; **5.22a:** ©McGraw-Hill Education/Dennis Strete; **5.23a, 5.24a, 5.25a, 5.26a:** ©Ed Reschke; **5.27a:** ©McGraw-Hill Education/Dennis Strete.

**Chapter 6**
**Opener:** ©SPL/Science Source/Science Source; **6.2a:** ©DLILLC/Corbis; **6.2b:** ©McGraw-Hill Education/Joe DeGrandis; **6.4:** ©Meckes/Ottawa/Science Source; **6.5a:** ©McGraw-Hill Education/Dennis Strete; **6.5b:** ©Dr. Richard Kessel & Dr. Randy K./Visuals Unlimited/Corbis; **6.5c:** ©Dr. Kessel & Dr. Kardon/Tissues and Organs/Visuals Unlimited, Inc; **6.6a:** ©McGraw-Hill Education/Dennis Strete; **(dark girl):** ©Tom & Dee Ann McCarthy/Corbis; **6.6b:** © McGraw-Hill Education/Dennis Strete; **(light boy):** ©Creatas/PunchStock RF; **6.7b:** ©CBS/Phototake; **6.8a–d:** ©McGraw-Hill Education/Joe DeGrandis; **6.11a–c:** ©McGraw-Hill Education/Dennis Strete; **6.12a:** ©NMSB/Custom Medical Stock Photo, Inc.; **6.12b:** ©Biophoto Associates/Science Source; **6.12c:** ©James Stevenson/SPL/Science Source; **6.13a:** ©SPL/Custom Medical Stock Photo, Inc; **6.13b,c:** ©John Radcliffe/Science Source.

**Chapter 7**
**Opener:** ©Univ. of Zurich/Science Source; **7.4a:** ©D.W. Fawcett/Visuals Unlimited; **7.4c:** ©Science VU/Visuals Unlimited; **7.4d:** ©Donald Fawcett/Visuals Unlimited; **7.5:** ©Robert Calentine/Visuals Unlimited; **7.8:** ©Ken Saladin; **7.10:** ©Biophoto Associates/Science Source; **7.11:** Courtesy of Utah Valley Regional Medical Center, Department of Radiology; **7.12:** ©Victor Eroschenko; **7.13:** ©McGraw-Hill Education/Joe DeGrandis; **7.17a:** ©Custom Medical Stock Photo, Inc.; **7.17b:** ©Howard Kingsnorth/Getty Images; **7.17c:** © Lester V. Bergman/Corbis; **7.17d:** ©Custom Medical Stock Photo, Inc.; **7.19:** ©SIU/Visuals Unlimited; **7.20a:** ©Michael Klein/Peter Arnold, Inc./Getty Images; **7.20b:** © Dr. P. Marzzi/Science Source; **7.20c:** ©Yoav Levy/Phototake.

**Chapter 8**
**Opener:** ©Mehau Kulyk/Science Photo Library/Science Source; **8.20:** ©McGraw-Hill Education/Bob Coyle; **8.34c:** ©NHS Trust/Tony Stone Images/Getty Images; **8.37(top, both):** ©David Hunt/specimens from the National Museum of Natural History, Smithsonian Institution; **8.37(bottom, both):** ©L. Bassett/

# INDEX

# N

# LEXICON OF BIOMEDICAL WORD ELEMENTS

**a-** no, not, without (atom, agranulocyte)

**ab-** away (abducens, abduction)

**acetabulo-** small cup (acetabulum)

**acro-** tip, extremity, peak (acromion, acromegaly)

**ad-** to, toward, near (adsorption, adrenal)

**adeno-** gland (lymphadenitis, adenohypophysis)

**aero-** air, oxygen (aerobic, anaerobe, aerophagy)

**af-** toward (afferent)

**ag-** together (agglutination)

**-al** pertaining to (parietal, pharyngeal, temporal)

**ala-** wing (ala nasi)

**albi-** white (albicans, linea alba, albino)

**algi-** pain (analgesic, myalgia)

**aliment-** nourishment (alimentary)

**allo-** other, different (allele, allograft)

**amphi-** both, either (amphiphilic, amphiarthrosis)

**an-** without (anaerobic, anemic)

**ana-** 1. up, build up (anabolic, anaphylaxis).
2. apart (anaphase, anatomy).
3. back (anastomosis)

**andro-** male (androgen)

**angi-** vessel (angiogram, angioplasty, hemangioma)

**ante-** before, in front (antebrachium)

**antero-** forward (anterior, anterograde)

**anti-** against (antidiuretic, antibody, antagonist)

**apo-** from, off, away, above (apocrine, aponeurosis)

**arbor-** tree (arboreal, arborization)

**artic-** 1. joint (articulation).
2. speech (articulate)

**-ary** pertaining to (axillary, coronary)

**-ase** enzyme (polymerase, kinase, amylase)

**ast-, astro-** star (aster, astrocyte)

**-ata, -ate 1.** possessing (hamate, corniculate).
2. plural of (stomata, carcinomata)

**athero-** fat (atheroma, atherosclerosis)

**atrio-** entryway (atrium, atrioventricular)

**auri-** ear (auricle, binaural)

**auto-** self (autolysis, autoimmune)

**axi-** axis, straight line (axial, axoneme, axon)

**baro-** pressure (baroreceptor, hyperbaric)

**bene-** good, well (benign, beneficial)

**bi-** two (bipedal, biceps, bifid)

**bili-** bile (biliary, bilirubin)

**bio-** life, living (biology, biopsy, microbial)

**blasto-** precursor, bud, producer (fibroblast, osteoblast, blastomere)

**brachi-** arm (brachium, brachialis, antebrachium)

**brady-** slow (bradycardia, bradypnea)

**bucco-** cheek (buccal, buccinator)

**burso-** purse (bursa, bursitis)

**calc-** calcium, stone (calcaneus, hypocalcemia)

**callo-** thick (callus, callosum)

**calori-** heat (calorie, calorimetry, calorigenic)

**calv-, calvari-** bald, skull (calvaria)

**calyx** cup, vessel, chalice (glycocalyx, renal calyx)

**capito-** head (capitis, capitate, capitulum)

**capni-** smoke, carbon dioxide (hypocapnia)

**carcino-** cancer (carcinogen, carcinoma)

**cardi-** heart (cardiac, cardiology, pericardium)

**carot-** 1. carrot (carotene). 2. stupor (carotid)

**carpo-** wrist (carpus, metacarpal)

**case-** cheese (caseosa, casein)

**cata-** down, break down (catabolism)

**cauda-** tail (cauda equina, caudate nucleus)

**-cel** little (pedicel)

**celi-** belly, abdomen (celiac)

**centri-** center, middle (centromere, centriole)

**cephalo-** head (cephalic, encephalitis)

**cervi-** neck, narrow part (cervix, cervical)

**chiasm-** cross, X (optic chiasm)

**choano-** funnel (choana)

**chole-** bile (cholecystokinin, cholelithotripsy)

**chondro-** 1. grain (mitochondria).
2. cartilage, gristle (chondrocyte, perichondrium)

**chromo-** color (dichromat, chromatin, cytochrome)

**chrono-** time (chronotropic, chronic)

**cili-** eyelash (cilium, superciliary)

**circ-** about, around (circadian, circumduction)

**cis-** cut (incision, incisor)

**cisterna** reservoir (cisterna chyli)

**clast-** break down, destroy (osteoclast)

**clavi-** hammer, club, key (clavicle, supraclavicular)

**-cle** little (tubercle, corpuscle)

**cleido-** clavicle (sternocleidomastoid)

**cnemo-** lower leg (gastrocnemius)

**co-** together (coenzyme, cotransport)

**collo-** 1. hill (colliculus).
2. glue (colloid, collagen)

**contra-** opposite (contralateral)

**corni-** horn (cornified, corniculate, cornu)

**corono-** crown (coronary, corona, coronal)

**corpo-** body (corpus luteum, corpora quadrigemina)

**corti-** bark, rind (cortex, cortical)

**costa-** rib (intercostal, subcostal)

**coxa-** hip (os coxae, coxal)

**crani-** helmet (cranium, epicranius)

**cribri-** sieve, strainer (cribriform, area cribrosa)

**crino-** separate, secrete (holocrine, endocrinology)

**crista-** crest (crista galli, mitochondrial crista)

**crito-** to separate (hematocrit)

**cruci-** cross (cruciate ligament)

**-cule, -culus** small (canaliculus, trabecula, auricular)

**cune-** wedge (cuneiform, cuneatus)

**cutane-, cuti-** skin (subcutaneous, cuticle)

**cysto-** bladder (cystitis, cholecystectomy)

**cyto-** cell (cytology, cytokinesis, monocyte)

**de-** down (defecate, deglutition, dehydration)

**demi-** half (demifacet, demilune)

**den-, denti-** tooth (dentition, dens, dental)

**dendro-** tree, branch (dendrite, oligodendrocyte)

**derma-, dermato-** skin (dermatology, hypodermic)

**desmo-** band, bond, ligament (desmosome, syndesmosis)

**dia-** 1. across, through, separate (diaphragm, dialysis). 2. day (circadian)

**dis-** 1. apart (dissect, dissociate).
2. opposite, absence (disinfect, disability)

**diure-** pass through, urinate (diuretic, diuresis)

**dorsi-** back (dorsal, dorsum, latissimus dorsi)

**duc** to carry (duct, adduction, abducens)

**dys-** bad, abnormal, painful (dyspnea, dystrophy)

**e-** out (ejaculate, eversion)

**-eal** pertaining to (hypophyseal, arboreal)

**ec-, ecto-** outside, out of, external (ectopic, ectoderm, splenectomy)

**ef-** out of (efferent, effusion)

**-el, -elle** small (fontanel, organelle, micelle)

**electro-** electricity (electrocardiogram, electrolyte)

**em-** in, within (embolism, embedded)

**emesi-, emeti-** vomiting (emetic, hyperemesis)

**-emia** blood condition (anemia, hypoxemia)

**en-** in, into (enzyme, parenchyma)

**encephalo-** brain (encephalitis, telencephalon)

**enchymo-** poured in (mesenchyme, parenchyma)

**endo-** within, into, internal (endocrine, endocytosis)

**entero-** gut, intestine (mesentery, myenteric)

**epi-** upon, above (epidermis, epiphysis, epididymis)

**ergo-** work, energy, action (allergy, adrenergic)

**eryth-, erythro-** red (erythema, erythrocyte)

**esthesio-** sensation, feeling (anesthesia, somesthetic)

**eu-** good, true, normal, easy (eupnea, aneuploidy)

**exo-** out (exopeptidase, exocytosis, exocrine)

**facili-** easy (facilitated)

**fasci-** band, bundle (fascia, fascicle)

**fenestr-** window (fenestrated)

**fer-** to carry (efferent, uriniferous)

**ferri-** iron (ferritin, transferrin)

**fibro-** fiber (fibroblast, fibrosis)

**fili-** thread (myofilament, filiform)

**flagello-** whip (flagellum)

**foli-** leaf (folic acid, folia)

**-form** shape (cuneiform, fusiform)

**fove-** pit, depression (fovea)

**funiculo-** little rope, cord (funiculus)

**fusi-** 1. spindle (fusiform).
   2. pour out (perfusion)

**gamo-** marriage, union (monogamy, gamete)

**gastro-** belly, stomach (gastrointestinal, digastric)

**-gen, -genic, -genesis** producing, giving rise to (pathogen, carcinogenic, glycogenesis)

**genio-** chin (geniohyoid, genioglossus)

**germi-** 1. sprout, bud (germinal, germinativum). 2. microbe (germicide)

**gero-** old age (progeria, geriatrics, gerontology)

**gesto-** 1. to bear, carry (ingest).
   2. pregnancy (gestation, progesterone)

**glia-** glue (neuroglia, microglia)

**globu-** ball, sphere (globulin, hemoglobin)

**glom-** ball (glomerulus)

**glosso-** tongue (glossopharyngeal, hypoglossal)

**glyco-** sugar (glycogen, glycolysis, hypoglycemia)

**gono-** 1. angle, corner (trigone).
   2. seed, sex cell, generation (gonad, oogonium, gonorrhea)

**gradi-** walk, step (retrograde, gradient)

**-gram** recording of (electrocardiogram, sonogram)

**-graph** recording instrument (sonograph, electrocardiograph)

**-graphy** recording process (sonography, radiography)

**gravi-** severe, heavy (gravid, myasthenia gravis)

**gyro-** turn, twist (gyrus)

**hallu-** great toe (hallux, hallucis)

**hemi-** half (hemidesmosome, hemisphere)

**-hemia** blood condition (polycythemia)

**hemo-** blood (hemophilia, hemoglobin, hematology)

**hetero-** different, other, various (heterozygous)

**histo-** tissue, web (histology, histone)

**holo-** whole, entire (holistic, holocrine)

**homeo-** constant, unchanging, uniform (homeostasis, homeothermic)

**homo-** same, alike (homologous, homozygous)

**hyalo-** clear, glassy (hyaline, hyaluronic acid)

**hydro-** water (dehydration, hydrolysis, hydrophobic)

**hyper-** above, above normal, excessive (hyperkalemia, hypertonic)

**hypo-** below, below normal, deficient (hypogastric, hyponatremia, hypophysis)

**-ia** condition (anemia, hypocalcemia, osteomalacia)

**-ic** pertaining to (isotonic, hemolytic, antigenic)

**-icle, -icul** small (ossicle, canaliculus, reticular)

**ilia-** flank, loin (ilium, iliac)

**-illa, -illus** little (bacillus)

**-in** protein (trypsin, fibrin, globulin)

**infra-** below (infraspinous, infrared)

**ino-** fiber (inotropic, inositol)

**insulo-** island (insula, insulin)

**inter-** between (intercellular, intervertebral)

**intra-** within (intracellular, intraocular)

**iono-** ion (ionotropic, cationic)

**ischi-** to hold back (ischium, ischemia)

**-ism 1.** process, state, condition (metabolism, rheumatism). **2.** doctrine, belief, theory (holism, reductionism, naturalism)

**iso-** same, equal (isometric, isotonic, isomer)

**-issimus** most, greatest (latissimus, longissimus)

**-ite** little (dendrite, somite)

**-itis** inflammation (dermatitis, gingivitis)

**jug-** to join (conjugated, jugular)

**juxta-** next to (juxtamedullary, juxtaglomerular)

**kali-** potassium (hypokalemia)

**karyo-** seed, nucleus (megakaryocyte, karyotype)

**kerato-** horn (keratin, keratinocyte)

**kine-** motion, action (kinetic, kinase, cytokinesis)

**labi-** lip (labium, levator labii)

**lacera-** torn, cut (foramen lacerum, laceration)

**lacrimo-** tear, cry (lacrimal gland, nasolacrimal)

**lacto-** milk (lactose, lactation, prolactin)

**lamina-** layer (lamina propria, laminar flow)

**latero-** side (bilateral, ipsilateral)

**lati-** broad (fascia lata, latissimus dorsi)

**-lemma** husk (sarcolemma, neurilemma)

**lenti-** lens (lentiform)

**-let** small (platelet)

**leuko-** white (leukocyte, leukemia)

**levato-** to raise (levator labii, elevation)

**ligo-** to bind (ligand, ligament)

**line-** line (linea alba, linea nigra)

**litho-** stone (otolith, lithotripsy)

**-logy** study of (histology, physiology, hematology)

**lucid-** light, clear (stratum lucidum, zona pellucida)

**lun-** moon, crescent (lunate, lunule, semilunar)

**lute-** yellow (macula lutea, corpus luteum)

**lyso-, lyto-** split apart, break down (lysosome, hydrolysis, electrolyte, hemolytic)

**macro-** large (macromolecule, macrophage)

**macula-** spot (macula lutea, macula densa)

**mali-** bad (malignant, malocclusion, malformed)

**malle-** hammer (malleus, malleolus)

**mammo-** breast (mammary, mammillary)

**mano-** hand (manus, manipulate)

**manubri-** handle (manubrium)

**masto-** breast (mastoid, gynecomastia)

**medi-** middle (medial, mediastinum, intermediate)

**medullo-** marrow, pith (medulla)

**mega-** large (megakaryocyte, hepatomegaly)

**melano-** black (melanin, melanocyte, melancholy)

**meno-** month (menstruation, menopause)

**mento-** chin (mental, mentalis)

**mero-** part, segment (isomer, centromere, merocrine)

**meso-** in the middle (mesoderm, mesentery)

**meta-** beyond, next in a series (metaphase, metacarpal)

**metabol-** change (metabolism, metabolite)

**-meter** measuring device (calorimeter, spirometer)

**metri-** 1. length, measure (isometric, emmetropic). 2. uterus (endometrium)

**micro-** small (microscopic, microcytic, microglia)

**mito-** thread, filament, grain (mitochondria, mitosis)

**mono-** one (monocyte, monogamy, mononucleosis)

**morpho-** form, shape, structure (morphology, amorphous)

**muta-** change (mutagen, mutation)

**myelo-** 1. spinal cord (poliomyelitis, myelin).
   2. bone marrow (myeloid, myelocytic)

**myo-, mysi-** muscle (myoglobin, myosin, epimysium)

**natri-** sodium (hyponatremia, natriuretic)

**neo-** new (neonatal, gluconeogenesis)

**nephro-** kidney (nephron, hydronephrosis)

**neuro-** nerve (aponeurosis, neurosoma, neurology)

**nucleo-** nucleus, kernel (nucleolus, nucleic acid)

**oo-** egg (oogenesis, oocyte)

**ob-** 1. life (aerobic, microbe). 2. against, toward, before (obstetrics, obturator, obstruction)

**oculo-** eye (oculi, oculomotor)

**odonto-** tooth (odontoblast, periodontal)

**-oid** like, resembling (colloid, sigmoid, ameboid)

**-ole** small (arteriole, bronchiole, nucleolus)

**oligo-** few, a little, scanty (oligopeptide, oliguria)

**-oma** tumor, mass (carcinoma, hematoma)

**omo-** shoulder (omohyoid, acromion)

**onycho-** nail, claw (hyponychium, onychomycosis)

**op-** vision (optics, myopia, photopic)

**-opsy** viewing, to see (biopsy, rhodopsin)

**or-** mouth (oral, orbicularis oris)